COMPREHENSIVE INORGANIC CHEMISTRY

PERGAMON PRESS

OXFORD · NEW YORK · TORONTO
SYDNEY · BRAUNSCHWEIG

Pergamon Press Ltd., Headington Hill Hall, Oxford
Pergamon Press (Aust.) Pty. Ltd., 19a Boundary Street, Rushcutters Bay, N.S.W. 2011, Australia
Vieweg & Sohn GmbH, Burgplatz 1, Braunschweig

First edition 1973

Library of Congress Cataloging in Publication Data

Main entry under title:
Comprehensive inorganic chemistry.

1. Chemistry, Inorganic. I. Bailar, John Christian, 1904– ed. II. Trotman-Dickenson, A. F., ed.
[DNLM: 1. Chemistry. QD 151.2 C737 1973]
QD151.2.C64 547 77–189736
ISBN 0–08–017275–X

Exclusive distributors in the Western Hemisphere
Compendium Publishers
Fairview Park, Elmsford, New York 10523

Printed in Great Britain by J. W. Arrowsmith Ltd., Bristol

ISBN 0 08 016988 0

PREFACE

THE Editorial Board of Comprehensive Inorganic Chemistry planned the treatise to fill a gap in the literature. There was no work that provided more information than could be found in single volumes but was not so large as to put it out of reach of all but a few central libraries.

The Editorial Board drew up and incorporated in instructions to authors a scheme that would make the best possible use of about five thousand pages. It was envisaged that the treatise would be of service to a wide range of readers many of whom would not be professional chemists. Convenience for all classes of reader was of paramount importance so that if a conflict arose between brevity and ease of use, the latter was preferred. Nevertheless the arrangement of the treatise is so systematic that such conflicts rarely occurred. The convenience of the reader has been further ensured by the adoption of a consistent arrangement of material within the chapters on the elements. The editors have been very gratified to observe that authors have not found the imposed pattern unduly restrictive. It has certainly helped to keep the accounts coherent and to preserve the intended balance between the chapters. The editors are very sensible to the effort that authors have made to collaborate.

The section of the book devoted to the survey of topics, particularly those relating to the transition elements, was a special interest of Sir Ronald Nyholm, whose death after most chapters were in proof saddened many chemists. We hope that those chapters which bear repeated evidence of his intellectual influence will be judged to be one of the many worthy memorials that he left behind him.

A. F. TROTMAN-DICKENSON

CONTENTS

CONTENTS OF VOLUME 1

CONTENTS OF VOLUME 3

CONTENTS OF VOLUME 4

For Chapter 45 see Volume 5.

CONTENTS OF VOLUME 5

LIST OF CONTRIBUTORS

Professor E. W. Abel,
Department of Chemistry,
University of Exeter,
Stocker Road,
Exeter, EX4 4QD

Dr. C. J. Adams,
Department of Inorganic Chemistry,
University of Oxford,
South Parks Road,
Oxford, OX1 3QR.

Professor K. W. Bagnall,
Department of Chemistry,
The University of Manchester,
Manchester, M13 9PL.

Dr. A. J. Downs,
Inorganic Chemistry Laboratory,
University of Oxford,
South Parks Road,
Oxford, OX1 3QR.

Professor E. A. V. Ebsworth,
Department of Chemistry,
University of Edinburgh,
West Mains Road,
Edinburgh, EH9 3JJ.

Dr. K. Jones,
Department of Chemistry,
The University of Manchester Institute of Science and Technology,
Sackville Street,
Manchester, M60 1QD.

Dr. J. A. Connor,
Department of Chemistry,
The University of Manchester,
Manchester, M13 9PL.

Dr. T. A. O'Donnell,
Department of Inorganic Chemistry,
University of Melbourne,
Parkville,
Victoria 3052,
Australia.

Professor E. G. Rochow,
Harvard University,
Department of Chemistry,
12 Oxford Street,
Cambridge,
Massachusetts 02138,
U.S.A.

Professor Dr. Max Schmidt,
Institut fur Anorganische Chemie,
Der Universität Würzburg,
8700 Würzburg,
Landwehr,
German Federal Republic.

Dr. W. Siebert,
Institut fur Anorganische Chemie,
Der Universität Würzburg,
8700 Würzburg,
Landwehr,
German Federal Republic.

Dr. J. D. Smith,
Department of Chemistry,
University of Sussex,
Falmer, Brighton,
Sussex, BN1 9QJ.

Dr. A. D. F. Toy,
Senior Scientist,
Eastern Research Center,
Stauffer Chemical Company,
Dobbs Ferry, New York 10522,
U.S.A.

Dr. J. J. Turner,
School of Chemistry,
The University,
Newcastle-upon-Tyne, NE1 7RH.

16. GERMANIUM

E. G. Rochow
Harvard University

1. THE ELEMENT

1.1. DISCOVERY AND HISTORY. OCCURRENCE

Germanium was discovered by Clemens A. Winkler in 1886[1], following the accurate prediction of its existence by Mendeleef in 1871 as part of his formulation of the Periodic Law and his construction of the Periodic Table[2]. Much has been made in the past of the "complete agreement" of Winkler's findings with Mendeleef's predictions, but a more critical appraisal making use of modern constants for the element and its compounds[3] shows that the correspondence is not as close as once believed (see Table 1). The observation by Winkler that elemental germanium is not attacked by hydrochloric acid, but *is* attacked by concentrated solutions of NaOH, is but the first of a long series of observations pointing toward a less metallic behavior for germanium than had been expected. The element actually resembles arsenic in many respects, and Winkler first thought he had found eka-antimony rather than eka-silicon. When we consider that a water solution of GeO_2 is acidic, the point is emphasized.

Winkler obtained his germanium from the mineral argyrodite, which is Ag_8GeS_6. The element also occurs in canfieldite, ($Ag_8(SnGe)S_6$), and in germanite, which is approximately $Cu_5(Cu,Fe)_6AsGeS_{12}$. Many other minerals (especially those of zinc, arsenic, tin, lead copper, and antimony) contain some germanium. Indeed, the element is extremely widespread in trace amounts: Papish demonstrated spectrographically that it is in almost all iron and zinc produced in the United States. The ionic potential (ratio of charge to radius) of germanium is so much like that of many other elements that germanium has very few distinctive minerals; it crystallizes with a dozen other elements in their pegmatite-type sulfide and arsenide minerals. Only the three special minerals mentioned above have enough Ge in them to have it appear in their formulas; some 16 other minerals contain 0.1% to 1% Ge, and 700 other minerals contain 0.0001% to 0.1% Ge[4]. Silicate rocks in general have the lowest Ge content. Coal ash from brown coal frequently contains Ge in the range 0.005% to 0.1%; the ash of some anthracites may run as high as 4% or even 7.5% Ge, thereby providing a possible source of the element[5]. Some spring waters from great depths in the earth contain Ge to the extent of 0.03 mg/l., but these are not practicable

[1] C. A. Winkler, *J. prakt. Chem.* (2) **34** (1886) 188; *Ber.* **19** (1886) 210.
[2] D. I. Mendeleef, *Liebig's Ann. Suppl.* **8** (1871) 133.
[3] A. Turk, H. Meislich, F. Brescia and J. Arents, *Introduction to Chemistry*, Academic Press, N.Y. (1968).
[4] W. Noddack and I. Noddack, *Freiburger wiss. Ges.* **26** (1937) 3/38, 21.
[5] *Gmelins Handbuch der anorganischen Chemie*, 8th ed., System No. 45, suppl. vol., pp. 1–27, Verlag Chemie, Weinheim (1958).

TABLE 1. PREDICTION OF THE PROPERTIES OF GERMANIUM

Property	"Eka-silicon" Predicted in 1871 by Mendeleef	Germanium Reported in 1886 by Clemens Winkler	Germanium Currently accepted
Atomic weight	72	72.32	72.59
Specific gravity	5.5	5.47	5.35
Melting point	High	—	947°C
Specific heat (cal/g deg)	0.073	0.076	0.074
Gram-atomic volume (cm^3)	13	13.22	13.5
Color	Dark gray	Grayish-white	Grayish-white
Valence	4	4	4
Reaction with acids and alkalis	Es will be slightly attacked by such acids as HCl, but will resist attack by such alkalis as NaOH	Ge is dissolved by neither HCl nor NaOH in dilute solution, but is dissolved by concentrated NaOH	Ge is dissolved by neither HCl nor NaOH in dilute solution, but is dissolved by concentrated NaOH
Boiling point of the tetraethyl derivative	160°C	160°C	185–187°C
Specific gravity of the dioxide	4.7	4.703	4.228
Specific gravity of the tetrachloride	1.9	1.887	1.8443
Boiling point of the tetrachloride	100°C	86°C	84°C

Adapted from Turk, Meislich, Brescia and Arents, *Introduction to Chemistry*, Academic Press, N.Y., 1968, by permission.

sources. On the cosmic scale, germanium occurs as a trace element in iron meteorites, in stony meteorites, in the sun, and in the stars[5].

The abundance of Ge in the igneous rocks of the earth is usually given as 7 g per metric ton, or $7 \times 10^{-4}\%$[6], so it is somewhat more rare than Ga and Pb (15 and 16 ppm, respectively) but more abundant than As, Be, B, and Br. The hydrosphere contains very little: some 0.00005 mg/1. The cosmic abundance is 50 atoms per 10^6 atoms of silicon[7].

1.2. PRODUCTION AND INDUSTRIAL USE

Minerals of high Ge content, such as germanite, usually are roasted first to convert the sulfide to oxides, and then are put through an acid or alkaline extraction process (the advantage of the alkaline process is that gallium may be recovered at the same time). Special methods are used for the recovery of Ge from coal and from zinc ore concentrates. The processes will be treated briefly here; full details and references are available in the Gmelin supplementary volume[8].

By the acid method, germanite or other roasted ore is treated with fuming HNO_3 or a mixture of H_2SO_4 and fuming HNO_3 until only a greenish-white powder is left. The

[6] T. Moeller, *Inorganic Chemistry*, p. 30, J. Wiley & Sons, N.Y. (1952).

[7] H. E. Suess and H. C. Urey, in *McGraw-Hill Encyclopedia of Science and Technology*, Vol. 4, p. 548, New York (1960).

[8] Gmelin, *loc. cit.*, pp. 27–34.

HNO_4 is fumed off, concentrated HCl is added, and the mixture is heated in a distilling flask fitted with a fractionating column and an inlet tube for chlorine. The GeO_2 slowly converts to $GeCl_4$ and distils over, leaving non-volatile chlorides behind. The purpose of the chlorine is to keep arsenic in the pentavalent state, because $AsCl_5$ is less volatile than $AsCl_3$. The $GeCl_4$ can be rectified further, or hydrolysed to GeO_2 and then put through the $HCl+Cl_2$ treatment again to purify it.

By the alkaline method, the finely powdered ore is digested with an equal weight of 50% aq. NaOH, evaporated to dryness, and then stirred with hot water, filtered and washed. The filtrate contains sodium salts of Ge, Ga, As, and Mo oxyacids and thioacids. It is neutralized with H_2SO_4 and acidified with HNO_3, precipitating As_2S_3 and sulfur. The filtrate is neutralized and NH_3 is added, precipitating a hydrous GeO_2. This can be filtered off and converted to $GeCl_4$ by the $HCl+Cl_2$ treatment described above. Gallium usually is recovered from the alkaline filtrate by electrolysis.

An alternative dry method involves heating the powdered original ore in an *oxygen-free* stream of N_2 at 800° for 5 hr to drive off sulfur and As_2S_3. The residue is then heated to 825° in a stream of NH_3 for 12 hr, whereupon GeS_2 is reduced to GeS and sublimes. This can be oxidized to GeO_2 with fuming HNO_3 and treated as above.

Germanium-containing zinc oxide can be mixed with powdered charcoal and NaCl, and then heated to sintering temperature to drive off $GeCl_4$ and other chlorides. The condensate is taken up in water, treated with H_2SO_4 to precipitate $PbSO_4$, and the filtrate treated with zinc dust to precipitate the Cu, Ge, and As. Ignition of the precipitate gives oxides which can be heated with conc. HCl and Cl_2 as above.

Germanium in coal presents a special problem because during the usual procedure for burning the coal the Ge is converted to volatile GeO and is lost from the ash. For this reason the best starting material for recovery of Ge from the coal is not the ash but the flue dust. This can be treated as an "ore", or simply heated with 20% HCl to drive off all of the volatile chlorides. The oily layer of the distillate is then fractionated in a stream of chlorine to recover the $GeCl_4$, and the water layer is treated with H_2S to precipitate As_2O_3, GeS_2, and most other sulfides. The precipitate is extracted with conc. NaOH, and the extract either neutralized and heated with a large excess of HCl and Cl_2 to distil $GeCl_4$, or made 12 N with H_2SO_4 and treated with H_2S to precipitate GeS_2.

Elementary Ge is usually prepared from the pure white dioxide by reduction. This has to be done very slowly, below 540°, if hydrogen is used as the reducing agent, because too much volatile brown GeO is lost at higher temperatures. A simple and effective laboratory method, developed by Dennis[9], consists in reducing GeO_2 with sugar charcoal (as a source of very pure carbon free of Si and metals) in a graphite crucible. The GeO_2 is ground with an equal volume of sugar charcoal and then mixed with a similar quantity of finely pulverized NaCl. This mixture is put into the bottom of a graphite crucible and overlaid with a layer of NaCl, then a thick layer of sugar charcoal, and finally more NaCl. The crucible is then heated slowly, preferably over its entire surface (as in an induction furnace). The GeO_2 is reduced, any fumes of GeO are caught by the charcoal layer, the NaCl melts and acts as a flux, and the globules of Ge (m.p. 947°, vs. m.p. 801° for NaCl) coalesce at the bottom of the melt. The cooled crucible is broken open and the salt dissolved away from the regulus with warm water. Germanium does not form a carbide, and the element is not attacked by molten NaCl.

To obtain ultrapure elementary Ge for semiconductor use, the tetrachloride usually is

[9] L. M. Dennis, *Zeit. anorg. allgem. Chem.* **174** (1928) 97; *J. Am. Chem. Soc.* **45** (1923) 2033.

purified extensively by distillation and then hydrolyzed to GeO_2. This is reduced slowly in H_2 (in graphite boats), melted to bars, and then purified by extensive zone-melting refining (see description under preparation of ultrapure silicon). The purity of the metalloid can be measured during the process by following its Hall-effect constant and its conductivity.

Nuclear Properties

The isotopes of Ge are listed in Table 2 (with nuclear masses given on the scale of $^{12}C = 12.00000$), and the nuclear behavior is summarized as of July 1966[10].

The natural mixture of isotopes has a cross-section of 2.23 barns for thermal neutrons; the cross-section becomes 5.13 barns for neutrons of 10.0 eV, and 4.47 barns for neutrons of 3800 eV energy. Neutron bombardment of the lighter isotopes of Ge results in the usual (n,γ) reaction to give the next heavier isotope, but ^{70}Ge undergoes an (n,p) reaction to give ^{70}Ga, and ^{72}Ge and ^{73}Ge react the same way. The ^{74}Ge isotope undergoes an (n,α) reaction to give ^{71}Zn. The isotopes of masses 74, 75, 76, 77 and 78 are fission products from the slow-neutron fission of ^{235}U.

Physical Properties

Germanium is a brittle metalloid which crystallizes in the cubic system, with a diamond lattice. There are no polymorphs. In the massive form its color is much like that of nickel; it is decidedly more yellowish than silicon. Since it is neither ductile nor malleable, its mechanical properties are difficult to measure with precision, but a list of selected physical properties is given in Table 3.

Because the classic investigations of Shockley and others on the properties of solid semiconductors (which led to the development of transistors, crystal diodes and many other electronic solid-state devices) were done with crystalline germanium, the electrical properties of the element have been studied more thoroughly than those of any other solid substance. The electrical conductivity and the Hall effect, in particular, have been investigated over an enormous range of compositions and of electrical and magnetic field strengths. Germanium of the very highest attainable purity is itself an intrinsic semiconductor; thermal promotion of electrons from a filled band to an empty conduction band is relatively easy (compared with that required in silicon or in diamond), and such promotion naturally increases the conductivity. Still easier access to conduction bands is provided by the deliberate introduction of new electron energy-level bands by adding controlled amounts of other elements. These impurities cause germanium to function also as an *extrinsic* semiconductor with just the right properties for the function in mind. The qualitative theory of such operation is easily understood: the introduction of B, Al, Ga and In causes substitution of some of the Ge atoms in the lattice with a Group III element, which can supply only three electrons to the binding network instead of four, and hence there is an electron deficiency (a "hole") in the structure. Migration of electrons from neighboring binding sites to the "holes" constitutes electrical conduction, and at the same time creates corresponding deficiencies elsewhere. Such introduction of Group III elements therefore makes an unfilled conduction band out of a previously filled one, and the resulting material is spoken of as *p*-type germanium because it contains "positive holes". Similarly, the introduction of P, As or Sb into the germanium crystal results in the replacement of some Ge atoms with Group V

[10] D. T. Goldman and J. R. Roeser, *Chart of the Nuclides*, Gen. Elect. Co. Ed. Div., Schenectady, N.Y. (1966).

TABLE 2. ISOTOPES OF GERMANIUM

Mass		Natural abundance, %	Cross-section, barns	Principal radiation	E, meV	Other radiation	E, meV	Disint. energy, meV	Half-life
^{65}Ge	—	0	—	β^{+}	3.7	γ	0.67, 1.72	6.4	1.5 m
^{66}Ge	—	0	—	β^{+}	1.3, 2.0	γ	0.38, 0.046	3.0	2.4 h
^{67}Ge	—	0	—	β^{+}	3.2, 2.3	γ	0.17, 0.92	4.4	19 m
^{68}Ge	—	0	—	no gamma	—	—	—	0.7	280 d
^{69}Ge	—	0	—	β^{+}	1.21, 1.61	γ	1.12, 0.58	2.23	37 h
^{70}Ge	69.9243	20.52	0.3	—	—	—	—	—	—
^{71}Ge	—	0	—	γ	0.175	iso.	trans.	0.198	0.02 s
						ec		0.23	11 d
^{72}Ge	71.9221	27.43	1.0	—	—	—	—	—	—
^{73}Ge	72.9235	7.76	14	upper isomer decays:				0.068	0.53 s
^{74}Ge	73.9219	36.54	0.2+0.3	(fission product of U-235)					
^{75}Ge	(f.p.)	0	—	iso. trans. to→					49 s
				β^{-}	1.20	γ	27.1	1.20	82 m
^{76}Ge	75.9214	7.76	0.1+0.1	(fission product of U-235)					
^{77}Ge	(f.p.)	0	—	β^{-}	2.9	iso. trans.→			59 s
				β^{-}	2.1	γ0.26		2.75	11.3 h
^{78}Ge	(f.p.)	0	—	β^{-}	0.71	γ0.28		0.99	88 m

elements having five electrons per atom instead of four, and the migration of these extra electrons under influence of an electric field constitutes the electrical conduction of *n*-type germanium. Junctions between *n*- and *p*-types of germanium function as rectifiers because they set up a barrier to conduction in one direction but not in the other: an a.c. field alternately sweeps out the carriers (when the situation is negative electrode/*p*-type/*n*-type/ positive electrode) and then pushes carriers across the junction (when the situation is positive

TABLE 3. PHYSICAL PROPERTIES OF ELEMENTAL GERMANIUM

Property	Constant	Ref.
Atomic weight	72.59	
Melting point	947.4°	d
Heat of fusion	98.3 ± 4.9 cal/g	b
Boiling point	2830°	d
Heat of vaporization, liquid	79.9 kcal/g-atom	d
Heat of sublimation	84 kcal/g-atom	b
Vapor pressure of liquid		
at 996°	10^{-4} torr	
at 1112°	10^{-3} torr	
at 1251°	10^{-2} torr	b
at 1421°	10^{-1} torr	
at 1635°	1 torr	
Specific heat C_p (cal/g-atom °C)	1.497 at 50°K	b
	3.309 at 100°K	
	4.453 at 150°K	
	4.992 at 200°K	
Thermal cond., cal/sec/cm²/°C	0.15	d
Entropy, std.	10.1 cal/g-atom °C	b
Debye temperature	362°K	d
Crystal lattice	diamond str., A^4 type	b
Thermal expansion at 25°	6.1×10^{-6}	d
Lattice constant		
at −253°	5.63 Å	
at 20°	5.6576 Å	b
at 840°	5.681 Å	
Interatomic distance	2.445 Å	b
Lattice energy	42 kcal/g-atom	b
Density D_4^{25}	5.323 g/cc	d
Atomic radius	1.26 Å	e
Ionic radii	Ge^{+} 0.73 Å	c
	Ge^{4+} 0.53 Å	
Atomic volume at 20°	13.57	b
Valency	1, 2 and 4	
Coordination nos.	12, 8, 4	
Electronegativity	2.01 Pauling	a
	2.02 electrostatic	a
Hardness		
Brinell	190 kg/mm²	b
Moh scale	6	
Bulk rigidity modulus	1.3×10^{12} dynes/cm²	d
Shear strength τ	1400 at 10,000 kg/cm²	b
	5700 at 50,000 kg/cm²	
Yield point		
at 523°	1.1 kg/mm²	b
at 614°	0.052 kg/mm²	
Elastic constants		
c_{11}	12.92×10^{11} dynes/cm²	b
c_{12}	4.79×10^{11} dynes/cm²	
c_{44}	6.70×10^{11} dynes/cm²	

TABLE 3—*Continued*

Property	Constant	Ref.
Elastic modulus E	8000 kg/mm^2	
Young's modulus Y_{111}	1.558	d
Torsion modulus F	3000 kg/mm^2	b
Poisson's number μ	0.32	
Compressibility $-\Delta v/v_o$		
at 30°	$14.11\times10^{-7}p$–$6.09\times10^{-12}p^2$	b
at 75°	$14.39\times10^{-7}p$–$6.96\times10^{-12}p^2$	b
Surface tension at m.p.	600 dynes/cm	b
Parachor	40.1	b
Velocity of sound at 20°	5.4×10^5 cm/sec	b
Optical reflectance		
at 8.7μ	35.4%	b
at 20μ	33.8%	
at 41μ	40.3%	
at 117μ	79%	
at 152μ	46%	
Refractive index, n_o	3.994	d
Refractive index at 2.0 microns	4.125	d
Refractive index, temperature dependence at 2.25 microns (dn/dT)	5.25×10^{-4}	d
Extinction coefficient	1.0 at 0.8μ	b
	0.1 at 1.2μ	
	10^{-3} at 1.6μ	
	10^{-6} at 2.0μ	
	10^{-7} at 2.2μ	
Magnetic properties:		
Specific susceptibility, $\times10^6$	−0.122 at 20°C	d
	−0.146 at −183°C	
Diamagnetic susceptibility, $\times10^6$	(Ge^{2+}) −16.8	
	(Ge^{4+}) −10	
	(Ge^{4-}) −100	
Electrochemical potentials:		
$Ge = Ge^{2+}+2e^-$	$E° = 0.0$ V	d
$Ge+5OH^- = HGeO_3^- +2H_2O+4e^-$	$E° = -1.0$ V	
$Ge+2H_2O = GeO_2+4H^+ +4e^-$	$E° = -0.15$ V	
$Ge+3H_2O = H_2GeO_3+4H^++4e^-$	$E° = 0.131$ V	
Electrical properties:		
Band gap at 0°K, eV	0.75	d
Intrinsic resistivity at 27°C, ohm-cm^2	47	
Dielectric constant, E	15.7	
Electron mobility drift, cm^2/−V/sec	3800 (3900)	
Hole mobility drift, cm^2/−V/sec	1800 (1900)	
Density of intrinsic electrons, n_i, at 300°K, per cc	2.4×10^{13}	
n_p (where k = Boltzmann's constant)	$3.1\times10^{32}T^3\exp(-0.785/kT)$	
Arc spectrum lines	2709, 2754, 2691, 2651.5, 2592, 2651, 3269 and 3039 Å	b
Polarographic waves	−1.45 and −1.70 V	b

[a] F. A. Cotton and G. Wilkinson, *Advanced Inorganic Chemistry*, 2nd ed., p. 103, Interscience, New York (1966).

[b] *Gmelins Handbuch der anorganischen Chemie*, 8th ed., System No. 45, supplementary volume, Verlag Chemie, Weinheim (1958).

[c] L. Pauling, *Z. Krist.* **67** (1928) 377.

[d] E. G. Rochow, *Ind. Eng. Chem.* **55** (1963) 32.

[e] R. T. Sanderson, *J. Chem. Phys.* **20** (1952) 535.

electrode/*p*-type/*n*-type/negative electrode). Such Ge diodes are cool, small, non-microphonic, rapid-operating devices which are very useful for radio demodulation and for switching circuits. The introduction of a *second* junction, either as a *pnp* or an *npn* arrangement, allows the center material to control much larger battery-induced currents between the two end materials, and so the arrangement (called a transistor) functions as an amplifier. Since the thickness of center material can be made very small (by diffusion techniques), the transit time for electrons can be correspondingly short, and the device will function at high

TABLE 4. IMPURITY ENERGY LEVELS IN GERMANIUM

Element as impurity	Electron donor (D) or acceptor (A)	Ionization energy gap, conductance band	eV, from valence band
B	A		0.0104
Al	A		0.0102
Ga	A		0.0108
In	A		0.0112
P	D	0.0120	
As	D	0.01257	
Sb	D	0.0097	
Li	D	0.0093	
Zn	A		0.031[a]
Ni	A		0.20
Co	A	0.3	0.25
Fe	A		0.25[a]
Cu	A–1		0.037[a]
	A–2		0.285–0.315
Au	A–1		0.15[a]
	A–2	0.20	
Pt	A–1		0.04
	A–2	0.2	

[a] Other (higher) values listed also.
Source: Gmelin, ref. 5, pp. 148, 149.

frequencies. Since it needs no cathode heater current, operates cold, and is small and mechanically nearly indestructible, the transistor has been welcomed into the electronics field and has been responsible for many of the new developments within that field.

Details of the effect on electrical conductivity brought about by controlled impurities in germanium, and of the dependence of conductivity on external factors, fill 330 pages of the supplementary volume of Gmelin (ref. 5, pp. 124–454) just in summary. Few of these facts are of interest to chemists, but the selected values for energy gaps (ionization energy in eV from nearest conductivity and valence bands) given in Table 4 may be useful.

1.3. CHEMICAL BEHAVIOR OF THE ELEMENT

Germanium does not absorb nitrogen, but adsorbs H_2, CO and CO_2, H_2O, alcohols, and oxygen- and halogen-containing compounds generally. When H_2O is adsorbed and the Ge subsequently is heated, H_2 is liberated and a film of oxide forms on the surface. Atomic hydrogen attacks Ge to form hydrides. Oxygen is adsorbed, but reaction with the underlying metalloid begins only at 575°; the rate of oxidation in terms of the quantity Q of O_2 taken up in time t follows the relation $Q = Q_\infty (1 - e^{-kt})$, where Q_∞ is the ultimate quantity

TABLE 5. REPRESENTATIVE INTERMETALLIC COMPOUNDS OF GE

Composition	Structure	Space group	Parameters			Measured density
			a	*b*	*c*	
LiGe	tetrag.	—	0.438	—	0.580	—
NaGe	monocl.	C^5_{2h}	1.233	0.670	1.142	3.09
KGe	cubic	—	1.278	—	—	2.78
RbGe	cubic	—	1.319	—	—	3.63
CsGe	cubic	—	1.367	—	—	4.28
ϵ-Cu_3Ge	rhomb.	D^4_{6h}	0.265	0.455	0.420	14.00
Mg_2Ge	cubic	O^5_h	0.639	—	—	3.09
$Ca_{33}Ge$	cubic	O^5_h	1.019	—	—	2.20
Ca_2Ge	rhomb.	D^{16}_{2h}	0.907	0.773	0.483	—
CaGe	rhomb.	D^{17}_{1h}	0.400	0.458	1.084	—
BaGe	rhomb.	D^{17}_{2h}	0.507	1.198	0.430	—
$BaGe_2$	cubic	O^7	1.452	—	—	4.28
Sc_5Ge_3	hexag.	D^3_{6h}	0.794	—	0.588	2.76
Y_5Ge_3	hexag.	D^3_{6h}	0.847	—	0.635	5.40
La_5Ge_3	hexag.	D^3_{6h}	0.896	—	0.680	—
Ce_5Ge_3	hexag.	D^3_{6h}	0.888	—	0.657	—
Pr_5Ge_3	hexag.	D^3_{6h}	0.880	—	0.660	—
Ni_5Ge_3	hexag.	D^3_{6h}	0.876	—	0.657	—
Sm_5Ge_3	hexag.	D^3_{6h}	0.865	—	0.649	—
Gd_5Ge_3	hexag.	D^3_{6h}	0.855	—	0.641	—
Tu_5Ge_3	hexag.	D^3_{6h}	0.831	—	0.623	—
Lu_5Ge_3	hexag.	D^3_{6h}	0.824	—	0.617	—
TiGe	rhomb.	C^1_{2v}	0.381	0.523	0.683	5.86
ZrGe	rhomb.	D^{16}_{2h}	0.708	0.390	0.540	—
$HfGe_2$	rhomb.	D^{17}_{2h}	0.382	1.500	0.378	—
$V_{11}Ge_8$	rhomb.	—	1.341	1.609	0.502	6.89
$NbGe_2$	hexag.	D^4_6	0.496	—	0.677	—
αTa_5Ge_3	tetrag.	D^{18}_{4h}	0.660	—	1.201	—
$TaGe_2$	hexag.	D^4_6	0.495	—	0.674	—
Cr_3Ge	cubic	O^3_h	0.461	—	—	7.28
Mo_3Ge	cubic	O^3_h	0.493	—	—	9.70
Mn_5Ge_3	hexag.	D^3_{6h}	0.7184	—	0.505	—
$Mn_{11}Ge_8$	rhomb.	—	1.322	1.583	0.509	7.38
Fe_3Ge	cubic	O^1_h	0.357	—	—	—
$FeGe_2$	tetr.	D^{18}_{4h}	0.590	—	0.494	7.70
$CoGe_2$	tetr.	D^{17}_{2v}	0.568	0.568	1.081	—
NiGe	rhomb.	D^{16}_{2h}	0.581	0.538	0.343	—
RuGe	cubic	T^4	0.485	—	—	—
RhGe	rhomb.	D^{16}_{2h}	0.570	0.648	0.325	9.70
PdGe	rhomb.	D^{16}_{2h}	0.6258	0.578	0.348	—
$OsGe_2$	tetr.	C^3_{2h}	0.899	0.309	0.769	11.10
IrGe	rhomb.	D^{16}_{2h}	0.628	0.561	0.349	—
PtGe	rhomb.	D^{16}_{2h}	0.609	0.573	0.371	—
$PtGe_2$	rhomb.	D^{12}_{2h}	0.619	0.577	0.291	—
Th_3Ge_2	tetr.	D^5_{4h}	0.797	—	0.417	10.48
U_5Ge_3	hexag.	D^3_{6h}	0.858	—	0.579	13.40
Pu_2Ge_3	hexag.	D^1_{6h}	0.3975	—	0.420	10.60
$GeAs_2$	rhomb.	D^9_{2h}	0.372	1.01	1.474	—
$GeSe_2$	rhomb.	—	1.299	0.694	2.213	4.56
Ge_2Te_3	hexag.	D^3_{3d}	0.432	—	0.530	—

of O_2 consumed after a very long time, and the rate constant k has a value of 0.034 min^{-1} at 615° and 0.87 min^{-1} at 703° [11].

Methyl and ethyl chlorides and bromides (and undoubtedly many other alkyl and some aryl halides) in the vapor phase attack heated germanium to form the corresponding organogermanium halides[12]. The powdered element burns in Cl_2 and Br_2 when warmed, forming the tetrahalide. Dry hydrogen chloride attacks the element at elevated temperatures to form $GeCl_4$ and impure $GeHCl_3$:

$$Ge+4HCl = GeCl_4+2H_2$$
$$Ge+3HCl = GeHCl_3+H_2$$

(pure $GeHCl_3$ is best made by the union of $GeCl_2$ and HCl; see Dennis[9]). Sulfur reacts with Ge only at red heat, but H_2S reacts at 400° to form GeS_2.

Water slowly dissolves a thin evaporated film of Ge on Cu if oxygen is present, presumably because of the appreciable solubility of GeO_2. A 3% aq. solution of H_2O_2 dissolves massive Ge slowly at 20°, but quite rapidly at 90°–100°. Concentrated sulfuric acid at 90° has a slight action on massive Ge (1% loss of weight in 1 week), and concentrated nitric acid attacks it only superficially. Aqueous solutions of NaOH and KOH have little effect, but fused NaOH, KOH, Na_2CO_3, Na_2O_2 and $Na_2B_4O_7$ dissolve all forms of Ge quickly, forming alkali germanates. A solution of NaOCl also dissolves powdered Ge, forming GeO_2. Solutions of amides in liq. NH_3 do not attack it.

Germanium appears to form no carbide, but it alloys with many metals and metalloids. A complete tabulation of intermetallic compounds (as of 1968)[13] lists 152 compositions, from which the representative metal germanides of Table 5 have been abstracted. The relation of Ge with Si is unique: a continuous series of solid solutions is formed, and the two liquids also form ideal mixtures. About eighty ternary intermetallic compounds are known, and phase diagrams for many of the binary and ternary systems have been reported[13].

1.4. BIOLOGICAL ACTIVITY

Germanium is a ubiquitous component of living organisms, but it has no structural function (as silicon has in some exoskeletons), nor is it a proven required trace element for plants or animals[14]. Sax[15] reports that germanium has a low order of toxicity. Its compounds in general are much less poisonous than those of lead and tin, but GeH_4 has a hemolytic effect and is dangerous at levels above 100 ppm. Germanium halides hydrolyze to hydrohalogen acids, of course, and can be biologically unpleasant for this reason. The other product of hydrolysis, GeO_2, becomes toxic only at the level of 300 to 600 mg per kilo of body weight, and some tolerance to it can apparently be acquired by repeated sublethal doses[15, 16].

11 R. B. Bernstein and D. Cubicciotti, *J. Am. Chem. Soc.* **73** (1951) 4112.

12 E. G. Rochow, *J. Am. Chem. Soc.* **69** (1947) 1729; *ibid.* (1948) 437; *ibid.* **72** (1950) 198.

13 G. V. Samsonov and V. N. Bondarev, *The Germanides* (in Russian), publ. by Metallurgy, Moscow (1968).

14 H. A. Schroeder and J. J. Balassa, Abnormal trace metals in man: germanium, in *J. Chron. Dis.* **20** (1967) 211–214.

15 N. I. Sax, *Dangerous Properties of Industrial Materials*, p. 736, Reinhold Publ. Corp., N.Y. (1962).

16 See summary of biological activity given by F. Rijkens in *Organogermanium Compounds: A Survey of the Literature from Jan. 1950 to July 1960* (Institute for Organic Chemistry of the T.N.O., Utrecht, Holland, 1960), and the subsequent publications of the Germanium Research Committee through that Institute.

Although not markedly toxic, germanium is of some interest in two widely separated areas of biology and physiology: its erythropoietic effect, and the possible bacteriocidal or fungicidal effect of organogermanium compounds. Stimulation of the formation of red blood cells (erythrocytes) by injections of GeO_2 was first reported in 1922[17], and caused a flurry of activity aimed toward the treatment of anemias. The effect seems to vary among individuals and with the kind of anemia being treated, and the results sometimes are temporary. In all cases the effect is less pronounced than with repeated small doses of arsenic trioxide, but of course the danger of poisoning also is less. Acting on the view that the low solubility of GeO_2 in water might be the reason for the marginal and inconsistent results, much higher levels of germanium in the form of the highly soluble dimethylgermanium oxide were adminstered to hamsters, and while these demonstrated the non-toxic behavior of $(CH_3)_2GeO$, the effect on erythrocyte levels was inconclusive[18].

The markedly poisonous nature of organolead compounds and the known bacteriocidal and fungicidal properties of some trialkyltin compounds have led to study of the effect of organogermanium compounds on microorganisms, but no utility has been found[16]. The marginal toxicity of most germanium compounds leads to the recurrent hope that some rather particular organogermanium structures will be found to be effective against invading microorganisms, or may restore balance to a pathogenic condition, without being dangerously toxic to a human host. So far, however, any such chemotherapeutic activity of germanium remains in the realm of hopeful speculation. On the average, each one of us ingests 1500 μg of germanium daily in our food[14], so we already have in our tissues an appreciable level of this element for which no function or decisive effect is known. Therefore any possible chemotherapeutic application would very likely require some specificity of structure (presumably in the direction of organogermanium chemistry), or some markedly higher local level of concentration of the element by methods as yet unknown.

1.5. ANALYTICAL DETERMINATION

Germanium can be detected by the blue line in its arc spectrum[19] at 4686 Å, by the band emission spectrum of its oxide in the visible range, and by the pale blue color of a flame into which a solution containing at least 10 mg of Ge per ml is sprayed. The decomposition of GeH_4 in a heated glass tube (as in the Marsh test for As) has also been used, but the GeH_4 must be generated by electrolysis of a dilute solution of NaOH containing the sample[20]. In solution, the presence of Ge can be determined by the precipitation of white GeS_2 from 12 N H_2SO_4 by addition of H_2S; the only other white sulfide is ZnS, and this precipitates only in alkaline solution. The presence of Ge in solution can also be recognized by the formation of a lemon-yellow Ge heteropolyacid when a 5% solution of ammonium molybdate is added in 0.15 to 0.34 N HNO_3. Various organic reagents (such as 9-phenyl-2,3,7-trioxy-6-fluoron) can also be used for the colorimetric estimation of Ge[21].

The quantitative determination of Ge can be accomplished gravimetrically by precipitation of GeS_2 from 6 N H_2SO_4, followed by thorough washing of the sulfide and then

[17] F. S. Hammett, J. E. Nowrey and J. H. Müller, *J. Exp. Medicine* **35** (1922) 173 and 507; J. H. Müller and M. S. Iszard, *Am. J. Med. Sci.* **163** (1922) 364; J. H. Müller, *J. Metab. Res.* **3** (1923) 181.

[18] E. G. Rochow and B. M. Sindler, *J. Am. Chem. Soc.* **72** (1950) 1218.

[19] J. Papish, F. M. Brewer and D. A. Holt, *J. Am. Chem. Soc.* **49** (1927) 3028; J. Papish, *Econ. Geol.* **23** (1928) 660 and **24** (1929) 470.

[20] S. A. Coase, *Analyst*, **59** (1934) 462.

[21] See *Gmelins Handbuch*, System No. 45, suppl. vol., pp. 460–463, Verlag Chemie, Weinheim (1958).

oxidation to GeO_2 by repeated treatment with 30% H_2O_2 before ignition and weighing[22]. Magnesium germanate, barium germanium tartrate and various forms of complexed germanium molybdates may also be precipitated as weighable forms of Ge[22]. Titration of Ge complexes of polyhydric alcohols such as mannite and glucose with phenolphthalein indicator also is possible. Various colorimetric methods, based mostly on molybdate complexes, have been used[22]. In mineral and nonvolatile inorganic samples, determination of Ge is best accomplished by arc emission spectroscopy, making use of the lines at 3039.1, 2754.6, 2709.6, 2651.2 and 2592.6 Å. By use of internal standards and previous calibration, determination down to the level of 0.0005% Ge is possible[19, 23]. The polarographic determination of Ge^{4+} at a mercury cathode in alkaline solution (1 M NH_4Cl+0.5 M NH_3; $E_c = -1.45$ and -1.70 V) or in acid solution (with the aid of NaH_2PO_2 as reducing agent) also is possible[24].

2. COMPOUNDS OF Ge(I)

2.1. MONOHYDRIDE, $(GeH)_x$

The action of moist air on alkali germanides, or better, the addition of NaGe to cold water[25], or the action of NH_4Br on NaGe in liq. NH_3[25], or the slow action of metallic Na on $GeHCl_3$ vapor at 60°[26], gives a brown amorphous solid of composition GeH which is insoluble in organic and inorganic solvents but is rapidly oxidized to GeO_2 by HNO_3 or H_2O_2. When dried in air it oxidizes with an explosive puff as soon as it is dry. In vacuum it evolves hydrogen slowly at 100° and decomposes explosively at 160° to form GeH_4 and finely divided Ge.

2.2. MONOHALIDES

Germanium monofluoride, GeF, is known only from the band emission spectrum from a discharge in GeF_4. The bond force constant of the molecule is calculated to be 3.92×10^5 dynes/cm[27].

Monomeric GeCl also is known from its band spectrum; the force constant is calculated to be 2.2×10^5 dynes/cm. A solid "subchloride", $(GeCl)_x$, is a high polymer of unknown structure made by passing $GeCl_4$ and H_2 through a hot tube at 1000°, or by reduction of an aq. solution of $GeCl_4$ with H_3PO_2. It is a dark brown solid which is stable up to 360° in vacuum but decomposes at 500° to give $GeCl_4$ and Ge; it is insoluble in common organic solvents, slightly soluble in H_2O (and slowly hydrolyzed), and oxidized rapidly by hot HNO_3. Boiling aq. KOH dissolves it very slowly:

$$2GeCl + 6KOH = 2K_2GeO_3 + 2KCl + 3H_2$$

The monobromide GeBr is known from its band spectrum, and the force constant is

22 See Gmelin, *loc. cit.*, pp. 465–467.

23 Wa. Gerloch and We. Gerlach, *Die chemische Emissionsspektralanalyse*, Part 2, pp. 63 and 155, Leipzig (1933).

24 I. M. Kolthoff and J. J. Lingane, *Polarography*, Vol. 2, p. 522, Interscience, N.Y. (1952).

25 L. M. Dennis and N. A. Skow, *J. Am. Chem. Soc.* **52** (1930) 2369; C. A. Kraus and E. S. Carney, *ibid.* **56** (1934) 765.

26 L. M. Dennis, W. R. Orndorff and D. L. Tabern, *J. Phys. Chem.* **30** (1926) 1050.

27 E. B. Andrews and R. F. Barrow, *Proc. Phys. Soc.* **A63** (1950) 185.

2.0×10^5 dynes/cm. No GeI is known, even from a band spectrum, but the force constant for a hypothetical Ge–I bond is calculated to be 1.46×10^5 dynes/cm. No oxide, nitride or sulfide of Ge in the +1 oxidation state is known.

3. COMPOUNDS OF Ge(II)

3.1. DIHYDRIDE, GeH_2 ("POLYGERMENE")

The action of conc. HCl on a mixture of powdered NaCl and CaGe (with exclusion of oxygen) produces a yellow solid polymer of the composition $(GeH_2)_x$. Hydrolysis of Mg_2Ge in equal volumes of alcohol and conc. HCl gives a less pure product; small amounts can also be made by the thermal decomposition of Ge_2H_6. The substance is readily oxidized by bromine to $GeBr_4$, and by conc. HNO_3 or H_2O_2 (in alcohol plus aq. NH_3) to GeO_2. Conc. aq. alkalies convert it (by complex reactions) to GeH_4 and alkali germanites. At 200° (in the absence of air) it decomposes slowly to H_2 and Ge, and rapidly at 300°. A sharp blow on the dry solid can cause explosive decomposition to Ge, with burning of the liberated hydrogen. Treatment with hot aq. HCl yields H_2, Ge, GeH_4, Ge_2H_6 and Ge_3H_8.

3.2. GERMANIUM DIFLUORIDE, GeF_2

Germanium is much more easily reduced to the +2 state than silicon, and the tetrahalides GeX_4 react with elementary Ge to form dihalides GeX_2. The action of GeF_4 on powdered Ge produces a white solid for which no physical constants are available, but the chemical properties show it to be quite certainly GeF_2 (see section on GeF_4). Better examples are found among the higher dihalides.

3.3. GERMANIUM DICHLORIDE, $GeCl_2$

The slow passage of $GeCl_4$ vapor (at low pressure, *ca.* 0.1 torr) over powdered Ge at only 300° produces a light-yellow solid $GeCl_2$, which condenses in the cooler parts of the tube[28]. The same product is obtained from the thermal decomposition of $GeHCl_3$ (which, however. is usually made by the combination of $GeCl_2$ and HCl):

$$GeHCl_3 \underset{20^\circ}{\overset{70^\circ}{\rightleftharpoons}} GeCl_2 + HCl$$

Excess H_2 and $GeCl_4$ vapor passed through a hot silica tube at 800° also produce some $GeCl_2$.

At room temperature $GeCl_2$ is stable but easily oxidized. It adds Cl_2 and Br_2 to form tetrahalides (although I_2 is likely to form GeI_2), and it adds HCl readily to form $GeHCl_3$, but strangely it does not seem to add alkyl halides to form organogermanium halides. It hydrolyzes to form yellow $Ge(OH)_2$ (which dehydrates to brown GeO) and HCl (which may react with excess $GeCl_2$). It disproportionates in vacuum at 1000° or more to $GeCl_4$ and elementary germanium.

[28] L. M. Dennis and H. L. Hunter, *J. Am. Chem. Soc.* **51** (1929) 1151; F. M. Brewer, *ibid.* **31** (1927) 1817.

3.4. GERMANIUM DIBROMIDE, $GeBr_2$

The reduction of $GeBr_4$ and $GeHBr_3$ with zinc[28], or the action of a deficiency of HBr on Ge at 400°, produces $GeBr_2$, a yellow solid which melts at 122° and is slightly soluble in 5 N aq. HBr. Such a solution forms white crystalline coordination compounds with $(CH_3)_4NBr$, $(CH_3)_4AsBr$ and $(CH_3)_3C_2H_5AsBr$[29]. Pure $GeBr_2$ disproportionates to $GeBr_4$ and Ge at 150°, and hydrolyzes to $Ge(OH)_2$. It reacts readily with dry HBr at 40° to form $GeHBr_3$[28].

3.5. GERMANIUM DIIODIDE, GeI_2

Low-temperature methods are preferred for the preparation of GeI_2. Treatment of freshly precipitated $Ge(OH)_2$ with hot conc. HI, followed by reduction of the liberated iodine with H_3PO_2 and then cooling of the solution, causes precipitation of flaky crystals of GeI_2. Similarly, powdered GeS may be extracted with warm conc. HI, the excess GeS filtered off, the solution cooled and the crystals of GeI_2 recovered. However, the simplest preparative method consists in the reduction of GeI_4 with a slight excess of H_3PO_2: 20 g of GeI_4 is dissolved in 10 ml of 57% HI+20 ml H_2O and 7.6 ml of 50% H_3PO_2 added with stirring. The mixture is refluxed, the red crystals of GeI_4 slowly being converted to bright-yellow GeI_2. The mixture is cooled to 10°, the GeI_2 filtered and washed with dil. HI, and dried in vacuum (yield 75%)[30].

Pure sublimed GeI_2 is an orange solid which crystallizes in the hexagonal system: $a = 4.249$, $c = 6.833$ Å, Ge–I distance = 2.99 Å[31]. At 544° to 643° it disproportionates

$$2GeI_2(s) \rightleftharpoons Ge(s) + GeI_4(g)$$

with a heat of sublimation $\Delta H° = 30.1$ kcal/mole, a free energy change $\Delta G° = 17.9$ kcal/mole, and an entropy change $\Delta S° = 41.0$ cal/mole deg[31]. The crystals do not oxidize in dry air, but hydrolyze slowly in moisture. They dissolve in liq. NH_3 (with formation of GeNH) and in aq. HI. Iodine in aq. KI+HCl oxidizes GeI_2 to GeI_4:

$$GeI_2 + I_2 \xrightarrow{HCl+KI} GeI_4\text{ (dissolved)} + 19.06\text{ kcal}^{31}$$

In the absence of HI, hydrolysis takes place during oxidation:

$$GeI_2 + I_3^- + 3H_2O = H_2GeO_3 + 4H^+ + 5I^- + 26.0\text{ kcal}^{31}$$

Many addition compounds of GeI_2 with amines and other nitrogenous bases are known[32].

3.6. GERMANIUM MONOXIDE, GeO

When GeO_2 is reduced to metal with carbon (q.v.) there is troublesome loss of volatile brown GeO, and the same can be said for any high-temperature reduction of Ge(IV) compounds with oxygen present. A mixture of Ge and GeO_2 powders heated to 1000° gives a yellow sublimate of amorphous GeO, and further heating at 650° gives dark-brown

[29] T. Karantassis and L. Capatos, *Compt. rend.* **199** (1934) 64.
[30] L. S. Foster, *Inorg. Syntheses*, Vol. 3, p. 63, McGraw-Hill, N.Y. (1950).
[31] W. L. Jolly and W. M. Latimer, *J. Am. Chem. Soc.* **74** (1952) 5752.
[32] T. Karantassis and L. Capatos, *Compt. rend.* **201** (1935) 74; D. A. Everest, *J. Chem. Soc.* **1952,** 1670.

crystalline GeO. Hydrolysis of germanium dihalides and GeS gives $Ge(OH)_2$, which dehydrates readily to GeO.

Crystalline GeO has a density of 1.83, a heat of formation (at 25°) of 6.0 kcal/mole, and a vapor pressure of 1.80 torr at 915°, 9.9 at 948° and 28.5 at 978°. The vapor-pressure equation is $4.57 \log p = -63{,}000/T + 57.9$[33], and the entropy content is $S_{298} = 56.1$ cal/mol deg. The molar magnetic susceptibility is -28.6×10^{-6}. Chemically, GeO disproportionates rapidly at 700° to $Ge + GeO_2$; the heat of this reaction is 54.8 kcal/mole and the entropy change is 42.0 cal/mol deg[33]. It is oxidized by a variety of reagents, among them silver perchlorate[33]:

$$GeO + 2AgClO_4 + 6HF = 2Ag + 2HClO_4 + H_2O + H_2GeF_6 + 66 \text{ kcal}$$

3.7. GERMANIUM(II) HYDROXIDE, $Ge(OH)_2$

Besides its derivation from other Ge(II) compounds, $Ge(OH)_2$ may also be prepared by the direct reduction of an acid solution of GeO_2. As done originally[34], GeO_2 is dissolved in aq. HCl (6 N) and reduced with H_3PO_2, whereupon yellow or red $Ge(OH)_2$ is precipitated by the addition of aq. NH_3. The product is washed with O_2-free water and dried. By a more recent procedure[31], 6 g of GeO_2 is dissolved in 30 ml of conc. NaOH, precipitated and redissolved by adding sufficient 6 N HCl, and then reduced with 45 ml of 50% H_3PO_2 at 100° over 5 to 6 hr. Neutralization of the cooled solution with NH_3 precipitates yellow $Ge(OH)_2$, which can be filtered under N_2. The color appears to be due to the presence of some GeO, because the material slowly dehydrates and darkens upon standing, and eventually becomes dark-brown GeO. The hydroxide is very easily oxidized to GeO_2; the standard potential for the reaction

$$Ge(OH)_2 + H_2O = GeO_2 + 2H^+ + 2e^-$$

is -0.118 V[31].

Because of the change in composition, the chemical properties of $Ge(OH)_2$ change with the age and color of the material. Freshly precipitated material dissolves readily in conc. HCl. The solubility of aged material in 4 N HCl is only 0.01 mole/l., but at high concentration of HCl the solubility increases markedly. Dilute aq. NaOH does not dissolve detectable amounts of $Ge(OH)_2$ if the base is less than 1 N, but 50% NaOH dissolves it by oxidizing it to sodium germanate, with evolution of hydrogen:

$$Ge(OH)_2 + 2NaOH = Na_2GeO_3 + H_2O + H_2$$

Latimer[35] gives the potential for the germanite–germanate oxidation in basic solution as -1.4 V.

3.8. GERMANIUM MONOSULFIDE, GeS

Although GeS was first made by reducing GeS_2 with Ge (Winkler) or with H_2 at 480° (Dennis), the simplest way to prepare it now is to precipitate GeS_2 (as by the action of H_2S

[33] W. Bues, H. v. Wartenburg, *Z. anorg. Chem.* **266** (1951) 281.

[34] L. M. Dennis and R. E. Hulse, *J. Am. Chem. Soc.* **52** (1930) 3553.

[35] W. M. Latimer, *Oxidation Potentials*, p. 136, Prentice-Hall, N.Y. (1938). The sign has been reversed in the sentence above, in keeping with present practice. For later information on Ge(II) compounds, see 2nd ed. of Latimer (1952), pp. 145–148.

on a solution of $GeCl_4$ in 6 N HCl) and reduce the fresh precipitate with excess H_3PO_2 by stirring on a steam bath until a clear, colorless solution results. The cooled solution is then saturated with H_2S and neutralized with aq. NH_3, with H_2S still flowing. The resulting GeS is filtered under N_2 and dried in N_2 over P_2O_5[36]. The resulting amorphous red-brown powder may be crystallized by sublimation in vacuum at 600°; the crystals are orthorhombic bipyramidal, with $a = 4.29$, $b = 10.42$ and $c = 364$ Å (space group D_{2h}^{16}). Each Ge atom is surrounded by six S atoms in a deformed octahedral structure, with Ge–S distances of 2.47, 2.64, 2.91 and 3.00 Å. The closest S–S approach is 3.55 Å[37]. The calculated density is 4.24 g/cc, and the molar susceptibility is -4.9×10^{-6}. The crystals appear black, but transmitted light through thin sections is red, and reflected light steel gray[37]. The powdered substance hydrolyzes slowly in moist air and rapidly in liq. H_2O to form $Ge(OH)_2$ and then GeO. It dissolves slightly in aq. NaOH, but apparently only if some GeS_2 is present[36]. The solubility product in liq. NH_3 is 9×10^{-6}, and no ammonolysis takes place. An excess of Na in liq. NH_3 dissolves GeS to form a reddish solution of sodium polygermanide, and addition of NH_4Br to this causes evolution of GeH_4[37].

3.9. GERMANIUM(II) SELENIDE AND TELLURIDE

When H_2Se is passed into an aq. solution of $GeCl_2$, a dark-brown precipitate of GeSe forms. This substance is insoluble in 5% aq. HCl, alcohol and ether; it oxidizes readily in air, but is stable in CO_2 and can even be prepared from the elements in a CO_2 atmosphere at 500°. It crystallizes in the tetragonal system with $a = 8.83$ and $c = 9.76$ Å. The measured density is 5.30, the calculated 5.266 g/cc. The melting point is about 667°. It dissolves in a solution of Br_2 in HCl, and H_2O_2 in alkaline solution oxidizes it to GeO_2 and Se. Aq. HNO_3 oxidizes it to GeO_2 and H_2SeO_3[38].

The system Ge–Te shows a eutectic at 375° and an incongruent melting point for the compound GeTe at 725°. The bond force constant is 2.9×10^5 dynes/cm, and the bond distance is 2.85 Å (calcd. from at. radii, 2.39 Å). The crystal structure is cubic, with $a = 5.99$ at 460° and 5.98 at 390° (like Ge and Te structures). At room temperature the structure is more complicated and probably is non-stoichiometric. It is metallic in appearance, with density 6.20. It reacts with conc. HCl, H_2SO_4 and H_2O when heated, but aqua regia attacks it in the cold, as does a mixture of H_2O_2 and HCl[38].

3.10. GERMANIUM(II) IMIDE AND NITRIDE

In liquid NH_3, GeI_2 is completely ammonolyzed to the imide GeNH:

$$GeI_2 + 3NH_3 = GeNH + 2NH_4I$$

The NH_4I can be washed from the product by repeated extraction with liq. NH_3, leaving a canary-yellow powder[39] which oxidizes rapidly in air to GeO_2 and hydrolyzes in oxygen-free water to $Ge(OH)_2$ and NH_3. When heated to 300° in vacuum, the imide is converted to a dark-brown nitride:

$$3GeNH = Ge_3N_2 + NH_3$$

---

[36] L. S. Foster, *Inorganic Syntheses*, Vol. 2, p. 102, McGraw-Hill, N.Y. (1946); D. A. Everest and H. Terry, *J. Chem. Soc.* **1950**, 2282.

[37] Gmelin, *loc. cit.*, pp. 536–539.

[38] Gmelin, *loc. cit.*, pp. 541–544.

[39] L. S. Foster, *Inorganic Syntheses*, Vol. 3, p. 63, McGraw-Hill, N.Y. (1950).

This nitride decomposes into the elements at 500°, but when heated in H_2 at 500° to 600° it forms Ge and NH_3[40]. The α form of Ge_3N_4 is hexagonal close-packed, $a = 8.202$ and $c = 5.941$ Å; the β form has the phenacite structure[40], with $a = 8.038$ and $c = 3.074$ Å.

3.11. GERMANIUM(II) COMPOUNDS OF PHOSPHORUS AND ARSENIC

By melting Ge and white P_4 in an evacuated glass tube, impure GeP is obtained; the heat of formation is only 6 kcal/mole. At 540° the substance decomposes according to the equation

$$4GeP = 4Ge + P_4 - 37 \text{ kcal}$$

There also are some hypophosphite–halide double salts which can be obtained by evaporating solutions of Ge(II) salts containing excess H_3PO_2 left from the reduction procedures described above. In particular $Ge(H_2PO_2)_2 \cdot GeCl_2$, $3Ge(H_2PO_2)_2 \cdot GeBr_2$ (m.p. 129°) and $3Ge(H_2PO_2)_2 \cdot GeI_2$ (m.p. 120°) have been crystallized[41]. These compounds are stable in dry air at 20°, but hydrolyze to $Ge(OH)_2$. Warm conc. H_2SO_4 and HNO_3 oxidize them rapidly to GeO_2.

If the reduction of GeO_2 with limited H_3PO_2 is carried out in hot conc. H_3PO_3 solution, in the absence of halide ions, the cooled solution yields crystals of yellow-green $GeHPO_3$. This substance begins to decompose at 230° without melting, and becomes black at 300°. It is soluble in warm aq. HCl, HBr and H_3PO_4, but insoluble in cold dil. H_2SO_4, CH_3COOH, HCOOH and organic solvents. Conc. HNO_3 oxidizes it to GeO_2, and alkaline solutions convert it to hydrous GeO[41].

The phase diagram for the system Ge–As shows the compounds $GeAs_2$ (m.p. 732°) and GeAs (m.p. 737°)[42]. The α-phase is made up of mixed crystals of As and $GeAs_2$ and is hexagonal in structure with $a = 3.701$ Å, $c = 10.71$ Å and $c/a = 2.894$; pure As crystallizes in a foreshortened hexagonal structure, with $a = 3.754$ Å, $c = 10.52$ Å and $c/a = 2.802$. Both GeAs and $GeAs_2$ show an enhanced diamagnetism; the values for atomic susceptibility at 20° are: pure As -6×10^{-6}, $GeAs_2$ -19×10^{-6}, GeAs -16×10^{-6} and pure Ge -13×10^{-6}. The alloying of As with Ge renders the As more stable to the air, and the compounds mentioned above remained unaltered for months[42].

3.12. ORGANIC DERIVATIVES OF Ge(II)

The action of C_6H_5Li on GeI_2 in organic solvents gives an orange-brown solution from which polymers of the composition $(C_6H_5)_2Ge$ have been isolated[43]. Gilman and Gerow[44] studied the reaction in detail, and noted that a complex of the composition $(C_6H_5)_3GeLi$ is first obtained, and this dissociates to products which are difficult to characterize. Bromination of the product, for example, gives $(C_6H_5)_3GeBr$ instead of the expected $(C_6H_5)_2GeBr_2$. Preparation of germanium(II) alkyls and aryls by other routes is difficult because germanium

[40] W. C. Johnson, *J. Am. Chem. Soc.* **52** (1930) 5160; S. N. Ruddlesden and P. Popper, *Acta Cryst.* **11** (1958) 465.

[41] D. A. Everest, *J. Chem. Soc.* **1952**, 1670; **1953**, 4117.

[42] Gmelin, *loc. cit.*, pp. 560 and 561.

[43] O. H. Johnson, *Chem. Revs.* **48** (1951) 259; L. Summers, *Iowa State Col. J. Sci.* **26** (1952) 292.

[44] H. Gilman and C. W. Gerow, *J. Org. Chem.* **23** (1958) 1582.

dihalides reduce mercury alkyls to metallic mercury, and organometallic reagents containing more active metals form complexes with the Ge(II) products[45].

4. COMPOUNDS OF Ge(III)

Germanium in an oxidation state of 3 is rare, but a few compounds of the type R_2Ge–GeR_2 (which some might consider compounds of divalent germanium, rather than trivalent) are known. The first of these is the supposed dimer $(GeH_2)_2$ which was obtained by the reaction of $NaGeH_3$ with bromobenzene in liquid ammonia[46]:

$$2NaGeH_3+2C_6H_5Br \xrightarrow{\text{liq. } NH_3} 2NaBr+2C_6H_6+(GeH_2)_2$$

The $(GeH_2)_2$ so obtained is not like the polygermene obtained by the action of conc. HCl on CaGe (*vide supra*), but is a white solid which disproportionates on warming to room temperature:

$$3(GeH_2)_2 = 2GeH_4+4GeH$$

The same white solid reacts with one equivalent of Na in liq. NH_3 to give a deep red solution supposedly of the composition NaH_2Ge–GeH_2Na (like analogous tin compounds)[46].

A supposed Ge_2Cl_6 has also been obtained, by the thermal decomposition of GeCl at 210° at only 0.05 to 0.1 torr. The white crystalline product dissolves in $GeCl_4$ and C_6H_6, melts at 30°, is oxidized by air and by H_2O_2, and dissolves in warm aq. KOH with evolution of H_2[47]. No Ge_2Br_6 or Ge_2I_6 has been reported.

One possible interpretation of the term "Ge(III)" would include compounds of the type Ge_2H_6 and Ge_2Cl_6. These are definitely of the structure X_3Ge–GeX_3, and so are compounds of tetracovalent germanium. For this reason, Ge_2H_6 is described below along with GeH_4 and the other hydrides of Ge(IV), Ge_2Cl_6 is included with the chlorides, and so on.

5. COMPOUNDS OF Ge(IV)

5.1. GERMANIUM(IV) HYDRIDES

Monogermane, GeH_4

This formerly was made by the action of acids on an alloy of magnesium with germanium[48]. If aqueous acids are used, large quantities of hydrogen are produced and must be separated from the mixture of germanium hydrides; furthermore, the yield of hydrides based on Ge is disappointing. Ammonia-system acids are more satisfactory: the action of NH_4Br on powdered Mg–Ge alloy (3 parts Ge+2 parts Mg fused at 800°) suspended in liq. NH_3 converts 60% to 70% of the Ge to the various hydrides, which are separated by distillation in a vacuum chain. Some GeH_4 also is obtained from the thermal

[45] G. Jacobs, *Compt. rend.* **238** (1954) 1825.

[46] S. N. Glarum and C. A. Kraus, *J. Am. Chem. Soc.* **72** (1950) 5398; C. A. Kraus, *J. Chem. Ed.* **29** (1952) 417.

[47] Gmelin, *loc. cit.*, p. 514.

[48] L. M. Dennis, R. B. Corey and R. W. Moore, *J. Am. Chem. Soc.* **46** (1924) 657; H. J. Emeléus and E. R. Gardner, *J. Chem. Soc.* **1938**, 1900.

decomposition of solid GeH, and also by the electrolysis of a solution of GeO_2 in H_2SO_4. More modern and practical methods consist in the reduction of $GeCl_4$ by $LiAlH_4$ in ether[49] (which requires only the simple separation of GeH_4 from ether vapor), and the still simpler and easier reduction of GeO_2 by $NaBH_4$ in water solution[50a, 50b]. By the last method, GeO_2 is dissolved in 1 M HBr and an excess of $NaBH_4$ solution (5 g in 100 ml H_2O) is added dropwise. The GeH_4 (with about 1% Ge_2H_6) is condensed from the stream of H_2 in a series of five traps cooled to $-196°$; the yield is 98%[50b].

Monogermane is a stable, colorless gas with physical properties listed in Table 6. It reacts with sodium dissolved in liq. NH_3 (with evolution of H_2) to form $NaGeH_3$, the same product obtained by splitting Ge_2H_6 with Na in the same medium. Treatment of the $NaGeH_3$ with NH_4Br in the same solvent then evolves pure GeH_4. Thermal decomposition (to $Ge+2H_2$) is very slow at 280°, but rapid at 375°; the deposited Ge is a catalyst for the further decomposition, so the rate is proportional to the 1/3 power of the pressure of GeH_4. Under appropriate conditions, the dissociation is a first-order reaction[51]. The photochemical decomposition (under Hg-arc irradiation, the GeH_4 being sensitized with Hg vapor) begins with stripping of one H atom and then a stepwise degradation of GeH_3 fragments[52].

In contrast to SiH_4 and SnH_4, GeH_4 does not ignite when it meets air. It can be mixed with pure oxygen, at low pressure, and the oxidation begins only slowly at 320°. In the range 230° to 330° the reaction produces H_2O and a white deposit of GeO_2, but at higher temperatures explosions may result, and brown Ge is deposited as the thermal dissociation of GeH_4 precedes the oxidation of its products.

In sharp distinction to SiH_4 (which is decomposed rapidly by even exceedingly dilute alkali solutions), GeH_4 is unaffected by 30% aq. NaOH, and of course is not hydrolyzed by the 1 M HBr in which it is made. It is a potent reducing agent; it reacts with aq. $AgNO_3$ to evolve hydrogen and precipitate a black mixture of Ag and Ge. Oxidizing agents convert it to GeO_2 and H_2O. It dissolves in liq. NH_3 to form a conducting solution believed to contain NH_4^+ and GeH_3^- ions, and the solution dissolves P_4 to form an ammonium phosphogermanide. Sodium and potassium in liq. NH_3 convert GeH_4 to $NaGeH_3$ and $KGeH_3$, white solids which are unstable at room temperature and decompose to the metal germanide and hydrogen. The substances $MGeH_3$ react in liq. NH_4 with CH_3Cl to form CH_3GeH_3, but the same substances react with CH_2Br_2 in a reductive way, not to yield $H_3GeCH_2GeH_3$[53]:

$$CH_2Br_2+2NaGeH_3+NH_3 \longrightarrow CH_3GeH_3+GeH_3NH_2+2NaBr$$

In the same solvent $NaGeH_3$ reduces aromatic halides to hydrocarbons, rather than forming $RGeH_2$[53]:

$$NaGeH_3+C_6H_5Br \longrightarrow C_6H_6+GeH_2+NaBr$$

---

49 A. E. Finholt, A. C. Bond, K. E. Wilzbach and H. I. Schlesinger, *J. Am. Chem. Soc.* **69** (1947) 2692.
50a T. S. Piper and M. K. Wilson, *J. Inorg. Nucl. Chem.* **4** (1957) 22.
50b J. E. Griffiths, *Inorg. Chem.* **2** (1963) 375.
51 H. J. Emeléus and H. H. G. Jellinek, *Trans. Faraday Soc.* **40** (1944) 93.
52 H. Romyn and W. A. Noyes, *J. Am. Chem. Soc.* **54** (1932) 4143; H. E. Mahncke and W. A. Noyes, *ibid.* **57** (1935) 456.
53 F. G. A. Stone, *Hydrogen Compounds of the Group IV Elements*, chap. 3, Prentice-Hall, N.J. (1962).

TABLE 6. PROPERTIES OF THE Ge(IV) HYDRIDES

Property	GeH_4	Ge_2H_6	Ge_3H_8	Ge_4H_{10}	Ge_5H_{12}
Melting point, °C	−164.8	−109	−105.6		
ΔH fusion, cal/mole	199.7				
Boiling point, °C	−88.1	29	110.5	176.9	234
ΔH vaporization, cal/mole	3360	6400	8000	9650	11,300
Vapor pressure					
at −127°C, torr	53.0				
at −115°	142.5				
at −105°	282.0				
at −95°	523.5				
at −68.1°		6.4			
at −38.9°		38.6			
at −10.2°		152.7			
at 0°		239.0			
at +18.8°		503.5			
at 25.3°			39.9		
at 45.0°			90.9		
at 80.9°			317		
at 99.5°			545		
Density at m.p., g/cc	1.52[c]	1.98	2.20		
Critical temp., °K	308	483	588	665	730
Critical pressure, atm	54.8	45.7	37.9	32.5	28.8
ΔS vap. (Trouton constant)	19.8	20.2	20.8	21.5	22.2
Molar heat capacity, cal. at m.p.	15				
Transition temps. in solid, °C	−210 −199.8 −196.5				
IR abs. bands, cm^{-1} [a]	2090 930.9 2114 819.3				
Raman frequencies, cm^{-1} [a]					
ν_1	2089				
δ_{12}	816				
ν_{234}	2106				
δ_{345}	920				
Heat of formation, kcal/mole	21.6	38.7			
Bond energy, kcal					
Ge–H	69.0				
Ge–Ge		37.9			
Ge–H bond distance, Å[b]	1.527				
Ge–Ge bond distance, Å[b]		2.41	2.41		
Index of refraction n					
at 5484 Å	1.00091				
at 5893 Å	1.00089				
Molar refraction, cm^3/mole at 5893 Å	13.35				
Surface tension at b.p., dynes/cm	15.80				
Parachor	108.5				

[a] See also spectroscopic data on GeH_4, GeD_4, etc., in F. G. A. Stone, *Hydrogen Compounds of the Group IV Elements*, pp. 74–76, Prentice-Hall, Englewood Cliffs, N.J. (1962).

[b] L. E. Sutton, *Interatomic Distances* (Spec. Publ. 11, Chemical Society, p. S-5).

[c] At −142°, not at m.p.

Digermane, Ge_2H_6

This is obtained as a by-product in the preparation of GeH_4 as described above, and also by the circulation of GeH_4 through a silent electrical discharge at low pressures[54]. The physical properties, insofar as they are known, appear in Table 6.

Digermane decomposes at 200° in a manner that fits the scheme[55]

$$GeH_6 \longrightarrow 2GeH_3$$
$$GeH_3+Ge_2H_6 \longrightarrow GeH_4+Ge_2H_5$$
$$Ge_2H_5 \longrightarrow GeH_2+GeH_3$$
$$GeH_2 \longrightarrow Ge+H_2$$
$$2GeH_2 \longrightarrow GeH_4+Ge$$

The activation energy for such pyrolysis is 33.7 kcal[55]. Digermane oxidizes more easily than GeH_4, but less readily than Si_2H_6; oxidation proceeds at 100°:

$$2Ge_2H_6+7O_2 \longrightarrow 4GeO_2+6H_2O$$

In liq. NH_3, a solution of Ge_2H_6 has a conductivity of 0.97×10^{-3} recip. ohm-cm at $-60°$ (1.35×10^{-3} at 47°), which is 10,000 times as large as that of the pure solvent and indicates formation of a salt-like $(NH_4)_2Ge_2H_4$. The solution dissolves P_4 in much the same way as does a solution of GeH_4[56].

The *higher germanes* also ensue from circulation of GeH_4 through an electric discharge[54], more Ge_3H_8 being obtained than Ge_2H_6. Some Ge_3H_8 also was obtained by the classical method of Dennis, Corey and Moore[48], and most of the physical properties of the substance in Table 6 come from a sample prepared that way. When exposed to the air, Ge_3H_8 soon changes to a white solid, but does not ignite. It does not dissolve in water, but appears to be oxidized by air in the water. It is immiscible with 33% aq. NaOH, and does not react with it. Like Ge_2H_6, Ge_3H_8 dissolves in CCl_4 and reacts with it, apparently to form $GeCl_4$.

The formulas Ge_4H_{10} and Ge_5H_{12} in Table 6 refer to mixtures of isomers in unknown proportions. The Ge_4H_{10} decomposes slowly above 50° and rapidly above 100° to give GeH_4 and liquid products[53], and Ge_5H_{12} decomposes at 100° to GeH_4 and a solid residue. At 350° all the higher hydrides become Ge and H_2.

5.2. GERMANIUM(IV) HALIDES

Germanium Tetrafluoride, GeF_4

This probably is obtainable from the elements, but is much more easily obtained in pure form by the thermal decomposition of barium fluorogermanate, $BaGeF_6$. This salt is made by adding $BaCl_2$ to a solution of GeO_2 in aq. HF; the white granular solid is washed with cold water and dried at 120°. When heated in a silica tube (copper is unsatisfactory), copious evolution of GeF_4 begins at 600°. The gas does not attack dry glass at 25°, nor silica glass at 700°, but copper and other base metals reduce it at elevated temperatures. The gas itself does not decompose at 1000° and 100 mm. When passed over reduced Ge at temperatures above 100°, or over crystalline Ge at 350°, a white volatile solid GeF_2 is

[54] J. E. Drake and W. L. Jolly, *Proc. Chem. Soc.* **1961**, 379.
[55] H. J. Emeléus and H. H. G. Jellinek, *Trans. Faraday Soc.* **40** (1944) 93.
[56] P. Royen, *Z. anorg. Chem.* **235** (1938) 324.

formed. This solid dissolves in water, and the solution reduces $KMnO_4$ and I_2; when H_2S is added to it, GeS precipitates[57].

The physical properties of GeF_4 are given in Table 7. Chemically, GeF_4 reacts with limited water to precipitate gelatinous GeO_2 and form HF. In a large excess of water, the gas dissolves and heat is evolved; addition of KOH precipitates K_2GeF_6, showing that H_2GeF_6 is formed in the solution. Hydrolysis of small amounts of GeF_4 vapor in the throat and lungs produces hoarseness and extreme irritation. When the vapor of GeF_4 (at 1 atm) is passed over heated $AlCl_3$, $MgCl_2$ or $FeCl_3$, complete exchange of halogen occurs, giving only $GeCl_4$ and no mixed halides.

Besides the two fluorogermanate salts already mentioned, many others can be precipitated or crystallized from water solution. The easiest method of preparation is to dissolve dried precipitated GeO_2 in an excess of 40% aq. HF, and then add the base and cool the solution. Many organic bases form fluorogermanates this way, as well as hydrazine, ammonia and metallic hydroxides.

Germanium Tetrachloride, $GeCl_4$

This is the oldest and best-known covalent compound of germanium. It was prepared in 1886 by Clemens Winkler, who found that the germanium in a solution of GeO_2 in a large excess of hydrochloric acid could all be recovered as $GeCl_4$ just by heating the solution. Finely divided Ge ignites in chlorine, and massive Ge heated in a stream of Cl_2 soon reaches incandescence and produces $GeCl_4$. The tetrachloride may also be made from GeO_2 by heating the dioxide to 500° in a stream of CCl_4 vapor carried by N_2. The reductive chlorination also is possible with $COCl_2$ or a mixture of CO and Cl_2 at 600°. Passing Cl_2 over a heated mixture of carbon and GeO_2 is not a satisfactory method; the best large-scale method is distillation from aq. HCl. Crude $GeCl_4$ can be freed from Cl_2 by blowing a stream of dry air through it (or by shaking it with Hg or with Hg_2Cl_2) and then purified by fractional distillation (see isolation of Ge from its ores, *supra*). The physical properties of pure $GeCl_4$ are given in Table 7. References 58–69 are a guide to the older literature.

Pure $GeCl_4$ is a colorless, refractive liquid with a sharp, penetrating odor. It fumes in moist air, and its vapor deposits a chalky coating of GeO_2 on glass surfaces which have not been dried in vacuum. The vapor is very stable (undecomposed at 950°) but very reactive toward alkali metals (mixing the vapors of $GeCl_4$ and K gives a rapid reaction and a blue chemiluminescence). It undergoes halogen exchange if treated with PI_3 or SbF_3 or $SnBr_4$. Hydrogen reduces the vapor to Ge at 600°, forming $GeCl_2$ as an intermediate product.

Liquid $GeCl_4$ reacts rather slowly with liquid water at the interface, depositing GeO_2. In liquid ammonia it forms the imide $Ge(NH)_2$, which decomposes on heating to form a nitride Ge_3N_4. The liquid is without action on H_2SO_4 or conc. aq. HCl, but reacts slowly with conc. aq. HNO_3 to become colored with oxides of nitrogen. In acetic anhydride $GeCl_4$ forms the tetra-acetate,

$$GeCl_4+4(CH_3CO)_2O = Ge(CH_3COO)_4+4CH_3COCl$$

but this cannot be isolated by distillation and is better prepared by a metathetic reaction with thallium monoacetate in acetic anhydride solution, precipitating TlCl. Primary and secondary amines react to form the amine hydrochloride and a substituted imide $Ge(NR)_2$, but tertiary amines such as $(C_2H_5)_3N$ form addition compounds, usually of the type

[57] L. M. Dennis and A. W. Laubengayer, *Zeit. f. Phys. Chem.* **130** (1927) 520; L. M. Dennis, *Z. anorg. Chem.* **174** (1928) 119.

TABLE 7. PROPERTIES OF THE GERMANIUM TETRAHALIDES GeX_4 [58–68]

Property	GeF_4	$GeCl_4$	$GeBr_4$	GeI_4
Melting point, °C	−15[f]	−49.5[a]	26	146
Crystal form and color			white octahedr.	orange cubic
Boiling point, °C	−36.5[c]	83.1	186	*ca.* 400
ΔH vaporization, kcal/mole	8.3[d]	7.35		20.1[d]
Entropy of vap., cal/mole-deg		20.65		37
Vapor pressure: at −62.7°, torr	75.9			
at −41.4°	452	1.0		
at 0°		25		
at 39.4°		161		
at 81.0°		708.5		
at 93.7°			58.4	
at 140.0°			232	
at 180.2°			655	
at 373°[b]				0.17
at 393°[b]				0.66
at 408°[b]				1.36
Critical temp., °K		548		
Critical pressure, atm		36.6		
Specific heat C_p, joule/g-deg	19.57[g]	0.699	23.34[g]	25.1[g]
Ratio of specific heats		1.42		
Standard entropy, cal/mole-deg	72.08	83.05	94.77	107.9
Density at 30°C	2.126[e]	1.844	2.1002	4.322[j]
Surface tension, dynes/cm at 30°		22.44	35.51	
Parachor		253.2	309.7	
Velocity of sound, m/sec, 30°		768.6		
Compressibility, $cm^2/dyne \times 10^{12}$		91.3		
Dielectric constant, 0°C		2.491		
Ge–X distance, Å	1.67	2.08	2.29	2.50
X–X distance, Å	2.73	3.39	3.82	4.09
Raman frequencies, cm^{-1}: ν_1	740	397	235	
δ_{12}		132	80	
ν_{234}		451	327	
δ_{345}		171	112	
Magnetic susceptibility, $\times 10^6$	−49[i]	−72.0		−174
Index of refraction: D line, 22.5°	695.4[h]	1.4644		
white light, 20°		1.4648	1.6296	
white light, 25°		1.4614	1.6270	
Molar refraction (Lorentz–Lorenz)	10.2	31.465		
Heat of formation, kcal/mole		165	78.5	
Bond energy, kcal/mole, Ge–X			63.5	48.1

[a] For α-form of $GeCl_4$; the metastable β-form melts at −52.0°. [b] Vapor pressure of solid which sublimes. [c] Sublimation temp. at 1 atm. [d] Heat of sublimation. [e] Density of liquid at 0° under pressure. [f] Melting point of solid under 3032 mm pr.; subs. sublimes. [g] In cal/mole-degree at 25°. [h] $(n-1)\times 10^6$ at Hg line 5461 Å. [i] Measured as liquid under pr., 20°; for gas -33 ± 0.01. [j] At 26°.

58 L. M. Dennis and F. E. Hance, *J. Am. Chem. Soc.* **44** (1922) 306.
59 L. M. Dennis and F. E. Hance, *Z. anorg. Chem.* **122** (1922) 275.
60 A. W. Laubengayer and D. L. Tabern, *J. Phys. Chem.* **30** (1926) 1048.
61 W. A. Roth and O. Schwartz, *Z. phys. Chem.* **134** (1928) 466.
62 R. N. Pease, *J. Am. Chem. Soc.* **43** (1921) 193.
63 M. E. Lear, *J. Phys. Chem.* **28** (1924) 889.
64 L. M. Dennis, W. R. Orndorf and D. L. Tabern, *J. Phys. Chem.* **30** (1926) 1052.
65 F. M. Brewer and L. M. Dennis, *J. Phys. Chem.* **31** (1927) 1537.
66 F. M. Jaeger, P. Terpstra and H. G. K. Westenbrink, *Proc. Acad. Amsterdam* **28** (1925) 747.
67 L. M. Dennis and A. W. Laubengayer, *Z. phys. Chem.* **130** (1927) 527.
68 L. M. Dennis, *Z. anorg. Chem.* **174** (1928) 119.

$GeCl_4 \cdot 4R_3N$. Alcohols appear to react only very slowly with $GeCl_4$ to form esters; since the reaction with water

$$GeCl_4+2H_2O = GeO_2+4HCl$$

is definitely reversible at high concentrations of HCl (*vide supra*), the reaction with alcohols probably soon comes to equilibrium. Hence germanium alkoxides are best made by dissolving sodium in the alcohol and then adding $GeCl_4$:

$$GeCl_4+4C_2H_5ONa = Ge(OC_2H_5)_4+4NaCl$$

Germanium tetrachloride is readily purified and stored, and so is the favorite starting material for preparing a wide variety of *organogermanium* compounds through the agency of Grignard reagents, lithium alkyls and sodium condensations:

$$4C_2H_5MgBr\ (\text{in } Et_2O)+GeCl_4 = (C_2H_5)_4Ge+4MgBrCl$$

$$C_6H_5MgCl\ (\text{in THF})+GeCl_4 = C_6H_5GeCl_3+MgCl_2$$

The subject of organogermanium chemistry is treated separately in a later section, but it may be said that ever since Clemens Winkler made the first organogermanium compound from it in 1887, $GeCl_4$ has been very important to the subject. It also is the common starting material for the preparation of addition compounds (with ethers and esp. cyclic and aromatic ethers, for example[69]) and of coordination compounds of the transition metals[70]. In this connection, it is interesting to note that although *silicon* forms very stable fluorosilicates containing the SiF_6^- ion, it forms no chlorosilicates M_2SiCl_6; *germanium*, on the other hand, not only forms fluorogermanates M_2GeF_6, but also chlorogermanates, such as Cs_2GeCl_6 [71].

Germanium Tetrabromide, $GeBr_4$

This is readily prepared from the elements by conducting Br_2 vapor (diluted with N_2) over powdered Ge heated to 220° and collecting the products in an ice-cooled trap, or by refluxing finely powdered Ge with liquid Br_2 for 4 hr, evaporating the excess bromine at 25°, removing the last Br_2 with HgCl, and then distilling the tetrabromide. Pure $GeBr_4$ forms colorless ice-like crystals which turn yellow in ultraviolet light and hydrolyze rapidly in moist air. Active chlorides (such as $C_6H_5PCl_2$) convert it to chlorobromides, and eventually to $GeCl_4$. However, there is no halogen exchange with $SnCl_4$. Active metals such as Mg strip the bromine from $GeBr_4$ at 25° to 40°, leaving black dispersed Ge.

Germanium Tetraiodide, GeI_4

This may also be prepared from the elements, but its purification by distillation or sublimation is difficult because of decomposition. An easier way is to heat GeO_2 and aq. HI together, and then to extract the GeI_4 into CCl_4 or C_6H_6 [72]. Thus GeO_2 is heated in a pressure flask with 20 equivalents of HI (sp. gr. 1.7) to 150° until the GeO_2 is dissolved, and then a stream of hot CO_2 is blown into the flask to vaporize any free iodine. Upon cooling,

[69] H. H. Sisler, H. H. Batey, B. Pfahler and R. Mattair, *J. Am. Chem. Soc.* **70** (1948) 3818, 3821.
[70] F. G. A. Stone, *New Pathways in Inorganic Chemistry*, p. 283, Cambridge Univ. Press (1968).
[71] A. W. Laubengayer, O. B. Billings, and A. E. Newkirk, *J. Am. Chem. Soc.* **62** (1940) 546.
[72] L. S. Foster and A. F. Williston, *Inorganic Syntheses*, Vol. 2, p. 112, McGraw-Hill, N.Y. (1946); H. Bauer and K. Burschkies, *Ber.* **66** (1933) 277.

TABLE 8. HYDRIDE–HALIDES OF GERMANIUM

Property	$GeHCl_3$	GeH_2Cl_2	GeH_3Cl	$GeHBr_3$	GeH_2Br_2	GeH_3Br	GeH_2I_2	GeH_3I
Melting point, °C	−71	−68	−52	−25	−15	−32	−44.9	−15
Boiling point, °C	75	69.5	28		89	52		
Vapor pressure, torr	26.5	36.9	185	1.9	6.5	73.8	0.1	20
at temp., °C	0.7	0.4	−5.0	0	0.6	−3.3	0	0
Other vapor pressures	a	b	d		f	h		
Vapor pressure equation		c	e		g	i		
Density, g/cc	1.93	1.90	1.75		2.80	2.34		
at temp., °C	0	−68	−52		0	29.5		
Ge–X distance, Å	2.114		2.147					
Ge–H distance, Å	1.55		1.52					
Ge–H bond force const., dynes/cm × 10^{-5}	2.7			2.7				
Raman freq., cm^{-1}								
δ_{12}	149			95				
δ_3	181			128				
ν_1	409			273				
ν_{23}	438			325				
δ_{45}	699			474				
ν_4	2159			2116				
Dipole moment, × 10^{18} esu		2.21	2.12					

[a] Vap. pr. 0.8 mm at −25°, 9.2 at −10°, 75.2 at 19.5°, 129.2 at 30.5°.
[b] Vap. pr. 2.1 mm at −41.8°, 19.3 at −11.5°, 110 at 20.7°, 258 at 40.2°.
[c] $\log p = -1742.7/t + 7.969$.
[d] Vap. pr. 11.3 mm at −52°, 88.9 at −20.7°, 345 at 8.8°.
[e] $\log p = -1527.4/t + 7.961$.
[f] Vap. pr. 1.2 mm at −26.6°, 28.3 at 21.7°, 102.4 at 43.2°.
[g] $\log p = -2461.9/t + 9.798$.
[h] Vap. pr. 8.1 mm at −44.6°, 43 at −16.4°, 145.4 at 10.9°, 265 at 24.5°.
[i] $\log p = -1614.7/t + 7.851$.

orange crystals of GeI_4 deposit. These crystals slowly turn brown and disintegrate in moist air, becoming eventually a brown paste of GeO_2 in HI. Dry GeI_4 decomposes at 440°. Aq. NaOH converts it to Na_2GeO_3 and NaI, and liq. NH_3 converts it to $Ge(NH_2)_2$. Warming with $SnCl_4$, $ZnCl_2$, $CdCl_2$, $HgCl_2$, $PbCl_2$, $AsCl_3$, $SbCl_3$ or $SbCl_5$ converts it to $GeCl_4$. Passing NH_3 through a solution of GeI_4 in CCl_4 gives $GeI_4{\cdot}8NH_3$, and in the same solvent organic amines react with GeI_4 to give addition compounds containing 4, 5, 6 or 10 molecules of amine[73].

A few mixed halides, such as GeF_3Cl, GeF_2Cl_2, $GeFCl_3$, $GeClBr_3$, $GeCl_2Br_2$ and $GeCl_3Br$ are known[74], with properties intermediate between those of the tetrahalides. Those containing chlorine and bromine disproportionate readily.

Halogermanes, GeH_aX_{4-a}

Many volatile hydride–halides of germanium are known, some of which are characterized as given in Table 8. Of these, the most important are the "haloforms" $GeHCl_3$ and $GeHBr_3$. Mention has already been made of the ready reaction of $GeCl_2$ with anhydr. HCl to form $GeHCl_3$; variations of this preparative method include passing $GeCl_4$ vapor and H_2 through a silica tube at 900°, and heating Ge powder in a stream of HCl (diluted with N_2) at 500°. Parallel methods may be used for preparing $GeHBr_3$, or GeS may be dissolved in 40% aq. HBr. A more difficult route involves the preparation of GeH_4 and then the halogenation of this by reaction with limited HCl or HBr in contact with crystals of $AlCl_3$ or $AlBr_3$. Warming $GeHCl_3$ to 75° causes it to return to $GeCl_2$ and HCl. The hydride nature of the hydrogen in $GeHCl_3$ is shown by dissolving the $GeHCl_3$ in alcohols, whereupon H_2 slowly is evolved and $GeCl_4$ and $GeCl_2$ remain. The reactions of the hydride–halides follow in other respects the reactions of both halides and hydrides[75], including preferential hydrolysis of the halogen:

$$2GeH_3Cl+H_2O = (GeH_3)_2O+2HCl$$

As in silicon chemistry, halogen exchange with chloroform and other carbon halides is noted:

$$GeHBr_3+CHCl_3 = GeHClBr_2+CHCl_2Br$$

Mixtures of $GeHCl_3$ and $GeHBr_3$ will also undergo halogen exchange to give $GeHClBr_2$ and $GeHCl_2Br$.

Some halogenated *digermanes* and *digermoxanes* are H_3GeGeH_2Cl (b.p. 88°), Ge_2Cl_6 (m.p. 41°), $Cl_3GeOGeCl_3$ (m.p. −60°, b.p. 70° at 13 mm, density 2.057 at 20°) and H_3GeGeH_3I (m.p. −17°).

5.3. GERMANIUM(IV) CHALCOGENIDES

Germanium Dioxide, GeO_2

This is the best-known solid compound of germanium. It is a stable, unreactive white powder which melts to a glass and can be used as a constituent of glasses which are more

[73] T. Karantassis and L. Capatos, *Compt. rend.* **193** (1931).

[74] W. E. Anderson, J. Sheridan and W. Gordy, *Phys. Rev.* (2) **81** (1951) 819; M. L. Delwaulle, *Compt. rend.* **234** (1952) 2361; G. S. Forbes and H. H. Anderson, *J. Am. Chem. Soc.* **66** (1944) 931.

[75] F. G. A. Stone, *Hydrogen Compounds of the Group IV Elements*, pp. 63–76, Prentice-Hall, Englewood Cliffs, N.J. (1962).

refractive than the corresponding silicate glasses[76]. It is readily obtained by the hydrolysis of $GeCl_4$, yielding a microcrystalline form of the hexagonal modification of GeO_2 (with a structure much like that of low quartz)[77]. This is often called the "soluble" form, because it dissolves in cold water to the extent of 4 g/l. The same crystalline modification also is obtained by holding GeO_2 glass at 1080° until it devitrifies. The only other modification of GeO_2 is the "insoluble" or tetragonal form, which is obtained by hydrothermal conversion of the "soluble" form (by heating with water in a bomb at 355° for 100 hr). The "insoluble" form is stable up to 1033°, where it inverts slowly to the "soluble" form (which melts at 1116°). The physical properties of both forms are given in Table 9. The "soluble" form of

TABLE 9. PHYSICAL PROPERTIES OF GeO_2 AND GeS_2

Property	Soluble GeO_2	Insoluble GeO_2	Vitreous GeO_2	GeS_2
Melting point, °C	1116	1086		*ca.* 800
Specific heat	[a]			
Crystal system	hexagonal	tetragonal	amorphous	orthorhomb.
Crystal habit	rhombohed.	prisms		
Crystal structural type	low quartz	rutile		
Unit cell, Å				
a	4.987	4.394		11.66
b				22.34
c	5.652	2.852		6.86
Space group	D_3^4	D_{4h}^{14}		C_{2v}^{19}
Ge–X dist., Å		1.86	1.65	2.19
Density, g/cc at 25°	4.228	6.239	3.637	2.942[b]
Index of refr.				
ω	1.695	1.99		
ϵ	1.735	2.05	1.607	
Inversion point, °C	1033	1033		
Solubility, g/l. H_2O, 25°	4.53	0	5.18	
Magnetic suscept., $\times 10^6$	-34.3			-53.3

[a] $C_p = 11.2+7.17\times10^{-3}\ T$ cal/mole.
[b] At 14°.

GeO_2 is dissolved and converted to H_2GeF_6 by 25 N HF, and is attacked by 12 N HCl and 5 N NaOH, but the "insoluble" form resists these reagents; it dissolves only in 10 times its weight of fused NaOH at 550°, or in 5 times its weight of fused Na_2CO_3 at 900° [77]. The vitreous form of GeO_2 has the same structure as fused silica, and reacts with aq. HF and HCl like the hexagonal crystal modification. The electrical conductivity of GeO_2 is about 10 times that of SiO_2, at temperatures up to 1100°.

Germanium Disulfide, GeS_2

As indicated above, the common hexagonal form of GeO_2 is very soluble in moderately concentrated aqueous acids. When 6 N H_2SO_4 is saturated with GeO_2 and then H_2S is introduced, a precipitate of white GeS_2 forms (sometimes in pearly platelets). This may be filtered, then washed with H_2S-saturated water, H_2S-saturated alcohol, and finally ether. The GeS_2 must be dried in a desiccator, otherwise it will oxidize and hydrolyze. The

[76] L. M. Dennis and A. W. Laubengayer, *J. Am. Chem. Soc.* **47** (1925) 1945; *J. Phys. Chem.* **30** (1926) 1510.
[77] A. W. Laubengayer and D. S. Morton, *J. Am. Chem. Soc.* **54** (1932) 2303.

physical properties are listed in Table 9. Pure GeS_2 dissolves in 12 N HCl, evolving H_2S. When heated in air it turns brown and melts to a dark, vitreous mass, evolving SO_2. In water it hydrolyzes slowly, with a prominent odor of H_2S. It is slightly soluble in liq. NH_3, and in that medium reacts with dissolved sodium to give Na_2S and a dispersion of elementary Ge; addition of more sodium converts the latter to sodium germanide, which reacts with NH_4Br to produce GeH_4. In absolute alcohol GeS_2 reacts with H_2S to form a polymeric white solid, $H_2Ge_2S_5$.

Germanium Diselenide, $GeSe_2$

This is produced when H_2Se is introduced into a solution of GeO_2 in 6 N HCl. It is an orange solid which melts at 707° (with decomposition). The density is 4.56 at 25°, and the crystal form is orthorhombic, with $a = 12.96$, $b = 6.93$, $c = 22.09$ Å. The crystals oxidize when heated in air, with separation of the selenium. Aqueous acids have little effect on the substance in the absence of an oxidizing agent, but Br_2 in 12 N HCl dissolves it. Conc. HNO_3 oxidizes it to GeO_2 and H_2SeO_3.

Tellurium forms alloys with Ge, and the phase diagram shows evidence of a compound GeTe, but no $GeTe_2$. The GeTe is metallic in appearance, with a density of 6.20. Conc. HCl, and H_2SO_4 and H_2O_2 have no action on the solid, but conc. HNO_3 attacks it and forms GeO_2.

5.3. OXYACID SALTS OF Ge(IV)

Acetate

The addition of $TlC_2H_3O_2$ to a solution of $GeCl_4$ in acetic anhydride gives a precipitate of TlCl and a solution of $Ge(C_2H_3O_2)_4$, from which the latter can be isolated by concentrating the solution at low pressure and then cooling it. The $Ge(C_2H_3O_2)$ separates in white needles, m.p. 156°. It is soluble in benzene and acetone, and almost insoluble in CCl_4 and ether. It hydrolyzes readily in moist air to GeO_2 and $HC_2H_3O_2$.

Oxalate

When excess GeO_2 is heated with oxalic acid, a syrupy liquid of the composition $H_2[Ge(C_2O_4)_3]$ is obtained. The NaOH titer of this corresponds to only three-quarters of the original oxalic acid. No crystalline compound can be isolated, but the liquid forms stoichiometric complexes with quinine and with strychnine.

Sulfate

When $GeCl_4$ is heated with SO_3 in a bomb tube at 160° for 12 hr, $S_2O_5Cl_2$ and an unstable sulfate of Ge(IV) are formed[78].

$$GeCl_4 + 6SO_3 = Ge(SO_4)_2 + 2S_2O_5Cl_2$$

Organogermanium Oxyacid Salts

The preparation of oxyacid derivatives of Ge(IV) is facilitated by the presence of alkyl and aryl groups linked to Ge, so that formates, acetates, chloroacetates, perfluoroacetates,

[78] E. Hayek and K. Hinterauer, *Monatsh.* **82** (1951) 205.

propionates and benzoates become possible. For preparation and properties, see the extensive publications of H. H. Anderson[79].

Alkoxides $Ge(OR)_4$

The action of sodium alkoxides (and of other alkoxides of active metals[80]) on $GeCl_4$, Ge_2OCl_6, etc., gives colorless liquid germanium alkoxides of pleasant odor. The physical properties of some representative compounds are given in Table 10[81]. These substances

TABLE 10. SOME ALKOXIDES OF GERMANIUM

Compound	M.p., °C	B.p., °C	At. pr. mm	n_D^{20}	d_4^{20}
$Ge(OCH_3)_4$	−18	145	760	1.4015	1.3257
$Ge(OC_2H_5)_4$	−72	184	741	1.4073	1.1395
$Ge(O\ n\text{-}C_3H_7)_4$		178	115	1.4200	1.0664
$Ge(O\ i\text{-}C_3H_7)_4$		109	30	1.4141	1.0245 at 25°
$Ge(O\ n\text{-}C_4H_9)_4$		143	8	1.4255 at 25°	1.0173 at 25°
$Ge(O\ i\text{-}C_4H_9)_4$		265	760		
$Ge(O\ sec\text{-}C_4H_9)_4$		137	54	1.4291 at 25°	1.0164 at 25°
$Ge(O\ tert\text{-}C_4H_9)_4$	44	224	760		1.0574 at 25°
$Ge(OC_6H_5)_4$		220	0.3		
$Ge(OSiPh_3)_4$	237				

Reference: V. F. Mironov and T. K. Gar, *Organic Compounds of Germanium* (in Russian), Academy of Sciences, U.S.S.R. Moscow (1967).

cannot be made in satisfactory yield directly from the alcohol and $GeCl_4$, probably because of reversibility of the alcoholysis. However, if a hydrohalogen acceptor such as NH_3 or pyridine is present, the alcoholysis proceeds satisfactorily. The germanium alkoxides are stable in dry air but are slowly hydrolyzed by moisture. Partial hydrolysis leads to polymeric alkoxygermanium oxides $(RO)_3GeOGe(RO)_3$, etc. The compound $Ge(OSiPh_3)_4$ is much more resistant to hydrolysis[82].

6. ORGANOGERMANIUM COMPOUNDS

6.1. GENERAL

Although germanium is a rather rare and little-known element, a large proportion of the research effort on it has been devoted to organogermanium chemistry. As a metalloid, germanium forms strong and durable bonds to carbon (bond dissociation energy $Ge–CH_3 = 59.1$ kcal/mole, $Ge–C_2H_5$ 56.7 kcal, $Ge–C_3H_7$ 56.8 kcal[83]), as well as to

[79] H. H. Anderson, *J. Am. Chem. Soc.* **73** (1951) 5798 and 5800; *ibid.* **74** (1952) 2370 and 2371; *ibid.* **79** (1957) 326; *J. Org. Chem.* **20** (1955) 536 and 900.

[80] D. C. Bradley, L. Kay and W. Wordlaw, *Chem. and Ind.* **1953**, 746; *J. Chem. Soc.* **1956**, 4916; *ibid.* **1958**, 3656.

[81] For references see F. Rijkens, *Organogermanium Compounds*, Inst. Org. Chem. T.N.O., Utrecht, Holland (1960), as well as ref. 80 above.

[82] V. Gutmann and A. Meller, *Monatsh.* **91** (1960) 519.

[83] H. A. Skinner, The strengths of metal-to-carbon bonds, *Advances in Organometallic Chemistry*, **2** (1964) 99, Academic Press, N.Y.

hydrogen, oxygen and the halogens. Hence there are very many varieties of organogermanium compounds, and the roster of individual compounds reached 2300 in 1967[84]. Obviously a complete account cannot be given here. All that can be done at present is to review the general methods of preparation, to list the physical properties of some representative compounds and to summarize the chemical behavior of these. For more detailed information, the reader is referred to several compilations and general textbooks of organometallic chemistry which have appeared[81, 84–87]. Current reviews appear in *Advances in Organometallic Chemistry* (Academic Press, N.Y., 1964–).

6.2. PREPARATIVE METHODS

1. *Alkylation of Ge halides, alkoxides, etc., by zinc alkyls and aryls:*

$$GeX_4+2R_2Zn = R_4Ge+2ZnX_2$$

This reaction, first used by Winkler in 1886, is fairly satisfactory for tetra-alkyls (aside from the unpleasant making and handling of zinc alkyls), but it is difficult to control for the purpose of making organogermanium halides, and so is not much used any more. Mercury alkyls can be used instead of those of zinc.

2. *Alkylation of Ge halides by Grignard reagents:*

$$GeCl_4+2RMgCl = R_2GeCl_2, \text{ etc.}$$

This method is so versatile and so readily controlled that it is the favorite for making most compounds of the type $RGeX_3$, R_3GeX and R_4Ge. It also is used for preparing those dialkyldichlorogermanes which cannot be obtained readily by direct synthesis. Mixed alkyls of the type RR′R″R‴Ge can also be made by successive alkylations with respective Grignard reagents, indicating the degree of flexibility which is possible. In the Grignard preparation of organogermanium halides, since halogen exchange between Ge and Mg is possible, it is important to use organomagnesium *bromides* with $GeBr_4$, and so on.

3. *Alkylation of Ge halides and pseudohalides by organolithium reagents:* It is principally a matter of choice whether organomagnesium or organolithium alkylating reagents be used, but sometimes there are advantages from the use of hydrocarbon solvents (possible with RLi), and sometimes no corresponding RMgX can be made (as was the case at one time when $CH_2{=}CHLi$ could be obtained but not $CH_2{=}CHMgX$).

4. *Sodium condensation reactions:* These are so difficult to control that they are seldom used.

5. *Direct synthesis:* This is principally a method for making dialkylgermanium dichlorides and dibromides, especially from the lower alkyl halides:

$$2CH_3Cl+Ge(+Cu \text{ catalyst at } 300°) = (CH_3)_2GeCl_2$$

---

[84] V. F. Mironov and T. K. Gar, *Organic Compounds of Germanium* (in Russian), Acad. Sciences, U.S.S.R. Moscow (1967). See also F. Glockling, *Chemistry of Germanium, Organic and Inorganic*, Academic Press, N.Y. (1969), which appeared after this account was written.

[85] *Gmelins Handbuch der anorganischen Chemie*, 8th ed., System No. 45, supplementary vol., Verlag Chemie, Weinheim (1958).

[86] R. Weiss, *Organometallic Compounds* (ed. by M. Dub), Vol. 11, *Compounds of Ge, Sn, and Pb*, Springer Verlag, N.Y. (1967).

[87] H. C. Kaufman, *Handbook of Organometallic Compounds*, Table for Group IVB, Van Nostrand, N.Y. (1961).

The reaction always produces some $RGeCl_3$ (and a trace of R_3GeCl) as well as R_2GeCl_2, but the proportion of $RGeCl_3$ usually is less than that obtained in the corresponding silicon reaction. The method is simple and inexpensive, where it is applicable, and complexes with ethers are avoided entirely. Either powdered fused germanium or the black powder obtained by reducing GeO_2 with H_2 may be used, and 10% Cu powder is appropriate as catalyst[88].

6. *Addition of Ge–H compounds to alkenes and alkynes:* The addition of $GeHCl_3$ (so readily obtained; see above) to terminal olefins provides a good example of this convenient reaction[89]:

$$GeHCl_3+CH_3(CH_2)_zCH=CH_2 = CH_3(CH_2)_zCH_2CH_2GeCl_3$$

The various additions are catalyzed by free-radical initiators, by ultraviolet light, by Pt on charcoal or by H_2PtCl_6, and by powdered Cu; each reaction has its own optimum conditions. In general, the additions are easier to conduct than those of the corresponding Si–H compounds, and $GeHCl_3$ is far more reactive than $SiHCl_3$ in such reactions. A tabulation of 109 organogermanium compounds prepared by such germane-to-olefin addition reactions, together with references and a discussion of the technique, is given in a 1966 review[90].

7. *Addition of alkali-metal derivatives of organogermanium compounds to unsaturated compounds:* This is a limited reaction, but a very handy one in some cases where a fourth organic group is to be added to germanium[91]:

$$(C_6H_5)_3GeLi+C_6H_{13}CH=CH_2 = (C_6H_5)_3GeCH_2CHLiC_6H_{13} \xrightarrow{H_2O} (C_6H_5)_3GeCH_2CH_2C_6H_{13}$$

$$(C_6H_5)_3GeK+HCHO = (C_6H_5)_3GeCH_2OK \xrightarrow{H_2O} (C_6H_5)_3GeCH_2OH$$

The reaction with carbonyl compounds does not always result in addition (as in the reaction with formaldehyde, above), but may result in decarbonylation:

$$2(C_6H_5)_3GeLi+CO(OC_2H_5)_2 = (C_6H_5)_3Ge\text{–}Ge(C_6H_5)_3+CO+2LiOC_2H_5$$

8. *Other methods:* Organodigermanes frequently are obtained by reductive coupling through the agency of Grignard reagents, and indeed the reaction of an excess of C_6H_5MgBr with $GeCl_4$ can produce a preponderance of either $(C_6H_5)_4Ge$ or $(C_6H_5)_3Ge\text{–}Ge(C_6H_5)_3$, depending on a slight change in the procedure[92]. These coupling reactions usually are explained by the formation of R_3GeMgX intermediates, although there is much controversy about whether germanium can form such Grignard-like compounds. Exchange reactions are quite common in germanium chemistry also, so that the interaction of $(C_6H_5)_3GeK$ and $(C_6H_5)_3SiCl$ gives not only $(C_6H_5)_3GeSi(C_6H_5)_3$, but also $(C_6H_5)_3GeGe(C_6H_5)_3$ and $(C_6H_5)_3SiSi(C_6H_5)_3$[93].

[88] E. G. Rochow, *J. Am. Chem. Soc.* **69** (1947) 1729; *ibid.* **72** (1950) 198.

[89] A. K. Fisher, R. C. West and E. G. Rochow, *J. Am. Chem. Soc.* **76** (1954) 5878.

[90] E. Y. Lukevits and M. G. Voronkov, *Organic Insertion Reactions of Group IV Elements*, Consultants Bureau, N.Y., Plenum Press (1966).

[91] H. Gilman and C. W. Gerow, *J. Am. Chem. Soc.* **77** (1955) 5740; *ibid.* **79** (1957) 342; *ibid.* **82** (1960) 4562.

[92] D. M. Harris, W. H. Nebergall and O. H. Johnson, *Inorganic Syntheses*, **5** (1957) 70–74.

[93] F. Rijkens, *Organogermanium Compounds*, p. 39, Germanium Research Committee, TNO, Utrecht (1960).

6.3. TETRA-ALKYLS AND TETRA-ARYLS, R_4Ge

The tetra-alkyls of germanium are somewhat more reactive than those of silicon, but are stable substances obtainable in great variety; hundreds of them (with similar and dissimilar R groups) are listed in the 1967 compendium prepared by Weiss[86]. The physical properties of a few are given in Table 11. The C–Ge bonds can be cleaved by bromine in CCl_4

$$(C_6H_5)_4Ge+Br_2 = (C_6H_5)_3GeBr+C_6H_5Br$$

or by hydrohalogen acids in the presence of aluminum halides (HF requires no AlF_3):

$$(CH_3)_4Ge+HBr(AlBr_3) = (CH_3)_3GeBr+CH_4$$
$$(C_2H_5)_4Ge+HF = (C_2H_5)_3GeF+C_2H_6$$

The R–Ge bond is moderately stable to aqueous acids, but HNO_3 in $(CH_3CO)_2O$ will cleave it, and so will $HClO_4$ in aq. C_2H_5OH. The relative rates of cleavage of $-C_6H_4OCH_3$ groups

TABLE 11. SOME TETRA-ALKYLS AND TETRA-ARYLS, R_4Ge[e]

Compound	Mp., °C	B.p., °C	At. pr. mm	n_D^{20}	d_{20}
$(CH_3)_4Ge$[a, d]	−88	43.5	760	1.3882	
$(C_2H_5)_4Ge$[e]	−92.5[b]	163.5	760	1.4428	0.9941
$(n\text{-}C_4H_9)_4Ge$		278	760	1.4571	0.934
$(i\text{-}C_4H_9)_4Ge$		135	17	1.4594	0.9374
$(C_6H_5)_4Ge$	235.7	>400	760		
$(CH_2{=}CH)_4Ge$		53	27	1.4676	1.040
$(CF_2{=}CF)_4Ge$		123	760	1.3662	1.7719
$(C_6H_5CH_2CH_2CH_2)_4Ge$		245	0.05	1.5704	1.106[c]
$(CH_3)_3C_6H_5Ge$		183	760	1.5045	
$(CH_3)_3C_6H_5CH_2Ge$		94	28	1.5140	1.1011
$(CH_3)_2(C_2H_5)_2Ge$		108	760	1.4221	0.9885
$(CH_3)_2(C_6H_5)_2Ge$		145	10	1.573	1.18
$CH_3(C_2H_5)_3Ge$		135	760	1.4328	0.9912
$(CH_3)_3CH{:}CH_2Ge$		70.6	735	1.4153	0.997
$(CH_3)_3CH_2CH{:}CH_2Ge$		101	760	1.4333	0.9952
$(C_2H_5)_3CH_2CH{:}CH_2Ge$		180	732	1.4594	1.0004
$(C_2H_5)_3C_5H_5Ge$		108	2	1.4598	0.9479

[a] Ge–C bond distance = 1.98 Å.
[b] Two crystal forms; m.p. of second form −89°.
[c] At 25°.
[d] Proton magnetic resonance shift τ = 9.87. For detailed information on N.M.R. spectra see M. L. Maddox, S. L. Stafford and H. D. Kaesz, *Adv. in Organometallic Chem.* 3 (1965) 1–179.
[e] N.M.R. constants in Maddox, Stafford and Kaesz, compilation cited above.

from Si, Ge, Sn and Pb in the reaction of $(C_2H_5)_3MC_6H_4OCH_3$ with $HClO_4$ were $1{:}36{:}10^5{:}10^8$, as measured in aq. ethanol.[94] Transfer of alkyl groups to Ge of $GeCl_4$ and $GeBr_4$ occurs in the presence of $AlCl_3$ or $AlBr_3$, and to $SnCl_4$ without the aluminum halide. Metallic Li or Na–K alloy will cleave $(C_6H_5)_4Ge$, but not $(C_4H_9)_4Ge$.

Some interest attaches to the *unsaturated* tetra-alkyls of germanium and their possible polymerization. Vinyl and allyl compounds of the type $R_nGeR'_{4-n}$ (where R′ is methyl or ethyl) have been polymerized at 6000 atm and 120°, using peroxide initiators[95]. Monovinyl and monoallyl compounds gave oily liquids, while the polyfunctional compounds

[94] C. Eaborn and K. C. Pande, *J. Chem. Soc.* **1960**, 1566.
[95] F. Rijkens, ref. 93, p. 29.

gave flexible or glassy solids. Readiness of polymerization of analogous .vinyl and allyl compounds of Group IV elements increases in the order Sn, Ge, C, Si[95]. Polymers may also be made by the addition of R_2MH_2 compounds (where M is a Group IV element) to unsaturated tetra-alkyls of the type $R_2MR'_2$, where R′ is an alkyl group with terminal unsaturation[95]. References to many polymerization experiments with olefinic R_4Ge compounds are given by Weiss[86].

6.4. ORGANOGERMANIUM HYDRIDES

Many alkyl and aryl derivatives of GeH_4 are known, and the properties of a few representative compounds are given in Table 12. They show most of the reactions of the

TABLE 12. SOME ORGANOGERMANIUM HYDRIDES[d]

Compound	M.p., °C	B.p., °C	At. pr., mm	n_D^{20}	d_{20}
$(CH_3)_3GeH$[d]		26	760		1.0128
$(CH_3)_2GeH_2$[c]	−149	6.5	744		
CH_3GeH_3[c]	−158	−23	760		
$(C_2H_5)_3GeH$[c]		120	760	1.4382	1.0043
$(C_2H_5)_2GeH_2$		72.5	740	1.4208	1.0378
$(C_3H_7)_3GeH$		183	760	1.4441	0.9694
$(C_3H_7)_2GeH_2$		126	760	1.4340	1.003
$(C_4H_9)_3GeH$		123	20	1.4508	0.9155
$C_4H_9GeH_3$		74	760	1.4200	1.022
$(C_6H_5)_3GeH$[c]	42.5[a]	128	0.03		
$(C_6H_5)_2GeH_2$[c]		93	1	1.5921	
$CH_2{=}CHCH_2GeH_3$		37	760	1.4315	1.0797
$(n\text{-}C_6H_{13})_3GeH$		123	0.5	1.4565	0.917[b]
$(n\text{-}C_6H_{13})_2GeH_2$		113	8	1.4522	0.9484
$n\text{-}C_6H_{13}GeH_3$		128	760	1.4350	0.9972
$(n\text{-}C_7H_{15})_3GeH$		182	17	1.4600	0.9108
$(n\text{-}C_7H_{15})_2GeH_2$		148	10	1.4543	0.9348
$n\text{-}C_7H_{15}GeH_3$		85	74	1.4390	0.9819
$n\text{-}C_8H_{17}GeH_3$		80	31	1.4422	0.9719

[a] α-form; β-form melts at 27°.
[b] At 25°.
[c] Observations on bond lengths in these compounds, together with some discussion of their chemical behavior, appear in F. G. A. Stone, *Hydrogen Compounds of the Group IV Elements*, chapter 3, Prentice-Hall, Englewood Cliffs, N.J. (1962).
[d] See also Monogermanes—their synthesis and properties, by J. E. Griffiths, *Inorg. Chem.* **2** (1963) 375. Vapor pressures for CH_3GeH_3, $(CH_3)_2GeH_2$ and $(CH_3)_3GeH$ are given there.

germanium hydrides themselves, moderated by the presence of R groups: hydrogen is replaced by halogen through the action of HX, reducing action is evident, hydrogen is expelled by $GeCl_4$ and SO_2Cl_2, etc., and addition to alkenes and alkynes occurs. It is interesting that lithium alkyls react in such a way as to metalate the germanium

$$(C_2H_5)_3GeH+RLi = (C_2H_5)_3GeLi+RH$$

whereas R_3SiH gives R_4Si and LiH.

6.5. ORGANOGERMANIUM HALIDES

Preparative methods 2, 3, 5, and 6 (see above) come into full play here, as well as halogen cleavage of tetra-alkyls and tetra-aryls. Hence very large numbers of compounds

R_3GeX, R_2GeX_2 and $RGeX_3$ are known, along with organogermanium halo-hydrides and organo-functional germanium halides. A full listing extends into the hundreds[86]; some constants for a few representative compounds are given in Table 13. Organogermanium halides hydrolyze *reversibly* (in sharp contrast to organosilicon halides) to give organogermanium oxides and hydroxides, and reaction with ammonia gives primary, secondary,

TABLE 13. SOME ORGANOGERMANIUM HALIDES[86] [a]

Compound	M.p., °C	B.p., °C	At. pr., mm	n_D^{20}	d_{20}
$(CH_3)_3GeF$		76	746	1.3863	1.230
$(CH_3)_3GeCl$	−15	97	730	1.4337	1.2493
$(CH_3)_3GeBr$		115	760	1.4660	1.5486
$(CH_3)_3GeI$		136	760	1.5159	
$(C_2H_5)_3GeF$		148	760	1.4206	1.1527
$(C_2H_5)_3GeCl$		174	760	1.4643	1.175
$(C_2H_5)_3GeBr$		191	760	1.4829	1.412
$(C_2H_5)_3GeI$		212	760	1.528	1.608
$(C_3H_7)_3GeF$	−27.5	203	760	1.4340	1.074
$(C_3H_7)_3GeCl$	−70	227	760	1.4641	1.10
$(C_3H_7)_3GeBr$	−47	242	760	1.4832	1.282
$(C_3H_7)_3GeI$	−38	259	760	1.5144	1.443
$(CH_2{=}CH)_3GeBr$		58	10	1.5057	
$(CH_3)_2C_2H_5GeCl$		99	760	1.4285	1.1763
$(C_4H_9)_2C_6H_5CH{=}CHGeCl$		140	0.4	1.540	1.1293
$(CH_3)_2GeF_2$		112	750	1.3743	1.5726
$(CH_3)_2GeCl_2$	−22	124	760	1.4600	1.4552
$(CH_3)_2GeBr_2$		153	746	1.5268	2.1163
$(C_2H_5)_2GeCl_2$		172.8	760		
$CH_3(C_2H_5)GeCl_2$		149	760	1.4660	1.4381
$(n\text{-}C_3H_7)_2GeCl_2$	0.5	183	760	1.4128	1.248
$(i\text{-}C_3H_7)_2GeCl_2$	−52	203	760	1.4738	1.264
CH_3GeF_3	38	96.5	751		
CH_3GeCl_3		110	727	1.4685	1.7053
CH_3GeBr_3		168	750	1.5770	2.6337
CH_3GeI_3	48–55	237	752		
$C_2H_5GeCl_3$		140	760	1.4750	1.6006
$C_3H_7GeCl_3$		163.5	756	1.4779	1.5146
$C_6H_5GeCl_3$		115	19	1.5702	1.6641
$CH_2{=}CHGeCl_3$		128.5	756	1.4815	1.6520
$CH_2{=}CHCH_2GeCl_3$		154	743	1.4928	1.5274
$Cl_3GeCH_2CH_2GeCl_3$	56	131	12		

[a] See also J. E. Griffiths, *Inorg. Chem.* **2** (1963) 375.

and tertiary amines (depending on conditions and size of R group). Lithium in THF brings about coupling of R_3GeBr and R'_3SiCl to give $GeSiR'_3$ compounds.

Organogermanium chlorides (and indeed $GeCl_4$ also) undergo an interesting and often useful reaction with diazomethane to replace chlorine ligands with chloromethyl groups, thereby establishing new Ge–C bonds[96]:

$$CH_3GeCl_3 + CH_2N_2 \longrightarrow CH_3(CH_2Cl)GeCl_2 + N_2$$

(yield 78%; b.p. 72° at 40 mm, $n_D^{25} = 1.4890$, $d_4^{25} = 1.642$).

This is an illustration of a general method for obtaining chloromethyl derivatives of

[96] D. Seyferth and E. G. Rochow, *J. Am. Chem. Soc.* **77** (1955) 907.

Ge and Si, in which a cold solution of CH_2N_2 in ether is added to the germanium chloride in the presence of a small amount of suspended copper powder. Nitrogen evolution takes place at $-60°$, and there are few by-products.

6.6. ORGANOGERMANIUM OXIDES

Considerable interest attaches to the possibility of making polymeric organogermanium oxides $(-R_2GeO-)_x$ analogous to the silicones $(-R_2SiO-)_x$, etc., so some attention has been given to the matter. When a mixture of $(CH_3)_2SiCl_2$ and $(CH_3)_2GeCl_2$ is hydrolyzed, the mixture of water-immiscible dimethylsiloxanes is found to contain no germanium; all of the germanium is found in the water layer. When $(CH_3)_2GeCl_2$ alone is added to water it dissolves, and evaporation of the solution in an oven at 105° leaves nothing behind. This indicates that the hydrolysis of $(CH_3)_2GeCl_2$ in pure water is reversible and incomplete, and that the partial hydrolysis products are water-soluble. A study of the hydrolytic dissociation[97] shows that a 0.6 M solution of $(CH_3)_2GeCl_2$ in water has a pH of 1.2, and that titration of this with 0.1 N NaOH produces a curve very much like that for the titration of HCl, with no precipitation of $(CH_3)_2GeO$ or $(CH_3)_2Ge(OH)_2$ even up to pH 12. The van't Hoff *i* factor is 5. This indicates that the hydrolytic dissociations

$$(CH_3)_2GeCl_2 \rightleftharpoons (CH_3)Ge^{++} + 2Cl^-$$

$$(CH_3)_2Ge^{++} + H_2O \rightleftharpoons (CH_3)_2GeOH^+ + H^+$$

and

$$(CH_3)_2GeOH^+ + H_2O \rightleftharpoons (CH_3)_2Ge(OH)_2(sol) + H^+$$

are taking place, and this was confirmed by precipitating sulfide, chromate, and thiocyanate salts of the $(CH_3)_2Ge^{++}$ ion[96]. Hence $(CH_3)_2GeCl_2$ behaves like a strong diprotic acid in water.

Extraction of a water solution of $(CH_3)_2GeCl_2$ with petroleum ether, followed by intensive drying of the ether layer before evaporation, leads to a cyclic tetramer $[(CH_3)_2GeO]_4$ which melts at 92°[98]. This dissolves readily in water (where it is present as monomeric units, probably $(CH_3)_2Ge(OH)_2$), and evaporation of the water solution yields a white polymer $[(CH_3)_2GeO]_x$ which melts at about 132° and appears fibrous but is revealed by electron microscopy[99] to consist of dendritic orthorhombic crystals. In the vapor phase at 200° and 100 mm, dimethylgermanium oxide exists as a trimer[98]; rapid quenching of the vapor produces unstable crystals which revert to the tetramer. The three forms of $(CH_3)_2GeO$ have different infrared spectra[98, 100], with Ge–C asymmetric stretching absorption at 598 cm^{-1}. The tetramer is non-toxic to hamsters, and causes little or no increase in the red blood cell count; both observations point toward no appreciable metabolism of the substance, and hence a biological stability[101].

*Diethyl*germanium oxide may also be made by hydrolyzing $(C_2H_5)_2GeCl_2$ prepared from Ge by the direct reaction of C_2H_5Cl[102]. The hydrolysis product is an oily mixture of polymers, less soluble in water than $(CH_3)_2GeO$ but decidedly soluble.

[97] E. G. Rochow and A. L. Allred, *J. Am. Chem. Soc.* **77** (1955) 4489.

[98] M. P. Brown and E. G. Rochow, *J. Am. Chem. Soc.* **82** (1960) 4166.

[99] E. G. Rochow and T. G. Rochow, *J. Phys. Colloid Chem.* **55** (1951) 9.

[100] M. P. Brown, R. Okawara and E. G. Rochow, *Spectr. Acta*, **16** (1960) 595.

[101] E. G. Rochow and B. M. Sindler, *J. Am. Chem. Soc.* **72** (1950) 1218. For erythropoietic effect of *inorganic* Ge compounds, see section on biological activity above.

[102] E. G. Rochow, *J. Am. Chem. Soc.* **72** (1950) 198.

6.7. ORGANOFUNCTIONAL ORGANOGERMANIUM COMPOUNDS

Besides the simple alkyl and aryl germanium hydrides, halides, and related derivatives, there are many organogermanium compounds which have organofunctionality of one or more types:

(a) Unsaturation, as in vinyl, allyl, and other alkenyl and alkynyl compounds, obtained by dehydrochlorination or by Grignard or related techniques.

(b) Halogen substitution, as in CH_2Cl–Ge compounds from diazomethane reactions, or CF_3–Ge compounds (from addition of CF_3I to GeI_2, etc.).

(c) Cyano-, acetoxy-, and alkoxy-alkyl derivatives, often obtained from the corresponding haloalkyl compounds, or by such specialized procedures as addition of acrylonitrile to Ge–H bonds. Fifty such compounds are listed (with properties and references) in the Weiss compilation[86].

6.8. OTHER ORGANOGERMANIUM COMPOUNDS

A long series of organogermanium *pseudohalides* has been obtained by the action of AgCN, AgCNS, AgCNO, etc., on organogermanium halides[103]. The properties of some representative compounds are given in Table 14.

TABLE 14. ORGANOGERMANIUM PSEUDOHALIDES

Compound	M.p., °C	B.p., °C	At. pr., mm	n_D^{20}	d_{20}
$(CH_3)_3GeCN$	38.5	150	760	—	
$(CH_3)_3GeCNS$		192	760	1.4960	
$(CH_3)_3GeN_3$	−65	138	760		
$(CH_3)_2Ge(N_3)_2$	−14	45.2	2		
$(C_2H_5)_3GeCN$	18	213	760	1.4509	1.111
$(C_2H_5)_3GeNCO$	−26.4	200	760	1.4519	
$(C_2H_5)_3GeNCS$	−46	252	760	1.517	1.184
$(C_2H_5)_2Ge(NCO)_2$	−32	135	52	1.4619	1.330
$(C_2H_5)_2Ge(NCS)_2$	16	114	1		1.356
$C_2H_5Ge(NCO)_3$	−31	139	52	1.4739	1.5344
$(n\text{-}C_3H_7)_3GeCN$	−13	116	10	1.544	1.041
$(n\text{-}C_3H_7)_3GeNCO$	−19	114	10	1.4575	1.055
$(n\text{-}C_3H_7)_3GeNCS$	−56	143	9	1.5063	1.105
$(n\text{-}C_4H_9)_3GeNCO$		109	2	1.4595	1.044
$(n\text{-}C_4H_9)_3GeNCS$		135	2	1.5039	1.071

Similarly, *organogermanium esters* of organic and inorganic acids may be made by reaction of silver or other heavy-metal salts of the acids with organogermanium halides, or by reaction of the latter with the acids themselves (or their anhydrides). Such esters act like organogermanium halides in their sensitivity to water and their solvolytic dissociations. Many are known (see ref. 86 for a listing); space permits only the physical constants for the trimethylgermanium esters and a few higher trialkylgermanium compounds in Table 15.

[103] H. H. Anderson, *J. Am. Chem. Soc.* **71** (1949) 1799; *ibid.* **72** (1950) 194; *ibid.* **73** (1951) 5439, 5798; *ibid.* **74** (1952) 1421, 2370, etc.; *ibid.* **79** (1957) 326.

TABLE 15. TRIALKYLGERMANIUM ESTERS OF INORGANIC ACIDS[86]

Compound	M.p., °C	B.p., °C	At. pr., mm	n_D^{20}	d_{20}
$[(CH_3)_3Ge]_3BO_3$		128	0.25	1.4607	1.1723
$(CH_3)_3GeCl_2PO_2$	61	dec. 100	subl. 1		
$[(CH_3)_3Ge]_3AsO_4$	46	130	1.5		
$[(CH_3)_3Ge]_3VO_4$	−18				
$[(CH_3)_2Ge]_2SO_4$	138	subl.			
$(CH_3)_3GeSO_3OSi(CH_3)_3$	93	dec.			
$[(CH_3)_3Ge]_2SeO_4$	147	135	1		
$(CH_3)_3GeClO_4$	5	91	2		
$[(C_2H_5)_3Ge]_2SO_4$	−4	165	3		
$(C_2H_5)_3GeSO_3CH_3$		280	760	1.4650	1.286
$[(n\text{-}C_3H_7)_3Ge]_2SO_4$		370	760		1.186
$[(i\text{-}C_3H_7)_3Ge]_2SO_4$		380	760	1.482	1.217
$(n\text{-}C_4H_9)_3GeSO_3C_2H_5$		337	760	1.4654	1.117

The *organogermanium oxides*[86] are best represented by the solid polymers of $(CH_3)_2GeO$ (*vide supra*) and by the bis-trialkylgermanium oxides and their aryl counterparts (which are liquids prepared by hydrolysis of the chlorides R_3GeCl, or by oxidation of the hydrides, or by hydrolytic cleavage reactions). The physical properties of a few such germanoxanes are given in Table 16, together with those of $(CH_3)_2GeO$ for comparison.

TABLE 16. ORGANOGERMANIUM OXIDES[86]

Compound	M.p., °C	B.p., °C	At. pr., mm	n_D^{20}	d_{20}
$(CH_3)_3GeOGe(CH_3)_3$	−61.1	137	730	1.4308	1.2086
$(C_2H_5)_3GeOGe(C_2H_5)_3$		131	20	1.4612	
$n\text{-}Pr_3GeOGen\text{–}Pr_3$	−55	175	14	1.4648	
$i\text{-}Pr_3GeOGei\text{–}Pr_3$		315 dec.	760	1.4836	1.112
$i\text{-}Pr_3GeOH$	−15	216	760	1.472	1.077
$n\text{-}Bu_3GeOGen\text{–}Bu_3$		173	1	1.4652	
$(n\text{-}C_5H_{11})_6Ge_2O$		157	0.06	1.4656	
$(n\text{-}C_6H_{13})_6Ge_2O$		210	0.04	1.4645	0.963 at 25°
$[(CH_3)_2GeO]_3$	133.4				
$[(CH_3)_2GeO]_4$	91	88	1		
$[(C_2H_5)_2GeO]_3$	19				
$[(C_2H_5)_2GeO]_4$	27.1	129	3		

Besides the organogermanium oxides which invoke purely Ge–O–Ge structures, it is quite easy to make *mixed organometallic oxides* which involve Ge–O–Si and Ge–O–Sn linkages. These result from the interaction of compounds of the type R_3SiOLi with R_3GeCl, or from the reaction of R_3SiCl with R_3GeOLi, and so on. The properties of some of these mixed oxides appear in Table 17. The oxygen linkage is reactive to several kinds of cleavage agent: phenyl lithium converts $(CH_3)_3GeOSi(CH_3)_3$ to $(CH_3)_3GeC_6H_5$ and $(CH_3)_6Si_2O$[86], and $AlCl_3$ converts it to $(CH_3)_3GeCl$ and $(CH_3)_3SiOAlCl_2$. Sulfur trioxide converts the same substance to the mixed sulfate $(CH_3)_3GeOSO_2OSi(CH_3)_3$[86]. The substance $(CH_3)_3GeOSi(CH_3)_3$ itself is said to be very toxic[86], which is surprising in that its fragments are not known to be.

TABLE 17. SOME ORGANOGERMANIUM MIXED OXIDES[86]

Compound	M.p., °C	B.p., °C	At. pr., mm
$(CH_3)_3GeOSi(CH_3)_3$	−68	117	723
$(CH_3)_3GeOSi(C_2H_5)_3$		33	1
$(C_2H_5)_3GeOSi(CH_3)_3$		77	16
$(CH_3)_2Ge[OSi(CH_3)_3]_2$	−61	54.5	11
$CH_3Ge[OSi(CH_3)_3]_3$		70	10
$(CH_3)_3GeOSn(CH_3)_3$		51	12

Closely related to the mixed oxides are the *organogermanium alkoxides*, which contain the linkage Ge–O–C and may be considered as alkoxide counterparts of the organogermanium halides. Indeed, they may readily be made from such halides by the action of sodium or lithium alkoxides, or by the action of the alcohols themselves if an HCl acceptor such as a tertiary amine also is used. Conversion reactions of organogermanium alkoxides with higher alcohols also are possible. The physical properties of some such alkoxides are given in Table 18.

TABLE 18. ORGANOGERMANIUM ALKOXIDES[86]

Compound	B.p., °C	At. pr., mm	n_D^{25}	d_{25}
$(CH_3)_3GeOCH_3$	88	753	1.401	1.075
$CH_3Ge(OCH_3)_3$	137	760	1.4053	1.264
$(C_2H_5)_3GeOCH_3$	163	760	1.4362[a]	1.068[a]
$(C_2H_5)_3GeOC_3H_7$	190	760	1.4388[a]	1.025[a]
$(C_2H_5)_3GeOCH_2C{\equiv}CH$	92	17	1.46009	1.095[a]
$(C_2H_5)_3GeOCH_2CH{=}CH_2$	85	19	1.4452[a]	1.041[a]
$(C_2H_5)_3GeOC_6H_5$	140	18	1.5102[a]	1.133[a]
$(C_2H_5)_3GeOCH_2CH_2OGe(CH_3)_3$	162	11	1.4640[a]	1.149[a]
$(C_4H_9)_3GeOCH_3$	136	16	1.4502[a]	0.988[a]
$(C_4H_9)_3GeOC_2H_5$	129	10	1.4481[a]	0.972[a]
$(C_4H_9)_3GeOCH{=}CH_2$	86	1	1.4580[a]	0.988[a]
$(C_4H_9)_3GeOC_6H_{11}$	180	16	1.4680[a]	0.996[a]

[a] At 20°C.

As expected, there also are many *organogermanium sulphides* and *mercaptides*. These are obtained by the reaction of organogermanium halides with sodium sulfide or mercaptide, or by reaction with H_2S in the presence of an HCl acceptor. Some compounds are listed in Table 19.

TABLE 19. ORGANOGERMANIUM SULFIDES AND MERCAPTIDES[86]

Compound	M.p., °C	B.p., °C	At. pr., mm	n_D^{20}	d_{20}
$(CH_3)_3GeSGe(CH_3)_3$	−27	68	12		
$[(CH_3)_2GeS]_3$	55.5	302	760		
$[(i\text{-}C_3H_7)_2GeS]_3$		119	1	1.551	1.327
$(CH_3)_3GeSeGe(CH_3)_3$		94	13		
$[(CH_3)_2GeSe]_3$	53				
$(C_2H_5)_3GeSCH_2C_6H_5$		131	1	1.549	1.139
$(C_2H_5)_3GeSC_4H_9$		120	4	1.4880	1.0546
$(C_2H_5)_3GeSC_6H_{13}$		109	1	1.488	1.029
$(C_2H_5)_3GeSC_6H_5$		113	1	1.553	1.153
$(C_6H_5)_3GeSH$	112				
$(C_6H_5)_3GeSCH_3$	87				

Similarly, *nitrogen derivatives* of organogermanium halides can be made by reaction with alkali–metal amides, or (in some cases) by reaction with ammonia or amines directly. Transamination is possible as a way of making one organogermanium amine from another, and even some organogermanium oxides can be converted to amines or amides by azeotropic distillation with an amine[86]. In general, organogermanium compounds of nitrogen and phosphorus and are sensitive to water, acids, and oxidizing agents. The properties of some representative compounds are given in Table 20. Related compounds containing the Ge–N–Sn and Ge–N–Pb linkages also are known[86], and can be prepared similarly.

TABLE 20. ORGANOGERMANIUM DERIVATIVES OF NITROGEN[86]

Compound	M.p., °C	B.p., °C	At. pr., mm	n_D^{20}	d_{20}
$[(CH_3)_3Ge]_3N$		60	2		
$[(CH_3)_3Ge]_2NH$		47	17		
$C_2H_5Ge[N(CH_3)_2]_3$	−46	106	34		1.049[a]
$C_2H_5Ge[N(C_2H_5)_2]_3$		118	12		1.108[a]
$Ge[N(C_2H_5)_2]_4$		109	2		1.215[a]
$[(C_4H_9)_2GeNH]_3$		172	0.004	1.4889	
$(C_4H_9)_2Ge[N(CH_3)_2]_2$		116	7	1.4605	1.001

[a] At 22°C.

Lastly, there are *organodigermanes* and *polygermanes*, made by the condensation of organogermanium halides, using alkali–metal acceptors for the halogen atoms. Mixed compounds containing Ge–Si, Ge–Sn, and Ge–Pb bonds also may be made, usually by the reaction of R_3GeK or R_3GeLi with the appropriate organometallic halide (or vice versa, using R_3GeCl and R_3SiK, etc.). Intercondensations such as that of R_3GeBr with R_3SiCl (using sodium) also are possible. Cleavage reactions of such polymetalloidal compounds provide the basis for much speculation about relative electronegativities, but the relative bond energies usually provide a more reliable clue to what happens. The physical properties of some organogermanes and related compounds are given in Table 21.

TABLE 21. ORGANODIGERMANES, POLYGERMANES AND RELATED COMPOUNDS[86] [c]

Compound	M.p., °C	B.p., °C	At. pr., mm	n_D^{20}	d_{20}
$(CH_3)_3GeGe(CH_3)_3$	−40	138	750	1.4564	
$(C_2H_5)_3GeGe(C_2H_5)_2$		62	0.007	1.4690	
$(C_2H_3)_3GeGe(C_2H_3)_2$		55	0.3	1.5217[a]	1.171[a]
$(C_6H_5)_3GeGe(C_6H_5)_3$	330	271[b]	1[b]		
$CH_3[Ge(CH_3)_2]_3CH_3$		44	0.05	1.4940	1.2311
$CH_3[Ge(CH_3)_2]_4CH_3$		82	0.8	1.5161	1.3094
$CH_3[Ge(CH_3)_2]_7CH_3$		97	0.2	1.5356	1.3806
$(C_2H_5)_3GeSi(C_2H_5)_3$		255	760	1.4860	0.9791[d]
$(C_6H_5)_3GeSi(C_2H_5)_3$	96				
$(C_6H_5)_3GeSi(C_6H_5)_3$	357				
$(C_2H_5)_3GeHgSi(C_2H_5)_3$		131	1.5		

[a] At 25°C.
[b] Sublimation conditions.
[c] Many more complicated organopolygermanes are listed in an article by H. Gilman entitled Catenated organic compounds of silicon, germanium, tin and lead, in *Adv. Organomet. Chem.* **4** (1966) 1–94.
[d] At 26°C.

7. COMPLEXES OF GERMANIUM(IV)

7.1. FLUOROGERMANATES AND CHLOROGERMANATES

As described in the section dealing with GeF_4, many fluorogermanate salts of organic and inorganic bases may be made by dissolving GeO_2 in 40% aq. HF and then adding the base[104]:

$$GeO_2+6HF = H_2GeF_6+2H_2O$$

$$H_2GeF_6+2C_6H_5NH_2 = (C_6H_5NH_3)_2GeF_6$$

Similarly, *chloro*germanates such as Cs_2GeCl_6 have been known many years[71]. Such halocomplexes of the form M_2GeX_6 constitute the simplest and earliest complexes of Ge(IV), and reveal much about germanium with a coordination number of six. The octahedral GeF_6 grouping is a dinegative ion, of course, but within the anion the bonding is much the same as that in the neutral SF_6 [105]. Alkali–metal fluorogermanates are soluble in water; the heavy-metal ones are not. The solubilities and densities of some nitrogen-base fluorogermanates[104] are given in Table 22.

TABLE 22. SOME NITROGEN-BASE FLUOROGERMANATES[104]

	d_{25}	H_2O	CH_3OH	C_2H_5OH
$(NH_4)_2GeF_6$	2.564	sol.	insol.	insol.
$(NH_3OH)_2GeF_6$	2.492	sol.[a]	sol.	sl. s.
$(N_2H_5)_2GeF_6$	2.406	sol.[a]	sl. s.	sl. s.
$(C_6H_5NH_3)_2GeF_6$	1.579	sol.[a]	sol.	s. hot
$(C_6H_5NH_2CH_3)_2GeF_6$	1.631	sol.[a]	sol.	sl. s.
$(C_6H_5NHMe_2)_2GeF_6$	1.548	sol.[a]	sol.	sol.

[a] With hydrolysis.

7.2. COMPLEXES OF TRANSITION METALS

The newer coordination chemistry of germanium deals largely with germane, chlorogermane, and organogermanium derivatives of π-complexes of the transition metals[106]. The presence of π acceptor ligands "conditions" the transition metal to form σ bonds to certain alkyl groups[107], and also to analogous groups containing other Group IV elements. A great many such compounds are known[106], some made by metathetical reactions involving a carbonyl metal anion:

$$(C_6H_5)_3GeCl+\pi\text{-}C_5H_5W(CO)_3^- = (C_6H_5)_3GeW(CO)_3\pi\text{-}C_5H_5+Cl^-$$

[104] L. M. Dennis, B. J. Staneslow and W. D. Forgeng, *J. Am. Chem. Soc.* **55** (1933) 4392.
[105] W. Hückel, *Structural Chemistry of Inorganic Compounds*, Vol. 2, p. 689, Elsevier Publ. Co., Amsterdam (1951).
[106] F. G. A. Stone, *New Pathways in Inorganic Chemistry* (ed. Ebsworth, Maddock and Sharpe), chapter 12, pp. 283–302, Cambridge Univ. Press (1968).
[107] J. Chatt and B. L. Shaw, *J. Chem. Soc.* **1959**, 705; J. Chatt, *Rec. Chem. Prog.* **21** (1960) 147.

This method can be used to attach four metal atoms to Ge, starting with $GeCl_4$. Other compounds are made by elimination reactions, such as

$$HGeCl_3+ClMn(CO)_5 = Cl_3GeMn(CO)_5+HCl$$

Another general method consists in the insertion of germanium *dihalides* into metal–metal bonds, giving M–Ge–M configurations:

$$GeI_2+Co_2(CO)_8 = (CO)_4Co(GeI_2)Co(CO)_4$$

Lithium derivatives of organogermanium compounds can also be used to react with halogen-bearing transition metals in suitable complexes:

$$2(C_6H_5)_3GeLi+[(C_2H_5)_3P]_2PtCl_2 = [(C_6H_5)_3Ge]_2Pt[P(C_2H_5)_3]_2+2LiCl$$

The structures of some transition metal–germanium compounds have been determined by X-ray crystallographic study, and these show the Ge atoms to be in distorted tetrahedral configuration, while the transition metals (here iron and manganese) show basically octahedral bonding[106]. In $(C_6H_5)_3GeMn(CO)_5$ the space group is P_1, and the Ge–Mn distance is 2.535±0.02 Å[108]. In $Cl_2Ge[Fe(CO)_2\pi\text{–}C_5H_5]_2$ the space group is $C\ 2/c$, and the Ge–Fe distance is 2.36±0.01 Å[109].

The square-planar Pt–Ge complexes are said to be stable up to 150°, but the triphenylgermanyl–palladium complex $[(C_2H_5)_3P]_2Pd[Ge(C_6H_5)_3]_2$ decomposes in solution at 20° [106]. The transition metal–germanium bonds are only moderately reactive: the compounds $(R_3P)_2Pt[Ge(C_6H_5)_3]_2$ are unaffected by air and water, but the Ge–Pt bonds are cleaved by I_2, HCl, and CCl_4. Reduction of Ge–Cl bonds with $NaBH_4$ is possible without destroying the Ge–Fe bond[106]:

$$Cl_2Ge[Fe(CO)_2\pi\text{–}C_5H_5]_2 \xrightarrow{NaBH_4} H_2Ge[Fe(CO)_2\pi\text{–}C_5H_5]_2$$

Related compounds containing organogermanium groups bonded to Cu, Ag, Au, Ti, Zr, Hf, Nb, Ta, Cr, W, and Mo have been collected by Weiss[86] from the patent literature (where they are of interest as catalytic agents), but the compositions and structures are not always clear. Some interest in such complexes as antiknock agents and as starting materials for thermal metal plating also is expressed.

[108] B. T. Kilbourne, T. L. Blundell and H. M. Powell, *Chem. Comms.* **1965**, 444.
[109] M. A. Bush and P. Woodward, *J. Chem. Soc.* **1967**, 1883.

17. TIN

E. W. Abel
University of Exeter

1. THE ELEMENT

1.1. HISTORICAL

In the book of Numbers in the Bible's Old Testament, tin is mentioned as a metal of value under the name *bedil*. The ancient Indian author Veda refers to tin as *trapu*. Objects made of tin have been found in the tombs of ancient Egypt[1], and the tin–copper alloy bronze has been used from ancient times. Caesar, recording the presence of tin in Britain, referred to it as *plumbum album* as also did Pliny, to distinguish it from lead, which was *plumbum nigrum*.

Up to the 12th century A.D., the tin deposits of Cornwall were the only large European source of tin[2]. Subsequently the mines of Saxony and Bohemia became important. Nowadays, however, the Continent of Europe is almost entirely dependent upon imported ores.

1.2. OCCURRENCE AND DISTRIBUTION

Tin is found in nature almost exclusively as the tin(IV) oxide known as cassiterite or tinstone. Small quantities of stannite, $Cu_2S.FeS.SnS_2$, are known, and occasionally small amounts of tin metal are found in nature along with gold.

Cassiterite is found as primary deposits interspersed in other rocks particularly as "reef-tin" in granite, or alternatively in secondary deposits as "stream-tin", mixed with large quantities of clay and sand.

Principal suppliers of tin are Malaysia, Bolivia, Indonesia, Congo, Siam and Nigeria. Further important deposits are worked in China, Australia, Rhodesia, South Africa, Alaska, United States, Chile and the United Kingdom.

Tin is estimated to be present in the earth's crust as $4 \times 10^{-3}\%$ by weight, and to be in sea water at a concentration of 0·003 g/ton.

1.3. EXTRACTION

Despite the 78·6% of tin in pure SnO_2, the tin ores for extraction of the metal often contain only a few per cent of tin. Initial concentration is carried out by flotation removal of lighter rocks, such as silica, and magnetic removal of certain impurities such as tungsten minerals. Roasting volatilizes sulphur and arsenic and oxidizes many metals

[1] A. H. Church, *Chem. News*, **36** (1877) 168.
[2] T. R. Holmes, *Ancient Britain*, p. 483, Oxford (1907).

present, such as bismuth, zinc, iron and copper, into oxides which can be washed out by acid.

Cassiterite is reduced by carbon in a blast or reverberatory furnace. The crude molten tin is usually contaminated with iron, which is removed by oxidation. (This used to be carried out by "poling" the molten tin with a fresh green wood pole, which on charring released gas vigorously, and stirred the tin into contact with atmospheric oxygen.)

Tin is extensively recovered from tin plate scrap, either by electrolysis, whereby the scrap is made the anode in a caustic soda cell, or by detinning with chlorine to produce tin(IV) chloride. Iron unlike tin is not readily attacked by dry chlorine.

1.4. INDUSTRIAL AND COMMERCIAL UTILIZATION OF TIN

The primary tin consumption for the U.S.A., Japan, U.K., Germany and France for 1968 totalled over 120,000 tons. This was utilized as tinplate (47%), solder (21%), bronze (8%), babbitt (bearings) (6%) and tinning (5%). The remaining 13% was utilized for a wide miscellany of alloys and chemicals.

In addition to the extensive use of tin metal itself and its alloys, various compounds of tin are used in industry and commerce. Tin(IV) oxide is used as an opacifier for vitreous enamels, in ceramic glazes and as a polishing powder. Tin chlorides are used to weight natural silk, stabilize soap perfumes, silver glass mirrors and as chemical reducing agents. Tin(II) fluoride finds extensive use as an additive to dentifrices[3].

Monobutyltin compounds stabilize plastic films, and dibutyl and other organotins are present in up to 1% by weight of tin in polyvinylchloride, as a stabilizer[4].

Tributyltin compounds are used in industrial fungicides, insecticides, anti-fouling paints and disinfectants; and triphenyltin derivatives are now manufactured in hundreds of tons for agricultural pesticide purposes[5].

1.5. ALLOTROPES OF TIN

Tin has two crystalline forms. Above 18° the stable form is the so-called β or white tin, and below 18° the stable form is the so-called α or grey tin. An often described γ form of tin has no basis in fact.

In the grey α-tin each atom has four tetrahedral bonds resulting in three-dimensional covalent bonding throughout the crystal. In this α-form tin has an analogous structure to the diamond form of carbon.

In β-tin each atom has four nearest neighbours at 3·016 Å in the form of a very flattened tetrahedron, and two further neighbours at 3·175 Å. Thus instead of tetrahedral co-ordination as in α-tin, the white β-form of tin has each tin atom in a somewhat distorted octahedral environment as shown in Fig. 1.

Tin metal, as normally utilized, is the white β-form, and although the transition temperature for β-tin to α-tin is about 18°, the conversion of white to grey tin does not

[3] *Tin in Your Industry*, Ed. W. T. Dunne, The Tin Industry Board, Malaysia.

[4] H. Verity Smith, *Organotin Stabilizers*, The Tin Research Institute, Greenford, Middlesex, England (1959).

[5] J. G. A. Luijten and G. J. M. van der Kerk, *A Survey of the Chemistry and Applications of Organotin Compounds*, The Tin Research Institute, Greenford, Middlesex, England (1952).

take place except at much lower temperatures. An exception to this, however, is when grey tin is present and catalyses conversion of β-tin to α-tin below 18°.

Once started, the change from β-tin to α-tin proceeds quite rapidly, and due to the very

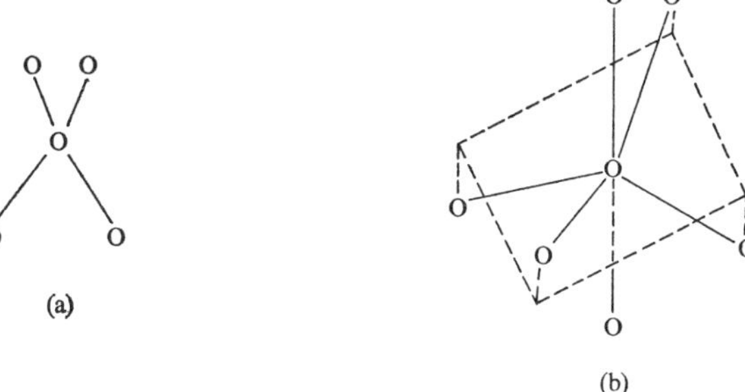

FIG. 1. (a) The tetrahedral environment of the tin atom in α-tin (cubic) and (b) the distorted octahedral environment of the tin atom in β-tin (tetragonal).

different crystalline forms and densities of the two allotropes, this conversion causes the tin to completely crumble and powder. The consequent destruction of tin objects caused this allotropic change to be known as tin disease, tin pest or tin plague.

1.6. ISOTOPES OF TIN

The isotopes of tin are listed in Table 1. None of the naturally occurring isotopes of tin is radioactive, and it is notable that no single isotope constitutes even one-third of natural tin.

TABLE 1. THE NATURAL AND ARTIFICIAL ISOTOPES OF TIN

Isotope	% abundance in natural tin	Half-life	Isotope	% abundance in natural tin	Half-life
^{108}Sn	—	0·167 days	^{119m}Sn	—	250 days
^{109}Sn	—	1080 sec	^{119}Sn	8·58	—
^{110}Sn	—	0·171 days	^{120}Sn	32·97	—
^{111}Sn	—	2100 sec	^{121m}Sn	—	> 400 days
^{112}Sn	0·95	—	^{121}Sn	—	1·05 days
^{113}Sn	—	118 days	^{122}Sn	4·71	—
^{114}Sn	0·65	—	^{123m}Sn	—	2370 sec
^{115}Sn	0·34	—	^{123}Sn	—	136 days
^{116}Sn	14·24	—	^{124}Sn	5·98	—
^{117m}Sn	—	14 days	^{125m}Sn	—	570 sec
^{117}Sn	7·57	—	^{125}Sn	—	9·9 days
^{118}Sn	24·01	—	^{126}Sn	—	3000 sec
			^{127}Sn	—	5400 sec

1.7. CHEMICAL PROPERTIES OF TIN METAL

Tin is virtually unattacked in air and water at ordinary temperatures, but at higher temperatures a film of oxide forms on the surface. Burning tin at white heat produces tin(II) oxide. Dilute acids only attack tin slowly to give tin(II) salts and hydrogen.

TABLE 2. PHYSICAL DATA FOR TIN
(Temperatures °C)

Melting point 232°; $\triangle H$ fusion 1·69 kcal/mole; $\triangle S$ fusion 3·35 cal/deg/mole
Boiling point 2687°; $\triangle H$ evaporation 69 kcal/mole

Temperature	700	1000	1200	1700	2300
Vapour pressure (mmHg)	$7 \cdot 4 \times 10^{-6}$	$8 \cdot 3 \times 10^{-3}$	0·17	30·6	638

Critical temperature	3460° (?)
Critical pressure	650 atm
Specific heat (cal/deg/mole)	6·30 (25°)
Heat of formation at 25°	$\triangle H^\circ$ = 0·00 (solid) $\triangle H^\circ$ = 72 kcal/mole (vapour)
Free energy of formation at 25°	$\triangle G^\circ$ = 0·00 (solid) $\triangle G^\circ$ = 64 kcal/mole (vapour)
Entropy at 25°	S° = 10·7 cal/deg/mole (solid) S° = 40·2 cal/deg/mole (vapour)
Thermal conductivity	0·157 (18°), 0·143 (200°), 0·078 (500°) cal/sec/cm/deg
Density	α 7·285, β 5·769 (−2·7% change in density on fusion)
Electrical resistivity at 20°	11×10^{-6} ohm cm; superconductor below 3·72°K

Ion	Sn^{+}	Sn^{2+}	Sn^{3+}	Sn^{4+}	Sn^{5+}	Sn^{6+}	Sn^{7+}	Sn^{8+}	Sn^{9+}
Ionization potential (eV)	7·332	14·6	30·7	46·4	91	(103)	(126)	(151)	(176)

Surface tension at 400° = 700±25 dynes/cm
Viscosity of liquid tin (centipoises) 1·88 (250°), 1·38 (400°), 0·87 (800°)

Temperature	−173	−73	10	20	
Form	α	α	α	β	liquid
Specific magnetic susceptibility $\times 10^{-6}$ cgs	−0·267	−0·289	−0·312	+0·026	−0·038

Ionic radii Sn^{2+} 1·02 Å; Sn^{4+} 0·71 Å
$\alpha Sn \rightarrow \beta Sn$ at 18° $\triangle H$ 0·6 kcal/mole $\triangle S$ 2·1 cal/deg/mole

TABLE 3. THERMODYNAMIC DATA FOR CERTAIN TIN SPECIES

Species	State	$\triangle H^\circ$ kcal/mole	$\triangle G^\circ$ kcal/mole	S° cal/deg/mole
α-Tin	crystal	0·6	1·1	10·7
β-Tin	crystal	0	0	12·3
Tin	gaseous	72	64	40·25
Sn^{+}	gaseous	242·6	—	—
Sn^{2+}	gaseous	581·4		
Sn^{2+}	aqueous	−2·39	−6·28	−5·9
Sn^{4+}	aqueous		0·65	
SnO	crystal	−68·4	−61·5	13·5
SnO_2	crystal	−138·8	−124·2	12·5
SnF_6^{2-}	aqueous	−474·7	−420	
$Sn(OH)_6^{2-}$	aqueous		−310·5	—
$SnCl_2$	crystal	−83·6	−72·2	29·3
$SnCl_4$	liquid	−130·3	−113·3	61·8
$SnBr_2$	crystal	−63·6	−59·5	34·9
$SnBr_4$	liquid	−97·1		
SnI_4	crystal	−34·4	−34·4	40·3
SnS	crystal	−18·6	−19·7	23·6
$Sn(SO_4)_2$	crystal	−393·4	−346·8	37·1

Concentrated nitric acid produces the hydrated tin(IV) oxide sometimes called metastannic acid. Tin evolves hydrogen with hot alkali solutions to form alkali stannates.

A number of dilute chemical solutions attack tin metal slowly, with consequent corrosion. Among these are aluminium chloride, aluminium potassium alum, ammonium sulphate, hydrochloric acid, hydrobromic acid, hydriodic acid, fluorine, chlorine, bromine, iodine, iron(III) chloride, potassium carbonate, potassium hydroxide, silver nitrate, zinc chloride, glycerol and maleic acid[6].

1.8. NUCLEAR MAGNETIC RESONANCE

The technique of nuclear magnetic resonance spectroscopy has been extensively utilized in tin chemistry.

There are three naturally occurring isotopes of tin which have non-zero nuclear magnetic moments—^{115}Sn (0·35% abundance), ^{117}Sn (7·67% abundance) and ^{119}Sn (8·68% abundance). In each of these isotopes the spin number (I) of the nucleus is one half; thus there are no quadrupole moments[7].

One of the most extensively studied effects of the presence of these three spin-active tin isotopes has been in organotin chemistry, where the presence of satellite bands on many

TABLE 4. CHEMICAL SHIFTS OF THE ^{119}Sn NUCLEAR MAGNETIC RESONANCE IN VARIOUS COMPOUNDS FROM THE ^{119}Sn RESONANCE IN $(CH_3)_4Sn$

Compound	Shift (δ) ppm
SnI_4 (CS_2 soln.)	1701
$SnSO_4$	909
$SnBr_4$	638
$SnCl_2$ (aqueous soln.)	521†
$SnCl_4$	150
$Sn(C_4H_9)_4$	12
$SnCl_3(C_4H_9)$	3
$(CH_3)_4Sn$	0
$(CH_3)_2SnCl_2$	−36
$(C_2H_5)_2SnCl_2$	−62
$(C_4H_9)_2SnCl_2$ (in acetone)	−71
$(C_4H_9)_2SnCl_2$ (in CS_2)	−114
$(C_2H_5)_3SnCl$	−151

† This figure is concentration dependent, and can be lowered by up to 200 ppm by the addition of hydrochloric acid, due presumably to the build up of $SnCl_3^-$ ion concentration.

of the proton magnetic resonances is due to the spin–spin interaction with tin isotopes of non-zero magnetic moments in about 17% of the molecules present. In common with many metal alkyls, the coupling constant for interacting nuclei separated by three bonds is usually greater than for nuclei separated by two bonds. Thus in tetraethyltin $J_{Sn\text{-}CH_2\text{-}CH_3}$ for the methyl protons is $\pm 71{\cdot}2$ c/s, and $J_{Sn\text{-}CH_2}$ for the methylene protons is $\pm 32{\cdot}2$ c/s.

[6] *Handbook of the Physicochemical Properties of the Elements*, Ed. G. V. Samsonov, Oldbourne Press, London (1968).

[7] J. W. Emsley, J. Feeney and L. H. Sutcliffe, *High Resolution Nuclear Magnetic Resonance Spectroscopy*, Vol. 2, p. 1082, Pergamon Press (1966).

In proton magnetic resonance spectra the satellite bands due to ^{117}Sn and ^{119}Sn can usually be well resolved.

In observing the nuclear magnetic resonance of the tin isotope atoms, however, the ^{119}Sn isotope is both slightly more sensitive to detection, and is also the most abundant isotope. Thus it is usual to observe the ^{119}Sn spectra in nuclear magnetic resonance investigations on tin.

The most common reference material to date for tin resonance measurements is the ^{119}Sn line in tetramethyltin. This is usually used as a capillary external reference. Table 4 shows some ^{119}Sn chemical shifts from the ^{119}Sn resonance in $(CH_3)_4Sn$.

From the table it will be seen that there are not different regions of resonance for Sn^{II} compounds and Sn^{IV} compounds. Large solvent effects in both chemical shifts and coupling constants involving the ^{119}Sn nucleus are extensively reported, and are assumed to reflect the different coordination tendencies of tin towards many solvents, and to other species present in solution[8].

Many useful relationships enable ^{119}Sn chemical shift data to be extensively applied; thus for example when chlorine, bromine or iodine are attached to tin, each makes a fixed contribution to the overall ^{119}Sn shift regardless of the other groups attached to tin.

The very large range over which the ^{119}Sn chemical shifts extend suggests that the paramagnetic contribution to the shielding is the major factor controlling the shielding of the tin nucleus.

1.9. MÖSSBAUER SPECTROSCOPY

In addition to being a most powerful probe of the structure and nature of solid tin containing materials in the laboratory, the Mössbauer effect is rapidly gaining usage in the field of analysis of mineralogical tin-containing samples[9]. A portable Mössbauer spectrometer weighing less than 7 lb is now available for this purpose.

Isomer shift values seem to follow the electron density in the 5*s* subshell of tin, and hence it is possible to distinguish between the formal oxidation states of tin(II) and tin(IV). If the α (grey) form of tin is taken as a standard, then all tin(IV) compounds fall below this and all tin(II) compounds are above[10].

In many minerals of the type $MSnO_n$, the relative oxidation states of M and Sn were often assigned intuitively, and frequently erroneously. Mössbauer spectroscopy has now been utilized to assign definitively an oxidation state of tin, and hence of M in these materials. Similarly the so-called di-organotins R_2Sn were first shown by Mössbauer spectroscopy to be polymeric derivatives of tin(IV), rather than tin(II) compounds.

Most Mössbauer measurements of ^{119m}Sn spectra are taken at very low temperature, but it has been noted that six-coordination of tin is usually accompanied by the presence of a measurably large resonance at ordinary temperatures[11].

A useful feature of ^{119m}Sn Mössbauer spectra is the quadrupole splittings, which can be informative on the environmental symmetry of tin in a compound. A non-zero value of field gradient is expected in any molecule which does not possess cubic symmetry, with an accompanying quadrupole splitting of the Mössbauer resonance.

[8] J. J. Burke and P. C. Lauterbur, *J. Am. Chem. Soc.* **83** (1961) 326.

[9] J. J. Zuckermann, in *Mössbauer Effect Methodology*, **3** (1967) 15.

[10] V. S. Shpinel, V. A. Bryukhanov and N. N. Delyagin, *Soviet Physics J.E.T.P.* **14** (1962) 1256.

[11] V. I. Goldanski, E. F. Makarov, R. A. Stukan, T. N. Sumarokova, V. A. Trukhtanov and V. V. Khrapov, *Doklady Akad. Nauk. S.S.S.R.* **156** (1964) 474.

Very curiously, there is an absence of this quadrupole splitting in many tin compounds possessing obvious chemical asymmetry at the tin atom. Such asymmetric tin compounds have been classified into two sets. In the first set, where tin is directly bonded to atoms possessing lone pairs of electrons (F, Cl, Br, I, O, S, N, etc.), there is usually the expected measurable quadrupole splitting. In the second set of asymmetric Sn(IV) compounds, where tin is bonded directly to atoms bereft of lone pair electrons (H, Li, C, Ge, Sn, Pb, etc.), the predicted quadrupole splitting is not observed[12].

TABLE 5. THE MÖSSBAUER SHIFTS OF TIN AND SOME TIN COMPOUNDS RELATIVE TO α-TIN

Compound	Shift (δ) relative to α-tin, mm/sec
Tin(II) chloride	2·40
Tin(II) bromide	2·10
Tin(II) iodide	2·00
Tin(II) fluoride (monoclinic)	1·45
Pyridine tin(II) dichloride	1·30
Tin(II) sulphide	1·25
Tin(II) fluoride (orthorhombic)	1·20
Tin(II) telluride	1·20
Tin(II) orthophosphate	1·10
Tin(II) oxide ((tetragonal)	0·70
β-Tin	0·60
Tin(II) oxide (orthorhombic)	0·60
α-Tin	0
Tin(IV) iodide	−0·20
Organometallic tin(IV) compounds	−0·30 to −1·93
Tin(IV) sulphide	−0·80
Tin(IV) bromide	−1·00
Tin(IV) chloride	−1·90
Tin(IV) oxide	−2·20
Tin(IV) fluoride	−2·50

From Table 5 it is tempting to infer that in grey (α) tin there is tin(IV) interbonded, and in white (β) tin there is tin(II) interbonded The suggestion receives chemical support, in that the action of concentrated hydrochloric acid upon grey tin and white tin produces respectively $SnCl_4.5H_2O$ and $SnCl_2.2H_2O$.

2. COMPOUNDS

2.1. ALLOYS

Tin is one of the very important alloying metals, and the number of different alloys are too numerous to mention. Small changes in composition are made to suit specific purposes, and some representative compositions are given in Table 6.

About 20% of the world tin production goes to make solders of different types. It is the tin present that confers the ability of the solder to stick metals together.

True bronzes are copper based tin alloys with a good combination of chemical resistance, mechanical strength and ease of manufacture.

[12] N. N. Greenwood, quoted in ref. 9.

Tin-rich babbitt alloys find very extensive use as bearing metals in marine engines and automobile engines.

TABLE 6. REPRESENTATIVE COMPOSITIONS OF SOME TIN ALLOYS

Alloy	Sn %	Cu %	Pb %	Sb %	Bi %	Other %
Rhine metal	97	3				
Tinfoil	88	3	8	1		
Woods metal (m.p. 70°C)	12·5		25		50	Cd 12·5
Fusible alloy	15		32		53	
Pewter	85	7		2	6	
Soft solder	50				50	
Babbitt	90	4		6		
White metal	5	1	75	19		
Gun metal (bronze)	10	90				
Speculum (bronze)	33	67				
Medal (bronze)	8	90				Zn 2
Phosphor (bronze)	10	79·7		9·5		P 0·8

Tin is commonly a constituent of easily fusible alloys used in fire warning apparatus and other safety devices.

2.2. TIN(II) BONDING

With a ground state configuration for tin of $5s^2 5p^2$ it can form covalent tin(II) compounds with use of two unpaired *p*-electrons. Should the $5s^2$ electrons take no part in compound formation, the bond angle in such an SnX_2 molecule would be 90°. Normally,

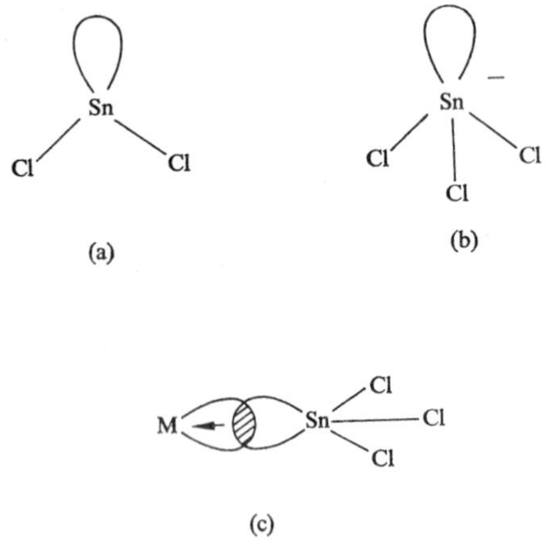

Sn Cl Cl

(a)

Sn Cl Cl Cl

(b)

M Sn Cl Cl Cl

(c)

FIG. 2. (a) $SnCl_2$ (gaseous), (b) $SnCl_3^-$, (c) M ← $SnCl_3$ complex, illustrating the directional lone pair of electrons in each case.

however, the *s*-electrons are incorporated into an sp^2 hybridized bond situation, and the resultant bond angle is about 120°. Thus, for example, tin(II) chloride in the gas phase has the structure illustrated in Fig. 2 (a), with two sp^2 orbitals forming covalent bonds to halogen and one directional lone pair of electrons. Figure 2 (b) shows the $SnCl_3^-$ ion in which tin(II) may be regarded as sp^3 hybridized, with the two tin chlorine bonds and the directional lone pair of $SnCl_2$, but with the coordination of a chloride ion into the vacant sp^3 orbital. Figure 2 (c) shows how the directional lone pair of $SnCl_3^-$ behaves as a donor ligand to metals, etc. In this latter case it is possible for the tin atom to act simultaneously as a π-electron acceptor, thus increasing the overall strength of the tin–acceptor atom bond[13].

With a second ionization potential of 14·63 eV tin easily looses two $5p$ electrons to form a dipositive ion of $5s^2$ configuration. The separation of this ground state ($5s^2$) of Sn^{2+} ion from the first excited state ($5s^1p^1$) is only 6·64 eV. In such circumstances it is possible to gain extra crystal field stabilization by s–p mixing. Such extra stabilization energy can only be achieved, however, by an unsymmetrical distortion of the environment of the ion[14].

The large size of the Sn^{2+} ion gives it usually an octahedral environment in materials containing smaller ions such as fluoride and oxide; and maximum distortion is to be expected in such compounds of tin(II). The extra stabilization, and hence the extent of distortion, will, however, fall rapidly with increasing tin–anion distances.

2.3. TIN(IV) BONDING

Hybridization of the available $5s$ and $5p$ orbitals of tin produces a suitable situation for the formation of four tetrahedral covalent bonds Many tin compounds such as the tin(IV) halides (except fluoride) and the organotins have this covalent mode of bonding.

Whereas tin(II) compounds do not appear to utilize the $5d$ orbitals except in the possible cases of π-bonding in donor–acceptor complexes, tin(IV) compounds make extensive use of $5d$ orbitals. Thus, for example, in complexes such as $(CH_3)_3SnCl.C_5H_5N$, the five-coordinate tin atom is sp^3d hybridized in a trigonal bipyramidal environment. Similarly in octahedral molecules or ions such as L_2SnCl_4 and $SnCl_6^{2-}$, the $5d$ orbitals of similar energy to the $5s$ and $5p$ orbitals are utilized to form sp^3d^2 hybrid orbitals for largely covalent bonds.

The Sn^{4+} ion (radius $\sim$ 0·74 Å) may be regarded as present in the lattices of compounds like tin(IV) oxide and various stannates. The $5s^0p^0$ configuration of the Sn^{4+} ion should give regular octahedral coordination for tin in ionic lattices, but there are notable tendencies towards a distortion from octahedral environments to tetrahedral environments, due possibly to the partially covalent (and hence tending towards tetrahedral) character of the bonding in the lattice.

Whilst these descriptions of ionic and covalent bonding in Sn(II) and Sn(IV) represent considerable oversimplification, they do allow an effort at classification of tin compounds.

2.4. ORGANOMETALLIC COMPOUNDS

Compounds of tin(II) possessing a tin–carbon bond are extremely rare, and many compounds previously characterized as dialkyl- and diaryltins are now known to be telomeric derivatives of tin(IV) of the type $(R_2Sn)_n$. Exceptions are 10,11-dihydrodibenzo-stannoepin[15] and bis(cyclopentadienyl)tin[16]. The structure of the 10,11-dihydrodibenzo-stannoepin is very likely to be as in Fig. 3. The structure of $(C_5H_5)_2Sn$ in the gas phase is

[13] J. D. Donaldson, *Progress in Inorganic Chemistry*, **8** (1967) 287.
[14] L. E. Orgel, *J. Chem. Soc.* (1959) 3815.
[15] H. G. Kuivila and O. F. Beumel, *J. Am. Chem. Soc.* **80** (1958) 3250.
[16] L. D. Dave, D. F. Evans and G. Wilkinson, *J. Chem. Soc.* (1959) 3684.

TABLE 7. BOND LENGTHS INVOLVING TIN

Bond	Bond length (Å)	Compound	Method
Sn–Sn	2·8099	α-Tin	X-ray diffraction[a]
Sn–Sn	3·022 3·181	β-Tin	X-ray diffraction[a]
Sn–H Sn–D	1·701 ±0·001	$SnHD_3$	Infrared spectroscopy[b]
Sn–H	1·700	CH_3SnH_3	Microwave spectroscopy[c]
Sn–H	1·785	SnH (short-lived species)	Infrared spectroscopy[d]
Sn–F	2·15 2·45 2·80	SnF_2 (orthorhombic)	X-ray diffraction[e]
SnF bridged terminal	2·08 1·88	SnF_4 (crystal)	X-ray diffraction[f]
SnF bridged	 2·15	$[(CH_3)_3SnF]_n$ (crystal)	X-ray diffraction[g]
Sn–Cl	2·42	$SnCl_2$ (gas)	Electron diffraction[h]
Sn–Br	2·55	$SnBr_2$ (gas)	Electron diffraction[h]
Sn–I	2·73	SnI_2 (gas)	Electron diffraction[h]
Sn–Cl bridged terminal	2·78 2·67	$[SnCl_2]_n$ (crystal)	X-ray diffraction[i]
Sn–Cl	2·31	$SnCl_4$ (gas)	Electron diffraction[j]
Sn–Br	2·44	$SnBr_4$ (gas)	Electron diffraction[k]
Sn–I	2·64	SnI_4 (gas)	Electron diffraction[k]
Sn–Cl	2·42 2·50 2·50	$\{C_6H_5CH:(NH_2^+)\}_2SnCl_6^{2-}$	X-ray diffraction[l]
Sn–Cl	2·43	Cs_2SnCl_6	X-ray diffraction[m]
	2·45	K_2SnCl_6	
	2·41	$(NH_4)_2SnCl_6$	
	2·39	Tl_2SnCl_6	
Sn–Br	2·64	Cs_2SnBr_6	X-ray diffraction[n]
	2·59	$(NH_4)_2SnBr_6$	
	2·59	Rb_2SnBr_6	
Sn–I	2·85	Cs_2SnI_6	X-ray diffraction[o]
	2·84	Rb_2SnI_6	
Sn–Cl Sn–C	2·32 2·19	CH_3SnCl_3	Electron diffraction[p]
Sn–Cl Sn–C	2·34 2·17	$(CH_3)_2SnCl_2$	Electron diffraction[p]
Sn–Cl Sn–C	2·37 2·19	$(CH_3)_3SnCl$	Electron diffraction[p]
Sn–Br Sn–C	2·45 2·17	CH_3SnBr_3	Electron diffraction[p]
Sn–Br Sn–C	2·48 2·17	$(CH_3)_2SnBr_2$	Electron diffraction[p]
Sn–Br Sn–C	2·49 2·17	$(CH_3)_3SnBr$	Electron diffraction[p]

TABLE 7—*Continued*

Bond	Bond length (Å)	Compound	Method
Sn–I	2·68	CH_3SnI_3	Electron diffraction[p]
Sn–C	2·17		
Sn–I	2·69	$(CH_3)_2SnI_2$	Electron diffraction[p]
Sn–C	2·17		
Sn–I	2·72	$(CH_3)_3SnI$	Electron diffraction[p]
Sn–C	2·17		
Sn–Cl	2·42	$C_5H_5N . (CH_3)_3SnCl$	X-ray diffraction[q]
Sn–C	2·18	$(CH_3)_4Sn$	Electron diffraction[r]
Sn–O	1·837	Sn–O (short-lived species)	Infrared spectroscopy[d]
Sn–O	1·98	$K_2Sn(OH)_6$	X-ray diffraction[s]
	1·93	$Na_2Sn(OH)_6$	
Sn–Mn	2·674	$(C_6H_5)_3SnMn(CO)_5$	X-ray diffraction[t]
Sn–Mn	2·627	$(C_6H_5)_3SnMn(CO)_4P(C_6H_5)_3$	X-ray diffraction[u]
Sn–Mn	2·70	$(C_6H_5)_2Sn\{Mn(CO)_5\}_2$	X-ray diffraction[v]
Sn–Fe	2·54	$Sn\{Fe(CO)_4\}_4$	X-ray diffraction[w]
Sn–Fe		$(CH_3)_4Sn_3\{Fe(CO)_4\}_4$	X-ray diffraction[x]
Interior	2·75		
Terminal	2·63		
Sn–Fe	2·49	$\{\pi\text{-}C_5H_5Fe(CO)_2\}_2SnCl_2$	X-ray diffraction[y]
Sn–Pt	2·54	$[Pt(SnCl_3)_5]^{3-}$	X-ray diffraction[z]
Sn–Ir	2·64	$(C_8H_{12})_2IrSnCl_3$	X-ray diffraction[α]

[a] J. Thewlis and A. R. Davey, *Nature*, **174** (1954) 1011 (α-tin), and W. B. Pearson, *Lattice Spacings and Structures of Metals and Alloys*, Pergamon Press (1957) (β-tin).
[b] G. R. Wilson and M. K. Wilson, *J. Chem. Phys.* **25** (1956) 784.
[c] D. R. Side, *J. Chem. Phys.* **19** (1951) 1605.
[d] G. Herzberg, *Molecular Spectra and Molecular Structure Infrared Spectra of Diatomic Molecules*, 2nd ed., Van Nostrand, New York.
[e] J. D. Donaldson and R. Oteng, *Inorg. and Nuclear Chem. Letters*, **3** (1967) 163.
[f] R. Höppe and W. Dahne, *Naturwiss.* **49** (1967) 254.
[g] H. C. Clark, R. J. O'Brien and J. Trotter, *J. Chem. Soc.* (1964) 2332.
[h] M. W. Lister and L. E. Sutton, *Trans. Faraday Soc.* **37** (1941) 406.
[i] R. E. Rundle and D. H. Olson, *Inorg. Chem.* **3** (1964) 596.
[j] R. L. Livingstone and C. N. R. Rao, *J. Chem. Phys.* **30** (1959) 339.
[k] M. Lister and L. E. Sutton, *Trans. Faraday Soc.* **37** (1941) 393.
[l] C. Kung-Du, L. Chao-Fa and T. You-Chi, *Acta Chim. Sin.* **25** (1959) 72.
[m] G. Engel, *Z. Krist.* **90** (1935) 341.
[n] J. A. A. Ketelaar, A. A. Rietdijk and C. H. Staveren, *Rec. Trav. Chim.* **56** (1937) 907.
[o] W. Werker, *Rec. Trav. Chim.* **58** (1939) 257.
[p] L. E. Sutton and H. E. Skinner, *Trans. Faraday Soc.* **40** (1944) 164.
[q] R. Huhne, *J. Chem. Soc.* (1963) 1524.
[r] L. O. Brockway and H. O. Jenkins, *J. Am. Chem. Soc.* **58** (1936) 2036.
[s] C. O. Björling, *Arkiv. Kemi*, **15** (1941) 100.
[t] H. P. Weber and R. F. Bryan, *Chem. Comms.* (1966) 443.
[u] R. F. Bryan, *J. Chem. Soc.* (*A*) (1967) 172.
[v] B. T. Kilbourn and H. M. Powell, *Chem. and Ind.* (1964) 1578.
[w] P. F. Lindley and P. Woodward, *J. Chem. Soc.* (*A*) (1967) 382.
[x] C. J. Fritchie, R. M. Sweet and R. Schunn, *Inorg. Chem.* **6** (1967) 749.
[y] J. E. O'Connor and E. R. Corey, *Inorg. Chem.* **6** (1967) 968.
[z] R. D. Cramer, R. V. Lindsey, C. T. Prewitt and U. G. Stolberg, *J. Am. Chem. Soc.* **87** (1965) 658.
[α] P. Porta, H. M. Powell, R. J. Mawby and L. M. Venanzi, *J. Chem. Soc.* (*A*) (1967) 455.

an angular sandwich molecule[17], but this may be different in the crystal, and could be different in the gas, solution and solid phases.

The literature of organometallic tin chemistry is now one of the largest for any element, but is fortunately particularly well documented and reviewed[18–21]. In this section discussion is confined to the tetra-organotins, as the chemistries of the extensive ranges of R_3SnX,

FIG. 3. An example of an organometallic derivative of tin(II) 10,11-dihydrobenzostannoepin.

R_2SnX and $RSnX_3$ compounds are more appropriately covered in those sections dealing with the properties of the functional groups X when attached to tin (*vide infra*).

The tetra-alkyl- and tetra-aryltins are either colourless liquids or white crystalline solids, which are stable to air and water; many of them are now available in commercial quantities. In addition to the symmetrical compounds R_4Sn where four identical alkyl or aryl groups are attached to tin in a tetrahedral configuration, many unsymmetrical organotins of type

TABLE 8. SOME EXAMPLES OF TETRA-ORGANOTINS

Compound	M.p. °C	B.p. °C/mm
$(CH_3)_4Sn$	−54	78/760
$(C_2H_5)_4Sn$	−136 to −146†	180/760
$(n\text{-}C_3H_7)_4Sn$	−109	111/10
$(iso\text{-}C_3H_7)_4Sn$		103/10
$(n\text{-}C_4H_9)_4Sn$		145/10
$(C_6H_5)_4Sn$	229	
$(CH_2{=}CH)_4Sn$		57/17
$(CH_2{=}CH-CH_2)_4Sn$		70/1·5
$(C_6H_5C{:}C)_4Sn$	174 decomp.	
$(CH_3)_3SnC_6F_5$		
$(CH_3)_3SnCF_3$		
$(m\text{-}CF_3{\cdot}C_6H_4)_4Sn$		
$(CH_3)_3SnC_6H_5$		208/760
$(CH_3)_2Sn(C_2H_5)_2$		144/760
$(C_6H_5{\cdot}CH_2)_2Sn(C_2H_5)(C_4H_9)$		209/9
$(C_6H_5)_3SnC{:}C-C{:}CSn(C_6H_5)_3$		

† Various crystal modifications.

R_3SnR', $R_2SnR'_2$, $R_2SnR'R''$, etc., have been synthesized. Table 8 contains a limited selection of tetra-organotins, illustrating the wide range of organic groups attachable to tin(IV).

[17] A. Almenningen, A. Haaland and T. Motzfeldt, *J. Organometallic Chem.* 7 (1967) 97.
[18] R. K. Ingham, S. D. Rosenberg and H. Gilman, *Chem. Revs.* 60 (1960) 459.
[19] W. P. Neumann, *Die Organische Chemie des Zinns*, Ferdinand Enke, Stuttgart (1967).
[20] *Organometallic Compounds*, Ed. M. Dub, 2nd ed. vol. II, Springer-Verlag, Berlin (1967).
[21] E. Krause and A. von Grosse, *Die Chemie der Metall-organischen Verbindungen*, Borntraeger, Berlin (1937).

Formation of Tin–Carbon Bonds

The different methods now available for the formation of tin–carbon bonds are classified below.

Use of Organomercurys and Organozincs

These methods are largely of historical importance, and some of the very earliest organotins were prepared from them. These methods still find application in special circumstances.

$$SnCl_2 + R_2Hg \longrightarrow R_2SnCl_2 + Hg$$

$$2SnCl_2 + 2RHgCl \longrightarrow R_2SnCl_2 + 2Hg + SnCl_4$$

$$SnCl_4 + 2R_2Zn \longrightarrow R_4Sn + 2ZnCl_2$$

$$SnCl_2 + R_2Zn \longrightarrow [R_2Sn]_n + ZnCl_2$$

$$Sn + RZnCl \longrightarrow R_4Sn$$

Use of Sodium and Sodium–Tin Alloys

These methods are also of historical interest, but have continued to receive considerable attention, as the corresponding reaction for lead has proved such a convenient source of organoleads. A virtually quantitative yield of tetra-ethyltin is reported from ethyl bromide and a tin alloy containing 14% of sodium and about 15% of zinc. It is important to note, however, that quantitative in this context means a maximum of 25% based on tin due to the composition (NaSn) of the alloy.

$$4RX + 4NaSn \longrightarrow R_4Sn + 4NaX + (3Sn\downarrow)$$

The reaction is actually known to be considerably more complex than indicated by this overall equation; and trialkyltin halides, dialkyltins and hexa-alkylditins are formed as by-products. Of recent interest has been the Wurtz type of reaction between organic halides and tin tetrahalides in the presence of sodium:

$$4RCl + SnCl_4 + 4Na \longrightarrow SnR_4 + 4NaCl$$

Direct Interaction of Alkyl and Aryl Halides with Metallic Tin

The requirements of large amounts of organotin halides for industry makes this a particularly attractive route, and considerable efforts have been made to perfect a process comparable to the Rochow Process for the organosilicon halides[22, 23]. Whilst the "direct process" reaction between tin and ethyl iodide has been known for well over a century, only recently have the right conditions for the formation of good yields from alkyl halides been determined.

$$CH_3Cl + Sn \xrightarrow{175^\circ} \underset{(6{\cdot}6\%)}{CH_3SnCl_3} + \underset{(39\%)}{(CH_3)_2SnCl_2} + \underset{(4{\cdot}6\%)}{(CH_3)_3SnCl}$$

In this particular reaction small amounts of iodomethane and triethylamine were the catalysts.

[22] H. Tokunaga, Y. Murayama and I. Kijima, Japanese Patent (1964) 24958; *C.A.* **62** (1965) 14, 726.
[23] Nitto Chemicals, French Patent (1965) 1,393,779; *C.A.* **63** (1965) 9985.

Use of Grignard Reagents and Organolithiums

This is the most important laboratory method for the formation of the tin–carbon bond and is probably still the most extensively used industrial process.

$$SnX_4+4LiR \longrightarrow SnR_4+4LiX$$

$$SnX_4+4RMgX \longrightarrow SnR_4+4MgX_2$$

The reactions are of wide applicability for a great range of organic groups. Steric hindrance affects these reactions considerably, and yields from secondary or tertiary aliphatic groups is poor; and only two tertiary butyl groups may be attached to a tin atom. An excess of the organolithium or Grignard reagent is normally used to avoid, as far as possible, the formation of organotin halides as by-products. In many cases the Grignard reagent is generated *in situ*, and the synthesis consists of running the organic halide into a mixture of tin tetrahalide and magnesium metal, often in a hydrocarbon solvent, rather than ether.

Miscellaneous Methods

In addition to the methods classified above, a number of other routes to tin–carbon bonds have been reported, and some of these are indicated in equation form below.

$$4R_3Al+3SnCl_4 \longrightarrow 3R_4Sn+4AlCl_3$$

$$4CH_2N_2+SnCl_4 \longrightarrow (ClCH_2)_4Sn+4N_2$$

$$4C_6H_5C{:}CNa+SnCl_4 \longrightarrow (C_6H_5C{:}C)_4Sn+4NaCl$$

$$CH_3Cl+SnCl_2 \longrightarrow CH_3SnCl_3$$

$$CH_3I+K_2SnO_2 \longrightarrow K[CH_3SnOO]+KI$$

$$C_6H_5N_2Cl+SnCl_2 \longrightarrow C_6H_5SnCl_3+N_2$$

Additional to the wide range of symmetrical tetra-alkyls and tetra-aryls of tin(IV), numerous compounds of general formula R_3SnR', $R_2SnR'R''$, etc., are known where different alkyl and aryl groups are attached to a single tin atom.

A representative selection of organotins is listed in Table 8. A number of very comprehensive listings of such compounds are available.

The existence of tin–carbon bonds is not restricted to linear species, and many rings are known in which the tin–carbon bond is part of the heterocyclic bonding system. In the main these are made by metathetical reactions or intermolecular hydride additions.

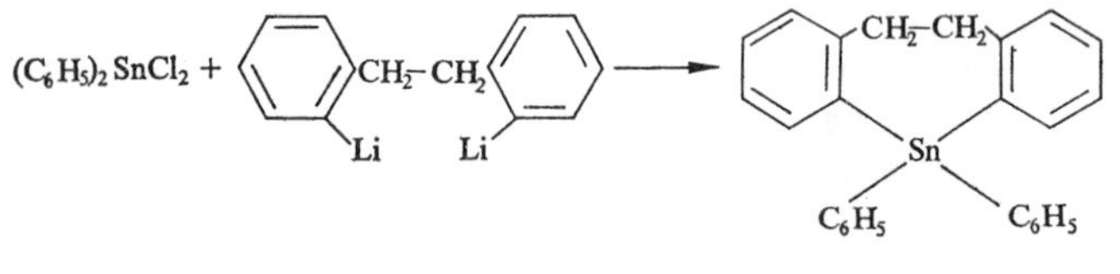

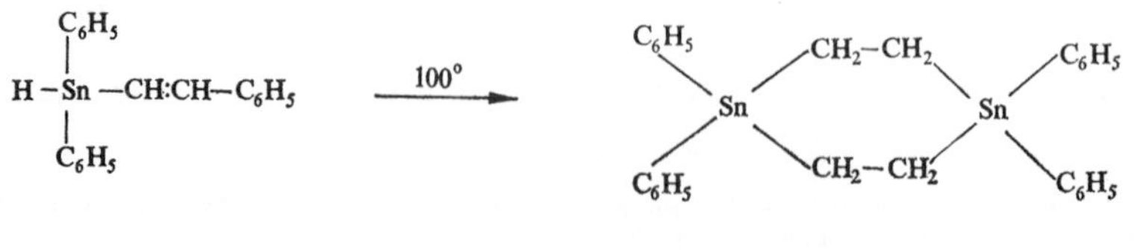

$(C_6H_5)_2SnCl_2$ + [2,2'-dilithiobibenzyl: $C_6H_4(Li)$–CH_2–CH_2–$C_6H_4(Li)$] ⟶ [cyclic $(C_6H_5)_2Sn$ bridging the two rings of C_6H_4–CH_2–CH_2–C_6H_4]

H–$Sn(C_6H_5)_2$–$CH{:}CH$–C_6H_5 $\xrightarrow{100^\circ}$ [1,4-distannacyclohexane ring: $(C_6H_5)_2Sn$ linked to $(C_6H_5)_2Sn$ by two CH_2–CH_2 bridges]

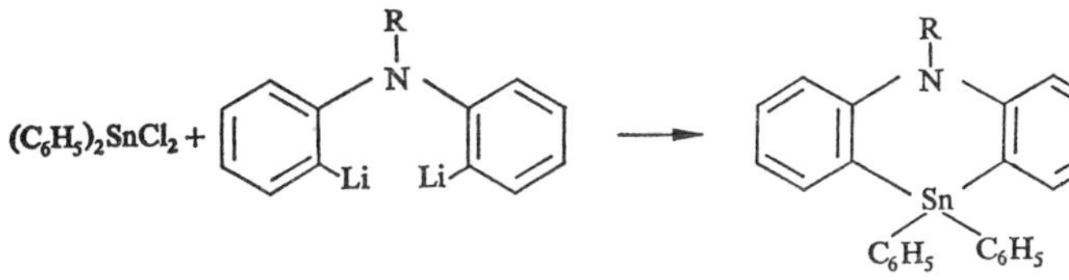

$$(C_6H_5)_2SnCl_2 + \text{[2-lithio-N-R-diphenylamine, Li Li]} \longrightarrow \text{[N-R, Sn}(C_6H_5)_2\text{ ring]}$$

Physical Properties of Organotins

The physical properties of the tetra-organotins greatly resemble the hydrocarbons of similar constitution, but with higher density, refractivity, etc. Thus the lower members of the tetra-alkyltins are volatile liquids which become waxy solids at long chain length, and the tetra-aryls are all high melting point solids.

Electron diffraction, microwave spectroscopy and X-rays have shown the four organic

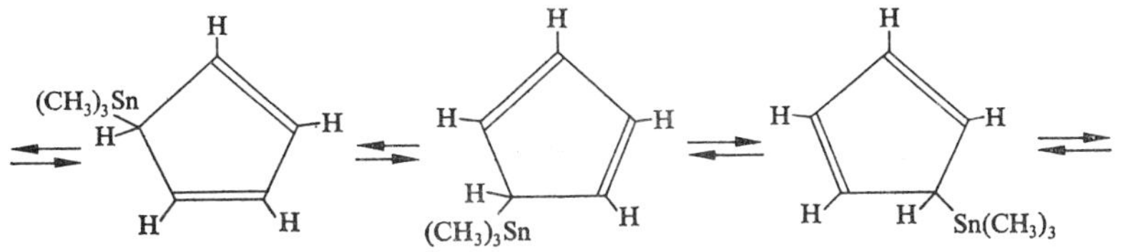

FIG. 4. Rapid movement of the trimethyltin group around the cyclopentadienyl ring in $C_5H_5Sn(CH_3)_3$.

groups in symmetrical organotins R_4Sn to be perfectly tetrahedral about the central tin atom.

The nuclear magnetic resonance spectra of organotins, in which a cyclopentadienyl group is attached to tin are of particular interest. They indicate that the σ-bond between tin and the carbon of the cyclopentadienyl group in, for example, $(CH_3)_3SnC_5H_5$ has a lifetime of less than 10^{-4} sec at ordinary temperatures, and that the tin atom wanders around the C_5 ring[24, 25].

Chemical Properties of Organotins

Here discussion is limited to those reactions of organotins in which the tin–carbon bond is involved Properties of the compounds of type R_3SnX, R_2SnX_2, etc., which are dependent upon the properties of the Sn–X bond rather than the organotin function of the molecule, are discussed under the various Sn–X sections below.

Organotins do not undergo hydrolysis or oxidation under ordinary conditions, but if ignited burn strongly with the emission of clouds of tin oxide.

Water and alcohol can sometimes bring about the fission of the tin–carbon bond in reactive organotins.

$$2C_6H_5{:}CSn(C_2H_5)_3 + H_2O \longrightarrow 2C_6H_5C{:}CH + [(C_2H_5)_3Sn]_2O$$
$$NC_5H_4.Sn(CH_3)_3 + ROH \longrightarrow C_5H_5N + (CH_3)_3SnOR$$

Mercaptans do not require especially active tin–carbon bonds to react with organotins, and under varying conditions virtually all types of tin–carbon bond can be converted to

[24] H. P. Fritz and C. G. Kreiter, *J. Organometallic Chem.* **1** (1964) 323.
[25] A. Davison and P. E. Rakita, *J. Am. Chem. Soc.* **90** (1968) 4479.

tin–sulphur bonds by thiols, and similarly for selenols. Elemental sulphur and selenium similarly break the tin–carbon bond

$$R_4Sn+R'SH \longrightarrow R_3SnSR'+RH$$
$$(C_6H_5)_4Sn+Se \longrightarrow (C_6H_5)_3SnSeC_6H_5$$

Under conditions of high temperature and pressure, hydrogen removes aryl groups from tin to deposit metal

$$(C_6H_5)_4Sn+H_2 \longrightarrow 4C_6H_6+Sn$$

The organic groups attached to tin vary considerably in their ease of fission by halogen. In increasing ease of removal these groups are butyl, propyl, ethyl, methyl, vinyl, phenyl, benzyl, allyl, cyanomethyl and $-CH_2COOR$, an order which applies to many other tin–carbon fission reactions besides those of halogens.

Tin will often exchange organic groups cleanly, in good yield, with another metal or non-metal, providing a useful synthetic route to these materials.

$$(C_6H_5CH_2)_3SnCl+4CH_3Li \longrightarrow 3C_6H_5CH_2Li+(CH_3)_4Sn+LiCl$$
$$(CH_2{:}CH)_4Sn+4C_6H_5Li \longrightarrow 4CH_2{:}CHLi+(C_6H_5)_4Sn$$
$$(C_6H_5)_4Sn+BCl_3 \longrightarrow C_6H_5BCl_2+(C_6H_5)_3SnCl$$

Among other reagents causing fission of tin–carbon bonds are hydrogen halides, mercury halides, bismuth halides, thallium chloride, arsenic halides, phosphorus halides, sulphuric acid, nitric acid, sulphur, sulphur dioxide, sulphuryl chloride and organic acids[18].

Organotins undergo proportionation reactions with tin tetrahalides to give the range of organotin halides, which are discussed in detail below.

$$nR_4Sn+(4-n)SnX_4 \longrightarrow 4R_nSnX_{4-n}$$
$$(n = 1, 2 \text{ and } 3)$$

2.5. HYDRIDES

Binary Hydrides

The two binary hydrides of tin that are known to date are SnH_4 and Sn_2H_6[26].

SnH_4 was originally prepared in very low yield by the treatment of tin–magnesium alloy with dilute acids. The lithium aluminium hydride reduction of tin(IV) chloride has made SnH_4 easily available; in the presence of a trace of oxygen this reaction gives 80–90% yields. The oxygen is believed to inhibit the decomposition of stannane to tin and hydrogen[27, 28]. Tin(II) chloride is reduced by aqueous sodium borohydride to 84% yield of SnH_4 under optimum conditions[29].

In addition to a good yield of SnH_4, the potassium borohydride reduction of stannite ion in solution produced the hydride distannane Sn_2H_6. This hydride may be distilled in a vacuum system without decomposition provided relatively low pressures are used, but if a vessel containing Sn_2H_6 is allowed to warm to room temperature it is completely decomposed[27].

[26] F. G. A. Stone, *Hydrogen Compounds of the Group IV Elements*, Prentice Hall (1962).
[27] W. L. Jolly, *Angew. Chem.* **72** (1960) 268; *J. Am. Chem. Soc.* **83** (1961) 335.
[28] H. J. Emeléus and S. F. A. Kettle, *J. Chem. Soc.* (1958) 2444.
[29] G. W. Schaeffer and M. Emilius, *J. Am. Chem. Soc.* **76** (1954) 1203.

From the infrared spectrum[30] of SnH_4, the estimated bond dissociation energy for Sn–H is 73·7 kcal/mole, which compares well with the value of 70·3 kcal/mole obtained as a mean dissociation energy for Sn–H from mass spectrometry measurements[31]. The thermochemical bond[32] energy of Sn–H has been obtained as 60·4 kcal/mole.

Analysis of the stretching vibrations[33] in $SnHD_3$ gives a value of the Sn–H distance as $1{\cdot}701 \pm 0{\cdot}001$ Å, in good agreement with the Sn–H bond distance of $1{\cdot}700 + 0{\cdot}015$ Å derived from the microwave spectrum[34] of CH_3SnH_3.

The nuclear magnetic resonance spectrum[35] shows a τ value of the protons in SnH_4 at 6·15, along with very large tin–proton coupling constants $^{119}Sn{-}H = 1931$ c/s and $^{117}Sn{-}H = 1846$ c/s.

All physical data on SnH_4 and SnD_4 are in full accord with a regular tetrahedral structure[36].

Stannane[37] melts at $-146°$ and has a normal boiling point of $-52{\cdot}5°$.

Stannane undergoes appreciable decomposition to tin and hydrogen even at ordinary temperatures, but about 100°C it is very rapidly decomposed. The decomposition is first order with respect to stannane. Somewhat remarkably, oxygen inhibits decomposition at pressures above 1mm, and stannane may be stored mixed with oxygen at room temperature. At low pressures oxygen catalyses the decomposition of stannane[38, 39].

Stannane is toxic and is intermediate between silane and germane in its chemical properties; it is unattacked by dilute acids and alkalis, but is decomposed by concentrated acid or alkali. It is a powerful reducing agent, and is rapidly decomposed by solutions of transitional metal salts. Stannane and hydrogen chloride evolve hydrogen to form stannyl chloride, H_3SnCl, which is very unstable, undergoing noticeable decomposition, even at $-70°$. (Contrast H_3GeCl stable at 20°C and H_3SiCl stable at 200°C.)

Sodium in liquid ammonia may be titrated against SnH_4 to produce H_3SnNa and H_2SnNa_2. It would appear, however, that these two sodium derivatives are only stable as amines, as removal of ammonia, even at $-63{\cdot}5°$, causes decomposition. H_3SnNa reacts with alkyl iodides in liquid ammonia to produce alkyl stannanes $RSnH_3$, and H_3SnNa with ammonium chloride regenerates SnH_4[28].

The SnH_3 radical may be trapped in an inert matrix at low temperature, and has been well characterized by its electron spin resonance spectrum[40].

Stannane reduces nitrobenzene in 94% yield and reduces benzaldehyde to benzyl alcohol virtually quantitatively. Only a 29% yield of isopropylamine was obtained from 2-nitropropane due to extensive decomposition of the stannane, catalysed by the amine[37].

Cobalt naphthenate di-tert-butyl peroxide, or palladium (10% on charcoal) or hexachloroplatinic acid are effective in catalysing the addition of stannane to olefins to form the corresponding tetra-alkyltins.

Stannane reacts with boron trifluoride to form tin(IV) fluoride quantitatively.

30 L. May and C. R. Dillard, *J. Chem. Phys.* **34** (1961) 694.
31 F. E. Saalfeld and H. J. Sues, *J. Inorg. and Nuclear Chem.* **18** (1961) 98.
32 S. R. Gunn and L. G. Green, *J. Chem. Phys.* **65** (1961) 779.
33 G. R. Wilkinson and M. K. Wilson, *J. Chem. Phys.* **25** (1956) 784.
34 D. R. Lide, *J. Chem. Phys.* **19** (1951) 1605.
35 H. D. Kaesz and N. Flitcroft, *J. Am. Chem. Soc.* **85** (1963) 1377.
36 I. W. Levin and H. Ziffer, *J. Chem. Phys* **43** (1965) 4023; I. W. Levin, *ibid.* **46** (1967) 1176.
37 G. H. Reifenberg and W. J. Considine, *J. Am. Chem. Soc.* **91** (1969) 2401.
38 K. Tamaru, *J. Phys. Chem.* **60** (1956) 610.
39 S. F. A. Kettle, *J. Chem. Soc.* (1961) 2569.
40 R. L. Moorehouse, J. J. Christiansen and W. Gordy, *J. Chem. Phys.* **45** (1966) 1751.

Organotin Hydrides

Any extensive study of the chemistry of the tin–hydrogen bond is extremely difficult in SnH_4 due to its inherent instability. In the case of the organotin hydrides, this difficulty is considerably allayed. The stability of the organotin hydrides increases with decreasing number of tin–hydrogen linkages in the molecule[19].

The organotin hydrides are almost invariably synthesized by the reduction of the corresponding organotin halide.

$$CH_3SnCl_3+LiAlH_4 \longrightarrow CH_3SnH_3$$
$$(C_2H_5)_2SnCl_2+LiAlH_4 \longrightarrow (C_2H_5)_2SnH_2$$
$$(C_6H_5)_3SnCl+LiAlH_4 \longrightarrow (C_6H_5)_3SnH$$
$$(CH_3)_3SnNa+NH_4Br \longrightarrow (CH_3)_3SnH$$
$$(C_4H_9)_3SnCl \xrightarrow{Al/Hg+H_2O} (C_4H_9)_3SnH$$

Examples of tin hydrides are known containing two or more tin atoms

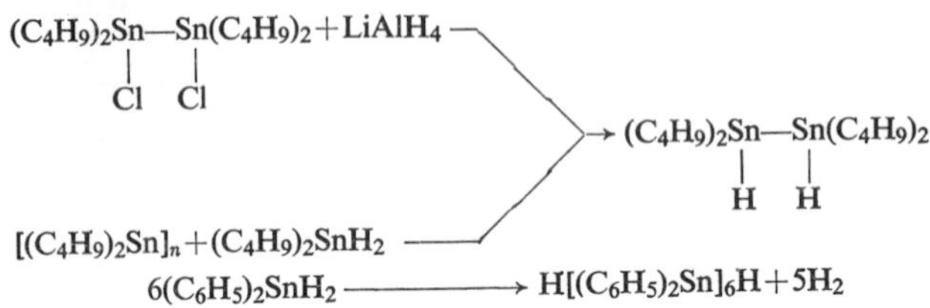

Table 9 shows some examples of organotin hydrides together with boiling points where recorded.

TABLE 9. ORGANOTIN HYDRIDES

Hydride	B.p. °/mm	Hydride	B.p. °/mm
SnH_4	−52/760	$(C_4H_9)_2SnH_2$	75–76/12
CH_3SnH_3	0/760	$(C_4H_9)_3SnH$	76–78/0·7
$(CH_3)_2SnH_2$	35/760	$C_6H_5SnH_3$	57–64/105
$(CH_3)_3SnH$	59/760	$(C_6H_5)_2SnH_2$	89–93/0·3
$C_2H_5SnH_3$	35/760	$(C_6H_5)_3SnH$	168–172/0·5
$(C_2H_5)_2SnH_2$	99/760	$(C_4H_9)_2SnH(Cl)$	
$(C_2H_5)_3SnH$	52/20	$(C_4H_9)_2SnH(OOCCH_3)$	
$(C_3H_7)_3SnH$	76–82/12	$(C_4H_9)_2Sn(H)$–$Sn(H)(C_4H_9)_2$	
$C_4H_9SnH_3$	99–101/760	$H[(C_6H_5)_2Sn]_6H$	

In general, pure samples of tri-organotin hydrides are only slightly decomposed after several months at ordinary temperature; but on the other hand butyltin trihydride deposits an extensive yellow precipitate after only a few hours.

The tin–proton coupling constants in the methyltin hydrides, CH_3SnH_3 ($J_{^{119}Sn-H}=1852$; $J_{^{117}Sn-H}=1770$); $(CH_3)_2SnH_2$ ($J_{^{119}Sn-H}=1758$; $J_{^{117}Sn-H}=1682$); $(CH_3)_3SnH$ ($J_{^{117}Sn-H}=1744$; $J_{^{117}Sn-H}=1664$) c/sec, show a steady fall from SnH_4 ($J_{^{119}Sn-H}=1931$; $J_{^{117}Sn-H}=1846$), with increasing methyl content. The extra methyl groups are believed to cause a reduction in the *s* character of the tin orbitals in the tin–hydrogen bond, and hence a reduction in the Sn–H coupling constant[41]. The dominant Fermi contact contribution to the coupling is proportional to the *s* character in the hybridized orbital used in the Sn–H bond

The reactions of the organotin hydrides are extremely diverse, and the organotin hydrides have found some use as specialist reducing agents in synthetic organic and inorganic chemistry. The following equations illustrate the diversity of organotin hydride reactions[42, 43].

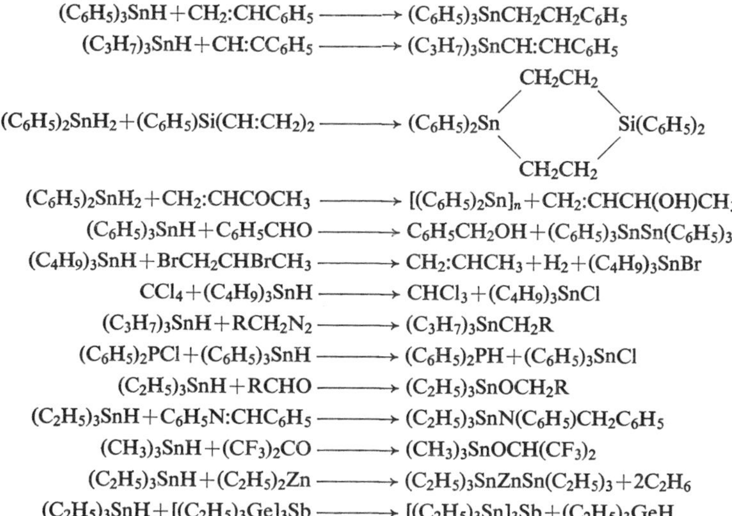

$$(C_6H_5)_3SnH + CH_2{:}CHC_6H_5 \longrightarrow (C_6H_5)_3SnCH_2CH_2C_6H_5$$
$$(C_3H_7)_3SnH + CH{:}CC_6H_5 \longrightarrow (C_3H_7)_3SnCH{:}CHC_6H_5$$
$$(C_6H_5)_2SnH_2 + (C_6H_5)Si(CH{:}CH_2)_2 \longrightarrow (C_6H_5)_2Sn\langle {}^{CH_2CH_2}_{CH_2CH_2} \rangle Si(C_6H_5)_2$$
$$(C_6H_5)_2SnH_2 + CH_2{:}CHCOCH_3 \longrightarrow [(C_6H_5)_2Sn]_n + CH_2{:}CHCH(OH)CH_3$$
$$(C_6H_5)_3SnH + C_6H_5CHO \longrightarrow C_6H_5CH_2OH + (C_6H_5)_3SnSn(C_6H_5)_3$$
$$(C_4H_9)_3SnH + BrCH_2CHBrCH_3 \longrightarrow CH_2{:}CHCH_3 + H_2 + (C_4H_9)_3SnBr$$
$$CCl_4 + (C_4H_9)_3SnH \longrightarrow CHCl_3 + (C_4H_9)_3SnCl$$
$$(C_3H_7)_3SnH + RCH_2N_2 \longrightarrow (C_3H_7)_3SnCH_2R$$
$$(C_6H_5)_2PCl + (C_6H_5)_3SnH \longrightarrow (C_6H_5)_2PH + (C_6H_5)_3SnCl$$
$$(C_2H_5)_3SnH + RCHO \longrightarrow (C_2H_5)_3SnOCH_2R$$
$$(C_2H_5)_3SnH + C_6H_5N{:}CHC_6H_5 \longrightarrow (C_2H_5)_3SnN(C_6H_5)CH_2C_6H_5$$
$$(CH_3)_3SnH + (CF_3)_2CO \longrightarrow (CH_3)_3SnOCH(CF_3)_2$$
$$(C_2H_5)_3SnH + (C_2H_5)_2Zn \longrightarrow (C_2H_5)_3SnZnSn(C_2H_5)_3 + 2C_2H_6$$
$$(C_2H_5)_3SnH + [(C_2H_5)_3Ge]_3Sb \longrightarrow [(C_2H_5)_3Sn]_3Sb + (C_2H_5)_3GeH$$

2.6. COMPOUNDS OF NITROGEN, PHOSPHORUS, ARSENIC, ANTIMONY AND BISMUTH

Tin forms binary compounds with each of these Group V elements.

Tin nitride Sn_3N_4 is obtained by thermal decomposition of the product of reaction from liquid ammonia and tin(IV) chloride[44], and also by atomization of a tin cathode in an atmosphere of nitrogen[45]. Sn_3N_4 is reported stable to water, but to form tin(IV) chloride and ammonium chloride on attack by hydrochloric acid.

Tin and red phosphorus react in sealed evacuated silica tubes at about 500° to produce tin phosphides.

Many stoichiometries for tin phosphides are reported in the earlier literature, but more recent investigations show the phase Sn_4P_3 to predominate, though other phases with higher phosphorus content such as SnP are also formed. A single crystal of Sn_4P_3 has trigonal symmetry. This gives a structure where two non-equivalent phosphorus atoms are both surrounded octahedrally by six tin atoms with a mean tin phosphorus distance of 2·7 Å. One type of tin atom is octahedrally surrounded by six phosphorus atoms, but the other

[41] J. R. Holmes and H. D. Kaesz, *J. Am. Chem. Soc.* **83** (1961) 3903.
[42] H. G. Kuivila, *Advances in Organometallic Chemistry*, **1** (1964) 47.
[43] N. S. Vyazankin, G. A. Rasuvaev and O. A. Kruglaya, *Organometallic Reviews*, **3** (1968) 323.
[44] R. Schwarz and A. Jeanmaire, *Chem. Ber.* **65** (1932) 1443.
[45] W. Janeff, *Z. Physik.* **142** (1955) 619.

type of tin atom is coordinated octahedrally by three phosphorus atoms and three tin atoms, with a tin–tin distance of above 3·25 Å[46].

A phase with this type of structure also crystallizes in the tin–arsenic system, but from phase analytical data the composition is near to Sn_3As_2. The substitutional solution of tin is believed to account for the deviation from the stoichiometric composition Sn_4As_3[47].

Tin becomes hard when alloyed with arsenic and there is a tendency for the alloys to crystallize. A full tin–arsenic phase diagram has been recorded[48].

The phases in the tin–antimony and tin–bismuth alloys have been investigated. Compounds of stoichiometry SnSb, Sn_4Sb_2, Sn_3Sb_2, $SnSb_2$ and Sn_3Bi have been claimed. The tin–antimony alloys form the basis of the so-called brittania metal and algiers metal.

Organotin Amines, Phosphines, Arsines, Stibines and Bismuthines

In addition to the binary compounds and alloys discussed above, there exist a large number of molecular compounds in which tin is bonded to the elements nitrogen, phos-

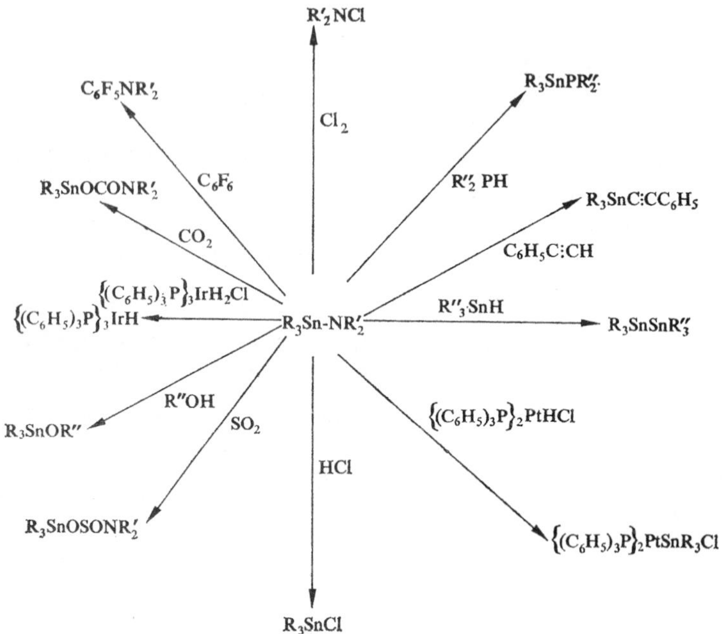

FIG. 5. Some reactions of the tin–nitrogen bond in organotin amines.

phorus, arsenic, antimony and bismuth. In these compounds it is possible to study the properties of discrete tin–Group V element single bonds.

Organotin nitrogen compounds are now very numerous[49], and tin derivatives of amines,

[46] O. Olofsson, *Acta Chem. Scand.* **21** (1967) 1659.
[47] G. Hagg and A. G. Hybinette, *Phil. Mag.* **20** (1935) 913.
[48] Q. A. Mansuri, *J. Chem. Soc.* (1923) 214.
[49] J. G. A. Luijten, F. Rijkens and G. J. M. van der Kerk, *Advances in Organometallic Chemistry*, **3** (1965) 413.

hydrazine, formamide, urea, carbodi-imide and carbamic acids, etc., have been reported. Most intensive investigations have been carried out on the organotin amines, and all types are known from $(R_3Sn)_3N$, where nitrogen is coordinated by three tin atoms, to $(R_2N)_4Sn$, in which one tin is coordinated by four nitrogens, and cyclic amines of the type $(R_2SnNR^1)_3$. All these organotin amines are very water sensitive.

Some syntheses of the organotin amines are outlined in equation form below, and some of the extensive reactions[50] of this class are illustrated in Fig. 5. It is obvious from Fig. 5 that the organotin amines are available as powerful synthetic intermediates for a number of important reactions[51, 52, 53].

$$SnCl_4 + 4LiNR_2 \longrightarrow Sn(NR_2)_4$$

$$Li_3N + 3R_3SnCl \longrightarrow (R_3Sn)_3N$$

$$R_3SnNR_2 + R_2''NH \longrightarrow R_3SnNR_2$$

$$(CH_3)_3SnN(CH_3)_2 + RNH_2 \longrightarrow [(CH_3)_3Sn]_2NR$$

Organotin phosphines, arsines, stibines and bismuthines are known, and are exemplified, together with some of the commoner organotin amines in Table 10.

TABLE 10. SOME EXAMPLES OF ORGANOTIN AMINES, PHOSPHINES, ARSINES, STIBINES AND BISMUTHINES

Compound	B.p. °/mm [m.p.]	Compound	B.p. °/mm [m.p.]
$(CH_3)_3SnN(CH_3)_2$	126/760	$(CH_3)_3SnAs(CH_3)$	
$(C_6H_5)_3SnN(CH_3)_2$	166/0·1	$(CH_3)_3SnAs(C_6H_5)_2$	136/0·5
$(C_6H_5)_2Sn\{N(CH_3)_2\}_2$	128/0·2	$\{(C_6H_5)_3Sn\}_2AsC_6H_5$	[115]
$Sn\{N(CH_3)_2\}_4$	51/0·15	$\{(C_6H_5)_3Sn\}_3As$	[216]
$CH_3Sn\{N(C_2H_5)_2\}_3$	92/0·1	$Sn\{As(C_6H_5)_2\}_4$	[70]
$\{(CH_3)_3Sn\}_2NCH_3$	64/3		
$\{(CH_3)_3Sn\}_3N$	133/20	$(C_2H_5)_3SnSb(C_6H_5)_2$	146/0·18
$\{(CH_3)_2SnNC_2H_5\}_3$	104/0·05	$(C_6H_5)_3SnSb(C_6H_5)_2$	[116]
		$\{(C_6H_5)_3Sn\}_2SbC_6H_5$	[120]
		$(C_6H_5)_2Sn\{Sb(C_6H_5)_2\}_2$	[150]
$(C_2H_5)_3SnP(C_2H_5)_2$	70/0·3	$\{(CH_3)_3Sn\}_3Sb$	[39]
$(CH_3)_3SnP(C_6H_5)_2$	147/0·7	$Sn\{Sb(C_6H_5)_2\}_4$	[75]
$\{(CH_3)_3Sn\}_2PCH_3$	90/3		
$(C_6H_5)_2Sn\{P(C_6H_5)_2\}_2$	[80]		
$\{(CH_3)_3Sn\}_3P$	137/3		
$\{(C_6H_5)_3Sn\}_3P$	[201]	$(C_2H_5)_3SnBi(C_6H_5)_2$	
$C_6H_5Sn\{P(C_6H_5)_2\}_3$	[117]	$\{(C_6H_5)_3Sn\}_2BiC_6H_5$	
$Sn\{P(C_6H_5)_2\}_4$	[107]	$\{(C_2H_5)_3Sn\}_3Bi$	
$[\{(C_6H_5)_3Sn\}_2P]_2$	[110]	$Sn\{Bi(C_6H_5)_2\}_4$	
$\{(C_6H_5)_2SnPC_6H_5\}_3$	[64]		

All of these compounds are basic in character and in contrast to the planar trisilylamines, the compounds $\{(CH_3)_3Sn\}_3X$, where X = N, P, As and Sb have all been shown to be pyramidal molecules[54].

[50] M. F. Lappert and B. Prokai, *Advances in Organometallic Chemistry*, **5** (1967) 225.
[51] K. Jones and M. F. Lappert, *Proc. Chem. Soc.* (1964) 22; *J. Organometallic Chem.* **3** (1965) 295.
[52] T. A. George and M. F. Lappert, *Chem. Comm.* (1966) 463.
[53] D. J. Cardin and M. F. Lappert, *Chem. Comm.* (1966) 506.
[54] R. E. Hester and K. Jones, *Chem. Comm.* (1966) 317.

2.7. OXIDES AND OXYGEN COMPOUNDS

Tin Oxides

Tin(II) oxide may be condensed from the vapour and trapped in an argon or nitrogen matrix at low temperature. In this form tin(II) oxide consists of Sn_2O_2 species which have a structure based on a planar (V_h) rhombus[55].

The common form of SnO is the black tetragonal orthorhombic modification (density 6·32). It is prepared by the hydrolysis of tin(II) salts to hydrated tin(II) oxide, with subsequent dehydration[56]. Special care has to be taken with reagents and with conditions to obtain pure samples[57]. This black oxide has a layer structure in which the tin atom lies at the apex of a square pyramid, the base of which is formed by the four nearest oxygen atoms (Sn–O = 2·21 Å). The Sn–Sn distance between the layers at 3·70 Å is so close to that of the Sn–Sn distance in tin metal, that some mode of tin–tin interaction cannot be ruled out[58].

A red orthorhombic modification of SnO is formed by heating suspensions of the white hydrous oxide. The red metastable form can be stabilized by the presence of about 1% phosphite, but if ammonia is used in the hydrolysis of the tin(II) chloride solutions, a red form of SnO is formed in the absence of stabilizing phosphite[59].

Tin(II) oxide undergoes rapid oxidation with incandescence to tin(IV) oxide upon heating around 300°. The action of heat upon tin(II) oxide in non-oxidizing conditions has yielded a very diverse range of reported results, and it is likely that different workers may have studied the oxide in differing states of purity[13]. On heating, tin(II) oxide disproportionates into tin(IV) oxide and tin. Whilst many intermediate oxides such as Sn_2O_3, Sn_3O_4 and Sn_5O_6 have been reported, it would appear that only Sn_3O_4 exists as a well-characterized discrete phase between SnO and SnO_2[60]. The formulation $Sn_2^{II}Sn^{IV}O_4$ for this intermediate oxide is not inconsistent with the Mössbauer spectrum.

Tin(II) oxide is amphoteric, dissolving in acids to give the tin(II) ion or anion complexes thereof; it dissolves in alkalis to form $Sn(OH)_3^-$ solutions.

Tin(IV) oxide occurs in nature as cassiterite, with a rutile type structure. It is also known in rhombic and hexagonal forms. Pure tin(IV) oxide is white and sublimes above 1800° without melting. It is not soluble in water, and is hardly attacked by acids or alkalis but dissolves easily on fusion with alkali hydroxides to form alkali stannates. SnO_2 is reduced to tin by heating with carbon or hydrogen, and forms tin(IV) chloride on heating in chlorine. Hydrous tin(IV) oxide is formed by hydrolysis of tin(IV) salt solutions. The so-called α-oxide (or α-stannic acid) formed by the slow, low temperature hydrolysis is readily soluble in acids and bases, showing the amphoteric nature of SnO_2. Rapid high temperature hydrolysis produces the so-called β-oxide (or β-stannic acid), which is exceedingly inert to solvents. Both α- and β-oxides have the rutile structure with absorbed water, and the somewhat considerable differences in their properties are believed due to differences in particle size and surface properties.

[55] J. S. Anderson, J. S. Ogden and M. J. Ricks, *Chem. Comm.* (1968) 1585.
[56] M. Baudler, in *Handbook of Preparative Inorganic Chemistry*, Ed. G. Baurer, 2nd ed., p. 736 (1963).
[57] W. Kwestroo and P. H. G. M. Vromans, *J. Inorg. and Nuclear Chem.* **29** (1967) 2187.
[58] W. J. Moore and L. Pauling, *J. Am. Chem. Soc.* **63** (1941) 1392.
[59] J. D. Donaldson, W. Moser and W. B. Simpson, *J. Chem. Soc.* (1961) 839; *Acta Cryst.* **16** (1963) A22.
[60] F. Lawson, *Nature*, **215** (1967) 955.

Hydroxy Compounds

Tin(II) hydroxide is not known. The hydrolysis[61] of tin(II) salt solutions under controlled conditions indicates the presence of the ions $SnOH^+$, $Sn_2(OH)_2{}^{2+}$ and $Sn_3(OH)_4{}^{2+}$, and it has been proposed that this has a ring structure as Fig. 6 (a).

Single crystals of the tin(II) oxide–hydroxide compound $3SnO.H_2O$ are obtained by the very slow hydrolysis of tin(II) perchlorate solution. In X-ray structural determination[62]

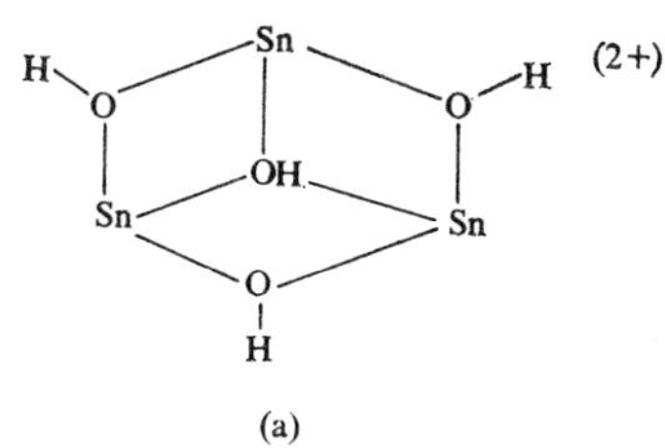

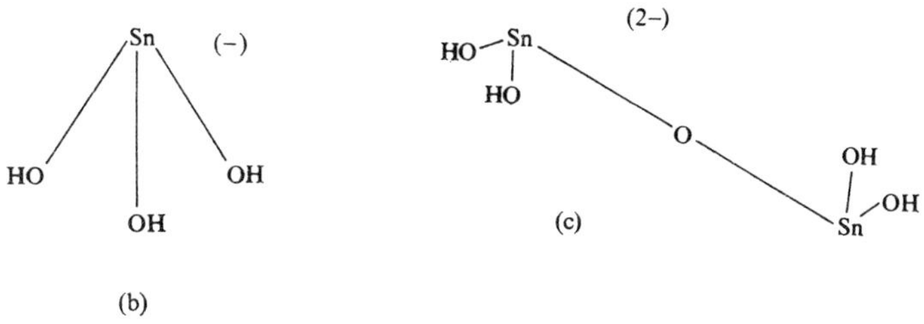

FIG. 6. Proposed structures of (a) $Sn_3(OH)_4^{2+}$, (b) $Sn(OH)_3{}^-$ and (c) $\{Sn_2O(OH)_4\}_2{}^-$.

this has been shown to contain Sn_6O_8 clusters, with six tin atoms at the corners of an octahedron, and an oxygen atom above each of the octahedron's eight faces, thus closely resembling the cluster structure of $Mo_6Cl_8^{4-}$. The formula is thus best written as $Sn_6O_8H_4$. This stoichiometry allows all of the oxygens in an infinite array of Sn_6O_8 clusters to be joined by hydrogen bonds.

It is interesting to note that the structure of the only well-characterized solid hydroxide phase, $Sn_6O_8H_4$, can be derived by the condensation and deprotonation of two of the cyclic ions $Sn_3(OH)_4^{2+}$, which are believed to predominate in solution.

Tin(II) oxide dissolves in alkalis to form the hydroxy-stannate(II) ion $Sn(OH)_3{}^-$, many salts of which have been isolated. The ion has the pyramidal structure illustrated in Fig. 6 (b). In addition to the $Sn(OH)_3{}^-$ ions the oxotetrahydroxyditin anions (Fig. 6 (c)) are known, and are partial dehydration products of the $Sn(OH)_3{}^-$ anions. Thus whereas $Ba\{Sn(OH)_3\}_2$ is the product of the interaction of barium hydroxide solution and $NaSn(OH)_3$ solution at 35°, at higher temperatures the salt $Ba\{Sn_2O(OH)_4\}$ of the dehydrated anion is formed[63].

[61] R. S. Tobias, *Acta Chem. Scand.* **12** (1958) 198.
[62] R. A. Howie and W. Moser, *Nature*, **219** (1968) 372.
[63] C. G. Davies and J. D. Donaldson, *J. Chem. Soc.* (*A*) (1968) 946.

Tin(IV) hydroxide is not known. Hydrolysis of tin(IV) salt solutions produces a white voluminous precipitate, which is not detectably crystalline, and may contain bonded water in a tin oxide gel. The freshly prepared material is called α-stannic acid, and is easily soluble in acid. Ageing produces β-stannic acid (or metastannic acid), which is more inert in acids, and has also developed an X-ray diffraction pattern characteristic of tin(IV) oxide. This metastannic acid may be obtained directly from tin and concentrated nitric acid. It remains open whether the differences between α- and β-stannic acids are solely due to differences in particle size.

The drying of tin(IV) oxide gel at 110° gives a reproducible product SnO_3H_2, which looses water stepwise between 110° and 600° to produce crystalline cassiterite[64].

Fusion of β-stannic acid with excess of alkaline hydroxide produces the crystalline salts

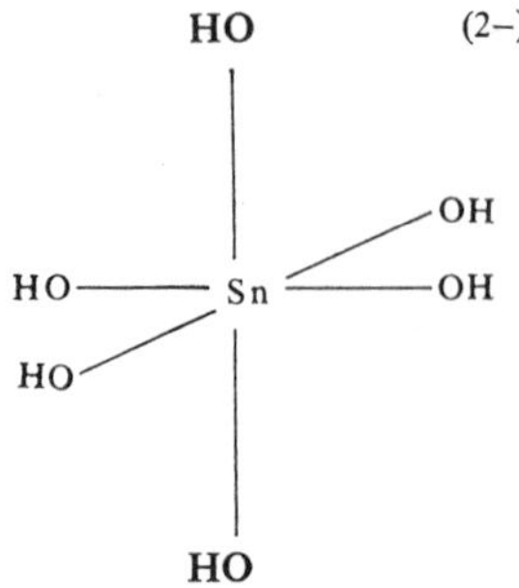

FIG. 7. Structure of the hexahydroxytin(IV) anion $Sn(OH)_6^{2-}$.

$M^I_2Sn(OH)_6$; and $Sn(OH)_6^{2-}$ ions are also well characterized in many other salts[65, 66]. The hexahydroxytin(IV) anions have the octahedral structure shown in Fig. 7, and the corresponding hexadeuteroxy compounds are known[67].

The hydrolysis of organotin halides gives rise to organotin hydroxides of different types, many of which, however, show a marked tendency to condense with loss of water[68, 69].

Trimethyltin hydroxide is an example of a well-characterized organotin hydroxide[70]. In the crystal it has the chain structure illustrated in Fig. 8 (a). In non-polar solvents a dimeric species $\{(CH_3)_3SnOH\}_2$ is present and is believed to have the dihydroxo-bridged structure of Fig. 8 (b).

Stannites and Stannates

The dehydration of trihydroxytin(II) anions to the oxotetrahydroxydistannites may be carried to completion on heating, to produce the stannites.

Many alkali and alkaline earth stannates have been reported and some have been well characterized. The action of heat upon the salt $K_2Sn(OH)6$ produces the hydrates $K_2SnO_3.H_2O$ and $3K_2SnO_3.2H_2O$ in addition to the anhydrous salt K_2SnO_3.

64 E. W. Giessekke, H. S. Gutowsky, P. Kirkov and H. A. Laitinen, *Inorg. Chem.* **6** (1967) 1294.
65 R. W. G. Wyckoff, *Am. J. Sci.* **15** (1928) 297.
66 R. L. Williams and R. J. Paces, *J. Chem. Soc.* (1957) 4143.
67 M. Maltese and W. J. Orville-Thomas, *J. Inorg. and Nuclear Chem.* **29** (1967) 2533.
68 J. G. A. Luijten, *Rec. Trav. Chim.* **85** (1966) 873.
69 R. Okawara and M. Wada, *Advances in Organometallic Chemistry*, **5** (1967) 137.
70 N. Kasai, K. Yasuda and R. Okawara, *J. Organometallic Chem.* **3** (1965) 172.

Three distinct anhydrous salt phases may be produced by heating together the alkali oxides M^I_2O and SnO_2 in different ratios and under varying conditions[71]. K_2SnO_4 is triclinic; K_2SnO_3 is orthorhombic with each tin atom almost at the centre of a deformed octahedron, with chains of octahedra sharing faces; $K_2Sn_3O_7$ is orthorhombic[72]. Strong

(a)

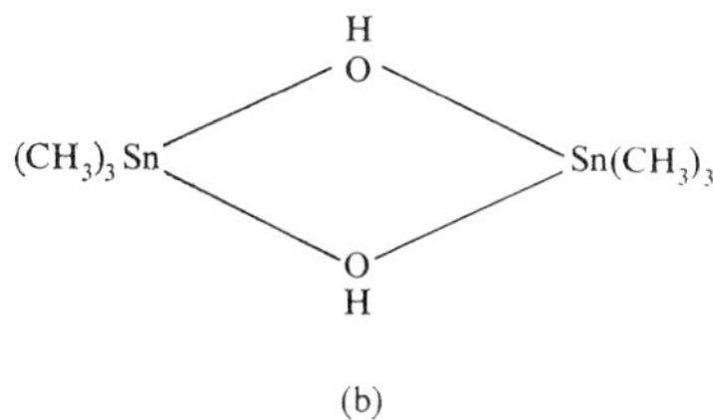

(b)

FIG. 8. (a) Polymeric structure of $(CH_3)_3SnOH$ in the crystal, (b) dimeric structure of $(CH_3)_3SnOH$ in solution.

heating of K_4SnO_4 converts it successively to K_2SnO_3, $K_2Sn_3O_7$ and eventually SnO_2 by loss of K_2O.

In addition to the alkali stannates, other tin(IV) stannates such as $CdSnO_3$, Ca_2SnO_4 and Cd_2SnO_4 have been fully characterized.

The stannates are usually white crystalline substances sensitive to carbon dioxide; they react vigorously with water[73, 74].

Peroxides and Perstannates

Many hydroperoxides of tin and perstannates with such formulations as $H_2Sn_2O_7$, $HSnO_4.3H_2O$, $KSnO_4.2H_2O$ and $K_2Sn_2O_7.3H_2O$ have been reported from the action of hydrogen peroxide upon solutions of tin(II) and tin(IV). The exact nature of these compounds is unknown, and in many cases they may be only hydrogen peroxide adducts[75, 76, 77].

Stable organotin hydroperoxides are well characterized. Thus trimethyltin hydroperoxide is formed from $(CH_3)_3SnOH$ and hydrogen peroxide, and is quite stable when pure.

[71] M. Tournoux, *Ann. Chim.* **9** (1964) 579.
[72] R. Hoppe, H. J. Roehrborn and H. Walker, *Naturwiss.* **51** (1964) 86.
[73] I. Morgenstern-Baradau, P. Poix and A. Michel, *Compt. rend.* **258** (1964) 3036.
[74] M. Troemel, *Naturwiss.* **54** (1967) 17.
[75] M. W. Spring, *Bull. Soc. Chim.* **1** (1889) 180.
[76] M. P. Pierron, *Bull. Soc. Chim.* (1950) 291.
[77] S. Tanatar, *Chem. Ber.* **38** (1905) 1184.

$(CH_3)_3SnOOH$ decomposes on warming to $(CH_3)_2SnO$, CH_3OH and oxygen[78]. Similarly, H_2O_2 on $\{(C_2H_5)_3Sn\}_2O$ produces $(C_2H_5)_3SnOOH{\cdot}H_2O_2$ at $-60°$, and at room temperature forms the explosive compound $(C_2H_2)_2Sn(OH)OOH$[79].

Organotin alkylhydroperoxides [80, 81, 82] have been synthesized by a number of methods. Exemplary of these is the action of alkylhydroperoxides upon organotinhydrides.

$$(C_4H_9^n)_3SnH + C_4H_9^tOOH \longrightarrow (C_4H_9^n)_3SnOOC_4H_9^t$$

Examples of the compounds $R'_{4-n}Sn(OOR)_n$ can be made in solution for $n = 1-4$, but only the species where $n = 1$ and $n = 2$ are isolable as pure materials. But these latter compounds such as $(C_4H_9^n)_3$ $SnOOC_4H_9^t$ and $(C_4H_9^n)_2Sn(OOC_4H_9^t)_2$ may be purified by vacuum distillation.

Dialkyltinoxyperoxides and bis(triaryl)tinperoxides $R_3SnOOSnR_3$ have been reported.

Organic Stannites and Stannates

Alkoxy compounds of both tin(II) and tin(IV) are known and may be regarded as the esters of the oxides SnO and SnO_2.

Tin(II) alkoxides have been prepared from either the action of tin(II) halides upon sodium alkoxides[83], or from the reaction between tin(II) halides and alcohols in the presence of triethylamine[84]. Both tin(II) methoxide and tin(II) ethoxide are very reactive compounds. They are hydrolysed rapidly to tin(II) oxide, but with limited water the bis{alkoxytin(II)} oxide ROSnOSnOR is formed.

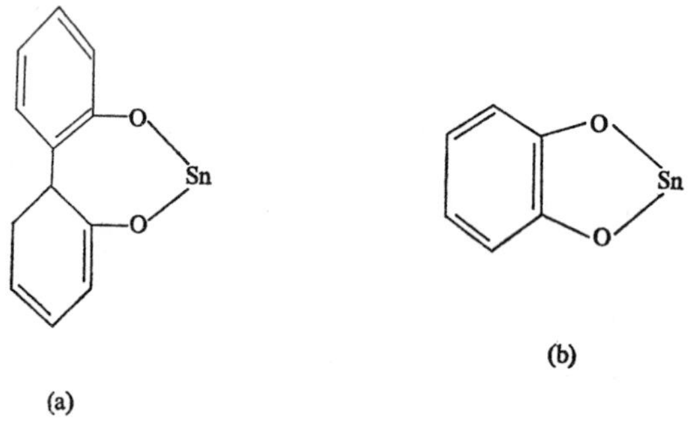

FIG. 9. (a) 2,2′-Biphenylenedioxytin(II) and (b) *o*-Phenylenedioxytin(II).

Another class of tin(II) organic oxygen derivatives are the tin(II) heterocycles formed by many dihydric phenols. Examples of this type of tin(II) oxygen ring system are 2,2′-biphenylenedioxytin(II) and *o*-phenylenedioxytin(II) illustrated in Fig. 9.

These tin(II) ring compounds are crystalline with remarkable thermal stability. They may be polymeric in the solid state with intermolecular tin–oxygen bridges, but in pyridine

78 R. L. Daunley and W. A. Aue, *J. Org. Chem.* **30** (1965) 3845.
79 Yu. A. Aleksandrov and V.A. Shushkov, *Zh. Obshch. Khim.* **35** (1965) 115.
80 A. J. Bloodworth, A. G. Davies and J. F. Graham, *J. Organometallic Chem.* **13** (1968) 351.
81 D. L. Alleston and A. G. Davies, *J. Chem. Soc.* (1962) 2465.
82 A. G. Davies and J. F. Graham, *Chem. and Ind.* (*London*) (1963) 1622.
83 E. Amberger and M. R. Kula, *Angew. Chem.* **75** (1963) 476; *Chem. Ber.* **96** (1963) 2556.
84 J. Morrison and H. M. Haendler, *J. Inorg. and Nuclear Chem.* **29** (1967) 373.

solution they are monomeric and may utilize an empty *p*-orbital to complex donor solvents such as pyridine[85, 86].

Tin(IV) organic oxygen compounds are of many different types, and their chemistry has been extensively investigated.

Tin(IV) Alkoxides

Tin tetrachloride and sodium ethoxide produce the double alkoxide $NaSn_2(OC_2H_5)_9$, and not $Sn(OC_2H_5)_4$. The tetra-alkoxides of tin are best produced by an alcohol exchange reaction with the dimeric isopropyl alcoholate of tin tetra-isopropoxide[87, 88].

$$\{Sn(OC_3H_7^i)_4C_3H_7^iOH\}_2 + 8ROH \longrightarrow 2Sn(OR)_4 + 10C_3H_7^iOH$$

The tin alkoxides of primary alcohols are involatile and are presumably polymeric, with bridging alkoxy groups, but the monomeric tertiary alkoxides distil easily at reduced pressure, e.g. $(C_4H_9{}^tO)_4Sn$ b.p. 99°/4 mm.

Organotin Oxides and Alkoxides

In addition to the tin(IV) alkoxides mentioned above, there have been synthesized many organotin compounds in which tin–oxygen bonds are present[18, 19]. The organostannoxanes are mostly obtained by the hydrolysis of the requisite halides, as exemplified below.

$$R_3SnX \xrightarrow{H_2O/OH^-} R_3SnOSnR_3$$

$$R_2SnX_2 \xrightarrow{H_2O/OH^-} R_2SnO$$

$$RSnX_3 \xrightarrow{H_2O/OH^-} RSn(O)OH$$

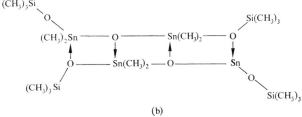

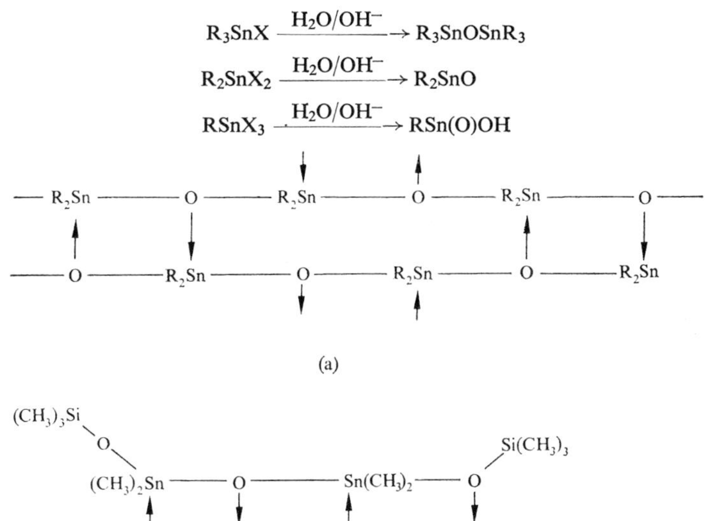

FIG. 10. (a) Proposed structure of $(R_2SnO)_n$ species, and (b) the structure of $\{(CH_3)_3SiO(CH_3)_2SnOSn(CH_3)_2OSi(CH_3)_3\}_2$.

---

85 J. J. Zuckermann, *J. Chem. Soc.* (1963) 1322.
86 G. T. Cocks and J. J. Zuckermann, *Inorg. Chem.* **4** (1965) 592.
87 D. C. Bradley, E. V. Caldwell and W. Wardlaw, *J. Chem. Soc.* (1957) 4775.
88 D. C. Bradley, *Progress in Inorganic Chemistry*, **2** (1960) 303.

$R_3SnOSnR_3$ species are monomeric and structurally analogous to the organic ethers, though they are believed to be considerably more basic[89, 90]. The dialkyl- and diaryltin oxides R_2SnO are formally analogous to the silicones, but bear no resemblance in physical properties, being semi-crystalline powders. They are believed to contain polymeric chains, in which coordination takes place as shown in Fig. 10 (a) to render each tin atom five-coordinate. This would suggest that all tin atoms are present in four-membered Sn–O rings[69].

This ladder type of structure has recently been confirmed[91] for the dimeric species $\{(CH_3)_2SiO(CH_3)_2SnOSn(CH_3)_2OSi(CH_3)_3\}_2$ as shown in Fig. 10 (b).

Little structural information is available about the so-called alkyl/arylstannonic acids RSn(O)OH, but these may have a ring structure, and undergo dehydration to polymers as indicated below.

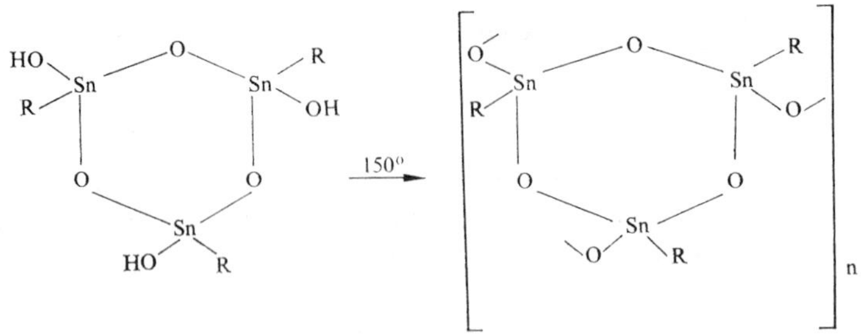

Organotin alkoxides have been synthesized in a variety of ways, as outlined below.

$$R_3SnOSnR_3 + R'OH \longrightarrow R_3SnOR'$$
$$R_2SnO + R'OH \longrightarrow R_2Sn(OR')_2$$
$$R_3SnCl + NaOR' \longrightarrow R_3SnOR'$$
$$R_3SnOR' + R''OH \longrightarrow R_3SnOR''$$
$$R_3SnNR_2 + R''OH \longrightarrow R_3SnOR''$$
$$R_3SnOR' + R''COOR''' \longrightarrow R_3SnOR'''$$

Some diols produce[92] cyclic alkoxides as illustrated in Fig. 11.

FIG. 11. The cyclic organotin alkoxides formed from $(C_4H_9)_2SnO$ and (a) 2,3-butanediol and (b) cyclohexanediol.

The tin–oxygen bonds of organotin oxygen compounds are often extremely reactive and undergo a wide variety of reactions as illustrated in Fig. 12.

89 E. W. Abel, D. A. Armitage and D. B. Brady, *Trans. Faraday Soc.* **62** (1966) 3459.

90 E. W. Abel, D. A. Armitage and S. P. Tyfield, *J. Chem. Soc.* (*A*) (1967) 554.

91 R. Okawara, N. Kasai and K. Yasuda, *Abstracts 2nd Intern. Symp. Organometallic Chem.*, p. 128, Wisconsin (1965).

92 W. J. Considine, *J. Organometallic Chem.* **5** (1966) 263.

Phenyltin trichloride and tropolone produce $C_6H_5Sn(tropolonato)_2Cl$, which is converted to $C_6H_5Sn(tropolonato)_3$ by reaction with sodium tropolonate. The complex $C_6H_5Sn(tropolonato)_3$ is monomeric, and is believed to contain hepta-coordinate tin[93].

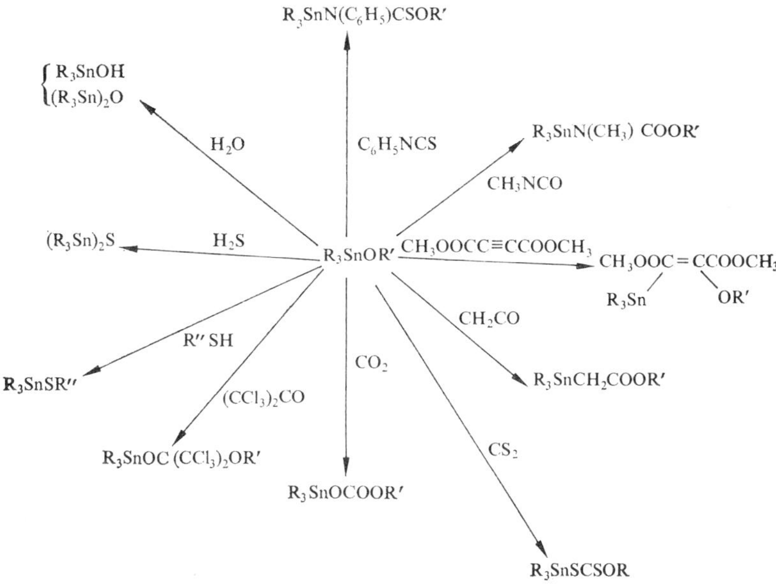

FIG. 12. Some reactions of organotin alkoxides.

In contrast to the corresponding alkoxides of tin, the oxinates $R_2Sn(Ox)_2$ are stable to moisture. From spectroscopic data these tin oxinates are believed to be chelate compounds of hexa-coordinate tin. The corresponding R_3SnOx types are believed to be penta-coordinate[94].

Carboxylates

Tin(II) carboxylates such as tin(II) formate[95] and tin(II) acetate[96] are made by dissolving tin(II) oxide in the requisite carboxylic acid. Tin(II) acetate is stable, and can be stored for some months in air, but over longer periods of time it undergoes slow oxidation. It can be sublimed in vacuum to give very pure $(CH_3COO)_2Sn$ at about 150°, but at higher temperatures decomposes to tin(II) oxide, carbon dioxide and acetone. Tin(II) acetate hydrolyses slowly over several hours in water, but the adduct $(CH_3COO)_2Sn.2CH_3COOH$ formed by dissolving tin(II) acetate in glacial acetic acid is unhydrolysed, even in boiling water.

Tin(II) acetate is a useful reducing agent, and can act under certain conditions both as a reducing and acetylating agent, thus for example a 75% yield of 1,4-diacetoxynaphthalene is obtained from 1,4-naphthaquinone and tin(II) acetate in 20 min.

[93] E. L. Meutterties and C. M. Wright, *J. Am. Chem. Soc.* **86** (1964) 5132; *Quart. Revs. (London)* **21** (1967) 123.

[94] K. Ramaiah and D. F. Martin, *Chem. Comm.* (1965) 130.

[95] J. D. Donaldson and J. F. Knifton, *J. Chem. Soc.* (1964) 4801.

[96] J. D. Donaldson, W. Moser and W. B. Simpson, *J. Chem. Soc.* (1964) 5942.

With the exception of tin(II) formate, the tin(II) carboxylates are fibrous in nature, and of low solubility, suggesting a linear polymeric structure which can break down on sublimation, or upon solution in donor solvents. The Mössbauer and infrared spectra of tin(II) carboxylates are both in accord with the presence of long chains of pyramidally coordinated tin atoms containing bridging carboxylate groups[97].

In addition to the simple tin(II) carboxylates, evidence has been presented for many more carboxylate species of tin(II); thus CH_3COOSn^+, $(CH_3COO)_2Sn$, $(CH_3COO)_7Sn_3^-$, $(CH_3COO)_5Sn_2^-$ and $(CH_3COO)_3Sn^-$ have all been reported in tin(II)–acetate solutions[98].

The tri-acetatostannates have been well characterized and isolated as pure salts of the alkali metals and ammonia, and also as the alkaline earth salts.

The discrete $(RCOO)_3Sn^-$ ions in the tricarboxylatostannates have distorted pyramidal

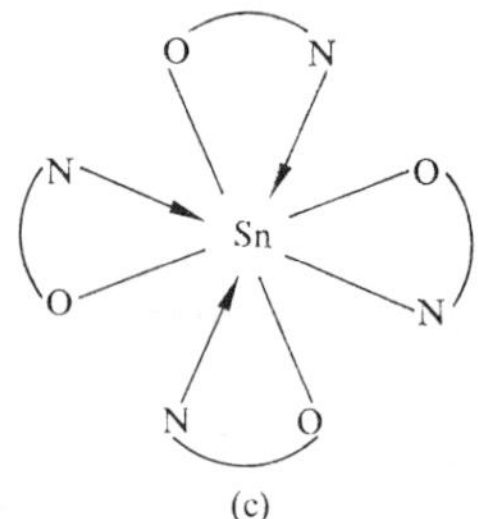

FIG. 13. (a) Five-coordinate tin in R_3Sn-oxinate, (b) six-coordinate tin in $R_2Sn(oxinate)_2$ and (c) eight-coordinate tin in $Sn(oxinate)_4$.

coordination about each tin atom as shown in Fig. 14(b). The more complex anions such as $(CH_3COO)_7Sn_3^-$ and $(CH_3COO)_5Sn_2^-$ may be regarded as having chain structures analogous to the tin(II) carboxylate chain illustrated in Fig. 14 (a).

Tin(IV) tetracarboxylates may be synthesized from anhydrous acetic acid and tin(IV) chloride, or from thallium(I) carboxylate and tin(IV) iodide. They are in marked contrast to the tin(II) carboxylates, in that they are exceedingly sensitive to hydrolysis by traces of moisture, but are soluble in most organic solvents[99, 100].

Ditin hexa-acetate $(CH_3COO)_6Sn_2$ is formed by the action of acetic acid upon hexaphenylditin at 120°. It is quite a stable compound, and is believed to have the structure illustrated in Fig. 15, where two bridging acetate groups are present in addition to the

[97] J. D. Donaldson and A. Jelen, *J. Chem. Soc.* (*A*) (1968) 1448.
[98] J. D. Donaldson and J. F. Knifton, *J. Chem. Soc.* (*A*) (1966) 332.
[99] M. Baudler, *Handbook of Preparative Inorganic Chemistry*, Ed. G. Brauer, Vol. II, 2nd ed., p. 747, Academic Press, New York.
[100] H. Schmidt, C. Blohm and G. Jander, *Angew. Chem.* *A***59** (1947) 233.

metal–metal bond. Hydrogen chloride at $-100°$ converts ditin hexa-acetate to ditin hexachloride[101].

The organotin(IV) carboxylates are of particular interest in that they appear to exist in more than one form. When trimethyltin hydroxide reacts with, for example, formic and acetic acids, the products are the very insoluble $(CH_3)_3SnOOCH$ and $(CH_3)_3SnOOCH_3$.

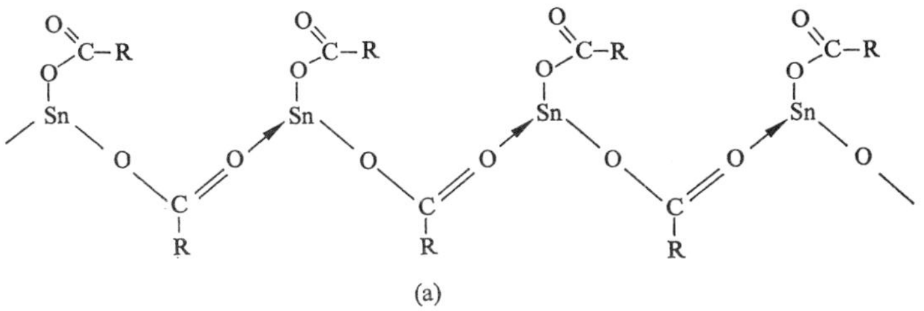

(a)

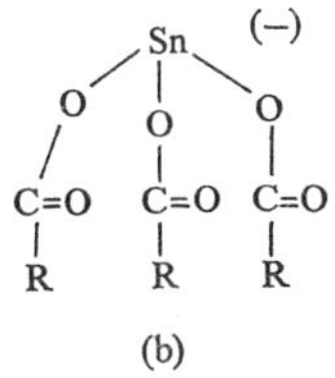

(b)

FIG. 14. Proposed structures of (a) tin(II) carboxylates and (b) the tricarboxylatotin anions.

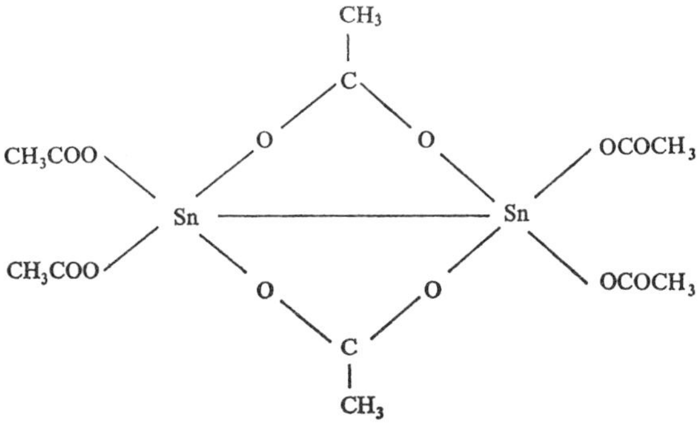

FIG. 15. Ditin hexa-acetate.

These insoluble products are believed to be associated by bridging carboxylate groups, either as linear or ring polymers[69].

If, however, the polymeric forms of $(CH_3)_3SnOOCH$ and $(CH_3)_3SnOOCCH_3$ are heated in cyclohexane at 100° for several hours, they are both converted to new soluble forms. A remarkable point of interest is that the soluble and insoluble forms of these organotin carboxylates are sublimable in high vacuum without interconversion[102].

[101] E. Wiberg and H. Behringer, *Z. anorg. Chem.* **329** (1964) 290.
[102] P. B. Simons and W. A. G. Graham, *J. Organometallic Chem.* **8** (1967) 479.

Tetra-ethylammonium trisoxalatostannate(IV) and tetra-ethylammonium trismalonato-stannate(IV) are made by the treatment of $\{(C_2H_5)_4N\}_2SnBr_6$ with silver oxalate and silver malonate respectively in acetone. These complex ions are not hydrolysed by water, and their infrared and Raman spectra have been measured and discussed[103].

2.8. TIN SALTS OF INORGANIC OXY-ACIDS

Numerous tin(II) and tin(IV) salts of inorganic oxy-acids, such as dithionates[106], carbonates, arsenates[104, 105], vanadates[104, 105], tungstates[104], borates[104], nitrates, sulphates, etc., have been reported in the literature. Very many of these, however, are ill characterized and of dubious stoichiometry. Some of the more authenticated compounds are discussed below.

Nitrates

A normal tin(II) nitrate is unknown, but dilute solutions of tin(II) nitrate are reasonably stable[107]. Such solutions may be made by nitrate metathesis from tin(II) chloride or sulphate, or alternatively by the dissolution of tin(II) oxide in dilute nitric acid. Efforts to evaporate such solutions to crystallization result in decomposition which may be violent. A number of explosions have been attributed to tin(II) nitrate systems, where the decomposition products are tin(IV) oxide and various nitrogen compounds such as hydroxylamine and nitrous oxide[108].

A basic tin(II) nitrate $Sn_3(OH)_4(NO_3)_2$ is made by the action of nitric acid upon a paste of hydrous tin(II) oxide. It is believed to be ionic and to contain the ion $Sn_3(OH)_4^{2+}$ illustrated in Fig. 6 (a). Basic tin nitrate is partly hydrolysed on dissolution in water. The presence of Sn(II) ions with NO_3^- makes thermal decomposition of the basic nitrate exceedingly violent, and $Sn_3(OH)_4(NO_3)_2$ may be classified as a high explosive[107].

Tin(IV) nitrate is formed by the action of N_2O_5, $ClNO_3$ or $BrNO_3$ upon tin(IV) chloride[109, 110]. The white volatile $Sn(NO_3)_4$ is believed to contain eight-coordinate tin, with each nitrate group behaving as a bidentate ligand. Tin(IV) nitrate reacts readily with aliphatic hydrocarbons with the formation of carboxylic acids, alkyl nitrates and nitro-alkanes; it reacts vigorously with diethyl ether. These reactions are attributed to the easy release of the very reactive NO_3 radicals.

$$Sn(NO_3)_4 \longrightarrow Sn(NO_3)_2 + 2NO_3$$

The compounds which release NO_3 radicals appear to be those in which the nitrate group is bonded in bidentate fashion. Thus the pyridine adduct $Sn(NO_3)_4(C_5H_5N)_2$, in which all of the nitrate groups are monodentate, is unreactive towards hydrocarbons and ether.

When $SnCl_4$ and N_2O_4 are mixed at $-10°$ in a stream of nitrogen, the compound

103 P. A. W. Dean, D. F. Evans and R. F. Phillips, *J. Chem. Soc.* (*A*) (1969) 363.
104 S. Prakash and N. R. Dahr, *J. Indian Chem. Soc.* **6** (1929) 587.
105 E. B. Maxted and N. J. Hassid, *J. Soc. Chem. Ind.* (*London*), **50** (1931) 399.
106 E. A. O'Connor, *Nature*, **139** (1937) 151.
107 J. D. Donaldson and W. Moser, *J. Chem. Soc.* (1961) 1996.
108 L. H. Milligan and G. R. Gillette, *J. Phys. Chem.* **28** (1924) 744.
109 W. Schmiesser, *Angew. Chem.* **67** (1955) 493.
110 C. C. Addison and W. B. Simpson, *J. Chem. Soc.* (1965) 598.

$SnCl_4.N_2O_4$ is formed as a pale yellow very hygroscopic powder. This is formulated as $NO^+[SnCl_4NO_3]^-$ in which a bidentate nitrate ligand produces an octahedral tin anion[111].

By a similar type of reaction between Cs_2SnCl_6 and liquid N_2O_5 the hexanitrate ion $Sn(NO_3)_6^{2-}$ was obtained as the caesium salt. The $Sn(NO_3)_6^{2-}$ may be presumed to be octahedral, with monodentate nitrate ligands[112].

A number of organotin nitrates such as $(CH_3)_3SnNO_3$ and $(CH_3)_2Sn(NO_3)_2$ are known[18, 113]. Structural information on these compounds to date is largely based upon spectroscopic data, and is conflicting. Organotin nitrates do however appear to contain metal coordinated nitrate groups, which may be mono- or bidentate, and in some cases bridging[69]. In the case of adducts such as $[(CH_3)_3Sn(NH_3)_2]^+NO_3^-$, however, there is non-coordinated nitrate ion present.

Triphenyltin nitrate has poor stability, and decomposes at quite low temperatures to nitrobenzene and diphenyltin oxide[114].

$$(C_6H_5)_3SnNO_3 \longrightarrow (C_6H_5)_2SnO + C_6H_5NO_2$$

Phosphates and Phosphites

The chemistry of tin phosphates is exceedingly complex, and very many stoichiometries have been reported[114, 115].

Tin(II) orthophosphate $Sn_3(PO_4)_2$ is reported from the interaction of sodium hydrogen phosphate upon tin(II) sulphate in sulphuric acid. Other tin(II) phosphates recorded are $SnHPO_4$, $Sn(H_2PO_4)_2$, $Sn_2P_2O_7$ and $Sn(PO_3)_2$.

Tin(IV) phosphates include $Sn_2O(PO_4)_2 10H_2O$, $Sn_2O(PO_4)_2$ and SnP_4O_7. Many complex phosphates such as $KSn(PO_4)_3$, $KSnOPO_4$ and $Na_2Sn(PO_4)_2$ are known[116].

The stoichiometry of the tin(II) and tin(IV) phosphates is controlled by the particular phosphate ion which is complexing to the tin(II) or the tin(IV) ion; thus a whole range of pyrophosphates and tripolyphosphates of tin have been detected, and there is a trend for tin(II) to complex long-chain polyphosphates, with average chain lengths up to fourteen[117].

Tin(IV) hypophosphite is remarkable in that it represents a class of compound containing tin(IV) and a strongly reducing anion. Considering the reduction potentials,

$$Sn^{4+} + 2e \longrightarrow Sn^{2+} \qquad (E = 0{\cdot}15V)$$
$$H_3PO_3 + 2H^+ \longrightarrow H_3PO_2 + H_2O \qquad (E = -0{\cdot}50V)$$

the very existence of $Sn(H_2PO_2)_4$ is surprising. The isolation of tin(IV) hypophosphite by bubbling oxygen through a solution of tin(II) oxide in hypophosphorous acid may be due to a combination of the low solubility of the product, and the often slow kinetics of hypophosphite reductions at ordinary temperatures[118].

In addition to the simple tin(IV) hypophosphite, a number of halogenohypophosphites

111 C. C. Addison and W. B. Simpson, *J. Chem. Soc.* (*A*) (1966) 775.
112 K. W. Bagnall, D. Brown and J. G. H. du Preez, *J. Chem. Soc.* (1964) 5523.
113 C. C. Addison, W. B. Simpson and A. Walker, *J. Chem. Soc.* (1964) 2360.
114 P. J. Shapiro and E. I. Becker, *J. Org. Chem.* **27** (1962) 4668.
115 E. Jablczynsky and W. Wieckowsky, *Z. anorg. Chem.* **152** (1926) 207.
116 J. W. Mellor, *Comprehensive Treatise in Inorganic and Theoretical Chemistry*, vol. VII, p. 481 Longmans, London (1927).
117 R. E. Mesmer and R. R. Irani, *J. Inorg. and Nuclear Chem.* **28** (1966) 493.
118 W. B. Simpson, *Chem. Comm.* (1967) 1100.

of tin(II) and tin(IV) such as $SnCl_4.Sn(H_2PO_2)_4$, $SnI_4.Sn(H_2PO_2)_4$, $SnBr_2.3Sn(H_2PO_2)_2$ and $SnI_2.3Sn(H_2PO_2)_2$ have been reported[119]

Perchlorates

The action of Cl_2O_3 upon tin(IV) chloride is reported to produce the tin(IV) perchlorates $Sn(ClO_4)_4.2Cl_2O_6$ and $Cl_2Sn(ClO_4)_2$, the structures of which are unknown, but which presumably must contain perchlorate groups coordinated to tin[120].

Organotin perchlorates such as $(CH_3)_3SnClO_4$ and $(CH_3)_2Sn(ClO_4)_2$ are better characterized, and are electrolytes in aqueous solution. Trimethyltin perchlorate (m.p. 128°) is volatile, and in the vapour the molecule is likely to have a bidentate perchlorate group attached to five-coordinate tin. In the crystal, however, the perchlorate group may have a bridging role in the structure[121].

Solutions of tin(II) perchlorate have been used in studies of tin complex formation, and one crystalline species has been isolated. Tin(II) perchlorate trihydrate is made by the action of tin(II) oxide upon 70% perchloric acid. $Sn(ClO_4)_2.3H_2O$ is very hygroscopic, melts at 240° and decomposes explosively at 250°. Mössbauer measurements suggest that one water molecule is bonded to tin, as in the tin(II) chloride dihydrate[122].

Sulphates

Tin(II) sulphate may be prepared by dissolving tin metal or a tin(II) salt in sulphuric acid. Pure tin(II) sulphate, however, is best made by the displacement of copper from a copper sulphate solution by metallic tin. Crystals of $SnSO_4$ are white and orthorhombic ($a = 8{\cdot}81$; $b = 7{\cdot}17$ and $c = 5{\cdot}35$ Å). Tin(II) sulphate has an excellent shelf life, and is strongly recommended as a pure source of Sn(II)[123].

The thermal decomposition of $SnSO_4$ yields sulphur dioxide and tin(IV) oxide, and is an example of the numerous internal oxidation–reduction reactions which are undergone by oxygen containing tin(II) compounds.

The solubility of tin(II) sulphate in water decreases steadily with rise in temperature (35·2 g/100 ml of solution at 20°, 22·0 g/100 ml of solution at 100°). It is also quite soluble in concentrated sulphuric acid.

The basic tin(II) sulphate $Sn_3(OH)_2OSO_4$ is obtained as a well-defined crystalline material by the action of ammonia solution upon tin(II) sulphate. This basic tin sulphate looses water in one stage at 230° to leave the tin(II) oxysulphate $Sn_3O_2SO_4$.

Tin(IV) sulphate as the dihydrate $Sn(SO_4)_2.2H_2O$ is obtained from hydrous tin(IV) oxide in hot dilute sulphuric acid. The colourless crystals are exceptionally hygroscopic and best stored in sealed ampoules[124]. Tin(IV) sulphate hydrolyses completely in water, with precipitation of hydrous tin(IV) oxide, but is nevertheless freely soluble in dilute sulphuric acid to give a clear solution.

119 D. A. Everest, *J. Chem. Soc.* (1951) 2903; *ibid.* (1954) 4698; *ibid.* (1959) 4149.
120 H. Schmeisser, *Angew. Chem.* **67** (1955) 493.
121 R. Okawara, B. J. Hathaway and D. E. Webster, *Proc. Chem. Soc.* (1963) 13.
122 C. G. Davies and J. D. Donaldson, *J. Inorg. and Nuclear Chem.* **30** (1968) 2635.
123 J. D. Donaldson and W. Moser, *J. Chem. Soc.* (1960) 4000.
124 M. Baudler, *Handbook of Preparative Inorganic Chemistry*, Ed. G. Brauer, Vol. I, 2nd ed., p. 744, Academic Press, New York.

Estimates of activity coefficients suggest that the dissolution of hydrous tin(IV) oxide in dilute sulphuric acid produces the $SnSO_4^{2+}$ ion in solution, with more concentrated sulphuric acid solutions causing further complexing to form $Sn(SO_4)_2$ and $H_2Sn(SO_4)_3$ in solution[125].

Tetramethyltin reacts with sulphuric acid to evolve methane and produce trimethyltin hydrogen sulphate $(CH_3)_3SnHSO_4$. The trialkyltin hydrogen sulphates and the dialkyltin dihydrogen sulphates behave as strong bases. It is most likely, however, that the cationic species present are protonated hydrogen sulphates, rather than stannonium ions[126].

Phenyltin compounds undergo cleavage in sulphuric acid to produce the hexa(hydrogen sulphate) stannic acid $H_2Sn(HSO_4)_6$ and its anions.

2.9. CHALCOGEN COMPOUNDS

Sulphides

The only three well-characterized sulphides of tin are SnS, Sn_2S_3 and SnS_2, though Sn_4S_5 has been claimed[127] by thermal decomposition of SnS_2. SnOS is reported from the action of sulphur upon tin(II) oxide, and Sn_2S_3O, $Sn_5O_6S_4$ and $Sn_3O_7S_2$ have been reported as hydrated materials from various aqueous precipitations[128, 129].

Tin(II) sulphide may be made by the direct combination of the elements, but this product is usually non-stoichiometric; it is usually obtained by the sulphide precipitation of tin(II) salts, whence the initially hydrated form is easily dehydrated. Crystalline SnS, m.p. 1153°K, b.p. 1500°K, density ~5, has a very distorted NaCl structure[130] in which each tin atom has four different near neighbour sulphur distances of 2·62Å, 2·68 Å (2), 3·27 Å (2) and 3·39 Å. The analogous $PbSnS_2$ which occurs in Bolivian ores as the mineral theallite has the same distorted structure, in marked contrast to the perfect NaCl lattice of PbS.

SnS is reduced to tin by hydrogen, and converts to tin(IV) oxide on heating in the air. It is virtually insoluble in water (solubility product $\sim 10^{-27}$ at 25°C), but it can be re-crystallized from tin(IV) chloride solutions as grey-blue metallic platelets.

Tin sesquisulphide has been reported as a constituent of Bolivian ores[131], and has been shown by X-ray work to be a distinct phase. The structure consists of infinite double rutile type strings of $Sn^{IV}S_6$ octahedra with Sn^{II} ions attached laterally[132].

Tin(IV) sulphide is prepared commercially as yellow platelets (density 4·51) by the action of tin foil upon flowers of sulphur, and is used as mosaic gold and may be purified by sublimation. It is precipitated from mildly acid solutions of tin(IV) salts by sulphide ion.

SnS_2 crystallizes with a CdI_2 structure[133]. When formed by precipitation from aqueous solution the SnS_2 is easily redissolved in warm hydrochloric acid, but the product of the high temperature dry synthesis is very insoluble.

125 C. H. Brubaker Jr., *J. Am. Chem. Soc.* **76** (1954) 4269; *ibid.* **77** (1955) 5265.
126 R. J. Gillespie, R. Kapoor and E. A. Robinson, *Canad. J. Chem.* **44** (1966) 1197.
127 Ya I. Gerasimov, E. V. Kruglova and N. D. Rozenblyum, *Zhur. Obsch. Khim., S.S.S.R.* **7** (1937) 1520.
128 F. W. Schmidt, *Chem. Ber.* **27** (1894) 2739.
129 P. Sisley and L. Meunier, *Bull. Soc. Chim.* **51** (1932) 939
130 W. Hofmann, *Z. Krist.* **92** (1935) 161
131 *Neues Jahrb. Mineral Monatsh.*, (1964) 64.
132 D. Mootz and H. Puhl, *Acta Cryst.* **23** (1967) 471.
133 A. F. Wells, *Structural Inorganic Chemistry*, 3rd ed., p. 520, Oxford (1962).

Thiostannates

The thiostannates may be made by a number of methods such as the action of alkali sulphide solutions upon tin(IV) sulphide or upon various stannate solutions. Well-crystallized species of definite stoichiometry are formed, which appear to be salts of the SnS_3^{2-} and SnS_4^{4-} ions. Invariably, however, the thiostannates crystallize highly hydrated, as for example in $Na_2SnS_3.8H_2O$ and $Na_4SnS_4.18H_2O$, and some water appears to be more strongly bound than the rest[134]. The structure of these compounds is unknown, but it has been suggested that the ions present may be either $[(HS)_3Sn(OH)_3]^{2-}$ or $[(H_2O)_3SnS_3]^{2-}$, and either $[(HS)_2Sn(OH)_2S_2]^{4-}$ or $[(H_2O)_2SnS_4]^{4-}$ rather than simply SnS_3^{2-} and SnS_4^{4-}.

In addition to the binary sulphides, other classes of compounds contain tin–sulphur bonds. Of these, the coordination complexes of thio-ethers with tin halides are discussed below in the chemistry of the tin halides.

Organotin Sulphides

The tin–sulphur covalent bond has been extensively studied, and a variety of synthetic methods[135] available for its formation are outlined below.

$$SnCl_4+NaSR \longrightarrow (RS)_4Sn$$

$$(C_6H_5)_3SnCl+AgS \longrightarrow (C_6H_5)_3SnSSn(C_6H_5)_3$$

$$(CH_3)_3SnOH+RSH \longrightarrow (CH_3)_3SnSR$$

$$R_2SnO+R'SH \longrightarrow R_2Sn(SR')_2$$

$$(R_3Sn)_2O+H_2S \longrightarrow (R_3Sn)_2S$$

$$(C_6H_5)_4Sn+S \longrightarrow (C_6H_5)_2SnS$$

$$(CH_3)_3SnSn(CH_3)_3+S \longrightarrow (CH_3)_3SnSSn(CH_3)_3$$

$$C_6H_5NCS+(C_2H_5)_3SnH \longrightarrow (C_2H_5)_3SnSCH{:}NC_6H_5$$

Some of the different types of organotin sulphur compounds are illustrated in Fig. 16. In all of these compounds, the tin(IV) atom is approximately tetrahedral, with the exception of Fig. 16 (f), which is polymeric[136], and in which tin is probably octahedral by intermolecular coordination. This polymer may be broken down by ligands such as amines to produce complexes like Fig. 16 (g)[137].

The base strength measurements suggest that the organotin sulphides such as $(CH_3)_3SnSSn(CH_3)_3$ are stronger bases than the simple organic sulphides. This would be in accord with little or no back-bonding from sulphur to tin, and tin's lower electronegativity[89,90].

Organotin sulphur compounds are fairly robust compounds, but undergo fission by halogens and halides. They are now manufactured on a large scale for industrial uses, especially as stabilizers for polyvinyl chloride and related plastics, where the $R_2Sn(SR')_2$ types are extensively used[4].

[134] E. E. Jelley, *J. Chem. Soc.* (1933) 1580.
[135] E. W. Abel and D. A. Armitage, *Advances in Organometallic Chemistry*, **5** (1967) 1.
[136] R. C. Poller, *Proc. Chem. Soc.* (1963) 312.
[137] R. C. Poller and J. A. Spillman, *J. Chem. Soc.* (*A*) (1966) 958.

Selenides

In addition to SnSe and $SnSe_2$, the phase diagram[138] for tin–selenium shows the formation of Sn_2Se_3 by a peritectic reaction at 650°. Sn_2Se_3 has a tetragonal type of lattice $a = 6{\cdot}77$ Å, $c = 5{\cdot}86$ Å[139].

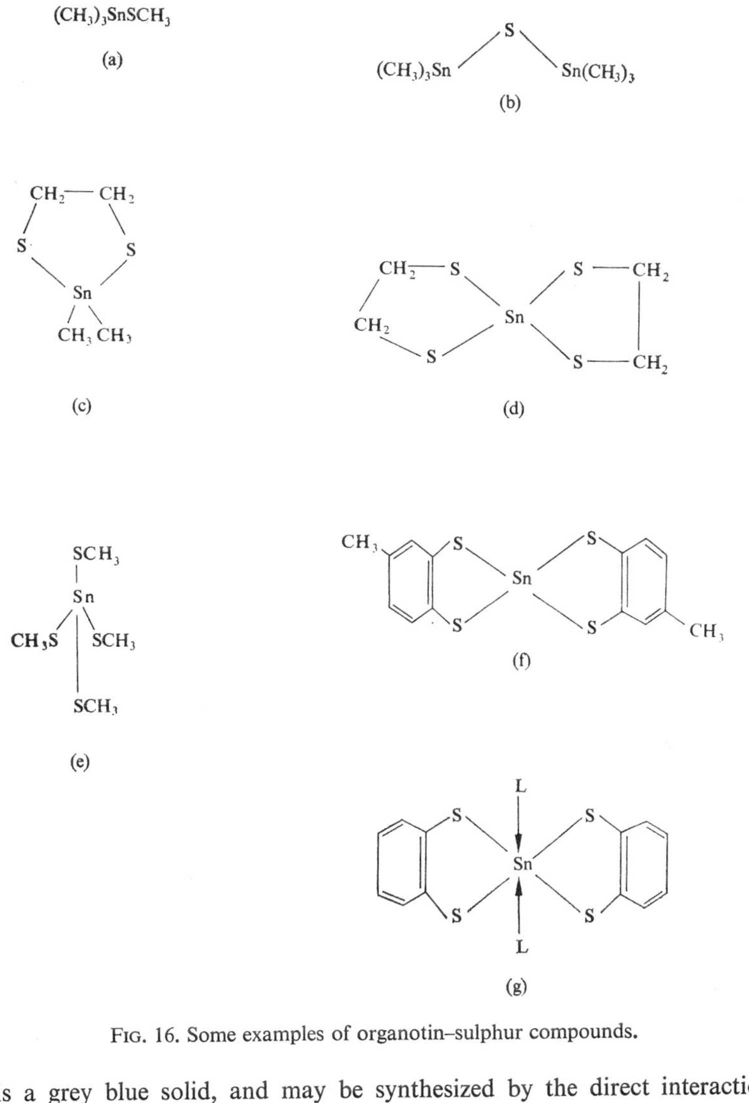

FIG. 16. Some examples of organotin–sulphur compounds.

SnSe is a grey blue solid, and may be synthesized by the direct interaction of the powdered elements, which starts at about 350°, and is extremely exothermic[140]. It melts at 861° and may be distilled without decomposition. Single crystals of considerable size

138 F. Laves and Y. Baskin, *Z. Krist.* **107** (1959) 377.
139 J. Willy, *Compt. rend.* **251** (1960) 1273.
140 Y. Matakuta, T. Yamamoto and A. Osazaki, *Mem. Fac. Sci. Kynsyu Univ.* *B***1** (1953) 98.

are grown with some ease[141]. Crystal data on SnSe are contradictory, giving either a rhombic[140] structure $a = 4{\cdot}33$ Å, $b = 3{\cdot}98$ Å, $c = 11{\cdot}18$ Å, or a cubic NaCl type of structure[142].

$SnSe_2$, m.p. 625°C, may be prepared by vacuum melting of stoichiometric quantities of the elements and as single crystals by the Bridgman method[143]. $SnSe_2$ has a CdI_2 structure $a = 3{\cdot}811$ Å and $c = 6{\cdot}137$ Å.

Both SnSe and $SnSe_2$ are semiconductors with specific resistances in the range 10^3 to 10^{-1} ohm cm, but electrical properties vary considerably with the history of heat treatment of the alloys. Impurities in SnSe behave somewhat unusually, and it is likely that antimony for example replaces tin atoms in the lattice up to 200°C, but above that replaces selenium in the lattice[144].

Both Sn_2Se_3 and $SnSe_2$ decompose on heating to SnSe and selenium. The molecules Sn_2Se_2 are detected by the mass spectrometer at high temperatures in the vapour, and these are believed to be tetrahedral clusters[145].

The selenothiostannates $SnSe_2S^{2-}$ and the selenostannates $SnSe_3^{2-}$ are analogous to the thiostannates, and though well characterized, are of unknown structure[146].

Organic Selenides

A variety of molecular compounds exist in which there are discrete tin–selenium bonds.

The sodium, potassium and magnesium salts of selenols react with tin tetrahalides to afford complete halogen replacement[147], and organotin selenolates have been made in aqueous solution from the selenols[148].

$$\left.\begin{matrix} RSeNa \\ RSeMgX \end{matrix}\right\} + SnX_4 \longrightarrow (RSe)_4Sn$$

$$R_3SnX + R'SeH \longrightarrow R_3SnSeR$$

The organoditin selenides are made by the action of the appropriate organotin halide upon sodium selenide[149, 150].

$$R_3SnCl + Na_2Se \longrightarrow R_3SnSeSnR_3$$

The organotin selenols R_3SnSeH have not been reported, but their lithium salts such as $(C_6H_5)_3SnSeLi$ are known[151].

Some representative organotin–selenium compounds are listed[135] in Table 11.

Tellurides

The heating together of tin and tellurium causes the formation of SnTe with incandescence. SnTe can be sublimed to give large crystals and is stable at high temperatures.

141 P. F. Wells, *J. Electrochem. Soc.* **113** (1966) 90.
142 L. S. Palatkin and V. V. Levin, *Doklady Akad. Nauk. S.S.S.R.* **96** (1954) 975.
143 O. Mitchell and H. Levinstein, *Bull. Am. Phys. Soc.* **6** (1959) 133.
144 D. M. Chizhikov and V. P. Shchastlivyi, *Selenium and Selenides*, Collets Press, London (1968).
145 R. Colin and J. Drowart, *Trans. Faraday Soc.* **60** (1964) 673.
146 F. C. Mathers and H. S. Rothrock, *Ind. Eng. Chem.* **23** (1931) 831.
147 H. J. Backer and J. B. G. Hurenkamp, *Rec. Trav. Chim.* **61** (1942) 802.
148 E. W. Abel, D. A. Armitage and D. B. Brady, *J. Organometallic Chem.* **5** (1966) 130.
149 I. Riudisch and M. Schmidt, *J. Organometallic Chem.* **1** (1963) 160.
150 M. Schmidt and H. Ruf, *Chem. Ber.* **96** (1963) 784.
151 H. Schumann, K. F. Thom and M. Schmidt, *J. Organometallic Chem.* **2** (1964) 361.

SnTe is deposited from tin(II) solutions by the action of hydrogen telluride[152], and similarly $SnTe_2$ is supposedly formed from hydrogen telluride and tin(IV) solutions[153]. The linear $SnTe_2$ molecule is analogous to CO_2 and is detected at high temperatures.

Melting SnTe under high pressure gives two solid polymorphs[154], as well as liquid around 117°K. The normally NaCl structure of SnTe is converted by a pressure of 18,000 atm to an orthorhombic form with a 360% increase in electrical resistivity[155].

TABLE 11. EXAMPLES OF MOLECULAR COMPOUNDS CONTAINING TIN–SELENIUM BONDS

$(C_6H_5Se)_4Sn$	$[(CH_3)_3Sn]_2Se$
$(p\text{-}CH_3\cdot C_6H_4Se)_4Sn$	$(CH_3Sn)_4Se_6$
$(\text{tert-}C_4H_9Se)_4Sn$	$(C_6H_5)_3SnSeLi$
$(p\text{-}Cl\cdot C_6H_4Se)_4Sn$	$(C_6H_5)_3SnSePb(C_6H_5)_3$
$(CH_3)_3SnSeC_6H_5$	$[(n\text{-}C_4H_9)_2SnSe]_n$

The organic tin–tellurium compounds have been little investigated, but the hexaphenylditintelluride $[(C_6H_5)_3Sn]_2Te$ and its lithium salt have been reported[156]

2.10. HALOGENO COMPOUNDS

Tin(II) Halides

Tin(II) fluoride is obtained as colourless monoclinic crystals by the evaporation of a solution of tin(II) oxide in 40% hydrofluoric acid[157, 158]. From solutions of tin(II) fluorides with other fluorides an orthorhombic form of pure tin(II) fluoride has been isolated[159]. In this form tin is in an irregular trigonal pyramidal environment, with two Sn–F bonds of 2·15 Å, and one of 2·45 Å[160]. Three other fluorines at about 2·80 Å make up an overall distorted octahedral environment. From Mössbauer spectroscopy the tin–fluorine bonds in the orthorhombic form are believed to be more covalent than in the monoclinic form. Tin(II) fluoride is very soluble in water, and is extensively utilized as a constituent of dentifrices[161–163]. The mixed halide SnFCl is obtained by action of hydrofluoric acid upon tin(II) chloride[164].

Tin(II) chloride is available commercially as the dihydrate, which is itself of interest. In crystalline $SnCl_2.2H_2O$ one molecule of water is lattice aquation[165], but the other is coordinated to the tin dichloride molecule, whereby the tin nominally attains an octet of electrons. Anhydrous tin(II) chloride is obtained from the dihydrate by the action of acetic

152 W. Biltz and W. Mecklenburgh, *Z. anorg. Chem.* **64** (1909) 226.
153 A. Brukl, *Montash.* **45** (1924) 471.
154 W. Klement and L. H. Cohen, *Science*, **154** (1966) 1176.
155 J. A. Kafalas and A. N. Mariano, *Science*, **143** (1964) 952.
156 H. Schumann, K. F. Thom and M. Schmidt, *J. Organometallic Chem.* **2** (1964) 361.
157 W. H. Nebergall, J. C. Muhler and H. G. Day, *J. Am. Chem. Soc.* **74** (1952) 1604.
158 G. Bergerhoff, *Acta Cryst.* **15** (1962) 509.
159 J. D. Donaldson, R. Oteng and B. J. Senior, *Chem. Comm.* (1965) 618.
160 J. D. Donaldson and R. Oteng, *J. Inorg. and Nuclear Chem. Letters*, **3** (1967) 163.
161 M. R. Mericle and J. C. Muhler, *J. Dental Research*, **43** (1964) 1227.
162 W. A. Zacherl and C. W. B. McPhail, *J. Can. Dental Assoc.* **31** (1965) 174.
163 V. Mercer and J. C. Muhler, *J. Oral Therap. Pharmacol.* **1** (1964) 141.
164 W. H. Nebergall, G. Baseggio and J. C. Muhler, *J. Am. Chem. Soc.* **76** (1954) 5353.
165 D. Grdenic and B. Kamenar, *Proc. Chem. Soc.* (1960) 312.

anhydride. Crystalline $SnCl_2$ has a layer structure in which there are Sn–Cl–Sn–Cl chains[166] The Sn–Cl bonds in the chain are 2·78 Å, and the pyramidal coordination of each metal is completed by a further chlorine at 2·67 Å. Considering the overall structure of $SnCl_2$ from layer to layer, tin may be regarded as nine-coordinate in this structure. Tin(II) bromide and tin(II) iodide are similar to $SnCl_2$ in the crystal, in that they do not contain

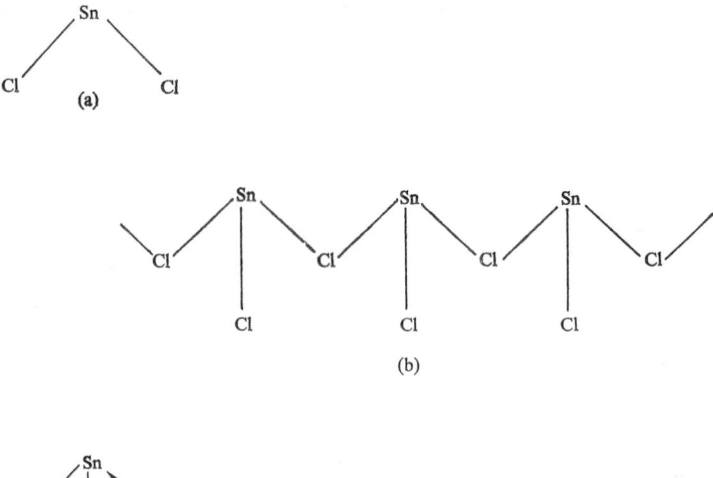

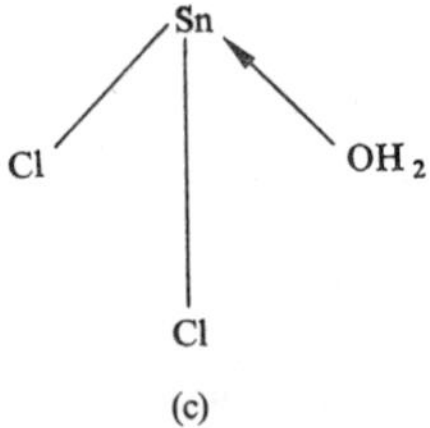

FIG. 17. Tin(II) chloride: (a) an angular discrete molecule in the gas phase; (b) the polymeric $(SnCl_2)_n$ chain present in crystalline $SnCl_2$; (c) coordination about tin in $SnCl_2{\cdot}2H_2O$, where only one water molecule is actually coordinated to tin.

discrete $SnBr_2$ and SnI_2 molecular units, but a polymeric layer structure; SnI_2 is conveniently made as brilliant red needles by heating iodine in 2N hydrochloric acid with tin[167].

In the gas phase $SnCl_2$, $SnBr_2$ and SnI_2 exist as discrete angular molecular species[168] (see Table 7).

The solubility of the tin(II) halides in water falls off markedly from $SnCl_2$ (84 g of $SnCl_2$ in 100 g of water at 0°C) to SnI_2 (0·96 g of SnI_2 in 100 g of water at 0°C).

Halogenotin(II) Anions

The halogenotin(II) acids H_2SnX_4 have been claimed both in solution and as crystalline solids in the older literature[169, 170]. Better characterized, however, are the many halogeno-

166 R. E. Rundle and D. H. Olson, *Inorg. Chem.* **3** (1964) 596.
167 W. Moser and I. C. Trevena, *J. Chem. Soc.* (*D*) (1969) 25.
168 M. W. Lister and L. E. Sutton, *Trans. Faraday Soc.* **37** (1941) 406.
169 S. W. Young, *J. Am. Chem. Soc.* **23** (1901) 450.
170 R. Wagner, *Chem. Ber.* **19** (1886) 896.

tin(II) salts such as $KSnF_3$, $Na_2Sn_2F_5$, $(NH_4)_4SnCl_6$, NH_4SnCl_3, K_4SnCl_6, $KSnBr_3$, K_2SnBr_4, K_4SnBr_6, $CsSn_2I_5$, $KSnI_3$, etc.[171–173].

Though the precise nature of many of these salts is unknown, information regarding the discrete ions present in some of these complexes show these materials to be lattice salts only.

Salts of the ions SnX_3 (where X=F, Cl, Br and I) have all been obtained from aqueous

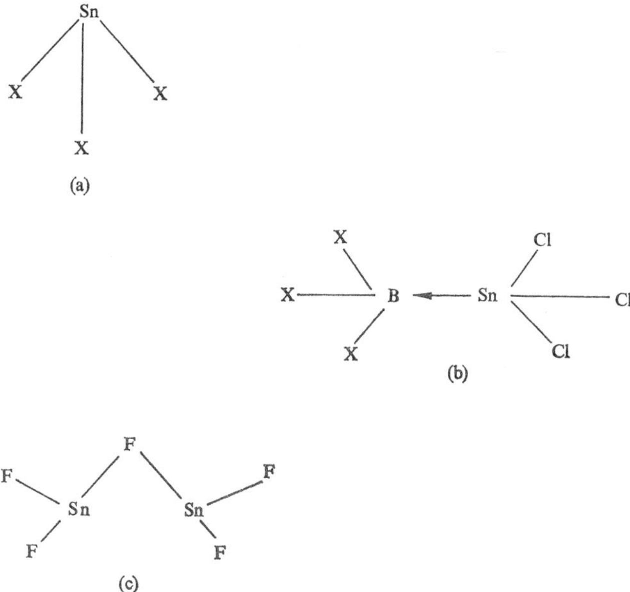

FIG. 18. (a) Pyramidal structure of the SnX_3^- ion due to the presence of a sterically active lone pair of electrons. (b) The complex ion $[X_3B.SnCl_3]^-$ illustrating coordination by the lone pair of electrons on $SnCl_3^-$. (c) Structure of the $Sn_2F_5^-$ anion.

solution. The SnX_3 ions are pyramidal, as shown in Fig. 18 (a), due to the presence of a sterically active non-bonding pair of electrons. This electron pair makes the trihalogenotin anions excellent ligands, resulting in a wide range of transitional metal complexes[174, 175]. In addition to being a σ-bonding donor group, the SnX_3 anion ligands are believed to be powerful π-acceptor ligands. Complexes which involve only the σ-donor characteristics are exemplified by the boron halide complex anions $[X_3B.SnCl_3]^-$, which are believed[176] to contain a boron–tin bond as shown in Fig. 18 (b). The formation and decomposition of some $SnCl_3$ complexes are exemplified in the following equations.

$$R_4AsSnCl_3+BX_3 \longrightarrow R_4As^+[X_3B.SnCl_3]^-$$
$$PtCl_4^{2-}+2SnCl_3^- \longrightarrow [Cl_2Pt(SnCl_3)_2]^{2-}+2Cl^-$$
$$[Cl_2Pt(SnCl_3)_2]^{2-}+2(C_6H_5)_3P \longrightarrow \{(C_6H_5)_3P\}_2PtCl_2+2SnCl_3^-$$
$$Pt(CO)_2Cl_2+2SnCl_3^- \longrightarrow [Cl_2Pt(SnCl_3)_2]_2^-+2CO$$

[171] E. Rimbach and K. Fleck, *Z. anorg. Chem.* **94** (1916) 139.
[172] W. Pugh, *J. Chem. Soc.* (1953) 2491.
[173] V. Auger and T. Karantassis, *Compt. rend.* **181** (1925) 665.
[174] J. F. Young, R. D. Gillard and G. Wilkinson, *J. Chem. Soc.* (1964) 5176.
[175] R. D. Cramer, R. V. Lindsey, C. T. Prewitt and U. G. Stolberg, *J. Am. Chem. Soc.* **87** (1965) 658.
[176] M. P. Johnson, D. F. Shriver and S. A. Shriver, *J. Am. Chem. Soc.* **88** (1966) 1588.

Despite the existence of many well-characterized salts, $M_2SnCl_4.H_2O$, $M^{II}SnCl_4.nH_2O$ and $M^{III}SnBr_4.nH_2O$, it is unlikely that any contain a discrete SnX_4^{2-} ion. A structural investigation of the salt $K_2SnCl_4.H_2O$ showed it to be a lattice salt $KCl.KSnCl_3.H_2O$ with the tin atom present in a distorted pyramidal $SnCl_3^-$ ion, with three more distant chlorine atoms in the lattice completing a roughly octahedral environment about tin[177].

The ions $Sn_2X_5^-$ (X = F and Cl) are well defined[178], thus $Sn_2F_5^-$ is formed in concentrated tin(II) solutions containing insufficient fluoride ions to complete the formation of SnF_3^-. $Na_2Sn_2F_5$ contains discrete $Sn_2F_5^-$ ions as shown in Fig. 18 (c). Two SnF_2 groups are linked by fluorine, with bridging Sn–F distances of 2·22 Å and terminal Sn–F bond lengths of 2·07 Å[179].

$Sn_2F_5^-$ anions are also present in salt melts, and similar tin(II) chloride melts with the alkali metal chlorides provide evidence for the presence of $Sn_2Cl_5^-$ ions, and also $Sn_3Cl_7^-$ ions[180].

SnF_3^- is considerably the strongest of the halogenotin(II) complexes, as illustrated[181] by the dissociation constants SnF_3^- $1{\cdot}1\times10^{-11}$, $SnCl_3^-$ 3×10^{-5}, $SnBr_3^-$ $4{\cdot}5\times10^{-2}$.

Coordination Complexes of Tin(II) Halides

Tin(II) fluoride, chloride and bromide all have Lewis acid capability to form complexes with suitable bases. Complexes of oxygen and nitrogen species are numerous, and those of sulphur somewhat rarer. Table 12 gives some examples of well-defined complexes of these three tin(II) halides[13].

TABLE 12. STOICHIOMETRIES OF SOME COORDINATION COMPLEXES OF TIN(II) HALIDES

SnF_2		$SnCl_2$		$SnBr_2$	
Water	(1:1)	Water	(1:1 and 1:2)	Pyridine	(1:2)
Dimethylsulphoxide	(1:1)	Dioxane	(1:1)	Aniline	(1:2)
Pyridine	(1:1 and 1:2)	Dimethylsulphoxide	(1:2 and 2:3)	Toluidine	(1:1 and 1:2)
		Dimethylformamide	(1:1)	Benzidine	(1:1)
		Pyridine N-oxide	(1:1 and 1:2)	Piperazine	(1:1)
		Pyridine	(1:1 and 1:2)	Tetramethylthiourea	(1:1)
		2,2′-Bipyridyl	(1:1)		
		Thiourea	(1:1, 1:2 and 2:5)		
		Tetramethylthiourea	(1:1)		

With six outer shell electrons the tin(II) halides can be expected to be monofunctional acceptors to complete a valence shell. Whilst this accounts for the many 1:1 complexes in Table 12, it cannot be an explanation for other well-characterized stoichiometries. Whilst many of these other stoichiometries may be analogous to $SnCl_2.2H_2O$, in which one water molecule is coordinated and the other is only lattice water, it cannot be ruled out that in certain cases tin may use[13] vacant *d* orbitals to form these $SnCl_2.(base)_n$ complexes, where $n > 1$.

177 H. Brasseur and A. de Rassenfosse, *Z. Krist.* **101** (1939) 389.
178 J. D. Donaldson and J. D. O'Donoghue, *J. Chem. Soc.* (1964) 271.
179 R. R. McDonald, A. C. Larson and D. T. Cromer, *Acta Cryst.* **17** (1964) 1104.
180 I. S. Marozor and Li Chih-Fa, *Zhur. Neorg. Khim.* **8** (1963) 708.
181 J. M. van den Berg, *Acta Cryst.* **14** (1961) 1002.

Tin(IV) Halides

Tin(IV) fluoride may be prepared by the action of anhydrous hydrogen fluoride upon tin(IV) chloride[182]. It is an extremely hygroscopic white crystalline solid which may be sublimed at about 700°. Although it is likely that gaseous SnF_4 has tetrahedral molecules with four-coordinate tin, solid SnF_4 differs markedly from the other three tin(IV) halides

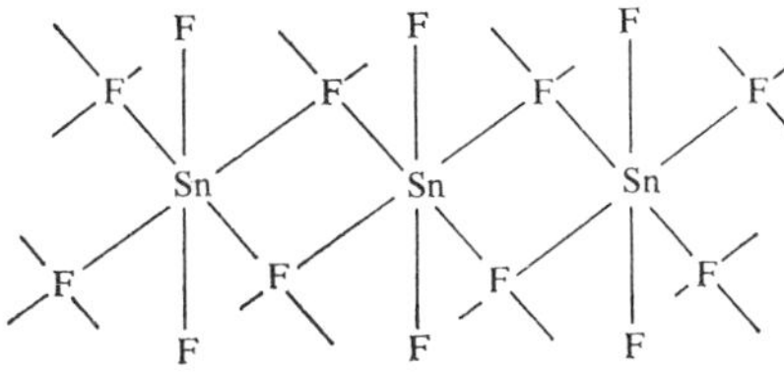

FIG. 19. The layer structure of crystalline SnF_4.

in that each tin atom is six-coordinate. Crystalline tin(IV) fluoride has a layer structure[183] as illustrated in Fig. 19. Each tin atom is attached to four fluorine atoms in the plane of the layer with Sn–F = 2·08 Å, and also to two terminal fluorine atoms above and below the layer plane with Sn–F = 1·88 Å.

Tin(IV) chloride, bromide and iodide are all conveniently made by the direct action of the halogen upon metallic tin. Some of their physical properties are described in Table 13.

TABLE 13. PROPERTIES OF Sn(IV) CHLORIDE, BROMIDE AND IODIDE

Property	$SnCl_4$	$SnBr_4$	SnI_4
Colour	Colourless	Colourless	Brown
Melting point, °C	−33·3	31	144
Boiling point, °C	114	205	348
$\triangle H$(fusion), kcal/mole	2·19	3·15	4·53
$\triangle H$(vap), kcal/mole	7·96	10·39	13·61
Density, g/ml	2·23 (20°C)	3·34 (35°C)	4·56 (20°C)
Sn–X bond length (Å)	2·31	2·44	2·64
Vapour pressure ($\log_{10}{}^{P}$) (T in °Kelvin)	$7 \cdot 59676 - \frac{1824 \cdot 9}{T}$	$7 \cdot 63048 - \frac{2270 \cdot 8}{T}$	$7 \cdot 66607 - \frac{2973 \cdot 9}{T}$
Critical temperature, °C	319		
Critical pressure, mmHg	28,080		
Specific heat, cal/deg/g	0·1476		
Dielectric constant	3·2		
Magnetic susceptibility (per mole × 10^{-6})	−115	−149	

All three of these tetrahalides are volatile molecular species, in which discrete SnX_4 molecules are present in the gas, liquid and solid phases.

Tin(IV) chloride forms a number of hydrates, but that stable at ordinary temperature and pressure is $SnCl_4.5H_2O$. Tin(IV) tetrahalides are all hydrolysed by aqueous alkali.

[182] O. Ruff and W. Plato, *Chem. Ber.* **37** (1904) 673.
[183] R. Hoppe and W. Dahne, *Naturwiss.* **49** (1967) 254.

There is a rapid exchange of halogens in mixtures of tin(IV) chloride, bromide and iodide, and Raman spectra have shown the presence of all possible molecules of formula $SnBr_nI_mCl_{4-m-n}$ in such a mixture[184].

Hexahalogenotin(IV) Anions

The anions SnX_6^{2-} (X = F, Cl, Br and I) may be obtained from solutions of the tetrahalides containing an excess of halide ion. They have been isolated as salts with a wide variety of cations M^+ and M^{2+}. The salts contain discrete cations and the SnX_6^{2-} anions which are octahedral. A crystal structure determination[185] of Na_2SnF_6 showed tin to be in a distorted octahedral environment of fluorines, with Sn–F bonds of 1·83, 1·92 and 1·96 Å.

The well-characterized salt formulated as K_3HSnF_8 is better written as $K_2SnF_6.KHF_2$, and consists of a lattice of K^+, SnF_6^{2-} and linear FHF ions[186].

The parent acids of hexachlorostannates[187, 188] and hexabromostannates[189] may be precipitated as $H_2SnCl_6.6H_2O$ and $H_2SnBr_6.8H_2O$ by the action of hydrogen halide upon a concentrated solution of the appropriate tin(IV) halide.

$(NH_4)_2SnCl_6$ is the "pink salt" mordant of the dyeing industry.

In addition to the hexahalogenostannates a number of other complex halogenostannates are known, in which other ligands besides halogen are coordinated to tin. These may be prepared in a number of ways, and have been extensively studied[190]. Thus increasing addition of alkali to aqueous solution of SnF_6^{2-} causes initial formation of $SnF_5(OH)^{2-}$, and then $SnF_4(OH)_2^{2-}$, which is present in *cis* and *trans* forms; finally $SnF_3(OH)_3^{2-}$ as *cis* and *trans* isomers is obtained. Examples of other mixed di-anions of this type are SnF_5Cl^{2-}, $SnF_4Cl_2^{2-}$ (*cis* and *trans*), $SnCl_3F_3^{2-}$ (*cis* and *trans*), $SnCl_5F^{2-}$, $SnBrF_5^{2-}$, $SnBrClF_4^{2-}$, $SnIF_5^{2-}$, $Sn(NCO)F_5^{2-}$, $Sn(NCO)_2F_4^{2-}$ (*cis* and *trans*), $Sn(NCS)_3F_3^{2-}$, $Sn(CN)F_5^{2-}$ and $Sn(OOCCF_3)F_5^{2-}$.

In addition, octahedral mono-anions are known where one of the ligands is neutral, as, for example, in $Sn(H_2O)F_5^-$, $Sn\{(CH_3)_2SO\}ClF_4^-$, $SnNCS\{(CH_3)_2SO\}F_4^-$ and $Sn\{(CH_6H_5)_3P\}F_5^-$.

It is unlikely that the pentahalogenotin(IV) anions SnX_3 exist in solution, as they would pick up a solvent molecule to complete six-coordination. The SnX_3^- anions are, however, known in crystals, and $SnCl_5^-$ has been shown to be a trigonal bipyramidal species[191].

Coordination Complexes of Tin(IV) Halides

All four tetrahalides of tin form addition compounds with certain electron donors (L), such as amines, cyanides, ethers, sulphides, ketones, phosphines, arsines, etc.[192]. The most usual stoichiometry of these compounds is L_2SnX_4, but the presence of 1:1 adducts has been demonstrated in solution[193], and the occasional 1:1 adduct has been reported as a

[184] C. Cerf and M. B. Delhaye, *Bull. Soc. Chim. France* (1964) 2818.
[185] Ch. Hebecker, H. G. von Schnering and R. Hoppe, *Naturwiss.* **53** (1966) 154.
[186] M. F. A. Dove, *J. Chem. Soc.* (1959) 3722.
[187] R. Engel, *Compt. rend.* **103** (1886) 213.
[188] K. Seubert, *Chem. Ber.* **20** (1887) 793.
[189] B. Rayman and K. Preis, *Ann. Chem. Liebigs.* **223** (1884) 324.
[190] P. A. W. Dean and D. F. Evans, *J. Chem. Soc.* (*A*) (1968) 1154.
[191] R. F. Bryan, *J. Am. Chem. Soc.* **86** (1964) 734.
[192] I. R. Beattie, *Quart. Revs.* **17** (1963) 382.
[193] D. P. N. Satchell and J. L. Wardell, *Proc. Chem. Soc.* (1963) 86.

solid. Some examples of tin(IV) halide complexes are given in Table 14, together with the stereochemistry, where known. In general, the stereochemistry of these octahedral complexes appears to be governed by steric factors, thus small ligands tend to give *cis* adducts, while larger sterically hindered ligands produce *trans* adducts.

The adducts of the more powerful donors give complexes such as (pyridine)$_2$SnCl$_4$ which are very insoluble[194], but with weaker ligands, complexes such as (tetrahydrofuran)$_2$SnCl$_4$ are soluble in solvents such as benzene, frequently with extensive dissociation[195].

Although crystalline adducts have not been isolated, there is evidence that 1:1 complexes are formed between tin(IV) chloride and aromatic hydrocarbons in solution[196].

Organotin Halides

The organotin halides are among the most widely investigated and used compounds of tin; and virtually the whole field of organotin chemistry is based upon the use of these materials as precursors[197]. The three basic types of organotin halide are R_3SnX, R_2SnX_2 and $RSnX_3$ and each class is known for a wide range of organic groups R, and all of the halogens with the sole exception of the type $RSnF_3$.

Many methods of synthesis are available, and all find their use for special cases[18]. The halogenation of tetra-organotins may be carried out by a number of reagents as outlined below, and possibly represents the most extensively used synthesis for the organotin halides.

$$R_4Sn + X_2 \longrightarrow R_3SnX + RX$$
$$R_3SnX + X_2 \longrightarrow R_2SnX_2 + RX$$
$$R_4Sn + HX \longrightarrow R_3SnX + RH$$
$$R_3SnX + HX \longrightarrow R_2SnX_2 + RH$$
$$(C_4H_9)_2Sn(CH=CH_2)_2 + AsBr_3 \longrightarrow (C_4H_9)_2SnBr_2 + (CH_2=CH)_2AsBr$$
$$(C_6H_5)_4Sn + BF_3 \longrightarrow (C_6H_5)_3SnF + C_6H_5BF_2$$
$$R_4Sn + SO_2Cl_2 \longrightarrow R_2SnCl_2 + 2RCl + SO_2$$

Another very widely used synthesis of the organotin halides involves the proportionation reaction between tetra-organotins and tin(IV) halides[198]. In these reactions, illustrated below, it is to be noted that there is conservation of the whole organic content of the reaction as the required organotin halide, in contrast to the losses of organic groups in the above equations.

$$3R_4Sn + SnX_4 \longrightarrow 4R_3SnX$$
$$R_4Sn + SnX_4 \longrightarrow 2R_2SnX_2$$
$$R_4Sn + 3SnX_4 \longrightarrow 4RSnX_3$$

These reactions take place in decreasing order of ease Cl > Br > I, with fluoride little utilized[199].

[194] S. T. Zenchelsky and P. R. Segatto, *J. Am. Chem. Soc.* **80** (1958) 4796.

[195] H. H. Sisler, E. E. Schilling and W. O. Groves, *J. Am. Chem. Soc.* **73** (1951) 426; H. H. Sisler, H. H. Batey, B. Pfahler and R. Mattair, *J. Am. Chem. Soc.* **70** (1948) 3821.

[196] J. J. Myher and K. E. Russell, *Canad. J. Chem.* **42** (1964) 1555.

[197] I. Ruidisch, H. Schmidbauer and H. Schumann, *Halogen Chemistry*, Vol. 2, p. 269, Ed. V. Gutmann, Academic Press (1967).

[198] G. E. Coates, M. L. H. Green and K. Wade, *Organometallic Compounds*, 3rd ed., Vol. 1, p. 430, Methuen (1967).

[199] E. V. van den Berghe and G. P. van der Kelen, *J. Organometallic Chem.* **6** (1966) 522.

The tin–tin bonds in hexa-organoditins and the cyclic di-organotins undergo fission with halogen to form the corresponding halides[19, 200].

$$R_3Sn.SnR_3+X_2 \longrightarrow 2R_3SnX$$
$$(R_2Sn)_n+nX_2 \longrightarrow nR_2SnX_2$$

The direct alkylation and arylation of tin halides with a variety of reagents is useful in certain cases, but usually leads to a mixture of products[197].

$$SnX_4+nRMgX \longrightarrow R_nSnX_{4-n}+nMgX_2$$
$$SnX_4+nRLi \longrightarrow R_nSnX_{4-n}+nLiX$$
$$SnX_2+R_2Hg \longrightarrow R_2SnX_2+Hg$$

There have been extensive investigations of the so-called "direct synthesis", from tin

TABLE 14. SOME COORDINATION COMPLEXES OF TIN(IV) HALIDES

Complex	Notes	Complex	Notes
$(Pyridine)_2SnF_4$		$\{(C_2H_5)_2S\}_2SnCl_4$	*trans*
$(CH_3)_3NSnF_4$		$\{(CH_3)_3N\}_2SnCl_4$	*trans*
$(CH_3CN)_2SnF_4$			
$(Tetrahydrofuran)_2SnF_4$		$(Pyridine)_2SnBr_4$	
		$\{(CH_3)_3N\}_nSnBr$	$n = 1$ or 2
$(Pyridine)_2SnCl_4$		$(Tetrahydrofuran)_2SnBr_4$	Soluble in chloroform, smells of tetrahydro-furan
$\{(CH_3)_3N\}_nSnCl_4$	$n = 1$ or 2		
$\{(CH_3)_2CO\}_2SnCl_4$	*cis*		
$(CH_3CN)_2SnCl_4$	*cis*, soluble in benzene	$(Pyridine)_2SnI_4$	Insoluble
$(POCl_3)_2SnCl_4$	*cis*	$\{(CH_3)_3N\}_2SnI_4$	Soluble in $(CH_3)_3N$
$(Tetrahydrofuran)_2SnCl_4$	*trans*		
$(Tetrahydrothiophen)_2SnCl_4$	*trans*		

metal and alkyl or aryl halides, and many organotin halides have been successfully synthesized in this way[201–204].

$$Sn+3CH_3Br \longrightarrow CH_3SnBr_3+SnBr_2$$
$$2Sn+3C_6H_5CH_2Cl \longrightarrow (C_6H_5CH_2)_3SnCl+SnCl_2$$
$$Sn+2RX \xrightarrow{\text{catalyst}} R_2SnX_2$$

In addition to the mononuclear organotin halides described above, there exist a very limited number of organoditin halides and organodistannoxane halides, as mentioned in Table 15.

With the exception of the fluorides, the organotin halides are liquids or low melting solids, and are mostly readily soluble in organic solvents, and are quite volatile.

200 W. P. Neumann and I. Pedain, *Ann. Chem. Liebigs.* **672** (1964) 34.
201 K. Kocheshkov, *Chem. Ber.* **61** (1928) 1659.
202 A. C. Smith and E. G. Rochow, *J. Am. Chem. Soc.* **75** (1953) 4103, 4105.
203 K. Shishido, Y. Takada and Z. Kinugawa, *J. Am. Chem. Soc.* **83** (1961) 538.
204 V. Oakes and R. E. Hutton, *J. Organometallic Chem.* **6** (1966) 133.

Weak intermolecular association in trimethyltin chloride is believed to be via chlorine bridges, as shown by the dependence of the position of the tin–chlorine stretching frequency

TABLE 15. SOME REPRESENTATIVE ORGANOTIN HALIDES

Compound	M.p. °	B.p. °/mm	Compound	M.p. °	B.p. °/mm
$(CH_3)_3SnF$	375 decomp.		$(C_6H_5)_2SnBr_2$	38	
$(CH_3)_3SnCl$	38	154/760	$C_6H_5SnI_3$		220/760 (decomp.)
$(CH_3)_3SnBr$	27	165/760	$(C_4H_9)_3SnCl$		147/5
$(CH_3)_3SnI$	3·4	170/760	$(CH_2=CH)_3SnCl$		60/6
$(CH_3)_2SnF_2$	360 decomp.		$(ClCH_2)_3SnCl$		140/5
$(CH_3)_2SnCl_2$	108	190/760	$(CH_3)_2(C_2H_5)SnCl$		180/760
$(CH_3)_2SnBr_2$	74	209/760	$(CH_3)(C_2H_5)SnCl_2$	52	
$(CH_3)_2SnI_2$	44	228/760	α-$C_{10}H_7SnCl_3$	78	
CH_3SnCl_3	46		$Cl(C_4H_9)_2Sn.Sn(C_4H_9)_2Cl$	112	
CH_3SnBr_3	53	211/760	$Br(C_4H_9)_2Sn.Sn(C_4H_9)_2Br$	104	
CH_3SnI_3	85		$Cl(C_2H_5)_2Sn.Sn(C_2H_5)_2Cl$	175	
$(C_6H_5)_3SnCl$	107	249/13·5	$Br(C_4H_9)_2SnOSn(C_4H_9)_2Br$	104	

upon the state of aggregation[205]. In general, however, organotin chlorides, bromides and iodides are unassociated in gas, liquid or crystal form.

In contrast, all of the organotin fluorides have strongly associated polymeric structures.

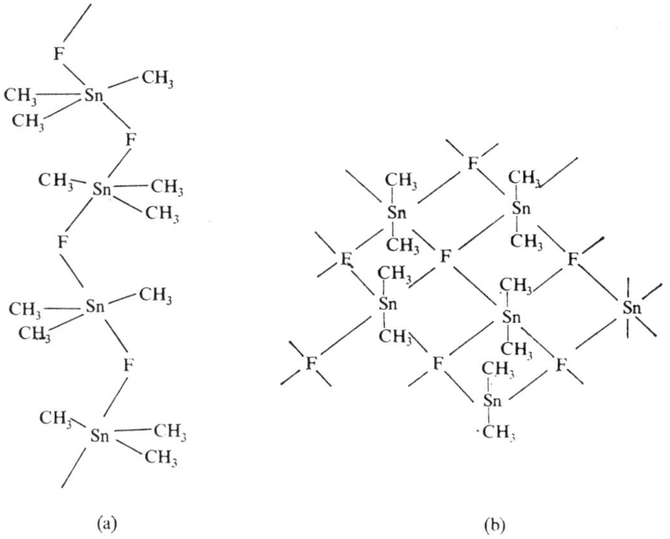

FIG. 20. (a) The chain structure of $(CH_3)_3SnF$ and (b) the sheet structure of $(CH_3)_2SnF_2$.

Thus, for example, trimethyltin fluoride has the chain structure shown in Fig. 20 (a), in which each tin atom is five-coordinate[206].

Dimethyltin difluoride forms a layer structure as shown in Fig. 20 (b), wherein SnF_2 units form a two-dimensional infinite sheet, and the two methyl groups attached to each tin

[205] H. Kriegsmann and S. Pischtschan, *Z. anorg. Chem.* **308** (1961) 212.

[206] H. C. Clark, R. J. O'Brien and J. Trotter, *Proc. Chem. Soc.* (1963) 85; *J. Chem. Soc.* (1964) 2332.

are present above and below the sheet, giving each tin atom a resultant octahedral coordination[207]. Such a structure is analogous to SnF_4 in Fig. 19 when terminal fluorine atoms are present instead of methyl groups.

Organotin halides are chemically very reactive, and the halogen may be replaced by a wide variety of other atoms or groups.

Hydrogen replaces halogens, upon treatment with various hydrides[19, 208, 209].

$$4R_{4-n}SnX_n + nLiAlH_4 \longrightarrow 4R_{4-n}SnH_n + nLiX + nAlX_3$$

(R = alkyl and aryl; X = Cl, Br and I; n = 1, 2 and 3)

$$R_nSnCl_{4-n} + R_2AlH \longrightarrow R_nSnH_{4-n} + R_2AlCl$$

(n = 1, 2 and 3)

The halogen of organotin halides is replaceable by a wide variety of metals and metalloids[197, 210].

$$R_3SnCl + Li \longrightarrow R_3SnLi$$

$$(CH_3)_3SnCl + NaMn(CO)_5 \longrightarrow (CH_3)_3SnMn(CO)_5$$

$$(CH_3)_2SnBr_2 + Na \longrightarrow \begin{cases} (CH_3)_2Sn.Sn(CH_3)_2 \text{ (each Sn bearing Na)} \\ (CH_3)_2SnNa_2 \end{cases}$$

$$(C_6H_5)_3SnCl + Ca \longrightarrow (C_6H_5)_3SnCaSn(C_6H_5)_3$$

$$(C_6H_5)_3SnCl + (C_6H_5)_3MK \longrightarrow (C_6H_5)_3SnM(C_6H_5)_3$$

(M = Si, Ge and Sn)

Similarly a range of non-metals replace halogen[211–213].

$$R_3SnCl + LiNR^1_2 \longrightarrow R_3SnNR^1_2$$

$$R_3SnCl + LiP(C_6H_5)_2 \longrightarrow R_3SnP(C_6H_5)_2$$

$$R_2SnX_2 + R'OH \longrightarrow R_2Sn(OR')_2$$

$$(CH_3)_2SnBr_2 + HSCH_2CH_2SH \longrightarrow (CH_3)_2Sn\langle S{-}CH_2{-}CH_2{-}S\rangle \text{ (ring)}$$

$$(CH_3)_3SnCl + NaN_3 \longrightarrow (CH_3)_3SnN_3$$

In addition to these reactions involving displacement of halogen, many organotin halides behave as Lewis acids, like the tin(IV) halides themselves. Many different types of coordination complex are known[214].

Although the tetra-organotins are unable to form stable complexes, in which tin has a coordination number greater than four, the introduction of one halogen atom on tin is

207 E. O. Schlemper and W. C. Hamilton, *Inorg. Chem.* **5** (1966) 995.
208 A. E. Finholt, A. C. Bond and H. I. Schlesinger, *J. Am. Chem. Soc.* **69** (1947) 1199.
209 W. P. Neumann, *Angew. Chem.* **73** (1961) 542.
210 F. G. A. Stone in *New Pathways in Inorganic Chemistry*, Ed. E. A. V. Ebsworth, A. Maddock and A. G. Sharpe, Cambridge University Press (1968).
211 G. J. M. van der Kerk, J. G. A. Luijten and M. J. Janssen, *Chimia.* **16** (1962) 10.
212 E. Wiberg and R. Rieger, German Patent 1,121,050 (1960).
213 E. W. Abel and D. B. Brady, *J. Chem. Soc.* (1965) 1192.
214 R. C. Poller, *J. Organometallic Chem.* **3** (1965) 321.

sufficient to produce Lewis acids which can complex with even relatively weak donor bases. Figure 21 (a) shows the stereochemistry of the complex (pyridine) $(CH_3)_3SnCl$[215, 216].

In addition to neutral ligands, the tri-organotin halides will take up halide ion[217] to form the complex ions $R_3SnX_2^-$; the trigonal bipyramidal structure of $(CH_3)_3SnBr_2^-$ is illustrated in Fig. 21 (b).

The Lewis acidity of the organotin halides increases considerably with higher halogen content, and the difference between their dipole moments in dioxane and hexane has been

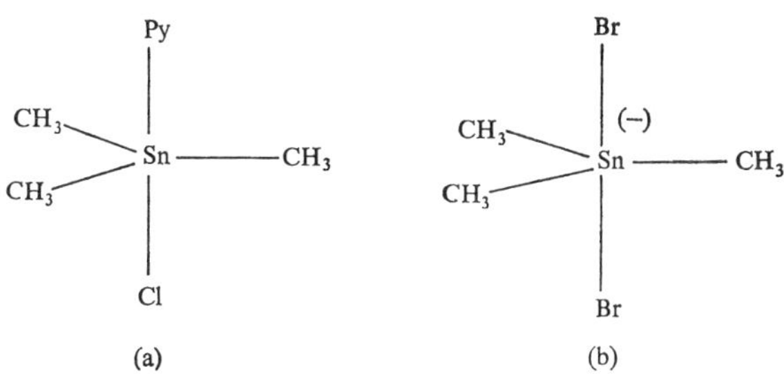

FIG. 21. Structures of (a) pyridine$(CH_3)_3SnCl$, and (b) the $(CH_3)_3SnBr_2^-$ cation.

used as a measure of complex forming ability to obtain the following series of descending Lewis acidity, $SnCl_4 > C_6H_5SnCl_3 > (C_4H_9)_2SnCl_2 > (C_6H_5)_2SnCl_2 > (C_6H_5)_3SnCl > (C_4H_9)_3SnCl$[218]. Thus the halides R_2SnX_2 react with many bases to form very stable 2:1 complexes, which contain octahedral six-coordinate tin. Examples of this type of complex are (pyridine)$_2(C_2H_5)_2SnCl_2$; α,α'-dipyridyl $(CH_3)_2SnCl_2$ and 1,10-orthophenanthroline R_2SnX_2.

The di-organotin dihalides coordinate with one or two halide ions[219] to give respectively trigonal bipyramidal ions such as $(CH_3)_2SnF_3^-$ or octahedral di-anions[220] like $(C_2H_5)_2SnCl_2^-$ and $(CH_3)_2SnBr_4^{2-}$.

Tin(IV) Oxyhalides

A variety of tin oxyhalide stoichiometries have been reported by heating partly hydrolysed tin halides, but all of these materials are rather ill defined[221–223].

The compounds of formula $SnOX_2$ (X = F, Cl, Br and I), however, are analytically well characterized, though of unknown structure to date.

$SnOF_2$ is prepared as outlined in the scheme below, and is reported to be polymeric, and to contain six-coordinate tin[224].

$$SnCl_4 \xrightarrow{ClF} SnF_2Cl_2 \xrightarrow{ClONO_2} F_2Sn(ONO_2)_2 \xrightarrow{heat} F_2SnO$$

[215] I. R. Beattie and G. P. McQuillan, *J. Chem. Soc.* (1963) 1519.
[216] R. Hulme, *J. Chem. Soc.* (1963) 1524.
[217] D. Seyferth and S. O. Grim, *J. Am. Chem. Soc.* **83** (1961) 1610.
[218] I. P. Goldshtein, E. N. Guryanova, E. D. Deleneskaya and K. A. Kocheshkov, *Doklady Akad. Nauk S.S.S.R.* **136** (1961) 1679.
[219] C. J. Wilkins and H. M. Haendler, *J. Chem. Soc.* (1965) 3175.
[220] O. A. Reutov, O. A. Ptytzina and M. D. Patrina, *Zhur. Obshch. Khim.* **28** (1958) 588.
[221] C. M. Carson, *J. Am. Chem. Soc.* **41** (1919) 1969.
[222] Davy, *Phil. Trans.* **102** (1912) 169.
[223] M. Randall and S. Murakami, *J. Am. Chem. Soc.* **52** (1930) 3967.
[224] K. Dehnicke, *Chem. Ber.* **98** (1965) 280.

$SnOCl_2$ is made directly by the action of Cl_2O upon tin(IV) chloride.

$$Cl_2O + SnCl_4 \longrightarrow SnOCl_2 + 2Cl_2$$

It is an isomorphous white powder, and a cryoscopic molecular weight determination suggests the trimeric formulation $(SnOCl_2)_3$, possibly as a cyclic stannoxane[225].

The passage of Cl_2O or ozone into molten tin(IV) bromide produces $SnOBr_2$ as a solid amorphous to X-rays[226]. Similarly, the iodide $SnOI_2$ is made from tin(IV) iodide and ozone; and from spectroscopic evidence it is believed to consist of discrete $SnOI_2$ molecules[227].

The oxyhalides were originally reported to form co-ordination compounds such as $SnOI_2$ pyridine, with a variety of bases. A careful re-investigation of the pyridine.$SnOCl_2$ system, however, has revealed that the complexes formed are those of tin(IV) chloride, with a stoichiometric quantity of tin(IV) oxide present[228].

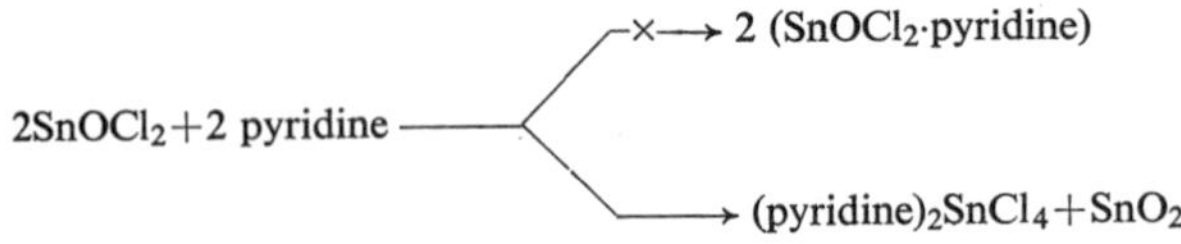

It would appear therefore that the tin(IV) oxyhalides undergo disproportionation to tin(IV) halide and tin(IV) oxide in the presence of base.

2.11. PSEUDOHALOGEN COMPOUNDS

Few simple tin pseudohalides are known, but there is an extensive chemistry of organotin pseudohalides[229], in which the synthesis and properties of the tin–pseudohalogen bonds have been studied.

Cyanides

The only cyanides of tin are the organometallic cyanides of tin(IV), which have a number of available syntheses[229].

$$(C_6H_5)_3SnOH + HCN \longrightarrow (C_6H_5)_3SnCN$$
$$(C_2H_5)_3SnOSn(C_2H_5)_3 + HCN \longrightarrow (C_2H_5)_3SnCN$$
$$R_2Sn[N(C_2H_5)_2]_2 + HCN \longrightarrow R_2Sn(CN)_2$$
$$(CH_3)_3SnI + AgCN \longrightarrow (CH_3)_3SnCN$$
$$(C_4H_9)_3SnCl + KCN \longrightarrow (C_4H_9)_3SnCN$$

Trimethyltin cyanide has a remarkable structure in which disordered CN groups are symmetrically disposed on either side of planar $(CH_3)_3Sn$ groups to produce polymeric chains containing trigonal bipyramidal five-coordinate tin.

Thus in the solid, the structure is intermediate between the cyanide and isocyanide forms, with Sn–C and Sn–N distances equal. The C–N bond is short (1·09 Å) indicating an

225 K. Dehnicke, *Z. anorg. Chem.* **308** (1961) 72.
226 K. Dehnicke, *Chem. Ber.* **98** (1965) 290.
227 K. Dehnicke, *Z. anorg. Chem.* **338** (1965) 279.
228 I. R. Beattie and V. Fawcett, *J. Chem. Soc. (A)* (1967) 1583.
229 J. S. Thayer and R. West, *Advances in Organometallic Chemistry*, **5** (1967) 169.

essentially ionic group (cf. CN in NaCN = 1·05 Å), compared to the covalent cyanides where C–N is about 1·16 Å[230].

The melting or disolution of $(CH_3)_3SnCN$ breaks up the polymeric structure, but there is no firm evidence whether the tin–carbon bonds remain intact to form $(CH_3)_3SnCN$, or whether they break to form the isocyanide $(CH_3)_3SnNC$. The formation of $(C_4H_9)_3SnNCS$ from tributyltin cyanide is taken as evidence for the isocyanide formulation, but this is not conclusive. Also iron pentacarbonyl suffers the displacement of a carbon monoxide group by trimethyltin cyanide, to give the substituted carbonyl $(CH_3)_3SnNC.Fe(CO)_4$, which is believed to be the isocyanide complex[231].

Azides

Tin(IV) azide is unknown, but from the interaction of tin tetrachloride and sodium azide, the hexa-azidostannate may be isolated[232].

$$NaN_3 + SnCl_4 \longrightarrow Na_2Sn(N_3)_6$$

Organotin azides[229, 233] and an azidochlorotin[234] are also known.

$$R_3SnCl + NaN_3 \longrightarrow R_3SnN_3$$
$$(R = CH_3, C_4H_9 \text{ and } C_6H_5)$$
$$SnCl_4 + ClN_3 \longrightarrow Cl_3SnN_3$$

The organometallic azides largely lack any sensitivity to shock, and are surprisingly thermally stable; thus triphenyltin azide decomposes around 300°C, to tetraphenyltin and nitrogen[235].

The azide group in R_3SnN_3 behaves as a Lewis base to form such complexes as $(C_6H_5)_3SnN_3.BBr_3$ and $(C_6H_5)_3SnN_3.SnCl_4$[236]. Such complexing is also believed to be

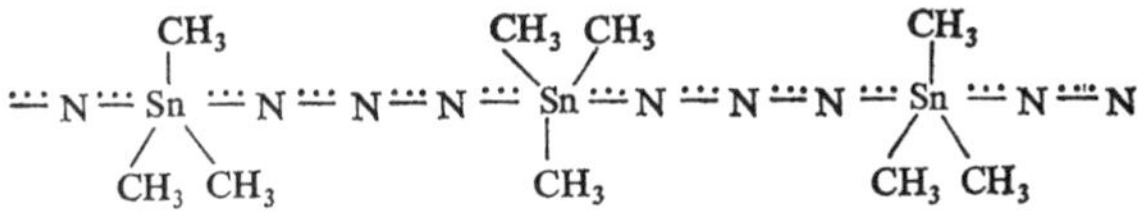

```
        CH3              CH3  CH3                  CH3
         |                 \  /                     |
 ···N···Sn···N···N···N···Sn···N···N···N···Sn···N···N
        / \                 |                     / \
     CH3  CH3              CH3                 CH3  CH3
```

FIG. 22. Trimethyltin azide polymer.

present in the solid azides, R_3SnN_3, where the tin atom is in a bipyramidal five-coordinate environment with azide groups as bridges between planar R_3Sn units. If, however, the azide is severely hindered sterically, such as in the compound $\{C_6H_5C(CH_3)_2CH_2\}_3SnN_3$, it is unassociated[237].

Tributyltin azide and acetylene dicarboxylic acid evolve hydrazine and form

230 E. Schlemper and D. Britton, *Inorg. Chem.* **5** (1966) 507.
231 D. Seyferth and N. Kahlen, *J. Am. Chem. Soc.* **82** (1960) 1080.
232 E. Wiberg and H. Michaud, *Z. Naturforsch.* **9b** (1954) 500.
233 M. F. Lappert and H. Pyzora, *Advances in Inorganic and Radio Chemistry*, **9** (1966) 133.
234 K. Dehnicke, *Angew. Chem., Intern. Edit.* **3** (1964) 142.
235 W. T. Reichle, *Inorg. Chem.* **3** (1964) 402.
236 J. S. Thayer and R. West, *Inorg. Chem.* **4** (1965) 114.
237 J. S. Thayer, *Organometallic Revs.* **1** (1966) 157.

$(C_4H_9)_3SnOOCC{:}CCOOSn(C_4H_9)_3$; with the acetylenedicarboxylic ester, however, nitrogen is retained with the formation of triazoles[238].

$$R_3SnN_3 + C_2H_5OOCC{:}CCOOC_2H_5 \longrightarrow \begin{array}{c} C_2H_5OOC{=}CCOOC_2H_5 \\ R_3Sn{-}N \quad N \\ N \end{array}$$

Isocyanates

$Si(NCO)_4$ and $Ge(NCO)_4$ are both known compounds[239], but there is no report to date of $Sn(NCO)_4$. There are, however, a number of covalent, volatile organotin(IV) isocyanates of the type R_3SnNCO and $R_2Sn(NCO)_2$[229, 233].

$$\left.\begin{array}{l} R_3SnOH \\ R_3SnOSnR_3 \end{array}\right\} + CO(NH_2)_2 \longrightarrow R_3SnNCO$$

The infrared spectrum of $(CH_3)_3SnNCO$ has been recorded and fully assigned[240]. The tin isocyanates can be assigned this formulation rather than the cyanate alternative with some confidence from spectroscopic results. No cyanates have been reported, and are presumably unstable with respect to the isocyanates, as in the very unstable organic cyanates. It may, however, be possible to isolate tin cyanates at low temperatures.

The organotin isocyanates are very sensitive to hydrolysis, and many partially hydrolysed isocyanates such as $(C_6H_5)_4Sn_2(NCO)_2(OH)_2$ and $(C_4H_9)_8Sn_4(NCO)_4O_2$ have been isolated.

The triphenyltin fulminate $(C_6H_5)_3SnCNO$ rearranges at 146–148° to the isocyanate $(C_6H_5)_3SnNCO$.

The carbon–nitrogen double bond of these isocyanates undergoes insertion of ammonia, amines, alcohols, etc., in the manner of the organic isocyanates[229, 233].

$$(C_4H_9)_3SnNCO + NH_3 \longrightarrow (C_4H_9)_3SnNH.CO.NH_2$$

Isothiocyanates and Isoselenocyanates

Although $Sn(SCN)_4$ has been briefly mentioned in the literature[241], the known chemistry of isothiocyanatotin(IV) compounds is restricted to organometallic isothiocyanates and complexions. A wide range of syntheses have been used to obtain the organotin(IV) isothiocyanates[229, 233].

$$(C_4H_9)_3SnCl + AgNCS \longrightarrow (C_4H_9)_3SnNCS$$
$$(CH_3)_2SnCl_2 + NaCNS \longrightarrow (CH_3)_2Sn(NCS)_2$$
$$(C_2H_5)_3SnOSn(C_2H_5)_3 + HNCS \longrightarrow (C_2H_5)_3SnNCS$$
$$(CH_3)_3SnNC + S \longrightarrow (CH_3)_3SnNCS$$
$$(C_4H_9)_3SnOSn(C_4H_9)_3 + NH_2CSNH_2 \longrightarrow (C_4H_9)_3SnNCS$$

These compounds are very stable thermally; thus $(C_4H_9)_3SnNCS$ withstands prolonged heating in vacuum at 180° without change.

238 J. G. A. Luijten and G. J. M. van der Kerk, *Rec. Trav. Chim.* **83** (1964) 295.
239 F. A. Miller and G. L. Carson, *Spectrochim. Acta* **17** (1961) 977.
240 J. S. Thayer and D. P. Strommer, *J. Organometallic Chem.* **5** (1966) 383.
241 R. Bock, *Z. anal. Chem.* **133** (1951) 110.

Spectroscopic data points to the isothiocyanate Sn—N=C=S structure[242], despite the very strong tendency for tin to bond to sulphur, and the normally weak tin–nitrogen bond. Similarly, the corresponding selenocyanates are believed to have the SnNCSe sequence of bonding[229].

Certain of the organotin isothiocyanates are sufficiently strong Lewis acids to form complexes, as do the tin(IV) halides. Thus in $(CH_3)_2Sn(NCS)_2$(bipy), the central tin atom becomes octahedral upon coordination of the 2,2′-bipyridyl. In the same way, the thio-

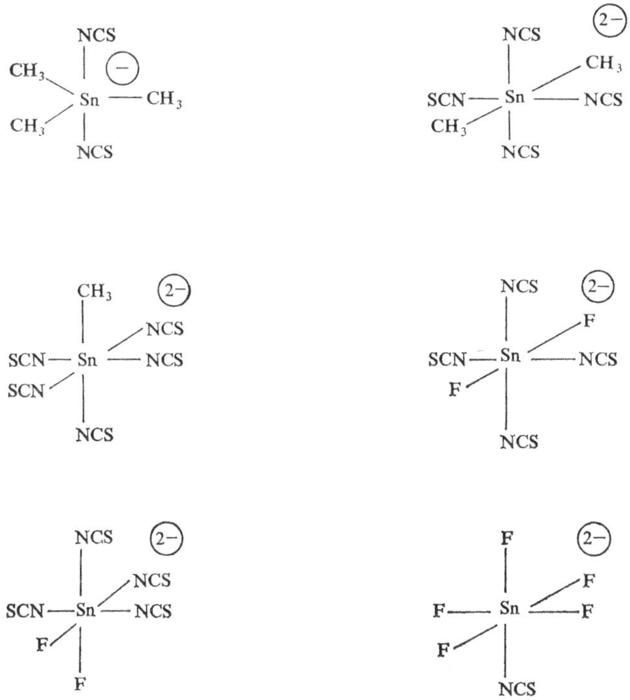

FIG. 23. Some isothiocyanato anions of tin(IV).

cyanate ion may behave as the Lewis base[243], and complex ions such as $[(CH_3)_3Sn(NCS)]_2^{-}$, $[(CH_3)_2Sn(NCS)_4]^{2-}$ and $[CH_3Sn(NCS)_5]^{2-}$ are known (Fig. 23).

The complex isothiocyanato-fluoride ions $F_{6-n}Sn(NCS)_n$ ($n = 1$, 2 and 3) have been characterized in solution, and *cis* and *trans* isomers recognized from ^{19}F nuclear magnetic resonance studies[190], where $n = 2$ and 3 (Fig. 23).

Tin(II) isothiocyanate is best made by controlled crystallization from a solution of tin(II) sulphate and sodium thiocyanate[244]. The white crystalline product is believed to contain predominantly N-bonded isothiocyanate groups, but extensive thiocyanato bridging is also suggested by infrared spectroscopy.

Thermal decomposition of $Sn(NCS)_2$ gives predominantly SnS at about 150°, a mixture, of SnS and SnS_2 at about 200°, and above this temperature predominantly SnS_2[244].

$Sn(NCS)_2$ is hydrolysed by water, and forms weak complexes with dialkyl ethers.

242 R. A. Cummins and P. Dunn, *Australian J. Chem.* **17** (1964) 411.
243 A. Cassol, R. Portanova and R. Barbieri, *J. Inorg. Nuclear Chem.* **27** (1965) 2275.
244 B. R. Chamberlain and W. Moser, *J. Chem. Soc.* (*A*) (1969) 354.

From potentiometric studies, the species $Sn(NCS)^+$, $Sn(NCS)_2$ and $Sn(NCS)_3^-$ are shown to be present in aqueous solution[245], and the salts $K_2Sn(NCS)_4.MeOH$, $KSn(SO_4)NCS$, $CsSn(NCS)_3$, $(CH_3)_4NSn(NCS)_3$ and $(C_2H_5)_4NSn(NCS)_3$ have been isolated, but no structural data is available for these compounds[244].

2.12. METAL–METAL BONDS INVOLVING TIN

Alkali Metal Derivatives

These compounds have proved very powerful intermediates in the preparation of a wide range of organotin compounds, and several methods have been investigated for their synthesis. Many of the syntheses are carried out in liquid ammonia[246–248].

$$R_3SnX+2Na \longrightarrow R_3SnNa+NaX$$
$$R_2SnX_2+4Na \longrightarrow R_2SnNa_2+2NaX$$
$$R_3SnSnR_3+2Na \longrightarrow 2R_3SnNa$$

Organotin alkali derivatives can, however, be prepared in solvents other than liquid ammonia, probably with greatest success in the case of lithium; thus $(C_6H_5)_3SnLi$ and $(CH_3)_3SnLi$ are obtained in about 80% yield from the corresponding halide and lithium in tetrahydrofuran[249]. In addition, the organotin lithiums can be prepared from the diphenyltins[250], or from tin(II) chloride[251].

$$\{(C_6H_5)_2Sn\}_n+C_6H_5Li \longrightarrow (C_6H_5)_3SnLi$$
$$3RLi+SnCl_2 \longrightarrow R_3SnLi+2LiCl_2$$

Conductivity measurements show triphenyltin sodium and trimethyltin sodium to be strong electrolytes in liquid ammonia[252, 253], but no solid structural information is available. It is possible that these materials have an electron deficient type of bonding in the crystals, similar to the alkali metal alkyls.

The organotin alkali metal compounds are air sensitive, converting to the oxides R_3SnONa, which hydrolyse to the hydroxide R_3SnOH or the oxide $(R_3Sn)_2O$.

The ionization of these compounds in solution may be formulated as $M^+SnR_3^-$, with the R_3Sn^- ion exactly analogous to the organic carbonion. Thus these powerfully nucleophilic ions complement the R_3Sn^+ cations present in solutions of the organotin halides, and where a synthesis is not possible with one, it can normally be accomplished by the other.

One specially interesting synthesis of organotin substituted alcohols involves the insertion of epoxides and other cyclic esters into the tin–alkali bond, and subsequent hydrolysis of the sodium alkoxide formed[254].

$$R_3SnNa+\overline{O.CH_2.CH}R \longrightarrow R_3SnCH_2CH(R)ONa$$

---

[245] A. M. Golub and V. M. Samoilenko, *Ukrain. Khim. Zhur.* **29** (1963) 789.
[246] R. H. Bullard and F. R. Holden, *J. Am. Chem. Soc.* **53** (1931) 3150.
[247] C. A. Kraus and W. N. Greer, *J. Am. Chem. Soc.* **47** (1925) 2568.
[248] C. A. Kraus and H. Eatough, *J. Am. Chem. Soc.* **55** (1933) 5014.
[249] H. Gilman and O. Marrs, quoted in ref. 18.
[250] G. Wittig, F. J. Meyer and G. Lange, *Ann. Chem. Liebigs.* **571** (1951) 167.
[251] H. Gilman and S. D. Rosenberg, *J. Am. Chem. Soc.* **75** (1953) 2507; *J. Org. Chem.* **18** (1953) 680 and 1554.
[252] C. A. Kraus and W. H. Kahler, *J. Am. Chem. Soc.* **55** (1933) 3537.
[253] C. A. Kraus and E. G. Johnson, *J. Am. Chem. Soc.* **55** (1933) 3542.
[254] H. Gilman, F. K. Cartledge and S. Y. Sim, *J. Organometallic Chem.* **4** (1965) 332.

Tin Bonded to Other Non-Transitional Metals

The organotin equivalents of Grignard reagents and the organotin equivalents of magnesium dialkyls are made according to the equations below[255].

$$C_2H_5MgBr+(C_6H_5)_3SnH \longrightarrow (C_6H_5)_3SnMgBr+C_2H_6$$

$$2(C_6H_5)_3SnCl+2Mg \xrightarrow{C_2H_5Br} \{(C_6H_5)_3Sn\}Mg$$

A similar displacement of hydrocarbon from complexed alkyls of cadmium and zinc produce metal–metal bonded species of these metals as shown in Fig. 24 (a) and (b)[256].

Tin–boron bonds have been formed by the metathesis of triethyltin lithium with the respective dimethylaminoboron chlorides to give species containing either one or two boron–tin bonds, as illustrated in Fig. 24 (c) and (d)[257].

The chemistry of the compounds containing tin bonded to another Group IV element, other than carbon or tin itself, is extensive. The two most useful synthetic methods have been the elimination of alkali halide, and the displacement of amine[258, 259].

$$R_3SnN(C_2H_5)_2+(C_6H_5)_3GeH \longrightarrow R_3SnGe(C_6H_5)_3$$

$$(C_6H_5)_3SnLi+(C_6H_5)_3SiCl \longrightarrow (C_6H_5)_3SnSi(C_6H_5)_3$$

Of particular interest are the compounds $\{(C_6H_5)_3M\}_4M'$ such as $\{(C_6H_5)_3Pb\}_4Sn$ and $\{(C_2H_5)_3Sn\}_4Ge$, for which almost all combinations of the Group IV elements are known for carbon to lead[260].

Tin–Tin Bonds

In addition to the tin–tin bonds present in the metallic tins, there exists a good range of molecular species which contain tin–tin bonds. A number of general methods have become apparent for the formation of the tin–tin bonds[261].

The tin equivalent of a Wurtz reaction between organotins and sodium yields the tin–tin bond[262].

$$2R_3SnCl+2Na \longrightarrow R_3Sn\text{–}SnR_3$$

The weak bonds between tin and oxygen and tin and nitrogen gives the opportunity for the displacement of amines[263] and alcohols[264] and even water[265] by tin hydrides, with a concomitant metal–metal bond formation.

$$R_3SnNR_2+HSnR_3'' \longrightarrow R_3SnSnR_3''+R_2'NH$$

$$R_3SnOR'+HSnR_3'' \longrightarrow R_3SnSnR_3''+R'OH$$

$$(R_3Sn)_2O+2HSnR_3 \longrightarrow 2R_3SnSnR_3+H_2O$$

255 C. Tamborski, F. E. Ford and E. Solonski, *J. Org. Chem.* **28** (1963) 181 and 237.
256 H. M. J. C. Creemers, *Hydrostannolysis*, T.N.O. Utrecht, Schotanus and Jens, Utrecht.
257 H. Nöth and K. H. Hermannsdörfer, *Angew. Chem.* **76** (1964) 377.
258 W. P. Neumann, B. Schneider and R. Sommer, *Ann. Chem. Liebigs.* **692** (1966) 1.
259 H. Gilman and S. D. Rosenberg, *J. Am. Chem. Soc.* **74** (1952) 531.
260 L. C. Willemsens and G. J. M. van der Kerk, *J. Organometallic Chem.* **2** (1964) 260.
261 H. Gilman, W. H. Atwell and F. K. Cartledge, *Adv. Organometallic Chem.* **4** (1966) 49.
262 H. Gilman and S. D. Rosenberg, *J. Org. Chem.* **18** (1953) 1554.
263 H. J. M. C. Creemers and J. G. Noltes, *Rec. Trav. Chim.* **84** (1965) 382.
264 A. K. Sawyer, *J. Am. Chem Soc.* **87** (1965) 537.
265 W. P. Neumann, B. Schneider and R. Sommer, *Ann. Chim. Liebigs.* **692** (1966) 1.

Tin hydrides lose hydrogen easily to form tin–tin bonds.

$$2R_3SnH \longrightarrow R_3SnSnR_3 + H_2$$

In addition, the metathetical elimination of alkali halides has proved useful for the non-symmetrical compounds.

$$R_3SnNa + ClSnR_3' \longrightarrow R_3SnSnR_3'$$

By these methods molecules containing up to nine sequential tin–tin bonds have been synthesized[19].

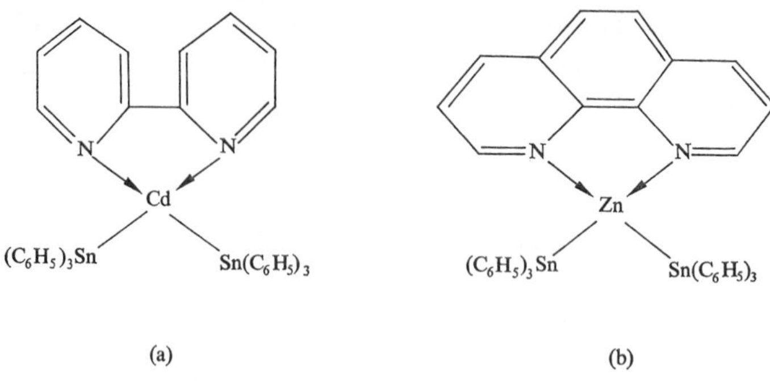

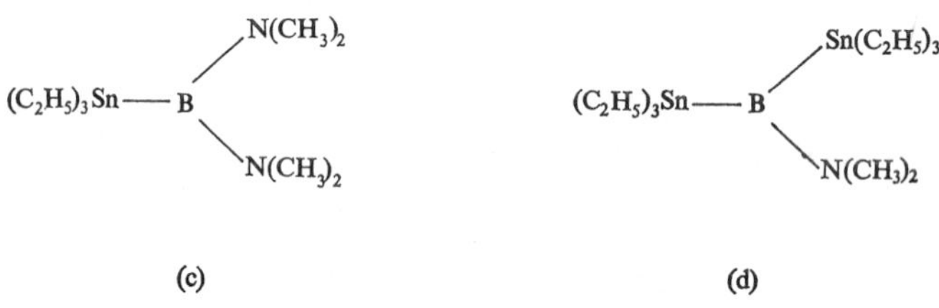

FIG. 24. Some examples of metal–metal bonded species containing (a) cadmium–tin bonds, (b) zinc–tin bonds, (c) and (d) tin–boron bonds.

Linear and Branched Polytins

Virtually all of the known compounds of this class are organotin derivatives, though a very few examples containing tin–halogen, tin–hydrogen and tin–oxygen bonds are known.

Ditins

A very large number of ditins are known, and have usually been made by one of the general reactions outlined above. The robustness of the tin–tin bond is very dependent upon the other substituent groups, thus hexamethylditin and hexa-ethylditin are both unstable in the presence of air, but hexaphenylditin is unaffected by oxygen in benzene solution[261]. The air oxidation of hexa-ethylditin is proposed to occur through a peroxide $(C_2H_5)_3SnOOSn(C_2H_5)_3$ to the oxide $(C_2H_5)_3SnOSn(C_2H_5)_3$ [266].

[266] Yu A. Alexandrov and N. N. Vyshinki, *Tr. Khim. i Khim. Tekhnol.* **4** (1962) 656.

The tin–tin bonds in hexa-arylditins do not undergo dissociation to give free radicals, as do the corresponding hexa-aryls of carbon.

The Raman spectrum of hexamethylditin has been analysed[267] to give a tin–tin stretching force constant of $1{\cdot}0 \pm 0{\cdot}1 \times 10^{-5}$ dynes/cm. Nuclear magnetic resonance spectra of di- and poly-tins have been interpreted to indicate an increasing percentage of *s* character in the tin–tin bond with increasing chain length of tin atoms[268]. Some of the strong ultraviolet absorptions in the hexa-organoditins are believed to be characteristic of the tin–tin bond[269].

The metal–metal bond in ditins undergoes a range of reactions which can be regarded as either fissions or insertions. In the latter type of reaction the two tin moieties remain in the same molecule.

Organotin halides are the products from the reaction of ditins with both halogen and hydrogen halide[18, 270].

$$R_3SnSnR_3 + X_2 \longrightarrow 2R_3SnX$$
$$R_3SnSnR_3 + 2HX \longrightarrow 2R_3SnX + H_2$$

In both cases reaction quantities and conditions must be carefully controlled in order to avoid fission of the tin–carbon bonds present.

Organic halides cause a number of different types of reaction; thus, for example, trifluoromethyliodide behaves as a halogen[271].

$$(CH_3)_3SnSn(CH_3)_3 + CF_3I \longrightarrow (CH_3)_3SnI + (CH_3)_3SnCF_3$$

Trityl chloride has an analogous behaviour to the reaction of hydrogen chloride above[272].

$$(C_2H_5)_3SnSn(C_2H_5)_3 + 2(C_6H_5)_3CCl \longrightarrow 2(C_2H_5)_3SnCl + (C_6H_5)_3C.C(C_6H_5)_3$$

These dehalogenations at carbon are extended in the action of hexa-ethylditin upon ethylene dibromide, where intramolecular removal of the halogen takes place[273].

$$(C_2H_5)_3SnSn(C_2H_5)_3 + BrCH_2CH_2Br \longrightarrow CH_2{:}CH_2 + 2(C_2H_5)_3SnBr$$

It is possible that such a dehalogenation reaction may have extensive usage in organic chemistry.

Hexa-organoditins take up sulphur to produce the bis(trialkyl/aryltin) sulphides $R_3SnSSnR_3$.

Organic peroxides[274], peranhydrides[275] and peroxy esters[276] cause the fission of tin–tin bonds to give the corresponding tin–oxygen compounds.

$$R_3SnSnR_3 + (\text{tert-}C_4H_9O)_2 \longrightarrow R_3SnO\text{–t-}C_4H_9$$
$$R_3SnSnR_3 + (C_6H_5COO)_2 \longrightarrow R_3SnOCOC_6H_5$$
$$R_3SnSnR_3 + C_6H_5COOO\text{–t-}C_4H_9 \longrightarrow R_3SnOCOC_6H_5 + R_3SnO\text{–t-}C_4H_9$$

267 M. P. Brown, E. Cartmell and G. W. A. Fowles, *J. Chem. Soc.* (1960) 506.
268 T. L. Brown and G. L. Morgan, *Inorg. Chem.* **2** (1963) 736.
269 W. Drenth, M. J. Janssen, G. J. M. van der Kerk and J. A. Vleigenthart, *J. Organometallic Chem.* **2** (1964) 265.
270 G. Tagliavini, S. Faleschini, G. Pilloni and G. Plazzongna, *J. Organometallic Chem.* **5** (1966) 136.
271 H. C. Clark and C. J. Willis, *J. Am. Chem. Soc.* **82** (1960) 1888.
272 G. A. Razuvaev, Y. I. Dergunov and N. S. Vyazankin, *Zhur. Obshchei. Khim.* **32** (1962) 2515.
273 G. A. Razuvaev, N. S. Vyazankin and Y. I. Dergunov, *Zhur. Obshchei. Khim.* **30** (1960) 1310.
274 W. P. Neumann and K. Rubsamen, *Chem. Ber.* **100** (1967) 1621.
275 G. A. Razuvaev, N. S. Vyazankin and O. A. Shchepetkova, *Zhur. Obshchei. Khim.* **30** (1960) 2498.
276 N. S. Vyazankin, G. A. Razuvaev and T. N. Brevnova, *Doklady Akad. Nauk. S.S.S.R.* **163** (1965) 1389.

Under special circumstances olefins[277] and acetylenes[278] will undergo an insertion type of reaction to the multiple bond by the addition of the two halves of a ditin.

$$CF_2{:}CF_2 + (CH_3)_3SnSn(CH_3\ _3 \longrightarrow (CH_3)_3Sn.CF_2CF_2.Sn(CH_3)_3$$

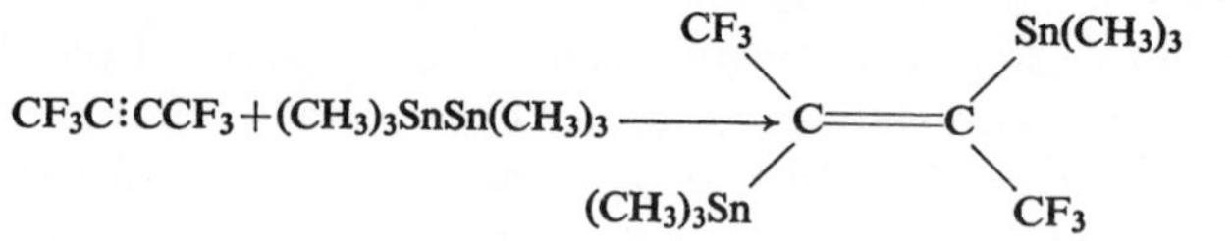

$$CF_3C{\vdots}CCF_3 + (CH_3)_3SnSn(CH_3)_3 \longrightarrow \begin{matrix} CF_3 & & Sn(CH_3)_3 \\ & C{=}C & \\ (CH_3)_3Sn & & CF_3 \end{matrix}$$

Hexachloroditin contains a tin–tin bond, but disproportionates below 0°C to tin(II) and tin(IV) chlorides[101].

Tri-, Tetra-, Penta- and Hexa-Tins

Whilst tin–tin bonds appear to be incapable of existing in the very long chains of the lighter Group IV elements, the ability to catenate is still considerable, and up to six adjacent tin–tin bonds are known, with even more in the cyclic species[19, 261]. The increasing number of tin–tin bonds lowers considerably the stability of many of these compounds, and renders them unstable in the atmosphere. Additional to the linear polytins are some branched species exemplified by $\{(C_6H_5)_3Sn\}_4Sn$ and $C_6H_5Sn\{Sn(C_2H_5)_3\}_3$ [279, 280].

$$(C_6H_5)_3SnLi \xrightarrow{SnCl_2} \{(C_6H_5)_3Sn\}Li \xrightarrow{(C_6H_5)_3SnCl} \{(C_6H_5)_3Sn\}_4Sn$$

$$C_6H_5SnH_3 \xrightarrow{3(C_2H_5)_3SnNR_2} C_6H_5Sn\{Sn(C_2H_5)_3\}_3 + 3R_2NH$$

Cyclic Polytins

Four-,[281] five-,[282] six- [282, 283] and nine-membered[284] rings are well characterized in this class.

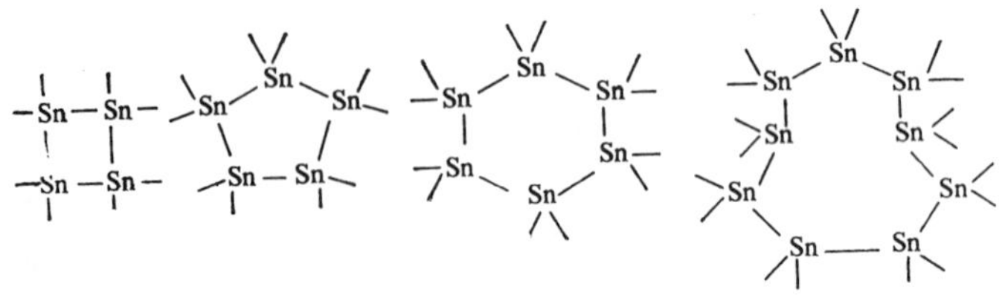

The synthetic methods[19] for formation of the tin–tin bonds, and the fission reactions of the tin rings largely parallel those for the linear polytins.

277 M. A. A. Beg and H. C. Clark, *Chem. and Ind.* (*London*) (1962) 140.
278 W. R. Cullen, D. S. Dawson and G. E. Styan, *J. Organometallic Chem.* 3 (1965) 406.
279 H. Gilman and F. K. Cartledge, *Chem. and Ind.* (*London*) (1964) 1231; *J. Organometallic Chem.* **5** (1966) 48.
280 H. M. J. C. Creemers, J. G. Noltes and G. J. M. van der Kerk, *Rec. Trav. Chim.* **83** (1964) 1284.
281 W. V. Farrar and H. A. Skinner, *J. Organometallic Chem.* 1 (1964) 434.
282 W. P. Neumann and K. Konig, *Ann. Chem. Liebigs.* **677** (1961) 1.
283 D. H. Olson and R. E. Rundle, *Inorg. Chem.* 2 (1963) 1310.
284 W. P. Neumann and J. Pedain, *Ann. Chem. Liebigs.* **672** (1964) 34.

$$(C_6H_5)_2SnH_2 \xrightarrow{\text{pyridine}} [(C_6H_5)_2Sn]_6$$

$$(\text{iso-}C_4H_9)_2Sn\{N(C_2H_5)_2\}_2+(\text{iso-}C_4H_9)_2SnH_2 \longrightarrow \{(\text{iso-}C_4H_9)_2Sn\}_9$$

$$(C_2H_5)_2SnH_2+(C_2H_5)_2SnCl_2 \xrightarrow{\text{pyridine}} \{(C_2H_5)Sn_2\}_9$$

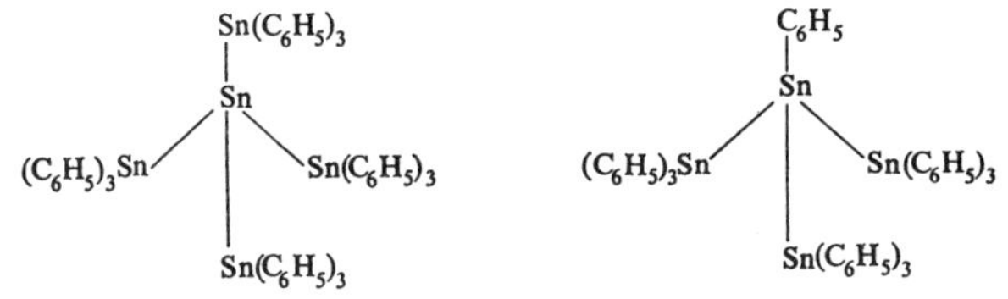

(a) central Sn bonded to four $Sn(C_6H_5)_3$ groups: $Sn(C_6H_5)_3$ (top), $(C_6H_5)_3Sn$ (left), $Sn(C_6H_5)_3$ (right), $Sn(C_6H_5)_3$ (bottom)

(b) central Sn bonded to C_6H_5 (top), $(C_6H_5)_3Sn$ (left), $Sn(C_6H_5)_3$ (right), $Sn(C_6H_5)_3$ (bottom)

FIG. 25. (a) Tetra(triphenyltin)tin and (b) phenyltri-(triphenyltin)tin.

Tin–Transitional Metal Bonds

Many transitional metals have now been incorporated into compounds containing tin–transitional metal bonds[210, 285]. The syntheses can be roughly divided into four main classes.

(a) Insertion Reactions

$$SnCl_2+Co_2(CO)_8 \longrightarrow (CO)_4Co\text{–}SnCl_2\text{–}Co(CO)_4$$

$$SnCl_2+[(C_6H_5)_3P]_2PtCl_2 \longrightarrow \{(C_6H_5)_3P\}_2Pt(SnCl_3)(Cl)$$

(b) Alkali Halide Metathesis

$$(C_6H_5)_3SnLi+\{(C_6H_5)_3P\}PtCl_2 \longrightarrow (C_6H_5)_3SnPt(Cl)\{P(C_6H_5)_3\}_2$$

$$(CH_3)_2SnCl_2+2NaRe(CO)_5 \longrightarrow (CO)_5ReSn(CH_3)_2Re(CO)_5$$

(c) Elimination Reactions

$$SnCl_2+\pi\text{-}C_5H_5Fe(CO)_2HgCl \longrightarrow \pi\text{-}C_5H_5Fe(SnCl_3)(CO)_2+Hg$$

$$Zr\{N(C_2H_5)_2\}_4+4(C_6H_5)_3SnH \longrightarrow Zr\{Sn(C_6H_5)_3\}_4+4(C_2H_5)_2NH$$

(d) Oxidative Addition Reactions

$$SnCl_4+\{(C_6H_5)_3P\}_2Ir(CO)Cl \longrightarrow \{(C_6H_5)_3P\}_2PIr(SnCl_3)(CO)Cl_2$$

$$(C_6H_5)_3SnCl+\{(C_6H_5)_3P\}_4Pt \longrightarrow (C_6H_5)_3SnPtCl\{(C_6H_5)_3P\}_2$$

The number of individual compounds that possess one or more tin–transitional metal bonds is very large, and the table only includes some representative examples.

Covalent metal–metal bonding in these compounds can be inferred from the synthetic

[285] N. S. Vyazankin, G. A. Razuvaev and O. A. Kruglaya, *Organometallic Revs.* (*A*), **3** (1968) 323.

methods and from chemical composition. Moreover, X-ray crystallographic studies of a number of compounds have shown directly the presence of the tin–metal bond[286–289].

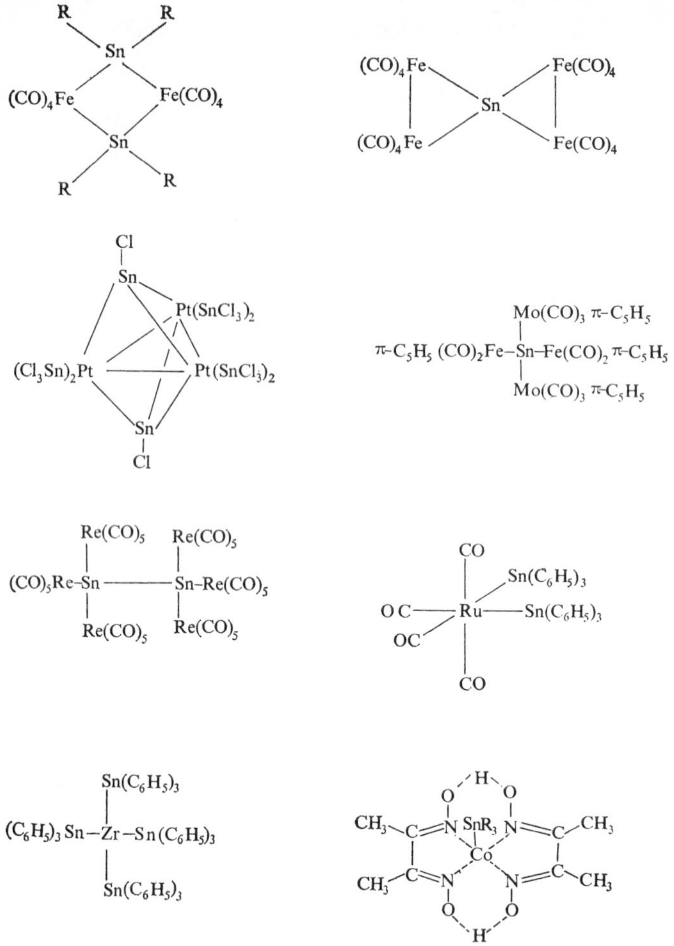

R R Sn (CO)$_4$Fe Fe(CO)$_4$ Sn R R

(CO)$_4$Fe Fe(CO)$_4$ Sn (CO)$_4$Fe Fe(CO)$_4$

Cl Sn Pt(SnCl$_3$)$_2$ (Cl$_3$Sn)$_2$Pt Pt(SnCl$_3$)$_2$ Sn Cl

Mo(CO)$_3$ π-C$_5$H$_5$ π-C$_5$H$_5$ (CO)$_2$Fe–Sn–Fe(CO)$_2$ π-C$_5$H$_5$ Mo(CO)$_3$ π-C$_5$H$_5$

Re(CO)$_5$ Re(CO)$_5$ (CO)$_5$Re-Sn——Sn-Re(CO)$_5$ Re(CO)$_5$ Re(CO)$_5$

CO Sn(C$_6$H$_5$)$_3$ OC—Ru—Sn(C$_6$H$_5$)$_3$ OC CO

Sn(C$_6$H$_5$)$_3$ (C$_6$H$_5$)$_3$ Sn–Zr–Sn(C$_6$H$_5$)$_3$ Sn(C$_6$H$_5$)$_3$

H O O CH$_3$ C=N SnR$_3$ N=C CH$_3$ Co CH$_3$ C=N N=C CH$_3$ O O H

FIG. 26. Some examples of metal sequences and metal clusters containing tin–transitional metal bonds.

Figure 26 illustrates some of the many metal sequences and metal clusters known to contain tin–transitional metal bonds.

A number of reagents cause fission of the tin–transitional metal bond[290, 291].

$$(CH_3)_3SnCo(CO)_4+I_2 \longrightarrow (CH_3)_3SnI+Co(CO)_4I$$
$$(CH_3)_3SnMn(CO)_5+(C_6H_5)_2PCl \longrightarrow (CH_3)_3SnCl+[(C_6H_5)_2PMn(CO)_4]_2+CO$$
$$(CH_3)_3SnMn(CO)_5+C_2F_4 \longrightarrow (CH_3)_3SnCF_2CF_2Mn(CO)_4$$

[286] H. P. Weber and R. F. Bryan, *Chem. Comms.* (1966) 443.
[287] R. F. Bryan, *Chem. Comms.* (1967) 355; *J. Chem. Soc* (*A*) (1967) 172.
[288] P. Woodward and P. F. Lindley, *J. Chem. Soc.* (*A*) (1967) 382.
[289] C. J. Fritchie, R. M. Sweet and R. Schunn, *Inorg. Chem.* **6** (1967) 749.
[290] E. W. Abel and G. V. Hutson, *J. Inorg. and Nuclear Chem.* **30** (1968) 2339
[291] H. C. Clark, J. D. Cotton and J. H. Tsai, *Inorg. Chem.* **5** (1966) 1582.

TABLE 16. COMPLEXES CONTAINING DISCRETE TIN–TRANSITIONAL METAL BONDS ARRANGED BY GROUPS

V	*Cr*	*Mn* $(C_6H_5)_3SnMn(CO)_5$ $(C_6H_5)_2Sn\{Mn(CO)_5\}_2$	*Fe* $\{C_5H_5Fe(CO)_2\}_4Sn$ $(CH_3)_4Sn_3Fe_4(CO)_{16}$	*Co* $(C_4H_9)_3SnCo(CO)_4$ $BrSn\{Co(CO)_4\}_3$	*Ni* $[C_5H_5(CO)Ni]_2SnCl_2$
Nb	*Mo* $(CO)_5MnSn(CH_3)_2Mo(CO)_3C_5H_5$	*Tc*	*Ru* $(CO)_4Ru\{Sn(C_6H_5)_3\}_2$ $[Ru(SnCl_3)_2(CO)_2Cl_2]^{2-}$	*Rh* $\{(C_6H_5)_3P\}_2(CO)_2RhSn(CH_3)_3$ $[Rh(SnCl_3)_2(CO)Cl]^{2-}$	*Pd*
Ta	*W* $C_5H_5(CO)_3WSn(CH_3)_3$	*Re* $(CH_3)_2Sn\{Re(CO)_5\}_2$ $[\{(CO)_5Re\}_3Sn]_2$	*Os*	*Ir* $(C_6H_5)_3P(CO)_3IrSn(C_6H_5)_3$ $[Ir_2Cl_6(SnCl_3)_4]^{4-}$	*Pt* $\{(C_6H_5)_3P\}_2ClPtSn(C_6H_5)_3$ $[Pt(SnCl_3)_2Cl_2]^{2-}$

In contrast, however, many reactions may be carried out on such molecules, whilst the metal–metal bond remains intact[292, 293].

$$(C_6H_5)_3SnMn(CO)_5 + Cl_2 \longrightarrow Cl_3SnMn(CO)_5 + C_6H_5Cl$$

$$Cl_2Sn\{Co(CO)_4\}_2 + RMgX \longrightarrow R_2Sn\{Co(CO)_4\}_2$$

Some complexes containing platinum–tin bonds function as hydrogenation catalysts[294], and although complexes in which the $SnCl_3^-$ ligand is attached to platinum can be isolated crystalline, in solution some dissociation takes place leaving vacant sites for coordination of olefins and hydrogen. Thus a highly reactive intermediate is produced, reverting to the original complex after hydrogen transfer to the olefin (or acetylene) has occurred. It is believed that the very high *trans*-activating ability of the $SnCl_3^-$ group is important here, and this *trans* effect is thought to originate mainly in the π-acceptor ability of the ligand These same platinum–tin complexes catalyse double bond migration in olefins.

292 R. D. Gorsich, *J. Am. Chem. Soc.* **84** (1962) 2486.
293 F. Bonati, S. Cenini, D. Morelli and R. Ugo, *J. Chem. Soc.* (*A*) (1966) 1052.
294 R. D. Crammer, E. L. Jenner, R. V. Lindsay and U. G. Stolberg, *J. Am. Chem. Soc.* **85** (1963) 1691

18. LEAD

E. W. Abel
University of Exeter

1. THE ELEMENT

1.1 HISTORICAL

The very easy extraction of lead from its ores made it one of the few metals used extensively from earliest times[1]. One of the earliest-dated specimens is a lead statue found in the Dardanelles on the site of the ancient city of Abydos. This is believed to date from 3000 B.C.

Lead was in common usage in Ancient Egypt for ornamental objects and solder, and lead salts were used to glaze pottery. The Hanging Gardens of Babylon were floored with sheet lead as a moisture retainer, and the Babylonians and other ancients used lead for caulking and for the fastening of iron bolts and hooks in bridges, houses and other stone buildings.

The most extensive use of lead by the ancients was for the manufacture of water pipes. The Romans produced lead pipes in standard diameters and regular 10-ft lengths. Many such pieces of pipe work have been recovered perfectly preserved in modern times from ruins in Rome and many other Roman sites throughout Europe. Over four centuries, it is estimated, the Roman Empire extracted and used six to eight million tons of lead, with a peak annual production of around sixty thousand tons[2].

1.2. OCCURRENCE AND DISTRIBUTION

The most important ore of lead is galena or lead glance PbS, which is widely distributed throughout the world[1]. Other ore minerals of lead are anglesite ($PbSO_4$), cerussite ($PbCO_3$), pyromorphite ($PbCl_2 . 3Pb_3(PO_4)_2$) and mimetesite ($PbCl_2 . 3Pb_3(AsO_4)_2$). Other lesser minerals are crocoite (or kallochrome, or red-lead ore) ($PbCrO_4$), wulfenite (or molybdenum lead spar, or yellow lead ore) ($PbMoO_4$) and stolzite ($PbWO_4$).

The main sources of lead ores currently worked are in the U.S.A., Australia, Mexico, Canada and the U.S.S.R., but substantial quantities are mined in Peru, Morocco, Yugoslavia, Germany, Spain, South Africa, Italy and Bolivia.

1.3. EXTRACTION

Lead ores are initially crushed and concentrated by flotation, when lighter impurities such as zinc sulphide are removed. At this stage ore concentrates in general contain about

[1] *Lead in Modern Industry*, p. 1, Lead Industries Association, New York (1952).
[2] *Nouveau traité de chimie minérale*, Vol. VIII, Part 3, p. 469, ed. P. Pascal, Masson et Cie, Paris (1963).

40% of lead by weight. The ore is roasted in air to convert the sulphide to oxide and sulphate, in which process extensive evolution of sulphur dioxide takes place.

After mixing with limestone and coke the roasted ore is smelted in blast furnaces to yield crude lead. Major impurities are gold, silver, copper, antimony, arsenic and bismuth. At this stage the lead is hard, due to the arsenic and antimony present. The lead is now melted and kept molten at a temperature below the melting point of copper, whereby the copper present crystallizes and can be skimmed out. If air is now blown into the molten lead, arsenic and antimony float out as oxides, and after this process the lead is referred to as soft lead[1].

Gold and silver may be removed from soft lead by the Parkes process. This involves the preferential extraction of silver and gold into added molten zinc, which then rises to the surface of the lead and after solidification can be skimmed off. The final impurity of a little zinc can be removed either by air oxidation, or more recently by evaporation.

This series of processes renders lead one of the purest of commercial metals, and the chemical specification for the purest commercial grade available in bulk requires 99·94% lead with no single impurity greater than 0·0025%, with the exception of bismuth which may be up to 0·05%.

Lead may be further refined electrolytically up to 99·995%, and finally zone refining. produces lead better than 99·9999% pure[3].

In the earth's crust lead is estimated to be present as $1 \cdot 6 \times 10^{-3}$% by weight, and to be in sea water at a concentration of 0·004 g/ton.

1.4. LEAD METAL

Lead has no allotropic modifications, and crystallizes with face-centred cubic structure (a = 4·9489 Å), with nearest neighbour lead–lead distance as 3·49 Å[4].

Lead, together with iridium and thallium, forms a group of close-packed metals isolated from the other close-packed metals, in that the interatom distances in iridium, thallium and lead are very large compared with those of neighbouring elements[5].

The outstanding physical properties of pure lead are its high density, softness, malleability, flexibility, low melting point, low strength and low elastic limit. These characteristics, together with its high corrosion resistance, are the basis of most applications of lead metal.

1.5. ISOTOPES

Table 1 illustrates the range of lead isotopes. Only four isotopes occur in nature and these are non-radioactive. Lead isotope analyses form an important tool in the calculation of geological ages in minerals in the region 1–6000 million years, and convenient tables have been prepared for this purpose. Only ^{204}Pb is known to be completely non-radiogenic. ^{206}Pb, ^{207}Pb and ^{208}Pb can be present (though not necessarily so) in minerals from radioactive decay[6].

[3] W. A. Tiller and J. W. Rutter, *Can. J. Phys.* **34** (1949) 96.
[4] H. P. Klug, *J. Am. Chem. Soc.* **68** (1946) 1493.
[5] A. F. Wells, *Structural Inorganic Chemistry*, 3rd ed., p. 975, Oxford University Press (1962).
[6] L. R. Stieff, T. W. Stern, S. Oshiro and F. E. Senftle, U.S. Geological Survey, 1959, Paper 334-A.

TABLE 1. THE STABLE AND RADIOACTIVE ISOTOPES OF LEAD

Isotope	Natural abundance (%)	Half-life
^{195}Pb		17 min
^{196}Pb		37 min
^{197m}Pb		42 min
^{198}Pb		2·4 hr
^{199m}Pb		12·2 min
^{199}Pb		90 min
^{200}Pb		21·5 hr
^{201m}Pb		61 sec
^{201}Pb		9·4 hr
^{202m}Pb		3·62 hr
^{202}Pb		*ca.* 3×10^5 years
^{203m}Pb		6·1 sec
^{203}Pb		52·1 hr
^{204m}Pb		66·9 min
^{204}Pb	1·40	
^{205}Pb		*ca.* 5×10^7 years
^{206}Pb	25·1	
^{207m}Pb		0·80 sec
^{207}Pb	21·7	
^{208}Pb	52·2	
^{209}Pb		3·30 hr
^{210}Pb RaD		19·4 years
^{211}Pb AcB		336 min
^{212}Pb ThB		10·64 hr
^{214}Pb RaB		26·8 min

1.6. NUCLEAR MAGNETIC RESONANCE

About one-fifth of the atoms in naturally occuring lead are present as ^{207}Pb, which has nuclear spin (I) of $\frac{1}{2}$, which enables nuclear magnetic resonance measurements to be carried out on lead and its compounds. Extremely large chemical shifts up to 16,000 ppm are encountered, which indicate a large paramagnetic contribution to shielding at the lead nucleus. The calculated values for paramagnetic shielding of ^{207}Pb nuclei are in good agreement with those observed experimentally[7]. Some ^{207}Pb chemical shifts are recorded in Table 2.

High resolution proton magnetic resonance spectra have been reported for organoleads and organolead hydrides[8]. In $(CH_3)_3PbH$ the τ value of the hydridic proton is 2·32, and the coupling constant for ^{207}Pb and the hydric proton is large, at 2379 cycles/sec. The τ value of the methyl protons in $(CH_3)_3PbH$ and $(CH_3)_4Pb$ are respectively 9·15 and 9·27.

In tetraethyl lead the observed coupling constant from ^{207}Pb to the methyl protons (125 cycles/sec) is larger than the coupling constant from ^{207}Pb to the methylene protons (41·0 cycles/sec), though notably the two constants are of opposite sign[9].

[7] L. E. Orgel, *Mol. Phys.* **1** (1958) 322.

[8] M. L. Maddox, S. L. Stafford and H. D. Kaesz, *Adv. Organometallic Chem.* **3** (1965) 12.

[9] J. W. Emsley, J. Feeney and L. H. Sutcliffe, *High Resolution Nuclear Magnetic Resonance*, Vol. 2, p. 688, Pergamon Press, Oxford (1966).

TABLE 2. SOME ^{207}Pb CHEMICAL SHIFTS RELATIVE TO METALLIC LEAD[10, 11]

Compound	Shift (δ) ppm	Compound	Shift (δ) ppm
Pb metal	0	$Pb(NO_3)_2$.solution	+14,400
PbO_2 (98% pure)	+6900	PbO (yellow)	+7400
$(CH_3COO)_2Pb.3H_2O$	+10,900	PbO (red)	+11,200
$(CH_3COO)_2Pb$ solution	+12,300	$Pb(ClO_4)_2$.solution	+14,100
$PbCl_2$.powder	+13,800	PbTe	+10,800
$PbSO_4$.powder	+15,200	PbS	+10,100
$Pb(NO_3)_2$.powder	+15,200	PbSe	+8700

1.7. ALLOYS

Lead is alloyed with many other metals, and some compositions of these alloys are represented in Table 3.

TABLE 3. REPRESENTATIVE COMPOSITIONS OF SOME LEAD ALLOYS†

Alloy	Pb %	Sn %	Sb %	As %	Cd %	Other %
Battery plate	94		6			
Magnolia metal	90		10			
Type metal	70	10	18			Cu 2
Aluminium solder (U.S.P. 1,333,666)	92				8	
Lead foil	87	12	1			
Marine babbitt	72	21	7			
Solder (plumber's)	67	33				
Solder (soft)	50	50				

† Most of these alloys are reported with considerable variations on these compositions.

1.8. CHEMICAL PROPERTIES OF METALLIC LEAD

Pure lead is bluish-white in colour and has a bright lustre. With air and water vapour the metal forms a thin surface film of oxycarbonate which protects the underlying metal. This film can easily be observed forming when freshly cut metal rapidly loses its lustre. Whilst bulk metal is only superficially oxidized, finely divided lead takes fire if dropped through the air.

The attack of lead by oxygenated water produces lead hydroxy compounds, which can result in lead poisoning by consumption of water carried in lead pipes. If the water contains small amounts of carbonate or silicate, however, attack of pipes becomes negligible.

10 L. H. Piette and H. E. Weaver, *J. Chem. Phys.* **28** (1958) 735.
11 J. M. Rocard, M. Bloom and L. H. Robinson, *Can. J. Phys.* **37** (1959) 522.

TABLE 4. PHYSICAL DATA FOR LEAD (TEMPERATURES °C)

Melting point 327°; ΔH (fusion) 1·22 kcal/mole; ΔS (fusion) 2·03 cal/deg/mole
Boiling point 1751°; ΔH (evaporation) 43·0 kcal/mole

Temperature	920	1000	1100	1250	1340
Vapour pressure, mm Hg	0·49	1·77	6·85	37·5	90·3

Specific heat (cal/deg/mole)	6·32 (25°)
Heat of formation at 25°	ΔH° = 0·00 (solid)
	ΔH° = 46·34 kcal (vapour)
Free energy of formation at 25°	ΔG° = 0·00 (solid)
	ΔG° = 38·47 kcal (vapour)
Entropy at 25°	S° = 15·51 cal/deg (solid)
	= 41·89 cal/deg (vapour)

Thermal conductivity 0·083 (0°), 0·076 (100°) cal/sec/cm/deg
Density 11·3415 g/cc at 20°; 10·69 at 327° (liquid); 11·00 at 327° (solid)
Electrical resistivity 19×10^{-6} ohms cm: superconductor below 7·23°K

Ion	Pb^{+}	Pb^{2+}	Pb^{3+}	Pb^{4+}	Pb^{5+}	Pb^{6+}	Pb^{7+}	Pb^{8+}	Pb^{9+}
Ionization potential, eV	7·415	15·03	31·93	39·0	69·7	(84)	(103)	(112)	(142)

Surface tension at 327° = 444 dynes/cm
Viscosity of liquid metal (centipoises) 3·2 (330°), 2·1 (440°), 1·2 (840°)

Temperature	−253	25	liquid
Specific magnetic susceptibility $\times 10^{-6}$ c.g.s.	−0·132	−0·111	−0·075

Ionic radii Pb^{2+} 1·21 Å; Pb^{4+} 0·84 Å

Both in the vapour and in solution in mercury, lead is monatomic.

With a potential of +0·130 V relative to the normal hydrogen electrode, it would be thought that lead should, in general, dissolve in dilute acids. The discharge of hydrogen upon lead does, however, have a very high overpotential. This, together with, in some cases the formation of insoluble coatings, affects profoundly the action of acids upon lead.

Sulphuric acid (unless perfectly anhydrous) does not dissolve lead, due to the insoluble lead sulphate coating. Lead is widely used for handling sulphuric acid, and is unattacked even by concentrated acid below 200°C.

Hydrochloric acid dissolves lead very slowly, but nitric acid dissolves lead rapidly with evolution of oxides of nitrogen and formation of lead nitrate.

Acetic acid rapidly attacks lead in the presence of oxygen.

The solubility of lead in pure water is low[12], and equilibration of lead with air-free water at 24° indicated a solubility of 311×10^{-6} g/litre.

[12] J. C. Pariaud and P. Archinard, *Bull Soc. Chim. France*, **19** (1952) 454.

TABLE 5. THERMODYNAMIC DATA FOR CERTAIN LEAD SPECIES

Species	State	ΔH° kcal/mole	ΔG° kcal/mole	S° cal/deg/mole
Lead	crystal	0	0	15·51
Lead (Pb)	gaseous	46·34	38·47	41·89
Lead (Pb_2)	gaseous	78		
Pb^+	gaseous	218·8		
Pb^{2+}	gaseous	566·9		
Pb^{2+}	aqueous	0·39	−5·81	5·1
Pb^{4+}	aqueous		72·3	
PbO (red)	crystal	−52·40	−45·25	16·2
PbO (yellow)	crystal	−52·07	−45·05	16·6
$Pb(OH)_2$	crystal	−123·0	−100·6	21
PbO_2	crystal	−66·12	−52·34	18·3
Pb_3O_4	crystal	−175·6	−147·6	50·5
PbF_2	crystal	−158·5	−148·1	29
PbF_4	crystal	−222·3	−178·1	35·5
$PbCl_2$	crystal	−85·85	−75·04	32·6
$PbBr_2$	crystal	−66·21	−62·24	38·6
PbI_2	crystal	−41·85	−41·53	42·3
PbI_3^-	aqueous		−48·55	
PbS	crystal	−22·54	−22·15	21·8
$PbSO_4$	crystal	−219·50	−193·89	35·2
PbSe	crystal	−18·0	−15·4	26·9
$PbSeO_4$	crystal	−148	−122	37
PbTe	crystal	−17·5	−18·1	27·6
$Pb(N_3)_2$	crystal	104·3	135·1	49·5
$Pb(NO_3)_2$	crystal	−107·35	−60·3	50·9
$Pb_3(PO_4)_2$	crystal	−620·3	−581·4	84·45
$PbHPO_3$	crystal	−234·5	−208·3	31·9
$PbCO_3$	crystal	−167·3	−149·7	31·3
PbC_2O_4	crystal	−205·1	−180·3	33·2
$Pb(C_2H_5)_4$	liquid	52		
$PbCrO_4$	crystal	−225·2	−203·6	36·5
$PbMoO_4$	crystal	−265·8	−231·7	38·5
$Pb(CH_3COO)_2$	crystal	−230·5		40
$PbSiO_3$	crystal	−258·8	−239·0	27

Bulk metal lead in contact with acidic solutions is a fairly good reducing agent, and with alkali it is quite a powerful reductor. A technique similar to that of the familiar Jones reductor with amalgamated zinc is also possible using lead in place of zinc. Although the reducing power of the lead reductor is less than the Jones zinc reductor, there are certain advantages in the selectivity of ions reduced[13].

1.9. INDUSTRIAL AND COMMERCIAL UTILIZATION[1]

World production of lead is about two million tons a year, but because lead is easily refined, and does not normally become contaminated in service, lead and lead-alloy scrap constitute an important factor in the lead market. It is estimated that over 80% of the lead

13 C. W. Sill and H. E. Peterson, *Anal. Chem.* **24** (1952) 1175.

TABLE 6. STANDARD REDUCTION POTENTIALS OF LEAD HALF REACTIONS AT 25°, REFERRED TO THE NORMAL HYDROGEN ELECTRODE[14]

Reduction		$E°$ in volts
$Pb^{2+}+2e^-$	$= Pb$	$-0{\cdot}126$
$HPbO_2^-+2e^-+H_2O$	$= Pb+3OH^-$	$-0{\cdot}54$
$PbCl_2(S)+2e^-$	$= Pb+2Cl^-$	$-0{\cdot}268$
$PbBr_2(S)+2e^-$	$= Pb+2Br^-$	$-0{\cdot}280$
$PbI_2(S)+2e^-$	$= Pb+2I^-$	$-0{\cdot}365$
$PbS(S)+2e^-$	$= Pb+S^{2-}$	$-0{\cdot}98$
$PbSO_4(S)+2e^-$	$= Pb+SO_4^{2-}$	$-0{\cdot}3563$
$Pb^{4+}+2e^-$	$= Pb^{2+}$	$\sim+1{\cdot}7$
$PbO_2(S)+2e^-+4H^+$	$= Pb^{2+}+2H_2O$	$+1{\cdot}455$
$PbO_2(S)+2e^-+4H^++SO_4^{2-}$	$= PbSO_4(S)+2H_2O$	$+1{\cdot}685$
$PbO_2(S)+2e^-+H_2O$	$= PbO(S, red)+2OH^-$	$+0{\cdot}248$

The $PbSO_4$–Pb couple is a common reference electrode whose potential has been determined as a function of temperature from 0° to 60°C. Similarly, data of the other half reactions that are involved in the lead acid storage battery are known accurately over a 0°–60°C temperature range[15].

used to manufacture batteries re-enters the market. The following list gives an approximate indication of some of the major consumers of lead: storage batteries (~35%), tetraethyl lead (~10%), cable covering (~10%), solder (~10%), red lead and litharge (~5%), building lead (~5%), caulking (~5%). Other lead uses are white lead, bearing metal, ammunition, type metal and radiation shields.

1.10. ANALYSIS

No truly specific reagent or test for lead exists. Thus unless the nature of the sample is known in advance, preliminary separations of lead are made. Useful separation procedures are well documented[16].

A number of the persistent lines of lead are suitable for spectrographic analysis by arc or spark spectra[17].

No single gravimetric method, of the great variety used in the determination of lead, is clearly the most satisfactory. The thermal stabilities of a number of the weighed precipitates used have been recorded, and such considerations are clearly important[18].

Among the best-known gravimetric precipitants for lead are sulphate, chromate, molybdate, phosphate, oxalate, anthranilate, oxinate, mercaptobenzothiazolate and phthalate.

[14] W. M. Latimer, *Oxidation Potentials*, 2nd ed., Prentice-Hall, New York (1952).

[15] H. S. Harned and W. J. Harmer, *J. Am. Chem. Soc.* **57** (1935) 33.

[16] T. W. Gilbert, in *Treatise in Analytical Chemistry*, Part II, Vol. 6, p. 69, ed. I. M. Kolthoff and P. J. Elving, Interscience, New York.

[17] L. H. Ahrens and S. R. Taylor, *Spectrochemical Analysis*, 2nd ed., p. 242, Addison-Wesley, Reading, Massachusetts.

[18] C. Duval, *Inorganic Thermogravimetric Analysis*, Elsevier, Amsterdam (1953).

Volumetric methods of lead estimation involving titration with molybdate, chromate, phosphoric acid, ferrocyanide, fluoride and iodide are all available, in addition to the complexometric titrations with reagents such as EDTA.

For the estimation of lead at extremely low concentrations polarographic methods are best utilized. Anodic dissolution polarography can estimate the lead content of 10^{-8} molar lead solutions to a precision of about 5%[19].

Very small traces of lead in water can be effectively collected by co-precipitation with mercury sulphide. After treatment with peroxide, the collected HgS is ignited and the mercury vaporized away. The residual lead can be tested with sodium rhodizonate. Lead at concentrations of 10^{-8} are detected easily by this method[20]. The method is also applicable to the analysis of reagents for traces of lead contaminant.

The presence of tetra-organolead compounds in fuels is detected by first decolorizing the fuel from added dye with activated charcoal. A drop of fuel is placed on a filter paper and irradiated under an ultraviolet lamp, hence evaporating fuel and decomposing the organolead. Lead is then detected by adding a drop of freshly prepared 0·1% solution of dithizone in chloroform. A deep red spot of lead dithizonate indicates presence of lead, and retention of the green dithizone colour indicates absence of the metal[16].

1.11. TOXICITY

All compounds of lead are toxic, and lead poisoning has been long known and exhaustively studied[21]. Nevertheless, lead poisoning still constitutes one of the most important industrial hazards[22].

Minute quantities of lead are ingested regularly by humans from contaminated food and drink, but such quantities are easily eliminated by normal body processes. Causes of harmful accumulations in this way are now very rare.

Serious lead intoxication is now most frequently encountered by inhalation of vapours or dusts of lead and lead compounds. Thus the processes of spraying or sanding lead paints are extremely hazardous without proper protection. The specified maximum allowable atmospheric pollution is 0·15 mg of lead per cubic metre of air.

Lead poisoning can be cured and recovery is usually 100%. The most common treatment now is the intravenous injection of the sodium calcium salt of ethylenediaminetetra-acetic acid, which results in an immediate ten- to thirty-fold increase in the urinary excretion of lead. The clinical symptoms of lead poisoning are relieved in this way in a short time.

2. COMPOUNDS

2.1. ORGANOLEAD COMPOUNDS

Tetraethyl-lead, together with tetramethyl-lead and the mixed ethylmethyl-leads, is made on a larger scale commercially than all other organometallic compounds together.

19 T. L. Marple and L. B. Rogers, *Anal. Chim. Acta*, **11** (1954) 574.
20 F. Feigl, *Spot Tests in Inorganic Analysis*, Elsevier, Amsterdam (1956).
21 H. B. Elkins, *The Chemistry of Industrial Toxicology*, 2nd ed., John Wiley, New York (1959).
22 *Occupational Lead Exposure and Lead Poisoning*, American Public Health Association Inc., New York (1943).

TABLE 7. BOND LENGTHS INVOLVING LEAD

Bond	Bond length (Å)	Compound	Method
Pb–Pb	3·49	Lead	X-ray diffraction[a]
Pb–Pb	2·88	$(CH_3)_3Pb–Pb(CH_3)_3$ (Gas)	Electron diffraction[b]
Pb–H	1·839	PbH (Short-lived species)	Infrared spectroscopy[c]
Pb–F	2·13	PbF_2 (Gas)	Electron diffraction[d]
Pb–Cl	2·46	$PbCl_2$	Electron diffraction[d]
	2·46	(Gas)	Electron diffraction[e]
Pb–Cl	2·50	Cs_2PbCl_4	X-ray diffraction[f]
Pb–Cl	2·43	$PbCl_4$ (Gas)	Electron diffraction[e]
PbBr	2·60	$PbBr_2$	Electron diffraction[d]
	2·60	(Gas)	Electron diffraction[e]
Pb–I	2·79	PbI_2	Electron diffraction[e]
	2·78	(Gas)	Electron diffraction[d]
Pb–C	2·303	$Pb(CH_3)_4$ (Gas)	Electron diffraction[g]
Pb–C	2·25	$(CH_3)_3Pb–Pb(CH_3)_3$ (Gas)	Electron diffraction[b]
Pb–O	1·922	PbO (Short-lived species)	Infrared spectroscopy[c]
PbO	2·30	PbO (Tetragonal)	X-ray diffraction[h]
PbO	{2·21, 2·49}	PbO (Rhombic)	Neutron diffraction[i, j]
PbS	2·394	PbS (Short-lived species)	Infrared spectroscopy[c]
PbS	2·962	PbS (Crystal)	X-ray diffraction[k]
PbS	2·75	$Pb(NS)_2NH_3$	X-ray diffraction[l]
PbN	2·27	$Pb(NS)_2NH_3$	X-ray diffraction[l]
PbS	{3·02, 3·10, 2·92}	$[(NH_2)_2CS]PbCl_2$	X-ray diffraction[m]
PbCl	{3·28, 3·17, 2·75}	$[(NH_2)_2CS]PbCl_2$	X-ray diffraction[m]
PbRe	2·77	$(C_6H_5)_3PbRe(CO)_5$	X-ray diffraction[n]
PbMo	2·90	$(C_6H_5)_3PbMo(CO)_3$–π-C_5H_5	X-ray diffraction[n]

[a] A. F. Wells, *Structural Inorganic Chemistry*, 3rd ed., p. 975, Oxford University Press (1962).

[b] H. A. Skinner and L. E. Sutton, *Trans. Faraday Soc.* **36** (1940) 1209.

[c] G. Herzberg, *Molecular Spectra and Molecular Structure: Infrared Spectra of Diatomic Molecules*, 2nd ed., van Nostrand.

[d] P. A. Akisin, V. P. Spiridonov and A. N. Khodankov, *Zhur. Fiz. Khim. SSSR*, **32** (1958) 1679.

[e] M. Lister and L. E. Sutton, *Trans. Faraday Soc.* **37** (1941) 393.

[f] G. Engel, *Z. Krist.* **90** (1935) 341.

[g] C. E. Wong and V. Schomaker, *J. Chem. Phys.* **28** (1958) 1007, but see Interatomic Distances Special Publication, No. 18, The Chemical Society, London.

[h] W. J. Moore and L. Pauling, *J. Am. Chem. Soc.* **63** (1941) 1392.

[i] J. Leciejewicz, *Acta Cryst.* **14** (1961) 66.

[j] M. I. Kay, *Acta Cryst.* **14** (1961) 80.

[k] E. Zeipel, *Arkiv. Mat. Astron. Fysik*, **25A** (1935) 1.

[l] J. Weiss and D. Neubauer, *Z. Naturforsch.* **13b** (1958) 459.

[m] M. Nardelli and G. Fava, *Acta Cryst.* **12** (1959) 727.

[n] Yu. T. Struchkov, K. N. Anisimov, O. P. Osipova, N. E. Kolobova and A. N. Nesmeyanov, *Dokl. Akad. Nauk SSSR*, **172** (1967) 107.

This renders it a fairly cheap chemical at about \$0·35 per pound. A quarter of a million tons is used annually in the U.S.A. alone as an antiknock agent in motor fuels[23–26].

Industrially, the tetra-alkyl-leads are normally made by the action of alkylchloride upon sodium–lead alloy[26], or by the electrolysis of alkylmagnesium halides using a sacrificial lead anode, where the anode is attacked by alkyl radicals or anions. On a laboratory scale, however, the action of a Grignard reagent upon lead(II) halide remains the favoured synthesis, especially of the alkyl compounds R_4Pb[27]. In the case of aryl compounds, R_6Pb_2 is formed and has to be treated with halogen to produce R_3PbX, which is then further treated with Grignard to give R_4Pb[24]. Some other methods utilized to form lead(IV)–carbon bonds are outlined below[24].

$$Pb+Li+C_6H_5Br \rightarrow (C_6H_5)_4Pb$$
$$(C_2H_5)_2Zn+PbCl_2 \rightarrow (C_2H_5)_4Pb$$
$$RLi+PbX_2 \rightarrow R_4Pb$$
$$R_3Al+PbX_2 \rightarrow R_4Pb$$

The lead(IV) alkyls and aryls are stable at ordinary temperatures, but release organic free radicals on heating. Subsequent reactions are complex and among the pyrolysis products of tetramethyl-lead are 2-methyl-2-propene, propylene, ethylene, hydrogen, methane and ethane.

The fission of the lead–carbon bond is caused by a number of reagents as outlined below[24].

$$R_4Pb+Na \xrightarrow{NH_3} R_3PbNa$$
$$(CH_3)_4Pb+AgNO_3 \longrightarrow (CH_3)_3PbNO_3$$
$$(C_2H_5)_4Pb+PCl_3 \longrightarrow C_2H_5PCl_2+(C_2H_5)_2PCl$$
$$R_4Pb+SOCl_2 \longrightarrow R_2PbCl_2+R_3PbCl$$

TABLE 8. SOME ORGANOLEAD(IV) COMPOUNDS

Compound	B.p., °C/mm Hg	M.p., °C
$(CH_3)_4Pb$	110° or 6/10	−27·5
$(C_2H_5)_4Pb$	78/10	
$(n\text{-}C_3H_7)_4Pb$	126/13	
$(iso\text{-}C_3H_7)_4Pb$	120/14	
$(C_6H_5)_4Pb$		225
$(C_2H_5)_3PbCH_2\text{–}CH{=}CH_2$	86/10	
$p\text{-}CH_3.C_6H_4Pb(CH_3)_3$	119/13	
$(CH_3)_2Pb(C_3H_7)_2$	72/10	
$(C_2H_5)_2Pb(CH_3)(C_3H_7)$	80/15	
$(C_2H_5)(CH_3)Pb(C_3H_7)(C_4H_9)$	103/13	
$(C_6H_5)_3Pb(C\ \ C)_2Pb(C_6H_5)_3$		187
$(C_6H_5)_3PbCH{:}C{:}CH_2$		
$(C_6F_5)_4Pb$		199

[23] L. C. Willemsens and G. J. M. Van der Kerk, *Investigations in the Field of Organolead Chemistry*, ILZRO, New York (1965).

[24] R. W. Leeper, L. Summers and H. Gilman, *Chem. Revs.* **54** (1954) 101.

[25] *Organometallic Compounds*, 2nd ed., Vol. II, p. 519, ed. M. Dub, Springer-Verlag, Berlin (1967).

[26] G. E. Coates, M. L. H. Green and K. Wade, *Organometallic Compounds*, 3rd ed., Vol. I, p. 482, Methuen, London (1967).

[27] M. Baudler, in *Handbook of Preparative Inorganic Chemistry*, 2nd ed., Vol. 1, p. 763, ed. G. Brauer, Academic Press, New York.

From various studies the following series of relative ease of cleavage by halogen or halides has been established for organic groups attached to lead: α-naphthyl, *p*-xylyl, *p*-tolyl, phenyl, methyl, ethyl, *n*-propyl, isobutyl, isoamyl, cyclohexyl (least easily cleaved)[24].

The only well-defined examples of lead(II) organometallic compounds are bis(cyclopentadienyl)lead(II) and its ring methyl derivative[28]. In the gas phase[29] this has a bent sandwich structure as shown in Fig. 1 (a). Such a structure is also proposed[30] to account for the solution spectra of $(C_5H_5)_2Pb$, where the two rings are believed bonded to two sp^2 lead hybrid orbitals, and a further sp^2 orbital contains a non-bonding pair of electrons. Such a bonding situation is in accord with the crystal structure of the orthorhombic form shown in Fig. 1 (b).

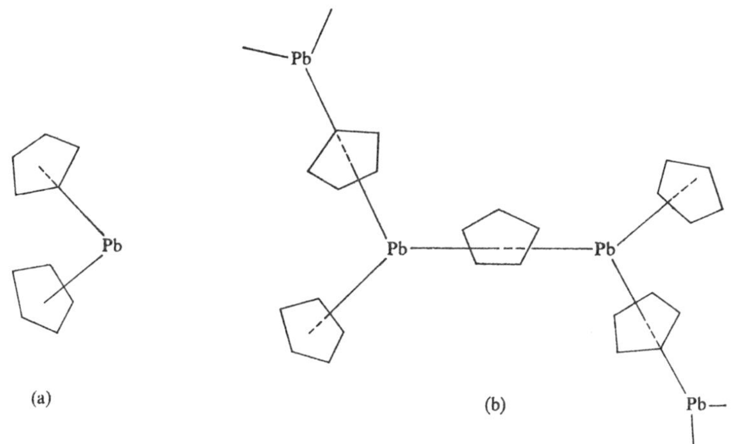

FIG. 1. The structure of $(C_5H_5)Pb$: (a) in the gas and solution and (b) in the orthorhombic crystal form.

Here each lead atom is attached to two bridging cyclopentadienyl groups and one terminal cyclopentadienyl group[31]. The lead to ring bonds are all at approximately 120°, and the polymer may be considered to arise by the lone pair of electrons on each monomer interacting with a cyclopentadienyl ring of another monomer.

A monoclinic form of $(C_5H_5)_2Pb$ is of unknown structure[31].

2.2. HYDRIDES

Lead hydride is the least well characterized of the Group IVB hydrides. It is formed along with hydrogen, on the electrolysis of dilute sulphuric acid with lead electrodes, and by the dissolution of lead–magnesium alloy in dilute acid[32]. The hydride of lead formed in small quantities is assumed to be PbH_4 (b.p. $\sim -13°C$)[33].

In alkali or weakly acid solutions, lead cathodes disintegrate at high current densities. This is believed due to the formation of the unstable hydride PbH_2 at the cathode, and at

[28] E. O. Fischer and H. Grubert, *Z. anorg. Chem.* **286** (1956) 237.
[29] A. Almenningen, A. Haaland and T. Motzfeldt, *J. Organometallic Chem.* **7** (1967) 97.
[30] L. D. Dave, D. F. Evans and G. Wilkinson, *J. Chem. Soc.* (1959) 3684.
[31] C. Panattoni, G. Bombieri and U. Croatto, *Acta Cryst.* **21** (1966) 823.
[32] F. Paneth and O. Norring, *Chem. Ber.* **53** (1920) 1693.
[33] F. G. A. Stone, *Hydrogen Compounds of the Group IV Elements*, Prentice-Hall (1962).

current densities over 10–50 mA/cm² the formation of PbH_2 was believed to be quantitative[34].

A lead hydride has been reported from the bombardment of lead films by hydrogen atoms[35].

Organolead Hydrides

The organohydrides of lead(IV) are not as robust as the corresponding compounds of tin, germanium and silicon, but are nevertheless sufficiently stable for an extensive study of the lead–hydrogen bond to be made.

Trialkyl-lead hydrides and dialkyl-lead dihydrides were made by reducing the corresponding chlorides with lithium aluminium hydride at low temperature (−78°)[36, 37, 38].

The exchange reaction between organotin hydrides and organolead salts has also proved a useful synthesis for organolead hydrides[39].

$$(n\text{-}C_4H_9)_3PbX + (C_6H_5)_3SnH \rightarrow (n\text{-}C_4H_9)_3PbH + (C_6H_5)_3SnX$$

Trimethyl-lead hydride (m.p. ∼ −106°) and triethyl-lead hydride (m.p. ∼ −145°) were found to decompose to the corresponding tetra-alkyl-lead, lead metal and hydrogen.

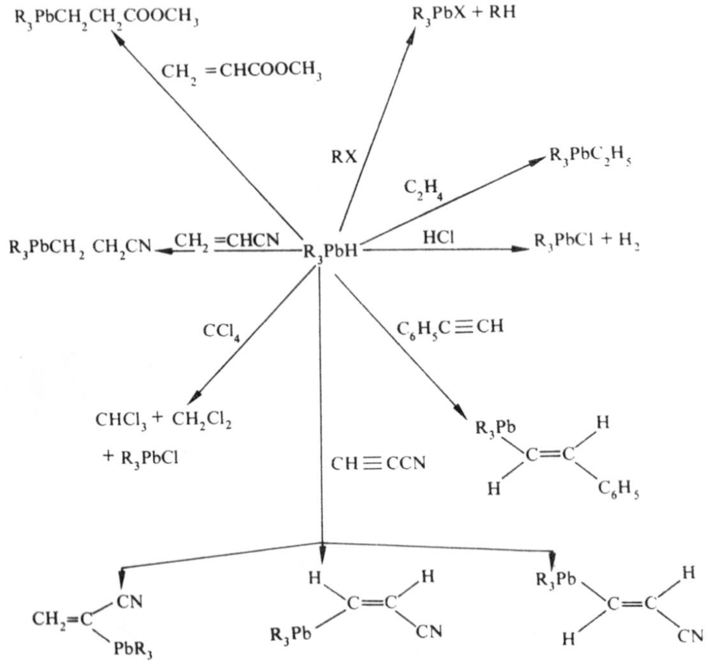

FIG. 2. Some reactions of the lead–hydrogen bond in trialkyl-lead hydrides.

34 H. W. Salzberg, *J. Electrochem. Soc.* **100** (1953) 146.
35 E. Pietsch and F. Seuferling, *Naturwiss.* **19** (1931) 574.
36 E. Amberger, *Angew. Chem.* **72** (1960) 494.
37 W. E. Becker and S. E. Cook, *J. Am. Chem. Soc.* **82** (1960) 6264.
38 W. P. Neumann and K. Kuhlein, *Angew. Chem.* **77** (1965) 808.
39 H. M. J. C. Creemers, A. J. Leusink, J. G. Noltes and G. J. M. van der Kerk, *Tetrahedron Letters* (1966) 3167.

Evolution of hydrogen begins on warming to about $-30°$ to $-20°$, but even at 0° the hydrides are not completely decomposed for several hours.

Some of the reactions of trialkyl-lead hydrides are outlined[40] in Fig. 2.

Very unstable trialkyl-lead borohydrides have been made from organolead alkoxides and diborane at $-78°$.

$$4R_3PbOCH_3+3(BH_3)_2 \rightarrow 4R_3PbBH_4+2HB(OCH_3)_2$$

These trialkyl-lead borohydrides react with methanol to release trialkyl-lead hydrides, and decompose at about $-30°$ to hexa-alkyldileads[41].

2.3. COMPOUNDS OF NITROGEN, PHOSPHORUS ARSENIC, ANTIMONY AND BISMUTH

Binary Compounds

Lead(II) azide is discussed with the pseudohalides, and lead nitrides are not known.

The phosphides of lead are not well characterized and appear to interconvert easily[42]. PbP_5 was obtained from a liquid ammonia solution of rubidium phosphide and lead nitrate. PbP_5 is spontaneously inflammable in air, and dissociates on heating with loss of phosphorus[43].

The action of phosphine upon alcoholic lead acetate solution produces Pb_3P_2 [44].

Arsine and lead(II) acetate form the very unstable arsenide Pb_3As_2[45].

In addition to these few binary compounds of lead with the Group VB elements, there is a considerable range of mineral compounds which contain lead, and either arsenic, antimony or bismuth, but these are ternary and quaternary materials such as sartorite $PbAs_2S_4$, dufrenoysite $Pb_2As_2S_5$, sperrylite $Pb_4As_2S_7$, sakharovite $PbBiSbS_4$ and semseyite $Pb_9Sb_8S_2$.

Organolead Amines, Phosphines, Arsines and Stibines

In addition to the binary compounds discussed above, there are a few examples of discrete lead–nitrogen[46], lead–phosphorus, lead–arsenic and lead–antimony bonds in various compounds, but they are considerably more rare than the corresponding compounds of tin, germanium and silicon.

Coordination compounds such as the di-ammine of diphenyl-lead dibromide contain discrete lead–nitrogen bonds, but such complexes are very unstable[47]. In the case of the diaryl-lead oxinates, however, the coordinate nitrogen–lead bond is quite stable[48].

The first organolead amine reported[49] was sec-$C_4H_9.NH.Pb(C_2H_5)_3$, prepared from triethyl-lead chloride and sodium sec-butylamide. Other amines of lead include $(CH_3)_3PbN(CH_3)Ge(CH_3)_3$ [50] and $(CH_3)_3PbN[Si(CH_3)_3]_2$ [51].

40 D. Seyferth and R. B. King, *Annual Surveys of Organometallic Chemistry*, **2** (1965) 184.
41 E. Amberger and R. Honigschmid-Grossich, *Chem. Ber.* **99** (1966) 1673.
42 *Nouveau traité de chimie minérale*, Vol. VIII, Part 3, p. 729, ed. P. Pascal, Masson et Cie, Paris (1963).
43 R. Bossuet and L. Hackspill, *Compt. Rend.* **157** (1913) 720.
44 A. Brukl, *Z. anorg. Chem.* **125** (1922) 252.
45 A. Brukl, *Z. anorg. Chem.* **131** (1923) 236.
46 J. G. A. Luijten, F. Rijkens and G. J. M. van der Kerk, *Adv. Organometallic Chem.* **3** (1965) 397.
47 P. Pfeiffer, P. Truskier and P. Disselkamp, *Chem. Ber.* **49** (1916) 2445.
48 R. Barbieri, G. Faraglia, M. Giustiniani and L. Roncucci, *J. Inorg. Nuclear Chem.* **26** (1964) 203.
49 D. O. Depree, U.S. Patent 2,893,857 (1959); *C.A.* **53** (1959) 18372.
50 I. Ruidisch and M. Schmidt, *Angew. Chem.* **76** (1964) 686.
51 O. Scherer and M. Schmidt, *J. Organometallic Chem.* **1** (1964) 490.

TABLE 9. SOME COMPOUNDS WITH LEAD BONDED TO EITHER NITROGEN, PHOSPHORUS, ARSENIC OR ANTIMONY

Compound	M.p., °C [B.p. °C]	Reference
$(C_6H_5)_2PbCl_2.4C_5H_5N$		a
$(C_6H_5)_2Pb(NO_3)_2.4C_5H_5N$		a
$(C_6H_5)_2PbBr_2.2NH_3$		a
$(C_6H_5)_2Pb(oxinate)_2$		b
$(C_2H_5)_2PbNHCH(CH_3)C_2H_5$		c
$(CH_3)_3PbN[Si(CH_3)_3]_2$	[85–87/3 mm]	d
$(CH_3)_3PbN(CH_3)Ge(CH_3)_3$	[49/2 mm]	e
$(C_2H_5)_3PbN$ CO CO	131	f
$(C_2H_5)_3PbN$ CO SO_2		g
$(C_2H_5)_3PbN$ CO NH CO	135	h
$(C_2H_5)_3PbNHSO_2CH_3$	97	h
$(C_2H_5)_3PbN(C_6H_5)SO_2CH_3$	116	h
$(C_6H_5)_3PbP(C_6H_5)_2$	100	i
$[(C_6H_5)_3Pb]_2PC_6H_5$	110	j
$[(CH_3)_3Pb]P$	47	k
$[(C_6H_5)_3Pb]P$	110	k
$(C_6H_5)_3PbP[Sn(C_6H_5)_3]_2$	172	l
$(C_6H_5)_3PbAs(C_6H_5)_2$	115	m
$[(CH_3)_3Pb]_3As$	45	k
$[(C_6H_5)_3Pb]_3As$	158	k
$(C_6H_5)_3PbSb(C_6H_5)_2$	115	m
$[(C_6H_5)_3Pb]_3Sb$	150	k

[a] P. Pfeiffer, P. Trusker and P. Disselkamp, *Ber.* **49** (1916) 2445.

[b] R. Barbieri, G. Faraglia, M. Guistiniani and L. Roncucci, *J. Inorg. Nucl. Chem.* **26** (1964) 203.

[c] D. O. DePree, U.S.P. 2,893,857 (1959); *Chem. Abs.* **53** (1959) 18372.

[d] O. Scherer and M. Schmidt, *J. Organometallic Chem.* **1** (1964) 490.

[e] I. Ruidisch and M. Schmidt, *Angew. Chem.* **76** (1964) 686.

[f] R. Heap and B. C. Saunders, *J. Chem. Soc.* (1949) 2983.

[g] W. B. Ligett, R. D. Closson and C. N. Wolf, U.S.P. 2,595,789 (1952); *Chem. Abs.* **46** (1952) 7701.

[h] B. C. Saunders, *J. Chem. Soc.* (1950) 684.

[i] H. Schumann, P. Schwabe and M. Schmidt, *J. Organometallic Chem.* **1** (1964) 366.

[j] H. Schumann, P. Schwabe and M. Schmidt, *Inorg. Nucl. Chem. Letters,* **2** (1966) 309.

[k] H. Schumann, A. Roth, O. Stelzer and M. Schmidt, *Inorg. Nucl. Chem. Letters,* **2** (1966) 311.

[l] H. Schumann, P. Schwabe and M. Schmidt, *Inorg. Nucl. Chem. Letters,* **2** (1966) 313.

[m] H. Schumann and M. Schmidt, *Inorg. Nucl. Chem. Letters,* **1** (1965) 1.

In addition to these ammines, many organolead sulphonamides, phthalimides and sulphimides contain lead–nitrogen bonds[52, 53, 54]. The sulphimides, phthalimides and sulphonamides are not hydrolysed by water, in marked contrast to the organolead amines.

The mono-, bi- and tri-lead phosphines exemplified by $(C_6H_5)_3PbP(C_6H_5)_2$ [55], $[(C_6H_5)_3P]_2PC_6H_5$ [56] and $[(C_6H_5)_3Pb]_3P$[57] are all crystalline solids which decompose readily upon warming. They are synthesized by the reaction of organolead halides with the requisite phosphine. The same method, along with the metathetical removal of sodium chloride from the organolead halide and the requisite sodium arsenide or stibnide, is used to prepare the organolead arsines and stibines in Table 9.

2.4. OXYGEN COMPOUNDS

Oxides of Lead

Lead(II) oxide (litharge) is the oxide of lead formed when lead is heated in the air. The industrial preparation involves blowing air into molten lead.

Two crystal forms of lead(II) oxide are known: the yellow orthorhombic form is the stable form above 488°C, and the red tetragonal form is stable at ordinary temperatures[58].

Ultrapure PbO is made by precipitation from lead acetate solution by ammonium hydroxide in polyethylene vessels. In such wet preparations of lead(II) oxide, the yellow orthorhombic form is first produced which undergoes transformation to the red tetragonal PbO. This transformation is particularly sensitive to impurities, and the presence of elements such as silicon, germanium, phosphorus, arsenic, antimony, selenium, tellurium, molybdenum and tungsten in concentrations as low as 10 ppm prevents the transformation. The use of polythene vessels for the preparation of ultrapure red lead(II) oxide is emphasized, because sufficient silica is released from glass vessels to prevent the yellow to red conversion[58, 59].

The so-called "black" lead(II) oxide is merely yellow or red PbO with a thin surface film of elemental lead.

In tetragonal lead(II) oxide[60], each lead atom has four oxygen near neighbours, all lying to one side, and all of the oxygen atoms lie between every other pair of lead layers as in Fig. 3.

The oxygen atoms in rhombohedral PbO have proved impossible to locate by X-ray intensity measurements[61], but their positions were determined by neutron diffraction[62]. The structure consists of a chain in which lead has two nearest oxygen neighbours (Pb–O = 2·21 Å). These chains are loosely bonded into layers by two Pb–O bonds (Pb–O = 2·49 Å).

The layer in rhombohedral PbO is a slightly puckered version of that in tetragonal PbO, where four equal Pb–O bonds (2·30 Å) replace the two short (2·21 Å) and two long (2·49 Å)

[52] H. McCombie and B. C. Saunders, *Nature*, **159** (1947) 491.
[53] R. Heap and B. C. Saunders, *J. Chem. Soc.* (1949) 2983.
[54] B. C. Saunders, *J. Chem. Soc.* (1950) 684.
[55] H. Schumann, P. Schwaben and M. Schmidt, *J. Organometallic Chem.* **1** (1964) 366.
[56] H. Schumann, P. Schwaben and M. Schmidt, *J. Inorg. Nuclear Letters*, **2** (1966) 309.
[57] N. H. W. Addink, *Nature*, **157** (1946) 764.
[58] W. Kwestroo and A. Huizing, *J. Inorg. and Nuclear Chem.* **27** (1965) 1951.
[59] W. Kwestroo, J. de Jonge and P. H. G. M. Vromans, *J. Inorg. and Nuclear Chem.* **29** (1967) 39.
[60] W. J. Moore and L. Pauling, *J. Am. Chem. Soc.* **63** (1941) 1392.
[61] A. Bystrom, *Arkiv. Kemi* **17B** (1943) No. 8.
[62] J. Leciejewicz, *Acta Cryst.* **14** (1961) 66.

Pb–O bonds. This close similarity of structures presumably allows small impurities to stabilize the yellow rhombic form at ordinary temperatures. The two crystal forms can be obtained simultaneously by heating lead carbonate at 400° in a melt of potassium and sodium nitrates, when over a period of one day red prisms of tetragonal PbO and yellow plates of rhombic PbO are formed along with orange needles of Pb_3O_4.

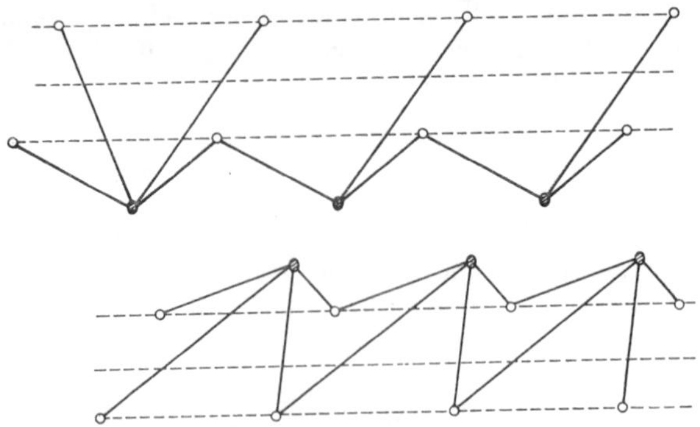

FIG. 3. Tetragonal lead(II) oxide.

Lead(IV) oxide can be obtained in a very pure form by the hydrolysis of lead tetra-acetate[63,64]. In this form it is particularly reactive and useful as an oxidizing agent in organic chemistry. More usually lead(IV) oxide is made by the hypochlorite oxidation of lead(II) acetate solutions[65], or by the action of nitric acid upon Pb_3O_4.

Formed in this way PbO_2 has the rutile structure[65], but another orthorhombic form of PbO_2 is known[66]. This is prepared by the anodic deposition during electrolysis of lead acetate, lead nitrate or sodium plumbite solutions under specific conditions. This so-called α-form of PbO_2 has the niobite (columbite) type of structure, which consists essentially of a hexagonal close-packed assemblage of oxygen atoms in which one-half of the octahedral holes are occupied by Pb^{4+} ions. This gives the Pb^{4+} ion six oxygen nearest neighbours, two at 2·16 Å, two at 2·17 Å and two at 2·22 Å[66].

Lead(IV) oxide is formed in nature as the somewhat rare mineral plattnerite, which is isomorphous with rutile and many other dioxides like SnO_2, MnO_2, ZrO_2 and ThO_2.

Lead (IV) oxide is difficult to obtain in a perfectly anhydrous state, as the temperature for complete drying initiates release of oxygen to form PbO and Pb_3O_4. If carefully heated in a stream of oxygen, however, completely anhydrous lead(IV) oxide can be made.

The action of heat decomposes lead(IV) oxide to oxygen (partial pressure of O_2 reaches one atmosphere at 344°C with a coarse-grained PbO_2) and to Pb and Pb_3O_4. Lead(IV) oxide is a powerful reducing agent but is not a peroxide. Rapid grinding with sulphur or red phosphorus produces inflammation, explaining the use of lead(IV) oxide in matches.

63 R. Kuhn and I. Hammer, *Chem. Ber.* **83** (1950) 413.

64 M. Baudler, in *Handbook of Preparative Inorganic Chemistry*, 2nd ed., Vol. 1, p. 757, ed. G. Brauer, Academic Press, New York.

65 A. Bystrom, *Arkiv. Kemi, Min. Geol.* **20A** (1945) No. 11.

66 A. I. Zaslavsky, Yu. D. Kondrashev and S. S. Tolkachev, *Dokl. Akad. Nauk SSSR*, **75** (1950) 559.

Lead(IV) oxide is an essential constituent of the lead–acid accumulator. A single cell of the battery consists of one electrode of lead sponge and the other of a mesh of lead which is impregnated with lead(IV) oxide. The electrolyte is strong sulphuric acid.

The two electrode reactions are:

$$PbSO_4 + 2e^- \rightarrow Pb + SO_4^{2-} \qquad E_0 = -0{\cdot}356 \text{ V}$$

$$PbO_2 + SO_4^{2-} + 4H^+ + 2e^- \rightarrow PbSO_4 + H_2O \qquad E_0 = 1{\cdot}685 \text{ V}$$

Acting in opposition the electrode reactions provide an overall potential of 2·041 V, but the potential delivered by the cell will depend upon temperature, the acid concentration and current flow.

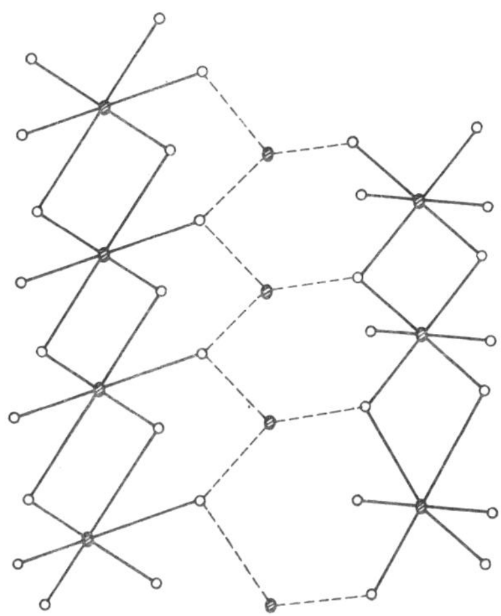

FIG. 4. Structure of Pb_3O_4.

Pb_3O_4 has a structure[67, 68, 69] which consists of chains of $Pb^{IV}O_6$ octahedra joined by opposite edges; the chains are linked by Pb(II) ions coordinated by three oxygen atoms pyramidally as in Fig. 4. Thus Pb_3O_4 can be looked upon as lead(II) plumbate, and nitric acid attacks Pb_3O_4 to, in effect, remove 2PbO and leave PbO_2.

The finely divided Pb_3O_4 (minium or red lead) used as a preservative and in paints is usually made by the action of heat upon PbO or $PbCO_3$ in a current of air. A macrocrystalline form is obtained by the slow crystallization of Pb_3O_4 from a mixed solution of $K_2Pb(OH)_4$ and $K_2Pb(OH)_6$ [70], or alternatively from PbO_2 and sodium hydroxide in the presence of water in a steel bomb at about 375° [71].

Pb_3O_4 evolves oxygen on heating, and the oxygen reaches a pressure of about 0·2 atm at about 550°. If, however, dissociation into oxygen and lead(II) oxide is prevented by raising the oxygen pressure over Pb_3O_4, it melts at about 830°.

67 S. T. Gross, *J. Am. Chem. Soc.* **65** (1941) 1107.

68 M. Straumanis, *Z. Phys. Chem.* **B52** (1942) 127.

69 A. Bystrom and A. Westgren, *Arkiv. Kemi, Min. Geol.* **16B** (1943) No. 14; *ibid.* **25A** (1947) No. 13.

70 M. Baudler, in *Handbook of Preparative Inorganic Chemistry*, 2nd ed., Vol. 1, p. 755, ed. G. Brauer, Academic Press, New York.

71 G. L. Clark, N. C. Schieltz and T. T. Quirke, *J. Am. Chem. Soc.* **59** (1937) 2305.

Much has been written of oxides of lead, other than Pb_3O_4, which lie between the compositions of PbO and PbO_2.

The differential thermal analysis of the decomposition of PbO_2, together with X-ray analysis, has shown two further oxide phases as illustrated below[72].

$$PbO_2 \xrightarrow{\text{about } 280°} \alpha\text{-}PbO_x \quad (x \sim 1{\cdot}66)$$
$$\alpha\text{-}PbO_x \xrightarrow{\text{about } 375°} \beta\text{-}PbO_y \quad (y \sim 1{\cdot}5)$$
$$\beta\text{-}PbO_y \xrightarrow{\text{above } 375°} Pb_3O_4$$

A monoclinic structure is proposed for α-$PbO_{1\cdot66}$, which is probably the Pb_7O_{11} and Pb_5O_8 of earlier workers. β-$PbO_{1\cdot5}$ has an orthorhombic structure, with composition Pb_2O_3.

Hydroxy Compounds

There appears to be only one crystalline lead oxide–hydroxide, which has been formulated as $3PbO.H_2O$[73]. Comparisons of the X-ray data with the corresponding tin compound suggest that this compound is better written as $Pb_6O_8H_4$, with a structure illustrated in Fig. 5, in which an octahedral Pb_6 cluster is contained in a cube of oxygen atoms, with hydrogen bonds between all oxygen atoms[74].

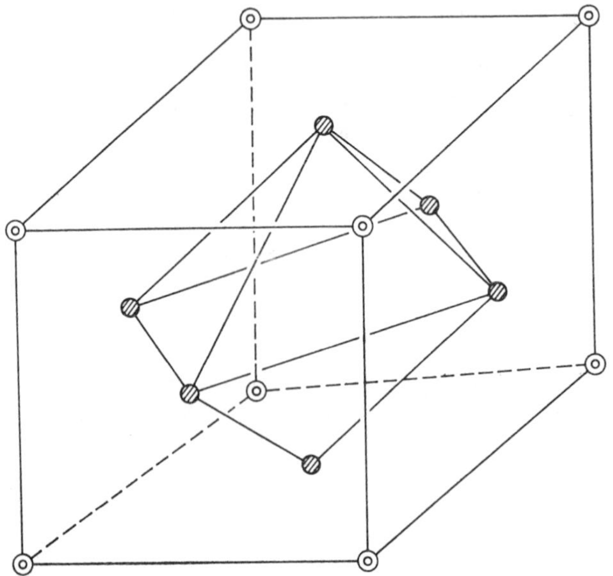

FIG. 5. Proposed cluster structure of the Pb_6O_8 units in lead oxide–hydroxide. ▨ = Pb; ◎ = O.

$Pb_6O_8H_4$ is obtained either by hydrolysis of lead(II) acetate solutions[75], or from reduced pressure evaporation of solutions of tetragonal PbO in large volumes of carbon dioxide free water[76]. Such a preparation and the proposed structure are in line with the hydroxy-lead(II)

72 G. Butler and J. L. Copp, *J. Chem. Soc.* (1956) 725.
73 G. Todd and E. Parry, *Nature*, **202** (1964) 386.
74 R. A. Howie and W. Moser, *Nature*, **219** (1968) 372.
75 F. C. Hentz and S. Y. Tyree, *Inorg. Chem.* **3** (1964) 844.
76 A. Olin, *Acta Chem. Scand.* **16** (1962) 983.

anions $Pb_3(OH)_4^{2+}$ and $Pb_6(OH)_6^{4+}$ which are reported in partially hydrolysed solutions of lead(II) salts. Such ions are likely to contain metal clusters.

When aqueous sodium hydroxide is added to a solution of lead nitrate, two basic nitrates are precipitated with stoichiometries $Pb(NO_3)_2 . Pb(OH)_2$ and $Pb(NO_3)_2 . 5Pb(OH)_2$. Even at pH = 12 (which is considerably past the equivalence point), there is no indication of precipitation of $Pb(OH)_2$ as such[77].

Many other alkali precipitations from solutions of lead(II) salts have various basic salt formulations, and are often well-characterized crystalline materials.

If a current of air is passed through a solution of lead(II) acetate during the addition of sodium hydroxide solution, the carbon dioxide content of the air is sufficient to precipitate all of the lead as hexagonal platelets with the formulation $3PbCO_3 . 2Pb(OH)_2$[78]. Another better-known basic carbonate of lead $2PbCO_3 . Pb(OH)_2$ is formed when lead sulphate or lead chloride is boiled with sodium carbonate solution.

The commercial product "white-lead" is of variable composition, but is essentially the $2PbCO_3 . Pb(OH)_2$ mentioned above. It is made by a number of processes, which give different physical characteristics such as particle size, which affect its use as a white paint pigment.

In the Dutch and German "dry" processes lead sheet is exposed to acetic acid vapour and carbon dioxide. Lead acetate, basic lead acetate and finally basic lead carbonate (white lead) appear to be formed sequentially. The so-called French, English and Hebrew "wet" processes involve the formation of a suspension or dough of basic lead acetate by dissolving lead(II) oxide in boiling lead acetate solution. The subsequent passage of carbon dioxide produces the basic lead carbonate. The formation of black lead sulphide in sulphur-containing atmospheres is a drawback in the use of $2PbCO_3 . Pb(OH)_2$ as a white pigment.

$2PbCO_3 . Pb(OH)_2$ is found as a mineral hydrocerussite (or plumbonacrite) in many parts of the world, and has also been obtained as a corrosion product of lead roofing[79].

The complex formation between Pb^{2+} and OH^- has been studied in perchlorate solutions and the following complex ions are noted[80, 81] to be present in solution before precipitation commences, $Pb(OH)_3^-$, $Pb_4(OH)_4^{4+}$, Pb_2OH^{3+}, $Pb_3(OH)_4^{2+}$, $PbOH^+$, $Pb_6(OH)_8^{4+}$, together with dissolved $Pb(OH)_2$.

The predominant ion present is $Pb_4(OH)_4^{4+}$, which is likely to have a cyclic bridged structure similar to that of $Sn_4(OH)_4^{4+}$.

The so-called metaplumbates $M_2^IPbO_3 . 3H_2O$ are made by fusing lead(IV) oxide with an excess of potassium hydroxide[82, 83]. These compounds are actually the alkali metal salts of the hexahydroxylead(IV) anion $Pb(OH)_6^{2-}$. The salts contain discrete $Pb(OH)_6^{2-}$ ions analogous to the corresponding $Sn(OH)_6^{2-}$ of tin(IV)[84, 85]. $Na_2Pb(OH)_6$ has also been made by the electrolytic oxidation of alkali hydroxide solutions of Pb(II), and also by the controlled alkali hydrolysis of lead(IV) acetate[86].

[77] J. L. Pauley and M. K. Testerman, *J. Am. Chem. Soc.* **76** (1954) 4220.
[78] H. Mauch and A. Brunhold, *Helv. Chim. Acta*, **60** (1957) 86.
[79] J. W. Mellor, *A Comprehensive Treatise on Inorganic and Theoretical Chemistry*, Vol. VII, p. 837, Longmans Green, London (1927).
[80] B. Carrell and A. Olin, *Acta Chem. Scand.* **14** (1960) 1999.
[81] A. Olin, *Acta Chem. Scand.* **14** (1960) 126 and 814.
[82] I. Bellucci and R. Parravano, *C.R. Accad. Lincei*, **14** (1905) 378.
[83] R. Scholder, in *Handbook of Preparative Inorganic Chemistry*, Vol. II, p. 1694, ed. G. Brauer, Academic Press, New York.
[84] R. L. Williams and R. J. Pace, *J. Chem. Soc.* (1957) 4143.
[85] M. Maltese and W. J. Orville-Thomas, *J. Inorg. and Nuclear Chem.* **29** (1967) 2533.
[86] A. Simon, *Z. anorg. Chem.* **177** (1929) 109.

Plumbites and Plumbates

As already mentioned in the hydroxide section above, alkaline solutions of lead contain a complex set of lead hydroxo cations; but at low lead concentrations the anion $Pb(OH)_3^-$ is present. This trihydroxyplumbite is also believed present when precipitated amphoteric lead "hydroxide" is dissolved in alkali. No evidence for the $Pb(OH)_4^{2-}$ anion has been found, though solutions of lead "hydroxide" in strong potassium hydroxide solution are often written as $K_2Pb(OH)_4$.

In addition to the hexahydroxyplumbates referred to above in the hydroxide section, there are two other types of plumbate. The alkali metal plumbates have the formulation $M_2^IPbO_3$. These are made by the carefully controlled dehydration of the $M_2^IPb(OH)_6$ salts of the alkali metals[86, 87]. If the dehydration is not carefully controlled below 110°, however, the $M_2^IPb(OH)_6$ decomposes with evolution of oxygen as well as water. Alternatively K_2PbO_3 along with $K_2Pb_3O_7$ can be made in a melt of KOH and PbO_2, and they are both isostructural with the corresponding stannates[88].

The metaplumbates of strontium and barium $M^{II}PbO_3$ are made by fusion of strontium or barium nitrates with lead(II) oxides[89]. The oxides of nitrogen released convert the lead(II) to lead(IV). Calcium orthoplumbate Ca_2PbO_4 is formed when calcium nitrate and lead(II) oxide are fused in the stoichiometry $Ca(NO_3)_2 . PbO$. The barium and strontium orthoplumbates are prepared by heating barium or strontium carbonates with lead(II) oxide in the stoichiometry $[MCO_3]_2 . PbO$; the necessary oxidation is caused by the atmosphere[90].

$BaPbO_3$ (density 8·30) has the ideal perovskite structure and $SrPbO_3$ (density 7·79) has a somewhat deformed orthorhombic perovskite structure[91].

Ba_2PbO_4 (density 7·34) is tetragonal with Pb–O distances 2·15 and 2·07 Å and BaO distances 2·66 and 2·88 Å [89].

Neither the metaplumbates nor the orthoplumbates can be regarded as containing discrete PbO_3^{2-} and PbO_4^{4-} ions respectively[90].

Oxy and Hydroxy Halides

The lead oxyhalides Pb_2OCl_2, $Pb_3O_2Cl_2$ and Pb_3OCl_4 occur in nature respectively as matlockite ($PbCl_2 . PbO$), mendipite ($PbCl_2 . 2PbO$) and penfieldite ($2PbCl_2 . PbO$).

Lead oxyfluorides of unknown composition are reported from the passage of fluorine over PbO_2, or steam over PbF_2. An oxyfluoride Pb_2OF_2 is reported to be tetragonal[92].

In addition to the oxychlorides occurring as minerals, the two compounds Pb_2OCl_2 and $Pb_3O_2Cl_2$ can be made in a number of ways, including the action of the requisite amount of alkali upon lead chloride solutions. $Pb_3O_2Cl_2$ melts without decomposition at 693°C, and the heat of formation from $PbCl_2$ and 2PbO is 9·24 kcal/mole[93].

The compositions Pb_2OBr_2, $Pb_3O_2Br_2$ and $Pb_5O_4Br_2$ have been reported for the lead oxybromides, but only Pb_2OI_2 for the oxyiodides[93].

[87] M. Baudler, in *Handbook of Preparative Inorganic Chemistry*, Vol. I, p. 758, ed. G. Brauer, Academic Press, New York.

[88] G. Foussassier, M. Tournoux and P. Hagemuller, *J. Inorg. and Nuclear Chem.* **26** (1964) 1811.

[89] R. Weiss and R. Faivre, *Compt. Rend.* **248** (1959) 106.

[90] *Nouveau traité de chimie minérale*, Vol. VIII, Part 3, p. 627, ed. P. Pascal, Masson et Cie, Paris (1963).

[91] R. Weiss, *Compt. Rend.* **246** (1958) 3073.

[92] A. Bystrom, *Arkiv. Kemi* **24A** (1947) No. 33.

[93] Reference 90, pp. 642 and 643.

The chlorite, chlorate and perchlorate together with the corresponding bromine and iodine compounds are known with various formulae $Pb_xO_yX_x$, but these are discussed below as oxy-salts, and should not be confused with the lead oxyhalides.

The lead(II) hydroxychloride Pb(OH)Cl occurs as the mineral laurionite, whose structure is closely related to that of lead(II) chloride. In Pb(OH)Cl, lead exhibits nine coordination, with five chlorine atoms (Pb–Cl = 3·23 Å) and one hydroxide group (Pb–OH = 2·67 Å) at the apices of a trigonal prism, and three further hydroxide groups in the equatorial plane beyond the centres of the prism faces[94].

An iodo-laurionite Pb(OH)I is made from a solution of lead acetate and warm lead iodide solution.

The lead(IV) oxychloride $PbOCl_2$ is prepared by the action of Cl_2O on a solution of lead(IV) chloride in carbon tetrachloride. $PbOCl_2$ (d. 5·04 g/cc) is pale violet when completely dry, and appears to be amorphous. It decomposes at 95° to lead(II) chloride and oxygen, and is very hygroscopic, undergoing hydrolysis to lead(IV) oxide and hydrogen chloride[95].

Organolead Oxides, Hydroxides and Alkoxides

The organolead hydroxides $R_2Pb(OH)_2$ and R_3PbOH are prepared by the alkaline hydrolysis of the corresponding halides. The corresponding oxides $R_3PbOPbR_3$ and $(R_2PbO)_n$ are prepared by the action of sodium upon the hydroxides. The oxides are extremely sensitive to water, undergoing rehydrolysis to the corresponding hydroxides[96, 97, 98].

The trialkyl-lead hydroxides and oxides are not easy to obtain completely pure, due to a marked tendency to decompose at room temperature. The triaryl-lead hydroxides, however, are much more stable[99].

The solvated $(CH_3)_2Pb^{2+}$ ion has been studied in aqueous solution, and it is concluded that in the $[(CH_3)_2Pb(H_2O)_n]^{2+}$ complex, the CH_3–Pb–CH_3 skeleton is linear. The equilibria outlined below are for the hydrated $(CH_3)_2Pb^{2+}$ cation under various conditions of acidity[100].

$$\underset{\text{pH} < 5}{(CH_3)_2Pb^{2+}} \underset{H^+}{\overset{OH^-}{\rightleftharpoons}} \underset{\text{pH 5–8}}{\left[(CH_3)_2Pb\langle{}^{OH}_{OH}\rangle Pb(CH_3)_2\right]^{2+}}$$

$$\text{pH 5–8 species} \quad \underset{OH^-}{\downarrow} \; \underset{H^+}{\uparrow} \quad \text{pH 8–10 species}$$

$$\underset{\text{pH} > 10}{[(CH_3)_2Pb(OH)_3]^-} \underset{H^+}{\overset{OH^-}{\rightleftharpoons}} \underset{\text{pH 8–10}}{(CH_3)_2Pb(OH)_2}$$

---

[94] H. Brasseur, *Bull. Soc. roy. sci. Liege*, **9** (1940) 166.
[95] K. Dehnicke, *Naturwiss.* **51** (1964) 535.
[96] Yu. A. Aleksandrov, T. G. Brilkina and V. A. Shushnov, *Tr. Khim. i. Khim. Tekhnol.* **2** (1959) 623; *C.A.* **56** (1962) 14314c.
[97] T. G. Brilkina, M. K. Safonova and V. A. Shushnov, *Zhur. Obschei. Khim.* **32** (1962) 2684; *C.A.* **58** (1963) 9112b.
[98] Yu. A. Aleksandrov, T. G. Brilkina and V. A. Shushnov, *Dokl. Akad. Nauk SSSR*, **136** (1961) 89.
[99] R. West, R. H. Baney and D. L. Powell, *J. Am. Chem. Soc.* **82** (1960) 6269.
[100] C. D. Freidline and R. S. Tobias, *J. Organometallic Chem.* **6** (1966) 535.

The ozonization of tetra-ethyl-lead at $-68°$ gives an unidentified product which decomposes at room temperature to produce $(C_2H_5)_3PbOH$, $[(C_2H_5)_2PbO]_n$, $(C_2H_5)_3PbOC_2H_5$ together with acetaldehyde and ethanol[101]. Good yields of the oxides $R_3PbOPbR_3$ are obtained by the action of ozone upon the hexa-organodileads R_3PbPbR_3.

With alcohols some alkyl-leads undergo fission.

$$R_4Pb + R'OH \rightarrow R_3PbOR' + RH$$

The alcohol reactivity falls off steadily from $R'OH = CH_3OH$ to $R'OH = C_4H_9OH$, which is unreactive[102].

The best preparation of organolead alkoxides involves the metathesis between sodium alkoxides and the organolead halides[103].

Trimethyl-lead methoxide $(CH_3)_3PbOCH_3$ is polymeric[104], as shown in Fig. 6. The properties of the lead–oxygen bond in the organolead alkoxides and the organolead oxides

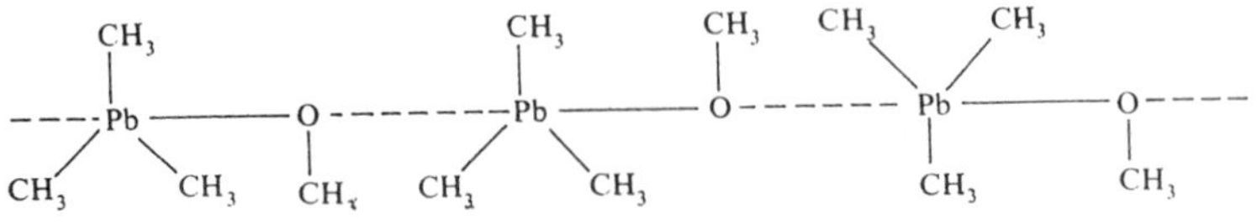

FIG. 6. Polymeric structure of $(CH_3)_3PbOCH_3$ in the crystal.

have been investigated and many reactions of the "insertion" type are known, as illustrated by the equations below.

$$(C_6H_5)_3PbOCH_3 + CS_2 \rightarrow (C_6H_5)_3Pb\text{–}S\text{–}\underset{\underset{S}{\|}}{C}\text{–}OCH_3$$

$$(C_4H_9)_3PbOCH_3 + Cl_3CCHO \rightarrow (C_4H_9)_3PbO\text{–}\underset{\underset{CCl_3}{|}}{CH}\text{–}OCH_3$$

$$(C_6H_5)_3PbOPb(C_6H_5)_3 + Cl_3CCHO \rightarrow (C_6H_5)_3PbO\underset{\underset{CCl_3}{|}}{C}HOPb(C_6H_5)_3$$

$$(C_6H_5)_3PbOCH_3 + C_6H_5NCS \rightarrow (C_6H_5)_3Pb\text{–}S\text{–}\underset{\underset{N\text{–}C_6H_5}{\|}}{C}\text{–}O\text{–}CH_3$$

$$(C_6H_5)_3PbOCH_3 + (CCl_3)_2CO \rightarrow (C_6H_5)_3PbO\overset{\overset{CCl_3}{|}}{\underset{\underset{CCl_3}{|}}{C}}OCH_3$$

Many of the complexes of lead such as lead(II) oxinate which are used in lead analyses presumably contain a lead–phenoxide type of linkage, but no structural data are available.

Triphenyl-lead 8-oxyquinolinate is reported to have lead four-coordinate in ethanol but five-coordinate in benzene solution (see Fig. 7 (a) and (b))[105]. In monophenyl-lead

101 Yu. A. Aleksandrov and N. G. Sheyanov, *Zhur. Obshchei. Khim.* **36** (1966) 953.
102 M. Pedinelli, R. Magri and M. Raudi, *Chem. Ind. (Milan)*, **48** (1966) 144.
103 A. G. Davies and R. J. Puddephatt, *J. Organometallic Chem.* **5** (1966) 590.
104 E. Amberger and R. Honigschmid-Grossich, *Chem. Ber.* **98** (1965) 3795.
105 L. Roncucci, G. Faraglia and R. Barbieri, *J. Organometallic Chem.* **6** (1966) 278.

tris(8-oxyquinolinate) only one of the nitrogen atoms can be quaternized by methyl iodide, suggesting the structure in Fig. 7 (c) in which lead is six-coordinate[106].

(a)

(b)

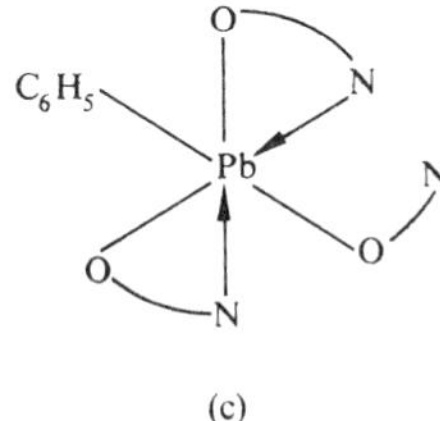

(c)

FIG. 7. (a) Triphenyl-lead 8-oxyquinolinate in ethanol solution. (b) Triphenyl-lead 8-oxyquinolinate in benzene solution. (c) Monophenyltris(8-oxyquinolinate). (ON represents the oxygen and nitrogen coordination points in 8-oxyquinolinate.)

Carboxylates

The best-known carboxylate of lead(II) is lead(II) acetate, $(CH_3COO)_2Pb.3H_2O$ (sugar of lead), made by dissolving lead(II) oxide in acetic acid. The salt is very soluble in water (1 g dissolves in about 1·6 ml of cold water or 0·5 ml of boiling water). Solutions are only partly ionized[107], which accounts for the strong solvent effect of acetates upon lead compounds. (Aqueous solutions of lead(II) acetate freely dissolve lead(II) oxide.)

$$Pb^{2+} + CH_3COO^- \rightarrow CH_3COOPb^+ \qquad K_1 = 145$$
$$Pb^{2+} + 2CH_3COO^- \rightarrow (CH_3COO)_2Pb \qquad K_2 = 810$$
$$Pb^{2+} + 3CH_3COO^- \rightarrow (CH_3COO)_3Pb^- \qquad K_3 = 2950$$

Lead acetate solutions absorb carbon dioxide from the air and precipitate insoluble carbonates.

Many other lead(II) carboxylates are known such as formate, propionate, butyrate, stearate, oxalate, tartrate, etc., many of which find industrial uses as lubricant additives, varnish driers, silk weighing, etc.[108].

[106] F. Huber and H. J. Haupt, *Z. Naturforsch.* **21b** (1966) 808.
[107] E. A. Burns and D. N. Hume, *J. Am. Chem. Soc.* **78** (1956) 3958.
[108] *The Merck Index*, 8th ed., Merck Sharpe and Dohm, Rahaway, New Jersey.

Lead(IV) acetate is best made by the action of glacial acetic acid upon Pb_3O_4. The colourless crystals are very sensitive to moisture, hydrolysing to lead(IV) oxide and acetic acid[109, 110].

It is very extensively used as a selective oxidizing agent in organic chemistry.

Other lead(IV) carboxylates such as propionate, butyrate and stearate are known.

Organolead(IV) carboxylates of general formulae R_3PbO_2CR', $R_2Pb(O_2CR')_2$ and $RPb(O_2CR')_3$ are known. The equations below indicate some of the syntheses available[111–115].

$$R_4Pb + R'COOH \rightarrow R_3PbO_2CR'$$
$$(C_2H_5)_4Pb + (CH_3COO)_4Pb \rightarrow (C_2H_5)_3PbO_2CCH_3$$
$$(C_2H_5)_6Pb_2 + (C_6H_5COO)_2 \rightarrow (C_2H_5)_3PbO_2CC_6H_5$$
$$R_3PbX + R'COOAg \rightarrow R_3PbO_2CR'$$
$$R_3PbOH + R'COOH \rightarrow R_3PbO_2CR'$$
$$R_2PbO + R'COOH \rightarrow R_2Pb(O_2CR')_2$$
$$R_2PbX_2 + R'COOAg \rightarrow R_2Pb(O_2CR')_2$$
$$R_2Pb(O_2CR')_2 + Hg(O_2CR')_2 \rightarrow RPb(O_2CR')_3$$
$$(CH_3COO)_4Pb + R_2Hg \rightarrow RPb(O_2CCH_3)_3$$
$$RPbO_2H + R'COOH \rightarrow RPb(O_2CR')_3$$

Very many organolead carboxylates have been investigated for potential commercial uses. Little structural information is available for these compounds, but the trimethyl-lead carboxylates are believed to be coordination polymers[116, 117] in which planar trimethyl-lead ions are joined by carboxylate groups as in Fig. 8, producing five-coordinate lead.

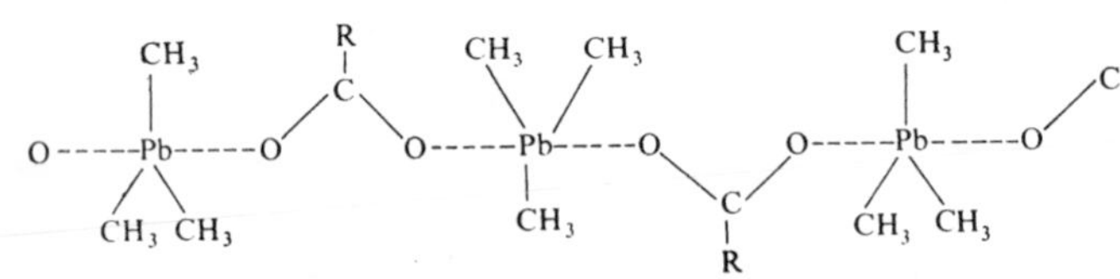

FIG. 8. Polymeric trimethyl-lead carboxylates.

Organolead peroxides of type R_3PbOOR and $R_3PbOOPbR_3$ exist and are reasonably stable[118, 119], but the peroxycarboxylates R_3PbO_3CR are very unstable and are not well characterized[120].

109 M. Baudler, in *Handbook of Preparative Chemistry*, 2nd ed., Vol. I, p. 767, ed. G. Brauer, Academic Press, New York (1963).
110 H. Mendel, *Rec. trav. Chim.* **59** (1940) 720.
111 H. Gilman, S. M. Spatz and M. J. Kobbezen, *J. Org. Chem.* **18** (1953) 1341.
112 B. C. Saunders and G. J. Stacey, *J. Chem. Soc.* (1949) 919.
113 R. Heap and B. C. Saunders, *J. Chem. Soc.* (1949) 2983.
114 M. M. Koton, *Zhur. Obshchei. Khim.* **11** (1941) 376.
115 K. A. Kocheshkov and E. M. Panov, *Izv. Akad. Nauk SSSR* (1955) 711.
116 R. Okawara and H. Sato, *J. Inorg. Nuclear Chem.* **16** (1961) 204.
117 M. J. Janssen, J. G. A. Luijten, G. J. M. van der Kerk, *Rec. trav. Chim.* **82** (1963) 90.
118 A. Rieche and J. Dahlmann, *Ann. Chem. Liebigs*, **675** (1964) 19.
119 Yu. A. Aleksandrov and T. G. Brilkina, *Dokl. Akad. Nauk SSSR*, **129** (1959) 321.
120 V. A. Shushnov and T. G. Brilkina, *Dokl. Akad. Nauk SSSR*, **141** (1961) 1391.

2.5. LEAD SALTS OF INORGANIC ACIDS

Sulphates, Selenates, Fluorosulphonates and Chlorosulphonates

Lead(II) sulphate is precipitated from lead(II) solutions by sulphate ions, and is one of the popular gravimetric precipitates for lead. Its increased solubility in concentrated sulphuric acid suggests formation of sulphato or bisulphato complexes of lead(II). Lead selenate is precipitated similarly from Pb^{++} and SeO_4^{--} mixing.

Lead(IV) sulphate may be prepared by a number of methods. Concentrated sulphuric acid on either lead(IV) fluoride[121] or lead(IV) acetate[122] produces $Pb(SO_4)_2$; and electrolysis of strong sulphuric acid with lead electrodes gives $Pb(SO_4)_2$ under carefully controlled conditions[123]. $Pb(SO_4)_2$ is stable in dry air, but is hydrolysed to PbO_2 by moisture[123]

If an acid solution of $Pb(SO_4)_2$ is treated with alkali sulphate, salts such as $K_2Pb(SO_4)_3$ are obtained. No structural data are available for these salts, but they are likely to contain the $[Pb^{IV}(SO_4)_3]^{2-}$ anion [124].

Lead(IV) acetate dissolves in concentrated sulphuric acid[125] to form hexa(hydrogensulphato)plumbic acid $H_2Pb(HSO_4)_6$, and the ions $[HPb(HSO_4)_6]^-$ and $[Pb(HSO_4)_6]^{2-}$. Lead(IV) acetate is also soluble in selenic, fluorosulphuric and chlorosulphuric acids, giving yellow solutions which may contain $H_2Pb(HSeO_4)_6$, $H_2Pb(FSO_3)_6$ and $H_2Pb(ClSO_3)_6$ respectively.

The organometallic sulphates of lead(IV) are stable crystalline compounds made by the action of sulphuric acid upon the requisite organometallic oxide or hydroxide.

Nitrates

Lead(II) nitrate is normally made by the action of nitric acid upon lead metal or lead(II) oxide. It is freely soluble in water, but contrary to the behaviour of most nitrates, it is incompletely dissociated in solution[126, 127]. The $PbNO_3^+$ cation has a formation constant of 15·1 at infinite dilution at 25°C.

$$Pb^{2+} + NO_3^- \rightarrow PbNO_3^+ \qquad (K = 15{\cdot}1)$$

The formation constant of the reaction at ionic strength 2·0 is approximately 2·6.

A number of basic nitrates of lead are known, and are discussed in the hydroxy section above.

Efforts to prepare lead(IV) nitrate or hexanitrato complexes of lead(IV) have, to date, been unsuccessful[128]. The organometallic nitrates of lead(IV) are, however, well known and crystalline specimens of such nitrates as $(C_2H_5)_3PbNO_3$, $(CH_3)_3PbNO_3$, $(C_6H_5)_3PbNO_3$, $(C_2H_5)_2Pb(NO_3)_2$, $(C_6H_5)_2Pb(NO_3)_2.2H_2O$ and $(p\text{-}CH_3.C_6H_4)_2Pb(NO_3)_2.3H_2O$ can be prepared by a variety of reactions.

$$(C_6H_5)_4Pb + HNO_3 \rightarrow (C_6H_5)_2Pb(NO_3)_2.2H_2O$$

$$(C_2H_5)_4Pb + HNO_3 \rightarrow (C_2H_5)_2Pb(NO_3)_2$$

$$(C_{12}H_{25})_3PbCl + AgNO_3 \rightarrow (C_{12}H_{25})_3PbNO_3$$

$$(C_6H_5)_4Pb + Cu(NO_3)_2 \rightarrow (C_6H_5)_3PbNO_3$$

---

[121] B. Brauner, *Z. anorg. Chem.* **7** (1894) 11.
[122] A. Hutchinson and W. Pollard, *J. Chem. Soc.* **69** (1896) 221.
[123] K. Elbs and F. Fischer, *Z. Electrochem.* **7** (1901) 343.
[124] R. J. Gillespie, R. Kapoor and E. A. Robinson, *Canad. J. Chem.* **44** (1966) 1197.
[125] R. J. Gillespie and E. A. Robinson, *Proc. Chem. Soc.* (1957) 145.
[126] G. H. Nancollas, *J. Chem. Soc.* (1955) 1458.
[127] H. M. Hershenson, M. E. Smith and D. N. Hume, *J. Am. Chem. Soc.* **75** (1953) 507.
[128] K. W. Bagnall, D. Brown and J. G. H. Du Preez, *J. Chem. Soc.* (1964) 5523.

The Raman spectrum of an aqueous solution of dimethyl-lead dinitrate suggests that the dimethyl-lead cation has undissociated nitrate ion as well as water in its coordination sphere[129].

The action of N_2O_4 on lead tetra-alkyls produces the compounds formulated as $[R_4Pb(NO)_2](NO_3)_2$, in which the octahedral complex ion $[R_4Pb(NO)_2]^{2+}$ is believed present. Although the stoichiometry of this series of compounds is well established, the structural formulation seems very dubious[130].

Oxyhalogen Salts

Lead(II) chlorite, chlorate, bromate, iodate and perchlorate are the known oxyhalogen salts of lead. $Pb(ClO_2)_2$ is made by the action of $Ba(ClO_2)_2$ solution upon lead nitrate solution[131]. The yellow crystals have a pseudotetragonal cell ($a = 4{\cdot}14$ Å, $c/a = 1{\cdot}51$)[132].

Lead(II) chlorite decomposes upon heating to a mixture of lead(II) chloride and lead(II) perchlorate, or alternatively to oxygen and lead(II) chloride depending upon the rate and intensity of heating.

Lead(II) chlorate is obtained as the hydrate $Pb(ClO_3)_2 . H_2O$ from the dissolution of lead carbonate in aqueous chloric acid. The anhydrous chlorate is got by careful heating of the hydrate at 110–150°, but at higher temperatures decomposition to lead(II) chloride, lead(IV) oxide, chlorine and oxygen occurs, and the breakdown can be explosive.

Solutions of lead acetate and potassium bromate precipitate the less soluble lead bromate. It is possible for the double salt $Pb(BrO_3)_2 . Pb(OCOCH_3)_2$ to be formed in this preparation, and this double salt explodes violently at about 165°, or even at room temperature upon percussion. Two fatal accidents have occurred by explosions of $Pb(BrO_3)_2$ due to the double salt impurity present, and lead(II) bromate prepared in this way should be treated with extreme care[133]. An alternative synthesis involves treatment of lead carbonate with bromic acid to form $Pb(BrO_3)_2 . H_2O$ [134].

Lead iodate precipitates as the anhydrous salt from lead(II) solutions and alkali iodate solutions. It is thermally more stable than the chlorate or bromate, and only decomposes at 250–300°.

The variation of the aqueous solubilities of lead(II) chlorate, bromate and iodate are remarkable. 1 ml of water at 18°C will dissolve 4·4 g of $Pb(ClO_3)_2$, $1{\cdot}3 \times 10^{-2}$ g of $Pb(BrO_3)_2$ and $1{\cdot}9 \times 10^{-5}$ g of $Pb(IO_3)_2$.

$Pb(ClO_4)_2 . 3H_2O$ (m.p. 84°) is made by the dissolution of lead carbonate in aqueous perchloric acid[135]. The monohydrate m.p. 153–155° is also known.

Phosphites and Phosphates

The phosphate of lead(IV) formulated as $Pb(H_2PO_4)_4$ was synthesized electrolytically from lead electrodes and strong phosphoric acid[136]. It is formed as white crystals which are stable in the air, and it behaves as a slow but fairly strong oxidizing agent, releasing iodine from potassium iodide.

129 C. E. Freidline and R. S. Tobias, *Inorg. Chem.* **5** (1966) 354.
130 B. Hetnaiski and T. Urbanski, *Tetrahedron*, **19** (1963) 1319.
131 G. Lasegue, *Bull. Soc. Chim. (France)*, **11** (1912) 884.
132 C. R. Levi and A. Scherillo, *Z. Krist.* **76** (1931) 431.
133 Victor, *Z. angew. Chem.* **40** (1927) 841; *C.A.* (1927) 3324.
134 C. F. Rammelsberg, *Ann. Physik. Chem.* **44** (1838) 566.
135 H. F. Roscoe, *Ann. Chem. Liebigs*, **121** (1862) 356.
136 K. Elbs and R. Nubling, *Z. Electrochem.* **9** (1903) 776.

The corresponding monoacid phosphate of lead(IV), $Pb(HPO_4)_2$, was made by the action of phosphoric acid upon lead(IV) acetate[137].

Lead(II) hypophosphite $Pb(H_2PO_2)_2$ was prepared by dissolution of lead(II) oxide[138] or lead carbonate[139] in hypophosphorous acid, or alternatively by metathesis between lead(II) nitrate and calcium hypophosphate[140]. It is slightly soluble in cold water to give a feebly acidic solution, but is much more soluble in hot water. On treatment of $Pb(H_2PO_2)_2$ with a strong solution of lead nitrate, it forms the nitrohypophosphite $Pb(H_2PO_2)NO_3$, which behaves as a detonant[140].

The neutral hypophosphite $Pb_2P_2O_6$ has been characterized by its infrared spectrum[141] over the region 4000–650 cm^{-1}.

The phosphite $PbHPO_3$ may be obtained by the metathesis of lead nitrate or lead acetate with alkalic phosphite[142], or alternatively by the partial neutralization of phosphorous acid solution with lead carbonate[143]. When $PbHPO_3$ is dissolved in a warm concentrated solution of phosphorous acid the acid phosphite $Pb(H_2PO_3)_2$ is obtained, and can be isolated as large transparent crystals[142].

Certain basic phosphites formulated as $2PbHPO_3.2PbO.H_2O$, $PbHPO_3.2PbO.\frac{1}{2}H_2O$ and $PbHPO_3.PbO.\frac{1}{2}H_2O$ have also been reported[139, 143, 144], though the stoichiometries may be dubious. In addition, halogenophosphites and nitratophosphite $PbPO_3F$ [145], $PbHPO_3.PbCl_2.H_2O$, $PbHPO_3.2PbBr_2.H_2O$ [146] and $PbHPO_3.Pb(NO_3)_2$ are claimed[142].

Melts of phosphorus(V) oxide and lead(II) oxide containing less than about 65% molar of lead(II) oxide form glasses, and are not discussed with the discrete phosphates of lead outlined below.

Lead(II) metaphosphates formulated as dimetaphosphate $Pb(PO_3)_2$, trimetaphosphate $\frac{1}{2}[Pb_3(PO_3)_6.3H_2O]$, tetrametaphosphate $Pb_2(PO_3)_4$ and hexametaphosphate $Pb_3(PO_3)_6$ have been claimed. Preparations usually involve the action of alkali metaphosphate upon soluble lead(II) salts, but conditions appear to be critical for obtaining the right product[147, 148, 149]. Fusion methods involving lead nitrate and phosphorus(V) oxide have also been used[150].

The three orthophosphates of lead are well characterized, and have been extensively studied.

$PbHPO_4$ exists in nature as the mineral monetite (density 5·66 g/cc), which is monoclinic (a = 4·66 Å, b = 6·64 Å, c = 5·77 Å, β = 83°), the unit cell containing two $PbHPO_4$ units[151].

137 A. Hutchison and W. Pollard, *J. Chem. Soc.* **69** (1896) 212.
138 H. Rose, *Ann. Physik. Chem.* **12** (1828) 288.
139 M. W. Lotz, *Ann. Chim. Phys.* [3] **43** (1865) 250.
140 E. von Herz, *Chem. Zentr.* (1919) 271.
141 D. E. C. Corbridge and E. J. Lowe, *J. Chem. Soc.* (1954) 493.
142 L. Amat, *Ann. Chim. Phys.* [6] **24** (1891) 315.
143 H. Rose, *Ann. Physik. Chem.* **9** (1827) 42 and 221.
144 L. M. Kebrich, U.S. Patent 2,483,469 (1949); *C.A.* **44** (1950) 2256.
145 D. E. C. Corbridge and E. J. Lowe, *J. Chem. Soc.* (1954) 4555.
146 R. Weinland and F. Paul, *Z. anorg. Chem.* **129** (1923) 243.
147 T. Fleitmann, *Ann. Physik. Chem.* **78** (1849) 253.
148 G. Knorre, *Z. anorg. Chem.* **24** (1900) 369.
149 A. Travers and Yu Kwong Chu, *Compt. Rend.* **198** (1934) 2169.
150 F. Warschauer, *Z. anorg. Chem.* **36** (1903) 137.
151 E. Bengtsson, *Arkiv. Kemi, Min. Geol.* **15B** (1941) No. 7.

$PbHPO_4$ is only slightly soluble in water, and disproportionates to the phosphate $Pb_3(PO_4)_2$ and phosphoric acid on boiling in water. $PbHPO_4$ is transformed to the pyrophosphate $Pb_2P_2O_7$ on heating. The decomposition starts at about 195° and probably proceeds via the tetrametaphosphate of lead.

$PbHPO_4$ may be synthesized from lead nitrate and phosphoric acid in the right conditions[152].

Phosphoric acid dissolves $PbHPO_4$ to form $Pb(H_2PO_4)_2$ [152]. This lead diacid phosphate is stable in the air but decomposed by water. A number of compounds of the type $Pb(HRPO_4)_2$, where R is CH_3, C_2H_5, C_3H_7, etc., are also known[153].

Lead orthophosphate $Pb_3(PO_4)_2$ may be prepared in a number of ways. $Pb_3(PO_4)_2$ is the product of the neutralization of $PbHPO_4$ with ammonia[154], and lead(II) chloride and Na_2HPO_4 give $Pb_3(PO_4)_2$ [155]. The solubility of lead orthophosphate in water is $1{\cdot}35 \times 10^{-4}$ g/l. at 20°. $Pb_3(PO_4)_2$ is dimorphic[156] with a transition point at 782°, and it melts at 1104°.

Various apatites of lead are known[156–160] with the general stoichiometry $3[Pb_3(PO_4)_2].PbX_2$ where X = F, Cl, Br, I and OH. These apatites of lead form a homogeneous crystallographic group in the same way as the better-known apatites of calcium. The chloroapatite $3[Pb_3(PO_4)_2].PbCl_2$ occurs in nature as the mineral pyromorphite (density 7·01 g/cc, dielectric constant 47·5 at λ = 75 cm).

A mixed hydroxyapatite of calcium and lead $Ca_5PB_5(PO_4)_6(OH)_2$ (density 5·39 g/cc) is hexagonal with a = 9·62 Å and c = 7·08 Å[161].

Many arsenites and arsenates comparable to the above phosphites and phosphates have been characterized[162].

Carbonates

Lead carbonate $PbCO_3$ occurs in nature as the mineral cerusite. It may be prepared by the action of alkali carbonate upon a solution of lead acetate or nitrate, but the preparation must be kept cold in order to avoid formation of the basic carbonate. Another method of preparation of the neutral carbonate involves shaking a cold solution of sodium carbonate with a suspension of a salt of lead which has a low solubility, but which is not as insoluble as lead carbonate itself[163]. Chloride, bromide and sulphate have been utilized in this context.

$PbCO_3$ as cerusite is orthorhombic and is isomorphous with aragonite $CaCO_3$, strontianite $SrCO_3$ and witherite $BaCO_3$.$PbCO_3$ has a density 6·58 g/cc and a unit cell a = 5·195 Å, b = 8·436 Å and c = 6·152 Å[164].

Lead carbonate appears to undergo slight decomposition on exposure to strong light and is decomposed to carbon dioxide and lead(II) oxide on heating; thus at 184°, for example, it

152 H. Alders and A. Stahler, *Chem. Ber.* **42** (1909) 2261.
153 J. Cavalier and E. Prost, *Bull. Soc. Chim.* [3] **23** (1900) 678.
154 J. J. Berzelius, *Ann. Chim. Phys.* **2** (1816) 258.
155 L. T. Fairhall, *J. Am. Chem. Soc.* **46** (1924) 1593.
156 A. Ferrari, *Gazz. chim. Ital.* **70** (1940) 457.
157 M. Amadori, *Gazz. chim. Ital.* **49** (1919) 38.
158 A. Ditte, *Compt. Rend.* **96** (1883) 846.
159 H. St. Claire Deville and H. Caron, *Ann. Chim. Phys.* **67** (1863) 451.
160 R. Klement, *Z. anorg. Chem.* **237** (1938) 161.
161 M. Muller, *Helv. chim. Acta*, **30** (1947) 2069.
162 *Nouveau traité de chimie minérale*, Vol. VIII, Part 3, pp. 740 et seq., ed. P. Pascal, Masson et Cie, Paris (1963).
163 W. Herz, *Z. anorg. Chem.* **72** (1911) 106.
164 H. E. Swanson and R. K. Fuyat, National Bureau of Standards U.S.A. Circular, No. 539, Vol. 2 (1953).

has a dissociation pressure of 10 mm of mercury[165]. Decomposition starts at about 130° and is complete at about 470° [166].

$PbCO_3$ has a solubility product in pure water of $3{\cdot}3 \times 10^{-14}$, but the solubility is much increased by the presence of carbon dioxide in the water, even in quite small amounts[167].

The halogenocarbonates $PbCl_2.PbCO_3$ and $PbBr_2.PbCO_3$ are isomorphous[168, 169], but the corresponding iodide could not be made. The chlorocarbonate $PbCl_2.PbCO_3$ is present in nature as the mineral phosgenite, and can be made by the action of phosgene upon lead hydroxide or by the action of carbon dioxide upon a solution of lead chloride. Natural and synthetic phosgenite have been shown to have identical crystal structures.

In the thermal decomposition of lead carbonate it is believed that the basic carbonates $PbO.PbCO_3$, $2PbO.PbCO_3$ and $3PbO.5PbCO_3$ are formed as intermediates[166, 170], but the very well-characterized basic carbonate is the hydroxycarbonate $Pb(OH)_2.2PbCO_3$, the mineral hydrocerusite. The basic carbonate $Pb(OH)_2.2PbCO_3$ can be made by the action of urea and water upon basic lead acetate. In a sealed tube at 130° the urea is converted to ammonium carbonate, which precipitates $Pb(OH)_2.2PbCO_3$ as fine hexagonal crystals[171]. The hydroxycarbonate $3PbCO_3.2Pb(OH)_2$ has been characterized by X-ray crystallography, and is made by passing air into an alkaline solution of lead acetate[172].

2.6. CHALCOGEN COMPOUNDS

Sulphides

Lead(II) sulphide crystallizes in a perfect NaCl structure[173] with Pb–S distance 2·97 Å, and is the only binary sulphide of lead. It is found in nature as large regular crystals of galena. It has a grey-blue metallic character and density about 7·6. It is best synthesized in a crystalline form by the action of thiourea upon sodium plumbite solution[174].

PbS is a *p*-type semiconductor when sulphur rich and an *n*-type semiconductor when lead rich. A conductivity of $1{\cdot}3 \times 10^{-4}\ \Omega^{-1}\ cm^{-1}$ is found for stoichiometric PbS [175]. Crystalline PbS is a remarkable radio detector, and was widely used in early crystal radio receivers.

Lead sulphide sublimes unchanged[176] and has a vapour pressure of about 20 mm of mercury at 1000°C, and boils at about 1200°C. It is exceedingly insoluble in water (solubility product about 8×10^{-28} at 25°C)[177], but it is attacked by strong acids.

Very many organolead compounds containing covalent lead–sulphur bonds are known

165 A. Colson, *Compt. Rend.* **140** (1905) 865.
166 A. Nicol, *Compt. Rend.* **226** (1948) 670.
167 W. Bottger, *Z. physik. Chem.* **46** (1903) 521.
168 L. G. Sillen and R. Petterson, *Naturwiss.* **32** (1944) 41; *Arkiv. Kemi, Min. Geol.* **21A** (1946) No. 13.
169 I. Oftedal, *Norsk. Geol. Tidsskr.* **24** (1945) 79.
170 G. A. Collins and A. G. Swan, *Canad. Min. Met. Bull.* **508** (1954) 533; *C.A.* **48** (1954) 12620f.
171 L. Bourgeois, *Compt. Rend.* **106** (1888) 1641.
172 H. Mauch and A. Brunhold, *Helv. Chim. Acta*, **40** (1957) 86.
173 B. Wasserstein, *Am. Mineral*, **36** (1951) 102; *C.A.* **47** (1953) 6824.
174 J. Emerson-Reynolds, *J. Chem. Soc.* **45** (1884) 162.
175 H. Hinterberger, *Z. Physik.* **119** (1942) 1.
176 R. Schenk and A. Albers, *Z. anorg. Chem.* **105** (1919) 164.
177 J. R. Goates, M. B. Gordon and N. D. Faux, *J. Am. Chem. Soc.* **74** (1952) 835.

with general formulae R_3PbSR', $R_2Pb(SR')_2$, $R_3PbSPbR_3$, etc., in which lead(IV) is present. Some synthetic routes to these types are outlined below[178].

$$(C_2H_5)_3PbOH + (NH_2)_2CS \rightarrow (C_2H_5)_3PbSPb(C_2H_5)_3$$
$$R_3PbOH + R'SH \rightarrow R_3PbSR'$$
$$(C_2H_5)_4Pb + S \rightarrow (C_2H_5)_3PbSPb(C_2H_5)_3$$
$$(C_6H_5)_3PbSLi + (C_6H_5)_3SnCl \rightarrow (C_6H_5)_3PbSSn(C_6H_5)_3$$

In addition to these linear compounds, the heterocyclic ring $(C_6H_5)_2\overline{PbSCH_2CH_2S}$ contains lead–sulphur bonds.

All these sulphur derivatives of lead(IV) are considerably less stable than the corresponding compounds of tin, germanium and silicon[178].

Lead, however, is unique in Group IVB in forming stable lead(II) thiolates. These are formed instantly as yellow crystalline precipitates by the direct action of thiols upon solutions of lead(II) salts. These lead mercaptides form yellow crystals with sharp melting points which are soluble in many organic solvents[179]. Whilst no structural data are available, these compounds appear to be covalent thiolates of lead(II).

Lead(II) mercaptides have been widely used as intermediates.

$$(CH_3S)_2Pb + CH_3COCl \rightarrow CH_3COSCH_3$$
$$(RS)_2Pb + BrCH_2CH_2COONa \rightarrow RSCH_2CH_2COONa$$

They are oxidized in the air or by iodine to produce the corresponding disulphides, but nitric acid produces the corresponding sulphonic acid.

$$(RS)_2Pb + O \rightarrow R_2S_2 + PbO$$
$$(RS)_2Pb + I_2 \rightarrow R_2S_2 + PbI_2$$

At about 200° the lead(II) mercaptides disproportionate to the organic sulphide and lead sulphide[179].

$$(RS)_2Pb \rightarrow R_2S + PbS$$

Selenides

Thermal and X-ray analysis, and other physical methods have shown that PbSe is the only compound formed in the lead–selenium system[180]. PbSe is a lead-grey compound, m.p. 1065°, with a cubic NaCl lattice (a = 6·12 to 6·15 Å from various determinations). PbSe occurs in nature as the mineral clausthalite, which is isomorphous with galena. Lead selenide is synthesized by the interaction of stoichiometric quantities of the elements at 1200–1250°, with subsequent annealing in argon at 800° [181]. It can also be made by the action of hydrogen selenide upon lead salts, but some free selenium is liberated during the reaction. Lead selenide is precipitated from a solution of lead acetate and selenourea in the presence of hydrazine. Purification of PbSe can be carried out by vacuum distillation at 10^{-6} mm Hg, or alternatively by zone refining. When PbSe crystals are grown from a melt containing a small excess of selenium or lead, the semiconductor (5×10^{-2} to 5×10^{-3} ohm cm) single crystals will have respectively hole or electron conductivity[180].

Films of lead selenide up to 1000 Å thick are completely oxidized to $PbSeO_3$ in air at 350° after 10 min; bulk samples oxidize more slowly. The photoconductivity of lead selenide is sensitive to the content of oxygen and other elements in the deposited film[180]. Photosensitivity is in the range 3–10 μ, and increases with reduction of temperature[182].

178 E. W. Abel and D. A. Armitage, *Adv. Organometallic Chem.* **5** (1967) 1.

179 E. E. Reid, *Organic Chemistry of Bivalent Sulphur*, Vol. 1, p. 151, Chemical Publ. Co., New York (1958).

180 D. M. Chizhikov and V. P. Shchastlivyi, *Selenium and Selenides*, Collets Press, London (1968).

181 N. Kh. Abrikosov and E. I. Elagina, *Dokl. Akad. Nauk SSSR*, **111** (1956) 353.

182 A. Roberts and J. Baines, *Phys. Chem. Solids*, **6** (1954) 184.

Few organic lead–selenium compounds are known, but organolead selenolates such as $(CH_3)_3PbSeR$ are known, as also are examples of the hexa-aryldilead selenides $(R_3Pb)_2Se$ [178]. The lithium salts $(C_6H_5)_3PbSeLi$ are formed in an interesting manner by the direct attack of selenium upon the lithium triaryl-lead.

$$(C_6H_5)_3PbLi + Se \rightarrow (C_6H_5)_3PbSeLi$$

Tellurides

PbTe is the sole telluride of lead and occurs in nature as the mineral altaite. It may be synthesized by high-temperature reaction of stoichiometric quantities of the elements, or by the action of tellurium powder on a boiling solution of lead salt.

PbTe melts at 860°. It behaves as a semiconductor of the *n*- and *p*-types depending upon which component is present in slight excess. PbTe displays photoconductivity at low temperatures[183].

The mixed telluride $Pb_xSn_{1-x}Te$ shows potential as a tuneable long-wavelength laser under variable temperature and pressure[184].

Organolead tellurides are very rare and appear to be currently restricted to $[(C_6H_5)_3Pb]_2Te$ (m.p. 129°) and its lithium salt $(C_6H_5)_3PbTeLi$ [178].

2.7. HALOGENO COMPOUNDS

Lead(II) Halides

Lead(II) fluoride is obtained as a white crystalline powder by dissolving lead carbonate in hydrofluoric acid and decomposing the hydrofluoride formed by rapid melting. Alternatively it is prepared from lead metal and anhydrous hydrofluoric acid at 160° under autogenous pressure.

Lead(II) fluoride is dimorphic, the low-temperature rhombic PbF_2 form having the $PbCl_2$ structure and the high-temperature (above 316°C) form having the cubic fluorite structure[185].

Lead(II) chloride, bromide and iodide are all prepared by addition of the requisite halide to solutions of lead(II) ion. Lead(II) chloride and bromide are white like the fluoride, and lead(II) iodide forms bright yellow very characteristic hexagonal plates. The solubilities of the lead(II) halides in water increase markedly on heating.

In crystalline $PbCl_2$ and $PbBr_2$ each lead atom is coordinated by nine halide ions, six of which lie at the apices of a trigonal prism, with the remaining three beyond the centres of the three prism faces. PbI_2 has the CdI_2 layer structure[186].

Electron diffraction measurements of the vapours of $PbCl_2$, $PbBr_2$ and PbI_2 show them as angular molecules, with bond lengths shown in Table 7. Some doubt, however, has been cast upon the existence of PbX_2 molecules in the vapour state[187].

Freezing point curves and conductimetric studies of mixtures of lead(II) halides show the existence of a variety of mixed halides such as PbFCl, PbFBr, PbFI, $PbCl_2.4PbF_2$ and $PbBr_2.4PbF_2$.

183 E. Schwartz, *Nature*, **162** (1948) 614.
184 J. O. Dimmack, I. Melngailis and A. J. Strausse, *Phys. Rev. Letters*, **16** (1966) 1193.
185 A. Bystrom, *Arkiv. Kemi*, **24A** (1947) No. 33.
186 A. F. Wells, *Structural Inorganic Chemistry*, 3rd ed., p. 902, Oxford University Press (1962).
187 P. A. Akisin, V. P. Spiridonov and A. N. Khodankov, *Zhur. Fiz. Khim. SSSR*, **32** (1958) 1679.

Lead(II) chlorofluoride is sparingly soluble in water, and is the precipitate deposited when a lead salt is added to a solution containing both fluoride and chloride ions. This precipitation of PbFCl is the basis of an important analytical method for the determination of fluoride. PbBrF is also only very sparingly soluble in water.

PbFCl and PbFBr have a tetragonal complex layer structure.

Lead(IV) Halides

Only lead(IV) fluoride and chloride are known. The non-existence of lead(IV) bromide and iodide can be put down to the reducing power of Br^- and I^- or, stated alternatively, the inability of bromine and iodine to oxidize lead(II).

Lead(IV) fluoride is obtained by the direct fluorination of lead(II) fluoride at about 300° [188]. It is formed as white needles which are very sensitive to moisture, immediately decolorizing due to the formation of brown PbO_2. PbF_4 has a tetragonal cell ($a = 4{\cdot}24$ Å and $c = 8{\cdot}03$ Å).

Lead(IV) chloride is a clear yellow highly refracting liquid which fumes in moist air, and under certain circumstances decomposes explosively to lead(II) chloride and chlorine. It is best prepared by the action of concentrated sulphuric acid upon pyridinium hexachloroplumbate, and stored under pure concentrated sulphuric acid at −80° in the dark[189].

Coordination Complexes of Lead Halides

An early report indicates[190] the following coordination complexes of lead(IV) chloride: $PbCl_4(NH_3)_2$; $PbCl_4(NH_3)_4$; $PbCl_4(CH_3NH_2)_4$; $PbCl_4(C_2H_5NH_2)_4$; $PbCl_4(C_3H_7NH_2)_4$; $PbCl_4(C_6H_5NH_2)_3$; $PbCl_4(C_5H_5N)_2$.

No structural information is available upon these complexes, but the L_2PbCl_4 types are likely to contain six-coordinate octahedral lead, and the L_4PbCl_4 types may contain eight-coordinate lead[191].

The complex bis(thiourea) $PbCl_2$ is of some interest as a coordination complex of lead(II) chloride. Each lead atom is in a distorted trigonal prism environment of four sulphur atoms (Pb–S = 3·02 Å) and two chlorine atoms (Pb–Cl = 2·75 Å). Lead may, however, be regarded as seven-coordinate rather than six-coordinate, as each lead atom also has a chlorine near the centre of a lateral face of the trigonal prism (Pb–Cl = 3·22 Å)[192].

Halogenolead(II) Ions

Solutions of lead(II) halides contain the PbX^+ ion in addition to the Pb^{2+} and X^- ions. The thermodynamics of these ion dissociations have been investigated by conductimetric methods[193].

$$Pb^{2+} + X^- \rightarrow PbX^+$$

X	ΔG at 25° kcal/mole	ΔH kcal/mole	ΔS cal/deg/mole
Cl	−2·18	4·38	22·0
Br	−2·02	2·88	18·4

188 H. von Wartemburg, *Z. anorg. Chem.* **244** (1940) 337.
189 W. Blitz and E. Meinecke, *Z. anorg. Chem.* **131** (1923) 1.
190 J. M. Mathews, *J. Am. Chem. Soc.* **20** (1898) 834.
191 E. L. Muetterties and C. M. Wright, *Quart. Revs.* (*London*), **21** (1967) 157.
192 M. Nardelli and G. Fava, *Acta Cryst.* **12** (1959) 727.
193 G. H. Nancollas, *J. Chem. Soc.* (1955) 1458.

Alkali fluorides and lead(II) fluoride form alkali fluoroplumbites. In the case of potassium the compound formed is K_4PbF_6, but in the cases of rubidium and caesium the compounds have stoichiometry $MPbF_3$ with perowskite structures[194].

The solubilities of lead(II) chloride, bromide and iodide decrease at first in the presence of halide ion, due to the common ion effect, but with further halide ions the solubility rapidly increases with the formations of complex ions.

From aqueous crystallizations and phase studies on mixed salt systems, a quite remarkable number of solid state stoichiometries have been obtained for the halogenoplumbites. Representative of these are KPb_2Cl_5, K_2PbCl_4, $CsPbCl_3$, $RbPbCl_3$, $KPbBr_3$, $KPbI_3$, Cs_4PbCl_6, K_4PbF_6, K_4PbCl_6 and K_4PbBr_6.

Spectroscopic measurements on $PbCl_2$–Cl^- solutions do not permit positive identification of $PbCl_3^-$ and $PbCl_4^{2-}$, but the limiting spectrum of $PbCl_2$ in 11 M hydrochloric acid is believed due to $PbCl_6^{4-}$ in solution[195].

It is likely that the PbX_3^- ions are pyramidal like the corresponding SnX_3^- ions. The pale yellow $KPbI_3.2H_2O$ is particularly well known, and is formed when warm aqueous solutions of lead nitrate and potassium iodide are mixed. The complex is somewhat weak, and renewed heating causes dissociation to potassium iodide and lead(II) iodide. Anhydrous white $KPbI_3$ is formed by storing the dihydrate over concentrated sulphuric acid, or by dissolving $KPbI_3.2H_2O$ in acetone (in which it is quite soluble), and precipitating with ether.

Anhydrous $KPbI_3$ is curious, in that when reattacked by water it immediately forms bright yellow lead(II) iodide rather than $KPbI_3.2H_2O$. This reaction is used for the detection of traces of water in gases or organic solvents. The stabilities of the halide complexes of lead[193,196] are in the order $I^- > Br^- > Cl^- \gg F^-$; no fluoride complexes are detectable in solution. It is believed that chlorine, bromine and iodine utilize *d*-orbitals for partial double bonding to lead in these complex ions[197].

In the salt $NH_4Pb_2Br_5$ the lead atom has two nearest bromine neighbours at 2·98 Å, two bromines at 3·16 Å and four bromines at 3·35 Å. It would appear to consist of discrete NH_4^+ and Br^- ions along with $PbBr_2$ molecules. The Br–Pb–Br angle in the $PbBr_2$ units is ~85·5°, and the Pb–Br bond length at 2·98 Å is about half-way between covalent and ionic distances for Pb–Br[198].

The salts K_4PbF_6, K_4PbCl_6 and K_4PbBr_6 are remarkable in having PbX_6^{4-} octahedral anions[199].

Halogenolead(IV) Ions

The hexafluoroplumbates M_2PbF_6 are prepared either by dissolution of alkali plumbate in hydrofluoric acid, or by neutralizing an alkali metal carbonate with a solution of PbF_4 in hydrofluoric acid.

The salt K_3HPbF_8 has been prepared in a number of ways, such as adding lead tetraacetate to a solution of potassium fluoride in hydrofluoric acid, or by dissolving a fusion of lead(IV) oxide and alkali in hydrofluoric acid. The well-defined crystals are made up of K^+, HF_2^- and PbF_6^- ions[200].

194 O. Schmitz-Dumont and G. Bergerhoff, *Z. anorg. Chem.* **283** (1956) 314.
195 G. P. Haight and J. R. Peterson, *Inorg. Chem.* **4** (1965) 1073.
196 R. E. Connick and A. D. Paul, *J. Am. Chem. Soc.* **80** (1958) 2069.
197 S. Ahrland, *Acta Chem. Scand.* **10** (1956) 723.
198 H. M. Powell and H. S. Tasker, *J. Chem. Soc.* (1937) 119.
199 G. Bergerhoff and O. Schmitz-Dumont, *Z. anorg. Chem.* **284** (1956) 10.
200 M. F. A. Dove, *J. Chem. Soc.* (1959) 3722.

The alkaline earth salts $MPbF_6$ present an interesting series of structural changes. $CaPbF_6$ contains no discrete complex ions, but consists of a superstructure of the ReO_3 type[201]. $SrPbF_6$ contains linear polymeric complex ions of the type $(PbF_5)_n^{n-1}$ along with F^- ions, and $BaPbF_6$ has discrete PbF_6^{2-} ions. The alkaline earth fluoroplumbates were made by the direct fluorination of the plumbates $MPbO_3$.

The hexachloroplumbate salts M_2PbCl_6 represent a convenient and stable source of lead(IV). A solution of H_2PbCl_6 is easily produced by the action of chlorine upon a fine suspension of lead(II) chloride in concentrated hydrochloric acid. Subsequent addition of solutions of either ammonium or potassium chloride produces the corresponding M_2PbCl_6 salt as lemon yellow crystals. The alkali hexachloroplumbates are isomorphous with the hexachlorostannates.

Organolead Halides

Compared with the other Group IVB organometallic halides, knowledge of organolead halides is somewhat sparse[24, 25, 26, 202]. The types R_3PbX and R_2PbX_2 are well characterized, but compounds of formula $RPbX_3$ are practically unknown.

The usual method for the preparation of organolead halides is by the action of hydrogen halides upon tetra-organoleads. Initial reaction involves fission of one carbon–lead bond, with loss of a further alkyl/aryl group under more vigorous conditions.

$$R_4Pb + HX \rightarrow R_3PbX + RH$$
$$R_3PbX + HX \rightarrow R_2PbX_2 + RH$$

The trialkyl- and triaryl-lead halides R_3PbX in which X is chlorine, bromine, or iodine have all been prepared in this way at low temperature in a solvent. At higher temperatures the principal product is the di-alkyl- or di-aryl-lead halide.

Halogenation by metal and other halides, and by the halogens themselves, has proved to be useful synthetic routes from the tetra-organoleads to the organolead halides in various special cases[203, 204].

$$(C_6H_5)_4Pb + 2AlCl_3 \rightarrow (C_6H_5)_2PbCl_2 + 2C_6H_5AlCl_2$$
$$R_4Pb + SOCl_2 \rightarrow R_3PbCl + RSOCl \qquad (R = \text{alkyl})$$
$$Ar_4Pb + SOCl_2 \rightarrow Ar_2PbCl_2 + Ar_2SO$$

Although at first appearing an unattractive route, the fission by halogens of the lead–lead bond in hexa-alkyl-dileads is in fact an economic route to the trialkyl-lead halides[202]. This is a result of the formation of the hexa-alkyl dileads as cheap by-products of the commercial synthesis of the antiknock tetra-alkyl-leads.

$$R_3Pb.PbR_3 + X_2 \rightarrow 2R_3PbX \qquad (R = \text{alkyl}, X = \text{Cl, Br and I})$$

The attractive proportionation reaction between tetra-organotins and tin(IV) tetrahalides for the preparation of organotin halides is of little use in organolead chemistry, as the low thermal stability of organolead halides and the lead(IV) halides does not allow the temperature range of these proportionations to be reached, except in a few special cases[205].

201 R. Hoppe and K. Blinne, *Z. anorg. Chem.* **293** (1958) 251.
202 I. Ruidisch, H. Schmidbauer and H. Schumann, *Halogen Chemistry*, Vol. 2, p. 269, ed. V. Gutmann, Academic Press (1967).
203 M. Geilen, J. Nasielski, J. E. Dubois and P. Fresnet, *Bull. Soc. Chim. Belg.* **73** (1964) 293.
204 R. Gelius, *Z. anorg. Chem.* **334** (1964) 72.
205 P. R. Austin, *J. Am. Chem. Soc.* **54** (1932) 3287.

One of the more important synthetic methods for the preparation of the organometallic halides of silicon, germanium, and tin is the "direct" reaction between organic halide and metal. All efforts to use this approach for the preparation of the organolead halides have been entirely unsuccessful.

Table 10 contains examples of some representative organolead halides. The fluorides are best prepared by the action of alkali fluorides upon organolead halides, or hydrogen fluoride upon alkyl-lead hydroxides[206].

$$(C_6H_5)_3PbBr + KF \rightarrow (C_6H_5)_3PbF + KBr$$
$$(CH_3)_3PbOH + HF \rightarrow (CH_3)_3PbF + H_2O$$

The sole examples of the halides $RPbX_3$ reported to date are obtained by partial alkylation of $CsPbCl_3$ with alkyl halide[207].

TABLE 10. SOME REPRESENTATIVE ORGANOLEAD HALIDES

Compound	M.p., °C	Compound	M.p., °C
$(CH_3)_3PbF$	305 decomp.	$(CH_3)_2PbCl_2$	155
$(CH_3)_3PbCl$	190 decomp.	$(C_6H_5)_2PbBr_2$	—
$(CH_3)_3PbBr$	133 decomp.	$(C_6H_5)_2PbI_2$	103
$(C_2H_5)_3PbI$	20	$(C_6H_5)_2(C_2H_5)PbCl$	147
$(CH_2{=}CH)_3PbCl$	119	$(C_6H_5)_2(CH_3)PbBr$	118

Definitive structural information on the organolead halides is extremely sparse, and structural determinations would be desirable. Most organolead halides have low solubilities in non-polar organic solvents, but dissolve more readily in strongly polar solvating solvents. The fluorides behave as ionic salts, but early suggestions that other organolead halides are also largely ionic have not been borne out. Thus trimethyl-lead chloride and tri-ethyl-lead chloride are non-conducting in tetramethylene sulphoxide[208].

Infrared data on the halides $(CH_3)_3PbX$ suggest a chain-like constitution involving halogen bridges between lead atoms[209].

An X-ray crystal structure determination of diphenyl-lead dichloride showed it to have a linear polymeric structure as shown in Fig. 9. There is an octahedral coordination about each lead atom, and all the chlorine atoms are bridging in character[210].

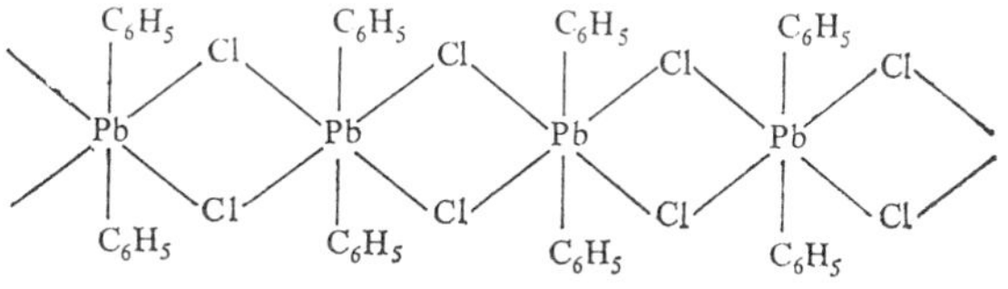

FIG. 9. Structure of $(C_6H_5)_2PbCl_2$ in the crystal.

[206] E. Krause and O. Schlotting, *Chem. Ber.* **58** (1925) 427.
[207] M. Lesbre, *Compt. Rend.* **200** (1935) 559; *ibid.* **210** (1940) 535.
[208] N. A. Matwiyoff and R. S. Drago, *Inorg. Chem.* **3** (1964) 337.
[209] E. Amberger and R. Honigschmidt-Grossich, *Chem. Ber.* **98** (1965) 3795.
[210] V. Busetti, M. Mammi and A. Del Pra, Int. Union of Cryst., 6th Int. Cong. Rome, 1963, Abstract A.73.

The monohalides R_3PbX have an unpleasant smell, and the lower trialkyl-lead halides are extremely sternutatory, and should be handled with care.

The organometallic halides are the starting point of most organometallic syntheses involving lead, and the halogen atom or atoms may be replaced by a great variety of other atoms and groups.

Borohydrides[211] and aluminohydrides[212] react with organolead halides to form the hydrides R_3PbH and R_2PbH_2, but due to the very low thermal stability of the products, such reactions must be carried out at very low temperatures.

$$(CH_3)_3PbBr + KBH_4 \rightarrow (CH_3)_3PbH$$
$$(CH_3)_2PbCl_2 + LiAlH_4 \rightarrow (CH_3)_2PbH_2$$

Metallation of the organolead halides by lithium, sodium or potassium yields the synthetically useful metal tri-organoleads, whose exact nature is presently not fully understood.

Under suitable conditions the action of sodium upon the organometallic lead halides produces a Wurtz type of reaction and the formation of the hexa-organodileads. The tri-organolead chlorides are reduced to the hexa-organodileads by aluminium in dilute potassium hydroxide solution[213].

In addition to the reactions involving the fission of the lead–halogen bond, the organolead halides undergo coordination reactions in which they remain intact, but undergo changes of stereochemistry due to the incoming ligand. $(CH_3)_3PbCl$ and $(C_2H_5)_3PbCl$ both form 1:1 complexes with tetramethylene sulphoxide, dimethylacetamide and dimethylformamide, in which the oxygen atoms bond to lead to give a trigonal bipyramidal structure with a planar tri-alkyl lead unit[208].

Diphenyl-lead dichloride complexes with such ligands as pyridine (1·2), dimethylsulphoxide (1:2), 1,1′-dipyridyl (1:1), and ortho-phenanthroline (1:1) to produce octahedral complexes[214] as illustrated in Fig. 10.

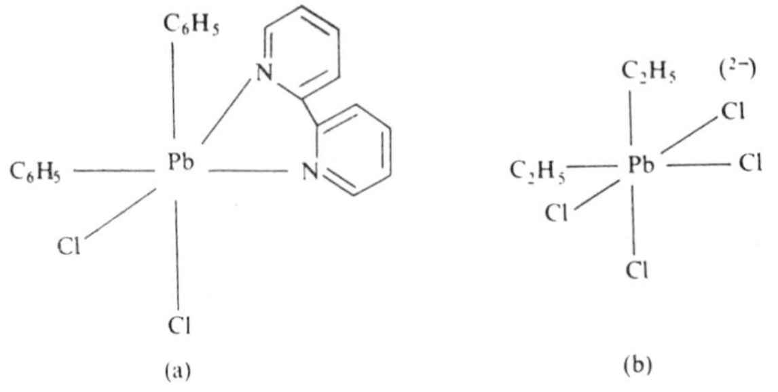

FIG. 10. Six-coordinate lead in organolead complexes. (a) $(C_6H_5)_2PbCl_2$.dipy. and (b) $[(C_2H_5)_2PbCl_4]^{2-}$.

Both tri-organolead halides and di-organolead dihalides complex with halide ions to produce trigonal bipyramidal and octahedral complex anions such as $[(C_6H_5)_3PbCl_2]^-$, $[(C_2H_5)_3PbCl_3]^{2-}$, $[(C_2H_5)_3PbCl_2]^-$ and $[(C_2H_5)_2PbCl_4]^{2-}$.

211 R. Duff and A. K. Holliday, *J. Chem. Soc.* (1961) 1679.

212 W. E. Becker and S. E. Cook, *J. Am. Chem. Soc.* **82** (1960) 6264.

213 G. A. Ruzuvaev, M. S. Fedotov, T. N. Zaichenko and K. Vuiskaya, *Sb. Stat. Obshchei. Khim.* **2** (1953) 1514.

214 K. Hills and M. C. Henry, *J. Organometallic Chem.* **3** (1965) 159.

2.8. PSEUDOHALIDES

Thiocyanates and Selenocyanates

Lead(II) thiocyanate is precipitated from lead(II) solutions by potassium thiocyanate solution; a very pure form is obtained from lead nitrate. It is light-sensitive, and it decomposes at about 190°C[215].

Lead(II) thiocyanate has a structure of particular interest. Each lead atom is eight-coordinate with two pairs of sulphur atoms (at 3·05 and 3·14 Å) and two pairs of nitrogen atoms (at 2·70 and 2·72 Å), all from different thiocyanate ions. Further, all thiocyanate ions are bridging, and each thiocyanate ion makes contact with four lead(II) ions—two at each end. The thiocyanate ion is effectively linear with S–C 1·53 Å and C–N 1·26 Å[216].

Lead(IV) isothiocyanate has been reported from the action of the acid upon lead tetra-acetate. $Pb(NCS)_4$ undergoes very ready disproportionation to thiocyanogen and lead(II) isothiocyanate, and has consequently been little investigated[217].

The corresponding organometallic derivatives of lead(IV), R_3PbNCS and $R_2Pb(NCS)_2$, are, however, much more stable and more extensively studied[218, 219]. The isothiocyanato linkage Pb–N=C=S is believed present in these complexes from spectroscopic evidence, and a full assignment of the infrared spectrum of $(CH_3)_3PbNCS$ has been made. In contrast, however, the selenocyanate $(C_6H_5)_3PbSeCN$ is believed to have the normal Pb–Se–C≡N attachment rather than the iso alternative[220].

Isocyanates

Only organometallic isocyanates of lead(IV) R_3PbNCO (R = CH_3 and C_2H_5) are known[218, 219].

$$(C_2H_5)_3PbCl + KOCN \rightarrow (C_2H_5)_3PbNCO$$

These are believed to contain metal–nitrogen rather than metal–oxygen bonds[218].

The triphenyl-lead fulminate $(C_6H_5)_3PbCNO$ is reported to rearrange to the isocyanate $(C_6H_5)_3PbNCO$ at 174° [221].

Cyanides

Lead(II) cyanide has been little studied[222, 223], and although a compound has been formulated as $Pb(CN)_2$, there is considerable doubt about its true nature. The action of hydrocyanic acid upon lead(II) salts has been variously reported as producing $Pb(CN)_2$ and $Pb(CN)_2PbO$.

Organometallic cyanides of lead(IV), however, are well known[112, 224, 225].

$$R_3PbCl + KCN \rightarrow R_3PbCN$$
$$R_3PbOH + HCN \rightarrow R_3PbCN$$
$$R_3PbOH + NaCN \rightarrow R_3PbCN$$

$(C_6H_5)_3PbCN$ is reported stable up to 250°C.

215 R. Barbieri, G. Faraglia and M. Guistiniani, *Ric. Sci. Rend.* (A) **34** (1964) 109.
216 J. A. A. Mokiolu and J. C. Speakman, *Chem. Comm.* (1966) 25.
217 H. P. Kaufmann and E. Kogler, *Chem. Ber.* **59** (1926) 178.
218 J. S. Thayer and R. West, *Adv. Organometallic Chem.* **5** (1967) 169.
219 M. F. Lappert and H. Pyzora, *Adv. Inorganic and Radio Chem.* **9** (1966) 133.
220 E. E. Aynsley, N. N. Greenwood and M. J. Sprague, *J. Chem. Soc.* (1965) 2395.
221 W. Beck and E. Schuierer, *Chem. Ber.* **97** (1964) 3517.
222 N. M. Gupta, *J. Soc. Chem. Ind.* **39A** (1920) 332.
223 S. Grundt, *Compt. Rend.* **185** (1927) 72.
224 H. H. Anderson, *J. Am. Chem. Soc.* **66** (1944) 934.
225 H. J. Emeléus and P. R. Evans, *J. Chem. Soc.* (1964) 510.

Azides

Azides of lead are known for both Pb(II) and Pb(IV) states.

Lead(IV) azide was made by the action of hydrazoic acid upon Pb_3O_4. The red solution of $Pb(N_3)_4$ fades rapidly and the product was not isolated, though nitrogen was evolved and $Pb(N_3)_2$ was deposited.

Highly explosive red crystals were deposited by the action of $(NH_4)_2PbCl_6$ upon sodium azide solution and, though not analysed, a lead(IV) complex azide was believed present[226].

The organometallic azides of lead(IV) are more stable and have been widely studied The types R_3PbN_3 and $R_2Pb(N_3)_2$ are obtained by the action of hydrazoic acid upon R_3PbOH or R_2PbO.

Triphenyl-lead azide melts without decomposition at 180°, and is the least thermally stable of the Group IV triphenylmetal azides; the products of decomposition are $(C_6H_5)_4Pb$ and nitrogen[218].

Lead(II) azide is dimorphic. The slow diffusion of lead nitrate and sodium azide produces a monoclinic (β) form ($a = 5{\cdot}09$ Å, $b = 8{\cdot}84$ Å and $c = 17{\cdot}51$ Å), whilst the α-form is precipitated on mixing solutions of alkaline azides and lead(II) salts, and is orthorhombic ($a = 6{\cdot}63$ Å, $b = 11{\cdot}31$ Å and $c = 16{\cdot}24$ Å). The β-form is converted to the α-form by the action of light, but may be preserved for years in the dark.

Both α- and β-forms of lead(II) azide are very sensitive to shock and to thermal decomposition, but the activation energy for decomposition of the β-form is considerably lower than that of the α-form, and at about 260°C the violence of detonation of the β-form is about twenty times greater than that of the α-form[227].

Lead(II) azide has a very low solubility (solubility product at 20°C $= 2{\cdot}58 \times 10^{-9}$), but is decomposed by dilute nitric and dilute hydrofluoric acid.

Lead(II) azide is an extraordinarily treacherous compound, and should be handled with great respect. Explosive decomposition is catalysed by small quantities of impurities.

Various basic azides such as $PbN_6.PbO$ and $(PbN_6)_3(PbO)_5$ have been characterized by X-rays from suspensions of lead azides and oxides[228].

2.9. METAL–METAL BONDS CONTAINING LEAD

Alkali Metal Derivatives

Alkali metal derivatives of the type R_3PbM are most stable for R = aryl and M = lithium, though many others are used *in situ* in reactions without being isolated[24, 26].

$$R_3Pb.PbR_3 + Na \xrightarrow{NH_3} R_3PbNa$$

$$(C_6H_5)_2PbBr_2 + Li \xrightarrow{NH_3} (C_6H_5)_2PbLi_2$$

$$PbCl_2 + LiR \longrightarrow R_3PbLi$$

$$R_4Pb + Na \xrightarrow{NH_3} R_3PbNa$$

No structural information is available for these compounds, but they are likely to have some ionic character $R_3Pb^{-\delta}$–$M^{+\delta}$.

226 H. Moller, *Z. anorg. Chem.* **260** (1949) 249.
227 C. S. Garner and A. S. Gomm, *J. Chem. Soc.* (1931) 2123.
228 W. Feitknecht and M. Sahli, *Helv. Chim. Acta*, **37** (1954) 1423 and 1431.

The alkali metal tri-organoleads have been widely utilized as synthetic intermediates and their reactions are exemplified in the lithium organoleads.

$(C_6H_5)_3PbLi$ is completely hydrolysed by water to Pb, LiOH and C_6H_6.

Displacement of the alkali metal by reactions with halides have proved very extensive.

$$(C_6H_5)_3PbLi + H_{4-n}CCl_n \rightarrow \{(C_6H_5)_3Pb\}_nCH_{4-n} \qquad n = 1, 2, 3 \text{ and } 4$$
$$(C_6H_5)_nPbLi_{4-n} + RX \rightarrow (C_6H_5)_nPbR_{4-n} \qquad n = 2 \text{ and } 3$$
$$(C_6H_5)_3PbLi + C_2Cl_6 \rightarrow (C_6H_5)_3PbCl + LiC_2Cl_5$$

Insertion reactions into the lithium–lead bond has realized a useful source of organolead functional derivatives[229].

$$(C_6H_5)_3PbLi + \overline{CH_2CH_2CH_2O} \rightarrow C_6H_5PbCH_2CH_2CH_2OLi$$
$$(C_6H_5)_3PbLi + \overline{CH_2CH_2S} \rightarrow (C_6H_5)_3PbCH_2CH_2SLi$$
$$(C_6H_5)_3PbLi + \overline{CH_2CH_2N}COR \rightarrow (C_6H_5)_3PbCH_2CH_2N(Li)COR$$
$$(C_6H_5)_3PbLi + X \rightarrow (C_6H_5)_3PbXLi \qquad X = S, Se \text{ and } Te$$

Compounds with Lead–Lead Bonds

In addition to the lead–lead bonded compounds discussed in detail below, there are many compounds known in which lead and another Group IV metal are bonded together[230].

$$(C_6H_5)_3GeLi + PbCl_2 \rightarrow [(C_6H_5)_3Ge]_4Pb$$
$$(C_6H_5)_3PbLi + GeCl_4 \rightarrow [(C_6H_5)_3Pb]_4Ge$$

As the least extensively catenating member of the Group IVB elements, the longest sequence of lead is a chain of three atoms.

The hexa-alkyl and hexa-aryl dileads are very numerous and many synthetic approaches are available[231].

$$(CH_3)_3PbCl + NH_4Pb(CH_3)_3 \rightarrow (CH_3)_3Pb.Pb(CH_3)_3$$
$$(C_6H_5)_3PbCl + Na \rightarrow (C_6H_5)_3Pb.Pb(C_6H_5)_3$$
$$(CH_3)_3PbH \xrightarrow{\Delta} (CH_3)_3Pb.Pb(CH_3)_3 + H_2$$
$$(CH_3)_2CO + NaPb + H_2SO_4 \rightarrow \{(CH_3)_2CH\}_3Pb.Pb\{CH(CH_3)_2\}_3$$
$$(C_6H_5)_3PbLi + (C_6H_5)_3PbCl \rightarrow (C_6H_5)_3Pb.Pb(C_6H_5)_3$$
$$(C_6H_{11})_3PbNa + Hg \rightarrow (C_6H_{11})_3Pb.Pb(C_6H_{11})_3$$
$$C_6H_5MgBr + PbX_2 \rightarrow (C_6H_5)_3Pb.Pb(C_6H_5)_3$$

The reaction of Grignard reagents upon lead(II) halides is believed to involve initially the formation of unstable lead(II) diaryls which polymerize and subsequently decompose to hexa-aryldileads and lead metal.

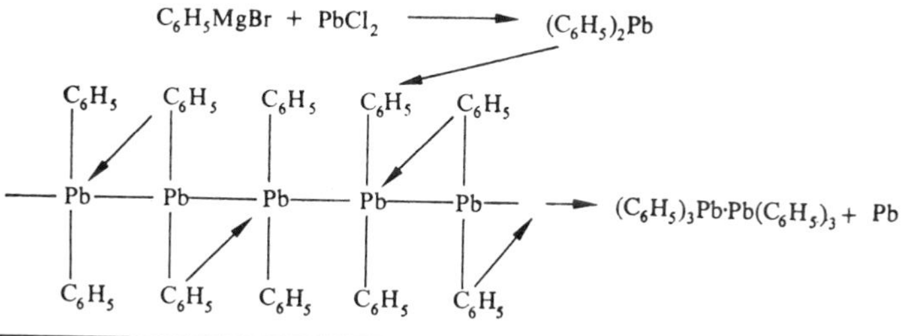

229 L. C. Willemsens and G. J. M. van der Kerk, *J. Organometallic Chem.* **4** (1965) 34.
230 L. C. Willemsens and G. J. M. van der Kerk, *J. Organometallic Chem.* **2** (1960) 260 et seq.
231 H. Gilman, W. H. Atwell and F. K. Cartledge, *Adv. Organometallic Chem.* **1** (1964) 90.

In addition to the compounds containing just one lead–lead bond, there exists $[(C_6H_5)_3Pb]_4Pb$, which contains four lead–lead bonds and six different lead–lead–lead sequences (Fig. 11). It is best made by the simultaneous hydrolysis and oxidation of triphenylphembyl lithium at low temperature[230].

The bright red product may be recrystallized from chloroform, but decomposes over a few days to hexaphenyldilead and lead.

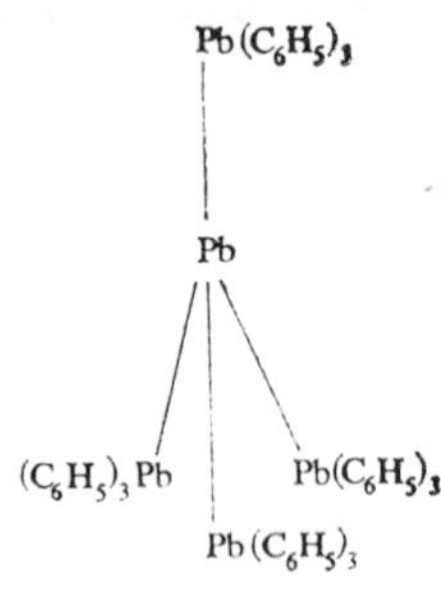

FIG. 11. Tetrakis (triphenyl-lead) lead.

All of the lead–lead bonds in $[(C_6H_5)_3Pb]_4Pb$ are broken by iodine to form $(C_6H_5)_3PbI$ and PbI_2.

Despite early statements to the contrary, the hexa-aryldileads do not dissociate in solution to radicals as do the hexa-arylethanes. An electron diffraction study of $(CH_3)_3Pb.Pb(CH_3)_3$ indicates a normal covalent structure, with Pb–Pb bond length 2·88 Å[232].

Lead–lead bonds undergo fission by a variety of agents, and some of the reactions are believed to have complex mechanisms. Some of the overall reactions are represented below.

$$R_6Pb_2+HX \rightarrow R_3PbX+PbX_2+RH$$
$$R_6Pb_2+X_2 \rightarrow R_3PX+R_2PbX_2+PbX_2$$
$$R_6Pb_2+AlCl_3 \rightarrow R_4Pb+PbX_2+RAlX_2$$
$$R_6Pb_2+AgNO_3 \rightarrow R_3PbNO_3+Ag$$
$$R_6Pb_2+KMnO_4 \rightarrow R_3PbOPbR_3$$
$$R_6Pb_2+C_2H_5ONa \rightarrow R_3PbOC_2H_5+Pb(OC_2H_5)_2$$
$$R_6Pb_2+RSSR \rightarrow R_3PbSR$$

Hexaphenyldilead disproportionates upon heating to tetraphenyl-lead and lead metal.

The well-characterized ions Pb_4^{4-} and Pb_9^{4-} may be regarded as containing lead–lead bonds. The structure of Pb_4^{4-} is tetrahedral and may be regarded as analogous to the P_4 moiety[233]. It is suggested that the Pb_9^{4-} ion takes the shape of a tricapped trigonal prism, with each lead atom virtually equidistant from the centre of the ion cluster[234].

232 H. A. Skinner and L. E. Sutton, *Trans. Faraday Soc.* **36** (1940) 1209.
233 R. E. Marsh and D. P. Shoemaker, *Acta Cryst.* **6** (1953) 197.
234 D. Britton, *Inorg. Chem.* **3** (1964) 305.

Lead–Transitional Metal Bonds

Whilst lead–transitional metal bonds have not been investigated to the same extent as the corresponding bonds of silicon, germanium and tin, there is a significant range of examples[235].

Organolead halides undergo metatheses with metal carbonyl anions.

$$(C_2H_5)_3PbCl + Fe(CO)_4^{2-} \rightarrow [(C_2H_5)_3Pb]_2Fe(CO)_4$$
$$(C_6H_5)_2PbCl_2 + Co(CO)_4^{-} \rightarrow (C_6H_5)_2Pb[Co(CO)_4]_2$$
$$(C_6H_5)_3PbCl + \pi\text{-}C_5H_5Cr(CO)_3^{-} \rightarrow (C_6H_5)_3PbCr(CO)_3(\pi\text{-}C_5H_5)$$

The lead–transitional metal bonds and sequences in these compounds are best illustrated in Fig. 12.

The lead–transitional metal bonds undergo rapid fission with halogens.

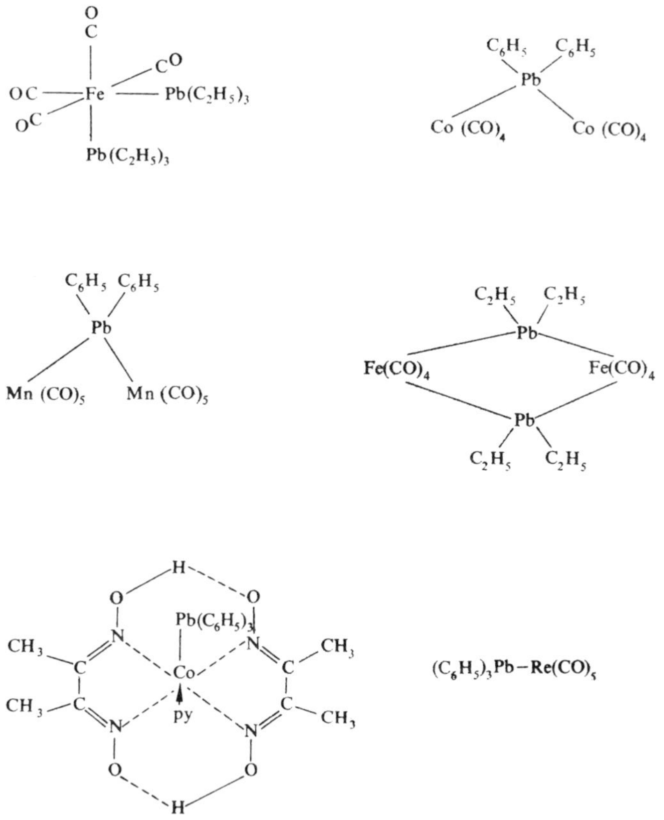

FIG. 12. Some compounds containing lead–transitional metal bonds.

---

[235] F. G. A. Stone, in *New Pathways in Inorganic Chemistry*, eds. E. A. V. Ebsworth, A. Maddock and A. G. Sharpe, Cambridge University Press (1968).

The lead–platinum bond has been quite extensively investigated[236] and the preparation and some reactions are illustrated in Fig. 13.

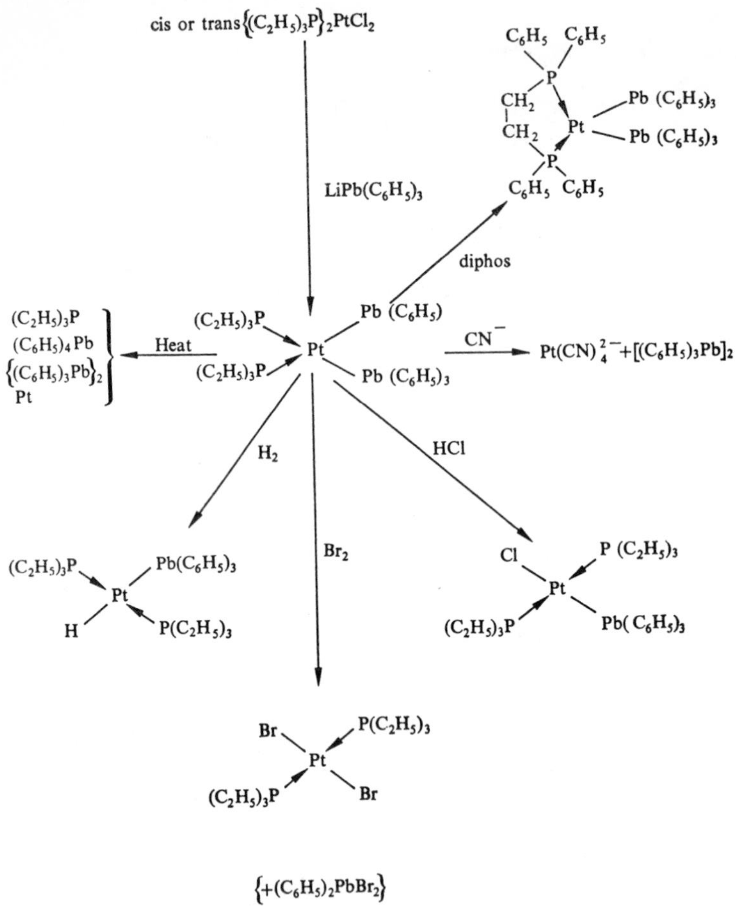

FIG. 13. Some reactions of the lead–platinum bond.

236 G. Deganello, G. Carturan and U. Belluco, *J. Chem. Soc.* (A) (1968) 2873.

19. NITROGEN

K. JONES
The University of Manchester Institute of Science and Technology

1. GENERAL INTRODUCTION

1.1. HISTORY

The history of nitrogen and its compounds dates back to early geological times and to the controversy of whether the earth's atmosphere then contained ammonia or nitrogen. Arabic alchemists are said to have known of salts of ammonia and nitric acid; however, it was not until 1772 that nitrogen was discovered and 1774 that free ammonia was prepared[1]. The preparation of urea from ammonium cyanate by Wöhler in 1828 played its part in helping to break down the "organic" theory[2]. Although this undoubtedly removed the barrier between organic and inorganic chemistry, in some respects the distinction has never been so rigid as today.

The importance of nitrogen compounds in the soil was only realized after about 1862, and it was 1886 before it was shown that nitrogen from the atmosphere was "fixed" by organisms inhabited in the nodules of roots on certain plants[3]. The sources of nitrogen fertilizers in use then were plant and animal wastes, Chile saltpetre, and ammonium sulphate—a byproduct of the coke and coal-gas industry.

The first synthetic processes involving nitrogen compounds were also important methods of nitrogen fixation. The Frank–Caro process (1895) for the production of calcium cyanamide is still operated in a modified form for a very small rapidly diminishing market, but the Birkeland–Eyde (1900) arc oxidation of nitrogen to nitric oxide and hence nitric acid is now obsolete. However, the first decade of this century saw the major developments in the industrial production of nitrogen compounds including the Raschig hypochlorite oxidation of ammonia to hydrazine (1907), the Ostwald catalytic oxidation of ammonia to nitric oxide for nitric acid production (1908) and the Haber catalytic synthesis of ammonia (1909), all of which are still operated today[4, 5]. Indeed, it is only recently that any alternative to the high-pressure catalytic synthesis of ammonia has seemed feasible, but with the recent preparation of stable molecular nitrogen complexes of certain transition metals, and the reported conversion of nitrogen to ammonia at room temperature and atmospheric pressure

[1] M. E. Weeks revised by H. M. Leicester, *Discovery of the Elements*, 7th edn., Journal of Chemical Education, Washington (1968), p. 197.

[2] N. V. Sidgwick, *The Organic Chemistry of Nitrogen*, 3rd edn., revised and rewritten by I. T. Millar and H. D. Springall, Oxford University Press, Oxford (1966), p. 415.

[3] W. D. Stewart, *Nitrogen Fixation in Plants*, Athlone Press, London (1966).

[4] D. W. F. Hardie and J. Davidson Pratt, *A History of the Modern British Chemical Industry*, Pergamon, Oxford (1966).

[5] C. Matasa and E. Matasa, *L'Industrie Moderne des Produits Azotes*, Dunod, Paris (1968) (in French).

in the presence of homogeneous molecular catalysts[6], a significant development in nitrogen fixation would seem to be imminent.

Finally, an historical section would not be complete without a mention of the considerable literature that is already available on the inorganic chemistry of nitrogen. Comprehensive texts include Mellor first published in 1928[7] and with two supplements as recent as 1964[8] and 1967[9], and *Gmelins Handbuch*[10] of 1936. A new series edited by Colburn[11] provides authoritative reviews of major areas of nitrogen chemistry up to the mid-sixties, and a bimonthly trade journal[12] indicates trends, supplies and prices of commercial materials. Most of us owe our introduction to nitrogen chemistry to one of the many inorganic texts[13–18], but especially to Yost and Russell's book[19] of 1944, and more recently the shorter monograph by Jolly[20] in 1964.

1.2. OCCURRENCE AND DISTRIBUTION

Nitrogen constitutes 78% by volume of the earth's atmosphere and occurs in considerable quantities in natural gas. The atmosphere also contains small amounts of ammonia, ammonium salts, nitrates and oxides of nitrogen produced by lightning discharge during thunderstorms as well as increased concentrations over cities due largely to automobile exhaust pollution. All soils contain up to 1% combined nitrogen, but the major natural deposits are as nitre (KNO_3) in India, and saltpetre ($NaNO_3$) in Chile and other desert regions of South America. Smaller amounts of combined-nitrogen are associated with coal and oil. A considerable amount is found *in vivo*, living organisms having an average combined-nitrogen content of about 16%.

As nitrogen is not very soluble in water, the oceans are not potential sources of combined nitrogen, although there is clearly sufficient to meet the demand of all marine life. Marine algae must obtain their nitrogen from that dissolved in ocean water, and this may be as low as 0.03 g/ton, of which only about 5% is in the form of nitrate[21].

[6] R. Murray and D. C. Smith, *Coordin. Chem. Rev.* **3** (1968) 429.

[7] J. W. Mellor, *Comprehensive Treatise on Inorganic and Theoretical Chemistry*, Vol. 8, *Nitrogen*, Longmans, London (1928).

[8] *Mellor's Comprehensive Treatise on Inorganic and Theoretical Chemistry*, Vol. 8, Suppl. I, *Nitrogen*, Part I, Longmans, London (1964).

[9] *Mellor's Comprehensive Treatise on Inorganic and Theoretical Chemistry*, Vol. 8, Suppl. II, *Nitrogen*, Part II, Longmans, London (1967).

[10] *Gmelins Handbuch der Anorganischen Chemie Stickstoff*, System-Nummer 4, Verlag Chemie, GmbH, Berlin (1936).

[11] *Developments in Inorganic Nitrogen Chemistry*, Vol. 1 (ed. C. B. Colburn), Elsevier, Amsterdam (1966).

[12] *Nitrogen, The Journal of World Nitrogen*, British Sulphur Corporation Ltd., 23 Upper Brook St., London, W1.

[13] F. A. Cotton and G. Wilkinson, *Advanced Inorganic Chemistry*, 2nd edn., Interscience, New York (1966), p. 323.

[14] R. B. Heslop and P. L. Robinson, *Inorganic Chemistry: A Guide to Advanced Study* (3rd edn.), Elsevier, Amsterdam (1967), p. 420.

[15] C. S. G. Phillips and R. J. P. Williams, *Inorganic Chemistry*, Oxford (1965).

[16] H. H. Sisler, in *Comprehensive Inorganic Chemistry* Vol. 5 (eds. M. C. Sneed and R. C. Brasted), Van Nostrand, New York (1956).

[17] H. Remy, *Treatise on Inorganic Chemistry*, Elsevier, Amsterdam (1956).

[18] T. Moeller, *Inorganic Chemistry, An Advanced Textbook*, Wiley, New York (1952).

[19] D. M. Yost and H. Russell, Jr., *Systematic Inorganic Chemistry*, Prentice-Hall, New York (1944).

[20] W. L. Jolly, *The Chemistry of Nitrogen*, Benjamin, New York (1964).

[21] D. F. Martin, *Marine Chemistry*, Vol. 1, Marcel Dekker, New York (1968).

At a time when the analysis of moon samples is awaited, it is appropriate to mention the extra-terrestrial sources of nitrogen compounds. Spectra of N_2, NH, NO and CN have been observed in giant superstars and comets, and recently ammonia spectra from interstellar dust clouds have been reported[22]. The atmosphere of Mars contains nitrogen, while that of Jupiter and probably Saturn contain ammonia[23].

1.3. PRODUCTION AND INDUSTRIAL USE

The need to develop industrial methods capable of even greater production capacity of "fixed" nitrogen containing materials for use as fertilizers has run parallel to the ever-increasing demand for food. Saltpetre can be used directly as a fertilizer, and in 1913 accounted for more than half of the world production of fixed nitrogen. Today, however, most of the total world production of nitrogen-containing compounds comes from synthetic routes starting with atmospheric nitrogen. Figure 1 summarizes the present-day route by which the principal nitrogen compounds are produced.

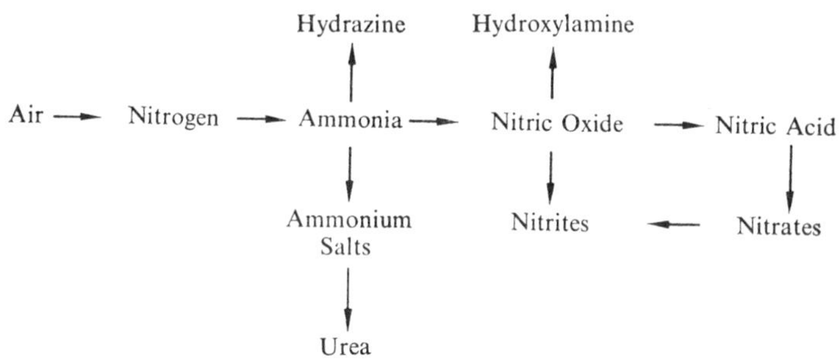

FIG. 1. Industrial routes to nitrogen compounds.

Although the manufacture of ammonia is the most important single use of elemental nitrogen, both gaseous and liquid nitrogen have found increasing application as an inert atmosphere and a cryogenic material. The demand for nitrogen is not expanding more rapidly than in the food industry where the gas is used during processing and in packaging to prevent oxidative deterioration, and liquid nitrogen is used for food freezing both in storage and during transport[24].

The principal methods of nitrogen manufacture are by liquefaction of air or removal of oxygen from air by combustion. The latter source is that normally used in the Haber process for the synthesis of ammonia[25]. This is the only nitrogen fixation process of major importance now in world-wide use, and involves the reaction of nitrogen and hydrogen at high temperatures and pressures in the presence of a catalyst according to the equation

$$N_2 + 3H_2 \xrightleftharpoons{\text{catalyst}} 2NH_3$$

[22] A. C. Cheung, D. M. Rank, C. H. Townes and W. J. Welch, *Nature*, **221** (1969) 917.
[23] F. H. Day, *The Chemical Elements in Nature*, Harrap, London (1963).
[24] M. Sittig, *Nitrogen in Industry*, Van Nostrand, New York (1965).
[25] R. Noyes, *Ammonia and Synthesis Gas*, Noyes Development Corporation, New Jersey (1967).

Hydrogen for the Haber process is commonly produced by the thermal decomposition, reforming or oxidation of hydrocarbons, and frequently in conjunction with the nitrogen purification process. Other sources of hydrogen, such as the electrolysis of water, etc., are equally acceptable, but depend on resources local to the production site.

A large quantity of ammonia is converted into a variety of ammonium salts by simple neutralization reactions, principally for use as fertilizers[26]. Both anhydrous and aqueous ammonia are also used directly as fertilizers by soil injection and spray techniques. Increasing amounts of ammonia are being converted into urea via ammonium carbamates:

$$2NH_3+CO_2 \rightarrow NH_2CO_2NH_4 \rightarrow NH_2CONH_2+H_2O$$

Urea contains a higher percentage of nitrogen than any other solid fertilizer, and also finds application in urea–formaldehyde resins as well as in melamine production.

The production of hydrazine by the Raschig process[27] only accounts for a modest amount of ammonia:

$$2NH_3+OCl^- \rightarrow N_2H_4+H_2O+Cl^-$$

Because of the high heat of combustion, one of the major uses of hydrazine and its derivatives, and potentially the largest, is as a component of rocket propellants. The rocket motor of the service module used during the recent successful moon landings was powered by such a fuel.

Large amounts of ammonia are oxidized via nitric oxide to nitric acid by the following reactions:

$$12NH_3+15O_2 \rightarrow 12NO+18H_2O$$
$$12NO+6O_2 \rightarrow 12NO_2$$
$$12NO_2+4H_2O \rightarrow 8HNO_3+4NO$$

Nitric acid is used principally for the production of nitrates as fertilizers, nitro-compounds as explosives and to a lesser extent for a multitude of purposes in the chemical industry[26]. Thus from these principal compounds it is possible to synthesize all nitrogen derivatives, the nature of which are to be described in subsequent sections.

It is probably apparent even from such a brief outline of the major industrial processes producing nitrogen compounds that they are related to such an extent with each other and also with the natural resources of a particular country that many factors enter into the planning of a viable production unit. As an indication of the current trend[28], one of the most modern sites is operated by Esso Chemie at Europoort in The Netherlands, and includes plants for the production of ammonia, nitric acid, urea and calcium ammonium nitrate. The ammonia plant is equal to the world's largest, producing 1500 short tons of liquid ammonia per day in a single production line. The synthesis gas is obtained by reforming natural gas from the nearby North Sea gasfields. The carbon dioxide produced in the shift conversion is removed and regenerated as feedstock for the urea plant. Ammonia is oxidized via nitric oxide to nitric acid from which ammonium nitrate is formed by neutralization. Coagulation of ammonium nitrate solution with limestone gives calcium ammonium nitrate, another useful fertilizer. The plant is operated with a high degree of automation, and is located conveniently for shipping.

[26] C. J. Pratt and R. Noyes, *Nitrogen Fertilizer Chemical Processes*, Noyes Development Corporation, New Jersey (1965).

[27] L. F. Audrieth and B. A. Ogg, *The Chemistry of Hydrazine*, Wiley, New York (1951).

[28] *Nitrogen*, No. 60 (1969) 34.

1.4. NUCLEAR PROPERTIES

1.4.1. Isotopes and Radioisotopes

Seven isotopes of nitrogen are known of which ^{14}N and ^{15}N are stable and occur in natural nitrogen-containing substances in the ratio 272:1. Natural abundances of stable isotopes vary slightly depending on the source of the element; the percentage abundances 99.635 and 0.365 for atmospheric nitrogen correspond to an atomic weight of 14.0067 based on the IUPAC scale ^{12}C = 12.000000 atomic mass units. The remaining five isotopes ^{12}N, ^{13}N, ^{16}N, ^{17}N and ^{18}N are radioactive, the longest lived being ^{13}N with a half-life of nearly 10 min, and since it is neutron deficient, charged particle accelerators are necessary to produce it in reasonable amounts. Even so, experiments can be performed near such equipment[29]. ^{13}N as $^{13}NH_3$ can be obtained by dissolving the target foils used in the radiochemical synthesis $^{12}C(d, n)^{13}N$, but a recent report[30] indicates that the product reacts as atomic nitrogen-13 (see section 2.12). The properties of nitrogen isotopes are listed in Table 1.

Thus as the half-lives of the radioisotopes are inconveniently short for most chemical experiments, tracer and isotope-effect studies are generally conducted using the stable ^{15}N isotope. Clearly, the scope and success of such investigations depend on the availability of efficient methods of separation and enrichment, the variety of labelled compounds which can be prepared and the analytical techniques to determine and monitor ^{15}N concentration.

1.4.2. Separation of ^{15}N from ^{14}N

(a) *By Chemical Exchange*[31]

If A' and B' are two molecules containing ^{14}N, and A'' and B'' are similar molecules containing ^{15}N, then the equilibrium constant for the reaction

$$A'_{(a)} + B''_{(b)} \rightleftharpoons A''_{(a)} + B'_{(b)}$$

can be expressed by

$$K = \frac{[Q_{A''}/Q_{A'}]_{(a)}}{[Q_{B''}/Q_{B'}]_{(b)}}$$

where $Q_{A''}$, etc., refer to the partition functions for the reaction between A'' and A', etc. The $Q_{'}/Q_{''}$ ratios can be calculated, and values greater than unity indicate that enrichment is possible. To obtain appreciable enrichment, a countercurrent system is desirable, and this is most effective if one phase is liquid and the other gaseous.

Thus for the exchange of gaseous ammonia with aqueous solutions of ammonium salts described by the equation

$$^{15}NH_3(g) + {}^{14}NH_4^+(aq) \rightleftharpoons {}^{14}NH_3(g) + {}^{15}NH_4^+(aq)$$

separations of up to 75% ^{15}N have been achieved, and with columns in series 99.9% enriched material can be obtained[32, 33]. The procedure is to pass ammonium nitrate solution down a packed column to a flask containing aqueous sodium hydroxide. In the flask,

29 J. Hudis, *The Radiochemistry of Carbon, Nitrogen, and Oxygen*, Natural Academy of Sciences—NRC, Nuclear Science Series, Washington (1960).

30 M. J. Welch, *Chem. Communs.* (1968) 1354.

31 *Separation of Isotopes* (ed. H. London), Newnes, London (1961).

32 W. Spindel, in *Inorganic Isotopic Syntheses* (ed. R. H. Herber), Benjamin, New York (1962), p. 74.

33 P. S. Baker, in *Survey of Progress in Chemistry*, Vol. 4 (ed. A. F. Scott), Academic Press, New York (1968), p. 69.

TABLE 1. PROPERTIES OF NITROGEN ISOTOPES[a, b]

		^{12}N	^{13}N	^{14}N	^{15}N	^{16}N	^{17}N	^{18}N
Natural abundance (%)				99.635	0.365			
Atomic weights $^{12}C = 12.000000$		12.018900	13.0057389	14.0030738	15.0001088	16.006089	17.008580	18.014266
Mass excess ($\triangle = M-A$) in MeV		17.36	5.345	2.8637	0.100	5.685	7.87	13.1
Half-life		0.0110 s	9.96 m			7.14 s	4.16 s	0.63 s
Type of decay		β^+ (100%) 3α (3.0%)	β^+			β^- α (0.0006%)	β^-, n	β^-
Major radiations and approximate energies in MeV with relative intensities (%)	β^+	16.4 (100%)	1.20 (100%)					
	β^-					10.24 (26%) 4.27 (68%)	8.68 (1.6%) 7.81 (2.6%) 4.1 (95%)	9.4 (100%)
	γ	0.511	0.511			2.75 (1%) 6.13 (69%) 7.11 (5%)	0.87 (3%) 2.19 (0.5%)	0.82 (59%) 1.65 (59%) 1.98 (100%) 2.47 (41%)
	α	0.195				1.7		
	n						0.40 (45%) 1.21 (45%) 1.81 (5%)	
Thermal neutron cross-section in barns				σ (n, p) 1.81	σ_c 2.4×10^{-5}			
Radiochemical methods of preparation		^{10}B (He, $3n$) ^{12}C (p, n) ^{14}N $(\gamma, 2n)$	^{10}B (α, n) ^{12}C (d, n) ^{12}C (p, γ) ^{13}C (p, n) ^{14}N (γ, n)	^{11}B (α, n) ^{16}O (d, α)	^{14}N (d, p) ^{14}N (n, α)	^{15}N (n, γ) ^{15}N (d, p) ^{16}O (n, p) ^{19}F (n, γ)	^{14}C (α, p) ^{15}N (t, p) ^{17}O (n, p) ^{18}O (γ, p)	^{18}O (n, p)

[a] C. M. Lederer, J. M. Hollander and I. Perleman, *Table of Isotopes*, 6th edn., Wiley, New York, 1968.
[b] C. J. L. Lock, in *Mellor's Comprehensive Treatise on Inorganic and Theoretical Chemistry*, Vol. VIII, Supplement II, Longmans, London, 1967, Section XXXV, p. 526.

ammonia gas is liberated and rises up the column and exchanges with the ammonium ions in the descending solution. As a result, ^{15}N tends to concentrate in the liquid phase while the gaseous ammonia contains relatively more of the unwanted lighter isotope. A counter-current system may take 2–3 weeks to reach equilibrium after which time the aqueous solution can be continuously bled off at such a rate so as not to disturb the equilibrium.

Similarly, 99% enrichment can be obtained from the process based on the exchange reaction

$$^{15}NO(g)+{}^{14}NO_3^-(aq) \rightleftharpoons {}^{14}NO(g)+{}^{15}NO_3^-(aq)$$

and this method is now used commercially. Nitric oxide is produced from nitric acid at the bottom of the column and converted back at the top according to the equations[34]

$$2HNO_3+3SO_2+2H_2O = 2NO+3H_2SO_4$$
$$2NO+1.5O_2+H_2O = 2HNO_3$$

Isotope separations based on chemical exchange are not restricted to gas–liquid counter-current methods. For example, high purity ^{15}N has been obtained by exchange between NH_4^+ adsorbed on Dowex X-12 resin and NH_4^+ in equilibrium with unionized NH_4OH in solution[35]. Since two chemically different species are involved, the isotope exchange factor is relatively large for the equilibrium

$$NH_4^+ + OH^- \rightleftharpoons NH_4OH$$

(b) *By Thermal Diffusion*[32]

Although $^{15}N_2$ can be separated by thermal diffusion, the quantity of separated isotope is relatively small. The separation in a thermal diffusion column containing a mixture of nitric oxide and nitrogen dioxide is assisted by the gas-phase equilibrium

$$^{14}NO_2(g)+{}^{15}NO(g) \rightleftharpoons {}^{15}NO_2(g)+{}^{14}NO(g)$$

(c) *Other Methods*

Many other methods of preparation have been reported for specific applications; even simple fractional distillation of nitric oxide provides an efficient method[36] for the simultaneous concentration of the rarer isotopes of both nitrogen and oxygen in the form $^{15}N^{18}O$. Methods applicable to specific nitrogen compounds will be mentioned in the appropriate section of this chapter dealing with such compounds.

1.4.3. Synthesis of Compounds Containing ^{15}N

The preparative route to a particular labelled compound is to some extent governed by the enriched starting materials that can be obtained directly from separation methods. Fortunately, many key nitrogen compounds are now available from commercial sources[37, 38] including N_2, NO, NO_2, NH_3, HNO_3, ammonium salts and nitrates, as 5% 30%, or 95% ^{15}N enriched materials. The degree of enrichment is likely to be dictated by the application for which the product is required.

[34] W. Spindel and T. I. Taylor, *J. Chem. Phys.* **23** (1955) 981; **24** (1956) 626; *Trans. NY Acad. Sci.* **19** (1956) 3.

[35] H. Spedding, J. E. Powell and H. J. Svec, *J. Am. Chem. Soc.* **77** (1955) 6125.

[36] B. B. McInteer and R. M. Potter, *Ind. Eng. Chem. Process Design Develop.* **4** (1965) 35.

[37] Office Nationale Industriel de l'Azote (ONIA), 40 avenue Hoche, Paris (VIIIe), France.

[38] Isomet Corporation, Palisades Park, New Jersey, USA.

The preparation of choice should preferably involve only one or two stages, and provide for quantitative conversion of all the ^{15}N in the enriched starting material into labelled product, which in turn should be easily separated in high yield from unreacted starting materials and byproducts. Clearly, methods requiring a gross excess of an enriched material should be avoided if possible, and to overcome these and other problems, syntheses which would not normally be used may be applicable. It is safe practice to carry out the preparation under identical conditions of scale and apparatus using unlabelled materials prior to the synthesis using enriched compound to establish the procedure for optimum yield. It is also necessary to consider ^{15}N recovery procedures should anything go wrong during the synthesis and also from the final product. Manipulative technique is simplified, however, by the fact that all ^{15}N enriched compounds prepared so far undergo negligible exchange with atmospheric nitrogen under normal conditions, in marked contrast to labelled hydrogen, carbon and oxygen compounds, where contact with atmospheric moisture and carbon dioxide must be avoided.

1.4.4. ^{15}N Analysis[32]

The mass spectrometer has proved to be the most useful instrument for isotopic analysis, and in general ^{15}N abundance ratios can be determined to a precision of $\pm 0.1\%$. Nitrogen gas is the most satisfactory material for mass spectrometric isotope analysis, as even mixtures of $^{14}N_2$, $^{14}N^{15}N$ and $^{15}N_2$ do not exchange at ordinary temperatures. Thus in most experiments involving ^{15}N, it is necessary to convert the combined nitrogen of the material to be analysed into elemental nitrogen. Inorganic compounds can either be reduced to ammonia directly, or oxidized to nitrate and then converted into ammonia by Devarda's alloy. Ammonia is readily oxidized to nitrogen by alkaline hypobromite. Procedures for converting organic nitrogen compounds into nitrogen are based on the methods of Dumas or Kjeldahl followed by oxidation, taking considerable precautions to avoid contamination by atmospheric nitrogen (see section 1.8).

Nitrous oxide can be used for mass-spectrometric ^{15}N analysis instead of nitrogen, and is useful when the labelling position is of particular interest. This arises because the thermal decomposition of labelled ammonium nitrate produces two different isotopic isomers of nitrous oxide, depending on where the ^{15}N originated, which can be distinguished by the mass spectrometer.

$$^{15}NH_4{}^{14}NO_3 \xrightarrow{\text{Thermal}} {}^{15}N^{14}NO + 2H_2O$$
$$^{14}NH_4{}^{15}NO_3 \xrightarrow[\text{decomposition}]{} {}^{14}N^{15}NO + 2H_2O$$

Caution should be exercised in the choice of reference standards for isotopic analysis as some compounds (salts produced as byproducts of nitric acid, for example) contain significantly lower than the normal natural abundance level of ^{15}N as a result of isotopic fractionation during the manufacturing process.

1.4.5. Uses of ^{15}N

Highly enriched $^{15}N_2$ may have nuclear reactor applications because of the low thermal neutron absorption cross-section compared to ordinary nitrogen[29], but the major use of ^{15}N has been in tracer work. Of particular mention is its use in mechanistic studies of chemical, biochemical and nitrogen-fixation processes.

1.4.6. Nuclear Magnetic Resonance[39, 40]

Although both of the stable isotopes of nitrogen have non-zero spin numbers, neither isotope is particularly favourable for n.m.r. investigation (Table 2). ^{14}N has a spin number $I = 1$ with consequent electric quadrupole moment, and very low n.m.r. sensitivity.

TABLE 2. DATA PERTINENT TO NUCLEAR MAGNETIC RESONANCE STUDIES WITH NITROGEN ISOTOPES[a, b]

	^{13}N	^{14}N	^{15}N
Natural abundance (%)		99.635	0.365
Spin number (I) in multiples of $h/2\pi$	$\frac{1}{2}$	1	$\frac{1}{2}$
Larmor frequency in MHz at 10^4 gauss		3.076	4.315
Electric quadrupole moment (Q) in multiples of $e \times 10^{-24}$ cm^2		7.1×10^{-2}	
Magnetic dipole moment (μ) in multiples of nuclear magnetons ($eh/4\pi Mc$)	± 0.3221	$+0.40361$	-0.28309
Relative sensitivity to ^1H for an equal number of nuclei:			
at constant field		1.01×10^{-3}	1.04×10^{-3}
at constant frequency		0.193	0.101

[a] E. A. C. Lucken, *NQR Constants*, Academic Press, New York, 1969.
[b] C. M. Lederer, J. M. Hollander and I. Perleman, *Table of Isotopes*, Wiley, New York, 1968.

In compounds where the nitrogen atom is in an asymmetric electronic environment, the spectra are affected by quadrupole broadening to such an extent that fine structure cannot be resolved. However, ^{14}N chemical shifts can be measured using broad-line equipment. Indeed, the discovery of chemical shift is associated with the work of Proctor and Yu[41], who found that the frequency of the ^{14}N resonance depended on the chemical compound chosen, and for NH_4NO_3 solution they detected two resonances separated by 1.0 kHz in a field of 10,500 gauss (*ca.* 300 ppm). ^{14}N gives rise to chemical shifts extending over a range of some 900 ppm (Table 3), and values can be correlated with the electronic environment at the nitrogen atom. Note that the symmetric ammonium ion has the highest chemical shift whereas structures of lower electronic symmetry at nitrogen such as the nitrite appear at lower chemical shift. The ammonium, nitrate and nitrite ions in aqueous solution are all used as reference standards for reporting chemical shifts.

In order to interpret observed chemical shift data, it is necessary to separate the diamagnetic and paramagnetic components. This has been carried out for NOX

[39] J. A. Pople, W. G. Schneider and H. J. Berstein, *High-resolution Nuclear Magnetic Resonance*, McGraw-Hill, New York (1959).

[40] J. W. Emsley, J. Feeney and L. H. Sutcliffe, *High-resolution Nuclear Magnetic Resonance Spectroscopy*, Pergamon Press, Oxford (1965), p. 1031.

[41] W. G. Procter and F. C. Yu, *Phys. Rev.* **81** (1951) 20.

TABLE 3. ^{14}N CHEMICAL SHIFTS DETERMINED BY BROAD-LINE NUCLEAR MAGNETIC RESONANCE SPECTROSCOPY QUOTED RELATIVE TO AQUEOUS NITRATE ION
(aq NH_4^+ at +353, aq NO_2^- at −232 ppm)

Compound	Shift in ppm	Compound	Shift in ppm
NOX		NO_2X	
NO^+	7	NO_2^+	125
NOF	−104	NO_2F	68
Me_2NNO	−158	NO_2Cl	63
MeONO	−188	N_2O_5/CCl_4	48
NO_2^- aq	−232	$ClONO_2$	44
NOCl	−223	$EtONO_2$	37
$O_2N\underline{N}O$	−302	$Me_2N\underline{N}O_2$	23
NOBr	−325	CF_3NO_2	23
CF_3SNO	−342	CCl_3NO_2	15
EtSNO	−412	CBr_3NO_2	13
CF_3NO	−424	$H_2N\underline{N}O_2$/ether	12
PhNO	−532	N_2O_4	11
Other NO compounds		$PhNO_2$	8
		$MeNO_2$	−2
$^-ONNO^-$ aq	−15	$ON\underline{N}O_2$	−67
$^-ONNO_2$ aq	−83		
N—F		N—S	
ONF_3	143	$H_2N—SO_3^-$ aq	293
N_2F_4	46	$H_3N^+—SO_3^-$ aq	285
NF_3	6	S_4N_4/dioxan	253
cis N_2F_2	−7	$XNSF_2$	220–245
$N_2F_3^+/HF$	−22	$S_3N_2O_2$	71
trans N_2F_2	−69	PhNSO	59
		$S_4N_3^+$	9 and −3
N–P in Et_2O			
$(NPF_2)_4$	298		
$(NPF_2)_3$	294		
$(NPCl_2)_4$	256		
$[NP(NCS)_2]_3$	253		
$(NPCl_2)_3$	247		

compounds[42, 43], and the graph illustrated in Fig. 2 shows that $-\sigma_p$ is inversely proportional to the energy of the $n \rightarrow \pi$ band in the ultraviolet-visible spectrum as predicted by Pople theory[39].

Although the principal causes of chemical shift differences are understood, the applications of ^{14}N chemical shifts have been mainly empirical. As an example, the characterization of metal–pseudohalide complexes raises the problem of which atom is bonded to the metal. For thiocyanates, where the point of attachment has been reliably determined by X-ray crystallography, it was found that sulphur-bonded complexes gave ^{14}N chemical shifts

[42] J. Mason and W. Van Bronswijk, *Chem. Communs.* (1969) 357.
[43] J. Mason, unpublished results.

0–20 ppm to low field for the free NCS^- resonance, whereas nitrogen-bonded complexes appear at 50–130 ppm to high field of NCS^-. Thus it is possible to predict whether related complexes contain metal–nitrogen or metal–sulphur bonds[44].

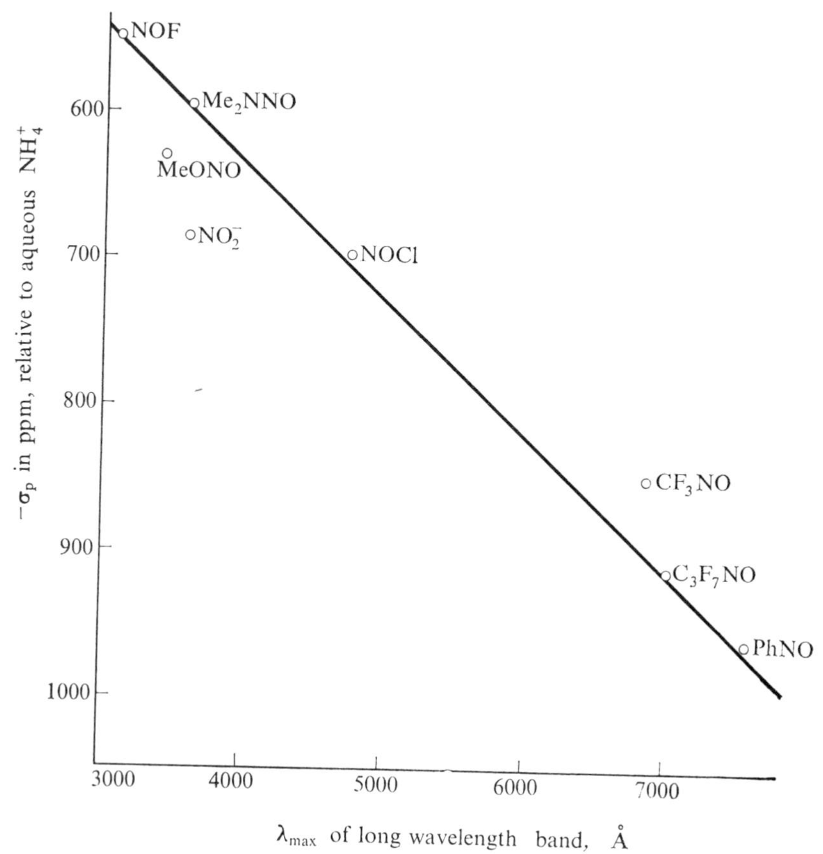

FIG. 2. Plot of the paramagnetic component of the observed shift versus the wavelength of the $n \rightarrow \pi$ band for NOX compounds.

The problem of ^{14}N quadrupole broadening and lack of fine structure can be overcome by using the ^{15}N isotope which has a spin number $I = \frac{1}{2}$. However, the very low n.m.r. sensitivity and the low natural abundance (0.365%) make ^{15}N isotope enrichment essential. It is for this reason that the technique is likely to be limited to specialized applications. For example, the structure of $S_4N_3^+$ illustrated in Fig. 3 is consistent with the spectrum obtained from a sample enriched to 97.2% ^{15}N [45].

[44] O. W. Howarth, R. E. Richards and L. M. Venanzi, *J. Chem. Soc.* (1964) 3335.
[45] M. Logan and W. Jolly, *Inorg. Chem.* **4** (1965) 1508.

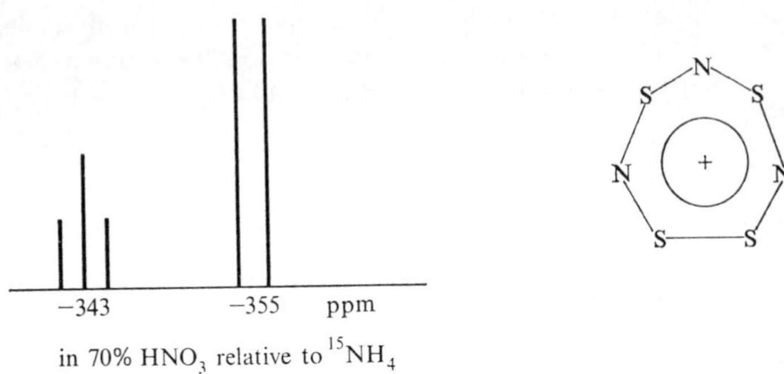

FIG. 3. The ^{15}N n.m.r. spectrum and structure of $S_4{}^{15}N_3^+$.

1.4.7. Nuclear Quadrupole Resonance[46, 47]

As a consequence of its spin number $I = 1$, the ^{14}N nucleus has an electric quadrupole moment. The coupling of this nuclear quadrupole moment with the field produced by electrons from neighbouring atoms causes a torque to be exerted on the molecule that in turn gives rise to hyperfine structure in the rotational spectrum. The magnitude of the quadrupole coupling constant eQq can be determined either from pure quadrupole spectra or from microwave spectra and related to the asymmetry of the electron charge cloud

TABLE 4. NUCLEAR QUADRUPOLE COUPLING CONSTANTS OF ^{14}N IN NITROGEN COMPOUNDS[a]

Molecule	eQq (MHz)
NH_3	-4.0842 ± 0.0003
NH_2D	-4.10
ND_3	-4.080 ± 0.003
NF_3	-7.09
NSF_3	1.19 ± 0.05
N_2O central N	-0.27
N_2O terminal N	-1.03 ± 0.10
HCN	-4.58
MeCN	-4.40
HC≡CCN	-4.20
ClCN	-3.63
MeNC	$+0.5$
HNCS	$+1.20$

[a] Landolt-Börnstein, *Numerical Data and Functional Relationships in Science and Technology*, New Series, Group II, Vol. 4, *Molecular Constants from Microwave Spectroscopy*, Springer-Verlag, Berlin (1967).

[46] E. A. C. Lucken, *NQR Constants*, Academic Press, New York (1969).

[47] H. Sillescu, in *Physical Methods in Advanced Inorganic Chemistry* (eds. H. A. O. Hill and P. Day), Wiley, London (1968), p. 434.

surrounding the nitrogen nucleus. Quadrupole coupling constants for some nitrogen compounds are listed in Table 4. For nitrous oxide, the central nitrogen atom is in a more symmetric electron environment than the terminal nitrogen atom corresponding to its lower coupling constant.

1.5. PHYSICAL PROPERTIES OF THE NITROGEN ATOM

It is usual when discussing the factors which contribute to the overall chemistry of an element to consider the fundamental atomic properties. Such properties fall broadly into two groups: (1) properties of the free nitrogen atom itself which can be measured or calculated directly, e.g., atomic weight or ionization energy, and (2) properties associated with concepts used to rationalize the behaviour of the nitrogen atom in chemically combined states, such as electro-negativity and electron affinity. Properties associated with the nitrogen nucleus have already been listed in Tables 1 and 2, where properties like nuclear spin can be classified as (1), while nitrogen chemical shift data presented in Table 3 would come under group (2). Other atomic nitrogen properties are presented in Table 5 and are concerned mainly with atomic dimensions and thermodynamic functions.

TABLE 5. PHYSICAL PROPERTIES OF THE NITROGEN ATOM

Property	Temperature (K)	Value
Atomic number		7
Atomic weight		14.0067
Metallic atomic radius		0.92 Å
Non-polar covalent radius		0.75 Å
Average electron density $D = 3Z/4\pi r^3$		3.95 electrons/$Å^3$
Atomic volume		15.95 cm^3
Crystal radius in N^{---}		1.71 Å
Crystal radius in NO_3^-		0.11 Å
Electro-negativity		2.93 (Sanderson)
		3.05 (Pauling)
Electron affinity		0.04 eV, 0.92 kcal/g atom
1st ionization potential		14.54 eV, 335.4 kcal/g atom
Average ionization energy per electron		1231 kcal/g atom
ΔH_f° (enthalpy of formation)	0	112.534 kcal/g atom
ΔH_f° (enthalpy of formation)	298.15	112.979 kcal/g atom
ΔG_f° (free energy of formation)	298.15	108.883 kcal/g atom
$H_{298}^\circ - H_0^\circ$ (enthalpy)	298.15	1.481 kcal/g atom
S° (entropy)	298.15	36.622 cal/deg g atom
C_p° (heat capacity)	298.15	4.968 cal/deg g atom

R. T. Sanderson, *Chemical Periodicity*, Reinhold, New York (1960).
D. D. Wagman, W. H. Evans, V. B. Parker, I. Halow, S. M. Bailey and R. H. Schumm, *Selected Values of Chemical Thermodynamic Properties*, NBS Technical Note 270-3, Washington (1968).

Parameters from nitrogen atomic spectra which are listed include the ground-state terms for nitrogen atoms and ions together with their corresponding ionization energies in Table 6 and the low-lying atomic energy terms in Table 7. In the nitrogen ^4S ground state, all three $2p$ electrons have parallel spins, while in the ^2D and ^2P states two of the $2p$ electrons are paired. These nitrogen atoms are produced during the dissociation processes and their correct assignment was important in determining the value of the dissociation energy of molecular nitrogen. In Table 6 one must note that N I refers to the ground state and N II to the singly ionized species. The ionization energies were calculated from the limiting wave numbers obtained from spectral terms.

TABLE 6. ELECTRONIC DESCRIPTION OF THE GROUND STATES OF FREE NITROGEN ATOM AND IONS TOGETHER WITH THE CORRESPONDING IONIZATION ENERGIES IN VARIOUS UNITS

N	Ground state	Spectroscopic state	Ionization energy		
			Wave numbers (cm^{-1})	Volts	kcal/g atom
I	$1s^22s^22p^3$	$^4S_{3/2}$	117 345	14.545	335
II	$1s^22s^22p^2$	$^3P_0$	238 846.7	29.605	683
III	$1s^22s^22p$	$^2P_0$	382 625.5	47.426	1 097
IV	$1s^22s^2$	$^1S_0$	624 851	77.450	1 784
V	$1s^22s$	$^2S_{1/2}$	789 532.9	97.863	2 256
VI	$1s^2$	$^1S_0$	4 452 800	551.93	12 730
VII	$1s$	$^2S_{1/2}$	5 379 860	663.82	15 300

C. E. Moore, *Atomic Energy Levels as derived from the Analyses of Optical Spectra*, Natl. Bur. Std. US Circ., 467, Vol. 1 (1949).

From another spectroscopic point of view, when materials are subjected to X-radiation, electrons from the innermost shells of an atom can be emitted and subsequently detected by high resolution magnetic or electric analysis. Thus electron spectroscopy for chemical analysis (ESCA)[48] is capable of determining the electronic levels of nitrogen in a molecule

TABLE 7. LOW-LYING ATOMIC ENERGY TERMS FOR NITROGEN

Spectroscopic state	Wave numbers (cm^{-1})	Electron volts	kcal/g atom
Ground state $^4S_{3/2}$	0	0	0
$^2D_{5/2}$	19 223	2.383	54.95
$^2D_{3/2}$	19 231	2.385	54.98
$^2P_{1/2,3/2}$	28 840	3.574	82.45

A. G. Gaydon, *Dissociation Energies and Spectra of Diatomic Molecules*, 3rd edn., Chapman & Hall, London (1968).

[48] K. Siegbahn, C. Nordling, A. Fahlman, R. Nordberg, K. Hamrin, J. Heldman, G. Johansson, T. Bergmark, S-E. Karlsson, I. Lindgren and B. Lindberg, *ESCA Atomic, Molecular and Solid State Structure Studied by Means of Electron Spectroscopy*, Nova Acta Regiae Societatis Scientiarum Upsaliensis, Ser. IV, Vol. 20, Uppsala (1967), p. 108.

to the precision set by the widths of the atomic levels themselves. Of particular use to the chemist is the comparison of values measured for the nitrogen 1*s* level in various compounds. For example, sodium azide $[Na^+(\overset{-}{N}{=}\overset{+}{N}{=}\overset{-}{N})]$, the central nitrogen atom of the azide ion, has a calculated charge of +0.64 while the other two nitrogen atoms have a charge of −0.72. The electron spectrum accordingly shows two nitrogen 1*s* lines of which the one of higher kinetic energy is twice the intensity of the other.

1.6. NITROGEN CHEMISTRY

Under the heading of General Introduction (section 1), this section attempts to indicate the scope of inorganic nitrogen chemistry and to outline the general trends and principles on which the subject can be rationalized. For this it will be convenient to list those concepts and fundamental properties associated with chemical combination and to define and illustrate such terms with examples from nitrogen chemistry for reference in the detailed discussion of each type of nitrogen compound in later sections.

Nitrogen will form bonds to most elements, perhaps the only exceptions being the noble gases; however, there is spectroscopic evidence for a diatomic xenon–nitrogen molecule of low stability[49]. Thus although nitrogen compounds of almost 100 elements need to be considered, the information is not evenly distributed over the Periodic Table. As so many of the most important nitrogen compounds contain only the light elements of the first short period, it is not surprising that nitrogen chemistry has centred around substances containing the elements hydrogen, carbon, nitrogen and oxygen, which include the organic compounds containing nitrogen[50], together with the hydrides, oxides and their related derivatives. Naturally these latter compounds will figure prominently in subsequent sections of this chapter. Apart from the halogen derivatives (see section 9), the nitrogen chemistry of the remaining elements will be considered individually in the chapter appropriate to the element concerned. However, as the majority of elements are metals, the nitrogen chemistry of metals as a group will be considered with reference to their complexes of nitrogen ligands of the inorganic or organic variety, and their salts of nitrogen anions such as amides, azides and nitrides. Mention will also be made to the considerable chemistry involving metal–nitrogen covalent bonds, which has largely emerged from developments in the organometallic chemistry of both transition and non-transition elements[51]. It is in such areas that the formal boundaries between organic, organometallic and inorganic chemistry become sufficiently blurred as to allow discussion of related compounds without embarrassment or apology to the purists. In fact, a survey of the known types of organic compounds containing nitrogen can suggest related structures containing other elements. Subsequent experiments based on such comparative exercises have led to the synthesis of novel compounds to which a great deal of recent experience in preparative inorganic chemistry bears witness. And perhaps this practical approach is no more appropriate than in nitrogen chemistry when one considers all that has been written, e.g., about the low chemical reactivity of molecular nitrogen, and this is equated with the facile reactions which have recently been

[49] J. H. Holloway, *Noble-gas Chemistry*, Methuen, London (1968), p. 44.

[50] N. V. Sidgwick, *The Organic Chemistry of Nitrogen*, 3rd edn., revised and rewritten by I. T. Millar and H. D. Springall, Oxford University Press, Oxford (1966).

[51] See *J. Organometal. Chem.*, 1963–date; *Organometal. Chem. Rev.*, 1966–date; *Annual Survey of Organometallic Chemistry*, 1964–date (eds. D. Seyferth and R. B. King), Elsevier, Amsterdam.

reported for the formation of nitrogen complexes in certain transition metal derivatives (see section 2.11).

The traditional exercise of vertical relationships between nitrogen and the other elements of Group V is probably not very meaningful except to emphasize their great differences. So few nitrogen compounds have isostructural phosphorus analogues that only isolated comparisons can be made[52]. It is more profitable to relate the chemistry of nitrogen to that of its neighbouring elements in the first short period, carbon and oxygen, the so-called horizontal relationship. For example, nitrogen is strongly electro-negative, forms hydrogen bonds and multiple bonds, is capable of catenation to a minor extent (the longest nitrogen chain so far reported[53] being N_8 in the organic derivative PhN=N—NPh—N=N—NPh—N=NPh), and is limited to a maximum covalency of four, all properties which would correctly position nitrogen between carbon and oxygen in the first row of the Periodic Table.

1.6.1. Bonding

The electronic structures of nitrogen compounds are generally rationalized on the basis of a completed octet of electrons around each atom in the molecules. From the electronic configuration $1s^2 2s^2 2p^3$, nitrogen has five electrons in the valence shell; hence the octet can be achieved by accepting or sharing three extra electrons from atoms or groups bonded to nitrogen. As nitrogen only accepts electrons from the most electro-positive elements,

TABLE 8. HOW NITROGEN CAN ACHIEVE AN OCTET OF ELECTRONS

Type	Example		
Electron gain	N^{---}	Li_3N	lithium nitride
Electron gain and sharing	HN^{--}	AgNH	silver imide
	H_2N^-	KNH_2	potassium amide
Electron sharing single bond	:N≤	NH_3	ammonia
Electron sharing double bond	⁄N=C<	R_2C=NR	imines
	⁄N=N⁄	RN=NR	diazenes
	⁄N=O	NOCl	nitrosyl chloride
Electron sharing triple bond	:N≡N:	N_2	nitrogen
	N≡C—C≡N	C_2N_2	cyanogen
	$(C≡N)^-$	KCN	potassium cyanide
Electron loss and sharing	NH_4^+	NH_4Cl	ammonium chloride
	NR_4^+		quaternary salts
	$N_2H_5^+$	$R_4NXN_2H_5Cl$	hydrazinium hydrochloride

bonding occurs predominantly by the formation of covalent bonds of the single or multiple variety, which may lead to ionic species by subsequent gain or loss of an electron. Examples of each type are included in Table 8. Ions formed by electron loss show nitrogen at its maximum coordination number of four, corresponding to the number of orbitals available in first short period elements.

[52] W. E. Dasent, *Nonexistent Compounds*, Edward Arnold, London (1965).
[53] P. A. S. Smith, *Open-chain Nitrogen Compounds*, Vol. 1, Benjamin, New York (1965), p. 10.

There also exist stable compounds such as nitric oxide and nitrogen dioxide in which there is an odd number of electrons and hence no possibility of achieving completed octets around all the atoms in the molecules. In such circumstances paramagnetism due to unpaired electrons can be detected, and bonding descriptions are usually rationalized in terms of molecular orbital theory. Thus, although electron counts for molecules are still noted, bonding is thought of in terms of the overlapping of atomic or hybrid orbitals, the effectiveness of such overlap being controlled by both symmetry and energy considerations. The usual way of obtaining molecular orbitals is by the linear combinations of atomic orbitals, the LCAO/MO method. Molecular orbitals formed by overlap of atomic orbitals along the molecular axis give rise to σ-bonds while those formed by overlap of atomic orbitals perpendicular to the molecular axis produce π-bonds. Nitrogen, like its neighbouring elements of the first short period, is capable of forming such π-bonds predominantly to itself, carbon, and oxygen of the p_π–p_π type as indicated in Table 8. While the majority of examples involving π-bonding will be of this type, mention will of course be made to p_π–d_π bonding (e.g. in Si–N compounds) which can occur with elements of the second and subsequent periods. Satisfactory explanations of the bonding in nitrogen compounds are provided either by the valence bond approach or molecular orbital theory, so both descriptions will be discussed for particular compounds in their appropriate sections[54].

Information about chemical bonds comes principally from two parameters, bond length as determined by diffraction or spectroscopic techniques[55], and bond strength from quantitative energy measurements during a bond-making or more frequently a bond-breaking process[56]. Theoretically one would expect to find a simple relationship between these two parameters, and undoubtedly short bonds are stronger than long bonds between the same elements. However, the nature of a particular bond will be dependent on the electronic configuration of the elements involved, the kind of orbital available and factors contributing to the bond polarity such as electro-negativity.

On the other hand, there is the problem of what is meant by bond strength. It is only with diatomic molecules that the bond dissociation energy can be defined unambiguously as the energy required to split the molecule into its component atoms. The bond dissociation energies of some nitrogen diatomic species are given in Table 9. For polyatomic molecules however, it is necessary to distinguish between two situations. The energy required to atomize a molecule can be determined, and by dividing this value by the number of bonds broken, the average bond energy $E(\text{N—X})$ results. Alternatively, bond strength could refer to the energy required to break one specific bond in the molecule in which case it is termed the bond dissociation energy $D(\text{NX}_2\text{—X})$. For example, consider the atomization of ammonia:

$$\text{NH}_3(\text{g}) = \text{N}(\text{g})+3\text{H}(\text{g}); \qquad \Delta H_{298} = +280.3 \text{ kcal/mol}$$

As three N–H bonds are broken, $E(\text{N—H}) = 280.3/3 = 93.4$ kcal/mol. However, this value is different from that for the removal of a single hydrogen atom from ammonia:

$$\begin{aligned} \text{NH}_3(\text{g}) &= \text{NH}_2(\text{g})+\text{H}(\text{g}); & D(\text{NH}_2\text{—H}) &= 104 \text{ kcal/mol} \\ \text{NH}_2(\text{g}) &= \text{NH}(\text{g})+\text{H}(\text{g}); & D(\text{NH—H}) &= 95 \text{ kcal/mol} \\ \text{NH}(\text{g}) &= \text{N}(\text{g})+\text{H}(\text{g}); & D(\text{N—H}) &= 81 \text{ kcal/mol} \end{aligned}$$

---

54 M. Green, in *Developments in Inorganic Nitrogen Chemistry*, Vol. 1 (ed. C. B. Colburn), Elsevier, Amsterdam (1966), p. 1.

55 *Tables of Interatomic Distances and Configuration in Molecules and Ions* (ed. L. E. Sutton), Chemical Society Special Publications 11 and 18, London (1958 and 1965).

56 L. Pauling, *The Nature of the Chemical Bond*, 3rd edn., Cornell University Press, Ithaca, New York (1960).

The sum of the bond dissociation energies is, of course, equal to the energy of atomization, but the regular decrease in dissociation energy is attributed to decreasing *s* character of the bonding orbitals.

TABLE 9. BOND DISSOCIATION ENERGIES IN SOME DIATOMIC SPECIES OF NITROGEN

Species	Bond dissociation energy eV	Bond dissociation energy kcal/mol
NH	3.21 ± 0.16	74
NH^+	3.7 ± 0.4	85
BN	4.0 ± 0.5	92
CN	7.75 ± 0.2	178
CN^+	4.7 ± 0.4	108
N_2	$9.760 + 0.005$	225.1
N_2^+	8.73 ± 0.01	201.4
NO	6.5 ± 0.01	149.9
NO^+	10.86 ± 0.04	250.4
NF	4 ± 1	92
MgN	1.5 ± 0.5	35 ± 10
AlN	3 ± 1	70
SiN	4.5 ± 0.4	104
PN	7.1 ± 0.05	164
NS	5.0 ± 0.7	115
NCl	4 ± 0.5	92
NSe	4.5 ± 1	104
NBr	2.8 ± 0.2	65
SbN	3.1 ± 0.5	71

A. G. Gaydon, *Dissociation Energies and Spectra of Diatomic Molecules*, 3rd edn., Chapman & Hall, London (1968), pp. 257–88.

Such information can be used to estimate other bond energies. For the atomization of hydrazine,

$$N_2H_4(g) = 2H(g) + 4H(g); \qquad \Delta H = 411.6 \text{ kcal/mol}$$

If we assume that the bond energy of the N–H bonds is the same as in ammonia, then the bond energy of the N–N bond $E(NH_2—NH_2) = 411.6 - (4 \times 93.4) = 38.0$ kcal/mol. The bond dissociation energy for the reaction

$$N_2H_4(g) = 2NH_2(g) \text{ gives } D(NH_2—NH_2) = 60 \text{ kcal/mol}$$

Clearly these values are significantly different, but by using the sum of the two appropriate dissociation energies D(NH—H) and D(N—H), a closer estimate to $D(NH_2—NH_2)$ can be obtained.

$$D(NH_2—NH_2) = 411.6 - [2 \times (95 + 81)] = 59.6 \text{ kcal/mol}$$

Thus bond energies calculated from thermochemical data and bond lengths determined by diffraction techniques for some common nitrogen bonds are listed in Table 10 together with the bond order for the particular bond in a specific compound. The bond order is defined as the number of electron pairs involved in bonding two atoms in a molecule, and represents the bond multiplicity. The data in Table 10 is reasonably consistent (compounds with anomalous values have been intentionally omitted) in that bond energy and bond order increase as bond length decrease, and can be used to estimate values of unrecorded bond energies or bond lengths.

TABLE 10. BOND ENERGIES AND BOND LENGTHS FOR SOME NITROGEN BONDS

Bond	Typical molecule	Bond order	Bond length[a] (Å)	Bond energy[b] (kcal/mol at 298 K)
N—H	NH	1	1.084	85
	NH_3	1	1.031	93.4
	NH_4^+	1	1.034	86
	HN_3	1	1.012	85
N—C	$MeNH_2$	1	1.47	69.7
N=C	$(MeC{=}NOH)_2$	2	1.29	147
N≡C	NCCN	3	1.16	197
	HCN	3	1.15	197
N—N	N_2H_4	1	1.45	38
	NH_2NO_2	1	1.40	67
	N_2O_4	1	1.75	52
N=N	MeN=NMe	2	1.24	99
	N_2O	2	1.29	119
	N_2^+	2.5	1.12	201.4
N≡N	N_2	3	1.098	225.1
N—O	NH_2OH	1	1.47	46
	$HO{-}NO_2$	1	1.36	57
	NO_3^-	1.33	1.23	91
	$MeNO_2$	1.5	1.22	103
	N_2O	1.5	1.19	147
	NO_2	1.75	1.197	112
	N_2O_4	1.75	1.18	
N=O	HONO	2	1.20	146
	NO_2^+	2	1.154	159
	NO	2.5	1.15	149.9
N≡O	NO^+	3	1.06	250.5

[a] *Tables of Interatomic Distances and Configuration in Molecules and Ions* (ed. L. E. Sutton), Chemical Society Special Publications 11 and 18, London (1958 and 1965).

[b] R. T. Sanderson, *J. Inorg. Nucl. Chem.* **30** (1968) 375.

1.6.2. Oxidation States

Nitrogen is capable of exhibiting all formal integral oxidation states from +5 to −3. By convention, the oxidation state of nitrogen in a particular molecule or ion containing fluorine, oxygen or hydrogen, is defined as the formal charge per atom of nitrogen if every fluorine, oxygen and hydrogen present carries a charge of −1, −2 and +1 respectively[57].

[57] C. K. Jorgensen, *Oxidation Numbers and Oxidation States*, Springer-Verlag, New York (1969).

Some examples are listed in Table 11. This definition relates nitrogen to fluorine—the most electro-negative element, and to oxygen and hydrogen—the constituent elements of water, the solvent for which most information is available. It is not in any way meant to restrict the application of the formal oxidation number concept to compounds only containing the elements mentioned. Since only oxygen and fluorine are more electro-negative than nitrogen,

TABLE 11. OXIDATION STATES OF NITROGEN COMPOUNDS

Oxidation state	Example
5	N_2O_5, HNO_3, NO_3^-
4	NO_2, N_2O_4
3	HNO_2, NO_2^-, NOF, NF_3
2	NO, N_2F_4
1	N_2O, $H_2N_2O_2$, $N_2O_2^{2-}$, HNF_2
0	N_2
−1	NH_2OH, H_2NF
−2	N_2H_4, $N_2H_5^+$
−3	NH_3, NH_4^+

it is with these elements that nitrogen can exhibit positive oxidation states although it should be realized that nitrogen in its positive states is not to be regarded as cationic nor in its negative states necessarily as an anion. Thus the concept of oxidation number to describe the oxidation state of an element is most useful in balancing equations, calculating the stoichiometry of reactions, and especially in conjunction with oxidation-reduction

TABLE 12. STANDARD REDOX POTENTIALS (pH = 0) FOR SOME NITROGEN COUPLES[a]

Electrode reaction	Equation	Couple	$E°$ (volts)
$N^{-II}+e = N^{-III}$	$N_2H_5^+ + 3H^+ + 2e = 2NH_4^+$	$N_2H_5^+/NH_4^+$	+1.275
$N^{-I}+e = N^{-II}$	$2NH_3OH^+ + H^+ + 2e = N_2H_5^+ + 2H_2O$	$NH_3OH^+/N_2H_5^+$	+1.42
$N^0+2e = N^{-II}$	$N_2 + 5H^+ + 4e = N_2H_5^+$	$N_2/N_2H_5^+$	−0.23
$N^0+e = N^{-I}$	$N_2 + 2H_2O + 4H^+ + 2e = 2NH_3OH^+$	N_2/NH_3OH^+	−1.87
$N^I+e = N^0$	$H_2N_2O_2 + 2H^+ + 2e = N_2 + 2H_2O$	$H_2N_2O_2/N_2$	+2.65
$N^{II}+e = N^I$	$2NO + 2H^+ + 2e = H_2N_2O_2$	$NO/H_2N_2O_2$	+0.71
$N^{III}+2e = N^I$	$2HNO_2 + 4H^+ + 4e = N_2O + 3H_2O$	HNO_2/N_2O	+1.29
$N^{III}+e = N^{II}$	$HNO_2 + H^+ + e = NO + H_2O$	HNO_2/NO	+1.00
$N^{IV}+e = N^{III}$	$N_2O_4 + 2H^+ + 2e = 2HNO_2$	N_2O_4/HNO_2	+1.07
$N^V+2e = N^{III}$	$HNO_3 + 2H^+ + 2e = HNO_2 + H_2O$	HNO_3/HNO_2	+0.94
$N^V+N^{II}+e = 2N^{III}$	$HNO_3 + NO + H^+ + e = 2NO_2^- + 2H^+$	HNO_3:NO/HNO_2	+0.49
$N^V+e = N^{IV}$	$2HNO_3 + 2H^+ + 2e = N_2O_4 + 2H_2O$	HNO_3/N_2O_4	+0.79
$3N^0+e = 3N^{I/3}$	$3N_2 + 2H^+ + 2e = 2HN_3$	N_2/HN_3	−3.1

[a] G. Charlot, D. Bézier and J. Courtot, *Selected Constants. Oxidation-reduction Potentials*, Pergamon, London (1958).

processes[58]. Redox reactions can be defined in terms of changing oxidation states, and if the values of the oxidation-reduction potentials are known, a clear picture of the redox solution chemistry of the element can be obtained. Apart from normal tabulations (Table 12), Latimer diagrams[59] as in Fig. 4 summarize the relationships between the various states, but these methods of presentation have largely been superseded by a graphical

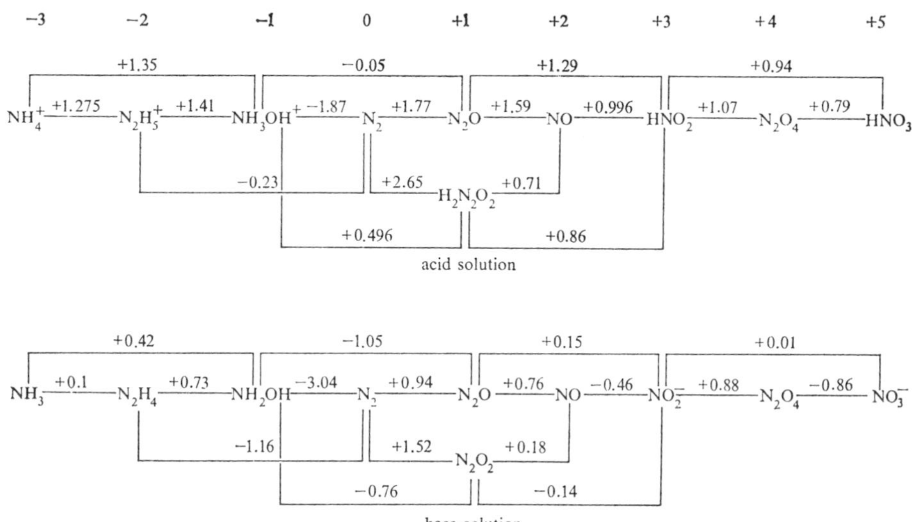

FIG. 4. Latimer diagrams for nitrogen species in aqueous solution.

method first suggested by Frost[60] and Ebsworth[61], and referred to as oxidation state diagrams (Fig. 5) by Phillips and Williams. These latter authors emphasize the use of such diagrams in their textbook, and take nitrogen as the example to illustrate the general principles of their application[62]. It will be shown that oxidation state diagrams for nitrogen can convey a good deal more information than can be conveniently presented by other methods.

The oxidation state diagram illustrated in Fig. 5 consists of a plot of the oxidation state versus the volt equivalent for various nitrogen species (molecules and ions containing the elements nitrogen, oxygen and hydrogen only) in aqueous acid solution. By definition, the volt equivalent of any nitrogen species, say HNO_3, is the product of its oxidation number and its redox potential measured relative to elemental nitrogen at 25°C.

$$E^\circ\ HNO_3/N_2 = +1.246\ V$$
$$N^V + 5e = N^0$$
$$VE = 5 \times 1.246 = +6.23\ V$$

[58] A. G. Sharpe, *The Principles of Oxidation and Reduction*, Royal Institute of Chemistry Monographs for Teachers Series, No. 2, London (1959).

[59] W. E. Latimer, *Oxidation Potentials*, Prentice-Hall, Englewood Cliffs, New Jersey (1952).

[60] A. A. Frost, *J. Am. Chem. Soc.* **73** (1951) 2680.

[61] E. A. V. Ebsworth, *Education in Chemistry* **1** (1964) 123.

[62] C. S. G. Phillips and R. J. P. Williams, *Inorganic Chemistry*, Oxford (1965), p. 314.

VE represents a reversible electrode potential corresponding to the free energy change (volts × 23.03 = kcal/mol) of the two nitrogen species involved according to the equation

$$HNO_3 + 2.5H_2 \rightarrow 0.5N_2 + 3H_2O$$
$$VE = (6.23 \times 23.06) = 144 \text{ kcal/mol}$$

Thus in Fig. 5, the values of *VE* in volts and their corresponding free energy values in kcal/mol are indicated for each nitrogen species, and hence the redox potential of any couple is simply the slope of the line joining the points corresponding to the two species involved.

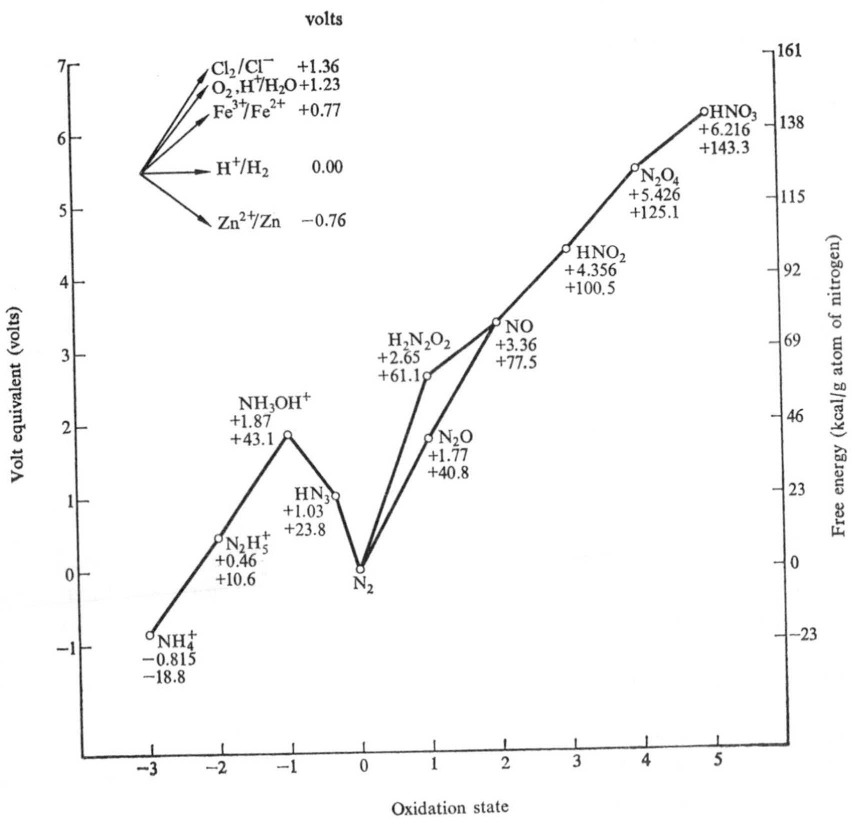

FIG. 5. The oxidation state diagram for nitrogen in acid solution.

The greater the slope, the stronger oxidizing or reducing the couple depending on whether the slope is positive or negative. So it follows that by comparing slopes of lines one can compare oxidizing and reducing properties not only within the nitrogen series but with other elements, and it is for this reason that a group of different sloped lines representing other redox couples are included near the upper left hand corner of the figure. For example, the N_2–$N_2H_5^+$ couple will be a thermodynamically stronger reducing agent than H^+–H_2, and likewise N_2–NH_3OH^+ will be stronger than Zn^{++}–Zn. A vertical downward movement on the diagram corresponds to an increase in stability and occurs with a decomposition reaction, e.g. $H_2N_2O_2$ decomposes readily into N_2O and H_2O. A similar situation arises in considering ammonium nitrate as the line joining NH_4^+ and nitric acid is bisected to give the position of ammonium nitrate at a point just above N_2O. The fact that N_2O can be conveniently prepared by heating ammonium nitrate is at least consistent with this observation. Thus when more than two species lie on the same line, an equilibrium is likely to result, whereas any species occurring at a convex point on the diagram should be thermodynamically unstable with respect to disproportionation into the neighbouring species.

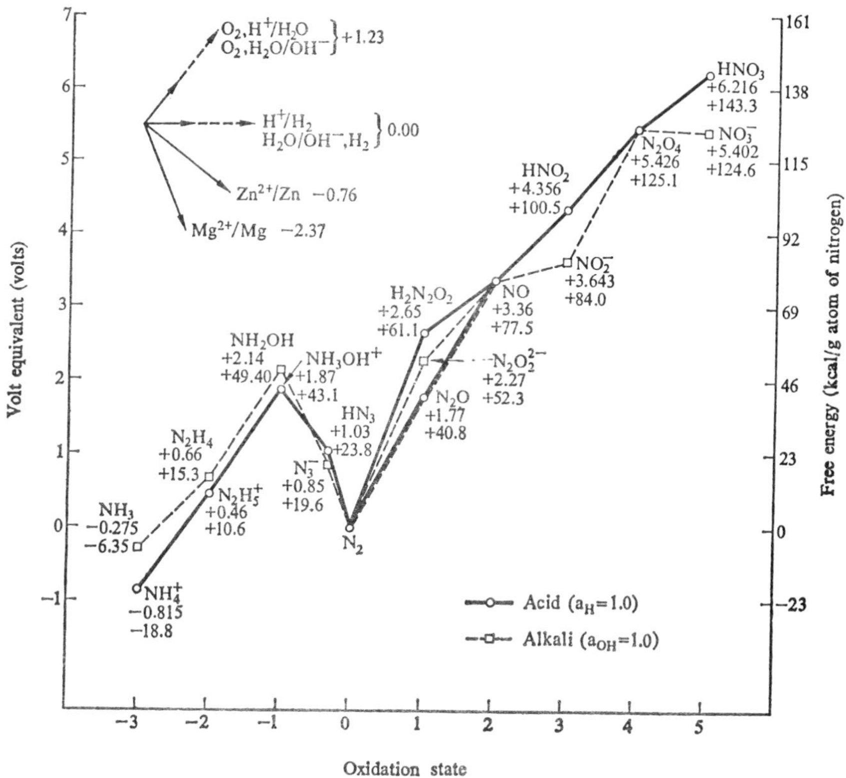

FIG. 6. The oxidation state diagram for nitrogen in acid and alkaline solution.

Reactions in alkaline solution can also be considered, and an oxidation state diagram constructed in a similar way. Figure 6 shows the oxidation state diagram for nitrogen in both acid and alkaline solutions, and includes the volt equivalents *VE* and *VE*(OH) together with corresponding free energy values for individual nitrogen species. *VE* and *VE*(OH) values differ for charged species due to the reaction

$$2H_2O+2e \rightarrow H_2+2OH^-$$

having a standard potential of 0.828 V on the standard hydrogen scale. This means that for fully ionized acids, the value of *VE*(OH) will be lower than *VE* by 0.828 V corresponding to a decrease of 19.1 kcal/mol. For example, $HNO_3 = 6.23-0.828 = 5.4$ V or

144−19.1 = 125 kcal/mol, the lower value being due to the neutralization reaction which occurs when nitric acid is formed in alkaline solution:

$$HNO_3+OH^- \rightarrow NO_3^-+H_2O$$

For weak acids, some energy is used to ionize the acid, and hence the $VE(OH)$ value will be higher than $VE-0.828$ to the extent of $(RT/F)\ \ln K_a$, where K_a is the dissociation constant for a monobasic acid, i.e.

$$VE(OH) = VE-[0.828+(RT/F)\ \ln K_a]$$

TABLE 13. THERMODYNAMIC FUNCTIONS AT 25°C FOR SELECTED NITROGEN SPECIES

Species	State	Formula weight	ΔH_f° (kcal/mol)	ΔG_f° (kcal/mol)	S° (cal/deg mol)
N	g	14.0067	112.979	108.883	36.622
N_2	g	28.0134	0	0	45.77
NH_3	g	17.0306	−11.02	−3.94	45.97
NH_3	aq		−19.09	−6.35	26.6
NH_4^+	aq	18.0386	−31.67	−18.97	27.1
N_2H_4	l	32.0453	12.10	35.67	28.97
N_2H_4	g		22.80	38.07	56.97
N_2H_4	aq		8.20	30.6	33
$N_2H_5^+$	aq	33.0532	−1.8	19.7	36
HN_3	l	43.0281	63.1	78.2	33.6
HN_3	g		70.3	78.4	57.09
HN_3	aq		62.16	76·9	34.9
NH_2OH	c	33.0300	−27.3		
NH_2OH	aq		−23.5	−5.60	40
NH_4OH	l	35.0460	−86.33	−60.74	39.57
NH_4HO_2	aq	51.0454	−69.99	−35.1	32.8
N_2H_5OH	g	50.0606	−49.0	−18.9	63
N_2H_5OH	aq		−60.11	−26.1	49.7
NO	g	30.0061	21.57	20.69	50.347
NO_2	g	46.0055	7.93	12.26	57.35
N_2O	g	44.0128	19.61	24.90	52.52
N_2O_3	g	76.0116	20.01	33.32	74.61
N_2O_4	g	92.0110	2.19	23.38	72.70
N_2O_4	l		−4.66	23.29	50.0
N_2O_5	c	108.0104	−10.3	27.2	42.6
N_2O_5	g		2.7	27.5	85.0
NO_2^-	aq	46.0055	−25.0	−8.9	33.5
NO_3^-	aq	62.0049	−49.56	−26.61	35.0
$N_2O_2^{2-}$	aq	60.0122	−4.1	33.2	6.6
HNO_2	aq	47.0135	−28.5	−13.3	36.5
HNO_3	l	63.0129	−41.61	−19.31	37.19
HNO_3	g		−32.28	−17.87	63.64
HNO_3	aq		−49.56	−26.61	35.0
$HNO_3 \cdot H_2O$	l	81.0282	−113.16	−78.61	51.84
$HNO_3 \cdot 3H_2O$	l	117.0589	−252.40	−193.91	82.93
$HN_2O_2^-$	aq	61.0202	−12.4	18.2	34
$H_2N_2O_2$	aq	62.0281	−15.4	8.6	52

D. D. Wagman, W. H. Evans, V. B. Parker, I. Halow, S. M. Bailey and R. H. Schumm, *Selected Values of Chemical Thermodynamic Properties*, NBS Technical Note 270-3, Washington (1968).

Bases in alkaline solution lie at slightly higher than the *VE* value, the increase being related to the acid dissociation constant of the conjugate acid of the base:

$$VE(\mathrm{OH}) = VE + (RT/F)\ln K_a$$

The free energies corresponding to the *VE*(OH) values for hydrides such as NH_3 or N_2H_4 are simply the standard free energies of formation from the elements. Clearly, neutral species such as N_2, N_2O, NO and N_2O_4 will be unaffected:

$$VE(\mathrm{OH}) = VE$$

Thus thermodynamic data for a large number of reactions can be calculated using the oxidation state diagram in Fig. 6 together with the heats of formation, free energies of formation and entropies of selected nitrogen species listed in Table 13. However, one should not forget the limitations of thermodynamic considerations which only sort out reactions into possible or impossible and does not take into account kinetic factors, i.e. the rate at which a possible reaction occurs or how they will affect the choice between two alternative reactions.

1.6.3. Stereochemistry

In common with other neighbouring first-row elements, the maximum coordination number for nitrogen is four, and thus the only regular stereochemical arrangements that it can exhibit are tetrahedral, pyramidal, planar, angular and linear. The shapes of most nitrogen molecules can be predicted using the Sidgwick–Powell[63] approach by allowing each pair of electrons on the central atom to take up the stereochemical position of maximum repulsion irrespective of whether the electrons are involved in bonding or not. The extension of this approach by Gillespie and Nyholm[64] is also appropriate for considering distortions produced by the electrostatic interactions between lone pairs and bonds. As the electrons of a lone pair lie closer to the atomic nucleus than those in a bond, electrostatic repulsion between electron pairs will be greatest in the sequence

loan pair:loan pair > loan pair:bond pair > bond pair:bond pair

For example, the lone pair:bond pair repulsion reduces the HNH bond angle in NH_3 to 108° from what is initially regarded as sp^3 hybrid orbitals with tetrahedral bond angles of 109° 28′.

Substituted ammonias of the type R_3N are in general also pyramidal. Thus molecules of the type RR′R″N should have non-superposable mirror images and hence exhibit optical activity. No such optical isomers have yet been isolated because inversion can readily take place, i.e. where one isomer reverts to the other by an umbrella type of movement and the nitrogen atom crosses over from one side of the plane N′ to the other N″ accompanied by the potential energy change illustrated in Fig. 7. For ammonia this potential energy barrier to inversion is only about 6 kcal/mol, and similar values would be expected for R_3N type molecules, whereas a barrier of the order of 20 kcal/mol would be necessary in order to make the rate of interconversion sufficiently slow for isolation of optical isomers to be possible[65].

[63] N. V. Sidgwick and H. M. Powell, *Proc. Roy. Soc.* (*London*) A, **176** (1940) 153.

[64] R. J. Gillespie and R. S. Nyholm, in *Progress in Stereochemistry*, Vol. 2 (ed. W. Klyne and P. B. D. de la Mare), Butterworths, London (1968), p. 261.

[65] R. E. Weston, *J. Am. Chem. Soc.* **76** (1954) 2645.

On the basis of this argument, one might expect all compounds containing nitrogen bonded to three groups to be pyramidal. However, in circumstances where the lone pair on nitrogen can participate in partial double bonding (π-bonding) with the neighbouring atom, even planar structures have been reported. For example, trisylylamine $(H_3Si)_3N$ is coplanar with Si–N bond lengths considerably shorter than a single bond (see section 4)[66]. Nitrogen is coplanar both in urea NH_2CONH_2 and in formamide $HCONH_2$ where the C–N bond lengths also suggest appreciable double bond character[67].

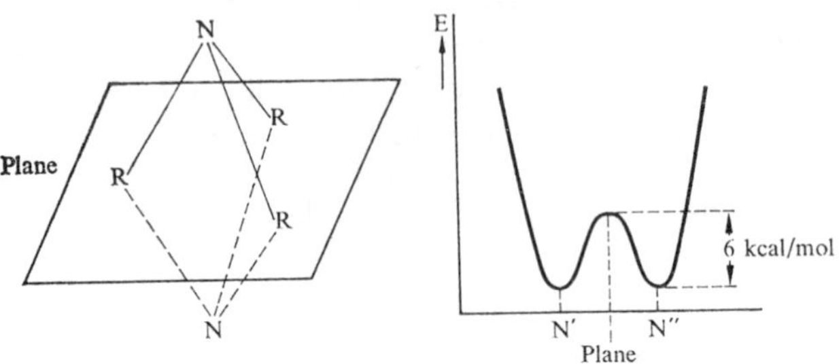

FIG. 7. Inversion of R_3N compounds.

1.7. BIOLOGICAL PROPERTIES OF NITROGEN

1.7.1. The Nitrogen Cycle

Nitrogen is found in many compounds which are vital to all forms of life including proteins, nucleic acids, co-enzymes, vitamins and hormones. Animals synthesize these complex molecules from organic intermediates available in food, while plants are able to utilize simple inorganic compounds which arise in the soil[23]. The concentration of nitrogen compounds in the soil, however, is low, and would quickly become depleted in the absence of some replenishing process involving the abundant source of the element available from the atmosphere[68]. This process whereby micro-organisms convert atmospheric nitrogen into simple nitrogen compounds on which the majority of plants can thrive is known as nitrogen fixation. The initial product of fixation is ammonia, from which further transformations can be easily rationalized.

The routes by which nitrogen pass between environment and living organisms is usually illustrated as the nitrogen cycle[69] in Fig. 8. Organisms thus play a critical role in the nitrogen cycle which maintains in the soil an adequate supply of nitrogen compounds essential to plant life. The extra addition of nitrogen compounds to the soil usually leads to both increased yields and better quality of many crops, but because nitrogen uptake is related to plant assimilation of other elements, commercial fertilizers are often compounded in definite ratios of available nitrogen, phosphorus and potassium. The choice of nitrogen compounds for use as fertilizer is quite wide, and includes ammonia, ammonium salts, nitrates, urea,

[66] K. Hedberg, *J. Am. Chem. Soc.* **77** (1955) 6491.

[67] A. F. Wells, *Structural Inorganic Chemistry*, 3rd edn., Oxford University Press, Oxford (1962).

[68] F. Call, in *Mellors Comprehensive Treatise on Inorganic and Theoretical Chemistry*, Vol. 8, Suppl. I, *Nitrogen*, Part I, Longmans, London (1964), Section XVII, p. 568.

[69] J. J. W. Baker and G. E. Allen, *The Study of Biology*, Addison-Wesley, Reading, Mass. (1967).

ureaform and calcium cyanamide, in a variety of physical forms as powders, crystals, pellets, prills and liquid. The actual physical form of a fertilizer is important from the view of storage and the mode of application[26].

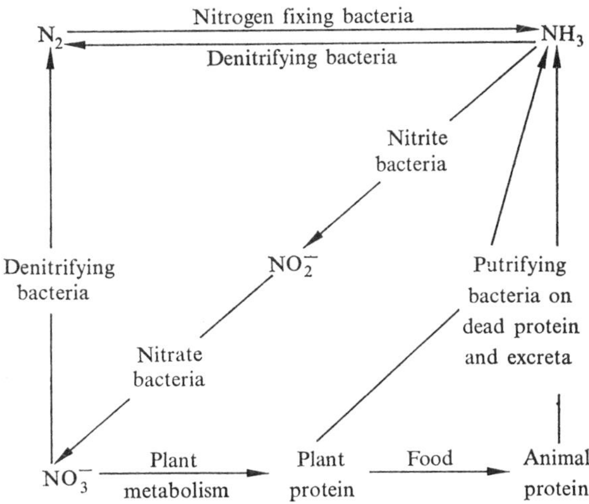

FIG. 8. The nitrogen cycle.

With regard to nitrogen uptake, fertilization by foliar sprays is now carried out on an increasing scale, but nitrogen is usually assimilated by plants in the form of ammonium or nitrate ions through the root system. Ammonia is rather immobile in the ground being held in the soil structure as the ammonium cation and tending to be released over a prolonged period. Nitrate, on the other hand, can move freely in the soil moisture and is taken up quickly by plants, although losses by leaching can be considerable. Calcium cyanamide is hydrolysed to give ammonia as also is urea, while guanyl urea and the urea–formaldehyde condensation product known as ureaform are slow-release fertilizers. Indiscriminate use of fertilizers can deplete the soil of other essential growth factors or accumulate undesirable byproducts, which emphasizes the need to select fertilizers to meet the specific needs of the particular soil and crop combination. The currently used methods of conserving land fertility by alternating grain with legumes or combining crop rotation with livestock farming will be substituted by balanced nitrogen fertilizers to ensure continuous utilization of all arable land.

1.7.2. Nitrogen Fixation

Micro-organisms from the genera *Clostridium*, *Azobacter* and *Azotobacter*, which are normally found in the soil, are capable of performing the complete process of nitrogen fixation quite independently, as also are certain species of aerobic, anaerobic and photosynthetic bacteria, actinomyces, yeasts and blue–green algae. The only plants which can take part in nitrogen fixation are those which bear root nodules (notably the legume family) inhabited by bacteria of the genus *Rhizobium*. This association between plant and bacteria is termed symbiosis, which means that in the presence of a fixed nitrogen source both can exist separately, but when fixed nitrogen is not available, neither can grow or fix nitrogen independently[3].

Research has been aimed at isolating the nitrogen-fixing enzyme system as cell-free extracts in a pure state[70]. It is known that the reduction of elemental nitrogen by the enzyme system nitrogenase occurs in conjunction with the related system hydrogenase to give ammonia as the product. No intermediate reduced nitrogen species have yet been isolated. The enzyme system required for nitrogen reduction involves at least two protein components which contain the elements iron and molybdenum[71], and is capable of other remarkable reductions. For example, azide can be reduced to ammonia, cyanide to methane and ammonia, and even acetylene has been reduced to ethylene. The proposed mechanisms of fixation have recently been reviewed[6] but are likely to be modified in the light of recent experience with the chemistry of nitrogen complexes (see section 2.11.3).

1.8. ANALYTICAL CHEMISTRY OF NITROGEN

The methods used in the identification, characterization and analysis of specific nitrogen compounds and ions will be discussed in each following section, and general reference will be made to the problems associated with their detection and determination in a variety of raw materials, processes and products. Thus in this section, only general principles behind the analytical chemistry of nitrogen will be indicated together with the routine methods of nitrogen determination currently used in analytical laboratories[72].

Nitrogen, and ultimately all nitrogen compounds are capable of being converted into molecular nitrogen, can be identified by its emission spectrum with a quartz spectrograph, and its purity determined by mass spectrometry. For detecting the presence of nitrogen compounds without regard to specific identification, oxidation to nitrate or reduction to ammonia followed by some suitable spot test[73] is usually adequate. Nitrogen in organic compounds is usually detected by Lassaigne fusion with sodium where the resulting cyanide is confirmed by the Prussian-blue test.

In general, the quantitative determination of nitrogen in unknown materials is also based essentially on simple conversions to nitrate, ammonia or nitrogen, which can be estimated by a range of conventional chemical analytical techniques including titrimetric[74,75], colorimetric[76,77], gravimetric[75] and gasometric methods[78]. However, the need to develop rapid methods for the analysis of compounds and mixtures associated with

[70] L. E. Mortenson, *Biochim. Biophys. Acta* **127** (1966) 18.

[71] L. E. Mortenson, in *Survey of Progress in Chemistry*, Vol. 4 (ed. A. F. Scott), Academic Press, New York (1968), p. 127.

[72] K. I. Vasu, in *Mellors Comprehensive Treatise on Inorganic and Theoretical Chemistry*, Vol. 8, Suppl. II, *Nitrogen*, Part II, Longmans, London (1967), Section XXXVII, p. 563.

[73] F. Feigl, *Spot Tests in Organic Analysis*, 7th edn., Elsevier, Amsterdam (1966); *Spot Tests in Inorganic Analysis*, Elsevier, Amsterdam (1958).

[74] A. J. Clear and M. Roth, in *Treatise on Analytical Chemistry*, Part II, Vol. 5 (eds. I. M. Kolthoff and P. J. Elving), Interscience, New York (1961), p. 217.

[75] A. F. Williams, in *Comprehensive Analytical Chemistry*, Vol. IC (eds. C. L. Wilson and D. W. Wilson), Elsevier, Amsterdam (1962), p. 203.

[76] G. Charlot, *Colorimetric Determination of Elements*, Elsevier, Amsterdam (1964).

[77] M. J. Taras, in *Colorimetric Determination of Nonmetals* (ed. D. F. Boltz), Interscience, New York (1958), p. 75.

[78] A. I. Vogel, *A Textbook of Quantitative Inorganic Analysis*, 3rd edn., Longmans, London (1961).

industrial processes has required the application of physical methods based on chromatography, thermal conductivity, mass spectrometry and absorption spectroscopy, the tendency being for continuous on-line analysis where possible.

The routine determination of nitrogen in organic materials is also as the element or ammonia, and the methods briefly outlined below are clearly applicable to the analysis of many nitrogen complexes of metals, organometallic derivatives, as well as some purely inorganic compounds.

The Dumas method[79] dating from 1831 is based simply on the thermal decomposition of organic material with a metal oxide in an atmosphere of carbon dioxide. The liberated nitrogen is then measured volumetrically over potassium hydroxide solution which absorbs the other combustion products and carbon dioxide carrier gas. The method is applicable with a wide range of samples, even inorganic compounds by adding some organic material such as benzoic acid, and has reached an advanced state of sophistication with the availability of an automatic commercial instrument[80]. For these reasons it is generally the method of choice.

The Kjeldahl method[79, 81], although not applicable to all organic nitrogen compounds, is still widely used in fertilizer and food analysis. The method involves digestion of organic nitrogen-containing material with sulphuric acid and potassium sulphate in the presence of a catalyst (HgO) at about 350°C. The nitrogen is converted to ammonia, and remains in the acid solution as ammonium sulphate. The acid digest is diluted and then made sufficiently basic to permit quantitative distillation of the ammonia. The ammonia is absorbed in standard acid and then back titrated with standard base. Developments have also been made in automating the method using the Technicon AutoAnalyzer[82].

In the ter Meulen method[79], organic nitrogen-containing material is pyrolysed in an atmosphere of hydrogen, and the vaporized products passed over Ni–Mg catalyst at 320°C to give ammonia which can be determined titrimetrically or colorimetrically depending on the amount of nitrogen present. This method is particularly good for determining traces of nitrogen and has found application in petroleum analysis.

The chemical composition of a substance, however, is often only an initial step to characterizing a new or known compound. The way in which the investigation will subsequently proceed depends largely on the equipment available and to some extent on the techniques in vogue at the time. Many nitrogen compounds which are stable in the gaseous or liquid phase can be identified, characterized and even analysed by the modern spectroscopic techniques applicable in organic chemistry of infrared, ultraviolet, n.m.r., mass spectrometry and gas–liquid chromatography. Detailed structural parameters of volatile materials can also be obtained from electron diffraction and other spectroscopic techniques. The investigation of solid materials, traditionally an inorganic activity, usually relies on finding suitable solvents in which to carry out spectroscopic examinations. However, although the results of such "sporting" techniques are adequate in identifying known substances and predicting structures for new compounds, the chemist is ultimately only satisfied when he can rationalize the chemistry of the material with structural parameters from X-ray crystallography.

[79] G. M. Gustin and C. L. Ogg, in *Treatise on Analytical Chemistry*, Part II, Vol. 11 (eds. I. M. Kolthoff and P. J. Elving), Interscience, New York (1961), p. 405.

[80] Coleman Instrument Co., Maywood, Illinois, USA.

[81] R. B. Bradstreet, *The Kjeldahl Method for Organic Nitrogen*, Academic Press, New York (1965).

[82] Technicon Instruments Co. Ltd., Hanworth Lane, Chertsey, Surrey, England.

2. ELEMENTAL NITROGEN

2.1. DISCOVERY AND HISTORY

Although it is difficult to say who first isolated nitrogen and clearly recognized it as a definite substance, Rutherford in Scotland is generally credited with the discovery[83], although Scheele in Sweden made similar independent observations about the same time. In 1772, oxygen from air was removed by combustion, then the soluble products were dissolved out, and the remaining gas was reported to extinguish flames and destroy life. This asphyxiating gas (German *Stickstoff*), which Scheele called "foul air", was later recognized to be an element by Lavoisier who called it *azote*, meaning without life; the prefix *az* still being used in chemical nomenclature as an indication of combined nitrogen in many of its compounds, e.g. azide, hydrazine, etc.

2.2. OCCURRENCE AND DISTRIBUTION

Nitrogen constitutes 78% by volume of the earth's atmosphere. The concentration of nitrogen gas decreases with increase in distance from the surface of the earth, and it has been claimed that perceptible fractionation of the two stable isotopes occurs due to gravitation[84]. Considerable quantities of nitrogen occur in natural gas and thermal springs, and it has even been detected in the atmosphere of the planet Mars.

2.3. PRODUCTION OF NITROGEN

The three principal methods of manufacturing nitrogen[85] are the low-temperature separation of air, the chemical deoxygenation of air and the dissociation of ammonia. The preparation of synthesis gas for the production of ammonia will be discussed in section 3.3.

2.3.1. Low-temperature Separation from Air

Gaseous and liquid nitrogen are produced by air separation plants based on the low-temperature fractional distillation of liquid air[86]. Modern gas liquefaction plants are based largely on the Linde process which revolutionized the industry in the early years of this century by applying the Joule–Thompson effect.

The main stages of the process involve (i) compression of air, (ii) purification of compressed air, (iii) cooling of compressed air, (iv) liquefaction by expansion and cooling, and (v) fractional distillation.

In the diagram illustrated in Fig. 9, filtered air enters the compressor cylinder *A* and the pressure is raised to 200 atm. After carbon dioxide and water have been removed in *B*, the heat produced by rapid compression is absorbed as it passes through the heat exchanger *C*. The compressed air is then allowed to expand suddenly at the orifice *D*, where it cools rapidly due to the Joule–Thompson expansion (an isenthalpic process). This cooled air

[83] M. E. Weeks, revised by H. M. Leicester, *Discovery of the Elements*, 7th edn., Journal of Chemical Education, Washington (1968).

[84] F. H. Day, *The Chemical Elements in Nature*, Harrap, London (1963).

[85] B. R. Brown, in *Mellor's Comprehensive Treatise on Inorganic and Theoretical Chemistry*, Vol. 8, Suppl. I, *Nitrogen*, Part I, Longmans, London (1964), pp.1–26.

[86] M. Sittig, *Nitrogen in Industry*, Van Nostrand, New Jersey (1965).

still at slight positive pressure enters the outer tube *E* surrounding the inner tube leading to *D*, and cools it further, producing such a low temperature that the incoming gases liquefy at *D* and collect in *F*. The cooled air leaving *E* can then be circulated back via the heat exchanger or direct to the cylinder, thus completing a partial cycle. The compressed air can also be liquefied in an expanding engine by an isentropic process. The liquid air is fractionated in various columns depending on whether gaseous or liquid products are required.

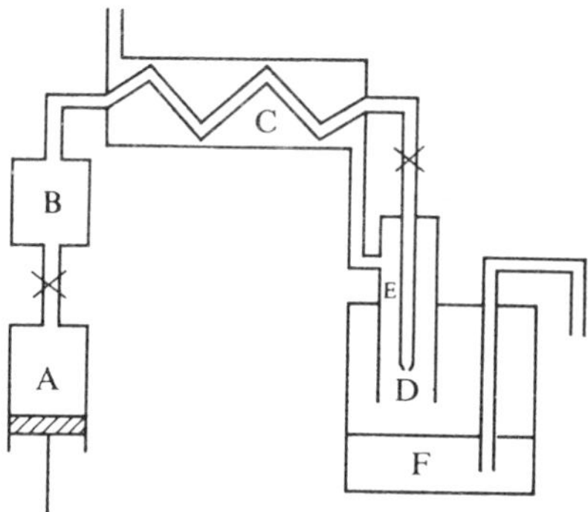

FIG. 9. The liquefaction of air.

Modern air separation plants operate very efficiently with a high degree of automation and are capable of producing several tons of liquid nitrogen per hour[87]. Laboratory units producing high purity liquid nitrogen at up to 4 litres per hour are also available[88].

2.3.2. Chemical Deoxygenation of Air

The obvious chemical method of removing oxygen from air is by combustion[89]. Usually the oxidizable substance is a carbonaceous material such as a hydrocarbon or coal. Quantities are calculated so that all the oxygen is converted to carbon dioxide which can be removed by scrubbing the gas under pressure with water or monoethanolamine. Residual water is removed by condensation or by passing through driers containing silica gel or activated alumina. Substantially pure nitrogen is thus obtained and used as inert atmosphere for steel- and metal-working.

2.3.3. Dissociation of Ammonia

For ammonia to be decomposed into nitrogen, the principal use of which is to prepare ammonia, must sound rather peculiar logic. However, ammonia is readily available in a pure state and reasonably cheap, and can be "cracked" directly simply by reversing the conditions required for its synthesis (see section 3.3). Ammonia at 600°C and pressures of 1–15 atm in the presence of a catalyst (usually a metal of the platinum group) is almost completely dissociated into nitrogen and hydrogen. The hydrogen can be partially or

[87] M. Sittig and S. Kidd, *Cryogenics—Research and Application*, Van Nostrand, New Jersey (1963), p. 100.
[88] H. E. Charlton and H. J. V. Charlton, *Trans. Inst. Chem. Engrs.* **41** (1963) CE 254.
[89] A. J. Moyes and R. F. Keyte, *Trans. Inst. Chem. Engrs.* **41** (1963) CE 248.
[90] Anon., *Nitrogen*, No. 5 (1960) 38.

totally removed by combustion with air, and this can occur concurrently with the dissociation reaction to maintain the temperature. After further treatment, nitrogen can be obtained pure if required, but the process is particularly useful when a reducing atmosphere is required.

2.3.4. Other Methods

One way of physically separating air into its components without resorting to low-temperature distillation is by molecular sieve[91]. The critical diameters of the atmosphere gases nitrogen, oxygen and argon are 3.0, 2.8 and 3.8 Å respectively. As natural and synthetic zeolites can be obtained with pore channels of specific dimensions, one with a cut-off at just below 3 Å would be suitable. The method has been shown to be capable of producing nitrogen 99.25% pure, but optimum conditions gave a product containing about 10% oxygen.

Other methods which have been suggested include separation by selectively permeable films, adsorption and diffusion. All these other methods could also be applied more directly to nitrogen–hydrogen mixtures from ammonia dissociation.

2.3.5. Storage and Handling of Nitrogen[86]

Although cylinders of compressed nitrogen are convenient for small-scale and intermittent use, the trend in recent years has been more towards the handling of large quantities as liquid. Small amounts of liquid can be stored in Dewar vessels of glass or copper insulated by a vacuum. Larger quantities are contained in Horton spheres which are made up of a stainless steel inner sphere surrounded by insulation and supported in a protective carbon steel shell. Twelve ton loads of liquid nitrogen can be transported by road and delivered as either gas or liquid.

The main precautions to be observed when handling liquid nitrogen and equipment that has been in contact with it are with regard to frostbite. Care should be taken to avoid being splashed especially when immersing an object at room temperature. Precautions should also be taken to avoid large areas of liquid nitrogen to be exposed to the atmosphere since it can condense oxygen in concentrations which can become hazardous in the presence of organic material. Adequate ventilation should be ensured when large amounts of evaporating liquid nitrogen are confined to enclosed places, as oxygen content of the air can be reduced with the danger of asphyxiation.

2.4. INDUSTRIAL USES OF NITROGEN

2.4.1. Chemical Applications

The manufacture of ammonia is the most important single use of elemental nitrogen[92]. In the Haber process, nitrogen and hydrogen react at high temperatures and pressures in the presence of a catalyst according to the equilibrium

$$N_2(g) + 3H_2(g) \xrightleftharpoons{\text{catalyst}} 2NH_3(g)$$

The process will be discussed in detail in section 3.3.

[91] Anon., *Chem. Week.* **92** (1963) 37.
[92] A. J. Harding, *Ammonia Manufacture and Uses*, Oxford University Press, Oxford (1959).

Important quantities of nitrogen have been used in calcium cyanamide production[93] which can be used as a fertilizer and was formerly an important route to ammonia. The Frank–Caro process involves burning limestone to produce lime which reacts with coke in an electric furnace giving calcium carbide as the main product. The finely ground carbide then reacts with pure nitrogen at 1100°C to give calcium cyanamide.

$$CaCO_3 \rightarrow CaO + CO_2$$
$$CaO + 3C \rightarrow CaC_2 + CO$$
$$CaC_2 + N_2 \rightarrow CaCN_2 + C$$

Related to this reaction is the Bucher process first described by Castner and which was formerly used for the manufacture of sodium cyanide by the reaction of nitrogen with red-hot charcoal and sodium.

$$Na_2CO_3 + 2C \rightarrow 2Na + 3CO$$
$$2Na + 2C \rightarrow Na_2C_2$$
$$Na_2C_2 + N_2 \rightarrow 2NaCN$$

The Birkeland–Eyde electric arc oxidation of atmospheric nitrogen to nitric oxide for nitric acid production was developed in Norway using low-cost hydroelectric power. At the present time, however, all such arc processes are uneconomical and obsolete. A large demonstration plant was built and operated during 1953 in an attempt to develop a direct nitrogen-fixation route to nitric oxide by the high-temperature combination of the elements in the "pebble-bed" or Wisconsin process[94], but this too proved uneconomic.

The use of nitrogen in metal nitride synthesis will be considered in section 4.2.

2.4.2. Cryogenic Applications[86, 87]

Liquid nitrogen is used in the laboratory as a convenient medium for cooling apparatus, condensing gaseous materials in vacuum apparatus, and for preparing "slush baths" by cooling a liquid to its melting point to provide a constant-temperature bath. Liquid air is no longer used in cryogenic work because of the potential inflammable and explosive hazards which can occur in the event of contamination by organic materials.

Many materials which are tough or pliable at ambient temperatures become brittle at liquid nitrogen temperatures and less likely to overheat or deteriorate during grinding operations. Other mechanical uses include shrink fitting where a slightly oversize component contracts sufficiently on cooling to −196°C to slide into position and be held firmly on expansion as it attains ambient temperature.

Liquid nitrogen is also finding increasing use in food-freezing processes and as a refrigerant for frozen and perishable foods during transit. Other modern applications are to be found in cryobiology and cryosurgery, while the largest single consumer of liquid nitrogen in the United States is the aerospace industry.

2.4.3. Inert Atmosphere Applications[86]

In the laboratory, dry nitrogen is a satisfactory substitute for argon in most of its applications as an inert atmosphere in dry boxes and glove bags. The food industry use nitrogen during preparation, packing and canning of a wide variety of food products to prevent spoilage by atmospheric oxygen. Nitrogen acts as a protective atmosphere for the

[93] Anon., *Nitrogen*, No. 22 (1963) 32.
[94] E. D. Ermenc, *Chem. Engng. Prog.* **52** (1956) 149.

bright hardening and annealing of steel, chilling in aluminium foundries and during arc welding. The molten tin bath used in the manufacture of float glass is also protected by nitrogen.

Chemical industries use nitrogen atmosphere either to avoid reaction with oxygen or to prevent the formation of dangerous explosive mixtures. Some examples of its application include the manufacture of reactive organic chemicals, acrylonitrile, polyamides, polyesters, lubricants and waxes. It is also used in the petroleum industry during fuel storage, polymerization processes and plastics fabrication.

Based on its properties as a non-reactive gas, nitrogen finds application as a fluidizing gas in powder technology, a propellant gas in aerosols, an inert pressure medium for liquids which are reactive with air and for the agitation of liquids.

2.5. LABORATORY PREPARATION OF NITROGEN

For most laboratory applications, nitrogen from cylinders can be used directly or after purification. However, laboratory methods of preparation may be appropriate if small quantities of high-purity nitrogen are required or when the conversion of a nitrogen compound to the element is required during an analytical procedure. Nitrogen prepared by chemical methods is free from inert gas impurities, which makes it particularly useful for spectroscopic purposes. Such methods include the following[95]:

(1) The reaction of ammonium ion and nitrite ion:

$$NH_4^+ + NO_2^- \rightarrow N_2 + 2H_2O$$

In practice, saturated sodium nitrite solution is added dropwise to hot saturated ammonium chloride solution, or saturated ammonium nitrite solution can be warmed directly.

(2) The reaction of ammonia and bromine:

$$8NH_3 + 3Br_2 \rightarrow N_2 + 6NH_4Br$$

The oxidation of ammonia is a general procedure, other oxidizing agents being dichromate (as in (3)), ozone, fluorine and manganese dioxide. Ammonia can also be thermally decomposed to nitrogen and hydrogen in the presence of a platinum catalyst.

(3) The decomposition of ammonium dichromate:

$$K_2Cr_2O_7 + (NH_4)_2SO_4 \rightarrow N_2 + Cr_2O_3 + K_2SO_4 + 4H_2O$$

The thermal decomposition of ammonium dichromate is an explosive reaction which can be controlled by the presence of sulphate, presumably as a result of slow diffusion in the solid state.

(4) The thermal decomposition of metal azides[96]:

$$2NaN_3 \xrightarrow{300^\circ} 2Na + 3N_2$$

Sodium or barium azide can be purified by recrystallization and careful drying, and if the forerun of nitrogen produced is discarded, then spectroscopically pure nitrogen can be obtained.

[95] Ref. 85, pp. 6–7.

[96] P. W. Schenk, in *Handbook of Preparative Inorganic Chemistry*, 2nd edn., Vol. 1 (ed. G. Brauer), Academic Press, London (1963), pp. 457–517.

2.6. PURIFICATION OF NITROGEN

Nitrogen prepared by laboratory methods from aqueous solutions may be purified by passing the gas through gas-washing bottles containing dilute sulphuric acid to remove ammonia, iron(II) sulphate solution to remove nitric oxide, chromium(II) or vanadium(II) solutions to remove oxygen, and, finally, concentrated sulphuric acid to remove water. The use of gas-washing bottles, however, tends to saturate the gas with water vapour, necessitating very efficient drying.

As most commercial sources of nitrogen are produced by the liquefaction of air, cylinders of compressed nitrogen usually contain inert gases (mainly argon), oil vapour, carbon dioxide, water and oxygen as impurities totaling not more than 0.5% depending on grade. In general, the low oxygen–low moisture grade nitrogen requires little or no purification. For most applications, the presence of inert gases is not important, but the complete removal of oxygen and water is often necessary. Water and carbon dioxide can be removed by passing the gas through Linde molecular sieve grade 4A followed by passage through a column of manganese(II) oxide supported on fireclay at 100°C to remove oxygen[97]. Oxygen can also be removed by passing the gas over copper at 400°C or preferably an active form[14] at 170°C, over a tungsten filament at 2300°C, or with hydrogen over a platinum catalyst. Any water present can be removed either by passing the gas through a trap cooled in liquid nitrogen at −196°C or through a desiccant train containing phosphorus pentoxide or anhydrous magnesium perchlorate. Small amounts of oxygen and water can be removed in one operation by bubbling the gas through a trap containing 25:75 sodium–potassium liquid alloy[95].

For spectroscopic purposes, nitrogen once purified should only be contained in closed systems previously baked out under high vacuum.

2.7. NUCLEAR ISOTOPES $^{14}N_2$, $^{14}N^{15}N$, $^{15}N_2$

Atmospheric nitrogen contains 0.365% of the heavier isotope ^{15}N, and the remaining 99.635% being the only other naturally occurring isotope ^{14}N. Thermal diffusion columns can be used to produce highly concentrated ^{15}N from nitrogen gas, but quantities produced in this way are small[98]. However, as previously mentioned (section 1.4.4), gaseous nitrogen is the form to which compounds are converted for mass-spectrometric isotope analysis, and hence the enriched material can be recovered as the element. Both $^{15}N_2$ and $^{15}N^{14}N$ are available commercially in at least 95% enriched material[99].

The mass spectrometry of N_2 has naturally received detailed attention[100], particularly in respect of calculating ^{15}N concentration from direct measurement of the ratio of masses 28:29 ($^{14}N_2^+$–$^{14}N^{15}N^+$) or 30:29 ($^{15}N_2^+$–$^{15}N^{14}N^+$). This is only possible with any accuracy because mixtures of $^{14}N_2$, $^{14}N^{15}N$, $^{15}N_2$ do not exchange isotopes at ordinary temperatures. Indeed, no evidence could be obtained for isotope exchange between $^{14}N_2$ and $^{15}N_2$ over

97 C. C. Addison and B. M. Davies, *J. Chem. Soc.* A (1969) 1822.

98 K. Clusius, *Helv. Chim. Acta* **33** (1950) 2134.

99 Office National Industriel de l'Azote (ONIA), 40 avenue Hoche, Paris (VIIIe), France; Isomet Corporation, Palisades Park, New Jersey, USA.

100 W. Spindel, in *Inorganic Isotopic Syntheses* (ed. R. H. Herber), Benjamin, New York (1962), pp. 81–118.

the temperature range 30–1000°C, and it was concluded that the lower limit activation energy for the exchange reaction

$$^{15}N + {}^{14}N^{14}N \rightarrow {}^{14}N + {}^{15}N^{14}N$$

must be between 14 and 31 kcal/mole[101].

The reaction of active nitrogen (see section 2.12) with labelled nitric oxide either at room temperature and 380°C yields $^{14}N^{15}N$ exclusively in amounts equivalent to the active nitrogen concentration[101]

$$^{14}N + {}^{15}NO \rightarrow {}^{14}N^{15}N + O$$

2.8. STRUCTURE AND BONDING

Solid nitrogen exists in two enantiotropic forms at atmospheric pressure with a transition point of 35.6 K [102]. At 20 K, the α form is cubic with a molecular lattice containing four molecules (internuclear distance 1.065 Å) per unit cell, whereas at 39 K β-nitrogen crystallizes in a hexagonal form. Recently, however, a third high-pressure modification γ-nitrogen has been reported to be tetragonal at 20.5 K and 4015 atm pressure[103]. Neutron diffraction indicates single neighbour distances of 1.1 Å for liquid nitrogen while gaseous nitrogen exists as a diatomic molecule with an internuclear distance of 1.097 Å as determined by microwave spectroscopy.

As the diatomic molecular structure is present in all three phases, the simplest description of the bonding in nitrogen is as three electron-pair bonds giving an octet of electrons on each atom[104]. Thus with six electrons involved in a triple bond, this leaves a lone pair at each end of the molecule (:N≡N:).

Molecular orbital theory can also supply a satisfactory description of the bonding in nitrogen which is consistent with most of its properties[105]. Molecular orbitals in their simplest form are considered to be linear combinations of atomic orbitals. This approximation is valid for valence orbitals which interact by the LCAO/MO method only if the orbitals involved have similar energies and the same symmetry, while the symmetry of the molecule requires that each molecular orbital be made up of equal contributions from the two atoms. Assuming the molecule to be lying on the z axis, the s orbitals of each nitrogen can overlap to form a σ-bonding and σ-antibonding molecular orbitals, and the p_z orbitals can do the same. The p_x orbitals of each nitrogen being perpendicular to the z axis overlap to form π-bonding and π-antibonding molecular orbitals parallel to the molecular axis. The p_y orbitals interact similarly. As each orbital is capable of containing two electrons, the electrons of nitrogen will fill the lowest energy orbitals first. Figure 10 shows the energy diagram for nitrogen, and can be summarized as

$$2[N(1s^2 2s^2 2p_x 2p_y 2p_z)] \rightarrow N_2(1s\sigma^b)^2(1s\sigma^*)^2(2s\sigma^b)^2(2s\sigma^*)^2(2p\sigma^b)^2(2p\pi^b)^4$$

The strikingly similar physical properties of the isoelectronic molecules, nitrogen and carbon monoxide, invites comparison of the bonding in view of their differing reactivity.

[101] R. A. Back and J. Y. P. Mui, *J. Phys. Chem.* **66** (1962) 1362.

[102] B. R. Brown, in ref. 85, pp. 27–149.

[103] R. L. Mills and A. F. Schuch, *Phys. Rev. Letters* **23** (1969) 1154.

[104] M. Green, in *Developments in Inorganic Nitrogen Chemistry*, Vol. I (ed. C. Colburn), Elsevier, Amsterdam (1966), pp. 1–71.

[105] C. J. Ballhausen and H. B. Gray, *Molecular Orbital Theory*, Benjamin, New York (1965).

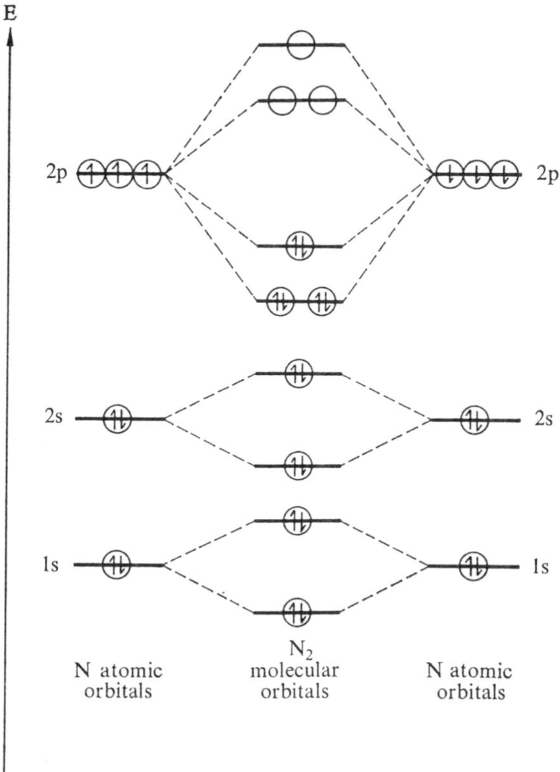

FIG. 10. Molecular orbital energy diagram for nitrogen.

A point that was not mentioned in the simple molecular orbital description above was that because the s and p_z atomic orbitals belong to the same symmetry species, they can be hybridized[106]. Hybridization will give rise to two new levels, σ_1 predominantly an s orbital with some p_z character, and σ_2 predominantly p_z with some s character, whose energies indicate that the major effect of hybridization is to increase the bonding power of σ_2. The energy level diagrams are illustrated in Fig. 11.

For carbon monoxide, however, the atomic orbitals of the two atoms which form the σ-bond are not equivalent. Figure 11 shows that the σ_1 of carbon interacts with σ_2 of oxygen to form the σ bond. Consequently the lone pair of carbon in :C≡O: will be in a σ_2 orbital and have largely p_z character, while the lone pair of the oxygen atom will be in σ_1 and have largely s character.

The bond order, calculated by subtracting the number of antibonding electrons from the number of bonding electrons and dividing by two, is three for the nitrogen molecule in agreement with the triple bond obtained by the simple octet rule. This is the maximum value as either the removal of one electron from a bonding orbital to form N_2^+ or the addition of an electron to an antibonding orbital to form N_2^- gives a decrease in the bond order

[106] H. H. Jaffé and M. Orchin, *Tetrahedron* **10** (1960) 212.

from 3 to $2\frac{1}{2}$ accompanied by appropriate increases in the bond length and decreases in bond dissociation energy.

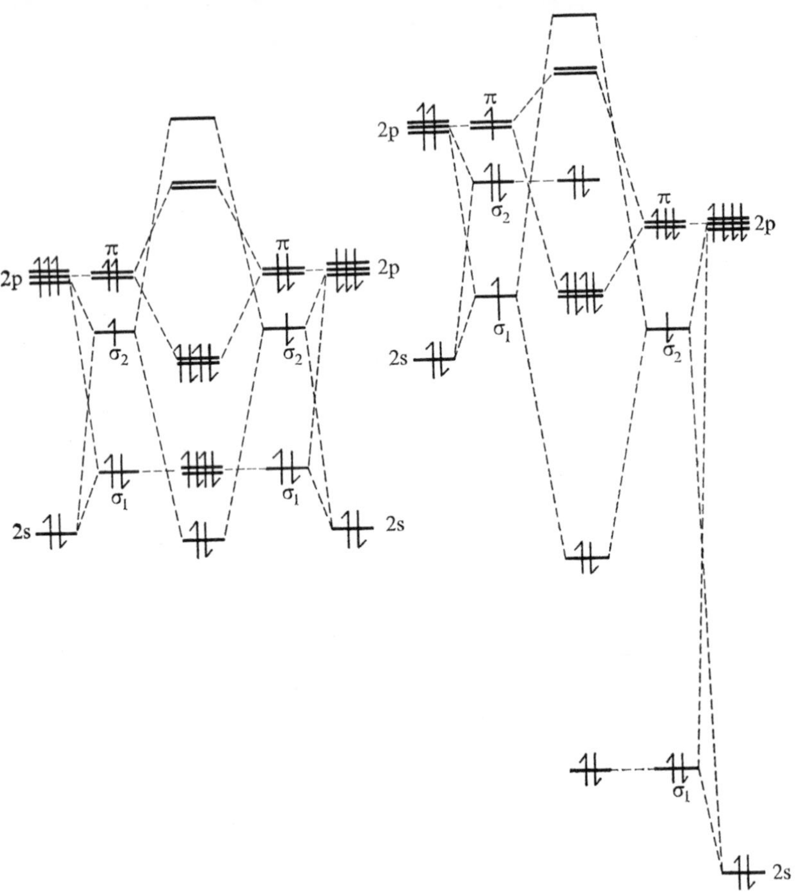

FIG. 11. Molecular orbital energy diagrams for nitrogen and carbon monoxide.

2.9. MOLECULAR CONSTANTS

Molecular constants for the nitrogen crystal modifications are tabulated in Table 14 and have been mentioned in section 2.8 in connection with bonding.

The principle source of molecular data is from spectroscopy[107]. Although the selection rule requires a change in dipole moment for infrared and microwave spectra to be observed,

[107] R. F. Barrow and A. J. Merer, in *Mellor's Comprehensive Treatise on Inorganic and Theoretical Chemistry*, Vol. 8, Suppl. II, *Nitrogen*, Part 2, Longmans, London (1967), p. 454.

TABLE 14. MOLECULAR CONSTANTS OF THE NITROGEN MOLECULE[102,103]

Crystal modification	α-nitrogen	β-nitrogen[a]	γ-nitrogen
Range of stability	<35.6 K	>35.6 K	>3500 atm
Temperature and pressure of X-ray structure determination	21 K/1 atm	50 K/1 atm	20.5 K/4015 atm
System	Cubic	Hexagonal	Tetragonal
Space group	$P2_13$	$P6_3/mmc$	$P4_2/mnm$
Symmetry	T^4	D^4_{6h}	D^4_{4h}
Lattice constants: a	5.667	3.93	3.957
c		6.50	5.109
c/a		1.654	1.29
Internuclear distance	1.065 Å		

[a] W. E. Streib, T. H. Jordan and W. N. Lipscomb, *J. Chem. Phys.* **37** (1962) 2962.

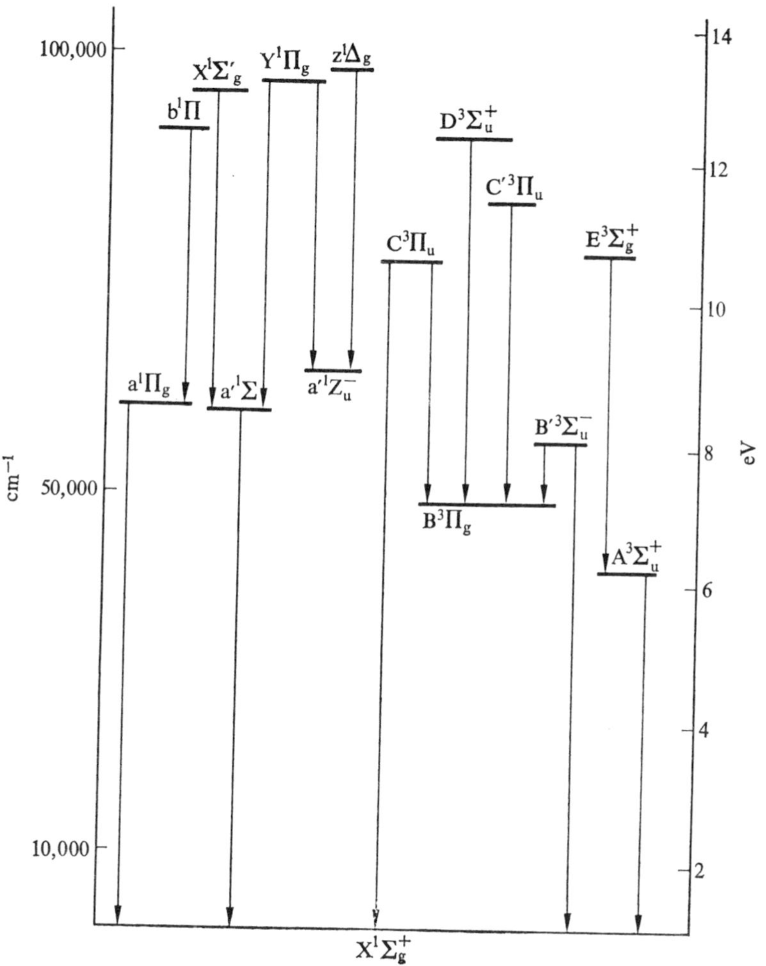

FIG. 12. Some N_2 band systems.

homonuclear molecules can always be suitably perturbed under pressure or in condensed phases. These absorptions agree with the value of the strongest Raman line at 2331 cm^{-1} and its overtones[108]. As regards electronic spectra, however, there is no shortage of data as more band systems are known for nitrogen than for any other molecule[107,109,110], and some of these are indicated in Fig. 12.

Electronic spectra arise from the promotion of an electron from one state to another and are observed when the electron returns to the original state. Thus for each excited state, an electronic configuration can be assigned and various parameters like the equilibrium bond length can be computed.

Values for the constants of the ground state of the nitrogen molecule, for spectroscopic purposes written

$$KK(2\sigma_g)^2(2\sigma_u)^2(1\pi_u)^4(3\sigma_g)^2X^1\Sigma_g^+,$$

result from the rotational analysis of the Lyman–Birge–Hopfield system and rotational and vibrational Raman spectral data:

$$G_v = 2358.07(v+\tfrac{1}{2})-14.188(v+\tfrac{1}{2})^2-0.0124(v+\tfrac{1}{2})^3$$
$$B_v = 1.9987-0.01781(v+\tfrac{1}{2})$$
$$r_e = 1.0976\ \text{Å}$$

The dissociation energy of N_2 has been the subject of considerable debate[109], but the value of 9.760±0.005 eV, 225.1 kcal/mol is now well established.

Three Rydberg series are known[107], the band heads being represented by the following equations.

The Worley–Jenkins series converging on $X^2\Sigma^+$ of N_2^+.

$$X\text{–}X:\ KK(\sigma_g)^2(\sigma_u)^2(\pi_u)^4(\sigma_g)(np\sigma_u),\ {}^1\Sigma_u^+$$
$$\nu_m = 125\,666-R[m+0.3450-0.1000/m-0.100/m^2]^2 \qquad m = 2, 3, \ldots, 26$$

Worley's third series, converging on $v = 1$ of $A^2\Pi_{\frac{1}{2}}$.

$$A\text{–}X:\ KK(\sigma_g)^2(\sigma_u)^2(\pi_u)^3(\sigma_g)(ns\sigma_g)^2,\ {}^1\Pi_u$$
$$\nu_m = 136\ 607-R/[m-0.0441-0.018/m]^2 \qquad m = 2, 3, \ldots, 6$$

The Hopfield series, which converges on $B^2\Sigma^+$.

$$B\text{–}X:\ KK(\sigma_g)^2(\sigma_u)(\pi_u)^4(\sigma_g)^2(ns\sigma_g),\ {}^1\Sigma_u^+$$

$$\nu_m = 151\,240-R/[m-0.092]^2 \qquad m = 3, 4, \ldots, 10,\ R = 109\ 735\ \text{cm}^{-1}$$

Five electronic states of the nitrogen molecule ion N_2^+ are known, $X^2\Sigma_g^+$, $A^2\Pi_u$, $B^2\Sigma_u^+$ $D(^2\Pi_g)$ and $C^2\Sigma_u^+$, and all the allowed transitions between these states have been observed[107].

---

[108] B. P. Stoicheff, *Adv. Spectrosc.* **1** (1959) 91.

[109] A. G. Gaydon, *Dissociation Energies and Spectra of Diatomic Molecules*, 3rd edn., Chapman & Hall, London (1968), p. 184.

[110] R. W. Nicholls, in *Physical Chemistry, An Advanced Treatise* (eds. H. Eyring, D. Henderson and W. Jost), Vol. III, *Electronic Structure of Atoms and Molecules* (ed. D. Henderson), Academic Press, New York (1969), p. 325.

The dissociation energy of N_2^+ ($X^2\Sigma_g^+$) has been calculated from the ionization potential of nitrogen and is related to the dissociation energy of N_2 by the equation

$$D_0(N_2^+) = D_0(N_2)+IP(N)-IP(N_2)$$
$$D_0(N_2^+) = 9.760+14.545-15.58$$
$$D_0(N_2^+) = 8.73\pm0.01 \text{ eV, } 201.4 \text{ kcal/mol}$$
$$r_e(N_2^+) = 1.118 \text{ Å}$$

2.10. PHYSICAL PROPERTIES

Physical constants for molecular nitrogen are given in Table 15 together with some specific values for temperature variable parameters while properties which are described by equations are listed in Table 16.

The adsorption of nitrogen[102] is being mentioned under physical properties although this is not intended to presume what mode of interaction between nitrogen and the adsorbent might be. Indeed, the terms physisorption and chemisorption were intended to indicate the ease or difficulty of removal of an adsorbate from an adsorbent surface, and are only adequate in describing extreme situations. Originally, the adsorption of nitrogen by various substances was a means of determining surface properties of the adsorbent, and useful in characterizing high surface-area materials. A variety of materials have been studied in this context, including silica gels, glass, carbon, minerals, salts, metal oxides and metals. In physisorption and surface area studies, nitrogen has given way to monatomic argon, but it is the sorption of nitrogen on metals that commands the interest from a catalytic viewpoint. The infrared absorption of nitrogen chemisorbed on a nickel surface[111] at 2202 cm^{-1} is only *ca.* 100 cm^{-1} lower than the Raman active band of nitrogen itself, and suggests that only one end of the nitrogen molecule is bonded to the adsorbent. Field emission microscopy has been used to study nitrogen adsorption on metal tips[112]. For iron, the nitrogen molecules are aligned with their axes perpendicular to the 111 face, and at 400–500°C it has been estimated that the heat of chemisorption to iron approximates to the heat of formation of iron nitride[113].

2.11. CHEMISTRY AND CHEMICAL PROPERTIES

2.11.1. Reactions of Nitrogen with the Elements

It is appropriate that the reaction of nitrogen with the first element of the Periodic Table, hydrogen, is the basis of nearly all nitrogen chemistry. Suffice it to mention here that the reaction to produce ammonia only proceeds at an appreciable rate at elevated temperatures and pressures in the presence of a catalyst (see section 3.3.1). Lithium is one of the few elements to enter into reaction with nitrogen at room temperature, but the formation of lithium nitride proceeds smoothly at elevated temperatures[114]:

$$6Li+N_2 \rightarrow 2Li_3N$$

[111] R. P. Eichens and J. Jacknow, *Proceedings of the 3rd International Congress on Catalysis*, N. Holland Pub. Co., Amsterdam (1965), p. 627.

[112] R. Brill, E. L. Richter and E. Ruch, *Angew. Chem. Int. Edn. Engl.* **6** (1967) 882.

[113] S. Brunauer, *Physical Adsorption of Gases and Vapours*, Oxford (1942).

[114] C. C. Addison and B. M. Davies, *J. Chem. Soc.* A (1969) 1822.

Although the other alkali metals also form nitrides, no reaction occurs between nitrogen and the element. Beryllium and the alkaline earth metals all react with nitrogen at elevated temperatures forming nitrides. Boron, aluminium, silicon and germanium also react with

TABLE 15. PHYSICAL CONSTANTS OF MOLECULAR NITROGEN[a]

Property	Temperature (K)	Value
Transition temperature (α-cubic to β-hexagonal)	35.61	−237.55°C
γ-Tetragonal triple point[b]	44.5	−228.65°C
γ-Tetragonal triple point pressure[b]		4650 atm
Melting point	63.14	−210.01°C
Boiling point	77.36	−195.79°C
Triple point	63.16	−209.99°C
Triple-point pressure		94.0 mm
Critical temperature	126.2	−146.95°C
Critical pressure		33.54 atm
Critical density[c]		0.311 g/cm³
ΔH° transition	35.61	54.7 cal/mol
ΔH fusion	63.14	172 cal/mol
ΔH evaporation	77.36	1.335 kcal/mol
S° (entropy)[d]	298	45.767 cal/deg mol
Increase in entropy per 100° rise		0.71 cal/deg mol
C_p° (heat capacity)[d]	298	6.96 cal/deg mol
Lattice energy		1.64 kcal/mol
Heat of dissociation[e]		225.1 kcal/mol
		9.760 ± 0.005 eV
Density of solid	20.6	1.0265 g/cm³
	63	0.8792 g/cm³
Density of liquid	70.38	0.8431 g/cm³
	76.21	0.8163 g/cm³
	82.57	0.7871 g/cm³
	89.38	0.7557 g/cm³
Density of gas	273.1/1 atm	1.25046 g/l
Viscosity of gas	290.16	1738.2×10^{-7} poise
Viscosity of liquid	64	2.10×10^{7} poise
	80	1.49×10^{7} poise
	100	9.40×10^{6} poise
	111.7	7.40×10^{6} poise
Surface tension of liquid	70	4.624 dynes/cm
Refractive index	293.16/1 atm	1.002941
Dielectric constant of gas	298.44/1.02 atm	1.00052
Dielectric constant of liquid	78.5	1.455
Thermal conductivity of gas	298.16/1 atm	5.71×10^{-5} cal/cm s K
Thermal conductivity of liquid	77.4	3.34×10^{-4} cal/cm s K
Magnetic susceptibility	293	−0.426
Ionization potential $N_2 \rightarrow N_2^+ + e^-$		15.65 eV
Molar volume	273.16/1 atm	22402.5 cm³/mol

[a] Ref 102 unless otherwise stated.

[b] Ref. 103.

[c] *Handbook of the Physicochemical Properties of the Elements* (ed. G. V. Samsonov), Oldbourne, London (1968).

[d] D. D. Wagman, W. H. Evans, V. B. Parker, I. Halow, S. M. Bailey and R. H. Schumm, *Selected Values of Chemical Thermodynamic Properties*, NBS Technical Note 270-3, Washington (1968).

[e] Ref. 109.

TABLE 16. PHYSICAL PROPERTIES OF MOLECULAR NITROGEN[a]

Vapour pressure of solid P

$\log P_{cm} = -381.6/T - 0.0062372T + 7.41105$

Vapour pressure of liquid P

$\log P_{cm} = -339.8/T - 0.0056286T + 6.71057$

K	°C	P	K	°C	P
47.0	−226.1	1 mm	77.3	−195.8	1 atm
54.0	−219.1	10	83.9	−189.2	2
59.1	−214.0	40	94.0	−179.1	5
63.4	−209.7	100	104.3	−169.8	10
72.2	−200.9	400	115.5	−157.6	20

Second Virial coefficient B[b]

$$(PV/nRT) = 1 + \frac{nB(T)}{V} + \dots$$

K	B	K	B
75	−274±5	200	−35.2
80	−242	300	− 4.2±0.5
90	−196	400	+ 9.0
100	−160±2	500	+16.9
125	−104	600	+21.3
150	− 71.5±1	700	+24.0

Heat capacity C_p°

$C_p^\circ = 6.83 + 0.0009T - 0.0000012T^{-2}$ over the range $T = 298\text{–}3000$ K

Density of liquid d

$d = 1.1604 - 0.00455T$

Density of gas ρ

$$\log \rho_T = 3.39858 - \frac{282.953}{T - 3.83}$$

Viscosity of gas η

$\eta = 1.377 \times 10^{-5} T^{3/2}/(102.7 + T)$

Surface tension γ

$\gamma = 11.68(1 - 0.00863T)$ dynes/cm

Molar volume V

$PV = nRT[1 - P(0.0022 + 19600/T^3)]$ over the range $P = 0.5\text{–}10$ atm

Solubility of gaseous nitrogen in water.

Volume at s.t.p. which dissolves in 1 volume of water when the partial pressure of nitrogen:

K	Volume	K	Volume
273	0.230	313	0.0119
283	0.183	323	0.0108
293	0.0165	333	0.0100
303	0.0133	343	0.0095

TABLE 16 (*cont.*)

Solubility of gaseous nitrogen at 1 atm in g/ml of solvent:

	K	g/ml		K	g/ml
MeOH	273	0.1532	95% EtOH	273	0.1053
	298	0.1645		298	0.1160
100% EtOH	273	0.1391	Acetone	273	0.1554
	298	0.1489		298	0.1816

Dissociation constant K_p

$K_p = (N)^2/(N_2)$

K	K_p	K	K_p
298	1×10^{-128}	4000	7×10^{-4}
500	1×10^{-74}	8000	281
1000	2×10^{-34}	16000	1.49×10^{5}
2000	4×10^{-14}		

[a] Ref. 102 unless otherwise stated.

[b] J. H. Dymond and E. B. Smith, *The Virial Coefficient of Gases—A Critical Compilation*, Oxford (1969).

nitrogen at high temperature to give nitrides, whereas cyanogen[115] (N≡C—C≡N) is formed when nitrogen reacts with incandescent coke. Of the transition metals, scandium, yttrium, the lanthanide elements, zirconium, hafnium, vanadium, chromium, molybdenum, tungsten, manganese, thorium, uranium and plutonium all react with nitrogen at elevated temperatures to form nitrides of the general formula MN[116, 117].

Reactions also occur between nitrogen and mixtures of elements. For example, hydrogen, carbon and nitrogen react slowly at temperatures above 1900 K to give hydrogen cyanide:

$$C(s)+\tfrac{1}{2}H_2(g)+\tfrac{1}{2}N_2(g) \rightarrow HCN(g)$$

This reaction is thought to have been associated with basic life processes[84] in connection with the equilibrium

$$NH_3(g)+C(s) \rightleftharpoons HCN(g)+H_2(g)$$

Many metal nitrides are also formed when metals react with ammonia[117] under conditions of temperature and pressure at which ammonia is at least partially dissociated.

2.11.2. Reactions of Nitrogen with Compounds

One might have expected the scope of reactions between nitrogen and compounds to be somewhat wider in view of the selection of reagents available, but until recently this was not

[115] T. K. Brotherton and J. W. Lynn, *Chem. Rev.* **59** (1959) 841.

[116] *Handbook of Physicochemical Properties of the Elements* (ed. G. V. Samsonov), Oldbourne, London (1968).

[117] R. T. Sanderson, *Chemical Periodicity*, Reinhold, New York (1960), p. 201.

so. One of the earliest reactions of nitrogen to be put to commercial use was the reaction with calcium carbide[118]:

$$CaC_2+N_2 = CaCN_2+C; \qquad \Delta H(298\text{ K}) = -72.7\text{ kcal/mol}$$

After heating to 700°C to initiate, the reaction is self-sustaining at 1100°C. Similarly, another feasible route to ammonia after hydrolysis of the product was the alkali cyanide process[119]

$$Na_2CO_3+4C+N_2 \xrightleftharpoons{\text{Fe catalyst}} 2NaCN+3CO; \qquad \Delta H(298\text{ K}) = 138.5\text{ kcal/mol}$$

There are also a few organic reactions in which nitrogen is involved; for example, a nitrogen-containing product is obtained when the benzene–sulphenium cation is generated in the presence of nitrogen[120]:

$$C_6H_5S^+ + N_2 \rightarrow [C_6H_5SN_2^+]$$

The exchange of labelled nitrogen with diazomethane has also been reported[121]:

$$CH_2N_2 + {}^{15}N_2 \rightarrow CH_2{}^{15}N_2 + N_2$$

Reactions of nitrogen with organometallic species were first reported[122,123] in 1964, and their investigation has continued with a view to establishing a basis for the catalytic fixation of nitrogen. In the Volpin reaction, nitrogen reacts at room temperature and modest pressures (< 150 atm) with a two-component system consisting of a transition metal halide (i.e. $TiCl_4$, $CrCl_3$, WCl_6) and an organometallic reagent (RLi, RMgX, R_3Al, etc.) in an organic solvent. In general, the reaction product is not isolated, but hydrolysed to ammonia in yields approaching 2 mole/mole of transition metal. Various mechanisms have been proposed[123], but all refer to the organometallic reagent as the reducing agent responsible for producing the transition metal in a low oxidation state.

Van Tamelen has reported 10–15% yields of ammonia without recourse to hydrolysis[124]. Titanium tetrachloride was added to potassium *tertiary* butoxide in diglyme under nitrogen to give bis(*tertiary*-butoxy)titanium dichloride.

$$TiCl_4 + 2Bu^tOK \xrightarrow{N_2} TiCl_2(OBu^t)_2$$

Addition of two moles of potassium resulted in the reduction of titanium(IV) to titanium(II) together with the slow evolution of ammonia over a period of several weeks. Recently, however, the fixation and reduction of nitrogen was achieved at room temperature and atmospheric pressure in an overall catalytic fashion[125].

118 M. L. Kastens and W. G. McBurney, *Ind. Eng. Chem.* **43** (1951) 1020.
119 J. E. Bucher, *J. Ind. Engng. Chem.* **9** (1917) 223.
120 D. C. Owsley and G. K. Helmkamp, *J. Am. Chem. Soc.* **89** (1967) 4558.
121 Yu. G. Borod'ko, A. E. Shilov and A. A. Shteinman, *Dokl. Chem. Proc. Acad. Sci. USSR* **168** (1966) 510.
122 M. E. Volpin and V. B. Shur, *Dokl. Chem. Proc. Acad. Sci. USSR* **156** (1964) 1102.
123 R. Murray and D. C. Smith, *Coordin. Chem. Rev.* **3** (1968) 429.
124 E. E. Van Tamelen, G. Boche, S. W. Ela and R. B. Fechter, *J. Am. Chem. Soc.* **89** (1967) 5707.
125 E. E. Van Tamelen, R. B. Fechter, S. W. Schneller, G. Boche, R. H. Greeley and B. Åkermark, *J. Am. Chem. Soc.* **91** (1969) 1551.

$$\begin{array}{ccccc} Ti(OR)_2 & \xrightarrow{N_2} & & & [Ti(OR)_2N_2]_n \\ \uparrow & & & & \downarrow \\ 2Na^+Np^- & \underset{2e}{\overset{2Na}{\rightleftharpoons}} & \text{(naphthalene)} & \underset{4e}{\overset{4Na}{\rightleftharpoons}} & 4Na^+Np^- \\ -2NaOR \uparrow 2Na & & & & \downarrow \\ Ti(OR)_4 & \xleftarrow[-4NaOR\,-2NH_3]{6ROH} & & & [\text{nitride}] \end{array}$$

Thus the overall process is expressed by the equation

$$N_2 + 6e^- + 6ROH \rightarrow 2NH_3 + 6OR^-$$

2.11.3. Molecular Nitrogen Adducts

(a) *Synthesis*

Undoubtedly the most exciting area of nitrogen chemistry at present is the work on nitrogen complexes of transition metals. It is well known that both iron(II) and molybdenum(V) are involved in the enzyme system responsible for natural nitrogen fixation (see section 1.7), and although it has been assumed by some that an initial stage in the fixation process might involve complex formation of the nitrogen molecule with the transition metal[123], until quite recently no such model compounds were known. In contrast to the forcing conditions required for combination with most elements, it is remarkable that molecular nitrogen can react with suitable substrates at ambient temperature and pressure to form stable adducts analogous to the complexes of the isoelectronic ligand carbon monoxide.

As with so many new discoveries, however, the first and several subsequent nitrogen complexes were obtained while pursuing other aims and were clearly not anticipated in those experiments where nitrogen was intended as an inert atmosphere. In 1965 Allen and Senoff were attempting to devise simple methods for the preparation of ruthenium amines when they found that hydrazine hydrate with potassium pentachloroaquaruthenite(III) or commercial ruthenium trichloride gave salts of yellow nitrogenpentamineruthenium(II)[126].

$$\begin{array}{l} K_2[Ru^{III}Cl_5(H_2O)] + N_2H_4 \cdot H_2O \rightarrow [Ru^{II}(NH_3)_5N_2]X_2 \quad X = Cl, Br, I, BF_4, PF_6 \\ \text{or } RuCl_3 \end{array}$$

The general formula of these complexes was proposed after taking into account their magnetic, conductivity, spectroscopic, analytical and chemical properties[127]. Of particular importance was the assignment of a very sharp infrared band in the region 2100–2170 cm^{-1} to the nitrogen–nitrogen stretching frequency, and this band appears to be characteristic for such complexes.

The following year, Collman and Kang reported[128] the preparation of chloronitrogenobis(triphenylphosphine)iridium(I)[129] from the reaction of Vaska's iridium compound and an acyl azide:

$$\begin{array}{c} Ph_3P \quad\; CO \\ \;\; Ir \\ Cl \quad\; PPh_3 \end{array} \xrightarrow[CHCl_3]{RCON_3} \begin{array}{c} Ph_3P \quad\; N_2 \\ \;\; Ir \\ Cl \quad\; PPh_3 \end{array} + RCONCO$$

[126] A. D. Allen and C. V. Senoff, *Chem. Communs.* (1965) 621.

[127] A. D. Allen and F. Bottomley, *Accounts Chem. Res.* **1** (1968) 360.

[128] J. P. Collman and J. W. Kang, *J. Am. Chem. Soc.* **88** (1966) 3459; J. P. Collman, M. Kubota, J-Y. Sun and F. Vastine, *ibid.* **89** (1967) 169.

[129] Available from Strem Chemicals Inc., 150 Andover St., Danvers, Mass. 01923, USA.

The first cobalt derivative was prepared by reacting cobalt acetylacetonate with diethylaluminium ethoxide in an atmosphere of nitrogen[130]:

$$Co(acac)_3 + 3Et_2AlOEt + 3Ph_3P + N_2 \rightarrow CoH(N_2)(PPh_3)_3$$

As no metal–hydrogen stretching frequency could be detected in the infrared spectrum, the product was initially formulated incorrectly as the nitride complex. An independent investigation using triisobutylaluminium reported a mixture of the nitride and the hydride/nitride complex[131]. The presence of hydride was confirmed by proton n.m.r.[131], and also when Sacco and Rossi discovered the equilibrium which can be quantitatively displaced in either direction under the conditions of room temperature and atmospheric pressure simply by changing the atmosphere[132].

$$CoH_3L_3 + N_2 \rightleftharpoons CoH(N_2)L_3 + H_2 \qquad L = PPh_3, PEtPh_2$$

More recently, the reaction of π-cyclo-octenyl-π-cyclo-octa-1,5-diene cobalt with tertiary phosphines under hydrogen and subsequently with nitrogen gives the complexes via the following equilibria[133]:

$$Co(C_8H_{13})(C_8H_{12}) + 2L \rightleftharpoons Co(C_8H_{13})L_2 + C_8H_{12}$$

$$Co(C_8H_{13})L_2 + L \rightleftharpoons [Co(C_8H_{13})L_3] \rightleftharpoons CoHL_3 + C_8H_{12}$$

$$CoHL_3 + N_2 \rightleftharpoons CoHN_2L_3$$

$$L = PPh_3, PMePh_2, PEtPh_2, PBuPh_2, PEt_2Ph, PBu_2Ph, PBu_3$$

Harrison and Taube investigated the behaviour of reduced solutions of ruthenium(III) in an atmosphere of nitrogen and characterized a binuclear ion[134]:

$$[Ru(NH_3)_5Cl]^{2+} \xrightarrow[0.1\,\text{M}\ H_2SO_4]{Zn/Hg} [Ru(NH_3)_5H_2O]^{2+}$$

$$\downarrow N_2;\ 10^{-3}\ \text{M soln.};\ \text{Zn removed}$$

$$[Ru(NH_3)_5N_2]^{2+} \xleftarrow{NH_4OH} \tfrac{1}{2}[(NH_3)_5RuN_2Ru(NH_3)_5]^{4+}$$

The binuclear complex shows no absorption in the region of the infrared spectrum associated with the N≡N stretching frequency but does exhibit a strong band in the Raman spectrum at 2100 cm^{-1} which shifts to 2030 cm^{-1} in the $^{15}N_2$ analogue[135], and is consistent with the predicted symmetrical structure[134]. A similar isotopic shift from 2145 to 2080 cm^{-1} has also been observed for the $^{14}N_2$ and $^{15}N_2$ complexes of $[Ru(NH_3)_5N_2][BF_4]_2$ [136].

Nitrogen complexes of osmium have now been prepared by hydrazine hydrate[137] and zinc amalgam[138] reduction methods, while iron compounds are formed via hydride

130 A. Yamamoto, S. Kitazume, L. S. Pu and S. Ikeda, *Chem. Communs.* (1967) 79.

131 A. Misono, Y. Uchida and T. Saito, *Bull. Chem. Soc. Japan* **40** (1967) 700; A. Misono, Y. Uchido, T. Saito and K. M. Song, *Chem. Communs.* (1967) 419; A. Misono, Y. Uchida, T. Saito, M. Hidai and M. Araki, *Inorg. Chem.* **8** (1969) 1.

132 A. Sacco and M. Rossi, *Chem. Communs.* (1967) 316.

133 M. Rossi and A. Sacco, *Chem. Communs.* (1969) 471.

134 D. E. Harrison and H. Taube, *J. Am. Chem. Soc.* **89** (1967) 5706; D. E. Harrison, H. Taube and E. Weissberger, *Science* **159** (1968) 320.

135 J. Chatt, A. B. Nikolsky, R. L. Richards and J. R. Sanders, *Chem. Communs.* (1969) 154.

136 J. Chatt, R. L. Richards, J. E. Fergusson and J. L. Love, *Chem. Communs.* (1968) 1522.

137 A. D. Allen and J. R. Stevens, *Chem. Communs.* (1967) 1147.

138 J. Chatt, G. J. Leigh and R. L. Richards, *Chem. Communs.* (1969) 515.

intermediates[139,140], and rhenium complexes are obtained by degradation of benzoylazo complexes[141] or reduction with zinc amalgam[141].

$$(NH_4)_2OsCl_6 + N_2H_4 \cdot H_2O \rightarrow [Os(NH_3)_5N_2]X_2 \qquad X = Cl, Br, I, ClO_4, BF_4 \text{ and } BPh_4$$

$$OsX_3L_3 + N_2 \xrightarrow[THF]{Zn/Hg} [OsX_2(N_2)L_3]$$

$X = Cl$, $L = PMe_2Ph$, PEt_2Ph, PPr^n_2Ph, PBu^n_2Ph, $PEtPh_2$, PEt_3

$X = Br$, $L = PMe_2Ph$, PEt_2Ph

$$FeCl_2 \cdot 2H_2O + 2L + NaBH_4 \xrightarrow[argon]{H_2 \text{ or}} FeH_2L_3$$

$$FeH_2L_3 + N_2 \rightarrow FeH_2N_2L_3 \qquad L = PEtPh_2, PBuPh_2$$

$$FeH_2N_2(PEtPh_2)_3 \rightleftharpoons H_2 + FeH(C_6H_4PEtPh)N_2(PEtPh_2)_2$$

$$FeH_2N_2(PEtPh_2)_3 + Et_3Al \rightarrow FeH_2N_2(PEtPh_2)_2 + Et_3AlPEtPh_2$$

$$FeHCl(Et_2PCH_2CH_2PEt_2)_2 + NaBH_4 + N_2 \xrightarrow{acetone} FeH(N_2)(Et_2PCH_2CH_2PEt_2)_2 + NaCl$$

$$ReCl_2(N{-}N{=}C(Ph){-}O)(PPh_3)_2 \xrightarrow[(MeOH)]{L} ReCl_2(N{=}NCOPh)L_3 \xrightarrow[(MeOH)]{L} ReCl(N_2)L_4 + PhCO_2Me + HCl$$

$L = PMePh_2$, PMe_2Ph, $Ph_2PCH_2PPh_2$, $Ph_2PCH{=}CHPPh_2$, $\frac{1}{2}(Ph_2PCH_2CH_2PPh_2)$

$$ReCl_2(PMe_2Ph)_4 \xrightarrow[N_2/THF]{Zn/Hg} trans\text{-}[ReCl(N_2)(PMe_2Ph)_4]$$

The first dinitrogen complex *cis*-$[Ru\,en_2(N_2)_2][BPh_4]_2$ was prepared by hydrazine reduction[142], and more recently infrared evidence has been presented[143] for the existence of $[Ru(NH_3)_4(N_2)_2]Br_2$.

(b) *Structure and Bonding*

The first molecular nitrogen complex to be prepared, $[Ru(NH_3)_5(N_2)]Cl_2$, has a disordered crystal, which means that the N_2 ligand randomly occupies any of the six octahedral positions around ruthenium in the unit cell[144]. Thus the X-ray structure determination showed ruthenium apparently surrounded by six ligands, each comprising five-sixths

139 A. Sacco and M. Aresta, *Chem. Communs.* (1968) 1223.
140 G. M. Bancroft, M. J. Mays and B. E. Prater, *Chem. Communs.* (1969) 687.
141 J. Chatt, J. R. Dilworth and G. J. Leigh, *Chem. Communs.* (1969) 687.
142 L. A. P. Kane-Maguire, P. S. Sheridan, F. Basolo and R. G. Pearson, *J. Am. Chem. Soc.* **90** (1968) 5295.
143 J. E. Fergusson and J. L. Love, *Chem. Communs.* (1969) 399.
144 F. Bottomly and S. C. Nyburg, *Chem. Communs.* (1966) 897.

NH_3 and one-sixth N_2 as illustrated in Fig. 13. Nitrogen is collinear with the metal within the error imposed by disorder, but no accurate bond lengths could be measured.

Fortunately, an X-ray structure determination has been completed on $CoH(N_2)(PPh_3)_3$ (Fig. 14)[145]. The structure is a trigonal bipyramid with the vacant site (due to low diffracting power of hydrogen) *trans* to the N_2 group.

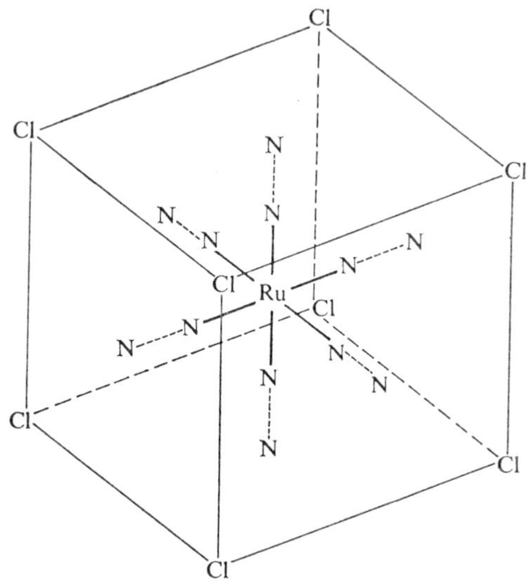

FIG. 13. The unit cell of $[Ru(NH_3)_5(N_2)]Cl_2$.

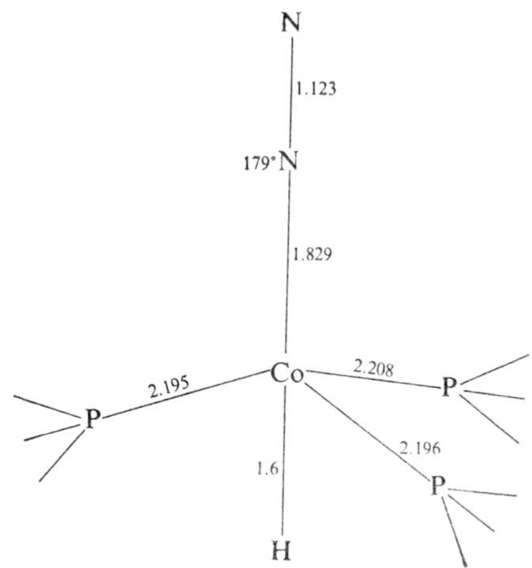

FIG. 14. The structure of $CoH(N_2)(PPh_3)_3$.

---

[145] J. H. Enemark, B. R. Davies, J. A. McGinnety and J. A. Ibers, *Chem. Communs.* (1968) 96; B. R. Davies, N. C. Payne and J. A. Ibers, *J. Am. Chem. Soc.* **91** (1969) 1240.

The parameters for two independent molecules were reported although the differences were not regarded as significant. Co—N—N is linear, while Co—N is comparable with Co—C for cobalt carbonyl compounds but significantly shorter than a typical Co—N bond distance of 1.96 Å in various cobalt amines. The N—N distance of 1.11 Å is comparable to 1.0976 Å in gaseous N_2 and 1.118 Å in N_2^+.

Prior to the formation of nitrogen complexes, attention was often drawn to the similarities in the physical properties of nitrogen and carbon monoxide, and in general the differences in the molecular orbital description of the bonding in the two molecules were regarded as significant in accounting for the marked variation in chemical reactivity[106]. Now that nitrogen complexes have been shown to have structural similarities with metal carbonyl derivatives, the simplest description of the bonding between nitrogen and the metal is exactly analogous to the situation in metal carbonyls, viz. the formation of a σ bond by donation of the lone-pair of electrons on one of the nitrogen atoms to a vacant orbital on the metal together with a π bond formed by the back donation of electrons from d orbitals on the metal to an empty antibonding π orbital on the nitrogen molecule. In carbon monoxide, the highest filled orbital contains the lone pair on the carbon atom which has predominantly p character and able to form a directional bond to metals. In nitrogen, however, the lone-pairs lie in σ orbitals of low energy similar to that of the oxygen atom in carbon monoxide, and are concentrated largely between the atoms[106]. Thus with the σbonding of nitrogen to the metal so ineffective, clearly the major contribution to the stability of nitrogen complexes must therefore be due to back-bonding from the metal into the lowest empty antibonding π orbital.

Consistent with this description of the bonding is the infrared spectral data. All mononuclear nitrogen complexes so far prepared have a strong sharp absorption band in the region 1919–2220 cm^{-1}, assigned to the nitrogen–nitrogen stretching frequency[123]. The absorption is sometimes split in solid phase spectra, but appears as a singlet for solutions. This range of values is lower than the Raman stretching frequency of 2330 cm^{-1} for molecular nitrogen, and suggests a slightly lower bond order in the complexes. Further lowering of the stretching frequency is observed for the binuclear complex $[(NH_3)_5RuN_2Ru(NH_3)_5]^{4+}$ consistent with increased π-bonding of the type Ru$\rightleftharpoons$N$=$N$\leftleftarrows$Ru, and an increase for the dinitrogen complex *cis*-$[Ru(en)_2(N_2)_2]^{2+}$ at 2220 and 2190 cm^{-1} over the mononitrogen complex at 2130 cm^{-1} suggests that these systems have to share the π-bonding ability of the metal.

(c) *Properties*

The properties so far examined have been mainly concerned with the characterization of these compounds as molecular nitrogen complexes. However, the incentive to find a simple transition metal complex to catalyse the reaction of atmospheric nitrogen with some hydrogen source (preferably water) to produce ammonia (comparable perhaps to the alchemist's search for philosophers stone) has restricted somewhat the field of investigation to reactions with reducing agents.

Although there have been several reports of molecular nitrogen complexes being reduced to ammonia, these have been shown by careful re-investigation to be due to the presence of impurities[146]. The reducing agents investigated to date (1969) have failed to reduce the

146 J. Chatt, R. L. Richards, J. R. Saunders and J. E. Fergusson, *Nature* **221** (1969) 551.

nitrogen molecule ligand in currently known complexes to ammonia, although perhaps even when it is achieved there may be some reluctance to rush into print.

The principal reactions that have been encountered so far lead to removal of the nitrogen ligand as molecular nitrogen or ligand exchange reactions[123,127], but no doubt a considerable chemistry awaits investigation.

2.12. ACTIVE NITROGEN

The glowing gas produced when an electric discharge is passed through molecular nitrogen at low pressure (0.1–2 mm) is commonly known as active nitrogen[102,147,148]. The name arises as a consequence of its high chemical reactivity when compared with ordinary nitrogen which may be attributed to the presence of ground state $N(^4S)$ atoms. These atoms have a relatively long lifetime as recombination involves a three-body collision:

$$N(^4S)+N(^4S)+M \xrightarrow{k} N_2+M$$

If M is a surface, k is the first order rate constant for heterogeneous recombination, and this mechanism may predominate at nitrogen pressures below 3 torr in systems of low atom concentration. Where M represents atomic or molecular nitrogen, a termolecular homogeneous association reaction, second order in active species, largely describes the rate of decay of active nitrogen above 3 torr. Both reactions are partially responsible for producing excited molecular states, and it is these excited molecules returning to the ground state that give rise to the emission of the first positive band system of the molecular nitrogen spectrum in the Lewis–Rayleigh afterglow:

$$N_2(B^3\Pi_g) \rightarrow N_2(A^3\Sigma_u^+)+h\nu$$

The nitrogen atom content of active nitrogen may be estimated from its reaction with nitric oxide by following visually the colour changes in a sequence of chemiluminescent association reactions which are known collectively as the "NO titration". With no added nitric oxide, active nitrogen in a flow system emits the characteristic yellow afterglow downstream from the discharge.

$$N(^4S)+N(^4S) \xrightarrow{M} N_2^* \rightarrow N_2+h\nu \quad \text{(yellow emission)} \tag{1}$$

Addition of NO at flow rates less than the available concentration of $N(^4S)$ atoms in the active nitrogen stream results in the very fast reaction

$$N(^4S)+NO \rightarrow N_2+O(^3P) \tag{2}$$

The oxygen atoms so formed may react with excess atomic nitrogen to produce excited nitric oxide molecules which emit the blue-coloured β and γ bands of NO:

$$N(^4S)+O(^3P) \rightarrow NO^* \rightarrow NO+h\nu \quad \text{(blue emission)} \tag{3}$$

With relatively small amounts of NO, reaction (3) occurs simultaneously with reaction (1), and the gas stream appears purple. When the flow rate of NO exceeds that of $N(^4S)$, reaction (2) being the fastest, eliminates all nitrogen atoms, hence reactions (1) and (3) are

[147] A. N. Wright and C. A. Winkler, *Active Nitrogen*, Academic Press, New York (1968).
[148] K. R. Jennings and J. W. Linnett, *Quart. Rev.* **12** (1958) 116.

no longer possible. Instead, excited nitrogen dioxide is formed which emits the greenish-yellow NO_2 bands associated with the "airglow":

$$NO+O(^3P) \rightarrow NO_2^* \rightarrow NO_2+h\nu \quad \text{(greenish-yellow)} \tag{4}$$

When NO is added at a flow rate equivalent to that of atomic nitrogen, both NO and $N(^4S)$ atoms are destroyed, and the slow light-producing association reactions (1), (3) and (4) cannot occur; hence no light is emitted. Thus, as the flow rate of NO is slowly increased, the colours change progressively through yellow, purple, blue, darkness and greenish yellow. Hence the flow rate of NO at the dark "end point" gives a measure of the concentration of $N(^4S)$ in the active nitrogen. The results of this relatively simple method agree well with estimations by other techniques[147].

The reactions of active nitrogen, however, not only involve nitrogen atoms. Indeed, metastable nitrogen molecules formed by three-body recombination reactions are capable of dissociating gases which are stable to direct attack by nitrogen atoms[149]:

$$N_2(A^3\Sigma_u^+)+CO_2(^1\Sigma_g^+) = N_2(X^1\Sigma_g^+)+CO(^1\Sigma_g^+)+O(^3P)$$
$$N_2(A^3\Sigma_u^+)+H_2O(^1A_1) = N_2(X^1\Sigma_g^+)+OH(^2\Pi)+H(^2S)$$

Although a number of chemical reactions of active nitrogen with other elements have been reported, the majority tend to result in the formation of nitrides. Even the reactions with hydrocarbons and organometallics are uneventful, and metal carbonyls generally only give rise to the emission spectra of the metal atoms. Recently, however, aqueous active nitrogen has been prepared, and its versatility as a reagent demonstrated[150] by the transformations $Fe^{II} \rightarrow Fe^{III}$ and $Fe^{III} \rightarrow Fe^{II}$ in 0.05 M H_2SO_4, $I^- \rightarrow I_3^-$ in neutral solution and $Ce^{IV} \rightarrow Ce^{III}$ in 0.5 M H_2SO_4.

Active nitrogen has also been encountered in other ways. Atomic ^{13}N was produced by the $^{12}C(d, n)^{13}N$ reaction using a 7 MeV deuteron beam on a carbon dioxide source[151]. The nitrogen produced was allowed to react with CO_2–N_2 mixtures at atmospheric pressure and the products analysed by radio-g.l.c. The product spectrum was explained by the following reaction scheme:

$$^{13}N^*+CO_2 \rightarrow {}^{13}NO+CO$$
$$^{13}N^*+N_2 \rightarrow {}^{13}NN^*+N$$
$$^{13}NN^*+CO_2 \rightarrow {}^{13}NNO+CO$$
$$^{13}NN^*+M \rightarrow {}^{13}NN+M^*$$

2.13. BIOLOGICAL PROPERTIES[152]

Although nitrogen is a vital element to all life, it cannot be assimilated by the higher animals nor by the majority of plants from its most abundant source as molecular nitrogen (see section 1.7). To mammals, atmospheric nitrogen is just a convenient inert carrier gas for respiratory oxygen and expiratory carbon dioxide. However, exposure to high-pressure air causes some nitrogen to dissolve in the blood and other body fluids. At very high pressures, nitrogen has a narcotic effect, apparently due to removal of oxygen from the nervous tissues. If the pressure is suddenly lowered, the dissolved nitrogen forms bubbles

149 I. M. Campbell and B. A. Thrush, *Chem. Communs.* (1967) 992.
150 N. N. Lichtin, S. E. Juknis, R. Melucci and L. Backenroth, *Chem. Communs.* (1967) 283.
151 M. J. Welch, *Chem. Communs.* (1968) 1354.
152 F. Call, in ref. 85, pp. 567–615.

which are released in the body, and can cause pains in the muscles and joints, partial paralysis, fainting and eventually death. Much attention has been paid to the problem of "diver's bends", and slow decompression is always practised by underwater swimmers.

Nitrogen, unlike isoelectronic carbon monoxide, is not toxic, but in high concentrations it is an asphyxiant. It is the principal constituent of "blackdamp" (87% N_2, 13% CO_2), a well-known hazard in coal-mines, but similar situations can be encountered in enclosed spaces where liquid nitrogen has been allowed to evaporate.

2.14. ANALYTICAL CHEMISTRY OF MOLECULAR NITROGEN[153, 154]

Nitrogen can generally be detected in any gaseous mixture with a quartz spectrograph without preliminary separation. This technique is also applicable for the determination of nitrogen in solids and in solutions by sparking under reduced pressure. The presence of larger amounts of nitrogen can be indicated by heating with magnesium:

$$3Mg + N_2 \rightarrow Mg_3N_2$$

Magnesium nitride is greenish yellow in colour, and is readily decomposed by water to give ammonia which can be identified by its odour or by a chemical test:

$$Mg_3N_2 + 6H_2O \rightarrow 2NH_3 + 3Mg(OH)_2$$

The determination of molecular nitrogen is most often encountered in connection with furnace or dry-box atmospheres and only occasionally as a product of reaction. In mixtures of gases, nitrogen is usually determined by difference after the other gases have been absorbed. The presence of inert gases would lead to high results; however, in such cases one would need to resort to absorbing nitrogen. One such method is the quantitative absorption by calcium carbide to form calcium cyanamide at 800°C.

$$CaC_2 + N_2 \xrightarrow{800^\circ C} CaCN_2 + C$$

Small amounts of nitrogen in the presence of other gases have been estimated by mass spectrometry, which is also used to detect and determine the impurities in cylinder nitrogen. Nitrogen dissolved in metals can be released by vacuum fusion techniques and, after removal of impurities, measured volumetrically.

3. AMMONIA

3.1. HISTORY OF AMMONIA

The penetrating odour of ammonia was known even in prehistoric times and was originally encountered in connection with ammonium chloride from urine and other animal wastes. Later, ammonia solutions were the only well-known soluble base, and were used in the fulling and dyeing of wool[155]. Notable dates in the history of ammonia were: 1661

153 A. J. Clear and M. Roth, in *Treatise on Analytical Chemistry*, Part II, Section A, Vol. 5 (eds. I. M. Kolthoff and P. J. Elving), Interscience, New York (1961), p. 217.

154 K. I. Vasu, in *Mellor's Comprehensive Treatise on Inorganic and Theoretical Chemistry*, Vol. 8, Suppl. II, *Nitrogen*, Part 2, Longmans, London (1967), pp. 563–673.

155 J. W. Mellor, *Comprehensive Treatise on Inorganic and Theoretical Chemistry*, Vol. 8, *Nitrogen*, Longmans, London (1928), Section 14.

when Boyle realized that ammonium chloride was composed of ammonia and hydrogen chloride; 1774 when Priestley distilled ammonium chloride with lime and collected ammonia over mercury; and 1785 when Berthollet decomposed ammonia by electric spark into its constituent elements and determined its stoichiometric composition. The early nineteenth century saw the production of ammonia liquor as a byproduct of the coal-gas industry[156] followed by development of synthetic routes to ammonia in the beginning of the twentieth century[157].

3.2. OCCURRENCE OF AMMONIA

Ammonia was believed to be present in the atmosphere during the early stages of the earth's evolution, but it is now thought that nitrogen and carbon dioxide predominated[156]. Ammonia can be detected in air, rain, snow, etc., originating from the immense quantities of ammonia which are constantly formed in nature by putrefaction of the protein from dead plants, animals and micro-organisms. Rivers and sea water may also contain appreciable amounts, particularly near the sea-bed as ammonium salts are found in the deposits obtained after evaporation of sea water[158]. The nitrogen-containing compounds in coal and oil are thought to originate from similar sources.

3.3. PRODUCTION OF AMMONIA

3.3.1. Synthetic Production from the Elements[159–162]

Virtually all ammonia is produced industrially by the reversible reaction between hydrogen and nitrogen at elevated temperatures and pressures in the presence of a catalyst according to the equation

$$N_2(g)+3H_2(g) \xrightleftharpoons{\text{catalyst}} 2NH_3(g); \qquad \triangle H(298\text{ K}) = -22.08\text{ kcal/mol}$$

Although Nernst and Yost[163] were the first to measure the improvement in the equilibrium yield under pressure, this process was developed by Haber[164] between the years 1905 and 1913, and illustrates how the application of thermodynamic principles can result in a major industrial process. Consider Le Chatelier's principle as applied to this equilibrium; the formation of ammonia will be favoured by high pressures (a consequence of the decrease in volume accompanying the formation of product from reactants) and low temperatures (a result of the exothermic nature of the reaction). However, when the temperature is lowered the rate at which the corresponding equilibrium mixture is attained also decreases.

[156] G. S. Cribb, in *Mellor's Comprehensive Treatise on Inorganic and Theoretical Chemistry*, Vol. 8, Suppl. I, *Nitrogen*, Part I, Longmans, London (1964), Section IV, pp. 250–8.

[157] G. H. Payn, in ref. 156, Part I, Section V, pp. 259–75.

[158] D. F. Martin, *Marine Chemistry*, Vol. 1, Marcel Dekker, New York (1968).

[159] C. Matasa and E. Matasa, *L'Industrie Moderne des Produits Azotes*, Dunod, Paris (1968), pp. 63–208.

[160] R. Noyes, *Ammonia and Synthesis Gas*, Noyes Development Corporation, New Jersey (1967).

[161] A. J. Harding, *Ammonia Manufacture and Uses*, Oxford University Press, London (1959).

[162] R. M. Stephenson, *Introduction to the Chemical Process Industries*, Reinhold, New York (1966), pp. 102–55.

[163] W. Nernst, W. Jost and G. Jellinek, *Z. Elektrochem.* **13** (1907) 52; **14** (1908) 373; *Z. anorg. Chem.* **57** (1908) 414.

[164] F. Haber and R. LeRossignol, *Ber.* **40** (1907) 2144; *Z. Elektrochem.* **14** (1908) 181.

Haber realized that the way to maintain an acceptable rate of equilibrium while keeping the temperature low enough to produce reasonable equilibrium concentrations of ammonia was to employ a catalyst. Finely divided iron containing one or more "promoters" was eventually found to be a most effective catalyst[165].

Many ammonia production plants have been built over the years throughout the world[166]. Probably no two plants are identical, but they are all in principle similar to the original Haber–Bosch process. What has changed dramatically, however, is the scale of operation, with the largest plants at present on stream being capable of producing 1500 tons of ammonia per day[167]. Such advancement represents major developments in the scale-up of high-pressure reactions and solution of the accompanying chemical engineering problems. For the purpose of considering the chemical basis of the main stages in ammonia production, the overall process can be described quite generally as in Fig. 15.

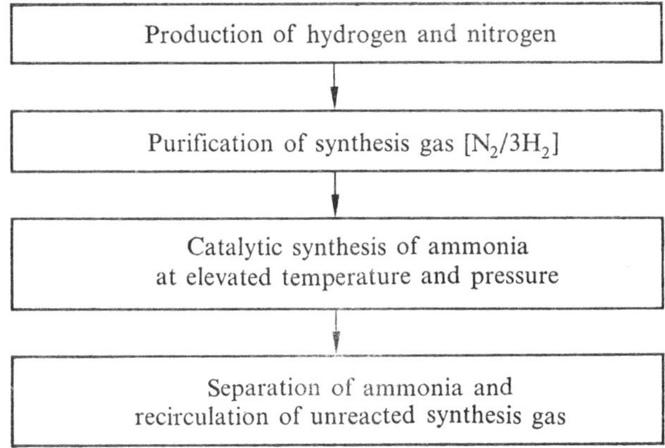

FIG. 15. General scheme for the production of ammonia.

(a) *Sources of Nitrogen and Hydrogen and Purification*

The mixture of hydrogen and nitrogen used as feedstock in the Haber process is appropriately known as synthesis gas, and for the successful operation of the process it is necessary that the mixture be of high purity. The only source of nitrogen is the atmosphere, from which it is obtained in a pure form either from the fractionation of liquid air[168] or the chemical deoxygenation of air[160]. Hydrogen may be available as a byproduct of some other process (such as catalytic cracking of petroleum naphtha in the production of petrol[160]) or from the electrolysis of water where electricity is cheap[167]; but when it is required specifically for ammonia synthesis, it is usually produced from hydrocarbon and carbonaceous fuels.

[165] A. Nielsen, *An Investigation on Promoted Iron Catalysts for the Synthesis of Ammonia*, Jul. Gjellerups Forlag, Copenhagen (1956).

[166] *World Fertilizer Atlas*, British Sulphur Corp. Ltd. (1967).

[167] Anon., *Nitrogen*, No. 60 (1969) 34.

[168] M. Sittig, *Nitrogen in Industry*, Van Nostrand, New Jersey (1965).

The incomplete combustion of a fuel such as coal, coke or charcoal gives "producer gas", the approximate composition of which is 25% CO, 70% N_2, 4% CO_2 together with smaller amounts of hydrogen, methane and oxygen. "Water gas" is prepared by blowing steam over coke or anthracite at bright red heat:

$$C+H_2O \rightleftharpoons CO+H_2; \qquad \Delta H(298) = +31.38 \text{ kcal/mol}$$

The carbon monoxide from both these processes can react with steam in the "shift reaction" to give more hydrogen:

$$CO+H_2O \xrightleftharpoons{\text{catalyst}} CO_2+H_2; \qquad \Delta H(298) = -9.84 \text{ kcal/mol}$$

Saturated hydrocarbons as found in natural gas, refinery gas and light naphtha readily decompose at high temperatures to methane which can undergo the process of steam reforming and subsequent shift reaction[159]:

$$CH_4+H_2O \rightleftharpoons CO+3H_2; \qquad \Delta H(298) = +49.27 \text{ kcal/mol}$$
$$CO+H_2O \rightleftharpoons CO_2+H_2; \qquad \Delta H(298) = -9.84 \text{ kcal/mol}$$

The heat for the endothermic steam-reforming reaction is produced by the introduction of a controlled amount of air causing partial oxidation of some of the gas stream, while at the same time providing nitrogen for the synthesis gas mixture[159]:

$$CH_4+\tfrac{1}{2}O_2 \rightleftharpoons CO+2H_2; \qquad \Delta H(298) = -8.50 \text{ kcal/mol}$$
$$CO+\tfrac{1}{2}O_2 \rightleftharpoons CO_2; \qquad \Delta H(298) = -63.64 \text{ kcal/mol}$$
$$H_2+\tfrac{1}{2}O_2 \rightleftharpoons H_2O; \qquad \Delta H(298) = -57.80 \text{ kcal/mol}$$

Thus, in effect, an equilibrium mixture is created between H_2, O_2, H_2O, CO, CO_2 and CH_4, nitrogen acting only as a diluent, and conditions of temperature, pressure, catalysts, residence times, etc., are chosen so that synthesis gas contaminated only by carbon dioxide is the predominant product[160].

(b) *Purification of Synthesis Gas*[159, 160]

Primarily, the purification of the synthesis gas stream is in order to maintain the life of the catalyst which would be "poisoned" by the presence of sulphur compounds, water, oxygen and oxides of carbon. Thus the oxidation of carbon monoxide in the shift reaction as well as being a further source of hydrogen is a necessary part of purification. However, sulphur compounds are removed with zinc oxide even prior to carbon monoxide conversion, as they would also poison the catalysts used in the shift reaction. Ammoniacal copper(I) solution is used to scrub out traces of carbon monoxide while active charcoal has been used for subsequent absorption of organic sulphur compounds. Carbon dioxide is removed under pressure by scrubbing with water, ethanolamine or hot potassium carbonate solution.

Final traces of impurities are sometimes removed by condensation at liquid nitrogen temperature and atmospheric or higher pressure. Indeed, in a process using a feedstock rich in hydrogen (>75%), there are advantages in using a low-temperature fractionation to separate the hydrogen[160]. In plants where cryogenic purification is omitted, inert gases accompanying nitrogen from air tend to concentrate in the synthesis reactor and in effect lower the rate of reaction. This necessitates periodic venting, although this vented gas provides an excellent enriched source for the separation of inert gases.

(c) *Reaction Conditions*[159]

The principle variable distinguishing the different processes is the operating pressure, of which a few are compared in Table 17.

TABLE 17. COMPARISON OF AMMONIA SYNTHESIS PROCESSES

Process	Pressure (atm)	Temperature (°C)	Catalyst	Ammonia (%)
Haber–Bosch	300–325	530–560	Double promoted iron	13–15
Fauser–Montecatini	280–300	520–550	Promoted iron	12–14
Mont Cenis	100–160	440–480	Iron cyanide	8–10
Kellogg	150–200	500–520	Promoted iron	11–12
Casale	500–900	500	Promoted iron	18–20
Claude	1000	495–600	Promoted iron	28–30

In general, pressures of about 300 atm have been preferred, but with the largest plants it has been necessary to operate at 200 atm due to compressor limitations[160]. For a particular pressure the choice of temperature has to be within the range in which the catalyst gives high activity, usually about 500°C. This is a convenient compromise between the low temperatures desirable for a high equilibrium concentration and the higher temperatures necessary for increased reaction rate.

A third variable is the space velocity[160]. Clearly, at a constant temperature, an increase in space velocity decreases the percentage of ammonia in the exit gas, but results in a substantial increase in the amount of ammonia produced per volume of catalyst. This leads to a fourth variable because at the space velocities normally used in ammonia synthesis, the maximum amount of ammonia in the exit gas occurs at the hydrogen–nitrogen ratio of 2.5:1.0 rather than the stoichiometric ratio of 3:1.

Ammonia is condensed from the exit gas as the liquid by water cooling and/or refrigeration, and the unreacted elements are compressed and recirculated.

(d) *Catalyst Systems*[169–171]

The iron catalyst first developed by Mittasch in 1910 is still the basis of catalyst systems in current use[172]. These are prepared when iron oxide, generally in the form of magnetite (Fe_3O_4) is sintered or fused with small amounts of one or more oxides Al_2O_3, K_2O, CaO and MgO, called "promoters". The mixture is then reduced either in the converter of an ammonia plant by slow heating with synthesis gas until the plant operating temperature is reached, or pre-reduced with pure hydrogen and the resultant catalyst stabilized to render it non-pyrophoric. Both methods reduce the oxide to iron metal in a suitably divided and activated state ready for use as the active catalyst. Thus "double promotion", i.e. the

[169] *Catalyst Handbook*, ICI, Wolfe Scientific Books, London (1970).

[170] W. G. Frankenburg, in *Catalysis*, Vol. 3 (ed. P. H. Emmett), Reinhold, New York (1955), pp. 171–263.

[171] C. Bokhoven, C. Van Heerden, R. Westrik and P. Zweitering, in *Catalysis*, Vol. 3 (ed. P. H. Emmett), Reinhold, New York (1955), pp. 265–348.

[172] A. Mittasch, *Z. Elektrochem.* **36** (1930) 569.

addition to iron oxide of both an acidic or amphoteric oxide such as Al_2O_3, and an alkaline oxide such as potassium oxide, increases the activity of the resultant catalyst over that for singly promoted catalysts. It is thought that in such systems the Al_2O_3 contributes to the generation and stabilization of porosity and of large surface, whereas K_2O promotes catalytic action even though its presence may reduce the free iron surface[167].

Decrease in the activity of a catalyst is always observed over a period of time, and in order to maintain efficiency and to prolong the life of such large amounts of catalyst as are involved with a synthesis plant, the pressure, temperature, residence time, etc., are systematically varied to maintain optimum economy. However, the catalyst can deteriorate more rapidly under severe operating conditions resulting in disintegration, while permanent poisoning, e.g. with sulphur compounds, can also necessitate total replacement.

The reaction at the catalyst site is exothermic as demonstrated by the rise in temperature of the gas stream as it proceeds through the reactor. Thus it is important to dissipate the heat produced particularly by the initial reaction of synthesis gas at the catalyst bed localized near the inlet port, while towards the end of the converter where the feedstock will be somewhat diluted with product and at slightly lower pressure, the catalyst will not be used to maximum efficiency. Synthesis converters are thus designed to give even reaction conditions throughout, and any excess heat produced is recovered to pre-heat the entering synthesis gas to the operating temperature of the converter.

(e) *Reaction Kinetics*

In the reversible reaction

$$N_2+3H_2 \underset{k_2}{\overset{k_1}{\rightleftharpoons}} 2NH_3; \qquad \Delta H(298) = -22.08 \text{ kcal/mol}$$

the net rate of reaction is the difference between the ammonia synthesis reaction rate k_1 and the ammonia decomposition reaction rate k_2. Equilibrium occurs when the net rate of reaction is zero, that is when $k_1 = k_2$. An increase in temperature will increase k_2 considerably and k_1 to a lesser extent because it is exothermic, while an increase in pressure will favour k_1. The effect of temperature and pressure on equilibrium volume of ammonia[173] can be seen in Table 18, related to $K_p = p^2_{NH_3}/p_{N_2} \cdot p^3_{H_2}$.

TABLE 18. VOLUME PERCENTAGE OF AMMONIA IN $N_2/3H_2$ EQUILIBRIUM MIXTURES

Pressure (atm)	Temperature (°C)					
	200	300	400	500	600	700
10	50.66	14.73	3.85	1.21	0.49	0.23
50	74.38	39.41	15.27	5.56	2.26	1.05
100	81.54	52.04	25.12	10.61	4.52	2.18
300	89.94	70.96	47.00	26.44	13.77	7.28
600	95.37	84.21	65.20	42.15	23.10	12.60
1000	98.29	92.55	79.82	57.47	31.43	12.87

173 A. T. Larsen, *J. Am. Chem. Soc.* **46** (1924) 367.

However, because these are not ideal gases, K_p does not remain constant for a given temperature and pressure above 50 atm, and the increase in the value of K_p at higher pressures indicates that the conversion is even more favourable under such conditions than would be predicted assuming perfect gases.

Emmett and Brunauer[174] first suggested that the rate-determining step in the synthesis reaction is the chemisorption of nitrogen on the surface of the catalyst. Although this step requires a considerable activation energy, the subsequent steps of hydrogenation on the surface of the catalyst and desorption of the ammonia require only small amounts of activation energy and thus occur at a fast rate. Similarly, in the decomposition of ammonia, the rapid adsorption and dehydrogenation of NH_3 on the surface of the catalyst is followed by slow desorption of nitrogen molecules from the surface of the catalyst. This hypothesis was used by Temkin and Pyzhev[175] to develop the rate equation

$$\frac{dP}{dt}\,NH_3 = k_1\,p_{N2}\left(\frac{p^3_{H_2}}{p^2_{NH_3}}\right) - k_2\left(\frac{p^2_{NH_3}}{p^3_{H_2}}\right)$$

The nitrogen adsorbed on the surface is in equilibrium with hydrogen and ammonia in the gas phase, and it is assumed that nitrogen adsorption is unaffected by the pressure of varying amounts of hydrogen and ammonia. The above equation was formulated with concentrations represented by partial pressures, but was later revised for use at high pressures by the introduction of fugacities and an additional correction factor[176]:

$$\frac{dP}{dt}\,NH_3 = k_1 f_{N_2}\left[\frac{f^3_{H_2}}{f^2_{NH_3}}\right]^{\alpha} - k_2\left[\frac{f^2_{NH_3}}{f^3_{H_2}}\right]^{1-\alpha} \frac{\exp\,(\overline{V}_s - \overline{V}_a)P}{RT}$$

The exponent α is <1 and is frequently taken as 0.5. $\overline{V}_s$ and $\overline{V}_a$ are the partial molar volumes of adsorbed nitrogen and of the transition state for adsorption respectively, and P is the total pressure; k_1 and k_2 are related to the fugacity equilibrium constant for the reaction by

$$k_1 = k_2 K$$

and if the exponential factor is donated by ψ, then the equation can be rearranged to give

$$r = k_2\left[\left(\frac{K_f f_{N_2} f^3_{H_2}}{f^2_{NH_3}} - 1\right)\left(\frac{f^2_{NH_3}}{f^3_{H_2}}\right)^{1-\alpha}\right]\psi$$

Recent evidence has confirmed that the chemisorption of nitrogen is the rate-determining step[177], that the rate of adsorption is equal to the rate of ammonia synthesis[178] and that the modified Temkin–Pyzhev equation correlates with the most recent P and T equilibrium data[179].

3.3.2. Small-scale Production of Ammonia

Compared to the scale on which the Haber process is operated, all other methods for the commercial production of ammonia are small scale. Prior to the Haber process, the principal synthetic route to ammonia was by the hydrolysis of cyanamides or cyanides.

[174] P. H. Emmett and S. Brunauer, *J. Am. Chem. Soc.* **56** (1934) 35.
[175] M. I. Temkin and V. Pyzhev, *Acta Physicochim. USSR* **12** (1940) 327.
[176] M. I. Temkin, *Zh. Fiz. Khim.* **24** (1950) 1312.
[177] C. Bokhoven, M. I. Gorgels and P. Mars, *Trans. Faraday Soc.* **55** (1959) 315.
[178] J. J. F. Scholten, J. A. Konvalinka and P. Zweitering, *Trans. Faraday Soc.* **56** (1960) 262.
[179] A. Nielsen, J. Kjaer and B. Hansen, *J. Catalysis* **3** (1964) 68.

Calcium cyanamide produced by the Frank–Caro process[168] can be hydrolysed with dilute sodium hydroxide solution in an autoclave above 100°C:

$$CaCN_2 + 3H_2O \rightarrow 2NH_3 + CaCO_3$$

Hydrolysis at 400–500°C of sodium cyanide, prepared by the Bucher process from coal, alkali and nitrogen, was considered as a potential route to ammonia even after the Haber process had become well established:

$$NaCN + 2H_2O \rightarrow HCO_2Na + NH_3$$

However, these methods are now of historic interest only.

Ammonia can be regenerated from ammonium salts simply by heating with a base. Alkaline earth metal oxides and hydroxides have been used with the naturally occurring ammonium chloride[156].

Ammonia can be released from a variety of organic waste materials. Such methods generally involve the residues of organic reactions where oxidation and hydrolysis of proteins release ammonia[156]. However, the principal source of non-synthetic ammonia is obtained when fuels such as coal, oil and peat are processed by distillation, gasification or hydrogenation[156]. Prior to the development of the Haber process, ammonia and ammonium sulphate were produced from coal as a byproduct of coke and coal-gas manufacture. Ammonia liquor collected from condensers and washers is steam distilled in the presence of lime, and the ammonia evolved is absorbed in sulphuric acid. Alternative methods of recovering ammonia simultaneously with sulphur, or as ammonium bicarbonate and phosphate, have been investigated in order to produce lower cost products[156]. Only part of the nitrogen in coal is recovered as ammonia and yields generally amount to less than 10 lb of ammonia per ton of coal. Thus with the low price of synthetic ammonia, the sale of byproduct ammonium sulphate only partially off-sets the cost of recovery.

3.3.3. Handling, Storage and Corrosion

Bulk ammonia is usually stored as cold liquid in insulated Horton spheres[161]. As the gas slowly volatilizes it can be refrigerated and returned to the tank or used directly in some ammonia-consuming process. It is transported as the liquid in cylinders or tank-wagons capable of withstanding the vapour pressure at ambient temperatures.

Although ammonia itself presents few corrosion problems, the conditions under which ammonia is prepared have required the development of special alloys for the construction synthesis reactors[162]. Liquid and gaseous ammonia can be stored or handled in equipment made from cast iron, steel, aluminium, polythene and natural or synthetic rubber. However, copper and its alloys, nickel or polyvinyl chloride should not be allowed into contact with either gaseous or liquid ammonia. Laboratory cylinder pipework and valves are best made from stainless steel.

From a safety consideration, it is not always realized that ammonia is flammable in certain concentrations in air, but since the high concentrations are seldom encountered, the relative explosion or fire hazard is small. Mercury may also react with ammonia to form explosive compounds; hence the use of mercury thermometers should be avoided. However, the greatest potential danger arises from accidental spillage or escape of ammonia leading to high concentrations which can cause severe irritation. Respirators should always be at hand in places where ammonia is handled.

3.4. USES OF AMMONIA

As the nitrogen fertilizer industry is based almost entirely on ammonia[180], its principal uses are in the production of ammonium salts (see section 5.2), and the oxidation of ammonia in the route to nitric acid and its many derivatives[181] (see section 18.3). Ammonium nitrate is related to both the fertilizer and explosive industries. Similarly, the production of urea, not only for fertilizer use but also for conversion to cyanuric acid and melamine, are new major ammonia users[159, 181].

Ammonia is used in the Solvay ammonia–soda process, where its properties as an alkaline material capable of easy regeneration finds application. Anhydrous ammonia is also used in the petroleum industry to neutralize acid constituents of oil, thus protecting the plant from corrosion without the formation of aqueous solutions or emulsions. A weak aqueous solution is also used as a household cleanser[182].

Since anhydrous ammonia can be transported as the liquid, it can be regarded as a portable source of nitrogen and hydrogen by ready dissociation, under conditions exactly the reverse to its formation.

The food and beverage industries use ammonia as nutriment in fermentation processes. The rubber industry use ammonia atmospheres for the vulcanization of certain hard rubber products, while natural and synthetic latex is stabilized during transportation and storage by ammonia[182]. The textile industry use ammonia in connection with the production of synthetic fibres such as cuprammonium rayon, and in the preparation of hexamethylenediamine for nylon-6,6 and caprolactam for nylon-6 [181]. The applications of ammonia cannot be concluded without reference to its considerable use in what organic chemists term ammonolysis reactions, leading to cyanides, amides, amines, nitriles, dye intermediates and urea[181, 183].

3.5. LABORATORY PREPARATION AND PURIFICATION OF AMMONIA

The laboratory preparation of ammonia from ammonium salts in strongly basic solution is usually only required for analytical purposes as pure synthetic NH_3 is commercially available in steel cylinders[184]. Passage of the gas through regenerated activated charcoal is sufficient to remove minor impurities such as oil vapour and carbon monoxide. The gas is then dried by passing over soda-lime, potassium hydroxide, sodium wire, and, finally, phosphorus pentoxide. The lack of reaction with P_2O_5 can be used as the indication of successful pre-drying. The gas that boils off the blue solution of liquid ammonia containing a little sodium is also perfectly dry[185]. Gaseous ammonia can be condensed to liquid by passing through traps cooled in liquid nitrogen. Alternatively, liquid ammonia can be withdrawn from an inverted cylinder into Dewar vessels in an efficient fume cupboard.

180 C. J. Pratt and R. Noyes, *Nitrogen Fertilizer Chemical Processes*, Noyes Development Corporation, New Jersey (1965).

181 M. Sittig, *Combine Hydrocarbons and Nitrogen for Profit*, Noyes Development Corporation, New Jersey (1967).

182 J. D. F. Marsh, in ref. 156, Section VII(b), p. 361.

183 H. Smith, *Organic Reactions in Liquid Ammonia*, Vol. I, Part 2, *Chemistry in Nonaqueous Ionizing Solvents* (eds. G. Jander, H. Spandau and C. C. Addison), Interscience, New York (1963).

184 The British Drug Houses Ltd., Poole, England.

185 P. W. Schenk, in *Handbook of Preparative Inorganic Chemistry*, Vol. I (ed. G. Brauer), Academic Press, New York (1963), p. 460.

To prepare a pure carbonate-free aqueous ammonia solution, cylinder ammonia is purified through activated charcoal and soda-lime and bubbled through well-boiled water in a container protected from the atmosphere by a soda-lime tube[185].

3.6. NUCLEAR ISOTOPES $^{15}NH_3$, $^{14}ND_3$, $^{15}ND_3$

Pure dry $^{15}NH_3$ is most conveniently prepared by treating a solution of an enriched ammonium salt with excess alkali (see section 1.4.3) and drying the final product over metallic sodium[186]:

$$^{15}NH_4^+ + KOH \rightarrow {}^{15}NH_3 + H_2O + K^+$$

As most nitrogen-containing compounds can be oxidized to nitrate or nitrite, the reduction to ammonia with Devarda's alloy (50% Cu, 45% Al, 5% Zn) in alkaline solution[187] provides a general method to recover enriched nitrogen as a useful starting material $^{15}NH_3$. As an alternative, the conversion of $^{15}N_2$ into $^{15}NH_3$ can be achieved by the formation of an ionic metal nitride and subsequent hydrolysis[188]:

$$3Ca + {}^{15}N_2 \rightarrow Ca_3{}^{15}N_2 \xrightarrow{6H_2O} 2{}^{15}NH_3 + 3Ca(OH)_2$$

Similarly, ammonia-d_3 can be prepared by the deuterolysis of an ionic metal nitride[157]:

$$Li_3N + 3D_2O \rightarrow ND_3 + 3LiOD$$
$$Mg_3N_2 + 6D_2O \rightarrow 2ND_3 + 3Mg(OD)_2$$

The method is also convenient for preparing the doubly labelled compound[190]:

$$Ca_2{}^{15}N_3 + 6D_2O \rightarrow 2{}^{15}ND_3 + 3Ca(OD)_2$$

$^{15}NH_3$, $^{14}ND_3$ and enriched ammonium salts are all available commercially[191].

3.7. STRUCTURE AND BONDING

The Sidgwick–Powell approach[192] to molecular shape predicts ammonia to be pyramidal. This result is arrived at by considering each of the three bonding pairs of electrons between nitrogen and the hydrogens together with the lone pair residual on nitrogen as occupying positions of mutual repulsion in a tetrahedral arrangement. The Nyholm–Gillespie extension to this approach[193] further suggests that the H–N–H angle is slightly smaller than the tetrahedral angle due to the greater repulsion between the lone pair on nitrogen and the bond-pairs than between the bond pairs themselves. These predictions are borne out by the structural parameters for gaseous ammonia as obtained from spectroscopic and electron

[186] W. Spindel, in *Inorganic Isotopic Synthesis* (ed. R. H. Herber), Benjamin Inc., New York (1962), p. 89.

[187] K. Clusius and H. Schumacher, *Helv. Chim. Acta* **43** (1960) 1562.

[188] K. Clusius, *Angew. Chem.* **66** (1954) 497.

[189] M. Baudler, in *Handbook of Preparative Inorganic Chemistry*, Vol. I (ed. G. Brauer), Academic Press, New York (1963), p. 137.

[190] K. Clusius and M. Huber, *Z. Naturforsch.* **10a** (1955) 556.

[191] Office Nationale Industriel de l'Azote (ONIA), 40 avenue Hoche, Paris (VIIIe), France. Isomet Corporation, Palisades Park, New Jersey, USA.

[192] N. V. Sidgwick and H. M. Powell, *Proc. Roy. Soc.* (*London*) A, **176** (1940) 153.

[193] R. J. Gillespie and R. S. Nyholm, in *Progress in Stereochemistry*, Vol. 2 (eds. W. Klyne and P. B. D. de la Mare), Butterworths, London (1968), p. 261.

diffraction data listed in Table 19. X-ray and neutron diffraction studies on crystalline ammonia also indicate a similar shape, but with significantly longer N–H bond lengths. Each nitrogen in the ammonia lattice has six equidistant nearest neighbour nitrogen atoms which by their position and distance indicate that the structure is further stabilized by hydrogen bonding[194].

TABLE 19. MOLECULAR CONSTANTS[a] OF AMMONIA AND AMMONIA-d_3

Property	NH_3	ND_3
X-ray diffraction data		
System	Cubic	Cubic
Space group	$P2_1,3$	$P2_1,3$
Lattice constant a	5.084 at −196°C 5.138 at −102°C	5.037 at −196°C 5.091 at −160°C
N–H bond length	1.13 Å	1.12 Å
N–N distance	3.380 Å	3.349 Å
N–N–N angles	118°, 71.6°	118°, 71.6°
Electron diffraction data		
d(N–H)	1.019 Å	1.020 Å
H–N–H bond angle	109.1°	106.1°
Fundamental vibration frequencies (cm^{-1})		
ν_1 (a_1)	3336.21	2419
ν_2 (a_1)	950.24 931.66 964.08	748.6 749.0
ν_3 (e)	3443.38	2555
ν_4 (e)	1627.77	1191.0
Parameters calculated from spectroscopic data		
d(N–H)	1.0173 Å	1.0155 Å
H–N–H bond angle	107.78°	107.59°
Height of pyramid	0.3670 Å	0.3687 Å
Inversion barrier	5.9 kcal/mol	
Dipole moment	1.46 D	1.50 D
Bond dissociation energies		
$D(NH_2–H)$	104 ± 2 kcal/mol	
D(NH–H)	88 ± 4 kcal/mol	
D(N–H)	88 ± 2 kcal/mol	
Ionization potential	10.154 eV	
Nuclear quadrupole coupling constant	−4.0842	−4.080

[a] Refs. 194, 195, 198, 199.

Thus although the bond angles in ammonia are a fine balance between electrostatic repulsive forces and hybridization energies, in fact the energy difference between the pyramidal and planar arrangement is small enough (5.9 kcal/mol) for the molecule to invert[195]. Figure 16 illustrates the resultant potential energy curve with two minima corresponding to the two equivalent nitrogen positions of equilibrium either side of the hydrogen atom plane. The dotted lines represent the potential energy of non-inverting molecules. The vibrational states are split into symmetric and antisymmetric levels, the effects of which can be detected in infrared and microwave spectra.

[194] J. Olovsson and D. H. Templeton, *Acta Cryst.* **12** (1959) 832.
[195] C. H. Townes and A. L. Schawlow, *Microwave Spectroscopy*, McGraw-Hill, New York (1955).

Self-consistent molecular orbital calculations[196,197] on the NH_3 molecule indicate that a small transfer of electrons leaves the hydrogen more positive, and there is clearly some hybridization of the nitrogen bonding orbitals.

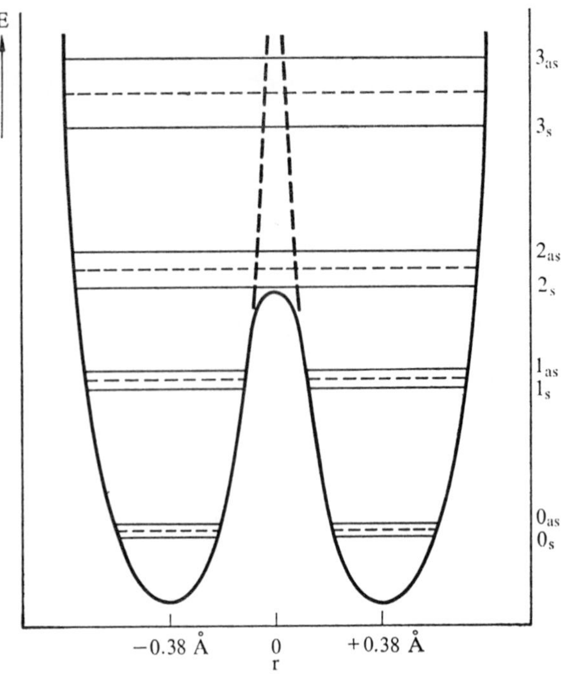

FIG. 16. The potential energy of ammonia as a function of the displacement of the nitrogen atom from the plane of the hydrogen atoms.

3.8. MOLECULAR CONSTANTS

Various molecular constants[198] of ammonia and deuteroammonia are listed in Table 19. The electronic spectrum of ammonia consists of five transitions and ionization continua in the far ultraviolet[199]. The infrared spectrum of ammonia is complicated by the large number of vibration–rotation interactions as well as by the inversion doubling previously mentioned[199]. The Raman spectrum is complex and the microwave spectrum contains some very intense lines also arising from transitions between the component levels of the inversion-doubled ground state[199]. The inversion frequency can be calculated from an equation of the type

$$\nu = \nu_0 - a[J(J+1)-K^2]+bK^2+\text{higher powers of } J \text{ and } K,$$

the origin of the system being at 23785.88 mHz. Such transitions have been used in the ammonia beam maser (microwave amplification by stimulated emission of radiation).

[196] H. Kaplan, *J. Chem. Phys.* **26** (1957) 1704.

[197] A. B. F. Duncan, *J. Chem. Phys.* **27** (1957) 423.

[198] J. D. F. Marsh, in ref. 156, Section VI, pp. 276–327.

[199] R. F. Barrow, A. J. Merer and A. Thorne, in *Mellor's Comprehensive Treatise on Inorganic and Theoretical Chemistry*, Vol. 8, Suppl. II, *Nitrogen*, Part II, Longmans, London (1967), Section XXXIV, pp. 484–95.

TABLE 20. PHYSICAL PROPERTIES OF AMMONIA[a]

Property	Temperature (K)	Value
Melting point	195.42	−77.74°C
Boiling point	239.74	−33.42°C
Triple point	195.42	−77.74°C
Triple-point pressure		45.58 mm
Critical temperature	405.6	132.4°C
Critical pressure		112.3 atm
Critical density		0.2362 g/cm³
ΔH fusion	195.42	1.3516 kcal/mol
ΔH vaporization	239.74	5.581 kcal/mol
ΔH_f (enthalpy of formation)	0	−9.34 kcal/mol [b]
ΔH_f (enthalpy of formation)	298	−11.02 kcal/mol [b]
Increase in enthalpy for each 100°C		2.06 kcal/mol
H_{298}–H_0 (enthalpy)	298	2.388 kcal/mol [b]
ΔG_f° (Gibbs energy of formation)	298	−3.94 kcal/mol [b]
S° (entropy)	298	45.97 cal/deg mol [b]
Increase in entropy for each 100°C		0.99 cal/deg mol
Entropy of vaporization	239.74	23.29 cal/deg mol
C_p° (heat capacity)	298	8.38 cal/deg mol
Density of solid	88.1	0.836 g/cm³
	194	0.817 g/cm³
Density of liquid	203	0.7253 g/cm³
	213	0.7138
	223	0.7020
	233	0.6900
	239	0.6826
	240	0.6814
	243	0.6776
	253	0.6650
	263	0.6520
	273	0.6386
	283	0.6247
	293	0.6103
	298	0.6028
	303	0.5952
Trouton's constant		23.56 cal/deg
Viscosity of gas	293	9.821×10^9 poise
Viscosity of liquid	239.5	0.254 centipoise
	247	0.230
	263	0.183
	269	0.170
	278	0.162
	283	0.152
	288	0.146
	293	0.141
	298	0.135
	303	0.138
Refractive index of liquid	289	1.325 (λ = 5899 Å)
Dielectric constant of gas	107	3.36
	161	3.42
	183	4.01
	194	4.83

[a] Refs. 198, 200.
[b] D. D. Wagman, W. H. Evans, V. B. Parker, I. Halow, S. M. Bailey and R. H. Schumm, *Selected Values of Chemical Thermodynamic Properties*, NBS Technical Note 270–3, Washington (1968).

TABLE 20 (*cont.*)

Property	Temperature (K)	Value
Dielectric constant of liquid	239	22
	278	18.94
	288	17.82
	298	16.90
	308	16.26
Surface tension of liquid	284.26	23.38 dyn/cm
	307.21	18.05
	332.14	12.95
Magnetic susceptibility	208	-1.115×10^{-6}
Thermal conductivity of gas	285.09/1.25–7.58 cm	5.51×10^{-5} g cal/cm s
Thermal conductivity of solid	169.26	0.00239 g cal/cm s
Specific conductance of liquid	201.7	$1.2\times10^{-7}\ \Omega^{-1}\ cm^{-1}$
	234.3	$1.97\times10^{-7}\ \Omega^{-1}\ cm^{-1}$
Electrical conductance of liquid		$1\times10^{-11}\ \Omega^{-1}\ cm^{-1}$

3.9. PHYSICAL PROPERTIES OF AMMONIA

Physical constants[200, 201] for ammonia are listed in Table 20 together with some specific values for those which vary with temperature, while properties which can be described by equations are listed in Table 21. Values which have been reported for physical properties of isotope isomers of ammonia (ND_3, $^{15}NH_3$) follow expected trends.

Perhaps the most significant effect on the physical properties of ammonia is that of hydrogen bonding which occurs in the condensed phases. The normal melting and boiling points (195.4 and 239.7 K) are much higher than would be expected for an unassociated hydride of a first short period element, and have been estimated to be 108 and 143 K respectively if hydrogen bonding were absent[202]. The breakdown of hydrogen bonding occurs to the extent of 26% on melting, 7% between the melting and boiling points and 67% on boiling. For the same reason, the thermodynamic properties are affected in that Trouton's constant is consistent with an associated solvent, and the enthalpy and entropy of vaporization are abnormally high. Properties associated with the excellent solvent properties of ammonia (see section 3.10.4) are a consequence of the high dipole moment and the dielectric constant.

The adsorption of gaseous ammonia on solids has been extensively studied[198]. Efficient adsorbents such as charcoal impregnated with various acidic materials are used in ammonia gas masks. The adsorption characteristics of ammonia at metal surfaces are also important in its synthesis and other catalytic reactions[198].

[200] J. Jander, *Chemistry in Anhydrous Liquid Ammonia*, Vol. I, Part I, of *Chemistry in Nonaqueous Ionizing Solvents* (eds. G. Jander, H. Spandau and C. C. Addison), Interscience, New York (1966).

[201] W. L. Jolly and C. J. Hallada, in *Non-aqueous Solvent Systems* (ed. T. C. Waddington), Academic Press, London (1965), pp. 1–45.

[202] J. D. F. Marsh, in ref. 156, Section VII(b), pp. 342–69.

TABLE 21. PHYSICAL PROPERTIES OF AMMONIA[a]

Vapour pressure of solid P

$\log P_{cm} = -1630.700/T + 10.00593$

Vapour pressure of liquid P

$\log P_{cm} = -1612.500/T - 0.012311T + 0.000012521T^2 + 11.83997$

K	°C	P	K	°C	P
164.06	−109.1	1 mm	254.45	−18.7	2 atm
181.26	−91.9	10	268.45	+4.7	5
193.96	−79.2	40	298.85	25.7	10
204.75	−68.4	100	323.25	50.1	20
228.95	−45.2	400	352.05	78.9	40
239.55	−33.6	1 atm	371.45	98.3	60

Second Virial coefficient B

$$(PV/nRT) = 1 + \frac{nB(T)}{V} + \ldots$$

K	B	K	B
293.4	−288	364.0	−154
313.6	−231	372.8	−139
323.0	−205	383.0	−137
333.0	−187	392.6	−128
343.7	−179	395.6	−118
351.4	−165		

Vapour density

L_p = mass of normal litre of NH_3 in g at p atm pressure

$L_p = 0.759913 + 0.011547p - 0.0000061p^2$

Surface tension of liquid γ

$\gamma = 23.41 - 0.3371t - 0.000943t^2 \quad (-75 < t < -39°C)$

Effect of pressure on melting point

$p = 16.29 + 310.22T + 1.33044T^2$

Specific heat

$C_p = 4.73(10^{-5})T^3$

[a] Ref. 198.

3.10. CHEMISTRY AND CHEMICAL PROPERTIES OF AMMONIA

3.10.1. Decomposition

The photochemical decomposition of ammonia has been studied with a view to establishing the mechanism including the following:

$$NH_3 + h\nu \rightarrow NH_2 + H$$
$$NH_3 + H + M \rightleftharpoons NH_4 + M$$
$$NH_4 + NH_2 \rightarrow 2NH_3 \text{ or } NH_3 + NH + H_2$$
$$2NH \rightarrow N_2 + H_2$$

Hydrazine is produced by both photochemical and by the glow discharge of ammonia, and decompositions in the presence of other substances have also been studied[203].

The decomposition of ammonia by electric discharge is not wholly attributable to ionization, and the detection of the NH radical in emission spectra when ammonia flows through a discharge tube is evidence for the mechanism:

$$NH_3 + e^- \rightarrow NH_3^* + e^-$$
$$NH_3 \rightarrow NH + H_2$$
$$NH + NH_3 \rightarrow N_2 + 3H_2$$

The presence of other materials with ammonia in glow or silent discharge can catalyse the decomposition to the elements or react to produce compounds which further interact.

The decomposition of ammonia to its elements at a solid surface has been widely investigated[203]. In particular the decomposition on iron and promoted iron catalysts has been mentioned in connection with ammonia synthesis (section 3.3) and nitrogen production (section 2.3.3). The platinum group metals are usually chosen as catalysts for the decomposition process, while other transition metals react with the nitrogen produced to form nitrides.

3.10.2. Oxidation of Ammonia[202]

The explosive limits of ammonia with air or oxygen depend not only on the temperature and pressure of the mixture but also on the means of ignition, size of open or closed vessel, etc. Explosive mixtures of ammonia in dry air at atmospheric temperature and pressure contain 16–27% by volume while the limits for ammonia with oxygen are 25–75%. These limits widen as the pressure is increased.

Stable ammonia–air flames cannot be produced by gas mixtures at room temperature, but with 19.4% NH_3 in air pre-heated to 150°C, a flame temperature of 1600°C can be obtained. Spectroscopic examination of such flames show bands due to NH_3, NH_2, NH, OH and O_2, while the combustion products contain traces of NH_3 and NO. Combustion of liquid ammonia with liquid oxygen in a rocket motor gives a temperature of 3060 K and an exhaust velocity of 1770 m/s. Liquid ammonia reacts with liquid oxygen at temperatures near the critical temperature (132.4°C) to form ammonium nitrate, ammonium nitrite and nitrogen.

The catalytic oxidation of ammonia to nitric oxide is important as the major step in the manufacture of nitric acid[159, 203]. For this process, it is the aim to promote the reaction:

$$4NH_3 + 5O_2 \rightarrow 4NO + 6H_2O; \qquad \Delta H(298) = -216.55 \text{ kcal/mol}$$

and suppress possible alternative reactions, some of which are:

$$4NH_3 + 3O_2 \rightarrow 2N_2 + 6H_2O; \qquad \Delta H(298) = -302.95 \text{ kcal/mol}$$
$$4NH_3 + 4O_2 \rightarrow 2N_2O + 6H_2O; \qquad \Delta H(298) = -263.85 \text{ kcal/mol}$$
$$4NH_3 + 7O_2 \rightarrow 4NO_2 + 6H_2O; \qquad \Delta H(298) = -271.15 \text{ kcal/mol}$$
$$4NH_3 + 6NO \rightarrow 5N_2 + 6H_2O; \qquad \Delta H(298) = -432.55 \text{ kcal/mol}$$
$$2NO \rightarrow N_2 + O_2; \qquad \Delta H(298) = -43.04 \text{ kcal/mol}$$
$$2NH_3 \rightarrow N_2 + 3H_2; \qquad \Delta H(298) = -22.04 \text{ kcal/mol}$$

Thus although the formation of nitric oxide is thermodynamically unfavourable with respect to nitrogen, by passing pre-heated ammonia–air mixture over a platinum catalyst at

[203] F. D. Miles, *Nitric Acid Manufacture and Uses*, ICI Ltd., Oxford University Press, London (1961).

temperatures above 650°C for a very short contact time, the formation of nitric oxide is favoured. Unfortunately, platinum has an appreciable vapour pressure at the reaction temperature, and although the less-volatile rhodium–platinum alloys are employed, the catalyst losses constitute a significant running cost which can only be partly offset by recovery from flue dust. All surfaces, in fact, catalyse the oxidation of ammonia, but most materials tend to produce nitrogen to such an extent as to make them unsuitable for this process.

It is now generally agreed that the oxidation occurs through collision of ammonia with a platinum surface almost completely covered with adsorbed oxygen. However, a large number of reaction intermediates have been isolated and there has been difference of opinion as to the initial nitrogen oxidation species. Mechanisms have been postulated based on the initial formation of three alternative species—imide, nitroxyl and hydroxylamine. The imide mechanism was first suggested by Raschig and developed by Zawadzki[204].

Reactions to give NO:

$$NH_3+O \rightarrow NH+H_2O$$
$$NH+O_2 \rightarrow HNO_2 \rightarrow NO+OH$$
$$2HNO_2 \rightarrow NO+NO_2+H_2O$$
$$HNO_2+O_2 \rightarrow HNO_4 \rightarrow NO+O_2+OH$$

Other side reactions:

$$NH+O \rightarrow HNO \rightarrow \tfrac{1}{2}N_2O+\tfrac{1}{2}H_2O$$
$$NH+NH_3 \rightarrow N_2H_4$$
$$2NH \rightarrow N_2+H_2$$
$$NH_3+HNO_2 \rightarrow N_2+2H_2O$$
$$2NO \rightarrow N_2+O_2$$
$$2OH \rightarrow H_2O+O$$

The nitroxyl theory of Andrussov[205] in which ammonia reacts initially with molecular oxygen is at variance with the generally accepted view that adsorbed oxygen is in the atomic form, but in fact, most of the reactions also appear in the other two mechanisms.

Reactions to give NO:

$$NH_3+O_2 \rightarrow HNO+H_2O$$
$$HNO+O_2 \rightarrow HNO_3$$
$$4HNO_3 \rightarrow 4NO+2H_2O+H_2$$

Other side reactions:

$$HNO+NH_3 \rightarrow N_2+H_2O+H_2$$
$$H_2+\tfrac{1}{2}O_2 \rightarrow H_2O$$

Hydroxylamine was proposed as the primary reaction product after it had been isolated from an ammonia oxidation reaction[206].

Reactions to give NO:

$$NH_3+O \rightarrow NH_2OH$$
$$NH_2OH+O_2 \rightarrow HNO_2+H_2O$$
$$HNO_2+O_2 \rightarrow HNO_4 \rightarrow NO+O_2+OH$$
$$4HNO_2 \rightarrow 4NO+O_2+2H_2O$$

---

204 J. Zawadzki, *Discuss. Farday Soc.* (1950) 140.
205 L. Andrussow, *Angew. Chem.* **63** (1951) 21.
206 M. Bodenstein, *Z. Elektrochem.* **47** (1941) 501.

Other side reactions:

$$NH_2OH+O \rightarrow HNO+H_2O$$
$$2HNO \rightarrow N_2O+H_2O$$
$$HNO_2+NH_3 \rightarrow N_2+2H_2O$$
$$HNO+NH_2OH \rightarrow N_2+2H_2O$$
$$2OH \rightarrow H_2O+O$$

The way in which the three primary products may be formed can be visualized as in Fig. 17. In this scheme, the ammonia becomes co-ordinated to an oxygen atom adsorbed on the catalyst via the nitrogen lone pair. Hydrogen of the ammonia may then hydrogen bond to the attached or other oxygen atoms adsorbed on the catalyst surface with resultant isomerization to hydroxylamine or dehydrogenation to nitroxyl, imine or even nitric oxide.

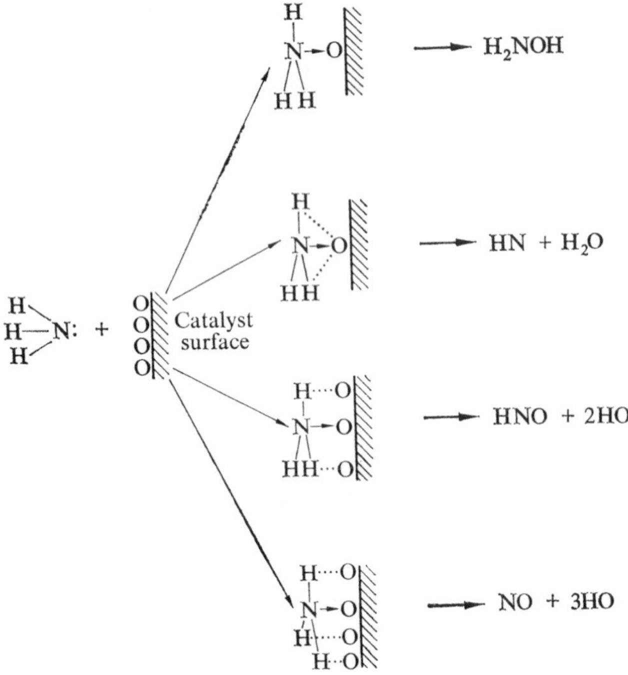

FIG. 17. Mechanism for the oxidation of ammonia.

3.10.3. Reactions of Ammonia with Elements and their Compounds[207]

The exchange reaction of ammonia with hydrogen or deuterium does not proceed at any appreciable rate unless catalysed by a metal such as platinum. Both the thermal and mercury photosensitized exchange reaction proceed by the chain mechanism via the initial monodeuterated product:

$$D_2 \rightleftharpoons 2D$$
$$NH_3+D \rightleftharpoons NH_2D+H$$
$$H+D_2 \rightleftharpoons HD+D$$

The reactions of ammonia with alkali and alkaline–earth metals will be discussed in section 3.10.4. Calcium, strontium and barium carbonates react with ammonia to form the

[207] J. D. F. Marsh, in ref. 156, Section VII(a), pp. 328–41.

corresponding cyanamide, whereas magnesium, zinc and lead carbonates are reduced to the metals:

$$CaCO_3+2NH_3 \rightarrow CaCN_2+3H_2O$$
$$3ZnCO_3+2NH_3 \rightarrow 3Zn+3H_2O+3CO_2+N_2$$

Many metal salts form adducts with ammonia (see section 3.10.6) which are often precursors of the corresponding amides, imides and nitrides (see section 4).

Boron hydrides and halides with ammonia form an extensive co-ordination and derivative chemistry as do the organo- and halo-derivatives of the other Group III elements.

Ammonia and carbon monoxide react under pressure at elevated temperature in the presence of catalysts to give formamide or hydrogen cyanide:

$$NH_3+CO \rightarrow HCONH_2$$
$$NH_3+CO \rightarrow HCN+H_2O$$

Hydrogen cyanide can also be formed by catalytic action of ammonia and air on hydrocarbons[162]:

$$2NH_3+2CH_4+3O_2 \rightarrow 2HCN+6H_2O$$

The reactions of ammonia with sulphur compounds give rise to an extensive N–S chemistry[208], and similarly phosphorus compounds give P–N derivatives[209]. Reactions of the halogens and interhalogen compounds with ammonia will be discussed in section 9.

3.10.4. Reactions of Liquid Ammonia

Liquid ammonia is probably the best-known non-aqueous solvent[200, 201, 210, 211]. Inevitably, its solution chemistry has been rationalized by analogy with aqueous chemistry. In

TABLE 22. SOME ANALOGOUS COMPOUNDS OF WATER AND AMMONIA

Aquo	Aquo-ammono	Ammono
H_2O		NH_3
KOH		KNH_2
Li_2O		Li_2NH
MgO		Mg_3N_2
H_2O_2	NH_2OH	N_2H_4
HNO_3		NH_2NO_2
N_2O		HN_3
$(HO)_2C{=}O$	$(NH_2)_2C{=}O$	$(NH_2)_2C{=}NH$
	$NH_2{-}C({=}O){-}OH$	
$(HO)_2SO_2$	NH_2SO_2OH	
	$(NH_2)_2SO_2$	
CO		HNC
CO_2	$HNCO$	$HNCNH$

[208] H. G. Heal, in *Inorganic Sulphur Chemistry* (ed. G. Nickless), Elsevier, Amsterdam (1968), p. 459.

[209] M. L. Nielsen, in *Developments in Inorganic Nitrogen Chemistry*, Vol. I (ed. C. B. Colburn), Elsevier, Amsterdam (1966), p. 307.

[210] V. Gutmann, *Coordination Chemistry in Non-aqueous Solutions*, Springer-Verlag, New York (1968), p. 38.

[211] G. W. A. Fowles, in *Developments in Inorganic Nitrogen Chemistry*, Vol. I (ed. C. B. Colburn), Elsevier, Amsterdam (1966), p. 522.

TABLE 23. SOLUBILITIES OF SALTS IN LIQUID AMMONIA[a]

Salt	Solubility in g per 100 g NH_3 at temperatures in °C			
	25	0	−33	Other temperatures
AgCl	0.83	0.28	0.215	
AgBr	5.92	2.35		
AgI	206.84	84.15		
$AgNO_3$	86.04			
AlI_3	10			
BF_3			~10	
$B_2H_6 \cdot 2PH_3$	~10			30.6/−75
H_3BO_3	1.92			
$BaCl_2$	0	0		
$BaBr_2$		0.017		
BaI_2		0.025		
$Ba(ClO_4)_2$			67	
$Ba(NO_3)_2$	97.22	17.88		
$CaCl_2$	0.08/19	0	0	
$CaBr_2$	0.03	0.009		
CaI_2		3.85		
$Ca(ClO_4)_2$			52.6	
$Ca(NO_3)_2$	80.22	45.13		
$CdCl_2$	0.02			
$Cd(NO_3)_2$	0			0.3/37
CsCl		0.381	0.36	
CsBr		4.38	5.48	
CsI			91.3	
CuI	21.0			
$Fe(ClO_4)_3$	18.5/15			
GeH_3K			133	
GeS			0.0314	
GeS_2			2.12	
$HgCl_2$			0.019	
KCl	0.04	0.132	0.13	
KBr	13.50	21.18	29.3	~45/−50
KI	182.0	64.81	94.1	
$KClO_3$	2.52			
$KBrO_3$	0.002			
KIO_3	0			
KCN			3.75	
KNCO	1.70			
KNH_2	3.6			
KNO_3	10.4	9.71		
K_2CO_3	0			
$K_2Cr_2O_7$	0.02/18			
K_2SO_4	0			
LiCl		1.43	0.538	
LiBr	0.169/20	0.038/10	1.32	
LiI			36.1	
$LiNO_3$	243.66			
Li_2SO_4	0			
$MgCl_2$	0			
$MgBr_2$	0.004			
MgI_2	0.156			
$Mg(ClO_4)_2$			38.0	

TABLE 23 (*cont.*)

Salt	Solubility in g per 100 g NH_3 at temperatures in °C			
	25	0	−33	Other temperatures
$MnCl_2$	0			
$MnBr_2$	0			
MnI_2	0.02			
NH_4F	0			
NH_4Cl	102.5	39.91	14.75	
NH_4Br	237.9	57.96	90.75	
NH_4I	368.5	76.99	71	
NH_4N_3		49	42	
NH_4HS	120			
NH_4ClO_4	137.93			
NH_4HCO_3	0			
NH_4NO_3	390		224	25/−82
NH_4OAc	253.2			
NH_4OCONH_2	0.15/20			0.1/60
NH_4SCN	312			
$(NH_4)_2CO_3$	0		0	
$(NH_4)_2HPO_4$	0			
$(NH_4)_2MoO_4$	0			
$(NH_4)_2S$	120			
$(NH_4)_2SO_3$	0			
$(NH_4)_2SO_4$			0	
$(NH_4)_2SeO_4$	0			
NaF	0.35			
NaCl	3.02	2.8	2.10	
NaBr	137.95	39.00	17.62	
NaI	161.9	132	51.0	
NaNCO	1.24/10	1.35		
NaSCN	205.50			
$NaNH_2$	0		3.9	
$NaNO_3$	97.60	127		
NaO_2			0.3	
NaOH	0			
Na_2CO_3	0			
Na_2CrO_4	0.05/18			
Na_2S			0.0017	
Na_2SO_3	0.088			
$Na_2S_2O_3$	0.17			
Na_2SO_4	0			
RbCl	0.22	0.289	0.27	
RbBr		18.23	25.0	
RbI		68.15	111.6	
$SrCl_2$	0			
$SrBr_2$		0.008		
SrI_2	0.308			
$Sr(NO_3)_2$	87.08	40.4		
$ZnCl_2$	0			
ZnI_2	0.10			
$Zn(NO_3)_2$		29.01		

[a] Ref. 200.

the same way that oxygen compounds are considered as being derived from water, so ammonia can be regarded as the parent solvent of the nitrogen system of compounds[212]. Thus the class of ammono compounds have direct analogues in the aquo series as indicated in Table 22. Comparisons of this type have found most successful application in preparative chemistry as structurally related compounds often react in similar ways.

Before discussing detailed properties and reactions involving liquid ammonia, it is useful to indicate the solvent properties of ammonia by comparison with the more familiar properties of water. The lower boiling point of ammonia (−33°C) compared to water (100°C) is indicative of the reduced amount of hydrogen bonding in liquid ammonia, but its influence is still seen by the fact that compounds capable of forming hydrogen bonds such as amines, alcohols and phenols, are miscible with ammonia in all proportions. The significant dielectric constant of ammonia ($\varepsilon = 23$ at −33°C), which although lower than water ($\varepsilon = 82$ at 25°C), means that ammonia is not only capable of dissolving a wide range of ionic materials (Table 23), but is superior to water in its solvent action towards organic compounds such as hexane, carbon tetrachloride, benzene, etc.[183].

The heats of solution of electrolytes in liquid ammonia are frequently more negative than in water as indicated in Table 24, consistent with ammonia being highly basic. Such

Table 24. Heats of Solution in Water and Ammonia[a] ($\triangle H°$ in kcal/mol)

Salt	H_2O (25°C)	NH_3 (25°C)	NH_3 (−33°C)
NaCl	1.02	−6.75	−1.57
KI	4.95	−9.44	−7.89
CsI	7.9	−5.27	
NH_4Cl	3.71	−8.23	−6.95
NH_4I	3.3	−16.10	−13.37
$Ba(NO_3)_2$	9.96	−15.31	

[a] Ref. 200.

basic solvent properties are particularly suitable for the study of reduction reactions[200]. Other solutes include the alkali metals which can dissolve in liquid ammonia reversibly, in contrast to the familiar violent reactions with water. Ammonia is also a stronger donor solvent than water as evidenced by the ease of formation of ammonia complexes with transition metal ions in aqueous solution. However, the self-ionization of water is considerably more extensive than for ammonia due to the fact that the O–H bond is more readily broken than the N–H bond. For this reason, hydrolysis reactions tend to be both more common throughout chemistry and more capable of reacting to completion than corresponding solvolysis reactions involving ammonia (ammonolysis).

212 E. C. Franklin, *The Nitrogen System of Compounds*, Reinhold, New York (1935).

(a) *Acid–Base Properties of Liquid Ammonia*[200, 201]

The acid–base properties of ammonia result largely from the self-ionization reaction

$$2NH_3 \rightleftharpoons NH_4^+ + NH_2^-; \qquad K_{-33^\circ} = [NH_4^+][NH_2^-] = \sim 10^{-30}$$

Thus although the conductivity of liquid ammonia is quite small, ammonia is a strong proton acceptor enabling most potential proton donors to act as an acid. Thus weak acids by aqueous standards such as acetic acid are almost completely dissociated in liquid ammonia similar to mineral acids in aqueous solution:

$$CH_3COOH + NH_3 \rightarrow NH_4^+ + CH_3COO^-$$

Likewise, conventionally very weak acids such as water or arsine can also form salts in liquid ammonia:

$$K^+ + NH_2^- + H_2O \xrightarrow{\text{liq. } NH_3} NH_3 + KOH$$

$$K^+ + NH_2^- + AsH_3 \xrightarrow{\text{liq. } NH_3} NH_3 + KAsH_2$$

This enhanced ionization is not meant to imply that weak acids become stronger, but because acid strength is relative to the solvent in which it is measured, it merely emphasizes the fact that the NH_4^+ ion, being symmetric and non-polar, is a weaker acid in liquid ammonia than is the asymmetric and polar H_3O^+ ion in water.

Alkali metal hydrides and oxides are stronger bases than the amide ion and hence are capable of removing protons from an ammonia molecule:

$$LiH + NH_3 \rightarrow LiNH_2 + H_2$$

$$Na_2O + NH_3 \rightarrow NaNH_2 + NaOH$$

(b) *Reactions in Liquid Ammonia*[200, 211]

By choosing salts of suitable solubility from Table 23, a large number of metathetical reactions can be carried out in liquid ammonia which would not be possible in water. Some examples are:

$$Ba(NO_3)_2 + 2AgCl \xrightarrow{\text{liq. } NH_3} BaCl_2 + 2AgNO_3$$

$$AgNO_3 + KNH_2 \xrightarrow{\text{liq. } NH_3} AgNH_2 + KNO_3$$

$$(NH_4)_2S + Ca(NO_3)_2 \xrightarrow{\text{liq. } NH_3} CaS + 2NH_4NO_3$$

Liquid ammonia has also long been used as a medium for coupling reactions, particularly in organometallic and metal carbonyl chemistry:

$$Ph_3GeNa + Me_3SnBr \xrightarrow{\text{liq. } NH_3} Ph_3GeSnMe_3 + NaBr$$

$$[Co(NH_3)_4][SCN]_2 + 2LiC_5H_5 \xrightarrow{\text{liq. } NH_3} [Co(NH_3)_4][C_5H_5]_2 + 2LiSCN$$

$$2Cr(CO)_6 + 2NaBH_4 \xrightarrow{\text{liq. } NH_3} Na_2[Cr_2(CO)_{10}]$$

Ammonia complexes of metal halides, $MX_n \cdot xNH_3$, often termed ammoniates, can be formed conveniently by dissolving the halide in liquid ammonia and evaporating off the excess solvent:

$$MX \xrightarrow{\text{liq. } NH_3} MX_n \cdot xNH_3$$

However, the interaction between metal halides and ammonia is often complicated by a further reaction involving the ammonolysis of one or more of the metal–halogen bonds:

$$MX \xrightarrow{\text{liq. } NH_3} MNH_2 + NH_4X$$

This reaction is one of the principal methods of preparing metal–nitrogen compounds.

(c) *Metal–Ammonia Solutions*[198,200,201,211,213]

Liquid ammonia is capable of dissolving the alkali metals, the alkaline-earth metals (except beryllium) and some of the lanthanide elements (particularly europium and ytterbium giving ions in the +2 oxidation state). All these elements are among the most electropositive, and it would appear that the factors responsible for a metal having a high oxidation potential (high solvation energy, low sublimation energy and low ionization energy) are the same as those responsible for solubility in ammonia. Some solubilities of alkali metals are listed in Table 25.

TABLE 25. SOLUBILITIES OF METALS IN LIQUID AMMONIA[a]

Metal	Temperature (°C)	g metal per 100 g NH_3	g atom metal per 100 g NH_3	mol NH_3/g atom metal
Li	0	11.3	1.631	3.60
	−33	10.9	1.566	3.75
	−63.5	10.7	1.541	3.81
Na	+22	22.0	0.956	6.14
	0	23.0	1.000	5.87
	−33	24.6	1.072	5.48
	−50	25.1	1.089	5.39
	−70	26.0	1.189	5.20
K	0	48.8	1.24	4.7
	−33	46.5	1.186	4.95
	−50	45.5	1.23	4.9
	−105	45.2		
Cs	−50	334	2.51	2.34
Ca	−35	33.6		

[a] Ref. 200.

Of their physical properties, dilute metal–ammonia solutions are characterized by their blue colour, while more concentrated solutions generally have a metallic or bronze appearance. The blue colour results from a very intense broad band that occurs around 15,000 Å, which for the alkali metals is similar in position, shape and extinction coefficient irrespective of which metal is studied. Dilute solutions are paramagnetic, but this decreases as solutions become more concentrated. As dilute metal–ammonia solutions become more

[213] W. L. Jolly, in *Progress in Inorganic Chemistry*, Vol. I (ed. F. A. Cotton), Interscience, New York (1959), p. 235.

concentrated, the conductivity decreases to a minimum then rises sharply to values approaching those of pure metals for solutions near saturation. When alkali metals dissolve in liquid ammonia, there is a considerable increase in the volume of solution, and vapour pressure measurements suggest considerable polymerization of the species in solution.

The physical properties of metal–ammonia solutions have been explained by regarding the process of dissolution as a dissociation giving solvated metal ions and solvated electrons, which in dilute solutions exist essentially independently. Thus although solvated metal ions are easy enough to visualize, solvated electrons are currently described in terms of the "cavity" model, which considers that the electron creates a large cavity for itself in the solvent while nearby solvent molecules orient themselves so as to position the hydrogen atoms nearest to the electron. An alternative description of solvated electrons is the "expanded metal" model which assumes that on dissolution the metal ion is separated from the electron by only a single layer of ammonia molecules. Both models are consistent with the anomalous density as a result of the space which the electron occupies, but the equivalent conductance values, and in particular the absorption spectra, are more consistent with a model involving solvated electrons essentially independent of the metal ions.

In concentrated metal–ammonia solutions, there are insufficient solvent molecules to co-ordinate both the metal ions and the electrons, so these solutions consist of solvated metal ions in an electron pool, similar to a metal with its electrons in a conductor band.

(d) *Reactions of Metal–Ammonia Solutions*[200,211,213]

Metal–ammonia solutions are stable over long periods, but decomposition to hydrogen and metal amide occurs slowly if impurities are present but rapidly with suitable catalysts such as transition metal salts:

$$2M+2NH_3 \xrightarrow{\text{liq. } NH_3} 2MNH_2+H_2$$

In view of the powerful reducing properties of metal–ammonia solutions, they are often regarded as convenient electron producing reagents:

$$M \xrightarrow{\text{liq. } NH_3} M^++e^- \qquad \text{(M, for example = alkali metal)}$$

Thus in the same way that partial ionic equations in aqueous chemistry are often used to describe individual reaction steps in an overall process, so the reactions of metal–ammonia solutions can be described as electron addition reactions which may also involve bond cleavage.

Electron addition without bond cleavage can be represented as

$$e^-+A \xrightarrow{\text{liq. } NH_3} A^- \quad \text{or} \quad M+A \xrightarrow{\text{liq. } NH_3} MA$$

For example, peroxides and superoxides can be prepared by bubbling oxygen through metal–ammonia solutions:

$$O_2+K \xrightarrow{\text{liq. } NH_3} KO_2$$

Similarly, the reduction of nitric oxide with sodium in liquid ammonia is thought to give the nitroside ion:

$$NO+Na \xrightarrow{\text{liq. } NH_3} NaNO$$

Other examples illustrate methods of preparing complexes and ions of metals in low oxidation states:

$$KMnO_4+K \xrightarrow{\text{liq. } NH_3} K_2MnO_4$$

$$Mn_2(CO)_{10}+2Na \xrightarrow{\text{liq. } NH_3} 2NaMn(CO)_5$$

$$K_2Ni(CN)_4+2K \xrightarrow{\text{liq. } NH_3} K_4Ni(CN)_4$$

$$[Pt(en)_2]I_2+2K \xrightarrow{\text{liq. } NH_3} [Pt(en)_2]+2KI$$

$$[Ir(NH_3)_5Br]Br+3K \xrightarrow{\text{liq. } NH_3} [Ir(NH_3)_5Br]+3KBr$$

$$[Pd\{o\text{-}C_6H_4(PEt_2)_2\}_2]Br_2+2Na \xrightarrow{\text{liq. } NH_3} [Pd\{o\text{-}C_6H_4(PEt_2)_2\}_2]+2NaBr$$

Electron addition accompanied by bond cleavage is described by the equations:

$$e^-+A-B \xrightarrow{\text{liq. } NH_3} A^{\cdot}+B^-$$

$$2e^-+A-B \xrightarrow{\text{liq. } NH_3} A^-+B^-$$

Some examples of this type of reaction include:

$$2NH_4Cl+2K \xrightarrow{\text{liq. } NH_3} 2NH_3+H_2+2KCl$$

$$2RC{\equiv}CH+2Na \xrightarrow{\text{liq. } NH_3} 2RC{\equiv}CNa+H_2$$

$$Fe(CO)_5+2Na \xrightarrow{\text{liq. } NH_3} Na_2[Fe(CO)_4]+CO$$

$$Ge_2H_6+2K \xrightarrow{\text{liq. } NH_3} 2KGeH_3$$

Unlike the above reactions in which the radical formed initially dimerizes, the triethyltin radical is stable in liquid ammonia:

$$Et_3SnCl+K \xrightarrow{\text{liq. } NH_3} Et_3Sn\cdot+KCl$$

In a few instances, the anions formed in such reactions undergo ammonolysis:

$$N_2O+NH_3+2K \xrightarrow{\text{liq. } NH_3} N_2+KOH+KNH_2$$

$$KNCO+NH_3+2K \xrightarrow{\text{liq. } NH_3} KCN+KOH+KNH_2.$$

3.10.5. Ammonia Hydrates and Aqueous Chemistry of Ammonia[198]

The extent to which ammonia in aqueous solution exists as ammonium hydroxide has long been discussed. Ammonia is very soluble in water in which it is involved in the equilibrium

$$NH_3+H_2O \rightleftharpoons NH_4^++OH^-; \qquad K = \frac{[NH_4^+][OH^-]}{[NH_3]} = 1.81\times10^{-5}\,(pK_b = 4.75)$$

This data suggests that the formation of ammonium hydroxide is improbable. The phase diagram of the ammonia–water system shows three eutectics and two compounds corresponding to the formulae $NH_3\cdot H_2O$, and $2NH_3\cdot H_2O$. Crystal structures confirm that neither substance contains NH_4^+, OH^- nor discrete NH_4OH molecules. The monohydrate has chains of H_2O linked by hydrogen bonds (2.76 Å), and cross-linked by NH_3 into a three-dimensional lattice by OH · · · N (2.78 Å) and O · · · H—N (3.21–3.29 Å). Aqueous ammonia

TABLE 26. PHYSICAL CONSTANTS FOR AMMONIA HYDRATES[a]

Property	Temperature (K)	$NH_3 \cdot H_2O$	$2NH_3 \cdot H_2O$
Crystal data	178		
System		Hexagonal	Orthorhombic
Space group			*Pbnm*
Lattice constants: *a*		11.21 ± 0.05	$a = b = 8.41 \pm 0.03$
c		4.53 ± 0.02	5.33 ± 0.02
Melting point		194.15 K	194.32 K
Density	205	0.89 g/cm³	0.82 g/cm³
ΔH fusion	194	2.352 kcal/mol	1.568 kcal/mol
ΔH_f (enthalpy of formation)	298	−86.33 kcal/mol	−102.94 kcal/mol
ΔG_f (free energy of formation)	298	−60.74 kcal/mol	−63.84 kcal/mol
S° (entropy)	298	39.57 cal/deg mol	63.94 cal/deg mol
C_p° (heat capacity)	298	37.02 cal/deg mol	59.08 cal/deg mol

[a] Refs. 198, 200.

TABLE 27. PHYSICAL PROPERTIES OF AQUEOUS AMMONIA SOLUTIONS AT 293 K [a]

Weight (%NH_3)	Molarity	NH_3 pressure (mm Hg)	Density (g/cm³)	Specific conductivity $\times 10^6$ mhos at 298 K	Viscosity (centipoise at 283 K)	Surface tension (dynes/cm)
0.404		3.1		529.0		
0.45	0.264		0.9960			72.55
1	0.584		0.9939			
1.042		8.4		820.7		
2	1.162	12.5	0.9895			
4	2.304	26.1	0.9811			
5.259		44.3		1275		
7.72	4.378		0.9661			65.74
8	4.534	60	0.9651			
8.8					1.435	
9.517		90.9		1142		
14.61	8.072		0.9409			62.15
16	8.796	156	0.9362			
18.61		245.6		625.4		
19.6					1.632	
27.20		516.3		285.7		
28	14.76	447	0.8980			
29.70	15.504		0.8890			55.58
30.97		686.7		196.5		
39.6					1.612	
47.45	23.302		0.8363			46.42
49.3					1.290	
61.16	28.192		0.7850			37.90
70.47	30.799		0.7443			32.99
80.95	33.198		0.6984			28.11
89.4					0.239	
90.81	34.826		0.6531			24.57
100	35.836		0.6103		0.152	22.03

[a] Ref. 200.

is probably hydrated in a similar way. The name "ammonium hydroxide" therefore is a misnomer since at present there is no evidence to suggest that undissociated NH_4OH exists in aqueous solution.

Some physical parameters of ammonia hydrates and aqueous ammonia solutions are listed in Tables 26 and 27. Vapour pressures and vapour pressure compositions have been determined for 0–100% aqueous ammonia over the range −50 to +210°C at pressures of 0.1–20 atm, and partial pressures can also be calculated from the empirical formula

$$\log_{10} p = \frac{7.6468t}{t+230} - \frac{0.7486t+236.68}{t+230} \times \log_{10}\frac{1000}{C} - 1.6$$

where p = partial pressure in mm, t = temperature in °C and c = concentration of ammonia in g NH_3 per 100 g solution.

3.10.6. Addition Compounds

The addition compounds of ammonia have been mentioned in connection with their formation during reactions in liquid ammonia (section 3.10.4.b). In ammoniates of the type $MX_n \cdot xNH_3$, the ammonia and halogen each occupy co-ordination positions round the metal in a molecular complex, and should be distinguished from metal amines in which ammonia molecules are co-ordinated round a metal ion. In this section we are primarily concerned with the former type. However, the interaction of metal halides with ammonia has been extensively studied[211] and formation of $MX_n \cdot xNH_3$ is frequently accompanied by ammonolysis which involves replacement of some or all of the halogen atoms by amino groups. Some metal halides are ammonolysed immediately, and in such cases ammoniates may only be considered to exist as transient intermediates. The course of such reactions has usually been considered to proceed through the sequence, metal halide (usually in solution) leading to ammoniated metal halide followed by stepwise base-catalysed elimination of hydrogen halide. The system is aided by solvents which can extract the ammonium halide as it forms, forcing the equilibrium reaction to completion.

Organic substituted ammonia (amines) can be employed to reduce or limit solvolysis reactions. Fully substituted (tertiary) amines, of course, prevent any further reaction after complex formation although primary and secondary amines can also have the same effect in certain stereochemical situations. A wide variety of organic-nitrogen mono-, di- and multi-dentate ligands including amines, diamines (ethylene-diamine) and heterocyclic bases (pyridine) to name a few, have found application[214], but it is only recently that amino derivatives of other elements (see section 4) have been considered as potential Lewis bases.

3.11. BIOLOGICAL PROPERTIES OF AMMONIA

As mentioned in section 1.7, ammonia is a key substance in the natural nitrogen fixation process. Formerly, most nitrogen fertilizers were based on ammonium salts, but more recently anhydrous ammonia and ammonia solutions have been applied directly. As regards the human metabolism of nitrogen compounds, ammonia occurs in all body fluids but is

[214] W. P. Griffith, in *Developments in Inorganic Nitrogen Chemistry*, Vol. I (ed. C. B. Colburn), Elsevier, Amsterdam (1966), p. 241.

difficult to differentiate from urea. Ammonia is excreted through the urine and together with urea accounts for the removal of waste nitrogen products[215].

Exposure to small concentrations of gaseous ammonia causes attack to the mucous membrane, and although it only acts as an irritant to the eyes and throat, exposure to high concentrations can induce oedema, haemorrhage and empyema. Some safety limits are indicated in Table 28.

TABLE 28. PHYSIOLOGICAL RESPONSE TO CONCENTRATIONS (IN ppm) OF AMMONIA IN AIR

Least detectable odour	53
Least amount causing immediate irritation to eyes	698
Least amount causing immediate irritation to throat	408
Least amount causing immediate coughing	1720
Maximum concentration allowable for prolonged exposure	100
Maximum concentration allowable for short exposure ½–1 hour	300–500
Dangerous concentration even for short exposure ½ hour	2500–4500

3.12. ANALYTICAL CHEMISTRY OF AMMONIA[216]

The following methods are applicable to the estimation of free ammonia to be distinguished from ammonium ion. Ammonia in sufficient quantity can easily be detected and identified by its odour, but the most common spot test is that with Nessler reagent. Other sensitive colorimetric methods of estimation are based on the blue colour formed by its reaction with phenol and hypochlorites, also the purple colour produced when ammonia solution is treated with chloramine-T followed by pyridine-pyrazolone reagent[217]. The latter method is particularly sensitive and applicable to coloured samples as the purple colour can be extracted into carbon tetrachloride and remains stable and reproducible.

A simple method for determining free ammonia in solution is by direct titration with an acid using methyl red as indicator[218]. Ammonia can also be oxidized by sodium hypobromite to nitrogen which can be measured volumetrically[219].

$$2NH_3 + 3NaOBr \rightarrow 3NaBr + H_2O + N_2$$

The principal impurities in cylinder ammonia are non-condensable gases, nitrogen, argon and neon together with moisture and traces of oil. An estimate of the impurity level in the atmosphere above liquid ammonia can be obtained by treating a sample in a gas burette with sulphuric acid and measuring the volume directly.

215 F. Call, in ref. 156, Part I, Section XVII, p. 604.

216 K. I. Vasu, in *Mellor's Comprehensive Treatise on Inorganic and Theoretical Chemistry*, Vol. 8, Suppl. II, *Nitrogen*, Part I, Longmans, London (1964), Section XXXVII, p. 616.

217 M. J. Taras, in *Colorimetric Determination of Nonmetals* (ed. D. F. Boltz), Interscience, New York (1958), p. 75.

218 N. H. Furman, in *Scott's Standard Methods of Chemical Analysis*, 6th edn., Vol. I, Van Nostrand, New York (1962), p. 736.

219 A. I. Vogel, *A Textbook of Quantitative Inorganic Analysis*, 3rd edn., Longmans, London (1961), p. 396.

4. AMIDES, IMIDES, AND NITRIDES

4.1. CLASSIFICATION OF COMPOUNDS

Although metal–nitrogen compounds will be discussed with the chemistry of each particular element, it is appropriate here to indicate the range and variety of such derivatives and to attempt a comparative study of these compounds. Thus, with the exception of hydrogen, nitrogen, oxygen and halogen compounds of nitrogen (discussion of which accounts for virtually the remainder of this chapter), the chemistry of compounds in which nitrogen is bonded to one or more of the remaining elements is outlined in this section. All of these elements, comprising essentially the metals and metalloids, have an electronegativity lower than nitrogen, which might imply an electronic displacement towards nitrogen and be useful in predicting possible reaction courses. Thus as oxides and hydroxides are regarded as being derived from the parent molecule water, so in the same way amides, imides and nitrides are related to ammonia[220]. In principle, these classes of compounds can be thought of as being formed by replacing one or more hydrogen atoms of ammonia by other elements. Clearly, any remaining hydrogen attached to nitrogen of the amide or imide produced could be substituted by some other element or group, as also could other valency positions on the element bonded to nitrogen. It is in this context that organo groups have played an important rôle in facilitating the synthesis of derivatives containing only one element–nitrogen bond[221], thus simplifying their chemical investigation without the complications of multi-functional reactants.

Simple metal amides $M(NH_2)_x$ of the alkali and alkaline–earth metals are quite stable, but amides of heavier elements such as silver can be explosive[222]. Very few imides are known, however, and lithium imide is probably the best characterized compound of its type. Analogous compounds derived from organic amines and imines are now known for many transition and non-transition metals and metalloids, but these, in general, are of quite different character to related conventional inorganic derivatives.

Metal nitrides can be conveniently divided into four classes[223]—salt-like, metallic, diamond-like and covalent. The salt-like ionic nitrides, which include compounds of lithium, alkaline–earth metals, together with some electropositive Group IIIA metals, are characterized by the presence of N^{3-} ions in the crystal lattice. The metallic nitrides, so named because of their physical similarity to refractory hard metals, are often called interstitial nitrides. They include compounds of transition metals, and are also related to borides, carbides, silicides and phosphides of such elements[223, 224]. Their structures are expanded metal lattices in which nitrogen atoms occupy interstitial positions which indicates why these compounds tend to exhibit non-stoichiometry and variable composition. The diamond-like nitrides of Group III elements boron, aluminium, gallium and indium have graphite- or wurtzite-type structures. The covalent nitrides, such as $(CN)_2$, PN, AsN, S_4N_4, are reasonably volatile materials and parent compounds to a large number of derivatives.

220 E. C. Franklin, *The Nitrogen System of Compounds*, Reinhold, New York (1935).

221 G. E. Coates, M. L. H. Green and K. Wade, *Organometallic Compounds*, 3rd edn., Methuen, London (1967).

222 J. Jander, *Chemistry in Anhydrous Liquid Ammonia*, Vol. 1, Part 1, of *Chemistry in Nonaqueous Ionizing Solvents* (eds. G. Jander, H. Spandau and C. C. Addison), Interscience, New York (1966).

223 B. R. Brown, in *Mellor's Comprehensive Treatise on Inorganic and Theoretical Chemistry*, Vol. 8, Suppl. I, *Nitrogen*, Part I, Longmans, London (1964), Section III, pp. 150–239.

224 P. Schwarzkopf and R. Kieffer, *Refractory Hard Metals*, Macmillan, New York (1953), pp. 223–60.

Organometallic derivatives of the type $(Me_3Sn)_3N$ and $[(Me_3Si)_2N]_2Cd$ would also be included in this group.

Thus, in review, the nitrogen chemistry of the elements can be summarized as follows. The alkali metals are capable of forming each class of nitrogen mentioned above, with lithium and sodium derivatives being important in organic chemistry. Alkaline-earth metals form amides and nitrides, but only magnesium and beryllium form derivatives of organic amines. The transition elements are predominantly useful as nitrides[223], but the synthesis and properties of amino derivatives of the elements titanium[225], zirconium[226], hafnium[226], vanadium[225], niobium[226], tantalum[226], chromium[227], molybdenum[228], tungsten[229], manganese[230], iron[231], cobalt[232], nickel[233], palladium[234], platinum[235], copper[236], silver[237], uranium[238] and thorium[226], have at least begun to be investigated. Zinc, cadmium and mercury form amides and unstable nitrides, and only a few organic derivatives have been reported[221]. The elements of Group III form covalent nitrides and organo-derivatives, and in particular the chemistry of boron nitrogen compounds[239,240] has received particular attention because of their relationship to organic systems, B–N being isoelectronic with C–C. From the Group IV elements, organic chemistry of nitrogen is a subject in itself[241], and organosilicon–nitrogen chemistry is advanced with many cyclic derivatives[242]. Germanium–nitrogen[243] closely resembles silicon, while organotin–nitrogen[244] compounds have turned up some unexpected properties which have led to the discovery of analogous situations in the chemistry of other elements. Phosphorus–nitrogen chemistry[245,246] is quite extensive in connection with phosphonitrilic halides and other condensed systems[247], but the nitrogen chemistry of arsenic and antimony have not received much attention[248],

225 E. C. Alyea, D. C. Bradley, M. F. Lappert and A. R. Sanger, *Chem. Communs.* (1969) 1064; A. R. Sanger, D.Phil. thesis, University of Sussex, 1969.

226 D. C. Bradley and M. H. Gitlitz, *J. Chem. Soc.* A (1969) 980.

227 E. C. Alyea, J. S. Basi, D. C. Bradley and M. H. Chisholm, *Chem. Communs.* (1968) 495.

228 D. A. Edwards and G. W. A. Fowles, *J. Chem. Soc.* (1961) 24.

229 B. J. Brisdon and G. W. A. Fowles, *J. Less-Common Metals* **7** (1964) 102.

230 T. A. George and M. F. Lappert, *Chem. Communs.* (1966) 463.

231 B. J. Bulkin and J. A. Lynch, *Inorg. Chem.* **7** (1968) 2654.

232 R. Bonnet, D. C. Bradley and K. J. Fisher, *Chem. Communs.* (1968) 886.

233 S. Otsuka, A. Nakamura and T. Yoshida, *Ann.* **719** (1969) 54.

234 R. L. Dutta and S. Lahiry, *J. Indian Chem. Soc.* **37** (1960) 789.

235 D. M. Roundhill, *Chem. Communs.* (1969) 567.

236 S. Miki and S. Yamada, *Bull. Chem. Soc. Japan* **37** (1964) 1044.

237 R. Nast and W. Danneker, *Ann.* **693** (1966) 1.

238 R. G. Jones, G. Karmas, G. A. Martin, Jr. and H. Gilman, *J. Am. Chem. Soc.* **78** (1956) 4285.

239 K. Niedenzu and J. W. Dawson, *Boron–Nitrogen Compounds*, Academic Press, New York (1965).

240 R. A. Geanangel and S. G. Shore, in *Preparative Inorganic Reactions*, Vol. 3 (ed. W. L. Jolly), Interscience, New York (1966), p. 123.

241 N. V. Sidgwick, *The Organic Chemistry of Nitrogen*, 3rd edn., revised and rewritten by I. T. Millar and H. D. Springall, Oxford University Press, Oxford (1966).

242 R. Fessenden and J. S. Fessenden, *Chem. Rev.* **61** (1961) 361.

243 M. Dub, *Organometallic Compounds*, Vol. 2, *Organic Compounds of Germanium, Tin, and Lead*, Springer-Verlag, Berlin (1961).

244 K. Jones and M. F. Lappert, *Organometal. Chem. Rev.* (1965) 295; also in *Organotin Compounds* Vol. 2 (ed. A. K. Sawyer), Marcel Dekker, New York (1971), Chapter 7.

245 E. Fluck, in *Topics in Phosphorus Chemistry*, Vol. 4 (eds. M. Grayson and E. J. Griffith), Interscience, New York (1967).

246 M. L. Nielsen, in *Developments in Inorganic Nitrogen Chemistry*, Vol. 1 (ed. C. B. Colburn), Elsevier, Amsterdam (1966), pp. 307–469.

247 R. A. Shaw, *Endeavour* **27** (1968) 74.

248 M. Dub, *Organometallic Compounds*, Vol. 3, *Organic Compounds of Arsenic, Antimony, and Bismuth*, Springer-Verlag, Berlin (1962).

In Group VI, sulphur–nitrogen chemistry is a flourishing field of research[249].

Finally, attention must be drawn to the problem of nomenclature. In organic chemistry, "amide" is the class name for compounds containing the group $—CONH_2$, whereas in inorganic parlance it refers to metal derivatives of ammonia, e.g. sodium amide is $NaNH_2$ and not $NaCONH_2$. On the other hand, metal amides can easily be thought of as amino derivatives of metals, and for this reason they are often referred to by this "organic" nomenclature, especially when naming cyclic derivatives. Some examples are given in Table 29.

TABLE 29. NAMES OF NITROGEN COMPOUNDS FO THE ELEMENTS

Formula	Name and alternatives
$NaNH_2$	Sodium amide
Ag_2NH	Silver imide
Li_3N	Lithium nitride
$Bu_3^nSiNH_2$	Tri-n-butylsilylamine, tri-n-butylsilylamide
$(Me_3Si)_2NH$	Bis(trimethylsilyl)amine, hexamethyldisilylamide
$(H_3Si)_3N$	Trisilylamine
$(HBNH)_3$	Borazine, borazole
$Ti(NMe_2)_4$	Tetrakis(dimethylamino)titanium, titanium tetradimethylamide
Me_3SnNEt_2	Diethylaminotrimethylstannane, trimethyltin diethylamide

4.2. METHODS OF PREPARATION

Solutions of alkali and alkaline–earth metals in liquid ammonia with a trace of catalyst (usually a simple transition-metal salt) produce the corresponding metal amide (1)[222]:

$$2Na+2NH_3 \xrightarrow[FeSO_4]{race} 2NaNH_2+H_2 \quad (1)$$

Another general method in liquid ammonia involves the precipitation of metal amides by adding potassium amide to liquid ammonia solutions of soluble salts of these metals as in (2)[222]:

$$CdCl_2+2KNH_2 \xrightarrow{liq.\ NH_3} Cd(NH_2)_2+2KCl \quad (2)$$

Similar amides can also be prepared by reaction of the corresponding metal–alkyl or metal–aryl with ammonia (3, 4)[222, 250]:

$$PhNa+NH_3 \rightarrow NaNH_2+C_6H_6 \quad (3)$$

$$Et_2Zn+2NH_3 \rightarrow Zn(NH_2)_2+2C_2H_6 \quad (4)$$

---

[249] H. G. Heal, in *Inorganic Sulphur Chemistry* (ed. G. Nickless), Elsevier, Amsterdam (1968), pp. 459–508; M. Becke-Goehring and E. Fluck, in *Developments in Inorganic Nitrogen Chemistry*, Vol. 1 (ed. C. B. Colburn), Elsevier, Amsterdam (1966), pp. 150–240.

[250] O. Schmitz-DuMont, *Rec. Chem. Progr.* **29** (1968) 13.

Metal amides containing organic substituents on the metal are not very common but have been prepared by method (5) comparable to (2), while metal amides derived from organic amines are obtained in ways (6–8)[243, 251] similar to (2) and (3):

$$Ph_3GeCl+KNH_2 \xrightarrow{\text{liq. } NH_3} Ph_3GeNH_2+KCl \quad (5)$$

$$C_4H_9Li+Me_2NH \rightarrow LiNMe_2+C_4H_{10} \quad (6)$$

$$TiCl_4+4LiNMe_2 \rightarrow Ti(NMe_2)_4+4LiCl \quad (7)$$

$$Me_3SnCl+LiNMe_2 \rightarrow Me_3SnNMe_2+LiCl \quad (8)$$

There are also particular instances when organometallic halides or oxides can react with simple amines to give the amino-metal derivative directly (9, 10):

$$Me_3SiCl+2Et_2NH \rightarrow Me_3SiNEt_2+Et_2NH \cdot HCl \quad (9)$$

$$BCl_3+6Me_2NH \rightarrow B(NMe_2)_3+3Me_2NH \cdot HCl \quad (10)$$

The organic groups attached to nitrogen can be varied by transamination reactions (11, 12)[251]:

$$Nb(NMe_2)_5+2Et_2NH \rightarrow (Me_2N)_3Nb(NEt_2)_2+2Me_2NH \quad (11)$$

$$Ti(NMe_2)_4+3Pr^n_2NH \rightarrow Me_2NTi(NPr^n_2)_3+3Me_2NH \quad (12)$$

Lithium imide is prepared by heating the amide (13), whereas organometallic imides are formed by exchange reactions (14):

$$2LiNH_2 \xrightarrow[\text{in vacuo}]{360°} Li_2NH+NH_3 \quad (13)$$

$$(Me_3Si)_2SO_4+3NH_3 \rightarrow (Me_3Si)_2NH+(NH_4)_2SO_4 \quad (14)$$

Imides have also been prepared by the thermal decomposition of dialkylamino-derivatives of niobium and tantalum (15)[252], and by reaction of amines with oxo-compounds of rhenium (16)[253].

$$Ta(NEt_2)_5 \rightarrow (Et_2N)_3Ta\!:\!NEt+Et_2NH+C_2H_4 \quad (15)$$

$$Cl_3Re(O)(EtPh_2P)_2+MeNH_2 \rightarrow Cl_3Re(:NMe)(EtPh_2P)_2+H_2O \quad (16)$$

The two main methods of forming binary compounds of nitrogen are from the elements[254, 255] (17)[256], and from the corresponding metal amide (18)[223]:

$$6Li+N_2 \rightarrow 2Li_3N \quad (17)$$

$$3Zn(NH_2)_2 \rightarrow Zn_3N_2+4NH_3 \quad (18)$$

Common modifications to these methods include the reduction of a metal oxide or halide in the presence of nitrogen (19–21), or the formation of a metal amide as an intermediate in the reaction between a metal and ammonia (22)[223]:

$$Al_2O_3+3C+N_2 \rightarrow 2AlN+3CO \quad (19)$$

$$2TiO_2+4TiC+3N_2 \rightarrow 6TiN+4CO \quad (20)$$

$$2ZrCl_4+N_2+4H_2 \rightarrow 2ZrN+8HCl \quad (21)$$

$$3Ca+6NH_3 \rightarrow [3Ca(NH_2)_2] \rightarrow Ca_3N_2+4NH_3 \quad (22)$$

---

[251] D. C. Bradley and I. M. Thomas, *J. Chem. Soc.* (1960) 3857.

[252] D. C. Bradley and I. M. Thomas, *Can. J. Chem.* **40** (1962) 1355.

[253] D. Bright and J. A. Ibers, *Inorg. Chem.* **8** (1969) 703.

[254] *Handbook of the Physicochemical Properties of the Elements* (ed. G. V. Samsonov), Oldbourne, London (1968).

[255] R. T. Sanderson, *Chemical Periodicity*, Reinhold, New York (1960), pp. 201–7.

[256] C. C. Addison and B. M. Davies, *J. Chem. Soc.* A (1969) 1822.

Most of these reactions are carried out at elevated temperatures and in some cases under pressure, and contrast with the facile formation of corresponding organometallic[257] (23) and metal complex nitride[258] (24) derivatives:

$$3Me_3SnCl + Li_3N \rightarrow (Me_3Sn)_3N + 3LiCl \qquad (23)$$

$$Cl_3Re(O)(EtPh_2P)_2 + 2NH_3 \rightarrow Cl_2Re(:N)(EtPh_2P)_2 + H_2O + NH_4Cl \qquad (24)$$

4.3. NUCLEAR ISOTOPES[259]

Labelled sodium amide has been prepared by heating sodium metal with enriched ammonia:

$$2Na + 2\,^{15}NH_3 \rightarrow 2Na^{15}NH_2 + H_2$$

A similar method can also be used to prepare $Ca(^{15}NH_2)_2$.

Calcium or magnesium react with nitrogen at elevated temperatures to give the nitride which can conveniently liberate ammonia by hydrolysis.

$$3Ca + {}^{15}N_2 \rightarrow Ca_3{}^{15}N_2$$

$$Ca_3{}^{15}N_2 + 6H_2O \rightarrow 3Ca(OH)_2 + 2\,^{15}NH_3$$

Labelled ammonia was used for the preparation of labelled trisilylamine in order to record the ^{15}N coupling in the nuclear magnetic resonance spectrum[260]:

$$3H_3SiCl + {}^{15}NH_3 + Et_3N \rightarrow (H_3Si)_3{}^{15}N + 3Et_3N \cdot HCl$$

4.4. USES[223, 261]

The applications of metal nitrides are based essentially on their physical properties of high melting points, extreme hardness, metallic conductivity and high chemical inertness. Aluminium nitride and silicon nitride, pure or in combination with the corresponding oxides or carbides, are used as refractory materials for crucibles, boats, thermocouple protecting tubes, crucible coatings and linings. Sintered parts containing these materials can be prefabricated in metal powder followed by nitriding in a nitrogen gas stream at high temperatures. Several metal nitrides have been used as heterogeneous catalysts in a variety of processes, one of the best known being iron nitride in the Fischer–Tropsch method of hydriding carbonyls. Boron nitride in the form with the zincblende structure, sometimes known as borazon, is about as hard as diamond, and silicon nitride bonded silicon carbide has also been used as an abrasive. Some III–V metal nitrides, AlN, GaN and InN, have applications as semiconductors. UN and ThN have been proposed as fuels and breeders in high-temperature nuclear reactors.

Metal amides are used as reducing agents while amino derivatives of some elements are useful intermediates in organic and organometallic syntheses.

257 W. L. Lehn, *J. Am. Chem. Soc.* **86** (1964) 305.
258 D. Bright and J. A. Ibers, *Inorg. Chem.* **8** (1969) 709.
259 W. Spindel, in *Inorganic Isotopic Synthesis* (ed. R. H. Herber), Benjamin, New York (1962), p. 74.
260 E. A. V. Ebsworth, G. Rocktäschel and J. C. Thompson, *J. Chem. Soc.* A (1967) 362.
261 M. Sittig, *Nitrogen in Industry*, Van Nostrand, Princeton, New Jersey (1965), pp. 70–76.

4.5. STRUCTURE AND BONDING

Descriptions of the bonding in metal–nitrogen compounds are based almost entirely on parameters obtained from structure determinations by X-ray diffraction techniques. The salt-like ionic nitrides all contain N^{3-} ions (~1.4 Å radius) in their crystal lattices. For example, lithium nitride Li_3N has a hexagonal structure, while many Group II nitrides are

TABLE 30. PHYSICAL PROPERTIES OF IONIC AND COVALENT METAL NITRIDES[a]

Metal nitride	Crystal system	Unit cell dimensions		Density	Melting point (°C)	$\Delta H_f^\circ(298)$ (kcal/g atom)
		a	*c*			
Li_3N	Hexagonal	3.658	3.882	1.28	548 d	47.17
Na_3N				1.7	d	
Cu_3N	Cubic	3.80		5.84		−17.8
Ag_3N	fcc	4.369			25 d	−61.0
Be_3N_2	Cubic $D5_3$	8.14			2200	−133.5
Mg_3N_2	bcc $D5_3$	9.93			271 d	−115.18
Ca_3N_2	Pseudo-hexagonal	3.533	4.11	2.62		102.6
	Cubic $D5_3$	11.38		2.54		
Sr_3N_2					1030	−91.4
Ba_3N_2						−89.9
Zn_3N_2	Anti-isomorphous with Mn_2O_3	9.74		6.4		24.06
Cd_3N_2	Anti-isomorphous with Mn_2O_3	10.79		7.67		38.6
Hg_3N					95/2 d	
BN	Hexagonal	2.504	6.661	2.34	3000	
BN	fcc zincblende	3.615		3.43		
AlN	Hexagonal wurtzite	3.11	4.975	3.05	2200	−57.4
GaN	Hexagonal wurtzite	3.19	5.18	6.10	600 *in vacuo*	24.9
InN	Wurtzite	3.53	5.70	6.89		4.6
LaN	NaCl	5.295				71.06
CeN	NaCl	5.01				−78
PrN	NaCl	5.15				
NdN	NaCl	5.14				
GdN	NaCl	4.99				
Si_3N_4	Hexagonal	13.16	8.72	3.2	1900	
Ge_3N_4	Orthorhombic	13.84	8.18 $b = 4.06$		900–1000 d	
UN_2	Fluorite	5.31				

[a] Ref. 223.

anti-isostructural with the oxides of these elements, i.e. the metal atoms occupy the positions of the oxygen atoms and the nitrogen those of the metal atoms in M_2O_3. Similarly, other nitrides with formulae corresponding to the usual valency of the metal in combination with nitrogen(III) can also be thought of as containing some ionic character. In Ge_3N_4

the metal has four tetrahedral nitrogen neighbours and nitrogen forms coplanar bonds to three metal atoms. Some lanthanide element nitrides of general formula MN which crystallize in the NaCl-type structure have magnetic moments which suggest M(III). Thus the metallic character shown by these nitrides is probably due to small deviations from stoichiometry such as vacancies among the N^{3-} anions[262]. UN_2 crystallizes as the fluorite structure and also seems to contain uranium(VI) and N^{3-} ions. Structural data on these and similar compounds are summarized in Table 30.

In metallic nitrides of general formulae MN, M_2N and M_4N, the nitrogen atoms occupy some or all of the interstices of cubic- or hexagonal-close-packed metallic lattices. These compounds are often referred to as hard metals with properties characteristic of alloys. But in spite of their metallic appearance and high conductivity, they are usually harder and melt at higher temperatures than metals[223]. Structural data on metallic nitrides are listed in Table 31.

TABLE 31. PHYSICAL PROPERTIES OF METALLIC NITRIDES[a]

Metal nitride	Crystal system	Unit cell dimensions Å		Density	Melting point (°C)	$\Delta H_f^\circ(298)$ (kcal/mol)	$\Delta G_f^\circ(298)$ (kcal/mol)	Hardness (Mohs' scale)
		a	*c*					
TiN	fcc NaCl	4.23		5.3	2950	−80.7	−75.4	8–9
ZrN	fcc NaCl	4.56		7.1	2980	−87.3		8+
HfN	fcc NaCl	4.51		13.6	2700	−78.3	−81.4	
VN	fcc NaCl	4.140		6.04	2050	−41.43	−35.06	9–10
$VN_{0.4}$	Hexagonal	2.835	4.541	5.98				
NbN	fcc NaCl	4.37		8.3	2300	−59		8+
	Hexagonal	2.952	11.25					
TaN	Hexagonal	5.191	2.906	13.7	3090	−58.1	−52.17	
Ta_2N	Hexagonal	3.041	4.907			−64.7		
CrN	fcc NaCl	4.14		6.1	d 1770	−28.3	−17.26	
Cr_2N	hcp	2.74	4.45			−26.3	19.545	
MoN	Hexagonal	5.725	5.608			−16.6		
Mo_2N	fcc	4.16		8.0	d			
WN	Hexagonal	2.893	2.826		d 600			
W_2N	fcc	4.12		12.0	d	−17		
Mn_3N_2	fcc	3.865						
Mn_2N	hcp	2.779	4.529	14.2				
Co_2N	Rhombic deformed hcp	2.842	4.330 *b* = 4.627					
Co_3N	Hexagonal	2.658	4.351					
Ni_3N	hcp	2.667	4.312		d 450			
ThN	fcc NaCl	5.21		11.5	2630			
Th_2N_3	La_2O_3	3.875	6.175					
UN	fcc	4.880		14.32	2800	−80		
U_2N_3	bcc	10.678		11.24	d	−71		
NpN	fcc NaCl	4.887		14.1				
PuN	fcc NaCl	4.906		14.2				

[a] Ref. 223.

[262] C. K. Jorgenson, *Oxidation Numbers and Oxidation States*, Springer-Verlag, New York (1969).

The nitrides of Group III elements, MN, have properties which tend towards covalency. Boron nitride can be prepared in two forms: one is related to graphite, being a lattice of hexagonal layers, and this form can be converted into material with the zincblende structure like that of diamond, with which it is both isoelectronic and isostructural. Relevant structural data is also listed in Table 30.

The structures of metal amides and imides have also been investigated (Table 32), but it is the organo-substituted derivatives which are of current interest and which reveal unique structural and bonding situations. Organoboron–nitrogen compounds show $(p \rightarrow p)\pi$-interaction in the B–N bonds of monomeric aminoboranes, and a consequence of this multiplicity is that free rotation about these bonds is hindered. This has been

TABLE 32. PHYSICAL PROPERTIES OF ALKALI AND ALKALINE EARTH METAL AMIDES AND IMIDES[a]

Compound	Crystal system	M.p. (°C)	Density	$\Delta H_f^\circ(298)$ (kcal/mol)	Solubility in liquid NH_3 (298)
$LiNH_2$	Tetragonal	374	1.18	−43.5	0
$NaNH_2$	Orthorhombic	208	1.39	−28.4	0.17 g per 100 g
KNH_2	fcc NaCl	338	1.64	−28.3	3.6 g per 100 g
$RbNH_2$	fcc NaCl	309	2.58	−25.7	Very soluble
$CaNH_2$	Tetragonal deformed CaCl (γ-NH_4Br struct.)				Very soluble
	fcc NaCl	262	3.43	−25.4	
$Ca(NH_2)_2$				−91.6	0
$Sr(NH_2)_2$				−83.2	0
$Ba(NH_2)_2$				−78.9	0
$Zn(NH_2)_2$			2.13	−38.8	0
$Cd(NH_2)_2$			3.05	−14.7	0
Li_2NH	Anti-fluorite		1.48		
CaNH	fcc				

[a] Refs. 222, 223.

demonstrated by ^1H n.m.r. studies on $Ph(Cl)BNMe_2$ and vinyl$(Br)BNMe_2$ which show non-equivalent methyl groups on nitrogen[263]. A wide selection of cyclic and condensed ring boron–nitrogen derivatives have also been prepared and characterized[221]. Fewer aluminium–nitrogen compounds are known[221, 264] but, none the less, $(PhAlNPh)_4$ has a unique structure based on a Al_4N_4 cube[265]. The dimethylamino derivative Me_2AlNMe_2 and its gallium and indium analogues are dimeric both in solution and in the vapour phase, but undergo transition to a glass at 50–80°C. These crystal-glass transitions are reversible,

[263] P. A. Barfield, M. F. Lappert and J. Lee, *Proc. Chem. Soc.* (1961) 421; *J. Chem. Soc.* A (1967) 362.

[264] J. K. Ruff, in *Developments in Inorganic Nitrogen Chemistry*, Vol. 1 (ed. C. B. Colburn), Elsevier, Amsterdam (1966), pp. 470–521.

[265] J. Idris Jones and W. S. McDonald, *Proc. Chem. Soc.* (1962) 366; T. R. R. McDonald and W. S. McDonald, *ibid.* (1963) 382.

TABLE 33. PHYSICAL PROPERTIES OF COVALENT NITROGEN COMPOUNDS AND SOME AMINO DERIVATIVES OF METALS AND METALLOIDS

Compound	M.p. (°C)	B.p. (°C/mm)	Bond lengths (Å)	Bond angles (°)	Ref.
$Be[N(SiMe_3)_2]_2$			Be–N 1.566 Si–N 1.726	N–Be–N 180 Si–N–Si 129.3	a
$(MeZnNPh_2)_2$				Zn–N–Zn 90	221
$(HBNH)_3$	−58	55/760	B–N 1.44		b
$(ClBNH)_3$	84				b
$(ClBNBu^t)_4$	248	sub 135/0.01			b
$B(NMe_2)_3$		147/760	B–N 1.43	B–N–C 123.9 N–B–N 120	b
$(Me_2AlN{=}CHMe)_2$			Al–N 1.96	Al–N–Al 94.6	221
$(PhAlNPh)_4$	>300		Al–N 1.93	Al–N–Al 90	221
$(Me_2GaNH_2)_2$	97				221
$MeNH_2$	−93.5	−6.3/760	C–N 1.474	C–N–H 112.1	c
Me_2NH	91.1	6.9/760	C–N 1.455	C–N–C 111.5	d
Me_3N	−117.1	2.8/760	C–N 1.47	C–N–C 109	c
$MeN{=}NMe$	−78	1.5/760	C–N 1.47	N–N–C 110	c
C_5H_5N	−42	115.5/760	C–N 1.33	C–N–C 117	c
$(CN)_2$	−34.4	−21.15/760	C–N 1.16	N–C–C 180	c
$(H_3Si)_3N$	−106	49/760	Si–N 1.738	Si–N–Si 119.6	269
$(Me_3Si)_3N$	70–71	215/760	Si–N 1.74	Si–N–Si 120	269
$(Me_3Si)_2NH$		126/760			269
$(H_3Ge)_3N$					269
$Ge(NMe_2)_4$	14	203/760			243
$(Me_3Ge)_2NH$		47/17			243
$(Me_3Ge)_3N$		60/2			243
$(Ph_3Ge)_3N$	163				243
$(Me_3Sn)_3N$		84/0.4			244
Me_3SnNMe_2		126/760			244
Me_3SnNEt_2		36/6			244
$Sn(NMe_2)_4$		51/0.15			244
$Me_3PbN(SiMe_3)_2$		85–7/3			243
$(PNCl_2)_3$		256/760	P–N 1.59	P–N–P 119 N–P–N 119	246
$(PNCl_2)_4$		328/760	P–N 1.57	P–N–P 131 N–P–N 121	246
Cl_2PNMe_2		151/760			b
$P(NEt_2)_3$		245/760			246
$As(NMe_2)_3$		36/2			b
S_4N_4	178		S–N 1.62	S–N–S 113 N–S–N 105	249
$S_4N_4H_4$	152		S–N 1.67	S–N–S 122 N–S–N 108	249
S_2N_2	23	explodes at 30°			249
$Fe[N(SiMe_3)_2]_3$			Fe–N 1.918 Si–N 1.731	Fe–N–Si 119.38 Si–N–Si 121.24	e

[a] A. H. Clark and A. Haaland, *Chem. Communs.* (1969) 912.
[b] K. Moedritzer, in *Inorganic Syntheses*, Vol. X (ed. E. L. Muetterties), McGraw-Hill, New York (1967), section 21, pp. 131–49.
[c] *Tables of Interatomic Distances and Configuration in Molecules and Ions* (ed. L. E. Sutton), Chemical Society Special Publications 11 and 18, London (1958 and 1965).
[d] B. Beagley and T. G. Hewitt, *Trans. Faraday Soc.* **64** (1968) 2561.
[e] D. C. Bradley, M. B. Hursthouse and P. F. Rodesiler, *Chem. Communs.* (1969) 14.

TABLE 33 (*cont.*)

Compound	M.p. (°C)	B.p. (°C/mm)	Bond lengths (M–N Å)	Bond angles (°)	Ref.
$Ti(NMe_2)_3$		d 100/0.01			225
$Ti(NMe_2)_4$		50/0.05			225
$Zr(NMe_2)_4$	70	80/0.05			226
$Nb(NMe_2)_5$		sub 100/0.1			226
$Ta(NMe_2)_5$		sub 110/0.1			226
$U(NEt)_4$		115–125/0.06			238
$(EtPh_2P)_2Cl_3ReNMe$			Re–N 1.685	Re–N–C 180	253
$(Et_2PhP)_3Cl_2ReN$			Re–N 1.79		258
$K_3\{[(H_2O)ClRu]_2N\}$			Ru–N 1.718	Ru–N–Ru 180	f
$K_2(Cl_5OsN)$			Os–N 1.614		g

[f] M. Ciechanowicz and A. C. Skapski, *Chem. Communs.* (1969) 574.
[g] L. A. Atovyman and G. B. Bokii, *J. Struct. Chem. (USSR)* **1** (1960) 501.

and the glasses are thought to consist of mixtures of cyclic oligomers and short-chain polymers[266].

The organic chemistry of nitrogen provides the ultimate in variety of compounds against which nitrogen derivatives of other elements can be compared[241]. Some parameters of organic nitrogen compounds are included in Table 33. Organosilicon–nitrogen compounds have attracted considerable attention[242], particularly since the molecular structure of trisilylamine $(H_3Si)_3N$, as determined by electron diffraction, was planar at nitrogen[267]. This result has been confirmed, and other related derivatives such as $(H_3Si)_2NN(SiH_3)_2$, $(H_3Si)_2NH$, $(H_3Si)_2NMe$, $(H_3Si)_2NBF_2$, $[(Me_3Si)_2N]_3Al$, all show similar stereochemistry at nitrogen[268]. With analogous carbon compounds being pyramidal, the most popular explanation for planarity at nitrogen is $(p \rightarrow d)\pi$-bonding interaction between nitrogen and silicon. The evidence for such interactions in silicon systems has recently been reviewed[269]. A similar explanation has been proposed for planar $(H_3Ge)_3N$ [268].

Little structural information is available on organotin–nitrogen compounds but $(p \rightarrow d)\pi$-interactions are not thought to be important. The chemistry of amino-tin derivatives can be rationalized on the basis of a weak polar Sn–N bond[244]. A wide variety of phosphorus–nitrogen compounds are known including cyclic derivatives[246], and like other second-row elements, the involvement of *d* orbitals in the P–N bond has been proposed.

Tetrasulphur tetranitride, the simplest nitride of sulphur, has an eight-membered cage structure in which the four sulphur atoms form a bisphenoid while the four nitrogen atoms are coplanar and occupy the corners of a square. A large number of other cyclic S–N systems are also known[249]. Structural parameters for covalent nitrides and their related organo derivatives are included in Table 33.

[266] O. T. Beachley and G. E. Coates, *J. Chem. Soc.* (1965) 3241.
[267] K. Hedberg, *J. Am. Chem. Soc.* 77 (1955) 6491.
[268] G. M. Sheldrick, personal communication.
[269] E. A. V. Ebsworth, in *Organometallic Compounds of the Group IV Elements*, Vol. 1, Part 1 (ed. A. G. MacDiarmid), Marcel Dekker, New York (1968), pp. 1–104.

4.6. PHYSICAL PROPERTIES

Some physical properties of metal amides, metal nitrides and a selection of representative amino derivatives of the elements are listed in Tables 30–33, together with structural parameters which have been commented on in section 4.5.

4.7. CHEMISTRY AND CHEMICAL PROPERTIES

The chemistry of metal nitrides is not very well developed. This is due to the fact that some nitrides are chemically quite inert, and those that do react tend to give hydrolysis or decomposition products:

$$2AlN + 3H_2O \rightarrow Al_2O_3 + 2NH_3$$
$$2VN + 3H_2SO_4 \rightarrow V_2(SO_4)_3 + 3H_2 + N_2$$

Many ternary metal nitrides, $MM'N_2$, etc., do not appear to have been investigated as to chemical reactivity and may well prove to be of interest.

Metal amides and imides are readily hydrolysed[222]:

$$NaNH_2 + H_2O \rightarrow NaOH + NH_3$$
$$Li_2NH + 2H_2O \rightarrow 2LiOH + NH_3$$

The metal amides can also be deamminated to the corresponding imide and nitride by heating, but in some instances further decomposition takes place:

$$Mg(NH_2)_2 \rightarrow Mg_3N_2 + 4NH_3$$
$$2LiNH_2 \rightarrow Li_2NH + NH_3$$
$$2NaNH_2 \rightarrow 2NaH + N_2 + H_2$$
$$2NaNH_2 \rightarrow 2Na + N_2 + 2H_2$$

Sodium amide can also be reduced with carbon at red heat:

$$NaNH_2 + C \rightarrow NaCN + H_2$$

The chemistry of alkali metal amides and a few substituted amino derivatives of metals and metalloids (predominantly main group elements) has been investigated sufficiently for some of the main lines of development to emerge. Tin is an element whose organometallic chemistry has received considerable recent attention[270] and although organotin–nitrogen compounds are among the more reactive, some of their reactions have since proved to be quite general and applicable to similar derivatives of other elements[244]. These reactions can be divided into the two types: substitution and addition:

$$\geqslant Sn—N\lessdot + XY \rightarrow \geqslant Sn—X + YN\lessdot$$
$$\geqslant Sn—N\lessdot + Z \rightarrow \geqslant Sn—Z—N\lessdot$$

Specific examples of these general reactions illustrate the variety of products that can be obtained and include the replacement of acidic, neutral and basic hydrogen, halogen

[270] W. P. Neumann, *Die Orgamische Chemie des Zinns*, Ferdinand Enke Verlag, Stuttgart (1967).

metathesis, transmetalation, dehydrohalogenation and insertion reactions (R and R′ can be any non-reacting organic group, usually Me, Et, Ph, etc.)[244]:

$$R_3SnNR_2 + EtOH \rightarrow R_3SnOEt + R_2NH$$
$$R_3SnNR_2 + PhC{\equiv}CH \rightarrow R_3SnC{\equiv}CPh + R_2NH$$
$$R_3SnNR_2 + R'_3SnH \rightarrow R_3SnSnR'_3 + R_2NH$$
$$R_3SnNR_2 + HMn(CO)_5 \rightarrow R_3SnMn(CO)_5 + R_2NH$$
$$R_3SnNR_2 + CF_2{=}CFCl \rightarrow R_3SnF + CFNR_2{=}CFCl$$
$$R_3SnNR_2 + Me_3SiCl \rightarrow R_3SnCl + Me_3SiNR_2$$
$$R_3SnNR_2 + (Ph_3P)_3IrHCl \rightarrow R_3SnCl \cdot R_2NH + (Ph_3P)_3IrH$$
$$R_3SnNR_2 + CS_2 \rightarrow R_3SnSCSNR_2$$
$$R_3SnNR_2 + PhNCO \rightarrow R_3SnNPhCONR_2$$

Similar reactions also occur with amino derivatives of transition metals of which titanium, zirconium, niobium and tantalum amides have been investigated most[225]. Clearly, a large area of metal nitrogen chemistry remains to be explored.

4.8. ANALYTICAL CHEMISTRY OF AMIDES, IMIDES AND NITRIDES[271]

Hydrolysis of simple amides gives ammonia and the metal hydroxide, both of which can be determined separately by titration with acid:

$$NaNH_2 + H_2O \rightarrow NaOH + NH_3$$

Similarly, imides and hydrolysable nitrides can also be dissolved in acid, and after making the solution strongly basic the ammonia can be distilled into standard acid and back-titrated with standard base.

Amino derivatives of metals and metalloids yield free amine on hydrolysis which can also by analysed by acid–base titrimetry. Alternatively, the conventional methods, used for determining organic nitrogen, of Dumas or Kjeldahl, are usually appropriate for amides derived from organic amines.

Metal nitrides which are soluble in acids present no problem in nitrogen analysis as the compound dissolves to give metal and ammonium salts. However, for metal nitrides which are resistant to chemical attack by wet methods, fusion or vacuum fusion in the presence of sodium peroxide may be necessary in order to obtain an estimate of nitrogen content.

5. AMMONIUM SALTS

5.1. OCCURRENCE[272, 273]

As ammonium salts are quite soluble in water, major terrestrial deposits which may have formed have long since been leached away. However, ammonium chloride is still found in places where natural combustion processes have been at work on nitrogen-containing

[271] K. I. Vasu, in *Mellor's Comprehensive Treatise on Inorganic and Theoretical Chemistry*, Vol. 8, Suppl. II, *Nitrogen*, Part II, Longmans, London (1967), Section XXXVII, p. 563.

[272] J. W. Mellor, *Comprehensive Treatise on Inorganic and Theoretical Chemistry*, Vol. 8, *Nitrogen*, Longmans, London (1928).

[273] *Mellor's Comprehensive Treatise on Inorganic and Theoretical Chemistry*, Vol. 8, Suppl. I, *Nitrogen*, Part I, Longmans, London (1964).

materials in the absence of water, e.g. in crevices in the vicinity of volcanoes, undoubtedly as a sublimation product. Ammonium sulphate has also been found in volcanic craters. Natural steam vents in Italy have been harnessed to yield considerable amounts of ammonium carbonate by the controlled interaction of carbon dioxide and ammonia present in the steam and water. Ammonium iodide can be sublimed from burning seaweed, and ammonium chloride was obtained from the combustion of animal waste also by sublimation. Apart from the byproduct ammonium salts from the coal-gas industry, all ammonium salts are now produced from synthetic ammonia.

5.2. PRODUCTION[274]

Ammonium fluoride is produced by neutralizing strong hydrofluoric acid with liquid ammonia until just alkaline and then filtering off the product[275]:

$$HF + NH_3 \rightarrow NH_4F$$

For the production of ammonium chloride, however, direct neutralization is only economical if chlorine or hydrochloric acid is available at an adjacent plant, in order to offset the costs due to corrosion problems. Thus the common method of producing fertilizer grade ammonium chloride is by dual-salt process[276]. Such a situation is present in the Solvay ammonia–soda process for producing sodium carbonate which can be described by the following reaction scheme (1)–(5):

$$CaCO_3 \xrightarrow{-CO_2} CaO \xrightarrow{+H_2O} Ca(OH)_2 \qquad (1)$$

$$NH_3 + H_2O + CO_2 \rightarrow NH_4HCO_3 \qquad (2)$$

$$NH_4HCO_3 + NaCl \rightarrow NaHCO_3 + NH_4Cl \qquad (3)$$

$$2NaHCO_3 \rightarrow Na_2CO_3 + CO_2 + H_2O \qquad (4)$$

$$2NH_4Cl + Ca(OH)_2 \rightarrow 2NH_3 + CaCl_2 + H_2O \qquad (5)$$

Modification of this process to produce ammonium chloride of suitable purity would also dispense with reactions (1) and (5) associated with ammonia recovery. After separation of the sodium bicarbonate in (3), more ammonia is added to the mother liquor, then salted out with sodium chloride and cooled to below 15°C at which temperature ammonium chloride crystallizes out. The product is centrifuged, washed and dried. Higher grade material can be produced by double decomposition[276]:

$$(NH_4)_2SO_4 + 2NaCl \rightarrow Na_2SO_4 + 2NH_4Cl$$

The ammonia–carbon dioxide–water system gives rise to a number of products which are often referred to collectively as ammonium carbonate[277] but which include, amongst others, the individual species $(NH_4)_2CO_3$, ammonium bicarbonate NH_4HCO_3 and ammonium carbamate $NH_2CO_2NH_4$. Ammonium bicarbonate free of these other compounds can be crystallized when an aqueous solution saturated with ammonia and carbon dioxide at 60°C is allowed to cool under a pressure of carbon dioxide. Ammonium carbonate

[274] C. J. Pratt and R. Noyes, *Nitrogen Fertilizer Chemical Processes*, Noyes Development Corporation, New Jersey (1965).
[275] D. M. McC. Steele, in ref. 273, Section VIII, pp. 370–7.
[276] N. L. Ross Kane, in ref. 273, Section IX, pp. 378–432.
[277] K. W. Allen, in ref. 273, Section XII, pp. 459–68.

can be formed when gaseous ammonia and carbon dioxide react in a humid atmosphere. At higher temperatures and pressures, ammonia and carbon dioxide form ammonium carbamate which is also in equilibrium with its dehydration product urea:

$$CO_2 + 2NH_3 \rightleftharpoons NH_2CO_2NH_4$$
$$NH_2CO_2NH_4 \rightleftharpoons NH_2CONH_2 + H_2O$$

Conditions favourable for the formation of commercially desirable urea hinge on the fact that the reaction velocity of the dehydration process proceeds at a reasonable rate only when the temperature is raised which likewise increases the decomposition pressure of the carbamate. Hence a higher pressure must also be used, and modern urea production units operate at 175–210°C and 170–400 atm [274].

Ammonium sulphate is produced principally from direct neutralization, but the gypsum process in which ammonium carbonate reacts with a suspension of calcium sulphate in water is still operated in countries having sources of calcium sulphate but no sulphur:

$$(NH_4)_2CO_3 + CaSO_4 \rightarrow CaCO_3 + (NH_4)_2SO_4$$

Direct neutralization is the only major commercial process for the production of ammonium nitrate either in solution or in the molten state[279]. Processes based on Chilean nitrate are nowadays much less important:

$$(NH_4)_2SO_4 + 2NaNO_3 \rightarrow 2NH_4NO_3 + Na_2SO_4$$

Crystalline diammonium hydrogen phosphate is produced from liquid ammonia and orthophosphoric acid[280]:

$$2NH_3 + H_3PO_4 \rightarrow (NH_4)_2HPO_4$$

5.3. APPLICATIONS AND USES[274, 281]

The principal use of ammonium salts is as fertilizers and as with all fertilizer salts, special attention must be paid not only to the chemical purity of the product, but also the physical characteristics of the material. Both pure compounds and mixtures (physical and chemical) are produced and applied as aqueous solutions or solids in the form of crystals, granules or prills, appropriate to the particular material. With solids, fairly even sized particles are desirable for it is the presence of "fines" that can lead to caking. This happens particularly with hygroscopic substances and can be troublesome from an application point of view, but in the case of ammonium nitrate could in certain circumstances constitute an explosion hazard[282]. For this reason, solid ammonium salt fertilizers are often coated with an inert material such as fine limestone or Kieselguhr, and water is eliminated as far as possible both before packing and during storage[283]. The factors influencing the choice of fertilizer for a particular application are varied and complex, but the nitrogen content of such materials is clearly favourably considered at the present time in view of the increased

278 F. Call, in ref. 273, Section XIV, pp. 473–505.
279 K. F. J. Thatcher, in ref. 273, Section XV, pp. 506–62.
280 W. H. Lee and M. F. C. Ladd, in ref. 273, Section XVI, pp. 563–6.
281 M. Sittig, *Combine Hydrocarbons and Nitrogen for Profit*, Noyes Development Corporation, New Jersey (1967).
282 T. Urbanski, *Chemistry and Technology of Explosives*, Vol. II, Pergamon Press, Oxford (1965).
283 A. J. Harding, *Ammonia Manufacture and Uses*, ICI Ltd., Oxford University Press, London (1959).

demand for fertilizers of high nitrogen content such as ammonium nitrate (35%), urea (46%) and the controlled-release fertilizer ureaform (38%).

Regarding the other uses of ammonium salts, all the halides are employed as catalysts[276, 284, 285], the fluoride being involved in isomerization, alkylation and cracking of hydrocarbons. The fluoride is also used for cleaning metal surfaces, as a solder flux, and for etching glass[275]. The main uses of ammonium chloride are in the manufacture of dry cell batteries and in galvanizing where the molten zinc bath is protected from atmospheric oxidation by a layer of ammonium chloride through which steel sheet and articles pass to ensure that they are free of surface oxide. Ammonium chloride is also a mordant in the dyeing industry[276].

Ammonium carbonate and related compounds when heated produce gaseous products leaving no residue, and such materials find application as blowing agents for sponge rubber and plastics formation, smelling salts and baking powder[277]. Urea, the dehydration product of ammonium carbamate, is used in urea–formaldehyde resins and in the production of melamine[274].

Ammonium sulphate like the chloride and the phosphate can be used as a fire-retardant and are constituents in aqueous fire-proofing solutions for cottons and cellulose fabrics. On the other hand, ammonium nitrate is an explosive which although not entirely satisfactory, because of its relatively cheap cost is widely used particularly for blasting purposes[279]. Being hygroscopic during storage, ammonium nitrate is somewhat unpredictable, and for this reason is often compounded with other recognized explosives. High purity ammonium nitrate is required in the preparation of nitrous oxide (see section 10) for use as an anaesthetic, and ammonium perchlorate is a solid rocket fuel[286].

5.4. LABORATORY PREPARATION

Ammonium halides are prepared by the straightforward neutralization of halogen acid and ammonia:

$$HX + NH_3 \rightarrow NH_4X$$

Ammonium fluoride can be sublimed from a mixture of ammonium chloride and excess sodium fluoride:

$$NH_4Cl + NaF \rightarrow NH_4F + NaCl$$

On further heating the product loses ammonia to leave ammonium hydrogen fluoride:

$$2NH_4F \rightarrow NH_4HF_2 + NH_3$$

Ammonium chloride is prepared by a number of methods all of which rely on sublimation for final purification[276]. The direct action of ammonia on bromine under water may be used to produce ammonium bromide, but care must be taken to ensure that an alkaline solution is maintained to avoid the formation of explosive nitrogen tribromide[284]:

$$8NH_3 + 3Br_2 \rightarrow 6NH_4Br + N_2$$

---

[284] J. D. Richards, in ref. 273, Section X, pp. 433–47.
[285] J. D. Richards, in ref. 273, Section XI, pp. 448–58.
[286] P. W. M. Jacobs and H. M. Whitehead, *Chem. Rev.* **69** (1969) 551.

Similarly, the iodide can also be formed by reacting the element with ammonium sulphide or ammonia and hydrogen peroxide[285]:

$$(NH_4)_2S + I_2 \rightarrow 2NH_4I + S$$
$$2NH_3 + H_2O_2 + I_2 \rightarrow 2NH_4I + O_2$$

Halogen oxyacid salts are usually prepared by neutralization methods, but as with the ammonium salts of all oxidizing anions, decomposition can occur unexpectedly, e.g. ammonium chlorate is unstable and has a tendency to explode without apparent cause:

$$HClO_3 + NH_3 \rightarrow NH_4ClO_3$$
$$HClO_4 + NH_3 \rightarrow NH_4ClO_4$$

Ammonium carbonate is prepared by bringing ammonia and carbon dioxide together in aqueous solution, but the product is invariably contaminated by ammonium bicarbonate and ammonium carbamate[277]. Urea, the dehydration product of ammonium carbamate, was originally prepared by Wöhler from ammonium cyanate which can also be related to the ammonia–carbon dioxide–water system[277]:

$$\begin{array}{ccccc}
NH_3 + CO_2 & \xrightarrow{H_2O} & NH_4HCO_3 & \xrightarrow{NH_3} & (NH_4)_2CO_3 \\
\downarrow & & \searrow^{NH_3} \text{ (from } CO_2\text{)} & & \downarrow -H_2O \\
[NH_2CO_2H] & \underset{-NH_3}{\overset{+NH_3}{\rightleftharpoons}} & & & NH_2CO_2NH_4 \\
\downarrow -H_2O & & & & \downarrow -H_2O \\
[HNCO \rightleftharpoons NCOH] & \underset{-NH_3}{\overset{+NH_3}{\rightleftharpoons}} & NCONH_4 & \rightleftharpoons & NH_2CONH_2
\end{array}$$

Ammonium carbonate is sometimes a convenient starting material for the preparation of other ammonium salts:

$$(NH_4)_2CO_3 + CaSO_4 \rightarrow (NH_4)_2SO_4 + CaCO_3$$
$$(NH_4)_2CO_3 + Ca_3(PO_4)_2 + 3SO_2 + 3O_2 + 2H_2O \rightarrow 2NH_4H_2PO_4 + 3CaSO_4 + CO_2$$

Ammonium hexafluorophosphate can be prepared either by the reaction of ammonium fluoride and phosphorus pentachloride or by the degradation of phosphonitrilic chloride with hydrogen fluoride[287]:

$$PCl_5 + 6NH_4F \rightarrow NH_4PF_6 + 5NH_4Cl$$
$$(PNCl_2)_n + 6nHF \rightarrow nNH_4PF_6 + 2nHCl$$

5.5. NUCLEAR ISOTOPES $^{15}NH_4$ AND ND_4

A number of ^{15}N ammonium salts are available commercially[288] and most could be readily prepared from ^{15}N ammonia. Singly labelled ^{15}N-urea can be prepared by Wöhler's reaction of $^{15}NH_4CNO$, but the most satisfactory method for fully labelled urea is by the

[287] W. Kwasnik, in *Handbook of Preparative Inorganic Chemistry*, 2nd edn., Vol. I (ed. G. Brauer), Academic Press, New York (1963), Section 4, pp. 150–271.

[288] Office Nationale Industriel de l'Azote (ONIA), 40 avenue Hoche, Paris (VIIIe), France; Isomet Corporation, Palisades Park, New Jersey, USA.

ammonolysis of diphenyl carbonate[289]. Deutero-ammonium salts are readily prepared by D_2O exchange reactions or directly by deuterolysis of metal nitrides[290]:

$$Mg_3N_2+8D_2O \rightarrow 2ND_4OD+3Mg(OD)_2$$

Magnesium ammonium orthophosphate is essentially insoluble (~0.5 g/l), and is used for the quantitative determination of radiophosphorus-32[291]. This salt can be precipitated from an orthophosphate solution by adding a slight excess of magnesia mixture (slightly acidic solution of magnesium and ammonium chlorides) followed by a large excess of concentrated ammonia:

$$^{32}PO_4^{3-}+Mg^{2+}+NH_4^+ \xrightarrow{H_2O} MgNH_4{}^{32}PO_4\cdot 6H_2O$$

5.6. STRUCTURE AND BONDING

The ammonium ion is tetrahedral with an overall radius of about 1.43 Å which compares with values of 1.33 and 1.48 Å for potassium and rubidium ions respectively[292], and indicates why many compounds of these ions are isostructural. However, where differences do occur, they are often attributed to hydrogen bonding[293]. For example, the structures of KHF_2 and NH_4HF_2 are of the CsCl type, but the HF_2^- ions are differently oriented with the NH_4^+ ion having four near F^- neighbouring ions in a tetrahedral arrangement with which it can form N–H–F bonds, whereas K^+ has all eight F^- neighbours equidistant[275]. Structural data on ammonium salts are listed in Table 34. In ammonium fluoride, where the strongest hydrogen bonds would be expected, the structure shows a completely ordered hexagonal arrangement of fluoride and ammonium ions. Each ammonium ion is surrounded tetrahedrally by four ammonium ions with each fluoride ion acting as hydrogen bond acceptor for four hydrogen bonds at the corners of the tetrahedra[294]. The remaining ammonium halides each have a high-temperature phase with the NaCl structure and lower temperature modifications with CsCl and tetragonal structures[292]. Explanations for the stability of these various phases have been discussed in terms of order–disorder phenomena. The lower transition temperatures are often referred to as lambda points, and are believed to correspond to transitions in which the restricted rotation of the ammonium ions become free. In ammonium perchlorate the ammonium group is apparently free to rotate[286], while ammonium hexafluorosilicate provides an example of dimorphism and disorder[294].

Ammonium sulphate undergoes a ferroelectric transition at −49.7°C, which has structurally been shown to involve the displacement of two independent ammonium ions in the structure which results in stronger hydrogen bonds in the ferroelectric phase (one in which there can occur a spontaneous polarization capable of reversal by the application of an electric field without causing dielectric breakdown of the material)[294].

Ammonium nitrate has a tendency to cake on storage which is due not only to its hygroscopic nature but also to breakdown of the particles as a direct result of changes in

[289] W. Spindel, in *Inorganic Isotopic Syntheses* (ed. R. H. Herber), Benjamin, New York (1962), pp. 74–118.

[290] M. Baudler, in *Handbook of Preparative Inorganic Chemistry*, 2nd edn., Vol. I (ed. G. Brauer), Academic Press, New York (1963), Section I, pp. 111–39.

[291] L. Lindner, in *Inorganic Isotopic Syntheses* (ed. R. H. Herber), Benjamin, New York (1962), pp. 143–92.

[292] A. F. Wells, *Structural Inorganic Chemistry*, 3rd edn., Oxford University Press, London (1962).

[293] G. C. Pimentel and A. L. McClellan, *The Hydrogen Bond*, W. H. Freeman, San Francisco (1960).

[294] W. C. Hamilton and J. A. Ibers, *Hydrogen Bonding in Solids*, Benjamin, New York (1968).

TABLE 34. STRUCTURAL DATA ON AMMONIUM SALTS[a]

Salt	Temperature range (°C)	Crystal system	Cell dimensions (Å)			Internuclear distance (Å)
			a	*b*	*c*	
NH_4F	> -27	Hexagonal wurtzite	4.39		7.02	N–F 2.66, H–F 1.66 N–H 1.04
NH_4HF_2		Orthorhombic Pseudotetragonal	8.426	8.18	3.69	N–H–F 2.80, H–F 1.184, F–F 2.32
NH_4Cl	> 185	NaCl				N–H 1.025
	-31 to 185	CsCl	3.866			N–H ~ 1
	< -31	CsCl				
NH_4Br	> 138	NaCl				
	-38 to 138	CsCl	4.041			N–H 1.025
	< -38	Tetragonal CsCl	4.034			
ND_4Br	> 125	NaCl				
	-58 to 125	CsCl	4.034			N–D 1.03, N–Br 3.47
	-104 to -58	Tetragonal CsCl	4.034			
	< -104	CsCl	3.981			
NH_4I	> -14	NaCl	7.244			
	-42 to -14	CsCl				
	< -42	Tetragonal CsCl	4.315			
ND_4I	at -65	Tetragonal CsCl	4.313			
NH_4HS		Tetragonal	6.01		4.01	
$(NH_2)_2SO_4$	> -50	Orthorhombic	5.98	10.62	7.78	
NH_4NO_3						
I ε	> 125	Cubic	4.40			
II δ	84 to 123	Tetragonal	5.75		5.00	
III γ	32 to 84	Orthorhombic	7.14	7.65	5.83	
IV β	-18 to 32	Orthorhombic	5.75	5.45	4.96	
V α	< -18	Hexagonal	5.75		15.9	
$(NH_4)_2SiF_6$	> 5	Trigonal				N–H 1.064
	< 5	Cubic				
NH_4ClO_4	> 240	Cubic	7.63			
	< 240	Orthorhombic	9.202	5.816	7.449	
$(NH_4)_3AlF_6$		Cubic	8.40			Al–F 1.66

[a] Refs. 275–80, 284–6.

the crystalline state. The phase transition at 32.1 °C is accompanied by a sudden expansion which may cause particle fracture and subsequent degradation during processing or storage, and in tropical countries the 84.2°C transition may also be involved[274].

5.7. PHYSICAL CONSTANTS OF THE AMMONIUM ION

The fundamental vibrational frequencies[295] of the various isotopic ammonium ions are given in Table 35 together with values for some thermodynamic functions[296, 297].

[295] K. Nakamoto, *Infrared Spectra of Inorganic and Coordination Compounds*, Wiley, New York (1963).
[296] D. D. Wagman, W. H. Evans, V. B. Parker, I. Halow, S. M. Bailey and R. H. Schumm, *Selected Values of Chemical Thermodynamic Properties*, NBS Technical Note 270-3, Washington (1968).
[297] J. D. F. Marsh, in ref. 273, Section VI, pp. 325–7.

TABLE 35. PHYSICAL CONSTANTS OF THE AMMONIUM ION

Vibrational frequencies[a]	$[^{14}NH_4]^+$	$[^{15}NH_4]^+$	$[ND_4]^+$	$[NT_4]^+$
ν_1	3040		2214	
ν_2	1680	1646	1215	976
ν_3	3145	1215	2346	2022
ν_4	1400	976	1065	913

Thermodynamic functions[b] for aqueous NH_4^+ in the standard state $m = 1$ aq at 298 K.

$\triangle H_f^\circ$ (enthalpy of formation)	−31.67 kcal/mol
$\triangle G_f^\circ$ (Gibbs energy of formation)	−18.97 kcal/mol
S° (entropy)	27.1 cal/deg mol
C_p° (heat capacity)	19.1 cal/deg mol

[a] Ref. 295. [b] Ref. 296.

TABLE 36. PHYSICAL PROPERTIES OF SOME AMMONIUM SALTS[a]

Salt	M.p. (°C)	B.p. (°C/mm)	d_4^{20}	Solubility in water (g per 100 g H_2O/°C)
NH_4F	d		1.009	100/0
NH_4HF_2	125	240	1.52	v. soluble
NH_4Cl	s 340	d 520	1.527	29.7/0, 37.2/20, 77.3/100, 87.3/115.6
NH_4ClO_3	102 explodes		1.80	28.7/0, 115/75
NH_4ClO_4			1.95	10.7/0, 42.5/85
NH_4Br	s 452	235 vac	2.429	97/25, 145.6/100
NH_4I	s 551	220 vac	2.514	154.2/0, 250.3/100
NH_4HS	118		1.17	128.1/0, d
$(NH_4)_2S$	d −18			v. soluble
NH_4HSO_3	s 150		2.03	267/0, 620/60
$(NH_4)_2SO_3 \cdot 6H_2O$	d 60–70	s 150	1.41	32.4/0, 60.4/100
NH_4HSO_4	146.9	d	1.78	100/0
$(NH_4)_2S_2O_3$	d 150		1.679	103.3/100
$(NH_4)_2S_2O_4$	d 120		1.982	58.2/0
NH_4N_3	160	s 134 explodes	1.346	20.16/30, 27.14/40
NH_4NO_2	60–70 explodes		1.69	v. soluble
NH_4NO_3	170.4	210/11	1.725	118.3/0, 871/100
$NH_4H_2PO_4$	190		1.803	22.7/0, 173.2/100
$(NH_4)_2HPO_4$	d 155		1.619	57.5/10, 106.0/100
NH_4HCO_3	107.5	s	1.58	18.4/15
$(NH_4)_2CO_3$	d 58		100/15	
NH_4PF_6	d 58		62/15	

[a] Refs. 275–80, 284–6.

5.8. PHYSICAL PROPERTIES

A few general physical properties of some ammonium salts are listed in Table 36. Selected values of the thermodynamic properties[296] for ammonium salts in the crystalline state (c), and for molar aqueous solutions (aq) are given in Table 37.

TABLE 37. THERMODYNAMIC PROPERTIES OF AMMONIUM SALTS[a]

Salt	State	ΔH°_{f0}	ΔH°_f	ΔG°_f	$H^\circ_{298}-H^\circ_0$	S°	C°_p
		(kcal/mol)				(cal/deg mol)	
NH_4F	c	−107.41	−110.89	−83.36	2.655	17.20	15.60
	aq		−111.17	−85.61		23.8	−6.4
NH_4HF_2	c	−187.94	−191.9	−155.6	4.243	27.61	25.50
	aq		−187.01	−157.15		49.2	
NH_4Cl	c		−75.15	−48.51		22.6	20.1
	aq		−71.62	−50.34		40.6	−13.5
NH_4ClO_3	aq		−55.4	−19.8		65.9	
NH_4ClO_4	c		−70.58	−21.25		44.5	
	aq		−62.58	−21.03		70.6	
NH_4Br	c		−64.73	−41.9		27	23
	aq		−60.72	−43.82		46.8	−14.8
NH_4BrO_3	aq		−51.7	−18.6		66.1	
NH_4I	c		−48.14	−26.9		28	
	aq		−44.86	−31.30		53.7	−14.9
NH_4OH	aq		−86.64	−56.56		24.5	−16.4
NH_4O_2H	aq		−69.99	−35.1		32.8	
NH_4HS	c		−37.5	−12.1		23.3	
	aq		−35.9	−16.09		42.1	
NH_4HSO_3	c		−183.7				
	aq		−181.34	−145.12		60.5	
$(NH_4)_2SO_3$	c		−211.6				
	aq		−215.2	−154.2		47.2	
NH_4HSO_4	c		−245.45				
	aq		−243.75	−199.66		58.6	−0.9
$(NH_4)_2SO_4$	c		−282.23	−215.56		52.6	44.81
	aq		−280.66	−215.77		58.6	−31.8
$(NH_4)_2S_2O_3$	aq		−219.2				
$(NH_4)_2S_2O_4$	c		−229.2				
	aq		−221.0	−138.6		77	−36
NH_4N_3	c		27.6	65.5		26.9	
	aq		34.1	64.3		52.7	
NH_4NO_2	c		−61.3				
	aq		−56.7	−27.9		60.6	−4.2
NH_4NO_3	c		−87.37	−43.98		36.11	33.3
	aq		−81.23	−45.58		62.1	−1.6
NH_4ONO_2	aq		−42.4				
$NH_4HN_2O_2$	aq		−44.1				
$(NH_4)_2N_2O_2$	aq		−67.4				
$NH_4H_2PO_4$	c		−346.8	−290.5		36.3	34.0
	aq		−342.9				
$(NH_4)_2HPO_4$	c		−376.1				43.5
	aq		−373.6				
$(NH_4)_3PO_4$	c		−401.8				
	aq		−394.0				
NH_4HCO_3	c		−203.0	−159.2		28.9	
	aq		−197.06	−159.23		48.9	
$NH_2CO_2NH_4$	c		−154.17	−107.09		31.9	
	aq		−150.4				
NH_4CNO	c		−72.75				
	aq		−66.6	−42.3		52.6	
$CO(NH_2)_2$	c		−79.56	−47.04		25	22.26
NH_4CNS	c		−18.8				
	aq		−13.40	3.18		61.6	9.5

[a] Ref. 296.

5.9. CHEMISTRY AND CHEMICAL PROPERTIES

Ammonium salts undergo slight hydrolysis in aqueous solution, but this does not usually restrict the solution chemistry:

$$NH_4^+ + H_2O \rightleftharpoons NH_3 + H_3O^+; \qquad K_{298} = 5.5\times 10^{-10}$$

Likewise, although most ammonium salts dissociate at elevated temperatures to give ammonia and the protonated anion, the process need not be regarded as a decomposition reaction, for subsequent recombination in the vapour phase is often a convenient method of purification[276]:

$$NH_4Cl(s) \rightleftharpoons NH_3(g) + HCl(g); \qquad \Delta H = 42.3 \text{ kcal/mol}; K_{298} = 10^{-16}$$

However, it would appear that salts containing small or highly charged anions polarize the ammonium cation so that dissociation occurs more readily than with salts containing large uninegative anions. In contrast, salts with oxidizing anions tend to decompose irreversibly. For example, in the same way that the ammonium ion can be oxidized to nitrogen and nitrogen oxides by nitric–hydrochloric acid mixture, so the violent high-temperature decomposition of ammonium nitrate is similar[279]:

$$2NH_4NO_3 \xrightarrow{300^\circ C} 2N_2 + O_2 + 4H_2O$$

The slow controlled decomposition of ammonium nitrate, however, provides the route to nitrous oxide (see section 10), and the thermal decomposition of ammonium perchlorate is also temperature dependent[286]:

$$NH_4NO_3 \xrightarrow{200\text{–}260^\circ C} N_2O + 2H_2O$$

$$4NH_4ClO_4 \xrightarrow{<300^\circ C} 2Cl_2 + 2N_2O + 3O_2 + 8H_2O$$

$$2NH_4ClO_4 \xrightarrow{>300^\circ C} Cl_2 + 2NO_2 + O_2 + 4H_2O$$

Ammonium chloride sublimes without melting and the vapour is completely dissociated into hydrogen chloride and ammonia. In its action on metals, however, ammonium chloride at, say, 250°C is considerably more reactive than the corresponding concentration of free hydrogen chloride. Also aqueous solutions of ammonium chloride react very much like dilute hydrochloric acid, properties which demonstrate the corrosive nature of the substance. An explanation for this is based on the theory that "onium" salts in the fused, solid or solution states, being acids in the Brönsted sense of the term, give rise to cations with a tendency to split off a proton:

$$NH_4Cl \rightleftharpoons Cl^- + NH_4^+ \rightleftharpoons NH_3 + H^+$$

Ammonium chloride can be oxidized to nitrosyl chloride and chlorine with strong oxidizing agents like nitric acid, and in a similar reaction potassium perchlorate is quantitatively reduced:

$$8HNO_3 + 6NH_4Cl \rightarrow 2NOCl + 2Cl_2 + 6N_2 + 3O_2 + 16H_2O$$
$$KClO_4 + 2NH_4Cl \rightarrow KCl + Cl_2 + N_2 + 4H_2O$$

From a synthetic point of view, the reaction of ammonium chloride and boron trichloride or one of its complexes gives B-trichloroborazine[276]:

$$3BCl_3 + 3NH_4Cl \rightarrow (ClBNH)_3 + 9HCl$$

Most reactions of ammonium fluoride and ammonium bifluoride are ones in which they act as fluorinating agents[275]:

$$NH_4F + SO_3 \longrightarrow NH_4SO_3F$$
$$NH_4F + P_2O_5 \longrightarrow NH_4PO_2F_2$$
$$3NH_4F + AsCl_3 \longrightarrow AsF_3 + 3NH_4Cl$$
$$6NH_4F + 3SiCl_4 \longrightarrow 2SiCl_3F + SiF_4 + 6NH_4Cl$$
$$NH_4F + BrF_3 \longrightarrow NH_4BrF_4$$
$$2NH_4HF_2 + MnF_2 \xrightarrow{\text{electrolyse}} (NH_4)_2MnF_6 + H_2$$

Ammonium iodide is somewhat similar in its reactions[285].

$$4NH_4I + SnO_2 \rightarrow SnI_4 + 4NH_3 + 2H_2O$$

Ammonium salts are important components in fused salt chemistry. For example, molten ammonium nitrate rapidly oxidizes and dissolves metals[279]. Ammonium sulphate and ammonium nitrate also form a large number of complex and double salts of interest in phase rule studies, ammonium salt purification processes and fertilizer compounding.

In the same way that alkali metals can be obtained in the free state by electrolysis of their salts, various attempts have been made to produce free "ammonium" by the reduction of ammonium salts. Electrolysis of a cold ammonium salt solution using a mercury cathode or treatment of this solution with an alkali–metal amalgam gives a product of ammonium in mercury known as ammonium amalgams:

$$NH_4^+ + NaHg_x \rightarrow (NH_4)Hg_x + Na^+$$

The amalgam is a stable solid at $-85°C$ which becomes a pasty material as it attains room temperature with accompanying decomposition to mercury, ammonia and hydrogen:

$$2(NH_4)Hg_x \rightarrow 2xHg + 2NH_3 + H_2$$

Amalgams of substituted ammonias can also be prepared by analogous methods[298].

Conventional electrolysis of aqueous ammonium halides can be represented by the following general equations:

$$2NH_4^+ + 2e^- \rightarrow 2NH_3 + H_2$$
$$2X^- \rightarrow 2X + 2e^-$$
$$2X + NH_4X \rightarrow NH_2X + 2HX$$
$$4NH_2X \rightarrow 2NH_4X + N_2 + X_2$$
$$2OH^- \rightarrow H_2O + O + 2e^-$$

Thus the final products are aqueous ammonia and hydrogen at the cathode, together with halogen, hydrogen halide, nitrogen and a little oxygen, are liberated at the anode[276, 284, 285].

[298] W. L. Jolly, *Inorganic Chemistry of Nitrogen*, Benjamin, New York (1964), p. 43.

5.10. BIOLOGICAL PROPERTIES[299]

The principal rôle of ammonium salts is as fertilizers, and the various physical and chemical characteristics of such materials are major factors taken into consideration in choosing a fertilizer for a particular application. Ammonium salts also play an important part in nitrogen metabolism, but this is not easy to define or analyse because of ammonia–ammonium salt–urea equilibria.

5.11. ANALYSIS OF AMMONIUM SALTS[300]

In the Formol titration, ammonium salts and organic amino acids react with formaldehyde to release acid which can be titrated with standard base:

$$4NH_4^+ + 6HCHO \rightarrow (CH_2)_6N_4 + 6H_2O + 4H^+$$

However, ammonium salts are generally detected and analysed as ammonia which can be released simply by making the solution alkaline. Ammonia can then be distilled into saturated boric acid, and the borate formed determined by titration with standard hydrochloric acid using bromocresol green–methyl red indicator.

Probably the most widely and commonly used method of estimating ammonium and ammonia in aqueous solution is by the colorimetric method using alkaline potassium mercuri–iodide commonly known as Nessler reagent. The reaction can be represented by the following equation, although the composition of the brown precipitate has also been given as $NHg_2I \cdot H_2O$:

$$2K_2HgI_4 + 2NH_3 \rightarrow NH_2Hg_2I_3 + 4KI + NH_4I$$

It is, of course, not necessary to know the exact stoichiometry of the reaction or even the species involved so long as Beer's law is obeyed for a known range of ammonia concentrations. An alternative colorimetric method of analysis is based on the reaction of ammonia with phenol and hypochlorites.

6. HYDRAZINE

6.1. PRODUCTION

The principal commercial method for producing hydrazine is the Raschig synthesis[301–4] of 1907 based on the partial oxidation of ammonia by hypochlorite in aqueous alkali (1):

$$2NH_3 + NaOCl \rightarrow N_2H_4 + NaCl + H_2O \qquad (1)$$

---

299 F. Call, in ref. 273, Section XVII, pp. 567–615.

300 K. I. Vasu, in *Mellor's Comprehensive Treatise on Inorganic and Theoretical Chemistry*, Vol. 8, Suppl. II, *Nitrogen*, Part II, Longmans, London (1967), pp. 563–673.

301 G. H. Hudson, R. C. H. Spencer and J. P. Stern, in ref. 300, pp. 69–114.

302 L. F. Andrieth and B. A. Ogg, *The Chemistry of Hydrazine*, Wiley, New York (1951).

303 F. Raschig, *Schwefel- und Stickstoffstudien*, Verlag Chemie, GmbH, Leipzig–Berlin (1924).

304 C. Matasa and E. Matasa, *L'Industrie Moderne des Produits Azotes*, Dunod, Paris (1968).

The overall process is stepwise and dependent on the rapid initial formation of chloramine which proceeds to completion even in the cold (2):

$$NH_3 + OCl^- \rightarrow NH_2Cl + OH^- \tag{2}$$

The reactions of chloramine that follow have since been shown[305] to produce hydrazine either by nucleophilic attack of ammonia (3) and subsequent neutralization (4), or by preliminary formation of the chloramide ion (5) followed by nucleophilic attack of ammonia (6):

$$NH_3 + NH_2Cl \rightarrow N_2H_5^+ + Cl^- \quad \text{slow} \tag{3}$$

$$N_2H_5^+ + OH^- \rightarrow N_2H_4 + H_2O \quad \text{fast} \tag{4}$$

$$NH_2Cl + OH^- \rightarrow NHCl^- + H_2O \quad \text{fast} \tag{5}$$

$$NHCl^- + NH_3 \rightarrow N_2H_4 + Cl^- \quad \text{slow} \tag{6}$$

However, a further reaction can occur between chloramine and the product hydrazine as in (7):

$$2NH_2Cl + N_2H_4 \rightarrow 2NH_4Cl + N_2 \quad \text{fast} \tag{7}$$

Clearly, reaction (7) is undesirable, particularly as it proceeds at a faster rate than the hydrazine producing reactions (3) and (4), or (5) and (6). It is now known that reaction (7) is catalysed by traces of metal ions, notably copper(II), even in concentrations of a few ppm. This can be inferred from the fact that reaction (7) does not occur when very pure reagents are used nor in the presence of complexing agents which inactivate metal ions in solution. Since the former method is somewhat impractical on an industrial scale, the addition of a glue-like material or gelatine to the reaction mixture as discovered by Raschig[303] is still used and serves to inhibit and suppress reaction (7) while in some way assisting the hydrazine-forming reactions.

The Raschig process as it operates industrially is continuous, and involves mixing an ammonia–hypochlorite (30:1) solution with a gelatine solution in the cold followed by rapid heating through a reactor at 150°C under pressure to avoid loss of ammonia. After a residence time in the reactor of only 1 second, the liquid contains about 0.5% by weight of hydrazine corresponding to about 60% conversion based on hypochlorite. Ammonia and steam are then stripped off in stages, concentrating the hydrazine solution through 2%, 10% and final distillation to give pure hydrazine hydrate $N_2H_4 \cdot H_2O$ (64% hydrazine) in overall yields of up to 70%[301,304]. Urea can also be used as a source of ammonia, otherwise the process is very similar requiring the presence of an inhibitor and rapid heating to give good yields[302]:

$$NH_2CONH_2 + NaOCl + 2NaOH \rightarrow N_2H_4 \cdot H_2O + NaCl + Na_2CO_3$$

New modifications of the Raschig process include the reaction of ammonia and chlorine in a ketone solvent[304,306] to give a diazacyclopropane, presumably via chloramine and

[305] G. Yagil and M. Anbar, *J. Am. Chem. Soc.* **84** (1962) 1797.
[306] *Chem. Engng. News* (1965), 5 July, p. 38.

ketimine, which can take up another mole of ketone under acid conditions giving a ketazine capable of acid hydrolysis under pressure to produce hydrazine sulphate:

$$2NH_3+Cl_2 \rightarrow NH_4Cl+NH_2Cl$$
$$NH_3+R_2CO \rightarrow R_2C{=}NH+H_2O$$
$$NH_4Cl+R_2C{=}NH+NH_3 \rightarrow NH_4Cl+R_2C\begin{matrix}NH\\ |\\ NH\end{matrix}$$
$$\downarrow R_2CO$$
$$2R_2CO+N_2H_6SO_4 \xleftarrow[H_2SO_4]{2H_2O} R_2C{=}N{-}N{=}CR_2+H_2O$$

Aqueous hydrazine and hydrazine hydrate can be concentrated to about 85% by further distillation. However, anhydrous hydrazine can only be obtained from aqueous media by dehydration with solid caustic soda or potash followed by distillation. Alternatively, hydrazine sulphate can be precipitated from dilute hydrazine solutions by treatment with dilute sulphuric acid, and after filtration and drying, reaction with liquid ammonia precipitates ammonium sulphate leaving anhydrous hydrazine after the ammonia has evaporated off the filtrate:

$$N_2H_4+H_2SO_4 \xrightarrow{H_2O} N_2H_6SO_4$$
$$N_2H_6SO_4+2NH_3 \xrightarrow{\text{liq. }NH_3} N_2H_4+(NH_4)_2SO_4$$

Other methods by which hydrazine is formed include the decomposition of ammonia and related derivatives by irradiation either in the presence or absence of inert gases krypton or xenon into amide radicals ($NH_2^{\cdot}$) which can dimerize[307]:

$$NH_3 \xrightarrow{h\nu} NH_3^{+}+e^-$$
$$NH_3^{+}+NH_3 \rightarrow NH_4^{+}+NH_2$$
$$NH_4^{+}+e^- \rightarrow NH_2^{\cdot}+H_2$$
$$NH_3 \xrightarrow{h\nu} NH_2^{+}+H^{\cdot}+e^-$$
$$N_2H_4+H^{\cdot} \rightarrow N_2H_3^{\cdot}+H_2$$
$$N_2H_4+H^{\cdot} \rightarrow NH_2^{\cdot}+NH_3$$

Similarly, in the direct oxidation of ammonia with molecular oxygen, hydrazine can be detected when the gas mixture is passed over the alkaline surface of activated metal oxides.

The majority of other reactions in which hydrazine is a major product are based on the reduction of compounds already containing N–N bonds including nitramide, nitrosohydroxylamine, nitrosoamines, nitrosoketones and organic azo compounds[301, 302, 308]. It has also been claimed that hydrazine is formed when carbonyl-forming metals such as nickel react with urea at 60–70°C [301]:

$$NH_2CONH_2 \xrightarrow{Ni} CO+N_2H_4$$

---

[307] N. J. Friswell and B. G. Gowenlock, in *Advances in Free-Radical Chemistry*, Vol. 2 (ed. G. H. Williams), Logos Press, Academic Press, London (1967), pp. 1–45.

[308] L. A. Raphaelson, in *Kirk–Othmer Encyclopedia of Chemical Technology*, 2nd edn., Vol. II, Interscience, New York (1966), pp. 164–96.

6.2. APPLICATIONS AND USES OF HYDRAZINE

Hydrazine is surprisingly stable in view of its endothermic nature:

$$N_2(g)+2H_2(g) \rightarrow N_2H_4(l); \qquad \triangle H_f^\circ = 12 \text{ kcal/mol}$$

It will burn in air or any other oxygen source with a considerable evolution of heat and conversion to low molecular weight products[302]:

$$N_2H_4(l)+O_2(g) \rightarrow N_2(g)+2H_2O(l); \qquad \triangle H = 148.6 \text{ kcal/mol}$$

Hence a major use of hydrazine and its derivatives is as a fuel for guided missiles and rockets[302]. When combined with an appropriate oxidizing agent such as liquid oxygen, hydrogen peroxide, nitric acid or fluorine, it gives high exhaust velocities and hence high specific impulse at relatively low combination temperatures. Its stability and good storage properties make it suitable for immediate use, and it can be rapidly ignited catalytically to give high exhaust velocities and hence high specific impulse at relatively low combination temperatures. The disadvantage of the relatively high melting point of hydrazine (2°C) is that at temperatures which prevail in the upper atmosphere it can solidify, and for this reason derivatives such as *NN*-dimethylhydrazine are used, either pure or as mixtures with hydrazine[301, 308].

Hydrazine salts of strongly oxidizing anions may also find application as explosives, e.g. solid hydrazine nitrate explodes more readily than ammonium nitrate[309]. Its affinity for oxygen has also been used to remove oxygen from high-pressure boiler systems, and similarly hydrazine prevents discoloration of amines, stabilizes lubrication oils and inhibits peroxide formation in glycol ethers, presumably by suppressing oxidative deterioration[301].

Hydrazine is attractive as a reducing agent because of its high available hydrogen content and because its oxidation product is nitrogen. Hydrazine will reduce a number of important metal salts to the element including silver, the traditional way of producing mirrors and metal coatings. Metals important as catalysts such as nickel can also be reduced in a finely divided form or coated on plastics[301, 308]. Hydrazine and its salts are also used as a reducing agent in organic syntheses[301] and in analytical chemistry[302]. The reducing properties also make hydrazine hydrohalides useful as fluxes for welding and soldering metals, particularly copper and its alloys, as no residue remains to initiate corrosion at the join[301]. A further use relying on reducing properties is as a fuel in the hydrazine–oxygen fuel cell[310] in which the overall reaction is:

$$N_2H_4+O_2 \rightarrow N_2+2H_2O; \qquad E^\circ = 1.56 \text{ V}$$

where the anode reaction is

$$N_2H_4+4OH^- \rightarrow N_2+4H_2O+4e^-; \qquad E^\circ = 1.16 \text{ V}$$

and the cathode reaction is

$$O_2+2H_2O+4e^- \rightarrow 4OH^-; \qquad E^\circ = 0.40 \text{ V}$$

High current densities can be obtained as the concentration polarization is preserved due to the high solubility of hydrazine in the electrolyte. A significant cost reduction in producing fuel cells would see their widespread application[308].

[309] T. Urbanski, *Chemistry and Technology of Explosives*, Vol. II, Pergamon Press, London (1965).

[310] S. S. Tomter and A. P. Anthony, *The Hydrazine Fuel Cell System in Fuel Cells*, Amer. Inst. of Chem. Engineers, New York (1963), pp. 22–31.

Applications arising from the difunctional nature of hydrazine include reactions with dibasic acids such as sebacic acid, to give polymeric amides that can be drawn into fibres, and it is also incorporated in some polymeric materials as a cross-linking agent[308]. The formation of free radicals can to some extent arise as a result of homolytic fission of the symmetrical diamine[307]. Free radicals formed during the decomposition or oxidation of hydrazine have been used as initiators to several polymerization processes, particularly in emulsion polymerization where hydrazine can also act as a blowing agent in the production of plastic foams[301]. Hydrazine derivatives find wide use as chemotherapeutic agents, particularly in the treatment of leprosy and tuberculosis[301].

6.3. LABORATORY PREPARATION OF HYDRAZINIUM SULPHATE, HYDRAZINE HYDRATE AND ANHYDROUS HYDRAZINE

Hydrazine sulphate is prepared by a small-scale Raschig process using either lime water or sodium ethylenediaminetetra-acetate to complex the traces of heavy metal present which catalyse the decomposition of the product[311]:

$$2NH_3+NaOCl+H_2SO_4 \rightarrow N_2H_6SO_4+NaCl+H_2O$$

The product is only slightly soluble in cold dilute acid and so can be filtered off and recrystallized from hot aqueous solution if necessary.

Hydrazine hydrate can be obtained by distillation of hydrazine sulphate in alkaline solution, and collecting the fraction with the boiling range 117–119°C at atmospheric pressure.

$$N_2H_6SO_4+2KOH \rightarrow N_2H_4 \cdot H_2O+K_2SO_4$$

Anhydrous hydrazine can be distilled in a stream of nitrogen after dehydrating hydrazine hydrate with solid sodium hydroxide or barium oxide. Alternatively, hydrazine salts can be cleaved with ammonia and the hydrazine separated by vacuum distillation.

$$N_2H_5Cl+NH_3 \rightarrow N_2H_4+NH_4Cl$$

The insolubility of ammonium sulphate in liquid ammonia also forms the basis of a separation to give hydrazine of high purity:

$$N_2H_6SO_4+2NH_3 \xrightarrow[\text{ammonia}]{\text{liquid}} N_2H_4+(NH_4)_2SO_4$$

6.4. NUCLEAR ISOTOPES $^{15}N_2H_4$

Because such a large excess of $^{15}NH_3$ is necessary to avoid side reactions; the Raschig synthesis is a particularly unfavourable process to carry out using enriched materials. However, by taking advanced precautions to ensure recovery of all ^{15}N byproducts, $^{15}N_2H_4$ can be obtained from the overall process[312]:

$$2\,^{15}NH_3+OCl^- \rightarrow {}^{15}N_2H_4+H_2O+Cl^-$$

---

311 P. W. Schenk, in *Handbook of Preparative Inorganic Chemistry*, 2nd edn., Vol. I (ed. G. Brauer), Academic Press, New York (1963), pp. 468–72.

312 J. W. Cahn and R. E. Powell, *J. Am. Chem. Soc.* **76** (1954) 2568.

Alternatively, the singly labelled compound can be prepared by adding excess $^{14}NH_3$ to the enriched ammonia–hypochlorite solution prior to heating[313]:

$$^{15}NH_3 + OCl^- \rightarrow {}^{15}NH_2Cl + OH^-$$

$$^{15}NH_2Cl + {}^{14}NH_3 + OH^- \rightarrow {}^{15}NH_2{}^{14}NH_2 + H_2O + Cl^-$$

The enriched hydrazine is distilled off with excess ammonia into dilute sulphuric acid and recrystallized as the sulphate. Anhydrous hydrazine can be obtained by shaking the labelled hydrazine sulphate with liquid ammonia at −45°C, filtering off the ammonium sulphate and then evaporating the ammonia solvent from the filtrate. N_2D_4 and related compounds have been used for spectroscopic purposes[314].

6.5. STRUCTURE AND BONDING

One of the significant comparisons between the ammono- and aquo- series of compounds is the structural and chemical comparison of hydrazine with hydrogen peroxide. As indicated by the high values for the melting point, boiling point and Trouton's constant, hydrazine is extensively associated through hydrogen bonding in the condensed phases, but monomeric in the gas phase[315]. Electron diffraction data[316] give the N–N–H angle as 112° and the N–N bond length as 1.45 Å, suggesting sp^3 hybridization at the nitrogen atoms. Thus it is reasonable to assume that the two nitrogen atoms are joined by a σ bond, rotation around which can give rise to one or other of the conformational isomers illustrated in Fig. 18.

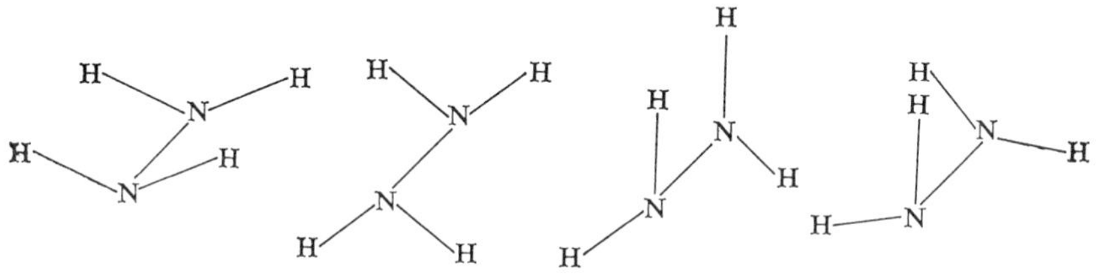

eclipsed *cis*	staggered *trans*	semi-eclipsed *half-cis*	gauche
C_{2v}	C_{2h}	C_2	C_2

FIG. 18. Conformations of hydrazine.

Clearly the large dipole moment of 1.85 Debye units eliminates the *trans* (C_{2h}) conformation, and the *gauche* form is usually considered to be the equilibrium conformation as both the eclipsed and semi-eclipsed conformations would involve coplanar repulsions[317]. *Ab*

[313] F. O. Rice and F. Scherber, *J. Am. Chem. Soc.* **77** (1955) 291.
[314] J. L. Durig, S. F. Bush and E. E. Mercer, *J. Chem. Phys.* **44** (1966) 4238.
[315] A. F. Wells, *Structural Inorganic Chemistry*, 3rd edn., Oxford University Press, Oxford (1962).
[316] Y. Morino, T. Iijima and Y. Murata, *Bull. Chem. Soc. Japan* **33** (1960) 46.
[317] A. Yamaguchi, I. Ichishima, T. Shimanouchi and S. Mizushima, *J. Chem. Phys.* **31** (1959) 843.

initio calculations of the barriers to internal rotation give values of 11.88 and 3.70 kcal/mol[318], and the fundamental frequencies, infrared and Raman bands have been assigned based on C_2 symmetry[314, 319].

Hydrazine being dibasic can give rise to two series of salts $N_2H_4 \cdot HA$ and $N_2H_4 \cdot 2HA$ which contain the $N_2H_5^+$ and $N_2H_6^{2+}$ ions respectively[301]. However, some confusion has arisen with salts of di- and poly-basic acids in part due to nomenclature. For example,

TABLE 38. CRYSTAL STRUCTURES OF HYDRAZINE AND SOME OF ITS SALTS[a]

Compound	Crystal system	Lattice parameters (Å)	Interatomic distances (Å)
N_2H_4	Monoclinic	$a = 3.56$, $b = 5.78$ $c = 4.53$, $\beta = 109.5°$	N–N = 1.46
$N_2H_4 \cdot H_2O$	Trigonal	$a = 4.873$, $c = 10.94$	N–O · · · H = 2.79, 3.11, 3.15 N–N = 1.447
$N_2H_6F_2$	Rhombohedral	$a = 5.43$, $\alpha = 38° 10'$	N–N = 1.42, N–H = 1.05 N–H–F = 2.62
N_2H_5Cl	Orthorhombic	$a = 12.49$, $b = 21.85$, $c = 4.41$	N–N = 1.45, N–H–N = 2.95 N · · · C = 3.12 and 3.4
$N_2H_6Cl_2$	Cubic	$a = 7.89$	N–N = 1.42
N_2H_5Br	Monoclinic	$a = 12.85$, $b = 4.54$, $c = 11.94$, $\beta = 110° 16'$	N–N = 1.45
$N_2H_5NO_3$	Monoclinic	$a = 11.23$, $b = 11.73$, $c = 5.17$, $\beta \sim 90°$	
$N_2H_6SO_4$	Orthorhombic	$a = 8.251$, $b = 9.159$, $c = 5.532$	N–N = 1.40, N–H = 2.79–2.99 N–H · · · O = 2.73 and 2.77 S–O = 1.48–1.49
$Li(N_2H_5)SO_4$ [b]	Orthorhombic	$a = 9.913$, $b = 8.969$ $c = 5.178$	N–N = 1.40, N–H = 1.02 Sulphate slightly distorted
$N_2H_5HC_2O_4$ [c]	Monoclinic	$a = 3.580$, $b = 13.321$ $c = 5.097$, $\beta = 102.62°$	N–N = 1.440 N–H = 0.998 N–H · · · N = 1.037 Short H bonds 2.448

[a] In ref. 301.
[b] V. M. Padmanabhan and R. Balasubramanian, *Acta Cryst.* **22** (1967) 532.
[c] In ref. 321.

$N_2H_6SO_4$, commonly known as hydrazine sulphate, contains the $N_2H_6^{2+}$ and SO_4^{2-} ions[320] and would be more precisely named as hydrazinium(+2) or hydrazonium sulphate, whereas $N_2H_6C_2O_4$ has been reported[321] to be hydrazinium monoxalate $N_2H_5^+$ $HC_2O_4^-$. In hydrazinium difluoride $N_2H_6F_2$, the $N_2H_6^{2+}$ ions have the fully staggered conformation when viewed along the N–N axes, and are linked to the fluoride ions entirely by hydrogen bonding[322]. As hydrogen bonding is not possible with the other halides, the dichloride

318 W. H. Fink, D. C. Pan and L. C. Allen, *J. Chem. Phys.* **47** (1967) 895.
319 F. G. Baylin, S. F. Bush and J. R. Durig, *J. Chem. Phys.* **47** (1967) 895.
320 I. Nitta, K. Sakurai and Y. Tomiie, *Acta Cryst.* **4** (1951) 289.
321 A. Nilsson, R. Liminga and I. Olovsson, *Acta chem. scand.* **22** (1968) 719.
322 M. L. Kronberg and D. Harker, *J. Chem. Phys.* **10** (1942) 309.

$N_2H_6Cl_2$ has a somewhat distorted fluorite structure[323]. The crystal structures of some simple hydrazine salts are given in Table 38. The name for hydrazine hydrate, however, would appear to be accurate, for in the crystal each water molecule is linked by six hydrogen bonds to discrete hydrazine molecules[324], and there is no evidence for structures based on $N_2H_5^+$ OH^- or $N_2H_6^{2+}$ O^{2-}.

Although in comparisons the similarities between hydrazinium and ammonium salts are frequently emphasized, there are often important structural differences. For example, $(NH_4)_2Zn(SO_4)_2 \cdot 6H_2O$ contains water of crystallization, some of which is co-ordinated to the metal, whereas $(N_2H_5)_2Zn(SO_4)_2$ is anhydrous implying that the hydrazinium groups may also be co-ordinated. X-ray structure determination indicates that this is so with the hydrazinium groups co-ordinated *trans* to octahedral zinc and participating in hydrogen bonding with bridging sulphate groups[325].

A large number of hydrazine complexes of general formula $M^{II}(N_2H_4)X_2$ in which the ligand acts as a bidentate chelate bridging between two metal atoms have been reported[326]. Very little structural information is available, but infrared studies indicate that the complex $[Hg(N_2H_4)Cl_2]_n$ has an extended chain structure. $Hg(N_2H_4)_2Cl_2$ is also known and may contain terminal hydrazine groups of C_2 local symmetry, bonded at one end only[327], but the formula can also be written $HgCl_2 \cdot 2N_2H_4$ suggesting the presence of hydrazine of crystallization. Other derivatives of hydrazine include the hydrazide ions $N_2H_3^-$ and $N_2H_2^{2-}$, for which infrared data on $[Zn(N_2H_3)_2]_n$ is consistent with a polymer chain of six-membered ring repeat units[328], while $[Hg_2(N_2H_2)]Cl_2$ has a layer lattice structure[327].

6.6. MOLECULAR CONSTANTS

Several infrared and Raman spectroscopic investigations have been carried out on hydrazine[314], and assignments are consistent with C_2 symmetry as indicated in Table 39. A variety of other spectroscopic data is also known for hydrazine including ultraviolet[301], microwave[329], nuclear magnetic resonance[330] and mass spectra[331]. Infrared spectra have been recorded on hydrazine complexes and salts[332], and the values for the band assigned to the N–N stretching frequency correlates well with the corresponding N–N bond length. In a series of related compounds, ν(N–N) can be used to assign the structure type[333] as the value is lower for unidentate N_2H_4 than for bridged N_2H_4, and $N_2H_5^+$ is lower than $N_2H_6^{2+}$, a trend which probably reflects the state of repulsion between the lone pairs.

323 J. Donohue and W. N. Lipscomb, *J. Chem. Phys.* **15** (1947) 115.
324 R. Liminga and I. Olovsson, *Acta Cryst.* **17** (1964) 1523.
325 C. K. Prout and H. M. Powell, *J. Chem. Soc.* (1961) 4177.
326 W. P. Griffiths, in *Developments in Inorganic Nitrogen Chemistry*, Vol. I (ed. C. B. Colburn), Elsevier, Amsterdam (1966), pp. 1–71.
327 K. Brodersen, *Z. anorg. allgem. Chem.* **290** (1957) 24.
328 J. Goubeau and U. Kull, *Z. anorg. allgem. Chem.* **316** (1962) 182.
329 T. Kojima, H. Hitawanka and T. Oka, *J. Phys. Soc. Japan* **13** (1958) 321.
330 B. E. Holder and M. P. Klein, *J. Chem. Phys.* **23** (1955) 1956.
331 V. H. Dibelev, J. L. Franklin and R. M. Reese, *J. Am. Chem. Soc.* **81** (1959) 68.
332 J. C. Decius and D. P. Pearson, *J. Am. Chem. Soc.* **75** (1953) 2436.
333 A. Braibanti, F. Dallavalle, M. A. Pellinghelli and E. Leporati, *Inorg. Chem.* **7** (1968) 1430.

TABLE 39. VIBRATIONAL FREQUENCIES AND BAND ASSIGNMENTS FOR HYDRAZINE[a]

Infrared gas phase (cm^{-1})	Raman liquid (cm^{-1})	Assignments		Infrared gas phase (cm^{-1})	Ramon liquid (cm^{-1})	Assignments	
3325		$\nu_1(A)$	$\nu(NH)$	3350	3336	$\nu_8(B)$	$\nu(NH)$
	3190	$\nu_2(A)$	$\nu(NH)$	3280	3273	$\nu_9(B)$	$\nu(NH)$
	3261			1587 }		$\nu_{10}(B)$	$\delta(HNH)$
1493	1628	$\nu_3(A)$	$\delta(HNH)$	1628 }			
1098	1111	$\nu_4(A)$	$\rho_r(NH_2)$	1275	1295	$\nu_{11}(B)$	$\rho_r(NH_2)$
	882	$\nu_5(A)$	$\nu(NN)$	966 }		$\nu_{12}(B)$	$\rho_w(NH_2)$
780	783	$\nu_6(A)$	$\rho_w(NH_2)$	933 }			
377		$\nu_7(A)$	$\rho_t(NH_2)$				

[a] Summarized in ref. 314.

TABLE 40. PHYSICAL PROPERTIES OF HYDRAZINE

Property	Temperature (K)	Value
Melting point	275.15	2.0°C
Boiling point	386.65	113.5°C/760 mm
Critical temperature	653.15	380°C
Critical pressure		145 atm
Critical density		0.231 g/cm^3
Vapour pressure of solid	273	2.60 mm
Vapour pressure of liquid	298	14.38 mm
	335	100 mm
Density of solid	268	1.146 g/cm^3
Density of liquid	298	1.00 g/cm^3
	323	0.982 g/cm^3
Viscosity	298	0.009 dyne s/cm^2
Surface tension	298	66.67 dyne cm^{-1}
	308	62.32 dyne cm^{-1}
Refractive index n_D	295	1.470
	308	1.46444
Dielectric constant	273	58.5
	298	51.7
Specific conductivity	273	$1.1–2.0 \times 10^{-16}\ \Omega^{-1}\ cm^{-1}$
	298	$2.3–2.8 \times 10^{-6}\ \Omega^{-1}\ cm^{-1}$
Dipole moment	291–298	1.83–1.85D
eQq (quadrupole coupling constant)		-4.09 ± 0.05
Trouton's constant		25.23 g cal/mol deg
ΔH fusion	275.95	3.025 kcal/mol
ΔH vaporization	386.65	10.2 kcal/mol
Heat of combustion		148.635 kJ/mol
Heat of solution (hydration)	298	−3.893 kcal/mol
Vapour pressure of liquid	293–387	$\log_{10} P_{mm} = 7.80687 - 1680.745t(°C) + 227.74$
Vapour pressure of vapour	413–653	$\log_{10} P_m = 9.40 - \frac{2814.9}{T(K)} - 0.006931T + 0.000003746T^2$
Density of liquid	275–386	$\rho = 1.0253\ (1 - 0.00085t)(t \text{ in } °C)$
Viscosity		$\log \eta = 536/T - 3.844$

6.7. PHYSICAL PROPERTIES

The physical properties of hydrazine are summarized in Table 40, and some properties of hydrazine salts are given in Table 41. Values for thermodynamic functions for hydrazine and its derivatives are given in Table 42. Other thermodynamic data which are pertinent to the chemistry of hydrazine are the standard heats of formation of the amino $NH_2^{\cdot}$ and hydrazino $N_2H_3^{\cdot}$ radicals at 44 ± 2 and 46 ± 5 kcal/mol respectively[334]. These values are conditional on the bond dissociation energies $D(NH_2–NH_2) = 60 \pm 4$ kcal/mol and $D(N_2H_3–H) = 76 \pm 7$ kcal/mol. From the heat of atomization of hydrazine of 411.6 kcal/mol, and assuming that the average N–H bond energy in hydrazine is the same as the average N–H bond energy in ammonia, the bond energy E(N–N) is 38 kcal/mol[307,335].

TABLE 41. PHYSICAL PROPERTIES OF SOME HYDRAZINE SALTS [a]

Compound	M.p. (°C)	B.p. (°C/mm)	Density	Solubility in water (g per 100 g/°C)
$N_2H_4 \cdot H_2O$	−51.7	118.5/760	1.0305/21	Inf.
$N_2H_6F_2$				
N_2H_5Cl	89	d 240		
$N_2H_6Cl_2$	198 (—HCl)	d 200	1.42/20	27.2/32
$N_2H_5ClO_4$	137 explodes		1.939/15	Fairly soluble
N_2H_5Br				
N_2H_5I	124–6			Soluble
$N_2H_5NO_3$ α	70.71			327.5/25
β	62.09			
$N_2H_5N_3$	75.4 explodes			
$N_2H_6SO_4$	254	d	1.37	3.415/25
$(N_2H_5)_2SO_4$	85			202.2/25
$(N_2H_5)_2S_4O_6$	d 85–87			
$N_2H_6(NO_3)_2$	d 104 fast d 80 slow			20.2/35
$(N_2H_5)_2HCO_3$	148			200/35

[a] In *Handbook of Chemistry and Physics*, 48th edn., Chemical Rubber Co., Cleveland, Ohio (1967).

6.8. CHEMISTRY AND CHEMICAL PROPERTIES

6.8.1. Decomposition

Although hydrazine is kinetically stable at room temperature and even at temperatures approaching its critical temperature (380°C), the positive free energy of formation, however,

[334] D. D. Wagman, W. H. Evans, V. B. Parker, I. Halow, S. M. Bailey and R. H. Schumm, *Selected Values of Chemical Thermodynamic Properties*, NBS Technical Note 270-3, Washington (1968).
[335] C. T. Mortimer, *Reaction Heats and Bond Strengths*, Pergamon Press, Oxford (1962), pp. 140–2.

TABLE 42. THERMODYNAMIC PROPERTIES OF HYDRAZINE AND ITS DERIVATIVES[a]

Compound	State	ΔH°_{f0}	ΔH°_{f298}	ΔG°_{f298}	$H^\circ_{298}-H^\circ_0$	S°	C°_p
		kcal/mol				(cal/deg mol)	
N_2H_4	g	26.18	22.80	38.07	2.743	56.97	11.85
	l		12.10	35.67		28.97	23.63
	aq		8.20	30.6			33
$N_2H_5^+$	aq		−1.8	19.7		36	16.8
$N_2H_4\cdot H_2O$	g		−49.0	−18.9		63	
	l		−58.01				
	aq		−60.11	−26.1		49.7	17.5
N_2H_5Cl	c		−47.0				
	aq		−41.8	−11.7		49.5	−15.8
$N_2H_5NO_3$	c		−60.13				
	aq		−51.41	−6.91		71	
N_2H_5Br	c		−37.2				
	aq		−30.8	−5.2		55.7	−17.1
$N_2H_5ClO_4$	c		−42.2				
	aq		−32.7	17.6		79.7	
$(N_2H_5)_2SO_4$	c		−229.2				
	aq		−221.0	−138.6		77	−36

[a] In ref. 334.

is indicative of potential instability. As the only stable products of hydrazine decomposition are ammonia, nitrogen and hydrogen, the varying amounts of each can often be expressed[302] in the form of a general equation (1):

$$3N_2H_4 \rightarrow 4(1-x)NH_3+(1+2x)N_2+6xH_2 \quad (1)$$

Thus the products of decomposition can be specified for a particular set of conditions simply by indicating the value of x. The overall general equation, however, represents the extent to which the individual reactions (2–4) occur.

$$3N_2H_4(g) \rightarrow 4NH_3(g)+N_2(g); \quad \Delta H = -37.4 \text{ kcal/mol} \quad (2)$$

$$N_2H_4(g) \rightarrow N_2(g)+2H_2(g); \quad \Delta H = -22.8 \text{ kcal/mol} \quad (3)$$

$$N_2H_4(g)+H_2(g) \rightarrow 2NH_3(g); \quad \Delta H = -44.8 \text{ kcal/mol} \quad (4)$$

As reactions which produce ammonia must be thermodynamically favoured under conditions where ammonia is stable, the presence of hydrogen would be expected to increase the stability of hydrazine as in (4). For example, when gaseous hydrazine is decomposed thermally at 250–310°C on silica, the products correspond to equation (2) ($x = 0$), whereas decomposition on platinum wire gives $x = 0.25$ [302].

Considerable evidence has accumulated to indicate that the gaseous decomposition of hydrazine involves free radicals[307]. Several mechanisms have been proposed[302] which can be separated into initiation, propagation and termination steps involving atomic hydrogen, imide, amide and hydrazyl radicals as intermediates in reactions of the type described below:

$$\left.\begin{array}{r} N_2H_4 \rightarrow 2NH_2^{\cdot} \\ N_2H_4 \rightarrow N_2H_3^{\cdot}+H^{\cdot} \end{array}\right\} \text{initiation}$$

$$\left.\begin{array}{r} N_2H_3^{\cdot}+N_2H_4 \rightarrow 2NH_3+N_2+H^{\cdot} \\ H^{\cdot}+N_2H_4 \rightarrow NH_3+NH_2^{\cdot} \\ NH_2^{\cdot}+N_2H_4 \rightarrow NH_3+N_2H_3^{\cdot} \\ H^{\cdot}+N_2H_4 \rightarrow H_2+N_2H_3^{\cdot} \end{array}\right\} \text{propagation}$$

$$\left.\begin{array}{r} 2H^{\cdot} \rightarrow H_2 \\ 2NH_2^{\cdot} \rightarrow N_2+2H_2 \\ 2NH_2^{\cdot} \rightarrow NH_3+NH^{\cdot} \\ NH^{\cdot}+N_2H_4 \rightarrow NH_3+N_2+H_2 \\ NH_2^{\cdot}+N_2H_3^{\cdot} \rightarrow NH_3+N_2+H_2 \\ H^{\cdot}+N_2H_3^{\cdot} \rightarrow N_2+2H_2 \\ 2N_2H_3^{\cdot} \rightarrow 2NH_3+N_2 \end{array}\right\} \text{termination}$$

Such decompositions are often used as sources of these radicals for further study. The amino radical $NH_2^{\cdot}$ can also be produced in reactions of HN_3, NH_2OH and Me_2NNH_2, whereas hydrazine is the principal source of the hydrazyl radical $N_2H_3^{\cdot}$. Reactions of $N_2H_3^{\cdot}$ in the above scheme reflect the tendency of the N–N bonded radical to produce N–N by the observations that dimerization to tetrazane has only been mentioned as a possibility at -178°C, and even thermodynamically favoured disproportionation is not suggested in favour of nitrogen and ammonia formation[302].

$$2N_2H_3^{\cdot} \longrightarrow NH_2NHNHNH_2$$

$$2N_2H_3^{\cdot} \longrightarrow N_2H_2 + N_2H_4$$

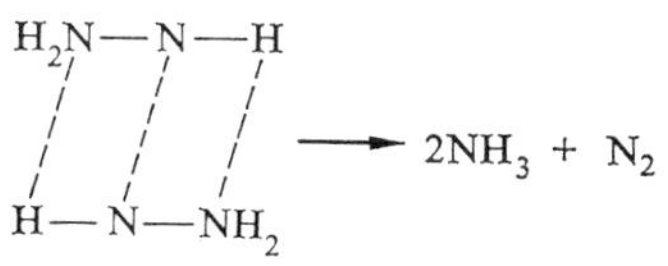

$$\begin{array}{c} H_2N—N—H \\ ⁄ \quad ⁄ \quad ⁄ \\ H—N—NH_2 \end{array} \longrightarrow 2NH_3 + N_2$$

6.8.2. Oxidation

Hydrazine–air mixture can be ignited above 130°C in a metal burner, whereas it is reported[302] that hydrazine–oxygen mixture with a platinum catalyst can ignite at only 30°C.

$$N_2H_4(l)+O_2(g) \rightarrow N_2(g)+2H_2O(l); \qquad \Delta H = -148.6 \text{ kcal/mol}$$

However, most oxidation reactions of hydrazine yield a mixture of ammonia and nitrogen, some give hydrazoic acid and a few nitrogen only. Such complex situations can only be rationalized for a number of idealized reactions in aqueous solution for which redox potentials can be compared as in equations (5)–(10) and in the oxidation state diagram illustrated in Fig. 19.

$$N_2H_5^+ \rightarrow NH_4^+ + \tfrac{1}{2}N_2 + H^+ + e^-; \quad E^\circ = -1.74\ V \tag{5}$$

$$N_2H_5^+ \rightarrow \tfrac{1}{2}NH_3 + \tfrac{1}{2}NH_4^+ + H^+ + 2e^-; \quad E^\circ = +0.11\ V \tag{6}$$

$$N_2H_5^+ \rightarrow N_2 + 5H^+ + 4e^-; \quad E^\circ = -0.23\ V \tag{7}$$

$$N_2H_4 + OH^- \rightarrow NH_3 + \tfrac{1}{2}N_2 + H_2O + e^-; \quad E^\circ = -1.16\ V \tag{8}$$

$$N_2H_4 + OH^- \rightarrow \tfrac{1}{2}N_3^- + \tfrac{1}{2}NH_3 + H_2O + 2e^-; \quad E^\circ = -0.92\ V \tag{9}$$

$$N_2H_4 + 4OH^- \rightarrow N_2 + 4H_2O + 4e^-; \quad E^\circ = -1.16\ V \tag{10}$$

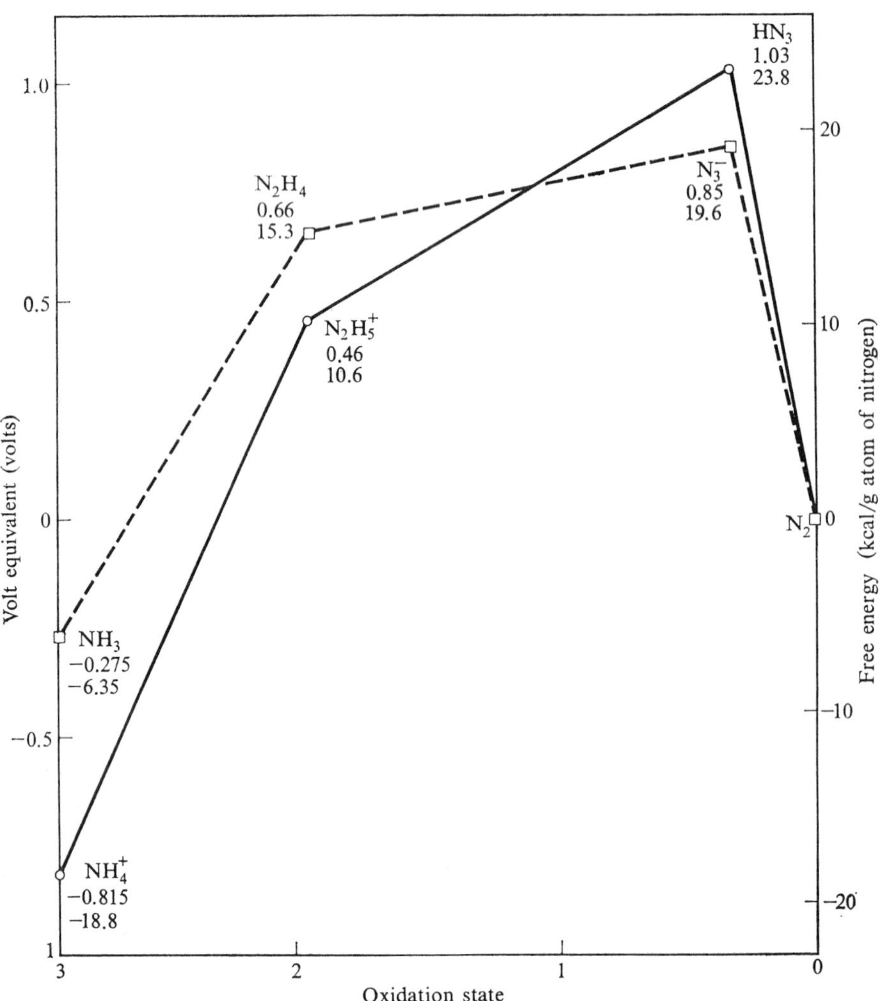

FIG. 19. Oxidation state diagram for hydrazine and related nitrogen species.

The shape of the graph in Fig. 19 suggests that disproportionation type reactions are possible. Several attempts have been made to classify oxidizing agents according to the equations (5)–(10), but invariably overall reactions are described by a combination of more than one equation. Experiments using $^{14}NH_2{}^{15}NH_2$ have been necessary to establish that $^{14}N^{15}N$ is formed by (7) and (10), but in (5) and (8) half of the nitrogen formed has a random distribution of the ^{15}N isotope[336]. Mechanisms consistent with this, and supported by other evidence, are generally free radical type involving the generation of diimide and hydrazyl radicals[302, 307]:

$$N_2H_4 \rightarrow N_2H_3^{\cdot} + H^{\cdot}$$

$$2N_2H_3^{\cdot} \rightarrow \underset{\text{Tetrazane}}{N_4H_6} \rightarrow N_2 + 2NH_3$$

$$N_2H_4 \rightarrow \underset{\text{Diimide}}{HN{=}NH} \rightarrow \underset{\text{Imide}}{2NH^{\cdot}}$$

$$2N_2H_2 \rightarrow \underset{\text{Tetrazene}}{N_4H_4} \rightarrow N_2 + N_2H_4$$

$$N_2H_2 + NH^{\cdot} \rightarrow \underset{\text{Triazene}}{H_3N_3} \rightarrow N_2 + NH_3$$

$$2N_2H_2 \rightarrow \underset{\text{Isotetrazene}}{NH{=}NNHNH_2} \rightarrow HN_3 + NH_3$$

6.8.3. Reactions of Anhydrous Hydrazine

The high dielectric constant of anhydrous liquid hydrazine suggests that it should be a reasonable solvent for many ionic compounds. These include alkali-metal halides, tetra-alkylammonium halides and many carboxylic acids which are completely dissociated in it, but a number of oxides, sulphates and carbonates are practically insoluble[337]. Complexes and solvates have been formed by dissolving metal halides, etc., in liquid hydrazine to give $M^{II}X_2 \cdot 2N_2H_4$ as well as hydrazinium double salts $(N_2H_5)_2M^{II}(SO_4)_2$ [326]. However, as a solvent for acid–base reactions, there is no advantage over the more readily available and safer liquid ammonia as their proton affinities are so similar.

The self-ionization reaction of pure liquid hydrazine is

$$2N_2H_4 \rightleftharpoons N_2H_5^+ + N_2H_3^-$$

Thus although many compounds are known to behave as strong acids, a few substances including the alkali metal hydrazides behave as bases. Sodium hydrazide NaN_2H_3 explodes violently in the presence of O_2 or when heated above 100°C. It was formerly prepared by dissolving sodium in liquid hydrazine in the absence of air, but even that method was unpredictably hazardous. Sodium hydrazide is now made from sodium amide or sodium hydride and used as a suspension in an ether solvent:

$$NaR + N_2H_4 \rightarrow NaN_2H_3 + RH \qquad (R = H, NH_2)$$

The reactions of sodium hydrazide with a number of organic compounds has been investigated[337], but apart from the neutralization reaction, the inorganic chemistry of metal hydrazides has not received much attention[336]:

$$NaN_2H_3 + N_2H_5Cl \rightarrow 2N_2H_4 + NaCl$$

---

[336] W. L. Jolly, *The Inorganic Chemistry of Nitrogen*, Benjamin, New York (1964).

[337] V. Gutmann, *Coordination Chemistry in Non-Aqueous Solutions*, Springer-Verlag, New York (1968), pp. 48–49.

More recently, however, organometallic derivatives of hydrazine have been prepared by methods appropriate to the formation of metal amides[339].

$$4Me_3SnNMe_2 + N_2H_4 \rightarrow (Me_3Sn)_2NN(SnMe_3)_2 + 4Me_2NH$$

There seems no reason why such derivatives should not be formed by elements other than boron[340], silicon[341] and tin[339].

6.9. BIOLOGICAL PROPERTIES[302]

In the nitrogen fixation process, no intermediates between nitrogen and ammonia have been isolated, and although a reductive pathway via diimine and hydrazine would seem to be preferred, there is little supporting evidence[342]. However, if either of these species are formed during fixation, it is likely that they remain closely associated with the enzyme until reduction to ammonia is complete.

Hydrazine is a dangerous though non-cumulative poison as it can inhibit certain enzyme systems causing reduced body metabolism. Skin contact with anhydrous hydrazine leads to caustic-like burns, and dissolves hair, while the vapour produces severe irritation of eyes and mucous membrane. However, some hydrazine salts and derivatives are used as chemotherapeutic agents[301].

6.10. ANALYSIS OF HYDRAZINE[301, 302]

Since hydrazine is a powerful reducing agent, tests with a wide range of oxidizing agents are only specific if other reducing agents like hydroxylamine are absent. The reaction of hydrazine with organic aldehydes to form coloured organic azines is often used as a spot test and is specific even in the presence of hydroxylamine and other nitrogen compounds. For example, the reaction with *p*-dimethylaminobenzaldehyde gives the corresponding azine which rearranges in the presence of strong acids to a quinoid structure which forms the basis of a colorimetric method of analysis:

$$2Me_2N{-}C_6H_4{-}CHO + N_2H_4 \longrightarrow Me_2N{-}C_6H_4{-}CH{=}N{-}N{=}CH{-}C_6H_4{-}NMe_2$$

$$\downarrow HCl$$

$$Me_2N{-}C_6H_4{-}CH{=}N{-}NH{-}CH{=}C_6H_4{=}N^{+}Me_2$$

As hydrazine is a weak base ($K_B = 8.5 \times 10^{-7}$ at 25°C), it can be determined together with its major impurity ammonia by titration with acid. Hydrazine alone can then be titrated in 3–5 M HCl with standard potassium iodate solution with chloroform to help

[338] Th. Kauffmann, *Angew. Chem. Int. Edn. Engl.* **3** (1964) 342.
[339] R. F. Dalton and K. Jones, unpublished observations.
[340] G. E. Coates, M. L. H. Green and K. Wade, *Organometallic Compounds*, 3rd edn., Methuen, London (1967).
[341] R. Fessenden and J. S. Fessenden, *Chem. Rev.* **61** (1961) 361.
[342] R. Murray and D. C. Smith, *Coord. Chem. Rev.* **3** (1968) 429.

detect the end point at which iodine is discharged. Although the overall reaction is described by the equation

$$N_2H_4+KIO_3+2HCl \rightarrow KCl+ICl+N_2+3H_2O$$

the process involves reduction of the iodate by hydrazine to iodine which is subsequently oxidized by additional iodate to iodine monochloride. This is known as the direct iodate method, the indirect method being based on the fact that both hydrazine and iodide ions reduce iodate to iodine which can be titrated with thiosulphate. Potassium bromate can also be used in place of iodate.

Volatile impurities in hydrazine are generally detected and determined by physical techniques, particularly by infrared spectroscopy on liquid or gaseous samples.

7. HYDROXYLAMINE

7.1. PRODUCTION AND INDUSTRIAL USE

In the conventional Raschig synthesis of hydroxylamine[343–6], nitrite is reduced with bisulphite and sulphur dioxide at 0°C to give hydroxylamido-*N*,*N*-disulphate which can be hydrolysed stepwise to hydroxylammonium sulphate. The nitrite solution is prepared by burning ammonia–air mixture over a platinum–rhodium catalyst grid to give NO and NO_2, and the gases absorbed in ammonium carbonate solution:

$$NO+NO_2+2(NH_4)_2CO_3+H_2O \rightarrow 2NH_4NO_2+2NH_4HCO_3$$
$$NH_4NO_2+2SO_2+NH_3+H_2O \rightarrow HON(SO_3NH_4)_2$$
$$HON(SO_3NH_4)_2+H_2O \rightarrow HONHSO_3NH_4+NH_4HSO_4$$
$$2HONHSO_3NH_4+2H_2O \rightarrow (HONH_3)_2SO_4+(NH_4)_2SO_4$$

Thus using ammonium salts, the process also yields a large amount of ammonium sulphate as saleable byproduct. In an effort to find a process giving much less or no byproducts, the catalytic reduction of nitric oxide has been suggested[343, 347]. For this process, nitric oxide is manufactured by passing ammonia–oxygen mixture and steam as diluent over platinum–rhodium catalyst at about 600°C:

$$4NH_3+5O_2 \rightarrow 4NO+6H_2O$$

Surprisingly, conditions can be controlled to give nitric oxide of greater than 99% purity in spite of the variety of other reactions which are thermodynamically favourable (see section 11). The reduction of nitric oxide is again platinum catalysed at low pH with $NO:H_2$ ratios greater than theoretical 1:1.5:

$$2NO+3H_2+H_2SO_4 \rightarrow (HONH_3)_2SO_4$$

An alternative reduction process is the electrolytic reduction of nitric acid usually in mixed sulphuric–hydrochloric acid using amalgamated lead electrodes[343, 345, 346]. The

343 K. G. Mason, in ref. 300, pp. 115–57.

344 C. Matasa and E. Matasa, *L'Industrie Moderne des Produits Azotes*, Dunod, Paris (1968), p. 580.

345 P. J. Baker, Jr., in *Kirk–Othmer Encyclopedia of Chemical Technology*, 2nd edn., Vol. II, Interscience, New York (1966), pp. 493–508.

346 D. M. Yost and H. Russell, Jr., *Systematic Inorganic Chemistry*, Prentice-Hall, New York (1944), pp. 90–98.

347 *Nitrogen*, No. 50 (1967) 27.

product is recovered by passing dry hydrogen chloride through the electrolysed solution and filtering off the precipitated hydroxylammonium chloride:

$$HNO_3 + 6H^+ + 6e^- \rightarrow HONH_2 + 2H_2O \xrightarrow{HCl} (HONH_3)Cl$$

The hydrolysis of aliphatic nitro compounds during reflux with strong mineral acids offers economic routes to hydroxylamine derivatives and carboxylic acids as useful by-products[343]:

$$2RCH_2NO_2 + H_2SO_4 + 2H_2O \rightarrow 2RCOOH + (HONH_3)_2SO_4$$

Such starting materials are readily obtained by vapour phase nitration of aliphatic hydrocarbons, e.g. 1,2-dinitroethane can be prepared from ethylene and dinitrogen tetroxide, and subsequent hydrolysis gives hydroxylammonium sulphate as the sole product in acid solution[343]:

$$CH_2{=}CH_2 + N_2O_4 \rightarrow O_2NCH_2CH_2NO_2 \xrightarrow[H_2O]{H_2SO_4} CO_2 + CO + (HONH_3)_2SO_4$$

Aqueous solutions of hydroxylamine are obtained from hydroxylammonium salts by ion exchange, while free hydroxylamine can be prepared by ammonolysis of hydroxylammonium sulphate with liquid ammonia. Insoluble ammonium sulphate can be filtered off and the solvent evaporated under reduced pressure to leave unstable solid hydroxylamine[343].

As part of its extensive organic chemistry[348], a major use of hydroxylamine or its salts is the preparation of cyclohexanone oxime in the manufacture of caprolactam which is an intermediate in the production of polyamides for synthetic fibres:

$$\text{C}_6\text{H}_5\text{OH} \xrightarrow[H_2]{Ni} \text{cyclohexanol} \xrightarrow{Zn} \text{cyclohexanone} \xrightarrow{HONH_2} \text{cyclohexanone oxime (NOH)} \xrightarrow{H_2SO_4} \text{caprolactam}\cdot H_2SO_4 \ (\text{NH·HSO}_4) \xrightarrow{2NH_3} \text{caprolactam (NH)}$$

Hydroxylamine and its salts in their rôle as reducing agents act as anti-oxidants in photographic developers, stabilizers of monomers and for reducing copper(II) to copper(I) in the dyeing of acrylic fibres. Hydroxylamine is an effective adsorbent for removing nitrous gases, but nitric oxide must first be oxidized to nitrogen dioxide. Hydroxylammonium sulphate on a silica-gel carrier is capable of performing both operations, and for this reason is used as an absorbent in combustion analysis[343].

7.2. LABORATORY PREPARATION OF HYDROXYLAMINE AND RELATED COMPOUNDS[349]

Hydroxylammonium phosphate is prepared in high yield and purity from the chloride and trisodium orthophosphate:

$$3(HONH_3)Cl + H_3PO_4 + 3NaOH \rightarrow (HONH_3)_3PO_4 + 3NaCl + 3H_2O$$

---

[348] N. V. Sidgwick, *The Organic Chemistry of Nitrogen*, 3rd edn., revised and rewritten by I. T. Millar and H. D. Springall, Oxford University Press, Oxford (1966).

[349] P. W. Schenk, in *Handbook of Preparative Inorganic Chemistry*, 2nd edn., Vol. I (ed. G. Brauer), Academic Press, New York (1963), pp. 500–11.

Free hydroxylamine can then be conveniently prepared by the thermal decomposition of tertiary hydroxylammonium phosphate under reduced pressure:

$$(HONH_3)_3PO_4 \rightarrow H_3PO_4 + 3NH_2OH$$

While the bulk of material distils over at 135–137°C/13 mm, the temperature and pressure must be monitored continually and not allowed to rise above 150°C or 30 mm for fear of explosion. The product which still contains water can be recrystallized from absolute alcohol at −18°C, rapidly suction-filtered and dried briefly in a vacuum desiccator over sulphuric acid. Hydroxylamine can also be prepared from hydroxylammonium chloride after reaction with sodium alkoxide in alcohol and separating free hydroxylamine by crystallization from the alcohol solution after filtering off the sodium chloride:

$$(HONH_3)Cl + NaOR \xrightarrow{ROH} NH_2OH + NaCl + ROH$$

Potassium hydroxylamine disulphonate can be prepared by the reaction of potassium nitrite and potassium metabisulphite:

$$KNO_2 + 2KHSO_3 \rightarrow HON(SO_3K)_2 + KOH$$

Hydrolysis of the product gives potassium sulphate and hydroxylammonium bisulphate which can be separated and recrystallized[349]:

$$HON(SO_3K)_2 + 2H_2O \rightarrow (HONH_3)HSO_4 + K_2SO_4$$

7.3. NUCLEAR ISOTOPES ND_2OD [350]

Attempts to prepare deuterohydroxylamine by successive exchanges of hydroxylamine with deuterium oxide were unsuccessful. However, deuterium oxide exchanges with tertiary hydroxylammonium phosphate to give the fully deuterated compound which can be thermally decomposed at 130°C/13 mm to give hydroxylamine-d_3.

$$2D_2O + (HONH_3)_3PO_4 \rightarrow 2H_2O + (DOND_3)_3PO_4 \xrightarrow{\Delta} D_3PO_4 + 3ND_2OD$$

7.4. STRUCTURE AND BONDING

Hydroxylamine can exist as two configurational isomers, usually termed *cis* and *trans*, together with numerous *gauche* conformations resulting from varying degrees of rotation of the OH group around the N–O bond, as illustrated in Fig. 20. An X-ray study of crystalline hydroxylamine detailed in Table 43 gave the N–O bond distance as 1.47 Å and other intermolecular hydrogen bonded N · · · O distances at 2.74, 3.07, 3.11 and 3.18 Å. Of these, the 2.74 Å bond corresponds to O–H · · · N and the 3.07 Å bond to O · · · H–N by analogy to the situation in oximes, while the third hydrogen atom of hydroxylamine remains uninvolved in hydrogen bonding which is taken as evidence consistent with the molecule having the *trans* configuration in the solid state[351]. Infrared spectra have also been fully assigned as in Table 44 based on hydroxylamine having C_s symmetry, and the presence of double bands indicates that the vapour phase probably contains a mixture of *cis* and *trans* isomers although the possibility of the *gauche* configuration being present cannot be excluded[352]. Using the molecular parameters computed from this spectroscopic data, N–O 1.46 Å, O–H 0.96 Å, N–H 1.01 Å, and angles HON 103°, HNO 105°, HNH 107°,

[350] R. E. Nightingale and E. L. Wagner, *J. Am. Chem. Soc.* **75** (1953) 4092.
[351] E. A. Meyers and W. N. Lipscomb, *Acta Cryst.* **8** (1955) 583.
[352] P. A. Gignère and I. D. Liu, *Can. J. Chem.* **30** (1952) 948.

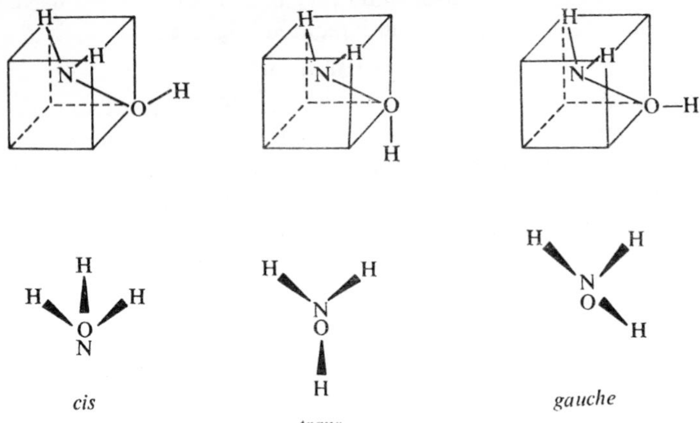

FIG. 20. Configurations of hydroxylamine.

TABLE 43. CRYSTAL STRUCTURES OF HYDROXYLAMINE AND ITS SALTS[a]

Compound	Crystal system	Lattice parameters (Å)	Interatomic distances (Å)
$HONH_2$	Orthorhombic	$a = 7.29$, $b = 4.39$, $c = 4.88$	N–O = 1.47 N···H–O = 2.74 N–H···O = 3.07 Other N···O = 3.11, 3.18
$(HONH_3)Cl$	Monoclinic	$a = 6.95$, $b = 5.95$, $c = 7.70$, $\beta = 120.8°$	N–O = 1.45 N–Cl = 3.16, 3.21, 3.23, 3.26
$(HONH_3)Br$	Monoclinic	$a = 7.29$, $b = 6.13$, $c = 8.04$, $\beta = 120.8°$	N–O = 1.45
$(HONH_3)ClO_4$	Orthorhombic[b]	$a = 7.52$, $b = 7.14$, $c = 15.99$	N–O = 1.41 Cl–O = 1.45

[a] In ref. 343.
[b] In ref. 356.

ab initio calculations of the barriers to internal rotation gave values of 1.16 and 11.95 kcal/mol corresponding to the *cis* and *trans* conformations[353].

The N–O bond length in hydroxylammonium halides as determined by X-ray diffraction[354] and spectroscopic techniques[355] is 1.45 Å, and 1.41 Å for hydroxylammonium perchlorate[356]. These values are in agreement with the N–O single covalent bond length

[353] W. H. Fink, D. C. Pan and L. C. Allen, *J. Chem. Phys.* **47** (1967) 895.
[354] B. Jerslev, *Acta Cryst.* **1** (1948) 21.
[355] D. L. Frasco and E. L. Wagner, *J. Chem. Phys.* **30** (1959) 1124.
[356] B. Dickens, *Acta Cryst.* B, **25** (1969) 1875.
[357] V. Schomaker and D. P. Stevenson, *J. Chem. Soc.* **63** (1941) 37.

TABLE 44. VIBRATIONAL FREQUENCIES AND BAND ASSIGNMENTS FOR HYDROXYLAMINE AND THE HYDROXYLAMMONIUM ION

Hydroxylamine[a]			Hydroxylammonium ion[b]	
Infrared (cm^{-1})	Assignments		Infrared (cm^{-1})	Assignment
3245, 3173	$\nu_1(A')$	ν(NH)		
			3080 (s)	OH stretch
			3025 (m)	NH stretch
2867	$\nu_2(A')$	ν(OH)		
			2795, 2760 (w)	NH stretch
			1876 (s)	
			1563 (s)	OH bend
1515	$\nu_3(A')$	δ(HNH)		
			1482 (s)	NH bend
1191	$\nu_4(A')$	δ(NOH)	1200, 1176 (s)	NH bend
912	$\nu_5(A')$	ν(NO)	1000 (s)	NO stretch
950	$\nu_6(A')$	$\rho_r(NH_2)$		
3302	$\nu_7(A'')$	ν(NH)		
867	$\nu_8(A'')$	$\rho_w(NH_2)$		
535	$\nu_9(A'')$	$\rho_t(NH_2)$		

[a] In ref. 352.
[b] In ref. 343.

calculated[357] by correcting the sum of the covalent radii of nitrogen and oxygen for differences in electro-negativities of the atoms, i.e.

$$(0.735+0.735)-[-0.08(\chi_N-\chi_O)] = 1.47-0.04 = 1.43\pm0.02 \text{ Å}.$$

However, the neutron diffraction study[358] of hydroxylammonium chloride gave an N–O bond length of 1.383 Å which is much shorter than the X-ray value of 1.47. The N–H, 1.044 Å, and O–H, 0.996 Å, bond lengths are unremarkable although hydrogen bonding to each chlorine is indicated, and the dihedral angle between the NOH plane and the ONH plane is 53.6°.

Only a few hydroxylamine complexes are known for which the bonding arrangements are illustrated in Fig. 21. Comparison of infrared spectra indicate a similarity between hydroxylamine complexes of cobalt(III) and platinum(II) with analogous amine complexes[359]. Thus it seems likely that the metal is bonded to nitrogen rather than oxygen.

[358] V. M. Padmanabhan, H. G. Smith and S. W. Peterson, *Acta Cryst.* **62** (1967) 928.
[359] Yu. Ya. Kharitonov and M. A. Sarukhanov, *Zh. Neorg. Khim.* **11** (1966) 2532.

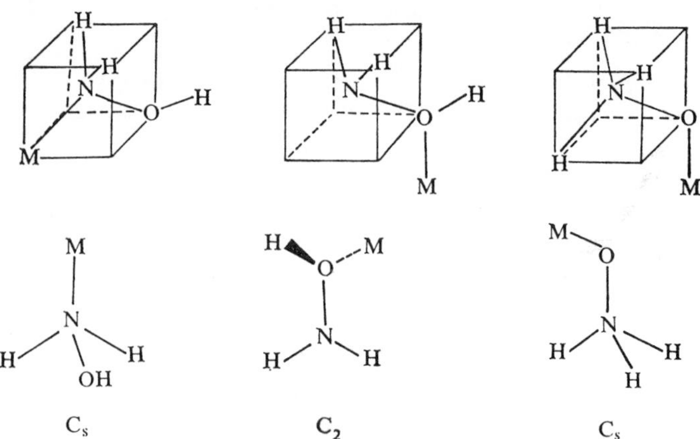

FIG. 21. Bonding arrangements in hydroxylamine complexes.

7.5. MOLECULAR CONSTANTS

The structural parameters of hydroxylamine and hydroxylammonium salts are given in Table 43. The infrared spectrum of hydroxylamine together with the principal bands assigned to the hydroxylammonium ion are listed in Table 44. Raman and infrared spectra have been recorded for hydroxylammonium halides and assigned[355] by considering the local symmetry of the molecule as $C_{3v}(H_3NO)$ and $C_s(OH)$.

7.6. PHYSICAL PROPERTIES

The values of some physical properties of hydroxylamine are listed in Table 45.

Such information is rather sparse on account of the fact that pure hydroxylamine is unstable, particularly in the liquid phase. Alkali metal halides and similar salts dissolve in liquid hydroxylamine which resembles water in its solvent properties. It is hygroscopic, forming reasonably stable solutions with water and with the lower alcohols, but has relatively low solubility in most other common organic solvents. Hydroxylamine can be regarded as water in which one of the hydrogens has been replaced by the more electronegative amide group, or as ammonia with a hydrogen replaced by the more electron-attracting hydroxyl group. Thus aqueous hydroxylamine solutions are less basic than either ammonia or hydrazine:

$$NH_2OH + H_2O \rightarrow HONH_3^+ + OH^-; \qquad K_b = 6.6 \times 10^{-9}$$

Aqueous solutions of hydroxylammonium salts are therefore less stable than corresponding ammonium salts, and undergo considerable hydrolysis in aqueous solutions to yield acidic solutions. The physical properties of a few hydroxylammonium salts are listed in Table 46.

Hydroxylamine, either anhydrous or in strong alkali, can undergo acidic ionization even

TABLE 45. PHYSICAL PROPERTIES OF HYDROXYLAMINE[a, b]

Property	Temperature (K)	Value
Melting point	305.20	32.05°C
Boiling point	339–340	56–57°C/22 mm
		142/760 mm (extrapolated)
Vapour pressure	305.15 (32°C)	5.3 mm
	320.35 (47.2°C)	10 mm
	337.75 (64.6°C)	40 mm
	350.65 (77.5°C)	100 mm
	372.35 (99.2°C)	400 mm
Density of liquid	306	1.204 g/cm³
ΔH_f° (heat of formation), cryst.	298	−27.3 kcal/mol
ΔH_f° (heat of formation), aq.	298	−23.5 kcal/mol
ΔG_f° (free energy of formation)	298	−5.60 kcal/mol
ΔH (heat of fusion)	305.2	3.94 kcal/mol
ΔH (heat of sublimation)		15.34 kcal/mol
ΔH (heat of solution)		−4.13 kcal/mol
ΔH (heat of hydrolysis)	293	1.96 kcal/mol
S° (entropy), aq.	298	40 cal/deg mol
S° (entropy), calc.	298	56.33 cal/deg mol
C_p° (heat capacity)	298	11.17 cal/deg mol
Molecular volume		27.4 cm³
Dissociation constant	293	1.07×10^{-8}
Dielectric constant		77.63–77.85
Proton affinity		211 kg cal

[a] Ref. 343.
[b] R. A. Back and J. Betts, *Canad. J. Chem.* **43** (1965) 2157.

though with a dissociation constant approaching 10^{-14} at an ionic strength of 0.72 it is a weaker acid than water itself:

$$2NH_2OH \rightarrow HONH_3^+ + NH_2O^-$$

$$NH_2OH + OH^- \rightarrow NH_2O^- + H_2O$$

Such properties are demonstrated by the formation of the unstable calcium salts $Ca(ONH_2)_2$ and $Ca(OH)(ONH_2)$ [345].

TABLE 46. PHYSICAL PROPERTIES OF HYDROXYLAMMONIUM SALTS[a]

Compound	M.p. (°C)	B.p. (°C)	Density d_4^{20}	Solubility in water (g per 100 g H_2O/°C)
$(HONH_3)Cl$	152 d		1.680	94.4/25
$(HONH_3)Br$			2.3514	
$(HONH_3)HSO_4$	57			100/25 [disproportionates to $(HONH_3)_2SO_4 + H_2SO_4$]
$(HONH_3)_2SO_4$	170 d			63.7/25
$(HONH_3)ClO_4$	88–89	120 d		

[a] In Ref. 345.

7.7. CHEMISTRY AND CHEMICAL REACTIONS

Anhydrous hydroxylamine is usually kept at 0°C as decomposition by internal oxidation-reduction to ammonia and nitrogen or nitrous oxide occurs even at room temperature. Hydroxylammonium salts derived from strong acids are more stable than the free base; nevertheless they do decompose near their melting point or above 200°C, often violently even when unconfined:

$$4HONH_3^+ \rightarrow N_2O + 2NH_4^+ + 3H_2O + 2H^+$$

Hydroxylammonium phosphate at lower temperatures splits off hydroxylamine which distils and can be separated.

The reaction of nitrous acid with the hydroxylammonium ion is thought to give nitroxyl which can decompose to nitrous oxide or dimerize to give hyponitrous acid[360]:

$$HONH_3^+ + HNO_2 \rightarrow 2HNO + H_3O^+$$

$$2HNO \rightarrow H_2O + N_2O$$

$$2HNO \rightarrow H_2N_2O_2$$

However, the main decomposition reaction in alkaline solution is

$$3NH_2OH \rightarrow NH_3 + N_2 + 3H_2O$$

while the reaction in acid solution gives mainly nitrous oxide:

$$4NH_2OH \rightarrow N_2O + 2NH_3 + 3H_2O$$

Thus a large number of reactions of hydroxylamine involve oxidation-reduction processes. Such properties seem appropriate when hydroxylamine is considered as the intermediate compound in relation to hydrogen peroxide and hydrazine. This comparison is even more clearly expressed in terms of the redox potentials associated with the following equations[361] and by the redox diagram illustrated in Fig. 22.

In acid solution:

$$2H_2O \rightarrow H_2O_2 + 2H^+ + 2e^-; \qquad E^\circ = 1.77\text{ V}$$

$$H_2O_2 \rightarrow O_2 + 2H^+ + 2e^-; \qquad E^\circ = 0.68\text{ V}$$

$$N_2H_5^+ + 2H_2O \rightarrow 2HONH_3^+ + H^+ + 2e^-; \qquad E^\circ = 1.42\text{ V}$$

$$NH_4^+ + H_2O \rightarrow HONH_3^+ + 2H^+ + 2e^-; \qquad E^\circ = 1.35\text{ V}$$

$$HONH_3^+ \rightarrow \tfrac{1}{2}N_2 + 2H^+ + H_2O + e^-; \qquad E^\circ = -1.87\text{ V}$$

$$2HONH_3^+ \rightarrow N_2O + 6H^+ + H_2O + 4e^-; \qquad E^\circ = -0.05\text{ V}$$

$$2HONH_3^+ \rightarrow H_2N_2O_2 + 6H^+ + 4e^-; \qquad E^\circ = -0.44\text{ V}$$

360 R. Nast and I. Föppl, *Z. anorg. Chem.* **263** (1950) 310.
361 W. E. Latimer, *Oxidation Potentials*, Prentice-Hall, Englewood Cliffs, New Jersey (1952).

In basic solution:

$$3OH^- \rightarrow HO_2^- + H_2O + 2e^-; \quad E^\circ = 0.87\ V$$
$$N_2H_4 + 2OH^- \rightarrow 2NH_2OH + 2e^-; \quad E^\circ = 0.74\ V$$
$$N_2H_4 + 4OH^- \rightarrow N_2 + 4H_2O + 4e^-; \quad E^\circ = -1.16\ V$$
$$2NH_2OH + 2OH^- \rightarrow N_2 + 4H_2O + 2e^-; \quad E^\circ = -3.04\ V$$
$$2NH_2OH + 4OH^- \rightarrow N_2O + 5H_2O + 4e^-; \quad E^\circ = -1.05\ V$$
$$2NH_2OH + 6OH^- \rightarrow N_2O_2^{2-} + 6H_2O + 4e^-; \quad E^\circ = -0.73\ V$$

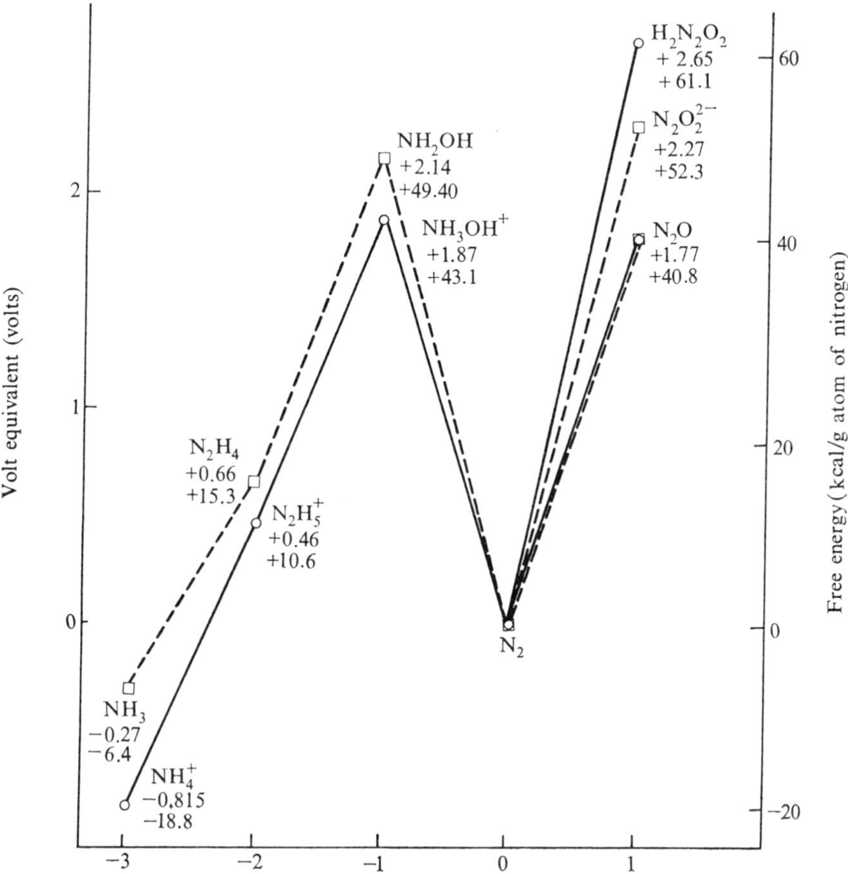

FIG. 22. Redox diagram for hydroxylamine and related species.

Thus, in general, hydrogen peroxide, hydroxylamine and hydrazine act as reducing agents in acidic solutions and as oxidizing agents in neutral or alkaline media. However, by varying the conditions even within this classification, different products predominate. For example, the reaction of silver bromide and hydroxylamine gives nitrogen and nitrous oxide, whereas mercurous nitrate gives mainly nitrous oxide:

$$2NH_2OH + 2AgBr \rightarrow 2Ag + N_2 + 2HBr + 2H_2O$$
$$2NH_2OH + 4AgBr \rightarrow 4Ag + N_2O + 4HBr + H_2O$$
$$2NH_2OH + 2Hg_2(NO_3)_2 \rightarrow 4Hg + N_2O + 4HNO_3 + H_2O$$

The investigation of the reduction of hydroxylamine with aqueous chromium(II) gave 1:2 stoichiometry and first-order reaction with the rate-determining step leading to the formation of $HONH_3^+$ [362].

$$2H^+ + 2Cr(H_2O)_6^{2+} + HONH_3^+ \rightarrow NH_4^+ + 2Cr(H_2O)_6^{3+} + H_2O$$

The participation of free-radicals ($NH_2^{\cdot}$) in such reactions is indicated by polymerization occurring when titanium(III) reacts with hydroxylamine in the presence of vinyl monomer[343].

A number of hydroxylamine complexes of transition metals have been prepared, particularly among the platinum metals, but the area is lacking in structural studies. The non-transition element derivatives of hydroxylamine are also mainly restricted to one area—that of the sulphuric acid derivatives[343, 363]. These include both N- and O-derivatives commonly described as derivatives of sulphuric (IUPAC) or sulphonic acids:

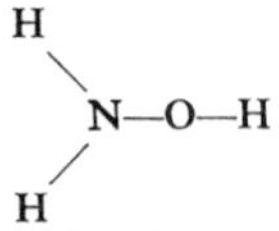

$$(H)(H)N—O—H$$

Hydroxylamine

$$(HSO_3)(H)N—O—H$$

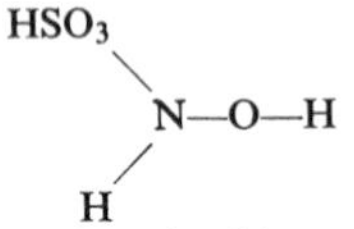

Hydroxylamido-*N*-sulphuric acid (I)

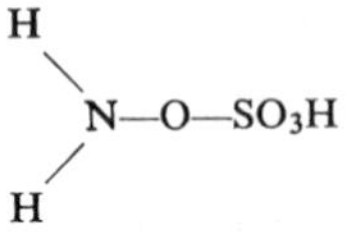

$$(H)(H)N—O—SO_3H$$

Hydroxylamido-*O*-sulphuric acid (II)

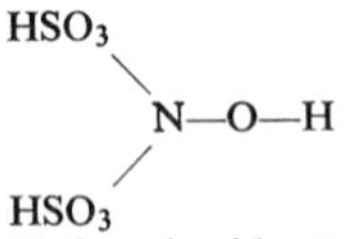

$$(HSO_3)(HSO_3)N—O—H$$

Hydroxylamido-*N,N*-disulphuric acid (III)

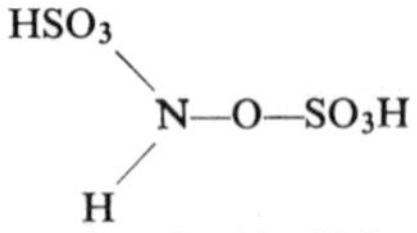

$$(HSO_3)(H)N—O—SO_3H$$

Hydroxylamido-*N,O*-disulphuric acid (IV)

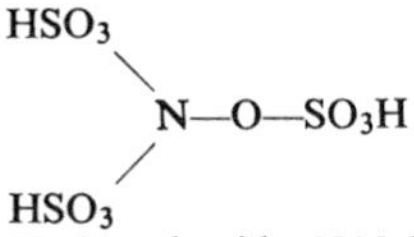

$$(HSO_3)(HSO_3)N—O—SO_3H$$

Hydroxylamido-*N,N,O*-trisulphuric acid (V)

A number of nitrosyl and nitroso derivatives can also be thought of as being derived from hydroxylamine. The compounds (I)–(V) above can be prepared conveniently by methods[363] outlined in equations below:

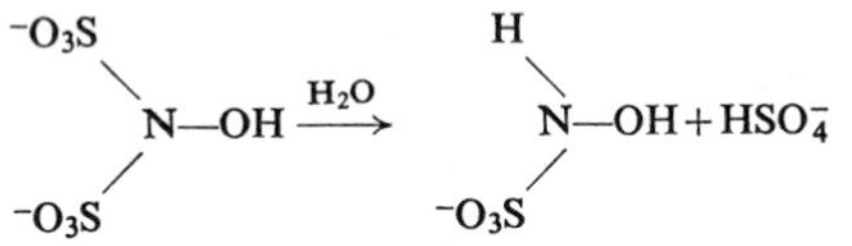

$$({}^-O_3S)({}^-O_3S)N—OH \xrightarrow{H_2O} (H)({}^-O_3S)N—OH + HSO_4^- \quad \text{(I)}$$

[362] W. Schmidt, J. H. Swinehart and H. Taube, *Inorg. Chem.* **7** (1968) 1984.

[363] K. W. C. Burton and G. Nickless, in *Inorganic Sulphur Chemistry* (ed. G. Nickless), Elsevier, Amsterdam (1968), pp. 607–67.

$$(HONH_3)HSO_4 + ClSO_3H \rightarrow HCl + H_2SO_4 + H_2NOSO_3H \rightleftharpoons H_3\overset{+}{N}OS\overset{-}{O_3} \qquad (II)$$

$$NO_2^- + SO_3H^- + SO_2 \rightarrow HON(SO_3)_2^{2-} \qquad (III)$$

$$({}^-O_3S)_2N{-}OSO_3^- \xrightarrow{H_2O} ({}^-O_3S)(H)N{-}OSO_3^- + HSO_4^- \qquad (IV)$$

$$3\,(KO_3S)_2N{-}OH + 2PbO_2 + KOH \rightarrow 2\,(KO_3S)_2N{-}OSO_3K + KNO_2 + 2Pb(OH)_2 \qquad (V)$$

A considerable organic chemistry of hydroxylamine centres around oxime formation, the Beckmann, rearrangement, and its use in organic synthesis as a reducing agent[334, 348]. Organometallic hydroxylamine derivatives have not received much attention, but tri-organosilyl chlorides react with hydroxylamine to give the silylamine oxide as evidenced by the characterization of the product from subsequent reaction with phenyl isocyanate[364]:

$$Me_3SiCl + NH_2OH \rightarrow [Me_3Si\overset{+}{N}{-}\overset{-}{O}H_2] \xrightarrow{2PhNCO} Me_3SiN(CONHPh)(CO_2NHPh)$$

7.8. BIOLOGICAL PROPERTIES

Hydroxylamine and its salts are moderately toxic substances which may cause both reversible and irreversible changes in the body associated with methaemoglobinemia[345]. Free hydroxylamine is known to be present in stagnant water due to bacteriological reduction of nitrite. For a long time its rôle in the nitrogen fixation process was considered to be important, but it is now thought to be the key intermediate[343].

7.9. ANALYSIS OF HYDROXYLAMINE[343, 365]

A very sensitive spot test for hydroxylamine is provided by the green colour produced by its reaction with 8-hydroxyquinoline, and it only fails in the presence of large amounts of hydrazine. Paper chromatography has also been used to separate various nitrogen containing compounds including hydroxylamine.

The fact that hydroxylamine is reduced by alkaline iron(II) sulphate, whereas hydrazine is not affected has been developed into a method for its determination by estimating the amount of ammonia evolved. Hydroxylamine hydrochloride can be determined readily by oxidation with iron(III) ammonium sulphate and the iron(II) produced titrated with

[364] U. Wannagat and J. Pump, *Monatsh. Chem.* **94** (1963) 141.

[365] A. J. Clear and M. Roth, in *Treatise on Analytical Chemistry*, Part II, Vol. 5 (eds. I. M. Kolthoff and P. J. Elving), Interscience, New York (1961), pp. 288–90.

permanganate, or by oxidation with excess potassium bromate in the presence of hydrochloric acid, and after the addition of potassium iodide, the iodine liberated by excess bromate titrated with thiosulphate. Appropriate reducing agents are titanium(III) chloride in excess followed by back-titration with iron(III) using thiocyanate as indicator.

8. HYDROGEN AZIDE AND AZIDES

8.1. PRODUCTION AND INDUSTRIAL USE

Sodium azide is a reasonably stable material which is produced on a commercial scale, and is the principal starting material for the preparation of all other azides[366–9]. It is obtained by passing nitrous oxide into fused sodamide. The melt acts as a vigorous reducing agent while the water formed is removed by immediate reaction with the excess sodium amide giving recoverable ammonia:

$$NaNH_2+N_2O \rightarrow NaN_3+H_2O$$
$$NaNH_2+H_2O \rightarrow NaOH+NH_3$$

Nitrous oxide, ammonia and sodium in liquid ammonia can react directly under pressure without prior isolation of the sodamide, and although the reaction is faster, the overall yield is lower due to the formation of elemental nitrogen[367]:

$$4Na+3N_2O+NH_3 \rightarrow NaN_3+3NaOH+2N_2$$

An alternative preparative route is the addition of powdered sodium nitrate to sodium amide at 175°C to give sodium azide in yields up to 65%:

$$3NaNH_2+NaNO_3 \rightarrow NaN_3+3NaOH+NH_3$$

As sodamide is difficult to produce industrially, a more convenient starting material is sodium oxide[370].

$$2Na_2O+N_2O+NH_3 \xrightarrow{110\text{–}190^\circ} NaN_3+3NaOH$$

In each process the products can be dissolved in water and sodium azide recrystallized from solution. Other methods of forming azides are from reactions of hydrazine with nitrous acid, nitrogen trichloride, or in low yields from a variety of oxidizing agents[369]:

$$N_2H_5^+ +HNO_2 \rightarrow HN_3+H^+ +2H_2O$$
$$N_2H_5OH+NCl_3+4OH^- \rightarrow N_3^- +3Cl^- +5H_2O$$

The major use of inorganic azide depends on the explosive nature of heavy metal azides, particularly lead azide. Lead azide can be precipitated from a mixture of aqueous azide and soluble lead salt solutions.

$$Pb(NO_3)_2+2NaN_3 \rightarrow Pn(N_3)_2+2NaNO_3$$

---

366 K. G. Mason, in ref. 300, pp. 1–15.

367 K. G. Mason, in ref. 300, pp. 16–68.

368 D. M. Yost and H. Russell, Jr., *Systematic Inorganic Chemistry*, Prentice-Hall, New York (1944), pp. 122–31.

369 H. M. Sisler, in *Comprehensive Inorganic Chemistry*, Vol. 5 (eds. M. C. Sneed and R. C. Brasted), Van Nostrand, New York (1956), pp. 41–52.

370 C. Matasa and E. Matasa, *L'Industrie Moderne des Produits Azotes*, Dunod, Paris (1968), p. 557.
Van Nostrand, New York (1956), pp. 41–52.

However, although precipitation methods appear to be attractive preparative routes from a separation point of view, the products are usually contaminated with trapped ions present in the solution. These trapped ions are thought to be a major contributing factor responsible for spontaneous detonation. This kind of contamination can clearly occur with all double decomposition reactions, while the neutralization of lead oxide with hydrazoic acid suffers the disadvantage of forming insoluble lead azide only on the surface of the lead oxide particles, in addition to the hazards of handling strong hydrazoic acid solutions. To overcome both of these problems the reaction of pure aqueous lead nitrite solution with an alcoholic solution of hydrogen azide has been proposed. The azide destroys the nitrite anion to give gaseous products while simultaneously forming insoluble lead azide as the only species remaining suspended in the solvent:

$$Pb(NO_2)_2 + 4HN_3 \rightarrow Pb(N_3)_2 + 2N_2 + 2N_2O + 2H_2O$$

Minor applications of azides include their incorporation of barium and strontium azides as getters in electric discharge tubes, alkali-metal azides in anti-corrosion solutions and the thermal decomposition of electro-positive metal azides in the production of foam rubber[367].

The wide use of lead azide for detonators in preference to other explosive materials such as mercury fulminate is based on reliability under a variety of adverse, in particular damp conditions rather than superior detonation characteristics.

8.2. LABORATORY PREPARATION OF HYDROGEN AZIDE AND AZIDES

Pure hydrogen azide is an extremely explosive material, but it can be formed by the action of molten stearic acid on sodium azide[366, 372], although its preparation should only be attempted in conjunction with extreme safety precautions. Because of the high incidence of explosions with pure hydrogen azide, it is expedient to work only with dilute solutions where possible. Aqueous solutions can be prepared in comparative safety by acidifying alkaline sodium azide solution with 40% sulphuric acid during the distillation of water to give a solution containing about 3% HN_3. Anhydrous ether solutions of HN_3 can also be prepared by dropwise addition of concentrated sulphuric acid to a mixture of ether and aqueous sodium azide during which time the bulk of the ether and HN_3 distils off. The ether solution can be dried over calcium chloride and redistilled if necessary.

Sodium azide is prepared in the laboratory by reacting dried nitrous oxide with sodium amide at 170–190°C:

$$2NaNH_2 + N_2O \rightarrow NaN_3 + NaOH + NH_3$$

Alternatively, hydrazine hydrate in ether reacts with nitrous acid formed by the breakdown of ethyl nitrite in the presence of sodium methoxide:

$$N_2H_4 \cdot H_2O + EtONO + NaOMe \rightarrow NaN_3 + EtOH + MeOH + 2H_2O$$

Ammonium azide can be prepared[367] from ammonia and nitrous oxide over $Ni\text{–}Al_2O_3$:

$$2NH_3 + N_2O \xrightarrow{Ni/Al_2O_3} NH_4N_3 + H_2O$$

---

371 P. Gray, *Quart. Rev.* **17** (1963) 441.

372 P. W. Schenk, in *Handbook of Preparative Inorganic Chemistry*, Vol. I (ed. G. Brauer), Academic Press, New York (1963), p. 472.

Potassium, rubidium or caesium azides can be prepared from metal carbonates[372] or hydroxides[373] and hydrogen azide produced *in situ* by acidifying sodium azide:

$$M_2CO_3+2HN_3 \rightarrow 2MN_3+H_2O+CO_2$$

Metal azides[367] and complex azides are also generally prepared from hydrogen azide or its sodium salt by various exchange reactions as illustrated below:

$$Ba(OH)_2+2HN_3 \longrightarrow Ba(N_3)_2+2H_2O$$
$$AlH_3+3HN_3 \longrightarrow Al(N_3)_3+3H_2$$
$$Me_2Be+2HN_3 \longrightarrow Be(N_3)_2+2MeH$$
$$(C_2H_5)_2Mg+2HN_3 \longrightarrow Mg(N_3)_2+2C_2H_6$$
$$LiBH_4+4HN_3 \longrightarrow LiB(N_3)_4+4H_2$$
$$B(N_3)_3+NaN_3 \longrightarrow NaB(N_3)_4$$
$$SnCl_4+6NaN_3 \longrightarrow Na_2Sn(N_3)_6+4NaCl$$
$$2Ph_4AsCl+CoSO_4+4NaN_3 \longrightarrow (Ph_4As)_2Co(N_3)_4+2NaCl+Na_2SO_4$$
$$3Ph_4AsCl+CrCl_3+6NaN_3 \longrightarrow (Ph_4As)_3Cr(N_3)_6+6NaCl$$
$$4Ph_4AsCl+CuSO_4+6NaN_3 \longrightarrow (Ph_4As)_4Cu(N_3)_6+4NaCl+Na_2SO_4$$
$$2Ph_4AsCl+K_2(PtCl_4)+4NaN_3 \xrightarrow{H_2O} (Ph_4As)_2Pt(N_3)_4\cdot H_2O+NaCl+2KCl$$
$$2Ph_4AsCl+H_2PtCl_6+6NaN_3 \longrightarrow (Ph_4As)_2Pt(N_3)_6+6NaCl+2HCl$$
$$Ph_4AsCl+K(AuCl_4)+4NaN_3 \longrightarrow (Ph_4As)Au(N_3)_4+4NaCl+KCl$$
$$Ph_4PCl+2CdSO_4+5NaN_3 \xrightarrow{H_2O} Ph_4P[Cd_2(N_3)_5]\cdot H_2O+2Na_2SO_4+NaCl$$

In a similar way, organometallic azides have also been synthesized[375]:

$$Ph_2BCl+NaN_3 \rightarrow Ph_2BN_3+NaCl$$
$$Me_6Al_2+4HN_3 \rightarrow 2MeAl(N_3)_2+4CH_4$$
$$Et_2AlCl+NaN_3 \rightarrow Et_2AlN_3+NaCl$$
$$(Me_3Si)_2NH+3HN_3 \rightarrow 2Me_3SiN_3+NH_4N_3$$
$$Bu^n_3SnCl+NaN_3 \rightarrow Bu^n_3SnN_3+NaCl$$
$$Ph_3PbOH+HN_3 \rightarrow Ph_3PbN_3+H_2O$$

Covalent azides are well-characterized derivatives[376]. Fluorine reacts with hydrazoic acid in a stream of nitrogen to give unstable fluorine azide:

$$2F_2+4HN_3 \rightarrow FN_3+N_2+NH_4F$$

Chlorine azide is also spontaneously explosive in both the condensed and solid phases, but can be prepared in carbon tetrachloride solution by mixing aqueous sodium azide with sodium hypochlorite over the organic solvent in a separating funnel, followed by acidification with acetic acid with vigorous mixing. After separating the organic layer, the safest way to obtain gaseous chlorine azide is to bubble nitrogen through the solution, and collect the nitrogen–chlorine azide mixture under reduced pressure:

$$NaOCl+NaN_3+2HOAc \rightarrow ClN_3+2NaOAc+H_2O$$

---

[373] A. D. Yoffe, in *Developments in Inorganic Nitrogen Chemistry*, Vol. I (ed. C. B. Colburn), Elsevier, Amsterdam (1966), pp. 72–149.
[374] W. Beck, E. Schuierer and K. Fedl, *Angew. Chem. Int. Edn. Engl.* **5** (1966) 249.
[375] J. S. Thayer, *Organometal. Chem. Rev.* **1** (1966) 157.
[376] K. Dehnicke, *Angew. Chem. Int. Edn. Engl.* **6** (1967) 240.

Analogous bromine and iodine compounds are also known, but are stable only at sub-zero temperatures[376].

From Group VI the reaction of sulphuryl chloride with sodium azide leads to unstable sulphuryl azide:

$$SO_2Cl_2+2NaN_3 \rightarrow SO_2(N_3)_2+2NaCl$$

Explosive azidosulphates are obtained readily by the reaction of hydrazinosulphuric acid with potassium nitrite:

$$H_2NHNSO_3H+KNO_2 \rightarrow N_3SO_3K+2H_2O$$

Hydrazoic acid and salts react with carbon disulphide to produce azidothiocarbonic acid derivatives from which free azidocarbonium disulphide can be formed[369]:

$$HN_3+CS_2 \rightarrow N_3CSSH$$
$$NaN_3+CS_2 \rightarrow N_3CSSNa$$
$$2N_3CSSNa+I_2 \rightarrow (N_3CSS)_2+2NaI$$

Phosphorus azides such as phosphonitrilic azide and a variety of organophosphorus azides are well known, while azides of nitrogen include unstable nitrosyl azide, *N*-azido-dimethylamine and related *N*-azidohexamethyldisilazane[367]:

$$(PNCl_2)_3+6NaN_3 \rightarrow [PN(N_3)_2]_3+6NaCl$$
$$(CF_3)_2PCl+LiN_3 \rightarrow (CF_3)_2PN_3+LiCl$$
$$NO \cdot HSO_4+NaN_3 \rightarrow NON_3+NaHSO_4$$
$$Me_2NCl+NaN_3 \rightarrow Me_2NN_3+NaCl$$
$$(Me_3Si)_2NCl+LiN_3 \rightarrow (Me_3Si)_2NN_3+LiCl$$

The nitrogen pseudohalogen derivatives N_3–N_3 and $N(N_3)_3$, which correspond to the chlorine analogues Cl_2 and NCl_3, are not known.

With regard to compounds containing carbon–azide bonds, a considerable organic chemistry has been investigated[377], and a few essentially inorganic compounds like carbonyl azide are also known[369]:

$$CO(NHNH_2)_2+2HNO_2 \rightarrow CO(N_3)_2+4H_2O$$

Dicyandiazide is obtained by the action of cyanogen bromide on sodium azide, and is one of a number of pseudohalogen type compounds[369]:

$$2CNBr+2NaN_3 \rightarrow (CNN_3)_2+2NaBr$$

8.3. NUCLEAR ISOTOPES OF HN_3 AND AZIDES

As no exchange occurs between the central and terminal nitrogen atoms of the azide ion, synthesis of compounds labelled in either or both positions is possible. Potassium

377 N. V. Sidgwick, *The Organic Chemistry of Nitrogen*, 3rd edn., revised and rewritten by I. T. Millar and H. D. Springall, Oxford University Press, Oxford (1966), pp. 488–502.

azide containing labelled terminal nitrogen atoms is prepared by the reaction of hydrazine hydrate and labelled ethyl nitrite in the presence of potassium methoxide[378]:

$$Na^{15}NO_2 + EtOH + HCl \rightarrow Et^{15}NO_2 + NaCl + H_2O$$

$$N_2H_4 \cdot H_2O \ + \ MeOK \ + \ Et^{15}NO_2 \longrightarrow K^{15}NN^{15}N$$

TABLE 47. STRUCTURES OF METAL AZIDES[a]

Compound	Crystal class	Cell constants (Å)	Bond lengths (Å)	
			$M-N_3$	N–N
LiN_3	Body-centred rhombohedral			
α-NaN_3				
β-NaN_3	Body-centred rhombohedral	$a = 5.488$, $\alpha = 38° 43'$	2.48	1.17
KN_3	Body-centred tetragonal	$a = 6.091$, $c = 7.056$	2.96	1.16
RbN_3	Body-centred tetragonal	$a = 6.36$, $c = 7.41$	3.11	1.13
CsN_3	Body-centred tetragonal	$a = 6.72$, $c = 8.04$	3.34	
NH_4N_3	Orthorhombic	$a = 8.93$, $b = 8.64$, $c = 3.80$	3.2 2.98	1.166
AgN_3	Orthorhombic	$a = 5.59$, $b = 5.91$, $c = 6.01$	2.79 2.56	1.16
CuN_3	Tetragonal	$a = 8.65$, $c = 5.59$	2.23	1.17
TlN_3	Body-centred tetragonal	$a = 6.23$, $c = 6.75$	2.98	
$Cu(N_3)_2$	Orthorhombic	$a = 9.23$, $b = 13.23$, $c = 3.07$		
$Cd(N_3)_2$	Orthorhombic	$a = 7.82$, $b = 5.46$, $c = 7.02$		
$Ca(N_3)_2$	Orthorhombic	$a = 11.62$, $b = 10.92$, $c = 5.66$		
$Sr(N_3)_2$	Orthorhombic	$a = 11.82$, $b = 11.47$, $c = 6.08$	2.63 2.77	1.12
$Ba(N_3)_2$	Monoclinic	$a = 6.22$, $b = 29.29$, $c = 7.02$, $\alpha = 105°$	2.937	
α-$Pb(N_3)_2$	Orthorhombic	$a = 6.63$, $b = 5.46$, $c = 16.25$	4.71	
β-$Pb(N_3)_2$	Monoclinic	$a = 5.09$, $b = 8.84$, $c = 17.51$, $\alpha = 90° 10'$	4.93	
γ-$Pb(N_3)_2$	Monoclinic	$a = 6.55$, $b = 10.51$, $c = 12.17$, $\alpha = 98° 30'$		
$[Co(NH_3)_5N_3](N_3)_2$	Orthorhombic	$a = 12.997$, $b = 8.031$, $c = 10.414$	1.943	1.145, 1.208 1.158 azide ion 1.172

[a] Refs. 367, 371, 373, 379.

[378] W. Spindel, in *Inorganic Isotopic Syntheses* (ed. R. H. Herber), Benjamin, New York (1962), p. 116.

Azide with only the central nitrogen atom labelled is most conveniently prepared from labelled nitrous oxide and calcium amide:

$$Ca(NH_2)_2 + {}^{15}NNO \xrightarrow{\text{dil. } H_2SO_4} HN^{15}NN \xrightarrow{KOH} KN^{15}NN$$

Alternatively, centrally labelled sodium azide can be synthesized by an organic route[367]:

$$p\text{-}NO_2C_6H_4NH_2 \xrightarrow{H^{15}NO_2} p\text{-}NO_2C_6H_4N^{15}N^+ \xrightarrow{Br_2} p\text{-}NO_2C_6H_4N^{15}NBr_3 \xrightarrow{NH_4OH} p\text{-}NO_2C_6H_4N^{15}NN \xrightarrow{NaOH} NaN^{15}NN$$

8.4. STRUCTURE AND BONDING

The structure of hydrogen azide as determined by electron diffraction methods shows that the three nitrogen atoms are collinear and the HNN bond angle is 112°. The two N–N distances differ considerably, $HN\text{–}N_2$ being 1.24 Å and $HN_2\text{–}N$ is 1.13 Å, and these values are comparable to those reported for organic azides[373]. Examination of metal azides by X-ray and neutron diffraction methods, some results of which are given in Table 47, reveal that discrete azide ions are also linear, but with both the N–N distances the same length of 1.16 Å.

Thus on the basis of bond lengths, a distinction can be made between ionic and covalent azides, and approximate bond orders can be estimated by comparison with the bond lengths in isoelectronic and other nitrogen species listed in Table 48. Also the higher symmetry

TABLE 48. BOND LENGTHS AND ANGLES IN NITROGEN AND RELATED SPECIES[a]

	–N–NN (Å)	–NN–N (Å)	Angles		–N–CX	–NC–X	Angles
HN_3	1.24	1.134	HNN 112.6	HNCO	1.21	1.17	HNC 128
MeN_3	1.24	1.10	CNN 120	MeNCO	1.19	1.18	CNC 125
$(NCN_3)_3$	1.26	1.11	CNN 113	CO_2		1.16	
MN_3	1.15	1.15		CO_2^+		1.18	
N_2	1.10			CS_2		1.55	
N_2^+	1.12			$M^+(NCO)^-$			
CH_2N_2	1.13			$M^{2+}(NCN)^{2-}$			
MeN=NMe	1.24						
FN=NF	1.25						
N_2O	1.29						
$H_2N\text{–}NH_2$	1.47						

[a] Refs. 367, 371, 373, 379.

associated with the azide ion means that both terminal nitrogen atoms are identical, whereas in covalent azides each of the three nitrogen atoms has a different electronic environment.

The bonding in hydrogen azide can be described by the valence bond approach and visualized in terms of resonance between the canonical structure illustrated in Fig. 23. The HNN bond angle and distances calculated on the basis of equal resonance between conventional electron pair structures (i) and (ii) agree well with reported values, while the non-pairing structure (iii) is an alternative approximation[379] predicing bond orders of about the correct magnitude and a bond angle consistent with the state of hybridization at nitrogen (bonded to hydrogen) being something between sp^2 and sp^3.

On the basis of molecular orbital theory, the sixteen valence electrons of hydrogen azide are distributed four into two lone pairs, six into three σ bonds, four into two π bonds,

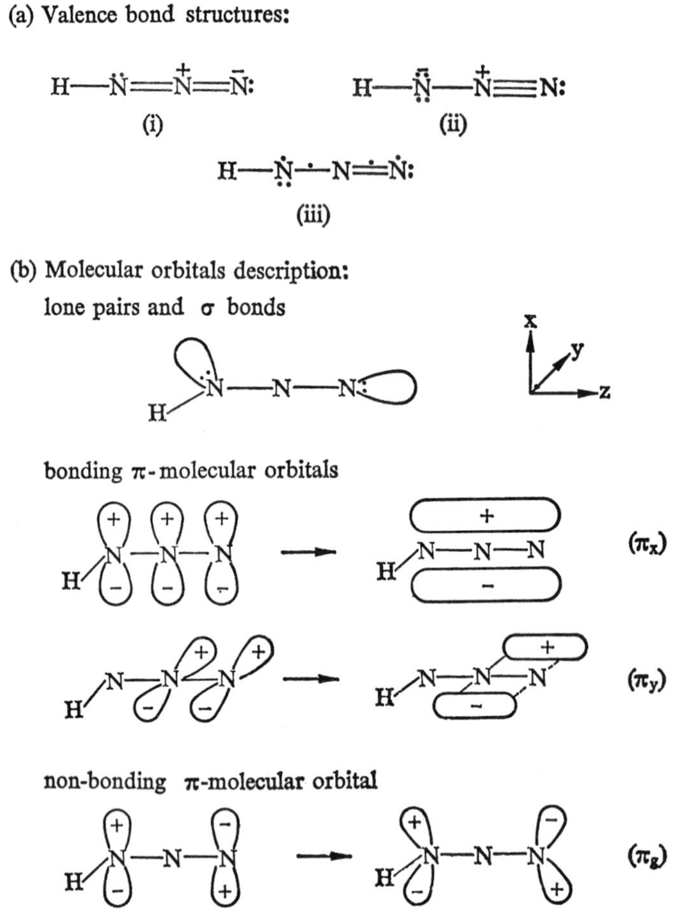

(a) Valence bond structures:

H—N̈=N⁺=N⁻: (i)   H—N̈⁻—N⁺≡N: (ii)

H—N̈—·—N=·=N: (iii)

(b) Molecular orbitals description:

lone pairs and σ bonds

bonding π-molecular orbitals

(π_x)

(π_y)

non-bonding π-molecular orbital

(π_g)

Fig. 23. Valence bond and molecular orbital descriptions of the bonding in hydrogen azide.

[379] M. Green, in *Developments in Inorganic Nitrogen Chemistry*, Vol. I (ed. C. B. Colburn), Elsevier, Amsterdam (1966), pp. 1–71.

and two into non-bonding orbitals as illustrated in Fig. 23. With the azide group of hydrogen azide being asymmetrical, the itr ogen atom bonded to hydrogen requires approximately sp^2 hybridization to account for the HNN bond angle, while the other two nitrogen atoms are sp hybridized.

The azide ion, however, is a symmetrical linear molecule, and as it contains sixteen electrons it is both isostructural and isoelectronic with the cynamide ion NCN^{2-}, the cyanate ion NCO^-, the fulminate ion CNO^-, the nitronium ion NO_2^+ and carbon dioxide. The valence bond canonical structures are illustrated in Fig. 24. However, the postulation

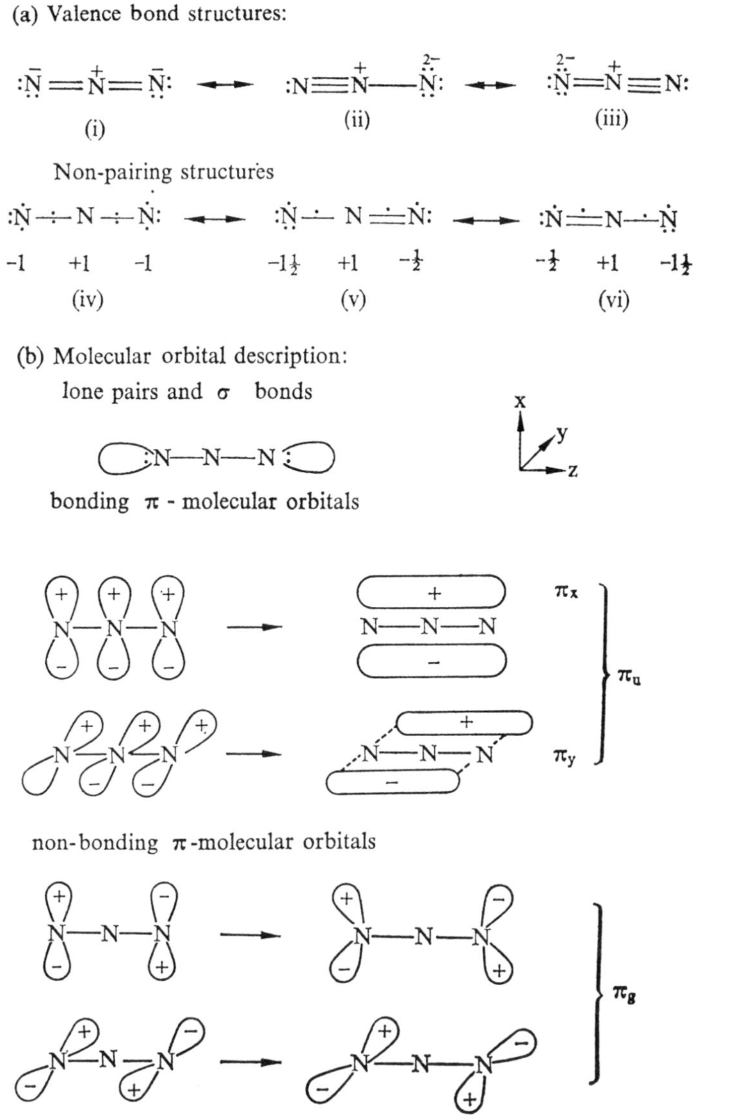

(a) Valence bond structures:

(i) (ii) (iii)

Non-pairing structures

−1 +1 −1 (iv)

−1½ +1 −½ (v)

−½ +1 −1½ (vi)

(b) Molecular orbital description:

lone pairs and σ bonds

bonding π - molecular orbitals

π_x, π_y, π_u

non-bonding π-molecular orbitals

π_g

FIG. 24. Valence bond and molecular orbital descriptions of the bonding in the azide ion.

of structures (ii) and (iii) with localized formal charges of -2 is somewhat extreme, and the description of the azide ion is perhaps better represented by non-pairing structures (iv)–(vi)[379].

The molecular orbital description of the azide ion also illustrated in Fig. 24 can be derived from first principles by considering the interactions of the available atomic orbitals of related symmetry. The symmetry of the orbitals of the central nitrogen atom can be simply assigned on the basis of the transformations which result from carrying out the symmetry operations of the point group to which the molecule belongs, namely $D_{\infty h}$. Likewise, linear combinations of related atomic orbitals on the terminal nitrogen atoms can also be assigned, and molecular orbitals are formed by combining orbitals of all three nitrogen atoms which transform alike, as indicated below:

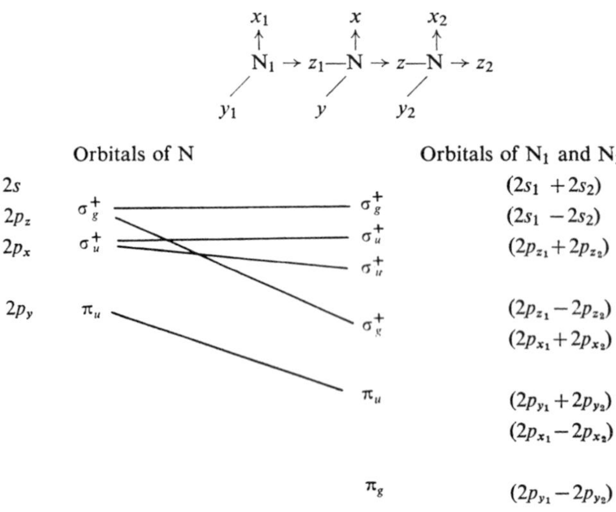

Thus in the ${}^1\Sigma_g^+$ ground state, the sixteen valence electrons occupy the lowest molecular orbitals of the energy level diagram illustrated in Fig. 25, giving the configuration $(1\sigma_g)^2(1\sigma_u)^2(2\sigma_g)^2(2\sigma_u)^2(1\pi_u)^4(1\pi_g)^4$, with the highest energy electrons in non-bonding orbitals.

The azide radical $N_3^{\cdot}$ is known in flash photolysis and molecular beam studies, but little has been deduced about its structure. Hence the linear structure and bonding description which has been proposed is based on the assumption that the removal of a further electron from the non-bonding orbital of the azide ion causes only slight alterations to internuclear distances and other bond parameters[371].

Finally, azides can act as bridging groups and although their structures have usually been assumed to be of the M—N—N—N—M′ type, there is, in principle no reason why structures of the type

```
                          M
M—N—N—M′                   \
     |        and           N—N—N
     N                     /
                          M′
```

should not also be considered[380].

[380] M. Straumanis and A. Cirulis, *Z. anorg. Chem.* **252** (1943) 9.

8.5. MOLECULAR CONSTANTS

The infrared and Raman spectra of hydrogen azide have been thoroughly investigated[381] and the assignment of five in-plane (A') vibrations and one out-of-plane (A'') vibration as listed in Table 49, is consistent with a planar molecule.

TABLE 49. FUNDAMENTAL FREQUENCIES IN HN_3 AND DN_3[a]

Vibration	Approximate description	Frequency (cm^{-1})			
		HN_3		DN_3	
		Gas	Solid	Gas	Solid
ν_1 (A')	H–N stretching	3336	3090	2480	2308
ν_2 (A')	N–N–N asym. stretching	2140	2162	2141	2155
ν_3 (A')	N–N–N sym. stretching	1274	1299	1183	1230
ν_4 (A')	H–N–H bending	1150	1180	955	977
ν_5 (A')	N–N–N bending	522	—	498	—
ν_6 (A'')	N–N–N bending	672	—	638	—

[a] D. A. Dows and G. C. Pimentel, *J. Chem. Phys.* **23** (1955) 1258.

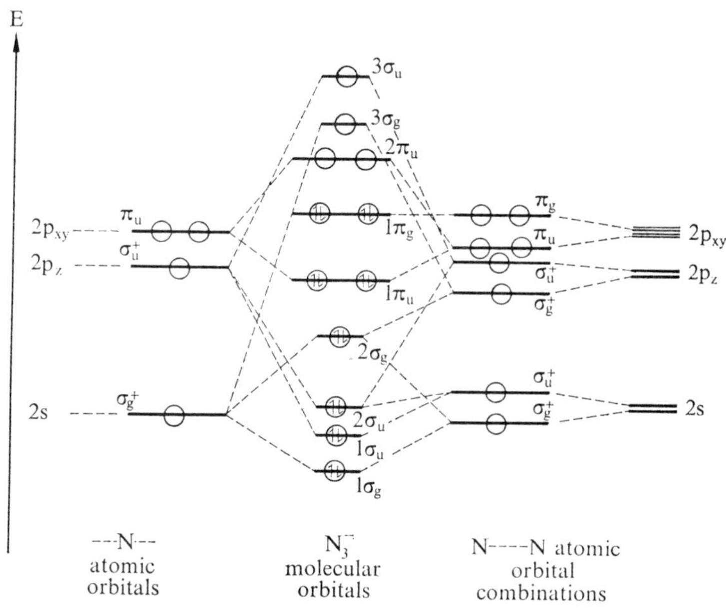

FIG. 25. The molecular orbital energy level diagram for the azide ion N_3^-.

[381] K. Nakamoto, *Infrared Spectra of Inorganic and Coordination Compounds*, Wiley, New York (1963), pp. 76, 176.

As the azide ion is a linear symmetrical triatomic group of $D_{\infty h}$ symmetry, the symmetric stretching frequency ν_1 is Raman active and infrared inactive while both the bending vibration ν_2 and the asymmetric stretching frequency ν_3 are infrared active and Raman inactive. However, values for all three modes can be derived from infrared spectra due to the presence of combination bands such as $\nu_3 - \nu_1$ or $\nu_1 + \nu_2$. Spectra of heavy metal and ionic azides which have been recorded are given in Table 50.

Although solid-phase infrared spectra often give rise to splitting due to the packing of the ions in the lattice, new bands related to ν_3 have been observed when alkali-metal azides were irradiated by ultraviolet light at low temperatures. The presence of transitory excited azide ions or charge transfer effects have been suggested as possible explanations[382].

Infrared spectra have also been most useful in the characterization of covalent azides in particular organometallic derivatives where the asymmetric stretching frequency is invariably characteristic[375], but the presence of the symmetric stretching frequency although theoretically active, is not always observed.

TABLE 50. INFRARED FREQUENCIES AND FORCE CONSTANTS OF THE AZIDE ION N_3^- IN METAL AZIDES[a]

	ν^1 (cm^{-1})	ν^2 (cm^{-1})	ν^3 (cm^{-1})	k_1	k^2	k_δ/l^2
				(10^5 dyne cm^{-1})		
NH_4N_3	1345	661, 650	2041	13.4	1.74	0.59
LiN_3	1369	635	2092	13.9	1.73	0.56
NaN_3	1358	639	2128	13.9	1.30	0.57
KN_3	1344	645	2041	13.4	1.74	0.58
RbN_3	1339	642	2024	13.2	1.78	0.57
CsN_3	1329	635	2062	13.3	1.45	0.56
$Ca(N_3)_2$	1381	638	2114	14.2	1.74	0.57
$Sr(N_3)_2$	1373	635	2096	14.0	1.75	0.56
$Ba(N_3)_2$	1354	650, 637	2123, 2083	14.2	1.97	0.55
TlN_3		634	2030			
AgN_3	1334	631	2073, 1960			
CuN_3	1337	615	2110			
$Hg_2(N_3)_2$	1322	647, 592	2080			
$Hg(N_3)_2$	1313	675, 644, 642, 584	2070			
α-$Pb(N_3)_2$	1325	628	2030, 1920			
β-$Pb(N_3)_2$	1352		2040			
Ph_3PbN_3			2030			

[a] Refs. 367, 373.

[382] H. A. Papazian, *J. Chem. Phys.* **32** (1960) 456; **34** (1961) 1614.

8.6. PHYSICAL PROPERTIES

Some related thermochemical properties of the azide radical and ions listed[371] in Table 51 can be used particularly in calculations predicting the course of decomposition reactions (see section 8.7.1). The electro-negativity calculated for the azide ion using this data lies close to that of the bromide ion.

TABLE 51. THERMOCHEMICAL PROPERTIES OF THE AZIDE RADICAL AND IONS

Property	Temperature (K)	Value
Enthalpy of formation		
$\triangle H_f$ (N_3^-, g)	298	34.8 ± 0.1 kcal/mol
$\triangle H_f$ ($N_3^{\cdot}$, g)	298	105 ± 3 kcal/mol
$\triangle H_f$ (N_3^+, g)	298	388 ± 5 kcal/mol
E (electron affinity) ($N_3^{\cdot}$)		70 ± 5 kcal/mol
I (ionization energy) ($N_3^{\cdot}$)		283 ± 6 kcal/mol
Electro-negativity $(E+I)/(130\ \text{kcal})$		2.71

The list of reasonably reliable values for the physical properties of hydrogen azide in Table 52 is somewhat incomplete due partly to the hazards and accompanying problems of

TABLE 52. PHYSICAL PROPERTIES OF HYDROGEN AZIDE[a]

Property	Temperature (K)	Value
Freezing point	~193	~ −80°C
Boiling point (estimated)	308.85	35.7°C
Vapour pressure		$\log_{10} p_{mm} = 8.198 - \frac{1643}{T}$
Density of liquid d_4^t	273–294	$d = \frac{1.126}{1+0.0013t}$
$\triangle H_{f0}^\circ$ (enthalpy of formation) (g)	273	71.82 kcal/mol
$\triangle H_f^\circ$ (enthalpy of formation) (g)	298	70.3 kcal/mol
$\triangle G_f^\circ$ (Gibbs energy of formation) (g)	298	78.4 kcal/mol
$H_{298}^\circ - H_0^\circ$ (enthalpy) (g)	298	2.599 kcal/mol
S° (entropy) (g)	298	57.09 cal/deg mol
C_p° (heat capacity) (g)	298	10.44 cal/deg mol
$\triangle H$ vaporization	285.5	7.3 kcal/mol
$\triangle H_{f0}^\circ$ (enthalpy of formation) (l)	298	64.4 kcal/mol
$\triangle G_{f0}$ (Gibbs energy of formation) (l)	298	78.2 kcal/mol
S° (entropy) (l)	298	33.6 cal/deg mol
$\triangle H_{f0}^\circ$ (enthalpy of formation), aq $H^+ + N_3^-$ at infinite dilution	298	65.5 kcal/mol
$\triangle H_{f0}^\circ$ (enthalpy of formation) of undissociated satd. soln.	298	61.9 kcal/mol
Heat of ionization	298	3.60 kcal/mol
Minimum heat of decomp. (calc.) of liq. HN_3	298	67 kcal/mol
Heat of solution of gas in H_2O	298	9.7 kcal/mol
Dipole moment		0.847 D

[a] Refs. 366, 371.

maintaining hydrogen azide in a pure state during their determination. For similar reasons, some properties are only discussed qualitatively. For example, the dielectric constant has not been measured, but as a number of salts dissolve in hydrogen azide to yield electrically conducting solutions, it must be fairly large.

Some physical properties of selected metal azides have been compiled in Table 53.

8.7. CHEMISTRY AND CHEMICAL PROPERTIES

8.7.1. Decomposition

Although the vast majority of azides undergo explosive decomposition, the slow decomposition of azides can also be brought about in a variety of ways. For example, hydrogen azide can be decomposed by thermal, photochemical, photolytic and by electric discharge methods[371]. Unlike explosive decomposition which results in the formation of the elements, slow decomposition can give rise to a variety of other compounds. For hydrogen azide these have been explained on the basis of the initial step in the decomposition being fission of the longest (consequently the weakest) N–N bond ($HN–N_2$) to give elemental nitrogen and the imine radical[373], probably in an excited state as dissociation into ground state products is forbidden by the correlation (spin conservation) rules and alternative modes of dissociation are thermodynamically less favourable:

$$\text{Allowed:} \quad HN_3(^1A') \rightarrow NH(^1\Delta)+N_2(^1\Sigma_g^+); \qquad \Delta H = 33\pm2 \text{ kcal/mol}$$

$$\text{Forbidden:} \quad HN_3(^1A') \rightarrow NH(^3\Sigma^-)+N_2(^1\Sigma_g^+); \qquad \Delta H = 5\pm2 \text{ kcal/mol}$$

$$\text{Allowed:} \quad HN_3(^1A') \rightarrow H(^2S_g)+N(^2\Pi_g); \qquad \Delta H = 85\pm3 \text{ kcal/mol}$$

Further reaction of the imine radical can lead to the formation of diimine or ammonia, and hydroxylamine or nitrous acid if water or oxygen are present. Diimine has been detected in mass spectrometric examination of the products after hydrogen azide was subjected to electric discharge[383], and is believed to be the active intermediate in the oxidation of hydrazine:

$$:NH+HN_3 \rightarrow HN{=}NH+N_2$$

$$:NH+2HN_3 \rightarrow NH_3+3N_2$$

$$:NH+H_2O \rightarrow NH_2OH$$

$$:NH+O_2 \rightarrow HONO$$

If the products of thermal decomposition of hydrogen azide at 1000°C under low-pressure conditions are collected on a surface cooled with liquid nitrogen, the stoichiometry of the process is

$$6HN_3 \rightarrow 7N_2+H_2+(NH)_4$$

[383] W. L. Jolly, *The Chemistry of Nitrogen*, Benjamin, New York (1964), p. 64.

TABLE 53. PHYSICAL PROPERTIES OF METAL AZIDES[a]

Azide	M.p., b.p. (°C/mm)	Density (d_4^{20})	Solubility in H_2O (g per 100 g H_2O/°C)	Enthalpy ΔH_f° cryst. (kcal/mol)	Dielectric constant K	Lattice energies (kcal/mol 298 K)	Specific resistance (Ω/cm)	Magnetic susceptibility (c.g.s. units)	Refractive index at = 5500 Å	Activation energy (kcal/mol)	Detonation velocity (m/s)
LiN_3	d 115–298		66.41/18 forms monohydrate	2.6		194				19	
NaN_3		1.8475/25	Trihydrate 28/0 35.6/100	5.1	6.4	175		-17×10^{-6}	1.38–1.52	34	
KN_3	350 *in vacuo*	2.04	107.1/16	−0.3	6.9	157		-25×10^{-6}	1.45, 1.66	49 *in vacuo*	
RbN_3	d ~ 310	2.788		−0.1		152					
CsN_3	310	4.65	224.2/0	−2.4		146					
NH_4N_3	160	1.346	20.16/30	26.8		175					
TlN_3	334		0.17/0	55.8	11.5	164	8×10^{12}			40	1480
AgN_3	252 297/760	5.1	Insol.	74.2	9.4	205	3×10^{8}		1.8	44 190°C 31 190°C	2170
CuN_3	exp	3.26	0.0007/20	67.2	9.3	227				27	2720
$Cu(N_3)_2$	exp 215	2.604	0.008/20	140.4							
$Hg_2(N_3)_2$	exp		0.025/20	141.5	4.6		High	-135×10^{-6}			2350
$Ca(N_3)_2$	exp 144–156		38.1/0	11.0		517					
$Sr(N_3)_2$				1.7	8.3	494					
$Ba(N_3)_2$		2.936		−5.3	7.7	469					
α-$Pb(N_3)_2$	exp 350	4.71	0.023/18	115.5		434	1.5×10^{13}				2300
β-$Pb(N_3)_2$		4.93		115.8					2.2		

[a] Refs. 367, 371, 373.

At $-196°C$ the blue paramagnetic substance $(NH)_4$ is unreactive and gives no X-ray diffraction pattern, but when it is allowed to attain a temperature between -150 and $-125°C$, it is transformed into colourless ammonium azide[384].

The decomposition reactions of metal azides are usually only explosive if the overall process is exothermic; however, since the elements are the products, the heat of decomposition corresponds to the standard enthalpy of formation of the salt (see Table 53). For example, the alkali metal azides are not explosive and decompose smoothly:

$$2NaN_3(s) \rightarrow 2Na(s) + 3N_2(g)$$

However, in some cases the nitride is formed:

$$3LiN_3 \rightarrow Li_3N + 4N_2$$

It is not clear whether the formation of nitride occurs by subsequent reaction between nitrogen and the metal or as a result of initial $MN\text{–}N_2$ bond fission as in covalent azide decompositions. This latter situation has been proposed for the photochemical decomposition of mercury(I) azide[385], and from this point of view such compounds are termed covalent solids.

The decomposition of heavy metal azides can be described in terms of excitation of the azide ion with subsequent promotion of an electron to the conduction band leaving the azide radical. The dissociation of the ground state azide radical ($^2\Pi_g$) into ground state products N and N_2 is forbidden by the correlation rules, while the allowed dissociation into excited states requires 62 kcal/mol and is therefore not important at normal temperatures. However, the interaction of two azide radicals is also allowed, and being an exothermic process, it is considered to be the important step in the decomposition of solid ionic azides:

$$N_3^- + h\nu \rightarrow N_3^{-*}$$

$$N_3^{-*} \rightarrow N_3 + e^-$$

$$N_3(^2\Pi_g) \rightarrow N(^4S_g) + N_2(^1\Sigma_g^+); \quad \Delta H = 7.5 \pm 3 \text{ kcal/mol}$$

$$N_3(^2\Pi_g) \rightarrow N(^2D) + N_2(^1\Sigma_g^+); \quad \Delta H = 62 \pm 3 \text{ kcal/mol}$$

$$2N_3(^2\Pi_g) \rightarrow 3N_2(^1\Sigma_g^+); \quad \Delta H = -210 \pm 6 \text{ kcal/mol}$$

The rapid exothermic decomposition of azide leading to explosive ignition or detonation has often been experienced but is less-well understood[371]. Apart from uniform or localized heating, the effects of electric current, impact, friction, ultrasonic and shock waves, exposure to intense light or ionizing radiation, can all cause azide to explode. The simplest interpretation which provides a relationship between each of these methods of initiation is on the basis of conversion to thermal energy. However, the various methods of initiation are now known to occur in more than one way and also to depend on other factors such as the variation in sensitivity of different azides, the presence of ionic and metallic impurities,

[384] F. O. Rice and T. A. Luckenbach, *J. Am. Chem. Soc.* **82** (1960) 2681.

[385] S. K. Deb and A. D. Yoffe, *Trans. Faraday Soc.* **55** (1959) 106; *Proc. Roy. Soc.* (*London*) A, **235** (1961) 106, 481.

and even the nature of the surface and physical size of the material itself. Nevertheless, the propagation of a "critical" condition is perhaps most easily visualized as proceeding by a self-heating mechanism in which the energy of decomposition brings about the decomposition of further material at an accelerating rate.

8.7.2. Reactions of Azides

Hydrogen azide dissolves in water to give a weakly acidic solution. From the measured dissociation constant, hydrazoic acid is about as strong as acetic acid:

$$HN_3(aq) \rightleftharpoons H^+(aq) + N_3^-(aq); \quad K = 1.8 \times 10^{-5}$$

$$pK_a = 4.77 \text{ at } 298 \text{ K}$$

$$\Delta H^\circ_{298} = 3.60 \text{ kcal/mol}$$

However, although the acid dissociates only slightly in aqueous solution, metals dissolve in it accompanied by the evolution of ammonia and nitrogen analogous on the ammonia–aquo basis to the action of nitric acid on metals.

$$Zn + 3HN_3 \rightarrow Zn(N_3)_2 + NH_3 + N_2$$

The basic dissociation constant of $H_2N_3^+$ has also been estimated from experiments involving the equilibration of hydrogen azide between chloroform and pure sulphuric acid.

$$HN_3 + H_2O \rightleftharpoons H_2N_3^+ + OH^-; \quad pK_a = -6.21$$

The oxidizing and reducing properties of azides in acidic and basic aqueous solutions can be summarized by the following couples:

In acid solution:

$$3NH_4^+ \rightleftharpoons HN_3 + 11H^+ + 8e^-; \quad E^\circ_{298} + 0.66 \text{ V}$$

$$NH_4^+ + N_2 \rightleftharpoons HN_3 + 3H^+ + 2e^-; \quad E^\circ_{298} + 1.82 \text{ V}$$

$$HN_3 \rightleftharpoons 3/2N_2 + H^+ + e^-; \quad E^\circ_{298} - 2.8 \text{ V}$$

In basic solution:

$$NH_3 + N_2H_4 + 7OH^- \rightleftharpoons N_3^- + 7H_2O + 6e^-; \quad E^\circ_{298} - 0.62 \text{ V}$$

$$N_3^- \rightleftharpoons 3/2N_2 + e^-; \quad E^\circ_{298} - 3.1 \text{ V}$$

However, the fractional oxidation numbers encountered with individual nitrogen atoms in azides can be confusing, and for this reason comparisons with other redox systems are more conveniently illustrated by oxidation state diagrams as discussed in section 1.6.2.

The principal inorganic reactions of hydrogen azide all involve breakdown of the N–N–N grouping and lead to the formation of simpler nitrogen species, often the element.

For example, the reduction of hydrogen azide can give various products but does not provide any useful preparative routes:

$$HN_3+6[H] \xrightarrow[\text{Pd}]{\text{colloidal}} NH_3+N_2H_4$$

$$HN_3+H_2 \xrightarrow[\text{black}]{\text{Pt}} NH_3+N_2$$

$$2HN_3+[O] \xrightarrow[\text{black}]{\text{Pt}} 3N_2+H_2O$$

$$3HN_3 \xrightarrow[\text{Ni}]{\text{Raney}} NH_3+4N_2$$

$$HN_3+3[H] \xrightarrow{\text{SnCl}_2\text{/HCl}} NH_4^++N_2$$

$$2HN_3+10[H] \xrightarrow{\text{Al/NaOH}} 3N_2H_4$$

Hydrogen azide and nitrous acid react quantitatively according to the equation:

$$HNO_2+HN_3 \rightarrow N_2O+N_2+H_2O$$

The course of this reaction has been determined by ^{15}N labelling experiments in which the nitrous acid nitrogen appeared quantitatively as the central atom in nitrous oxide while the central nitrogen of the azide group always resulted as elemental nitrogen. The reaction of hydrogen azide with strong nitric acid gives nitrogen as the chief product with varying amounts of N_2O, NO, N_2O_3 and NO_2 presumably formed from further reactions of nitric oxide with excess nitric acid[366].

$$HN_3+HNO_3 \rightarrow N_2+2NO+H_2O$$

The action of sulphuric acid on hydrogen azide gives some hydroxylamine, while hydrogen chloride gives ammonium chloride:

$$HN_3+H_2O \xrightarrow{\text{sulphuric acid}} N_2+NH_2OH$$

$$3HN_3+HCl \rightarrow NH_4Cl+4N_2$$

In general, the azide ion (N_3^-) in metallic azides behaves like a halide ion, and being both isoelectronic and isostructural with cyanate (CNO^-) and isocyanate (NCO^-), it is regarded as a true pseudohalide. The azide ion has a cylindrical shape[371] about 5.1 Å long overall with a diameter of 3.5 Å and its position in the spectrochemical series has been estimated to lie between water and fluoride[386]. These favourable steric properties and greater ligand strength over chloride, whose ionic diameter is 3.62 Å and ligand strength less than fluoride, makes azide a most useful competitive ligand in reactions of transition metal complexes[387].

Some typical reactions of metallic azides are illustrated by the following equations:

$$NaN_3+H_2 \xrightarrow{\text{colloidal Pd}} NaNH_2+N_2$$

$$2N_3^-+3H_2 \xrightarrow{\text{Al–Hg}} 2NH_3+2N_2$$

$$2NaN_3+2ICl \xrightarrow{0^\circ} 2NaCl+I_2+3N_2$$

$$NaN_3+NOCl \rightarrow NaCl+N_2O+N_2$$

---

[386] P. J. Staples and M. Tobe, *J. Chem. Soc.* (1960) 4812.

[387] W. P. Griffith, in *Developments in Inorganic Nitrogen Chemistry*, Vol. I (ed. C. B. Colburn), Elsevier, Amsterdam (1966), pp. 241–306.

Another general reaction of the azide ion which has been extensively investigated is the oxidation of azide by iodine in acid solution in the presence of a catalyst such as sodium thiosulphate:

$$2N_3^- + 2H^+ + I_2 \rightarrow 2HI + 3N_2$$

Reactions which are less easy to formulate include the ability of azide to inhibit catalase activity in the catalase–iron redox system, while from a preparative point of view, $Na^{14}CN$ can be obtained in good yields when sodium azide and barium [^{14}C] carbonate are carefully heated in an atmosphere of nitrogen.

In contrast, halogen azides are hydrolysed in basic solution, and this provides the basis of a convenient method for their analysis by oxidation of arsenite:

$$XN_3 + 2OH^- \rightarrow N_3^- + OX^- + H_2O; \qquad X = \text{halogen}$$

$$AsO_3^{3-} + XN_3 + 2OH^- \rightarrow AsO_4^{3-} + N_3^- + X^- + H_2O$$

Finally, mention should also be made concerning the rôle of azides as reagents in organic chemistry in which the group can take part in a wide variety of substitution and addition reactions as well as the cyclization and acidic reactions of hydrogen azide[366, 367].

8.8. BIOLOGICAL PROPERTIES[388]

Hydrogen azide has an unbearably pungent odour, inhalation of which causes dizziness, headache and strong irritation of the mucous membranes. It is classed as a deadly non-cumulative poison, and concentrations in air of even less than 1 ppm are considered very dangerous. Sodium azide can cause cardiac irregularities and lowering of blood pressure, although non-lethal doses can confer some protection against the action of body X-irradiàtion.

8.9. ANALYSIS OF AZIDES[367]

Silver azide precipitated from neutral or weak nitric acid solutions resembles silver chloride except that it is a sensitive explosive but stable to light. Azide solutions give a blood-red colour with ferric salts which is used as a colorimetric method for its determination. Azides react with cerium(IV) in acid solution:

$$2Ce^{4+} + 2N_3^- \rightarrow 3N_2 + 2Ce^{3+}$$

The nitrogen liberated can be measured volumetrically or excess cerium(IV) can be added and back titrated with standard ferrous ammonium sulphate. Another method for the determination of lead azide is based on the conversion to hydrogen azide with dilute sulphuric acid which can then be distilled into excess cerium(IV) solution which is then titrated with standard sodium oxalate. Excess permanganate in the presence of manganese(II) will also oxidize azide in acid media to nitrogen and can be back titrated iodometrically.

388 F. Call, in *Mellor's Comprehensive Treatise on Inorganic and Theoretical Chemistry*, Vol. 8, Suppl. I, *Nitrogen*, Part I, Longmans, London (1964), Section XVII, p. 568.

9. NITROGEN–HALOGEN COMPOUNDS

9.1. CLASSES OF NITROGEN–HALOGEN COMPOUNDS

In this section, only molecules containing a nitrogen–halogen bond will be considered[389]. Unfortunately, the formulae of compounds containing nitrogen, halogen and other elements are not always explicit in describing which atoms are bonded together. For example, nitrosyl chloride is usually formulated NOCl, whereas the constituent atoms are bonded O=N—Cl, while NSF thiazyl fluoride is not a sulphur analogue of the nitrosyl halides as the halogen is attached to sulphur N≡S—F. Further confusion results from the poor systematization of the nomenclature of such derivatives, particularly those containing nitrogen, fluorine and oxygen. For example, there is no systematic name for the class of compounds F—N=N(O)—R, and of the various trivial and reasonable names which have been used to describe F—N=N(O)—Me, perhaps methyl *N'*-fluorodiimide *N*-oxide is least ambiguous.

Attention is drawn to these problems of formulation and chemical nomenclature in this field because of the current attention being given to such materials as high-energy oxidizing agents[390]. Indeed, it is in this context of investigating the properties of such compounds as potential ingredients of advances chemical propellants that so much new information has become available. Clearly the fluorine derivatives have received most attention[391–3] because of the position of $—NF_2$ when substituent oxidizer groups are arranged in decreasing order of oxidizing power:

$$-F,\ -OF,\ -NF_2,\ -ClF_4,\ -O,\ -ClF_2,\ -NO_3,\ -ClO_4,\ -NO_2$$

Apart from the binary nitrogen–halogen series of compounds, the other classes of compounds which will be considered are the haloamines[394] HNX_2 and H_2NX, the oxohalides[395–7] including nitrosyl halides[398, 399] XNO, nitryl halides XNO_2, trifluoramine oxide[390] ONF_3, related ions NF_2O^-, $NFO_2^=$, $HNFO_2^-$, $NF_2O_2^-$ and a few sulphur derivatives[400] of NF_2 such as NF_2SF_5. Numerous organic N–X derivatives are also well known[391, 401]. The binary nitrogen–halogen compounds can be further subdivided on the basis of the number of nitrogen atoms per molecule. This is primarily relevant to fluorine–nitrogen derivatives which include the mononuclear NF_4^+ ion, NF_3, NF_2X and possibly

[389] J. D. Richards, in *Mellor's Comprehensive Treatise on Inorganic and Theoretical Chemistry*, Vol. 8, Suppl. II, *Nitrogen*, Part II, Longmans, London (1967), Section XXXIII, pp. 409–39.

[390] E. W. Lawless and I. C. Smith, *Inorganic High-energy Oxidizers*, Edward Arnold, London (1968).

[391] C. B. Colburn, in *Advances in Fluorine Chemistry*, Vol. 3 (eds. M. Stacey, J. C. Tatlow and A. G. Sharpe), Butterworths, London (1962), p. 92.

[392] A. V. Pankratov, *Russian Chem. Rev.* **32** (1963) 157.

[393] J. K. Ruff, *Chem. Rev.* **67** (1967) 665.

[394] A. V. Fokin and Yu. M. Kosyrev, *Russian Chem. Rev.* **35** (1966) 791.

[395] J. W. George, in *Progress in Inorganic Chemistry*, Vol. 2 (ed. F. A. Cotton), Interscience, New York (1960), p. 33.

[396] C. Woolf, in *Advances in Fluorine Chemistry*, Vol. 5 (eds. M. Stacey, J. C. Tatlow and A. G. Sharpe), Butterworths, London (1965), pp. 1–30.

[397] R. Schmutzler, *Angew. Chem. Int. Edn. Engl.* **7** (1968) 440.

[398] C. J. Hoffman and R. G. Neville, *Chem. Rev.* **62** (1962) 1.

[399] D. M. Yost and H. Russell, *Systematic Inorganic Chemistry*, Prentice-Hall, New York (1946), p. 41.

[400] O. Glemser and M. Fild, in *Halogen Chemistry*, Vol. 2 (ed. V. Gutmann), Academic Press, London (1967), pp. 1–30.

[401] J. P. Freeman, in *Advances in Fluorine Chemistry*, Vol. 6 (eds. J. C. Tatlow, R. D. Peacock and H. H. Hyman), Butterworths, London (1970), p. 287.

NFX_2, dinitrogen compounds difluorodiazine N_2F_2, tetrafluorohydrazine N_2F_4, and related ions N_2F^+, $N_2F_3^+$, nitrosodifluoramine NF_2NO and halogen azides[402] (see section 8) as polynuclear species.

9.2. PREPARATION

9.2.1. Nitrogen–Halogen Compounds

Nitrogen trifluoride was first prepared[403] in 1928 by the electrolysis of molten ammonium hydrogen fluoride, and this is the basis even of the present-day commercial production. Direct fluorination methods have also been investigated[390] and the reaction with ammonia can be controlled in the presence of copper to give NF_3 in yields up to 60%:

$$2NH_3+3F_2 \rightarrow 6HF+N_2$$

$$4NH_3+3F_2 \xrightarrow{Cu} NF_3+3NH_4F$$

Nitrous oxide can be fluorinated or electrolysed in anhydrous hydrogen fluoride solution, but both reactions give rise to oxyfluoride byproducts:

$$N_2O+2F_2 \xrightarrow{400\text{–}700^\circ} NF_3+NOF$$

Nitrogen trifluoride is often a product in the reactions of other nitrogen–fluorine derivatives, e.g.

$$2N_3F+2OF_2 \xrightarrow{UV} 2NF_3+2N_2+O_2$$

and also of the fluorination of nitrogen-containing materials, but direct combination of the elements has only been observed under glow discharge conditions:

$$N_2+F_2 \xrightarrow[10\text{ mm},\ -196^\circ]{5\text{ kV, }10\text{ mA}} NF_3$$

Nitrogen trichloride is formed almost quantitatively by the action of excess chlorine or hypochlorous acid with ammonium ion in acid solution:

$$NH_4Cl+3Cl_2 \rightarrow NCl_3+4HCl$$

The reaction can be carried out in chloroform to give solutions which are stable for several days. The handling of dilute solutions containing less than 18% NCl_3 is not dangerous, but above this concentration it is an exceedingly explosive substance. Thus in order to avoid the build up of dangerous concentrations or the handling of bulk quantities when NCl_3 is used on a large scale in the bleaching and sterilizing of flour, it is prepared by electrolysing an acid solution of ammonium chloride at pH 4. The gaseous product is swept out of the cell diluted with air for immediate use and its production simply controlled electrically as required[389].

An equilibrium is thought to exist[404] when bromine reacts in liquid ammonia at -78°, as a pale reddish-yellow colour is observed with solutions of low concentration while a

[402] K. Dehnicke, *Angew. Chem. Int. Edn. Engl.* **6** (1967) 240.

[403] O. Ruff, J. Fischer and F. Luft, *Z. anorg. allgem. Chem.* **172** (1928) 417.

[404] J. Jander, *Chemistry in Anhydrous Liquid Ammonia*, Vol. 1, Part I, of *Chemistry in Nonaqueous Ionizing Solvents* (eds. G. Jander, H. Spandau and C. C. Addison), Interscience, New York (1966), p. 374.

reddish-violet solid of composition $NBr_3 \cdot 6NH_3$ together with ammonium bromide remains when the solvent is evaporated at $-78°$:

$$3NH_2Br + 4NH_3 \rightleftharpoons NBr_3 \cdot 6NH_3$$

Decomposition of this material is slow at $-70°$, rapid at $-50°$ and explosive when warmed to room temperature.

The compound commonly known as nitrogen triiodide is also an ammonia complex formed when iodine, potassium triiodide or potassium dibromiodide is added to ammonia[405]:

$$3I_2 + 5NH_3 \rightarrow NI_3 \cdot NH_3 + 3NH_4I$$

$$3KIBr_2 + 5NH_3 \rightarrow NI_3 \cdot NH_3 + 3NH_4Br + 3KBr$$

The use of finely divided iodine, formed by precipitation when alcoholic iodine solution is poured into water, can give a product containing less ammonia than indicated by the above formula[406]. All forms of the material are extremely dangerous, the brown–black powder explodes at the slightest touch and sometimes even under water. Thus detonation is so unpredictable that storage of the material is impossible.

Of the mixed halogen–nitrogen compounds, chlorodifluoramine has been most thoroughly investigated as an intermediate for the synthesis of potential high energy NF_2 compounds. It was first characterized[407] in 1959 as a product of decomposition when difluoramine–boron trichloride complex was allowed to warm to room temperature:

$$HNF_2 + BCl_3 \xrightarrow{-130°} HNF_2 \cdot BCl_3 \xrightarrow{-80 \text{ to } 25°} \underset{(50\%)}{ClNF_2} + HCl + Cl_2 + BF_3 + BCl_3$$

$ClNF_2$ is now prepared by the hypochlorite method[390] in which aqueous NaOCl (5–10% NaOH saturated with chlorine) reacts at pH 6 with difluoramine, *N,N*-difluoroureas or *N,N*-difluorosulphonamide (prepared by aqueous fluorination of sulphamide):

$$HNF_2 + NaOCl \xrightarrow{0°} ClNF_2 + NaOH$$

$$NH_2CONF_2 + NaOCl \rightarrow ClNF_2$$

$$NH_2SO_2NF_2 + NaOCl \xrightarrow{5°C} ClNF_2 \quad (70\%)$$

Quantitative yields can be obtained by using HgO instead of NaOH to displace the Cl_2/H_2O equilibrium:

$$HNF_2 + HOCl \xrightarrow[HgO]{0°} ClNF_2 + H_2O$$

Other methods of preparing $ClNF_2$ include:

$$HNF_2 + Bu^tOCl \rightarrow ClNF_2 \quad (100\%)$$

$$4NaCl + NaN_3 + F_2 \xrightarrow{0°} ClNF_2$$

$$N_3F + Cl_2 \xrightarrow{50°–80°} ClNF_2$$

---

405 U. Engelhardt and J. Jander, *Fortschr. Chem. Forsch.* **5** (1966) 663.

406 P. W. Schenk, in *Handbook of Preparative Inorganic Chemistry*, 2nd edn., Vol. 1 (ed. G. Brauer), Academic Press, New York (1963), pp. 457–517.

407 R. C. Petry, *J. Am. Chem. Soc.* **82** (1960) 2400.

Bromodifluoramine, which is less stable than the chlorine analogue, was prepared in 1960 by the hypohalite method[390]:

$$HNF_2+Br_2(aq) \xrightarrow{HgO} BrNF_2+HBr$$

Dichlorofluoramine is explosive in the liquid state and was prepared in 1960 by the reaction[408]:

$$ClF+NaN_3 \xrightarrow[\text{Ni tube}]{-10° \text{ to } 0°} Cl_2NF$$

The reaction of BrF with NaN_3 gave an unidentified product, but did not appear to be Br_2NF. No mixed halides containing iodine have as yet been characterized[390].

Tetrafluorohydrazine was first identified[390, 409] in 1957 among the thermal decomposition products of nitrogen trifluoride, and its preparation is based on the thermal reduction of NF_3 by copper, arsenic, antimony or bismuth:

$$2NF_3+2M \rightarrow N_2F_4+2MF$$

NF_3 can also be reduced through hot fluidized carbon[410], and although the yields of N_2F_4 can be high, there is a separation problem from fluorocarbon contaminants:

$$4NF_3+C \xrightarrow{440°} 2N_2F_4+CF_4$$

Mixtures of nitrogen trifluoride and nitric oxide also react when passed through a heated nickel tube for short residence times giving moderate yields of N_2F_4:

$$2NF_3+2NO \xrightarrow[600°]{\text{Ni tube}} N_2F_4+2NOF$$

The other major production route to N_2F_4 is from difluoramine generated by the urea method[390]. HNF_2 can be oxidized quantitatively with hypochlorite at pH 12, or reversibly with aqueous acidic ferric chloride solution, N_2F_4 being the only gaseous product evolved from the aqueous phase:

$$2HNF_2+NaOCl \xrightarrow{\text{pH } 12} N_2F_4+NaCl+H_2O$$

$$2HNF_2 \underset{Fe^{2+}/H_2SO_4}{\overset{Fe^{3+}/HCl}{\rightleftharpoons}} N_2F_4$$

It should of course be pointed out that the formation of tetrafluorohydrazine implies the presence of the stable difluoroamine free radical ($\cdot NF_2$) with which there is ready equilibrium[411]:

$$N_2F_4 \rightleftharpoons 2\cdot NF_2$$

Difluorodiazine N_2F_2 was first identified[412] in 1942 as a thermal decomposition product of N_3F, but it is now most conveniently prepared[413] in almost quantitative yields by the thermal decomposition of the $KF \cdot HNF_2$ complex:

$$KF+HNF_2 \xrightarrow{-80°} KF\cdot HNF_2 \xrightarrow{20°} N_2F_2+KHF_2$$

---

408 B. Sukornick, R. F. Stahl and J. Gordon, *Inorg. Chem.* **2** (1963) 875.
409 C. B. Colburn and A. Kennedy, *J. Am. Chem. Soc.* **80** (1958) 5004.
410 J. R. Gould and R. A. Smith, *Chem. Engng. News* **38** (1960) 85.
411 C. B. Colburn, *Chem. in Britain* **2** (1966) 336.
412 J. F. Haller, thesis, Cornell University, Ithaca, New York (1942).
413 E. A. Lawton, D. Pilipovich and R. D. Wilson, *Inorg. Chem.* **4** (1965) 118.

Small quantities of N_2F_2 can be prepared in good yield by the fluorination of sodium azide, and it is also an isolatable product of the following reactions:

$$NH_4F \cdot HF \xrightarrow{\text{electrolysis}} N_2F_2 \qquad (5\text{–}10\%)$$

$$NF_3 + NH_3 \xrightarrow{\text{Cu reactor}} N_2F_2 \qquad (5\%)$$

$$NF_3 + Hg(g) \xrightarrow[\text{discharge}]{\text{electric}} N_2F_2 \qquad (15\%)$$

$$HNF_2 + KF(aq) \xrightarrow{\text{pH } 8.6} N_2F_2 \qquad (75\%)$$

These methods give mixtures of both the *cis*- and *trans*-isomers, but in the following methods only the *trans*-isomer is found[390]:

$$N_2F_4 + AlCl_3 \xrightarrow{-80° \text{ to } -112°} N_2F_2 \qquad (48\%)$$

$$N_2F_4 + MCl_2 \rightarrow N_2F_2 \qquad (M{=}Sn, Co, Fe, Mn, Ni)$$

$$N_2F_4 \cdot SbF_5 + NOCl \rightarrow N_2F_2 \qquad (90\%)$$

The two isomers are thermally interconvertible and *trans*-N_2F_2 isomerizes[390] into the *cis*-form in >90% yields with a trace of fluorine in a stainless steel reactor at 75°.

The preparation of halogen azides FN_3, ClN_3 and BrN_3 has been described[402] in section 8 under azides, but mention is included here in view of their similarity to other nitrogen–halogen compounds.

Considerable effort has been spent in searching for and investigating binary fluorine–nitrogen ions[390]. Salts of the tetrafluoroammonium cation NF_4^+ have been synthesized by the following methods and their presence demonstrated by ^{19}F n.m.r.:

$$NF_3 + F_2 + AsF_5 \xrightarrow[\text{discharge}]{\text{electric glow}} NF_4^+AsF_6^-$$

$$NF_3 + F_2 + SbF_5 \xrightarrow{200°/100 \text{ atm}} NF_4^+SbF_6^-$$

$$N_2F_3^+AsF_6^- + F_2 \rightarrow NF_4^+AsF_6^-$$

Although the NF_2^+ ion is observed as the most intense peak in the NF_3 mass spectrum, no stable salts have yet been prepared.

Fluorodiazonium N_2F^+ salts have been made[390] by abstraction of a fluoride ion from *cis*-difluorodiazine with BF_3 or AsF_5:

$$cis\text{-}N_2F_2 + BF_3 \xrightarrow[50°]{150 \text{ atm}} N_2F^+BF_4^-$$

$$cis\text{-}N_2F_2 + AsF_5 \xrightarrow{-196 \text{ to } 25°} N_2F^+AsF_6^-$$

An $N_2F_3^+$ salt is believed to be formed[390] when N_2F_4 reacts with the $AsF_3 \cdot SbF_5$ complex[414]. The product is stable and has the composition NF_2SbF_5, which may be formulated $N_2F_4 \cdot 2SbF_5$ or $N_2F_3^+SbF_6^- \cdot SbF_5$, in which the SbF_5 group is either complexed to $N_2F_3^+$ or in a bridged arrangement as $Sb_2F_{11}^-$. Similarly, 1:1 and 1:2 complexes of N_2F_4 with AsF_5 are known[390] and appear to be $N_2F_3^+ \cdot AsF_6^-$ and $N_2F_3^+ \cdot As_2F_{11}^-$.

414 J. K. Ruff, *J. Am. Chem. Soc.* **87** (1965) 1140.

9.2.2. Haloamines

Difluoramine HNF_2 is an unstable explosive material[415] which was first reported[416] in 1931 as a product from the electrolysis of ammonium bifluoride. It can be made in 40% yield by the reaction of N_2F_4 with thiophenol or butyl mercaptan[417]:

$$N_2F_4 + 2PhSH \xrightarrow[4\text{ hr}]{50^\circ} 2HNF_2 + PhSSPh$$

However, it is most conveniently prepared[390] by the hydrolysis of *N,N*-difluorourea obtained from the aqueous fluorination of urea:

$$NH_2CONH_2(aq) \xrightarrow{F_2/N_2} \underset{(70\%)}{NF_2CONH_2} \xrightarrow{c.\ H_2SO_4} \underset{(100\%)}{HNF_2}$$

Monofluoramine H_2NF was also claimed to be a minor product of the electrolysis of ammonium bifluoride, but this does not seem to have been confirmed[390].

Dichloramine $HNCl_2$ has not been isolated pure but it can be formed in solution by the reaction of aqueous ammonia with hypochlorite in acid solution at about pH 5:

$$NH_3 + 2OCl^- + H^+ \rightarrow HNCl_2 + H_2O + OH^-$$

If the reaction is carried out at pH 8.5, the product is monochloramine H_2NCl:

$$NH_3 + OCl^- \xrightarrow{pH\ 8.5} H_2NCl + OH^-$$

Monochloramine can be isolated by distillation at high vacuum followed by drying and condensation or by fractional distillation from the reaction of chlorine with excess gaseous ammonia[406]:

$$Cl_2 + 2NH_3 \rightarrow H_2NCl + NH_4Cl$$

Ethereal solutions of monobromamine H_2NBr and dibromamine $HNBr_2$ can be prepared by the reaction of bromine and ammonia in ether at low temperatures[404]. Iodamine has also been postulated as an intermediate in the iodine–ammonia reaction[405].

Several attempts have been made without success[390] to prepare compounds containing fluorammonium ions NH_3F^+, or $NH_2F_2^+$. However, low temperature 1:1 complexes of HNF_2 with KF or RbF may contain the bifluoramide ion $FHNF_2^-$:

$$HNF_2 + KF \rightarrow KF \cdot HNF_2 \rightarrow K^+[HFNF_2]^-$$

9.2.3. Oxohalides

The nitrosyl halides[406] FNO, ClNO and BrNO, are all usually prepared by direct halogenation of nitric oxide:

$$2NO + X_2 \rightarrow 2XNO$$

The preparations have each been the subject of much study. The mechanism of the FNO synthesis is thought to be

$$F_2 + NO \rightarrow FNO + F$$

$$F + NO \rightarrow FNO^* \xrightarrow{(M)} FNO$$
$$\downarrow$$
$$FNO + h\nu$$

[415] E. A. Lawton and J. Q. Weber, *J. Am. Chem. Soc.* **85** (1963) 3595.
[416] O. Ruff and L. Staub, *Z. anorg. allgem. Chem.* **198** (1931) 32.
[417] J. P. Freeman, A. Kennedy and C. B. Colburn, *J. Am. Chem. Soc.* **82** (1960) 5304.

FNO can also be prepared by fluorination of NO with a metal fluoride such as AgF_2:

$$NO + AgF_2 \rightarrow FNO + AgF$$

The NO–Cl_2 reaction is kinetically third order[418], one mechanism for which assumes the formation of the NO dimer as an intermediate:

$$\text{Fast equilibrium:} \qquad 2NO \rightleftharpoons (NO)_2$$

$$\text{Rate-determining step:} \quad (NO)_2 + Cl_2 \rightarrow 2ClNO$$

$$\frac{d[ClNO]}{dt} = k[NO]^2[Cl_2]$$

The NO–Br_2 reaction is also third order and perhaps better represented as an equilibrium in that BrNO cannot be obtained pure as it decomposed reversibly to NO and Br_2:

$$2NO + Br_2 \rightleftharpoons 2BrNO$$

Nitryl fluoride FNO_2 is prepared[397] by fluorination of NO_2 or sodium nitrate with elemental fluorine or a metal fluoride such as CoF_3:

$$2NO_2 + F_2 \rightarrow 2FNO_2$$

$$NaNO_2 + F_2 \rightarrow FNO_2 + NaF$$

$$NO_2 + CoF_3 \xrightarrow{300^\circ} FNO_2 + CoF_2$$

The NO_2–F_2 reaction is believed to proceed:

$$\text{Rate-determining step:} \quad NO_2 + F_2 \rightarrow FNO_2 + F$$

$$\text{Fast reaction:} \quad F + NO_2 + M \rightarrow FNO_2 + M$$

Nitryl chloride $ClNO_2$ can be produced[406] by the interaction of chlorosulphonic acid and anhydrous nitric acid:

$$ClSO_3H + HNO_3 \rightarrow ClNO_2 + H_2SO_4$$

Similarly, any reactions generating NO_2^+ from nitric acid under anhydrous conditions in the presence of chlorine will produce $ClNO_2$:

$$HNO_3 + 2H_2SO_4 \rightarrow NO_2^+ + H_3O^+ + 2HSO_4^-$$

$$\downarrow \text{HCl}$$

$$ClNO_2 + H_2O + 2H_2SO_4$$

$$N_2O_5 + PCl_5 \rightarrow 2ClNO_2 + POCl_3$$

Nitryl bromide $BrNO_2$ is reported[404] to exist in equilibrium with bromine and NO_2–N_2O_4 in a mixture of these reagents.

Trifluoramine oxide ONF_3 was reported[419] in 1966 but had previously been synthesized in 1961. It was obtained[390] by electric discharge of NF_3–O_2–OF_2 mixtures at $-196°$, flame

418 I. C. Hisatsune and L. Zaforte, *J. Phys. Chem.* **73** (1969) 2980.
419 N. Bartlett and S. P. Beaton, *Chem. Communs.* (1966) 167.

fluorination of NO or photochemical fluorination of NOF at 25°. ONF_3 is now made in stoichiometric yields by fluorination of NOF with iridium hexafluoride[420] whereas OsF_6 or PtF_6 give only trace amounts:

$$3NOF + 2IrF_6 \xrightarrow{20^\circ} ONF_3 + 2NO^+IrF_6^-$$

Attempts have been made to prepare salts of the ions NOF_2^-, $NO_2F^=$, HNO_2F^- and $NO_2F_2^-$, but no conclusive evidence is available[390].

The majority of compounds containing sulphur, nitrogen and halogen do not contain an N–X bond[400, 421]. However, a number of NF_2, NF and NCl derivatives are known and their preparations are exemplified by the following equations:

$$N_2F_4 + S_2F_{10} \xrightarrow{140^\circ} 2SF_5NF_2 \ (60\%)$$

$$N_2F_2 + N_2F_4 + SF_4 \xrightarrow[17\text{ hr}]{100^\circ} SF_5NF_2 \ (100\%)$$

$$N_2F_4 + CF_3SF_3 \xrightarrow{UV} trans\text{-}CF_3SF_4NF_2 \ (7\%)$$

$$N_2F_2 + N_2F_4 + SO_2 \xrightarrow[21\text{ hr}]{100^\circ} FSO_2NF_2 \ (83\%)$$

$$(FSO_2)_2NH + F_2/He \xrightarrow{25^\circ} (FSO_2)_2NF$$

$$NSF + Cl_2 \xrightarrow{CsF} ClN{=}SF_2$$

Finally, the compound nitrosodifluoramine NF_2NO has been identified[422] as the intense dark blue material which collects on a liquid nitrogen cold finger when the gaseous products formed by heating 10:1 mixtures of NO and N_2F_4 to 300° are condensed. The compound dissociates on vaporization to give NO and N_2F_4 in a 2:1 ratio.

9.3. STRUCTURE AND BONDING

Structures which have been determined for nitrogen–halogen and oxygen–nitrogen–halogen compounds are detailed in Tables 54 and 55 respectively.

In general, their shapes are as would be predicted on the basis of the Sidgwick–Powell approach. The NF_3 molecule has similar symmetry to ammonia[423] and its bonding is readily described in terms of molecular orbitals. The decrease in FNF bond angle as compared to HNH, although usual, is surprising when one considers that fluorine is much larger than hydrogen. N_2F_4 has a skew configuration[424] similar to hydrazine, due presumably to the repulsion between the nitrogen lone-pairs in an sp^3 environment. Dissociation results in the $\cdot NF_2$ free radical whose nineteen electrons can be contained in the resonance structures illustrated in Fig. 26. On a molecular orbital description, the unpaired electron probably lies in a $2b_1$ antibonding orbital[423], also illustrated in Fig. 26.

420 N. Bartlett, J. Passmore and E. J. Wells, *Chem. Communs.* (1966) 213.

421 H. G. Heal, in *Inorganic Sulphur Chemistry* (ed. G. Nickless), Elsevier, Amsterdam (1968), p. 459.

422 C. B. Colburn and F. A. Johnson, *Inorg. Chem.* **1** (1962) 715.

423 M. Green, in *Developments in Inorganic Nitrogen Chemistry*, Vol. 1 (ed. C. B. Colburn), Elsevier, Amsterdam (1966), pp. 1–71.

424 R. K. Bohn and S. H. Bauer, *Inorg. Chem.* **6** (1967) 304.

TABLE 54. STRUCTURAL PARAMETERS[a] OF NITROGEN–HALOGEN COMPOUNDS

Compound	Shape	Point group	Bond lengths (Å)			Bond angles		
			N–X	N–H	N–H	XNX	NNX	HNX
NF_3	Pyramidal	C_{3v}	1.37			103		
N_2F_4	Skew	C_2	1.39	1.48			102.5	
							105	
							dihedral angle = 66	
$\cdot NF_2$	Bent	C_{2v}	1.365			103		
cis-N_2F_2	Planar	C_{2v}	1.41	1.21			114	
trans-N_2F_2	Planar	C_{2h}	1.40	1.22			106	
HNF_2	Pyramidal	C_2	1.400		1.026	102.9		99.8 (microwave)
			1.37		1.08	104.3		103.5 (electron diffraction)
NCl_3	Pyramidal	C_{3v}						
$HNCl_2$	Pyramidal	C_2	1.75			106		102
H_2NCl	Pyramidal	C_2	1.76			HNH = 106.8		102
$NI_3 \cdot NH_3$ [b]			2.14, 2.15, 2.30 (bracketed)					

[a] Refs. 389, 390.
[b] H. Hartl, H. Bärnighausen and J. Jander, *Z. anorg. allgem. Chem.* **357** (1968) 225.

TABLE 55. STRUCTURAL PARAMETERS[a] OF OXYGEN–NITROGEN–HALOGEN COMPOUNDS

Compound	Shape	Point group	Bond lengths (Å)		Bond angles	
			N–X	N–O	XNO	ONO
ONF_3	Tetrahedral	C_{3v}	1.48	1.15		
FNO	Bent	C_s	1.52	1.13	110	
ClNO	Bent	C_s	1.98	1.14	113	
BrNO	Bent	C_s	2.14	1.15	117	
FNO_2	Planar	C_{2v}	1.35	1.23		125
$ClNO_2$	Planar	C_{2v}	1.84	1.20		131
NO	Linear	$C_{\infty v}$		1.15		
NO_2^-	Bent	C_{2v}		1.24		

[a] Ref. 390.

Both the isomers of difluorodiazine N_2F_2 have N–F bond lengths longer[425] than those of NF_3 and an N–N distance shorter than in MeN=NMe. It has been in situations like this that the Linnett approach to the theory of chemical bonding has been most useful[426]. The basis of this approach is the assignment of electrons into spin sets rather than electron pairs. For structures containing one or more atoms capable of catenation, a number of combinations would be possible. On this basis N_2F_2 would have four satisfactory electronic structures as illustrated in Fig. 27. The structural parameters clearly demonstrate the importance of the mirror image resonance structures. Furthermore, its dissociation product, the NF molecule, would by this description have a formal charge of +1 on the fluorine atom which may indicate why it dimerizes so readily.

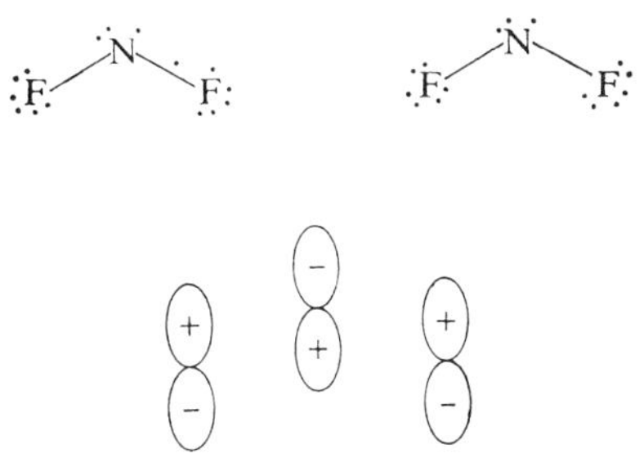

$2b_1$ antibonding orbital

FIG. 26. Resonance structures of $\cdot NF_2$.

425 R. K. Bohn and S. H. Bauer, *Inorg. Chem.* **6** (1967) 309.
426 J. W. Linnett, *The Electronic Structure of Molecules*, Wiley, New York (1964).

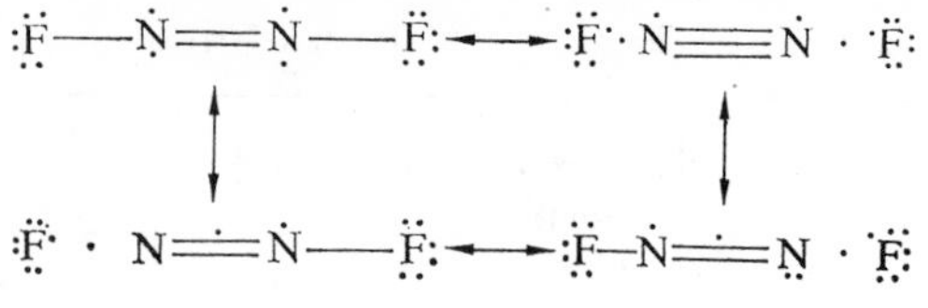

:F—N=N—F: ⟷ :F· N≡N · F:

↕ ↕

:F · N=N—F: ⟷ :F—N=N · F:

FIG. 27. Linnett structures of N_2F_2.

The nitrosyl halides contain a long nitrogen–halogen bond while the N–O bond length is about the same as in nitric oxide. FNO with eighteen electrons is best described in terms of a Linnett non-pairing structure illustrated in Fig. 28 rather than the formal F—N=O. A similar situation is present in ClNO.

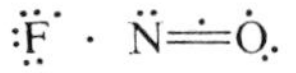

:F · N=O.

FIG. 28. Linnett structure of FNO.

The nitryl halides FNO_2 and $ClNO_2$, which contain twenty-four electrons, are isoelectronic and isostructural with the nitrate ion and, as indicated in Table 55, the NO_2 groups are similar in shape and size to the nitrite ion. In FNO_2 the N–F bond length is similar to that in NF_3, which implies a full two-electron bond. Assignment of three electrons to each N–O bond satisfies the electron count and places formal charges of $+1$ on nitrogen and $-\frac{1}{2}$ on each oxygen as illustrated in Fig. 29, but the N–Cl distance in $ClNO_2$ is lengthened slightly suggesting a greater contribution from the alternative structure.

FIG. 29. Linnett structures of FNO_2 and $ClNO_2$.

9.4. MOLECULAR CONSTANTS

As the majority of nitrogen–halogen compounds are covalent, much of the available data for calculating molecular constants arises from information given by methods applicable in organic chemistry, in particular the spectroscopic techniques of infrared, n.m.r. and mass spectrometry. Infrared spectra of N–X species have been widely used as a means of indicating the presence of either a particular structural feature or a particular compound[390]. Full assignments of the bands in some nitrogen–fluorine derivatives have led to the calculation of structural parameters. With regards to n.m.r. data, ^{14}N n.m.r. can theoretically be recorded for all N–X compounds but in fact experimental difficulties have prevented this until quite recently[427, 428]. ^{15}N n.m.r. can only be considered when circumstances warrant the expense incurred by the synthesis of such compounds with highly enriched ^{15}N materials. However, much information has been obtained through ^{19}F n.m.r.

[427] J. (Banus) Mason and W. Van Bronswijk, *Chem. Commun*s. (1969) 357.
[428] L.-O. Andersson, J. (Banus) Mason and W. Van Bronswijk, *J. Chem. Soc.* A (1970) 296.

of fluorine–nitrogen species[429], the spectra of which are relatively easy to obtain, frequently simple and can often establish unequivocal structures. Some ^{19}F and ^{14}N chemical shifts of nitrogen–fluorine species relative to $CFCl_3$ and the saturated aqueous nitrite ion respectively are given in Table 56.

TABLE 56. ^{19}F AND ^{14}N CHEMICAL SHIFTS OF NITROGEN–FLUORINE SPECIES

Compound	^{19}F shift relative to $CFCl_3$	^{14}N shift relative to aqueous nitrite ion
FNO	−479	+128
FNO_2	−393	+297
NF_4^+ (in HF)	−215	
$NF_2{=}N\underline{F}^+$ (in SO_2)	−192	
$N\underline{F}_2{=}NF^+$ (in SO_2)	−144	
NF_3 (l)	−142	+238
$ClNF_2$	−141	
NF_3 (g)	−138	
cis-N_2F_2	−129	
N_2F^+ (in HF)	−103	
trans-N_2F_2	−88	
SF_5NF_2	−66	
NF_2NO	−65	
N_2F_4 (l)	−58	
N_2F_4 (g)	−53	
HNF_2	+6	

TABLE 57. DIPOLE MOMENTS[a] AND BOND ENERGIES[b] OF NITROGEN–HALOGEN COMPOUNDS

Compound	Dipole moment (*D*)	*E*(N–X)	*D*(N–X) (kcal/mol)	*D*(N–N)
NF_3	0.234	66.4	57.1	
			71	
N_2F_4	0.26		71	20.3
cis-N_2F_2	0.18			
trans-N_2F_2	0.00		68.2	107.9
HNF_2	1.92			
FNO	1.81		55.4	
FNO_2	0.47		45	
NCl_3	0.6			
ClNO	0.42			
$ClNO_2$	0.42			
$ONNF_2$				10.1

[a] Ref. 430. [b] Ref. 390.

[429] W. S. Brey, Jr. and J. B. Hynes, in *Fluorine Chemical Reviews*, Vol. 2 (ed. P. Tarrant), Marcel Dekker, New York (1968), p. 111.

The interpretation of fragmentation patterns in mass spectra has also led to useful predictions concerning the chemistry of nitrogen–fluorine compounds[409], but the generally low intensities of such species[391] due to the reluctance of fluorine to lose an electron and form a positive ion is a major limitation[392].

Dipole moments[430] have been measured for a number of N–X compounds which are listed in Table 57. In comparing the values of NF_3 and NH_3, 0.234 and 1.47 *D* respectively, the large difference is explained on the basis that in NF_3 the N–F bond moments are so oriented as to almost cancel the lone-pair–nitrogen moment, whereas the bond and lone-pair–nitrogen moments of NH_3 are additive. Likewise the strength of N–F bonds in terms of bond energies have been discussed, and again the distribution in NF_3 is believed to be different to that in ammonia[390]. Some other values are also included in Table 57 for comparison.

9.5. PHYSICAL PROPERTIES

The physical properties of nitrogen–halogen compounds which are easy to prepare and handle have been well investigated, but the unstable and explosive materials are less well documented. Values for these properties are presented in groups of related compounds in Tables 58–64.

9.6. CHEMISTRY AND CHEMICAL PROPERTIES

9.6.1. Nitrogen–Halogen Compounds

Nitrogen trifluoride is a remarkably unreactive material whose properties are more like those of inert CF_4 than of its other immediate neighbour highly reactive OF_2. This may to some extent be attributed to its thermal stability with respect to the elements; however, at elevated temperatures, NF_3 acts as a fluorinating agent towards a variety of metals and metalloids and this provides a method of preparing tetrafluorohydrazine[390]:

$$2NF_3 + 2M \rightarrow N_2F_4 + 2MF$$

NF_3 is slowly hydrolysed by aqueous base at 100° or aqueous HCl at 133°, but is resistant to strong acids and neutral aqueous solutions. It reacts with aqueous iodide, conventional reducing agents such as $LiAlH_4$, and sparking with reducing agents such as H_2, NH_3, CH_4, CO, H_2S or H_2O can lead to violent explosions[389]:

$$4NF_3 + 3LiAlH_4 \rightarrow 4NH_3 + 3LiAlF_4$$

$$2NF_3 + 3H_2O \xrightarrow{\text{spark}} 6HF + NO + NO_2$$

Although weak associations have been reported[431] between NF_3 and BF_3 at −125° and for NF_3 and BCl_3 at −100°, it is clear that the unshared electron pair of NF_3 does not contribute much Lewis base character.

Nitrogen trichloride undergoes photochemical decomposition to the elements particularly in the presence of chlorine, while detonation can be initiated in several ways including thermal and mechanical shock[389]. The reactions of NCl_3 with nitrogen oxides are complex, but the main reaction with NO at −80° is thought to be

$$NCl_3 + 2NO \rightarrow N_2O + NOCl + Cl_2$$

---

430 A. L. McClellan, *Tables of Experimental Dipole Moments*, W. H. Freeman, London (1963).
431 A. D. Craig, *Inorg. Chem.* **3** (1964) 1628.

TABLE 58. PHYSICAL PROPERTIES OF NITROGEN TRIFLUORIDE[a] AND TRIFLUORAMINE OXIDE[b]

Property	Temperature (K)	NF_3	Temperature (K)	ONF_3
Melting point	66.35	−206.8°C	113.15	−160°C
Boiling point	144.15	−129.0°C	185.65	−87.6°C
Triple point	66.35	−206.8°C		
Critical temperature	233.85	−39.3°C	243.65	29.5°C
Critical pressure		44.7 atm		63.5 atm
$\triangle H$ fusion	66.35	0.095 kcal/mol		
$\triangle H$ vaporization	144.15	2.77 kcal/mol	185.65	3.85 kcal/mol
$\triangle H_f^\circ$ (enthalpy of formation)	0	−28.43 kcal/mol		
	298	−29.8 kcal/mol		
H_{298}–H_0 (enthalpy)	298	2.827 kcal/mol	298	3.282 kcal/mol
$\triangle G_f^\circ$ (Gibbs energy of formation)	298	−19.9 kcal/mol		
S° (entropy) (g)	144.15	62.29 cal/deg mol		
S° (entropy) (l)	144.15	35.2 cal/deg mol		
S° (entropy)	298	54.5 cal/deg mol	298	66.50 cal/deg mol
C_p° (heat capacity)			298	16.21 cal/deg mol
Density of solid				0.593 g/cm³
Density of liquid	66.35	1.92 g/cm³		
Trouton's constant		19.94		20.7
Vapour pressure	$\log p_{mm} = 4.646 - 662T$		$\log p_{mm} = 180/T - 10.39$	

[a] Refs. 390, 398.
[b] W. B Fox, J. S. McKenzie, E. R. McCarthy, J. R. Holmes, F. R. Stahl and R. Juurik, *Inorg. Chem.* 7 (1968) 2064.

NCl_3 is readily hydrolysed but its reaction with chlorite even in aqueous solution can be used for the preparation of chlorine dioxide[389]:

$$NCl_3+3H_2O \rightarrow NH_3+3HOCl$$
$$NCl_3+3H_2O+6NaClO_2 \rightarrow 6ClO_2+3NaCl+3NaOH+NH_3$$
$$2NCl_3+6NaClO_2 \rightarrow 6ClO_2+6NaCl+N_2$$

With the uncertainty as to the nature of nitrogen tribromide, little is known of its chemistry[389]. Nitrogen triiodide–ammonia complex has maximum stability to hydrolysis at pH 10, but it readily reacts with acids, base or aqueous sulphite[389]:

$$NI_3\cdot NH_3+HCl \rightarrow [NI_3+NH_4Cl] \xrightarrow{4HCl} 3ICl+2NH_4Cl$$
$$NI_3\cdot NH_3+4NaOH \rightarrow 2NaI+NaNO_2+NH_3+2H_2O$$
$$NI_3\cdot NH_3+3Na_2SO_3+3H_2O \rightarrow 3Na_2SO_4+3HI+2NH_3$$

TABLE 59. PHYSICAL PROPERTIES[a] OF DIFLUORAMINE AND CHLORODIFLUORAMINE

Property	Temperature (K)	HNF_2	$ClNF_2$
Melting point	157.15	-116°C	
Boiling point	249.55	-23.6°C	
	206.15		-67°C
Critical temperature	403.15	130°C	
Critical pressure	·	75 atm	
Critical density		$1.424-0.00202t$ g/cm³	
ΔH vaporization	249.55	5.940 kcal/mol	4.350 kcal/mol
ΔH_f (enthalpy of formation)	0	-16.1 kcal/mol	3.2 kcal/mol
H_{298}–H_0 (enthalpy)	298	25.582 kcal/mol	
$S°$ (entropy)	298	60.40 cal/deg mol	
C_p° (heat capacity)	298	10.37 cal/deg mol	
Trouton's constant		23.7	21.0
Vapour pressure	145–250	$\log p_{mm} = 8.072+1.298/T$	
Density of liquid		$1.424-0.00202t$ g/cm³	

[a] Ref. 394.

The reactions of chlorodifluoramine that have been studied as part of the advanced oxidizer research programme have recently been summarized[390]. Many reactions give quite complex mixtures, some of which are more easily achieved by analogous reactions of N_2F_4–Cl_2 mixture. Photolytic reactions of $ClNF_2$ or their reactions with nucleophiles frequently give N_2F_4 as the only product while its reaction with n-alkyl mercury compounds leads to the formation of difluoraminoalkanes[391]:

$$3ClNF_2+R_2Hg \rightarrow RNF_2+2RHgCl+RCl+N_2F_4$$

Reactions of other mixed halogen–nitrogen compounds only lead to decomposition products.

There have been several attempts to classify the numerous reactions of tetrafluorohydrazine[391, 392]. Many reactions involve $\cdot NF_2$ radicals formed by dissociation at elevated temperatures.

Examples[390] include additions to unsaturated systems, reactions with other radicals or processes involving hydrogen abstraction:

$$N_2F_4 \rightleftharpoons 2\cdot NF_2 \xrightarrow{R_fCF=CF_2} R_fCF(NF_2)CF_2NF_2$$

$$N_2F_4 \rightleftharpoons 2\cdot NF_2 \xrightarrow{2NO} 2NONF_2$$

$$N_2F_4 \rightleftharpoons 2\cdot NF_2 \xrightarrow[50^\circ\ 4\ hr]{2PhSH} 2HNF_2 + PhSSPh$$

TABLE 60. PHYSICAL PROPERTIES OF NITROGEN TRICHLORIDE[a], DICHLOROFLUOROAMINE[b] AND BROMODIFLUORAMINE[b]

Property	Temperature (K)	NCl_3	Cl_2NF	$BrNF_2$
Melting point		233 K −40°C		
Boiling point		344 K 71°C	263–273 K −10 to 0°C	237 K −36°C
ΔH_f (enthalpy of formation)	298	−54.7 kcal/mol		
Density of liquid	293	1.65 g/cm³		

[a] Ref. 395.
[b] Ref. 390

TABLE 61. PHYSICAL PROPERTIES[a] OF TETRAFLUOROHYDRAZINE AND THE DIFLUOROAMINO FREE RADICAL

Property	Temperature (K)	N_2F_4	$\cdot NF_2$
Melting point	108.65	−164.5°C	
Boiling point	200.15	−73°C	
Triple point	104.85	−168.3°C	
Critical temperature	309.15	36°C	
ΔH vaporization	200.15	3.710 kcal/mol	
ΔH_f (enthalpy of formation)	0	0.88 kcal/mol	10.92 kcal/mol
	298	−1.7 kcal/mol	10.3 kcal/mol
$H_{298}-H_0$ (enthalpy)	298	3.710 kcal/mol	2.527 kcal/mol
ΔG_f (Gibbs energy of formation)	298	19.4 kcal/mol	13.8 kcal/mol
S° (entropy)	298	71.96 cal/deg mol	59.71 cal/deg mol
C_p° (heat capacity)	298	18.9 cal/deg mol	9.80 cal/deg mol
Density of liquid	173	1.5 g/cm³	
	200	1.397 g/cm³	
Density of gas	294	0.00444 g/cm³ at 1 atm	
Vapour pressure		$\log p_{mm} = -692/T + 6.33$	

[a] Refs. 390, 398.

Irradiation of the $N_2F_4/{}^{\cdot}NF_2$ equilibrium mixture can give an excited $^{\cdot}NF_2$ radical as well as NF and F species[390].

$$\begin{array}{lcl} N_2F_4 \rightleftharpoons 2{}^{\cdot}NF_2 & \xrightarrow{SO_2} & FSO_2NF_2 \\ \quad\downarrow & & \\ \quad NF_2^{*} & \xrightarrow{SF_4} & SF_5NF_2 \\ \quad\downarrow & & \\ \quad NF+F & \xrightarrow{RNO} & RN(O){=}NF \\ & \xrightarrow{CO} & FCONF_2 \end{array}$$

TABLE 62. PHYSICAL PROPERTIES[a] OF *cis*- AND *trans*-DIFLUORODIAZINE

Property	*cis*-N_2F_2	*trans*-N_2F_2
Melting point	<78.15 K <-195°C	101.15 K -172°C
Boiling point	167.45 K -105.7°C	161.75 K -111.4°C
Critical temperature	272 K -1°C	260 K -13°C
Critical pressure	70 atm	55 atm
ΔH vaporization	3.670 kcal/mol	3.400 kcal/mol
ΔH_f (enthalpy of formation)	16.6 kcal/mol	19.6 kcal/mol
Vapour pressure	$\log p_{mm} = 7.675-803.0/T$	$7.470-742.3/T$

[a] Ref. 398.

N_2F_4 can act as a strong fluorinating agent under a wide range of conditions and hence unlikely to occur by way of a common mechanism[390]:

$$N_2F_4+SiH_4 \xrightarrow{25^\circ} SiF_4+N_2+H_2$$

$$N_2F_4+Li \xrightarrow{-80 \text{ to } 250^\circ} LiF+Li_3N$$

$$N_2F_4+S \xrightarrow{110 \text{ to } 140^\circ} SF_4+SF_5NF_2$$

The hydrolysis of N_2F_4 under various conditions leads to the formation of HF together with NO at 60°, N_2 and NO_3^- at 133°, and N_2O and NO_2^- at 133° in basic solution. Furthermore, N_2F_4 can behave as a reducing agent in its reaction with oxygen difluoride:

$$2N_2F_4+OF_2 \rightarrow 3NF_3+NOF$$

Finally, reactions in which a fluoride ion is abstracted from N_2F_4 lead to stable salts[414] which are thought to be the $N_2F_3^+$ ion:

$$N_2F_4+AsF_5 \rightarrow N_2F_3^+AsF_6^-$$

Attempts to carry out addition reactions across the nitrogen–nitrogen double bond of *cis*- and *trans*-difluorodiazine have so far been unsuccessful. All that can be said is that the

TABLE 63. PHYSICAL PROPERTIES OF NITROSYL HALIDES

Property	Temperature (K)	FNO [a]	ClNO [b]	BrNO [b]
Melting point		140.65 K	213.55 K	
		−132.5°C	−59.6°C	
Boiling point		213.25 K	266.75 K	~273 K
		−59.9°C	−6.4°C	~0°C
Triple point			−211.65K/38.6 mm	
			−61.5°C/38.6 mm	
Critical temperature			440.65 K	
			167.5°C	
Critical pressure			90 atm	
Critical density			0.47 g/cm³	
$\triangle H$ fusion	213.55		1.430 kcal/mol	
$\triangle H$ vaporization	213.25	4.607 kcal/mol	6.0 kcal/mol	
	266.75		6.16 kcal/mol	
$\triangle H_f^\circ$ (entropy of formation)	0	−15.33 kcal/mol	12.81 kcal/mol	21.86 kcal/mol
$\triangle H_f^\circ$ (entropy of formation)	298	−15.9 kcal/mol	12.36 kcal/mol	19.64 kcal/mol
H_{298}–H_0 (enthalpy)	298	2.557 kcal/mol	2.716 kcal/mol	2.785 kcal/mol
$\triangle G_f$ (Gibbs energy of formation)	298	−12.2 kcal/mol	15.79 kcal/mol	19.70 kcal/mol
S° (entropy)	298	59.27 cal/deg mol	62.52 cal/deg mol	65.38 cal/deg mol
C_p° (heat capacity)	298	9.88 cal/deg mol	10.68 cal/deg mol	10.87 cal/deg mol
Density of solid		1.719 g/cm³		
Density of liquid	213.25	1.326 g/cm³	1.592 g/cm³	
	213.25	$1.919-0.00278T$ g/cm³		
Trouton's constant		21.6		
Vapour pressure		100 mm at −88.8 °C	100 mm at −46.3 °C	
		200 mm at −79.2 °C	200 mm at −34.0 °C	
Specific conductivity	263		$6\times10^{-7}\ \Omega^{-1}\ cm^{-1}$	
Dielectric constant	263		19.7	
	253.5		21.4	
	246		22.5	

[a] Refs. 393, 396, 397.
[b] Refs. 389, 395.

TABLE 64. PHYSICAL PROPERTIES OF NITRYL FLUORIDE[a] AND NITRYL CHLORIDE[b]

Property	Temperature (K)	FNO_2 [a]	Temperature (K)	$ClNO_2$ [b]
Melting point	107.15	−166°C	128.15	−145°C
Boiling point	200.65	−72.5°C	257.25	−15.9°C
Critical temperature	349.45	76.3°C		
$\triangle H$ vaporization		4.32 kcal/mol		5.0 kcal/mol
$\triangle H_f^\circ$ (enthalpy of formation)			0	4.29 kcal/mol
$\triangle H_f^\circ$ (enthalpy of formation)	298	−19 kcal/mol	298	3.00 kcal/mol
$H_{298}–H_0$ (enthalpy)	298	2.755 kcal/mol	298	2.915 kcal/mol
$\triangle G^\circ$ (Gibbs energy of formation)	298	−8.88 kcal/mol	298	13.0 kcal/mol
S° (entropy)	298	62.2 cal/deg mol	298	65.02 cal/deg mol
C_p° (heat capacity)	298	11.9 cal/deg mol	298	12.71 cal/deg mol
Density of solid		1.92 g/cm³		
Density of liquid	172.15	1.571 g/cm³	193	1.5 g/cm³
	200.65	1.494 g/cm³	257	1.41 g/cm³
		$2.143-0.00323T$		
Trouton's constant		21.2		
Viscosity	172.15	0.00572 poise	193	0.00616 poise
	200.65	0.00460 poise		
Vapour pressure	172.15	113.8 mm	193	16.1 mm
Surface tension	168.65	27.6 dyne/cm		

[a] Refs. 393, 396, 397.
[b] Ref. 395.

cis-form exhibits greater reactivity than the *trans*-isomer as evidenced[432] by its reaction over a period of 2 weeks with glass to form SiF_4 and N_2O.

Reactions of binary fluorine–nitrogen ions[390] include those of the fluorodiazonium ion as the hexafluoroarsenate $N_2F^+AsF_6^-$ in HF solution:

$$N_2F^+AsF_6^- + NaF \xrightarrow{HF} NaAsF_6 + cis\text{-}N_2F_2$$

$$N_2F^+AsF_6^- + H_2O \longrightarrow N_2O + HAsF_6 + HF$$

$$2N_2F^+AsF_6^- + 2NO \xrightarrow{25^\circ} 2NOAsF_6 + N_2 + cis\text{-}N_2F_2$$

$$2N_2F^+AsF_6^- + 2O_2 \longrightarrow 2O_2AsF_6 + 2N_2 + F_2$$

Reactions of the ionic species thought to be $N_2F_3^+$ with reducing agents have been carried out in liquid SO_2 or AsF_3 and lead to the formation of *trans*-N_2F_2:

$$N_2F_3^+SbF_6^- \cdot SbF_5 + 2(C_5H_5)_2Fe \rightarrow N_2F_2 + 2(C_5H_5)_2FeSbF_6$$

$$N_2F_3^+SbF_6^- \cdot SbF_5 + 2NOCl \rightarrow N_2F_2 + 2NOSbF_6 + Cl_2$$

$$N_2F_3^+SbF_6^- \cdot SbF_5 + I_2 \rightarrow N_2F_2 + [2I^+] + 2SbF_6^-$$

9.6.2. Haloamines

Difluoramine forms weak complexes with both Lewis acids such as boron halides, and Lewis bases like ethers, and can thus be described as amphoteric. These complexes are not usually very stable and either decompose or dissociate[433]:

$$HNF_2 + BCl_3 \xrightarrow{-80^\circ} BCl_3 \cdot HNF_2 \xrightarrow[\text{to } 25^\circ]{\text{warming}} ClNF_2 + BCl_3 + BF_3 + HCl + Cl_2 + \text{white solid}$$

$$HNF_2 + BF_3 \xrightleftharpoons{-56 \text{ to } -23^\circ} BF_3 \cdot HNF_2$$

The relative base strengths of difluoramines[431] decrease in the order

$$EtNF_2 > MeNF_2 \sim CD_3NF_2 > HNF_2 > ClNF_2 > N_2F_4 \sim CF_3NF_2 \sim NF_3$$

Complexes of difluoramine with basic alkali metal fluorides[434], $MF \cdot HNF_2$, are thought to be hydrogen-bonded complexes of the type $M^+F^- \cdot HNF_2$, although some evidence suggests $M^+ \cdot FH\text{–}NF_2^-$. Thermal decomposition of such complexes gives *cis*- and *trans*-isomers of difluorodiazine in good yield:

$$2MF \cdot HNF_2 \rightarrow FN{=}NF + 2MF \cdot HF$$

The redox reaction of HNF_2 with base is described by the following equations and any additional fluoride ion which is formed over the stoichiometric amount probably results from the hydrolysis of N_2F_2 or N_2F_4:

$$2HNF_2 + 2NaOH \rightarrow N_2F_2 + 2NaF + 2H_2O$$

$$4HNF_2 + 4NaOH \rightarrow N_2F_4 + N_2 + 4NaF + 4H_2O$$

---

[432] C. B. Colburn, F. A. Johnson, A. Kennedy, K. McCallum, L. C. Metzger and C. O. Parker, *J. Am. Chem. Soc.* **81** (1959) 6397.

[433] R. C. Petry, *J. Am. Chem. Soc.* **82** (1960) 2400.

[434] H. E. Dubb, R. C. Greenough and E. C. Curtis, *Inorg. Chem.* **4** (1965) 648.

HNF_2 is reduced rapidly and quantitatively with aqueous hydrogen iodide[435] while its oxidation with aqueous ferric solution to give tetrafluorohydrazine is thought to proceed via NF_2^-:

$$HNF_2+4HI \longrightarrow NH_4F+2I_2+HF$$

$$2HNF_2+2Fe^{3+} \xrightarrow{H_2O} N_2F_4+2Fe^{2+}+2H^+$$

However, it would appear that more than one of these three reaction types complex formation, redox and ionic, can occur simultaneously[390].

Monochloramine is an important intermediate in the Raschig synthesis of hydrazine from the reaction of ammonia with hypochlorite (see section 6.1). In alkaline solution, H_2NCl decomposes to nitrogen and ammonia by way of hydroxylamine intermediate. Presumably the initial step is NH_2OH formation by nucleophilic substitution followed by rapid decomposition[436]:

$$H_2NCl+OH^- \rightarrow NH_2OH+Cl^-$$

$$H_2NCl+2NH_2OH+OH^- \rightarrow N_2+NH_3+3H_2O+Cl^-$$

Reaction of monochloramine with tertiary amines and phosphines provides convenient routes to substituted hydrazonium chlorides and aminophosphonium chlorides[437]:

$$H_2NCl+R_3N \rightarrow (R_3NNH_2)Cl$$

$$H_2NCl+R_3P \rightarrow (R_3PNH_2)Cl$$

9.6.3. Oxohalides

Nitrosyl fluoride FNO readily reacts with many elements forming metal fluorides and with a variety of metal fluorides to form nitrosyl salts[396, 397]:

$$FNO+M \rightarrow MF+NO$$

$$FNO+AsF_5 \rightarrow NOAsF_6$$

The reaction of FNO with iridium hexafluoride also leads to the formation of trifluoramine oxide[420]:

$$3FNO+2IrF_6 \rightarrow 2NOIrF_6+ONF_3$$

Like the other nitrosyl halides, FNO reacts with hydroxylic compounds to give nitrites:

$$XNO+ROH \rightarrow RONO+HX$$

In the reaction with water, however, the nitrous acid formed initially reacts further to give nitric acid and nitric oxide:

$$XNO+H_2O \rightarrow HNO_2+HX$$

$$3HNO_2 \rightarrow HNO_3+2NO+H_2O$$

Thus the overall mixture is a potent solvent for metals, being similar in composition to aqua regia (HNO_3–HCl):

$$4XNO+3H_2O \rightarrow HNO_3+HNO_2+2NO+4HX$$

---

[435] E. A. Lawton and J. Q. Weber, *J. Am. Chem. Soc.* **81** (1959) 4755.
[436] M. Anbar and G. Yagil, *J. Am. Chem. Soc.* **84** (1962) 1790.
[437] W. Jolly, *Inorganic Chemistry of Nitrogen*, Benjamin, New York (1964).

The reaction of nitrosyl halides in basic solutions is also similar[389]:

$$4XNO + 6NaOH \rightarrow NaNO_3 + NaNO_2 + 4NaX + 2NO + 3H_2O$$

FNO reacts with SO_2 to give nitrososulphuryl fluoride[438] which can be stored at $-78°$ and acts as stabilized FNO by decomposing into its constituent compounds on warming to 19°:

$$FNO + SO_2 \rightleftarrows FSO_2NO$$

Equilibrium constants for the thermal decomposition of nitrosyl chloride have been determined:

$$2ClNO \rightleftharpoons 2NO + Cl_2$$

At elevated temperatures, however, the rate of decomposition is faster than would be expected due to the formation of chlorine atoms by the reaction

$$NO + Cl_2 \rightleftharpoons ClNO + Cl$$

which can then cause further decomposition of ClNO by the reverse process. ClNO also reacts rapidly and reversibly with other halogens[389]:

$$2ClNO + I_2 \rightleftharpoons 2NO + 2ICl$$

The oxidation of ClNO leads to decomposition into nitrogen dioxide and chlorine:

$$ClNO + \tfrac{1}{2}O_2 \rightarrow NO_2 + \tfrac{1}{2}Cl_2$$

Liquid nitrosyl chloride is a useful ionizing solvent[439] for other nitrosyl salts (see section 11.7), and can act as a medium for preparing nitrosyl derivatives of metal halide anions:

$$ClNO \rightleftharpoons NO^+ + Cl^-$$

$$ClNO + FeCl_3 \xrightarrow[ClNO]{liq.} NO^+FeCl_4^-$$

Nitrosyl bromide like ClNO also decomposes reversibly even at room temperature (~7% at 1 atm pressure of BrNO), and a rapid reversible reaction occurs when chlorine is present[389]:

$$2BrNO \rightleftharpoons 2NO + Br_2$$

Nitryl fluoride FNO_2 reacts with most metals to form both the metal fluoride and oxide, whereas some Group VI metals form the oxyfluoride, while almost all non-metals give nitryl salts[397]:

$$\left.\begin{array}{l} 2FNO_2 + Zn \rightarrow ZnF_2 + 2NO_2 \\ NO_2 + Zn \rightarrow ZnO + NO \end{array}\right\}$$

$$2FNO_2 + Cr \rightarrow CrO_2F_2 + 2NO$$

$$\left.\begin{array}{l} 3FNO_2 + B \rightarrow BF_3 + 3NO_2 \\ FNO_2 + BF_3 \rightarrow NO_2BF_4 \end{array}\right\}$$

Nitryl chloride $ClNO_2$ decomposes at about 120° according to the following equations:

$$ClNO_2 \rightarrow NO_2 + Cl$$

$$Cl + ClNO_2 \rightarrow NO_2 + Cl_2$$

---

438 W. Kwasnik, in *Handbook of Preparative Inorganic Chemistry*, 2nd edn., Vol. I (ed. G. Brauer), Academic Press, New York (1963), Section 4, pp. 150–271.

439 V. Gutmann, in *Halogen Chemistry*, Vol. 2 (ed. V. Gutmann), Academic Press, London (1967), p. 399; V. Gutmann, *Coordination Chemistry in Non-aqueous Solutions*, Springer-Verlag, New York (1968).

The reaction of $ClNO_2$ with NO leads to two other nitrogen species:

$$ClNO_2 + NO \rightarrow NO_2 + ClNO$$

Hydrolysis of nitryl chloride gives nitric and hydrochloric acids as final products formed when the initial products nitrous and hypochlorous acids interact, whereas ammonolysis in liquid ammonia yields chloramine and ammonium nitrite:

$$ClNO_2 + H_2O \rightarrow [HOCl + HNO_2] \rightarrow HNO_3 + HCl$$
$$ClNO_2 + 2NH_3 \rightarrow H_2NCl + NH_4NO_2$$

Other nitryl salts are formed by the reaction of $ClNO_2$ with metal chlorides in liquid chlorine as solvent[389]:

$$ClNO_2 + SbCl_5 \xrightarrow{\text{liq. } Cl_2} NO_2^+SbCl_6^-$$

9.7. BIOLOGICAL PROPERTIES

Nitrogen trifluoride is a toxic, colourless gas with little odour at concentrations of 500 ppm in air but has a pungent odour at higher concentrations. Tetrafluorohydrazine also causes considerable respiratory distress and is highly toxic[390]. Thus, because so few details have been reported, it is expedient to regard all nitrogen–halogen compounds as toxic and to be handled out of contact in closed systems.

9.8. ANALYSIS

The majority of nitrogen–halogen compounds are characterized by normal research laboratory procedures of micro-analysis of one or more of the elements present together, with an investigation of some spectroscopic[440] and other physical properties. The techniques used to handle volatile nitrogen–halogen compounds are similar to those used for other reactive halogen compounds[441].

10. NITROUS OXIDE

10.1. PRODUCTION AND INDUSTRIAL USE

The industrial manufacture of nitrous oxide is essentially a scale-up of the well-known laboratory method involving the thermal decomposition of ammonium nitrate[442]:

$$NH_4NO_3 \rightarrow N_2O + 2H_2O$$

As has been mentioned previously (section 5.3), ammonium nitrate is a high explosive and, consequently, various precautions are taken to reduce or preferably eliminate the possibility of detonation during the decomposition process. For this purpose, catalysts can be employed, but the process can be controlled by leading a concentrated aqueous ammonium nitrate solution into a reactor at 275°C.

440 R. F. Barrow and A. J. Merer, in ref. 389, Section XXXIV, pp. 454–525.

441 R. D. Peacock, in *Advances in Fluorine Chemistry*, Vol. 4 (eds. M. Stacey, J. C. Tatlow and A. G. Sharpe), Butterworths, London (1965), p. 31.

442 I. R. Beattie, in ref. 389, p. 158.

Alternative methods have been suggested including the reaction of hydroxylamine and aqueous nitrite, but the oxidation of ammonia by oxygen in the presence of a catalyst yields such a variety of products in various amounts depending on the experimental conditions that it might be possible to optimize on nitrous oxide as the principal product. The mechanism is complex, but N_2O can be envisaged as being formed by one or other of the following routes:

$$\left.\begin{aligned} NH_3+O &\rightarrow NH_3O \\ NH_3O+O &\rightarrow HNO+H_2O \\ 2HNO &\rightarrow N_2O+H_2O \end{aligned}\right\}$$

$$\left.\begin{aligned} NH_3+O &\rightarrow NH_3O \\ NH_3O+O_2 &\rightarrow HNO_2+H_2O \\ NH_3O+HNO_2 &\rightarrow N_2O+2H_2O \end{aligned}\right\}$$

$$\left.\begin{aligned} NH_3+O &\rightarrow NH+H_2O \\ NH+O &\rightarrow HNO \\ 2HNO &\rightarrow N_2O+H_2O \end{aligned}\right\}$$

Added to these are many reactions producing NO, N_2 and other byproducts. However, in spite of the obvious complexity, up to 88% nitrous oxide can be obtained from ammonia oxidation using manganese catalysts containing iron or bismuth at 200–300°C, and 71% nitrous oxide from MnO–BiO catalysts at 200°C.

A small concentration of nitrous oxide is also found in the atmosphere which is thought to originate either from the decomposition of fixed nitrogen compounds in the soil, bacterial metabolism, or by the interaction of nitrogen and oxygen atoms produced by photochemical decomposition of ozone[443].

The principal use of nitrous oxide is as an anaesthetic for which it is available in a high state of purity in cylinders. It also has a high solubility in cream, and hence can be used as the propellant gas in forming whipped-cream products[443].

10.2. LABORATORY PREPARATION AND PURIFICATION OF NITROUS OXIDE

Potassium dinitrososulphite, which can be prepared by passing a fast stream of nitric oxide through potassium sulphite solution, decomposes in weakly acid solution to give nitrous oxide[444]:

$$K_2SO_3+2NO \rightarrow K_2SO_3(NO)_2 \rightarrow K_2SO_4+N_2O$$

Alternatively, ammonium nitrate decomposes exothermically at 170°C to give nitrous oxide:

$$NH_4NO_3 \rightarrow N_2O+2H_2O$$

Since the reaction can be explosive, care must be taken to ensure that the temperature does not rise above 250°C during the reaction, and also to prevent condensed water returning into the molten salt[444].

[443] I. R. Beattie, in ref. 389, p. 189.

[444] P. W. Schenk, in *Handbook of Preparative Inorganic Chemistry*, 2nd edn., Vol. 1 (ed. G. Brauer), Academic Press, London (1963), p. 484.

The gas can be purified by passing through 4M potassium hydroxide solution followed by liquefaction and fractional distillation at low temperature. Pure nitrous oxide in steel cylinders can be used directly or after further purification by low temperature distillation.

Nitrous oxide is also formed when nitramide or hyponitrites decompose, during the reduction of nitrites and nitrates, and in the reaction of nitrite with hydroxylamine or hydrazoic acid[443]:

$$NH_2NO_2 \xrightarrow{\Delta} N_2O + H_2O$$

$$Na_2N_2O_2 + H_2O \xrightarrow{\Delta} N_2O + 2NaOH$$

$$2HNO_2 + SO_2 \rightarrow N_2O + H_2O + SO_3$$

$$HNO_2 + NH_2OH \rightarrow N_2O + 2H_2O$$

$$HNO_2 + HN_3 \rightarrow N_2O + N_2 + H_2O$$

10.3. NUCLEAR ISOTOPES OF N_2O

The various labelled isotope isomers $^{15}N^{14}NO$, $^{14}N^{15}NO$, and $^{15}N_2O$ may all be conveniently synthesized by the thermal decomposition of the appropriately labelled form of ammonium nitrate[445]:

$$^{15}NH_4NO_3 \rightarrow {}^{15}NNO + 2H_2O$$

$$NH_4{}^{15}NO_3 \rightarrow N^{15}NO + 2H_2O$$

$$^{15}NH_4{}^{15}NO_3 \rightarrow {}^{15}N_2O + 2H_2O$$

The temperature of decomposition should be maintained at 200 ± 20°C as appreciable amounts of nitrogen and other oxides are formed at higher temperature or explosions can occur.

Nitrous oxide enriched in ^{18}O has been prepared by adding a solution of sodium azide in ^{18}O-enriched water to a solution of nitrite–^{18}O also in ^{18}O-enriched water at pH 2.5, then purified by storing first over solid NaOH then concentrated sulphuric acid[446]:

$$HN_3(aq) + HN^{18}O_2(aq) \rightarrow N_2{}^{18}O + N_2 + H_2O$$

Alternatively, the reaction of hydroxylamine hydrochloride and nitrous acid[446] formed by bubbling dry hydrogen chloride through sodium nitrite in ^{18}O-enriched water yields nitrous oxide–^{18}O:

$$NH_2OH(aq) + HN^{18}O_2(aq) \rightarrow N_2{}^{18}O + 2H_2O$$

10.4. STRUCTURE AND BONDING

The crystal structure of N_2O and CO_2 are so similar that their shapes were erroneously considered to be the same. Early diffraction data clearly indicated that N_2O was linear, but the atoms could not be individually located because the scattering power of nitrogen and oxygen are so similar[443]. Furthermore, now that the asymmetrical NNO structure has been confirmed by a variety of spectroscopic methods, the X-ray and neutron diffraction studies

[445] W. Spindel, in *Inorganic Isotopic Syntheses* (ed. R. H. Herber), Benjamin, New York (1962), p. 87.
[446] I. Dostrovsky and D. Samuel, in *Inorganic Isotopic Syntheses* (ed. R. H. Herber), Benjamin, New York (1962), p. 133.

are consistent with a disordered structure in that N_2O molecules are randomly oriented NNO and ONN. Further evidence for this disorder is given by the extrapolated entropy of N_2O at 0 K being 1.14 e.u. close to the calculated value $R \ln 2 = 1.38$ for random orientation rather than zero for a perfectly ordered crystal[447].

The NO and NN distances[448] indicate bond orders of roughly 1.5 and 2.5 respectively; electronic structures can be rationalized by postulating approximately equal contributions from each of the two resonance structures:

$$\overset{+1}{:N{\equiv}}\overset{}{N}{-}\overset{-1}{\ddot{\underset{..}{O}}}: \qquad \overset{-1}{:N}{=}\overset{+1}{\underset{..}{N}}{=}\dot{\underset{.}{O}}:$$

However, the bonding situation in N_2O is most simply represented by the Linnett non-pairing structure[449]:

$$\overset{-\frac{1}{2}}{:N}{=}\overset{+1}{N}{-}\overset{-\frac{1}{2}}{\underset{..}{O}}:$$

A similar description can be deduced from molecular orbital theory in which all the bonding and non-bonding orbitals are filled while the anti-bonding levels are empty. Such a situation seems to confer extra stability and arises with sixteen-electron triatomic systems of which N_2O is an example together with other isoelectronic species NO_2^+, CO_2, N_3^-, NCO^-, CN_2^{2-}, etc.

10.5. MOLECULAR CONSTANTS

Molecular constants for nitrous oxide are listed in Table 65. The electronic spectrum consists of a number of diffuse bands and several Rydburg series, the two lowest giving ionization potentials in agreement with the photo-ionization value[450]. One ultraviolet

TABLE 65. MOLECULAR CONSTANTS OF NITROUS OXIDE[443, 448, 450]

Property	Value
Diffraction data	
System	
Space group	T^4-P2_13
NN distance	1.126 Å
NO distance	1.186 Å
Fundamental vibration frequencies	
ν_1	1276.522 cm^{-1}
ν_2	589.193 cm^{-1}
ν_3	222.745 cm^{-1}
Dipole moment	0.166 D
Ionization potential	12.9 eV
eQq (nuclear quadrupole coupling constant): $\underline{N}NO$	−1.03 MHz
$N\underline{N}O$	−0.27 MHz

447 W. Jolly, *Inorganic Chemistry of Nitrogen*, Benjamin, New York (1964).

448 L. E. Sutton, *Tables of Interatomic Distances and Configurations in Molecules and Ions*, Chemical Society, London (1958).

449 M. Green, in *Developments in Inorganic Nitrogen Chemistry*, Vol. 1 (ed. C. Colburn), Elsevier, Amsterdam (1966), p. 47.

450 R. F. Barrow and A. J. Merer, in *Mellor's Comprehensive Treatise on Inorganic and Theoretical Chemistry*, Vol. 8, Suppl. II, *Nitrogen*, Part 2, Longmans, London (1967), p. 503.

band-system of the gaseous N_2O^+ ion is also known, and e.s.r. investigation of the γ-irradiation of solid nitrous oxide at 77K indicates[451] the presence of at least three radicals one of which is N_2O^+. The infrared spectra of the various nitrogen and oxygen isotope isomers of nitrous oxide have also been fully investigated[450, 452], while the presence of two strong Raman lines corresponding to the fundamental frequencies ν_1 and ν_3 indicates that nitrous oxide cannot have a centre of symmetry[450].

10.6. PHYSICAL PROPERTIES

Physical constants for nitrous oxide are listed in Table 66.

TABLE 66. PHYSICAL PROPERTIES OF NITROUS OXIDE[443]

Property	Temperature (K)	Value
Melting point	182.29	−90.86°C
Boiling point	184.67	−88.48°C
Critical temperature	309.65	36.5°C
Critical pressure		71.7 atm
Critical density		0.452 g/cm³
Vapour pressure of solid	162.80	100 mm
Vapour pressure of gas	215.15	5 atm
Density of liquid	184.67	1.2257 g/cm³
	309.65	0.45 g/cm³
Density of gas	273.15	1.9084 g/atm
$\triangle H$ fusion	182.29	1.563 kcal/mol
$\triangle H$ vaporization	184.67	3.956 kcal/mol
$\triangle H^\circ_{f0}$ (enthalpy of formation)	0	20.435 kcal/mol
$\triangle H_{f0}$ (enthalpy of formation)	298	19.61 kcal/mol
$\triangle G^\circ_f$ (free energy of formation)	298	24.90 kcal/mol
$H^\circ_{298}-H^\circ_0$ (enthalpy)	298	2.284 kcal/mol
S° (entropy)	298	52.52 cal/deg mol
C°_p (heat capacity)	298	9.19 cal/deg mol
Viscosity	290	1441.3×10^{-7} g/cm s
	296.1	1470.2×10^{-7} g/cm s
Refractive index (λ = 5462.3 Å)		1.0005078
Dielectric constant	70	2.023
Surface tension	298	0.552 dynes/cm
	223	14.39 dynes/cm
Magnetic susceptibility		-0.43×10^{-6}
Dielectric constant	70	2.023
Second Virial coefficient B $(PV/nRT) = 1+\frac{nB(T)}{V}+\ldots$	240	−219
	260	−181
	280	−151
	300	−128
	320	−103
	360	−85
	400	−68

[451] D. R. Smith and W. A. Seddon, *Chem. Phys. Letters* 3 (1969) 640.
[452] K. Nakamoto, *Infrared Spectra of Inorganic and Coordination Compounds*, Wiley, New York (1963), p. 80.

The adsorption of nitrous oxide on various solids has been investigated as well as solubility studies in water and organic solvents[443]. The hydrate $N_2O \cdot 6H_2O$ has a clathrate structure, but no oxygen exchange was detected using $H_2{}^{18}O$ over a wide pH range[453].

10.7. CHEMISTRY AND CHEMICAL PROPERTIES

10.7.1. Decomposition[443]

The thermal decomposition of nitrous oxide results in the formation of the elements at an appreciable rate in the temperature range 585–850°C by a unimolecular process involving the fission of the weaker bond:

$$N_2O \rightarrow N_2 + \tfrac{1}{2}O_2$$

As the activation energy for the process is 59 kcal/mol, the energetics contravene the spin conservation rule requiring zero change in total electron spin in elementary processes:

$$N_2O \rightarrow N_2 + O[^1D]; \qquad \Delta H^\circ = 85.1 \text{ kcal/mol}$$
$$N_2O \rightarrow N_2 + O[^3P]; \qquad \Delta H^\circ = 39.7 \text{ kcal/mol}$$

The presence of other molecules either act as catalysts or inhibitors, while photochemical decomposition also requires consideration of nitric oxide formation in spite of both its contradiction of the spin conservation rule and unfavourable energy requirements.

$$N_2O \rightarrow NO + \tfrac{1}{2}N_2; \qquad \Delta H^\circ = 115.1 \text{ kcal/mol}$$

10.7.2. Reactions

Compared to other oxides of nitrogen, nitrous oxide is relatively unreactive and inert towards halogens and alkali metals at room temperature. With hydrogen however, the main reaction is:

$$N_2O + H_2 \rightarrow N_2 + H_2O; \qquad \Delta H^\circ = 75 \text{ kcal/mol}$$

The ignition temperature for H_2–N_2O gas mixtures is lower than that for hydrogen in air or oxygen, and it is more exothermic than the H_2–O_2 reaction due to the lower energy requirement to dissociate nitrous oxide (45 kcal/mol) as opposed to molecular oxygen (60 kcal/mol). Hence the process is not simply dependent on the dissociation of nitrous oxide and the following scheme has been suggested[454]:

$$\left.\begin{array}{c} 2N_2O \rightarrow N_2O^* + N_2O \\ N_2O + H_2 \rightarrow N_2O^* + H_2 \\ N_2O \rightarrow N_2 + O \\ O + H_2 \rightarrow OH + H \end{array}\right\} \text{Initiation}$$

$$\left.\begin{array}{c} H + N_2O \rightarrow OH + N_2 \\ OH + H_2 \rightarrow H_2O + H \end{array}\right\} \text{Propagation}$$

$$2H + M \rightarrow H_2 + M \qquad \text{Termination}$$

453 R. Bonner and J. Bigeleisen, *J. Am. Chem. Soc.* **74** (1952) 4944.
454 H. W. Melville, *Proc. Roy. Soc.* A, **146** (1934) 737.

Explosion limits are extremely wide, and even nitrous oxide diluted with up to 79% nitrogen can form explosive mixtures with hydrogen. Also the H_2–O_2 reaction is sensitized by nitrous oxide, e.g. when oxygen is added to N_2O–H_2 mixtures, instantaneous ignition or an autocatalytic reaction resulting in an explosion occurs.

Nitrous oxide is only slowly reduced by $TiCl_3$ to ammonia and by $CrCl_2$ to nitrogen. Sodium borohydride, however, does not react with nitrous oxide by itself, but a number of transition metal complexes, particularly cobalt(I) derivatives, can act as catalysts for the reduction to nitrogen[455].

$$N_2O+2H \rightarrow N_2+H_2O$$

A further recent development in this area has been the reduction of N_2O with nitrogentris-(triphenylphosphine)cobalt hydride and tris(triphenylphosphine)cobalt trihydride in the presence of excess triphenylphosphine[456]. Results have been interpreted in terms of oxidation of triphenylphosphine to triphenylphosphine oxide and the evolution of hydrogen and nitrogen as well as formation of nitrogen complexes in solution suggesting the following reaction:

$$H_3Co(PPh_3)_3+Ph_3P+N_2O \rightarrow H(N_2)Co(PPh_3)_3+Ph_3PO+H_2$$

When carbon, phosphorus, sulphur or metals whose oxides have high heats of formation are heated in nitrous oxide, they ignite and often burn more vigorously than in air, forming the oxide and nitrogen:

$$M+N_2O \rightarrow MO+N_2$$

Similarly, with atomic chlorine, the intermediate oxide decomposes to give the elements as final products[443]:

$$Cl+N_2O \rightarrow [ClO]+N_2$$
$$2[ClO] \rightarrow Cl_2+O_2$$

Thus the reactions of nitrous oxide in its rôle as an oxidizing agent involve large free energy changes but have higher activation energies than the corresponding reactions involving molecular oxygen.

Probably the most remarkable reaction of nitrous oxide is with molten sodium or potassium amide at about 200°C to yield azides (see section 8.1):

$$2KNH_2+N_2O \rightarrow KN_3+KOH+NH_3$$

10.8. BIOLOGICAL PROPERTIES

Inhaled in small amounts, nitrous oxide induces intoxication and hysterical excitement often accompanied by convulsive laughter ("laughing gas"), whereas in larger amounts it acts as a narcotic. As it is not capable of supporting respiration, in anaesthesia it is administered as a mixture with oxygen and is particularly used in dentistry and gynaecology. With lower concentrations of nitrous oxide in oxygen, it can be used as an analgesic. Prolonged anaesthesia may lead to undesirable side-effects[457].

The rôle of nitrous oxide in natural nitrogen fixation processes is not understood, but it is a product and substrate of bacterial metabolism[458].

[455] R. G. S. Banks, R. J. Henderson and J. M. Pratt, *Chem. Communs.* (1967) 387.
[456] L. S. Pu, A. Yamamoto and S. Ikeda, *Chem. Communs.* (1969) 189.
[457] F. Call, in *Mellor's Comprehensive Treatise on Inorganic and Theoretical Chemistry*, Vol. 8, Suppl. I, *Nitrogen*, Part I, Longmans, London (1964), p. 606.
[458] B. A. Fry, *The Nitrogen Metabolism of Micro-organisms*, Methuen, London (1955).

10.9. ANALYSIS OF NITROUS OXIDE[459]

Thermal decomposition of nitrous oxide in the presence of platinum or palladium catalyst will liberate oxygen which can be detected with alkaline pyrogallol reagent, but is not specific to this particular oxide. The mass spectrometer is probably the most reliable technique for detecting small quantities of gas. However, infrared absorption spectroscopy is a more accessible technique, and depending on other gases present, the intensity of one of the infrared bands at 590, 1285 and 2224 cm^{-1} can be used to estimate the nitrous oxide content of gas mixture. In connection with the use of nitrous oxide as an anaesthetic, its estimation in admixture with oxygen and ether can be carried out by the acoustic gas analyser which, as its name implies, is based on the variation of the velocity of sound according to the density of the gas. Hence by pre-calibrating for various mixtures, nitrous oxide content can be continuously monitored. If nitrogen is also present, oxygen has to be determined separately by means of an oxygen analyser for the four components to be estimated.

Combustion methods include reduction over a palladium catalyst followed by volumetric determination of the nitrogen:

$$N_2O + H_2 \rightarrow N_2 + H_2O$$

Alternatively, nitrous oxide can be reduced over heated copper gauze:

$$N_2O + Cu \rightarrow CuO + N_2$$

or as another variation to eliminate interference from carbon dioxide:

$$N_2O + CO \rightarrow N_2 + CO_2$$

11. NITRIC OXIDE

11.1. PRODUCTION AND INDUSTRIAL USE

From an industrial point of view, nitric oxide is probably the most important oxide of nitrogen[460]. It is produced by the catalytic oxidation of ammonia[461, 462] at 800–960°C:

$$4NH_3 + 5O_2 \xrightarrow{\text{Pt-Rh}} 4NO + 6H_2O; \qquad \triangle H(298) = -216.55 \text{ kcal/mol}$$

This, and the alternative reactions which are possible in ammonia oxidation, have been considered in more detail in section 3.10.2

The direct reaction of nitrogen and oxygen requires high temperatures before equilibrium conditions are such as to bring any appreciable yields of nitric oxide. The Birkeland–Eyde electric arc oxidation of atmospheric nitrogen was developed during the 1920's in countries where hydroelectric power was cheap, but the process is now obsolete[460]. More recently the combustion of air–natural gas mixtures in the "pebble-bed" or Wisconsin process can produce temperatures of over 2000°C, and although the process is feasible, it is not at present economic[463]. The equilibrium concentrations of NO for each process

459 K. I. Vasu, in *Mellor's Comprehensive Treatise on Inorganic and Theoretical Chemistry*, Vol. 8, Suppl. II, *Nitrogen*, Part 2, Longmans, London (1967), p. 628.

460 I. R. Beattie, in ref. 459, Section XXII, p. 158.

461 I. R. Beattie, in ref. 459, Section XXV, p. 216.

462 G. C. Bond, *Catalysis by Metals*, Academic Press, New York (1962), p. 456.

463 E. D. Ermenc, *Chem. Engng. Progress* **52** (1956) 149.

are 4% and 2% respectively, and the gas mixture has even to be cooled rapidly to prevent any further loss by decomposition (see section 11.7.1).

Clearly, the fate of most nitric oxide is oxidation to nitrogen dioxide almost as soon as it is formed, and the principal industrial application of both oxides is as intermediate oxidation products in the manufacture of nitric acid[464]. The presence of nitric oxide in the atmosphere can obviously occur through lightning discharge and it is also present in automobile exhausts[465], hence subsequent oxidation and dissolution produces nitric acid in very low concentration in rain water. However, although this could be considered as providing a beneficial source of fixed nitrogen, the build-up of concentrations of nitrous fumes in city atmospheres is likely to be a considerable health hazard. Nitric oxide is also present in the upper atmosphere, and important in photochemical reactions arising from solar radiation[466].

11.2. LABORATORY PREPARATION AND PURIFICATION OF NITRIC OXIDE[467]

Nitric oxide may be prepared from aqueous solution by the reaction between nitrous acid and iodine or ferrocyanide:

$$2KNO_2+2KI+2H_2SO_4 \rightarrow 2NO+2K_2SO_4+I_2+2H_2O$$

$$KNO_2+K_4Fe(CN)_6+2CH_3COOH \rightarrow NO+K_3Fe(CN)_6+2CH_3COOK+H_2O$$

When dilute sulphuric acid is added dropwise on to sodium nitrite, the nitrous acid formed initially decomposes to give a steady stream of nitric oxide:

$$6NaNO_2+3H_2SO_4 \rightarrow 4NO+2H_2O+3Na_2SO_4+2HNO_3$$

The gas is purified by passing through 90% sulphuric acid followed by 50% potassium hydroxide then dried by passing through a trap cooled in solid CO_2–ether mixture, and condensed over phosphorus pentoxide cooled with liquid nitrogen. On fractional distillation, the middle cut is collected and redistilled to give pure NO.

Dry nitric oxide can be obtained directly simply by heating the solid mixture of chromium(III) oxide with nitrate and nitrite, preferably in the presence of calcined iron(III) oxide[468]:

$$3KNO_2+KNO_3+Cr_2O_3 \rightarrow 4NO+2K_2CrO_4$$

Nitric oxide is also formed in conjunction with other gases during the reduction[461] of nitric acid, nitrates and nitrites, or nitrosyl derivatives:

$$8HNO_3+3Cu \longrightarrow 2NO+3Cu(NO_3)_2+4H_2O$$

$$2Ba(NO_2)_2+I_2 \xrightarrow{225°C} 2NO+Ba(NO_3)_2+BaI_2$$

$$2NOCl+2I^- \longrightarrow 2NO+I_2+2X^-$$

---

[464] C. J. Pratt and R. Noyes, *Nitrogen Fertilizer Chemical Process*, Noyes Development Corporation, New Jersey (1965).

[465] W. B. Innes and K. Tsu, in *Kirk–Othmer Encyclopedia of Chemical Technology*, 2nd edn., Vol. 2, Interscience, New York (1963), p. 814.

[466] J. Heicklen and N. Cohen, in *Advances in Photochemistry*, Vol. 5 (eds. W. A. Noyes, Jr., G. S. Hammond and J. N. Pitts), Interscience, New York (1968), pp. 157–328.

[467] P. W. Schenk, in *Handbook of Preparative Inorganic Chemistry*, Vol. I (ed. G. Brauer), Academic Press, New York (1963), p. 460.

[468] R. A. Ogg, Jr. and J. D. Ray, *J. Am. Chem. Soc.* **78** (1956) 7993.

Nitrosonium salts react with hydroxylic solvents:

$$NO^+ + OH^- \rightleftharpoons HNO_2 \rightleftharpoons NO_2^- + H^+$$

Thus the equilibrium is only displaced to the left in strongly acidic media. Nitrosonium hydrogen sulphate is formed therefore by dissolution of nitrites or dinitrogen trioxide in sulphuric acid[469]:

$$N_2O_3 + H_2SO_4 \rightarrow 2NO^+ + 3HSO_4^- + H_3O^+$$

although hydrolysis occurs in solutions containing less than 80% H_2SO_4.

11.3. NUCLEAR ISOTOPES OF NITRIC OXIDE[470]

^{15}NO is commercially available or can be prepared by the reduction of nitrogen dioxide, nitrites or nitrates with mercury and sulphuric acid:

$$6\,{}^{15}NO_2 + 4Hg + 2H_2SO_4 \rightarrow 4\,{}^{15}NO + Hg_2(NO_3)_2 + 2HgSO_4 + 2H_2O$$

Conversion of ammonia into nitric oxide can be achieved most conveniently without proceeding via nitrogen by direct oxidation with permanganate in an autoclave for 7–8 hr at 170–180°C to nitrate which can be reduced to nitric oxide as before. By passing a mixture of oxygen and nitrogen through a high-tension electric arc, a mixture of N_2O_3 and N_2O_4 is formed which can also be reduced to NO as above. This method has been used in the preparation of ^{15}NO and $N^{18}O$ by using the appropriate enriched elements[471].

11.4. STRUCTURE AND BONDING

X-ray crystallography[472] shows that solid nitric oxide consists of a loosely bound dimeric $(NO)_2$ species which may have a rectangular shape. The residual entropy and other physical properties[461] also suggest that complete randomization of the two orientations occurs, and although the mean distance between the two NO units is about 2.4 Å, it is not possible to tell whether they are oriented in the same direction or head to tail as illustrated in the structure:

```
N . . . . . . O
|             |   1.10 Å
O . . . . . . N
    2.38 Å
```

The blue-coloured liquid phase is also asssociated and infrared studies[473] provide some evidence for N–N bonds implying a *cis*-configuration with the structure:

```
·Ṅ—N·
 ‖   ‖
 O   O
```

However, nitric oxide is normally encountered as a gas, in which phase it is colourless and monomeric with little tendency to associate via electron pairing.

469 C. C. Addison and J. Lewis, *Quart. Rev.* (9) (1955) 115.
470 W. Spindel, in *Inorganic Isotopic Synthesis* (ed. R. H. Herber), Benjamin, New York (1962), p. 100.
471 I. Dostrovsky and D. Sammuel, in *Inorganic Isotopic Synthesis*, (ed. R. H. Herber), Benjamin, New York (1962).
472 W. J. Dulmange, E. A. Meyers and W. N. Lipscomb, *Acta Cryst.* **6** (1953) 760.
473 W. G. Fateley, H. A. Bent and B. Crawford, *J. Chem. Phys.* **31** (1959) 204.

Using molecular orbital theory, the electronic structure of nitric oxide in the ground state is usually written[474]

$$NO\ {}^2\Pi\ (1\sigma^+)^2(2\sigma^+)^2(3\sigma^+)^2(1\pi)^4(4\sigma^+)^2(5\sigma^+)^2(2\pi)^1$$

Molecular orbital calculations suggest[475] that the 3σ, 4σ and 5σ orbitals are bonding, non-bonding and anti-bonding respectively, with the last electron entering into an antibonding π orbital and accounting for the paramagnetism. Furthermore this theory predicts the bond order to be 2.5, and that it should not be too difficult to remove an electron to form the NO^+ ion with a shorter and stronger bond than NO itself. In fact, each of these predictions is correct, the bond length of NO (1.14 Å) lies between that of a double (1.18 Å) and a triple bond (1.06 Å), the ionization potential of 9.25 eV is appreciably lower than for similar molecules (N_2, O_2 and CO are 15.6, 12.1 and 14.0 eV respectively) and the stretching frequency of the NO^+ ion in nitrosyl salts (2150–2400 cm^{-1}) is higher than nitric oxide itself (1888 cm^{-1}). In fact, salts of the nitrosyl cation, or the nitrosonium ion as it is alternatively named, such as the perchlorate, bisulphate and tetrafluoroborate, are well-characterized crystalline solids which are isomorphous with corresponding hydroxonium and ammonium salts[469]. The NO^+ species is also isoelectronic with N_2, CO and CN^-, each of which can be considered by molecular orbital theory as having an electronic structure containing no antibonding π electrons and a bond order of 3. Thus it is not surprising that a wide range and variety of NO complexes of transition metals (see Chapter 46) analogous to metal carbonyls are known. In most cases, NO transfers an electron to the metal, and the NO^+ so formed further donates a lone pair of electrons from nitrogen to the metal to give compounds like $Fe(NO)_2(CO)_2$ and $Co(NO)(CO)_3$ which are stable and isoelectronic with $Ni(CO)_4$.

The valence bond approach is also worthy of mention, for although NO is often represented by the resonance forms

$$\cdot\overset{+}{N}-\overset{-}{O} \leftrightarrow \cdot N{=}O \leftrightarrow \overset{-}{N}{=}\overset{+}{O}\cdot \leftrightarrow N{\equiv}O\cdot$$

it is adequately described by the Linnett non-pairing structure[474]:

$$:\dot{N}\dot{=}\dot{O}:$$

while the cation NO^+ is

$$[:N{\equiv}O:]^+$$

Similarly, the anion NO^- would on the same basis have the structure

$$[:N{=}O:]^-$$

however, most compounds in which these ions were thought to be present are now known to contain the anion $N_2O_2^{2-}$ [476].

[474] M. Green, *Developments in Inorganic Nitrogen Chemistry*, Vol. I (ed. C. B. Colburn), Elsevier, Amsterdam (1966), p. 28.

[475] H. Brion, C. Moser and M. Yamazaki, *J. Chem. Phys.* **30** (1959) 673.

[476] N. Gee, D. Nicholls and V. Vincent, *J. Chem. Soc.* (1964) 5897.

11.5. MOLECULAR CONSTANTS

Various molecular constants of nitric oxide are listed in Table 67.

The electronic absorption spectrum of NO extends from just below 2300 Å to the vacuum ultraviolet region. Transitions between the ground electronic state and the $A^2\Sigma^+$, $B^2\Pi_r$, $C^2\Pi$ and $D^2\Sigma^+$ states are usually referred to as γ, β, δ and ε bands respectively and have been extensively investigated[466, 477]. Infrared and Raman spectra of various isotope isomers of nitric oxide as gas, liquid, solid and matrix trapped molecule samples also indicate the presence of *cis* or *trans* dimer (ONNO) forms[477]. However, the fundamental bands are also split due to an interaction between the angular and spin momentum vectors ($L = 1$, $S = \frac{1}{2}$) which may couple $L+S$ or $L-S$ dividing the $^2\Pi$ ground state into

TABLE 67. MOLECULAR CONSTANTS OF NITRIC OXIDE[a]

Property	Value
X-ray diffraction data[b]	
System	Monoclinic
Space group	$P2_1/a$
Lattice constants	$a = 6.68$, $b = 3.96$, $c = 6.55$, $\alpha = 127.9°$
N–O bond length	1.10 Å
N · · · O intermolecular distance	2.38 Å
Electron diffraction data[c]	
N–O bond length	1.15 Å
Fundamental vibration frequency	1876 cm^{-1}
Force constant	1.59×10^6 dynes/cm
Moment of inertia	16.47×10^{-40} g/cm^2
Dipole moment	0.15 *D*
Bond dissociation energy	6.50 eV, 149.9 kcal
First ionization potential	9.25 eV
Magnetic susceptibility (293 K)	1.46×10^{-3} c.g.s. units
Magnetic moment (296 K)	1.837 BM

[a] Ref. 461.
[b] Ref. 472.
[c] L. E. Sutton, *Tables of Interatomic Distances and Configurations in Molecules and Ions*, Chemical Society, London (1958).

an upper paramagnetic state $^2\Pi_{3/2}$ and a lower diamagnetic state $^2\Pi_{\frac{1}{2}}$. Interaction with the rotational motion of the molecule also causes the upper state to split into four sub-levels giving rise to three absorption lines with the usual selection rule $m_s = \pm 1$, and in fact they are each observed to be further split into three components due to the interaction of the spin of the nitrogen nucleus[461]. The energy difference between the paramagnetic and diamagnetic $^2\Pi$ states is only about 350 cal/mol and as kt is considerably greater than this value at room temperature and comparable down to even quite low temperatures, it is this

[477] R. F. Barrow and A. J. Merer, in ref. 459, p. 507.

phenomenon that gives rise to the temperature dependence of the magnetic moment. The hyperfine splitting of the electron spin resonance spectrum of NO is also explained on the basis of a strong interaction between the moment of the unpaired electron and the rotation of the molecule, and also leads to the conclusion that about 60% of the spin density is concentrated on the nitrogen atom[478]. Microwave spectral data have been used to obtain values for magnetic hyperfine structure constants, the nuclear quadrupole resonance moment and the dipole moment[477].

The various dissociative ionization processes have also been studied in great detail and their corresponding appearance potentials are consistent with the value of 6.50 ± 0.01 eV for the dissociation energy of nitric oxide. Photo-ionization and electron impact studies are in agreement that the ionization potential for the formation of NO^+ is 9.25 ± 0.02 eV[20].

11.6. PHYSICAL PROPERTIES

Physical constants for nitric oxide are listed in Table 68.

11.7. CHEMISTRY AND CHEMICAL PROPERTIES

11.7.1. Decomposition of Nitric Oxide

As the free energy change in the formation of nitric oxide from the elements is large and positive:

$$N_2 + O_2 \rightarrow 2NO; \qquad \Delta G = +20 \text{ kcal/mol}$$

the isolation of a reasonable yield of product requires not only a high temperature but also a rapid rate of cooling of the equilibrium mixture.

The equilibrium constant of the reaction $K = [NO]^2/[N_2][O_2]$, which at 298 K is 5.27×10^{-31}, has been calculated for a wide range of conditions, but their experimental confirmation has been difficult.

Although many products have been reported for the thermal decomposition of nitric oxide, the principal reaction to be considered is that reverting to the elements:

$$2NO \rightarrow N_2 + O_2$$

The results of several investigations indicate that at 1130–1330°C the decomposition is described by a second-order homogeneous reaction, whilst at higher temperatures reactions involving atomic nitrogen and nitrogen dioxide become important[461]. Heterogeneous decomposition of nitric oxide on platinum and platinum–rhodium wire above 1000°C is bimolecular with respect to NO and is retarded proportionally to the oxygen concentration. Photochemical decomposition of nitric oxide has also been widely investigated, and the products vary to some extent with the state of excitation from which the nitric oxide is decomposed[466].

11.7.2. Oxidation

Undoubtedly the most important reaction of nitric oxide is its oxidation to nitrogen dioxide[460]:

$$2NO + O_2 \rightarrow 2NO_2$$

---

[478] R. Beringer, E. Rawson and A. Henry, *Phys. Rev.* **94** (1954) 343.
[479] G. R. A. Johnson, in ref. 459, p. 544.

TABLE 68. PHYSICAL PROPERTIES OF NITRIC OXIDE

Property	Temperature (K)	Value
Melting point	109.49	−163.6°C
Boiling point	121.36	−151.8°C
Critical temperature	179.15	−94°C
Critical pressure		65 atm
Critical density		0.52 g/cm^3
ΔH fusion	109.49	0.5495 kcal/mol
ΔH vaporization	121.36	3.292 kcal/mol
ΔH_f (enthalpy of formation)[b]	0	21.45 kcal/mol
ΔH_f (enthalpy of formation)	298	21.57 kcal/mol
ΔG_f (Gibbs energy of formation)	298	20.69 kcal/mol
S° (entropy)	298	50.347 cal/deg mol
C_p° (heat capacity)	298	7.133 cal/deg mol
Density of solid	20	1.57 g/cm^3
	78	1.556 g/cm^3
(calculated from X-ray)	98	1.46 g/cm^3
Density of liquid	110.15	1.332 g/cm^3
	113.65	1.306 g/cm^3
	117.15	1.277 g/cm^3
	119.55	1.227 g/cm^3
Density of gas	293	1.3402 g/l
Viscosity of liquid	120	843.6×10^7 g/cm s
	200	1371.2×10^7 g/cm s
	280	1837.6×10^7 g/cm s
Thermal conductivity	120	2.580 g cal/cm s
	300	6.189 g cal/cm s
Virial coefficient[c] B	121.72	−224.4
$(PV/nRT) = 1+\frac{nB(T)}{V}+\ldots$	153.06	−107.9
	274.00	−22.8
	277.60	−26.2 (c = 1400)
	310.94	−19.0 (c = 1900)

Vapour pressure of solid P

$$\log P_{\text{cm}} = \frac{-867}{T} + 0.00076T + 9.05125$$

Vapour pressure of liquid P

$$\log P_{\text{cm}} = \frac{-776}{T} + 0.002364T + 8.562128$$

K	°C	P	K	°C	P
88.6	−184.55	1 mm	121.4	−151.75	1 atm
94.9	−178.25	10 mm	128.0	−145.1	2 atm
101.4	−171.75	40 mm	137.4	−135.7	5 atm
107.1	−166.05	100 mm	146.8	−127.3	10 atm
116.3	−156.85	400 mm			

[a] Ref. 461.

[b] D. D. Wagman, W. H. Evans, V. B. Parker, I. Halow, S. M. Bailey and R. H. Schumm, *Selected Values of Chemical Thermodynamic Properties*, NBS Technical Note 270-3, Washington (1968).

[c] J. H. Dymond and E. B. Smith, *The Virial Coefficient of Gases—A Critical Compilation*, Oxford (1969).

Work on this classical termolecular reaction was first carried out by Bodenstein[480] in 1918, and extensive investigations which have been reported since that date confirm that it follows a simple third-order rate law over a wide range of experimental conditions:

$$-d(NO)/dt = k[NO]^2[O_2]$$

However, a further unusual feature of this reaction is that the rate constant decreases with increasing temperature. This can be accounted for in the mechanism which postulates the initial formation of dimer:

$$2NO \rightleftharpoons N_2O_2$$

$$N_2O_2+O_2 \xrightarrow{k''} 2NO_2$$

Since the equilibrium constant K for the dimerization step should decrease with increasing temperature, the process can now be described by the rate law:

$$-d[NO]/dt = k'K[NO]^2[O_2]$$

where K is the equilibrium constant for the dimerization and $k'K$ is synonymous with k. Thus the negative temperature coefficient is explained on the basis that since K should be expected to decrease with temperature, it is not unreasonable to expect $k'K$ to do the same, particularly with a very low energy of activation.

An alternative mechanism[481] involves the species NO_3, together with the possibility of the pernitrite–nitric oxide complex. Like nitrogen trioxide, dinitrogen pentoxide has also been detected by infrared spectroscopy[482] during nitric oxide–oxygen reactions which may lend support for the scheme:

$$NO+O_2 \rightleftharpoons NO_3$$

$$NO_3+NO \rightarrow 2NO_2$$

$$NO_3+NO_2 \rightarrow N_2O_5$$

$$N_2O_5+NO \rightarrow 3NO_2$$

11.7.3. Reactions of Nitric Oxide

Being the simplest stable molecule with an odd number of electrons, nitric oxide has been well investigated with regard to its reactivity towards atoms, free radicals and other paramagnetic species. Many of these reactions have provided unique systems for study by gas kinetics[466, 483] and the results of which have led to an understanding of their mechanisms.

The use of nitric oxide in the monitoring of atomic nitrogen has already been mentioned in section 2.12, while chemiluminescence due to the reaction

$$NO+O \rightarrow NO_2+h\nu$$

has likewise been used for both the qualitative and quantitative estimation of atomic oxygen[484]. The termolecular reaction

$$NO+O+M \rightarrow NO_2+M$$

is also well known and first order in each reactant.

[480] M. Bodenstein, *Helv. Chim. Acta* **18** (1935) 743.
[481] J. D. Ray and R. A. Ogg, Jr., *J. Chem. Phys.* **26** (1957) 984.
[482] R. A. Ogg, Jr., *J. Chem. Phys.* **18** (1950) 770.
[483] A. F. Trotman-Dickenson and G. S. Milne, *Tables of Bimolecular Gas Reactions*, National Standard Reference Data Series in National Bureau of Standards, 9, Washington (1969).
[484] F. Kaufman, in *Progress in Reaction Kinetics*, Vol. I (ed. G. Porter), Pergamon, Oxford (1961), p. 1.

Of similar major importance is the reaction of NO with hydrogen atoms produced by electric discharge[485] to give nitroxyl HNO, which has also been prepared by the flash photolysis of nitroalkanes or isoamyl nitrite. Nitroxyl has also been postulated as an intermediate in several industrial processes (see section 3.10.2), and its infrared spectrum in an argon matrix at 20°C has been recorded[486], as well as the value of 48.6 kcal/mol for the bond strength[487] D(H–NO).

In the reactions of NO with the halogens, both atomic and molecular species need to be considered and their mechanisms are among those generalized as follows:

$$NO + X_2 \rightarrow X_2NO$$
$$NO + X_2NO \rightarrow 2XNO$$
$$NO + X_2 \rightarrow XNO + X$$
$$NO + X \rightarrow XNO^*$$
$$XNO^* \rightarrow XNO + h\nu$$
$$XNO^* + M \rightarrow XNO + M$$

Although the NO–F_2 reaction gives FNO predominantly, some ONF_3 can be obtained by photochemical fluorination of NO even at room temperature[488]. The nitric oxide–chlorine system has been most thoroughly investigated as both radical and molecular halogen reactions are important contributing factors in the overall mechanism[466, 489]. Less is known about the NO–Br_2 reaction, but with iodine the mechanism is thought[466] to be

$$NO + I \rightleftharpoons INO$$
$$INO + I \rightarrow NO + I_2$$

Likewise, the reaction of NO with HI

$$NO + 6HI \rightarrow NH_4I + H_2O + 5/2I_2$$

is thought to involve the following steps[490]:

$$NO + HI \rightarrow HNO + I$$
$$HNO + HI \rightarrow H_2 + INO$$
$$HNO + I_2 \rightarrow HI + INO$$
$$INO + I \rightarrow NO + I_2$$

Reactions with other halogen compounds are equally complex. Nitryl chloride reacts rapidly with NO[491], but it is uncertain whether this reaction involves oxygen or chlorine transfer.

$$NO + ClNO_2 \rightarrow NO_2 + ClNO$$

---

485 M. A. A. Clyne and B. A. Thrush, *Trans. Faraday Soc.* **57** (1961) 1305.
486 H. W. Brown and G. C. Pimentel, *J. Chem. Phys.* **29** (1958) 883.
487 M. J. Y. Clement and D. A. Ramsay, *Can. J. Phys.* **39** (1961) 205.
488 E. W. Lawless and I. C. Smith, *Inorganic High-energy Oxidizers*, M. Dekker, New York (1968), p. 68.
489 P. G. Ashmore and M. S. Spencer, *Trans. Faraday Soc.* **55** (1959) 1868.
490 J. L. Holmes and E. V. Sundaram, *Trans. Faraday Soc.* **62** (1966) 910.
491 E. C. Freiling, H. S. Johnston and R. A. Ogg, *J. Chem. Phys.* **20** (1952) 327.

Nitric oxide and hydrogen chloride condense to a deep blue liquid which may contain the complex NOHCl. With xenon fluorides, the reactions with NO have been followed by mass spectrometry which indicates[492] stepwise removal of fluorine:

$$XeF_4+NO \rightarrow XeF_3+FNO$$
$$XeF_2+NO \rightarrow XeF+FNO$$

Nitric oxide can, however, be oxidized with iodine pentoxide[461]:

$$10NO+2I_2O_5 \rightarrow 2I_2+5N_2O_4$$
$$10NO+3I_2O_5 \rightarrow 3I_2+5N_2O_5$$

Halogenated methyl radicals add to NO to form nitroso compounds[466] of which, perhaps, the best known[493] is the blue gas trifluoronitrosomethane CF_3NO:

$$CF_3+NO \rightarrow CF_3NO$$
$$CFCl_2+NO \rightarrow CFCl_2NO$$
$$CF_2H+NO \rightarrow CF_2HNO \rightarrow CF_2{=}NOH$$

However, with diboron tetrahalides breakdown of the unstable adducts occurs[494, 495]:

$$3B_2Cl_4+3NO \xrightarrow{-40°C} 3B_2Cl_4\cdot NO \rightarrow B_2(NO)_3\cdot BCl_3+BCl_3$$
$$\downarrow$$
$$B_2(NO)_3+BCl_3$$
$$3B_2F_4+6NO \longrightarrow 4BF_3+3N_2O+B_2O_3$$
$$6B_2F_4+6NO \longrightarrow 8BF_3+3N_2+2B_2O_3$$

The primary step in the reaction of nitric oxide with ozone is a bimolecular second-order reaction[466]:

$$NO+O_3 \rightleftharpoons NO_2+O_2; \qquad \Delta H(298) = -47.8 \text{ kcal/mol}$$

with possible secondary reactions:

$$2NO_2+O_3 \rightarrow N_2O_5+O_2$$
$$2NO+O_2 \rightarrow 2NO_2$$

Similarly, the gas phase oxidation of NO with HNO_3 also gives nitrogen dioxide[496]:

$$NO+2HNO_3 \rightarrow 3NO_2+H_2O$$

The reactions of NO with the oxides of sulphur have been investigated in view of their relevance to the chamber process for the manufacture of sulphuric acid[497], and losses occur when nitrous oxide is produced:

$$2NO+SO_2 \rightarrow N_2O+SO_3$$
$$NO+2SO_3 \rightarrow (SO_3)_2NO$$
$$2NO+2SO_2+2H_2O \rightarrow 2H_2SO_3NO \rightarrow H_2SO_3(NO)_2 \rightarrow N_2O+H_2SO_4$$
$$+$$
$$H_2SO_3$$

---

492 H. S. Johnston and R. Woolfolk, *J. Chem. Phys.* **41** (1964) 269.
493 J. Banus, *J. Chem. Soc.* (1953) 3755.
494 A. K. Holliday and A. G. Massey, *J. Inorg. Nucl. Chem.* **18** (1961) 108.
495 A. K. Holliday and F. B. Taylor, *J. Chem. Soc.* (1962) 2767.
496 J. H. Smith, *J. Am. Chem. Soc.* **69** (1947) 1741.
497 T. J. P. Pearce, in *Inorganic Sulphur Chemistry* (ed. G. Nickless), Elsevier, Amsterdam (1968), p. 543.

The intermediate dinitrososulphites can be prepared as follows[467]:

$$2NO + K_2SO_3 \rightarrow K_2SO_3(NO)_2$$

$$6NO + Na_2S_2O_4 + 2NaOH \rightarrow 2N_2O + Na_2SO_3 \cdot N_2O_2 + Na_2SO_4 + H_2O$$

Nitrogen compounds which react with NO include ammonia as in the side-reactions during NO production by ammonia oxidation[464]:

$$2NH_3 + 3NO \rightarrow 5/2N_2 + 3H_2O$$

NO at 20 atm pressure is also taken up by solutions of nitrosonium salts forming the deep blue $N_2O_2^+$ cation: which is thought to be responsible for the colour during the chamber process for the production of sulphuric acid[498]:

$$NO^+HSO_4^- + NO \rightarrow N_2O_2^+HSO_4^-$$

Nitrogen trichloride reacts with nitric oxide[499] and at $-150°C$ the overall reaction is

$$NCl_3 + 3NO \rightarrow 2NOCl + N_2O + Cl$$

but at $-80°C$ the main reaction is represented by

$$NCl_3 + 2NO \rightarrow N_2O + NOCl + Cl_2$$

Reaction of nitric oxide with carbon compounds, in particular organic-free radicals, have been reviewed[466]. The explosive reaction with carbon disulphide has received much attention[461] because of the associated light emission. The reaction of sodium oxide with NO gives sodium hydronitrite initially which then decomposes[500]:

$$4Na_2O + 4NO \xrightarrow{100°C} 4Na_2NO_2 \rightarrow 2Na_2O + 2NaNO_2 + Na_2N_2O_2$$

With alkali metal hydroxides, both N_2O and N_2 are formed[501]:

$$4NO + 2MOH \rightarrow N_2O + 2MNO_2 + H_2O$$

$$6NO + 4MOH \rightarrow N_2 + 4MNO_2 + 2H_2O$$

Sodium reacts with NO directly or in liquid ammonia to give $(NaNO)_x$ [502]; which on the evidence of its X-ray diffraction pattern is thought to be different to ordinary sodium hyponitrite and possibly may be *cis*-$Na_2N_2O_2$. Infrared spectra of the product of reaction of lithium with NO in an argon matrix suggests LiON rather than LiNO [503]. A solution of hyponitrous acid is obtained when a stream of NO is passed through an ether solution of lithium aluminium hydride[504].

[498] F. Seel, B. Ficke, L. Riehl and E. Vo'lkl, *Z. Naturforsch.* **86** (1953) 607.
[499] W. A. Noyes, *J. Am. Chem. Soc.* **53** (1931) 2137.
[500] E. Zintl and H. H. Baumbach, *Z. anorg. allgem. Chem.* **198** (1931) 88.
[501] E. Barnes, *J. Chem. Soc.* (1931) 2605.
[502] H. Gehlen, *Ber.* **72B** (1939) 159.
[503] W. L. S. Andrews and G. C. Pimentel, *J. Chem. Phys.* **44** (1966) 2361.
[504] P. Karrer and R. Schwyzer, *Rec. Trav. Chim.* **69** (1950) 474.

The reaction of nitric oxide with transition metal compounds[505] is an important route to nitrosyl complexes which will be discussed in Chapter 46. Classes of compounds containing the nitrosyl group including metal nitrosyls $[Ru(NO)_4]$, nitrosyl carbonyls $[Co(NO)(CO)_3]$, nitrosyl halides $[Fe(NO)_2I]$, nitrosyl pseudohalides $[Fe(NO)(CO)_5]^{2-}$ and organometallic (principally cyclopentadienyl) nitrosyl derivatives[506]

$$[(\pi\text{-}C_5H_5)_2Mn_2(NO)_3]$$

have been extensively investigated.

11.8. ANALYSIS OF NITRIC OXIDE

The classic qualitative test for NO is the brown ring test which is demonstrated by the interaction of NO with aqueous iron(II) solution to give $[Fe(NO)]^{2+}$. A standard quantitative estimation is based on the assumption that nitric and nitrous acid solutions produce pure NO when shaken with concentrated sulphuric acid which can be measured volumetrically in a Lunge nitrometer. Combustion methods based on the reaction of NO with carbon monoxide lead to the volumetric measurement of nitrogen after absorbing other residual gases[507].

$$2NO+2CO \rightarrow 2CO_2+N_2$$

Of the other chemical methods, NO can be absorbed directly in an excess of acid permanganate solution and the residual permanganate reduced with excess iron(II) solution which is then back-titrated with standard permanganate.

$$10NO+6KMnO_4+9H_2SO_4 \rightarrow 3K_2SO_4+6MnSO_4+10HNO_3+4H_2O$$

Most other methods depend on the conversion of NO to nitrites:

$$4NO+O_2+4KOH \rightarrow 4KNO_2+2H_2O$$

$$2NO+HNO_3+3H_2SO_4 \rightarrow 3NOHSO_4 \xrightarrow{3H_2O} 3HNO_2+3H_2SO_4$$

and subsequent determination either by titration with permanganate:

$$5HNO_2+2KMnO_4+3H_2SO_4 \rightarrow K_2SO_4+2MnSO_4+5HNO_3+3H_2O$$

or colorimetrically[508] by a diazo coupling reaction such as Griess reagent (sulphanilic acid and α-naphthylamine).

Of the physical techniques, infrared spectroscopy is applicable and the bands at 1850 and 1925 cm^{-1} useful to estimate the concentration of NO in the gas phase. However, mass spectrometry is probably the best method of determining NO content, and a good example of its potential is its ability to monitor NO concentrations in the exhausts of internal combustion engines[509].

[505] B. F. G. Johnson and J. A. McCleverty, in *Progress in Inorganic Chemistry*, Vol. 7 (ed. F. A. Cotton), Interscience, New York (1966), p. 277.

[506] W. P. Griffith, in *Advances in Organometallic Chemistry*, Vol. 7 (eds. F. G. A. Stone and R. West), Academic Press, New York (1968), p. 211.

[507] K. I. Vasu, in ref. 459, pp. 563–673.

[508] M. J. Taras, in *Colorimetric Determination of Non-metals* (ed. D. F. Boltz), Interscience, New York (1958).

[509] R. D. Craig, in *Modern Aspects of Mass Spectroscopy* (ed. R. I. Reed), Plenum, New York (1968).

12. DINITROGEN TRIOXIDE

12.1. PREPARATION

The history of dinitrogen trioxide has been traced back[510, 511] even to before 1816 when Gay-Lussac proposed its existence. However, pure N_2O_3 can only be obtained as a pale-blue solid or as an intense blue liquid just above its freezing point (−100.1°C). At higher temperatures dissociation becomes important:

$$\underset{\text{Blue}}{N_2O_3} \rightleftharpoons \underset{\text{Colourless}}{NO} + \underset{\text{Brown}}{NO_2}$$

$$\underset{\text{Brown}}{2NO_2} \rightleftharpoons \underset{\text{Colourless}}{N_2O_4}$$

Thus as the temperature of the liquid N_2O_3 rises towards 0°C, it becomes tinged green due presumably to the colour combination of blue N_2O_3 with the pale-yellow colour of liquid N_2O_4 containing a trace of NO_2. This explanation is consistent with the fact that the dissociation of N_2O_4 (like N_2O_3) only occurs at temperatures above its melting point (−11°C). In the gas phase, N_2O_3 is largely dissociated as evidenced by the brown colour of NO_2 particularly above the boiling point of N_2O_4 (21.3°C).

Dinitrogen trioxide is usually prepared by condensing equimolar amounts of nitric oxide and nitrogen dioxide (in equilibrium with dinitrogen tetroxide) at about −20°C.

$$2NO + N_2O_4 \rightarrow 2N_2O_3$$

The nitrogen dioxide can of course be prepared *in situ* simply by introducing the calculated quantity of oxygen to bring about the controlled oxidation of half of the nitric oxide:

$$4NO + O_2 \rightarrow 2N_2O_3$$

It can also be readily prepared[512] by adding 1:1 nitric acid dropwise onto arsenic(III) oxide warmed to about 70°C:

$$2HNO_3 + 2H_2O + As_2O_3 \rightarrow N_2O_3 + 2H_3AsO_4$$

Alternatively, passing sulphur dioxide into fuming nitric acid gives a solution of nitrosyl hydrogen sulphate from which N_2O_3 can be obtained simply by hydrolysis[512]:

$$SO_2 + HNO_3 \rightarrow NOHSO_4$$

$$2NOHSO_4 + H_2O \rightarrow N_2O_3 + H_2SO_4$$

There is, however, a problem when it comes to drying the product, for although phosphorus pentoxide is capable of dehydrating N_2O_3, it can also react with its dissociation products[513]:

$$P_2O_5 + N_2O_4 \rightarrow P_2O_5 \cdot 2NO + O_2$$

---

510 I. R. Beattie, in *Progress in Inorganic Chemistry*, Vol 5. (ed. F. A. Cotton), Interscience, New York (1963), pp. 1–26.

511 I. R. Beattie, in ref. 459, Section XXVI, pp. 241–5.

512 P. W. Schenk, in *Handbook of Preparative Inorganic Chemistry*, Vol. I (ed. G. Brauer), Academic Press, New York (1963), p. 487.

513 E. M. Stoddard, *J. Chem. Soc.* (1938) 1459; (1945) 448.

12.2. STRUCTURE AND BONDING

X-ray data on single crystals of dinitrogen trioxide at $-115°C$ suggest that the structure is disordered[514], and this may have to do with the phase transition that occurs at $-125°C$. Matrix trapped molecules of N_2O_3 do not simplify the structural situation either, for both the nitrito and nitroso-nitro isomeric forms have been identified spectroscopically[515]:

More recently, electron diffraction data indicate that the molecule is planar with non-equivalent nitrogen atoms (Fig. 30) in agreement with the nitroso-nitro structure previously proposed on the basis of microwave results[517].

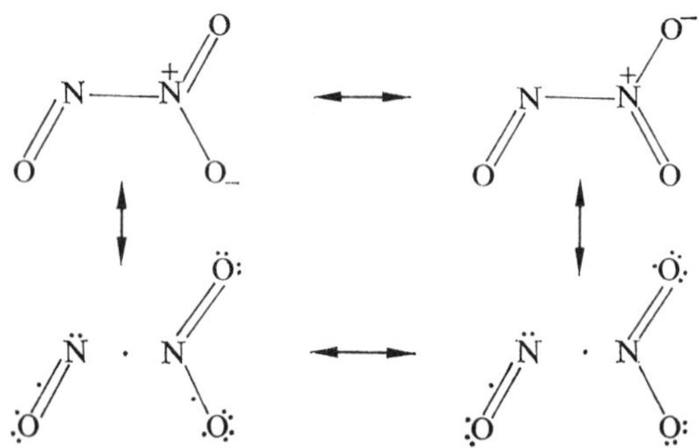

FIG. 30. The nitroso-nitro structure of dinitrogen trioxide.

Some molecular constants are given in Table 69.

The N–N bond length in N_2O_3 (1.86 Å) is unusually long when compared with the conventional N–N single bond length as in hydrazine (1.45 Å), although such a situation is not unique in nitrogen chemistry. In this connection, the bonding in N_2O_3 is thought to be closely related to the situation in N_2O_4 (see section 13) where calculations[518] involving 8π and 10σ electrons lead to a bond order of 0.76 consistent with the observed bond length. Other electronic structures which have been proposed to represent the bonding in N_2O_3 range from the Linnett non-pairing structures also illustrated in Fig. 30, to the π-only descriptions which are supposed to involve overlap only between the lone pairs on each nitrogen[519].

[514] T. B. Reed and W. N. Lipscomb, *Acta Cryst.* **6** (1953) 781.
[515] W. G. Fateley, H. A. Bent and B. Crawford, *J. Chem. Phys.* **31** (1959) 204.
[516] A. H. Brittain, A. P. Cox and R. L. Kuczkowski, *Trans. Faraday Soc.* **65** (1969) 1963.
[517] R. L. Kuczkowski, *J. Am. Chem. Soc.* **87** (1965) 5259.
[518] R. D. Brown and R. D. Harcourt, *Proc. Chem. Soc.* (1961) 217.
[519] J. Mason, *J. Chem. Soc.* (1959) 1288.

Mention should also be made to the fact that molecular orbital calculations together with X-ray photoelectron spectroscopy show that the anion in Angeli's salt ($Na_2N_2O_3$) has the structure[520]:

$$O{=}N{-}N\begin{matrix} O^- \\ O^- \end{matrix}$$

TABLE 69. MOLECULAR CONSTANTS OF DINITROGEN TRIOXIDE[a]

Property	Value
X-ray diffraction data	
System	Tetragonal
Space group	D_4^{10} $I4_12$
Lattice constant	$a = 16.4$ Å, $c = 8.86$ Å
Unit cell	32 molecules
Electron diffraction data[b]	
N–N bond length	1.864 Å
Microwave spectra[c]	
N–N bond length	1.85±0.03 Å
Infrared spectra	
N–N bond length	2.08 Å
N–O bond length	1.12 Å
$N\begin{matrix} O \\ O \end{matrix}$ bond length	1.18 Å
^{14}N n.m.r. data (satd. aq. $NO_2^- = 0$)[d]	
$\underline{N}ONO_2$	−70 ppm
$NO\underline{N}O_2$	+165 ppm
Magnetic susceptibility	$\sim -0.2 \times 10^{-6}$
Bond dissociation energy	9.5 kcal/mol

[a] Ref. 510.
[b] Ref. 516
[c] Ref. 517.
[d] Ref. 527.

12.3. PHYSICAL PROPERTIES

Gas-phase equilibrium constants and related thermodynamic functions have been determined[521] for the dissociation reaction

$$N_2O_3(g) \rightleftharpoons NO(g) + NO_2(g)$$

Thus because of this reaction many physical constants cannot be measured, while the values for some among the brief list in Table 70 are only approximate.

At temperatures around its freezing point, however, it is likely that N_2O_3 behaves as a pure compound, but even here other complications occur. It has a tendency to supercool,

[520] D. N. Hendrickson, J. M. Hollander and W. L. Jolly, *Inorg. Chem.* **8** (1969) 2642.
[521] I. C. Hisatsune, *J. Phys. Chem.* **65** (1961) 2249.

and some confusion has arisen between the freezing point (−100.7°C) and the nearby eutectic (33.7 wt.% NO in NO_2) temperature[522] at −106.2°C. At the other end of the liquid phase, dissociation clearly makes it impossible to record a true boiling point[523], for as liquid N_2O_3 warms up, NO is evolved together with the N_2O_3 vapour pressure leaving a mixture of NO_2 and N_2O_3 in the liquid phase. Values which have been quoted for the boiling point usually refer to the temperature at which the total vapour pressure over the mixture is equal to one atmosphere. Vapour pressure changes for liquid mixtures of nitric oxide and dinitrogen tetroxide vary considerably with composition in the region $NO_{1.5}$ as the vapour above such mixtures is almost pure nitric oxide which has such a low solubility.

TABLE 70. PHYSICAL PROPERTIES OF DINITROGEN TRIOXIDE[a]

Property	Temperature (K)	Value
Melting point	172.45	−100.7°C
Boiling point		between −40°C and +3°C
Transition temperature	148	−125°C
Density of solid	78	1.782 g/cm³
	158	1.694 g/cm³
Density of liquid	253	1.476
	275	1.447
ΔH°_{f0} (enthalpy of formation)	0	21.628 kcal/mol [b]
ΔH°_f (enthalpy of formation) (g)	298.15	20.01 kcal/mol [b]
ΔH°_f (enthalpy of formation) (l)	298.15	12.02 kcal/mol [b]
ΔG°_f (Gibbs energy of formation)	298.15	33.32 kcal/mol [b]
S° (entropy)	298.15	74.61 cal/deg mol [b]
C°_p (heat capacity)	298.15	15.68 cal/deg mol [b]
Vapour pressure equation	$\log_{10} P_{mm}$ =	$8.95-f(x)T$
	x	$f(x)$
	1.525	1496
	1.6	1583
	1.8	1696
	2.0	1783
For the dissociation reaction[c]	$N_2O_3(g) \rightleftharpoons$	$NO(g)+NO_2(g)$
ΔH°_{298} (enthalpy of dissociation)	298.15	9.69 kcal/mol
ΔG°_{298} (Gibbs energy of dissociation)	298.15	−0.38 kcal/mol
S° (entropy of dissociation)	298.15	33.2 cal/deg mol
K_{eq} (dissociation constant)	298.15	1.91 atm

[a] Ref. 510.
[b] D. D. Wagman, W. H. Evans, V. B. Parker, I. Halow, S. M. Bailey and R. H. Schumm, *Selected Values of Chemical Thermodynamic Properties*, NBS Technical Note 270-3, Washington (1968).
[c] Ref. 521.

The infrared spectra of N_2O_3 including the ^{15}N isomers in the various phases have been recorded and assignments based on the nitroso-nitro structure suggested[524]. Six of the fundamental bands have been observed in the Raman spectra of N_2O_3 solution[525]. Visible and

522 I. R. Beattie, S. W. Bell and A. J. Vosper, *J. Chem. Soc.* (1960) 4796.
523 I. R. Beattie and A. J. Vosper, *J. Chem. Soc.* (1960) 4799.
524 I. C. Hisatsune and J. P. Devlin, *Spectrochim. Acta* **17** (1961) 218.
525 I. C. Hisatsune and J. P. Devlin, *Spectrochim. Acta* **16** (1960) 401.

near ultraviolet spectra have also received attention[519]. The blue colour due to a band at 5400–5900 λ is usually assigned to an $n \rightarrow \pi^*$ electronic transition between a lone-pair electron on the nitroso-nitrogen and an antibonding π^* orbital, while the band in the near ultraviolet at ~6500 λ is associated with a π–π^* transition involving the N–N bond. The ^{14}N n.m.r. spectrum also offers a direct demonstration[526] of the structure with a resonance at +165 ppm (relative to saturated aqueous nitrite ion at room temperature) due to the nitro-nitrogen, and the nitroso-nitrogen gives a line at −70 ppm. The low field position of this latter resonance is explained on the premiss that the mixing of the readily accessible $n_N \rightarrow \pi^*$ excited states with the ground state by the magnetic field de-shields the ^{14}N nucleus.

12.4. REACTIONS

One of the main problems in studying the reactions of a substance like N_2O_3 is one of carrying out adequate ancillary experiments to determine the effects of its dissociation products on the reaction under investigation. This is further complicated by their participation in exchange processes:

$$^{14}NO + ^{15}NO_2 \rightleftharpoons ^{15}NO + ^{14}NO_2; \qquad K_{298} = 0.96 \pm 0.02$$

Exchange even between NO and solid N_2O_3 at −118°C is known to occur[527], although gas–liquid and vapour phase equilibria are of interest in ^{15}N isotope concentration[528].

The reactions involving water support the idea of N_2O_3 being the formal anhydride of nitrous acid:

$$2HNO_2 \rightarrow N_2O_3 + H_2O$$

Indeed, the reaction of N_2O_3 or an equimolar mixture of NO and NO_2 in aqueous alkaline solution result in conversion to nitrite:

$$\begin{array}{l} N_2O_3 + 2OH^- \rightarrow 2NO_2^- + H_2O \\ (NO + NO_2) \end{array}$$

Gas phase reactions lead to rather more complex equilibria which are involved in nitric acid formation[529]:

$$N_2O_3 \rightleftharpoons NO + NO_2$$
$$NO + NO_2 + H_2O \rightleftharpoons 2HNO_2$$
$$3HNO_2 \rightarrow HNO_3 + 2NO + H_2O$$

The reactions of N_2O_3 with concentrated acids such as sulphuric, selenic, perchloric and tetrafluoroboric, provide preparative routes to the corresponding nitrosyl salts, $NO^+HSO_4^-$, $NO^+HSeO_4^-$, $NO^+ClO_4^-$, $NO^+BF_4^-$:

$$N_2O_3 + 3H_2SO_4 \rightarrow 2NO^+ + H_3O^+ + 3HSO_4^-$$

Reference should also be made to the solubility of N_2O_3 in other non-aqueous solvents particularly organic solvents such as benzene, toluene, carbon tetrachloride and chloroform, as well as its rôle in liquid N_2O_4 reactions[530] which produce nitric oxide (see section 13).

526 L.-O. Andersson and J. Mason, *Chem. Communs.* (1968) 99.
527 E. Leifer, *J. Chem. Phys.* **8** (1940) 301.
528 W. Spindel, in *Inorganic Isotopic Synthesis* (ed. R. H. Herber), Benjamin, New York (1962), p. 97.
529 A. Klemenc, *Z. anorg. allgem. Chem.* **280** (1955) 100.
530 C. C. Addison, *Chemistry in Liquid Dinitrogen Tetroxide*, Vol. III, Part 1, of *Chemistry in Nonaqueous Ionizing Solvents* (eds. G. Jander, H. Spandau and C. C. Addison), Pergamon, Oxford (1967), p. 55.

12.5. ANALYSIS

Dinitrogen trioxide can be detected and determined even in the presence of its dissociation products NO and NO_2, by infrared and ultraviolet spectroscopic methods[510]. Frequently, however, only an estimate of the total nitrogen oxide content is required, and methods based on their absorption as nitrite or nitrate followed by colorimetric analysis are employed[531]. Such methods are commonly used for analysis of "nitrous fumes"[532] which usually refers to mixtures of NO and NO_2 and other oxides that exist in equilibrium with them (principally N_2O_3 and N_2O_4). However, the full analysis of such mixtures often requires a combination of techniques including physical equipment such as gas chromatographs using helium carrier gas and thermal conductivity detectors, or mass spectrometers[531].

13. NITROGEN DIOXIDE AND DINITROGEN TETROXIDE

13.1. PRODUCTION AND INDUSTRIAL USE

Nitrogen dioxide and dinitrogen tetroxide are discussed together because under ordinary conditions of temperature and pressure both species occur in the presence of each other in a state of equilibrium[533-5]:

$$2NO_2 \rightleftharpoons N_2O_4$$

Thus N_2O_4 having the higher molecular weight is the low-temperature form, and NO_2 the dissociation product at higher temperatures. Their equilibrium concentrations at various temperatures are indicated in Table 71.

Like other molecules with an odd number of electrons (free radicals), the monomer NO_2 is coloured, whereas the dimeric form N_2O_4 is colourless, hence the intensity of the red–brown colour in the gas phase can be used to monitor the monomer concentration[535].

TABLE 71. EQUILIBRIUM CONCENTRATIONS OF NO_2 AND N_2O_4 AT ORDINARY TEMPERATURES

Temperature (°C)	Phase	Equilibrium concentration	
		NO_2	N_2O_4
−11.2	Solid	0	100
−11.2	Liquid	0.01	99.99
21.15	Liquid	0.1	99.9
21.15	Gas	15.9	84.1
135	Gas	99	1

[531] K. I. Vasu, in ref. 459, p. 633.

[532] *Methods for the Detection of Toxic Substances in Air*, Booklet No. 5, *Nitrous Fumes*, Ministry of Employment and Productivity, HM Factory Inspectorate, HMSO, London (1969).

[533] C. C. Addison, *Chemistry in Liquid Dinitrogen Tetroxide*, Vol. III, Part I, *Chemistry in Nonaqueous Ionizing Solvents* (eds. G. Jander, H. Spandau and C. C. Addison), Pergamon, London (1967).

[534] I. R. Beattie, in ref. 459, Section XXVII, pp. 246–68.

[535] P. Gray and A. D. Yoffe, *Chem. Rev.* **55** (1955) 1069.

On an industrial scale, NO_2–N_2O_4 is produced by the oxidation of nitric oxide during the manufacture of nitric acid[536]:

$$2NO(g)+O_2(g) \rightarrow 2NO_2(g)$$

Thus the principal use of NO_2–N_2O_4 in forming nitric acid[536, 537] is described by its reaction with water according to the overall equation

$$3NO_2(g)+H_2O(l) \rightleftharpoons 2HNO_3(aq)+NO(g)$$

13.2. PREPARATION OF NO_2/N_2O_4

The equilibrium mixture NO_2–N_2O_4 is usually prepared on a laboratory scale either by reduction or decomposition of nitric acid, or more conveniently by the thermal decomposition of a heavy metal nitrate. Rigorously dried lead nitrate can be heated in a steel bomb[533] or a tube furnace[538] under a slow stream of oxygen which helps prevent contamination by N_2O_3:

$$2Pb(NO_3)_2 \rightarrow 4NO_2+2PbO+O_2$$

The evolved gases are passed through a condenser to remove traces of nitric acids then over phosphorus pentoxide to complete the drying process. The gas is condensed at -78°C and can be further purified by fractional distillation, rejecting the first fraction and collecting the remainder in individual ampoules which are then sealed. Pure dinitrogen tetroxide freezes as a colourless solid, whereas a pale-green solid results if traces of moisture are present, colour being a sensitive test of purity[533].

Other methods which have been used to prepare nitrogen dioxide include the oxidation of nitric oxide:

$$2NO+O_2 \rightarrow 2NO_2$$

the reaction of dinitrogen trioxide and dinitrogen pentoxide:

$$N_2O_3+N_2O_5 \rightarrow 4NO_2$$

the reaction of nitric acid with sulphur dioxide:

$$2HNO_3+SO_2 \rightarrow 2NO_2+H_2SO_4$$

the reaction of nitric acid and phosphorus pentoxide:

$$4HNO_3+2P_2O_5 \rightarrow 4NO_2+O_2+4HPO_3$$

the reaction of nitric acid and copper:

$$Cu+4HNO_3 \rightarrow 2NO_2+Cu(NO_3)_2+2H_2O$$

the reaction of nitrosyl chloride with silver nitrate:

$$NOCl+AgNO_3 \rightarrow 2NO_2+AgCl$$

the reaction of nitrosonium hydrogen sulphate and potassium nitrate:

$$NOHSO_4+KNO_3 \rightarrow 2NO_2+KHSO_4$$

---

[536] I. R. Beattie, in ref. 459, Section XXII, pp. 158–72.

[537] C. Matasa and E. Matasa, *L'Industrie Moderne des Produits Azotes*, Dunod, Paris (1968), p. 460.

[538] P. W. Schenk, in *Handbook of Preparative Inorganic Chemistry*, 2nd edn., Vol. I (ed. G. Brauer), Academic Press, New York (1963), p. 488.

Such methods may be convenient and adequate for some purposes, but the presence of other non-metal compounds generally leads to impurities in the product.

13.3. NUCLEAR ISOTOPES OF NITROGEN DIOXIDE[539]

The oxidation of enriched nitric oxide with excess molecular oxygen is rapid and complete at room temperature:

$$^{15}NO+\frac{1}{2}O_2 \rightarrow {}^{15}NO_2$$

Similarly, $N^{18}O^{16}O$ and $N^{18}O_2$ can be prepared by treating $N^{18}O$ with normal and ^{18}O-enriched oxygen respectively.

Based on the thermal decomposition of heavy metal nitrates, $^{15}NO_2$ has been prepared by heating $K^{15}NO_3$ with PbO_2, or dissolving $K^{15}NO_3$ in 85% phosphoric acid and adding copper metal:

$$M(^{15}NO_3)_2 \rightarrow 4\,^{15}NO_2+2MO+O_2$$

Nitrogen dioxide enriched in ^{18}O can also be prepared[540] by heating the lead nitrate that is formed from lead chloride and $KN^{18}O_3$. The nitrogen oxides collected at $-196°C$ are vaporized and allowed to react with oxygen to ensure complete oxidation of the nitric oxide to the nitrogen dioxide. NO_2 is then re-condensed at $-78°C$, and the excess oxygen and nitrogen pumped off.

13.4. STRUCTURE AND BONDING

The solid phase consists entirely of N_2O_4 molecules and shows two modifications known as the monoclinic form[541] which is unstable with respect to the cubic form[542]. The structures of the molecules in both phases which have been determined by X-ray diffraction are similar and are in agreement with a planar structure with a centre of symmetry:

```
O       O
 \     /
  N—N
 /     \
O       O
```

although infrared evidence suggests that non-planar $O_2N–NO_2$ and even $ONONO_2$ structures are present at $-196°C$ and $-269°C$ respectively[543]. Indeed, nitrosonium nitrate $NO^+NO_3^-$ has been prepared by oxidation of NO at $-196°C$ and it has been suggested[544] that the stabilities of the N_2O_4 isomers decrease in the order

$$O_2NNO_2 > NO^+NO_3^- > ONONO_2.$$

Structural evidence for N_2O_4 in the liquid phase is less direct, but it is thought that more than one form exists and in particular one through which the formation of ion pairs $[NO^+][NO]_3^-$ can be accounted[533]. Electron diffraction study of the vapour at $-20°C$

[539] W. Spindel, in *Inorganic Isotopic Syntheses* (ed. R. H. Herber), Benjamin, New York (1962), p. 101.
[540] I. Dostrovsky and D. Samuel, in *Inorganic Isotopic Syntheses* (ed. R. H. Herber), Benjamin, New York (1962), p. 134.
[541] P. Groth, *Acta chem. scand.* **17** (1963) 2419.
[542] J. S. Broadley and J. M. Robertson, *Nature* **164** (1949) 915.
[543] I. C. Hisatsune and J. P. Devlin, *J. Chem. Phys.* **31** (1959) 1130.
[544] L. Parts and J. Y. Miller, Jr., *J. Chem. Phys.* **43** (1965) 136.

supports the planar structure[545] although there seems to be no obvious reason why there should not be free rotation of the NO_2 groups about the long N–N bond.

The V-shaped structure of nitrogen dioxide in the gas phase has been confirmed by a variety of spectroscopic methods and by electron diffraction[546] giving the N–O bond length of 1.19 Å and the ONO bond angle of 134°. Using Linnett-type valence bond structures, electronic configurations which maintain a complete octet of electrons round each atom are possible for the seventeen-electron NO_2 molecule[547] as shown in Fig. 31.

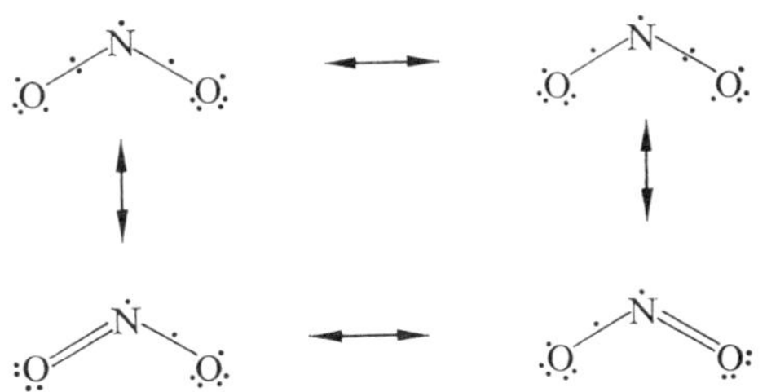

FIG. 31. Valence bond structures of NO_2.

The bonding can also be described in terms of molecular orbital theory, and NO_2 provides a good example to illustrate the procedure by which molecular orbitals can be constructed for simple molecules[548]. Since NO_2 is angular, the orbitals of the molecule must be characterized according to the operations of the point group which describes the molecular shape C_{2v} for which the character table is:

C_{2v}	E	$C_2(z)$	$\sigma_v(xz)$	$\sigma_v(yz)$
A_1	1	1	1	1
A_2	1	1	−1	−1
B_1	1	−1	1	−1
B_2	1	−1	−1	1

The NO_2 molecule is fixed in the co-ordinate system so as to be situated say in the *xz* plane as illustrated in Fig. 32.

By carrying out the operations of the point group $C_{2v}[C_2(z), \sigma_v(xz)$ etc.] on each atomic orbital of nitrogen ($2s$, $2p_z$, etc.) and then on linear combinations of the orbitals on the oxygen atoms [$(s+s)$, $(s-s)$, (p_z+p_z), etc.], the transformation properties of each orbital can be calculated and assigned to an irreducible representation (a_1, a_2, etc., corresponding

[545] R. G. Snyder and I. C. Hisatsune, *J. Chem. Phys.* **26** (1957) 960.

[546] G. R. Bird, *J. Chem. Phys.* **25** (1956) 1040.

[547] M. Green, in *Developments in Inorganic Nitrogen Chemistry*, Vol. I (ed. C. B. Colburn), Elsevier, Amsterdam (1966), pp. 1–71.

[548] C. J. Ballhausen and H. B. Gray, *Molecular Orbital Theory*, Benjamin, New York (1965), p. 76.

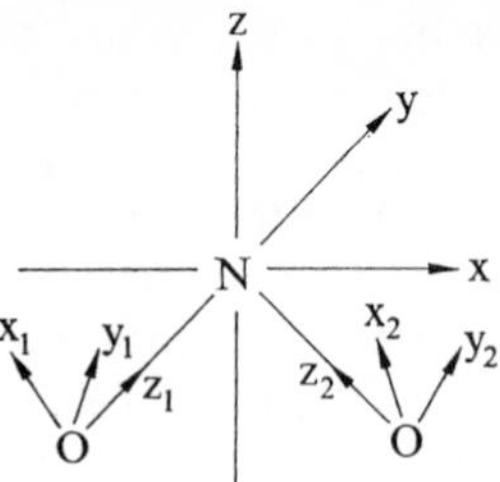

FIG. 32. The NO_2 molecule.

to A_1, A_2 in the character table). Thus having assigned the available atomic orbitals to the appropriate symmetry type, the molecular orbitals for NO_2 can be constructed simply by combining those atomic orbitals with the same symmetry. These molecular orbitals can then be presented in the form of an energy-level diagram as in Fig. 33 into which electrons

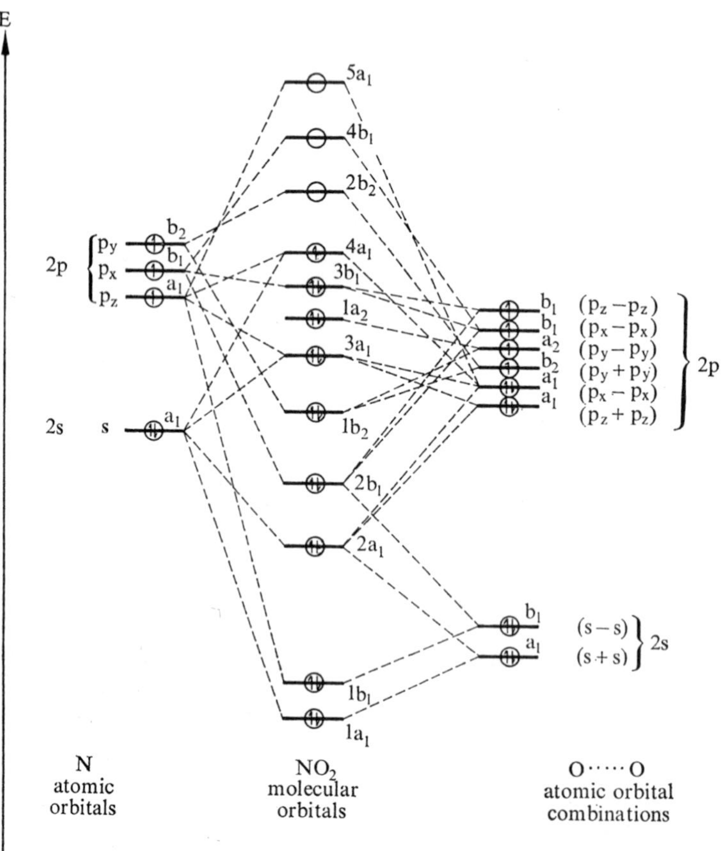

FIG. 33. Molecular orbital diagram for NO_2.

are allocated to the lowest molecular orbitals to give the ground state configuration of the molecule:

$$^2A_1 \ (1a_1)^2(1b_1)^2(2a_1)^2(2b_1)^2(1b_2)^2(3a_1)^2(1a_2)^2(3b_1)^2(4a_1)^1$$

Thus although the $1a_2$, $3b_1$ and $4a_1$ levels are all fairly close together, the unpaired electron is assigned to the $4a_1$ orbital if one considers the large splitting that is observed in the e.s.r. spectrum. This is produced by an unpaired *s* electron interacting with the spin of the ^{14}N nucleus, and both $1a_2$ and $3b_1$ are made up of *p* functions, whereas the $4a_1$ level is composed partly of nitrogen and oxygen 2*s* orbitals. The first excited state has the configuration:

$$^2B_1 \ (1a_1)^2(1b_1)^2(2a_1)^2(2b_1)^2(1b_2)^2(3a_1)^2(1a_2)^2(3b_1)^1(4a_1)^2$$

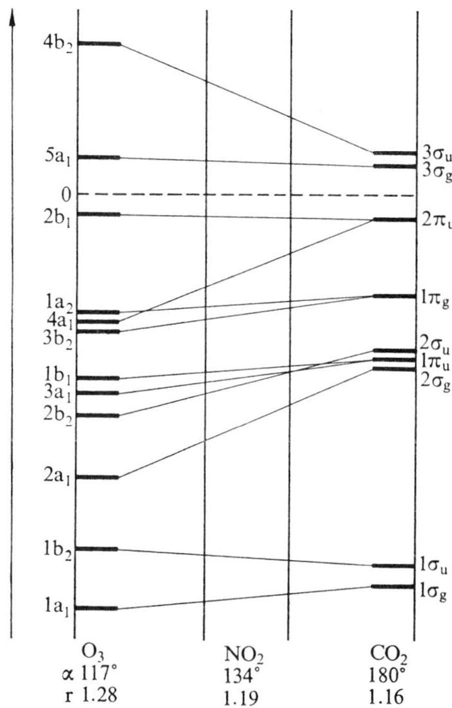

FIG. 34. Ozone, nitrogen dioxide and carbon dioxide correlation diagrams.

and the absorption band in the NO_2 spectrum at 23,000 cm^{-1} is usually assigned to the $^2A_1 \rightarrow {}^2B_1$ transition[547, 548].

A diagram similar to that in Fig. 33 can be arrived at by correlating the molecular orbital diagrams of structurally related compounds. From this point of view, NO_2 can be considered as the compromise situation between ozone, a C_{2v} molecule with smaller angle and longer bonds, and carbon dioxide, a linear molecule with point group $D_{\infty h}$ and shorter bonds. The correlation diagram in Fig. 34 indicates the approximate energies of NO_2 molecular orbitals simply by joining the appropriate corresponding orbitals of O_3 and CO_2 (note that the different notation for CO_2 is due to it having a centre of symmetry). These correlation diagrams can also be used to estimate the bond angles of low-lying excited states while the bond length can be inferred by considering to what extent antibonding orbitals are involved[549].

With regard to the description of the bonding in N_2O_4, this would appear to be less straightforward[533]. There is general agreement that the molecule is planar, has a long N–N bond and all the N–O bonds are of equal length. Planarity in this situation implies a restriction to rotation about the N–N bond which normally would be expected to result from steric restraint or multiple bonding. However, there would not appear to be any steric hindrance, particularly in the gas phase, and a bond length longer than that of a normal N–N single bond (N–N in hydrazine is 1.47 Å) is not consistent with multiple bonding either. Thus this unusual situation has led to the proposal of various novel descriptions of the bonding. The usual structure is a resonance hybrid of the canonical forms illustrated in Fig. 35.

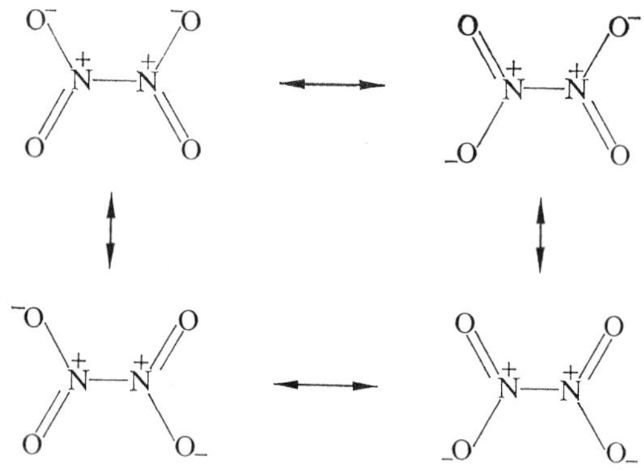

FIG. 35. Valence bond structures for N_2O_4.

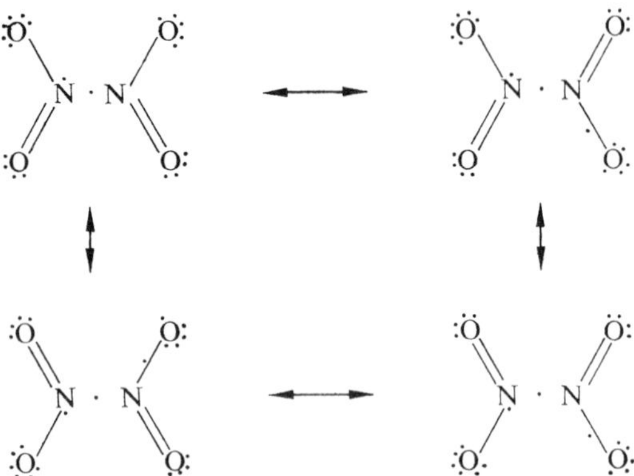

FIG. 36. Linnett non-pairing structures for N_2O_4.

---

[549] A. D. Walsh, *J. Chem. Soc.* (1953) 2266.

Here the long N–N bond is explained on the basis of repulsion due to formal positive charges lying on adjacent atoms, but it does not offer a satisfactory explanation for planarity. Similarly, resonance between the Linnett non-pairing structures in Fig. 36 also provides a simple picture of N_2O_4 which nicely explains the long one-electron N–N bond but fails to account for the planarity[547].

More recently it was suggested that the central bond is a π bond with little or no σ character[550]. However, this is not now thought to be correct, and calculations[551] suggest

TABLE 72. MOLECULAR CONSTANTS OF N_2O_4

Property	Value	
X-ray diffraction data[a]		
System	Monoclinic (unstable)	Cubic (stable)
N–N bond length	1.75 Å	1.64 Å
N–O bond length	1.21 Å	1.17 Å
ONO bond angle	135°	126°
Electron diffraction data[b] on gaseous N_2O_4		
N–N bond length	1.75 Å	
N–O bond length	1.18 Å	
ONO bond angle	133.7°	
Fundamental vibration frequencies[c] for liquid N_2O_4	(cm^{-1})	
ν_1 (a_g)	1379.6	
ν_2 (a_g)	808	
ν_3 (a_g)	260	
ν_4 (a_u)	(~50)	
ν_5 (b_{1g})	1712	
ν_6 (b_{1g})	482	
ν_7 (b_{1u})	429	
ν_8 (b_{2g})	622	
ν_9 (b_{2u})	1748	
ν_{10} (b_{2u})	381	
ν_{11} (b_{3u})	1262	
ν_{12} (b_{3u})	750	
Enthalpy of homolytic dissociation	12.9 kcal/mol	
Enthalpy of heterolytic dissociation	7.87 kcal/mol	
Entropy of homolytic dissociation	42 cal/deg mol	
Entropy of heterolytic dissociation	−80.1 cal/deg mol	
Dipole moment	0.55 *D*	
Dielectric constant	2.42	
Magnetic susceptibility (293)	2.52×10^{-7} c.g.s. units g^{-1}	
Ultraviolet absorption maxima	343 mμ	
Refractive index (5893 Å)	1.420	

[a] Ref. 542.
[b] Ref. 545.
[c] Ref. 552.
[d] Ref. 533.

550 C. A. Coulson and J. Duchesne, *Bull. Acad. Belg. Cl. Sci.* **43** (1957) 522.
551 M. Green and J. W. Linnett, *Trans. Faraday Soc.* **57** (1961) 10.

that although the main part of the N–N bond is σ in character, there is a fairly large π contribution from orbitals which lie in the plane of the molecule (similar to combinations of the $3b_2$ orbitals of NO_2) and not from the more usual orbitals perpendicular to the plane.

13.5. MOLECULAR CONSTANTS

Some molecular constants for N_2O_4 are listed in Table 72, while values for NO_2 and NO_2^+ are compared in Table 73 together with some relevant spectroscopic[552] and thermodynamic data[553]. Electron spin resonance spectra have been recorded for NO_2 in various matrices[554].

TABLE 73. MOLECULAR CONSTANTS[a] OF NO_2 AND NO_2^+

Property	NO_2	NO_2^+
Structural parameters		
N–O bond length	1.197 Å	1.15 Å
ONO bond angle	134° 15′	180°
Fundamental vibration frequencies[b]		
ν_1 (a_1)	1322.5 cm^{-1}	
ν_2 (a_1)	749.7 cm^{-1}	
ν_3 (b_1)	1617.75 cm^{-1}	
Magnetic susceptibility	28.2×10^{-2} c.g.s. units g^{-1}	
ESR hyperfine splitting (NO_2 trapped in argon)	58 oersted	
Ionization potential	11 eV	
Dipole moment	0.39 *D*	
Thermodynamic functions[c]		
ΔH°_{f0} (enthalpy of formation (0 K)	8.6 kcal/mol	276.6 kcal/mol
ΔH°_{f0} (enthalpy of formation) (298 K)	7.93 kcal/mol	277.4 kcal/mol
$H^\circ_{298}-H_0$ (enthalpy)	2.438 kcal/mol	
ΔG_{f0} (Gibbs energy of formation) (298 K)	12.26 kcal/mol	
S° (entropy)	57.35 cal/deg mol	
C°_p (heat capacity)	8.89 cal/deg mol	

[a] Refs. 533–5.
[b] Ref. 552.
[c] Ref. 553.

13.6. PHYSICAL PROPERTIES

Some physical data for NO_2 have already been mentioned in Table 73, but so few experimentally determined values are known because of their need to be measured at temperatures above 200°C to ensure complete dissociation of N_2O_4. Values of physical constants for N_2O_4 are listed in Table 74.

[552] R. F. Barrow and A. J. Merer, in ref. 459, p. 515.
[553] D. D. Wagman, W. H. Evans, V. B. Parker, I. Halow, S. M. Bailey and R. H. Schumm, *Selected Values of Chemical Thermodynamic Properties*, NBS Technical Note 270-3, Washington (1968).
[554] P. W. Atkins and M. C. R. Symons, *The Structure of Inorganic Radicals*, Elsevier, Amsterdam (1967), p. 127.

However, only the solid phase consists of pure N_2O_4, and values for properties of the liquid refer to material containing the equilibrium quantity of NO_2. This probably makes no difference just above the melting point but is likely to be significant towards the boiling point, accompanying the colour change from very pale yellow to deep red–brown over the same temperature range. The solvent properties of liquid N_2O_4 will be discussed further in section 13.7.2.

TABLE 74. PHYSICAL PROPERTIES OF DINITROGEN TETROXIDE [a]

Property	Temperature (K)	Value
Freezing point	261.95	−11.2°C
Boiling point	294.30	21.15°C
Critical temperature	158.2	−114.95°C
Critical pressure		99.96 atm
Critical density		0.570 g/cm^3
Critical volume		165.3 cm^3/mol
$\triangle H$ fusion	261.95	3.502 kcal/mol
Entropy of fusion	261.95	13.37 cal/deg mol
$\triangle H$ vaporization	294.30	9.110 kcal/mol
Entropy of vaporization	294.30	30.96 cal/deg mol
Specific heat	261.95	32.65 cal/deg mol
Specific heat	294.30	33.90 cal/deg mol
Cryoscopic constant		3.64°C depression/kg N_2O_4
Ebullioscopic constant		1.37°C elevation/kg N_2O_4
Thermodynamic properties[b]		
$\triangle H^\circ_{f0}$ (enthalpy of formation) (g)	0	4.49 kcal/mol
$\triangle H_{f0}$ (enthalpy of formation) (g)	298	2.19 kcal/mol
$H_{298}-H_0$ (enthalpy) (g)		3.918 kcal/mol
$\triangle H_{f0}$ (enthalpy of formation) (l)	298	−4.66 kcal/mol
$\triangle G_f$ (Gibbs energy of formation) (g)	298	23.38 kcal/mol
$\triangle G_f$ (Gibbs energy of formation) (l)	298	23.29 kcal/mol
S_0 (entropy) (g)	298	72.70 cal/deg mol
S_0 (entropy) (l)	298	50.0 cal/deg mol
C_{p0} (heat capacity) (g)	298	18.47 cal/deg mol
C_{p0} (heat capacity) (l)	298	34.1 cal/deg mol
Vapour pressure of solid	240.3–261.9	
	$\log_{10} P = 9.58149 - 2460.000/T + 7.61700/T^3 - 1.51335T^2/10^5$	
Vapour pressure of liquid	261.9–294.9	
	$\log_{10} P = 8.00436 - 1753.000/T - 11.8078/10^4 + 2.0954T^2/10^6$	
	248	70 mm
	261.95	139.8 mm
	272.31	266 mm
	283	454 mm
	288	569.4 mm
	294.30	760 mm
Density of solid	78	1.979 g/cm^3
Density of liquid	253–294.30	$d = 1.4927 - 2.235t/10^3 - 2.75t^2/10^6$
	253	1.5364 g/cm^3
	258	1.5256 g/cm^3
	263	1.5147 g/cm^3
	273	1.4927 g/cm^3
	283	1.4702 g/cm^3
	293	1.4469 g/cm^3

TABLE 74 (*cont.*)

Property	Temperature (K)	Value
Viscosity of liquid	253–294.3	$\log_{10} \eta = 400/T - 1.742$
	253	0.687 poise
	258	0.641 poise
	263	0.599 poise
	273	0.527 poise
	283	0.468 poise
	293	0.420 poise
Electrical conductivity of liquid	248–294.3	$\log \kappa = -1267/T - 8.260$
	248	$4.3 \times 10^{-14}\ \Omega^{-1}\ \text{cm}^{-1}$
	290	$2.36 \times 10^{-13}\ \Omega^{-1}\ \text{cm}^{-1}$

[a] Ref. 533. [b] Ref. 553.

13.7. CHEMISTRY AND CHEMICAL PROPERTIES

13.7.1. Reactions of NO_2

It is unlikely that perfectly pure NO_2 can be formed at atmospheric pressure, for the temperature required to dissociate N_2O_4 also coincides with the onset of thermal decomposition of NO_2 which becomes significant above 150°C and complete[536] at about 600°C:

$$N_2O_4 \rightleftharpoons 2NO_2 \rightleftharpoons 2NO + O_2$$

At low partial pressures, photochemical decomposition occurs and can be studied under conditions where the recombination of nitric oxide and oxygen is so slow as to be negligible. The proposed mechanism[555] is:

$$NO_2 + h\nu \rightarrow NO_2^*$$
$$NO_2^* + NO_2 \rightarrow 2NO + O_2$$

Low concentrations of NO_2 in hydrogen and oxygen–air mixtures depress considerably the ignition temperatures and effect the explosion limits[556]. The overall reaction between NO_2 and hydrogen is given as

$$NO_2 + H_2 \rightarrow NO + H_2O$$

for which the following mechanism has been proposed[557]:

$$H_2 + NO_2 \rightarrow H + HNO_2$$
$$H + NO_2 \rightarrow NO + OH$$
$$OH + H_2 \rightarrow H_2O + H$$
$$OH + NO_2 + M \rightarrow HNO_3 + M$$
$$OH + NO + M \rightarrow HNO_2 + M$$

555 R. G. W. Norrish, *J. Chem. Soc.* (1929) 1158.
556 A. C. Egerton and J. Powling, *Proc. Roy. Soc.* A, **193** (1948) 172.
557 P. G. Ashmore and B. P. Levitt, *Trans. Faraday Soc.* **53** (1957) 945.

Such reactions are also thought to be involved in the reactions between water and NO_2 in the production of nitric acid[536]:

$$2NO_2+H_2O \rightarrow HNO_3+HNO_2$$
$$3HNO_2 \rightarrow HNO_3+2NO+H_2O$$

It is not surprising therefore that NO_2 in the presence of those compounds with which it is in equilibrium is strongly corrosive towards metals.

Other oxidations involving nitrogen dioxide[558] include the reactions with halogen compounds:

$$NO_2+2HCl \rightarrow NOCl+H_2O+\tfrac{1}{2}Cl_2$$

At higher temperatures the nitrosyl halide decomposes:

$$NO_2+2HCl \rightarrow NO+H_2O+Cl_2$$
$$NO_2+2HBr \rightarrow NO+H_2O+Br_2$$

Likewise, a variety of bimolecular gas phase reactions involving NO_2 have been studied[559]:

$$NO_2+NH_3 \rightarrow HNO_2+NH_2$$
$$NO_2+F_2 \rightarrow NO_2F+F$$
$$NO_2+Cl_2O \rightarrow NO_2Cl+OCl$$
$$NO_2+ClO_2 \rightarrow NO_3+OCl$$
$$NO_2+CO \rightarrow NO+CO_2$$

Thus the description of nitrogen dioxide as a fairly reactive free radical is well illustrated by dimerization, combination with other free radicals[560], addition, and abstraction reactions, with inorganic and organic molecules[561].

13.7.2. Reactions of Liquid Dinitrogen Tetroxide[533]

Homolytic dissociation of N_2O_4 always occurs to a small extent in the liquid phase:

$$N_2O_4(l) \rightleftharpoons 2NO_2(l); \qquad \Delta H_{diss} = 19.5 \text{ kcal/mol}$$

but of the various alternative heterolytic dissociation processes, it is perhaps surprising that the nitronium and nitrite ions have never been recognized as free ions in liquid N_2O_4. In fact the chemistry of the N_2O_4 solvent system is rationalized on the basis of self-ionization to give nitrosyl and nitrate ions according to the equation

$$N_2O_4 \rightleftharpoons NO^++NO_3^-$$

analogous to the ionization of water or liquid ammonia:

$$2H_2O \rightleftharpoons H_3O^++OH^-$$
$$2NH_3 \rightleftharpoons NH_4^++NH_2^-$$

---

558 J. H. Thomas, in *Oxidation and Combustion Reviews*, Vol. I (ed. C. F. H. Tipper), Elsevier, Amsterdam (1965), p. 137.

559 A. F. Trotman-Dickenson and G. S. Milne, *Tables of Bimolecular Gas Reactions*, NBS-9, Washington (1969).

560 Y. Rees and G. H. Williams, in *Advances in Free-radical Chemistry*, Vol. 3 (ed. G. H. Williams), Logos Press, London (1969), p. 199.

561 A. V. Topchiev (translated by C. Matthews), *Nitration of Hydrocarbons and other Organic Compounds*, Pergamon, London (1959).

Thus although the free ions in liquid N_2O_4 cannot be detected by methods such as infrared or Raman spectroscopy due to their low concentration, the large difference between the molar polarization and molar refraction is attributed[562] to the presence of polar constituents in the liquid phase which are not present in either solid or gaseous states. These polar constituents are likely to be ion pairs rather than free ions because of the low dielectric constant of the solvent, and although ionic reactions will still occur, they do so as a result of collisions between ion pairs rather than free ions. Consequently, simple metal salts are insoluble in pure liquid N_2O_4, but do dissolve in mixtures of N_2O_4 diluted with solvents of higher dielectric constant[533]. Thus the addition of nitromethane ($\varepsilon = 37$) produces conducting solutions in which the self-ionization of N_2O_4 is enhanced to such an extent that dissociation is complete in solvents like perchloric or sulphuric acids:

$$N_2O_4 + 3H_2SO_4 \rightarrow NO^+HSO_4^- + HNO_3 + SO_3 + H_3O^+HSO_4^-$$

Likewise a similar situation can be achieved by the addition of a donor solvent (such as ethyl acetate, diethyl ether, diethylnitrosamine, etc.). This arises from the electron-deficient nature of N_2O_4 and its tendency to form molecular addition compounds with donor molecules[533], in particular oxygen and nitrogen bases bonded to the nitrogen(s) of N_2O_4:

$$nD + N_2O_4 \rightleftharpoons D_n \cdot N_2O_4 \rightleftharpoons (D_nNO)^+NO_3^-$$

The wide variety of reactions in which liquid N_2O_4 can participate are summarized briefly below, although it should not be assumed that all such reactions are necessarily unique to the liquid phase. Acid–base reactions are described by the neutralization equation

$$NO^+ + NO_3^- \rightleftharpoons N_2O_4$$

Thus in liquid N_2O_4, nitrosyl chloride reacts with silver nitrate[563]:

$$NOCl + AgNO_3 \xrightarrow{N_2O_4} AgCl + N_2O_4$$

comparable with the corresponding reactions in water or liquid ammonia:

$$HCl + KOH \xrightarrow{H_2O} KCl + H_2O$$
$$NH_4Cl + NaNH_2 \xrightarrow{NH_3} NaCl + 2NH_3$$

Similarly, compounds that produce "basic" solutions in liquid N_2O_4 are typified by alkylammonium nitrates which introduce excess nitrate into the system. Thus the reaction of zinc in these solutions illustrates the amphoteric behaviour of zinc compounds[564] in N_2O_4:

$$Zn + 2EtNH_3 \cdot NO_3 + 2N_2O_4 \rightarrow [EtNH_3]_2[Zn(NO_3)_4] + 2NO$$

again directly analogous to reactions in water and liquid ammonia:

$$Zn + 2NaOH + 2H_2O \rightarrow Na_2[Zn(OH)_4] + H_2$$
$$Zn + 2NaNH_2 + 2NH_3 \rightarrow Na_2[Zn(NH_2)_4] + H_2$$

---

562 C. C. Addison, H. C. Bolton and J. Lewis, *J. Chem. Soc.* (1951) 1294.
563 C. C. Addison and R. Thompson, *J. Chem. Soc.* (1949) S211.
564 C. C. Addison and N. Hodge, *J. Chem. Soc.* (1954) 1138.

In contrast, the nitrosonium compounds are the "solvo-acids" and increase the concentration[533] of the NO^+ ion in liquid N_2O_4. Thus using tin as example,

$$Sn + 2NOCl \xrightarrow{N_2O_4} SnCl_2 + 2NO$$

again analogous to the corresponding reactions in water and liquid ammonia:

$$Sn + 2HCl \xrightarrow{H_2O} SnCl_2 + H_2$$
$$Sn + 2NH_4Cl \xrightarrow{NH_3} SnCl_2 + 2NH_3 + H_2$$

Silver, zinc, mercury and the alkali metals react quite readily when brought into contact with liquid N_2O_4. For example, sodium requires continual stirring to expose clean metal surface if the reaction is to proceed to completion:

$$Na + N_2O_4 \rightarrow NaNO_3 + NO$$

At room temperature nitric oxide is evolved, but at lower temperatures or under pressure the formation of blue–green N_2O_3 is apparent[533]. The reaction is again directly analogous to those in water and liquid ammonia:

$$2Na + 2H_2O \rightarrow 2NaOH + H_2$$
$$2Na + 2NH_3 \rightarrow 2NaNH_2 + H_2$$

However, a number of other metals (Cd, Mn, Co, Cu, In, U) are only soluble if the liquid N_2O_4 is diluted with a solvent of higher dielectric constant such as nitromethane. The mixture becomes more reactive by enhancement of the self-ionization process of N_2O_4:

$$Cu + 3N_2O_4 \xrightarrow{MeNO_2} Cu(NO_3)_2 \cdot N_2O_4 + 2NO$$

Similarly, the presence of donor solvents like ethyl acetate, diethyl ether, dimethyl sulphoxide, etc., in liquid N_2O_4 form the basis of excellent solvent systems for metals giving a variety of products including some stable solvate compounds containing dinitrogen tetroxide of crystallization[533]:

$$Ca + 2N_2O_4 \xrightarrow{EtOAc} Ca(NO_3)_2 + 2NO$$
$$Mg + 3N_2O_4 \xrightarrow{EtOAc} Mg(NO_3)_2 \cdot N_2O_4 + 2NO$$
$$Bi + 3N_2O_4 \xrightarrow{Me_2SO} Bi(NO_3)_3 \cdot 3Me_2SO + 3NO$$
$$Cu + N_2O_4 \xrightarrow{AcOH} Cu(OAc)_2$$
$$Cu + 5N_2O_4 \xrightarrow{5\text{–}30\%\ PhCN} Cu(NO_3) \cdot 2PhCN \cdot 4N_2O_4 + NO$$

Solvolysis reactions in liquid N_2O_4 are reactions of a salt MX with the solvent in which the anion X is totally or partially replaced by the nitrate ion, being the anion characteristic of the medium:

$$MX + N_2O_4 \rightarrow MNO_3 + NOX$$

Many such reactions have been reported including the following selection[533]:

$$R_4NCl + N_2O_4 \rightleftharpoons R_4N \cdot NO_3 + NOCl$$
$$ZnCl_2 + N_2O_4 \rightleftharpoons Zn(NO_3)_2 + 2NOCl$$
$$CaO + 2N_2O_4 \rightarrow Ca(NO_3)_2 + N_2O_3$$
$$Na_2O_2 + N_2O_4 \rightarrow 2NaNO_3$$
$$NaOH + N_2O_4 \rightarrow NaNO_3 + HNO_2$$

This reaction provides an excellent route to anhydrous metal nitrates from the corresponding halides, particularly where X = Br or I as NOX decomposes preventing the possible formation of nitrosyl compounds[533]:

$$TiI_4+4N_2O_4 \rightleftharpoons Ti(NO_3)_4+4NO+2I_2$$

The oxidation properties of N_2O_4 towards inorganic compounds generally involve the donation of oxygen atoms rather than the removal of electrons from metal ions. This is well illustrated by the unusual salts that have been prepared by the oxidation of sodium hyponitrite[565] with N_2O_4:

$$Na_2N_2O_2 \xrightarrow{\text{rapid}} \beta\text{-}Na_2N_2O_3 \xrightarrow{\text{slow}} Na_2N_2O_5 \xrightarrow[100^\circ]{\text{slow}} Na_2N_2O_6$$

$$\left.\begin{matrix} NH_2OH \\ EtNO_3 \\ NaOEt \end{matrix}\right\} \rightarrow \alpha\text{-}Na_2N_2O_3 \xrightarrow{\text{rapid}} Na_2N_2O_4 \xrightarrow{\text{slow}} Na_2N_2O_6$$

Accordingly, the reaction of N_2O_4 with cobalt forms cobalt(II) rather than cobalt(III) compounds, and its selective oxidizing properties in organic chemistry[566] are unique, particularly its ability to restrict oxidation to a single stage when further oxidation would appear to be possible.

Dinitrogen tetroxide reacts readily with water according to the overall equation

$$3N_2O_4+2H_2O \rightarrow 4HNO_3+2NO$$

Several mechanisms, more than one of which may be applicable, have been proposed to account for the absorption stage in nitric acid production[536] (see section 18.3) including the following:

$$N_2O_4(g) \rightarrow N_2O_4(l)$$

$$N_2O_4(l)+H_2O(l) \rightarrow HNO_3(l)+HNO_2(l)$$

$$2HNO_2(l) \rightarrow HNO_3(l)+NO(l)$$

$$N_2O_4 \rightarrow NO^++NO_3^-$$

$$N_2O_4+H_2O \rightarrow H^++NO_3^-+HNO_2$$

$$2HNO_2 \rightarrow H_2O+NO+NO_2$$

However, in spite of this ready reaction, simple hydrated salts do not dehydrate completely when immersed in liquid N_2O_4 (e.g. $MgSO_4\cdot 7H_2O \rightarrow MgSO_4\cdot 2H_2O$), and the selective reactivity may be related to possible differences in the bonding of the water[567].

Metal carbonyls react with N_2O_4 under mild conditions and provide simple preparative routes to metal nitrates and their N_2O_4 solvates[533]:

$$Co_2(CO)_8+8N_2O_4 \rightarrow 2Co(NO_3)_2\cdot 2N_2O_4+8CO+4NO$$

$$Fe(CO)_5+4N_2O_4 \rightarrow Fe(NO_3)_3\cdot N_2O_4+5CO+3NO$$

$$Mn_2(CO)_{10}+N_2O_4 \rightarrow Mn(CO)_5NO_3+[Mn(CO)_x(NO)_y]$$

---

565 C. C. Addison, G. A. Gamlen and R. Thompson, *J. Chem. Soc.* (1952) 338.
566 B. O. Field and J. Grundy, *J. Chem. Soc.* (1955) 1110.
567 C. C. Addison and D. J. Chapman, *J. Chem. Soc.* (1965) 819.

The reaction of N_2O_4 with organometallic compounds like tetramethyltin is explosively violent even at $-80°$ and the reagents require dilution with an inert solvent[566]:

$$Me_4Sn + 2N_2O_4 \xrightarrow{EtOAc} Me_2Sn(NO_3)_2$$

Finally, the reaction of N_2O_4 with the electron acceptor molecule boron trifluoride has received much attention. The reagents react both in 1:1 and 1:2 ratio to give stoichiometric solid products which have been formulated as containing the nitronium ion[567]:

$$N_2O_4 \xrightarrow{BF_3} NO_2^+[BF_3 \cdot NO_2]^- \xrightarrow{BF_3} NO_2^+[F_3BONOBF_3]^-$$

Thus the overall scheme for N_2O_4 dissociation can be considered as in the following scheme[533]:

$$2NO_2 \rightleftharpoons N_2O_4 \rightleftharpoons [NO_2^+ + NO_2^-] \xrightarrow[\text{transfer}]{\text{O atom}} NO^+ + NO_3^-$$

$$[NO_2^+ + NO_2^-] \xrightarrow{BF_3} NO_2^+ [BF_3 \cdot NO_2]^-$$

$$NO^+ + NO_3^- \xrightarrow{\text{donor solvents}} D_n \cdot N_2O_4 \quad (D_n \cdot NO)^+NO_3^-$$

13.8. BIOLOGICAL PROPERTIES[570]

The toxic properties of nitrogen dioxide are well known, but the presence of dangerous concentrations in everyday situations is not widely appreciated. NO_2 is the most dangerous component of "nitrous fumes", the name given to mixtures of oxides of nitrogen of variable composition. These mixtures can arise as a result of many industrial processes involving combustion but particularly electric-arc processes, electric-arc welding and oxy-acetylene welding. Hazards occur when such operations are carried out in confined, unventilated spaces. It is not considered safe to work for prolonged periods in an atmosphere containing >5 ppm NO_2, and short periods at higher concentrations can lead to symptoms such as inflammation of the lower respiratory tract and a marked elevation in the blood platelet count. Continued inhalation of concentrations >100 ppm has led to death.

13.9. ANALYSIS OF NO_2–N_2O_4[571]

In large enough amounts, nitrogen dioxide is readily recognized by its characteristic red–brown colour. It can also be identified in equilibrium with its dimer by spectroscopic methods[552]; NO_2 gives infrared absorption at 1626 cm^{-1}. Ultraviolet spectroscopy can also be used because dinitrogen tetroxide absorbs up to about 390 $m\mu$, and nitrogen dioxide in the 390–500 $m\mu$ region. Nitrogen dioxide gives a positive test with Griess reagent due

568 C. C. Addison, W. B. Simpson and A. Walker, *J. Chem. Soc.* (1964) 2360.

569 R. W. Sprague, A. B. Garrett and H. H. Sisler, *J. Am. Chem. Soc.* **82** (1960) 1059.

570 F. Call, in *Mellor's Comprehensive Treatise on Inorganic and Theoretical Chemistry*, Vol. 8, Suppl. I, *Nitrogen*, Part I, Longmans, London (1964), pp. 604–9.

571 K. I. Vasu, in *Mellor's Comprehensive Treatise on Inorganic and Theoretical Chemistry*, Vol. 8, Suppl. II, *Nitrogen*, Part II, Longmans, London (1967), p. 632.

572 M. J. Taras, in *Colorimetric Determination of Nonmetals* (ed. D. F. Boltz), Interscience, New York (1958).

to the presence of nitrite in solution, which can also form the basis of a colorimetric method of analysis[572]. The application of titrimetric methods is somewhat similar to those described for nitric oxide. Nitrogen dioxide can be oxidized with excess cerium (IV) which is back-titrated with sodium oxalate:

$$NO_2 + Ce^{4+} + H_2O \rightarrow NO_3^- + Ce^{3+} + 2H^+$$

Reference should also be made, however, to the analysis of nitrous fumes. This usually refers to mixtures of nitric oxide, nitrogen dioxide and other oxides that exist in equilibrium with them (principally N_2O_3 and N_2O_4). Their toxicity has directed considerable attention to the detection and estimation of small amounts of such materials[573]. Thus in this connection, and frequently in other situations, only the total nitrogen oxide content is required, and this can be obtained by oxidation with alkaline hydrogen peroxide to nitrate which can be determined colorimetrically by the phenoldisulphonic acid method. This can be preceded by absorption in Griess–Ilosvay reagent to obtain a separate estimation of NO_2 content if required. For full analysis of such nitrogen oxide mixtures, mass spectrometry provides the quickest most reliable method. However, more recently, gas chromatographic techniques have been developed which are capable of separating mixtures containing oxides of both nitrogen and carbon as well as elemental nitrogen using helium as carrier gas and thermal conductivity cell detectors[574].

14. DINITROGEN PENTOXIDE

14.1. PREPARATION

Dinitrogen pentoxide is usually prepared in the laboratory by the dehydration of concentrated nitric acid with phosphorus pentoxide[575, 576]:

$$2HNO_3 + P_2O_5 \rightarrow N_2O_5 + 2HPO_3$$

Considerable care must be exercised during this reaction, as the procedure[577] involves either the dropwise addition of concentrated HNO_3 (d 1.525 g/cm^3) onto a large excess of P_2O_5 at −10°C followed by slow distillation at about 35°C, or by adding P_2O_5 in one portion to nitric acid previously cooled to −78°C then allowing the mixture to attain room temperature slowly[578]. The principal alternative method is by the oxidation of nitrogen dioxide with ozone:

$$2NO_2 + O_3 \rightarrow N_2O_5 + O_2$$

---

[573] *Methods for the Detection of Toxic Substances in Air*, Booklet No. 5, *Nitrous Fumes*, Ministry of Employment and Productivity, HM Factory Inspectorate, HMSO, London (1969).

[574] J. Janak and M. Rusek, *Chemické Listy* **48** (1954) 397.

[575] I. R. Beattie, in ref. 571, pp. 269–77.

[576] P. W. Schenk, in *Handbook of Preparative Inorganic Chemistry*, Vol. I (ed. G. Brauer), Academic Press, New York (1963), p. 489.

[577] N. S. Gruenhut, M. Goldfrank, M. L. Cushing and G. V. Caesar, *Inorganic Syntheses*, Vol. III (ed. L. F. Andrieth), McGraw-Hill, New York (1950), p. 78.

[578] G. V. Caesar and M. Goldfrank, *J. Am. Chem. Soc.* **68** (1962) 372.

The difficulties encountered when attempting to dry NO_2 make it preferable[579] to oxidize dry NO with dry O_2 first, and when this reaction is complete then passing O_2–O_3 through. Another variation is bubbling ozonized oxygen through liquid N_2O_4.

N_2O_5 is also formed in the Birkland–Eyde arc process for the fixation of nitrogen and also when chlorine, phosphoric oxychloride or nitryl chloride reacts with silver nitrate[575]:

$$NO_2Cl + AgNO_3 \rightarrow N_2O_5 + AgCl$$

The electrolysis of N_2O_4 dissolved in nitric acid also affords oxidation[575] to N_2O_5:

$$2NO_3^- + N_2O_4 \rightarrow 2N_2O_5 + 2e^-$$

14.2. STRUCTURE AND BONDING

Dinitrogen pentoxide is a colourless crystalline solid which is stable in diffuse light below 8°C, but decomposes on exposure to sunlight or on warming even to room temperature. Above ~ -78°C, the crystal structure of N_2O_5 as determined by X-ray diffraction is consistent with the fully ionic nitronium nitrate $NO_2^+NO_3^-$ lattice in which the nitronium ion is linear and symmetrical with NO bond lengths (1.154Å) comparable to isoelectronic and isostructural carbon dioxide (C–O = 1.163 Å), while the nitrate group is both symmetrical and planar (N–O = 1.24 Å) as usual. Raman and infrared spectra[581] also indicate the presence of nitronium and nitrate ions. At lower temperatures, however, a molecular structure, $O_2N—O—NO_2$, similar to that found by electron diffraction for the gaseous phase[582] is present. The central NON bond angle is close to 180° and the bonding can be described by a Linnett non-pairing structure in Fig. 37.

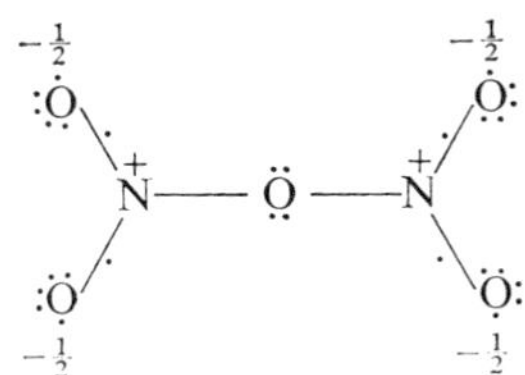

FIG. 37. Non-pairing valence bond configuration for molecular N_2O_5.

Valence bond structures for the NO_2^+ ion (also isoelectronic and isostructural with N_3^- and CN_2^{2-} ions) together with Linnett non-pairing structures[583] are illustrated in Fig. 38.

Alternatively, in terms of molecular orbital theory, the ground state configuration is

$$(1\sigma_g)^2(1\sigma_u)^2(2\sigma_g)^2(2\sigma_u)^2(1\pi_u)^4(1\pi_g)^4$$

in which the sixteen electrons are contained in filled bonding and non-bonding orbitals while the anti-bonding levels remain empty[583].

579 R. E. Nightingale, A. R. Downie, D. L. Rothenberg, B. Crawford and R. A. Ogg, Jr., *J. Phys. Chem.* **58** (1954) 1047.
580 E. Grison, K. Eriks and J. L. de Vries, *Acta Cryst.* **3** (1950) 290.
581 R. F. Barrow and A. J. Merer, in ref. 571, p. 520.
582 L. E. Sutton, *Tables of Interatomic Distances and Configurations in Molecules and Ions*, Chemical Society, London (1958).
583 M. Green, in *Developments in Inorganic Nitrogen Chemistry*, Vol. I (ed. C. B. Colburn), Elsevier, Amsterdam (1966), p. 40.

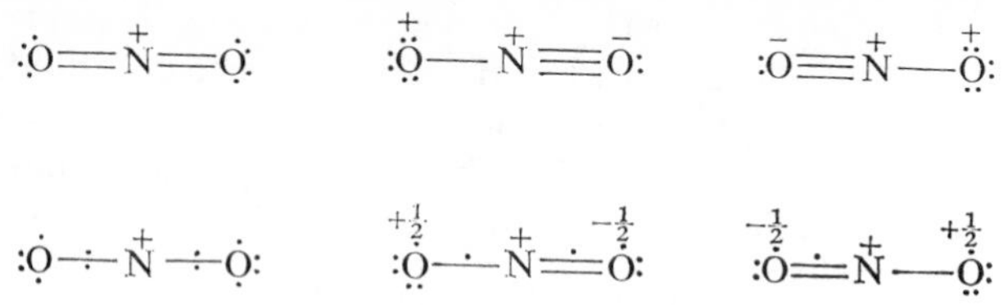

$:\dot{O}=\overset{+}{N}=\dot{O}:\qquad :\overset{+}{\ddot{O}}-\overset{+}{N}\equiv\overset{-}{O}:\qquad :\overset{-}{O}\equiv\overset{+}{N}-\overset{+}{\ddot{O}}:$

$:\dot{O}\div\overset{+}{N}\div\dot{O}:\qquad :\overset{+\frac{1}{2}}{\ddot{O}}\cdot\overset{+}{N}\dot{=}\overset{-\frac{1}{2}}{O}:\qquad :\overset{-\frac{1}{2}}{O}\dot{=}\overset{+}{N}-\overset{+\frac{1}{2}}{\ddot{O}}:$

FIG. 38. Valence bond configuration for NO_2^+.

14.3. MOLECULAR CONSTANTS

A few molecular constants for N_2O_5 in the crystalline phase ($NO_2^+NO_3^-$) and the molecular form are included in Table 75.

TABLE 75. MOLECULAR CONSTANTS OF N_2O_5

Property	Value	
X-ray diffraction data[a] on $NO_2^+NO_3^-$		
Space group	$D_{6h}^4-C_6/mmc$	
Lattice constants	$a = 5.41$ Å, $c = 6.57$ Å	
N_2O_5 structure [b]		
Terminal N–O	1.2 Å	
Central N–O	1.3–1.4 Å	
Dipole moment[c]	1.40 D	
	NO_2^+	NO_3^-
N–O bond length	1.154 Å	1.24
Raman spectrum[d]	1400 cm^{-1}	1050 cm^{-1}
	538 cm^{-1}	824 cm^{-1}
	2375 cm^{-1}	1413 cm^{-1}
		722 cm^{-1}

[a] Ref. 580.
[b] Ref. 582.
[c] A. L. McClellan, *Tables of Experimental Dipole Moments*, W. H. Freeman, San Francisco (1963).
[d] Ref. 581.

14.4. PHYSICAL PROPERTIES

Some values for the physical constants of N_2O_5 are listed in Table 76, principally thermodynamic data[584]. Spectroscopic measurements have been recorded for N_2O_5 in both gaseous and solid phases and also in solution[581].

[584] D. D. Wagman, W. H. Evans, V. B. Barker, I. Halow, S. M. Bailey and R. H. Schumm, *Selected Values of Chemical Thermodynamic Properties*, NBS Technical Note 270-3, Washington (1968).

TABLE 76. PHYSICAL PROPERTIES[a] OF N_2O_5

Property	Temperature (K)	Value
Sublimation temperature	305.6	32.4°C
Density	78	2.175 g/cm³
(calculated)	213	2.14 g/cm³
	288	2.05 g/cm³
Thermodynamic functions[b]		
N_2O_5 (cryst.)		
ΔH (enthalpy of sublimation)	305.6	13.6 kcal/mol
S (entropy of sublimation)	305.6	44.5 cal/deg mol
ΔH_f° (enthalpy of formation)	298	−10.3 kcal/mol
ΔG° (Gibbs energy of formation)	298	27.2 kcal/mol
S° (entropy)	298	42.6 cal/deg mol
C_p° (heat capacity)	298	34.2 cal/deg mol
N_2O_5 (g)		
ΔH_f° (enthalpy of formation)	0	5.7 kcal/mol
ΔH_f° (enthalpy of formation)	298	2.7 kcal/mol
$H_{298}-H_0$ (enthalpy)	298	4.237 kcal/mol
ΔG_f° (Gibbs energy of formation)	298	27.5 kcal/mol
S° (entropy)	298	85.0 cal/deg mol
C_p° (heat capacity)	298	20.2 cal/deg mol
Vapour pressure equation	$\log_{10} P_{mm} = 1244/T + 34.1 \log_{10} T - 85.929$	
	236.4	1 mm
	256.5	10 mm
	270.3	40 mm
	280.6	100 mm
	297.6	400 mm
	305.6	760 mm

[a] Ref. 575. [b] Ref. 584.

14.5. REACTIONS

Although N_2O_5 is colourless, absorption occurs at ~3800 Å in the ultraviolet region, and this has been assigned to photodecomposition according to the equation[581]

$$N_2O_5 \rightleftharpoons N_2O_4 + O(^3P)$$

The thermal decomposition of N_2O_5 has played an important rôle in the theory of reaction rates[575]:

$$2N_2O_5 \rightarrow 2N_2O_4 + O_2$$

The decomposition rate is first order:

$$-d[N_2O_5]/dt = k[N_2O_5]$$

and the overall reaction has been described by the following scheme[585]:

$$N_2O_5 \rightleftharpoons NO_2 + NO_3$$
$$NO_2 + NO_3 \rightarrow NO_2 + O_2 + NO$$
$$NO + N_2O_5 \rightarrow 3NO_2$$

585 R. A. Ogg, Jr., *J. Chem. Phys.* **15** (1947) 337.

Labelled NO_2 undergoes rapid exchange[586] with N_2O_5:

$$^{14}N_2O_5 + {}^{15}NO_2 \rightleftharpoons {}^{15}N_2O_5 + {}^{14}NO_2$$

and many gas phase reactions of N_2O_5 depend on initial dissociation to NO_2 and NO_3 with the latter then further acting as an oxidizing agent.

Dinitrogen pentoxide is the formal anhydride of nitric acid and being deliquescent it is very readily hydrated:

$$N_2O_5 + H_2O \rightarrow 2HNO_3$$

With hydrogen peroxide, pernitric acid is also formed[587]:

$$N_2O_5 + H_2O_2 \rightarrow HNO_3 + HNO_4$$

In anhydrous solvents like nitric, perchloric, sulphuric and phosphoric acids, N_2O_5 dissociates providing a convenient source of nitronium ions and a suitable preparative route to nitronium salts[588]:

$$N_2O_5 + 3H_2SO_4 \rightarrow 2NO_2^+ + H_3O^+ + 3HSO_4^-$$

$$N_2O_5 + 3HClO_4 \rightarrow 2NO_2^+ + H_3O^+ + 3ClO_4^-$$

$$2N_2O_5 + H_3O^+ClO_4^- \rightarrow NO_2^+ClO_4^- + 3HNO_3$$

$$N_2O_5 + 2SO_3 \rightarrow [NO_2^+]_2S_2O_7^{2-}$$

$$N_2O_5 + FSO_3H \rightarrow NO_2^+FSO_3^- + HNO_3$$

The strong oxidizing action of N_2O_5 can cause violent reactions with reducing agents including metals and organic substances to give nitrates and/or oxides[575]:

$$N_2O_5 + Na \rightarrow NaNO_3 + NO_2$$

$$N_2O_5 + NaF \rightarrow NaNO_3 + FNO_2$$

$$N_2O_5 + I_2 \rightarrow I_2O_5 + N_2$$

15. NITROGEN TRIOXIDE

The existence of the nitrogen trioxide radical NO_3 has for long been assumed judging by its frequent postulation in nitrogen oxide reaction mechanisms. For example, dinitrogen pentoxide undergoes thermal decomposition by a homogeneous first-order gas-phase reaction (see section 14). However, the reaction is not unimolecular as was initially thought, and one mechanism which is consistent with the data involves the preliminary equilibrium[589]

$$N_2O_5 \rightleftharpoons NO_2 + NO_3$$

followed by

$$NO_3 + NO_2 \rightarrow NO_2 + O_2 + NO$$

$$NO_3 + NO \rightleftharpoons 2NO_2$$

---

[586] A. R. Amell and F. Daniels, *J. Am. Chem. Soc.* **74** (1952) 6209.
[587] R. Schwartz, *Z. anorg. allgem. Chem.* **256** (1947) 3.
[588] D. R. Goddard, E. D. Hughes and C. K. Ingold, *J. Chem. Soc.* (1950) 2559.
[589] R. A. Ogg, Jr., *J. Chem. Phys.* **5** (1937) 873.

Similarly, NO_3 is involved in the rate-determining step to account for the products of the NO_2–methane reaction[590]:

$$2NO_2 \rightleftharpoons NO + NO_3$$

$$NO_3 + CH_4 \rightarrow HNO_3 + CH_3$$

$$CH_3 + NO_2 \rightarrow CH_3NO_2$$

However, in the N_2O_5 catalysed decomposition of ozone, the steady state of NO_3 can be high enough for its spectrum to be recorded as an extremely intense banded absorption in the region 5000–7100 Å [591]. Indeed its presence in ozone can be detected by a colour change from the violet–blue of ozone to a more intense pure bright blue. Thus the mechanism for the N_2O_5–O_3 reaction involves the equilibria which are important in N_2O_5 formation[592]:

$$N_2O_5 \rightleftharpoons NO_2 + NO_3$$

$$NO_2 + O_3 \rightarrow NO_3 + O_2$$

$$NO_3 + NO_3 \rightarrow 2NO_2 + O_2$$

Nitrogen trioxide has also been identified by e.s.r. in electron irradiated $NaNO_3$ crystals[593]. It is usually assumed to have a planar symmetrical structure in the ground state, and its electronic structure can be described by Linnett non-pairing structures[594] as illustrated in Fig. 39.

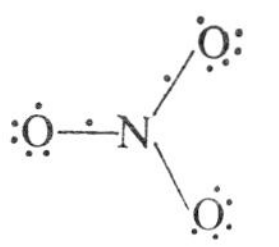

FIG. 39. Linnett non-pairing structure of NO_3.

A molecular orbital description suggests that the unpaired electron lies in an energy state composed only of orbitals from oxygen[595].

An isomeric form with the asymmetric pernitrite OONO structure has also been detected in the reaction of nitric oxide and oxygen at low pressure[596]:

$$O_2 + NO \rightleftharpoons OONO$$

[590] J. H. Thomas, in *Oxidation and Combustion Reviews*, Vol. I (ed. C. F. H. Tipper), Elsevier, Amsterdam (1965), p. 137.

[591] E. J. Jones and O. R. Wulf, *J. Chem. Phys.* **5** (1937) 873.

[592] H. S. Johnson and F. Leighton, Jr., *J. Am. Chem. Soc.* **75** (1953) 3612.

[593] R. Adde, *Compt. rend.* **264** (1967) 1905.

[594] M. Green, in *Developments in Inorganic Nitrogen Chemistry*, Vol. I (ed. C. B. Colburn), Elsevier, Amsterdam (1966), p. 55.

[595] A. D. Walsh, *J. Chem. Soc.* (1953) 2301.

[596] W. A. Guillory and H. S. Johnson, *J. Chem. Phys.* **42** (1965) 2457.

16. HYPONITROUS ACID AND HYPONITRITES

16.1. PREPARATION

Hyponitrous acid crystallizes as white deliquescent plates[597] which are explosively unstable, detonating even when rubbed with a glass rod. It can be prepared reasonably pure by treating silver hyponitrite with anhydrous hydrogen chloride in ether, and after the silver chloride has been filtered off, the ether can be evaporated off from the filtrate[598].

$$Ag_2N_2O_2+2HCl \xrightarrow{\text{ether}} H_2N_2O_2+2AgCl$$

The metal salts are generally prepared in variable yield by reduction methods, sodium amalgam being a favourite reducing agent:

$$2NaNO_3+8Na/Hg+4H_2O \rightarrow Na_2N_2O_2+8NaOH+8Hg$$
$$2AgNO_3+2NaNO_2+4Na/Hg+2H_2O \rightarrow Ag_2N_2O_2+2NaNO_3+4NaOH+4Hg$$

^{15}N isotopically labelled hyponitrites have also been synthesized by this method. Similarly, if bulk sodium metal is allowed to disintegrate in pyridine under a hydrogen atmosphere, the sodium–pyridine adduct can act as a reducing agent towards nitric oxide:

$$2Na+2NO \xrightarrow[\text{benzene}]{\text{pyridine}} Na_2N_2O_2$$

The hyponitrites of the alkaline earth metals have each been obtained by reducing the corresponding nitrate with magnesium amalgam[599]:

$$Ca(NO_3)_2+4Mg/Hg+4H_2O \rightarrow CaN_2O_2+4Mg(OH)_2+4Hg$$

Other methods of preparation include the condensation of hydroxylamine (prepared from the hydrochloride and sodium alkoxide), alkyl nitrite and sodium alkoxide[597]:

$$NH_2OH+RO{-}N{=}O \rightarrow H_2N_2O_2+ROH$$
$$H_2N_2O_2+2NaOR \rightarrow Na_2N_2O_2+2ROH$$

The silver salt has also been prepared by way of hydroxylamine disulphonate and sodium hydroxylamine monosulphonate[598]:

$$NaNO_2+NaHSO_3 \rightarrow HONHSO_3Na+NaOH$$
$$2HONHSO_3Na+2KOH+2AgNO_3 \rightarrow Ag_2N_2O_2+2NaHSO_3+2KNO_3+2H_2O$$

Mention should also be made to the isomer nitramide NH_2NO_2 which is formed by acid hydrolysis of the nitrocarbamate ion followed by ether extraction[598]:

$$EtOCONH_2+EtONO_2 \rightarrow EtOCONHNO_2+EtOH$$
$$EtOCONHNO_2 \xrightarrow{2KOH} K_2(OCONNO_2)+EtOH+H_2O$$
$$K_2(OCONNO_2)+H_2SO_4 \rightarrow NH_2NO_2+CO_2+K_2SO_4$$

---

597 M. N. Hughes, *Quart. Rev.* **22** (1968) 1.

598 P. W. Schenk, in *Handbook of Preparative Inorganic Chemistry*, Vol. I (ed. G. Brauer), Academic Press, New York (1963), pp. 492–8.

599 H. Block, in ref. 571, pp. 407–8.

16.2. STRUCTURE AND BONDING

The presence of an N=N double bond is suggested by both chemical and physical properties of hyponitrites, and hence the question of *cis–trans*-isomerism arises[597]:

It is now known that the *trans*-isomer is usually isolated from normal preparative methods. Evidence for this includes the assignment of three fundamental vibrations which are both infrared and Raman active consistent with a structure belonging to the C_{2h} point group[600]. Infrared and n.m.r. studies also suggest that the free acid has a symmetrical structure[597].

There is some infrared evidence that the so-called alkali metal "nitrosyls" (formed by the action of NO on alkali metals or in liquid ammonia) are dimeric and have the *cis*-hyponitrite configuration[601].

The red nitrosopentammines of cobalt(III), formerly regarded as monomeric $[Co(NH_3)_5NO]^{2+}$, have been shown to be dimeric with bridging hyponitrito groups in a *cis* planar configuration[602]

$$[(NH_3)_5Co{-}\underset{\displaystyle O}{\underset{|}{N}}NO{-}Co(NH_3)_5]^{4+}$$

Their infrared spectra also show $\nu(NO)$ to be in the region 1040–1195 cm^{-1} similar to the spectra of other complexes where bridging $N_2O_2^{2-}$ groups are suspected[603], e.g., $K_6[(CN)_5Co(N_2O_2)Co(CN)_5]$ and iron and ruthenium nitrosyls $[(NO)_3M(N_2O_2)M(NO)_3]$.

In the absence of structural parameters, a full discussion of the bonding in hyponitrites is not possible. However, the strong absorption in the ultraviolet region at 248 mμ (ε = 3980) is characteristic of compounds containing a N=N bond[604], which suggests that the bonding in the hyponitrite ion is adequately described by the structure

:Ö:⁻–N̈=N̈–:Ö:⁻ (trans)

Crystalline nitramide, however, is quite different[605]:

H₂N —1.40 Å— N (1.18 Å to O; O–N–O 129°) with two O

600 G. E. McGraw, D. L. Bernitt and I. C. Hisatsune, *Spectrochim. Acta* A, **23** (1967) 25.
601 J. Goubeau and K. Laitenberger, *Z. anorg. allgem. Chem.* **78** (1963) 320.
602 B. F. Hoskins, F. D. Whillans, D. H. Dale and D. C. Hodgkin, *Chem. Communs.* (1969) 69.
603 J. B. Raynor, *J. Chem. Soc.* A (1966) 997.
604 C. C. Addison, G. A. Gamlen and R. Thompson, *J. Chem. Soc.* (1952) 338.
605 C. A. Beevers and A. F. Trotman-Dickenson, *Acta Cryst.* **10** (1957) 34.

16.3. PHYSICAL PROPERTIES

Due, no doubt, to the instability of hyponitrous acid, few of its physical constants have been recorded[597]. Perhaps the most important fact is that it reacts as a weak dibasic acid, and the values for the ionization constants are $K_1 = 9\times10^{-8}$ and $K_2 = 1\times10^{-11}$:

$$H_2N_2O_2 \overset{K_1}{\rightleftharpoons} H^+ + HN_2O_2^- \overset{K_2}{\rightleftharpoons} 2H^+ + N_2O_2^-$$

From the temperature variation of the dissociation constants the first and second enthalpies of ionization for hyponitrous acid have been estimated[606] to be 3.7 and 6.1 kcal/mol. This compares with the calorimetric value (11.1 kcal/mol) for the sum of the two[607]. The enthalpy of formation of aqueous $HN_2O_2^-$ and $H_2N_2O_2$ is -12.4 and -15.4 kcal/mol respectively[608].

It is not usual to regard nitrous oxide as the anhydride of hyponitrous acid because the equilibrium

$$N_2O(g) + H_2O(l) \rightleftharpoons H_2N_2O_2(aq)$$

lies completely to the left. Indeed, it has been calculated that an equilibrium pressure of 10^{27} atm of N_2O would be necessary in order to obtain a 0.001 M aqueous solution of hyponitrous acid[609]. Organic solvents are particularly useful in $H_2N_2O_2$ chemistry and the partition coefficients between aqueous solution and benzene, diethyl ether and di-n-butyl ether, $K = [H_2N_2O_2]_{solvent}/[H_2N_2O_2]_{aqueous}$, are 3.6×10^{-4}, 1.14 and 7.7×10^{-2} respectively[597].

16.4. CHEMICAL PROPERTIES

The decomposition of hyponitrous acid in aqueous solution leads principally to nitrous oxide and water. At pH 4–14 the predominant unstable species in hyponitrous acid decompositions is the hydrogen hyponitrite ion

$$HN_2O_2^- \rightarrow N_2O + OH^-$$

However, at pH 1–3 and 25°C, $H_2N_2O_2$ has a half-life of about 16 days which decreases to orders of minutes if the temperature or acidity is increased. Under such conditions the following mechanism is thought to be relevant:

$$H^+ + H_2N_2O_2 \underset{}{\overset{\text{fast}}{\rightleftharpoons}} \left[\begin{array}{c} H\diagdown\overset{+}{O}\diagup H \\ \diagdown N{=}N\diagdown \\ O\diagdown H \end{array}\right] \overset{\text{slow}}{\rightleftharpoons} H_2O + N_2O + H^+$$

However, the fact that decomposing hyponitrous acid solutions cause polymerization of added acetonitrile suggests the presence of free radical intermediates. The OH and NO_2 radicals are likely to be involved because nitrophenols are formed when benzene is added,

[606] M. N. Hughes and G. Stedman, *J. Chem. Soc.* (1963) 1239.

[607] W. M. Lattimer and H. W. Zimmerman, *J. Am. Chem. Soc.* **61** (1939) 1550.

[608] D. D. Wagman, W. H. Evans, V. B. Parker, I. Halow, S. M. Bailey and R. H. Schumm, *Selected Values of Chemical Thermodynamic Properties*, NBS Technical Note 270-3, Washington (1968).

[609] D. M. Yost and H. Russell, Jr., *Systematic Inorganic Chemistry*, Prentice-Hall, New York (1944), pp. 52–57.

and in fact, the process is initiated by OH radicals in the presence of oxidation transfer agents M according to the scheme[597]:

$$M+OH \rightarrow M^{+}+OH^{-}$$
$$M^{+}+H_2N_2O_2 \rightarrow M+HONNO+H^{+}$$
$$HONNO \rightarrow N_2O+OH$$
$$HONNO+H_2N_2O_2 \rightarrow 2NO+N_2+H_2O+OH$$
$$NO+H_2N_2O_2 \rightarrow HNO_2+N_2+OH$$
$$HNO_2+H_2N_2O_2 \rightarrow [ONON{=}NOH] \rightarrow NO_2+N_2+OH$$
$$NO_2+H_2N_2O_2 \rightarrow H^{+}+NO_3^{-}+N_2+OH$$
$$OH+HNO_2 \rightarrow H_2O+NO_2$$
$$OH+NO_2 \rightarrow H^{+}+NO_3^{-}$$

The reaction of nitrous acid with hyponitrous acid in dilute solution has also been studied separately[610] and is essentially described by the equation

$$HNO_2+H_2N_2O_2 \rightarrow HNO_3+N_2+H_2O$$

although at higher concentrations some N_2O is formed.

Hyponitrites are relatively stable towards reduction, as witnessed by their formation in the presence of strong reducing agents like sodium amalgam, whereas oxidation to nitrite or nitrate occurs readily. For example, the oxidation with iodine provides the basis of a convenient analytical method[611]:

$$H_2N_2O_2+3I_2+3H_2O \rightarrow HNO_3+HNO_2+6HI$$

The oxidation of $Na_2N_2O_2$ with liquid N_2O_4 to form $Na_2N_2O_x$ (where x = 3, 4, 5, 6) has been considered[604] in section 13. However, in the presence of an inert solvent of high dielectric constant such as nitromethane, the reaction goes to completion rapidly:

$$2N_2O_4+N_2O_2^{2-} \rightarrow ONON{=}NONO+2NO_3^{-}$$
$$\downarrow$$
$$N_2+2NO_2$$

The mechanism based on electrophilic attack by NO^{+} on the oxygen atoms of the hyponitrite anion is further supported by the similar reaction of nitrosonium perchlorate with hyponitrite in nitromethane suspension[612]:

$$2NO^{+}ClO_4^{-}+N_2O_2^{2-} \rightarrow ONON{=}NONO+2ClO_4^{-}$$
$$\downarrow$$
$$N_2+2NO_2$$

Perhaps, however, the most important aspect of hyponitrite redox chemistry is the possibility that it is an intermediate in the nitrogen cycle[597]. In particular, it is the ammonia–nitrite oxidation step that is thought to proceed via hydroxylamine and hyponitrous acid although the latter could result from dimerization of nitroxyl NOH.

Sodium hyponitrite decomposes on heating to 260°C *in vacuo* or 335°C at atmospheric pressure to give sodium oxide, nitrite and nitrate together with nitrogen. It is partially

[610] M. N. Hughes and G. Stedman, *J. Chem. Soc.* (1963) 2824.
[611] G. Ferrani, *Ann. Chim.* (*Italy*) **48** (1958) 322.
[612] M. N. Hughes and H. G. Nicklin, *Chem. Communs.* (1969) 80.

hydrolysed in water or dilute acids to N_2O, NO_4 and N_2, but oxidized by permanganate. Alkali-metal hyponitrites also decompose thermally[613]:

$$MNO_2 \rightarrow MO + N_2O$$
$$3MNO_2 \rightarrow 2MO + M(NO_2)_2 + 2N_2$$

in an analogous manner to the silver salt:

$$Ag_2N_2O_2 \rightarrow Ag_2O + N_2O$$
$$3Ag_2N_2O_2 \rightarrow 2Ag_2O + 2AgNO_2 + 2N_2$$

Finally, nitramide like isomeric hyponitrous acid is also a weak acid in aqueous solution:

$$NH_2NO_2 \rightleftharpoons H^+ + NHNO_2^-; \qquad K = 2.6 \times 10^{-7}$$

and decomposes by a base-catalysed reaction to nitrous oxide and water:

$$NH_2NO_2 + B \xrightarrow{\text{slow}} NHNO_2^- + BH^+$$
$$NHNO_2^- \xrightarrow{\text{fast}} N_2O + OH^-$$

17. NITROUS ACID AND NITRITES

17.1. PRODUCTION AND INDUSTRIAL USE

Nitrous acid is an important intermediate in the N_2O_4–H_2O reaction during the production of nitric acid (see section 18.3), but it cannot be isolated during this process[614].

$$N_2O_4 + H_2O \rightleftharpoons HNO_2 + HNO_3$$

For although the salts and esters of nitrous acid are fairly stable, the free acid is usually used and prepared *in situ* by acidification of aqueous nitrite solutions below room temperature due to its instability with respect to[614, 615]:

$$3HNO_2 \rightleftharpoons H_3O^+ + NO_3^- + 2NO$$

However, it is the reverse of such equilibria that are operative in industrial processes whereby metal nitrites are obtained from the absorption of nitrous fumes in basic solutions of metal hydroxides, carbonates, sulphides, etc.:

$$NO + NO_2 + 2NaOH \rightarrow 2NaNO_2 + H_2O$$

For sodium nitrite, industrially the most important nitrite, although the purity of the raw product from this process is often less than 60%, better than 99% purity can be obtained simply by four recrystallizations. Alkali-metal nitrites can also be obtained by the thermal decomposition of readily available nitrates:

$$2NaNO_3 \xrightarrow{\Delta} 2NaNO_2 + O_2$$

---

613 T. M. Oza and V. T. Oza, *J. Phys. Chem.* **60** (1956) 192.
614 H. Block, in ref. 571, Section XXXII, pp. 353–407.
615 D. M. Yost and H. Russell, Jr., *Systematic Inorganic Chemistry*, Prentice-Hall, New York (1944), pp. 58–70.

However, the process can be carried out at lower temperatures and with better yields by the addition of a reducing agent such as coke or iron, etc.:

$$2NaNO_3 + C \rightarrow 2NaNO_2 + CO_2$$

Sodium nitrite is used industrially[614, 616] for the synthesis of hydroxylamine and its derivatives, organic nitrites and nitro-compounds, and particularly in diazotization reactions leading to azo dyes and pharmaceuticals. Inorganic nitrites are also components of textile bleaching agents, heat transfer media, corrosion inhibitors, blowing agents, monomer stabilizers and commercial explosives.

17.2. LABORATORY PREPARATION

Nitrous acid is prepared simply by the acidification of a nitrite with any moderately strong or mineral acid at below room temperature. An aqueous solution free of any dissolved salts is conveniently obtained using an exchange reaction which produces an insoluble salt[614]:

$$Ba(NO_2)_2 + H_2SO_4 \rightarrow 2HNO_2 + BaSO_4$$
$$AgNO_2 + HCl \rightarrow HNO_2 + AgCl$$

Pure nitrous acid has also been prepared by low-temperature ion exchange of sodium nitrite in aqueous glycol dimethylether[617]. Where the presence of salts is of no consequence, sodium nitrite is acidified with hydrochloric acid cooled in a salt-ice bath[614].

Commercial sodium nitrite can be easily purified by recrystallization or prepared by one of the following methods:

$$2NaNO_3 \xrightarrow{\Delta} 2NaNO_2 + O_2$$
$$NaNO_3 + Na_2Fe_2O_4 + 2NO \rightarrow 3NaNO_2 + Fe_2O_3$$
$$2NaOH + N_2O_3 \rightarrow 2NaNO_2 + H_2O$$

Similarly, silver nitrite, the other convenient starting reagent for introducing the nitrite group by halide replacement reactions, is best obtained as the sparingly soluble salt from the exchange reaction[614]:

$$AgNO_3 + NaNO_2 \rightarrow AgNO_2 + NaNO_3$$

Also, dinitrogen trioxide in alkaline solutions, as indicated above for sodium nitrite, gives practically pure nitrite solutions. Thus these reagents provide convenient laboratory methods of synthesis for most nitrites as illustrated by the following examples:

$$Li_2O + N_2O_3 \rightarrow 2LiNO_2$$
$$KCl + AgNO_2 \rightarrow KNO_2 + AgCl$$
$$2NH_3 + H_2O + N_2O_3 \rightarrow 2NH_4NO_2$$
$$Ba(OH)_2 + NO + NO_2 \rightarrow Ba(NO_2)_2 + H_2O$$

Such methods are also applicable to the formation of transition metal complexes:

$$[Co(NH_3)_5H_2O]^{3+} + HNO_2 \rightarrow [Co(NH_3)_5ONO]^{2+} + H^+ + H_2O$$

---

616 C. Matasa and E. Matasa, *L'Industrie Moderne des Produits Azotes*, Dunod, Paris (1968), p. 567.
617 C. S. Scanley, *J. Am. Chem. Soc.* **85** (1963) 3888.

17.3. NUCLEAR ISOTOPES OF NITROUS ACID AND NITRITES

The availability of $Na^{15}NO_2$ free of nitrate contamination is a vital reagent for introducing ^{15}N into a very large number of organic compounds. Nitrogen-15 can be oxidized in a high voltage electric arc to give a mixture of nitrogen oxides, of which the higher oxides can be reduced to nitric oxide with mercury and sulphuric acid. The oxidation by permanganate in an autoclave at 180–190°C gives nitrate which can be similarly reduced. If the subsequent oxidation of ^{15}NO over sodium hydroxide solution is controlled by slow addition of oxygen, the enriched nitrogen dioxide reacts with the excess nitric oxide as it is formed producing dinitrogen trioxide $^{15}N_2O_3$ which dissolves in the caustic soda to give labelled nitrite:

$$^{15}NO+\tfrac{1}{2}O_2 \rightarrow {}^{15}NO_2$$
$$^{15}NO+{}^{15}NO_2 \rightarrow {}^{15}N_2O_3$$
$$^{15}N_2O_3+2NaOH \rightarrow 2Na^{15}NO_2+H_2O$$

Free sodium nitrite–^{15}N can be obtained by crystallization after concentration by evaporation in a platinum dish[618].

Alkali metal nitrites can also be prepared by heating a nitrate with metallic lead, but the product is usually contaminated with nitrate:

$$Na^{15}NO_3+Pb \rightarrow Na^{15}NO_2+PbO$$

Oxygen-18-enriched nitrites can be prepared by simple exchange reactions between the normal nitrite and ^{18}O-enriched water. For example sodium nitrite–^{18}O can be prepared simply by dissolving dry unlabelled sodium nitrite in ^{18}O-enriched water acidified with perchloric acid to pH 4.9. Silver nitrite–^{18}O can then be precipitated out when silver perchlorate, also in ^{18}O-enriched water, is added to the enriched nitrite solution[619]:

$$NaNO_2+2H_2{}^{18}O \rightarrow NaN^{18}O_2+2H_2O$$
$$AgClO_4+NaN^{18}O_2 \rightarrow AgN^{18}O_2+NaClO_4$$

17.4. STRUCTURE AND BONDING

Three structures can be envisaged for nitrous acid, two of which would be related as *cis* and *trans* isomers as illustrated in Fig. 40.

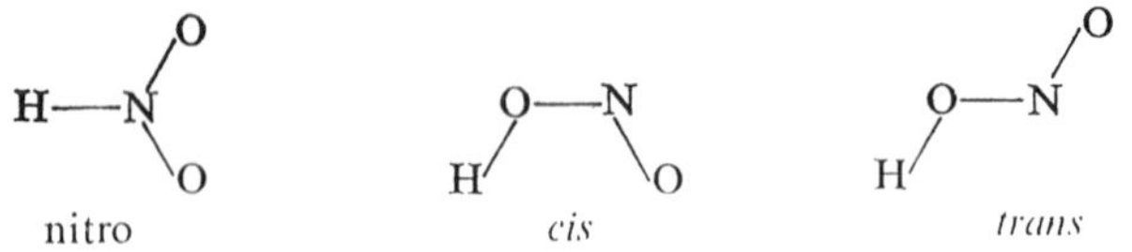

FIG. 40. Possible structures for nitrous acid.

That hydrogen is bonded to oxygen rather than nitrogen is indicated by magnetic susceptibility measurements[614], while infrared spectra[620] show absorptions which are assigned to both *cis* and *trans* configurations, and which also suggest that the *trans* isomer is more

[618] W. Spindel, in *Inorganic Isotopic Syntheses* (ed. R. H. Herber), Benjamin, New York (1962), p. 95.

[619] I. Dostrovsky and D. Samuel, in *Inorganic Isotopic Syntheses* (ed. R. H. Herber), Benjamin, New York (1962), p. 134.

[620] R. F. Barrow and A. J. Merer, in ref. 571, p. 480.

stable than *cis* by about 500 cal/mol[621]. The probable structural parameters for nitrous acid[614] are shown in Fig. 41.

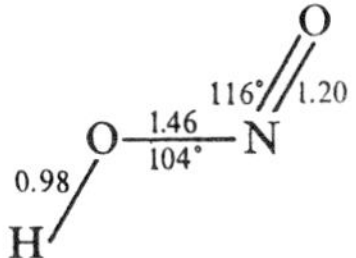

Fig. 41. Structures for nitrous acid and the nitrite ion.

Crystal data on some ionic metal nitrites[614] are listed in Table 77, and in all cases the nitrite ion has an angular shape with N–O distances reported between 1.13 to 1.23 Å and the ONO bond angle between 116° and 132°.

Table 77. Crystal Structures of Ionic Nitrites[a]

Compound	Crystal system	Lattice parameters (Å)
$LiNO_2 \cdot H_2O$	Monoclinic	$a = 3.31$, $b = 14.10$, $c = 6.36$, $\beta = 105°$
$NaNO_2$	Orthorhombic	$a = 3.55$, $b = 5.56$, $c = 5.37$
KNO_2	Monoclinic (distorted $NaNO_2$-type)	$a = 4.45$, $b = 4.99$, $c = 7.31$, $\beta = 114° 50'$
$CsNO_2$	Monometric	$a = 4.34$
$AgNO_2$	Orthorhombic	$a = 3.505$, $b = 6.14$, $c = 5.16$
$NaAg(NO_2)_2$	Orthorhombic	$a = 7.89$, $b = 10.62$, $c = 10.83$
$Ba(NO_2)_2$	Hexagonal	$a = 7.05$, $b = 17.66$
$2Ba(NO_2)_2 \cdot TlNO_2$	Rhombic	$a = 13.48$, $b = 17.94$, $c = 4.958$

[a] Ref. 614.

The electronic structure of nitrous acid is more or less as would be expected, and the N–O bond lengths indicated in Fig. 41 are consistent with bond orders of 1 and 2. The nitrite ion, however, contains eighteen electrons for which two resonance forms seem likely as well as the alternative non-pairing structure[622] illustrated in Fig. 42.

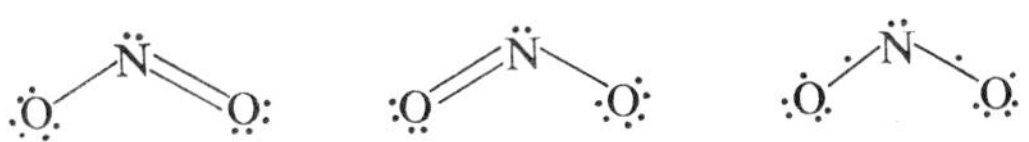

Fig. 42. Electronic structures of NO_2^-.

The ONO bond angle is consistent with sp^2 hybridization at the nitrogen atom, but the N–O bond lengths are closer to a bond order of 2 than 1.5.

621 A. P. Altshuller, *J. Phys. Chem.* **61** (1957) 251.

622 M. Green, in *Developments in Inorganic Nitrogen Chemistry*, Vol. I (ed. C. B. Colburn), Elsevier, Amsterdam (1966), p. 42.

The molecular orbital description of the NO_2^- ion is

$$(1a_1)^2(1b_2)^2(2a_1)^2(1b_1)^2(3a_1)^2(2b_2)^2(1a_2)^2(3b_1)^2(4a_1)^2$$

Four of these orbitals are bonding and five non-bonding suggesting individual N–O bond orders of 2. However, both these descriptions are approximations and overlook the fact that overlap can vary considerably and also the slight anti-bonding character of some levels.

Nitrite is also capable of bonding to metal atoms in other ways[623] as indicated in Fig. 43.

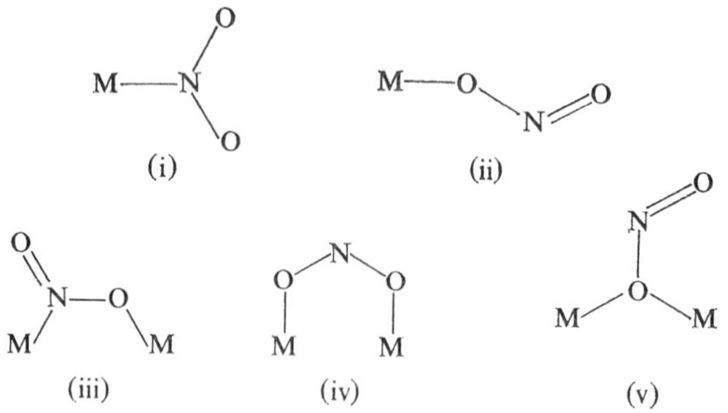

(i) (ii) (iii) (iv) (v)

FIG. 43. Bonding of the nitrite group to metals.

The formation of nitro (i) and nitrito (ii) complexes such as $[Co(NH_3)_5NO_2]^{2+}$ and $[Co(NH_3)_5ONO]^{2+}$ is well established. Evidence for such structures is largely spectroscopic because differences in the bond orders and symmetries of the nitrite group in such configurations are reflected in their infrared spectra[624]. In general, nitrito complexes are less stable than the corresponding nitro complexes, and isomerization usually occurs quite readily. The rate of isomerization from nitrito to the nitro form is faster in solution than in the solid state although the absence of exchange between the —ONO group and the solvent or added nitrite ion implies that the group does not leave the co-ordination sphere of the metal during isomerization[625]:

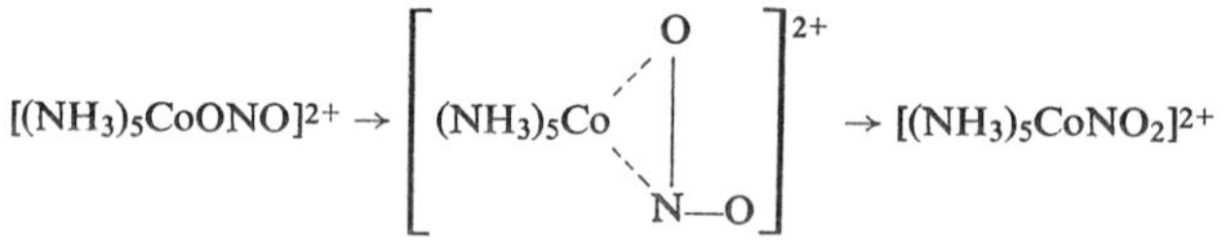

$$[(NH_3)_5CoONO]^{2+} \rightarrow \left[(NH_3)_5Co \begin{matrix} O \\ | \\ N—O \end{matrix} \right]^{2+} \rightarrow [(NH_3)_5CoNO_2]^{2+}$$

Of the bridging structures (iii)–(v) illustrated in Fig. 43, structure (iii) has been confirmed for the compound $[Nien_2(ONO)]BF_4$ which forms poymer chains of Ni—ON(O)Ni units[626].

[623] W. P. Griffith, in *Developments in Inorganic Nitrogen Chemistry*, Vol. I (ed. C. B. Colburn), Elsevier, Amsterdam (1966), p. 268.

[624] K. Nakamoto, *Infrared Spectra of Inorganic and Coordination Compounds*, J. Wiley, New York (1963), p. 151.

[625] R. K. Murman and H. Taube, *J. Am. Chem. Soc.* **78** (1956) 4886.

[626] M. G. B. Drew, D. M. L. Goodgame, M. A. Hitchman and D. A. Rogers, *Chem. Communs.* (1965) 477.

TABLE 78. THERMODYNAMIC FUNCTIONS[a] OF NITROUS ACID AND NITRITE ION

Property	Temperature (K)	*cis*-HNO_2	*trans*-HNO_2	Equilibrium HNO_2	Undissociated HNO_2 standard state (m = 1 aq)	NO_2^- in standard rate (m = 1 aq)	Units
ΔH°_{f0} (enthalpy of formation)	0	−17.12	−17.68				kcal/mol
ΔH_{f0} (enthalpy of formation)	298	−18.64	−19.15	−19.0	−28.5	−25.0	kcal/mol
$H-H_0$ (enthalpy)	298	2.608	2.652				kcal/mol
ΔG_f (Gibbs energy of formation)	298	−10.27	−10.82	−11.0	−13.3	−8.9	kcal/mol
S (entropy)	298	59.43	59.54	60.7	36.5	33.5	cal/deg mol
C_p (heat capacity)	298	10.70	11.01	10.9		−23.3	cal/deg mol

[a] Ref. 627.

17.5. MOLECULAR PROPERTIES

The infrared and Raman spectra of nitrous acid have been recorded in solution. Infrared and ultraviolet absorption spectroscopy have been important in demonstrating that the nitrous acid molecule exists in equilibrium in the gas phase[627]

$$NO(g)+NO_2(g)+H_2O(g) \rightleftharpoons 2HNO_2(g); \quad K(298) = 1.74$$

Spectra of *cis* and *trans* HNO_2 and DNO_2 have been recorded. The non-linear structure of the nitrite ion has also been confirmed by the appearance of fundamental frequencies ν_1, ν_2 and ν_3 in both infrared and Raman spectra at 1335, 830 and 1250 cm^{-1} respectively[624].

17.6. PHYSICAL PROPERTIES

Nitrous acid is unknown as a pure substance in condensed phases, but in the gas phase it can be detected in the absence of dissociation products when equilibrium is rapidly established[614]. Solutions of nitrous acid have a limited useful life due to the decomposition

$$3HNO_2 \rightarrow H_3O^+ + NO_3^- + 2NO$$

Consequently few physical constants for HNO_2 have been measured. Some thermodynamic data[627] are given in Table 78.

In aqueous solution, HNO_2 is a weak monobasic acid [pK_a(291) = 3.3] which forms reasonably stable salts with electro-positive elements. Ionic nitrites have high solubility in water[615], and the properties of aqueous solutions such as densities, viscosities and vapour pressures have been investigated[614]. Phase diagrams for metal nitrite–water systems in the presence of related substances have also been recorded together with thermodynamic data on fused salt mixtures containing sodium nitrite. Some physical properties of ionic nitrites are given in Table 79.

TABLE 79. PHYSICAL PROPERTIES OF IONIC NITRITES[a]

Property	NH_4NO_2	$LiNO_2$	$NaNO_2$	KNO_2	$AgNO_2$
Melting point (°C)	s 32–3		282–4	441	
Decomposition point (°C)	60–70		320		
Transition temperature (°C)			160–2	47	
Density d_0^{20} (g/cm^3)			2.144		
Heat of solution (kcal/mol)		2.38	−3.57	−3.40	10.07
Heat of formation (kcal/mol)		94.16	85.06	87.66	

[a] Ref. 614.

17.7. CHEMISTRY AND CHEMICAL PROPERTIES

In aqueous solutions, HNO_2 is best regarded[628] as hydroxylated NO^+ implying that OH^- can be replaced by other bases to give NOX. Highly acidic solutions favour the

627 D. D. Wagman, W. H. Evans, V. B. Parker, I. Halow, S. M. Bailey and R. H. Schumm, *Selected Values of Chemical Thermodynamic Properties*, NBS Technical Note 270-3, Washington (1968).

628 T. A. Turney and G. A. Wright, *Chem. Rev.* **59** (1959) 497.

formation of NO^+ or its hydrated equivalent ($H_2NO_2^+$), while in dilute acid the unionized species HONO is present.

$$HONO + H^+ \rightleftharpoons NO^+ + H_2O; \qquad K(293) = 2\times10^{-7}$$

This sort of equilibrium has been demonstrated by the exchange reaction

$$H_2{}^{18}O + HONO \rightleftharpoons H^{18}ONO + H_2O$$

However, aqueous solutions of nitrous acid are also unstable and decompose rapidly when heated. The following mechanism has been suggested:

$$4HNO_2 \rightleftharpoons N_2O_4 + 2NO + 2H_2O; \qquad \text{rapid}$$
$$N_2O_4 \rightarrow NO^+ + NO_3^-; \qquad \text{rate determining}$$
$$NO^+ + H_2O \rightleftharpoons HNO_2 + H^+$$

The decomposition of ammonium nitrite solution can also be regarded as similar to that of nitrous acid and ammonia in aqueous solution:

$$NH_4NO_2 \rightarrow NH_4^+ + NO_2^-$$
$$NH_4^+ \rightarrow NH_3 + H^+$$
$$2H^+ + NO_2^- \rightarrow H_2O + NO^+$$
$$NO^+ + NH_3 \rightarrow NH_3NO^+ \rightarrow N_2 + H^+ + H_2O$$

Explosive decomposition occurs[629] with solid NH_4NO_2:

$$NH_4NO_2 \rightarrow N_2 + 2H_2O$$

Silver nitrite is also susceptible to disproportionation during thermal decomposition[614, 615]:

$$AgNO_2 \rightleftharpoons Ag + NO_2$$
$$AgNO_2 + NO_2 \rightleftharpoons AgNO_3 + NO$$

The redox aqueous solution chemistry of nitrites is best described[630] by the following equations:

In acid solution:

$$HNO_2 + H_2O \rightleftharpoons NO_3^- + 3H^+ + e^-; \qquad E^\circ\ -0.94\ V$$
$$NO + H_2O \rightleftharpoons HNO_2 + H^+ + e^-; \qquad E^\circ\ -0.99\ V$$
$$N_2O + 3H_2O \rightleftharpoons 2HNO_2 + 4H^+ + 4e^-; \qquad E^\circ\ -1.29\ V$$

In alkaline solution:

$$NO_2^- + 2OH^- \rightleftharpoons NO_3^- + H_2O + 2e^-; \qquad E^\circ\ -0.01\ V$$
$$NO + 2OH^- \rightleftharpoons NO_2^- + H_2O + e^-; \qquad E^\circ\ +0.46\ V$$
$$N_2O + 6OH^- \rightleftharpoons 2NO_2^- + 3H_2O + 4e^-; \qquad E^\circ\ -0.15\ V$$

Oxidation of nitrous acid by atmospheric oxygen proceeds by oxidation of nitric oxide formed by decomposition of the acid, and the resulting dinitrogen tetroxide reacting with water:

$$4HNO_2 \rightarrow N_2O_4 + 2NO + 2H_2O$$
$$2NO + O_2 \rightarrow N_2O_4$$
$$N_2O_4 + H_2O \rightarrow HNO_2 + HNO_3$$

---

629 T. Urbański, *Chemistry and Technology of Explosives*, Vol. II, Pergamon, Oxford (1965), p. 491.
630 P. Gray and A. D. Yoffe, *Chem. Rev.* **55** (1955) 1069.

The reaction with hydrogen peroxide also produces nitrate, but the mechanism probably involves the formation of peroxynitrous acid HOONO, isomeric with nitric acid, as an intermediate:

$$HNO_2+H^+ \rightleftharpoons NO^++H_2O$$
$$H_2O_2+NO^+ \rightarrow HOONO+H^+ \rightarrow NO_3^-+2H^+$$

As such mixtures act as initiators of polymerization reactions, it is thought that peroxynitrous acid can decompose by way of free radicals[631]:

$$HOONO \rightarrow {}^{\cdot}OH+NO_2^{\cdot}$$

Nitrous acid and nitrites are readily reduced[615] to a variety of nitrogen species, even to hyponitrous acid with tin(II) and to ammonia with H_2S, but more commonly to NO and N_2O:

$$SO_2+2HNO_2 \rightarrow 2NO+H_2SO_4$$
$$H_2O+2SO_2+2HNO_2 \rightarrow N_2O+2H_2SO_4$$

Hydrazine salts lead to the formation of azides which also react with HNO_2 to give nitrous oxide[632]:

$$N_2H_5^++HNO_2 \rightarrow HN_3+2H_2O+H^+$$
$$HN_3+HNO_2 \rightarrow N_2O+N_2+H_2O$$

The mechanism of this latter reaction has been investigated using ^{15}N substituted azides, and the route of each nitrogen atom traced[633]:

$$\overset{*}{N}NN^-+\overset{\dagger}{N}O^+ \rightleftharpoons \overset{*}{N}NN\overset{\dagger}{N}O \rightarrow \overset{*}{N}N+N\overset{\dagger}{N}O$$

With hydroxylamine, nitrous oxide is also obtained but the route is different[615] under acid and neutral conditions[634]:

$$NH_2OH+H\overset{*}{N}O_2 \xrightarrow{\text{neutral}} HON{=}\overset{*}{N}{-}OH+H_2O$$
$$\swarrow \qquad \searrow$$
$$\overset{*}{N}NO+H_2O \qquad N\overset{*}{N}O+H_2O$$

$$NH_2OH+H\overset{*}{N}O_2 \xrightarrow{\text{acid}} HONH\overset{*}{N}{=}O+H_2O$$
$$\downarrow$$
$$N\overset{*}{N}O+H_2O$$

Other reactions of nitrites include[614]:

$$NaNO_2+NH_2CONH_2 \longrightarrow NaCNO+N_2+2H_2O$$
$$NaNO_2+NO_2 \xrightarrow{450^\circ C} NaNO_3+NO$$
$$NaNO_2+Na_2O \longrightarrow Na_3NO_3$$
$$2NaNO_2+2Na \xrightarrow{\text{liq. } NH_3} Na_4N_2O_4 \ (Na_2NO_2)$$
$$2Ba(NO_2)_2+I_2 \rightleftharpoons Ba(NO_3)_2+2KI+2NO$$
$$2HNO_2+2HI \longrightarrow I_2+2NO+2H_2O$$
$$HNO_2+H_2O+I_2 \longrightarrow HI+HNO_3$$

---

631 E. Halfpenny and P. L. Robinson, *J. Chem. Soc.* (1952) 928.
632 P. Biddle and J. H. Miles, *J. Inorg. Nucl. Chem.* **30** (1968) 1291.
633 K. Clusius and H. Knopf, *Ber.* **89** (1956) 681.
634 A. A. Bothner-By and L. Friedman, *J. Chem. Phys.* **20** (1952) 459.

17.8. BIOLOGICAL PROPERTIES[635]

Nitrite is a principal intermediate in the nitrogen cycle and important in both nitrification and denitrification processes. In the former, nitrite is regarded as the precursor to nitrate, while the latter can occur by reduction of nitrite to a volatile material or by its reaction with amino groups giving elemental nitrogen.

Nitrites are dangerously poisonous materials and show symptoms which resemble those of cyanide poisoning. Fatal accidents have occurred through the use of sodium nitrite in place of sodium chloride to season food.

17.9. ANALYSIS OF NITROUS ACID AND NITRITES[636]

Nitrous acid is never encountered in the pure state and is stable only in cold aqueous solutions:

$$3HNO_2 \rightleftharpoons H_3O^+ + NO_3^- + 2NO$$

However, all nitrites react in acid solution, by intermediate formation of nitrous acid, with primary aromatic amines to form diazonium compounds which are capable of coupling with amino or hydroxy derivatives to form azo dyes. For example, α-naphthylamine can be diazotized and coupled with sulphanilic acid to give the red azo dye α-naphthylamine-*p*-azobenzene-*p*-sulphonic acid.

Titrimetric procedures involve oxidation with permanganate according to the equation

$$5NO_2^- + 2MnO_4^- + 6H^+ \rightarrow 5NO_3^- + 2Mn^{2+} + 3H_2O$$

although it is recommended that the nitrite solution be mixed with excess permanganate prior to acidification, and the permanganate back-titrated iodometrically to avoid any losses due to nitrous acid decomposition.

Small amounts of nitrite in the presence of nitrate can be determined by reaction with sulphamic acid by measuring the volume of nitrogen evolved:

$$NaNO_2 + HSO_3NH_2 \rightarrow NaHSO_4 + N_2 + H_2O$$

18. NITRIC ACID AND NITRATES

18.1. DISCOVERY AND HISTORY

Nitric acid was certainly known by the alchemists of the thirteenth century who prepared it by heating a mixture of copper sulphate, alum and saltpetre for use in refining precious metals. In 1776 Lavoisier proved that nitric acid contained oxygen, and in 1783 Cavendish showed that the dew produced by the electric sparking of humid air contained nitric acid and also demonstrated that it contained nitrogen and oxygen. The complete chemical composition was established in 1816 by Gay-Lussac and Berthollet[637].

635 F. Call, in *Mellor's Comprehensive Treatise on Inorganic and Theoretical Chemistry*, Vol. 8, Suppl. I, *Nitrogen*, Part I, Longmans, London (1964), Section XVII, pp. 567–615.

636 K. I. Vasu, in *Mellor's Comprehensive Treatise on Inorganic and Theoretical Chemistry*, Vol. 8, Suppl. II, *Nitrogen*, Part II, Longmans, London (1967), p. 643.

637 F. D. Miles, *Nitric Acid Manufacture and Uses*, ICI Ltd., Oxford University Press, London (1961).

From 1825 for almost a century, the saltpetre beds of Chile provided the world demand for nitrate. As the nineteenth century relied on natural resources, so the twentieth century has seen the necessity of synthetic routes from readily available starting materials. The investigations of Ostwald into the catalytic oxidation of ammonia at about the time the Haber process for the synthetic production of ammonia was being developed have provided the basis of the nitric acid industry today[637, 638].

18.2. OCCURRENCE OF NITRIC ACID AND NITRATES

A very large amount of nitric acid is produced by lightning and other electric discharges in the atmosphere and washed down on to the land by rain. It has been suggested that apart from providing land with combined nitrogen, that some may have led to the formation of alkali metal nitrate, deposits, particularly in dry climates. The extensive deposits in the caliche beds of the desert region of Chile have been exploited since the early nineteenth century. Chile saltpetre is sodium nitrate containing some potassium nitrate and sodium iodate. The caliche layer is usually 2–4 m thick just below the surface and can contain up to 70% $NaNO_3$ although much of the ore now mined contains less than 10% nitrates[639]. The chemical composition also suggests that the origin of the deposits may have been sea-weed washed up from the sea followed by the action of nitrifying bacteria in a dry climate[637].

18.3. PRODUCTION AND INDUSTRIAL USE OF NITRIC ACID AND NITRATES

Nitric acid was formerly made[637, 638, 640] from the reaction of sodium nitrate recrystallized from Chile saltpetre with concentrated sulphuric acid:

$$2NaNO_3 + H_2SO_4 \rightarrow Na_2SO_4 + 2HNO_3$$

The nitric acid was distilled from the mixture after rejecting the forerun containing hydrochloric acid and nitrosyl chloride resulting from chloride impurity. Excess sulphuric acid was added forming some bisulphate to maintain a fluid melt, but this also produced frothing due to the water vapour released during pyrosulphate formation:

$$NaNO_3 + H_2SO_4 \rightarrow NaHSO_4 + HNO_3$$
$$2NaHSO_4 \rightarrow Na_2S_2O_7 + H_2O$$

Nowadays, the principal route to nitric acid is based on the oxidation of nitric oxide to nitrogen dioxide and subsequent absorption in water[637, 639, 641, 642]:

$$2NO + O_2 \rightarrow 2NO_2$$
$$3NO_2 + H_2O \rightarrow 2HNO_3 + NO$$

---

638 D. W. F. Hardie and J. Davidson Pratt, *A History of the Modern British Chemical Industry*, Pergamon, Oxford (1966).

639 C. J. Pratt and R. Noyes, *Nitrogen Fertilizer Chemical Processes*, Noyes Development Corporation, Pearl River, New York (1965).

640 T. H. Chilton, *Strong Water Nitric Acid: Its Sources, Methods of Manufacture and Uses*, MIT Press, Cambridge, Mass. (1968).

641 R. G. Halford, in ref. 636, pp. 278–95.

642 C. Matasa and E. Matasa, *L'Industrie Moderne des Produits Azotes*, Dunod, Paris (1968), pp. 419–671.

Nitric oxide can be produced directly by the oxidation of atmospheric nitrogen but neither the Birkeland–Eyde arc oxidation[637] nor the Wisconsin thermal process[643] are economical at the present time. Thus, the catalytic oxidation of ammonia (see section 3.10.2) is the only method currently used[639]. Under suitable conditions platinum–rhodium catalyst[644] has a high specificity in promoting the oxidation to nitric oxide rather than the thermodynamically more favourable reaction to nitrogen:

$$4NH_3+5O_2 \rightarrow 4NO+6H_2O; \qquad \Delta H = -229.4 \text{ kcal/mol}$$
$$4NH_3+3O_2 \rightarrow 2N_2+6H_2O; \qquad \Delta H = -312.1 \text{ kcal/mol}$$

This reaction can be carried out at atmospheric or higher pressures using air or pure oxygen. The oxidation of NO to NO_2 (see section 11.7.2) is also favoured by high pressures as well as low temperatures, with the conversion time being proportional to the square of the pressure. Higher pressures also improve the absorption of nitrogen dioxide in water giving stronger nitric acid[639]. Some of the reactions involved include the following:

$$2NO_2+H_2O \rightarrow HNO_3+HNO_2$$
$$2HNO_2 \rightarrow H_2O+NO_2+NO$$
$$3NO_2+H_2O \rightarrow 2HNO_3+NO$$

The exothermic NO_2–H_2O reaction is carried out in towers by counter-current absorption giving dilute acid from which the excess NO_2 is bleached out with air. The gases leaving the towers usually contain low concentrations of nitrogen oxides which may require further treatment before being allowed to enter the atmosphere[639]. Most of the energy requirements of a modern HNO_3 unit can be supplied by steam generated by the plant.

Nitric acid of 50–70% strength is produced by commercial processes operating at atmospheric and higher pressures. However, because of azeotrope formation between nitric acid and water, simple distillation at atmospheric pressure does not increase the strength of weak acid above 68.5%. The usual method of concentrating nitric acid is by counter-current dehydration using 95% sulphuric acid giving 96–98% nitric acid and 65% sulphuric acid which can be concentrated by evaporation.[637] An alternative and cheaper method is that in which concentrated magnesium nitrate solution and weak nitric acid together with 80–90% nitric acid are fed into a stripping column[639]. The nitric acid vapours evolved are further fractionated, condensed and part of the product returned to the stripping column. The weak magnesium nitrate solution can be concentrated by thermal and vacuum evaporation. The use of chemical dehydration agents can be avoided if liquid N_2O_4 is used in the absorption step. 98–99% HNO_3 is obtained when liquid N_2O_4 is heated to 70°C at 50 atm for about 4 hr with theoretical quantities of air and dilute nitric acid[639].

The production of nitrates is probably the principal use of nitric acid[642]. Ammonium nitrate is manufactured in large quantities by the neutralization reaction for use as fertilizer and explosive[645] (see section 5). Sodium nitrate can also be readily obtained from weak nitric acid and soda ash[639]:

$$2HNO_3+Na_2CO_3 \rightarrow 2NaNO_3+CO_2+H_2O$$

or from sodium chloride[646]:

$$4HNO_3+3NaCl \rightarrow 3NaNO_3+H_2O+Cl_2+NOCl$$
$$3NOCl+2Na_2CO_3 \rightarrow NaNO_3+3NaCl+2NO_2+2CO_2$$

---

643 I. R. Beattie, in ref. 636, p. 183.

644 G. C. Bond, *Catalysis by Metals*, Academic Press, New York (1962), p. 456.

645 K. F. J. Thatcher, in *Mellor's Comprehensive Treatise on Inorganic and Theoretical Chemistry*, Vol. 8, Suppl. I, *Nitrogen*, Part I, Longmans, London (1964), pp. 506–62.

646 E. D. Crittenden, in *Fertilizer Nitrogen* (ed. G. Sauchelli), Reinhold, New York (1964), pp. 331–43.

Chile saltpetre is still mined, and the sodium nitrate is extracted from the caliche by leaching the crushed ore at 40°C with water. The solution is cooled to 15°C in heat exchangers then to 5°C by liquid ammonia, and the sodium nitrate slurry separated in centrifuges. Solar evaporation plants are also used to recover nitrate from residues[639].

Potassium nitrate is made by the double decomposition reaction

$$KCl + NaNO_3 \rightarrow KNO_3 + NaCl$$

Alkaline earth metal nitrates and some compound fertilizers are also produced by similar methods. For example, nitric acid can replace sulphuric acid in the treatment of phosphate rock to give nitrophosphates according to the generalized equation[639]

$$Ca_{10}(PO_4)_6F_2 + 14HNO_3 \rightarrow 3Ca(H_2PO_4)_2 + 7Ca(NO_3)_2 + 2HF$$

Large quantities of nitric acid are used for the nitration of organic compounds in the production of commercial explosives (nitroglycerine, nitrocellulose, trinitrotoluene, cyclotrimethylenetrinitramine) and numerous organic intermediates[639, 647]. The oxidizing properties of HNO_3 are manifest in its application as a rocket propellant[642] and in many organic[647] and inorganic[642] oxidation reactions. Most metals can be made to dissolve in nitric acid, and even the noble metals are attacked in the presence of hydrochloric acid (aqua regia). Stainless steel, silicon iron and aluminium resist high-strength acid, but alloy steels are necessary for dilute solutions. Stainless steel is often used for modern plant construction and fluorocarbon resins are inert to all strengths of acid and being flexible can be used for gaskets, seals and linings[639]. Nitrates are also incorporated in explosives (KNO_3 in gunpowder) and have many minor applications, but their principal use is as fertilizers[639, 648].

18.4. LABORATORY PREPARATION AND PURIFICATION

Nitric acid is usually available from commercial sources and so laboratory preparations are not often required, but because of its continual decomposition in the presence of light or heat it is necessary to purify it for certain purposes. Technical grade nitric acid can be rendered halogen-free by distilling over a small amount of silver nitrate and using the middle-cut[649]. Anhydrous nitric acid can be obtained by distilling acid of the highest concentration available over P_2O_5 or with pure H_2SO_4, in an all-glass vacuum distillation apparatus (no grease) in the absence of light. The product distills over at 36–38°C/20 mm pressure as a colourless liquid which can be redistilled from pure H_2SO_4 with a stream of ozonized oxygen passing through the liquid and stored below its freezing point[650].

The preparative methods appropriate to the nitrates of individual elements will be discussed in their respective chapters, but mention will be made here to four reagents from which most nitrates can be prepared[651, 652]. Aqueous nitric acid reacts with most metals

[647] M. Sittig, *Combine Hydrocarbons and Nitrogen for Profit*, Noyes Development Corporation, New Jersey (1967), pp. 59–98.

[648] *Chemical Fertilizers* (ed. G. Fauser), Pergamon, Oxford (1968).

[649] P. W. Schenk, in *Handbook of Preparative Inorganic Chemistry*, Vol. I (ed. G. Brauer), Academic Press, New York (1963), p. 460.

[650] S. A. Stern, J. T. Mullhaupt and W. B. Kay, *Chem. Rev.* **60** (1960) 185.

[651] B. O. Field and C. J. Hardy, *Quart. Revs.* (*London*) **18** (1964) 361.

[652] C. C. Addison and N. Logan, in *Preparative Inorganic Reactions*, Vol. I (ed. W. L. Jolly), Interscience, New York (1964), p. 141.

and their oxides or carbonates to form metal nitrate hydrates. Liquid N_2O_4 (see section 13.7.2), particularly in the presence of solvents of higher dielectric constant, dissolve a number of metals to give anhydrous metal nitrates[653]:

$$Ca + 2N_2O_4 \xrightarrow{MeNO_2} Ca(NO_3)_2 + 2NO$$

Liquid N_2O_5 also reacts with metal oxides and chlorides to give addition compounds which can be thermally decomposed *in vacuo* to anhydrous metal nitrates[651]:

$$TiCl_4 + 4N_2O_5 \rightarrow Ti(NO_3)_4 + 2N_2O_4 + 2Cl_2$$

One of the most powerful nitrating agents is $ClONO_2$ nitrosyl hypochlorite (often referred to as chlorine nitrate $ClNO_3$), which reacts with most metals[654] or metal chlorides explosively at room temperature but can be controlled at temperatures between −40 and −70°C:

$$SnCl_4 + 4ClONO_2 \rightarrow Sn(NO_3)_4 + 4Cl_2$$

Perhaps the least familiar nitrate compounds are those of the halogens, and these have been prepared by the following methods[651, 655]:

$$F_2 + NaNO_3 \longrightarrow FONO_2 + NaF$$

$$2ClO_2 + 2N_2O_5 \longrightarrow 2ClONO_2 + N_2O_4 + 4O_2$$

$$BrF_3 + 3N_2O_5 \xrightarrow[-30\text{ to }-50^\circ]{CCl_3F} Br(NO_3)_3 + 3FNO_2$$

$$ICl_3 + 3ClONO_2 \xrightarrow{-70^\circ} I(NO_3)_3 + 3Cl_2$$

18.5. NUCLEAR ISOTOPES OF NITRIC ACID AND NITRATES

Nitrogen can be oxidized in a high-voltage discharge to nitrogen dioxide which is then dissolved in a slight deficiency of dilute potassium hydroxide solution:

$$2^{15}NO_2 + 2KOH \rightarrow K^{15}NO_3 + K^{15}NO_2 + H_2O$$

Over a period of a few hours, the resulting acid solution causes the nitrite present to disproportionate:

$$3H^{15}NO_2 \rightarrow H^{15}NO_3 + 2^{15}NO + H_2O$$

The presence of excess oxygen at 50°C converts the oxides of nitrogen completely into nitrate within a few hours, from which $K^{15}NO_3$ can be crystallized out, or alternatively concentrated sulphuric acid is added and $H^{15}NO_3$ obtained by distillation[656].

$^{15}NH_3$ can be oxidized directly to nitrate with aqueous permanganate in an autoclave at 170–180°C for 7–8 hr. The enriched product is separated from permanganate decomposition products, and after addition of sulphuric acid, $H^{15}NO_3$ distilled off. The labelled nitric acid can be converted into any of its salts by simple neutralization reactions[656].

653 C. C. Addison, *Chemistry in Liquid Dinitrogen Tetroxide*, Vol. III, Part I, *Chemistry in Nonaqueous Ionizing Solvents* (eds. G. Jander, H. S. Spandau and C. C. Addison), Pergamon, Oxford (1967).

654 M. Schmeisser and K. Brandle, *Angew. Chem.* **73** (1961) 388.

655 S. M. Williamson, in *Preparative Inorganic Reactions*, Vol. I (ed. W. L. Jolly), Interscience, New York (1964), p. 246.

656 W. Spindel, in *Inorganic Isotopic Syntheses* (ed. R. H. Herber), Benjamin, New York (1962), p. 99.

Nitric acid-d is prepared by the reaction of nitrogen dioxide with ozone in deuterated water[657]:

$$2NO_2+O_3 \rightarrow N_2O_5+O_2$$
$$N_2O_5+D_2O \rightarrow 2DNO_3$$

Oxygen-18-enriched nitric acid is prepared by exchange between nitric acid and ^{18}O-enriched water[658]. The potassium salt can then be obtained by neutralizing the acid with a solution of potassium hydroxide also in ^{18}O-enriched water. After crystallization, the product can be further enriched by heating with more enriched water and fuming nitric acid in a sealed tube at 70°C for 38 hr. After the removal of the volatiles by vacuum distillation, the potassium nitrate remaining contains up to 84% ^{18}O.

18.6. STRUCTURE AND BONDING

In general the nitrate ion is planar (D_{3h} symmetry) with all N–O bond distances close to 1.22 Å. This can be represented in valence bond terms by resonance structures based on those illustrated in Fig. 44.

FIG. 44. Valence bond and non-pairing structures for NO_3^-.

Molecular orbitals can also be constructed for the nitrate ion on the basis of three σ bonds using sp^2 hybrid orbitals and the p_z orbitals of nitrogen and the three oxygens combining to form a π orbital containing two electrons. Thus twenty-four electrons are contained in bonding and non-bonding orbitals only which is a stable configuration for a tetratomic species, similar to sixteen-electron triatomic and ten-electron diatomic molecules[659].

In nitric acid, however, the hydrogen is covalently bonded to one of the oxygen atoms of the nitrate group. Structural parameters[660, 661] for HNO_3 and FNO_3 are illustrated in Fig. 45.

FIG. 45. The structure of HNO_3 and FNO_3.

657 H. L. Crespi, in *Inorganic Isotopic Syntheses* (ed. R. H. Herber), Benjamin, New York (1962), p. 38.

658 I. Dostrovsky and D. Samuel, in *Inorganic Isotopic Syntheses* (ed. R. H. Herber), Benjamin, New York (1962), p. 135.

659 M. Green, in *Developments in Inorganic Nitrogen Chemistry*, Vol. I (ed. C. B. Colburn), Elsevier, Amsterdam (1966), p. 54.

660 V. Luzzati, *Acta Cryst*. 4 (1951) 120.

661 L. Pauling and L. O. Brockway, *J. Am. Chem. Soc.* **59** (1937) 13.

From these data, the electronic configuration is probably best represented by the non-pairing structures in Fig. 46 analogous to nitryl halides.

FIG. 46. Linnett non-pairing structures for covalent $XONO_2$ molecules.

Covalent metal nitrates[662, 663] present the opportunity for the nitrate group to demonstrate the different ways it can bond to metals as illustrated in Fig. 47.

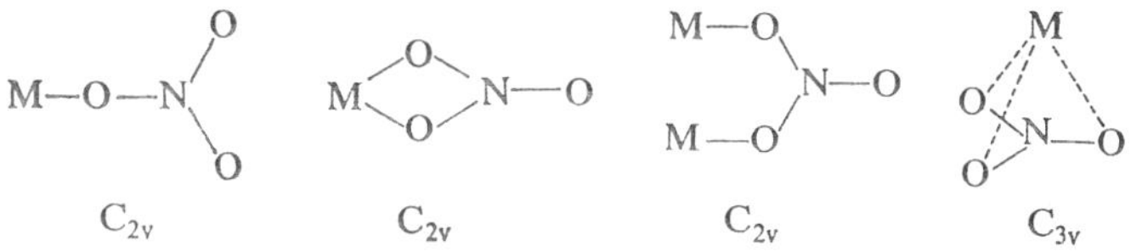

FIG. 47. Bonding possibilities in covalent metal nitrates.

For example, the nitrate groups act as unidentate ligands[653] in $(CO)_5MnONO_2$, bidentate ligands in uranyl nitrate[664] and bridging groups in basic beryllium nitrate[665], the symmetry in each case being at best C_{2v}. An example of a compound with the terdentate C_{3v} structure is not known, but each of the oxygen atoms in half the nitrate groups of the α-form of copper(II) nitrate are involved in bonding to a copper atom[666].

Thus covalent bonding can be easily recognized by infrared spectroscopy[667] due to the lowering of the symmetry of the nitrate group. The greater stability of covalent metal nitrates over non-metal nitrates may be due to back donation of electrons from the metal to empty orbitals of the nitrate group.

18.7. MOLECULAR CONSTANTS

Pure anhydrous nitric acid exists only in the solid phase, and X-ray analysis of single crystals indicates[660] that the unit cell is monoclinic. In contrast, the hydrates $HNO_3 \cdot H_2O$ and $HNO_3 \cdot H_2O$ are orthorhombic[668]. Electron diffraction[669] and microwave data[670] for

662 C. C. Addison and N. Logan, in *Advances in Inorganic Chemistry and Radiochemistry*, Vol. 6 (eds. H. J. Emeleus and A. G. Sharpe), Academic Press, New York (1964), p. 72.

663 C. C. Addison and D. Sutton, in *Progress in Inorganic Chemistry*, Vol. 8 (ed. F. A. Cotton), Interscience, New York (1967), pp. 195–286.

664 J. E. Fleming and H. Lynton, *Chem. Ind.* (*London*) (1960) 1416.

665 C. C. Addison and A. Walker, *Proc. Chem. Soc.* (1961) 242.

666 S. C. Wallbank and W. E. Addison, *J. Chem. Soc.* (1965) 2925.

667 K. Nakamoto, *Infrared Spectra of Inorganic and Coordination Compounds*, Wiley, New York (1963), p. 92.

668 L. Bouttier, *Compt. rend.* **228** (1949) 1419.

669 O. Redlich and L. E. Nielsen, *J. Am. Chem. Soc.* **65** (1943) 654.

670 D. J. Millen and J. R. Morton, *J. Chem. Soc.* (1960) 1523.

gaseous HNO_3 have also been reported, and the structural parameters show some deviations from the crystal data as indicated in Table 80.

TABLE 80. MOLECULAR CONSTANTS FOR NITRIC ACID AND THE NITRATE ION

X-ray diffraction data[a] for HNO_3

Crystal system	monoclinic
Symmetry	$P2_1/a\text{-}C_{2h}^5$
Lattice parameters	$a = 16.23$, $b = 8.57$, $c = 6.71$, $B = 90°$
Molecules/unit cell	16

Structural parameters of HNO_3	X-ray (Å)	Electron diffraction[b] (Å)	Microwave spectra[c] (Å)
N–O bond length	1.24	1.22	1.206
N–OH bond length	1.30	1.41	1.405
ONO bond angle	134°	$130 \pm 5°$	130°
ON–OH bond angle	114°	$115 \pm 2.5°$	$113 \pm 5°$

Infrared data[d]	HNO_3 (cm^{-1})	DNO_3 (cm^{-1})	Moments of inertia[c]	HNO_3 (cm^{-1})	DNO_3 (cm^{-1})
ν_1	886	888	I_A	38.85368	38.97481
ν_2	1320	1313	I_B	41.77966	44.68656
ν_3			I_C	80.74801	83.76847
ν_4	1710	1685	Δ	0.11467	0.10710
ν_5	583	543			
ν_6	765	764			

N–O bond length in nitrates
From X-ray data 1.22–1.24 Å

Infrared data for the nitrate ion[e] (cm^{-1})

ν_1 (A_1')	NO stretch	R active	1050
ν_2 (A_2')	Out of plane rock	IR active	832
ν_3 (E')	Sym. and asym. stretch	IR and R active	1360
ν_4 (E')	Sym. and asym. deform	IR and R active	720

[a] Ref. 660. [b] Ref. 669. [c] Ref. 670. [d] Ref. 667. [e] Ref. 650.

The spectroscopic properties of nitric acid have received much attention[671] in particular as a means of establishing what species are present in nitric acid solutions[672, 673]. Satisfactory assignments of the fundamental frequencies were made[671] for the HNO_3 molecule by treating the OH group as of C_s symmetry and the NO_2 group as of C_{2v}. For the nitrate ion,

[671] R. F. Barrow and A. J. Merer, in *Mellor's Comprehensive Treatise on Inorganic and Theoretical Chemistry*, Vol. 8, Suppl. II, *Nitrogen*, Part II, Longmans, London (1967), p. 476.

[672] W. H. Lee, in *The Chemistry of Non-Aqueous Solvents*, Vol. III (ed. J. J. Lagowski), Academic Press, New York (1967), pp. 151–89.

[673] P. E. Curry, in ref. 671, pp. 296–313.

TABLE 81. PHYSICAL PROPERTIES OF NITRIC ACID[a, b, c]

Property	Temperature (K)	Value
Melting point	231.55	−41.6°C
Boiling point	356.15	82.6°C
ΔH fusion	231	2.503 kcal/mol
ΔH vaporization	356	9.426 kcal/mol
ΔH_f (enthalpy of formation) (g)	0	−29.94 kcal/mol
ΔH_f (enthalpy of formation) (g)	298	−32.28 kcal/mol
ΔH_f (enthalpy of formation) (l)	298	−41.61 kcal/mol
ΔH_f (enthalpy of formation) ($m = 1$) (aq)	298	−49.56 kcal/mol
$H_{298}-H_0$ (enthalpy) (g)	298	2.815 kcal/mol
ΔG_f (Gibbs energy of formation) (g)	298	−17.87 kcal/mol
ΔG_f (Gibbs energy of formation) (l)	298	−19.31 kcal/mol
ΔG_f (Gibbs energy of formation) ($m = 1$) (aq)	298	−26.61 kcal/mol
$S°$ (entropy) (g)	298	63.64 cal/deg mol
$S°$ (entropy) (l)	298	37.19 cal/deg mol
$S°$ (entropy) ($m = 1$) (aq)	298	35.0 cal/deg mol
C_p° (heat capacity) (g)	298	12.75 cal/deg mol
C_p° (heat capacity) (l)	298	26.26 cal/deg mol
C_p° (heat capacity) ($m = 1$) (aq)	298	−20.7 cal/deg mol
Trouton's constant		21.2 cal/deg
Density of solid	231.5	1.895 g/cm^3
Density of liquid d	243–305	$d = 1.5492 - 0.00183t$ g/cm^3
	273	1.549 g/cm^3
	298	1.504 g/cm^3
	313	1.477 g/cm^3
Vapour pressure of liquid	273	14.7 mm
	298	57 mm
	323	215 mm
	353	625 mm
Viscosity of liquid	273	10.92 centipoise
	298	7.46 centipoise
Surface tension	273	43.56 dyne cm
	293	41.15 dyne cm
	313	37.76 dyne cm
Velocity of sound	289	1425 m/s
Refractive index ($\lambda = 5461$)	278	1.4030
Specific conductance	293	$3.72 \times 10^{-2}\ \Omega^{-1}\ cm^{-1}$
Dielectric constant	287	50 ± 10

[a] Ref. 672. [b] Ref. 673. [c] Ref. 676.

electronic and infrared spectra have been studied in molten alkali nitrates and solids, and the allowed fundamentals have also been observed in Raman solution spectra. Values of bands recorded for alkali metal nitrates are in close agreement with calculated[650] figures listed in Table 80.

18.8. PHYSICAL PROPERTIES

The values of physical properties listed in Table 81 refer predominantly to liquid nitric acid in which self-dissociation occurs to a greater extent than any other pure liquid[672]. The presence of ionic species is supported by the high electrical conductivity[673] and also by the appearance of weak lines at 1050 cm^{-1} and 1400 cm^{-1} when the Raman spectrum is recorded on pure nitric acid just after melting[671]. Further spectroscopic studies[671, 674] have established that such species result from self-ionization:

$$2HNO_3 \rightleftharpoons H_2O+N_2O_5 \rightleftharpoons H_2O+NO_2^++NO_3^-$$

followed by further ionization of H_2O in HNO_3:

$$H_2O+HNO_3 \rightleftharpoons H_3O^++NO_3^-$$

The single ^{15}N n.m.r. peak which appears at the average position of NO_2^+, NO_3^- and HNO_3 suggests that rapid exchange occurs between such species[675]. Thermodynamic properties are also listed[676] in Table 81.

The concentration of water formed during the dissociation equilibrium of nitric acid is low[677]; however, HNO_3 is miscible with water in all proportions and can be extracted from 6–16 M aqueous solutions into benzene or toluene as the hemi-hydrate $2HNO_3 \cdot H_2O$. Some properties of the two crystalline hydrates, HNO_3H_2O (Raman spectra of which show that it exists as hydroxonium nitrate[678] $H_3O^+NO_3^-$) and $HNO_3 \cdot 3H_2O$ are given in Table 82, and the existence of a dimer hydrate $2HNO_3 \cdot 3H_2O$ has also been confirmed[679]. A few acid salts such as $NH_4NO_3 \cdot 2HNO_3$ and $KNO_3 \cdot HNO_3$ which may be thought of either as

TABLE 82. PHYSICAL PROPERTIES OF NITRIC ACID HYDRATES[a, b]

Property	Temperature (K)	$HNO_3 \cdot H_2O$ value	$HNO_3 \cdot 3H_2O$ value
Melting point	235.47	−37.68°C	
Melting point	254.68		−18.47°C
Boiling point	392.15	119.8°C	
ΔH_f (enthalpy of formation) (l)	298	−113.16 kcal/mol	−252.40 kcal/mol
ΔG_f (Gibbs energy of formation) (l)	298	−78.61 kcal/mol	−193.91 kcal/mol
$S°$ (entropy) (l)	298	51.84 cal/deg mol	82.93 cal/deg mol
$C_p^°$ (heat capacity)	298	43.61 cal/deg mol	77.71 cal/deg mol
ΔH (enthalpy of dilution to infinite dilution)		−4.732 kcal/mol	−2.123 kcal/mol

[a] Ref. 673.
[b] Ref. 676.

674 C. K. Ingold and D. J. Millen, *J. Chem. Soc.* (1950) 2612.
675 R. A. Ogg, Jr. and J. D. Ray, *J. Chem. Phys.* **25** (1956) 1285.
676 D. D. Wagman, W. H. Evans, V. B. Parker, I. Halow, S. M. Bailey and R. H. Schumm, *Selected Values of Chemical Thermodynamic Properties*, NBS Technical Note 270-3, Washington (1968).
677 G. Charlot and B. Tremillon, *Chemical Reactions in Solvents and Melts*, Pergamon, Oxford (1969), p. 227.
678 D. J. Millen and E. Vaal, *J. Chem. Soc.* (1956) 2913.
679 T. G. Berg, *Acta chem. scand.* **8** (1954) 374.

containing HNO_3 of crystallization or solvated anions are known[680]. However, the evolution of heat which occurs when nitrates are dissolved in HNO_3 is probably due to the formation of such species. Thus although nitric acid is a good ionizing solvent for electrolytes in general, salts are only sparingly soluble unless they produce NO_2^+ or NO_3^- ions.

Physical properties of nitrates of the elements allow a rather arbitrary division into ionic and covalent compounds. In this context the term nitrate is applied to compounds containing nitrate ions, whereas the prefix nitrato refers to materials in which the group is covalently bonded through oxygen[651]. Ionic metal nitrates form hydrated compounds, have high solubility in water and give rise to an extensive aqueous solution chemistry. However, the relatively low melting points of the anhydrous alkali and alkaline–earth metal nitrates and their eutectic mixtures, together with their ease of handling, have made such materials ideal for study as molten salts[681].

18.9. CHEMISTRY AND CHEMICAL PROPERTIES

The decomposition of nitric acid occurs in both the liquid and the gas phase according to the general equation[672]

$$2HNO_3 \rightleftharpoons 2NO_2+H_2O+\tfrac{1}{2}O_2$$

Like so many decomposition reactions of compounds containing nitrogen and oxygen, the various steps in the proposed mechanisms are further complicated by the possibility of setting up other equilibria. One mechanism is based on the dehydration of nitric acid[682]:

$$2HNO_3 \rightarrow N_2O_5+H_2O$$

followed by the normal first order decomposition of dinitrogen pentoxide[683]:

$$N_2O_5 \rightarrow NO_2+NO_3$$

$$NO_2+NO_3 \rightarrow NO_2+O_2+NO$$

$$NO+N_2O_5 \rightarrow 3NO_2$$

An alternative mechanism involves OH radicals[684]:

$$HNO_3 \rightarrow NO_2+OH$$

$$OH+HNO_3 \rightarrow NO_3+H_2O$$

$$NO_3+NO_2 \rightarrow NO_2+O_2+NO$$

$$NO+NO_3 \rightarrow 2NO_2$$

A similar mechanism was also proposed for the decomposition of nitric acid under flash photolysis conditions in which the NO_3 radical was identified as an intermediate[685].

680 V. Gutmann, *Coordination Chemistry in Non-Aqueous Solutions*, Springer-Verlag, New York (1968), pp. 75–76.
681 D. Inman and S. H. White, *Annual Reports Chem. Soc.* **62** (1965) 107.
682 C. Frejacques, *Compt. rend.* **232** (1951) 2206.
683 R. A. Ogg, Jr., *J. Chem. Phys.* **15** (1947) 337.
684 H. S. Johnson, L. Foering and R. J. Thompson, *J, Phys. Chem.* **57** (1953) 390.
685 D. Husain and R. G. W. Norrish, *Proc. Roy. Soc.* (*London*) A, **273** (1953) 165.

The decomposition of liquid nitric acid is much faster than gas-phase reactions at corresponding temperatures, and considerable effort has been directed to finding additives which may act as inhibitors during storage. However, although the addition of 1% lead dioxide to colourless concentrated nitric acid prevents decomposition to the extent that it remains colourless for some time, water has the greatest inhibiting effect, which suggests that dehydration mechanisms are important[686].

The thermal decomposition of metal nitrates by heating provides a method of preparing nitrites of the more electropositive elements (see section 17.1):

$$2NaNO_3 \rightarrow 2NaNO_2 + O_2$$

The temperature at which decomposition commences varies for different compounds[651, 687] and further decomposition can proceed simultaneously:

$$2KNO_3 \rightarrow K_2O + N_2 + 5/2O_2$$

Heavy metal nitrates decompose directly to the oxide presumably because the corresponding nitrites are thermally so much less stable:

$$2Pb(NO_3)_2 \rightarrow 2PbO + 4NO_2 + O_2$$

Silver nitrate decomposes directly to the metal:

$$2AgNO_3 \rightarrow 2Ag + 2NO_2 + O_2$$

The thermal and explosive decomposition of ammonium nitrate has been discussed elsewhere[645, 688] (see section 5). Covalent halogen nitrates ($XO—NO_2$) are thermally unstable even at room temperature[655, 689].

In the presence of suitable reducing agents, the oxygen released during solid state nitrate decompositions can give rise to explosive oxidations (e.g.: gunpowder). Thus nitrates in the solid and molten states are strong oxidizing agents, whereas neutral or alkaline aqueous solutions of the nitrate ion are weaker oxidizing agents than nitric acid. The redox properties of nitrate in aqueous acidic and basic solutions are summarized in the Ebsworth diagrams in section 1.6.2, and by which the course of reaction between nitrate and other nitrogen species can be predicted. Some other redox reactions of nitric acid are illustrated by the following equations[690]:

$$2HNO_3 + 3SO_2 \rightarrow 2NO + 2SO_3 + H_2SO_4$$

$$HNO_3 + 3HCl \rightarrow NOCl + Cl_2 + 2H_2O$$

$$2HNO_3 + 3H_2AsO_3 \rightarrow 2NO + 3H_3AsO_4 + H_2O$$

---

[686] G. D. Robertson, D. M. Mason and W. H. Corcoran, *J. Phys. Chem.* **59** (1955) 683.
[687] M. W. Lister, *Oxyacids*, Oldbourne, London (1965), p. 107.
[688] T. Urbanski, *Chemistry and Technology of Explosives*, Vol. II, Pergamon, Oxford (1965).
[689] C. Woolf, in *Advances in Fluorine Chemistry*, Vol. 5 (eds. M. Stacey, J. C. Tatlow and A. G. Sharpe), Butterworths, London (1965), pp. 1–30.
[690] M. G. B. Wright and P. Evans, in ref. 671, pp. 314–52.

Nitric acid in general behaves as a typical strong acid, and its reactions with metals, oxides, carbonates and other anions constitute important methods of producing nitrates[642, 648, 690]. Metals are usually attacked more rapidly by dilute rather than concentrated nitric acid from which oxides of nitrogen are normally liberated but seldom according to stoichiometry[690]. Pure nitric acid, however, is considered as an acceptor solvent on the basis of its reactions with donor compounds and its inability to solvate metal ions[680]. However, in concentrated sulphuric acid, the solution chemistry of HNO_3 is dominated by the presence of the nitronium ion[672]:

$$HNO_3 + H_2SO_4 \rightleftharpoons NO_2^+ + H_3O^+ + 2HSO_4^-$$

Such solutions are of importance in aromatic nitration reactions[672, 688, 691], on which a considerable part of the chemical industry is based[637, 642, 647]. Crystalline nitronium salts have been prepared[692], isolated and fully characterized[672].

18.10. BIOLOGICAL PROPERTIES[693]

Nitrates are the most important group of fertilizers and a source of fixed nitrogen which can be taken up by plants rapidly. This may be related to the high solubility of nitrates in water, and also to the requirement of cations such as potassium in the soil. However, nitrate can be lost by reduction to free nitrogen by soil bacteria in denitrification processes, although other organisms are involved in replacing it by nitrification of ammonia.

Liquid nitric acid is dangerous to the body as it penetrates and destroys tissue, the extent of the damage being proportional to the contact time. Nitric acid vapours, or as an aerosol, are particularly toxic, giving symptoms similar to those of nitrous fumes which may only occur after the time of exposure. Contact even to low concentrations for short periods of time should be avoided.

18.11. ANALYSIS OF NITRIC ACID AND NITRATES[694]

A very sensitive spot test for nitrate is the yellow colour produced when it reacts with diphenylamine under short ultraviolet radiation. Nitrates can also be reduced with Devarda's alloy in alkaline solution to ammonia which can be detected or determined by a variety of methods as previously described in section 3.12. A gasometric method for the determination of either organic or inorganic nitrates is available, based on the reduction of nitrate to nitric oxide by mercury in concentrated sulphuric acid:

$$2KNO_3 + 4H_2SO_4 + 3Hg \rightarrow 2NO + K_2SO_4 + 3HgSO_4 + 4H_2O$$

Even a gravimetric method is known using "nitron" (1,4-diphenyl-3,5-endanilo-4,5-dihydro-1,2,4-triazole); but many ions are known to interfere.

691 *The Chemistry of the Nitro and Nitroso Group* (ed. H. Feuer), Interscience, New York (1969).

692 D. R. Goddard, E. D. Hughes and C. K. Ingold, *J. Chem. Soc.* (1950) 2559.

693 F. Call, in *Mellor's Comprehensive Treatise on Inorganic and Theoretical Chemistry*, Vol. 8, Suppl. I, *Nitrogen*, Part I, Longmans, London (1964), pp. 567–615.

694 K. I. Vasu, in *Mellor's Comprehensive Treatise on Inorganic and Theoretical Chemistry*, Vol. 8, Suppl. II, *Nitrogen*, Part II, Longmans, London (1967), pp. 563–673.

Of the numerous colorimetric methods available for the determination of nitrate, the majority seem to be based on nitration of phenols, reactions with amines, or reduction to nitrite followed by a diazotization and coupling reaction.

The analysis of nitric acid poses several problems of its own, which arise mainly from its volatility, reversible decomposition on exposure to light or heat, and high reactivity. Mixtures with sulphuric acid are also encountered through nitrating mixtures and in the manufacture of sulphuric acid.

20. PHOSPHORUS

ARTHUR D. F. TOY
Stauffer Chemical Company, New York

1. ELEMENTAL PHOSPHORUS

1.1. DISCOVERY

Bones, guano, excrement, and fish have been used as fertilizer since antiquity. The major element in these substances responsible for the increase in crop yield was not discovered until 1669. This discovery has been credited to Hennig Brandt, an impoverished German merchant who sought to become rich by converting base metal to gold. According to the account by G. W. von Leibniz[1], Brandt made the discovery while carrying out alchemical experiments with urine. The substance he obtained glowed in the dark and burst into flame when exposed to air; it was subsequently named phosphorus, a term derived from φώς, light, φορέω, I bear.

The presence of phosphorus in the vegetable kingdom was first detected by B. Albino in 1688, as recorded in his Dissertato de Phosphoro Liquido et Solido[1]. Almost a hundred years later, in 1769–70, phosphorus as an essential ingredient in the bones of both men and animals[1] was recognized by J. G. Gahn and C. W. Scheele. The first discovery of phosphorus in a mineral (pyromorphite, a lead phosphate) was made by Gahn in 1779[2]. Subsequently, T. Bergman and J. L. Proust found phosphorus also in the mineral apatite. The elucidation of the elemental nature of phosphorus and the composition of phosphoric anhydride however was credited to L. Lavoisier[1, 3].

1.2. OCCURRENCE AND DISTRIBUTION

For a century after its discovery, the only source of phosphorus was urine. As the demand for phosphorus and phosphates increased, guano and bones became the main source. In fact, the need for bones in England was so great in the early 1800s that battlefields of Europe were turned to as the source of supply[2]. Bone ash from South America was also a source of phosphate. Bone contains approximately 23% mineral matter of which calcium phosphate calculated as $Ca_3(PO_4)_2$ is 87%, and calcium carbonate 12%[4]. Most of the bones were converted to phosphate fertilizers by treatment with sulfuric acid.

The use of minerals as sources of phosphate began in England in the early 1840s.

[1] J. W. Mellor, *Comprehensive Treatise on Inorganic and Theoretical Chemistry*, Vol. 8, pp. 729–36, Longmans, Green and Co. Inc. (1947).

[2] W. H. Waggaman, *Phosphoric Acid, Phosphates and Phosphatic Fertilizers*, 2nd ed., p. 3, Reinhold Publishing Corp., New York (1952).

[3] E. Farber, *History of Phosphorus*, U.S. National Museum Bulletin, No. 240 (1965), pp. 177–200.

[4] Ref. 2, pp. 30–32.

Since only low grade coprolite was available, explorations were undertaken in other countries for richer sources. Phosphatic minerals are quite abundant on the earth. According to Clarke and Washington[5], phosphorus is more abundant than carbon and less than chlorine in the ten mile crust of the earth, hydrosphere and atmosphere. They estimated that phosphorus is twelfth in the list of elements which make up 99.6% of the crust of the earth, or 0.142%. Phosphorus is even found in meteorites of extraterrestrial origin in the form of iron and nickel phosphides.

In *Dana's System of Mineralogy*[6] many naturally occurring crystalline phosphate minerals are listed along with their accepted names. However, only the apatite family of minerals is of industrial importance. Apatites have hexagonal crystalline structure and include fluorapatite, $Ca_5(PO_4)_3F$, chlorapatite, $Ca_5(PO_4)_3Cl$, and hydroxylapatite, $Ca_5(PO_4)_3(OH)$. There is also the carbonate–apatite, $Ca_{10}(PO_4)_6(CO_3).H_2O$ which is not a true apatite. Apatite deposits are found in many parts of the world. The largest deposit which is actively mined is in the Kola Peninsula, near Kirovsk[7], in the Soviet Arctic part of Scandinavia.

The world's largest source of phosphate, however, is the amorphous phosphate rock or phosphorite, of which the largest known deposits are found in the United States in the states of Florida, Tennessee, Utah, Idaho, Wyoming, Montana, North and South Carolina, Kentucky, Virginia and Arkansas. Important deposits are also found in the countries of Algeria, Tunisia, Egypt and Morocco. Lesser deposits are in the islands in the Indian and Pacific oceans, which include those of the Ocean, Christmas and Nauru Islands and the islands of the Marshall, Pellew, and Society group. Less important deposits which are actively mined are located in Australia, Japan, New Zealand, Belgium, France and England[8].

Origin and Nature of Phosphorite Deposits

The large deposits of phosphorites over the world are believed to have originated from the small quantity of phosphate normally present in granite rocks in the earth's surface. Through weathering and leaching, it found its way into the soil, spring water, and eventually to the sea where it was absorbed and concentrated in the shells, bones, and tissue of marine organisms. Remains of these organisms accumulated in the bottom of the sea along with calcium carbonate, and the latter, due to its higher solubility, gradually leached away. The phosphate which remained formed into compact phosphatic nodules. Over the years, due to the uplift of the ocean floor and other geological changes, some of the phosphatic sediments became deposits in dry land[1].

All of the phosphorites found in the mineral deposits have approximately the same $CaO : P_2O_5 : F$ ratio as the mineral fluorapatite: a typical analysis for F is between 2 to 4%. Their physical character varies from hard and flint-like to soft and plastic. Nearly all contain iron and aluminum oxides, carbonates, silicates, and organic matters. The phosphate content, expressed as $Ca_3(PO_4)_2$ (or BPL, bone phosphate of lime), varies from about 65% in Florida pebbles to about 75% in the high-grade deposits of Morocco, Algeria and Tunisia. Brown rocks of Tennessee and oolitic phosphatic material of Western

[5] F. W. Clarke and H. S. Washington, *Proc. Am. Nat. Acad.* **8** (1922) 108.

[6] E. Palache, H. Berman and C. Frondel, *Dana's System of Mineralogy*, 7th ed., pp. 654–1016, John Wiley and Sons Inc., New York (1951).

[7] F. S. Noyes, *Min. and Met.* **25** (1944) 495–506.

[8] Ref. 2, p. 37.

States also have phosphate analysis falling within this range. The crystalline apatite of Kola Peninsula, upon concentration by flotation, yields material up to 88% BPL. Apatite, in spite of its high analysis, is not always competitive with phosphorite on an equal basis for conversion to superphosphate fertilizers by treatment with sulfuric acid. Crystalline apatite in its reaction with sulfuric acid requires about three times as long for completion than the porous phosphorite[9]. This factor is not significant when the material is used for the electric furnace production of phosphorus.

The world production of phosphate rock in 1967 is estimated to be 100 million short tons. In the United States the production is about 39 million short tons with Florida accounting for about 78%, the Western States for 14%, and Tennessee and North Carolina the remainder[10].

1.3. PRODUCTION

Methods

The original process for the production of elemental phosphorus, as described by Robert Boyle in 1680, was based on the method of Brandt[1]. It involved the distillation of a large quantity of partly digested urine to the consistency of a thick syrup. Fine white sand was added and the mixture heated in a retort first gently to remove the volatiles, and then very intensely to produce phosphorus which distilled over and collected under water. When bone ash was used as phosphatic raw material, the process consisted of treatment with sulfuric acid to produce phosphoric acid which was then concentrated and heated with coke in a retort to produce phosphorus.

Production of phosphorus by heating a mixture of silica, coke and phosphate rock was first proposed by Auberton and Boblique in 1867 and the use of the electric furnace for heating such a mixture was proposed by J. B. Readman in 1890[11]. The basic method for producing elemental phosphorus today, except for engineering improvements, is essentially that of the method originated by Readman. Lower grade phosphate sand contaminated with clay is concentrated by washing to an average P_2O_5 content of 28 to 30%. Higher grade sand of 26–28% P_2O_5 content is used directly in combination with washed sand. These fine phosphatic grains are compacted or "nodulized," and then sintered into fused agglomerates having an average diameter of $\frac{1}{2}$ to $1\frac{1}{2}$ in. The nodules, which are mechanically very strong and resist abrasion even at high temperature, are then mixed with silica and coke particles of similar size. Such a mixture is called "furnace burden". A typical furnace burden has a SiO_2/CaO ratio of 0.8 to 1.2 and a P_2O_5/C ratio of 2.3 to 2.6.

The electric furnace for the reaction is lined on the bottom and sides with thick carbon blocks while the top is lined with refractory bricks. In a 65–70 MW furnace the three carbon electrodes which supply power for the reaction weigh about 25 tons each. They go through an almost air-tight seal in the furnace roof into the reaction zone and are supported in a manner which enables them to move vertically automatically depending on the power requirements of the fluctuating furnace conditions. The temperature in the reaction zone is 1400–1500°. The phosphorus produced vaporizes and rises with carbon monoxide and entrained dust through the space between the furnace burden

[9] Ref. 2, p. 122.

[10] R. W. Lewis, *U.S. Dept. of Interior, Bureau of Mines, Mineral Industry Survey*, Dec. 1967.

[11] Ref. 2, p. 132.

particles, and the mixture then passes next through a Cottrell electrostatic precipitator where most of the dust is removed. The phosphorus vapor is then cooled, condensed, and collected under water. The CO gas is recovered for use as fuel in the sintering operation. By-product calcium silicate is drawn off from the bottom of the furnace as a molten liquid. The iron phosphide or "ferrophos" formed from the iron impurities present in the phosphate ore is also drawn off as a melt.

Many proposals have been made for modifying the phosphorus production process. They have varied from the use of intimately mixed powdered carbon, phosphate rock and silica for promoting easier reaction to the use of a fluid bed as the reactor[12–15]. All of the proposals are chemically feasible, but thus far none has been able to overcome the consequent engineering problems well enough to replace the electric furnace process.

Chemistry

The mechanism of phosphate reduction is quite complex and there is no complete agreement among investigators on the exact path of each step in the reaction sequence. The overall reaction is generally represented by the following simplified equation:

$$Ca_3(PO_4)_2 + 3SiO_2 + 5C \longrightarrow 3CaSiO_3 + 5CO + P_2$$

Some proposed mechanisms for the reaction are listed below:

Phosphide mechanism[16]:

$$5Ca_3(PO_4)_2 + 40C \longrightarrow 5Ca_3P_2 + 40CO$$
$$3Ca_3(PO_4)_2 + 5Ca_3P_2 \longrightarrow 24CaO + 8P_2$$

Acid displacement mechanism:

$$2Ca_3(PO_4)_2 + 6SiO_2 \longrightarrow 6CaSiO_3 + P_4O_{10}$$
$$P_4O_{10} + 10C \longrightarrow 2P_2 + 10CO$$

CO reduction mechanism:

$$Ca_3(PO_4)_2 + 5CO \longrightarrow 3CaO + 5CO_2 + P_2$$
$$5CO_2 + 5C \longrightarrow 10CO$$

The phosphide theory is generally regarded for thermodynamic reasons as unlikely. A recent study by Dorn[17] on the equilibrium of the system CaO, $Ca_4P_2O_9$, $Ca_3P_2O_8$, Ca_3P_2, CaC_2, CO and P_2, showed that calcium phosphate is reduced by carbon directly to Ca_3P_2 only at a P_2 pressure of greater than 10^4 atm. At lower pressures only CaO is formed.

The acid displacement mechanism has considerable experimental support. According to this mechanism, fused silica behaves as an acid at high temperatures and it displaces P_4O_{10} from $Ca_3(PO_4)_2$. Molten $CaSiO_3$ [18] is formed as a by-product. The expelled P_4O_{10} is reduced by carbon to give elemental phosphorus along with carbon monoxide. These combined reactions are achieved in the temperature range 1250–1500°. In the absence

12 S. Tour and L. Burgess, U.S. 2,907,637, Oct. 6, 1959.
13 W. C. Schreiner and D. E. Loudon, U.S. 3,026,181, Mar. 20, 1962, to M. W. Kellogg Co.
14 W. J. Darby, U.S. 3,335,094, Aug. 8, 1967, to Tennessee Valley Authority.
15 I. I. Ablichenkov and M. V. Miniks, *Khim. Prom.* **42** (1) (1966) 37–40.
16 H. H. Franck and H. Füldner, *Z. anorg. u. allgem. Chem.* **204** (1932) 97–139.
17 F. Dorn, Knapsack-Griesheim A-G. Knapsack, Germany, private comm.
18 R. B. Burt and J. C. Barber, *Production of Elemental Phosphorus by the Electric-Furnace Method*, Chemical Engineering Report No. 3, pp. 3–4, Tennessee Valley Authority, Wilson Dam, Alabama (1952).

of silica, thermal decomposition of $Ca_3(PO_4)_2$ to P_4O_{10} and CaO occurs at above 1800°. Ross, Mehring and Jones[19] found that when pure tricalcium phosphate mixed with one-fifth of its weight of carbon is heated in a reducing atmosphere, 96% of the phosphorus content is volatilized at 1400° and 100% at 1550°.

The decomposition reaction is shown as follows:

$$Ca_3(PO_4)_2+5C \longrightarrow 3CaO+P_2+5CO$$

requires 418,900 cal with a decomposition temperature of 1385°. For the same reaction in the presence of silica, as shown in the following equation, the decomposition temperature is reduced to 1190° and the heat requirement is 365,350 cal.

$$Ca_3(PO_4)_2+3SiO_2+5C \longrightarrow 3CaSiO_3+P_2+5CO$$

The formation of calcium silicate flux reduces the fusion temperature of the reaction mixture and changes the environment of the reaction. The molten slag also serves as a fluid medium in which the overall reaction proceeds more readily than is possible with the solid system.

In proposing the CO reduction mechanism, Dorn[17] questioned the validity of the acid displacement theory. From published thermochemical data, he calculated that the P_4O_{10} pressure over $Ca_3P_2O_8+3SiO_2$ melt is only 10^{-18} atm. His own experiments have shown that in the reaction of calcium phosphate with carbon in the presence of CO, the tricalcium phosphate is reduced by the CO. The P_2 formed volatilizes off with the by-product CO_2, leaving behind solid CaO. The CO_2 is subsequently reduced by C to CO to continue the reaction. In this process CO gas diffuses into the phosphate particle and the P_2 and CO_2 diffuse out of the particle. One support for the theory is that the reduction also goes, though at a reduced rate, even when the carbon is not in direct contact with the phosphate particle. The heat requirement for the described overall reaction was calculated as 398,000 cal/mole P_2. The observations made by Ross, Mehring and Jones[19] on the direct reduction of phosphate by carbon at 1400–1500° may be interpreted as support for the CO reduction mechanism.

Dorn has also studied the reduction of the $CaO–P_2O_5–SiO_2$ melt system. It was shown that a graphite rod partially immersed in the melt has a higher consumption just *above* the melt surface rather than beneath the melt surface. This observation supports the fact that the reduction of calcium phosphate/calcium silicate melt occurs predominately at the melt–vapor boundary via the CO reduction mechanism.

Ershov[20] believes that calcium phosphate is not reduced in an electric furnace until melted. This is based on the analysis of samples of furnace burden taken at different heights from a furnace during overhaul. He found that the P_2O_5 content of the phosphorite decreased from that of the original phosphorite only after it was melted. From his data, he proposed the following mechanism: At the high temperature zone, the phosphorite begins to melt and dissolves some SiO_2. The liquid phase flows down and wets and dissolves solid lumps of phosphorite. The dissolved calcium phosphates then react with the coke which decreases the P_2O_5 content of the melt. The phosphorus produced is vaporized.

In an industrial electric furnace, the yield of elemental phosphorus based on the phosphorus content in the burden is approximately 88–92%. Some of the phosphorus is

[19] W. R. Ross, A. L. Mehring and R. N. Jones, *Ind. Eng. Chem.* **16** (1924) 563–66.
[20] V. A. Ershov, *Khim. Prom.* **42** (4) (1966) 283–4.

lost in the calcium silica slag. More of the phosphorus is lost as "ferrophos". Power consumption averages 13,200 kWh or 11,357,557 kcal per ton of phosphorus.

1.4. ALLOTROPES OF PHOSPHORUS AND P_2 VAPOR

White Phosphorus

The most common form of phosphorus is white phosphorus, a solid obtained by the condensation of phosphorus vapor under water. Impurities such as arsenic and hydrocarbons are usually present. For industrial purposes, many methods of purification are used. One method proposed involves the extraction of molten phosphorous with polyphosphoric acid followed by decantation and washing with water. After filtering through a column of activated charcoal and Fuller's earth, the final product contains only 0.005% inorganic and 0.009% organic impurities[21]. Phosphorus may also be purified by redistillation in the presence of an absorbent such as activated carbon[22]. Purified phosphorus is white or colorless depending on the size of the crystals, and has a waxy appearance not unlike that of paraffin. Since it ignites in air with the self-ignition temperature of 34°, it is usually transported under water. Ordinary white phosphorus is the *a* or high temperature variety with a specific gravity of 1.8232. It is insoluble in water, but very soluble in CS_2 (89.8 g/100 g sat. soln. at 10°)[23], PCl_3, $POCl_3$, liquid SO_2, and liquid NH_3, slightly soluble in alcohol, ether, benzene, xylene, methyl iodide, glycerine, and acetic acid. It crystallizes from CS_2 in rhombic dodecahedral[24]. X-ray diffraction shows that it is cubic with large unit cells containing 56 molecules of P_4 and a lattice constant of 7.17 Å. It is a non-conductor. It has a melting point of 44.1° and a boiling point of 287°.

Beta-phosphorus is obtained by the conversion of the α form at $-76.9°$ under atmospheric pressure; or at 64.4° under 11,600 atm[25]. It consists of hexagonal crystals with a specific gravity of 1.88. Its other properties are similar to those of the α form. The heat of transformation of the α to the β forms is equal to -3.8 ± 0.2 kcal/mole P_4 at $-76.9°$ [26].

P_2 is formed to an appreciable amount when P_4 is heated to 800°. The P_2 in turn dissociates to P and the equilibrium of this latter dissociation has been studied in the vapor phase over UP–UO_2. The dissociation energy of P_2 is calculated as to have a most probable experimental value of 114 ± 6 kcal/mole[27]. The heat capacity of P_2 vapor at 25° is 7.63 cal/°C mole; that for P_4 vapor is 16.0 cal/°C mole[28].

Red Phosphorus

Commercial red phosphorus is prepared by heating white phosphorus at about 400° for several hours. Iodine, sulfur or sodium may be used as a catalyst. The residual white phosphorus is removed by wet grinding and boiling with sodium carbonate solution. Red phosphorus under normal temperature and humidity conditions reacts very slowly with

21 J. Cremer, H. Kribbe and F. Rodis, Ger. 1,143,794, Feb. 21, 1963, to Knapsack-Griesheim A/G.

22 N. D. Talanov, A. D. Mikhailin and A. M. Ezhova, U.S.S.R. 167,835, Feb. 5, 1965.

23 W. F. Linke, *Seidell's Solubilities Inorganic and Metal Organic Compounds*, 4th ed., Vol. II, p. 1243, Am. Chem. Soc., Washington, D.C. (1965).

24 Ref. 2, p. 5.

25 P. W. Bridgman, *J. Am. Chem. Soc.* **36** (1914) 1344.

26 H. J. Rodewald, *Helv. Chim. Acta,* **43** (1960) 878–85.

27 K. A. Gingerich, *J. Chem. Phys.* **44**(4) (1966) 1717–18.

28 F. D. Rossini, D. D. Wagman, W. H. Evans, S. Levine and I. Jaffe, *Selected Values of Thermodynamic Properties,* Nat. Bur. Standards (U.S.) Cir. 500, **72** (1952) 570.

water vapor and O_2 in air. This reaction, catalyzed by metal impurities such as Fe and Cu, results in the formation of phosphine and oxyacids of phosphorus and may be stabilized with 1% by weight of magnesium oxide or sodium aluminate.

Many forms of red phosphorus have been reported[29]. Pauling and Simonetta[30] have suggested that the structure of red phosphorus may be the result of cleavage of one of the P_4 tetrahedron bonds followed by polymerization into the following molecular chain:

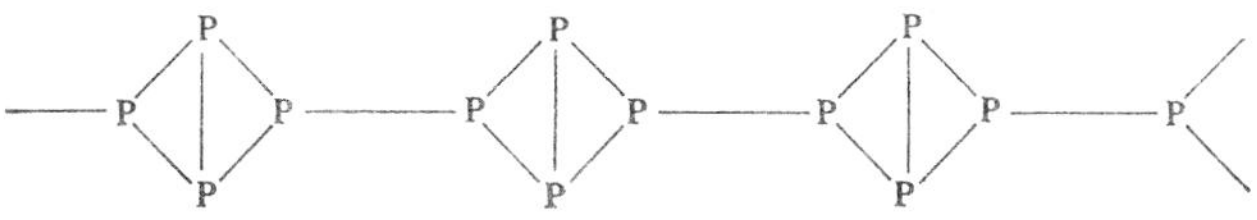

The conditions of formation, chain structures of variable length and different terminal groups (such as the catalyst used, or O and OH groups from stray oxygen or water and other impurities) probably account for some of the forms reported in the literature[31].

Red phosphorus as produced by commercial processes is almost entirely amorphous (average density = 2.16 g/cm^3). Amorphous red phosphorus may also be produced by ultraviolet radiation of white phosphorus. The latter reaction is responsible for the nodules of red phosphorus seen on the surface of solid sticks of white phosphorus after long storage and exposure to light or the precipitation of red phosphorus from a solution of phosphorus in PCl_3. Photochemical production of red phosphorus is believed to be through the dissociation of P_4 to P_2 molecules which recombine to form polymeric red phosphorus.

Rubenstein and Ryan[32] have studied the preparation of red phosphorus from 300–610° X-ray diffraction data, diffuse reflection spectra, and fluorescence emission data showed only two allotropic forms of red phosphorus. Phosphorus when heated at 420° and lower is deep red and often metallic, while heating higher than 436° gives forms varying in color from deep maroon to orange-red. The two allotropic forms observed were prepared by heating at 570° for 16 hr and 170 hr, respectively. They correspond most to forms IV and V of Roth, DeWitt, and Smith[29].

The most well-characterized red or violet phosphorus is the so-called Hittorf's phosphorus[33]. It was prepared originally by heating phosphorus in contact with lead at a temperature range of 500–600°. The phosphorus which dissolved in the lead while hot separated out on cooling as small dark reddish-violet crystals. Various literature references report Hittorf's phosphorus as form V of red phosphorus and with a crystal structure which is rhombohedral or triclinic and density of 2.34 g/cm^3. Thurn and Krebs[34] reported that Hittorf's phosphorus crystallizes in monoclinic form with space group $P2/c$ ($a = 9.21$, $b = 9.15$, $c = 22.59$ Å, and $\beta = 106.1°$); $Z = 84$. The structure is based on two units: a P_8 group and a P_9 group. The units are linked, each through two further P atoms, to form infinite, tube-like structures of pentagonal cross-sections. The tubes are packed parallel to one another to form layers. A double layer consists of two interlocking systems held together by van der Waals' forces.

The enthalpies of transition at 25° were reported by O'Hare and Hubbard[35] for the following processes:

[29] W. L. Roth, T. W. DeWitt and A. J. Smith, *J. Am. Chem. Soc.* **69** (1947) 2881.
[30] L. Pauling and M. J. Simonetta, *Chem. Phys.* **20** (1952) 29.
[31] M. Ya. Kraft and V. P. Parina, *Dokl. Akad. Nauk SSSR*, **77** (1951) 57.
[32] M. Rubenstein and F. M. Ryan, *J. Electrochem. Soc.* **113**(10) (1966) 1063–7.
[33] M. Hittorf, *Phil. Mag.* (4) **31** (1865) 311.
[34] H. Thurn and H. Krebs, *Angew. Chem. Intern. Ed.* **5**(12) (1966) 1047–8.
[35] P. A. G. O'Hare and Ward N. Hubbard, *Trans. Faraday Soc.* **62**(10) (1966) 2709–15.

$$\text{P}(\alpha\text{-white}) \longrightarrow \text{P(amorphous red)}, \quad \Delta H_t = -3.9 \pm 0.4 \text{ kcal/g-atom};$$
$$\text{P}(\alpha\text{-white}) \longrightarrow \text{P(red V)}, \quad \Delta H_t = -3.9 \pm 1.5 \text{ kcal/g-atom};$$
$$\text{P(red IV)} \longrightarrow \text{P(red V)}, \quad \Delta H_t = -1.2 \pm 1.0 \text{ kcal/g-atom}.$$

Black Phosphorus

There are four forms of black phosphorus reported in the literature. The crystalline orthorhombic form was originally prepared by Bridgman[25, 36] by heating white phosphorus to 200° under a pressure of 12,000 atm. It has a density of 2.69 g/cm^3 and melts at about 610°. Under very high pressure, this orthorhombic form is reported to undergo reversible transition to successively denser rhombohedral and cubic forms[37]. The amorphous form[38] with a density of 2.25 g/cm^3 transforms to the orthorhombic form at higher temperature and pressure (see Fig. 1).

In the conversion of white phosphorus to black phosphorus by heat and pressure, soluble impurities present in industrial grade yellow phosphorus inhibit the polymorphous

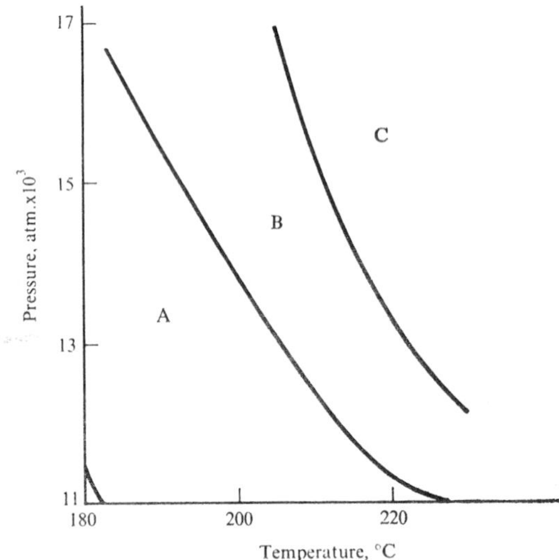

FIG. 1. Regions of temperature and pressure corresponding to the formation of black phosphorus. A: formation of amorphous black phosphorus. B: formation of amorphous black phosphorus followed by transition to crystalline variety. C: immediate formation of crystalline black phosphorus[38]. (Reprinted by permission of John Wiley & Sons Inc.)

transformation. Studies with purified white phosphorus showed that the m.p. of black phosphorus rises smoothly with increase in pressure, and is about 1000° at 18,000 kg/cm^2 [39]. No new modification is formed up to 20,000 kg/cm^2 and 1000° [40].

[36] P. W. Bridgman, *J. Am. Chem. Soc* **38** (1916) 609.

[37] E. C. Corbridge, *Topics in Phosphorus Chemistry*, **3** (1966) 57–394.

[38] R. B. Jacobs, *J. Chem. Phys.* **5** (1937) 945.

[39] V. P. Butuzov, S. S. Boksha, and M. G. Gonikberg, *Dokl. Akad. Nauk SSSR*, **108** (1956) 837–40.

[40] V. P. Butuzov, S. S. Boksha, *Rost Kristallov, Akad. Nauk SSSR, Inst. Krist., Doklady Soveshchaniya* (1956), 311–19 (Pub. 1957); *Growth of Crystals*, Repts. 1st Conf. Moscow (1956), 245–51 (Pub. 1958)/Eng. translation.

Patz[41, 42] reported that the transformation of colorless phosphorus into black modification at a pressure of 13,868 kg/cm^2 starts at 201°, 20,802 kg/cm^2 at 183°, 27,736 kg/cm^2 at 164°, and 34,670 kg/cm^2 at 145°. At a pressure of 41,600 kg/cm^2 only a few minutes were necessary.

Orthorhombic black phosphorus may be made under atmospheric pressure using mercury as catalyst. Purified white phosphorus is distilled into an ampule and heated in the presence of mercury and seed crystals of black phosphorus: 3 days at 280° increase over 1 day to 380°, and hold at 380° for 3 days. The mercury is removed by amalgamation with lead[43].

The black modification of phosphorus can also be prepared in the highly crystalline orthorhombic form from a solution of phosphorus in liquid bismuth[44]. The crystals obtained after dissolving the bismuth matrix with 1 : 1 nitric acid are generally needle

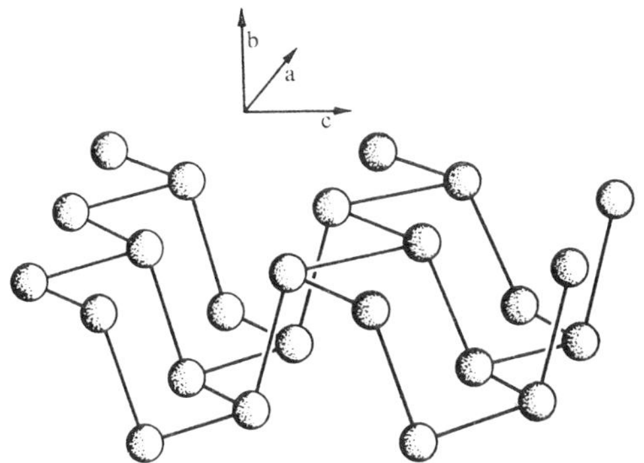

FIG. 2. A part of an infinite puckered layer of phosphorus atoms parallel with the (010) plane in the structure of black phosphorus[44].

shape with the needle axis coincident with the crystallographic a-axis. A complete single-crystal study by Brown and Rundqvist verified the earlier work of Hultgren *et al.*[45]. The cell dimensions are $a = 3.3136 \pm 0.0005$, $b = 10.478 \pm 0.001$ and $c = 4.3763 \pm 0.0005$ Å. Space group is *Cmca*; $Z = 8$. The main atomic arrangement of orthorhombic black

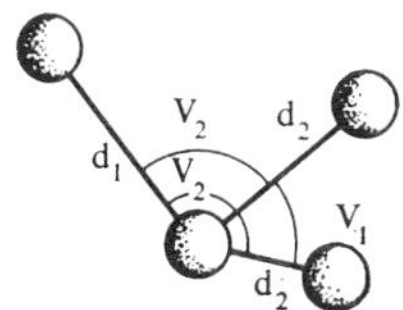

FIG. 3. Representation of bond distances and bond angles in the structure of black phosphorus: $d_1 = 2.244$ Å, $d_2 = 2.224$ Å, $v_1 = 96.34°$, $v_2 = 102.09°$ [44].

phosphorus is composed of puckered layers parallel with the 010 plane (Fig. 2). The phosphorus atoms have close neighbors within the same layer with the bond distances and angles denoted in Fig. 3[44].

[41] K. Pätz, *Abhandl. deut Akad. Wiss. Berlin, Kl. Chem., Biol. u. Geol.* (1955) No. 7, 93–9 (Pub. 1957).

[42] K. Pätz, *Z. anorg. u. allgem. Chem.* **299** (1959) 297–30.

[43] H. Krebs, *Inorganic Synthesis*, Vol. VII, p. 60, ed. J. Kleinberg, McGraw-Hill Book Co., New York (1963).

[44] A. Brown and S. Rundqvist, *Acta Cryst.* **19**(4) (1965) 684–5.

[45] R. Hultgren, N. S. Gingrich and B. E. Warren, *J. Chem. Phys.* **3** (1935) 351.

In the lattice of the orthorhombic black phosphorus, up to 74% of the atoms can be substituted by arsenic[46]. The material exhibits a flakiness somewhat similar to that of mica and graphite. It is a semiconductor of electricity with an electrical resistivity of 0.711 ohm cm^{-3} at 0°[47]. From 300 to 700°K, resistivity can be fitted by the expression $\rho_{av} = 4.6 \times 10^{-3} \exp (0.35/2kT)$ ohm-cm[48]. At high temperature, it is an intrinsic semiconductor with a gap width of 0.33 eV[46].

Black orthorhombic phosphorus is the least reactive form of the elemental phosphorus;

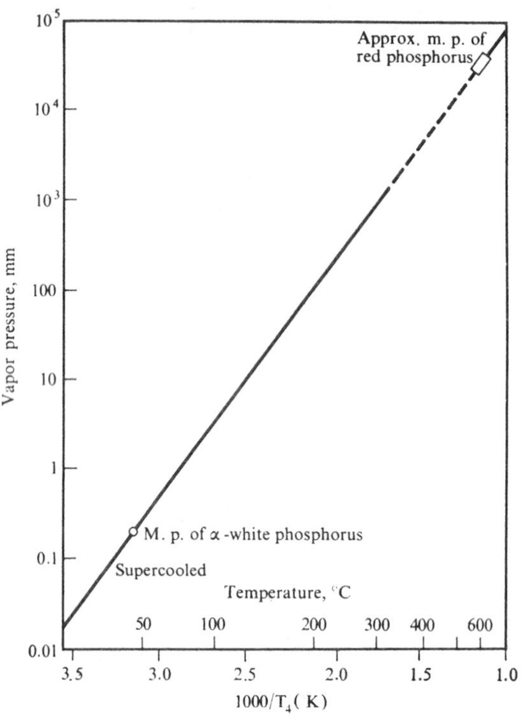

FIG. 4. Vapor pressure of liquid phosphorus. Liquid does not exist in the region represented by the broken portion of the line due to rapid red phosphorus formation[52]. (Reprinted by permission of John Wiley & Sons Inc.)

completely stable in air. It can be ignited with a match only with difficulty. It sublimes in a nitrogen atmosphere at 490° [25]. The heat of sublimation is 38.7±0.2 kcal/mole P_4 at 25° [26].

By means of the curves C_p vs. T and C_p vs. $\ln T$, respectively, the enthalpy change of crystalline orthorhombic black phosphorus between 0 and 298.15°K, $H = 882.6 \pm 1.5$ cal/mole, and the absolute entropy at 298.15°K, $S = 5.457 \pm 0.010$ cal/deg/mole were calculated. At 20–37°K, the heat capacity of phosphorus is proportional to $T^{2.3}$ and at 13–20°K to $T^{2.7}$, where T is the absolute temperature[49].

The rhombohedral form of black phosphorus is detected by high-pressure X-ray

[46] H. Krebs, W. Holz and K. H. Worms, *Chem. Ber.* **90** (1957) 1031–7.

[47] P. W. Bridgman, *Proc. Nat. Acad. Sci.* **21** (1935) 109.

[48] D.Warschauer, *J. Appl. Phys.* **34**(7) (1963) 1853–60.

[49] I. E. Paukov, P. G. Strelkov, V. V. Nogteva and V. I. Belyi, *Dokl. Akad. Nauk SSSR*, **162**(3) (1965) 543–5.

diffraction technique when orthorhombic black phosphorus is subjected to a pressure of 83–102 kb. It is of the A7 (arsenic) type of structure with $a = 3.377$ Å, $c = 8.806$ Å.

The cubic form of black phosphorus, $a = 2.377$ Å, shows up in the X-ray pattern when the pressure is increased to around 124 kb. The transition of the cubic and the rhombohedral forms is reversible at 111 ± 9 kb. When the pressure is returned to atmospheric pressure, only the orthorhombic form is shown[50].

Liquid Phosphorus

The same liquid is obtained no matter whether white, red or black phosphorus is melted or the vapor is condensed[51]. On rapid cooling, white phosphorus is obtained. The vapor pressure curve of liquid phosphorus is shown in Fig. 4[51]. X-ray diffraction studies[53] of liquid phosphorus indicate that the phosphorus atoms are presented in the form of symmetrical P_4 tetrahedral with a P–P distance of 2.25 Å. The density (d = g/cm^3) and viscosity (η in centipoises) of pure liquid phosphorus are given by the following equation[49].

$$d = 1.7862 - (9.195 \times 10^{-4})t \text{ for } 280° > t > 10°$$

$$\log \eta = -1.3879 = 514.4/(273.2 + t) \text{ for } 140° > t > 20°$$

Water dissolves to the extent of 3.6 mg H_2O/g P_4 in the temperature range of 25–45°: mercury dissolves 0.29 mg per g P_4 at 25°[52].

The physical properties and physical constants of α-white phosphorus are summarized in Table 1.

TABLE 1. PHYSICAL PROPERTIES AND PHYSICAL CONSTANTS OF α-WHITE PHOSPHORUS[51, 54–57]

Atomic weight	30.9738
Crystal structure	Cubic system lattice constant 7.17 Å
Molecular unit	P_4
Dielectric constant at 20° and 1 atm	4.1 cgse
Electrical resistivity	1×10^{11} ohm-cm at 11°
Density	1.828 g/cm^3
Atomic volume of solid	16.9 cm^3
Melting point	44.1°
Heat of fusion	$\Delta H_{317 \cdot 26} = 600 \pm 3$ cal/mole P_4
Boiling point	280.5°
Critical temperature	695°
Critical pressure	82.2 atm
Surface tension (liquid)	43.09 dynes/cm at 78.3° 35.56 dynes/cm at 132.1°
Index of refraction	1.8244 for D line at 29.2°

50 J. C. Jamieson, *Science*, **139** (1963) 1291.

51 T. D. Farr, *Phosphorus, Properties of the Element and Some of Its Compounds*, Chemical Engineering Report No. 8, pp. 2–17, TVA, Wilson Dam, Alabama (1950).

52 J. R. van Wazer, *Phosphorus and Its Compounds*, Vol. 1, pp. 101–23, Interscience Publishers Inc., New York (1958).

53 C. D. Thomas and N. S. Gingrich, *J. Chem. Phys.* **6** (1938) 659–65.

54 From Atomic Weight Commission of IUPAC in 1961, *Pure and Applied Chem.* **5** (1962) 255–92.

55 T. Moeller, *Inorganic Chemistry*, p. 557, John Wiley and Sons Inc., New York (1952).

56 F. Albert Cotton and G. Wilkinson, *Advanced Inorganic Chemistry*, pp. 372–3, Interscience Publishers Inc., New York (1962).

57 D. M. Yost and H. Russell, Jr., *Systematic Inorganic Chemistry*, pp. 159–60, Prentice-Hall Inc., New York (1944).

TABLE 1—*Continued*

Heat capacity	$C_p = 13.615 + 2.872 \times 10^{-2}T$ cal/mole P_4/deg where T is 273.16° to 317.26°K C_p at 25° = 22.18 cal/mole/deg C_p at 44.1° = 22.73 cal/mole/deg
Sublimation pressure	T, °C 20 25 30 35 40 P_{mm} 0.025 0.043 0.072 0.089 0.122
Thermodynamics of sublimation	$P_4(\alpha) \rightleftharpoons P_4(g)$ $\Delta C_p^\circ = 5.515 - 28.210 \times 10^{-3}T - 2.98 \times 10^5 T^{-2}$ $\Delta H^\circ = 12{,}080 + 5.515T - 14.105 \times 10^{-3}T^2 + 2.98 \times 10^5 T^{-1}$ $\Delta F^\circ = 12{,}080 + 4.396T - 12.699T \log T + 14.105 \times 10^{-3}T^2 + 1.49 \times 10^5 T^{-1}$ $\Delta S^\circ = 1.119 + 12.699 \log T - 28.210 \times 10^{-3}T + 1.49 \times 10^{-5}T^{-2}$
Reference thermodynamic constants (298.16°K) for sublimation of white phosphorus are:	
	$\Delta H^\circ = 13{,}470$ cal/mole $\Delta F^\circ = 5{,}784 \pm 8$ cal/mole $\Delta S^\circ = 25.78 \pm 0.03$ cal/mole/deg
Ionization potentials	
3rd	30.15 eV
5th	65.0 eV
Electronegativity	2.19
Radii	
ionic	1.85 (P^{3-}) Å
covalent (for trivalent state)	1.10 Å
Bond energy[58]	
P–P	18.9 kcal/mole
Heat of combustion[59]:	
White phosphorus	710.2±1.0 kcal/mole P_4
Amorphous red phosphorus	703.2±0.5 kcal/mole P_4
Crystalline red phosphorus (form IV)	697.7±0.4 kcal/mole P_4
Enthalpies of transition[59]:	
White to amorphous red phosphorus	−7.0±1.1 kcal/mole P_4
White to crystalline red phosphorus (form IV)	−12.5±1.1 kcal/mole P_4
Mean vaporization coefficient[60] (−20 to +20°)	0.57

Levine and Stull[61] had calculated the thermodynamic functions of gaseous P_4 between 0° and 1000°K with the use of known spectroscopic data. The values for 298.15°K were $S^\circ = 66.893$, $C_p = 16.051$ and $-(G^\circ - H^\circ_{298.15})/T = 66.893$ cal/deg/mole.

1.5. CHEMISTRY AND CHEMICAL PROPERTIES

Atomic Structure

The electronic structure of phosphorus consists of 15 electrons in the ground state distributed as $1s^2\ 2s^2\ 2p^6\ 3s^2\ 3p_x\ 3p_y\ 3p_z$. The atomic energy levels for phosphorus are shown as follows[62]:

[58] Ref. 55, p. 569.
[59] W. S. Holmes, *Trans. Faraday Soc.* **58** (1962) 1916–25.
[60] H. Ramthun and I. N. Stranski, *Z. Elektrochem.* **61** (1957) 819–26.
[61] S. Levine and D. R. Stull, *Ill. State Acad. Science* **56** (1963) 88–9.
[62] R. F. Hudson, *Structure and Mechanism of Organo-Phosphorus Chemistry*, pp. 1–3, Academic Press, London (1965).

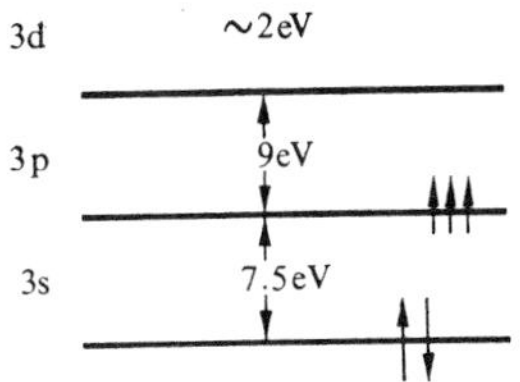

The gaseous P_4 molecule shows four phosphorus atoms arranged in a regular tetrahedron with a P–P distance of 2.21 ± 0.02 Å and a moment of inertia about any of its three major axes of 2.51×10^{-40} g/cm^2 [63]. The nuclear spin of $_{15}P^{31}$ is $\frac{1}{2}(h/2\pi)$, with zero electric quadrupole moment, and therefore suited for interpretation of the electronic structure of its molecules or molecule ions by nuclear magnetic resonance data. In a 10 kG field $_{15}P^{31}$ resonates at 17.24 Mc/s[64].

Pauling and Simonetta calculated that the bonding atomic orbitals in P_4 are 98% 3*p*, with only 2% of 3*s* and 3*d* character. Consequently the bonds in P_4 are bent. The P_4 molecule has P–P–P bond angles of 60°, and since an unstrained P–P–P bond angle should

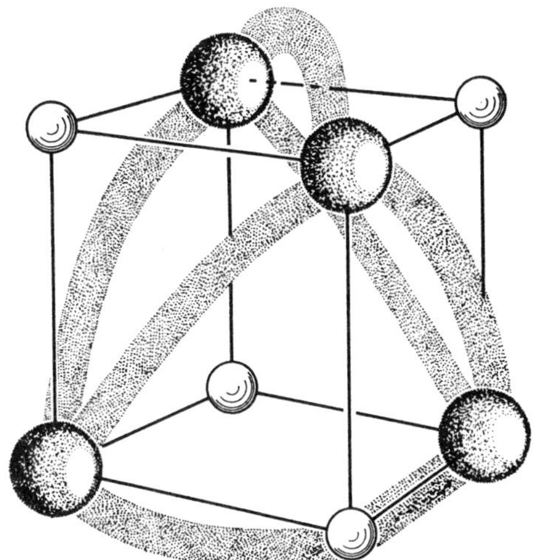

FIG. 5. A sketch demonstrating the manner in which the valence-shell electron density is distributed in three dimensions[66].

be 90° this represents a strain energy of 22.8 kcal/mole of P_4, and accounts for its high reactivity.

Hart, Robin and Kuebler[66], though their study of the electronic spectrum of P_4, have calculated that strong pi bonds are present in the molecule and that their presence leads to a resonance which vitiates the concept of "bend bonds" and "strain energy" in P_4. Based on their calculations, an electron density map shows that regions of highest density are situated symmetrically above each of the triangular faces of the tetrahedral P_4 molecules

63 L. R. Maxwell, S. B. Hendricks and V. M. Mosley, *J. Chem. Phys.* **3** (1925) 699.
64 Ref. 52, pp. 3–4.
65 L. Pauling and M. Simonetta, *J. Chem. Phys.* **20** (1952) 29.
66 R. R. Hart, M. B. Robin and N. A. Kuebler, *J. Chem. Phys.* **42**(10) (1965) 3631–8.

with each of the high density regions connected to the other three by lower density streamers, as shown in Fig. 5.

General Types of Phosphorus Compounds and Their Molecular Structures

A phosphorus atom forms many types of compounds. Some of the general types with the orbitals used in the formation of the bonds and their geometry are shown in Table 2 [56, 62, 67].

TABLE 2

Number of bonds or valency	Orbital or hybrid	Directional properties and geometry		Example
3	p (lone pair present)		pyramidal	PH_3
4	sp^3 (in compounds with PO and PS bond d orbitals of phosphorus may be used in $d\pi-p\pi$ bonding)		tetrahedral	PH_4^+ $OP(OH)_3$, $POCl_3$
5	sp^3d	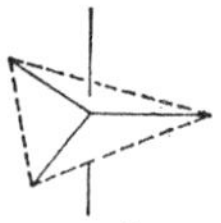	trigonal bipyramid	PCl_5, PF_5
6	sp^3d^2	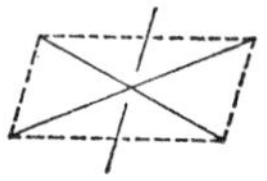	octahedral	PF_6^-

Phosphorus has empty *d* orbitals of fairly low energy, so if the atom to which phosphorus donates electrons has the same symmetry as the empty *d* orbitals in the phosphorus, this usually results in back donation. An over-all multiple bond character may result from this phenomenon.

1.6. BIOLOGICAL PROPERTIES

White phosphorus is very poisonous, with a lethal dose of approximately 0.1 g[68], and the maximum allowable concentration of white phosphorus is 0.1 mg/m³ of air[69]. It is absorbed in the liver and in the blood. The chronic disease contracted by workers in factories where yellow phosphorus is carelessly handled is known as necrosis of the bones, especially of the teeth and jaw. It is known also by the common name of "phossy-jaw" disease. With the advent of strict safety and hygienic control in phosphorus factories

[67] H. R. Allcock, *Heteroatom Ring Systems and Polymers*, p. 21, Academic Press, New York (1967).
[68] Ref. 2, p. 5.
[69] M. B. Jacobs, *The Analytical Toxicology of Industrial Inorganic Poisons*, p. 576, Interscience Publishers Inc., New York (1967).

this disease is no longer common. Since one entry of phosphorus to the body is via decayed teeth, periodic dental check-ups are required for all workers and no open cavities are permitted.

The first aid treatment for white phosphorus taken internally consists of emptying the stomach by induced vomiting or stomach pump, or by drinking a glass of water containing $\frac{1}{4}$ g of cupric sulfate. The cupric sulfate is effective because it coats the phosphorus[70] with a layer of copper phosphide.

Phosphorus must never be permitted to come in contact with skin or clothing and prolonged exposure to phosphorus vapor should be avoided. Phosphorus removal after accidental contact is facilitated by application of a dilute cupric sulfate solution which converts finely divided phosphorus particles to black copper phosphide and coats the larger particles with copper phosphide and renders it non-igniting.

1.7. NUCLEAR PROPERTIES

Phosphorus has only one stable isotope, P^{31}. The data on the six radioactive isotopes of phosphorus are summarized in Table 3[71, 72].

TABLE 3

Isotope	Half-life	Type of decay	Major radiations: Approximate energies (MeV) and intensities	Principal means of production	Decay product
${}_{15}P^{28}$	0.28 sec	β^+, no α	β^+ 11.0 max. γ 0.511 (200%, $\gamma^\pm$), 1.780 (75%), 2.6, 4.44 (10%), 4.9, 6.1, 6.7, 7.0, 7.6 (5%)	$Si^{28}(p, n)$	${}_{14}Si^{28}$
${}_{15}P^{29}$	4.45 sec	β^+	β^+ 3.95 max. γ 0.511 (200%, $\gamma^\pm$), 1.28 (0.8%), 2.43 (0.2%)	Si^{28} (d, n)	${}_{14}Si^{29}$
P^{30}	2.50 min	β^+	β^+ 3.24 max. γ 0.511 (200%, $\gamma^\pm$), 2.23 (0.5%)	Al^{27} (α, n) S^{32} (d, α) Si^{29} (p, γ)	${}_{14}Si^{30}$
P^{32}	14.28 days	β^-	β^- 1.710 max. average β energy: 0.69 colorimetric 0.70 ion ch	P^{31} (n, γ) S^{34} (d, α) S^{32} (n, p)	${}_{16}S^{32}$
P^{33}	24.4 days	β^-	β^- 0.248 max. γ no γ	S^{33} (n, p) Cl^{37} (γ, α)	${}_{16}S^{33}$
P^{34}	12.4 sec	β^-	β^- 5.1 max. γ 2.13 (25%), 4.0 (0.2%)	Cl^{37} (n, α) S^{34} (n, p)	${}_{16}S^{34}$

The isotopes P^{28}, P^{29} and P^{34} have half-lives too short to be of interest in isotopic synthesis. Isotopes P^{32} and P^{33} are useful for labelling purposes[73]. Of the two, P^{32} is

[70] J. R. van Wazer, *Encyclopedia of Chemical Technology*, Ed. R. E. Kirk and D. F. Othmer, Vol. 10, p. 456, The Interscience Encyclopedia, Inc., New York (1953).

[71] D. Strominger, J. M. Hollander and G. T. Seaborg, *Revs. Mod. Phys.* **30** (1958) 585.

[72] M. Lederer, J. M. Hollander and I. Perlman, *Table of Isotopes*, 6th ed., pp. 10–11, John Wiley and Sons Inc., New York (1968).

[73] R. H. Herber, *Inorganic Isotopic Synthesis*, p. 143, W. A. Benjamin Inc., New York (1962).

commercially available. It is produced by nuclear reactor irradiation with either (1) $P^{31}(n, \gamma)P^{32}$ or (2) $S^{32}(n, p)P^{32}$. Reaction (1) is carried out by the action of thermal neutrons on red phosphorus as the target. No further chemical treatment is required if one allows possible interfering activities such as Si^{31}, Al^{28} with short-lived periods to decay. One disadvantage of this process is the relatively low specific activity that can be obtained. The highest specific activity of P^{32} available as red phosphorus is reported to be 550 mc/g = 0.55 μc/μg.

The production of P^{32} by method (2) uses the very high fast-neutron flux near the core of the reactor. It gives an essentially carrier-free product with specific activities as high as 1 to 5 mc/μg. After a short cooling period only S^{35} remains which may then be removed by distillation at 500°. If the sulfur is purified prior to irradiation other purification steps are seldom necessary[73]. The P^{32} residue may be recovered by dissolution in 0.05 N HCl and radiometrically[74] the yield of the separation approaches 99%.

Red phosphorus, P^{32} and $KH_2P^{32}O_4$ are the two unprocessed forms of radioisotopic phosphorus available commercially.

1.8. USES

Most of the phosphorus produced is converted to phosphoric acids and their various salts. Their chemistry and application will be covered in their respective sections.

Elemental white phosphorus because of its ease of self-ignition in air was originally used for starting a fire. In the late 1780s the "ethereal match" consisted of a slip of paper tipped with white phosphorus sealed in a glass tube. When a fire was needed, the glass tube was broken to admit air to oxidize the phosphorus which ignited the paper. Later developments led to the use of wooden splints with a tip of white phosphorus, chalk or diatomaceous earth, glue and sulfur or paraffin. Potassium chlorate was subsequently also introduced into the formulation. When such a match is struck, frictional heat causes phosphorus to ignite by air. Friction also causes the rapid oxidation of phosphorus and organic binder by potassium chlorate. The combined heat of these exothermic reactions rapidly ignites the sulfur which then transmits the flame to the wooden splint. The inert diatomaceous filler serves the purpose of regulating and moderating the combustion reactions.

Due to the high toxicity of white phosphorus, which exacted a fearful toll of lives among the early matchmakers, its use for matches was banned. This use started in Finland in 1872, and was taxed out of existence in the United States in 1913. White phosphorus was replaced by the much less toxic red phosphorus. This change led to the development of the "safety match" as represented by the book matches. In such matches, a head contains approximately 50% potassium chlorate, 10% animal glue binder, 5% sulfur and the remainder being mostly ground glass on a tip of a paper splint previously dipped in paraffin. A separate friction striking surface has as basic ingredients approximately 50% red phosphorus as the igniting agent, 15% glue as the binder, 30% glass powder as friction agent and 5% zinc oxide or calcium carbonate as stabilizer for the red phosphorus[75].

White phosphorus is used extensively as a chemical warfare agent. When oxidized in an excess of air, it forms phosphorus oxide which appears as a dense white smoke with

[74] F. de la Cruz Castilla, C. Suarez and G. Domingues, *Energie Nucl.* (*Madrid*) **6**(22) (1962) 4–15.

[75] J. R. van Wazer, *Phosphorus and Its Compounds*, Vol. II, pp. 1943–4, Interscience Publishers Inc., New York (1961).

TABLE 4. COMPOSITION OF WELL-CHARACTERIZED PHOSPHIDES[78]

Li_3P	Be_3P_2											$B_{13}P_2$ BP	
Na_3P	Mg_3P_2											AlP	
	Ca_3P_2 CaP	ScP	Ti_3P Ti_5P_3 TiP	V_3P VP	Cr_3P CrP	Mn_3P Mn_2P MnP	Fe_3P Fe_2P FeP FeP_2	Co_2P CoP CoP_3	Ni_3P Ni_5P_2 $Ni_{12}P_5$ Ni_2P Ni_5P_4 NiP NiP_2 NiP_3	Cu_3P CuP_2	Zn_3P_2 ZnP_2*	GaP	
		YP	ZrP*	NbP NbP_2	Mo_3P Mo_4P_3 MoP MoP_2		Ru_2P RuP RuP_2	Rh_2P Rh_4P_3 RhP_2 RhP_3	Pd_3P Pd_7P_3 PdP_2 PdP_3	AgP_2	Cd_3P_2 CdP_2 CdP_4	InP	SnP
		LaP	HfP*	TaP TaP_2	W_3P WP WP_2*	Re_2P	OsP_2	Ir_2P IrP_2 IrP_3	Pt_5P_2 PtP_2				

LaP	CeP	PrP	NdP		SmP		GdP	TbP	DyP	HoP	ErP		

$ThP_{0 \cdot 7}$ Th_3P_4		UP U_3P_4 UP_2	Np_3P_4	PuP

* There is more than one phase with or near this composition.

very high obscuring power. It is used also in incendiary bombs and shells as a self-igniting agent. Phosphorus-containing munition shells when exploded have the additional property of raining small particles of burning phosphorus which stick tenaciously to the clothing and skin of the enemy.

The sensitivity of phosphorus to ignition is also utilized for amusement purposes, e.g. in caps for cap pistols. These caps have one section of potassium chlorate and the other of red phosphorus, sulfur and calcium carbonate. When the two sections are struck together in a cap pistol, a small explosion results.

Other uses for elemental phosphorus are as poisons in baits for rodents and as ingredients for incendiaries.

2. PHOSPHIDES

Phosphorus reacts with most of the metallic elements at elevated temperature to form phosphides. Each element forms one or more crystalline phosphides of a specific metal-to-phosphorus ratio. Some ratios correspond to the classical valency rule and some do not. Literature references on phosphides are rather numerous. With the introduction of modern phase-analytical methods, many members of this class of compounds are well characterized. The properties of some metallic phosphides were summarized by Farr[76] in 1950. Van Wazer[77] reviewed the subject in 1958 and more recently (1965) Aronsson, Lundström and Rundqvist[78] critically reviewed the preparation, properties and crystal chemistry of various phosphides. The structural chemistry of the phosphides was reviewed by Corbridge[79]. In Table 4 are listed the compositions of well-characterized phosphides compiled by Aronsson, Lundström and Rundqvist[78]. Readers are referred to the above references for detailed information and additional references.

2.1. METHODS OF PREPARATION

1. By the direct reaction of the elements heated in vacuum or in an inert atmosphere: For elements which form several stable phosphides of different metal-to-phosphorus ratio, the correct ratio of the elements must be used. For elements which form only one phosphide, an excess of phosphorus, either red or white, may be used and the excess phosphorus then removed by distillation. Other sources of phosphorus are the phosphorus-rich phosphides which when heated decompose to the lower phosphides with the liberation of phosphorus.

2. By the reduction of metal phosphate with carbon at high temperature: An example of this method is the preparation of calcium phosphide, Ca_3P_2, by the reduction of calcium phosphate with carbon.

3. By the heating of the metal or metal halide, or metal sulfide, with phosphorus

[76] T. D. Farr, *Phosphorus, Properties of the Element and Some of Its Compounds*, Chemical Engineering Report No. 8, pp. 69–74, Tennessee Valley Authority, Wilson Dam, Alabama (1950).

[77] J. R. van Wazer, *Phosphorus and Its Compounds*, Vol. I, pp. 123–75, Interscience Publishers Inc., New York (1958).

[78] B. Aronsson, T. Lundström and S. Rundqvist, *Borides, Silicides and Phosphides*, John Wiley & Sons Inc., New York (1965).

[79] D. E. C. Corbridge, *Topics in Phosphorus Chemistry*, 3 (1966) 57–394.

halide or phosphorus hydride: This method may be illustrated by the preparation of boron phosphide by the heating of B_2S_3 with PH_3 at 1200° to 1400°.

4. By the high temperature electrolysis of molten salt baths containing the metal oxide and an alkali phosphate: This was reported to be the only successful method for the preparation of W_3P.

2.2. CLASSES, PROPERTIES AND CRYSTAL STRUCTURES

Phosphides may be divided into two general classes:

(1) Those which are reactive and easily undergo hydrolysis.

(2) Those which are metallic in nature and do not undergo hydrolysis easily.

The alkali and alkaline earth phosphides, as well as the phosphides of the lanthanides

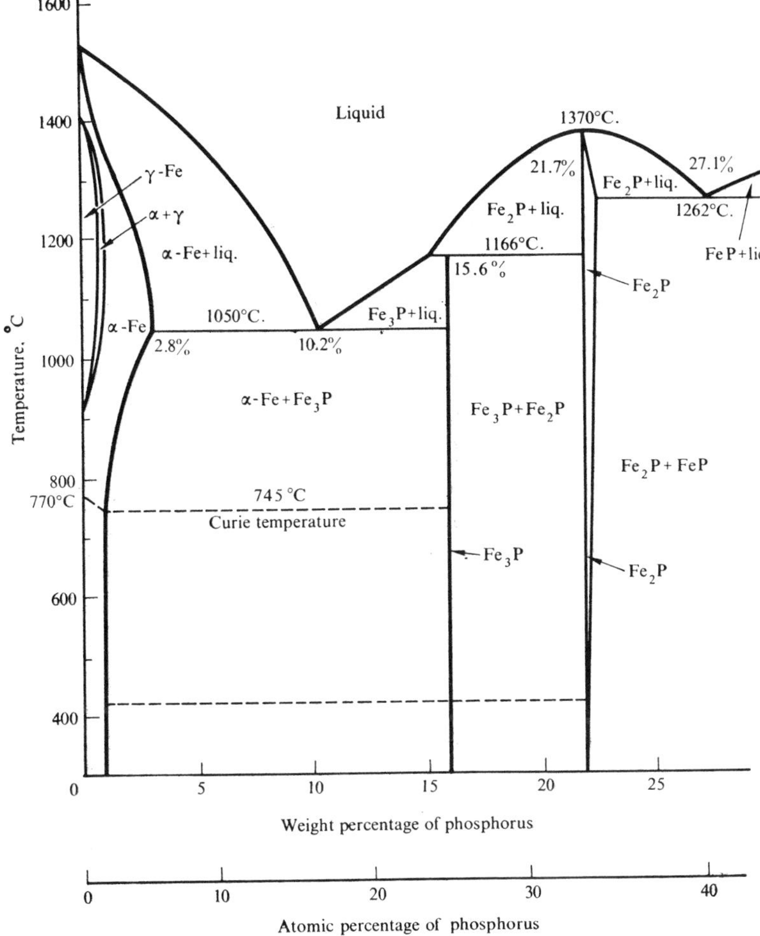

FIG. 6. Phase diagram of the iron–phosphorus system[77]. (Reprinted by permission of John Wiley & Sons Inc.)

TABLE 5

Phosphides	References	Color, crystal system, structure type	Molecular data dimension in Å		Methods of preparation	Physical and chemical properties
Li_3P	a, b, c	Reddish-brown hexagonal	$a = 4.273$ $c = 7.595$	Li–Li = 2.53 Li–P = 2.64 P–P = 4.26	Li+P	Density at 25° = 1.43 g/cm³. Reacts with water to give phosphine
LiP	e	Black needles with metallic luster	$a = 5.55$ $b = 4.98$ $c = 10.19$	$\beta = 117.1°$	Li_3P+red P	
Na_3P	a, c	Hexagonal	$a = 4.990$ $c = 8.815$	Na–P = 3.09 Na–Na = 2.93 P–P = 4.98	Na+P	Density at 25° = 1.74 g/cm³. Reacts with water to give phosphine
KP_{15}	f	Transparent, ruby flat needles, triclinic	$a = 22.74$ $b = 9.69$ $c = 7.21$	$\alpha = 116.7°$ $\beta = 97.5°$ $\gamma = 90.0°$	K+red P at 300–20°	Odorless, stable in air, not attacked by water, slowly attacked by concd. oxidizing acids
Be_3P_2	a, b, c	Brown cubic anti-Mn_2O_3 type	$a = 10.17$	P–P = 3.59 Be–P = 2.20	Be+red P	Density at 25° = 2.25 g/cm³. Reacts with water to give phosphine
Mg_3P_2	b, c	Cubic anti-Mn_2O_3 type	$a = 12.03$		Mg+red P	
Ca_3P_2	a, b, c	Reddish-brown crystals tetragonal	$a = 5.44$ $c = 6.59$		Ca+red P $Ca_3(PO_4)_2$+carbon	Density at 15° = 2.51 g/cm³ stable to 1250° under inert atm. Reacts with moist air to give phosphine, incandescent with O_2 at 300°
β–ZrP	b, c	Hexagonal	$a = 3.684$ $c = 12.554$		$2Zr+P_4 \rightarrow 2ZrP_2$ $4ZrP_2(s) \rightarrow 4ZrP(s)+P_4(g)$	Hard metallic $\beta\text{-}Zr_{0.99}P \underset{900°}{\overset{>1400°}{\rightleftharpoons}} \alpha\text{-}Zr_{0.93}P+P$
Zr_3P	g	Ti_3P type	$a = 10.7994 \pm 0.0003$ $c = 5.3545 \pm 0.0003$		$ZrP_{1.2}$+Zr	
ZrP_2	g	$PbCl_2$ type	$a = 6.4940 \pm 0.0005$ $b = 8.7434 \pm 0.0005$ $c = 3.5135 \pm 0.0003$		P_4+Zr	
HfP	b, c	Hexagonal cubic	$a = 3.651$ $c = 12.37$		Hf+P	
VP	b, c	Hexagonal NiAs type	$a = 3.18$ $c = 6.22$		Electrolysis of metaphosphate melt containing V_2O_5	Hard metallic

TABLE 5—*Continued*

Phosphides	References	Color, crystal system, structure type	Molecular data dimension in Å	Methods of preparation	Physical and chemical properties
GeP	h	SiAs type	$a = 15.14$ $\beta = 101.1°$ $b = 3.638$ $Z = 12$ $c = 9.19$	Ge+P at 900° $\Delta H = -37$ kcal	
SiP_2	h	$GeAs_2$ type	$a = 13.97$ $Z = 8$ $b = 10.08$ $c = 3.436$	Si+P at 900°	
Mo_3P	a, b, c	Grey-black $a = V_3S$-type structure	$a = 9.794$ $c = 4.827$	Crystallize from melt of MoP alloy	Density at 25° = 9.07 g/cm³ Chemically inert
W_3P	c	Tetragonal isostructural with Fe_3P	$a = 9.890$ $c = 4.808$	Electrolysis of fused $(NaPO_3)_n$ containing WO_3 and NaCl	
MnP	a, c, i	Orthorhombic	$a = 5.258$ $b = 3.172$ $c = 5.918$	Electrolysis in electrolyte compn. of NaCl, $Na_4P_2O_7$, $Na_2B_4O_7$, $NaPO_3$ and MnO_2 at 925°	M.p. 1193°C, sp.gr. 5.0
Fe_2P	b, c	Hexagonal	$a = 5.865$ $c = 3.456$	$P_{red}+2Fe \rightarrow Fe_2P$ $\Delta H_{903} = -34.5$ kcal/mole	
FeP	a, b, c	Orthorhombic MnP type	$a = 5.191$ $b = 3.099$ $c = 5.792$	$P_{red}+Fe \rightarrow FeP$ $\Delta H_{903} = -25$ kcal/mole	Density at 25° = 6.07 g/cm³ $4FeP(s) \rightarrow 2Fe_2P(s)+P_2(g)$ $\Delta H_{av} = 80.2$ kcal
FeP_2	a, b, c	Orthorhombic	$a = 4.791$ $b = 5.654$ $c = 2.719$	$2P_{red}+Fe \rightarrow FeP_2$ $\Delta H_{903} = -34$ kcal/mole	$2FeP_2(s) \rightarrow 2FeP(s)+P_2(g)$ $\Delta H_{av} = 67$ kcal
OsP_2	b	Grey-black FeS_2 type	$a = 5.098$ $b = 5.898$ $c = 2.918$	Os+P	Decomposes to metal and phosphorus above 1200°C
CoP_3	a, b	$CoAs_3$ type	$a = 7.706$	$Co+3P_{red} \rightarrow CoP_3$ $\Delta H = -52$ kcal/mole	Density at 25° = 4.26 g/cm³ $2CoP_3(s) \rightarrow 2CoP(s)+P_4(g)$ $\Delta H_{av} = 71.6$ kcal
Rh_2P	a, b, c	Cubic system	$a = 5.498$	Rh_4P_3+heat	Very hard, chemically inert
Rh_4P_3	b	Grey-black	$a = 11.662$ $b = 3.317$ $c = 9.994$	Rh+P	Very hard, chemically inert
Ir_2P	b	Lavender color cubic	$a = 5.543$	IrP_2+heat	Very hard, chemically inert. Decomposed by alkali fusion

TABLE 5—*Continued*

Phosphides	References	Color, crystal system, structure type	Molecular data dimension in Å	Methods of preparation	Physical and chemical properties
IrP_2	b, c	Grey-black FeAsS type	$a = 5.746$ $b = 5.791$ $\beta = 111° 60'$ $c = 5.850$	Ir+exc. P_4	
Ni_3P	a, c	Tetragonal	$a = 8.954$ $c = 4.386$	$3Ni+P_{red} \rightarrow Ni_3P$ $\Delta H_{903} = -48.4$ kcal/mole	Density at 25° = 7.7 g/cm³ Decomposes at 960° to form Ni_5P_2
NiP_2	j	Black lustrous rod shape crystals pyrite-type	$a = 5.4706 \pm 0.0002$ P–P = 2.12 ± 0.03 Ni–P = 2.290 ± 0.01	Ni+P 1200° at 65 kb	Density = 4.70 g/cm³
NiP_3	a, b	$CoAs_3$ type	$a = 7.819$	Ni+P or Ni–Sn alloy+P. Sn_3P_4 removed by dissolution in HCl	Density at 18° = 4.19 g/cm³ $4NiP_3(s) \rightarrow 4NiP_2(s)+P_4(g)$ $\Delta H_{av} = 43$ kcal
Cu_3P	a, b, c	Metallic luster hexagonal Cu_3AS type	$a = 6.95$ $c = 7.15$	$3Cu+P_{red} = Cu_3P$ $\Delta H = -32$ kcal/mole at 888–903°K, heating CuP_2; electrolysis of sodium phosphate melt containing CuO	Conducts electricity, brittle and dense, feebly attacked by HCl, readily oxidized by air
Zn_3P_2	b, c, k, l	Iron-grey-metallic tetragonal	$a = 8.11$ $c = 11.47$	Zn salt+PH_3: ZnP_2+heat; $Zn+P_{red} \rightarrow Zn_3P_2$ $\Delta H_{298°K} = -39.5 \pm 5$ kcal/mole	Conductor of electricity $Zn_3P_2 \xrightarrow{620–820°K} 3Zn(g)+1/2P_4(g)$ toxicity = 10–11 mg/kg to hens 8–10 mg/kg to ducks and geese 10–12 mg/kg to pigs
ZnP_2	c	Red (tetragonal) Black (monoclinic)	$a = 5.08$ $c = 18.59$ $a = 8.85$ $\beta = 102° 3'$ $b = 7.29$ $c = 7.56$	Zn+P	
Cd_3P_2	b, c	Iron-grey metallic tetragonal	$a = 8.76$ $c = 12.31$	Cd salt+PH_3	Good conductor of electricity
CdP_2	b, c	Orange-red tetragonal	$a = 5.29$ $c = 19.74$	Cd+P	Does not react with water, reacts with acid to give PH_3
$B_{13}P_2$	c, m	B_4C type	$a = 5.984$ $c = 11.850$	B+P at 1600° at 100 atm.	

TABLE 5—*Continued*

Phosphides	References	Color, crystal system, structure type	Molecular data dimension in Å	Methods of preparation	Physical and chemical properties
BP	b, c, n, o, p	Cubic	a = 4.538 cube edge 4.538 B–P distance 1.964	ZnP_2+BBr_3 at 900°; $B+P$; $B_2S_3+PH_3$ at 1200–1400°; BCl_3+H_2 over P_{red} at 500°	$B+P_{violet}$ at 22° = -29 ± 2 kcal/mole Standard entropy = 6.4 ± 0.1 cal/°K mole, stable to boiling water, decomposed by boiling conc. alkali $BP+H_2 \xrightarrow{1000°} B_5P_3$
AlP	a, b, q	Cubic	a = 5.451 Al–Al distance = 3.84 Al–P = 2.34	P+Al	Pure AlP insol. in cold and boiling water, dissolves in dilute HCl to give PH_3, can act as p or n semiconductors
GaP	b, c, r	Cubic ZnS type	a = 5.4506	$Ga+PCl_3 \rightarrow 2GaP+3GaCl_2$; Ga+P	
InP	b, s, t, u	Cubic ZnS type	a = 5.8688	In+P $In+P_4O_{10}$ $In+ZnP_2$	Exhibits transistor properties, dissociation temp. = $1015\pm4°$
LaP	b, c	Cubic	a = 6.02	La+P	Reacts with water to give PH_3
Th_3P_4	a, b, c	Greyish-black metallic cubic	a = 8.618 Th–P = 2.98 P–P = 3.20 space group 143d	Th+P	Decomposes in cold HCl soln. to liberate PH_3 density at 25° = 8.44
UP	e, v	Greyish-black metallic cubic	a = 5.589	$U+PH_3$; UO_2+C+PH_3	Decompd. to free metal below m.p. density at 25° = 10.16 g/cm³
PuP	c, w	Cubic	a = 5.644	$PuH_3+PH_3 \xrightarrow{400–600°} PuP+3H_2$	

For refs. to table see p. 412.

[a] T. D. Farr, *Phosphorus, Properties of the Element and Some of Its Compounds*, Chemical Engineering Report No. 8, pp. 69–74. Tennessee Valley Authority, Wilson Dam, Alabama (1950),
[b] J. R. van Wazer, *Phosphorus and Its Compounds*, Vol. I, pp. 123–75. Interscience Publishers Inc., New York (1958).
[c] B. Aronsson, T. Lundström and S. Rundqvist, *Borides, Silicides and Phosphides*, John Wiley & Sons Inc., New York (1965).
[d] D. E. C. Corbridge, *Topics in Phosphorus Chemistry*, **3** (1966) 57–394.
[e] K. Langer and R. Juza, *Naturwissenschaften*, **54**(9) (1967) 225
[f] H. G. von Schnering and H. Schmidt, *Angew. Chem. Intern. Ed. Eng.* **6**(4) (1967) 356.
[g] T. Lundström, *Acta Chem. Scand.* **20**(6) (1966) 1712–14.
[h] T. Wadsten, *Acta Chem. Scand.* **21**(2) (1967) 593–4.
[i] D. H. Baker, Jr., *Trans. Met. Soc. AIME* **239**(5) (1967) 755–6.
[j] P. C. Donohue, T. A. Bither and H. S. Young, *Inorg. Chem.* **7**(5) (1968) 998–1001.
[k] M. N. Mirianashvili, *Sb. Tr., Gruz. Zootekh-Vet, Ucheb-Issled. Inst.* **35** (1965) 415–17.
[l] A. R. Venkitaraman and P. K. Lee, *J. Phys. Chem.* **71**(8) (1967) 2676–83.
[m] P. E. Grayson, J. T. Buford and A. F. Armington, *Electrochem. Technol.* **3**(11–12) (1965) 338–9.
[n] J. Cueilleron and F. Thevenot, *Bull. Soc. Chim. Fr.* (1966) (**9**) 2763–4.
[o] W. Kischio, *Z. anorg. u. allg. Chem.* **349**(3–4) (1967) 151–7.
[p] J. Cueilleron and F. Thevenot, *Bull. Soc. Chim. Fr.* (1965) (**2**) 402–4.
[q] W. E. White and A. H. Bushey, *Inorganic Synthesis*, Vol. IV, pp. 23–25. Ed. J. C. Bailar, Jr., McGraw-Hill Book Co., New York (1963).
[r] G. Sanjiv Kamath and D. Bowman, *J. Electrochem. Soc.* **114**(2) (1967) 192–5.
[s] M. Kuisl, *Angew. Chem. Intern. Ed. Engl.* **6**(2) (1967) 177.
[t] A. Addamiano, U.S. 3,379,502, April 23, 1968, to General Electric Co.
[u] Y. A. Ugai and L. A. Bityutskaya, *Simp. Protsessy Sin. Rosta Krist. Plenok Poluprov. Mater, Tezisy Dokl. Novosibirsk*, (1965) 42–3.
[v] M. Allbutt, A. R. Junkison and R. F. Carney, *Proc. Brit. Ceram. Soc.* No. **7** (1967) 111–26.
[w] O. L. Kruger, J. B. Moser and B. Wrona, U.S. 3,282,656, Nov. 1, 1966, to U.S. Atomic Energy Commission.

and other electropositive metals, are in general very reactive and hydrolyze in water to give phosphines.

The phosphides of the transition metals constitute the largest and the most studied class of phosphides. These phosphides and the phosphides of the more noble metals are characterized by hardness, high melting point, high thermal and electric conductivity, metallic luster and resistance to attack by dilute alkaline or acids[79]. Some of this class of phosphides, however, will react with hot nitric acid.

The properties of the phosphides are intimately associated with the electronic structure of the metal component. However, minute impurities also produce drastic changes in the properties, especially the electrical properties. Differences in homogeneity and porosity (high melting materials are made by sintering) account for further discrepancies[78].

Phosphides can also be classified as either "phosphorus rich" or "metal rich". The structure of the phosphorus-rich phosphides is that of polymerized phosphorus atoms around the metal atoms. The metal-rich phosphides may be described in terms of co-ordination polyhedra of metal atoms enclosing phosphorus atoms. The immediate coordination sphere of the metal atoms contains both metal and phosphorus atoms[79].

In Table 5 are listed some examples of phosphides; their method of preparation, crystal structure, molecular data and some physical and chemical properties.

Of the metal–phosphorus systems studied, one of the more important ones is that of iron–phosphorus. Since the iron-phosphides "ferrophos" are produced as by-products in the commercial production of elemental phosphorus, they represent the largest volume of metal phosphides produced. A phase diagram[77] for this system is shown in Fig. 6.

2.3. APPLICATIONS

"Ferrophos" until recently was added to the iron smelting process used to manufacture high-grade steel. In this process, phosphorus in the "ferrophos", along with the phosphorus impurity present originally in the iron, is oxidized and reacted with the calcium and magnesium oxides in the lining of the furnace to form slag. Such phosphate-containing slag is useful as fertilizer. The phosphorus and iron values in ferrophos are thus recovered.

Phosphorus in the concentration of 0.1 to 0.3% as an alloying element in iron greatly increases its strength and corrosion-resistance. Another interesting application for phosphorus in steel is the preventing of sticking of steel sheets together when several sheets are pack rolled[80].

Phosphorus has a marked effect in improving the hot and cold roll characteristics, softening, recrystallization, and grain growth of copper. Alloys of copper with 2–6% phosphorus can be hot-rolled at 450–650° down to 0.021 in. in thickness. This rolling is carried out with light passes which favors the breaking of the copper–copper phosphide eutectic[80].

One application for reactive phosphides depends on their property of releasing the highly toxic phosphines when reacted with moisture. Thus aluminum phosphide is used in formulations for fumigants[81]. Zinc phosphide has been used as the poison ingredient

[80] J. R. van Wazer, *Phosphorus and Its Compounds*, Vol. II, pp. 1823–55, Interscience Publishers Inc., New York (1961).

[81] L. Hüter, U.S. 2,826,486, March 11, 1958, to Deutsche Gold und Silber Scheideanstalt vormals Roessler.

in baits for rodents[82]. Zinc phosphide was reported to have a toxicity (mg/kg) to hens of 10–11, to ducks and geese of 8–10, and to pigs of 10–12[83]. For comparison, the toxicity (mg/kg) of HCN to birds is 0.1 and to rabbits 4.

Calcium phosphide, when reacted with water, releases spontaneously flammable phosphines. This property makes it an important ingredient in certain types of navy sea flares.

Phosphides of tantalum, tungsten, and niobium are resistant to oxidation at high temperature. It has been suggested that nose cones of space vehicles made of these metals be coated with layers of the respective phosphides to protect against too rapid oxidation upon re-entry and passage through the atmosphere[84].

Many phosphides are semiconductors. The energy gap E in eV for BP = 4.5, AlP = 2.5, GaP = 2.25, InP = 1.25[78]. Thus some of these phosphides are proposed as ingredients for transistors.

3. PHOSPHORUS HYDRIDES AND PHOSPHONIUM COMPOUNDS

Investigations on phosphorus hydrides began in the latter part of the eighteenth century. The best known members of this class of compound are phosphine, PH_3, and biphosphine, P_2H_4. The compound PH has been characterized only by spectroscopic means. Some of the higher phosphines have been isolated, but not fully characterized.

3.1. THE PH MOLECULE

The PH molecule does not exist at room temperature. It is formed by the reaction of phosphorus vapor with hydrogen and detected by the prominent spectral band at 3400 Å. This corresponds to the $^3\Pi \rightarrow {}^3\Sigma$ transition with a ground state of $^3\Sigma^-$ The PH distance has been found to be 1.43 Å or a single bond[85]. Peyton[86] in his spectroscopic study of the green chemiluminescence from an atomic reaction between H atom and P vapor, postulated that besides the $^3\Sigma^-$ ground state, there is also a new electronic $^1\Sigma^+$ state. Spectra of PH molecules are also observed in the shock wave decomposition of argon-diluted PH_3 [87].

3.2. PHOSPHINE

Preparation

There are many methods for the preparation of phosphine. Convenient preparative methods for the laboratory are:

1. By the hydrolysis of a metal phosphide such as aluminum phosphide or calcium phosphide. An example of this method is that described by Baudler, Ständeke and Dobbers[88] for the preparation of several hundred grams quantity of phosphine at one time. It involves the hydrolysis of calcium phosphide, and the by-products P_2H_4 and higher phosphines

[82] J. H. Krieger, *Agr. Chemicals* **7** (No. 4) (1952) 46–48, 135–43.

[83] M. N. Mirianashrili, *Sb. Tr., Gruz. Zootekh-Vet, Ucheb-Issled. Inst.* **35** (1965) 415–17.

[84] H. A. Wilhelm and J. H. Witte, U.S. 3,318,246, May 9, 1967, to U.S. Atomic Energy Commission.

[85] J. R. van Wazer, *Phosphorus and Its Compounds*, Vol. I, pp. 179–93, Interscience Publishers Inc., New York (1958).

[86] M. Peyron, *U.S. Dept. Com. Office Tech. Serv.* (1961) A. D. 271740.

[87] H. Guenebaut and B. Pascat, *Compt. Rend.* **255** (1962) 1741–3.

[88] M. Baudler H. Ständeke and J. Dobbers, *Z. anorg. u. allgem. Chem.* **353** (1967) 122–6.

are simultaneously decomposed thermally into more phosphine along with phosphorus and hydrogen.

2. By the pyrolysis of lower acids of phosphorus[89] such as phosphorous acid at 205–210°; this process is described in ref. 89.

3. By the action of a KOH solution on PH_4I[87]. This method produces very pure phosphine.

For the industrial production of phosphine, white phosphorus is hydrolyzed in the presence of sodium hydroxide; sodium hypophosphite is obtained as a co-product. It is also produced by the electrolytic reduction of white phosphorus. This process with its many variations has been reported in the patent literature[90].

Phosphine may also be formed by the direct combination of gaseous hydrogen and phosphorus. The yield increases with an increase in pressure and a decrease in temperature. This reaction proceeds too slowly to be of practical use[91]. The preparation of phosphine by the reduction of phosphorus trichloride with lithium hydride[92] or lithium aluminum hydride[93] has also been reported.

Structure and Physical Properties of Phosphine

Phosphine has a pyramidal structure with a H–P–H bond angle of 93° 5′—close to the theoretical value for *p*-bonding. The unshared pair of electrons is thus predominantly *s* character. It has a P–H ristance of 1.419 Å and a dipole moment of 0.55*D*. The average bond energies $E_{P\text{-}H} = 76.4$ kcal/mole.

In the mass spectrometer, the most abundant ions and their appearance potentials are:

PH^+ ($m/e=32$) at 13.1 ± 0.2 eV, PH_3^+ ($m/e=74$) at 10.4 ± 3 eV,
PH_2^+ ($m/e=33$) at 14.0 ± 0.2 eV and P^+ ($m/e=1$) at 16.0 ± 31 eV[94].

In the solid state, phosphine has four modifications[95]. The temperature (°C) and $\Delta H°$ (cal/mole) for various phase changes are as follows:

$$\begin{array}{lccccccccc} \text{Temp.} & -242.81 & & -223.67 & & -185.00 & & -133.75 & & -89.72 \\ PH_{3(s\gamma)} & \rightleftarrows & PH_{3(s\alpha)} & \rightleftarrows & PH_{3(s\beta)} & \rightleftarrows & PH_{3(s\delta)} & \rightleftarrows & PH_{3(l)} & \rightleftarrows PH_{3(g)} \\ \Delta H° & 19.6 & & 185.7 & & 115.8 & & 270.4 & & 3{,}489 \end{array}$$

At −170°, phosphine is a face centered cubic with $a = 6.31$ Å[96].

Chemical Properties and Toxicity

A solution of phosphine in water (solubility 26 cc in 100 cc H_2O at 17°) is neither basic nor acidic by pH measurements. The acid constant is about 10^{-29} and the base constant about 10^{-25} [97]. Phosphine undergoes gradual decomposition in water, forming

[89] S. D. Gokhale and Wm. L. Jolly, *Inorganic Synthesis*, Vol. IX, pp. 56–8. Ed. S. Y. Tyree, McGraw-Hill Book Co., New York (1967).

[90] I. Gordon *et al.*, U.S. 3,109,785 to 3,109,795, Nov. 5, 1963, to Hooker Chemical Corp.

[91] D. M. Yost and H. Russel, Jr., *Systematic Inorganic Chemistry of the Fifth- and Sixth-Group Non-Metallic Elements*, pp. 234–53, Prentice-Hall Inc., New York (1944).

[92] Siemens-Schuckertwerke A-G, Neth. Appl. 6,504,634, Nov. 29, 1965.

[93] S. R. Gunn and L. G. Green, *J. Phys. Chem.* **65** (1961) 779–83.

[94] M. Halmann, *Topics in Phosphorus Chemistry*, **4** (1967) 71.

[95] M. C. Sneed and R. C. Brasted, *Comprehensive Inorganic Chemistry*, Vol. V, p. 114, D. Van Nostrand Co. Inc., New Jersey (1956).

[96] D. E. C. Corbridge, *Topics in Phosphorus Chemistry*, **3** (1966) 63–363.

[97] F. A. Cotton and G. Wilkinson, *Advanced Inorganic Chemistry*, pp. 377–8, Interscience Publishers Inc., New York (1962).

phosphorus, hydrogen and a solid of the approximate composition of P_2H. A hydrate, $PH_3 \cdot 5.9H_2O$, has been reported[85].

Phosphine burns readily in air (lower explosive limit is 1.79 vol. % in air)[98]. However, in the gas phase oxidation with oxygen, there are upper and lower critical pressures with an intermediate range of explosive oxidation. This has been shown to be a branch chain reaction and is affected by presence of impurities. The product of the oxidation is various oxy-acids of phosphorus and water[94]. Phosphine is stable at room temperature or below, but does undergo thermal decomposition to elemental phosphorus (red) and hydrogen. Phosphine also photodissociates. With flash photolysis at above 1950 Å, PH and PH_2 radicals are detected spectroscopically[99]. Red phosphorus and hydrogen are also obtained in this decomposition reaction[100].

One of the hydrogens in phosphine may be replaced by an alkali metal. Thus $NaPH_2$ may be prepared by the reaction of phosphine with triphenylmethyl sodium, $(C_6H_5)_3CNa$, in an ether solution. From the industrial standpoint, the most important reaction of phosphine is that with formaldehyde in an aqueous hydrochloric acid solution:

$$PH_3+4CH_2OH+HCl \longrightarrow (HOCH_2)_4PCl$$

The product, tetrakis (hydroxymethyl) phosphonium chloride[101], is the major ingredient in the formulation with urea–formaldehyde or melamine–formaldehyde resins for the permanent flameproofing of cotton cloth[°°2].

Phosphine is extremely poisonous. The recommended maximum concentration in the atmosphere for 8 hr of exposure is 0.3 ppm by volume. Short exposure tolerance is 50–100 ppm for 1 hr and the immediately fatal concentration is 400 ppm. The garlic odor threshold is about 2 ppm by volume[98]. Symptoms of phosphine poisoning are anorexia, abdominal pain, headache, nausea, running nose, cough and depression. Because of its toxicity, phosphine has been used as a grain fumigant (27,000 lb used in 1964)[98]. Phosphine generated *in situ* from AlP is also used for this purpose (see section on phosphides). For protection against phosphine, gas mask canisters containing specially impregnated activated charcoal are used. Impregnating agents such as $K_2Cr_2O_7$ and Ag_2CrO_4 are described in the patent literature[103].

Phosphine in air may be determined by metering a known volume of air through a tube packed with silica gel impregnated with $AgNO_3$. The quantity (at the ppb level) of phosphine is determined by the length of black color formed in the tube as compared to a known standard[104]. Phosphine may also be indicated (in the 0.1 to 4 parts/liter gas range) by the reduction of a gold salt to produce a blue-violet color of colloidal gold[105].

3.3. PHOSPHONIUM HALIDES

Phosphine reacts with hydrogen halides to form the corresponding phosphonium halides of which the best known is phosphonium iodide, a beautiful colorless tetragonal

[98] E. J. Kloos, L. Spinetti and L. D. Raymond, *U.S. Bur. Mines Inform. Cir.* No. 8291 (1966) 7.
[99] D. Kley and K. H. Welge, *Z. Naturforsch.* **20a** (1965) 124–31.
[100] R. W. G. Norrish and G. A. Oldershaw, *Proc. Roy. Soc. (London)* **A262** (1961) 1–7.
[101] A. Hoffman, *J. Am. Chem. Soc.* **43** (1921) 1684.
[102] W. Reeves and J. Guthrie, *Ind. Eng. Chem.* **48** (1956) 64.
[103] H. Heidrich and W. Lemke, Ger. 1,084,137, June 23, 1960.
[104] J. P. Nelson and A. J. Milun, *Anal. Chem.* **29** (1957) 1665–6.
[105] K. Grosskopf, *Draeger-Hefte*, **255** (1964) 21–6.

crystalline compound. Phosphonium iodide is also prepared by the reaction of phosphorus with iodine in carbon disulfide solution. The carbon disulfide is then evaporated and the residue of P_2I_4 and phosphorus obtained upon treatment with water under CO_2 atmosphere forms the phosphonium iodide[106]. The compound has a vapor pressure of 50 mm at 20° and 760 mm at 62.5°. The melting point is 18.5° (under its own vapor pressure) and the heat of fusion is 12,680 cal/mole. It is easily purified by sublimation. The space group and unit cell data are: space group $P4/nmm$, $z = 2$, $a = 6.34$ Å and $c = 4.62$ Å. NMR data showed a P–H distance of 1.42 Å[96]. The calculated total lattice energy is 14.2±5 kcal/mole[107].

TABLE 6. PHYSICAL PROPERTIES OF PH_3

Density PH_3(g) (air = 1) = 1.1829; 1.529 g/liter at 0° and 1 atm
Sp.gr. (D) of PH_3(l) at T (°K) = D = 0.744+0.0005952 (186.8−T)[a]
Surface tension of liquid at −105.9° = 22.783 dynes/cm[a]
Vapor pressure equation of liquid phosphine =
$\log p \text{ cm} = 9.73075 - 1.7853 \times 10^{-2}T + 2.9135 \times 10^{-5}T^2 - 1.0273 \times 10^3 T^{-1}$ [b]
Vapor pressure PH_3 at m.p. of CS_2 = 171 mm[c]
Colorless gas at r.t. and normal pressure[b]
B.p. = −87.8° [b]
M.p. = −133.8° [b]
Heat of vaporization (kcal) = 3.49
Heat of fusion (kcal)[b] = 0.27
Critical temperature = +51° [b]
Critical pressure = 64 atm[b]
Solubility, water: 26 cc in 100 cc at 17°[d]; cyclohexanol: 286 cc in 100 cc at 25° [b]
Autoignition temperature = 104° to 122°F[e]
Heat of formation (kcal) = 2.29[b]
Entropy $S°_{298}$ = 50.35 cal/deg[a]
Electron affinity = 0.5 eV[f]
Far u.v. adsorption: intense absorption bands of PH_3(g) at 1910 Å[g]
NMR chemical shift from H_3PO_4 = 238 ppm[h]

[a] D. M. Yost and H. Russel, Jr., *Systematic Inorganic Chemistry of the Fifth- and Sixth-Group Non-Metallic Elements*, pp. 234–53, Prentice-Hall Inc., New York (1944).
[b] J. R. van Wazer, *Phosphorus and Its Compounds*, Vol. I, pp. 179–93, Interscience Publishers Inc., New York (1958).
[c] S. R. Gunn and L. G. Green, *J. Phys. Chem.*, **65** (1961) 779–83.
[d] T. D. Farr, *Phosphorus, Properties of the Element and Some of Its Compounds*, Chemical Engineering Report No. 8, pp. 75–77, TVA, Wilson Dam, Alabama (1950).
[e] E. J. Kloos, L. Spinetti and L. D. Raymond, *U.S. Bur. Mines Inform. Cir.* No. 8291 (1966) 7.
[f] F. B. Saalfeld and H. J. Svec, *Inorg. Chem.* 3 (10) (1964) 1442–3.
[g] M. Halmann, *J. Chem. Soc.* (1963) 2853–6.
[h] V. Mark, C. Dungan, M. Crutchfield and J. van Wazer, *Topics in Phosphorus Chemistry*, **5** (1967) 227–457.

Both phosphonium bromide and chloride are gases at room temperature under 1 atm pressure. Phosphonium bromide is also tetragonal with the unit cell dimension of $a = 6.042$ Å and $c = 4.378$ Å[108]. The calculated total lattice energy is 147.3±5 kcal/mole[107].

Phosphonium chloride has a m.p. of 28.5° under its own vapor at 46 atm. pressure where

[106] J. B. Work, *Inorganic Synthesis*, Vol. II, pp. 141–4, Ed. W. C. Fernelius, McGraw-Hill Book Co., New York (1946).
[107] T. C. Waddington, *Trans. Faraday Soc.* **61**(12) (1965) 2652–5.
[108] V. Scatturin, P. L. Bellon and E. Frasson, *Atti est. veneto sci lettere ed arti, Classe sci. mat. e nat.* **14** (1965–6) 67–73.

its heat of fusion is 12,680 cal/mole. Its sublimation point is $-28°$, critical temperature 48.8–50.1° and critical pressure 74.2–75.0 atm[91].

Other phosphonium compounds such as PH_4BF_3Cl, PH_4BCl_4, PH_4BBr_4 and PH_4SO_3Cl have also been reported in the literature[109, 110].

3.4. DIPHOSPHINE AND HIGHER PHOSPHINES

Diphosphine, P_2H_4, formed as a minor co-product in the generation of phosphine by the hydrolysis of calcium phosphide, is separated from phosphine by freezing. It is a liquid which boils at 56° (vapor pressure at 25° = 200 mm[93]) and freezes at $-99°$. The specific gravity is 1.02 at 12°. It is quite unstable and decomposes to give PH_3 and polymeric liquid and solid phosphorus hydrides of variable composition[96]. The calculated value for $D(H_2P\text{–}PH_2) = 74.1$ kcal/mole[111].

A new phosphorus hydride, P_2H_2, has been reported by Fehlner[112]. It is identified by mass spectrometry from the products of the pyrolytic decomposition of P_2H_4. The triphosphine P_3H_5 is isolated as a white solid also by the decomposition of diphosphine. Baudler and her co-workers[113] have identified a series of higher chain and cyclic phosphines by mass spectrometry from the higher boiling fraction of the product from the hydrolysis of Ca_3P_2 or by the disproportionation of P_2H_4.

4. PHOSPHORUS HALIDES AND PHOSPHORUS PSEUDOHALIDES

Phosphorus combines with halogens in various ratios to form phosphorus halides. Phosphorus monohalides PCl, PBr, and PI are unstable under normal conditions and are detected only in spectral bands. Phosphorus dihalides, PX_2, which are actually X_2PPX_2, are represented by the known compounds P_2F_4, P_2Cl_4 and P_2I_4. The best known of phosphorus halides, the trihalides, are PF_3, PCl_3, PBr_3 and PI_3. Phosphorus pentahalides with the exception of the pentaiodide are all known. There exist also certain phosphorus polyhalides containing more than five halogens per phosphorus atom. Various pseudohalides as well as mixed halides have also been prepared and characterized.

4.1. PHOSPHORUS DIHALIDES (TETRAHALODIPHOSPHINES[114])

Tetrafluorodiphosphine

Phosphorus difluoride or tetrafluorodiphosphine is the newest member of the tetrahalodiphosphine family to be discovered. Parry *et al.*[115] prepared it in high yield by coupling

[109] T. C. Waddington and F. Klanberg, *J. Chem. Soc.* (1960) 2332–8, 2339–43.
[110] T. C. Waddington and A. J. White, *J. Chem. Soc.* (1963) 2701–7.
[111] F. B. Saalfeld and H. J. Svec, *Inorg. Chem.* 3(10) (1964) 1442–3.
[112] T. P. Fehlner, *J. Am. Chem. Soc.* **88**(8) (1966) 1819–21; *ibid.* **88**(11) (1966) 2613–14.
[113] M. Baudler, H. Ständeke, M. Borgardt, H. Strabel and J. Dobber, *Naturwissenschaften*, **53**(4) (1966) 106.
[114] D. S. Payne, *Topics in Phosphorus Chemistry*, **4** (1967) 85–144.
[115] R. W. Rudolph, R. C. Taylor and R. W. Parry, *J. Am. Chem. Soc.* **88**(16) (1966) 3729–34.

two PF_2 groups together under reduced pressure according to the following equation:

$$2PF_2I+2Hg \longrightarrow P_2F_4+Hg_2I_2$$

Available data indicated that the compound undergoes hydrolysis reaction as shown in the following equation:

$$2P_2F_4+H_2O \longrightarrow 2PHF_2+F_2POPF_2$$

Tetrachlorodiphosphine

Tetrachlorodiphosphine, Cl_2PPCl_2, may be obtained (in low yield) by an electric discharge passing through a mixture of PCl_3 and hydrogen under reduced pressure[114] or by a microwave discharge on PCl_3 at 1–5 mm pressure.

The heat of formation of Cl_2PPCl_2 as estimated from mass spectral data is $\Delta H_f = -106$ kcal/mole, and the P–P bond energy is 58 kcal/mole. NMR spectra, using a frequency of 24,288 Mc/s, give a single peak with a chemical shift of -155 ppm relative to 85% H_3PO_4[116].

Tetrachlorodiphosphine is a colorless liquid with a m.p. of $-28°$. It decomposes on standing at room temperature to PCl_3 and a nonvolatile solid. Hydrolysis of P_2Cl_4 in basic solution has been studied by anion exchange chromatography. This reaction gives 50% P_2H_4 and 50% $P_2(OH)_4$. The latter product then reacts with water to give H_3PO_2 and H_3PO_3[117].

The presence of a lone pair of electrons on each phosphorus atom in P_2Cl_4 permits it to act either as a mono- or difunctional Lewis base. It reacts with nickel carbonyl to form a series of complexes. In the presence of a large excess of $Ni(CO)_4$ at 0° under CO pressure, a yellow white solid $P_2Cl_4 \cdot 2Ni(CO)_4$ is formed which decomposes when the CO pressure is released[118].

Tetraiododiphosphine

Many methods are reported in the literature for the preparation of tetraiododiphosphine: (1) by heating a mixture of iodine with red phosphorus at 180–190°; (2) by heating phosphorus(III) iodide with red phosphorus in n-butyl iodide[114]; (3) by the reaction of iodine with white phosphorus dissolved in carbon disulfide solutions. Evaporation of this last solution produces orange crystalline solid in high yield.

The heat of formation of P_2I_4, $\Delta H_{f298 \cdot 16}$, is -19.76 kcal/mole P_2I_4(s)[119]. The compound has a m.p. of 125.5°[120]. It is very soluble in carbon disulfide, slightly soluble in benzene, ethyl ether and liquid sulfur dioxide, and may be conveniently recrystallized from chlorobenzene. The crystal structure is triclinic with a P–P distance of 2.21 Å, P–I bond length of 2.475 Å. The dipole moment in carbon disulfide is 0.45 *D*.

Tetraiododiphosphine in CS_2 solution reacts with oxygen to form phosphorus(III) iodide and an unstable yellow polymer of variable composition of phosphorus iodide and oxygen. Its reaction with sulfur in CS_2 gives $P_2I_4S_2$. Most of the reaction of P_2I_4, however, results in the cleavage of the P–P bond[121]. Thus, hydrolysis in water gives a mixture of oxyacids of phosphorus [including $(HO)_2(O)PP(O)(OH)_2$] along with phosphines. Reaction with bromine splits the P–P bond to produce $PBrI_2$ in about 90% yield[122].

[116] A. A. Sandoval and H. C. Moser, *Inorg. Chem.* **2** (1963) 27–9.
[117] J. H. Wiersma and A. A. Sandoval, *J. Chromatog.* **20**(2) (1965) 374–7.
[118] C. B. Lindahl, *Nucl. Sci. Abstr.* **18**(19) (1964) 4717.
[119] T. D. Farr, *Phosphorus, Properties of the Element and Some of Its Compounds*, Chemical Engineering Report No. 8, pp. 79–84, TVA, Wilson Dam, Alabama (1950).
[120] M. Baudler, *Z. Naturforsch.* **13b** (1958) 266–7.
[121] T. Moeller and J. E. Huhecy, *J. Inorg. Nucl. Chem.* **24** (1962) 315–19.
[122] A. H. Cowley and S. T. Cohen, *Inorg. Chem.* **4**(8) (1965) 1221–2.

4.2. PHOSPHORUS TRIHALIDES AND PHOSPHORUS TRIPSEUDOHALIDES

Phosphorus Trifluoride

Phosphorus trifluoride (trifluorophosphine) is prepared by the fluorination of phosphorus trichloride with AsF_3, ZnF_2 or CaF_2. The method described in ref. 123 involved the use of zinc fluoride as the fluorinating agent. Other fluorinating agents such as benzoyl fluoride may also be used[124]. The preparation of PF_3 by the reaction of red phosphorus with liquid HF in a closed system has also been reported[125].

The heat of reaction between CaF_2 and PCl_3 obtained at 356.6° in a fluid bed reactor and corrected to 25° was found to be $\Delta H = -2.361$ kcal/mole PCl_3[126]. The standard values of formation at 0° for PF_3 are $\Delta H_0 = -226.03$ kcal/mole, $\Delta F_0 = -242.36$ kcal/mole, $S_0 =$ 54.78 e.u. and P–F bond energy is 119.4 kcal/mole[147].

Phosphorus trifluoride is a colorless gas. It forms complexes with transition metals similar to those formed by CO. Like CO it is also very poisonous due to the formation of a complex with blood hemoglobin. It hydrolyzes slowly in moist air, more rapidly in water and very rapidly in alkaline solution.

Pace and Petrella[128] observed two solid transitions, one at 83.75°K with a heat of transition of 59.9 cal/mole and the other at 110.56°K with a heat of transition of 552.3 cal/mole. The heat of fusion at the triple point temperature of 121.85°K was reported to be 223.9 cal/mole. The heat of vaporization at b.p. (171.77°K) is 3.485 kcal/mole compared to the value of 3.94 kcal/mole reported in the earlier literature. The vapor pressure of the liquid is given by the equation:

$$\log P_{mm} = -606.1714/T + 2.98575 \log T - 0.26322$$

The density of the liquid is given as:

$$d = 2.332 - 0.004407\ T$$

The entropy of the gas in the standard state at b.p. is calculated to be 58.48 cal/mole from calorimetric data.

Phosphorus Trichloride

Phosphorus trichloride is industrially the most important of phosphorus halides. It is prepared on a large scale by the chlorination of an excess of white phosphorus dissolved in a refluxing phosphorus trichloride solution. In the laboratory preparation, it is more convenient to replace white phosphorus with red phosphorus[129]. Other methods of preparation are only of academic interest.

The standard heats of information are $\Delta H^\circ_{f(l)} = -73.3$ kcal/mole and $\Delta H^\circ_{f(g)} =$ -65.5 kcal/mole[114]. The vapor pressure is expressed as:

$$\log P_{mm} = 7.472 - 1600/T \text{ (10 mm at 22°, }-50\text{ mm at 6°, 100 mm at 21°)}$$

The mean bond-dissociation energy D(P–Cl) is 76.2 kcal/mole[130].

[123] A. A. Williams, *Inorganic Synthesis*, Vol. V, pp. 95–97, Ed. T. Moeller, McGraw-Hill Book Co., New York (1952).

[124] F. Seel, K. Ballreich and W. Peters, *Chem. Ber.* **92** (1959) 2117–22.

[125] E. L. Muetterties and J. E. Castle, *J. Inorg. and Nuclear Chem.* **18** (1961) 148–53.

[126] H. C. Duus and D. P. Mykytink, *J. Chem. Eng. Data* **9**(4) (1964) 585–8.

[127] A. Finch, *Rec. Trav. Chim.* **24**(3) (1965) 424–8.

[128] E. L. Pace and R. V. Petrella, *J. Chem. Phys.* **36** (1962) 2991–4.

[129] M. C. Forbes, C. A. Roswell and R. N. Maxson, *Inorganic Synthesis*, Vol. II, pp. 145–7, Ed. W. C. Fernelius, McGraw-Hill Book Co., New York (1946).

[130] C. T. Mortimer, *Pure Appl. Chem.* **2** (1961) 71–6.

TABLE 7. PROPERTIES OF PHOSPHORUS HALOGEN AND PHOSPHORUS PSEUDOHALOGEN COMPOUNDS

Compounds	Physical state at 25°C	M.p. °C	B.p. °C	Density g/ml	Refractive index	NMR Chemical ppm shift from H_3PO_4	Interatomic distance Å P–X	Bond angle X–P–X	Heat of vaporization kcal/mole	Critical temp. °C	Critical pressure atm	Refs.
PF_3	Colorless gas Poisonous	−151.5°	−101.8			−97	1.546	104°	3.94	−2.05	42.69	b, e, f
PCl_3	Colorless liquid	−93.6	76.1	1.5751 at 20°	N_D^{20} = 1.515	−219.4	2.039 (0.0014)	100.27°	7.28	285.5		a, b, d, e, f, g
PBr_3	Colorless liquid	−41.5	173.2	2.852 at 15°	N_D^{20} = 1.6903	−227	2.18	101.5°	9.28			b, e, f, g, h
PI_3	Red hexagonal crystals	61.2	> 200 (decomp.)			−178	2.43 (0.04)	102°				b, e, f, g
PF_2Cl	Colorless gas	−164.8	−47.3						4.30	89.17	44.61	b, g, 1
$PFCl_2$	Colorless gas	−144.0	13.85				P–F = 1.55 P–Cl = 2.02	102°	5.95	189.84	49.3	b, g, 1
PF_2Br	Colorless liquid	−133.8	−16.1	2.181 at 0°					5.57	113°		b, g, h, i
$PFBr_2$	Colorless liquid	−115.0	78.4	2.1810 at 0°					7.62	254°		b, g
$P(CN)_3$	White needles	~ 200°	Sublimes 160–180			135.7	P–C = 1.79 (0.03)					e, f, j
PCl_2Br						−228						f
PCl_2I						−221						f
$PClI_2$						−208						f
PBr_2I						−224						f
$PBrI_2$						−208						f
$PF_2(NCO)$		*ca.* −108	12.3	*ca.* 1.444	N_D^{20} = 1.3700	−130.6						f, 1
$PCl_2(NCO)$		−99	104.4	1.513 at 31°		−165.7						f
$PF(NCO)_2$		−55.0	98.7	1.475	N_D^{20} = 1.4678	−127.9			9.25			f, 1
$PCl(NCO)_2$		−50	134.6	1.505		−128						f, 1
$PBr(NCO)_2$						−126.7						f
$P(NCO)_3$		−2	169.3	1.439 at 25°	N_D^{20} = 1.5352	−96.4			11.9			f, 1, k

TABLE 7—*Continued*

Compounds	Physical state at 25° C	M.p. °C	B.p. °C	Density g/ml	Refractive index	NMR Chemical ppm shift from H_3PO_4	Interatomic distance Å P–X	Bond angle X–P–X	Heat of vaporiz-ation kcal/mole	Critical temp. °C	Critical pressure atm	Refs.
$PCl_2(NCS)$		148	−76			−155.3						f, k
$PF_2(NCS)$		90.3	−95									k
$PBr_2(NCS)$						−152.8						f
$PCl(NCS)_2$						−114.0						f
$PBr(NCS)_2$						−111.5						f
$P(NCS)_3$		< −20	265			−85.6						f, k
P_2F_4						−226						f
P_2Cl_4	Colorless oily liquid	−28	*ca.* 180 (decomp.)	1.701 at 0°		−155						a, b, f
P_2I_4	Pale orange triclinic crystals	125.5	Decomp.	d_4^{20} = 4.178 ±0.002		−170						a, c, f
PF_5	Colorless gas	−93.7	−84.5			35.1	1.551 (0.001)	90° and 120°	4.11			e, f, h, l
PCl_5	Greenish white tetragonal crystals	167 (919 mm)	160 (subl.)	~1.6			(Tetrahedral) 1.97 (Octahedral) 2.04 (plane) 2.08 (apical)					b, g
PBr_5	Reddish-yellow rhombohedral crystals	<100 (decomp.)	106 (decomp.)			−229±1 (in CS_2)	[PBr_4] P–Br =2.2Å					a
PF_3Cl_2	Colorless gas	−130 to −125	7.1						5.66			h
PF_3Br_2 Covalent Form	Gas	−20	+16									h

[a] D. S. Payne, *Topics in Phosphorus Chemistry*, **4** (1967) 85–144.
[b] T. D. Farr, *Phosphorus, Properties of the Element and Some of Its Compounds*, Chemical Engineering Report No. 8, pp. 79–84. TVA, Wilson Dam, Alabama (1950).
[c] M. Baudler, *Z. Naturforsch.* **13b** (1958) 266–7.
[d] F. Seel, K. Ballreich and W. Peters, *Chem. Ber.* **92** (1959) 2117–22.
[e] D. E. Corbridge, *Topics in Phosphorus Chemistry*, **3** (1966) 57–396.
[f] V. Mark, C. H. Dungan, M. M. Crutchfield and J. R. van Wazer, *Topics in Phosphorus Chemistry*, **5** (1967) 227–457.
[g] J. R. van Wazer, *Phosphorus and Its Compounds*, Vol. I, pp. 184–261, Interscience Publishers, Inc., New York (1958).
[h] R. Schmutzler, *Advances in Fluorine Chemistry*, Vol. 5, pp. 34–265, Ed. M. Stacey, J. C. Tatlow and A. G. Sharpe, Butterworth, London (1965).
[i] T. Moeller, *Inorganic Chemistry*, pp. 478–624, John Wiley and Sons, Inc., New York (1952).
[j] P. A. Staats and H. W. Morgan, *Inorganic Synthesis*, Vol. VI, pp. 84–87, Ed. E. G. Rochow, McGraw-Hill Book Co., New York (1960).
[k] H. H. Anderson, *Silicon-Sulfur-Phosphates*, pp. 235–7, IUPAC Colloquium Münster (West), 2–6 Sept. 1954, Verlag Chemie, Weinheim, Germany.
[l] K. W. Hansen and L. S. Bartell, *Inorg. Chem.* **4** (12) (1965) 1775–6.

The predominant reaction of phosphorus trichloride in a mass spectrometer is the formation of the positive ion PCl_2^+ at 12.5 eV. Other ions formed are PCl_3^+ at 12.2 eV, PCl^+ at 17.5 eV and P^+ at 22.1 eV[131].

Phosphorus trichloride molecule has a pyramidal structure of C_{3v} symmetry. The dipole moment measured in carbon tetrachloride is 0.8 ± 0.1 D[114]. Sobhanadri[132] calculated the dipole moment from molecular orbitals and concluded that the lone pair of electrons contributes to the dipole moment and that the calculations indicate about 8% hybridization for PCl_3. The critical density and critical temperature are 0.520 and 290.0°, respectively[133]. Various other physical properties are listed in Table 7.

Phosphorus trichloride is soluble in ether, benzene, chloroform, carbon disulfide, carbon tetrachloride and sulfur dioxide. The P–Cl bond is highly polar and the phosphorus atom is quite susceptible to nucleophilic attack. It hydrolyzes violently in water, forming HCl and phosphorous acid. When the hydrolysis is carried out under controlled conditions, especially in ice-cold sodium carbonate solution, various mixtures of lower oxoacids of phosphorus are formed.

Reaction of phosphorus trichloride with a stoichiometric quantity of phenol gives triphenyl phosphite:

$$PCl_3 + 3C_6H_5OH \longrightarrow P(OC_6H_5)_3 + 3HCl$$

The direct reaction of phosphorus trichloride with aliphatic alcohols gives dialkyl phosphonate:

$$PCl_3 + 3ROH \longrightarrow (RO)_2\overset{\overset{\displaystyle O}{\|}}{P}H + RCl + 2HCl$$

When the reaction with alcohol is carried out in the presence of a tertiary amine as HCl acceptor, trialklyl phosphite results.

$$PCl_3 + 3ROH + 3R'_3N \longrightarrow (RO)_3P + 3R'_3N{\cdot}HCl$$

Phosphorus trichloride is an important intermediate for the formation of organic phosphorus compounds containing the carbon to phosphorus bond, e.g.:

$$PCl_3 + RCl + AlCl_3 \longrightarrow [RPCl_3]^+[AlCl_4]^- \xrightarrow{H_2O} RPOCl_2{}^{134}$$

$$2PCl_3 + RH + O_2 \longrightarrow RP(O)Cl_2 + POCl_3 + HCl^{135}$$

$$PCl_3 + 3RMgBr \longrightarrow R_3P + 3MgBrCl^{136}$$

$$PCl_3 + 3C_6H_5Na \longrightarrow (C_6H_5)_3P + 3NaCl$$

$$PCl_3 + C_6H_6 \xrightarrow{\Delta} C_6H_5PCl_2 + HCl^{137}$$

$$PCl_3 + CH_4 \xrightarrow[O_2\ \text{catalyst}]{\Delta} CH_3PCl_2 + HCl^{138}$$

The phosphorus atom in phosphorus trichloride is a good electron donor and forms purely

131 M. Halmann, *Topics in Phosphorus Chemistry*, **4** (1967) 49–80.
132 J. Sobhanadri, *Proc. Natl. Inst. Sci. India*, **Pt. A26** (1960) 110.
133 L. A. Nisel'son, U. V. Mogucheva and T. D. Sokolova, *Zh. Neorgan. Khim.* **10**(3) (1965) 592–5.
134 A. M. Kinnear and E. A. Perren, *J. Chem. Soc.* (1952) 3437.
135 J. O. Clayton and W. L. Jensen, *J. Am. Chem. Soc.* **70** (1948) 3880.
136 K. D. Berlin, T. H. Austin, M. Peterson and M. Nagabhushanam, *Topics in Phosphorus Chemistry*, **1** (1964) 17–50.
137 A. Toy, R. S. Cooper, U.S. 3,029,282, April 10, 1962, to Victor Chemical Works.
138 J. A. Pianfetti and L. D. Quin, *J. Am. Chem. Soc.* **84** (1962) 851–4.

σ-bonded complexes with boron halides. An example of such a complex is $PCl_3 \cdot BBr_3$, a solid of m.p. 42°[114]. This compound undergoes a halogen exchange reaction on standing to form phosphorus tribromide and boron trichloride.

Phosphorus trichloride is quite toxic and the danger limit is above 4 mg per cubic meter of air. It attacks the respiratory tract and causes a sensation of suffocation and inflammation of the lungs with frothy bloodstained expectoration. A concentration of 600 ppm is lethal in a few minutes. Concentrations of 50–80 ppm are harmful after exposure of 30–60 min and a concentration of 0.7 ppm is the maximum that can be tolerated for exposure of several hours. The maximum allowable concentration is considered to be 0.5 ppm[134]. Accidental contact with skin calls for thorough washing with water.

The important uses for phosphorus trichlorides are as indicated above: (1) as an intermediate for phosphite esters, such as triphenyl phosphite, trimethyl phosphite, tris-betachloroethyl phosphite and diethyl phosphonate $\left[\begin{array}{c} O \\ \| \\ (C_2H_5O)_2PH \end{array}\right]$, which are all compounds of industrial importance; (2) as a chlorinating agent for converting alkyl alcohols to alkyl chlorides, and organic acids to organic acid chlorides; (3) as intermediate for organophosphorus compounds containing the carbon to phosphorus bonds; (4) as an intermediate for phosphorous acid and metal phosphites; (5) as an intermediate for the preparation of PCl_5, $POCl_3$, $PSCl_3$ and the phosphorus pseudohalogens.

Phosphorus Tribromide

Phosphorus tribromide is prepared by the bromination of an excess of phosphorus in phosphorus bromide as the solvent. Other inert solvents such as benzene or carbon tetrachloride may also be used. Their use, however, necessitates fractionation of the final reaction mixture[140].

Phosphorus tribromide is a colorless liquid. Its vapor pressure is represented by the equation:

$$\log p_{\mathrm{mm}} = 7.53 - \frac{2076}{T} \quad (10 \text{ mm at } 48°;\ 100 \text{ mm at } 84°)$$

The standard heats, free energies, and equilibrium constants of formation at 25°[119] are as follows:

$$\Delta H_f°(\mathrm{l}) = -47.5 \text{ kcal/mole},\ \Delta H_f°(\mathrm{g}) = -35.9 \text{ kcal/mole};$$
$$\Delta F_f°(\mathrm{g}) = 41.2 \text{ kcal/mole},\ \log K_f = 30.19$$

Other miscellaneous properties include:

$$C_p(\mathrm{g}) = 18{,}154 + 2.045 \times 10^{-3}T - 0.153 \times 10^5 T^{-2}$$

where T is 298–800°K, and entropy of vaporization, $\Delta S_{446.4} = 20.79$ e.u[119]. The mean bond-dissociation energy D(P–Br) is 61.7 kcal/mole[130]. Other physical constants are shown in Table 7.

Based on electron diffraction or microwave spectroscopy data, the molecule of phosphorus tribromide is pyramidal with a Br–P–Br angle close to 100°. This structure has been confirmed also by Raman spectral studies[114].

[139] M. B. Jacobs, *The Analytical Toxicology of Industrial Inorganic Poisons*, pp. 576–89, Interscience Publishers Inc., New York (1967).

[140] J. F. Gay and R. N. Maxson, *Inorganic Synthesis*, Vol. II, pp. 147–51, Ed. W. C. Fernelius, McGraw-Hill Book Co., New York (1946).

The chemical properties of phosphorus tribromide follow closely those of phosphorus trichloride except for the difference in reactivity due to the difference in the degree of polarity of the phosphorus halogen bonds. This applies to the action of nucleophilic reagents on the phosphorus atom as well as to the hydrolysis reaction. (Hydrolysis under controlled conditions yields polyacids of complex structures.) The hydrolysis of phosphorus tribromide in acid solution to give HBr and H_3PO_3 involves an energy change of -67.2 ± 0.6 kcal/mole[114]. The oxidation reaction with gaseous oxygen is not so easily controlled as in the case of PCl_3 and results in an explosive reaction with the formation of phosphorus pentoxide and bromine.

Phosphorus Triiodide

Phosphorus triiodide, a hexagonal red crystalline material, is prepared by the direct action of iodine on white phosphorus in carbon disulfide. The heat of formation is reported to be $\Delta H^\circ_{f298\cdot16} = -10.9$ kcal/mole[119]. Alternative methods of synthesis involve the action of HI or metal iodide on phosphorus trichloride.

The molecule of phosphorus triiodide is pyramidal in shape with very low polarity of the phosphorus–iodine bond. This is consistent with the observed zero dipole moment of the compound in CS_2 solution. The chemical properties of phosphorus iodide are not as well studied as those of the chloride or the bromide. In general, the products of hydrolysis are quite similar to those from the chloride and the bromide, the major difference being that a greater percentage of phosphine is formed from PI_3 as are compounds with a P–P linkage[114].

4.3. PHOSPHORUS TRI-MIXED HALIDES

Phosphorus tri-mixed halides are formed by the rearrangement of a mixture of the pure trihalides:

$$PBr_3+PCl_3 \rightleftharpoons PClBr_2+PCl_2Br$$

The reaction reaches equilibrium in about 1 to 1.5 hr[141]. Existence of the specific compounds is detected by NMR data as well as by Raman and infrared studies. The fluorine containing halides are sufficiently stable for isolation. The chloride-bromides or chloride-iodides have not been definitely isolated as pure compounds[114].

The phosphorus difluoride halides, PF_2Cl, PF_2Br and PF_2I, are prepared by the reaction shown in the following equation[142]:

$$(CH_3)_2NPF_2+2HX \longrightarrow PF_2X+(CH_3)_2NH\cdot HX$$

All of these compounds undergo decomposition reaction fairly rapidly and may be stored only at $-196°$. The physical properties of some phosphorus mixed fluoride halides are listed in Table 7.

4.4. PHOSPHORUS TRIPSEUDOHALIDES AND TRI-MIXED HALIDE PSEUDOHALIDES

Phosphorus tripseudohalides are in general prepared by exchange reaction between phosphorus trichloride and the metal pseudohalides.

[141] M. L. Delwaulle and M. Bridoux, *Compt. Rend.* **248** (1959) 1342–4.

[142] J. G. Morse, K. Cohn, R. W. Rudolph and R. W. Parry, *Inorganic Synthesis*, Vol. X, pp. 153–6, Ed. E. L. Muetterties, McGraw-Hill Book Co., New York (1967).

Phosphorus Tricyanide

This compound is prepared by the action of phosphorus trichloride on silver cyanide in the presence of solvents such as chloroform, carbon tetrachloride or benzene[143]:

$$PCl_3 + 3AgCN \longrightarrow P(CN)_3 + 3AgCl$$

To avoid possible explosion in this reaction, silver cyanide precipitated from slightly acidic silver nitrate solution and stored in a dark bottle is recommended[144].

Pure phosphorus tricyanide is a white needle-like crystalline compound which sublimes at 160–180° and melts close to 200°. It is quite sensitive to water, forming hydrogen cyanide, phosphorous acid and a yellow phosphorus compound. It is, however, stable at room temperature to dry oxygen, air or ozone[145].

The crystals of phosphorus tricyanide are tetragonal with $z = 16$, $a = 14.00$ and $c = 10.81$ Å. The molecules are pyramidal and exhibit C_{3v} symmetry[146].

Phosphorus Triisocyanate

The compound $[P(NCO)_3]$ may be prepared by the action of a suspension of lithium isocyanate [LiNCO from $CO(NH_2)_2$ and $LiCO_3$] in benzene or phosphorus trichloride[147]. It may also be prepared by treating NaOCN with PCl_3 in liquid SO_2 solvent. It is soluble in acetone, benzene and chloroform. Colton and St. Cyr reported that the structure is iso (P–NCO) and not normal (POCN)[148].

The isocyanate groups react with alcohols to form the corresponding carbamates
H O
| ||
$>$ P–N–C–OR and with ammonia or amines to form the phosphorus containing ureas
H O H
| || |
$>$ P–N–C–N–R. The latter type of compounds have been reported to be useful for imparting flame resistance to cotton[149]. The pure compound polymerizes on standing.

Phosphorus Triisothiocyanate, $P(NCS)_3$

This compound is prepared by the action of ammonium or potassium thiocyanate on phosphorus trichloride in CH_3CN or liquid SO_2 as solvents. The isostructure is indicated by the infrared spectra and molar refraction[150]. It has a m.p. $< -20°$ and b.p. 265°. When heated, it forms a polymer which is stable[151].

Phosphorus Tri-mixed Halide Pseudohalides

Phosphorus triisocyanate and triisothiocyanate also form phosphorus tri-mixed halide-isocyanates and halide-isothiocyanates. The mixed chloride-isocyanates and chloride-isothiocyanates, $PCl_2(NCO)_2$, $PCl(NCO)_2$, $PCl_2(NCS)$ are prepared by the reaction of

143 P. A. Staats and H. W. Morgan, *Inorganic Synthesis*, Vol. VI, pp. 84–7, Ed. E. G. Rochow, McGraw Hill Book Co., New York (1960).

144 T. D. Smith and P. G. Kirk, *Chem. and Ind.* **11** (1967) 1909.

145 J. Gonbeau, H. Haeberle and H. Ulmer, *Z. anorg. u. allgem. Chem.* **311** (1961) 110–16.

146 K. Emerson and D. Britton, *Acta Cryst.* **17**(9) (1964) 1134–9.

147 L. H. Jenkins and D. S. Sears, U.S. 2,873,171, 10 Feb. 1959, to Virginia-Carolina Chemical Corp.

148 E. Colton and L. St. Cyr, *J. Inorg. and Nuclear Chem.* **7** (1958) 424–5.

149 H. C. Fielding and F. Nyman, Brit. 966,525, 12 Aug. 1964, to Imperial Chemical Industries.

150 D. B. Sowerby, *J. Inorg. Nucl. Chem.* **22** (1961) 205–12.

the silver isocyanate or thioisocyanate with an excess of phosphorus trichloride. The fluoride-isocyanates and fluoride-isothiocyanates $PF_2(NCO)$, $PF(NCO)_2$, $PF_2(NCS)$ are prepared by the fluorination of the $P(NCO)_3$ or $P(NCS)_3$ with antimony trifluoride[151].

4.5. PHOSPHORUS PENTAHALIDES

The known phosphorus pentahalides are PF_5, PCl_5 and PBr_5.

Phosphorus Pentafluoride

Phosphorus pentafluoride is prepared in a high yield by the reaction of PCl_5 with CaF_2 at 300–400° [152]. The heat of formation as determined by fluorination of α-white phosphorus is $\Delta H^\circ_{f298}(g) = -381.4 \pm 0.8$ kcal/mole. The average energy of the P–F bonds was calculated as $\Delta E = -109.7$ kcal per bond[153].

Electron diffraction studies showed that PF_5 is a trigonal bipyramid with a significant difference between the lengths of the axial and equatorial bonds: P–F(eq) = 1.534±0.004 Å and P–F(ax) = 1.577±0.005 Å and an average P–F of 1.551±0.001 Å. Calculations by Cotton[154] predict that the axial bonds are weaker than the equatorial ones.

At room temperature, PF_5 is a colorless gas hydrolyzed immediately by water. It is stable in dry glass at 250° and can be liquefied under 46 atm pressure at 16° [155].

Phosphorus pentafluoride strongly complexes with amines, ethers, nitrates, sulfoxides and organic bases. It is a strong catalyst in ionic polymerization reactions such as the polymerization of tetrahydrofuran to form elastomers[152].

Phosphorus Pentachloride

Chlorination of phosphorus trichloride under controlled conditions results in the formation of phosphorus pentachloride[156]. It is a greenish-white tetragonal crystal, soluble in carbon disulfude and benzoyl chloride. It may be recrystallized from carbon tetrachloride solution.

Electron diffraction studies showed that in the vapor state, the molecule is in the form of a trigonal bipyramid with the phosphorus atom approximating sp^3d hybridization. In the solid state, the crystals have an ionic structure corresponding to $(PCl_4)^+$ $(PCl_6)^-$ with sp^3-hybridization and sp^3d^2-hybridization for the cation phosphorus and anion phosphorus respectively[157]. In the tetrahedral cation, the P–Cl bond is 1.97 Å while in the octahedral anion the planar P–Cl bond is 2.04 Å and the apical P–Cl bond is 2.08 Å. In the gaseous state, the apical P–Cl bond distance is also greater than the distance of the equatorial bonds[158].

[151] H. H. Anderson, *Silicon-Sulfur-Phosphates*, pp. 235–7, IUPAC Colloquium Münster (West), 2–6 Sept. 1954, Verlag Chemie, Weinheim, Germany.

[152] E. L. Muetterties, T. A. Bither, M. W. Farlow and D. D. Coffman, *J. Inorg. and Nuclear Chem.* **16** (1960) 52–9.

[153] P. Gross, C. Hayman and M. C. Stuart, *Trans. Faraday Soc.* **62** (10) (1966) 2716–18.

[154] F. A. Cotton, *J. Chem. Phys.* **35** (1961) 228–31.

[155] R. Schmutzler, *Advances in Fluorine Chemistry*, Vol. 5, pp. 34–265, Eds. M. Stacey, J. C. Tatlow and A. G. Sharpe, Butterworth, London (1965).

[156] R. N. Maxson, *Inorganic Synthesis*, Ed. H. S. Booth, Vol. I, pp. 99–100, McGraw-Hill Book Co., New York (1939).

[157] J. R. van Wazer, *Phosphorus and its Compounds*, Vol. I, pp. 184–261, Interscience Publishers Inc., New York (1958).

[158] D. E. Corbridge, *Topics in Phosphorus Chemistry*, **3** (1966) 57–396.

Phosphorus pentachloride dissociates in the vapor state:

$$PCl_5(g) \rightleftharpoons PCl_3(g) + Cl_2$$

and follows the equation:

$$\log K_{mm} = -20{,}000/(457T) + 1.75 \log T + 6.66.$$

At 160°, the vapor of PCl_5 is 13.5% dissociated. In the exchange of the chlorine atoms of phosphorus pentachloride with radioactive chlorine in carbon tetrachloride solution, it was shown that three equatorial chlorine atoms undergo rapid exchange and the two apical chlorines exchange more slowly[157].

The reaction of phosphorus pentachloride with water is very violent, forming hydrogen chloride and phosphoric acid. However, with a controlled quantity of water, phosphorus oxychloride is formed:

$$PCl_5 + H_2O \longrightarrow POCl_3 + 2HCl$$

Phosphorus pentachloride has been used as a chlorinating agent in organic chemistry:

$$ROH + PCl_5 \longrightarrow POCl_3 + RCl + HCl$$

One interesting reaction is that with ammonium chloride to form various polymeric forms of phosphonitrilic chloride (see section on phosphonitrilic chloride):

$$PCl_5 + NH_4Cl \longrightarrow (PNCl_2)_3\text{–to higher polymers} + 3HCl$$

Phosphorus pentachloride also forms additional compounds with certain metal halides Examples of such compounds are $PCl_5.BCl_3$, $2PCl_5.GaCl_3$ and $PCl_5.AlCl_3$. This reaction results in the transfer of a chloride ion to the acceptor molecule: e.g., $[PCl_4]^+ [AlCl_4]^-$.

Phosphorus Pentabromide

Pure phosphorus pentabromide is obtained by the reaction of phosphorus tribromide with an excess of bromine in carbon disulfide solution. It may be recrystallized from nitrobenzene solution.

Phosphorus pentabromide decomposes completely in the vapor phase to $PBr_3 + Br_2$. In the solid state, it has the structure of $[PBr_4]^+ Br^-$. In CH_3CN solution, phosphorus pentabromide is a conductor with species PBr_4^+ and PBr_6^-. The anion is stabilized by solvation while the cation is relatively stable[158]. The force constants according to the Urey–Bradley field method are $k_1 = 2.19 \times 10^5$ dynes/cm, $k_2 = 0.18 \times 10^5$ dynes/cm, $\gamma_3 = 0.028 \times 10^5$ dynes/cm[159]. It acts as a brominating agent for converting organic acids to acyl bromides. It also forms addition compounds with some metal bromides, e.g., $AuBr_3PBr_5$ or $[PBr_4]^+ [AuBr_4]^-$, $[PBr_4]^+ [BBr_4]^-$, $[PBr_4]^+ [SnBr_5]^-$ [114].

4.6. PHOSPHORUS MIXED PENTAHALIDES

Phosphorus mixed pentahalides are prepared by the addition of one halogen to the phosphorus trihalide of a second halogen. The best known compounds are the mixed fluoride-halides and fluoride-bromides, e.g., PF_4Cl, PF_3Cl_2, PF_2Cl_3, $PFCl_4$, PF_3Br_2, PF_2Br_3, $PFBr_4$ [125].

The compounds PF_3Cl_2, PF_2Cl_3, $PFCl_4$ all appear to have a basic trigonal bipyramid structure. The spectrum of $PFCl_4$ shows the F atom occupying the axial site; the F atoms

[159] H. Gerding and P. C. Nobel, *Rec. trav. chim.* 77 (1958) 472–8.

in PF_2Cl_3 also assume axial positions, while in PF_3Cl_2 the structure is C_{2v}, in which there are one equatorial and two axial F atoms[160]. Hückel-type molecular orbital calculations by van der Voorn and Drago showed that the equatorial *p* orbitals are more electronegative than the axial orbitals; this is because the phosphorus orbital is concentrated in the equatorial orbitals. This gives rise to stronger equatorial than axial bonds[161]. The vapor-state electric dipole moment was found to be 0.68 ± 0.02 *D* for PF_3Cl_2, 0 for PF_2Cl_3 and 0.21 *D* for $PFCl_4$. The differences are attributed to the differences in bonding between equatorial and axial positions of the trigonal bipyramid[162].

In the phosphorus fluorochloride series, the last member to be prepared was PF_4Cl. This was accomplished by the low temperature controlled fluorination of the molecular form of PF_3Cl_2 by SbF_3 [163] (b.p. $= -43.4°$, m.p. $= -132 \pm 3°$, heat of vaporization $= 5.16$ kcal/mole). It decomposes on storage at room temperature slowly to PF_3 and PCl_2F_3.

The mixed phosphorus fluoride–bromide PF_3Br_2 is a solid with a m.p. $-20°$ and a b.p. *ca.* 106°, and is prepared by the addition of Br_2 to PF_3. The compound $PFBr_4$ is prepared by passing a bromine–nitrogen mixture into liquid PBr_2F or its CCl_4 solution at $-75°$. The compound is ionic $[[PBr_4]^+F^-]$ at room temperature, m.p. 87° (decomp.)[164].

4.7. PHOSPHORUS POLYHALIDES

Phosphorus chlorides of P : Cl ratio greater than 1 : 5 have not been well characterized. Phosphorus polybromides PBr_7 (red crystals), PBr_9 and PBr_{17} (m.p. 14.5°) have been reported. Crystalline phosphorus poly mixed halides of seven or more halogen atoms per phosphorus have also been reported in the literature. Examples of such compounds are PCl_6I, PCl_5BrI, PCl_3Br_4, $PClBr_5I$, PBr_6I[157]. The crystal structure of PCl_6I as determined by X-ray diffraction shows a mixture of symmetrical tetrahedral $[PCl_4^+]$ cations and linear $[ClICl]^-$ anion. Many of the phosphorus poly mixed halides have not as yet been well characterized.

5. PHOSPHORYL HALIDES, PHOSPHORYL PSEUDOHALIDES, PHOSPHORYL HALIDE-PSEUDOHALIDES, PYROPHOSPHORYL HALIDES AND POLYPHOSPHORYL HALIDES

5.1. PHOSPHORYL HALIDES

Preparation

The most important member of this class of compound is phosphoryl chloride which is prepared commercially by the two methods shown in the following equations:

$$2PCl_3 + O_2 \longrightarrow 2POCl_3$$
$$P_4O_{10} + 6PCl_5 \longrightarrow 10POCl_3$$

Pure oxygen reacts rapidly with PCl_3 at 20–50°; air is not so effective as an oxidation agent

[160] J. E. Griffiths, R. P. Carter, Jr. and R. R. Holmes, *J. Chem. Phys.* **41**(3) (1964) 863–76.
[161] P. C. van der Voorn and R. S. Drago, *J. Am. Chem. Soc.* **88**(14) (1966) 3255–60.
[162] R. R. Holmes and R. P. Carter, *J. Chem. Phys.* **43**(5) (1965) 1645–9.
[163] R. P. Carter, Jr. and R. R. Holmes, *Inorg. Chem.* **4**(5) (1965) 738–9.
[164] L. Kolditz and K. Bauer, *Z. anorg. u. allgem. Chem.* **302** (1959) 230–6.

The reaction of PCl_5 with P_4O_{10} is generally carried out by chlorinating a slurry of P_4O_{10} in PCl_3. In this manner the handling of two reactive solids is avoided. The PCl_5 formed *in situ* reacts instantaneously with P_4O_{10}, first in the excess PCl_3 and then in the $POCl_3$ product as solvent for the reaction. The standard heat, free energy and equilibrium constants of formation for $POCl_3$ at 25° are:

$$-\Delta H_f°(\text{vapor}) = 141.5 \text{ kcal/mole}, \quad -\Delta H_f°(\text{liquid}) = 151.0 \text{ kcal/mole}$$

$$-\Delta F_f°(\text{vapor}) = 130.3 \text{ kcal/mole}, \quad \log K_f(\text{vapor}) = 95.508$$[165].

Phosphoryl bromide can be prepared by methods analogous to those for $POCl_3$. An alternative method reported in the patent literature[166] consists of bubbling gaseous HBr through a mixture of $POCl_3$ in the presence of $AlCl_3$ catalyst at about 80°. When a deficiency of HBr is used, $POCl_2Br$ may be isolated from the reaction mixture. The method described in ref. 167 is that of the reaction between P_4O_{10} and PBr_5.

Phosporhyl fluoride is generally prepared by the fluorination of $POCl_3$ with metallic fluorides such as lead, zinc, silver, or sodium fluoride. When sodium fluoride is used, the reaction can be carried out in such nonaqueous solvents as CH_3CN by heating under atmospheric pressure[168]. When benzoyl fluoride is used as the fluorinating agent, $POCl_2F$ and $POClF_2$, in addition to POF_3, are also obtained[169]. Phosphoryl fluoride may also be prepared directly by the action of MgF_2 on $Mg_2P_2O_7$ at above 750° [170]. The phosphoryl mixed fluoride–chloride and mixed fluoride–bromide may also be prepared by the fluorination of $POCl_3$ or $POBr_3$, respectively, with SbF_3.

The compound $POCl_2Br$ in addition to the process of reacting $POCl_3$ with HBr in the presence of $AlCl_3$ catalyst can be prepared by the bromination of $C_2H_5POCl_2$. In this reaction ethyl bromide is the by-product. Phosphoryl chloride–dibromide, $POClBr_2$, has been reported to be prepared by the action of dry HBr on $POCl_3$ at 400–500° [171]. Kuchen and Ecke[172] isolated $POCl_2Br$ and $POClBr_2$ from a mixture resulting from heating PBr_3 and PCl_5 followed by reaction with bromine and subsequent heating with P_4O_{10}. In contrast to the mixed phosphorus trihalides, mixed phosphoryl chloride–bromide does not form by the reorganization of $POCl_3$ and $POBr_3$ at room temperature[171].

Structure and Properties

Phosphoryl halides, as shown by electron diffraction, microwave studies, and Raman spectra, have a tetrahedral structure with the phosphorus atom surrounded by four other atoms. Nagarajan and Mueller[173] showed that the chemical shift of ^{31}P NMR in the series OPF_3, OPF_2Cl, $OPFCl_2$ and $OPCl_3$; when the halogen atoms are substituted by less negative atoms, the ^{31}P resonance signal is shifted to a lower field. In the series $OPCl_3$, $OPCl_2Br$, $OPClBr_2$ and $OPBr_3$, the reverse is true. This is explained as due to the decreasing

[165] T. D. Farr, *Phosphorus, Properties of the Element and Some of Its Compounds*, Chemical Engineering Report No. 8, T.V.A., Wilson Dam, Alabama (1950).

[166] A. A. Asadorian and G. A. Burk, U.S. 2,941,864, 21 June 1960, to Dow Chemical Co.

[167] H. S. Booth and C. G. Seegmiller, *Inorganic Synthesis*, Vol. II, pp. 151–2, Ed. W. C. Fernelius, McGraw-Hill Book Co., New York (1946).

[168] C. W. Tullock and D. D. Coffman, *J. Org. Chem.* **25** (1960) 2016–19.

[169] F. Seel, K. Ballreich and W. Peters, *Chem. Ber.* **92** (1959) 2117–22.

[170] J. Berak and I. Tomczak, *Roczniki Chem.* **39**(12) (1965) 1761–7.

[171] J. R. van Wazer, *Phosphorus and Its Compounds*, Vol. I, pp. 245–64, Interscience Publishers Inc., New York (1958).

[172] W. Kuchen, H. Ecke and H. G. Beckers, *Z. anorg. u. allgem. Chem.* **313** (1961) 138–43.

[173] G. Nagarajan and A. Mueller, *Z. Naturforsch.* **b21**(6) (1966) 505–7.

TABLE 8. PROPERTIES OF PHOSPHORYL HALIDES AND PHOSPHORYL PSEUDOHALIDES

	Physical state at 25°	M.p. °C	B.p. °C	Heat of vaporization kcal/mole	Density g/cm³	Interatomic distances Å	Bond angle X–P–X	Crit. temp. °C	Crit. press. atm	NMR Chem. shift ppm from H_3PO_4	Vapour pressure mm	Refs.
POF_3	Colorless gas	−39.1 (785 mm)	−39.7	$\Delta H_{sub} = 9.08$ $\Delta H_{vap} = 5.52$		P–O = 1.56 P–F = 1.52	∠FPF = 107°	73.3	41.8	35.5	1. Sublimation: $\log P = -1984.7/T + 11.3755$ ($T = 188°$ to 233°K) 2. Liquid: $\log P = -1207/T + 8.0524$ ($T = 233°$ to 253°K)	a, d
$POCl_3$	Colorless liquid	1.25	105.1	8.06	1.645 at 25°	P–O = 1.58 P–Cl = 2.02	∠ClPCl = 106°	329		2.2	t°C 40.5 63.6 82.9 Pmm 76 190 380	a, b, d
$POBr_3$	Solid	55	191.7	$\Delta H_{464.9} = 9.08$		P–O = 1.41 ± 0.07 P–Br = 2.06 ± 0.03				102.9	Entropy of vaporization $\Delta S_{464.9} = 19.53$ e.u.	a, d
POF_2Cl	Colorless gas	−96.4	3.1 ± 0.1	6.08	1.6555 at 0°	P–P = 1.55 P–Cl = 2.01 P–F = 1.51	∠FPCl = ∠FPF = ∠ClPCl = 106°	150.6	43.4	15	$\log P = -1328.3/T + 7.6904$ ($T = 233°$ to 288°K)	a, b, d
$POFCl_2$	Colorless liquid	−80.1	52.9 ± 0.1	7.40	1.5931 at 0°	P–O = 1.54 P–Cl = 1.94 P–F = 1.50	Same as POF_2Cl			0	$\log P = -1618.23/T + 7.8440$ ($T = 243°$ to 362°K)	a, d
POF_2Br	Colorless liquid	−84.8	31.6 ± 0.1	7.09	2.099 at 0°						$\log P = -1550/T + 7.9662$ ($T = 220°$ to 305°K)	a
$POFBr_2$	Colorless liquid	−117.2	110.1 ± 0.1	7.52	2.568 at 0°						$\log P = -1642.9/T + 7.1687$ ($T = 300°$ to 384°K)	a
$POCl_2Br$		10–11	52.3 at 39 mm									c
$POClBr_2$		31	49° at 12 mm									c
$PO(NCO)_3$	Colorless liquid	5.0	193.1	13.41	1.510 at 25°					40.9		b, d
$PO(NCS)_3$	Colorless liquid	13.8	300.1 106–107 at 0.5 mm	14.82	1.484 at 25°					61.9		b, d

[a] T. D. Farr, *Phosphorus, Properties of the Element and Some of Its Compounds*, Chemical Engineering Report No. 8, TVA, Wilson Dam, Alabama (1950).
[b] J. R. van Wazer, *Phosphorus and Its Compounds*, Vol. I, pp. 245–64. Interscience Publishers Inc., New York (1958).
[c] W. Kuchen, H. Ecke and H. G. Beckers, *Z. anorg. u. allgem. Chem.* **313** (1961) 138–43.
[d] V. Mark, C. Dungan, M. Crutchfield and J. van Wazer, *Topics in Phosphorus Chemistry*, **5** (1967) 228–457.

effect of the screening of the P nucleus with increasing electronegativity of the ligand as well as the double bonding character of the P–F bond.

The mean bond dissociation energies calculated from thermochemical data for $O{=}PX_3$ are $D(P{=}O) = 129.8$ kcal/mole for POF_3, 127.5 kcal/mole for $POCl_3$, 124.9 kcal/mole for $POBr_3$, i.e., an increase in $D(P{=}O)$ occurs with an increase in the electronegativity of the halide[174]. Other properties of phosphoryl halides are listed in Table 8.

Reactions of Phosphoryl Chloride

Phosphoryl chloride (phosphorus oxychloride) is at present the only phosphoryl halide produced in a large volume. Its major use is as an intermediate for the synthesis of phosphate esters. Phosphorus oxychloride reacts with an excess of phenols at elevated temperature usually in the presence of catalyst such as $MgCl_2$ or $AlCl_3$ to form triaryl phosphates:

$$3ArOH + POCl_3 \longrightarrow (ArO)_3P = O + 3HCl$$

Its reaction with aliphatic alcohols to form neutral trialkyl phosphate, $(RO)_3P{=}O$, must be carried out under carefully controlled conditions or in the presence of an acid acceptor in order to prevent the attack of the neutral ester by HCl. Neutral chloroalkyl phosphates are formed by the reaction of $POCl_3$ with alkylene epoxides:

$$3\overset{\diagup O \diagdown}{CH_2CH_2} + POCl_3 \xrightarrow[TiCl_4]{} (ClCH_2CH_2O)_3P = O$$

Phosphorus oxychloride forms addition compounds such as $POCl_3.BCl_3$, $POCl_3.BBr_3$, $POCl_3.AlCl_3$, $POCl_3.TiCl_4$ and $POCl_3.MoCl_5$. In addition, in complexes such as $Cl_3POSbCl_5$, X-ray analysis indicates that coordination occurs through the oxygen atom: $Cl_3PO \rightarrow SbCl_5$. This mechanism is confirmed by the lowering of the $P{=}O$ stretching frequency in the IR[175]. Addition compounds with $MgBr_2$, $MgBr_2.2POCl_3$ and $MgBr_2.3POCl_3$, are formed by the reaction of $POCl_3$ with $MgBr_2$ in a molar ratio of $\geqq 2:1$ at room temperature. Coordination in these compounds is also through the oxygen as donor atom as evidenced by the decrease in $\sim 30\ cm^{-1}$ absorption in the $P{=}O$ frequency of the adducts. The compound $PSCl_3$ without the $P{=}O$ group does not react analogously with $MgBr_2$ [176]. Experimental data on other addition compounds of $POCl_3$ do not so clearly indicate whether the oxygen or chlorine atom of $POCl_3$ acts as the electron donor in the complex formation[171]. In a system such as $OPCl_3$–$SbCl_5$, two modes of coordination are possible:

$$OPCl_2^+SbCl_6^- \xleftarrow{(a)} OPCl_3 + SbCl_5 \xrightarrow{(b)} Cl_5Sb \leftarrow OPCl_3$$

As has already been shown, the coordination through the oxygen (b) has been proven to be more significant. Further evidence on this is that the formation of a coordinate bond from the oxygen to the metal weakens the back-donation of electrons from the oxygen to the phosphorus and thus reduces the $P{=}O$ bond order, as indicated by the strong bathochromic shift of the $P{=}O$ frequency[177].

Gutmann has studied the acid and base reaction in $POCl_3$ as a non-aqueous solvent.

[174] C. T. Mortimer, *Pure Appl. Chem.* **2** (1961) 71–6.
[175] D. E. C. Corbridge, *Topics in Phosphorus Chemistry*, **3** (1966) 63–363.
[176] M. Baudler, G. Fricke and H. I. Oezdemir, *Z. anorg. u. allgem. Chem.* **339**(5–6) (1965) 262–73.
[177] H. Teichmann and G. Hilgetag, *Angew. Chem., Intern. Ed.* **6**(12) (1967) 1013–23.

Experimental evidence indicates that $POCl_3$ may ionize as shown in the following equation:

$$POCl_3 \rightleftharpoons POCl_2^+ + Cl^-$$

or

$$2POCl_3 \rightleftharpoons POCl_2^+ + POCl_4^-$$

Thus in $POCl_3$ solvent, $AlCl_3$ would be a Lewis acid and pyridine a Lewis base[171].

$$AlCl_3 + POCl_3 \rightleftharpoons POCl_2^+ + AlCl_4^-$$

$$C_5H_5N + POCl_3 \rightleftharpoons C_5H_5N \cdot POCl_2^+ + Cl^-$$

Phosphoryl chloride addition compounds have been used for the separation of closely related metals, e.g., the 1 : 1 addition product of $POCl_3$ with $NbCl_5$ and $TaCl_5$ may be separated by fractional distillation to recover the Nb and Ta values[178]. The insolubility of $AlCl_3.POCl_3$ in organic solvents renders $POCl_3$ useful for the complete removal of $AlCl_3$ from Friedel–Crafts reaction mixtures[179].

5.2. PHOSPHORYL PSEUDOHALIDES

Phosphoryl isocyanate and phosphoryl isothiocyanate as well as the mixed halide–isocyanate and halide-isothiocyanate are known.

Phosphoryl isocyanate, $OP(NCO)_3$, is prepared by the action of $POCl_3$ on AgNCO in benzene solvent[180]. It may also be prepared by the oxidation of $P(NCO)_3$ with anhydrous liquid SO_3 in anhydrous liquid SO_2 solvent[181].

Phosphoryl isothiocyanate, $OP(NCS)_3$, can be prepared either by the reaction of $POCl_3$ with AgNCS, KCNS or NaCNS in benzene solvent[180, 182] or by the reaction of $POCl_3$ with NH_4SCN in MeCN solvent[183]. Trimethylsilanyl isothiocyanate may also be used as the source of the isothiocyanate moiety. The by-product in this case is $(CH_3)_3SiCl$[184]. The iso-structure of the isothiocyanate is indicated by molar refraction while the structure of the isocyanate is not so definite[180].

Both the isothiocyanate and isocyanate can be isolated as pure compounds, but they polymerize on standing or heating. The isothiocyanate reacts with $1,4\text{-}C_6H_4(NH_2)_2$ and $H_2N(CH_2)_6NH_2$ to form phosphorus-containing polymers[182]. The isocyanate has been claimed in the patent literature[185] to impart flame-and-glow resistance to cellulosic textiles.

5.3. PHOSPHORYL HALIDE-PSEUDOHALIDES

The mixed halide–isothiocyanates such as $OPF_2(NCS)$, $OPF(NCS)_2$, $OPCl_2(NCS)$, $OPCl(NCS)_2$ and the mixed halide–isocyanates such as $OPF_2(NCO)$, $OPFCl(NCO)$, $OPCl_2(NCO)$ have also been reported in the literature.

The compound $OPF_2(NCS)$ is prepared by the reaction of KSCN with pyrophosphoryl

178 W. Scheller and H. Abegg, U.S. 2,936,214, 10 May 1960, to Ciba Ltd.
179 J. Dye, *J. Am. Chem. Soc.* **70** (1948) 2595.
180 H. H. Anderson, *Silicon, Sulfur, Phosphates*, pp. 235–7, IUPAC Colloquium, Münster, 2–6 Sept. 1954.
181 Imperial Chemical Industries Ltd., Belg. 612,359, 5 July 1962.
182 C. N. Kenney, Brit. 883,488, 29 Nov. 1961, to Imperial Chemical Industries Ltd.
183 B. S. Green, D. B. Sowerby and K. J. Wihksne, *Chem. and Ind.* (1960) 1306–7.
184 B. Anders and H. Malz, Ger. 1,215,144, 28 April 1966, to Farbenfabriken Bayer AG.
185 R. L. Holbrook, U.S. 2,898,180, 4 Aug. 1959, to Olin Mathieson Chemical Corp.

tetrafluoride at room temperature, while $OPF(NCS)_2$ is obtained through the disproportionation of $OPF_2(NCS)$ at around 65° [186]. For $OPF_2(NCO)$ (b.p. 69°) the method of preparation involves the action of KOCN on $F_2P(O)OP(O)F_2$ [187].

The mixed chloro-isothiocyanate can be obtained by the ligand exchange between $OPCl_3$ and $OP(NCS)_3$ at 130–150°. Thus $OPCl_2(NCS)$ and $OPCl(NCS)_2$ are detected in the equilibrium mixture by means of the NMR spectra[188]. The compound $OPCl_2(NCS)$ can also be obtained by the reaction of an excess of $POCl_3$ with AgNCS in benzene solvent[180].

In the mixed phosphoryl halide–isocyanate series, $OPCl_2(NCO)$ (b.p.$_{50}$ 60 – 64°) is prepared in good yield by the reaction of PCl_5 with $H_2NCOOC_2H_5$ [189, 190]. The product is unstable and partially decomposes on storage at room temperature. Fluorination of $OPCl_2(NCO)$ with SbF_3 gives $OPF_2(NCO)$ (b.p. 68–68.5°, n_D^{25} 1.3381, d_{25} 1.5899) and $OPFCl(NCO)$ (b.p. 101 – 103°, n_D^{25} 1.4024). When $OPF_2(NCO)$ is reacted with CH_3OH in CH_2Cl_2 solvent, a high yield of $OPF_2(NHCO_2CH_3)$ is obtained. The reaction with amines gives $OPF_2(NHCONR_2)$[189]. Reaction of $OPCl_2NCO$ with the alcohols and phenols also gives the corresponding carbamates[190, 191] and $OPCl_2(NHCOOH)$ is obtained by the interaction of $OPCl_2(NCO)$ with one mole of H_2O added in the form of concentrated HCl[192].

5.4. PYROPHOSPHORYL HALIDES AND POLYPHOSPHORYL HALIDES

Pyrophosphoryl fluoride, $F_2P(O)OP(O)F_2$, is prepared by the dehydration of two moles of $HOP(O)F_2$ with P_2O_5. It has a b.p. of 71°. The symmetrical structure is consistent with NMR data[193].

The original methods for the preparation of pyrophosphoryl chloride, i.e., by the oxidation of phosphorus trichloride or by the controlled hydrolysis of phosphoryl chloride, resulted only in a 10 to 15% yield. Grunze[194] was able to isolate it in 30% yield by the fractionation of a reaction mixture of P_4O_{10} and an excess of $POCl_3$ which had been heated in a sealed tube for 48 hr at 200°. When the reaction with P_4O_{10} is carried out with PCl_5 using a $PCl_5 : P_4O_{10}$ ratio of 4 : 1, a lower temperature of 105° may be used to give a 59% yield of pyrophosphoryl chloride[195]. The yield of pyrophosphoryl chloride is increased to greater than 90% by the reaction of one mole of PCl_5 with 2 moles of $HOPO(Cl_2)$ (prepared from partial hydrolysis of $POCl_3$ or from the sealed-tube reaction of H_3PO_4 with $POCl_3$). An equimolar quantity of $POCl_3$ is formed in this reaction[196].

$$PCl_5 + 2HOP(O)Cl_2 \longrightarrow Cl_2(O)POP(O)Cl_2 + POCl_3 + 2HCl$$

[186] H. W. Roesky and A. Müller, *Z. anorg. u. allgem. Chem.* **353** (1967) 265–9.
[187] H. W. Roesky, *Angew. Chem. Intern. Ed.* **6**(1) (1967) 90.
[188] E. Fluck, H. Binder and F. L. Goldmann, *Z. anorg. u. allgem. Chem.* **338**(1–2) (1965) 58–62.
[189] S. J. Kuhn and G. A. Olah, *Can. J. Chem.* **40** (1962) 1951–4.
[190] Armour and Co., Brit. 877,671, 20 Sept. 1961.
[191] A. V. Kirsanov and M. S. Marenets, *Zhur. Obshchei Khim.* **31** (1961) 1607–11.
[192] A. V. Kirsanov and L. P. Zhuravleva, *Dopovidi Akad. Nauk Ukr. RSR* (1960) 929–31.
[193] E. A. Robinson, *Can. J. Chem.* **40** (1962) 1725.
[194] H. Grunze, *Z. anorg. u. allgem. Chem.* **296** (1958) 63–72.
[195] P. C. Crofts, I. M. Downie and R. B. Heslop, *J. Chem. Soc.* (1960) 3673–5.
[196] H. Grunze, *Z. Chem.* **2** (1962) 313.

Pyrophosphoryl chloride has also been obtained by the heating of $HOP(O)Cl_2$ with an excess of $POCl_3$ at 90° [197], or by the vapor phase reaction of $CH_3OP(O)Cl_2$ with an excess of $POCl_3$ at above 150° [198].

Pyrophosphoryl chloride is a fuming liquid with a b.p. = 90° at 12 mm, m.p. = −165°, and d_4^{25} = 1.821 g/cm^3. It is extremely sensitive to attack by atmospheric moisture and decomposes partially at its boiling point at the range of 210–215°. The NMR chemical shift is 10 ppm from H_3PO_4.

The symmetrical pyrophosphoryl difluoride dichloride $O[P(O)ClF]_2$ has been isolated from a reaction mixture obtained by heating $POCl_2F$ with P_4O_{10} for 48 hr at 200°. It has a b.p. of 36° at 12 mm, n_D^{20} 1.3860 and d_{20}^{20} 1.7918. The symmetrical nature of the compound was confirmed by conversion to $O[P(O)(NEt_2)F]_2$ by reaction with diethyl amine and to $C_2H_5OP(O)ClF$ by reaction with ethyl alcohol[199].

Various polyphosphoryl chlorides have also been reported in the literature. Grunze[194] heated $POCl_3$ with P_4O_{10}, and after the distillation of $POCl_3$ and $P_2O_3Cl_4$ from the reaction mixture, obtained a glassy residue with a composition of (PO_2Cl) from which $(PO_2Cl)_3$ was isolated. Presumably the latter trimeric compound is a cyclic phosphoryl chloride:

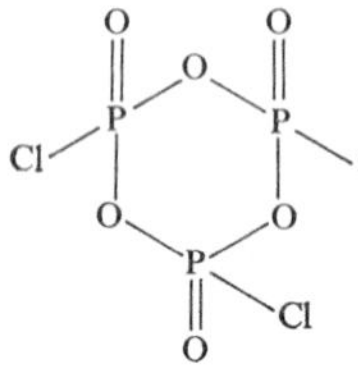

A product with an empirical composition of PO_2Cl has also been obtained by the treatment of $(PNCl_2)_3$ with an excess of Cl_2O. Such a product can be heated to 300° without decomposition[200].

6. THIOPHOSPHORYL HALIDES AND THIOPHOSPHORYL PSEUDOHALIDES

6.1. THIOPHOSPHORYL HALIDES

The preparation of thiophosphoryl halides, SPX_3 (phosphorus sulfohalides), is analogous to those for the corresponding phosphoryl halides.

Thiophosphoryl Fluoride, SPF_3

This compound is prepared by the fluorination of $PSBr_3$ with antimony fluoride[201]. It can also be prepared from $PSCl_3$ by fluorination with KSO_2F[202]. Another method is

[197] M. Becke-Goehring and E. Fluck, Ger. 1,098,495, 5 June 1959, to Chemische Fabrik Joh. A. Benckiser G.m.b.H.

[198] A. D. F. Toy and J. E. Blanch, U.S. 3,034,862, 15 May 1962, to Stauffer Chemical Co.

[199] C. Stolzer and A. Simon, *Chem. Ber.* **94** (1961) 1976–9.

[200] K. Dehnicke, *Chem. Ber.* **97**(12) (1964) 3358–62.

[201] H. S. Booth and C. A. Seabright, *Inorganic Synthesis*, Vol. II, pp. 153–4, Ed. W. C. Fernelius, McGraw-Hill Book Co., New York (1946).

[202] F. Seel, K. Ballreich and R. Schmutzler, *Chem. Ber.* **95** (1962) 199–202.

the fluoride–chloride exchange reaction between $PSCl_3$ and NaF in tetramethylene sulfone solvent at 170° under atmospheric pressure (53% conversion)[203].

Thiophosphoryl fluoride reacts with $(CH_3)_2NH$ to give $(CH_3)_2NP(S)F_2$ [204]. The reaction is not clear-cut since other phosphorus-containing products are also formed.

Thiophosphoryl Chloride

A convenient laboratory method for the preparation of $PSCl_3$ is by heating an intimately mixed powdered phosphorus pentachloride with powdered phosphorus pentasulfide[205]:

$$P_2S_5 + 3PCl_5 \longrightarrow 5PSCl_3$$

Commercially it is prepared by the addition of sulfur to PCl_3 in presence of a catalyst such as $AlCl_3$. In this reaction the mechanism assumes the formation of $AlCl_3.PCl_3$ complex as the intermediate[206]:

$$PCl_3 + AlCl_3 \longrightarrow AlCl_3.PCl_3$$
$$AlCl_3.PCl_3 + S \longrightarrow PSCl_3 + AlCl_3$$

Other materials such as $FeCl_3$, activated charcoal, and Na_2S_5 [207] are also active catalysts.

The heat of formation of $PSCl_{3(l)}$ at 25° is −79.2 kcal/mole. The P=S group has a dissociation energy D of 61 kcal/mole when $H_f^\circ(s) = 57$ kcal/mole or 70 kcal/mole when $H_f^\circ(s) = 66$ kcal/mole[208].

Spectral studies on $PSCl_3$ as well as other thiophosphoryl halides are in agreement with X-ray and electron diffraction studies that it has a tetrahedral structure with the phosphorus atom surrounded by three chlorine atoms and a sulfur atom[209]. The characteristic infrared adsorption frequency for the P=S group of $PSCl_3$ is 750 cm^{-1}, and is affected when the chlorine atoms are replaced by other groups[210].

Thiophosphoryl chloride is a colorless fuming liquid having two allotropic forms: the α forms which solidify at −40.8°, and the α form which solidifies at −36.2°. It is soluble in benzene, carbon tetrachloride, carbon disulfide and $CHCl_3$, and hydrolyzes slowly in water yielding orthophosphoric acid, hydrochloric acid and hydrogen sulfide[206]:

$$PSCl_3 + 4H_2O \longrightarrow H_3PO_4 + 3HCl + H_2S$$

It reacts with aqueous sodium hydroxide to give the thiophosphate Na_3PO_3S[211]. The reaction with one mole of alcohol gives $ROPSCl_2$ but its reaction with a second mole of alcohol for the formation of $(RO)_2\overset{\overset{S}{\|}}{P}Cl$ is very poor and not clear cut. It does react with

[203] C. W. Tullock and D. D. Coffman, *J. Org. Chem.* **25** (1960) 2016–19.

[204] R. G. Cavell, *Can. J. Chem.* **46** (1968) 613–21.

[205] D. R. Martin and W. M. Duvall, *Inorganic Synthesis*, Vol. IV, pp. 73–4, Ed. J. C. Bailar, Jr., McGraw-Hill Book Co., N.Y. (1953).

[206] T. Moeller, H. J. Birch and N. C. Nielsen, *Inorganic Synthesis*, Vol. IV, pp. 71–73, Ed. J. C. Bailar, Jr., McGraw-Hill Book Co., N.Y. (1953).

[207] N. N. Mel'nikov, Y. A. Mandel'baum and V. N. Volkov, *Zhur. Priklad. Khim.* **31** (1958) 938–40.

[208] M. F. Mole and J. C. McCoubrey, *Nature*, **202**(4931) (1964) 450–1.

[209] J. R. van Wazer, *Phosphorus and Its Compounds*, Vol. I, pp. 245–64, Interscience Publishers, Inc., N.Y. (1958).

[210] E. M. Popov, T. A. Mastryukova, N. P. Rodionova and M. I. Kabachnik, *Zhur. Obshchei Khim.* **29** (1959) 1998–2006.

[211] S. K. Yasuda and J. L. Lambert, *Inorganic Synthesis*, Vol. V, pp. 102–4, Ed. T. Moeller, McGraw-Hill Book Co., N.Y. (1957).

alcoholates or an excess of the alcohols in the presence of a base to form the neutral ester $(RO)_3P{=}S$.

The complex $SPCl_3.AlCl_3$ according to ^{31}P–NMR and conductivity measurements appears to exist in CH_2Cl_2 solution as the chloro complex $SPCl_2^+ AlCl_4^-$ [212]. Speculation is that some neutral complexes with other thiophosphoryl compounds are coordinated through the sulfur atom of the P=S group, as indicated by an appreciable bathochromic shift of the P=S frequency due to the sulfur–metal coordination. Teichmann and Hilgetag[213] advanced the theory that the thiophosphoryl sulfur is a typical "soft base" and reacts preferentially with soft acid, i.e., the sub-group B-metals, halogens and sp^3 hybridized carbon and that it is largely inert to "hard acids" such as protons, $\rangle$C=O, and tetrahedral phosphorus.

Thiophosphoryl Bromide

This compound is prepared by adding P_2S_5 to freshly prepared PBr_5 (from red phosphorus and liquid bromine) and the mixture heated in the water bath for 2 hr[201] or by the heating of PBr_3 with S under nitrogen at 130° [214]. The product is a yellow solid soluble in such organic solvents as carbon disulfide, chloroform and ether. It is attacked only slowly by cold water.

Thiophosphoryl Iodide

Baudler *et al.*[215] reported the preparation of PSI_3 by the reaction of PI_3 with sulfur in CS_2 solvent for 3 to 4 days at 10–15° in the absence of light. It is obtained as brick-red to brown-red leaves from CS_2 and can be kept at $-20°$, but decomposes at room temperature in the presence of light or when it is heated to its m.p. of 48°. Raman and infrared measurements show the characteristic frequencies of 309, 321, and 673 cm^{-1}.

Thiophosphoryl Mixed Halides

Thiophosphoryl fluoride–chloride and fluoride–bromide can be prepared by the fluorination of $PSCl_3$ or $PSBr_3$ with SbF_3. An interesting method is by the decomposition of N,N-diethyl amide of difluorophosphoric acid, $(C_2H_5)_2NP(S)F_2$, with HCl to give $P(S)ClF_2$ (80% yield) and with HBr to give $P(S)BrF_2$ (33% yield)[216].

Thiophosphoryl chloride–bromide can be obtained by the fractionation of a reaction mixture resulting from the heating of PBr_3 with PCl_5 followed by addition of red phosphorus and then heated with sulfur[214]. The compound $P(S)Cl_2Br$ has a b.p. of 42 – 44° at 13 mm, and the $P(S)ClBr_2$ has a b.p. of 67° at 13–14 mm.

6.2. THIOPHOSPHORYL PSEUDOHALIDES AND THIOPHOSPHORYL HALIDE-PSEUDOHALIDES

Thiophosphoryl isocyanate, $PS(NCO)_3$, and isothiocyanate, $PS(NCS)_3$, as well as the mixed halide–isocyanate and mixed halide–isothiocyanate are also known. The compound

[212] L. Maier, *Z. anorg. u. allgem. Chem.* **345**(1–2) (1966) 29–34.
[213] H. Teichmann and Hilgetag, *Angew. Chem. Intern. Ed.* **6**(12) (1967) 1013–1126.
[214] W. Kuchen, H. Ecke and H. G. Beckers, *Z. anorg. u. allgem. Chem.* **313** (1961) 138–43.
[215] M. Baudler, G. Fricke and K. Fichtner, *Z. anorg. u. allgem. Chem.* **327**(3–4) (1964) 124–7.
[216] A. Mueller, H. G. Horn and O. Glemser, *Z. Naturforsch.* **b20**(12) (1965) 1150–5.

TABLE 9. PROPERTIES OF THIOPHOSPHORYL HALIDES AND THIOPHOSPHORYL PSEUDOHALIDES

	Physical state at 25°	M.p. °C	B.p. °C	Thermodynamic data	Inter-atomic distances, Å	Bond angle X–P–X	Crit. temp. °C	Crit. press. (atm)	NMR Chemical ppm from H_3PO_4	Specific gravity	Refs.
PSF_3	Colorless gas	−148.8	−52.2	$\Delta H_{221 \cdot 0} = 4.68$ $S_{221 \cdot 0} = 21.18$ e.u. $H_{vap} = 4.730$ kcal/mole	P–S = 1.85 P–F = 1.51	99.5°	72.8	37.7	32.4		b, c
$PSCl_3$	Colorless liquid	−35	125	$C_p°_{298} = 20.99$ cal/deg $S°_{298} = 79.245$ cal/mole/°K	P–S = 1.94 P–Cl = 2.01	107°			−28.8	1.6271 at 25°	b, d, e
$PSBr_3$	Yellow cubic crystals	37.8	212 (decomp.) 125–130 at 25 mm	$C_p°_{298} = 22.69$ cal/deg $S°_{298} = 89.0$ cal/mole/°K	P–S = 1.89 ± 0.06 P–Br = 2.13 ± 0.03	106±3°			111.8	2.85 at 17°	a, d, e
PSF_2Cl	Colorless gas	−155.2	6.3	$H_{vap} = 5.703$ kcal/mole			166	40.9	50.0	1.484 at 0°	a, c, e
$PSFCl_2$	Colorless liquid	−96.0	64.7	$H_{vap} = 6.893$ kcal/mole						1.590 at 0°	a, e
PSF_2Br		−136.9	35.5	$H_{vap} = 6.775$ kcal/mole	P–S = 1.87 ± 0.05 P–F = 1.45 ± 0.08 P–Br = 2.14 ± 0.04	106±3°			28.6	1.94 at 0°	a, c, e
$PSFBr_2$		−75.2	125.3	$H_{vap} = 8.351$ kcal/mole	P–S = 1.87 ± 0.05 P–F = 1.50 ± 0.1 P–Br = 2.18 ± 0.03					2.390 at 0°	a, e
$PS(NCO)_3$	Colorless liquid	8.8	215	$H_{vap} \sim 16$ kcal/mole						1.538 at 25°	a, e
$PS(NCS)_3$			121–3 at 0.3 mm						9.3		e

[a] J. R. van Wazer, *Phosphorus and Its Compounds*, Vol. I, pp. 245–64, Interscience Publishers Inc., New York (1958).
[b] T. D. Farr, *Phosphorus, Properties of the Element and Some of Its Compounds*, Chemical Engineering Report No. 8, T.V.A. Wilson Dam, Alabama (1950).
[c] H. G. Horn and A. Mueller, *Z. anorg. u. allgem. Chem.* **346**(5–6) (1966) 266–71.
[d] V. Mark, C. Dungan, M. Crutchfield and J. van Wazer, *Topics in Phosphorus Chemistry*, **5** (1967) 231–448.
[e] *Gmelins Handbuch der Anorganischen Chemie*, 8th ed., Part 16–C, pp. 585–603, Verlag Chemie GmbH (1965).

$PS(NCO)_3$ is obtained by the addition of sulfur to $P(NCO)_3$ at 140° [217]. It is stable when formed but polymerizes to a solid on heating. The compound $PS(NCS)_3$ is prepared by the reaction of $PSCl_3$ with NH_4SCN in SO_2 solvent at −30° [218],

$$SPCl_3 + 3NH_4SCN \xrightarrow[-30°]{SO_2} SP(NCS)_3 + 3NH_4Cl$$

or in CH_3CN as solvent at room temperature[219].

The thiophosphoryl fluoride diisothiocyanate, $SPF(NCS)_2$, is obtained by the reaction of $Cl_2P(S)F$ with KSCN (b.p. 46–47° at 1 mm, d_4^{20} 1.4827, n_D^{20} 1.6390). It is a colorless liquid with a sharp odor and powerful lachrymatory properties. It decomposes gradually on storage and hydrolyzes slowly in water[220]. The compounds $SPF(NCS)_2$ and $SPF_2(NCS)$ are prepared by the fluorination of $SP(NCS)_3$ with SbF_3 under reduced pressure[221].

Liquid exchange reactions between $SPBr_3$ and $SP(NCS)_3$ result in the formation of $SPBr_2(NCS)$ and $SPBr(NCS)_2$ as indicated by the new lines in the NMR spectra[218]. The equilibrium state for this ligand exchange reaction is reached after heating at 150° for 156 hr.

Compositions with various ratios of phosphorus–sulfur–halogen, other than those described, have also been reported in the literature. Many of these compounds are as yet not well characterized.

The properties of some thiophosphoryl halides and thiophosphoryl pseudohalides are listed in Table 9.

7. PHOSPHORUS OXIDES

The oxidation of white phosphorus takes place by a chain reaction between a lower and upper limit of oxygen pressures. At the lower limit, the oxygen chain carriers are destroyed by impact with the walls and at the upper limit triple collision occurs and O_3 is formed[222]. The net result is the formation of many intermediate oxidation products as represented by the many oxides reported in the literature. Of the oxides, phosphorus pentoxide only is commercially important. Phosphorus trioxide has been the subject of many recent investigations. Much less is known, however, on phosphorus tetroxide and other oxides such as P_4O_7, P_4O_8, and P_4O_9.

7.1. PHOSPHORUS MONOXIDE

Phosphorus forms a diatomic molecule, PO, with oxygen. It is unstable under ordinary conditions and is detected in the vapor state only in spectral bands. Cordes and Warkehr[223] showed that the systems correspond to $^2\Sigma \rightarrow X^2\pi$ and $^2\pi \rightarrow X^2\pi$ transitions. They reported the dissociation energy of PO as 5.30 eV. Other workers calculated the PO

[217] H. H. Anderson, *Silicon, Sulfur, Phosphates*, pp. 235–7, IUPAC Colloquium, Münster, 2–6 Sept. 1954.
[218] E. Fluck, H. Binder and F. L. Goldmann, *Z. anorg. u. allgem. Chem.* **338**(1–2) (1965) 58–62.
[219] B. S. Green, D. B. Sowerby and K. J. Wihksne, *Chem. & Ind.* (1960) 1306–7.
[220] Zh. M. Ivanova, E. A. Stukalo and G. I. Derkach, *Zhurnal. Obshchei Khimii*, **37**(5) (1967) 1144–7
[221] H. W. Roesky and Z. Muller, *Z. anorg. u. allgem. Chem.* **353** (1967) 266–9.
[222] H. Cordes and W. Witschel, *Z. Physik. Chem.* **46**(1/2) (1965) 35–48.
[223] H. Cordes and E. Warkehr, *Z. Physik. Chem.* **46**(1/2) (1965) 26–34.

dissociation energy as 6.2 eV[224] and 6.8 eV[225]. The P–O distance is 1.447 Å and the excitation energy for the molecule is 143 kcal/mole. The dissociation is to oxygen and an excited phosphorus atom[226].

A polymeric $(PO)_n$ has also been reported. It is formed by the electrolysis of $POCl_3$ in $(C_2H_5)_3N.HCl$ at 0° or by the reaction of $POBr_3$ with magnesium in C_2H_5OH solution. The product is a brownish amorphous–crystalline substance insoluble and stable in water but decomposes in an aqueous alkali solution with liberation of PH_3. It oxidizes to P_2O_5 in an oxygen atmosphere at 300° [227].

7.2. PHOSPHORUS TRIOXIDE

Phosphorus trioxide, P_2O_3 or P_4O_6, is prepared by the controlled oxidation of phosphorus at a pressure of 90 mm with air enriched to contain 75% of total oxygen[226]. A more recent method is by the oxidation of phosphorus with N_2O at 550–600° under 70 torr. A yield of 50% of P_4O_6 is obtained[228]. The usual contamination of 1 to 2% of dissolved white phosphorus is removed by conversion to the less soluble and less volatile red phosphorus by u.v. radiation followed by distillation *in vacuo*.

Phosphorus trioxide has a tetrahedral symmetry of the point group T_d. This is confirmed by two strong absorption bands at 639 and 911 cm^{-1} together with seven Raman frequencies[229]. The phosphorus atoms are located at the corners of a regular tetrahedron and each of the six oxygens located between two phosphorus atoms in the plane of symmetry passing through the trigonal axes[230]:

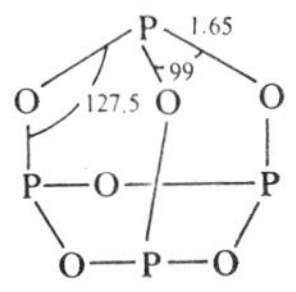

The P–P distance is 2.95±0.03 Å.

Phosphorus trioxide is a colorless liquid at ordinary temperature and has a m.p. of 23.9° and b.p. of 175.4°. The standard enthalpy of formation ΔH_f° of P_4O_6 (crystal) is calculated to be −392 kcal/mole [taking ΔH_f° P_4O_{10} (crystal) as −713 kcal/mole]. The heat of sublimation is 16 kcal/mole and the entropy of vaporization = P_4O_6 (liq.)→P_4O_6 (gas): $\Delta S_{448 \cdot 56}$ = 21.0 cal mole/deg[231]. The bond energy = (P–O) is calculated to be 86 kcal/mole[232]. The dielectric constant is 3.2 at 22° and the surface tension is 36.6 dynes/cm at 34.3°.

The compound hydrolyzes in an excess of water forming H_3PO_3. Upon heating to

[224] J. Berkowitz, *J. Chem. Phys.* **30** (1959) 858–60.

[225] C. V. V. S. N. K. Santharam and P. Tiruvenganna Rao, *Indian J. Phys.* **37** (1963) 14–17.

[226] J. R. van Wazer, *Phosphorus and Its Compounds*, Vol. I, pp. 266–86, Interscience Publishers Inc., N.Y. (1958).

[227] H. Spandau and A. Beyer, *Naturwissenschaften*, **46** (1959) 400.

[228] E. Thilo and D. Heinz, Ger. 1,172,241, 18 June 1964, to Deutsche Akademie der Wissenschaften zu Berlin.

[229] T. A. Sidorov, *Trudy Fiz. Inst. im. P.N. Lebedeva, Akad. Nauk. SSSR* **12** (1960) 225–73.

[230] D. E. C. Corbridge, *Topics in Phosphorus Chemistry*, **3** (1966) 71–81.

[231] T. D. Farr, *Phosphorus, Properties of the Element and Some of Its Compounds*, Chemical Engineering Report No. 8, pp. 18–25, T.V.A., Wilson Dam, Alabama (1950).

[232] S. B. Hartley and J. C. McCoubrey, *Nature*, **198** (1963) 476.

above 210°, it disproportionates to red phosphorus and a phosphorus oxide of the approximate composition of P_nO_{2n}[231]. It displaces CO from $Ni(CO)_4$ forming successively coordinates containing one to four $Ni(CO)_3$ groups. The compound P_4O_6 $[Ni(CO)_3]_4$ is crystalline. Reaction of P_4O_6 with B_2H_6 results in the symmetrical cleavage of B_2H_6 with the formation of $H_3BP_4O_6BH_3$ [233–5].

7.3. PHOSPHORUS PENTOXIDE

Preparation, Structure and Properties

When phosphorus burns in the presence of an excess of oxygen in air, the product formed is P_4O_{10}, commonly called phosphorus pentoxide. One commercial process for collecting the fine white smoke is by condensing it continuously on a fluidized bed of granular P_4O_{10} at 129° into free flowing dustless small beads[236]. The P_4O_{10} thus obtained is the hexagonal or "H" form, the product of commerce.

The free energy of formation of P_4O_{10} is −644.8 kcal/mole and the standard entropy at 298.15°K is 54.66±0.1 cal/mole/deg. The heat capacity at 298.15°K $C_p = 50.60$ cal/deg. Other molar thermodynamic properties at 298.15°K are $(H–H_0^\circ) = 8117$ cal, $(H–H_0^\circ)/T =$ 27.22 cal/deg and $-(G–H_0^\circ)/T = 27.47$ cal/deg[237]. The value of enthalpy of formation of hexagonal P_4O_{10} at 25° has also been reported as −713.2 kcal/mole[238]. The mean bond energy E(P–O) and dissociation energy $\bar{D}$(P–O) were calculated as 86 and 138 kcal/mole respectively[232].

The P_4O_{10} molecule in the vapor state with a T_d symmetry has the arrangement of atoms as shown in the following diagram[230]. It has been estimated that there is essentially

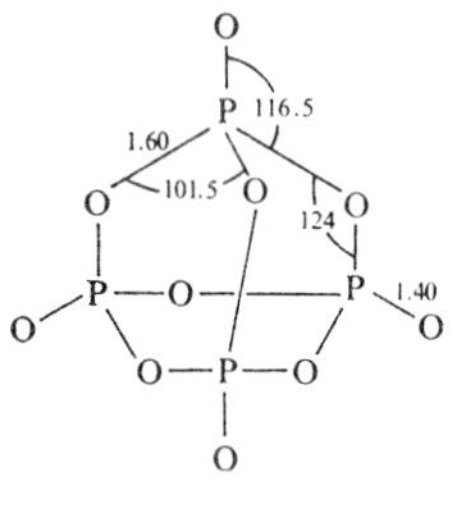

no π bonding between the phosphorus and the shared oxygen and about 1.5–2.0 π-bond between the phosphorus and the unshared oxygen in addition to the δ-bond[226]. The crystals are uniaxial positive with the indexes of refraction in sodium light: $n_w = 1.469$ and $n_\varepsilon = 1.471$, both values ±0.002[231]. It has a density of 2.30.

There are two other crystalline modifications of phosphorus oxide: the O-form and a stable orthorhombic form designated as the O′-form.

The O-form is prepared by heating the H-form in a closed system for 2 hr at 400°. It is built of rings of ten PO_4 tetrahedra linking together to form a continuous three-

[233] J. G. Riess and J. R. van Wazer, *J. Am. Chem. Soc.* **88** (10) (1966) 2166–70.
[234] G. Kodama and H. Kondo, *J. Am. Chem. Soc.* **88**(9) (1966) 2045–6.
[235] J. G. Riess and J. R. van Wazer, *J. Am. Chem. Soc.* **88**(10) (1966) 2339–40.
[236] G. I. Klein, R. E. Newby and L. B. Post, U.S. 3,100,693, 13 August 1963, to Stauffer Chemical Co.
[237] R. J. L. Andon, J. F. Counsell, H. McKerrell and J. F. Martin, *Trans. Faraday Soc.* **59**(492) (1963) 2702–5.
[238] W. S. Holmes, *Trans. Faraday Soc.* **58** (1962) 1916–25.

dimensional structure[230]. The unit cell is orthorhombic with $a = 16.3$ Å, $b = 8.12$ Å, and $c = 5.25$ Å. The indexes of refraction in sodium light are $n_\alpha = 1.545$, $n_\beta = 1.578$, $n_\gamma = 1.589$, all values ± 0.002. The density is 2.72[231].

The O′-form which is also orthorhombic is prepared by heating the H-form in a closed system for 24 hr at 450°. The product consists of horny aggregates of relatively large orthorhombic crystals with a structure composed of infinite corrugated sheets parallel to [100] plane. The indices of refraction in sodium light are $n_w = 1.599$ and $n_\varepsilon = 1.624$, both values ± 0.002. The density calculated from the indices of refraction is 2.89. It melts slowly at 580°±5° to give a very viscous liquid with a vapor pressure of 720 mm at 600° [231].

The conversion of H-form to the O- or O′-form is facilitated by condensing the H-form vapor on a seed bed of either the O- or O′-form, in the presence or absence of 0.2 to 0.3% moisture[239, 240].

The H-form absorbs moisture at a fast rate and dissolves in water with a hissing sound. The O′-form reacts less rapidly with water but more rapidly than the O-form. The final product in the presence of excess water in all cases is orthophosphoric acid. Under controlled conditions, the H-form hydrolyzes to form principally tetrametaphosphoric acid[241]. The reaction of P_4O_{10} with 100% H_2O_2 gives peroxomonophosphoric acid and lower polyphosphoric acids but no tetrametaphosphoric acid since cyclic phosphates are rapidly decomposed by water in the presence of much H_2O_2 [242].

Applications

The largest use for phosphoric pentoxide is the direct hydration into phosphoric acid. A very important use is as the intermediate for the preparation of triethyl phosphate. This is accomplished via the very elegant reaction of P_4O_{10} with ethyl ether to form ethyl polyphosphates which, on subsequent pyrolysis and distillation, disproportionate into triethyl phosphate.[243]. Phosphorus pentoxide is used also for the synthesis of the mixed mono- and dialkyl-phosphoric acids by direct reaction with alcohol. This reaction involves the cleavage of the POP bonds with ROH:

$$P_4O_{10} + 6ROH \longrightarrow 2RO\overset{\overset{O}{\|}}{P}(OH)_2 + 2(RO)_2\overset{\overset{O}{\|}}{P}OH$$

For this reaction, the generation of P_4O_{10} *in situ* by blowing air through a solution of white phosphorus in the alcohol at 150°F has been claimed in the patent literature[244]. Phosphorus pentoxide is also used as a desiccant[245] and as a dehydrating agent in organic reactions. For the latter purpose, phosphorus pentoxide was, until recently, employed extensively in the synthesis of methyl methacrylate.

Phosphorus pentoxide reacts with ammonia to give a white powdery solid of unknown structure containing both P–N–P as well as $PONH_4$ nitrogen. This product has been employed in flameproofing compositions. Phosphorus pentoxide is used as catalyst in

239 F. McCollough, Jr., U.S. 3,034,860, 15 May 1962, to Stauffer Chemical Co.
240 W. F. Tucker, U.S. 2,907,635, 5 October 1959, to Monsanto Chemical Co.
241 M. Shima, K. Hamamoto and S. Utsumi, *Bull. Chem. Soc. Japan*, **33** (1960) 1386–9.
242 P. W. Schenk and K. Dommain, *Z. anorg. u. allgem. Chem.* **326**(3–4) (1963) 139–51.
243 D. C. Hull and J. R. Snodgrass, U.S. 2,407,279, 1946, to Eastman Kodak Co.
244 R. W. Malone, U.S. 3,167,577, 26 January 1965, to Esso Research and Engineering Co.
245 F. Trusell and H. Diehl, *Anal. Chem.* **35** (1963) 674–7.

the air blowing of asphalt [246] and when supported on kieselguhr acts as catalyst for the polymerization of isobutylenes[247].

7.4. PHOSPHORUS TETROXIDE AND OTHER OXIDES

Phosphorus tetroxide $(PO_2)_n$ with a molecular weight ranging from 293 to 721 is best prepared by the heating of P_4O_6 in an evacuated tube at 200–250° for several days. It is separated from the by-product, red phosphorus, by fractional sublimation. The product is a lustrous transparent crystal with a density of 2.54 at 23°. It is deliquescent in air and very soluble in water[226].

Thilo, Heinz, and Jost[248] in their study of the thermal decomposition of P_4O_6 obtained two phosphorus (III/V) oxides, rhombohedral mixed crystals of P_4O_9 and P_4O_8 (α-form) and monoclinic mixed crystals of P_4O_8 and P_4O_7 molecules (β-form). The P_4O_9, P_4O_8, and P_4O_7 molecules are structurally similar to P_4O_{10} but lack respectively one, two or three of the terminal oxygens. These mixed oxides can also be obtained by reaction between P_4O_{10} and elemental phosphorus in an inert gas atmosphere. The relationship of the structure of these oxides with that of P_4O_{10} and P_4O_6 is shown below[249]:

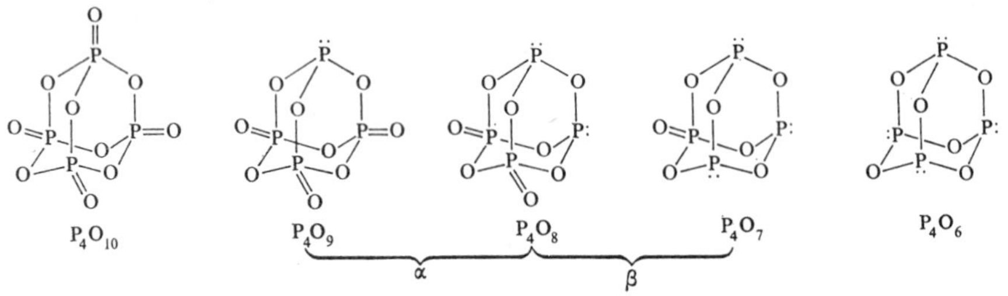

P_4O_{10} P_4O_9 P_4O_8 P_4O_7 P_4O_6

α β

TABLE 10. PHYSICAL PROPERTIES OF PHOSPHORUS OXIDES

Compound	M.p. °C	B.p. °C	Heat of vaporization (kcal/mole)	Heat of fusion (kcal/mole)	Heat of sublimation (kcal/mole)	Density	Ref.
P_4O_6	23.8	175.4	—	—	—	2.135 at 21°	a
P_4O_{10}[H]	420	340 360 (sublime)	16.2	6.5	22.7	2.3	a, b
P_4O_{10}(O]	562	605	18.7	7.7	36.4	2.72	a, b
P_4O_{10}[O′]	580	605	18.7	5.2	33.9	2.89	a, b
$[PO_2]_2$	sublimes	—		—		2.54 at 23°	a

[a] J. R. van Wazer, *Phosphorus and Its Compounds*, Vol. I, pp. 266–86. Interscience Publishers Inc., New York (1958).

[b] T. D. Farr, *Phosphorus, Properties of the Element and Some of Its Compounds*, Chemical Engineering Report No. 8, pp. 18–25, TVA, Wilson Dam (1950).

246 E. K. Brown and R. A. Burge, U.S. 2,886,506, 12 May 1959, to Standard Oil of Indiana.

247 T. A. Kolesnikova, O. I. Lapitskaya and T. N. Lanina, *Tr. Boshkirsk. Nauchn. Issled. Inst. po Pereabotke Nefte*. No. 5 (1960) 176–80.

248 E. Thilo, D. Heinz and K. H. Jost, *Angew. Chem. Intern. Ed.* 3(3) (1964) 232.

249 D. Heinz, *Z. anorg. u. allgem. Chem.* **336** (1965) 137.

Schenk and Vietzke[250, 251] obtained a mixture of two peroxides by exposing a mixture of P_4O_{10} and oxygen to a corona discharge. One product is violet in color and radical-like. It is stable at room temperature; on heating it is converted to the colorless P_4O_{11} which is stable up to 130° and hydrolyzes according to:

$$P_4O_{11}+4H_2O \longrightarrow H_4P_2O_8+H_4P_2O_7$$

8. PHOSPHORUS SULFIDES, OXYSULFIDES AND RELATED COMPOUNDS

Phosphorus forms with sulfur four well-characterized crystalline compounds, P_4S_{10}, P_4S_7, P_4S_5 and P_4S_3. Of these, P_4S_{10} and P_4S_3 are produced commercially. The melting point diagram for the sulfur–phosphorus system is shown in Fig. 7[252].

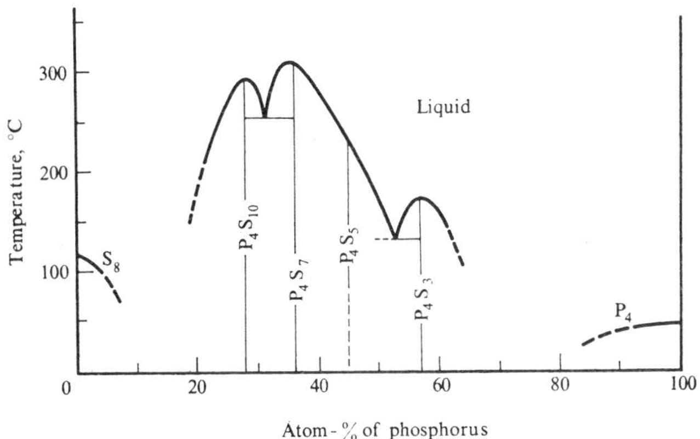

FIG. 7. Phosphorus–sulfur system.[252] (Reprinted by permission of John Wiley & Sons Inc.)

The structures of these sulfides may be treated as insertion and addition of the correct amount of sulfur to the phosphorus tetrahedral. They are shown[253] in Fig. 8.

8.1. PHOSPHORUS SESQUISULFIDE, P_4S_3

Phosphorus sesquisulfide is prepared by heating a stoichiometric mixture of white phosphorus and sulfur at above 180° under an inert atmosphere. It is purified by distillation at 420° under ordinary pressure or by recrystallization from toluene. In an industrial scale, to moderate the highly exothermic reaction, phosphorus is added continuously to sulfur. The final purified product is a yellow rhombic crystal which dissolves 31.2 g at 17° and 15.4 g at 111° in 100 g of toluene. The structure of phosphorus sesquisulfide

[250] P. W. Schenk and H. Vietzke, *Angew. Chem.* **74** (1962) 75.
[251] P. W. Schenk and H. Vietzke, *Z. anorg. u. allgem. Chem.* **326**(3–4) (1963) 152–69.
[252] J. R. van Wazer, *Phosphorus and Its Compounds*, Vol. I, pp. 289–309, Interscience Publishers Inc., New York (1959).
[253] D. E. C. Corbridge, *Topics in Phosphorus Chemistry*, **3** (1966) 81–5.

is shown in Fig. 8. The entropy at 298.15°K for P_4S_3 is 48.60 cal/deg[254]. The melted P_4S_3 has a specific gravity of 1.7953 g/cm³ and a surface tension of 22.86 ergs/cm², both

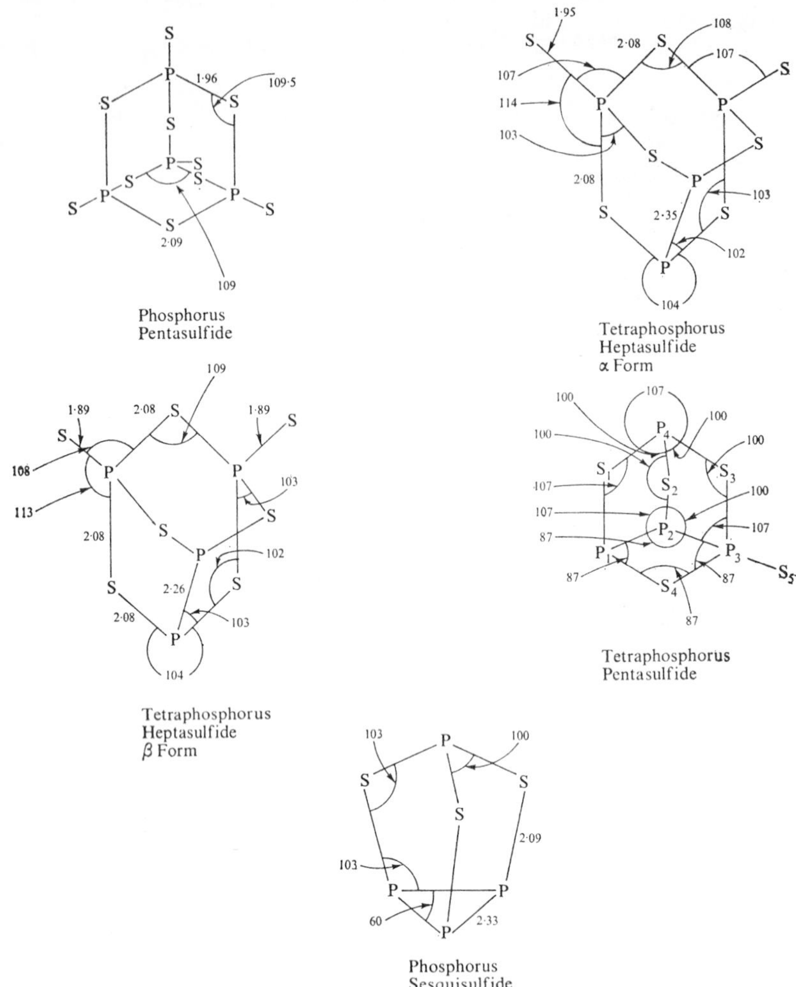

FIG. 8. Structures of phosphorus sulfides.[253] (Reprinted by permission of John Wiley & Sons Inc.)

measured at 172.2°. The parachor was calculated as 329.8[255]. Other physical properties are listed in Table 11.

Under ordinary conditions phosphorus sesquisulfide is stable and unaffected by exposure to the atmosphere or water at room temperature. It decomposes gradually in boiling water but is unaffected by cold hydrochloric or sulfuric acid. It reacts rapidly with an alkaline solution to form sulfide ions, PH_3, H_2, hypophosphite, and a phosphite[256]. From

[254] H. Clever, E. F. Westrum, Jr. and A. W. Cordes, *J. Phys. Chem.* **69**(4) (1965) 1214–19.

[255] E. I. Krylov, *Izvest., Vysshikh Ucheb. Zavedenii, Khim. i Khim. Teknol.* **3** (1960) 223–5.

[256] D. M. Yost and H. Russell, Jr., *Systematic Inorganic Chemistry of the Fifth- and Sixth-Group Nonmetallic Elements*, pp. 182–90, Prentice Hall Inc., N.Y. (1944).

this reaction mixture a monothiophosphite has been isolated by anion exchange column as the diammonium salt, $(NH_4)_2PHO_2S$[257]. Dry P_4S_3 reacts with liquid ammonia to give a solid of the composition $(NH_4)_2[P_4S_3(NH_2)_2]$. The $-NH_2$ groups are presumably

TABLE 11. PROPERTIES OF PHOSPHORUS SULFIDES AND OXYSULFIDE

	P_4S_{10}	P_4S_7	P_4S_5	P_4S_3	$P_4O_6S_4$	Ref.
B.p., °C	514±1	523	—	408	295	a
M.p., °C	288±2	308±3	170–220	174	102	a
Physical state	yellow	nearly colorless pale yellow	bright yellow	yellow		a
Crystalline form	triclinic	monoclinic	monoclinic	orthorhombic	tetragonal	a, b
Density	2.09 (17°)	2.19 (17°)	2.17 (25°)	2.03 (17°)		c
Solubility, g/100g solvent						c
CS_2 (0°)	0.182	0.005		27.0	S	
CS_2 (17°)	0.222	0.029	~10	100	S	
C_6H_6 (17°)	—	—	—	2.5	S	
C_6H_6 (80°)	—	—	—	11.1	S	
Heat of vaporization, kcal/mole	21.3±0.6	22.4±0.6	—	15.2±0.4	—	d

[a] J. R. van Wazer, *Phosphorus and Its Compounds*, Vol. I, pp. 289–309, Interscience Publishers Inc., New York (1958).
[b] A. H. Cowley, *J. Chem. Ed.* **41** (10) (1964) 530–4.
[c] D. M. Yost and H. Russell, Jr., *Systematic Inorganic Chemistry of the Fifth and Sixth-Group Non-metallic Elements*, pp. 182–90, Prentice-Hall Inc., New York (1944).
[d] R. Foerthmann and A. Schneider, *Naturwissenschaften* **52** (13) (1965) 390–1.

attached to the pyramidal P atoms via the lone pair. It can be converted back to P_4S_3 to the extent of 60% when heated to 150–180°. By alternate treatment with NH_3 and heating finally to 275° it gives $P_4(NH)_3$.

Phosphorus sesquisulfide reacts with sulfur in CS_2 solution to give P_4S_5. The reaction with I_2 in CS_2 is more complex, yielding $P_4S_3I_8$, $P_4S_3I_2$ as well as PI_3 and P_4S_7[252].

Phosphorus sesquisulfide ignites in air at above 100°, and glows with luminescence when oxidized in air at 40–50°, having a higher and lower limit of oxygen pressure for the appearance of the glow[252].

One major use for phosphorus sesquisulfide is in combination with potassium chlorate, sulfur and lead dioxide for match tips. Zinc oxide is usually added as a stabilizer and kieselguhr and powdered glass added to moderate the combustion reaction. The whole composition is held together by glue and resin[258].

8.2. TETRAPHOSPHORUS PENTASULFIDE, P_4S_5

This compound is formed by the action of diffuse light on a CS_2 solution of sulfur and P_4S_3 in the presence of iodine catalyst[256]. It recrystallizes as bright yellow crystals from carbon disulfide. The structure is shown in Fig. 8. At above 170° it decomposes

[257] J. D. Murray, G. Nickless and F. H. Pollard, *J. Chem. Soc.* **A**(1967) 1726–30.
[258] J. C. Morales, Span. 281,960, 24 Jan. 1963.

(reversibly) to P_4S_3 and P_4S_7, and for this reason it does not appear in the phase diagram. The crystallographic data for P_4S_5 are as follows: space group $= P2$; $a = 6.412 \pm 0.003$ Å; $b = 10.903 \pm 0.005$ Å; $c = 6.694 \pm 0.003$ Å; and $\beta = 111.66 \pm 0.1°$ [259].

8.3. TETRAPHOSPHORUS HEPTASULFIDE, P_4S_7

Several methods have been reported for the preparation of P_4S_7. The most direct methods are by heating 2P+3S in a glass tube or by heating 4P+7S+5%P_4S_3. The P_4S_3 and P_4S_{10} formed as impurities are removed by extraction with carbon disulfide. Carbon disulfide is also a good solvent for recrystallization of P_4S_7 [256]. Crystals of P_4S_7 are also obtained by reacting white phosphorus and sulfur in CS_2 in the presence of iodine catalyst and the mixture allowed to stand a few days in daylight. Mixture of P_2I_4 and sulfur in CS_2 solvent kept for five days in diffused daylight at room temperature also gives large crystals of P_4S_7[260].

The monoclinic unit cell of α-P_4S_7 contains molecules having the structure shown in Fig. 8. It has the following crystallographic data: space group $= P2_1/n$; $a = 8.895 \pm 0.004$Å; $b = 17.44 \pm 0.015$ Å; $c = 6.779 \pm 0.004$ Å; and $\beta = 92.73 \pm 0.1°$ [259].

A new phosphorus sulfide obtained from a melt of the composition P_4S_5–$P_4S_{6.9}$ with an approximate composition of $P_4S_{6.5}$ was shown by X-ray analysis to be mainly the β-P_4S_7 with structure shown also in Fig. 8. The deficiency in the sulfur analysis is due to a small fraction of the product which contains one less terminal sulfur atom[253].

Of all the crystalline phosphorus sulfides, tetraphosphorus heptasulfide reacts the most rapidly with moisture, giving off hydrogen sulfide. Hydrolysis either in alkaline or acidic solution gives mostly phosphoric and phosphorous acid along with a little phosphine. Reaction of P_4S_7 with ethanol in CS_2 gives mostly $(C_2H_5O)_2P(S)SH$ and $(C_2H_5O)_2PP(OC_2H_5)SH$. The latter decomposes further to $(C_2H_5O)_2P(S)SC_2H_5$ and $(C_2H_5O)_2P(S)H$[261].

The reaction of P_4S_7 with liquid ammonia at $-33°$ gives $(NH_4)_3PS_3$, $(NH_4)_2PS_3NH_2$, $(NH_4)_3P_3S_3(NH)_3$, and NH_4P_2SN in 1 : 2 : 1 : 1 ratio[262].

8.4. PHOSPHORUS PENTASULFIDE, P_4S_{10}

This is the most well known of all phosphorus sulfides and commercially the most important. It is manufactured by the addition of phosphorus and sulfur continuously under an inert atmosphere to a reactor containing molten P_4S_{10}. The temperature is maintained at greater than 300° and the last addition of phosphorus is made after all the sulfur is added. The ratio of the reactants is maintained at close to the theoretical 4P to 10S[263, 264]. The product is purified by distillation under atmospheric pressure.

Phosphorus pentasulfide exhibits a triclinic crystal structure with two P_4S_{10} per unit cell. The structure as shown in Fig. 8 has essentially the same tetrahedral *Td* ($\bar{4}3m$) symmetry

[259] A. Vos, R. Olthof, F. van Bolhuis and R. Botterweg, *Acta Cryst.* **19**(5) (1965) 864–7.

[260] H. Falius, *Naturwissenschaften*, **50** (1963) 126.

[261] P. and E. Steger, *Angew. Chem.* **76**(8) (1964) 344.

[262] H. Behrens and K. Kinzel, *Z. anorg. u. allgem. Chem.* **299** (1959) 241–51.

[263] J. Cremer, *Chem. Inqr. Tech.* **37**(7) (1965) 705–9.

[264] H. Niermann, H. Harnisch and J. Cremer, Ger. 1,195,282, 24 June 1965, to Knapsack-Griesheim A.G.

as the H–form of P_4O_{10} [253]. It has the following crystallographic data: space group = $P1$; $a = 9.072$; $b = 9.199$; and $c = 9.236$, all ± 0.003 Å; $\alpha = 92.58 \pm 0.07°$; $\beta = 100.90 \pm 0.1°$; and $\gamma = 110.18 \pm 0.07°$ [259].

The reactivity of P_4S_{10} with alcohols depends on the rate of cooling of the P_4S_{10} through the phase transition range of 280–60°. The more rapid the rate of cooling the higher is the rate of reactivity[265].

The initial decomposition temperature of P_4S_{10} is quite close to the b.p. of 513–515°. Extensive decomposition occurs even at 600°, as measured by the molecular weights at that temperature. The initial decomposition is believed to follow the equilibrium below:

$$P_4S_{10} \rightleftharpoons P_4S_7 + 3S$$

Upon condensation of the vapor, the equilibrium shifts back to the left[260]. Very rapid condensation of the vapor on an extremely cold surface results in a green solid which turns yellow when the temperature is allowed to go above −100°. It is speculated that this color may be due to the freezing of some free radicals[252].

The vapor pressure of P_4S_{10} is given according to the following equation:

$$\log p_{\text{mm}} = -4940/T + 9.17$$

On hydrolysis, phosphorus pentasulfide gives mostly hydrogen sulfide and phosphoric acid:

$$P_4S_{10} + 16H_2O \longrightarrow 4H_3PO_4 + 10H_2S$$

Presumably intermediates of thiophosphoric acids are first formed. At 100°, P_4S_{10} on hydrolysis in 1.0 M hydrochloric acid goes rapidly to the monophosphate. When the hydrolysis is carried out in a 1 M NaOH solution at 100°, after 1 hr, almost 75% of the phosphorus is present as the mono- and dithiophosphates[267].

The most important reaction of phosphorus pentasulfide is that with alcohols or phenols for the formation of dialkyl or diaryl phosphorodithioic acids:

$$8ROH + P_4S_{10} \longrightarrow 4(RO)_2P(S)SH + 2H_2S$$

Zinc and barium salts of these acids are used extensively as additives to lubricating oil for improving the extreme pressure properties. They also act as antioxidants, detergents and

$$(CH_3O)_2P(=S)SH + \begin{matrix} CHC(=O)-OC_2H_5 \\ \| \\ CHC(=O)-OC_2H_5 \end{matrix} \longrightarrow (CH_3O)_2P(=S)-S-\begin{matrix} CHCOOC_2H_5 \\ | \\ CH_2COOC_2H_5 \end{matrix}$$

$$(C_2H_5O)_2P(=S)SH + Cl_2 \longrightarrow (C_2H_5O)_2P(=S)Cl + HCl + S$$

$$(C_2H_5O)_2P(=S)Cl + NaO-C_6H_4-NO_2 \longrightarrow (C_2H_5O)_2P(=S)O-C_6H_4-NO_2 + NaCl$$

parathion

---

265 S. Robota, U.S. 3,023,086, 27 February 1962, to Hooker Chemical Corp.
266 G. Perot, *Chimie et Industrie*, **87** (1962) 78–86.
267 G. Nickless, F. H. Pollard and D. E. Rogers, *J. Chem. Soc.* (**A**) (1967) 1721–6.

corrosion inhibitors. Some short chain alkyl or cresylic phosphorodithioic acids and their sodium or ammonium salts are used as flotation collectors to separate such minerals as lead and zinc sulfide from the contaminating gangue.

A very important application for dimethyl and diethyl phosphorodithioic acid is as intermediate for the organic phosphorus insecticides. It may be used directly or it may be first converted to the acid chloride for subsequent reactions as shown above.

Phosphorus pentasulfide reacts with priminary amines to give $(RNH)_2P(S)SH$ or $(RNH)_3PS$ depending on the reactant ratios and the temperature[268].

Phosphorus pentasulfide is toxic and the maximum allowable concentration is 1 mg per cubic meter of air. Since it hydrolyzes to H_2S and H_3PO_4, methods for its determination are based on the analysis of the two products of decomposition[269].

8.5. MISCELLANEOUS PHOSPHORUS SULFIDES

The compound P_2S_3 has been reported in the literature. Pitochelli and Audrieth obtained a product with such empirical composition by the action of PCl_3 on H_2S. Extraction and crystallization from CS_2 showed that the product is a mixture of P_4S_3 and P_4S_7. So if P_2S_3 were formed initially, it was unstable at room temperature and underwent disproportionation in CS[270]. Foerthmann and Schneider[271] reported a new sulfide, P_4S_2, with a m.p. of 46° in their studies of the S–P phase diagram. They obtained it as crystalline needles by crystallizing from a mixture of P_4S_2 and white phosphorus at room temperature.

Phosphorus and sulfur also form a solid solution at below 100°. A partially crystalline polymeric phosphorus monosulfide $(PS)_x$ is prepared by the reaction of $PSBr_3$ with magnesium. The formation of $(PS_x)_n$ is by the action of SCl_2 or S_2Cl_2 on H_3PS_4[268].

8.6. PHOSPHORUS OXYSULFIDES

Phosphorus oxysulfides of the following empirical formula have been reported: $P_2O_3S_2$, $P_2O_2S_3$, $P_4O_4S_3$ and $P_6O_{10}S_5$[252]. The best known member of this series, $P_2O_3S_2$, can be obtained by heating P_4O_6 and sulfur in a sealed tube at 160° or by heating a mixture of three moles of P_2O_5 with two moles of P_2S_5. It may be purified by distillation under vacuum to give colorless hygroscopic crystals[272] that melt at about 102° and boil at 295°. However, on each redistillation only about 50–60% of the compound is recovered in the distillate. The rest is removed as polymeric residue.

Electron diffraction studies and Raman data show that the vapor of $P_4O_6S_4$ has a structure similar to that of P_4O_{10} or P_4S_{10}, i.e., it may be regarded as derived from the P_4O_{10} structure with the four terminal oxygens replaced by four sulfur atoms. The crystalline solid, however, belongs to the tetragonal system[253].

Tetraphosphorus tetrathiohexaoxide, $P_4O_6S_4$, reacts with liquid ammonia in a sealed tube at room temperature to yield $(NH_4)_2(PO_2SNH_2)$ and $PS(NH_2)_3$[273].

[268] A. H. Cowley, *J. Chem. Ed.* **41**(10) (1964) 530–4.

[269] M. B. Jacobs, *The Analytical Toxicology of Industrial Inorganic Poisons*, pp. 585–6, Interscience Publishers Inc., New York (1967).

[270] A. R. Pitochelli and L. F. Audrieth, *J. Am. Chem. Soc.* **81** (1959) 4458–60.

[271] R. Foerthmann and A. Schneider, *Naturwissenschaften*, **52**(13) (1965) 390–1.

[272] J. C. Pernert and J. H. Brown, *Chem. Eng. News*, **27** (1949) 2143.

[273] H. Behrens and L. Huber, *Chem. Ber.* **93** (1960) 1137–43.

Hydrolysis of $P_4O_6S_4$ at 100° in 1.0 M hydrochloric acid gives after 1 hour only the monophosphate. When the hydrolysis is carried out with 1 M NaOH solution at 100°, less than 10% of the total phosphorus is the monophosphate, the main products being the monothiophosphate and two anions tentatively identified as

$$\left[\begin{array}{ccccc} & S & & O & \\ & \| & & \| & \\ O- & P & -O- & P & -O \\ & | & & | & \\ & O & & O & \end{array}\right]^{-4} \quad \text{and} \quad \left[\begin{array}{ccccc} & S & & S & \\ & \| & & \| & \\ O- & P & -O- & P & -O \\ & | & & | & \\ & O & & O & \end{array}\right]^{-4}$$ [267]

8.7. PHOSPHORUS SELENIDES AND TELLURIDES

The best characterized phosphorus selenide is P_2Se_5, prepared by heating a stoichiometric mixture of the elements for 12 hr at 450°. It is a black-purple amorphous solid. An attempt to recrystallize it from CCl_4 was not successful[274]. On heating it decomposes to phosphorus and selenium.

Phosphorus pentaselenide reacts with liquid ammonia. Depending on the temperature used, products such as $(NH_4)_2[PSe_3NH_2]$, $[NH_4]_3[PSe_4]$ and $[NH_4]_2$ $[PSe_2(NH_2)NH]$ are obtained. The latter compound at higher temperature deammoniates to other phosphorus–selenium–nitrogen compounds[275].

In its reaction with alkyl amines, P_2Se_5 forms compounds such as $(RNH)_3$ P $=$ Se and $(RNH)_2P(Se)Se^-RNH_3^+$ [276]. When P_2Se_5 reacts with alcohols, the expected $(RO)_2P(Se)SeH$ which forms is unstable and can be isolated only as its potassium salt $(RO)_2P(Se)SeK$ or its chromium(III) complexes $Cr[Se_2P(OR)_2]_3$ [274].

Tetraphosphorus triselenide, P_4Se_3, is prepared in 55–60% yield by refluxing a mixture of white phosphorus with selenium in heptane in the presence of charcoal. It has a m.p. of 245–246° [277]. When it reacts with liquid ammonia, it gives a dark red compound $[NH_4]_2[P_4Se_3(NH_2)_2]$ that when heated to 20° loses NH_3 to yield $[NH_4]_2[P_4Se_3NH]$. Continued heating to 90° yields $[NH_4][P_4Se_3NH_2]$ and finally at 200° it goes back to P_4Se_3 [275].

Phosphorus tritelluride, P_2Se_3, is prepared by the heating of tellurium with white phosphorus at 320° in a sealed tube. It is a black solid which decomposes slowly in moist air to give phosphine[252].

9. PHOSPHORUS–NITROGEN COMPOUNDS

Phosphorus forms many compounds with nitrogen. They may be regarded as nitrogen derivatives of the oxyacids of phosphorus. By such classification, the fundamental chemical relationships which characterize many of these compounds become more apparent[278, 279].

The organic N-substituted derivatives of phosphorus–nitrogen compounds prepared from amines are in general more stable and easier to prepare than the corresponding ammono

[274] M. V. Kudchadker, R. A. Zingaro and K. J. Irgolic, *Can. J. Chem.* **46** (1968) 1415–24.
[275] H. Behrens and G. Haschka, *Chem. Ber.* **94** (1961) 1191–9.
[276] R. C. Melton and R. A. Zingaro, *Can. J. Chem.* **46** (1968) 1425–8.
[277] K. Irgolic, R. A. Zingaro and M. Kudchadker, *Inorg. Chem.* **40**(10) (1965) 1421–3.
[278] L. F. Audrieth, R. Steinman and A. D. F. Toy, *Chem. Rev.* **32** (1943) 99–108.
[279] L. F. Audrieth, *Chem. Eng. News*, **25** (1947) 2552.

CHART 1

Nitrogen Derivatives of Phosphorous Acid[278 279]

aquo-ammono derivatives:

$$\mathrm{HP(O)(OH)_2} \longrightarrow \mathrm{H_2NP(O)(H)\text{-}OH} \longrightarrow \mathrm{(H_2N)_2P(O)H} \longrightarrow \mathrm{(H_2N)_3P}$$

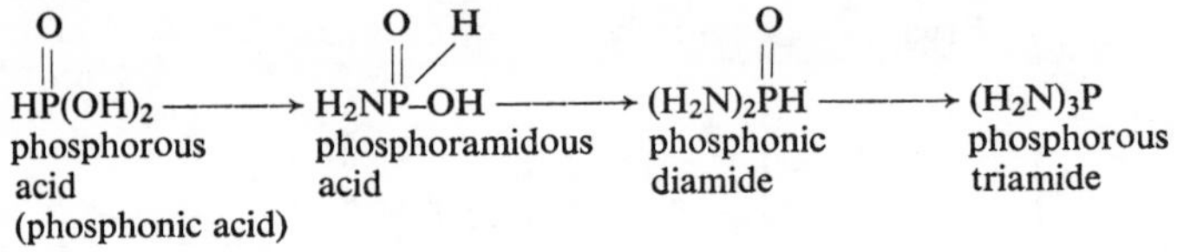

phosphorous acid (phosphonic acid) → phosphoramidous acid → phosphonic diamide → phosphorous triamide

ammono derivatives:

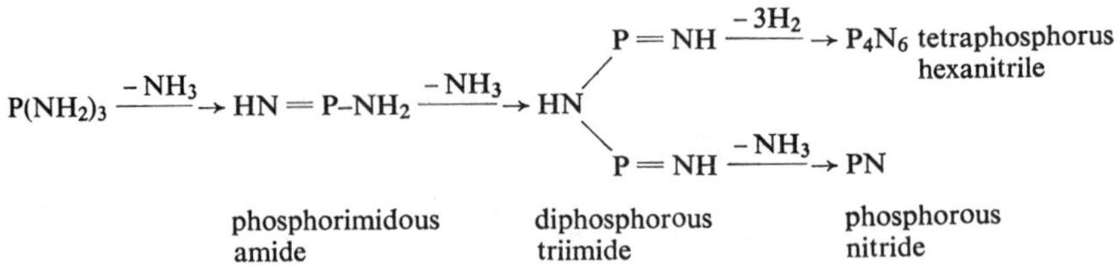

$$\mathrm{P(NH_2)_3} \xrightarrow{-\mathrm{NH_3}} \mathrm{HN{=}P\text{-}NH_2} \xrightarrow{-\mathrm{NH_3}} \mathrm{HN(P{=}NH)_2}$$

$$\mathrm{HN(P{=}NH)_2} \xrightarrow{-3\mathrm{H_2}} \mathrm{P_4N_6} \text{ tetraphosphorus hexanitrile}$$

$$\mathrm{HN(P{=}NH)_2} \xrightarrow{-\mathrm{NH_3}} \mathrm{PN}$$

phosphorimidous amide — diphosphorous triimide — phosphorous nitride

Nitrogen Derivatives of Phosphoric Acid

aquo-ammono derivatives:

$$\mathrm{OP(OH)_3} \longrightarrow \mathrm{OP(NH_2)(OH)_2} \longrightarrow \mathrm{OP(NH_2)_2(OH)} \longrightarrow \mathrm{OP(NH_2)_3}$$

phosphoric acid → phosphoramidic acid → phosphorodiamidic acid → phosphoric triamide

phosphoric acid $\xrightarrow{-2\mathrm{NH_3}}$ $\mathrm{N[P(O)(OH)_2]_3}$ nitridotri-phosphoric acid

phosphoramidic acid $\xrightarrow{-\mathrm{NH_3}}$ $\mathrm{HN[P(O)(OH)_2]_2}$ diphosphorimidic acid $\xrightarrow{-\mathrm{H_2O}}$ $\mathrm{HN[P(O)(OH)_2]\text{-}P(O)(OH)\text{-}NH[P(O)(OH)_2]}$ triphosphordiimidic acid

phosphorodiamidic acid $\xrightarrow{-\mathrm{NH_3},\ +\mathrm{H_2O}}$ $\mathrm{[NP(OH)_2]_x}$ metaphosphimic acid (phosphonitrile acid)

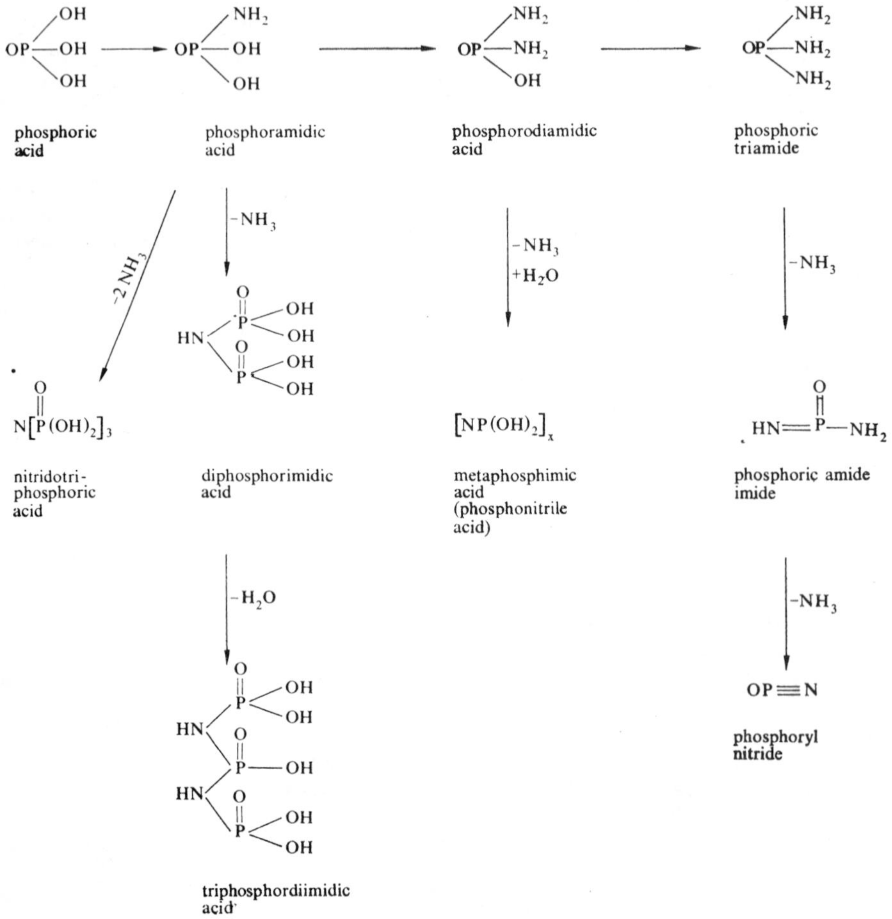

phosphoric triamide $\xrightarrow{-\mathrm{NH_3}}$ $\mathrm{HN{=}P(O)\text{—}NH_2}$ phosphoric amide imide $\xrightarrow{-\mathrm{NH_3}}$ $\mathrm{OP{\equiv}N}$ phosphoryl nitride ammono derivatives:

$$P(NH_2)_5 \xrightarrow{-NH_3} HN{=}P(NH_2)_3 \xrightarrow{-NH_3} [NP(NH_2)_2]_x \xrightarrow{-NH_3} N{\equiv}P{=}NH \xrightarrow{-NH_3} P_3N_5$$

phosphorus pentamide; phosphoimidic triamide; phosphonitrilamide; phospham; triphosphorus pentanitride

derivatives prepared from ammonia. Some of the compounds listed on chart 1 are, at present, unknown. With minor exceptions, the organic derivatives will not be included in this chapter. A review on phosphorus–nitrogen compounds by E. Fluck[280] is recommended as reference for the subject.

9.1. NITROGEN DERIVATIVES OF PHOSPHOROUS ACID

The reaction of phosphorus trichloride with a saturated solution of ammonia in chloroform at $-78°$ forms as product a mixture of phosphorous triamide and ammonium chloride. Phosphorous triamide is unstable at room temperature and attempts at isolation from ammonium chloride have not been successful. It is obtained, however, as an adduct with boron hydride, $(NH_2)_3PBH_3$, by the ammoniation of $F_3P{\cdot}BH_3$ [280].

The polymeric phosphorimidous amide, $(HN{=}P{-}NH_2)_n$, is prepared by the addition of phosphorus trichloride to ammonia in ethereal solution at $-20°$. By-product ammonium chloride is removed by washing with liquid ammonia in which the desired product is only difficultly soluble. Phosphorimidous amide is unstable at room temperature, decomposing into a brick-red solid of indefinite composition.

Phosphorous nitride, PN, is formed when an excess of PCl_3 is added to liquid ammonia. The NH_4Cl by-product may be removed by washing with liquid ammonia or by heating with diethylamine in chloroform. In the latter process, ammonium chloride is converted to chloroform soluble diethylamine hydrochloride and volatile ammonia. Phosphorous nitride is isolated as a yellow amorphous powder which is insoluble in chloroform or liquid ammonia[280].

From the reaction of PCl_3 with CH_3NH_2, Holmes *et al.* isolated an interesting crystalline compound of the composition $P_2N_3(CH_3)_3$ (m.p. 122.0–122.8°, b.p. 303–304°/739 mm). Molecular weight measurements and NMR data indicated that the compound has a tetrahedral structure analogous to that of phosphorous oxide, P_4O_6.

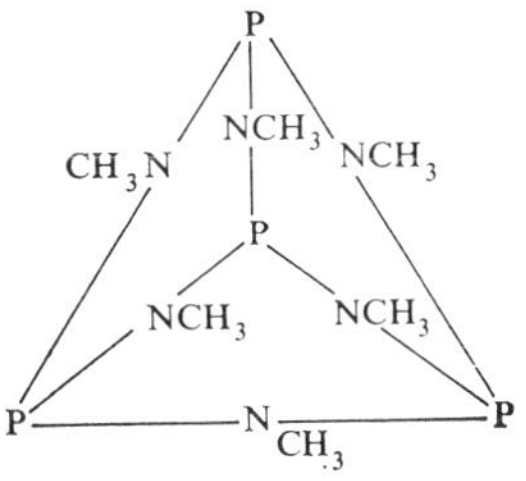

P; CH_3N; NCH_3; NCH_3; P; NCH_3; NCH_3; P; N; CH_3; P

It undergoes many reactions via the phosphorus atoms such as with O_2 or S analogous to those reactions of phosphorous oxide.

[280] E. Fluck, *Topics in Phosphorus Chemistry*, **4** (1967) 293–466.

9.2. NITROGEN DERIVATIVES OF PHOSPHORIC ACID

The sodium salts of both phosphoramidic acid, $H_2NP(O)(OH)_2$ [281], and phosphorodiamidic acid, $(H_2N)_2P(O)OH$, were first prepared by Stokes by the saponification of the corresponding phenyl esters with strong caustic[282]. The phenyl esters, $(C_6H_5O)_2P(O)NH_2$ and $C_6H_5OP(O)(NH_2)_2$, were in turn prepared by the ammoniation of the pure $C_6H_5OPOCl_2$ or $(C_6H_5O)_2POCl$. An alternative procedure for the phenyl esters is the ammoniation of the reaction mixture $C_6H_5OH–POCl_3–C_5H_5N$ prepared using the optimum ratio selected for the desired amide. The pure amides are then separated by crystallization from selective solvents[283].

The free phosphoramidic acid, $H_2NP(O)(OH)_2$, is obtained from the sodium salt by first converting to the lead salt followed by treatment with H_2S. It has also been obtained directly from the phenyl ester by catalytic hydrogenation[280]. The compound is a white solid soluble in water with the dissociation constants of $K_1 = 1.2 \times 10^{-3}$ and $K_2 = 2.1 \times 10^{-8}$ or $pK_1 = 4.6$ and $pK_2 = 7.7$. At room temperature it decomposes slowly (more rapidly at 100°) into ammonium polyphosphates having the same analytical composition as that of the parent compound.

The sodium salt of phosphoramidic acid is stable and resists hydrolysis in an aqueous solution at a pH above 7. It undergoes thermal condensation when heated under vacuum at 210° into the sodium salt of diphosphorimidic acid, $Na_4P_2O_6(NH)$, with the evolution of one mole of ammonia. Under more drastic conditions (450° for seven days), three moles of sodium phosphoramidate condense with the evolution of two moles of ammonia to form sodium nitridotriphosphate[280].

$$3(NaO)_2P(O)NH_2 \longrightarrow 2NH_3 + N[P(O)(ONa)_2]_3$$

Phosphorodiamidic acid, $(H_2N)_2P(O)OH$, is obtained in high purity by the treatment of the silver salt, $AgOP(O)(NH_2)_2$, with HBr. The silver salt in turn is derived from the sodium salt prepared either by the saponification of the phenyl ester or by the hydrolysis of phosphoric triamide, $OP(NH_2)_3$, in aqueous sodium hydroxide.

The dissociation constant for phosphorodiamidic acid is $K = 1.2 \times 10^{-5}$. The compound

[281] R. Klement, *Inorganic Synthesis*, Vol. XI, pp. 100–101, Ed. E. G. Rochow, McGraw-Hill Book Co., New York (1960).

[282] H. N. Stokes, *Am. Chem. J.* **15** (1893) 198; *ibid.* **16** (1894) 123.

[283] L. F. Audrieth and A. D. F. Toy, *J. Am. Chem. Soc.* **63** (1941) 2117–19.

reacts with atmospheric moisture to form ammonium hydrogen phosphoramidate, $NH_4O\overset{O}{\overset{\|}{P}}(OH)NH_2$. In an aqueous solution it is converted eventually to diammonium phosphate, $OP(OH)(ONH_4)_2$.

Anhydrous sodium phosphorodiamidate, $NaO\overset{O}{\overset{\|}{P}}(NH_2)_2$, undergoes thermal condensation at above 155–163° with evolution of ammonia into a polymeric compound containing the phosphorus imide, $-\overset{O}{\overset{\|}{P}}-\overset{H}{\overset{|}{N}}-\overset{O}{\overset{\|}{P}}-$, linkages; the degree of condensation increases with the temperature and time of the reaction[280].

Several silver salts of phosphorodiamidic acid have been described by Stokes[282]. The mono-silver salt, $AgO\overset{O}{\overset{\|}{P}}(NH_2)_2$, when treated with KOH forms $KO\overset{O}{\overset{\|}{P}}(NHAg)_2$ which on heating in hot water disproportionates into a brown salt having a proposed structure of $AgOP(O)(NAg_2)_2$. This last compound detonates readily.

Over the years it has been assumed that the reaction product of $POCl_3$ with ammonia contains as the desired product, phosphoric triamide:

$$OPCl_3+6NH_3 \longrightarrow OP(NH_2)_3+3NH_4Cl$$

The isolation of the triamide in the pure form, however, was not achieved until the development of the elegant method of Klement and Koch[284, 285] which by treatment of the reaction mixture with diethyl amine in $CHCl_3$ solution converts the NH_4Cl to the chloroform soluble diethylammonium chloride and volatile ammonia. The chloroform insoluble $OP(NH_2)_3$ is then isolated by filtration.

Phosphoric triamide crystallizes as colorless needles from methanol. It dissolves in water with only slight dissociation. In dilute caustic, it gives the sodium salt of phosphorodiamidic acid described above. On prolonged contact with atmospheric moisture it hydrolyzes to ammonium hydrogen phosphoramidate, $OP(OH)(ONH_4)NH_2$.

When phosphoric tetramide is heated, progressive condensation occurs with the elimination of ammonia. Imidodiphosphoric tetramide, $(NH_2)_2P(O)NHP(O)(NH_2)_2$, is first formed, then diimidotriphosphoric tetramide, $(NH_2)_2P(O)NHP(O)NHP(O)(NH_2)_2$. Finally, at 450–500°, the highly polymeric phosphoryl nitride, $(OPN)_n$, is formed. The compound $(OPN)_n$ loses 50% of its weight when heated for 4 hr at 600° and 84% when heated for 3 hr at 800° [286]. The polymeric residue left after heating at 800° has a spatial structure in which one nitrogen atom is bonded to three phosphorus atoms[287].

The compound phospham, NPNH, an insoluble and amorphous polymeric product, has been known since 1811. It was originally prepared by heating the reaction product of PCl_5

[284] R. Klement and O. Koch, *Chem. Ber.* **87** (1954) 333.

[285] R. Klement, *Inorganic Synthesis*, Vol. VI, pp. 108–10, Ed. E. G. Rochow, McGraw-Hill Book Co., New York (1960).

[286] T. D. Averbukh, N. P. Bakina and L. V. Alpatova, *Vysokomol. Soedin.* **8**(10) (1966) 1754–9.

[287] M. Ya Koroleva and V. G. Dubinin, *Zh. Prikl. Spektrosk.* **5**(3) (1966) 344–8.

and NH_3. Subsequently it has been prepared by the thermal decomposition of the trimeric or tetrameric phosphorus nitrile amide:

$$[NP(NH_2)_2]_{3-4} \xrightarrow{\Delta} [(NPNH)_{3-4}]_n + 3\text{–}4\ NH_3$$

The phosphorus atoms in these products are believed to be joined by nitrogen and imido bridges[280]. Thermal decomposition of phospham leads to triphosphorus pentanitride (P_3N_5) and phosphorus mononitride (PN).

Becke-Goehring and Scharf[288] obtained from the reaction of PCl_5 and an excess of liquid ammonia under anhydrous conditions a compound of the structure:

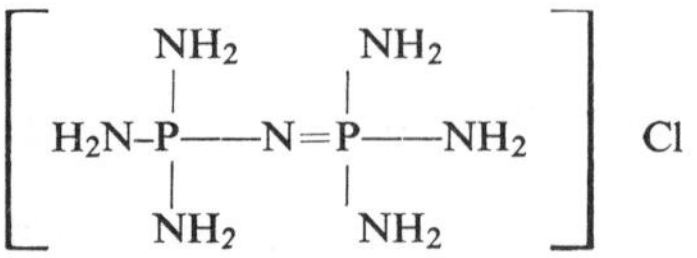

$$\left[H_2N\text{–}\underset{NH_2}{\overset{NH_2}{P}}\text{—}N{=}\underset{NH_2}{\overset{NH_2}{P}}\text{—}NH_2 \right] Cl$$

The NH_4Cl by-product formed in this reaction was removed by the method of diethylamine in chloroform solution.

9.3. NITROGEN DERIVATIVES OF POLYPHOSPHORIC ACIDS

Tetramide of pyrophosphoric acid, $(NH_2)_2P(O)\text{–}O\text{–}P(O)(NH_2)_2$, is obtained by the reaction of pyrophosphoryl chloride with ammonia:

$$Cl_2P(O)\text{–}O\text{–}P(O)Cl_2 + 8NH_3 \longrightarrow (NH_2)_2P(O)\text{–}O\text{–}P(O)\ (NH_2)_2 + 4NH_4Cl$$

Ammonium chloride by-product is removed by extraction with liquid ammonia[289].

Amido derivatives of polyphosphates are obtained by the ammonolysis of the polyphosphate P–O–P bonds. For example, amidotriphosphate is formed by the action of aqueous ammonia on ammonium trimetaphosphate:

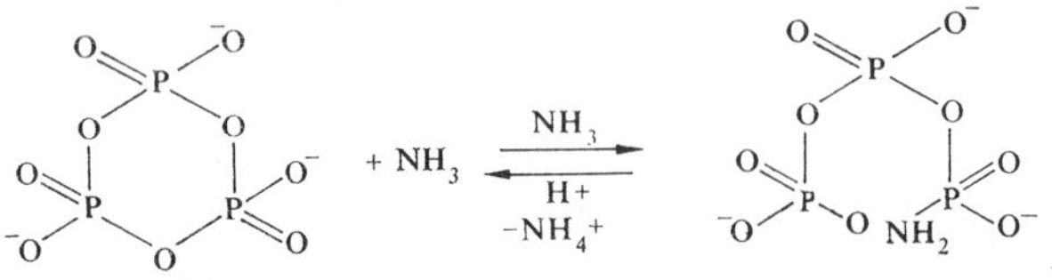

$$\text{(trimetaphosphate ring)} + NH_3 \underset{\substack{H+ \\ -NH_4^+}}{\overset{NH_3}{\rightleftharpoons}} \text{(amidotriphosphate)}$$

This reaction is reversible. Acidification regenerates the trimetaphosphate ion[290].

A water-insoluble polyphosphoric amide prepared by the thermal condensation of the reaction product of $POCl_3$ and ammonia at 150–200° (NH_4Cl removed by washing with water) has been found to be a good flame retardant for polyurethane foam[291].

9.4. METAPHOSPHIMIC ACIDS

Both tri- and tetrametaphosphimic acids and their salts were first prepared by Stokes[292] by the hydrolysis of the trimeric and tetrameric phosphonitrile chlorides:

288 M. Becke-Goehring and B. Scharf, *Z. anorg. u. allgem. Chem.* **353** (1967) 320–3.
289. R Klement and L. Benek, *Z. anorg. u. allgem. Chem.* **287** (1956) 12.
290 W. Feldman and E. Thilo, *Z. anorg. u. allgem. Chem.* **328** (1964) 113.
291 Pittsburgh Plate Glass Co., Brit. 918,636, 13 Feb. 1963.
292 H. N. Stokes, *Am. Chem. J.* **17** (1895) 275.

$$(PNCl_2)_3 \xrightarrow{HOH} (PO(OH)NH)_3$$

$$(PNCl_2)_4 \xrightarrow{HOH} (PN(OH)_2)_4 \longrightarrow (PO(OH)NH)_4$$

Subsequent modification on the preparation of trisodium trimetaphosphimate 1-hydrate involves the hydrolysis of trimeric phosphonitrile chloride in a dioxane solution with an aqueous solution of sodium acetate at 45–55° . The tripotassium trimetaphosphimate is similarly prepared by the use of an aqueous solution of potassium acetate[293]. Trimetaphosphimic acid is then obtained by the acidification of the potassium salt with perchloric acid and the insoluble potassium perchlorate removed by filtration. Upon concentration of the filtrate, the trimetaphosphimic acid is isolated as the dihydrate $H_3(PO_2NH)_3{\cdot}2H_2O$, which transforms to the anhydrous form at 33°. It is a strong acid with three replaceable hydrogens giving a single deflection in electrometric titration curves. The mono-, di- or trisodium salts may be prepared by progressive neutralization[294].

The potassium, rubidium, cesium and ammonium salts of tetrametaphosphimic acid have also been prepared and characterized in the various hydrated and anhydrous forms. The geometric isomerism of tetrametaphosphimate ion allows the existence of two species of the hydrated potassium salts and two modifications of the anhydrous compound[295].

Wanek and Thilo studied the hydrolysis of sodium trimetaphosphimate under a stream of moist air at 135° and found that the ring opens to form amidodiimidotriphosphate, which under these conditions is very unstable and hydrolyzes quickly to form Na_2HPO_4 [296].

Under a severe condition of alkaline hydrolysis, Stokes obtained from $(PNCl_2)_3$ also the sodium amidodiimidotriphosphate

$$(NaO)_2P(O){-}NH{-}P(O)(ONa){-}NH{-}P(O)(ONa)NH_2$$

Acid hydrolysis of sodium trimetaphosphimate was found by Quimby *et al.*[298] to

[293] M. L. Nielsen and T. J. Morrow, *Inorganic Synthesis*, Vol. VI, pp. 97–9, Ed. E. G. Rochow, McGraw-Hill Book Co., New York (1960).

[294] M. L. Nielsen, *Inorganic Synthesis*, Vol. VI, pp. 79–80, Ed. E. G. Rochow, McGraw-Hill Book Co., New York (1960).

[295] K. Lunkwitz and E. Steger, *Z. anorg. u. allgem. Chem.* **358** (1968) 111–124.

[296] W. Wanek and E. Thilo, *Z. Chem.* 7(3) (1967) 108.

[297] H. N. Stokes, *Am. Chem. J.* **18** (1896) 629–63.

[298] A. Narath, F. H. Lohman and O. T. Quimby, *J. Am. Chem. Soc.* **78** (1956) 4493–4.

proceed largely through intermediate ring compounds with one, two and three oxygen atoms replacing successively the imide linkages of the ring:

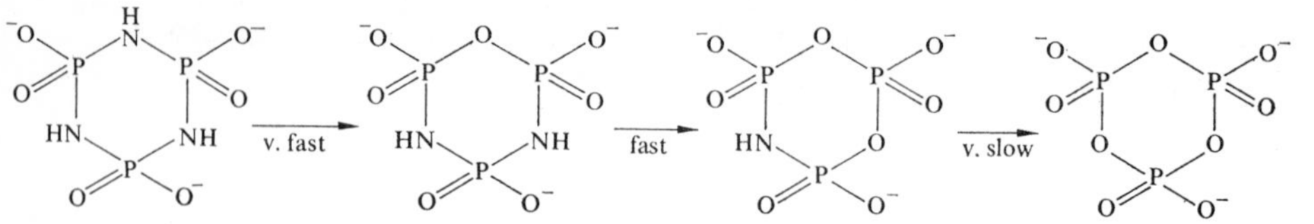

v. fast → fast → v. slow

In this acid hydrolysis, chain imidophosphates were never found in large amount. The P–O–P linkages in the ring are formed by the elimination of NH_3 between P–OH and P–NH_2 groups. The ring compounds containing both the imide, $P\!-\!\overset{H}{N}\!-\!P$ and P–O–P bridges are converted to the linear imidophosphate by treatment with 30% NaOH. Cleavage occurs at the P–O–P linkage while the $P\!-\!\overset{H}{N}\!-\!P$ bridge appears to be inert at the pH above 11[299]. Thus:

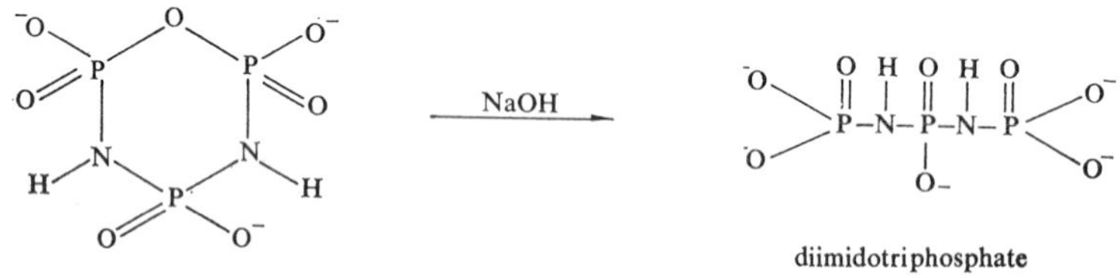

NaOH →

diimidotrimetaphosphate

diimidotriphosphate

9.5. NITROGEN DERIVATIVES OF THIOPHOSPHORIC ACID

Sodium salt of phosphoramidothioc acid, $Na_2PO_2S(NH_2)$, is obtained by the saponification of the diphenyl ester, $(C_6H_5O)_2\overset{S}{\overset{\|}{P}}NH_2$. Sodium phosphorodiamidothioate, $NaPOS(NH_2)_2$, is obtained by hydrolysis of phosphorothioic triamide $SP(NH_2)_3$ in dilute NaOH.

A mixture of diammonium phosphoramidothioate, $(NH_4)_2PO_2S(NH_2)$, and ammonium phosphorodiamidothioate, $NH_4POS(NH_2)_2$, is obtained by the reaction of $PSCl_3$ with aqueous solution of ammonia. In this reaction the yield of the diamido compound is favored by the use of a higher concentration of ammonia. The free acids of both of the mono and diamido compounds have not been isolated[280].

Phosphorothioic triamide, $SP(NH_2)_3$, is obtained along with ammonium chloride when $PSCl_3$ is treated with an excess of ammonia. The same procedure as used in the isolation of phosphoric triamide which involves diethylamine in chloroform is employed for the separation of the desired product from ammonium chloride[300].

Phosphorothioic triamide, a colorless crystalline rhombohedric solid, is unstable in moist air, decomposing to the diammonium salt of phosphorothioic acid, $(NH_4)_2HPO_3S$.

[299] O. T. Quimby and A. Narath, *Inorganic Synthesis*, Vol. VI, pp. 104–6, Ed. E. G. Rochow, McGraw-Hill Book Co., New York (1960).

[300] R. Klement, *Inorganic Synthesis*, Vol. VI, pp. 111–12, Ed. E. G. Rochow, McGraw-Hill Book Co., New York (1960)

When heated it loses ammonia, and at 800° the final product is triphosphorus pentanitride, P_3N_5[280], instead of the anticipated phosphorothioic nitride, SPN, the thio-analog of OPN from the corresponding thermal decomposition of $OP(NH_2)_3$. The polymeric phosphorothioic nitride, $(SPN)_n$, is however obtained via an entirely different route, that of the ammonolysis of P_4S_{10} [280]:

$$P_4S_{10}+4NH_4Cl \longrightarrow 4(SPN)_n+6H_2S+4HCl$$

10. PHOSPHONITRILIC HALIDES (HALOPHOSPHAZENES)

10.1. PHOSPHONITRILIC CHLORIDES

Phosphonitrilic chlorides with a general formula of $(PNCl_2)_n$ were first discovered in 1834 by J. von Liebig[301] in very low yield from the reaction of ammonia with PCl_5. Later Stokes[302] in the reaction of NH_4Cl with PCl_5 (heated in a sealed tube at 150–200°) produced and isolated various phosphonitrilic chlorides ranging from the trimer to the hexamer and an impure heptamer in addition to a polymeric residue. Stokes also found that the trimer when heated in a sealed tube at 250° polymerized to a rubbery polymer, the first inorganic rubber.

During the last few decades, the search for a new inorganic type of polymer provided a strong impetus for research on phosphonitrilic chemistry. The first review on the subject appeared in 1943[303]. More recent surveys are by Allcock[304, 305], Fluck[306], Paddock[307], Shaw[308], and Yvernault and Casteignau[309].

Mono- and Dimeric Adducts

Phosphonitrilic chloride does not exist in the monomeric form, except possibly as an adduct. The tertiary amine adduct has been reported in the literature[303] as $PNCl_2.2NR_3$. The monomeric formula is assumed because the adducts, whether obtained from the trimeric or tetrameric phosphonitrilic chlorides, have the same melting point.

Adducts such as $POCl_3.PNCl_2$ and $PNCl_2.2PCl_5$ have also been reported. The compound $POCl_3.PNCl_2$ has probably the structure of $Cl_2P(=O)–N=PCl_3$. One method for its preparation[310] is by the reaction of PCl_3 and PCl_5 with hydroxylammonium chloride:

$$2PCl_3+PCl_5+2[H_3NOH]Cl \longrightarrow Cl_3P=N–P(=O)Cl_2+4HCl+NH_4Cl+POCl_3$$

It is a colorless crystal, m.p. 35.7°, b.p. 102°/1 mm. A super-cooled melt has a refractive

[301] J. von Liebig, *Ann. Chem. Pharm.* **11** (1834) 139.
[302] H. N. Stokes, *Am. Chem. J.* **19** (1897) 782.
[303] L. F. Audrieth, R. Steinman and A. D. F. Toy, *Chem. Rev.* **32** (1943) 109.
[304] H. R. Allcock, *Chem. and Eng. News*, 22 April (1968) 68–81.
[305] H. R. Allcock, *Heteroatom Ring Systems and Polymers*, Academic Press, New York (1967).
[306] E. Fluck, *Topics in Phosphorus Chemistry*, **4** (1967) 293–466.
[307] N. L. Paddock, *Quart. Rev. (London)*, **18** (1964) 168–210.
[308] R. A. Shaw, B. W. Fitzsimmons and B. C. Smith, *Chem. Rev.* **62** (1962) 247.
[309] T. Yvernault and G. Casteignau, *Bull. Soc. Chim. de France*, **4** (1966) 1469–93.
[310] M. Becke-Goehring and E. Fluck, *Inorganic Synthesis*, Vol. VIII, pp. 92–94, Ed. H. F. Holtzclaw Jr., McGraw-Hill Book Co., New York (1966).

index $n_D^{25} = 1.5313$. This compound is sensitive to moisture and soluble in benzene, nitrobenzene and sym-tetrachloroethane.

The $PNCl_2.2PCl_5$ adduct has actually the structure of $[Cl_3P{=}N{-}PCl_3]\,[PCl_6]$. It may be prepared by the reaction of PCl_5 with NH_4Cl in the presence of $POCl_3$ as solvent in which case $PNCl_2.POCl_3$ is obtained as a by-product[311]. A more straightforward method of synthesis is that of Becke-Goehring and Fluck[312]:

$$3PCl_5+NH_4Cl \longrightarrow [Cl_3P = N{-}PCl_3]\,[PCl_6]+4HCl$$

This procedure involves the heating of the reactants in sym-tetrachloroethane solvent under vacuum at 80° until most of the PCl_5 is consumed. The product is isolated as needle-like pale greenish-yellow crystals, m.p. 310–315°. Reaction of $[Cl_3P{=}N{-}PCl_3]\,[PCl_6]$ with more NH_4Cl[306] results in the formation of $[Cl_3P{=}N{-}PCl_2{=}N{-}P{-}Cl_3]\,[PCl_6]$. This compound has the composition equivalent to the adduct of the dimeric phosphonitrilic chloride with 2 moles of phosphorus pentachloride:

$$(PNCl_2)_2.2PCl_5$$

Cyclic Polymers

Preparation. The preparation of cyclic and polymeric phosphonitrilic chlorides today is based essentially on the original reaction described by Stokes. One basic modification is that of Schenck and Römer[313] who found that the use of a solvent to carry out the reaction at around 140–150° improves the yield of the distillable cyclic polymers. Sym-tetrachloroethane with a b.p. of 146.3° has been the solvent of choice. However, chlorobenzene and chlorotoluene are also effective solvents.

The general procedure[314] for the preparation of the trimeric and tetrameric phosphonitrilic chloride is based on the following equations:

$$3\,PCl_5+3NH_4Cl \longrightarrow (PNCl_2)_3+12HCl$$
$$4\,PCl_5+4NH_4Cl \longrightarrow (PNCl_2)_4+16HCl$$

A slurry of ammonium chloride (about 40 to 50% excess) is heated with PCl_5 in sym-tetrachloroethane under reflux until the evolution of HCl ceases. The excess ammonium chloride is removed by filtration and the filtrate then concentrated by distilling off the solvent under reduced pressure. The buttery residue left is distilled quickly under reduced pressure to obtain a mixture of the trimer, tetramer and a little higher cyclic polymers in the distillate. The trimer and tetramer may then be separated by fraction distillation and recrystallization. The heat of formation of $(PNCl_2)$ is -178.1 ± 3 kcal/mole[315].

One modification of the above procedure is by the preparation of NH_4Cl *in situ* in chlorobenzene solvent by the reaction of gaseous HCl and NH_3. Ammonium chloride of the particle size of 5μ is thus obtained. Such fine NH_4Cl increases the yield of the cyclic polymers[316].

Another modification is by the addition of PCl_5 to a refluxing suspension of NH_4Cl in sym-tetrachloroethane during a 7-8 hr period. The yield of the trimer is thus increased

311 G. Barth-Wehrenalp and A. Kowalski, U.S. 2,975,028, 14 March 1961, to Pennsalt Chemicals Corp.
312 Ref. 310, pp. 94–97.
313 R. Schenck and G. Römer, *Ber.* **57B** (1924) 1343.
314 M. L. Nielsen and G. Cranford, *Inorganic Synthesis*, Vol. VI, pp. 94–97, Ed. E. G. Rochow, McGraw-Hill Book Co., New York (1960).
315 C. T. Mortimer, *Pure and App. Chem.* **2**(1–2) (1961) 71–76.
316 Albright and Wilson, Brit. 1,017,375, 19 Jan. 1966.

at the expense of higher linear polymer, $(PNCl_2)_n.PCl_5$. From the petroleum solvent soluble fraction of this reaction mixture, Paddock[317] and his co-workers, using a combination of fractional extraction, fractional distillation *in vacuo*, and fractional crystallization, have been able to obtain individual cyclic polymers from the trimer to the octamer.

The mechanism of formation of $[NPCl_2]_n$ is believed to proceed through the intermediate cations of the type $[Cl_3P=NPCl_3]^+$ and $[Cl_3P=N(PCl_2=N)_nPCl_3]^+$ [318, 319], as evidenced by the fact that trimeric, tetrameric and higher cyclic polymers of phosphonitrilic chlorides have been obtained directly from compounds containing these cations by further reaction with NH_4Cl.

The intermediate products from the reaction of PCl_5 and NH_4Cl preceding the actual formation of the cyclic phosphonitrilic chlorides have been summarized by Fluck as shown in Chart 2[306]. Some reactions of these intermediates as well as the alternative routes to their synthesis are included in this chart.

CHART 2 [306]

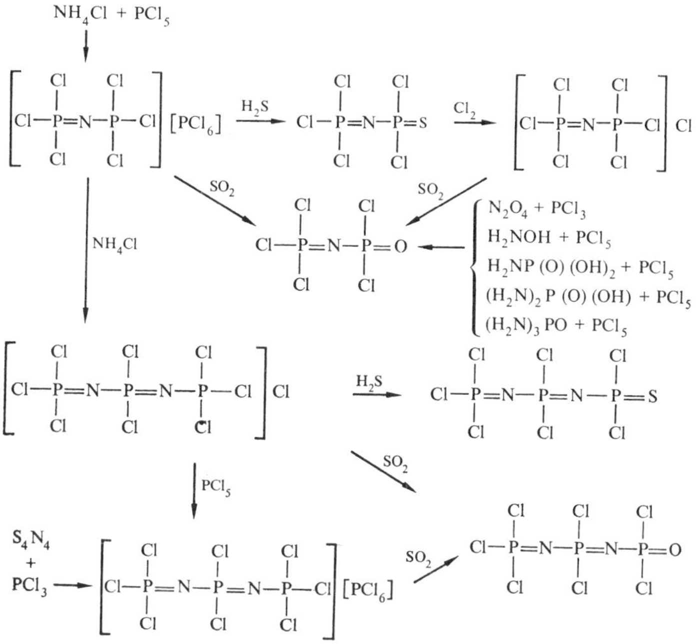

$NH_4Cl + PCl_5 \rightarrow [Cl_3P{=}N{-}PCl_3][PCl_6]$

$[Cl_3P{=}N{-}PCl_3][PCl_6] \xrightarrow{H_2S} Cl_3P{=}N{-}PCl_2{=}S \xrightarrow{Cl_2} [Cl_3P{=}N{-}PCl_3]Cl$

$[Cl_3P{=}N{-}PCl_3][PCl_6] \xrightarrow{SO_2} Cl_3P{=}N{-}PCl_2{=}O \xleftarrow{SO_2} [Cl_3P{=}N{-}PCl_3]Cl$

$Cl_3P{=}N{-}PCl_2{=}O \leftarrow \begin{cases} N_2O_4 + PCl_3 \\ H_2NOH + PCl_5 \\ H_2NP(O)(OH)_2 + PCl_5 \\ (H_2N)_2P(O)(OH) + PCl_5 \\ (H_2N)_3PO + PCl_5 \end{cases}$

$[Cl_3P{=}N{-}PCl_3][PCl_6] \xrightarrow{NH_4Cl} [Cl_3P{=}N{-}PCl_2{=}N{-}PCl_3]Cl \xrightarrow{H_2S} Cl_3P{=}N{-}PCl_2{=}N{-}PCl_2{=}S$

$[Cl_3P{=}N{-}PCl_2{=}N{-}PCl_3]Cl \xrightarrow{SO_2} Cl_3P{=}N{-}PCl_2{=}N{-}PCl_2{=}O$

$[Cl_3P{=}N{-}PCl_2{=}N{-}PCl_3]Cl \xrightarrow{PCl_5} [Cl_3P{=}N{-}PCl_2{=}N{-}PCl_3][PCl_6]$

$S_4N_4 + PCl_3 \rightarrow [Cl_3P{=}N{-}PCl_2{=}N{-}PCl_3][PCl_6] \xrightarrow{SO_2} Cl_3P{=}N{-}PCl_2{=}N{-}PCl_2{=}O$

Structure. The structure of trimeric[320] phosphonitrilic chloride (Fig. 9) as determined by X-ray diffraction studies is a six-membered ring of alternating phosphorus and nitrogen atoms in a plane with two chlorine atoms attached to each phosphorus atom. The structure of the tetramer is similar except that the ring is puckered[321]. The ten-membered ring structure of the pentamer has again a planar configuration[322].

[317] L. G. Lund, N. L. Paddock, J. E. Proctor and H. T. Searle, *J. Chem. Soc.* (1960) 2542–7.
[318] M. Becke-Goehring and W. Lehr, *Z. anorg. u. allgem. Chem.* **327**(3–4) (1964) 128–38.
[319] E. Kobayashi, *Nippon Kagaku Zasshi*, **87**(2) (1966) 135–41.
[320] A. Wilson and D. F. Carroll, *J. Chem. Soc.* (1960) 2548–52.
[321] J. R. van Wazer, *Phosphorus and Its Compounds*, Vol. I, pp. 309–28, Interscience Publishers Inc. New York (1958).
[322] A. W. Schlueter and R. A. Jacobson, *J. Am. Chem. Soc.* **88**(9) (1966) 2051.

In the skeletal structure of the phosphonitrilic system, phosphorus is tetracoordinated and nitrogen is divalent. Four of the five outer-shell electrons of phosphorus are used to form sigma bonds with the nitrogen and chlorine atoms. Two of the five L-shell electrons of nitrogen are used for the sigma bonds to the phosphorus atoms while two lone-pair

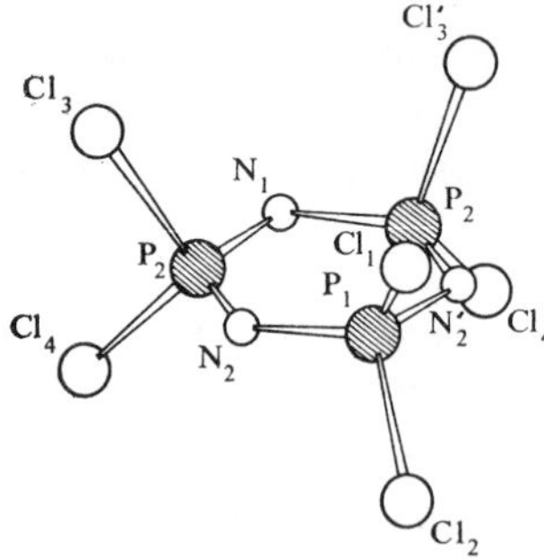

FIG. 9. The $(PNCl_2)_3$ molecule. The atoms P_1, Cl_1, Cl_2, and N_1 lie on a mirror plane,[320] (Reprinted by permission of the Chemical Society).

electrons are probably retained in the third sp^2 hybrid lobe. The remaining one electron each for phosphorus and nitrogen probably interact to form $d\pi$–$p\pi$ bonds by overlap of phosphorus $3d$ orbitals with nitrogen $2p$ or $2sp^3$ orbitals[304].

After a study of the theoretical and experimental evidence on the structure of the

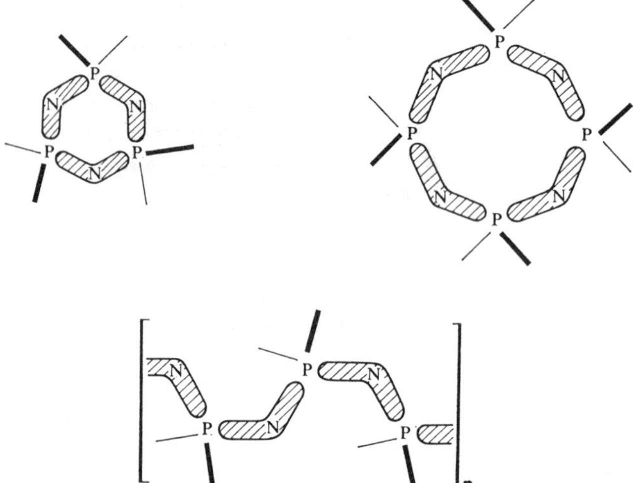

FIG. 10. Phosphorus atoms interrupt the discrete "islands" of pi character[304]. (Reprinted by permission of the American Chemical Society.)

phosphonitrilic ring system, Allcock[304] has concluded that the rings are not truly "pseudo-aromatic", except when very negative substituents are present. For most other substituents an "island-type" pi structure is favored. That is, the rings are stabilized by skeletal pi-bonding, and the pi bonds are interrupted at each phosphorus atom (see Fig. 10).

The $\gamma_{P:N}$ of N=P–N in cyclic compounds is assigned to the region of

1170–1438 cm^{-1} [323]. The trimer has a strong band at 1220 cm^{-1} and the tetramer has a strong band at 1315 cm^{-1} [324]. The phosphorus–nitrogen bond energy, $E(P=N-P)$ for $(PNCl_2)_3$ is 153 kcal/mole[315]. The NMR chemical shift relative to 85% H_3PO_4 is −20 ppm for the trimer and +7 ppm for the tetramer[317].

Polymerization. Cyclic phosphonitrilic chlorides polymerize to a colorless rubbery polymer when heated at 250°. The reaction is exothermic with $-\Delta H_{(polym)}$ decreasing with increasing ring size from 1.39 kcal/mole for the trimer to 0 kcal for the heptamer. The polymer is linear with a degree of polymerization of at least 15,000. It is soluble in such solvents as benzene or tetrahydrofuran. At down to −63°, it is still an elastomer and does not melt below 300° [304].

The polymerization reaction is reversible; depolymerization back to the cyclic oligomers occurs at above 350° and to the extent of 76–82% at 600° [325]. The phosphorus–chlorine bonds in the polymer are hydrolytically unstable. This causes the polymer to degrade slowly when exposed to atmospheric moisture.

The mechanism of the polymerization process is not entirely clear. The facts that peroxides do not seem to have an effect on the rate of polymerization and that no free radicals were detected by electron spin resonance techniques during polymerization seem to rule out the free radical mechanism. Allcock[304] favors the ionic mechanism. Compounds such as metals and carboxylic acids which facilitate the removal of the chloride ions from phosphorus are polymerization catalysts. Also conductance and capacitance measurements on equilibrating systems demonstrate the presence of ionic and polarizable species in the mixture[326]. The initiation step, therefore, probably involves the ionization of chloride ion from the phosphorus.

Derivatives. Phosphonitrilic chlorides undergo nucleophilic and electrophilic substitution reactions as well as salt and complex formation. Halogen replacements in the trimer and tetramers by hydroxide ions result in the formation of metaphosphimic acids:

$$[NPCl_2]_{3-4}+6H_2O \longrightarrow [NP(OH)_2]_{3-4}$$

In the reaction with other nucleophilic reagents such as alkoxides, phenoxides, primary or secondary amines, and organic metallic reagents, it is possible to obtain the fully or partially substituted appropriate derivatives. For the partially substituted derivatives, halogen replacement can take place geminally or nongeminally and *cis* or *trans*. Amines such as methylamine, dimethylamine or piperidine which are strongly basic, substitute nongeminally. Ammonia and tertbutylamine substitute geminally. Phenoxide ion substitutions are predominantly non-geminal. Electron supply from the amino group, steric factor, and structure of the transition states all probably play a part in the mechanism of these substitutions[304].

Of the cyclic phosphonitrilic halides, the tetramer reacts with n-propylamine faster than the trimer. High polymers of phosphonitrilic chlorides are also highly reactive. It is suggested that this reactivity is due to the fact that the more reactive species have greater flexibility in the approach to the transition state.

Of other substitution reactions, Friedel–Crafts arylation reactions with benzene in the presence of aluminum chloride are geminal while those with organometallic reagents are completely random[304].

[323] R. A. Chittenden and L. C. Thomas, *Spectrochim. Acta* **22**(8) (1966) 1449–63.

[324] R. A. Shaw, *Chem. and Ind.* (1959) 59.

[325] M. Yokoyama and S. Konya, *Kogyo Kagaku Zasshi*, **69**(9) (1966) 1835–6.

[326] H. R. Allcock and R. J. Best, *Can. J. Chem.* **42**(2) (1964) 447–55.

The alkoxy derivatives undergo a thermal rearrangement to the N-alkyl derivative:

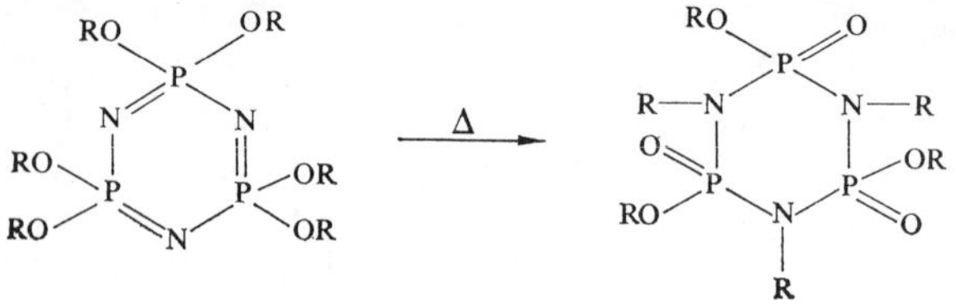

and the bishydroxy derivative rearranges to metaphosphimic acid:

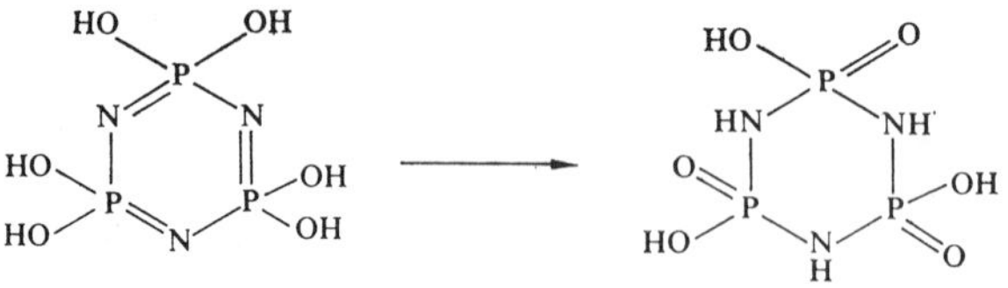

Applications. Because of their tendency to lose chlorine atoms upon exposure to atmospheric moisture, polymeric phosphonitrilic chlorides have not found practical applications. The cyclic phosphonitrilic compounds, in which all of the chlorine atoms are replaced by organic groups, e.g., $[NP(C_6H_5)_2]_4$, do not polymerize to a high molecular weight polymer.

Allcock has been successful in replacing all of the chlorine atoms in a noncross-linked high polymer $(PNCl_2)_n$ ($n \approx 15{,}000$) from the polymerization of pure $(PNCl_2)_3$ or $(PNCl_2)_4$. The trifluoroethoxy and phenoxy derivatives which he prepared are unaffected after immersion for several months in concentrated aqueous sodium hydroxide solution. These polymers are transparent, flexible and may be converted into films. One major drawback to them is their tendency to depolymerize to low-molecular weight cyclic homologs when heated near 200°.

The hydroquinone derivatives of the mixed trimeric and tetrameric phosphonitrilic chlorides may be cross-linked into flame-retardant polymers. The polymer of the hydroquinone derivative, in combination with the synthetic rubber Hycar, has been reported to be an excellent adhesive[327]. The hexabis (ethyleneimino) derivative

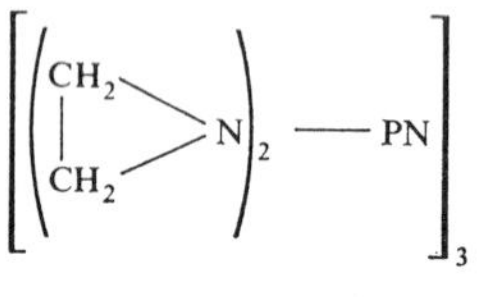

$$\left[\left(\begin{matrix}CH_2\\ |\\ CH_2\end{matrix}\!\!>\!N\right)_2 - PN\right]_3$$

has been found to be useful as a chemosterilant for houseflies and other insects[304]. The propyl ester is employed industrially as an additive type of flame retardant for viscose rayon. It is incorporated inside the fiber by addition to the viscose rayon dope prior to the extrusion process[328].

[327] R. G. Rice, B. H. Geib and J. R. Hooker, U.S. 3,219,915, 23 Nov. 1965, to General Dynamics Corp.
[328] L. E. A. Godfrey, Summaries of Papers Presented at 39th Annual Meeting, Textile Research Institute, Princeton, New Jersey, 9–11 April 1969, pp. 77–80, FMC, Austrian Patent Application A8560/67.

10.2. OTHER PHOSPHONITRILIC HALIDES

Phosphonitrilic Bromides

These compounds can be prepared by the action of NH_4Br on PBr_5 in a reaction analogous to that for phosphonitrilic chlorides. John and Moeller[329] found that a comparable reaction involving PBr_3, Br_2 and NH_4Br in sym-tetrachloroethane gives a somewhat better yield of the trimer and tetramer:

$$n\text{PBr}_3+n\text{Br}_2+n\text{NH}_4\text{Br} \longrightarrow (\text{PNBr}_2)_n+4n\text{HBr} \qquad (n = 3 \text{ or } 4)$$

This reaction takes about ten days for completion. After the removal of excess NH_4Br by filtration and the solvent by evaporation under reduced pressure, the residue obtained is extracted with benzene. From the benzene extracts, after evaporation of benzene, a dark semi-crystalline mass is obtained. Upon sublimation, the white crystalline product obtained contains about 92% trimer and 8% tetramer. The trimer is separated from the tetramer by fractional crystallization from anhydrous heptane.

Coxon and Sowerby[330] in their study of the reaction between PBr_5 and NH_4Br in sym-tetrachloroethane found besides the lower homologs also a small quantity of the hexameric compounds in the form of white needle crystals with a m.p. of 153–154°. Mass-spectroscopic studies indicated that the cyclic polymers were contaminated with mixed chloride-bromides due to halogen exchange reactions with the solvent sym-tetrachloroethane. This difficulty was eliminated when 1,2-dibromoethane was used as the solvent.

The solubilities in anhydrous n-heptane and petroleum ether (b.p. 90° to 110°) at 25° are for the trimer, 1.45 g and 2.30 g; for tetramer, 0.15 and 0.27 g per each 100 g of solvent respectively.

The P–N stretching frequencies are 1175 cm^{-1} for the trimer and 1272 cm^{-1} for the tetramer.

Trimeric and tetrameric phosphonitrilic bromides polymerize at near 250° in a manner similar to the chloride analogs[305]. Herring[331] also obtained the $(NPBr_2)_n$ polymer by the reaction of NaN_3 with PBr_3 at 165–167°:

$$\text{PBr}_3+\text{NaN}_3 \longrightarrow (\text{NPBr}_2)_n+\text{NaBr}+\text{N}_2$$

Phosphonitrilic Fluorides

Both trimeric and tetrameric phosphonitrilic fluorides are prepared by the fluorination of the corresponding chlorides with NaF in acetonitrile or nitrobenzene as solvents[332, 333]. Alternative fluorinating agents such as KSO_2F [334] or a mixture of KF and SO_2 [335, 336] have also been used. By the latter procedure, the cyclic polymers from the trimer to the undecamer were prepared and characterized. Another method of preparation for the

[329] K. John and T. Moeller, *Inorganic Synthesis*, Vol. VII, pp. 76–79, Ed. J. Kleinberg, McGraw-Hill Book Co., New York (1963).

[330] G. E. Coxon and D. B. Sowerby, *J. Chem. Soc.* (A) (1967) 1566–8.

[331] D. L. Herring, *Chem. and Ind.* (*London*) (1960) 717–18.

[332] R. Schmutzler, *Inorganic Synthesis*, Vol. IX, pp. 76–78, Ed. S. Y. Tyree, Jr., McGraw-Hill Book Co., New York (1967).

[333] T. Moeller and F. Tsang, *Inorganic Synthesis*, Vol. IX, pp. 78–79, Ed. S. Y. Tyree, Jr., McGraw-Hill Book Co., New York (1967).

[334] F. Seel and J. Langer, *Angew. Chem.* **68** (1956) 461.

[335] H. T. Searle, Brit. 895,969, 9 May 1962, to Albright and Wilson Ltd.

[336] A. C. Chapman, N. L. Paddock, D. H. Paine and H. T Searle, *J. Chem. Soc.* (1960) 3608–14.

CHART 3. PHYSICAL CONSTANTS OF PHOSPHONITRILIC CHLORIDES

	M.p. °C	Crystallographic data	Density d_{20}(g cc^{-1})	B.p. °C°	Bond distances and angles	Ref.
$(PNCl_2)_3$	112.8	a = 14.15Å b = 6.20Å c = 13.07Å	1.99	256	P–N ≃ 1.6Å P–Cl = 2.04Å ∠ ClPCl ≃ 100° ∠ PNP ≃ 120°	b, i
$(PNCl_2)_4$	122.8	a = 10.82Å c = 5.95Å	2.18	328.5	P–N = 1.68Å P–Cl = 2.00Å ∠ PNP = 123° ∠ NPN = 117° ∠ ClPCl = 105° 30′	b, i
$(PNCl_2)_5$	41.3	a = 19.37Å b = 15.42Å c = 6.23Å	2.02	224 at 13 mm		b i
$(PNCl_2)_6$	92.3	triclinic a = 10.6Å b = 10.7Å c = 11.4Å α = 93.5° β = 90° γ = 117°	1.96	262 at 13 mm		b, c
$(PNCl_2)_7$	8 – 12		1.890	293 at 13 mm		b i
$(PNCl_2)_8$	57 – 58	a = 24.7Å b = 6.2Å c = 20.4Å	1.99			b
$(PNBr_2)_3$	192	a = 14.43Å b = 13.36Å c = 6.63Å	3.182		P–N = 1.51 – 1.54Å P–Br = 2.08 – 2.22Å ∠ PNP = 122° – 126° ∠ NPN = 116° – 117° ∠ BrPBr = 102° – 103°	d j
$(PNBr_2)_4$	202	a = 11.18Å c = 6.29Å	3.439			d
$P_3N_3Cl_5Br$	122.5 – 123.5	a = 14.24Å b = 6.28Å c = 13.00Å	2.27			h j

CHART 3—*Continued*

	M.p. °C	Crystallographic data	Density d_{20}(g cc^{-1})	B.p. °C	Bond distances and angles	Ref.
$P_3N_3Cl_4Br_2$	132.5 - 135	a = 14.27Å b = 6.34Å c = 13.02Å	2.44			h
$P_3N_3Cl_2Br_4$	167.5 - 169	a = 14.29Å b = 6.48Å c = 13.33Å	2.84			
$(PNF_2)_3$	26	orthorhombic a = 6.948Å b = 12.190Å c = 8.723Å	2.237	50.9	P–N = 1.546Å 1.563Å 1.572Å P–F = 1.521Å ∠ PNP = 119.6° 121.1° ∠ NPN = 119.5° 119.3°	h e g
$(PNF_2)_4$	28 - 29	monoclinic a = 7.40Å b = 13.83Å c = 5.16Å β = 109.5°	2.239 (solid)	89.7	P–N = 1.51±0.02Å P–F = 1.51±0.02Å ∠ PNP = 147.2±1.4° ∠ NPN = 122.7±1.0° ∠ FPF = 99.9±1.0°	f g
$(PNF_2)_5$	-50		1.825	120		a
$(PNF_2)_6$	-45 5		1.841	147		a
$(PNF_2)_7$	-61		1.849	170		a
$(PNF_2)_8$	-16		1.856	192		a
$(PNF_2)_9$	-78		1.859	214		a
$P_4N_4Cl_2F_4$	-12.1		1.8742 at 13.5°	105.8		i

[a] T. Yvernault and G. Casteignau, *Bull. Soc. Chim. de France*, **4** (1966) 1469–93.

[b] L. G. Lund, N. L. Paddock, J. E. Proctor and H. T. Searle, *J. Chem. Soc.* (1960) 2542–7.

[c] M. Becke-Goehring and W. Lehr, *Z. anorg. u. allgem. Chem.* **327** (3–4) (1964) 128–38.

[d] K. John and T. Moeller, *Inorganic Synthesis*, Vol. VII, pp. 76–79, Ed. J. Kleinberg, McGraw-Hill Book Co., New York (1963).

[e] R. Schmutzler, *Inorganic Synthesis*, Vol. IX, pp. 76–78, Ed. S. Y. Tyree, Jr., McGraw-Hill Book Co., New York (1967).

[f] T. Moeller and F. Tsang, *Inorganic Synthesis*, Vol. IX, pp. 78–79, Ed. S. Y. Tyree, Jr., McGraw-Hill Book Co., New York (1967).

[g] T J. Mao, R. D. Dresdner and J. A. Young, *J. Am. Chem. Soc.* **81** (1959) 1020.

[h] R. B. Rice, L. W. Daasch, J. R. Holden and E. J. Kohn, *J. Inorg. and Nucl. Chem.* **5** (1958) 190–200.

[i] D. M. Yost and H. Russell, Jr., *Systematic Inorganic Chemistry of the Fifth- and Sixth-Group Non-metallic Elements*, pp. 108–12, Prentice-Hall Inc., New York (1944).

[j] *Gmelins Handbuch der anorganischen Chemie*, System-number 16, *Phosphorus*, Part C, 8th ed. pp. 541–67, Verlag Chemie, Weinheim (1965).

mixture of the trimer and tetramer involves the fluorination of P_3N_5 with CF_3SF_5 or NF_3 at 700° [337].

Trimeric phosphonitrilic fluoride is a colorless liquid with a heat of vaporization $\Delta H_{vap} = 7600$ cal/mole, $\Delta S_{vap} = 23.6$ e.u. The P–N stretching frequency is 1297 cm^{-1} [336]. Tetrameric phosphonitrilic fluoride is also a colorless liquid having a heat of vaporization $\Delta H_{vap} = 8.9$ kcal/mole, $\Delta S_{vap} = 26.4$ e.u. The P–N stretching frequency is 1419 cm^{-1} [336]. Unlike the corresponding chloride, the P–N ring $(PNF_2)_4$ is planar.

The trimer is less reactive than the tetramer. It may be washed with water. The trimer undergoes thermal polymerization in a sealed tube at 350° to a rubbery material. This polymer, however, hydrolyzes readily and depolymerizes partially when heated under vacuum[332].

Phosphonitrilic Mixed Halides

Trimeric phosphonitrilic mixed chloride–bromides are also known[338]. Thus $P_3N_3Cl_5Br$ (m.p. 122.5–123.5°), a white crystalline solid, is prepared by the reaction of PCl_5, NH_4Br and Br_2. When 3 moles of PBr_3 and 2 moles of PCl_5 are reacted with an excess of NH_4Cl, $P_3N_3Cl_4Br_2$ is obtained (m.p. 132.5–135.0). Reaction of one mole of PCl_3 and 1 mole of Br_2 with excess NH_4Br yields $P_3N_3Cl_2Br_4$ (m.p. 167.5–169°).

An interesting compound is $P_4N_4Cl_2F_6$, a derivative of the tetrameric $(PNCl_2)_4$ and yet it is obtained by the heating of a mixture of PbF_2 with the trimeric $(PNCl_2)_3$ [339]. It has a m.p. of −12.1°, b.p. 105.8° and a heat of vaporization of 8750 cal/mole.

The physical properties of various phosphonitrilic halides are summarized in Chart 3.

11. LOWER OXYACIDS OF PHOSPHORUS

Various oxyacids of phosphorus and their structural relationship to each other are listed on the following page.

In addition, there are the lower oxyacids of phosphorus in which each molecule contains two or more phosphorus atoms in the same or different oxidation states.

11.1. HYPOPHOSPHOROUS ACID AND HYPOPHOSPHITES

Preparation. Salts of hypophosphorous acid are prepared by heating white phosphorus in an aqueous solution of an alkali or alkaline earth hydroxide. The reaction is fairly complicated and by-products such as hydrogen, phosphite and phosphine are obtained. The main reactions appear to be:

$$P_4 + 4OH^- + 4H_2O \longrightarrow 4H_2PO_2^- + 2H_2$$
$$P_4 + 4OH^- + 2H_2O \longrightarrow 2HPO_3^{2-} + 2PH_3$$

The calcium salt of hypophosphite is water soluble and serves as an easy means of separating the hypophosphite from the insoluble calcium phosphite. A typical preparation

337 T. J. Mao, R. D. Dresdner and J. A. Young, *J. Am. Chem. Soc.* **81** (1959) 1020.
338 R. B. Rice, L. W. Daasch, J. R. Holden and E. J. Kohn, *J. Inorg. and Nuclear Chem.* **5** (1958) 190–200.
339 D. M. Yost and H. Russell, Jr., *Systematic Inorganic Chemistry of the Fifth- and Sixth-Group Nonmetallic Elements*, pp. 108–112, Prentice-Hall Inc., New York (1944).

involves the heating of white phosphorus at 80–90° in the presence of an aqueous solution of a mixture of NaOH–$Ca(OH)_2$. The yield of NaH_2PO_2 is in the order of 55–9%[340]. Improvement in the yield (73% based on the phosphorus consumed) is claimed by the use of a 5% excess of finely divided phosphorus and the reaction carried out at 35–40° [341].

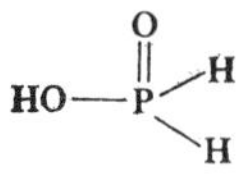

hypophosphorous acid

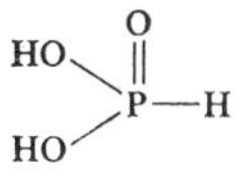

phosphorous acid

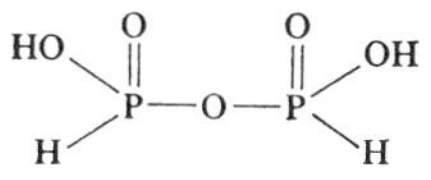

pyrophosphorous acid

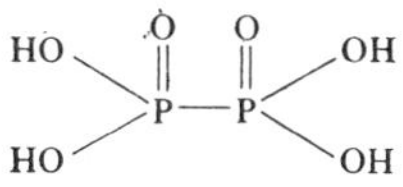

hypophosphoric acid

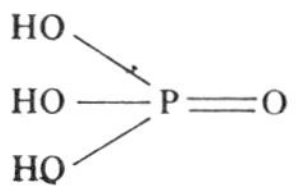

orthophosphoric acid

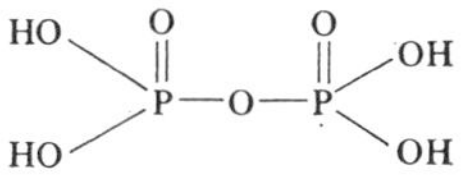

pyrophosphoric acid

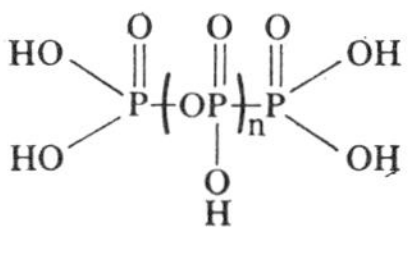

polyphosphoric acid

$(HPO_3)_n$ metaphosphoric acid

A similar modification carried out at 45–50° uses an emulsion of phosphorus to increase the reactive surface and amyl alcohol as catalyst to shorten the reaction time. The use of lower temperature for the reaction is claimed to have the effect of suppressing the formation of phosphine and phosphite with a resulting increase in the yield of hypophosphite[342].

Free hypophosphorous acid is generally prepared from its calcium salt by treatment

[340] E. M. Morgunova and T. D. Averbukh, *Zh. Prikl. Khim.* **40**(2) (1967) 174–84.

[341] H. Harnisch and F. Rodis, Ger. 1,112,054, 3 Aug. 1961, to Knapsack-Griesheim A.G.

[342] R. R. Pahud, Swiss 322,980, 30 Aug. 1957, to La Fonte electrique, S.A.

with an equivalent amount of sulfuric acid or oxalic acid[343]. It is not obtained by evaporation of an aqueous solution, since it would undergo oxidation to phosphorous and phosphoric acids as well as disproportionation to phosphine and phosphorous acid. Hypophosphorous acid is slightly soluble in ethyl ether and it has been found practical to extract it from an aqueous solution with ethyl ether in a continuous liquid–liquid extraction apparatus. The acid thus obtained is a colorless liquid which crystallized in cooling[344].

Hypophosphorous acid is formed also by the oxidation of PH_3 with iodine in water. This reaction has to be carried out under controlled conditions and in a weak acid medium. In the presence of excess iodine and strong acid medium, the oxidation reaction continues to the formation of phosphorous acid[345].

Hypophosphite is also formed by the oxidation of PH_3 in an aqueous solution of NaClO in accordance with the following proposed reaction scheme[346]:

$$OCl^- + H_3O^+ \rightleftharpoons HOCl + H_2O \text{ (fast)}$$

$$PH_3 + HOCl \rightarrow [PH_3O] + H^+ + Cl^- \text{ (rate detg).}$$

$$[PH_4O] + OCl^- \rightarrow H_3PO_2 + Cl^- \text{ (very fast)}$$

Structure. The hypophosphite anion of ammonium hypophosphite[347] has been shown to be a distorted tetrahedron with the phosphorus atom in the center, two oxygen atoms in two corners and two hydrogen atoms in the other two corners. The bond distances are P–O = 1.51 Å, P–H about 1.5 Å, and the bond angles: H–P–H = 92°, O–P–O = 120°. The O–P–O bond angle of the anion of the magnesium salt is quite smaller, being 109°. Infrared studies show that the hypophosphite anion of the potassium, sodium and ammonium salts exhibit C_{2v} symmetry while that of the calcium salt indicates a C_1 symmetry[348].

Properties. Pure hypophosphorous acid[347] is a white crystalline solid that melts at 26.5° and easily supercools to a viscous liquid. It is a monobasic acid with a dissociation constant of 8.0×10^{-2}, and $pk_1 = 1.1$ at 20–25°, n_D^{20} 1.4601, d_4^{20} 1.479. A water solution of hypophosphorous acid is not oxidized by atmospheric oxygen but will decompose when heated to above 140° to give phosphine, phosphoric acid and hydrogen. To avoid decomposition, the crystalline acid should not be heated to above 50°. In alkaline solution, the decomposition follows the equation:

$$H_2PO_2^- + OH^- \longrightarrow HPO_3^= + H_2$$

and the rate of decomposition increases with the OH^- concentration[349].

The alkali and alkaline earth salts, as are also most of the heavier metal salts of hypophosphite, are quite soluble in water. The hypophosphite ion is a reducing agent and its characteristic reduction reactions are those with silver nitrate to give black, metallic silver, cupric salts to give cuprous salts and metallic copper[349], the dichromate to Cr^{+++}[350]. The oxidation potential for the half cell reaction:

[343] M. Halmos, *Acta Univ. Szegediensis, Acta Phys. et Chem N.S.* **2** (1956) 85–86.
[344] K. V. Nikonorov and E. A. Gurylev, *Ivest. Akad. Nauk. SSSR Ser. Khim.* **3** (1967) 683.
[345] P. Svehla, *Collection Czech. Chem. Commun.* **31** (1966) 4712–17.
[346] J. J. Lawless and H. T. Searle, *J. Chem. Soc.* (1962) 4200-5.
[347] J. R. van Wazer, *Phosphorus and Its Compounds*, Vol. I, pp. 355–418, Interscience Publishers Inc., New York (1958).
[348] R. W. Lovejoy and E. L. Wagner, *J. Phys. Chem.* **68**(3) (1964) 544–50.
[349] D. M. Yost and H. Russell, Jr., *Systematic Inorganic Chemistry of the Fifth- and Sixth-Group Nonmetallic Elements*, pp. 191–8, Prentice-Hall Inc., New York (1944).
[350] K. Pang and S. H. Lin, *J. Chinese Chem. Soc.* (*Taiwan*) **7** (1960) 75–80.

$$H_3PO_2 + H_2O \longrightarrow H_3PO_3 + 2H^+ + 2e^-$$

is estimated to be 0.59 V[349].

Besides salt formation through the acidic hydrogen, hypophosphorous acid also reacts with aldehydes through the P–H bonds[351]:

$$H_3PO_2 + CCl_3CCHO \xrightarrow{36^\circ} Cl_3CCH(OH){-}P(=O)(H){-}OH$$

$$\xrightarrow{80\text{–}100^\circ} (Cl_3CCH(OH){-})_2P(=O)OH$$

Analogous compounds $H_2C(OH){-}P(=O)(H){-}OH$ and $(H_2C(OH){-})_2P(=O)OH$ are obtained in reactions with formaldehyde[352].

For analytical purposes, hypophosphite separates cleanly from phosphites, phosphates, hypophosphates and diphosphites, etc., by gradient elution and anion exchange chromatographic technique at a pH of 6.8 (NH_4OAc–KCl buffer system)[353, 354], using Dowex 1–X8 resin or by the use of a cellulose powder packed column and eluting with an aqueous ammoniacal iso-propyl and tert-butyl alcohol solution[355].

Applications. The hypophosphite of the greatest industrial importance is the sodium salt ($NaH_2PO_2 . H_2O$). Its main application is as reducing agent for the electroless plating of a nickel coating on metallic parts such as steel or aluminum. One proposal[356] for the mechanism of this plating reaction is as follows:

$$H_2PO_2^- + H_2O \longrightarrow HPO_3^{--} + 2H^+ + H^-$$
$$Ni^{++} + 2H^- \longrightarrow Ni^\circ + H_2$$
$$H^+ + H^- \longrightarrow H_2$$

This plating process has been commercialized by the General American Transportation Corporation under the trademark of "Kanigen." It has been used as a relatively inexpensive and convenient method for plating of the inside of mild steel tank cars and containers used for holding corrosive materials. It is also used for the nickel plating of intricately shaped objects such as valves and pump parts and chemical processing equipments needing corrosion resistant surfaces. The nickel coating is amorphous and contains 6–15% of P depending on the concentration of the hypophosphite solution used. The hardness of the coating is increased by heating (1 hr at 400° increases the hardness from 49 Rockwell C to 70 Rockwell C), which results in the precipitation of Ni_3P in the nickel–phosphorus lattice[357].

The plating solution contains as chief ingredients sodium hypophosphite and nickel

351 K. V. Nikonorov, E. A. Gurylev and F. F. Fakhrislamova, *Izv. Akad. Nauk. SSSR Ser. Khim.* (1966)(6) 1095.
352 J. Horak and V. Ettel, *Collection Czechoslov. Chem. Communs.* **26** (1961) 2401–9.
353 F. H. Pollard, D. E. Rogers, M. T. Rothwell, and G. Nickless, *J. Chromatog.* **9** (1962) 227–30.
354 F. H. Pollard, G. Nickless and M. T. Rothwell, *J. Chromatog.* **10**(2) (1963) 212–14.
355 F. H. Pollard, G. Nickless and J. D. Murray, *J. Chromatog.* **27**(1) (1967) 271–2.
356 R. M. Lukes, *Plating*, **51** (1964) 969–71.
357 E. F. Jungslager, *Tijdschr. oppervlakte Tech. Methalen*, **9**(1) (1965) 2–9.

sulfate. In addition, it also contains auxiliary reagents such as complexing agent (e.g., lactates) to prevent the precipitation of nickel phosphite, accelerators (e.g., malates and succinates) to speed up the rate of nickel deposition, and stabilizers (e.g., lead salts) to prevent the random decomposition of the bath[358].

Hypophosphite electroless nickel plating is also applicable to certain plastic parts. In this case a palladium pre-coating on the plastic surface is needed to initiate the plating of nickel which then becomes autocatalytic. The electroless nickel coating thus obtained serves then as the electroconducting surface for the eventual chrome-plating of the plastic piece.

In addition to Ni–P coatings, hypophosphite plating process is also useful for the application of Co–P[359], and tin[360] coatings.

A very old use for hypophosphite is as an ingredient for tonic compositions. For this purpose sodium, manganese[361], calcium[362], and methionine[363] salts are claimed in the patent literature.

Because of its reducing properties, hypophosphorous acid and its salts are claimed as anti-oxidant for the prevention of discoloration during the cooking of modified alkyd resins from polyols and monobasic carboxylic acids[364]. They are also used as a light and heat stabilizer for polyamides[365]; as light stabilizer for polyvinyl chlorides[366]; and as stabilizer to prevent the decomposition of polyurethane foam during its preparation[367].

11.2. PHOSPHOROUS ACID AND PHOSPHITES

Preparation. Phosphorous acid is usually prepared by the hydrolysis of PCl_3:

$$PCl_3+3H_2O \longrightarrow H_3PO_3+3HCl$$

One industrial method involves the spraying of PCl_3 under nitrogen into a reactor containing an excess of water in the form of steam at 185–190°. The product is then sparged of residual H_2O and HCl at 166° with the aid of nitrogen. Crystalline H_3PO_3 obtained by this method contains $< 0.05\%$ Cl^-, $< 1.4\%$ PO_4^{3-} and $< 1.5\%$ H_2O. Small-scale laboratory[369] preparation is carried out by adding water to an ice bath cooled solution of PCl_3 in CCl_4. The aqueous layer separated is removed of excess water and HCl by heating under vacuum at 60°. The purity of the crystalline acid obtained is 99.5%. Phosphorous acid can also be prepared by the direct hydration of P_4O_6 obtained from the controlled oxidation of phosphorus[370].

[358] J. R. van Wazer, *Phosphorus and Its Compounds*, Vol. II, pp. 1888–95, Interscience Publishers Inc., New York (1961).

[359] J. S. Judge, J. R. Morrison and D. E. Speliotis, *J. Electrochem. Soc.* **113**(6) (1966) 547–51.

[360] L. H. Shipley, U.S. 3,303,029, 7 Feb., 1967, to Shipley Co., Inc.

[361] T. E. T. Weston, Brit. 1,062,899, 22 March 1967.

[362] F. J. Meyer and W. Wirth, Ger. 954,726, 20 Dec. 1956, to Farbenfabriken Bayer A.G.

[363] A. Polgar, Fr. M 3733, 10 Jan. 1966.

[364] K. Yoshitomi, M. Nagakura and K. Matsunuma, Japan 12,997, 24 June 1965, to Nisshin Oil Mills Ltd.

[365] P. V. Papero, U.S. 3,242,134, 22 March 1966, to Allied Chemical Corp.

[366] M. Schuler, Swiss 326,175, 31 Jan. 1958, to Lonza Elektrzitatswerke and Chemische Fabriken A.G.

[367] Imperial Chemical Industries Ltd., Fr. 1,391,335, 5 March 1965.

[368] J. Cremer, U. Thuemmer, F. Schulte and H. Harnisch, Ger. 1,206,406, 9 Dec. 1965, to Knapsack Griesheim A.G.

[369] D. Voigt and F. Gallais, *Inorganic Synthesis*, Vol. IV, pp. 55–8, Ed. J. C. Bailar, McGraw-Hill Book Co., New York (1953).

[370] Proctor and Gamble Co., Belg. 701,548, 19 July 1967.

Structure. X-ray diffraction studies on the magnesium salt, $MgHPO_3.6H_2O$, show he phosphorous atom to be surrounded tetrahedrally by one hydrogen atom and three oxygen atoms.

In an aqueous solution, the ionic radius of $HPO_3^{=}$ of NaH_2PO_3 and K_2HPO_3 in the concentration range of 1 g-equivalent per 10–100 moles of water at 25° is 1.33–1.52 Å. The radius decreases with increase in dilution, possibly due to the compression of H_2O molecules in the vicinity of $HPO_3^{=}$ [371].

NMR spectra of both hydrogen and phosphorus of a phosphite salt solution exhibit 1–1 spin–spin splitting, thus indicating that the hydrogen atom is attached directly to the phosphorus atom[372].

Properties. Since phosphorous acid has a structure of $HP(O)(OH)_2$ rather than $P(OH)_3$, it is a dibasic acid. At 18°, the first ionization constant is 5.1×10^{-2}, the second ionization constant is 1.8×10^{-7}. The heat of fusion of the crystalline acid is 3.07 kcal/mole, and the heat of solution in an excess of water is −0.13 kcal/mole[347].

Phosphorous acid is an extremely hygroscopic white crystalline solid, m.p. 73–74°. The specific gravity of the liquid is D_4^{76} 1.597 [369].

The potential constants for $PO_3^{\equiv}$ calculated on the basis of the Urey–Bradley type potential field showed a stretching force constant of 4.808, bending force constant of 0.095 and repulsion force constant of 1.193 (all $\times 10^5$ dynes/cm)[373].

Thermal decomposition of phosphorous acid is fairly complex, but basically it involves self-oxidation-reduction reactions which result in the formation of phosphoric acid, phosphine, hydrogen and some red phosphorus. This decomposition reaction is quite rapid at 250–275° [347].

Atmospheric oxygen does not oxidize phosphorous acid except in the presence of I_2 and light. Phosphorous acid, as a reducing agent, reduces Ag^+ to metallic silver, reacts with hot concentrated H_2SO_4 to give H_3PO_4 and SO_2. It is also oxidized in aqueous solution to the phosphates by halogens, mercuric chloride and other oxidizing agents. The oxidation potential of the couple, H_3PO_3, H_3PO_4, in acid solution is estimated to be 0.28 V and in alkaline solution the potential $HPO_3^{=}$, $PO_4^{\equiv}$ is 1.12 V[347].

Derivatives. Because of its dibasic nature, phosphorous acid forms two series of salts, the normal and the acid phosphites. Compounds such as $(NH_4O)HP(O)(OH)$, $(NH_4O)_2P(O)H.H_2O$, $LiOP(O)H(OH)$, $(LiO)_2P(O)H$ are known. The sodium and potassium salts in addition to the normal and acid salts: $Na_2PHO_3.5H_2O$, $NaH_2PO_3.2.5H_2O$, K_2HPO_3 and KH_2PO_3 also form adducts such as $Na_2HPO_3.2H_3PO_3.0.5H_2O$, $K_2HPO_3.2H_3PO_3$ [347], $2KH_2PO_3.H_3PO_3$ and $KH_2PO_3.H_3PO_3$ [374]. By specific electric conductivity measurements at various concentrations, it was found that the adducts of the potassium salts exist in aqueous solution just the same as in the solid phase. Their formation increases with increase in the concentration with negligible temperature effect at the range of 25–45° [375].

The monobasic salts, such as $Mg(H_2PO_3)_2$, $Ca(H_2PO_3)_2$ and $Sr(H_2PO_3)_2$, are prepared in greater than 99% purity by the addition of the respective carbonate or hydroxide to an excess of molten H_3PO_3 at 80°. The excess H_3PO_3 is then removed by washing with ethyl

[371] M. Ebert, *Collection Czech. Chem. Commun.* **31**(2) (1966) 481–8.
[372] Shigeru Ohashi, *Topics in Phosphorus Chemistry*, **1** (1964) 113–89.
[373] K. V. Rajalakshmi, *Current Sci.* (*India*), **31** (1962) 329.
[374] M. Ebert and A. Muck, *Collection Czech. Chem. Commun.* **28** (1963) 257–61.
[375] M. Ebert and J. Cipera, *Chem. Zvesti* **19**(9) (1965) 679–83.

alcohol or acetone[376]. For the preparation of such compounds as $NH_4H_2PO_3$ and $(NH_4)_2HPO_3.H_2O$, the neutralization is carried out in methyl or ethyl alcohol or in acetone as solvent[377].

Esters of phosphorous acid, both the tri $[P(OR)_3]$ and dialkyl esters, $[(RO)_2P(O)H]$, are usually produced from the reaction of PCl_3 with the corresponding alcohols under proper conditions, e.g.:

$$PCl_3+3ROH \longrightarrow (RO)_2\overset{\overset{\displaystyle O}{\|}}{P}H+RCl+2HCl$$

$$PCl_3+3ROH+3R_3N \longrightarrow (RO)_3P+3R_3N.HCl$$

Triaryl phosphites are prepared by the direct action of PCl_3 with phenols while the diaryl esters, $(ArO)_2\overset{\overset{\displaystyle O}{\|}}{P}H$, are best prepared by the rearrangement reaction of phosphorous acid with triaryl phosphites at 100–160° [378]:

$$3ArOH+PCl_3 \longrightarrow (ArO)_3P+3HCl$$

$$2(ArO)_3P+H_3PO_3 \longrightarrow 3(ArO)_2\overset{\overset{\displaystyle O}{\|}}{P}H$$

Long chain dialkyl esters have also been prepared by direct esterification of phosphorous acid[379]:

$$H_3PO_3+2ROH \longrightarrow (RO)_2\overset{\overset{\displaystyle O}{\|}}{P}H+2H_2O$$

An interesting reaction of phosphorous acid is that with NH_4Cl and aqueous formaldehyde to form the sequestrant, nitrilotrimethylenephosphonic acid:[380]:

$$NH_4Cl+3H_2CO+3H_3PO_3 \longrightarrow N[CH_2\overset{\overset{\displaystyle O}{\|}}{P}(OH)_2]_3+3H_2O+HCl$$

The trialkyl phosphites, especially the long chain ones, are readily prepared by the ester exchange reaction between the alcohol and triphenyl phosphite[381]:

$$(C_6H_5O)_3P+3ROH \longrightarrow (RO)_3P+3C_6H_5OH$$

For quantitative determination of lower trialkyl and dialkyl esters in the presence of each other, advantage is taken of the fact that trialkyl phosphite hydrolyzes only slowly in an alkaline alcohol solution but rapidly in an acidic alcohol solution to the dialkyl phosphonate, and that dialkyl phosphonate hydrolyzes immediately in an alkaline alcohol solution. In a mixture of trialkyl phosphite and dialkyl phosphonate, the latter can thus be titrated in an alkaline alcohol solution before the trialkyl phosphite begins to hydrolyze[382].

Applications. The important applications of phosphorous acid are based on its salts and esters. Most of the applications depend on the reducing property of the trivalent

[376] Z. Dlouhy, M. Ebert and V. Vesely, *Collection Czech. Chem. Communs.* **24** (1959) 2801–2.
[377] M. Ebert, *Collection Czech. Chem. Communs.* **24** (1959) 3348–52.
[378] E. N. Walsh, U.S. 2,984,680, 16 May 1961, to Stauffer Chemical Co.
[379] K. A. Petrov, E. E. Nifant'ev, R. G. Gol'tsova, M. A. Belaventsev and S. M. Korneev, *Zh. Obshch. Khim.* **32** (1962) 1277–9.
[380] K. Moedritzer and R. R. Irani, *J. Org. Chem.* **31** (1966) 1603–7.
[381] K. A. Petrov, E. E. Nifant'ev, A. A. Shchegolev, M. M. Butilov and I. F. Rebus, *Zh. Obshch Khim.* **33**(3) (1963) 899–901.
[382] D. N. Bernhart and K. H. Rattenbury, *Anal. Chem.* **28** (1956) 1765–6.

phosphorus atom in these compounds. One use is as stabilizers for polyvinyl chloride and other polymers. When used with other stabilizing ingredients, they prevent the discoloration of the polymers by heat or u.v. light. Some of the more widely used phosphites for this purpose are triphenyl phosphite, trinonylphenyl phosphite and tri-iso-octyl phosphite. The organic groups attached to the phosphorus atom, depending on their nature, improve the compatibility of the stabilizer to the polymer system. Specific organic groups also have special synergistic action for certain systems. For example, triphenyl phosphite is most widely used for polyvinyl chloride, while trinonylphenyl phosphite is an excellent stabilizer for the GR–S polymer.

11.3. PYROPHOSPHOROUS ACID AND POLYPHOSPHOROUS ACID

In contrast to pyrophosphoric and polyphosphoric acids, all of which may be obtained by the thermal condensation of orthophosphoric acid, it is not possible to prepare pyrophosphorous or polyphosphorous acid by the analogous thermal condensation of phosphorous acid. Thermal dehydration of phosphorous acid at 80° and 10^{-3} mm does not yield pyrophosphorous acid after 8 hr. Pyrophosphorous acid has to be prepared by other than the thermal condensation route. One such route is the reaction PCl_3 with H_3PO_3 [383]:

$$5H_3PO_3 + PCl_3 \rightleftharpoons 3H_4P_2O_5 + 3HCl$$

This reaction is reversible and it necessitates the complete removal of HCl.

Attempts at analogous reaction to produce polyphosphorous acid had not been successful. Apparently, if polyphosphorous acid were formed, it was too unstable and underwent disproportionation reactions since only phosphoric acid and products of lower degree of oxidation were obtained[384].

Disodium pyrophosphite may be prepared by the dehydration of monosodium phosphite, $NaH_2PO_3.2.5H_2O$, under vacuum at 150°. Similarly, the diammonium, dipotassium, calcium, strontium, barium and lead pyrophosphites are prepared from the respective acid orthophosphites[347]. When the dehydration of NaH_2PO_3 is carried out at 300–400°, the reaction leads to complex oxidation and rearrangement products[385]. It is not possible to form polyphosphites by the dehydration of the acid phosphite[386].

Pyrophosphite hydrolyzes slowly to the orthophosphite in neutral solution and more rapidly in either strong acid or strong base solution. The approximate half-lives at 30° are pH 5–7, 1000 hr; pH 4, 60 hr; pH 3, 8 hr; pH 1.5, 5 min[347].

```
     H   H
     |   |
O = P–O–P = O
     |   |
     O–  O–
```

11.4. HYPOPHOSPHORIC ACID AND HYPOPHOSPHATES

Preparation. The common method for the preparation of hypophosphate is by controlled oxidation of elemental phosphorus. A procedure proposed by Genge, Nevett and Salmon[387]

383 J. P. Ebel and F. Hossenlopp, *Bull. Soc. Chim. France*, **8** (1965) 2219–2; *ibid.* 2221–8.
384 Ref. 383, pp. 2229–33.
385 B. Krukowska-Fulde, *Roczniki Chem.* **35** (1961) 1203–10.
386 D. Grant, D. S. Payne and S. Skledar, *J. Inorg. Nucl. Chem.* **26**(12) (1964) 2103–11.
387 J. A. R. Genge, B. A. Nevett and J. E. Salmon, *Chem. and Ind., (London)* (1960) 1081–2.

produces the disodium salt $Na_2H_2P_2O_6.6H_2O$ in greater than 99.5% purity. This procedure involves the use of a 35×3.5 cm column cooled internally using a 1.5 cm diameter tube, and externally using a 6.0 cm diameter tube, and packed with alternate layers of glass beads (0.3 to 0.4 cm diameter) and 35 g of red phosphorus supported on porcelain chips and glass wool. The red phosphorus is first cleaned by digesting with 10% HCl, followed by washing. A solution of 135 g of sodium chlorite in 750 cc of water is passed through the column at a rate of about 150 cc/hr, keeping the effluent temperature at 15–18°. The effluent is then adjusted to a pH of 5.2 with NaOH. Upon cooling to 0°, $Na_2H_2P_2O_6.6H_2O$ crystallizes out. If the pH were adjusted to 10, $Na_4P_2O_6.10H_2O$ would crystallize out. After two recrystallizations from water, 45 g of disodium salt of high purity is obtained. A similar procedure described in *Inorganic Synthesis*[388] reported a yield of approximately 30% of $Na_2H_2P_2O_6.6H_2O$ based on the weight of sodium chlorite consumed.

The disodium salt is also obtained by the oxidation of red phosphorus with a cold solution of calcium hypochlorite. In this procedure, the $Ca_2P_2O_6.2H_2O$ formed is acidified with H_2SO_4 to precipitate $CaSO_4.2H_2O$. The filtrate obtained upon treatment with sodium acetate forms crystals of $Na_2H_2P_2O_6.6H_2O$[389, 390]. Oxidation of red phosphorus with H_2O_2, followed by adjusting the pH of the solution to 5.2 with NaOH, also produces $Na_2H_2P_2O_6.6H_2O$[391]. Other oxidants such as $KMnO_4$, NaClO, NaBrO, or acid solutions of $K_2S_2O_8$, $NaBrO_3$ have also been used. By-products from these oxidation reactions are phosphite, orthophosphate, pyrophosphate and polyphosphate[392]. Since phosphorus sulfides such as P_4S_3, P_4S_5 and P_4S_7 and P_2I_4 also contain P–P bonds, they too produce hypophosphates as part of the product-mix on oxidative hydrolysis[393, 394].

Hypophosphoric acid, $H_2P_2O_6{\cdot}2H_2O$, is obtained by passing a solution of $Na_2H_2P_2O_6{\cdot}6H_2O$ through a column of cation exchange resin[387]. The anhydrous acid is obtained by the dehydration of the dihydrate *in vacuo* over P_2O_5 for a period of two months.

The ester of hypophosphate is prepared by the reaction of sodium dialkyl phosphite with SO_2Cl_2[395]:

$$2\,(RO)_2\overset{\overset{O}{\|}}{P}Na + SO_2Cl_2 \longrightarrow (RO)_2\overset{\overset{O}{\|}}{P}-\overset{\overset{O}{\|}}{P}(OR)_2 + SO_2 + 2NaCl$$

Structure. The tetrasodium salt, $Na_4P_2O_6{\cdot}10H_2O$, is monoclinic, space group $C2/c$, with 4 mols per unit cell. The cell dimensions are $a = 16.97$, $b = 6.97$, $c = 14.50$ Å, $\beta = 116°$. The disodium salt $Na_2H_2P_2O_6{\cdot}6H_2O$, has also 4 mols per unit cell, space group $C2/c$ with unit cell dimensions of $a = 14.12$, $b = 7.00$, $c = 12.71$ Å, $\beta = 115.4°$. The trisodium salt, $Na_3HP_2O_6{\cdot}9H_2O$, belongs to space group Pn with 4 mols per monoclinic unit cell. The cell dimensions are $a = 26.96$, $b = 6.00$, $c = 9.26$ Å and $\beta = 110°$ [396].

X-ray diffraction pattern, NMR studies, infrared and Raman spectra, all support the anion structure of

[388] E. Leininger and T. Chulski, *Inorganic Synthesis*, Vol. 4, pp. 68–71, Ed. J. C. Bailar, Jr., McGraw-Hill Book Co., New York (1953).

[389] W. G. Palmer, *J. Chem. Soc.* (1961) 1079–82.

[390] M. Ishaq and Badar-ud-Din, *Pakistan J. Sci. Ind. Res.* 7(1) (1964) 17–19.

[391] N. Yoza and S. Ohashi, *Bull. Chem. Soc. Japan*, **38**(8) (1965) 1408–9.

[392] H. Remy and H. Falius, *Z. anorg. u. allgem. Chem.* **306** (1960) 211–15.

[393] H. Falius, *Z. anorg. u. allgem. Chem.* **326**(1–2) (1963) 79–88.

[394] M. Baudler and G. Fricke, *Z. anorg. u. allgem. Chem.* **319** (1963) 211–29.

[395] J. Michalski and T. Modro, *Roczniki Chem.* **36** (1962) 483–8.

[396] D. E. C. Corbridge, *Acta Cryst.* **10** (1957) 85.

```
     O O
     || ||
  -O-P-P-O-
     | |
    -O O-
```

Properties. In the absence of moisture both $H_4P_2O_6$ and $H_4P_2O_6.2H_2O$ are stable at 0–5°. The anhydrous acid begins to melt at 73° without a sharp m.p.[397] It undergoes rearrangement and self-oxidation-reduction at room temperature into $H(HO)(O)POP(O)(OH)_2$, $H_4P_2O_7$, and $[(HO)_2P]_2O$[398]. The dissociation constants for hypophosphoric acid are K_1 = about 6×10^{-3}, K_2 = about 1.5×10^{-3}, $K_3 = 5.4\times10^{-8}$ and $K_4 = 0.93\times10^{-10}$ at 25° [372].

The hypophosphate anion is very stable to alkali hydroxide and does not decompose even when heated at 200° with 80–90% NaOH. It hydrolyzes slowly (half-life in 1 N acid at 25° = 180 days), however, in an acid solution:

```
   O O                       O        O
   || ||          H+         ||       ||
HOP-POH+H2O ---------> HO-P-H+HOP-OH
   | |                       |        |
  HO OH                     HO       OH
```

The rate of hydrolysis is quite rapid in 4 N HCl (1 hr completely changed).

11.5. MISCELLANEOUS LOWER OXYACIDS OF PHOSPHORUS

A series of lower oxyacids of phosphorus containing two or more phosphorus atoms of the same or different oxidation state in each molecule were reported by Blaser and Worms. The chemistry of these compounds has recently been reviewed by Ohashi[372]. These compounds are of such structures that they are difficult to name by present rules for nomenclature of phosphorus compounds. The structural formula and the abbreviated notations originally proposed by Blaser and Worms[399] will be used in this section. These notations use the phosphorus and oxygen linkage which comprised the skeleton of the compound. The oxidation state of the phosphorus atom is indicated by the corresponding number.

```
               H H
               O O
 2 2           | |
 P-P acid, H-P-P-H
               || ||
               O O
```

This acid is prepared by the hydrolysis of a carbon disulfide solution of P_2I_4 at 0°. The barium salt, $BaH_2P_2O_4$, is precipitated from the aqueous solution. The acid undergoes disproportionation reaction to form phosphine and phosphorus or phosphoric acids. However, it is stable in an aqueous solution at pH 7. It oxidizes by oxygen readily in an aqueous solution to $\overset{2\ 4}{\text{P–P}}$ and $\overset{4\ 4}{\text{P–P}}$ acids. The structure of the $\overset{2\ 2}{\text{P–P}}$ acid is assumed from the method of preparation and chemical properties.

[397] H. Remy and H. Falius, *Naturwissenschaften*, **43** (1956) 177.
[398] H. Remy and H. Falius, *Chem. Ber.* **92** (1959) 2199–2205.
[399] B. Blaser and K. H. Worms, *Z. anorg. u. allgem. Chem.* **300** (1959) 225–8.

```
                   H H
                   O O
 2 4               | |
 P–P acid,       HP–POH
                   ‖ ‖
                   O O
```

Hydrolysis of PBr_3 or PI_3 in a water suspension of sodium hydrogen carbonate at 0–5° produces the sodium salt of $\overset{2}{\mathrm{P}}$–$\overset{4}{\mathrm{P}}$ acid along with some sodium hypophosphate. By recrystallization from water, $Na_3P_2HO_5.12H_2O$ is obtained. The $\overset{2}{\mathrm{P}}$–$\overset{4}{\mathrm{P}}$ anion is stable in alkaline solution but when heated with 80% NaOH, it converts to the $\overset{4}{\mathrm{P}}$–$\overset{4}{\mathrm{P}}$ anion with liberation of hydrogen. In an acid solution it hydrolyzes readily at 25° into 2 moles of phosphorous acid[372]:

```
     H H                     H
     O O                     O
     | |                     |
 H–P–P–OH  +  H2O ——→ 2H–P–OH
     ‖ ‖                     ‖
     O O                     O
```

The $\overset{2}{\mathrm{P}}$–$\overset{4}{\mathrm{P}}$ anion undergoes oxidation with bromine to the hypophosphate in an acid solution but to the pyrophosphate in a $NaHCO_3$ solution.

$(\overset{3}{\mathrm{P}})_6$ **ring acid,**

(six-membered ring of P atoms; substituents: OH, O, O, OH, OH, O, HO, O, O, O, OH, OH)

When red phosphorus is oxidized in a solution of potassium hypochlorite and potassium hydroxide, a very small yield of $(\overset{3}{\mathrm{P}})_6$ ring acid is obtained as the crystalline potassium–sodium salt (80–85% K and 16–18% Na).

It hydrolyzes in acidic solution to the phosphite and a minor amount of hypophosphite and phosphate. Hydrolysis in alkaline solution produces the phosphite and some $\overset{2}{\mathrm{P}}$–$\overset{4}{\mathrm{P}}$ anion[400]. The cyclic puckered six-membered ring structure was determined by the X-ray diffraction study of the cesium salt. Other supporting evidences for the structure are (1) infrared spectra which show no P–H bonds and (2) pH titration curve which shows only one inflection point[372].

```
                          H     H H
                          O     O O
 3   4 4                  |     | |
 P–O–P–P acid,          H–P–O–P–P–OH
                          ‖     ‖ ‖
                          O     O O
```

This acid is the anhydride of phosphorous and hypophosphoric acids, and it is prepared by boiling an aqueous solution of sodium pyrophosphite and sodium hypophosphate:

[400] B. Blaser and K. H. Worms, *Z. anorg. u. allgem. Chem.* **300** (1959) 237–49.

```
O⁻  O⁻        O⁻O⁻ 1              O⁻  O⁻O⁻        O⁻
|   |         |  |                |   |  |        |
H–P–O–PH + ⁻O–P–P–O⁻ ———→ H–P–O–P–P–O⁻ + H–P–O⁻
  ||  ||      || ||               ||  || ||       ||
  O   O       O  O                O   O  O        O
```

As expected from the structure, it hydrolyzes readily in 0.5 N NaOH solution to the phosphite and hypophosphate[372].

```
         H H H
         O O O
4 3 4    | | |
P–P–P acid, HO–P–P–P–OH
         || || ||
         O  O  O
```

The pentasodium salt of this acid, $Na_5P_3O_8.14H_2O$, is obtained in low yield by the oxidation of $\overset{3}{P}_6$ ring anion in a solution of $KHCO_3$ with iodine. It is also one of the very minor products of oxidation of red phosphorus with an aqueous solution of sodium hypochlorite[372].

```
           H     H
           O     O
3      5   |     |
P–O–P acid, H–P–O–P–OH
           ||    ||
           O     O
```

This acid, being the anhydride of phosphorous and phosphoric acid may be prepared by several methods, e.g.:

```
O   O       O               O   O       O
||  ||      ||              ||  ||      ||
H–P–O–P–H + ⁻O–P–O⁻ ———→ HP–O–P–O⁻ + H–P–O⁻
|   |       |               |   |       |
O⁻  O⁻      O⁻              O⁻  O⁻      O⁻

           O                   O   O
           ||                  ||  ||
POCl3 + HP–O⁻ + 2H2O ———→ H–P–O–P–O⁻ + 3HCl + H⁺
           |                   |   |
           O⁻                  O⁻  O⁻

           O                   O   O
           ||                  ||  ||
PCl3 + ⁻O–P–O⁻ + 2H2O ———→ H–P–O–P–O⁻ + 3HCl + H⁺
           |                   |   |
           O⁻                  O⁻  O⁻
```

It is also a product of the spontaneous rearrangement of anhydrous hypophosphoric acid[398].

The structure of the compound is indicated by the methods of preparation and the reactions with oxidizing agents. Two trisodium salts, the tetrahydrate and octahydrate, are known. These salts are stable in the solid state and in neutral solution, but hydrolyze in acid and alkaline solutions to the phosphite and phosphate[372].

$(-\overset{4}{P}-\overset{4}{P}-O-)_2$ ring acid,

```
      OH  HO
    O\ |   | //O
       P———P
      /     \
    O         O
      \     /
       P———P
     /\\   //\
   HO  O O    OH
```

This acid is formed by the dehydration of 2 moles of tetrapotassium hypophosphate with acetic anhydride. Attempted preparation by thermal dehydration of the sodium or potassium acid hypophosphate was not successful. The crystalline tetraguanidium salt is obtained by treating an aqueous solution containing the $\overset{4}{P}-\overset{4}{P}-O_2$ ring anion at 0° with guanidium chloride. The ring anion is hydrolyzed quantitatively at 150° with 60% KOH to the hypophosphate anions. Under milder conditions of hydrolysis, the pyrohypophosphate anion, $\overset{4}{P}-\overset{4}{P}-O-\overset{4}{P}-\overset{4}{P}$ is formed.

$\overset{4}{P}-\overset{4}{P}-O-\overset{4}{P}-\overset{4}{P}$ acid,

```
      H H   H H
      O O   O O
      | |   | |
   HO-P-P-O-P-P-OH
      ‖ ‖   ‖ ‖
      O O   O O
```

The hexasodium salt of this acid is obtained in very high yield by the hydrolysis of the tetrasodium salt of the ring acid $\overset{4}{P}-\overset{4}{P}-O_2$ in a 1.5 N NaOH solution at 25°. The anion $\overset{4}{P}-\overset{4}{P}-O-\overset{4}{P}-\overset{4}{P}$ is quite stable in alkaline solution but does hydrolyze quantitatively to the hypophosphate anion when heated at 150° with 60% NaOH. Hydrolysis in a hot acid solution results in the degradation to phosphorous acid and phosphoric acid with hypophosphoric acid as the intermediate[372].

$\overset{5}{P}-O-\overset{4}{P}-\overset{4}{P}$ acid,

```
      H   H H
      O   O O
      |   | |
   HOP-O-P-P-OH
      ‖   ‖ ‖
      O   O O
```

Since this acid is the anhydride of phosphoric and hypophosphoric acid, it can be prepared by the reaction of $POCl_3$ with hypophosphoric acid:

```
             H H                       H     H H
   Cl        O O                       O     O O
   |         | |                       |     | |
  ClPCl+HO-P-P-OH+2H2O ———→ HOP-O—P-P-OH+3HCl
   ‖         ‖ ‖                       ‖     ‖ ‖
   O         O O                       O     O O
```

It is also prepared in good yield by the oxidation of $\overset{4}{P}-\overset{3}{P}-\overset{4}{P}$ acid with bromine in a solution of $KHCO_3$. The anion hydrolyzes to the phosphate and hypophosphate anions when heated with 60% KOH. Similar results are obtained when the hydrolysis is carried out in dilute mineral acid.

The lower oxyacids of phosphorus are of interest only from the academic standpoint. No practical applications have been developed for them.

12. PHOSPHORIC ACIDS

12.1. ORTHOPHOSPHORIC ACID

Preparation. Orthophosphoric acid is one of the most important of phosphorus compounds. It was prepared originally by the "wet process" which involved the treatment of calcined bones with sulfuric acid. A large quantity of it is still prepared by this process except that now the more abundant phosphate rock has replaced bone as the phosphatic raw material. The major reactions of calcium phosphate and the combined calcium fluoride (as occur in fluorapatite) with sulfuric acid are shown in the following formalized equations:

$$Ca_3(PO_4)_2+3H_2SO_4+6H_2O \longrightarrow 3CaSO_4.2H_2O+2H_3PO_4$$
$$CaF_2+H_2SO_4+H_2O \longrightarrow CaSO_4.2H_2O+2HF$$

The relatively insoluble $CaSO_4.2H_2O$ is filtered off and the dilute phosphoric acid, containing 30–32% P_2O_5, is then concentrated by evaporation. The acid obtained in this manner is quite impure but of sufficient quality for most fertilizer preparations. Further purification by removal of residual F^- (as the insoluble Na_2SiF_6), iron, aluminum, and other metallic impurities present originally in the phosphate ore results in an acid of adequate purity suitable for sodium phosphates used in detergent applications. Engineering modifications and process changes in the wet process have made possible the preparation of phosphoric acid of 50 to 54% P_2O_5 content prior to concentrating by evaporation.

One interesting new development in the wet process involves the acidulation of phosphate rock with hydrochloric acid. The phosphoric acid thus generated is first extracted from the water solution away from the soluble calcium chloride by-product and other impurities by means of organic solvents such as butyl or amyl alcohols. It is then recovered from the organic solvent by extraction with water[401]. If sufficient numbers of extraction stages are used in this process, phosphoric acid of high purity is obtained. Recent reviews of the wet process phosphoric acid technology have been compiled by Slack[402].

Commercially, all pure phosphoric acid suitable for application in food products is manufactured by the hydration of phosphoric anhydride obtained by the oxidation of elemental phosphorus:

$$P_4+5O_2 \longrightarrow P_4O_{10}$$
$$P_4O_{10}+6H_2O \longrightarrow 4H_3PO_4$$

In this process, the phosphoric anhydride formed is hydrated immediately in the reactor systems where phosphorus is burned. The acid obtained, known as "thermal acid", is then treated with hydrogen sulfide to remove arsenic impurity as the insoluble arsenic sulfide. After filtration, the acid is diluted to 75, 80 or 85% H_3PO_4 for commercial applications.

401 A. Baniel, R. Blumberg, A. Alon, M. El-Roy and D. Goniadski, *Chem. Eng. Progress*, **58** (11) (1962) 100–4.

402 A. V. Slack, *Phosphoric Acid*, Vol. I, Part I and Part II, Marcel Dekker, Inc., New York (1968).

Properties. Solid crystalline orthophosphoric acid can be obtained on a laboratory scale by removal of free water from commercial 85% H_3PO_4 and the anhydrous liquid acid allowed to crystallize[403]. It has a m.p. of 42.35° and forms a hemihydrate with a m.p. of 29.32°. Pure phosphoric acid crystallizes only slowly and is quite easy to supercool into a glass. The glassy transition point is −121°. The heat of formation of crystalline

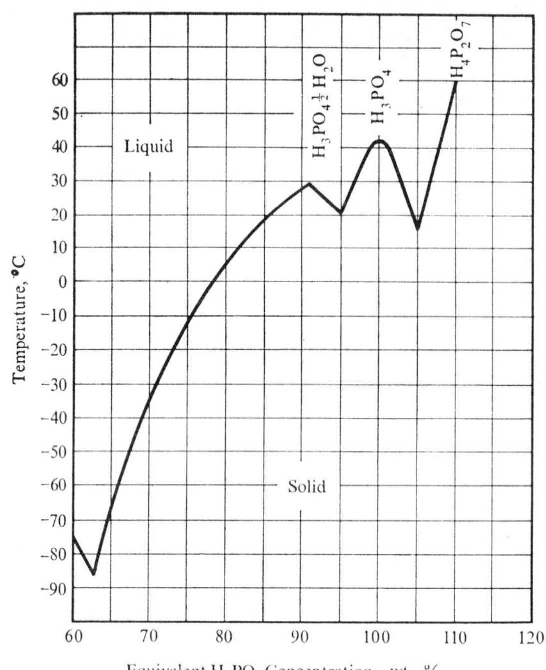

FIG. 11. H_3PO_4–H_2O system[405]. (Reprinted by permission of John Wiley & Sons Inc.)

H_3PO_4 is $\Delta H_f^\circ = -306.2$ kcal/mole. For the pure liquid H_3PO_4: $\Delta H_f^\circ = -303.57$ kcal/mole, $\Delta F_f^\circ = -268.73$ kcal/mole, $S^\circ = 38.3$ e.u.[404]. The liquid–solid phase diagram for the phosphoric acid–H_2O system from 0 to 11% equivalent H_3PO_4 is shown in Fig. 11[405].

The equation for the vapor pressure of an aqueous solution of phosphoric acid is:

$$\log P_{mm} = -A/T - 4.9373 \log T + 1.639 \times 10^{-5} T + 4.874 \times 10^{-8} T^{-2} + B$$

The vapor pressure equation constants, vapor pressure, heat of vaporization for selected concentrations of H_3PO_4 are summarized in Table 12[404].

The density, boiling point, specific heat, electric conductivity and approximate viscosity of phosphoric acid at selected concentration and temperature are shown in Table 13[404, 406].

403 A. G. Weber and G. B. King, *Inorganic Synthesis*, Vol. I, pp. 101–3, Ed. H. S. Booth, McGraw-Hill Book Co., New York (1939).

404 T. D. Farr, *Phosphorus, Properties of the Element and Some of Its Compounds*, Chemical Engineering Report No. 8, pp. 26–51, T.V.A., Wilson Dam, Alabama (1950).

405 C. Y. Shen and C. F. Callis, *Preparative Inorganic Reactions*, Vol. 2, pp. 139–67, Ed. W. L. Jolly, Interscience Publishers Inc., New York (1965).

406 J. R. van Wazer, *Phosphorus and Its Compounds*, Vol. I, pp. 479–91, Interscience Publishers Inc., New York (1958).

TABLE 12

Weight % H_3PO_4	Vapor pressure constants		Vapor pressure mm Hg at		Heat of vaporization cal/mol H_2O vaporized	
	A	B	25°	100°	$\Delta H_{298.16}$	$\Delta H_{373.16}$
29.80	2943.8	23.4092	21.19	684	10,560	9840
50.20	2981.4	23.4275	16.61	565	10,730	10,010
70.90	3066.2	23.3837	7.80	303	11,120	10,400
80.14	3145.0	23.3422	3.86	170	11,480	10,760

Phosphoric acid is tribasic. The dissociation constants at 25° are:

$$K_1 = 0.7107\times10^{-2},\ K_2 = 7.99\times10^{-8},\ \text{and}\ K_3 = 4.8\times10^{-13}$$

The heats and free energies of ionization for the three dissociations are:

	$\Delta H_{25°}$	$\Delta F_{25°}$
1st	−1.88 kcal	+2.90 kcal
2nd	+0.99 kcal	+9.82 kcal
3rd	+3.5 kcal	+16.0 kcal

At elevated temperature, phosphoric acid is quite reactive toward most metals and reactive oxides. For concentrating strong phosphoric acid at high temperature, carbon equipment is recommended. Reduction of phosphoric acid with reducing agents such as carbon or hydrogen occurs only at above 350–400°.

Raman pattern for 100% H_3PO_4 showed four lines (cm^{-1}):

$$W_1 = 910,\ W_2 = 359,\ W_3 = 1079,\ \text{and}\ W_4 = 496$$[404].

Structure. The crystalline acid and its crystalline hydrates consist of tetrahedral PO_4 connected by hydrogen bonds. These hydrogen bonds are retained in concentrated solutions and are responsible for their viscous nature. In more dilute solution (54% H_3PO_4), the phosphate ions are hydrogen bonded to the liquid water rather than other PO_4 ions. The structure of anhydrous phosphoric acid is shown below[406, 407]:

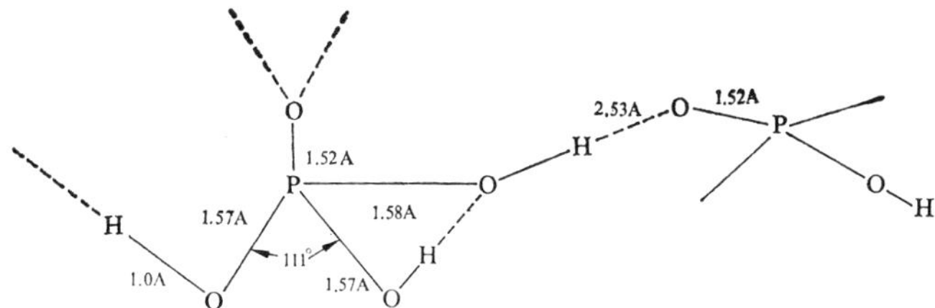

As indicated above, three oxygen atoms of the PO_4 group are bonded covalently to hydrogen atoms which in turn are hydrogen-bonded to oxygen atoms of the neighboring PO_4 group. The fourth oxygen, even though it is hydrogen bonded, is shorter in P–O distance than that of P–OH. When a phosphoryl oxygen atom is involved in intermolecular

[407] D. E. C. Corbridge, *Topics in Phosphorus Chemistry*, **3** (1966) 220–1.

TABLE 13

Concentration		Density 25° (g/ml)	B.p. °C	F.p. °C	Viscosity 20° (cp)	Specific heat 20–120° (cal/g)	Electric conductivity at 25°		Heat of formation $\Delta H^{\circ}_{298.16}$ kcal/mole H_3PO_4
Wt. % H_3PO_4	Wt. % P_2O_5						Specific L (mhos)	Equivalent Δ (mhos)	
10	7.24	1.0523	100.2	−2.1	1.3	0.939	0.0617	19.2	−309.10
30	21.73	1.1794	101.8	−11.8	2.6	0.798	0.1836	17.0	−308.66
50	36.22	1.3334	108	−44	5.7	0.656	0.2324	11.4	−307.78
75	54.32	1.5725	135	−17.5	20.0	0.542	0.1373	3.80	−306.38
85	61.57	1.6850	158	21.1	47.0	0.493	0.907	2.07	−305.56

H-bonding, it usually shows a lowering of the phosphoryl stretching frequency together with an increase in absorption intensity[407].

Applications. Most phosphoric acids produced commercially are converted into various salts. They will be discussed in their respective sections. The free acid itself, however, has many industrial applications. One important use is for metal treatment such as in phosphatizing, in which steel or iron surface is protected by a rust-resistant coating of tertiary zinc or manganese phosphates and secondary ferrous phosphate. This treatment is carried out by applying to the metal surface a solution of phosphoric acid containing zinc or manganese ions[408]. Zinc phosphatizing gives a light gray, smooth, slate-like textured surface which is about 0.000025 in. thick. Such surface serves as an excellent corrosion-resistant undercoating for paint or enamel. Manganese phosphatizing results in a relatively coarse black absorbent surface about 0.0002 in. thick. It is applied on bolts, nuts, screws and machine parts to protect them against corrosion.

Phosphoric acid solution containing nitric acid (about 95 parts of 85% H_3PO_4 and 4–5 parts of 68% HNO_3 and additives) is very effective for the chemical polishing of aluminum articles to give them mirror-like finishes[409]. One theory for this action is that the nitric acid oxidizes the rough aluminum surface to a porous aluminum oxide film which is then removed by dissolution in phosphoric acid as aluminum phosphates. The diffusion of aluminum phosphate is faster from the microscopic mountains than the valleys on the rough aluminum surface. This results in the selective removal of the mountains, thus leaving a smooth surface. This process is known commercially as the "bright dip process."

A solution of 75% phosphoric acid, alone or with added sulfuric acid and other additives, is used as a bath for the electropolishing of stainless steel articles of regular or irregular shapes such as forks, knives, spoons and surgical instruments to give them bright finishes[410]. Inhibited phosphoric acid in 5% concentration is very effective for removal of mill scales and rust in new power generating boilers and water-formed insoluble deposits in old boilers[411].

For food applications, phosphoric acid imparts the sour and tart taste to cola drinks (0.057 to 0.084% of 75% H_3PO_4) and root beer drinks (0.013% of 75% H_3PO_4). As animal feed supplement, phosphoric acid is added to molasses. This addition, besides supplying extra phosphorus, has the added advantage of reducing the viscosity and stickiness of the molasses and thus permits easier handling. As fertilizer, phosphoric acid has been used directly by addition to irrigation water.

12.2. PYROPHOSPHORIC, POLYPHOSPHORIC AND METAPHOSPHORIC ACIDS

When orthophosphoric acid is condensed by elimination of water between two or more molecules, polyphosphoric acid is formed as a complex mixture of linear molecules of various chain length. These chains consist of PO_4 tetrahedrons joined by a shared oxygen atom at the corner of the tetrahedra.:

[408] R. M. Burns and W. W. Bradley, *Protective Coatings for Metals*, 3rd ed., Reinhold Publishing Corp. New York (1967).
[409] J. R. van Wazer, *Phosphorus and Its Compounds*, Vol. II, pp. 1880–3, Interscience Publishers Inc. (1961)
[410] P. M. Strocchi, D. Sinigaglia and B. Vicentini, *Electrochim. Metal*, 1(1) (1966) 107–11.
[411] T. E. Purcell and S. F. Whirl, Paper presented at the Midwest Power Conference, April 5–7, 1950.

TABLE 14. COMPOSITION OF THE STRONG PHOSPHORIC ACIDS[412]

Total P_2O_5 (%) *	$\frac{[P_2O_5]}{[H_2O]}$	Percentage composition in terms of the constituent polyphosphoric acids †														High-poly.	Tri-meta-	Tetra-meta-
		1	2	3	4	5	6	7	8	9	10	11	12	13	14			
67.4[a]	0.263	100.0																
68.7[a]	0.279	99.7	0.33															
70.4[a]	0.302	96.2	3.85															
71.7[a]	0.321	91.0	8.86	tr.														
73.5[a]	0.352	77.1	22.1	0.79														
73.9[b]	0.360	73.6	25.1	1.34														
75.7[b]	0.394	53.9	40.7	4.86	0.46													
77.5[a]	0.438	33.5	50.6	11.5	2.68	0.74	tr.											
79.1[a]	0.481	22.1	46.3	20.3	7.82	2.26	1.02	0.34										
80.5[b]	0.523	13.8	38.2	23.0	13.0	6.86	3.38	1.67	1.03	0.22								
81.0[b]	0.542	12.2	34.0	22.7	14.6	8.42	4.36	2.27	1.41	0.56	tr.							
81.2[b]	0.549	10.9	32.9	22.3	15.0	9.36	5.41	2.85	1.75	0.97	0.36	0.05						
82.4[a]	0.594	7.32	23.0	19.3	15.9	12.3	8.21	5.73	3.89	2.52	1.36	0.91	0.14			tr.		
84.0[a]	0.667	3.92	11.8	12.7	12.0	10.5	8.97	7.99	6.62	5.63	4.54	3.72	3.03	2.46	1.68	6.63		
85.0[b]	0.717	2.28	6.36	7.32	8.01	8.17	7.67	7.22	6.93	6.42	5.89	5.27	4.69	3.99	3.83	16.9		
85.3[a]	0.736	1.87	4.73	6.33	6.58	6.66	6.71	6.36	6.11	5.88	5.46	5.07	4.90	4.64	4.38	25.6		
86.1[b]	0.787	1.46	2.81	3.74	4.43	4.52	4.77	4.79	4.93	4.67	4.54	4.67	4.63	4.38	4.17	43.5	0.17	
87.1[b]	0.860	0.83	1.81	2.17	2.53	3.09	3.39	3.46	3.33	3.55	3.47	3.45	3.52	3.26	3.24	61.1	tr.	
87.9[b]	0.920	0.50	0.82	1.56	1.76	1.72	2.03	2.13	2.26	2.07	2.26	2.06	2.20	1.99	2.30	76.4	0.42	0.11
89.4[b]	1.066	1.88	1.52	0.77	0.61	0.62	0.68	0.54	0.71	0.86	1.03	0.98	1.16	1.23	1.37	86.8	1.17	0.41

* [a] Prepared by dehydration of orthophosphoric acid; [b] prepared by dissolving phosphoric oxide in orthophosphoric acid.

† 1 = Ortho-, 2 = pyro, 3 = tri-, 4 = tetraphosphate, etc.; high-poly. = material retained by resin and includes the 15-phosphoric acid when this was separated.

```
O     O  O
||    || ||
HO-P-(O-P-)nP-OH
|     |  |
O     O  O
H     H  H
```

$n = 0$ for pyrophosphoric acid
$n = 1$ for tripolyphosphoric acid
$n = 2$ for tetrapolyphosphoric acid

At equilibrium, the exact constituent composition of a condensed liquid phosphoric acid depends on its P_2O_5 content. Thus, acid mixtures of the same P_2O_5 content have identical compositions whether obtained by removal of water from orthophosphoric acid or by the solution of P_2O_5 in it. Compositions of strong condensed phosphoric acids with a P_2O_5 concentration up to 89.4% have been determined for their constituent polyphosphoric acids. The method for this analysis employs ion exchange chromatography using a gradient-elution technique supplemented by ascending one- and two-dimensional paper chromatography. The resin recommended is of the strongly basic quaternary ammonium type. A sample of the polyphosphoric acid is first neutralized at −5° with 0.5 N NaOH and then passed through the ion exchange column at 15°. By such a technique, in addition to orthophosphoric acid, polyphosphoric acids of 2 to 14 phosphorus atoms have been separated. These data are shown in Table 14. The high poly portion indicated on the table is the polymeric material retained in the ion exchange column. It may be hydrolyzed to lower phosphoric acids by 6 N HCl[412].

The theoretical P_2O_5 value for H_3PO_4 is 72.4%. As can be seen from Table 14, at this range of P_2O_5 concentration liquid phosphoric acid is no longer pure H_3PO_4. Pure H_3PO_4, not contaminated with other phosphoric acids, is found only at between 0 and about 67.4% P_2O_5 concentration. The composition corresponding to 72.4% P_2O_5 contains 12.7% pyrophosphoric acid. This means that the liquid phosphoric acid of 72.4% P_2O_5 concentration contains some free water to counterbalance the presence of pyrophosphoric acid[413]. Upon crystallization, however, the equilibrium of the 72.4% P_2O_5 acid shifts to the formation of pure crystalline orthophosphoric acid.

12.3. SUPERPHOSPHORIC ACID

As shown in Fig. 11, the melting point of the eutectic equivalent to 105–106% H_3PO_4 is 16°. A phosphoric acid of this approximate composition (75–77% P_2O_5) is of industrial importance. Known as "superphosphoric acid", it permits the easy handling and shipping of a concentrated phosphoric acid in the liquid form[414]. "Superphosphoric acid", prepared from wet process phosphoric acid, has a P_2O_5 content of 69.5 to 70.6%. This change in the composition is due to the presence of metallic impurities present in wet process phosphoric acid.

12.4. PYROPHOSPHORIC ACID

Pyrophosphoric acid is the only member of the linear polyphosphoric acid series which is readily obtained in a crystalline form. It has a theoretical P_2O_5 content of 79.8%.

[412] R. F. Jameson, *J. Chem. Soc.* (1959) 752–9.
[413] Ref. 406, p. 749.
[414] T. A. Blue, *Chemical Economics Handbook* (1966) 761.5075B-761.5075E.

However, syrupy liquid phosphoric acid of this concentration has only 42.5% $H_4P_2O_7$, the remainder being orthophosphoric acid (17.2%), tripolyphosphoric acid (25%), tetrapolyphosphoric acid (10.5%) and higher chain polyphosphoric acids[415].

When liquid acid of 79.8% P_2O_5 content is allowed to crystallize, a white solid pyrophosphoric acid (Form I), the common form, is obtained (m.p. = 54.3°). When crystalline Form I is heated in a sealed tube for several hours at about 50°, it is converted to Form II with a m.p. of 71.5°. Form II is the stable form at room temperature[416]. The ionization constants of $H_4P_2O_7$ at 18° are $K_1 = 1.4 \times 10^{-1}$, $K_2 = 1.1 \times 10^{-2}$, $K_3 = 2.1 \times 10^{-7}$, $K_4 = 4.1 \times 10^{-10}$. At 25°, the values for K_3 and K_4 are 2.7×10^{-7} and 2.4×10^{-10} respectively[404]. The heat of formation of crystalline pyrophosphoric acid is −538.0 kcal/mole and of the liquid −545.9 kcal/mole. The heat of fusion at 61° is 2.20 kcal/mole and the entropy of fusion is 6.587 cal/mole/deg[404].

The rate of hydrolysis of 0.100 molal $H_4P_2O_7$ prepared by dissolution of the crystalline acid has been determined at 30.0° and 60.0°. The half-life for the first order conversion of pyrophosphoric acid to orthophosphoric acid is given in Table 15[417].

TABLE 15. HYDROLYSIS OF PYROPHOSPHORIC ACID

% strength $\frac{g\ H_4P_2O_7}{g\ H_2O + g\ H_4P_2O_7} \times 100$	Molality	Temp. °C	$t_{\frac{1}{2}}$ (hr)
1.75	0.100	30.0	183
1.75	0.100	60.0	7.7
8.25	0.505	30.0	39.0
26.5	2.02	30.0	9.8
49.9	5.58	30.0	142 min

12.5. METAPHOSPHORIC ACIDS

Polymetaphosphoric Acids

Strong phosphoric acid forms an azeotrope containing 92% P_2O_5 and has a boiling point of 864° at 760 mm Hg pressure[418]. When glacial phosphoric is heated for a short time at 300°, a water-soluble acid which hydrolyzes to the pyrophosphoric acid at room temperature is obtained. This is called meta-acid-I and is generally considered a dimer $(HPO_3)_2$. Heating of glacial H_3PO_4 or meta-acid-I at 320° for 1 hr or 218° for 20 hr has been reported to give a water-insoluble crystalline meta-acid-II[404]. When meta-acid-I is heated to above 400°, a transparent water-soluble modification called meta-acid-III is formed. Heating of meta-acid-III for several hours at red heat produces a meta-acid-IV which disintegrates in water with a crackling sound. Since meta-acid-IV is difficultly soluble in water, a turbid dispersion of fragments in water is obtained. The polymeric metaphosphoric acids described above hydrolyze very slowly to orthophosphoric acid. The hydration proceeds in two steps, one of depolymerization and one of hydration.

[415] Ref. 406, pp. 619–20.

[416] C. Y. Shen and D. R. Dyroff, *Preparative Inorganic Reactions*, Vol. 5, p. 171, Ed. W. L. Jolly, Interscience Publishers, New York (1968).

[417] A. K. Nelson, Unpublished results, Stauffer Chemical Company.

[418] Ref. 406, p. 773.

Cyclic Metaphosphoric Acids

In addition to the poly-metaphosphoric acid described above, there are also trimeta- and tetrametaphosphoric acids both having a cyclic ring structure. In the elution of strong phosphoric acid through an ion exchange column (Table 14), small quantities of both trimeta- and tetrametaphosphoric acid are detected. It is believed that they are products of hydrolysis of the more complicated ring and branched chain polymers present in the strong phosphoric acid[412].

The best way to prepare trimetaphosphoric acid is by removal of the sodium ions from sodium trimetaphosphate (to be discussed under metaphosphates) by the use of a cation exchange resin. Tetrametaphosphoric acid can be made by adding slowly the hexagonal crystalline form of P_4O_{10} to ice water. To avoid further hydrolysis, the product is neutralized to a pH of 7 shortly after it is prepared. A 75% yield of the tetrametaphosphate (based on the P_4O_{10} used) is obtained[419]. The formation of tetrametaphosphoric acid from P_4O_{10} requires the scission of two P–O–P linkages in the P_4O_{10} tetrahedron:

$$P_4O_{10} \xrightarrow{2H_2O} H_4P_4O_{12}$$

Applications

For industrial applications, only superphosphoric acid and polyphosphoric acid of 82–84% P_2O_5 concentration are important. Superphosphoric acid's effectiveness depends on its high acid concentration, its easy-to-handle characteristics and its pyrophosphoric acid content. It is used in large scale for the preparation of high-strength phosphatic fertilizers. The pyrophosphoric acid content is especially beneficial in superphosphoric acid produced from wet process acids which is contaminated with metallic ions such as iron, aluminum, and magnesium. These metallic ions usually cause sludge formation when phosphoric acid is ammoniated for use as fertilizer. The pyrophosphate effectively complexes the metallic ions and holds them in solution.

Polyphosphoric acids of 82–84% P_2O_5 concentration find application as catalyst in the petroleum industry for alkylation, dehydrogenation, polymerization and isomerization reactions. They are particularly effective when supported in solid kieselguhr or diatomaceous earth. The catalytic activity of the acids depends largely on their hydrogen ions. One advantage of polyphosphoric acids is that they retain their acid characteristics at a much higher temperature than most other acids. Many of the reactions using polyphosphoric acids as catalyst are carried out at around 200°. To suppress the dehydration (and lowering of activity) of the polyphosphoric acid under reaction conditions, wherever conditions permit, 2–10% of water is introduced with the reactants[420].

[419] R. N. Bell, L. F. Audrieth and O. F. Hill, *Ind. Eng. Chem.* **44** (1952) 570.

[420] W. H. Waggaman, *Phosphoric Acid, Phosphate and Phosphatic Fertilizers*, 2nd ed., pp. 566–8, Longmans, Green and Co. Inc., New York (1947).

13. ALKALI METAL ORTHOPHOSPHATES

All of the hydrogens in orthophosphoric acid are replaceable with metal ions. The general method for the preparation of alkali metal phosphates is to react orthophosphoric acid with an amount of alkali metal hydroxide necessary to give a concentrated solution of the desired compositions. With few exceptions, the desired compound precipitates as crystals on cooling. Formation of specific hydrates of any compound from a solution depends on the temperature range at which the crystals are precipitated. Mono-, di-, and trialkali metal phosphates and their various hydrates as well as some double salts can thus be prepared.

Preparative methods for alkali metal phosphates with the limiting compositions of their saturated solutions at a specific temperature have been reviewed recently by Shen and Callis[421]. The phase-equilibria of alkali metal orthophosphates in water have been summarized in recent surveys[422, 423] and the chemistry of sodium phosphates has been reviewed by Quimby[424].

13.1. LITHIUM ORTHOPHOSPHATES

Lithium phosphates represent the unusual compounds in the alkali metal orthophosphate series. In contrast to other alkali metal orthophosphates which are all quite water soluble, $Li_3PO_4.12H_2O$ is insoluble, and Li_3PO_4 is only soluble to the extent of 0.03 g per 100 g of solution at 25°. Dilithium phosphate does not appear in the phase diagram; and monolithium phosphate is known, but it does not form any hydrate.

The crystal structure of $Li_3PO_4.12H_2O$ is trigonal with a crystal density of 1.645. The crystal structure of Li_3PO_4 is orthorhombic and the crystal density is 2.537[425]. In Li_3PO_4, the PO_4 tetrahedra is regular with a P–O bond length of 1.55 ± 0.02 Å and the O–P–O angle of $109.7 \pm 2°$. The Li ions coordinate tetrahedrally to the oxygen at an average distance of Li–O $= 1.96 \pm 0.03$ Å[426].

13.2. SODIUM ORTHOPHOSPHATES

Crystalline sodium phosphates which are stable at between 25° to 100° are[422, 426]:

NaH_2PO_4	Na_2HPO_4	Na_3PO_4
$NaH_2PO_4.H_2O$	$Na_2HPO_4.2H_2O$	$Na_3PO_4.1/2H_2O$
$NaH_2PO_4.2H_2O$	$Na_2HPO_4.7H_2O$	$Na_3PO_4.6H_2O$
$2NaH_2PO_4.Na_2HPO_4.2H_2O$	$Na_2HPO_4.8H_2O$	$Na_3PO_4.8H_2O$
$NaH_2PO_4.Na_2HPO_4$	$Na_2HPO_4.12H_2O$	$Na_3PO_4.ca12H_2O$
$NaH_2PO_4.H_3PO_4$		

Preparation. The hemisodium salt, $NaH_2PO_4.H_3PO_4$, is prepared by mixing aqueous solutions of H_3PO_4 and NaH_2PO_4. Upon concentration under vacuum, the crystals

421 C. Y. Shen and C. F. Callis, *Preparative Inorganic Reactions*, Vol. 2, pp. 139–67, Ed. W. L. Jolly, Interscience Publishers (1965).

422 B. Wendrow and K. A. Kobe, *Chem. Revs.* **54** (1954) 891-924.

423 J. R. van Wazer, *Phosphorus and Its Compounds*, Vol. 1, pp. 491–559, Interscience Publishers (1958).

424 O. T. Quimby, *Chem. Rev.* **40** (1947) 141–76.

425 J. R. van Wazer, *Kirk-Othmer Encyclopedia of Chemical Technology*, 2nd ed., Vol. 15, pp. 236–9, Interscience Publishers (1968).

426 D. E. C. Corbridge, *Topics in Phosphorus Chemistry*, **3** (1966) 168–71.

separated are ground fine and washed with Et_2O at 20° to remove any excess H_3PO_4. The compound is stable in humid weather to its fusion point. In a dry atmosphere, it dehydrates at 105° to condensed phosphates[427].

Both mono- and disodium phosphates are generally prepared on an industrial scale in the United States by the use of sodium carbonate as source of sodium. For the preparation of trisodium phosphate by crystallization from a solution, most of the third sodium atom has to come from the more expensive sodium hydroxide. The CO_2 from sodium carbonate cannot be removed from solution at above a pH of 8.

In the preparation of disodium phosphate, both the concentration of reactants and the temperature of reaction have to be carefully controlled. Too high a concentration or temperature results in the formation of tetrasodium pyrophosphate as a co-product. As will be shown later, the latter's presence in disodium phosphate, even in the order of 1–2%, is detrimental for its use as a cheese emulsifier. Anhydrous Na_3PO_4 can be prepared by calcining a solution containing $3Na_2O$ and $1P_2O_5$ at a temperature above 400° [421].

The most complex system of the sodium salts is the trisodium phosphate. Commercial crystalline trisodium phosphate dodecahydrate contains an excess of sodium hydroxide and is believed to consist of two isomorphous alkaline mixed salts having the formula $Na_3PO_4.1/4NaOH.12H_2O$ and $Na_3PO_4.1/7NaOH.12H_2O$[425]. Trisodium phosphate dodecahydrates having a phosphate to NaOH ratio different than those shown above have also been reported in the literature.

Sodium compounds, other than sodium hydroxides, also enter into complex formation with trisodium phosphate hydrates. Examples of such compounds are:[428]

$4(Na_3PO_4.11H_2O)NaNO_2$
$7(Na_3PO_4.11H_2O)NaNO_2$
$5(Na_3PO_4.11H_2O)NaMnO_4$
$7(Na_3PO_4.11H_2O)NaMnO_4$
$4(Na_3PO_4.11H_2O)NaOCl$
$5(Na_3PO_4.11H_2O)NaCl$
$4(Na_3PO_4.11H_2O)NaNO_3$
$Na_3PO_4.NaBO_2.18H_2O$

These compounds may be prepared by crystallizing from a liquor having the proper Na_2O–P_2O_5 ratio containing an excess of the added salt. In the presence of more than one added salt, the order of preference for complex formation is as follows:

Complex-forming salt	m.p. °C
NaOH	74
$NaMnO_4$	d70
NaOCl	62
NaCl	61
$NaNO_3$	60

If sufficient quantity of one salt is present to form the complex, any salt of lowei order in the series is found to be uncombined in the product.

Structure. IR spectra of the sodium phosphate solids and Raman spectra of their solutions showed the PO_4 group to have an approximate tetrahedral configuration[426].

427 A. Norbert, *Rev. Chim. Minerale*, **3**(1) (1966) 1–59.
428 R. N. Bell, *Ind Eng. Chem.* **41** (1949) 2901–5.

Properties. The solubility of the mono-, di-, and trisodium orthophosphates and their hydrates is shown in Fig. 12[423]. Hemisodium orthophosphate and trisodium phosphate (both the hemihydrate and the anhydrous form) dissolve incongruently in water. A water solution of these compounds upon evaporation yields crystals of another species.

Measurements by adiabatic colorimetry yielded heat capacity values at 298.15°K

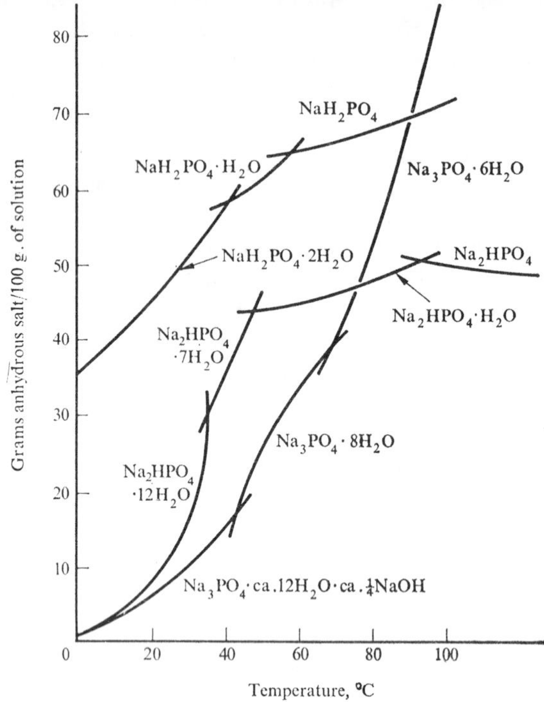

FIG. 12. Solubility of some sodium phosphates[423]. (Reprinted by permission of John Wiley & Sons Inc.)

(cal/deg) for $NaH_2PO_4 = 27.93$, $Na_2HPO_4 = 32.34$, and $Na_3PO_4 = 36.68$. The entropies (cal/deg) calculated for these compounds are $S_{298.15} = 30.47, 35.97$, and 41.54 respectively[429].

Applications. Industrial applications for the sodium phosphates depend on the chemical properties of these compounds. For example, monosodium phosphate is a water-soluble solid acid and many applications for it depend on this property. In water solution it is a mild phosphatizing agent on steel surfaces and provides an undercoating for paints such as in painted metal furniture where high quality protective coating from phosphoric acid–zinc salt phosphatizing is not necessary. The acidic property of monosodium phosphate is also effectively utilized in effervescent laxative tablets and in boiler water treatments for pH adjustments.

One large application for disodium phosphate is as a buffering agent. At 25°, pH buffers containing equimolar concentration of Na_2HPO_4 and KH_2PO_4, at a concentration ranging from 0.01 to 0.2 molar, are represented by the equation:

$$pH = 7.169 + 4.78c - 3.287(c)^{1/2}$$

where c is the total molar concentration of KH_2PO_4 plus Na_2HPO_4 [423].

[429] R. J. L. Andon, J. F. Counsell, J. F. Martin and C. J. Mash, *J. Appl. Chem.* **17** (1967) 65–70.

Another important use for disodium phosphate is as an emulsifier for the manufacture of pasteurized American processed cheese from aged cheddar cheese. The mechanism of this emulsifying action is not completely understood, but addition of about 1.8 to 2.0% by weight of disodium phosphate during processing prevents the separation of fat globules in processed cheese even when the latter is heated to a melt. Disodium phosphate used for this purpose must not contain any pyrophosphate. Presence of pyrophosphate impurity even at less than 1% level results in processed cheese which hardens upon ageing and which also won't melt when heated.

Disodium phosphate has many other applications in food products. It is added at about 0.1% to evaporated milk to maintain the correct calcium to phosphate balance. This prevents gelation of the milk during processing and subsequent storage. It is added at about 5% level to the sodium chloride brine solution used in the pickling of ham. The added phosphate reduces the exudation of juices during cooking which results in a more tender and juicy ham. Starch is chemically modified by heating with disodium phosphate. Such starch even though containing only very little phosphate has the desirable property of forming very stable cold water gel. Addition of 0.75 to 1.10% by weight of disodium phosphate to farinaceous products raises the pH to slightly on the alkaline side of neutrality. This treatment provides "quick-cooking" breakfast cereals.

Trisodium phosphate is highly alkaline in an aqueous solution as the result of the hydrolysis:

$$PO_4^{-3}+H_2O \rightleftharpoons HPO_4^{-2}+OH^-$$

Many of its applications depend on this property. Industrial hard surface (e.g., painted surfaces) cleaners and scouring powders are formulated with trisodium phosphate. The high alkalinity saponifies the oxidized oil on the painted surface or the grease in the kitchen sink to render them easy for removal.

The sodium hypochloride complex of trisodium phosphate is also strongly alkaline (pH of 1% solution = 11.8), and it has the added property of releasing active chlorine when wetted. In its use in scouring powder and automatic dishwashing formulations the high alkalinity provides cleansing action and the released chlorine gives bleaching action for stain removal as well as bactericidal action for sanitizing[430].

13.3. POTASSIUM ORTHOPHOSPHATES

The well-characterized potassium orthophosphates obtained by crystallization from saturated solutions include[421].:

$KH_5(PO_4)_2$	K_2HPO_4	$K_3PO_4.3H_2O$
KH_2PO_4	$K_2HPO_4.3H_2O$	$K_3PO_4.7H_2O$
$KH_2PO_4.2K_2HPO_4.H_2O$	$K_2HPO_4.6H_2O$	$K_3PO_4.9H_2O$[423]
$KH_2PO_4.3K_2HPO_4.2H_2O$		

Preparation. Hemi-potassium phosphate, $KH_5(PO_4)_2$, similar to the sodium analog, is prepared by mixing aqueous solutions of KH_2PO_4 and H_3PO_4 together and then concentrating under vacuum to obtain the crystalline compound. Any residual H_3PO_4 is removed by washing with diethyl ether at 20°. The $KH_5(PO_4)_2$ is not hygroscopic and may also be prepared by the reaction of H_3PO_4 with KCl at 20° in an open container[427].

[430] Product Report IPR-4, Stauffer Chemical Company.

Mono-, di-, and tripotassium phosphates can be prepared by neutralizing phosphoric acid with calculated quantity of potassium hydroxide to give a solution of the desired composition. Concentration of the solution followed by cooling results in the crystallization of the desired salt. Monopotassium phosphate, which has a solubility of 20 g per 100 g of the solution at 20°, is thus obtained.

Alternative procedures have also been reported for the preparation of monopotassium phosphate. Most of these methods proposed the use of the cheaper KCl as the source of potassium:

$$H_3PO_4 + KCl \longrightarrow HCl + KH_2PO_4$$

The HCl formed may be removed by anion exchange resins[431], long chain organic tertiary amines[432], or by distillation using octane as carrier[433]. Product obtained via these routes contains minor amount of KCl.

The compound $K_2HPO_4.3H_2O$ is prepared by the neutralization of 1 mole of H_3PO_4 with 2 moles of KOH. Since it has a solubility of over 60 g (based on the anhydrous salt) per 100 g of solution at 20°, it is necessary to concentrate the solution to a high degree before crystallization occurs on cooling. Water removal from this solution is usually carried out under reduced pressure at not higher than 50–60° to avoid possible pyrophosphate formation.

Tripotassium phosphate of commerce is the anhydrous form. It is prepared by calcining at 300–400° the reaction mixture from 3 moles of potassium hydroxide and 1 mole of phosphoric acid.

Structure. The structure of monopotassium phosphate has been investigated quite extensively. Raman data on a 35% aqueous solution show the following frequency values (cm^{-1}): $w_1 = 883$, $w_2 = 402$, $w_3 = 1069$, and $w_4 = 513$[434]. At room temperature crystalline monopotassium phosphate has a tetragonal form with each PO_4 tetrahedron linked to four others through hydrogen bonding at each of its corners. The lattice parameters as determined by X-ray diffraction are $a = 7.4528 \pm 0.0004$, and $c = 6.9683 \pm 0.0004$ Å.

Properties. The principal coefficients of thermal expansion for the range 25–150° for monopotassium phosphate follow the equations:

$$\alpha_a = 10.10 \times 10^{-6} + 21.68 \times 10^{-8} t - 2.62 \times 10^{-10} t^2$$
$$\alpha_c = 28.73 \times 10^{-6} + 17.65 \times 10^{-8} t + 1.83 \times 10^{-10} t \text{ (435)}$$

At a low temperature just above the Curie point of −151°, the tetragonal cell contracts and the hydrogen bonds are shorter than at normal temperature. However, the tetrahedral symmetry about the phosphorus atom is retained. At a temperature below the Curie end point the crystal structure is changed to that of orthorhombic and this change is associated with an ordering of the H atoms and the distortion of the PO_4 group. Monopotassium phosphate has piezoelectric and ferroelectric properties and the loss of ferroelectric properties (and an increase in the dielectric constant as well as the heat capacity) at the Curie point is associated with the above reversible tetragonal to orthorhombic transition at lower temperatures[426].

[431] S. Suzuki, M. Shimoyamada, H. Akabayashi, Y. Ogawa and T. Iwatsuki, Japan 14,293, 8 July 1965, to Missan Chemical Industries, Ltd.

[432] W. E. Clifford, French 1,443,004, 17 June 1966, to Kaiser Aluminum and Chemical Corp.

[433] M. Israel, R. Blumberg, A. M. Baniel and P. Melzer, Israel 21,072, 22 Dec. 1966.

[434] T. D. Farr, *Phosphorus, Properties of the Element and Some of Its Compounds*, Chemical Engineering Report 8, pp. 4 and 61–7, T.V.A., Wilson Dam, Alabama (1950).

[435] D. B. Sirdeshmukh and V. T. Deshpande, *Acta Crystallogr.* **22**(3) (1967) 438–9.

The electric conductivity of a single crystal of KH_2PO_4 has been measured at 50–180°. The conductivity along the two axes perpendicular to the *c* axis is approximately 1.5 times the value of the conductivity in the *c* direction at $< 110°$ and approximately the same above this temperature[436].

Applications. Monopotassium phosphate crystals can be grown to several inches long and because of its piezoelectric effect, which at room temperature is straightforward (the e.m.f. is proportional to the distortion force with essentially no hysteresis), it has been used in submarine sonar systems and other applications based on piezoelectricity[423].

The main application of dipotassium phosphate is as a buffering agent to maintain the pH of around 9 (pH of 1% solution = 8.9) such as in the car radiator coolants to prevent corrosion[437]. It is also used in special fertilizers and fermentation nutrients.

Tripotassium phosphate is an ingredient in some recipes for the polymerization of styrene-butadiene rubber where it acts as an electrolyte to regulate the rate of polymerization and also to control the stability of synthetic latex formed. Another application of tripotassium phosphate depends on its alkalinity (pH of 1% solution = 11.8). A water solution of it absorbs H_2S from gas streams. Such a solution may be regenerated by heating to drive off the absorbed H_2S[438].

13.4. AMMONIUM ORTHOPHOSPHATES

The following ammonium phosphates are present in the solid phases in the system $(NH_4)_2O.P_2O_5.H_2O$ from 0 to 75° [422]:

$NH_4H_5(PO_4)_2.H_2O$	$(NH_4)_2HPO_4$	$(NH_4)_3PO_4.3H_2O$
$NH_4H_5(PO_4)_2$	$2(NH_4)_2HPO_4.(NH_4)_3PO_4$	
$NH_4H_2PO_4$		

The industrially important ammonium phosphates are $NH_4H_2PO_4$ and $(NH_4)_2HPO_4$.

Preparation. Monoammonium phosphate is prepared by passing ammonia gas into an 80% solution of phosphoric acid. The composition of the product is controlled by maintaining a pH range of 3.8–4.5. Crystals of monoammonium phosphate separate out on cooling.

Diammonium phosphate is prepared by the addition of two moles of ammonia to one mole of 80% phosphoric acid solution. The correct NH_3 to H_3PO_4 ratio may also be controlled by maintaining the reaction mixture of pH 8. To prevent dissociation, the addition of the final quantity of ammonia is carried out at less than 50°. Drying of the crystals is similarly carried out at the low temperature range.

Structure. The crystal system of $NH_4H_2PO_4$ is body-centered tetragonal with 4 molecules per unit cell and lattice dimensions of $a = 7.48$ Å, $c = 7.56$ Å[434]. Crystals of $NH_4H_2PO_4$ exhibit piezoelectric properties. While not ferroelectric, it resembles KH_2PO_4 in that it also undergoes a transition from tetragonal to orthorhombic symmetry at low temperature (−125°)[426].

Properties. Of the ammonium phosphates, both $(NH_4)_3PO_4$ and the double salt $(NH_4)_3PO_4.2(NH_4)_2HPO_4$ are unstable at room temperature. They lose ammonia to

436 L. B. Harris and G. J. Vella, *J. Appl. Phys.* **37**(11) (1966) 4294.

437 R. F. Monroe and A. J. Maciejewski, U.S. 3,291,741, 13 Dec. 1966, to Dow Chemical Company.

438 H. W. Wainwright, G. P. Egleson, C. M. Brock, J. Fischer and A. E. Sands, *Ind. Eng. Chem.* **45** (1953) 1378–84.

TABLE 16. PROPERTIES OF MONO- AND DIAMMONIUM PHOSPHATES

	Crystal system	Refractive index *D* line	Sp. gravity d_4^{20}	Solubility g/100g soln	Dissociation pressure P_{NH_3}	Mean specific heat cal/g/deg	pH of 0.1M soln
$NH_4H_2PO_4$	Body centred tetragonal	$n_w = 1.525$ $n_\epsilon = 1.479$	1.803	18.2 to $0.449t$ ($t = 0$ to $110°$)	0.05 mm at 125°	0.3089 ($t = 0$ to $99.6°$)	4
$(NH_4)_2HPO_4$	Monoclinic		1.619	$36.28+0.218t$ ($t = 10$ to $80°$)		0.3408 ($t = 0$ to $99.6°$)	7.8

form diammonium phosphate. However, in the high ammonia region of the $(NH_4)_2O \cdot P_2O_5 \cdot H_2O$ system, triammonium phosphate occurs at up to 60° and $(NH_4)_3PO_4 2(NH_4)_2HPO_4$ double salt is stable at an even higher temperature. The thermal decomposition of mono-, di-, and triammonium phosphates has been studied by thermogravimetric and differential thermal analyses. In this test the minimum temperatures for the onset of decomposition are 30°, 140° and 170° for the tri-, di- and monoammonium phosphates respectively[439].

The vapor pressure of a saturated solution of $NH_4H_2PO_4$ is expressed by the equation

$$\log p_{mm} = -2240/T + 8.862$$

where T is 292° to 363°K.

The specific equation is: $p = 0.2773 + 7.30 \times 10^{-4}t$ cal/g $NH_4H_2PO_4$/degC at the temperature range of -25 to 100°. The pH of a 1% solution is 4.5. The heat of formation for $NH_4H_2PO_{4(c)}$, $\Delta H^\circ_{f(298 \cdot 16)}$, is -346.75 kcal/mole; free energy of formation $\Delta F^\circ_{f(298 \cdot 16)}$ is -290.46 kcal/mole and the entropy, $S^\circ_{297 \cdot 1}$, is 36.32 ± 0.1 cal/mole $NH_4H_2PO_4$/deg. The heat of neutralization of liquid H_3PO_4 with gaseous NH_3 to form $NH_4H_2PO_4$ is $\Delta H = -32.19$ kcal. Some other properties of $NH_4H_2PO_4$ are listed in Table 16[434].

The dissociation pressure of $(NH_4)_2HPO_4$ is represented by the equation, $\log p_{mm} = -4211/T + 12.040$, where T is 353° to 398°K; the vapor pressure equation of a saturated solution is $\log p_{mm} = -2240/T + 8.807$, where T is 292° to 328°K. The partial pressure of NH_3 over a saturated solution is $p_{mm} = 1$ at 25°, 9 at 50°, and 19 at 60°. The pH of 1% solution is 8.0. The heat of formation of $(NH_4)_2HPO_{4(c)}$ is $\Delta H^\circ_{f(298 \cdot 16)} = -376.12$ kcal/mole. The heat of neutralization of liquid H_3PO_4 with gaseous NH_3 to form $(NH_4)_2HPO_4$ is $\Delta H = -51.45$ kcal. The thermodynamics of the dissociation: $(NH_4)_2HPO_4 \rightarrow NH_4H_2PO_{4(c)} + NH_{3(g)}$ are $\Delta C_p = -0.81$ cal (at 49.8°), $\Delta H = 19.26$ kcal. Other properties of $(NH_4)_2HPO_4$ are listed in Table 16[434].

Applications. Mono- and diammonium phosphate are used interchangeably for many applications. They are both used as fertilizers and nutrients in fermentation broths to supply phosphorus and nitrogen. For fertilizer application, phosphorus promotes development of plant roots. It is also essential for flower development and fruit production. Nitrogen is essential for the growth of the vegetative organs. It also greatly increases the utilization of fertilizer phosphate by the plant, especially during the early growth period[439].

One very important application for ammonium phosphates is as a flame retardant agent for cellulosic materials. This action depends on the property of ammonium phosphates to dissociate to NH_3 and H_3PO_4 when heated. The phosphoric acid generated catalyzes the decomposition of cellulose to char (carbon) which is slow burning and also smothers the flame. Without the acid catalyst, cellulose decomposes on heating first to liquid fragments which on further heating convert to flammable volatiles[441].

A dilute solution of diammonium phosphate with an initial pH of 7.85, upon boiling, evolves NH_3 and in about 2.5 hr the pH drops to about 5.78. This property is used for the precipitation of colloidal wool dyes on wool fabrics. Wool dyes remain dispersed

[439] E. V. Margulis, L. I. Beisekeeva, N. I. Kopylov and M. A. Fishman, *Zh. Prikl. Khim.* **39**(10) (1966) 2364–6.

[440] J. R. van Wazer, *Phosphorus and Its Compounds*, Vol. II, pp. 1534–8, Interscience Publishers, New York (1961).

[441] R. W. Little, *Flameproofing Textile Fabrics*, Reinhold Publishing Corp., New York (1957).

under alkaline conditions, but precipitate rapidly and unevenly on the fabric under acidic conditions. The gradual change of the pH in the dye bath, as controlled by the decomposition of $(NH_4)_2HPO_4$, permits the gradual deposition of the dye. This makes for level dyeing of the cloth with good penetration[442].

14. ALKALINE EARTH PHOSPHATES

14.1. GENERAL PREPARATIVE METHODS

Alkaline earth phosphates in general are prepared by the reaction of alkaline earth hydroxides or oxides with phosphoric acid. Monoalkaline earth phosphates are crystallized from a solution containing an excess of phosphoric acid. Dialkaline earth phosphates are obtained as crystalline precipitates by the addition of two equivalents of the oxide, hydroxide or carbonate[443, 444] to slightly more than 1 mole of phosphoric acid. Trialkaline earth phosphates are prepared by the reverse addition of phosphoric acid to a slurry of alkaline earth hydroxides.

Another general method for the preparation of these compounds is by precipitation from the metathetical reaction between $(NH_4)_2HPO_4$ or other water-soluble dialkali-metal orthophosphates and the respective water-soluble alkaline earth salts. The temperature at which crystallization takes place determines whether hydrates or anhydrous forms of the compounds are to be obtained. Lower temperature ranges favor the higher hydrates and higher temperature ranges favor the anhydrous forms. Examples of some compounds prepared by the above method are listed in Table 17[445–7].

TABLE 17[447]

Compounds	Special requirements on temperature of crystallization, °C
$Mg(H_2PO_4)_2$	Between 10–130
$Mg(H_2PO_4)_2.2H_2O$	Between 10–130
$Mg(H_2PO_4).4H_2O$	Below 58
$Ca(H_2PO_4)_2$	Evaporate at 125–130
$Ca(H_2PO_4)_2.H_2O$	Evaporate at 35–40
$Ca_{10}(PO_4)_6(OH)_2$	Heat to boiling
$CaHPO_4$	Heat to 100–110
$CaHPO_4.2H_2O$	Below 40
$Sr(H_2PO_4)_2$	Below 60
$Sr(H_2PO_4)_2.H_2O$	Evaporate under vacuum
$Ba(H_2PO_4)_2$	Evaporate at 25
$Ba_3(PO_4)_2$	Reaction at 100

442 J. N. Dalton and J. P. Ploubides, U.S. 2,590,847, April 1952, to Pacific Mills.

443 A. T. Jensen and J. Rathlev, *Inorganic Synthesis*, Vol. 4, pp. 18–19, Ed. J. C. Bailar, Jr., McGraw-Hill Book Co., New York (1953).

444 A. G. Knapsack-Griesheim, Neth. Appl. 6,607,318, 19 Dec. 1966.

445 J. R. van Wazer, *Kirk-Othmer Encyclopedia of Chemical Technology*, 2nd ed., Vol. 15, pp. 246–64, Interscience Publishers (1968).

446 R. W. Mooney and M. A. Aia, *Chem. Rev.* **61** (1961) 433–59.

447 C. Y. Shen and C. F. Callis, *Preparative Inorganic Reactions*, Vol. II, pp. 148–53, Ed. W. L. Jolly, Interscience Publishers, New York (1965).

Special conditions are sometimes necessary for the preparation of some other alkaline earth phosphates from an aqueous solution. These compounds and the conditions for their preparation are shown in Table 18[446, 447].

TABLE 18[446, 447]

Compounds	
$Mg_3(PO_4)_2$	$MgCl_2$, KH_2PO_4 in stoichiometric ratio mixed in NaOH solution
$Mg_3(PO_4)_2.22H_2O$	20% $MgSO_4$ add to 2 vol. of 9% Na_3PO_4 and filter
$Mg_3(PO_4)_2.8H_2O$	Precipitate of $Mg_3(PO_4)_2.22H_2O$ left standing for 2–5 weeks with mother liquor
$Mg_3(PO_4)_2Mg(OH)_29H_2O$	Hydrolysis of $Mg_3(PO_4)_2$ in 0.0l N NaOH at 37°
$CaHPO_4$ $2H_2O$[a]	$Na_2HPO_4.2H_2O+CaCl_2.6H_2O \longrightarrow CaHPO_42H_2O+2NaCl+6H_2O$
$Ca_8H_2(PO_4)_6.5H_2O$	$CaHPO_4.2H_2O$ in 0.5 M CH_3COONa at 40°
β-$Ca_3(PO_4)_2$	$Ca(NO_3)_2+Na_2HPO_4+1\%$ Mg^{++} or Mn^{++} at 70°
β-$SrHPO_4$	$(NH_4)_2HPO_4$ added to Sr^{++} below 25°
α-$SrHPO_4$	$(NH_4)_2HPO_4$ added to Sr^{++} above 50°
$Sr_6H_3(PO_4)_5.2H_2O$	$Sr(NO)_2$ soln added to a KH_2PO_4 solution which is adjusted to pH 7.7 at 25°
$Sr_3(PO_4)_2.4H_2O$	$Sr(NO_3)_2$ added to excess K_2HPO_4 and KOH at 10°
$Sr_3(PO_4)_2$ (large crystals)	Slow hydrolysis of $SrHPO_4$ in Soxhlet
$Sr_{10}(OH)_2(PO_4)_6$	$Sr(NO_3)_2+(NH_4)_2HPO_4$ to pH 12 with ethylene diamine. Filter and heat precipitate at 950°
$Ba_3(PO_4)_2$	$BaCl_2+KH_2PO_4$ in stoichiometric ratio in NaOH solution

[a] A. T. Jensen and J. Rathlev, *Inorganic Synthesis*, Vol. IV, pp. 19–21, Ed. J. C. Bailar, Jr., McGraw-Hill Book Co., New York (1953)

Of the alkaline earth phosphates, the calcium salts are the most important and have received the most study[446, 448, 449].

14.2. MONO- AND DICALCIUM PHOSPHATES

One important property of the mono- and dicalcium phosphates (and the analogous magnesium phosphates) is that they have incongruent water solubility, i.e., in water the solid phase is in equilibrium with a solution more acidic than the solid phase composition.

Monocalcium Phosphate

As shown in the CaO–H_2O–P_2O_5 phase diagram (Fig. 13) monocalcium phosphate decomposes in the presence of water partially to the more basic dicalcium phosphate and phosphoric acid. The extent of this reaction increases with the amount of water and temperature. Since commercial monocalcium phosphate monohydrate is prepared either by total evaporation of the aqueous reaction mixture of hydrated lime and phosphoric acid or by crystallization from the aqueous system, it is thus always contaminated with various amounts of dicalcium phosphate. A typical product from the crystallization process contains 5.8% dicalcium phosphate while those from the total evaporation process contain up to 8–9% of dicalcium phosphate.

In the commercial production of monocalcium phosphate, extreme care is needed to

[448] T. D. Farr, *Phosphorus, Properties of the Element and Some of Its Compounds*, Chemical Engineering Report No. 8, pp. 52–60, T.V.A., Wilson Dam, Alabama (1950).

[449] J. R. van Wazer, *Phosphorus and Its Compounds*, Vol. I, pp. 510–40, Interscience Publishers, New York (1958).

avoid presence of residual phosphoric acid. Free phosphoric acid renders the material hygroscopic and the absorbed water along with the acid catalyze the decomposition to more free phosphoric acid and dicalcium phosphate. Monocalcium phosphate contaminated with free phosphoric acid is thus an unstable product. In fact, a portion of the

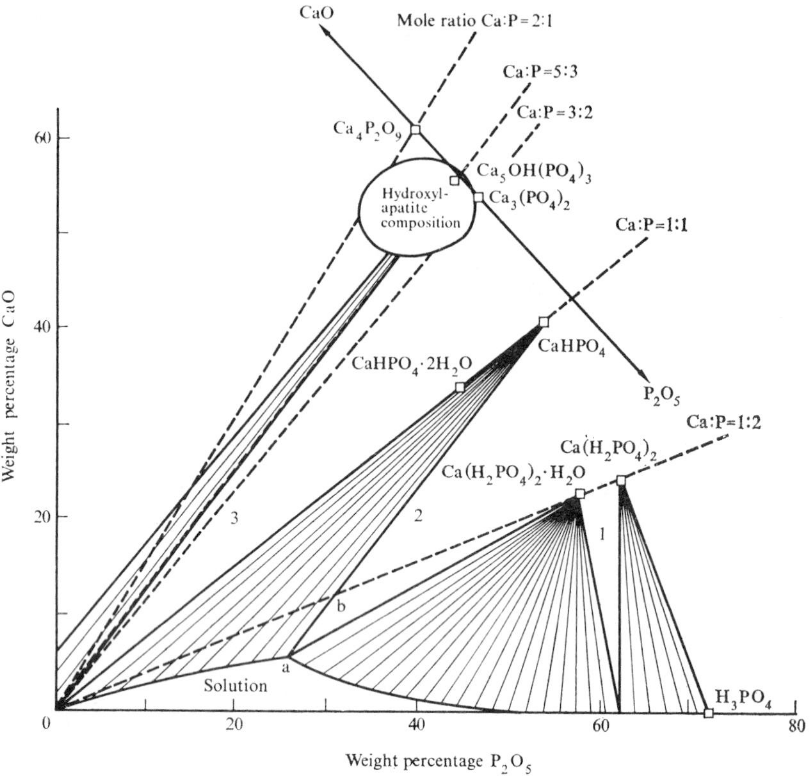

FIG. 13. $CaO–H_2O–P_2O_5$ phase diagram.[449] (Reprinted by permission of John Wiley & Sons Inc.)

dicalcium phosphate impurity present in industrial monocalcium phosphate is the result of intentional addition of an excess of lime for the removal of residual free phosphoric acid.

Dicalcium Phosphate

In an aqueous system, it decomposes to the more basic calcium hydroxyl apatite and free phosphoric acid. This decomposition is also favored by an excess of water and high temperature. It goes through an intermediate stage, that of the formation of octacalcium phosphate $[Ca_8H_2(PO_4)_6.5H_2O]$ and free phosphoric acid[450]. The decomposition of $CaHPO_4.2H_2O$ in an aqueous solution is also accelerated by the presence of fluoride ions. The final product in this case always contains fluoroapatite, $Ca_5(PO_4)_3F$[451].

[450] W. E. Brown, J. P. Smith, J. R. Lehr and A. W. Frazier, *Nature*, **196**, No. 4859 (1962) 1048–55.
[451] D. H. Booth and R. V. Coates, *J. Chem. Soc.* (1961) 4914–21.

Thermogravimetric, IR and NMR studies indicated that the water of hydration in $CaHPO_4.2H_2O$ is in two states, one loosely bound and the other tightly bound, with some mobility of water molecules between these two states[452]. Crystalline $CaHPO_4.2H_2O$ loses both of its water of hydration at a moderately elevated temperature and/or on storage. Dehydration is accelerated by the presence of moisture, as for example, when it is formulated as the polishing agent in a toothpaste. The dehydration and subsequent growth of the anhydrous dicalcium phosphate crystals result in the solidification of the paste to a hard rock-like mass.

There has been speculation that the dehydration of $CaHPO_4.2H_2O$ to $CaHPO_4$ in the presence of a little moisture is related to the decomposition of $CaHPO_4.2H_2O$ in a water slurry to hydroxyl apatite and free phosphoric acid. The formation of $CaHPO_4$ is postulated to be from the reaction of free phosphoric acid and hydroxyl apatite resulting from the decomposition of $CaHPO_4.2H_2O$. It is known that the reaction between phosphoric acid and finely divided hydroxyl apatite is reasonably rapid at room temperature[449].

Recent kinetic studies, however, show that the dehydration of $CaHPO_4.2H_2O$ in a water vapor atmosphere takes place in a single step. After an induction period, which could be long if the water vapor is low, the reaction rate increases and then reaches a maximum[453].

The instability of $CaHPO_4.2H_2O$ as prepared by the neutralization of H_3PO_4 with a slurry of lime at 38–40° is an important industrial problem. Two methods of stabilization have been developed and are in actual use. The first method involves the addition of a small quantity of pyrophosphate ion to the compound. This is carried out by adding a few per cent of tetrasodium pyrophosphate to the slurry of dicalcium phosphate during the manufacturing process[454]. The second method is for stabilizing $CaHPO_4.2H_2O$ from dehydrating in a moist toothpaste formulation and it is carried out by adding 2–3% of trimagnesium phosphate to the $CaHPO_4.2H_2O$ as a dry mix[455]. The mechanism of stabilization by both the pyrophosphate ion and trimagnesium phosphate is not well understood. However, in the hydrolysis of $CaHPO_4.2H_2O$ through the sequence to octacalcium phosphate to hydroxyl apatite and free phosphoric acid stages, it has been found that as little as 0.001 N of magnesium ion inhibits the further hydrolysis of octacalcium phosphate[450]. If the speculation that the dehydration of $CaHPO_4.2H_2O$ is through the mechanism of hydrolysis first to octacalcium phosphate then to hydroxyl apatite and phosphoric acid and subsequent reaction of hydroxyl apatite with phosphoric acid to form anhydrous $CaPHO_4$ were true, then the inhibition of hydrolysis of octacalcium phosphate by Mg^{++} would interrupt the above reaction sequence and thus account for the inhibition of dehydration of moist $CaHPO_4.2H_2O$ by $Mg_3(PO_4)_2$.

The decomposition of $Ca(H_2PO_4)_2.H_2O$ and $CaHPO_4.2H_2O$ is also important in the manufacture of fertilizers. Superphosphate is ammoniated to convert the water-soluble $Ca(H_2PO_4)_2.H_2O$ into citrate-soluble dicalcium phosphate and water-soluble mono- or diammonium phosphate, e.g.:

$$Ca(H_2PO_4)_2.H_2O+2NH_3 \longrightarrow CaHPO_4+(NH_4)_2HPO_4+H_2O$$

If the temperature of this reaction is allowed to go above 40°, the $CaHPO_4$ formed is

452 J. Fraissard, A. de S. Dupin and A. Boulle, *Compt. Rend.* **261**(23) (Group 6) (1965) 5040–3.
453 P. Dugleux and A. de S. Dupin, *Bull. Soc. Chim. France*, **1** (1967) 144–50.
454 H. V. Moss and M. G. Kramer, U.S. 2,287,699, June 1942, to Monsanto Chemical Co.
455 G. A. McDonald and D. Miller, U.S. 2,018,410, 1935, to Victor Chemical Works.

decomposed further to the insoluble hydroxyl apatite, an undesirable product for fertilizer applications[456].

Other properties of mono- and dicalcium phosphates are summarized in Table 19[448]. The solubility product data listed are calculated thermodynamically.

TABLE 19. PROPERTIES OF MONO- AND DICALCIUM PHOSPHATES[448, 449]

	Crystal system	Density	Refractive index	Solubility product K 25°
$Ca(H_2PO_4)_2.H_2O$	Triclinic pinacoidal	2.220 g/cm³ at 16°	$n_\alpha = 1.496$ $n_\beta = 1.515$ $n_\gamma = 1.527$	$a_{Ca^{2+}}.a^2_{H_2PO_4^{1-}} = 7.19 \times 10^{-2}$ ($pK_{sp} = 1.14$)
$Ca(H_2PO_4)_2$	Triclinic	2.546 g/cm³ at 16°	$n_\alpha = 1.547$ $n_\beta = 1.580$ $n_\gamma = 1.601$	
$CaHPO_4.2H_2O$	Monoclinic prismatic, biaxial positive Lattice dimension: $a = 10.47$Å $b = 15.15$Å $c = 6.37$Å $\beta = 150° 8'$	2.306 g/cm³ at 16.5°	$n_\alpha = 1.539$ $n_\beta = 1.544$ $n_\gamma = 1.549$	$a_{Ca^{2+}}.a_{HPO_4^{2-}} = 2.18 \times 10^{-7}$ ($pK_{sp} = 6.66$)
$CaHPO_4$	Triclinic pinacoidal	2.892 g/cm³ at 16°	$n_\alpha = 1.587$ $n_\beta = 1.615$ $n_\gamma = 1.640$	2.18×10^{-7}. Solubility in acid solution decreases with rising temperature

14.3. OCTACALCIUM PHOSPHATE AND TRICALCIUM PHOSPHATE

Octacalcium phosphate, $Ca_8H_2(PO_4)_65H_2O$, and hydroxyl apatite, $Ca_5(PO_4)_3OH$, mentioned above have received extensive studies. The reason for this is that hydroxyl apatite is now recognized as the most important of the inorganic compounds in the structure of bone[457] and teeth and that octacalcium phosphate may be a precursor to it.

Octacalcium Phosphate

As shown in Table 18, pure octacalcium phosphate is prepared by the controlled hydrolysis of $CaHPO_4.2H_2O$ in a sodium acetate buffer at 40° [458]. Its structure is related to hydroxyl apatite but the two compounds are not isostructural. Octacalcium phosphate has the following unit cell dimensions: $a = 19.87$ Å, $b = 9.63$ Å, $c = 6.87$ Å, $\alpha = 89°17'$, $\beta = 92°13'$, $\gamma = 108°57'$[450]. It has a pK_{sp} of 46.9 at 25°. Infrared spectra show a broad band near 3500 cm^{-1} and strong bands at 1000 and 1100 cm^{-1}. Octacalcium phosphate has also two characteristic bands (865 and 910 cm^{-1}) which are useful in identifying it in the presence of hydroxyl apatite. Two other major bands for the compound occur at 599 and 559 cm^{-1} [459]. Boiling of octacalcium phosphate in water results in gradual hydrolysis to hydroxyl apatite[450].

[456] W. H. Waggaman, *Phosphoric Acid, Phosphates and Phosphatic Fertilizers*, 2nd ed., pp. 321–8, Reinhold Publishing Corp., New York (1952).

[457] V. K. Lamer, *J. Phys. Chem.* **66** (1962) 973–8.

[458] W. E. Brown, J. R. Lehr, J. P. Smith and A. W. Frazier, *J. Am. Chem. Soc.* **79** (1957) 5318.

[459] B. O. Fowler, E. C. Moreno and W. E. Brown, *Arch. Oral. Biol.* **11**(5) (1966) 477–92.

Tricalcium Phosphate

Commercial tricalcium phosphate, as prepared by the addition of phosphoric acid to a slurry of hydrated lime, consists mostly of the amorphous form of hydroxyl apatite $Ca_5(PO_4)_3OH$. There is, however, a true tricalcium phosphate $Ca_3(PO_4)_2$. The β-form occurs in nature as the mineral whitlockite and contains 6.8 at.% magnesium. Synthetically, as shown in Table 17, it may be prepared by the reaction of a water solution of $Ca(NO_3)_2$ with that of Na_2HPO_4 in the presence of Mg^{++} or Mn^{++} as stabilizer. The β-$Ca_3(PO_4)_2$, when heated, undergoes the following progression:

$$\beta\text{-}Ca_3(PO_4)_2 \xrightarrow{1180^\circ} \alpha\text{-}Ca_3(PO_4)_2 \xrightarrow{1430^\circ} \alpha'\text{-}Ca_3(PO_4)_2$$

The refractive indices of β-$Ca_3(PO_4)_2$ are $n_\alpha = 1.620$ and $n_\gamma = 1.622$, while those of α-$Ca_3(PO_4)_2$ are (biaxial positive): $n_\alpha = 1.588$ and $n_\gamma = 1.591$. The solubility in citrate solution for β-$Ca_3(PO_4)_2$ is 20 to 50%, while that of α-$Ca_3(PO_4)_2$ is 85 to 98%. Infrared spectra of β-$Ca_3(PO_4)_2$ have been studied by Mooney *et al.*[460]

Hydroxyl apatite is generally represented by the formula $Ca_5(PO_4)_3OH$. However, by variations in the methods of synthesis, compounds having Ca/P mole ratio of from 1.41 to 1.75 as shown on Table 20[447] are obtained and which all have nearly the same X-ray patterns.

TABLE 20. CALCIUM HYDROXYL APATITE[447]

Mole ratio Ca/P	Method of preparation
1.41	Dilute $CaCl_2$ plus excess of dilute Na_2HPO_4 at 25°
1.50	$Ca(OH)_2$ added to H_3PO_4 to phenolthalein end point or by slow hydrolysis of $CaHPO_4.2H_2O$ (gets good crystals)
1.61	$Ca(OH)_2$ added to dilute H_3PO_4 to phenolthalein end point and then bo led
1.67	$Ca(OH)_2$ added to dilute H_3PO_4 then neutralized at boiling
1.75	Freshly precipitated "tricalcium phosphate" plus lime

The method described in *Inorganic Synthesis* for the preparation of $Ca_5(PO_4)_3OH$ involves the use of an aqueous NH_4OH solution to maintain a sufficiently high pH (pH = 12). Under high pH conditions, the PO_4^{3-} concentration is not exceeded by that of HPO_4^{2-} and thus avoids the formation of $Ca_8H_2(PO_4)_6.5H_2O$[461]. The reaction is shown in the following equation:

$$5Ca(NO_3)_2 + 3(NH_4)_3PO_4 + NH_4OH \longrightarrow Ca_5(OH)(PO_4)_3 + 10NH_4NO_3$$

The precipitate obtained is heated to 800°. It shows a structure corresponding to hexagonal prisms which terminate in pyramids. The $Ca_5(PO_4)_3OH$ crystal is uniaxial negative with index of refraction- $n_\epsilon = 1.644$ and $n_\omega = 1.651$ and lattice dimensions of $a = 9.422$ Å, $c = 6.883$ Å. Infra studies on hydroxyl apatite show strong bands at 1000–1100 cm^{-1} and medium intensity bands at 631, 601 and 501 cm^{-1} [458]. Hydroxyl apatite has a specific gravity of 3.21 and decomposes at 1400°. The solubility product constant is $K_{sp} = 1.53 \times 10^{-112}$ at 25° [448, 462].

Hydroxyl apatite may be regarded as the parent member of a whole series of calcium

[460] R. W. Mooney, S. Z. Tuma and R. L. Goldsmith, *J. Inorg. and Nuclear Chem.* **30** (1968) 1669–75

[461] E. Hayek and H. Newesley, *Inorganic Synthesis*, Vol. VII, pp. 63–5, Ed. J. Kleinberg, McGraw-Hil Book Co. (1963).

[462] R. A. Young and J. C. Elliott, *Arch. Oral. Biol.* **11**(7) (1966) 699–707.

phosphates with related structures. This series includes fluorapatite, $Ca_5(PO_4)_3F$, chlorapatite, $Ca_5(PO_4)Cl$, and bromapatite, $Ca_5(PO_4)_3Br$. All these apatites have the same hexagonal structure with the halogens[463] replacing the hydroxyl groups in the crystal lattice. By a diffusion process at 1200°, a 0.4 mm diameter single crystal of synthetic chlorapatite was changed into hydroxyl apatite while maintaining the single crystal form. The lattice parameters for fluorapatite are $a = 9.364$ Å, $b = 6.879$ Å, unit cell volume = 521.3 Å^3, and for chlorapatite $a = 9.634$ Å, $c = 6.783$ Å, and unit cell volume = 545.2 Å^3 [462].

Applications

Calcium phosphates are the only alkaline earth phosphates with important industrial applications. Very crude monocalcium phosphate $Ca(H_2PO_4)_2.H_2O$ as prepared by the reaction of phosphate mineral with sulfuric acid is known as superphosphate:

$$2Ca_5(PO_4)_3F + 7H_2SO_4 + H_2O \longrightarrow 7CaSO_4 + 3Ca(H_2PO_4)_2 . H_2O + 2HF$$

In admixture with the $CaSO_4$ by-product it is used as fertilizer. A purer grade of monocalcium phosphate fertilizer is the "triple superphosphate" prepared by the reaction of calcium phosphate mineral with H_3PO_4:

$$Ca_5(PO_4)_3F + 7H_3PO_4 + 5H_2O \longrightarrow 5Ca(H_2PO_4)_2 . H_2O + HF$$

As mineral supplement for animal feed, "stock food" grade of crude dicalcium phosphate is prepared by reacting pasty hydrated lime with 75–80% H_3PO_4. Such product contains anhydrous dicalcium phosphate as the main component along with monocalcium phosphate, tricalcium phosphate and unreacted lime.

Defluorinated phosphate rock is also used for animal feed purposes. It is prepared by heating phosphate mineral, $Ca_5(PO_4)_3F$, to 1200–1400° in the presence of water vapor and added H_3PO_4, Na_2SO_4 and SiO_2. The finished product is essentially α-$Ca_3(PO_4)_2$ and in order to meet prescribed specification it must not have more than 100 ppm F for each per cent of phosphorus[464, 465].

Calcium phosphates for human consumption or "food grade" are prepared from high quality phosphoric acid and lime. In the 1850s, monocalcium phosphate monohydrate $Ca(H_2PO_4).H_2O$ was introduced as the solid acid to react with $NaHCO_3$ to generate CO_2 for leavening in baking. Since then, its applications have greatly broadened. It is used also as the acid in effervescent tablets, as additive in phosphated flour, and as bread improver to stimulate the growth of yeast in yeast leavened products.

Anhydrous monocalcium phosphate is hygroscopic and was regarded for many years as unsuitable for use as a leavening acid. It was found, however, that when $Ca(H_2PO_4)_2$ is crystallized from a mother liquor containing added minor quantity of K^+, Na^+, and Al^{3+}, upon filtration, the minute crystals are wetted with the mother liquor containing the acid phosphates of these metals. When these crystals are dried and heat treated at 210–220°, the process resulted in the covering of each $Ca(H_2PO_4)_2$ crystal with a contiguous glassy coating of the added metal polyphosphates[466]. Such product then becomes an extremely useful baking acid in that it has a delayed reaction with $NaHCO_3$. Less CO_2 is lost in the dough mixing stage and thus allows more leavening action to occur during the baking process. The

[463] D. E. C. Corbridge, *Topics in Phosphorus Chemistry*, **3** (1966) 171–95.
[464] Ref. 456, pp. 376–406.
[465] Smith-Douglas Co., Brit. 902,361, 1 Aug. 1962.
[466] J. R. Schlaeger, U.S. 2,160,232, May 1939, to Victor Chemical Works.

coated $Ca(H_2PO_4)_2$ also permits the formulation of more storage stable self-rising flours and prepared baking mixes in which the leavening ingredients are premixed with the flour.

For $CaHPO_4.2H_2O$, the major application is as polishing agent in toothpaste formulations. It cleans teeth to provide a glossy surface on the tooth enamel without being too abrasive. Anhydrous dicalcium phosphate, on the other hand, is quite abrasive. It is used only in combination with the less abrasive $CaHPO_4.2H_2O$ or tricalcium phosphate for specialty toothpaste formulations in which extra abrasiveness is required as in the case of "smoker's toothpaste" for stain removal. Dicalcium phosphates are also used in pharmaceutical tablets as supplement for calcium and phosphorus.

Commercial tricalcium phosphate has particle size of the order of few microns in diameter. It is added at 1 to 2% level to coat hygroscopic salt or sugar particles so that they remain free flowing even in humid weather. It has also been claimed as a component along with organic bromides and zinc chloride in flame retardant formulations for polyacrylonitrile[467].

15. OTHER INORGANIC ORTHOPHOSPHATES

Metal ions other than alkali metal and alkaline earth metals also form orthophosphate salts. Mixed metal orthophosphates are also known. The three general methods for their preparation are: (1) high temperature solid state reaction: (2) precipitation of the less soluble phosphate from solution; and (3) crystallization from an equilibrium solution.

The procedure of choice depends on the nature of the compound. High temperature reactions are limited to the anhydrous tertiary phosphates and are carried out below the fusion point of the compound. The precipitation method is widely used because the majority of the orthophosphates are of limited solubility in an aqueous medium. For this method, phosphoric acid or water-soluble alkali phosphates are reacted with a water-soluble metal salt bearing the cation. However, in order to obtain duplicable products, exact conditions of reaction must be followed. There are a multiplicity of cases of polymorphism in many compounds and different crystalline varieties of the same compound are obtained under very nearly the same conditions[468].

The crystallization process gives definite compounds in accordance with the phase diagram[469]. In order to grow large crystals of a specific composition and crystal structure, sometimes it is necessary to heat the aqueous system under high temperature and pressure, that is by the hydrothermal process.

15.1. TRIPLY CHARGED METAL ION ORTHOPHOSPHATES

The best known of metal orthophosphates, other than the alkali-metal and alkaline earth phosphates, are the tertiary boron, aluminum and iron phosphates.

Boron Phosphate

This compound is prepared by heating stoichiometric quantity of the reaction product of

[467] F. J. Lowes, Jr., U.S. 3,271,343, 6 Sept. 1966, to Dow Chemical Co.

[468] F. d'Yvoire, *Bull. Soc. Chim. France* (1962) 1243–6.

[469] C. Y. Shen and C. F. Callis, *Preparative Inorganic Reactions*, Vol. II, pp. 153–64, Ed. W. L. Jolly, Interscience Publishers, New York (1965).

B_2O_3 and H_3PO_4 to 400–500°. It may also be obtained by adding H_3BO_3 to H_3PO_4 at 25° and then crystallized from a solution containing more than 48% P_2O_5[469].

The crystal structure of BPO_4 is similar to that of SiO_2. Each phosphorus atom is surrounded by a tetrahedron of oxygen atoms, and the boron atom is also tetrahedrally coordinated by oxygens. The crystals of BPO_4, therefore, consist of alternate tetrahedra linked together to form a continuous three-dimensional structure. Since each bond is formed by the sharing of an oxygen atom between a phosphorus and a boron atom, the compound may be thought of as a mixed anhydride of P_2O_5 and B_2O_3[470].

Boron phosphate is thermally stable, showing no appreciable vapor pressure until above 1000°. Some studies have been made to use it as catalyst supports. It is used also in ceramics.

Aluminum Phosphates

In the aluminum phosphate system, the following compounds are known:

$AlH_3(PO_4)_2.3H_2O$

$AlH_3(PO_4)_2.H_2O$

$Al(H_2PO_4)_3.1.5H_2O$

$Al(H_2PO_4)_3$

$Al_2(HPO_4)_3.3H_2O$

$AlH(HPO_4)_2.3H_2O$

$AlPO_4.3.5H_2O$

$AlPO_4.2H_2O$

$AlPO_4$

All of these compounds may be crystallized from solution under equilibrium conditions.

The acidic aluminum phosphate, $AlH_3(PO_4)_2.H_2O$, is isolated as crystals by evaporating at about 100° a solution of alumina in phosphoric acid having a $P_2O_5:Al_2O_3$ ratio of about 3.5[471]. The crystals appear as rectangular platelets or as hexagonal lamallae. The tri-hydrate, $AlH_3(PO_4)_2.3H_2O$, is obtained by allowing a concentrated solution of P_2O_5 and Al_2O_3 (ratio between 6 and 2.6) to evaporate in air at ambient temperature. The crystals formed belong to the rhombohedral system and appear as platelets in the form of equilateral triangles or triangular pyramids with a density of 2.192[471].

Monoaluminum phosphate, $Al(H_2PO_4)_3$, is reported to exist in at least three forms. The classical variety referred to by d'Yvoire[472] as type C is obtained by evaporating in air at 95° a solution of alumina in phosphoric acid having a $P_2O_5:Al_2O_3$ ratio of $\geqq 6$ (long needle crystals) or under vacuum at room temperature (hexagonal prisms). These crystals have a rhombohedral lattice. Two other varieties of $Al(H_2PO_4)_3$ referred to by d'Yvoire as A and B are identified (not isolated) as components in the thermal decomposition product of acid aluminum phosphates, $AlH_3(PO_4)_2{\cdot}3H_2O$ and $AlH_3(PO_4)_2.H_2O$. The particular variety obtained depended on the nature and physical state of the starting compound and dehydrating conditions[472].

[470] J. R. van Wazer, *Phosphorus and Its Compounds*, Vol. I, pp. 550–9, Interscience Publishers Inc., New York (1958).

[471] F. d'Yvoire, *Bull. Soc. Chim. France* (1961) 2283–90.

[472] F. d'Yvoire, *Bull. Soc. Chim. France* (1961) 2277–82.

Upon heating, $Al(H_2PO_4)_3$ loses water, forming at between 290–490° the acid triphosphate $H_2(AlP_3O_{10})$ which on further heating to 900° yields the crystalline cyclic tetrametaphosphate and a long chain polyphosphate.

Not too much is known on the compound $Al_2(HPO_4)_3$. It is obtained as an amorphous precipitate containing various water of hydration when an acidic solution of aluminum sulfate and ammonium phosphate is boiled. It precipitates from solution only over a narrow range of acid concentration[471].

The dihydrate, $AlPO_4{\cdot}2H_2O$, occurs naturally in two forms: as the orthorhombic variscite and as the monoclinic metavariscite. Both forms can be prepared synthetically by boiling a solution of alumina in phosphoric acid with a $P_2O_5{:}Al_2O_3$ ratio of 2.7. Variscite is formed from dilute solutions, metavariscite from concentrated solutions, and a mixture of the two varieties from solutions of medium concentration.

In the preparation of other hydrates of $AlPO_4$, it is quite easy to obtain different results from apparently identical conditions. This is due to the readiness of the Al_2O_3–P_2O_5–H_2O system to form metastable solutions which are probably colloidal in nature. Such systems can undergo changes with time and/or with minor changes in the preparative methods[473].

Anhydrous $AlPO_4$ can be obtained by the thermal dehydration of the hydrates or by crystallization from solution at above about 130°. Large crystals may be grown by heating in an autoclave a saturated solution of $NaAlO_2$ in H_3PO_4 very slowly from 132 to 315°[474].

Tertiary aluminum phosphate, like boron phosphate, is isostructural with SiO_2. The phosphorus atom is surrounded tetrahedrally by four oxygen atoms. It has been found to exist in at least six different crystalline modifications that are structurally parallel to some of the known forms of silica[475–7].

$$SiO_2 \left\{ \begin{array}{l} \text{Quartz} \xrightleftharpoons[867^\circ]{} \text{Tridymite} \xrightleftharpoons[1470^\circ]{} \text{Cristobalite} \xrightleftharpoons[1713^\circ]{} \text{melt} \\ \beta \xrightleftharpoons[573^\circ]{} \alpha \quad \beta \xrightleftharpoons[117^\circ]{} \alpha \xrightleftharpoons[163^\circ]{} \alpha_2 \quad \beta \xrightleftharpoons[220^\circ]{} \alpha \end{array} \right.$$

$$AlPO_4 \left\{ \begin{array}{l} \text{Berlinite} \xrightleftharpoons[705^\circ]{} \text{Tridymite analog} \xrightleftharpoons[1025^\circ]{} \text{Cristobalite analog} \xrightleftharpoons[>1600^\circ]{} \text{melt} \\ \beta \xrightleftharpoons[586^\circ]{} \alpha \quad \beta \xrightleftharpoons[93^\circ]{} \alpha_1 \xrightleftharpoons[130^\circ]{} \alpha_2 \quad \beta \xrightleftharpoons[210^\circ]{} \alpha \end{array} \right.$$

Despite the similarity between SiO_2 and $AlPO_4$, the attempts to obtain solid solutions by heating the cristobalite forms of the two compounds together have not been successful[473].

Aluminum phosphates as impurities in the wet process phosphoric acid are highly objectionable in that when wet process acid is ammoniated the insoluble aluminum phosformed tie up the phosphate value and reduce its availability to the plants as fertilizers. These precipitates also clog up fertilizer equipment. Pure aluminum phosphates are, however, quite useful. Monoaluminum phosphate in phosphoric acid solution is applied to steel plates used in electrical transformer construction. Upon heat treatment, it forms a surface coating which minimizes the occurrence of eddy currents in the transformer during its

[473] F. d'Yvoire, *Bull. Soc. Chim. France* (1961) 1762–76.
[474] J. M. Stanley, *Ind. Eng. Chem.* **46** (1954) 1684–9.
[475] D. E. C. Corbridge, *Topics in Phosphorus Chemistry*, **3** (1966) 195–208.
[476] D. Schwarzenbach, *Naturwissenschaften*, **52**(12) (1965) 343–4.
[477] D. Schwarzenbach, *Z. Krist.* **123**(3–4) (1966) 161–85.

operation. Aluminum acid phosphates, due to the formation of thermally stable aluminum phosphates, and crystalline $AlPO_4$ upon heating to a high temperature, are used also as binder in refractories[478-9].

Iron Phosphates

These compounds are quite similar to the aluminum phosphates. There are two varieties of $Fe(H_2PO_4)_3$ (A and B) which are isomorphous with the A and B forms of $Al(H_2PO_4)_3$. Formation of $Fe(H_2PO_4)_3$ is through the evaporation at 95° of a concentrated solution of $FeCl_3$ in H_3PO_4 in which the $P_2O_5:Fe_2O_3$ ratio is between 4 and 7. Variety A, in the shape of rhombohedra, is favored when crystals are allowed to form without agitation. Variety B, in the shape of leaves, is obtained when the solution is kept at 95° and subjected to successive crystallization and redissolution with frequent agitation[472].

Both varieties of $Fe(H_2PO_4)_3$ are very hygroscopic. They absorb water and undergo hydrolysis to give $FeH_3(PO_4)_2{\cdot}2.5H_2O$ and a very acidic solution[472].

Upon heating to above about 480° $Fe(H_2PO_4)_3$ undergoes decomposition similar to $Al(H_2PO_4)_3$ with the formation of a mixture of crystalline cyclic tetrametaphosphate and a long chain polyphosphate[472]. The similarity between $Fe(H_2PO_4)_3$ and $Al(H_2PO_4)_3$ is also shown by the formation of a solid solution when Al_2O_3 and Fe_2O_3 are dissolved in H_3PO_4. The substitution of Fe in the C variety of $Al(H_2PO_4)_3$ has been observed for ratio Fe/(Fe+Al) less than about 0.5[472].

The industrial applications of iron phosphates are fairly limited. $FePO_4{\cdot}2.5H_2O$ formed by precipitation from the reaction of ferric chloride with an alkali metal orthophosphate in an aqueous solution is used as mineral supplement to supply iron.

15.2. MIXED ALKALI METAL—OTHER METAL ORTHOPHOSPHATES

Many crystalline orthophosphates containing two or more cations have been prepared. The best known of these is $NH_4MgPO_4{\cdot}6H_2O$ which is insoluble in an ammoniacal solution and has been used for the quantitative gravimetric determination of phosphate.

Two compounds of recent commercial importance are $NaAl_3H_{14}(PO_4)_8{\cdot}4H_2O$[480] and $Na_3Al_2H_{15}(PO_4)_8$. These compounds can be precipitated as crystals from cooling of a hot concentrated phosphoric acid solution containing the proper $Na_2O:Al_2O_3$ ratio. Their crystal structures have not been determined, but both are manufactured industrially on a large scale. The compound $NaAl_3H_{14}(PO_4)_8{\cdot}4H_2O$ was the first and the more important of the two to be commercialized. Its main use is as a leavening acid for baking. Commercially, it is used in combination with the heated treated or "coated" anhydrous monocalcium phosphate. The main advantage of such a combination is that in the dough system it has the proper rate of reaction with $NaHCO_3$ to release CO_2 gas in a proper sequence suitable for the chemical and physical changes during the baking process to yield superior baked goods. Also, its lack of reacticity with $NaHCO_3$ at room temperature when mixed with flour permits its use in self-rising flours and prepared cake mixes in which the baking powder ingredients are premixed with other baking ingredients before packaging. Such packages are quite

[478] I. L. Rashkovan, L. N. Kuzminskaya and V. A. Kopeikin, *Izv. Akad. Nauk SSSR, Neorgan, Materialy*, **2**(3) (1966) 541–9.

[479] V. M. Medvedeva, A. A. Medvedev and I. V. Tananaev, *Izv. Akad. Nauk SSSR, Neorgan, Materialy*, **1**(2) (1965) 211–17.

[480] G. A. MacDonald, U.S. 2,550,490, 24 April 1951, to Victor Chemical Co.

stable even when stored under adverse conditions of high temperature, high humidity and over-long storage.

It may be of interest to note here that the pure compound, $NaAl_3H_{14}(PO_4)_8.4H_2O$, is quite hygroscopic and difficult to handle in commercial equipment. The solution to this problem is by the exchange of some of the H^+ on the surface of the crystals with K^+, thus rendering the product non-hygroscopic without affecting its chemical properties for baking purposes[481–2].

16. PYROPHOSPHATES

16.1. GENERAL METHODS OF PREPARATION

The common method for the preparation of metal pyrophosphates is by thermal condensation of the acid orthophosphates:

$$2MH_2PO_4 \longrightarrow M_2H_2P_2O_7+H_2O$$
$$2M_2HPO_4 \longrightarrow M_4P_2O_7+H_2O$$

For thermally unstable pyrophosphates, e.g., $(NH_4)_4P_2O_7$ [483], the preferred method is the direct neutralization of a solution of pure $H_4P_2O_7$ at low temperature with the hydroxide of the desired cation. For multiple charged metal pyrophosphates that are extremely insoluble in water, many are prepared as precipitates by the reaction of a soluble salt of the respective cation with $Na_4P_2O_7$ in an aqueous solution.

Various inorganic pyrophosphates and their methods of preparation are summarized in a recent review by Shen and Dyroff[484].

16.2. GENERAL STRUCTURE AND PROPERTIES

The structure of the pyrophosphate anion as determined by X-ray studies consists of two PO_4 tetrahedra bridged by a mutual oxygen atom.

```
O         O         O
 \      /   \      /
O—P            P—O
 /               \
O                 O
```

The size of the central P–O–P angle varies with different metal pyrophosphates. Most of the crystalline salts have a bent nonlinear configuration.

Pyrophosphates have a great tendency to occur in polymorphic phases and this polymorphism may be due to three kinds of structure arrangements: (a) linear ions with P–O–P angle $= 180°$, (b) non-linear ions with P–O–P angle $< 180°$, or (c) statistically linear ions derived from suitable arrangement of nonlinear ions. Thus the P–O–P for α-$Mg_2P_2O_7$ = 144°, and for β-$Mg_2P_2O_7$ = 180° [485].

The compound $Zr_2P_2O_7$, which received detailed X-ray studies, has a P–O–P angle of

[481] J. E. Blanch and F. McCullough, Jr., U.S. 3,205,073, 7 Sept. 1965, to Stauffer Chemical Co.

[482] L. B. Post and J. E. Blanch, U.S. 3,411,872, 19 Nov. 1968, to Stauffer Chemical Co.

[483] C. Swanson and F. McCollough, *Inorganic Synthesis*, Vol. VII, pp. 65–67, Ed. J. Kleinberg, McGraw-Hill Book Co., New York (1963).

[484] C. Y. Shen and D. R. Dyroff, *Preparative Inorganic Reactions*, Vol. V, pp. 157–89, Ed. W. L. Jolly, Interscience Publishers (1968).

[485] D. E. C. Corbridge, *Topics in Phosphorus Chemistry*, **3** (1966) 232–47.

180°, and the cations coordinated sixfold by oxygen. The crystals are cubic with four molecules per unit cell. Isomorphous with $Zr_2P_2O_7$ in the cubic series is SiP_2O_7 where the silicon atom is also surrounded by six oxygen atoms. The latter is a rare example of an oxygen-containing compound of silicon having an oxidation number of +4 where the silicon is sixfold coordinated by oxygen.

Neutral anhydrous pyrophosphates are, in general, thermally quite stable. When heated, however, many do undergo polymorphic phase changes until melted. One of the exceptions is $Al_4(P_2O_7)_3$ which is formed by the heating of $Al_2(HPO_4)_3$ at 400°. When heated to 1000°, $Al_4(P_2O_7)_3$ decomposes to the phosphocristobalite $AlPO_4$ and P_2O_5 and at 1300–1800° $AlPO_4$ decomposes into Al_2O_3 and P_2O_5, plus a residue with an Al_2O_3/P_2O_5 ratio of 1/2.3[486].

16.3. AMMONIUM PYROPHOSPHATES

Following are the known ammonium pyrophosphates:[487]

$(NH_4)_4P_2O_7$

$(NH_4)_4P_2O_7.H_2O$

$(NH_4)_3HP_2O_7$

$(NH_4)_3HP_2O_7.H_2O$

$(NH_4)_2H_2P_2O_7$

The morphological and optical properties of these compounds are listed in Table 21.

One general method for the preparation of these compounds is by the ammoniation of an aqueous solution of pure $H_4P_2O_7$. Calculated quantity of NH_3 is added at below 10° followed by evaporation under vacuum or precipitation by addition of ethanol to obtain crystalline products.

Tetraammonium pyrophosphate monohydrate, $(NH_4)_4P_2O_7.H_2O$, is obtained as a precipitate by the addition of ethanol to a concentrated aqueous solution, pH 6.5 at 0°. Anhydrous $(NH_4)_4P_2O_7$ is obtained by precipitation with ethanol at above 25°. The monohydrate dehydrates readily upon exposure to the atmosphere. The anhydrous form loses NH_3 slowly in the open, even at room temperature, and this loss is accelerated at above 50% relative humidity.

The compound $(NH_4)_3HP_2O_7.H_2O$ is precipitated from its aqueous solution, pH 6, with alcohol at below 55°. It is non-hygroscopic and stable at room temperature at relative humidity above 45%. The anhydrous form may be obtained by dehydration of $(NH_4)_3HP_2O_7.H_2O$ at above 55° or by crystallization from an aqueous solution at above 55°.

Diammonium pyrophosphate $(NH_4)_2H_2P_2O_7$ forms dimorphs. The monoclinic form is stable at room temperature or below and may be crystallized from an aqueous solution, pH 3.4, at below room temperature by the addition of ethanol. The orthorhombic form is obtained by the same method except at a higher temperature. At room temperature, the orthorhombic form changes rapidly to the monoclinic form which is stable and non-hygroscopic.

[486] V. M. Medvedeva, A. A. Medvedev and I. V. Tananaev, *Izv. Akad. Nauk SSSR, Neorgan, Materialy*, **1**(2) (1965) 211–17.

[487] A. W. Frazier, J. P. Smith and J. R. Lehr, *J. Agr. Food Chem.* **13**(4) (1965) 316–22.

TABLE 21. MORPHOLOGICAL AND OPTICAL PROPERTIES OF SOME PYROPHOSPHATES

Compound	Space group	Crystal system	Lattice constants				Density g/cm³	Refractive indices	Ref.
			a (Å)	b (Å)	c (Å)	β			
$Na_4P_2O_7 \cdot 10H_2O$	$C2/c$	Monoclinic	16.95	6.94	14.81	112°	1.83		b
$Na_2H_2P_2O_7 \cdot 6H_2O$	$C2/c$	Monoclinic	14.11	7.03	13.50	117.6°	1.85		b
$Na_3HP_2O_7 \cdot 9H_2O$	$P2_1/a$	Monoclinic	8.59	31.65	6.13	113.7°	1.78		b
$K_2H_2P_2O_7 \cdot 0.5H_2O$	$C2/c$	Monoclinic	17.92	7.01	14.27	120.7°	2.27		b
β-$Ca_2P_2O_7$	$P4_1$	Tetragonal	6.66		23.86		3.1		b
α-$Ca_2P_2O_7$	$P2_1/n$	Monoclinic	12.66	8.543	5.315	90.3°	2.947 (at 22.5°)		c
$(NH_4)_4P_2O_7$	$C2/c$	Monoclinic	11.77	6.51	13.63	105°	1.61	$\alpha = 1.514$ $\beta = 1.520$ $\gamma = 1.521$	a
$(NH_4)_4P_2O_7 \cdot H_2O$		Monoclinic				104°	1.54	$\alpha = 1.496$ $\beta = 1.498$ $\gamma = 1.503$	a
$(NH_4)_3HP_2O_7$	$P2_1/c$	Monoclinic	6.63	19.74	7.09	110.8°	1.73	$\alpha = 1.518$ $\beta = 1.521$ $\gamma = 1.531$	a
$(NH_4)_3HP_2O_7 \cdot H_2O$	$P\bar{1}$	Triclinic	9.13	6.33	9.96	77.2°	1.66	$\alpha = 1.494$ $\beta = 1.509$ $\gamma = 1.517$	a
$(NH_4)_2H_2P_2O_7$ γ-form, stable at room temp.	$P2_1/c$	Monoclinic	7.36	11.26	9.06	108.5°	1.89	$\alpha = 1.505$ $\beta = 1.521$ $\gamma = 1.556$	a
$(NH_4)_2H_2P_2O_7$ α-form, stable above room temp.		Orthorhombic					1.81	$\alpha = 1.495$ $\beta = 1.505$ $\gamma = 1.521$	a

[a] A. W. Frazier, J. P. Smith and J. R. Lehr, *J. Agr. Food Chem.* **13**(4) (1965) 316–22.
[b] D. E. C. Corbridge, *Acta Cryst.* **10** (1957) 85.
[c] C. Calvo, *Inorg. Chem.* **7**(7) (1968) 1345–51.

16.4. SODIUM PYROPHOSPHATES

The following sodium pyrophosphates are known[488]:

$Na_4P_2O_7.10H_2O$	$Na_3HP_2O_7$
$Na_4P_2O_7$	$Na_2H_2P_2O_7.6H_2O$
$Na_3HP_2O_7.9H_2O$	$Na_2H_2P_2O_7$
$Na_3HP_2O_7.H_2O$	$NaH_3P_2O_7$

Preparation and Properties. Anhydrous tetrasodium pyrophosphate is prepared by dehydration of Na_2HPO_4 at 500° for 5 hr[489]. Thermal-gravimetric studies showed that actual condensation occurred at 330° to 340° [488]:

$$2Na_2HPO_4 \longrightarrow Na_4P_2O_7 + H_2O$$

It can be easily recrystallized from water. The hydrate, $Na_4P_2O_7.10H_2O$, is obtained at between −0.4° and 79° while the anhydrous form $Na_4P_2O_7$ is obtained at above 79°. Anhydrous $Na_4P_2O_7$ is reported as having five crystalline modifications with one form only stable at room temperature[490]:

$$Na_4P_2O_7\ \text{V} \xrightleftharpoons{400°} \text{IV} \xrightleftharpoons{510°} \text{III} \xrightleftharpoons{520°} \text{II} \xrightleftharpoons{545°} \text{I} \xrightleftharpoons{985°} \text{melt.}$$

Crystallographic data are summarized in Table 21.

The solubility of $Na_4P_2O_7.10H_2O$ in water is equivalent to approximately 9 g of the anhydrous salt per 100 g of solution at 25°. A 1% solution has a pH of 10.2 and the compound is stable to hydrolysis with no noticeable hydrolysis after 60 hr at 70°. The hydrate has a m.p. 79.5° and the crystals are biaxial positive with $\alpha = 1.450$, $\beta = 1.453$ and $\gamma = 1.460$ (±0.002).

The anhydrous crystal is also biaxial positive with $\alpha = 1.475$, $\beta = 1.477$ and $\gamma = 1.496$ (±0.002)[489] and a heat capacity $C_{p298.15°K} = 47.36$ cal/deg and entropy $S_{298.15K} = 52.63$ cal/deg[491].

Trisodium pyrophosphate is prepared by acidification of an aqueous solution of $Na_4P_2O_7$ with HCl at ≦ 35° followed by crystallization of the resulting $Na_3HP_2O_7.9H_2O$ and its separation from NaCl[492]. The monohydrate is formed in solution at about 30°. Heating of $Na_3HP_2O_7.H_2O$ for several days at 150° results in the formation of the anhydrous form.

Disodium dihydrogen pyrophosphate is formed when NaH_2PO_4 is heated to 210° for 12 hr[489]:

$$2NaH_2PO_4 \longrightarrow Na_2H_2P_2O_7 + H_2O$$

In the commercial process, the conversion is carried out in a rotary-type continuous converter at 225–250° under a controlled atmosphere of water vapor to prevent metaphosphate formation. The hexahydrate, $Na_2H_2P_2O_7.6H_2O$, exists in the solution up to about 27° where it is converted to the anhydrous form.

488 L. Steinbrecher and J. F. Hazel, *Inorg. Nucl. Chem. Letters*, **4** (1968) 559–62.

489 R. N. Bell, *Synthesis*, Vol. III, pp. 98–101, Ed. L. F. Audrieth, McGraw-Hill Book Co., New York (1950).

490 T. D. Farr, *Phosphorus, Properties of the Element and Some of Its Compounds*, Chemical Engineering Report No. 8, pp. 39–40, 58–59, T.V.A., Wilson Dam, Alabama (1950).

491 R. J. L. Andon, J. F. Counsell, J. F. Martin and C. J. Mash, *J. Appl. Chem.* **17**(3) (1967) 65–70.

492 I. L. Gofman, K. S. Zotova, I. P. Khudolei and D. N. Shevchenko, U.S.S.R. 175,493, 9 Oct. 1965, *Chem. Abstr.* **64** (1966) 6156.

Monosodium pyrophosphate is also known. It is prepared by mixing equimolar proportions of $Na_2H_2P_2O_7$ and $H_4P_2O_7$ in an aqueous solution at 0°. Upon evaporation of the solution under vacuum, the crystalline $NaH_3P_2O_7$ separated has a m.p. of 185° [493]. It is very soluble in water. Addition of alcohol or acetone to a concentrated aqueous solution causes decomposition which results in the precipitation of $Na_2H_2P_2O_7$.

Applications. Tetrasodium pyrophosphate was introduced in the 1930's as a builder for soap. It is a good disperser for lime soap which formed in hard water. Since the advent of sodium tripolyphosphate and synthetic detergents, the use of $Na_4P_2O_7$ as a soap builder has greatly decreased. However, it is still used as a builder in some cleaning formulations[494].

Pyrophosphate ions form a gel when reacted with a soluble calcium salt in water solution. This property makes $Na_4P_2O_7$ useful as an ingredient for starch type instant pudding which requires no cooking[495]. Tetrasodium pyrophosphate is also used as a stabilizer for hydrogen peroxide employed in textile bleaching. This application depends on the ability of $Na_4P_2O_7$ to sequester heavy metal cations which normally catalyze the decomposition of peroxides. Other applications for $Na_4P_2O_7$ are for water treatment and as disperser for clays.

The main application for $Na_2H_2P_2O_7$ is as a leavening acid in baking. Its major advantage is that in a dough or batter system it has relatively little action with $NaHCO_3$ until heat is applied. This property permits commercial bakeries to prepare a large quantity of dough or batter at one time. Such mixtures are stable until heated in the oven during the baking to the finished products.

Most of the commercially produced $Na_2H_2P_2O_7$ contains added minor amounts of K^+, Ca^{++} and Al^{+++}. These additives permit the controlled retardation of the rate of reaction of $Na_2H_2P_2O_7$ with $NaHCO_3$ in the baking system. The reason for the slow rate of reaction of even the pure $Na_2H_2P_2O_7$ with $NaHCO_3$ at room temperature in the bake-mix is not well understood since their reaction in an aqueous solution is quite rapid. One theory proposed is that the retardation is due to the formation of a thin coating of $Ca_2P_2O_7$ on the surface of the $Na_2H_2P_2O_7$ particles from the reaction of the latter with Ca^{++} in the bake-mix. To support this theory, it has been found that the slow reaction of $Na_2H_2P_2O_7$ with $NaHCO_2$ in a bake-mix is even slower (1) in the presence of added soluble calcium salt and (2) when $Na_2H_2P_2O_7$ used is milled finer. In actual baking applications, $Na_2H_2P_2O_7$ is used largely as the leavening acid for cake doughnuts, cakes and refrigerated biscuits. One major disadvantage of $Na_2H_2P_2O_7$ as a baking acid is that it imparts an astringent aftertaste to the baked goods.

16.5. POTASSIUM PYROPHOSPHATES

The best known of potassium pyrophosphates is $K_4P_2O_7$ which is formed by the dehydration of K_2HPO_4 at 350–400°. It is very soluble in water (187 g per 100 g H_2O at 25°). In a saturated solution, the solid in equilibrium with the solution is $K_4P_2O_7.3H_2O$. The monohydrate $K_4P_2O_7.H_2O$ is also known. Both $K_4P_2O_7$ and $K_4P_2O_7.H_2O$ are hygroscopic while $K_4P_2O_7.3H_2O$ is hygroscopic at above a relative humidity of 45%[496].

$K_2H_2P_2O_7$ is difficult to prepare by thermal dehydration of KH_2PO_4 since the rate

[493] A. Norbert and C. Dautel, *Compt. Rend.*, Ser. C, **262**(21) (1966) 1534–6.

[494] *Oil, Paint and Drug Reporter*, 28 Aug. 1967, p. 7.

[495] M. H. Kennedy and M. P. Castagna, U.S. 2,607,692, 19 Aug. 1952, to Standard Brands, Inc.

for dehydration to form $(KPO_3)_n$ is about as rapid unless the reaction is carried out under a humid atmosphere. The preferred preparative method for $K_2H_2P_2O_7$ is by the acidification of a $K_4P_2O_7$ solution with HCl to a pH of 4.5 and then alcohol added to precipitate the product.

Only $K_4P_2O_7$ is of commercial importance. It was introduced in the middle 1950's as a builder for light duty liquid detergents. It is used at about 25% by weight of the total formulation which contains approximately 50% water. The effectiveness of $K_4P_2O_7$ as a builder depends on the pyrophosphate ion. Even though it is not as effective a builder as tripolyphosphate or long chain polyphosphate ions, it has the advantage of having a high degree of hydrolytic stability (half-life of several years at 50°). Tetrapotassium pyrophosphate is more expensive than $Na_4P_2O_7$ but it is still widely used since its high water solubility permits easy formulation in liquid detergents[494].

16.6. CALCIUM PYROPHOSPHATE

When $CaHPO_4.2H_2O$ is dehydrated thermally, DTA and TGA studies show the formation of amorphous $Ca_2P_2O_7$ at 360–450°. Continuation of heating results in an exotherm at 530° due to the conversion of the amorphous $Ca_2P_2O_7$ to the crystalline γ-$Ca_2P_2O_7$ [497]. Further heating converts the γ-form to the β-form at 750° [490] and to the α-form at 1171–1179? [498]. The α-$Ca_2P_2O_7$ melts at 1352°.

Steger and Kassner[499] have recently measured the IR spectra of α-$Ca_2P_2O_7$, β-$Ca_2P_2O_7$ and γ-$Ca_2P_2O_7$ along with those of other pyrophosphates The indices of refraction of α-calcium pyrophosphate are n_α = 1 584, n_β = 1 599, and n_γ = 1.605[490]. The crystals of β-$Ca_2P_2O_7$ are uniaxial and positive, with n_ω = 1 630 and n_ε = 1 639.

Calcium pyrophosphate, because of its low water solubility and chemical inertness, is used as an abrasive for a Sn^{++} and F^- containing toothpaste For this purpose a mixture of γ-$Ca_2P_2O_7$ and β-$Ca_2P_2O_7$ forms has been found to have the desired abrasiveness and yet is still sufficiently compatible with the added fluoride and stannous ions in the formulation. To improve the compatibility of $Ca_2P_2O_7$ to Sn^{++} and F^- in toothpaste formulations even upon prolonged storage, various modifications have been introduced during the conversion of $CaHPO_4.2H_2O$. Examples of such modifications are by the addition of 750–4000 ppm of soluble alkali metal ions to the $CaHPO_4.2H_2O$ before dehydration[500] or by the dehydration in presence of added steam[501].

17. TRIPOLYPHOSPHATES

17.1. SODIUM TRIPOLYPHOSPHATE

Preparation The common commercial method for the preparation of sodium tripolyphosphate (pentasodium tripolyphosphate, $Na_5P_3O_{10}$) is by heating at a controlled

[496] J. R. van Wazer, *Phosphorus and Its Compounds*, Vol. I, pp. 617–33, Interscience Publishers Inc., New York (1958).

[497] J. G. Rabatin, R. H. Gale and A. E. Newkirk, *J. Phys. Chem.* **64** (1960) 491–3.

[498] J. A. Parodi, R. L. Hickok, W. G. Segelken and J. R. Cooper, *J. Electrochem. Soc.* **112**(7) 688–92.

[499] E. Steger and B. Kassner, *Z. anorg. u. allgem. Chem.* **355** (1967) 131–44.

[500] E. Saunders and F. H. Wright, Can. 758,508, 9 May 1967, to Monsanto Company.

[501] T. J. Dolan and F. H. Wright, Can. 758,507, 9 May 1967, to Monsanto Company.

temperature range an intimate mixture of powdered $2Na_2HPO_4$ and NaH_2PO_4:

$$2Na_2HPO_4+NaH_2PO_4 \longrightarrow Na_5P_3O_{10}+2H_2O$$

Anhydrous sodium tripolyphosphate occurs in two crystalline forms. Form II, the low temperature variety, is obtained at about 400°. It is usually converted to Form I, the thermodynamically stable phase and the high temperature variety, by heating at 500–550°. The phase transition temperature of Form II to Form I is 417°.

Reverse transition of Form I to Form II is difficult. It can be done by heating a sample for about a month in a sealed tube in the presence of 0.1% $Na_5P_3O_{10}.6H_2O$ at 409° to 320° [502].

Laboratory preparation of $Na_5P_3O_{10}$ can be carried out by heating any mixture of sodium phosphate salts with a Na_2O:P_2O_5 ratio of 5 : 3 to a clear melt (850–900°) and the melt cooled quickly to a glass. Subsequently, tempering of this glass at above 470° results in the formation of crystalline Form I[503]. Form II is formed from annealing the glass for at least 2 hr at 550°, and then cooling rapidly in air. The solid undergoes a crystalline phase transition at 150–100° and disintegrates into a powder of Form II[504].

The hydrate of sodium tripolyphosphate, $Na_5P_3O_{10}.6H_2O$, is formed by the addition of either Form I or II to water. Both anhydrous forms are metastable in water and cannot be recrystallized from water. The hexahydrate can also be obtained by the hydrolysis of $(NaPO_3)_3$ in a strong alkaline solution[505]:

$$(NaPO_3)_3+2NaOH+5H_2O \longrightarrow Na_5P_3O_{10}.6H_2O$$

Structure. Sodium tripolyphosphate anion consists of triply condensed PO_4 tetrahedra (Fig. 14)[506]. Titration studies show that the acid has one weak hydrogen ion in each

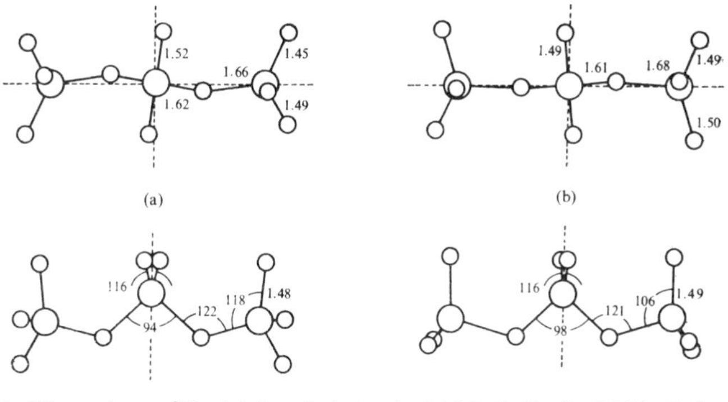

(a) (b)

FIG. 14. Dimensions of the triphosphate ion in (a) $Na_5P_3O_{10}$–I; (b) $Na_5P_3O_{10}$–II[506].

end group and one strong hydrogen ion in each of all three phosphorus atoms. The dimensions of the tripolyphosphate ion are shown in Fig. 14[506].

The major difference between the structures of Form I and Form II is in the ionic

[502] G. W. Morey, *J. Am. Chem. Soc.* **80** (1958) 775.

[503] E. Thilo, *Advances in Inorganic Chemistry and Radiochemistry*, Vol. IV, pp. 1–66, Eds. H. J. Emeleus and A. G. Sharpe, Academic Press Inc., New York (1962).

[504] R. N. Bell, *Inorganic Synthesis*, Vol. III, pp. 101–6, Ed. L. F. Audrieth, McGraw-Hill Book Co. Inc., New York (1950).

[505] C. Y. Shen, *Ind. Eng. Chem., Prod. Res. Develop.* **5**(3) (1966) 272–6.

[506] D. E. C. Corbridge, *Topics in Phosphorus Chemistry*, **3** (1966) 232–66.

coordination of the sodium ions. In Form II, all sodium ions are coordinated octahedrally by oxygen while in Form I some sodium ions are surrounded by only four oxygen atoms[506].

Properties. Anhydrous $Na_5P_3O_{10}$ melts incongruently at 662° into a liquid plus crystalline $Na_4P_2O_7$. Further heating to 865° results in a complete melt which when chilled quickly forms a glass. As mentioned earlier, tempering of this glass, depending on the temperature conditions, converts it back to either Form I or Form II.

The heat capacity and entropy of sodium tripolyphosphates at 298.15°K are for Form I: C_p = 78.16 cal/deg, S = 91.25 cal/deg; Form II: C_p = 77.2 cal/deg, S = 87.37 cal/deg;

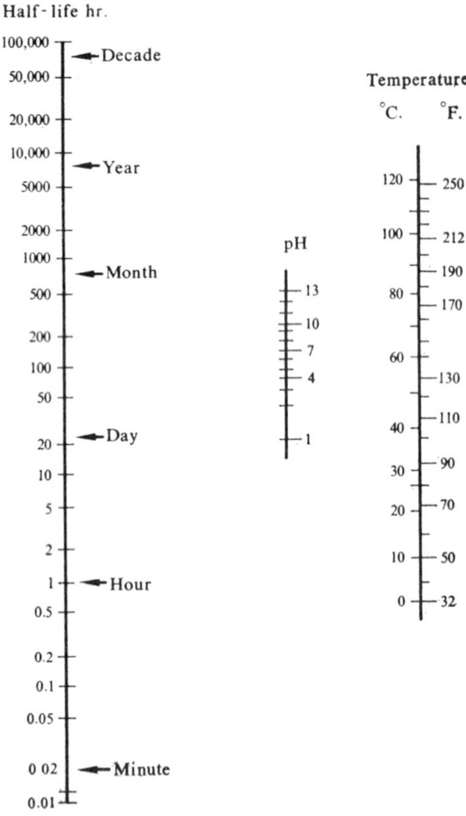

FIG. 15. Half-life of the tripolyphosphate ion as a function of pH and temperature. (Data refer to tetramethylammonium tripolyphosphate dissolved in 0.65 N tetramethylammonium bromide.)[508] (Reprinted by permission of John Wiley & Sons Inc.)

and $Na_5P_3O_{10}.6H_2O$: C_p = 137.1 cal/deg, S = 146.1 cal/deg[507]. The density of sodium tripolyphosphate in Form I = 2.62 g/cm³; Form II = 2.57 g/cm³.

The solubility of $Na_5P_3O_{10}$ in water is a function of temperature. At room temperature, $Na_5P_3O_{10}.6H_2O$ dissolves to an equivalent of about 13 g $Na_5P_3O_{10}$/100 g of solution. Both anhydrous Forms I and II are more soluble than $Na_5P_3O_{10}.6H_2O$. However, in the temperature range of 0 to 100°, the saturating crystal phase is the hexahydrate. Because Form I forms the hexahydrate crystals very rapidly, its high initial solubility in water

[507] R. J. L. Andon, J. F. Counsell, J. F. Martin and C. J. Mash, *J. Appl. Chem.* (1967) (17) 65–70.

drops almost immediately to that of the hexahydrate. Form II, on the other hand, forms the hexahydrate at a much slower rate, so it has an immediate water solubility of about 32 g per 100 g of solution. This solubility decreases slowly with the formation of $Na_5P_3O_{10}.6H_2O$, and after 15–20 min it drops to that of the equilibrium solubility of hexahydrate. The greater ease of hydration of Form I is attributed to the presence of the four-coordinated sodium ions in the structure. Such a feature in other related compounds is known to be associated with a strong affinity for water.

The solubility characteristics of $Na_5P_3O_{10}$ described above applies only to fairly pure $Na_3P_5O_{10}$. The rate of $Na_5P_3O_{10}.6H_2O$ formation is greatly retarded by impurities such as other phosphates normally present in commercial $Na_5P_3O_{10}$. For example, in the presence of small amount of glassy metaphosphates, the amount of dissolved $Na_5P_3O_{10}$ does not decrease to that of the equilibrium solubility of $Na_5P_3O_{10}.6H_2O$ even after many hours.

The rapid conversion of Form I into $Na_5P_3O_{10}.6H_2O$ crystals is a serious problem in detergent manufacturing where sodium tripolyphosphate is used as a builder. Under certain conditions these hexahydrates could cement together and plug the detergent manufacturing equipment. As a consequence, additives and/or the correct ratio of Form I and Form II in the anhydrous material must be carefully controlled.

A 1% solution of $Na_5P_3O_{10}$ has a pH of 9.7. The rate of hydrolysis of the tripolyphosphate depends on the temperature and the pH of the solution. The products of hydrolysis are a mole each of pyrophosphate and orthophosphate. At its natural pH of 9.7, approximately 50% is hydrolyzed in 6 hr at 100° and 15% in 60 hr at 70° [504]. The nomograph showing the half-life of the tripolyphosphate anion as a function of temperature and pH is shown in Fig. 15[508]. The rate of hydrolysis in alkaline solution is accelerated by the presence of sodium ions or other cations which form stronger complexes than do sodium ions with the tripolyphosphate anion. However, these accelerating effects are minor compared to that of the effect of temperature and pH.

Crystalline $Na_5P_3O_{10}.6H_2O$, upon storage in a sealed system at room temperature, undergoes self-hydrolysis mainly as follows:

$$2Na_5P_3O_{10}.6H_2O \longrightarrow Na_4P_2O_7 + 2Na_3HP_2O_7 + 11H_2O$$

At 100°, hydrolysis is principally:

$$Na_5P_3O_{10}.6H_2O \longrightarrow Na_3HP_2O_7 + Na_2HPO_4 + 5H_2O$$

At above 140°, the products of hydrolysis recondense to $Na_5P_3O_{10}$ Form II and water[506].

Applications. Crystalline sodium tripolyphosphate was first described in 1895 by Schwartz[508]. It was not until the 1940s that it was introduced commercially as a builder for sodium alkylbenzenesulfonate containing synthetic detergents. The total production in 1967 in the United States alone exceeds 2.2 billion pounds.

For detergent application, the commercial product is usually a mixture of Form I and Form II in a ratio determined by the needs of the specific detergent manufacturer. In addition, commercial products also contain a few per cent of $Na_4P_2O_7$. The latter is the result of using a slight excess of Na_2O over the theoretical Na_2O/P_2O_5 ratio of 5 : 3 during the manufacturing process for the purpose of avoiding the formation of any insoluble sodium metaphosphate. Presence of even a minor amount of insoluble metaphosphate

[508] J. R. van Wazer, *Phosphorus and Its Compounds*, Vol. I, pp. 638–659, Interscience Publishers Inc., New York (1958).

in the product would cause turbidity in a water solution—highly undesirable for detergent applications.

The two anhydrous forms of $Na_5P_3O_{10}$ can be differentiated by their X-ray diffraction pattern or IR spectra so that their ratio in a commercial mixture can thus be determined. For industrial purposes, however, a simpler and more rapid, though not quite as accurate, method of determination is desired. The accepted procedure is a thermal test developed at Proctor and Gamble and is known as Temperature Rise Test. This procedure is based on the fact that in a water–glycerine mixture, Form I hydrates more rapidly than Form II. The test, when carried out under the directions given, percentage of Form I can be calculated by the following equation[509]:

$$\% \text{ Form I} = 4[(\text{temp. rise in °C}) - 6]$$

Data from this test have been calibrated against X-ray diffraction data. The quantity of $Na_4P_2O_7$ normally present in the commercial mixture does not interfere with this test. Heats of solution for Form I and Form II are −16.1 and −14.0 kcal/mole respectively.

Sodium tripolyphosphate, when used as a builder for synthetic detergents, constitutes approximately 45% by weight of each formulation. Through hydration in the synthetic detergent manufacturing process, except for a few per cent of degradation products, all of the original $Na_5P_3O_{10}$ is converted to $Na_5P_3O_{10}.6H_2O$.

The major function of $Na_5P_3O_{10}$ as a builder depends on its water-softening action through complexing or sequestering of Ca^{++} and Mg^{++} in hard water. At 0 ionic strength, the formation constants for some tripolyphosphate complexes are as follows[503]:

$NaP_3O_{10}^{4-}$	pK = 2.8
$CaP_3O_{10}^{3-}$	pK = 8.1
$MgP_3O_{10}^{3-}$	pK = 8.6

The stability of these complexes is greater than the corresponding complexes of the pyrophosphates:

$NaP_2O_7^{3-}$	pK = 2.3
$CaP_2O_7^{2-}$	pK = 6.8
$MgP_2O_7^{2-}$	pK = 7.2

Another important function of sodium tripolyphosphate in detergent is the property to suspend and peptize dirt particles. The tripolyphosphate anion is also important for its ability to lower the critical micelle concentration of the detergent.

Compared to its use as builder for detergents, other uses of $Na_5P_3O_{10}$ become minor. However, large tonnages are consumed annually as deflocculent for solid slurries to reduce the amount of water needed as in the cases of cement manufacturing[510], oil well drilling mud formulations, and kaolin clay mining.

17.2. OTHER TRIPOLYPHOSPHATES

Potassium tripolyphosphate, $K_5P_3O_{10}$, may be prepared by a method analogous to that used for the sodium compound:

$$2K_2HPO_4 + KH_2PO_4 \longrightarrow K_5P_3O_{10} + 2H_2O$$

---

[509] J. D. McGilvery, *ASTM Bull.* No. 191 (1933) 45–48.

[510] W. B. Chess, D. N. Bernhart and R. Gates, *Rock Products*, **10** (1961) 121–2.

Quantitative differential thermal analysis shows that this condensation reaction occurs at 200–300° with $H = 32.4$ kcal/mole[511]. However, in the actual manufacturing process, the temperature range employed is 325–400°.

Potassium tripolyphosphate melts incongruently at 614.5° with the formation of crystalline $K_4P_2O_7$ and a melt. Total liquefication occurs at about 940° [503].

Crystalline $K_5P_3O_{10}$ is hygroscopic and extremely soluble in water. When exposed to atmospheric moisture, it absorbs 2 moles of water to form the α-form of the dihydrate. Another dihydrate, the β-form, is obtained by precipitation from an aqueous solution with methanol[503].

Tripolyphosphate salts of multiple charged cations are also known. These are made by precipitation from a solution of sodium tripolyphosphate with the respective polyvalent cations. Many crystalline salts thus obtained contain also the sodium ions. They are of the types $Na_3M^{++}P_3O_{10}$, $NaM_2^{++}P_3O_{10}.aq$. These salts are slightly soluble in water but in most cases dissolve in excess of tripolyphosphate. The low solubility of $NaZn_2P_3O_{10}.9H_2O$ is used for the rapid identification of tripolyphosphate[503].

18. POLYPHOSPHATES, CYCLIC METAPHOSPHATES AND ULTRAPHOSPHATES

The polyphosphate anion may be represented by the general formula:

$$({}^{-}O)_2P(=O)-\left[O-P(=O)(O^{-})\right]_n-O-P(=O)(O^{-})_2$$

When $n = 0$, the compound is pyrophosphate; $n = 1$, tripolyphosphate. When n is very large, the sodium salts are represented by three high molecular weight crystalline compounds known as $NaPO_3$–II (Maddrell's salt), $NaPO_3$–III (low temperature variety of Maddrell's salt), and $NaPO_3$–IV (Kurrol's salt). The potassium salt is represented by the high molecular weight crystalline potassium polyphosphate known as potassium-Kurrol's salt, or potassium metaphosphate. In addition, there are also the sodium polyphosphate glasses, or Graham's salts, which are not pure compounds but mixtures of polyphosphates having a wide distribution of molecular weights.

The high molecular weight compounds described above, though commonly called metaphosphates, belong actually to the polyphosphate system since each molecule contains the two end groups. However, as n approaches infinity, the total composition of the compound is close to that of $(NaPO_3)_n$ or the metaphosphate.

True metaphosphates have the exact formula of $(NaPO_3)_n$ or the middle groups only in the polyphosphate anion formula. These are cyclic compounds and the first well-characterized member in the series is the sodium trimetaphosphate, $(NaPO_3)_3$, known as $NaPO_3$–I, or Knorre's salt.

Another class of condensed phosphate is the ultraphosphate. These are polymeric phosphates containing branched chains, e.g.:

[511] E. Calvert, M. Gambino and M. L. Michel, *Bull. Soc. Chim. France* (1965)(6) 1719–21.

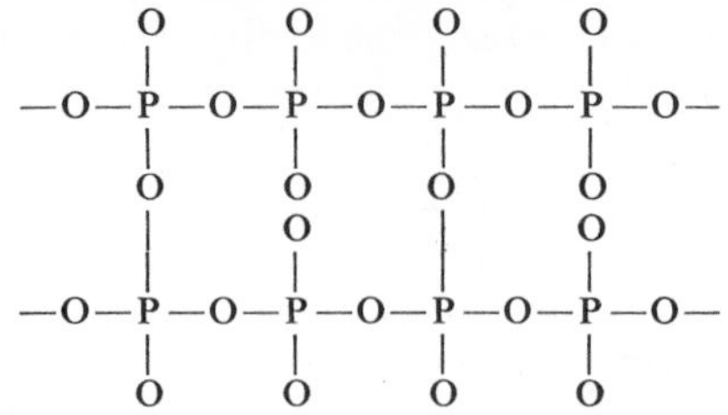

```
     O      O      O      O
     |      |      |      |
—O—P—O—P—O—P—O—P—O—
     |      |      |      |
     O      O      O      O
     |      O      |      O
     |      |      |      |
—O—P—O—P—O—P—O—P—O—
     |      |      |      |
     O      O      O      O
```

Linear sodium polyphosphates have a ratio $Na_2O/P_2O_5 = >1$ and $\leqq 2$; cyclic sodium metaphosphates $Na_2O/P_2O_5 = 1$, and the ultraphosphates $Na_2O/P_2O_5 = >1$.

18.1. TETRAPOLYPHOSPHATES

The next member in the linear chain polyphosphate series from tripolyphosphate is tetrapolyphosphate. It is much more difficult to obtain by thermal condensation method than tripolyphosphate. Heating to over 550° the Pb or Ba monophosphates with metal-to-phosphorus ratio calculated for tetrapolyphosphate has been reported to yield the pure tetrapolyphosphates.

Sodium tetrapolyphosphate cannot be prepared in a pure form by thermal methods. It is prepared by opening the ring of sodium tetrametaphosphate with aqueous NaOH at temperatures up to 40°:

$$(NaPO_3)_4 + 2NaOH \longrightarrow Na_6(P_4O_{13}) + H_2O$$

This ring cleavage reaction takes advantage of the fact that hydrolysis of cyclic metaphosphate is strongly catalyzed by OH^- (and H^+) and that the chain phosphates are sufficiently stable in an alkali solution to permit isolation. The product obtained is amorphous but can be converted to the crystalline tetrapolyphosphate salt of other cations, e.g., hexaguanidinium[512] tetrapolyphosphate and ammonium tetrapolyphosphate, $(NH_4)_6P_4O_{13}.6H_2O$.

The linear nature of the tetrapolyphosphate anion has been demonstrated by NMR studies on a solution of crystalline ammonium tetrapolyphosphate. The acid of tetrapolyphosphate has four strongly and two weakly dissociated hydrogen ions.

Tetrapolyphosphate anion is quite stable in neutral or alkaline solution but not as stable as tripolyphosphate anion over the whole pH range. Its maximum stability at 65.5° is at pH 10, and this stability decreases with decreasing pH. The hydrolytic cleavage starts at the end of the chain forming first the mono- and tripolyphosphate[513]. At pH $\leqq 3$, cyclic trimetaphosphate is additionally formed. There are no evidences for symmetric splitting into two moles of pyrophosphates[514].

The alkaline earth tetrapolyphosphates are considerably more soluble than the corresponding pyro- and tripolyphosphates. The stability of the tetrapolyphosphate complexes of these metal ions is also greater than those of pyro- and tripolyphosphate. This is as one might expect from the increasing number of tetrahedra available in the tetrapolyphosphate for complex bonding[515].

[512] O. T. Quimby and F. P. Krause, *Inorganic Synthesis*, Vol. V, pp. 97–102, Ed. T. Moeller, McGraw-Hill Book Co. (1957).

[513] E. Thilo, *Advances in Inorganic Chemistry and Radiochemistry*, Vol. IV, pp. 1–66, Eds. H. J. Emeleus and A. G. Sharpe, Academic Press Inc., New York (1962).

[514] W. Wieker, *Z. anorg. u. allgem. Chem.* **355** (1967) 20–29.

[515] J. I. Watters and R. Machen, *J. Inorg. and Nucl. Chem.* **30** (1968) 2163–6.

18.2. HIGHER POLYPHOSPHATES

Longer linear polyphosphates above tetrapolyphosphates are more difficult to obtain in a pure form. The mineral Troemelite which occurs as crystalline plates or fibers formed by tempering a melt of the correct $CaO : P_2O_5$ ratio at between 920–990° has been shown to be calcium hexapolyphosphate[513, 516].

Long chain polyphosphates normally occur as mixtures of molecules of various chain length. As the linear chains become very long, the composition of each chain approaches that of $(NaPO_3)_n$ and their properties become so similar that for practical purposes the mixture behaves as a pure compound. Also in these long chains, the substitution of one for another during crystal growth is not difficult since one chain goes through several

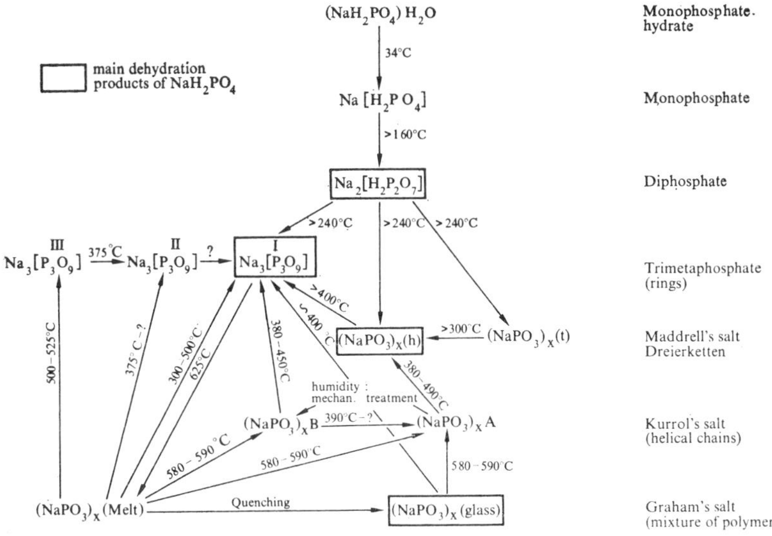

FIG. 16. Interrelationship of metaphosphates[518]. (Reprinted by permission of Verlag Chemie, GmbH.)

unit cells of the crystal and the exact chain length has little effect on the lattice parameter of the crystals. For these reasons several high molecular weight polyphosphates are obtained as crystalline substances.

Since each long chain molecule contains a large number of metaphosphate $(NaPO_3)$ middle groups, with only two end groups, the overall composition is very near that of the metaphosphate. They are, therefore, usually referred to as metaphosphates.

Both cyclic and long chain sodium metaphosphates can be made by thermal dehydration of monosodium phosphate. They can also be made by tempering of the polyphosphate glass of $Na_2O : P_2O_5$ ratio of 1. The specific compound to be formed depends on the temperature of heating or tempering. The interrelationship of these compounds and conditions for their preparation have been summarized by Thilo[517, 518] as shown in Fig. 16.

516 C. Y. Shen and D. R. Dyroff, *Preparative Inorganic Reactions*, pp. 157–222, Ed. W. L. Jolly, Interscience Publishers, New York (1968).

517 D. E. C. Corbridge, *Topics in Phosphorus Chemistry*, **3** (1966) 232–66.

518 E. Thilo, *Angew. Chem. Intern. Ed.* **4**(12) (1965) 1061–71.

18.3. GLASSY SODIUM METAPHOSPHATE (GRAHAM'S SALT)

Preparation. This glassy polyphosphate was first described by Thomas Graham in 1833. It is now manufactured industrially by heating NaH_2PO_4 at above 620° to a melt and then cooling rapidly[519]. Another industrial process involves the direct reaction of a spray of aqueous sodium hydroxide with the vapor from the burning of elemental phosphorus, the resultant sodium polyphosphate melt is withdrawn from the reaction zone and cooled rapidly over a cooled metal surface[520]. The product solidifies into a glass with about 90% of its content in the form of a mixture of high molecular weight polyphosphates, the remaining 10% being various cyclic metaphosphates[518]. Other glassy metaphosphates with a somewhat higher $Na_2O : P_2O_5$ ratio than Graham's salt are also known as "hexametaphosphate." This term (which is now proven incorrect) was given by Fleitman[521], who in 1849 assigned to the glassy sodium metaphosphate $(NaPO_3)_x$ which he studied a degree of polymerization of six.

The best known of commercial glassy metaphosphates goes by the registered trademark of "Calgon" (from calcium gone). It contains approximately 67.8% P_2O_5 or a $Na_2O : P_2O_5$ ratio of slightly over 1.1 : 1. Its molecular weight is in the range of 1500 to 2000 or a degree of polymerization of 15 to 20[522].

Properties. Graham's salt, as shown in Fig. 16, when annealed, depending on conditions, converts to sodium Kurrol's salt, Maddrell's salt, or cyclic trimetaphosphate. They are all crystalline compounds. In this transition, the PO_3^- units are cleaved and rearranged into new chains or cyclic rings. The cations have some influence on the crystalline arrangement of the final product. For example, annealing of metaphosphate glasses of bivalent cations such as Cu, Mg, Ni, Co, Fe and Mn convert them ultimately to the cyclic tetrametaphosphates[518].

In an aqueous solution, the maximum degree of dissociation of the polyphosphate, like that of polyphosphoric acid, is in the order of 30%. When a solution of a salt of a cation is added to a solution of an acid, corresponding to Graham's salt, the pH of the solution decreases due to the exchange of the previously undissociated hydrogen ions by the cations. It has been found that bivalent metal ions form stronger bonds with the polyphosphate anion than do the monovalent cations. Suggestions have been made that one of the reasons for the tendency of polyphosphate to ion-exchange is caused by the ability of the PO_4 tetrahedra in the chain to rotate freely about the P–O–P bonds. This mobility permits the cations to be linked to more than one PO_3^- in the form of relatively stable chelates. A support for this hypothesis is the observation that a phosphate chain the length of several phosphorus atoms is needed for firm combination with Ca ions[518].

Sodium polyphosphate is quite stable at room temperature at neutral pH. It hydrolyzes more rapidly in acid solution and at high temperatures. The hydrolysis is through the removal of the monophosphate from the ends of the chains. Degradation of the polyphosphate chain is also catalyzed by presence of cations. This action is due to the chelation of the cation on the PO_4 tetrahedra which causes the latter to become positive and thus become more susceptible to combination with OH^- ions or with water molecules[518].

[519] R. N. Bell, *Inorganic Synthesis*, Vol. III, pp. 103–6, Ed. L. F. Audrieth, McGraw-Hill Book Co., New York (1950).

[520] K. W. H. Dribbe, H. Harnisch and J. Cremer, U.S. 3,393,043, 16 July 1968, to Knapsack-Griesheim Aktiengesellschaft.

[521] T. Fleitman, *Ann. Physik.* **78** (1849) 233, 338.

[522] *Calgon In Industry*, Hagan Chemical and Controls Inc., Pittsburg, Penn. (1959).

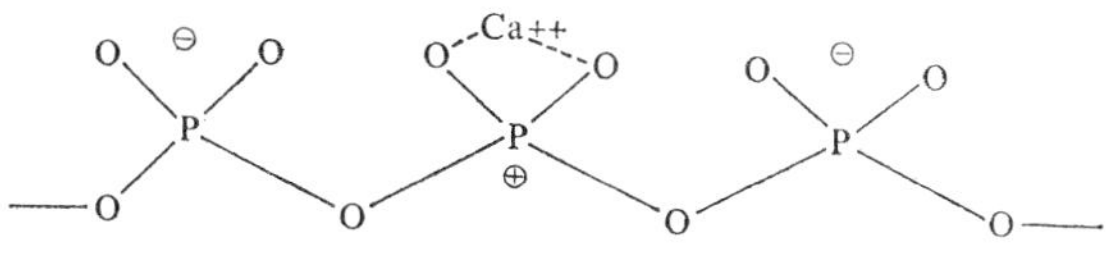

Applications. A large use for sodium metaphosphate glass, as represented by Calgon®, is for water treatment. This application depends largely on the ability of the polyphosphate anion to sequester the many cations such as Ca^{++}, Mg^{++} and Fe^{++} present in hard water and convert them into stable soluble complexes. Cations thus complexed are unavailable for the precipitation of soap in hard water.

Minute quantities (1 to 2 ppm) of glassy sodium metaphosphate in water also inhibit the crystal growth of calcium carbonate. This is known as the "threshold treatment" and prevents the deposition of scales in boilers. One theory for this inhibitory effect is that the growing faces of the calcium carbonate crystal are covered by the adsorbed polyphosphate molecule ions and this slows down the crystal growth. Upon long standing, large crystals of calcium carbonate do form, but they are grossly distorted due to the adsorbed polyphosphate ions[523]. Presence of dissolved glassy sodium metaphosphate (40–60 ppm) also inhibits the corrosion of cleaned iron surfaces in boilers and iron pipes by the formation of a protective film.

The polyvalent anion of glassy sodium metaphosphate has also dispersive action on finely divided solids. This property is used for reducing the viscosity of clay, oil well drilling mud, and various types of pigments, e.g., titanium dioxide in paper coating.

18.4. SODIUM POLYPHOSPHATES (MADDRELL'S SALT AND SODIUM KURROL'S SALT)

As shown in Fig. 16, there are two forms of Maddrell's salt obtained by the thermal condensation of NaH_2PO_4. The molecular weight of Maddrell's salt increases when the condensation is carried out under increasing water vapor pressure and dehydration time[524].

Each form of Maddrell's salt has its own X-ray pattern but their physical and chemical properties are very similar. They are referred to in the literature as Form II (high temperature Maddrell) and Form III (low temperature Maddrell). Both forms are practically insoluble in water. The structure of Form II is shown in Fig. 17.

Commercial insoluble sodium metaphosphate (IMP) consists of a mixture of these two forms and about 2–3% of $Na_2H_2P_2O_7$. Because of its low water solubility, the mixture is used as a polishing agent in dentifrice which contains soluble fluorides as tooth decay preventative. Insoluble sodium metaphosphate can solubilize a limited amount of calcium, as for example from the tooth enamel. Dentifrice formulations using IMP therefore generally contain small amounts of calcium phosphate such as the γ-form of calcium pyrophosphate as the calcium sacrificing agent.

Sodium Kurrol's salt is prepared by tempering $(NaPO_3)_x$ glass at 580–590°. It is a fibrous crystalline material referred to in the literature also as $NaPO_3$–IV or Form IV sodium metaphosphate. When added to water it swells into a gum. Upon standing

[523] B. Raistrich, *Discussions Faraday Soc.* (1949) 234–7.
[524] N. M. Dombrovskii, *Vysomolekul. Soedin* **8**(1) (1966) 38–41.

or boiling in water, it eventually dissolves into a viscous solution. As shown in Fig. 17, the anion of the sodium Kurrol's salt consisted of a spiral chain.

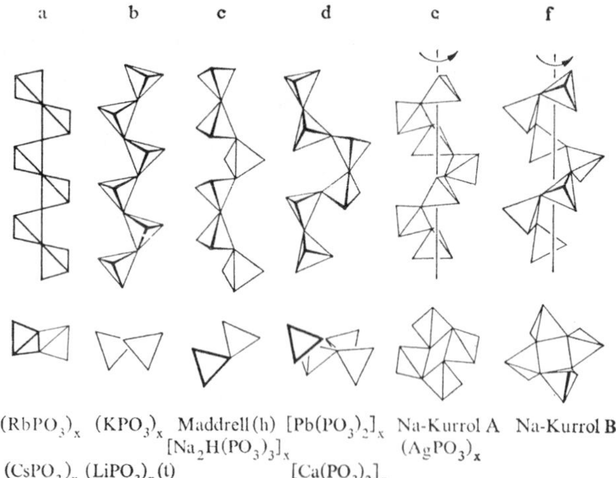

FIG. 17. The six known types of anion chains present in high molecular weight crystalline polyphosphates. Above: vertical projections; below: projections along the chain axis[518]. (Reprinted by permission of Verlag Chemie, GmbH.)

18.5. POTASSIUM POLYPHOSPHATE (POTASSIUM KURROL'S SALT)

Potassium dihydrogen phosphate dehydrates rapidly into potassium Kurrol's salt at 205–230°. Because of the high cost of K^+ from KOH, many proposals have been made for the preparation of potassium metaphosphate by the direct action of KCl on phosphoric acid[525, 526]:

$$KCl+2H_3PO_4 \rightleftharpoons KH_5(PO_4)_2+HCl$$

$$KH_5(PO_4)_2 \rightleftharpoons KH_2PO_4+H_3PO_4$$

$$KH_2PO_4 \rightleftharpoons (KPO_3)_n+H_2O$$

Via the above route, a product with 0–0.5 Cl content has been reported[527].

The degree of polymerization of potassium metaphosphate increases with the temperature and length of time of heating. Heating at 260° for 215–280 hr the degree of polymerization is 2300; at 445° for 115–240 hr, the degree of polymerization is 11,700. The common product has a molecular weight range of 250,000 to several million[528].

Potassium polymetaphosphate is practically insoluble in water. However, it goes into a viscous solution in the presence of other alkali metal cations such as Na^+. This phenomenon is attributed to the exchange of the Na^+ for the K^+ and thus disturbs the compact structure of the crystal to permit dissolution by water. An interesting example of this dissolution through ion exchange is found in the case of a mixture of potassium

[525] A. Norbert, *Rev. Chim. Minerale*, 3(1) (1966) 1–59.

[526] J. Martin and J. R. Andrieu, U.S. 3,378,340, 16 April 1968, to Société Chimiques pour l'Industrie et l'Agriculture.

[527] S. I. Vol'fkovich, M. V. Lykov, A. S. Cherepanova, G. V. Kryukov, E. G. Polievktova and V. V. Doronin, *Zh. Prikl. Khim.* 39(1) (1966) 3–7.

[528] J. R. van Wazer, *Phosphorus and Its Compounds*, Vol. I, pp. 601–800, Interscience Publishers Inc., New York (1958).

metaphosphate and sodium metaphosphate. Even though both salts individually are practically insoluble in distilled water, the mixture in water dissolves each other through the mutual exchange of the cations[528].

The long chain molecules of potassium metaphosphate can also be cross-linked into a rubbery mass by kneading in a solution of $CaCl_2$ and NaCl.

Industrial application of potassium metaphosphate is still fairly limited. One use is in combination with $Na_4P_2O_7$ for sausages to improve the texture and homogeneity of ground meat [529]. It also reduces separation of fat and loss of meat juices during cooking and thus reduces shrinkage. The use of potassium metaphosphate in food products while prevalent in Germany has not been approved by the Food and Drug Administration in the United States.

18.6. AMMONIUM POLYPHOSPHATE

A finely divided crystalline ammonium polyphosphate precipitates when the product of ammoniation of polyphosphoric acid containing 83–84% P_2O_5 is added to water. The precipitate contains 35–40% of the original phosphorus in the polyphosphoric acid. It consists of long linear chains of the general formula $(NH_4)_{n+2}P_nO_{3n+1}$ in which n is 50 or higher[530].

Dried crystalline product prepared in this manner appears as stubby rods ($n_\alpha = 1.485$, $n_\gamma = 1.500$) and is stable at room temperature. It maintains the same crystal structure even after heating for 48 hr at 105°. It melts at 355°. Its solubility in water is in the order of 0.1 g per 100 cc after 24 hr at 25°[530].

Ammonium polyphosphate cannot be prepared by the direct thermal dehydration of $NH_4H_2PO_4$ since such reaction is accompanied by the loss of NH_3. Shen, Stahlheber and Dyroff[531] have recently described a successful thermal condensation process which involves the use of urea as the source of ammonia and dehydrating agent:

$$H_2N\overset{\overset{\displaystyle O}{\|}}{C}NH_2 + H_3PO_4 \longrightarrow (1/x)(NH_4PO_3)_x + CO_2 + NH_3$$

The product (Form I) obtained by heating an equal molar mixture of $NH_4H_2PO_4$ and urea under an NH_3 atmosphere at 280° for 16 hr is the same as the product obtained by the ammoniation of polyphosphoric acid described earlier. Four other crystalline modifications of ammonium polyphosphate (Forms II, III, IV and V) have been obtained by the treatment of Form I under various temperature conditions. Form IV, obtained in a mixture with glassy impurities by cooling rapidly a melt from heating Form I to 450–470°, has an X-ray diffraction pattern similar to that of potassium Kurrol's salt.

The solubility of ammonium polyphosphate, though slight in water, is also increased by ion exchange with alkali metal ions as described for Maddrell's salt and potassium Kurrol's salt.

The original development of ammonium polyphosphate by the Tennessee Valley Authority was as a superconcentrated fertilizer. It releases slowly the nitrogen and phosphorus values in the soil. It is now also used in formulations for flame retardants.

529 H. Huber and K. Vogt, U.S. 2,852,392, 16 Sept. 1958, to Chemische Werke Albert.
530 A. W. Frazier, J. P. Smith and J. R. Lehr, *J. Agr. Food Chem.* **13**(4) (1965) 316–22.
531 C. Y. Shen, N. E. Stahlheber and D. R. Dyroff, *J. Am. Chem. Soc.* **91** (1969) 62–67.

18.7. CYCLIC METAPHOSPHATES

Cyclic metaphosphates from trimeta to octameta have been observed from chromatographic study of a solution of Graham's salt after the linear polyphosphates have been removed by precipitation. With the exception of heptameta, all members from trimeta to octameta have been either isolated in minor quantities from Graham's salt or prepared by specific methods.

Trimetaphosphates

Sodium trimetaphosphate exists in three polymorphic forms (Forms I, I′, I″).[517] The common and stable form is $(NaPO_3)_3$–I or Form I and is known also as Knorre's salt. All three forms may be made by tempering the sodium polyphosphate melt under specific temperature conditions[518, 519] (see Fig. 16). Form I is obtained in high purity by the use of a tempering temperature range of 300–500°[532]. It can also be made by the direct condensation of NaH_2PO_4. Thermal gravimetric studies showed this reaction to occur at the temperature range of 320–380° [533]. A high quality product, not contaminated with Maddrell's salt, is also obtained by the condensation of NaH_2PO_4 in the presence of a small percentage of ammonium nitrate[534].

Trimetaphosphate anion has been shown by X-ray structural analysis to contain a six-membered ring of alternating phosphorus and oxygen atoms. In the acid form, all hydrogen atoms are equivalent and strongly acidic. NMR studies on the solution show only one peak corresponding to the middle groups in the chain polyphosphate. Raman spectra indicate the ion in solution to be a flat ring. In crystalline $LiK_2P_3O_9.H_2O$, however, the ring is a puckered chair shape[517].

At 299.15°K, $Na_3P_3O_9$ has $C_p = 62.00$ cal/deg and $S = 68.47$ cal/deg[535].

Sodium trimetaphosphate is water-soluble, but has no calcium sequestering power. It converts rapidly to sodium tripolyphosphate ($Na_5P_3O_{10}.6H_2O$) when heated at 100° in the presence of an excess of sodium hydroxide.

Crystals of the 6-hydrate, $Na_3P_3O_9.6H_2O$, are triclinic rhombohedra, biaxial negative with $\alpha = 1.433$, $\beta = 1.442$, and $\gamma = 1.446$ ($\pm.002$). This hydrate effloresces at room temperature and loses water rapidly at 50°[519].

Other salts of trimetaphosphate, such as La, Ce, Pr, are prepared by the addition of a solution of the respective metal chlorides to an aqueous solution of $Na_3P_3O_9.6H_2O$[536]. The salt $Ca_3(P_3O_9)_2.10H_2O$ is also prepared from $Na_3P_3O_9$ by the addition of Ca^{++}. It is very soluble in water and reacts with NH_4^+ to form a compound of the approximate composition $NH_4CaP_3O_9$ [537].

Sodium trimetaphosphate reacts with methyl and ethyl alcohols in a water alkali solution at room temperature to form methyl and ethyl tripolyphosphate. It reacts with ethylene glycol to produce the 2-hydroxyethyl monophosphate[538]. One industrial application related to these reactions is the phosphorylation of starch with sodium trimeta-

[532] G. Heymer, J. Cremer and H. Harnisch, U.S. 3,389,958, 25 June 1958, to Knapsack Aktiengesellschaft.
[533] L. Steinbrecher and J. F. Hazel, *Inorg. Nucl. Chem. Letters*, **4** (1968) 559–62.
[534] H. A. Rohlfs and H. Schmidt, U.S. 3,393,974, 23 July 1968, to Chemische Werke Albert.
[535] R. J. L. Andon, J. F. Counsell, J. F. Martin and C. J. Mach, *J. Appl. Chem.* (1967) (17) 65–70.
[536] O. A. Serra and E. Giesbrecht, *J. Inorg. Nucl. Chem.* **30** (1968) 793–9.
[537] W. Feldmann and I. Grunze, *Z. anorg. u. allgem. Chem.* **360** (1968) 225–30.
[538] W. Feldmann, *Ber.* **100** (1967) 3850–60.

phosphate. Such modified starch though containing very little phosphate, has the property of forming stable cold water gels[539]. The property of $Na_3P_3O_9$ to cleave readily into $Na_5P_3O_{10}.6H_2O$ in the presence of aqueous NaOH has been used for the *in situ* generation of $Na_5P_3O_{10}.6H_2O$ in detergents[540].

Tetrametaphosphates

Pure tetrametaphosphates are formed when dihydrogen phosphates of Al, Cu^{++}, Mg, Ni^{++}, Co^{++}, Mn^{++}, Fe^{++}, Zn and Cd are dehydrated at 400–500°. Alternatively, they are also formed when glasses resulting from cooling of the melts of the above condensation products are tempered at 400–500°.

Sodium tetrametaphosphate can be prepared by the reaction of Cu or Zn tetrametaphosphate with Na_2S in an aqueous solution[518]. A more convenient method is the selective splitting of the P_4O_{10} tetrahedra by the controlled hydrolysis of the hexagonal form of P_4O_{10} at 15°. The solution obtained is neutralized to pH 7.0 with 30% NaOH, and NaCl then added. Upon standing overnight at below 25°, $(NaPO_3)_4.10H_2O$ crystallizes out.

Sodium tetrametaphosphate also forms a tetrahydrate which has two polymorphic forms. The low temperature form is stable at below approximately 54° and the high temperature modification which forms above 54° does not revert to the low temperature form on cooling. At 100°, the hydrates lose all water to give the anhydrous compound[541].

Tetrametaphosphate ion has an eight-membered cyclic ring structure. It has been suggested that the observed existence of two crystalline varieties of the same salts is due to the two different configurations of the tetrametaphosphate rings (e.g., boat and chair)[517].

Thermally, $(NaPO_3)_4$ is less stable than $(NaPO_3)_3$. Heating at 400–500° converts it to the smaller ring compound.

In 1% NaOH and at 100°, $(NaPO_3)_4$ hydrolyzes completely in about 4 hr. The products of hydrolysis are tripolyphosphates and orthophosphate via the intermediate linear tetrapolyphosphate. Linear tetrapolyphosphate can be obtained by the controlled hydrolysis of $(NaPO_3)_4$ at below 40° with aqueous NaOH. Precipitation with ethanol gives a syrupy sodium tetrapolyphosphate layer which is converted to the crystalline hexaguanidinium 1-hydrate salt upon addition of guanidinium chloride in water[512].

The rate of cleavage of P–O–P bond in tetrapolyphosphate increases about threefold for every 10° rise in temperature. As a consequence the reactions on the degradation of tetrapolyphosphate and the opening of the tetrametaphosphate ring are competitive[512].

Higher Membered Ring Metaphosphates

Pentametaphosphate and hexametaphosphate of Na, Ag, and Ba, containing water of crystallization, have been obtained in gram quantities from fractionation of the cyclic-metaphosphate fraction of Graham's salt prepared with a Na : P ratio of $\leqq$ 1.25 : 1[542].

Lithium hexametaphosphate is prepared by heating a mixture of Li_2O/P_2O_5 in a ratio near 7:5 to a temperature of 275°. The product contains only hexametaphosphate with some pyrophosphate. A water solution of it upon passing through a strong acid ion exchange resin and neutralized with Na_2CO_3 to pH 5 to 6 gives a precipitate of sodium hexametaphosphate.$6H_2O$ when CH_3OH is added.

[539] R. W. Kerr and F. C. Cleveland, U.S. 2,884,413, 28 April 1959, to Corn Products.
[540] C. Y. Shen, *JAOCS*, **7** (1968) 510–16.
[541] R. N. Bell, L. F. Audrieth and O. F. Hill, *Ind. Eng. Chem.* **44** (1952) 568–72.
[542] E. Thilo and U. Schuelke, *Z. anorg. u. allgem. Chem.* **341** (5–6) (1965) 293–307.

The compound $Na_6P_6O_{18}.6H_2O$ appears as monoclinic needles and becomes anhydrous when heated to above 120° without breaking the ring[543]. The anion ring has been shown to consist of six PO_4 tetrahedra connected at the corners[544].

In the cyclic metaphosphate system, the stability of the ring to cleavage by NaOH increases rapidly as the size of the ring increases. Thus, pentametaphosphate and hexametaphosphate rings are much more stable than the tri- and tetrametaphosphate rings, e.g., half-life on cleavage with 0.1 NaOH (hours): ring size (P atoms) 3 = 4.5, 4 = 150, 5 = 200 and 6 = 1000[518].

Lead octametaphosphate is formed by heating $Pb_2(P_4O_{12}) \cdot 4H_2O$ to 300–350°. The reaction product is contaminated with about 30% of high molecular weight lead polyphosphate. The sodium salt, $Na_8(P_8O_{24}).6H_2O$, is obtained from lead salt by treatment with $Na_2S.9H_2O$ in water. It precipitates from an aqueous solution as needles upon addition of ethanol.

Sodium and potassium octametaphosphates are stable in aqueous solution for many weeks at neutral pH. They hydrolyze in 0.4N NaOH at 40°. The primary product of hydrolysis is octapolyphosphate of which a great portion hydrolyzes further to the lower polyphosphates before the rest of the octametaphosphate is cleaved[545].

18.8. ULTRAPHOSPHATES

This class of compound is condensed polymeric phosphates having a cation to phosphorus ratio of less than one. Each molecule contains varying numbers of PO_4 tetrahedra some of which are linked to the other PO_4 tetrahedra through three oxygen atoms, i.e., a branched chain structure. In contrast to linear polyphosphates in which the PO_4 tetrahedra are connected through two oxygen atoms and the bonds are resonance stabilized, the branched chain ultraphosphates are not resonance stabilized[518]:

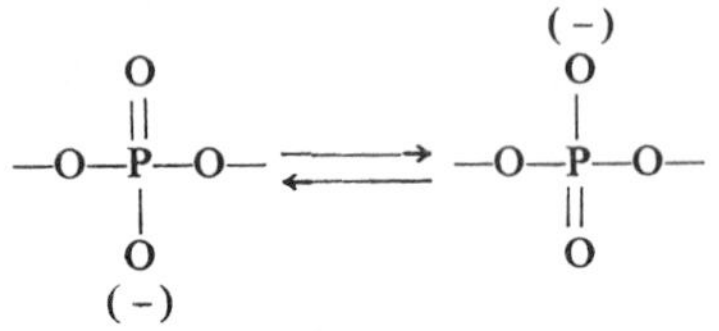

```
                          (-)
       O                   O
       ‖                   |
  —O—P—O—   ⇄   —O—P—O—
       |                   ‖
       O                   O
      (-)
```

Resonance stabilization

```
       O
       ‖
  —O—P—O—
       |
       O
       |
  —O—P—O—
       ‖
       O
```

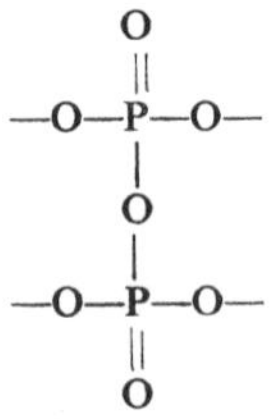

No resonance stabilization

The tertiary PO_4 tetrahedra in ultraphosphate are therefore more favored for nucleophilic

[543] E. J. Griffith and R. L. Buxton, *Inorg. Chem.* **4** (1965) 549–52.
[544] K. H. Jost, *Acta Cryst.* **19**(4) (1965) 555–60.
[545] U. Schuelke, *Z. anorg. u. allgem. Chem.* **360** (1968) 231–46.

attack by OH^-. The net result is that the triply linked PO_4 tetrahedra of ultraphosphate is much more easily hydrolyzed than the linear polyphosphates.

The structures of ultraphosphates are endless depending on the cation to P ratio. An example of ultraphosphate in the acid form is "azeotropic phosphoric acid" which boils at 864° at 760 mm Hg pressure. It has a P_2O_5 content of 92.4%[528].

19. CONDENSED PHOSPHATES WITH OXY ANIONS OF OTHER ELEMENTS

Orthophosphates, in addition to the ability to form homopolymers through sharing of oxygen atoms between the PO_4 tetrahedra, form copolymers also with oxy anions of other elements. There are two general types of such copolymers. One consists of PO_4 tetrahedra joined with other oxy anion tetrahedra, XO_4, in a random fashion through P–O–X linkages. Such copolymers are represented by phosphate-silicates, phosphate-sulfates, phosphate-vanadates and phosphate-arsenates. The other consists of definite compounds having specific ratios of phosphate anion to the other oxy anions. The acid forms of such compounds are called heteropoly acids. When the other oxy anion is molybdate, compounds with P:Mo ratios of 1:12, 1:11, 1:10, 2:18 and 2:17 have been reported[546].

The best known of the heteropoly acids is that with a phosphate:oxy anion ratio of 1:12 and is known as 12-heteropoly acid. They are represented by compounds such as 12-molybdophosphoric acid, 12-tungstophosphoric acid and their salts. There are also compounds containing mixed oxy anions such as 12-molybdovanadophosphoric acids and their salts.

19.1. HETEROPOLYMERS OF PHOSPHATE-SILICATE, PHOSPHATE-SULFATE, PHOSPHATE-VANADATE, AND PHOSPHATE-ARSENATE

These copolymers have been reviewed recently by Ohashi[547]. They are prepared by melting together the acids, salts or acid anhydrides of the oxy acids. The oxy acids which form copolymers with phosphates can by themselves form copolymers. Their tetrahedral anions are shown below:

```
      O⁻                 O                 O⁻                 O⁻
      |                  ||                |                  |
  ⁻O—Si—O⁻          ⁻O—S—O⁻           ⁻O—V—O⁻           ⁻O—As—O⁻
      |                  ||                ||                 ||
      O⁻                 O                 O                  O
   Silicate           Sulfate           Vanadate           Arsenate
```

Copolymers of the above oxy acid anions with phosphates are usually glasses containing P–O–X linkages. When they are dissolved in water, in all cases cleavage occurs first at the P–O–X bond.

Phosphate-Silicate

In the phosphate-silicate copolymer system, the calcium salt is prepared by fusing together calcium metaphosphate [prepared by fusing of $Ca(H_2PO_4)_2$] with the desired amount of

[546] F. A. Cotton and G. Wilkinson, *Advanced Inorganic Chemistry*, 2nd ed., pp. 941–6, Interscience Publishers (1966).

[547] S. Ohashi, *Topics in Phosphorus Chemistry*, **1** (1964) 189–239.

silica gel at 980–1000°. Glassy copolymers with a SiO_2 content of 12.5% or below are cleer and completely, though slowly, soluble in water. Products with a SiO_2 content of more than 12.5% are turbid in appearance and not all of the SiO_2 content is soluble in water. Dissolution of calcium phosphate silicate glass in water is more rapid in the presence of cation exchange resin (Dowex 50, H form). The solution becomes saturated when the P:Si ratio reaches about 4 and from this data it has been postulated that a compound having a P:Si ratio = 4, e.g., $Ca_2SiP_4O_{14}$, may be present in the glass along with calcium polyphosphates of various degrees of polymerization[547].

In the sodium phosphate silicate system, when a mixture of sodium metaphosphate and sodium metasilicate is melted together in a 1:1 molar ratio, no copolymer forms. The reaction proceeds with the formation of trisodium phosphate and α-cristobalite:

$$NaPO_3 + Na_2SiO_3 \longrightarrow Na_3PO_4 + SiO_2$$

Also, no copolymers containing P–O–Si linkages are present in the $NaPO_3$–SiO_2 system with a P:Si ratio of 0.5 to 4. In this system, the phosphates are present mostly as trimetaphosphate along with pyro and tripolyphosphate and very little orthophosphate. The SiO_2 content is present mostly as α-cristobalite and little α-tridymite.

The presence of P–O–Si linkages in sodium phosphate silicate glasses is assumed from indirect evidence. When glasses with P:Si ratios greater than 6 (both in the $NaPO_3$–Na_2SiO_3 and $NaPO_3$–SiO_2 systems) are dissolved in water, polyphosphates of wide molecular weight distribution are obtained. The average chain length of the polyphosphates increases with the increase in the P:Si ratio. It is assumed then that the formation of these polyphosphates is the result of cleavage of the P–O–Si linkages present in the phosphate-silicate glasses. Such linkages are known to hydrolyze easily in water[547].

Phosphate-Sulfate

Copolymers of sodium phosphate-sulfate, containing P–O–S bonds, can be prepared only in melts in which the molar ratio $(P_2O_5+SO_3) > Na_2O$. The nature of the starting material is unimportant[548]. Such copolymers may be prepared by the absorption of SO_3 on pyro-, tri-, or polyphosphates in the temperature range of 400–450°. The resultant melt is then chilled rapidly into a glass. Alternatively, P_4O_{10} and varying amounts of sodium pyro- or tripolyphosphate may react with sodium sulfate at about 400° to form a melt which is then cooled to a glass[547].

Thilo and Blumenthal prepared sodium phosphate-sulfate glasses by adding various quantities of either sodium trimetaphosphate or Graham's salt to molten $Na_2S_2O_7$ at about 400°. Sulfur trioxide is evolved only when the heating is prolonged or when a large excess of $Na_2S_2O_7$ is used[548].

No P–O–S linkages are formed in melts of high Na_2O content. For example, when an equimolar mixture of $NaHSO_4$ and Na_2HPO_4 are heated together, Na_2SO_4 and polyphosphates with an average chain length corresponding to the residual sodium content are obtained. Sulphur trioxide is evolved at $\geqq 550°$.

Phosphate-sulfate copolymers can be isolated in solution from contaminating ionic sulfates and polyphosphate by means of anion exchange resin (the chloride form). The ionic sulfate is eluted alone and quantitatively by 0.2 N KCl. The phosphate-sulfate is eluted with 3 N HCl while the polyphosphate is eluted only by concentrated acid[548].

[548] E. Thilo and G. Blumenthal, *Z. anorg. u. allgem. Chem.* **348** (1966) 77–88.

The copolymers formed by the reaction of sodium polyphosphate and $Na_2S_2O_7$ with a molar ratio of $Na_2O/(P_2O_5+SO_3) = <1$, when dissolved in water, have been assigned the following structure:

$$\left[\begin{matrix} \text{O O} \\ \text{—OSOPO—} \\ \text{O O} \end{matrix} - - - - - \begin{matrix} \text{O O} \\ \text{POSO—} \\ \text{O O} \end{matrix}\right]^{n-} Na_n^+$$

This assignment is based on results of hydrolysis studies of the phosphate-sulfate fraction eluted from the anion exchange resin column. Terminal $NaSO_3$ group in the phosphate-sulfate copolymer hydrolyzes at 60° with a half-life of 7.5 hr at pH = 4.5 and 11.7 hr at pH = 13. One equivalent of a strongly acidic HSO_4^- ion and one weakly acidic POH end group are formed for each mole of sulfate hydrolyzed off. In contrast, the P–O–P bond in the polyphosphates portion hydrolyzes at 60° with a half-life of 4.4 days at a pH of 5, and 18.5 days at a pH of 8, while the S–O–S bond in the disulfate hydrolyzes with a half-life of 0.12 min in the pH range of 3–14[548].

Phosphate-Vanadate

Copolymers of phosphate-vanadate containing P–O–V linkages are presumed to form when glassy sodium polyphosphate with Na/P ratio of 1 is heated to a melt with various amounts of V_2O_5 and Na_2CO_3 at 900°. Rapid cooling of the melt results in glasses having colors which vary from dark brown to yellow as the P/V ratio is increased. At a P/V ratio of 3, about 19% of the total vanadium is reduced to V^{IV}. At a P/V ratio above 50, the total percentage of vanadium reduced to V^{IV} is 37–38%.

Assuming that P–O–V bonds hydrolyze immediately upon dissolving in water, glasses with P/V ratio of 1 have probably essentially an alternate phosphate-vanadate structure. This is shown by the fact that the phosphate moieties in the hydrolysate contain 89% ortho-, only 11% pyro- and no higher polyphosphates. As expected, glasses of higher P/V ratio have increasingly more longer chain polyphosphates in the hydrolysate.

Phosphate-Arsenate

Glassy copolymers of phosphate-arsenate are obtained by melting a mixture of NaH_2PO_4 and NaH_2AsO_4 at 68° and the melt then cooled rapidly. The distribution of the arsenic atoms in the chain of the copolymer is approximately at random with a slight preference for As–O–P bonds. Since the P–O–As and As–O–As bonds are easily hydrolyzed, the copolymers are converted in a water solution to monoarsenic and ortho- or polyphosphate mixtures.

In the case of the crystalline phosphate-arsenate copolymer, obtained by annealing of the glass at 470° for 24 hr, Thilo[549] has shown that it crystallizes with the structure of Maddrell's salt. The arsenic atom is preferentially located in the XO_4 tetrahedra forming the "noses" of the "Dreierketten" (Fig. 17). Since these crystalline copolymers resemble in structure crystalline Maddrell's salts, they are only very slowly soluble in water. Also similar to Maddrell's salt, they dissolve in a solution of alkali metal salts other than sodium. This is due to the ion-exchange phenomenon which disturbs the close pack structure of the crystal.

Potassium phosphate-arsenate forms crystalline compounds having six membered rings

[549] E. Thilo, *Angew. Chem. Intern. Ed.* **4** (1965) 1061–71.

analogous to potassium trimetaphosphate. These trimeric rings contain one or two arsenate anions:

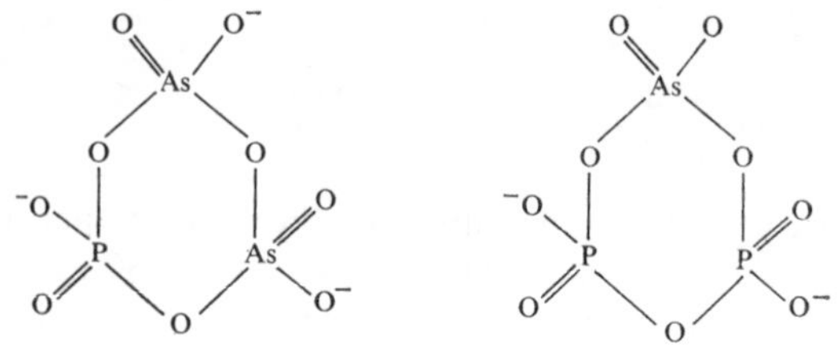

Obtained only as mixtures, their presence is deduced from the fact that P–O–As or As–O–As bonds are easily hydrolyzed while the P–O–P bond is rather stable to hydrolysis. Paper chromatographic analysis of the hydrolysate of potassium metaphosphate-arsenate shows that as the arsenic content is increased, the percentage of orthophosphate increases and the pyro- and trimetaphosphate decreases. This is in line with the proposed structures for the cyclic phosphate-arsenate.

19.2. 12-HETEROPOLY ACIDS AND THEIR SALTS

The term heteropoly acids has been designated for definite compounds of specific structures. Compounds containing anions of oxy acids such as phosphoric, arsenic, boric, or silicic acids and anions of other oxy acids such as molybdic and tungstic acids are examples of heteropoly acids. This section concerns itself only with examples of the phosphorus-containing 12-heteropoly acids.

When phosphoric acid is reacted with ammonium molybdate in a nitric acid solution, a yellow precipitate of ammonium 12-molybdophosphate forms. It is a complex compound having a phosphorus nucleus with a formula of $(NH_4)_3H_4[P(Mo_2O_7)_6].H_2O$. Since it is only soluble to the extent of 0.03 g/100 g H_2O at 15°, and no corresponding precipitates form with pyro- or metaphosphate ions, it is used for the qualitative determination of soluble orthophosphate. The free acid of the compound, 12-molybdophosphoric acid, $H_7[P(Mo_2O_7)_6].10H_2O$, is soluble in water and many of its normal salts such as silver, mercury and guanidinium salts are known[550].

Heteropoly acid of phosphorus with tungsten is also prepared by the direct combination of the components followed by crystallization. For example, the 12-tungstophosphoric acid, $H_7[P(W_2O_7)_6]\ xH_2O$ is prepared by dissolving sodium tungstate ($Na_2WO_4.2H_2O$) and disodium phosphate in boiling water and the solution acidified with concentrated hydrochloric acid:

$$12Na_2WO_4+Na_2HPO_4+26HCl+H_2O \rightarrow H_7[P(W_2O_7)_6]\cdot xH_2O+26NaCl$$

The precipitated acid is purified by the addition of ether. The water-insoluble heteropoly acid-ether complex being insoluble in both ether or water is washed free of impurities with dust-free air through the solution. The free acid which is extremely soluble in water is obtained as very heavy white octahedral crystals. It is easily reduced by organic reducing

[550] D. M. Yost and H. Russell, Jr., *Systemic Inorganic Chemistry of the Fifth- and Sixth-Group Nonmetallic Elements*, pp. 232–3, Prentice-Hall Inc., N.Y. (1944).

vapors or light to a blue color. Reoxidation is easily accomplished by heating with water.

The sodium salt is soluble in water while the potassium and ammonium salts are comparatively insoluble. 12-tungstophosphoric acid finds use as precipitant for proteins, alkaloids and certain amino acids[551].

Mixed molybdovanadophosphoric acids and their salts are also known. Three of these heteropoly acids have the following formula:

$$H_4[PMo_{11}VO_{40}]\cdot 34H_2O,\ H_5[PMo_{10}V_2O_{40}]\cdot 32H_2O,\ \text{and}\ H_6[PMo_9V_3O_{40}]\cdot 34H_2O.$$

11-molybdo-1-vanadophosphoric acid, $H_4[PMo_{11}VO_{40}]\cdot 34H_2O$, is prepared using Na_2HPO_4, $NaVO_3$ and $Na_2MoO_4.2H_2O$ in the molar ratio of 1:1:11. The water solution of disodium phosphate and sodium metavanadate is acidified to a red color with concentrated H_2SO_4. A water solution of $Na_2MoO_4.2H_2O$ is added, and then the whole reaction mixture is acidified with concentrated H_2SO_4. Upon cooling the heteropoly acid is extracted

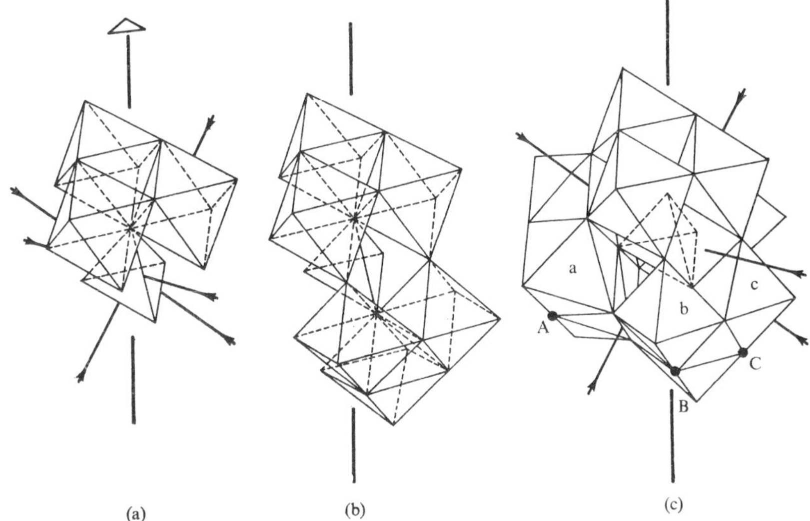

FIG. 18. Arrangement of (a) one group and (b) two groups of three WO_6 octahedra relative to the central PO_4 tetrahedron, and (c) the complete anion $PW_{12}O_{40}^{3-}$.[555]

with ethyl ether. The heteropoly acid etherate is freed of ether by sparging with air. The orange solid residue obtained is recrystallized from water. The crystalline acid effloresces slowly at room temperature which accounts for the slight variations in the amount of water of crystallization from sample to sample.

The 10-molybdo-2-vanadophosphoric acid, $H_5[PMo_{10}V_2O_{40}].32H_2O$, is similarly prepared using Na_2HPO_4:$NaVO_3$:$Na_2MoO_4.2H_2O$ in the molar ratio of 1 : 4 : 10. The solid obtained from the etherate upon crystallization from water yields large red crystals. It dissolves in ether to form an oil which is ether insoluble. It is soluble and forms a single phase, however, in alcohols, ethyl acetate, dimethyl sulfoxide and tetrahydrofuran.

The preparation of 9-molybdo-3-vanadophosphoric acid $H_6[PMo_9V_3O_{40}].34H_2O$, is carried out in a same manner, using a Na_2HPO_4:$NaVO_3$:$Na_2MoO_4.2H_2O$ in the ratio of 1 : 6 : 4.5.

[551] J. C. Bailar, Jr , *Inorganic Synthesis*, Vol. I, pp. 132–3, Ed. H. S. Booth, McGraw-Hill Book Co. 939).

Triammonium 11-molybdo-1-vanadophosphate, $(NH_4)_3H[PMo_{11}VO_{40}].3.5H_2O$, and triammonium 10-molybdo-2-vanadophosphate, $(NH_4)_3H_2[PMo_{10}V_2O_{40}].7.5H_2O$, are insoluble in water. They are prepared as precipitates by the addition of a NH_4Cl solution to a solution of the corresponding free acid. The sodium 11-molybdo-1-vanadophosphate, $Na_4[PMo_{11}VO_{40}]\cdot 8H_2O$, however, is very soluble in water. It can be prepared by passing a solution of the acid through the sodium form of Dowex 50–X8 ion exchange resin and the effluent evaporated to dryness[552].

The structure of crystalline 12-tungstophosphoric acid, $H_3[PW_{12}O_{40}]5H_2O$, was studied by Keggin by X-ray technique[553]. It was found that in the anion, a PO_4 tetrahedron is located at the center of a structure surrounded by twelve WO_6 octahedra. These octahedra have shared corners and shared edges but no shared faces. It may be considered as four groups of three WO_6 octahedra. In each group there is one oxygen atom shared by all three octahedra. The orientation of the four groups is such that the four triply shared oxygen atoms are located at the corners of the central tetrahedron—that of the PO_4 anion[554, 555]. In such an arrangement, there are large spaces between the atoms which accounts for the existence of hydrates containing large numbers of water molecules, e.g., $H_3[PW_{12}O_{40}]\cdot 29H_2O$. Such hydrates can be dehydrated without causing any important structural changes.

Tsigdinos and Hallada have proposed that the molybdovanadophosphoric acids they prepared have anions also of the structure analogous to that proposed by Keggin for 12-tungstophosphoric acid[552].

20. HALOACIDS OF PHOSPHORUS AND THEIR SALTS

This class of compounds may be regarded as derived from substituting the OH groups in orthophosphoric acid with halogens. The best known and well-characterized compounds in this series are the fluoro derivatives.

20.1. FLUOROPHOSPHORIC ACIDS AND FLUOROPHOSPHATES

Monofluorophosphoric Acids and Monofluorophosphates

Fluorophosphoric acids interconvert with relative ease. Their relationship to each other and to orthophosphoric acid is shown below:

$$HO\overset{\overset{\displaystyle O}{\|}}{\underset{\underset{\displaystyle H}{\underset{\displaystyle O}{|}}}{P}}OH \underset{H_2O}{\overset{HF}{\rightleftarrows}} HO\overset{\overset{\displaystyle O}{\|}}{\underset{\underset{\displaystyle F}{|}}{P}}OH \rightleftarrows HO\overset{\overset{\displaystyle O}{\|}}{\underset{\underset{\displaystyle F}{|}}{P}}F \underset{2H_2O}{\overset{4HF}{\rightleftarrows}} HPF_6$$

Based on the above equilibrium, monofluorophosphoric acid is formed to a considerable amount when HF is reacted with concentrated orthophosphoric acid[556]. It can also be

552 G. A. Tsigdinos and C. J. Hallada, *Inorg. Chem.* 7(3) (1968) 437–41.

553 J. F. Keggin, *Proc. Roy. Soc.* (*London*), **144A** (1934) 75–100.

554 H. B. Jonassen and S. Kirschner, *The Chemistry of Coordination Compounds*, pp. 472–86, Ed. J. C. Bailar, Jr., Reinhold Publishing Corp., New York (1956).

555 A. F. Wells, *Structural Inorganic Chemistry*, 3rd ed., p. 450, Oxford, Clarendon Press (1962).

556 W. Lange, *Inorganic Synthesis*, Vol. II, pp. 155–6, Ed. C. W. Fernelius, McGraw-Hill Book Co., New York (1946).

prepared by the hydrogenation of $(C_6H_5O)_2P(O)F$ in tetrahydrofuran solvent for 10 hr at room temperature in the presence of $PtO_2.2H_2O$ catalyst[557]. The preferred preparative method is the conversion of Na_2PO_3F by the acid form of a cation exchange resin[558]. Commercial monofluorophosphoric acid is prepared by the reaction of 69% aqueous HF and phosphoric anhydride:

$$P_2O_5+2HF+H_2O \longrightarrow 2H_2PO_3F$$

It contains usually 15–20% each of H_3PO_4 and HPO_2F_2[559].

Monofluorophosphoric acid is a colorless viscous oily liquid, strongly resembles H_2SO_4 and is completely soluble in water. Potentiometric titration shows it to be a dibasic acid and the ionization constants have been estimated to be $K_1 = 0.28$ and $K_2 = 1.58 \times 10^{-5}$ [560]. The neutralization points for a 0.05 N solution are pH 3.5 and pH 8.5[559]. Due to intramolecular hydrogen bonding, F---H–O, it acts as a monobasic acid in aqueous solution by conductivity measurements[558]. The density of the 100% acid is $D_4^{25} = 1.818$[560]. Little decomposition occurs even when heated to 185° under reduced pressure. However, it is not distillable. It does not attack glass in the absence of water. At 30° it solidifies and at −70° it becomes a rigid glass.

Sodium monofluorophosphate is prepared by heating anhydrous sodium trimetaphosphate with anhydrous sodium fluoride in a platinum dish at 800° until complete fusion of the mixture. The product obtained contains about 90% Na_2PO_3F. It is purified by dissolving in water and removing the sodium pyrophosphate contaminant as the insoluble silver pyrophosphate by the addition of Ag_2PO_3F. (Silver monofluorophosphate is prepared from Na_2PO_3F by reaction with $AgNO_3$.) Sodium monofluorophosphate is then obtained as crystals upon concentration of the aqueous solution. It melts at about 625°. It is very soluble in water and hydrolyzes to the orthophosphate and fluoride slowly in neutral solution and more rapidly in acid solution[561]. The pH of 2% solution is about 7.2[559].

The major application of sodium monofluorophosphate is as an additive in dentifrice formulation for caries inhibition[562]. Tests have also shown that when Na_2PO_3F is added at a level of 40 ppm F equivalent to the drinking water of Syrian hamsters, a reduction of dental caries in these animals is observed. Since the PO_3F^- ion apparently is not hydrolyzed appreciably in the animal body, the fluorine therefore does not have to be in the free ionic form to be active as dental caries inhibitor. The results on dental caries inhibition by Na_2PO_3F are comparable to those obtained by NaF at the same F level. Calculated on the basis of F content, the toxicity of Na_2PO_3F is only about 2/5 to 1/3 that of NaF to rats[563].

Diammonium fluorophosphate is formed in equal amount with monoammonium difluorophosphate when NH_4F is heated with P_4O_{10} at about 135°. It is obtained as an insoluble residue when the reaction product is extracted with absolute alcohol to remove ammonium difluorophosphate. Upon recrystallization from water, the monohydrate, $(NH_4)_2PO_3F.H_2O$, occurs as rectangular prisms which effloresces in air at room temperature

[557] K. B. Boerner, C. Ztoelzer and A. Simon, *Ber.* **96** (1963) 1328–34.

[558] J. Neels and H. Grunze, *Z. anorg. u. allgem. Chem.* **360** (1968) 284–92.

[559] W. E. White and C. Pupp, *Kirk-Othmer Encyclopedia of Chem. Technol.*, 2nd ed., Vol. IX, pp. 635–826, Interscience Publishers (1966).

[560] J. R. van Wazer, *Phosphorus and Its Compounds*, Vol. I, pp. 801–26, Interscience Publishers, New York (1958).

[561] O. F. Hill and L. F. Audrieth, *Inorganic Synthesis*, Vol. III, pp. 106–17, Ed. L. F. Audrieth, McGraw-Hill Book Co., New York (1950).

[562] E. Saunders and T. Schiff, U.S. 3,308,029, 7 March 1967, to Monsanto Chemical Company.

[563] K. L. Shourie, J. W. Hein and H. C. Hodge, *J. Dental Research*, **29** (1950) 529–33.

and becomes anhydrous at 105° [557]. On heating to 220°, $(NH_4O)_2POF$ loses NH_3 accompanied by condensation and reorganization reactions to difluoro- and monofluoropolyphosphates and polyphosphates[558].

Alkali acid monofluorophosphates such as $NaHPO_3F$, $KHPO_3F$ and NH_4HPO_3F are obtained by dissolving the corresponding dialkali metal monofluorophosphate and H_2PO_3F in water[558].

Mixed alkali metal monofluorophosphates are obtained by evaporation of a solution of the two dialkali monofluorophosphates. In this manner $NaKPO_3F$, $NaNH_4PO_3F.H_2O$ and KNH_4PO_3F are isolated[558]. Many metal monofluorophosphates are prepared by metathesis reaction from Ag_2PO_3F and the metal chlorides. Examples of compounds prepared in this manner are $CdPO_3F.8H_2O$, $MnPO_3F.3H_2O$, $NiPO_3F.7H_2O$, $Cr_2(OP_3F)_3.18H_2O$ and $Fe_2(PO_3F)_3.12H_2O$[568].

Difluorophosphoric Acid and Difluorophosphates

Commercial difluorophosphoric acid, HPO_2F_2, is prepared by the treatment of phosphoric anhydride with HF according to the following equation:

$$P_4O_{10}+6HF \longrightarrow 2HPO_2F_2+2H_2PO_3F$$

It is separated from monofluorophosphoric acid by distillation under reduced pressure: b.p.= 116° at 760 mm or 51° at 100 mm. The vapor pressure at between 22–108° follows the equation:

$$\log p = -421.195/T - 0.478214 \log T + 0.0137121T$$

At between 51.6–93°,

$$\log p = 7.333 - 1732\ T.$$ [559, 565]

It has a m.p. of $-96.5\pm1°$ and density of d_4^{25} 1.583. The heat of vaporization is 7925 cal/g mole. When dissolved in water it behaves as a monobasic acid with a neutralization end point at pH 7.3. A neutralized solution, however, slowly turns acidic due to hydrolysis to monofluorophosphoric acid and hydrofluoric acid[559].

Many industrial applications have been suggested for difluorophosphoric acid in the patent literature. These include its use as catalysts for isomerization and polymerization of hydrocarbons. At the present its actual industrial applications are limited.

Salts of difluorophosphoric acid are well known and the most widely studied ones are those of the alkali metals. As indicated earlier, ammonium difluoride phosphate is prepared as a co-product with ammonium monofluorophosphate when a mixture of P_4O_{10} and NH_4F is heated to 135°. It is isolated by extraction with boiling absolute alcohol and recrystallized quickly from water[556].

Sodium difluorophosphate may be prepared by heating $NaHF_2$ with P_4O_{10} through several temperature stages until 300–320° when all of the $NaHF_2$ is decomposed[566]. Another procedure is by the melting of $NaPF_6$ and $(NaPO_3)_n$:

$$NaPF_6+2/n(NaPO_3)_n \longrightarrow 3NaPO_2F_2$$

Quite pure $NaPO_2F_2$ is thus obtained[567].

[564] E. B. Singh and P. C. Sinha, *J. Indian Chem. Soc.* **41** (1964) 407–10, 411–14.

[565] *Gmelin Handbuch der Anorganischen Chemie*, 8th ed., Part C, p. 408, Verlag Chemie, G.m.b.H. (1965).

[566] I. G. Ryss and V. B. Tul'chinskii, *Zh. Neorgan. Khim.* **7** (1962) 1313–15.

[567] H. Jonas, Ger. 813,848, 17 Sept. 1951, to Farbenfabriken Bayer.

Potassium difluorophosphate is prepared analogously by heating KPF_6 with $(KPO_3)_n$ to 400°:

$$KPF_6+2/n(KPO_3)_n \longrightarrow 3KPO_2F_2$$

The melt obtained upon cooling contains 97.5% KPO_2F_2 which may be further purified by recrystallization from 1:1 mixture of isopropanol–water at between 65° and −5°. It has a m.p of 255°±2° and a density of 2.442 g/ml[568].

The crystals of potassium difluorophosphate have an orthorhombic structure ($a=8.03_9$, $b=6.20_5$, $c=7.63_5$). The difluorophosphate anion has the dimensions of P–O = 1.470±0.005Å, P–F = 1.575±0.005Å, ∠FPF = 97.1°±0.4°, ∠OPO = 122.4°±0.4° [569].

Hexafluorophosphoric Acid and Hexafluorophosphates

Commercial concentrated hexafluorophosphoric acid is prepared by reacting P_2O_5 with anhydrous HF in the ratio of 1 to 12:

$$P_2O_5+12HF \longrightarrow 2HPF_6+5H_2O$$

The product contains besides HPF_6 also an equilibrium amount of the products of hydrolysis of HPF_6 which includes mono- and difluorophosphoric acids and HF.

When the solution is cooled, the hexahydrate, $HPF_6 \cdot 6H_2O$, crystallizes out, having a m.p. 31.5° [560]. The anhydrous acid may be prepared by the reaction of PF_5 with HF in low boiling inorganic liquids such as SO_2[559]. On standing at room temperature, it decomposes back to HF and PF_5.

Ammonium and alkali metal hexafluorophosphate can be prepared directly from the reaction of the chloride salt of the cation and PCl_5 in an excess of liquid HF in a silver or aluminum reactor:

$$NaCl+PCl_5+6HF \longrightarrow NaPF_6+6HCl$$

The hexafluorophosphate formed precipitates readily from the anhydrous HF solution[560]. Patent literature has also claimed the preparation of hexafluorophosphate by heating a mixture of alkali metal fluoride, elemental phosphorus and anhydrous HF at 180–220° under autogenous pressure[570]. Dry alkali metal hexafluorophosphates are also formed by heating metal fluorides with PF_5 in a closed system. This is due to the high electronegativity of fluorine which permits PF_5 to combine easily with a fluoride ion to form PF_6^-.

The solubility characteristics of hexafluorophosphates resemble those of the perchlorates. Potassium and tetramethylammonium hexafluorophosphate are only moderately soluble in water (KPF_6 = 3.56 g/100 solution at 0° and 38.3 g/100 g solution at 100°)[559]. The sodium, ammonium and alkaline earth salts, however, are very soluble. These salts undergo hydrolysis even at ice temperature when the concentration approaches saturation[560]. They are, however, stable to dilute alkali. When heated, they dissociate with the liberation of PF_5 and this reaction serves as a convenient method for the laboratory preparation of PF_5.

It is of interest to note that KPF_6 is rated as practically non-toxic; the minimal lethal dose to female albino rats by intraperitoneal injection is 1120 mg/kg[559]. The density of KPF_6 is 2.55 g/ml and the m.p. is 575° (with decomposition).

568 V. B. Tul'chinskii, I. G. Ryss and V. I. Zubov, *Zh. Neorg. Khim.* **11**(12) (1966) 2694–6.
569 R. W. Harrison, R. C. Thompson and J. Trotter, *J. Chem. Soc.* **A**(1966) (12) 1775–80.
570 A. W. Jache and S. Kongpricha, U.S. 3.380,803, 30 April 1968, to Olin Mathieson Chemical Corp.

The PF_6^- anion has the phosphorus atom present as an sp^3d^2 hybrid. The salts MPF_6 where M is K, Cs, or NH_4, crystallize with a NaCl type packing of M^+ and octahedral PF_6^-.

In the crystal structure of $NaPF_6.H_2O$, the PF_6^- octahedral is distorted. The four equatorial fluorine atoms with sodium ions as nearest neighbors have a P–F distance of 1.58 Å. The apical fluorine atoms with water molecules as nearest neighbor have the P–F distance 1.73 Å[571].

20.2. CHLOROPHOSPHOSPHORIC ACID AND CHLOROPHOSPHATES

Dichlorophosphoric Acid and Dichlorophosphates

Chlorophosphoric acids are much more difficult to prepare and isolate than the fluorophosphoric acids. Of the possible chlorophosphoric acids only dichlorophosphoric acid has been prepared. One method of preparation involves the hydrolysis of $Cl_2\overset{O}{\overset{\|}{P}}O\overset{O}{\overset{\|}{P}}Cl_2$ with one mole of water at −60° to −70° [572]:

$$Cl_2\overset{O}{\overset{\|}{P}}O\overset{O}{\overset{\|}{P}}Cl_2 + H_2O \longrightarrow 2HO\overset{O}{\overset{\|}{P}}Cl_2$$

Dichlorophosphoric acid is also formed by the controlled hydrolysis of $POCl_3$. Product thus obtained is usually contaminated with unreacted $POCl_3$ and polyphosphoryl chlorides[573]. Heating of a mixture of $POCl_3$ and H_3PO_4 in a 2 : 1 ratio at 80–150° gave a 60% yield of $HOPOCl_2$ along with condensation products[574].

Dichlorophosphoric acid crystallizes as needles with a m.p. of −18° and $d_4^{25} = 1.686$ g/ml. At room temperature it is a clear mobile liquid stable in the absence of air and soluble in organic solvents. When distilled under reduced pressure, the distillate contains pyrophosphoryl chloride and a little $POCl_3$. Distillation under normal pressures resulted in the loss of HCl and the formation of $POCl_3$ and polymeric condensed chlorophosphoric acids[575].

Further hydrolysis of HPO_2Cl_2 with a little H_2O does not result in the formation of monochlorophoric acid, only H_3PO_4 and $POCl_3$ [576].

Dichlorophosphoric acid precipitates as the nitron and brucine salt at −15° [576]. Metal salts such as the dichlorophosphates of Be, Al, Ga and In, and Fe^{+++} are formed as precipitates when the anhydrous chlorides of the metals are dissolved in an excess of $POCl_3$ and a stream of Cl_2O diluted with inert gas is introduced into the solution. In this reaction, to avoid gel formation, the temperature is allowed to go to about 80°. Infrared spectra indicated $Be(PO_2Cl_2)_2$ to consist of Be^{+2} and $PO_2Cl_2^-$ while the other compounds are of polymeric structure with O bridges[577].

Manganous dichlorophosphate forms a stable adduct with ethyl acetate, $Mn(PO_2Cl_2)_2(CH_3COOC_2H_5)_2$. Its crystal structure indicates that the Mn octahedrally

[571] R. W. G. Wyckoff, *Crystal Structures*, 2nd ed., pp. 570–1, Interscience Publishers, New York (1965).
[572] H. Grunze and E. Thilo, *Angew. Chem.* **70** (1958) 73.
[573] H. Grunze, *Z. anorg. u. allgem. Chem.* **313** (1961) 316–22.
[574] H. Grunze, *Monatsber. Deut. Akad. Wiss, Berlin*, **5**(10) (1963) 636–41.
[575] H. Grunze, *Z. anorg. u. allgem. Chem.* **298** (1958) 152–63.
[576] J. Goubeau and P. Schulz, *Z. anorg. u. allgem. Chem.* **294** (1958) 224–32.
[577] H. Mueller and K. Dehnicke, *Z. anorg. u. allgem. Chem.* **350**(5–6) (1967) 231–6.

coordinated to four O atoms of $PO_2Cl_2^-$ ions and two carbonyl oxygens of the $CH_3COOC_2H_5$ [578].

Miscellaneous Halophosphates

Dibromophosphoric acid cannot be prepared by reactions analogous to those for the preparation of difluorophosphoric acid and dichlorophosphoric acid. It can be isolated as a sparingly soluble nitron salt, $C_{20}H_{16}N.HOPOBr_2$ in 85% yield when a nitron–acetic acid–water solution is added to $POBr_3$ dissolved in acetone[579]. The ethyl acetate adducts of the calcium and magnesium salts, $Ca(PO_2Br_2)_2.2CH_3COOC_2H_5$ and $Mg(PO_2Br_2)_2.2CH_3COOC_2H_5$, are formed as monoclinic needles when the metal oxides are added to a solution of $POBr_3$ in $CH_3COOC_2H_5$ [580].

Dichlorophosphorothioic acid is also isolated as the difficultly soluble nitron complex, $C_{20}H_{16}N_4.HOPSCl_2$, which has a m.p. of 182°. The hydrolysis of $PSCl_3$ to $HOPSCl_2$ is carried out in an acetone solution. In this reaction the PCl bond is split while the PS bond remains intact[581].

21. PEROXYPHOSPHATES AND PHOSPHATE PEROXYHYDRATES

21.1. PEROXYPHOSPHATES

Orthophosphates form permonophosphates and perdiphosphates analogous to the better known persulfates:

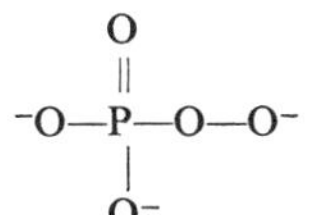

Peroxymonophosphate anion

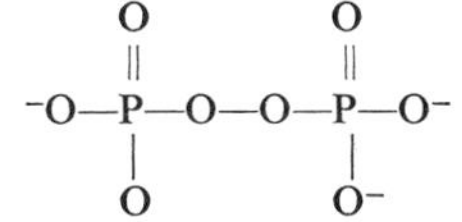

Peroxydiphosphate anion

Perdiphosphates

Potassium perdiphosphate is prepared by the anodic oxidation of potassium orthophosphate in the presence of KF and K_2CrO_4. A typical solution for oxidation contains 302.2 g KH_2PO_4, 198 g KOH (equivalent to a molar composition of $2K_2HPO_4$ plus $3K_3PO_4$), 120 g of KF and 0.355 K_2CrO_4. The solution is electrolyzed at about 14° with platinum electrodes at a current density of less than 0.15 A/cm^2 at the anode. By-product permonophosphate decomposes when the electrolyzed solution is allowed to stand overnight. The yield of crystalline $K_4P_2O_8$ is in the order of 80%[582].

Other metal salts of perdiphosphoric acid such as the sparingly soluble $Ba_2P_2O_8$, $Zn_2P_2O_8$, $Pb_2P_2O_8$ are prepared from an aqueous solution by the metathesis of $K_4P_2O_8$ with $BaCl_2$, $ZnSO_4$ or $Pb(NO_3)_2$ respectively[582].

Lithium perdiphosphate is obtained from the potassium salt as the tetrahydrate, $Li_4P_2O_8.4H_2O$. It is converted to the tetramethylammonium salt by the use of a column

[578] J. Danielsen and S. E. Rasmussen, *Acta. Chem. Scand.* **17**(7) (1963) 1971–9.
[579] H. Grunze and G. U. Wolf, *Z. anorg. u. allgem. Chem.* **329**(1–2) (1964) 56–67.
[580] H. Grunze and K. H. Jost, *Z. Naturforsch.* **20b**(3) (1965) 268.
[581] H. Grunze and M. Meisel, *Z. Naturforsch.* **18b**(8) (1963) 662.
[582] F. Fichter and E. Gutzwiller, *Helv. Chim. Acta* **11** (1928) 323–37.

containing a tenfold excess of the ion exchange resin Dowex 50–X4 charged with tetramethylammonium chloride.

Free perdiphosphoric acid, $H_4P_2O_8$, has not been isolated and characterized. Its third and fourth dissociation constant, however, has been measured at 25° by pH titration of the tetramethylammonium salt with HCl. The values obtained upon extrapolation to infinite dilution are $K_3 = 6.6 \pm 0.3 \times 10^{-6}$, $K_4 = 2.1 \pm 1 \times 10^{-8}$. The first and second dissociation constant could not be measured by this method, but were estimated to be $K_1 = \simeq 2$ and $K_2 \simeq 3 \times 10^{-1}$ [583].

At room temperature, the perdiphosphate anion, $P_2O_8^{4-}$, is quite stable in neutral or basic solutions. In acidic solution, it undergoes rapid hydrolysis to permonophosphoric acid and the rate of hydrolysis increases with increase in hydrogen ion concentration:

$$H_4P_2O_8 + H_2O \longrightarrow H_3PO_5 + H_3PO_4$$

Perdiphosphate anion forms complexes with alkali metal cations as does the pyrophosphate anion. However, with alkaline earth cations such as Mg^{++}, its complex is much weaker than the corresponding pyrophosphate anion complex.

Permonophosphate

Permonophosphoric acid, as indicated above, can be prepared by the hydrolysis of perdiphosphoric acid. It is also produced by the anodic oxidation of the orthophosphate, a procedure in which a larger yield is obtained in an acidic solution. For the manufacture of a crude product containing permonophosphoric acid, patent literature has described alternative procedures which involved the reaction of P_2O_5 with aqueous solution of H_2O_2 [584] or by the heating of 85% H_3PO_4 and P_2O_5 with 95.3% H_2O_2 [585].

The acid ionization constants of permonophosphoric acid have been determined by spectrophotometric technique (UV absorption) and found to be $K_1 = 8 \times 10^{-2}$, $K_2 = 3 \times 10^{-6}$ and $K_3 \simeq 2 \times 10^{-13}$ at 25° or $pK_1 = 1.1$, $pK_2 = 5.5$ and $pK_3 = 12.8$ (the peroxide proton) as compared to H_3PO_4, $pK_1 = 2.1$, $pK_2 = 7.1$ and $pK_3 = 12.3$ [586].

Permonophosphoric acid hydrolyzes in a water solution in the presence of perchloric acid to hydrogen peroxide and phosphoric acid. The hydrolysis is first order in per-acid and the rate increases with added $HClO_4$. At all $HClO_4$ concentration and temperature studied, the hydrolysis follows the rate law:

$$\frac{-d[H_3PO_5]}{dt} = k_1[H_3PO_5]$$

where k_1 is pseudo-first order rate constant dependent on both the temperature and acidity of the solution[586].

Permonophosphoric acid is a strong oxidizing agent and liberates iodine at once from acidified KI solution. Under the same conditions, perdiphosphoric liberates iodine much more slowly and can be kept in such a dilute solution for a relatively long period of time.

Both perdiphosphate and permonophosphate can be determined polarographically with an error of $\leq 2\%$ in a supporting electrolyte of 0.1M H_3PO_4 or H_2SO_4. Permonophosphate yields two reduction waves. It can be determined in an alkaline solution from

583 M. M. Crutchfield and J. O. Edwards, *J. Am. Chem. Soc.* **82** (1960) 3533.

584 E. W. Heiderich, U.S. 2,765,216, 2 Oct. 1956, to E. I. du Pont de Nemours & Co.

585 F. Beer and J. Muller, Ger. 1,096,339, 5 Jan. 1961, to Deutsche Gold and Silber-Scheideanstalt vorm Roessler.

586 C. J. Battaglia and J. V. Edwards, *Inorg. Chem.* **4** (1965) 552–8.

the first reduction wave (about −0.2 V) even in the presence of $P_2O_8^{4-}$ since the latter is not reduced in an alkaline solution. In 0.1 or N KOH, the first reduction wave of $PO_5^{\equiv}$ (0.07–0.8 millimole/l) will tolerate a tenfold amount of $P_2O_8^{4-}$ (587).

One suggested use for perphosphoric acid is for the detoxification of hydrogenation catalysts which have been poisoned by small concentration of CS_2 or other sulfides[588].

21.2. PEROXYHYDRATES OF ORTHO-, PYRO-, TRIPOLY AND TRIMETAPHOSPHATES

Hydrogen peroxide forms a number of peroxyhydrates with alkaline or neutral ortho-, pyro-, tripoly or trimetaphosphates. These compounds are analogous to the hydrates with H_2O_2 replacing all or part of the H_2O in the hydrates. The general method of synthesis involves the solution of the anhydrous salt in an aqueous solution of selected concentration of H_2O_2.

Peroxyhydrates of Orthophosphates

In the Na_3PO_4–H_2O_2–H_2O system at 0°, solution concentrations of 5.4% H_2O_2 and 4.9% Na_3PO_4 are in equilibrium with two solid phases: $Na_3PO_4.12H_2O$ and $Na_3PO_4.H_2O_2$. At solution concentration of 26.7–28.0% H_2O_2 and 15.4–16.7% Na_3PO_4, the two solid phases are $Na_3PO_4.H_2O_2$ and $Na_3PO_4{\cdot}4.5H_2O_2$. When the concentration of H_2O_2 is $\geqq$ 45.2%, decomposition occurs. At −20°, the system in the range of concentrations studied above, only $Na_3PO_4.5H_2O_2$ is found in the solid phase. Decomposition occurs when the concentration of H_2O_2 is $\geqq$ 56.9%[589].

The monoperoxyhydrate, $Na_3PO_4.H_2O_2$, has a decomposition temperature of 70°, while $Na_3PO_4{\cdot}4.5H_2O_2$ has a m.p. of 65° and a decomposition temperature of about 120°[589].

Mixed peroxyhydrate–hydrates of Na_3PO_4 are formed when solutions of the proper $H_2O_2 : H_2O$ ratio are allowed to react with anhydrous Na_3PO_4 suspended in an organic solvent such as $CHCl_3$ at 15–20°. Compounds such as $Na_3PO_4.8H_2O.2H_2O_2$ and $Na_3PO_4.4H_2O_2.2H_2O$ are thus prepared[590]. The standard heat of formation of $Na_3PO_4.4H_2O_2.2H_2O(s)$ is -815.08 ± 0.52 kcal/mole[591].

For the system Na_2HPO_4–H_2O_2–H_2O at 0°, solution concentrations of 27.0% H_2O_2 and 42.2% Na_2HPO_4 are in equilibrium with the two solid phases, $Na_2HPO_4.12H_2O$ and $Na_2HPO_4{\cdot}1.5H_2O_2$. When the H_2O_2 concentration is $\geqq$ 53.6%, the solid peroxyhydrate obtained is $Na_2HPO_4{\cdot}2.5H_2O_2$ [589]. The decomposition temperature of $Na_2HPO_4.1.5H_2O_2$ is 110°.

Peroxyhydrate of Pyrophosphates

Pyrophosphates also form peroxyhydrates. Compounds such as $Na_4P_2O_7.3H_2O_2$, $K_4P_2O_7.3H_2O_2$, and $(NH_4)_2H_2P_2O_7.H_2O_2.2H_2O$ are obtained as crystalline products. In the system $Na_4P_2O_7$–H_2O_2–H_2O at 0°, the phase boundary $Na_4P_2O_7.10H_2O$–$Na_4P_2O_7.6H_2O_2$ occurs at about the liquid phase concentration region of 32.7–52.1% H_2O_2 and 37.0–33.3%

587 A. Vaskelis, *Lietuvos TSR Moklu Akad. Darbai*, Ser. B, No. 4 (1962) 41–52.

588 J. R. van Wazer, *Phosphorus and Its Compounds*, Vol. I, pp. 821–4, Interscience Publishers Inc., New York (1958).

589 E. A. Ukraintseva, *Izv. Sib. Otd., Akad. Nauk SSSR, Ser. Khim. Nauk* (1963) (1) 14–24.

590 V. Habernickel, Ger. 1,066,190, 1 Oct. 1959, to Henkel & Cie. G.m.b.H.

591 E. A. Ukraintseva, *Izv. Sib. Otd. Akad. Nauk SSSR, Ser. Khim. Nauk* (1966) (2) 153–5.

$Na_4P_2O_7$. The compound $Na_4P_2O_7.6H_2O_2$ obtained melts at 59° and decomposes at 75° [589].

The peroxyhydrates of $Na_4P_2O_7$, containing 1 to 3 H_2O_2, have been promoted as bleaching agents. They can be prepared by mixing the anhydrous powdered $Na_4P_2O_7$ with the proper amount of 45–75% H_2O_2 at below 60° and the paste obtained dried at 100° to a solid which is then crushed to a fine powder[592–94]. The standard heat of formation of $Na_4P_2O_7.3H_2O_2(s) = -913.93 \pm 0.08$ kcal/mole[591].

Peroxyhydrates of Tripolyphosphates and Trimetaphosphate

When sodium tripolyphosphate reacts with H_2O_2 and water in the mole ratio of 1 : 1 : 5 at $< 40°$ in an inert organic suspending agent such as benzene, crystalline $Na_5P_3O_{10}.H_2O_2.5H_2O$ is formed. This compound has been suggested as an additive to bleaches to be used with detergents[595].

The peroxyhydrate of sodium trimetaphosphate is $Na_3P_3O_9.H_2O_2$. Magnesium silicate or the sodium salt of ethylenediamine tetraacetic acid have been recommended as stabilizers for peroxyhydrates[596].

General Properties of Peroxyhydrates. All peroxyhydrates liberate H_2O_2 when dissolved in water. The difference between the peroxyhydrate salt and the corresponding nonperoxy salts can be shown by EPR spectra. When these compounds are irradiated with UV light at −196°, the peroxyhydrate gives a signal of the radical HO_2. Owing to the stabilizing effect of the lattice, the HO_2 radicals are more stable in these compounds than in H_2O_2. They have a life of about 30 min at −80° but decompose quickly at 20°.

22. THIOPHOSPHATES (PHOSPHOROTHIOATES)

The oxygen atoms in oxyphosphorus compounds may be replaced from one to all by sulfur atoms. However, the methods of synthesis for the thiophosphorus compounds do not depend on this direct replacement reaction. An exception is that of the replacement of the oxygen in P=O with sulfur from P=S in the preparation of organophosphonothionic chlorides, as for example[597]:

$$5ClCH_2\overset{\overset{\large O}{\|}}{P}Cl_2 + P_2S_5 \longrightarrow 5ClCH_2\overset{\overset{\large S}{\|}}{P}Cl_2 + P_2O_5$$

$$ClCH_2\overset{\overset{\large O}{\|}}{P}Cl_2 + PSCl_3 \longrightarrow ClCH_2\overset{\overset{\large S}{\|}}{P}Cl_2 + POCl_3$$

Inorganic thiophosphates are in general prepared from sulfur-containing phosphorus compounds. They are relatively unstable in water, hydrolyzing to the corresponding oxy-compound with liberation of hydrogen sulfide. Hydrolysis is more rapid in an acid solution, especially with compounds having several sulfur atoms attached to the same phosphorus atom.

592 R. T. Russell, Brit. 990,172, 28 April 1965, to Albright and Wilson.
593 A. V. Yanush and A. S. Farberov, *Khim. Prom. Ukr.* (5) (1966) 5–6.
594 T. Kawasaki, Japan 22,553, 24 Oct. 1963, to Takeda Chemical Industries Ltd.
595 V. Habernickel, Ger. 1,048,265, 8 Jan. 1959, to Henkel & Cie. G.m.b.H.
596 W. Burger and J. Heidelmann, Ger. 1,065,124, 10 Sept. 1959, to Kraemer and Flammer K.-G.
597 E. Uhing, K. Rattenbury and A. D. F. Toy, *J. Am. Chem. Soc.* **83** (1961) 2299–2303.

Tetrathiophosphates

Thio(ortho)phosphates, M_3PS_4, can be prepared by heating the metal, phosphorus and sulfur together in a sealed tube. Alternatively, they can be formed by the reaction of the metal phosphide with sulfur. Thus PBS_4 is obtained by heating a stoichiometric mixture of boron, phosphorus and sulfur or by heating a mixture of 1 mole of boron phosphide and 4 moles of sulfur in the absence of water vapor and oxygen.

An orthorhombic modification of BPS_4 is formed at 450–500°. It consists of one-dimensional infinite chain molecules with lattice dimensions $a = 5.60 \pm 0.06$, $b = 5.25 \pm 0.06$ and $c = 9.04 \pm 0.06$ Å and interatomic distances B–S 1.89 Å, P–S 2.16 Å and B–P 2.63 Å. It is a hygroscopic, colorless, crystalline substance which decomposes in air with evolution of H_2S. Another modification is the monoclinic form which is produced at 650–700°. It crystallizes in brownish plates which are fairly stable in air. It has unit cell dimensions of $a = 10.38$, $b = 6.05$, $c = 6.69$ Å, and $\beta = 75°$. The orthorhombic form can be converted to the monoclinic form by annealing at 650° [598].

Another example of MPS_4 is $AlPS_4$, prepared by heating AlP in a sulfur atmosphere at 650°. It is a hygroscopic solid, decomposing in air with evolution of H_2S. Its structure consists of Al and P atoms surrounded by sulfur atoms[599]:

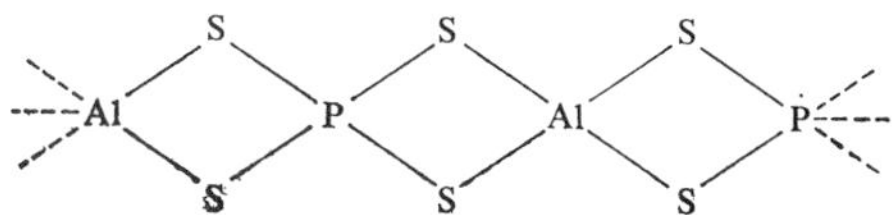

Another example, the sodium salt of tetrathiophosphoric acid, $Na_3PS_4 . 8H_2O$, is formed by the addition of P_4S_{10} to a saturated solution of Na_2S.

The free acid, tetrathiophosphoric acid, H_3PS_4, is unstable even at below room temperature. Schmidt and Wieber[600] claimed to have prepared it in an aqueous solution by the reaction of Na_3PS_4 with HCl and the compound stabilized by sudden chilling in acetone cooled to −78°. At 0°, H_3PS_4 decomposes with evolution of H_2S into H_3PO_3S[601].

Trithiophosphates

The action of P_2S_5 on a solution of NaOH saturated with H_2S at 20° produces a mixture of sodium thiophosphates. Sodium trithiophosphate, $Na_3PS_3O . 11H_2O$, is obtained by dissolving the above reaction mixture in a 10% Na_2S solution and fractionally crystallized at 4° [601, 602]. The crystalline product obtained has an orthorhombic lattice with 4 mols per unit cell and lattice parameters of 12.40, 14.0 and 9.02 Å. Measurements by infrared indicate that the $P \rightarrow O$ fundamentals occur at ~ 1000 cm^{-1} and the $P \rightarrow S$ probably at ~ 550 cm^{-1} [603].

The barium salts of trithiophosphoric acid, $Ba(POS_3)_2 . 6H_2O$, and other insoluble trithiophosphates such as the lead or silver salts are obtained by the reaction of the sodium salts with the corresponding metal cations[601].

598 A. Weiss and H. Schaefer, *Z. Naturforsch.* **18b**, No. 1 (1963) 81–82.
599 A. Weiss and H. Schafer, *Naturwissenschaften*, **47** (1960) 495.
600 M. Schmidt and M. Wieber, *Z. anorg. u. allgem. Chem.* **326** (3–4) (1963) 182–5
601 R. Klement, *Z. anorg. Chem.* **253** (1947) 237–48.
602 P. Tribodet, *Rev. Chim. Minerale*, **2**(2) (1965) 321–5.
603 G. Tribot and P. Tribodet, *Ind. Chim. (Paris)* **52**(579) (1965) 291–9; (581) 369–78.

Dithiophosphates

Trisodium dithiophosphate, $Na_3PS_2O_2.11H_2O$, is prepared by the partial hydrolysis of the mixed tri- and dithiophosphates obtained by the reaction of P_2S_5 in aqueous NaOH solution. This hydrolysis is carried out at 36° for 1 hr. The dodecahydrate salt melts at 48° in its water of crystallization. It is converted to the decahydrate $Na_3PS_2O_2.10H_2O$ at 58° and to the anhydrous form by drying *in vacuo* at 150° in a nitrogen atmosphere[602].

Disodium dithiophosphate, $Na_2HPS_2O_2.H_2O$, is obtained by the acidification of $Na_3PS_2O_2.11H_2O$ followed by drying *in vacuo* at 2–3°. Tribodet claimed that when this compound is heated at 72–137°, it converts to the tetrasodium trithiopyrophosphate, $Na_4P_2S_3O_4.2H_2O$, with evolution of H_2S. Further heating to 200–260° results in the formation of $Na_4P_2SO_6$ [602]

The two sulfur atoms in dithiophosphate are shown by iodine oxidation in an alkaline solution to be bonded differently to the phosphorus atom. One of the bonds is semi-polar. This suggests that the structure for $Na_2HPS_2O_2$ is $(NaO)_2(HS)P \rightarrow S$[603].

Monothiophosphates

The monothiophosphate is the most stable of the thiophosphates to hydrolysis. Trisodium monothiophosphate is obtained by the hydrolysis of $PSCl_3$, in a sodium hydroxide solution at 103–109°:

$$PSCl_3+6NaOH \longrightarrow Na_3PO_3S+3NaCl+3H_2O$$

It is separated from by-product NaCl as $Na_3PO_3S.12H_2O$ by fractional precipitation from an aqueous solution with anhydrous methanol. The anhydrous salt obtained by stirring the hydrate in anhydrous methanol is a white crystalline solid[604].

An alternative procedure for the preparation of anhydrous sodium monothiophosphate is by the reaction of Na_2CS_3 with $(NaPO_3)_3$. The product forms rapidly in a yield of 84–90% when the reaction is carried out at 550° under nitrogen atmosphere or *in vacuo*. Thermogravimetric studies indicated that the actual reaction is between $(NaPO_3)_3$ and Na_2S resulting from the thermal decomposition of Na_2CS_3 [605].

Anhydrous Na_3PO_3S is very stable at room temperature and not affected by moisture up to a relative humidity of 31%. It is quite soluble in water, dissolving 79.2 g per liter of saturated solution at 18°. The 12-hydrate liberates H_2S slowly at room temperature and decomposes rapidly at 60°.

Aqueous solution of the sodium salt decomposes only very slowly when stabilized with added sodium carbonate. It has been reported, however, that the rate of hydrolysis in strongly alkaline solution is increased from 2.5×10^{-4} to 2.3×10^{-3} min^{-1} upon addition of 30 mg silicone grease per 100 ml of the solution [606]. In strongly acidic solution, titration with iodine leads to the formation of the disulfide $HO_3PSSPO_3H^{=}$[604]. Infrared studies of the $PO_3S^{\equiv}$ ion from several salts give a frequency of 438 cm^{-1} for γ_{P-S}. The force constant for the P–S bond is 2.68 millidynes/A. It is reported that the resonance effects in $PO_3S^{\equiv}$ involves largely the P–O bonds[607].

The disodium monothiophosphate, $Na_2HPO_3S\cdot6.5H_2O$, is prepared by acidifying

[604] S. K. Yasuda and J. Lambert, *Inorganic Synthesis*, Vol. V, pp. 102–4, Ed. T. Moeller, McGraw-Hill Book Co., New York (1957).

[605] A. Lamotte, M. Porthault and J. C. Merlin, *Bull. Soc. Chim. France* (1965) (4) 915–19.

[606] S. Akerfeldt, *Nature*, **200**(4909) (1963) 881–2.

[607] E. Steger and K. Martin, *Z. anorg. u. allgem. Chem.* **308** (1961) 330–6.

an aqueous solution of $Na_3PO_3S.12H_2O$ with one mole of conc. HCl at < 10°. The product is precipitated in the cold by the addition of ethanol. It decomposes with evolution of H_2S upon standing in air or when heated to 40° [608].

Potassium monothiophosphate is prepared by passing a solution of $Na_3PO_3S.12H_2O$ through the H^+ form of an ion exchange resin and the eluate adjusted immediately to pH 4.5 with 2N KOH. All of the operations are carried out under ice cooling. After dilution with CH_3OH and H_2O and allowing to stand at −10°, KH_2PO_3S precipitates[609]. Other salts, M_2HPO_3S, are similarly prepared.

Monothiophosphoric acid (phosphorothioic acid), H_3PO_3S, is the sole product left when $Ba_3(PS_4)_2$ is acidified with H_2SO_4, the $BaSO_4$ precipitate formed removed by filtration and the filtrate allowed to stand at 0°. Under these conditions, all higher thio acids which formed also in the original reaction decomposed with the evolution of H_2S. Monothiophosphoric acid decomposes only to an extent of 10% to H_3PO_4 and H_2S when it is allowed to stand at −2° for a period of 18 weeks[601]. In a more precise study, Dittmer and Ramsay[610] showed that the hydrolysis of H_3PO_3S in an aqueous solution at constant ionic strength has a maximum rate at pH ~3.0 and pH ~8.0. The minimum rate is at pH ~7.0 and pH ~0.30.

As indicated earlier, titration of strongly acidified Na_3PO_3S with iodine resulted in the formation of $HO_3PSSPO_3H^=$. Similarly H_3PO_3S is oxidized to $HO_3PSSPO_3H^=$ by ferricyanide in a mole to mole ratio. The kinetics of the reduction of ferricyanide ion by phosphorothioate are found to be 2nd order with respect to the reactants in the pH range of 5.8 to 7.8 at the temperature range of 15° to 40° and the ionic strength $\mu = 0.035$ to 0.06. The 2nd order rate constant for the reduction of ferricyanide ion by P–S– shows a maximum at pH 6.2 at 26° and ionic strength $\mu = 0.06$.[611]

The industrial application of phosphorothioates are limited. A solution of a mixture of sodium thiophosphates along with sodium sulfide is used as a depressant in mining operations for the separation of molybdenite from copper sulfide ores by flotation. Such a product is prepared by the reaction of P_2S_5 with an aqueous solution of sodium hydroxide. It is known as Nokes reagent, after one of the inventors[612].

608 M. Schmidt and R. R. Waegerle, *Naturwissenschaften*, **50**(21) (1963) 662–3.
609 G. Ladwig, *J. prakt. Chem.* **27**(3–4) (1965) 117–32.
610 D. C. Dittmer and O. B. Ramsay, *J. Org. Chem.* **28** (1963) 1268–72.
611 H. Neumann, I. Z. Steinberg and E. Katchalski, *J. Am. Chem. Soc.* **87**(17) (1965) 3841–8.
612 C. M. Nokes, C. G. Quigley and R. T. Pring, U.S. 2,492,936, 27 Dec. 1949.

21. ARSENIC, ANTIMONY AND BISMUTH

J. D. Smith
University of Sussex

The elements of Group V are sometimes known as pnictides. The chemistry of nitrogen and phosphorus has been worked out in great detail: the remaining elements—arsenic, antimony and bismuth—have been less thoroughly studied. As they form a closely related group, they are considered together. Interest centres on the trend from non-metallic to metallic properties with increasing atomic weight. Thus there are many parallels between phosphorus and arsenic, but considerably fewer between phosphorus and bismuth, which is a typical B metal like tin or lead. Arsenic and antimony are important largely because of their intermediate or metalloid character.

In the following sections the Group V element is given the symbol E. Halogen atoms are denoted by X and other electro-negative atoms or groups by Y or Z. Electro-positive (metallic) atoms are denoted by the symbol M.

1. THE ELEMENTS

1.1. ARSENIC

The yellow pigment orpiment (As_2S_3) and red realgar (As_4S_4) were known to Greek alchemists, who probably also succeeded in isolating the metal. The German Dominican scholar Albertus Magnus (1193–1280) described the preparation of arsenic by heating orpiment with soap, but the alchemists probably thought of the metal as a kind of mercury. The relation between As^{III} oxide and metallic arsenic, and the similarity between arsenic and tin, were established in the early eighteenth century. Arsenic compounds were also known in early Chinese civilizations both as pigments and poisons, and were used to kill mice and insects. The chief uses have remained as insecticides and weed-killers and in alloys with lead and copper. More recently, arsenic compounds have found important applications in semiconductor devices.

Arsenic minerals are widely distributed in many countries of the world: a few examples follow. The commonest minerals are those, e.g. mispickel, in which arsenic is combined both with iron or nickel and with sulphur.

Sulphides: realgar (As_4S_4); orpiment (As_2S_3)
Oxides: claudetite or arsenolite (As_2O_3)
Arsenides: mispickel (FeAsS); nickel glance (NiAsS); chloanthite ($NiAs_2$); löllingite ($FeAs_2$); niccolite (NiAs)
Arsenates: pharmacolite ($CaHAsO_4,2H_2O$); erythrite ($Co_3(AsO_4)_2,8H_2O$)

Small amounts of native arsenic are found. Traces occur widely in soil, mineral water and in the human body.

Extraction

Ores such as mispickel may be heated in the absence of air, when arsenic sublimes:

$$FeAsS \rightarrow FeS + As$$

Arsenic trapped in the residue may be released by roasting in air, when arsenic trioxide (As_2O_3) volatilizes and condenses in the flue system. It may be purified by further sublimation. The oxide may be obtained directly from arsenic-containing ores by roasting them in air, and also as a by-product in the smelting of copper or lead. It may be converted to a wide range of arsenic compounds (section 6.2), or heated with coke, when it is reduced to the metal.

Arsenic, obtained by one of the processes described above, is contaminated by oxide and, perhaps, by sulphide. Further purification can be effected by resubliming from more charcoal, and, finally, by subliming in hydrogen. The most difficult impurities to remove are antimony and Group VI elements such as sulphur. Since ultra-pure arsenic (impurities less than 1 in 10^6) is an important commercial material in the semiconductor industry, considerable effort has been put into its purification[1]. Some of the more successful methods appear to be (a) by sublimation from solutions in lead (much of the sulphur remains as lead sulphide), (b) by crystallization from molten arsenic at high pressures, and (c) by conversion to arsine, which can be scrubbed by alkali and dried and then pyrolysed at about 600°C to the elements.

Allotropes and Structures

Relationships between the various modifications of arsenic are shown in Scheme 1. The stable form at 25°C is grey or metallic arsenic, which has a rhombohedral structure (*Strukturbericht* symbol A7)[3]. Atoms are linked into sheets which are fitted together so

SCHEME 1. Allotropes of arsenic[2].

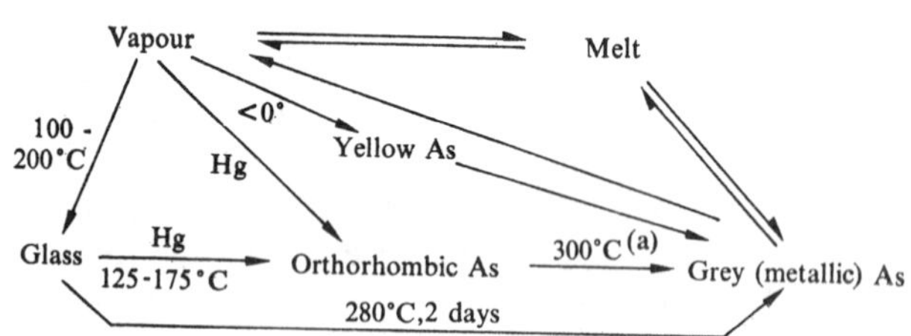

[a] At much lower temperature (*ca.* 200°) in the presence of mercury.

that each atom has six neighbours (Fig. 1). Three on one side are at a distance r_1, and three on the other side are slightly further away at a distance r_2. Similar structures are adopted by antimony and bismuth: as the atomic weight increases, the ratio r_2/r_1 becomes more

[1] R. K. Willardson and H. L. Goering (eds.), *Compound Semiconductors*, Vol. I, *Preparation of III–V Compounds*, Reinhold, New York (1962).

[2] H. Stöhr, *Z. anorg. allgem. Chem.* **242** (1939) 138; H. Krebs, W. Holz and K. H. Worms, *Chem. Ber.* **90** (1957) 1031; H. Krebs and R. Steffen, *Z. anorg. allgem. Chem.* **327** (1964) 224.

[3] W. B. Pearson, *A Handbook of Lattice Spacings and Structures of Metals and Alloys*, Pergamon, Vol. 1 (1958), Vol. 2 (1967).

nearly unity (Table 1), showing the tendency towards higher co-ordination numbers in the more metallic elements. Arsenic vapour at 1 atm consists mainly of tetramers As_4 at 800°C, and dimers As_2 at 1750°C. The As_4 molecules comprise tetrahedra of atoms (cf. P_4) and the As–As distance is 243.5±0.4 pm [4].

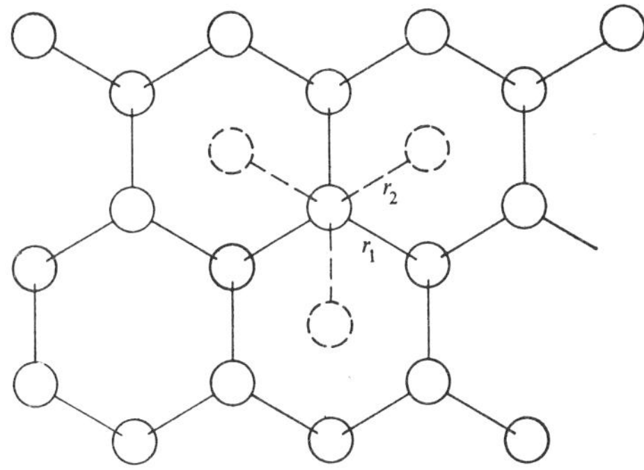

FIG. 1. Plan of a sheet of atoms in rhombohedral arsenic. Atoms in the adjacent sheet (dotted) are placed over holes in the sheet beneath. Each atom has six neighbours, three at a distance r_1 and three at r_2.

TABLE 1. NEAREST NEIGHBOURS IN RHOMBOHEDRAL ARSENIC, ANTIMONY AND BISMUTH

	r_1 (pm)	r_2 (pm)	r_2/r_1
As (295.6 K)	250.4	313.6	1.25 [a]
Sb (298 K)	290.8	335.5	1.154 [b]
Bi (298 K)	307.1	352.9	1.149 [c]

[a] From data by J. B. Taylor, S. L. Bennett, and R. D. Heyding, *J. Phys. Chem. Solids* **26** (1965) 69 and in ref. 3.
[b] C. S. Barrett, P. Cucka and K. Haefner, *Acta Cryst.* **16** (1963) 451.
[c] P. Cucka and C. S. Barrett, *Acta Cryst.* **15** (1962) 865.

Yellow arsenic, obtained by sudden cooling of arsenic vapour, is assumed to consist of As_4 molecules. It can be recrystallized from carbon disulphide (solubility: 11 g in 100 ml at 46°C; 4 g at 0°C; 0.8 g at −80°C) but is rapidly converted to grey arsenic in light, and very easily oxidized[5]. Above 300°C, arsenic vapour condenses as grey arsenic, but, at 100–200°C, glassy modifications result. These presumably consist of disordered cross-linked chains formed at temperatures low enough to prevent ready reorganization to the sheets of the crystalline grey form. The amorphous arsenic has a lower density (4730–5180 kg m^{-3}) and higher diamagnetic magnetic susceptibility (-2.89×10^{-10} m^3 mol^{-1}) than the rhombohedral form (see Table 5). The shortest As–As distance is 249 pm.

[4] Y. Morino, T. Ukaji and T. Ito, *Bull. Chem. Soc. Japan* **39** (1966) 64.
[5] G. Brauer, *Handbook of Preparative Inorganic Chemistry*, Academic Press (1963).

A third crystalline modification, made from the vapour in the presence of mercury, is orthorhombic and analogous to black phosphorus, with which it forms mixed crystals. Orthorhombic arsenic consists of chains of atoms linked into double layers: each atom has two nearest neighbours and one slightly further away. The bonds between layers are weaker than in grey arsenic.

TABLE 2. ISOTOPES OF ARSENIC ($Z = 33$)

Isotope	Half-life	Decay mode[a]	Major radiation (MeV)[b]	Production
^{69}As	15 m	β^+		^{70}Ge(p, 2n)
^{70}As	52 m	β^+		^{70}Ge(p, n); ^{70}Ge(d, 2n)
^{71}As	62 h	EC (70%) β^+ (30%)		^{69}Ga(α, 2n); ^{70}Ge(d, n)
^{72}As	26 h	β^+ (78%) EC (22%)	β^+: 2.50 (56%) γ: 0.835 (78%)	^{69}Ga(α, n)
^{73}As	76 d	EC		^{73}Ge(d, 2n); ^{73}Ge(p, n)
^{74}As	18 d	β^+ (29%) β^- (32%) EC (39%)	β^+: 0.91 (26%) γ: 0.596 (60%)	^{71}Ga(a, n); ^{74}Ge(d, 2n)
^{75}As [c]	stable	Natural abundance 100%; mass 74.9216		
^{76}As [c]	26.5 h	β^-	β^-: 2.41 (30%); 2.97 (56%) γ: 0.56 (44%)	^{75}As(n, γ)
^{77}As	39 h	β^-	β^-: 0.68 (94%)	^{76}Ge(n, γ)^{77}Ge(β^-)
^{78}As	91 m	β^-		^{81}Br(n, α); ^{78}Se(n, p)
^{79}As	9 m	β^-		^{82}Se(n, α)^{79}Ge(β^-) ^{80}Se(n, pn); ^{80}Se(γ, p)
^{80}As	15 s	β^-		^{80}Se(n, p)
^{81}As	33 s	β^-		^{82}Se(n, pn); ^{82}Se(γ, p)

[a] β^+, positron emission; β^- electron emission; EC, electron capture.
[b] Only those with intensities greater than 20% of initial disintegrations are given. All β^+ emitters give 0.511 MeV γ-radiation. For complete list, including radiations from shorter lived isotopes, see ref. 6.
[c] Other nuclear properties are as follows:

	^{75}As	^{76}As
Spin	3/2	2
Magnetic dipole moment (nuclear magnetons)	1.435	−0.903
Electric quadrupole moment × 10^{28} (m^2)	0.27	
N.m.r. frequency for 1 T field (MHz)	7.292	

Isotopes

Isotopes of arsenic[6] are described in Table 2. Only one isotope, ^{75}As, is stable. This has a spin of 3/2, which means that useful information can be obtained from n.m.r. The sensitivity, compared with the proton at the same field, is 0.025,[7] so successful ^{75}As spectra can be

[6] C. M. Lederer, J. M. Hollander and I. Perlman, *Table of Isotopes*, 6th edn., Wiley (1957); B. J. Wilson (ed.), *Radiochemical Manual*, Radiochemical Centre, Amersham (1966).

[7] Data of Varian Associates given by J. W. Emsley, J. Feeney and L. H. Sutcliffe, *High Resolution NMR Spectroscopy*, Pergamon (1966).

obtained only from concentrated solutions. Coupling constants involving ^{75}As can sometimes be derived from measurements on resonances of other magnetic nuclei. Best spectra are given by symmetrical species, e.g. AsF_6^-, since in these quadrupole relaxation effects are minimal. The quadrupole moment of ^{75}As is utilized in nuclear quadrupole spectroscopy. (Resonances are in the range 60–120 MHz.) ^{72}As and ^{74}As are useful positron sources, and ^{76}As and ^{77}As are β-emitters suitable for tracer experiments and activation analysis (as little as 10^{-10} g can be detected). The high energy β-particles from ^{76}As can readily be distinguished from other radiations.

1.2. ANTIMONY

The sulphide stibnite, Sb_2S_3, was mentioned in a papyrus from the sixteenth century BC; it was a constituent of the eye-paint used by oriental women. The Greeks and Chaldeans knew that "lead"—presumably antimony—could be obtained from stibnite, and in a vase dated 3000 BC and now in the Louvre there is antimony which is almost pure. The alchemists made extensive studies of antimony compounds, but they, too, did not distinguish between antimony and lead. Antimony was later used in gold refining, as it combines with base metals to form dross, and alloys with tin were used for type-metal. This use continues today. Antimony compounds as medicants were described in a *Treatise on Antimony* published by Nicholas Lémery in 1707.

Occurrence and Extraction

Stibnite is the most important ore. It occurs in large quantities in China and in parts of South America and Europe. It appears to be a hydrothermal deposit and often is found in veins of granite. Other sulphide ores are ullmannite NiSbS, livingstonite ($HgSb_4S_8$), tetrahedrite (Cu_3SbS_3), wolfsbergite ($CuSbS_2$) and jamesonite ($FePb_4Sb_6S_{14}$). Very small amounts of antimony are found native and small amounts occur as the oxides valentinite and senarmontite (Sb_4O_6), cervantite (Sb_2O_4) and stibiconite (Sb_2O_4,H_2O).

Rich stibnite ores are heated, when the sulphide melts and separates from other rock. The metal is obtained by heating the sulphide with iron:

$$Sb_2S_3 + 3Fe \rightarrow 2Sb + 3FeS$$

Poorer ores are roasted in air, when the oxide Sb_4O_6 sublimes. It may be collected and reduced to the metal by heating with charcoal or anthracite, using an alkali metal salt, e.g. carbonate or sulphate, as a flux. Antimony obtained by these methods is up to 99% pure.

Metal of 99.999% purity is available commercially[1]. Antimony from a smelting process is converted to a crystallizable derivative, e.g. $2HSbCl_6,9H_2O$. After purification, this may be hydrolysed to the oxide, and the metal made by reduction in a stream of pure hydrogen. Further purification is achieved by zone melting, which removes nickel, lead, silver and copper, or by vacuum distillation. The most difficult impurities to remove are those which form azeotropes or solid solutions with antimony, e.g. arsenic, tin and bismuth. Arsenic may be removed by zone refining in the presence of small quantities (*ca.* 0.1%) of aluminium. The ternary Al–As–Sb eutectic is lower melting than antimony, and so arsenic may be moved in the liquid zone to one end of an ingot.

Allotropy[8]

The stable form of antimony at 25°C has the same rhombohedral structure as arsenic and bismuth (see Fig. 1, p. 549). The lattice parameters and interatomic distances have been determined with great accuracy (see Table 1). At high pressure (5×10^9 N m^{-2}), rhombohedral antimony transforms to a primitive cubic lattice in which each atom is surrounded by six others at a distance of 296 pm. At still higher pressures, the metal becomes hexagonal-close-packed with an interatomic distance of 328 pm. These changes represent the transition from a metalloid structure, with co-ordination number of 3(+3), to a fully metallic structure with a co-ordination number of 12. Attempts to prepare modifications corresponding to yellow or orthorhombic arsenic do not seem to have been successful, but several amorphous forms of antimony have been made. "Yellow" and "black" antimony, from liquid stibine and oxygen at low temperatures, appear to contain some hydrogen, and "explosive antimony", made by the electrolysis of antimony trichloride in hydrochloric acid, always contains residual chlorine. Both these forms are more reactive chemically than rhombohedral antimony and are rapidly and exothermically transformed into the stable form, often as a result of slight heating or mechanical stress.

Isotopes

About thirty isotopes are known[6], and the most important are described in Table 3. Two, ^{121}Sb and ^{123}Sb, occur naturally, and both n.m.r. and n.q.r. measurements are

TABLE 3. ISOTOPES OF ANTIMONY ($Z = 51$)

Isotope[a]	Half-life	Decay[b]	Production
^{112}Sb	0.9 m	β^+, EC	^{112}Sn(p, n)
^{113}Sb	6.4 m	EC, β^+	^{112}Sn(d, n); ^{114}Sn(p, 2n)
^{114}Sb	3.3 m	β^+, EC	^{114}Sn(p, n); ^{115}Sn(p, 2n)
^{115}Sb	31 m	EC (67%), β^+ (33%)	^{114}Sn(d, n); ^{116}Sn(p, 2n); ^{113}In(α, 2n)
^{116}Sb	16 m	EC (72%), β^+ (28%)	Daughter ^{116}Te; ^{115}In(α, 3n)
^{117}Sb	2.8 h	EC (97%), β^+ (3%)	^{115}In(α, 2n)
^{118}Sb	3.5 m	EC, β^+	Daughter ^{118}Te; ^{115}In(α, n)
^{119}Sb	38 h	EC	^{121}Sb(p, 3n)^{119}Te(EC); ^{119}Sn(p, n); ^{118}Sn(d, n)
^{120}Sb	5.8 d	EC	^{119}Sn(d, n); ^{120}Sn(d, 2n)
^{121}Sb [d]	Stable	Abundance 57.25%; mass 120.9038	
^{122}Sb [c, d]	2.8 d	β^- (97%), EC (3%)	^{121}Sb(n, γ)
^{123}Sb [d]	> 10^{16} y	Abundance 42.75%; mass 122.9041	
^{124}Sb [c]	60.4 d	β^-	^{123}Sb(n, γ)
^{125}Sb [c]	2.71 y	β^-	^{124}Sn(n, γ)^{125m}Sn(β^-)

[a] Several metastable nuclear states are also known. In addition, the following isotopes are found in fission products; all decay by β^--emission (half-lives in parentheses). ^{126}Sb (12.5 d; 19.0 m); ^{127}Sb (93 h); ^{128}Sb (10.8 m; 8.6 h); ^{129}Sb (4.3 h); ^{130}Sb (33 m; 7.1 m); ^{131}Sb (26 m); ^{132}Sb (2.1 m); ^{133}Sb (4.2 m).

[b] β^+, positron emission; β^-, electron emission; EC, electron capture.

[c] Radiations with intensities greater than 20% of initial disintegrations (MeV). ^{122}Sb: β^-, 1.42 (63%), 1.99 (30%); γ, 0.57 (66%). ^{124}Sb: β^-, 0.61 (51%), 2.31 (23%); γ, 0.60 (98%), 1.69 (48%). ^{125}Sb: β^-, 0.124 (30.7%), 0.295 (40%); γ, 0.427 (31%).

[d] Other nuclear properties (refs. 6 and 7):

[8] S. S. Kabalkina and V. P. Mylov, *Soviet Physics Doklady* **8** (1964) 917; H. Krebs, F. Schultze-Gebhardt and R. Thees, *Z. anorg. allgem. Chem.* **282** (1955) 177.

TABLE 3 (*cont.*)

	^{121}Sb	^{122}Sb	^{123}Sb
Spin	5/2	2	7/2
Magnetic dipole moment (nuclear magnetons)	3.342	−1.904	+2.535
N.m.r. for 1 T field (MHz)	10.19		15.518
N.m.r. sensitivity (relative to ^{1}H at constant field)	0.16		0.046
Electric quadrupole moment × 10^{28} (m^2)	−0.8	+0.47	−1.0

possible. These nuclei provide probes for the obtaining of structural information in favourable circumstances, but spectra are often complicated and difficult to interpret. Nuclear quadrupole resonances are in the range 30–120 MHz. ^{121}Sb has an excited state at 37.2 keV which can be used in Mössbauer spectroscopy. Isomer shifts for ^{121}Sb in antimony compounds parallel those for ^{119}Sn in tin compounds, ranging from about -6 mm s^{-1} in SbF_3 in which the Sb^{III} has a lone pair, to $+12$ mm s^{-1} in $KSb^{V}F_6$ in which the lone-pair electron density has been largely removed. (The isomer shift in InSb is taken as zero[9].) ^{124}Sb and ^{122}Sb are useful tracers. Both are obtained by n, γ reactions on the naturally occurring isotopes, and quantities as small as 10^{-9} g of ^{122}Sb may be detected in neutron activation analysis. A mixture containing radioactive ^{124}Sb and beryllium is a neutron source giving 1.6×10^6 neutrons per second per curie with energy 24.8 keV.

1.3. BISMUTH

The earliest references to bismuth suggest that, by the end of the fifteenth century in Germany, it was extracted from its ores by heating them with charcoal, and was used as a base for lacquer or gold on decorative articles such as caskets or chests. Demand for bismuth increased with the development of printing, and it became a usual constituent of alloys for type-metal. Some writers as early as the sixteenth century distinguished bismuth from lead, antimony and tin, but others confused these elements for a further 200 years.

Small amounts of the metal occur native with lead, silver or cobalt ores. Other minerals are the oxide Bi_2O_3, the sulphide Bi_2S_3 and the basic carbonate $(BiO)_2CO_3$. Extraction of the metal is based on standard reactions. Thus sulphide ores are roasted to oxides, which are reduced by iron or carbon.

Bismuth can also be obtained as a byproduct in the extraction of other metals. It collects either in flue dust from smelting processes (since the oxide Bi_2O_3 is slightly volatile) or, with lead, in anode slimes from copper refining. In the extraction of lead[10], bismuth appears in the smelted metal. It must be removed, as it weakens the resistance of the lead to corrosion. It also makes the lead difficult to roll and unsuitable for the manufacture of tetraethyl-lead. Bismuth may be separated from large amounts of lead either electrolytically or by the addition of calcium– or magnesium–lead alloys. The base metals, the bismuth,

[9] (a) S. L. Ruby, G. M. Kalvius, G. B. Beard and R. E. Snyder, *Phys. Rev.* **159** (1967) 239; (b) T. Birchall and B. Della Valle, *Chem. Communs.* (1970) 675.

[10] The author acknowledges helpful comments from Mr. H. C. Wesson of the Lead Development Association.

TABLE 4. ISOTOPES OF BISMUTH ($Z = 83$) [a]

Isotope	Half-life	Decay	Production	Isotope	Half-life	Decay	Production
^{199}Bi	24 m	EC	Protons/Pb	^{209}Bi [c]	$>2\times10^{18}$ y	Natural abundance 100%	
^{200}Bi	35 m	EC	Protons/Pb	^{210}Bi (RaE) [e]	5 d	β^- (>99%) α (10^{-4}%)	^{209}Bi(n, γ) Desc. ^{226}Ra
^{201}Bi	1.85 h	EC	Protons/Pb	^{211}Bi (AcC) [e]	2.15 m	α (>99%)[d]	Desc. ^{227}Ac
^{202}Bi	95 m	EC	Daughter ^{202}Po	^{212}Bi (ThC) [e]	60.6 m	β^- (0.27%) β^- (64%)[d] α (36%)	Desc. ^{228}Th
^{203}Bi	11.8 h	EC	^{206}Pb(p, 4n)				
^{204}Bi	11.2 h	EC	^{206}Pb(p, 3n) ^{203}Tl(α, 3n) ^{204}Pb(d, 2n)				
^{205}Bi	15.3 d	EC	^{206}Pb(d, 3n) ^{209}Bi(p, 5n) ^{205}Po(EC)	^{213}Bi [e]	47 m	β^- (98%) α (2%)	Desc. ^{233}U
^{206}Bi	6.3 d	EC [b]	^{206}Pb(d, 2n)	^{214}Bi (RaC) [e]	19.7 m	β^- (99%) α (0.02%)	Desc. ^{226}Ra
^{207}Bi	28 y	EC [b]	^{207}Pb(p, n)	^{215}Bi	7 m	β^-	Desc. ^{227}Ac
^{208}Bi	3.68×10^5 y	EC	^{209}Bi(n, 2n)				

[a] For more details see refs. 6 and 7.

[b] EC, electron capture. Energies (MeV) and intensities (as % initial disintegrations) of major radiations (only those of intensity >20% given). ^{206}Bi: γ, 0.18 (21%), 0.34 (25%), 0.52 (40%), 0.54 (31%), 0.803 (99%), 0.88 (68%), 1.72 (33%). ^{207}Bi: γ, 0.57 (98%), 1.06 (67%).

[c] Mass 208.9806. Other nuclear properties: spin, 9/2; magnetic dipole moment, 4.039 BM; electric quadrupole moment, -0.34×10^{-28} m^2. N.m.r. for 1 T field, 6.842 MHz; n.m.r. sensitivity (relative to proton at constant field), 0.137.

[d] Major radiations (MeV). ^{211}Bi: γ, 0.35 (13%). ^{212}Bi: 1.52 (5%), 2.25 (54%), and several γ. ^{214}Bi: β^-, 1.0 (23%), 1.51 (40%), and several others to 3.26 (19%). Several γ to 2.43.

[e] For members of radioactive decay series, radiations from daughters must be considered. Major radiations (MeV) from bismuth isotopes as follows. ^{210}Bi: α, 5.06 (10^{-4}%); β, 1.17. ^{211}Bi: α, 6.273(17%), 6.617 (83%), γ, 0.35 (13%). ^{212}Bi: α, 6.04 (25%), 6.08 (10%); β^-, 1.52 (5%), 2.25 (54%), and several γ. ^{213}Bi: α, 5.87; β^-, several to 1.39. ^{214}Bi: α, 5.51; β^-, 1.0 (23%), 1.51 (40%) and several others to 3.26 (19%); several γ to 2.43.

and some lead give a high melting alloy, which can be separated from the molten lead. Subsequently, the bismuth can be recovered by chlorination, when volatile bismuth trichloride sublimes. The metal can then be obtained electrolytically.

Commercial bismuth has a purity of about 99%. It is used as a constituent of a wide range of alloys (see Table 15A). Much purer bismuth, up to 99.9999%, is also available and is used mainly in the semiconductor industry. This very pure bismuth may be obtained by reduction of the oxide made by heating recrystallized bismuth nitrate. The metal is distilled in high vacuum and zone refined.

Isotopes

Some of the more important isotopes are shown in Table 4. The longest lived isotope, ^{209}Bi, is, for all practical purposes, the final product of the $4n+1$ radioactive decay series, and bismuth isotopes occur in the three other decay series. A number of ^{209}Bi nuclear

TABLE 5. PHYSICAL PROPERTIES OF THE ELEMENTS[a]

		As	Sb	Bi
Atomic weight (^{12}C = 12.0000)[b]		74.9216	121.75	208.9806
Melting point (°C)		817	630	271
Latent heat of fusion at melting point (kJ mol^{-1})		21.3	19.9	10.8
Vapour pressure: 10^{-3} atm at (°C)[c]		356	734	882
10^{-1} atm at (°C)		421	952	1047
10^{-1} atm at (°C)		502	1247	1266
Boiling point (°C)		616	1635	1580
Density (kg m^{-3})		5780	6680	9800
Hardness		3.5		2–2.5
Contraction on freezing (%)[d]		10	0.8	−3.35
10^6 coefficient expansion (K^{-1})		6	8–11	13.4
Thermal conductivity (W m^{-1} K^{-1})			17.6	7.9
Electrical resistance ($\mu\Omega$ m)(s)		0.40/20°C	0.42/20°C	1.16/20°C
(s)		2.00/800°[d]	1.54/500°	2.60/250°
(l)		3.90/820°	1.13/630°	1.28/271°
10^3 temperature coefficient resistance (0–100°) (K^{-1})			5.1	4.2
Photoelectric work function (V)		5.1	4.1	4.4
Thermoelectric e.m.f. (relative to Pt; cold jn. 0°) V			4.89/100°	−7.34/100°
			10.14/200°	−13.57/200°
10^{10} magnetic susceptibility (m^3 mol^{-1})		+0.067	−10.1	−36.4
Energies characteristic X-ray spectra (kV)	K	11.9	30.4	90.1
	L	1.52	4.69	16.4
	M	—	0.94	4.01
	N	—	0.15	0.96
K emission lines (pm)	α_1	117.98	47.483	16.558
	α_2	117.58	47.026	16.073
	β_1	105.73	41.707	14.234
	β_2	104.49	40.793	13.647
Absorption edge (pm)		104.47	40.692	13.706

[a] C. J. Smithells (ed.), *Metals Reference Book*, 4th edn., Butterworths (1967).
[b] N. N. Greenwood, *Chem. Brit.* **6** (1970) 119.
[c] atm = 101.325 kN m^{-2}.
[d] W. Klemm and H. Niermann, *Angew. Chem. Int. Edn. Engl.* **2** (1963) 523; W. Klemm, H. Spitzer and H. Niermann, *Angew. Chem.* **72** (1960) 985.

quadrupole resonances in various compounds have been reported in the range 20–55 MHz. Bismuth has found applications in the atomic energy industry since the metal is low melting and ^{209}Bi is a strong neutron absorber.

1.4. PHYSICAL PROPERTIES

Some properties of the rhombohedral forms of arsenic, antimony and bismuth are shown in Table 5 and thermodynamic data are given in Table 6.

By comparison with the transition metals, arsenic, antimony and bismuth are volatile and low melting; antimony and bismuth have high entropies of fusion. The melting points decrease with increasing atomic weight, but antimony and bismuth have much higher

TABLE 6. THERMODYNAMIC PROPERTIES OF ARSENIC, ANTIMONY AND BISMUTH[a]

	ΔH_f° (kJ mol^{-1})	ΔG_f° (kJ mol^{-1})	$H^\circ_{298}-H^\circ_0$ (kJ mol^{-1})	S° (J mol^{-1} K^{-1})	C_p (J mol^{-1}K^{-1})	Vapour pressure at b.p. (atm)[b]
As (s)	0	0	5.129	35.1	24.6	
As (g)	302.5	261.1	6.196	174.1	20.8	
As_2 (g)	222.2	172.0	9.418	239.3	35.0	5.5×10^{-4} [c]
As_4 (g)	143.9	92.5		313.8		1.0
Sb (s)	0	0	5.899	45.69	25.2	
Sb (g)	262.3	222.2	6.196	180.2	20.8	0.0735
Sb_2 (g)	235.6	187.0	9.874	254.8	36.4	0.775
Sb_4 (g)	205.0	141.4	18.41	351		0.151
Bi (s)	0	0	6.425	56.73	25.5	
Bi (g)	207.1 [d]	168.2	6.196	186.9	20.8	0.395
Bi_2 (g)	219.7 [d]	172.4	10.259	273.6	37.0	0.604
Bi_4 (g)	241.0 [d]					

[a] Except where indicated, refs. 11 and 12.
[b] atm = 101.325 kN m^{-2}.
[c] D. R. Stull and G. C. Sinke, *Thermodynamic Properties of Elements*, Adv. Chem. Series, **18** (1956).
[d] F. J. Kohl, O. M. Uy and K. D. Carlson, *J. Chem. Phys.* **47** (1967) 2667.

boiling points than arsenic. Hence arsenic has no liquid range at normal pressure but antimony and bismuth have long liquid ranges. This may be related to the generalization already noted that, whereas in the various forms of arsenic each atom interacts strongly with three neighbours, in antimony and bismuth interactions are spread more evenly over a larger number of neighbours. (Compare the interatomic distances in the rhombohedral forms given in Table 1 and the greater diversity of structures shown by arsenic in comparison with antimony and bismuth.) Thus, although more energy is needed to atomize arsenic than bismuth, less is needed to form As_4 molecules (in which each arsenic atom keeps three neighbours) than to form Bi_4 molecules. At the normal boiling point, E_4

[11] D. D. Wagman, W. H. Evans, I. Halow, V. B. Parker and R. H. Schumm, *Selected Values of Chemical Thermodynamic Properties*, NBS Technical Note 270-1 (1965).

[12] R. Hultgren, R. L. Orr, P. D. Anderson and K. K. Kelly, *Selected Values of Thermodynamic Properties of Metals and Alloys*, Wiley (1963).

molecules dominate in arsenic vapour, whereas they are largely dissociated in the vapours of antimony and bismuth. The formation of E_4 molecules by phosphorus, arsenic, antimony, and bismuth is not easy to explain in terms of valence theory. Delocalized bonds within the tetrahedra must be considered; a system of six localized two-centre bonds is not adequate. Many of the possible mixed molecules, e.g. $E^1_nE^1_{4-n}$ or E^1E^2, have been characterized by spectroscopic or molecular beam techniques. It is possible that all possible combinations are stable species in the gas phase. Those that have been identified show systematic trends in atomization energies, as the following data[13] indicate.

Sb_2, 289; BiSb, 249; Bi_2, 195.
Sb_4, 844; $BiSb_3$, 797; Bi_2Sb_2, 726; Bi_3Sb, 657; Bi_4, 587 (kJ mol^{-1}).

The Raman spectrum of arsenic at 900°C is very similar to that of phosphorus at 500°C and assignments have been made as follows[14]:

421 vw (cm^{-1})	As_2	250 m	As_4 (ν_2)
400 vvw	As_4 ($2\nu_3$)	200 w	As_4 (ν_3)
340 s	As_4 (ν_1)	~40 vvw	As_4 ($\nu_2 - \nu_3$)

The values of thermal and electrical properties in Table 6 are for polycrystalline materials. They vary slightly according to the method of sample preparation. Properties of single crystals of the rhombohedral forms are, of course, strongly anisotropic, since atomic interactions are much weaker perpendicular to the sheets of atoms (along the *c*-axis) than parallel to them. The anisotropy is greatest for arsenic (Table 7) where interactions between sheets are weakest, but is still important for antimony and bismuth. Thus the coefficient of expansion of arsenic is much greater along *c* than along *a*; though *c* expands rapidly on heating, *a* is unchanged between 20° and 400°C. The explanation seems to be that, at room temperature, the valence angles are forced out by the packing of the sheets, beyond the angles in many molecules. As the temperature is raised and the distance between the layers increases, the valence angles contract. The effect is to compensate for the increased As–As distance so that the cell parameter *a*, parallel to the sheets, is almost unchanged. Similar features are revealed by measurements of the thermal expansion parallel to, and perpendicular to, the zigzag chains in selenium and tellurium[15].

As is also shown in Table 6, bismuth contracts on melting, a fact which leads to important uses in alloys. At the same time, the electrical conductivity increases (in contrast, the conductivity of metals usually decreases on melting). These changes are associated with

Table 7. Anisotropy of Arsenic

	Along *c*	Along *a*
Coefficient of expansion (20–400°C) (K^{-1})	47.2×10^{-6}	0
Resistivity at 295 K ($\mu\Omega$ m) $\times 10^6$	0.356	0.255
Temperature coefficient resistivity (K^{-1}) $\times 10^3$	4.5	4
Magnetic susceptibility (m^3 mol^{-1}) $\times 10^{10}$	5.58	−2.69

J. B. Taylor, S. L. Bennett and R. D. Heyding, *J. Phys. Chem. Solids* **26** (1965) 69.

[13] F. J. Kohl and K. D. Carlson, *J. Am. Chem. Soc.* **90** (1968) 4814.
[14] G. A. Ozin, *Chem. Communs.* (1969) 1325.
[15] W. Klemm, *Proc. Chem. Soc.* (1958) 329.

an increased co-ordination number in molten, compared with solid, bismuth. In other words, molten bismuth is more metallic than the solid. Arsenic expands on melting and its conductivity falls. Antimony is intermediate between arsenic and bismuth: its conductivity rises, showing that the melt is more metallic than the solid, but the effect on the inter-atomic distance is sufficient only partly to cancel the normal expansion on melting. Further non-metallic features of solid bismuth and antimony are their low thermal and electrical conductivities compared with other metals, and their high diamagnetism. This falls rapidly near the melting point. The high negative thermoelectric e.m.f. of bismuth relative to platinum is also unusual. The magnetic properties of single crystals of arsenic (Table 7) are strikingly anisotropic; paramagnetism along *c* practically cancels diamagnetism along *a* in polycrystalline samples.

1.5. CHEMICAL PROPERTIES

The elements do not react with water or oxygen at 20°C and so are stable in air. At higher temperatures, however, they react with most metals (see section 2), and many non-metals (not, however, with hydrogen since the hydrides are endothermic). Tervalent derivatives (Scheme 2) are usually produced, except for antimony, when some oxidation to Sb^V is common. Since elements are, in general, easier to purify than compounds, many

SCHEME 2. Reactions of the elements.

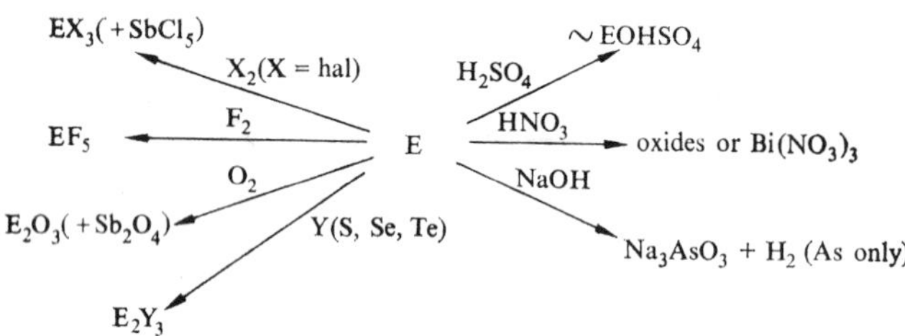

compounds of arsenic, antimony and bismuth required in high purity are made from stoichiometric quantities of ultra-pure elements, usually by reactions at elevated temperatures in quartz apparatus. The elements are not affected by dilute acids, but they all dissolve in concentrated sulphuric acid to form complex sulphates. Aqua regia oxidizes antimony to hexachloroantimonate(V) ($SbCl_6^-$). Arsenic dissolves in fused alkali, but antimony and bismuth are not much affected.

2. ALLOYS AND INTERMETALLIC COMPOUNDS

Pure arsenic, antimony and bismuth have few uses, but alloys with other metals have been made and exploited since the earliest times. The use of the Group V elements, e.g. in hardening lead to make it suitable for shot, or in increasing the resistance of copper to corrosion, has been largely empirical, but more recent research on intermetallic systems, stimulated by applications in the electronics industry, has revealed much of interest concerning chemical bonding. Structural data on intermetallic compounds is given in Table 8. A number of more comprehensive critical compilations, including detailed phase diagrams and crystallographic data, are available[3,16]. Thermodynamic data on intermetallic compounds is not very plentiful. For example, heats of formation have been measured by various techniques, but it has not always been possible to be sure that equilibrium has been reached. Values of −150 to −300 kJ per g atom of element E have been obtained for alkali metal compounds and rather more negative values for compounds with the alkaline earth metals. Compounds between Group V elements and transition metals have less negative heats of formation (0 to −60 kJ per g atom of E)[12].

2.1. INTERMETALLIC COMPOUNDS WITH PRE-TRANSITION ELEMENTS

The electro-negativity differences between the alkali or alkaline earth metals and the Group V elements are such that intermetallic compounds may be expected to show "salt-like" characteristics described by structures such as $Na_3^+Sb^-$. Though ionic formulations are useful in the interpretation of the chemical reactions of the intermetallic compounds, they are not entirely adequate for the solid phases themselves. These are characterized by high co-ordination numbers and interatomic distances between like atoms comparable with those in metals. However, strong interactions between unlike atoms are indicated by the high melting points of some intermetallic compounds. For example, Na_3Bi melts at 840°, sodium at 98° and bismuth at 271°C.

Intermetallic compounds with formulae M_3E (where M = Li, Na, K, Rb, Cs and E = As, Sb, Bi) are formed for all combinations of alkali and heavier Group V elements (Table 8). Sodium arsenide has the anti-tysonite (LaF_3) structure. Equal numbers of arsenic and sodium atoms (Fig. 2) make hexagonal nets as in boron nitride. The remaining sodium atoms are arranged in layers on either side of these nets. Each arsenic has five neighbours at the corners of a trigonal bipyramid (three at 294 and two at 299 pm) and six

TABLE 8. CRYSTAL STRUCTURES[a] OF INTERMETALLIC COMPOUNDS

Compound	Structure	Compound	Structure	Compound	Structure
Li_3As	Na_3As ($D0_{18}$)	Li_3Sb	Na_3As, Li_3Bi [b]	Li_3Bi	($D0_3$)
LiAs	Monoclinic			LiBi	CuAu ($L1_0$)
Na_3As	($D0_{18}$)	Na_3Sb	Na_3As ($D0_{18}$)	Na_3Bi	Na_3As ($D0_{18}$)
		NaSb	LiAs	NaBi	CuAu ($L1_0$)
K_3As	Na_3As ($D0_{18}$)	K_3Sb	Na_3As ($D0_{18}$)	K_3Bi	Na_3As ($D0_{18}$)
		KSb	LiAs		
				KBi_2	$MgCu_2$ (C15)
Rb_3As	Na_3As ($D0_{18}$)	Rb_3Sb	Na_3As, Li_3Bi [b]	Rb_3Bi	~NaTl (B32)
				$RbBi_2$	$MgCu_2$ (C15)
		Cs_3Sb	~ NaTl (B32)	Cs_3Bi	~NaTl (B32)
				$CsBi_2$	$MgCu_2$ (C15)
Mg_3As_2	Mn_2O_3 ($D5_3$), La_2O_3 [b]	Mg_3Sb_2	La_2O_3 ($D5_2$)	Mg_3Bi_2	La_2O_3 ($D5_2$)
		Ca_3Sb_2 [c]		Ca_3Bi_2 [c]	
				$CaBi_3$ [c]	
		Sr_3Sb_2 [c]		Sr_3Bi_2	
				$SrBi_3$	$AuCu_3$ ($L1_2$)

[16] M. Hansen and K. Anderko, *Constitution of Binary Alloys*, 2nd edn., McGraw Hill (1958). Supplement by R. P. Elliott (1961).

TABLE 8 (*cont.*)

Compound	Structure	Compound	Structure	Compound	Structure
		Ba_3Sb_2 [c]		Ba_3Bi_2	
				$BaBi_3$	$CuTi_3$ ($L6_0$)
		M_4Sb_3 [d]	Th_3P_4 ($D7_3$)	M_4Bi_3 [d]	Th_3P_4 ($D7_3$)
MAs [d]	NaCl (B1)	MSb [d]	NaCl (B1)	MBi [d]	NaCl (B1)
		MSb_2 [d]	$LaSb_2$		
$ThAs_2$	$PbCl_2$ (C23)	$ThSb_2$	Cu_2Sb (C38)	$ThBi_2$	Cu_2Sb (C38)
UAs_2	Cu_2Sb (C38)	USb_2	Cu_2Sb (C38)	UBi_2	Cu_2Sb (C38)
		Ti_3Sb	Cr_3Si (A15) [e]	Ti_3Bi	Tetragonal
TiAs	TiP(B_i), NiAs [f]	TiSb	NiAs (B8)		
$TiAs_2$	Orthorhombic	$TiSb_2$	$CuAs_2$ (C16)		
Zr_3As	Ti_3P	Zr_3Sb	Fe_3P	Zr_3Bi	
		Zr_2Sb	Hexagonal		
ZrAs	TiP(B_i) [f]	Zr_5Sb_3	Mn_5Si_3		
$ZrAs_2$	$PbCl_2$ (C23)			$ZrBi_2$	Orthorhombic
Hf_3As	Ti_3P	Hf_3Sb	Fe_3P		
HfAs	TiP(Bi) [f]				
$HfAs_2$	$PbCl_2$ (C23)	$HfSb_2$	$TiAs_2$ [b]	$HfBi_2$	
V_3As [g]	Cr_3Si (A15)	V_3Sb [g]	Cr_3Si (A15)	V_3Bi	Cr_3Si (A15)
VAs	MnP (B31)	VSb	NiAs (B8)		
VAs_2	$NbAs_2$	VSb_2	$CuAl_2$ (C16)		
Nb_3As	Ti_3P	Nb_3Sb	Cr_3Si (A15)	Nb_3Bi	
NbAs	Tetragonal [f]	Nb_5Sb_4	Ti_5Sb_4		
$NbAs_2$	Monoclinic	$NbSb_2$	$NbAs_2$		
Ta_3As	Ti_3P	Ta_3Sb	Cr_3Si (A15)		
TaAs	NbAs	Ta_5Sb_4	Ti_5Sb_4		
$TaAs_2$	$NbAs_2$	$TaSb_2$	$NbAs_2$		
Cr_2As	Cu_2Sb (C38)				
CrAs [g]	MnP (B31)	CrSb [g]	NiAs (B8)		
		$CrSb_2$	FeS_2 (C18) [h]	$CrBi_2$	
MoAs	MnP (B31)				
$MoAs_2$	$NbAs_2$				
WAs_2	$NbAs_2$				
Mn_3As	Orthorhombic				
Mn_2As	Cu_2Sb (C38)	Mn_2Sb	Cu_2Sb (C38)		
MnAs	NiAs, MnP [i]	MnSb	NiAs (B8)	MnBi	NiAs (B8)
Fe_2As	Cu_2Sb (C38)	Fe_3Sb_2	Ni_2In ($B8_2$)		
FeAs	MnP (B31) [f]	FeSb	NiAs (B8)		
$FeAs_2$	FeS_2 (C18) [h]	$FeSb_2$	FeS_2 (C18) [h]		
RuAs	MnP (B31)	RuSb	MnP (B31)		
$RuAs_2$	FeS_2 (C18) [h]	$RuSb_2$	FeS_2 (C18) [h]	$RuBi_3$	$NiBi_3$
$OsAs_2$	FeS_2 (C18) [h]	$OsSb_2$	FeS_2 (C18) [h]		
CoAs [g]	MnP (B31) [i]	CoSb [g]	NiAs (B8)	CoBi	
$CoAs_2$	$CoSb_2$	$CoSb_2$	Monoclinic		
$CoAs_3$	($D0_2$)	$CoSb_3$	$CoAs_3$ ($D0_2$)		
Rh_2As	CaF_2 (C1), $PbCl_2$ (C23) [b]				
RhAs	MP (31)	RhSb	MnP (B31)	RhBi	NiAs (B8)
$RhAs_2$	$Cnob_2$ B	$RhSb_2$	$CoSb_2$	$RhBi_2$	$CoSb_2$ [b]
$RhAs_3$	$CoSAs_3$ ($D0_2$)	$RhSb_3$	$CoAs_3$ ($D0_2$)	$RhBi_3$	$NiBi_3$
		IrSb	NiAs (B8)		
$IrAs_2$	$CoSb_2$	$IrSb_2$	$CoSb_2$	$IrBi_2$	$CoSb_2$
$IrAs_3$	$CoAs_3$ (D 0_2)	$IrSb_3$	$CoAs_3$ ($D0_2$)	$IrBi_3$	$NiBi_3$
Ni_5As_2 [g]	Hexagonal	Ni_3Sb [g]	Cu_3Ti [b]		

TABLE 8 (*cont.*)

Compound	Structure	Compound	Structure	Compound	Structure
NiAs	(B8)	NiSb	NiAs (B8)	NiBi	NiAs (B8)
$NiAs_2$	FeS_2 (C18) [h, b]	$NiSb_2$	FeS_2 (C18) [h]	$NiBi_3$	Orthorhombic
$NiAs_3$	~$CoAs_3$ ($D0_2$)				
Pd_3As [g]	Fe_3P				
Pd_5As_2		Pd_5Sb_3	Ni_2In ($B8_2$)	Pd_5Bi_3	Ni_2In ($B8_2$)
Pd_2As	Monoclinic[b]				
		PdSb	NiAs (B8)	PdBi	
$PdAs_2$	FeS_2 (C2) [h]	$PdSb_2$	FeS_2 (C2) [h]	$PdBi_2$	Monoclinic[b]
		PtSb	NiAs (B8)	PtBi	NiAs (B8)
$PtAs_2$	FeS_2 (C2) [h]	$PtSb_2$	FeS_2 (C2) [h]	$PtBi_2$	FeS_2 (C2) [h, b]
Cu_3As	Cu_3P ($D0_{21}$) [b]	Cu_3Sb	Li_3Bi, Cu_3Ti [b]		
Cu_5As_2	Tetragonal[j]	Cu_5Sb_2 [j]			
Cu_2As	Hexagonal	Cu_2Sb	(C38)		
AgAs [k]		Ag_3Sb	Cu_3Ti (A_e)	$AgBi_2$	
		$AuSb_2$	FeS_2 (C2) [h]	Au_2Bi	$MgCu_2$ (C15)
Zn_3As_2 [l]	Zn_3P_2 ($D5_9$)	$ZnMg_2Sb_2$	La_2O_3		
$ZnAs_2$	Orthorhombic	ZnSb	CdSb (B_e)		
Cd_3As_2 [l]	Zn_3P_2 ($D5_9$)	Cd_3Sb_2	Monoclinic		
		CdSb	(B_e)		
$CdAs_2$	Tetragonal				
AlAs	ZnS (B3) [m]	AlSb	ZnS (B3) [m]		
GaAs	ZnS (B3) [m]	GaSb	ZnS (B3) [m]		
InAs	ZnS (B3) [m]	InSb	ZnS (B3) [m]	InBi	LiOH (B10)
				In_2Bi	Ni_2In ($B8_2$)
		Tl_7Sb	($L2_2$)		
GeAs	GaTe				
$GeAs_2$	Orthorhombic				
Sn_4As_3	Cubic				
SnAs	NaCl (B1)	SnSb	NaCl (B1)		

[a] Structures are denoted by a common representative. The nomenclature of the *Strukturbericht*, widely used in this field, is also given.

[b] Several modifications are known. See refs. 3 and 16.

[c] These compounds are deduced from phase diagrams.

[d] M = Sc, Y, lanthanide or actinide element. Rare earth diantimonides are synthesized at 7×10^9 N m^{-2} (N. L. Eatough and H. T. Hall, *Inorg. Chem.* **8** (1969) 1439.

[e] There appear to be two closely related forms of Ti_3Sb: one cubic and one tetragonal. The tetragonal form may be stabilized by impurities; the cubic form is superconducting below 5.8 K. Several less well-characterized titanium compounds have been reported.

[f] Relations between TiP, NbAs, MnP and NiAs structures are discussed by H. Boller and E. Parthé, *Acta Cryst.* **16** (1963) 1095.

[g] Several other compounds of V, Cr, Mo, Co, Pd and several mixed Fe–Co–Ni compounds have been described.

[h] The marcasite structure of FeS_2 is denoted by C18 and the pyrites structure by C2.

[i] Many of these compounds are polymorphic and show both NiAs and MnP structures. Considerable deviations from stoichiometry are common. See text.

[j] Several other Cu–As and Cu–Sb compounds are described. Some of these are superstructures related to hexagonal close packing. The Ag–Sb system is similar. Bismuth and copper form a system with a simple eutectic and no intermediate phases.

[k] AgAs is stable only at high temperatures.

[l] These compounds are polymorphic; a second form is described as the Zn_3As_2 structure.

[m] The zincblende structure is denoted by B3. These compounds change to metallic (white tin or NaCl) structures at high pressures. The high pressure form of InSb is a superconductor; T_c, 1.6 . . . 2.1 K. S. Minomura and H. G. Drickamer, *J. Phys. Chem. Solids* **23** (1962) 451.

other sodium atoms (at 330 pm) form a trigonal prism. There are two sorts of sodium. One sort, *a*, in the AsNa net, has three arsenic neighbours at 294 pm and six sodium neighbours at 330 pm. The other sort of sodium, *b*, has four arsenic neighbours at the

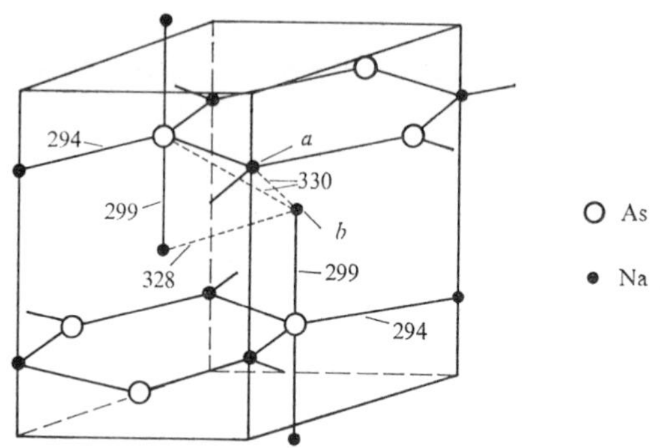

FIG. 2. The hexagonal structure of Na_3As. (Distances in pm.)

corners of a distorted tetrahedron (one at 299 and three at 330 pm) and six sodium neighbours, three at 328 and three at 330 pm. All these Na–Na distances are less than that in sodium metal (371.6 pm). Thus the structure shows high co-ordination numbers characteristic of metals. Similar conclusions can be drawn from the cubic structure of lithium bismuthide (Li_3Bi) (Fig. 3), which is more compact than sodium arsenide. Each bismuth atom has eight lithium neighbours, *a* or *b*, at the corners of a cube and six, *c*, at the corners of an octahedron. There are two sorts of lithium. One sort, *a* or *b*, has six lithium neighbours at 335, four lithium neighbours at 291 and four bismuth neighbours at 291 pm. The other sort, *c*, has eight lithium neighbours at 291 and six bismuth neighbours at 335 pm. Again, the co-ordination numbers are high and the interatomic distances are comparable with those in the metals (e.g. Li–Li 304 pm in metallic lithium). In the compounds of the heavier alkali metals, e.g. Cs_3Bi, there is some interchange of alkali and Group V metal

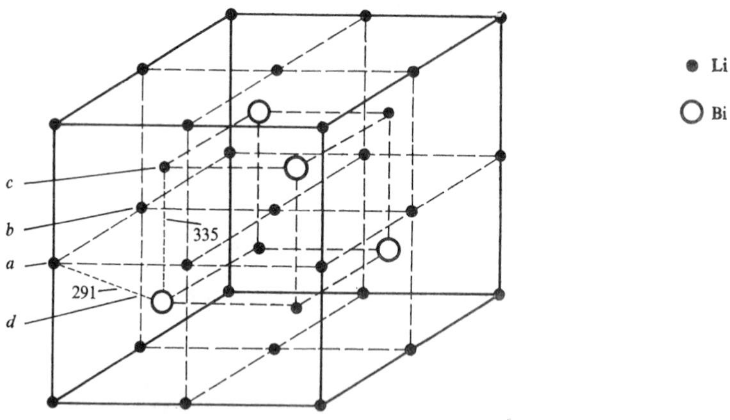

FIG. 3. The structure of Li_3Bi. (Distances in pm.)

atoms. Thus in Fig. 3 the positions *b* and *c* are occupied by caesium atoms and the positions *a* and *d* by randomly distributed caesium and bismuth atoms[17]. Several ternary antimonides, e.g. K_2CsSb (see Table 12), have related structures.

Some of the alkali and Group V metal systems give compounds ME (Table 9). The arsenic and antimony compounds have parallel infinite spirals of arsenic or antimony atoms linked by bonds to the alkali metals. If the structures are represented by the formula $M_n^+(E_n)^{n-}$, the spirals $(E_n)^{n-}$ are isoelectronic with those of elemental selenium or tellurium. The bismuth compounds LiBi and NaBi have typical alloy structures, and are superconductors at low temperatures (LiBi below 2.47, NaBi below 2.22 K)[18].

TABLE 9. INTERATOMIC DISTANCES IN COMPOUNDS ME AND ME_2
(Shortest distances in metals in parentheses)

Compound	E–E (pm)	M–E (pm)	M–M(pm)	Ref.
LiAs	245, 247 (250)	278	300–378 (304)	a
NaSb	286, 284.5 (291)	331	344–445 (372)	a
KSb	281, 288 (291)	364	396–505 (454)	b
KBi_2	336 (307)	394	411 (454)	c
$RbBi_2$	340 (307)	399	416 (496)	c
$CsBi_2$	345 (307)	405	422 (532)	c

[a] D. T. Cromer, *Acta Cryst.* **12** (1959) 36, 41.
[b] E. Busmann and S. Lohmeyer, *Z. anorg. allgem. Chem.* **312** (1961) 53.
[c] Ref. 3.

Potassium, rubidium and caesium also form well-defined compounds ME_2, which have the $MgCu_2$ structure. These also show high co-ordination numbers (16 for M and 12 for E). The E–E bond lengths are greater than those in the metals (a reflection of the higher co-ordination number in the alloy), but the M–M bond lengths are strikingly shorter than those in the alkali metals.

The compound Na_3Bi shows metallic conduction, but the antimony compounds are semiconductors with energy gaps as follows[19]:

n-type: Na_3Sb, 1.1; K_3Sb, 1.1; $(Na,K)_3Sb$, 1.0 eV
p-type: Rb_3Sb, 1.0; Cs_3Sb, 1.6 eV

Caesium antimonide is one of the most efficient photoemitters known. At 450 nm, the quantum efficiency is 20–30%, i.e., one electron is emitted for every 3–5 photons. The compounds MBi_2 become superconducting below the temperatures T_c: KBi_2, 3.58; $RbBi_2$, 4.25; $CsBi_2$, 4.75 K. These compounds are unusual in having much higher atomic volumes than most other superconductors with the $MgCu_2$ structure.

The intermetallic compounds M_3E are partly soluble in molten salts, e.g. LiCl–LiF or NaCl–NaI, and cryoscopic measurements suggest that there is some ionization, e.g. to Na^+, Bi^{3-} in dilute solution. The compounds ME are less soluble. Sodium antimonide

[17] J. P. Suchet, *Acta Cryst.* **14** (1961) 651.
[18] B. T. Matthias and J. K. Hulm, *Phys. Rev.* **87** (1952) 799.
[19] W. E. Spicer, *Phys. Rev.* **112** (1958) 114; K. H. Jack and M. M. Wachtel, *Proc. Roy. Soc.* (*London*) **A**, **239** (1957) 46.

(NaSb) appears to give the species Sb_3^{3-} besides Sb^{3-}; the Sb–Sb chains are thus partly preserved in solution[20]. The alkali metal compounds are readily oxidized and are hydrolysed by water or acid. The arsenic and antimony compounds yield hydrides (section 4); in this they are "salt-like", with the Group V element formally in oxidation state −3.

The best characterized compounds with the alkaline earth elements have the formula M_3E_2. Those for which the structures have been found have either the Mn_2O_3(C-M_2O_3) or La_2O_3(A-M_2O_3) structures. Several are dimorphic, and each of these structures is stable over a particular temperature range. The heats of formation[12] and melting points are high. Chemically, these compounds behave like the alkali metal compounds and give good yields of hydrides on hydrolysis. Some cubic phases ME_3 have also been characterized, though the complete phase diagrams of a number of systems with alkaline earth metals do not appear to be well established. The compounds $BaBi_3$ (T_c, 5.69 K), $SrBi_3$ (T_c, 5.62 K) and $CaBi_3$ (T_c, 1.7 K) are superconductors.

The compounds of scandium, lanthanides and actinides are of three main types. Those with composition ME have the sodium chloride structure, and those with composition M_4E_3 the body-centred thorium phosphide ($D7_3$) structure, in which each M atom is surrounded by eight E atoms. In thorium arsenide ($ThAs_2$), each thorium atom has nine arsenic neighbours.

Thorium arsenides or antimonides have low thermal conductivities and high electrical conductivities and may have possibilities as thermoelectric materials for use at high temperatures, provided that sufficient control over impurities can be exercised (Table 10). Material with composition $Th_3(As_{0.9}Sb_{0.1})_4$ has a figure of merit Z of about 0.5×10^{-3} K^{-1} at 800°C. See section 7.3 for definition of Z and a fuller discussion of thermoelectric properties.

TABLE 10. PROPERTIES OF THORIUM ARSENIDES AND ANTIMONIDES

	Th_3P_4	Th_3As_4	Th_3Sb_4	$Th_3(As_{0.9}Sb_{0.1})_4$
Thermoelectric power (μV K^{-1})	−223	−160	−52.5	−120
Electrical conductivity (Ω^{-1} m^{-1}) $\times 10^{-2}$	1.5	545	3370	~500
Thermal conductivity (W m^{-1} K^{-1})	27.6	5.44	7.48	4.30

C. E. Price and L. H. Warren, *J. Electrochem. Soc.* **112** (1965) 510.

2.2. COMPOUNDS WITH TRANSITION METALS

Arsenic, antimony and bismuth often give complex phase diagrams with transition elements, and many intermediate phases have been characterized. Table 8 shows that transition metals with similar electronic structures usually give intermetallic compounds of similar stoichiometries. Large groups of compounds have similar structures, and some of these are described below.

Most compounds ME have either the hexagonal nickel arsenide structure (Fig. 4) or the closely related orthorhombic manganese phosphide structure. Transitions between

[20] M. S. Foster, C. E. Crouthamel, D. M. Gruen and R. L. McBeth, *J. Phys. Chem.* **68** (1964) 980; M. Okada, R. A. Guidotte and J. D. Corbett, *Inorg. Chem.* **7** (1968) 2118.

these structures are common. In nickel arsenide, each nickel atom has six arsenic neighbours at the corners of an octahedron, and each arsenic atom has six nickel neighbours at the corners of a trigonal prism. The Ni–As distance is 243 pm. There are

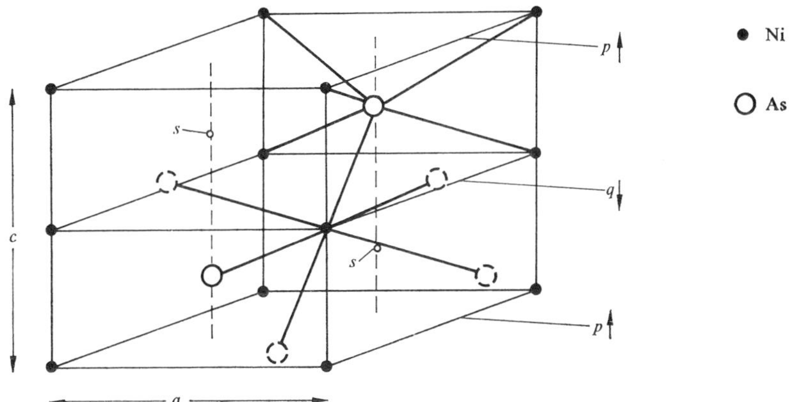

FIG. 4. The hexagonal nickel arsenide structure. The arsenic atoms shown by broken lines are in adjacent unit cells.

two further important features. First, the structure may be described as a hexagonal-close-packed array of arsenic atoms with nickel atoms in octahedral holes. If some of these holes are unfilled, or if other than octahedral holes are occupied, the stoichiometry can deviate widely from ME. For example, vacancies in layers marked *q* in Fig. 4 are evident in $Ni_{0.95}As$; in the extreme, if all atoms in layer *q* are absent, the structure becomes that of cadmium iodide. Extra M atoms at points *s* in Fig. 4, each surrounded by five E atoms at the corners of a trigonal bipyramid, give stoichiometries such as $Co_{1.3}Sb$ to $Co_{1.03}Sb$ or Pd_5Sb_3. In the extreme, if all the sites *s* are occupied, as in In_2Bi, the structure becomes that of Ni_2In. The nickel arsenide structure is thus a very flexible one, accommodating considerable deviations from the ME stoichiometry. Secondly, the distance between nickel atoms in adjacent layers in nickel arsenide is 252 pm, comparable with that in nickel metal (249 pm). There is thus some indication of Ni–Ni bonding, confirmed by the observation that the coefficient of expansion perpendicular to the principal axis is eight times that along it. Metal–metal bonding is also shown by the axial ratios (c:a) found for compounds ME with the nickel arsenide structure. Low axial ratios indicate M–M bonding. In general, axial ratios of arsenides are greater than those of antimonides and bismuthides.

Many transition metal atoms have unpaired electrons and are therefore magnetic centres. The short M–M distances in nickel arsenide structures facilitate interactions between metal atoms, and so magnetic behaviour may be complicated. Some compounds are antiferromagnetic (Table 11). For example, neutron diffraction studies show that in CrAs or CrSb the magnetic moments are aligned, as shown on the right of Fig. 4, with those in layers *p* and *q* opposed. Other compounds are ferromagnetic, e.g. MnAs below 400 K or MnSb below 587 K. Susceptibilities of single crystals are anisotropic, and the anisotropy is strongly temperature-dependent. Thus the easy direction of magnetization in MnSb is below 520 K in the basal plane and above 520 K along the principal axis. The compound

MnBi passes over to a defect Ni_2In structure above 630 K. Below this temperature it is ferromagnetic with a strong coercive force; it thus has possibilities as a material for permanent magnets[23].

TABLE 11. ELECTRICAL AND MAGNETIC PROPERTIES OF INTERMETALLIC COMPOUNDS[a]

Néel temperatures (K)
UAs, 128; UAs_2, 283; USb, 213; USb_2, 206; UBi, 290; UBi_2, 183
Cr_2As, 393; CrAs, 823; CrSb, 723; Mn_2As, 580; Fe_2As, 351.
Energy gaps (eV) *in semiconductors*
$FeAs_2$, 0.2; $FeSb_2$, 0.05 [b]; RuPAs, 0.8; $RuAs_2$, 0.8; $RuSb_2$, 0.3 [c]; RuSbSe, 0.35; RuSbTe, 0.5 [c]; OsAsS, 1.3 [c]
RhAsS, 1.2 [c]
PdPAs, 0.45; $PdAs_2$, 0; $PdSb_2$, 0 [d]; $PtAs_2$, 0.5; $PtSb_2$, 0.05; PtBi2, 0
Critical temperatures for superconductivity (K)
Ti_3Bi, 5.8; Zr_3Bi, 2.35; V_3Sb, 0.8; Nb_5Sb_4, 8.60; $MoAs_2$, 0.41; Mo_3Sb_7, 2.31; Re_2Bi, 2.2
RuSb, 1.27; $RuBi_3$, 4.1; RhAs, 0.58; RhBi 2.0 . . . 2.2; $RhBi_3$, 3.2; $RhBi_4$, 2.70[e]
NiBi, 4.25; $NiBi_3$, 4.06; Pd_5As_2, 0.46; Pd_2As, 0.6; PdSb, 1.5; $PdSb_2$, 1.25; PdBi, 3.74; α-$PdBi_2$, 1.7; β-$PdBi_2$, 4.25; PdSbSe, 1.0; PdSbTe, 1.2 [e]; PdBiSe, 1.0 [e]; PdBiTe, 1.2 [e]; PtSb, 2.1; PtBi, 1.21 . . . 2.4; β-$PtBi_2$, 0.15; $PtBi_{0\cdot1-1}Sb_{0\cdot9-0}$, 1.21 . . . 2.05 [e]; PtBiSe, 1.45 [e]; PtBiTe, 1.15 [e]
Cu_3Sb, 1.3 . . . 1.8; Cu_2Sb, 0.085; CuBi, 1.3 . . . 1.4; $AgBi_2$, 2.8 . . . 3.0; $AuSb_2$, 0.58; Au_2Bi, 1.84; In_2Bi, 5.6

[a] Except where indicated from ref. 21.
[b] W. D. Johnston, R. C. Miller and D. H. Damon, *J. Less-Common Metals* **8** (1965) 272.
[c] F. Hulliger, *Nature* **198** (1963) 1081; **201** (1964) 381.
[d] F. Hulliger, *Nature* **200** (1963) 1064.
[e] B. W. Roberts, *Superconductive Properties* in ref. 22.

Compounds with stoichiometry ME_2 usually crystallize with the pyrites or marcasite structures, or with the distorted marcasite structure shown by $CoSb_2$. The pyrites structure is derived from the NaCl structure by retaining the cations and replacing the anions by E_2 groups disposed along threefold axes. Marcasite is orthorhombic. In both structures, M atoms are octahedrally co-ordinated by six E atoms and E_2 groups have six octahedrally disposed M neighbours. As_2 or Sb_2 groups may be replaced by combinations E^1E^2, e.g. AsSb or AsS, and a large number of ternary compounds are thus possible. Some of these are shown in Table 12[3,21]. If the metalloid atoms are randomly distributed over possible sites, the symmetrical pyrites or marcasite structures may be retained. In other cases, where E^1E^2 groups are ordered, less symmetrical structures, e.g. those of NiSbS or FeAsS, result. Since the E^1E^2 group (E^1 = As, Sb, Bi; E^2 = S, Se, Te) has one more electron then E^1_2, arsenoselenides, for example, of a given group of elements have structural and electrical properties like those of the diarsenides of elements one place to the right in the Periodic Table. Thus the co-ordination round M is tetragonally distorted in löllingites, compounds

[21] F. Hulliger, *Structure and Bonding* **4** (1968) 83.
[22] J. H. Westbrook (ed.), *Intermetallic Compounds*, Wiley (1967).
[23] J. S. Kouvel, *Magnetic Properties* in ref. 22.

ME_2 of the iron group (M = Fe, Ru, Os) with eighteen valence electrons. The compounds ME_2 of the cobalt group (M = Co, Rh, Ir) and the compounds ME^1E^2 of the iron group, all with nineteen valence electrons, have monoclinic structures in which M atoms are grouped in pairs with especially short M–M distances. The compounds ME_2 of the nickel group (M = Ni, Pd, Pt) and ME^1E^2 of the cobalt group (with twenty valence electrons) have the much more symmetrical pyrites or ullmanite structures. These generalizations are related to the occupancy of the d_{xy}, d_{xz}, d_{yz} orbitals of the metal. Table 12 shows that symmetrical structures are favoured when E^1 and E^2 are similar in size.

TABLE 12. SOME TERNARY DERIVATIVES

FeS_2 (*marcasite*), $CoSb_2$ *or* FeAsS (*arsenopyrite*) *structures*[a]									
18 [b]	RuAsP	RuAsSb	OsAsP						
19 [b]	FeAsS	FeAsSe	FeAsTe	FeSbS	FeSbSe	FeSbTe			
	RuAsS	RuAsSe	RuAsTe	RuSbS	RuSbSe	RuSbTe			
	OsAsS	OsAsSe	OsAsTe	OsSbS	OsSbSe	OsSbTe			
	RhAsSb	IrAsSb							
FeS_2 (*pyrites*) *or* NiSbS (*ullmannite*) *structures*[c]									
20 [b]	CoAsS	CoAsSe							
	RhAsS	RhAsSe	RhAsTe	RhSbS*	RhSbSe	RhSbTe	RhBiS*	RhBiSe	RhBiTe
	IrAsS	IrAsSe	IrAsTe	IrSbS	IrSbSe*	IrSbTe*	IrBiS*	IrBiSe*	IrBiTe
	PdAsSb	PtAsP							
21 [b]	NiAsS	NiAsSe		NiSbS*	NiSbSe*			NiBiSe*	
	PdAsS*	PdAsSe*		PdSbS*	PdSbSe*			PdBiSe*	PdBiTe
	PtAsS				PtSbSe*	PtSbTe		PtBiSe*	PtBiTe*
MgAgAs *and antifluorite structures*[d]									
	MgLiAs	MgAgAs	ZnLiAs*	ZnNaAs*	ZnAgAs*				
	MgLiSb	MgNiSb	MgCuSb	MnCoSb*	MnNiSb	MnCuSb			
	CoTiSb	CoVSb	CdCuSb	FeTiSb	FeVSb	NiTiSb	NiVSb		
	MgLiBi	MgNiBi	MgCuBi						
	Li_5SiAs_3*	Li_5GeAs_3*	Li_5TiAs_3						
$AlCu_2Mn$ *structure*[e]									
	Na_2KSb	K_2CsSb	Ni_2MgSb	Ni_2MnSb	Cu_2MnSb				
$CuFeS_2$ (*chalcopyrite*) *structure*[f]									
	$ZnSiAs_2$	$ZnGeAs_2$	$ZnGeAs_2$	$CdGeAs_2$	$CdSnAs_2$				

[a] The marcasite, $CoSb_2$ and FeAsS structures are closely related. Compounds in this section have the FeAsS or $CoSb_2$ structures except for RuAsP, RuAsSb and OsAsP (FeS_2).

[b] Number of valence electrons.

[c] Compounds with the NiSbS structure (a less symmetrical form of pyrites) are shown by an asterisk.

[d] Compounds with the antifluorite structure are marked by an asterisk. The MgAgAs structure is an antifluorite superstructure.

[e] A superstructure of body-centred cubic packing, closely related to the antifluorite or Li_3Bi structures (see Fig. 3).

[f] The $CuFeS_2$ structure is the same as that of zincblende (B3) with zinc atoms replaced alternately by copper and iron. $ZnSnAs_2$ has the ZnS (B3) structure, i.e. zinc and tin atoms are disordered. It forms mixed crystals with InAs. Other related ternary sulphides, e.g. Cu_3ES_3, are known; some are described in section 7.6.

The compounds ME_2 may show either metallic conductivity or, if the number of valence electrons is twenty or less, semiconductivity (see Table 11). In general, the energy gap between valence and conduction bands increases (a) with increasing atomic weight of the element M, and (b) with decreasing atomic weight of the element E. Substitution of a Group VI for a Group V element also increases the energy gap. All these trends correspond to gradation from metallic to covalent–ionic chemical bonding. It is interesting to note that the elements of the platinum and gold groups form particularly strong complexes with heavier metal donors: they are "class (b)" acceptors[24]. Their arsenides, antimonides and bismuthides are thus compact, and the distances between heavy metal atoms are brought within the range for which superconductivity is possible (see Table 11).

Phases of composition M_2E are also widely distributed. The compound Au_2Bi is anti-isomorphous with the alkali metal compounds already discussed. More commonly, the compounds M_2E crystallize with the tetragonal Cu_2Sb structure in which layers of M atoms

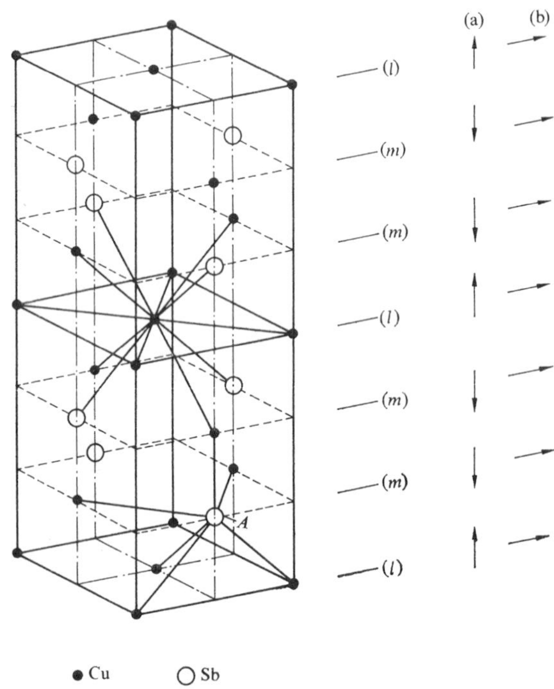

FIG. 5. The structure of Cu_2Sb.

alternate with (slightly puckered) layers of both M and E atoms (Fig. 5). Each E atom has nine neighbours (six of the nine bound to atom *A* are shown in Fig. 5: the other three are in adjacent unit cells). The M atoms in layers (*m*) have nine neighbours (five E and four M) distributed like the neighbours of atom *A*. The M atoms in layers (*l*) have twelve neighbours (eight M and four E) making a dodecahedron. Co-ordination numbers are high and bonding between like atoms is apparent. Phases with this structure show interesting and often complicated magnetic properties. For example, in Mn_2Sb at room temperature, there is long range coupling of spins in two interpenetrating lattices (Fig. 5 (a)). Since the manganese atoms in layers (*l*) and (*m*) are not equivalent, there is only partial cancellation of spins, and the lattice is ferrimagnetic. The closely related alloys Cr_2As, Mn_2As,

[24] S. Ahrland, J. Chatt and N. R. Davies, *Quart. Revs. London* **12** (1958) 265.

Cu_2Sb, $Mn_{2-x}A_xSb$ (A = V, Co, Cu), $Mn_2Sb_{1-x}Z_x$ (Z = Ge, As) are antiferromagnetic with spins oriented over two unit cells as indicated in Fig. 5 (b). The alloy $Mn_{2-x}Cr_xSb$ ($x \leqslant 0.003$) may be obtained in a state intermediate between the cases (a) and (b) with a spiral orientation of magnetic moments.

The compounds ME_3, formed by the elements cobalt, rhodium and iridium, crystallize with the cubic (skutterudite) $CoAs_3$ structure, in which groups of four E atoms are arranged in squares. Each M has six E neighbours at the corners of an octahedron and each E has four neighbours (two E and two M) at the corners of a distorted tetrahedron.

2.3. COMPOUNDS WITH POST-TRANSITION ELEMENTS

Arsenic, antimony and bismuth interact with the post-transition metals to form substances ranging from intermetallic phases with properties like those of the elements, to well-defined groups of compounds when the electro-negativities of the constituent elements differ considerably. Several such groups of compounds are discussed in later sections.

Compounds with Zinc and Cadmium

Arsenides of composition M_3As_2, MAs_2 are formed by zinc and cadmium (Table 13). The compounds M_3As_2 are dimorphic, with one form like the arsenides of the calcium group. They have been investigated as semiconductors: the cadmium compound has an

TABLE 13. PROPERTIES OF ZINC AND CADMIUM DERIVATIVES

	M.p. (°C)	Dissociation pressure (p in mmHg)	Enthalpy change (kJ mol^{-1})	Energy of gap (eV)
$ZnAs_2$	768 [a]	$\log p = 12300/T + 15.0$ [b]	78.6	0.92 [c]
Cd_3As_2		$\log p = 6600/T + 9$ [d]	418	0.5 [e]
$CdAs_2$		$\log p = 7100/T + 11$ [b]	46	1.1 [e]
CdSb		$\log p = 7200/T + 10$ [d f]	137.9	0.51

[a] Under arsenic pressure.
[b] For $MAs_2(s) \rightarrow \frac{1}{3}M_3As_2(s) + \frac{1}{3}As_4(g)$.
[c] V. J. Lyons, *J. Phys. Chem.* **63** (1959) 1142.
[d] For $Cd_3As_2(s) \rightarrow 3Cd(g) + \frac{1}{2}As_4(g)$.
[e] V. J. Lyons and V. J. Silvestri, *J. Phys. Chem.* **64** (1960) 266.
[f] For $CdSb(s) \rightarrow Cd(g) + Sb(s)$. J. V. Silvestri, *J. Phys. Chem.* **64** (1960) 826; W. I. Turner, A. S. Fischler and W. E. Reese, *J. Electrochem. Soc.* **106** (1959) 206C.

exceptionally high electron mobility and has possibilities as an Ettingshausen cooler (see p. 576). The orthorhombic compounds ZnSb and CdSb, also semiconductors, have distorted diamond-type lattices, with each atom surrounded by four others (three the same and one different) at the corners of a tetrahedron. The shortest interatomic distances are like those in the metals[25],

e.g. in CdSb: Cd–Cd, 299; Cd–Sb, 280–291; Sb–Sb, 281 pm
in ZnSb: Zn–Zn, 258; Zn–Sb, 266–274; Sb–Sb, 281 pm
Cf. Cd–Cd in metal 298, Zn–Zn in metal 266 pm (sixfold co-ordination)

[25] K. E. Almin, *Acta chem. scand.* **2** (1948) 400.

Bismuth and mercury are miscible, but there is no evidence for compound formation; solid bismuth and cadmium are largely insoluble in each other, and bismuth and zinc form a simple eutectic.

III–V Compounds

Arsenic, antimony and bismuth give compounds ME with the elements of the aluminium group. The cubic zincblende structure predominates at normal pressures (see Table 8). The only exceptions are thallium compounds, TlSb and TlBi, which have the caesium chloride structure, and indium bismuthide InBi, which has the structure shown in Fig. 6. Each indium atom has four bismuth neighbours at the corners of a tetrahedron, and

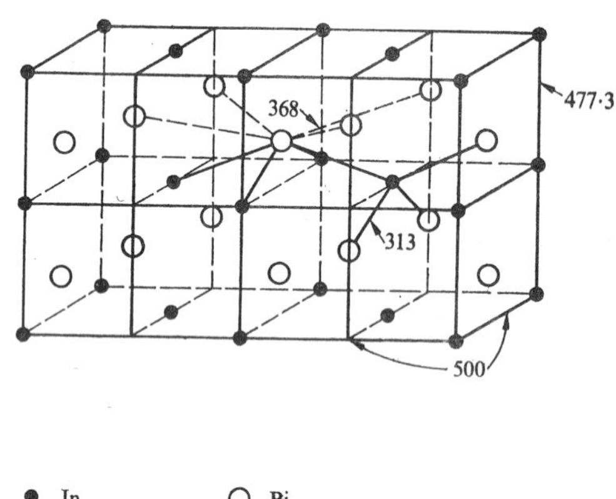

FIG. 6. The tetragonal structure of indium bismuthide (W. P. Binnie, *Acta Cryst.* **9** (1956) 686). (Distances in pm.)

each bismuth has four indium neighbours in a square to one side. Because of their very extensive applications in the electronics industry, the compounds between arsenic or antimony and aluminium, gallium or indium, often called III–V compounds, are some of the most widely studied of all substances. Very full accounts of their preparation[1] and properties[26–28] have been published.

The principal method for making III–V compounds is by the interaction of stoichiometric quantities of the elements in sealed tubes at elevated temperatures. Most applications demand extremely stringent control of impurities. These are inevitably introduced during preparation, and the procedure adopted for one application may be quite unsuitable for another. Some of the problems are indicated here in a general way; full experimental details may be found in specialized works:

(a) There may be impurities in starting materials. These are minimal in modern commercially available metals.

[26] R. K. Willardson and A. C. Beer, *Semiconductors and Semimetals*, Academic Press. Several volumes in this series are devoted to III–V compounds.

[27] C. Hilsum and A. C. Rose-Innes, *Semiconducting III–V Compounds*, Pergamon (1961).

[28] *Gallium Arsenide*, Institute of Physics and Physical Society (1967); O. Madelung, *Physics of III–V Compounds*, Wiley (1964).

(b) Several of the compounds have appreciable dissociation pressures at their melting points (Table 14). These must be taken into account in controlling stoichiometry. For example, gallium arsenide is made in a sealed quartz ampoule containing gallium at one

TABLE 14. PROPERTIES OF III–V COMPOUNDS[a, b e]

	AlAs	AlSb	GaAs	GaSb	InAs	InSb
Lattice constant a (pm) at 18.0°C	566.0 [c]	613.55	565.34	609.54	605.84	647.88
Bond length $\sqrt{3}a/4$ (pm)	245.0	265.5	244.9	263.9	262.1	280.0
Melting point (°C)	1740 [c]	1080	1240	712	943	533
Approximate dissociation pressure at m.p. (atm)	1		0.9		0.3	
ΔH°_{f298} (kJ mol^{-1})	−146.5	−104.7	−83.7	−44	−54.4	−33.5
S°_{298} (J mol^{-1} K^{-1})		63	63	75	75.8	86.2
ΔG°_{f298} (kJ mol^{-1})		−102	−77.5	−37.6	−50.3	−25.1
$\Delta H_{atom298}$ (kJ mol^{-1}) [d]	774	694	663	584	600	538
Energy of gap E_g (eV)	2.25 [c]	1.62	1.39	0.67	0.36	0.17
Electron mobility (m^2 V^{-1} s^{-1}) μ_n		0.02	0.85	0.4	3.3	7.8
Hole mobility (m^2 V^{-1} s^{-1}) μ_p		0.042	0.042	0.14	0.046	0.075
Refractive index		3.18	3.30	3.74	3.42	3.96
Dielectric constant		11	12.5	15	14	17

[a] N. N. Sirota in Vol. IV, of ref. 26.
[b] Ref. 27.
[c] W. Kischio, *Z. anorg. allgem. Chem.* **328** (1964) 187. Melting point under 1 atm.
[d] Heats of atomization calculated using heats of atomization of the elements in ref. 11.
[e] Boron arsenide (BAs) is similar to the other III–V compounds. It has the zincblende structure with $a = 477.7$ pm. Above ~1000°C it loses arsenic with formation of an orthorhombic lower arsenide $B_{5-7}As$. Boron arsenides are very inert chemically but dissolve in molten sodium hydroxide. (J. A. Perri, S. LaPlaca and B. Post, *Acta Cryst.* **11** (1958) 310; F. V. Williams and R. A. Ruehrwein, *J. Am. Chem. Soc.* **82** (1960) 1330.)

end and arsenic at the other. The gallium is slowly heated to the melting point of gallium arsenide and the arsenic to 605°C so that the arsenic pressure is maintained at 0.9 atm (Table 14). The ingot of gallium arsenide is allowed to solidify from one end, so that impurities are somewhat segregated. Indium arsenide may be made similarly. The antimonides of gallium and indium have low dissociation pressures and may be made in ampoules without differential heating.

(c) Impurities may be incorporated from the boat or other container. Quartz, aluminium oxide or graphite are commonly used for containers, but none is wholly free from disadvantages. For example, silicon may be incorporated from quartz into molten gallium arsenide:

$$2Ga_{(GaAs)} + SiO_2 \rightarrow 2GaO + Si_{(GaAs)}; \qquad 3GaO \rightarrow Ga_2O_3 + Ga$$

This reaction is minimized by working at temperatures as low as possible and by careful control of temperature gradients to prevent transport of the oxide GaO to cooler regions where it can disproportionate[29].

[29] J. E. Wardill and D. J. Dowling, *Chem. Brit.* **5** (1969) 226.

(d) Oxygen and water must be rigorously excluded, as oxygen is difficult to remove from the product. Hydrogen, helium or argon may be used as inert atmospheres.

(e) The III–V compounds expand on freezing (e.g. volume changes: GaSb, 7%; InSb, 12%), and allowance must be made for this so that the container does not break.

Some of the difficulties associated with preparations at high temperatures and pressures are avoided if the III–V compounds can be made at lower temperatures, e.g. by elimination of methane from adducts such as Me_3In,SbH_3 or Me_3Ga,AsH_3, or by passing hydrogen and arsenic trichloride vapour over heated gallium. The latter is converted to the monochloride GaCl which reacts with arsenic vapour formed by reduction of the trichloride:

$$6GaCl + As_4 \rightarrow 4GaAs + 2GaCl_3$$

Large crystals are difficult to prepare by this process.

Purification of III–V compounds may be effected by zone melting, but for arsenides, the pressure of arsenic must be controlled to maintain the stoichiometry. The purity may be assessed[1] by emission or mass spectroscopy, colorimetry, radioactivation analysis or by examination of electrical properties (see below).

Properties of III–V Compounds

The lattice constants and bond lengths are known with very great accuracy, and accurate electron density maps have been constructed. In contrast to the other intermetallic compounds discussed in the previous sections, co-ordination numbers in the III–V compounds are low (four), and the M–M and E–E distances are much larger than in the elements. Thus the bonding is between unlike atoms. The zincblende structures lack centres of symmetry, so the [111] and [$\bar{1}\bar{1}\bar{1}$] directions are inequivalent (Fig. 7). Careful examination of X-ray intensities suggests that at (111) faces, M atoms are bound to three E atoms

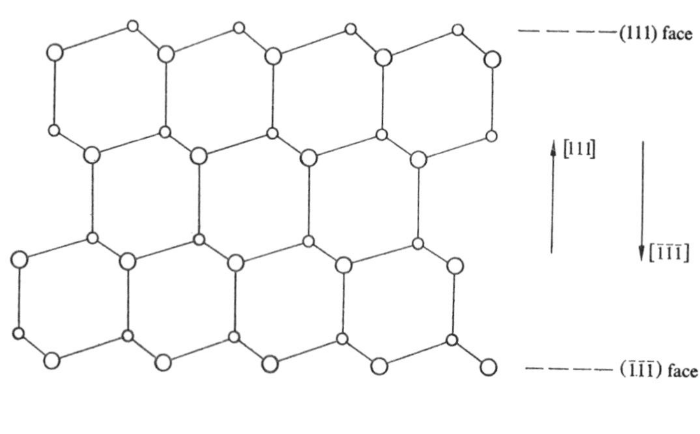

FIG. 7. Projection of the zincblende structure.

within the crystal, and at ($\bar{1}\bar{1}\bar{1}$) faces, E atoms are bound to three M atoms within the crystal. The inequivalence of the [111] and [$\bar{1}\bar{1}\bar{1}$] directions shows in the rates of chemical reactions. Thus etching agents such as $HF–HNO_3$ attack (111) faster than ($\bar{1}\bar{1}\bar{1}$) faces. Crystals of indium antimonide grow more rapidly along [$\bar{1}\bar{1}\bar{1}$], where lone pairs of E atoms are exposed, than along [111]. The polarity of the [111] directions gives piezoelectric effects which have been observed in InSb at low temperatures[27]

The low co-ordination numbers and open structures account for the expansion on freezing and the consequent fall in the melting point as the pressure increases. Dissociation pressures are not certainly known at present. The heats of formation have been obtained, e.g. by molten tin solution calorimetry; arsenides are 20–40 kJ mol^{-1} more exothermic than antimonides. Since there is a difference of about 40 kJ mol^{-1} in the heats of atomization of arsenic and antimony, the heats of atomization of the arsenides, which may be taken as a measure of the strength of the bonding, are 60–80 kJ mol^{-1} more than the heats of atomization of the antimonides. The stronger bonding in arsenides results in part from smaller M–As compared with M–Sb distances. Variations in properties which depend on the tightness of the bonding, such as hardness, melting point and the energy gap E_g between valence and conduction bands (Table 14) are thus readily understood.

The importance of the III–IV compounds arises mainly from their electrical properties. These depend on the number of carriers capable of transferring charge, and the characteristics of the lattice, which determine the energy gap and the mobilities. For a given compound, the number of carriers can in principle be controlled, e.g. by variation from exact stoichiometry or by addition of impurities. Many of the measured electrical properties depend on the net effects of negative carriers (electrons) and positive carriers (holes). If the effects of positive carriers dominate, the substance shows a positive Hall effect coefficient and the semiconductor is *p*-type; if the effects of negative carriers dominate, the Hall effect coefficient is negative and the substance is an *n*-type semiconductor. By careful addition of impurities it is possible to make III–V compounds either *p*- or *n*-type, and many devices depend on this possibility. The most commonly added impurities are the acceptors—zinc, cadmium or magnesium—which give *p*-type semiconductors, or the donors—sulphur, selenium or tellurium—which give *n*-type semiconductors. Transition metal atoms or elements of Group IV, e.g. Ge, Sn, Pb, may be either donors or acceptors. Impurities are sometimes distributed evenly by zone levelling, in which a narrow molten zone is passed along a cylindrical crystal or ingot successively in opposite directions. The main difficulties in the development of the III–V compounds arise from unwanted impurities from starting materials or contamination during preparation, and the full potential of these compounds is far from realized. It is, of course, much easier to prepare an element, e.g. germanium, with an impurity content of less than 1 ppm than to prepare a compound with a similar impurity level.

In spite of these difficulties, research on III–V compounds has been pursued because applications of semiconductors are limited by lattice properties as well as by impurities. The wide variations in energy gap and mobilities afforded by the III–V compounds increases their potential compared with silicon and germanium, provided difficulties in preparation can be overcome. Some possibilities are indicated very briefly. Above a certain temperature which depends on E_g, the majority of carriers are derived from electrons excited from the valence to the conduction band. It is then impossible to make devices which depend on *n–p* junctions. The maximum operating temperature for a silicon junction is 250°C (E_g 1.09 eV) and for germanium is 100°C (E_g 0.66 eV). Table 14 shows that gallium arsenide should be effective at higher temperatures (up to 450°C) and aluminium antimonide should be usable to 500°C; the indium compounds InAs, InSb are suited for junctions operating at low temperatures. In certain cases, gallium arsenide, with its high mobility and low dielectric constant, may be preferable to germanium, e.g. for variable capacity diodes. Gallium arsenide may also be used for microwave rectifiers, switching diodes (these have high on–off speeds), and tunnel diodes. The Hall effect can be exploited in a range of devices to measure magnetic fields, or in multipliers. The Hall effect coefficient R_H is given by $R_H = \mu/\sigma$, where μ is the carrier mobility and σ the electrical conductivity. The power transferred from an input circuit to a load connected to the Hall electrodes is proportional to the square of the mobility. Indium antimonide is thus 400 times as efficient as germanium ($\mu_n = 0.4\ m^2\ V^{-1}\ s^{-1}$). The III–V compounds have suitable energy gaps and high carrier mobilities and so show important photoelectric and electroluminescent effects. For gallium arsenide, the band structure is such that, above a certain current, the electroluminescent radiation becomes coherent, and a laser, with electrical rather than optical pumping, can be made. This is not as good as other solid-state lasers, but improvement and development may be possible. Another application is in solar batteries. It is calculated that in outer space, where the ultraviolet radiation of the sunlight is not absorbed, the optimum energy gap for a battery material is about 1.6 eV. Clearly, GaAs and AlSb are superior to silicon, which is commonly used.

Most of the applications so far have been made with gallium arsenide and indium arsenide and antimonide. The aluminium compounds could be useful for some applications, but these are limited by (a) their hydrolysis in moist air, and (b) the fact that, at their high melting points, containers are readily attacked, and so impurity contents are difficult to regulate.

Alloys with Germanium, Tin and Lead

The germanium arsenides are GeAs, m.p. 737°C, and $GeAs_2$, m.p. 732°C. The diarsenide has a layer structure which, from a chemical point of view, is intermediate between those of the two elements[30]. Thus each germanium atom has four arsenic neighbours and each arsenic atom makes three short bonds and three longer, as in arsenic metal. The shortest As–As distances are 250 and 315 pm (cf. p. 549). Tin gives two compounds with arsenic. The compound SnAs has the NaCl structure and may be represented as $Sn^{2+}Sn^{4+}As_2^{3-}$; the compound Sn_4As_3 is similar. The tin arsenides are superconductors at low temperatures[31] (T_c: SnAs, 3.41 . . . 3.65 K; Sn_4As_3, 1.16 . . . 1.19 K). The cubic SnSb is a component of some bearing alloys (Table 15A). These require a hard constituent in a softer matrix. In the tin-based alloys the hard compounds SnSb and Cu_6Sn_5 are dispersed in a tin–copper–antimony matrix, and in the lead-based alloys SnSb is in a Pb–SnSb eutectic.

Lead and antimony form a system with a simple eutectic, without intermediate phases, and the only intermediate phase in the lead–bismuth system is unstable at its melting point. The lead–bismuth eutectic is an important superconducting alloy (critical temp., 7.3. . .8.8 K; critical magnetic field at 4.2 K, 1.5 T; critical current density, $10^6\ A\ m^{-2}$). The ternary eutectic Sb–Sn–Bi and the quaternary eutectic Sb–Sn–Bi–Cd are important commercially because of their low melting points. They form the basis of the "fusible alloys", used for safety plugs in boilers and fire sprinklers, low-temperature solders, numerous applications in the plastics industry, and for cores in electroforming or in tubes which are to be bent. The fusible metal cores are readily removed by melting. Antimony and bismuth alloys are also used in printing. Alloys with more than 47% Bi or more than 75% Sb expand on freezing: those with less antimony or bismuth, and more tin or lead, contract.

30 J. H. Bryden, *Acta Cryst.* **15** (1962) 167.
31 S. Geller and G. W. Hull, *Phys. Rev. Letters* **13** (1964) 127.

TABLE 15A. COMMERCIAL ALLOYS [a]

Deoxidized arsenical copper	As, 0.35%	
Tough pitch arsenical copper	As, 0.4%	
Lead shot	As, 0.3–1%	
Antimonyl lead[b]	Sb, 0.25–8%	
Regulus metal (lead base)[c]	Sb, 8–20%	
Printing alloys (lead based)		
Electroplating metal	Sb, 2–5; Sn, 2–4%	
Slug-casting metal	Sb, 10–13; Sn, 2–5%	
Monotype metal	Sb, 15–19; Sn, 6–10%	
Cast-type metal	Sb, 17–27; Sn, 9–18%	
Brittania metal (tin base)	Sb, 5–10; Cu, 1–3%	
Die casting metal (tin base)	Sb, 8; Cu 8%	
Bearing alloys[d]		
Lead-base whitemetal	Sb, 15; Sn, 5–10%	
Tin-base whitemetal	Sb, 5–10; Cu 2%	
Fusible alloys[e]		
Binary eutectic	Bi, 55.5; Pb, 44.5%	m.p., 124°
Cerromatrix[f]	Bi, 48; Sb, 9; Sn, 14.5; Pb, 28.5%	m.p.[g], 103s, 227°l
Rose's alloy	Bi, 50; Sn, 22; Pb, 28;	m.p.[g], 96s, 110°l
Wood's alloy	Bi, 50; Sn, 12.5; Pb, 25; Cd, 12.5%	m.p.[g], 70s, 72°l
Cerrolow 117 [f]	Bi, 44.7; Sn, 8.3; Pb, 22.6; Cd, 5.3; In, 19.1%	m.p., 47°

[a] C. J. Smithells (ed.), *Metals Reference Book*, 4th edn., Butterworths, London (1967).
[b] For cable coverings, *ca.* 1% Sb; pipes, *ca.* 6% Sb; battery plates, *ca.* 9% Sb.
[c] Improved hardness and corrosion resistance.
[d] P. G. Forrester, *Babbit Alloys for Plain Bearings*, Tin Research Institute (1963).
[e] *Fusible Alloys containing Tin*, Tin Research Institute (1967).
[f] Registered trade mark of Cerro de Pasco Corporation, New York, and Mining Chemical Products Ltd., London. Cerromatrix has a particularly high tensile strength.

[g] s solidus; l liquidus (°C).

With suitable constituents there is little volume change on freezing, and so very sharp images can be taken by solid casts. Tin-based alloys with antimony can also take sharp images because of the formation of rigid shells of the compound SnSb. Table 15A shows only a very small selection of commercially available alloys. Materials with a wide range of melting points and other physical properties have been made.

Antimony–Arsenic and Antimony–Bismuth Alloys

Antimony and bismuth are completely miscible in both solid and liquid, and the solid alloys have the same rhombohedral structure as the elements. In bismuth the valence and conduction bands overlap by 0.0184V, so the element is a metallic conductor; but addition of a few per cent of antimony separates the bands so that the alloy becomes semiconducting. The energy separation rises to a maximum at about 12 atomic % Sb and then falls. Alloys with more than 40 atomic % Sb are metallic. The changes in band structure are revealed by measurements of resistivities and Hall coefficients, which show anomalies for materials with an energy gap near to the maximum. Similar electrical properties are found for arsenic–antimony alloys: those with 9–40 atomic % As are semiconducting between 240 and 310 K[32].

[32] A. L. Jain, *Phys. Rev.* **114** (1959) 1518; S. Tanuma, *J. Phys. Soc. Japan* **14** (1959) 1246; **16** (1961) 2349; K. Tanaka, *ibid.* **20** (1965) 1374; M. Ohyama, *ibid.*, p. 1538.

Alloys of bismuth with a few per cent of antimony show large galvanothermomagnetic effects, and attempts have been made to utilize these in solid-state cooling devices[33]. When a current I_y, perpendicular to a field H_z, is passed along a conductor (Fig. 8), the

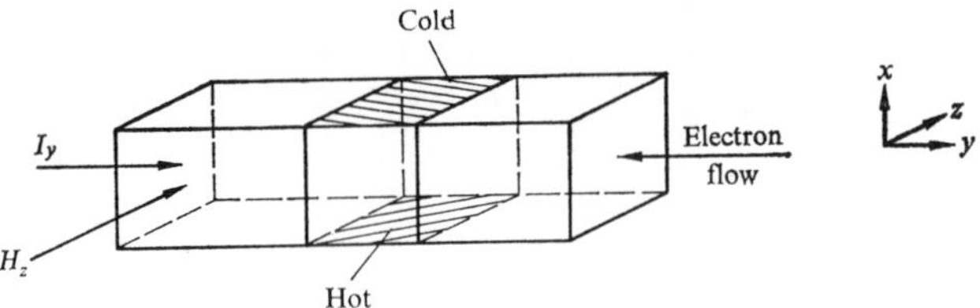

FIG. 8. The Ettingshausen effect.

electrons, which constitute the current, are deflected by the magnetic field. This results in (a) an opposing electric field along x perpendicular to both current and magnetic field (Hall effect), and (b) a thermal gradient along x (Ettingshausen effect). For most conductors, the Ettingshausen effect is small. However, when the current is carried by approximately equal numbers of electrons and holes, both positive and negative carriers are deflected in the same direction by the magnetic field. The net Hall effect is small, and the effect of the magnetic field is to produce a large thermal gradient. In any device to utilize this effect, the current must be carried through electrodes which are themselves good electrical and thermal conductors, and which therefore tend to diminish the thermal gradient. However, if the material used has a low thermal conductivity and if the conductor is long and thin, as shown in the figure, a large thermal gradient may be maintained at the centre (shaded). Equal numbers of holes and electrons are most easily obtained in semiconductors with small energy gaps or in metals where overlap of valence and conduction bands is small. Lattice thermal conductivity is smallest in structures with heavy atoms, and can be further reduced by limited isomorphous replacements with lighter atoms. Bismuth–antimony alloys seem on these criteria to be suitable materials, and those with about 3% Sb are best. The cooling may be enhanced by shaping the block of thermomagnetic material or by using single crystals, e.g. for Bi–Sb the optimum directions are with the current along the threefold axis, the heat flow along the twofold axis and the field perpendicular to both these directions. The effect is temperature dependent and bismuth–antimony Ettingshausen coolers are most efficient in the range from 77 to 200 K. At 77 K fields provided by a small permanent magnet are suitable. Further improvement may be possible by doping to control numbers and mobilities of carriers, and by alloying to cut down thermal conductivity.

3. GENERAL ASPECTS OF THE CHEMISTRY OF ARSENIC, ANTIMONY AND BISMUTH

3.1. ATOMIC PROPERTIES

Ionization Potentials

Values of ionization potentials from atomic spectra are given in Table 15B. The first

[33] T. C. Harman and J. M. Honig, *Semiconductor Products* (July 1963), p. 19; S. R. Hawkins, C. F. Kooi, K. F. Cuff, J. L. Weaver, R. B. Horst and G. M. Enslow, *Adv. Cryogenic Engineering* **9** (1964) 367.

ionization potentials, corresponding to the removal of electrons from the outer p orbitals, decrease with increasing atomic number, as in most groups of the Periodic Table. However, the second, third and fourth ionization potentials of bismuth are slightly greater than those of antimony.

Ionization potentials are illustrated further in Fig. 9, which shows, A, the energy required to remove the first three electrons ($s^2p^3 \rightarrow s^2$), and, B, the energy to remove the next two ($s^2 \rightarrow s^0$). The line B for removal of the fourth and fifth electrons, instead of falling smoothly

TABLE 15B. IONIZATION POTENTIALS[a]

Configuration[b]	Ground state	As		Sb		Bi	
		cm^{-1}	kJ mol^{-1}	cm^{-1}	kJ mol^{-1}	cm^{-1}	kJ mol^{-1}
E $(n-1)d^{10}ns^2np^3$	$^4S_{1\frac{1}{2}}$	79 165	947	69 700	833.6	58 790	703.1
E^+ $(n-1)d^{10}ns^2np^2$	$^3P_0$	150 290	1798	133 327	1592	134 600	1609
E^{2+} $(n-1)d^{10}ns^2np^1$	$^2P_{0\frac{1}{2}}$	228 670	2734	204 248	2450	206 180	2466
E^{3+} $(n-1)d^{10}ns^2$	$^1S_0$	404 369	4834	356 156	4255	365 500	4370
E^{4+} $(n-1)d^{10}ns^1$	$^2S_{0\frac{1}{2}}$	505 136	6040	449 300	5400	451 700	5403
E^{5+} $(n-1)d^{10}$	$^1S_0$	1 028 800	12300	868 140	10420	712 000	8520

[a] C. E. Moore, *Atomic Energy Levels*, NBS Circular 467 (1952, 1958).
[b] As, $n = 4$ (atomic number $Z = 33$); Sb, $n = 5$ ($Z = 51$); Bi, $n = 6$ ($Z = 83$).

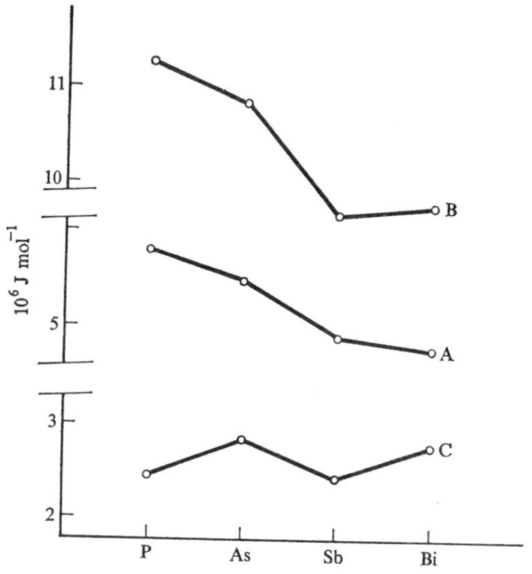

FIG. 9. A, sum of first three ionization potentials. B, sum of fourth and fifth ionization potentials. C, $s^1 \rightarrow d^1$ promotion energies for E^{4+}.

from phosphorus to bismuth, shows a pronounced step. The 4*s* electrons in arsenic penetrate within the $3d^{10}$ shell and are thus held more firmly than expected by extrapolation from phosphorus where the 3*d* shell is empty. Similarly, the 6*s* electrons in bismuth penetrate within the $4f^{14}$ shell and so are held more firmly than expected by extrapolation from antimony where the 4*f* shell is empty. A similar "alternation" shows in line *A*, and in plots of some other electronic parameters, such as the $s^1 \rightarrow d^1$ promotion energies for E^{4+} shown in line C [34].

Sizes

Although the size of an atom depends on its environment, it is useful to have some radius for each element which indicates roughly its size relative to other elements. Values given for arsenic, antimony and bismuth are as follows:

	As	Sb (pm)	Bi
From metal	125	145	153
Tetrahedral	118	136	146

Radii from metals are half the shortest internuclear distances (p. 549), and refer to situations in which bonds to nearest neighbours make angles just over 90°. Tetrahedral radii are derived from compounds with the zincblende (p. 570) or pyrites structures; internuclear distances in these compounds are well reproduced by addition of appropriate tetrahedral radii[35]. The radii of arsenic, antimony and bismuth in most other compounds may be taken as within the limits shown above.

3.2. BINARY COMPOUNDS

The ionization potentials discussed in the previous section show that the first three electrons may become involved in chemical bonding much more easily than the next two. Involvement of more than five electrons in bonding requires prohibitively large amounts of energy. There are thus two well-established series of compounds, in which the elements E show oxidation states III (sections 4–10) and V (sections 11–15). In a few compounds, arsenic, antimony, and bismuth show oxidation states less than three (sections 4.3 and 5.5), and in others the Group V elements appear as both E^{III} and E^{V} (section 16). Compounds of As^{V} and Bi^{V} are much less well known than the corresponding compounds of Sb^{V}. This has been attributed[34] to the effects of electron penetration (see previous section). Analogous effects appear in the chemistry of the elements S–Po and the halogens.

There are still many gaps in the thermodynamic data on compounds of arsenic, antimony and bismuth. Some free energies of formation (per gram atom of E) of binary compounds have been plotted in Fig. 10.

---

[34] R. S. Nyholm, *Proc. Chem. Soc.* (1961) 276; W. E. Dasent, *Non-existent Compounds*, Edward Arnold, London (1965).

[35] L. Pauling, *The Nature of the Chemical Bond*, 3rd edn., Cornell University Press, Ithaca, NY (1960), Chs. 7 and 11.

It is clear that the elements arsenic, antimony and bismuth are very closely related with rather small variations in stability within a series of compounds with a given element. Mean bond energies $\bar{H}$ (Table 16) have been calculated where the necessary data is available,

$$-3\bar{H}(\text{As—H}) = -\Delta H_f^\circ\text{As(g)} - 3\Delta H_f^\circ\text{H(g)} + \Delta H_f^\circ\text{AsH}_3\text{(g)}$$

For a given element E, E–X bond energies decrease with increasing atomic weight of X.

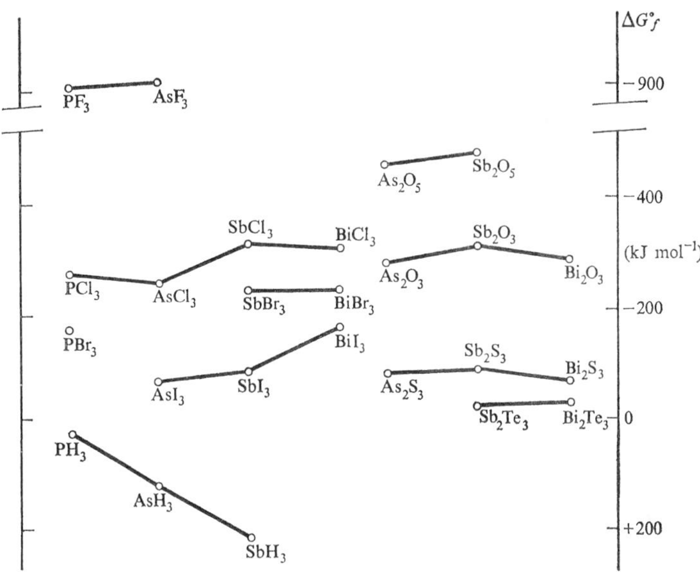

FIG. 10. Free energies of formation per gram atom E.

For compounds with a given element X, there are quite small changes in E–X bond energies for the series arsenic, antimony and bismuth. Pauling[35] electro-negativities, derived empirically from bond energies, reflect the similarity of the three elements. Allred and Rochow[36] electro-negativities, calculated from the force on electrons at distances from the nuclei equal to the covalent radii, decrease from arsenic to bismuth:

Electro-negativities	As	Sb	Bi
Pauling	2.0	1.9	1.9
Allred and Rochow	2.20	1.82	1.67

Bond dissociation energies for diatomic molecules, derived from spectroscopy, have also been collected in Table 16. There is a general decrease in bond dissociation energy D(E–X) for the series (E =) arsenic, antimony and bismuth.

[36] A. L. Allred and E. G. Rochow, *J. Inorg. Nucl. Chem.* **5** (1958) 264.

TABLE 16. BOND ENERGIES AND BOND DISSOCIATION ENERGIES

Mean bond energies (kJ mol^{-1}) [a]					
AsO	276				
AsH	296	SbH	257	BiH	194 [h]
AsC [b]	229	SbC	215	BiC	143
AsF [c]	484				
AsCl	306	SbCl [c]	313	BiCl	279
AsBr	256	SbBr	264	BiBr	233
AsI	195	SbI	194	BiI	180
Bond dissociation energies of diatomic molecules (kJ mol^{-1}) [d]					
As_2	380	Sb_2	289	Bi_2	195
As_2^+	250				
AsO	470	SbO	390	BiO	339 [e]
AsO^+	740			BiS	312
AsN	480	SbN	300	BiSe	276
				BiTe	228
				BiH	240
		SbF	430	BiF	255
		SbCl	350	BiCl	298 [f]
		SbBr	310	BiBr	264
				BiI	215 [g]

[a] For compounds EX_3. From data in subsequent tables and ref. 11. See footnote c for data for compounds EX_5.

[b] For methyl compounds.

[c] As^V–F, 385; Sb^V–Cl, 253 kJ mol^{-1}.

[d] Except where indicated from A. G. Gaydon, *Dissociation Energies and Spectra of Diatomic Molecules*, 3rd edn., Chapman & Hall, London (1968). Many of these values are of low accuracy and so they have been rounded to the nearest 10 kJ mol^{-1}.

[e] O. M. Uyz and J. Drowart, *Trans. Farad. Soc.* **65** (1969) 3221.

[f] D. Cubicciotti, *J. Phys. Chem.* **71** (1967) 3067.

[g] D. Cubicciotti, *Inorg. Chem.* **7** (1968) 211.

[h] Upper limit from appearance potentials, ref. 49.

3.3. CHEMISTRY IN AQUEOUS SOLUTION

Most binary compounds of arsenic, antimony and bismuth either react with water or are insoluble in it. The chemistry of aqueous solutions of these elements is the chemistry of oxyions, or complexes with ligands other than water: species $E(H_2O)_m^{n+}$ are probably rarely formed. For many purposes, however, the actual species in solution do not matter and equations may be written with non-committal symbols such as As^{III} aq. or Sb^V aq. The basicity of the metal increases from arsenic to bismuth. Thus there is a well-developed series of arsenites and arsenates; arsenic usually appears in anionic species. Bismuth gives a series of compounds which can formally be considered as Bi^{III} salts, but these do not give clear solutions in water except in high acid concentration, when complex species are present. On dilution, basic salts, which formally contain the bismuthyl ion BiO^+, precipitate. These compounds, though often with simple stoichiometry, rarely have simple structures; usually bismuth–oxygen frameworks (oxycations) have anions interposed. In alkaline solution, bismuth precipitates as bismuth hydroxide, $Bi(OH)_3$. Antimony is intermediate: it forms both stable oxyanions, in antimonates and antimonites, and oxycations, in basic salts.

Relationships between the oxidation states are shown in Fig. 11, which is like those given by Phillips and Williams[37], with data for acid solution plotted with reference to $E^\circ = 0$

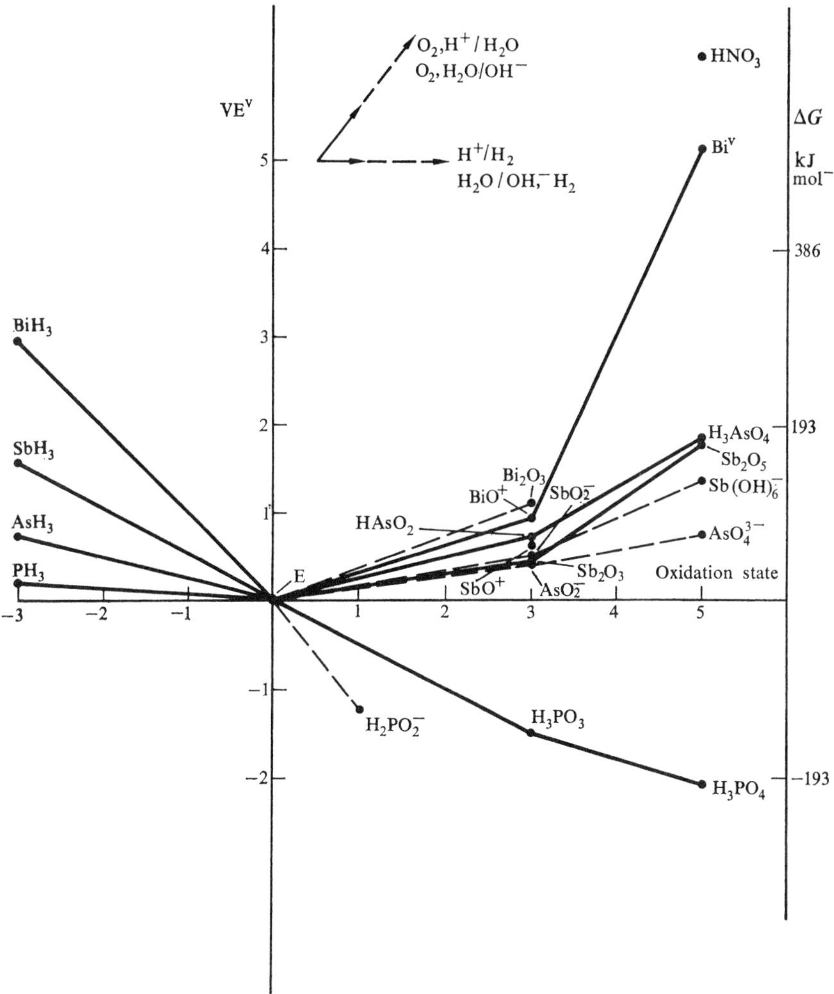

FIG. 11. Oxidation state diagram for arsenic, antimony and bismuth. Species present in solution are not necessarily those indicated. For example, AsO_2^- represents $H_2AsO_3^-$ or $As(OH)_4^-$. See sections 6.3 and 12.1–3.

for H^+ (a = 1)/H_2 and data for alkaline solution with reference to $E^\circ = 0$ for $H_2O/H_2,OH^-$ (a = 1). Oxidation-reduction data [$E^\circ = 0$ for H^+ (a = 1)/H_2] are also given in Table 17. The similarity between the three elements, and especially between arsenic and antimony, is apparent. The Bi^V–Bi^{III} couple is strongly oxidizing, able to oxidize water to oxygen. As^V and Sb^V are weak oxidizing agents, but stronger in acid

[37] C. S. G. Phillips and R. J. P. Williams, *Inorganic Chemistry*, 2 volumes (1965 and 1966).

TABLE 17. OXIDATION POTENTIALS[a]

	V
1. $As(s)+3H^{+}+3e^{-} \rightleftharpoons AsH_3(g)$	−0.239 [b]
2. $HAsO_2+3H^{+}+3e^{-} \rightleftharpoons As(s)+2H_2O$	+0.247$_5$
3. $\frac{1}{2}AS_2O_3(s)+3H^{+}+3e^{-} \rightleftharpoons As(s)+3/2H_2O$	+0.234
4. $H_3AsO_4+2H^{+}+2e^{-} \rightleftharpoons HAsO_2+2H_2O$	+0.559
5. $AsO_2^{-}+2H_2O+3e^{-} \rightleftharpoons As(s)+4OH^{-}$	−0.68
6. $AsO_4^{3-}+2H_2O+2e^{-} \rightleftharpoons AsO_2^{-}+4OH^{-}$	−0.67
7. $Sb(s)+3H^{+}+3e^{-} \rightleftharpoons SbH_3(g)$	−0.51
8. $SbO^{+}+2H^{+}+3e^{-} \rightleftharpoons Sb(s)+H_2O$	+0.212
9. $\frac{1}{2}Sb_2O_3(s)+3H^{+}+3e^{-} \rightleftharpoons Sb(s)+3/2H_2O$	+0.152
10. $Sb^{V}+2e^{-} \rightleftharpoons Sb^{III}$ (6 M HCl)	+0.818
11. $\frac{1}{2}Sb_2O_5(s)+2H^{+}+2e^{-} \rightleftharpoons \frac{1}{2}Sb_2O_3(s)+H_2O$	+0.671 [c]
12. $SbO_2^{-}+2H_2O+3e^{-} \rightleftharpoons Sb+4OH^{-}$	+0.66 [c]
13. $Sb^{V}+2e^{-} \rightleftharpoons Sb^{III}$ (10 M KOH)	−0.589
14. $Bi(s)+3H^{+}+3e^{-} \rightleftharpoons BiH_3(g)$	−0.97 [b]
15. $BiOCl(s)+2H^{+}+3e^{-} \rightleftharpoons Bi(s)+H_2O+Cl^{-}$	+0.160
16. $BiO^{+}+2H^{+}+3e^{-} \rightleftharpoons Bi(s)+H_2O$	+0.32 [c]
17. $Bi^{V}+2e^{-} \rightleftharpoons Bi^{III}$ (0.5 M H^{+}; I = 2 M)	+2.1 [d]
18. $\frac{1}{2}Bi_2O_3+3/2H_2O+3e^{-} \rightleftharpoons Bi(s)+3OH^{-}$	−0.46

[a] Except where indicated from ref. 38.
[b] Calculated from data of ref. 11.
[c] W. M. Latimer, *Oxidation Potentials*, 2nd edn., Prentice-Hall (1952).
[d] M. H. Ford-Smith and J. J. Habeeb, *Chem. Communs.* (1969) 1445.

than in alkaline solution. Many of these E^{III}–E^{V} oxidation-reduction reactions proceed cleanly and quantitatively, and so form the basis of wet methods of analysis (p. 583). By comparison, the high oxidation states of nitrogen are more strongly oxidizing, and those of phosphorus less strongly oxidizing than arsenic, antimony, or bismuth (Fig. 11). The greater basicity of bismuth, compared with antimony and arsenic, is shown in Fig. 11 by the lower line for acid, compared with alkaline, solution. For arsenic, the line for alkaline solution is lower, and for antimony points for acid and alkaline conditions are close. It must, however, be remembered that the positions representing various oxidation states on such diagrams depend on the precise conditions—particularly on the presence of ligands forming complexes or precipitates. Neither of the oxidation states III or V for the elements arsenic, antimony and bismuth disproportionates in solution.

The elements themselves are noble in that they are weaker reducing agents than hydrogen. They are stable with respect to disproportionation and so differ from phosphorus (Fig. 11). There are thus no reactions for the heavier elements of Group V corresponding to the reaction of phosphorus with alkali to give phosphine and hypophosphite (Chapter 20). The Sb_2O_3–Sb couple has been used in the "antimony electrode" which is suitable for the measurement of pH in the range from 3 to 8. The electrode is simple and robust, and usually

[38] L. G. Sillén and A. E. Martell, *Stability Constants of Metal Ion Complexes*, Chem. Soc. Special Publication, 17 (1964); L. G. Sillén, A. E. Martell, E. Högfeldt and R. H. Smith, *ibid.*, 25 (1971); D. D. Perrin, *Pure Appl. Chem.* **20** (1969) 133.

there is sufficient oxide on the surface of the antimony. It cannot be used in the presence of complexing agents like tartrates, or with ions of metals more noble than antimony.

3.4. ANALYSIS[39-41]

Arsenic and antimony are readily determined volumetrically. For example, the reaction between As^{III} oxide and iodine is complete in buffers of sodium bicarbonate, borax–boric acid or Na_2HPO_4–NaH_2PO_4:

$$\tfrac{1}{2}As_2O_3+I_2+H_2O \rightarrow \tfrac{1}{2}As_2O_5+2H^++2I^-$$

Other convenient oxidizing agents for As^{III} are potassium bromate, potassium iodate and potassium permanganate. Similar reactions occur with antimony compounds, but precipitation of basic salts, which may occur in all except the strongest acid, must be prevented by addition of tartrate, which forms strong complexes with Sb^{III} (section 6.4). Only the most powerful oxidizing agents will oxidize Bi^{III} to Bi^V (Fig. 11) and so volumetric methods applicable to arsenic and antimony cannot be used. Bi^{III} can, however, be titrated with ethylenediaminatetraacetic acid (H_4–EDTA). The very high formation constant for the complex $[Bi(EDTA)]^-$, 10^{28}, shows that it forms even in solutions of very low pH, where less strongly complexed cations do not interfere[40]. $[Bi(EDTA)]^-$ shows strong absorption in the ultraviolet, and so concentrations as low as 10^{-6} M can be estimated spectrophotometrically.

TABLE 18. GRAVIMETRIC METHODS

Element[a]	Precipitate	Compound weighed	Ignition temperature (°C)
As^{III}	As_2S_3 from 9 M HCl	As_2S_3	105
As^V	$MgNH_4AsO_4 \cdot 6H_2O$	$Mg_2As_2O_7$	800–900
As^V	$NH_4UO_2AsO_4 \cdot xH_2O$ from acetic acid	U_3O_8	800
Sb^{III}	Sb_2S_3 from 2.5 M HCl	Sb_2S_3	280–300 in CO_2
Sb^{III}	$Sb(C_6H_5O_3)$ [b] from tartrate soln.	$Sb(C_6H_5O_3)$	100–105
Bi^{III}	BiOI by boiling soln. containing Bi^{III} and I^-	BiOI	105–110 (BiOCl similarly but above 325)
Bi^{III}	$BiPO_4$	$BiPO_4$	380–900
Bi^{III}	Cupferron[c] complex	Bi_2O_3	960
Bi^{III}	Bi by reduction with H_2CO or by electrodeposition	Bi	105

[a] The element E must be in the appropriate oxidation state before the determination.
[b] Pyrogallol.
[c] *N*-nitroso-*N*-phenylhydroxylamine.

[39] I. M. Kolthoff and P. J. Elving, *Treatise on Analytical Chemistry*, Interscience, London, Part II, Vol. 8 (1963), Vol. 12 (1965).
[40] G. Schwarzenbach and H. Flaschka, *Complexometric Titrations*, Methuen, London (1969), pp. 306, 308, 309.
[41] F. D. Snell, C. T. Snell and C. A. Snell, *Colorimetric Methods of Analysis*, 3rd edn., Van Nostrand, Princeton, Vol. II (1954), Vol. IIA (1959).

Gravimetric methods for arsenic, antimony, and bismuth are listed in Table 18. The methods for antimony are not very satisfactory, and so volumetric estimations are usually preferred. Small amounts of the Group V elements may be determined colorimetrically[41]. Some examples are as follows. The wavelengths given indicate strong absorptions: the wavelength used in any particular estimation depends on the absorption from any interfering species:

As: as molybdenum blue, 840 nm; as diethylthiocarbamate, 560 nm

Sb: as Rhodamine B complex, 550 nm; as SbI_4^-, 330 nm

Bi: as BiI_4^-, 330 nm; as thiourea complex, 420 nm; as dithizonate, 500 nm; as diethylthiocarbamate, 370–400 nm; with EDTA, 265 nm. (In some of these determinations interference from antimony may be prevented by complexing with tartrate)

Arsenic can often be separated from complex mixtures containing heavy metals by distillation as chloride in HCl below 108°C. Germanium tetrachloride is the only similar compound to distil under these circumstances. Arsenic may also be separated as arsine by boiling the substance under test with zinc and acid. Some classical procedures are given in Table 19. In a more modern example, arsenic in hydrocarbons has been determined by adsorption on to silica gel and removal of organic matter by a mixture of sulphuric, nitric and perchloric acids. (This mixture is commonly used for the digestion of organo-arsenic compounds[39].) The arsenic is then reduced by zinc, and the arsine is absorbed in silver diethylthiocarbamate and determined spectrophotometrically[42]. Alternatively, the arsine may be converted to arsenate by iodine in bicarbonate buffer. Excess iodine is removed, and the arsenate treated with ammonium molybdate and reduced by hydrazine sulphate.

TABLE 19. ESTIMATION OF ARSINE AND STIBINE[a]

Marsh	AsH_3,SbH_3 decomposed by heat to metallic mirror. As, but not Sb, oxidized by NaOCl
Gutzeit	AsH_3,SbH_3 absorbed on paper impregnated with $AgNO_3$ or $HgCl_2$. Can be made quantitative by comparison with standard stains for known samples
Fleitmann	AsH_3 generated in alkaline soln (Al–NaOH); SbH_3 not formed

[a] See also W. A. Campbell, *Chem. Brit.* **1** (1965) 198.

The intensely coloured "molybdenum blue" may be determined colorimetrically. Antimony can also be separated from heavy metals as the chloride. Thus, when whitemetals (Table 15A) are treated with dry chlorine, tin and antimony are volatilized quantitatively and may be separately determined in the distillate[43]. All three metals—arsenic, antimony and bismuth—may be precipitated as sulphides using hydrogen sulphide or thioacetamide in acid solution. This forms the basis of their separation in classical analysis (section 7.4).

Traces of arsenic, antimony and bismuth may be estimated by emission spectrography or by neutron activation analysis[1].

[42] G. W. Powers, R. L. Martin, F. J. Piehl and J. M. Griffin, *Anal. Chem.* **31** (1959) 1589.
[43] E. Bishop, *Analyst* **78** (1953) 61.

4. HYDRIDES EH_3

Besides an earlier account[44], a critical review[45] gives the most satisfactory methods for laboratory preparations of the hydrides and lists many physical properties.

4.1. PREPARATIONS

Arsine (AsH_3), and stibine (SbH_3), have been obtained by the usual methods for the preparation of covalent hydrides. The most important of these are as Table 20 shows.

TABLE 20. PREPARATION OF HYDRIDES[45]

	Reagents	% yield	Ref.
AsH_3	$NaOH–As_2O_3–KBH_4$	59 [a]	b
AsD_3	$AsCl_3–LiAlD_4$ (−90°)	83 [c]	d
SbH_3	$C_4H_4O_7SbK \cdot \frac{1}{2}H_2O$ [e]–KBH_4	51 [a]	b
SbD_3	$SbCl_3–LiAlD_4$ (−90°)	82 [c]	d
BiH_3	$BiCl_3–LiAlH_4$ (−100°)	1 [c]	f

[a] Based on BH_4^- in equation $4E(OH)_4^- + 3BH_4^- + 7H^+ \rightarrow 4EH_3 + 3H_3BO_3 + 7H_2O$. The borohydride is the most expensive starting material.
[b] W. L. Jolly and J. E. Drake, *Inorg. Synth.* 7 (1963) 34.
[c] Based on ECl_3 converted to hydride or deuteride.
[d] E. Wiberg and K. Moedritzer, *Z. Naturforsch.* **12b** (1957) 123.
[e] Potassium tartratoantimonite (III) (see section 6.4).
[f] E. Amberger, *Chem. Ber.* **94** (1961) 1447.

Reductions with Lithium Aluminium Hydride or Potassium Borohydride

Lithium aluminium hydride must be used in diethyl or dibutyl ether. It is important to remove the gaseous products from the reaction mixture, usually by condensing them in liquid nitrogen, to prevent further reactions, e.g. with formation of $LiAl(AsH_2)_4$ or metallic antimony. Good yields based on arsenic or antimony trichlorides are readily obtained, but use of the hydride reducing agent is inefficient unless conditions are chosen carefully[45]. Potassium borohydride may be used in aqueous solution, with less possibility of attack on grease by solvent.

Solvolysis of Arsenides or Antimonides

Arsenic or antimony is alloyed with sodium, magnesium or zinc, usually by fusing proportions of the elements to give $M_3^{I}E$ or $M_3^{II}E_2$, and the arsenides or antimonides are treated, either with dilute sulphuric acid, or with ammonium bromide in liquid ammonia. The hydrides produced are contaminated with less volatile lower hydrides (section 4.3), but these are easily separated in a vacuum line.

[44] D. T. Hurd, *An Introduction to the Chemistry of the Hydrides*, Chapman & Hall, London (1952), p. 129.
[45] W. L. Jolly and A. D. Norman, *Preparative Inorganic Reactions* **4** (1968) 1.

Arsine and stibine are also formed by the electrolytic reduction of solutions containing As^{III} or Sb^{III} [45]. Since the hydrides are poisonous, their formation can constitute a hazard in plants using electrolytic processes on arsenic or antimony compounds, and, less obviously, where arsenic and antimony may be impurities in other metal electrodes, e.g. in lead batteries. It has been suggested that copper salts suppress the formation of hydrides.

Bismuthine is too unstable for preparations in high yield; but small amounts can be made by the methods described above. It must be stored in liquid nitrogen, and any measurements at room temperature must be completed in a few minutes.

TABLE 21. PROPERTIES OF THE TRIHYDRIDES[a]

	AsH_3	SbH_3	BiH_3
Melting point (°C)	−116.9	−88	
Latent heat of fusion (kJ mol^{-1})	1.20		
Boiling point (°C)	−62.5	−18.4	16.8 extrap.
Latent heat of evaporation (kJ mol^{-1})	16.75	21.0	25.2 (−43°C)
Vapour pressure (mmHg)	35 (−111.6°C) [b]	81 (−63.5°C) 224 (−45.2°C) [c]	4.6 (−78°C)
$H^\circ_{298}-H^\circ_0$ (kJ mol^{-1}) [d]	10.20	10.47	
ΔH°_f (kJ mol^{-1}) [d]	66.44	145.1	277.8 [e]
C°_p (J mol^{-1} K^{-1}) [d]	38.07	41.05	
S° (J mol^{-1} K^{-1}) [d]	222.7	232.7	
ΔG°_f (kJ mol^{-1}) [d]	68.91	147.7	
Ionization potential (kJ mol^{-1}) [f]	101		
Density (kg m^{-3})	1622 (−63°)	2150	
Bond length (pm) [g]	151.9±0.2 [m]	170.7±0.25 [n]	
HEH bond angle (°) [g]	91.8±0.3	91.3±0.3	
Vibrational frequencies (cm^{-1}):			
ν_1 (a_1)	2116 [h, i]	1891 [j, k]	
ν_2 (a_2)	905	781.5	
ν_3 (e)	2123	1894	
ν_4 (e)	1003	831	
Dipole moment (D)	0.22 [l]	0.116 [l]	
Nuclear quadrupole coupling Constant e^2Qq (MHz)	^{75}As: −160.1±0.4 [m]	^{121}Sb: 458.7±0.8 [n] ^{123}Sb: 586.0±0.8	
N.m.r. (ppm) (TMS = 10.0) [o]	8.50	8.62	

[a] Except where indicated, data from ref. 45.
[b] Log p_{mmHg} = $-1403.3/T - 9.45935 \log T + 0.008037T + 28.82835$.
[c] Log p_{mmHg} = $-1446.34/T - 3.1200 \log T + 1.48 \times 10^{-5} p + 16.0522$
[d] Ref. 11.
[e] Lower limit, from appearance potential.
[f] Ref. 46.
[g] Ref. 47.
[h] V. M. McConaghie and H. H. Nelson, *Phys. Rev.* **75** (1949) 633.
[i] AsD_3: ν_1, 1523; ν_2, 660; ν_3, 1529; ν_4, 714 cm^{-1}.
[j] W. H. Haynie and H. H. Nelson, *J. Chem. Phys.* **21** (1953) 1839.
[k] SbD_3: ν_1, 1359; ν_2, 561; ν_3, 1362; ν_4, 592 cm^{-1}.
[l] C. C. Loomis and M. W. P. Strandberg, *Phys. Rev.* **81** (1951) 798; cf. PH_3, 0.55 D.
[m] G. S. Blevins, A. W. Jache and W. Gordy, *Phys. Rev.* **97** (1955) 685; e^2Qq, −165.9±0.4 MHz in $^{75}AsD_3$.
[n] A. W. Jache, G. S. Blevins and W. Gordy, *Phys. Rev.* **97** (1955) 680; e^2Qq, 465.4±0.8 MHz in $^{121}SbD_3$, 592.8±0.8 MHz in $^{123}SbD_3$.
[o] Infinite dilution in CCl_4. Single line, narrowed on cooling by increased quadrupole relaxation from E nucleus. E. A. V. Ebsworth and G. M. Sheldrick, *Trans. Faraday Soc.* **63** (1967) 1071.

4.2. PROPERTIES

Thermodynamic and structural properties are given in Table 21. The melting points increase, and the volatilities decrease with increasing molecular weight. There is no evidence for association in the liquid. The hydrides are endothermic compounds, i.e. their heats of formation are positive, but this is largely a reflection of large heats of atomization of the elements: the E–H bond is moderately strong. The bond energies decrease from arsine to bismuthine. The hydrides are thus kinetically stable; they will decompose to the elements provided a mechanism is available. Care must be taken, for example, to keep the gases away from flames used for sealing ampoules. At room temperature, decomposition of arsine is slow, but it proceeds at a measurable rate on an arsenic, or, better, antimony surface at 300°C. The decomposition is first order in arsine[48]. A mixture of AsH_3 and AsD_3 gives some HD, but a mixture of AsH_3 and D_2 does not; exchange must therefore occur only between hydrogens bound to arsenic. Stibine is considerably less stable. Samples to be kept more than a day or two should be stored in liquid nitrogen. The extreme instability of bismuthine has been mentioned. The quantitative decomposition of stibine, but not arsine, to the elements can be initiated in a calorimeter by a small heated filament, and this provides a convenient method for the measurement of the energy of decomposition ($-\Delta E_f^\circ$). Mixtures of other hydrides (including AsH_3) and stibine are also quantitatively decomposed, making possible the determination of a large number of energies of decomposition. The strongest peaks in the mass spectra of the trihydrides are from MH^+ ions[49]; there are also peaks from MH_3^+, MH_2^+ and M^+. Along the series AsH_3, SbH_3, BiH_3, the relative abundances of the parent ions decrease, corresponding with weakening M–H bonds, and the relative abundances of M^+ increase, reflecting the decreasing ionization potential of the elements. The thermal decomposition of the hydrides is of historical interest in two ways. It formed the basis of Marsh's test (Table 19), which effectively decreased the use of a popular poison. Also, the detection of bismuthine, from the reaction of acid with an alloy of magnesium and thorium-C (^{212}Bi), was one of the first applications of radioactive tracers.

The standard entropies have been derived from spectral data, and, in the case of arsine, from specific heat measurements. These reveal a transition at 35 K associated with the onset of free rotation in the solid. The gaseous molecules are pyramidal (C_{3v}) and the bond lengths and angles have been found from microwave measurements and vibrational fine structure. The bond angles are close to 90°, smaller than in the trihalides, about the same as in phosphine (93.3°) but much smaller than in ammonia. The bond angles suggest that in arsine and stibine the metals use mainly *p* orbitals for bonding, so the lone pairs are largely *s*, but the quadrupole coupling constants show appreciable *s* involvement in the bonds. The molecules have very small dipole moments. The four fundamental vibrational frequencies, all active in both the infrared and Raman spectra, have been assigned in hydrides, deuterides and tritides. At low resolution, the peaks associated with the two M–H stretching frequencies overlap.

[46] F. E. Saalfeld and M. V. McDowell, *Inorg. Chem.* **6** (1967) 96.

[47] L. E. Sutton (ed.), *Interatomic Distances*, Chem. Soc. Special Publication 11 (1958). Supplement, *ibid.* 18 (1965).

[48] K. Tamaru, *J. Phys. Chem.* **59** (1955) 777.

[49] F. E. Saalfeld and H. J. Svec, *Inorg. Chem.* **2** (1963) 46, 50.

Arsine and stibine are soluble in organic solvents and slightly soluble in water. They are poor donors and very few complexes with Lewis acids are known (see section 10.1). Arsine is much more reluctant than phosphine or ammonia to accept protons to form arsonium salts. Infrared frequencies assigned to $[AsH_4]^+$ have been detected in the solid formed from equimolar mixtures of arsine and hydrogen bromide at 82 K, but they rapidly disappear on warming. No stibonium $[SbH_4]^+$ salts have been described.

The hydrides are good reducing agents (Fig. 11 and Table 17) and so are destroyed by most aqueous inorganic oxidizing agents. They are oxidized in air. The black precipitates formed with silver nitrate have been used for the detection of small quantities of the gases.

4.3. LOWER HYDRIDES

The compounds As_2H_4, As_2H_5 and Sb_2H_4 have been made only in very low yields, and so information about them is limited[50]. The catenated compounds are isolated as by-products in the preparation of the trihydrides from metal arsenides or antimonides or from borohydride reductions. They may also be made by passing the trihydrides through a silent electric discharge in an ozonizer. Attempts to make the lower hydrides by the reduction of As^{III} compounds by mild reducing agents such as stannous chloride, gave only arsenic metal. It seems that As_2H_4 and As_2H_5 form only in the presence of excess trihydride.

The lower hydrides decompose rapidly at room temperature with formation of the trihydrides and non-stoichiometric E_2H. There is very little structural data but it has been possible[49] to record the mass spectra of As_2H_4 and Sb_2H_4. The heats of formation (lower limits) of the lower hydrides, deduced from appearance potential measurements, are As_2H_4 148 and Sb_2H_4 239 kJ mol^{-1}, and the M–M bond energies (upper limits) are As–As 187 and Sb–Sb 128 kJ mol^{-1}.

From mixtures of arsenides and phosphides, or arsenides and nitrides, ternary hydrides AsH_2PH_2, AsH_2NH_2, $HAs(NH_2)_2$ and $As(NH_2)_3$ have been identified. Organometallic hydrides are described in section 9.4 and silyl and germyl derivatives in section 9.6.

5. HALIDES EX_3

The trihalides of arsenic, antimony and bismuth have been described in several reviews[51].

5.1. PREPARATION

Two main methods are available. (a) Stoichiometric quantities of the elements are heated in a sealed tube, or the halogen, diluted with nitrogen, is passed over the heated metal. Excess halogen is pumped away and the halide purified by distillation or sublimation in the absence of moisture. (b) Trioxides are heated with the hydrogen halide. For arsenic

[50] W. L. Jolly, L. B. Anderson and R. T. Beltrami, *J. Am. Chem. Soc.* **79** (1957) 2443.
[51] (a) J. W. George, *Progr. Inorg. Chem.* **2** (1960) 33; (b) D. S. Payne, *Quart. Revs.* (*London*) **15** (1961) 173; (c) L. Kolditz, *Adv. Inorg. Chem. Radiochem.* **7** (1965) 1, and in V. Gutmann (ed.), *Halogen Chemistry*, Academic Press, London (1967), Vol. 2, p. 115.

TABLE 22. PROPERTIES OF TRIHALIDES

	AsF_3	$AsCl_3$	$AsBr_3$	AsI_3
Preparation (see text)	b [a]	b [b]	a, b [b α]	b [c]
Melting point (°C)	−5.95 [d]	−16.2 [d]	31.2 [b]	140.4 [e]
ΔH_{fus} (kJ mol^{-1})	10.4 [d]		17.1 [f]	21.8 [e]
Boiling point (°C)	62.8 [d]	103.2 [b]	221 [f]	371 [e]
Vapour pressure	g			h
$H^\circ_{298}-H^\circ_0$ (g) [i] (kJ mol^{-1})	14.29	17.31	19.12	20.22
ΔH°_{f298} [i] (kJ mol^{-1})	−956.5	−305.0	−197.0	−58.2
S°_{298} [i] (J mol^{-1} K^{-1})	181.2	207.5		277.4
ΔG°_{f298} [i] (kJ mol^{-1})	−909.2	−256.9		−78.9 [j]
$\Delta H^\circ_{298\,sub}$ [i] (kJ mol^{-1})	35.6	46.4	67.8	95.0 [e]
$\Delta S^\circ_{298\,sub}$ [i] (J mol^{-1} K^{-1})	108	119.6		177 [e]
C_p [i] (l) (J mol^{-1} K^{-1})	126.6			140.2 [e]
C_p [i] (g) (J mol^{-1} K^{-1})	65.6	75.7	79.2	82.4 [e]
Density (kg m^{-3})		2200 [b]	3400/25°C [f]	
Conductivity (Ω^{-1} m^{-1} 298 K)	2.4×10^{-3} [d]	10^{-5}/292 K [k]	1.6×10^{-5} [f]	
Dielectric constant		12.8/293 K [l]	9.3/308 K [f]	
10^{10} magnetic susceptibility[y] (m^3 mol^{-1})		−9.1	−13.3	−17.8

	SbF_3	$SbCl_3$	$SbBr_3$	SbI_3
Preparation (see text)	b [m]	a, b [b]	a [b α]	a [c α]
Melting point (°C)	292 [d]	73 [b]	97 [f]	171 [n]
ΔH_{fus} (kJ mol^{-1})				22.8
Boiling point (°C)		223 [b]	288 [f]	401 [o]
Vapour pressure ($\log p_{mmHg}$)				$-3412/T+7.944$ [o]
$H^\circ_{298}-H^\circ_0$ (g) [i] (kJ mol^{-1})		17.86	19.77	20.79
ΔH°_{f298} [i] (kJ mol^{-1})	−915.4	−382.0	−259.4	−100.4 [n]
S°_{298} [i] (J mol^{-1} K^{-1})		184.1	207.1	215.5
ΔG°_{f298} [i] (kJ mol^{-1})		−323.8	−239.3	−99.2 [j]
$\Delta H^\circ_{298\,sub}$ [i] (kJ mol^{-1})		68.2	64.8	101.7
$\Delta S^\circ_{298\,sub}$ [i] (J mol^{-1} K^{-1})		153.5	165.7	182.8
C°_p (s) [i] (J mol^{-1} K^{-1})		107.9		97.5
C°_p (g) [i] (J mol^{-1} K^{-1})		76.7	80.2	81.3
Density (kg m^{-3})		3140/293 K [b]	4150 [b]	4920/295 K [b]
Conductivity (Ω^{-1} m^{-1} 298 K)	1 [d]		10^{-3}/373 K [f]	2×10^{-2} [z]
Dielectric constant			20.9/373 K [f]	
$10^{10}\times$ magnetic susceptibility[y] (m^3 mol^{-1})		−10.4	−14.0	−18.5

	BiF_3	$BiCl_3$	$BiBr_3$	BiI_3
Preparation (see text)	b [p]	a, b [b]	a, b [q]	a [r]
Melting point (°C)	725–30 [d]	233.5 [s]	219 [t]	408.6 [u]
ΔH°_{fus} (kJ mol^{-1})		23.9 [s]	21.8 [t]	39.2 [v]
Boiling point (°C)		441 [s]	462 [q]	542 extrap. [u]
Vapour pressure ($\log p_{mmHg}$)		$8.547-4032/T$ [s]	w	$8.170-4310/T$ [u]

TABLE 22 (*cont.*)

$H^\circ_{298}-H^\circ_0$ (kJ mol^{-1})		24.9 [s]	20.5 [q]	21.4 [v]
ΔH_{f298} (kJ mol^{-1})		−379.1 [s]	−276.1 [q]	−150 [v]
S_{298} (J mol^{-1} K^{-1})		174.5 [s]	190.4 [q]	224.7 [v]
ΔG°_{f298} (kJ mol^{-1})		−314.6 [s j]	−247.7 [q j]	−148.7 [v j]
$\Delta H^\circ_{298\,sub}$ (kJ mol^{-1})		114.2 [s]	115.5 [q]	134.3 [v]
$\Delta S^\circ_{298\,sub}$ (J mol^{-1} K^{-1})		182.8 [s]	181.6 [q]	183.3 [v]
C_p [i] (s) (J mol^{-1} K^{-1})		107	108.6	
C_p [i] (l) (J mol^{-1} K^{-1})		136.3	157.5 [t]	151 [n]
Density (kg m^{-3})		4750 [b]	5720 [f]	5640
Maximum conductivity (Ω^{-1} m^{-1}) [x]		58/700 K	36/700 K	31/800 K
$10^{10}\times$ magnetic susceptibility (m^3 mol^{-1}) [y]		−12.75	−17.1	−20.2

[a] C. J. Hoffman, *Inorg. Synth.* **4** (1953) 150.
[b] Ref. 5, pp. 596, 608, 621.
[c] J. C. Bailar, *Inorg. Synth.* **1** (1939) 103, 104.
[d] R. D. W. Kemmitt and D. W. A. Sharp, *Adv. Fluorine Chem.* **4** (1965) 142.
[e] D. Cubicciotti and H. Eding, *J. Phys. Chem.* **69** (1965) 2743.
[f] Z. E. Jolles, *Bromine and its Compounds*, E. Benn, London (1966), pp. 222 ff.
[g] Log $p_{mmHg} = 61.3797-4149.78/T-18.2640 \log T$ [d].
[h] Log $p_{mmHg} = 30.148-4897/T-7.0 \log T$ [e].
[i] Except where indicated, from ref. 11.
[j] Calculated from other data here and in ref. 11.
[k] L. H. Anderson and I. Lindqvist, *Acta chem. scand.* **9** (1965) 79.
[l] V. Gutmann, *Z. anorg. allgem. Chem.* **266** (1951) 331.
[m] F. A. Andersen, B. Bak and A. Hillebert, *Acta chem. scand.* **7** (1953) 236.
[n] D. Cubicciotti and H. Eding, *J. Phys. Chem.* **69** (1965) 3621.
[o] B. L. Bruner and J. D. Corbett, *J. Inorg. Nucl. Chem.* **20** (1961) 62.
[p] B. Aurivillius, *Acta chem. scand.* **9** (1955) 1206.
[q] D. Cubicciotti, *Inorg. Chem.* **7** (1968) 208.
[r] G. W. Watt, W. W. Hakki and G. R. Choppin, *Inorg. Synth.* **4** (1953) 114.
[s] D. Cubicciotti, *J. Phys. Chem.* **70** (1966) 2410; **71** (1967) 3066.
[t] L. E. Topol and L. D. Ransom, *J. Phys. Chem.* **64** (1960) 1339.
[u] D. Cubicciotti and F. J. Keneshea, *J. Phys. Chem.* **63** (1959) 295.
[v] D. Cubicciotti, *Inorg. Chem.* **7** (1968) 211.
[w] Shown graphically by D. Cubicciotti and F. J. Keneshea, *J. Phys. Chem.* **62** (1958) 999.
[x] L. F. Grantham and S. J. Yosim, *J. Phys. Chem.* **67** (1963) 2506.
[y] Ref. 52.
[z] For liquid at melting point, G. Fischer, *Helv. Phys. Acta* **34** (1961) 827.
[α] P. M. Druce and M. F. Lappert, *J. Chem. Soc.* (A) (1971) 3595.

trifluoride, which is easily hydrolysed, it is convenient to prepare hydrogen fluoride in the presence of concentrated sulphuric acid:

$$As_2O_3+3CaF_2+3H_2SO_4 \rightarrow 3AsF_3+CaSO_4+3H_2O$$

For halides which are less readily hydrolysed, concentrated aqueous solutions of the hydrogen halide may be used. Trioxides may also be converted to trichlorides by a mixture of sulphur and chlorine or by sulphur monochloride; tribromides may be made similarly. Bromides EBr_3 and iodides EI_3 may be conveniently made from chlorides using boron tribromide or iodide.

The trihalides are all available commercially. References to laboratory preparations are given in Table 22, which includes physical properties and thermodynamic data.

[52] M. Prasad, C. R. Kanekar and L. N. Mulay, *J. Chem. Phys.* **19** (1951) 1051, 1440.

5.2. PROPERTIES

The compounds are liquids or solids at room temperature, and, with the exception of the fluorides of antimony and bismuth, the volatility decreases with increasing molecular weight. The molecules in the gas phase are all pyramidal, with bond angles close to 100° (Table 23). Accurate X-ray measurements on single crystals have been made in some cases. The bond lengths and angles of the gaseous molecules are largely preserved in the solids, though usually individual molecules are distorted, giving two long and one short intramolecular distances (e.g. 251 and 246 pm in $SbBr_3$; intermolecular Sb–Br distance, 369 pm). The molecules form layers in which the halogens are roughly close-packed, but the arrangements differ slightly from compound to compound. According to powder photographs, bismuth iodide has a structure in which bismuth ions occupy octahedral holes in a close-packed array of iodide ions, i.e. intramolecular and intermolecular Bi–I bond lengths are equal. Alternatively, the structure may be described as largely ionic with the lone pair in Bi^{III} stereochemically inactive (see section 10.3).

It is clear from the melting points and volatility shown in Table 22 that in antimony trifluoride there are especially strong intermolecular attractions. Like a number of other trifluorides of heavy metals, bismuth trifluoride crystallizes with the YF_3 structure, in which the metal has eight nearest neighbours and a ninth only slightly further away, and in which there is no recognizable molecule. The more compact structure of bismuth trifluoride, compared with the other halides of arsenic, antimony and bismuth, is reflected in its high density, high melting point and low volatility.

The liquid trihalides of arsenic and antimony have low electrical conductivities when pure. Arsenic trichloride has been suggested as a solvent for n.m.r. spectroscopy[53]. Molten

TABLE 23. STRUCTURAL AND SPECTROSCOPIC PROPERTIES OF TRIHALIDES

	F	Cl	Br	I
Structural parameters[a, b]				
As–X (pm)	171.2±0.5(*M*)	216.1±0.4(*M*)	233±1(*E*)	255.7±0.5[c]
∠XAsX (°)	96.0±0.5(*E*)[d]	98.4±0.5(*M*)(*E*)[d]	99.7(*E*)[d]	100.2±0.4
Sb–X (pm)	203[e]	232.5±0.5(*M*)	251±2	271.9±0.2[f]
∠XSbX (°)	88	99.5±1.5	97±2	99.1
Bi–X (pm)		248±2(*E*)	263±2(*E*)	
∠XBiX (°)		100±6	100±4	
Vibrational spectra (cm^{-1})				
AsX_3 ν_1 (a_1)	700[g]	411[h, i]	272[j, i]	216[j, k]
ν_2 (a_1)	340	195	130	94
ν_3 (e)	643	380	287	221
ν_4 (e)	272	159	100	70
SbX_3 ν_1 (a_1)		364[h, l]	252[h, m]	177[j]
ν_2 (a_1)		164	103	89
ν_3 (e)		320	243	147
ν_4 (e)		134	83	71
BiX_3 ν_1 (a_1)		280[n]	196[j, n]	145[j]
ν_2 (a_1)			104	90
ν_3 (e)		262, 254	169	115
ν_4 (e)			89	71

[53] E. G. Brame, R. C. Ferguson and G. J. Thomas, *Anal. Chem.* **39** (1967) 517.

TABLE 23 (*cont.*)

	F	Cl	Br	I
Nuclear quadrupole spectra [o]				
^{75}As [p] $\|e^2Qq\|$(MHz)	−236.2(*M*) [a]	157.9 [p]	127.1 [p]	58.67
X(^{35}Cl, ^{81}Br, ^{127}I) [q] (MHz)		25	172	207, 396
^{121}Sb $\|e^2Qq\|$ (MHz)		376.9 [r]	320.0 [r]	169.4 [p]
^{123}Sb $\|e^2Qq\|$ (MHz)		480.4 [r]	407.9 [r]	
η		0.159	0.082	
X(^{35}Cl, ^{81}Br, ^{127}I) [q] (MHz)		20	140	175, 255
^{209}Bi $\|e^2Qq\|$(MHz)		318.8/299 K	340.5/300 K	
η		0.555	0.553	
X(^{35}Cl, ^{127}I) [q] (MHz)		17		110, 201
Dipole moments[s] (D)AsX_3	2.815	2.1 ± 0.3	1.7	0.96 ± 0.4
Dipole moments (D) SbX_3		3.9 ± 0.4	2.8 ± 0.4	1.58 ± 0.4

[a] For gaseous molecules, by electron diffraction *E* or microwave spectroscopy *M*; except where indicated from ref. 47.

[b] Crystal structure determinations: $AsBr_3$, J. Trotter, *Z. Krist.* **122** (1965) 230° AsI_3, J. Trotter, *Z. Krist.* **121** (1965) 81; SbF_3, A. J. Edwards, *J. Chem. Soc.* A (1970) 2751; $SbCl_3$, I. Lindqvist and A. Niggli, *J. Inorg. Nucl. Chem.* **2** (1956) 345; $SbBr_3$, D. W. Cushen and R. Hulme, *J. Chem. Soc.* (1962) 2218; BiF_3, B. Aurivillius, *Acta chem. scand.* **9** (1955) 1206 (powder); $BiCl_3$ and $BiBr_3$, G. M. Wolten and S. W. Mayer, *Acta Cryst.* **11** (1958) 739 (powder); BiI_3, J. Trotter and T. Zoebel, *Z. Krist.* **123** (1965) 67.

[c] Y. Morino, T. Ukaji and T. Ito, *Bull. Chem. Soc. Japan* **39** (1966) 71.

[d] Y. Morino, K. Kuchitsu and T. Moritani, *Inorg. Chem.* **8** (1969) 867.

[e] In crystal, errors uncertain.

[f] A. Almenningen and T. Bjorvatten, *Acta chem. scand.* **17** (1963) 2573.

[g] J. A. Evans and D. A. Long, *J. Chem. Soc.* A (1968) 1688.

[h] P. W. Davis and R. A. Oetjen, *J. Molec. Spectrosc.* **2** (1958) 253.

[i] T. M. Loehr and R. A. Plane, *Inorg. Chem.* **8** (1969) 73.

[j] On solids: T. R. Manley and D. A. Williams, *Spectrochim. Acta* **21** (1965) 1467, 1773.

[k] M. A. Hooper and D. W. James, *Aust. J. Chem.* **21** (1968) 2379.

[l] J. K. Wilmshurst, *J. Molec. Spectrosc.* **5** (1960) 343.

[m] J. C. Evans, *J. Molec. Spectrosc.* **4** (1960) 435.

[n] R. P. Oertel and R. A. Plane, *Inorg. Chem.* **8** (1969) 1188. Solid spectra complicated.

[o] Except where indicated, at 77 K. Data from E. A. C. Lucken, *Nuclear Quadrupole Coupling Constants*, Academic Press, New York (1969), and references therein. Coupling constants, e^2Qq; asymmetry parameter, η. (*M*) from microwave spectra.

[p] η not known, assumed to be zero.

[q] Several resonances in solid, approximate positions given here. For ^{127}I, transitions correspond to $\pm\frac{1}{2} \leftrightarrow \pm\frac{3}{2}$, $\pm\frac{3}{2} \leftrightarrow \pm\frac{5}{2}$. Coupling constants (MHz) for iodides (η in parentheses): AsI_3, 1330.2 (0.189); SbI_3, 895.8 (0.565); BiI_3, 682.2 (0.274).

[r] At 31°C; T-C. Wang, *Phys. Rev.* **99** (1955) 566.

[s] P. Kisliuk, *J. Chem. Phys.* **22** (1954) 86.

bismuth halides have higher conductivities which reach maxima 125–250° above their melting points. Solid antimony and bismuth iodides are semiconductors.

Spectroscopic Properties

Vibrational spectra (Table 23) for gaseous and liquid halides show four fundamentals all both infrared and Raman active for molecules with C_{3v} symmetry. Similar features appear in spectra from solutions in carbon tetrachloride, carbon disulphide or molten paraffins. Spectra from solutions in donor solvents show shifts attributed to complex formation, and E-donor and E–X vibrations may be confused. Interactions between molecules, which lower the symmetry at E from C_{3v}, must be taken into account in interpreting the spectra of solids. Assignments of spectra of arsenic compounds are reasonably well established, but it is difficult to be sure of the correct allotment of the very similar symmetric and asymmetric stretching vibrations (ν_1 and ν_3). The assignments given in Table 23 for some of the antimony and bismuth halides are not so well established.

Nuclear quadrupole resonances (for ^{35}Cl, ^{81}Br and ^{127}I), and the coupling constants for ^{127}I, decrease along the series of halides AsX_3, SbX_3 and BiX_3. There are usually two or three closely spaced resonances, since all the halogen atoms are not usually equivalent in the solids. Nuclear quadrupole coupling constants relating to the atoms E are given in Table 23. For arsenic trichloride, the coupling constant from nuclear quadrupole resonance measurements on the solid is similar to that from microwave spectra on the vapour. For antimony and bismuth halides, the non-zero asymmetry parameters are consistent with a molecular symmetry lower than C_{3v} in the solid.

5.3. REACTIONS

Halogenations[54]

The fluorides of arsenic and antimony are important reagents in preparative inorganic chemistry for the conversion of non-metal chlorides to fluorides[51, 55, 56]:

$$[PCl_4]^+[PCl_6]^- + AsF_3 \rightarrow [PCl_4]^+[PF_6]^- + AsCl_3$$
$$SiCl_4 + SbF_3 \rightarrow SiCl_3F,\ SiCl_2F_2,\ SiClF_3$$
$$CCl_3\cdot CCl_3 + SbF_3 \rightarrow CCl_2F\cdot CCl_2F$$

Arsenic trifluoride is very rapidly hydrolysed and is toxic, but can be easily and safely manipulated in a vacuum line. It would be preferred for the preparation of a high boiling fluoride, since arsenic trichloride (b.p. 130°) can be distilled off. Antimony trifluoride is preferred for making low boiling fluorides, which can be removed readily from antimony trichloride (b.p. 223°). It is often advantageous to add a small amount of chlorine or antimony pentachloride[54, 55], but this is, of course, not possible when the substrate is easily oxidized.

Arsenic and antimony trifluorides, though less strongly oxidizing than fluorine, can still be reduced by strong reducing agents. Fluorophosphines are obtained from the fluorides EF_3 and chloro- or iodo- phosphines with an electro-negative group attached to phosphorus, but phosphoranes result from many other chlorophosphines and phosphorus sulphides:

$$3CF_3PCl_2 + 2SbF_3 \rightarrow 3CF_3PF_2 + 2SbCl_3$$
$$3PhPCl_2 + 4SbF_3 \rightarrow 3PhPF_4 + 2Sb + 2SbCl_3$$
$$3R_3PS + 2SbF_3 \rightarrow 3R_3PF_2 + Sb_2S_3$$
$$Me_2P(S)P(S)Me_2 + 6SbF_3 \rightarrow 6Me_2PF_3 + 2Sb + 2Sb_2S_3$$

The potentiality of a fluorinating agent EF_n with respect to a chloride E′Cl depends on the difference in free energies of formation per gram atom halogen ($\Delta G'/n$) of EF_n and ECl_n. The values in Table 22 show that antimony trifluoride is thermodynamically a better

[54] E. L. Muetterties and C. W. Tullock, *Preparative Inorganic Reactions* **2** (1965) 237.

[55] R. Schmutzler, *Inorg. Synth.* **9** (1967) 63, and in V. Gutmann (ed.), *Halogen Chemistry*, Academic Press, London (1967), Vol. 2, p. 31; A. L. Henne, *Organic Reactions* **2** (1944) 49; D. R. Martin, *Inorg. Synth.* **4** (1953) 134.

[56] H. J. Emeléus, P. M. Spaziante and S. M. Williamson, *Chem. Communs.* (1969) 768.

fluorinating agent than arsenic trifluoride ($\Delta G'/n$: SbF_3, about +180; AsF_3, +217 kJ mol^{-1}. Compare CF_4, 203; SiF_4, 238; NaF, 157; AgF, 75 kJ mol^{-1}). For the conversion of arsenic or antimony chlorides to fluorides, hydrogen fluoride, ammonium fluoride, or potassium fluorosulphonate have been used. Arguments similar to that above show that arsenic tribromide has potential as a brominating agent, e.g.

$$6PbO + 4AsBr_3 \rightarrow 6PbBr_2 + As_4O_6$$

The driving force for this reaction is in the formation of the strong As–O bonds from the weaker As–Br bonds (cf. Fig. 10).

Pseudohalides may be made by reactions similar to the halogenations described above. Thus arsenic trichloride reacts with silver cyanide to give the tricyanide $As(CN)_3$ (section 9.3). The radical $(CF_3)_2NO^{\cdot}$, which in its reactions resembles a halogen, reacts with $AsBr_3$, AsI_3, $BiBr_3$, BiI_3 and BiOI with quantitative formation of the tris(trifluoromethylnitroxy)-derivatives $[(CF_3)_2NO]_3E$. The bismuth compound may also be made directly from metallic bismuth and $(CF_3)_2NO^{\cdot}$. Arsenic trichloride can be converted into $[(CF_3)_2NO]_3As$ by reaction with mercury(II) bistrifluoromethylnitroxide; with $(CF_3)_2NO^{\cdot}$ in the presence of iodine, the partially substituted compound $[(CF_3)_2NO]_2AsCl$ is obtained[56].

Solvent Properties

The conductivities of the halides AsF_3, $AsCl_3$, $AsBr_3$, $SbCl_3$ and $SbBr_3$ are markedly increased by substances which can donate halide ions, e.g. KF, Me_4NCl, TlBr, or accept them, e.g. SbF_5, $FeCl_3$, $AlBr_3$. Further, halide ions have very high mobilities in the liquid trihalides. This has led to suggestions[57] that the solvents ionize, e.g.

$$2AsX_3 \rightleftharpoons AsX_2^+ + AsX_4^-$$

From the conductivities of the pure solvents, the ionic products $[EX_2]^+[EX_4]^-$ must be $< 10^{-15}$ mol^2 l^{-2}. Evidence given in support of self-ionization is as follows (cf. ref. 133).

(a) The conductivity of a solution of ferric chloride in arsenic trichloride, to which a solution of tetramethylammonium chloride is added, reaches a minimum at the end point. Such conductimetric titrations can be made with a variety of halide donors and acceptors.

(b) Potentiometric titrations using silver–silver chloride electrodes can be made on similar systems.

(c) Solid compounds with EX_4^- ions, e.g. $[Me_4N][AsCl_4]$, can be isolated (see section 10.3). It has been suggested that certain solids (e.g. $AsSbCl_8$) may contain EX_2^+ ions. Structural determinations on some of these would be worthwhile.

(d) Complexes such as $Tl(AlBr_4)$ can be prepared by precipitation from $TlBr–SbBr_3$ and $AlBr_3–SbBr_3$.

Though it is clear that ionic species are formed when some substances dissolve in arsenic and antimony halides, many halide transfer reactions could proceed by molecular mechanisms without requiring self-ionization of the pure solvent. More work is needed on species in solution.

The ready exchange of halogen between halides EX_3 and KX^* provides a convenient method for the preparation of labelled EX_3^* [58].

[57] I. Lindqvist, *Acta chem. scand.* **9** (1955) 73, 79; G. Jander and J. Weiss, *Z. Elektrochem.* **61** (1957) 1275; G. Jander and K. Günther, *Z. anorg. allgem. Chem.* **297** (1958) 81; **298** (1959) 241; **302** (1960) 155.

[58] M. F. A. Dove and D. B. Sowerby, in V. Gutmann (ed.), *Halogen Chemistry*, Academic Press, London (1967).

Formation of Mixed Halides

Although a number of mixed halides of phosphorus are known, the compounds of the heavier elements are much less easily isolated. The halides AsF_3 and $AsCl_3$ are immiscible below 19°C, but the ^{19}F magnetic resonance spectrum of a mixture of arsenic trichloride and an excess of arsenic trifluoride at room temperature shows three signals[59]. The strong absorption at 40.4 ppm (from $CFCl_3$) is assigned to AsF_3, and the weaker signals at 49 and 67.4 ppm are assigned to AsF_2Cl and $AsFCl_2$. Equilibrium constants for the exchange reactions can be found from the peak areas:

$$2AsF_3 + AsCl_3 \rightleftharpoons 3AsF_2Cl; \qquad K \sim 5\times 10^{-3}$$
$$2AsCl_3 + AsF_3 \rightleftharpoons 3AsFCl_2; \qquad K \sim 5\times 10^{-3}$$

Similar values are obtained from mass spectral studies. It has also been suggested that Raman spectra of mixtures of $AsCl_3$ and $AsBr_3$ show the presence of mixed halides $AsCl_2Br$ and $AsClBr_2$. Because of the rapid establishment of equilibria such as those above, it is probable that most mixed halides cannot be isolated as pure substances, but the compound SbI_2Br (m.p. 88°) is apparently obtained by the elimination of ethyl bromide from the organo-metallic compound $EtSbI_2Br_2$ (see section 14.3).

Reactions with Alcohols and Mercaptans

The trihalides react with alcohols, often in the presence of bases, or with sodium derivatives NaOR, to give arsenite and antimonite esters $E(OR)_3$ (see Table 31, p. 610)[60, 61]:

$$AsCl_3 + 3MeOH + 3Me_2NPh \rightarrow As(OMe)_3 + 3[Me_2NPhH]Cl$$
$$SbCl_3 + 3NaOSiEt_3 \rightarrow Sb(OSiEt_3)_3 + 3NaCl$$
$$SbCl_3 + 3Bu^tOH + 3NH_3 \rightarrow Sb(OBu^t)_3 + 3NH_4Cl$$
$$AsCl_3 + 3PhOH \rightarrow As(OPh)_3 + 3HCl$$

Halide esters, AsX_2Y, AsY_2X (X = hal; Y = OR) can be made similarly, or from the esters $E(OR)_3$ and alkyl or acyl halides:

$$AsCl_3 + 2NaOEt \longrightarrow AsCl(OEt)_2 + 2NaCl$$
$$AsCl_3 + EtOH \xrightarrow{CO_2} AsCl_2OEt + HCl$$
$$Sb(OEt)_3 + MeI \longrightarrow [SbMe(OEt)_3]I \rightarrow (EtO)_2SbI + EtOMe$$
$$As(OPr)_3 + MeCOCl \longrightarrow (PrO)_2AsCl + MeCOOPr$$

The halide esters can be detected in mixtures of AsX_3 and AsY_3 (X = hal; Y = OR). Nuclear magnetic resonance measurements show that the X and Y groups exchange rapidly at room temperature, but distribution of X and Y between possible species is not random, and formation of the halide esters AsX_2Y, AsY_2X is favoured. Several of these can be isolated as pure substances by distillation of appropriate mixtures of AsX_3 and AsY_3 (compare previous section). In mixtures of arsenic and antimony compounds, transfer of alkoxide to arsenic seems to be preferred:

$$AsCl_3 + Sb(OEt)_3 \rightarrow AsCl(OEt)_2 + SbCl_2OEt$$

---

[59] J. K. Ruff and G. Paulett, *Inorg. Chem.* **3** (1964) 998.

[60] W. Herrmann, in *Methoden der Organischen Chemie*, 4th edn. (Houben Weyl), G. Thieme Verlag, Stuttgart (1963), Vol. 6, Part 2, p. 363.

[61] K. Moedritzer, *Inorg. Synth.* **11** (1968) 181.

Reactions with Amines

The trihalides react with secondary amines, with alkali-metal amides, or with compounds having silicon–nitrogen or tin–nitrogen bonds to give amido derivatives[62]:

$$AsCl_3 + 6Me_2NH \rightarrow As(NMe_2)_3 + 3[Me_2NH_2]Cl$$

$$SbCl_3 + 3LiNMe_2 \rightarrow Sb(NMe_2)_3 + 3LiCl$$

$$Et_2NSiMe_3 + AsCl_3 \rightarrow Et_2NAsCl_2 + Me_3SiCl$$

$$SbF_3 + 3Me_2NSnMe_3 \rightarrow Sb(NMe_2)_3 + 3Me_3SnF$$

Derivatives of bulky amines are often conveniently made by transamination:

$$As(NMe_2)_3 + Bu_2NH \rightarrow As(NBu_2)_3 + 3NMe_2H$$

Like the alkoxyhalides, most of the amidohalides, e.g. $AsX(NR_2)_2$, AsX_2NR_2, are stable with respect to reorganization and so can be isolated as pure compounds. An exception is $AsF(NMe_2)_2$ which easily reorganizes to AsF_2NMe_2 and $As(NMe_2)_3$. Mixtures of trisamido and trisalkoxy compounds give a random distribution of OR, NR_2 groups over all possible species[63].

TABLE 24. AMIDO DERIVATIVES

	M.p. (°C)	B.p. [a] (°C)	Vapour pressure (log p_{mmHg})	N.m.r. (ppm rel. to TMS)
$As(NMe_2)_3$ [b]	−53	57/10	$8.289 - 2391/T$	−2.53 [c]
$MeAs(NMe_2)_2$ [b]	−62	148	$7.648 - 2046/T$	
Me_2AsNMe_2 [b]	−110	108	$7.881 - 1917/T$	
$ClAs(NMe_2)_2$ [d]		69/10		−2.48 [f]
Cl_2AsNMe_2 [e]		50/10		−2.38 [f]
$Sb(NMe_2)_3$ [c]		33/0.45		−2.70

[a] Pressures in mmHg; where none shown at 760 mmHg (101.325 kN m^{-2}).
[b] K. Moedritzer, *Chem. Ber.* **92** (1959) 2637.
[c] N.m.r. on neat liquid: K. Moedritzer, *Inorg. Chem.* **3** (1964) 609.
[d] Ref. 64 b.
[e] G. A. Olah and A. A. Oswald, *Can. J. Chem.* **38** (1960) 1428, 1431.
[f] N.m.r. in CCl_4: H.-J. Vetter, *Naturwissenschaften* **51** (1964) 240.

Properties of some amido compounds are given in Table 24, and their chemical reactions in Scheme 3. The amido compounds closely resemble the arsenite and antimonite esters (section 6.4). Compounds such as R_2AsNR_2 are important intermediates, since the As–N bonds can easily be broken by a variety of reagents.

[62] E. W. Abel, D. A. Armitage and G. R. Willey, *J. Chem. Soc.* (1965) 57; T. A. George and M. F. Lappert, *J. Chem. Soc.* A (1969) 992.

[63] K. Moedritzer and J. R. van Wazer, *Inorg. Chem.* **3** (1964) 139; H. C. Marsmann and J. R. van Wazer, *J. Am. Chem. Soc.* **92** (1970) 3969.

SCHEME 3. Amido compounds.[64]

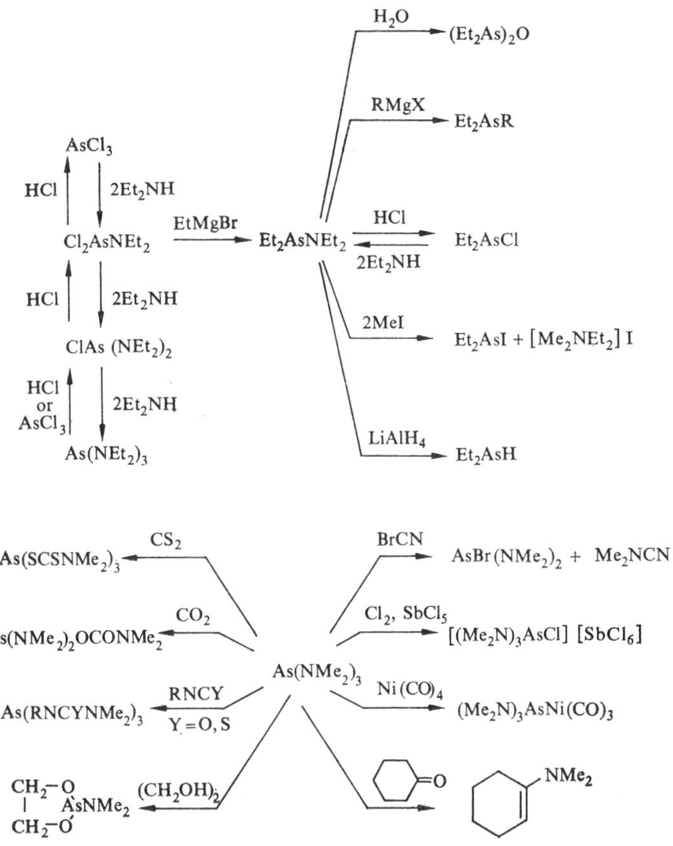

Arsenic trihalides and primary amines give imido-derivatives (Table 25). Some of these may also be made by transaminations.

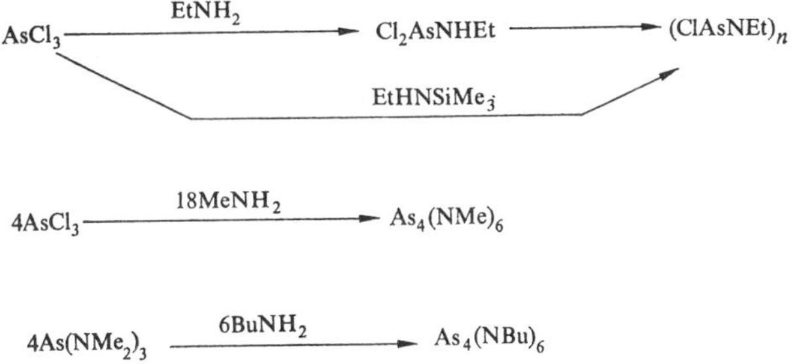

$$4As(NMe_2)_3 \xrightarrow{6BuNH_2} As_4(NBu)_6$$

[64] (a) A. Tzschach and W. Lange, *Z. anorg. allgem. Chem.* **326** (1964) 280; (b) H.-J. Vetter and H. Nöth, *ibid.* **330** (1964) 233; (c) G. Oertel, H. Malz and H. Holtschmidt, *Chem. Ber.* **97** (1964) 891; R. H. Anderson and R. H. Cragg, *Chem. Communs.* (1970) 425; H. Weingarten and W. A. White, *J. Org. Chem.* **31** (1966) 4041.

The compounds $(XAsNR)_n$ are slightly volatile liquids or solids, which in benzene show molecular complexities between 2 and 4. The methyl derivative ClAsNMe is trimeric in the solid, but it is not in all cases clear whether the non-integral values obtained for *n* are due to decompositions or to equilibria between various species. The butyl compounds seem to show lower values for *n* than compounds with less bulky substituents.

TABLE 25. IMIDO DERIVATIVES

$As_4(NMe)_6$: m.p. 113–116°C. $As_4(NPr^i)_6$: m.p. 47–51°C. $As_4(NBu)_6$: b.p. 197–201°C/0.1 mmHg [a]
$As_4(NPr)_6$. $As_4(NCH_2Ph)_6$: m.p. 63–66°C. $As_4(NPh)_6$ m.p. 265°C. $As_4(4\text{-}MeC_6H_4N)_6$ [b] m.p. 260°C
$Sb_4(NPh)_6$. $Sb_4(2\text{-}MeC_6H_4N)_6$: m.p. 86°C [c]

[a] H-J. Vetter, H. Nöth and W. Jaher, *Z. anorg. allgem. Chem.* **328** (1964) 144; H-J. Velter, H. Nöth and U. Hayduk, *Z. anorg. allgem. Chem.* **331** (1964) 35.
[b] D. Hass, *Z. anorg. allgem. Chem.* **325** (1963) 139.
[c] D. Hass, *Z. anorg. allgem. Chem.* **332** (1964) 287.

The imido compounds $As_4(NR)_6$ (these formulae are shown by molecular weights in benzene) have the same structure as the oxide As_4O_6 [65] (section 6.1) or the phosphorus imide $P_4(NMe)_6$. The phosphorus and arsenic compounds react differently with methyl iodide. Whereas the phosphorus compound gives a simple adduct, the arsenic derivative loses one NMe group:

$$As_4(NMe)_6 + 3MeI \rightarrow As_4(NMe)_5I_2 + Me_4NI$$

The product can be further degraded with methyl iodide or arsenic triiodide. Similar progressive degradation of the As_4N_6 cage is effected by hydrogen chloride:

$$As_4(NMe)_6 \xrightarrow{3HCl} As_4(NMe)_5Cl_2 \xrightarrow{3HCl} 4/3(MeNAsCl)_3 \xrightarrow{12HCl} 4AsCl_3$$

Attempts to make the t-butyl analogue of $As_4(NMe)_6$ yielded instead the compound $(Bu^tNHAsNBu^t)_2$, m.p. 80–85°C, which is assumed to have the cyclic structure [1].

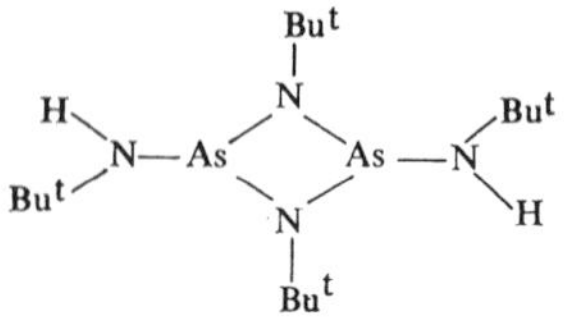

[1]

The antimony compounds $Sb_4(NR)_6$ (R = Ph, $4\text{-}MeC_6H_4$) are obtained from alkoxy-derivatives and amines:

$$4Sb(OEt)_3 + 6RNH_2 \rightarrow Sb_4(NR)_6 + 12EtOH$$

The reaction is reversible and the equilibrium is shifted from left to right by removal of the volatile ethanol.

[65] J. Weiss and W. Eisenhuth, *Z. Naturforsch.* **22b** (1967) 454; *Z. anorg. allgem. Chem.* **350** (1967) 9.

5.4. OXIDE HALIDES

The trihalides of arsenic, antimony and bismuth are readily hydrolysed and, if they are to be kept pure, are best handled in a drybox or vacuum line. The hydrolysis is usually reversible, and the halides, e.g. arsenic triiodide or bismuth trifluoride, can often be recovered quantitatively by precipitation from concentrated aqueous solutions of hydrogen halides.

Arsenic trichloride is not miscible with water at $AsCl_3:H_2O$ mole ratios of $>1:10$. The Raman spectra[66] of concentrated ($As:H_2O$ between 0.1 and 0.05) solutions of arsenic trichloride in water suggest the formation of the complex $H_2O,AsCl_3$, which is probably in equilibrium with ionic species:

$$H_2O,AsCl_3+H_2O \rightarrow H_3O^+ + [HOAsCl_3]^-$$

Evidence for compounds with less than three chlorine atoms per arsenic is not very strong, but small concentrations of such compounds may be present in some solutions of arsenic trichloride. At $AsCl_3:H_2O$ ratios of about 1:20, As^{III} oxide precipitates; the species in solution at this stage is the arsenious acid $As(OH)_3$ (section 6.3).

Arsenic trioxide dissolves in arsenic trifluoride to form homogeneous solutions[67], the viscosity of which increases dramatically with increasing oxide content (e.g. from

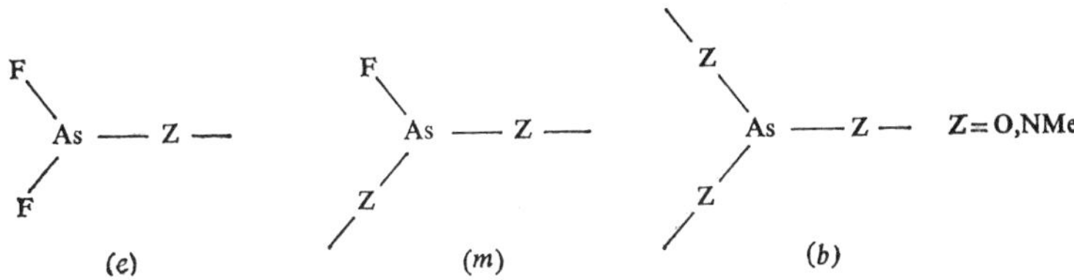

10^{-3} kg m^{-1} s^{-1} for AsF_3 to 10^4 kg m^{-1} s^{-1} for $5AsF_3.6As_2O_3$ at 20°C). The ^{19}F n.m.r. spectra of fluoride–oxide mixtures show a random mixture of polymeric species with end (*e*), middle (*m*), and branching (*b*) groups, as well as the free AsF_3 (*n*). Individual species cannot be isolated. The activation energies for fluorine exchange and for viscous flow, obtained from the temperature dependence of the ^{19}F n.m.r. spectra and the viscosity, are the same, suggesting that these processes have the same mechanism. Similar polymeric species have been postulated for $As_4(NMe)_6$–AsF_3 mixtures. In this case, distribution of polymeric species is not random. Proportions of end and middle groups are more than expected, showing that chains predominate. Equilibrium constants (27°C) are as follows:

Z	$K_1 = nm/e^2$	$K_2 = eb/m^2$	*K* for random distribution 0.33
O	0.33	0.31	
NMe	0.1	0.05	

Polymeric species are also shown by cryoscopic, ebullioscopic, and magnetic resonance measurements on As_2O_3–$AsCl_3$, As_2O_3–$AsBr_3$, As_2O_3–$As(OMe)_3$ or As_2O_3–$As(NMe_2)_3$ mixtures[68]. Arsenic oxide fluoride AsOF has been reported as a product from the reaction

[66] T. M. Loehr and R. A. Plane, *Inorg. Chem.* **8** (1969) 73.

[67] J. R. van Wazer, K. Moedritzer and D. W. Matula, *J. Am. Chem. Soc.* **86** (1964) 807; M. D. Rausch, J. R. van Wazer and K. Moedritzer, *ibid.*, p. 814.

[68] E. Thilo and P. Flögel, *Z. anorg. allgem. Chem.* **329** (1964) 244; K. Moedritzer and J. R. van Wazer, *Inorg. Chem.* **4** (1965) 893.

of arsenic trioxide and trifluoride in a sealed tube at 320° but has not been fully characterized[69].

Water is a non-electrolyte in antimony trichloride.[70] With excess water, however, antimony halides give oxide halides. The compounds SbOX have structures with sheets, $[Sb_6O_6Cl_4]^{2+}$, of antimony atoms linked by oxygen and halogen bridges, interleaved with layers of halide ions. Bismuth oxide halides (Table 26) similarly have layer structures[71].

TABLE 26. BISMUTH OXIDE HALIDES

	BiOCl	BiOBr	BiOI
Bond lengths[a] (pm): Bi–O	231	232	233
Bi–X [b]	307, 349	318, 404	336, 488
Decomposition temperature (°C) [c]	575	560	300
Infrared absorption (cm^{-1}) [c]	528, 285	520, 265	487, 249
Solubility product (log K_s) [d]	−34.9	−34.8	
ΔH_f° (kJ mol^{-1})	−367 [e]	−297	

[a] Ref. 71.
[b] Bi–X distance within layer, followed by shortest Bi–X distance to next layer.
[c] R. Bonnaire, *Compt. rend.* **266B** (1968) 1415.
[d] S. Ahrland and I. Grenthe, *Acta chem. scand.* **11** (1957) 1111. $K_s = [Bi^{3+}][X^-][OH^-]^2$.
[e] For BiOCl: ΔH_f°, −367 kJ mol^{-1}; S°, 121 J mol^{-1} K^{-1}; ΔG_f°, −322 kJ mol^{-1} K^{-1}. Ref. 11.

Within a layer, each bismuth has four oxygen and four chlorine neighbours, but in the chloride and bromide, the bonds to halogens in adjacent layers are little longer than those to halogens within the layers. The low solubility of the oxide halides means that they are precipitated when solutions of the trihalides EX_3 in concentrated acids HX are diluted. A series of complex oxide halides of bismuth can be made by fusing the compounds BiOX with halides of Groups I and II. These have tetragonal structures in which metal–oxide layers (MO_2M) are interleaved with halide layers (X). Several repeating sequences parallel to the *c*-axis have been found, and some examples are as follows. More complicated sequences are also known[71].

...$\{(MO_2M)X\}(MO_2M)X$...	$LiBi_3O_4Cl_2$, $Cd_2Bi_2O_4Br_2$
...$\{(MO_2M)XX\}(MO_2M)XX$...	BiOX
...$\{(MO_2M)X(MO_2M)XX\}$...	$SrBi_3O_4Cl_3$
...$\{(MO_2M)X(MX)X\}(MO_2M)X$...	$Ca_{1.25}Bi_{1.5}O_2Cl_3$

5.5. LOWER HALIDES

All the elements, arsenic, antimony and bismuth, form compounds in which the formal oxidation state of the Group V element is less than three. Arsenic forms an iodide, As_2I_4,

[69] H. J. Emeléus and M. J. Dunn, *J. Inorg. Nucl. Chem.* **27** (1965) 269.
[70] J. R. Atkinson, E. C. Baughan and B. Dacre, *J. Chem. Soc.* A (1970) 1377.
[71] M. Edstrand, *Acta chem. scand.* **1** (1947) 178. Most of the structural work on oxide halides has been by L. G. Sillén. It is summarized in *Naturwissenschaften* **28** (1940) 396; **30** (1942) 318, and by A. F. Wells in *Structural Inorganic Chemistry*, 3rd edn., Oxford (1962), pp. 390, 673.

like the corresponding phosphorus compound, but it disproportionates rather easily. The antimony compound Sb_2I_4 is even more unstable. Bismuth, in common with other heavy metals, forms metal clusters, several of which have been characterized by spectroscopic and crystallographic studies.

The Compounds E_2I_4

When arsenic and iodine are heated in a sealed tube at 260°C in octahydrophenanthrene, red crystals of the iodide, As_2I_4 (m.p. 137°C), separate. These can be recrystallized from carbon disulphide at −20°C, and are stable up to 150°C in an inert atmosphere. As_2I_4 is readily oxidized. It disproportionates easily, e.g. in warm carbon disulphide and on hydrolysis. Disproportionation is quantitative at 400°C:

$$3As_2I_4 \rightarrow 4AsI_3 + 2As$$

At lower temperatures, and in carbon disulphide at 20°C, there is evidence for formation of solids with approximate composition As_4I [72].

The unstable iodide Sb_2I_4 can be detected by e.m.f. or vapour pressure measurements on solutions of antimony in molten antimony triiodide at 230°C, but only antimony and antimony triiodide are obtained when the solutions are cooled.

The Compounds BiX

Early work suggesting formation of compounds BiX_2 has been discounted[73]. Formation of the species BiX can be shown in the vapour in equilibrium with Bi–BiX_3 mixtures and consistent thermodynamic data has been obtained.

TABLE 27. THERMODYNAMIC DATA FOR HALIDES BiX [a]

	ΔH_{298} for reaction Bi(s)+$\frac{1}{2}BiX_3$(g) = BiX(g) (kJ mol^{-1})	$H^\circ_{298}-H^\circ_0$ (kJ mol^{-1})	S° (J mol^{-1})	Consolute temperature[b] (K)
BiCl	117	9.8	263 (298 K)	1053
BiBr	102.5	10.0	301 (935 K)	811
BiI	101.7	10.5	307 (913 K)	731

[a] D. Cubicciotti, *J. Phys. Chem.* **71** (1967) 3066; *Inorg. Chem.* **7** (1968) 208, 211.
[b] S. J. Yosim, A. J. Darnell, W. G. Gehman and S. W. Mayer, *J. Phys. Chem.* **63** (1959) 230; S. J. Yosim, L. D. Ransom, R. A. Sallach and L. E. Topol, *J. Phys. Chem.* **66** (1962) 28.

The phase diagrams of Bi–BiX_3 systems show in each case a region where two liquids are in equilibrium over a large range of temperature and composition. These liquids become miscible only at a high temperature (Table 27). For Bi–$BiCl_3$ (Fig. 12) the two liquids are in equilibrium with a solid phase of composition approximately BiCl at about 320°C, but

[72] M. Baudler and H.-J. Stassen, *Z. anorg. allgem. Chem.* **343** (1966) 244; **345** (1966) 182.
[73] J. D. Corbett, *J. Am. Chem. Soc.* **80** (1958) 4757.

equilibrium is reached only very slowly. The solid phase has been isolated[74] as orthorhombic crystals, by heating bismuth–bismuth trichloride mixtures to 325°C and cooling them during several weeks to 270°C, and removing excess bismuth trichloride by sublimation, or

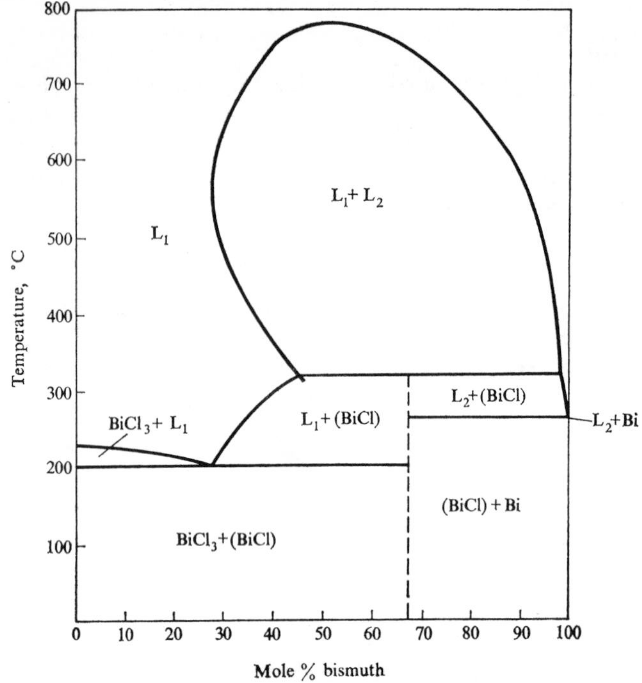

FIG. 12. Phase diagram for Bi–$BiCl_3$ from *J. Phys. Chem.* **63** (1959) 230.

extraction with benzene. The diamagnetic "bismuth monochloride" which remains has a structure with bismuth clusters (Fig. 13) and two different complex anions. It is thus $(Bi_9^{5+})_2$ $[BiCl_5]_4^{2-}[Bi_2Cl_8]^{2-}$ or $BiCl_{1.167}$. It is stable in vacuum up to 200°C, but disproportionates

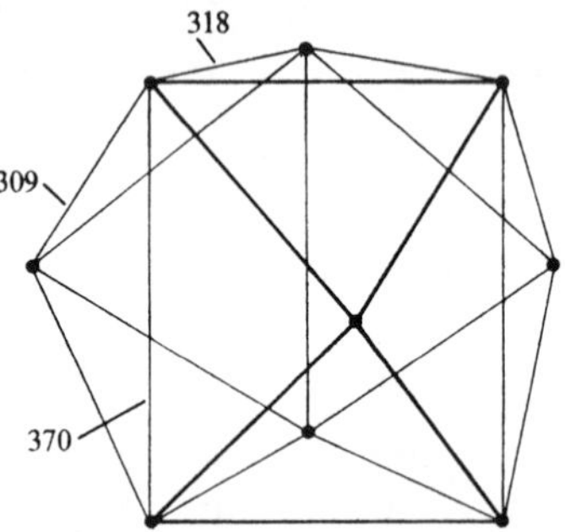

FIG. 13. Idealized (D_{3h}) structure for the Bi_9^{5+}, a trigonal prism with further bismuth atoms above each face. In solid $BiCl_{1.167}$ the cluster is considerably distorted by the unsymmetrical environment of the anions. (Average distances in pm.)

[74] A. Hershaft and J. D. Corbett, *Inorg. Chem.* **2** (1963) 979; R. M. Friedman and J. D. Corbett, *Chem. Communs.* (1971) 422.

at higher temperatures. It is readily hydrolysed to the oxide chloride, and it disproportionates in the presence of ligands which complex strongly with bismuth trichloride.

There are apparently related bismuth clusters in the compounds $Bi_5[AlCl_4]_3$ and $Bi_8[AlCl_4]_2$, which are obtained by reduction of a mixture of bismuth and aluminium trichlorides with bismuth in sodium tetrachloraluminate as solvent. Electronic spectra show that the same bismuth clusters Bi_5^{3+} and Bi_8^{2+} are present in other solvents of low pCl, e.g. $3NaCl/AlCl_3$ (eutectic) or $KCl/2ZnCl_2$; in very dilute solutions of bismuth in bismuth trichloride, which have a higher pCl, Bi^+ (presumably solvated) and Bi_3^+ predominate. Species with bismuth in low oxidation states are listed in Table 28, together with details of their characteristic spectra. Spectra of Bi^+ show a similar pattern of absorptions in various (molten salt) solvents. They have been interpreted in terms of *p–p* transitions between various states of the p^2 configuration, modified by the crystal field of the anions,

TABLE 28. BISMUTH CLUSTERS IN HALIDE MELTS

	Probable shape	No. bonding *p* electrons	Electronic spectra[a] (nm)	Ref.
Bi^+	—		900 800 **690** **663** **585**	b
Bi_3^+	?			c
Bi_5^{3+}	D_{3h}	12	875 790 540 455 **390** 345 305	c
Bi_8^{2+}	D_{4d} (antiprismatic)	22	530 450 350	d
Bi_9^{5+}	D_{3h} (Fig. 13)	22		

[a] Strongest absorptions in bold.
[b] N. J. Bjerrum, C. R. Boston and G. P. Smith, *Inorg. Chem.* **6** (1967) 1162, 1172.
[c] J. D. Corbett, *Inorg. Chem.* **7** (1968) 198.
[d] N. J. Bjerrum and G. P. Smith, *Inorg. Chem.* **6** (1967) 1968.

e.g. $[AlCl_4]^-$ or $[ZnCl_4]^{2-}$, by arguments similar to those used for the interpretation of *d–d* transitions. Stable clusters seem to be those with a maximum number of nearest neighbours. The distribution of species can be rationalized by considering the effects of counter-ions. Thus Bi_8^{2+} is found with the compact ions $[AlCl_4]^-$ or $[Al_2Cl_7]^-$ but Bi_9^{5+}, with a higher charge density, requires larger, more polarizable anions, e.g. $BiCl_5^{2-}$ or $Bi_2Cl_8^{2-}$. If the *s* electrons are assumed not to take part in the bonding (see p. 577), the clusters are electron-deficient, and molecular orbital schemes like those for boron hydrides or carboranes give satisfactory accounts of the observed electronic spectra[75a].

Coloured solutions obtained by oxidizing antimony with peroxodisulphate in fluorosulphonic acid, or with arsenic pentafluoride in liquid sulphur dioxide, may contain antimony clusters[75b].

6. OXIDES E_2O_3

Extraction processes for arsenic, antimony and bismuth, using oxidizing conditions, yield the oxides E_2O_3, which sublime away from involatile ash. They can be purified by

[75] (a) J. D. Corbett and R. E. Rundle, *Inorg. Chem.* **3** (1964) 1408; (b) R. C. Paul, K. K. Paul and K. C. Malhotra, *Chem. Communs.* (1970) 453; (c) P. A. W. Dean and R. J. Gillespie, *ibid.*, p. 853.

further sublimations, by recrystallization from hydrochloric acid, or by conversion to the anhydrous chlorides which can be distilled before hydrolysis back to the oxides.

6.1. PROPERTIES AND STRUCTURES

There is little recent structural work on these compounds, but it is clear that all three oxides E_2O_3 exist in several modifications (Scheme 4).

SCHEME 4. The oxides E_2O_3.

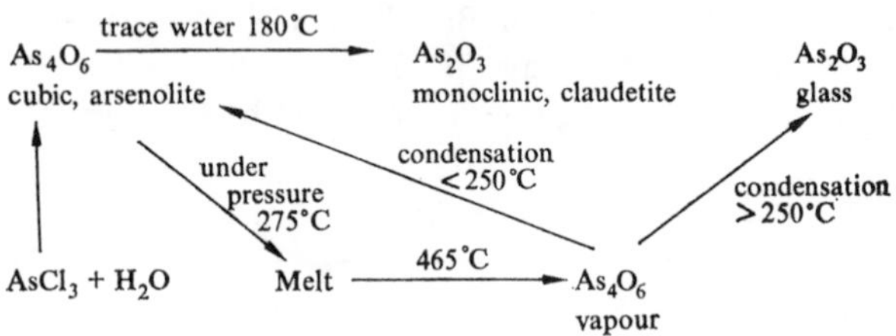

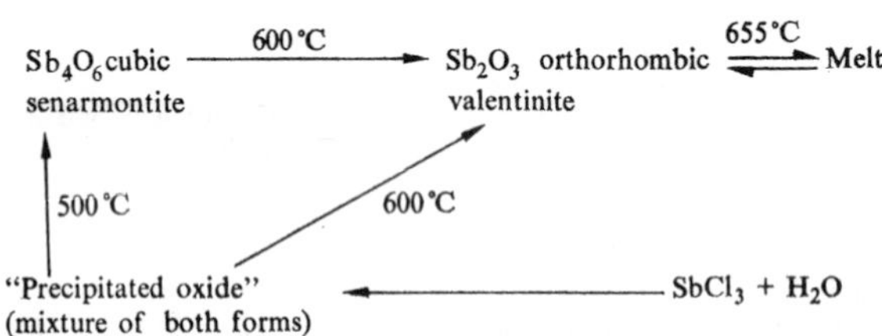

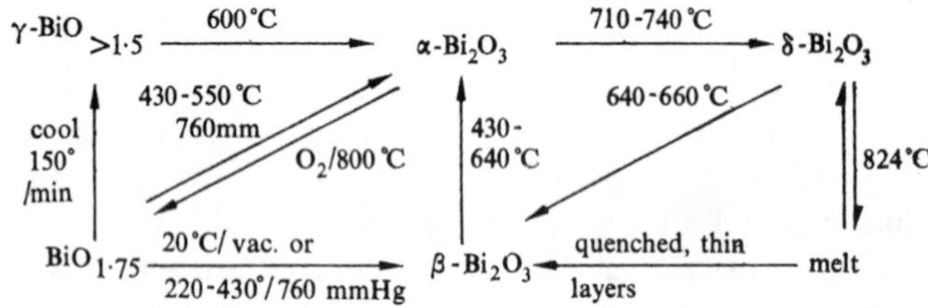

The molecules As_4O_6, with a structure [2] like that of P_4O_6, are found in the cubic form of the solid (arsenolite), in nitrobenzene solution and in the vapour below 800°C. At 1800° the vapour consists of molecules As_2O_3. Bond lengths and angles in As_4O_6 are given in Table 29.

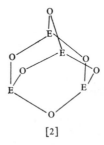

[2]

TABLE 29. PROPERTIES OF OXIDES E_2O_3

	As_2O_3 cubic	As_2O_3 monoclinic	Sb_2O_3 cubic	Sb_2O_3 orthorhombic	Bi_2O_3 monoclinic
Melting point (°C)	278 [a]	312 [a]		655 [b]	824 [c]
Boiling point (°C)	465 [b]		1425 [b]		
Vapour pressure[d] (mmHg/°C)	3.44/231 26.7/275	1.31/231 17/275	0.41/550 20.8/750 (liq.)	7.6/650	
ΔH_f° [e] (kJ mol^{-1})	−656.9	−654.8	−720.5	−708.5	−574
S° [e] (J mol^{-1} K^{-1})	107	117	110	123	151.5
ΔG_f° [e] (kJ mol^{-1})	−576.3	−577.4	−634	−626.5	−493.8
C_p [e] (J mol^{-1} K^{-1})	95.6			101.3	113.5
Density (kg m^{-3})	3890 [a]	4230(I) [a] 4020(II)	5200 [b]	5790 [b]	9200 [h]
E–O bond length (pm)	180±2 [f]	180 [a]	200 [f]		(i) 208–229 [h] (ii) 248–280
Angles OEO (°)	100±1.5 [f]		97 [f]	79, 92, 100 [g]	
(°)	126±3 [f]		130 [f]	115, 129 [g]	
Vibrational and nuclear quadrupole resonance spectra	i		i, j		
Magnetic susceptibilities (m^3 mol^{-1})×10^{10} [k]	−5.21		−6.95		−10.1

[a] Ref. 76.
[b] Ref. 5, pp. 600, 615.
[c] Ref. 78. Double oxides, e.g. $Bi_2O_3MoO_3$, are also known.
[d] J. H. Schulman and W. C. Schumb, *J. Am. Chem. Soc.* **65** (1943) 878; W. B. Hincke, *ibid.* **52** (1930) 3869.
[e] Ref. 11. Values based on E_2O_3. For As_2O_3: $\Delta H^\circ_{298\,\mathrm{sub}}$, 52.4 kJ mol^{-1}; $\Delta S^\circ_{298\,\mathrm{sub}}$, 83.2 J mol^{-1} K^{-1}.
[f] Ref. 47.
[g] M. J. Buerger and S. B. Hendricks, *J. Chem. Phys.* **5** (1937) 600.
[h] The five-co-ordinate Bi^{III} (see text) has three Bi–O distances in range (i) and two in range (ii). The six-co-ordinate Bi^{III} has three Bi–O distances in range (i) and three in range (ii). G. Malmros, *Acta chem. scand.* **24** (1970) 384.
[i] Assignments have been made as follows (I. R. Beattie, K. M. S. Livingston, G. A. Ozin and D. J. Reynolds, *J. Chem. Soc.* A (1970) 449):

T_d	t_2	a_1	t_2	e	a_1	t_2	e	t_2
As_4O_6(g)		556	492 m	409 w	381 s	253 m	184 m	99 m
Senarmontite	717 w	452 m	376 m	359 vw	256 vs	193 m	121 m	87 m

The Raman spectrum of arsenolite is like that of the vapour.
[j] Nuclear quadrupole coupling constants (77 K): ^{121}Sb, 554.8; ^{123}Sb, 707.1 MHz. R. G. Barnes and P. J. Bray, *J. Chem. Phys.* **23** (1955) 1177.
[k] Ref. 52.

There are apparently[76] two monoclinic forms of arsenious oxide (claudetite I and II), in which alternate arsenic and oxygen atoms are linked into sheets making open macromolecular structures (cf. orpiment, p. 613). The glassy form of arsenious oxide is similar, but the macromolecular structure is irregular. In all the structures, pyramidal AsO_3 groups

[76] K. A. Becker, K. Plieth and I. N. Stranski, *Progr. Inorg. Chem.* **4** (1962) 1.

share oxygen atoms, but the linking of the pyramids is different in the various modifications. The cubic and monoclinic forms have very similar stabilities with respect to the elements. Thus formation of cubic arsenolite by crystallization from aqueous solutions appears to be dominated by kinetic factors, since claudetite is the thermodynamically stable form. Interconversion of various forms of As^{III} oxide is slow, since rearrangements with breaking of many As–O bonds are involved. Changes are catalysed by water; presumably As–O bonds are protonated and the AsOH units can reform into a new pattern more easily than the original AsOAs units[76].

The molecules Sb_4O_6 are found in the cubic senarmontite and in the vapour below 1000°C. In the orthorhombic valentinite, alternate antimony and oxygen atoms are linked into bands (cf. sheets in claudetite), making a macromolecular structure. The cubic form appears to be the more stable below 570°C, but orthorhombic valentinite is metastable for long periods[77]. As with As^{III} oxide, the macromolecular structure has the higher density. The powder obtained by precipitation from a neutral aqueous solution is valentinite; that from aqueous sodium carbonate is mainly valentinite, with a small amount of senarmontite. Again, kinetic factors seem to be important in determining which modification is formed. The slow interconversions between the forms are consistent with transformations between molecular and macromolecular structures.

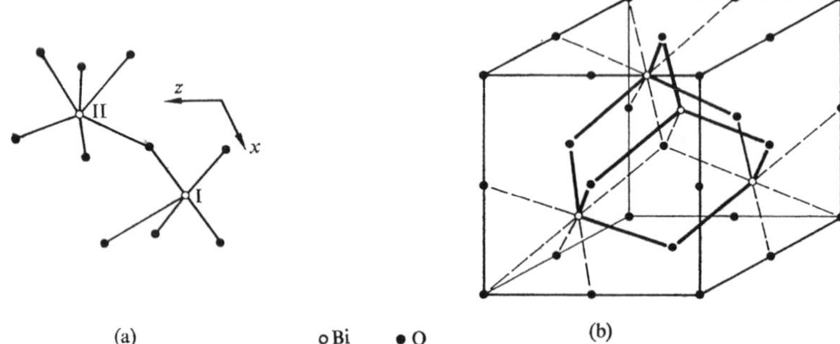

FIG. 14. (a) The co-ordination polyhedra in α-Bi_2O_3. (b) The fluorite structure of δ-Bi_2O_3 (one-quarter of the oxygen atoms are missing).

Bi^{III} oxide is polymorphic[78]. The most stable modification at 20°C (α-Bi_2O_3) has a complex monoclinic structure with layers of bismuth atoms and oxygen atoms parallel to y and z. There are two sorts of bismuth (Fig. 14). One (Bi_I) has five oxygen neighbours at the

O
O—Bi—O
O
:
O

[3]

[77] E. J. Roberts and F. Fenwick, *J. Am. Chem. Soc.* **50** (1928) 2125.

[78] G. Gattow and H. Schröder, *Z. anorg. allgem. Chem.* **318** (1962) 176; G. Gattow and D. Schütze, *ibid.* **328** (1964) 44.

corners of a distorted octahedron with the bismuth lone pair occupying the sixth corner [3]. Two more oxygen atoms are weakly bound. The other sort of bismuth (Bi_{II}) has six oxygen neighbours at the corners of an octahedron, distorted to suggest that here also the lone-pair affects the stereochemistry. The co-ordination polyhedra are linked to give tunnels parallel to z into which the Bi^{III} lone pairs project.

The cubic δ-form of Bi_2O_3, stable above 717°C, has a fluorite defect structure in which oxygen vacancies are randomly distributed; each bismuth atom has eight neighbours. The relation between the macromolecular structure of δ-Bi_2O_3 and the molecular structures of As^{III} and Sb^{III} oxides is shown in Fig. 14b. If the oxygen at the centre of the cube is missing and the surrounding atoms are drawn in regularly, discrete E_4O_6 molecules result.

The β-modification and several oxygen-rich forms (Scheme 4) are related to the high temperature δ-form. Some of the oxygen-rich forms are stabilized by small amounts of elements such as B, Al, Tl, Si, Ge, P, As, V, Mn, Fe, Co, and so may be double oxides pMO$_n$, qBi$_2$O$_3$. For example, the bismuth germanium oxide $Bi_{12}GeO_{20}$ has a body-centred structure (space group $I23$)[79]. Germanium atoms make four bonds to oxygen at the corners of a tetrahedron and bismuth atoms are like the Bi_I atoms of α-Bi_2O_3. Five oxygen atoms form an incomplete octahedron [3] (Bi–O, 207–264 pm) and two further oxygen atoms are more weakly bound (Bi–O, 308, 317 pm). The oxide is strongly optically active and piezoelectric along the threefold axes. Double oxides may be made by fusing Bi^{III} oxide and oxides of Ca, Sr, Ba, Cd or Pb. These have $(BiO)_n$ layers, as in the oxide halides (p. 600), interleaved with M^{II} ions. Other double oxides have formulae $M^1M^2_4Bi_6O_{18}$, where M^1 = Na, K, Ca, Sr, Ba, Pb, and M^2 = Ti, Nb, Ta.

6.2. REACTIONS

The oxides are starting materials for the preparation of many other compounds of arsenic, antimony and bismuth. Reactions of As^{III} oxide are shown in Scheme 5; Sb^{III}

SCHEME 5. Reactions of As^{III} oxide.

$$
\begin{array}{ccccc}
AsX_3 & \xleftarrow{HX} & As_2O_3 & \xrightarrow{S} & As_2S_3 \\
AsH_3 & \xleftarrow{NaOH/KBH_4} & As_2O_3 & \xrightarrow{NaOH/RX} & RAsO_3Na_2\text{(R alkyl)} \\
R_3As & \xleftarrow{R_3Al} & As_2O_3 & \xrightarrow{ROH} & As(OR)_3 \\
H_3AsO_4 & \xleftarrow{HNO_3} & As_2O_3 & \xrightarrow{NaOH/ArN_2Na} & ArAsO_3Na_2\text{(Ar = aryl)}
\end{array}
$$

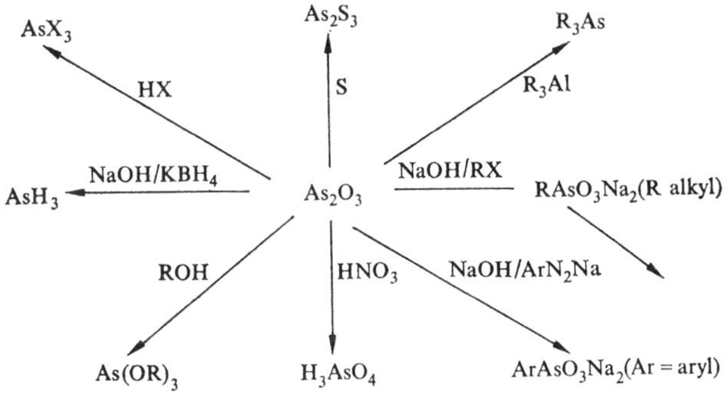

oxide is similar. Bi^{III} oxide is more basic. It is insoluble in alkali but dissolves in acid to give bismuth salts (section 8). As^{III} and Sb^{III} oxides are used in analytical chemistry as mild

[79] S. C. Abrahams, P. B. Jamieson and J. L. Bernstein, *J. Chem. Phys.* **47** (1967) 4034.

reducing agents, and in preparative chemistry for the conversion of transition metal halides to oxyhalides[80]:

$$3MX_n + E_2O_3 \rightarrow 3MX_{n-2}O + 2EX_3$$

The volatile halide EX_3 is readily removed. In this reaction, Sb^{III} oxide has been most frequently used: the reaction is of wide application. Thiohalides may be made, using Sb^{III} sulphide.

As^{III} oxide has been used as a "fining agent" in glass manufacture. Small quantities may be incorporated into the Si–O frameworks, and redox reactions between As^{III} and As^{V} assist in the control of small bubbles of oxygen in the molten glass. Under certain conditions, As^{III} oxide may be a useful decolorizer by assisting in redox reactions involving transition metal (e.g. iron) oxide impurities[81].

6.3. AQUEOUS CHEMISTRY: ARSENITES AND ANTIMONITES

As^{III} oxide is not very soluble in water (0.104 mol As_2O_3 per kg water at 25°C). The solubility decreases with increasing HCl concentration, becoming a minimum (0.075 mol per kg water) at about 3 M HCl and then increasing[82]. Presumably at high HCl concentrations species with As–Cl bonds are formed (section 10.3). In neutral or acid solution, the predominant species is probably $As(OH)_3$. On crystallization, the oxide As_2O_3 is obtained.

TABLE 30. RAMAN SPECTRA OF SOLUTIONS OF ARSENIOUS ACID (cm^{-1}) [a]

$As(OH)_3$	C_{3v}	ν_1, 710(a_1); ν_3, 655(e)
$[AsO(OH)_2]^-$	C_s	ν_1, 790(a_1); ν_5, 610(a''); ν_2, 570(a'); $\nu_{3,4,6}$, 370–320
$[AsO_2(OH)]^{2-}$	C_s	810, 770, 670–520
AsO_3^{3-}	C_{3v}	ν_1, 752(a_1); ν_3, 680(e); $\nu_{2,4}$ 340

[a] T. M. Loehr and R. A. Plane, *Inorg. Chem.* **7** (1968) 1708.

Samples of As_2O_3, which have been precipitated from HCl and sublimed, are suitable for use as a primary standard in quantitative analysis. As^{III} oxide is much more soluble in basic solutions, probably giving species $[AsO(OH)_2]^-$, $[AsO_2(OH)]^{2-}$ and AsO_3^{3-}, but not AsO_2^- (Table 30). There seems to be no evidence for formation of polymerized species, e.g. $As_2O_5^{4-}$, $As_3O_6^{3-}$, in basic solutions for As^{III} concentrations of 0.6–5 M. As–O stretching absorptions appear at 750–790 cm^{-1} and As–OH absorptions vary from 710 cm^{-1} in $As(OH)_3$ to 570 cm^{-1} in $[AsO(OH)_2]^-$. The Raman spectra show that arsenious acid does not have a structure like phosphorous acid $HPO(OH)_2$ with an As–H bond. The difference between phosphorus and arsenic is consistent with the reluctance of arsenic to reach oxidation state V and with the weaker As–H, compared with P–H, bonds. Proton n.m.r. spectra of $As(OH)_3$ show only one peak, indicating that all the protons are equivalent on the time scale of the experiment. Arsenious acid is weak (pK 9.2)[38], in accord with Pauling's rules for the structure $As(OH)_3$.

[80] P. C. Crouch, G. W. A. Fowles, I. B. Tomkins and R. A. Walton, *J. Chem. Soc.* A (1969) 2412, and references therein. D. Britnell, G. W. A. Fowles and R. Mandyczewsky, *Chem. Communs.* (1970) 608.

[81] Information about the use of arsenic compounds may be obtained from the Arsenic Development Committee, 26 rue La Fayette, Paris, 9e.

[82] A. B. Garrett, O. Holmes and A. Laube, *J. Am. Chem. Soc.* **62** (1940) 2024; K. H. Gayer and A. B. Garrett, *ibid.* **74** (1952) 2353.

When alkaline solutions of As^{III} oxide are evaporated, salts known as arsenites are obtained. These are not very well characterized, but structures appear to consist of metal ions and pyramidal AsO_3 groups linked into sheets or chains like those of the macromolecular forms of As^{III} oxide[83]. Yellow silver arsenite (contrast the arsenate which is brown) and copper arsenite (Scheele's green) may be precipitated from neutral solutions. Sodium arsenite has been used as an insecticide in cattle and sheep dips, as a long-lasting herbicide, and as a fungicide[81].

Sb^{III} oxide is very slightly soluble in water and dilute acid (3×10^{-5} mol Sb_2O_3 per kg water at 25°C) and rather more soluble in alkali (5×10^{-4} mol Sb_2O_3 per kg 0.1 M NaOH)[82]. The species in solution are not well established, but the following data have been given[38] from solubility measurements:

	log K
$\frac{1}{2}Sb_2O_3 + \frac{3}{2}H_2O \rightarrow Sb(OH)_3$	~ −4
$\frac{1}{2}Sb_2O_3 + OH^- + \frac{3}{2}H_2O \rightarrow [Sb(OH)_4]^-$	−2.06
$\frac{1}{2}Sb_2O_3 + H^+ \rightarrow SbO^+ + \frac{1}{2}H_2O$	−3.11

Antimonites are not well characterized. The compounds MSb_2O_4 (M = Mg, Mn^{II}, Fe^{II}, Co^{II}, Ni^{II}) and $NiAs_2O_4$ are isostructural with the lead oxide Pb_3O_4.

6.4. ARSENITE AND ANTIMONITE ESTERS

The preparation of these compounds has been described in section 5.3. Other routes to them include the following.

(a) Transesterification:

$$As(OEt)_3 + ROH \rightarrow As(OR)_3 + 3EtOH$$

(b) From oxides and alcohols. The water is removed by a drying agent such as calcium oxide or as an azeotrope. The method is not suitable for the more volatile esters because the water cannot be satisfactorily removed by distillation:

$$As_2O_3 + 6ROH \rightarrow 2As(OR)_3 + 3H_2O \qquad \text{(R alkyl, aryl)}$$

(c) By isomerization of alkylarsonates (section 14.4). This reaction is the reverse of the Arbuzov reaction, and the contrast between arsenites and phosphites is striking:

$$MeAsO(OMe)_2 \rightarrow As(OMe)_3 \quad (52\%)$$

Mixed esters $E(OR^1)_n(OR^2)_{3-n}$ may be made by transesterification or from halide esters $E(OR^1)_nCl_{3-n}$ and alkoxides $NaOR^2$. The mixed esters tend to reorganise at high temperatures:

$$3Sb(OBu)(OEt)_2 \xrightarrow{150\,°C} Sb(OBu)_3 + 2Sb(OEt)_3$$

The arsenite and antimonite esters are colourless volatile materials (Table 31), soluble in organic solvents and very readily hydrolysed by water to alcohols and oxides E_2O_3.

[83] J. W. Menary, *Acta Cryst.* **11** (1958) 742.

They are assumed to have the C_{3v} configuration. At 220°C, tricyclohexyl antimonite decomposes giving $C_6H_{11}^{\cdot}$ and $C_6H_{11}O^{\cdot}$ radicals[84].

TABLE 31. ARSENITE AND ANTIMONITE ESTERS

	M.p. (°C)	B.p./mmHg [a]		M.p. [a] (°C)	B.p./mmHg [a]
$As(OMe)_3$ [b]		128–130	$Sb(OMe)_3$ [c]	123–4	
$As(OEt)_3$ [b]		59/14	$Sb(OEt)_3$ [c]		94/10
$As(OCH_2)_3CMe$	41–2		$Sb(OPr)_3$ [d]		115/15
$As(OPh)_3$		212/4	$Sb(OPh)_3$	100	241/4
$As(OSiPh_3)_3$	189–90		$Sb(OSiEt_3)_3$		170/3
$As(OEt)_2Cl$		64–65/20	$Sb(OEt)_2Cl$ [c]		119–20/20
$As(OEt)Cl_2$		58–9/45	$Sb(OEt)Cl_2$ [c]	70	114–18/10
$As(SEt)_3$		90/1	$Sb(SEt)_3$		167–70/4

[a] Refs. 60, 61. Boiling point at 760 mmHg (101 325 N m^{-2}) where no pressure indicated.
[b] For $As(OMe)_3$: ΔH_f°, -591 kJ mol^{-1}; vapour pressure, $\log p_{mmHg} = -2200/T+8.358$; $\overline{D}$ (As–OMe), 262 kJ mol^{-1}. For $As(OEt)_3$: ΔH_f°, -706 kJ mol^{-1}°; vapour pressure, $\log p = -2570/T+8.831$. $\overline{D}$ (As–OEt), 270 kJ mol^{-1}. (T. Charnley, C. T. Mortimer and H. A. Skinner, *J. Chem. Soc.* (1953) 1181.)
[c] See also C. Russias, F. Damm, A. Deluzarche and A. Maillard, *Bull. Soc. chim. France* (1966) 2275; O.D. Dubrovina, *Chem. Abstr.* **51** (1957) 6534.
[d] ^{121}Sb nuclear quadrupole coupling constant (77 K), 560.7 MHz (cf. Tables 23 and 29). R. G. Barnes and P. J. Bray, *J. Chem. Phys.* **23** (1955) 1177.

Thioesters $E(SR)_3$ and halide thioesters $EX(SR)_2$, $EX_2(SR)$ may be made from mercaptans by methods like those used for analogous oxygen compounds[60]. As^{III} *NN*-diethyl-dithiocarbamate $As(S_2CNEt_2)_3$, from arsenic trichloride and sodium diethyl dithiocarbamate, has a structure in which each arsenic atom makes three bonds (As–S, 235 pm) at 90°, and three longer bonds (As–S, 280–290 pm), so that the configuration round arsenic is that of a distorted antiprism[85].

The antimony compounds react with alkoxides NaOR to give the derivatives $Na^+[Sb(OR)_4]^-$. With Grignard reagents, the organometallic compounds R_3E are obtained. Arsenite and antimonite esters may be oxidized by bromine (section 12.4).

The stereochemical consequences of the lone pair are of interest in the compounds $M^+[E(OR)_4]^-$, which may be compared with the halide complexes described in section 10.3.

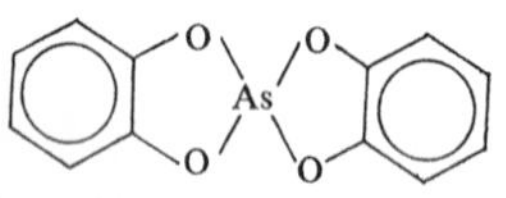

[4]

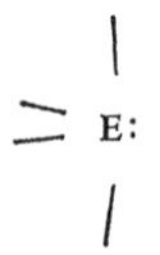

[5]

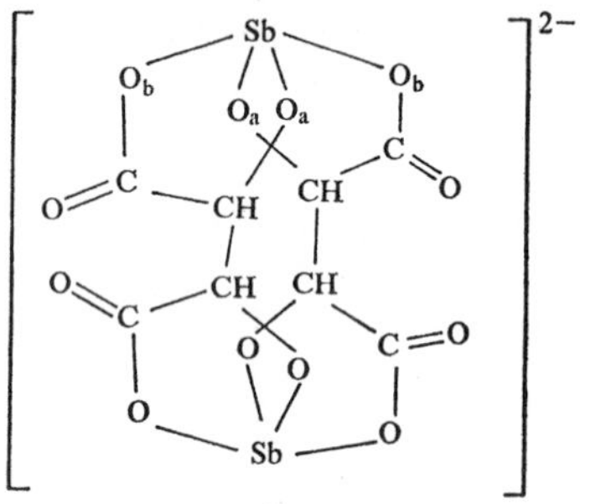

[6]

[84] L. Golder, A. Deluzarche and A. Maillard, *Bull. Soc. chim. France* (1961) 1805.
[85] M. Colapietro, A. Domenicano, L. Scaramuzza and A. Vaciago, *Chem. Communs.* (1968) 302.

In potassium di-*o*-phenylenedioxyarsenite [4] each arsenic makes four bonds to the corners of a distorted trigonal bipyramid [5], with equatorial bonds, opposite the lone pair, 180.7 and axial bonds 199.5 pm [86]. A similar configuration at antimony is found in antimony hydrogen bis(thioglycollate) $Sb(SCH_2CO_2)_2H$, which has Sb–O (axial) 228, Sb–S (equatorial) 242.8 pm.

TABLE 32. CRYSTAL DATA ON TARTRATOANTIMONITES $M_2[Sb_2(C_4H_2O_6)_2]nH_2O$

M	*n*	Tartrate	Space group	Piezoelectricity[a]	Sb–O a (pm) [6]	Sb–O b (pm) [6]	Ref.
NH_4, Rb	3	d-	$C222_1$	*pp*	202	218	b
NH_4	4	Racemic	$C2/c$	None	204	214	c
K	3	Racemic	$Pca2_1$	*p*	201	220	d
$Fe(C_{12}H_8N_2)_3$	8	d-			194	214	e

[a] *pp*, greater than quartz; *p*, less than quartz. G. A. Kiosse, N. I. Golovastikov and N. V. Belov, *Soviet Physics Crystallography* **9** (1964) 321.
[b] K salt isomorphous. G. A. Kiosse, N. I. Golovastikov, A. V. Ablov and N. V. Belov, *Dokl. Akad. Nauk. SSSR* **177** (1967) 329.
[c] Rb salt isomorphous. G. A. Kiosse, N. I. Golovastikov and N. V. Belov, *Dokl. Akad. Nauk. SSSR* **155** (1964) 545.
[d] B. Kamenar, D. Grdenić and C. K. Prout, *Acta Cryst.* **26B** (1970) 181.
[e] $C_{12}H_8N_2$ *o*-phenanthroline. D. H. Templeton, A. Zalkin and T. Ueki, *Acta Cryst.* **21** (1966) A154.

Potassium tartratoantimonite(III) (tartar emetic) is one of the commonest and most useful compounds of antimony and often the best source of Sb^{III} ions in solution. The older name for this compound, potassium antimonyl tartrate, is misleading because it implies SbO^+ ions, for which there is no evidence. In the d- and racemic tartrates (Table 32), both hydroxyl and carboxyl groups of the tartrate are bound to the antimony to make dimeric anions (structure [6]): hydrogen-bonded water molecules and cations fill the remaining space. An acid solution is obtained when potassium tartratoantimonite is dissolved in water. The Hg^{II} salt is precipitated from aqueous solutions: in alkali it disproportionates to Hg^0 and antimonate(V) [87].

6.5. BISMUTH–OXYGEN SPECIES IN SOLUTION

Bismuth hydroxide dissolves in acid, giving solutions which contain Bi^{III} ions. Such solutions remain clear only at high acid concentrations. As the hydrogen ion concentration is reduced, oxy-salts are precipitated. Before precipitation occurs, a series of polymeric cations can be detected in solution. The best characterized ion is $[Bi_6(OH)_{12}]^{6+}$, which has a structure (Fig. 15) like $Ta_6Cl_{12}^{2+}$ with a Bi_6 octahedron joined by oxygen bridges. The Bi–Bi distance is 370 pm (cf. 307, 353 in the metal) and the shortest Bi–O distance is 233 pm. The totally symmetric band in the Raman spectrum at 177 cm^{-1} is more intense than that at 450 cm^{-1}. It has been argued that, if the intensity of the Raman bands must be

[86] A. C. Skapski, *Chem. Communs.* (1966) 10; I. Hansson, *Acta chem. scand.* **22** (1968) 509.
[87] E. Chinoporos and N. Papathanasopoulos, *J. Phys. Chem.* **65** (1961) 1643.

attributed only to Bi–O bonds, the mode involving mainly stretching (that with higher energy) should be more intense than the mode involving mainly bending, and that the intensity of the low energy absorption must thus be attributed to Bi–Bi interactions.

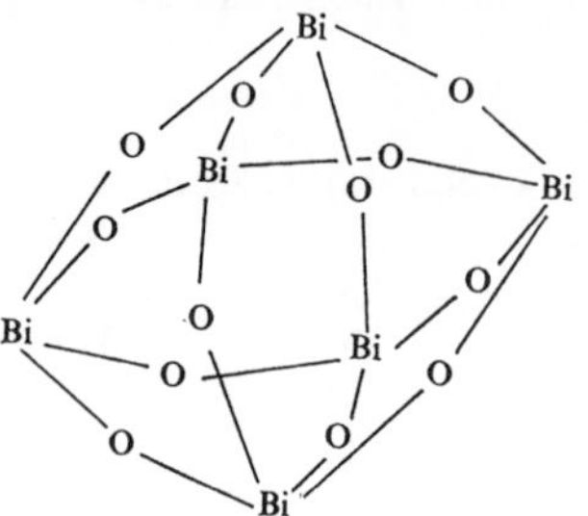

FIG. 15. The ion $[Bi_6(OH)_{12}]^{6+}$.

7. SULPHIDES, SELENIDES AND TELLURIDES

Sulphides of arsenic and antimony were used as pigments for centuries. With the development of organic dyes, their importance decreased, but recently interest in the compounds of antimony and bismuth with the Group VI elements has revived. The structures of the sulphides, selenides and tellurides show an interesting gradation from the small molecules of the arsenic sulphides to the chain-like molecules of the antimony compound Sb_2S_3 and the sheet-like molecules of bismuth selenide and telluride.

7.1. PREPARATION

The compounds may be made fusing together stoichiometric quantities of the elements in sealed quartz tubes at 500–900°C. Selenides and tellurides are almost always made in this way, and very pure elements are available. Crystals usually form as the tubes are cooled. Sometimes, for example, with As_2Se_3, the product is amorphous; in this case crystals may be obtained by slow sublimation at 330°C. Realgar, As_4S_4, may be recrystallized from carbon disulphide or benzene.

The compounds E_2X_3 (X = S, Se, Te) may also be obtained by heating the oxide with the element X:

$$2As_2O_3+9S \rightarrow 2As_2S_3+3SO_2$$

The amorphous sulphides may be made by precipitation from acidified solutions of E^{III} with hydrogen sulphide. This precipitation is quantitative at pH 1–2.

7.2. STRUCTURES

Realgar, As_4S_4, is molecular in both the solid and the vapour: the dimensions of the molecule [7] are given in Table 33. Above 550°C some dissociation to As_2S_2 is observed. α-Dimorphite, As_4S_3, has a structure [8] like the phosphorous compounds P_4S_3, P_4Se_3. Orpiment vapour consists of molecules As_4S_6, with the same structure [2] as As_4O_6. In

TABLE 33. BOND DISTANCES AND INTERBOND ANGLES IN ARSENIC SULPHIDES

	As_4S_4(v) [a]	As_4S_4(s) [a]	As_4S_3(s) [b]	As_4S_6(v) [a]	As_2S_3(s) [c]
As–X (X = S, Te) (pm)	223±2	221	221±1	225±2	224
As–As (pm)	249±4	254	245±1		325 [d]
∠As S As	101±4	102	106±0.4	100±2	102(S_I), 94.5(S_{II}) [e]
∠S As S	93	93	98±0.5	114±2	99
∠S As As	100	102	102±1		

[a] Ref. 47.
[b] H. J. Whitfield, *J. Chem. Soc.* A (1970) 1800.
[c] N. Morimoto, *Mineral J. (Japan)* **1** (1954) 160.
[d] Non-bonding.
[e] S_I within spirals, S_{II} between spirals.

the monoclinic solid, alternate arsenic and sulphur atoms, instead of forming oligomers as in realgar, form spirals (parallel to *c*) linked into sheets (parallel to 010) similar to, but not identical with, those in claudetite. Each arsenic is bound to three sulphur neighbours, making an AsS_3 pyramid with bond angles at As considerably less than in the vapour. Each sulphur makes two bonds to arsenic: the sulphur atoms within (S_I) and between (S_{II}) the spirals are inequivalent. Arsenic selenide is similar, but in arsenic telluride[89], though there are again chains linked into sheets, half the arsenic atoms are octahedrally co-ordinated by tellurium {Te–As (3 co-ordinate) 274; Te–As (6 co-ordinate) 286 pm}.

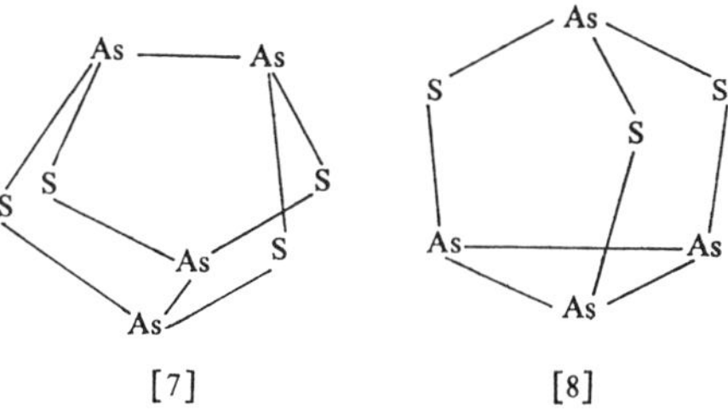

[7] [8]

The stibnite structure, adopted by Sb_2S_3, Sb_2Se_3 and Bi_2S_3 (and Bi_2Se_3 at high pressure) consists of a complex network (Fig. 16). The units (i) are linked into infinite bands, as in (ii), and the bands are linked as in (iii) and (iv). Internuclear distances are given in Table 34. All five atoms in unit (i) are inequivalent. It is interesting that one type of antimony atom (see arrows) has a square pyramidal environment, with the antimony slightly below the base of the square. Compare structure [3]. This is the configuration predicted by assuming sp^3d^2 hybridization of orbitals on antimony, with a lone pair occupying one co-ordination position. Antimony has this stereochemistry in several ternary sulphide minerals and in pentaphenylantimony (section 14.1). The bond angles at antimony are very close to 90° as in other compounds with non-metals.

[88] H. A. Levy, M. D. Danford and P. A. Agron, *J. Chem. Phys.* **31** (1959) 1458; V. A. Maroni and T. G. Spiro, *J. Am. Chem. Soc.* **88** (1966) 1410; *Inorg. Chem.* **7** (1968) 183.
[89] G. J. Carron, *Acta Cryst.* **16** (1963) 338.

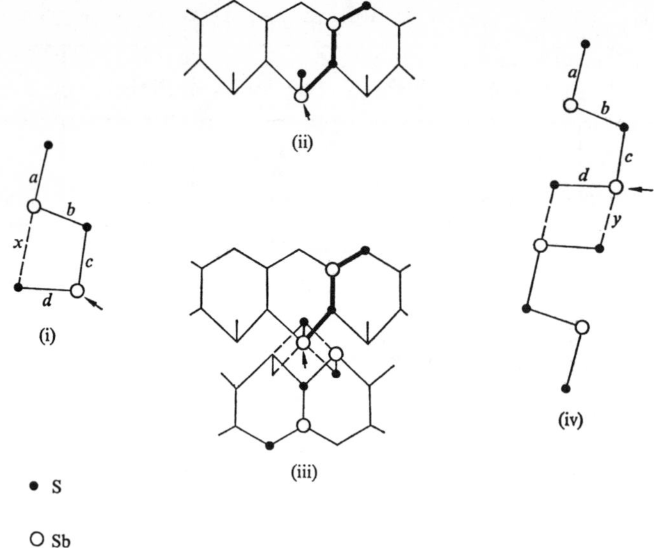

FIG. 16. The stibnite structure.

TABLE 34. BOND DISTANCES AND INTERBOND ANGLES IN STIBNITE STRUCTURES[a]

	a	*b*	*c*	*d*	*x*	*y*	*ab*	*bc*	*cd*	Ref.
Sb_2S_3	257	258	268	249	315	282	87.7	99.4	90.2	b
Sb_2Se_3	266.5	265.8	277.7	257.6	322	298	86.6	98.9	90.1	c

[a] See Fig. 16. Distances in pm, angles in degrees.
[b] S. Sćavničar, *Z. Krist.* **114** (1960) 85.
[c] N. W. Tideswell, F. H. Kruse and J. D. McCullough, *Acta Cryst.* **10** (1957) 99.

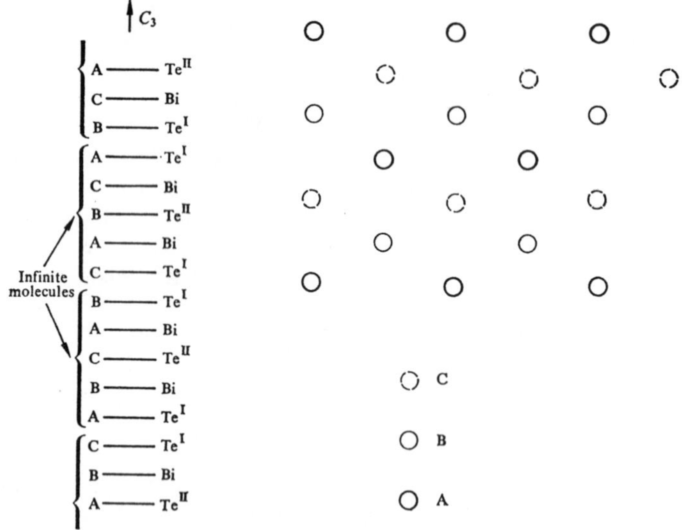

FIG. 17. The structure of bismuth telluride.

In bismuth telluride the molecules are infinite sheets. Atoms are in layers, in an extremely elongated rhombohedral arrangement shown in plan and elevation in Fig. 17. The molecules are perpendicular to the principal axis and weak forces hold them together (compare the bond lengths in Table 35). Bismuth selenide and antimony telluride have similar structures.

TABLE 35. STRUCTURES AND PROPERTIES OF COMPOUNDS $Bi_2Te_{(3-x)}Se_x$, Bi_2Te_2S AND Sb_2Te_3

	Bi_2Te_3		Bi_2Te_2Se	Bi_2TeSe_2	Bi_2Se_3	Bi_2Te_2S	Sb_2Te_3
x	0	$\frac{1}{2}$	1	2	3		
Bi–X_I (X = S, Se, Te) (pm)	306 [a]		305 [a]	296 [a]	286 [a]	311 [b]	295 [c]
Bi–X_{II} (pm)	324		309	308	310	306	316
X_I–X_{II} (pm)	363		367	360	354	370	361
Seebeck coefficient S [d] (μV K^{-1})	212 *p*	290 *p*	0	135 *n*	70 *n*		
Electrical conductivity [d] (Ω^{-1} m^{-1}) (κ) $\times 10^{-4}$	7.3	2.0	0.85	8.3	19.5		
Thermal conductivity [d] (W m^{-1} K^{-1}) (λ)	1.98	1.19	1.56	1.45	2.70		
Figure of merit (Z) $\times 10^3$ (K^{-1}) [d, e]	1.66	1.46	0	1.05	0.35		

[a] S. Nakajima, *J. Phys. Chem. Solids* **24** (1963) 479.
[b] Calculated from data by J. R. Wiese and L. Muldawer, *J. Phys. Chem. Solids* **15** (1960) 13, and from ref. 3.
[c] Calculated from data by S. A. Semiletov, *Soviet Physics Crystallography* **1** (1956) 317.
[d] I. Cadoff in ref. 22.
[e] See text.

7.3. PROPERTIES

The sulphides, selenides and tellurides are highly coloured solids. Some of their physical properties are given in Table 36.

As^{III} sulphide and selenide, like As^{III} oxide, readily form glasses if samples of the liquids are quenched, and some of these glasses are important commercially. The compounds of antimony and bismuth form glasses less readily[90]. Extensively glassy regions are also observed in the systems Si(or Ge)–As–Te and Ge–Sb–Se: the glassy regions are more extensive in the silicon than in the germanium system. The glasses, which have high softening points (up to 500°C), are of some importance, since some are transparent over a large part of the infrared spectrum. The molecular species in these glasses are not very well established[90].

The solids E_2X_3 are semiconductors with energy gaps which decrease in the series; for a given X, arsenic > antimony > bismuth compound; for a given E, sulphide > selenide > telluride. Either *n*- or *p*-type properties may be obtained by controlled deviations from exact stoichiometry. For bismuth telluride an excess of bismuth gives *p*-type properties, perhaps because some bismuth atoms occupy tellurium sites, and an excess of tellurium gives *n*-type properties. Lead, as impurity, gives *p*-type, and copper, silver and bromine

[90] H. Krebs, *Angew. Chem. Int. Edn. Engl.* **5** (1966) 544.

n-type, semiconductivity. As first obtained from the melt, bismuth telluride is slightly non-stoichiometric with an excess of bismuth. Replacement of tellurium by sulphur or selenium gives substances of composition $Bi_2Te_{3-x}X_x$ (X = S, Se). As x increases from 0 to 1, the X atoms first occupy Te_{II} sites (note the decrease in Bi–X_{II} distance in Table 35), leading to a tightening of the structure, an increase in the bond angles (e.g. ∠BiSBi 90° in Bi_2Te_2S), and an increase in the energy gap. As x increases further, Te_I atoms are replaced and the energy gap decreases. Bismuth atoms cannot now occupy the (smaller) anion sites, and so the substance changes to an *n*-type semiconductor. Replacement of bismuth by antimony is also possible, and solid solutions are stable over considerable composition ranges. Up to 45% of the antimony atoms in Sb_2Te_3, and up to 25% of the bismuth atoms in Bi_2Te_3, can be replaced by indium. The antimony compounds remain *p*-type, but the Bi_2Te_3 compounds become *n*-type semiconductors[91].

TABLE 36. PROPERTIES OF SULPHIDES, SELENIDES AND TELLURIDES

	As_4S_4	As_2S_3	Sb_2S_3	Bi_2S_3	As_2Se_3
Melting point (°C) [a]	320		546	850	
Boiling point (°C)	565	723			
Density (kg m^{-3}) [b]	3560	3490	4610	6780	4800 [c]
ΔH°_{f298} (kJ mol^{-1}) [d]	−285	−169.0	−174.9	−143.1	
S°_{289} (J mol^{-1} K^{-1}) [d]		163.6	182.0	200.4	
ΔG°_{f289} (kJ mol^{-1}) [d]		−168.6	−173.6	−140.6	
C_p (J mol^{-1} K^{-1}) [d]		116.2	119.8	122.1	
E_g (eV) [e]			1.7	1.3	2.1 [c]
Magnetic susceptibility (m^3 mol^{-1}) × 10^{10} [g]		−8.8	−10.8	−15.4	
Spectra			[h]		

	Sb_2Se_3	Bi_2Se_3 [f]	As_2Te_3	Sb_2Te_3	Bi_2Te_3 [d, f]
Melting point (°C) [a]	612	706	360	620	580
Density (kg m^{-3}) [b]	5810	7500	6250		
ΔH°_{f298} (kJ mol^{-1}) [d]		−149.8		−56.5	−77.4
S°_{298} (J mol^{-1} K^{-1}) [d]				234	261
ΔG°_{f298} (kJ mol^{-1}) [d]				−55.2	−77.0
C_p (J mol^{-1} K^{-1}) [d]					120.4
E_g (eV) [e]	1.3	0.35	~1	0.3	0.15

[a] Refs. 5 and 16.
[b] Ref. 3.
[c] S. A. Dembovskii and A. A. Vaipolin, *Soviet Physics Solid State* **6** (1964) 1388.
[d] Ref. 11. For Bi_2Te_3: $H^\circ_{298}-H^\circ_0$, 30.9 kJ mol^{-1}.
[e] N. B. Hannay (ed.), *Semiconductors*, Reinhold, New York (1959), p. 54. For Bi_2Te_3, mobilities: 0.125 *n*, 0.051 *p*, m^2 s^{-1} V^{-1}.
[f] ΔH°_f: Bi_2Te_2Se, −112; Bi_2TeSe_2, −128 kJ mol^{-1}. S. Misra and M. B. Bever, *J. Phys. Chem. Solids* **25** (1964) 1233.
[g] Ref. 52.
[h] Nuclear quadrupole coupling constants, 0°C; ^{121}Sb, 298.7; ^{123}Sb, 380.7 MHz. η, 0.008. T-C. Wang, *Phys. Rev.* **99** (1955) 566.

[91] A. J. Rosenberg and A. J. Strauss, *J. Phys. Chem. Solids* **19** (1961) 105.

Systems involving In, Tl, Sb, Bi, Se and Te have been widely studied, particularly because of their thermoelectric properties, used in solid-state refrigerators. If, in the arrangement of Fig. 18, the junction J_1 is at temperature T_1 and junctions J_2, J_3 at temperature T_2, the emf developed is given by $E_{AB} = (S_A - S_B)(T_1 - T_2)$. The Seebeck

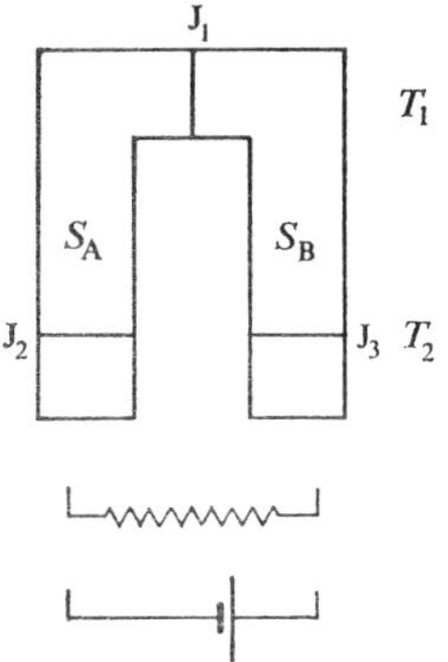

FIG. 18. The thermoelectric effect.

coefficients S may be positive (p-type) or negative (n-type). The materials A and B may be chosen so that they have similar electrical conductivities κ and thermal conductivities λ but Seebeck coefficients which are of opposite sign. To use the arrangement as a refrigerator, a current is passed from the battery, so that heat is absorbed at junction J_1, and released at J_2 and J_3. The thermoelectric heat flow is in addition to the heat flow associated with thermal conductivity and to the resistive heating due to the passage of current. The heat balance shows that the transfer of heat from the junction at temperature T_1 to those at T_2 is most efficient when the "figure of merit", Z, given by $Z = S^2\kappa/\lambda$, is maximum. A good material for use in a thermoelectric refrigerator thus has a low thermal conductivity λ, a high electrical conductivity κ and a high Seebeck coefficient S. Metals have high thermal conductivities and so are unsuitable. For semiconductors, the quantities κ and S vary in opposite senses with the carrier concentration, and it can be shown that $S^2\kappa/\lambda$ is maximum when the carrier concentration is $\sim 10^{25}\ m^{-3}$. Thermal conductivity, which involves transmission of lattice vibrations, is curtailed in complex or disordered structures, in structures with heavy atoms, and in structures with highly anharmonic vibrations, i.e. with atoms of unequal masses. The selenides and tellurides of bismuth may thus have particularly large thermoelectric figures of merit. Moreover, the optimum carrier concentration can be achieved by deviations from exact stoichiometry or by doping. Some data for the Bi–Te–Se system is shown in Table 35. The maximum figure of merit for bismuth telluride alloys (*ca.* $3 \times 10^{-3}\ K^{-1}$) is achieved in $(Bi,Sb)_2Te_3$. The figure of merit is strongly temperature dependent: for bismuth telluride alloys the maximum is at about room temperature, and so they are particularly suitable for refrigerators. They have been used in situations where elaborate pumps are undesirable.

7.4. REACTIONS

Sulphides, selenides and tellurides are stable in air at room temperature, but realgar is easily oxidized in light. All compounds are oxidized on heating in air, and by halogens, giving the chalcogenohalides (section 7.5).

The sulphides are only slightly soluble in water: solubility products have been given[38] as 10^{-93} for Sb_2S_3, and 10^{-96} mol^5 l^{-5} for Bi_2S_3. The sulphides are precipitated in the copper group of classical analysis. Arsenic and antimony sulphides dissolve in 0.5 M alkali with formation of oxy- and thio-anions, e.g. SbS_3^-. Bismuth sulphide is insoluble in alkali, but, unlike mercuric sulphide, it dissolves in nitric acid, probably with formation of nitrate complexes. Bismuth is not precipitated as sulphate, unlike lead, but, with ammonia, forms white gelatinous bismuth hydroxide (unlike copper and cadmium which form ammonia complexes). The hydroxide is readily reduced to bismuth metal. Arsenic sulphide is oxidized, by nitric acid or by hydrogen peroxide, to arsenate, but it is much less soluble in hydrochloric acid than antimony sulphide; this is exploited for separation of arsenic and antimony.

7.5. SULPHIDE, SELENIDE AND TELLURIDE HALIDES

By fusing mixtures of the compounds $E_2X_3^1$ (X^1 = S, Se, Te; E = Sb, Bi) and the halides EX_3^2 (X^2 = Cl, Br, I), ternary compounds EX^1X^2, analogous to the oxide halides (section 5) are obtained. They are stable in air and do not react with dilute acids. The

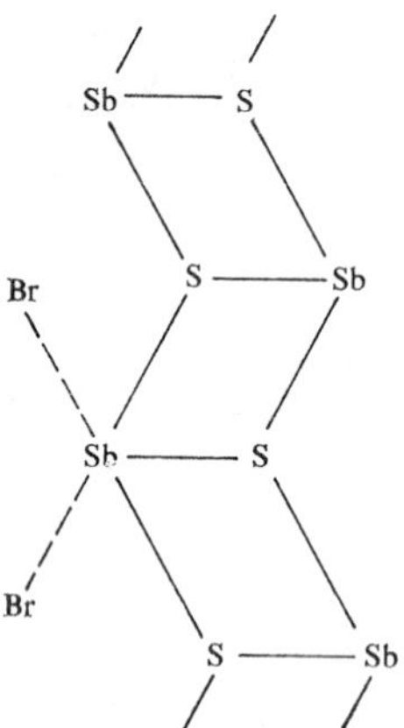

FIG. 19. The band structure of Sb^{III} sulphide bromide.

sulphide and selenide halides have structures with parallel pleated bands $(EX^1)_n^{n+}$ (Fig. 19) interspersed by halide ions. In SbSBr each antimony atom has three sulphur neighbours (one at 249 and two at 267 pm) and two bromine neighbours at 294 pm in a square pyramidal configuration. (Compare Sb_2S_3, Fig. 16 and structure [3] (p. 606).) The telluride halides BiTeBr and BiTeI have layer structures related to Bi_2Te_3[92].

There are large regions of glass formation in the ternary system As–S–Br, and glasses with widely differing softening points may be obtained. Thus, a glass with composition $AsS_{1.8}Br_{2.5}$ has a softening point of -60°C. As the bromine content decreases, the

[92] E. Dönges, *Z. anorg. allgem. Chem.* **263** (1950) 112, 280; **265** (1951) 56; G. D. Christofferson and J. D. McCullough, *Acta Cryst.* **12** (1959) 14.

softening point rises (e.g. to 90° for $AsS_{2.7}Br_{0.6}$ and 200° for AsS_2), as the divalent sulphur atoms are able to link together structures of high molecular weight[90].

7.6. TERNARY SULPHIDES, SELENIDES AND TELLURIDES

Many ternary sulphides are known as minerals. Others, with melting points between 400° and 600°C, can be made by fusing appropriate quantities of the binary sulphides. Compounds with the marcasite, arsenopyrite, pyrites and ulmanite structures have been given in Table 12 (p. 567). Further ternary compounds are described in Table 37, which is not comprehensive.

TABLE 37. TERNARY SULPHIDES, SELENIDES, AND TELLURIDES

$LiBiS_2$, $NaBiS_2$, $KBiS_2$, $NaBiSe_2$, $KBiSe_2$	NaCl structure; statistical occupation of cation sites[a]
$AgAsSe_2$, $AgAsTe_2$, $AgSbSe_2$, $AgSbTe_2$	NaCl structure[b]
$AgBiS_2$, $AgBiSe_2$	NaCl at high temperatures. Rhombohedral structure at 20°C, but distortions from NaCl very small. Bi–Se, 201 pm; Bi–S, 293 pm [b]
$TlSbTe_2$, $TlBiTe_2$	Rhombohedral like $AgBiS_2$. Sb–Te 311 pm; Bi–Te 319 pm [c]
$TlAsS_2$ (lorandite)	$(AsS_2)_n$ chains. As–S 223, 233 pm [d]
$CuSbS_2$ (wolfsbergite)	$(SbS_2)_n$ chains linked to double sheets by Cu. Double sheets joined by weaker Sb–S bonds[e]
$PbAs_2S_4$ (scleroclase)	Two kinds of $(AsS_2)_n$ chains As_I–S, 214, 256 (twice), 292 (twice) pm. As_{II}–S, 236, 259 (twice), 287 (twice) pm [f]
$FeSb_2S_4$ (berthierite)	$(SbS_2)_n$ chains Sb–S 250 pm [g]
$HgSb_4S_8$ (livingstonite)	$(SbS_2)_n$ double chains linked by S_2 groups or by Hg. Sb–S 247–315 pm [h]
$FePb_4Sb_6S_{14}$ (jamesonite)	SbS_3 groups linked to Sb_3S_7 units[h]
Ag_3AsS_3 (xanthoconite)	AsS_3 pyramids joined to double sheets by Ag atoms. As–S 223–227 pm [i]
$Ag_5S(SbS_3)$ (stephanite)	SbS_3 pyramids Sb–S 247 pm [j]

[a] O. Glemser and M. Filcek, *Z. anorg. allgem. Chem.* **279** (1955) 321; G. Gattow and J. Zemann, *ibid.* p. 324. $MSbE_2$ (M = Li, Na, K, Cs; E = S, Se) are described by S. I. Berul', N. P. Luzhnaya and Ya. G. Finkel'shtein, *Russ. J. Inorg. Chem.* **13** (1968) 662.

[b] J. H. Wernick, S. Geller and K. E. Benson, *J. Phys. Chem. Solids* **4** (1958) 154; **7** (1958) 240; S. Geller and J. H. Wernick, *Acta Cryst.* **12** (1959) 46. Lattice constants *a* are as follows: $AgSbSe_2$, 578.6; $AgSbTe_2$, 607.8; $AgBiSe_2$, 583.2; $AgBiTe_2$, 615.5 pm..

[c] E. F. Hockings and J. G. White, *Acta Cryst.* **14** (1961) 328.

[d] A Zemann and J. Zemann, *Acta Cryst.* **12** (1959) 1002.

[e] J. H. Wernick and K. E. Benson, *J. Phys. Chem. Solids* **3** (1957) 157.

[f] Y. Iitaka and W. Nowacki, *Acta Cryst.* **14** (1961) 1291.

[g] M. J. Buerger and T. Hahn, *Am. Miner.* **40** (1955) 226; *Chem. Abstr.* **50** (1956) 9239.

[h] M. J. Buerger and N. Niizeki, *Z. Krist.* **109** (1957) 129, 161.

[i] P. Engel and W. Nowacki, *Acta Cryst.* **24B** (1968) 77. Another modification of Ag_3AsS_3 is known as proustite.

[j] B. Ribár and W. Nowacki, *Acta Cryst.* **26B** (1970) 201.

The alkali metal and silver compounds, with the sodium chloride structure, are probably best described as double sulphides $M^IE^{III}S_2$, but, in some of the other compounds, anions based on ES_3 pyramids may be distinguished. These anions may be discrete as in Ag_3AsS_3, joined into chains, as in $TlAsS_2$, or small rings, as in $FePb_4S_6S_{14}$. In a number of ternary sulphides, e.g. in $FeSb_2S_4$, $PbAs_2S_4$, and $HgSb_4S_8$, the element E has the square pyramidal environment derived from an octahedron in which a lone pair occupies one co-ordination site (compare Sb_2S_3, section 7.2, and the complex halides, section 10.3).

Several of the ternary sulphides have been investigated as semiconductors, and energy gaps of 0.2–1 eV have been reported. In the silver compounds $AgAsSe_2$, $AgAsTe_2$, $AgSbSe_2$ and $AgSbTe_2$, the normal *p*-type properties are enhanced by excess silver and changed to *n*-type by addition of zinc, cadmium or manganese. All four compounds have the same sodium chloride structure and mixtures of compounds yield a complete series of solid solutions. The thermoelectric powers of about 200 $\mu V\ K^{-1}$ are comparable with those of systems based on bismuth telluride (Table 35).

8. SALTS OF ANTIMONY AND BISMUTH

Antimony and bismuth are sufficiently basic to form compounds which can formally be considered as salts. Halides (section 5) and sulphides (section 7) have already been discussed. Salts of oxyacids are considered in this section. There is, in general, little structural work or quantitative information. Antimony and bismuth salts are usually hygroscopic and react with water to give rather insoluble basic compounds, unless ligands are present which form strong complexes (section 10). Only basic salts of weak acids can be obtained.

8.1. PERCHLORATES

Bismuth perchlorate, $Bi(ClO_4)_3,5H_2O$, is well established. It dissolves in water to give a clear solution containing bismuth oxocations (section 6.5). The basic salts are not readily precipitated.

8.2. NITRATES

Bismuth nitrate pentahydrate may be crystallized from a solution of Bi^{III} oxide or carbonate in concentrated nitric acid. In the solid, each bismuth is co-ordinated by three unsymmetrical bidentate nitrate groups as well as three molecules of water. The metal thus has the high co-ordination number often found in hydrated salts of the heavier elements. Raman spectra of aqueous solutions containing various ratios of nitrate to bismuth and also of $Bi(NO_3)_3,5H_2O$–KNO_3 melts have been interpreted as showing that both the bidentate nitrate groups and also co-ordinated water persist in solution in a series of complexes with one to four nitrates per bismuth[93]. A basic salt $BiONO_3$ precipitates when solutions of the pentahydrate are diluted. The composition of the solid varies with the reaction conditions[94]. Basic salts, obtained by heating the pentahydrate, have been formulated with oxocations:

$$Bi(NO_3)_3,5H_2O \xrightarrow{50\text{–}60°C} [Bi_6O_6]_2(NO_3)_{11}(OH),6H_2O \xrightarrow{77\text{–}130°C} [Bi_6O_6](NO_3)_6,3H_2O \xrightarrow{400\text{–}450°C} \alpha\text{-}Bi_2O_3$$

[93] R. P. Oertel and R. A. Plane, *Inorg. Chem.* 7 (1968) 1192.

In the hemihydrate $BiONO_3,\frac{1}{2}H_2O$, the bismuth atoms are arranged in the centrosymmetric cluster [9] with Bi–Bi distances 318–342 pm [94] (compare 307, 353 pm in the metal).

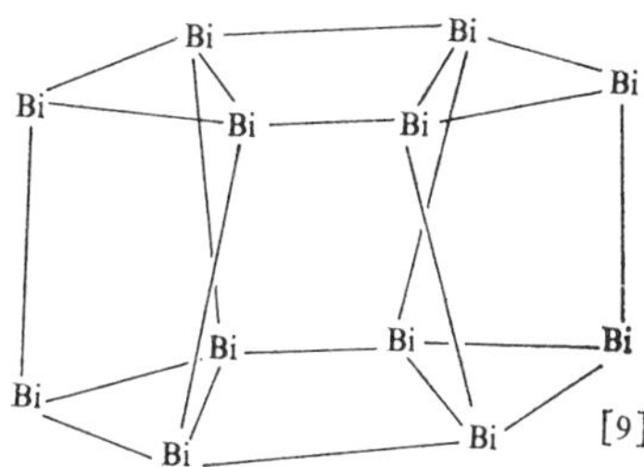

[9]

It does not appear to be possible to make the anhydrous nitrate from the pentahydrate, either by dehydration over sulphuric acid or by treatment with nitrogen dioxide, which often removes lattice water. With nitrogen dioxide, an adduct, probably $Bi(NO_3)_3,N_2O_4$, is formed. This decomposes to oxide nitrates:

$$Bi(NO_3)_3,N_2O_4 \xrightarrow{200°C} Bi_2O(NO_3)_4 \xrightarrow{415°C} Bi_4O_5(NO_3)_2$$

Infrared spectra suggest that the nitrate groups are bidentate and covalent in $Bi_2O(NO_3)_4$, but probably ionic in $Bi_4O_5(NO_3)_2$, which may thus have a structure like the oxide halides (section 5.4). Treatment of the pentahydrate $Bi(NO_3)_3,5H_2O$ with nitrogen pentoxide yields an adduct which has been formulated as $[NO_2]^+[Bi(NO_3)_4]^-$ [95]. Complex nitrate ions are found also in the compounds $M_3Bi_2(NO_3)_{12},24H_2O$ (M = Mg, Zn, Co, Ni). These are isomorphous with the cerium compound $[Mg(H_2O)_6]_3[Ce(NO_3)_6]_26H_2O$ in which each cerium is co-ordinated by six bidentate nitrate groups so that the oxygen atoms make an irregular icosahedron[96].

Antimony reacts with nitrogen dioxide in dimethylsulphoxide to give the basic salt $SbO(NO_3),(CH_3)_2SO$. Under the same conditions bismuth gives $Bi(NO_3)_3,3(CH_3)_2SO$, and arsenic does not react.

8.3. SULPHATES

The compounds $Bi_2(SO_4)_3,nH_2O$, $KBi(SO_4)_2$ and $Sb_2(SO_4)_3$ have been described. They are easily hydrolysed to basic salts.

Cryoscopic and electrical conductivity measurements[97] on dilute solutions of As^{III} oxide in 100% sulphuric acid have been interpreted as showing the formation of species such as $As(OH)(HSO_4)_2$ and $[As(OH)(SO_4H)]^+$. At higher concentrations, polymeric species are probably formed and $(AsO)_2SO_4$ precipitates. In oleum, the species $As(HSO_4)_3$, $[(HSO_4)_2As]_2O$ and $[(HSO_4)_2As]_2SO_4$ with ··· AsOSO ··· frameworks seem probable.

[94] G. Gattow and D. Schott, *Z. anorg. allgem. Chem.* **324** (1963) 31; G. Gattow and G. Kiel, *ibid.* **335** (1965) 61; *Naturwissenschaften* **55** (1968) 389.

[95] G. C. Tranter, C. C. Addison and D. B. Sowerby, *J. Inorg. Nucl. Chem.* **30** (1968) 97; D. K. Straub, H. H. Sisler and G. E. Ryschkewitsch, *ibid.* **24** (1962) 919.

[96] A. Zalkin, J. D. Forrester and D. H. Templeton, *J. Chem. Phys.* **39** (1963) 2881.

[97] R. J. Gillespie and E. A. Robinson, *Can. J. Chem.* **41** (1963) 450.

8.4. PHOSPHATES

Antimony phosphate, $SbPO_4$, may be obtained from Sb^{III} oxide and phosphoric acid. It has a layer structure in which antimony atoms are linked by PO_4 tetrahedra. The co-ordination round antimony is one-sided [5] with the lone pairs on the outside of the layers. Sb–O distances are 198–219 pm [98].

Bismuth phosphate is of some importance as a carrier for the separation of plutonium in the atomic energy industry. Since it is insoluble in moderately strong nitric and sulphuric acids (solubility product $[Bi^{3+}][PO_4^{3-}] = 10^{-23}$ $mol^2\,l^{-2}$ in HNO_3)[38], it may be precipitated under conditions where uranium is strongly complexed with nitrate. The bismuth phosphate dissolves in concentrated nitric acid and so may be readily recycled.

8.5. BISMUTH FORMATE AND ACETATE

In bismuth formate[99a], HCOO groups link bismuth atoms into sheets, so that each bismuth has three oxygen neighbours at 238, three at 252 and three at 277 pm. If the six shortest Bi–O distances are considered, the co-ordination polyhedron round bismuth

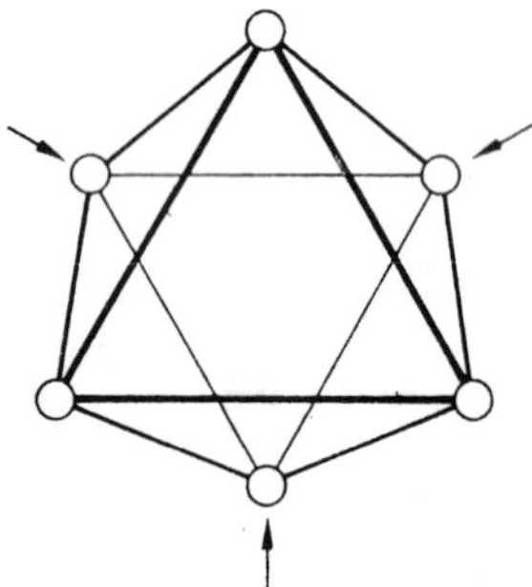

FIG. 20. The distorted octahedron round bismuth in $Bi(O{\cdot}CO{\cdot}H)_3$. Three more oxygen atoms above those arrowed, together with the six shown, make a face-centred trigonal prism.

(Fig. 20) is a distorted octahedron with one face larger than the one opposite (compare α-Bi_2O_3, p. 606). If three more Bi–O distances are considered also, the atoms round bismuth form a trigonal prism with extra atoms over the rectangular faces. A similar environment is found in bismuth fluoride (section 5.2).

Bismuth oxide acetate CH_3CO_2OBi is said to have a structure like the oxide chloride (section 5.4)[99b].

9. ORGANOMETALLIC COMPOUNDS

As long ago as 1760, Cadet heated As^{III} oxide with potassium acetate and obtained a spontaneously inflammable liquid which was a mixture of cacodyl, $Me_2AsAsMe_2$, and

98 B. Kinberger, *Acta chem. scand.* **24** (1970) 320.
99(a) C.-I. Stålhandske, *Acta chem. scand.* **23** (1969) 1525; (b) B. Aurivillius, *ibid.* **9** (1955) 1213.

cacodyl oxide, $Me_2AsOAsMe_2$. Since then, many thousands of organometallic derivatives of arsenic, antimony and bismuth have been made. The arsenic compounds are known best: their chemistry is similar to that of the analogous compounds of phosphorus. Antimony and bismuth compounds are less thoroughly studied. The field is well documented[101]. Good recent reviews have covered arsenic[100a] and antimony and bismuth[100b] derivatives. Work before 1937 is described by Krause and von Grosse, and there is a comprehensive non-critical summary of the literature from 1937 to 1964 which gives very full details[102].

9.1. TERTIARY ARSINES, STIBINES, AND BISMUTHINES, R_3E

Preparation

Although C–As, C–Sb and C–Bi bonds can be made by a variety of methods, the most important for the preparation of the symmetrical compounds R_3E is the replacement of all three groups X in the compounds EX_3 (X = halogen, OR, NR_2) by organic groups. The most widely used reagents for this purpose are Grignard and organolithium reagents, organoaluminium compounds and sodium–alkyl halides (Wurtz reaction):

$$AsCl_3 + MeMgI \xrightarrow{Bu_2O} Me_3As$$

$$As(OEt)_3 + PhMgBr \xrightarrow{Et_2O} Ph_3As$$

$$MeCH{:}CHLi + SbCl_3 \xrightarrow{Et_2O} (MeCH{:}CH)_3Sb$$

$$BiCl_3 + Et_6Al_2 \xrightarrow{Et_2O} Et_3Bi$$

$$AsCl_3 + PhCl + Na \xrightarrow{C_6H_6} Ph_3As$$

Alkyl, aryl, vinyl, allyl, ethynyl, perfluorophenyl and carborane (e.g. $PhC_2B_{10}H_{10}$—) derivatives can be made by these reactions. Unsymmetrical compounds $R^1R^2_2E$, $R^1R^2R^3E$, can be obtained from intermediates R^2_2EX (X = halogen or NR_2) and R^2R^3EX.

Oxides E_2O_3 can also be used as starting materials. The compounds R_3E are obtained with an excess of trialkylaluminium in hexane:

$$Et_6Al_2 + Sb_2O_3 \rightarrow 2Et_3Sb$$

Yields fall in the series Sb > As > Bi, and with increasing size of alkyl group. If zinc or mercury alkyls are available, they may be used to convert the halides EX_3, or the metals, into trialkyl derivatives. Pure alkyls can also be made by reduction of the halides R_3ECl_2 (section 14.3).

Properties and Structural Data

Methyl and phenyl compounds are described in Table 38; more extensive data is available in other compilations[102]. The arsines, stibines and bismuthines are colourless,

[100] (a) W. R. Cullen, *Adv. in Organomet. Chem.* **4** (1966) 145; M. Fild and O. Glemser, *Fluorine Chem. Rev.* **3** (1969) 129. (b) G. E. Coates, M. L. H. Green and K. Wade, *Organometallic Compounds*, Methuen, London (1967), Vol. 1, Chap. 5.

[101] G. O. Doak and L. D. Freedman, *Organometallic Compounds of Arsenic, Antimony and Bismuth*, Wiley (1970).

[102] (a) E. Krause and A. von Grosse, *Die Chemie der Metallorganischen Verbindungen*, Bornträger, Berlin (1937); (b) M. Dub (ed.), *Organometallic Compounds: Methods of Synthesis, Physical Constants and Chemical Reactions*, Vol. 3, *Compounds of Arsenic, Antimony and Bismuth*, Springer, Berlin (1968).

TABLE 38. PROPERTIES OF THE COMPOUNDS R_3E

	Me_3As	Me_3Sb	Me_3Bi
Melting point (°C) [a]	−87	−62	−86
Boiling point (°C) [a]	50	80	109
Vapour pressure [a] (log p_{mmHg})	$7.394-1456/T$	$7.707-1697/T$	$7.71-1840/T$
ΔH_f° (l) (kJ mol^{-1}) [b]	-14.8 ± 5	-0.4 ± 13	158 ± 8
ΔH_f° (g) (kJ mol^{-1}) [b]	15.3 ± 5	31 ± 13	193 ± 8
Mean bond energy (kJ mol^{-1}) [b]	229 ± 6	215 ± 6	143 ± 5
Bond dissociation energy D (Me_2E–Me) (kJ mol^{-1}) [b]	263[m]	238	184
Ionization potential (eV) [c]	8.3	8.0	
Vibrational spectra [d]:			
ν_1 (a_1)	568	513	460
ν_2 (a_1)	223	188	171
ν_3 (e)	583	513	460
ν_4 (e)	239	188	171
E–C distance (pm)	195.9 ± 1 [e]		
CEC angle (°)	96 ± 3 [f]		
Nuclear quadrupole coupling constant e^2Qq (MHz)	^{75}As: −203.15 [e]		
Dipole moment (D) [e]	0.86		

	Ph_3As	Ph_3Sb	Ph_3Bi
Melting point (°C) [a]	61	55	78
ΔH_f° (s) (kJ mol^{-1}) [b]	292 ± 8	330 ± 12	470 ± 8
ΔH_f° (g) (kJ mol^{-1})	392 ± 10	437 ± 13	580 ± 13
Mean bond energy (kJ mol^{-1})	267 ± 10	244 ± 10	177 ± 10
Ultraviolet spectra [g]: λ_{max} (nm)	248	256	248
$10^{-4}\,\varepsilon_{max}$ (nm)	1.23	1.17	1.29
E–C distance (pm)	196–199 [h]		224 [i]
CEC angle (°)	102		94
Dipole moment (D) [j]	1.23	0.77	0
Nuclear quadrupole coupling constant $\lvert e^2Qq\rvert$ (MHz)	^{75}As: 194.6, 193.6 [k]	^{121}Sb: 510.4 (η, 0.03), 500.7 (η, 0.07, 650.7 (η, 0.02) [k]	^{207}Bi: 669.1 (η, 0.09) [l]

[a] Ref. 102b and references therein. mmHg = 133.32 N m^{-2}.
[b] H. A. Skinner, *Adv. Organomet. Chem.* **2** (1964) 49.
[c] R. E. Winters and R. W. Kiser, *J. Organomet. Chem.* **10** (1967) 7.
[d] E–C skeleton only, nomenclature as for EH_3,EX_3. E. J. Rosenbaum, D. J. Rubin and C. R. Sandberg, *J. Chem. Phys.* **8** (1940) 366.
[e] D. R. Lide, *Spectrochim. Acta* **15** (1959) 473.
[f] Assumed.
[g] H. H. Jaffé, *J. Chem. Phys.* **22** (1954) 1430.
[h] In $(4\text{-}MeC_6H_4)_3As$, $(2,5\text{-}Me_2C_6H_3)_3As$. J. Trotter, *Can. J. Chem.* **41** (1963) 14; *Acta Cryst.* **16** (1963) 1187.
[i] D. M. Hawley and G. Ferguson, *J. Chem. Soc.* A (1968) 2059.
[j] M. J. Aroney, R. J. W. Le Fèvre and J. D. Saxby, *J. Chem. Soc.* (1963) 1739.
[k] η for Ph_3As not known; assumed to be zero. R. G. Barnes and P. J. Bray, *J. Chem. Phys.* **23** (1955) 407, 1177.
[l] H. G. Robinson, H. G. Dehmelt and W. Gordy, *Phys. Rev.* **89** (1953) 1305.
[m] S. J. W. Price and J. P. Richard, *Can. J. Chem.* **48** (1970) 3209.

volatile liquids or solids, with latent heats and Trouton constants much as for organic compounds. In general, volatilities decrease with increasing size of R and with increasing atomic weight of E. Except for the methyl and trifluoromethyl derivatives, however, the effect on the volatility of the element E is small (compare $(CH_2{:}CH)_3As$, b.p. 146°C; $(CH_2{:}CH)_3Sb$, 160°C; $(CH_2{:}CH)_3Bi$, 179°C). Thermodynamic properties are not known with very great accuracy for two reasons. First, the organometallic compounds burn in oxygen to complex products, such as various oxides, metal and soot, as well as carbon dioxide and water, and this entails rather uncertain corrections to heats of combustion. Secondly, the heats of formation of the products are themselves not known very well. Nevertheless, it seems that the compounds become more endothermic in the series from arsenic to bismuth, and the mean bond energy decreases with increasing atomic weight of E. Thus the compounds R_3Bi may decompose photolytically or when heated, and, like the lead and mercury compounds, R_4Pb and R_2Hg, may be used as sources of radicals $R^{\cdot}$ (e.g. $Ph^{\cdot}$). Bond dissociation energies, determined kinetically, for removal of a single methyl are greater than the mean bond energy, especially for the bismuth compounds. Trends in bond energies are shown in mass spectra. Thus the most abundant peaks in the 70 eV mass spectra of the arsenic and antimony compounds are Me_2As^+, Me_2Sb^+, $PhAs^+$ and $PhSb^+$, but the ion Bi^+ predominates in the spectrum of both Me_3Bi and Ph_3Bi. By contrast, the strongest peaks in the mass spectra of tertiary amines and phosphines, R_3N and R_3P, are often those of the parent ions R_3E^+ [103]. Mass spectra of trifluoromethyl compounds CF_3EX_2 show important peaks attributed to the transfer of fluorine from carbon to the element E.

The molecules R_3E are pyramidal in both vapour and solid, but reliable structural data is very incomplete. As–C bond lengths in tertiary arsines seem to be in the range 196–199 pm, but wider variations (191–206 pm) are found in compounds where the arsenic is part of a five- or six-membered ring. Very few Sb–C or Bi–C bond lengths have been measured in compounds R_3E. The CEC angles seem to be significantly smaller in triphenylbismuth than in triarylarsines, probably indicating less involvement of the lone pair of bismuth in the E–C bonds. The angle in Me_3As has been assumed to be 96° (cf. 99.1° in Me_3P). In the crystalline derivatives of triphenyl-arsenic, -antimony and -bismuth, the rings are twisted by angles ϕ from the planes bisecting the angles between the E–C bonds. In solution (probably), and in $(4\text{-}MeC_6H_4)_3As$ or $(2,5\text{-}Me_2C_6H_3)_3As$, the angles ϕ for each aryl group are the same so the symmetry is C_3 (propeller), but, more usually, the three aryl groups are twisted by different angles, and there is no molecular symmetry in the crystal.

Molecules $R^1R^2R^3E$, with a pyramidal configuration at E, lack symmetry even in solution, and so can be resolved into optically active isomers, provided that there is a sufficiently high energy barrier between the enantiomers. For tertiary amines, the barrier to inversion is too small for resolution, but optically stable phosphines, arsines and stibines may be isolated, e.g. by stereospecific reactions of optically active onium salts or by resolution of distereoisomeric transition metal complexes[104]. Several compounds $HO_2C{\cdot}C_6H_4{\cdot}ER^1R^2$ have also been resolved. Barriers to inversion of about 200 kJ mol^{-1}

[103] R. G. Kostyanovskii and V. V. Yakshin, *Izvest. Akad. Nauk SSSR Ser. Khim.* (1967) 2363; D. E. Bublitz and A. W. Baker, *J. Organomet. Chem.* **9** (1967) 383; R. C. Dobbie and R. G. Cavell, *Inorg. Chem.* **6** (1967) 1450, *J. Chem. Soc.* A (1968) 1406; A. T. Rake and J. M. Miller, *ibid.* (1970) 1881.

[104] V. I. Sokolov and O. A. Reutov, *Russ. Chem. Rev.* **34** (1965) 1; G. Kamai and G. M. Usacheva, *ibid.* **35** (1966) 601; B. Bosnich and S. B. Wild, *J. Am. Chem. Soc.* **92** (1970) 459.

have been measured in arsines. Optically active arsines may be converted to optically active sulphides and may be racemized by mineral acids. The rate depends on the concentration of anions as well as on the pH, and the reaction may involve five- or six-co-ordinate intermediates: either would lead to loss of activity[105].

$$R_3As \underset{-HX}{\overset{HX}{\rightleftharpoons}} R_3HAsX \underset{-X^-}{\overset{X^-}{\rightleftharpoons}} [R_3AsX_2H]^-$$

Diarsines (section 9.2) are less stable optically and the racemization of PhMeAsAsMePh has been studied by temperature-dependent n.m.r. [106].

Organometallic compounds R_3E (R alkyl) are readily oxidized by air and must be manipulated in a nitrogen atmosphere, but phenyl derivatives are more easily handled. Controlled reactions with oxidizing agents and with alkyl halides are described in sections 14.1 and 14.3. Tertiary alkyl and aryl compounds are not hydrolysed by water or acid, but electronegative organic groups (e.g. CF_3—, C_6F_5—) are quantitatively cleaved by concentrated aqueous alkali. The formation of CF_3H provides a useful method for the estimation and analysis of trifluoromethyl derivatives[107].

A very large number of complexes of tertiary arsines and stibines with transition metal compounds have been isolated[108]: no attempt is made to describe them here. In addition, many derivatives of polydentate arsines (e.g. [10], [11], [12]) and related compounds (e.g. [13]) have been characterized:

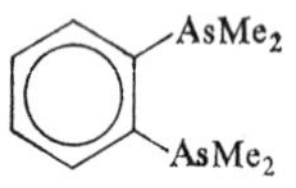

[10]

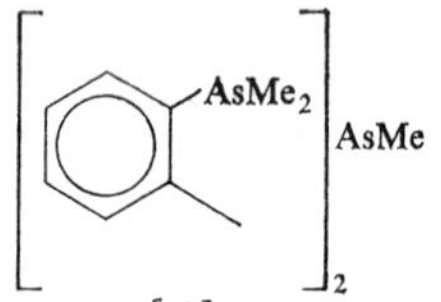

[11]

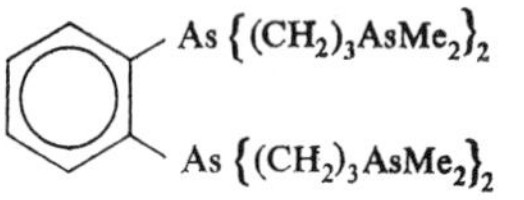

[12]

The bond systems of the arsines are usually preserved in transition metal complexes, but sometimes unexpected rearrangements accompany complex formation[109]:

$$\text{Diarsine}[10] + [Ni(H_2O)_6]^{2+} \xrightarrow[\text{diethylene glycol}]{\text{reflux}} [Ni\{\text{Triarsine }[11]\}\{\text{Diarsine }[10]\}]^{2+}$$

$$Me_2AsC{=}CAsMe_2CF_2CF_2 + Fe_3(CO)_{12} \longrightarrow$$

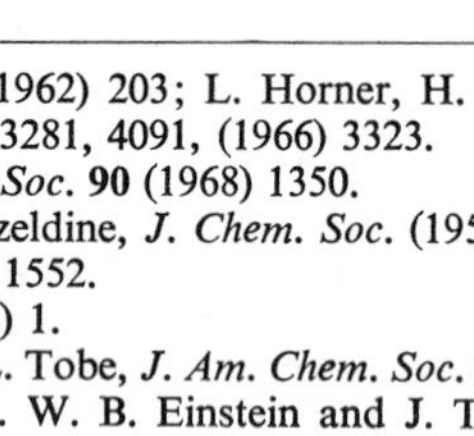

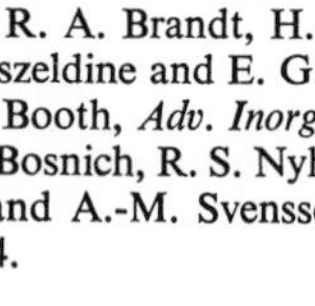

[13]

[105] L. Horner and H. Fuchs, *Tetrahedron Letters* (1962) 203; L. Horner, H. Winkler and E. Meyer, *ibid.* (1965) 789; L. Horner and W. Hofer, *ibid.* (1965) 3281, 4091, (1966) 3323.

[106] J. B. Lambert and G. F. Jackson, *J. Am. Chem. Soc.* **90** (1968) 1350.

[107] G. R. A. Brandt, H. J. Emeléus and R. N. Haszeldine, *J. Chem. Soc.* (1952) 2552; H. J. Emeléus, R. N. Haszeldine and E. G. Walaschewski, *ibid.* (1953) 1552.

[108] G. Booth, *Adv. Inorg. Chem. Radiochem.* **6** (1964) 1.

[109] B. Bosnich, R. S. Nyholm, P. J. Pauling and M. L. Tobe, *J. Am. Chem. Soc.* **90** (1968) 4741; F. W. B. Einstein and A.-M. Svensson, *ibid.* **91** (1969) 3663; F. W. B. Einstein and J. Trotter, *J. Chem. Soc.* A (1967) 824.

Arsenic and antimony may be incorporated in organic heterocyclic compounds[110]: for arsenic, systems with four- to ten-membered rings are known, and for antimony, five-, six- and seven-membered rings. Some of the known basic ring systems are shown. Corresponding antimony compounds can be named by substituting "stib" for "ars" and, for the six-membered rings, "antimon" for "arsen".

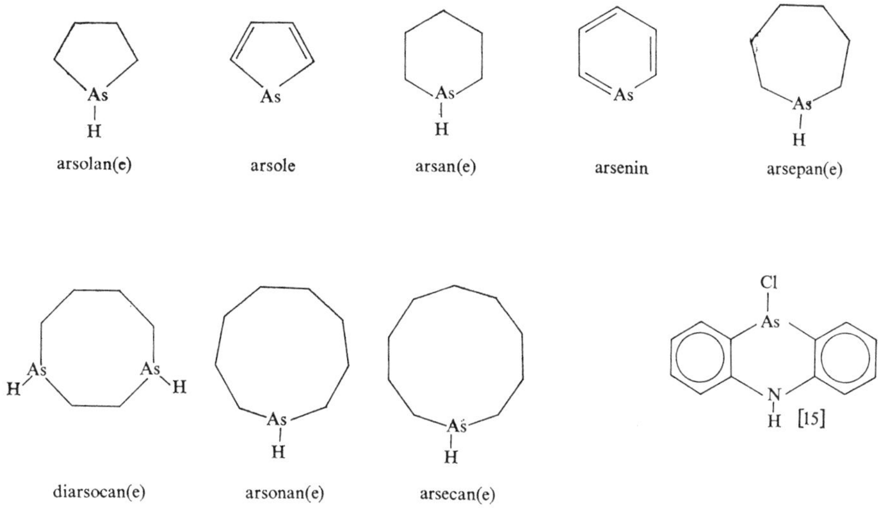

arsolan(e) arsole arsan(e) arsenin arsepan(e)

diarsocan(e) arsonan(e) arsecan(e) [15]

The commonest method for preparation of the ring systems is by reaction between a dihalide $RAsCl_2$ and a dilithio- or di-Grignard reagent. The molecular structures of

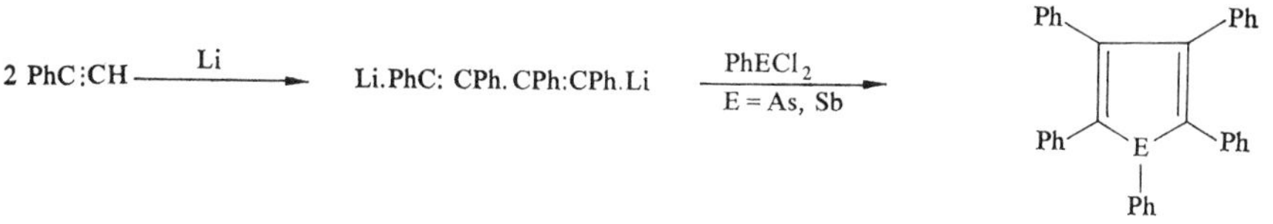

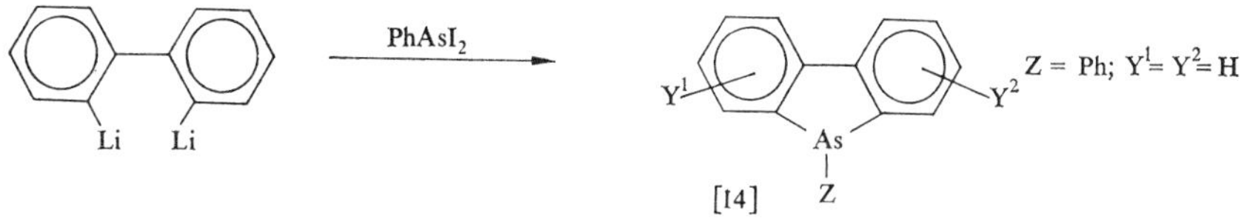

2 PhC:CH —Li→ Li.PhC: CPh.CPh:CPh.Li —$PhECl_2$, E = As, Sb→

—$PhAsI_2$→ [14] Z = Ph; $Y^1 = Y^2 = H$

9-phenyl-9-arsafluorene ([14] Z = Ph; Y′ = Y^2 = H)[111] and 10-chloro-5,10-dihydrophenarsazine [15] have been found. In the former, the bonds to arsenic have a pyramidal configuration, with CAsC angles of 88° and 98° and As–C bond lengths 198 and 202 pm. The arsafluorene ring is flat, so only unsymmetrically substituted compounds ($Y^1 \neq Y^2$)

[110] F. G. Mann, *The Heterocyclic Derivatives of Phosphorus, Arsenic, Antimony and Bismuth*, 2nd edn., Wiley-Interscience (1970).

[111] D. Sartain and M. R. Truter, *J. Chem. Soc.* (1963) 4414.

can be resolved into optically active isomers. In the phenarsazine [15] the As–C bonds are slightly shorter than expected by comparison with other compounds (see Table 41, p. 633); this has been taken to imply some interaction between the lone pairs of arsenic and nitrogen with the π-electrons of the aromatic rings.

9.2. COMPOUNDS WITH E–E BONDS

Cacodyl had already been mentioned. The compounds R_2AsAsR_2 may be made as shown in Scheme 6. Though it seems that reactions (ii) and (iv) could be used to prepare

SCHEME 6. The chemistry of diarsines.

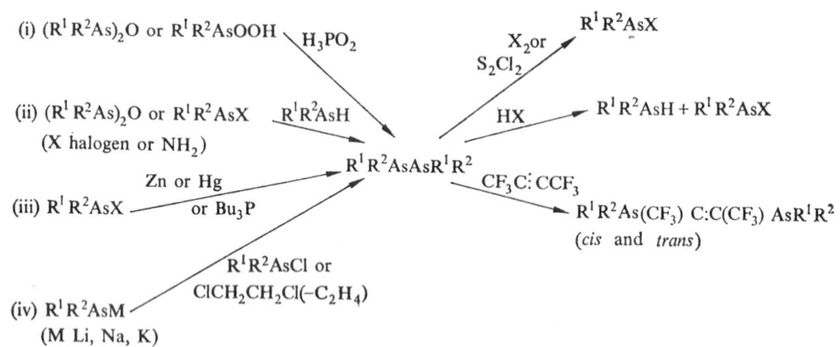

unsymmetrical diarsines, e.g. $R^1R^2AsAsR^3R^4$, mixtures of $(R^1R^2As)_2$ and $(R^3R^4As)_2$ are usually obtained instead. A number of antimony compounds made by similar reactions have been described, but their chemistry is less well explored than that of the arsenic compounds. Very little is known about dibismuthines.

The diarsines and distibines are easily oxidized; the more volatile compounds are spontaneously inflammable in air. The products depend on the reaction conditions, and in the presence of water various arsinic and arsonic acids may be isolated. Other reactions of the diarsines are shown in Scheme 6.

Some physical properties are as follows[102] (structural information is sparse):

$(CF_3)_2AsAs(CF_3)_2$: b.p. 106°C	$(CF_3)_2SbSb(CF_3)_2$: b.p. 34°C/14mm
$Me_2AsAsMe_2$: m.p. −1°C; b.p. 78°C	$Me_2SbSbMe_2$: m.p. 17°C; b.p. 76°C/6 mm
$Ph_2AsAsPh_2$: m.p. 127°C	$Ph_2SbSbPh_2$: m.p. 125°C

The compounds $(RAs)_n$, which can be made from organoarsenic dihalides or from arsonic acids (Scheme 7), are often polymeric, with both crystalline and amorphous

SCHEME 7. The chemistry of polyarsines $(RAs)_n$.

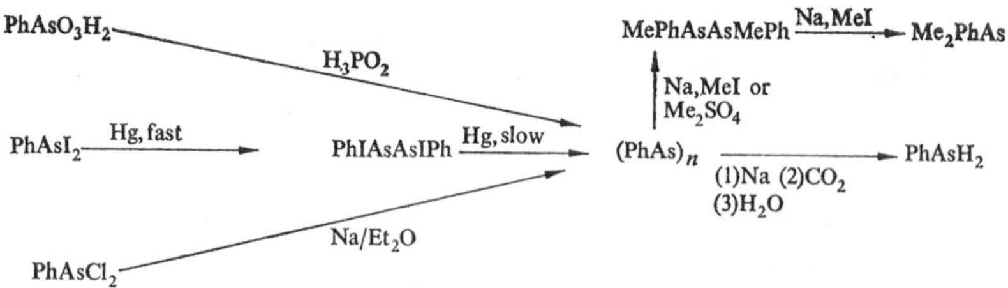

forms. In some of the crystalline forms there are oligomers, with molecular complexities which depend on the group R. Thus arsenomethane, $(MeAs)_n$, crystallizes in yellow and red forms, with the red form more stable at room temperature. The yellow form consists of pentamer molecules with a puckered As_5 ring. In the crystalline form of arsenobenzene, $(PhAs)_6$, the arsenic atoms form a six-membered puckered ring (chair form). Other oligomers which have been detected are $(CF_3As)_4$ (in the vapour), $(CF_3As)_5$ and $(C_6F_5As)_4$ (in solution and vapour)[112]. The average bond distances in the solid arsines conceal considerable variations which are outside the limits of experimental error (Table 39). The low symmetry of the oligomers results from puckering of the rings so that the carbon and arsenic atoms

TABLE 39. STRUCTURAL DATA FOR THE COMPOUNDS $(RAs)_n$

	As–As (pm)	∠CAsAs (°)	As–C (pm)	∠AsAsAs (°)	Ref.
$(MeAs)_5$	242.8	96.9	195	101.8	a
$(PhAs)_6$	245.6	100.1	196.6	91	b

[a] J. H. Burns and J. Waser, *J. Am. Chem. Soc.* **79** (1957) 859.
[b] K. Hedberg, E. W. Hughes, and J. Waser, *Acta Cryst.* **14** (1961) 369.

adjacent to a particular As–As bond are in a staggered configuration. The 60 MHz n.m.r. spectra[113] of $(MeAs)_5$ and $(CF_3As)_5$ in solution above -130°C show three resonances with relative intensities 2:2:1, suggesting that a rapid motion gives the rings time-averaged planes of symmetry. Small amounts of a variety of linear and cyclic methylpolyarsines are apparent in mixtures of $(MeAs)_5$ and $Me_2AsAsMe_2$, and three- and four-membered cyclic polyarsines, as well as pentamers, can be detected in the gas phase. Interconversions and equilibria between various polyarsines are complicated[114]. Several structurally similar cyclic phosphines have been characterized, e.g. $(MeP)_5$, $(CF_3P)_4$, $(CF_3P)_5$, $(PhP)_5$ and $(PhP)_6$, but there is no convincing explanation for the molecular complexities found for these cyclic compounds of phosphorus and arsenic.

The cyclic polyarsines are less readily oxidized than the diarsines R_2AsAsR_2. They form coloured derivatives with sodium, which are useful intermediates in further syntheses (Scheme 7). Red or black amorphous compounds $(RSb)_x$ have been described.

A wide variety of substituted arsenobenzenes have been made for chemotherapeutic use, though interest in this field has abated with the discovery of antibiotics. The effectiveness of $(AsPh)_5$ is limited by its low solubility in water, so attention has been directed to more soluble derivatives, especially those with substituents OH, NR_2 and SO_3H in the benzene rings. The most common method for preparation is by the reduction of the corresponding arsonic acid by hypophosphorous acid or sodium dithionite. Unsymmetrical

[112] J. Donohue, *Acta Cryst.* **15** (1962) 708; A. H. Cowley, A. B. Burg and W. R. Cullen, *J. Am. Chem. Soc.* **88** (1966) 3178; M. Green and D. Kirkpatrick, *J. Chem. Soc.* A (1968) 483.

[113] E. J. Wells, R. C. Ferguson, J. G. Hallett and L. K. Peterson, *Can. J. Chem.* **46** (1968) 2733; C. L. Watkins, L. K. Krannich and H. H. Sisler, *Inorg. Chem.* **8** (1969) 385. P. S. Elmes, S. Middleton and B. O. West, *Austr. J. Chem.* **23** (1970) 1559.

[114] F. Knoll, H. C. Marsmann and J. R. van Wazer, *J. Am. Chem. Soc.* **92** (1969) 4986.

compounds, i.e. those with several differently-substituted rings, can be made from appropriate mixtures of arsonic acids. Many of these substances are described as highly coloured powders. Some are crystalline, with sharp melting points. The properties of others, e.g. Salvarsan, 3-amino-4-hydroxyarsenobenzene, structure [16], vary with the method of preparation, suggesting that they are polymeric, with molecules of various complexities. There seems to be little recent structural work[102].

Arsenomethane $(MeAs)_5$ reacts with metal carbonyls in several ways[115]: the cyclic molecule may function as a ligand which is tridentate, as in [17], or monodentate, as in [18].

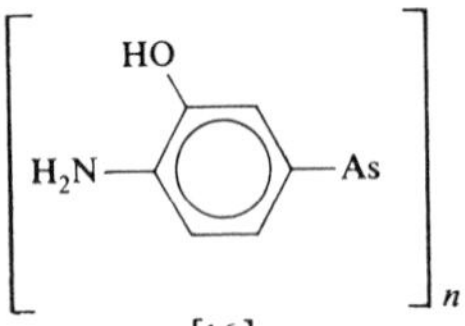

HO
H_2N—⬡—As]$_n$

[16]

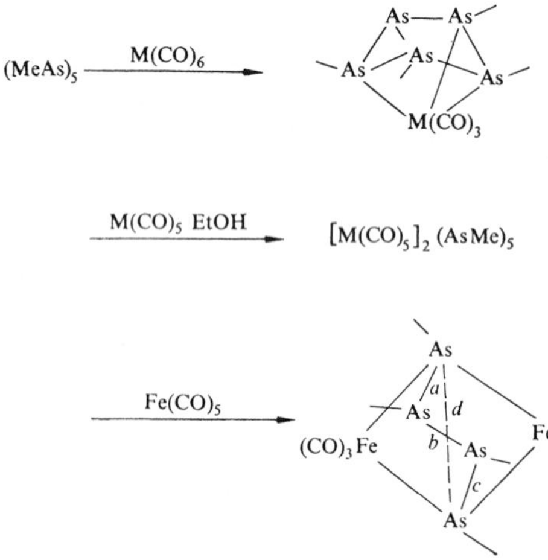

$(MeAs)_5$ —$M(CO)_6$→ [18] As—As, As, As, As, $M(CO)_3$ [17] M=Cr,Mo,W

—$M(CO)_5$ EtOH→ $[M(CO)_5]_2(AsMe)_5$ [18] M=Cr,Mo,W

—$Fe(CO)_5$→ As, As, As, As, $(CO)_3Fe$, $Fe(CO)_3$, a, b, c, d [19]

In the complex [19] the As_5 ring is not preserved (As–As: 245.3 pm (*a*, *c*); 239.1 (*b*); 288.8 (*d*))[116]. Among the products of the reaction between arsenomethane and cobalt carbonyl is the complex $As_3Co(CO)_3$ [20], in which one of the arsenic atoms of an As_4 tetrahedron has been replaced by the $Co(CO)_3$ group. A related compound, $As_2[Co(CO)_3]_2$,

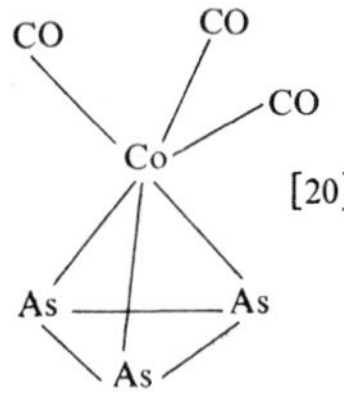

CO CO CO
Co
As As
As

[20]

[115] P. S. Elmes and B. O. West, *Coord. Chem. Rev.* **3** (1968) 279.

[116] B. M. Gatehouse, *Chem. Communs.* (1969) 948; P. S. Elmes, P. Leverett and B. O. West, *ibid.* (1971) 747.

can be isolated as a red liquid from cobalt carbonyl and arsenic trichloride in tetrahydrofuran. Replacement of arsenic atoms in As_4 by $Co(CO)_3$ results in a shortening of As–As bonds compared with As_4 (As–As, 243.5 pm).

$$As_3Co(CO)_3\text{: As–As, } 237.2 \pm 0.5 \text{ pm}$$

$$As_2[Co(CO)_3][Co(CO)_2PPh_3]\text{: As–As, } 227.3 \pm 0.3 \text{ pm}^{117}$$

9.3. THE COMPOUNDS R_2EX, REX_2 (X = HALOGEN)

Alkyl and some aryl halides react with metallic arsenic, particularly in the presence of copper, to give complex mixtures which contain the compounds R_2AsX and $RAsX_2$ and often R_3As, AsX_3 and As_2X_4 as well. The proportions of the various compounds obtained depend on the exact experimental conditions, but, provided that the products can be separated, e.g. by distillation, the reaction is useful preparatively:

$$\text{MeBr, As, Cu} \xrightarrow{350\text{–}370°\text{C}} \underset{\text{b.p. } 51°\text{C}/42 \text{ mm}}{Me_2AsBr\ (34\%)} + \underset{\text{b.p. } 89°\text{C}/41 \text{ mm}}{MeAsBr_2\ (62\%)} \quad \text{[Ref. 118]}$$

$$\text{MeCl, As, Cu} \xrightarrow{350\text{–}370°\text{C}} Me_2AsCl\ (20\%) + MeAsCl_2\ (80\%) \quad \text{[Ref. 118]}$$

$$CF_3I\text{, As, } I_2 \xrightarrow{220°\text{C}} (CF_3)_3As\ (78\%),\ (CF_3)_2AsI\ (13\%),\ CF_3AsI_2\ (4\%) \quad \text{[Ref. 107]}$$

Antimony reacts similarly, but yields of the organometallic compounds are lower; perhaps because they are thermally unstable at the temperatures required for formation. Only very low yields of bismuth compounds can be made this way[118]. These reactions are examples of the "direct synthesis" of organometallic compounds.

The halides can also be made by reduction of arsonic acids by sulphur dioxide in the presence of the hydrogen halide[119], by reorganization reactions, or by loss of alkylhalide from the compounds R_3EX_2 (section 14.3).

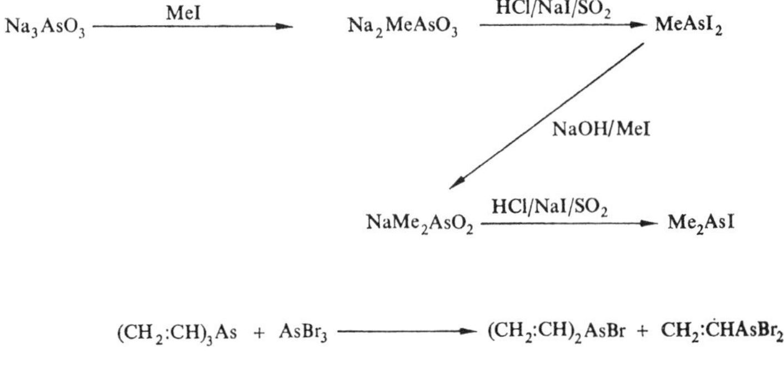

$$Me_3SbCl_2 \longrightarrow Me_2SbCl + MeCl$$

In the preparation from R_3EX_2, alkyl iodides are eliminated most readily and fluorides often with difficulty. In many cases, the five-co-ordinate bromides and iodides cannot be made.

The addition of arsenic trichloride to acetylene in the presence of aluminium trichloride yields a mixture known as lewisite, developed, but not used, as a war gas in 1918. The main

117 A. S. Foust, M. S. Foster and L. F. Dahl, *J. Am. Chem. Soc.* **91** (1969) 5631, 5633.
118 L. Maier, *Inorg. Synth.* **7** (1963) 82.
119 I. T. Millar, H. Heaney, D. M. Heinekey and W. C. Fernelius, *Inorg. Synth.* **6** (1960) 113, 116.

product is *trans*-2-chlorovinyldichloroarsine HCCl:CH·$AsCl_2$, but other compounds such as $(HCCl{:}CH)_2AsCl$ and $(HCCl{:}CH)_3As$ are also produced. The reorganization reactions are probably catalysed by the aluminium chloride.

The halogen derivatives R_2EX, REX_2 are volatile liquids or solids (Table 40) with properties similar to those of the compounds R_3E and the trihalides EX_3. Volatilities decrease from fluorides to iodides and with increasing size of R; trifluoromethyl compounds

TABLE 40. THE COMPOUNDS R_2EX, REX_2 (X = HALOGEN OR H) [a]

X	F	Cl	Br	I	H
Me_2AsX	75	107	127	154	36
Me_2SbX		157/750 [b]	107/90 [b]	80/17 [b]	61 [c]
$MeAsX_2$	76	131	181	128/16 [d]	2
$MeSbX_2$		115/60 [b]	*42* [b]	*110* [b]	41 [c]
$(CF_3)_2AsX$	25	46	59/745	14/54	19
$(CF_3)_2SbX$		~88	113	16/8	
CF_3AsX_2		71	118/745	100/48	−12.5/753
CF_3SbX_2			34/2.5	*6*	
Ph_2AsX	157	*39* [e]	*55* [e]	*38* [e]	161/20
Ph_2SbX		*68*			116/0.5
Ph_2BiX			*156*		f
$PhAsX_2$	110/48	117/10	~105/10		*−47*
$PhSbX_2$		*60*		*69*	*−38* [c]
$PhBiX_2$		*72*		*180*	f

[a] Ref. 102. Boiling points (°C) at pressures (mmHg) indicated; at 760 mmHg where no pressure shown. Melting points in italics.

[b] G. T. Morgan and G. R. Davies, *Proc. Roy. Soc.* **110** (1926) 523. H. Hartman and G. Kühl, *Z. anorg. allgem. Chem.* **312** (1961) 186.

[c] Decomposed at 20°C to di- or poly-stibines; boiling points extrapolated from vapour pressure data.

[d] Ref. 119.

[e] W. R. Cullen and J. Trotter, *Can. J. Chem.* **39** (1961) 2602.

[f] There have been unsuccessful attempts to make $PhBiH_2$, Ph_2BiH (E. Wiberg and K. Mödritzer, *Z. Naturforsch.* **12b** (1957) 132) but the methyl compounds $MeBiH_2$ (b.p. (extrap.) 72°C) and Me_2BiH (b.p. (extrap.) 103°C) have been described (E. Amberger, *Chem. Ber.* **94** (1961) 1447).

are more volatile than the methyl analogues. Pure compounds can be isolated from mixtures only if boiling points are sufficiently different: thus the iodides $(CF_3)_2AsI$, CF_3AsI_2 are easily separated, but the corresponding fluorides are not. Fluorides and chlorides are usually colourless: bromides and iodides are usually yellow or brown.

The molecules of the mono- and di-halides, like those of the compounds R_3E and EX_3, are pyramidal. Several have been investigated (Table 41) by X-ray methods or electron diffraction, and have As–C and As–X bond lengths similar to those in the tertiary arsines and trihalides. The CAsC bond angles in the phenyl derivatives are about 105°, but a smaller value was assumed in early electron diffraction work on the methyl compounds. There is considerable variation in XAsC angles. The microwave spectrum of the difluoride $MeAsF_2$ yields values of the quadrupole coupling constant ($e^2Qq_{cc} = -220+4$ MHz) and asymmetry parameter ($\eta = (q_{bb}-q_{aa})/q_{cc} = -1.2\pm0.05$) which are consistent with greater *s*-character in the As–F bond than the As–C bonds, as expected from electronegativities. In the solid cyanides Me_2AsCN (m.p. 30°C) and $MeAs(CN)_2$ (m.p. 127°C), as in $As(CN)_3$, there is evidence for interaction between a nitrogen atom in each molecule and an arsenic atom in a neighbouring molecule. This interaction probably accounts for the high melting points of the cyanides compared with the chlorides Me_2AsCl and $MeAsCl_2$, which are liquid at room temperature.

TABLE 41. STRUCTURAL DATA ON HALIDES AND CYANIDES

	Method[a]	As–C[b]	As–X[b]	∠CAsC[b]	∠CAsX[b]	Ref.
$MeAsF_2$	*M*	192±10	174±6	—	95.9±0.3	c
Me_2AsCl	*E*	198	218±4	96[d]	98±3	e
Ph_2AsCl	*X*	197±4	226±2	105±2	96±1	f
[15]	*X*	191.7±0.7	230.1±0.4	97±0.4	96.1±0.2	g
Me_2AsBr	*E*	198	234±4	96[d]	96±3	e
Ph_2AsBr	*X*	199±4	240±1	105±2	95±1	h
Me_2AsI	*E*	198	252±4	96[d]	98±4	e
$MeAsI_2$	*X*		254±1	—		i
Ph_2AsI	*X*		253			j
Me_2AsCN	*X*	193±3	201±3	105±2	91±1.5	k
$MeAs(CN)_2$	*X*	200±4	198±3		96±2	l
$As(CN)_3$	*X*		193±5		90.5±1.5	l

[a] *X*, X-ray; *E*, electron diffraction; *M*, microwave.
[b] Distances in pm. Angles in degrees.
[c] L. J. Nugent and C. D. Cornwell, *J. Chem. Phys.* **37** (1962) 523.
[d] Assumed value.
[e] H. A. Skinner and L. E. Sutton, *Trans. Faraday Soc.* **40** (1943) 164.
[f] J. Trotter, *Can. J. Chem.* **40** (1963) 1590.
[g] 10-chloro-5,10-dihydrophenarsazine. A. Camerman and J. Trotter, *J. Chem. Soc.* (1965) 730.
[h] J. Trotter, *J. Chem. Soc.* (1962) 2567.
[i] N. Camerman and J. Trotter, *Acta Cryst.* **16** (1963) 922.
[j] J. Trotter, *Can. J. Chem.* **41** (1963) 191.
[k] N. Camerman and J. Trotter, *Can. J. Chem.* **41** (1963) 460.
[l] E. O. Schlemper and D. Britton, *Acta Cryst.* **20** (1966) 777.

The halogen compounds are the starting materials for a wide variety of compounds in which the R_2As and RAs groups are preserved. Some reactions of the arsenic compounds are illustrated in Schemes 3 and 6–10. The antimony and bismuth compounds are like those of arsenic, but there is much less information about structures and physical properties.

9.4. HYDRIDES R_2EH, REH_2

The chemistry of the hydrides R_2AsH and $RAsH_2$ is summarized in Scheme 9. The arsenic compounds (Table 40) are reasonably well characterized, and many of them may be made without difficulty, provided that oxygen is rigorously excluded from reaction mixtures. The antimony compounds are much less well known and readily decompose, with formation of hydrogen and ill-defined solids. Alkylarsines may be made from $NaAsH_2$ and alkyl halides, or by reduction of the halides $RAsCl_2$ with lithium aluminium hydride. Low yields by this last route have been attributed to formation of stable Al–As bonds: borohydrides or related reducing agents may be more efficient. Phenylarsines are made by

SCHEME 8. Reactions of halides R_2AsX and $RAsX_2$ (X = Cl, Br, I).

Starting material	Reagent	Product
R_2AsX	AgF	R_2AsF
	AgCN	R_2AsCN
	R'MgX etc.	$R'R_2As$
	$NaOSiMe_3$	$R_2AsOSiMe_3$
	$(EtS)_2Pb$	R_2AsSEt
	NaOPh	R_2AsOPh
	$NaMn(CO)_5$	$[R_2AsMn(CO)_4]_2$
$RAsX_2$	NH_4F	$RAsF_2$
	AgSCN	$RAs(SCN)_2$
	R'MgX	R'RAsX
	$R'OH/Et_3N$	$RAs(OR')_2$
	H_2S	RAsS
	$R'N_2HSO_4$ / $CuCl_2$	$R'RAsO_2H$

SCHEME 9. Reactions of hydrides.

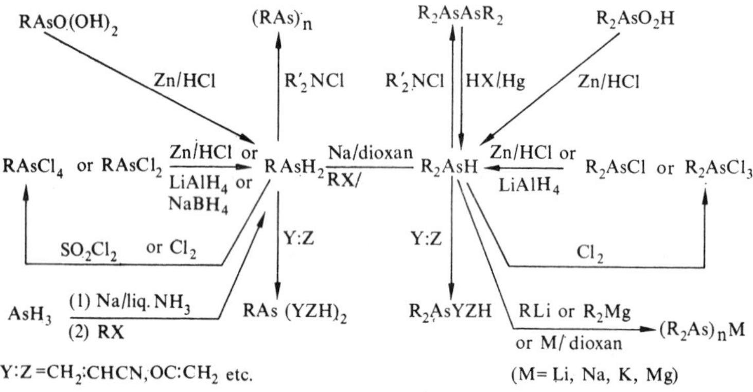

reduction of the corresponding arsonic or arsinic acids, and perfluoroalkylarsines are best made from the iodides or the diarsines by reduction with hydrogen iodide and mercury[120].

$$CF_3AsI_2 \text{ or } (CF_3As)_n \xrightarrow{HI/Hg} CF_3AsH_2$$

$$(CF_3)_2AsI \text{ or } [(CF_3)_2As]_2 \xrightarrow{HI/Hg} (CF_3)_2AsH$$

The arsines and stibines do not react with water. They are readily oxidized, e.g. by air or by halogens. In this last reaction, intermediates RAsHX have been detected[121], but these readily give hydrogen halides and polyarsines.

$$MeAsH_2 \xrightarrow{I_2} MeAsHI \rightarrow (MeAs)_5 + HI$$

The hydrogen atoms in REH_2, R_2EH can be replaced by alkali or alkaline earth metals by treatment with organometallic reagents, and the resulting metal derivatives, e.g. $Mg(AsPh_2)_2$, $LiAsBu^t_2$ may be used in further syntheses. The insertion reaction with species with double bonds, Y:Z (Scheme 9) is also of wide application. The compounds (Ph_2AsMMe_2; M = Al, Ga, In) have been made.

[120] R. G. Cavell and R. C. Dobbie, *J. Chem. Soc.* A (1967) 1308.
[121] A. L. Reingold and J. M. Bellema, *Chem. Communs.* (1969) 1058.

9.5. OXYGEN DERIVATIVES

Oxygen derivatives of As^{III} are as follows:

R_2AsOH, R^1R^2AsOH	$(R_2As)_2O$	$RAs(OH)_2$	$RAsO$
arsinous acids hydroxyarsines	oxybis(secondary arsines)	arsonous acids	oxoarsines

A number of arylarsinous acids have been made by reduction of the corresponding arsinic acids (section 14.4) or by alkaline hydrolysis of oxybis(secondary arsines). The free acids are not well characterized.

$$R_2AsOOH \xrightarrow{SO_2/I^-} R_2AsOH \xleftarrow{OH^-} (R_2As)_2O$$

Many arsinous acids are known as esters $R^1_2AsOR^2$, $R^1R^2AsOR^3$, which are readily obtained as volatile liquids and solids from halides R_2AsCl and sodium alkoxides (Scheme 8), or by heating oxides $(R_2As)_2O$ with alcohols in the presence of copper sulphate:

$$(R^1_2As)_2O + R^2OH \rightarrow R^1_2AsOR^2$$

The properties of the arsinite esters are intermediate between those of the arsines (section 9.1) and arsenites (section 6.4). They show no tendency to rearrange by Arbuzov-type reactions. A number of antimony compounds have been described.

The oxides $(R_2E)_2O$ (E = As, Sb) are liquids or crystalline solids, and appear to be momomeric in solution. The alkyl compounds are made by hydrolysis of the halides R_2EX; aryl compounds may also be prepared from halides and diazonium compounds (Scheme 10). The molecule of oxybis(diphenylarsine) has bond lengths As–C, 190; As–O, 167 pm. The AsOAs bond angle is 137°, and the bond angles at arsenic are close to 100° [123].

Oxygen derivatives with one organic group attached to arsenic are similar to those with two. Thus arylarsonous acids are made by reduction of the arsonic acids (section 14.4) or by hydrolysis of halides. Alkylarsonous acids are not well characterized.

$$RAsO(OH)_2 \xrightarrow{SO_2/HX} RAsX_2 \xrightarrow{OH^-} RAs(OH)_2$$

Many hydrated arsonous acids and oxoarsines have been isolated by careful control of reaction conditions[102]. Some of these derivatives are crystalline, but others are insoluble in organic solvents and amorphous. They may contain oligomeric[63] or polymeric frameworks of As–O bonds, but there is little precise structural work. The compound $\{MeCH(AsO)_2\}_2$ may have a realgar-like structure [7]. Oxostibines are also known (Scheme 10) and may be useful intermediates in syntheses.

Arsonous esters $RAs(OR)_2$ are readily made by treatment of the halides $RAsX_2$ with alkoxides, or with alcohols and tertiary amines, or by heating the oxoarsines with alcohols (sometimes with copper sulphate). As with the arsenites (section 6.4) and the arsinites, derivatives with involatile alcohols can be made by transesterification. The arsonites are volatile with pesticidal properties.

122 M. Becke-Goehring and K. Sommer, *Z. anorg. allgem. Chem.* **370** (1969) 31.
123 W. R. Cullen and J. Trotter, *Can. J. Chem.* **41** (1963) 2983.

SCHEME 10. Reactions of oxygen compounds[101, 102, 122]

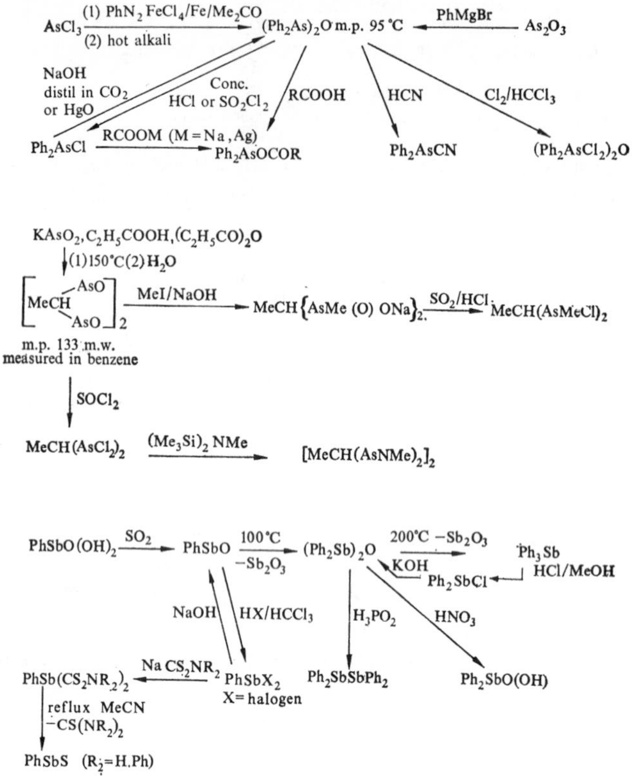

$AsCl_3$ —(1) $PhN_2FeCl_4/Fe/Me_2CO$, (2) hot alkali→ $(Ph_2As)_2O$ m.p. 95 °C ←PhMgBr— As_2O_3

$(Ph_2As)_2O$ —Conc. HCl or SO_2Cl_2→ Ph_2AsCl; Ph_2AsCl —NaOH distil in CO_2 or HgO→ $(Ph_2As)_2O$

$(Ph_2As)_2O$ —RCOOH→ $Ph_2AsOCOR$; Ph_2AsCl —RCOOM (M = Na, Ag)→ $Ph_2AsOCOR$

$(Ph_2As)_2O$ —HCN→ Ph_2AsCN

$(Ph_2As)_2O$ —$Cl_2/HCCl_3$→ $(Ph_2AsCl_2)_2O$

$KAsO_2, C_2H_5COOH, (C_2H_5CO)_2O$ —(1) 150 °C (2) H_2O→ $[MeCH(AsO)_2]_2$ m.p. 133 m.w. measured in benzene

$[MeCH(AsO)_2]_2$ —MeI/NaOH→ $MeCH\{AsMe(O)ONa\}_2$ —SO_2/HCl→ $MeCH(AsMeCl)_2$

$[MeCH(AsO)_2]_2$ —$SOCl_2$→ $MeCH(AsCl_2)_2$ —$(Me_3Si)_2NMe$→ $[MeCH(AsNMe)_2]_2$

$PhSbO(OH)_2$ —SO_2→ $PhSbO$ —100 °C, $-Sb_2O_3$→ $(Ph_2Sb)_2O$ —200 °C, $-Sb_2O_3$→ Ph_3Sb

Ph_3Sb —HCl/MeOH→ Ph_2SbCl —KOH→ $(Ph_2Sb)_2O$

$PhSbO$ —HX/$HCCl_3$→ $PhSbX_2$ (X = halogen); $PhSbX_2$ —NaOH→ $PhSbO$

$(Ph_2Sb)_2O$ —H_3PO_2→ $Ph_2SbSbPh_2$

$(Ph_2Sb)_2O$ —HNO_3→ $Ph_2SbO(OH)$

$PhSbX_2$ —$NaCS_2NR_2$→ $PhSb(CS_2NR_2)_2$ —reflux MeCN, $-CS(NR_2)_2$→ $PhSbS$ (R_2 = H, Ph)

Amido groups may sometimes be cleaved from arsenic by alcohols or phenols, or by aqueous alkali.

$$MeAs(NMe_2)_2 + PhOH \rightarrow MeAs(OPh)_2$$
$$R_2AsNR_2 + OH^- \rightarrow (R_2As)_2O$$

However, the oxide $(Me_2As)_2O$ and the ester Me_2AsOPh react with sodamide to form the unstable amide Me_2AsNH_2, which decomposes to the compound $(Me_2As)_3N$. The corresponding antimony compound may also be made[124].

Sulphur compounds $R^1_2AsSR^2$, $(R^1_2As)_2S$, $R^1As(SR^2)_2$, RAsS can be prepared by routes similar to those for oxygen analogues. Fungicidal properties of many esters $R^1As(SR^2)_2$ have been investigated[102]. The substance made from methylarsine and sulphur, or from

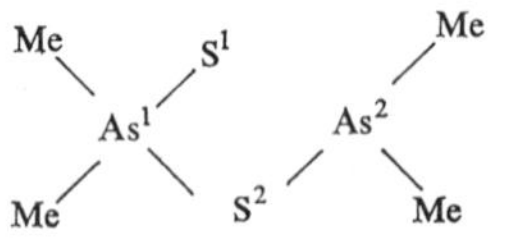

Me, Me (on As^1); S^1, S^2; Me, Me (on As^2)

[21]

[124] O. J. Scherer, J. F. Schmidt and M. Schmidt, *Z. Naturforsch.* **19b** (1964) 447.

dimethylarsonic acid (section 14.4) and hydrogen sulphide, is dimethylarsino-dimethyldithioarsinate [21]. The bond lengths are As^1S^1, 207.5; As^1S^2, 221.4; As^2S^2, 227.9±0.7 pm. The bond angles at As^1 are between 101 and 116° and at As^2 between 96 and 99° [125].

9.6. DERIVATIVES OF SILICON, GERMANIUM, TIN AND LEAD[126, 127]

The compounds $(R_3E^1)_nER^1_{3-n}$, $R_mE^1(ER^1_2)_{4-m}$ {R, R^1 = H, alkyl, aryl; E^1 = Si, Ge, Sn; E = As, Sb, Bi} have been compared with the organometallic compounds described in sections 9.1 and 9.4 in attempts to assess the rôle of *d* orbitals in EE^1 bonds. The organosilicon, -germanium and -tin compounds are also useful synthetic intermediates.

The most important preparative routes are as follows.

(a) Halides R_3E^1X react with alkali derivatives of arsenic, antimony and bismuth in liquid ammonia, or organic solvents such as hexane or ether.

$$Me_3SiCl + Li_3Sb \rightarrow (Me_3Si)_3Sb$$

Alkali metal derivatives of Group IV elements, e.g. triphenylstannyllithium, react with organo-arsenic and -antimony halides, but these reactions are often complicated by halogen–metal exchange, so that low yields of the desired compounds with EE^1 bonds are obtained.

(b) Halides R_3E^1X react with organo-arsenic and -antimony hydrides in the presence of tertiary amines:

$$Ph_3SnCl + Ph_2AsH \xrightarrow{Et_3N} Ph_3SnAsPh_2$$

(c) Stibines and bismuthines $(R_3E^1)_3E$ (E = Sb: E^1 = Si, Ge, Sn; E = Bi: E^1 = Sn) can be made from triethyl-stibine or -bismuthine and the hydrides R_3E^1H:

$$3Ph_3SnH + BiEt_3 \rightarrow (Ph_3Sn)_3Bi + 3EtH$$

(d) E^1–N bonds may be cleaved by hydrides:

$$Me_3SnNMe_2 + Ph_2AsH \rightarrow Me_3SnAsPh_2 + NMe_2H$$

(e) The unstable germyl compounds $(H_3Ge)_3E$ (E = As, Sb) may be made from germyl bromide and the silyl compounds $(H_3Si)_3E$.

Ternary hydrides $(H_3E^1)_mEH_{3-m}$ have been made by the above reactions, or from the hydrides E^1H_4 and EH_3 in electric discharges, and have been characterized by spectroscopic methods[127]. N.m.r. spectra show resonances at τ 6 to 6.5 due to E^1H protons and at about τ 9 due to EH protons.

The molecules of trisilyl-phosphine, -arsine and -stibine are pyramidal (contrast planar trisilylamine), and infrared spectra of other compounds $(R_3E^1)_3E$ have been assigned by assuming C_{3v} symmetry (Table 42). The participation of *d* orbitals has been invoked to account for chemical properties but physical evidence for π-bonding is slight.

The stability of compounds with E–E^1 bonds—both thermal and towards chemical attack—seems to be greatest when E and E^1 are about the same size. The ternary hydrides,

[125] N. Camerman and J. Trotter, *J. Chem. Soc.* (1964) 219.
[126] H. Schumann, *Angew. Chem. Int. Edn. Engl.* **8** (1969) 937.
[127] J. E. Drake and C. Riddle, *Quart. Rev.* **24** (1970) 263, and references therein.

TABLE 42. DERIVATIVES OF SILICON, GERMANIUM, TIN AND LEAD[a]

	M.p. (°C)	B.p. (°C/mmHg)	Infrared spectra[b] (cm^{-1})		
			ν_{as} (*e*)	ν_s (*a*)	
$(H_3Si)_3As$ [c]		120 [d]	362 [e]	346	
$(H_3Ge)_3As$	−84 [f]		263 [f]	245	90, 78
$(H_3Si)_3Sb$ [g]		225 [d]	309 [e]	308	
$(H_3Ge)_3Sb$	−60 [f]		216 [f]	209	68
$(Me_3Si)_3As$		82/4 [h]			
$(Me_3Ge)_3As$		67/0.1	275	256	
$(Me_3Sn)_3As$		99/1	233	211	
$(Me_3Pb)_3As$	44		208	182	
$(Me_3Si)_3Sb$	0 [i]				
$(Me_3Ge)_3Sb$	12		229 [i]	176	
$(Me_3Sn)_3Sb$	39				
$(Me_3Ge)_3Bi$		115/0.1	210 [j]	157	

[a] Except where shown, data from refs. 126 and 127.
[b] Peaks mainly from vibrations of E_3^1E (C_{3v}) skeleton.
[c] As–Si, 235.5±0.1 pm; SiAsSi, 93.8±0.2° by electron diffraction. B. Beagley, A. G. Robiette and G. M. Sheldrick, *J. Chem. Soc.* A (1968) 3006.
[d] Boiling point at 760 mmHg, extrapolated from vapour pressure data. For $(H_3Si)_3As$, $\log p_{mmHg} = -2143/T + 8.333$. For $(H_3Si)_3Sb$, $\log p_{mmHg} = -1670/T + 6.041$. E. Amberger and H. D. Boeters, *Chem. Ber.* **97** (1964) 1999.
[e] D. C. McKean, *Spectrochim. Acta* **24A** (1968) 1253.
[f] Data refer to triple points and Raman spectra. $(H_3Ge)_3As$ and $(H_3Ge)_3Sb$ decompose slowly at 25°C to germane and involatile products. E. A. V. Ebsworth, D. W. H. Rankin and G. M. Sheldrick, *J. Chem. Soc.* A (1968) 2828.
[g] Sb–Si, 255.7 pm. ∠SiSbSi, 88.6°. D. W. H. Rankin, A. G. Robiette, G. M. Sheldrick, B. Beagley and T. G. Hewitt, *J. Inorg. Nucl. Chem.* **31** (1969) 2351.
[h] A. B. Bruker, L. D. Balashova and L. Z. Soborovskii, *Dokl. Akad. Nauk SSSR Otdel Khim. Nauk* **135** (1960) 843.
[i] E. Amberger and R. W. Salazar, *J. Organomet. Chem.* **8** (1967) 111.
[j] I. Schuman-Ruidisch and H. Blass, *Z. Naturforsch.* **22b** (1967) 1081.

like the hydrides E^1H_4 and EH_3, are very readily oxidized by air, and are rapidly decomposed by halogens and compounds HX (X = halogen, OH, SH).

$$GeH_3I + AsH_3 \xleftarrow{HI} (GeH_3)_3As \xrightarrow{H_2O} (GeH_3)_2O + AsH_3$$

Alkyl- and aryl-substituted derivatives are also oxidized by air.

$$Ph_3GeAsPh_2 \xrightarrow{O_2} Ph_3GeOAs(O)Ph_2 \xleftarrow{Et_3N} Ph_3GeCl + Ph_2AsO\cdot OH$$

Reactions of dimethyl(trimethylsilyl)arsine (Scheme 11) may involve cleavage of the Si–As bond or addition to unsaturated systems[128]. The products reflect the polarity Si^+–As^-. The trimethylstannyl compound $Me_3SnAsMe_2$ is similar, but less reactive: for example, there is no reaction with water or alcohols. The secondary arsine $Me_3SiAsHMe$ readily loses methylarsine[129]: the diarsine $Me_2Si(AsHMe)_2$ gives a cyclic oligomer $(Me_2SiAsMe)_4$.

$$Me_3Si(AsHMe) \rightarrow (Me_3Si)_2AsMe + MeAsH_2$$

$$Me_2Si(AsHMe)_2 \rightarrow \tfrac{1}{4}(Me_2SiAsMe)_4 + MeAsH_2$$

[128] E. W. Abel and S. M. Illingworth, *J. Chem. Soc.* A (1969) 1094; *J. Organomet. Chem.* **17** (1969) 161.
[129] E. W. Abel and J. P. Crow, *J. Organomet. Chem.* **17** (1969) 337.

SCHEME 11. Reactions of dimethyl(trimethylsilyl)arsine.

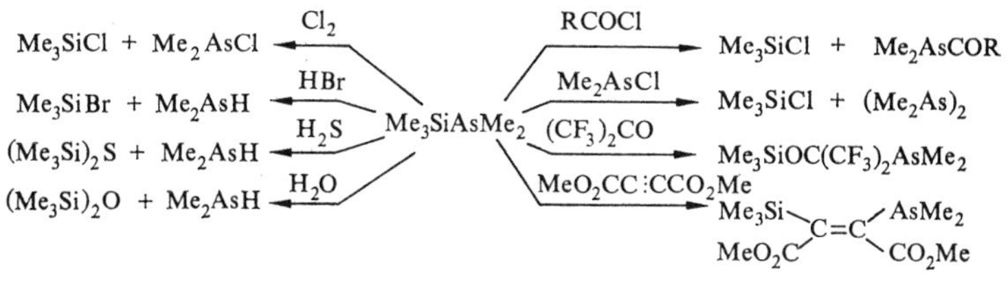

$$Me_3SiCl + Me_2AsCl \xleftarrow{Cl_2} Me_3SiAsMe_2 \xrightarrow{RCOCl} Me_3SiCl + Me_2AsCOR$$

$$Me_3SiBr + Me_2AsH \xleftarrow{HBr} Me_3SiAsMe_2 \xrightarrow{Me_2AsCl} Me_3SiCl + (Me_2As)_2$$

$$(Me_3Si)_2S + Me_2AsH \xleftarrow{H_2S} Me_3SiAsMe_2 \xrightarrow{(CF_3)_2CO} Me_3SiOC(CF_3)_2AsMe_2$$

$$(Me_3Si)_2O + Me_2AsH \xleftarrow{H_2O} Me_3SiAsMe_2 \xrightarrow{MeO_2CC{:}CCO_2Me} (Me_3Si)(MeO_2C)C{=}C(AsMe_2)(CO_2Me)$$

The organosilicon and organotin di- and tri-arsines $R_mE^1(ER^1_2)_{4-m}$ give complexes with transition metal derivatives[130].

10. COMPLEXES OF TERVALENT ARSENIC, ANTIMONY, AND BISMUTH

The compounds EX_3 (X = hal., R, H, etc.) can form several series of complexes. Electrons from the lone pair on E^{III} may be donated to suitable acceptor molecules, or the element E may expand its co-ordination number by accepting electrons from donor molecules.

10.1. THE MOLECULES EX_3 AS DONORS

Complexes with main group acceptors are usually weak and are formed most readily when X = R, H. Donor strength decreases in the series $PX_3 > AsX_3 > SbX_3 > BiX_3$. Stronger complexes are formed with late transition metal atoms (Class (b) acceptors)[24,108,130]. Even antimony trichloride reacts with iron and nickel carbonyls to give the compounds $Fe(CO)_3(SbCl_3)_2$ and $Ni(CO)_3SbCl_3$ [131].

TABLE 43. COMPLEXES OF R_3E AND EH_3

Me_nAsH_{3-n}, BH_3	Stability decreases as n decreases. Me_3Sb,BH_3 dissociated at 20°C [a]
Ph_3As,BH_3	No Ph_3Sb,BH_3 or Ph_3Bi,BH_3 obtained[b]
Ph_3Bi,SO_3	As and Sb derivatives decompose to Ph_3EO and SO_2 [b]
Ph_3E,SO_2	Coloured solutions decompose (rate increases As < Sb < Bi) to Ph_2SO_2 and $PhESO_2$ [c]
R_3SbHgX_2 (X = Cl, Br, I)	In refluxing THF, $R_2SbCl + RHgCl$ formed[d]

[a] F. G. A. Stone and A. B. Burg, *J. Am. Chem. Soc.* **76** (1954) 386; A. P. Lane and A. B. Burg, *ibid.* **89** (1967) 1040; F. Hewitt and A. K. Holliday, *J. Chem. Soc.* (1953) 530. H. C. Miller, N. Miller and E. L. Muetterties, *J. Am. Chem. Soc.* **85** (1963) 3885; *Inorg. Chem.* **3** (1964) 1456.

[b] M. Becke-Goehring and H. Thielemann, *Z. anorg. allgem. Chem.* **308** (1951) 33.

[c] S. I. A. El Sheikh and B. C. Smith, *Chem. Communs.* (1968) 1474.

[d] G. Deganello, G. Dolcetti, M. Giustiniani and U. Belluco, *J. Chem. Soc.* A (1969) 2138.

[130] E. W. Abel, R. Honigschmidt-Grossich and S. M. Illingworth, *J. Chem. Soc.* A (1968) 2623; E. W. Abel, J. P. Crow and S. M. Illingworth, *ibid.* (1969) 1631.

[131] G. Wilkinson, *J. Am. Chem. Soc.* **73** (1951) 5502.

SCHEME 12. Reactions of Me_3As,BH_3.

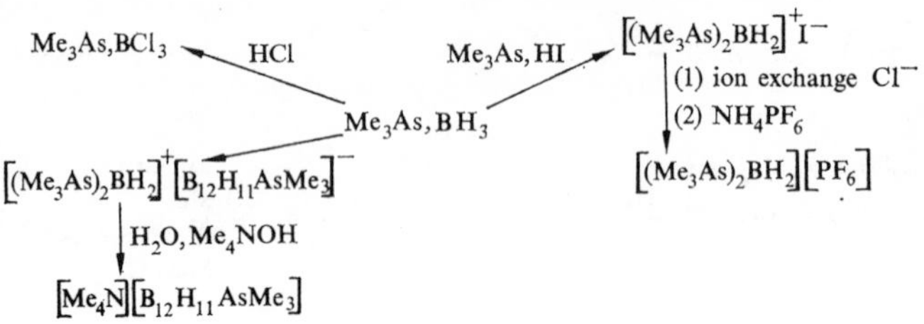

Me_3As,BCl_3 ← HCl — Me_3As,BH_3 — Me_3As,HI → $[(Me_3As)_2BH_2]^+I^-$

$[(Me_3As)_2BH_2]^+I^-$ → (1) ion exchange Cl^- (2) NH_4PF_6 → $[(Me_3As)_2BH_2][PF_6]$

Me_3As,BH_3 → $[(Me_3As)_2BH_2]^+[B_{12}H_{11}AsMe_3]^-$

$[(Me_3As)_2BH_2]^+[B_{12}H_{11}AsMe_3]^-$ → H_2O, Me_4NOH → $[Me_4N][B_{12}H_{11}AsMe_3]$

Reactions of the adduct Me_3As,BH_3 (Table 43) are shown in Scheme 12. The sparingly soluble salts of $[(Me_3As)_2BH_2]^+$ can be rapidly recrystallized from warm water. The adducts R_2AsH,BH_3 (R = Me, CF_3) readily lose hydrogen to give a series of stable compounds $(R_2AsBH_2)_n$ (n = 3, 4). The methyl compounds are unchanged at 200°C, but they are less stable thermally and more reactive than the phosphorus analogues. The monomeric Me_2SbBH_2, from Me_4Sb_2 and B_2H_6 at 100°C, is also unchanged at 200°C: the difference in molecular complexity between the corresponding arsenic and antimony compounds is strange.

Several other adducts are described in Table 43.

TABLE 44. COMPLEXES OF THE COMPOUNDS EX_3

Complex		Ref.
$2SbCl_3, C_6H_6$	Infrared and ^{35}Cl n.q.r. spectra recorded	a
$2SbCl_3, C_{10}H_8$	Trigonal bipyramidal configuration at Sb. Sb–Cl (*trans*) [b], 236.7° Sb–Cl (*cis*) [b], 234.7 pm	c
$SbCl_3$, L; $SbCl_3$, 2L[1]	L = amine; L[1] = phosphine. $Me_3As,SbCl_3$ made	d
$SbCl_3, C_6H_5NH_2$	Trigonal bipyramidal configuration at Sb. Sb–Cl (*trans*) [b], 251.6; Sb–Cl (*cis*) [b], 232.9 pm; Sb–N, 253 pm	e
$SbCl_3$, L; $SbCl_3$2L	L = oxygen donor. Examples: $AsCl_3,Me_3PO$; $SbCl_3,Me_3PO$; $SbCl_3,2Me_3PO$; $SbCl_3,Me_2SO_2$; $SbCl_3,2SOMe_2$; $SbCl_3,2Me_2CO$; $AsCl_3,POCl_3$; $SbCl_3,2POCl_3$	f
$SbCl_3, 2Ph_3AsO$	Configuration [3] at Sb [b]: Sb–Cl (*trans*), 259; Sb–Cl (*cis*), 220 pm	f
$2SbI_3$, (1,4-dithiane)	Configuration [3] at Sb [b]. . . . $S[(CH_2)_2]_2S(SbI_3)_2S[(CH_2)_2]_2S(SbI_3)_2$ . . . chains Sb–S, 327, 333 pm; Sb–I, longer than in SbI_3	g
$SbI_3, 3S_8$	Sb–I, 274.7; ∠ISbI 96.5°	h
$Sb(CF_3)_3$; C_5H_5N	M.p. 39°C. Dissociated in vapour	i

[a] L. W. Daasch, *Spectrochim. Acta* **15** (1959) 726; T. Okuda, A. Nakao, M. Shiroyama and H. Negita, *Bull. Chem. Soc. Japan* **41** (1968) 61.
[b] See text. *Cis* and *trans* with respect to donor.
[c] R. Hulme and J. T. Szymanski, *Acta Cryst.* **25B** (1969) 753. $C_{10}H_8$ = naphthalene.
[d] R. R. Holmes and E. F. Bertaut, *J. Am. Chem. Soc.* **80** (1958) 2980, 2983.
[e] R. Hulme and J. C. Scruton, *J. Chem. Soc.* A (1968) 2448.
[f] Ref. 132.
[g] T. Bjorvatten, *Acta chem. scand.* **20** (1966) 1863.
[h] T. Bjorvatten, O. Hassel and A. Lindheim, *Acta chem. scand.* **17** (1963) 689.
[i] J. W. Dale, H. J. Emeléus, R. N. Haszeldine and J. H. Moss, *J. Chem. Soc.* (1957) 3708.

[132] I. Lindqvist, *Inorganic Adduct Molecules of Oxo Compounds*, Springer, Berlin (1963).

10.2. THE MOLECULES EX_3 AS ACCEPTORS

When X is electronegative (e.g. hal), the molecules EX_3 are good acceptors (Table 44), and a wide range of complexes EX_3L, EX_3L_2 (where L is a monodentate ligand) have been isolated. Little detailed thermodynamic data relating to complex formation is available, but it seems clear that, to oxygen donors, antimony trichloride is a weaker acceptor than antimony pentachloride (section 15.3)[132]. More antimony trihalide complexes are known than arsenic or bismuth trihalide complexes, but a quantitative comparison of complex formation is not at present possible.

Structurally, the halide complexes are of interest because of the part played by the lone pair in determining the stereochemistry. In the 1:1 complexes, the lone pair occupies an equatorial position in a trigonal bipyramidal configuration (compare [5], p. 610, or [25], p. 643) with the donor L in an axial position. In the 1:2 complexes, the lone pair occupies a octahedral position, so that a distorted square pyramidal environment for the antimony results (compare [3], p. 606, or [27], p. 643). In each case Sb–X bonds *trans* to the donor L are longer than those *cis* (Table 44). In the weak complex $SbI_3,3S_8$, bonding between the sulphur and antimony triiodide molecules seems to involve mainly S–I interactions. The antimony iodide molecules are linked by weak bonds (Sb–I 385 pm) as in antimony triiodide itself.

Trialkyl- or triaryl-arsines or -stibines do not appear to form complexes with electron donors. The effect of the electro-negative trifluoromethyl groups shows in the formation of the weak complex between the stibine $(CF_3)_3Sb$ and pyridine.

Complexes of antimony trichloride with aromatic hydrocarbons are of two types. In the first, e.g. the naphthalene adduct in Table 44, the hydrocarbon behaves as an olefin, donating π-electron density to one co-ordination position at the antimony. There is considerable distortion of the naphthalene skeleton, with a shortening of the complexed C⋯C bond. In the second type, e.g. in the complexes with perylene [22] or phenanthrene, in the halide as solvent, the hydrocarbon appears to complex the $[SbCl_2]^+$ ion. Polycyclic hydrocarbons thus behave as weak electrolytes in antimony trichloride[133]. The cations may be readily protonated or be oxidized to radical ions by oxygen (compare section15.3).

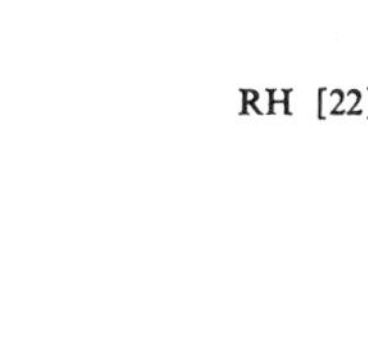

[22]

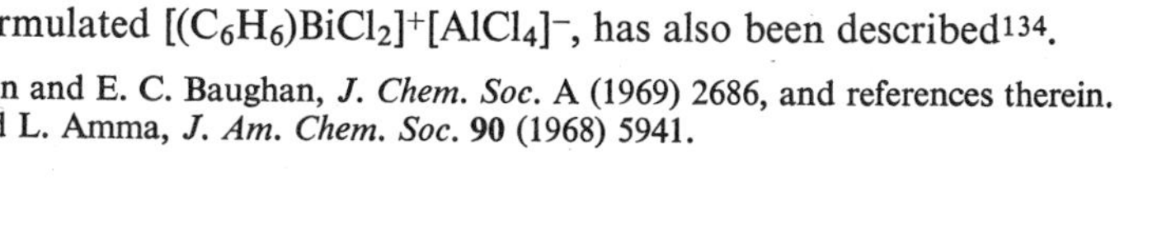

$$\text{RH [22]} + 2\,SbCl_3 \longrightarrow [RH, SbCl_2]^+ + [SbCl_4]^-$$

$$[RH, SbCl_2]^+ \xrightarrow{O_2} [RH^{\cdot}]^+ + \tfrac{1}{2}\,SbCl_3 + \tfrac{1}{2}\,SbOCl$$

$$[RH, SbCl_2]^+ \xrightarrow{HCl} [RH_2]^+ + SbCl_3$$

A compound, formulated $[(C_6H_6)BiCl_2]^+[AlCl_4]^-$, has also been described[134].

133 P. V. Johnson and E. C. Baughan, *J. Chem. Soc.* A (1969) 2686, and references therein.
134 Th. Auel and L. Amma, *J. Am. Chem. Soc.* **90** (1968) 5941.

10.3. COMPLEX HALIDES

The compounds EX_3 are strong acceptors towards halide ions (section 5.3). This section deals mainly with structural aspects of the resulting complex ions.

Complex Fluorides

Arsenic trifluoride reacts with alkali metal fluorides, except those of sodium and lithium, to form complexes MEF_4 [135]. The stability, with respect to decomposition to fluorides MF and arsenic trifluoride, increases with increasing atomic weight of M, as expected from lattice energy considerations. Dissociation pressures at 100°C are: $KAsF_4$, 0.18, $CsAsF_4$, 0.07 atm. The solids may contain AsF_4^- ions, isostructural with SeF_4, or more complicated anions. They dissolve in arsenic trifluoride, but AsF_4^- ions cannot be detected with certainty. Fluorarsenites are very hygroscopic and are rapidly hydrolysed by water.

TABLE 45. E–X BOND LENGTHS IN COMPLEXES HALIDES (pm)

		Reference
$NaSbF_4$	~[23] *a*, 193; *b*, 203, 208; *c*, 219, 251	A. Byström, S. Bäcklund and K. A. Wilhelmi, *Ark. Khem.* **6** (1953) 77
$KSbF_4$	[24] 200–230	A. Byström, S. Bäcklund and K. A. Wilhelmi, *Ark. Khem.* **4** (1952) 175
K_2SbF_5	[27] 4×202, 208	A. Byström and K. A. Wilhelmi, *Ark. Khem.* **3** (1951) 461
$CsSb_2F_7$	201, 208, 222, 238	A. Byström and K. A. Wilhelmi, *Ark. Khem.* **3** (1951) 373
$C_5H_6NSbCl_4$	[26] *a*, 238; *b*, 263; *c*, 313 (± 1)	S. K. Porter and R. A. Jacobson, *J. Chem. Soc.* A (1970) 1356
$[NH_4]_2SbCl_5$	[27] *a*, 236; *b*, 258, 269	M. Webster and S. Keats, *J. Chem. Soc.* A (1971) 298.
$[NH_4]_2SbBr_6$	279.5±6: small deviation from O_h	See section 16.2
$[C_5H_5NH]_5[Sb_2Br_9]Br_2$	[29] *a*, 263; *b*, 300	S. K. Porter and R. A. Jacobson, *J. Chem. Soc.* A (1970) 1359
NH_4BiF_4	219–286 (±3)	B. Aurivillius and C.-I. Lindblom, *Acta chem. scand.* **18** (1964) 1554
$C_6H_8NBiBr_4$	[26] *a*, 264; *b*, 283, 308; *c*, 297, 327 (±2)	B. K. Robertson, W. G. McPherson, and E. A. Meyers, *J. Phys. Chem.* **71** (1967) 3531
$C_6H_8NBiI_4$	[26] *a*, 288; *b*, 311, 331; *c*, 309, 345 (± 1)	
$[C_5H_{12}N]_2BiBr_5$	[28] *a*, 267.1; *b*, 285.7; *c*, 301.9, 312.5 (±1)	W. G. McPherson and E. A. Meyers, *J. Phys. Chem.* **72** (1968) 532
$[Me_2NH_2]_3BiBr_6$	284.0: small deviation from O_h	W. G. McPherson and E. A. Meyers, *J. Phys. Chem.* **72** (1968) 3117
$BiCl_{1.167}$ $[Bi_2Cl_8]^{2-}$	[23] *a*, 260; *b*, 261; *c*, 286 (± 3)	See p. 602. Bond lengths for half $BiCl_5^-$ ions: the others are distorted
$[BiCl_5]^-$	[27] *a*, 262; *b*, 274 (± 3)	
$Cs_3Bi_2I_9$	[29] *a*, 294; *b*, 324	A. Nyström and O. Lindqvist, *Acta chem. scand.* **21** (1967) 2570

135 E. L. Muetterties and W. D. Phillips, *J. Am. Chem. Soc.* **79** (1957) 3686.

A series of complex fluorides of antimony (Table 45) can be isolated from solutions of alkali metal carbonates and the appropriate amounts of antimony trioxide in hydrofluoric acid. Species in solution are not well established, but a series of complex ions is

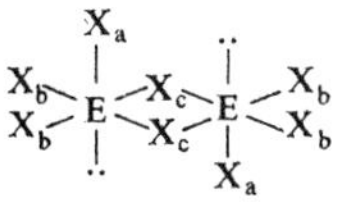

[23] 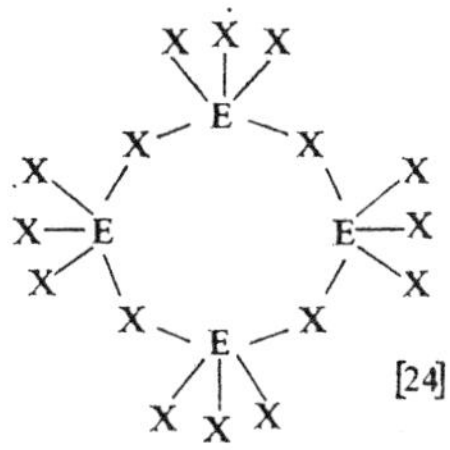 [24]

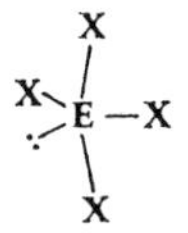

[25] 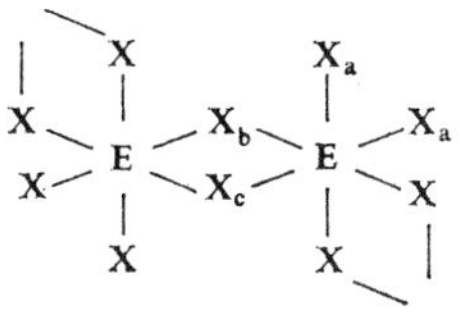 [26]

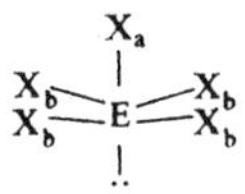

[27] 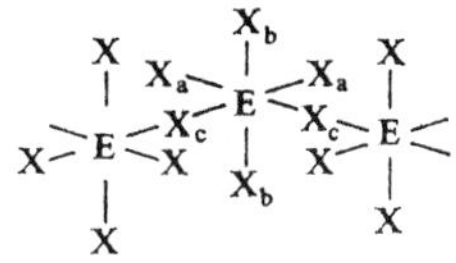 [28]

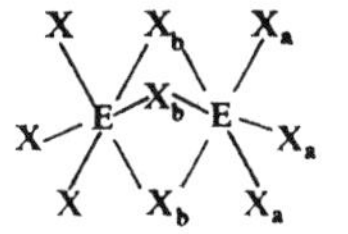

[29]

found in the solids. The nature of these ions depends on the counterions. Thus $[SbF_4]^-$ groups are said to be arranged in pairs in $NaSbF_4$ (distorted [23]), and in groups of four in $KSbF_4$ [24]. In $NaSbF_4$ and in $CsSb_2F_7$ each antimony atom forms four bonds to fluorine shorter than 230 pm, but in $KSbF_4$ and in K_2SbF_5 each antimony atom forms five such bonds. In the salts MSb_4F_{13} (M = K, Rb, Cs, NH_4, Tl) apparently only three bonds to each antimony atom are shorter than 230 pm. In all cases, the co-ordination round antimony is one-sided, showing that the lone pair is stereochemically active, but the precise distribution of valencies depends on how fluorine atoms are shared. In general, bonds to bridging (shared) fluorine atoms are longer than those to non-bridging fluorine atoms. The possibilities for fluorine bridging between antimony atoms are complicated, and further work both on solids and solutions is needed completely to rationalize the bonding in complex fluorides. The problem may be complicated by incorporation of Sb–O bonds in solids isolated from aqueous solution.

Bismuth trifluoride is rather insoluble in water, and complex fluorides of bismuth are not well known. In the complex NH_4BiF_4 each bismuth atom has nine fluorine neighbours (as in BiF_3); nets of BiF_4 co-ordination polyhedra are linked to nitrogen atoms by hydrogen bonds.

Complex Chlorides

Arsenic trichloride and dry hydrogen chloride apparently do not interact, but Raman spectra of arsenic trichloride in aqueous hydrochloric acid indicate the presence of species other than $AsCl_3$. These may be extracted into ether, but not benzene, and seem to be $AsCl_4^-$ ions, presumably combined with solvated protons. Vibrational spectra are given in Table 46. Assignments have been made on the basis of the following symmetries: C_{2v} for

TABLE 46. VIBRATIONAL SPECTRA OF HALIDE COMPLEXES

	Spectra (cm^{-1}) and assignments	Ref.
$HCl/AsCl_3/Bu_2O$	408, 383, 309, 256, 230, 192, 157 (infrared and Raman) 290 (Raman only)	a
$[Et_4N]^+[AsCl_4]^-$	385, 326, 302, 282	b
$[Bu_4N]^+[AsBr_4]^-$	275, 254, 231, 221	b
$[Et_4N]^+[SbCl_4]^-$	332, 293, 313, 248	b
$[Bu_4N]^+[SbBr_4]^-$	218, 203, 194, 181	b
$[Bu_4N]^+[SbI_4]^-$	171, 106, 154	b
$[Et_4N]^+[BiCl_4]^-$	299, 283, 266, 249	b
$[Bu_4N]^+[BiBr_4]^-$	189, 179, 170	b
$[Bu_4N]^+[BiI_4]^-$	137, 134, 127, 114	b
$[NH_4]_2^+[SbCl_5]^{2-}$	445, 300, 285, 255, 230, 180 90 (infrared and Raman) 420 (Raman only)	c
$[Co(NH_3)_6]^{3+}[SbCl_6]^{3-}$	267, 214, 178	d
$[Co(NH_3)_6]^{3+}[BiCl_6]^{3-}$	259, 222, 175	d

[a] J. E. D. Davies and D. A. Long, *J. Chem. Soc.* A (1968) 1761.

[b] G. Y. Ahlijah and M. Goldstein, *J. Chem. Soc.* A (1970) 326, 2590. Spectra from solutions and solids are similar except for $[Bu_4N][SbI_4]$.

[c] H. A. Szymanski, R. Yelin and L. Marabella, *J. Chem. Phys.* **47** (1967) 1877.

[d] T. Barrowcliffe, I. R. Beattie, P. Day and K. Livingston, *J. Chem. Soc.* A (1967) 1810. See also E. Martineau and J. B. Milne, *ibid.* (1970) 2971; C. J. Adams and A. J. Downs, *Chem. Communs.* (1970) 1609.

ECl_4^- [25] (compare $SeCl_4$), C_{4v} for ECl_5^{2-} [27] and O_h for ECl_6^{3-}, but it is difficult to be sure of assignments based on solids, or on solutions in which the ions may be strongly solvated. In solid $C_5H_5NHSbCl_4$, the anions have infinite chains of antimony atoms linked by double chlorine bridges [26]. The bond distances show that the chains may be considered as made of $SbCl_4^-$ ions linked by weaker ($Sb–Cl_c$) bonds (Table 45). It is likely that these are broken in solvating media. The $SbCl_5^{2-}$ ions has also been isolated in crystalline derivatives. It has a distorted octahedral structure [27] in which the shortest bond is opposite the lone pair. Alternatively, the configuration of the ion may be described as square pyramidal with the antimony atom slightly below the base of the pyramid (compare with the sulphides, sections 7.2 and 7.5). The vibrational spectra have been assigned assuming C_{4v} symmetry. The $SbCl_6^{3-}$ ion, like the isoelectronic $TeCl_6^{2-}$, is symmetrical; the lone pair does not occupy a co-ordination position.

Chloride complexes of bismuth have been extensively studied in aqueous solution, and information about individual species has been obtained from ultraviolet and Raman spectroscopy (Table 47). The structures of these ions are assumed to be those of the antimony analogues, with lone pairs stereochemically active in the four- and five-co-ordinate species, but not in $BiCl_6^{3-}$. Stability constants have been determined under a

TABLE 47. CHLORIDE COMPLEXES OF BISMUTH(III)

	Ultraviolet[a]			Ultraviolet[a]		Raman $\bar{\nu}_{max}$ (cm^{-1})		
	λ_{max} (nm)	$10^4\ \varepsilon_{max}$		λ_{max} (nm)	$10^4 \varepsilon_{max}$	ps	dm	dw[b]
Bi^{3+}	222	1.10	$BiCl_3$	300	0.49	307	272	120
$BiCl_2^+$	238	0.68	$BiCl_4^-$	317	0.95	293	256	115
$BiCl^{+}$	255	0.4	$BiCl_5^{2-}$	327	1.61	280	235	115
			$BiCl_6^{3-}$			263	220	110

[a] L. Newman and D. N. Hume, *J. Am. Chem. Soc.* **79** (1957) 4576.
[b] p, polarized; d, depolarized; s, strong; m, medium; w, weak. R. P. Oertel and R. A. Plane, *Inorg. Chem.* **6** (1967) 1960.

variety of conditions[38]. For comparison, values obtained for halide complexes of bismuth are given in Table 48. The stability of a given complex BiX_n increases as X changes from chloride to iodide: bismuth may thus be said to show class (b) [24] acceptor properties. The ions $BiCl_5^{2-}$ and $Bi_2Cl_8^{2-}$ have been found in $BiCl_{1.167}$ (section 5.5). Many crystalline chloride complexes have been isolated[135]; in particular, large tervalent cations, e.g. $Co(NH_3)_6^{3+}$, form insoluble compounds with ions ECl_6^{3-} [136a]. The $BiCl_5^{2-}$ ion has structure [27] and $Bi_2Cl_8^{2-}$ [23] is rather similar to $Sb_2F_8^{2-}$, except that the bridge bonds are symmetrical. In both $BiCl_5^{2-}$ and $Bi_2Cl_8^{2-}$, lone pairs occupy co-ordination positions. The absorption of solutions of halide complexes is used in the spectrophotometric estimation of bismuth (section 3.4).

Organometallic complex ions $[Ph_nBiX_{4-n}]^-$ have been described[136b].

TABLE 48. STABILITY CONSTANTS FOR BISMUTH HALIDE COMPLEXES

NaClO$_4$, 2 M; H$^+$, 1 M; 20 °C $\qquad \beta_n = \dfrac{[BiX_n^{3-n}]}{[Bi^{3+}][X^-]^n}$

n	log β_n 1	2	3	4	5	6
Cl	2.36	3.5	5.35	6.10	6.72	6.56
Br	2.26	4.45	6.30	7.70	9.28	9.38
I				14.95	16.8	18.8
SCN	1.15	2.26		3.41		4.23

S. Ahrland and I. Grenthe, *Acta chem. scand.* **11** (1957) 1111; **15** (1961) 932, and ref. 38.

[136] (a) H. Remy and L. Pellens, *Chem. Ber.* **61** (1928) 862; W. Pugh, *J. Chem. Soc.* (1954) 1385; (b) G. Faraglia, *J. Organomet. Chem.* **20** (1969) 99.

Bromide Complexes

Bromide complexes of As^{III} have not been extensively studied. Many bromide complexes of Sb^{III} and Bi^{III} have been isolated, particularly as salts of organic bases[136a, 137]. Though the stoichiometry has been established by analysis, structural work has been begun only recently (Table 45). The reaction between pyridine and antimony tribromide, in hydrogen bromide, yielded a crystalline compound with bromide ions and complex ions $Sb_2Br_9^{3-}$ in which two $SbBr_6$ octahedra share a common face [29]. The lone pair of the Sb^{III} does not occupy a co-ordination position. The same is true in the ions $Sb^{III}Br_6^{3-}$ and $Bi^{III}Br_6^{3-}$, which deviate only slightly from O_h symmetry. In $(C_6H_8N)BiBr_4$, and the corresponding iodide, the ions are linked as in [26], with a distorted octahedral co-ordination round bismuth and the longest bonds opposite the shortest, and in $(C_5H_{12}N)_2BiBr_5$, the distorted octahedra round bismuth are linked by single bromide bridges [28]. Again, the Bi^{III} lone pairs are largely stereochemically inactive in the solid. The interactions linking $[EBr_n]^{(n-3)-}$ units may be broken in solution.

Formation constants have been given in Table 48.

Iodide Complexes

The most studied complexes are those of bismuth. Individual species have been identified[137] and formation constants given. Many crystalline antimony and bismuth iodides have been obtained, but there has been little structural work. In $Cs_3Bi_2I_9$, the $Bi_2I_9^{3-}$ ion consists of two distorted octahedra which share a common face [29].

Complex halide ions may be present in the many ternary halides of the Group V elements with zinc, cadmium and mercury[138a].

10.4. COMPLEXES WITH ORGANIC LIGANDS

Many complexes of arsenic, antimony and bismuth with organic ligands are known; no doubt many more may be made. Some of the most important are with hydroxy-acids or with polyols. For example, As^{III}, like boron, forms a series of complexes with glycols and sugars, and tartrate is a good complexing ion for Sb^{III} (section 6.4). A few stability constants have been measured [38]. In the complex oxalate $K_3Sb(C_2O_4)_3, 4H_2O$, all six oxygen atoms bound to any one antimony lie on the same side; this provides a dramatic example of a stereochemically active Sb^{III} lone pair [138b].

11. HALIDES EX_5

Pentafluorides of all three elements are known; the only other well-documented pentahalide is antimony pentachloride, but a number of compounds with more than one halogen have been isolated[51, 54]. Why arsenic fails to form a pentachloride is not clear. Thermodynamic quantities relevant to the formation of pentahalides from trihalides are shown in Scheme 13 and values for a number of systems are presented in Table 49. The mean bond energies in the pentahalides (h_7) are smaller than in the corresponding trihalides (h_6). Since ΔS for the reaction $AsCl_3 + Cl_2 \rightarrow AsCl_5$ is almost certainly negative, the enthalpy

[137] J. R. Preer and G. P. Haight, *Inorg. Chem.* **5** (1966) 656; G. P. Haight and L. Johansson, *ibid.* **7** (1968) 1255; A. J. Eve and D. N. Hume, *ibid.* **6** (1967) 331.

[138] (a) H. Puff and J. Berg, *Z. anorg. allgem. Chem.* **343** (1966) 259, and references therein. (b) M. C. Poore and D. R. Russell, *Chem. Communs.* (1971) 18.

SCHEME 13. Formation of pentahalides from trihalides.

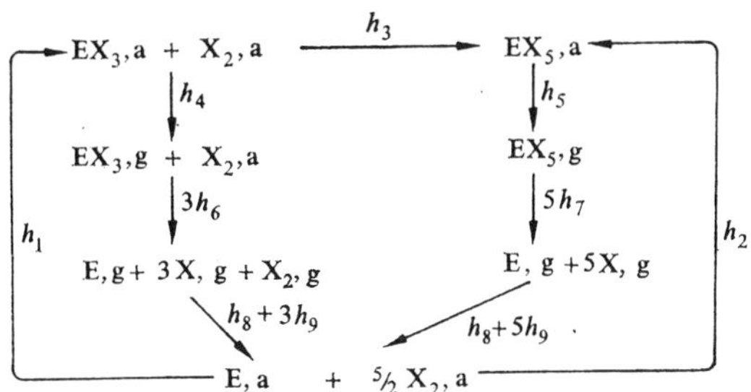

a standard state at 298 K
h enthalpy change at 298 K

TABLE 49. THERMODYNAMIC QUANTITIES RELATING TO THE FORMATION OF PENTAHALIDES FROM TRIHALIDES (kJ mol^{-1}, 298 K) [a, b]

MX_3	h_1	h_2	h_3	h_4	h_5	h_6	h_7	h_8	h_9
PCl_3	−311	−463	−152	32	64	319	265	−313	−124
$AsCl_3$	−305			46		306		−301	−124
$SbCl_3$	−382	−440	−58	68	46	313	253	−259	−124
AsF_3	−958	−1238	−280	36	0	484	385	−301	−76

[a] See Tables 22 and 50 for references.
[b] See Scheme 13.

change, h_3, must outweigh $T\Delta S$ for the reaction to be feasible. For reaction between antimony trichloride and chlorine, $T\Delta S$ at 298 K is −32.2 kJ mol^{-1}. If $T\Delta S$ for the $AsCl_3$–Cl_2 reaction is similar and if the heat of sublimation of $AsCl_5$ is 50 kJ mol^{-1}, formation of $AsCl_5$ is possible if h_2 is more negative than −338 kJ mol^{-1}, i.e. if the As^{V}–Cl bond energy is at least 242 kJ mol^{-1}. Such a value is not unreasonable compared with other halides (Table 49). Many attempts to make arsenic pentachloride, however, have failed. The freezing points of mixtures of $AsCl_3$–Cl_2, $AsBr_3$–Br_2, $SbBr_3$–Br_2 show simple eutectics, and vapour pressure measurements likewise give no evidence for formation of compounds[139]. The exchange of radiochlorine between chlorine and arsenic trichloride is much slower than the reaction between chlorine and phosphorus trichloride, and appears to involve a complex mechanism without even transient formation of $AsCl_5$ [140]. Phosphorus pentachloride gives the ions $[PCl_4]^+$,$[PCl_6]^-$ in acetonitrile. The arsenic analogues of both are known, but conductivity and Raman measurements on $AsCl_3$–Cl_2–MeCN mixtures show no evidence for ions[141].

Bi^{V} is probably too strongly oxidizing to form stable compounds with halide ions, other than fluoride.

139 W. E. Dasent, *J. Chem. Educ.* **34** (1957) 535; R. R. Holmes, *J. Inorg. Nucl. Chem.* **19** (1961) 363.
140 J. H. Owen and R. E. Johnson, *J. Inorg. Nucl. Chem.* **2** (1956) 260.
141 I. R. Beattie and K. M. Livingston, *J. Chem. Soc.* A (1969) 859.

11.1. PREPARATION AND PROPERTIES

The pentafluorides are made by the reaction between fluorine and the elements or the oxides. Bismuth pentafluoride requires a high temperature (*ca.* 600°C), but antimony and arsenic pentafluorides are made more easily. Antimony pentafluoride may be made from antimony pentachloride and anhydrous hydrogen fluoride, but replacement of the final chlorine is slow[142].

TABLE 50. PROPERTIES OF PENTAHALIDES EX_5

	AsF_5	SbF_5	BiF_5	$SbCl_5$
Preparation	As or As_2O_3|F_2 [a]	Sb_2O_3|F_2 [b]	Bi|F_2|600° [c]	$SbCl_3$|Cl_2 [d]
Melting point (°C)	−79.8 [e]	8.3 [e]	151.4 [e]	4 [d]
Latent heat fusion (kJ mol^{-1})	11.5 [f]			
Boiling point (°C)	−52.8 [f]	141 [e]	230 [e]	140 dec [d]
Latent heat evaporation (kJ mol^{-1})	21 [e]	45 [e]	62.5 [e]	
Trouton constant (kJ mol^{-1} K^{-1})	95 [g]	105 [g, h]	124 [g, e]	
Vapour pressure (mmHg) at (K)	70|−87.4 [f, i] 386|−65.2	14|48.5 [h, j]	k	6|51 [d] 14|68 [d]
Density (kg m^{-3}) at (K)	2330|−53 [f]	3110|25 [h]	5400|25 [e]	2350|21 [d]
ΔH_f° (kJ mol^{-1})	−1238 [l]			−440 [m]
S° (J mol^{-1} K^{-1})	326.3 [l]			301.2 [m]
ΔG_f° (kJ mol^{-1})	−1172 [l]			−350 [m]
Bond lengths (pm): eq.	171·1±0·5 [r]			229 [n]
ax.	165·6±0·4			234
^{19}F n.m.r. spectra[p, q]	−11.3	+6.83(1); +26.2(2); +52(2)		
N.q.r. spectra, $\|e^2Qq\|$ (MHz)				^{121}Sb, 84.6[q]; ^{123}Sb, 107.9

[a] F. Seel and O. Detmer, *Z. anorg. allgem. Chem.* **301** (1959) 113.
[b] A. A. Woolf and N. N. Greenwood, *J. Chem. Soc.* (1950) 2200.
[c] J. Fischer and E. Rudzitis, *J. Am. Chem. Soc.* **81** (1959) 6375.
[d] Ref. 5, p. 610.
[e] R. D. W. Kemmitt and D. W. A. Sharp, *Adv. Fluorine Chem.* **4** (1965) 142.
[f] O. Ruff, A. Braida, O. Bretschneider, W. Menzel, and H. Plaut, *Z. anorg. allgem. Chem.* **206** (1932) 59.
[g] Calculated from other data here. "Normal" Trouton constant about 96 kJ mol^{-1} K^{-1}.
[h] C. J. Hoffmann and W. L. Jolly, *J. Phys. Chem.* **61** (1957) 1574.
[i] Solid: $\log p_{mmHg} = 10{,}952 - 1692/T$; liquid: $\log p = 7.845 - 1094/T$.
[j] $\text{Log}\ p_{mmHg} = 8.567 - 2364/T^\circ$. mmHg = 133.33 N m^{-2}.
[k] $\text{Log}\ p_{mmHg} = 9.340 - 3250/T^\circ$.
[l] P. A. G. O'Hare and W. N. Hubbard, *J. Phys. Chem.* **69** (1965) 4358.
[m] Ref. 11. For $SbCl_5$, g: H_{298}–H_0, 26.5; ΔH_f°, −394.0 kJ mol^{-1}; S°, 402; C_p, 121 J mol^{-1} K^{-1}.
[n] For crystal, S. Ohlberg, *J. Am. Chem. Soc.* **81** (1959) 811. The Sb–Cl (apical) distance in the gas has been given as 243 pm.
[o] Ppm relative to CF_3COOH. Relative areas in parentheses. Above 80°C SbF_5 has one peak. E. L. Muetterties and W. D. Phillips, *J. Am. Chem. Soc.* **81** (1959) 1084.
[p] ^{75}As and ^{121}Sb n.m.r. spectra have been given by E. D. Jones and E. A. Uehling, *J. Chem. Phys.* **36** (1962) 1691.
[q] 249 K. R. F. Schneider and J. V. DiLorenzo, *J. Chem. Phys.* **47** (1967) 2343. Spectra below 210 K are more complicated. See text.
[r] F. B. Clippard and L. S. Bartell, *Inorg. Chem.* **9** (1970) 805.

142 L. Kolditz and H. Daunicht, *Z. anorg. allgem. Chem.* **302** (1959) 230.

The greater volatility of the pentahalides compared with the trihalides (Table 50) is associated with the zero dipole moments of the pentahalide molecules. Arsenic pentafluoride, like sulphur hexafluoride, is very volatile, a reflection of a low heat, rather than a low entropy, of evaporation. In antimony pentafluoride, however, association between molecules is clearly shown (a) by the high viscosity (0.46 N s m^{-2} at 20°C, about that of glycerol. Compare AsF_3, 1.17×10^{-3} N s m^{-2} at 8.3°C), (b) by molecular weight measurements on the vapour $\{(SbF_5)_3$ at 152°C and $(SbF_5)_2$ at 252°C$\}$, (c) by the high Trouton constant, (d) by the mass spectrum and (e) by vibrational and n.m.r. spectra (see below). Bismuth pentafluoride also has a high Trouton constant—probably indicating association. The solid appears to have the uranium pentafluoride structure, with infinite chains of *trans*-bridged BiF_6 octahedra, but more detailed crystal data would be interesting. Sb^V chloride is molecular both in the liquid and in the solid down to −63°C. The n.q.r. spectrum suggests that all the antimony atoms are equivalent. Below −63°C, however, the n.q.r. and Raman spectra become more complicated, and several different environments for antimony and chlorine can be detected. A halogen-bridged structure, like that in antimony pentafluoride, is possible for the low-temperature form. There is no evidence for formation of ionic species in either antimony pentafluoride or pentachloride (cf. PCl_5): the pentafluoride has a low electrical conductivity ($1.2 \times 10^{-6}\ \Omega^{-1}\ m^{-1}$ at 298 K).

TABLE 51. VIBRATIONAL SPECTRA OF PENTAHALIDES

	cm^{-1} [a]
AsF_5 [b, c]	I.r. (g): 787 vs, 400 s (a_2''); 811 vs, 369 s, 123 w (e') Raman (l): 733 vs, 642 w (a_1'); 809 w, 366 vw (e'); 388 w (e'')
$SbCl_5$ [b, d]	I.r. (l): 371 vs, 154 m (a_2''); 395 vs, 68 vw (e') Raman (l): 357 vs, 307 m (a_1'); 397 w, 177 m, 68 m (e'); 165 w (e'')
SbF_5 [e]	I.r. (l): 742 s, 705 s, 669 s, 450 w, 310 w, 200 w Raman (l): 718, 670 s, 349 w, 268 mw, 231 mw, 189 mw, 140 vw, 116 vw
BiF_5 [e]	I.r. (s): 627 s, 450 m, 220 m Raman (s): 595, 570, 255, 167, 101
$SbCl_4F$ [e]	I.r. (s): 446 mw, 416 s, 400 m, 371 vs, 348 m, 174 m, 158 ms, 140 sh Raman (s): 489 vw, 446 vw, 394 m, 370 wsh, 349 s, 274 w, 230 vw, 204 w, 178 w, 165 mw, 154 m, 138 mw, 128 msh, 115 w, 98 mw, 94 msh, 60 vw, 53 vw, 40 vw

[a] s, strong; m, medium; w, weak; v, very; sh, shoulder.
[b] In D_{3h}: a_1', Raman polarized; a_2'', i.r.; e', i.r. and Raman depolarized; e'', Raman depolarized.
[c] L. C. Hoskins and R. C. Lord, *J. Chem. Phys.* **46** (1967) 2402.
[d] G. L. Carlson, *Spectrochim. Acta* **19** (1963) 1291. The gas-phase Raman spectrum is described by I. R. Beattie and G. A. Ozin, *J. Chem. Soc.* A (1969) 1691.
[e] For detailed assignments, see I. R. Beattie, K. M. S. Livingston, G. A. Ozin and D. J. Reynolds, *J. Chem. Soc.* A (1969) 958.

Vibrational spectra of arsenic pentafluoride and antimony pentachloride have been interpreted in terms of D_{3h} symmetry (Table 51), but the fact that the ^{19}F n.m.r. spectrum of AsF_5 shows only one line, even at the lowest temperature observable, suggests that the barrier to pseudorotation—which interchanges axial and equatorial halogens—is small. Attempts have been made to calculate energy barriers, but several puzzling features of these systems remain[143]. For antimony and bismuth pentafluorides, interpretation of vibrational spectra is less simple, since the site symmetry at the heavy atom in the liquid or solid is less than that in the isolated molecule. The ^{19}F n.m.r. spectra of a supercooled sample of antimony pentafluoride, or a solution in perfluorocyclobutane, suggested that the molecules were associated by *cis*-Sb–F–Sb bridges [30] [144]. There are thus three types of fluorine: bridging (a), *trans* to bridging (b), and *cis* to bridging (c). The vibrational spectra are

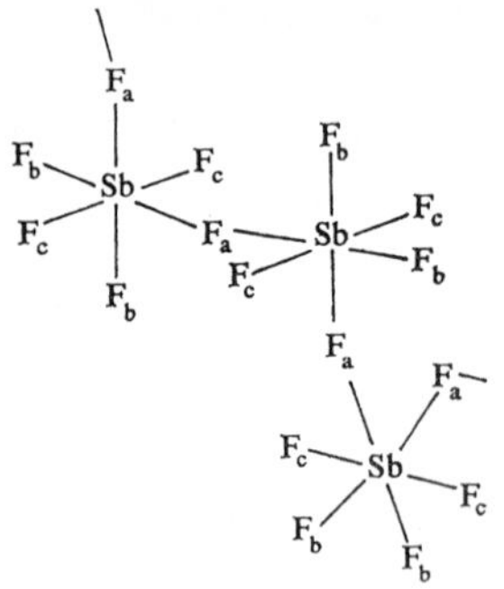

[30]

consistent with this interpretation, since three high frequencies associated with the SbF_4 residue suggest that this has C_{2v} symmetry (*cis*-bridging) rather than D_{4h} (*trans*-bridging). Association persists up to 125°C, but at this temperature the fluorine atoms interchange sufficiently quickly for the n.m.r. spectrum at 40 MHz to show only one peak. For bismuth pentafluoride, the Raman data confirm that the bridges linking octahedrally co-ordinated bismuth atoms are *trans*.

SCHEME 14. Reactions of pentahalides.

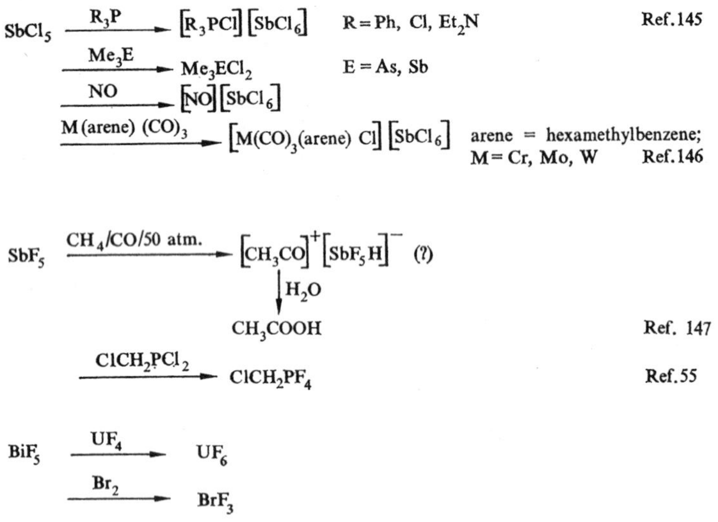

$SbCl_5 \xrightarrow{R_3P} [R_3PCl][SbCl_6]$ R = Ph, Cl, Et_2N Ref. 145

$SbCl_5 \xrightarrow{Me_3E} Me_3ECl_2$ E = As, Sb

$SbCl_5 \xrightarrow{NO} [NO][SbCl_6]$

$SbCl_5 \xrightarrow{M(arene)(CO)_3} [M(CO)_3(arene)Cl][SbCl_6]$ arene = hexamethylbenzene; M = Cr, Mo, W Ref. 146

$SbF_5 \xrightarrow{CH_4/CO/50\ atm.} [CH_3CO]^+[SbF_5H]^-$ (?) $\xrightarrow{H_2O} CH_3COOH$ Ref. 147

$SbF_5 \xrightarrow{ClCH_2PCl_2} ClCH_2PF_4$ Ref. 55

$BiF_5 \xrightarrow{UF_4} UF_6$

$BiF_5 \xrightarrow{Br_2} BrF_3$

---

143 R. R. Holmes and R. M. Deiters, *J. Am. Chem. Soc.* **90** (1968) 5021.

144 T. K. Davies and K. C. Moss, *J. Chem. Soc.* A (1970) 1054.

145 J. K. Ruff, *Inorg. Chem.* **2** (1963) 813.

146 M. H. B. Stiddard and R. E. Townsend, *J. Chem. Soc.* A (1969) 2355.

147 H. Hogeveen, J. Lukas and C. F. Roobeek, *Chem. Communs.* (1969) 920; W. Bracke, W. J. Cheng, J. M. Pearson and M. Szwarc, *J. Am. Chem. Soc.* **91** (1969) 203.

11.2. REACTIONS

The chemistry of the pentahalides is dominated by their tendency to form complexes (section 15). Other aspects of their chemistry have been neglected, but the halides EX_5 may be used as oxidizing agents (Scheme 14). Like other non-metal halides, they react with water to give oxides or oxide halides, and with alkoxides and related metal derivatives[60].

$$SbCl_5 \xrightarrow{NaOR} Sb(OR)_5 \xrightarrow{NaX} [Na][Sb(OR)_5X] \qquad (X = Cl, OR)$$

Antimony pentachloride, like phosphorus pentachloride, dissociates in the vapour (above 200°C) to the trichloride and chlorine. The same reaction is partly responsible for the exchange of antimony between Sb^{III} chloride and Sb^{V} chloride in carbon tetrachloride at 50–80°C. A second mechanism, perhaps involving a chlorine bridged intermediate Sb_3Cl_{13}, is probably also involved[148].

11.3. MIXED HALIDES

Many derivatives with more than one halogen are conceivable, but few have been fully characterized (Table 52). It has been suggested[51c] that the mixed halides may, like PCl_5, exist in ionic and molecular forms (i.e. $[EX_4]^+[EX^1_6]^-$ and $EX_2X^1_3$), but more structural work is needed.

The addition of small amounts of antimony pentachloride to the pentafluoride gives a dramatic decrease in viscosity and increase in electrical conductivity. These changes have

TABLE 52. MIXED PENTAHALIDES

	Preparation	Properties	Structure	Ref.
$[AsCl_4]^+[AsF_6]^-$	AsF_3–$AsCl_3$–Cl_2 $AsCl_3$–ClF_3	M.p. 130° dec., subl. 70°/10^{-3} mmHg	Confirmed by vibrational spectra[a]	b, c
$[AsCl_4]^+[SbCl_6]^-$	$AsCl_3$–$SbCl_5$–Cl_2		Confirmed by vibrational spectra[a]	d
$SbCl_2F_3$	$SbCl_5$–ClF–80°	B.p. 91–93°/1 mmHg; m.p. 67–70°C		e
$SbCl_4F$	AsF_3–$SbCl_5$	M.p. 83°C	F-bridged tetramers in solid	f, g
$SbCl_4N_3$	R_3SiN_3–$SbCl_5$	M.p. 131°C (may explode)	Azide-bridged dimers	h

[a] See sections 15.1 and 15.4.
[b] L. Kolditz, *Z. anorg. allgem. Chem.* **280** (1955) 313; H. M. Dess, R. W. Parry and G. L. Vidale, *J. Am. Chem. Soc.* **78** (1956) 5730. $[AsBr_4]^+[AsF_6]^-$ has not been isolated, but may be an intermediate in the reaction of AsF_3 with bromine water.
[c] J. Weidlein and K. Dehnicke, *Z. anorg. allgem. Chem.* **337** (1965) 113.
[d] I. R. Beattie, T. Gilson, K. Livingston, V. Fawcett and G. A. Ozin, *J. Chem. Soc.* A (1967) 712.
[e] K. Dehnicke and J. Weidlein, *Z. anorg. allgem. Chem.* **323** (1963) 267.
[f] L. Kolditz, *Z. anorg. allgem. Chem.* **289** (1957) 128.
[g] H. Preiss, *Z. Chem.* **6** (1966) 350. Sb–Cl 223–232 pm. Sb–F 205, 218 pm.
[h] N. Wiberg and K. H. Schmid, *Chem. Ber.* **100** (1967) 741; *Angew. Chem.* **79** (1967) 938.

---

148 F. B. Barker and M. Kahn, *J. Am. Chem. Soc.* **78** (1956) 1317; K. R. Price and C. H. Brubaker, *Inorg. Chem.* **4** (1965) 1351.

been attributed to the breaking of fluorine bridges between pentafluoride molecules, and the formation of ionic species such as $[Sb_nF_{5n+1}]^-$. As more pentachloride is added, the conductivity decreases and the viscosity rises again, as the ionic species are converted to the tetrameric $SbCl_4F$, which, like antimony pentafluoride, has *cis*-fluorine bridges between antimony atoms. The addition of still more pentachloride results in a fall in viscosity, as the polymeric species in solution are diluted with monomeric $SbCl_5$ molecules. It is probable that a variety of ionic and polymeric species are present in SbF_5–$SbCl_5$ mixtures: some of these may have been isolated in a series of adducts made many years ago by Ruff[149].

The viscosity and n.m.r. spectrum of material of composition SbF_3Cl_2 suggest that this may have a fluorine-bridged structure, like that of antimony pentafluoride[150]. The mixed halide (Swartz reagent) is often a more effective fluorinating agent than antimony trifluoride (cf. section 5.3), provided that yields are not lowered by oxidations[54].

$$SOCl_2 \xrightarrow{SbF_3Cl_2} SOF_2$$

$$POCl_3 \xrightarrow{SbF_3Cl_2} POFCl_2$$

$$CCl_2{:}CCl{\cdot}CCl{:}CCl_2 \xrightarrow{SbF_3Cl_2} CF_3CCl{:}CClCF_3 \xrightarrow{KMnO_4/KOH} 2CF_3COOH \quad \text{[Ref. 151]}$$

11.4. OXIDE HALIDES

Reports on oxide halides are confusing[152].

Thus fluorine was said to react with a mixture of As^{III} chloride and As^{III} oxide to give a compound $AsOF_3$, b.p. $-25.6°C$. In later work, however, the oxide fluorides $AsOF_3$, $SbOF_3$ and SbO_2F were obtained as hygroscopic, amorphous, non-volatile solids, insoluble in organic solvents.

$$[AsCl_4]^+[AsF_6]^- \xrightarrow{ClONO_2} [As(ONO_2)_4]^+[AsF_6]^- \xrightarrow{160\text{–}170} AsOF_3$$

Oxide halides molecules are said to be associated by AsO · · · As interactions.

12. OXIDES E_2O_5 AND RELATED OXYACIDS

The oxides E_2O_5 are much less well-characterized than the oxides E_2O_3, and there is little information about structures and physical properties. Ternary compounds, which may be described as double oxides, or as arsenates, antimonates and bismuthates, are better known.

12.1. ARSENIC(V) OXIDE AND ARSENATES

The reaction between As^{III} oxide and concentrated nitric acid yields a solution from which either of two solids may be recrystallized, depending on the temperature.

$$As_2O_3 + HNO_3 \rightarrow \begin{matrix} 2H_3AsO_4, H_2O \text{ (recryst. } < 30°C) \\ \text{or} \\ As_2O_5, 5/3H_2O \text{ (recryst. } > 100°C) \end{matrix} \xrightarrow{>170°C} As_2O_5$$

---

[149] N. E. Aubrey and J. R. van Wazer, *J. Inorg. Nucl. Chem.* **27** (1965) 1761; O. Ruff, *Chem. Ber.* **42** (1909) 4021.

[150] E. L. Muetterties, W. Mahler, K. J. Packer and R. Schmutzler, *Inorg. Chem.* **3** (1964) 1298, and references therein.

[151] A. L. Henne and P. Trott, *J. Am. Chem. Soc.* **69** (1947) 1820.

[152] J. Weidlein and K. Dehnicke, *Chem. Ber.* **98** (1965) 3053; *Z. anorg. allgem. Chem.* **342** (1966) 225; G. Mitra, *J. Am. Chem. Soc.* **80** (1958) 5639.

The hydrate $2H_3AsO_4,H_2O$ has a structure in which double sheets of AsO_4^{3-} tetrahedra and water molecules are linked by hydrogen bonds. It is similar to the hydrate $2H_3PO_4,H_2O$, with which it forms mixed crystals. It loses water at 100°C and is changed to the solid $As_2O_5,5/3H_2O$. This has a structure in which infinite band-like molecules $(H_5As_3O_{10})x$ are joined by short hydrogen bonds (O · · · O 240–245 pm) (Fig. 21). Two of the arsenic atoms in the repeating unit have the usual tetrahedral co-ordination

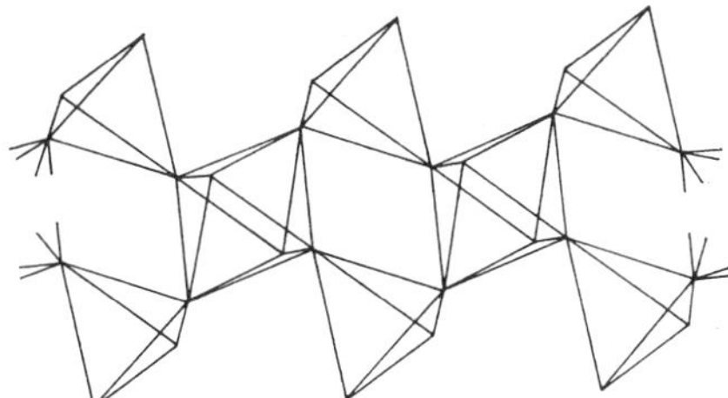

FIG. 21. The linked AsO_4 tetrahedra and AsO_6 octahedra in As_2O_5, $5/3H_2O$.

(As–O, 169.3 pm): the other atom, however, is six-co-ordinate (As–O, 182 pm). Both solids yield analytically pure As_2O_5 (ΔH_f°, −924.9; ΔG_f°, −782.4 kJ mol^{-1}; S^0, 105.5; C_p, 116.5 J mol^{-1} K^{-1}), when heated to constant weight at 350°C. This forms mixed crystals $(As,P)_2O_5$ which may contain up to 50 mol % P_2O_5, and it has been suggested that the structure may contain two different kinds of arsenic—only one of which may be replaced by phosphorus[153].

Arsenic acid is tribasic (at 298 K: pK_1, 2.2; pK_2, 6.9; pK_3, 11.5[38]) with similar dissociation constants to phosphoric acid. Alkali metals give two series of acid salts; like the acid phosphates, these lose water on heating.

$$NaH_2AsO_4 \rightarrow NaAsO_3 + H_2O$$

Salts of bi- and ter-valent ions are insoluble in water (Table 53). Solutions which lead to the formation of copper or chromium arsenates inside wood fibres have been used as preservatives. The insoluble arsenates are not easily leached, and so their effect is long-lasting[81].

TABLE 53. SOLUBILITY PRODUCTS OF ARSENATES[a]

M^{n+}	S[b]	M^{n+}	S[b]	M^{n+}	S[b]
Mg^{2+}	−19.7	Mn^{2+}	−28.7	Cr^{3+}	−20.1
Ca^{2+}	−18.2	Fe^{3+}	−20.2	Al^{3+}	−15.8
Sr^{2+}	−18.0	Co^{2+}	−28.1	Bi^{3+}	−9.4
Ba^{2+}	−50.1	Ni^{2+}	−25.5	$UO_2^{2+}K^+$	−22.6
Zn^{2+}	−27.9	Cu^{2+}	−35.1	$UO_2^{2+}Na^+$	−21.9
Cd^{2+}	−32.7	Pb^{2+}	−35.4	$UO_2^{2+}NH_4^+$	−23.8
Ag^+	−22.0				

[a] Ref. 38

[b] $S = \log_{10} [M^{n+}]_3[AsO_4^{3-}]_n$ ($n = 1, 2$), $\log_{10} [M^{3+}][AsO_4^{3-}]$, or $\log_{10} [M^{2+}][M'^{+}][AsO_4]$

[153] A. Winkler and E. Thilo, *Z. anorg. allgem. Chem.* **339** (1965) 71; **337** (1965) 149; **346** (1966) 92.

The structures of many naturally occurring and synthetic arsenates have been determined (Table 54). One group has isolated AsO_4^{3-} tetrahedra linked by cations or by systems of hydrogen bonds. The tetrahedra are often distorted, with As–O distances from 165 to 175 pm. The shorter bonds probably have order greater than one; the longer bonds are associated with hydrogen atoms (As–OH). The distortion of the AsO_4 tetrahedra has been studied by ^{75}As n.q.r. [154]. Movement within the hydrogen bond system of the dihydrogen arsenates MH_2AsO_4 (M = K, Rb_3, Cs, NH_4) provides a mechanism for ferroelectric properties.

TABLE 54. STRUCTURES OF ARSENATES

	Ref.	
$2H_3AsO_4, H_2O$	a	H Worzala, *Acta Cryst.* **24B** (1968) 987
$As_2O_5, 5/3H_2O$	b	K.-H. Jost, H. Worzala and E. Thilo, *Acta Cryst.* **21** (1966) 808
$(LiAsO_3)_x$	c	W. Hilmer and K. Dornberger-Schiff, *Acta Cryst.* **9** (1956) 87
$(NaAsO_3)_x$	c	F. Liebau, *Acta Cryst.* **9** (1956) 811
β-$K_3As_3O_9$	d, e	I. Grunze, K. Dostál and E. Thilo, *Z. anorg. allgem. Chem.* **302** (1959) 221
KH_2AsO_4; $AgAsO_4$	a, e	L. Helmoltz and R. Levine, *J. Am. Chem. Soc.* **64** (1942) 354
$CaHAsO_4, H_2O$ (haidingerite)	a	H. Binas, *Z. anorg. allgem. Chem.* **347** (1966) 133
$CaHAsO_4, 2H_2O$ (phamacolite)	a	M. Calleri and G. Ferraris, *Acta Cryst.* **25B** (1969) 1544
$M^IM^{II}AsO_4$; $M^{II}_{0.5}M^{IV}_{0.5}AsO_4$	f	R. Klement and P. Kresse, *Z. anorg. allgem. Chem.* **310** (1961) 53; R. Klement and H. Haselbeck, *ibid.* **334** (1964) 27
$Cu_3(AsO_4)_2$	a	S. J. Poulsen and C. Calvo, *Can. J. Chem.* **46** (1968) 917
$Cu_2(AsO_4)(OH), 3H_2O$ (euchroite)	a	J. J. Finney, *Acta Cryst.* **21** (1966) 437
$Cu_3AsO_4(OH)_3$ (clinoclase)	a	S. Ghosh, M. Fehlmann and M. Sundaralingam, *Acta Cryst.* **18** (1965) 777
$BiAsO_4$	a, g	R. C. L. Mooney, *Acta Cryst.* **1** (1948) 163
$BAsO_4$; $AlAsO_4$	h	F. Dachille and L. S. Dent Glasser, *Acta Cryst.* **12** (1959) 820
QH_3AsO_4	a, i	M. Currie and J. C. Speakman, *J. Chem. Soc.* A (1969) 1648

[a] AsO_4 tetrahedra linked by cations and/or hydrogen bonds.
[b] See text and Fig. 21.
[c] See text and Fig. 22.
[d] Structure based on powder data, mixed crystal formation and chemical evidence.
[e] Like phosphorus analogues.
[f] Structures based on powder photography.
[g] Scheelite ($CaWO_4$) structure.
[h] These substances are polymorphic: forms analogous to quartz and cristobalite are known.
[i] QH^+ = 1-methyl-2-quinolonium

N O+
Me H

[154] A. P. Zhukov, L. S. Golovchenko and G. K. Semin, *Izvest. Akad. Nauk SSSR Ser. Khim.* (1968) 1399; *Chem. Communs.* (1968) 854.

In other arsenates, the AsO_4^{3-} tetrahedra are joined by the sharing of corners. Cyclic $As_3O_9^{3-}$ ions have been claimed in the potassium salt β-$K_3As_3O_9$. Chains (Fig. 22), with a repeating trimeric unit, are found in sodium meta-arsenate, and diopside-like chains, with a repeating dimeric unit, appear in lithium arsenate. Both these substances show fibre-like properties.

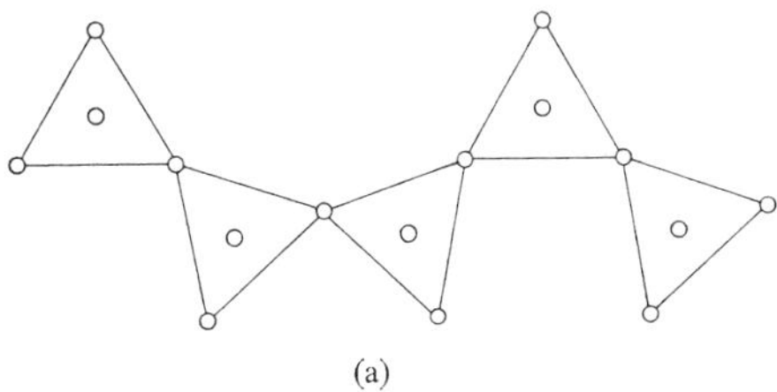

(a)

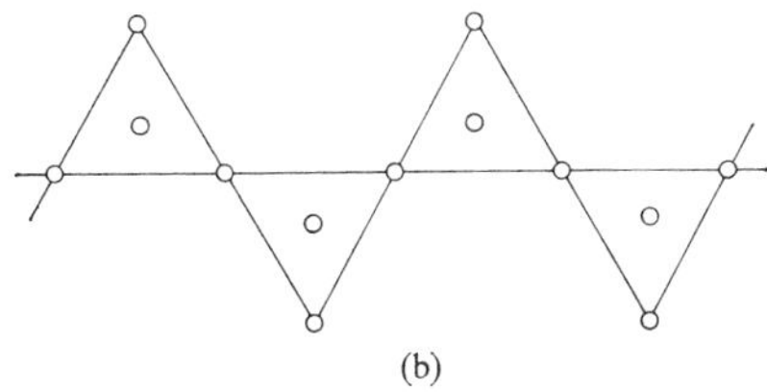

(b)

FIG. 22. (a) The repeating trimeric units of $NaAsO_3$. (b) The repeating dimeric units of $LiAsO_3$.

Synthetic arsenates and antimonates have been suggested as ion-exchange materials[154].

In a general way, the structural chemistry of the arsenates resembles that of phosphates and silicates. Some isomorphous replacement of arsenic by phosphorus is possible, but not all arsenates have the same structures as the corresponding phosphates. Condensed As–O frameworks are much less common than phosphate and silicate frameworks, and As–O–As bonds are apparently broken in aqueous solution[153]. Vibration frequencies assigned[156] to tetrahedral ions are as follows:

	ν_1	ν_2	ν_3	ν_4
AsO_4^{3-}	813	342	813	402
AsS_4^{3-}	386	171	419	216
SbS_4^{3-}	366	156	380	178

[155] M. Qureshi and V. Kumar, *J. Chem. Soc.* A (1970) 1488; M. Qureshi, R. Kumar and H. S. Rathore, *ibid.*, p. 272, 1986; M. Abe, *Bull. Chem. Soc. Japan* **42** (1969) 2683.

[156] H. Siebert, *Z. anorg. allgem. Chem.* **275** (1954) 225.

12.2. ANTIMONY(V) OXIDE AND ANTIMONATES

Sb^V oxide is obtained as a pale yellow powder by hydrolysis of antimony pentachloride with dilute ammonia, followed by dehydration at 275°C[5]. Care must be taken to avoid loss of oxygen, with reduction of some of the antimony to Sb^{III} (see section 16.1). Although Sb^V oxide is not very well characterized, thermodynamic data have been given (ΔH_f°, −974; ΔG_f°, −830 kJ mol^{-1}; S°, 125 J mol^{-1} K^{-1}). Several hydrates have been described, but it would be interesting to have comparative structural information on arsenic and antimony compounds.

Antimony is more often co-ordinated to six oxygen atoms than to four, and early X-ray work established the similarity between the sodium salts $NaSbF_6$ and $NaSb(OH)_6$. "Antimonic acid", obtained by passing the sparingly soluble alkali salts $MSb(OH)_6$ through an ion-exchange column in the acid form, gives a titration curve characteristic of a strong acid[157]. The pK_a is given[38] as 2.5. The solution probably contains $Sb(OH)_6^-$ ions, but there is evidence for condensed species in concentrated solutions at low pH [158].

TABLE 55. STRUCTURES OF ANTIMONY(V) OXIDES

$M^{n+}[Sb(OH)_6]_n^-$ [a]		J. Beintema, *Rec. trav. chim.* **56** (1937) 931; N. Scherewelius, *Z. anorg. allgem. Chem.* **238** (1938) 241
$LiSbO_3$		M. Edstrand and N. Ingri, *Acta chem. scand.* **8** (1954) 1021
Li_3SbO_4	NaCl superstructure	G. Blasse, *Z. anorg. allgem. Chem.* **326** (1963) 44
$NaSbO_3$, $KSbO_3$	Ilmenite	G. Blasse, *J. Inorg. Nucl. Chem.* **26** (1964) 1191
$M^{II}Sb_2O_6$ (M = Mg, Fe, Co, Ni, Zn)	Trirutile [b]	
$MnSb_2O_6$	Columbite	
$Zn_7Sb_2O_{12}$	Spinel	H. Saalfeld, *Acta Cryst.* **16** (1963) 836
$M^{III}SbO_4$ (M = Al, Cr, Fe, Rh, Ga)	Rutile (statistical)	K. Brandt, *Arkiv. Kemi Miner. Geol.* **17A** (1943), No. 15

[a] M^{n+} = $[Ni(H_2O)_6]^{2+}$, $[Mg(H_2O)_6]^{2+}$, $[Cu(H_2O)_3(NH_3)_3]^{2+}$, Na^+, Ag^+.
[b] A rutile superstructure in which the repeating unit corresponds to three unit cells of rutile. The columbite structure is related.

Solid antimonates have structures based on SbO_6 octahedra, with Sb–O distances 197–206 pm. These structures (Table 55) are commonly shown by ternary oxides, and so there is no reason why the antimony compounds should not be similarly described. In the rutile structure of $M^{III}SbO_4$ and in the spinel $Zn_7Sb_2O_{12}$, Sb^V and other metal ions are statistically distributed over octahedral holes in the lattice of oxide ions. In other structures, e.g. the ilmenite structure of $NaSbO_3$ or the trirutile structure of $FeSb_2O_6$, the antimony atoms are ordered in SbO_6 octahedra which may share corners or edges. In the compounds $LiSbO_3$ and $MnSb_2O_6$, there are strings of octahedra with shared edges (Fig. 23).

Oxidations involving antimonates in aqueous solution are discussed in section 16.3.

[157] D. V. S. Jain and A. K. Banerjee, *J. Inorg. Nucl. Chem.* **19** (1961) 177.
[158] G. Jander and H.-J. Ostmann, *Z. anorg. allgem. Chem.* **315** (1962) 241, 250.

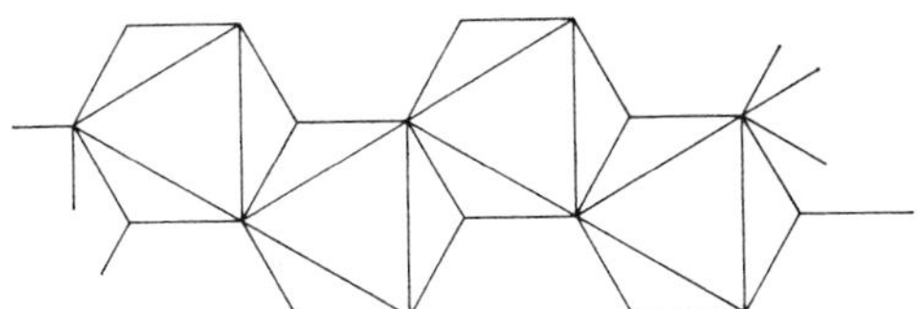

FIG. 23. SbO_6 octahedra with shared edges in $LiSbO_3$.

12.3. BISMUTH(V) OXIDE AND BISMUTHATES

Oxidizing agents such as chlorine or peroxydisulphide give brown or black precipitates with alkaline solutions of Bi^{III} oxide. These precipitates are taken to be ill-defined higher oxides. More definite brown compounds have been obtained by heating mixtures of alkali or alkaline earth, oxides with Bi^{III} oxide in oxygen, but structural work is still very sparse [159]. It has been claimed that sodium bismuthate has the same ilmenite structure as sodium antimonate $NaSbO_3$. Bismuthates are useful, strong, oxidizing agents, particularly in acid solution[160]; in phosphoric acid, insoluble and readily removed bismuth phosphate is formed. Glycols may be converted to compounds with carbonyl functions, and manganese (e.g. in the analysis of steel) may be quantitatively oxidized to permanganate.

12.4. ALKOXIDES

As^V alkoxides or arsenate esters have formulae $(RO)_3AsO$ and $(RO)_5As$. Sb^V alkoxides are of the second type, $(RO)_5Sb$, in keeping with the greater tendency of antimony to expand its co-ordination beyond four. The alkoxides may be made by methods similar to those used to obtain E^{III} alkoxides (antimonite esters, section 6.4), e.g. from halides and alkali alkoxides, by oxidation of compounds $(RO)_3E$, or by transesterification in a solvent such as benzene which forms azeotropes with low-boiling alcohols[60, 161].

$$SbCl_5 + NaOMe \rightarrow Sb(OMe)_5 \text{ (subl. 180°C/0.01 mm)}$$

$$2(MeO)_3As + Cl_2 \rightarrow (MeO)_3AsO \text{ (b.p. 70–75°C/3–5 mm), } MeOAsCl_2, MeCl$$

$$(EtO)_3Sb + Br_2 \rightarrow (EtO)_3SbBr_2 \xrightarrow{NaOEt} (EtO)_5Sb \text{ (b.p. 134°C/0.01 mm)}$$

$$(EtO)_5Sb + PhOH \rightarrow (PhO)_5Sb \text{ (m.p. 45°C)} + EtOH$$

$$2(MeO)_3AsO + NaOMe \rightarrow Na[O_2As(OMe)_2] + As(OMe)_5 \text{ (b.p. 70°C/11 mm)}$$

$$AgAsO_4 + Me_3SiCl \rightarrow (Me_3SiO)_3AsO \text{ (b.p. 76–77°C/1.2 mm)}$$

The alkoxides are moisture-sensitive, volatile substances. Because of their thermal instability they are most readily purified by crystallization, but methyl and ethyl derivatives may be distilled at reduced pressure.

There is evidence from cryoscopic measurements that the lower Sb^V alkoxides are associated in solution[161]. The low volatility of $Sb(OMe)_5$ may also indicate association in the solid. The compound $SbCl_4OEt$ is dimeric in the solid [31], with Sb–Cl 234 and

[159] R. Scholder, K-W. Ganter, H. Gläser and G. Merz, *Z. anorg. allgem. Chem.* **319** (1963) 375; B. Aurivillius, *Acta chem. scand.* **9** (1955) 1219.

[160] L. F. Fieser and M. Fieser, *Reagents for Organic Syntheses*, Wiley, New York (1967).

[161] A. Maillard, A. Deluzarche, J.-C. Marie and L. Havas, *Bull. Soc. chim. France* (1965) 2962.

Sb–O 209, 216 pm [162]. Reactions of As^{V} and Sb^{V} alkoxides have been extensively studied, especially by Kolditz and by Hass: much of this work is summarized in ref. 51c. In particular, reactions with fluorides yield derivatives

$$EF_n(OR)_m(OH)_l \ (l+m+n = 5) \quad \text{and} \quad M[EF_n(OR)_m(OH)_l] \ (l+m+n = 6),$$

and reactions with amines yield compounds with As–N frameworks. With phenyl magnesium bromide, the arsenic compound $(MeO)_3AsO$ yields tetraphenylarsonium bromide.

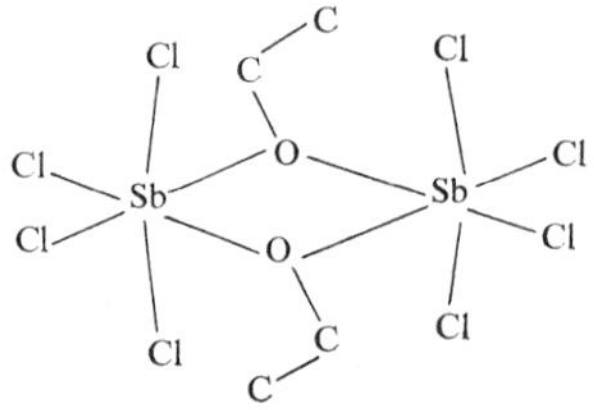

[31]

13. SULPHIDES E_2S_5

As^{V} sulphide As_2S_5 is not well characterized. It may be obtained[5] by passing hydrogen sulphide into an ice-cold solution of arsenate in concentrated hydrochloric acid and carefully dried. It easily decomposes, e.g. in hot water, to As^{III} sulphide and sulphur. Sb^{V} sulphide is said to be made similarly from more dilute hydrochloric acid: its solubility in the concentrated acid is attributed to the ready formation of $Sb^{V}Cl_6^-$ ions (section 15.1).

Both sulphides exchange sulphur rapidly with hydrogen sulphide in the presence of base by formation of anionic species[163]. These may be obtained in crystalline compounds by dissolving the pentasulphides in ammonium or alkali sulphides and cooling the resulting solutions. Depending on the precise reaction conditions, a variety of compounds with

TABLE 56. THIOARSENATES(V) AND THIOANTIMONATES(V)

		Ref.
Na_3AsS_4, $8H_2O$ [a]	In $(NH_4)_3AsS_4$, AsS_4 tetrahedra[b] with As–S 222 pm	c
Na_3AsO_3S, $12H_2O$ [a]	Infrared spectrum AsO_3S^{3-} has peaks at 791, 441, 390, 378, 312 cm^{-1}	
Li_3AsO_3S, $4H_2O$ •		d
$Na_3AsO_2S_2$, $11H_2O$		e
Na_3SbS_4, $9H_2O$	SbS_4 tetrahedra[b] with Sb–S 237 pm	f

[a] Preparation described ref. 5, p. 603.
[b] For infrared frequencies, see p. 655.
[c] H. Schaefer, G. Schaefer and A. Weiss, *Z. Naturforsch.* **18b** (1963) 665.
[d] K. Martin and E. Steger, *Z. anorg. allgem. Chem.* **345** (1966) 306.
[e] F. Remy and H. Guérin, *Bull. Soc. chim. France* (1968) 2327.
[f] A. Grund and A. Preisinger, *Acta Cryst.* **3** (1950) 363.

162 H. Preiss, *Z. anorg. allgem. Chem.* **362** (1968) 24.
163 J. R. Mickelsen, T. H. Norris and R. C. Smith, *Inorg. Chem.* **5** (1966) 911.

tetrahedral ions $[ES_nO_{4-n}]^{3-}$ may be isolated. Some of these are described in Table 56. Thioarsenates may also be formed by oxidation of arsenites with species containing S–S bonds. They are converted by oxidizing agents to arsenates and sulphates[164].

$$AsO_3^{3-} \xrightarrow{S_2O_3^{2-} \text{ or } S_xO_6^{2-}} AsO_3S^{3-} + SO_3^{2-}$$

$$AsO_3^{3-} \xrightarrow{S_x^{2-}} AsO_3S^{3-} + S^{2-}$$

$$AsO_3S^{3-} + 4I_2 + 10OH^- \longrightarrow AsO_4^{3-} + SO_4^{2-} + 8I^- + 5H_2O$$

The existence of Sb^V sulphide has been doubted[9b].

14. ORGANOMETALLIC COMPOUNDS

Attempts to make hydrides EH_5 of arsenic, antimony and bismuth have been unsuccessful, probably because the E–H bonds are insufficiently strong to compensate for the promotion of the Group V element to the E^V valence state (cf. Table 49). Organic derivatives ER_5 and halides $ER_{5-n}X_n$ are known[101, 102]. These and related compounds in which the Group V element is formally in oxidation state V are described in this section.

14.1. PENTA-ALKYLS AND -ARYLS R_5E

Pentaphenylantimony (m.p. 169°C) can be made from the reaction of phenyl-lithium with the chlorides Ph_3SbCl_2 or Ph_4SbCl, or, in lower yield, directly from phenyl-lithium and antimony pentachloride in ether. The arsenic and bismuth analogues Ph_5As and Ph_5Bi are obtained from the chlorides Ph_3ECl_2 and phenyl-lithium at −70°C. The ease of preparation and the thermal stabilities of the penta-alkyls and -aryls reflect the reluctance of arsenic and bismuth to reach oxidation state V. Thus penta-alkyl derivatives of arsenic and bismuth have not been described, but methyl-, ethyl-, vinyl- and propenyl-antimony compounds are known. Penta-aryls are also accessible from tosylimines, made from triaryl compounds and chloramine-T in dry alcohol or methyl cyanide.

$$Ph_3As \xrightarrow{MeC_6H_4SO_2NClNa} Ph_3As{:}N \cdot SO_2C_6H_4Me \xrightarrow{2PhLi} Ph_5As$$

The alkyl and aryl compounds of antimony and arsenic are usually colourless liquids or solids, and pentamethylantimony is appreciably volatile. Pentavinylantimony is described as a green oil. The molecules ER_5 are of interest as species involving five-co-ordination. Thus pentamethylantimony (Table 57) appears from its vibrational spectrum to be a well-behaved trigonal bipyramidal species: the correct selection rules for D_{3h} are observed. Crystalline pentaphenylarsenic is isomorphous with pentaphenylphosphorus, in which the molecules have a trigonal pyramidal configuration round phosphorus. In crystalline pentaphenylantimony[165], however, the molecules are best described in terms of a square pyramid. One bond (211.5+0.5 pm) is appreciably shorter than the other four (average 221.6±0.7 pm), but quite small distortions of the CSbC angles (96–106°) would make a description in terms of a distorted trigonal bipyramid more appropriate. Square pyramidal and trigonal bipyramidal configurations have very similar energies, and may interconvert

[164] M. Schmidt and R. R. Wägerle, *Chem. Ber.* **96** (1963) 3293; *Z. anorg. allgem. Chem.* **330** (1964) 48.

as a result of small changes in intermolecular forces; the precise shape of the molecules in the crystal is not necessarily maintained in solution. The fluxional characteristics of the molecules R_5E, allowing easy exchange of axial and apical groups (cf. AsF_5), are shown in the 60 MHz n.m.r. spectra of Me_5Sb and $(4\text{-}MeC_6H_4)_5E$ (E = As, Sb), which show only one line at $-60°$ [150,169]. It is not clear how the dipole moments of Ph_5As (1.32 D) and Ph_5Sb (1.59 D) in benzene are to be interpreted.

TABLE 57. PROPERTIES OF PENTAMETHYLANTIMONY[a]

M.p., $-19°C$; vapour pressure, 8 mm/25°C; May explode on distillation at 1 atm.[168]

Vibrational spectra[b]	ν_1 (a'_1)	ν_2 (a'_1)	ν_3 (a''_2)	ν_4 (a''_2)	ν_5 (e')	ν_6 (e')	ν_7 (e')	ν_8 (e'')
Raman (cm^{-1})	493	414			514	199	104	239
Infrared (cm^{-1})			456	213	516	195	108	

Ultraviolet spectrum: λ_{max}, 238 nm; log ε_{max}, 3.7; λ_{max}, 250 nm; log ε_{max}, 3

[a] A. J. Downs, R. Schmutzler and I. A. Steer, *Chem. Communs.* (1966) 221.
[b] Absorptions attributed to Sb–C skeleton; symmetry species in D_{3h}.

Reactions of pentaphenylantimony are shown in Scheme 15. The thermal decomposition to triphenylantimony and biphenyl appears to be an intramolecular process, but the photolysis in benzene gives complicated products from radical intermediates[166]. The formation of tetraphenylstibonium salts is common when large anions are available: contrast the behaviour of chlorine and iodine. Bromine at $-70°C$ gives the stibonium salt $[Ph_4Sb][Br_3]$, which decomposes at 20°C to Ph_3SbBr_2 and PhBr. The phenyl groups of ^3H-substituted phenyl-lithium exchange only slowly with those of pentaphenylarsenic[167], but the salts $LiEPh_6$ (E = Sb, Bi) are easily formed. Similar compounds may account for the easy intermolecular exchange of alkyl groups between R_5Sb molecules[168]. The penta-alkyl- but not the penta-aryl-antimony compounds are hydrolysed by water. Biphenyl derivatives have been studied in some detail[169, 170]; some reactions of the arsenic compounds are given in Scheme 15. The ions $As(bb')_2^+$ and $As(bb')_3^-$, with unsymmetrical biphenylyl groups (bb' = 4-methyl-2,2'-biphenylyl), have been resolved into optically active isomers, but not the unsymmetrical penta-aryl [32] with R = Ph, Y = Me, Z = H). An interesting feature of mass spectra of the binuclear compound [33] is the particularly intense peak due to the doubly charged positive ion; this probably rearranges to the structure [34] which has only paired electrons and favourable charge separation.

[165] A. L. Beauchamp, M. J. Bennett and F. A. Cotton, *J. Am. Chem. Soc.* **90** (1968) 6675.
[166] K. Shen, W. E. McEwen and A. P. Wolf, *J. Am. Chem. Soc.* **91** (1969) 1283.
[167] H. Daniel and J. Paetsch, *Chem. Ber.* **101** (1968) 1451.
[168] H. A. Meinema and J. G. Noltes, *J. Organomet. Chem.* **22** (1970) 653.
[169] G. Wittig and D. Hellwinkel, *Chem. Ber.* **97** (1964) 769; D. Hellwinkel and G. Kilthau, *Ann. Chem. Liebigs* **705** (1967) 66; *Chem. Ber.* **101** (1968) 121; D. Hellwinkel and C. Wünsche, *Chem. Communs.* (1969) 1412.
[170] D. Hellwinkel and M. Bach, *J. Organomet. Chem.* **17** (1969) 389, **20** (1969) 273.

SCHEME 15. Reactions of penta-aryl compounds.

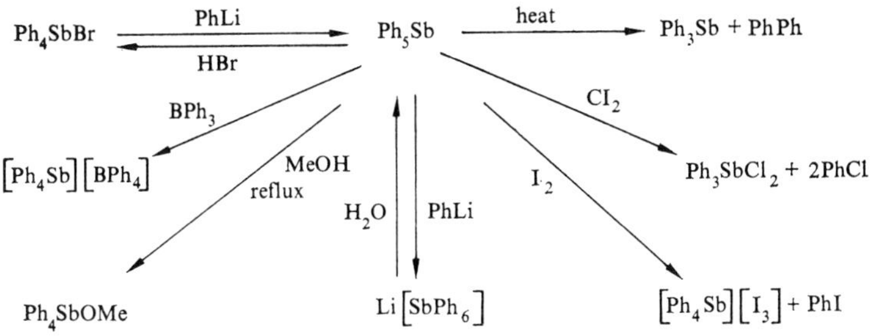

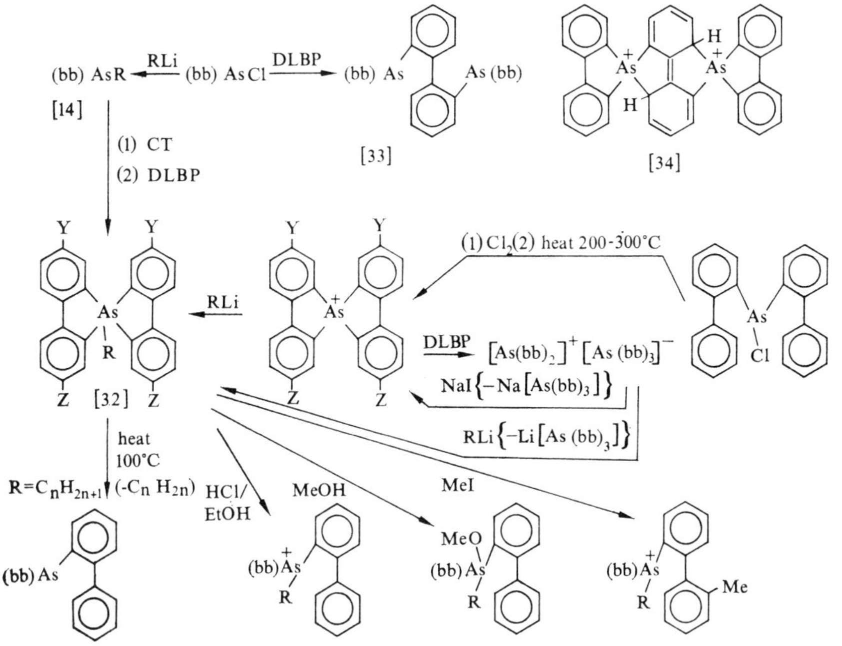

$Ph_4SbBr \underset{HBr}{\overset{PhLi}{\rightleftharpoons}} Ph_5Sb \xrightarrow{heat} Ph_3Sb + PhPh$

$Ph_5Sb \xrightarrow{BPh_3} [Ph_4Sb][BPh_4]$; $Ph_5Sb \xrightarrow{MeOH,\ reflux} Ph_4SbOMe$; $Ph_5Sb \underset{H_2O}{\overset{PhLi}{\rightleftharpoons}} Li[SbPh_6]$; $Ph_5Sb \xrightarrow{I_2} [Ph_4Sb][I_3] + PhI$; $Ph_5Sb \xrightarrow{Cl_2} Ph_3SbCl_2 + 2PhCl$

(bb) AsR ←RLi— (bb) AsCl —DLBP→ (bb) As–biphenyl–As (bb) [33]

[14] (bb) AsR —(1) CT, (2) DLBP→ [32]

[34]

[32] ←RLi— spiroarsonium cation ←(1) Cl_2 (2) heat 200–300°C— (biphenylyl)₂AsCl

spiroarsonium cation —DLBP→ $[As(bb)_2]^+[As(bb)_3]^-$; NaI{–Na[As(bb)₃]}; RLi{–Li[As(bb)₃]}

[32] —heat 100°C, R=C_nH_{2n+1} (–C_nH_{2n})→ (bb)As–biphenylyl

[32] —HCl/EtOH→ $(bb)\overset{+}{A}s(R)$–biphenylyl; —MeOH→ MeO(bb)As(R)–biphenylyl; —MeI→ $(bb)\overset{+}{A}s(R)$–biphenylyl–Me

CT, $MeC_6H_4SO_2NClNa$; DLBP, 2,2′-dilithiobiphenyl; (bb), 2,2′-biphenylyl, 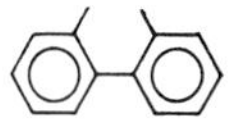 Y = Z = H

14.2. YLIDES[171]

The reactions between phenyl-lithium and arsonium salts $[R_3AsCHR_2]^+Br^-$ give a series of ylides comparable with the phosphorus analogues, and represented by the resonance structures

$$R^1_3As{=}CR^2_2 \leftrightarrow R^1_3\overset{+}{A}s{-}\overset{-}{C}R^2_2 \qquad [35]$$

Many are coloured. In some cases, sodium in liquid ammonia, or aqueous alkali, may be used in place of phenyl-lithium.

$$[Ph_3AsMe]Br \xrightarrow{PhLi} Ph_3\overset{+}{A}s{-}\overset{-}{C}H_2$$

$$[Ph_3AsCH_2COPh]Br \xrightarrow{Na/liq.\ NH_3} Ph_3\overset{+}{A}s{-}\overset{-}{C}H{\cdot}COPh$$

Other routes to ylides are as follows[171,172].

$$\mathrm{Ph_3AsCl_2 + H_2CR^2_2 \rightarrow Ph_3\overset{+}{As}{-}\overset{-}{C}R^2_2}$$

$$\mathrm{(Ph_4C_5^{-})\overset{+}{N}_2 + Ph_3E \longrightarrow Ph_3E{=}C_5Ph_4 \longleftrightarrow Ph_3E^{+}{-}(C_5Ph_4^{-}) + N_2} \quad [37]$$

$$\mathrm{(C_{13}H_8^{-})\overset{+}{As}Ph_3} \quad [36]$$

Trimethylmethylenearsorane $\mathrm{Me_3\overset{+}{As}-\overset{-}{C}H_2}$ (colourless, m.p. 33–35°C) is made indirectly from the trimethylsilyl compound[173].

$$\mathrm{Me_3As \xrightarrow{Me_3SiCH_2Cl} [Me_3AsCH_2SiMe_3]Cl \xrightarrow{BuLi} Me_3\overset{+}{As}{-}\overset{-}{C}HSiMe_3 \xrightarrow{Me_3SiOH} Me_3\overset{+}{As}{-}\overset{-}{C}H_2}$$

The thermal and chemical stability of the ylides depends on the ability of the groups R^1 and R^2 to accommodate the charges in the polar structure [35]. Thus in most crystalline ylides R^2 is unsaturated, e.g. CN, SO_2Ph, CO_2Me. The ylide [36] is unchanged at its m.p. (188–190°C) and heating under reflux with alcoholic alkali is required for its hydrolysis. Trimethylmethylenearsorane is decomposed above 60°C. Its n.m.r. spectrum at 60 MHz shows two types of protons (signals at τ 10.19 $\{CH_2\}$ and τ 9.19 $\{CH_3\}$), which exchange rapidly at room temperature in methanol.

$$\mathrm{Me_3\overset{+}{As}{-}\overset{-}{C}H_2 \underset{-H^+}{\overset{H^+}{\rightleftharpoons}} [Me_3AsCH_3]^+}$$

Reactions of the arsenic ylides are very similar to the phosphorus analogues; the arsenic compounds are the more basic.

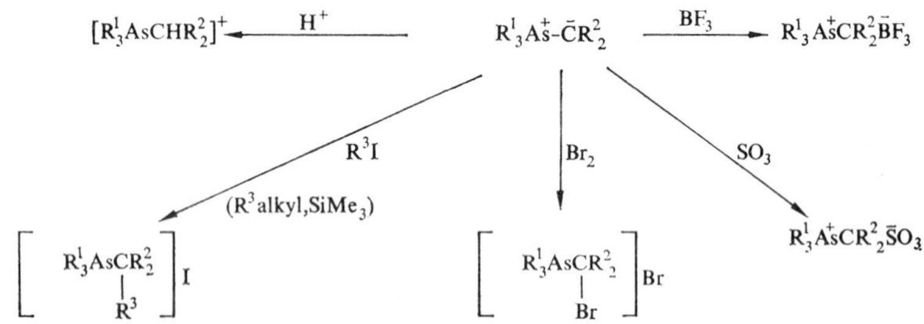

Arsenic ylides react with carbonyl compounds to give (a) normal Wittig products, or (b) arsines and products from rearrangement of oxirans.

$$\mathrm{R^1_3\overset{+}{As}{-}\overset{-}{C}R^2_2 + O{=}CR^3_2 \longrightarrow R^1_3\overset{+}{As}{-}CR^2_2{-}CR^3_2{-}\overset{-}{O} \longrightarrow}$$

$$\mathrm{(a)\ R^1_3AsO + R^2_2C{:}CR^3_2}$$

$$\mathrm{(b)\ R^1_3As + R^2_2C{-}CR^3_2\ (epoxide,\ {-}O{-}) \longrightarrow etc.}$$

[171] A. W. Johnson and H. Schubert, *J. Org. Chem.* **35** (1970) 2678 and references therein.
[172] D. Lloyd and M. I. C. Singer, *Chem. and Ind.* (*London*) (1967) 510, 787; *Chem. Communs.* (1967) 1042.
[173] N. E. Miller, *Inorg. Chem.* **4** (1965) 1458; H. Schmidbaur and W. Tronich, *ibid.* **7** (1968) 168.

The relative importance of these two reactions depends on the groups R^1 and R^2. For example, ethylidenefluorene can be obtained in 91 % yield from the ylide [36] and acetaldehyde.

Little is known about antimony and bismuth ylides. Several have been postulated as intermediates, and the compounds [37] E = Sb, Bi have been isolated. The blue bismuth compound is more like the unstable pyridinium tetraphenylcyclopentadienylide than like the colourless or yellow cyclopentadienylides of phosphorus, arsenic and antimony.

Imines[174]

The compounds $R^1_3\overset{+}{E}-\overset{-}{N}R^2$, isoelectronic with ylides, have been postulated as intermediates in the catalytic conversion of isocyanates to carbodiimides by triphenylarsine oxide. Several arsenic compounds have been isolated; their reactions parallel those of ylides, but they are less stable than phosphinimines.

$$Ph_3As \xrightarrow{ClNH_2} \left[Ph_3AsNH_2\right]Cl \xrightarrow[\text{liq. } NH_3]{NaNH_2} Ph_3\overset{+}{As}.\overset{-}{N}H$$

$$Ph_3AsO + PhCO.NCO \xrightarrow{-CO_2} Ph_3\overset{+}{As}.\overset{-}{N}.COPh \xleftarrow{NEt_3} Ph_3AsBr_2 + H_2NCOPh$$

$$Ph_3\overset{+}{As}.\overset{-}{N}.COPh \xrightarrow{200^\circ C} Ph_3AsO + PhCN$$

$$Ph_3\overset{+}{As}.\overset{-}{N}.COPh \xrightarrow{H^+ \text{ or } OH^-} Ph_3AsO + PhCONH_2$$

$$Ph_3\overset{+}{As}.\overset{-}{N}.COPh \xrightarrow{MeI} \left[Ph_3AsN{:}CPhOMe\right]I + \left[Ph_3As.NMe.COPh\right]I$$

$$Bu^t_2AsNHLi \xrightarrow{Me_3SiCl} \xrightarrow{BuLi} Bu^t_2AsNLiSiMe_3 \xrightarrow{MeCl} Bu^t_2Me\overset{+}{As}.\overset{-}{N}SiMe_3 \xrightarrow{MeI} \left[Bu^t_2MeAsNMeSiMe_3\right]I$$

14.3. HALIDES

The Compounds R_4EX

These may be made by two methods.

(1) The quaternization reaction between the compounds R_3E and alkyl halides may be rapid and quantitative even at low temperatures.

$$Me_3As + MeI \rightarrow [Me_4As]I$$

Quaternization becomes more difficult (a) in the series from arsines to bismuthines (trimethylbismuthine and methyl iodide do not react), and (b) as methyl groups are replaced by phenyl or trifluormethyl. In some cases, the onium salt $[Me_3O][BF_4]$ is a powerful reagent for quaternizations.

$$Ph_3Sb \xrightarrow{[Me_3O][BF_4]/\text{liq. } SO_2} [Ph_3SbMe][BF_4] \xrightarrow{NaI} [Ph_3SbMe]I$$

[174] O. J. Scherer and W. Janssen, *J. Organomet. Chem.* **16** (1969) P69; P. Frøyen, *Acta chem. scand.* **23** (1969) 2935.

(2) Grignard reagents react with compounds R_3EO or R_3ECl_2. Tetra-arylonium compounds can be made by this means.

$$Ph_3AsO + PhMgBr \rightarrow Ph_4AsOMgBr \xrightarrow{HCl} [Ph_4As][HCl_2]$$

$$Ph_3SbCl_2 + PhMgBr \rightarrow [Ph_4Sb]Br$$

Bromides and iodides are formulated as arsonium or stibonium salts. The halide ions may be replaced by a variety of anions, and some of the resulting onium compounds are soluble in water. The least soluble compounds are usually those with large anions (cf. the separation of $[Ph_4As][HCl_2]$ rather than $[Ph_4As]Cl$ above). Tetraphenylarsonium or tetraphenylstibonium derivatives are used for the isolation of complex anions, e.g. in analysis or for crystallographic studies. The As–C distances in Ph_4As^+ and Me_4As^+ have been given in the range 185–195 pm. In both cases, the carbon atoms are disposed very nearly at the corners of a tetrahedron, but the tetraphenylarsonium ion sometimes has no overall symmetry in solids because of twisting of the phenyl groups about As–C bonds. Arsonium and stibonium salts [abcdE]X contain asymmetric centres[104,105]. Optically active isomers of several halides such as benzylmethylphenylpropylarsonium bromide have been isolated, but they racemize easily, especially in non-polar solvents. Sulphates and perchlorates are more stable optically, and it may be that the ready racemization of the halides proceeds via fluxional five- co-ordinate or ionic species.

$$\text{abcdAsX} \rightleftharpoons [\text{abcdAs}]^+\ X^- \rightleftharpoons \tfrac{1}{2}\,[\text{abcdAs}]^+ \left[\text{(a,b,c,d)As}X_2\right]^-$$

Solvents such as alcohols, water and acids stabilize the arsonium halides and slow down racemization. Resolution of stibonium salts may also be prevented by fast racemization reactions; asymmetric tetrahedral tin compounds are likewise exceedingly difficult to resolve.

Tetraphenylstibonium fluoride is more soluble in carbon tetrachloride than in water, and can be quantitatively extracted from aqueous media. The compounds Me_4SbF, Me_4SbOH and Ph_4SbOH (see Table 61) have molecular structures [175].

The Compounds R_3EX_2

Of the remaining organometallic halides, the most easily obtained have formulae R_3EX_2. Tertiary arsines or stibines are oxidized by halogens or by sulphuryl chloride.

$$(CF_3)_3As + Cl_2 \longrightarrow (CF_3)_3AsCl_2$$

$$Ph_3E + SO_2Cl_2 \xrightarrow[\text{toluene}]{0°C} Ph_3ECl_2 \qquad (E = As, Sb)$$

Fluorides are usually made from chlorides.

$$(C_6H_5CH_2)_3ECl_2 + AgF \longrightarrow (C_6H_5CH_2)_3EF_2 \qquad (E = As, Sb)$$

[175] K. D. Moffett, J. R. Simmler and H. A. Potratz, *Anal. Chem.* **28** (1956) 1356; H. E. Affsprung and H. E. May, *ibid.* **32** (1960) 1164; H. Schmidbaur, J. Weidleim and K.-H. Mitsche, *Chem. Ber.* **102** (1969) 4136.

The most easily prepared compounds are those with the lighter Group V elements, which least readily lose alkyl halides.

$$Me_3SbBr_2 \xrightarrow{180^\circ C} MeBr + Me_2SbBr$$

$$Ph_3Bi + I_2 \xrightarrow[MeCN]{-35^\circ C} Ph_3BiI_2 \xrightarrow{20^\circ C} PhI + Ph_2BiI$$

$$(CF_3)_3SbBr_2 \xrightarrow{20^\circ C} CF_3Br + (CF_3)_2SbBr$$

In crystals, the dihalides show a trigonal bipyramidal configuration at the Group V elements (Table 58) and, as usual in five-co-ordinate species[150], the most electro-negative groups, here the halogens, occupy apical positions. In organic solvents, the compounds

TABLE 58. STRUCTURAL DATA ON DIHALIDES

	E–X (pm)	E–C (pm)		E–X (pm)	E–C (pm)
Me_3SbCl_2 [a]	249		Ph_3SbCl_2 [c]	245.8, 250.9±0.4	216±30
Me_3SbBr_2 [a]	263	213	Ph_3BiCl_2 [d]	257±20	212±8
Me_3SbI_2 [a]	288		$[Ph_3AsBr][IBr_2]$ [e]	226.3±0.5	191±1
$(ClCH{:}CH_2)_3SbCl_2$ [b]	245	215			

[a] A. F. Wells, *Z. Krist.* **99** (1938) 367.
[b] Yu. T. Struchkov and T. L. Khotsyanova, *Dokl. Akad. Nauk SSSR* **91** (1953) 565.
[c] T. N. Polynova and M. A. Porai-Koshits, *J. Struct. Chem.* **7** (1966) 691.
[d] D. M. Hawley and G. Ferguson, *J. Chem. Soc.* A (1968) 2539.
[e] R. S. McEwen and G. A. Sim, unpublished results (see ref. 177).

are monomeric. Mixtures of dihalides Me_3EX_2, Me_3EY_2 in benzene show only peak in one, the n.m.r. spectrum at room temperature, implying that the Me_3E moieties and halogens are exchanging rapidly. At lower temperatures, exchange is slowed and separate signals due to Me_3EX_2, Me_3EY_2 and Me_3EXY are detected[176]. The mixed halides cannot, however, be isolated by distillation.

The compounds Ph_3BiX_2 and Ph_3SbX_2 are not dissociated in methyl cyanide, but Ph_3AsCl_2 and Ph_3AsBr_2 are weak electrolytes, giving ions at low dilutions. The addition of halogens to the dihalides has been studied in detail: the large increase in conductivity after the addition of one equivalent has been attributed to the formation of polyhalides, e.g. $[Ph_3SbBr]I_3$. Some of these have been isolated[177].

The dihalides R_3EX_2 can often be recrystallized from concentrated aqueous solutions containing X^- ions, but they are hydrolysed in dilute solution. For example, when trimethylantimony dichloride is passed through an ion-exchange column, the hydroxide $Me_3Sb(OH)_2$ may be recovered from the extract and recrystallized from acetone. The

[176] G. C. Long, C. G. Moreland, G. O. Doak and M. Miller, *Inorg. Chem.* **5** (1966) 1358; C. G. Moreland, M. H. O'Brien, C. E. Douthit and G. C. Long, *ibid.* **7** (1968) 834.

[177] A. D. Beveridge and G. S. Harris, *J. Chem. Soc.* (1964) 6076; A. D. Beverage, G. S. Harris and F. Inglis, *J. Chem. Soc.* A (1966) 520.

halogens X may be replaced by a variety of other groups by reactions such as the following:

$$Me_3SbBr_2 + AgY \rightarrow Me_3SbY_2 \quad (Y = F, NO_3, \tfrac{1}{2}CO_3, \tfrac{1}{2}SO_4, \tfrac{1}{2}CrO_4, \tfrac{1}{2}C_2O_4)$$

$$Ph_3SbO + YH \rightarrow Ph_3SbY_2 \quad (Y = MeCO_2, HCO_2, NO_3)$$

Infrared spectra suggest that the anhydrous compounds Me_3SbY_2 are five-co-ordinate species. The species formed in aqueous solution have not been completely identified, but there is some evidence that a planar Me_3Sb skeleton is preserved, perhaps in ions such as $Me_3Sb(OH_2)_2^{2+}$ or $Me_3Sb(OH)(OH_2)^+$ with anions Y^-. Reactions of trimethylantimony dichloride are shown in Scheme 16. Tristifluoromethylantimony dichloride $(CF_3)_2SbCl_2$ reacts with water to give mono- and di-hydrates (e.g. [38]), which may be formulated as

SCHEME 16. Reactions of trimethylantimony dichloride[178, 179].

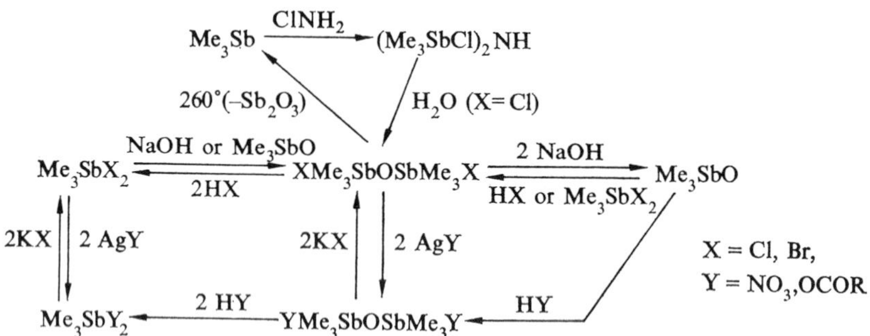

species with six-co-ordinate antimony. Derivatives of tristrifluoromethylantimonic acid (e.g. [39]) may be isolated as pyridinium salts. The contrast between the weakly acidic methyl and strongly acidic trifluoromethyl compounds is notable.

$$(CF_3)_3SbCl_2 \xrightarrow{2H_2O} \underset{[38]}{[H_3O][(CF_3)_3SbCl_2(OH)]} \xrightarrow[C_5H_5N]{\text{excess } H_2O}$$

$$\underset{[39]}{[C_5H_5NH][(CF_3)_3Sb(OH)_3]} \xrightarrow{HX} [C_5H_5NH][(CF_3)_3SbX_3] \quad (X = Cl, Br)$$

The Compounds R_2EX_3 and REX_4

The alkyl derivatives R_2EX_3 and REX_4 are much less stable than the dihalides described in the previous section. They are made by the reaction of halogens with the tervalent mono- or di-halides, but easily lose alkylhalides, often below room temperature.

$$Me_2SbCl + Cl_2 \rightarrow Me_2SbCl_3 \rightarrow MeSbCl_2 + MeCl$$

$$MeSbCl_2 + Cl_2 \rightarrow MeSbCl_4 \rightarrow SbCl_3 + MeCl$$

Aryl derivatives, e.g. Ph_2SbCl_3 and $PhSbCl_4$ (m.p. 60–65°C), are much more stable thermally. They may be made from the stibinic or stibonic acids and aqueous hydrochloric acid.

$$Ph_2SbO(OH) \xrightarrow{HCl} Ph_2SbCl_3$$

$$PhSbO(OH)_2 \xrightarrow{HCl} PhSbCl_4$$

---

[178] M. Shindo and R. Okawara, *J. Organomet. Chem.* **5** (1966) 537.
[179] R. L. McKenney and H. H. Sisler, *Inorg. Chem.* **6** (1967) 1178.

The trichloride isolated (m.p. 176°C) is, however, a hydrate, with the molecular structure [40] and dimensions as follows[180]: Sb–Cl_a, 233.5±0.6; Sb–Cl_b, 248.1±0.6; Sb–C, 215.8±2; Sb–O, 220.5±2 pm; ∠$OSbCl_b$, 85±0.6°; ∠OSbC, 81±2°. The environment of

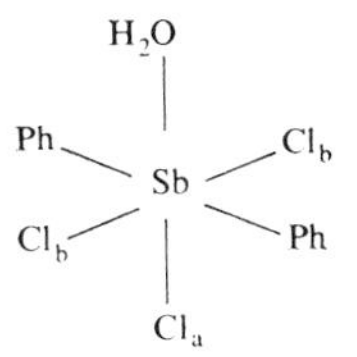

[40]

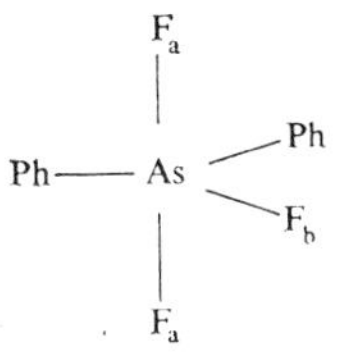

[41]

the antimony is very like that in the $SbCl_5$-adducts (section 15.3). The arsenic fluoride Me_2AsF_3 (m.p. 85°C) is associated, perhaps by fluorine bridges, and the phenyl derivative [41] is sufficiently stable dynamically to give two signals in the ^{19}F n.m.r. spectrum, a doublet from atoms F_a and a triplet from atoms F_b [150].

14.4. OXYGEN, SULPHUR, AND SELENIUM DERIVATIVES

The main groups of compounds are as follows:

R_3EY	$R_2E(Y)YH$	$RE(Y)(YH)_2$	$R_{5-n}E(YH)_n$
e.g. Me_3AsO, trimethylarsine oxide	e.g. $Me_2As(O)OH$, dimethylarsinic acid	e.g. $MeAsO(OH)_2$, methylarsonic acid	e.g. Ph_4SbOH, hydroxytetraphenylantimony(V)

(E = As, Sb, Bi; Y = O, S, Se)

In general, the oxygen compounds are the best known, though many sulphur and selenium derivatives have been made. The arsenic compounds are like the phosphorus analogues. The antimony compounds are formally similar, but, because of the tendency of antimony to form five or six bonds, the species in aqueous solutions of similar arsenic and antimony compounds may be quite different. Little is known about bismuth compounds.

Arsine and Stibine Oxides

The oxides are most commonly made by hydrolysis of the dihalides R_3EX_2, or by oxidation of the organometallic derivatives R_3E. Reaction with air is not usually easily controlled, and better oxidizing agents are mercuric oxide in alcohol, for the more easily oxidized compounds, or hydrogen peroxide—the most versatile reagent of all.

$$Me_3Sb + HgO \rightarrow Me_3SbO + Hg$$
$$Ph_3As + H_2O_2 \rightarrow Ph_3AsO + H_2O$$

Trialkyl sulphides and selenides can easily be made by refluxing arsines or stibines with sulphur or selenium.

$$Et_3As + S \rightarrow Et_3AsS$$

---

180 T. N. Polynova and M. A. Porai-Koshits, *J. Struct. Chem.* **8** (1967) 92.

Sulphides have also been made from oxides or dibromides.

$$Ph_3AsO \xrightarrow{CS_2} Ph_3AsS \xleftarrow{MeOH/NH_3/H_2S} Ph_3AsBr_2$$

The negative charge localized on the oxygen results in an interesting series of hydrogen-bonded compounds, as well as many complexes with Lewis acids (section 10.2) and transition metal compounds. Trialkyl-arsine and -stibine oxides are hygroscopic, and care is needed in the isolation of anhydrous compounds. Triphenylarsine oxide crystallizes as a monohydrate. The water, which is lost at about 70°C, is held by hydrogen bonds making centrosymmetric dimers [42]. As in the derivatives R_3E, the phenyl groups are twisted so that the arsine oxide has no symmetry in the crystal. "Short" hydrogen bonds have been found in triphenylarsine oxide–hydrogen chloride [43], in which there is no direct bond

Ph$_3$AsO------O
86·8° 93·2°
O------OAsPh$_3$

[42]

Ph$_3$AsO ---H—X

[43]

O=AsR$_2$CH$_2$ CH$_2$R$_2$As =O
H H
O O
O C O
H H
O=AsR$_2$CH$_2$ CH$_2$R$_2$As =O

[44]

between chlorine and arsenic. The n.m.r. absorption of the phenyl protons is at τ 2.14 in $CDCl_3$ (compare Ph_3AsO, τ 2.40), and the infrared spectrum shows strong absorption at 2750–2000 cm^{-1} but not at 888 cm^{-1} (ν As=O in Ph_3AsO, Table 59). The dipole moment is high (9.2 D). The ion $[Ph_3AsOH]^+$, which absorbs at 2800–3000 cm^{-1}, can be isolated in trihalides, which are strong electrolytes in methyl cyanide, but, from the reaction between triphenylarsine oxide, hydrogen bromide and mercuric bromide, the salt

$$[Ph_3AsOHOAsPh_3]_2[Hg_2Br_6]$$

is obtained. This also has a short hydrogen bond in the cation. Hydrogen bonds have been suggested in the adducts of triphenylarsine oxide with hydrogen peroxide, and nitric

TABLE 59. THE COMPOUNDS $R_3E(Y)$

	R_3AsO	R_3AsS	R_3AsSe	R_3SbO	R_3SbS	R_3SbSe
M.p. R = Et (°C) [a]	116	120	130	145 [g]	118	124
M.p. R = Ph (°C) [a]	194	164	125–30	222 [c]	112	
ν E=Y (cm^{-1}) [i]	880–90 [d]	472–85 [e]	336–60 [f]	650–680 [g, h]	421–440 [b,g]	270–300 [g]

[a] Ref. 102.
[b] M. Shindo, Y. Matsumura and R. Okawara, *J. Organomet. Chem.* **11** (1968) 299.
[c] G. H. Briles and W. E. McEwen, *Tetrahedron Letters* (1966) 5299.
[d] K. A. Jensen and P. H. Nielsen, *Acta chem. scand.* **17** (1963) 1875.
[e] R. A. Zingaro, R. E. McGlothin and R. M. Hedges, *Trans. Faraday Soc.* **59** (1963) 798.
[f] R. A. Zingaro and A. Merijanian, *Inorg. Chem.* **3** (1964) 580.
[g] G. N. Chremos and R. A. Zingaro, *J. Organomet. Chem.* **22** (1970) 637.
[h] Stibine oxides also show absorption at 450–480 cm^{-1}.
[i] Shifted to lower frequencies in complexes.

acid, and in the spirocyclic adduct $C(CH_2 \cdot Ph_2As{:}O)_4 \cdot 2H_2O_2$ [44]. The structural data (Table 60) shows that the As–C distances are like those in phenylarsonium salts and that longer As–O distances are associated with shorter hydrogen bonds.

TABLE 60. COMPLEXES OF TRIPHENYLARSINE OXIDE AND ARSINIC ACIDS

	As–O (pm)	O · · · H · · · X (pm)	As–C (pm)	∠CAsC (°)	∠CAsO (°)	Ref.
Ph_3AsO, H_2O [42]	164.4±0.7	281, 278	190.7±0.9	108	110.9	a
Ph_3AsO, HCl [43]	169.9±1.4	283.6±1.7	189.5			b
Ph_3AsO, HBr [43]	171.2±1.2	303.4±1.2	189.5			b
$[Ph_3AsOHOAsPh_3]^+$	168±2	240±3				c
$(Ph_3AsOHgCl_2)_2$	166±2		191±5	109 [e]	110	d
$(Ph_3AsO)_2HgCl_2$	169±3		192±4	108	111	d
$(Ph_3AsO)_2SbCl_3$	165		193	108.4	110.3	d
$Me_2AsO \cdot OH$	162±3	257	191±4	109.9	109.4	f
$Bu_2AsO \cdot OH$	167±1	247±2	195±2	109	110	g

[a] G. Ferguson and E. W. Macaulay *J. Chem. Soc.* A (1969) 1.
[b] G. Ferguson and E. W. Macaulay, *Chem. Communs.* (1968) 1288.
[c] G. S. Harris, F. Inglis, J. M. McKechnie, K. K. Cheung and G. Ferguson, *Chem. Communs.* (1967) 442.
[d] Ref. 132.
[e] Average; considerable distortion of Ph_3AsO.
[f] J. Trotter and T. Zobel, *J. Chem. Soc.* (1965) 4466.
[g] M. R. Smith, R. A. Zingaro and E. A. Meyers, *J. Organomet. Chem.* **20** (1969) 105.

Hydrates of trialkyl- or triaryl-antimony oxides are less well characterized. The compound $Me_3Sb(OH)_2$ (p. 671) is a non-electrolyte, but it is not known whether the crystals contain five co-ordinate species or hydrogen-bonded complexes.

The Acids $RE(O)(OH)_2$ and $R_2E(O)OH$

Arsonic and stibonic acids may be made by several methods.

(a) Primary alkyl halides may react with sodium arsenite[182].

$$EtBr + AsO_3^{3-} \rightarrow EtAsO_3^{2-} + Br^-$$

(b) Arylarsonic acids are best made from diazonium compounds[183]. There are several variations in procedure which are sometimes known by the following names:

Bart reaction $PhN_2^+Cl^- + As(ONa)_3 \rightarrow PhAsO_3Na_2 + NaCl + N_2$

Béchamp $HOPh + H_3AsO_4 \rightarrow 4\text{-}HOC_6H_4AsO_3H_2 + H_2O$

Rosenmund

$$2\text{-}BrC_6H_4CO_2H + As(OK)_3 \rightarrow 2\text{-}(K_2O_3As)C_6H_4CO_2K + KBr$$

Scheller $PhN_2^+HSO_4^- + AsCl_3 \xrightarrow[\text{(2) } H_2O]{\text{(1) ROH/trace CuCl}} PhAsO_3H_2 + N_2$

[181] G. V. Howell and R. L. Williams, *J. Chem. Soc.* A (1968) 117; G. C. Tranter, C. C. Addison and D. B. Sowerby, *J. Organomet. Chem.* **12** (1968) 369; J. Ellermann and D. Schirmacher, *Angew. Chem. Int. Edn. Engl.* **7** (1968) 738.

[182] C. K. Banks, J. F. Morgan, R. L. Clark, E. B. Hatlelid, F. H. Kahler, H. W. Paxton, E. J. Cragoe, R. J. Anders, B. Elpern, R. F. Coles, J. Lawhead and C. S. Hamilton, *J. Am. Chem. Soc.* **69** (1947) 927.

[183] G. O. Doak and H. G. Steinman, *J. Am. Chem. Soc.* **68** (1946) 1987, 1989, 1991; C. S. Hamilton and J. F. Morgan, *Org. Reactions*, **2** (1944) 415.

Hundreds of arylarsonic acids have been made[101, 102]. Stibonic acids made by the last reaction may be freed from inorganic antimony compounds by converting them to the pyridinium salts $[C_5H_5NH][RSbCl_5]$ which can be recrystallized from HCl/alcohol.

(c) Dihalides may be hydrolysed and oxidized with hydrogen peroxide.

$$CF_3AsI_2 \xrightarrow{H_2O_2} CF_3AsO(OH)_2$$

Arsinic acids and stibinic acids are made by analogous reactions.

$$R^1N_2Cl + R^2AsCl_2 \longrightarrow R^1R^2AsOOH$$

$$Me_2SbCl \xrightarrow{H_2O_2} Me_2SbOOH$$

Arsonic and arsinic acids are usually colourless crystalline solids, only slightly soluble in ether and non-polar solvents, but more soluble in alcohol and water. Like the alkyl carboxylic acids, the dialkylarsinic acids crystallize as centrosymmetric hydrogen-bonded dimers (Table 60). The acids are rather weak (e.g. $MeAsO(OH)_2$: pK_1, 3.61; pK_2, 8.25; $PhAsO(OH)_2$: pK_1, 3.47; pK_2, 8.48), and cacodylic acid $Me_2AsO(OH)$ is amphoteric (pK_a, 6.27; pK_b, 12.4 at 25°C)[38]. The trifluoromethyl compounds $CF_3AsO(OH)_2$ and $(CF_3)_2AsO(OH)$ are completely ionized in aqueous solution and are probably intermediate in strength between hydrochloric and nitric acids. They are thus similar to the trifluoromethyl derivatives of phosphorus and sulphur.

Arylstibonic acids are rather insoluble in both polar and non-polar solvents and molecular weight measurements show the presence of associated species. The nature of the polymerization is not known. It is probable that in alkaline solution the ions $[RSb(OH)_5]^-$ are formed.

Arsonic and stibonic acids are starting materials for the preparations of many organometallic compounds. Conversions to hydrides and halides have been mentioned. The As–C or Sb–C bonds are very stable. For example, it is possible to effect nitrations, diazotizations, Sandmeyer reactions and oxidations of methyl substituents in the aromatic rings of arylarsonic acids without removing the arsonic acid function. Cacodylic acid is unaffected by strong oxidizing agents, e.g. nitric acid or permanganate. Bistrifluoromethylarsinic acid, however, liberates trifluoromethane with strong alkali.

Methylarsonic acid and its ammonium, alkali, alkaline earth and titanium salts are used as herbicides. The chromium, iron and mercuric salts and other ammonium ferric alkylarsonates are used in horticultural and agricultural fungicides and bactericides[81]. Arylarsonic acids, especially derivatives of arsanilic acid $4\text{-}NH_2C_6H_4AsO_3H_2$, were important historically as early examples of compounds developed by planned research as chemotherapeutic agents. They were used in the treatment of syphilis and sleeping sickness. Cacodylic acid, which is more stable, less easily assimilated and less poisonous, was used in treatment of malaria. The acid $2,4\text{-}Cl_2C_6H_3AsO_3H_2$ is used as a wood preservative.

A number of azo-arsonic acids, e.g. arsenazo(III) [45], may be used for the preparation of transition metal complexes for spectrophotometric estimations. The Zr^{IV}, Th^{IV},

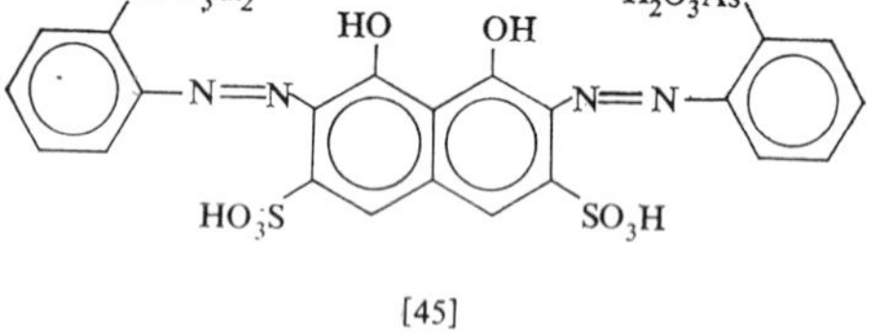

AsO$_3$H$_2$ HO OH H$_2$O$_3$As
N=N N=N
HO$_3$S SO$_3$H

[45]

UO_2^{II}, Sn^{IV} and Bi^{III} salts of methylarsonic acid are precipitated at pH 1. As the acidity is decreased, metal ions are precipitated in the following order[184]: Ti^{IV}, Sn^{II}, Hg^{I}; In^{III}, Al^{III}, Y^{III}; Ce^{III}, Cu^{II}, Hg^{II}; La^{III}, Mn^{II}, Fe^{II}, Co^{II}, Ag^{I}, Cd^{II}, Pb^{II}; Be^{II}, Zn^{II}, Ni^{II}; Fe^{III}; Mg^{II}. Dibutylarsinic acid has been suggested as a complexing agent for solvent extraction of metal ions.

Hydroxy and Alkoxy Derivatives of Antimony(V)

Tetraphenylstibonium bromide may be hydrolysed, first with acid and then with ammonia, to yield the hydroxy compound Ph_4SbOH, which can be recrystallized from aqueous methanol. This has a molecular structure in the solid, with the hydroxy group occupying an axial position in the trigonal bipyramid round each antimony (Table 61) (cf. ref. 150).

TABLE 61. HYDROXY- AND ALKOXY-DERIVATIVES OF ANTIMONY(V)

	M.p. (°C)	Sb–O	Sb–C	SbC	Ref.
Ph_4SbOH		204.8	221.8	213.1	a
Ph_4SbOMe	131	206.1	219.9 ± 1.4	212.0 ± 1	b
$Ph_3Sb(OMe)_2$	101	203 ± 0.8		212.0 ± 1	b

[a] A. L. Beauchamp, M. J. Bennett and F. A. Cotton, *J. Am. Chem. Soc.* **91** (1969) 297.

[b] K. W. Shen, W. E. McEwen, S. J. La Placa, W. C. Hamilton and A. P. Wolf, *J. Am. Chem. Soc.* **90** (1968) 1718.

The hygroscopic alkoxide Ph_4SbOMe, made from sodium methoxide and the bromide Ph_4SbBr, has a similar structure in the solid. In solution, however, axial and equatorial phenyl groups probably exchange rapidly, since the 60 MHz n.m.r. spectrum of the *p*-tolyl compound $(4\text{-}MeC_6H_4)_4SbOMe$ at −60° in $CDCl_3$–CCl_4 shows only one 4-Me peak.

The slow methanolysis of methoxytetraphenylantimony occurs by an ionic mechanism.

$$Ph_4SbOMe + OMe^- \rightarrow [Ph_4Sb(OMe)_2]^- \xrightarrow{\text{slow}} Ph_3Sb(OMe)_2 + Ph^-$$

This reaction contrasts with methanolyses of tetra-arylphosphonium salts, which yield ethers and triarylphosphine oxides, and shows the greater tendency of antimony to maintain co-ordination numbers greater than four[185]. The alkoxides $R_3Sb(OR)_2$ are easily obtained from the halides R_3SbX_2 and the alkoxides RONa. With chelating alkoxides, the organoantimony(V) halides yield compounds $R_nSbCl_{4-n}(YY)$ (YY = oxinate, acac). Most of these are monomeric in benzene, and probably contain six-co-ordinate antimony[186].

[184] R. Pietsch, *Mikrochim. Acta* (1960) 539; (1962) 1124.

[185] W. E. McEwen, G. H. Briles and B. E. Giddings, *J. Am. Chem. Soc.* **91** (1969) 7079.

[186] H. A. Meinema, E. Rivarola and J. G. Noltes, *J. Organomet. Chem.* **17** (1969) 71; **16** (1969) 257.

15. COMPLEXES OF ARSENIC(V), ANTIMONY(V) AND BISMUTH(V)

Hundreds of complexes of the pentahalides EX_5 have been described and much structural information is available. The complexes have important applications in preparative chemistry.

15.1. HEXAFLUORO-ARSENATES(V), -ANTIMONATES(V) AND -BISMUTHATES(V)

A full range of salts $M^{n+}[EF_6]_n$ can be made, either by heating together the fluorides MF_n and EF_5, or by treating a mixture of oxides with a powerful fluorinating agent, such as bromine trifluoride.

$$AgO + As_2O_3 + BrF_3 \longrightarrow AgAsF_6$$

$$BiF_3 \xrightarrow{BrF_3/F_2} BrF_2BiF_6 \xrightarrow{AgBrF_4} AgBiF_6$$

Hexafluorobismuthates(V) are less well known than the analogous arsenic and antimony compounds.

TABLE 62. STRUCTURAL CHEMISTRY OF HEXAFLUORO-ARSENATES(V) AND -ANTIMONATES(V), MEF_6

M	Li	Na	Ag	K	Tl	Rb	Cs
$MAsF_6$	R_1	R_1	C_2	R_2	R_2	R_2	R_2
$MSbF_6$	R_1	C_1	T	T	R_2	R_2	R_2

R_1, rhombohedral $LiSbF_6$ structure. C_1, face-centred-cubic $NaSbF_6$ structure. C_2, cubic $CsPF_6$ structure. R_2, rhombohedral $KOsF_6$ structure. T, tetragonal $KNbF_6$.

The crystal structures (Table 62) show almost symmetrical octahedral anions. With small cations, e.g. Li^+, Na^+, the structures are related to that of sodium chloride, with discrete EF_6^- octahedra in anion positions, so that each cation has six fluorine neighbours. With larger cations, e.g. Rb^+, Cs^+, the structures resemble that of caesium chloride, and each cation has twelve fluorine neighbours[187].

The ions EF_6^- have been characterized by infrared and, where suitable solvents are available, by n.m.r. spectroscopy, and the absorptions (Table 63) remain roughly consistent from compound to compound. In many solids, however, the site symmetries at the heavy atoms are less than O_h and this leads to a splitting of modes degenerate in O_h and to transitions forbidden in O_h. The ^{19}F n.m.r. spectra of solutions containing AsF_6^- often show four broad lines from coupling between the ^{75}As nucleus ($I = 3/2$, Table 2, p. 550) and fluorine. At low temperatures, the spectra may collapse to a single line because of increased ^{75}As quadrupolar relaxation[188].

[187] R. D. W. Kemmitt, D. R. Russell and D. W. A. Sharp, *J. Chem. Soc.* (1963) 4408.
[188] K. J. Packer and E. L. Muetterties, *Proc. Chem. Soc.* (1964) 147; S. Brownstein, *Can. J. Chem.* **47** (1969) 605.

TABLE 63. PROPERTIES OF COMPLEX HALIDE IONS

EX_6^-			AsF_6^-	SbF_6^-	$AsCl_6^-$ [e]	$SbCl_6^-$	$SbBr_6^-$
Bond lengths (pm)			177 [a, b]	188 [b, c]			255 [d]
Ultraviolet absorption λ_{max} (nm)					329 [e]	272 [f]	
Vibrational spectra (cm^{-1})	ν_1	a_{1g} [g]	685 [h]	668 [h]	337 [i]	329 [i]	
	ν_2	e_g	576	558	289	280	
	ν_3	t_{1u}	699	669	333	348	
	ν_4	t_{1u}	392	350	220	182	
	ν_5	t_{2g}	372	294	202	170	
^{19}F n.m.r. (ppm relative to CF_3COOH) [j]			−18.1	+32.3			
J(E–F) (Hz) (E = ^{75}As, ^{121}Sb)			930	1843			
^{35}Cl nuclear quadrupole coupling[k] constant (MHz)						47.6	

$E_2F_{11}^-$		$As_2F_{11}^-$ [46]	$Sb_2F_{11}^-$
^{19}F n.m.r. (ppm relative to $CFCl_3$)	Multiplet (a) [l]	21 [m]	89 [n]
	Doublet of doublets (b)	48	109
	Quintet (c)	85	131

$AsCl_4^+$

Vibrational spectra (cm^{-1}) [g, i]: ν_1 (a_1), 409; ν_2 (e), 149; ν_3 (t_2), 492; ν_4 (t_2), 186
^{35}Cl nuclear quadrupole coupling constant[k] 73.9 MHz

[a] In $KAsF_6$. J. A. Ibers, *Acta Cryst.* **9** (1956) 967.
[b] Similar E–F distances in [MeCO][SbF_6] and in pyridinium salts. R. F. Copeland, S. H. Conner and E. A. Meyers, *J. Phys. Chem.* **70** (1966) 1288.
[c] In $LiSbF_6$. J. H. Burns, *Acta Cryst.* **15** (1962) 1098.
[d] In $[C_6H_7N_4]_2[Sb^VBr_6][Br_3]$. The $SbBr_6^-$ ion is distorted to D_{4h} by a complicated charge transfer system involving cations and both sorts of anions (cf. section 16.2). S. L. Lawton and R. A. Jacobson, *Inorg. Chem.* **7** (1968) 2124.
[e] Ref. 192 and footnotes Table 52.
[f] H. M. Neumann, *J. Am. Chem. Soc.* **76** (1954) 2611.
[g] In O_h: a_{1g}, e_g, t_{2g}, Raman active; t_{1u}, i.r. active.
In T_d: a_1, e, t_2, Raman active; t_2, i.r. active.
[h] In solid $CsAsF_6$ and $LiSbF_6$. G. M. Begun and A. C. Rutenberg, *Inorg. Chem.* **6** (1967) 2212. Similar results on solutions in dimethylformamide have been obtained by J. A. Evans and D. A. Long, *J. Chem. Soc.* A (1968) 1688.
[i] $AsCl_6^-$ from solid Et_4NAsCl_6; $SbCl_6^-$ from $C_5H_5NH \cdot SbCl_6$ in nitromethane. $AsCl_4^+$ in $[AsCl_4][SbCl_6]$. I. R. Beattie, T. Gilson, K. Livingston, V. Fawcett and G. A. Ozin, *J. Chem. Soc.* A (1967) 712.
[j] E. L. Muetterties and W. D. Philips, *J. Am. Chem. Soc.* **81** (1959) 1084.
[k] At 79 K; average from several compounds; asymmetry assumed to be zero. J. V. DiLorenzo and R. F. Schneider, *Inorg. Chem.* **6** (1967) 766.
[l] See [46].
[m] R. A. W. Dean and R. J. Gillespie, *Chem. Communs.* (1969) 990.
[n] J. Bacon, P. A. W. Dean and R. J. Gillespie, *Can. J. Chem.* **47** (1969) 1655.

Ions $[E_nF_{5n+1}]^-$ may be formed in the presence of excess halide EF_5. The ion $[As_2F_{11}]^-$ [46] is characterized as its tetrabutylammonium salt in SO_2ClF solvent; $Sb_2F_{11}^-$ is known as the caesium salt, and in several solvents such as liquid sulphur dioxide. This is a good solvent for reactions of antimony pentafluoride, with which it forms a 1:1 complex (see Table 64).

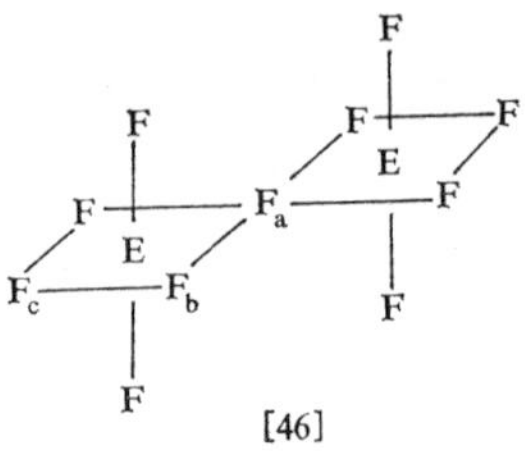

[46]

The $Sb_3F_{16}^-$ ion has been identified in the complex $[Br_2][Sb_3F_{16}]$: the Sb–F bond lengths are 183 (terminal) and 203 (bridge) pm, and the SbFSb angle 147°. The bridging fluorines are *trans* in the solid, but *cis* in solution.

Reactions of antimony pentafluoride in protonated solvents, and substituted hexafluoro-arsenates(V) and -antimonates(V) are referred to in section 15.5.

The reaction between Sb^V fluoride and sodium chloride yields compounds, e.g. Na[*cis*-SbF_2Cl_4], with both chlorine and fluorine in the anion. Similar anions

$$[(CF_3)_3AsF_3]^-, [(CF_3)_2AsF_4]^-$$

can be isolated as caesium salts from the reaction between trifluoromethylarsenic(V) fluorides and caesium fluoride[189].

15.2. ADDUCTS OF ARSENIC AND ANTIMONY PENTAFLUORIDES

The fluorides EF_5 form many adducts of low volatility with non-metal fluorides (A–F) [190] (Table 64). These usually involve transfer of fluoride, giving in the extreme, ionic structures.

$$\text{A–F} \xrightarrow{EF_5} [A]^+[EF_6]^- \xrightarrow{EF_5} [A]^+[Sb_nF_{5n+1}]^-$$

Evidence for ions has been taken from i.r. and n.m.r. spectra (Table 63). Full structure determinations, by X-ray crystallography, of solid adducts show that ionic formulations are nearly, but not completely, adequate. Some of the compounds examined show shorter A–F(Sb) and longer (A)F–Sb distances than would be expected for $[A][EF_6]$, and so a complete description of the solid must include a small component A–F–EF_5 (Fig. 24).

BrF_3, SbF_5	or	$[BrF_2][SbF_6]$	SbF_a, 184 (av.); SbF_b, 191 pm BrF_b, 229 BrF_c, 169
XeF_2, $2SbF_5$	or	$[XeF][Sb_2F_{11}]$	SbF_a, 183 (av.); SbF_b, 193; $SbF_{b'}$, 198, 205 XeF_b, 235 XeF_c, 184

The solids probably ionize in suitable solvents.

189 S. S. Chan and C. J. Willis, *Can. J. Chem.* **46** (1968) 1237; U. Müller, K. Dehnicke and K. S. Vorres, *J. Inorg. Nucl. Chem.* **30** (1968) 1719.

190 M. Webster, *Chem. Revs.* **66** (1966) 87.

TABLE 64. ADDUCTS OF ARSENIC AND ANTIMONY PENTAFLUORIDES

		Ref.
$RCOF, EF_5$	Prepared for wide range of R. Evidence for both $[RCO][EF_6]$ and $RFCO, EF_5$ in solution. Short C–C (137.8 pm) in solid $[MeCO][SbF_6]$. Ready decarbonylation when R secondary or tertiary	a
$R[SbF_6]$	From RF, SbF_5 or $RCOF, SbF_5$. Most stable when R tertiary, e.g. Bu^t. $MeF–SbF_5$ powerful methylating agent for Me_2E' ($E' = O, S$) → $Me_3E'^+$ and ArH → ArMe (Ar = aryl)	b
F F; R^1, R^3, R^2, R^4; F $[SbF_6]$	From (F F; R^1, R^3, R^2, R^4; F F) $+SbF_5$. With H_2O gives (O; R^1, R^3, R^2, R^4; F F) and (O; R^1, R^3, R^2, R^4; O)	c
Me_3SnEF_6	From Me_3SnBr, $AgEF_6$/liq. SO_2. F-bridged structures suggested	d
$[NF_4][EF_6]$	From NF_3, F_2, EF_5. Evidence from spectra for ionic structure. $N_2F_4, 2SbF_5$; $N_2F_2, 2SbF_5$ also described	e
$[A][EF_6]$	$A^+ = NO_2^+$, F_2NO^+, from AF,EF_5; $A^+ = NO_2{}^+$, NO^+ from oxide A, E_2O_3, BrF_3. Various complexes from reactions between nitrogen oxides, EF_5	f
$MeCN,AsF_5$	Doubt about species in MeCN. Possibly $[EF_4(MeCN)_2][EF_6]$ at low concentrations	g
Me_2PF_3, SbF_5; $EtPF_4, SbF_5$	$(NPF_2)_n, 2SbF_5$ (n = 3–6) also described	h
AsF_3,SbF_5; SbF_3,AsF_5	$[Sb_2F_{11}]^-$ detected by n.m.r. in solutions of SbF_5 in AsF_3.SbF_3 insoluble in SbF_5	i
$[VO_2][SbF_6]$	Adducts $[VO_2]\,[Sb_nF_{5n+1}]$ from VO_2F or VO_2Cl and SbF_5. $TiO_2[SbF_6]_2$ made from $TiOCl_2$ and SbF_5	j
$[O_2][FE_6]$	From O_2F_2, EF_5. From vibrational spectra and powder photographs, like $NOEF_6$	k
$[A][EF_6]_n$	Polymeric cations of S, Se, Te, by oxidation of elements with EF_5. $A = S_4^{2+}, S_8^{2+}, S_{16}^{2+}, Se_4^{2+}, Te_4^{2+}$. Similar species obtained in other highly acidic solvents	l
SO_2,SbF_5	Molecular structure in solids: Sb–O, 213; Sb–F, 185 pm	m
$SO_2, 2SbF_5$	In 1:2 adduct, oxygen *cis* to fluorine bridge	
$[E^1X_3F][EF_6]$	E^1 = S, Se, Te; SO.X = F, Cl. Crystalline adducts. Some evidence for ionic structures from vibrational spectra, and cryoscopic and conductimetric measurements	n, u
$[A][E_nF_{5n+1}]$	Polymeric halogen cations. $[Br_3][AsF_6]$ from O_2AsF_6, Br_2; $[Br_2][Sb_3F_{16}]$ from SbF_5, BrF_5, Br_2. Iodine and SbF_5 give blue $I(SbF_5)_2$ and brown $ISbF_5$. It has been suggested that these may contain ions I_2^+ and I_3^+	o
$[Me_2X][SbF_6]$(X = Cl, Br) $[Me_2X][Sb_2F_{11}]$	From MeX, SbF_5 in liquid SO_2. Raman and n.m.r. suggest bent cations	p
$[A][EF_6]$	From halogen fluorides AF and EF_5. AF = 2ClF (complex dissociated at 20°), ClF_3, ClF_5, BrF_3 [t], IF_7. Vibrational spectra interpreted in terms of ionic formulae	q, u
$[ClO_2][EF_6]$	From ClO_2F and EF_5.ClO_2 displaced by NO, NO_2	r
$[A][EF_6]$	From xenon fluorides AF and EF_5. AF = XeF_2, $2XeF_2$	s
$[XeF][Sb_2F_{11}]$ [t]	XeF_6, $2XeF_6$	

[a] G. A. Olah, S. J. Kuhn, W. S. Tolgyesi and E. B. Baker, *J. Am. Chem. Soc.* **84** (1962) 2733; G. A. Olah, W. S. Tolgyesi, S. J. Kuhn, M. E. Moffatt, I. J. Bastien and E. B. Baker, *ibid.* **85** (1963) 1328; F. P. Boer, *ibid.* **90**(1968) 6706; L. Lunazzi and S. Brownstein, *ibid.* **91**, 3034.

TABLE 64 (*cont.*)

[b] G. A. Olah, E. B. Baker, J. C. Evans, W. S. Tolgyesi, J. S. McIntyre and I. J. Bastien, *J. Am. Chem. Soc.* **86** (1964) 1360; G. A. Olah, J. R. DeMember and R. H. Schlosberg, *ibid.* **91** (1969) 2113; G. A. Olah and M. B. Comisarow, *ibid.*, p. 2955.

[c] V. D. Shteingarts, Yu. V. Pozdnyakovich and G. G. Yakobson *Chem. Communs.* (1969) 1264.

[d] H. C. Clark and R. J. O'Brien, *Proc. Chem. Soc.* (1963) 113.

[e] J. P. Guertin, K. O. Christie, A. E. Pavlath and W. Sawodny, *Inorg. Chem.* **5** (1966) 1921; **6** (1967) 533; W. E. Tolberg, R. T. Rewick, R. S. Stringham and M. E. Hill, *ibid.*, p. 1156; J. K. Ruff, *Inorg. Chem.* **5** (1966) 1791.

[f] See ref. 190. E. E. Aynsley, G. Hetherington and P. L. Robinson, *J. Chem. Soc.* (1954) 1119; A. A. Woolf and H. J. Emeléus, *ibid.* (1950) 1050; A. A. Woolf, *ibid.*, p. 1053. W. B Fox, C. A. Wamser, R. Eibeck, D. K. Huggins, J. S. MacKenzie and R. Juurik, *Inorg. Chem.* **8** (1969) 1247. R. D. Peacock and I. L. Wilson *J. Chem. Soc.* A (1969) 2030.

[g] L. Kolditz and I. Beierlein, *Z. Chem.* **7** (1967) 469.

[h] E. L. Muetterties and W. Mahler, *Inorg. Chem.* **4** (1965) 119; T. Chivers and N. L. Paddock, *J. Chem. Soc.* A (1969) 1687.

[i] Ref. 144.

[j] J. Weidlein and K. Dehnicke, *Z. anorg. allgem. Chem.* **348** (1966) 278.

[k] A. R. Young, T. Hirata and S. I. Morrow, *J. Am. Chem. Soc.* **86** (1964) 20; J. Shamir, J. Binenboym and H. H. Classen, *ibid.* **90** (1968) 6223.

[l] J. Barr, R. J. Gillespie and P. K. Ummat, *Chem. Communs.* (1970) 264; R. J. Gillespie and J. Passmore, *ibid.* (1969) 1333; J. Barr, D. B. Crump, R. J. Gillespie, R. Kapoor and P. K. Ummat, *Can. J. Chem.* **46** (1968) 3607; J. Barr, R. J. Gillespie, R. Kapoor and G. P. Pez, *J. Am. Chem. Soc.* **90** (1968) 6855.

[m] J. W. Moore, H. W. Baird and H. B. Miller, *J. Am. Chem. Soc.* **90** (1968) 1358; P. A. W. Dean and R. J. Gillespie, *ibid.* **91** (1969) 7260.

[n] See ref. 190. M. Brownstein, P. A. W. Dean and R. J. Gillespie, *Chem. Communs.* (1970) 9; M. Brownstein and R. J. Gillespie, *J. Am. Chem. Soc.* **92** (1970) 2718.

[o] O. Glemser and A. Šmalc, *Angew. Chem. Int. Edn. Engl.* **8** (1969) 517; A. J. Edwards, G. R. Jones, and R. J. C. Sills, *Chem. Communs.* (1968) 1527; R. D. Kemmitt, M. Murray, V. M. McRae, R. D. Peacock, M. C. R. Symons and T. A. O'Donnell, *J. Chem. Soc.* A (1968) 862.

[p] G. A. Olah and J. R. DeMember, *J. Am. Chem. Soc.* **92** (1970) 720, 2563; **91** (1969) 2113.

[q] I. Sheft, A. F. Martin and J. J. Katz, *J. Am. Chem. Soc.* **78** (1956) 1557; K. O. Christie and W. Sawodny, *Inorg. Chem.* **8** (1969) 212; **6** (1967) 313, 1783; K. O. Christie and D. Pilipovich, *ibid.* **8** (1969) 391; A. J. Edwards and G. R. Jones, *J. Chem. Soc.* A (1969) 1467.

[r] M. Schmeisser and W. Fink, *Angew. Chem.* **69** (1957) 780; K. O. Christie, C. J. Schack, D. Pilipovich and W. Sawodny, *Inorg. Chem.* **8** (1969) 2489.

[s] N. Bartlett and F. O. Sladky, *J. Am. Chem. Soc.* **90** (1968) 5316; F. O. Sladky, P. A. Bulliner, N. Bartlett, B. G. de Boer and A. Zalkin, *Chem. Communs.* (1968) 1048; V. M. McRae, R. D. Peacock and D. R. Russell, *ibid.* (1969) 62.

[t] See discussion of crystal structures in text.

[u] F. Seel and O. Detmar, *Z. anorg. allgem. Chem.* **301** (1959) 113.

Fluoride abstraction from non-metal fluorides is a very general method for the preparation of new cations, particularly where isoelectronic analogues of the cations are known. Examples in Table 64 are NF_4^+, RCO^+, Bu^{t+}, O_2^+, FCl_2^+, ClF_2^+, IF_6^+. Many of the adducts react with potassium fluoride.

$$SF_4, SbF_5 + KF \rightarrow KSbF_6 + SF_4$$

Such displacements do not necessarily imply ionic structures for the adducts. The volatility of the displaced ligand is an important factor in shifting the equilibrium to the right. Many similar reactions are known; some of these involve oxidation-reduction.

$$SF_4AsF_5 + SbF_5 \rightarrow SF_4, SbF_5 + AsF_5$$

$$2(C_5H_5)_2Fe + N_2F_4, 2SbF_5 \rightarrow 2[(C_2H_5)_2Fe][SbF_6] + N_2F_2$$

The arsenic and antimony pentafluoride adducts may be used in place of the non-metal fluorides in chemical reactions: e.g. the antimony pentafluoride adduct of bromine trifluoride is said to be a better fluorinating agent, as well as more easily handled, than the very corrosive bromine trifluoride. Thus even refractory oxides, e.g. ThO_2, Al_2O_3, CaO, are converted to fluorides. Oxygen is liberated quantitatively from organic compounds.

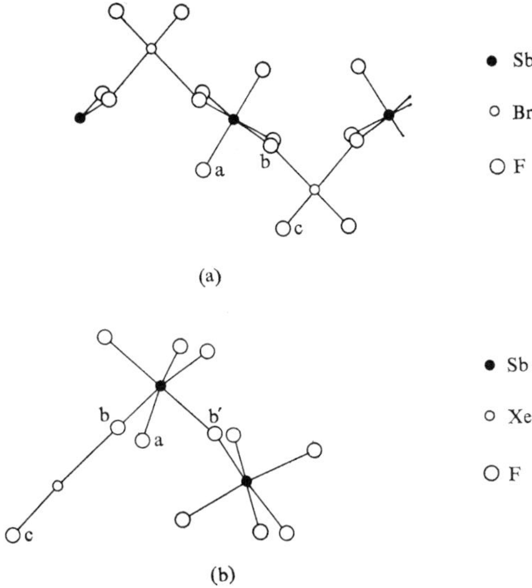

FIG. 24. The structures of (a) $[BrF_2]$, $[SbF_6]$, and (b) $[XeF]$, $[Sb_2F_{11}]$.

15.3. ADDUCTS OF ANTIMONY PENTACHLORIDE

The number of complexes between antimony pentachloride and electron donors is very large indeed[190]. The list in Table 65 is far from comprehensive, and concentrates on those adducts for which structural and thermodynamic information has been obtained. Many involve halogen transfer, and others, for which the structures are not known, may do so. Often the main interest is in the cations: e.g. the adduct written as $[Ph_2CN{:}CPh_2]^+[SbCl_6]^-$ has a cation isoelectronic with tetraphenylallene. As with fluorine analogues (section 15.2) simple valence structures are not always adequate. Thus the structure of the compound $ICl_3,SbCl_5$ is intermediate between the ionic $[ICl_2][SbCl_6]$ and the chlorine bridged $\cdots Cl–ICl_2–Cl–SbCl_4 \cdots$. Many adducts, however, have molecular structures: e.g. in $POCl_3,SbCl_5$, the donor atom is oxygen, and a slightly distorted octahedral $OSbCl_5$ arrangement results. Careful structural work, mainly by Lindqvist (Table 66), shows that, as the Sb–O bond length increases, the octahedron becomes more distorted as the chlorine atoms move towards the oxygen[132]. The Sb–Cl bond lengths (233–236 pm) are almost the same as in antimony pentachloride. A similar slightly distorted octahedral array is found in adducts with nitrogen donors. The sulphur–nitrogen ring is considerably flattened in the S_4N_4 adduct, compared with the free donor, but S_2N_2 is little distorted on co-ordination.

191 H. Kainer and K. H. Hausser, *Chem. Ber.* **86** (1953) 1563; G. E. Blomgren and J. Kommandeur, *J. Chem. Phys.* **35** (1961) 1636; E. A. C. Lucken, *J. Chem. Soc.* (1962) 4963; M. Kinoshita and H. Akamatu, *Bull. Chem. Soc. Japan* **35** (1962) 1040, 1137.

192 C. D. Schmulbach, *Inorg. Chem.* **4** (1965) 1232; I. Lindqvist and G. Olofsson, *Acta chem. scand.* **13** (1959) 1753.

TABLE 65. ADDUCTS OF ANTIMONY PENTACHLORIDE

	Ref.
Chlorine donors	
ICl_3	C. G. Vonk and E. W. Wiebenga, *Rec. trav. chim.* **78** (1959) 913; *Acta Cryst.* **12** (1959) 859
E^1Cl_4 (E^1 = S, Se, Te)	I. R. Beattie and H. Chudzynska, *J. Chem. Soc.* A (1967) 984
NOCl, NO_2Cl	F. Seel and Th. Gössl, *Z. anorg. Chem.* **263** (1950) 253; R. C. Paul, D. Singh and K. C. Malhotra, *J. Chem. Soc.* A (1969) 1396
R_3PCl_2 (R = Ph, NEt_2, Cl) Me_3ECl_2 (E=As, Sb)	A. Schmidt, *Chem. Ber.* **202** (1969) 380; **101** (1968) 4015; *Z. anorg. allgem. Chem.* **362** (1968) 129
ROCl	J.-M. de Carpentier and R. Weiss, *Chem. Communs.* (1968) 596
RCl	H. M. Nelson, *J. Phys. Chem.* **66** (1962) 1380; F. P. DeHaan, M. G. Gibby, and D. R. Aebersold, *J. Am. Chem. Soc.* **91** (1969) 4860
Ph_2C:NC·Ph_2Cl	B. Samuel and K. Wade, *J. Chem. Soc.* A (1969) 1742
Oxygen and sulphur donors	
R_3PO, $(RO)_3PO$, Cl_3PO R_2SO, R_2SO_2, Cl_2SO, Cl_2SO_2, R_2CO, $R_2CO_2R^1$, $(Me_2N)_2CO$, $RCONMe_2$, RCHO	Ref. 132. V. Gutmann, A. Steininger, and E. Wychera, *Monatsh. Chem.* **97** (1966) 460; G. Oloffson, *Acta chem. scand.* **21** (1967), 2143, 2415; **22** 377; R. C. Paul, H. R. Singal, and S. L. Chadha, *J. Chem. Soc.* A (1969) 1849
H_2O, ROH, R_2O, RSH, R_2S	E. Klages, H. Meuresch and W. Steppich, *Ann. Chem. Leibigs* **592** (1955) 81; F. Klages, A. Gleissner and R. Ruhnau, *Chem. Ber.* **92** (1959) 1834
Nitrogen donors	
C_5H_5N, R_3N	R. R. Holmes, W. P. Gallagher, and R. P. Carter, *Inorg. Chem.* **2** (1963) 437
RCN·RN_3	J. Goubeau, E. Allenstein and A. Schmidt, *Chem. Ber.* **97** (1964) 884
S_4N_4, S_2N_2	A. J. Banister and J. S. Padley, *J. Chem. Soc.* A (1969) 658; R. L. Patton and W. L. Jolly, *Inorg. Chem.* **8** (1969) 1389

TABLE 66. BOND LENGTHS AND ANGLES IN ANTIMONY PENTACHLORIDE ADDUCTS $DSbCl_5$

D [a]	Me_3PO	$HCONMe_2$	$SeOCl_2$	Me_2SO_2	Ph_2SO	$POCl_3$	S_4N_4 [b]	MeCN [c]	S_2N_2 [d]
Sb–O (pm)	194	205	208	212	216	217	217	223	228
∠Cl–Sb–O	89	87	86	86	86	85	89	84	83

[a] Ref. 132, p. 76.
[b] D. Neubauer and J. Weiss, *Z. anorg. allgem. Chem.* **303** (1960) 28.
[c] H. Binas, *Z. anorg. allgem. Chem.* **352** (1967) 271.
[d] R. L. Patton and K. N. Raymond, *Inorg. Chem.* **8** (1969) 2426.

Besides crystallographic evidence on the relative strengths of antimony pentachloride complexes, there is information from calorimetry and measurements of equilibrium constants. In 1,2-dichloroethane, the complexes $DSbCl_5$ become more dissociated, i.e. weaker, in the series:

(D =) $HCONMe_2$ > Me_2SO > Et_2POCl > Et_2O > H_2O > $PhPOCl_2$ > $PhPOF_2$ > MeCN
$POCl_3$ > $SOCl_2$ > PhCOCl > MeCOCl

The heats of dissociation decrease in the series:

(D =) $C_5H_5N > Me_2SO > HCONMe_2 > (MeO)_3PO > Ph_2POCl > Et_2O > Me_2CO > MeCN$
$SeOCl_2 > POCl_3 > SOCl_2 > SO_2Cl_2$
(D =) $Ph_2SeO > Ph_3AsO > Me_3PO > Me_2SO > Et_2S > Me_2CO > MeCOOEt > Et_2CO$
$Et_2O > Me_2SO_2 > Ph_2SO_2 > POCl_3 > SeOCl_2 > SOCl_2$

It seems that, towards pyridine, antimony pentachloride is about as good an acceptor as boron trichloride. The formation of complexes may be accompanied by oxidation-reduction. For example, thianthrene [47], some aromatic amines, or hydrocarbons such as perylene [22], react with antimony pentachloride in carbon tetrachloride to give coloured substances, which contain radical cations: little has been done to characterize the corresponding antimony-containing species[191] (cf. section 10.2).

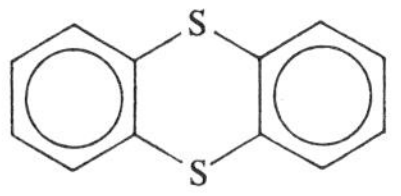

[47]

Many adducts between antimony pentachloride and alcohols, ethers, ketones, mercaptans and thioethers, have been made. These react with hydrogen chloride to give compounds formulated as onium salts. At higher temperatures the compounds $ROSbCl_4$ may be obtained from alcohol adducts.

$$R_2O,\ SbCl_5 \underset{-HCl}{\overset{HCl}{\rightleftharpoons}} [R_2OH][SbCl_6] \xrightarrow{CH_2N_2} [R_2OMe][SbCl_6]$$

$$[ROH_2][SbCl_6] \underset{HCl}{\overset{-HCl}{\rightleftharpoons}} ROH,\ SbCl_5 \underset{HCl}{\overset{-HCl}{\rightleftharpoons}} ROSbCl_4$$

Azide and cyanide adducts of antimony pentachloride, and adducts of the compounds R_3ECl_2 (E = P, As, Sb), show similar reactions.

$$MeN_3,\ SbCl_5 \xrightarrow[-N_2]{HCl} [H_2C{:}NH_2][SbCl_6]$$

$$HCN,\ SbCl_5 \xrightarrow{2HCl} [HClC{:}NH_2][SbCl_6] \xrightarrow{EtOH} [H(EtO)C{:}NH_2][SbCl_6]$$

$$Me_3PCl,\ SbCl_6 \xrightarrow{H_2O} Me_3PO\cdot SbCl_5 \xrightarrow[-40^\circ C]{HCl} [Me_3POH][SbCl_6]$$

$$\xrightarrow{MeOH} [Me_3POMe][SbCl_6]$$

$$\xrightarrow{NaN_3} [Me_3PN_3][SbCl_6] \xrightarrow{NaN_3} [Me_3PN_3][SbCl_5N_3]$$

It has been suggested that antimony pentachloride may form complexes by chloride donation, forming $SbCl_4^+$, but at present the evidence is not very clear. Presumably such complexes would require powerful chloride acceptors.

15.4. COMPLEXES OF ARSENIC PENTACHLORIDE

Although arsenic pentachloride has not yet been made, a number of its complexes have been isolated by the following reactions[192]:

$$AsCl_3 + Cl_2 + M'Cl_n \rightarrow [AsCl_4][M'Cl_{n+1}],\ \text{e.g. } M' = Al, Ga, Sb$$
$$AsCl_3 + Cl_2 + Q^+Cl^- \rightarrow Q^+AsCl_6^-,\ \text{e.g. } Q^+ = Et_4N^+$$
$$AsCl_3,\ xD + Cl_2 \rightarrow AsCl_5,\ yD + [x-y]D,\ \text{e.g. } D = R_3PO$$

The $AsCl_5$ can thus act as a donor, giving $AsCl_4^+$, or as an acceptor, giving $AsCl_6^-$. The $AsCl_4^+$ ion, which is isoelectronic with $GeCl_4$ or $GaCl_4^-$, is found in the mixed halide $AsCl_2F_3$ (section 11.3), or in adducts with strong halide acceptors, e.g. $AlCl_3$, $GaCl_3$, $SbCl_5$. It is presumed to be tetrahedral and some of its properties are given in Table 63. The $AsCl_6^-$ ion (Table 63) is formed in the presence of strong chloride donors. There is some evidence that the complex $AsCl_5PCl_5$, which is difficult to obtain pure, is $[PCl_4][AsCl_6]$. The structure reflects the weaker chloride acceptor properties of phosphorus compared with antimony pentachloride.

15.5. HALIDE COMPLEXES IN WATER AND OTHER PROTIC SOLVENTS

Hexafluoroarsenate(V) ions are hydrolysed only with difficulty in water. They may thus be quantitatively precipitated from aqueous solutions by nitron or tetraphenylarsonium cations. Colourless crystals of $HAsF_6 \cdot 6H_2O$ may be obtained by the evaporation of a solution made by passing hexafluoroarsenates down a cation-exchange column in the acid form. Hexa-fluoroantimonate(V) and -bismuthate(V)[193] ions are hydrolysed more readily.

The hydrates $HSbCl_6$, nH_2O ($n = 4\frac{1}{2}$, 2, 1) have been described, but $SbCl_6^-$ ions can be preserved only in solutions with high concentrations of HCl. At lower HCl concentrations, partially hydrolysed species $[Sb(OH)_nCl_{6-n}]^-$ predominate.

M (HCl)	12	8	6	5
n for main species	0	1	2	3

The hydrolysis of $SbCl_6^-$ in the pH range 2–12 is complicated. A base- and buffer-catalysed reaction probably involves S_N2 displacements in six-co-ordinate Sb^V species, and a pH independent reaction may be unimolecular[194].

A good deal of effort has been devoted to the characterization of species in anhydrous hydrogen fluoride and fluorosulphuric acid: $HF–SbF_5$ or $FSO_3H–SbF_5$ have been called "magic acids" or "superacids".

Hydrogen fluoride and antimony pentafluoride are miscible in all proportions. The electrical conductivity rises to a maximum at 3.5 mol % of SbF_5 and falls rapidly at higher SbF_5 concentrations. This behaviour and the details of the ^1H and ^{19}F n.m.r. spectra can be understood in terms of the reactions:

$$2HF + SbF_5 \rightleftharpoons H_2F^+ + SbF_6^-; \qquad SbF_6^- + nSbF_5 \rightleftharpoons [Sb_{n+1}F_{5n+6}]^-$$

Like other ions formed from solvent by proton transfer, the H_2F^+ and SbF_6^- ions have exceptionally high mobilities (H_2F^+ 0.035, and SbF_6^- 0.02 Ω^{-1} m^2 mol^{-1} at 0°)[195]. Sulphur trioxide and antimony pentafluoride react to give viscous adduct, formulated as polymeric SbF_4SO_3F. This is a strong acid in fluorosulphuric acid.

$$SbF_4SO_3F + 2HSO_3F \rightarrow H_2SO_3F^+ + [SbF_4(SO_3F)_2]^-$$

Various other species $\{SbF_n(SO_3F)_{5-n}\}$, which in fluorosulphuric acid give ions $[SbF_n(SO_3F)_{6-n}]$, have been characterized: their acidity increases as n increases from 1 to 3 [196].

[193] G. T. Burstein and G. A. Wright, *Nature* **221** (1969) 169.

[194] H. M. Neumann and R. W. Ramette, *J. Am. Chem. Soc.* **78** (1956) 1848; S. B. Willis and H. M. Neumann, *ibid.* **91** (1969) 2925.

[195] R. J. Gillespie and K. C. Moss, *J. Chem. Soc.* A (1966) 1170.

[196] R. J. Gillespie and R. A. Rothenbury, *Can. J. Chem.* **42** (1964) 416; R. C. Thompson, J. Barr, R. J. Gillespie, J. B. Milne and R. A. Rothenbury, *Inorg. Chem.* **4** (1965) 1641; R. J. Gillespie, K. Ouchi and G. P. Pez, *ibid.* **8** (1969) 63; P. W. A. Dean and R. J. Gillespie, *J. Am. Chem. Soc.* **92** (1970) 2362.

The acid solvents EF_5–HF and EF_5–HSO_3F dissolve metals with the liberation of hydrogen[197]. They have also been used for the protonation of weak organic bases such as nitriles, ketones and hydrocarbons[198] and for the preparation of radical ions for e.s.r. studies. An inorganic example is the paramagnetic Cl_2^+ made from chlorine fluoride. There has been some important work on the reactions of carbonium ions.

16. MIXED VALENCY COMPOUNDS AND MECHANISMS OF REDOX REACTIONS

Antimony sometimes appears as Sb^{III} and Sb^V in the same compound[199].

16.1. ANTIMONY TETROXIDE Sb_2O_4 AND RELATED COMPOUNDS

When antimony trioxide is heated in air at 700–1000°C, a compound, Sb_2O_4, is formed. This can crystallize in two forms (α and β); in each there are two quite distinct environments for antimony. Sb^V atoms have six oxygen neighbours at the corners of a slightly distorted octahedron and Sb^{III} atoms have the usual one-sided environment [48] (cf. structure [5], p. 610). The α and β modifications differ only in the packing of the Sb^VO_6 and $Sb^{III}O_4$ polyhedra. Antimony tetroxide is more compact and has a greater density than the trioxide (Table 67). It is also very stable and is unchanged in air at 900°C. The impure solid obtained by roasting stibnite in air has been used for colouring glass yellow.

TABLE 67. PROPERTIES OF ANTIMONY TETROXIDE Sb_2O_4

Thermodynamic properties[a]	ΔH_f°, −907; ΔG_f°, −795.8 kJ mol^{-1}; S°, 127; C_p°, 114 J K^{-1} mol^{-1}
Density (β-form)	6730 kg m^{-3}
Bond lengths and angles (β-form)[b] (see [48])	Sb^{III}–O_a, 203.2; Sb^{III}–O_b, 221.8 (±0.9) pm O_a–Sb–O_a, 87.9°; O_b–Sb–O_b, 148° (±0.4); Sb^V–O, mean 196.8±0.9 pm

[a] Ref. 11.
[b] D. Rogers and A. C. Skapski, *Proc. Chem. Soc.* (1964) 400; *Chem. Communs.* (1965) 611.

The oxides $Sb^{III}Nb^VO_4$, $Sb^{III}Ta^VO_4$, $Bi^{III}Nb^VO_4$, $Bi^{III}Ta^VO_4$ are isostructural with the α-form which also occurs as the mineral cervantite. The oxide $Bi^{III}Sb^VO_4$ is isostructural with β-Sb_2O_4. A compound $Sb^{III}Sb_2^VO_6OH$, and not very well-defined colourless oxides and sulphides of As^{III} and As^V, have also been described.

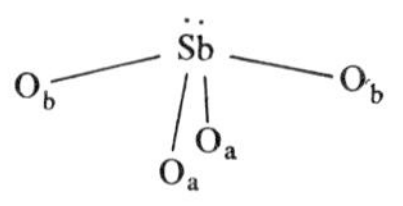

[48]

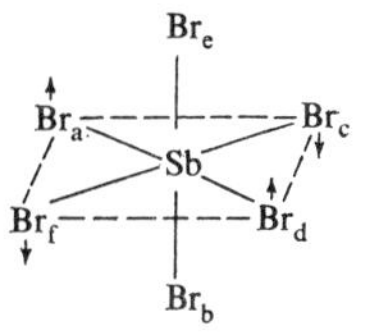

[49]

197 A. F. Clifford, H. C. Beachell and W. M. Jack, *J. Inorg. Nucl. Chem.* **5** (1957) 57.
198 G. M. Kramer, *J. Am. Chem. Soc.* **91** (1969) 4819; G. A. Olah, G. Klopman and R. H. Schlosberg, *ibid.*, p. 3261; G. A. Olah and P. J. Szilagy, *ibid.*, p. 2949; G. A. Olah and Mihai Calin, *ibid.* **90** (1968) 4763; G. A. Olah and M. B. Comisarow, *ibid.*, p. 5033; G. Olah and T. E. Kiovsky, *ibid.*, p. 4666.
199 M. B. Robin and P. Day, *Adv. Inorg. Chem. Radiochem.* **10** (1967) 247, and references therein.

16.2. COMPLEX HALIDES

Intensely coloured complex halides of empirical formulae M_2SbX_6 (e.g. M = Rb, Cs, R_4N) crystallize from solutions with appropriate amounts of Sb^{III}, Sb^{V} and MX in the 12 M acid HX. These compounds form mixed crystals with the corresponding tin compounds M_2SnX_6, and have only slightly distorted K_2PtCl_6 (antifluorite) structures in which M and EX_6^{n-} ions can be distinguished. Although the E^{III} and E^{V} atoms are in rather similar environments, they remain distinct in the ground state for the following reasons[199]:

(a) The compound are diamagnetic. Sb^{IV} (s^1) would be paramagnetic.

(b) Samples from radioactive $^{125}Sb^{III}$ do not yield equal amounts of $^{125}Sb^{III}$ and $^{125}Sb^{V}$ on dissolution as would be the case if all the antimony atoms were equivalent.

(c) The Sb^{III} atoms in M_2SbCl_6 may be replaced by Bi^{III}, In^{III} or Tl^{III} with very little change in structure.

(d) In the crystal of $(NH_4)_2SbBr_6$ there are two environments for antimony.

$$Sb^{III}Br_6^{3-}(O_h):\ Sb^{III}\text{–}Br,\ 279.5\pm0.6\ \text{pm}$$
$$Sb^{V}Br_6^{-}(D_{2d}):\ Sb^{V}\text{–}Br,\ 256.4\pm0.6\ \text{pm}$$

The environment of the Sb^{V} atoms is only slightly distorted from octahedral. Thus in structure [49] $\angle Br_aSbBr_b = 93.7$; $\angle Br_aSbBr_c = 90.2°$.

(e) Infrared and Raman spectra of Cs_2SbCl_6, $Cs_2(Sb,Bi)Cl_6$ show bands attributed to both $E^{III}Cl_6^{3-}$ (Table 46) and $E^{V}Cl_6^{-}$ (Table 63).

(f) For compounds $M_2Sb_xSn_{1-x}Cl_6$ (M = NH_4, Cs), $Cs_2In^{III}_{\frac{1}{2}-y}Sb^{III}_ySb^{V}_{\frac{1}{2}}Cl_6$, the intensity of the absorption in the visible spectrum is proportional to the concentration of $Sb^{III}Sb^{V}$ pairs, and a quantitative account of this intensity is given if it is assumed that delocalization of the 5*s* electrons of Sb^{III} to Sb^{V} is less than 1% in the ground state. The absorption is thus due to charge transfer.

$$Sb^{III}(5s^2)+Sb^{V}(5s^0) \xrightarrow{h\nu} Sb^{IV}(5s^1)+Sb^{IV}(5s^1)$$

The width of the absorption is accounted for by the changes in dimensions on excitation[200].

The mixed valency salts are semiconductors. The electronic properties of materials with more than 10% antimony, i.e. enough to form a continuous path through the lattice, are consistent with a mechanism in which $Sb^{IV}Cl_6^{2-}$ formation from Sb^{III}–Sb^{V} pairs is followed by migration of holes (migration of $Sb^{IV}Cl_6^{2-}$ through the $Sb^{III}Cl_6^{3-}$ ions).

16.3. MECHANISMS OF REDOX REACTIONS

There have been several kinetic studies of oxidations of solutions of E^{III} by various reagents (Table 68). Rate equations are sometimes rather complicated and, even for the same substrates, may differ for analogous arsenic, antimony and bismuth compounds. There may be several species of both E^{III} and oxidant in solution, and the concentrations of these may be pH dependent.

Oxidations by E^{V} have not been well studied. In the oxidation of iodide by Sb^{V}, two iodide ions are involved in the activated complex to give (by elimination of I_2) the change of 2 in oxidation state. A number of oxidations by Bi^{V} are, however, zero order in

[200] L. Atkinson and P. Day, *J. Chem. Soc.* A (1969) 2423, 2432.

reductant: the rate-determining step seems to involve oxidation of water by Bi^V possibly to give a chelated peroxide intermediate[202].

$$Bi^V\begin{matrix}\diagup OH\\ \diagdown OH\end{matrix} \; -2H^+ \longrightarrow Bi^{III}\begin{matrix}\diagup O\\ \diagdown O\end{matrix} \xrightarrow{2X^-} Bi^{III}\begin{matrix}\diagup OH\\ \diagdown OH\end{matrix} + X_2$$

TABLE 68. KINETIC STUDIES ON MECHANISMS OF REDOX REACTIONS

		Ref.
As^{III}, I_2, pH 8.2–9.2	$\dfrac{-dAs^{III}}{dt} = \{k[I_3^-][H_2AsO_3^-]+k^1[I_2]\}[H_2AsO_3^-]$	201
As^{III}, Br_2	Zero order in bromine	202
Sb^{III}, Br_2 or Cl_2	$\dfrac{-dSb^{III}}{dt} = k[X_2][H^+]^x[Sb^{III}]$ $x = 1$ for X = Cl; x fractional for X = Br	202
Sb^{III}, Ce^{IV} Sb^{III}, $Fe(CN)_6^{3-}$	Kinetics complicated by presence of various species of oxidant or reductant	203
Sb^V, I^-	Most important term in rate equation $\dfrac{dSb^V}{dt} = k[I^-]^2[H^+]^2[Sb^V]$	202, 204
Bi^V, X	Zero order in X for X = Cl^-, Br^-, I^-, Fe^{2+}, $IrCl_6^-$, SCN^-	205
As^{III}, As^V in HCl Sb^{III}, Sb^V in HCl	Kinetics show complicated dependence on HCl	206, 207
Sb^{III}, Sb^V in CCl_4	See p. 651	148

The exchange between Sb^{III} and Sb^V, i.e. oxidation of Sb^{III} by Sb^V, has been studied by spectrophotometry and radioactive tracers[58], in both water and carbon tetrachloride. In aqueous solution, the exchange is slow in alkali, or in acid solution containing anions (e.g. SO_4^{2-}) which are poor ligands for antimony, but faster in the presence of chloride. The kinetics are complicated by reactions involving several partially hydrolysed species of Sb^V, e.g. $[Sb(OH)_nCl_{6-n}]^-$. Activated complexes with two bridging chlorine atoms have been postulated to give the required two-electron change.

The author thanks Mrs. M. E. Smith and Mrs. J. Moncur for typing the manuscript and Dr. K. J. Alford and several other colleagues at the University of Sussex for helpful comments.

201 D. C. Johnson and S. Bruckenstein, *J. Am. Chem. Soc.* **90** (1968) 6592.
202 J. J. Habeeb and M. H. Ford-Smith, unpublished results.
203 S. K. Mishra and Y. K. Gupta, *J. Inorg. Nucl. Chem.* **30** (1968) 2991; *J. Chem. Soc.* A (1970) 260; L. Meites and R. H. Schlossel, *J. Phys. Chem.* **67** (1963) 2397.
204 A. Bahsoun and J. Lefebvre, *Bull. Soc. chim. France* (1970) 881.
205 M. H. Ford-Smith and J. J. Habeeb, *Chem. Communs.* (1969) 1445.
206 L. L. Anderson and M. Kahn, *J. Am. Chem. Soc.* **66** (1962) 886.
207 A. Turco, *Gazz. chim. ital.* **83** (1953) 231; C. H. Cheek, N. A. Bonner and A. C. Wahl, *J. Am. Chem. Soc.* **83** (1961) 80; N. A. Bonner and W. Goishi, *ibid.*, p. 85, and references therein; H. M. Neumann and H. Brown, *ibid.* **78** (1956) 1843.

22. OXYGEN

E. A. V. EBSWORTH
University of Edinburgh

J. A. CONNOR
The University of Manchester

and

J. J. TURNER
The University, Newcastle-upon-Tyne

1. OXYGEN

1.1. DISCOVERY OF OXYGEN[1]

By the middle of the seventeenth century it was appreciated that air contained a component associated with breathing and burning. In the first theory of burning to become widely accepted, this component was called phlogiston. When something burned, it was believed to release phlogiston to the air around it. If the burning took place in a sealed system, it stopped after a time because the air in the system became saturated with phlogiston. There were serious difficulties about this interpretation. For instance, metals such as tin gain in weight when they burn. However, the theory was widely accepted until the end of the eighteenth century.

Both the experimentalists generally credited with the discovery of oxygen, Joseph Priestly and Carl Wilhelm Scheele, were believers in the phlogiston theory. Indeed, Scheele called his experiments "... proofs that Heat or Warmth consists of Phlogiston and Fire Air". Scheele obtained oxygen, which he called fire air, by heating nitrates, mercuric oxide, or manganese dioxide in retorts to the ends of which bladders had been fixed; Priestly heated mercuric oxide with a magnifying glass and collected the gas over water. Both found that the gas they had obtained would support combustion better than does common air; and after some experiments with mice Priestly ventured to breath some of it himself, with very pleasant results. Priestly isolated what he called "dephlogisticated air" on 1 August 1774; after further experiments he wrote about his results to Sir John Pringle, the President of the Royal Society, in March 1775, and his letter was read before the Society on 23 March, while a detailed account of his experiments was published in the same year in the second volume of his book *Experiments and Observations on Different Kinds of Air*. Scheele worked at about the same time, but he had difficulties with his publishers (as others have

[1] M. E. Weeks, *The Discovery of the Elements*, 6th edn., published by the *Journal of Chemical Education*, New York (1956); D. McKie, *Antoine Lavoisier*, Gollancz, London (1935); J. G. Gillam, *The Crucible*, Robert Hale, London (1954); *The Collected Papers of C. W. Scheele* (translated by L. Dobbin), Bell, Edinburgh (1931).

had since then), and his book, *Chemische Abhandlung von der Luft und dem Feuer*, did not appear until 1777.

Neither Priestly nor Scheele, however, seems fully to have understood the significance of their discoveries. When Priestly was in Paris in October 1774, he mentioned some of his results to the distinguished French Academician, Antoine Lavoisier, who was interested in combustion. Lavoisier was also in touch with Scheele. He repeated and extended Priestly's experiments, and began to consider his results in the light of the deficiencies of the phlogiston theory. He burned tin in a sealed vessel, and showed that after combustion the weight of the vessel was effectively unchanged as long as the seal was not broken; thus the increase in weight of the tin on combustion could not be derived from outside the vessel. If the tin had lost phlogiston, then the phlogiston must have negative mass. From these and other experiments he concluded that when the metal burned what really happened was that it combined with something in the air; that "something" was the gas discovered by Priestly and Scheele, which Lavoisier called "principe oxygine". His view was not absolutely correct. He thought he was talking of a principle of acidity, for "oxygine" comes from the Greek word *oxus*, meaning sharp, and hence acid. But it was Lavoisier's penetration of mind, coupled with the experiments of himself, of Priestly, and of Scheele, that led to the collapse of the phlogiston theory and to the development of modern chemistry.

1.2. GENERAL

Oxygen is the eighth element in the Periodic Table. The electronic structure can be represented in terms of one-electron wave functions as $1s^2 2s^2 2p^4$; the ground state and some of the lower excited state terms are given in Table 1, with their energies above the ground state, and some other important properties of oxygen are given in Table 2. Detailed calculations of the wave function for the oxygen atom have been made using SCF and other methods[2].

Oxygen forms compounds with all the elements of the Periodic Table except for the lightest rare gases. These compounds could in principle be formed if oxygen were to lose electrons (forming cations), to gain electrons (forming anions) or to share electrons (forming bonds). The first four ionization potentials and the first two electron affinities are given in Table 2. These values show that the loss of electrons from oxygen is a process requiring much energy. Compounds are known in which oxygen is formally cationic, such as $O_2^+PtF_6^-$, but in all known cases the cations are polyatomic, and in general the chemistry of oxygen is not cationic. The doubly charged anion O^{-2} is a common species, even though its formation from the gaseous atom involves substantial absorption of energy; in ionic oxides the lattice energies are very high and more than compensate for the energy of formation of O^{-2}. This must also explain why the singly charged O^- is only known as an unstable species in irradiated solids and in the gas phase; in oxides it is unstable with respect to disproportionation to O_2 and O^{-2}. The coordination numbers of O^{-2} ions in oxides are set out in Table 2, and are considered in more detail in section 3.3.

Oxygen can form two additional bonds either by forming two σ-bonds with other atoms or groups, as in $(CH_3)_2O$, or by forming a σ-bond and a π-bond with the same other atom

[2] A. L. Merts and M. D. Torrey, *J. Chem. Phys.* **39** (1963) 694; C. C. J. Roothaan and P. S. Kelly, *Phys. Rev.* **131** (1963) 1177; E. Clementi, *J. Chem. Phys.* **40** (1964) 1944.

Table 1. Atomic Energy Levels of Oxygen

Electron configuration	State	J	Energy above ground state (cm^{-1})
$2s^22p^4$	$2p^4$ ^3P	2	0.0
		1	158.5
		0	226.5
	$2p^4$ ^1D	2	15 867.7
	$2p^4$ ^1S	0	33 792.4
$2s^22p^3(^4S^0)3s$	$3s$ $^5S^0$	2	73 767.81
$2s^22p^3(^4S^0)3s$	$3s$ $^3S^0$	1	76 794.69
$2s^22p^3(^4S^0)3p$	$3p$ ^5P	1	86 625.35
		2	86 627.37
		3	86 631.04
$2s^22p^3(^4S^0)3p$	$3p$ ^3P	2	88 630.84
		1	88 630.30
		0	88 631.00
$2s^22p^3(^4S^0)4s$	$4s$ $^5S^0$	2	95 476.43
$2s^22p^3(^4S^0)4s$	$4s$ $^3S^0$	1	96 225.5
$2s^22p^3(^4S^0)3d$	$3d$ $^5D^0$	4	97 420.24
		3, 2	97 420.37
		2, 1, 0	97 420.50
Ionization: $2s^22p^3$	$2p^3$ $^4S^0$	3/2	109 836.7

Data from C. E. Moore, *Atomic Energy Levels*, NBS Circular 467 (1949).

Table 2. Some Physical Properties of the Oxygen Atom

Ionization potentials [a] (eV)	1st, 13.614; 2nd, 35.146; 3rd, 54.934; 4th, 77.394
Electron affinities (eV)	1st, 1.478 ± 0.002 [b]; 2nd, ($O_g \rightarrow O_g^{-2}$), −7.8 ± 0.3
Atomic weight (C^{12} scale) [d]	15.9994
Atomic radius [e]	0.73 Å
Ionic radius [d]	1.39 ± 0.004 Å
Van der Waals radius [f]	~1.50 Å
Electronegativity [e]	3.46
Coordination numbers at oxygen: [g]	
(i) In ionic or near-ionic compounds	2 (e.g. SiO_2), 3 (e.g. rutile), 4 (e.g. ZnO), 6 (e.g. MgO), 8 (e.g. Na_2O)
(ii) In molecular compounds	1 (e.g. CO), 2 (e.g. H_2O), 3 (e.g. Me_2OBF_3), 4 (e.g. Be_4OAc)

[a] C. E. Moore, *Atomic Energy Levels*, NBS Circular 467, 1949.
[b] R. S. Berry, J. C. Mackie, R. L. Taylor and R. Lynch, *J. Chem. Phys.* **43** (1965) 3067.
[c] From thermochemical cycles: M. F. C. Ladd and W. H. Lee, *Acta Cryst.* **13** (1960) 959.
[d] A. E. Cameron and E. Wickers, *J. Am. Chem. Soc.* **84** (1962) 4175.
[e] R. T. Sanderson, *Chemical Periodicity*, Reinhold (1960).
[f] A. Bondi, *J. Phys. Chem.* **68** (1964) 441. The value is about the same for −O− and for =O.
[g] See section 3.

or group, as in $\rangle C{=}O$. Species in which oxygen forms just one σ-bond with another group (e.g. OH) have been detected spectroscopically, but they are free radicals and are not normally stable under chemical conditions. For once, all scales of electronegativity agree that oxygen is a very electronegative element. The electronegativity depends on the orbitals and the electron configuration; calculations have been made of the parameters associated with the valence state of oxygen, based on spectroscopic measurements[3]. As expected, the double bonds are shorter than the single bonds, and they have higher energies and stretching frequencies. Bonds of intermediate order are found in many compounds, including oxyanions such as RCO_2^-, CO_3^{-2}, NO_3^-, or SO_4^{-2}; relationships between bond length, bond order and stretching force constants have been described for BO [4], CO [5], NO [6], SiO [7], PO [7], SO [8] and ClO [7] bonds. Like fluorine, oxygen is a ligand which tends to promote oxidation of other elements to which it is bound (cf. $Os^{VIII}O_4$).

Bonds from 2-coordinated oxygen are usually considered as formed from (roughly) sp^3 hybrid orbitals, leading to bond angles at oxygen near the tetrahedral value. This leaves two lone pairs, which are also regarded as being in roughly sp^3-orbitals. Thus in each of the compounds H_2O, Me_2O and F_2O, the angles are near 109° (Table 3). However, oxygen can also use its lone pairs to form either intermolecular σ-bonds or intramolecular π-bonds additional to the normal σ-bonds.

Additional σ-bonds. Despite its high electronegativity, oxygen is a lone pair donor. Compounds like Me_2O form complexes with acceptors such as BF_3, and values for the energies of some donor–acceptor bonds involving oxygen are given in Table 4, with some values for other elements for comparison. Water is a well-known donor ligand in transition metal chemistry; oxygen is also a hydrogen bond acceptor, as in carboxylic acid dimers and in ice. All these interactions involve the lone pairs. In ice and in basic beryllium acetate, the angles at the 4-coordinated oxygen atoms are roughly tetrahedral, which also implies that the lone pairs are in approximately sp^3-orbitals; similarly, in H_3O^+, which is isoelectronic with NH_3, the angle is near 109°. In compounds, where it forms two σ-bonds (e.g. Me_2O), oxygen is a hard (class A) base, though in compounds like $Me_2C{=}O$ it has some "soft" character. Water comes near the "small Δ" end of the spectrochemical series; water and OH^- have small *trans*-effects, and in the nephelauxitic series water and OH^- come close to F^- as ligands with the smallest effect.

Internal π-bonding. If an attached atom or group Q has empty orbitals of π-symmetry relative to the Q–O σ-bond, these will overlap the lone pair orbitals at oxygen, and the overlap may lead to the formation of a donor π-bond. Thus in carbon monoxide the CO bond is of higher order than 2, and the strength of the bond can be attributed at least partly to an interaction of the form $(O^+{\equiv}C^-)$. In compounds in which oxygen is forming two or more σ-bonds, donor π-bonding like this will affect bond angles at oxygen. The donor π-overlap will be greatest when the lone pairs are in pure *p*-orbitals. If one lone pair is in a pure *p*-orbital, the two σ-bonds and the other lone pair must be built from one *s*- and two *p*-orbitals. This would lead (if all three σ-orbitals are equivalent) to an angle of 120°. If both lone pairs are in pure *p*-orbitals, the σ-bonds must be built from *sp*-orbitals,

[3] G. Pilcher and H. A. Skinner, *J. Inorg. Nucl. Chem.* **24** (1962) 937.
[4] J. Krogh-Moe, *Acta Chem. Scand.* **17** (1963) 843.
[5] J. P. Fackler and D. Coucouvanis, *Inorg. Chem.* **7** (1968) 181.
[6] Yu. Ya. Kharitonov, *Izv. Akad. Nauk SSSR, Otdel Khim. Nauk* 1962, 1953.
[7] E. A. Robinson, *Can. J. Chem.* **41** (1963) 3021.
[8] P. Haake, W. B. Miller and D. A. Tyssee, *J. Am. Chem. Soc.* **86** (1964) 3577.

TABLE 3. ANGLES AT OXYGEN

	Angle	How measured	Phase	Reference
In species QOZ				
1. Neither Z nor Q π-acceptors:				
H_2O	104.52°	vib.	vap.	a
F_2O	103.1 ± 0.05°	μ wave	vap.	b
$(CH_3)_2O$	111.5 ± 1.5°	ED	vap.	c
CH_3OH	109° ± 3°	ED	vap.	c
RbOH	180°	μ wave	vap.	d
2. Q, π-acceptor; Z, not:				
SiH_3OCH_3	120.6 ± 0.9°	ED	vap.	e
3. Q and Z both π-acceptors:				
Cl_2O	110.8 ± 1°	ED	vap.	f
Cl_2O_7	118.6 ± 0.7°	ED	vap.	g
$SiH_3OC_6H_5$	121 ± 1°	ED	vap.	h
$(SiH_3)_2O$	144.1 ± 0.9°	ED	vap.	i
SiOSi in silicates	140–180°	X-ray	solid	j
$(GeH_3)_2O$	126.5 ± 0.3°	ED	vap.	k
$[O_3POPO_3]^{-4}$	133.5°	X-ray	solid	l
$[O_3SOSO_3]^{-2}$	124°	X-ray	solid	m
$[O_3CrOCrO_3]^{-2}$	115°	X-ray	solid	n
$[Cl_5MOMCl_5]^{-4}$ (M = Ru, Re)	180°	X-ray	solid	o
4. Q_3O^+:				
H_3O^+	112°	X-ray	solid	p
$(ClHg)_3O^+$	120°	X-ray	solid	q

[a] W. S. Benedict, N. Gailar and E. K. Plyler, *J. Chem. Phys.* **24** (1956) 1139.
[b] Y. Morino and S. Saito, *J. Mol. Spectrosc.* **19** (1966) 435.
[c] K. Kimura and M. Kubo, *J. Chem. Phys.* **30** (1959) 151.
[d] C. Matsuma and D. R. Lide, *J. Chem. Phys.* **50** (1969) 71.
[e] C. Glidewell, D. W. H. Rankin, A. G. Robiette and G. M. Sheldrick (to be published).
[f] J. D. Dunitz and K. Hedberg, *J. Am. Chem. Soc.* **72** (1950) 3108.
[g] B. Beagley, *Trans. Faraday Soc.* **61** (1965) 1821.
[h] C. Glidewell, D. W. H. Rankin, A. G. Robiette, G. M. Sheldrick, B. Beagley and J. M. Freeman, *Trans. Faraday Soc.* **65** (1969) 2621.
[i] A. Almenningen, O. Bastiansen, V. Ewing, K. Hedberg and M. Traetteberg, *Acta Chem. Scand.* **17** (1963) 2455.
[j] D. W. J. Cruickshank, *J. Chem. Soc.* 1961, 5486; D. W. J. Cruickshank, H. Lynton and G. A. Barclay, *Acta Cryst.* **15** (1962) 493.
[k] C. Glidewell, D. W. H. Rankin, A. G. Robiette, G. M. Sheldrick, B. Beagley and S. Cradock, *J. Chem. Soc.* (A), 1970, 315.
[l] D. M. Macarthur and C. A. Beevers, *Acta Cryst.* **10** (1957) 428.
[m] H. Lynton and M. R. Truter, *J. Chem. Soc.* 1960, 5112.
[n] C. A. Brystrom and K. A. Wilhelmi, *Acta Chem. Scand.* **5** (1951) 1003.
[o] A. M. Mathieson, D. P. Mellor and N. C. Stephenson, *Acta Cryst.* **5** (1952) 185; J. C. Morrow, *Acta Cryst.* **15** (1962) 851.
[p] C. E. Nordman, *Acta Cryst.* **15** (1962) 18.
[q] S. Šćavničar and D. Grdenić, *Acta Cryst.* **8** (1955) 275.

TABLE 4. GAS-PHASE DISSOCIATION ENTHALPIES FOR SOME MOLECULAR COMPLEXES (kcal mol^{-1})

Complex	ΔH	Complex	ΔH
Me_2OBF_3	13.3	Me_2SGaMe_3	~8
Et_2OBF_3	10.9	Me_2SBH_3	5.2
$THF \cdot BF_3$	13.4	Me_3NBF_3	b
Me_2OBMe_3	a	Me_3NAlMe_3	b
Me_2OAlMe_3	b	Me_3NGaMe_3	21
Me_2OGaMe_3	9.5	Me_3PBF_3	18.9
Me_2OBH_3	a		

Data from F. G. A. Stone, *Chem. Rev.* **58** (1958) 101.
[a] Too unstable to study.
[b] Too stable to determine.
[c] For dissociation into Me_2S and B_2H_6.

and the angle at oxygen will be 180°. These possibilities, which are represented below in valence-bond terms, are extremes: intermediate angles might be expected, deriving from a balance between π-bonding (widening the angle, and removing electrons from oxygen) and charge distribution (which will tend to keep electrons on the oxygen atom):

$^{-}Q \Leftarrow \overset{+}{O} - Z$ (bent)　　　$^{-}Q \Leftarrow \overset{+2}{O} \Rightarrow Z^{-}$

Internal π-bonding will lead to shorter and stronger bonds, and should weaken the donor properties of the oxygen atom; the shortness and strength of bonds between oxygen and boron[9], silicon[10] or transition elements[11] have been accounted for in terms of this type of interaction. Several compounds are known in which oxygen is bound to one or two π-acceptors and in which the angle at oxygen is unusually wide (see Table 3). A similar argument could be used to explain why angles in Q_3O^+ might be nearer 120° than 109°; π-interactions between silicon and Q would be greatest if the lone pair at Q were in a pure *p*-orbital. However, this argument should be used with some caution; the angle at oxygen in RbOH is 180°, yet there is unlikely to be significant π-bonding between rubidium and oxygen.

Oxygen is known in a variety of formal oxidation states, from +2 to −2. Of these, the positive states are (by definition) only found when oxygen is bound to a more electronegative element—which must be fluorine—or forms part of a cation such as O_2^+. In other oxidation states greater than −2 the oxygen atom concerned must either be bound to at least one other oxygen atom or form part of a cation or free radical. A potential diagram for the redox chemistry of oxygen in aqueous solution is given in Table 5.

[9] C. A. Coulson and T. W. Dingle, *Acta Cryst.* **B24** (1968) 153.
[10] D. W. J. Cruickshank, *J. Chem. Soc.* 1961, 5486.
[11] F. A. Cotton and R. M. Wing, *Inorg. Chem.* 4 (1965) 867.

TABLE 5. REDUCTION POTENTIALS OF OXYGEN

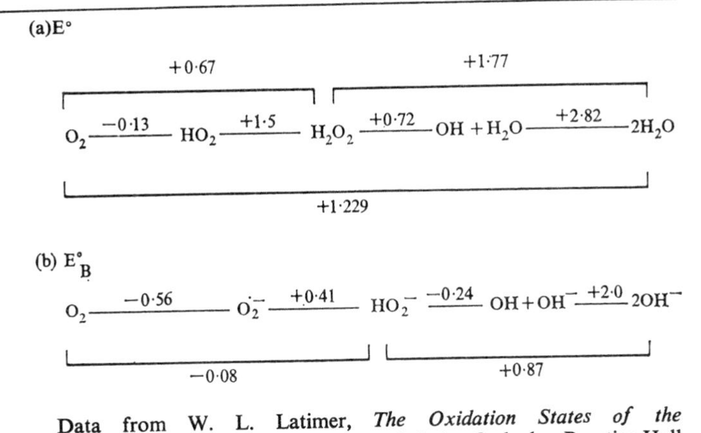

Data from W. L. Latimer, *The Oxidation States of the Elements and their Potentials in Aqueous Solution*, 2nd edn., Prentice-Hall (1952).

1.3. ISOTOPES OF OXYGEN

Unstable Isotopes

The known radioactive isotopes of oxygen are all artificial; they are listed in Table 6 with representative nuclear reactions by which they have been made. The half-lives are all so short that the isotopes are unsatisfactory for tracer work, though some tracer studies[12] have been made with O^{15}.

TABLE 6. RADIOACTIVE ISOTOPES OF OXYGEN

Isotope	Formation	$t\frac{1}{2}$, sec	Decay	Reference
O^{13}	$O^{16}(He^3, He^6)O^{13}$			a
O^{14}	$C^{12}(He^3, n)O^{14}$	73	β^+	b
O^{15}	$O^{16}(He^3, \alpha)O^{15}$	122	β^+	b
O^{19}	$O^{18}(n, \gamma)O^{19}$	29.4	β^-	b
O^{20}	$O^{18}(t, p)O^{20}$	14	β^-	b

[a] G. W. Butler, US Atomic Energy Comm. UCRL-17783, CFSTI, 1967. *CA* **68** (1968) 64756n.
[b] R. L. Heath, in *Handbook of Chemistry and Physics*, Chemical Rubber Co., New York (1967).

Stable Isotopes

By far the most abundant isotope of oxygen is O^{16}, but natural oxygen (both element and compounds) contains small amounts of O^{17} and O^{18} (Table 7). The proportions of

[12] C. T. Dollery and J. B. West, *Nature* **187** (1960) 1121.

these isotopes depend to a significant extent upon the source of oxygen. Natural processes such as the evaporation of water lead to some fractionation because of the influence of molecular weight upon physical properties; there are small differences in the proportion of O^{18} in water from different natural sources, and indeed the proportion of O^{18} in ocean water varies with the depth[13]. There is also some isotopic fractionation in chemical cycles in nature that involve oxygen. The variations in apparent atomic weight of natural oxygen may be as great[14] as ± 0.0003.

TABLE 7. STABLE ISOTOPES OF OXYGEN

Isotope	Mass [a]	Natural abundance [a]
	(C^{12} = 12.000 000 0)	
O^{16}	15.994 915	99.7587
O^{17} [b]	16.999 134	0.0374
O^{18}	17.999 160	0.2039

[a] W. H. Johnson and A. O. Nier, *Handbook of Physics*, McGraw-Hill, 2nd edn., 1967, pp. 9–63.

[b] Nuclear spin = 5/2; nuclear magnetic moment = 1.8930 nuclear magnetons; nuclear quadrupole moment = $-4e \times 10^{-27}$ cm^2. (From J. A. Pople, W. G. Schneider and H. J. Bernstein, *Nuclear Magnetic Resonance*, McGraw-Hill, 1959.)

Many different processes have been described for producing molecular oxygen or oxygen compounds enriched in O^{17} or O^{18}; some make use of the slightly different equilibrium constants for compounds of O^{16} and O^{17} or O^{18}, but most depend on some physical property of O_2 or oxygen compounds such as H_2O or N_2O. In practice the most important processes are the fractional distillation of water and (for high enrichments) the thermal diffusion of oxygen gas; the gas fed into the column for separation by thermal diffusion may be obtained by the electrolysis of water that has already been enriched in O^{18} by fractional distillation. Heavy water of enrichment up to 98 atom-% O^{18} and 20 atom-% O^{17} is commercially available, as is oxygen gas 99% in O^{18} or 90–95% in O^{17}. These are the most commonly used starting materials; of course it is possible to produce enriched O_2 by the electrolysis of enriched water. Samples of oxygen and its compounds are usually analysed for isotopic composition by mass spectrometry; if compounds are involatile, their oxygen must be converted into some volatile species first.

O^{18}. Oxygen-18 has been extensively used to study mechanisms of reactions, and in particular of hydrolysis. The exchange rates of many common oxyacids and oxyanions have been studied quantitatively or semi-quantitatively[15]; the structures of simple molecules containing oxygen have been determined by microwave spectroscopy[16] with the help of samples enriched in O^{18}, and vibrations involving the movement of oxygen atoms have been identified by the extent to which they shift in samples enriched in O^{18}. In the infrared

13 M. Dole, *J. Gen. Physiol.* **49** (1965) 5.
14 A. E. Cameron and E. Wickers, *J. Am. Chem. Soc.* **84** (1962) 4175.
15 D. Samuel, in *Oxygenases* (ed. O. Hayashi), Academic Press, New York (1962).
16 M. C. L. Gerry, J. C. Thompson and T. M. Sugden, *Nature* **211** (1966) 846.

spectra of adducts of molecular oxygen with compounds of transition metals, there is a band near 800–900 cm^{-1} that is assigned to the (O–O) stretching mode. In $(tbuNC)_2Ni(O_2)$ and $(tbuNC)_2Pd(O_2)$ this band splits into three when the complex is prepared from O_2 containing 25 atom-% O^{18}: the three components are due to stretching modes in $L_2M(O_2^{16})$, $L_2M(O^{16}O^{18})$ and $L_2M(O_2^{18})$; the relative intensities and the differences in frequency between the bands confirm this assignment[17]. A similar approach has led to the identification in matrices at low temperatures of active species containing more than one oxygen atom.

The preparation of O^{18} in naturally occurring systems has been much studied in connection with climatology and palaeoclimatology. The variation in O^{18} content of arctic snow over the year helps to establish the origin of the water from which the snow was formed[18]. In palaeoclimatology it is the oxygen isotopic composition of carbonate in fossils that is measured[19, 20]. The variation with temperature of the equilibrium constant for the reaction

$$H_2O^{18} + 1/3(CO_3^{16})^{-2} \rightleftharpoons H_2O^{16} + 1/3(CO_3^{18})^{-2}$$

has been determined accurately. Certain fossil shell-fish contain carbonate rock. The temperature at which this carbonate was precipitated can be determined from the isotopic distribution of the carbonate in the shell, using the equation

$$t = 16.5 - 4.3\delta + 0.14\delta^2$$

where t is the temperature in °C and δ is a measure of the difference between the O^{18} content of the sample and of the standard working gas. This use of the O^{18} content of carbonate is sometimes referred to as the carbonate thermometer.

In practice, matters are not so simple[19]. For the information obtained to have any meaning, certain conditions must be fulfilled. The carbonate must have been precipitated in isotopic equilibrium with the surrounding water; the isotopic composition of that water must be known; and there must have been no subsequent exchange. Though some shell-fish (e.g. coral) precipitate carbonate that is not in equilibrium with the surrounding water, in molluscs the equilibrium is maintained. For shells from open oceans it is usually assumed that the O^{18} content of the surrounding water was not very different from that of ocean water nowadays; for isolated oceans this may well not be so because of such things as evaporation, and allowance must be made for this (perhaps by using the O^{18} content of the phosphate of the shell). But it is with the third condition that the most serious difficulties arise. It is hard enough to make sure that the oxygen from the carbonate in the shell is extracted and analysed without exchange; it is even harder to make sure that nothing untoward has happened to the sample between fossilization and analysis. Fortunately there are some internal tests that can be applied. For example, in most, but not all, fossil shells studied the ratio of $(O^{18}:O^{16})$ to Sr^{+2} is the same as it is in oceans at present; therefore shells in which this ratio is very different may be supposed to have undergone "diagenetic" changes (in other words they are useless for isotopic analysis). However, there is always the nagging possibility that a particular sample may be atypical—some molluscs, for instance, do not deposit carbonate uniformly throughout the year—and it is rare that

17 K. Hirota, M. Yamamoto, S. Otsuka, A. Nakamura and Y. Tatsuno, *Chem. Communs.* 1968, 533.
18 L. Aldaz and S. Deutsch, *Earth Planet Sci. Letters* **3** (1967) 267.
19 H. A. Lowenstam, in *Problems in Palaeoclimatology* (ed. A. E. M. Nairn), Wiley (1964), p. 227.
20 R. Bowen, *Palaeotemperature Analysis*, Elsevier, Amsterdam (1966).

enough good samples from a particular area are available for a proper statistical analysis. Some claims about palaeotemperature may have been based on very slender evidence derived from the carbonate thermometer; but when all is said and done the remarkable thing is that such an extrapolation to the temperatures of past ages is even possible.

O^{17}. The isotope O^{17} is important despite its small natural abundance because it is the only stable isotope of oxygen with nuclear spin, and so is the only one that can be used to study oxygen n.m.r. or oxygen hyperfine interactions in electron resonance spectroscopy[21]. Unfortunately, however, the spin of 5/2 is associated with a large quadrupole moment, leading to broad n.m.r. lines because of the relatively rapid relaxation of the oxygen nuclei. Chemical shifts are difficult to determine, and coupling is often not observed at all. However, broadening due to quadrupolar relaxation has itself been used to give information about the field gradient at the oxygen nucleus; lines are sharper when oxygen is bound to an electropositive element, even in a high oxidation state, because then the oxygen atom approximates to the ion O^{-2} in which the field gradient at the nucleus is zero.

Some values for O^{17} chemical shifts are given in Table 8; they were measured relative to H_2O^{17} (usually as external standard), and some of the uncertainties are substantial. It is clear from these and other values that oxygen forming two σ-bonds gives resonance in the range -250 to 0 ppm from H_2O^{17} (+ve to high field), but that oxygen forming a double bond gives resonance in the range -1400 to -250 ppm. The dichromate ion[22], for instance, gives two resonances, the one (stronger and sharper) at -1129 ppm due to the terminal, and the other (weaker and broader) at -345 ppm due to the bridging oxygen atoms. Similarly[23], aqueous acetone gives a single O^{17} resonance in the $\rangle C{=}O$ region, at -523 ppm, indicating that the molecule dissolves unchanged; formaldehyde gives a resonance at -51 ppm, in the "single-bond" region, so it is concluded that the species present is the hydrated form $H_2C(OH)_2$. Acetaldehyde gives two peaks, the one at -550 ppm associated with the unchanged molecule CH_3CHO, and the other at -67 ppm associated with the hydrated form $CH_3CH(OH)_2$. Exchange between these two species and the solvent is slow on the n.m.r. timescale.

For doubly-bound oxygen there is a correlation between the chemical shift and the frequency of the first ultraviolet/visible absorption band[24]. Furthermore, a correlation has been reported between bond length, bond order and O^{17} chemical shifts in some Cr(VI)–O species[25]. The full range of validity of the latter correlation has still to be established.

Ion hydration has been extensively studied by O^{17} n.m.r. spectroscopy. For certain paramagnetic ions there are "contact shifts" which with Co^{+2} and Ni^{+2} lead to the observation of separate resonances due to water in the hydration shell of the cation and to the solvent itself. For certain diamagnetic ions which exchange their hydration water slowly with the solvent, separate O^{17} resonances for solvent and for solvating water have been observed (Table 9). There are other cases where exchange on other grounds is known to be slow, but where, none the less, the only O^{17} resonance observed is that due to the solvent. The addition of Co^{+2} or Dy^{+3} ions to the solution causes the solvent resonance to shift to low field through the "contact shift" mechanism, and so can expose the resonance due to the hydrated diamagnetic cation; this device has also been used to separate the

[21] B. L. Silver and Z. Luz, *Quart. Rev.* **21** (1967) 458.
[22] B. N. Figgis, R. G. Kidd and R. S. Nyholm, *Can. J. Chem.* **43** (1965) 145.
[23] P. Greenzaid, Z. Luz and D. Samuel, *J. Am. Chem. Soc.* **89** (1967) 749.
[24] B. N. Figgis, R. G. Kidd and R. S. Nyholm, *Proc. Roy. Soc.* A, **269** (1962) 469.
[25] R. G. Kidd, *Can. J. Chem.* **45** (1967) 605.

TABLE 8. CHEMICAL SHIFTS FOR O^{17}, RELATIVE TO H_2O^{17} (ppm)

Transition element systems		Main group systems			Carbon–oxygen systems [c]	
Na_3VO_4 [a]	−571±4	H_2O_2 (30% aq.)	a	−187±5	MeOH	+37
Na_2CrO_4 [a]	−835±5	K_2CO_3 (7 M [b])	a	−192±4	EtOH	−6
K_2MoO_4 [a]	−540±2	HNO_3 (100%)	a	−414±3	tBuOH	−70
Na_2WO_4 [a]	−420±2	$POCl_3$ (l)	a	−216±2	Furan	−241
$NaMnO_4$ [a]	−1219±8	H_3PO_4 (l)	c	−80	CH_3CHO	−595
$NaTcO_4$ [a]	−749±7	$SOCl_2$ (l)	a	−292±2	Acetic acid	−254
$NaReO_4$ [a]	−569±4	SO_2Cl_2 (l)	a	−304±3	CH_3COOCH_3	−355 (C=O)
RuO_4 [a]	−1119±10	H_2SO_4 (conc.)	c	−140		−137 (OMe)
OsO_4 [a]	−796±3	Na_2SeO_4 (3 M [b])	a	−204±12	$(CH_3CO)_2O$	−393 (C=O)
		H_6TeO_6 (aq.)	g	−120		
CrO_2Cl_2 (l) [d]	−1460±8	$NaClO_3$ (2.4 M [b])	a	−287±3		−259 (>O)
$Na_2Cr_2O_7$ [d]	−1129 (term.)	$HClO_4$ (60% aq.)	c	−288	di-tBu_2O_2	−269
	−345 (br.)	$NaClO_4$ (2.4 M [b])	a	−288±5	$MeONH_2$	−35
UO_2^+ aq. [f]	−1115±2	$NaBrO_3$ (2.4 M [b])	a	−297±5	tBuONO	−838 (N=O)
$Fe(CO)_5$ [e] (l)	−388±8	$XeOF_4$	h	−313±2		−513 (>O)
$Ni(CO)_4$ [e] (l)	−362±8	$Xe(OH)_2$ [h] (aq.)	i	−278±2		

[a] Data for aqueous solutions, unless otherwise stated, from B. N. Figgis, R. G. Kidd and R. S. Nyholm, *Proc. Roy. Soc.* A, **269** (1962) 469.

[b] 0.1 N aqueous alkali.

[c] Data for pure liquids from H. A. Christ, P. Diehl, H. R. Schneider and H. Dahn, *Helv. Chim. Acta* **44** (1961) 865.

[d] B. N. Figgis, R. G. Kidd and R. S. Nyholm, *Canad. J. Chem.* **43** (1965) 145.

[e] R. Bramley, B. N. Figgis and R. S. Nyholm, *Trans. Faraday Soc.* **58** (1962) 1893; R. G. Kidd, *Canad. J. Chem.* **45** (1967), 605.

[f] S. W. Rabideau, *J. Phys. Chem.* **71** (1967) 2747.

[g] Z. Luz and I. Pecht, *J. Am. Chem. Soc.*, **88** (1966), 1152.

[h] J. Shamir, H. Selig, D. Samuel and J. Reubin, *J. Am. Chem. Soc.* **87** (1965) 2359.

[i] J. Reuben, D. Samuel, H. Selig and J. Shamir, *Proc. Chem. Soc.* 1963, 270.

resonances of water and of phosphoric acid[26]. The hydration of ions or complexes of Be^{+2}, Al^{+3}, Ga^{+3}, VO^{+2}, Tl^{+3}, Co^{+2}, Ni^{+2} and Cu^{+2} has been studied using these and similar techniques. Finally, O^{17} n.m.r. affords information about H bonding and the structure of aqueous solutions; O^{17} resonances, like proton resonances, are shifted to low field by hydrogen bonding[27].

The e.s.r. spectra of organic compounds containing O^{17} have been important in relation to calculations of the electronic structures of these compounds, and particularly of heterocyclic compounds. The hyperfine interaction constants in O^{17}-labelled $Mn(acac)_3$ has been used[28] as the basis for an estimate of the covalent character of the metal–oxygen bonds. The e.s.r. spectrum of a sample of CF_3OOCF_3 that had been photolysed in the

Table 9. Coupling Constants Involving O^{17} (Hz)

Species	Coupling nuclei	Coupling constants	Reference
ClO_4^-	$Cl^{35}–O^{17}$	85.5 ± 0.5	a
$XeOF_4$	$Xe^{129}–O^{17}$	692 ± 10	b
H_2O	$H^1–O^{17}$	79 ± 2	c
$(MeO)_3P$	$P^{31}–O^{17}$	160	d
Cl_3PO	$P^{31}–O^{17}$	225	d
MnO_4^-	$Mn^{55}–O^{17}$	30	e
Me_2CO	$C^{13}–O^{17}$	22	e

[a] M. Alei, *J. Chem. Phys.* **43** (1965) 2904.
[b] J. Shamir, H. Selig, D. Samuel and J. Reuben, *J. Am. Chem. Soc.* **87** (1967) 2359.
[c] A. E. Florin and M. Alei, *J. Chem. Phys.* **47** (1967) 4268.
[d] H. A. Christ, P. Diehl, H. R. Schneider and H. Dahn, *Helv. Chim. Acta* **44** (1961) 865.
[e] M. Broze and Z. Luz, *J. Phys. Chem.* **73** (1969) 1600.

presence of small amounts of O_2 enriched ($\sim 30\%$) in O^{17} showed that a free radical had been formed with three different sets of O^{17} hyperfine splitting constants, and it was concluded[29] that the radical was the trioxide species CF_3COOO. Hyperfine structure in the e.s.r. spectrum of $[(H_3N)_5Co(O_2)Co(NH_3)_5]^{+5}$ due to O^{17} in the peroxy-bridge provided the first direct evidence that the unpaired electron is associated with the bridge as well as with the cobalt nuclei[30].

1.4. OCCURRENCE AND EXTRACTION

Oxygen is the most abundant element[31]. In combined form it makes up 46.60% by weight of the rocks of the earth's crust, and (as the element) $20.946 \pm 0.002\%$ by volume of

[26] J. A. Jackson and H. Taube, *J. Phys. Chem.* **69** (1965) 1844.
[27] Z. Luz and G. Yagil, *J. Phys. Chem.* **70** (1966) 554.
[28] Z. Luz, B. L. Silver and D. Fiat, *J. Chem. Phys.* **46** (1967) 469.
[29] R. W. Fessenden, *J. Chem. Phys.* **48** (1968) 3725.
[30] J. A. Weil and J. K. Kinnaird, *J. Phys. Chem.*, **71** (1967) 3341.
[31] *The Production of Oxygen, Nitrogen and the Inert Gases*, British Oxygen Co. (1967).

dry air at sea level. Nowadays the atmosphere is the primary source of oxygen for large-scale commercial production of the gas or liquid, but chemical processes have been devised and extensively used in the past. Most are based on substances which combine with oxygen at one temperature and release it on heating. The old Brin process made use of the reaction between barium oxide and oxygen at 600°:

$$BaO + 0.5O_2 \underset{\text{low pressure}}{\overset{\text{high pressure}}{\rightleftharpoons}} BaO_2$$

The barium oxide is heated with air that has been freed of CO_2, water vapour, organic matter and dust; the pressure is lowered, and pure oxygen is evolved. A modification of the process has recently been devised[32] in which use is made of the fact that the melting point of BaO_2 (450°) is much lower than that of BaO (1923°). Several similar processes have been proposed for making pure oxygen, which has also been made electrolytically, but almost all oxygen is now produced by the liquefaction and fractional distillation of air. Methods for doing this differ in detail. The Linde double column consists of one fractionating column operating at about atmospheric pressure, and another operating at 5–6 atm. At the higher pressure nitrogen boils at −170°, and so may be condensed using liquid oxygen from the column at lower pressure as refrigerant. After initial compression and cooling by water, the air is further cooled by heat exchange with gases leaving the columns; the cooling by expansion that follows may take place under Joule–Thompson conditions, by doing external work, or by a combination of the two. In the production of oxygen gas, the use of heat exchangers called regenerators is of considerable importance[33].

Alternative sources of oxygen for enclosed systems have become of increasing importance in the past few years with the development of space craft and nuclear submarines. Schemes have been devised for producing oxygen from CO_2 by means of algae, or by catalytic hydrogenation followed by electrolysis of the water formed[34]. Oxygen "candles" have been proposed for carrying oxygen; these consist of salts of oxy-anions which are both thermally unstable and rich in oxygen, such as $LiClO_4$, $NaClO_3$ or KO_2. A lithium perchlorate candle has been described consisting of 84.82% by weight of $LiClO_4$, 10.94% Mn and 4.24% of Li_2O_2; the solid is compressed to a density of 2.32 g/cc, and the oxygen available on heating is equivalent, volume for volume, to that from liquid oxygen[35].

In the laboratory[36], oxygen may be made by heating potassium permanganate or other salts thermally unstable and rich in oxygen, but commercial oxygen can be purified by successive treatment with $KMnO_4$, with KOH, and with concentrated H_2SO_4. The best method of making pure oxygen in the laboratory is to catalyse the decomposition of 30% H_2O_2 with thin nickel sheet suspended by a platinum wire; electrolysis may also be used.

32 S. A. Guerrieri, US pat. 3,310,381, 1966 (*Chem. Abstr.* **66** (1967) 106659z).

33 R. L. Shower and L. C. Matsch, *Adv. Petrol Chem. Refining* **9** (1964) 1.

34 P. J. Hannan, R. L. Shuler and C. Patouillet, NASA Acc. No. N63-14930, 1962 (*Chem. Abstr.* **62** (1965) 2006a); NASA Acc. No. N64-10551, 1963 (Report No. AD 420927) (*Chem. Abstr.* **61** (1964) 16488d).

35 C. S. Coe, *Chem. Engng. Progress, Symp. Ser.* **60** (1964) 161.

36 G. Brauer (ed.), *Handbook of Preparative Inorganic Chemistry*, 2nd edn., Academic Press, New York (1963), p. 334.

1.5. STRUCTURAL PROPERTIES

Electronic Structure

The electronic structure of the oxygen molecule can be represented very simply as O=O. This representation indicates that the O–O bond should be short and strong (as it is—see Table 10) but offers no explanation for the paramagnetism of the molecule. To account for this property in terms of simple valence-bond theory, it is necessary to invoke

TABLE 10. SOME MOLECULAR PROPERTIES OF O_2 IN ITS GROUND STATE

		Reference
Bond dissociation energy	$D = 5.114 \pm 0.002$ eV	a
Bond length	$= 1.207\ 398$ Å (from u.v. spectrum)	b
	$= 1.207\ 41 \pm 0.000\ 02$ Å (from the e.s.r. spectrum)	c
Bond stretching-force constant	$k = 11.409 \times 10^{-5}$ dyne cm^{-1}	b
Ionization potential (adiabatic)	$= 12.075 \pm 0.01$ eV	d
Electron affinities: 1st	$= 0.43 \pm 0.01$ eV	e
'Double'	$= -6.7 \pm 0.6$ eV (for $O_2(g) \rightarrow O_2^{-2}(g)$)	f
Polarizability: $\alpha_{xx} = \alpha_{yy}$	$= 1.2 \times 10^{-24}$	g
α_{zz}	$= 2.4 \times 10^{-24}$	
α_0	$= 1.6 \times 10^{-24}$	
Spin–orbit coupling constant λ	$= 1.985$ cm^{-1}	h

[a] P. Brix and G. Herzberg, *Can. J. Phys.* **32** (1954) 110.
[b] G. Herzberg, *Spectra of Diatomic Molecules*, 2nd edn., van Nostrand (1950).
[c] M. Tinkham and M. W. P. Strandberg, *Phys. Rev.* **97** (1955) 951.
[d] K. Watanabe, *J. Chem. Phys.* **26** (1957) 542.
[e] J. L. Pack and A. V. Phelps, *J. Chem. Phys.* **44** (1966) 1870.
[f] L. A. D'Orazio and R. H. Wood, *J. Chem. Phys.* **69** (1965) 2558.
[g] Landolt-Börnstein, *Zahlenwerte und Functionen aus Physik-Chemie-Astronomie-Geophysik und Technik*, Sechste Auflage, 1 Band, 3 Teil, p. 510.
[h] K. Kayama and J. L. Baird, *J. Chem. Phys.* **46** (1967) 2604.

three-electron bonds[37] or to use the "double quartet" approach[38]. In simple molecular orbital terms, the paramagnetism is explained by the two unpaired electrons in the anti-bonding π_g-orbital[39]. The configuration is

$$(\sigma_g 1s)^2(\sigma_u 1s)^2(\sigma_g 2s)^2(\sigma_u 2s)^2(\sigma_g 2p)^2(\pi_u 2p)^4(\pi_g 2p)^2.$$

Although in B_2, C_2 and N_2 the $(\sigma_g 2p)$ level is above the $(\pi_u 2p)$ level because of interaction with $(\sigma_g 2s)$, in O_2 the order is as shown. More detailed calculations using the LCAO SCF method with configurational interaction[40] gives good agreement with the observed value for the total energy, but the agreement between the observed and the calculated values for the dissociation energy is less satisfactory.

The photo-electron spectrum of molecular oxygen excited using HeI radiation (21.23 eV) gives at least four and possibly five bands, all with vibrational progressions. The adiabatic ionization potentials and observed vibration frequencies are given in Table 11A.

[37] L. Pauling, *The Nature of the Chemical Bond*, 3rd edn., Cornell University Press, Ithaca, NY.
[38] J. W. Linnett, *The Electronic Structure of Molecules*, Methuen, London (1964).
[39] J. N. Murrell, S. F. A. Kettle and J. M. Tedder, *Valence Theory*, Wiley, New York (1965).
[40] K. Ohno, in *Middle U.V.: Its Science and Technology* (ed. A. E. S. Green), Wiley, New York (1966).

The band at lowest I.P. corresponds to removal of an anti-bonding π_g electron producing O_2^+ in its electronic ground-state $^2\Pi_g$; as would have been expected, the vibration frequency is greater than in O_2. The next band, with a long vibrational progression, corresponds to removal of a π-bonding electron giving $O_2^+(^4\Pi_u)$; peaks associated with the production of $O_2^+(^2\Pi_u)$ may also occur in this region. The next band corresponds to removal of a σ-bonding electron giving $O_2^+(^4\Sigma_g)$, and the fifth band may correspond to the formation of O_2^+ $(^4\Sigma_u)$ or $O_2^+(^2\Sigma_g^-)$. In each case the electron removed is bonding, as shown by the drop in vibration frequency on ionization[40a].

TABLE 11A. PARAMETERS FROM THE PHOTO-ELECTRON SPECTRUM OF O_2

Adiabatic I.P., eV	Vibration frequency excited (0–1), cm^{-1}	State of O_2 formed
12.07	1780	$^2\Pi_g$
16.12	1010	$^4\Pi_u$
a	2887	$^2\Pi_u$
18.7	1090	$^4\Sigma_g$
20.29	1130	$^4\Sigma_u$ or $^2\Sigma_g^-$

[a] The (0–0) transition of this series was not identified.
Data from D. W. Turner *et al.*, *Molecular Spectroscopy*, Wiley, New York, 1970.

Molecular and Crystal Structure

The O–O distance in gaseous O_2 has been determined very accurately by analysis of the Schumann–Runge bands in the u.v. and also by microwave spectroscopy (see Table 10). The O–O distances in liquid and solid oxygen have not been accurately determined, but the diffraction measurements for the three solid forms are consistent with (O–O) bonded distances close to those for free O_2, as is the small change in (O–O) stretching frequency with change in phase.

TABLE 11B. SOLID O_2-STRUCTURES

Form	Class	Space group	Unit cell dimensions	Molecules in unit cell	Reference
α	Monoclinic	$C2/m(C_{2h}^3)$	$a = 5.403 \pm 0.005$ $b = 3.429 \pm 0.003$ $c = 5.086 \pm 0.005$	2	a
β	Rhombohedral	$R3m(D_{3h}^5)$	$a = 4.210 \pm 0.07$ $\alpha = 46° 16' \pm 9'$	3	b
γ	Cubic	$P_m3n(O_h^3)$	$a = 6.83 \pm 0.05$	8	c

[a] C. S. Barrett, L. Meyer and J. Wassermann, *J. Chem. Phys.* **47** (1967) 592.
[b] R. A. Alikhanov, *Soviet Physics JETP* **18** (1964) 556; E. M. Hörl, *Acta Cryst.* **15** (1962) 845.
[c] T. A. Jordan, W. E. Streib, H. W. Smith and W. N. Lipscomb, *Acta Cryst.* **17** (1964) 777.

[40a] D. W. Turner, C. Baker, A. D. Baker and C. R. Brundle, *Molecular Spectroscopy*, Wiley, New York, 1970.

The structure of crystalline γ-oxygen (the highest temperature form)[41] is very like that of α-F_2. There are eight molecules in the unit cell: two are roughly spherically disordered, on $m3(T_h)$ sites, and six are cylindrically disordered on $\bar{4}2m(D_{2d})$ sites, with the $\bar{4}$ (S_4) axis perpendicular to the molecular axis. This phase is soft and transparent, and less dense than the others; its structure is regarded as more like that of liquid oxygen than like that of the

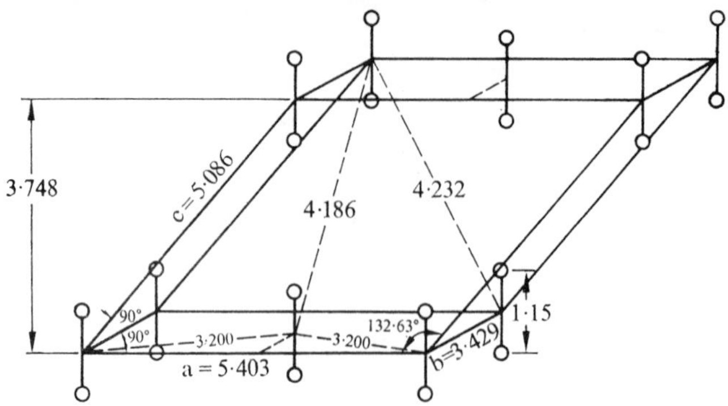

FIG. 1. The structure of α-oxygen; monoclinic $C2/m$. Atom centres and intermolecular distances are indicated; molecules are centred at 000 and $\frac{1}{2}\frac{1}{2}0$, with O–O bonds approximately normal to the (O–O) plane. (Reproduced with permission from C. S. Barrett, L. Meyer and J. Wasserman, *J. Chem. Phys.* **47** (1967) 592.)

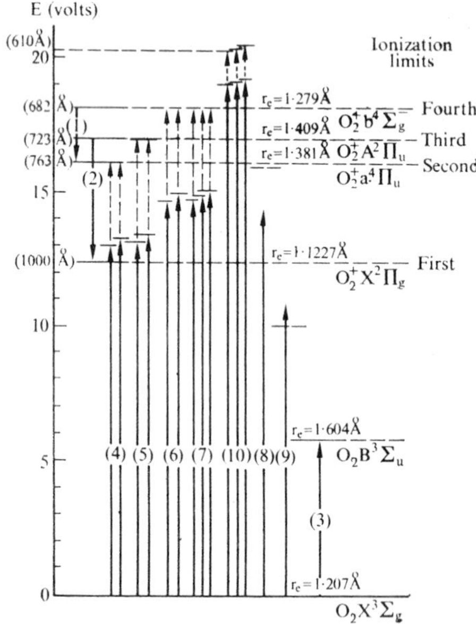

FIG. 2. Energy level diagram of the O_2 molecule. (Reproduced with permission from G. L. Weissler and Po Lee, *J. Opt. Soc. Am.* **42** (1952) 200.)

---

41 T. A. Jordan, W. E. Streib, H. W. Smith and W. N. Lipscomb, *Acta Cryst.* **17** (1964) 777.

β-form. This is consistent with the relatively small heat and volume changes on melting (smaller than the corresponding changes associated with the transition of β to α—see Table 15c)[42, 43], and with the closely similar vibrational spectra of γ-O_2 and liquid O_2. The β-form is rhombohedral[42, 43] and the α-form is monoclinic (Table 11B). The structure of the α-form is given in Fig. 1; the O_2 molecules lie parallel to one another, and with their molecular axes perpendicular to the O–O planes[43].

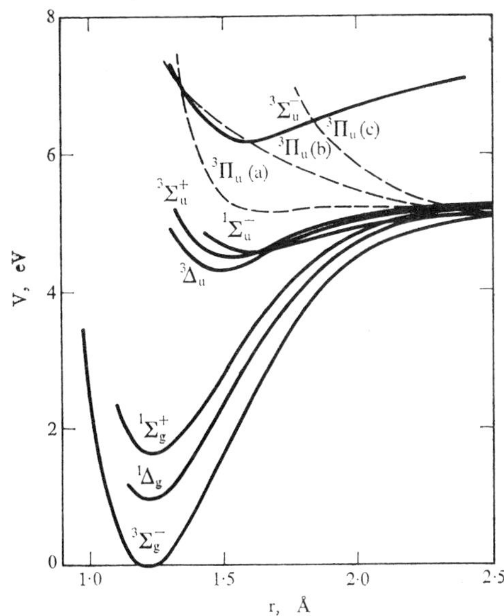

FIG. 3. Potential energy as a function of internuclear distance for states of the O_2 molecule. The dotted lines show results of different calculations for the $^3\pi_u$ state. (Reproduced with permission from D. H. Volman, *Advances in Photochemistry* **1** (1963) 46.)

1.6. SPECTROSCOPIC PROPERTIES

Electronic Spectra

The spectrum of isolated O_2. The ground state[44] of O_2 is $^3\Sigma_g^-$. The states $^1\Delta_g$ and $^1\Sigma_g^+$ are derived from the same electron configuration; transitions to each of these states from the ground state are formally forbidden, but are observed as very weak bands in the absorption of O_2 at ~8000 cm^{-1} ($^1\Delta_g \leftarrow {}^3\Sigma_g^-$) and ~13,200 cm^{-1} ($^1\Sigma_g^+ \leftarrow {}^3\Sigma_g^-$). Transitions to the next states involve excitation of a π_u-electron to a π_g-orbital. Fragments of a very weak band series due to the transition $^3\Delta_u \leftarrow {}^3\Sigma_g^-$ were detected by Herzberg[45] close to the weak band series for the transition $^3\Sigma_u^+ \leftarrow {}^3\Sigma_u^-$ near 36,000 cm^{-1} (the Herzberg bands) and the even weaker series[46] associated with the transition $^1\Sigma_u^- \leftarrow {}^3\Sigma_g^-$ near 37,000 cm^{-1}. All three of these transitions are forbidden. The strong Schumann–Runge bands near 50,000 cm^{-1} are

42 R. A. Alikhanov, *Soviet Physics JETP* **18** (1964) 556.
43 C. S. Barrett, L. Meyer and J. Wassermann, *J. Chem. Phys.* **47** (1967) 592.
44 G. Herzberg, *The Spectra of Diatomic Molecules*, 2nd edn., Van Nostrand, New York (1950).
45 G. Herzberg, *Can. J. Phys.* **31** (1953) 657.
46 P. Brix and G. Herzberg, *Can. J. Phys.* **32** (1954) 110.

TABLE 12. PARAMETERS FROM THE ELECTRONIC SPECTRA OF O_2. STATES ARE FOR O_2 UNLESS OTHERWISE STATED

Electronic state	*p*-electron configuration	r_e (Å)	T_e (cm^{-1})	ω_e (cm^{-1})	$\omega_e x_e$ (cm^{-1})	$\omega_e y_e$ (cm^{-1})	Reference
$^2\Pi_g(O_2^+)$	$\sigma_g^2\pi_u^4\pi_g^1$	1.1227		1876.4	16.53		
$^3\Sigma_u^-$	$\sigma_g^2\pi_u^3\pi_g^3$	1.60_4	49 357.6	700.36	8.002_3	-0.3753_5	a, b
$^1\Sigma_u^-$	$\sigma_g^2\pi_u^3\pi_g^3$	1.597	36 678.9_1	650.4_9	17.03_6	0.105_6	c
$^3\Sigma_u^+$	$\sigma_g^2\pi_u^3\pi_g^3$	[1.42]	36 096	$819._0$	22.5_0		a
$^1\Sigma_g^+$	$\sigma_g^2\pi_u^4\pi_g^2$	1.227 65_0	13 195.222_1	1432.687_4	13.9500_8	−0.01075	a
$^1\Delta_g$	$\sigma_g^2\pi_u^4\pi_g^2$	1.2155	7918.1	$1509._3$	$12._9$		a
$^3\Sigma_g^-$	$\sigma_g^2\pi_u^4\pi_g^2$	1.207 39_8	0	1580.361_3	12.0730	0.0546	a

r_e = equilibrium internuclear distance; T_e = electronic energy above the ground state; ω_e = harmonic fundamental vibrational frequency; $\omega_e x_e$ = vibrational constant for second-order anharmonic term; $\omega_e y_e$ = vibrational constant for third-order anharmonic term.

[a] G. Herzberg, *The Spectra of Diatomic Molecules* 2nd edn., van Nostrand, New York (1950).

[b] P. Brix and G. Herzberg, *Canad. J. Phys.* **32** (1954) 110.

[c] G. Herzberg, *Canad. J. Phys.* **31** (1953) 657.

derived from the allowed transition $^3\Sigma_u^- \leftarrow {}^3\Sigma_g^-$; they converge to a limit at 57,128 cm^{-1} and a continuum to about 70,000 cm^{-1}. At higher energies the spectrum becomes complex, and fragments of several Rydberg series have been detected[47] (Fig. 2).

Band systems associated with all these transitions except $^3\Delta_u \leftarrow {}^3\Sigma_g^-$ have been analysed, and the derived molecular constants are given in Table 12; promotion of an electron from π_u to π_g leads to a lengthening of the O–O bond and a decrease in the vibration frequency, as simple theory predicts. On the other hand, ionization strengthens the bond by removing an antibonding electron. In the spectrum of solid α-oxygen at 4°K, the structure of the bands indicate that the molecular constants for the gas and for α-O_2 are very similar but not exactly the same[48]. In γ-O_2 and β-O_2 the bands are broader than in the liquid or in α-O_2 [49]. There are further splittings of the vibrational bands in the spectrum of the crystal[48, 49].

Potential energy curves have been calculated for all the bound states, and for some of the repulsive ones (Fig. 3).

The lower states dissociate to two ^3P oxygen atoms, but the $^3\Sigma_u^-$ state gives one ^3P and one ^1D atom. There is clear evidence for predissociation in the Schumann–Runge system[50]; the relevant repulsive states seems to be $^3\Pi_g$, but the details of the process are not clear.

Bands involving intermolecular interactions. In the spectra of condensed or compressed oxygen there are bands that are not due to transitions of isolated O_2 molecules[51]; some of these features lie under the banded spectra of the O_2 transitions[52]. They are attributed to transitions in molecular complexes (perhaps short-lived collision complexes), but there must be more than one type of intermolecular process involved, for in compressed oxygen the intensities of different bands vary in different ways as the pressure is changed or as foreign gases are added. The band near 15,800 cm^{-1} is attributed to the (0–0) transition associated with simultaneous excitation of two $^3\Sigma_g^-$ oxygen molecules to the $^1\Delta_g$ state[53]. Systems of $^1\Delta_g$ molecules show emission at this wavelength due to "energy pooling" of two molecules[54]:

$$2O_2(^1\Delta_g) \rightarrow 2O_2(^3\Sigma_g^-) + h\nu(15{,}800 \text{ cm}^{-1})$$

Similar simultaneous excitation of two molecules gives bands near 21,000 cm^{-1} [(0–0) transition of $^1\Delta_g + {}^1\Sigma_g^+ \leftarrow 2{}^3\Sigma_g^-$] and 26,500 cm^{-1} [(0–0) transition of $2{}^1\Sigma_g^+ \leftarrow 2{}^3\Sigma_g^-$]. In O_2 gas, the band at 7890 cm^{-1} is induced by the presence of a foreign gas, such as CO_2; the band at 9370 cm^{-1}, however, is only induced by increasing the pressure of O_2, and so is assigned to a transition in a collision complex of O_2 molecules[52].

The strong bands largely responsible for the blue colour of liquid oxygen are due to (0–0), (0–1) and (0–2) transitions from the $2{}^1\Delta_g \leftarrow 2{}^3\Sigma_g^-$ and the $^1\Delta_g + {}^1\Sigma_g^+ \leftarrow 2{}^3\Sigma_g^-$ bands mentioned above[53].

Vibrational Spectra

The fundamental of O_2 is forbidden in the infrared, though it has been detected in the infrared spectra of oxygen gas at high pressures, of liquid oxygen, and of all three forms

47 G. L. Weissler and Po Lee, *J. Opt. Soc. America* **42** (1952) 200.
48 A. F. Prikhotko, T. P. Ptukha and L. I. Shanskii, *Optics Spectrosc.* **22** (1967) 203.
49 A. Landau, E. J. Allan and H. L. Welsh, *Spectrochim. Acta* **18** (1962) 1.
50 R. D. Hudson and V. L. Carter, *Can. J. Chem.* **47** (1969) 1840.
51 V. I. Dianov-Klokov, *Optics Spectrosc.* **20** (1966) 530.
52 R. M. Badger, A. C. Wright and R. F. Whitlock, *J. Chem. Phys.* **43** (1965) 4345.
53 E. A. Ogryzlo, *J. Chem. Educ.* **42** (1965) 647.
54 S. J. Arnold, N. Finlayson and E. A. Ogryzlo, *J. Chem. Phys.* **44** (1966) 2529.

of solid oxygen; the vibration is Raman active, and has been observed in the Raman spectra of all five phases of elementary O_2 (Table 13). The infrared spectra of the different phases of solid oxygen also show broad bands or shoulders to high frequencies of the fundamental band. In α-O_2, the fundamental is very sharp, and there is a broad band centred at 1605 cm^{-1}; the sharp peak is assigned to vibrations of molecules near a lattice imperfection, and the broad band to a combination of ν with lattice modes[55]. Vibrational

TABLE 13. VIBRATIONAL SPECTRA OF O_2

Phase	Infrared (cm^{-1})		T (°K)	Raman (cm^{-1})		T (°K)
Gas	1556	[a]	? r.t.	1554.7±1 [d]		r.t.
Liquid	1557 3102	(ν) [b] (2ν)	55	1552.5	(ν) [e]	77
Solid, γ	1552 1610 sh	(ν) [b] (? $\nu+l$)	44	1552.5	(ν) [e]	48
Solid, β	1552.4 1596 br 1618 br	(ν) [c] ($\nu+l_1$) ($\nu+l_2$)	26–32	46 1552.5	(l) (ν) [e]	43
Solid, α	1548.4 1591 br 1671	(ν) [c] ($\nu+l_1$) ($\nu+l_2$)	15–20	44 78 1552.5	(l_2) (l_2) (ν) [e]	22

ν = fundamental; l_1, l_2 = librational lattice modes.
[a] M. F. Crawford, H. L. Welsh and J. L. Locke, *Phys. Rev.* **75** (1949) 1607.
[b] A. L. Smith and H. L. Johnston, *J. Chem. Phys.* **20** (1952) 1972.
[c] B. R. Cairns and G. C. Pimentel, *J. Chem. Phys.* **43** (1965) 3432.
[d] F. Rasetti, *Phys. Rev.* **34** (1929) 367.
[e] J. E. Cahill and G. E. Leroi, *J. Chem. Phys.* **51** (1969) 97.

lattice modes were observed[56] in the Raman spectra of solid α-O_2 (at 44 cm^{-1} and 78 cm^{-1}) and β-O_2 (46 cm^{-1}). The strong winging on the Rayleigh line in the Raman spectra of solid γ-O_2 and liquid O_2 is associated with external motion; the torsional motion is highly hindered in each phase, and the r.m.s. torsional amplitude increases[56] from 9° for α-O_2 to 18° for liquid O_2.

Microwave and e.s.r. Spectra of O_2 $^3\Sigma_g^-$

The O_2 molecule has no permanent dipole moment, but its magnetic moment, associated with electron spin, is coupled to the total rotational angular momentum. This leads to a splitting of the K-levels into three non-degenerate rotational states:

$$J = K+1,\ J = K \text{ and } J = K-1.$$

Transitions between levels where $\Delta J = \pm 1$ are allowed and give rise to weak absorption

[55] B. R. Cairns and G. C. Pimentel, *J. Chem. Phys.* **43** (1965) 3432.
[56] J. E. Cahill and G. E. Leroi, *J. Chem. Phys.* **51** (1969) 97.

in the microwave region of the spectrum[57]. At atmospheric pressure the lines are broad, but fine structure has been studied at low pressures[58, 59].

The e.s.r. spectrum is very complicated. It arises because of the interaction of the molecular magnetic moment with the magnetic field; the spectrum is influenced by second-order coupling with the orbital motion of the electron and by a rotation-induced magnetic moment[60]. Analysis of the spectrum gives values for the internuclear distance and the spin–orbit coupling constant. There has been disagreement about some features of the spectrum and their interpretation[61].

TABLE 14. MAGNETIC AND RELATED PROPERTIES

Magnetic susceptibility	
Gas [a]	
$\chi_g T = 3115 \pm 7 \times 10^{-5}$ c.g.s. units at 293°K	
Liquid [b]	
$\chi_g(T-\theta') = C'$, where $\theta' = -71°$; $C' = 38{,}600 \times 10^{-6}$	
Solid[c–e]	
γ roughly as for liquid ($\theta' \sim -60$)	
β $\chi = 160 \times 10^{-6}$ c.g.s. units at 43°K	
$= 127 \times 10^{-6}$ c.g.s. units at 24°K	
α $14 < T < 20°$K, $\chi = 49.3[1+0.00276+(T-14.50)^2] \times 10^{-6}$	
$1.5 < T < 4.2°$K, $\chi = 50.7[1-0.01719+(T-2.906)] \times 10^{-6}$	
Verdet constants	
O_2 gas (0°, 1 atm; 5460 Å)	5.986×10^{-6} min oe^{-1} cm^{-1} [f]
O_2 liquid (5461 Å)	0.776×10^{-2} min oe^{-1} cm^{-1} at 90.1° [g]
	0.945×10^{-2} min oe^{-1} cm^{-1} at 63.4°
Kerr constants	
O_2 gas (0°, 1 atm; $\lambda = 6500$ Å)	$B = 6.94 \times 10^{-12}$ c.g.s. units [h]
O_2 liquid (5461 Å)	$B = 22.38 \times 10^{-9}$ c.g.s. units [i]

[a] A. Burris and C. D. Hause, *J. Chem. Phys.* **11** (1943) 442.
[b] H. Kamerlingh Onnes and E. Oosterhuis, *Comm. Leiden* 1913, 132e.
[c] E. Kanda, T. Hasada and A. Otsubo, *Physica*, **20** (1954) 131.
[d] A. S. Borovik-Romanov, *Zh. Exptl. Teoret. Fiz.* **21** (1951) 1303. For α-oxygen, χ is effectively constant between 4.2 and 14K.
[e] A. S. Borovik-Romanov, M. P. Orlova and P. G. Strelkov, *Doklady Akad. Nauk SSSR* **99** (1954) 699.
[f] L. R. Ingersoll and D. H. Lieberberg, *J. Opt. Soc. America* **44** (1954) 566.
[g] F. Gaume, *Comptes rendus* **232** (1951) 2304.
[h] W. M. Breazeale, *Phys. Rev.* **48** (1935) 237.
[i] R. Guillien, *Comptes rendus* **200** (1935) 1840.

1.7. MAGNETIC PROPERTIES

Elementary oxygen is paramagnetic in all phases. The susceptibility of the gas at 1 atm and 20°C corresponds closely with that calculated from the "spin-only" formula[62]; at

57 M. W. P. Strandberg, C. Y. Meng and J. G. Ingersoll, *Phys. Rev.* **75** (1949) 1524.
58 R. S. Anderson, W. V. Smith and W. Gordy, *Phys. Rev.* **82** (1951) 264.
59 M. Tinkham and M. W. P. Strandberg, *Phys. Rev.* **97** (1955) 937.
60 M. Tinkham and M. W. P. Strandberg, *Phys. Rev.* **97** (1955) 951.
61 K. D. Bowers, R. A. Kamper and L. D. Lustig, *Proc. Roy. Soc.* A, **251** (1959) 565; L. K. Keys, *J. Phys. Chem.* **70** (1966) 3760; A. Carrington, D. H. Levy and T. A. Miller, *J. Phys. Chem.* **71** (1967) 2372.
62 A. Burris and C. D. Hause, *J. Chem. Phys.* **11** (1943) 442.

TABLE 15A. BULK PHYSICAL PROPERTIES OF O_2

		Reference
Gas		
Density (1 atm, 0°) = 1.429 00 g l^{-1}		a
Compressibility $\frac{(pv)_{p=0}}{(pv)_{p=1\ \text{atm}}}$ (0°) = 1.000 92$_4$		b
Modified van der Waals equation:		
$p = [RT/(v-b_0+b_1p)-a/v-\alpha]^2$	$R = 0.003\ 666$	c
between 0°, 1 atm, and	$b_0 = 0.002\ 05$	
200°, 400 atm	$b_1 = 0.000\ 000\ 9$	
	$a = 0.002\ 72$	
	$\alpha = 0.000\ 60$	
Virial coefficients		
For $PV_m = RT\left(1+\frac{B}{V_m}+\frac{C}{V_m^2}\right)$, V_m = molar volume		c′
$B = -21.89$ cm^3 mol^{-1}	$T = 273.16$°K	
$C = 1230$ cm^6 mol^{-2}	$P < 135$ atm	
Critical temperature, T_c	154.78±0.03°K	d
Critical pressure, P_c	50.14±0.01 atm	d
Critical density, ρ_c	0.408 g cm^{-3}	e
Critical compressibility, Z_c	$(RT_c/P_cV_c) \sim 3.3$	d
Pitzer's acentric factor, ω	0.019	f
Boyle temperature, T_B	150°C	g
Specific heats (1 atm, 300°): C_p	0.2199 cal g^{-1}	h
C_v	0.1575 cal g^{-1}	h
Standard enthalpy, $H^\circ_{298.15} - H^\circ_0$	2.0746 kcal mole^{-1}	i
Standard entropy, $S^\circ_{298.15}$	49.003 cal mol^{-1} deg^{-1}	i
Viscosity (1 atm, 0°)	1.9192×10^{-4} poise	a
Thermal conductivity (1 atm, 0°)	5.867×10^{-5} cal cm^{-1} sec^{-1} °K^{-1}	a
Refractive index (1 atm, 0°; 5897.4 Å)	1.000 275 09	j
Dielectric constant (1 atm, 0°C; 24 000 MHz)	1.000 531 0	k
Velocity of sound (1 atm, 0°)	315.12 m sec^{-1}	l

[a] J. Hilsenrath *et al.*, NBS circular 564 (1955).

[b] Gmelin, *Handbuch der Anorganische Chemie, Sauerstoff*, Lieferung 3, System-nummer 3′ (1958), p. 395.

[c] J. B. Goebal, *Z. phys. Chem.* **47** (1904) 471.

[c′] J. H. Dymond and E. B. Smith, *The Virial Coefficients of Gases*, Oxford University Press, Oxford (1969).

[d] H. J. Hoge, *J. Nat. Bureau Standards*, **44** (1950) 321.

[e] The critical density is variously given as 0.38 g cm^{-3} and 0.4299 g cm^{-3}. This value is from A. V. Voronel, A. V. Chashkin, V. A. Popov and V. G. Sinikin, *Zh. Eksp. Teor. Fiz.* **45** (1963) 828.

[f] H. W. Cooper and J. C. Goldfrank, *Hydrocarbon Processing* **46** (12) (1967) 141.

[g] L. Holborn and J. Otto, *Z. Physik.* **23** (1924), 77.

[h] *Properties of Materials at Low Temperature* (*Phase* 1) (ed. V. J. Johnson), Pergamon Press, New York (1961).

[i] From NBS Technical Note 270-3 (1968). Values over a range of temperature are given in ref. a.

[j] R. Ladenburg and G. Wolfsohn, *Z. Phys.* **79** (1932) 42.

[k] L. Esson and K. D. Froome, *Proc. Phys. Soc.* **B64** (1951) 862.

[l] D. Bancroft, *Am. J. Phys.* **24** (1956) 355.

TABLE 15B. PROPERTIES OF THE LIQUID

	Reference
Boiling point: 90.1777°K	a
Vapour pressure: $\log_{10}p_{mm} = 28.901\,847 - 10.534\,702\log_{10}T + 0.019\,329\,48T - \frac{667.855\,76}{T} + \frac{1937.538}{T^2}$	a
Density, ρ (g/cc) $= (1.5656 - 4.896\times10^{-3}T + 2.136\times10^{-6}T^2) + p(-0.533\,27 + 0.028\,24T)\times10^{-4}$ (Range: 65–90°K; p up to 150 kg cm^{-2}); T in °K; p in kg cm^{-2}	b
$\rho = 1.142$ g cm^{-2} at 90°K	c
Viscosity: $\ln(10^5\eta) = 3.490 + \frac{0.756\rho}{(1.6144 + 0.632\times10^{-4}p - \rho)}$ at 89.7° 11 kgf cm^{-2}, $\eta = 0.001\,99$ poise	d
Specific heats: At 56.95°K $C_p = 12.76$ cal mole^{-1} °K^{-1}; At 90.33°K $C_p = 12.99$ cal mole^{-1} °K^{-1}	e
Surface tension: Between 71°K and 90°K, $\gamma_T = 36.479 - 0.259T$	f
At 70°, 18.35; at 90°, 13.23 dyne cm^{-2}	g
Thermal conductivity at 73.16°K, 20 atm, = 1.721 mW cm^{-1} °K^{-1}	c
Dielectric constant: At 54.33°K $\varepsilon = 1.594$; At 90.14°K $\varepsilon = 1.4837$ } In this range, $P = \frac{\varepsilon-1}{\varepsilon+2}\times\frac{1}{\rho} = 0.1211 \pm 0.05\%$	c
Adiabatic compressibility: At 60.5° 0.608×10^{-10} cm^2 dyn^{-1}; At 90.0° 1.050×10^{-10} cm^2 dyn^{-1}	h
Refractive index (5461 Å)[1]: At 65.0°K 1.2483; At 90.3°K 1.2243	i
Velocity of sound (535 kHz): At 60.5° 1128.9 m sec^{-1}; At 90.5° 902.3 m sec^{-1}	h

[a] R. Muijlwijk, M. R. Moussa and V. Van Dijk, *Physica* **32** (1966) 805. For measurements so precise as these, the thermometry becomes important; this reference discusses the small differences between precise measurements of the same parameter in different laboratories.
[b] A. Van Itterbeek and O. Verbeke, *Cryogenics* **1** (1960) 77.
[c] *Properties of Materials at Low Temperatures (Phase I)* (ed. V. J. Johnson), Pergamon Press, New York (1961).
[d] W. Grevendonk, W. Herreman, W. de Pesseroey and A. de Bock, *Physica* **40** (1968) 207.
[e] W. F. Giauque and H. L. Johnston, *J. Am. Chem. Soc.* **51** (1929) 2300.
[f] P. Waldon, *Z. phys. Chem.* **65** (1909) 129.
[g] E. Kanda, *Bull. Chem. Soc. Japan* **12** (1937) 469.
[h] A. Van Itterbeek, A. de Back and L. Verhaegen, *Physica* **15** (1949) 624.
[i] H. E. Johns and J. O. Wilhelm, *Canad. J. Res.* **A15** (1937) 101.

higher temperatures the Curie law is obeyed with agreement between observed and calculated values of about 1.5%, and agreement is still good at low temperatures[63]. Similarly, the susceptibility at high pressures is given by $\chi(T+\theta') =$ const., where θ' depends on the pressure[64]; at low temperatures and high pressures Curie's law is not obeyed[65]. The susceptibilities of liquid oxygen and of γ-O_2 obey the Curie–Weiss law with a negative value for θ'; both α-O_2 and β-O_2 show antiferromagnetic properties. In α-O_2, neutron diffraction

[63] H. Kamerlingh Onnes and E. Oosterhuis, *Communs. Leiden* **134d** (1913) 31.
[64] M. Kanzler, *Ann. Phys.* **36** (1939) 38.
[65] H. R. Woltjer, C. W. Coppoolse and E. C. Wiersma, *Communs. Leiden* **201** (1929) 35.

TABLE 15C. SOLID OXYGEN

Phase changes

Change	Temperature (°C)	ΔH (cal mole^{-1})	ΔV (cm^3 mole^{-1})
Melting point (triple point)	54.3496 [a]	106.3 ± 0.3 [b]	0.918 ± 0.02 [c]
γ–β	43.800 [c]	177.6 ± 0.5 [b]	1.08 ± 0.05 [c]
β–α	23.886 ± 0.005 [c]	22.4 ± 0.1 [b]	

Vapour pressure equation [d]

γ-O_2: $\log_{10} p_{mm} = \dfrac{-478.4}{T} + 12.40 - 2.055 \log_{10} T$

β-O_2: $\log_{10} p_{mm} = \dfrac{-452.2}{T} + 2.061 - 4.78 \log_{10} T$

α-O_2: $\log_{10} p_{mm} = \dfrac{-454.7}{T} + 1.794 + 5.085 \log_{10} T$

	γ-O_2	β-O_2	α-O_2
Density: measured (g cm^{-3})	1.30 [e]	1.395 [e]	1.426 [f]
from crystallography	1.334 [g]	1.495 [h]	1.53 [i]

Specific heat (cal mol^{-1} °K^{-1})[b] 11.03 at 45.90°K; 5.42 at 25.02°K; 9.34×10^{-4} at 4°
11.06 at 52.12°K; 10.73 at 42.21°K; 4.40 at 22.24°
(cal gm^{-1})[j] $C/T = 1.46 \times 10^{-5}\, T^2$ below 4°K

[a] R. Muiklwijk, N. R. Moussa and H. Van Dijk, *Physica* **32** (1966) 805.
[b] W. F. Giauque and H. L. Johnston, *J. Am. Chem. Soc.* **51** (1929) 2300.
[c] J. A. Jahnke, *J. Chem. Phys.* **47** (1967) 336.
[d] Landolt-Börnstein, *Zahlenwerte und Functionen aus Physik-Chemie-Astronomie-Geophysik und Technik*, Sechste Auflage, II Band, 2 Teil, pp. 5, 10, 15.
[e] L. Vegard, *Nature* **136** (1935) 720.
[f] J. Dewar, *Proc. Roy. Soc.* **73** (1904) 251.
[g] T. A. Jordan, W. E. Streib, H. W. Smith and W. N. Lipscomb, *Acta Cryst.* **17** (1964) 777.
[h] E. M. Hörl, *Acta Cryst.* **15** (1962), 845.
[i] C. S. Barrett, L. Meyer and J. Wassermann, *J. Chem. Phys.* **47** (1967) 592.
[j] *Properties of Materials at Low Temperatures* (*Phase* I) (ed. V. J. Johnson), Pergamon Press, New York (1961).

indicates that at 4°K the magnetic vectors of the individual molecules are aligned so that the vector of each molecule, perpendicular to the molecular axis, is opposed to the vectors of its eight nearest neighbours[68]. There may be a further antiferromagnetic transition[69] at

[66] H. Kamerlingh Onnes and A. Perrier, *Communs. Leiden* **116** (1910) 1.
[67] A. S. Borovik-Romanov, M. P. Orlova and P. G. Strelkov, *Doklady Akad. Nauk SSSR* **99** (1954) 699.
[68] C. S. Barrett, L. Meyer and J. Wassermann, *J. Chem. Phys.* **47** (1967) 592.
[69] W. H. Lien and N. E. Phillips, *J. Chem. Phys.* **34** (1961) 1073.

temperatures below 15°K. There is no evidence from neutron diffraction for long-range magnetic order in β-O_2, but some features of the diffraction patterns are consistent with short-range antiferromagnetism[70]. Intermolecular magnetic interaction in the liquid phase is strong, for the susceptibility increases at a given temperature on dilution with liquid nitrogen, and in very dilute solution approaches the "spin-only" value[71]. (Tables 14, 15A, 15B, 15C and 16.)

TABLE 16. SOLUBILITY OF O_2 [a]

1 cm^3 water dissolves 0.0308 cm^3 O_2 (measured at s.t.p.) at 20°, 0.0208 cm^3 at 50° and 0.0177 cm^3 at 80°. The concentration in ppm of O_2 in contact with air saturated with water vapour (total pressure one atmosphere) is given by

$$C = 14.161 - 0.3943t + 0.007\,714t^2 - 0.000\,064\,6t^3 \quad (t \text{ in } ^\circ\text{C})$$

The solubility is lower in solution of electrolytes, bases, or acids, but goes through a minimum in sulphuric acid and is nearly as great in 96% H_2SO_4 as in water. At 25° under a pressure of atmosphere O_2, 1 cm^3 of each of the following solvents dissolves the stated volume of O_2 (measured at 1 atm and 25°):

CCl_4	0.302
benzene	0.223
acetone	0.280
diethyl ether	0.455

[a] W. F. Linke, *Solubilities of Inorganic and Metal-Organic Compounds*, Vol. II, American Chemical Society (1965).

1.8. REACTIONS OF O_2

General

The bond in O_2 is very strong, but oxygen forms strong bonds with many other elements, and most elements and many compounds are thermodynamically unstable to O_2. Living organisms are unstable with respect to oxidation by O_2 to carbon dioxide and water; most metals are unstable with respect to oxidation by O_2 to an oxide. In aqueous solution O_2 is thermodynamically a powerful oxidizing agent. In practice, however, few of these systems react with O_2 under normal conditions; most of the processes are kinetically controlled. Activation energies for the oxidation of organic systems are often high; a metal may become protected by a thin film of oxide. Such metastable systems may of course be activated; under some conditions, the chain reaction between H_2 and O_2, with explosion limits depending on composition and other conditions, can be set off by a catalyst. Indeed, much of the interest in the chemistry of O_2 has been associated with different ways of activating O_2: converting it into atoms, using catalysts or by means of radiation. Oxygen, after all, is by far the most abundant oxidizing agent, and the cheapest. Oxidations based on molecular oxygen are the primary sources of energy in many living organisms; catalysts called oxygenases which can use O_2 in oxidizing cycles are associated with systems (oxygen carriers) that take up oxygen from the atmosphere (or from the lungs of the organism) and release it in the organism, so that the release of energy can be controlled. These aspects are discussed in subsequent sections. Despite the kinetic barriers to the reactions of O_2, liquid oxygen reacts explosively with organic matter and with many metals, which makes the handling of the material hazardous.

[70] M. F. Collins, *Proc. Phys. Soc.* **89** (1966) 415.
[71] A. Perrier and H. Kamerlingh Onnes, *Communs. Leiden* **139d** (1914) 37.

The process of "apparently uncatalysed oxidation of a substance exposed to the oxygen of the air" has been defined as autoxidation[72]; species that autoxidize include unsaturated organic systems, transition metals in low oxidation states and many free radicals. Oxygen, with its triplet ground state, is in condition to combine with other molecules with paramagnetic ground states or low-lying excited states. Indeed, many autoxidations involve free radicals. Some are sensitized by other species, not themselves oxidized; the wide variety of processes and systems cannot be discussed in detail here.

Oxygenation in Solution

Oxygen carriers and oxygenases. Oxygen carriers[73], which combine reversibly with molecular oxygen, maintain the level of molecular oxygen in the blood of animals and reptiles; oxygenases[74] are the catalysts that enable organisms to obtain energy by oxidation processes involving molecular oxygen. Molecular oxygen is not very soluble in blood, but the presence of oxygen carriers enables the blood of reptiles to contain 5–10%, and that of mammals 15–30%, of its own volume of O_2 in equilibrium with the atmosphere at room temperature[75]. The best-known natural oxygen carrier is haemoglobin, in which the oxygen is associated with Fe(II); the stereochemical arrangement of the combined O_2 is still a matter of controversy (see section 7). A number of systems containing Fe(II) react with O_2; indeed, many oxygenases contain iron. However, in most of these systems the products of the reaction are Fe(III) and H_2O_2 or water, and though the iron may be reduced back to Fe(II), the result of the reaction is the reduction of O_2 and the oxidation of iron or some other substrate. The remarkable thing about haemoglobin is not that it combines with O_2, but that it releases the combined O_2 reversibly. Other oxygen-carrying systems found in nature contain iron or copper; haemocyanin, the oxygen-carrier in snail's blood, contains Cu(I). Complexes of other transition metals, such as cobalt or iridium, have been shown to combine reversibly with molecular oxygen.

There is an enormous variety of oxygenases and related enzymes; like the oxygen carriers, most contain some transition element such as iron, copper or molybdenum. Among the enzymes associated with oxidation by molecular oxygen the following types have been distinguished[76]:

(1) Oxygenases proper, in whose operations both atoms of the O_2 molecule are transferred to the substrate. Among them is pyrocatecholoxidase, which catalyses the oxidation of catechol:

OH OH $\xrightarrow{O_2^{18}}$ O^{18} OH OH O^{18}

[72] N. Uri, in *Autoxidation and Antioxidants* (ed. W. O. Lundberg), Interscience, New York, **1** (1961) 55.

[73] L. H. Vogt, H. M. Faigenbaum and C. E. Wiberley, *Chem. Rev.* **63** (1963) 269; C. Manwell, *Ann. Rev Physiol.* **22** (1960) 191.

[74] O. Hayaishi (ed.), *Oxygenases*, Academic Press, New York (1964).

[75] J. Krog, in *Oxygen in the Animal Organism*, Pergamon Press, New York (1964).

[76] S. Fallab, *Z. Naturw.-med. Grundlagenforschung* **1** (1963) 333.

(2) Hydroxylases, such as monophenol oxidase, which transfer only one oxygen atom from O_2 to substrate:

$$\text{Me}_2\text{C}_6\text{H}_3\text{OH} + O_2^{18} + XH_2 \rightarrow \text{Me}_2\text{C}_6\text{H}_2(\text{OH})(O^{18}H) + X + H_2O^{18}$$

(3) Electron transferases, such as laccase, in the course of whose action both atoms of O_2 are reduced to water:

$$\text{HO-C}_6\text{H}_4\text{-OH} \xrightarrow{O_2^{18}} \text{O=C}_6\text{H}_4\text{=O} + 2H_2O^{18}$$

Some of the enzymic systems, such as the cytochrome oxidase system, are extremely complex and imperfectly understood. The formation of Fe(IV) has been proposed as part of the mechanism of operation of some iron-containing oxygenase systems[77].

Autoxidation and homogeneous catalysis. In the homogeneous oxidation of metal ions in aqueous solution[78, 79] at least two types of mechanism have been proposed. In one of these oxygen is reduced to water by a series of one-electron steps:

$$O_2+H^++e^- \rightarrow HO_2\cdot \quad \text{estimated}^{77}\ E^\circ = -0.32\ \text{V}$$
$$HO_2\cdot+H^++e^- \rightarrow H_2O_2 \quad \text{estimated}\ E^\circ = +1.68\ \text{V}$$
$$H_2O_2+e^- \rightarrow OH^-+HO\cdot \quad \text{estimated}\ E^\circ = +0.80\ \text{V}$$
$$HO\cdot+H^++e^- \rightarrow H_2O \quad \text{estimated}\ E^\circ = +2.74\ \text{V}$$

As is obvious from the E° values, the first of these steps is energetically unfavourable, whereas the rest all proceed with release of energy. Moreover, the direct reduction of O_2 to H_2O_2 is also energy-releasing:

$$O_2+2H^++2e^- \rightarrow H_2O_2 \quad E^\circ = 0.68\ \text{V}.$$

In the alternative mechanism, the first step involves transfer of two electrons, giving H_2O_2 directly and avoiding the energetically awkward formation of $HO_2\cdot$. In some cases there is evidence to indicate that the second mechanism operates. The autoxidation of V(II), for example, produces polynuclear V(III) species whose formation can be understood in terms of the steps below:

$$V^{+2}+\tfrac{1}{2}O_2 \rightarrow VO^{+2}$$
$$VO^{+2}+V^{+2} \rightarrow [V^{III}OV^{III}]^{+4}$$

Measurements of rates of autoxidation of transition metals clearly imply that in most cases studied the transfer of electrons from metal to O_2 takes place within some complex in which O_2 is bound to the metal, though some of the evidence is indirect; the transfer of

[77] P. George, in *Oxidases and Related Systems* (ed. T. E. King, H. S. Mason and M. Morrison), Wiley, New York (1965), p. 3.

[78] S. Fallab, *Angew. Chem., Int. Edn.* **6** (1967) 496.

[79] H. Taube, *J. Gen. Physiol.* **49** (1965) 29.

protons to combined O_2 from water or OH groups also bound to the metal may be important too.

The autoxidation of organic compounds in the liquid phase is catalysed by a variety of metal ions. The process that has been most extensively studied is the oxidation of olefins; this is catalysed by ions of such metals as[80] iron, cobalt, nickel, copper or zinc. The catalysed process occurs through a radical mechanism, the initiation of which is brought about in most cases by the interaction of olefin hydroperoxide with the metal ion. The redox behaviour of the metal has an important influence in the course of the reaction, and, indeed, for some metals the chief function of molecular oxygen may be to reoxidize the metal, which itself acts as the primary oxidizing agent and is reduced in oxidizing the substrate. In other systems, however, there is evidence that the oxidizing species is a complex of O_2 with the metal; some, but not all, of the metal–O_2 complexes that have been isolated are effective oxidants[81].

Chemisorption and Heterogeneous Catalysis

Oxygen is chemisorbed by almost all metals, by many metal oxides and by a variety of other materials; this chemisorption is involved in the heterogeneous catalysis of the reactions of molecular oxygen[82, 83]. Heats of adsorption of oxygen on metals are high; adsorption is rapid, and it is usually supposed that oxygen is present on the surface as monatomic species. This view is generally supported by results of low-energy electron diffraction[84]. The surface of nickel on which oxygen has been adsorbed shows some of the properties of a surface oxide[85]. A study[86] by mass spectrometry of the flash desorption of O_2 chemisorbed on tungsten ribbon indicates that only monatomic oxygen species or WO_n (n = 1–3) are evolved, so it seems unlikely that significant concentrations of diatomic oxygen species (O_2 or O_2^-) are present on the surface. On many metals (among them tungsten and silver)[87] at least two modes of chemisorption have been recognized; on silver it has been suggested that one mode involves the adsorption of peroxy-species, and experiments with $^{18}O_2$ were taken as confirming this view, but it now seems that insufficient allowance was made for the retention of unlabelled oxygen in the silver metal catalyst even after degassing at 500° under vacuum[88]. The inability of the oxygenated silver surface to catalyse the *ortho–para*-hydrogen conversion also implies the absence of such paramagnetic species as O_2. On the other hand, silver does catalyse the oxidation of cumene to cumene hydroperoxide[89].

Chemisorption on metal oxides[90] is closely connected with the presence of lattice defects or with metals that can easily be oxidized or reduced. Defects induced by radiation may be associated with chemisorption of O_2. Furthermore, some e.s.r. signals associated with

80 A. Chalk and F. Smith, *Trans. Faraday Soc.* **53** (1957) 1214.

81 E. A. V. Ebsworth and J. A. Connor, *Advances in Inorganic Chemistry and Radiochemistry* (ed. H. J. Emeléus and A. G. Sharpe), Academic Press, New York, **6** (1964) 279.

82 D. O. Hayward and B. N. W. Trapnell, *Chemisorption*, 2nd edn., Butterworths (1964).

83 J. M. Thomas and W. T. Thomas, *Introduction to Principles of Heterogeneous Catalysis*, Academic Press, New York (1967).

84 A. J. Pignocco and G. E. Pellissier, *J. Electrochem. Soc.* **112** (1965) 1188.

85 M. W. Roberts and B. R. Wells, *Disc. Faraday Soc.* **41** (1966) 162.

86 B. McCarroll, *J. Chem. Phys.* **46** (1967) 863.

87 J. H. Singleton, *J. Chem. Phys.* **47** (1967) 73.

88 Y. L. Sandler and W. M. Hickam, *Proc. 3rd Int. Congr. Cat.* **1** (1964) 227.

89 J. H. deBoer, *Discussion, 2nd Int. Conf. Surface Activity*, Butterworths, **II** (1957) 337.

90 F. S. Stone, *Adv. Catalysis* **13** (1962) 1.

91 J. H. Lumford and J. P. Jayne, *J. Chem. Phys.* **44** (1966) 1487.

lattice defects disappear in many systems when oxygen is adsorbed[91]. The chemisorbed species may be O_2, O_2^-, $O\cdot$, O^- or O^{-2}, and the formation of such polymers as O_4 or O_4^- has been suggested[92, 93]; the electrons necessary for the formation of the reduced species must come, of course, from the lattice. In all cases, however, oxygen will behave as an electron sink on adsorption; for this reason, adsorption of oxygen on oxides which are *n*-type semiconductors will be limited, whereas on *p*-semiconductors the same limitation will not hold. The e.s.r. spectra of zinc[91, 94] or titanium[95] oxides on which O_2 has been chemisorbed have been interpreted as showing the presence of O_2^- and perhaps O_2, but none of the signals assigned to these species was observed[95] for O_2 chemisorbed on MoO_3

TABLE 17. HETEROGENEOUS CATALYSIS OF SOME REACTIONS OF O_2

Other reactant(s)	Important product(s)	Catalyst
H_2	Water	Metals [a]
CO	CO_2	Cu_2O, NiO, etc. [b]
SO_2	SO_3	V_2O_5 or Pt [a, b]
NH_3	N oxides	Pt or Rh
C_2H_4	Ethylene oxide	Ag [a, b]
Propylene	Acreolin	Cu_2O [b]
Propylene + NH_3	Acrylonitrile	Bi(III)/Mo(VI) oxides [d]
Olefins	Dienes, aldehydes, acids, CO, CO_2	Bi(III)/Mo(VI) oxides [b, e]
Benzene	Maleic anhydride	V_2O_5 [b]
C_2H_4 + HCl	Ethylene dichloride	$CuCl_2$ [c]
CH_4 + H_2O	Formaldehyde	Cu or Ag [a]

[a] G. C. Bond, *Catalysis by Metals*, Academic Press (1962).
[b] J. M. Thomas and W. T. Thomas, *Introduction to Principles of Heterogeneous Catalysis*, Academic Press (1967).
[c] E. F. Edwards and T. Weaver, *Chem. Eng. Prog.* **61** (1965) 21. This is important because the product may be cracked to give vinyl chloride.
[d] C. R. Adams and T. J. Jennings, *J. Catalysis* **3** (1964) 549.
[e] It has been suggested that dehydrogenation is specifically associated with reaction with lattice oxygen.

or NiO. On MgO that had previously been γ-irradiated, adsorption of O_2 was associated with the appearance of a signal in the e.s.r. spectrum that was assigned to O_2^-, and with O_2 enriched in O^{17} the signal showed hyperfine structure that could be interpreted in terms of a mixture of $O^{18}O^{17}$ and O_2^{17} in the proportions required by the statistical distribution of O^{18} and O^{17}. There appeared to be no oxygen exchange with the lattice over several months at room temperature[96]. When adsorption is associated with rapid exchange between $^{16}O_2$ and $^{18}O_2$ it has been argued that the O–O bonds must be broken on adsorption, but more recently it has been suggested that such exchange could occur through the formation of

92 P. Chan and A. Steinmann, *Surface Sci.* **5** (1966) 267.
93 K. Hirota and M. Chono, *Sci. Papers Int. Phys. Chem. Res. Tokyo* **58** (1964) 115.
94 R. J. Kokes, *Proc. 3rd Int. Congr. Cat., Amsterdam* **1** (1964), 484.
95 P. F. Cornaz, J. H. C. Van Hoof, F. J. Pluijm and G. C. A. Schuit, *Disc. Faraday Soc.* **41** (1966) 290.
96 A. J. Tench and P. Holroyd, *Chem. Communs.* 1968, 471.
97 G. K. Boreskov, *Disc. Faraday Soc.* **41** (1966) 263.

O_4 or O_4^- species[93] on the surface. Formation of O^{-2} should lead to exchange between the metal oxide and the oxygen gas. Many metal oxides which have been preheated in O_2 are effective catalysts for the exchange between $^{16}O_2$ and $^{18}O_2$ only at high temperatures; the exchange has the same rate parameters as the exchange between the O_2 gas and lattice oxide ions, and is presumed to involve lattice oxide. On the other hand, oxides preheated under vacuum are effective catalysts for $^{16}O_2/^{18}O_2$ exchange at low temperatures, under conditions where exchange with lattice oxide does not occur[97]. The rates of isotopic exchange on many metal oxides have been determined, and the kinetic parameters discussed in terms of possible mechanisms of exchange[98]. The unusual ease of exchange of O_2 with Cu_2O has been related to the unusual structure of the oxide[90].

Catalytic oxidation at the surfaces of metal oxides may take place through direct interaction between the adsorbed species and the material to be oxidized; alternatively, the substrate may be oxidized by lattice oxide ions, with reduction of the catalyst, which is then reoxidized by oxygen. The oxidation of SO_2 or of hydrocarbons on a catalyst of V_2O_5 is believed to follow the latter mechanism[83]; on the other hand, experiments using labelled oxygen[83] indicate that lattice oxide is not involved in the oxidation of CO with O_2 on NiO. Some examples of reactions of molecular oxygen catalysed by a variety of solid materials are given in Table 17. The unique oxidation of ethylene to ethylene oxide at a silver catalyst is remarkable, since most olefins on all other catalysts give aldehydes, acids or (ultimately) CO and CO_2.

Photochemical Oxygenation: Reactions of Singlet O_2

Electronically excited O_2 molecules are not efficiently produced by radiation at frequencies less than 50 000 cm^{-1}. Many reactions of molecular oxygen, however, particularly those with unsaturated or aromatic organic compounds, may be induced photochemically, either with or without a sensitizer; compounds that act as sensitizers are usually dyes or other strong absorbers of light (e.g. rose bengal, eosin or chlorophyll). The system initially absorbs energy from radiation by electronic excitation of either the substrate or the sensitizer; this energy must lead to the production of some active intermediate. Various mechanisms have been proposed[99]:

(a) Free radicals may be produced by dissociation of the electronically excited substrate. The photochemical oxidation of aldehydes, for example, is believed to take place by initial formation of radicals[100]. Such processes in which free radicals are involved are called "type-1 photochemical oxygenations".

(b) Reactions involving only excited molecules are called type-2 processes. The species initially excited by radiation may be either the substrate or the sensitizer. In either case, the state formed is usually the first excited singlet state, here labelled $^1S_1$. This can transfer energy to O_2 with conservation of spin:

$$^1S_1 + {}^3O_2 \rightarrow {}^3S_1 + {}^1O_2$$

where $^3S_1$ is the lowest triplet state of S (assumed to have a singlet ground state). Singlet O_2 can only be formed in this way[99] if the energy gap $^1S_1 \rightarrow {}^3S_1$ is greater than the gap $^3O_2 \rightarrow {}^1O_2$, which for $O_2(^1\Delta_g)$ is 22.6 kcal $mole^{-1}$; if the energy transferred is greater than

[98] E. R. S. Winter, *J. Chem. Soc.* A, 1968, 2889.
[99] K. Gollnick, *Adv. Photochem.* (Wiley) **6** (1968) 1.
[100] M. Niclause, J. Lemaire and M. Letort, *Adv. Photochem.* (Wiley) **4** (1966) 25.

37 kcal mole^{-1}, some $O_2(^1\Sigma_g^+)$ may also be produced, until at a value of 49 kcal mole^{-1} about ten times as much $O_2(^1\Sigma_g^+)$ as $O_2(^1\Delta_g)$ may be formed. If the excited sensitizer or substrate is in an excited triplet state, however, singlet O_2 can be produced with the other species (substrate or sensitizer) in its ground state:

$$^3S_1 + {}^3O_2 \rightarrow {}^1S_0 + {}^1O_2$$

This is because the interaction of molecules with spin vectors of magnitude 1 can give a singlet collision complex, which can then dissociate to form singlet O_2 and the substrate or sensitizer in its singlet ground state[101]. It is also possible that the active intermediate might be the excited substrate itself, produced by the initial absorption of radiation; alternatively, the excited substrate or sensitizer might form a longer-lived activated complex (or moloxide) with O_2. Recently it has proved possible to study O_2 containing substantial amounts of singlet O_2; this species is formed by the action of an electrodeless discharge on oxygen gas[102], by the reaction between hydrogen peroxide and hypochlorite or chlorine in aqueous solution[103], or by the decomposition at $-78°$ of the adduct of triphenyl phosphite with ozone[104]. The product in the last case has been shown to contain $O_2(^1\Delta_g)$ by its e.p.r. spectrum[105]. The reactions of singlet O_2 from these different sources are very like the photochemically induced reactions of O_2, not only in products but also (in some cases) in kinetics. It therefore seems very likely that singlet O_2 is the active intermediate in photochemical oxygenations of type 2. Of the two mechanisms described above for the production of singlet O_2, the latter is by far the more important; the triplet senstitizer or substrate may be formed by intersystem crossing.

The products of many photochemical oxygenations depend on the solvent and the conditions, and are not always the same in sensitized and in unsensitized processes; with 1,4-dienes, O_2 may behave as a dienophile to give products of what is effectively a Diels–Alder reaction. Peroxides, hydroperoxides or substituted allyl alcohols may be prepared in this way. Unsaturated sulphur compounds or sulphides are oxidized photochemically by O_2 in the presence of a sensitizer; amines are also oxidized, but it is not clear what the primary products are[99].

The O_2 Electrode[106]

The electrochemistry of oxygen is of great and increasing importance, impinging as it does on batteries and fuel cells, on corrosion and on the analytical chemistry of oxygen. The reversible potential for the reaction

$$O_2 + 4H^+ + 4e^- \rightleftharpoons 2H_2O$$

(see Table 5, p. 691) has been calculated from thermodynamic data as +1.229 V, and the reversible rest potential has been observed for platinum that has been anodized and heated in pure O_2 or treated with HNO_3; such an electrode is probably covered with a uniform film of oxide. The electrode behaves reversibly so long as the current drawn is less than

101 J. S. Griffith, *Oxygen in the Animal Organism* (ed. F. Dickens and E. Neil), Pergamon Press, New York (1964).

102 R. P. Wayne, *Adv. Photochem.* (Wiley) **7** (1969) 311.

103 C. S. Foote, S. Wexler, W. Ando and R. Higgins, *J. Am. Chem. Soc.* **90** (1968) 975.

104 R. W. Murray and M. L. Kaplan, *J. Am. Chem. Soc.* **90** (1968) 4161.

105 E. Wassermann, R. W. Murray, M. L. Kaplan and W. A. Yagen, *J. Am. Chem. Soc.* **70** (1968) 4160.

106 T. P. Hoar, *The Electrochemistry of Oxygen*, Interscience (1968).

40 μA cm^{-2}, and it follows that the mechanism for the reduction of O_2 under these conditions must be precisely the reverse of the mechanism for O_2 evolution. It is believed that in the reversible reduction of O_2 the O–O bond is broken before electron transfer. The reversible potential value has also been observed for bright platinum electrodes in acid solutions, but here the process is irreversible and the mechanism is different: the system in this case has been described as a polyelectrode, and in the reduction of O_2 under these conditions peroxide is an intermediate. The rest potential of an oxygen electrode on other noble metals rarely corresponds to the reversible potential, and depends very much on the nature of the metal surface under O_2 (about which little is known); for this reason the electrochemical behaviour of O_2 is closely related to chemisorption.

The overvoltage for evolution of O_2 is substantial. Oxygen electrodes are very easily polarized, and the polarization depends on the nature of the electrode and on the current density. The mechanism for the evolution of O_2 from acid solutions on platinum is believed to involve the formation of adsorbed OH radicals, giving oxygen atoms, which then combine to give O_2 molecules. The mechanism in alkaline solution may be different. Cathodes at which O_2 is reduced are also very easily polarized. On platinum in acid solutions the slow step is the formation of O_2^- on the surface of the electrode from absorbed O_2; this is protonated and followed by further reduction and protonation to give H_2O_2. The overvoltage for reduction of O_2 in alkali is lower, and under these conditions it has been suggested that the intermediate is HO_2^-.

Particular impetus to the study of the oxygen electrode has come from the development of the H_2/O_2 fuel cell as a source of energy for many systems, including space craft[107]. In a fuel cell the reaction between hydrogen and oxygen is used to provide electrical energy; a hydrogen electrode forms the anode and an oxygen electrode the cathode. It is important to prevent diffusion of the electrode gases through the electrolyte, as they may then react chemically without the production of electrical energy; for this reason the gases are usually allowed to diffuse through porous electrode materials (such as nickel or carbon) on which some catalytic material such as platinum is held. The electrolyte solvent may be aqueous, but salt melts (with anions that are not easily oxidized or reduced) have been widely used for high-temperature cells. The electrochemical behaviour of the oxygen electrode in fused salts is complicated. In fused nitrates, for example, reversible systems have been described and E° values determined, but the electrode appears to behave under some conditions as a one-electron and under others as a two-electron system[108, 109]. It seems that in silica-free and thoroughly dried nitrate melts the concentration of oxide ion at 229° cannot be appreciable[109].

1.9. REMOVAL OF O_2

Oxygen may have to be removed from various systems for a variety of reasons; there are very many ways of doing this. Gases may have to be purified of the last traces of oxygen for the study of systems very sensitive to oxygen; this may be achieved using physical adsorption[111], by adding hydrogen and treating the gas with a catalyst to remove O_2 as water[110], or by treatment with alkali metals (liquid or vapour) or with the amalgams of

107 K. R. Williams (ed.), *An Introduction to Fuel Cells*, Elsevier (1966).
108 R. N. Kust, *J. Phys. Chem.* **69** (1964) 3662.
109 P. G. Zambonin and J. Jordan, *J. Am. Chem. Soc.* **91** (1969) 2225.
110 B. Schaub and J. P. Molin, *Vide* **20** (1965) 281.
111 H. C. Kauchmann and E. F. Yeudall, US Pat. 3,169,845 (1965).

aluminium, calcium or magnesium (which are said to reduce the residual oxygen content to 10^{-25} vol.%)[112]. The level of O_2 in N_2 can be reduced to $\sim 10^{-5}$ vol.% by passing the gas over heated copper[113]. The radical anion $C_6N_4^-$, made from $(NC)_2C{=}C(CN)_2$ and an alkali metal, is a very good oxygen scavenger[114]. Oxyanions of sulphur in low oxidation states (e.g. SO_3^{-2}, $S_2O_4^{-2}$) are effective in removing oxygen from aqueous systems[115]; hydrazine has been proposed as a deoxygenator for boiler plant water[116], and oxygen can be removed from closed hot water systems by electrolysis with a steel cathode and an aluminium anode[117].

1.10. ANALYSIS

For O_2

In a mixture of gases, molecular oxygen may be determined by gas chromatography, or by magnetic methods, which make use of the fact that O_2 is one of the few common gases that is paramagnetic[118]. Alternatively, the gas mixture may be treated with a solution that absorbs O_2, such as alkaline pyrogallol or 20% sodium dithionite containing sodium anthraquinone β-sulphonate. In aqueous systems O_2 is determined by cathodic reduction using a polarograph or some other suitable electrode system[119], or by oxidation of Mn(II) in alkaline solution (the Winkler method). The Mn(III), which is formed quantitatively, oxidizes iodide quantitatively to iodine when the solution is acidified[120]. There are special problems associated with the determination of oxygen in water in industrial systems[121].

For O in Compounds of Oxygen

Oxygen in organic compounds may be determined by pyrolysis in a stream of nitrogen; the gases evolved are treated with carbon at high temperatures, when all the oxygen is converted into CO. The CO may then be estimated either as a reducing agent, by treatment with I_2O_5 or (after oxidation) as CO_2. This is usually called the Unterzaucher method; it is not always satisfactory, but recent modifications have made it much more reliable[122]. Oxygen in compounds may also be determined by activation analysis[123].

1.11. TOXIC EFFECTS OF AN EXCESS OF O_2

Oxygen is essential to animal life; but too much oxygen can be fatal. Some of the ill effects of hyperoxia were noticed by Lavoisier, but oxygen toxicity has become a far more

112 E. Steinmetz, K. W. Lange and K. K. G. Schmitz, *Chem. Ing. Tech.* **36** (1964) 1103.

113 G. Brauer, *Handbook of Preparative Inorganic Chemistry*, 2nd edn., Academic Press, New York (1963), p. 458.

114 S. I. Weissman, US Pat. 3,222,385 (1965).

115 V. Kadlec, J. Wuensch and A. Brodsky, Czech Pat. 116,230 (1965).

116 H. Anders, *Zucker* **19** (1966) 552.

117 V. L. Loser, *Energetik* **13** (1965) 5.

118 H. H. Willard, L. L. Merritt and J. A. Dean, *Instrumental Methods of Analysis*, 4th edn., Van Nostrand, New York (1965).

119 I. M. Kolthoff and J. J. Lingane, *Polarography*, 2nd edn., Interscience, New York (1952); J. P. Payne and D. W. Hill, *Oxygen Measurements in Blood and Tissue, and their Significance*, Churchill, London (1966).

120 H. A. Laitinen, *Chemical Analysis*, McGraw-Hill, New York (1960).

121 British Standards, BS 2690, part 2.

122 R. Belcher, G. Ingram and J. C. Majer, *Talanta* **16** (1969) 881.

123 J. R. Vogt and W. D. Ehmann, *Radiochimia Acta* **4** (1965) 24.

serious problem in recent years with the increasing use of oxygen gas at high partial pressure to relieve hypoxia, and with the extension of man's activities to regions (such as space or the depths of the sea) where he must rely on artificial atmospheres.

The lungs of guinea-pigs, when subjected to hyperoxia for an extended period, show adverse effects that include oedema, hepatization, pleural effusion, extravasion of red blood corpuscles, haemorrhage, atelectasis and consolidation to a degree where the lung fails to function, oxygen is not supplied to the organism, and the animal dies from what amounts to hyperoxic anoxia[124]. Adult humans are more resistant than are small laboratory animals to these adverse effects of an excess of oxygen, but continued exposure to oxygen at partial pressures greater than 0.75 atm leads to lung damage in a time that depends inversely on the oxygen pressure but is very dependent on the particular subject. Continued exposure to oxygen at partial pressure of 1 atm or more leads to[124, 125] fever, vomiting, substernal pain and decreased vital capacity, and oxygen at 3–5 atm precipitates convulsions resembling those associated with grand mal epilepsy[126]. Retrolental fibroplasia in infants, induced by O_2 in excess, may result in blindness[124].

The ways in which an excess of O_2 operates on the organism are extremely complex. Enzyme systems may be poisoned either directly (by reaction with O_2) or by peroxides formed as a result of other reactions of O_2; however, no effects on isolated enzyme systems are fast enough to explain the speed of onset of convulsions associated with O_2 poisoning in intact animals[127]. High pressures of O_2 saturate the blood haemoglobin and increase the amount of oxygen dissolved in blood and other body fluids, and so influence indirectly the amount of CO_2 dissolved and the general balance of buffering and of gas transport in the body[126]. There is evidence that the endocrine system is involved in some aspects of oxygen poisoning[124], but the processes and mechanisms are extremely complex and do not yet seem to be fully understood.

1.12. PRODUCTION AND USES OF O_2

Oxygen is produced on a very large scale, and the amount produced is continually rising; in the United States, 10^6 tons were produced in 1966. Of this, over half went into the production of steel. The use of O_2 as a direct oxidant in processes involving organic systems is also growing[128]. On a smaller scale, oxygen is important in high-temperature processes such as welding; as a refrigerant, liquid oxygen is less popular than liquid nitrogen because of the hazards associated with liquid oxygen, but in some cases its higher boiling point gives it some advantages. The development of space flight has given a very strong impulse to the associated development of the technology of liquid oxygen. Finally, it would be impossible to maintain life under extremely abnormal conditions—in space, under water or in some clinical states—if O_2 gas were not freely available.

[124] J. W. Bean, in *Oxygen in the Animal Organism*, Pergamon Press, New York (1964), p. 455.
[125] A. P. Morgan, *Anaesthesiology* **29** (1968) 570.
[126] C. J. Lambertsen, in *Handbook of Physiology*, American Physiological Society, New York (1965), Section III, Vol. 2, p. 1027.
[127] H. C. Davies and R. E. Davies, *ibid.*, p. 1047.
[128] M. Sittig, *Combining Hydrocarbons and Oxygen for Profit*, Noyes Development Co., Princeton (1968).

2. OXYGEN ATOMS AND IONS

2.1. ATOMIC OXYGEN

Oxygen atoms play an important part in reactions in the earth's atmosphere[129], and have been extensively studied in the laboratory; but, as with many extremely reactive species, there are serious problems associated with producing and detecting them. In most of the early studies, oxygen atoms were generated by passing a microwave or an electric discharge through oxygen gas[130]. Unfortunately, however, oxygen atoms are not the only active species formed under these conditions; oxygen molecules are excited electronically, and the presence of impurities containing hydrogen leads to the formation of other very reactive radicals[131]. In recent studies, some of these difficulties have been avoided by passing the discharge through argon containing a very small proportion of very pure oxygen[132, 133]; it is hoped that under these circumstances oxygen atoms are the only active species formed. The thermal decomposition of ozone[134] or nitrous oxide give oxygen atoms, which are also produced when nitrogen atoms (themselves formed by discharging nitrogen) react with nitric oxide[135]:

$$N+NO \rightarrow N_2+O.$$

Photolysis of oxygen also gives oxygen atoms, but excited O_2 molecules may also be formed, and the reactions of the two may not be easy to distinguish[133]; the photolysis of N_2O, however, produces oxygen atoms, together with N_2, which is inert[135]. The photolysis of NO_2, giving O and NO, is less satisfactory[136, 137]. All these methods give $O(^3P)$ atoms; photolysis of ozone, or photolysis of N_2O, CO_2 or O_2 with radiation of higher energy, gives[138, 139] $O(^1D)$ or even $O(^1S)$, while $O(^5S)$ may have been detected in the products of a microwave discharge in O_2 by molecular beam magnetic resonance[140].

Oxygen atoms have been detected and estimated using chemical and physical techniques[133]. Chemically, the best-established technique is the NO_2 titration. Oxygen atoms react very fast indeed with NO_2, forming NO [136]:

$$O+NO_2 \rightarrow O_2+NO$$

The NO thus produced reacts much more slowly with any excess of oxygen atoms to reform NO_2, and this reaction produces a strong glow:

$$NO+O \rightarrow NO_2+h\nu$$

Therefore NO_2 is added to the system until the glow is sharply extinguished; then all the oxygen atoms are consumed in the very fast first process and $[O]=[NO_2]$. Absorption and emission spectroscopy have proved useful, but the emission spectra of certain species

129 *Disc. Faraday Soc.* **37** (1964); *Can. J. Chem.* **47** (1969) 1703.
130 F. Kaufman, *Prog. Reaction Kinetics* **1** (1961) 1.
131 F. Kaufman and J. R. Kelso, *Disc. Faraday Soc.* **37** (1964) 26.
132 M. A. A. Clyne, D. J. McKenny and B. A. Thrush, *Trans. Faraday Soc.* **61** (1965) 2701.
133 H. I. Schiff, *Can. J. Chem.* **47** (1969) 1905.
134 F. Kaufman and J. R. Kelso, *J. Chem. Phys.* **40** (1964) 1162.
135 D. G. Williamson and K. D. Bayes, *J. Phys. Chem.*, **73** (1969), 1232.
136 R. J. Cvetanovic, *J. Chem. Phys.* **23** (1955) 1203.
137 A. A. Westerberg and N. deHaas, *J. Chem. Phys.* **50** (1969) 707.
138 P. Warneck and J. O. Sullivan, *Ber. Bunsensges. Phys. Chem.* **72** (1968) 159.
139 R. A. Young, G. Black and T. S. Slanger, *J. Chem. Phys.* **49** (1968) 4758, 4769.
140 G. O. Brink, Report AD 643526 (1966) (*Chem. Abstr.* **67** (1967) 59232p).

TABLE 18. METHODS OF PRODUCTION AND SOME REACTIONS OF OXYGEN ATOMS

State	Methods of production	Some reactions studied
^3P	Electric or microwave discharge; Hg-sensitized photolysis of N_2O; photolysis of O_2, N_2O, NO_2; shock tubes; flames	With O [a], O_2 [b], NO [c], NO_2 [d], H_2 [e], Cl_2 [f], hydrocarbons, [g, h] acetylene (in matrices) [i], P_4 [j], PH_3 [j], S [k], Se [k], C_3O_2 [l], CCl_4 [m], H_2S [n], $(CN)_2$ [o], CN [p], ClCN [p'], ICN [q], B_5H_9 [r], CO [l], COS [s], CS_2 [s], Cl_2O [t]
^1D	Photolysis of O_3, CO_2, N_2O (in absence of Hg); microwave discharge	With O_2 [u], N_2 [v], CO_2 [w], N_2O [x], H_2O [y]
^1S	Photolysis of O_2 [z], CO_2 [aa], N_2O [bb]	Quenched by H_2, O_2, CO, CO_2, H_2O, N_2O, not by rare gases or N_2

[a] F. Kaufman, *Prog. Reaction Kinetics* **1** (1961) 1.
[b] M. F. R. Mulcahy and D. J. Williams, *Trans. Faraday Soc.* **64** (1968) 59.
[c] A. McKenzie and B. A. Thrush, *Chem. Phys. Letters* **1** (1968) 681.
[d] F. S. Klein and J. T. Herron, *J. Chem. Phys.* **41** (1964), 1285.
[e] A. A. Westenberg and N. DeHaas, *J. Chem. Phys.* **50** (1969) 2512.
[f] H. Niki and B. Weinstock, *J. Chem. Phys.* **47** (1967) 3249.
[g] R. Cvetanović, *Adv. Photochem.* (Wiley) **1** (1963) 115.
[h] C. Mitchell and J. P. Simons, *J. Chem. Soc.* (B) 1968, 1005.
[i] G. C. Pimentel, *Proc. Am. Petrol. Inst.* **41** (1961) 189.
[j] P. B. Davies and B. A. Thrush, *Proc. Roy. Soc.* A, **302** (1967) 243.
[k] F. Cramarossa, E. Molinari and B. Roio, *J. Phys. Chem.* **72** (1968) 84.
[l] H. Van Weyssenhoff, S. Dondes and P. Harteck, *J. Am. Chem. Soc.* **84** (1962) 1526.
[m] A. Y.-M. Ung and H. I. Schiff, *Canad. J. Chem.* **40** (1962) 486.
[n] E. L. Merryman and A. Levy, *J. Air Poll. Contr. Ass.* **17** (1967) 800.
[o] D. W. Setser and B. A. Thrush, *Proc. Roy. Soc.* A, **288** (1965) 275.
[p] J. C. Boden and B. A. Thrush, *Proc. Roy. Soc.* A, **305** (1968) 107.
[p'] P. B. Davies and B. A. Thrush, *Trans. Faraday Soc.* **64** (1968) 1836.
[q] J. F. Grady, C. G. Freeman and L. F. Phillips, *J. Phys. Chem.* **72** (1968) 743.
[r] J. P. Rosenkrans, *Diss. Abstr.* **25** (1964) 1608.
[s] A. A. Westerberg and N. deHaas, *J. Chem. Phys.* **50** (1969) 707.
[t] C. G. Freeman and L. F. Phillips, *J. Phys. Chem.* **72** (1968) 3025.
[u] R. A. Young, G. Black and T. G. Slanger, *J. Chem. Phys.* **49** (1968) 4758.
[v] R. G. W. Norrish and R. P. Wayne, *Proc. Roy. Soc.* A, **288** (1965) 200.
[w] N. G. Moll, D. R. Clutter and W. E. Thompson, *J. Chem. Phys.* **45** (1966) 4469.
[x] P. Warnek and J. O. Sullivan, *Ber. Bunsensges. Phys. Chem.* **72** (1968) 159.
[y] W. D. McGrath and R. G. W. Norrish, *Proc. Roy. Soc.* A, **234** (1958) 317.
[z] S. V. Filseth and K. H. Welge, *J. Chem. Phys.* **51** (1969) 839.
[aa] F. Stuhl and K. H. Welge, *Canad. J. Chem.* **47** (1969) 1870.
[bb] G. Black, T. G. Slanger, G. A. St. John and R. A. Young, *Canad. J. Chem.* **47** (1969) 1873.

are not observed in some systems. Catalytic probes, heated by the recombination of atoms on their surfaces, and special pressure gauges inside which atoms recombine, were extensively employed at one time, but at present the physical techniques most widely used in the measurement of concentrations of oxygen atoms are mass spectrometry and e.s.r.[133]; gas chromatography is very extensively used to analyse reaction products. Mass spectrometry, particularly when used to give appearance potentials, is a most powerful technique. Of the atomic species of oxygen that are usually studied, only the ground state O(^3P)

gives an e.s.r. signal (six lines at room temperature, due to transitions between different M_J states of the $^3P_2$ and $^3P_1$ levels)[141]. These signals may be calibrated using the e.s.r. spectrum of O_2. The excited states $O(^1D)$ and $O(^1S)$ have so far only been identified by their emission spectra and possibly [for $O(^1D)$ and $O(^5S)$] by molecular beam magnetic resonance; but indirect evidence comes from the processes in which they are formed and by which they react, for their reactions are different from those of $O(^3P)$. Some of the reactions of oxygen atoms that have been studied are set out in Table 18. A particularly interesting example is the reaction between $O(^1D)$ and CO_2, which has been shown by vibrational spectroscopy and by matrix techniques[142] to give the unstable radical CO_3. Many of the reactions are associated with strong after-glows, and some of the atmospheric glows are also associated with the reactions of oxygen atoms[143].

TABLE 19. SOME REACTIONS OF IONIZED SPECIES DERIVED FROM OXYGEN OR O_2

Reactions of O^- [a]		
	$O^- + N_2 + He$	$\rightarrow N_2O^- + He$
	$O^- + 2CO_2$	$\rightarrow CO_3^- + CO_2$
	$O^- + NO_2$	$\rightarrow O_2^- + NO$
	$O^- + NO_2$	$\rightarrow NO_2^- + O$
	$O^- + O_3$	$\rightarrow O_3^- + O$
Reactions of O_2^- [a]		
	$O_2^- + 2CO_2$	$\rightarrow CO_4^- + CO_2$
	$O_2^- + SO_2$	$\rightarrow SO_2^- + O_2$
Reactions of O^+ [b]		
	$O^+ + N_2$	$\rightarrow NO^+ + N$
	$O^+ + O_2$	$\rightarrow O + O_2^+$
	$O^+ + CO_2$	$\rightarrow CO + O_2^+$
Reactions of O_2^+ [b]		
	$O_2^+ + NO$	$\rightarrow NO^+ + O_2$
	$O_2^+ + Na$	$\rightarrow O_2 + Na^+$
	$O_2^+ + Na$	$\rightarrow NaO^+ + O$

[a] E. E. Ferguson, *Canad. J. Chem.* **47** (1969) 1815.
[b] W. L. Fite, *Canad. J. Chem.* **47** (1969) 1797.

2.2. ION–MOLECULE REACTIONS

The reactions of ionized species derived from oxygen atoms and molecules have been extensively studied in recent years[144]. The ions are produced in flames, in mass spectrometers or in molecular beams; some representative reactions are set out in Table 19.

[141] A. A. Westenberg and N. deHaas, *J. Chem. Phys.* **40** (1964) 3087.
[142] E. Weissberger, W. H. Breckenridge and H. Taube, *J. Chem. Phys.* **47** (1967) 1764.
[143] R. A. Young, *Can. J. Chem.* **47** (1969) 1927.
[144] C. F. Giese, *Adv. Chem. Ser.* **58** (1966) 20; J. F. Paulson, *ibid.*, p. 28; A. V. Phelps, *Can. J. Chem.* **47** (1969) 1784; W. L. Fite, *ibid.*, p. 1797; E. E. Ferguson, *ibid.*, p. 1815.

3. OXIDES AS A CLASS

3.1. INTRODUCTION

The formation of an element oxide M_xO_y from the reaction between the element M and molecular oxygen is a commonly observed process. Many metallic elements in a high state of purity react spontaneously with oxygen. This is in some respects a remarkable phenomenon when it is realized that the production of oxide ions from molecular oxygen is a strongly endothermic process [ΔH_f° (O^{2-}, g) = 220 kcal mole^{-1}]. The same is true of the energetics of oxidation of the element M [e.g. ΔH_f° (Fe^{3+}, g) = 1365 kcal mole^{-1}]. The driving force in the process is the high lattice energy of the element oxide.

It is possible to classify oxides under two general headings: acid–base character and structure.

In order to achieve satisfactory order within this section, the term oxide will be considered in its most rigorous sense. Thus only non-charged compounds of the type M_xO_y and $X_aY_b \ldots O_n$ will be considered and, specifically, oxy-anions such as carbonate, silicate, molybdate and tungstate, for example, will be ignored, as also will be oxy-halides, hydroxides and oxy-chalcogenides.

3.2. ACID–BASE CHARACTER OF SIMPLE OXIDES

The description of an oxide in terms of its acid–base character allows for a subdivision into four classes: acidic, basic, amphoteric and neutral.

An *acidic* oxide is one which, when dissolved in water, affords an acidic solution. For example phosphorus(V) oxide, P_4O_{10}, is completely hydrolysed in water to orthophosphoric acid,

$$P_4O_{10}+6H_2O \rightarrow 4H_3PO_4 \tag{1}$$

A *basic* oxide is one which, when dissolved in water, affords an alkaline solution. For example, calcium oxide, CaO, affords calcium hydroxide in aqueous solution,

$$CaO+H_2O \rightarrow Ca(OH)_2 \tag{2}$$

An *amphoteric* oxide is one which, under different conditions, exhibits the properties of both an acidic and a basic oxide. For example, antimony(III) oxide, Sb_4O_6, will react with both acids and bases,

$$Sb_4O_6+12H^+ \rightarrow 4Sb^{3+}+6H_2O \tag{3}$$
$$Sb_4O_6+nOH^- \rightarrow 4[Sb(OH)_6] \tag{4}$$

A *neutral* oxide is essentially inert and unreactive towards water. For example, carbon monoxide and nitrous oxide, which are almost insoluble in water, are classed as neutral oxides. This class is the smallest of the four.

Among the non-transition metal, or main group elements, the oxides of the most electropositive elements are basic while those of the more electronegative elements are acidic. A comparison between Na_2O and Cl_2O exemplifies this well:

$$Na_2O+H_2O \rightarrow 2NaOH \quad \text{basic oxide} \tag{5}$$
$$Cl_2O+H_2O \rightarrow 2HClO \quad \text{acidic oxide} \tag{6}$$

Between these two extremes there is a gradation in behaviour leading to an amphoteric oxide, although it is not usually possible to place a clear boundary between the three classes.

Na_2O	MgO	Al_2O_3	SiO_2	P_4O_{10}	SO_3	Cl_2O
B	B	AB	A	A	A	A

A = acidic B = basic AB = amphoteric

Within any one group, the acid character of any particular oxide M_xO_y, tends to decrease as the atomic number of the element increases.

N_2O_3	P_4O_6	As_4O_6	Sb_4O_6	Bi_2O_3
A	A	A	AB	B

In the case of an element, M, which forms more than one (stoichiometric) oxide and, particularly, where an element can exist in a number of different formal oxidation states, then the oxide of the highest oxidation state is more acidic than that of the lowest oxidation state. This may be expressed in an alternative form by stating that for the oxide M_xO_y the acidic character of the oxide decreases as the ratio y/x decreases. For example, among the oxides of chlorine, acidity decreases in the order

$$Cl_2O_7 > ClO_3 > ClO_2 > ClO > Cl_2O$$

and among the oxides of manganese,

$$Mn_2O_7 > MnO_2 > Mn_2O_3 > MnO$$

Although this qualitative classification is apparently quite straightforward, problems immediately arise when we inquire more closely into the reasons for the particular behaviour of any particular oxide. To a certain extent the acidic or basic character of an oxide may be explained in terms of the ionization behaviour of the element concerned. No satisfactory answer can be obtained by asking why Na_2O does not dissolve in water according to the hypothetical equation

$$Na_2O+H_2O \rightarrow 2H^+ + 2NaO^- \tag{7}$$

or why, conversely, Cl_2O does not behave according to

$$Cl_2O+H_2O \rightarrow 2Cl^+ + 2OH^- \tag{8}$$

The appropriate physical data are shown in Table 20.

TABLE 20

	First ionization potential	Electron affinity
Na	118	17
Cl	300	83

Values in kcal $mole^{-1}$ to nearest whole number.

These figures merely reflect that it is easier to ionize sodium than chlorine and, conversely, that the electron affinity of chlorine is greater than that of sodium. Consequently, the oxides will be expected to react according to eqns. (5) and (6) rather than as indicated by eqns. (7)

and (8). If the question is presented in terms of the contrast between the known behaviour of P_4O_{10} [eqn. (1)] and the hypothetical equation

$$\tfrac{1}{2}P_4O_{10}+5H_2O \rightarrow 2P^{5+}+10OH^- \qquad (9)$$

then the differences become much more apparent. The ionization energy of the P^{5+} ion (4075 kcal $mole^{-1}$) would appear to make reaction (9) very unfavourable unless there is a correspondingly greater solvation energy.

There remains the problem of amphoteric oxides. On the basis of the preceding argument we would expect the energy balance as expressed by electron affinity and ionization potential to be closer. It is difficult to know how much validity can be given to such a procedure, and so the appropriate figures are presented in Table 21.

TABLE 21. IONIZATION POTENTIALS AND ELECTRON AFFINITY OF THE ELEMENTS Na–Cl

Element	I_1 (eV)	E (eV)	Ion	$\sum_0^n I_n$ (eV)	Oxide
Na	5.14	0.54	Na^+	5.14	Na_2O
Mg	7.64	−0.22	Mg^{2+}	22.6	MgO
Al	5.98	0.20	Al^{3+}	53.2	Al_2O_3
Si	8.15	1.36	Si^{4+}	103.1	SiO_2
P	10.48	0.71	P^{5+}	176.8	P_4O_{10}
S	10.36	2.04	S^{4+}	116.1	SO_2
Cl	13.01	3.62	Cl^+	13.01	Cl_2O

I_1, first ionization potential.
E, electron affinity (R J. Zollweg, *J. Chem. Phys.* **50** (1969) 4251).

Table 21 shows that there is an increase in E from magnesium to aluminium which might be expected from the amphoteric behaviour of Al_2O_3, but otherwise the data are not particularly helpful. We must, therefore, seek another explanation.

We consider two possible alternative paths for the dissolution in water of an oxide M_xO_y and define a number n, such that $n = 1$ when $x = 2$, and $n = 2$ when $x = 1$. One path could be

$$\frac{1}{x}M_xO_y+\frac{y}{x}H_2O \rightarrow M^{(ny)+}(aq)+nyOH^-(aq) \qquad (10)$$

The species on the right-hand side of eqn. (10) can be referred to the standard states of their constituents,

$$M^{(ny)+}(aq)+nyOH^-(aq) \rightarrow M(s)+\frac{ny}{2}H_2(g)+\frac{ny}{2}O_2(g) \qquad (11)$$

The other path could be

$$\frac{1}{x}M_xO_y+\frac{y}{x}H_2O \rightarrow nH^+(aq)+MO_z^{n-}(aq)+uH_2O \qquad (12)$$

in which we have written $z = n(y+1)/2$ and $u = n(y-1)/2$. As before, the species on the right-hand side of eqn. (12) can be referred to the standard states of their constituents,

$$nH^+(aq)+MO_z^{n-}(aq)+uH_2O \rightarrow M(s)+\frac{ny}{2}H_2(g)+\frac{ny}{2}O_2(g) \qquad (13)$$

By combining the two processes [e.g. eqns. (10) and (12)] through their common origin and common terminus [eqns. (11) and (13)] we obtain the scheme shown in Fig. 4.

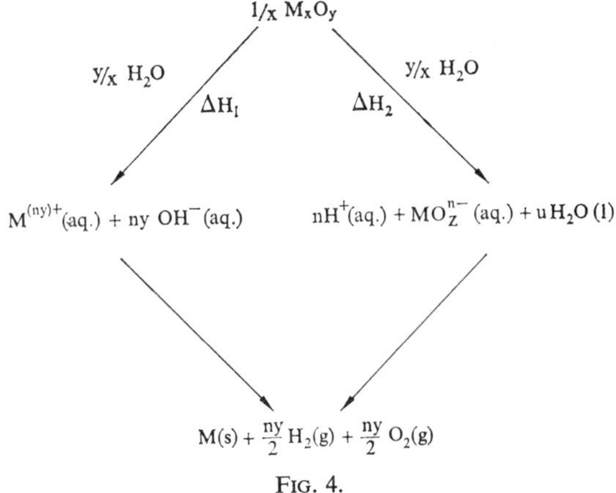

FIG. 4.

The following standard enthalpies (ΔH_f°) are known:

$$\tfrac{1}{2}H_2(g) \rightarrow H^+(aq) \quad \Delta H_f^\circ = 0.0 \text{ kcal mole}^{-1}$$
$$\tfrac{1}{2}H_2(g)+\tfrac{1}{2}O_2(g)+e \rightarrow OH^-(aq) \quad = -54.97$$
$$H_2(g)+\tfrac{1}{2}O_2(g) \rightarrow H_2O(l) \quad = -68.32$$

For the purposes of the following argument we shall take values of enthalpy to the nearest whole kilocalorie. From Fig. 4 we obtain the equation

$$\Delta H_1+A+55ny = \Delta H_2+B+34n(y-1)$$

whence
$$\delta = \Delta H_2-\Delta H_1 = A+21ny+34n-B \tag{14}$$

where $A = -\Delta H_f^\circ\,[M^{(ny)+}, aq]$ and $B = -\Delta H_f^\circ\,[MO_z^{n-}, aq]$.

Values of the part expression, $Q = 21ny+34n$ are:

n	y	ny	Q (kcal mole^{-1})
1	1	1	55
2	1	2	110
1	3	3	97
2	2	4	152
1	5	5	139

For the typical example of an oxide M_2O_3:

$$x = 2 \therefore n = 1$$
$$y = 3 \therefore Q = 97 \text{ kcal mole}^{-1}$$

Thus
$$\delta = A+97-B$$

Q is always positive, so that if $A+Q > B$, δ is positive, with the result that dissolution of M_xO_y will lead to an alkaline solution.

Conversely, if $B > A+Q$, δ is negative, with the result that dissolution of M_xO_y will lead to an acidic solution.

Let us consider as an example, the behaviour of SO_2. Following our treatment, we may describe two equations to represent the dissolution of SO_2 in water:

$$SO_2+2H_2O \rightarrow 2H^+(aq)+SO_3^{2-}(aq)+H_2O(l)$$

or
$$\rightarrow S^{4+}(aq)+4OH^-(aq)$$

Whereas the value ΔH_f° (SO_3^{2-}, aq) = −152 kcal mole^{-1} is known exactly, the value of ΔH_f° (S^{4+}, aq) is not known. It is possible to make an approximate estimate of this unknown quantity by using the following equations[145]:

$$\Delta G^\circ_{solv} = -\frac{164Z^2}{r+0.72} \quad (15)$$

$$S^\circ = 1.5R \ln M - \frac{270Z}{(r+x)^2} + 37 \quad (16)$$

For S^{4+}, $r = 0.37$Å; $Z = 4$; $x = 2$; $M = 32$.

On substitution of these values in eqns. (15) and (16) we obtain

$$\Delta G^\circ_{solv} = -2406 \text{ kcal mole}^{-1}$$
$$S^\circ = -112 \text{ cal deg}^{-1} \text{ mole}^{-1}$$

Now

$$\Delta G = \Delta H - T\Delta S$$

so that at 300°K, $\Delta H^\circ_{solv} = -2443$ kcal mole^{-1} and ΔH_f° (S^{4+}, g) = 2750 kcal mole^{-1}, so that we obtain

$$\Delta H_f^\circ (S^{4+}, aq) = 307 \text{ kcal mole}^{-1}$$

We may now evaluate eqn. (14):

$$\delta = -307+152-152$$
$$= -307 \text{ kcal mole}^{-1}$$

which indicates that on thermodynamic grounds alone, SO_2 should dissolve in water to give an acidic solution.

The approximations and inaccuracies inherent in the calculation of ΔH_f° [$M^{(ny)+}$, aq] from eqns. (15) and (16) are well documented, but it is generally conceded that they may provide values of ΔH°_{solv} which are correct to better than ±50 kcal mole^{-1}. It is clear that the dominant contributor to ΔH°_{solv} is the free energy of solvation, ΔG°_{solv}. In Table 22, we present representative values of $-\Delta G^\circ_{solv}$ calculated from eqn. (15).

TABLE 22. VALUES OF $-\Delta G^\circ_{solv}$ (kcal mole^{-1})

$Z =$	1	2	3	4	5	6
$r = 0.2$	178	712	1602	2850	4450	6608
0.3	161	644	1449	2756	4025	5796
0.4	146	584	1317	2336	3650	5256
0.5	134	537	1206	2150	3350	4824
0.6	124	496	1118	1984	3105	4664
0.7	115	461	1035	1846	2875	4140

The major factor in ΔH_f° [$M^{(ny)+}$, g] is the total (overall) ionization potential, ΣI_{ny}, for the process M(g) → $M^{(ny)+}$(g). This is clearly demonstrated by data for the iso-electronic ions of the second short period (Table 23). The table also includes the value of the effective ionic radius (r^{VI}Å), of the 6-coordinate cation[146].

[145] D. A. Johnson, *Some Thermodynamic Aspects of Inorganic Chemistry*, Cambridge University Press (1968).

[146] R. D. Shannon and C. T. Prewitt, *Acta Cryst.* **B25** (1969) 925.

TABLE 23. HEATS OF FORMATION ΔH_f° (g), OVERALL IONIZATION POTENTIAL $\Sigma_0^n I_n$ AND EFFECTIVE IONIC RADIUS r_+^{VI} OF SOME SECOND-ROW ELEMENTS

	Na^+	Mg^{2+}	Al^{3+}	Si^{4+}	P^{5+}	S^{6+}	Cl^{7+}
ΔH_f° (g) (kcal mole^{-1})	146	561	1310	2492	4168	6654	9472
$\Sigma_0^n I_n$ (kcal)	118	523	1228	2378	4075	6379	9431
r_+^{VI} (Å)	1.02	0.72	0.52	0.40	0.35	0.30	0.26

From the data in Tables 22 and 23 it is clear that as the positive charge on an ion increases and the size of the ion decreases, it will become more unlikely that the enthalpy of hydration of the gaseous cation, ΔH_f° (M^{n+}, g) can cancel out the ionization energy. In other words, we expect that the value of A in eqn. (14) will become negative as the charge, $(ny)^+$, on M increases. Consequently we can define three situations as follows:

(1) ny small	δ positive	$\Delta H_1 < \Delta H_2$
(2)	$\delta \sim 0$	$\Delta H_1 \simeq \Delta H_2$
(3) ny large	δ negative	$\Delta H_1 > \Delta H_2$

These three situations correspond exactly to what are termed respectively as basic, amphoteric and acidic oxides in the preceding discussion.

An oxide M_xO_y will have basic character when the value of y is small, or, alternatively, when the radius of $M^{(ny)+}$ is large; however, the radius of an ion decreases as the value of (ny) increases.

Table 24 contains values of the effective ionic radii of the 6-coordinate ions M^{3+} and M^{5+} of the elements of Group V. In the light of what has just been established, it is clear that Bi_2O_3 must be a more basic oxide that P_2O_5.

TABLE 24. EFFECTIVE IONIC RADII OF 6-COORDINATE IONS OF THE GROUP V ELEMENTS (Å)

	r^{VI} (M^{3+})	r^{VI} (M^{5+})
P	0.44	0.35
As	0.58	0.50
Sb	0.76	0.61
Bi	1.02	0.74

In this way we succeed in providing a fairly firm foundation for exploring the empirical facts of our previous discussion. To show the validity of the distinction we have made on thermodynamic grounds, let us consider the case of Al_2O_3. This system is exceptional in that all the relevant chemical thermodynamic data are known from experiment, and are as follows:

$$\Delta H_f^\circ (Al^{3+}, aq) = -127 \text{ kcal mole}^{-1}$$

$$\Delta H_f^\circ (AlO_2^-, aq) = -220 \text{ kcal mole}^{-1}$$

From eqn. (14) and using $Q = 97$ kcal mole^{-1}, $\delta = +4$ kcal mole^{-1}.

This fine difference between the enthalpies of the two possible processes suggests that although $\Delta H_1 < \Delta H_2$ and that therefore Al_2O_3 will behave as a basic oxide, a small change in the pH of the medium will cause the appearance of acidic character in Al_2O_3. This is in good agreement with the known amphoteric character of alumina. In the case of BeO, the value of δ is $+15$ kcal mole^{-1}.

We might, therefore, expect that amphoteric oxides will occur for intermediate values of (ny) where the ionic radius of M is within certain defined limits. Table 25 contains information which is relevant to the dissolution of Group III element sesquioxides M_2O_3, in water.

TABLE 25

M	r^{VI} (M^{3+}) (Å)	ΔH_f° (M^{3+}, aq) (kcal mole^{-1})	ΔH_f° (MO_2^-, aq) (kcal mole^{-1})	δ (kcal mole^{-1})
B	0.23	~ +200	−185	−288
Al	0.53	−127	−220	+4
Ga	0.62	−51	?	
In	0.79	−25	?	
Tl	0.88	−47	?	

r^{VI} (M^{3+}) is the radius of the 6-coordinate M^{3+} ion.

From Table 25 we can estimate that Ga_2O_3 will be amphoteric if ΔH_f° (GaO_2^-, aq) is approximately -146 kcal mole^{-1}. The corresponding figures for In_2O_3 and Tl_2O_3 are -122 and -50 kcal mole^{-1} respectively. In practice, the oxides of indium and thallium are purely basic, so that it is reasonable to conclude that the values of ΔH_f° (MO_2^-, aq) are much more positive than shown.

There remains the question of what factors contribute most decisively to ΔH_f° (MO_z^{n-}, aq). Unfortunately, the use of a thermochemical cycle such as that shown in Fig. 5, contains two unknown quantities ΔH_f° (MO_z^{n-}, g) and ΔH_{solv}° (MO_z^{n-}, g). One is therefore confined to an inspection of known values of ΔH_f° (MO_z^{n-}, aq) for a clarification of the problem. Table 26 contains a selection of such data for some simple oxyanions.

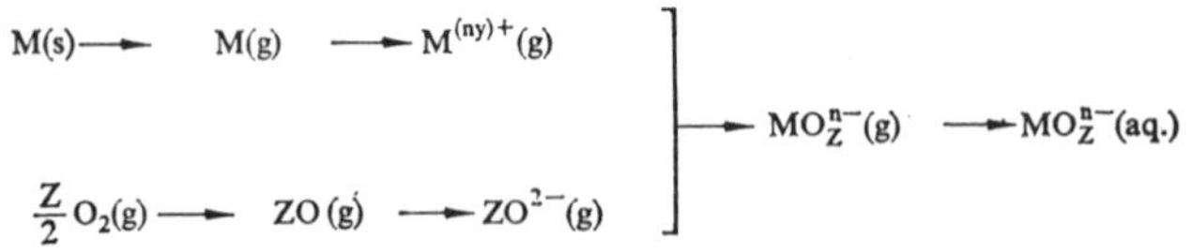

$$
\left.\begin{array}{l} M(s) \longrightarrow M(g) \longrightarrow M^{(ny)+}(g) \\ \frac{Z}{2}O_2(g) \longrightarrow ZO(g) \longrightarrow ZO^{2-}(g) \end{array}\right] \longrightarrow MO_Z^{n-}(g) \longrightarrow MO_Z^{n-}(aq.)
$$

FIG. 5.

With the exception of the halogen oxyanions, for which the values of ΔH_f° (MO_z^{n-}, aq) are uniformly small, it is clear that the enthalpy of formation increases as the value of z

increases. A similar change is noted with increasing values of n. Within any one periodic group, the value of ΔH_f° does not appear to change very markedly, thus:

M	ΔH_f° (MO_3^{2-}, aq)	M	ΔH_f° (MO_3^-, aq)
S	−152 kcal mole^{-1}	Cl	−24 kcal mole^{-1}
Se	−122	Br	−20
Te	−142	I	−53

In general, therefore, $-\Delta H_f^\circ$ (MO_z^{n-}, aq) increases as the values of n and y increase. This is in complete agreement with our earlier conclusions. There appears to be no correlation between ΔH_f° (MO_z^{n-}, aq) and the value of $r[M^{(ny)+}]$, although this is not true for the correlation with the thermochemical "radii" derived by Kapustinski which show that $-\Delta H_f^\circ$ (MO_z^{n-}, aq) increases with increasing thermochemical radius.

TABLE 26. ENTHALPIES OF FORMATION, ΔH_f° (MO_z^{n-}, aq) FOR SOME SIMPLE OXYANIONS OF NON-METALS (kcal mole^{-1})

BO_2^-	−185			NO_2^-	−25	O_3^-	−11		
BO_3^{3-}	−251*	CO_3^{2-}	−162	NO_3^-	−50				
								ClO^-	−26
AlO_2^-	−220			PO_2^{3-}	−144*			ClO_2^-	−15
				PO_3^{3-}	−233	SO_3^{2-}	−152	ClO_3^-	−24
				PO_4^{3-}	−305	SO_4^{2-}	−217	ClO_4^-	−31
								BrO^-	−22
		GeO_3^{2-}	−198*	AsO_3^{3-}	−175*	SeO_3^{2-}	−122	BrO_3^-	−20
				AsO_4^{3-}	−212	SeO_4^{2-}	−143		
								IO^-	−26
						TeO_3^{2-}	−142	IO_3^-	−53
				SbO_4^{3-}	−215*	TeO_4^{2-}	−165*	IO_4^-	−35

The values marked (*) are approximations based on ΔH_f° (H_nMO_z, aq).

It may therefore be concluded that the chemical behaviour of simple oxides in aqueous solution is satisfactorily explained in terms of the chemical thermodynamic properties of the system.

3.3. OXIDES AS A CLASS: STRUCTURES

We may distinguish four broad classes of structure among oxides; there are (a) infinite three-dimensional lattices, (b) layer lattices, (c) chain lattices and (d) molecular units.

We shall deal briefly with each of these in turn. However, it should be pointed out that this topic is treated in much greater detail than is warranted here by Wells[147] (especially chapters 11 and 12) and by Wyckoff[148].

[147] A. F. Wells, *Structural Inorganic Chemistry*, 3rd edn., Clarendon Press, Oxford (1962).

[148] R. W. G. Wyckoff, *Crystal Structures*, 2nd edn., Interscience, Wiley, New York (1963), Vols. 1, 2 and 3.

Infinite Three-dimensional Lattices

(a) *Simple Oxides*, M_xO_y

This class includes the majority of known oxide structures in which it is not possible to distinguish the presence of finite structural units. Following Wells, it is possible to subdivide this class according to the formula type M_xO_y to which a particular oxide belongs and then further to account for differences in the coordination numbers of M and O within the lattice. The generic name of the lattice type and a single example of each is given in Table 27.

TABLE 27

Formula type	Coordination number M	Coordination number O	Generic name	Example
MO_3	6	2	Rhenium trioxide	ReO_3
MO_2	8	4	Fluorite	UO_2
	6	3	Rutile	TiO_2
	4	2	Silica	GeO_2
M_2O_3	6	4	Corundum	V_2O_3
	*		Lanthanide-A (trigonal)	La_2O_3
	*		Lanthanide-B (monoclinic)	Gd_2O_3
	6	4	Lanthanide-C (cubic)	In_2O_3
MO	6	6	Sodium chloride	SrO
	4	4	Zinc blende	ZnO
	4	4	Wurtzite	BeO
	4	4	Platinum sulphide	CuO
M_2O	4	8	Anti-fluorite	Na_2O
	2	4	Cuprite	Cu_2O

* The metal ion in these systems is 7-coordinate.

Among the oxides of the transition metals three factors may, individually or in concert, cause departures from a regularly ordered structure. These are[149]:

(1) The formation of metal–metal bonds which, for example, lead to distorted rutile structures for a number of transition dioxides such as VO_2, MoO_2 and ReO_2.

(2) The presence of a covalent contribution to the metal–oxygen interaction which results in a contraction of one or more metal–oxygen bonds in a lattice.

(3) The effect of the ligand field of the (close-packed) oxide ion environment of the metal ion.

Even if the coordination number (CN) of the metal atom remains constant, the structure of the oxide will change considerably as the oxidation state of the metal changes; thus, to choose a simple well-known example, in the V–O system we have the situation summarized in Table 28.

[149] J. D. Dunitz and L. E. Orgel, *Adv. Inorg. Radiochem.* **2** (1960) 1.

This relatively simple picture of transition metal oxide systems belies their true complexity. More detailed studies show that in many metal–oxygen phase diagrams there are definite phases corresponding to intermediate oxides[150], many of which are non-stoichiometric. Thus in the case of the Mo–O_2 system the following phases lie between MoO_2 (distorted rutile) and MoO_3 (layer structure: see below); Mo_4O_{11} ($MoO_{2.75}$), Mo_8O_{23} ($MoO_{2.875}$) and Mo_9O_{26} ($MoO_{2.88}$). These are all of the general formula Mo_nO_{3n-1} and have structures which are related to that of ReO_3. Again, in the V–O system there is a series of six oxides intermediate between V_2O_3 and VO_2 which are of the type V_nO_{2n-1} (n = 3–7) and have structures related to that of TiO_2.

TABLE 28

Oxide	VO	V_2O_3	VO_2
Structure	NaCl	Al_2O_3	Distorted rutile
CN of V	6	6	6
CN of O	6	4	2

Among oxides of the type MO and MO_2, where a number of different structures are possible in which the metal is either 4-, 6- or 8-coordinate, the concept of the minimum radius ratio can be useful in determining the type of structure present (Table 29).

TABLE 29

Coordination polyhedron	Coordination number	$(r_M/r_O)_{min}$
Tetrahedron	4	0.225
Octahedron	6	0.414
Cube	8	0.732

This suggests that, for example, a dioxide MO_2, will adopt the 8-coordinated fluorite structure only if the ratio $r_M/r_O(r[O^{2-}] = 1.40$ Å) is greater than 0.732; should the radius ratio lie between 0.414 and 0.732, the compound MO_2 will have the 6-coordinated rutile structure[151], and if $0.225 < r_M/r_O < 0.414$, then the metal ion will be in a tetrahedral environment as in quartz. This criterion works fairly well in practice as a guide to the type of stereochemistry to be expected, but its use in a strictly quantitive manner is circumscribed both by the crudeness of the simple model adopted and by the uncertainties in the values of the ionic radii employed.

150 A. D. Wadsley, *Helv. Chim. Acta Fascic. Extraord. A. Werner* **207** (1966).
151 D. B. Rogers, R. D. Shannon, A. W. Sleight and J. L. Gillson, *Inorg. Chem.* **8** (1969) 841.

(b) *Complex Oxides*, $X_aY_b \ldots O_n$

In this class it is implied that within the lattice it is not possible to distinguish the presence of discrete complex ions such as SiO_4^{4-} or WO_6^{6-}. It is possible to rationalize the structures of complex oxides by considering them as being derived from simple oxides as shown in Table 30.

TABLE 30

Simple oxide	Ratio M/O	Degree of polymerization	Total cation charge	Complex oxide	Example
MO	1	$(MO)_2$	4	XYO_2	$NaFeO_2$
$[M_3O_4]$	0.75		8	XY_2O_4	$MgAl_2O_4$
			8	X_2YO_4	Mg_2SiO_4
M_2O_3	0.67		6	XYO_3	$FeTiO_3$
		$(M_2O_3)_2$	12	X_3YO_6	Ba_3WO_6
MO_2	0.5	$(MO_2)_2$	8	XYO_4	$MgWO_4$
		$(MO_2)_3$	12	XY_2O_6	$MgNb_2O_6$

It should not be implied from this that the complex oxide necessarily has the same structure as the simple oxide. This is clearly shown by considering next the coordination or packing of the cations and anions in these lattices.

In an array of close-packed oxygen atoms there are two types of hole, which are known as tetrahedral (4-coordinate) and octahedral (6-coordinate) holes, from the number of nearest-neighbour contacts which can be made by a particle of suitable size placed in such a hole. Clearly, the type of structure adopted by a particular complex oxide will depend markedly upon the size of the ions to be incorporated in the lattice. Consequently we expect considerable differences between lattices containing Be^{2+} (0.31 Å) and Ba^{2+} (1.35 Å), especially since in the latter case the barium ion is comparable in size with O^{2-} (1.40 Å).

We shall deal briefly in turn with the structures of complex oxides following the order given in Table 30, column 5.

XYO_2. In many oxides of this type one of the ions is an alkali metal ion and so we may formulate the class more specifically as $M^IM^{III}O_2$ complex oxides. Examples of XYO_2 oxides which do not contain an alkali metal are $CuCrO_2$ and $CuFeO_2$. Complex oxides of this type have a structure of the NaCl type in which both X and Y occupy octahedral holes in a close-packed oxide lattice. The simple NaCl structure suffers two types of distortion in XYO_2 oxides. In the one, extension parallel to a body diagonal or edge leads to a rhombohedral structure (e.g. $NaInO_2$), while in the other compression in the same direction leads to a tetragonal structure (e.g. $LiInO_2$). It is clear, however, that both modifications preserve the NaCl superstructure which is found for simple oxides, MO (see above).

XY_2O_4 *and* X_2YO_4. This class includes many important minerals and is of interest because it introduces the spinel structure. The necessity of formulating both types of oxide is shown when we consider the three most commonly quoted examples of complex oxide in this category, as in Table 31.

The fractions shown in Table 31 indicate the proportion of tetrahedral or octahedral holes which are occupied by the ions X and Y in the oxide lattice. The reader should consult Wells[147] (p. 505) for a characteristically critical summary of the distinction between the olivine and phenacite structures: suffice it to say that arguments based on differences in ionic radii are not acceptable.

TABLE 31

Oxide	Generic name	Tetrahedral		Octahedral	
		X	Y	X	Y
Be_2SiO_4	Phenacite	$\frac{1}{2}$	$\frac{1}{2}$		
Mg_2SiO_4	Olivine		$\frac{1}{8}$	$\frac{1}{2}$	
$MgAl_2O_4$	Spinel—normal	$\frac{1}{8}$			$\frac{1}{2}$
$MgFe_2O_4$	Spinel—inverse		$\frac{1}{4}$	$\frac{1}{8}$	$\frac{1}{4}$

Turning to the spinel structures, it is possible to distinguish three kinds of spinel compounds on the basis of oxidation state of the ions X and Y, thus:

$X^{II}Y^{III}_2O_4$	2:3	$CoAl_2O_4$	$NiFe_2O_4$
$X^{IV}Y^{II}_2O_4$	4:2	$TiZn_2O_4$	$SnCo_2O_4$
$X^{VI}Y^{I}_2O_4$	6:1	$MoAg_2O_4$	WNa_2O_4

As implied in Table 31, there are two types of spinel structure known as normal and inverse spinels. These differ in the disposition of the X and Y ions among the tetrahedral and octahedral holes in the close-packed oxide lattice. In the examples just quoted, $CoAl_2O_4$ is a normal spinel ($X^{II}Y_2^{III}O_4$), while $NiFe_2O_4$ is an inverse spinel [$Y^{III}(X^{II}Y^{III})O_4$]. Certain elements form binary oxides, M_3O_4, which have the spinel structure; thus Mn_3O_4 and Co_3O_4 have the normal structure, whilst Fe_3O_4 has the inverse structure.

The fraction of Y atoms in tetrahedral holes λ, may be used to describe spinels which are neither normal ($\lambda = 0$) nor inverse ($\lambda = 0.5$), e.g. $MnFe_2O_4$ has $\lambda = 0.1$. Calculations indicate that 4:2 spinels will have the inverse structure and all such spinels so far examined confirm this. On the other hand, similar calculations suggest that 2:3 spinels will have the normal structure in the absence of crystal field effects. As we have already seen, a number of 2:3 spinels are inverse, and this suggests that crystal field effects may be important in determining the preference of a metal ion for a particular (tetrahedral or octahedral) environment. Calculations of the difference in stabilization energy acquired by ions in an octahedral environment E_O, compared to that for the same ion in a tetrahedral environment E_T, provides in $(E_O - E_T)$ a measure of the preference of the ion for an octahedral site in a close-packed oxide lattice. This procedure has been used with success to explain the occurrence of inverse 2:3 spinels, especially those containing Fe^{3+} which is shown to have $(E_O - E_T) \sim 0$ and consequently to be capable of incorporation in a tetrahedral hole.

Although most spinels possess cubic symmetry, a small number show tetragonal distortions. These distortions may usually be traced to d^4 and d^9 metal ions in an octahedral environment or else to d^3, d^4, d^8 and d^9 metal ions in a tetrahedral environment. Thus Mn^{3+} (d^4) in $Mn^{2+}(Mn^{3+})_2O_4$ will be in a distorted octahedral environment with the result that Mn_3O_4 is a tetragonally distorted normal spinel.

XYO_3. When two-thirds of the octahedral holes in a cubic close-packed oxide lattice are occupied, the structure of the oxide M_2O_3 will be that of corundum (α-Al_2O_3). If the cations are of two different types, X and Y, then each will occupy one-third of the octahedral holes. Depending upon the pattern of occupation of the octahedral sites, one may distinguish two types of XYO_3 structure. The most common form, known by the generic name ilmenite ($FeTiO_3$), is commonly exhibited by complex oxides of the $M^{II}TiO_3$ type in which M^{2+} is small. The other form, exemplified by $LiSbO_3$, has a slightly distorted hexgonal close-packed oxide lattice. For further details see Wells[147] (pp. 127 ff.).

The most common XYO_3 structure is known as the perovskite ($CaTiO_3$) structure. This structure prevails when the cation X is comparable in size with the oxide ion; this results in a lattice of close-packed layers made up of $[XO_3]$ units in an infinite array. The smaller Y cations occupy the octahedral holes in the close-packed $[XO_3]$ lattice. The cation X is 12-coordinate. The ideal perovskite structure is cubic, but the occurrence of distorted variants is common. These distorted perovskites have considerable technological importance because of their dielectric and magnetic properties. There are three classes of perovskite compounds thus:

			r (X^{n+}) (Å)	r (Y^{n+}) (Å)
$X^IY^VO_3$	1:5	$KNbO_3$	1.60	0.64
$X^{II}Y^{IV}O_3$	2:4	$BaZrO_3$	1.60	0.72
$X^{III}Y^{III}O_3$	3:3	$LaGaO_3$	1.32	0.62

Typical of the distortions suffered by perovskites are those of $BaTiO_3$:

Distortion	Temperature range (°K)
Rhombohedral	193
Orthorhombic	193–278
Tetragonal	278–393
Cubic	393–1733
Hexagonal	1733–1885 (melting point)

The first three forms are ferroelectric.

There are also a number of complex oxides of the type X_2YO_4 (e.g. Sr_2TiO_4) which have the K_2NiF_4 structure which is in turn closely related to the perovskite structure (e.g. $SrTiO_3$). Whereas in the XYO_3 perovskite structure the X cation has twelve nearest neighbours, in the X_2YO_4–K_2NiF_4 structure the cation X has only nine nearest neighbours.

X_3YO_6. Complex oxides in this class, which may be more exactly described as having the formula $X_2Y'Y''O_6$ (following Wells[147]), have the cubic cryolite $[(NH_4)_3AlF_6]$ structure or a distorted variant of this structure. As in the XYO_3 class, the close-packed unit is XO_3 and the remaining cations occupy octahedral holes in the lattice. An example of a compound having the $X_2Y'Y''O_6$ constitution is Ba_2CaWO_6: the Ca^{2+} and W^{6+} ions occupy octahedral holes in the BaO_3 close-packed lattice.

XYO_4. While many compounds of this general type are known in which the total formal cation charge is +8, the majority have structures containing discrete YO_4 anions and therefore do not concern us here (iodates, molybdates, tantalates, silicates, phosphates, etc.). Many tungsten-bearing ores are of the type XWO_4, in which both X^{2+} and the W^{6+} cations occupy one quarter each of the octahedral holes in an hexagonal close-packed oxide lattice. Examples of the cation X^{2+} are Fe (ferberite), Mn (hübnerite), Co and Ni. In many of these, the WO_6 octahedron is more or less severely distorted[152].

XY_2O_6. Finally, mention must be made of the niobite structure. As in the preceding section, XYO_4, one-half of the octahedral holes in a close-packed oxide lattice are occupied by the cations. Most of the complex oxides having this structure are of the type $X^{2+}Y_2^{5+}O_6$ and consequently examples are restricted to the larger cations of Groups V^A and V^B. This structural type, together with the trirutile structure (e.g. $ZnSb_2O_6$) and a hexagonal modification of trirutile (e.g. $PbSb_2O_6$), are common among complex oxides of antimony of the type $X^{2+}Sb_2O_6$. The trirutile structure is also encountered for certain tantalum compounds such as $CoTa_2O_6$.

Bronzes. Brief mention must be made of this type of complex oxide of which, undoubtedly, the tungsten bronzes are the most familiar[153]. When the complex oxide Na_2WO_4 is reduced it is possible to isolate and identify a number of phases, Na_xWO_3 ($x < 1$), which lie between $W^{VI}O_3$ (distorted ReO_3 structure) and the hypothetical NaW^VO_3 (perovskite). The structure and physical properties of these phases change in a regular manner with the sodium content. These bronzes have a metallic lustre and a wide range of colours; they are electrical conductors. Perhaps the simplest representation of the constitution of the tungsten bronzes is $M^I_xW^VW^{VI}_{1-x}O_3$. Complex oxides of this type are discussed in much greater detail elsewhere[154, 150].

No attempt has been made in all that has been said to describe in any depth the enormous variety that is to be found among complex oxides, even within the broad general categories we have mentioned. Specifically we have excluded all compounds containing oxy-anions in which there is a high proportion of covalent character in the bonds between a formal cation and a number of oxide ions.

Much of the current interest in mixed metal oxide systems derives from the interest of the electronic industries in exploiting the electrical and magnetic properties of these substances.

Layer Lattices

A small group of simple oxides have layer structures. The compounds concerned are MoO_3, Re_2O_7, SnO, PbO and As_2O_3.

In MoO_3 the molybdenum atom is octahedrally coordinated by six oxygen atoms. Each MoO_6 octahedron shares two adjacent edges with its neighbours and, in addition, the octahedra are linked through corners.

152 K. S. Vorres, *J. Chem. Educ.* **39** (1962) 566.
153 *Bull. soc. chim. France* 1965, 1051–215.
154 R. Ward, *Prog. Inorg. Chem.* **1** (1959) 465.

In Re_2O_7, chains of ReO_6 octahedra are linked through corners. These chains are connected through ReO_4 tetrahedra to form double layers, every tetrahedron sharing one oxygen with each of two octahedra. In this way each octahedron is linked to two other octahedra and to two other tetrahedra through corners.

In the low-temperature monoclinic form of As_2O_3, a hexagonal net of arsenic atoms is joined up through oxygen atoms to give (twelve-membered) hexagonal rings which are puckered.

Both SnO and the tetragonal form of PbO have the same structure in which the metal atom is coordinated by four oxygen atoms all of which lie to one side of the metal. It has been suggested that the fifth coordination position is occupied by a pair of electrons (inert-pair effect). Inter-layer contancts occur through the metal atoms.

Chain Lattices

A small number of simple oxides have chain structures. The compounds concerned are CrO_3, Sb_2O_3, SO_3, SeO_2 and HgO.

In contrast to its molybdenum and tungsten analogues, CrO_3 has a comparatively low melting point (197°). The structure of the solid consists of infinite chains of tetrahedra in which each chromium atom is coordinated by four oxygen atoms, two of which are shared with two other chromium atoms. The chains are held together by weak (van der Waals) forces. This structure is similar to that of one of the polymorphs of solid SO_3. In both molecules the M–O (in-chain) distance is longer than the out-of-chain (unshared) distance.

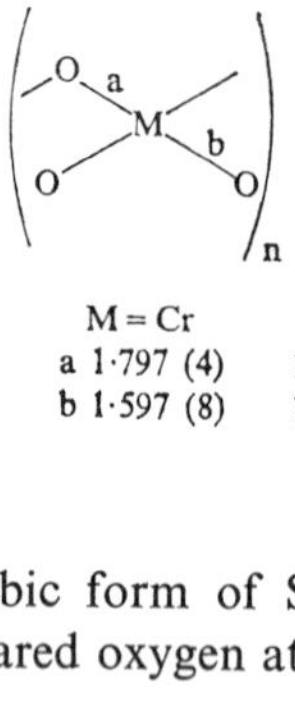

	M = Cr	S
a	1·797 (4)	1·61
b	1·597 (8)	1·41

The low-temperature orthorhombic form of Sb_2O_3 has an interesting double-chain structure in which there are no unshared oxygen atoms.

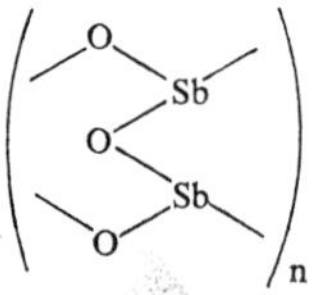

The structure of solid SeO_2 consists of infinite chains in which the selenium atom is trigonally coordinated by three oxygen atoms, one of which is unshared (Se–O = 1.73(8) Å) while the other two (shared) oxygen atoms are included in the chain (Se–O = 1.78(3) Å).

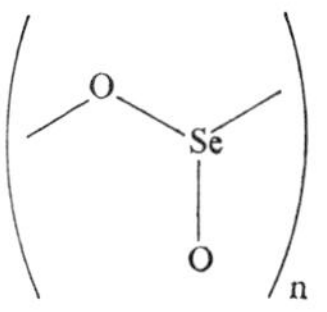

$$\left(\text{—O—Se(=O)—} \right)_n$$

Both the orthorhombric and hexagonal forms of HgO consist of infinite $[Hg–O]_n$ chains. In the former the chains are planar zigzag whilst in the latter they are helical. The coordination around mercury is strongly distorted octahedral.

Molecular Units

In this group we include all those simple oxides, in whatever phase, which have discrete (finite) molecular structural units. Thus, on the one hand, we have gaseous oxides such as CO_2 and Cl_2O while, on the other hand, there are oxides such as OsO_4 and Tc_2O_7 which have a molecular structure in the solid state.

Once again, this class is large and shows very considerable variation among structures of its members. The majority of molecular oxides are formed by the smaller non-metallic elements. It is possible to classify these oxides according to their formula type (Table 32).

TABLE 32

Formula	O/M	Examples
M_2O	0.5	H_2O, N_2O, S_2O, Cl_2O
MO	1	CO, NO
M_2O_3	1.5	N_2O_3, P_4O_6, As_4O_6, Sb_4O_6
MO_2	2	CO_2, NO_2, SO_2, ClO_2
M_2O_5	2.5	N_2O_5, P_4O_{10}
M_2O_7	3.5	Cl_2O_7, Mn_2O_7 (?), Tc_2O_7
MO_4	4	RuO_4, OsO_4

Within each class as defined by the ratio O/M there may be a variety of structures. The diatomic oxides are, of course, linear. Triatomic oxides M_2O and MO_2 are either linear or bent (V-shaped) thus:

Linear	CO_2, N_2O
V-shaped	H_2O, NO_2, S_2O, SO_2, Cl_2O, ClO_2

The most straightforward distinction between these two groups is based upon the total number of valence-shell electrons in the compound. Thus the oxides which are linear have a total of sixteen valence-shell electrons [CO_2: 4+6+6; N_2O: 5+5+6]. As the total number of valence electrons increases, the angle about the central atom decreases (see Table 33).

As the number of atoms in the molecule increases, so does the variety of structural types. The tetratomic SO_3 has a planar triangular structure in the gas phase, but in the solid phase a boat-shaped trimer is found for the orthorhombic modification. The pentatomic molecule N_2O_3 is thought to exist in two forms in the solid state, one of which is believed to be ONONO. The gaseous forms of the lower oxides of phosphorus, arsenic and antimony, $(M_2O_3)_2 \equiv M_4O_6$, have a structure based on a tetrahedron of M atoms. The volatile metal oxides, MO_4 (M = Ru, Os), have a tetrahedral structure.

TABLE 33

Compound	Number of valence electrons	Angle (°)
CO_2	16	180
NO_2	17	134
O_3	18	117
SO_2	18	119
ClO_2	19	117
Cl_2O	20	111

Dinitrogen tetroxide, the dimer of NO_2, has the structure of $O_2N–NO_2$. The molecule is planar. The "pentoxides" of the Group V elements show similar deviations between nitrogen and the other members of the group as found for the "trioxides". N_2O_5 in the gas phase probably has the structure $O_2N \cdot O \cdot NO_2$, while P_4O_{10} has a structure based on a tetrahedron of phosphorus atoms as found in P_4O_6 and in elementary white phosphorus.

TABLE 34

Oxide	Vapour phase	Solid phase
O_3	Molecular: $O\hat{O}O$ 116.8°	
SO_2	Molecular: $O\hat{S}O$ 119.5°	Molecular
SeO_2	Molecular: $O\hat{Se}O$ 125°	Chain lattice: tetragonal
TeO_2		Three-dimensional rutile lattice: tetragonal
PoO_2		Three-dimensional: tetragonal
		Three-dimensional: cubic

Summary

It will have become clear that the structures of oxides show a great deal of variety. While all complex oxides, $X_aY_b \ldots O_n$ have infinite three-dimensional lattice structure, simple oxides, M_xO_y, may have either a three-dimensional lattice structure, or a layer structure, or a chain structure, or else a molecular structure.

In order to demonstrate the type of variation found among the structures of simple oxides, we present examples from certain homologous series of oxides (Tables 34 and 35).

TABLE 35

Oxide	Solid phase structure
CrO_3	Chain lattice
MoO_3	Layer lattice
WO_3	Three-dimensional distorted ReO_3 lattice

3.4. ALLOTROPY OF OXIDES

Many simple oxides, particularly among those of the heavier non-transition metals, crystallize in a number of different allotropic forms. A well-known example of this behaviour is provided by SiO_2 which has three different allotropic forms, each of which occurs as an α- (low temperature) and β- (high temperature) modification.

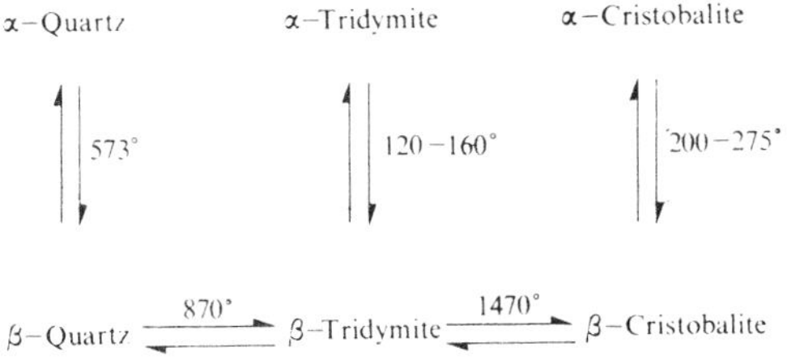

Only α-quartz is thermodynamically stable at ordinary temperatures, all the other forms being metastable.

Other examples of allotropic oxides are M_2O_3 (M = P, As, Sb, Bi); MO_2 (Ge, Se, Te, Po); SO_3 and P_2O_5. This list is not intended to be exhaustive, and for further information the reader should consult Wyckoff[148].

3.5. THERMODYNAMIC CONSIDERATIONS

Following our earlier explanations, there still remain a number of fundamental questions about metal oxide systems. Many of these, in the absence of sufficient reliable thermo-

TABLE 36. BOND ENERGY TERMS (kcal mole^{-1})

	M=C	Source	M=Si	Source
E (M–O)	86	Organic	111	SiO_2 (c)
E (M=O)	192	CO_2 (g)	153	SiO_2 (g)
E (M≡O)	257	CO (g)	192	SiO (g)

dynamic data, cannot be given an adequate answer. For example, we may ask why is CO_2 a gaseous molecule under normal conditions whereas SiO_2 under the same conditions is a crystalline solid?

To answer this question by referring merely to the availability of *d*-orbitals on silicon is simply to avoid the issue and certainly does not in itself answer the question; thus, as we have just seen, O_3 and SO_2 have very similar structures.

The question is really to be phrased in the following terms: Why does carbon form a double bond to oxygen in CO_2 but silicon forms a formal single bond to oxygen in SiO_2? It is possible in this instance and in a certain restricted number of other cases[145] to provide an adequate answer by considering the relative bond energies in the actual and the hypothetical molecules.

In simple organic molecules, E (C–O) (e.g. ethers) is roughly $\frac{1}{2}E$ (C═O) (e.g. in aldehydes or ketones where a typical value for E (C═O) would be ~177 kcal mole^{-1}). From Table 36 it is seen that E (C–O) is less than $\frac{1}{2}E$ (C═O) in CO_2. On the other hand, E (Si–O) is much more than $\frac{1}{2}E$ (Si═O). On this simple basis it is easy to show that a silica-like structure for CO_2 is thermodynamically unstable to the gaseous molecule.

Some experimentally determined bond energy values are given in Table 37 (all values in kcal mole^{-1}).

TABLE 37

(a) Single bonds										
B–O	125	C–O	86	N–O	48	O–O	34	F–O	51
		Si–O	111	P–O	88	S–O	65	Cl–O	49
		Ge–O	86	As–O	79				
(b) Double bonds									
		C═O	177	N═O	142	O═O	119		
		Si═O	153	P═O	123	S═O	125		
(c) Triple bonds									
		C≡O	257	N≡O	151				
		Si≡O	192						

3.6. GEOMETRICAL EFFECTS

In their recent compilation of an empirical set of effective ionic radii in oxide and fluoride structures, Shannon and Prewitt[146] have drawn attention to the small but significant variation of the radius of the oxide ion O^{2-}, with its coordination number. The appropriate figures are as shown in Table 38.

TABLE 38

Coordination number	r (O^{2-}) Å	Range Å
2	1.35	1.349±5
3	1.36	1.357±10
4	1.38	1.378±4
6	1.40	1.396±9
8	1.42	

Using these values it is possible to calculate lattice parameters of ionic oxides with considerable accuracy in many cases.

The structures of molecular oxides show considerable variations in the M–O bond length according to the compound studied. In many cases these differences may be correlated with M–O bond order. While extensive compilations of M–O distances are to be found in Wells[147] and elsewhere, we present here (Table 39) a selection of element–oxygen bond lengths in simple oxide systems.

TABLE 39. ELEMENT–OXYGEN DISTANCES IN BINARY COMPOUNDS (Å)

B	C	N	O	F
1.205 BO	1.128 CO 1.162 CO_2	1.187 N_2O 1.150 NO 1.192 NO_2	1.207 O_2 1.278 O_3	1.418 F_2O
	Si 1.509 SiO	P 1.448 PO 1.638 P_4O_6 1.429 P_4O_{10}	S 1.465 S_2O 1.493 SO 1.432 SO_2 1.43 SO_3	Cl 1.700 Cl_2O 1.571 ClO 1.473 ClO_2 1.405 Cl_2O_7 1.703
	Ge 1.650 GeO Sn 1.838 SnO 2.05 SnO_2 Pb 1.922 PbO	As 1.78 As_4O_6	Se 1.61 SeO_2	Br 1.65 BrO

4. WATER

4.1. PHYSICAL PROPERTIES

Water is one of the most familiar substances on earth. It plays an essential part in the chemical basis of life, and is the commonest solvent for solution studies of inorganic systems. The influence of water on the chemical behaviour of aqueous systems is too large a subject for discussion here; this section will be concerned with the physical properties and structure of water in its various phases[155, 156].

155 N. E. Dorsey, *Properties of Ordinary Water Substance*, Reinhold, New York (1940).

156 D. Eisenberg and W. Kauzmann, *The Structure and Properties of Water*, Oxford University Press, Oxford (1969).

TABLE 40. PROPERTIES OF WATER

Dimensions of the free molecule

		Reference
Internuclear distance, $\bar{r}_e$	0.9572 ± 0.0003 Å	a
Angle	$104.52 \pm 0.05°$	a
Dipole moment (298–484°K)	1.84–1.85 Debyes	b
Mean quadrupole moment, $\bar{Q}$	$-5.6 \pm 1.0 \times 10^{-26}$ e.s.u. cm^2	c
Mean polarizability, $\bar{\alpha}$	1.444×10^{-24} cm^3	c

Crystal parameters of crystalline ice polymorphs [c, d, e]

Form	Crystal system	Space group	Molecules per unit cell	H atom arrangement
Ice Ih	Hexagonal	$P6_3/mmc$	4	D
Ic	Cubic	$F\bar{4}3m$	8	D
II	Rhombohedral	$R\bar{3}$	12	O
III	Tetragonal	$P4_12_12$	12	D
V	Monoclinic	$A2/a$	28	D
VI	Tetragonal	$P4_2/nmc$	10	D
VII	Cubic	$Im3m$	2	D
VIII	Cubic*	$Im3m$	2	? O [d]
IX	Rhombohedral	$P4_12_12$	12	O

* Structure incomplete.

Vibrational constants for the water molecule (cm^{-1}) [a]

$\omega_1 = 3832.17$ $x_{11} = -42.576$ $x_{12} = -15.933$

$\omega_2 = 1648.4$ $x_{22} = -16.813$ $x_{13} = -165.824$

$\omega_3 = 3942.53$ $x_{33} = -47.566$ $x_{23} = -20.332$

Vibrational frequencies of water in different phases (cm^{-1})[c, e, f]

Mode	Vapour	Liquid	Ice I	Ice II*	Ice V
ν_T	—	~60, ~190	~60, ~225	~150	169
ν_L	—	685 (vbr)	~840	800	~730
δ(HOH), ν_2	1594.59	1650	1650	1690	1680
ν_A	—	2130	2270	2225	2210
$2\nu_2$	3151.4		~3130	3000, 3225	
νO,H (ν_1, ν_3)	3656.65 (ν_1) 3755.79 (ν_3)	~3250, ~3400, ~3600	3220, ~3350	3230, shs to higher frequency	3250 3440

vbr = very broad; shs = shoulders. * Many submaxima observed in this region 400–800 cm^{-1} in the infrared. ν_T are hindered translational lattice modes; ν_L is the librational lattice mode; and ν_A is the "association" band; ν_1 is the a_1 OH stretching mode, ν_2 the bend and ν_3 the b_1 stretch. Bands in the spectra of liquid water and of the disordered forms of ice (I and V in the table) are in general very broad, and the maxima are not always well defined; the spectra of other forms of ice differ in detail, save for ice Ih and ice Ic, where no differences were detected.

Force constants [c]

$k_{\bar{r}} = 8.54 \times 10^5$ dynes cm $k_a = 0.761$

$k_{\bar{r}}' = -0.101$ $k_{\bar{r}a} = 0.228$

TABLE 40 (*cont.*)

Ultraviolet absorption of water

	Vapour [g]	Liquid [h]	Ice [i]
λ_{max} (Å)	1655	1470	(1570)
ε (mol^{-1} cm^{-1})	3200	1700	

The value for ice is estimated for the onset of continuous absorption. The measurements above were made in the region > 1300 Å; there is a broad continuum for water vapour between 1150 Å and 1430 Å, on which diffuse bands are superimposed, and diffuse bands have been observed in the region 500–1000 Å [g, j].

N.m.r. spectrum of water

H^1 chemical shift = −0.60 ppm (vapour), −5.18 ppm (liquid) [k] relative to CH_4 gas; −4.32 ppm in NH_3(l) [k] relative to Me_4Si external standard

O^{17} chemical shift (H_2O^{17} vapour–H_2O^{17} liquid) = +36 ppm [l]

J (H–O^{17}) = 79±2 Hz for water vapour [l]

Ionization potentials [c] (eV)

1st 12.62 3rd 16.3±0.3
2nd 14.5±0.3 4th 18.0±0.5
Proton affinity [m] = 164±4 kcal $mole^{-1}$

Thermodynamic properties [c, m, n]

ΔH_f° (0°K)	−57.102 kcal $mole^{-1}$	C_p° (298°K)	8.025 cal $mole^{-1}$
ΔH_f° (298°K)	−57.796 kcal $mole^{-1}$	D (H–OH) (0°K)	117.8 kcal $mole^{-1}$
ΔG_f° (298°K)	−54.634 kcal $mole^{-1}$	E (H–O)	110.6 kcal $mole^{-1}$
S° (298°K)	45.104 cal^{-1} $°K^{-1}$ $mole^{-1}$		

Thermodynamics of ice–ice transitions [c]

		t (°C)	p (kbar)	ΔV (cm^3 mol^{-1})	ΔS (e.u.)	ΔH (cal mol^{-1})
From ice I	to ice II	−35	2.13	−3.92	−0.76	−180
I	III	−22	2.08	−3.27	0.4	94
II	III	−24	3.44	0.26	1.22	304
II	V	−24	3.44	−0.72	1.16	288
III	V	−17	3.46	−0.98	−0.07	−17
V	VI	0.16	6.26	−0.70	−0.01	−4
VI	VII	81.6	22	−1.05	~0	~0
VI	VIII	~5	~21	~0.0	~−1.01	−282

Critical temperature, T_c 647.30°K
Critical pressure, P_c 218.3 atm
Critical volume, V_c 59.1±0.5 cm^3 mol^{-1}
Van der Waals constants for water vapour [b]:
a 5.464, *b* 0.03049, when P is in atmospheres and V in litres
Virial coefficients [p] at 573.16°K:
B −112.9 cm^3 mol^{-1}, C −3470 cm^6 mol^{-2}

TABLE 40 (*cont.*)

Properties for liquid water [q]	
Density at 0°C	0.999 841 g cm^3
4°C	0.999 973 g cm^3
Density of ice I at 0°C	0.916 71 g cm^3
Absolute refractive index for Na *D*-line at 20°C	1.333 35
Kerr constant for Na *D*-line at 17°C	0.0363 cm volt^{-2}
Verdet constant for Na *D*-line	$(131.1-0.004\,00t-0.000\,400t^2)$ $\times 10^{-4}$ min gauss^{-1} cm^{-1} ($4° < t < 98°$C)
Thermal conductivity at 0°C	0.001 348 cal sec^{-1} cm^{-2} (°C cm^{-1})$^{-1}$
Surface tension at 18°C	73.05 dynes cm
Viscosity at 20°C	$0.010\,019 \pm 0.000\,003$ poise
Ionic dissociation constant [c] *K* at 25°C	$[H^+][OH^-]/[H_2O] = 1.821_4 \times 10^{-16}$ mol l^{-1}
Ionic concentrations [c] $[H^+] = [OH^-]$ at 25°C	1.004×10^{-7} mol l^{-1}
D.c. conductivity [c] at 20°C	5.7×10^{-8} ohm^{-1} cm^{-1}
Dielectric constant [c], ε_0	$87.740-0.400\,08t+9.398\times10^{-4}t^2$ $-1.410\times10^{-6}t^3$ (t = 0–100°C)
Specific magnetic susceptibility χ_g at 20°C	$-719.92 \pm 0.11 \times 10^{-9}$ c.g.s.u.

[a] W. S. Benedict, N. Gailar and E. K. Plyler, *J. Chem. Phys.* **24** (1956) 1139.
[b] *Chemical Rubber Handbook*, 48th edn. (1968–9).
[c] D. Eisenberg and W. Kauzmann, *The Structure and Properties of Water*, Oxford University Press, Oxford (1969).
[d] E. Whalley, *Ann. Rev. Phys. Chem.* **18** (1966) 205.
[e] J. E. Bertie and E. H. Whalley, *J. Chem. Phys.* **40** (1964) 1637, 1647.
[f] J. Schiffer and D. F. Hornig, *J. Chem. Phys.* **49** (1969) 4150.
[g] K. Watanabe and M. Zelikoff, *J. Opt. Soc. America* **43** (1953) 753.
[h] R. A. Verrall and W. A. Senior, *J. Chem. Phys.* **50** (1969) 2746.
[i] K. Dressler and O. Schnepp, *J. Chem. Phys.* **33** (1960) 270.
[j] S. Bell, *J. Mol. Spectrosc.* **16** (1965) 205.
[k] W. G. Schneider, H. J. Bernstein and J. A. Pople, *J. Chem. Phys.* **28** (1958) 601; T. Birchall and W. L. Jolly, *J. Am. Chem. Soc.* **87** (1965) 3007.
[l] A. E. Florin and M. Alei, *J. Chem. Phys.* **47** (1967) 4268.
[m] NBS Technical Note 270-3 (1968); J. L. Beauchamp and S. E. Buttrill, *J. Chem. Phys.* **48** (1968) 1783.
[n] T. L. Cottrell, *The Strengths of Chemical Bonds*, 2nd edn., Butterworths (1958).
[p] J. H. Dymond and E. B. Smith, *The Virial Coefficients of Gases*, Oxford University Press, Oxford (1969).
[q] N. E. Dorsey, *Properties of Ordinary Water Substance*, Reinhold, New York (1940).

The composition of water was established by the early work of Cavendish, Lavoisier, Gay-Lussac and Humbolt. The structure of the free molecule has been determined very accurately (Table 40), but the structures of the various forms of ice and of liquid water are less well understood. The unusual physical properties of liquid water—its high melting point and latent heat of vaporization, the increase in its density from 0° to 4°, the unusual spectroscopic and transport properties—indicate clearly that there is extensive hydrogen bonding in the liquid; this persists in the solid.

There are at least nine different crystalline forms of ice[156, 157]; each is apparently based on tetrahedrally coordinated oxygen, the oxygen atoms each being bound to two hydrogen atoms by "normal" bonds and to two others by hydrogen bonds. In hexagonal ice I, sometimes called ice Ih, the dimensions of the water molecule are close to those for the free molecule. Pauling showed that if the individual water molecules are preserved, and if each is involved in forming four hydrogen bonds with the hydrogen atoms between pairs of oxygen atoms, ice at 0°K should have residual entropy because of possible disorder in the

[157] E. Whalley, *Ann. Rev. Phys. Chem.* **18** (1967) 205; S. W. Rabideau, E. D. Finch, G. P. Arnold and A. I. Bowman, *J. Chem. Phys.* **49** (1968) 2514.

arrangement of the hydrogen atoms; the residual entropy calculated on this basis[156] is 0.8145 ± 0.0002 e.u., as against an experimental value of 0.82 ± 0.15 e.u. The other forms of ice are stable under different conditions of temperature and pressure (Fig. 6); in at least three the arrangement of the hydrogen atoms appears to be ordered.

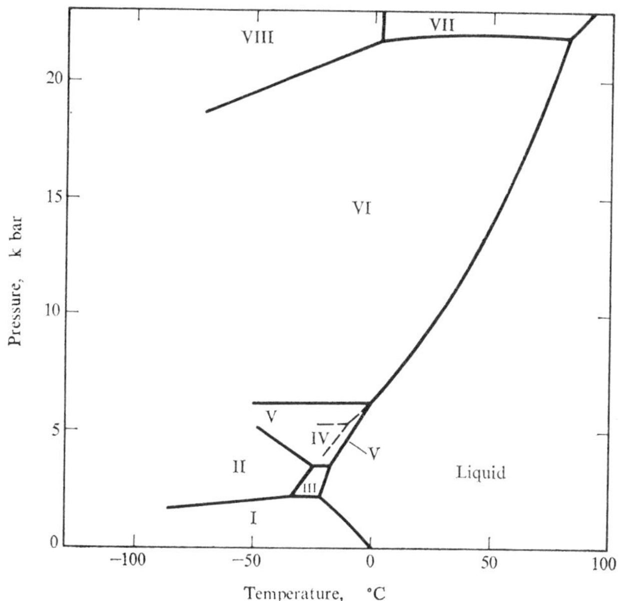

Fig. 6. Phase diagram for H_2O. The field for metastable ice IV, indicated by dashed lines, was mapped only for D_2O. (Reproduced with permission from D. Eisenberg and W. Kauzmann, *The Structure and Properties of Water*, Oxford University Press, 1969, p. 93.)

The structure of liquid water has been investigated by X-ray diffraction, by spectroscopy and by calculation of thermodynamic properties based on different models. Interpretation of diffraction patterns suggests that each water molecule has over a period of time an average of rather more than four nearest neighbours; the average drops a little with increasing temperature. The details of the "instantaneous" structure are controversial. As a consequence of measurements of dielectric relaxation, it now seems unlikely that liquid water consists of a whole range of units from monomer through dimer, trimer and so on up to very highly polymerized species. Models, known as mixture models, have been proposed in which a limited number of distinct species are in equilibrium with one another; this equilibrium may be maintained by the simultaneous making and breaking of a large number of hydrogen bonds (the "flickering-cluster" picture). The simplest mixture models, "two-state models", postulate monomeric species in equilibrium with high polymers; the monomeric water molecules have been described as analogous to certain hydrates, with the free molecules as interstitial species in a lattice of water polymers. An alternative approach involves the postulate that the hydrogen bonds are distorted, rather than broken and reformed, as a result of molecular motion in the liquid. Certain features of the infrared spectrum, notably the breadth of the band due to OH stretching and the unexpectedly large intensity of the overtone of the bending mode, have been explained in terms of collisional distortion[158]. None of these models is completely satisfactory as yet, but most of the properties of liquid water may be understood at least qualitatively on the base of the two-state model.

The state of water in different solvents depends on the nature of the solvent concerned. In non-polar solvents, it has been shown by classical physical techniques and by spectroscopy that even in solutions approaching saturation the concentrations of dimers or polymers of water are small; in polar solvents, aggregates are formed[159]. Association in the vapour is indicated by calculations based on the virial coefficients[156] and by spectroscopic methods[160].

The strength of the hydrogen bonds in water has been variously estimated at between 1 and 12 kcal (mole-bond)$^{-1}$; at present most estimates lie in the range 2–4 kcal (mole-bond)$^{-1}$. The ordered structure of liquid water is of the greatest importance in connection with solvation and with life chemistry; it is probably responsible for the enormous mobility of the proton in liquid water, which takes place through a cooperative "proton-switch" mechanism.

In the past few years a material called "anomalous water" or "polywater" has been obtained by condensing water vapour at partial pressures less than the saturated pressure in quartz capillaries in the absence of air[161]. The liquid obtained in this way is more dense, less volatile and more viscous than ordinary water, and its coefficient of thermal expansion is much higher. The structure and even the composition of anomalous water is uncertain. Structures proposed include tetrameric, tetrahedral units, not unlike P_4O_6 but with only two "terminal" protons, and a polymeric structure with symmetrical hydrogen bonds, giving O–H–O units analogous to the bifluoride ion[162, 163]. Preliminary studies of the mass, i.r. and n.m.r. spectra, however, revealed no anomalies[164]. It now seems almost certain that the anomalous properties are due to the presence of impurities[165] and that pure "anomalous" water does not exist.

Calculations of the properties of the water molecule have been made by the SCF MO method, using both Gaussian and Slater functions; the properties derived from these calculations correlate quite well with experimental values[166].

Some physical and structural properties of water are collected in Table 40.

[158] J. Schiffer and D. F. Hornig, *J. Chem. Phys.* **49** (1969) 4150.
[159] S. D. Christian, A. A. Taha and B. W. Gajh, *Quart. Rev.* **24** (1970) 20.
[160] W. A. P. Luck and W. Ditter, *Ber. Bunsensger. Phys. Chem.* **70** (1966) 1113.
[161] B. V. Derjaguin, N. V. Churaev, N. N. Fedyakin, M. V. Talaev and I. G. Ershova, *Bull. Acad. Sci. USSR* 1967, 2095.
[162] R. W. Bolander, J. L. Kassner and J. T. Zung, *Nature* **221** (1969) 1233.
[163] E. R. Lippincott, R. R. Strombers, W. H. Grant and G. L. Cessac, *Science* **164** (1969) 1482.
[164] E. Willis, G. R. Rennie, C. Smart and B. Pethica, *Nature* **222** (1969) 159.
[165] P. Barnes, I. Cherry, J. L. Finney and S. Petersen, *Nature* **230** (1971) 31.
[166] S. Aung, R. M. Pitzer and S. I. Chan, *J. Chem. Phys.* **49** (1968) 2071.

4.2. CHEMICAL PROPERTIES

Oxygen in water is in its lowest oxidation state; the oxidation potential is high (see Table 5, p. 691). On many surfaces there is a substantial overvoltage for both oxidation and reduction of water; these overvoltages reflect a kinetic resistance of water to oxidation or reduction, and so the range of redox processes that can be studied in aqueous solution is large. Because of its high dielectric constant and strong solvating properties, water is a good solvent for many ionic and polar species; the hydrogen bond energy of water itself, however, means that substances do not dissolve in water unless they are strongly solvated. As with other "structured" solvents, solutes may be classified as "structure making" or "structure breaking", depending on their effect on the structure of the solvent; the different types may be distinguished by n.m.r. or by study of such physical properties as viscosity.

The importance of the self-ionization of water in the chemistry of aqueous systems is enormous:

$$2H_2O \rightleftharpoons H_3O^+ + OH^-$$

The H_3O^+ ion, which is pyramidal, is, of course, more extensively solvated in aqueous solution; species such as $H_5O_2^+$ have been identified crystallographically, and aggregates $H^+(H_2O)_n$ have been detected in the mass spectrometer. Coordinated water may also behave as an acid, and, indeed, the system $H_2O \cdot BF_3$ is exceptionally acidic. Reactions involving the hydrolysis of halides, or anhydrides, and of cations cannot be described here; suffice it to say that water can be regarded as the parent of the whole series of hydroxy-compounds ROH, and as a base because of the lone pairs of electrons at oxygen.

5. OXYGEN FLUORIDES

5.1. INTRODUCTION

The oxygen fluorides have been the object of considerable study and possess unusual chemical and physical properties. At this time of writing (July 1969) the following species have been mentioned in the literature:

$$OF_2,\ OF,\ O_2F_2,\ O_2F,\ O_3F_2,\ O_4F_2,\ O_5F_2,\ O_6F_2$$

By no means has the preparation and characterization of all these species been unequivocally demonstrated. In fact the only ones for which the evidence seems completely reliable are:

$$OF_2,\ OF,\ O_2F_2,\ O_2F,\ O_4F_2$$

For the others, O_3F_2 is almost certainly a mixture of O_2F_2 and O_4F_2 (which is itself in equilibrium with $2O_2F$), and the evidence for O_5F_2 and O_6F_2 is based only on the ratio of oxygen to fluorine following decomposition.

The compounds are powerful fluorinating and oxidizing agents and, but for OF_2, are only stable well below room temperature. In 1963, Streng[167] described the chemistry of the non-radical compounds (OF_2, O_2F_2, "O_3F_2", O_4F_2) in a comprehensive review article. More recently Turner[168] has considered the physicochemical properties of the compounds and included a description of the preparation and spectroscopic observation of the radicals

[167] A. G. Streng, *Chem. Revs.* **63** (1963) 607.
[168] J. J. Turner, *Endeavour* **27** (1968) 42.

OF and O_2F. A very useful summary of properties of oxygen fluorides is contained in a recent book by Lawless and Smith[169].

In what follows an attempt is made to consider the oxygen fluorides in turn, but there will be considerable overlap; for example, the radical O_2F seems to be present to some extent in all condensed oxygen fluoride systems.

5.2. OXYGEN DIFLUORIDE, OF_2

Preparation

Although the first preparation[170] of an oxygen fluoride involved electrolysis of molten, slightly moist, KF–HF to give OF_2, the most common preparative method[171] depends on the reaction of fluorine gas with 2% caustic soda solution:

$$2F_2+2NaOH \rightarrow OF_2+2NaF+H_2O$$

If the concentration of base is not carefully controlled OF_2 is lost via the secondary reaction:

$$OF_2+2OH^- \rightarrow O_2+2F^-+H_2O$$

Full spectroscopic studies on OF_2 demand consideration of isotopic molecules, and Reinhard and Arkell[172] prepared $^{17}OF_2$ and $^{18}OF_2$ by the electrolysis of HF containing up to 0.5% enriched water.

Physical Properties

At room temperature OF_2 is a colourless, very poisonous gas; it decomposes at temperatures above 200°C, and freezes at −223.8°C to give a pale yellow liquid. Table 41 summarizes most of the known physical properties.

The most recent thermochemical determination[173] of the heat of formation of OF_2 allows an accurate estimate of the *mean* bond energy in OF_2, but to obtain a reliable value for the *dissociation* bond energy (i.e. ΔH for $OF_2 \rightarrow OF+F$) is a more difficult problem (see below).

Oxygen difluoride is quite stable at room temperature in *dry* glass vessels.

Some Chemical Properties

Although less reactive than fluorine, OF_2 is a powerful oxidizing agent, sometimes fluorinating and sometimes—particularly in aqueous solution—donating oxygen. Streng[167] and Lawless and Smith[169] have provided a comprehensive catalogue of reactions of OF_2, and here we merely outline some characteristic ones.

Most metals react with OF_2, particularly on warming, to give the corresponding fluoride. Metallic oxides can also produce the fluoride, e.g. CrO_3 forms the fluoride on

[169] E. W. Lawless and I. C. Smith, *Inorganic High-energy Oxidisers*, Arnold, London, and Dekker, New York (1968).

[170] P. Lebeau and A. Damiens, *Compt. rend.* **185** (1927) 652.

[171] For example, G. H. Cady, *J. Am. Chem. Soc.* **57** (1935) 246.

[172] R. R. Reinhard and A. Arkell, *Int. J. Appl. Radiation Isotopes* **16** (1965) 498.

[173] R. C. King and G. T. Armstrong, *J. Res. Natl. Bur. Std. U.S.* **72A** (1968) 113.

gentle warming, whereas CaO only reacts on strong heating. With non-metallic solids, fluorides and oxyfluorides are obtained, e.g. S, Se, Te give among other products SO_2, SF_4, SeF_4 and TeF_4.

TABLE 41. PHYSICAL PROPERTIES OF OXYGEN DIFLUORIDE, OF_2

Melting point	−223.8°C (49.4°K)
Boiling point	−145.3°C (127.9°K)
Decomposition temperature	~200°C
Heat of vaporization	2.650 kcal mole^{-1} at b.p.
Vapour pressure: vapour	$\log p$ (mmHg) $= 7.2242 - \frac{555.42}{T}$ (195°C to −145°C)
liquid	T (°K) 60.213 67.535 76.883 89.236 106.317 127.856
	p (mmHg) 0.01 0.10 1.0 10.0 100.0 760.0
Critical temperature	−58.0±0.1°C (215.2±0.1°K)
Critical pressure	48.9 atm
Heat of formation (ΔH_f°)	5.86±0.38 kcal mole^{-1}
Average bond energy	44.72 kcal mole^{-1}
Bond dissociation energy D (F–OF) [a]	38.0±3.7, 42.7±4.1 kcal mole^{-1}
Density: vapour	2.41 mg^{-1} cc at n.t.p.
liquid	$d = 2.190 - 0.005\,23T$ g cc^{-1}
Ionization potential	315±4 kcal mole^{-1} (13.7±0.2 eV)
Ultraviolet spectrum	Gas and liquid very similar—continuum with possible maxima at 4210 Å, 3580 Å and 2940 Å
liquid: λ (Å)	3000 4000 5000 5800
ε (l mole^{-1} cm^{-1}) [b]	0.48 0.055 0.012 0.0012
ε (l mole^{-1} cm^{-1}) [c]	0.061 0.021 0.008
Dipole moment	0.2–0.3 D
Infrared maxima (cm^{-1}) (fundamentals) [d]	461 (s, doublet, $\nu_2[A_1]$); 831 (vs, PQR, $\nu_3[B_1]$); 928 (s, doublet, $\nu_1[A_1]$)
Force constants (mdyn [f] Å^{-1})	$f_{OF} = 3.95$; $f_{OF-OF} = 0.81$; $f_{FOF} = 0.72$; $f_{OF-FOF} = 0.14$
Structure of gas (microwave) [e]	r_e (O–F) $= 1.4053 \pm 0.0004$ Å
	θ_e (F–O–F) $= 103°\ 4' \pm 3'$
Chemical shift	^{19}F [g] (v. $CFCl_3$ in ppm): gas, −248±1; liquid, −249±1; in SF_6 solution −274
	^{17}O [h] (v. $H_2{}^{17}O$ in ppm); liquid, −830

Unless stated otherwise, data from A. G. Streng, *Chem. Revs.* **63** (1963) 607 and/or E. W. Lawless and I. C. Smith, *Inorganic High-energy Oxidisers*, Arnold, London, and Dekker, New York (1968).

[a] M. A. A. Clyne, R. T. Watson, *Chem. Phys. Letters*, in press; M. C. Lin and S. H. Bauer, *J. Am. Chem. Soc.* **91** (1969) 7737.

[b] F. I. Metz, J. W. Nebgen, W. B. Rose and F. E. Welsh, unpublished data.

[c] A. G. Streng and L. V. Streng, *J. Phys. Chem.* **69** (1965) 1079.

[d] J. W. Nebgen, F. I. Metz and W. B. Rose, *J. Mol. Spectrosc.* **21** (1966) 99; these authors investigated the Fermi resonance between ν_1 and $2\nu_2$ first considered in the microwave work of ref. e. See D. J. Gardiner and J. J. Turner, *J. Mol. Spectrosc.* **38** (1971) 428 for liquid Raman data.

[e] Y. Morino and S. Saito, *J. Mol. Spectrosc.* **19** (1966) 435.

[f] L. Pierce, N. Di Cianni and R. H. Jackson, *J. Chem. Phys.* **38** (1963) 730; the force constants are calculated to be consistent with vibrational spectra and centrifugal distortion constants in microwave.

[g] See, for example, J. W. Nebgen, W. B. Rose and F. I. Metz, *J. Mol. Spectrosc.* **20** (1966) 72.

[h] I. J. Solomon, J. N. Keith, A. J. Kacmarek and J. K. Raney, *J. Am. Chem. Soc.* **90** (1968) 5408.

Oxygen difluoride and hydrogen sulphide explode on mixing at room temperature: with xenon, XeF_4 and xenon oxyfluorides are produced: with atomic hydrogen, even at 77°K, HF, H_2O and H_2O_2 are formed.

OF_2 and F_2 form cryogenic solutions on mixing in all proportions and do not react up to 300°C. On warming OF_2 with Cl_2, Br_2 and I_2 explosions occur to give largely halogen fluorides plus perhaps Cl_2O and I_2O_5. There is *some* evidence for ClOF, which would be the only example of a mixed oxygen halide of general formula XOY; experiments said to provide evidence for this species include the passage of Cl_2 and OF_2 through a copper tube at 300°C, the action of ultraviolet light on mixtures of Cl_2 and OF_2 and the reaction of ClF and OF_2 at 25°C [169].

Oxygen difluoride is slightly soluble in water but a slow hydrolysis occurs:

$$OF_2+H_2O \rightarrow O_2+2HF$$

In aqueous solutions of HCl, HBr and HI free halogen is liberated; with H_2S colloidal sulphur is precipitated and reactions which demonstrate oxidation in aqueous solution include:

$$MnSO_4 \rightarrow MnO_2;\ FeSO_4 \rightarrow Fe_2(SO_4)_3;\ Cr^{3+} \xrightarrow{OH^-} CrO_4^{2-}$$

Properties Relevant to Reaction Mechanism and the OF Radical

Until recently it was widely accepted that the thermal decomposition of OF_2, which proceeds overall according to

$$OF_2 \rightarrow \tfrac{1}{2}O_2+F_2,$$

involved the initial simple dissociation[174]

$$OF_2+M \rightarrow OF+F+M \qquad (1)$$

and that the reactions

$$OF+OF \rightarrow O_2+2F \qquad (2)$$
$$2F+M \rightarrow F_2+M \qquad (3)$$

proceed with zero activation energy; thus the overall rate constant k_{ov} should equal k_1 the second-order rate constant for (1). However recent work, particularly using shock-tubes[175, 176], has demonstrated the importance of the reverse of (2) and (3). In the most recent work[176]:

$$k_{ov} = 10^{15.8}e^{-37,800/RT}\ \text{cm}^3\ \text{mol}^{-1}\ \text{sec}^{-1}\ (T < 1000°\text{K})$$
$$k_1 = 10^{17.3}e^{-(42,500\pm4100)/RT}\ \text{cm}^3\ \text{mol}^{-1}\ \text{sec}^{-1}$$
$$k_2 = 10^{12.10\pm0.12}\ \text{cm}^3\ \text{mol}^{-1}\ \text{sec}^{-1}$$

Thus, applying RRKM theory, $D(\text{FO–F}) = 42.7$ kcal mole^{-1} and hence combining with $\Delta H_f^\circ(OF_2)$, $D(\text{OF}) = 48.3$ kcal mole^{-1}.

The photochemical behaviour is also best interpreted on the basis of OF and F. For example, Schumaker *et al.*[177] studied the photochemical decomposition between 15° and 45°C to give F_2 and O_2 and deduced the mechanism:

$$OF_2 \xrightarrow[3650\ \text{Å}]{h\nu} F+OF$$

$$F+F+M \longrightarrow F_2+M$$
$$OF \longrightarrow \tfrac{1}{2}O_2+\tfrac{1}{2}F_2$$

---

174 W. Koblitz and H. J. Schumaker, *Z. physik. Chem.* **B25** (1934) 283.

175 J. A. Blauer and W. C. Solomon, *J. Phys. Chem.* **72** (1968) 2307; W. C. Solomon, J. A. Blauer and F. C. Jaye, *J. Phys. Chem.* **72** (1968) 2311.

176 M. C. Lin and S. H. Bauer, *J. Am. Chem. Soc.*, **91** (1969) 7737.

177 R. Gatti, E. Staricco, J. E. Sicre and H. J. Schumaker, *Z. physik. Chem.* (*Frankfurt*) **35** (1962) 343.

They also concluded that any reaction

$$F + OF_2 \longrightarrow OF + F_2 \tag{4}$$

must have an activation energy of at least 15 kcal mole^{-1}, but it has been pointed out[175] that they neglected any contribution from the reverse of (4).

Chemical demonstration that OF_2 reacts photochemically as OF plus F was provided by Franz and Neumayr[178], following Schumaker. The photolysis of OF_2/SO_3, $OF_2/S_2F_2O_6$, OF_2/SO_2 gas mixtures with light of wavelength 3650 Å (which is absorbed only by OF_2) gave several compounds of F, S and O. For example, with OF_2 and SO_3 the only product was $FS(O_2)OOF$ implying addition of F and OF. More recently Solomon *et al.*[179] have repeated the OF_2/SO_3 photolysis experiments using ^{17}O labelled compounds. The products were analysed by ^{17}O nuclear magnetic resonance and the results can be summarized:

$$^{17}OF_2 + SO_3 \xrightarrow{h\nu} FS(O_2)O^{17}OF$$
$$OF_2 + S^{17}O_3 \xrightarrow{h\nu} FS(^{17}O_2)^{17}OOF$$

thus demonstrating the presence of OF as an intermediate.

Liquid OF_2 and low-temperature solutions of OF_2 in freons[180] show e.s.r. absorptions after exposure to light. This free radical is, however, certainly O_2F not OF, and is possibly formed by some such reaction as

$$OF_2 \xrightarrow{h\nu} OF + F$$
$$OF + OF_2 \rightarrow O_2F + F_2$$

The properties of the O_2F radical will be discussed following the section on the chemistry of O_2F_2.

5.3. OXYGEN MONOFLUORIDE, OF

The previous section demonstrated the evidence for OF as an intermediate in OF_2 chemistry. Attempts to observe this species in the gas phase or in solution by such techniques as flash photolysis and e.s.r. have until very recently all failed. Further *indirect* evidence was obtained by first Schumaker and then Kirshenbaum[181] by studying the photolysis (3650 Å) of gaseous F_2/O_3 mixtures at a variety of temperatures. At 120°K the products were O_2F_2, OF_2 and O_2 consistent with a mechanism:

$$F_2 \rightarrow 2F$$
$$F + O_3 \rightarrow O_2 + OF$$
$$OF + OF \rightarrow O_2F_2$$
$$OF + F \rightarrow OF_2$$
$$OF + F_2 \rightarrow OF_2 + F$$

The only current *proof* of OF is provided by the matrix isolation work of Arkell *et al.*[182] (Although the radical was claimed[183a] to result from the photolysis of OF_2 in CCl_3F at 77°K, the product was almost certainly[183b] O_2F.) In Arkell's work a gaseous mixture of OF_2 in argon (or nitrogen) of composition 1 in 40 was slowly deposited on a CsBr window maintained at 4°K in a special cold cell employing liquid helium as coolant. The infrared

178 G. Franz and F. Neumayr, *Inorg. Chem.* **3** (1964) 921.

179 I. J. Solomon, A. J. Kacmarek and J. K. Raney, *J. Phys. Chem.* **72** (1968) 2262.

180 F. I. Metz, F. E. Welsh and W. B. Rose, *Adv. Chem. Ser.* **54** (1966) 202.

181 E. Staricco, J. E. Sicre and H. J. Schumaker, *Z. physik. Chem. (Frankfurt)* **31** (1962) 385; A. D. Kirshenbaum, *Inorg. Nucl. Chem. Letters* **1** (1965) 121.

182 A. Arkell, R. R. Reinhard and L. P. Larson, *J. Am. Chem. Soc.* **87** (1965) 1016.

spectrum of the solid deposit showed sharp absorptions due to the OF_2 isolated in the argon matrix. Photolysis of this deposit with a 1000 W ultraviolet lamp, with appropriate filters, resulted in the development of an infrared band at 1028.5 cm^{-1} and concurrent decrease in the intensity of the OF_2 absorption. On allowing the deposit to warm to approximately 45°K, the 1028.5 cm^{-1} band disappeared and the intensity of the OF_2 bands returned to their original value; in some experiments weak absorptions due to O_2F_2 were observed on warming. On photolysis of an $^{18}OF_2$, $^{16}OF_2$ mixture two bands appeared at 1028.5 cm^{-1} and 997.4 cm^{-1}. Clearly photolysis of OF_2 generates OF and F, the latter escaping from the matrix cage of inert gas atoms—the 1028.5 cm^{-1} band is due to OF. On allowing warm-up to 45°K, the matrix becomes "diffusing" and the F and OF recombine, the small amount of O_2F_2 observed being due to the recombination $OF+OF \rightarrow O_2F_2$. More recently Arkell[184] has obtained a substantial increase in the quantity of OF produced by the addition of N_2O to the matrix—fluorine atoms produced photolytically from OF_2 can react with N_2O to give OF plus N_2. With F_2 and N_2O in argon, photolysis gave OF and OF_2 and this allows a lower limit to be put on the OF bond energy of approximately 40 kcal $mole^{-1}$ [185].

In very recent work Clyne and Watson[185a] have obtained mass spectral evidence for OF by the reaction of F atoms with O_3. The appearance potential of OF^+ from OF is 1.6_5 eV lower than its appearance potential from OF_2. This allows an estimate of $D(\text{FO–F})$ [38.0 ± 3.7 kcal $mole^{-1}$] and $D(\text{OF})$ [51.4 ± 4.1 kcal $mole^{-1}$].

The bond energy of the OF radical has been a subject of some controversy. Early electron impact studies[186] gave a value of 28.8 ± 0.5 kcal $mole^{-1}$ (corrected for the most recent values of other relevant thermodynamic data); matrix work suggested ~ 56 (from frequency of OF vibration) and > 40 (from $F+N_2O$ reaction). However it does now seem that the best experimental value is

$$D(\text{OF}) = 50 \pm 6 \text{ kcal mole}^{-1}.$$

Theoretical estimates have ranged from 45 to 56 kcal $mole^{-1}$; the most recent SCF calculations[187a] give a value of 69 (+7–18) kcal $mole^{-1}$.

5.4. DIOXYGEN DIFLUORIDE, O_2F_2

Preparation

This compound was first prepared by Ruff and Menzel[188] by passage of an electric discharge through a gaseous oxygen–fluorine mixture at low pressure and with the discharge vessel maintained at liquid air temperature. Photolysis of liquid mixtures[189] of oxygen and fluorine produced red and yellow crystals believed to be O_3F_2 and O_2F_2.

The common current method of preparation employs the electric discharge technique and Table 42 outlines the conditions for O_2F_2 and for the higher oxygen fluorides reported[169].

183a F. Neumayr and N. Vanderkooi, Jr., *Inorg. Chem.* **4** (1965) 1234.
183b F. E. Welsh, F. I. Metz and W. B. Rose, *J. Molec. Spectrosc.* **21** (1966) 249.
184 A. Arkell, *J. Phys. Chem.* **73** (1969) 3877.
185 J. S. Ogden and J. J. Turner, *J. Chem. Soc.* A, 1967, 1483.
185a M. A. A. Clyne and R. T. Watson, *Chem. Phys. Letters*, in press.
186 V. H. Dibeler, R. M. Reese and J. L. Franklin, *J. Chem. Phys.* **27** (1957) 1296.
187 G. Glockler, *J. Chem. Phys.* **16** (1948) 604; W. C. Price, T. R. Passmore and D. M. Roessler, *Disc. Faraday Soc.* **35** (1963) 207; M. Green and J. W. Linnett, *J. Chem. Soc.* 1960, 4959.
187a P. A. G. O'Hare and A. C. Wahl, *J. Chem. Phys.*, **53** (1970) 2469.
188 O. Ruff and W. Menzel, *Z. anorg. allgem. Chem* **211** (1933) 204; **217** (1934) 85.
189 S. Aoyama and S. Sakuraba, *J. Chem. Soc. Japan* **59** (1938) 1321.

It will be noted that the lower the temperature and the milder the discharge conditions, the higher the oxygen to fluorine ratio.

More recently Streng and Streng[190] have employed a discharge of OF_2 and O_2 rather than F_2 and O_2.

Most reports of properties of O_2F_2 describe a melting point of ~109°K for the reddish solid. Goetschel *et al.*[191] have very recently prepared O_2F_2 by the irradiation of liquid O_2–F_2 at 77°K with 3 MeV bremsstrahlung. Higher oxides produced in this preparation were decomposed by pumping at 195°K to give oxygen and O_2F_2; the latter was a pale *yellow* solid with a sharp melting point at 119°K. These workers believe that the previous preparations of O_2F_2 have resulted in substantial higher oxide impurities thus lowering the melting point.

TABLE 42. CONDITIONS FOR PREPARATION OF OXYGEN FLUORIDES

O/F compound	Temp. of discharge vessel (°K)	Gas		Power
		O_2:F_2 ratio	Pressure (mmHg)	
O_2F_2	90	1:1	12±5	25–30 mA, 2.1–2.4 kV
O_3F_2	77	3:2	12±5	20–25 mA, 2.0–2.2 kV
O_4F_2	60–77	2:1	5–15	4.5–4.8 mA, 0.8–1.3 kV
O_5F_2	60–77	5:2	0.5–8	4–6 W
O_6F_2	60–77	3:1	0.5–8	4–6 W

Levy and Copeland[192] have observed that O_2F_2 can be formed by the photolysis of *gaseous* O_2–F_2 mixtures at −42°C (no reaction at 0°C)—this reaction will be considered further under O_2F.

Physical Properties

Following Goetschel, O_2F_2 at low temperature is a pale yellow solid which melts at 119°K to give a yellow liquid. The compound is thermally unstable at room temperature, the rate of decomposition to O_2 and F_2 increasing rapidly above about 150°K.

Table 43 summarizes most of the important physical properties of O_2F_2. Certain unusual properties are of particular interest:

(1) The O–O bond length is surprisingly short, being almost the same as in O_2 itself; the O–F bond length is considerably *longer* than in OF_2.

(2) The infrared spectrum shows an O–O stretching vibration at 1306 cm^{-1}, which though lower in frequency than the corresponding mode in O_2F (1499 cm^{-1}) leads to a similar O–O force constant.

(3) The ^{19}F chemical shift is a very long way to low field of (i.e. less shielded than) $CFCl_3$.

[190] A. G. Streng and L. V. Streng, *Inorg. Nucl. Chem. Letters* **2** (1966) 107.
[191] C. T. Goetschel, V. A. Campanile, C. D. Wagner and J. N. Wilson, *J. Am. Chem. Soc.* **91** (1969) 4702.
[192] J. B. Levy and B. K. W. Copeland, *J. Phys. Chem.* **69** (1965) 408; **72** (1968) 3168.

(4) The activation energy for the thermal decomposition (17.3 kcal mole^{-1}) is perhaps assignable to FOOF → F+OOF. Mass spectral studies suggest a value of 18 kcal mole^{-1} for this step.

Chemical Properties

Dioxygen difluoride is a particularly powerful oxidizing agent. Because of the instability of O_2F_2, reactions must be carried out under cryogenic conditions and even then explosions

TABLE 43. PHYSICAL PROPERTIES OF DIOXYGEN DIFLUORIDE, O_2F_2

Melting point [a]	−154°C (119°K)
Boiling point (theoretical, from extrapolation)	−57°C (216°K)
Decomposition: $-\frac{d(O_2F_2)}{dt} = [O_2F_2] \times 10^{12.4} e^{-1700/4.57T}$	
Activation energy for thermal decomposition	17.3 kcal mole^{-1}
Heat of vaporization	4.583±0.10 kcal mole^{-1}
Vapour pressure	$\log p$ (mmHg) = 7.515 − 1000/T (T < 173°K)
Heat of formation (ΔH_f°)	4.73±0.30 kcal mole^{-1}
D (FO–OF) [b]	103.5±5 kcal mole^{-1}
D (F–OOF) [b]	≈ 18 kcal mole^{-1}
Ionization potential [b]	300–310 kcal mole^{-1} (13.0–13.4 eV)
Density (liquid)	2.074–0.002 91T g cc^{-1} (117–186°K)
Ultraviolet spectrum (solution in freon 13, at 77°K)	λ (Å) 3500 4000 4500 5000 6000 7000 ε (l mole^{-1} cm^{-1}) 13.13 8.02 2.52 0.63 0.10 0.10
Dipole moment	1.44±0.04 D

Infrared maxima (solid) (cm^{-1}) [c]

$^{16}O_2F_2$	$^{18}O_2F_2$	
1306	1239	O–O stretch
621	595	O–F sym. stretch
615	586	O–F asym. stretch
457	444	O–O–F asym. bend
369	362	O–O–F sym. bend
205	—	torsion

Force constants [c] (mdyn Å^{-1})

f_{OF}	1.36	f_{OO}	10.25
f_{OOF}	1.17	f_r	0.455
f_{OO-OF}	1.18	f_{OF-OOF}	0.114

Structure of gas (microwave)
r (O–O) = 1.217±0.003 Å
r (O–F) = 1.575±0.003 Å
OOF angle = 109° 30′±30′
dihedral angle = 87° 30′±30′

Chemical shift: ^{19}F (v. $CFCl_3$ in ppm) liquid, −865±1 [d]; dil. solution in CF_3Cl, −825±10 [e]
^{17}O (v. $H_2{}^{17}O$ in ppm) liquid, −647 [f]

Unless otherwise stated, data from A. G. Streng, *Chem. Revs.* **63** (1963) 607 and/or E. W. Lawless and I. C. Smith, *Inorganic High-energy Oxidisers*, Arnold, London, and Dekker, New York (1968).

[a] C. T. Goetschel, V. A. Campanile, C. D. Wagner and J. N. Wilson, *J. Am. Chem. Soc.* **91** (1969) 4702.

[b] T. J. Malone and H. A. McGee, Jr., *J. Phys. Chem.* **69** (1965) 4338; **70** (1966) 316.

[c] K. R. Loos, C. T. Goetschel and V. A. Campanile, *Chem. Communs.* (1968) 1633; *J. Chem. Phys.* **52** (1970) 4418; D. J. Gardiner, N. J. Lawrence and J. J. Turner, *J. Chem. Soc.* A (1971) 400.

[d] J. W. Nebgen, F. I. Metz and W. B. Rose, *J. Am. Chem. Soc.* **89** (1967) 3118.

[e] N. J. Lawrence, J. S. Ogden and J. J. Turner, *Chem. Communs.* (1966) 102; *J. Chem. Soc.* A (1968) 3100.

[f] I. J. Solomon, J. K. Raney, A. J. Kacmarek, R. G. Maguire and G. A. Noble, *J. Am. Chem. Soc.* **89** (1967) 2015.

frequently occur. A reactant can be condensed on to solid O_2F_2 at 77°K and slowly be allowed to warm to the reaction temperature; liquid O_2F_2 can be slowly dropped on to liquid or solid reactants; reactions with gases can be studied by bubbling the gases through liquid O_2F_2 or its solution in freons. Streng[167, 193] has provided a comprehensive list of reactions, and more recent work is briefly described by Lawless and Smith[169].

Some violent reactions with largely uncharacterized products involve liquid O_2F_2 with organic compounds, solid NH_3 (110°K), ice (130–140°K), S (90°K), charcoal, platinum coated with PtF_4 (160°K). With liquid NO_2F and liquid N_2F_2, O_2F_2 mixes in all proportions, but it is insoluble in liquid NF_3.

The overall result of many reactions with compounds or low fluorides is conversion into high fluorides. Streng[193] has established the stoichiometry of several such reactions and knowing the appropriate heats of formation has calculated the heats of reaction. For example:

	ΔH (kcal mole^{-1})
$O_2F_2+ClF \rightarrow ClF_3+O_2$	-30.1
$O_2F_2+BrF_3 \rightarrow BrF_5+O_2$	-46.1
$O_2F_2+SF_4 \rightarrow SF_6+O_2$	-121.5
$O_2F_2+\frac{1}{2}HCl \rightarrow \frac{1}{2}ClF_3+\frac{1}{2}HF+O_2$	-45.7
$O_2F_2+\frac{1}{3}HBr \rightarrow \frac{1}{3}BrF_5+\frac{1}{3}HF+O_2$	-58.9
$O_2F_2+\frac{1}{4}H_2S \rightarrow \frac{1}{4}SF_6+\frac{1}{2}HF+O_2$	-108.2

In those reactions with the smallest ΔH values it has been possible to obtain strong evidence for intermediate addition compounds. For example, if the reaction of ClF and O_2F_2 is carried out above 140°K, there is a violent reaction following the above equation; however, at lower temperatures (119–130°K), and with slow addition of ClF, a violet compound of formula O_2ClF_3 is obtained:

$$O_2F_2+ClF \rightarrow O_2ClF_3$$

This compound can also be produced by the slow addition of Cl_2 to O_2F_2 at 130°K, by the reaction of O_2F_2 and HCl at 130–140°K and by the ultraviolet irradiation of liquid ClF_3 under a pressure of approximately 2 atm of O_2 at 195°K. The solid violet compound is stable at 195°K and is readily soluble in ClF, O_2F_2, ClF_3, HF; the conductivity of non-aqueous HF is unchanged on addition of O_2ClF_3. The most likely structures are

$$\underset{\text{(i)}}{\text{F—}\underset{\underset{\displaystyle F}{|}}{\text{Cl}}\text{—OOF}} \quad \text{or} \quad \underset{\text{(ii)}}{\text{FO—}\underset{\underset{\displaystyle F}{|}}{\text{Cl}}\text{—OF}}$$

In view of the evidence (see below) that O_2F_2 behaves as F+OOF but not FO+OF, (i) is the more probable. Other similar compounds proposed include O_2BrF_5 and O_2SF_6, the evidence for the latter being sketchy.

With the avid fluoride ion acceptors BF_3, PF_5, AsF_5 and SbF_5, dioxygen difluoride reacts to form O_2BF_4 and O_2MF_6 (M = P, As, Sb):

$$O_2F_2+BF_3 \rightarrow O_2BF_4+\tfrac{1}{2}F_2$$
$$O_2F_2+MF_5 \rightarrow O_2MF_6+\tfrac{1}{2}F_2$$

---

193 A. G. Streng, *J. Am. Chem. Soc.* **85** (1963) 1380.

These solids are paramagnetic and almost certainly are best formulated as $O_2^+BF_4^-$ and $O_2^+MF_6^-$ (infrared, Raman and X-ray powder evidence). The boron compound has been intensively investigated. It is only slowly decomposed in the absence of moisture at 0°C, but it is an extremely active oxidizing agent; for example, benzene inflames on addition of a small particle of O_2BF_4 at room temperature. Hydrolysis of O_2BF_4 produces O_2, O_3, F_2 and HBF_4.

Finally, O_2F_2 and xenon react to give XeF_2 via an unstable yellow intermediate which may be $XeOF_2$ [194].

Properties Relevant to Reaction Mechanism and the OOF Radical

As mentioned above, thermal decomposition and mass spectral studies are consistent with the step FOOF → F+OOF. Recent infrared matrix isolation work on O_2F_2 co-condensed from a low-temperature bath with a stream of argon or nitrogen or CO_2 at 15°K provides evidence for O_2F and O_4F_2 [195].

Electron spin resonance studies on either liquid or solid O_2F_2 or solutions of O_2F_2 in freons at low temperature show the presence of a high concentration of free radical[196]. The spectra (except for solids where free rotation is prevented) consist of a symmetrical doublet with $g \sim 2.004$ and $A \sim 12$ gauss. These spectra are identical[197] to that obtained by electron irradiation of liquid CF_4 containing traces of O_2 and certainly ascribable to O_2F (see below).

Direct chemical evidence that O_2F_2 behaves as F+OOF has only been obtained fairly recently. At −183°C, O_2F_2 and C_3F_6 in $CClF_3$ reacted smoothly to give a reasonable yield of a fraction containing $CF_3CF(OOF)CF_3$ and $CF_3CF_2CF_2(OOF)$ [198] (cf. O_2F_2 with ClF, above).

Solomon *et al.*[199] have investigated the O_2F_2/BF_3 reaction using ^{18}F as a tracer; the data are consistent with the following reaction scheme:

$$O_2F_2 \rightarrow F+OOF$$
$$F+F \rightarrow F_2$$
$$F+B^{18}FF_2 \rightarrow {}^{18}F+BF_3$$
$$OOF+B^{18}FF_2 \rightleftharpoons O_2B^{18}FF_3$$

The equilibrium in the final step was demonstrated by boron exchange of $B^{18}FF_2$ with $O_2^{10}BF_4$ at room temperature. It should also be mentioned here that O_4F_2, which is believed to be in equilibrium with $2O_2F$, reacts with BF_3 much more readily than O_2F_2 to give the same product, O_2BF_4.

Thus there is strong evidence that O_2F_2 reacts initially as F+OOF.

5.5. DIOXYGEN MONOFLUORIDE, O_2F

The above evidence suggests that O_2F exists at low temperature as a relatively stable entity. It should be noted that the observation of doublet e.s.r. spectra in condensed oxygen

[194] S. A. Morrow and A. R. Young, *Inorg. Chem.* **4** (1965) 759.
[195] D. J. Gardiner, N. J. Lawrence and J. J. Turner, *J. Chem. Soc.* A (1971) 400.
[196a] P. H. Kasai and A. D. Kirshenbaum, *J. Am. Chem. Soc.* **87** (1965) 3069.
[196b] F. E. Welsh, F. I. Metz and W. B. Rose, *J. Molec. Spectrosc.* **21** (1966) 249.
[197] R. W. Fessenden and R. H. Schuler, *J. Chem. Phys.* **44** (1966) 434.
[198] I. J. Solomon, A. J. Kacmarek, J. N. Keith and J. K. Raney, *J. Am. Chem. Soc.* **90** (1968) 6557.
[199] J. N. Keith, I. J. Solomon, I. Sheft and H. H. Hyman, *Inorg. Chem.* **7** (1968) 230.

fluorides (OF_2, O_2F_2, "O_3F_2", O_4F_2) does not prove the existence of O_2F but merely O_nF since ^{16}O has no magnetic moment. However, Fessenden and Schuler[197] examined the e.s.r. spectra produced on electron irradiation of liquid CF_4 containing traces of $^{16}O_2$, $^{16}O^{17}O$ and $^{17}O_2$. The ^{17}O nucleus has a spin of 5/2, and the presence of $^{16}O_2F$, $^{16}O^{17}OF$, $^{17}O^{16}OF$ and $^{17}O_2F$ was unequivocally demonstrated; the spectral parameters are given in Table 44, and it is now accepted that the same species is present in the condensed O/F systems. In addition, Metz *et al.*[183b] concluded from e.s.r. work that O_2F was present in ^{17}O enriched pure liquid O_2F_2.

TABLE 44. PHYSICAL PROPERTIES OF DIOXYGEN MONOFLUORIDE, O_2F

Ionization potential [a]	290 ± 5 kcal mole^{-1} (12.6 ± 0.2 eV)				
Dissociation bond energies [a]:					
D (O–OF)	85 kcal mole^{-1}				
D (OO–F)	≈ 18 kcal mole^{-1}				
Infrared maxima [b] (cm^{-1})	$^{16}O^{16}OF$	$^{16}O^{18}OF$	$^{18}O^{16}OF$	$^{18}O^{18}OF$	
	1495.0	(1453.9)	(1453.9)	1411.7	O–O stretch
	584.5	581.2	563.4	560.1	O–F stretch
	376.0	—	—	266.6	O–O–F bend
Force constants [b] (mdyn Å^{-1})	f_{OO} 10.50;	f_{OF} 1.32;	f_{OOF} 0.52;	f_{OO-OF} 0.300;	f_{OF-OOF} 0.019
Electron spin resonance data:					
(i) [c] $g = 2.0038$					
Hyperfine constants:	^{19}F 12.83 gauss (36.0 Mc/s)				
	^{17}O 22.17 and 14.50 gauss (62.2 and 40.7 Mc/s)				
(ii) [d] $g_1 = 2.0080$, $g_2 = 2.0008$, $g_3 = 2.0022$					
Hyperfine tensors (^{19}F)	$A_1 \pm 102.8$ gauss (288.4 Mc/s)				
	$A_2 \mp 50.2$ gauss (141.1 Mc/s)				
	$A_3 \mp 14.0$ gauss (39.2 Mc/s)				
Isotropic constant	$A - 12.8$ gauss (36.0 Mc/s)				

[a] T. J. Malone and H. A. McGee, Jr., *J. Phys. Chem.* **69** (1965) 4338; **70** (1966) 316 (electron impact studies) the value for *D* (OO–F) assumes that *D* (F–OOF) and *D* (OO–F) are equal. The value for *D* (O–OF) assumes the same and in addition that *D* (OF) = 50 kcal mole^{-1}.

[b] Data are quoted from P. N. Noble and G. C. Pimentel, *J. Chem. Phys.* **44** (1966) 3641—photolysis of F_2 in O_2 (1:250) at 20°K. The $^{16}O^{18}F$ and $^{18}O^{16}F$ O–O stretching vibrations are calculated, using the above force constants, to be separated by only 2–3 cm^{-1} and are not resolved.

[c] R. W. Fessenden and R. H. Schuler, *J. Chem. Phys.* **44** (1966) 434.

[d] F. J. Adrian, *J. Chem. Phys.* **46** (1967) 1543.

Arkell[200] and Spratley *et al.*[201a] have obtained convincing infrared spectral evidence for O_2F by photolysis of several low temperature (4°K or 20°K) matrix mixtures, e.g. F_2 plus O_2 very dilute in argon. Presumably

$$F_2 \xrightarrow{h\nu} 2F$$

$$F + O_2 \rightarrow FO_2$$

and the presence of O_2F was proved by the use of $^{16}O_2$, $^{16}O^{18}O$ and $^{18}O_2$. For example, the band assigned to the O–F stretching vibration could be resolved under high resolution[201b]

[200] A. Arkell, *J. Am. Chem. Soc.* **87** (1965) 4057.

[201a] R. D. Spratley, J. J. Turner and G. C. Pimentel, *J. Chem. Phys.* **44** (1966) 2063.

[201b] P. N. Noble and G. C. Pimentel, *J. Chem. Phys.* **44** (1966) 3641.

into four components, due to $^{16}O^{16}OF$, $^{16}O^{18}OF$, $^{18}O^{16}OF$ and $^{18}O_2F$ (see Table 44). On diffusion (cf. matrix work on OF) the narrow absorption at 1499 cm^{-1} changed to a broad feature at $\sim$1510 cm^{-1} and the rest of the spectrum remained unchanged; this is probably because diffusion results in the formation of $(O_2F)_2$ which is a very weakly bonded dimer (see below).

More recently Adrian[202] has obtained e.s.r. spectra for O_2F by the matrix technique. These spectra are rather better than the original ones of Kasai and Kirshenbaum[196] from O/F compounds in solid CF_3Cl. One interesting feature of Adrian's work is the use of "forbidden" lines in which the electron spin transition is accompanied by a nuclear spin flip to obtain the relative signs of the fluorine hyperfine splitting tensor components. Theoretical work suggests that A_1 is negative, which means that the isotropic coupling constant is negative, -36.0 Mc/s.

Since O_2F can be formed[201] at 4°K by the reaction $F+O_2 \rightarrow O_2F$, the activation energy for this reaction must be close to zero. Levy and Copeland[192], during work on the oxygen inhibition of the H/F reaction, came to similar conclusions. As mentioned previously, at -42°C the photolysis of gaseous O_2 and F_2 gives O_2F_2 as product, presumably via

$$F_2 \xrightarrow{h\nu} 2F$$
$$F+O_2+M \rightarrow O_2F+M$$
$$O_2F+F+M \rightarrow O_2F_2+M$$

Clearly O_2F must be a relatively stable entity. [At 0°C the oxygen inhibition of the H_2/F_2 reaction does not involve what might be thought the obvious mechanism,

$$H+O_2F \rightarrow HF+O_2$$

since the step

$$H_2+O_2F \rightarrow HF+O_2+H$$

seems to be predominant.]

The reaction of O_2 with fluorine atoms is also presumably responsible for the very high concentrations of O_2F radicals (e.s.r. intensity evidence) which result from photolysis of liquid OF_2/O_2 mixtures[203].

5.6. ENERGETICS AND NUCLEAR MAGNETIC RESONANCE

Before considering the chemistry of other oxygen fluorides two general topics will be considered.

Energetics of O/F Species

So far values for various energy terms have been quoted. This seems an appropriate place to point out that they are not all self-consistent.

From Table 45 we obtain:

$$\text{FOOF} \rightarrow \text{FO}+\text{OF} \quad \Delta H = 52 \text{ kcal mole}^{-1} \quad (4)-2\times(3) \text{ cf. } (6)$$
$$\text{OF} \rightarrow \text{O}+\text{F} \quad \Delta H = 25 \text{ kcal mole}^{-1} \quad \tfrac{1}{2}[(4)-(6)] \text{ cf. } (3)$$

Clearly one or more of ΔH_f° (OF_2), ΔH_f° (O_2F_2), D (F–OF), D (F–OOF), D (FO–OF) must

202 F. J. Adrian, *J. Chem. Phys.* **46** (1967) 1543.
203 N. J. Lawrence, J. S. Ogden and J. J. Turner, *J. Chem. Soc.* A, 1968, 3100.

be wrong, but it is not possible to say which. However, it *is* clear that the O–O bonds in FOOF and OOF are particularly strong and the O–F bonds weak.

TABLE 45. SUMMARY OF SOME ΔH VALUES

	ΔH (kcal mole^{-1})	Method
(1) FOF → F+O+F	90	Calorimetry
(2) FOF → F+OF	40	Electron impact and kinetics
(3) OF → O+F	50	(1)–(2)
(4) FOOF → F+O+O+F	152	Calorimetry
(5) FOOF → F+OOF	18	Electron impact
(6) FOOF → FO+OF	103	Electron impact
(7) OOF → OO+F	18	Electron impact
(8) OOF → O+O+F	137	(7)+D (O–O)

Free Radicals and Nuclear Magnetic Resonance

It is well known that paramagnetic species in solution can affect the position and line widths of n.m.r. signals; usually, of course, n.m.r. signals from free radicals themselves are too broad to observe. Of particular relevance to oxygen fluoride work is the paper by de Boer and Maclean[204] on the effect of paramagnetic species exchanging with the molecule being studied. For example, in the equilibrium

$$\text{FOOF}+\text{OOF} \rightleftharpoons \text{FOO}+\text{FOOF}$$

the dependence of the ^{19}F chemical shift of O_2F_2 on O_2F can be expressed by

$$\frac{\Delta H}{H} = -\frac{g\beta}{g_N\beta_N} \times \underline{a} \times \frac{S(S+1)}{3kT} \times f_p \left[1+\frac{f_d \tau_p^2 \underline{a}^2/4}{1+2\tau_p T_{1e}^{-1}}\right]^{-1} \tag{5}$$

where $g\beta$ and $g_N\beta_N$ are the gyromagnetic ratios in Bohr magnetons of the electron and nucleus N (^{19}F in this case), $\underline{a}$ is the hyperfine interaction constant, S is the nuclear spin of N ($\frac{1}{2}$ for ^{19}F), f_p, f_d are the fractions of para- and dia-magnetic species respectively, τ_p is the lifetime of the free radical and T_{1e} is the electron longitudinal relaxation time. For this equilibrium eqn. (5) becomes

$$\frac{\Delta H}{H} = -\frac{g\beta}{g_N\beta_N} \times \frac{a}{4kT} \times \frac{f_p}{1+X}, \tag{6}$$

where X is a positive quantity. In the "slow exchange" case, X will be very large and hence $\Delta H/H = 0$, i.e. the chemical shift is unaffected by O_2F, although the line width may be affected. In the other extreme of $X = 0$, substituting for the physical constants and using the value -36 Mc/s for $\underline{a}$ [202], we have

$$\frac{\Delta H}{H} \sim 10^5\left(\frac{3f_p}{T}\right) \quad \text{(ppm)} \tag{7}$$

[204] E. de Boer and C. MacLean, *J. Chem. Phys.* **44** (1966) 1335.

Thus at, say, 150°K, $\Delta H/H \sim 2000 f_p$. Lawrence *et al.*[203] observed no change in the ^{19}F chemical shift of O_2F_2 on increasing the concentration of O_2F by an order of magnitude on addition of OF_2—presumably f_p was very small. However, Solomon *et al.*[205] in their studies on O_3F_2 (probably $O_2F_2+O_4F_2$ ($\rightleftharpoons 2O_2F$), see below) observed that the ^{19}F signal of O_2F_2 was very sensitive to temperature. *But* the explanation for this phenomenon must be more complex than the *simple* equilibrium suggested above since the ^{17}O n.m.r. signal seems to be *independent* of temperature. Substitution of the physical constants, $S = 5/2$ and $a \sim 50$ Mc/s, suggests a much *greater* ^{17}O variation with temperature. Conceivably a species $O_2F \ldots F'O'O''F''$ in which a (F′ or F″) is much greater than a (O′ or O″), is responsible for the shifts.

5.7. TETRAOXYGEN DIFLUORIDE, O_4F_2, AND THE EQUILIBRIUM $O_4F_2 \rightleftharpoons O_2F$

We shall first consider the preparation of O_4F_2 and some of its properties before returning to a discussion of the equilibrium with the O_2F radical.

Preparation

Table 42 (p. 753) has already outlined the discharge conditions for the preparation[206] of O_4F_2; the compound has also been obtained by electrical discharge in OF_2/O_2 mixtures[190], and by radiolysis of liquid O_2/F_2 mixtures[191].

The possibility that O_4F_2 was simply O_2F_2 with dissolved O_2 was eliminated by demonstrating that O_2 can be pumped off O_2F_2; the possibility of dissolved O_3 was eliminated by showing that O_3 can be extracted into liquid O_2, whereas O_4F_2 in liquid O_2 shows no trace of ozone[206].

Tetraoxygen difluoride is an even more powerful oxidizing agent than O_2F_2, and is considerably less thermally stable, decomposing ultimately into $2O_2$ and F_2, via O_2F_2 and O_2.

Physical Properties

The dark red–brown solid melts at −191°C (82°K) to give a similarly coloured liquid. The vapour pressure is "less than 1 mm at 90°K" [206].

Electron spin resonance spectra[196a, 207] of O_4F_2 show *very* intense signals due to the O_2F radical. Thus although it has been possible to obtain *some* physical data for O_2F, data for O_4F_2 must be treated with caution, and there has been no really systematic investigation of physical properties. However, since there is very strong evidence[205] that O_3F_2 is a mixture of O_4F_2 and O_2F_2, and several physical properties of O_3F_2 have been measured (see Table 46) it may be possible to estimate O_4F_2 properties from Tables 43 and 46—the reader is left to do this for himself.

Chemical Properties

Again there is virtually no information, but the reader can make what use seems appropriate of the descriptions of the properties of O_2F_2 (above) and O_3F_2 (below).

205a I. J. Solomon, J. K. Raney, A. J. Kacmarek, R. G. Maguire and G. A. Noble, *J. Am. Chem. Soc.* **89** (1967) 2015.

205b I. J. Solomon, J. N. Keith, A. J. Kacmarek and J. K. Raney, *J. Am. Chem. Soc.* **90** (1968) 5408.

206 A. V. Grosse, A. G. Streng and A. D. Kirshenbaum, *J. Am. Chem. Soc.* **83** (1961) 1004; A. G. Streng, *Can. J. Chem.* **44** (1966) 1476.

207 A. D. Kirshenbaum and A. G. Streng, *J. Am. Chem. Soc.* **88** (1966) 2434.

The Equilibrium $O_4F_2 \rightleftharpoons 2O_2F$

The ready reaction of O_4F_2 with BF_3 to give O_2BF_4 does underline its behaviour as $2O_2F$.

The e.s.r. spectrum of O_4F_2 ($\sim$3% by volume) in *solid* CF_3Cl at 77°K has been interpreted on the basis of $\sim$5 mole% concentration of O_2F [196a]. Goetschel *et al.*[191] have reconsidered this and propose an equilibrium constant at about 80°K of 8×10^{-5} in mole fraction units. This equilibrium constant is only meaningful if the *solid* CF_3Cl readily permits diffusion, otherwise O_2F radicals may be trapped in a cage of CF_3Cl molecules. Note that in the matrix isolation studies the O_2F infrared band at 1499 cm^{-1} only disappeared after warming to the diffusion temperature (40°K). It seems reasonable to assume that at temperatures close to the melting point of CF_3Cl (92°K), it is readily diffusing. Assuming from analogous dimerizations that the standard entropy change is $\sim$15 e.u., the authors calculate a standard enthalpy change of $\sim$3 kcal mole^{-1}.

Fessenden and Schuler[197] concluded that at about 10^{-3} molar in liquid CF_4 ($\sim$100°K) the "radical does not exist largely as the dimer". These two e.s.r. observations are not irreconcilable since the great dilution in CF_4 would drive the equilibrium to the right.

To date no one has published data on the n.m.r. spectrum of O_4F_2 by examining O_4F_2. However, Solomon *et al.*[205] believe that they have obtained ^{19}F and ^{17}O spectra for O_4F_2 by examining liquid O_3F_2 (see below). The position of the ^{19}F signal was a sensitive function of temperature, ascribable presumably to some exchange process probably more complex than just $O_4F_2 \rightleftharpoons 2O_2F$. The situation is complicated, however, by the decomposition of O_2F, probably into O_2F_2 and O_2, and, in fact, Solomon concludes that between 83° and 110°K the concentration of radicals *decreases* with temperature.

It is clear that because of the great difficulty in handling these compounds, the equiibrium is not yet fully understood.

5.8. TRIOXYGEN DIFLUORIDE, O_3F_2

At various times so far it has been suggested that O_3F_2 is really a mixture of O_4F_2 and O_2F_2. Before considering the evidence for this conclusion we shall first outline the preparation and properties of this "compound".

Preparation

Trioxygen difluoride is prepared by the reactions of O_2 and F_2 in the electric discharge (see Table 42) or OF_2 and F_2 in the discharge[190]. The assignment of the formula O_3F_2 to the compound depends only on the ratio of O_2:F_2 consumed and the ratio of O_2:F_2 obtained on thermal decomposition.

Physical Properties

Bearing in mind that O_3F_2 is probably a mixture of O_2F_2 and O_4F_2, some of the physical properties described by Streng[167] are outlined in Table 46.

Chemical Properties

Not surprisingly, O_3F_2 is a potent oxidizing agent and is unstable at temperatures above its melting point although it can be stored indefinitely at 77°K in dark, dry Pyrex. It is soluble in OF_2, O_2F_2 and O_3 although the latter solution can readily explode.

Some idea of its reactivity is demonstrated by dropping liquid O_3F_2 on to solid anhydrous ammonia at 90°K when an instantaneous flame and explosion results; its behaviour is similar with a whole range of compounds. There is some evidence for the formation of a purple species (O_2ClF_3?) on reaction of O_3F_2 and ClF at 77°K—at 90°K an explosion occurs.

"O_3F_2" or "$O_2F_2+O_4F_2$"

Solomon *et al.*[205] have carefully examined the ^{19}F and ^{17}O n.m.r. spectra of liquid O_3F_2. The ^{17}O spectrum consisted of three lines, one at −647 ppm (w.r.t. $H_2{}^{17}O$) and a pair of lines of equal intensity at −971 and −1512 ppm; the first line was easily attributable to O_2F_2 by performing experiments with O_2F_2 alone. It is difficult to assign two lines of equal intensity to a molecule of such structure as FOOOF. Solomon believes this pair is due to O_4F_2, i.e. FO′O″O″O′F, one line arising from O′ nuclei and one from O″ nuclei. Moreover, the ^{19}F spectrum of O_3F_2 was an asymmetrical line resolved with difficulty into two overlapping lines—one presumably due to O_2F_2 and one due to O_4F_2. The position of these ^{19}F lines, as described previously, was a sensitive function of temperature, probably due to some exchange mechanism with O_2F. Because of the apparent lack of temperature sensitivity of the ^{17}O signals, it is not possible to decide on a mechanism at this stage.

TABLE 46. PHYSICAL PROPERTIES OF TRIOXYGEN DIFLUORIDE, O_3F_2

Melting point	−190°C (83°K)
Boiling point	−60°C, decomposes (213°K)
Decomposition temperature	above m.p.
Heat of vaporization	4.581 ± 0.2 kcal mole^{-1}
Vapour pressure	$\log p\ (\text{mmHg}) = 6.1343 - \frac{675.57}{T}$ (79–114°K)
Heat of formation (ΔH_f°)	6.24 ± 0.75 kcal mole^{-1}
Density of liquid	$d = 2.357 - 0.006\,76T$ g cc^{-1}

Data from A. G. Streng, *Chem. Revs.* **63** (1963) 607.

Metz *et al.*[208] have also examined the ^{19}F n.m.r. spectra of liquid O_3F_2. They did not resolve the signal into two components and concluded that O_3F_2 could *not* be a mixture of O_2F_2 and O_4F_2. They did, however, on theoretical grounds reject the structure FOOOF and suggested that O_3F_2 is O_2F_2 plus "interstitial oxygen molecules" on the grounds that the only decomposition products of O_3F_2 are O_2F_2 and O_2. These workers object to the O_4F_2/O_2F_2 mixture theory on the grounds that the chemical shift of *pure* O_2F_2 is insensitive to temperature and that if Solomon's ^{19}F doublet were due to O_2F_2 and O_4F_2 the O_2F_2 signal should stay in the same place as the temperature was raised. However, as described previously, this insensitivity of O_2F_2 observed by Metz and Lawrence *et al.*[203] is probably due to the very low concentration of O_2F; in Solomon's work the O_2F concentration was probably very high.

[208] J. W. Nebgen, F. I. Metz and W. B. Rose, *J. Am. Chem. Soc.* **89** (1967) 3118.
[209] T. J. Malone and H. A. McGee, *J. Phys. Chem.* **71** (1967) 3060.

Malone and McGee[209] have examined the variation in fragment intensity pattern with temperature in the mass spectra of O_3F_2 and O_3F_2/O_2F_2 mixtures using a cryogenic inlet system. They concluded from this variation and appearance potential data that the OF^+ ion came from OF_2 which was itself a decomposition product of O_3F_2. They propose that O_3F_2 is best described as a loose adduct of the radicals OF and O_2F:

$$2FO_2{\cdot}OF \xrightarrow{I} 2FO_2 + 2OF \begin{cases} \xrightarrow[\text{fast}]{II} O_2F_2 + 2FO_2 \xrightarrow[\text{slow}]{III} O_2F_2 + O_2F_2 + O_2 \\ \xrightarrow[IV]{} 2OF_2 + 2O_2 \end{cases}$$

Such a mechanism explains the substantial concentration of O_2F (e.s.r. evidence) since III is slow. Moreover the authors believe the formation of O_2F_2 and O_2 would "predominate over the formation of OF_2 and O_2".

Malone and McGee point out that Linnett structures can be drawn for O_3F_2, analogous to O_3 in its first excited state, and that these structures exhibit the basic features of OF and O_2F joined by a weak, single electron bond.

Solomon *et al.*[205b] have objected to both Metz *et al.* and Malone and McGee. Firstly, they have repeated their ^{19}F work on O_3F_2 and are convinced that there are *two* signals. Secondly, they can find no n.m.r. evidence for OF_2 on allowing O_3F_2 to slowly decompose in a n.m.r. tube. Thirdly, experiments involving the reaction of BF_3 with O_4F_2, O_3F_2 and O_2F_2 confirm that, on the basis of the ratio of BF_3 consumed to F_2 formed, the stoichiometry of the reaction with O_3F_2,

$$2O_3F_2+3BF_3 \rightarrow 3O_2BF_4+\tfrac{1}{2}F_2$$

is consistent with the sum of the reactions with O_2F_2 and O_4F_2

$$O_2F_2+BF_3 \rightarrow O_2BF_4+\tfrac{1}{2}F_2$$
$$O_4F_2+2BF_3 \rightarrow 2O_2BF_4$$

If O_3F_2 were O_2F_2+"interstitial O_2", then presumably O_2 would be released; if it were $FO_2{\cdot}OF$, then it is arguable that since O_2F_2 and O_4F_2 behave as sources of O_2F (see above), $FO_2{\cdot}OF$ would act as FO_2+OF and the OF released would decompose to O_2+F_2.

On balance, therefore, the present author supports the O_2F_2/O_4F_2 mixture hypothesis.

The absence of a molecular FOOOF compound is not surprising on energetic grounds. O_2F_2 and O_4F_2 are readily formed by the addition of fluorine atoms to O_2 *molecules* in the discharge or other systems, the ratio of O_2F_2 to O_4F_2 depending on the ratio of F_2 to O_2 in the gas mixture.

$$F_2 \rightarrow 2F,\ F+O_2 \rightarrow O_2F;\ F+O_2F \rightarrow FO_2F,\ O_2F+O_2F \rightarrow O_4F_2$$

Molecular O_3F_2 would require oxygen *atom* production with its large energy requirement and, conversely, would probably be very unstable.

This, of course, does not prevent the formation of FOOOF under other conditions, and in fact Arkell[200] obtained *some* evidence for this species in his O_2F matrix experiments. Since $^{16}OF_2$ was produced during photolysis of $^{18}OF_2$ in $^{16}O_2$ he concluded that an exchange mechanism involving O_3F_2 might be important:

$$^{18}OF_2+^{16}O_2 \rightarrow F^{18}O^{16}O^{16}OF \rightarrow ^{16}OF_2+^{16}O^{18}O$$

Clearly such possibilities merit further exploration.

5.9. PENTAOXYGEN DIFLUORIDE, O_5F_2, AND HEXAOXYGEN DIFLUORIDE, O_6F_2

Preparation

These two species have been reported by Streng and Grosse[210] to be formed by electric discharge at low temperature and under mild discharge conditions in O_2/F_2 gas mixtures of appropriate composition. The products were analysed by allowing them to decompose on warming and estimating the oxygen to fluorine ratio.

Properties

O_5F_2 is a reddish-brown liquid at 90°K, similar to O_4F_2; at 77°K, O_5F_2 forms an oil. Although stable at 77°K it decomposes at 90°K giving, finally, O_2 and F_2. The e.s.r. spectrum is said to be substantially different from that of O_4F_2.

O_6F_2 is a crystalline solid at 60°K with a metallic lustre. Slow warming to 90°K results in decomposition to lower oxygen fluorides and ozone; fast warming to 90°K leads to explosion; the compound even exploded on occasions when illuminated by a flashlight.

Attempts to prepare higher oxygen fluorides failed, and the authors suggest that if any are prepared it will be at temperatures below 60°K.

If the species exist as FOOOOOF and FOOOOOOF, the authors make the interesting suggestion that hydrogen atom reaction at low temperature *could* lead to cyclic O_5 and O_6 molecules.

Much more information is required about these species before molecular formulae can be determined.

5.10. OTHER OXYGEN FLUORIDE POSSIBILITIES

During Arkell's[200] matrix work on O_2F he observed bands at 1503 and 1512 cm^{-1} on photolysis of F_2 in O_2 at 4°K. These were partially resolved sidebands on the side of the 1496 cm^{-1} band which was assigned to the O–O stretching vibration of O_2F. Arkell has suggested that these other two bands are due to O_3F and O_4F, but there is no firm evidence available yet.

Some support for the existence of such species is provided by Goetschel's[201] electron irradiation preparative work. Certain highly unstable products were obtained in small concentration, and the authors suggest the possible presence of O_4F or O_6F_2. As well as the reaction of O_2F_2 with BF_3 to give O_2BF_4 there was *some* evidence for less stable BF_4^- salts, conceivably $O_4^+BF_4^-$ (from O_4F+BF_3?) and $O_6^+BF_4^-$ (from O_6F+BF_3?). Clearly these suggestions are highly speculative at this stage, and it is obvious that oxygen fluorine chemistry is not yet "all tied up".

5.11. COMPOUNDS OF CF_3, FLUORINE AND OXYGEN

It would be possible to compare the properties of the oxygen fluorides with a wide range of species. However it is instructive to consider a single class of compounds which might be expected to have similar properties to oxygen fluorides—at least structurally—the CF_3 derivatives. Recently the field of C/F/O chemistry has expanded considerably and many interesting new compounds have been made. In what follows we shall concentrate simply on those species which can be theoretically derived from the corresponding O/F species by substitution of one or more fluorine atoms by the CF_3 group. The compounds of particular interest are:

$$CF_3OF,\ CF_3O,\ CF_3OOCF_3,\ CF_3OOO,\ CF_3OOF,\ CF_3OOOCF_3,\ CF_3OOOOCF_3$$

Trifluoromethyl Hypofluorite, CF_3OF

This is a stable colourless gas prepared[211] by the reaction of fluorine in the presence of silver(II) fluoride catalyst with CO, CO_2, COF_2 or methanol vapour. Some physical properties are listed in Table 47.

TABLE 47. SOME PHYSICAL PROPERTIES OF CF_3OF

Melting point [a]	$-215°C$ (58°K)
Boiling point [a]	$-95°C$ (178°K)
Heat of formation [b]	-184.0 ± 2.5 kcal mole^{-1}
D (CF_3O–F) [c]	43.5 ± 0.5 kcal mole^{-1}

[a] C. J. Hoffmann, *Chem. Revs.* **64** (1964) 91.
[b] D. R. Stull, *JANAF Chemical Tables* (1964).
[c] J. Czarnowski, E. Castellano and H. J. Schumacher, *Chem. Communs.* (1968) 1255.

There has also been a thorough investigation of the vibrational spectrum of CF_3OF [212].

The important feature about CF_3OF is that it behaves as $CF_3O + F$ not as $CF_3 + OF$. For example,

$$2CF_3OF + N_2F_4 \rightarrow 2CF_3ONF_2 + F_2$$

$$CF_3OF + OSO_2 \rightarrow CF_3OOSO_2F$$

Photochemical behaviour also seems to be dependent on an initial step to give $CF_3O + F$. For example, as well as COF_2, CF_3OOCF_3 results from the gas phase photolysis of CF_3OF alone[213]. The existence of CF_3O as intermediate is also inferred from the photochemical reaction of F_2 with COF_2 to give CF_3OF at room temperature[214].

In spite of this indirect evidence there is as yet no positive identification of the CF_3O radical. For instance, the photolysis[215] of CF_3OF in argon at 15°K with a variety of photolytic sources gives only COF_2—presumably because the CF_3O intermediate readily decays spontaneously to $COF_2 + F$ or because the radical is more photosensitive than the parent CF_3OF.

The Trifluoromethoxide Ion, OCF_3^-

The use of fluoride salts, especially caesium, for the preparation of several $CF_3/O/F$ type compounds suggests a mechanism involving a fluoro anion. In fact the reaction[216] of a

210 A. G. Streng and A. V. Grosse, *J. Am. Chem. Soc.* **88** (1966) 169.
211 See C. J. Hoffmann, *Chem. Revs.* **64** (1964) 91.
212 P. M. Wilt and E. A. Jones, *J. Inorg. Nucl. Chem.* **30** (1968) 2933.
213 C. I. Merrill and G. H. Cady, Second International Symposium on Fluorine Chemistry, *J. Am. Chem. Soc.* 1962, 414.
214 P. J. Aymonino, *Proc. Chem. Soc.* 1964, 341.
215 R. D. Clarke, A. J. Rest and J. J. Turner, unpublished observations.
216 M. E. Redwood and C. J. Willis, *Can. J. Chem.* **43** (1965) 1893.

heavy alkali metal fluoride (MF), suspended in CH_3CN at 20°C, with COF_2 gas leads to the isolation of $M^+(OCF_3)^-$, which anion is isoelectronic with BF_4^-. Other, less stable, salts of this kind which have been prepared include $OCF_2CF_3^-$, $OCF_2CF_2CF_3^-$ and $OCF(CF_3)_2^-$.

It is noteworthy that although F^- is easily obtained, CF_3^- does not exist, and although OF^- is not a stable entity, OCF_3^- is.

Perfluorodimethyl Peroxide, CF_3OOCF_3

Preparation. On heating equimolar quantities of CF_3OF and COF_2 to about 290°C and then cooling to room temperature, a substantial conversion to CF_3OOCF_3 resulted[217]. The compound can also be produced by heating F_2 and CO_2 (2:1 ratio) to 325°C and then cooling and also by passing F_2 and CO (3:2) through a reactor containing a catalyst composed of copper ribbon coated with fluorides of silver.

Properties. CF_3OOCF_3 is a stable colourless gas (cf. FOOF) at room temperature with a boiling point ~ -37°C. The compound begins to decompose at about 225°C.

In its chemistry CF_3OOCF_3 behaves as $(CF_3O)_2$ not CF_3+OOCF_3. In fact there is no evidence for the $OOCF_3$ radical. The dissociation into CF_3O radicals is greatly helped by photolysis with ultraviolet light.

There is little structural information on this compound but the infrared spectrum[218] does suggest that the C–O and O–O bonds are ordinary single bonds; certainly there is no evidence for the very strong O–O bond found in O_2F_2.

Trifluoromethyl Trioxide, CF_3OOO

Vanderkooi and Fox[219] studied the e.s.r. spectra of photolysed solutions of CF_3OF and CF_3OOCF_3 in NF_3 as solvent at low temperature. They assigned the six line spectra produced in both cases to CF_3OO; a freely rotating CF_3 group would produce a 1:3:3:1 quartet, and the authors explained the doublet of triplets observed on the basis of restricted rotation. (^{19}F hyperfine splittings = 6.72 and 0.54 gauss.) More recently, Fessenden[220] has examined the photolysis of CF_3OOCF_3 (and $CF_3{}^{17}OOCF_3$) in NF_3 solution at -196°C in the presence of traces of oxygen and ^{17}O enriched oxygen. These results demonstrate that the radical produced is not CF_3OO, but CF_3OOO with $g = 2.00373 \pm 0.00001$ and ^{17}O hyperfine constants 23.3, 14.0 ($CF_3O\underline{O}\underline{O}$), 3.59 ($CF_3\underline{O}OO$) gauss. The point is also made that the g value is markedly different from other RO_2 radicals and provides a further reason for rejecting CF_3OO. However, the two larger ^{17}O hyperfine constants and the g value are very close to those of O_2F. Presumably, therefore, the CF_3O radical produced on photolysis immediately reacts with an O_2 molecule to give $(CF_3O)O_2$ in precisely the same way as fluorine atoms react with O_2 to give O_2F. It is tempting therefore to postulate that the terminal O–O band in CF_3OOO is likely to be very strong and that when comparing the chemistry of fluorine with CF_3 species the unit of comparison should be CF_3O rather than CF_3.

[217] R. S. Porter and G. H. Cady, *J. Am. Chem. Soc.* **79** (1957) 5628.

[218] A. J. Arvia and P. J. Aymonino, *Spectrochim. Acta* **18** (1962) 1299; D. W. Wertz and J. R. Durig, *J. Mol. Spectrosc.* **25** (1968) 467.

[219] N. Vanderkooi and W. B. Fox, *J. Chem. Phys.* **47** (1967) 3634.

[220] R. W. Fessenden, *J. Chem. Phys.* **48** (1968) 3725.

Bis **(Trifluoromethyl) Trioxide, CF_3OOOCF_3**

Preparation. Although this compound can be obtained by the direct fluorination[221a] of salts of trifluoroacetic acid, it is best prepared[221b] by the reaction of OF_2 with COF_2 over a caesium fluoride catalyst. The postulated mechanism for this reaction,

$$COF_2 + MF \rightarrow M^+OCF_3^-$$
$$OF_2 + OCF_3^- \rightarrow CF_3OOF + F^-$$
$$CF_3OOF + OCF_3^- \rightarrow CF_3OOOCF_3 + F^-$$
$$CF_3OOF + COF_2 \rightarrow CF_3OOOCF_3$$

has received support from ^{17}O tagging experiments[221c].

Properties. CF_3OOOCF_3 is a stable material with a melting point of -138°C and a normal boiling point of -16°C. It begins to decompose in glass at about 70°C to give CF_3OOCF_3 and O_2 along with trace amounts of COF_2 and SiF_4.

CF_3OOOCF_3 reacts with N_2F_4, SF_4, SO_2, etc., to give simple CF_3O products (CF_3ONF_2, $CF_3OSF_4OCF_3$, etc.) which can also be obtained by starting with CF_3OOCF_3 or CF_3OF [222]. However, the reactions are cleaner and more controllable than corresponding reactions with CF_3OF and proceed with higher yield and at lower temperature.

Electron spin resonance experiments similar to those with CF_3OF and CF_3OOCF_3 in NF_3 demonstrated[222] the production of the same radical. It seems unlikely, in view of its chemical behaviour, that CF_3OOO results from CF_3 loss from CF_3OOOCF_3 but rather that CF_3O is first produced and this then reacts with traces of O_2.

A further important point is that the infrared spectrum[222] suggests that the C–O and O–O bonds are normal single bonds.

This might at first sight seem rather surprising since the CF_3OOO radical appears to have a very strong terminal O–O bond. In other words since FOOF and OOF both have strong O–O bonds, so should CF_3OOOCF_3 and CF_3OOO. However, if we again consider the unit CF_3O, the molecule which should possess the strong O–O bond is $CF_3O\cdot\cdot O{=}O\cdot\cdot OCF_3$ rather than CF_3OOOCF_3. Unfortunately the only report of $CF_3OOOOCF_3$ is that it may be produced as an impurity in the preparation of the trioxide[221a]. Similarly, the only published information on CF_3OOF whose properties would be particularly interesting for a comparison of CF_3, CF_3O and F groups, is the chemical shift of the OOF fluorine: -291 ppm v. $CFCl_3$. This is the same as the chemical shift of FSO_2OOF, and is a long way to high field of O_2F_2 (-825 ppm). It is tempting to suggest that CF_3OOF is analogous to FOF (i.e. CF_3O–O–F), but that the fluorine resonance of CF_3OOOF, if it is prepared, will be to very low fields.

5.12. SOME COMMENTS ON THE BONDING IN OXYGEN FLUORIDES

Table 48 lists some relevant physical properties. It is important to emphasize that the use of force constants to predict bond energies is not reliable; note particularly that the F_2 force constant is greater than that for Cl_2 but that F_2 has a weaker bond than Cl_2. Generally,

[221a] P. G. Thompson, *J. Am. Chem. Soc.* **89** (1967) 4316.
[221b] L. R. Anderson and W. B. Fox, *J. Am. Chem. Soc.* **89** (1967) 4313.
[221c] Private communication from I. J. Solomon to W. B. Fox.
[222] W. B. Fox, private communication; R. P. Hirschmann, W. B. Fox and L. R. Anderson, *Spectrochim. Acta* A **25** (1969) 811.

TABLE 48. SOME PHYSICAL PROPERTIES OF OXYGEN FLUORIDES AND RELATED SPECIES

	Bond length [a] (Å)			Force constants [b] (mdyne/Å)				Bond dissociation energies [c]	
	r_{OH}	r_{OF}	r_{OO}	f_{OH}	f_{OF}	f_{OCl}	f_{OO}	($\Delta H°$ kcal mole^{-1})	
OH	0.97			7.1				$OH \rightarrow O+H$	102
OF					5.42			$OF \rightarrow O+F$	50
OCl	1.57					6.4 [d]		$OCl \rightarrow O+Cl$	64
O_2			1.21				11.43	$O_2 \rightarrow 2O$	119
OH_2	0.96			7.66				$HOH \rightarrow H+O+H$	2×111
								$HO+H$	120
OF_2		1.41			3.95			$FOF \rightarrow F+O+F$	2×45
OCl_2						2.75 [e]		$ClOCl \rightarrow Cl+O+Cl$	2×54
O_2H				6.46			6.2 [f]	$OOH \rightarrow O_2+H$	47
								$O+O+H$	166
								$O+OH$	64
O_2F					1.32		10.50	$OOF \rightarrow O_2+F$	18
								$O+OF$	87
O_2Cl						1.29	9.7 [g]		
O_2H_2	0.95		1.48	7.4			4.6	$HOOH \rightarrow H+O_2+H$	2×68
								$H+OOH$	88
								$H+O+O+H$	256
								$HO+OH$	52
O_2F_2		1.575	1.22		1.36		10.25	$FOOF \rightarrow FO+OF$	52
								$F+O_2F$	18
								$F+O+O+F$	152
H_2	(r = 0.75)			(f = 5.13)				$H_2 \rightarrow 2H$	104
F_2	(r = 1.42)			(f = 4.45)				$F_2 \rightarrow 2F$	38
Cl_2	(r = 1.99)			(f = 3.20)				$Cl_2 \rightarrow 2Cl$	58

[a] Values from readily available books except for O_2F_2 (see Table 43) and OCl, A. Carrington, P. N. Dyer and D. H. Levy, *J. Chem. Phys.* **47** (1967) 1756.

[b] Values from readily available books or this chapter except for those stated.

[c] Values from NBS circulars *Selected Values of Thermodynamic Properties*, 270-1 (1965), 270-2 (1966); data for O/F compounds from previous tables.

[d] Based on ν 995 cm^{-1} (L. Andrews and J. I. Raymond, *J. Chem. Phys.* **55** (1971) 3087).

[e] M. M. Rochkind and G. C. Pimentel, *J. Chem. Phys.* **42** (1965) 1361.

[f] D. E. Milligan and M. E. Jacox, *J. Chem. Phys.* **38** (1963) 2627.

[g] A. Arkell and I. Schwager, *J. Am. Chem. Soc.* **89** (1967) 5999. (N.B.—This is OOCl not OClO and is produced by photolysis of OClO in argon at 4°K.)

however, substantial differences in force constants imply considerable differences in bond energy in the same direction. It is also important to realize that bond dissociation energies do not necessarily bear any relation to the "intrinsic" energy of a bond. For example, in the reaction $HOOH \rightarrow H+O_2+H$ one cannot infer an intrinsic bond energy of 68 kcal mole^{-1} for the O–H bond since ΔH for the reaction is much affected by the formation of an O–O double bond.

The O–X Bond in OX, OX_2

The bond lengths, force constants and bond energy terms indicate that the O–H bond has the same properties in OH and OH_2. This is not surprising since there is no possibility of π-bonding or "non-bonding electron repulsion". By contrast the striking difference in the O–Cl bond in OCl and OCl_2 suggests the presence of π-bonding in OCl which would have a nominal bond order of $1\frac{1}{2}$. Linnett's[223] double-quartet structure for OCl

```
 ∘ x      x ∘
   x O x ∘ Cl x
 ∘ x    ∘  x ∘
```

is consistent with this picture since the formal charge of $+\frac{1}{2}$ on the Cl atom is acceptable.

The bond energies and force constants for OF and OF_2 suggest a slightly stronger O–F bond in OF. Simple molecular orbital theory predicts a bond order of $1\frac{1}{2}$ for OF, but the formal change of $+\frac{1}{2}$ on the fluorine atom in the Linnett structure is not allowed.

The O–X Bond in O_2X, O_2X_2

There is *some* evidence that the O–H bond in O_2H is slightly weaker than the presumably ordinary single bond in OH, OH_2 and O_2H_2. There is, however, no doubt that the O–F and O–Cl bonds in O_2F, O_2F_2 and O_2Cl are considerably weaker than the nominally single bonds in OF_2 and OCl_2. It is interesting to note that Rochkind and Pimentel[224] obtained *some* matrix evidence for ClOOCl which was described as a weak dimer of OCl, i.e. the O–Cl bond was comparable with that in the OCl radical.

The O–O Bond in O_2, O_2X and O_2X_2

It is clear that the O–O bonds in O_2F, O_2F_2 and O_2Cl are practically the same as the double bond in O_2 itself. In O_2H the force constant suggests an O–O bond between O_2 and O_2H_2 in properties—the O–O force constant in O_3 is 7.3 mdyn Å^{-1}. Interestingly, if spectroscopic evidence[224] is reliable, the O–O bond in O_2Cl_2 is very weak.

Any successful bonding model must be able to explain these phenomena plus the observations on CF_3 compounds. Simple valence-bond theory can make good sense of the difference between, say, O_2H_2 and O_2F_2:

$$\underset{\text{(i)}}{\text{X—O—O—X}} \leftrightarrow \underset{\text{(ii)}}{\text{X}^- \ \text{O=O}^+\text{—X}} \leftrightarrow \underset{\text{(iii)}}{\text{X—O}^+\text{=O} \ \text{X}^-}$$

When X = F, (ii) and (iii), which increase the O–O bond strength and decrease the O–X bond strength, can make a substantial contribution because of the electronegativity of fluorine. When X = H, this does not happen, and structure (i) predominates. Similarly, FOOOOF is predicted to have weak O–F bonds, strong terminal O–O bonds and a weak central O–O bond. However, FOOOF is also predicted to be a stable entity, and "strongish" central bonds are predicted for CF_3OOCF_3 and ClOOCl. It is, of course, possible to say that whether a molecule adopts structure (iv) or (v)

$$\underset{\text{(iv)}}{\text{X}\cdot\cdot\text{O=O}\cdot\cdot\text{X}} \qquad \underset{\text{(v)}}{\text{X}\overline{\cdot\cdot}\text{O}\cdot\cdot\text{O}\overline{\cdot\cdot}\text{X}}$$

[223] J. W. Linnett, *The Electronic Structure of Molecules*, Methuen, London, and Wiley, New York (1964).
[224] M. M. Rochkind and G. C. Pimentel, *J. Chem. Phys.* **46** (1967) 4481.

depends on relative bond strengths. Rochkind and Pimentel[224] argued that the weak dimer [i.e. (v)] structure for ClOOCl is due to the high bond energy of the OCl radical. It is difficult to argue this way in the case of CF_3OOCF_3 since there is no reason why the C–O bond in the OCF_3 radical should be any stronger than a normal single bond. One difference between the CF_3 group and the halogen atoms is that there is no possibility of π-bonding to oxygen, and this may be important.

Linnett[223] has discussed the structure of O_2F_2 and concludes that the preponderance of structures (vi) and (vii) is responsible for the long O–F and short O–O bonds.

$$\overset{x}{—}F \circ \mathring{O} \overset{x}{\underset{x}{—}} \mathring{O} \circ F \overset{x}{—} \qquad \overset{x}{—}F \circ \overset{x}{\mathring{O}} = \mathring{O} \; x \; F—$$

(vi) (vii)

These structures are particularly favoured because they place formal charges of $-\frac{1}{2}, \frac{1}{2}, \frac{1}{2}, -\frac{1}{2}$ (vi), and $-\frac{1}{2}, \frac{1}{2}, \frac{1}{2}, -\frac{1}{2}$ (vii) on the four atoms. It is possible that in O_2Cl_2 there is a considerable contribution from

$$\dot{Cl} \overset{x}{—} O \cdot \overset{x}{O} — Cl —$$

(viii)

and its mirror image; in (viii) the formal charges are $\frac{1}{2}$, 0, $-\frac{1}{2}$, 0 [which is acceptable in this case but forbidden in O_2F_2 because of the $\frac{1}{2}$ on the end halogen], and there is a weak O–O and strong O–Cl bonds. For O_2H_2 the only really likely structure is

$$—H—O—O—H—$$

Linnett structures are readily drawn for O_4F_2, suggesting a loose O_2F dimer, but in the case of O_3F_2, the only satisfactory[208, 209] Linnett structures imply $O_2F \cdot OF$. The method can be extended to O_2X species with results in broad agreement with experimental observation. In particular a structure involving a one-electron O–O bond in O_2Cl places a formal charge of $-1\frac{1}{2}$ on the terminal oxygen which is not acceptable. Presumably structures involving a strong O–O bond then become dominant.

A quite different model was introduced by Jackson[225] and subsequently extended by Pimentel *et al.*[226]. In the species O_2X (or O_2X_2) the X lone electron is considered to interact with one of (or both) the antibonding π-orbitals of the O_2 molecule—a [p (or s)–π^*] σ-bond. The more electron density flows from X to O_2, the weaker the O_2 bond and the stronger the O–X bond. Consequently, in O_2F, because of electronegativity differences, it is not surprising that O–F is weak and O–O strong. In O_2H, the O–O bond order will be approximately $1\frac{1}{2}$ because of the non-interacting O_2 π system. This idea can be extended to systems

[225] R. H. Jackson, *J. Chem. Soc.* 1962, 4585.

where the non-interacting π-antibonding orbital has either *two* (XNO) or *no* (XCO) electrons[226]. In particular, a detailed molecular orbital study[227] on FNO supports such a simplified model. In O_2F_2 the model predicts strong O–O and weak O–F bonds and *similarly* for O_2Cl, contrary to the experimental evidence.

Other theoretical treatments have been suggested[228] but it is probably too much to hope that any one simple model will explain such a diversity of experimental fact.

6. HYDROGEN PEROXIDE[229]

6.1. HISTORY

Hydrogen peroxide was first reported[230] by Thenard in 1818. He treated barium peroxide with acid, and after removing other ions from the solution by precipitation he obtained an aqueous solution of hydrogen peroxide; he removed water by evaporation under reduced pressure, and was able in this way to prepare H_2O_2 that was almost anhydrous. He found that his product evolved oxygen when treated with a catalyst such as MnO_2, and from the volume of gas given off he concluded that the compound must contain twice as much oxygen as does water.

6.2. SYNTHESIS

Oxygen in hydrogen peroxide is in the oxidation state (-1); the compound may be obtained either by oxidizing oxygen from the (-2) state, or by reducing molecular oxygen, and processes based on reactions of both of these types have been used on an industrial scale. The oxidative route starts from aqueous sulphuric acid or sulphate, which is oxidized electrolytically at high current densities between platinum electrodes. The oxidative step may be represented:

$$2HSO_4^- = [O_3S{\cdot}OO{\cdot}SO_3]^{-2}+2H^++2e^-$$

The perdisulphate formed is hydrolysed in acid solution:

$$[O_3S\text{–}OO\text{–}SO_3]^{=}+2H_2O = 2HSO_4^-+H_2O_2$$

Hydrogen peroxide may be removed from the hydrolysate with steam, and sulphate is regenerated. The conditions must be carefully controlled to avoid side-reactions leading to the formation of O_2 rather than H_2O_2. The earliest process for manufacturing hydrogen peroxide was based on the reduction of atmospheric oxygen with barium metal, giving barium peroxide, from which (as Thenard found) hydrogen peroxide may be obtained by treatment with acid. Since 1945 processes have been developed that make use of the autoxidation of certain organic substrates; atmospheric oxygen is reduced to H_2O_2, and the oxidized substrate may then be reduced to its original state quickly and cheaply, often by catalytic

[226] R. D. Spratley and G. C. Pimentel, *J. Am. Chem. Soc.* **88** (1966) 2394; J. S. Shirk and G. C. Pimentel, *ibid.* **90** (1968) 3349.

[227] S. D. Peyerimhoff and R. J. Buenker, *Theor. Chim. Acta* **9** (1967) 103.

[228] S. J. Turner and R. D. Harcourt, *Chem. Communs.* (1967) 4; N. J. Lawrence, thesis, Cambridge (1967).

[229] W. L. Schumb, C. N. Satterfield and R. L. Wentworth, *Hydrogen Peroxide*, Reinhold (1955).

[230] L. J. Thenard, *Ann. chim. phys.* **8** (1818) 306.

hydrogenation. The only reagents consumed are atmospheric oxygen and hydrogen gas. One widely used substrate is 2-ethylanthraquinol[231]:

$$\text{2-ethylanthraquinol} \xrightarrow{O_2} H_2O_2 + \text{2-ethylanthraquinone}; \quad \text{2-ethylanthraquinone} \xrightarrow{H_2,\ \text{catalyst}} \text{2-ethylanthraquinol}$$

The primary product of all these processes is aqueous H_2O_2, which is an important industrial material. The liquid can be concentrated to *ca.* 98% by weight by fractional distillation at low pressure; very pure and anhydrous H_2O_2 is obtained by fractional crystallization or by extraction with organic solvents. The same methods may be used to produce anhydrous H_2O_2 in the laboratory from commercial aqueous H_2O_2; hydrogen peroxide is not normally synthesized in the laboratory although the preparation of D_2O_2 on a laboratory scale from D_2O and $K_2S_2O_8$ has been described[232], and H_2O_2 [233] has been prepared[234] from hydroxylamine and molecular oxygen in alkaline perchlorate solution at room temperature.

The greatest care is needed when H_2O_2 is purified by extraction with organic solvents; explosive compounds are all too easily formed, as are mixtures that may detonate.

6.3. MOLECULAR STRUCTURE AND SPECTRA

The structure of H_2O_2 is represented by HOOH; the isomeric H_2O:O has not been detected (but see section 7). The four atoms are not normally coplanar. In crystalline H_2O_2 (space group[235] $D_4^8P4_32{,}2$; four molecules in the unit cell) the dihedral angle is just over 90°, and each oxygen atom is linked by hydrogen bonds to three other molecules. In the vapour the vibration–rotation spectrum can be accounted for[236] in terms of a dihedral angle of 111.5° (though an earlier study[237] led to a value of 119.8°); in the peroxy-hydrate of sodium oxalate, $Na_2C_2O_4 \cdot H_2O_2$, the H_2O_2 molecule is constrained[238] into a planar configuration. Thus the dihedral angle is easily deformed. The other molecular parameters

231 *Industrial Chemist* 1959.
232 M. Schulz, A. Rieche and K. Kirschke, *Chem. Ber.* **100** (1967) 370.
233 F. Fehér, *Ber.* **72** (1939) 1789.
234 M. Anbar, Z. Baruch and D. D. Meyerstein, *Int. J. Appl. Radiation Isotopes* **17** (1966) 256.
235 W. R. Busing and H. A. Levy, *J. Chem. Phys.* **42** (1965) 3054.
236 R. M. Hunt, R. A. Leacock, C. W. Peters and K. Hecht, *J. Chem. Phys.* **42** (1965) 1931.
237 R. L. Redington, W. B'Olson and P. C. Cross, *J. Chem. Phys.* **36** (1962) 1311.
238 B. F. Pedersen and B. Pedersen, *Acta Chem. Scand.* **18** (1964) 1454.

(see Table 49) are much less sensitive to environment, though d (O–O) may be rather less in the crystal than in the vapour.

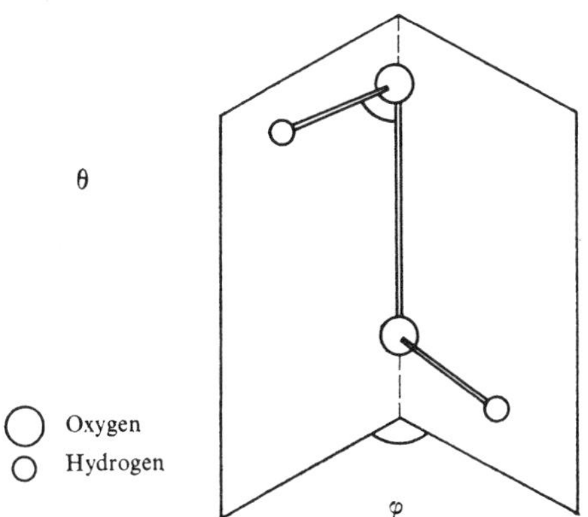

FIG. 7. Diagrammatic representation of the arrangement of the atoms in H_2O_2. $<\theta = 94.8°$, $<\varphi$ (dihedral angle) $= 111.5°$ for vapour.

The fundamental vibration frequencies are relatively easily identified, with the exception of the torsional mode ν_4. Gaseous H_2O_2 gives a complicated vibration–rotation spectrum between 11 and 680 cm^{-1}; all the bands observed have been accounted for by an analysis using torsional barriers[236] $V_{cis} = 2460$ cm^{-1} and $V_{trans} = 386$ cm^{-1}, and a similar analysis has been used for the spectrum of gaseous D_2O_2. The low-frequency spectra of condensed phases are harder to account for in detail; in a matrix of N_2, ν_4 has been given the value[239] of 385 cm^{-1}. The ultraviolet absorption of H_2O_2 increases with decreasing wavelength; for aqueous solutions there is a maximum just above 2000 Å that has been assigned to a charge transfer transition[240].

In terms of simple theory, H_2O_2 may be described as a saturated molecule, with single bonds between the hydrogen and the oxygen atoms, and between the two oxygen atoms. The σ-bonds from oxygen are described as between pure p and sp^3 in character. A number of more sophisticated calculations of the electronic structure have been made, using[241] LCAO SCF MO and other[242] methods; particular attention has been paid to the origin and magnitude of the torsional barrier, for H_2O_2 is the simplest molecule to show restriction of rotation about a single bond.

6.4. OTHER PHYSICAL PROPERTIES

The pure liquid is syrupy and colourless or nearly so. Like water, it is extensively associated, and the Trouton constant indicates that this association persists even near the

[239] E. Catalano and R. H. Sanborn, *J. Chem. Phys.* **38** (1963) 2273.
[240] J. Jortner and G. Stein, *Bull. Res. Co. Israel* **6 A** (1957) 239.
[241] W. H. Fink and L. C. Allen, *J. Chem. Phys.* **46** (1967) 2261.
[242] S. J. Turner and R. D. Harcourt, *Chem. Communs.* 1967, 4.

boiling point. Hydrogen peroxide is miscible in all proportions in the liquid phase with water, with which it forms a compound $H_2O_2 \cdot 2H_2O$ that is congruently melting and almost completely dissociated in the liquid phase. The vapour–pressure curve for mixtures of H_2O_2 and H_2O shows no maxima or minima. Hydrogen peroxide is soluble in many organic solvents, including alcohols, ethers, esters and amines (see note about hazards). Other physical properties are given in Table 49.

TABLE 49. PHYSICAL PROPERTIES OF H_2O_2

Molecular properties and structure

	Vapour (vib.-rot. spectrum)	Crystal (neutron diffraction) [c]
r(O–O), Å	1.475 [a]	1.453 ± 0.007
r(OH), Å	0.950 [a]	0.988 ± 0.005 (1.008 ± 0.005 after thermal correction)
<OOH	94.8° [a]	102.7 ± 3°
Dihedral angle	111.5 ± 0.5 [b]	90.2 ± 0.6°

Fundamental vibration frequencies (cm^{-1})	Vapour (IR) [a, b]	Liquid [d]	Crystal (−190°) (IR) [d]
ν(OH), ν_1	3599	3400 (IR, R)	3285
δ(OH), ν_2	(1380)	1400 (R)	1407
ν(O–O), ν_3	880	880 (IR, R)	882
Torsion, ν_4	242.8, 370.7	?	? ? 690
ν(OH), ν_5	3608	3400 (IR, R)	3192
δ(OH), ν_6	1266	? 1350 (IR)	1385

Rotational barriers [b]	$V_{trans} = 386 \pm 4$ cm^{-1}; $V_{cis} = 2460$ cm^{-1}
Ultraviolet absorption (aqueous solution) [e]	$\lambda_{max} = 2165 \pm 5$Å, $\varepsilon_{max} = 460$
Dipole moment (vapour; Stark effect) [f]	2.1–2.3 D

Thermodynamic properties and bond strengths

Formation thermodynamics [g]:	
ΔG°_f, 298	−25.24 kcal $mole^{-1}$
ΔH°_f, 298	−32.58 kcal $mole^{-1}$
S°	55.6 cal $mole^{-1}$ $°K^{-1}$
C_p (solid) (273°K > T > 220°K)	$0.0722 + 0.001\,239T$ cal gm^{-1}
C_p (liquid), 0–27°C	0.628 cal gm^{-1}
D (HO–OH) (298°)	48.8 kcal $mole^{-1}$ [h]
D (H–OOH) (298°)	89.6 kcal $mole^{-1}$ (electron impact) [h]
Ionization potential [h]	10.92 ± 0.05 eV

Other bulk properties	
Melting point	$-0.43°$
Boiling point	$150.2°$
Vapour pressure [i]	$\log p_{mm} = -\frac{3\,140.594\,30}{T} + 9.827\,689 - 0.007\,436\,0T + 0.000\,004\,151\,6T^2$
ΔH_{fus}, 0°	87.84 cal gm^{-1}
ΔH_{subl}, 0°	457.8 cal gm^{-1}
ΔH_{vap}, $1 \rightarrow v$, 0°	370.2 cal gm^{-1}
Trouton's constant	26.5 cal $mole^{-1}$ $°K^{-1}$
Density: solid	1.6434 gm cm^{-3} at $-4.45°$
liquid [j], ρ	$1.5970 + 0.000\,078\,4T - 0.00\,000\,197\,0T^2$
Mean coefficient of cubical expansion for liquid	(0–25°) 7.9×10^{-4} $°C^{-1}$ (25–96°) 8.58×10^{-4} $°C^{-1}$
Viscosity (20°)	0.01249 poise
Surface tension (20°)	80.4 dyne cm^{-1}
Dielectric constant, ε	$84.2 - 0.62t + 0.0032t^2$ (t in °C)
Refractive index, n_D^{20}	1.4077_4
Polarizability, $\alpha \times 10^{24}$	2.30 cm^3 $molecule^{-1}$
Specific conductance (99.9%)	3.9×10^{-7} ohm^{-1} cm^{-1}
Magnetic susceptibility, χ_g	-0.50×10^{-6} c.g.s. e.m.u. g^{-1} at 10°C
Verdet constant (5461 Å)	13.32×10^{-3} min oe^{-1} cm^{-1}

Data from W. C. Schumb, C. N. Satterfield and R. L. Wentworth, *Hydrogen Peroxide*, Reinhold (1955), unless otherwise stated.

[a] R. L. Redington, W. B. Olson and P. C. Cross, *J. Chem. Phys.* **36** (1962) 1311.

[b] R. M. Hunt, R. A. Leacock, C. W. Peters and K. T. Hecht, *J. Chem. Phys.* **42** (1965) 1931.

[c] W. R. Busing and H. A. Levy, *J. Chem. Phys.* **42** (1965) 3054.

[d] R. L. Miller and D. F. Hornig, *J. Chem. Phys.* **34** (1961) 265.

[e] In alkaline solution; assigned to HO_2^-; aqueous H_2O_2 gives no maxima at $\lambda > 2000$ Å; J. Jortner and G. Stein, *Bull. Res. Co. Israel* **6***A* (1957) 239.

[f] J. T. Massey and D. R. Bianco, *J. Chem. Phys.* **22** (1954) 442.

[g] NBS Technical Note 270-3 (1968).

[h] S. N. Foner and R. L. Hudson, *J. Chem. Phys* **36** (1962) 2676, 2681. P. A. Giguère, *J. Chem. Phys.* **30** (1959) 322.

[i] A. L. Tsykalo and A. G. Tabachnikov, *Teplofiz. Svoistva Veshchestv, Akad. Nauk Ukr. SSR, Respub. Mezhvedom* (1968) 120.

[j] A. L. Tsykalo and A. G. Tabachnikov, *Teor. Eksp. Khim.* **2** (1966) 837.

6.5. CHEMICAL PROPERTIES

Hydrogen peroxide has an extensive redox chemistry, and also shows hydroxylic properties, derived from the two OH groups. The redox properties will be considered first.

The compound is stable with respect to its elements (see Table 49), but is unstable with respect to decomposition into water and molecular oxygen in the solid, the liquid or the vapour phases, or in aqueous solution:

$$H_2O_2(l) = H_2O(l) + 1/2O_2(g) \quad \Delta G°_{298} = -27.92 \text{ kcal mole}^{-1}$$

This instability with respect to disproportionation is reflected in the standard potentials:

$$O_2 + 2e^- + 2H^+ = H_2O_2 \qquad E° = 0.693 \text{ V}$$

$$O_2 + H_2O + 2e^- = OH^- + O_2H^- \qquad E_B^° = -0.084 \text{ V}$$

$$H_2O_2 + 2e^- + 2H^+ = 2H_2O \qquad E° = 1.76 \text{ V}$$

$$HO_2^- + 2e^- + H_2O = 3OH^- \qquad E_B^° = 0.87 \text{ V}.$$

In the vapour phase, decomposition, a first-order process, is associated with dissociation into radicals such as OH or OOH; photolysis or decomposition of H_2O_2 by X-rays have been shown by e.s.r. to produce radicals like these[243]. In all phases decomposition is very sensitive to homogeneous or heterogeneous catalysis by metals, metal oxides or metal ions, and in homogeneous systems it is probable that peroxy-compounds of metals are formed as intermediates. In the absence of catalysts the pure liquid is stable indefinitely at room temperature and may be distilled at low pressures without decomposition, but concentrated solutions are usually stabilized by the addition of compounds such as sodium stannate or 8-hydroxyquinoline, which inhibit catalysts. These very concentrated solutions are normally stored and handled in apparatus made from very pure (>99.5%) aluminium, stainless steel, glass, porcelain, or plastic such as polyvinyl chloride, polyethylene or PTFE.

Since oxygen in H_2O_2 is in an intermediate oxidation state (−1), hydrogen peroxide can act either as an oxidizing or as a reducing agent; the appropriate potentials are given above. Its behaviour is very much influenced by kinetic factors, for its reactions are not simple electron-transfer processes. Some clearly involve free radicals; indeed, the system Ti^{+3}/H_2O_2 has been studied extensively in recent years in connection with free radical oxidations of organic compounds. The reactions of Fenton's reagent (a mixture of H_2O_2 and Fe^{+2}) are also free radical oxidation processes. Other reactions may proceed by two-electron transfer steps; in all systems involving transition metal ions the formation of metal peroxy-complexes may be important. The mechanisms and the products of the reactions of H_2O_2 in aqueous solution are very sensitive to pH, as illustrated by the system H_2O_2/Cr(III)/Cr(VI). The potentials for Cr(VI)/Cr(III) are:

$$CrO_4^{-2}+3e^-+8H^+ \rightarrow Cr^{+3}+4H_2O \qquad E^\circ = 1.33\ V$$
$$CrO_4^{-2}+3e^-+4H_2O \rightarrow Cr(OH)_3+5OH^- \qquad E^\circ_B = -0.13\ V.$$

Thus in acid solution H_2O_2 is capable of oxidizing Cr(III) to Cr(VI), and of reducing Cr(VI) to Cr(III). It is the reduction that predominates; but in alkaline solution chromate is reduced to the Cr(V) species $Cr(O_2)_4^{-3}$, which decomposes to regenerate chromate. Aqueous H_2O_2 exchanges oxygen with water immeasurably slowly[244] in the presence of $HClO_4$, and extremely slowly in the presence of hydroxide ion or of sulphuric acid; exchange is faster with water containing nitric acid, and kinetic analysis suggests that peroxynitric acid is formed by reaction between H_2O_2 and N_2O_5 in the solution. The small but measurable exchange[245] between H_2O_2 and H_2O [247] both in evolved O_2 and in residual H_2O_2 in the presence of such species as Fe^{+2} or ClO^- has also been explained in terms of intermediate formation of unstable peroxy-species.

Besides these reactions, hydrogen peroxide oxidizes a number of organic compounds, and is useful in the preparation of epoxides and of sulphones; α-halo-hydroperoxides have been prepared by treatment of olefins with halogens and hydrogen peroxide[246, 247]. It is also an intermediate in many biological oxidation processes.

As a hydroxy-compound, hydrogen peroxide can, in principle, act as an acid or as a base. For the acid dissociation[247] of H_2O_2 in water, the equilibrium favours the formation of H_3O^+:

243 B. G. Ershov and A. K. Pikaev, *Izvest. Akad. Nauk SSSR, Ser. Khim.* 1964, 922.
244 M. Anbar and S. Guttman, *J. Am. Chem. Soc.* **83** (1961) 2035.
245 M. Anbar, *J. Am. Chem. Soc.* **83** (1961) 2031.
246 K. Weissermel and M. Lederer, *Chem. Ber.* **96** (1963) 77.
247 J. Jortner and G. Stein, *Bull. Res. Co. Israel* **6A** (1957) 239.

$$H_2O_2+H_2O = H_3O^++OOH^- \quad pK_a = 11.85\pm0.10 \text{ at } 19°.$$

This shows that H_2O_2 is rather more acidic than water. Derivatives of H_2O_2 analogous to hydroxides and oxides—that is, hydroperoxides and peroxides—are discussed in the next section. Hydrogen peroxide is much less basic than water ("at least one millionth" less basic[248]); the lower conductivity of, for example, perchloric acid in H_2O_2/H_2O than in water may be connected with the difficulty of protonating H_2O_2, which perhaps interferes with proton mobility[249]. The ion $H_3O_2^+$ may be formed[250] in solutions of H_2O_2 in HBF_4, but its salts have not been characterized. The specific conductivity of 99.9% H_2O_2 is very low, so self-ionization is not extensive.

Other "hydroxylic" reactions of H_2O_2 follow those of water. Acid anhydrides gives peroxy-acids:

$$RCOOCOR+H_2O_2 = RCOOOH+RCOOH$$

aldehydes[251] gives peroxy-aldols:

$$HOOH + \begin{matrix} R \\ H \end{matrix}\!\!>\!C{=}O = \begin{matrix} R \\ H \end{matrix}\!\!>\!C\!<\!\!\begin{matrix} OOH \\ OH \end{matrix}$$

Again like water, hydrogen peroxide forms adducts with many salts and compounds such as triphenyl phosphine. These adducts are presumably held together by hydrogen bonds. The crystal structures of some have been determined, and in each case are consistent with the presence of molecules of H_2O_2 within the crystal lattice.

6.6. ANALYTICAL CHEMISTRY

Hydrogen peroxide is normally estimated with potassium permanganate, which it reduces quantitatively. The strengths of solution of hydrogen peroxide are often expressed in terms of "volumes", which means "the volume of gaseous oxygen at n.t.p. produced by one volume-unit of liquid". Thus 100-volume H_2O_2 gives 100 ml (at n.t.p.) of O_2 for each millilitre of liquid completely decomposed, meaning that each millilitre of liquid must contain about 9 mmol H_2O_2, so that "100-volume" H_2O_2 must contain about 30% by weight of H_2O_2.

6.7. USES OF HYDROGEN PEROXIDE

Hydrogen peroxide is of considerable importance in connection with the catalysis of polymerization reactions and in the production of epoxides; it is used to oxidize organic nitrogen compounds and organic and inorganic sulphur compounds, and in the synthesis of organic peroxides and peroxyacids and of sodium perborate, all of which are significant commercial products. Hydrogen peroxide has been extensively used to bleach many different materials, among them cellulose, textiles, leather, furs, human and animal hair,

248 A. G. Mitchell and W. F. K. Wynne-Jones, *Trans. Faraday Soc.* **52** (1954) 824.
249 M. Kilpatrick, *Can. J. Chem.* **37** (1959) 163.
250 R. W. Alder and M. C. Whiting, *J. Chem. Soc.* 1964, 4707.
251 E. G. Sander and W. P. Jenks, *J. Am. Chem. Soc.* **90** (1968) 4377.

tripe, sausage-skin and headstones; it has been used as a source of oxygen in rocket fuels and in submarines, and as an aerating agent. In dilute form it is a mild disinfectant and antiseptic. In qualitative and quantitative analysis its specific colour-reaction with certain metals is useful, and it has been used in pregnancy testing.

Toxicity. In low concentrations H_2O_2 is not physiologically dangerous, but in high concentrations it is; it attacks the eyes and skin.

7. PEROXIDES AND RELATED COMPOUNDS

This section is concerned with other compounds containing at least one pair of linked oxygen atoms, but no oxygen chains with more than two atoms. They are of several types:

(1) Salts of the ion O_2^+.
(2) Ionic superoxides, containing the ion O_2^-.
(3) Ionic peroxides, containing the ion O_2^{-2}, and hydroperoxides, containing the ion O_2H^-.
(4) Solid, non-stoichiometric peroxides.
(5) Compounds containing the peroxy-group linked by directed bonds to other atoms.
(6) Oxygen carriers.

7.1. DIOXYGENYL SALTS

Though the electronic spectrum of the O_2^+ ion was analysed many years ago, the first salt of this species, $O_2^+PtF_6^-$, was not reported[252] until 1962. It was obtained by fluorinating platinum in glass or silica apparatus; since then, dioxygenyl salts of the anions MF_6^- (M = P, As, Sb), SnF_6^{-2} and BF_4^- have been prepared from the highest fluoride of the appropriate element by one or more of the following routes:

(1) Treatment with a mixture of fluorine and oxygen, either at high temperature[253] or in daylight[254]:

$$O_2+\frac{1}{2}F_2+AsF_5 \rightarrow O_2^+AsF_6^-$$

(2) Treatment with OF_2 at temperatures greater than 150°:[253]

$$2OF_2+AsF_5 \rightarrow O_2^+AsF_6^-+\frac{3}{2}F_2$$

(3) Treatment with dioxygenyl difluoride at low (~ −190°) temperatures[255, 256]:

$$O_2F_2+BF_3 \rightarrow O_2^+BF_4^-+\frac{1}{2}F_2$$

The compounds are all paramagnetic; some properties of the O_2^+ ion are given in Table 50. The fluoroborate and the fluorophosphate decompose slowly at 0°, but the hexafluoroarsenate and antimonate are stable under an inert atmosphere at temperatures up to 100°. All the compounds are powerful oxidizing and fluorinating agents. The fluoroantimonate

[252] N. Bartlett and D. H. Lohmann, *Proc. Chem. Soc.* 1962, 115.
[253] J. B. Beal, C. Pupp and W. E. White, *Inorg. Chem.* **8** (1969) 828.
[254] J. Shamir and J. Binenboym, *Inorg. Chim. Acta* **2** (1968) 37.
[255] A. R. Young, T. Hirata and S. I. Morrow, *J. Am. Chem. Soc.* **86** (1964) 20.
[256] D. V. Bantov, V. F. Sukhoverkhov and Yu. N. Mikhailov, *Izv. Sib. Otdel. Akad. Nauk SSSR, Ser. Khim. Nauk* 1968, 84.

is soluble in excess of SbF_5 and the fluoroarsenate reacts with NO_2, giving the nitronium salt of the same anion; all the salts react very violently with water, giving oxygen and ozone.

TABLE 50. SOME PROPERTIES OF THE IONS O_2^+, O_2^- AND O_2^{-2}

	O_2^+	O_2^-	O_2^{-2}
d (O–O) (Å)	1.17±0.17 [a]	1.32–1.35 [b]	1.48–1.49 [c]
ν (O–O) (cm^{-1})	1858 (O_2AsF_6) [d]	1145 (KO_2) [e]	738, 794 (Na_2O_2) [f]
μ_{eff} (BM) (300°K)	1.57 [g]	2.04 [h]	—
g_{av}	1.9980±0.0002 [i]	2.058 (Na_2O_2) [j]	—

[a] N. Bartlett and D. H. Lohmann, *Proc. Chem. Soc.* 1962, **115**. This agrees within error with the value 1.1227±0.0001 from spectroscopy (see Table 12, p. 702).

[b] F. Halverson, *Phys. Chem. Solids* **23** (1962) 207.

[c] N.-G. Vannerberg, *Progr. Inorg. Chem.* (ed. F. A. Cotton) **4** (1963) 125.

[d] J. Shamir, J. Binenboym and H. H. Claasen, *J. Am. Chem. Soc.* **90** (1968) 6223. The frequency is not significantly different for O_2SbF_6 dissolved in liquid SbF_5.

[e] J. A. Creighton and E. R. Lippincott, *J. Chem. Phys.* **40** (1964) 1779. This frequency varies between 1120 and 1145 cm^{-1} in alkali metal halide lattices (J. Rolfe, W. Holzen, W. E. Murphy and H. J. Bernstein, *J. Chem. Phys.* **49** (1968) 963).

[f] For the anhydrous compound. For $Na_2O_2 \cdot 8H_2O$, ν (O–O) is 842 cm^{-1} (J. C. Evans, *J. Chem. Soc. D*, 1969, 682).

[g] N. Bartlett and I. P. Beaton, *Chem. Communs.* 1966, 167.

[h] L. Pauling, *The Nature of the Chemical Bond*, 3rd edn., Cornell University Press, Ithaca, New York (1960), p. 351.

[i] J. Shamir and J. Binenboym, *Inorg. Chim. Acta* **2** (1968) 37.

[j] In KCl, g_{av} = 2.1140 (P. W. Atkins and M. C. R. Symons, *The Structure of Inorganic Radicals*, Elsevier, Amsterdam (1967), p. 111).

7.2. IONIC SUPEROXIDES[257]

Superoxides of the alkali metals may be prepared by direct reaction between the metal and oxygen (except for lithium), or by oxygenation of solution of the metal in liquid ammonia; superoxides are also formed by dissociation of hydrogen peroxide adducts of metal hydroperoxides. The superoxide ion is present in the crystal lattice; those superoxides whose structures have been studied in detail are polymorphic, but in all the bond length in O_2^- is greater than in O_2, consistent with an electronic structure in which there is one more π_g^* electron than in O_2. Some structural parameters are given in Table 50. Superoxides are coloured and paramagnetic; changes in magnetic susceptibility and colour appear to be associated with phase changes. Besides the superoxides, mixed peroxides–superoxides (such as $K_4(O_2)(O_2)_2$) are known.

When heated, superoxides decompose into the metal oxide and molecular oxygen; the electron affinity of O_2 is barely exothermic (see Table 10, p. 698), and the ion is stabilized by the lattice energy. Superoxides are powerful oxidizing agents; they react with water to give oxygen and the unstable ion HO_2^-:

$$2MO_2(s)+H_2O \rightarrow 2H^+ + 2HO_2^- + O_2(g).$$

257 N.-G. Vannerberg, *Prog. Inorg. Chem.* (ed. F. A. Cotton), Interscience, **4** (1963) 125.

258 L. Andrews, *J. Chem. Phys.* **50** (1969) 4288.

They react with CO_2 to give oxygen and the metal carbonate through an intermediate peroxycarbonate.

Molecules of alkali metal superoxides, trapped in matrices at low temperatures, have recently been detected by infrared spectroscopy.[258]

7.3. IONIC PEROXIDES[257]

Peroxides containing the ion O_2^{-2} are formed by the alkali metals and the alkaline earth metals; they are formally derived from H_2O_2 by deprotonation. They may be prepared by oxygenation of the metal or of solutions of the metal in liquid ammonia, or by treatment of the metal or some compound with H_2O_2. Ionic peroxides are white, and contain the O_2^{-2} ion, which is diamagnetic and isoelectronic with F_2; some structural parameters are given in Table 50. The additional π_g^* electron leads to a longer bond in O_2^{-2} than in O_2^-, and to a substantially lower stretching frequency. Ionic peroxides decompose when heated into the metal oxide and molecular oxygen; they react with acids to give H_2O_2, and with water to give H_2O_2 and HO_2^-; they are, of course, powerful oxidizing agents. The reaction with CO_2, liberating oxygen, is of some importance in maintaining artificial atmospheres:

$$K_2O_2 + CO_2 \rightarrow K_2CO_3 + \tfrac{1}{2}O_2.$$

Peroxides form a large number of hydrates and of peroxyhydrates (which contain H_2O_2 of crystallization). Ionic hydroperoxides, containing the ion O_2H^-, are also known and are formally related to ionic peroxides.

7.4. NON-STOICHIOMETRIC PEROXIDES[259]

Treatment of solutions of salts of thorium or of some actinide elements with H_2O_2 leads to the formation of rather ill-defined solids containing metal ions, peroxide, other anions and water. It has been suggested that some at least of these compounds contain layers of metal ions and of peroxy-groups, with the other species held between the layers; the proportion of the various components depends on the conditions of preparation of the sample concerned and upon its previous history.

7.5. PEROXIDES CONTAINING DIRECTED BONDS

There is an enormous range of compounds in which peroxy- or similar groups are directly bound to atoms other than hydrogen. The peroxy-group may replace an oxygen atom as a bridge between two other groups, giving compounds such as $Me_3SnOOSnMe_3$, $K_2[O_3SOOSO_3]$ or $F_3CO{\cdot}OO{\cdot}COCF_3$; it may replace the oxygen atom of a hydroxy-group in a hydroxide, an alcohol or an acid, giving such compounds as Me_3SnOOH, C_4H_9OOH or $CH_3CO{\cdot}OOH$; it may replace an oxygen atom linked to another atom by a formal double bond, as in the ion $[HCr(O)_2(O_2)_2]^-$. The compounds of these different

[259] J. A. Connor and E. A. V. Ebsworth, *Advances in Inorganic Chemistry and Radiochemistry* (ed. H. J. Emeléus and A. G. Sharpe), Wiley (1964), Vol. 6.

classes are vastly different in structure and in chemistry[260–262]. Many, as the organic peroxides, are easily recognized, but it is sometimes very hard to distinguish between solids containing bound peroxide and others that contain H_2O_2 of crystallization. Chemical tests, of which the best-known is the Riesenfeld–Liebhafsky test (based on oxidation of buffered aqueous I^-), have been devised to distinguish between true peroxides and peroxy-hydrates, but these tests have not proved reliable. They depend on differences in reactivity between H_2O_2 itself and bound peroxide, and so cannot distinguish between a peroxy-hydrate and a compound containing bound peroxide that is hydrolysed very rapidly. Recently, criteria have been based on n.m.r.[263], or e.s.r.[264], and on Raman spectroscopy[265], where the (O–O) stretching mode of H_2O_2 has proved particularly useful.

In general, peroxides may be prepared by reduction of molecular oxygen (autoxidation) or ozone, or by treatment of an appropriate compound with H_2O_2 or an ionic peroxide:

$$O_2 + R_3B \rightarrow RB(O_2R)_2 + RB(OR)(O_2R)$$

$$O_2 + (Ph_3P)_4Pt \rightarrow (Ph_3P)_2Pt(O_2) + 2Ph_3P$$

$$C_6H_4(CO)_2O + H_2O_2 \longrightarrow C_6H_4(CO.OOH)(CO.OH)$$

Almost all preparative routes start from some compound already containing the O–O group; an important exception is the preparation of peroxydisulphate, $[O_3S \cdot O_2 \cdot SO_3]^{-2}$, by the anodic oxidation of acid sulphate solution.

7.5.1. Stereochemistry

The geometry of the peroxy-group in these compounds is very varied. In ether-type peroxides QOOQ the following types of structure may be distinguished (Fig. 8).

Planar, trans-QOOQ. This arrangement is found in the perdisulphate ion[266] and in the paramagnetic species[267] $[(H_3N)_5Co(O_2)Co(NH_3)_5]^{+5}$, which from the short (O–O) distance (1.31 Å) has been described as a superoxo-complex.

Non-planar, QOOQ. This is the type of structure adopted by H_2O_2. Dihedral angles vary widely, from 146° in diamagnetic[268] $[(H_3N)_5Co(O_2)Co(NH_3)_5]^{+4}$ (described as a peroxo-complex because the (O–O) distance is 1.47 Å) to 64° in the cyclic anion[269] $[B_2(O_2)_2(OH)_4]^{=}$.

Near-planar, cis QOOQ. The paramagnetic cobalt complexes[270]

$$[en_2Co(\mu\text{-}O_2)(\mu\text{-}NH_2)Coen_2]^{+4}$$

[260] A. G. Davies, *Organic Peroxides*, Butterworths, London (1962).
[261] G. Sosnovsky and J. H. Brown, *Chem. Rev.* **66** (1966) 528.
[262] E. G. E. Hawkins, *Organic Peroxides*, Spon, London (1961).
[263] T. M. Connor and R. E. Richards, *J. Chem. Soc.* 1958, 289.
[264] I. F. Franchuk, *Teor. Eksperim. Khim.* **1** (1965) 531.
[265] W. P. Griffith and J. D. Wickens, *J. Chem. Soc.* A, 1968, 397.
[266] W. H. Zachariasen and R. C. L. Mooney, *Z. Krist.* **88** (1934) 63.
[267] R. E. Marsh and W. P. Schaeffer, *Acta Cryst.* **24** (1968) 246.
[268] W. P. Schaeffer, *Inorg. Chem.* **7** (1968) 725.
[269] A. Hansson, *Acta Chem. Scand.* **15** (1961) 934.
[270] U. Thewalt and R. Marsh, *J. Am. Chem. Soc.* **89** (1967) 6364.

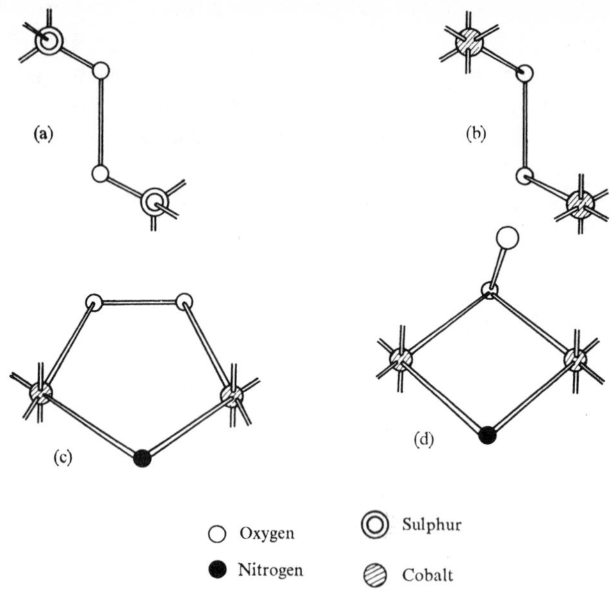

FIG. 8. Configurations in peroxides.

(a) *Trans*, planar in S_2O_8 = (K salt).
(b) Skew, non-planar in $[(NH_3)_5Co(O_2)Co(NH_3)_5]^{+4}$.

(c) *Cis*, near-planar, in

O—O
$[en_2Co$ $Coen_2]^{+4}$
NH_2

(d) Isomeric, in

OH
O
$[en_2Co$ $Coen_2]^{+4}$
NH_2

All from X-ray studies.

adopt this configuration; these compounds, with *d* (O–O) 1.31–1.32 Å, are described as superoxo-complexes.

Isomeric Q_2O:O. The only compound so far proved to have this type of structure is a salt of the red, diamagnetic cation[270]

OH
O
$[en_2Co$ $Coen_2]^{+4}$
NH_2

In solution this cation is formed from a diamagnetic species believed to be

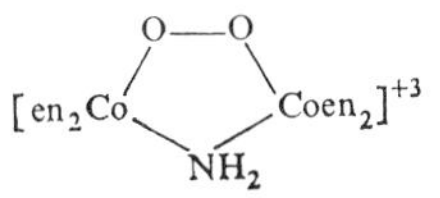

$$\left[en_2Co\begin{matrix} O{-}O \\ \diagup \qquad \diagdown \\ \diagdown \qquad \diagup \\ NH_2 \end{matrix}Coen_2\right]^{+3}$$

which isomerizes when the peroxy-bridge is protonated; the activation energy for isomerization is of the order of 15–20 kcal $mole^{-1}$, and the process is reversible[271].

In peroxy-derivatives in which the two oxygen atoms are bound to a single metal, the two metal–oxygen distances in all compounds so far studied are about equal, though there are some small deviations from symmetry that may be significant. Compounds of this type are so far only known for transition elements (though the vibrational spectra of alkali metal peroxide and superoxide molecules trapped in matrices at low temperatures suggest that the MOO systems here too may form isoceles triangles[258]). The peroxy-group is in effect doubly bound to the metal; the bonding has been discussed in terms of donation to the metal[260] from π- and π^*-orbitals of O_2^{-2}, and also by using the equivalent orbital approach[272]. In most peroxy-complexes the (O–O) distance is in the range 1.4–1.5 Å, though it is rather shorter in the paramagnetic cobalt complexes, but in one series of compounds the (O–O) bond length is remarkably sensitive to small changes in the rest of the molecule. In the species *trans*-$ClIr(Ph_3P)_2(CO)(O_2)$, *trans*-$IIr(Ph_3P)_2(CO)(O_2)$ and $(diphos)_2Ir(O_2)^+BF_4^-$, the (O–O) distances are respectively[273–274] 1.30 ± 0.03 Å, 1.51 ± 0.03 Å and 1.66 ± 0.03 Å.

Other Physical Properties

The (O–O) bond dissociation energy has been determined in a number of ether-type organic peroxides; the values are in the range 30–50 kcal $mole^{-1}$. In H_2O_2 and in many peroxy-complexes of transition metals the (O–O) stretching modes are found[265] near 800–900 cm^{-1}. The infrared spectra of adducts of O_2 with compounds such as $(Ph_3P)_2Ni$ also contain bands in this region that have been shown by isotopic substitution to be associated with the (O–O) group and are assigned to the (O–O) stretching mode[275]. It is remarkable that in the three iridium complexes described above the band assigned to ν (O–O) is almost unchanged in frequency, despite the very large change in the (O–O) distance.

Chemical Properties

Organic peroxides decompose readily to give free radicals, and have been extensively studied and used in connection with the initiation and progress of free radicals reactions in organic chemistry. They are important in many industrial oxidation processes, including the production of epoxides. Peroxy-disulphuric acid and its salts are very important as

[271] M. Mori and J. A. Weil, *J. Am. Chem. Soc.* **89** (1967) 3732.
[272] W. P. Griffith, *J. Chem. Soc.* 1964, 5248.
[273] S. J. LaPlaca and J. A. Ibers, *J. Am. Chem. Soc.* **87** (1965) 2581; J. A. McGinnety, R. J. Doedens and J. A. Ibers, *Inorg. Chem.* **6** (1967) 2243.
[274] J. A. McGinnety and J. A. Ibers, *Chem. Communs.* 1968, 235.
[275] K. Hirota, M. Yamamoto, S. Otsuka, A. Nakamura and Y. Tatsuno, *Chem. Communs.* 1968, 533.

unusually powerful oxidizing agents in inorganic chemistry[276, 277]. Many transition metal peroxides are dangerously explosive, and relatively little is known about their chemistry. The oxygen adduct $(Ph_3P)_2Pt(O_2)$ reacts with ketones[278] to give peroxy-derivatives:

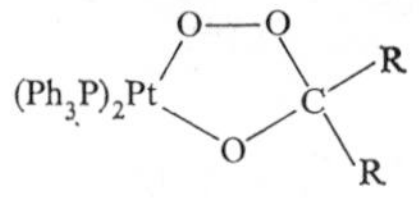

O—O
R
$(Ph_3P)_2Pt$ C
O
R

the reaction with CO_2 leads to the formation of a carbonato-complex, but a peroxy-carbonato species is apparently formed first, and the reaction with CS_2 may also give a peroxy-species[279]. The peroxy-compounds of transition metals (Table 51) are important in analytical chemistry (many of them are coloured), in connection with the catalytic decomposition of H_2O_2, and as models for natural oxygen-carriers.

TABLE 51. SOME OF THE ELEMENTS FORMING PEROXY-COMPOUNDS

Ionic superoxides	Alkali metals, alkaline earths ? Ti, Zr, Hf; Fe; Co, Ag, etc.
Ionic peroxides and hydroperoxides	Alkali metals, alkaline earths; Zn–Hg
Solid non-stoichiometric	Rare earths, Th, Pa
"Ether"-type peroxides	Be, B, Al, Ga; C–Pb; F, ? Cl
"alcohol"-type hydroperoxides	
Peroxyacids, peroxy salts	? B, C, ? N, P, S; V, Nb, Ta; Cr, Mo, W; Mn; U
Peroxy complexes (neutral or cations)	Ti, Zr, Hf; V, Nb, Ta; Cr, Mo, W; Re; ? Fe, Ru; Co, Rh, Ir; Ni, Pd, Pt; Cu; U
Oxygen carriers	Fe, Co, Ir, Cu

7.6. OXYGEN CARRIERS[280]

Certain compounds combine reversibly with molecular oxygen. The naturally bearing oxygen carriers include haemoglobin and haemocyanin; the simplest synthetic oxygen carrier is Vaska's compound, *trans*-$ClIr(Ph_3P)_2CO$. These compounds, when oxygenated, are closely related to peroxy-derivatives of transition elements; indeed, they really represent a special class of peroxides rather than a class of compounds that is quite distinct. While the oxygen adduct of Vaska's compound gives up its oxygen reversibly, the oxygen adduct of the very similar *trans*-$IIr(Ph_3P)_2CO$ does not, so the iodide cannot be regarded as an oxygen carrier. The configuration of oxygen in the oxygen adduct of haemoglobin is still unknown.

7.7. RELATIONSHIPS BETWEEN BOND LENGTH AND BOND ORDER IN PEROXIDES AND RELATED COMPOUNDS

If bond energies are plotted against bond lengths for the simple O_2^+, O_2, O_2^- and O_2^{-2} peroxy-species, a smooth curve is obtained[281]; a smooth curve is also obtained if (O–O)

[276] D. A. House, *Chem. Rev.* **62** (1962) 185.
[277] J. Burgess, *Ann. Rep. Chem. Soc.* **65A** (1968) 398.
[278] R. Ugo, F. Conti, S. Cenini, R. Mason and G. Robertson, *Chem. Communs.* 1968, 1498.
[279] P. J. Hayward, D. M. Blake, C. J. Nyman and G. Wilkinson, *Chem. Communs.* 1969, 987.
[280] L. H. Vogt, H. M. Faigenbaum and S. Wiberley, *Chem. Rev.* **63** (1963) 269.

stretching frequencies are plotted against d (O–O) for O_2 in its ground and excited states[282], as well as for O_2^- and O_2^{-2}. However, the very wide variation in d (O–O) in complex peroxides, coupled with the apparent constancy in ν (O–O), makes these relationships of doubtful value in assessing (O–O) bond orders. The reactivity of bound O_2 has been discussed[283] in terms of its resemblance to electronically excited O_2.

8. OZONE AND RELATED SPECIES

8.1. OZONE

During the past decade two factors have contributed to the increased attention paid by chemists and others to the behaviour of ozone; these factors are, on the one hand, chemical aeronomy and, on the other, the recognition of the strong oxidizing power of ozone.

A summary of physical data relating to ozone is given in Table 52. In the solid state, ozone is blue–black in colour, melting to a deep blue liquid and boiling to give a blue, thermally unstable gas.

Ozone, O_3, is normally prepared in the laboratory by passing an electric discharge through oxygen gas. Machines for the production of ozone, known as ozonizers, are marketed commercially. More generally, ozone is formed whenever oxygen is subjected to intense radiation, whether photochemical or electrical: thus studies have been carried out on ozone produced by pulse radiolysis[284], flash photolysis[284], mercury-sensitized photolysis[285] and ^{60}Co-γ radiolysis[286] of oxygen.

The growing interest in physical phenomena of the upper atmosphere (thermosphere), together with problems associated with atmospheric pollution in large cities[287], has produced a large volume of fundamental work on the rates of reaction of atoms with simple molecules[288]. Many of these reactions occur with concomitant emission of light (chemiluminescence). Examples of these fundamental rate studies which involve ozone as one component include the following:

$H+O_3 \rightarrow OH+O_2$	ref. 289	(1)
$N+O_3 \rightarrow NO+O_2$	290	(2)
$SO+O_3 \rightarrow SO_2+O_2$	291	(3)
$O_3+NO \rightarrow O_2+NO_2$	292	(4)
$M+O_3 \rightarrow M+O_2+O$	293	(5)
$O+O_3 \rightarrow 2O_2$	294	(6)
$H_2S+O_3 \rightarrow SO_2+H_2O$	295	(7)

281 D. G. Tuck, *J. Inorg. Nucl. Chem.* **26** (1964) 1525.
282 J. C. Evans, *J. Chem. Soc.* D, 1969, 682.
283 R. McWeeny, R. Mason and A. D. C. Towl, *Disc. Faraday Soc.* **47** (1969) 20.
284 C. J. Hochanadel, J. A. Ghormley and J. W. Boyle, *J. Chem. Phys.* **48** (1968) 2416.
285 R. J. Fallon, J. T. Vanderslice and E. A. Mason, *J. Phys. Chem.* **64** (1960) 505.
286 J. T. Sears and J. W. Sutherland, *J. Phys. Chem.* **72** (1968) 1166.
287 R. P. Wayne and J. N. Pitts, *J. Chem. Phys.* **50** (1969) 3644.
288 M. Nicolet, *Disc. Faraday Soc.* **37** (1964) 7 *et seq.*
289 D. Garvin, H. P. Broida and H. J. Kostkowski, *J. Chem. Phys.* **32** (1960) 880.
290 L. F. Phillips and H. I. Schiff, *J. Chem. Phys.* **36** (1962) 1509.
291 C. J. Halstead and B. A. Thrush, *Proc. Roy. Soc.* **295** (1966) A, 380.
292 P. N. Clough and B. A. Thrush, *Trans. Faraday Soc.* **65** (1969) 23.
293 W. M. Jones and N. Davidson, *J. Am. Chem. Soc.* **84** (1962) 2868.
294 W. Kaufmann and J. R. Kelso, *J. Chem. Phys.* **47** (1967) 4541.
295 R. D. Cadle and M. Ledford, *Air Water Pollution* **10** (1966) 25.

TABLE 52. PHYSICAL DATA FOR OZONE

Molecular weight, M	48
Melting point (°K)	80.5
Boiling point (°K)	161.3
Critical data:	
Temperature, T_c (°K)	261.1
Pressure, P_c (atm)	54.6
Volume, V_c (cc mole^{-1})	147.1
Density:	
Gas at 273°K, 760 mmHg (g l^{-1})	2.144
Liquid at 77.4°K, 760 mmHg (g cc^{-1})	1.614
Solid at 77.4°K, 760 mmHg (g cc^{-1})	1.728
Surface tension at 90°K (dyne cm^{-1})	38.4
Viscosity at 90°K (centipoise)	1.57
Dielectric constant at 90°K	4.75 ± 0.02
Dipole moment, D	0.58
Ionization potentials (eV):	
1	12.3 ± 0.1
2	12.52 ± 0.05
3	13.52 ± 0.05
4	$16.4 - 17.4$
5	19.2 ± 0.1
Electron affinity (kcal mole^{-1})	$+44 \pm 10$
Enthalpy of formation:	
ΔH_f° (g) (kcal mole^{-1})	+34.1
ΔH_f° (aq) (kcal mole^{-1})	+32.2
Free energy of formation, ΔG° (g)	+39.0
Entropy, S° (cal mole^{-1} deg^{-1})	57.1
Molecular dimensions:	
r (O–O) (Å)	1.278(3)
α (OÔO) (deg)	116° 45′
Redox potentials (V):	
(a) $O_3 + 2H^+ + 2e \rightarrow O_2 + H_2O$	+2.07
(b) $O_3 + H_2O + 2e \rightarrow O_2 + 2OH^-$	+1.24
Point group, C_{2v}	
Vibrational frequencies (cm^{-1}):	
ν_1 IR, Raman (pol)	1110
ν_2 IR, Raman (pol)	705
ν_3 IR	1043
Force constants (mdyn Å^{-1})	
f_d 5.70 f_{α/d^2}	1.28
f_{dd} 1.52 $f_{\alpha\alpha/d^2}$	0.33

From studies of the kinetics of thermal decomposition of ozone in the gas phase, it is generally agreed that the process[296] occurs in two stages[297] as shown in eqns. (5) and (6). The photochemical decomposition of ozone,

$$O_3 \xrightarrow{h\nu} O_2 + O \tag{8}$$

has been studied in detail[298].

The reactions of ozone with more complicated systems provide examples of its considerable oxidizing power. The redox potential, E_0, of ozone is such as to place it in a class

[296] S. W. Benson and A. E. Axworthy, *J. Chem. Phys.* **42** (1965) 2614.
[297] M. F. R. Mulcahy and D. J. Williams, *Trans. Faraday Soc.* **64** (1968) 59.
[298] R. G. W. Norrish and R. P. Wayne, *Proc. Roy. Soc.* A, **288** (1965) 200, 361.

with molecular fluorine, the hydroxyl radical and atomic oxygen, all of which are very strong oxidizing agents:

$O_3 + CH_4 \rightarrow CO + CO_2 + HCOOH + H_2O$	ref. 299	(9)
$+ CO_2 \rightarrow CO_3$	300	(10)
$+ R_3SiH \rightarrow R_3SiOH + (R_3Si)_2O$	301	(11)
$+ HX \rightarrow HOX + O_2$	302	(12)
$+ XeO_3 \rightarrow XeO_6^{4-}$	303	(13)
$+ NH_3 \rightarrow NH_4O_3 + NH_4NO_3$	304	(14)
$+ NO_2 \rightarrow NO_3 + N_2O_5 + NO$	305	(15)

The reactions of ozone with organic compounds have been reviewed[306–308]. Particular interest attaches to the reaction with alkenes. The end product of the reaction

$$O_3 + \gt C{=}C\lt \;\longrightarrow\; \gt C\langle O, O{-}O\rangle C\lt \qquad (16)$$

is termed an ozonide (but see following section). The ozonide may be destroyed (Zn/H_2O; $MeOH/I^-$), leaving the organic product which is either an aldehyde or a ketone. This reaction is useful as a means of determining the position of unsaturated linkages in organic compounds, for example:

$$\text{Butene-1} \quad CH_3CH_2CH{=}CH_2 \xrightarrow{O_3} CH_3CH_2CHO + HCHO \qquad (17)$$

$$\text{Butene-2} \quad CH_3CH{=}CHCH_3 \xrightarrow{O_3} 2CH_3CHO \qquad (18)$$

The exact mechanism of the reaction leading to the formation of the ozonide, eqn. (16), is the subject of controversy; however, the current view[309] appears to favour the initial formation of a π-complex between an ozone molecule and an olefin molecule. Ozone may be regarded as a 1,3-dipolar molecule.

An interesting contrast has been shown[310] in the rate of reaction between ozone and the simple olefins ethylene, propylene and butene-2, on the one hand, and their perfluorinated analogues, on the other hand. This difference has been taken to illustrate the effect of fluorination upon the C=C bond strength of olefins.

299 F. J. Dillemuth, D. R. Skidmore and C. C. Schubert, *J. Phys. Chem.* **64** (1960) 1496.
300 N. G. Moll, D. R. Clutter and W. E. Thompson, *J. Chem. Phys.* **45** (1965) 4469.
301 L. Spialter and J. D. Austin, *Inorg. Chem.* **5** (1966) 1975. J. D. Austin and L. Spialter, *Adv. Chem. Ser.* **77** (1968) 26.
302 I. Schwager and A. Arkell, *J. Am. Chem. Soc.* **89** (1967) 6006.
303 T. M. Spittler and B. Jaselskis, *J. Am. Chem. Soc.* **88** (1966) 2942.
304 I. J. Solomon, K. Hattori, A. J. Kacmarek, G. M. Platz and M. J. Klein, *J. Am. Chem. Soc.* **84** (1962) 34.
305 W. B. De More and N. Davidson, *J. Am. Chem. Soc.* **81** (1959) 5869.
306 P. S. Bailey, *Chem. Rev.* **58** (1958) 925.
307 A. T. Menyailo and M. V. Pospelov, *Russ. Chem. Rev.* **36** (1967) 284.
308 F. D. Gunstone, *Educ. Chem.* **5** (1968) 166.
309 S. Fliszár and J. Carles, *J. Am. Chem. Soc.* **91** (1969) 2637; R. W. Murray, *Accounts Chem. Res.* **1** (1968) 313.
310 J. Heicklen, *J. Phys. Chem.* **70** (1966) 477.

Various theoretical studies have been made of the ozone molecule. Both *ab initio* SCF MO[311] and INDO[312] calculations produce estimates of the bond angle, $\widehat{OOO}$ which are in satisfactory agreement with the experimental value. Ozone is a 24(18) electron molecule, isoelectronic with NO_2^- and NOCl.

A number of different methods are available for the quantitative estimation of ozone. The simplest chemical method involves the oxidation of KI to give iodine which can then be titrated with thiosulphate. The reaction

$$O_3+2KI+H_2O \rightarrow I_2+KOH+O_2 \tag{19}$$

is quantitative. Ozone can also be determined photometrically by measurement at the maximum absorption in the ultraviolet region[313] (λ_{max} 2537 Å; ε 135 cm^{-1}). This band in the ultraviolet spectrum of ozone is known as the Hartley band.

Largely as a consequence of its oxidizing power, ozone produces technological problems by causing rubbers to crack and the general deterioration of textile fibres. Extensive studies have been carried out to determine ways in which such deleterious effects can be inhibited by treatment of the substances with chemical agents such as diamines.

In addition to the simple reactions shown in eqns. (1)–(7), some of which are chemiluminescent, the reactions between ozone and CO, CS_2 and a wide range of organic compounds give rise to chemiluminescence[314], which in certain cases may be very intense.

8.2. OZONIDES

In the previous section it was noted that ozone is a 24(18) electron molecule. Addition of one electron to this entity produces the ozonide ion, O_3^-, which is isoelectronic with the 33(19) electron molecule ClO_2. The angle ($\widehat{OClO}$) in ClO_2 is 117° [315]; in O_3^- the angle ($\widehat{OOO}$) is 100° [316].

The literature on ozonides has been reviewed on a number of occasions[317–320]. Ozonides have been intensively investigated in recent years as potential sources of oxygen because of the aerospace industrial interest in non-regenerative air revitalization agents and propellant materials of high specific impulse.

Although the parent acid, HO_3, is unknown, ozonide salts of the alkali metals[321], the alkaline earth metals[322] and both the ammonium[323] and tetramethylammonium[321] cations

[311] S. D. Peyerimhoff and R. J. Buenker, *J. Chem. Phys.* **47** (1967) 1953.

[312] J. A. Pople, D. L. Beveridge and P. A. Dobosh, *J. Chem. Phys.* **47** (1967) 2026.

[313] W. B. De More and O. F. Raper, *J. Phys. Chem.* **68** (1964) 412.

[314] R. L. Bowman and N. Alexander, *Science* **154** (1966) 1454.

[315] R. F. Curl, J. L. Kinsey, J. G. Baker, J. C. Baird, G. R. Bird, R. F. Heidleberg, T. M. Sugden, D. R. Jenkins and C. R. Kenney, *Phys. Rev.* **121** (1961) 1119.

[316] L. V. Azárov and I. Corvin, *Proc. Nat. Acad. Sci. US* **49** (1963) 1.

[317] N. G. Vannerberg, *Prog. Inorg. Chem.* **4** (1962) 125.

[318] A. W. Petrocelli and R. V. Chiarenzelli, *J. Chem. Educ.* **39** (1962) 557.

[319] I. I. Vol'nov, *Usp. Khim.* **34** (1965) 2111.

[320] I. I. Vol'nov, *Peroxides, Superoxides and Ozonides of Alkali and Alkaline Earth Metals* (English translation), Plenum Press, New York (1966).

[321] I. J. Solomon, A. J. Kacmarek, J. M. McDonough and K. Hattori, *J. Am. Chem. Soc.* **82** (1960) 5640.

[322] I. I. Vol'nov, S. A. Tokareva, V. N. Belevskii and G. P. Pilipenko, *Izvestia Akad. Nauk SSSR, Ser. Khim.* 1967, 416.

[323] I. J. Solomon, K. Hattori, A. J. Kacmarek, G. M. Platz and M. J. Klein, *J. Am. Chem. Soc.* **84** (1962) 34.

have been described. The reaction between ozone and an alkali metal hydroxide in solution was first described in 1866 [324], when it was noted that a red colour was produced in the solution. Subsequently it was found that the reaction between a powered alkali metal hydroxide and gaseous ozone in an oxygen carrier at room temperature or below, produces the ozonide of the metal, and the presence of the O_3^- anion was established[325].

A detailed study of the mechanism of formation of potassium ozonide, KO_3, using $K^{18}OH$, has shown that the overall stoichiometry of the reaction is given by the equation[326]

$$5O_3+2KOH \rightarrow 5O_2+2KO_3+H_2O \quad (1)$$

where $H^\circ = -158$ kcal mole^{-1}. The reaction by which ozonide salts in general are formed is catalysed by ammonia[327], and the ozonide, once formed, is often isolated by extraction into liquid ammonia. Alkali metal ozonides are soluble in liquid ammonia, methylamine and dimethylformamide[321]. Although ozonides of all the alkali metals are known, lithium ozonide is known only in the form of its ammonia adduct, $LiO_3 \cdot 4NH_3$ [327]. Calculations of the heat of formation, ΔH_f°, and the lattice energy of the hypothetical LiO_3 have suggested[328] that it will not be stable with respect to decomposition to a lower oxide in the free (non-solvated) state.

The thermal stability of an ozonide is largely determined by the cation present. Thus among the alkali metals and the alkaline earth metals, the following order of decreasing thermal stability is observed:

$$Cs > Rb > K > Na > Li$$
$$Ba > Sr > Ca$$

which is in agreement with calculations of the thermodynamic stability of these systems[322, 329]. The enthalpy of formation of various ozonides has been measured, and the results shown in Table 53 obtained.

TABLE 53. ENTHALPY OF FORMATION OF SOLID OZONIDES (kcal mole^{-1})

$LiO_3 \cdot 4NH_3$	-135 ± 5	Ref. 327	$Ca(O_3)_2$	-69	Ref. 322
KO_3	-62.1 ± 0.9	330	$Sr(O_3)_2$	-68	322
Me_4NO_3	-49.5 ± 4.5	321	$Ba(O_3)_2$	-67	322

$\Delta H_f^\circ (O_3^-, g) = -11 \pm 10$ (ref. 328).

In the solid state, ozonides of the alkali and alkaline earth metals are coloured. Both NaO_3 and KO_3 are dichroic, while crystallographic studies have shown that both have a body-centred tetragonal unit cell (*I4/mcm*) with the following lattice parameters:

324 C. F. Schönbein, *J. Prakt. Chem.* **45** (1866) 469.

325 I. A. Kazarnovskii, G. P. Nikolskii and T. A. Abletsova, *Doklady Akad. Nauk SSSR* **64** (1949) 69.

326 I. I. Vol'nov, V. N. Chamova and E. I. Latysheva, *Izvestia Akad. Nauk SSSR, Ser. Khim.* 1967, 1183.

327 A. J. Kacmarek, J. K. McDonough and I. J. Solomon, *Inorg. Chem.* **1** (1962) 659.

328 R. L. Wood and L. A. D'Orazio, *J. Phys. Chem.* **69** (1965) 2562.

329 G. P. Nikolskii, Z. A. Bagdasar'yan and I. A. Kazarnovskii, *Doklady Akad. Nauk SSSR* **77** (1951) 69.

	a	*c*	Ref.
NaO_3	11.61	7.66	331
KO_3	8.597	7.080	316

The structure of KO_3 is related to that of KN_3. A monoclinic form of KO_3 related to KNO_2 has been described[332]. From this data the lattice energy and Madelung constant of KO_3 have been calculated[328] for two possible models of the charge distribution of the ozonide ion thus:

Model *a* charge distribution $O^{-\frac{1}{3}}O^{-\frac{1}{3}}O^{-\frac{1}{3}}$
Model *b* charge distribution $O^{-\frac{2}{3}}O^{+\frac{1}{3}}O^{-\frac{2}{3}}$

	Model a	*Model b*
Lattice energy (kcal mole^{-1})	166.7	178.4
Madelung constant	2.1012±0.03%	2.2554±0.08%

This data has been used to calculate the electron affinity of ozone (44 ± 10 kcal mole^{-1}) and thermodynamic functions ($\Delta H°$, $\Delta G°$) of other alkali metal ozonides.

When ozone is passed over ammonia at $-130°$, ammonium ozonide, NH_4O_3, is produced[323]. From a study of the infrared spectrum of NH_4O_3 at temperatures below $-90°$, the following values of the fundamental frequencies of the O_3^- ion were obtained[333] as ν_1 1260, ν_2 800, ν_3 1140 cm^{-1}. The values of the structural parameters of the ozonide ion obtained from spectroscopic[333] and crystallographic[316] measurements together with values calculated[334] by comparison with NO_2, NO_2^- and O_3 are presented in Table 54. The corresponding values for ozone are provided for comparison

TABLE 54. STRUCTURAL PARAMETERS OF O_3^- ION

Method	*r* (O–O) Å	($\widehat{OOO}$) (deg)	Ref.
X-ray	1.19	100	316
Infrared	1.22	~100	333
Calculation	1.34	110±5	334
O_3 microwave	1.278	116.8	335

In addition to the methods mentioned above, the ozonide ion may be produced in solution as a result of pulse radiolysis[336], flash photolysis[337] or γ-irradiation[338] of such substances

330 G. P. Nikolskii, L. J. Kazarnovskaya, Z. A. Bagdasar'yan and I. A. Kazarnovskii, *Doklady Akad. Nauk SSSR* **72** (1950) 713.

331 S. N. Tokareva, M. S. Dobrolyubova and S.Z. Makarov, *Khim. Perekisnykh Soedin. Akad. Nauk SSSR, Inst. Obshch i Neorg. Khim.* 1963, 188; *Chem. Abs.* **60** (1964) 12884a.

332 I. J. Solomon and A. J. Kacmarek, *J. Phys. Chem.* **64** (1960) 168.

333 K. Herman and P. A. Giguere, *Can. J. Chem.* **43** (1965) 1946.

334 P. Smith, *J. Phys. Chem.* **60** (1956) 1471.

335 R. H. Hughes, *J. Chem. Phys.* **24** (1956) 131.

336 G. Czapski, *J. Phys. Chem.* **71** (1967) 1683.

337 L. Dogliotti and E. Hayon, *J. Phys. Chem.* **71** (1967) 2511.

338 K. Tagaya and T. Nogaito, *J. Phys. Soc. Japan* **23** (1967) 70.

as H_2O_2, the persulphate ion, $AgNO_3$ and $KClO_4$. In these experiments the presence of the ozonide ion has been established from its characteristic visible absorption which shows λ_{max} at 4300 Å. This value should be compared with that found in the ozonide salts[323], all of which show a maximum absorption at ~4500 Å, together with five subsidiary maxima at the following approximate wavelengths; 5020, 4850, 4660, 4330, 4300 Å. The spectra of these salts all show a slight solvent dependence.

The ozonide ion is paramagnetic. The moment of KO_3 (μ) was found[325] to be 1.63 BM. The e.s.r. spectra of several ozonides have been studied[317] in order to determine the g-value of the unpaired electron and the following results have been obtained[325, 339].

	g-value		g-value
NaO_3	2.012	NH_4O_3	2.0119
KO_3	2.0124	Me_4NO_3	2.0144

From these data it has been concluded that there is considerable spin–orbit coupling in the ozonide ion.

The application of e.s.r. spectroscopy in a study[340] of the hydrolysis of KO_3 has shown that, although the overall process may be described[325] by the equation

$$4KO_3+2H_2O \rightarrow 4KOH+5O_2 \qquad (2)$$

free hydroxyl radicals, ·OH, are produced as intermediates in the reaction thus:

$$KO_3+H_2O \rightarrow KOH+O_2+\cdot OH \qquad (3)$$

Thermal decomposition of ozonide salts[319, 341] has been studied by a variety of methods, particularly by d.t.a. The reaction is complex, but in its most simple form may be described by the equations

$$MO_3 \rightarrow MO_2+O \qquad (4)$$

$$MO_3+O \rightarrow MO_2+O_2 \qquad (5)$$

so that the overall process is described by

$$2MO_3 \rightarrow 2MO_2+O_2+11.6 \text{ kcal} \qquad (6)$$

Further thermal decomposition of the superoxide, MO_2, according to

$$2MO_2 \rightarrow M_2O_2+O_2 \qquad (7)$$

can also occur, so that the general decomposition reaction is[342]

$$2MO_3 \rightarrow M_2O_2+2O_2 \qquad (8)$$

339 A. D. McLachlan, M. C. R. Symons and M. G. Townsend, *J. Chem. Soc.* 1959, 952.
340 I. A. Kazarnovskii, N. P. Lipikhin and N. N. Bubnov, *Izvestia Akad. Nauk SSSR, Ser. Khim.* 1966, 2247.
341 C. B. Riolo and T. F. Soldi, *Chim. Ind. (Milan)* **48** (1966) 846.
342 E. I. Sokovnin, *Izvestia Akad. Nauk SSSR, Ser. Khim.* 1963, 181.

The reaction between potassium superoxide, KO_2, and ozone in oxygen in a fluorocarbon solvent at $-100°$ to $-60°$, leads[343] to the formation of KO_3 by a process similar to that described by the reverse of eqn. (6). The peculiar advantage of this method is that it avoids the use of liquid ammonia.

9. OTHER SPECIES CONTAINING O_3 AND O_4 GROUPS

9.1. O_4

In 1924 G. N. Lewis explained the drop in paramagnetic susceptibility of O_2 on condensation and on cooling by postulating[344] the formation of $(O_2)_2$ dimers, with a heat of formation of 128 cal mole^{-1}; since then his suggestion has been supported by evidence drawn from measurements of magnetic susceptibility[345], from electronic spectroscopy[346] and from a preliminary study of the crystal structure of oxygen[347]. Thorough analysis of the structures of α-, β- and γ-O_2, however[348–350], shows that dimers are not present in these phases; the antiferromagnetism has been accounted for in terms of long- and short-range magnetic ordering[351], which probably persists in the liquid; thus most of the evidence on which the suggestion of dimerization was based has lost its force. There are electronic transitions in all phases that are clearly associated with co-operative effects, but in the vapour and the liquid phases it is not clear whether the polymolecular species is longer lived than a collision complex[352]. Calculations based on the second virial coefficient, using a Lennard-Jones potential[353], suggest that the concentration of dimers in O_2 at 300°K and 1 atm pressure is 6.4×10^{-4} dimers mole^{-1}, but experiments with a mass spectrometer indicate that the concentration is much lower[354]. It should be noted that in gaseous argon or krypton under the same conditions the concentration of dimers is higher.

9.2. CHARGED O_4 SPECIES

The molecule ions O_4^+ and O_4^- have both been detected by mass spectrometry[355, 356], and the e.s.r. spectra of irradiated nitrates at low temperatures contain signals that have been assigned[357] to O_4^-. In the gas phase the heats of formation of the two ions are[356]:

$$O_2^+(g)+O_2(g) = O_4^+(g) \quad \Delta H°\,(298) = -10 \text{ kcal mole}^{-1}$$

$$O_2^-(g)+O_2(g) = O_4^-(g) \quad \Delta H°\,(298) = -13.55 \pm 0.16 \text{ kcal mole}^{-1}$$

---

[343] I. I. Volnov, S. A. Tokareva, V. I. Klimanov and G. P. Pilipenko, *Izvestia Akad. Nauk SSSR, Ser. Khim.* 1966, 1267.
[344] G. N. Lewis, *J. Am. Chem. Soc.* **46** (1924) 2027.
[345] L. N. Mulay and L. K. Keys, *J. Am. Chem. Soc.* **86** (1964) 4489.
[346] V. I. Dianov-Klokov, *Optics Spectrosc.* **20** (1966) 530
[347] L. Vegard, *Nature* **136** (1935) 720.
[348] C. S. Barrett, L. Meyer and J. Wassermann, *J. Chem. Phys.* **47** (1967) 592.
[349] R. A. Alikhanov, *Soviet Physics JETP* **18** (1964) 556.
[350] T. A. Jordan, W. E. Streib and W. N. Lipscomb, *Acta Cryst.* **17** (1964) 777.
[351] M. F. Collins, *Proc. Phys. Soc.* **89** (1966) 415.
[352] R. P. Blickensderfer and G. E. Ewing, *J. Chem. Phys.* **47** (1967) 331.
[353] D. E. Stogryn and J. Hirschfelder, *J. Chem. Phys.* **31** (1959) 1531; **33** (1960) 942.
[354] T. A. Milne and F. T. Greene, *J. Chem. Phys.* **47** (1967) 3668.
[355] R. E. Voshall, J. G. Pack and A. V. Phelps, *J. Chem. Phys.* **43** (1965) 990.
[356] D. C. Conway and L. E. Nesbitt, *J. Chem. Phys.* **48** (1968) 509.
[357] R. Kikuchi, T. Nogaito, K. Tagaya and K. Matsumo, *Mem. Inst. Sci. Ind. Res. Osaka Univ.* **24** (1967) 53.

The rate constant for the former reaction has been determined[358]. The heats of formation are in reasonable agreement with calculations by the modified Hückel method assuming that the molecule ions are planar.

9.3. HYDROGEN SUPEROXIDES[359]

The formation of superoxides H_2O_3 and H_2O_4 has been postulated to explain the delay in oxygen evolution when H_2O_2 is oxidized by potassium permanganate; the compounds may also be formed in H–O flames and in irradiated aqueous H_2O_2. Some of the observations have been questioned, and for others there are other possible explanations. There is much evidence relating to the possible formation of hydrogen superoxides from studies at low temperatures of the products of the action of an electric discharge on H_2O or H_2O_2, or of the reaction between ozone or oxygen and hydrogen atoms. On warming the products to 160°K, heat and oxygen are evolved; the molar ratio of evolved O_2 to residual H_2O_2 is said for the system hydrogen atoms/liquid ozone to be 1:1, and it has been suggested that the evolved oxygen comes from decomposition of H_2O_4. The melting point of the condensate is lower than the lowest point in the H_2O/H_2O_2 eutectic; the X-ray and electron diffraction patterns are different from those given by mixtures of H_2O and H_2O_2, the product is diamagnetic below 160°K (indicating the absence of radicals) and there are bands in the infrared spectra that have been assigned to modes of H_2O_4 because no other plausible assignment could be devised. This evidence has been vehemently contested[360]. The spectroscopic assignments have been questioned[361]; other explanations have been put forward for most of the observations. If H_2O_3 or H_2O_4 is formed, it is clear that they are unstable at normal temperatures; either or both may be formed at low temperatures, but the evidence so far available is by no means decisive.

9.4. OTHER TRIOXIDES AND TETROXIDES

When t-butyl hydroperoxide is treated with lead tetraacetate in CH_2Cl_2 at low temperatures, a substance is formed which appears stable at temperatures below −30°, but which evolves oxygen rapidly at higher temperatures; it is believed to be di(t-butyl) trioxide[362]. An earlier claim to have prepared the compound[363] has been discounted[364]. From the constancy of the e.s.r. signal assigned to t-butyl peroxy-radicals (formed by photolysis of di-t-butyl peroxycarbonate at −70°) in the temperature range −110° to −85°, it is suggested[365] that in that temperature range there is appreciable dimerization to form di(t-butyl peroxide). Neither of these compounds is stable enough to have been isolated in a pure form.

358 D. A. Dundas, P. Kerbarle and A. Good, *J. Chem. Phys.* **50** (1969) 805.
359 M. Venugopalan and R. G. Jones, *Chem. Rev.* **66** (1966) 133; *The Chemistry of Dissociated Water Vapour and Related Systems*, Interscience (1968).
360 P. A. Giguère, *J. Am. Chem. Soc.* **90** (1968) 7181.
361 K. Herman and P. A. Giguère, *Can. J. Chem.* **46** (1968) 2649.
362 P. D. Bartlett and P Günther, *J. Am. Chem. Soc.* **88** (1966) 3288.
363 N. A. Milas and F. G. Arzoumanidis, *Chem. Ind.* 1966, 66.
364 R. D. Youssefyeh and R. W. Murray, *Chem. Ind.* 1966, 1531.
365 P. D. Bartlett and G. Guaraldi, *J. Am. Chem. Soc.* **89** (1967) 4799.

The perfluoroalkyl trioxides CF_3OOOCF_3 and $CF_3OOOC_2F_5$, on the other hand, have been isolated and characterized by infrared, n.m.r. and mass spectrometry, and by measurements of vapour density. The compounds were prepared by the direct fluorination of metal salts of trifluoracetic acid[366] and by the reaction between dioxygen difluoride and carbonyl fluoride[367]. Besides these, the trioxide radical CF_3OOO has been identified by e.s.r. spectroscopy[368] among the photolysis products of CF_3OOCF_3 (see section 5.11, under *Bis* (trifluoromethyl trioxide, CF_3OOOCF_3).

366 P. G. Thompson, *J. Am. Chem. Soc.* **89** (1967) 4316.
367 L. R. Anderson and W. B. Fox, *J. Am. Chem. Soc.* **89** (1967) 4313.
368 R. W. Fessenden, *J. Chem. Phys.* **48** (1968) 3725.

23. SULPHUR

MAX SCHMIDT and WALTER SIEBERT
University of Würzburg

1. THE ELEMENT

1.1. HISTORY

Sulphur in its free state was well known to the ancients. In Assyrian texts (Assurbanipal, 668–626 B.C.) sulphur is often mentioned as the "product of the riverside", which was described to have a yellow and sometimes black colour. This black colour of sulphur certainly originated from contents of asphalt contained in it. Originally sulphur was used for medicinal-religious purposes to purify ills and ward off evils. To the early Hebrews sulphur symbolized God's wrath, which is illustrated in various parts of the Old Testament (e.g. destruction of Sodom and Gomorrah).

Sulphur was supposedly long known in China and India. However, there is a lack of information to support this assumption. More information about the use of sulphur is available from the Greeks. Homer (9th century B.C.) mentioned the "pest-averting sulphur" and the "divine and purifying fumigations"[1]. Thucydides reported the application of sulphur by the tribe of Bootier, who in 424 B.C. destroyed a city's wooden wall with a burning mass of coal, sulphur and tar.

The first theoretical studies of sulphur and its compounds were originated by the alchemists who discovered that mercury could be "fixed" with sulphur to form a solid product. Further information was obtained when many of the "compounds" known at that time were subjected to the newly developed technique of distilling. Thereby it was discovered that various products contained sulphur. On this knowledge the alchemists based their remarkable theories which lasted for nearly fifteen centuries[2].

Besides the application of sulphur for "gunpowder" (12th century in China), the most outstanding result of the occidental alchemists was the synthesis of sulphuric acid from vitriol or pyrites as well as from sulphur. Then in the 18th century Lavoisier opened the door to modern chemistry with the conclusion that sulphur does not consist of various "elements" but represents one element.

1.2. THE OCCURRENCE OF SULPHUR AND ITS COMPOUNDS

Sulphur is widely distributed over the earth. The average sulphur content of the outer crust amounts to 0·048%, which puts sulphur in 15th place in the abundance of the elements.

[1] M. E. Weeks, Discovery of the elements, published by the *Journal of Chemical Education*, Easton, Pa., 1968.

[2] *Gmelins Handbuch der Anorganischen Chemie*, 8. Auflage, Systemnummer 9, S. 1, Verlag Chemie, Weinheim (1953).

Besides oxygen and silicon, sulphur is the most abundant element in minerals. Since sulphur occurs in various sulphur compounds as well as in elemental form, it seems convenient to discuss its occurrence with respect to its oxidation state.

Elemental Sulphur

Elemental sulphur is found in various deposits of sedimental or volcanic origin all over the earth. The most important occurrences are in Louisiana and Texas in the U.S.A. Before 1903, when the Frasch process was first introduced, sulphur mined in Sicily held a near-monopoly position in world production. Both the Gulf of Mexico and the Sicilian deposits[3] are of sedimental origin (bacterial reduction of sulphates), whereas in Japan, the Philippines, Peru, Chile and the west of North America occurrences of volcanic origin are found.

Hydrogen Sulphide

Hydrogen sulphide occurs in small amounts almost everywhere, owing to bacterial decomposition of proteins and sulphates (e.g. *Spirillus desulphuricans* reduces $CaSO_4$ to CaS; water and carbon dioxide react with CaS to form $CaCO_3$ and H_2S). The large amounts of H_2S found in the natural gas of Lacq (France) and Alberta (Canada) contain about 15% and up to 34% H_2S, respectively[3]. In the past decade production of sulphur from these natural sources has become very important.

Sulphur Oxides

The natural occurrence of sulphur oxides in the atmosphere is limited to sulphur dioxide, which originates from gases and springs in volcanic areas. Most of the sulphur dioxide now found in the atmosphere has been generated by combustion of coal and oil products containing more or less amounts of sulphur (up to 5%). Since, however, these energy sources are used in millions of tons per year, SO_2 released into the atmosphere also amounts to millions of tons. This pollution represents a threat to all life on earth. It has been estimated that in critical areas of high pollution (e.g. cities such as Los Angeles, New York, London) the sulphur content of the energy sources should be restricted to 0·3–0·5%, whereas in areas with fewer sources of pollution a higher content (1–2%) might be allowed[4]. Certainly, the pollution control restrictions on the sulphur content of fuels slowly begins to influence the production pattern of coal and oil.

Solid Mineral Sulphides[5]

Since sulphur reacts with almost all elements (exceptions are nitrogen, iodine and the noble gases), the formation of compounds with electropositive as well as electronegative sulphur is quite common in nature. Besides the sulphates, the sulphides represent a large amount of sulphur on earth. Of all possible sulphides, the transition metal sulphides are the most important minerals for industrial use. In Western Europe pyrites, FeS_2, provide the raw material for SO_2 as well as for iron production. The total pyrites deposits in the earth

[3] *Ullmans Encyklopädie der technischen Chemie*, 3. Auflage, Band 15, S. 370, Urban & Schwarzenberg, München–Berlin (1964).

[4] *Chem. and Eng. News*, March 10 (1969) 46.

[5] *Gmelins Handbuch der Anorganischen Chemie*, 8. Auflage, Systemnummer 9, Teil A, S. 69, Verlag Chemie, Weinheim (1953).

have been estimated at 1100 million tons. An additional 200 million tons will be obtained as by-products of copper, zinc and lead mining. Because of the differential content of Cu, Zn, Pb, Co, Ni, Mn, Bi, Ag, Au, As and Sb the iron sulphides exhibit variable colours (mostly grey to greenish).

Spain is considered to possess the biggest supply in pyrites (about 500 million tons) and ranks today in second position behind Japan in its mining. In comparison, the pyrite deposits of the U.S.A. and Canada amount only to 32 and 52 million tons, respectively. Most pyrite ores contain about 45% sulphur.

Sulphur Containing Ions in Aqueous Solutions

As pointed out before, the earth's crust contains about 0·048% sulphur, which occurs mostly in heavy metal sulphides. Whenever oxidation conditions have been prevalent, deposits of slightly soluble alkaline earth sulphates have been formed. The more soluble sulphates have been carried away by rain and ground water, resulting in more or less high concentrations in rivers, lakes and oceans. The average S-content of the oceans amounts to 0·0884%. Of the oceanic salts, 7·7% are sulphates. The Dead Sea and the Great Salt Lake contain about 25% and 20% salt with only 0·24% and 6·7% SO_4^{2-}, respectively. Some other lakes (e.g. Ebaity Lake near Omsk) reportedly contain about 25% salt, of which one-third is sulphate.

The H_2S content of lakes, rivers and springs is mainly a function of the temperature and the acidity of the water. It has been observed that the concentration of H_2S increases whereas the concentration of sulphate decreases going from the surface to the bottom of the oceans.

Solid Mineral Sulphates

The predominant cations of the mineral sulphates besides Fe, Cu, Pb and Zn are K, Na, NH_4, Mg, Ca and Al, which usually do not form sulphides. Less frequent are the cations Sr, Ba, Mn, U, Ni, V, Co and Ag. The amount of sulphur in sulphate minerals varies with the number of cations; thus anhydrite, $CaSO_4$, contains about 23% S, whereas chalkophyllite, $Cu_{18}Al_2[(OH)_9SO_4AsO_4]_3.36H_2O$, contains only 2·9% S. Because of the similar tetrahedral arrangement the sulphate is replaced in many cases by PO_4^{3-} and AsO_4^{3-} ions. The change in the negative charges is compensated by simultaneous entry of the higher valent cations in the lattice. In general, sulphate minerals are distributed all over the earth.

Organic Sulphur Compounds

The occurrence of sulphur bonded to carbon is restricted to coal and oil deposits. Two theories attempt an explanation of the origin of the sulphur: it could originate from proteins in the material from which coal was formed or from inorganic sulphur compounds, converted into organic sulphur derivatives. The latter assumption is based on calculations that the protein sulphur content of plants could not have been so high as to explain the amount of sulphur found today. Furthermore, if the sulphur in the oil originated entirely from the material from which the oil was formed, the content of nitrogen compounds should be higher as observed. Common sulphur compounds of the oil are the following: mercaptans, unsymmetrical sulphides, cyclic thiophanes, thiophenes and dimethyl sulphide. The overall sulphur content of coal and oil depends on the various locations. Oil contains up to 4% sulphur, whereas sulphur coal may be richer in sulphur owing to the pyrites present.

1.3. PRODUCTION OF SULPHUR

From Deposits of Elemental Sulphur

Today most of the elemental sulphur from sulphur-containing ores is processed by the Frasch process. It is the only method by which sulphur can be produced without mining the ore. However, the applicability of the Frasch process is restricted to certain geological structures similar to the salt domes of the Gulf of Mexico. These salt domes, measuring 100 m to a few kilometres in diameter, are covered by the "cap-rock", a combination of anhydrite and calcite. The sulphur-bearing calcite is covered by calcite cap-rock and overlying sediments, which have to be impermeable. In the Frasch process the sulphur in the calcite is melted by a combination of superheated water and steam (160°, 16 atm) and forced to the surface by air pressure (20–25 atm). For storage, the liquid sulphur is poured into large basins and allowed to cool. The purity of the sulphur reaches 99·5%. The underground thermal efficiency of the Frasch process is actually only about 5%, due to the tremendous heat losses to the barren formations, to the overlying sediment and to the bleedwater. In addition, the characteristics of the sulphur-bearing formation are very important for the Frasch process[6]. Usually the action radius of one well amounts to 50–80 m and yields about 300 tons per day. The border of profitability lies around 13 m^3 of superheated water per ton of sulphur. Therefore the process not only depends on favourable geological formations but also on cheap energy sources (natural gas) and water supply. The production of one well usually reaches 100,000 tons per year, though this may be greatly exceeded. The plant "Boling Dome" of the Texas Gulf Sulphur Company produced 50 million tons in the period 1933–63.

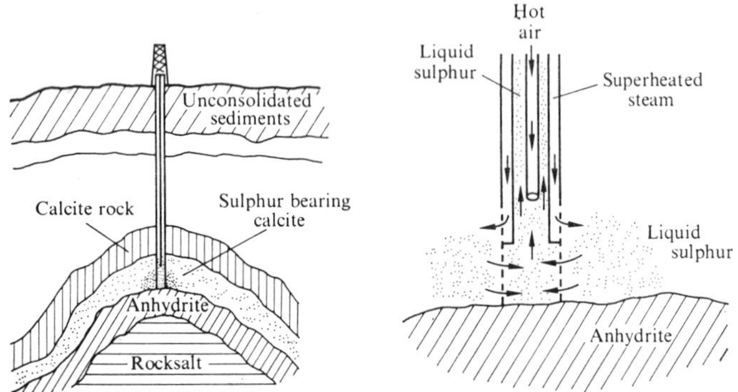

FIG. 1. The Frasch process.

From Hydrogen Sulphide of Natural Gases

The production of sulphur from H_2S plays an important role, because H_2S in industrial and natural gases has to be removed before these gases can be used for industrial purposes. If the H_2S content of the gas is not very high, it is eliminated by the so-called "wet process". Being regenerated from the absorption solution, the more concentrated H_2S is subjected to catalytic oxidation with oxygen (Claus process).

$$2H_2S+O_2 \longrightarrow \tfrac{1}{4}S_8+2H_2O \qquad \Delta H^\circ = -106 \text{ kcal mole}^{-1}$$
$$\Delta G = -74 \text{ kcal } (400^\circ K)$$

[6] R. C. Brasted, *Comprehensive Inorganic Chemistry*, p. 9, D. Van Nostrand Co., Princeton, N.J.

The redox reaction is very temperature-dependent. Below 350° the reaction does not proceed fast enough; therefore porous catalysts (e.g. Al_2O_3, Fe_2O_3) are used. The pores of the catalyst have to possess a certain diameter, since very small pores easily get clogged up with condensed sulphur. Charcoal is a very efficient catalyst at low temperature, but it is easily loaded up with sulphur. From time to time the sulphur must be removed by solvent extraction.

The original Claus process was developed in 1882–3. By now many improvements have been worked out. In the I.G. Claus process H_2S is reacted with a stoichiometric amount of oxygen in combustion chambers (Claus vessel), which produces about 70% sulphur. The remaining SO_2 and H_2S of the first reaction is then passed through two reaction ovens that are loaded with catalysts. This finally results in an overall yield of 92–94%.

$$2H_2S+SO_2 \xrightarrow{300^\circ} \tfrac{3}{8}S_8+2H_2O \qquad \Delta H = -35{\cdot}7 \text{ kcal}$$

Recently[7] a modified Claus process has been reported to yield 97% sulphur (two reaction ovens) and >98% (three reaction ovens). The importance of the Claus process for the production of sulphur from H_2S increases with increasing production of natural gas and oil. In 1967 about one-third of the world production of sulphur was obtained from H_2S sources[7].

1.4. INDUSTRIAL USE OF SULPHUR

Since the time of "alchemy" sulphur has attracted the interest of many investigators because of its unusual physical and chemical properties. When sulphur became readily available in a high purity state by the Frasch and Claus processes, efforts were made to use it not only as a reagent but also to stabilize its polymeric forms. This has been attempted with organic as well as inorganic materials[8]. Among the latter the polyvalent elements of phosphorus, arsenic, antimony and selenium have been studied. The arsenic–sulphur system proved to be reasonably stable and permitted the fabrication of products such as lenses, prisms, tubes and fibres. However, organic compounds, e.g. polysulphides and polymeric sulphides, are more promising modifiers for polymeric sulphur. They tend to stabilize the plasticized form. Mixtures of ethylene polysulphides with sulphur have been studied. In general, the attempts to stabilize sulphur by means of various additives have not been satisfactory. Therefore, the synthesis of polysulphides of organic and inorganic origin containing smaller sulphur units seems more promising. Recently[9], it has been demonstrated that phenylene sulphide polymers may be prepared by the direct reaction of sulphur and benzene.

The use of sulphur in industrial processes is manifold[10]. Most of it is consumed for sulphuric acid production (>80%). This fact has influenced the production patterns from the raw materials, e.g. pyrites, which are directly converted into sulphur dioxide by roasting. Elemental sulphur obtained from the Frasch or Claus process dominates world production, as is shown in Table 1.

The dominant industrial position[11] of elemental sulphur is due, among other factors, to its easy storage and low shipping costs in comparison with other sources. Thus, nearly one million tons of sulphur were shipped from U.S.A. to Europe in 1960. The consumption of

[7] S. Peter and H. Woy, *Chem. Ing. Techn.* **41** (1969) 1.
[8] M. D. Barnes, in B. Meyer (Ed.), *Elemental Sulphur*, p. 356, Interscience, New York (1965).
[9] M. Schmidt, in F. G. A. Stone and W. A. G. Graham (Eds.), *Inorganic Polymers*, p. 98, Academic Press, New York (1962).
[10] R. C. Brasted, *Comprehensive Inorganic Chemistry*, Vol. 8, p. 34, D. Van Nostrand Co., Princeton, N.J.
[11] K. Möbius, *Chemiker Ztg.* **88** (1964) 99.

TABLE 1. FREE WORLD PRODUCTION OF SULPHUR[a]

Source	Production	Percentage
Elemental sulphur		
U.S. (Frasch)	5225	25
Mexico (Frasch)	1600	8
W. Canada	1475	7
France	1475	7
U.S.	1075	5
Rest of world	1075	5
	11,925	57
Pyrites	5800	28
Smelter gas and other	3150	15
Total sulphur in all forms, 1964	20,875	100

[a] Thousand long tons equivalent.

sulphur, apart from sulphuric acid production, stems from its use in the production of vulcanized rubber, enamels and cement. It also goes into the production of explosives, matches, vermilion and ultramarine. Large quantities of sulphur are required for dyes and carbon disulphide production as well as for fine chemicals, insecticides and pharmaceutical products.

The consumption of sulphuric acid in a country reflects the standard of living, since sulphuric acid is an essential for products such as fertilizers and polyamides. Sulphur is also used in the production of petroleum, iron, cellulose, steel, titanium and other pigments. Table 2 reflects the U.S. consumption of sulphur in 1964. The proportions are similar in other countries.

TABLE 2. U.S. CONSUMPTION OF SULPHUR IN ALL FORMS[a]

Sulphuric acid for	
Fertilizer	3·100
Industrial	3·200
Non-acid sulphur	1·150
Total consumption	7·450

[a] Thousand long tons sulphur equivalent.

The use of sulphur in modified forms as a highway marking material ("sulphur paints") is a new area of application. These paints are obtained when sulphur is mixed with the appropriate amount and kind of modifier and heated above the melting point of sulphur.

1.5. PURIFICATION OF SULPHUR IN LABORATORY SCALE

Elemental sulphur usually contains varying amounts of impurities which influence its properties considerably[12] (for 25 commercial forms of sulphur see ref. 13). The choice of the best purification method for a given use of the element depends on the character and the

[12] *Gmelins Handbuch der Anorganischen Chemie*, 8. Auflage, Systemnummer 9, Teil A, S. 512–516, Verlag Chemie, Weinheim (1953).

[13] B. Meyer, in B. Meyer (Ed.), *Elemental Sulfur, Chemistry and Physics*, pp. 74–75, Interscience, New York (1965).

tolerable concentration of the interfering impurity. Some general impurity problems are as follows: first, sulphur contains organic contaminations. The organic compounds react slowly with the element, producing predominantly soluble compounds which cannot easily be separated. Second, liquid sulphur dissolves appreciable amounts of hydrogen sulphide which interferes with many reactions and changes the physical behaviour of the melt drastically. Third, sulphur often contains traces of selenium, tellurium and arsenic, especially if it was produced from minerals. The removal of these is difficult because of the similarity between the chalcogens and since arsenic forms a soluble sulphide. Fourth, sulphur is usually exposed to air and moisture, and reacts with both slowly, especially when exposed to light. Many purification methods have been proposed[12], most of them undesirable. The most common method is recrystallization from carbon disulphide or other solvents. This is very unsatisfactory since the crystals (although very beautiful) contain not only considerable amounts of solvent inclusions, but also H_2S and SO_2.

The distillation of sulphur also cannot be recommended. Sulphur reacts with air at high temperature, and many impurities distil together with the element. In addition, sulphur reacts with glass containers at its boiling point. The acid washing of sulphur (especially of the melt) reduces the organic impurity content, but it is difficult to remove the remaining acid traces and moisture.

A mechanical filtering through glass wool helps to separate solid sediments from liquid sulphur, but during the process sulphur is saturated with atmospheric gases and easily contaminated with dust.

Chromatographic separation of sulphur impurities is undesirable because of the high reactivity of sulphur on and with the chromatographic column, which is probably enhanced by the allotropic conversions of the element. Sulphur–sulphur bonds are sensitive towards all hitherto-known column materials. Useful purification methods are (resulting in reproducible viscosity measurements over the whole temperature range of the melt):

(1) Sulphur is boiled 48 hr with 2% MgO. The liquid is then filtered and the process repeated twice. The organic carbon content of this material is 0·001%. This method is therefore more efficient than double distillation (organic carbon content 0·002%) or sublimation (carbon content 0·003%) in removing organic matter, but has two drawbacks: first, an appreciable amount of magnesium salts and their impurities is introduced, and second, measurable quantities of hydrogen sulphide and sulphur dioxide remain dissolved. Vacuum distillation of sulphur after MgO treatment improves the quality considerably[14].

(2) The following method yields 99·999% pure sulphur[15]: The element is heated with sulphuric acid to 150°C, and nitric acid is added slowly for 6 hr. The liquid is then cooled, washed, melted and refluxed under an inert gas atmosphere. The sample is cooled and transferred into an ampoule. Water is added, the ampoule sealed in an inert gas atmosphere and heated to 125°C. The elution process is repeated three times, and the sample finally stored in an evacuated ampoule. The carbon content of such a sample is around 0·00085%[16].

(3) The following method is simple and efficient[17]: A quartz heater (700°C) is inserted into liquid sulphur. The carbon impurities reach the heater through diffusion and decompose either by forming volatile products or by precipitation on the heater as carbon. If the heater is cleaned daily for one week and then the sulphur is distilled from the container under vacuum, the carbon content of this material can be as low as 0·0009%.

[14] R. F. Bacon and R. Fanelli, *J. Am. Chem. Soc.* **65** (1943) 639.
[15] T. J. Murphy, W. S. Clabaugh and R. Gilchrist, *J. Res. Nat. Bur. Std.* **69A** (1960) 355.
[16] F. Fehér, K. H. Sauer and H. Morien, *Z. anal. Chem.* **192** (1963) 389.
[17] H. von Wartenberg, *Z. anorg. allg. Chem.* **297** (1958) 226.

(4) Zone melting[18] has also been proposed for simple and efficient purification of sulphur, but the ultimate power of this method has not yet been demonstrated, perhaps because of the analytical problems of the determination of very small carbon contents. The best present technique for quantitative determination of carbon involves the burning of the sulphur and dissolving the evolved carbon dioxide in barium hydroxide, followed by potentiometrical determination[16]. The purity of sulphur can be tested in several ways. Clean sulphur is pale yellow and does not smell. Small amounts of hydrogen sulphide and of organic impurities change the viscosity of liquid sulphur and lead to irreproducible results. Oil and as little as 0·5% organic dust cause sulphur to become dark when heated and cooled.

1.6. ALLOTROPE MODIFICATIONS OF SULPHUR

Sulphur atoms are characterized by a pronounced tendency to form sulphur–sulphur–σ bonds. This ability for catenation is only surpassed by carbon. It leads to the formation of many different chain-like sulphur compounds. Their properties and structures reveal a more or less close genetic connection with elementary sulphur itself. Elementary sulphur is usually regarded as the most simple substance that contains sulphur–sulphur bonds. But the word "simple" in this connection can only mean the stoichiometric composition of sulphur, which, of course, is indeed the most simple one can think of. This naturally holds for every chemical element. But it is worth while to realize the very important fact that—with the exception of the noble gases—all the chemical elements under normal conditions are indeed chemical compounds. There is—in spite of a traditional but severely misleading definition—no real difference between chemical compounds composed from different atoms and chemical compounds composed from equal atoms. From this point of view elementary sulphur is a most complicated system. At ordinary temperature only one single compound of sulphur "with itself" is thermodynamically stable: the crown-shaped cyclo-octasulphur, S_8. But just by varying the temperature within comparatively narrow ranges in the system S_x values for x from 1 to about 10^6 can be realized! That means that at least one million compounds of sulphur can be prepared, even more: they really do exist, but most of them only in very complicated temperature-dependent equilibria systems which we are far from really understanding. It seems that future research must bring a drastic increase of our knowledge of the element sulphur, which despite its tremendous theoretical and practical importance is still so poorly characterized.

Sulphur is unequalled among the elements in the confusion which exists in the nomenclature. Many criteria have been used to name the various forms and preparations of sulphur, with the result that a systematic nomenclature simply does not exist. There is not even one allotrope for which one single name is commonly used. Even worse, the same name has, in more than one case, been applied to two different forms. A rather recent review on the problems of nomenclature[19] attempts to resolve this peculiar situation. Its conclusions are strongly recommended and will be applied in this discussion of the allotropy of sulphur.

Crystalline Forms

In solid sulphur we have to distinguish two kinds of allotropy: first, the intramolecular allotropy, which accounts for the different molecular species formed by chemical bonding of

[18] F. Fehér and H. D. Lutz, *Z. anorg. allg. Chem.* **334** (1964) 235.

[19] I. Donohue and B. Meyer, in B. Meyer (Ed.), *Elemental Sulfur, Chemistry and Physics*, pp. 1–11, Interscience, New York (1965).

sulphur atoms, and second, the intermolecular allotropy, which accounts for the different structural arrangements of the molecules in crystals. Compared with other elements sulphur shows a large number of polymorphs. Some thirty-odd modifications are mentioned in the literature[20]. Quite a few of these may not bear a critical re-examination, but on the other hand, at least three new ones have recently been added to the list. Sulphur as a chalcogen element with an outer electronic shell complete but for two electrons (in the ground state the term symbol is $^3P_2$, indicating that the two unpaired electrons are situated in different *p*-orbitals; this explains why sulphur is divalent: sulphur atoms will form only two covalent bonds with adjacent sulphur atoms in the oxidation number zero) should indeed manifest a very large number of *a priori* possible molecular forms. It can in principle form linear molecules of arbitrary extent, which may close to form ring molecules; otherwise they become infinitely long chains. A great many of these hypothetical molecular conformations will be ruled out on stereochemical grounds; others do not fit into a crystalline packing. In principle, however, all ring molecules can be packed in a crystallographic way without penetrating each other. Only those linear chains with straight axes (which are linear in the crystallographic sense of the word) can be arranged in a crystalline packing. Therefore, with the exception of the "non-straight" types of chain, molecular lattices can be formed with any of the molecular types. Since the orientation in space of the covalent bonds depends on the orientation of the next nearest bond in the chain, the permissible molecules will have a staggered shape. In other words, the expected molecular conformations have a low degree of symmetry. A most sophisticated theory of possible modifications of elementary sulphur has been published recently[21]. All the interesting speculations of earlier workers are therein reviewed masterfully. We therefore can restrict ourselves to the relatively few experimentally secured compounds of sulphur with itself—that is, cyclo-octasulphur, S_8, cyclohexasulphur, S_6, cycloheptasulphur, S_7, cyclodecasulphur, S_{10}, and cyclododecasulphur, S_{12}, besides the crystalline forms of polycatenasulphur, S_∞.

Cyclo-octasulphur, S_8

Orthorhombic sulphur, S_α. The most important variety of sulphur is the orthorhombic form, S_α. It consists of S_8 molecules in the form of staggered eight-membered rings with a mean S–S bond length of 2·037 Å, a mean S–S–S angle of 107° 48′, and a mean dihedral angle of 99° 16′ [22]. For other structural data see Table 3. Orthorhombic sulphur is formed by all other modifications of sulphur on standing, because it is the only stable form at room temperature. Commercial roll sulphur and flowers of sulphur produce identical Hall patterns, which confirm the identity of their crystalline structures—both S_α [23].

Milk of sulphur is described in the literature as an amorphous modification of sulphur. The X-ray powder diagram of freshly prepared milk of sulphur, however, revealed its S_α-type of crystalline structure[23]. Figure 2 demonstrates the unit cell and a packing drawing of S_α. An excellent review concerning its structure is given by Donohue[24].

Monoclinic sulphur, S_β. If a sulphur melt crystallizes, monoclinic crystals form. Such crystals can be grown to very big specimens. Below 95·4°C the crystals transform into the

[20] B. Meyer, in ref. 19, p. 71.

[21] F. Tuinstra, *Structural Aspects of the Allotropy of Sulfur and Other Divalent Elements*, Waltman, Delft (1967).

[22] B. E. Warren and J. T. Burwell, *J. Chem. Phys.* **3** (1935) 6.

[23] S. R. Das, Colloquium der Sektion für Anorganische Chemie der Internationalen Union für Reine und Angewandte Chemie, p. 103, Münster/Westf., September 1954, Verlag Chemie, Weinheim (1955).

[24] J. Donohue, in ref. 19, pp. 13–43.

orthorhombic (α) form with a heat transition of 96 cal/mole, but the quenched crystals can be maintained at room temperature up to a month. The structure of this "high" temperature form of sulphur was unknown until very recently[24]. Some doubt has been raised whether

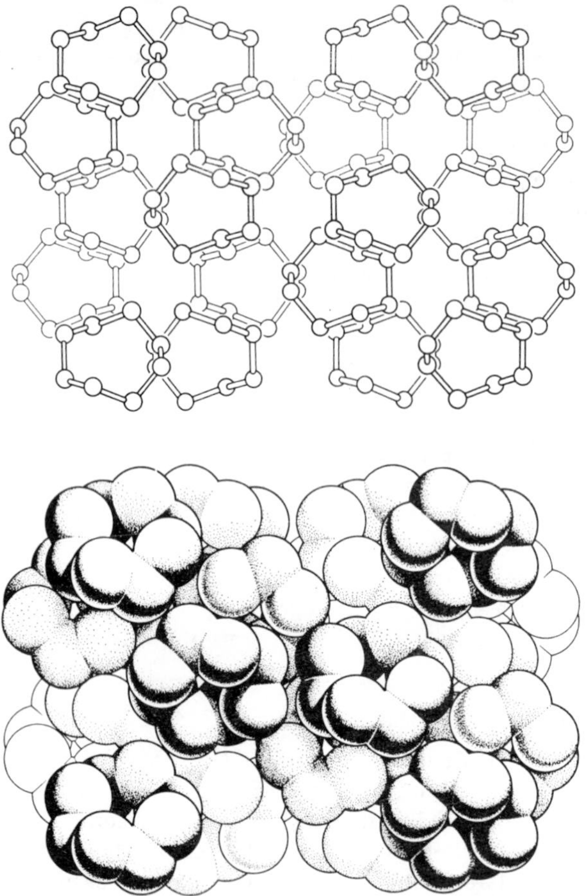

FIG. 2. Unit cell of S_α projected down the *a*-axis, and packing drawing of the same view.

the molecular unit was S_8. However, recent studies indicate that the structure is similar to that of orthorhombic S_α, except for the presence of disordered sites, in which the molecules, at random, can assume a normal or an inverted position[25]. Some mechanical properties of S_β have been studied, and heat capacity, heat conductivity and some electric constants are available. The infrared spectrum is also known[26]. For crystallographic data see Table 3.

The enantiotropic transition $S_\alpha \rightleftharpoons S_\beta$ between the claimed transition point of 96·5°C

[25] D. E. Sands, *J. Am. Chem. Soc.* **87** (1965) 1395.

[26] B. Meyer, in E. Nickless (Ed.), *Inorganic Sulphur Chemistry*, p. 241, Elsevier, Amsterdam, London, New York (1968).

and the melting point has formed the subject of many investigations[27]. They cannot be discussed here in detail. Instead, the very instructive review on phase transition rate measurements by Thackray is recommended to the reader who is interested in phase transitions[28].

Mother of pearl-like sulphur, S_γ. Monoclinic prismatic crystals of S_γ are formed by slow cooling of a sulphur melt heated above 150°C or by chilling of hot sulphur solutions in alcohols, hydrocarbons and carbon disulphide. Its stability is still disputed. For the crystallographic data of the "sheared penny roll" crystals see ref. 24 and Table 3. The melting point of S_γ is given as 106·8°C. It transforms into S_β and/or into S_α. These transformation processes have been carefully studied[28]. Like S_α and S_β, S_γ consists of staggered S_8 rings.

Insufficiently identified forms of cyclo-octasulphur. A great number of other allotropes of cyclo-octasulphur have been described[26]. Their identification is either incomplete or doubtful.

Cyclohexasulphur, S_6

A six-membered sulphur ring can be prepared according to the methods of Aten or Engel[29]. Concentrated hydrochloric acid is mixed with cooled concentrated thiosulphate solutions. The resulting mixture of S_8 and S_6 is then extracted with toluene or benzene. On cooling of the extract, cyclohexasulphur crystallizes in typical orange-coloured rhombohedral crystals. The mechanism of the formation of this thermodynamically unstable ring molecule seems to be a sort of reversal of the "sulphite degradation" of sulphur chains (see page 818)—that is, a stepwise building up of chains, followed by ring closure via an intramolecular S_N2 reaction[30].

A clear-cut kinetically controlled synthetic route for the preparation of cyclohexasulphur (and also other hitherto-unknown thermodynamically unstable sulphur rings) has been worked out recently[31]. It is the reaction of sulphanes with chlorosulphanes under suitable experimental conditions, according to

$$H_2S_x + Cl_2S_y \longrightarrow S_{x+y} + 2HCl \qquad (x+y = 6)$$

for instance,

$$H_2S_2 + Cl_2S_4 \longrightarrow S_6 + 2HCl$$

The thermodynamically unstable cyclohexasulphur as pure crystals or in pure solution can be stored for extended periods of time. However, in the presence of small amounts of impurities, it decomposes very fast. It is sensitive towards visible light. Also chemically it is much more reactive than S_8 (with nucleophiles it reacts about 10^4 times faster![30]). For crystal data see Table 3 and ref. 24. Only preliminary UV and IR spectra of S_6 are available[32]. The molecules of cyclohexasulphur are very efficiently packed (Fig. 3). The crystal

27 S. R. Das, *Sci. and Cult.* **4** (1939) 11.
28 M. Thackray, in ref. 19, pp. 45–69.
29 R. M. Engel, *Compt. Rend.* **112** (1891) 886; A. H. Aten, *Z. physik. Chem.* **88** (1914) 321.
30 P. D. Bartlett, R. E. Davies and E. F. Cose, *J. Am. Chem. Soc.* **83** (1961) 103.
31 M. Schmidt, in ref. 19, p. 327.
32 J. Berkowitz and W. Chupka, *J. Chem. Phys.* **40** (1964) 287; *ibid.* **48** (1968) 1300.

has a unit cell of 18 atoms, and a density of 2·21 g/cm³, the highest density of any known modification, including the only thermodynamically stable orthorhombic form, S_8.

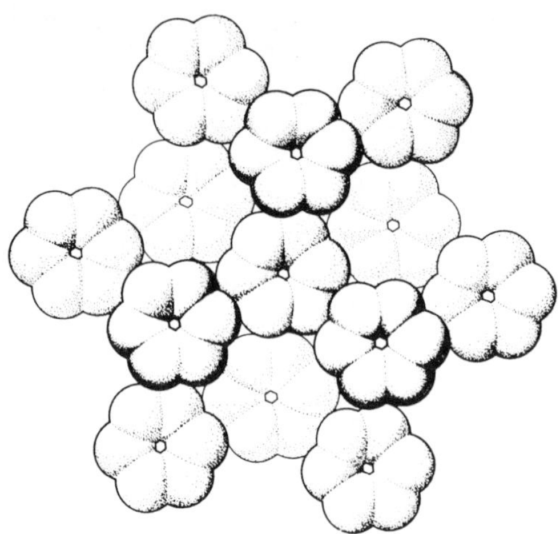

FIG. 3. Packing drawing of cyclohexasulphur, projected down the *c*-axis.

Cycloheptasulphur, S_7

Odd-numbered low-molecular sulphur rings have been regarded as "improper rings"[33] in theoretical considerations of possible modifications of elemental sulphur. So, a seven-membered ring because of geometrical reasons cannot exist with equal bond angles, dihedral angles and bond distances and therefore with equal energy content of all seven sulphur atoms (if a planar configuration is ruled out for obvious reasons). In spite of these theoretical predictions, cycloheptasulphur recently could be synthesized[34] by the kinetically controlled reaction of bis(π-cyclopentadienyl)titanium(IV)pentasulphide with dichlorodisulphane, S_2Cl_2, according to

$$(C_5H_5)_2TiS_5+S_2Cl_2 \longrightarrow (C_5H_5)_2TiCl_2+S_7$$

(The same reaction with SCl_2 in 87% yields very pure cyclohexasulphur[34].) From toluene, S_7 can be crystallized in the form of very long needles. Their intensely yellow colour, in contrast to S_8, does not disappear by cooling to liquid air temperature. S_7 melts reversibly at 39°C, polymerizes at about 45°C, shows a low viscosity again at about 115°C and polymerizes a second time at the familiar temperature of 159°C. Visible light induces a rapid transformation into S_8, via polymeric forms, as does storing at room temperature. In the cold it is stable for weeks. X-rays also decompose S_7 at ordinary temperature, but satisfactory single-crystal pictures could be taken at −80°C. A complete structure determination

---

[33] F. Tuinstra, *J. Chem. Phys.* **46** (1967) 2741.

[34] M. Schmidt, B. Block, H. D. Block, H. Köpf and E. Wilhelm, *Angew. Chem.* **80** (1968) 660.

of S_7 is not yet finished; for the already-known data see Table 3. The mass spectrum of cycloheptasulphur confirms its composition[35].

TABLE 3. STRUCTURAL DATA FOR THE RING MODIFICATIONS OF SULPHUR

Molecule	Modification	Cell dimensions (Å)	Cell content	Space group	Point group	Parameters
S_6 ring[a]	rhombohedral S_ρ	$a = 10{\cdot}82$ $c = 4{\cdot}28$	18 atoms	*R*3	D_{3d}	$d = 2{\cdot}057 \pm 0{\cdot}018$ Å $\beta = 102{\cdot}2 \pm 1{\cdot}6°$ $\gamma = 74{\cdot}5 \pm 1{\cdot}5°$[c]
S_7 ring[b,d]	orthorhombic	$a = 21{\cdot}77$ $b = 20{\cdot}97$ $c = 6{\cdot}09$	112 atoms	*Cmma*		$\alpha = \beta = \gamma = 90°$
S_8 ring[a]	orthorhombic S_α	$a = 10{\cdot}46$ $b = 12{\cdot}87$ $c = 24{\cdot}49$	128 atoms	*Fddd*	D_{4d}	$d = 2{\cdot}060 \pm 0{\cdot}002$ Å $\beta = 108{\cdot}0 \pm 0{\cdot}5°$ $\gamma = 98{\cdot}7 \pm 0{\cdot}5°$[c]
S_8 ring[a]	monoclinic S_β	$a = 11{\cdot}04$ $b = 10{\cdot}98$ $c = 10{\cdot}92$	48 atoms	$P2_1/c$	D_{4d}	$d = 2{\cdot}063$ Å $\beta = 96{\cdot}7°$
S_8 ring[a]	monoclinic S_γ	$a = 13{\cdot}88$ $b = 13{\cdot}12$ $c = 9{\cdot}26$	64 atoms	$P2/n$	D_{4d}	$\beta = 91{\cdot}7°$
S_9 ring[36]						
S_{10} ring[b]	monoclinic	$a = 12{\cdot}7$ $b = \sim 7$ $c = \sim 10$				
S_{12} ring[a]	orthorhombic	$a = 4{\cdot}73$ $b = 9{\cdot}10$ $c = 14{\cdot}57$	24 atoms	*Pnnm*	D_{3d}	$d = 2{\cdot}005$ Å $\beta = 106{\cdot}5°$ $\gamma = 87{\cdot}5°$[c]

[a] F. Tuinstra, *Structural Aspects of the Allotropy of Sulphur and the Other Divalent Elements*, Waltman, Delft (1967).
[b] M. Schmidt, B. Block, H. D. Block, H. Köpf and E. Wilhelm, *Angew. Chem.* **80** (1968) 660.
[c] Dihedral angle.
[d] I. Kawada and E. Hellner, *Angew. Chem.* **82** (1970) 390.

Cyclodecasulphur, S_{10}

The hitherto-unknown decameric sulphur rings S_{10} could recently be prepared by the kinetically controlled reaction of bis(π-cyclopentadienyl)titanium(IV) pentasulphide with sulphurylchloride[34], according to

$$2(C_5H_5)_2TiS_5 + 2SO_2Cl_2 \longrightarrow 2(C_5H_5)_2TiCl_2 + S_{10} + 2SO_2$$

From carbon disulphide it crystallizes in characteristic rhombic plates of intensely yellow colour. Like S_6, S_{10} does not show a sharp melting or decomposition point, but polymerizes at about 60°C. It is very sensitive towards visible light, and can only be stored under cooling (about −40°C) for longer periods of time. The structure of cyclodecasulphur has not yet been determined. Its mass spectrum confirms the composition[35].

Cyclododecasulphur, S_{12}

The above-mentioned method[31] for the synthesis of new sulphur rings, which is the first preparative method for the kinetically controlled "direct" synthesis of compounds between equal atoms—that is, new element modifications—has made possible a preparative synthesis

[35] U. I. Zahorszky, *Angew. Chem.* **80** (1968) 661.

of the new sulphur modification S_{12} [36a]. This allotrope is not only of special interest because its properties do convincingly refute a very popular prognosis[37], but also because of its surprisingly high stability. The light yellow needles show the highest melting point (with decomposition) of all the known sulphur modifications: 148°C. Its solubility in the usual solvents for sulphur is unexpectedly low. The mass spectrum[38] as well as the structure determination[39] confirm the composition. The reactivity of cyclododecasulphur lies between S_8 and S_6, but more on the side of S_8.

Cyclononasulphur, S_9

This has been prepared recently [36].

Polycatenasulphur

Radiation and heat lead to homolytic S–S bond scission which is followed by polymerization or chain formation. If a viscous sulphur melt ($T > 160$°C) is quenched in air, in ice water, in dry ice or in liquid nitrogen, high yields of a plastic material are obtained. The plastic solid can be purified by washing with CS_2 and it can be stretched into fibres. During the stretching two phases form. The phases can clearly be observed[40]. This solid, with or without the inclusions of S_α and before or after stretching, has been widely studied and analyzed. For various portions of it the terms polymeric sulphur (with different Greek letters), fibrous, plastic, elastic and many other names have been employed. A reproducible method for preparing long fibres is: sulphur heated for 5 min to 300°C and then poured into ice water in a thin stream. Long fibres form which are then stretched under water. The fibres obtained in this way are highly elastic and can be stretched 10 to 15 times their length over more than a dozen cycles. The stretched fibres display a typical X-ray pattern, not unlike stretched rubber. The geometry and the packing of this form seem to consist of sulphur helices with about three and one-half atoms per turn. The CS_2 insoluble solid is commerciably available in several forms as "super sublimation sulphur", "Crystex" or under other trade names. Since the insoluble forms slowly transform into S_8 and therefore consist of mixtures with ill-defined, unreproducible and time-dependent properties, more low-temperature studies of the pure metastable solids should be undertaken in the future.

A very competent review on the structure of polycatenasulphur is given by Tuinstra[21].

Molten Sulphur

At 119·25°C, cyclo-octasulphur melts to a yellow liquid even less viscous than water. The physical properties of liquid sulphur are so unusual that they have attracted the interest of many research workers over a long period. Some recent reviews survey this work[26, 41, 42]. At the transition temperature or floor temperature of 159°C, liquid sulphur abruptly changes into a very viscous material that cannot be poured from containers. This extraordinary rise in viscosity by a factor of 2000 within a very narrow temperature

[36] M. Schmidt and E. Wilhelm, *Chem. Commun.* (1970) 1111.
[36a] M. Schmidt and E. Wilhelm, *Angew. Chem.* **78** (1966) 1020; *Chem. Ber.* **101** (1968) 381.
[37] L. Pauling, *Proc. Nat. Acad. Sci. U.S.* **35** (1949) 495.
[38] J. Buchler, *Angew. Chem.* **78** (1966) 1021.
[39] A. Kutoglu and E. Hellner, *Angew. Chem.* **78** (1966) 1021.
[40] A. E. B. Presland and A. Jaskolska, *Nature*, **208** (1965) 1088.
[41] W. J. Macknight and A. V. Tobolsky, in ref. 19, pp. 95–107.
[42] J. A. Poulis and C. H. Massen, in ref. 19, pp. 109–123.

range is accompanied by a change of colour from light yellow to dark red. Almost all other physical properties—for instance, the specific heat, density, electric conductivity, velocity of sound and so on—show a discontinuity at this transition temperature[42]. As the most famous example, Fig. 4 demonstrates the viscosity behaviour of liquid sulphur.

This peculiar behaviour of liquid sulphur is undoubtedly caused by a polymerization process. So, the system of elementary sulphur may be regarded as a model example of a

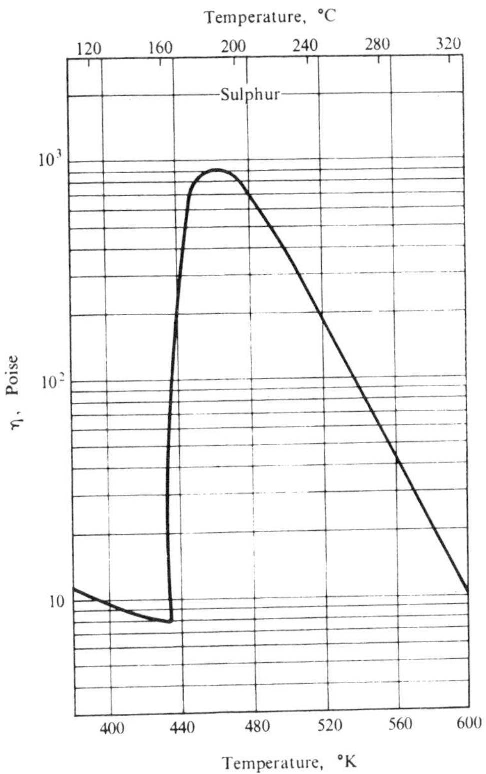

FIG. 4. Viscosity–temperature curve for liquid sulphur.

heat-induced polymerization. In spite of quite a number of qualitative and semi-quantitative explanations of the viscosity behaviour[43], it was not until 1959 that Tobolsky and Eisenberg[44] were able to describe it over the entire liquid range including the most important region of transition; this became possible by introducing two equilibrium constants, one for initiation and the other for propagation of the polymerization.

The relationship between cyclo-octasulphur and the very highly polymeric sulphur chains is thereby regarded as a completely reversible reaction with S_8 as the "monomer". The first step of the reaction is a thermally induced homolytic fission of a sulphur–sulphur bond in the ring, thus forming the biradical catenaoctasulphur S_8. These biradicals by polymerization can form longer radical chains, or can, as of course also can the longer chains, attack cyclo-octasulphur under ring opening, again extending the chain by eight

[43] M. Schmidt, *Inorganic Macromolecular Reviews*, **1** (1970) 101.
[44] A. V. Tobolsky and A. J. Eisenberg, *J. Am. Chem. Soc.* **81** (1959) 780.

sulphur atoms. The only species present in the system according to this hypothesis are S_8 rings as "monomers" and long chains, obviously thought of as being composed of multiple of eight atoms.

$$\text{cyclo-}S_8 \rightleftharpoons \text{catena-}S_8;\ [\text{catena-}S_8] = K[\text{cyclo-}S_8]$$

$$(\text{catena-}S_8)_n + \text{cyclo-}S_8 \rightleftharpoons (\text{catena-}S_8)_{n+1}$$

$$[(\text{catena-}S_8)_n] = K[\text{cyclo-}S_8] + (K_3[\text{cyclo-}S_8])^{n-1}$$

The number-average degree of polymerization P can be expressed as

$$P = \frac{1}{1-K_3[\text{cyclo-}S_8]}$$

and the total concentration of S_8 units in monomers and polymers is

$$[S_8] = [\text{cyclo-}S_8] + K[\text{cyclo-}S_8]/(1-K_3[\text{cyclo-}S_8])^2$$

The thermal constants for these two processes have been given as[41]:

$$\Delta H^\circ = 32{\cdot}8 \text{ kcal/mole};\ \Delta S^\circ = 23 \text{ cal/deg mole}$$

$$\Delta H_3^\circ = 3{\cdot}1 \text{ kcal/mole};\ \Delta S_3^\circ = 4{\cdot}6 \text{ cal/deg mole}$$

This hypothesis describes many properties of liquid sulphur below and above the viscosity maximum very well by assuming that, at the melting point, the liquid consists of free S_8 units, which at 159°C polymerize to an average chain length of over 10^5 S_8 units. At still higher temperatures a steady depolymerization is assumed. Many properties indicate, however, that liquid sulphur is in reality still a much more complicated system. The melting process is a good example of this: it has long been known that sulphur has a "natural" and an "ideal" melting point[45]. A new component forms in the liquid phase which, as a solute, causes a depression of the melting point. The search for this component, "S_π", has, in spite of much work, not yet led to a definite and experimentally proved answer. There are some indications that it might consist of catena-S_8[46]. It can be separated from cyclo-S_8 by dissolving in CS_2 at −78°C, where S_π is only poorly soluble. The assumption of at least up to 4% catena-S_8 in the melt is, however, complicated by the ESR evidence that only irrelevant amounts of free radicals exist below 159°C[47]. However, short chains might stabilize the free electrons and therefore only give a very broad ESR signal. Also measurements of the composition of the sulphur vapour indicate the presence of other low-molecular species than S_8 (see page 811) in the melt. The influence of light (UV radiation leads to S–S– scission) can influence the composition and therefore the properties of liquid sulphur, as can traces of impurities—for instance, H_2S [48]. The first experimental proof that the above simple model of the equilibrium polymerization of liquid sulphur does not give a completely correct picture of the molecular complexity of sulphur melts is the isolation of cyclododecasulphur from such melts[49]. Pure S_{12} could be isolated from rapidly cooled melts (after heating for some minutes to different temperatures between 140 and 400°C). The amount is very small—about 0·1%—but practically independent of the time and temperature of heating. According to this, in liquid sulphur there is besides cyclo-octasulphur—already below the transition temperature—present at least one other species

[45] M. Schmidt, in F. G. A. Stone and W. A. G. Graham (Eds.), *Inorganic Polymers*, p. 103, Academic Press, New York, London (1962).
[46] P. W. Schenk and U. Thümmler, *Z. anorg. allg. Chem.* **315** (1962) 217.
[47] D. M. Gardner and E. K. Fraenkel, *J. Am. Chem. Soc.* **78** (1956) 3279.
[48] K. T. Wiewiorowsky and F. J. Touro, *J. Phys. Chem.* **70** (1966) 234 and 239.
[49] M. Schmidt and H. D. Block, *Angew. Chem.* **79** (1967) 944.

of low molecular weight, namely S_{12}. Its concentration in the equilibrium mixture is not yet known. Most of the S_{12} will not "survive" the cooling and the time required for working up the solid—being thermodynamically unstable with respect to S_8. Also the isolation of less than 1% S_{12} from more than 99% S_8 is experimentally difficult and will never be quantitative. The fact that it is present at all—only after storage of the cooled melt at room temperature for a couple of days it disappeared completely—leads to the conclusion that the molecular composition of sulphur melts is more complex than is usually thought. Many other possible sulphur rings of low molecular weight might well take part in complicated equilibrium reactions but disappear rapidly during cooling and working up. It therefore seems necessary to plan and pursue new and probably quite sophisticated physical and chemical studies to learn more about the molecular complexity of liquid sulphur.

Sulphur Vapour

Sulphur vaporizes already at 100°C to an appreciable amount, and its vapour pressure reaches 1 atm at 444·6°C. This makes the region between 100°C and 1000°C the most interesting for experimental research. The vapour phase has a very complex composition which was only recently realized. In spite of much work over a long period of time, it was assumed that only even-numbered molecules (first only S_2 and S_8, later also S_6 and then S_4) were constituents of the vapour. This completely erroneous assumption is still to be found in many modern textbooks on inorganic chemistry. The results of recent mass spectroscopy demonstrated clearly that sulphur vapour also contains odd-numbered molecules. For an excellent review on this development of our knowledge see ref. 50. The analysis of vapours with the help of mass spectroscopy is somewhat complicated by the processes in the ion source, where part of the vapour is not only ionized but also fragmented. At temperatures above 2500°K and at pressures below 10^{-5} mmHg, sulphur atoms dominate. Below this the very wide pressure and temperature region down to 400°C and a pressure of several mmHg belongs to the S_2 molecule. At lower temperatures, however, the vapour equilibration is very slow, and the vaporization from solids leads only to metastable, unequilibrated systems in which the composition strongly depends on the solid; equilibrium in such metastable vapours can be obtained by catalysts such as aluminium[51]. The usual sources for vapours of various compositions are Knudsen cells, containing S_8, S_6 or S_∞, to give S_8, S_6 or S_7 vapour in high yield, and double furnaces for unsaturated vapour with higher S_6, S_3 and especially S_2 concentrations. Unsaturated vapours can also be produced by the thermal decomposition of sulphides such as CdS, FeS, HgS, etc. An exceedingly clean and elegant electrolytic vapour source has been developed by Rickert. From it precisely measured amounts of sulphur can be released[52]. The combination of this vapour source with mass spectrometer technique revealed the best present knowledge of the composition of sulphur vapour between 200°C and 400°C[53]. Table 4 shows the partial pressures of saturated sulphur vapour at different temperatures, Table 5 the thermodynamic properties of the gaseous sulphur molecules derived from this research. According to this, the enthalpies of formation of a sulphur–sulphur bond increases from about 51 kcal/mole in S_2 to about 62 kcal/mole in S_6. Molecules with more than six atoms all have about the

[50] J. Berkowitz, in ref. 19, pp. 125–159.
[51] J. Berkowitz and W. Chupka, *J. Chem. Phys.* **48** (1968) 1300.
[52] H. Rickert and N. Birks, *Ber. Bunsenges. Physik.-Chem.* **67** (1963) 97.
[53] D. Detry, J. Drowart, P. Goldinger, H. Keller and H. Rickert, *Z. physikal. Chem.* **55** (1967) 314, and *Adv. in Mass Spectrometry*, **4** (1968) 499.

TABLE 4. PARTIAL PRESSURES IN SATURATED SULPHUR VAPOUR (ATM)[a]

Molecule	Temperature, °K				
	473	523	573	623	673
S_2	$1{\cdot}40\times10^{-6}$	$2{\cdot}60\times10^{-5}$	$2{\cdot}68\times10^{-4}$	$1{\cdot}90\times10^{-3}$	$9{\cdot}40\times10^{-3}$
S_3	$1{\cdot}70\times10^{-7}$	$3{\cdot}38\times10^{-6}$	$3{\cdot}66\times10^{-5}$	$2{\cdot}68\times10^{-4}$	$1{\cdot}34\times10^{-3}$
S_4	$1{\cdot}65\times10^{-7}$	$3{\cdot}04\times10^{-6}$	$3{\cdot}25\times10^{-5}$	$2{\cdot}15\times10^{-4}$	$1{\cdot}04\times10^{-3}$
S_5	$1{\cdot}56\times10^{-5}$	$1{\cdot}72\times10^{-4}$	$9{\cdot}64\times10^{-4}$	$4{\cdot}20\times10^{-3}$	$1{\cdot}43\times10^{-2}$
S_6	$5{\cdot}50\times10^{-4}$	$3{\cdot}60\times10^{-3}$	$1{\cdot}60\times10^{-2}$	$5{\cdot}25\times10^{-2}$	$1{\cdot}37\times10^{-1}$
S_7	$3{\cdot}28\times10^{-4}$	$2{\cdot}63\times10^{-3}$	$1{\cdot}27\times10^{-2}$	$4{\cdot}55\times10^{-2}$	$1{\cdot}26\times10^{-1}$
S_8	$1{\cdot}89\times10^{-3}$	$1{\cdot}02\times10^{-2}$	$3{\cdot}64\times10^{-2}$	$9{\cdot}70\times10^{-2}$	$2{\cdot}14\times10^{-1}$

[a] I. Drowart, P. Goldfinger, D. Detry, H. Rickert and H. Keller, *Advances in Mass Spectrometry*, **4** (1967) 506.

TABLE 5. THERMODYNAMIC PROPERTIES OF THE GASEOUS SULPHUR MOLECULES[a]

A. Measured reaction, enthalpies, and entropies

Molecule	Equilibrium	Temperature, °K	$\Delta H_T°$, kcal/mole	$\Delta S_T°$, cal/deg mole
S_2	$2S(rh.) \longrightarrow S_2(g)$	460–670	$28{\cdot}08\pm0{\cdot}35$	$32{\cdot}6$
S_3	$2S_3(g) \longrightarrow 3S_2(g)$	566–669	$26{\cdot}6\ \pm2{\cdot}0$	$(37{\cdot}6\pm3{\cdot}0)$
S_4	$S_4(g) \longrightarrow 2S_2(g)$	615	$28{\cdot}2\ \pm2{\cdot}0$	$(36{\cdot}7\pm2{\cdot}5)$
S_5	$2S_5(g) \longrightarrow 5S_2(g)$	565–620	$95{\cdot}5\ \pm4{\cdot}0$	$(111{\cdot}4\pm8{\cdot}0)$
S_6	$\frac{3}{4}S_8(g) \longrightarrow S_6(g)$	435–625	$6{\cdot}26\pm0{\cdot}33$	$7{\cdot}6\pm0{\cdot}3$
S_7	$\frac{7}{8}S_8(g) \longrightarrow S_7(g)$	435–625	$5{\cdot}77\pm0{\cdot}31$	$7{\cdot}2\pm4{\cdot}2$
S_8	$S_8(g) \longrightarrow 4S_2(g)$	460–625	$96{\cdot}8\ \pm2{\cdot}2$	$110{\cdot}2\pm4{\cdot}2$

B. Standard entropies and heats of sublimation

Molecule	$S_T°$ cal/deg mole	$\Delta H_0°$ sub., kcal/mole	
		Ref a	Refs. 32, 43
S_2	$59{\cdot}6$ (565°K)[42]	$30{\cdot}86\pm0{\cdot}35$	$30{\cdot}81$
S_3	$(71{\cdot}8\pm1{\cdot}5)$ (618°K)	$33{\cdot}3\ \pm1{\cdot}3$	$33{\cdot}1$
S_4	$(83{\cdot}9\pm2{\cdot}5)$ (615°K)	$33{\cdot}5\ \pm2{\cdot}2$	$31{\cdot}1$
S_5	$(94{\cdot}3\pm4{\cdot}0)$ (592°K)	$29{\cdot}1\ \pm2{\cdot}5$	$28{\cdot}4$
S_6	$101{\cdot}6\pm3{\cdot}3$ (512°K)	$24{\cdot}8\ \pm2{\cdot}7$	$25{\cdot}1$
S_7	$116{\cdot}9\pm3{\cdot}8$ (512°K)	$28{\cdot}1\ \pm2{\cdot}7$	$27{\cdot}8$
S_8	$126{\cdot}8\pm4{\cdot}2$ (530°K)	$26{\cdot}0\ \pm2{\cdot}7$	$25{\cdot}3$

[a] I. Drowart, P. Goldfinger, D. Detry, H. Rickert and H. Keller, *Advances in Mass Spectrometry*, **4** (1967) 506.

same bond energy. Why, then, do we find such drastic differences in stability towards light, heat and nucleophiles with the hitherto-isolated and characterized ring molecules S_6, S_7, S_8, S_{10} and S_{12} in the solid state as well as in solution at room temperature? Question upon question is still open in sulphur chemistry, especially in the chemistry of the element itself.

Chilled Sulphur Vapour

If sulphur vapour is trapped below −200°C, a coloured solid is formed which gives a complex EPR spectrum[54]. The composition, colour and behaviour of these metastable

[54] A. Chatelain and J. Buttet, in ref. 19, pp. 209–239.

deposits depend on the vapour temperature, vapour pressure, condensation speed and condensation temperature. Only under very special conditions is it possible to isolate S_2, which is blue, as also is its vapour[55].

1.7. NUCLEAR PROPERTIES OF SULPHUR[56]

The second element of the VIth main group of the Periodic Table with the atomic number 16 has ten known isotopes. Four are stable, the remaining six are radioactive. As found in nature, sulphur has a mean atomic mass of 32·064±0·003, with carbon as reference at 12·00000. The distribution between the four stable isotopes is (because of different cycles of sulphur in nature there are some fractionation effects and therefore different isotope abundance ratios) approximately as follows:

95·1% ^{32}S, 0·74% ^{33}S, 4·2% ^{34}S and 0·016% ^{36}S

Some data on the radioactive isotopes are presented in Table 6. The best known of the radioactive isotopes is ^{35}S, which is made on a commercial scale usually through the process $^{35}Cl(n, p)^{35}S$. Thus it can be obtained in an isotopically pure form. It has very

TABLE 6. RADIOACTIVE ISOTOPES OF SULPHUR

Isotope	Mass	Half-life	Decay modes (MeV)
^{29}S	—	—	3·60, 3·86, 5·35 and 5·59
^{30}S	29·9847	1·35±0·1 sec	4·30, 4·98
^{31}S	30·97960	2·6 sec	3·87
^{35}S	34·96903	87 days	0·167
^{37}S	36·9710	5·1 min	1·6, 4·8
^{38}S	37·9712	2·87 hr	1·1, 1·88

good characteristics for tracer studies and is used extensively for this purpose. Its β-radiation is low and it must therefore be counted using scintillation techniques. For different reasons not to be discussed here[56] NMR and NQR spectroscopy of sulphur isotopes with the present available technique is not yet promising and practicable.

1.8. PHYSICAL PROPERTIES OF SULPHUR

Many important physical properties of sulphur have already been discussed in sections 1.5 to 1.7. The treatment of other properties seems to be too specialized in the framework of this review on sulphur and its compounds. References to recent reviews will guide the interested reader to the original literature: Phase transition rate measurements[57], properties of polymeric sulphur[58], physical properties of liquid sulphur[59], mechanical properties of sulphur[60, 61], high pressure behaviour of sulphur[62], the electrical and photoconductive

[55] L. Brewer, G. D. Brabsen and B. Meyer, *J. Chem. Phys.* **42** (1965) 1385.

[56] G. Nickless, in G. Nickless (Ed.), *Inorganic Sulphur Chemistry*, p. 241, Elsevier, Amsterdam, London, New York (1968).

[57] M. Thackray, in B. Meyer (Ed.), *Elemental Sulphur, Chemistry and Physics*, p. 45, Interscience, New York (1965).

[58] W. J. Macknight and A. V. Tobolsky, in ref. 57, pp. 95–107.

[59] J. A. Poulis and C. H. Massen, in ref. 57, pp. 109–123.

[60] J. M. Dale and A. C. Ludwig, in ref. 57, pp. 161–177.

[61] J. M. Dale and A. C. Ludwig, *The Sulphur Institute Journal*, **5**, No. 2 (1969) 2.

[62] R. E. Harries and G. Jura, in ref. 57, pp. 179–183.

properties[63] of S_8, electron paramagnetic resonance studies of unstable sulphur forms[64], vibrational spectra of elemental sulphur[65], electronic spectrum and electronic states of S_2 [66], and liquid solutions of sulphur[67].

1.9. CHEMISTRY AND CHEMICAL PROPERTIES OF SULPHUR

Sulphur is to be regarded as the "typical" element of the VIth main group of the Periodic Table[68]. These elements are but two electrons short of inert gas configuration. This fact together with the knowledge of some other characteristics (covalent radius in S_α = 1·04 Å, first ionization potential = 10·357 eV, electron affinity for two electrons = −3·44 eV, electronegativity in the Pauling scale = 2·5, $E°_{298}$ for $H_2S \rightleftharpoons S+2H^++2e^-$ = −0·14 V) allows a fairly good prediction of the general chemical behaviour of sulphur. The element will try to achieve the argon atom structure and thus the oxidation number −2 by either the gain of or the sharing with two electrons from other atoms. Only with very electropositive metals will it be able to form predominantly ionic bonds and thus dinegative anions S^{--}. Those sulphur compounds with negatively charged sulphur atoms, by nature, never can act as oxidants in chemical reactions independently of their ionic or covalent bond character; they are more or less strong reducing compounds.

Regarding the very high ionization potential already for the first electron, positively sulphur ions are not to be expected under "usual" conditions of chemical reactions. Positive oxidation states are very common in sulphur chemistry, the most important of which are +4 and +6. As is to be expected, the bonding in these positive oxidation states is primarily covalent. The compounds are either volatile or anionic. The most important compounds of this type contain oxygen as partner of sulphur, but halogen compounds are also very important. In lower oxidation states than +6, these compounds show both oxidizing and reducing properties. In the +6 state they are, by nature, only oxidizing agents.

Sulphur is a very reactive element. It can react under suitable conditions directly with most other elements except the noble gases, iodine and molecular nitrogen. Compounds of sulphur with iodine and especially with nitrogen are also well known.

Sulphur reacts not only with most of the elements but also with innumerable inorganic and organic compounds. The products of those reactions are described in the following sections of this review, as far as they are of general interest. For a formal systematic survey of reactions of sulphur with elements and compounds the reader is referred to *Gmelins Handbuch*[69]. Such a systematic survey must be read with much care, because it is suited to support an old prejudice in chemistry that "sulphur" (or any other element, except the noble gases) is something definite. It is not! As already mentioned in Section 1.6, "sulphur" is a most complicated system of chemical compounds between equal atoms. By nature, the innumerably possible allotropes of this element must, because of the different energy content of their atoms, show different chemical behaviour. This statement holds for

[63] W. E. Spear and A. R. Adams, in ref. 57, pp. 185–207.
[64] A. Chatelain and J. Buttet, in ref. 57, pp. 209–239.
[65] H. L. Strauss and J. A. Greenhouse, in ref. 57, pp. 241–249.
[66] R. F. Barrow and R. P. du Parcq, in ref. 57, pp. 251–263.
[67] R. L. Scott, in ref. 57, pp. 337–355.
[68] M. Schmidt, *Anorganische Chemie*, I, p. 80, Bibliographisches Institut, Mannheim (1967).
[69] *Gmelins Handbuch der Anorganischen Chemie*, 8. Auflage, Systemnummer 9, Verlag Chemie, Weinheim (1953).

every chemical element. If we, for instance, say that sulphur does not react directly with nitrogen, this is right and wrong at the same time: it is right as long as we are aware of the fact that those statements usually are made and are right only under the condition that the "elements" in question are in their thermodynamic stable "state". It is wrong if we regard as an "element" the sum of single atoms, characteristic for this element. Nitrogen atoms do react with sulphur very well (do sulphur atoms also react with nitrogen, N_2?). "Nitrogen does not react with sulphur." Statements such as this will only be overcome if we increasingly realize the often very complicated "compound character" of "elements" under normal conditions.

In the case of sulphur, unfortunately, only very few studies have been undertaken to distinguish the chemical properties of different allotropes. Therefore, the reader should be aware of the fact that the designation "sulphur" in connection with chemical reactions always means its stable form prevailing under the conditions in which the chemical reactions take place. This very often will be the case with cyclo-octasulphur and with the complicated mixtures of sulphur allotropes in the melt or in the vapour phase. Much work will have to be done in the future to clear up this very unsatisfactory situation. Because of interesting studies within the last ten years a first differentiation is possible: between reactions of atomic sulphur and reactions of compounds of sulphur with itself (mostly cyclo-octasulphur).

Reactions of Sulphur Atoms

Divalent sulphur atoms possess chemical reactivity closely resembling that of a large variety of methylenes and other polyatomic biradicals, but also that of oxygen atoms. Reactions of the latter one (created by electrical discharges) have been studied very intensively already since the early 1930's, but it was not until photochemical sources of oxygen atoms became available that reliable quantitative information was obtained on the nature and rate of oxygen atom reactions with other, mostly organic, molecules.

Sulphur atoms can, of course, be created thermally. But they are of no particular value since at the necessarily high temperatures the sulphur-containing reaction products would be far too unstable. Also recoil sulphur atoms from nuclear reactions (1.7) are not useful for mechanistic and preparative studies of their reactions, which can best be carried out with photochemically produced sulphur atoms. Only recently have convenient photochemical methods been developed for generating sulphur atoms in specific electronic states and studying the reactivities of these species with various types of bond system. For a recent review see ref. 70.

The outer valence electrons of sulphur atoms (as also oxygen or selenium atoms) tend to form electronic states in which the two spin vectors are either parallel (triplet state) or antiparallel (singlet state). In elementary processes the total spin of the system is conserved. Therefore, the chemical reactions associated with the singlet and triplet states of sulphur atoms can be quite different, since the requirement of spin-conservation may dictate different routes for the reaction. When excited to their lowest metastable singlet level, sulphur atoms undergo insertion reactions with hydrogen-containing saturated compounds in a concerted type addition process. In their ground (triplet) state they exhibit an entirely different behaviour in that they either attack C–H bonds abstractively or do not react at all. Their reactions with unsaturated compounds (olefins or acetylenes) are characterized by the initial

[70] O. P. Strausz and H. E. Gunning, in A. V. Tobolsky (Ed.), *The Chemistry of Sulfides*, pp. 23–43, Interscience, New York (1968).

formation of cyclic adducts which in some cases are readily stabilized by collision and in others are inherently unstable.

From the electronic configuration of the sulphur atom the spectroscopic states $3(^3P_2)$, $3(^3P_1)$, $3(^3P_0)$, $3(^1D_2)$ and $3(^1S_0)$ result. Transitions among these states are optically forbidden. The excited $^1D_2$ and $^1S_0$ states are metastable, with long radiative lifetimes, and consequently may undergo bimolecular reactions. The ground state is the triplet (^3P) and the lowest-lying excited level is the $^1D_2$ which is located 26·4 kcal above the ground state. Only these two states have significance in current studies.

Table 7 presents the low-lying electronic states of the sulphur atom[71]. The most valuable sulphur atom source is at present the ultraviolet photolysis of carbonyl sulphide,

TABLE 7. LOW-LYING ELECTRONIC STATES OF THE SULPHUR ATOM

Term	Energy cm^{-1}	kcal/mole
$^3P_2$	0·0	0·0
$^3P_1$	396·8	1·136
$^3P_0$	573·6	1·639
$^1D_2$	9239·0	26·40
$^1S_0$	22,181·4	63·39

COS[72]. It gives rise to carbon monoxide and elemental sulphur. The primary photolytic step involves scission into CO and a $^1D_2$ excited sulphur atom to the extent of at least 74%:

$$COS + h\nu \longrightarrow CO + S(^1D_2)$$

The remaining fraction may give a ground state atom:

$$COS + h\nu \longrightarrow CO + S(^3P)$$

The quantum yield of about 2 dictates that sulphur atoms produced in these reactions abstractively attack carbonyl sulphide according to

$$S(^1D) + COS \longrightarrow S_2 + CO$$
$$S(^3P) + COS \longrightarrow S_2 + CO$$

in exothermic processes (55 and 29 kcal for ground state products). Through an electronic deactivation by carbon dioxide it is possible to produce triplet atoms conveniently if the photolysis is carried out in the presence of a large excess of CO_2. CO_2 is chemically completely inert with respect to sulphur atoms.

An alternative method for generating triplet sulphur atoms is the $Hg(^3P_1)$ photosensitization of COS itself:

$$Hg + h\nu(2537\ \text{Å}) \longrightarrow Hg(^3P_1)$$
$$Hg(^3P_1) + COS \longrightarrow CO + Hg + S(^3P)$$

The quantum efficiency of this process is the same as that of the direct photolysis. In both cases the carbon monoxide produced is an excellent monitor of the production of atomic sulphur.

The photolysis of carbonyl sulphide has also been studied under flash conditions. None of the intermediates of atomic sulphur polymerization that are formed shows any detectable reactivity with respect to organic molecules.

[71] H. E. Gunning, in B. Meyer (Ed.), *Elemental Sulphur, Chemistry and Physics*, pp. 265–300, Interscience, New York (1965).

[72] K. S. Sidhu, I. E. Csizimadia, O. P. Strausz and H. E. Gunning, *J. Am. Chem. Soc.* **88** (1966) 2412.

Among other sources of sulphur atoms are the photolysis of carbon disulphide at wavelengths shorter than 2100 Å, of ethylene episulphide (2200–2600 Å) and of SPF_3 (2100–2300 Å) The first two reactions afford triplet atoms, the latter singlet sulphur atoms.

When the photolysis of COS or SPF_3 is carried out in the presence of paraffins, the corresponding mercaptan (the isomeric distribution of products is statistical, the *H/D* kinetic isotope effect is unity!) is the one and only addition product formed. The mechanism is a concerted single-step insertion process with the sulphur atom reactive only in its excited singlet state. The ground state is inert to paraffins. Analogous insertion processes of excited singlet sulphur atoms have also been observed with methylsilane, trimethylsilane, tetramethylsilane and diborane[70].

With olefins, sulphur atoms lead to three types of isomeric product: episulphides, alkenyl mercaptans and vinylic mercaptans. According to the experimental conditions and the olefin, the insertion of singlet atoms or the stereospecific addition of triplet sulphur atoms can play an important role. A detailed description of the many recent studies of these reactions is outside the scope of this book. This holds also for the rather complex reaction of photochemically produced sulphur atoms with acetylenes[70].

Unfortunately, practically no systematic studies have been undertaken on the reactions of sulphur atoms with inorganic compounds. This would seem to be a very promising field for further research.

Reactivity of Sulphur towards Hydrocarbons

The reaction of organic compounds with elemental sulphur, which is the complicated mixture of compounds of sulphur "with itself" that prevails under the reaction conditions at the relatively high temperatures, cannot be dealt with in this book on inorganic chemistry. Despite their tremendous practical interest, very little is known about them. Much more work has to be done before this intriguing field of chemistry can be described in a short and rational manner. Only some hints to the recent literature can be given here.

The dehydrogenation of alkanes is illustrated by the classical laboratory preparation of hydrogen sulphide involving pyrolysis of a mixture of sulphur and paraffin wax[73]. In such reactions alkanes are dehydrogenated to olefins with attendant formation of hydrogen sulphide. The olefins produced may then react with the excess sulphur, forming thiophens, polysulphides and various other compounds. At still higher temperatures, complete sulphuration occurs, giving carbon disulphide and hydrogen sulphide. For the mechanistic aspects of those reactions see ref. 74. Also the reaction of sulphur with alkenes cannot be described here in detail—it is widely regarded as part of organic chemistry. Reference should be made to the review of Bateman and Moore on this field of olefin chemistry[75].

Nucleophilic Degradation of Sulphur–Sulphur Bonds in Elemental Sulphur and Catenated Sulphur Compounds

Reactions of elemental sulphur under the usual experimental conditions with different reaction partners very often are degradation reactions of sulphur–sulphur bonds, because "elemental sulphur" is a compound composed from sulphur atoms, naturally linked together by S–S bonds, mostly cyclo-octasulphur, S_8. The sulphur atoms in such compounds—that

[73] J. Vorga and P. Benedels, *Magy. Kem. Folvoirat.* **56** (1959) 36.

[74] R. E. Davies, in G. Nickless (Ed.), *Inorganic Sulphur Chemistry*, pp. 85–133, Elsevier, Amsterdam, London, New York (1968).

[75] L. Bateman and C. G. Moore, in N. Kharash (Ed.), *Organic Sulfur Compounds*, Vol. 1, pp. 210–228, Pergamon Press (1961).

is, in elemental sulphur and in many chain-like sulphur compounds—mostly react as electrophilic centres; they are susceptible towards nucleophilic attack via a S_N2 mechanism.

Many reactions of sulphur formerly thought to be quite diverse may be rationally explained and understood from the unifying viewpoint that they are all Lewis acid–Lewis base interactions. This valuable working hypothesis makes the chemistry of elemental sulphur and the many compounds with sulphur–sulphur bonds in the molecule easy to understand and, what is more important, easy to predict[76]. This hypothesis of a stepwise degradation of sulphur–sulphur bonds may be demonstrated in the example of the formation of thiosulphate from sulphite ions and cyclo-octasulphur.

It has long been known that elemental sulphur dissolves in boiling aqueous sodium sulphite solutions with the formation of sodium thiosulphate. Indeed, this reaction has been used for the preparation of $Na_2S_2O_3$. For a long time the literature did not offer an indication as to the actual course of this reaction. In most textbooks on inorganic chemistry it still is formulated as

$$Na_2SO_3+S \longrightarrow Na_2S_2O_3$$

That this formulation of the familiar redox reaction does not describe the true reaction course (the ideal aim of every mechanistic study), but merely represents the starting materials and the end product, follows from the fact that sulphur, under the conditions of the thiosulphate synthesis, is present not as atomic sulphur but in the S_8 ring form. A more realistic formulation of the thiosulphate formation therefore has to be written as

$$S_8+8Na_2SO_3 \longrightarrow 8Na_2S_2O_3$$

If the thiosulphate formation actually did proceed according to this equation, then it would be a 9th order reaction, which, of course, is impossible. It is clear, therefore, that this "simple" reaction must proceed through intermediate stages. The first step is a nucleophilic attack of the base SO_3H^- (or SO_3^{--}, depending on the pH value of the solution) on the "acid" cyclo-octasulphur (all eight sulphur atoms in this ring are, by nature, electronically equivalent; there is no electrophilic centre for this first reaction step). Thereby a sulphur–sulphur bond of the ring is broken and a new S–S bond between that attacking SO_3H^- anion and the attacked sulphur atom is formed. This bimolecular reaction between the sulphur ring and the reductant or nucleophile is rate-determining for the overall reaction. It results in the formation of an open sulphur chain anion, terminated at one end by a SO_3H group, that is the anion of octasulphane monosulphonic acid, $H_2S_9O_3$:

$$S_8\ (\text{ring}) + SO_3H^- \longrightarrow {}^-S{-}S{-}S{-}S{-}S{-}S{-}S{-}S{-}SO_3H$$

This acid then is degraded in a series of seven very fast steps by sulphite:

$$S_8SO_3H^-+SO_3^{--} \longrightarrow S_7SO_3H^-+S_2O_3^{--}$$
$$S_7SO_3H^-+SO_3^{--} \longrightarrow S_6SO_3H^-+S_2O_3^{--}$$
$$S_6SO_3H^-+SO_3^{--} \longrightarrow S_5SO_3H^-+S_2O_3^{--}$$
$$S_5SO_3H^-+SO_3^{--} \longrightarrow S_4SO_3H^-+S_2O_3^{--}$$
$$S_4SO_3H^-+SO_3^{--} \longrightarrow S_3SO_3H^-+S_2O_3^{--}$$
$$S_3SO_3H^-+SO_3^{--} \longrightarrow S_2SO_3H^-+S_2O_3^{--}$$
$$S_2SO_3H^-+SO_3^{--} \longrightarrow SSO_3H^- +S_2O_3^{--}$$

[76] M. Schmidt, in ref. 71, pp. 301–326.

Thus, application of knowledge drawn from the experimental study of the formerly unknown sulphane monosulphonic acids to the reaction of elemental sulphur with sulphite led to the first experimental proof for part of an ingenious concept put forward by Foss already in 1950 on theoretical grounds[77] (which, in principle, goes back to a very old hypothesis on the cleavage of sulphur–sulphur bonds, put forward already in 1876[78]). Foss was right in his prediction of the first step of this reaction (a similar first step has also been found on the basis of kinetic measurements for the reactions of sulphur with tertiary phosphines[79] and cyanide[80]), but not in his conclusion that "the sulphur chains bearing sulphite groups at one end must, of course, be pictured as unstable intermediates only". In contrast to this, the then-unknown sulphane monosulphonic acids are stable and can be studied.

According to this scheme, eight molecules of thiosulphate must arise from one S_8 ring. In contrast to the former belief, this reaction does not proceed by simple addition of sulphur atoms to sulphite ions, the driving force for which is the desire of the central atom to achieve the coordination number four; rather it is a cleavage process of the S_8 ring by sulphite ions. The reaction proceeds quantitatively if sulphur and excess sodium sulphite are boiled for some time in weakly alkaline solution. The thiosulphate may be determined by direct iodometric titration, after the excess sulphite is complexed with formalin. In the cold, however, practically no reaction occurs between sulphur and sulphite. This is not primarily a question of reaction rate, as was formerly thought, but can be explained by the fact that sulphur is far too hydrophobic to be able to react with sulphite in aqueous solution. The reason for the extremely slow reaction at room temperature is that sulphur is only poorly wetted. If, however, the sulphur is first dissolved in an organic solvent such as chloroform or carbon tetrachloride and then an aqueous solution of excess sodium sulphite, along with a sufficient amount of a second organic solvent such as acetone or methanol to homogenize the aqueous and organic phases, is added, then all the sulphur reacts with the sulphite within a few seconds quantitatively to form thiosulphate. This method for the reaction of sulphur with sulphite to form thiosulphite, which may then be iodometrically determined, offers a most convenient determination of elementary sulphur or of sulphur solutions in organic solvents, as are often found in the rubber industry[81].

The same mechanism seems to hold for many reactions of elemental sulphur with a variety of nucleophiles—that is, reductants—that are oxidized by sulphur under suitable conditions. Examples for those reactions are the reactions of sulphur with cyanide ions, leading to thiocyanate[81], with sulphide ions, leading to polysulphides[82], with arsenite ions, leading to thioarsenates[83], with lithiumorganyls and Grignard reagents, leading to mercaptides[84] with sodium phenylacetylide[85], sodium bis-trimethylsilylamide[86], a series of organic compounds of germanium, tin and lead[87], etc., always leading to the expected products in stepwise degradation reactions.

One more reaction of this type will be described in this connection for two reasons:

77 O. Foss, *Acta Chem. Scand.* **4** (1950) 404.
78 R. Schiller and R. Otto, *Chem. Ber.* **9** (1876) 1637.
79 P. D. Bartlett and G. J. Merguerian, *J. Am. Chem. Soc.* **78** (1956) 3710.
80 P. D. Bartlett and R. E. Davies, *J. Am. Chem. Soc.* **80** (1958) 2513.
81 M. Schmidt and G. Talsky, *Z. Anal. Chem.* **166** (1959) 274.
82 M. Schmidt and G. Talsky, *Chem. Ber.* **92** (1959) 1526.
83 M. Schmidt and R. Wägerle, *Z. anorg. allg. Chem.* **330** (1964) 48.
84 R. Wägerle, Thesis, Univ. Munich (1960).
85 M. Schmidt and V. Potschka, *Naturwiss.* **50** (1963) 302.
86 M. Schmidt and O. Scherer, *Z. Naturforsch.* **18b** (1963) 317.
87 H. Schumann and M. Schmidt, *Angew. Chem.* **77** (1965) 1049.

first, it has been studied very carefully by Bartlett *et al.*[74]. They could show exactly how the first step of the overall reaction is a bimolecular Lewis acid–Lewis base interaction between sulphur and the phosphine. Secondly, this reaction is one of the all too few examples in which a drastic difference in reaction rates between different sulphur allotropes has been sought and found.

The reaction

$$8(C_6H_5)_3P + S_8 \longrightarrow 8(C_6H_5)_3PS$$

leading quantitatively to the phosphine sulphide, is strictly second order, first order in each reactant. The reaction rate is markedly dependent upon the solvent. This dependency proves the formation of charged intermediates from uncharged species. This requires a high solvation energy for stability in solution. Solvents of high ionizing and solvating power can greatly stabilize the ions, and thus lower the free energy of activation and hence increase the rate of the reaction. The study of the salt effects also fits into this picture. So, the formation of phosphine sulphides from phosphines and cyclo-octasulphur can be formulated as:

$$\text{cyclo-}S_8 + R_3P \longrightarrow {}^{-}S\text{—}S\text{—}S\text{—}S\text{—}S\text{—}S\text{—}S\text{—}S\text{—}\overset{\oplus}{P}R_3$$

$$^{-}S\text{-}S\text{-}S\text{-}S\text{-}S\text{-}S\text{-}S\text{-}S\text{-}^{+}PR_3 + R_3P \longrightarrow {}^{-}S\text{-}S\text{-}S\text{-}S\text{-}S\text{-}S\text{-}S\text{-}^{+}PR_3 + R_3PS$$

$$^{-}S\text{-}S\text{-}S\text{-}S\text{-}S\text{-}S\text{-}S\text{-}^{+}PR_3 + R_3P \longrightarrow {}^{-}S\text{-}S\text{-}S\text{-}S\text{-}S\text{-}S\text{-}^{+}PR_3 + R_3PS$$

$$^{-}S\text{-}S\text{-}S\text{-}S\text{-}S\text{-}S\text{-}^{+}PR_3 + R_3P \longrightarrow {}^{-}S\text{-}S\text{-}S\text{-}S\text{-}S\text{-}^{+}PR_3 + R_3PS$$

$$^{-}S\text{-}S\text{-}S\text{-}S\text{-}S\text{-}^{+}PR_3 + R_3P \longrightarrow {}^{-}S\text{-}S\text{-}S\text{-}S\text{-}^{+}PR_3 + R_3PS$$

$$^{-}S\text{-}S\text{-}S\text{-}S\text{-}^{+}PR_3 + R_3P \longrightarrow {}^{-}S\text{-}S\text{-}S\text{-}^{+}PR_3 + R_3PS$$

$$^{-}S\text{-}S\text{-}S\text{-}^{+}PR_3 + R_3P \longrightarrow {}^{-}S\text{-}S\text{-}PR_3 + R_3PS$$

$$^{-}S\text{-}S\text{-}^{+}PR_3 + R_3P \longrightarrow 2R_3PS$$

The first step of this reaction, that is the ring opening, is the rate-determining step. Because of the different energy content of different sulphur rings ("ring strain"), the reaction rate must depend on the size of the rings. Indeed, the authors found that cyclohexasulphur, S_6, in benzene solution reacts at a rate 25,000 times faster than S_8 at about 7°C (photolytically produced polycatenasulphur reacts immeasurably fast with the phosphine). More recent studies included also cyclododecasulphur in this series[88]. Kinetic measurements at different temperatures in toluene revealed the following activation energies for the ring opening: S_6 = 4·3 kcal mole^{-1}, S_8 = 14·5 kcal mole^{-1} and S_{12} = 9·5 kcal mole^{-1} (S_{10} and S_7 have not yet been studied). These results (which are in accordance with the determination of the heat of combustion of the different sulphur rings) clearly prove a drastic ring strain in the thermodynamically unstable sulphur rings S_6 and S_{12} (the latter, by the way, may also be obtained from S_6 photolytically in solution[88]).

A rather peculiar reaction of sulphur shall close the series of nucleophilic degradation reactions. Formally, it may be regarded as "quite normal"—the true course of this reaction is not yet known. Heating of elemental sulphur (S_8) with sodium nitrite in dimethylformamide to about 80°C quantitatively yields analytically pure and anhydrous sodium thiosulphate[89], according to the overall equation

$$S_8 + 8NaNO_2 \longrightarrow 4Na_2S_2O_3 + 4N_2O$$

---

[88] M. Schmidt and G. Knippschild, Thesis, Univ. Würzburg (1968).
[89] M. Schmidt and R. Wägerle, *Angew. Chem.* **70** (1958) 594.

If this reaction follows the above-mentioned type, then the first part would be the formation of a hitherto-unknown "thio nitric acid" salt, $NaNO_2S$, by stepwise sulphur degradation:

$$S_8+8NaNO_2 \longrightarrow 8NaNO_2S$$

The latter salt decomposes, forming thiosulphate and dinitrogen oxide:

$$2NaNO_2S \longrightarrow Na_2S_2O_3+N_2O$$

The above example of nucleophilic degradations of sulphur molecules may serve to demonstrate the importance of this reaction type in sulphur chemistry; sulphur–sulphur bonds are present not only in elemental sulphur but in a very large number of sulphur compounds which can all be regarded as derivatives of the sulphanes (Section 2.1). Cleavage reactions of such S–S bonds can be dealt with in this review only in connection with sulphonic acids. The detailed study of such cleavage reactions has many theoretical, synthetic, biochemical and industrial interests. It is, by nature, closely related to all studies on the nature of chemical bonding in sulphur compounds with the still-unanswered question of the participation of *d*-orbitals, etc. An exhaustive treatment of this very important problem in sulphur chemistry is outside the scope of this book; reference may be made to more competent reviews[90, 91].

A very valuable interpretative bibliography on the problem of scissions of sulphur–sulphur bonds is found in ref. 92. Some special aspects of it are described in ref. 93.

The hypothesis of the nucleophilic degradation of sulphur–sulphur bonds has been proved as a very useful tool in sulphur chemistry within the last 15 years. Further refinement, however, is still urgently needed for a deeper understanding and at the same time for a more accurate prediction of reactions in sulphur chemistry. A very sophisticated refinement has been put forward by Davies[74], who relates the relative nucleophilic (or "thiophilic") power towards sulphur of quite a considerable number of Lewis bases in a semi-quantitative manner in his "oxibase scale". But, in spite of its physical and partly even mathematical "appearance", it is not the "break through" sulphur chemists are looking forward to and are working for. The same also holds for a rather qualitative and speculative working hypothesis that has been developed and put forward by one of the authors of this review[76, 94, 95]. It is based on the assumption (derived not from theoretical considerations of sulphur–sulphur bonds, but from many experimental studies of sulphur with nucleophiles) of a delocalized electron system in sulphur–sulphur bonds and with this on the assumption that such bonds are not simple σ-bonds, but are reinforced by partial multiple bond character using empty *d*-orbitals. This qualitative hypothesis permits a very good interpretation without any contradiction of the many experimental findings known to date. It still needs many essential quantitative refinements which will surely not be simple to work out. It is to be hoped, however, that above and beyond experimental findings this working hypothesis—in spite of its partly speculative character—can contribute to a deeper understanding of the many unsolved problems by stimulating fruitful critical discussions and, what is far more important, by the conception and realization of new experiments.

[90] D. W. J. Cruickshank and B. C. Webster, in ref. 74, pp. 7–47.
[91] F. A. Gianturco, *J. Chem. Soc.*, Sect. A (1969) 1293.
[92] N. Kharash and A. J. Parker, *Quarterly Reports on Sulfur Chemistry*, **1** (1966) 285–378.
[93] J. L. Kice, *Accounts of Chemical Research*, **1** (1968) 58–64.
[94] M. Schmidt, *Angew. Chem.* **73** (1961) 394.
[95] M. Schmidt, *Österr. Chem. Ztg.* **64** (1963) 236.

Electrophilic Degradation of Sulphur–Sulphur Bonds

The scission of sulphur–sulphur bonds may occur homolytically as is most probably the case in melts and vapour of elemental sulphur; it may be caused by many nucleophiles heterolytically, as has been shown above (page 819); and it may further be induced by strong electrophiles. In the latter case sulphur (or chain-like sulphur compounds) does not act as a Lewis acid as was the case previously, but as a Lewis base. Many electrophilic scissions of sulphur–sulphur bonds are known (see for instance refs. 92 and 93). But there is not a simple and widely accepted theory or hypothesis, on the exact course of the reaction as is the case in nucleophilic scissions. Two examples demonstrate this sort of reaction· degradation of elemental sulphur by hydrogen iodide and by boron iodides.

Elemental sulphur in an anhydrous medium reacts with hydrogen iodide quantitatively, forming iodine and hydrogen sulphide:

$$S_x + 2xHI \longrightarrow xI_2 + xH_2S$$

This reaction may be of some importance in analytical chemistry because it offers the possibility of differentiation between chemically bound sulphur and cyclic allotropes of the element (the latter only react when irradiated by visible light; only open chains are attacked by HI in the dark) in complicated reaction mixtures[96]. In contrast to hydrogen iodide, the strong Lewis acid boron iodide, BI_3, and its organic derivatives RBI_2 and R_2BI react not only with open sulphur chains or chain-like sulphur compounds (such as sulphanes), but also with the cyclic allotropes (with different reaction rates). So, boron iodide already under very mild conditions attacks cyclo-octasulphur under ring opening and formation of the new boron–sulphur ring system trithiadiborolane[97], besides elemental iodine:

```
                              S——S
                              |    |
2BI3+3/8S8 ———→ 2I2+I–B    B–I
                              \  /
                               S
```

The same holds for alkyl and aryl boron di-iodides (see Section 2.11) (resulting in the B-substituted ring).

By the same sort of electrophilic sulphur–sulphur bond fission, a disulphane derivate is formed from the Lewis acid n-dibutyl boron iodide, $(Bu)_2BI$, and cyclo-octasulphur[88, 98].

$$2(Bu)_2BI + \tfrac{1}{4}S_8 \longrightarrow (Bu)_2B\text{–}S\text{–}S\text{–}B(Bu)_2 + I_2$$

This latter reaction can be followed kinetically in dilute organic solutions. The rate-determining step obviously is, as also with the nucleophilic attack on S–S bonds, the ring opening. Therefore, a comparison of the reaction rates with cyclohexa-, cyclo-octa and cyclododecasulphur must reveal the different ring strains in those allotropes. This is indeed the case: a clear decrease in rate is found in the expected order $S_6 > S_{12} > S_8$. (S_7 and S_{10} have not yet been studied[88].)

1.10. INTERCHALCOGEN COMPOUNDS (OXIDES EXCLUDED)

The analogous behaviour of sulphur and selenium in forming ring structures (of quite different stability) poses the question: "To what extent can selenium atoms enter sulphur

96 M. Schmidt and D. Eichelsdörfer, *Z. anorg. allg. Chem.* **330** (1964) 113 and 122.
97 M. Schmidt and W. Siebert, *Chem. Ber.* **102** (1969) 2752.
98 W. Siebert, E. East and M. Schmidt, *J. Organometal. Chem.* **23** (1970) 329.

rings, or vice versa?" This interesting question was previously considered in 1890[99], but only recently has considerable interest again been focused on the problem[100, 101]. Sulphur–selenium mixtures were melted together at rather high temperatures. The products obtained thereby were extracted by carbon tetrachloride and then crystallized. The crystalline samples have been analysed by mass spectrometry. All members of the octa-atomic series S_xSe_{8-x} have been identified by the characteristic group of peaks produced by the natural isotopic distribution of sulphur and selenium. A theoretical computation of the relative spectral peak intensities has confirmed the identity of this octameric series of sulphur–selenium compounds[102]. The mass spectral fragmentation patterns of the sulphur–selenium system show significant amounts of peaks corresponding to hexa-atomic ions. Whether these ions come from fragmentated octa-atomic species or are indicative of the real presence of hexa-atomic (or lower) species in the crystalline samples is impossible to deduce from the spectra.

The octa-atomic species identified by mass spectroscopy may not necessarily be unique molecules. Ring isomerism is possible when more than one atom of selenium is introduced into the S_8 ring. Thus: S_6Se_2—three isomers, S_5Se_3—five isomers, S_4Se_4—eight isomers, etc.[103].

The first studies for a preparative synthesis of mixed sulphur–selenium ring molecules under kinetically controlled conditions were carried out by the reaction of hydrogen selenide with dichloro disulphane in dilute solutions at rather low temperatures[104]. The primary product seems to be a polymeric chain from sulphur and selenium atoms, formed according to

$$xH_2Se + xCl_2S_2 \longrightarrow 2xHCl + \text{-(S-S-Se-)}_x$$

This interchalcogen chain "unrolls" easily to the surprisingly stable octa-atomic species S_6Se_2 and S_5Se_3 in the expected ratio. They may easily be separated by fractional crystallization into beautiful crystals of bright orange colour (deepening with the selenium content). These new interchalcogen compounds have been compared with cyclohexa-, cyclo-octa- and cyclododecasulphur on the one side and with cyclo-octaselenium on the other side with respect to their reactivity towards triphenyl phosphine in the well-known nucleophilic degradation of the chalcogen rings (page 819). The following order was found for the rate of ring opening by the nucleophile $S_6 > Se_8 > S_{12} > Se_3S_5 > Se_2S_6 > S_8$[105].

Also the sulphur–tellurium system has been studied recently[106]. The carbon disulphide extract of a high-temperature melt of sulphur and tellurium has been examined by mass spectroscopy. In contrast to the sulphur–selenium system, in this case only within the octa-atomic species was one tellurium-containing ion detected, namely, S_7Te. No reports on a kinetically controlled synthesis (or experiments with this aim) of sulphur–tellurium compounds are known (a report on an adduct of S_7Te with SnI_4 must be taken with critical care[107]). No species of higher molecularity than eight are observed, nor are there any

99 W. Muthmann, *Z. Krist.* **17** (1890) 336.
100 R. Cooper and J. V. Culka, *J. Inorg. Nucl. Chem.* **27** (1965) 755.
101 J. Berkowitz, in B. Meyer (Ed.), *Elemental Sulphur, Chemistry and Physics*, pp. 125–159, Interscience, New York (1965).
102 R. Cooper and J. V. Culka, *J. Inorg. Nucl. Chem.* **29** (1967) 1217.
103 J. V. Culka, Thesis, Univ. of Melbourne (1969).
104 M. Schmidt and E. Wilhelm, Unpublished.
105 G. Knippschild, Thesis, Univ. Würzburg (1968).
106 R. Cooper and J. V. Culka, *J. Inorg. Nucl. Chem.* **29** (1967) 1877.
107 L. Hawes, *Nature, Lond.* **198** (1963) 1267.

detectable species containing more than one tellurium atom. Synthetic studies should serve to elucidate the hitherto relatively unknown situation in the sulphur–tellurium situation.

The same holds also for the selenium–tellurium system. The crystalline products obtained by solvent extraction of the high-temperature reaction mixture of elemental selenium and tellurium have been analysed by mass spectrometry. Only a fragmentation pattern due to the ion Se_5Te^+ has been observed. No octa-atomic species or hexa-atomic species of a greater tellurium content than 1 have been detected[108].

Much more synthetic work will have to be done before a good and satisfactory survey on interchalcogen compounds between atoms of sulphur, selenium and tellurium can be given.

1.11. BIOLOGICAL ACTIVITIES OF SULPHUR AND ITS COMPOUNDS[109, 110]

Sulphur

Elemental sulphur, because of its hydrophobic character, is only slightly soluble in animal or human organs. However, sulphur reacts relatively easily with proteins, yielding H_2S. Presumably sulphur is attacked by thio amino acids, since the sulphur-containing proteins were found to be most reactive. Elemental sulphur has been used in medicine as well as in cosmetics. It causes after oral application a stimulation of the peristalsis by the formation of H_2S. Colloidal sulphur is more active than crystalline sulphur on the skin. It may cause irritation, especially to the eyelid.

Sulphanes

Hydrogen sulphide is as toxic as HCN and it can cause lethal poisoning in a very short time. By its unpleasant, strongly repellent odour H_2S usually prevents inhalation for longer periods of time in contrast to HCN. Poisonings with H_2S in small amounts are not dangerous, but one can get used to a low H_2S concentration, which then of course will cause serious defects. Early signs of H_2S inhalation are coughing, vomiting, headache, dizziness and weakness. Hydrogen sulphide is rapidly oxidized in the organism, and therefore it cannot be accumulated. It has been demonstrated that when inhalation of a lethal dose occurs no H_2S is found in the exhalation air. The formation of the so-called "sulphhaemoglobin" by H_2S presumably requires simultaneously an oxidation of haemoglobin. In contrast to "methaemoglobin" (caused by CO poisoning), "sulphhaemoglobin" cannot be converted into the normal haemoglobin.

Fortunately the intensive odour of H_2S warns people to leave areas poisoned with hydrogen sulphide. Use is often made of this effect by adding H_2S to heating gas so that leaks can be readily detected. H_2S is just noticeable at about 0·025 ppm; at 3–5 ppm the smell is very unpleasant, and at 20–30 ppm hard to stand. Surprisingly, at 200 ppm the smell is less intensive and at 700 ppm unconsciousness or lethal poisoning may happen before it is noticed.

The sulphanes H_2S_x ($x = 2$–8) all exhibit H_2S odour, since the sulphanes are thermodynamically unstable with respect to their decomposition products H_2S and sulphur. H_2S_2

[108] R. Cooper and J. V. Culka, *J. Inorg. Nucl. Chem.* **31** (1969) 685.

[109] *Ullmanns Encyklopädie der technischen Chemie*, 3. Auflage, Band 15, S. 380, 423, 464, 502, 530, 556, Urban & Schwarzenberg, München–Berlin (1964).

[110] *Gmelins Handbuch der Anorganischen Chemie*, 8. Auflage, Systemnummer 9, Teil A, S. 503, Verlag Chemie, Weinheim (1953).

and H_2S_3 as well as the higher sulphanes smell like camphor and irritate the mucous membranes of the nose, eyes and mouth. The camphor-like odour is usually predominant.

Sulphur Dioxide

Since SO_2 is the most important pollutant of the air in areas with and without industry, it has been studied intensively. Irritation of the mucous membrane occurs rapidly so people are forced to leave rooms with SO_2 concentrations over 100 ppm. Short exposures to 400–500 ppm SO_2 may be lethal. Poisoning occurs mainly in the upper respiratory tract by swelling of the mucous membranes and spasm of the bronchia muscles. 10–20 ppm of SO_2 can be easily detected by smell and by taste. It has been suggested that the maximum of SO_2 concentration should not exceed 5 ppm (= 13 mg SO_2/m^3) in areas where people work. In combination with dust, SO_2 exhibits a stronger effect on people, which actually may be caused by SO_3 formed by oxidation of SO_2. A further escalation of dangerous effects occurs when dust and SO_2 combine with fog to build up the so-called smog (smog disasters in London 1952 and 1962).

Animals such as rabbits, rats or mice are less sensitive to SO_2, since concentrations of 10 ppm given over a period of 90 days did not result in any symptoms. In contrast, plants are quite sensitive to SO_2. A concentration of 1–2 ppm caused damage within a few hours, presumably by inhibiting the photosynthesis. Thus the steadily increasing air pollution by SO_2 threatens not only man but also vegetation.

Sulphuric Acid and Its Salts

Concentrated sulphuric acid causes severe destruction on contact, since in contrast to conc. HNO_3 it does not build up a protective layer on the skin but reaches into deeper layers of the skin. SO_3 gas gives results similar to SO_2, an irritation of the respiratory tract. Local destruction of teeth by SO_3 gas has also been observed. Concentrations below 1 mg/m^3 are detectable neither by smell nor by taste. However, most people cannot stand 5 mg SO_3/m^3. Irritation of the mucous membranes causes strong coughing.

Sulphur Halogen Compounds

The poisonousness of most of the sulphur compounds containing Cl–S bonds is due to the ease of hydrolysis, thereby forming two acids. Chlorosulphonic acid hydrolyses to H_2SO_4 and HCl and both acids will cause severe damage to the skin and mucous membranes. The symptoms are similar to those originated by SO_2 and SO_3. Sulphur hexafluoride is totally intoxic: however, some of its synthesis by-products, the lower sulphur fluorides (e.g. SF_4, S_2F_{10}), are extremely toxic. The fluoride ion is responsible for the additional danger of a sulphur fluoride hydrolysis in the organism, since it easily combines with Ca^{2+} ions to form insoluble CaF_2. Thereby the equilibrium between Na, K and Ca ions in the cell is disturbed.

Carbon Sulphur Compounds

Besides the numerous organosulphur compounds, only a few inorganic compounds are known, e.g. COS, $SCCl_2$ and CS_2. Carbon oxisulphide was found to be more poisonous than H_2S. It causes death in a very short time when the concentration of COS amounts to 0·1%. In contrast to COS, carbon disulphide represents a technical product which is synthesized in large quantities. Its use as solvent for the rubber industry and for artificial silk production often leads to severe poisoning of workers. CS_2 is easily absorbed through the skin, especially between the fingers. Most of the poisoning, however, is caused by inhalation of CS_2. Unfortunately people working with CS_2 get used to it very soon. Part of the inhaled CS_2 is oxidized and eliminated as sulphate through the kidneys. The danger of CS_2 stems mainly from its solubility in lipoids. Thus it will be easily accumulated in the central nervous system. Concentrations of 1000–3000 ppm (0·1–0·3%) inhaled for 30 to 60 min may be lethal. Frequent inhalations of 300–500 ppm create very complex pathological symptoms. Animal tests have shown that CS_2 causes severe polyneuritis.

2. SULPHUR COMPOUNDS

Elemental sulphur is a very complicated system composed of very many compounds of sulphur "with itself". There is indeed no real reason for a division of chemical compounds composed by equal atoms and compounds composed from different atoms—the inherent problem is always the same: how do atoms interact with each other, and why do they interact as they do interact? It is just by tradition that we stick to the surpassed historical separation of an element and its compounds (which only holds for the noble gases) in this treatment of comprehensive inorganic chemistry.

2.1. SULPHANES

The most simple (from the stoichiometric point of view) binary compounds of sulphur, those which only contain hydrogen besides sulphur, are named sulphanes. All sulphanes can, by nature, only contain two hydrogen atoms and, therefore, must have the composition H_2S_x. In contrast to earlier belief (branchings in the chain!) all sulphanes are just unbranched sulphur chains, terminated at both ends by hydrogen atoms. They are not only the acid parent compounds of the metal polysulphides. Also all the many covalent compounds containing sulphur–sulphur bonds, such as for instance materials of the thiokol-type, sulphane-sulphonic acids, vulcanized rubber and so forth, may all be regarded as derivatives of the sulphanes. So, the sulphanes link together in a very close genetic connection, elemental sulphur on one side, with all the many compounds containing sulphur–sulphur bonds in their molecules on the other side. The great variety of all these genetically connected materials do, in spite of many quite different other properties, all manifest similarities imposed on them by the nature of the sulphur–sulphur bond. So, for instance, all the many different polymers containing chains of sulphur atoms inherently must and do react as electrophiles towards "thiophiles".

Free Sulphanes, H_2S_x

This very interesting class of compounds has been known since the 18th century, but has been neglected for a long period of time. The main reason for this is probably the fact that the extremely sensitive sulphanes are difficult to prepare, to purify, and to handle. Our knowledge, therefore, of these polymeric sulphur compounds is based mainly on a fascinating series of ingenious papers by Fehér and his co-workers which may be regarded as outstanding examples of preparative inorganic chemistry. For a historical review see refs.

111 and 112. Bloch and Hoehn first isolated reasonably pure sulphanes[113]. They prepared various sodium sulphides by fusing crude $Na_2S.9H_2O$ with various amounts of sulphur. The resulting aqueous polysulphide solutions were poured into dilute hydrochloric acid, which was stirred and kept at $-10°C$ (the reaction is exothermic). The separated "crude oil" was dried with $CaCl_2$. All glass vessels used were pretreated with HCl gas because of the pronounced alkali sensitivity of the sulphanes. The "crude oil", a yellow oily mixture of different sulphanes, has the viscosity of concentrated sulphuric acid and an odour reminiscent of sulphur monochloride and camphor. Alkali, even the alkali content of usual glass surfaces, catalyses the decomposition of the higher sulphanes into H_2S and polymeric sulphur. The composition of the quite heavy crude oil (its density varies from 1·625 to 1·697 at room temperature) varies with the sulphur content of the starting material.

Fehér and Laue[114] studied the influence of different factors on yield and composition of the crude oil. They were able to work out a simple method for preparing it conveniently in large preparative amounts. Its average sulphur content always is higher than that of the polysulphide solution due to disproportion reactions such as:

$$2H_2S_3 \longrightarrow H_2S + H_2S_5$$

By this method it is not possible to prepare a sulphane mixture with an average sulphur content below $H_2S_{4\cdot5}$ to $H_2S_{4\cdot0}$. The best yields of crude oil (more than 80%) are obtained from $Na_2S_{4\cdot4-4\cdot7}$; its sulphur content is then $H_2S_{6\cdot04-6\cdot5}$. The authors determined the density, dynamic viscosity and index of refraction of the crude oil as a function of its average composition.

Bloch and Hoehn succeeded in preparing reasonably pure H_2S_2 and H_2S_3 by vacuum distillation of their crude oil. Fehér and Baudler[115] showed by a careful study of the Raman frequencies of the different sulphanes that the crude oil does not contain appreciable amounts of H_2S_2, H_2S_3 or elemental sulphur, but only H_2S_4, H_2S_5, H_2S_6 and H_2S_7. Its overall composition varies from $H_2S_{4\cdot5}$ to $H_2S_{6\cdot5}$.

H_2S_2 and H_2S_3 are formed according to the results from higher sulphanes in the crude oil by cracking during distillation. By developing a more efficient distillation-cracking apparatus it became possible to obtain H_2S_2 and H_2S_3 in reasonably good yields[116]. However, the Raman spectrum indicated a contamination of H_2S_3 by H_2S_2 and H_2S_4. Pure H_2S_3 then was obtained by "thin layer distillation". It disproportionates under the influence of light into H_2S_2 and H_2S_4. H_2S_5 and H_2S_6 are also found as disproportionation products in small yields[116].

Pentasulphane, H_2S_5, and hexasulphane, H_2S_6, have also been prepared from the crude oil by thin layer distillation. They are deep yellow, viscous liquids of irritating odour, soluble in benzene. Some apparative improvements for the classical synthesis of sulphanes were reported in 1956[117].

A preparation of sulphanes from anhydrous polysulphides is in principle possible but

[111] M. Schmidt, in F. G. A. Stone and W. A. G. Graham (Eds.), *Inorganic Polymers*, p. 103, Academic Press (1962).

[112] K. W. C. Burton and P. Machmer, in G. Nickless (Ed.), *Inorganic Sulphur Chemistry*, pp. 336–364, Elsevier, Amsterdam, London, New York (1968).

[113] I. Bloch and F. Hoehn, *Chem. Ber.* **41** (1908) 1961.

[114] F. Fehér and W. Laue, *Z. anorg. allg. Chem.* **288** (1956) 103.

[115] F. Fehér and M. Baudler, *Z. anorg. allgem. Chem.* **258** (1949) 132.

[116] F. Fehér and M. Baudler, *Z. anorg. allg. Chem.* **254** (1948) 289.

[117] F. Fehér, W. Laue and G. Winkhaus, *Z. anorg. allg. Chem.* **288** (1956) 113.

without practical advantages[118]. To some extent this also holds for the electrochemical preparation of sulphanes[119]. Whereas an anodic oxidation of hydrogen sulphide failed for this purpose, a cathodic reduction of sulphur dioxide is possible. The primary reduction product is $H_2S_2O_4$, which in strong acid medium decomposes into sulphurous acid plus sulphanes.

The most important method for the preparation of higher sulphanes (and also the higher chlorosulphanes)[120] uses as basis the long-known condensation reaction between a sulphur–hydrogen group and a sulphur–chlorine group. According to the simple scheme

$$\text{—S}\boxed{\text{H + Cl}}\text{S—} \longrightarrow \text{HCl} + \text{—S—S—}$$

we thereby observe the formation of a new sulphur–sulphur bond by elimination of hydrogen chloride. Those condensation reactions may be symbolized by the following equations:

$$H_2S_x + Cl_2S_y \longrightarrow 2\,HCl + S_{x+y}$$

$$\text{H—S—S—}\boxed{\text{H + Cl}}\text{—S—S—Cl} \longrightarrow \text{H—S—S—S—S—Cl + HCl}$$

$$\text{H—S—S—S—S—}\boxed{\text{Cl + H}}\text{—S—S—H—} \longrightarrow \text{H—S—S—S—S—S—S—H + HCl}$$

$$\text{H—S—S—S—S—S—}\boxed{\text{H + Cl}}\text{—S—S—Cl} \longrightarrow \text{H—S—S—S—S—S—S—S—S—Cl + HCl}$$

tc.

The equation

$$H\text{–}S_x\text{–}Cl \longrightarrow HCl + S_x$$

is an idealized formulation of the formation of elemental sulphur from sulphanes and chlorosulphanes. It is a redox reaction whereby a chain $-S_x-$ of an overall oxidation number -2 is oxidized by a chain $-S_y-$ of an overall oxidation number $+2$, the latter one being reduced during the reaction. Elemental sulphur is formed besides hydrogen chloride.

Such a formulation, of course, does not say anything on the molecular weight of the sulphur so formed. The course of the reaction will widely depend on the reaction conditions —solvent, concentrations, temperature and so forth. In any case, the overall reaction must proceed via single steps, as is exemplified by the above-formulated equations for the first steps of the redox reaction between H_2S_2 and Cl_2S_2. Pure sulphur can only be formed thereby if an unsymmetrical chain molecule with suitable geometric orientation undergoes an intramolecular cyclization process:

$$H\text{–}S_x\text{–}Cl \longrightarrow HCl + S_x$$

The sulphur atoms of such an intermediate unsymmetrically substituted sulphur chain have an average oxidation number of zero, that is the oxidation number of elemental sulphur—a rather unique case with a main group element.

The preparation of sulphanes (or chlorosulphanes, respectively) may be expressed by the following general formulation:

$$aH_2S_x + bS_yCl_2 \longrightarrow (a-b)H_2S_z + 2bHCl$$

[118] F. Fehér and R. Berthold, *Z. anorg. allg. Chem.* **290** (1957) 251.
[119] F. Fehér, F. Schliep and H. Weber, *Z. Elektrochem.* **57** (1953) 916.
[120] F. Fehér and W. Kruse, *Z. anorg. allg. Chem.* **293** (1958) 302.

where $z = \frac{xa+yb}{(a-b)}$, $x = 1, 2, 3, \ldots$, $y = 0, 1, 2, 3, \ldots$ if $a > b$ sulphanes and $b > a$ chlorosulphanes are prepared.

From kinetic considerations, assuming that all the molecules have the same reactivity (which is not correct), the final molecular distribution of the products in the mixture formed can be calculated[112]. For a sulphane molecule of the general composition $H_2S_{w(x+y)+x}$ the molecular distribution is a $(1-b/a)^2(b/a)^w$, where $w = 0, 1, 2, 3, \ldots$.

This can be illustrated for the case of the synthesis of hexasulphane from the reaction of excess disulphane with dichlorodisulphane, which yields a mixture of the sulphane homologs H_2S_6, H_2S_{10}, H_2S_{14}, H_2S_{18}, . . . of an average chain length H_2S_z:

$$aH_2S_2+Cl_2S_2 \longrightarrow (a-1)H_2S_z+2HCl$$

relative to one mole of Cl_2S_2, $x = 2$, $y = 2$, and the general formula of the sulphane molecule is H_2S_{4w+2}.

The main reaction is thus:

$$H_2S_2+Cl_2S_2+S_2H_2 \longrightarrow H_2S_6+2HCl$$

followed by various secondary reactions such as:

$$H_2S_2+Cl_2S_2+H_2S_6 \longrightarrow H_2S_{10}+2HCl$$
$$H_2S_6+Cl_2S_2+H_2S_6 \longrightarrow H_2S_{14}+2HCl$$
$$H_2S_6+Cl_2S_2+H_2S_{10} \longrightarrow H_2S_{18}+2HCl$$
$$H_2S_2+Cl_2S_2+H_2S_{10} \longrightarrow H_2S_{14}+2HCl, \text{ etc.}$$

For the situation where the sulphane–chlorosulphane ratio is 2, the final molecular distribution of the products is calculated as 50% H_2S_2, 25% H_2S_6, 12·5% H_2S_{10}, 6·25% H_2S_{14} and 3·12% H_2S_{18}, etc. Thus a crude oil would be obtained containing 50% of hexasulphane, after removal of the volatile unreacted disulphane by vacuum distillation. However, increasing the molar ratio H_2S_x/Cl_2S_y will decrease the proportion of higher sulphanes in the mixture, so that with a sufficiently large excess of the initial sulphane reactant it is possible to obtain the intended primary product in a relatively pure state. In practice, a ratio of about 20:100 is applied.

Kinetic considerations of the composition of such reaction mixtures are based on the assumption that all molecules in such mixtures have the same reactivity. This is not true and also is not to be expected. Those condensation reactions between sulphanes and chlorosulphanes have to be looked at, as was already mentioned above, as redox reactions between sulphur chains of an overall oxidation number of -2, $H\text{–}S_x^{-2}\text{–}H$, and of $+2$, $Cl\text{–}S_x^{+2}\text{–}Cl$. The longer the sulphur chain in a given sulphane, the more the average oxidation number of the single sulphur atoms approaches the value of zero, either from -2 or from $+2$. Now, the smaller the difference between the formal oxidation numbers of the reaction partners (that is, the longer the sulphur chains are), the slower the redox reaction will be. In fact the reactivity of the sulphanes decreases rapidly with increasing chain length so that considerably less of the higher sulphanes are produced than is theoretically predicted.

Naturally, it is essential that only volatile sulphanes are used, which restricts the choice to hydrogen sulphide and disulphane (to a much lesser extent also to trisulphane). To obtain relatively pure sulphanes it is necessary to work under anhydrous conditions at low temperatures.

The reaction of hydrogen sulphide with chlorine instead of chlorosulphanes is a special case with $x = 1$ and $y = 0$ in the above general formulation. In aqueous solution this reaction leads to sulphur and hydrogen chloride:

$$Cl_2+H_2S \longrightarrow 2HCl+(1/x)S_x$$

This has been studied under anhydrous conditions at −100°C[121] and in the vapour phase[122]. The formation of sulphanes has not been observed under these conditions. With liquid hydrogen sulphide, however, chlorine reacts under proper conditions forming a "new" crude oil with a composition approximating H_2S_4 [123]. In contrast to the normal crude oil, it contains considerable amounts of H_2S_2 and H_2S_3. These most important sulphanes may be isolated from it by a simple vacuum distillation on a preparative scale without cracking. The most important intermediate of this process is HSCl, formed according to the following:

$$H_2S+Cl_2 \longrightarrow HSCl+HCl$$

HSCl reacts either with hydrogen sulphide to form disulphane:

$$HSCl+H_2S \longrightarrow H_2S_2+HCl$$

or with already-formed sulphanes yielding higher sulphanes:

$$HSCl+H_2S_x \longrightarrow H_2S_{x+1}+HCl$$

The process may be carried out continuously (higher sulphanes are insoluble in liquid hydrogen sulphide, which is rather remarkable) and therefore is most suitable for preparing sulphane mixtures. Bromine reacts in an analogous manner to chlorine. Apart from these above-mentioned preparative methods there are various other reactions which are reported to form sulphanes—for instance, the reaction of hydrogen sulphide with molten sulphur[124]. Also the reduction of aqueous sulphurous acid with hypophosphoric acid[125] and the acid decomposition of thiosulphate[126] yields sulphanes. The latter method is reported to form a sulphane mixture of average composition H_2S_{354} (?).

Properties of Sulphanes

The relationship of the sulphanes as the most simple binary sulphur compounds with elemental sulphur is very close. The similarity of many chemical and physical properties logically will increase rapidly with increasing chain length of the sulphane. In a molecule

$$\overset{+1}{H}-\overset{-2}{S_x}-\overset{+1}{H}$$

the two chain terminating hydrogen atoms have an oxidation number of +1, whereas the compensating oxidation number −2 is distributed along the sulphur chain. The longer the chain the higher becomes the average oxidation number of each sulphur atom—from −2 asymptotically approaching zero with the extension of the chain and by this logically approaching elemental sulphur itself. The most marked difference, of course, must be found between sulphur and the first member of the sulphane series—that is, monosulphane or hydrogen sulphide with $x = 1$. This is indeed the case. The difference in oxidation

121 A Stock, *Chem. Ber.* **53** (1920) 837.
122 P. W. Schenk and S. Sterner, *Monatsh. Chem.* **80** (1949) 117.
123 H. Dersin, Diplomarbeit, Univ. Munich (1958).
124 T. K. Wiewicrowski and F. J. Touro, in A. V. Tobolsky (Ed.), *The Chemistry of Sulphides*, pp. 9–21, Interscience, New York (1968).
125 O. von Deines, *Liebigs Ann. Chem.* **440** (1924) 213.
126 F. Fehér and H. J. Berthold, *Z. anorg. allg. Chem.* **267** (1951) 251.

numbers here is 2—no real close similarity between sulphur and hydrogen sulphide is found, which, of course, is also due to the fact that H_2S does not contain a sulphur–sulphur bond, characteristic for all the other sulphanes and their many formal derivatives. These are the main reasons for the extraordinary behaviour and characteristics of the first member of the homologous series of the sulphanes, namely monosulphane or hydrogen sulphide. It justifies a separate treatment of this very important compound, H_2S, that will be given at the end of the treatment of the higher sulphanes.

Physical properties. The well-characterized sulphanes H_2S_2 to H_2S_8 are yellow liquids, with the exception of disulphane, which is colourless. The lower members of the series up to a chain length of 4 to 6 may be purified by vacuum distillation. Sulphanes synthesized by the above condensation reactions in general do not need any further purification.

The physical constants measured, e.g. molar volume and molar refraction, are linear functions of chain within the homologous series[127]. The relation

$$\left.\begin{array}{l}\text{Molar volume}\\ \text{Molar refraction}\end{array}\right\} = 2a+(n-2)b$$

is followed accurately, a being the contribution of the end group, $-$SH, and b the increment for each inner sulphur atom. For the molar volume $a = 24{\cdot}8$ and $b = 16{\cdot}4$, and for the molar refraction $a = 8{\cdot}9$ and $b = 8{\cdot}6$, see Table 8.

TABLE 8. DENSITY, MOLAR VOLUME, INDEX OF REFRACTION AND MOLAR REFRACTION OF THE SULPHANES[a] [128]

Sulphane	d_{20}	V	ΔV	n_{20}	R	ΔR
H_2S_2	1·334	49·6	—	1·631	17·7	—
H_2S_3	1·491	65·9	16·3	1·729	20·2	8·5
H_2S_4	1·582	82·3	16·4	1·791	34·9	8·7
H_2S_5	1·644	98·7	16·4	1·836	43·6	8·7
H_2S_6	1·688	115·2	16·5	1·867	52·2	8·6
H_2S_7	1·721	131·6	16·4	1·893	60·9	8·7
H_2S_8	1·747	148·0	16·4	1·912	69·5	8·6

[a] F. Fehér, W. Laue and E. Winkhaus, *Z. anorg. allg. Chem.* **290** (1957) 52.

The constancy of the increments b in the above-mentioned relation offers strong support for the view that in the sulphanes the sulphur atoms are linked to form unbranched chains. This also is proved by the Raman spectra of the sulphanes, excited by the green mercury line (5461 Å): the characteristic S–S stretching frequencies at about 400 to 500 cm^{-1} and the S–S–S bending frequencies at about 150 to 200 cm^{-1} are excellent aids in the identification and characterization of the different members of this homologous series. The Raman spectra further prove the absence of considerable amounts of physically dissolved sulphur. For the spectra see Table 9.

Although Raman spectroscopy is very useful for analysing and characterizing sulphanes, it does not give any information on the valence electron system of the sulphur chain. A deepening colour of sulphanes with increasing chain length is observed, and the ultraviolet and visible absorption spectra of several series of chain-like sulphur compounds have been interpreted in terms of the Kuhn electron gas model with a delocalized electron cloud in overlapping sulphur $3d$ orbitals[129]. However, the evidence for a participation of $3d$ orbitals

[127] F. Fehér, W. Laue and G. Winkhaus, *J. anorg. allg. Chem.* **290** (1957) 52.
[128] F. Fehér and G. Winkhaus, *Z. anorg. allg. Chem.* **288** (1956) 123.
[129] F. Fehér and H. Münzner, *Chem. Ber.* **96** (1963) 1131.

TABLE 9. CHARACTERISTIC RAMAN SPECTRA OF POLYSULPHANES. ALL FREQUENCIES IN cm^{-1} [a]

H_2S	H_2S_2	H_2S_3	H_2S_4	H_2S_5	H_2S_6	H_2S_7	H_2S_8
				149(3)	149(3)	149(3)	149(2)
			185(7)	184(4)	185(2)	185(2)	185(2)
		210(8)	229(2)	217(2)	217(2)	217(2)	217(2)
				242(2)	242(0)	242(0)	242(0)
			450(4)	439(1)	439(1)	439(0)	439(2)
				464(5)	464(9)	464(10)	464(10)
	509(9)	483(9)	483(9)	485(8)	484(10)	484(9)	484(8)
	883(2*d*)	862(2*d*)	862(2*d*)	861(3*d*)	862(1*d*)	862(1*d*)	862(1*d*)
1236							
2615							
2632	2509(2*d*)	2502(3*d*)	2501(2*d*)	2498(3*d*)	2500(5*d*)	2500(5*d*)	2500(5*d*)

[a] K. W. C. Burton and P. Machmer, in G. Nickless (Ed.), *Inorganic Sulphur Chemistry*, p. 353, Elsevier, Amsterdam, London, New York (1968).

in sulphur–sulphur bonding still is inconclusive and further work needs to be carried out in this field.

A study of the viscosities of the sulphanes indicates again that they are in fact straight chain members of a homologous series[130]. However, mixtures of sulphanes give higher viscosity values than would be predicted from a consideration of the viscosities of the pure compounds, and because of this it is not possible by such measurements to predict accurately the composition of oils obtained from condensation reactions. No simple rules were found to account for this anomalous behaviour, but the viscosity can be correlated to the square of the average molecular weight[131]. In general, the viscosity of sulphane mixtures increases with the width of the distribution curve. Physically dissolved sulphur lowers the viscosity of those mixtures, the effect being only weak in the case of the lower sulphanes, but becoming more important as the chain length increases. he evaporation enthalpy of Tthe sulphanes H_2S_2–H_2S_5 were measured by calorimetric methods at room temperature[132]. These studies revealed a linear relationship between the sulphane chain length and the evaporation enthalpy.

The vapour pressures of the sulphanes at 20°C were estimated by utilizing the fact that the amount of cooling of liquids during their evaporation is a function of the vapour pressure. With the aid of the Clausius–Clapeyron equation it is possible to construct, for a set of standards, a curve of the cooling as a function of the vapour pressure by calculation using experimental values for the evaporation enthalpies and boiling points. With this calibration curve it is then possible to estimate the vapour pressure of the sulphanes at 20°C, from the experimental values of the cooling during evaporation. The boiling points of the various sulphanes and the variation in vapour pressure with temperatures were also evaluated using the Clausius–Clapeyron equation. From this experimental data the critical temperatures and pressures were calculated (Table 10).

Some thermochemical studies and calculations on the sulphanes revealed that the energies of the sulphur–sulphur bonds and of the sulphur–hydrogen bonds are almost

[130] F. Fehér, W. Kruse and W. Laue, *Z. anorg. allg. Chem.* **292** (1957) 203.
[131] P. Flory, *J. Am. Chem. Soc.* **62** (1940) 157.
[132] F. Fehér and G. Hitzemann, *Z. anorg. allg. Chem.* **294** (1958) 50.

TABLE 10. CRITICAL TEMPERATURES AND PRESSURES OF THE SULPHANES[a]

Formula H_2S_n	P_{20}	B.p.$_{760}$	T_c	P_c	L_s	K_{Tr}
H_2S_2	87·7	70	572	58·3	7497	21·9
H_2S_3	1·4	170	738	50·6	9327	21·0
H_2S_4	0·035	240	855	43·1	11261	21·9
H_2S_5	0·0012	285	930	38·4	13340	23·9

P_{20} Vapour pressure (mm) at 20°C.
B.p.$_{760}$ Boiling point (°C, calculated) at normal pressure.
T_c Critical temperature (°K).
P_c Critical pressure (atm).
L_s Evaporation enthalpy at the boiling point, normal pressure (cal/mole).
K_{Tr} Trouton's constant (cal g^{-1} $mole^{-1}$).
[a] K. W. C. Burton and P. Machmer, in G. Nickless (Ed.), *Inorganic Sulphur Chemistry*, p. 353, Elsevier, Amsterdam, London, New York (1968).

constant for all sulphanes. This furnished further proof that the sulphanes consist of unbranched sulphur chains[112]. Although from those thermodynamic considerations the sulphanes are unstable at room temperature with respect to hydrogen sulphide and sulphur, they are metastable because the decomposition reaction requires a rather high activation energy of 25 kcal/mole (for disulphane[133]). This is analogous to the alkanes, the essential difference being that, because of the bonded hydrogen atoms on the carbon chains, the alkanes are more resistant to attack by decomposition catalysts.

The problem of a potential barrier to inner rotation of disulphane (and also longer sulphur chains) cannot be dealt with here in detail[112]. There are two possible interpretations for the hindered internal rotation. First by a double bond character of the sulphur–sulphur bond, produced by a $(p \rightarrow d)\pi$ overlap, or second alternatively as a result of the repulsion of the p_z-lone pair electrons of the sulphur atoms. Sulphanes are soluble in anhydrous organic solvents such as ether, benzene, carbon disulphide, etc. (decreasingly with increasing chain length), but are insoluble in aqueous acids (except H_2S; in neutral and still faster in alkaline solutions they are decomposed completely and very rapidly). The higher sulphanes are considerably more acidic than hydrogen sulphide[134].

Analytical determination and chemical properties of sulphanes. For chemical analysis of sulphanes, degradation reactions by nucleophiles may be used, followed by determination of the reaction products. So sulphanes react with an excess of sulphite quantitatively to form sulphide and thiosulphate, according to

$$H_2S_x + (x-1)SO_3^{--} \longrightarrow H_2S + (x-1)S_2O_3^{--}$$

The formed hydrogen sulphide is precipitated as CdS and then determined iodometrically. Its amount is a direct measure for the amount of the sulphanes. Excess sulphite in the filtrate is removed by adding formalin, so that the thiosulphate formed can also be determined iodometrically. The ratio sulphide/thiosulphate is a measure for the chain length of the sulphane[135]. The method of the chain degradation by cyanide ions

$$H_2S_x + (x-1)CN^- \longrightarrow H_2S + (x-1)SCN^-$$

[133] F. Fehér and H. Weber, *Z. Elektrochem.* **61** (1957) 285.
[134] G. Schwarzenbach and A. Fischer, *Helv. Chim. Acta*, **43** (1960) 1365.
[135] M. Schmidt and G. Talsky, *Z. analyt. Chem.* **166** (1959) 274.

is similar to the sulphite method, but in this case the thiocyanate formed can be easily determined colorimetrically as $Fe(SCN)_3$. For some other gravimetric or volumetric analytical methods for the quantitative determination of sulphanes see ref. 112.

All the hitherto-known chemical analysis methods inherently show a severe handicap: they only allow a determination of the absolute amount of the sulphanes present, and the average chain length of the mixture, but not of the exact stoichiometric composition of the mixture (newer studies have revealed that "pure" sulphanes in most cases really are more or less complicated sulphane mixtures). This handicap makes kinetically controlled reactions with sulphanes of exactly known composition very difficult, as they are necessary for instance for the synthesis of thermodynamically unstable sulphur rings. Therefore, the suitability of NMR spectroscopy was tested for the problems connected with a good analysis of sulphane mixtures and of pure sulphanes[136]. This method proved to be particularly effective; it will greatly facilitate future investigations of the chemistry of the sulphanes.

A differentiation of different sulphanes by NMR spectroscopy depends on two important conditions: (1) The exchange rate of protons between individual species (via hydrogen bridges) has to take place slowly enough so that the mean residence time of individual hydrogen atoms at sulphur atoms is sufficiently far above the relaxation period of the experiment. It is known that this condition is not verified in mixtures of water and hydrogen peroxide. They show only one common proton signal. The observation of the H–S–D coupling in the hydrogen sulphide molecule[137] showed already, however, that this exchange proceeds slowly enough in H_2S. (2) The distinguishability of proton signals of sulphanes requires a sufficiently high substituent effect of the sulphur atoms newly added during growth of the chain. It can be predicted that this effect will become too small above a certain chain length. Both conditions are realized. The spectrum of sulphane mixtures gives an entire series of well-resolved and partly highly displaced proton signals of different intensity. Each of these signals can be attributed to a certain pure sulphane. High-purity carbon disulphide solutions show besides a solvent shift identical signals. The assignment of the individual signals can therefore simply be solved by comparison spectra of pure sulphanes with or without solvents. The signal of H_2S has been selected as an internal standard. Its relative position with respect to tetramethylsilane is known[137]. Whereas sulphanes from one to six sulphur atoms easily can be found in the spectra, the proton signals of the higher sulphanes differ only by very small amounts, probably below the detection limit: They appear in almost identical positions of the spectrum (within $\pm 0{\cdot}5$ c/s). Spectrometers with a higher power of resolution probably will enable a further differentiation of those higher sulphanes. The relative position of the signals of individual sulphanes permits not only a qualitative proof, but also, by means of the relative area ratios, a quantitative statement concerning the composition of a mixture of sulphanes. Table 11 shows the chemical shifts of the proton signals of sulphanes.

Until now the chemistry of the sulphanes has been only poorly characterized. A main reason for this fact is to be seen in the severe difficulties of preparation, separation and handling of these compounds in rather pure state and even in analytical difficulties, encountered until quite recently. Those experimental difficulties will be overcome more and more; this will broaden up considerably the interesting field of the chemistry of the sulphanes. Only a few of their reactions have hitherto been used for practical (synthetic

[136] H. Schmidbaur, M. Schmidt and W. Siebert, *Chem. Ber.* **97** (1964) 3374.
[137] H. Schmidbaur and W. Siebert, *Chem. Ber.* **97** (1964) 2090.

TABLE 11. CHEMICAL SHIFTS OF THE PROTON SIGNALS OF SULPHANES IN VARIOUS SOLVENTS[c]

	In CS_2		In H_2S_2		In crude sulphane	
	a	b	a	b	a	b
H_2S	0	− 51·0	0	− 93·0	0	− 99·0
H_2S_2	−112·0	−163·0	−116·0	−209·0	−109·0	−208·0
H_2S_3	−192·5	−244·5	−198·5	−291·5	−194·0	−293·0
H_2S_4	−195·0	−246·0	−200·5	−293·5	−195·0	−294·0
H_2S_5	−202·5	−253·5	−207·5	−300·5	−202·0	−301·0
H_2S_6	−207·5	−258·5	−210·7	−303·7	−206·0	−305·0

[a] All values in c/s at 60 kc/s. Negative values indicate the position of the signals at lower field strengths, referred to the standard hydrogen sulphide. Limits of error ±2 and ±1 c/s with small differences. Temperature 35±2°.

[b] As in (a), but against the external standard of tetramethylsilane.

[c] M. Schmidt, in B. Meyer (Ed.), *Elemental Sulphur*, p. 333, Interscience, New York (1965).

or analytical) uses, as for instance their condensation with chlorosulphanes for the synthesis of either higher sulphanes and chlorosulphanes[112], respectively, or of thermodynamically unstable sulphur rings; also the analytically useful degradation of sulphane chains by sulphite and other nucleophiles have already been mentioned above[135]. The mechanism may be symbolized, for the example of the cyanide degradation of tetrasulphane, as

$$\begin{array}{llll} \text{H–S–S–S–SH} & +\text{CN}^- \longrightarrow & \text{H–S–S–S–H} & +\text{SCN}^- \\ \text{H–S–S–S–H} & +\text{CN}^- \longrightarrow & \text{H–S–S–H} & +\text{SCN}^- \\ \text{H–S–S–H} & +\text{CN}^- \longrightarrow & \text{H–S–H} & +\text{SCN}^- \\ \hline \text{H}_2\text{S}_4+3\text{CN}^- & \longrightarrow & \text{H}_2\text{S}+3\text{SCN}^- & \end{array}$$

The reaction of sulphanes with sulphur trioxide, leading to sulphonic acids of sulphanes, will be dealt with on page 886, those with boron halides as Lewis acids in Section 2.11. For some reactions of sulphanes with halogenated organic compounds see ref. 112. Only recently the sulphanes have been proved as a useful tool in synthetic organic chemistry: 1·3-diketones in organic solvents saturated with hydrogen chloride react with sulphanes under formation of 1·2-dithiolium ions[138–140]:

$$R_1\text{–C(OH)=C(}R_2\text{)–C(=O)–}R_3 + H_2S_x + HCl \longrightarrow [\text{1,2-dithiolium}(R_1,R_2,R_3)]^{\oplus}\,Cl^- + 2H_2O + \frac{x-2}{8}\,S_8$$

Hydrogen Sulphide

As already mentioned, the first member of the series of the sulphanes, H_2S, deserves a special treatment, not only because of its great practical interest, but also because of the

[138] H. Schulz, Thesis, Univ. Würzburg (1969).

[139] M. Schmidt and H. Schulz, *Chem. Ber.* **101** (1968) 277.

[140] M. Schmidt and H. Schulz, *Z. Naturforsch.* **23b** (1968) 1540.

fact that hydrogen sulphide differs in many of its physical and chemical properties much more from elemental sulphur on the one side and the higher sulphanes on the other side than do the sulphanes with S–S bonds in their molecules. Hydrogen sulphide as the only thermodynamically stable binary sulphur–hydrogen compound frequently occurs in nature, but usually only in small concentrations (produced by different bacteria from sulphates or organic sulphur compounds)—for instance, in springs. In natural gas, however, the H_2S content may well exceed 10% (Laqu, France)[141].

The usual method for preparing hydrogen sulphide is either the long-used decomposition of ferrous sulphide by hydrochloric acid in a Kipp apparatus

$$FeS + 2HCl \longrightarrow FeCl_2 + H_2S$$

or the heating of a mixture from sulphur and paraffin wax[142]. For most laboratory uses the rather considerable amounts of impurities of those products do no harm. Purer products are obtained by hydrolysis of the sulphides of alkali and alkaline earth metals, especially of aluminium sulphide[143]. The purest hydrogen sulphide is made directly from the elements at 600°C[144]

$$(1/x)S_x + H_2 \rightleftharpoons H_2S + 4{\cdot}8 \text{ kcal}$$

The colourless gas with its well-known unpleasant odour (already in very high dilution!) is extremely poisonous, similar to hydrogen cyanide! Because it is heavier than air (1 litre $H_2S = 1{\cdot}5392$ g), special care must be taken for useful exhaust systems when working with hydrogen sulphide. The gas may easily be condensed to a colourless liquid. There is, in contrast to water, no considerable hydrogen bonding affecting the physical properties of hydrogen sulphide. Some of its properties are shown in Table 12.

TABLE 12. SOME PHYSICAL PROPERTIES OF HYDROGEN SULPHIDE[145]

Melting point	−85·60°C
ΔH (fusion)	0·5676 kcal/mole
Boiling point	−60·75°C
ΔH (vaporization)	4·463 kcal/mole
Transition temperatures	−169·6°C and −145·0°C
Critical temperature	100·4°C
Critical pressure	89 atm
Heat of formation	+4·80 kcal/mole
Density	0·993 g/cm^3 (−60°C)
Dielectric constant	~10 (−60°C), ~6 (0°C)
Dipole moment	$1{\cdot}10 \times 10^{-18}$ μ (gas phase)

The hydrogen sulphide molecules have a bent configuration with bond angle varying from 90° in the liquid phase to 92°20′ in the gas phase. The H–S bond distance in the gas phase (ground state) is 1·35 Å. In the solid state there seem to exist four modifications of hydrogen sulphide. It is probable that H_2S crystallizes in the fluorite system[112]. Hydrogen sulphide is soluble in water to a considerable extent. One volume of H_2O dissolves 4·65 volumes H_2S at 0°C, 3·44 at 10°C, and 2·61 at 20°C. Such solutions act as two-basic weak

[141] H. Laurien and G. G. Wedekind, *Erdöl u. Kohle* **14** (1961) 223.
[142] F. Seel, *Angew. Chem.* **68** (1956) 739.
[143] A. Klemenc, *Die Behandlung und Reindarstellung von Gasen*, Georg Thieme, Leipzig (1938).
[144] A. Klemenc and O. Bankowski, *Z. anorg. allg. Chem.* **208** (1932) 366.
[145] H. Remy, *Lehrbuch der Anorganischen Chemie*, Band I, 9. Auflage, S. 802, Akademische Verlagsgesellschaft, Leipzig (1957).

acids, with $pK_{A1} = 6{\cdot}88$ and $pK_{A2} = 14{\cdot}15$[146]. They are easily oxidized, mostly to elemental sulphur, even by air. Hydrogen sulphide may be ignited in air and burns with a light blue flame

$$H_2S + 1{\cdot}5O_2 \longrightarrow H_2O + SO_2$$

There is considerable practical interest and therefore a very comprehensive technical literature dealing with the hydrogen sulphide–oxygen system as well as with the problems connected with a catalytic oxidation of H_2S [147]; those systems serve on the one hand for the purification of technical gases from small amounts of H_2S, and on the other hand for the production of sulphur and sulphur dioxide by application of the Claus method, which involves the burning of hydrogen sulphide in air. Many other oxidants also can oxidize the negatively charged sulphur in H_2S, either to the oxidation number of zero or to higher oxidation numbers, depending on the relative strength of the oxidants; many redox reactions with hydrogen sulphide do, however, only proceed in the presence of at least catalytic amounts of water (or at rather high temperatures).

The same statement also holds for the reaction of H_2S with many metals, forming the corresponding metal sulphides, in spite of the usually very high tendency of formation of such metal sulphides. These salts of the acid hydrogen sulphide will be treated in this treatise on comprehensive inorganic chemistry in connection with the different metals. Only their formation in connection with the use of H_2S for analytical purposes will briefly be described in the section on sulphur and its compounds (2.16). Just one brief mention, however, will be made of an interesting study of the low-temperature reaction between alkali metals and solid hydrogen sulphide[148]. From this study it is concluded that complete transference of the valence electron from the metal atom to a hybrid orbital of the sulphur atom takes place, forming the ion H_2S^- as an intermediate.

Salts of Sulphanes (Polysulphides)[134, 149]

Alkali and alkaline-earth metal polysulphides have been known for a long time. They may be prepared from the elements directly, from the elements (or metal amides and sulphur) in liquid ammonia (which still seems to be the best synthetic method), from metal sulphides and sulphur in the melt, in water or in alcohol, also from metal hydrosulphides or metal hydroxides with sulphur, etc.

Most of these methods do not yield pure products, but only mixtures of different polysulphides. Under anhydrous conditions, however, a complete series of potassium polysulphides could be prepared: K_2S_2, K_2S_3, K_2S_4, K_2S_5 and K_2S_6. For the preparation of X-ray pure substances, liquid ammonia is the best solvent. Without any doubt, the polysulphides contain S_x^{--} anions, which are unbranched chains of sulphur atoms. This is proved by X-ray studies as well as by determinations of molar volumes and molar refractions. Older hypotheses of branched structures for polysulphides no longer can be maintained. Polysulphides of organic bases have been prepared[150] by dissolving amine and sulphur in a nonpolar solvent and slowly passing hydrogen sulphide into the mixture.

146 M. Widher and G. Schwarzenbach, *Helv. Chim. Acta*, **47** (1964) 266.

147 L. Gmelin, *Handbuch der Anorganischem Chemie*, 8. Auflage, Band IXb, Abt. 1, S. 47, and Band IXa, Abt. 2, S. 241, Verlag Chemie, Weinheim (1953).

148 J. E. Bennet, B. Mile and A. Thomas, *Chem. Comm.* **7** (1966) 182.

149 M. Schmidt, in F. G. A. Stone and W. A. G. Graham (Eds.), *Inorganic Polymers*, p. 103, Academic Press, New York, London (1962).

150 H. Krebs, E. F. Weber and H. Balters, *Z. anorg. allg. Chem.* **275** (1954) 147.

The least soluble polysulphide separates. The polysulphide formation is enhanced by the opening of the S_8 ring caused by the amine sulphide, or rather the S^{--} ion. Heptasulphides and hexasulphides of different amines were prepared in this way in the crystalline state. They are yellow to orange in colour and quite unstable. By the same method also polysulphides of diamines were prepared[151]. However, since they are much more stable, they could simply be prepared in water or alcohol–water mixtures by adding sulphur to the amine solution and saturating it with hydrogen sulphide. The salts crystallize out. Tri-, tetra, penta-, hexa-, and heptasulphides of different diamines were obtained in this way. Water-soluble polysulphides in aqueous solution always are decomposed into a complicated mixture of different polysulphides. The conditions in such polysulphide solutions are easy to understand with the hypothesis of the nucleophilic degradation of sulphur–sulphur bonds. It is well known that elemental sulphur dissolves readily in alkali sulphide solutions with formation of alkali polysulphides. Why are such polysulphide solutions stable at all? On the one hand it is known that the sulphanes are extremely unstable towards traces of alkali, so that they only can be prepared by careful pouring of the polysulphide solution into a large excess of cold hydrochloric acid, and never in the reverse way, as with other acids or with hydrogen sulphide. On the other hand, polysulphide solutions react by hydrolysis to form strongly alkaline solutions, which can arise only from the fact that, in spite of the alkaline reaction, the solution must contain free sulphanes (arising from the reaction $S_x^{--}+2HOH \rightarrow H_2S_x+2OH^-$). This apparent contradiction may be understood by the following concept (in principle already explained on page 818). The elemental sulphur dissolves in the sulphide solutions because the eight-ring S_8 is easily cleaved by the strong thiophilic S^{--} or SH^- ions according to

$$S_8\ (\text{ring}) + S^{--} \rightarrow {}^-S{-}S{-}S{-}S{-}S{-}S{-}S{-}S{-}S^-$$

with the formation of a nonasulphide ion, which then is degraded to shorter chains. At the same time, longer chains are formed from the S^{--} ions, such as S_2^{--}, S_3^{--} and so on, which themselves can engage further in the decomposition process. One may consider the processes in aqueous polysulphide solutions as a continuous "sulphide decomposition" of polysulphide solutions and as a simultaneous polysulphide synthesis of sulphides, which can be schematically represented as

$$\begin{aligned}
{}^-S{-}S{-}S{-}S{-}S{-}S{-}S{-}S{-}S^- + S^{--} &\longrightarrow {}^-S{-}S{-}S{-}S{-}S{-}S{-}S{-}S^- + S_2^{--}\\
{}^-S{-}S{-}S{-}S{-}S{-}S{-}S{-}S^- + S_2^{--} &\longrightarrow {}^-S{-}S{-}S{-}S{-}S{-}S{-}S^- + S_3^{--}\\
{}^-S{-}S{-}S{-}S{-}S{-}S{-}S^- + S_3^{--} &\longrightarrow {}^-S{-}S{-}S{-}S{-}S{-}S^- + S_4^{--}\\
{}^-S{-}S{-}S{-}S{-}S{-}S^- + S_4^{--} &\longrightarrow {}^-S{-}S{-}S{-}S{-}S^- + S_5^{--}
\end{aligned}$$

etc.

Since the reactions, as indicated, proceed in discrete steps, the degradation of nonasulphanes proceeds through octa-, hepta-, hexa and pentasulphane, etc., parallel with the synthesis of disulphanes through tri-, tetra-, pentasulphane, etc., whereby a dynamic equilibrium is set up. We have, therefore, in an aqueous polysulphide solution of the type which might be formed by the action of alkali sulphide solution on elemental sulphur, no statically existing compounds in the presence of each other but rather a complicated

[151] H. Krebs and K. H. Müller, *Z. anorg. allg. Chem.* **281** (1955) 187.

dynamic equilibrium involving sulphides of various chain lengths. Naturally, only unbranched sulphur chains may take part in those equilibria. Those solutions, then, may be nominated "pseudostable". This interpretation explains the fact that in no instance could an aqueous solution of polysulphides of definite chain length be obtained by treatment of stoichiometric quantities of sulphur and sulphide, and from which by subsequent treatment with acid a pure sulphane could be obtained. In every case, only a mixture of different sulphanes was found[149].

Polysulphides of some heavy metals such as Pb, Zn, Cd, Hg, Cu, As, Sb and Bi have been reported[152]. These mostly amorphous materials are formed by the reaction of metal thiophenolates with elemental sulphur in the presence of amines. Their physical and chemical properties have not yet been studied in detail. Three recent results on transition metal polysulphides will, however, be briefly mentioned, because they show the existence of surprisingly stable chains of five or four, respectively, sulphur atoms, chelating a central metal atom. A new field of heterocyclic sulphur metal rings seems to open up with these studies.

Already in 1903 the most interesting sulphane complex $(NH_4)_2PtS_{15}$ was isolated[153]. The authors were aware of the fact that S_5^{--} groups were probably present, but they did

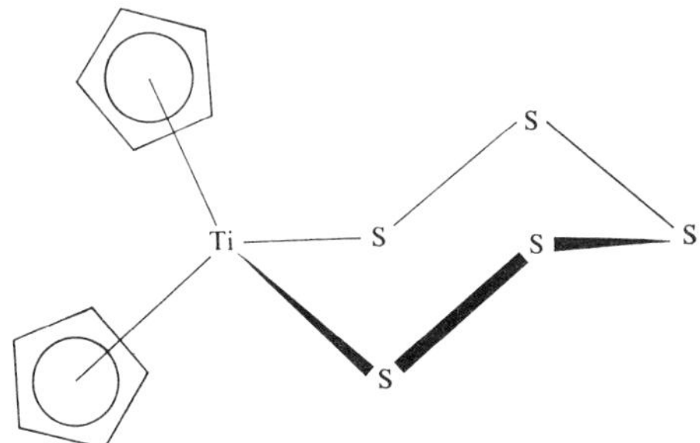

FIG. 5. Conformation of $(C_5H_5)_2TiS_5$.

not propose a structure for this rather unusual compound. The structure was only recently revealed[154] as such of a platinum(IV) complex ion in which the central atom octahedrally is surrounded by three S_5 groups; three six-membered rings composed of five sulphur atoms and one platinum atom are connected by this single metal atom that is common to all three rings that are present in the chair conformation. Quite recently the synthesis of the complex has been improved and some of its chemical reactions have been studied[155]. One ligand reaction, the attack of the sulphane chelate rings by cyanide ions, has been investigated in detail. The overall stoichiometry of this reaction is

$$PtS_{15}^{--} + 17CN^- \longrightarrow Pt(CN)_4^{--} + 13SCN^- + 2S^{--}$$

Early in the reaction a new sulphane chelate is formed, PtS_{10}^{--}. This new complex consists of planar platinum(II) surrounded by two pentasulphide chelating ligands, that is from two six-membered rings that have one platinum atom in common. A similar pentasulphane

[152] H. Krebs, H. Fassbänder and F. Jorgens, *Chem. Ber.* **90** (1957) 425.
[153] K. A. Hofmann and F. Höchtlein, *Chem. Ber.* **36** (1903) 3090.
[154] P. E. Jones and L. Katz, *Chem. Comm.* (1967) 842.
[155] A. E. Wickenden and R. A. Krause, *Inorg. Chem.* **8** (1969) 779.

derivative that is extremely stable (with a reversible melting point of 202°C far above the floor temperature of elemental sulphur it is the thermally most stable covalent ring, containing sulphur–sulphur bonds, hitherto known) has been synthesized by the following methods:

$$(C_5H_5)_2TiCl_2+(NH_4)_2S_5 \longrightarrow (C_5H_5)_2TiS_5+2NH_4Cl$$
$$2(C_5H_5)_2Ti(SH)_2+S_8 \longrightarrow 2(C_5H_5)_2TiS_5+3H_2S$$
$$(C_5H_5)_2Ti(SH)_2+Cl_2S_3 \longrightarrow (C_5H_5)_2TiS_5+2HCl$$

This interesting bis-π-cyclopentadienyl-titanium(IV)-pentasulphide exists in a fixed conformation that is shown in Fig. 5[156–158]. The first chelate of tetrasulphane with a transition metal has been synthesized very recently by the reaction[159]

$$(C_5H_5)_2MoCl_2+(NH_4)_2S_5 \longrightarrow (C_5H_5)_2MoS_4+\tfrac{1}{8}S_8+2NH_4Cl$$

This five-membered ring from four sulphur and one molybdenum atoms also is surprisingly stable. Its most probable structure is shown in Fig. 6.

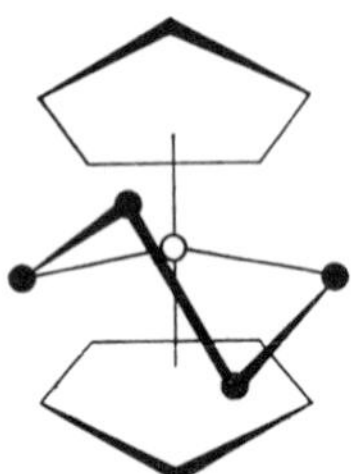

FIG. 6. Bis-π-cyclopentadienyl-molybdenum(IV)-tetrasulphide.

Monomeric and Polymeric Organic Sulphane Derivatives

Organic sulphane derivatives have been known for a long time, but their structure has been the subject of considerable controversy, some authors preferring an unbranched structure, others a branched structure, and some even cyclic structures. Today, however, we possess overwhelming evidence, experimental and theoretical, that such branched structures are not capable of existence[149]. The most important organosubstituted sulphanes are disulphanes of the general formula R–S–S–R. They may be prepared by a number of redox reactions described earlier[160] from mercaptanes, by alkylation or arylation of alkali disulphides, etc.[149]. Organic trisulphanes, tetrasulphanes, pentasulphanes, and hexasulphanes can be obtained from mercaptanes and the corresponding chlorosulphanes via HCl elimination[149]. Except for the disulphanes, which are higher boiling than the corresponding thio ethers, the yellow oils (or solids) cannot be distilled without decomposition. They are very toxic and characterized by an extremely unpleasant odour. The extraordinary difficulty of preparing higher organic sulphanes in the pure state may explain

156 H. Köpf, B. Block and M. Schmidt, *Chem. Ber.* **101** (1968) 272.
157 H. Köpf and B. Block, *Chem. Ber.* **102** (1969) 1504.
158 H. Köpf, *Chem. Ber.* **102** (1969) 1509.
159 H. Köpf, *Angew. Chem.* **81** (1969) 332.
160 A. Schöbert and A. Wagner, *Methoden der organischen Chemie* (Houben-Weyl), 4th ed., Vol. 9, p. 59, Georg Thieme, Stuttgart (1955).

some severe discrepancies in the literature regarding their properties (it is very difficult to decide among solutions of sulphur in lower sulphanes, sulphane mixtures and pure sulphanes). Reactions dealing with the scission of the sulphur–sulphur bond in these compounds have been reviewed[161]. For some physical properties of organic sulphane derivatives see ref. 149. Sulphur chains or also only single sulphur atoms can link together bi- or trifunctional hydrocarbon groups, thus forming polymeric sulphur compounds[162]. In these typical borderline cases between inorganic and organic chemistry one has to differentiate between mono- and polysulphides and between aliphatic and aromatic hydrocarbon groups. Aliphatic polysulphides of the thiokol type have the general composition $(C_xH_{2x}S_y)_n$ and are very well known and also widely used in practice. In nearly all of the technical polymers of this type $x = 2$ and $y = 4$; one hydrogen atom thereby may be replaced by other organic groups of a different kind. With $x = 2$ and varying y from 1 to 8, rubber-like insoluble polymers may be obtained, for example $(CH_2S_5)_n$. With $x = 1$ and $y = 1$ one ends up with the familiar polythioformaldehydes $(CH_2S)_n$. They may be regarded as intermediates between pure hydrocarbon chains and pure sulphur chains. For a long time besides different rather scarcely characterized high polymeric forms the only known low molecular species was the trimeric molecule, trithiane. By the reaction of methylene chloride with sodium sulphide under suitable conditions a well-defined tetrameric and pentameric thioformaldehyde could be synthesized, $(CH_2S)_4$ and $(CH_2S)_5$ [163, 164]. The tetrameric form, which is isoelectronic with cyclo-octasulphur, may easily be polymerized to a highly

$$n\,C_6H_5{-}Cl + nS_x \longrightarrow n\,HCl + [{-}C_6H_4{-}S_x{-}]_n$$

$$n\,Cl{-}C_6H_4{-}Cl + nS_x \longrightarrow n\,HCl + [{-}C_6H_3(Cl){-}S_x{-}]_n$$

$$[{-}C_6H_3(Cl){-}S_x{-}]_n + nS_y \longrightarrow n\,HCl + [{-}C_6H_3(S_y{-}){-}S_x{-}]$$

161 A. J. Parker and N. Kharash, *Chem. Revs.* **59** (1959) 584.
162 M. Schmidt, *Inorganic Macromolecular Reviews* **1** (1970) 101.
163 M. Schmidt, K. Blättner, P. Kochendörfer and H. Ruf, *Z. Naturforsch.* **21b** (1966) 622.
164 M. Russo, *Polymer Letters* **3** (1965) 455.

crystalline polymeric $(CH_2S)_n$ of melting point 247°C. Aliphatic polysulphides may easily be prepared by a nucleophilic displacement of chlorine by sulphide, that is via a simple polycondensation. This is incomparably more difficult with aromatic compounds because of the pronounced unreactivity of aromatically bound chlorine. But aromatic polysulphides attract the interest of polymer chemists because a phenylene linkage in the backbone of a polymer chain or network should be a desirable moiety that imparts regularity and rigidity to the structure[165, 166]. In this connection[162] the behaviour of elemental sulphur towards chlorinated aromatic compounds was studied. The following equations demonstrate how elemental sulphur and monochlorobenzene do react with each other under elimination of hydrogen chloride, whereby polymeric phenylene polysulphides are formed. The average number of sulphur atoms between the phenylene groups depends on the amount of sulphur in the reaction mixture and may vary from 2 to about 6 or 7.

This reaction indeed takes place quantitatively within 50 hr, if the reaction mixture is heated to 225°C in an autoclave. The polymers are soluble in carbon disulphide and melt at about 100°C; they are miscible with liquid sulphur. In a radial mechanism, of course, not only para-linked but also ortho- and meta-linked phenylene polysulphides are formed.

With dichlorobenzene at the same temperature a similar reaction takes place

$$nC_6H_4Cl_2 + nS_z \longrightarrow nHCl + (C_6H_4ClS_z)_n$$

The polymer thus obtained contains exactly half of the original chlorine, the other half being eliminated as hydrogen chloride. This chlorophenylene–polysulphide polymer is still soluble in carbon sulphide. By heating it at 260°C together with more sulphur a further elimination of HCl takes place, leading quantitatively according to

$$(C_6H_4ClS_z)_n + nS_y \longrightarrow (C_6H_4S_zS_y)_n + nHCl$$

to a two-dimensional polymer which is insoluble in carbon disulphide. Of course, the same product may also and more easily be prepared in one step by heating the starting materials directly at temperatures above 250°C. With x and y about 2, the polymers are stable up to 350°C. It is obvious that by the reaction of mixtures of mono- and dichlorobenzenes with sulphur and thus combining the two reactions, polymers of intermediate properties because of only partial crosslinking may be obtained. These radical reactions of sulphur are of a more general type and are not restricted to aromatic halogene compounds. For instance, the principle also works with benzene sulphonic acid—in this case, of course, sulphurous acid is set free instead of hydrogen chloride. According to this a one-step synthesis of those polymers is also possible

$$n\,C_6H_6 + nS_x + nH_2SO_4 \longrightarrow 2nH_2O + nSO_2 + [\!-C_6H_4-S_x-]_n$$

directly from benzene, sulphuric acid and sulphur. The amount of crosslinking in this case may be influenced by variation of the ratio benzene : sulphuric acid from 1 : 1 to 1 : 2. Those polymeric sulphane derivatives will be worth further studies from a theoretical point of view as well as (even much more) from a practical point of view[162].

[165] A. D. Macallum, *J. Org. Chem.* **13** (1948) 154.
[166] R. W. Lenz and C. E. Handlorits, *J. Polymer Sci.* **43** (1960) 167.

2.2. SULPHUR HALIDES

Significant differences between first and second row elements and their compounds are also found in the series of sulphur halogen compounds. Thus sulphur fluorides behave quite differently in their reactivity and stability from that of the other sulphur halides. In general the multivalent fluorides of sulphur exhibit a higher stability than the lower fluorides. Sulphur chlorides behave just the opposite way. Bromine and iodine compounds possess very weak S–X bonds; actually no definite sulphur iodine compound has been reported[167].

Sulphur and Fluorine

Sulphur forms stable compounds with fluorine in each of its typical valency states 2, 4 and 6. In contrast, the other halides do not show a tendency to combine with sulphur for SX_4 and SX_6 compounds, which is due to their smaller electronegativities and larger covalent radii in comparison to fluorine. The following binary sulphur fluorides have been prepared by various methods: S_2F_2, SF_2, $S{=}SF_2$, SF_4, S_2F_{10}, SF_6 and derivatives of XSF_5. Since the lower fluorides are highly reactive and disproportionate easily, they are difficult to analyse. This may explain the controversy about the structure of some of these compounds, which had been until recently in literature.

S_2F_2, Disulphur difluoride. This compound is synthesized by reacting sulphur with silver fluoride at 125° in a rigorously dried apparatus. Although the preparation of S_2F_2

TABLE 13. STRUCTURAL DATA OF DISULPHUR DIHALIDES

Compound	*d*(S–X) (Å)	*d*(S–S) (Å)	< S–S–X (°)	Dihedral angle	Ref.
S_2F_2	1·635±0·01	1·888±0·01	108·3±0·5	87·9± 1·5	(a)
S_2Cl_2	2·07 ±0·01	1·97 ±0·03	107 ±2·5	82·5±12	(b)
S_2Br_2	2·24 ±0·02	1·98 ±0·04	105 ±3	83·5±11	(b)

[a] R. L. Kuozkowski, *J. Am. Chem. Soc.* **86** (1964) 3617.
[b] E. Hirota, *Bull. Chem. Soc. Japan*, **31** (1958) 138.

has been claimed in various publications within the past four decades, it was first synthesized in high purity by Seel[168] and coworkers. The colourless liquid does not attack carefully dried glass. It isomerizes to thiothionyl fluoride in the presence of alkali metal fluorides. S_2F_2 is readily oxidized by N_2O_4 to form nitrosyl fluorosulphate. Physical data of S_2F_2 are given in Tables 13 and 14.

SF_2, Sulphur difluoride. Sulphur difluoride has been reported[169]; however, the available analytical data are insufficient to confirm the compound.

$S{=}SF_2$, Thiothionyl fluoride. As already mentioned, S_2F_2 is unstable in the presence of alkali metal fluoride and rearranges as follows:

$$\text{F–S–S–F} \xrightarrow{\text{KF}} \text{S=SF}_2$$

It has also been prepared according to the following:

$$2KSO_2F + S_2Cl_2 \longrightarrow S{=}SF_2 + 2KCl + 2SO_2$$

[167] H. L. Roberts, in G. Nickless (Ed.), *Inorganic Sulphur Chemistry*, p. 420, Elsevier, Amsterdam, London, New York (1968).
[168] F. Seel and R. Budenz, *Chimia* **17** (1963) 355; *ibid.* **22** (1968) 79.
[169] G. H. Cady, in H. J. Emeleus and A. G. Sharpe (Eds.), *Advances in Inorganic and Radiochemistry*, p. 120, Academic Press, New York (1961).

S=SF_2 represents the only compound having the thiothionyl structure. Hydrolysis in aqueous alkali yields sulphur and thiosulphate, whereas in acidic media various oxyacids are formed. Similar to other compounds having the X=S moiety, S=SF_2 reacts with electrophilic reagents[170]:

$$F_3B\text{–}SF_4 + S \xleftarrow{BF_3} |S\text{–}SF_2| \xrightarrow{HCl} Cl\text{–}S\text{–}S\text{–}Cl + HF$$

$$SO_2, OSF_2, O_2SF_2 \xleftarrow{O_2} |S\text{–}SF_2| \xrightarrow{HF} SF_4 + H_2S$$

The structure[171] of S=SF_2 has been established by microwave and by infrared spectroscopy.

SF_4, Sulphur tetrafluoride. SF_4 has been reported in literature since 1905. However, Silvey and Cady carried out the first unambiguous synthesis and characterization[172] by reacting a thin sulphur film on a cool surface with elemental fluorine:

$$S + 2F_2 \longrightarrow SF_4$$

ClF_3 and IF_5 have also been used successfully as a fluorinating agent for sulphur. The most convenient method was established by Tullock and coworkers[173], which proceeds in a polar solvent such as acetonitrile:

$$3SCl_2 + 4NaF \longrightarrow S_2Cl_2 + SF_4 + 4NaCl$$

The structure of SF_4 has been elucidated by a study of its vibrational[174], NMR[175] and microwave[176] spectra. Vibrational data first indicated that the structure was that of a trigonal bipyramid (C_{2v} symmetry) with one equatorial atom replaced by a lone pair of electrons (sp^3d-hybrid). From the microwave spectrum it was found, however, that deviations

TABLE 14. PHYSICAL DATA OF SULPHUR FLUORIDES

	S_2F_2	S=SF_2	SF_4	SF_6	S_2F_{10}
M.p., °C	−133	−165	−121	subl.	−52·7
B.p., °C	15	−10·6	−38	−63·8	+30
Density (g/cm³)			1·919 (200°K)	1·722 (−20°C)	2·08 (0°C)
Heat of formation (kcal/mole)			6·32	0·26	
Dipole moment (Debye)			0·632		
Trouton's constant			27·1	22	23·0
ΔH(S–F) (kcal/mole)			~78	~58	

occur from the ideal trigonal bipyramid ($<F_e\text{–}S\text{–}F_e = 101°33'$, $<F_a\text{–}S\text{–}F_a = 173°04'$). The deviation from the 120° trigonal angle in the equatorial plane is remarkable. It has been suggested to explain this fact by considering sp^2-hybridization in the equatorial plane and *pd*-hybridization for the two axial fluorines. Both axial fluorines are slightly bent towards the equatorial fluorine. Physical data are given in Table 14. Since SF_4 is extremely

170 F. Seel and D. Görlitz, *Z. anorg. allg. Chem.* **327** (1964) 32.
171 R. D. Brown, G. P. Pez and M. F. O'Dwyer, *Australian J. Chem.* **18** (1965) 627.
172 G. A. Silvey and G. H. Cady, *J. Am. Chem. Soc.* **72** (1950) 3624.
173 C. W. Tullock, F. S. Fawcett, W. C. Smith and D. D. Coffman, *J. Am. Chem. Soc.* **82** (1960) 539.
174 R. E. Dodd, H. L. Roberts and L. A. Woodward, *Trans. Faraday Soc.* **52** (1956) 1052.
175 E. L. Muetterties and W. D. Phillips, *J. Am. Chem. Soc.* **81** (1959) 1084.
176 W. M. Tolles and W. D. Gwinn, *J. Chem. Phys.* **36** (1962) 1118.

reactive, it has been successfully applied as a fluorinating agent for both inorganic and organic compounds[177], as is demonstrated by the following equations:

$$C_6H_5\text{–}\overset{O}{As}(OH)_2 + 3SF_4 \longrightarrow C_6H_5AsF_4 + 3OSF_2 + 2HF$$
$$CH_3\text{–}CO\text{–}CH_3 + SF_4 \longrightarrow CH_3\text{–}CF_2\text{–}CH_3 + OSF_2$$

SF_4 decomposes readily in the presence of moisture:

$$SF_4 + H_2O \longrightarrow 2HF + OSF_2 \xrightarrow{H_2O} SO_2 + 2HF$$

Towards certain inorganic fluorides, SF_4 behaves as a Lewis base, thereby forming addition compounds: these are listed in approximate order of decreasing stability:

$SbF_5.SF_4$, $AsF_5.SF_4$, $IrF_5.SF_4$, $BF_3.SF_4$, $PF_5.SF_4$, $GeF_4.2SF_4$ and $AsF_3.SF_4$.

They were first[178] formulated as donor acceptor complexes, $F_4S.BF_3$. Seel[179], however, suggested on the basis of solid infrared spectra that $SF_4.BF_3$ should be regarded as $SF_3^+BF_4^-$. SF_4 can be liberated by other nonvolatile fluorides:

$$BF_3.SF_4 + NaF \longrightarrow NaBF_4 + SF_4$$

It was observed that SF_4 also reacts as electron acceptor, since in pyridine a stable 1 : 1 complex is formed. The acceptor capability of SF_4 is further demonstrated by the formation of $[(CH_3)_4N][SF_5]$ and $CsSF_5$ species[180]. Strong fluorinating agents oxidize SF_4 to SF_6 or derivatives of it:

$$SF_4 + F_2 \longrightarrow SF_6$$
$$2SF_4 + ClF_3 \xrightarrow{200^\circ} SF_6 + SF_5Cl$$
$$Cl_2 + SF_4 + CsF \longrightarrow SF_5Cl + CsCl$$

Interesting insertion reactions of SF_4 occur with certain peroxides:

$$F_5S\text{–}O\text{–}O\text{–}SF_5 + SF_4 \longrightarrow F_5S\text{–}O\text{–}SF_4\text{–}O\text{–}SF_5$$

The two F_5S–O groups are in *cis*-position. A similar reaction has been observed with hypofluorites[181]:

$$F_5S\text{–}OF + SF_4 \longrightarrow F_5S\text{–}O\text{–}SF_5$$

The oxidation of SF_4 by oxygen proceeds slowly. However, it may be catalysed by NO_2:

$$2SF_4 + O_2 \longrightarrow 2F_4SO$$

In certain cases halogen exchange occurs accompanied by a redox reaction:

$$3SF_4 + BCl_3 \longrightarrow 4BF_3 + 3SCl_2 + 3Cl_2$$

Intermediates could not be observed since the energy required of fluorine exchange amounts only to 4–5 kcal/mole. An interesting reaction takes place between diphenyldisulphide and AgF_2, which yields an organic derivative of SF_4:

$$(C_6H_5S)_2 \xrightarrow{AgF_2} 2C_6H_5\text{–}SF_3$$

This method[182] does not work with aliphatic disulphides below n-butyl. Aliphatic

[177] W. C. Smith, *Angew. Chem. Intern. Ed. Engl.* **1** (1962) 467.
[178] N. Bartlett and P. L. Robinson, *Chem. Ind.* (1956) 1351.
[179] F. Seel and O. Detner, *Z. anorg. allg. Chem.* **301** (1959) 113.
[180] E. L. Muetterties, *J. Am. Chem. Soc.* **82** (1960) 1082.
[181] G. Pass and H. L. Roberts, *Inorg. Chem.* **2** (1962) 1052.
[182] W. A. Sheppard, *J. Am. Chem. Soc.* **84** (1962) 3058.

derivatives have been obtained by Muetterties[183], when perfluoro-olefins were reacted with SF_4 in the presence of CsF:

$$CF_3\text{–}CF{=}CF_2+SF_4 \longrightarrow CF_3\text{–}CF_2\text{–}CF_2\text{–}SF_3$$

SF_6, Sulphur hexafluoride. SF_6 was first prepared by Moissan[184] by burning sulphur in an atmosphere of fluorine and removing the lower fluorides by pyrolysis at 400°. This method is still used for commercial synthesis of SF_6, which is a colourless, odourless, tasteless nontoxic and nonflammable gas under ordinary conditions. It possesses a very low solubility in water. Sulphur hexafluoride deserves special attention because of its unusual industrial applications. It exhibits excellent insulating properties for high voltage apparatus even at relatively low pressure and it is stable to silent electric discharge. SF_6 remains unchanged even at 500°, which again demonstrates its unique inactivity. No reaction was found with metals such as copper, aluminium and steel under dry conditions at 110°. Silver is not attacked at the melting point of hard glass. Phosphorus and arsenic distil undecomposed in an atmosphere of SF_6, and NH_3 does not react at dull red heat. Boiling sodium attacks SF_6 to yield Na_2S and NaF. However, sodium dissolved in dimethylglyoxim already reacts at −78°. Hydrogen chloride does not react at high temperatures, but SF_6 decomposes in H_2S atmosphere to give HF and sulphur. The chemical inertness of SF_6 can be explained by its highly symmetrical structure of an octahedron, represented by a sp^3d^2 hybridization with *d*-orbitals reduced to one-half of its original size. Direct attack of nucleophiles (e.g. OH^-, CN^-) on the sulphur atom in the covalently saturated sulphur hexafluoride can only take place by extensive electronic rearrangement.

Fluorine exchange, as it has been observed in SF_4, does not proceed in SF_6 at all. However, SF_6 and its derivatives show a considerable reactivity towards electrophilic compounds such as $AlCl_3$ and SO_3 [185].

$$SF_6+2SO_3 \longrightarrow 3SO_2F_2$$

The differences in the chemical reactivity towards nucleophilic and electrophilic reagents may also indicate that SF_6 is kinetically stable rather than thermodynamically. Support for this can be obtained by calculating the free energy of hydrolysis of SF_6, which indeed proves its thermodynamic instability:

$$SF_6(g)+3H_2O(g) \longrightarrow SO_3(g)+6HF(g) \quad G^\circ = -48 \text{ kcal/mole}$$

From the S–F bond energy in SF_6 (58 kcal/mole) and in SF_4 (78 kcal/mole) the reverse order of reactivity would be expected. Other physical data of SF_6 are listed in Table 14. Oxidation of SF_6 occurs, when a mixture of SF_6/O_2 is heated by the microwave discharge[186]. The reaction also can be initiated by the electrical explosion of extremely small masses of platinum[187], thereby predominantly forming F_4SO.

S_2F_{10}, Disulphur decafluoride. S_2F_{10} is obtained as a by-product of the SF_6 synthesis. The reduction of F_5SCl represents a more convenient synthesis of the compound:

$$2F_5SCl+H_2 \xrightarrow{h\nu} S_2F_{10}+2HCl$$

---

[183] R. M. Rosenberg and E. L. Muetterties, *Inorg. Chem.* **1** (1962) 756.
[184] H. Moissan and Lebeau, *Compt. Rend.* **130** (1900) 865, 984.
[185] J. R. Case and F. Nyman, *Nature*, **193** (1962) 473.
[186] H. J. Emeleus and B. Tittle, *J. Chem. Soc.* (1963) 1644.
[187] B. Siegel and P. Breisacher, *J. Inorg. Nucl. Chem.* **31** (1969) 675.

The disulphur decafluoride differs remarkably in its reactivity from the apparently inert SF_6. On heating it undergoes disproportionation at 150° to form SF_6 and SF_4.

$$S_2F_{10} \xrightarrow{150^\circ} SF_6+SF_4$$

The powerful redox potential of S_2F_{10} and its facile rearrangement account for the extreme toxicity of S_2F_{10}.

$$F_5S\text{–}SO_2F \xleftarrow{SO_2} \boxed{S_2F_{10}} \xrightarrow{KI} I_2 \text{ and var. prod.}$$

$$F_5SBr \xleftarrow{Br_2} \boxed{S_2F_{10}} \xrightarrow{NH_3} NSF_3$$

However, S_2F_{10} is quite resistant towards hydrolysis in acidic and alkaline media. This is due to the fact that substitution of fluorine in SF_6 by the SF_5 moiety does not cause any change in the S–F bond distance (Table 15).

TABLE 15. BOND DISTANCES IN F_5SR COMPOUNDS

Compound	*d*(S–F) (Å)	*d*(S–R) (Å)	Ref.
SF_6	1·56 ±0·02	—	(a)
$F_5S\text{–}SF_5$	1·56 ±0·02	2·21 ±0·03	(b)
$F_5S\text{–}Cl$	1·58 ±0·01	2·030±0·002	(c)
$F_5S\text{–}Br$	1·597±0·002	2·190±0·002	(d)
$F_5S\text{–}OF$	1·53 ±0·04	1·66 ±0·05	(e)
$F_5SO\text{–}OSF_5$	1·56 ±0·02	1·66 ±0·05	(f)

[a] H. Braune and S. Knore, *Z. Phys. Chem.* **321** (1933) 297.
[b] R. B. Hervey and S. H. Bauer, *J. Am. Chem. Soc.* **75** (1953) 2840.
[c] R. Kewley, K. S. R. Murty and T. M. Sudgen, *Trans. Faraday Soc.* **56** (1960) 1732.
[d] E. W. Neuman and A. W. Jache, *J. Chem. Phys.* **39** (1963) 596.
[e] R. A. Crawford, F. Dudley and K. W. Hedberg, *J. Am. Chem. Soc.* **81** (1959) 5287.
[f] F. Dudley, G. H. Cady and D. F. Eggers, *J. Am. Chem. Soc.* **78** (1956) 1553.

Derivates of SF_6. As is shown above, F_5SBr can be prepared by direct bromination of S_2F_{10}. A more general approach to these compounds is represented by the oxidation of SF_4:

$$SF_4+ClF \xrightarrow{380^\circ} F_5SCl$$

$$SF_4+CsF+Cl_2 \xrightarrow{110^\circ} F_5SCl+CsCl$$

F_5SCl is stable up to 400°, while F_5SBr already decomposes at 150° to form SF_6, SF_4 and Br_2. Some physical data of sulphur pentafluorides are given in Table 16. The remarkable difference of these compounds to SF_6 and S_2F_{10} is demonstrated by the ease of hydrolysis, which may be reflected by an electrophilic attack of the supposedly positive chlorine or bromine atom.

$$F_5\overset{-}{S}\text{–}\overset{+}{Cl}+H_2O \longrightarrow F_5S^-+HOCl+H^+$$

$$F_5S^- \longrightarrow SF_4+F^-$$

$$SF_4+3H_2O \longrightarrow SO_3^{2-}+4HF+2H^+$$

Support for this mechanism stems from the fact that F_5SBr hydrolyses more rapidly than F_5SCl (because of its higher positive charge) and that with benzene the halobenzenes are formed rather than $C_6H_5SF_5$. The compounds F_5SX (X=Cl, Br) can be used for the synthesis of organic and inorganic derivatives of SF_6, as is shown in the following equations:

$$2F_5SBr+O_2 \longrightarrow F_5S\text{–}O\text{–}O\text{–}SF_5+Br_2$$

$$F_5SCl+N_2F_4 \longrightarrow F_5S\text{–}NF_2$$

$$F_5SCl+C_2H_4 \longrightarrow F_5S\text{–}CH_2\text{–}CH_2\text{–}Cl \longrightarrow F_5S\text{–}CH{=}CH_2$$

Similarly $F_5S\text{–}C{\equiv}CH$ has been obtained[188]. Numerous other organic compounds have been prepared. However, space limitations prevent further discussion of this interesting field of sulphur fluorine chemistry.

TABLE 16. PHYSICAL DATA OF INORGANIC SULPHUR PENTAFLUORIDES

Compound	F_5SCl	F_5SBr	$(F_5S)_2O$	$(F_5SO)_2$	F_5SNF_2	$(F_5S)_2$	F_5SOF
M.p., °C	−64	−79	−118	−95·4		−52·7	−86
B.p., °C	−21	3·1	31	49·4	−18	30·0	−35·1

Direct fluorination of OSF_2 yields $F_5S\text{–}OF$, sulphur pentafluorohypofluorite, which reacts with F_4SO to form $F_5S\text{–}OO\text{–}SF_5$. Both compounds, $F_5S\text{–}OF$ and $F_5S\text{–}OO\text{–}SF_5$, undergo facile decomposition into radicals and they are powerful oxidizing agents[189].

$$OSF_2+F_2 \longrightarrow OSF_4 \xrightarrow{F_2} F_5S\text{–}OF$$

Thionyl tetrafluoride has a trigonal bipyramidal structure (the oxygen atom is in the plane). Recently $F_5S\text{–}OCl$ has been prepared from OSF_4 and Cl–F[190].

Thionyl Fluorides

The thionyl fluorides, OSF_2, OSFCl, OFS–OM, are derivatives of the sulphurous acid, in which one or two OH groups have been replaced by Cl or F. The long-known OSF_2 was first prepared by reacting $OSCl_2$ with ZnF_2. However, better methods are available:

$$OSCl_2+SbF_3 \xrightarrow{SbF_5} OSF_2+SbCl_3$$

$$OSCl_2+NaF \xrightarrow{CH_3CN} OSFCl+NaCl$$

$$OSFCl+NaF \xrightarrow{CH_3CN} OSF_2+NaCl$$

The structure of OSF_2 as well as of the other thionyl halides consists of a trigonal pyramide with tetrahedral arrangement of the four bonding pairs of electrons. Oxygen, two fluorine and one lone pair of electrons represent the four substituents on the sulphur atom.

Thionyldifluoride hydrolyses slowly to SO_2 and HF. It can be used for replacing Cl or OH by fluorine in silicon compounds[191]. When KF is dissolved in SO_2, KO–SOF is formed[192]. This salt may be regarded as a monosubstituted derivative of sulphurous acid.

188 F. W. Hoover and D. D. Coffman, *J. Org. Chem.* **21** (1964) 3567.
189 C. I. Merrill and G. H. Cady, *J. Am. Chem. Soc.* **83** (1961) 298; *ibid.* **85** (1963) 909.
190 C. J. Schack, R. D. Wilson, J. S. Muirhead and S. N. Cohz, *J. Am. Chem. Soc.* **91** (1969) 2907.
191 R. Müller and V. Mross, *Z. anorg. allg. Chem.* **334** (1965) 86.
192 F. Seel and L. Riehl, *Z. anorg. allg. Chem.* **282** (1955) 293.

Sulphuryl Fluorides

Several routes lead to the synthesis of O_2SF_2: oxidation of SO_2 with fluorine, replacement[193] of chlorine in O_2SCl_2 and heating barium fluorosulphate[194]. Sulphuryl difluoride has a distorted tetrahedral structure and possesses a short S–O bond (1·405 Å) in comparison with SO_3 or O_2SCl_2. Similarly the S–F distance is also remarkably short, which indicates exceptionally strong bonds (Table 17). The formation of O_2SF_2 from SO_3 and the "inert" SF_6 exhibits its high tendency of formation.

Sulphuryl chlorofluoride is obtained by the reaction of O_2SCl_2 with O_2SF_2. Several other sulphuryl fluorides have been prepared. BF_3 and SO_3 yield $FSO_2–O–SO_2–O–SO_2$ [195]. This compound is also formed by heating a mixture of $KBF_4.4SO_3$ and liquid SO_3 or by thermal decomposition of $KBF_4.4SO_3$ at 100°. Peroxysulphuryl difluoride is prepared by the reaction between SO_3 and F_2:

$$2SO_3+F_2 \longrightarrow FSO_2\text{–}OO\text{–}O_2SF$$

Both fluorine atoms are in identical environments according to NMR studies. Pyrosulphuryl difluoride, $FSO_2–O–SO_2F$, may be obtained by heating CaF_2 with two moles of SO_3, resulting in $Ca(SO_3F)_2$, which reacts with SO_2 to form $Ca(S_2O_5F)_2$. Treating the latter with sulphuric acid yields $S_2O_5F_2$. Fluorosulphonic acid can also be used for the synthesis of disulphuryl difluoride:

$$FSO_3H+As_2O_3 \longrightarrow F_2S_2O_5+As_2O_3F_2$$

The hydrolysis of sulphuryl fluoride proceeds slowly in aqueous sodium hydroxide, faster in alcoholic potash media. Ammonolysis yields sulphamide, $O_2S(NH_2)_2$. Hexasulphuryl fluoride, $S_6O_{17}F_2$, has been characterized by ^{19}F NMR and IR spectra[196].

TABLE 17. PHYSICAL DATA OF SULPHURYL HALIDES

Compound	O_2SF_2	O_2SCl_2	O_2SFCl	O_2SFBr
M.p., °C	−120	−54	−125	−86
B.p., °C	−55	69	7	41
Density (g/cm³)		1·66 (20°)		
d(S–X) (Å)	1·530±0·003	1·99±0·02		
d(S–O) (Å)	1·405±0·003	1·43±0·02		

Fluorosulphonic Acid

Fluorosulphonic acid was first prepared by Thorpe and Kirmann in 1892 by the reaction of SO_3 with anhydrous HF. Replacement of chlorine in chlorosulphonic acid by inorganic fluorides, e.g. NH_4F or KF, or reaction of these fluorides with oleum also yields fluorosulphonic acid. This acid possesses remarkable properties in theoretical as well as industrial respect. Five reviews [197–201] have already been published within this decade, which may

[193] C. W. Tullock and D. D. Coffman, *J. Org. Chem.* **25** (1960) 2016.
[194] F. Fehér, in G. H. Brauer (Ed.), *Handbook of Preparative Inorganic Chemistry*, Vol. 1, Academic Press (1963).
[195] H. A. Lehmann and L. Kolditz, *Z. anorg. allg. Chem.* **272** (1956) 1.
[196] K. Stopperka and V. Grove, *Z. anorg. allg. Chem.* **359** (1968) 30.
[197] A. A. Woolf, in E. A. V. Ebsworth, A. G. Maddock and A. G. Sharpe (Eds.), *New Pathways in Inorganic Chemistry*, p. 327, Cambridge University Press (1968).
[198] R. C. Thompson, in G. Nickless (Ed.), *Inorganic Sulphur Chemistry*, p. 587, Elsevier, Amsterdam, London, New York (1968).
[199] A. Engelbrecht, *Angew. Chem. Intern. Ed.* **4** (1965) 641.
[200] G. H. Cady, in ref. 168, p. 123.
[201] R. J. Gillespie, *Accounts of Chem. Research*, **1** (1968) 202.

reflect the interest on and the importance of this acid. Therefore the following will only deal with the most interesting features of FSO_3H.

Fluorosulphonic acid is a colourless fuming liquid, which represents the strongest simple acid so far known. It reacts with rubber and wood and at elevated temperatures it dissolves sulphur, lead, mercury and tin. The structure is, analogous to H_2SO_4, of a tetrahedral arrangement of the four ligands around the sulphur atom. The anion SO_3^- also exhibits tetrahedral structure (*C3v* symmetry) according to infrared and Raman studies. Physical data are listed in Table 18. The hydrolysis of FSO_3H is a complex reaction, part of the acid hydrolysis depending on the rate of its addition to water and on the temperature. Water is simultaneously protonated, leaving the FSO_3^- anion, which then undergoes slow hydrolysis. When water is added to an excess of FSO_3H, a stable equilibrium is achieved:

$$H_3O^+ + FSO_3^- \rightleftharpoons HF + H_2SO_4$$

The oxidation properties of FSO_3H are due to a small amount of SO_3, which is always present in its solutions. This may also account for the sulphonation of benzene, when benzene is dissolved in the acid. The unique properties of FSO_3H are demonstrated by its abilities to solve electrolytes and stabilize cationic species, which are not stable in less acidic (more basic) media.

$$KCl + FSO_3H \longrightarrow K^+ + FSO_3^- + HCl$$
$$KClO_4 + FSO_3H \longrightarrow K^+ + FSO_3^- + HClO_4$$

Both reactions clearly indicate the high acidity of fluorosulphonic acid by complete solvolysis of salts derived from other strong acids. Unsaturated organic or basic (e.g. having a lone pair of electrons on N or O) systems are protonated in fluorosulphonic acid. It was found that FSO_3H represents an excellent medium for NMR studies on organic molecules (e.g. carbonium ions). When iodine is mixed with $F_2S_2O_6$ in fluorosulphonic acid, iodofluorosulphates are obtained[202].

$$3I_2 + F_2S_2O_6 \longrightarrow 2I_3^+FSO_3^-$$
$$5I_2 + F_2S_2O_6 \longrightarrow 2I_5^+FSO_3^-$$

However, I_5^+ is unstable at 25° and decomposes to I_3^+ and I_2. $[I^+][FSO_3^-]$ could not be prepared; all attempts yielded I_2^+ species. At high concentrations $[I_2^+][FSO_3^-]$ undergoes disproportionation to form $I[FSO_3]_3$.

TABLE 18. PHYSICAL PROPERTIES OF FSO_3H AND $ClSO_3H$

Compound	FSO_3H	$ClSO_3H$
M.p., °C	−88·98	−80
B.p., °C	163	155
Density (g/cm³)	1·7264	1·77
	1·7292	(18°)
Heat of formation (kcal/mole)	191	
Dielectric constant	120–150	
Trouton's constant	19·8	

Several inorganic fluorides exhibit acid behaviour: AuF_3, TaF_5, PbF_4 and SbF_5. From detailed conductive studies it has been learned that actually two species are present in solutions of SbF_5 and FSO_3H: a monomeric $H[SbF_5(FSO_3)]$ and a dimeric acid

202 R. J. Gillespie and J. B. Milne, *Inorg. Chem.* **5** (1966) 1577.

$H[(SbF_5)_2(FSO_3)]$. The former is a weak acid (pK: $3{\cdot}7\times10^{-3}$), while the latter is a fully dissociated acid:

$$H[(SbF_5)_2FSO_3]+FSO_3H \longrightarrow [(SbF_5)_2FSO_3]^-+FSO_3H_2^+$$

The mobility of $H_2SO_3F^+$ in this acid solution, like SO_3F^- in base solutions, is abnormally high. From low-temperature ^{19}F NMR studies information is obtained regarding the structure of this anionic species[203]. Two oxygens of FSO_3^- are coordinated to two SbF_5 molecules to build the complex anion. There are several other acids known of the general formula $H[SbF_{5-x}(FSO_3)_{1+x}]$ (x = 0, 1, 2, 3), which have been prepared from SbF_5, FSO_3H and SO_3.

Fluorosulphonic acid has been also used as fluorinating agent in inorganic and organic chemistry. Numerous fluorides have been synthesized according to the following:

$$B(OH)_3+3FSO_3H \longrightarrow BF_3+3H_2SO_4$$

In addition to its fluorinating properties it was found that FSO_3H can be used as catalyst in alkylation-, acylation-, polymerization- and hydro-fluorination reactions. Furthermore, it has been applied to the isomerization of hydrocarbons, to the production of substituted pyridins, to certain condensation reactions and to the preparation of derivatives of perfluoro aromatic acids. A list of more than 20 references regarding the use of FSO_3H in the above listed reactions is given in ref. 198.

Compounds Containing Sulphur, Nitrogen and Fluorine

Strictly speaking, this subsection should be listed under 2.5, "Sulphur and Nitrogen". However, since most of the compounds described below exhibit exceptional features due to the fluorine atoms, it is useful to discuss them within the scope of the Sulphur–Fluorine subsection. A combination of S, N and F atoms demands for the majority of compounds sulphur in its oxidation state +4 and +6. We will therefore restrict our discussion to these classes of compound in terms of acyclic and cyclic species.

The field of sulphur–nitrogen–fluorine chemistry was first explored by Glemser about fifteen years ago. Since then five reviewing articles on this subject have been published[204–208]. *Acyclic Compounds Containing Sulphur in the Oxidation State +4.* These compounds may be derived from SF_4 by replacing fluorine by amine. Hence different types of compound are possible: $FS{\equiv}N$, $F_2S{=}NR$, $F_3S{-}NR_2$ and $FOS{-}NR_2$. Thiazyl fluoride has been prepared by several routes: among them are:

$$NF_3+3S \xrightarrow{400^\circ} F{-}S{\equiv}N+S{=}SF_2$$

$$Hg(N{=}SF_2)_2 \xrightarrow[\text{vac.}]{100^\circ} 2F{-}S{\equiv}N+HgF_2$$

The latter reaction represents the most convenient synthesis, since $Hg(N{=}SF_2)_2$ can be easily stored and decomposed[209]. The FSN molecule is bent ($<$ N–S–F 116°52′) and it has

[203] R. C. Thompson, J. Barr, R. J. Gillespie, J. B. Milne and R. A. Rothenbury, *Inorg. Chem.* **4** (1965) 1641.

[204] O. Glemser, *Angew. Chem.* **75** (1963) 697.

[205] O. Glemser and M. Fild, in V. Gutmann (Ed.), *Halogen Chemistry*, Vol. 2, p. 1, Academic Press, London (1967).

[206] O. Glemser and U. Biermann, *Nachr. Acad. Wiss., Göttingen, II. Math.-Phys. Kl.* **5** (1968) 65.

[207] O. Glemser, *Endeavour*, **28** (1969) 86.

[208] S. M. Williamson, *Prog. Inorg. Chem.* **7** (1966) 39.

[209] O. Glemser R. Mews and H. W. Roesky, *Chem. Ber.* **102** (1969) 1523.

the following interatomic distances: d(S–F) 1·646 Å and d(S–N) 1·446 Å. The bond order of S–N amounts to 2·4. Thiazyl fluoride hydrolyses easily to form in the first step O=S=NH, *cis*-thionyl imid, which is then cleaved, yielding hydroxylamine and "sulphoxylic acid". The latter decomposes to sulphite and sulphur. Finally, $S_3O_6^{2-}$, monosulphane disulphonate, and $S_6O_6^{2-}$, tetrasulphane disulphonate, are formed, detectable as less soluble tetraphenylphosphonium salts.

Thiazyl fluoride trimerizes to give $(NSF)_3$, a cyclic six-membered system containing localized double bonds.

Iminosulphur difluoride, $RN{=}SF_2$, is obtained by ammonolysis of SF_4:

$$3RNH_2 + SF_4 \longrightarrow RN{=}SF_2 + 2(RNH_3)F$$

Further reaction with amine results in sulphur di-imides:

$$RN{=}SF_2 + 3RNH_2 \longrightarrow RN{=}S{=}NR + 2(RNH_3)F$$

Iminosulphur difluoride is also formed by the reaction of SF_4 with nitrile:

$$RC{\equiv}N + SF_4 \longrightarrow R{-}CF_2{-}N{=}SF_2$$

An interesting preparative approach to $RN{=}SF_2$ has been carried out by converting silicon–nitrogen bonds into the favourable Si–F unit[210]:

$$(R_3Si)_2NR + SF_4 \longrightarrow 2R_3SiF + RN{=}SF_2$$

The mercury derivative of $HNSF_2$ represents a valuable starting material not only for the preparation of NSF but also for various other compounds. Thus its reaction with halogens produces the less stable $XN{=}SF_2$ compounds. $Hg(NSF_2)_2$ is prepared from HgF_2 and $FCO{-}NSF_2$. Its structure has been recently elucidated[211].

$$HgF_2 + 2FCO{-}N{=}SF_2 \longrightarrow Hg(NSF_2)_2 + OCF_2$$

Perfluoroalkyl iminosulphur difluoride can be oxidized by elemental fluorine or N_2F_4 according to the following scheme:

$$2R_fN{=}SF_2 + 3N_2F_4 \xrightarrow{\text{u.v.}} 2F_5S{-}NF_2 + 2R_fF + 3N_2$$
$$R_fN{=}SF_2 + 2F_2 \longrightarrow R_fFN{-}SF_5$$

Replacing only one fluorine in SF_4, amidosulphur trifluorides are obtained, in which the R_2N group occupies an equatorial position. The reaction of thionyl fluoride with amine results in the formation of $R_2N{-}SOF$:

$$2R_2NH + OSF_2 \longrightarrow R_2N{-}SOF + (R_2NH_2)F$$

Cyclic Iminosulphur Fluorides with Sulphur in the Oxidation State +4. When S_4N_4 is reacted with AgF_2, tetrameric $(NSF)_4$ is obtained, whereas polymerization of NSF always results in trimeric $(NSF)_3$. Gaseous decomposition products of S_4N_4 form with AgF_2 in CCl_4 yellow-green crystals of the composition $S_3N_2F_2$. The compound exists in two polymorphic crystalline forms, which rapidly decompose above 100°. From hydrolysis experiments the conclusion was drawn that $S_3N_2F_2$ has probably a linear structure: F–S–N=S=N–S–F. A cyclic structure[212] as it is found in $S_3N_2Cl_2$ might just as well be possible for $S_3N_2F_2$. However, the compound's solubility in nonpolar solvents makes the

210 G. C. Demitras and A. G. MacDiarmid, *Inorg. Chem.* **6** (1967) 1903.
211 B. Krebs, E. Meyer-Hussein, O. Glemser and R. Mews, *Chem. Commun.* (1968) 1578.
212 A. Zalkin, T. E. Hopkins and D. H. Templeton, *Inorg. Chem.* **5** (1966) 1767.

linear covalent structure more likely. The physical properties of some of the compounds described above are listed in Table 19.

TABLE 19. PHYSICAL PROPERTIES OF SULPHUR–NITROGEN–FLUORINE COMPOUNDS

Compound	N=S–F	$N_3S_3F_3$	$N_4S_4F_4$	N=SF_3	FN=SF_2	$N_2S_3F_2$
M.p., °C	−89	74·2	153 (decomp.)	−72		83
B.p., °C	0·4	92·5		27·1	−6·7	
Density (g/cm³)			2·326	1·92		
Heat of vaporization (kcal/mole)	6·05			5·52	5·76	
Trouton's constant	22·1			22·5	21·6	
Heat of formation (kcal/mole)	+30			−95		

The formation of $(NSF)_4$ can only be achieved from a compound which contains already the eight-membered backbone of N and S atoms. This and the fact that polymerization of NSF exclusively yields $(NSF)_3$ must originate from structural aspects. An X-ray study[213] of $(NSF)_4$ showed that the molecule consists of a corrugated ring, which differs in shape from S_4N_4. Surprisingly two different S–N bond lengths were found: 1·66 Å (corresponding to a bond order of 1·4) and 1·54 Å (corresponding to a N–S double bond). Hence there is little or no delocalization in $(NSF)_4$. The ^{19}F NMR spectrum consists only of one signal, which appears in the region of the SF_6 signal. This is due to the equivalent electronical environment of all fluorine atoms. Some other examples of bond order–bond length relationship in N–S compounds are given below.

TABLE 20. RELATIONSHIP BETWEEN N–S DISTANCE AND BOND ORDER

Compound	Bond length (Å)	Bond order	References
SN^+	1·25	3·0	D. Chapman and D. C. Waddington, *Trans. Faraday Soc.* **58** (1962) 1679
NSF	1·416	2·7	H. Richert and O. Glemser, *Z. anorg. allgem. Chem.* **307** (1961) 328
SN	1·496	2·5	V. Schomaker and D. P. Stevenson, *J. Am. Chem. Soc.* **63** (1941) 37
$(NSF)_4$	1·54	2·0	G. A. Wiegers and A. Vos, *Acta Cryst.* **16** (1963) 152
α-$(NSOCl)_3$	1·564	1·9	G. A. Wiegers and A. Vos, *Proc. R. Soc.* (1962) 387
$(NSCl)_3$	1·605	1·7	*Ibid.*
S_4N_4	1·63	1·5	W. H. Kirchhoff and E. B. Wilson, *J. Am. Chem. Soc.* **85** (1963) 1726
$(NSF)_4$	1·66	1·4	G. A. Wiegers and A. Vos, *Acta Cryst.* **16** (1963) 152
$(NSH)_4$	1·67	1·4	D. Chapman and T. D. Waddington, *Trans. Faraday Soc.* **58** (1962) 1679
H_2NSO_3H	1·74	1·0	V. Schomaker and D. P. Stevenson, *J. Am. Chem. Soc.* **63** (1941) 37

Trithiazyl trifluoride, which can also be obtained by action of AgF_2 on $(SNCl)_3$ in CCl_4, exhibits only one ^{19}F NMR signal (near the SF_6 signal). From this fact the conclusion was drawn that $(NSF)_3$ contains fixed double bonds. No X-ray study of $(NSF)_3$ has been reported, but it may be assumed that its structure is similar to $(NSCl)_3$ (chair conformation, all Cl in axial positions).

[213] C. A. Wiegers and A. Vos, *Acta Cryst.* **16** (1963) 152.

Acyclic Compounds Containing Sulphur in the Oxidation State +6. Similarly, as has been shown for SF_4, the fluoride atoms may be replaced in SF_6. Theoretically numerous possibilities for new types of compound are possible and a large number have been already verified. The following classes of compound have been prepared[207]:

$F_5S{-}NR_2$ ($R_2{=}H_2$, Cl_2, CF_2, CO, CS, SF_2, SCl_2, etc.)
$F_5S{-}NFR$ ($R{=}CF_3$, C_2F_5, SO_2F)
$F_4S{=}NR$ ($R{=}CF_3$)
F_4SRR' ($R{=}CF_3$, OCF_3, $R'{=}NF_2$)
$F_2RS{\equiv}N$ ($R{=}F$, OR, NR_2)
$F_2S({=}NR)_2$ ($R{=}CF_3$, C_2F_5, SF_5)
$F_2S(O)NR$ ($R{=}H$, F, Cl, COF, CN, CF_3, C_6F_5, SO_2F)
$FS(R')NR(O)$ ($R{=}C_6H_5$, $R'{=}NH{-}CH_2COOH$)
$FSNR_2O_2$ ($R_2{=}H_2$, F_2, CO, SO, SF_2, SCl_2)
$(FSO_2)_2NR$ ($R{=}H$, F, Cl, COCl, CH_3)
$FSO_2NR{-}NR{-}SO_2F$ and $FSO_2NR{-}SO_2{-}NRSO_2F$ ($R{=}H$)

The most interesting compound of these classes is undoubtedly thiazyl trifluoride, F_3SN, which can be obtained in large quantities[214] by the action of AgF_2 on NSF.

$$N{\equiv}SF + 2AgF_2 \longrightarrow N{\equiv}SF_3 + 2AgF$$

The structure is different from the octahedral arrangement of SF_6. All fluorine atoms occupy equivalent positions on sulphur as it is indicated by one single ^{19}F NMR signal. However, in stronger solutions it splits up into three signals with roughly equal intensities. This has been explained by spin–spin interactions between nitrogen and fluorine nuclei. The isomer of thiazyl trifluoride, $FN{=}SF_2$, has been prepared[215] in low yield by the reaction of F_2 with $Hg(NSF_2)_2$ at $-80°$. The chemical properties of NSF_3 are related to those of SF_6, though SF_6 is sp^3d^2 and NSF_3 is sp^3 hybridized. The molecule has the symmetry C_{3v} and is isosteric to OPF_3 and $FClO_3$. Thiazyl trifluoride exhibits a stability similar to SF_6. Above 400° reaction with sodium takes place, thereby forming Na_2S, NaF and nitrogen. Boiling aqueous NaOH hydrolyses NSF_3 to fluoride and amidosulphonic acid. Hydrogen fluoride adds to the triple bond to yield $F_5S{-}NH_2$. A colourless compound is formed from BF_3 and NSF_3, which has the composition $NSF_3.BF_3$[216]. Infrared measurements and molecular weight determinations indicate that the gaseous phase consists of an equimolar mixture of BF_3 and NSF_3. The unusual increase of the S–N frequency of about 200 cm^{-1} has been explained by the increase of the bond order to 2·9. Since the spectrum of the liquid compound in the near infrared does not resemble those of alkali metal boron fluorides, the ionic structure $(NSF_2)^+(BF_4)^-$ has been excluded. At present the adduct is believed to be a donor acceptor compound, $F_3B{-}N{\equiv}SF_3$.

Recently it was found that BCl_3 reacts with NSF_3 at room temperature to form a yellow crystalline substance of the composition BNS_2Cl_6, which proved to be an interesting ionic compound:

$$2NSF_3 + BCl_3 \longrightarrow (NS_2Cl_2)^+(BCl_4)^- + \tfrac{1}{2}N_2 + \tfrac{3}{2}Cl_2 + 2BF_3$$

The structure of this compound is discussed in a later section (page 908).

[214] L. C. Duncan and G. H. Cady, *Inorg. Chem.* **3** (1964) 1045.
[215] O. Glemser, R. Mews and H. W. Roesky, *Chem. Commun.* **16** (1969) 914.
[216] O. Glemser, H. Richert and H. Haeseler, *Angew. Chem.* **71** (1959) 524.
[217] F. Seel and O. Dettmar, *Z. anorg. allg. Chem.* **301** (1959) 113.

Substitution products of NSF_3 may be obtained by reaction of NSF_3 with secondary amine. Sulphur tetrafluoride forms a SF_6-derivative, which hydrolyses to $F_5S{-}NH_2$, SO_2 and HF.

$$NSF_3+2R_2NH \longrightarrow N{\equiv}SF_2NR_2+(R_2NH_2)F$$
$$NSF_3+SF_4 \longrightarrow F_5S{-}N{=}SF_2$$
$$F_5S{-}NSF_2+H_2O \longrightarrow F_5S{-}NSO+2HF$$
$$F_5S{-}NSO+H_2O \longrightarrow F_5S{-}NH_2+SO_2$$
$$F_5S{-}NH_2 \longrightarrow NSF_3+2HF$$

When SF_5Cl is reacted with CN-containing compounds, the products of the type $F_5S{-}NCRCl$ are obtained. Two interesting isomers have been prepared by substitution of fluorine in $F_5S{-}NF_2$. The entering substituent occupies in $F_2N{-}SF_4{-}CF_3$ a *cis*-position[218], whereas in $F_2N{-}SF_4{-}COF$ it occupies a *trans*-position[214].

Compounds containing N, S, O and F are derived from $HNS(O)F_2$, which is formed[219] from OSF_4 and NH_3 at $-35°$. On heating, $HN{=}S(O)F_2$ loses HF, yielding a gummy product of the composition $(-NSOF-)_n$. This polymeric sulphanuric fluoride is isoelectronic with phosphorusnitril fluoride, $(-N{=}PF_2-)_n$. No cyclic compound has been isolated. Various compounds of the type $RN{=}S(O)F_2$ and $RN{=}S(O)FR'$ have been described[205]. Replacement of four fluorine atoms in F_5SNR_2 by oxygen leads to the class of sulphonyl fluoride amide compounds. Thus $H_2N{-}SO_2F$, fluorosulphonyl amide, is obtained[206] from $S_2O_5F_2$ and NH_3. The two hydrogen atoms in H_2NSO_2F may be substituted by various groups, e.g. Cl, F, SF_2, PCl_3, CCl_2, etc.[207]. Thus F_2NSO_2F is formed from SO_2 and N_2F_4.

$$SO_2+N_2F_4 \xrightarrow{\text{u.v.}} FSO_2NF_2+\tfrac{1}{2}N_2F_2$$

Furthermore, $(FSO_2)_2NH$ is prepared from $FSO_2{-}NCO$ and FSO_3H. It easily forms salts, which can be used for various synthetic approaches:

$$(FSO_2)_2NAg+Cl_2 \longrightarrow (FSO_2)_2NCl+AgCl$$
$$(FSO_2)_2NAg+(CH_3)_3SnCl \longrightarrow (FSO_2)_2N{-}Sn(CH_3)_3+AgCl$$

The reaction of $FSO_2{-}N{=}S{=}O$, N-sulphinylfluoro sulphonylimide, with fluorine yields $FSO_2{-}NSF_2$, fluorosulphonyl iminosulphur-oxy difluoride[220].

A cyclic fluorine compound containing sulphur in the oxidation state +6 is obtained when α-sulphanuric chloride is treated with KF in acetonitrile[221].

Sulphur and Chlorine

In contrast to the large variety of fluorine compounds of di-, tetra- and hexavalent sulphur, the corresponding chlorine compounds exist in a smaller number. A series of binary chlorides S_xCl_2 (x = 1 to 8) is known and has been characterized by ultraviolet[222] and Raman spectra[223]. Disulphur dichloride is readily obtained by the reaction of chlorine with sulphur as a golden-yellow liquid. Further chlorination results in the formation of SCl_2, a cherry-red liquid, which can be distilled in the presence of PCl_5 ($\sim 1\%$). The first

218 A. L. Logothetis, N. G. Sausen and R. J. Shozda, *Inorg. Chem.* **2** (1963) 173.
219 G. W. Parshall, R. Cramer and R. E. Forster, *Inorg. Chem.* **1** (1962) 677.
220 H. W. Roesky and D. P. Babb, *Inorg. Chem.* **8** (1969) 1733.
221 T. Moeller and A. Ouchi, *J. Inorg. Nucl. Chem.* **28** (1966) 2147.
222 F. Fehér and Munzer, *Chem. Ber.* **96** (1963) 1150.
223 F. Fehér and S. Ristic, *Z. anorg. allg. Chem.* **293** (1958) 307.

experimental proof for the existence of higher chlorosulphanes was provided by Fehér[224] and Baudler in 1952. Chilling a hot vapour mixture of S_2Cl_2 and hydrogen results in an orange-yellow viscous oil and a light-yellow solid of the approximate composition of Cl_2S_{20-24} and Cl_2S_{100}, respectively. These reaction mixtures consisted of chlorosulphanes with different chain lengths, but not—as one would expect—of a solution of sulphur in lower chlorosulphanes. A more convenient approach to higher chlorosulphanes was carried out by Fehér and his school according to the following reaction[225]:

$$2Cl_2S_x + H_2S_y \longrightarrow Cl_2S_{2x+y} + 2HCl$$

By careful addition of H_2S, H_2S_2, H_2S_3 or H_2S_4 to a high excess of SCl_2 at $-80°$ the chlorosulphanes Cl_2S_x ($x = 3$ to 8) may be prepared in excellent yields and sufficient purity. These chlorosulphanes are yellow to orange-yellow viscous liquids of an irritating odour. Physical data of chlorosulphanes are given in Table 21. They are thermally and hydrolytically unstable. The hydrolysis of the chlorosulphanes proceeds rapidly in the first step. Presumably $S(OH)_2$ and $S_2(OH)_2$ are formed, which decompose further to yield different products (e.g. H_2S, SO_2, H_2SO_3, H_2SO_4, $H_2S_xO_6$).

TABLE 21. PHYSICAL DATA OF SULPHUR CHLORIDES

Compound	SCl_2	S_2Cl_2	S_3Cl_2
M.p. (°C)	−122	−82	−46
B.p. (°C)	59·6	137·1	54/0·18 mn
Density (g/cm³)	1·622 (15°)	1·6733 (25°)	2·63 (20°)
Heat of formation (kcal/mole)	−11·99 (25°)	−14·31 (18°)	

Oxidation of SCl_2 produces $OSCl_2$ and O_2SCl_2; with fluorine, SF_4 and SF_6 are formed. The following reactions have already been discussed in a previous section (page 844):

$$3SCl_2 + 4NaF \longrightarrow SF_4 + S_2Cl_2 + 4NaCl$$

$$2KSO_2F + S_2Cl_2 \longrightarrow S{=}SF_2 + 2KCl + 2SO_2$$

Atomic nitrogen reacts with both SCl_2 and S_2Cl_2 to yield in the first step NSCl, which undergoes further reaction with S_2Cl_2[226].

$$2NSCl + S_2Cl_2 \longrightarrow S_3N_2Cl_2 + SCl_2$$

This thiadithiazyl chloride consists of a cyclic five-membered cation $S_3N_2Cl^+$ and chloride as anion (page 907). The formation of Cl_2S "adducts" has been reported with $AlCl_3$, $FeCl_3$ and $SbCl_5$[227], formulated as $AlCl_4^- SCl^+$. With RSCl similar compounds have been obtained. The chlorine in S_xCl_2 may be replaced by different nucleophiles:

$$2RSO_2Na + ClS_3Cl \longrightarrow RSO_2\text{–}S_3\text{–}SO_2R + 2NaCl$$

$$2RSH + S_xCl_2 \longrightarrow R_2S_{x+2} + 2HCl$$

$$S_xCl_2 + Hg(SCN)_2 \longrightarrow S_x(SCN)_2 + HgCl_2$$

Additions to double bonds have been studied. Thus C_2H_4 reacts with SCl_2 to yield

[224] F. Fehér, in ref. 194.
[225] F. Fehér, K. Naused and H. Weber, *Z. anorg. allg. Chem.* **290** (1957) 52.
[226] J. J. Smith and W. L. Jolly, *Inorg. Chem.* **4** (1965) 1006.
[227] N. S. Nabi and M. A. Khaleque, *J. Chem. Soc.* (1965) 3626.

$(ClCH_2–CH_2)_2S$ (mustard gas)[228]. In contrast to SF_4, SCl_4 is unstable above $-30°$ and decomposes to give SCl_2 and Cl_2. The organic derivatives $RSCl_3$ exhibit a slightly higher stability[229].

Thionyl Chloride

Thionyl chloride has a trigonal pyramidal structure. The compound is readily prepared by different methods[224]:

$$S_2Cl_2+SO_3 \longrightarrow OSCl_2+SO_2+\tfrac{1}{8}S_8$$
$$SCl_2+SO_3 \longrightarrow OSCl_2+SO_2$$
$$2P+2SO_2+5Cl_2 \longrightarrow 2OSCl_2+2OPCl_3$$

Thionyl chloride is quite reactive and begins to decompose above its boiling point (78°) into S_2Cl_2, SO_2 and Cl_2, which illustrates its use both as an oxidizing and chlorinating agent in organic chemistry. Since $OSCl_2$ reacts vigorously with water, it has been successfully applied to the preparation[230] of anhydrous metal halides:

$$CrCl_3{\cdot}6H_2O+6OSCl_2 \longrightarrow CrCl_3+6SO_2+12HCl$$

Thionyl chloride undergoes self-ionization to form $OSCl^+$ and Cl^-, which is responsible for its conductivity. Heteropolar substances are only slightly soluble in $OSCl_2$, in contrast to homopolar compounds, of which some increase the conductivity by electrolytical dissociation[231]. Substituted ammonium salts show the strongest electrolytical properties in $OSCl_2$. The solvolysis of iodides proceeds as follows:

$$2I^-+OSCl_2 \rightleftharpoons OSI_2+2Cl^-$$
$$2OSI_2 \rightleftharpoons SO_2+S+2I_2$$

Some chlorine/fluorine exchange reactions have been studied[232]:

$$OSCl_2+NaF \xrightarrow[-NaCl]{} OSFCl \xrightarrow{NaF} OSF_2+NaCl$$

TABLE 22. PHYSICAL DATA OF THIONYL HALIDES

Compound	OSF_2	OSFCl	$OSCl_2$	$OSBr_2$
M.p. (°C)	−110	−110 to −139	−101	−50
B.p. (°C)	−44	12	76	140; 48/20 mn
Density (g/cm³)	3·0		1·675 (0°)	2·685 (20°)
Dipole moment (Debye)	1·618		1·58	
Trouton's constant	22·6		21·2	
Heat of vaporization (kcal/mole)	5·18		7·48	
Heat of formation (kcal/mole)			50·2	
d(S–X) (Å)	1·585±0·001		2·07±0·03	2·27±0·02
d(S=O) (Å)	1·412±0·001		1·45±0·02	1·45 (assumed)

Sulphuryl Chloride

The compound represents the acid chloride of sulphuric acid. It is industrially prepared by oxidation of SO_2 with chlorine. Both chlorine atoms can be replaced by OH (hydrolysis)

228 L. A. Wiles and Z. S. Ariyan, *Chem. Ind.* (1962) 2102.
229 J. B. Druglass, K. R. Brower and F. T. Martin, *J. Am. Chem. Soc.* **73** (1951) 5787.
230 J. H. Freeman and M. L. Smith, *J. Inorg. Nucl. Chem.* **7** (1958) 224.
231 V. Gutmann in V. Gutmann (Ed.), *Halogen Chemistry*, Vol. 2, p. 399, Academic Press, London, New York (1967).
232 C. W. Tullock and D. D. Coffman, *J. Org. Chem.* **25** (1960) 2016.

and by NH_2 (ammonolysis)[233]. Because of its dissociation at higher temperature into SO_2 and Cl_2, it has been used in organic chemistry to introduce Cl or $ClSO_2$ into various systems. Gutmann[234] has studied its properties as a solvent and found that O_2SCl_2 parallels $OSCl_2$ with respect to its self-dissociation and its ability to solve covalent halides.

$$O_2SCl_2 \rightleftharpoons O_2SCl^+ + Cl^-$$
$$SbCl_3 + O_2SCl_2 \rightleftharpoons SbCl_4 + O_2SCl^+$$

Some other oxyhalogen compounds of hexavalent sulphur, trisulphuryl chloride, $S_3O_8Cl_2$, and pyrosulphuryl chloride, $S_2O_5Cl_2$, have been prepared according to the following scheme:

$$3SO_3 + CCl_4 \xrightarrow{80^\circ} ClSO_3\text{–}SO_2\text{–}SO_3Cl + OCCl_2$$
$$Cl_2S_3O_8 \xrightarrow{116^\circ} Cl_2S_2O_5 + SO_3$$

Physical data of O_2SCl_2 are listed in Table 17.

Chlorosulphonic Acid

The properties of chlorosulphonic acid are less exciting than those of FSO_3H. The acid is readily prepared from PCl_5 and H_2SO_4:

$$PCl_5 + H_2SO_4 \longrightarrow OPCl_3 + ClSO_3H + HCl$$

Its acidity may be compared with the acidity of fuming H_2SO_4. Therefore $ClSO_3H$ has been used as sulphonating agent as well as H_2O acceptor. The chlorine also can be substituted by SH, thereby yielding thiosulphuric acid[235]. This sulphane–sulphonic acid is very unstable and decomposes to SO_3 and H_2S. Technically chlorosulphonic acid is prepared from SO_3 and HCl. As a mixture with 40–50% SO_3 it was used in warfare to screen areas with smoke.

Sulphur Bromides and Iodides

Since the Br–S bond is very reactive, only few bromine compounds are stable enough under ordinary conditions. Iodosulphur compounds have not yet been isolated though several attempts have been made. Bromosulphanes, S_xBr_2, have been synthesized by Fehér[236] using similar techniques already applied for the formation of chlorosulphanes. The purification of S_xBr_2 is difficult, since these compounds are sensitive to heat. The best approach to obtain pure bromosulphanes starts with pure chlorosulphanes, which are reacted with HBr:

$$S_xCl_2 + 2HBr \longrightarrow S_xBr_2 + 2HCl$$

The bromosulphanes S_xBr_2 ($x = 2$ to 8) are dark red-orange viscous liquids, which in contrast to sulphanes, organic sulphanes and chlorosulphanes exhibit a decreasing intensity of colour with increasing chain length. The chemical behaviour of the bromosulphanes has not yet been evaluated. Thionyl bromide is obtained by the reaction of HBr or KBr with $OSCl_2$. Exchange studies have been carried out on the thionyl halides. A complete and rapid exchange of sulphur occurs between $OSCl_2$ and $OSBr_2$ in liquid SO_2 at $-50°$. In

[233] M. Goehring, *Ergebnisse und Probleme der Chemie der Schwefel-Stickstoff-Verbindungen*, Academie Verlag, Berlin (1957).

[234] V. Gutmann, *Monatsh.* **85** (1954) 393.

[235] M. Schmidt, *Z. anorg. allg. Chem.* **289** (1957) 141.

[236] F. Fehér and G. Rempe, *Z. anorg. allg. Chem.* **281** (1955) 161.

contrast to the relatively stable thionyl bromide, sulphuryl bromide does not exist under normal conditions. Similarly, bromosulphonic acid is also unstable at room temperature; it decomposes to Br_2 and SO_2. The acid has been prepared from SO_3 and HBr in liquid SO_2.

2.3. OXIDES OF SULPHUR

In compounds with oxygen, sulphur can because of the drastic difference in the electronegativity of these two homolog elements, only accept positive oxidation numbers. Binary sulphur–oxygen compounds are of enormous technical importance. In spite of many claims in the older literature[237] only two compounds are well defined and stable under normal conditions, namely sulphur dioxide and sulphur trioxide with the oxidation numbers +4 and +6 for sulphur. Disulphur monoxide, S_2O, is also known, but is thermodynamically unstable. (The formal oxidation number +1 of sulphur in this compound does not justify its denomination as a "lower" oxide, because the central sulphur atom in this "thiosulphur dioxide" still has the oxidation number +4; the second sulphur atom is, as also the oxygen atom, a ligand with the oxidation number −2.) There exists an extensive literature on sulphur monoxide, SO [238], with the oxidation number +2 for the sulphur atom. This very often described and intensively studied compound simply does not exist in the usual sense of the word. By physical methods it may be traced as a very short living intermediate during reduction of SO_2, SO_3 and $SOCl_2$, or during oxidation of sulphur, H_2S, CS_2 and COS, but it is unstable at all temperatures.

Sulphur Monoxide, SO

SO is formed when SO_2 is reduced with sulphur vapour in a glow discharge. According to mass-, UV- and ESR-absorption, SO is in a Σ ground state like the formal analog molecules O_2 and S_2 [239, 240]. The internuclear distance is 1·48 Å, the dipole moment 1·55 *D*, the force constant 7·94 mdyne/Å and the bond order 1·98 [241]. The SO valency vibration occurs at 1123·7 cm^{-1}. The molecular heat and standard entropy were calculated from spectroscopic data to be $C_p^\circ = 7{\cdot}22$ cal deg^{-1} $mole^{-1}$ and $S^\circ = 53{\cdot}07$ cal deg^{-1} $mole^{-1}$ (25°C)[237]. The dissociation energy and enthalpy of formation were disputed for a long time. The predissociation and mass spectroscopy investigations of the vaporization equilibria of inorganic sulphides and oxides show clearly that $D(SO) = 123{\cdot}5$ kcal/mole and $\Delta H_0^\circ = +1{\cdot}6$ kcal/mole[242]. The chemical properties of the very short living radical SO are not yet known in any detail (this holds even more for the rather hypothetical dimeric sulphur monoxide, S_2O_2)[237].

Disulphur Monoxide, S_2O

This thermodynamically unstable and only very poorly characterized compound is reported to form according to the following reactions (never in 100% yield and also never in a pure state):

[237] P. W. Schenk and R. Steudel, in G. Nickless (Ed.), *Inorganic Sulphur Chemistry*, pp. 369–416, Elsevier, Amsterdam, London, New York (1968).
[238] P. W. Schenk and R. Steudel, *Angew. Chem.* **77** (1965) 437.
[239] A. Carrington and D. M. Levy, *J. Phys. Chem.* **71** (1967) 2.
[240] J. M. Daniels and P. B. Dorain, *J. Chem. Phys.* **45** (1966) 26.
[241] T. Amano, E. Hiroito and Y. Morino, *J. Phys. Soc. Japan* **22** (1967) 299.
[242] R. Steudel and P. W. Schenk, *Z. physik. Chem.*, N.F. **43** (1964) 33.

$$3SO \longrightarrow S_2O + SO_2$$
$$2SO + S_2 \longrightarrow 2S_2O$$
$$SOCl_2 + \text{Metal S} \longrightarrow S_2O + \text{Metal } Cl_2 \text{ [243]}$$
$$SOCl_2 + H_2S \longrightarrow S_2O + 2HCl \ (200\text{–}360^\circ C)$$
$$S_2Cl_2 + \text{Metal O} \longrightarrow S_2O + \text{Metal } Cl_2 \ (100\text{–}400^\circ C)$$
$$3xS_x + \text{Metal O} \longrightarrow S_2O + \text{Metal S} \ (250\text{–}400^\circ C)\text{[244]}$$

Analysis of its microwave spectrum showed that S_2O has a similar structure to SO_2. The arrangement of the atoms is SSO with a valence angle of 118°. The interatomic distances are $d(SS) = 1{\cdot}884$ Å and $d(SO) = 1{\cdot}465$ Å; the dipole moment is 1·47 *D*. The S–S bond is a pure double bond. The bond order of the sulphur–oxygen bond is 1·80. A resonance formulation has therefore to be used to describe the molecule:

The enthalpy of formation of S_2O is still uncertain. For some spectroscopic data see ref. 237. The chemistry of disulphur monoxide, which is not stable in condensed phases, is only very poorly characterized, as for instance its thermal decomposition into sulphur and sulphur dioxide. Reports on its reactions in solution should at present only be read with severe scepticism[245]. The same holds for all reports on the so-called polysulphur oxides, $(S_xO)_y$, which have not yet been characterized in a satisfactory manner as definite chemical species[237]. Another oxide of sulphur, disulphur trioxide, S_2O_3, is very of en to be found in the literature and even in simple textbooks; in spite of this it does not exist. The reaction of elemental sulphur with sulphur trioxide, as well as the solvation of sulphur in sulphuric acid, leads to an until now not readily understood system of intensively coloured sulphur compounds[246] but not to a definite sulphur oxide of the composition S_2O_3.

Sulphur Dioxide, SO_2

Sulphur dioxide is formed in numerous redox reactions of sulphur compounds[247]. It may be prepared by the combustion of sulphur or sulphur compounds (e.g. H_2S) with lower oxidation numbers as +4, by the decomposition of sulphites, thiosulphates or dithionates, or by the reduction or thermal decomposition of SO_3 or sulphates, etc.

Technically, this very important oxide is prepared (1) by the combustion of sulphur or hydrogen sulphide, (2) by roasting of metal sulphides (in particular of pyrite, FeS_2), (3) by reduction with coal or the thermal decomposition of sulphates of calcium, magnesium and iron, (4) from thermal cracking of waste sulphuric acids, and (5) from the reduction of sulphur trioxide with sulphur[248]. Most of the prepared sulphur dioxide goes into the production of sulphuric acid. In addition to this, SO_2 is used in the sulphochlorination of hydrocarbons as bleach, disinfectant or preservative. In liquid form it is used for refrigeration and as a solvent.

[243] P. W. Schenk and R. Steudel, *Z. anorg. allg. Chem.* **342** (1966) 253.

[244] A. R. V. Murthy, *Z. anorg. allg. Chem.* **330** (1964) 245.

[245] R. A. V. Murthy, in *Annales du Génie Chimique*, p. 79, Édouard private, Toulouse (1968).

[246] H. Lux and E. Böhm, *Chem. Ber.* **98** (1965) 3210.

[247] L. Gmelin, *Handbuch der Anorganischen Chemie*, Schwefel B, Lieferungen I, II, III, Verlag Chemie, Weinheim (1953), (1960) and (1963).

[248] *Ullmanns Encyclopädie der technischen Chemie*, 3. Auflage, Band 15, Urban & Schwarzenberg, München–Berlin (1964).

For laboratory scale, sulphur dioxide is available in steel cylinders (vapour pressure at 25°C = 3·9 atm). It is separated from technical roasting gases by cooling, or from more dilute gases by dissolving, adsorption or absorption at normal temperature, followed by a heating process. The most common impurities in such samples are variable amounts of SO_3, CO_2, H_2O, O_2 and N_2. Washing and suitable drying removes those impurities with the exception of carbon dioxide. Also very "pure" SO_2 (prepared by heating concentrated sulphuric acid with metals such as mercury or copper[237]) is reported to contain small amounts of water, in spite of drying over P_4O_{10}[249].

The sulphur dioxide molecule may be described by the following five resonance formulations:

I ⟷ II ⟷ III ⟷ IV ⟷ V

The percentage of individual resonance structures is not yet known for certain, but structure III represents best the ground state of SO_2. The bond angle is 119·5° [250], the S–O bond distance is 1·432 Å[251]. The bond order is 1·93 [250]. The bond distance lies between a double

TABLE 23. PHYSICAL PROPERTIES OF SULPHUR DIOXIDE

Melting point	−75·48°C[a]
ΔH fusion	1·769 kcal/mole[a]
Boiling point	−10·02°C[a]
ΔH vaporization	5·955 kcal/mole[a]
V.p. equation for liquid	$\log P = +12{\cdot}07540 - 1867{\cdot}52T^{-1} - 0{\cdot}015865T + 0{\cdot}000015574T^2$
V.p. equation for solid	$\log p = 10{\cdot}45 - 1850T^{-1}$ [b]
Critical temperature	157·5°[d]
Critical pressure	77·7 atm[a]
Heat capacity	9·51 cal deg^{-1} mole^{-1} (25°C)[a]
Entropy	95·40 cal deg^{-1} mole^{-1} (25°C)[a]
Heat of formation	$\Delta H^\circ_\sigma = -70{\cdot}96$ kcal/mole (25°C)[c]
Bond dissociation energies	D(OS–O) = 131 kcal/mole[d], 125 kcal/mole[e]
Density	1·46 g/cm^3 (−10°C)[a]
Solubility	3937 ml SO_2(g) in 100 g H_2O (20°C),[a] 1877 ml at 100°C
Dielectric constant	17·27 (−16·5°C)[a]
Dipole moment	1·16D (25°C)[a]
Fundamental vibration frequencies	$\nu_1 = 1151$ cm^{-1}, $\nu_2 = 518$ cm^{-1}, $\nu_3 = 1362$ cm^{-1} [a]

[a] Ref. 237.
[b] Landolt-Börnstein, *Zahlenwerte und Funktionen*, Band II, Teile 2a und 2b, Springer, Berlin (1960) and (1962).
[c] V. Schumaker and D. P. Stevensen, *J. Am. Chem. Soc.* **62** (1940) 1270.
[d] I. Dubois and A. Groetsch, *Bull. Soc. Roy. Sci. Liege*, **33** (1964) 833.
[e] R. Steudel and P. W. Schenk, *Z. Physik. Chem.* NF **43** (1964) 33.

and a triple bond. Molecular orbital calculations showed that all the electrons of the valence shell, with the exception of the 2*s* and 3*s* electrons, are found in bonding orbitals. The calculated bond order amounts to 2·66, which is indicated by the contribution of the

[249] G. H. Weinreich, M. Jufresa and M. L. Salvat, in ref. 245, p. 151.
[250] W. Moffit, *Proc. Roy. Soc.* (*London*) **200** (1950) 409.
[251] M. H. Sirveitz, *J. Chem. Phys.* **19** (1951) 938.

resonance structures IV and V. The effective charge on sulphur (+0·35) and on the oxygen atoms (−0·18) determines the polarity of the molecule[252]. The dipole moment in the gas phase is 1·62*D*[253]. The extent of association in the liquid state is still disputed[237]. Sulphur dioxide crystallizes in the orthorhombic form. The space group is C_{2v}^{17}-*Aba* with four molecules per unit cell. The lattice constants are a = 6·07, b = 5·94, c = 6·14 Å; the molecular parameters: *d*(SO) = 1·43 Å, *d*(OO) = 2·46 Å, bond angle (OSO) = 119±2° [254].

At room temperature sulphur dioxide is a colourless poisonous gas with a sharp odour (maximum safe concentration 5 ppm). It condenses to a clear liquid which yields white crystals on further cooling. SO_2 is thermodynamically stable. Table 23 demonstrates some physical properties of this compound.

TABLE 24. INFRARED SPECTRUM OF GASEOUS SULPHUR DIOXIDE[a] [237]

No.	Assignment	Band	Relative intensity	No.	Assignment	Band	Relative intensity
1	ν_2	517·69	455	7	$\nu_2+\nu_3$	1875·55	6·0
2	$\nu_3-\nu_2$	844·93	0·55	8	$2\nu_1$	2295·88	5·5
3	ν_1	1151·38	565	9	$\nu_1+\nu_3$	2499·55	20·0
4	ν_3	1361·76	1000	10	$2\nu_3$	2715·46	0·2
5	$3\nu_2$	1535·06	0·1	11	$2\nu_1+\nu_2$	2808·32	0·8
6	$\nu_1+\nu_2$	1665·07	0·1	12	$2\nu_1+\nu_3$	3629·61	0·8

[a] P. W. Schenk and R. Steudel, in G. Nickless (Ed.), *Inorganic Sulphur Chemistry*, p. 373, Elsevier, Amsterdam, London, New York (1968).

The infrared spectrum of gaseous sulphur dioxide is seen in Table 24. The thermal (only above 1000°C) and photochemical decomposition of sulphur dioxide is reported to proceed via the unstable sulphur monoxide according to

$$SO_2 \xrightarrow{h\nu} SO_2^* \begin{cases} \nearrow SO + O \\ \searrow^{SO_2} SO + SO_3 \end{cases}$$

Since SO very rapidly disproportionates, the observed end products of the decomposition reactions are sulphur, oxygen and sulphur trioxide.

The most important chemical reaction of sulphur dioxide is its oxidation to sulphur trioxide, according to the equilibrium reaction

$$SO_2+\tfrac{1}{2}O_2 \rightleftharpoons SO_3 \qquad \Delta H_0^\circ = -22{\cdot}85 \text{ kcal/mole}$$

The equilibrium constant

$$K_p = \frac{p(SO_3)}{p(SO_2).(O_2)^{\frac{1}{2}}}$$

decreases markedly with increasing temperature (ln K_p at 800°C = 3·49, at 1100°C = −0·52)[255]. For maximum oxidation it is necessary to work at low temperatures, and the reaction rate must be increased by the use of catalysts. The SO_2–air mixture at 400–450°C is led over platinum or V_2O_5 contact catalysts. After the first contact, the gas is freed from

[252] S. P. Ionov, M. A. Porai and G. V. Tsintsadze, *Soobshch. Akad. Nauk. Gruz. SSR* **35** (1964) 559.
[253] J. E. Boggs, C. M. Crain and J. E. Whiteford, *J. Phys. Chem.* **61** (1957) 482.
[254] B. Post, R. S. Schwartz and J. Fankuchen, *Acta Cryst.* **5** (1952) 372.
[255] R. J. Lovejoy, J. H. Colwell and G. D. Halsey, *J. Chem. Phys.* **36** (1962) 612.

SO_3 by scrubbing with sulphuric acid. Then it is diluted with further air and led over a second contact. Hence a yield of 99·5% SO_3 or more is obtained, containing a minimum quantity of sulphur dioxide[208]. The technical catalysts used at present have an overall composition $V_2O_5.nK_2O.nSO_3$ (on supports as kieselguhr, xeolite, etc.). The following reaction scheme is given for the SO_2 oxidation at those catalysts[256, 257]:

$$SO_2+2V^{5+}+O^{2-} \longrightarrow SO_3+2V^{4+}$$
$$2V^{4+}+\tfrac{1}{2}O_2 \longrightarrow 2V^{5+}+O^{2-}$$

Much less suited as catalysts are activated charcoal, ferric oxide and numerous metal oxide mixtures[258]. In the presence of water the oxidation takes place even at room temperature. However, only under certain conditions or in the presence of some catalysts does the reaction take place with sufficient velocity[248]. The reaction of sulphur dioxide with water leads to the complicated system "sulphurous acid" (cf. page 878). Only the solid gas hydrate $SO_2.6H_2O$ will be mentioned here. It is formed as a cage compound from sulphur dioxide and water at 0°C (enthalpy of formation = −8·0 kcal/mole from gaseous SO_2 and ice) in cubic crystals of ideal composition $SO_2.5\frac{2}{3}H_2O$. At 7·1°C the dissociation pressure reaches 1 atm[259]. The redox reaction between sulphur dioxide and hydrogen sulphide also is of considerable technical importance. In the absence of water there is no reaction at all at room temperature (neither in the gaseous nor liquid forms nor in solutions in CCl_4). The reaction takes place in the presence of water or at high temperature. In the former case we deal with reactions of sulphurous acid (page 878). In the second case an endothermic reaction takes place with marked velocity above 300°C with the formation of sulphur and water:

$$SO_2+2H_2S \rightleftharpoons 3/xS_x+2H_2O$$

This technical reaction is catalytically speeded up, even at lower temperatures, by quartz, bauxite, metal sulphides, etc.[248, 258]. Also of technical importance is the reduction of sulphur dioxide with CO, COS and CS_2. Equilibria reactions such as

$$SO_2+2CO \rightleftharpoons 1/xS_x+2CO_2$$
$$SO_2+2COS \rightleftharpoons 3/xS_x+2CO_2$$
$$SO_2+CS_2 \rightleftharpoons 3/xS_x+CO_2$$

play a role in the preparation of sulphur from roasting gases[248, 258].

The reactions of sulphur dioxide with ammonia and amines are often extremely complicated and lead to a wide variety of sulphur–nitrogen–oxygen–hydrogen compounds, the most important of which will be dealt with in Section 2.5. For a most competent review see ref. 260. The oxidation of sulphur dioxide by nitric oxides[248] was of considerable technical importance in the lead chamber process for the preparation of sulphuric acid. The presence of water plays a great part in these reactions of acids of sulphur and their derivatives. N_2O_4 and N_2O_5 oxidize liquid SO_2 even at 25°C, forming SO_3, whereas NO

256 G. K. Boreskov, in ref. 245, pp. 221–230.

257 A. R. Glueck, F. P. B. Holroyd and C. N. Kenney, in ref. 245, p. 215.

258 L. Gmelin, *Handbuch der Anorg. Chemie*, Schwefel B, Lieferungen I, II, III, Verlag Chemie, Weinheim (1953), (1960) and (1963).

259 H. R. Müller and M. v. Stackelberg, *Naturwissenschaften*, **39** (1952) 20.

260 M. Goehring, *Ergebnisse und Probleme der Chemie der Schwefelstickstoffverbindungen*, Akademie Verlag, Berlin (1957).

reacts with SO_2 only at high temperatures[249]. With hot metal surfaces sulphur dioxide usually reacts under formation of metal sulphides besides metal oxides (due to secondary processes also sulphur and sulphates may be formed)

$$3M+SO_2 \longrightarrow MS+2MO$$

The reversion of such a reaction at high temperatures is used, for example, in the technical preparation of crude copper[261]. For the reduction of aqueous SO_2 solutions by certain metals or alloys to dithionites see page 881. For the many reactions of sulphur dioxide as an electron donor or acceptor with metal halides we must refer to a standard handbook; the same holds for the various reactions with metal oxides[258].

Sulphur dioxide is the most important gas in the atmospheric impurities in closely populated areas. It is not only formed in the manufacture of sulphur and its compounds, but in much higher quantities in the burning of coal and heating oils which practically always contain considerable amounts of sulphur compounds. In low concentrations it is not harmful to humans and animals, except under special unfortunate conditions such as smog. In the case of plants, however, concentrations of 1–2 ppm SO_2 already cause acute danger. The extraction of sulphur dioxide from exhaust gases of different types, therefore, is a most important process today and a very challenging task to chemists and technicians[258, 262].

Liquid sulphur dioxide, like liquid ammonia, is an interesting "water-like" solvent. For recent reviews see refs. 263–267. It dissolves many inorganic and especially organic compounds, often forming crystalline solvates, and is useful in various synthetic reactions. The electrical conductivity of the pure solvent is slight, but the solutions often are better conductors. The dissociation and the ionic reactions are more difficult to interpret than in aquo- and ammono-systems, since SO_2 provides a proton-free medium. The conductivity of liquid sulphur dioxide is of the same order of magnitude as in the case of water. For solubilities in liquid SO_2, electrical conductivity of such solutions, solvolysis reactions therein, double decompositions, and redox reactions see also ref. 237. For a theory upon the solvent action of liquid SO_2 it would be necessary to explain the following phenomena: the low electrical conductivity, the conductivity of solutions of electrolytes, the chemical reactivity, the isotope exchange processes, the electrode potentials, and the colour change of indicators. Such a comprehensive theory does not yet exist[237]. Obviously, much more work will have to be pursued before the real nature of solutions and reactions in liquid sulphur dioxide will be fully understood.

Sulphur Trioxide

Sulphur trioxide is prepared technically on a huge scale by the catalytic oxidation of sulphur dioxide (page 871) and converted to sulphuric acid. It is commercially available as a liquid, which contains additives to prevent polymerization, or as a "fuming" sulphuric acid ("oleum"), i.e. dissolved in sulphuric acid with a usual concentration of about 25–65% SO_3 in H_2SO_4. Sulphur trioxide is stable in the gaseous, liquid, and solid states.

261 L. Davignon, *Compt. Rend.* **262** (1966) 1380.
262 M. D. Thomas, *J. Air Pollution Control Assoc.* **14** (1964) 517.
263 L. F. Audrieth and J. Kleinberg, *Non-aqueous Solvents*, Wiley, New York (1953).
264 P. J. Elving and J. M. Markowitz, *J. Chem. Educ.* **37** (1960) 75.
265 V. Gutmann, *Quart. Rev. Chem. Soc.* **10** (1956) 451.
266 T. H. Norries, *J. Phys. Chem.* **63** (1959) 383.
267 H. H. Sisler, *Chemistry in Non-aqueous Solvents*, Reinhold, New York (1961).

It must be kept in mind, however, that experimentally it is extremely difficult to prepare, handle and study real chemically pure SO_3—in spite of its being so familiar to every chemist. So, for example, commercial liquid SO_3 is "stabilized" by different additives, and two of the hitherto-known three solid "modifications" (α, β) are in reality "compounds" between large amounts of SO_3 and small amounts of "impurities" such as water. A tabellaric representation of some of the more important physical data, therefore, would be rather confusing; it seems more reasonable in this case to give those data in the connection with the description of the different known "forms" of sulphur trioxide. A simple valence bond symbolization for the SO_3 molecule would result in a sextet of electrons around the central sulphur atom

Such a formulation, of course, cannot represent a physical reality, but can only demonstrate the strong potential electrophilic nature of the molecule, a typical Lewis acid. In the absence of other Lewis bases as reaction partners, the isolated molecules (in the gas phase—that is, at sufficiently high temperatures) will "satisfy themselves" by interactions of "free" electrons from the ligand–oxygen atoms with vacant sulphur orbitals via $(p \rightarrow p)\pi$ and $(p \rightarrow d)\pi$ bonds. As a result, we only can circumscribe the reality by a number of resonance formulations such as

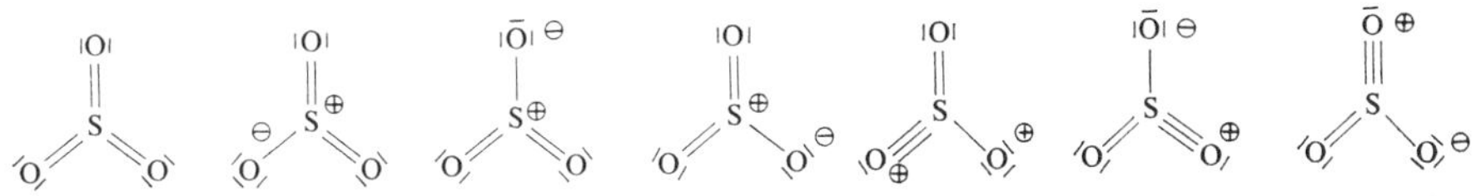

By lowering the temperature this intramolecular stabilization is overcome by an intermolecular interaction between free electrons of oxygen atoms from one SO_3 molecule with vacant orbitals at sulphur from another molecule—that is, via polymerization. The smallest polymeric unit is the trimeric sulphur trioxide, consisting of a six-membered ring of (alternating) three sulphur and three oxygen atoms (γ–SO_3). Higher polymeric forms are chain-like molecules with some impurities as end groups of the chains. The colourless gas SO_3 is mainly monomeric. The molecules are planar with a symmetry D_{3h} [255]. The dipole moment in this trigonal planar molecule is, by nature, zero.

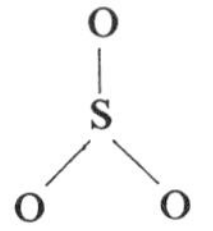

$d(SO) = 1{\cdot}43$ Å
bond angle (OSO) = 120°

The force constant of the S–O vibration (K = 10·77 mdyne/Å) leads to a bond order of 2·0 [250] in the sp^2 hybridized molecules. At room temperature only very pure SO_3 can be kept in the liquid state for a longer period of time without additives. Already 10^{-3} mole% of water catalyse the polymerization to β–SO_3. Its dielectric constant is 3·11 (18°C). This liquid form is a mixture of monomeric and trimeric sulphur trioxide[268]. Commercially

[268] K. Stopperka, *Z. Chem.* 6 (1966) 153.

available liquid SO_3 is stabilized by additives (up to 2%), such as BCl_3, $SOCl_2$, $TiCl_4$, derivatives of sulphonic acids, methyl siloxane, etc.[269, 270, 248]. The reason of the stabilizing effect is still disputed[257].

Solid sulphur trioxide exists in three (colourless) modifications:

(1) γ–SO_3, or ice-like SO_3. On cooling, pure liquid SO_3 solidifies at 16·86°C to this SO_3 modification. The orthorhombic crystals consist of cyclic S_3O_9 units. The space group is *Pbn* with $Z = 4$, $a = 5{\cdot}2$, $b = 10{\cdot}8$, $c = 12{\cdot}4$ Å. The puckered rings consist of SO_4 tetrahedra. The S–O distance in the ring is 1·626 Å, and for the ligand oxygen it is axial 1·371 Å and equatorial 1·430 Å. The valence angles in the ring are: O–S–O 99°, and S–O–S 121° [237]. It is possible that some monomeric SO_3 is still present in the lattice (mixed crystals).

S_3O_9

(2) β–SO_3 consists of polymeric molecules of unknown magnitude with a helical chain structure. It forms glistening needle-like crystals by the spontaneous polymerization of liquid or γ–SO_3. (γ–SO_3 is thermodynamically unstable with respect to the other modifications.) Polymerization to β–SO_3 is catalysed by traces of water, and probably proceeds by an ionic mechanism. Low temperature favours the formation and precipitation of polysulphuric acids, according to the equilibrium

$$H_2S_nO_{3n+1}+SO_3 \rightleftharpoons H_2S_{n+1}O_{3n+4}(\text{liquid} \rightleftharpoons H_2S_{n+1}O_{3n+4}(\text{solid})$$

Above 30°C no solid polymers are formed. This modification is not well defined; it probably consists of a mixture of polymers, which melts incompletely at 32–45°C. This, however, is not a real melting process but a slow depolymerization to liquid SO_3. Nevertheless, a constant equilibrium vapour pressure is measured after a few hours[237]. β–SO_3 is prepared by the cooling to 0°C of an incompletely dry sample of liquid SO_3, warming to 20°C and removing the remaining liquid and a part of the solid by vacuum distillation. β–SO_3 also is called asbestos-like sulphur trioxide. It forms monoclinic crystals of space group C^5_{2h}–$P2_{1/c}$. The cell constants are $a = 6{\cdot}20$ Å, $b = 4{\cdot}06$ Å (axis of the needle-like crystals), $c = 9{\cdot}31$ Å, $\beta = 109°50'$. The structure is made up of chains lying in the direction of the needle axis. They consist of a succession of SO_4 tetrahedra, arranged in a spiral. The S–O bond distance of 1·01 Å in the chain is longer than in the branched structure[258, 271, 237].

(3) α–SO_3 has the lowest vapour pressure of the three modifications and at the same time also the highest melting point (62°C). This most stable form at room temperature consists of SO_3 chains as β–SO_3. The chains are partially linked together, thus forming a layer structure[272]. The enthalpy of formation (25°C) is −94·45 kcal/mole for the gas, −104·67 kcal/mole for the liquid, −107·45 kcal/mole for the β modification, and −110·52 kcal/mole for the α modification. The standard entropy (25°C) is $S° = 61{\cdot}24$ cal/deg mole; the heat of polymerization $3SO_{3(g)} \rightarrow S_3O_{9(g)} = -30$ kcal/mole (0°K); the energy of dissociation,

[269] D. C. Abercromby, R. A. Hyne and P. F. Tiloy, *J. Chem. Soc.* (1963) 5832.
[270] E. E. Gilbert, in ref. 245, pp. 42–47.
[271] R. Pascard and C. Pascard-Billy, *Acta Cryst.* **18** (1965) 830.
[272] E. S. Scott and L. F. Audrieth, *J. Chem. Educ.* **31** (1954) 174.

$D_0(O_2S\text{–}O) = 81{\cdot}85$ kcal/mole[273]. For other tabulated thermodynamic functions of sulphur trioxide see ref. 274. The vapour pressure of liquid SO_3 at 25°C is 265 Torr, and is given by the following interpolation formulae:

$$\log p\ (\text{mm}) = 6{\cdot}6570 - 0{\cdot}1549 \times 10^3 T^{-1} - 0{\cdot}33165 T^{-2} \quad T \text{ in °K (0–45°C)}$$
$$\log p\ (\text{mm}) = 7{\cdot}8663 - 1158{\cdot}9/(t + 188{\cdot}00) \quad t \text{ in °C (17–50°C)}$$
$$\log p\ (\text{atm}) = 4{\cdot}2719 - 945{\cdot}78/(t + 180) \quad t \text{ in °C (80–200°C)}$$

From the second equation after interpolation the boiling point is found to be $44{\cdot}45 \pm 0{\cdot}15$°C. The molar heat of vaporization falls continuously from 10·5 kcal/mole at 20°C to 9·6 kcal/mole at 50°C. The critical temperature and pressure are $217{\cdot}7 \pm 0{\cdot}2$°C and $80{\cdot}8 \pm 0{\cdot}3$ atm respectively. Pure SO_3 solidifies at $16{\cdot}86 \pm 0{\cdot}02$°C. The vapour pressure of the γ modification is given by:

$$\log p\ (\text{mm}) = 12{\cdot}2346 - 2{\cdot}9160 \times 10^3 T^{-1} \quad (T \text{ in °K})$$

and for the β modification by:

$$\log p\ (\text{mm}) = 12{\cdot}5615 - 3{\cdot}0401 \times 10^3 T^{-1} \quad (T \text{ in °K})$$

The enthalpy of melting and sublimation of γ–SO_3 is 2·27 kcal/mole (16·86°C) and 13·45 kcal/mole. The vapour pressure curves for liquid and β–SO_3 cross at 30·45°C. This triple point should indicate the co-existence of melting SO_3 and β–SO_3. However, the depolymerization is rather hindered and β–SO_3 melts only very slowly above 31°C. The enthalpies of melting and sublimation are 3·21 kcal/mole (30·54°C) and 13·91 kcal/mole, respectively. A corresponding triple point for the co-existence of melting SO_3 and α–SO_3 is found at 62·2°C. The vapour pressure of the α modification is 76·0 Torr at 24·8°C and obeys the equation

$$\log p\ (\text{mm}) = 13{\cdot}9 - 3580 T^{-1} \quad (T \text{ in °K})$$

The enthalpies of melting and sublimation of the α-form are 6·09 kcal/mole (62·2°C) and 15·91 kcal/mole (51·6°C)[268, 275, 276]. For the viscosity of liquid sulphur trioxide in the temperature range 17–50°C the following equation applies[276]:

$$\log \eta\ (\text{cp}) = -1{\cdot}3726 - 404{\cdot}82 T^{-1} + 2{\cdot}6583 \times 10^5 T^{-2} \quad (T \text{ in °K})$$

The density of liquid SO_3 is 1·9255 g/cm³ at 20°C, 1·8335 g/cm³ at 40°C, and of γ–SO_3 at 10°C 2·29 g/cm³ [258].

Solubility of sulphur trioxide. SO_3 is very soluble in liquid sulphur dioxide and in sulphuryl chloride. In carbon disulphide[277] and in carbon tetrachloride[278] it dissolves at low temperatures, but above 0°C by-products are slowly formed:

$$CS_2 + SO_3 \longrightarrow COS + 1/x S_x + SO_2$$
$$CCl_4 + 2SO_3 \longrightarrow COCl_2 + S_2O_5Cl_2.$$

With water and most other compounds a fast reaction takes place. CF_2Cl_2 is, at low temperatures, also a suitable solvent. At 25°C pure SO_3 is soluble in CH_2Cl_2 and in C_2Cl_4, forming stable solutions. Nitromethane can be used below 0°C, but at higher temperatures decomposition occurs.

273 *Selected Values of Chemical Thermodynamic Properties*, Part I, U.S. Department of Commerce, NBS Circular 500, Part I.
274 G. Nagarajan, E. R. Lippincott and J. M. Stutman, *J. Phys. Chem.* **69** (1965) 2017.
275 J. M. Colwell and G. D. Halsey, *J. Phys. Chem.* **66** (1962) 2179, 2182.
276 R. A. Hyne and P. F. Tiley, *J. Chem. Soc.* (1961) 2348.
277 R. Steudel, *Z. anorg. allg. Chem.* **346** (1966) 255.
278 H. A. Lehmann and G. Ladwig, *Z. anorg. allg. Chem.* **284** (1956) 1.

Infrared spectrum. The four fundamental vibrations of sulphur trioxide are $\nu_1 = 1068\ \text{cm}^{-1}$, $\nu_2 = 495\ \text{cm}^{-1}$, $\nu_3 = 1391\ \text{cm}^{-1}$ and $\nu_4 = 529\ \text{cm}^{-1}$. Apart from the fundamental vibrations, the overtones and combination bands occur weakly at $2773\ \text{cm}^{-1}$ $(2\nu_3)$ and $2443\ \text{cm}^{-1}$ $(\nu_1+\nu_3)$. ν_1 is infrared-inactive, indicating the planar configuration of the molecule; it is found from the Raman spectrum to be $1068\ \text{cm}^{-1}$. S_3O_9 absorbs at 856, 1228 and $1510\ \text{cm}^{-1}$. In an xenon matrix and as a pure condensate monomeric SO_3 absorbs as in the gas phase, except that ν_2 is displaced slightly to a lower wave number ($464\ \text{cm}^{-1}$) and ν_1 becomes weakly infrared-active, with a sharp band at $1070\ \text{cm}^{-1}$. This indicates that the condensed SO_3 molecules in their own crystal field become partially distorted from the D_{3h} configuration[277, 255].

Chemical reactions of sulphur trioxide. SO_3 is one of the most reactive inorganic compounds. Sulphur being in the highest possible oxidation state, SO_3 must always act as an oxidizing agent. The potential sextet of electrons at the central atoms makes SO_3 at the same time to one of the most active Lewis acids.

According to the equation

$$SO_2+\tfrac{1}{2}O_2 \rightleftharpoons SO_3;\ \Delta H_0^\circ = -22{\cdot}85\ \text{kcal/mole}$$

SO_3 is stable only at room temperature or at very high pressures. Once formed, SO_3, in the absence of catalysts, decomposes completely into SO_2 and oxygen (in this connection see also page 862) only at about 1200°C. Below 1000°C the equilibrium is stationary. In the presence of platinum the decomposition begins already at 430°C and is complete at 1000°C [258].

For a summary of the reactions of sulphur trioxide with various nonmetals and metals see ref. 258. Addition reactions to compounds with donor oxygen, sulphur, nitrogen and phosphorus atoms normally lead to derivatives of sulphuric acid; they will be dealt with on pages 872–911. For a general summary see ref. 237.

Polysulphur peroxides. Higher oxides of sulphur of the type S_2O_7, S_3O_{11}, SO_4, etc., are described in the older literature. They do not exist. However, solid polymeric compounds of the composition SO_{3-4} may be prepared in almost continuous succession[279, 280, 281]. The compounds are derived from polymeric β-SO_3 by the substitution of oxygen bridges by peroxide bridges:

$$-\overset{O}{\underset{O}{S}}-O-\left(\overset{O}{\underset{O}{S}}-O-O-\overset{O}{\underset{O}{S}}-O\right)_n-\overset{O}{\underset{O}{S}}-O-O-$$

The individual members of the polysulphur peroxide are of non-uniform composition except for the end members $(SO_3)_n$ and $(SO_4)_n$. The –O– and –O–O– groups are statistically distributed. Those compounds are prepared as solid colourless condensates from mixtures of SO_2 or SO_3 with oxygen exposed to silent electric discharges. The polymeric compounds may be hydrolysed to H_2SO_5 and H_2SO_4, with H_2O_2 and oxygen as secondary products.

2.4. OXYACIDS OF SULPHUR

In this treatment of the oxyacids of sulphur we will only deal with such substances that really exist, either in the free state or in solution or in the form of definite salts. A critical

[279] M. Schmidt, in ref. 245, pp. 15–23.
[280] U. Wannagat and J. Rademacher, *Z. anorg. allg. Chem.* **286** (1956) 81.
[281] U. Wannagat and R. Schwarz, *Z. anorg. allg. Chem.* **286** (1956) 180.

review of the literature clearly shows that this is not the case with the so-called "sulphoxylic acid" (in its different (!) forms) and with "thiosulphurous acid". In spite of a rather large amount of literature (not seldom of a contradictory nature) on the "compounds" H_2SO_2 and $H_2S_2O_2$, these molecules, at best, may be regarded as speculated intermediates in some complicated reactions of sulphur chemistry. For a recent review on the two "acids" see ref. 282.

For a long time the reducing agent "Rongalite" was regarded as the most familiar "derivate" of sulphoxylic acid. Its structure, however, is clearly that of an organic sulphinic acid, hydroxymethane sulphinic acid[283].

$$HO\text{–}CH_2\text{–}S\begin{array}{l} /\!\!/ O \\ \backslash ONa \end{array}$$

The formulation of compounds such as SCl_2 and S_2Cl_2 as derivatives of unknown acids must be regarded as pure formalism without any heuristic or even didactic value (what about the formerly unknown higher chlorosulphanes $S_3Cl_2 \ldots S_8Cl_2$?), as long as there is no experimental proof for such connections. The realization of these facts makes the picture of the oxyacids of sulphur somewhat clearer. It may be further simplified by a hitherto

TABLE 25. OXYACIDS OF SULPHUR[a]

	H_2O	H_2O_2	H_2S	H_2S_x
SO_3	$H_2SO_4 \xrightarrow{SO_3} H_2S_2O_7$	$H_2SO_5 \xrightarrow{SO_3} H_2S_2O_8$	$H_2S_2O_3 \xrightarrow{SO_3} H_2S_3O_6$	$H_2S_{x+1}O_3 \xrightarrow{SO_3} H_2S_{x+2}O_6$
	$SO_2 \downarrow$			
	$H_2S_2O_6$			
	$SO_3 \uparrow$			
SO_2	$H_2SO_3 \xrightarrow{SO_2} H_2S_2O_5$			
	$SO \downarrow$			
	$H_2S_2O_4$			

[a] M. Schmidt, *Anorganische Chemie*, I, Bibliographisches Institut, Mannheim (1967).

unconventional classification of the remaining (that is, really existing) acids. They may all formally be regarded (and very often also really be synthesized according to this scheme) as reaction products of the oxides SO_2 and SO_3 on the one side, with water, hydrogen peroxide and sulphanes on the other side. Table 25 demonstrates this classification.

Derivatives of Water

Sulphuric Acid, H_2SO_4

The Lewis base H_2O adds easily to the Lewis acid SO_3, thus forming a complex that, by migration of a proton, is stabilized as the monosulphonic acid of water, $HO\text{–}SO_3H$. This sulphuric acid is the most important compound of sulphur and at the same time the

[282] D. Lyons and G. Nickless, in G. Nickless (Ed.), *Inorganic Sulphur Chemistry*, pp. 509–518, Elsevier, Amsterdam, London, New York (1968).

[283] M. R. Truter, *J. Chem. Soc.* (1955) 3064.

most important mineral acid at all. At the present time, the total world production of sulphuric acid is in excess of 40 million tons per year.

$$H_2O + SO_3 \longrightarrow [H_2O\text{–}SO_3] \longrightarrow HO\text{–}SO_2\text{–}OH$$

It enters into the manufacture of a very wide range of substances, although rarely forming part of the final substances. 100% pure sulphuric acid is a colourless oily liquid. It is highly associated because of very strong intermolecular hydrogen bonding. The sulphur atoms are surrounded symmetrically by four oxygen atoms. The SO_4^{--} ion can be circumscribed by the following resonance formulations:

$$[O^-\text{–}S(=O)(O^-)\text{–}O]^{--} \longleftrightarrow [O^-\text{–}S(=O)(O^-)\text{–}O^-]^{--} \longleftrightarrow [O\text{=}S(O^-)(O^-)\text{–}O^-]^{--} \longleftrightarrow [O^-\text{–}S(=O)(O^-)\text{–}O^-]^{--}$$

In addition to these formulae, because of the short S–O bond distance of 1·51 Å one has to assume some back bonding of the $(p \rightarrow d)\pi$ type. Table 26 gives some physical data of pure H_2SO_4 [284].

TABLE 26. SOME PHYSICAL CONSTANTS OF SULPHURIC ACID[285]

Freezing point	10·371°	
Boiling point	279·6°C	
Viscosity (cp)	24·54	25°
Density (d^{25})	1·8269	25°
Dielectric constant	100	25°
Specific conductance (ohm^{-1} cm^{-1})	$1{\cdot}0439 \times 10^{-2}$	25°
Heat capacity (cal deg^{-1} g^{-1})	0·3373	25°
Heat of fusion (cal mole^{-1})	2560	10·37°

Sulphuric acid is readily miscible with water in any concentrations. The heat of hydration is extremely high (about 210 kcal/mole at infinite dilution), so that addition of water to concentrated H_2SO_4 may be dangerous because of an explosive-like splashing of the mixture (the acid must be added under stirring to water, and never vice versa!). Definite hydrates are formed with a considerable volume-contraction. The most important of these hydrates are (melting point and heat of fusion in parentheses) $H_2SO_4.H_2O$ (+8.5°C, 4646 cal/mole), $H_2SO_4.2H_2O$ (−39·47°C, 4360 cal/mole), $H_2SO_4.3H_2O$ (−36·39°C, 5736 cal/mole) and $H_2SO_4.H_2O$ (−28·27°C, 7322 cal/mole)[286]. Commercial sulphuric acid contains about 96% H_2SO_4. Sulphuric acid and water form a constant boiling mixture when the sulphuric acid concentration is 98·3% and consequently this concentration has the minimum vapour pressure at any given temperature. The vapour of such an acid contains H_2SO_4, H_2O and SO_3. 100% sulphuric acid, therefore, cannot be distilled without partial

[284] M. Schmidt, *Anorganische Chemie*, I, Bibliographisches Institut, Mannheim (1967).

[285] R. Gillespie, in ref. 296, p. 563.

[286] *Ullmanns Encyclopädie der technischen Chemie*, 3. Auflage, Band 15, Urban & Schwarzenberg, München–Berlin (1964).

decomposition. Sulphuric acid dissolves sulphur trioxide easily. The very important practical system H_2O–H_2SO_4–SO_3 has been studied very carefully. It cannot be described here in any detail. For detailed data on melting points, densities, vapour pressures and vapour compositions, boiling points and viscosities of the system in dependence of its stoichiometrical composition see ref. 286.

Preparation of sulphuric acid. The raw materials for H_2SO_4 manufacture are sulphur dioxide, air (oxygen) and water. For the detailed technical processes see refs. 285, 287. Only the fundamental chemical reactions underlying the manufacture processes can be described in our connection. For the conversion of sulphur dioxide (the gas mixtures usually contain 5–10% SO_2) to sulphuric acid two processes have been used: the chamber process, which is now obsolescent, and the "contact" or catalytic process, which is now mainly used (exclusively for the preparation of very concentrated sulphuric acid). In the chamber process oxides of nitrogen are used to oxidize the sulphur dioxide. The chemical reactions are complex involving the formation of intermediate compounds which are subsequently decomposed by water to form sulphuric acid and regenerate the nitrogen oxides. On the other hand, the contact process is relatively simple, involving only the direct oxidation of SO_2 by air to SO_3 by means of a catalyst and subsequent absorption of the trioxide in sulphuric acid containing a small amount of water. The present view on the mechanism of the chamber process[288] will be given only in a systematic scheme of formulae (the hot starting gas mixture enters first the water-sprayed "Glover-tower", then successively some lead chambers or ceramic towers, and at the end the H_2SO_4-sprayed "Gay Lussac-tower").

Glover tower reactions:

$$SO_2+H_2O \rightleftharpoons SO_3H^-+H^+$$

$$NO^++SO_3H^- \longrightarrow NOSO_3^-+H^+$$

$$NO^++NOSO_3^- \longrightarrow 2NO+SO_3$$

$$SO_3+H_2O \longrightarrow H_2SO_4$$

Net reactions:

$$2NO^++SO_2+2H_2O \longrightarrow 2NO+H_2SO_4+2H^+$$

Chamber reactions:

$$NO+\tfrac{1}{2}O_2 \longrightarrow NO_2 \quad \text{(gas phase reaction)}$$

$$\left.\begin{array}{l} NO_2+NO+H_2O \longrightarrow 2ON.OH \\ SO_2+H_2O \longrightarrow H_2SO_3 \end{array}\right\} \text{surface reactions}$$

$$2ON.OH+2H^+ \longrightarrow 2NO^++2H_2O$$

$$H_2SO_3 \longrightarrow SO_3H^-+H^+$$

$$NO^++SO_3H^- \longrightarrow NO.SO_3^-+H^+$$

$$NO^++NO.SO_3 \longrightarrow 2NO+SO_3$$

$$SO_3+H_2O \longrightarrow H_2SO_4$$

Net reaction:

$$SO_2+\tfrac{1}{2}O_2+H_2O \longrightarrow H_2SO_4$$

Gay Lussac-tower reactions:

$$\tfrac{1}{2}[NO_2+NO+H_2O] \longrightarrow ON.OH$$

$$ON.OH+H^+ \longrightarrow NO^++H_2O$$

Net reaction:

$$\tfrac{1}{2}[NO_2+NO]+H^+ \longrightarrow NO^++\tfrac{1}{2}H_2O$$

---

287 T. J. P. Pearce, in ref. 282, pp. 535–560.

288 F. Seel and H. Meier, *Z. anorg. allg. Chem.* **274** (1953) 197.

The essential feature of the contact process for the manufacture of sulphuric acid is the production of sulphur trioxide by the catalytic oxidation of sulphur dioxide in the 400–600°C temperature range. This has, in principle, already been described on page 862. For further technical details, especially for thermodynamic relationships, converter design, catalysts and their poisoning, the kinetics of the oxidation on different catalysts, etc., see refs. 286, 287. The absorption of the so formed sulphur trioxide is a problem by itself. The only practical medium to absorb sulphur trioxide is concentrated sulphuric acid. H_2SO_4 of strength less than 98·3% has a vapour pressure of water which increases with decreasing acid strength. If weak acid is used to absorb SO_3, the SO_3 in the gas phase combines with water vapour from the weak acid to form a sulphuric acid fog which cannot be absorbed because the particles lack the mobility of molecules and have insufficient mass to be trapped by impingement. On the other hand, more concentrated sulphuric acid exerts a vapour pressure of sulphur trioxide and consequently the extent to which SO_3 can be scrubbed out of converted gas is limited by the vapour pressure of SO_3 exerted by the sulphuric acid. In practical systems one or more scrubbing towers are used in series each with its own acid circulating system and coolers to remove the heat of absorption of SO_3 in sulphuric acid. In the first tower, oleum which has a substantial partial pressure of SO_3 may be circulated to absorb most of the sulphur trioxide. However, in the last tower it is important to scrub the gas with optimum strength just specified to achieve maximum absorption of SO_3. Continuous addition of weak acid or water to the acid circulating in the final scrubbing tower is necessary to maintain the strength of the acid entering the tower constantly at the optimum. In the most modern plants when oleum is not required only one scrubber is now installed.

Chemical properties of sulphuric acid. In aqueous solution, sulphuric acid is a strong two basic Broensted acid. Its apparent degree of dissociation is 51% in 1 N solution, and 59% in 0·1 N solution[289]. The dissociation of sulphuric acid clearly is to be observed in two steps. In solutions of middle concentration the first hydrogen-ion is dissociated completely. The dissociation of the second one obeys the usual law

$$\frac{[SO_4^{--}]x[H^+]}{[HSO_4^-]} = K = 1{\cdot}29\times10^{-2} \qquad (18°C)$$

Two series of salts, therefore, are known: "acid" sulphates or hydrogen sulphates, M^IHSO_4, and "normal" sulphates, $M^I_2SO_4$. In the solid state, "acid" sulphates are only known from alkali metals; they may be prepared according to

$$Na_2SO_4+H_2SO_4 \longrightarrow 2NaHSO_4$$
$$NaCl+H_2SO_4 \longrightarrow NaHSO_4+HCl$$

They are easily soluble in water. Heating above the melting point leads to the formation of "pyrosulphates":

$$2NaHSO_4 \longrightarrow H_2O+Na_2S_2O_7$$

"Normal" sulphates, $M^I_2SO_4$, may be easily prepared by either one of the following methods: solution of the metal in sulphuric acid, neutralization of the acid with a metal oxide or hydroxide, decomposition of salts from volatile acids by sulphuric acid metathetic reactions, and oxidation of sulphides or sulphites. Most sulphates are easily soluble in water; scarcely soluble are the sulphates of lead, the alkaline earth metals, and to some extent also silver

[289] H. Remy, *Lehrbuch der Anorganischen Chemie*, Band I, 9. Auflage, Akademische Verlagsgesellschaft, Leipzig (1957).

sulphate. In dilute aqueous solution sulphuric acid is only a very poor oxidant. Such solutions do not react with non-metals and with metals with a redox potential more negative than hydrogen (the more electropositive metals are dissolved with evolution of hydrogen, forming sulphate solutions). Rise of temperature and much more so increasing concentration increase the oxidative power considerably; hot concentrated sulphuric acid, therefore, is a strong oxidant. From phosphorus, for instance, it will be reduced to sulphur, from carbon, sulphur and many metals to sulphur dioxide. Of considerable technical importance is the behaviour of sulphuric acid towards iron and lead. In dilute solutions, iron is readily dissolved; however, by a passivation process it becomes resistant towards highly concentrated acid (and solutions of SO_3 in the acid—that is, "oleum") that can be stored, shipped and handled in iron containers. Lead by dilute sulphuric acid (up to about 75%) is only attacked at the surface because of the insolubility of $PbSO_4$. In higher concentrated acid lead sulphate becomes considerably more soluble, especially on heating. Therefore, hot concentrated sulphuric acid cannot be handled in lead vessels.

Sulphuric acid as a solvent system. Sulphuric acid is one of the best-known non-aqueous protonic solvents[285]. Because of its high dielectric constant it is generally a good solvent for electrolytes, but because of its highly associated nature it tends to be a rather poor solvent for non-electrolytes. However, this does not necessarily mean that it is a poor solvent for organic compounds, since very many organic compounds are protonated or form strongly hydrogen-bonded complexes with sulphuric acid and are, therefore, soluble. The rather high specific conductivity of 100% H_2SO_4 is due to the ions $H_3SO_4^+$ and HSO_4^- formed in the autoprotolysis of the solvent:

$$2H_2SO_4 \rightleftharpoons H_3SO_4^+ + HSO_4^-$$

The rather extensive autoprotolysis of sulphuric acid shows that despite its very high acidity, sulphuric acid is also appreciably basic. The ions produced by autoprotolysis are of fundamental importance in the chemistry of any protonic solvent and, primarily because they determine acid–base behaviour in the solvent and, in particular, they limit the acid–base range that is accessible in the solvent. In sulphuric acid any solute that produces HSO_4^- ions may be regarded as a base, and any solute that produces $H_3SO_4^+$ ions as an acid. The $H_3SO_4^+$ ion is the strongest possible acid and the HSO_4^- ion is the strongest possible base of the sulphuric acid solvent system. Sulphuric acid is also slightly self-dissociated into sulphur trioxide and water:

$$H_2SO_4 \rightleftharpoons H_2O + SO_3$$

Hence it is important to understand the behaviour of both these solutes in sulphuric acid. Water is nearly completely ionized as a base:

$$H_2O + H_2SO_4 \rightleftharpoons H_3O^+ + HSO_4^-$$

Sulphur trioxide is completely converted to disulphuric acid, $H_2S_2O_7$. This acid is ionized as a moderately weak acid:

$$H_2S_2O_7 + H_2SO_4 \rightleftharpoons H_3SO^+ + HS_2O_7^-.$$

Thus since the ions $H_3SO_4^+$ and HSO_4^- are in equilibrium as a consequence of the autoprotolysis reaction it follows that the ions H_3O^+ and $HS_2O_7^-$ must also be in equilibrium:

$$2H_2SO_4 \rightleftharpoons H_3O^+ + HS_2O_7^-.$$

This is called the ionic self-dissociation reaction. The complete self-dissociation reaction in the sulphuric acid solvent system can be described then by the above equations. The equilibrium constants at 25°C are:

$$K = [H_3SO_4^+][HSO_4^-] \qquad 2{\cdot}7\times 10^{-4}$$
$$K = [H_3O^+][HS_2O_7^-] \qquad 5{\cdot}1\times 10^{-5}$$
$$K = [H_3SO_4^+][HS_2O_7^-]/[H_2S_2O_7] \qquad 1{\cdot}4\times 10^{-2}$$
$$K = [H_3O^+][HSO_4^-]/[H_2O] \qquad 1$$

The experimental methods for the investigation of solutions in sulphuric acid are mainly cryoscopy and electrical conductivity besides Raman, infrared, ultraviolet, visible absorption and NMR spectroscopy. For details see ref. 285.

Bases in sulphuric acid. Basic solutions may be formed in a number of different ways, e.g. (1) from metal sulphates, (2) by solvolysis of salts of weak acids, (3) by protonation of compounds containing a lone pair of electrons and (4) by dehydration reactions. Only a few examples will be given here very briefly; for more details on this interesting and rather new field of inorganic chemistry the reader must be referred to a more competent review[285].

(1) Hydrogen sulphates are fully ionized strong bases and are the direct analogs of the hydroxides in water

$$KHSO_4 \longrightarrow K^+ + HSO_4^-$$

(2) A number of salts of other acids (if they are not insoluble as, for example, AgCl, $CuBr_2$, $AlCl_3$, etc.) undergo complete solvolysis and thus give rise to strongly basic solutions. Solvolysis occurs because the other acids in sulphuric acid are either exceedingly weak (as $HClO_4$), or do not behave as acids at all but react as bases (as H_3PO_4 and HNO_3), e.g.

$$NH_4ClO_4 + H_2SO_4 \longrightarrow NH_4^+ + HSO_4^- + HClO_4$$
$$KNO_3 + H_2SO_4 \longrightarrow K^+ + HSO_4^- + HNO_3$$

(3) A very large number of substances behave as bases forming their conjugate acids by the addition of a proton. Thus almost any organic molecule with a potentially basic site, such as a lone pair of electrons, or an unsaturated system, dissolves in sulphuric acid with the formation of its conjugate acid. Many ketones, carboxylic acids, esters, amines and amides dissolve in sulphuric acid in this manner, e.g.

$$(CH_3)_2CO + H_2SO_4 \longrightarrow (CH_3)COH^+ + HSO^-$$
$$CH_3COOH + H_2SO_4 \longrightarrow CH_3COOH_2^+ + HSO_4^-$$

Even triphenylamine and triphenylphosphine behave as strong bases in sulphuric acid. Water and phosphoric acid provide examples of inorganic substances that behave as strong bases:

$$H_2O + H_2SO_4 \longrightarrow H_3O^+ + HSO_4^-$$
$$H_3PO_4 + H_2SO_4 \longrightarrow H_4PO_4^+ + HSO_4^-$$

(4) Many oxy- and hydroxy-compounds are dehydrated in sulphuric acid. The simplest reactions of this type are:

$$XOH + 2H_2SO_4 \longrightarrow XSO_4H + H_3O^+ + HSO_4^-$$
$$X_2O + 3H_2SO_4 \longrightarrow 2XSO_4H + H_3O^+ + HSO_4^-$$

Very often the hydrogen sulphate XSO_4H is ionized so that the general equations may be written in the form:

$$XOH+2H_2SO_4 \longrightarrow X^++H_3O^++2HSO_4^-$$

$$X_2O+3H_2SO_4 \longrightarrow 2X^++H_3O^++3HSO_4^-$$

A number of new cations have been prepared by reactions of this type, the most familiar of which is probably the nitronium ion. Solutions of nitric acid, metal nitrates and dinitrogen pentoxide in sulphuric acid have long been known to be efficient reagents in aromatic nitration. The old assumption of the formation of the nitronium ion, NO_2^+, as the active species has been proved correct by detailed studies in the sulphuric acid solvent system. The ion is formed according to

$$N_2O_5+3H_2SO_4 \longrightarrow 2NO_2^+ +H_3O^++3HSO_4^-$$

$$KNO_3+3H_2SO_4 \longrightarrow NO_2^+ +K^++H_3O^++3HSO_4^-$$

A number of stable salts of the nitronium ion have been prepared[285]. Acyl ions and carbonium ions may also be prepared in the sulphuric acid solvent system. They cannot be dealt with in this connection.

Acids in the sulphuric acid system. Until now, besides the rather weak acids $H_2S_2S_7$, $H_2S_3O_{10}$ and FSO_3H, only one strong acid has been found in the H_2SO_4 solvent system: tetrahydrogensulphato boric acid, $HB(HSO_4)_4$. Solutions of this acid can be prepared by dissolving boric acid, or boric oxide, in oleum instead of sulphuric acid, in which case the H_3O^+ ion is removed by the reaction:

$$H_3O^++SO_3 \longrightarrow H_3SO_4^+$$

and the overall reaction in the case of boric acid is

$$H_3BO_3+3H_2S_2O_7 \longrightarrow H_3SO_4^+ +B(HSO_4)_4^- +H_2SO_4$$

For some other inorganic solutes in H_2SO_4 see ref. 285.

Miscellaneous reactions in sulphuric acid. (1) *Oxidation.* At low temperature 100% sulphuric acid is only a moderately good oxidizing agent. In the oxidation process, sulphuric acid is reduced to sulphur dioxide according to the equation

$$5H_2SO_4+2e^- \longrightarrow SO_2+2H_3O^++4HSO_4^-$$

Electron spin resonance studies have shown that radical cations are formed by the oxidation of hydrocarbons such as anthracene, etc., in sulphuric acid[290, 291]. The concentration of the radical ion that is formed must, however, be very small[292]. Iodine is only slightly soluble in sulphuric acid and gives a pale violet coloured solution. A slow reaction with the formation of I_5^+ and I_3^+ ions seems to take place[293]. Selenium and tellurium dissolve in sulphuric acid to give green and red solutions respectively. In contrast to earlier assumptions (formation of $SeSO_3$ and $TeSO_3$ [294]) it has been established that the elements are oxidized to the new cationic species Se_8^{2+} and Te_4^{2+} according to the equations[295]:

[290] S. I. Weissmann, E. deBoer and J. J. Conradi, *J. Chem. Phys.* **26** (1956) 963.
[291] A. Carrington, F. Dravieko and M. C. R. Symons, *J. Chem. Soc.* (1959) 947.
[292] V. Gold and T. L. Tye, *J. Chem. Soc.* (1952) 2172.
[293] J. Arotsky, H. C. Mishra and M. C. R. Symons, *J. Chem. Soc.* (1961) 12.
[294] K. W. Bagnall, *The Chemistry of Selenium, Tellurium and Polonium*, Elsevier, Amsterdam (1966).
[295] J. Barr, R. J. Gillespie, R. Kapoer and K. C. Malhotra, *Can. J. Chem.* **42** (1968) 149.

$$8/xSe_x + 5H_2SO_4 \longrightarrow Se_8^{2+} + 2H_3O^+ + 4HSO_4^- + SO_2$$

$$4/xTe_x + 5H_2SO_4 \longrightarrow Te_4^{2} + 2H_3O^+ + 4HSO_4^- + SO_2$$

The green solution of selenium becomes yellow on heating or on addition of sulphur trioxide or another oxidizing agent such as selenium dioxide or persulphate as a consequence of the further oxidation of Se_8^{2+} to Se_4^{2+}.

(2) *Sulphonation.* Very many organic compounds give stable solutions in sulphuric acid from which they can often be quantitatively recovered by pouring the sulphuric acid solution onto ice. This is at least partly due to the fact that they are generally protonated and the protonated group deactivates the rest of the molecule towards sulphonation. Thus, if an aromatic molecule contains non-basic or very weakly basic activating groups, in addition to the protonated group, or if the protonated group is sufficiently separated from the aromatic ring, the compound will often protonate and dissolve, but will also rapidly sulphonate at the aromatic ring. For example, benzoic acid in the protonated form is quite stable, but *m*-methoxy benzoic acid is sulphonated completely in 1 hr, and *o*- and *p*-methoxybenzenes are sulphonated in 4–5 hr[296].

(3) *Hydrolysis.* It is at first sight somewhat surprising to find that, despite the fact that sulphuric acid is such a strongly dehydrating medium, occasionally hydration occurs in which a solute extracts water from the solvent, leaving an excess of sulphur trioxide. This is, for example, the case with the anhydrides of acetic acid and benzoic acid[297]:

$$(RCOO)_2O + 3H_2SO_4 \longrightarrow 2RCOOH_2^+ + HS_2O_7^- + HSO_4^-$$

Disulphuric Acid (and Higher Polysulphuric Acids)

Sulphuric acid is, as already mentioned, the most important compound of sulphur. It is the reaction product of one mole of water with one mole of sulphur trioxide, and therefore may be regarded as monosulphonic acid of water. Only a few arbitrarily chosen features of the properties and reactions of this extremely important compound could be dealt with in the foregoing section of this review on oxyacids of sulphur—otherwise a whole book on it would have resulted. Water can react not only with one mole SO_3 but also with two moles, thus forming the disulphonic acid of water:

$$HO_3S\text{–}OH + SO_3 \longrightarrow HO_3S\text{–}O\text{–}SO_3H$$

Hitherto, this disulphonic acid or "pyrosulphuric acid", $H_2S_2O_7$, could not be isolated in the free state. It does, however, certainly exist in solutions of sulphur trioxide in sulphuric acid ("oleum")[285]. Alkali metal salts of this acid may easily be prepared by heating hydrogensulphates above their melting point[284], such as

$$2NaHSO_4 \longrightarrow Na_2S_2O_7 + H_2O$$

The ion $S_2O_7^{--}$ is constructed from two SO_4-tetrahedra with a common oxygen atom as a common corner. Theoretically, the S–O–S chain in disulphuric acid may be elongated infinitely—always by the addition of the Lewis acid sulphur trioxide to a Lewis basic S–OH end group of the chain—forming an infinite number of polysulphuric acids $H_2S_xO_{3x+1}$. Neither definite members of this theoretical series of acids nor definite salts have been isolated until now; intensive studies, however, certainly will reveal new and interesting results in this field of research, which is still widely open. Mixtures of polysulphuric acids

296 M. S. Newman and N. C. Deno, *J. Am. Chem. Soc.* **73** (1951) 3651.

297 R. H. Flowers, R. J. Gillespie and J. V. Oubridge, *J. Chem. Soc.* (1956) 607.

will probably be important constituents of solutions of sulphur trioxide in sulphuric acid. Also the α- and β-modification of sulphur trioxide may be regarded as very high molecular forms of such acids.

Dithionic Acid, $H_2S_2O_6$

Sulphuric acid and disulphuric acid (including the polysulphuric acids) can be regarded as reaction products of sulphur trioxide with water. Therefore, the oxidation number of the sulphur atoms in these compounds is +6. The next lower oxidation state of sulphur is +5. It is formally realized in dithionic acid and its salts. Dithionic acid is unknown, hitherto, in the pure state. However, relatively concentrated aqueous solutions can be prepared by treatment of solutions of the barium salt with the correct amount of sulphuric acid:

$$BaS_2O_6 + H_2SO_4 \longrightarrow H_2S_2O_6 + BaSO_4$$

Normal salts of the acid are quite stable at room temperature and are well characterized. They are commonly prepared by oxidation of the corresponding metal sulphite or of sulphur dioxide. Thus with a suspension of manganese dioxide in sulphurous acid[298] at 0°C the reaction is:

$$2MnO_2 + 3H_2SO_3 \longrightarrow MnSO_4 + MnS_2O_6 + 3H_2O$$

The solution is treated with excess barium ions and the sulphate precipitated removed and the excess barium ions are then neutralized with sulphuric acid. The salt $BaS_2O_6.2H_2O$ may then be crystallized from solution. Alternatively, ferric oxide may be used when the following reactions take place:

$$Fe_2^{III}O_3 + 3SO_2 \longrightarrow Fe_2^{III}(SO_3)_3$$

$$Fe_2^{III}(SO_3)_3 \longrightarrow Fe^{II}SO_3 + Fe^{II}S_2O_6$$

Acid salts of dithionic acid are hitherto unknown. Dithionic acid is a truly intermediate state between sulphuric acid and sulphurous acid. In its stable state—in solution or in stable salts—however, it cannot be regarded as a "sulphonic–sulphinic acid of water", as might be anticipated from the systematic description expressed by Table 25. For the existence of an unsymmetrical form with a S–O–S bridge in the molecule see ref. 299. The dithionite ion has the structure[300]

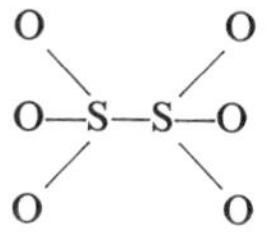

```
O         O
  \      /
O—S—S—O
  /      \
O         O
```

S–S = 2·15–2·16 Å
S–O = 1·45 Å
S-S-O = 103°

The oxygen atoms are symmetrically placed around each sulphur atom. The molecule is free to rotate around the S–S bond (cf. the abnormal long S–S distance!). The ion has D_{3d} symmetry[301]. Dithionic acid is a rather strong acid with the two ionization constants apparently very close[282]. Dilute solutions of the acid are relatively stable, but on concentration decomposition takes place at temperatures as low as 50°C. The overall reaction

[298] *Inorganic Synthesis*, II (1946) 167.
[299] K. Steinle, Diplomarbeit, Univ. Munich (1959).
[300] E. Stanley, *Acta Cryst.* **9** (1956) 897.
[301] W. G. Palmer, *J. Chem. Soc.* (1961) 1552.

of this decomposition confirms the right position of dithionic acid in the system of oxyacids of sulphur, represented by Table 25:

$$H_2S_2O_6 \longrightarrow H_2SO_4 + SO_2$$

For a reasonable mechanism of this decomposition reaction see ref. 299. Most dithionites are very soluble in water, including the barium salt. The solutions are stable up to the boiling point. However, at 200°C the salts break down to give the metal sulphate and sulphur dioxide (a reverse reaction could not yet be realized[299]). Dithionites are relatively stable towards oxidation, but very strong oxidizing agents such as the halogens, dichromate and permanganate will bring about oxidation to sulphate. Likewise, only very powerful reducing agents attack them, sodium amalgam, for example, leading to sulphite and dithionite.

Sulphurous Acid, H_2SO_3

In dithionates, the sulphur atoms have the formal oxidation number +5. The derivate or reaction product of water with the second thermodynamically stable oxide of sulphur, sulphur dioxide, must by nature exhibit an oxidation number of +4 for the involved sulphur atoms. This is indeed the case in the monosulphinic acid of water, HO–SO_2H. This sulphurous acid is not yet known in the free state, nor does it exist in aqueous solution. Sulphur dioxide gas dissolves fairly readily in water (at 15°C, 45 vol. SO_2 in 1 vol. water) to give a rather complicated reaction mixture. This reaction mixture is called "sulphurous acid" by tradition without having a definite composition. A spectroscopic study[302] of this complicated mixture (the composition is strongly dependent on concentration, temperature and pH-value[303] revealed) that it mainly consists of physically dissolved (and loosely hydrated) sulphur dioxide and of the ions H_3O^+, $S_2O_5^{--}$ and HSO_3^-; SO_3^- and $HOSO_2^-$ concentrations are so low that they are just at the borderline of spectroscopic detection. Undissociated sulphurous acid, H_2SO_3, may be assumed as a short-lived intermediate in the acidification of sulphites at low temperatures in ethers[304], but does not exist in detectable amounts in aqueous solutions of sulphur dioxide that are called "sulphurous acid". It is, therefore, a rather academic question which structure of the two possible ones

$$(HO)_2S{=}O \qquad (HO)(O{=})S(=O)(OH)$$

is the correct one. (The well-known existence of two different organic "derivates" of sulphurous acid, the alkyl sulphites and the alkyl sulphonates, does not allow a reasonable conclusion to the open question. It is nothing but a pure formalism to compare the unknown acid with those organic compounds.)

"Sulphurous acid" gives rise to two series of salts, both of which are well characterized. The normal or "neutral" sulphites contain the ion SO_3^{--}, and the acid or hydrogen sulphites the ion HSO_3^- (the latter ones are only known in aqueous solution, but not in the solid state). Normal sulphites are most conveniently prepared in two stages: reaction of sulphur

302 A. Simon and K. Waldmann, *Wiss. z. Techn. Hochschule Dresden*, **5** (1955–6), Heft 3.
303 K. A. Kolbe and K. C. Hellwig, *Ind. Eng. Chem.* **47** (1955) 1116.
304 M. Schmidt and B. Wirwoll, *Z. anorg. allg. Chem.* **303** (1960) 184.

dioxide with a metal hydroxide leads to the acid sulphite, and treatment of its solution with the corresponding carbonate gives the neutral salt[303]:

$$NaOH+SO_2 \longrightarrow NaHSO_3$$

$$NaHSO_3+Na_2CO_3 \xrightarrow{\text{boiling}} 2Na_2SO_3+H_2O+CO_2.$$

The HSO_3^- ion which is only stable in aqueous solution[304] has an unsymmetrical structure with the hydrogen atom directly bound to the central sulphur atom[305]. The structure of the sulphite ion, SO_3^{--}, in solution as well as in the solid state, is pyramidal with C_{3v} symmetry similar to that of the isoelectronic chlorate ion[306]

S–O = 1·39 Å

The ionic equilibria at 25°C and dissociation constants for "sulphurous acid" are given by the equations[307]

$$\text{"}H_2SO_3\text{"} \rightleftharpoons H^+ + HSO_3^- \qquad K_A = 1{\cdot}6\times10^{-2}$$

$$HSO_3^- \rightleftharpoons H^+ + SO_3^{--} \qquad K_A = 1{\cdot}0\times10^{-7}$$

From this it can be seen that aqueous SO_2 solutions "apparently" only are rather weak acids and that the solutions mostly contain HSO_3^- ions. Solutions of normal sulphites, therefore, must react clearly alkaline by the reaction

$$SO_3^{--}+H_2O \longrightarrow HSO_3^- + OH^-$$

All normal sulphites are nearly insoluble in water, except those of the alkali metals and of ammonium, whereas all known hydrogen sulphites are easily soluble and cannot be isolated in substance because of the equilibrium

$$2HSO_3^- \rightleftharpoons S_2O_5^{--}+H_2O$$

that goes to the right by increasing the concentration; only disulphites, therefore, can be isolated by evaporation of aqueous hydrogen sulphite solutions. In their general reactions, sulphites and hydrogen sulphites are moderately strong reducing agents yielding either dithionate or sulphate upon reaction. This reducing behaviour under all pH conditions may be seen in the redox potentials[284]

$$SO_2.xH_2O \rightleftharpoons SO_4^{--}+4H^+ +(x-2)H_2O+2e^- \qquad E_0 = 0{\cdot}18\text{ V}$$

$$SO_3^{--}+2OH^- \rightleftharpoons SO_4^{--}+H_2O+2e^- \qquad E_0 = +0{\cdot}93\text{ V}$$

Already the oxygen in the air can oxidize sulphur of the oxidation number +4 to the oxidation number +6 (sulphuric acid or sulphates). The rate of this reaction is strongly dependent on the pH-value of the solutions. This oxidation is catalysed by manganese dioxide and by such reducing agents as iron(II) and arsenite[308]. Oxidation by hydrogen peroxide also takes place easily and without the need of catalysts although various oxyacid

[305] A. Simon and K. Waldmann, *Z. anorg. allg. Chem.* **281** (1955) 113.
[306] W. H. Zachariasen and H. E. Buckley, *Phys. Rev.* **37** (1931) 1925.
[307] R. C. Brasted, *Comprehensive Inorganic Chemistry*, Vol. 8, p. 113, Van Nostrand, New York (1961).
[308] A. M. Koganowskii and P. N. Taran, *Ukr. Chim. Zh.* **21** (1955) 472.

anions speed the reaction considerably[309]. The familiar titrimetric quantitative determination of "sulphurous acid" and of sulphites is also based on this characteristic reducing behaviour:

$$SO_3^{--}+H_2O+I_2 \longrightarrow SO_4^{--}+2H^++2I^-.$$

As well as being a reducing agent, sulphurous acid can act as an oxidizing agent, depending, by nature, on the reaction partner. Thus with sodium amalgam dithionite is obtained[310] while various formic acid derivatives are oxidized to oxalate, the sulphite going to thiosulphate. Thiosulphate is also obtained when solutions of the alkali metal sulphites are treated with elemental sulphur[311]. The reduction of "sulphurous acid" with hydrogen sulphide leading to Wackenroder's solution will be dealt with briefly on page 889. Reactions of sulphurous acid with nitrogen-containing compounds frequently lead to derivatives containing partially oxidized sulphite as sulphonic acid groups[312]. Simultaneous oxidation to sulphate may take place as in the reaction with hydroxylamine which yields both sulphamic acid, NH_2SO_3H, and sulphuric acid. The sulphite ion, SO_3^{--}, is an excellent complexing agent and is known in many complexes, mainly with metals of periodic group VIII. Each individual sulphite group usually occupies only one coordination position although several may coordinate to the same metal atom. Coordination is through the sulphur atom[313]. Typical complexes containing the SO_3^{--} ion are $[Ru^{I}(NH_3)_5SO_3]^-$ [314], $[Cr_2(H_2O)_6(OH)_2SO_3]^{++}$ [315] and $Na_2[Pt(NH_3)_3(SO_3)_2]$[316]. Complexes are also known containing the hydrogen sulphite group, e.g. $[Ru^{II}(NH_3)_4(SO_3H)_2]$[317].

Disulphurous Acid, $H_2S_2O_5$

Disulphurous acid, as sulphurous acid, is a product of the reaction between water and sulphur dioxide. The oxidation number of the sulphur atoms, therefore, must be +4. The free acid is known neither in the free state, nor in solution—as is also the case with sulphurous acid. Therefore, of course, nothing is known on its structure. Its situation in Table 25 would suggest its nature as disulphinic acids of water, $HO_2S{-}O{-}SO_2H$. The $S_2O_5^{--}$ ion, however, surprisingly and also in contrast to disulphuric acid, does not contain a S–O–S bond, but a direct sulphur–sulphur bond. As such it is unsymmetrical[318], and a detailed analysis of the bonding has revealed that the most favourable hybridization requires that two different S–O distances should be present, as is indeed observed:

```
 O  O
 |  |
O–S1–S2–O
     |
     O
```

$S_1{-}S_2 = 2{\cdot}205$ Å
$O{-}S_1 = 1{\cdot}499$ Å
$O{-}S_2 = 1{\cdot}431, 1{\cdot}472$ Å

Disulphites are prepared by concentration of aqueous hydrogen sulphite solutions

$$2HSO_3^- \rightleftharpoons S_2O_5^{--}+H_2O$$

[309] P. M. Mader, *J. Am. Chem. Soc.* **80** (1958) 2634.
[310] A. W. Weng and G. L. Putnam, *Trans. Ind. Inst. Chem. Engrs.* **3** (1948–50) 35.
[311] F. Foerster and R. Vogel, *Z. anorg. allg. Chem.* **155** (1926) 161.
[312] D. S. Brackman and W. C. E. Higginson, *J. Chem. Soc.* (1953) 3896.
[313] M. M. Gurin, *Dokl. Akad. Nauk SSSR* **50** (1945) 201.
[314] K. Gleu and W. Breuel, *Z. anorg. allg. Chem.* **235** (1938) 211.
[315] S. G. Shuttlewerth, *J. Am. Leather Chemists Assoc.* **47** (1952) 387.
[316] A. Binz and E. Haberland, *Ber.* **53** (1920) 2030.
[317] K. Gleu, W. Breuel and K. Rehm, *Z. anorg. allg. Chem.* **235** (1938) 201.
[318] I. Lindquist and M. Morsell, *Acta Cryst.* **10** (1957) 406.

Acidification of their solutions results in the formation of hydrogen sulphites and sulphur dioxide again. The chemistry of the disulphite ions is largely that of the normal sulphites and hydrogen sulphites in aqueous solutions, from which they may also be prepared just by treating with an excess of sulphur dioxide.

Dithionous Acid, $H_2S_2O_4$

The lowest positive oxidation number that sulphur can assume in its oxyacids or derivatives thereof, respectively, is +3, the oxidation number in dithionous acid. The acid is known neither in the free state nor in stable aqueous solution. Freshly prepared solutions (by treatment of "sulphurous acid" with zinc amalgam) decompose by disproportionation:

$$2S_2O_4^{--} + H_2O \longrightarrow 2HSO_3^- + S_2O_3^{--}$$

Normal salts of the acid, however, are well characterized. They are stable when kept anhydrous, while aqueous solutions decompose much more slowly than those of the free acid. Sodium dithionite, $Na_2S_2O_4$, finds wide application as an industrial reducing agent. It is prepared by reduction of sodium sulphite in various ways, the most common being by zinc or sodium amalgams[310, 319, 320, 321] or by electrolytic reductions. Corresponding laboratory procedures are available. Thus the reaction with zinc amalgam follows the equation

$$2HSO_3^- + \text{"}H_2SO_3\text{"} + Zn \longrightarrow ZnSO_3 + S_2O_4^{--} + 2H_2O$$

Excess sulphite is removed with limewater and the dithionite obtained as the dihydrate by precipitation with sodium chloride[322]. All stages of the reaction must be carried out in the absence of oxygen. The same dihydrate is also the product when sodium amalgam reacts with sulphur dioxide solutions containing up to 20% alcohol[323]. The anhydrous salt is obtained either by vacuum drying or by alcohol dehydration. Sodium dithionite has also been found in the product of the reaction between sodium borohydride and sodium hydrogensulphite[324], while up to 84% yields of zinc dithionite have been claimed by the reaction of zinc powder with liquid sulphur dioxide in the presence of a trace of water[325].

The structure of the dithionite ion, after some controversy in the literature[282], has been confirmed by X-ray crystallographic studies on $Na_2S_2O_4.2H_2O$ [326]. The crystals of this salt are monoclinic of space group $P2/c$ and have approximately symmetry C_{2v}. The shape of the ion is·

```
    S ——— S
   / \   / \
  O   O O   O
```

S–S = 2·389 Å
S–O = 1·1496 Å, 1·515 Å

The two planes are almost parallel with an angle of 100° between the SO_2 plane and the S–S bond. The extremely (!) long bond S–S of 2·389 Å together with the unusual shape of the molecule with eclipsed SO_2 groups has been explained by the suggestion that each sulphur atom forms two *pd* hybrid orbitals one of which it uses to bond to the other atom. The question arises whether this long S–S distance is not the "real" sulphur–sulphur bond

319 D. M. Yost and H. Russell, *Systematic Inorganic Chemistry*, p. 354, Prentice-Hall, New York (1944).
320 K. F. Andryushchenko, *Ukr. Khim. Zh.* **29** (1963) 125.
321 C. C. Patel and M. R. A. Rao, *Proc. Natl. Acad. Sci. India*, **15** (1949) 115.
322 K. Jellinek, *Z. anorg. allg. Chem.* **70** (1911) 93.
323 H. Ostertag and Y. Choissen, *Compt. Rend.* **242** (1956) 1732.
324 G. S. Panson and C. E. Weill, *J. Inorg. Nucl. Chem.* **15** (1960) 184.
325 K. Uchigasuki, *Kogyo Kaysku Zasshi* **61** (1958) 670.
326 J. D. Dunitz, *Acta Cryst.* **9** (1956) 579.

single distance, whereas all the many considerably shorter S–S bonds are shortened by some sort of back-bonding effects such forming six multiple bond participations. The reaction studied with the dithionite ion may well be explained by assuming that the rather unstable ion splits according to the equation

$$S_2O_4^{--} \rightleftharpoons 2SO_2^-$$

to form two radical ions. These may then undergo further reduction or oxidation. Air oxidation studies show that in alkaline solutions at 30–60°C the oxidation is half order with respect to $S_2O_4^-$ suggesting that the above formulated fission really occurs. Recent electron spin resonance studies have confirmed the presence of the radical SO_2^-. It is present to the extent of one part in three thousand[327]. Dithionic acid is supposed to be a strong acid and ionizes at 25°C according to the equilibria[307].

$$H_2S_2O_4 \rightleftharpoons HS_2O_4 + H^+ \qquad K_A = 4{\cdot}5 \times 10^{-1}$$
$$HS_2O_4 \rightleftharpoons H^+ + S_2O^{--} \qquad K_A = 3{\cdot}5 \times 10^{-3}$$

The hydrolysis has been studied in some detail under both acid and alkaline conditions[328]. The acid hydrolysis follows second order kinetics[320] and is given by

$$2S_2O_4^{--} + H_2O \longrightarrow S_2O_3^{--} + 2HSO_3^-$$

while in excess alkali, sulphide is produced:

$$3Na_2S_2O_4 + 6NaOH \longrightarrow 5Na_2SO_3 + Na_2S + 3H_2O.$$

Decomposition of dithionite solutions can also occur quantitatively via a disproportionation reaction

$$2S_2O_4^{--} \longrightarrow S_2O_3^{--} + 2HSO_3^-$$

This reaction will take place in the absence of air and even in solid dithionites if a little moisture is present. Heating solid dithionites leads to the reaction

$$S_2O_4^- \longrightarrow S_2O_3^{--} + SO_3^{--} + SO_2.$$

In the case of the sodium salt this reaction occurs violently at 190°C [329]. In their normal reactions dithionites are strong reducing agents, this being consistent with their unsaturation with respect to oxygen[330]. Thus, they will reduce H_2O_2, MnO_4^-, I_2, IO_3^- and molecular oxygen. Many metal ions such as Cu^I, Ag^I, Pb^{II}, Sb^{III}, Bi^{III} are reduced to the metals while the TiO^{2+} ion is reduced to Ti^{III}.

Derivatives of Hydrogen Peroxide

The Lewis base water reacts with the Lewis acid sulphur trioxide in the molar ratio 1 : 1 and 1 : 2, forming a monosulphonic acid of water, H_2SO_4, and a disulphonic acid of water, $H_2S_2O_7$. According to Table 25, this is also to be expected from the Lewis base hydrogen peroxide, which in many respects is very similar to water. This formal analogy is indeed correct: we know a monosulphonic acid of hydrogen peroxide, H_2SO_5, and a disulphonic acid of hydrogen peroxide, $H_2S_2O_8$.

[327] S. Lynn, R. E. Rinker and W. H. Concoran, *J. Phys. Chem.* **68** (1964) 2363.
[328] E. M. Marshak, *Khim. Naukai Prom.* **2** (1957) 524.
[329] E. Schulek and L. Moros, *Magy. Kem. Fdyoirat*, **63** (1957) 41.
[330] H. Stamm and M. Goehring, *Angew. Chem.* **58** (1945) 52.

Peroxomonosulphuric Acid, H_2SO_5

The monosulphonic acid of hydrogen peroxide may be prepared in the free state by the reaction of chlorosulphonic acid with hydrogen peroxide under exclusion of water[331]:

$$HO_3SCl + HO\text{–}OH \longrightarrow HCl + HSO_3\text{–}OOH$$

It is a colourless solid with melting point 45°C. It should be handled with great care because of the danger of explosions[332]! This anhydrous acid is without any practical importance. Its aqueous solutions are stronger oxidizing agents than hydrogen peroxide on the one side and peroxodisulphonic acid on the other side. As an intermediate in the preparation of hydrogen peroxide from peroxodisulphonic acid and water it is rather important[333]. Aqueous solutions of the acid may be prepared from peroxodisulphates and sulphuric acid or from concentrated hydrogen peroxide and sulphuric acid (not in pure form but in mixtures with sulphuric acid).

Peroxodisulphuric Acid, $H_2S_2O_8$

This disulphonic acid of hydrogen peroxide[333] has the structure:

$$HO_3S\text{–}O\text{–}O\text{–}SO_3H \qquad \begin{array}{l} O\text{–}O = 1{\cdot}31\ \text{Å} \\ S\text{–}O = 1{\cdot}50\ \text{Å} \end{array}$$

In pure form it is a colourless solid with melting point 65°C (with decomposition). It is soluble in water to any extent. The solutions (as also the free acid) are strong oxidizing agents, and are not very stable, the stability decreasing with decreasing pH-value. It is—as also are the salts—prepared by the anodic oxidation of HSO_4^- ions under suitable conditions:

$$2HSO_4^- \longrightarrow H_2S_2O_8 + 2e^-$$

The acid is hydrolysed by heating, forming hydrogen peroxide via peroxymonosulphuric acid. This reaction is by far the most important reaction of these peroxo compounds of sulphur:

$$\begin{array}{l} H_2S_2O_8 + H_2O \longrightarrow H_2SO_5 + H_2SO_4 \\ H_2SO_5 + H_2O \longrightarrow H_2SO_4 + H_2O_2 \\ \hline H_2S_2O_8 + 2H_2O \longrightarrow 2H_2SO_4 + H_2O_2 \end{array}$$

All the hitherto-known salts of peroxodisulphuric acid are easily soluble in water. The most important are $(NH_4)_2S_2O_8$ and $K_2S_2O_8$. They are strong oxidizing agents.

Salts of a peroxotetrasulphuric acid can be prepared by three different methods[334]:

$$\begin{array}{l} 2KO_2 + 4SO_3 \longrightarrow K_2S_4O_{14} + O_2 \\ Na_2O_2 + 4SO_3 \longrightarrow Na_2S_4O_{14} \\ K_2S_2O_8 + 2SO_3 \longrightarrow K_2S_4O_{14} \end{array}$$

In these salts (that are of no practical importance) two $-SO_3^-$ ions are linked together by a –O–O– group. In all hitherto-described derivatives of hydrogen peroxide, the sulphur atoms of the sulphur oxyacids have an oxidation number of +6. Theoretically, also sulphinic acids of hydrogen peroxide could exist, as for instance a monosulphinic acid,

331 J. D'Ans and W. Friederich, *Z. Elektrochem.* **17** (1911) 849.

332 J. O. Edwards, *Chem. Eng. News*, **33** (1955) 3336.

333 *Ullmanns Encyclopädie der Technischen Chemie*, 3. Auflage, Band 15, Urban & Schwarzenberg, München–Berlin (1964).

334 M. Schmidt and H. Bipp, *Z. anorg. allg. Chem.* **303** (1960) 201.

HOO–SO_2H, which with the oxidation number of +4 for the sulphur atom would be an isomer of the familiar sulphuric acid. Experiments to isolate this interesting compound have not yet been successful, obviously because of a very fast rearrangement into sulphuric acid[335].

Derivatives of Sulphanes

In the same manner as water and hydrogen peroxide, the sulphanes also behave as Lewis bases towards the Lewis acid sulphur trioxide, thus forming monosulphonic acids, $H_2S_xO_3$, and disulphonic acids, $H_2S_xO_6$.

Sulphane Monosulphonic Acids, $H_2S_xO_3$

The sulphanes are unbranched sulphur chains, terminated by hydrogen atoms. The first member of this homolog series, hydrogen sulphide, for different reasons (cf. page 826) deserves special treatment. This also holds, naturally, for its reaction product with sulphur trioxide, the sulphane monosulphonic acid, HS–SO_3H, known for more than 250 years under the name thiosulphuric acid, $H_2S_2O_3$, but only in the form of some salts.

Thiosulphuric Acid, $H_2S_2O_3$

Despite a great number of experiments, for a very long time it was not possible to prepare free thiosulphuric acid or stable solutions of it. Instability was regarded as one of the characteristics of this acid[336]. The course as well as the products of its decomposition (acidifying of aqueous thiosulphate solutions) was not known. The usual formulation in textbooks

$$H_2S_2O_3 \longrightarrow S + H_2O + SO_2$$

only gives a very rough description of this decomposition. Depending on the conditions, besides sulphur (partially in the form of cyclohexa sulphur) and sulphur dioxide, hydrogen sulphide, higher sulphanes, sulphuric acid and polythionates are also formed thereby[336–338]. In the anhydrous state, the acid could be produced in ether solution as a dietherate, $H_2S_2O_3 . 2(C_2H_5)_2O$, by the reaction of sodium thiosulphate with hydrogen chloride in diethyl ether at −78°C:

$$Na_2S_2O_3 + 2HCl \longrightarrow 2NaCl + H_2S_2O_3$$

Small catalytic amounts of water must be added to the reaction mixture. The decomposition of the anhydrous acid proved surprisingly simple: on warming it decomposes quantitatively below 0°C to form hydrogen sulphide and sulphur trioxide, according to

$$H_2S_2O_3 \longrightarrow H_2S + SO_3$$

Under anhydrous conditions the reaction products do not enter a redox reaction but are stable in contact with each other. This decomposition is completely analogous to the thermal decomposition of sulphuric acid at temperatures above about 330°C:

$$H_2SO_4 \longrightarrow H_2O + SO_3$$

---

335 M. Schmidt and P. Bornmann, *Z. anorg. allg. Chem.* **331** (1964) 92.
336 M. Schmidt, *Z. anorg. allg. Chem.* **289** (1957) 147.
337 F. Foerster and H. Umbach, *Z. anorg. allg. Chem.* **217** (1934) 175.
338 H. D. Block, Diplomarbeit, Univ. Würzburg (1967).

and opens up an easy way for the preparation of thiosulphuric acid: stoichiometric amounts of sulphur trioxide and hydrogen sulphide under anhydrous conditions react at $-78°C$ in ether to form quantitatively $H_2S_2O_3$:

$$H_2S+SO_3 \longrightarrow H_2S_2O_3$$

This reaction may be extended to mercaptans and thiophenols, thus forming for the first time free alkyl or aryl thiosulphuric acids[339]

$$RSH+SO_3 \longrightarrow RSSO_3H$$

where R is an alkyl or an aryl group. Without a solvent, or in nonpolar solvents such as fluorochloromethanes, hydrogen sulphide and sulphur trioxide at low temperatures form a white crystalline Lewis adduct, $H_2S.SO_3$, which is isomeric with thiosulphuric acid and is easily decomposed *in vacuo* even at low temperatures into its components H_2S and SO_3. According to its formation and decomposition reactions, thiosulphuric acid must be regarded as the monosulphonic acid of monosulphane, and SO_3 may be regarded as the anhydride of sulphuric acid and at the same time as the "ansulphhydride" of thiosulphuric acid. Similarly, chlorosulphonic acid then is the acid chloride of sulphuric acid as well as of thiosulphuric acid, depending upon whether it is reacting with water or with hydrogen sulphide. So, a third way is possible for the preparation of anhydrous thiosulphuric acid, thiolysis of chlorosulphonic acid:

$$HO_3S\text{–}Cl+H\text{–}SH \longrightarrow HO_3SSH+HCl$$

By this method the acid can be synthesized free of any solvent at all[340].

The stable salts of thiosulphuric acid contain the ion $S_2O_3^{--}$; this ion is derived from the sulphate ion by the replacement of one ligand oxygen atom by a sulphur atom (naturally also in the oxidation number -2, whereas the central sulphur atom—as also in sulphuric acid—has an oxidation number of $+6$):

S O
\ /
S S–S distance = 1·98 Å[341]
/ \
O O

Only normal but no acid salts of thiosulphuric acid are known hitherto. They may easily be prepared by boiling aqueous sulphite solutions with elemental sulphur according to

$$SO_3^{--}+1/xS_x \longrightarrow S_2O_3^{--}.$$

For the mechanism of this reaction see page 818. Another method of preparation is the oxidation by iodine of a mixture of sulphides and sulphites:

$$S^{--}+SO_3^{--}+I_2 \longrightarrow S_2O_3^{--}+2J^-.$$

Of technical importance for the preparation of thiosulphates also is the oxidation of polysulphides by air, e.g.

$$Na_2S_5+\tfrac{3}{2}O_2 \longrightarrow Na_2S_2O_3+\tfrac{3}{x}S_x.$$
$$CaS_2+\tfrac{3}{2}O_2 \longrightarrow CaS_2O_3$$

Most thiosulphates are easily soluble in water (except those of lead, silver, thallium^{I}, and

339 M. Schmidt and G. Talsky, *Chem. Ber.* **94** (1961) 1352.
340 M. Schmidt and G. Talsky, *Chem. Ber.* **92** (1959) 1526.
341 O. Foss, *Advances in Inorg. Chem. and Radiochem.*, Vol. 2 (1960).

barium) and form well-shaped crystals with water of crystallization, e.g. $Na_2S_2O_3.5H_2O$. Thiosulphates of transition metals tend to the formation of complexes. This is of considerable practical importance in photography where sodium thiosulphate is used to dissolve unreacted silver bromide from emulsion, according to the reaction

$$AgBr+3Na_2S_2O_3 \longrightarrow Na_5[Ag(S_2O_3)_3]+NaBr$$

Thiosulphate ion is a reducing agent of moderate strength, as is indicated in the couple

$$2S_2O_3^{--} \rightleftharpoons S_4O_6^{--}+2e^-; E^\circ_{298} = 0{\cdot}17 \text{ V}$$

Oxidation of thiosulphate by iodine, forming tetrathionate and iodide, is perhaps the best-known reaction of the thiosulphates. It is the basis of all the very many titrimetric iodometric quantitative determinations of oxidizing agents:

$$2S_2O_3^{--}+I_2 \longrightarrow S_4O_6^{--}+2I^-.$$

Oxidation to sulphite is complicated by the inherent reducing power of sulphite. Accordingly, sulphate is the common oxidation product of thiosulphate when strong oxidizing agents are employed. A typical example for such a reaction is the oxidation of thiosulphates by chlorine, of technical importance in the use of thiosulphates as "antichlorine" in the bleaching industry when an excess of chlorine in the fibres is destroyed by $S_2O_3^{--}$:

$$S_2O_3^{--}+4Cl_2+5H_2O \longrightarrow 2HSO_4^-+8H^++8Cl^-.$$

In the course of the iodometric oxidation of thiosulphate to tetrathionate the average oxidation number of sulphur is brought from +2 to +2·5, whereas in the oxidation to sulphate by chlorine it rises from +2 to +6; in the oxidation of a given amount of thiosulphate, therefore, the consumption of chlorine is eight times that of iodine!

Higher Sulphane Monosulphonic Acids

The reaction of hydrogen sulphide with sulphur trioxide can be extended further. SO_3 forms an acid not only with water, but also with hydrogen peroxide. This holds for both the thio analog of water, H_2S, and the thio analog of H_2O_2, disulphane. According to the reaction

$$HSSH+SO_3 \longrightarrow HSSSO_3H$$

disulphane and sulphur trioxide react in ether at low temperatures to form the acid $H_2S_3O_3$ quantitatively. Its constitution is analogous to that of peroxomonosulphuric acid, H_2SO_5. The O–O group of this compound is replaced by a S–S group[342].

Monosulphane monosulphonic acid, $H_2S_2O_3$, and disulphane monosulphonic acid, $H_2S_3O_3$, are the first two members of the sulphur acid series sulphane monosulphonic acids, $H_2S_xO_3$. They are formed according to

$$HS_xH+SO_3 \longrightarrow HS_xSO_3H$$

H_2S and H_2S_2 yield $H_2S_2O_3$ and $H_2S_3O_3$, as already mentioned. H_2S_3 forms trisulphane monosulphonic acid, $H_2S_4O_3$. H_2S_4 forms tetrasulphane monosulphonic acid, $H_2S_5O_3$, H_2S_5 forms pentasulphane monosulphonic acid, $H_2S_6O_3$, and H_2S_6 forms hexasulphane monosulphonic acids, $H_2S_7O_3$. The latter one is the highest member of this series known at the present time[342, 343]. Sulphane monosulphonic acids may also be prepared by the

342 M. Schmidt, *Z. anorg. allg. Chem.* **289** (1957) 158.
343 M. Schmidt and H. Dersin, *Z. Naturforsch.* **14b** (1959) 735.

reaction of chlorosulphonic acid with sulphanes[344, 345]. These acids are stable in etheral solutions at low temperatures. At room temperatures they are decomposed more or less quickly, depending on the chain length. Water and especially aqueous alkali decompose the acids very rapidly, thiosulphate, sulphur dioxide and elemental sulphur being the main decomposition products. With excess sulphite ions, sulphane monosulphonic acids react very fast in a stepwise mechanism, forming thiosulphate:

$$S_xO_3^{--}+(x-2)SO_3^{--} \longrightarrow (x-1)S_2O_3^{--}$$

These degradation reactions are part of the reaction of elemental sulphur with sulphite ions (see page 818) in which sulphane monosulphonic acids are intermediates. A similar degradation of the sulphane monosulphonic acids occurs with many nucleophiles—for example, with cyanide ions, whereby thiocyanate and thiosulphate are formed:

$$S_xO_3^{--}+(x-2)CN^- \longrightarrow S_2O_3^{--}+(x-2)SCN^-$$

Again, this reaction is part of the sulphur degradation with cyanide ions.

The structure of the sulphane monosulphonic acids has not yet been established by physical methods, but follows from their formation from sulphur trioxide and the linear sulphanes as well as from their oxidation to linear polythionic acids, $H_2S_xO_6$. They obviously are built up from unbranched skewed sulphur chains with a $-SO_3H$ group on one end and a hydrogen atom on the other end. Hexasulphane monosulphonic acid, as an example, therefore appears to have the structure

```
HO3S      S       S       SH
    \    / \     / \     /
     S      S       S
```

These sulphane monosulphonic acids, by nature, are derivatives of hydrogen sulphide insofar as they still contain the characteristic $-SH$ group. With this terminating group these acids can undergo condensation reactions with ClS-groups, forming new sulphur–sulphur bonds with elimination of hydrogen chloride. Typical reactions of this type leading to polythionic acids will be described in the next section of this review.

Sulphane Disulphonic Acids, $H_2S_xO_6$

Compounds of the composition $H_2S_xO_6$ have been known for a long time in the form of some salts that are rather unstable in aqueous solution. This holds for the ions $S_3O_6^{--}$, $S_4O_6^{--}$, $S_5O_6^{--}$ and $S_6O_6^{--}$. They have been designated as polythionic acids[346]. Those compounds were found in several complex mixtures of sulphur oxyacids, but for a long time could neither be isolated in substance nor be understood in their chemical behaviour and composition. The situation only was changed completely with the finding of a close genetic connection between elemental sulphur, the sulphanes, the sulphane monosulphonic acids and the polythionic acids[346]. It could be shown that the polythionic acids are in reality the disulphonic acids of the sulphanes. This nomenclature should replace the established and customary older names, at least in cases in which the genetic relationship of the polythionic acids to the class of sulphanes and their monosulphonic acids is emphasized.

344 F. Fehér, J. Schotten and B. Thomas, *Z. Naturforsch.* **13b** (1958) 624.
345 M. Schmidt and G. Talsky, *Angew. Chem.* **70** (1958) 312.
346 M. Schmidt, *Z. anorg. allg. Chem.* **289** (1957) 193.

(This nomenclature would also express the fact that dithionic acid, $H_2S_2O_6$, as an oxidation product of sulphur dioxide, has chemically nothing to do whatsoever with the real "polythionic acids" that are reduction products of sulphur dioxide. The fact that dithionic acid in some modern textbooks, e.g. ref. 347, is denominated as the first member of the series $H_2S_xO_6$ can only be explained—and excused—by the erroneous nomenclature that does not allow any reasonable conclusion on the structure and reactions of the named compounds.)

The older literature on these polythionic acids is literally overwhelming. The problems connected with their chemical nature, reactions, formation and decomposition for a very long time was one of the great puzzles in inorganic chemistry. The fact that the very many publications not infrequently are contradictory shows that there existed no real understanding of the essential nature of these compounds. The very confusing literature up to 1952 has been masterfully collected and reviewed[348]. Sulphane disulphonates are formed with different yields and purity mainly in the following ways:

(1) Interaction of chlorosulphanes with the ions HSO_3^- and $S_2O_3^{--}$. These reactions have been studied in detail for quite a long time[349]. The conclusions from these studies are that the chlorosulphanes first are hydrolysed to intermediates such as $S_2(OH)_2$, S_2^+, S_2O; these intermediates of unknown composition should then react with the sulphite or thiosulphate ions, respectively. From today's knowledge, however, this hypothesis no longer can be accepted; the synthetic very useful reactions must be regarded as condensations between the free chlorosulphanes and the acid ions[350], e.g.

$$^-O_3SH + Cl\text{–}S\text{–}Cl + HSO_3^- \longrightarrow {}^-O_3S\text{–}S\text{–}SO_3^- + 2H^+ + 2Cl^-$$

$$^-O_3SH + Cl\text{–}S_2\text{–}Cl + HSO_3^- \longrightarrow {}^-O_3S\text{–}S\text{–}S\text{–}SO_3^- + 2H^+ + 2Cl^-$$

$$^-O_3S\text{–}SH + Cl\text{–}S\text{–}Cl + HS\text{–}SO_3^- \longrightarrow {}^-O_3S\text{–}S\text{–}S\text{–}S\text{–}SO_3^- + 2H^+ + 2Cl^-$$

$$^-O_3S\text{–}SH + Cl\text{–}S\text{–}S\text{–}Cl + HS\text{–}SO_3^- \longrightarrow {}^-O_3S\text{–}S\text{–}S\text{–}S\text{–}S\text{–}SO_3^- + 2H^+ + 2Cl^-$$

The redox reactions between thionyl chloride and sulphuryl chloride with thiosulphuric acid lead to the formation of sulphane disulphonates (polythionates) according to the same principal scheme[351].

(2) Interaction of hydrogen sulphide with excess sulphur dioxide in aqueous solution (Wackenroder's liquid). In aqueous solution hydrogen sulphide and sulphur dioxide react with each other to form an extremely complex mixture of sulphur–oxygen–hydrogen compounds, mainly sulphane-, disulphane-, trisulphane- and tetrasulphane-disulphonic acids (trithionic, tetrathionic, pentathionic and hexathionic acids) in varying amounts, depending strongly on the reaction conditions. This "Wackenroder's liquid" still represents a "classical" problem in inorganic chemistry[348, 352, 353, 354]. There is, hitherto, no experimentally secured mechanism known for the formation of sulphur and the sulphane disulphonic acid. On the basis of a very careful study the conclusion can be drawn that the first stable product in the complicated mixture is disulphane disulphonic acid (tetrathionic acid, $H_2S_4O_6$); the other sulphane disulphonic acids are then formed by side- or secondary reactions which can

347 E. S. Gould, *Inorganic Reactions and Structure*, Holt, Rinehart and Winston, New York (1962).
348 M. Goehring, *Fortschr. Chem. Forsch.* **2** (1952) 444.
349 M. Goehring and H. Stamm, *Z. anorg. allg. Chem.* **250** (1942) 56.
350 M. Schmidt and Th. Sand, *Z. anorg. allg. Chem.* **330** (1964) 179.
351 M. Schmidt and Th. Sand, *Chem. Ber.* **97** (1964) 282.
352 E. Blasius and W. Burmeister, *Z. anorg. allg. Chem.* **268** (1959) 1.
353 E. Blasius and R. Krämer, *J. Chromatog.* **20** (1965) 367.
354 H. Stamm, M. Becke-Goehring and M. Schmidt, *Angew. Chem.* **72** (1960) 34.

be understood from the chemical behaviour of the sulphane disulphonic acids towards sulphite and thiosulphate ions. A detailed description of the various and often contradictory hypotheses is out of the scope of this book. The, at present, most convincing hypothesis[355] will be sketched by giving some brutto formulae:

$$\mathrm{HS^-+O\underset{O}{\overset{O}{S}}\!-\!\underset{O}{\overset{OH^-}{S}} \longrightarrow HS\!-\!\underset{O}{\overset{O}{S}}OH+\underset{O}{\overset{OH^-}{S}}}$$

$$\mathrm{HO\underset{O}{\overset{O}{S}}S^-+\underset{OH}{\overset{OH}{S}} \longrightarrow HO\underset{O}{\overset{O}{S}}\!-\!S\!-\!SOH+OH^-}$$

$$\mathrm{HO\underset{O}{\overset{O}{S}}\!-\!S\!-\!SOH+\underset{O}{\overset{O}{S}}OH^- \longrightarrow HO\underset{O}{\overset{O}{S}}\!-\!S\!-\!S\!-\!\underset{O}{\overset{O}{S}}OH+OH^-}$$

Side reaction:

$$\mathrm{HO\underset{O}{\overset{O}{S}}\!-\!S\!-\!SOH+S\underset{O}{\overset{O}{S}}OH^- \longrightarrow HO\underset{O}{\overset{O}{S}}\!-\!S\!-\!S\!-\!S\!-\!\underset{O}{\overset{O}{S}}OH+OH^-}$$

Formulation of sulphur (schematically):

$$\mathrm{HOSOH+HS^- \longrightarrow HOSSH+OH^-}$$

$$x\mathrm{HOSSH \longrightarrow S_{2x}}+x\mathrm{H_2O}$$

The consequence that in the presence of thiosulphate ions the main product of Wackenroder's reaction is not disulphane disulphonate (tetrathionate), but trisulphane disulphonate (pentathionate), had been proved experimentally. Another important consequence of this hypothesis is the demand that the elemental sulphur formed in the course of the reaction does not stem from the reaction

$$\mathrm{2H_2S+SO_2 \longrightarrow \tfrac{3}{8}S_8+2H_2O}$$

It also could be proved experimentally. But still more sophisticated chemical and physical methods will have to be worked out and applied before a definite and final description of the interesting reaction of hydrogen sulphide and sulphur dioxide in aqueous solution can be given.

(3) Oxidation of thiosulphates with oxidants such as I_2, Cu^{2+}, ICN, BrCN, $S_2O_8^{--}$, NO_2^-, or H_2O_2, or the anode. When thiosulphate is treated with weak oxidizing agents, the main product of the reactions is disulphane disulphonate(tetrathionate). The by far best-known example of such a reaction is the quantitative oxidation of thiosulphate by iodine[356]:

$$\mathrm{2S_2O_3^{--}+I_2 \longrightarrow S_4O_6^{--}+2I^-}.$$

Despite the extensive use of this reaction in analytical chemistry, its mechanism still remains unexplained in real detail. The same oxidation reaction is also reported to take place with the oxidants Fe^{III}, Au^{III}, and MnO_4^{--} ions[357, 358].

(4) Disproportionation of sulphurous acid and its derivatives at higher temperatures[348].

(5) Interaction of thiosulphates with acids in the presence of special catalysts such as As or Sb salts[348]. These special reactions especially are not at all understood at the present time; they urgently deserve further experimental and theoretical studies.

[355] K. Steinle, Thesis, Univ. Munich (1962).
[356] E. Abel, *Z. anorg. allg. Chem.* **269** (1952) 207.
[357] K. Lar and G. Singh, *J. Ind. Chem. Soc.* **33** (1956) 668.
[358] D. V. R. Rao and S. Pani, *J. Sci. Ind. Res.* **15B** (1956) 667.

(6) Specific synthetic methods for some salts: (a) Monosulphane disulphonates (trithionates). The potassium salt $K_2S_3O_6$ may be prepared in satisfactory yield and purity by the interaction of thiosulphate with sulphur dioxide in water[359]. The mechanism of this redox reaction is not yet known.

(b) Disulphane disulphonates (tetrathionates). They may easily be obtained in a pure state by the oxidation of thiosulphates by iodine[359]. A second method is interesting not only because of its good synthetic results but also out of theoretical reasons: many experiments aiming at the formation of sulphur chains in which an inner sulphur atom forms one or two linkages to oxygen failed[360].

$$-S_x-\underset{\downarrow \atop O}{\overset{O \atop \uparrow}{S}}-S_y-$$

Compounds of such a configuration are extremely unstable and therefore easily split off sulphur dioxide already at low temperatures, as has been found with sulphuryl thiocyanate:

$$NCS\underset{| \atop O}{\overset{O \atop |}{S}}SCN \longrightarrow SO_2 + NCSSCN$$

This also holds for the reaction of sulphuryl chloride with thiosulphates[351] which results in practically quantitative yields of pure disulphane disulphonates and sulphur dioxide:

$$HO_3SS-H + Cl-\underset{O}{\overset{O}{S}}-Cl + H-SSO_3H \longrightarrow HO_3S-S-\underset{O}{\overset{O}{S}}-S-SO_3H + 2\,HCl$$

$$HO_3S-S-\underset{O}{\overset{O}{S}}-S-SO_3H \longrightarrow HO_3S-S-S-SO_3H + SO_2$$

(c) Trisulphane disulphonates (pentathionates). Pure salts, e.g. $K_2S_5O_6.1H_2O$, are obtained under carefully controlled experimental conditions by the acid decomposition of thiosulphates in the presence of As_2O_3[359].

(d) Tetrasulphane disulphonates (hexathionates). They are best prepared by a redox reaction between thiosulphates and nitrous acid, again under carefully controlled conditions. The mechanism of this reaction is not known[359]. The following physical properties of the potassium sulphane disulphonates $K_2S_3O_6$, $K_2S_4O_6$, $K_2S_5O_6$ and $K_2S_6O_6$ have been determined recently: solubility in hydrochloric acid, crystal habitus[359], ultraviolet spectra in aqueous solution[361], infrared spectra[362], polarographic behaviour[363], dielectric constant[364] and behaviour on anion exchangers[365].

[359] M. Schmidt and Th. Sand, *J. Inorg. Nucl. Chem.* **26** (1964) 1165.
[360] M. Schmidt and D. Eichelsdörfer, *Z. anorg. allg. Chem.* **319** (1963) 350.
[361] M. Schmidt and Th. Sand, *J. Inorg. Nucl. Chem.* **26** (1964) 1173.
[362] M. Schmidt and Th. Sand, *J. Inorg. Nucl. Chem.* **26** (1964) 1179.
[363] M. Schmidt and Th. Sand, *J. Inorg. Nucl. Chem.* **26** (1964) 1185.
[364] M. Schmidt and Th. Sand, *J. Inorg. Nucl. Chem.* **26** (1964) 1189.
[365] M. Schmidt and Th. Sand, *Z. anorg. allg. Chem.* **330** (1964) 188.

Synthesis of free sulphane disulphonic acids. The discovery of the sulphane monosulphonic acids opened up a number of new and especially clear-cut ways into the field of sulphane disulphonic acid chemistry. The monosulphonic acid reacts in etheral solution with sulphur trioxide to form anhydrous sulphane disulphonic acids in a simple reaction in quantitative yields[366]:

$$HO_3SS_xH + SO_3 \longrightarrow HO_3SS_xSO_3H$$

The following acids have been prepared by this method: $H_2S_3O_6$, $H_2S_4O_6$, $H_2S_5O_6$, $H_2S_6O_6$, $H_2S_7O_6$ and $H_2S_8O_6$. One can just as well start from the sulphanes as from their monosulphonic acids; the corresponding amount of sulphur trioxide must then be used—that is, $H_2S_x : SO_3 = 1:2$:

$$HS_xH + 2SO_3 \longrightarrow HO_3SS_xSO_3H$$

The sulphane disulphonic acids thereby are formed in a clearly defined manner, free from side products or other impurities.

A third route for the formation of sulphane disulphonic acids was found in the oxidation of the monosulphonic acids with iodine in aqueous medium:

$$HO_3SS_xH + I_2 + HS_xSO_3H \longrightarrow HO_3SS_{2x}SO_3H + 2HI$$

This method (for thiosulphates, of course, known for a long time) yields the highest members of the sulphane disulphonic acid series hitherto known as well-defined individuals. There is a suggestion[367] that the sulphane disulphonate series extends up to and including the hydrophilic Oden sulphur sols. These appear to be sodium salts, $Na_2S_xO_6$, with x from 50 to 100. The stability of these salts with respect to liberation of sulphur decreases up to x about 20 and then increases as the properties approach those of the sulphur sols. The probable existence of higher sulphane disulphonic acids—up to a chain length of 12—also in Wackenroder's liquid, has been postulated[368]. The following acids have been synthesized by this oxidation method: $H_2S_4O_6$, $H_2S_6O_6$, $H_2S_8S_6$, $H_2S_{10}O_6$, $H_2S_{12}O_6$ and $H_2S_{14}O_6$. Stoichiometric amounts of chlorine instead of iodine as oxidant permit the preparation of these acids in an anhydrous medium[369].

Two further ways also lead from sulphane monosulphonic acids to sulphane disulphonic acids, namely condensation reactions with chlorosulphanes[370, 371]

$$HO_3SS_x\text{–}H + Cl\text{–}S_y\text{–}Cl + H\text{–}S_xSO_3H \longrightarrow HO_3SS_{2x+y}SO_3H + 2HCl$$

and with chlorosulphuric acid[370, 372]

$$HO_3SS_x\text{—}\boxed{H + Cl}\text{—}SO_3H \longrightarrow HO_3SS_xSO_3H + HCl$$

Structure of the sulphane disulphonates. The structure of the until recently only known four sulphane disulphonates has been a matter of controversy for a long time. The structure of monosulphane disulphonate was generally accepted as $^-O_3S\text{–}S\text{–}SO_3^-$, but the problem remained as to whether further sulphur atoms extended the chain or added on to the central

366 M. Schmidt, *Z. anorg. allg. Chem.* **289** (1957) 175.
367 E. Weitz, K. Gieles, J. Singer and B. Alt, *Chem. Ber.* **89** (1956) 2365.
368 R. Barbieri and M. Bruno, *J. Inorg. Nucl. Chem.* **14** (1960) 148.
369 M. Schmidt and H. Dersin, *Z. Naturforsch.* **14b** (1959) 735.
370 F. Fehér, J. Schotten and B. Thomas, *Z. Naturforsch.* **13b** (1958) 624.
371 M. Schmidt and B. Wirwoll, *Z. anorg. allg. Chem.* **303** (1960) 184.
372 M. Schmidt and G. Talsky, *Angew. Chem.* **70** (1958) 312.

sulphur atom. Thus alternative structures are possible already with disulphane disulphonate:

$$\begin{array}{c}\phantom{{}^-O_3S-}S\\\phantom{{}^-O_3S-}\uparrow\\{}^-O_3S{-}S{-}SO^-\end{array}\qquad {}^-O_3S{-}S{-}S{-}SO_3^-$$

Many authors preferred structures with branched sulphur chains, mostly on the ground that the sulphur atoms of di-, tri- and tetrasulphane disulphonates, which are so readily given off by the action of basic reagents such as sulphite ions, cyanide ions, etc., must be bonded differently from the others and therefore could not be part of unbranched chains. Today, however, we know with certainty that the sulphane disulphonic acids possess an unbranched structure as the sulphanes and as sulphur itself (these different compounds of sulphur are genetically closely related and linked together by the sulphane monosulphonic acids) not only from their chemical behaviour but also from a number of physical measurements such as determination of refraction, viscosity and electrical conductivity[373], of the Raman spectra[374] and of the *Kα* X-ray fluorescence[375]. The most convincing arguments for the unbranched structure are structure determinations of the salts by X-ray methods, carried on by Foss, who also wrote an excellent review on chain-like sulphur compounds[376]. According to this determination the sulphane disulphonate ions consist of two distorted $S_2O_3^{--}$ tetrahedra joined by a common corner ($S_3O_6^{--}$), a covalent bond ($S_4O_6^{--}$), a sulphur atom ($S_5O_6^{--}$) or a disulphane group ($S_6O_6^{--}$). As a typical example trisulphane disulphonate is shown:

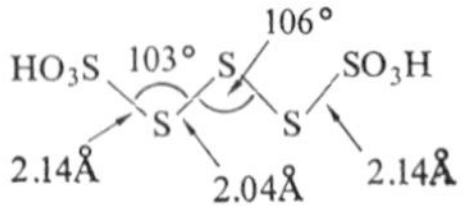

HO₃S–S–S–S–SO₃H (angles 103°, 106°; bond lengths 2.14Å, 2.04Å, 2.14Å)

Also there can be no doubt that the higher sulphane disulphonic acids have an analogous structure; unbranched skewed zig-zag chains of sulphur atoms terminated by SO_3H groups. Two types of sulphur–sulphur bonds occur in the sulphane disulphonates, namely, between divalent sulphur atoms in the middle of the chains and between one divalent and one sulphonate sulphur atom at the ends. The bond length of this terminal bond is 2·14 Å. The middle bonds have within the errors the same length as the sulphur–sulphur bonds in orthorhombic sulphur, 2·04 Å, which is also the value found for organic sulphane derivatives. The difference in length between the two types of bond indicates, apart from a possible effect of different hybridization of σ-bond orbitals at divalent and sulphonate sulphur, that bonds between divalent sulphur atoms possess some $(p \rightarrow d)\pi$-bond character, or, what seems to be less probable, that the terminal bonds are longer than single bonds.

Reactions of sulphane disulphonic acids. Aqueous solutions of sulphane disulphonates in alkaline or neutral solutions are unstable (except $S_3O_6^{--}$ and $S_4O_6^{--}$); the instability also in acid solutions is rapidly increasing with the length of the sulphur chain (higher sulphane disulphonates are extremely sensitive towards traces of alkali). The decomposition reactions are very complex and result in reaction mixtures of sulphur with different oxyacids, not hitherto known in detail. Research on the chemical reactions of sulphane disulphonates has been concentrated for a long time on their reactions with nucleophilic partners. Our present view of those reactions will be described here in some detail on the example of the

373 H. Hertlein, *Z. physikal. Chem.* **19** (1896) 287.
374 M. Eucken and J. Wagner, *Acta Phys. Austriaca*, **1** (1946) 339.
375 A. Faessler and M. Goehring, *Naturwissenschaften*, **39** (1952) 169.
376 O. Foss, *Advances in Inorg. Chem. Radiochem.* **2** (1960) 237.

interaction of sulphane disulphonates and sulphite. A brief summary of a qualitative working hypothesis on the nature of the sulphur–sulphur bond[377] must be given first, since it is part of the basis of the assumed reaction mechanism of the sulphane disulphonates.

Multiple bond components in sulphur–sulphur bonds. Sulphur–sulphur bonds mostly are formulated and regarded as simple single σ-bonds, according to a proposal made by Pauling in 1949[378]. Every sulphur atom must then still possess two free electron pairs. The *s*-pair is distributed around the nucleus in spherical symmetry, the *p*-pair is concentrated on a 90° space axis (atomic nucleus in the intersection point). The hindrance of free rotation about the S–S bond and also the dihedral angles in sulphur chains and rings are then traced to a Coulomb repulsion of the free *p*-electron pairs of neighbouring sulphur atoms. However, this hypothesis does not satisfactorily explain many experimental results about the nucleophilic degradation of sulphur–sulphur bonds[377]. They are much easier to understand by the assumption that the sulphur atom not only has the possibility but even shows a strong tendency to overcome the octet with involvement of its *d*-orbitals. With the multiplicity of compound formation of the element with itself, this strong tendency must perforce result in sulphur–sulphur bonds usually being other than single bonds. It is assumed that *d*-orbitals must participate in the required multiple bond components. An interpretation of "free" *p*-electrons with empty *d*-orbitals can be qualitatively explained from two different extremes: the participating orbitals must be correspondingly hybridized in order to create appropriate symmetry conditions, or the p_x^2 electrons coupled in the ground state must be uncoupled in order to make possible a higher bonding ability (e.g. $H—\bar{S}—\bar{S}—H \rightarrow H—\bar{S}{\equiv}\bar{S}—H$). Both processes require the expenditure of energy. However, this energy expenditure obviously is overcompensated by the resulting greater bond strength. The cause of the phenomenon is only of secondary importance for the hypothesis of a multiple bond content between sulphur atoms as a simple explanation of many experimental observations, as long as the hypothesis is possible in principle. For this reason it will not be further discussed in this connection.

Naturally "free" electron pairs cannot be considered in isolated fashion. Nevertheless only the *p*-electrons will be included in the following considerations because the experimental findings already can be qualitatively described with this simplification. For a more quantitative treatment one would naturally have to consider also the "free" *s*-pair. The possible expansion of the principle of the hypothesis will only be indicated here, but will not be considered in the formula outline employed for reasons of clarity and simplification.

The fact that sulphur–sulphur bonds are so easily attacked by nucleophilic partners proves that the sulphur atoms in their natural environment show electrophilic behaviour and thus a tendency to overcome the octet of electrons. However, this tendency need not yet necessarily lead to a multiple bond content. The presence of such a component is concluded from the behaviour of sulphur and chain-like sulphur compounds towards Lewis acids—that is, electrophilic partners. If reaction takes place at all (stable adducts could never be isolated at all), then a cleavage of the sulphur–sulphur bond takes place immediately as in the reaction of S_8 with the Lewis acid SO_3. Likewise the oxidation of sulphur with oxygen never leads to the much-sought S_8O_8 (or even S_8O_{16}), but always directly to monomeric sulphur oxides—in other words, to a cleavage of sulphur–sulphur bonds.

From such findings it is concluded that the "free" *p*-pairs of each sulphur atom are not

377 M. Schmidt, in B. Meyer (Ed.), *Elemental Sulphur Chemistry and Physics*, pp. 301–326, Interscience, New York (1965).

378 L. Pauling, *Proc. Nat. Acad. Sci. U.S* **35** (1949) 495.

actually free and available as such for donor bonding to Lewis acids. They are rather involved in the particular sulphur–sulphur bond in question and strengthen this bond over and beyond the usual single bond with participation of unoccupied *d*-orbitals. This state of affairs formally may be symbolized in the example of cyclo-octasulphur as a sort of resonance formulations (the *s*-pairs are given as non-participating for purposes of simplification), whereby each sulphur atom in the formal sense has a shell of ten electrons:

In this outline the π-electrons in elemental sulphur no longer can be regarded as localized. This hypothesis explains easily quite a number of physical and chemical properties of sulphur[377]. If those considerations are correct, then the conclusions drawn heretofore from physical sulphur–sulphur distance determinations must be re-examined. The distance of about 2·04 Å found in the element naturally at first was regarded as the real "single bond distance". However, according to this new hypothesis, the actual single bond distance may be somewhat longer. In this connection it seems appropriate to refer to the "unusually" long –S–S– distance of 2·39 Å in the relatively unstable dithionite ion $S_2O_4^{--}$, a value which exceeds by 0·5 Å the heretofore shortest measured distance of 1·89 Å in diatomic sulphur and so convincingly demonstrates the "softness" of a sulphur atom. The distance in cyclo-octasulphur is only about 0·15 Å longer than in thiosulphate, but about 0·35 Å shorter than in $S_2O_4^{--}$. The large difference of 20–25% from the measured –S–S– distances makes the requirement of a delocalized electron system seem readily possible.

According to this concept the required multiple bond components in the sulphane disulphonic acids can be symbolized (with the example of tetrasulphane disulphonic acid, $H_2S_6O_6$) as

$$\underset{\alpha}{HO_3S}\overset{\rightarrow}{—}\underset{\beta}{S}\overset{\rightarrow}{—}S\overset{\rightarrow}{—}S\overset{\rightarrow}{—}\underset{\beta}{S}—\underset{\alpha}{SO_3H} \qquad \underset{\alpha}{HO_3S}\overset{\leftarrow}{—}\underset{\beta}{S}\overset{\leftarrow}{—}S\overset{\leftarrow}{—}S\overset{\leftarrow}{—}\underset{\beta}{S}—\underset{\alpha}{SO_3H}$$

Here it should be noted that the two terminal (α) sulphur atoms are not components of the sulphane chain in which these compounds are based. With their oxidation state +6 they are the centre of a (somewhat distorted) tetrahedron with three oxygen atoms at the three corners and the terminal sulphur atom at the fourth corner. The sulphur chain with the two β-atoms at its end has altogether the oxidation state −2.

According to the bonding concept mentioned, the individual sulphur atoms in such sulphane derivatives (from the trisulphane disulphonic acid, $H_2S_5O_6$ onward) are no longer electronically equally surrounded. Thus they must differ from each other in their electrophilic behaviour. By means of this simple conception it is possible to understand the –S–S– bond distance found in the ions

$$S_3O_6^{--},\ S_4O_6^{--}, S_5O_6^{--} \text{ and } S_6O_6^{--}$$

and also to understand their reactions with nucleophilic partners. In the $S_2O_3^{--}$ tetrahedron of the thiosulphate ion the certainly polar S–S– bond of 1·98 Å is shorter through a strong multiple bonding component than the "ordinary" sulphur–sulphur bond. If now the "single bonding" sulphur atom is connected to a second sulphur atom, as is the case in the sulphane disulphonic acids, then the "free" electrons of this atom must serve to reinforce two bonds. This involves an enlargement of the original bond distance. However, the positive sulphur atom in the centre of the now more distorted tetrahedron will surely not return any additional electrons. This is governed by the experimentally measured two different bond distances in sulphane disulphonates: the spacing between the first (α) and second (β) sulphur atom is at 2·11 Å, appreciably greater than in thiosulphate and also longer than that between the other sulphur atoms of the chain, a constant $\sim$2·04 Å (as in elemental sulphur).

By "superposition" of the two "border formulations" for sulphane disulphonic acids a 9-shell formally results for both β-atoms, but for the remaining atoms of the chain a 10-shell of electrons results. This formulation is only intended to illustrate that the sulphur atoms in such chains are no longer surrounded in the same manner electronically. This means then that a nucleophilic attack upon these chains takes place preferentially at electrophilic centres and thus in a definite order rather than being statistically random. This requirement of an ordered stepwise degradation results from the experimental observations.

One knows that sulphane disulphonates in aqueous solution spontaneously react with a series of nucleophilic substances already at room temperature—for instance, with SH^-, CN^-, $AsO_3H_2^-$ and SO_3H^-. As a random example, the hydrogen sulphite ion, HSO_3^-, will be considered in this connection.

The gross equation for the "sulphite degradation" of the sulphane disulphonates is

$$H_2S_xO_6+(x-3)H_2SO_3 \longrightarrow H_3S_3O_6+(x-3)H_2S_2O_3$$

In the presence of sulphite in excess, this degradation proceeds spontaneously and quantitatively. It has been employed for a long time for the analysis of the sulphane disulphonates. However, the explanation of this degradation was never quite clear. The original assumption interpreted the reaction by stating that the sulphur atom in the sulphite ion has such a strong tendency to enter the coordinate 4-bond condition that it takes a sulphur atom away from the sulphane disulphonate and thereby becomes thiosulphate. This mechanically hardly comprehensible assumption was thoroughly disproven in 1949[379]. In the reaction of disulphane disulphonate with ^{35}S– labelled sulphite the entire activity was found not in the thiosulphate formed, but rather in the monosulphane disulphonate:

$$S_4O_6^{--} + {}^{35}SO_3H^- \not\longrightarrow S_3O_6^{--} + S\,{}^{35}SO_3H^-$$

$$S_4O_6^{--} + {}^{35}SO_3H^- \longrightarrow {}^-O_3S—S—{}^{35}SO_3^- + S_2O_3H^-$$

This finding can only be understood on the basis of a nucleophilic attack of the SO_3H^- ions on one of the two equivalent sulphur atoms of the disulphane disulphonic acid in which, in a S_N2 mechanism, the more weakly nucleophilic (equals more strongly acidic) thiosulphate ion is displaced from one of the two β-sulphur atoms by the more strongly nucleophilic sulphite ion:

$$\begin{array}{l} {}^-O_3S—S—\vdots S—SO_3^- \\ \quad + \quad \overset{*}{S}O_3H^- \end{array} \longrightarrow {}^-O_3\overset{*}{S}—S—SO_3^- + {}^-O_3S—SH$$

[379] J. A. Christiansen and W. Drost Hansen, *Nature*, **164** (1949) 759.

Since in the case of the disulphane disulphonic acid one is dealing with a symmetrical disulphane derivative, every hypothesis based upon a nucleophilic attack or on "ionic displacement reactions" must naturally lead to the same result. Differences can only start to appear in trisulphane derivatives (or unsymmetrical disulphane derivatives). Here indeed the problem sets in. The previous formulations assume that one molecule of sulphane disulphonate reacts with one mole of sulphite in such a way that there is always obtained one mole of thiosulphate and one mole of sulphane disulphonate shortened by one sulphur atom, this continuing until there remains, in addition to the $S_2O_3^{--}$, the monosulphane disulphonate $S_3O_6^{--}$ which is not degraded by sulphite. Using the example of tetrasulphane disulphonate, the following steps seem to be involved in the reaction:

$$S_6O_6^{--} + SO_3^{--} \rightleftharpoons S_5O_6^{--} + S_2O_3^{--}$$
$$S_5O_6^{--} + SO_3^{--} \rightleftharpoons S_4O_6^{--} + S_2O_3^{--}$$
$$S_4O_6^{--} + SO_3^{--} \rightleftharpoons S_3O_6^{--} + S_2O_3^{--}$$
$$\overline{S_6O_6^{--} + 3SO_3^{--} \longrightarrow S_3O_6^{--} + 3S_2O_3^{--}}$$

The gross reaction can only be observed quantitatively in the presence of an appreciable excess of sulphite. Stoichiometric amounts of sulphite or especially an insufficient amount of sulphite lead to complicated mixtures of different sulphane disulphonates. The formation of these difficult to analyse mixtures was attributed to the presence of equilibria as formulated above. The presence of such equilibria was also used as the basis for the interpretation of kinetic studies of reactions of several sulphane disulphonates with nucleophilic agents. Since the analytical methods employed cannot distinguish fundamentally between the products formed according to the above equations, such kinetic investigations lose their significance.

The formulation according to which a nucleophilic sulphite ion displaces the more weakly nucleophilic thiosulphate ion from the γ-atom of a sulphane disulphonate ion with a disulphane disulphonate which is poorer by one sulphur atom can no longer be maintained. The above formulated equations cannot be realized experimentally (the same also holds for the higher sulphane disulphonates!). Thus, $S_2O_3^{--}$ and $S_4O_6^{--}$ are proven not to be in equilibrium with $S_5O_3^{--}$ and SO_3^{--}. Therefore the reactions must be explained differently. According to the above-mentioned hypothesis, the β-atoms of a sulphane disulphonic acid are the electrophilic centres of such molecules. An attack by the nucleophilic sulphite ion will take place there and form one mole of monosulphane disulphonate, $S_3O_6^{--}$, and the residue $S_xSO_3^{--}$ via a transition state

$$\left[\begin{array}{c} HO_3S\text{–}S\text{–}S_x\text{–}SO_3H \\ \uparrow \\ SO_3H \end{array}\right]^-$$

Thus, according to this hypothesis the sulphane monosulphonate ion $S_xSO_3^{--}$ is displaced from the β-atom, and not a thiosulphate ion from the γ-atom. For the sulphane monosulphonic acids an electron distribution can be symbolized as follows:

$$H\text{—}\underline{\vec{S}}\text{—}\underline{\vec{S}}\text{—}\underline{\vec{S}}\text{—}\underline{\vec{S}}\text{—}\underline{\vec{S}}\text{—}SO_3H$$

The electrophilic centre in such compounds is the terminal α-atom which in contrast to the formal electron decet of the other chain atoms possesses only an octet of electrons.

Accordingly, the attack by the nucleophilic sulphite ion must take place at this α-atom, and the more weakly (than SO_3H^-) nucleophilic $HO_3S{-}S_x^-$ ion must be displaced from the most electrophilic α-atom via the transition state

$$\left[\begin{array}{c} HO_3S—S_x —SH \\ \uparrow \\ SO_3H \end{array}\right]^-$$

and by the reaction

$$H_2S_zO_3+H_2SO_3 \longrightarrow H_2S_{z-1}O_3+H_2S_2O_3$$

The sulphite degradation of the sulphane disulphonates does not proceed via the above-formulated equilibria reactions but must be formulated by the following steps (with $H_2S_6O_6$ as example):

$$H_2S_6O_6+H_2SO_3 \longrightarrow H_2S_3O_6+H_2S_4O_3$$
$$H_2S_4O_3+H_2SO_3 \longrightarrow H_2S_3O_3+H_2S_2O_3$$
$$H_2S_3O_3+H_2SO_3 \longrightarrow H_2S_2O_3+H_2S_2O_3$$

As already emphasized, the analytical methods employed cannot distinguish in principle between the two different reaction paths, since in the final analysis they all are based on the sum of the oxidation numbers of the sulphur atoms, which are, of course, the same in each case. According to this interpretation, the sulphite ion attacks the β-sulphur atom not only in the case of the disulphane disulphonic acid $H_2S_4O_6$ (for which both hypotheses must perforce lead to the same conclusion) but also in all higher sulphane disulphonic acids. The β-atoms always represent the electrophilic centres of such sulphur chains. The basic sulphite ion displaces the $HO_3S{-}S_x^-$ ion from it, this ion being more weakly nucleophilic than the HO_3S^- ion. The nucleophilic character of $HO_3S{-}S_x^-$ decreases with increasing values of x (acid strength increasing correspondingly). Naturally the decrease is more marked going from $x = 1$ to $x = 2$ than in going from $x = 2$ to $x = 3$. With increasing chain length the differences become even smaller. In long sulphur chains there will be a point at which they can no longer be detected. However, in the sulphane disulphonic acids isolated until now, and also in the primary reaction products of the S_8 ring with Lewis bases, they are still sufficiently marked to determine the course of the reactions with "normal" nucleophilic partners. The decreasing, but up to a chain length of about 6 to 8 atoms still clearly apparent, influence of an additional sulphur atom in the chain upon the electronic surroundings of the chain-terminating positively polarized hydrogen atom is convincingly shown by the characteristic proton resonance signals of the sulphanes.

The earlier interpretations of the sulphite degradation of sulphane disulphonates can no longer be maintained in view of the newer findings and considerations. According to these, the reaction of a sulphite ion with a sulphane disulphonate ion does not give thiosulphate and a sulphane disulphonate shorter by one sulphur atom. The first reaction product always is a monosulphane disulphonate, $S_3O_6^{--}$, and a sulphane monosulphonate $HO_3S{-}S_x^-$. Likewise in all other reactions of the sulphane disulphonates with nucleophilic partners, which fundamentally attack at the β-atom, the primary reaction products are sulphane monosulphonates. Since such sulphane monosulphonic acids were not known before 1956 (except for $H_2S_2O_3$), the versatile reactions of the sulphane disulphonates with basic materials naturally could not be interpreted correctly. In this special instance, the sulphane monosulphonic acids represent a kind of "missing link". The earlier interpretation of the sulphane disulphonates reaction cannot serve to explain their "instability" in the presence of a little sulphite or thiosulphate ("thiosulphate catalysis"). Trisulphane and tetrasulphane disulphonate are—as is also true of the higher members of this series—unstable in the presence of very small amounts of SO_3^{--} and $S_2O_3^{--}$. According to the earlier formulations, $S_4O_3^{--}$ and $S_2O_3^{--}$ should form from $S_5O_6^{--}$ and SO_3^{--}. However, these are stable in the presence of one another. The actual reaction product is a very complex mixture of sulphane sulphonates. These experimental findings are readily understandable under the assumption that the primary products of these reactions are sulphane monosulphonates. These are very unstable and correspondingly reactive. If they are not decomposed in ordinary fashion (as stronger electrophilic agents than the disulphonates; by excess sulphide, which is lacking in these cases) they can change to sulphane disulphonates either through oxidation or through condensation (splitting off sulphide). The sulphide formed in the condensation can, as a very good nucleophilic agent, likewise have a degrading effect. Thus the formation of complicated mixtures from pure sulphane disulphonates with a little sulphite or thiosulphite is readily understandable and need not be described in detail in this connection.

The same explanation holds for the "instability" of sulphane disulphonates which are not perfectly pure. The almost always present contaminations are, because of the usual preparative procedures, precisely SO_3^{--} and $S_2O_3^{--}$. The "autodecomposition" in such cases is a consequence of the primary formation of sulphane monosulphonates by the contaminants. Genuinely pure sulphane disulphonates are relatively quite stable in the presence of one another. The reaction of sulphane disulphonates with sulphite was treated in such detail because it is typical of the reaction of water-soluble sulphur chains with nucleophilic partners. It can also be applied in an analogous way to the reactions with other bases such as $AsO_3H_2^-$, CN^- and SH^-. Such reactions can then be readily understood. For the analytical chemistry of sulphur as well as of other compounds of sulphur see the most recent review of Blasius *et al.*[380].

2.5. COMPOUNDS CONTAINING SULPHUR AND NITROGEN

Sulphur Nitrides

This section will deal with compounds consisting of nitrogen and sulphur. It will also include a brief discussion of the thiothiazyl halides, which actually belong to the class of nitrogen–sulphur–halogen compounds. However, since the thiothiazyl cation has the characteristics of a nitrogen–sulphur compound, it should be discussed together with the sulphur nitrides. Several reviews on sulphur–nitrogen chemistry are available[381–383]. The simplest N–S compound is represented by the "thiazyl" S≡N, the monomeric sulphur nitride. Since NS is the thio-analog of NO and therefore possesses an unpaired electron, it polymerizes easily to form $(SN)_2$, $(SN)_4$ and, under certain conditions, $(SN)_x$.

Tetrasulphur Tetranitride, $(SN)_4$

This compound is obtained by several reactions. However, the yields are poor in all cases. The most convenient procedure uses a CCl_4 solution of S_2Cl_2 saturated with chlorine

[380] E. Blasius, G. Horn, A. Knöchel, J. Münch and H. Wagner, in ref. 282, pp. 201–230.

[381] M. Becke-Goehring in H. J. Emeleus and A. G. Scharpe (Eds.), *Advances in Inorganic Chemistry and Radiochemistry*, Vol. 2, p. 169, Academic Press, New York (1960).

[382] M. Becke-Goehring and E. Fluck, in C. B. Colburn (Ed.), *Developments in Inorganic Nitrogen Chemistry*, Vol. 1, p. 150, Elsevier, Amsterdam, London, New York (1966).

[383] H. G. Heal, in G. Nickless (Ed.), *Inorganic Sulphur Chemistry*, p. 459, Elsevier, Amsterdam, London, New York (1968).

and ammonia as starting materials[384]. By heating NH_4Cl with S_2Cl_2 at 160°, S_4N_4 is formed in a 26% yield according to the following scheme:

$$6S_2Cl_2+4NH_4Cl \longrightarrow S_4N_4+S_8+16HCl$$

The puzzle of the ring formation has yet to be solved. However, it has been learnt through several studies that a five- and a seven-membered ring system, $S_3N_2Cl_2$ and S_4N_3Cl, are intermediates in route to the eight-membered ring, S_4N_4. Recently it was found[385] that S_4N_3Cl can be enlarged to tetrasulphur tetranitride by means of $Al(N_3)_3$:

$$3S_4N_3Cl+Al(N_3)_3 \longrightarrow 3S_4N_4+AlCl_3+3N_2$$

On the other hand, S_4N_4 reacts with S_2Cl_2 to form S_4N_3Cl:

$$3S_4N_4+2S_2Cl_2 \longrightarrow 4S_4N_3Cl$$

These two equations illustrate a rare reaction cycle of contracting and enlarging ring systems in inorganic chemistry. Some of the physical properties of S_4N_4 are given in Table 28.

TABLE 27. INTERATOMIC DISTANCES AND BOND ANGLES IN S_4N_4

		Electron diffraction data	Crystal measurements
[structure of S_4N_4]	d(S–N) (Å)	1.62	1.62
	∢N–S–N (°)	112	113
	∢S–N–S (°)	106	105
	d(S–S) (Å)		2.58

TABLE 28. PHYSICAL PROPERTIES OF SULPHUR NITRIDES

Compound	M.p. (°C)	Density (g/cm³)	Approx. decomp. temp. (°C)
S_4N_4	178·2[a]	2·24[b]	206[c]
S_2N_2			30[d]
$(SN)_x$	130	219[e]	130
S_4N_2	23[f]	1·71[f]	100[f]
$S_{11}N_2$	150–155		>150
$S_{15}N_2$	137		
$S_{16}N_2$	122		

[a] M. H. M. Arnold, J. A. C. Hugill and J. M. Hutson, *J. Chem. Soc.* (1936) 1648.
[b] F. P. Burt and F. L. Usher, *Proc. Roy. Soc.* **A85** (1911) 84.
[c] J. P. Koettnitz, *Z. Elektroch.* **34** (1928) 770.
[d] M. Goehring, *Ergebnisse u. Probleme der Chemie der Schwefel-Stickstoff-verbindungen*, p. 145, Academie-Verlag, Berlin (1957).
[e] L. Gmelin, *Handbuch der Anorganischen Chemie*, Schwefel B, 3, S. 1537, Verlag Chemie, Weinheim (1963).
[f] *Ibid.*, S. 1535.

S_4N_4 is a yellow-orange compound at room temperature, but darkens at higher temperatures. Since S_4N_4 decomposes upon rapid heating or striking to form N_2 and sulphur, it should be handled with care. Its enthalpy of formation is +128·8 kcal/mole. The

[384] *Gmelins Handbuch der Anorganischen Chemie*, Systemnummer 9, Teil B, S. 1537, Verlag Chemie, Weinheim (1963).
[385] M. Becke-Goehring and G. Magin, *Z. Naturforsch.* **20b** (1965) 493.

cage-like structure of S_4N_4 has been found[386]. It represents nearly a spheric molecule—every atom fitting exactly inside a closed shell. The S–N distances are all alike within the ring and correspond to a bond order of 1·5. The S–S distance of 2·58 Å is considerably shorter than the sum (3·7 Å) of van der Waals' radii. This fact suggests that the whole system contains delocalized π-electrons, which may be responsible for the abnormally high diamagnetic susceptibility[387]. Additional proof for the π-electron system stems from ESR studies[388]. In THF below 0° a nine-line spectrum was found with a g-value of 2·0006—due to the $S_4N_4^-$ anion—in which the unpaired electron is delocalized over the entire S_4N_4 ring. Above 0° the radical decomposes to give a one- and a two-nitrogen radical, and, upon further reduction, another four-nitrogen radical appears, probably an S_xN^- species. The dipole moment amounts to $0{\cdot}72 \times 10^{-18}$ esu in CS_2. The ^{14}N chemical shift for S_4N_4 is $+485 \pm 20$ ppm from saturated aqueous nitrite ion. It is much nearer the shifts for singly bonded S–N compounds (530–540 ppm) than it is to the range (200–300 ppm) observed for thiazenes such as $S_4N_3^+$ ion, despite the double bonding in S_4N_4. This high shielding can be explained in terms of the high symmetry (D_{2d}) of the near-spherical S_4N_4, compared with the flatter thiazenes[389]. The reduction of S_4N_4 with $SnCl_2$ or dithionite results in $(SN\text{–}H)_4$. Silver difluoride in CCl_4 fluorinates it to form $(F\text{–}SN)_4$ [390]. However, chlorination under mild conditions yields $(Cl\text{–}SN)_3$. Aqueous hydrogen halides form S_4N_3X[391], but when HI is used in excess, a complete destruction of S_4N_4 takes place.

$$S_4N_4 + 4HCl \longrightarrow S_4N_3Cl + NH_4Cl + Cl_2$$
$$S_4N_4 + 12HI \longrightarrow 4S + 4NH_3 + 6I_2$$

Upon hydrolysis, NH_3 is formed in addition to various sulphur compounds. Hydrolysis of S_4N_4 is catalysed by fluoride ions in neutral media[392].

Ammonolysis of tetrasulphur tetranitride yields a red compound $S_4N_4.2NH_3$ of unknown structure. This product is also obtained by the reaction of $OSCl_2$ with liquid ammonia. Oxidation with air converts $S_4N_4.2NH_3$ into an ionic cluster-like species, $S_4N_5O^-$, tetrasulphur tetranitride oxide imide. Several salts have been prepared[393, 394]: NH_4^+ (yellow), Tl^+, Ag^+, Na^+, K^+. S_4N_4 behaves as an inorganic diene, as is shown by the formation of $S_4N_4.4C_5H_6$, $S_4N_4.2C_7H_{10}$ and $S_4N_4.2C_7H_8$ with cyclopentadiene, norbornene and bicycloheptadiene, respectively[395].

Thiophiles such as CN^- or triphenylphosphine[396] react with S_4N_4 to form smaller ring substrates. The ruby-red compound $(SN)_3\text{—}N\text{=}P(C_6H_5)_3$ has been described. Its structure, based upon analysis, infrared spectra and data on hydrolysis, is unique. The cage structure of S_4N_4 can be easily bent open by Lewis acids[397], e.g. SO_3, BX_3, $SnCl_4$ or $SbCl_5$, which

[386] D. Clark, *J. Chem. Soc.* (1952) 1615.
[387] R. C. Brasted, *J. Chem. Soc.* (1965) 2297.
[388] R. A. Meinzer and R. J. Myers, ref. by W. L. Jolly in A. V. Tobolsky (Ed.), *The Chemistry of Sulphides*, p. 6, Interscience Publishers, New York, London, Sydney (1968).
[389] J. Mason, *J. Chem. Soc.* **A1969**, 1567.
[390] O. Glemser, *Preparative Inorganic Reactions*, Vol. I, p. 227, Interscience Publishers, New York, London, Sydney (1964).
[391] A. G. MacDiarmid, *J. Am. Chem. Soc.* **78** (1956) 3871.
[392] H. W. Roesky, O. Glemser and A. Hoff, *Chem. Ber.* **101** (1968) 1219.
[393] M. Becke-Goehring and K. Erhard, *Naturwissenschaften* **56** (1969) 415.
[394] R. Steudel, *Z. Naturforsch.* **24b** (1969) 934.
[395] M. Becke-Goehring and D. Schläfer, *Z. anorg. allg. Chem.* **356** (1968) 234.
[396] E. Fluck, M. Becke-Goehring and G. Dehoust, *Z. anorg. allg. Chem.* **312** (1961) 60.
[397] W. L. Jolly and K. J. Wynne, *Inorg. Chem.* **6** (1967) 107.

form coloured donor–acceptor complexes. Thus, for the compound $S_4N_4.SbCl_5$ the structure has been elucidated, showing that the complex is formed via an N–Sb bond[398].

When S_4N_4 reacts with nickel, cobalt or palladium dichloride in alcoholic media, the cage system decomposes into smaller units and simultaneously is reduced. Complexes of the type $M(HN_2S_2)_2$, which all exhibit intensive colours, are obtained. With $NiCl_2$, for instance, $Ni(HN_2S_2)_2$ is formed in addition to NiS_5N_3H and NiS_6N_2. Ni- and Co-carbonyls were found to react with S_4N_4 in alcoholic solvents in the same manner.

The platinum complex has a planar structure in which the two hydrogens occupy the *cis*-position[399].

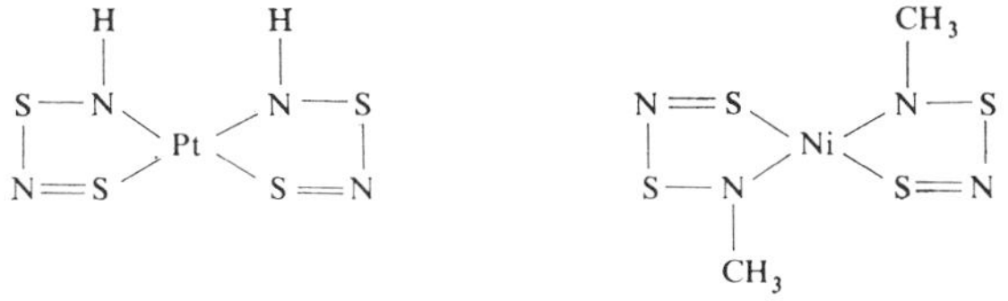

In contrast, the dimethyl derivative of $Ni(HN_2S_2)$, obtained by causing $Ni(HN_2S_2)_2$ to react with CH_3J, possesses a *trans*-configuration[400].

Disulphur Dinitride, S_2N_2

This compound can be prepared in an evacuated apparatus[401] by passing S_4N_4 vapour through silver wool heated up to 300°. It readily dissolves in organic solvents such as THF and dioxane. It represents the most dangerous of all $(NS)_n$ compounds. Decomposition takes place upon striking or warming above 30°. Infrared studies indicate a flat, four-membered ring with alternating sulphur and nitrogen atoms[402]. Traces of alkali or potassium cyanide catalyse its dimerization to S_4N_4. The reactions of S_2N_2 with NH_3, S_2Cl_2 and aqueous alkali result in the formation of the same products as form when S_4N_4 is used.

Solutions of S_2N_2 react with antimony pentachloride (in excess) to form a di adduct $S_2N_2(SbCl_5)_2$ which can further react with S_2N_2 to form the mono adduct[403]. This is a reversible process. The physical and chemical properties of these compounds indicate that the S_2N_2 ring structure is maintained intact. The mono adduct reacts irreversibly with S_2N_2 to form both the previously characterized $S_4N_4SbCl_5$ and, in lower yields, a less reactive material $(S_4N_4SbCl_5)_n$. Reactions of BF_3 and BCl_3 with S_2N_2 yield the corresponding adducts, which resemble in their physical and chemical properties the $S_2N_2SbCl_5$ adducts[403]. The reactions between tetrasulphur tetranitride[404] and chlorides of (a) manganese(II) and cobalt(II); (b) zinc(II), chromium(III), iron(III) and zirconium(IV); (c) antimony; and (d) beryllium in thionyl chloride solution have been found to give products of the types (a) $SNMCl_2$, (b) $S_2N_2MCl_n$, (c) $S_3N_3SbCl_n$ and (d) S_2N_2OBeCl (M = metal).

398 D. Neubauer and J. Weiss, *Z. anorg. allg. Chem.* **303** (1960) 28.
399 J. Lindquist and J. Weiss, *J. Inorg. Nucl. Chem.* **6** (1958) 184.
400 J. Weiss and M. Ziegler, *Z. anorg. allg. Chem.* **322** (1962) 184.
401 M. Goehring and D. Voigt, *Z. anorg. allg. Chem.* **285** (1956) 181.
402 J. R. W. Warn and D. Chapman, *Spectrochim. Acta*, **22** (1966) 1371.
403 L. Patton and W. L. Jolly, *Inorg. Chem.* **8** (1969) 1384, 1392.
404 A. J. Bannister and J. S. Padley, *J. Chem. Soc.* **A1969**, 658.

Polymeric Sulphur Nitride, $(SN)_x$

This compound is reported to have a high degree of electron delocalization. The product may be obtained[405] by leaving S_2N_2 in an evacuated desiccator at room temperature for 30 days.

Tetrasulphur Dinitride, S_4N_2

The compound can be prepared by heating S_4N_4 with S_8 in CS_2 at 120°. The dinitride can be made in 42%-yields by a curious reaction:

$$Hg_5(NS)_8 + 4S_2Cl_2 \xrightarrow{CS_2} 4S_4N_2 + 3HgCl_2 + Hg_2Cl_2$$

S_4N_2 is a red-brown liquid with an unpleasant odour. It decomposes within a few hours at room temperature. The position of nitrogen in the structure of S_4N_2 is not yet known. Chemical evidence—mainly from the products of hydrolysis—allows one to assume that it has an unsymmetrical structure, a 1,3-dinitrido-cyclohexasulphur.

Nitrides Derived from Cyclo-octasulphur, $S_{15}N_2$, $S_{16}N_2$

According to the condensation reaction, both compounds have been prepared[406] from S_2Cl_2, SCl_2 and S_7NH:

$$S_xCl_2 + 2S_7NH \longrightarrow S_7N\text{–}S_x\text{–}NS_7 + 2HCl$$

The yellow crystalline substances are quite stable at room temperature and dissolve easily in CS_2, but are less soluble in other organic solvents. An attempt to prepare $S_7N\text{–}NS_7$ from $Hg(NS_7)_2$ and iodine resulted in polymeric material $(S_{11}N)_8$. Recently an interesting new sulphur nitride $S_{11}N_2$ has been reported[407], which has been prepared by using the Ruggli–

$$S_6(NH)_2 + S_5Cl_2 \xrightarrow{-2\,HCl} S_{11}N_2$$

Ziegler dilution technique. The formula of the compound is based on S and N analysis and on the mass spectrum, which exhibits $S_{11}N_2^+$ as heaviest ion. The compound decomposes at its melting point (m.p. 150–155°).

Thio Trithiazyl Cation, $S_4N_3^+$

The structure of this cation, which has been known[408] since 1880, has recently been elucidated[409, 410] by X-ray diffraction studies. An elegant ^{15}N NMR study by Jolly[411] confirms the nearly planar seven-membered cation. Due to the spin 1/2, a relatively sharp

405 M. Goehring, *Ergebnisse und Probleme der Chemie der Schwefel-Stickstoffverbindungen*, p. 20, Academie-Verlag, Berlin (1957).
406 M. Becke-Goehring, H. Jenne and V. Rekalic, *Chem. Ber.* **92** (1959) 855, 1237.
407 H. G. Heal and M. S. Shahid, *Chem. Commun.* (1969) 1064.
408 E. Demarcay, *Compt. Rend.* **91** (1880) 854.
409 R. F. Kruh, A. W. Cordes, R. M. Lawrence and R. G. Goforth, *Acta Cryst.* **14** (1961) 1306.
410 J. Weiss, *Angew. Chem.* **74** (1962) 216.
411 N. Logan and W. L. Jolly, *Inorg. Chem.* **4** (1965) 1508.

triplet and a doublet in the ratio 1 : 2 were observed, indicating two differently bonded nitrogens in the ring system. Thio trithiazyl chloride is readily obtained when S_4N_4 is heated with S_2Cl_2 in CS_2:

$$3S_4N_4+2S_2Cl_2 \longrightarrow 4[S_4N_3]^+Cl^-$$

It explodes with a blue luminescence upon being heated in air, but it is not sensitive to shock. Due to the ionic structure of the compound, it is insoluble in solvents with a small dielectric constant. The anion may be exchanged by treating the chloride with concentrated nitric or sulphuric acid[408]. S_4N_3Cl is enlarged to S_4N_4 when it reacts with $Al(N_3)_3$[385]. A hydroxide, S_4N_3OH, is derived from the chloride by careful hydrolysis. More drastic hydrolysis with alkali metal hydroxides leads to the formation of ammonia, sulphite, thiosulphate and some sulphide.

The interatomic distances of the cation are shown in Table 29. The measured S–S bond length of 2·06 Å differs insignificantly from the S–S bond (2·04 Å) in S_8. On the basis

TABLE 29. INTERATOMIC DISTANCES AND BOND ANGLES[410] IN THE ION $S_4N_3^+$

Structure	Distances (Å)		Angles (°)	
(2) S; (2) N, N (1); (3) S, S (1); N—S (3) (4)	S_1–S_4	2.06	S_1–S_4–N_3	113.1
	S_1–N_1	1.52	S_4–N_3–S_3	149.1
	S_2–N_2	1.54	N_3–S_3–N_2	119.2
	S_3–N_3	1.57	S_3–N_2–S_2	135.0
	S_2–N_1	1.54	N_2–S_2–N_1	116.7
	S_3–N_2	1.60	S_2–N_1–S_1	155.4
	S_4–N_3	1.56	N_1–S_1–S_4	110.1

of the bond distances, it has been assumed that the electron system is delocalized, e.g. the positive charge is distributed over the entire ring system. $S_4N_3NO_3$ is monoclinic and crystallizes in the space group C_{2h}^5–$P2_1/c$ with lattice constants a = 5·81 Å, b = 10·42 Å, c = 12·47 Å, β = 108°. The elementary cell contains four formula units.

Sulphur Imides

Imides contain the bivalent group =N–H which may be bonded to carbon, phosphorus or sulphur. The discussion of sulphur imides in this section will be limited to cyclic compounds containing the elements S, N and H. The acyclic imides, in general, have to be stabilized with more electronegative elements such as oxygen or fluorine. Depending on the ring size the cyclic imides represent derivatives of S_8 or S_6. As we have already seen (page 899), up to four sulphur atoms of S_8 can be replaced by nitrogen to form S_4N_4. Reduction of the latter yields $S_4N_4H_4$, tetrasulphur tetra-imide. Theoretically this compound should be obtained when ammonia reacts with SCl_2 in equal amounts. However, despite numerous studies, this has not yet been established. In a non-polar solvent, S_4N_4 is formed whereas in polar solvents such as dimethylformamide, various imides are obtained. Thus, starting with 170 g S_2Cl_2 and the corresponding amount of NH_3, the following products have been isolated:

32·0 g S_8	0·98 g 1,3-$S_6(NH)_2$	0·08 g 1,3,5-$S_5(NH)_3$
15·4 g S_7NH	2·3 g 1,4-$S_6(NH)_2$	0·32 g 1,3,6-$S_5(NH)_3$
	0·82 g 1,5-$S_6(NH)_2$	

In no case have compounds with adjacent NH groups been detected, probably because such substrates once formed lose molecular nitrogen, leaving behind an unstable sulphane, H_2S_6. Recently Heal reported an easy preparation of S_7NH and $S_6(NH)_2$ in similar proportions and fairly good yields by mixing a CS_2 solution of S_2Cl_2 with aqueous ammonia[412]. When LiN_3 reacts with S_2Cl_2, S_4N_4 is formed in an inert solvent. However, using a polar organic solvent[413] the reaction yields $S_6(NH)_2$ and $S_7(NH)$.

Heptasulphurimide, S_7NH

Macbeth and Graham[414] first isolated S_7NH from the reaction product of S_2Cl_2 and NH_3, but the exact formula of this compound was assigned 30 years later. The preparation of S_7NH is preferably carried out in dimethylformamide by having chlorosulphanes, S_xCl_2, react with ammonia[415]. S_7NH is a stable compound, which melts at 113·5° without decomposing. Slight amounts of sulphur depress the melting point; the eutectic mixture with sulphur melts at 91·5° and contains 46% sulphur. The proton of S_7NH is acidic[416] and can be replaced by various organic and inorganic moieties. These reactions have been studied mainly by Becke-Goehring.

$$S_7NH + SO_3 \longrightarrow S_7NSO_3H \quad [417]$$
$$S_7NH + CH_2O \longrightarrow S_7N\text{–}CH_2OH \quad [418]$$
$$S_7NH + BCl_3 \longrightarrow S_7N\text{–}BCl_2 + HCl \quad [419]$$

Steric hindrance prevents dimerization of $S_7N\text{–}BCl_2$, since R_2NBX_2 easily dimerizes when R represents a small group (e.g. CH_3). BBr_3 reacts similarly to form S_7NBBr_2. However, BI_3 was found[419] to decompose S_7NH. This may be well explained in terms of a redox reaction between BI_3 and S_7NH, which should yield elemental iodine and compounds containing B–S and B–N bonds. The authors have studied a similar reaction between S_8 and BI_3, which leads to elemental iodine and $(IB)_2S_3$ (page 924). Various other substitution reactions of S_7NH have been reported as indicated in the following equations:

$$2S_7NH + Hg(CH_3COO)_2 \longrightarrow 2CH_3COOH + Hg(NS_7)_2 \text{ (yellowish)}$$
$$S_7NH + Ph_3CNa \longrightarrow Ph_3CH + S_7NNa \text{ (olive green)}$$
$$2S_7NH + (Me_3Si)_2NH \longrightarrow NH_3 + 2Me_3Si\text{–}NS_7 \quad [420]$$

The last equation represents the first S_7N-derivative in the organometallic field, which seems to be a promising area for further exploration. Recently[421] it was found that S_7NH undergoes a redox reaction with concentrated sulphuric acid, which leads to the formation of free radicals. The ESR spectrum of the red solution consists of five equidistant lines with the ratio of intensities 1 : 2 : 3 : 2 : 1. The formation of the radical cations occurs in the following way:

$$2S_7NH + H_2SO_4 \longrightarrow S_7N\text{–}NS_7 + 2H_2O + SO_2$$

412 H. G. Heal, *J. Inorg. Nucl. Chem.* **29** (1967) 1538.
413 F. Fehér and P. Junkes, *Z. Naturforsch.* **21b** (1966) 592.
414 A. K. Macbeth and H. Graham, *Proc. Roy. Irish Acad.* **36B** (1923) 31.
415 M. Becke-Goehring, H. Jenne and E. Fluck, *Chem. Ber.* **91** (1958) 1947.
416 B. A. Olsen and F. P. Olsen, *Inorg. Chem.* **8** (1969) 1736.
417 M. Goehring, in ref. 405, p. 109.
418 A. Meuwsen and F. Schlossnagel, *Z. anorg. allg. Chem.* **271** (1953) 226.
419 H. G. Heal, *J. Chem. Soc.* (1962) 4442.
420 M. Becke-Goehring, *Angew. Chem.* **73** (1961) 589.
421 P. Machmer, *Z. Naturforsch.* **24b** (1969) 1056.

The diamagnetic $S_{14}N_2$ is further oxidized to the radical cations.

$$2S_7N\text{–}NS_7 + 3H_2SO_4 \longrightarrow 2S_7N\text{–}\overset{+}{N}S_7 + 2HSO_4^- + H_2O + SO_2$$

Similarly S_4N_4 also forms free radicals in concentrated sulphuric acid.

Hexasulphur Di-imide, $S_6(NH)_2$

The hexasulphur di-imides are colourless, crystalline compounds which are as stable as S_7NH. The three isomers are formed when SCl_2 reacts with NH_3 (page 903). Separation of the products can be achieved by column chromatography. The 1,5-isomer's structure has been elucidated by X-ray technique[422]. The compound forms a puckered eight-membered ring similar to that of S_8 and $S_4N_4H_4$. The lattice constants are a = 7·386 Å,

TABLE 30. INTERATOMIC DISTANCES AND BOND ANGLES IN 1,5–$S_6(NH)_2$

	Distances (Å)		Angles (°)	
S_1 S_2 S_3	S_1–S_2	2.04	N_1–S_1–S_2	112.3
H–N_1 NH	S_2–S_3	2.05	S_1–S_2–S_3	109.5
S_1 S_3	N_1–S_1	1.62	S_2–S_3–N_2	107.2
S_2	N_2–S_2	1.68	S_1–N_1–S_1	120.3
			S_3–N_2–S_3	117.3

b = 8·69 Å, c = 12·828 Å. Since the hexasulphur di-imides are not readily available in large quantities, their reactivity in respect to the hydrogen displacement at the nitrogen atom has been studied only sporadically. Some organic derivatives have been prepared[423, 424]. The reaction of 1,5-$S_6(NH)_2$ with S_2Cl_2 gives rise to linear polymers of the general formula $HNS_6N\text{–}(S_2\text{–}NS_6N)_n\text{–}S_2\text{–}NS_6NH$ (n = 0, 1, 2, 3), the first four members of which have been isolated by chromatography[425].

Pentasulphur Tri-imides, $S_5(NH)_3$

Heal[426] first isolated these compounds by using the chromatography technique. By reducing S_4N_4 with hydrazine[427], the 1,3,5-isomer could be prepared. Structural assignment of the isomers was based upon the dipole moment. Recently an X-ray study confirmed the puckered ring for the 1,3,5-isomer. Some derivatives have been synthesized by the reaction of methylamin with S_2Cl_2[421].

Tetrasulphur Tetra-imide, $S_4(NH)_4$

Reduction of S_4N_4 under mild conditions[428] with $SnCl_2$ or dithionite yields $S_4(NH)_4$ at about 60%. It forms small colourless crystals which are insoluble in water but readily

422 J. C. Van de Grampel and A. Vos, *Acta Cryst.* **25** (1969) 611.
423 H. G. Heal and J. Kane, *J. Chem. Eng. Data*, **10** (1965) 386.
424 E. M. Tingle and F. P. Olsen, *Inorg. Chem.* **8** (1969) 1741.
425 H. G. Heal and J. Kane, in ref. 383, p. 489.
426 H. G. Heal and J. Kane, *Nature*, **203** (1964) 971.
427 H. Garcia-Fernandez, *Compt. Rend.* **260** (1965) 6107.
428 G. Brauer, *Handbook of Preparative Inorganic Chemistry*, Vol. 1, p. 411, Academic Press, New York, London (1963).

soluble in pyridine, presumably through hydrogen bonding. The X-ray study confirms a puckered ring analog to S_8. However, the position of the hydrogen is not exactly known. Diamagnetic susceptibility gives no evidence for delocalized double bonds.

Though S_6 is easily prepared by two different methods (page 806), only one di-imide has been reported[429]. The reduction of S_4N_4 gave besides $S_4N_4H_4$ a substance corresponding

TABLE 31. PHYSICAL DATA OF SULPHUR IMIDES, $S_n(NH)_{8-n}$

Compounds n = 0–4	M.p. (°C)	Symmetry classes	Dipole moments	Magnetic susceptibility (X)
S_8	114·5	$D_{4d} = S_{8v}$[a]	0[b]	$-121·6$ to $128·2 \times 10^{-6}$[c]
S_7NH	113·5[d]		1·28[e]	
1,3-$S_6(NH)_2$	130[f]	C_s	1·28[e]	
1,4-$S_6(NH)_2$	133[f]	C_1	1·23[e]	
1,5-$S_6(NH)_2$	155[f]	C_{2v}	1·74[e]	
1,3,5-$S_5(NH)_3$	124[e]		2·8[e]	
1,3,6-$S_5(NH)_3$	131[e]		1·0[e]	
$S_4(NH)_4$	145[g]	C_{4v}		-88×10^{-6}[h]

[a] S. C. Abrahams, *Acta Cryst.* **8** (1955) 661.
[b] S. Dobinsky, *Bull. Acad. Polon. Sci.* **A** (1932) 239.
[c] Gmelin, *Handbuch der Anorganischen Chemie*, Schwefel A, 3, S. 677, Verlag Chemie, Weinheim (1953).
[d] H. Tavs, J. Schultze-Steinen and J. E. Colchester, *J. Chem. Soc.* (1963) 2555.
[e] H. G. Heal and J. Kane, *Nature*, **203** (1964) 971.
[f] H. G. Heal, *Nature*, **199** (1963) 371.
[g] H. Garcia-Fernandez, *Compt. Rend.* **260** (1965) 6107.
[h] M. Becke-Goering, *Chem. Ber.* **80** (1947) 110.

TABLE 32. STRUCTURAL DATA OF $S_n(NH)_{8-n}$

Compound	d(S–S) Å	d(S–N) Å	S–N–S	S–S–S	N–S–S (°)
1,4–$S_6(NH)_2$[a]	2·04	1·73	118	107	110
1,5–$S_6(NH)_2$[b]	2·04	1·65	118·8	107·2	109·7
$S_4(NH)_4$[c]		1·67	122·2		108·4
S_7NH[b]	2·05	1·68	113·3	102·7	N–S–N
	2·11			103·8	109·0
	1·90			114·4	

[a] J. C. Van de Grampel and A. Vos, *Rec. Trav. Chim.* **84** (1965) 599.
[b] J. Weiss, *Z. anorg. allgem. Chem.* **305** (1960) 190.
[c] R. L. Sass and J. Donohue, *Acta Cryst.* **11** (1958) 497.

to $S_4(NH)_2$ (m.p. 0° to 10°). Organic derivatives[430] of the 1,4-tetrasulphur di-imide $(RN)_2S_4$ have been prepared by causing primary amines to react with S_2Cl_2 in an inert solvent. To date no attempt has been made to synthesize imides of S_{10} and S_{12} (page 807). Sulphur di-imide, HN=S=NH, and diaminosulphur di-imide, $(H_2N)_2S(NH)_2$, are thermally not stable; however, derivatives are known. When S,S dimethylsulphur di-imide is treated with potassium amide in liquid ammonia, a salt is formed[431]:

$$(CH_3)_2S(=NH)_2+3KNH_2 \longrightarrow K_3[NS(NH)_3].NH_3+2CH_4$$

[429] H. Garcia-Fernandez, *Bull. Soc. Chim. France* (1959) 760.
[430] M. Becke-Goehring and H. Jenne, *Chem. Ber.* **92** (1959) 1149.
[431] R. Appel and B. Ross, *Angew. Chem.* **80** (1968) 561.

An interesting eight-membered ionic N–S ring is obtained, when N,N′ di-bromo dimethyl-sulphur di-imide is reacted with disulphides. The di-cation contains alternating four S and four N atoms. Six alkyl groups are attached to the sulphur atoms, resulting to sulphonium–sulphur atoms[432]. The bromide ions represent the anions in this salt.

2.6. COMPOUNDS COMPOSED OF SULPHUR, NITROGEN AND HALOGEN

In general, the stability of compounds containing N,S and halogen decreases with increasing molecular weight of the halogen. The number of compounds to be discussed here is limited to the chlorine compounds, since we already have lined out the general features of sulphur–nitrogen–fluorine chemistry on pages 851–5. Bromine and iodine derivatives of sulphur–nitrogen compounds described in the literature are not yet sufficiently characterized. Two interesting compounds of the composition $S_3N_2Cl_2$ and BSN_2Cl_6 have been reported. The former has been mentioned already; it is an intermediate *en route* to S_4N_4. When a hot solution of S_4N_4 in S_2Cl_2 is allowed to cool, rust brown crystals are formed[433]. This product is also obtained from the action of nitrosyl chloride or thionyl chloride on S_4N_4 in a polar solvent[434]:

$$S_4N_4+2NOCl \longrightarrow S_3N_2Cl+\tfrac{1}{2}S_2Cl_2+2N_2O$$

The structure of $S_3N_2Cl_2$ has been elucidated by an X-ray diffraction study[435]. The compound is a salt with the cation $(S_3N_2Cl)^+$, which consists of a slightly puckered five-membered ring. Structural data are given below.

TABLE 33. INTERATOMIC BOND DISTANCES AND BOND ANGLES IN $S_3N_2Cl_2$

Structure	Bond	Bond length (Å) (± 0.005)	Angle	Bond angles (°) (± 0.3)
$Cl_{(1)}$–$S_{(2)}$–$S_{(3)}$, $N_{(1)}$, $N_{(2)}$, $S_{(1)}$ ring (+); Cl^-	S_1-N_1	1.617	$N_1-S_1-N_2$	106.3
	S_1-N_2	1.543	$S_2-N_1-S_1$	118.0
	N_1-S_2	1.581	$N_1-S_2-S_3$	97.8
	N_2-S_3	1.615	$S_2-S_3-N_2$	95.6
	S_2-S_3	2.136	$S_3-N_2-S_1$	120.7
	S_2-Cl_1	2.168	$N_1-S_2-Cl_1$	106.5
			$S_3-S_2-Cl_1$	100.1

$S_3N_2Cl_2$ crystallizes in the monoclinic space group $P2_1$ with the cell dimensions a = 6·546 Å, b = 8·600 Å, c = 5·508 Å and β = 102°37′. The Cl anion and its closest approach are to S_1 at 2·90 Å, S_3 at 2·93 Å and S_2 at 3·04 Å. If a plane is drawn through the three sulphur atoms, the Cl_1 atom is 2·09 Å above the plane, N_1 is 0·188 Å below the plane and the opposing N_2 atom is 0·140 Å above the plane. Recently[436] a compound BNS_2Cl_6 has been prepared by the reaction of NSF_3 with BCl_3 (page 854). Its structure has been elucidated. The salt crystallizes in the monoclinic space group C^5_{2h}–$P2_1/c$ with the cell dimensions

432 R. Appel, D. Hänssgen and W. Müller, *Chem. Ber.* **101** (1968) 2855.

433 A. Meuwsen, *Chem. Ber.* **65** (1932) 1724.

434 M. Goehring, *Ergebnisse und Probleme der Chemie der Schwefel-Stickstoffverbindungen*, Academie-Verlag, Berlin (1957).

435 A. Zalkin, T. E. Hopkins and D. H. Templeton, *Inorg. Chem.* **5** (1966) 1767.

436 O. Glemser, B. Krebs, J. Wegner and E. Kindler, *Angew. Chem.* **81** (1969) 568.

$a = 6{\cdot}441$ Å, $b = 16{\cdot}008$ Å, $c = 9{\cdot}864$ Å and $\beta = 103{\cdot}3°$. The compound was found to be planar with *cis*-configuration (C_{2v} symmetry). Structural data are listed below.

TABLE 34. INTERATOMIC BOND DISTANCES AND BOND ANGLES IN $(Cl_2S_2N)^+BCl_4^-$

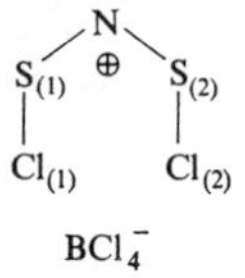

	Bond length (Å)		Bond angles (°)	
$[Cl_{(1)}-S_{(1)}-N^{\oplus}-S_{(2)}-Cl_{(2)}]$ BCl_4^-	Cl_1-S_1	1.985	Cl_1-S_1-N	112.0
	Cl_2-S_2	1.985	Cl_2-S_2-N	110.5
	S_1-N	1.532	S_1-N-S_2	149.2
	S_2-N	1.537		

The structure of the BCl_4 anion was not known before; the assumed tetrahedral conformation has been confirmed. The B–Cl distances vary slightly: 1·833, 1·835, 1·845 and 1·856 Å; the tetrahedral angles range between 109·0° and 109·9°. The salt is very hygroscopic and decomposes easily under evolution of BCl_3. Decomposition of the compound at 80° yields N_2, S_2Cl_2, SCl_2 and BCl_3.

Thio trithiazyl chloride has already been discussed in the subsection of sulphur nitrides (page 902). It consists of a seven-membered cation $S_4N_3^+$, which obviously represents a stable arrangement. Within the row of the cations S^+, $Cl_2S_2N^+$, $S_3N_2Cl^+$, $S_4N_3^+$ and $S_5N_4^+$ the first and last members are not known as cations of salts. Recently the thiazyl cation NS^+ has been prepared from N≡S–F and AsF_5 [437]. Besides the above-described ionic, some covalent sulphur nitrogen chlorides have been reported. Thiazyl chloride, NSCl, polymerizes easily to trimeric $(NSCl)_3$. On heating to 110° in high vacuum, NSCl may be obtained as a greenish-yellow gas. When S_4N_4, dissolved in CCl_4, is treated with chlorine, $(NSCl)_3$ is obtained in large yellow needles[438]. This is surprising since fluorination of S_4N_4 yields $(NSF)_4$. It has been suggested that during chlorination the $(NS)_4$-ring is cracked into small fragments, which finally build the six-membered ring. The compound is stable in dry atmosphere; it explodes weakly on sudden heating. Its good solubility in organic solvents indicates the covalent character of the compound. The structure has been elucidated by an X-ray study.

TABLE 35. INTERATOMIC DISTANCES AND BOND ANGLES IN $(NSCl)_3$

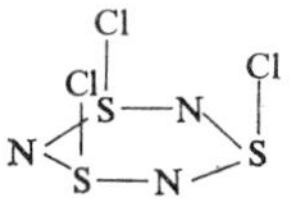

	Bond distances (Å)		Bond angle (°)	
$(NSCl)_3$ ring (S—N six-membered ring, Cl on each S)	S–N	1.61	S–N–S	123.8
	S–Cl	2.15	N–S–N	113.4
			N–S–Cl	113.8

In the nearly flat ring all N–S bonds are of the same length, which indicates a delocalized π-electron system. When $(NSCl)_3$ is heated with SO_3 to 140°, α-sulphanuric chloride is formed. By hydrolysis and alkoholysis the $(NSCl)_4$ system is decomposed to NH_3, SO_2 and HCl. With liquid ammonia, HN=S=NH is supposedly formed.

437 O. Glemser and W. Koch, *Angew. Chem.* **83** (1971) 145.
438 M. Goehring in ref. 434, p. 155.

2.7. COMPOUNDS CONTAINING SULPHUR, NITROGEN, OXYGEN AND HYDROGEN

The compounds described in this section belong to different classes in which sulphur may occur in the oxidation states of 2, 4 and 6. Since there is such a large store of information available, we can give only a brief discussion of the main features of these compounds. Recent reviews[439–441] provide detailed information about the compounds containing S, N, O and H.

Sulphur–Nitrogen Oxides

Trisulphur Dinitrogen Dioxide, $S_3N_2O_2$

The compound has been prepared by the reaction of $SOCl_2$ with ammonia[442] as well as by the reaction of S_4N_4 with thionyl chloride[443]. If the latter reaction is carried out at room temperature in a polar solvent such as nitromethane, tetrasulphur dinitride, thiodiazyl chloride and chlorosulphanes are obtained in addition to the pale yellow $S_3N_2O_2$:

$$S_4N_4+2OSCl_2 \longrightarrow S_3N_2O_2+2Cl_2+S_2N_2+S$$

The formation of S_4N_2, $S_3N_2Cl_2$ and chlorosulphanes stems from the reactions of the intermediates S, N_2S_2 and Cl_2. $S_3N_2O_2$ is readily oxidized by SO_3, thus forming $S_3N_2O_5$:

$$S_3N_2O_2+3SO_3 \longrightarrow S_3N_2O_5+3SO_2$$

By labelling SO_3 with ^{35}S it was found that SO_3 is not built in the molecule but merely acts as an oxidizing agent. Moist air converts $S_3N_2O_2$ to SO_2 and S_4N_4. The structure of $S_3N_2O_2$ has been elucidated[444]. In an NMR study[445] two equivalent nitrogens were observed.

TABLE 36. BOND DISTANCES AND BOND ANGLES IN $S_3N_2O_2$

Structure	Distances (Å)		Angles (°)	
O=S(2)–N(1)–S(1)–N(2)–S=O	S_1-N_1	1.69	$N_1-S_1-N_2$	95.3
	S_2-N_1	1.58	N_1-S_2-O	115.3
	S_2-O	1.37	$S_1-N_1-S_2$	120.0

The compound is monoclinic and crystallizes in the space group C^6_{2h}–$C2/h$ with the lattice constants of $a = 6{\cdot}84$ Å, $b = 4{\cdot}56$ Å, $c = 16{\cdot}52$ Å and $\beta = 97°$.

Trisulphur Dinitrogen Pentoxide, $S_3N_2O_5$

When S_4N_4 or $S_3N_2O_2$ are treated with SO_3, the cyclic pentoxide is formed. By using $^{35}SO_3$, it was shown that only two labelled sulphur atoms are built in the molecule.

$$S_4N_4 + 6\overset{*}{S}O_3 \longrightarrow 2\ S\begin{matrix} \diagup\!\!\!\!= N-\overset{*}{S}O_2 \\ \diagdown\!\!\!\!= N-\overset{*}{S}O_2 \end{matrix}\!\!>\!O \ + \ 2\,SO_2 + 2\,\overset{*}{S}O_2$$

[439] M. Becke-Goehring, in F. A. Cotton (Ed.), *Progress in Inorganic Chemistry*, Vol. 1, p. 268, New York (1959).

[440] M. Becke-Goehring and E. Fluck in C. B. Colburn (Ed.), *Developments in Inorganic Nitrogen Chemistry*, Vol. 1, pp. 152–234, Elsevier, Amsterdam, London, New York (1966).

[441] K. W. W. Burton and G. Nickless in G. Nickless (Ed.), *Inorganic Sulphur Chemistry*, p. 608, Elsevier, Amsterdam, London, New York (1968).

[442] W. L. Jolly and M. Becke-Goehring, *Inorg. Chem.* **1** (1962) 76.

[443] M. Becke-Goehring and J. Heinze, *Z. anorg. allg. Chem.* **272** (1953) 297.

[444] J. Weiss, *Z. Naturforsch.* **16b** (1961) 477.

[445] M. Becke-Goehring and E. Fluck, in ref. 440, p. 220.

The formation of $2H_3NSO_3H$ and SO_2 upon hydrolysis proves the structure of the compound[446].

Di-isothiazylsulphoxide, $(S{=}N)_2SO$

The compound is obtained[447] according to

$$Hg_5(NS)_8+4SOCl_2 \longrightarrow 4(S{=}N)_2SO+3HgCl_2+Hg_2Cl_2.$$

The stability of the yellow-orange oil may be compared with that of S_4N_2; it hydrolyses to form SO^{2-}, $S_2O_3^{2-}$ and NH_3 in a 1:1:2 ratio.

Sulphuryl Azide

The compound may be obtained as a colourless oil by causing SO_2Cl_2 to react with NaN_3[448].

Amido and Imido Derivates of Sulphurous Acid

Thionylamides and Thionylhydroxide Sulphurous Acid

These are analogous with respect to the isostery of OH and NH_2. Therefore, it is not surprising that thionylamides are fairly unstable. Early reports claiming the synthesis of $OS(NH_2)_2$ proved to be erroneous. In a later study[449] a reaction product of NH_3 and $OSCl_2$ having a molar ratio of S:N = 1:2 was obtained. However, this compound may have been the ammonium salt of the thionyl imide, $[NH_4][N{=}S{=}O]$. In contrast, the N-alkylated thionylamides are well known. The N-alkylated monoamides may be obtained through the reaction of SO_2 with primary or secondary amines[450]. In the solid state the structure $R_2N^+HSO_2^-$ is verified[451].

Thionylimide, HNSO

When NH_3 and $OSCl_2$ react in the vapour phase, monomeric colourless HNSO is formed (m.p. −94°). At −70° thionylimide polymerizes to give a glassy material of a yellow-red to brown colour[452]. Since the infrared spectrum exhibits a N–H vibration, the structure of the polymer should be of the $-NH[-SO-NH-]_nSO-$ type. Upon heating, the compound depolymerizes, and in the presence of small amounts of water it yields S_4N_4.

$$6HNSO+2H_2O \longrightarrow S_4N_4+NH_4HSO_4$$

The reaction of $OSCl_2$ with NH_3 proceeds entirely differently in organic solvents such as $CHCl_3$. A violet-red substance is obtained in the presence of a HCl-acceptor[453].

$$OSCl_2+NH_3+CaO \longrightarrow HO\text{–}SN+CaCl_2+H_2O$$

This compound represents the isomer of thionylimide:

$$HO-\overline{S}=\overline{N}| \text{ or } HO-\overline{N}=\overline{S}|.$$

---

[446] M. Goehring, H. Hohenschutz and J. Ebert, *Z. anorg. allg. Chem.* **276** (1954) 47.
[447] A. Meuwsen and M. Lösel, *Z. anorg. allg. Chem.* **271** (1953) 222.
[448] T. Curtins and F. Schmidt, *Chem. Ber.* **55** (1922) 1576.
[449] O. Gehrig, Ph.D. thesis, Heidelberg (1953).
[450] A. Michaelis and O. Storbeck, *Ann.* **274** (1893) 187.
[451] G. Zinner, *Arch. Pharm.* **291/63** (1958) 7.
[452] W. P. Schenk and E. Krone, *Angew. Chem.* **73** (1961) 762.
[453] M. Becke-Goehring, R. Schwarz and W. Spiess, *Z. anorg. allg. Chem.* **293** (1957) 294.

With $OSCl_2$ the acid chloride Cl—S≡N is obtained, which easily trimerizes:

$$3ClS{\equiv}N \longrightarrow (ClSN)_3$$

When sulphur is heated in air with $S_4N_4H_4$, a red solid of the composition $(OSNH)_4$ is formed[454].

Sulphur Di-imide

The replacement of both oxygens in SO_2 by the imine moiety will lead to HN═S═NH. However, this compound is not known, but derivatives of it have been identified. Thus, RN═S═NR (R = C_6H_5, C_4H_9)[455] have been prepared. Also several imidosulphur difluorides have been reported (page 852).

Amino, Imino and Nitrido Derivatives of Sulphuric Acid

In contrast to the nitrogen–oxygen compounds of sulphur in the oxidation state +4, the corresponding compounds derived from sulphuric acid exhibit a higher stability and a larger variety of reaction possibilities. Many of the compounds have been already prepared in the nineteenth century[456], which demonstrates the early interest in this part of sulphur chemistry.

Amides of H_2SO_4

Sulphamic acid, NH_2SO_3H, may be considered a member of a series of aquo-ammono acids. It is readily obtained by different methods. The free acid was first prepared by Raschig[457] according to the following scheme:

$$H_2NOH+SO_2 \longrightarrow H_2NSO_3H$$

Other methods are as follows:

$$SO_3+NH_3 \longrightarrow H_2NSO_3H$$
$$FSO_3H+NH_3 \longrightarrow H_2NSO_3H+HF$$

A commercial synthesis starts with urea and fuming sulphuric acid:

$$OC(NH_2)_2+H_2S_2O_7+H_2O \longrightarrow CO_2+H_2NSO_3H+NH_4SO_4H$$

The structure of free sulphamic acid is best represented by the arrangement $H_3\overset{+}{N}{-}SO_3^-$, in which three oxygens and one nitrogen occupy four tetrahedral positions around sulphur. The various bond distances and bond angles are given below.

TABLE 37. BOND DISTANCES AND BOND ANGLES IN SULPHAMIC ACID[458]

Structure	Bond distances (Å)		Bond angles (°)	
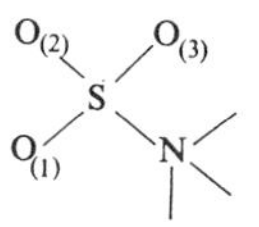 $O_{(2)}$, $O_{(3)}$, $O_{(1)}$ around S; S–N	S-N	1.764 ± 0.020	N-S-O_1	103.2
	S-O_1	1.421 ± 0.021	N-S-O_2	102.9
	S-O_2	1.452 ± 0.022	N-S-O_3	103.5
	S-O_3	1.445 ± 0.022	O_1-S-O_2	113.4
			O_2-S-O_3	114.7
			O_1-S-O_3	117.3

[454] E. Fluck and M. Becke-Goehring, *Z. anorg. allg. Chem.* **292** (1957) 229.
[455] D. Cramer, *J. Org. Chem.* **26** (1961) 3476.
[456] A. Claus, *Chem. Ber.* **4** (1871) 504.
[457] F. Raschig, *Chem. Ber.* **39** (1906) 245.
[458] R. L. Sass, *Acta Cryst.* **13** (1960) 320.

The sulphur–nitrogen bond distance in the sulphamate ion (1·60 Å) is considerably shorter than in the free acid, which fact has been attributed to π-bonds using *d*-orbitals. Replacing the OH group by halogen yields the acid halides of sulphamic acid. $H_2N–SO_2Cl$ has been prepared according to the following[459]:

$$Cl–CN+SO_3 \longrightarrow Cl–SO_2–NCO \xrightarrow{H_2O} H_2N–SO_2Cl+CO_2$$

This acid chloride can be converted by means of KF in acetonitrile into FSO_2NH_2, which has also been obtained by ammonolysis of $F_2S_2O_5$. Sulphamates of the general formula $H_2NSO_2OR(R = CH_3, C_2H_5)$ may be synthesized from the chloride $H_2N–SO_2Cl$ and alcoholate[460]. An easy access to the N-alkylated sulphamic acids is achieved by using Raschig's method:

$$R_2NOH+SO_2 \longrightarrow R_2NSO_3H$$

When sulphamic acid is exposed to oxidizing agents such as chlorine, bromine or chlorate, N_2 and H_2SO_4 are formed.

$$2H_2NSO_3H+KClO_3 \longrightarrow KCl+H_2O+2H_2SO_4+N_2$$

PCl_5[461] reacts with H_2NSO_3H to give $ClSO_2–NPCl_3$, which hydrolyses to form sulphamic and phosphoric acid. Anodic oxidation of potassium sulphamate results in the formation of potassium azodisulphonate[462], $KO_3S—N{=}N—SO_3K$.

The diamide of sulphuric acid was first prepared by Traube[463]. It can be obtained either by ammonolysis of SO_3 or O_2SCl_2. With the latter compound several other products are formed: ammonium salt of sulphimide, imidosulphamide and long-chain sulphurylimido-amides. Similar results are observed when SO_3 is used[464]. The structure of sulphamide has been elucidated. The bond distances and angles are given in Table 38.

TABLE 38. INTERATOMIC DISTANCES AND BOND ANGLES IN SULPHAMIDE

	Bond distances (Å)		Bond angles (°)	
$O_{(2)}$, $O_{(1)}$, S, $N_{(1)}$, $N_{(2)}$	$S–O_1$	1.391 ± 0.008	$O_1–S–O_2$	119.4 ± 0.8
	$S–N_1$	1.600 ± 0.009	$N_1–S–N_2$	112.1 ± 0.7
	$O–O_2$	2.402 ± 0.016	$O_1–S–N_1$	106.6 ± 0.5
	$N_1–N_2$	2.654 ± 0.016	$O_1–S–O_2$	106.2 ± 0.5
	$O_1–N_1$	2.401 ± 0.011		
	$O_1–N_2$	2.394 ± 0.011		

As indicated by the shortness of the interatomic distances, the N–S bond exhibits certain characteristics of a double bond. A discussion of this matter has been published by Cruickshank[465].

Imides of H_2SO_4

Replacement of two hydrogens in NH_3 by the SO_3H moiety results, theoretically, in the

459 R. Appel and G. Berger, *Chem. Ber.* **91** (1958) 1339.
460 R. Appel and W. Senkpiel, *Angew. Chem.* **70** (1958) 504.
461 O. Wallach and T. Huth, *Chem. Ber.* **8** (1875) 317.
462 A. Krettler and W. Teske, *Angew. Chem.* **71** (1959) 69.
463 W. Traube, *Chem. Ber.* **25** (1892) 2472.
464 R. Appel and W. Huber, *Chem. Ber.* **89** (1956) 386.
465 D. W. Cruickshank, *J. Chem. Soc.* (1961) 5486.

formation of the free imidosulphuric acid, which is unstable. However, salts of this acid are known in large number. Ammonia reacts with SO_3 to form the triammonium salt:

$$2SO_3+4NH_3 \longrightarrow NH_4[N(SO_3NH_4)_2]$$

The structure of the dipotassium salt has been elucidated[466]. For comparison the structural data of the pyrosulphate ion are included in Table 39.

TABLE 39. BOND DISTANCES AND BOND ANGLES IN THE IMIDODISULPHATE AND THE DISULPHATE IONS

	Bond distances (Å)	Bond angles (°)
	$HN(SO_3)_2^{2-}$ in $K_2(O_3S)_2NH$	
H		
O N O		
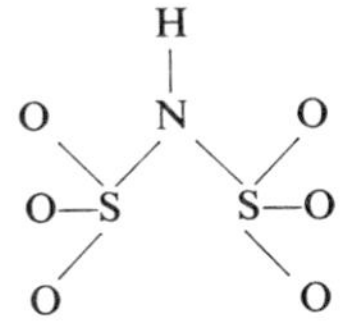 O–S S–O	S–O 1·453	O–S–O 113
O O	S–N 1·662	O–S–N 105·5 S–N–S 125·5
	$O(SO_3)^{2-}$ in $K_2(O_3S)_2O$	
O O O		
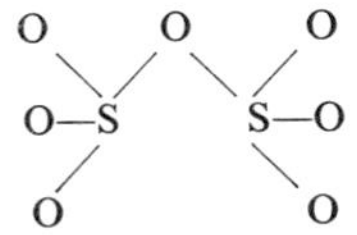 O–S S–O	S–O 1·437	O–S–O 114
O O	$S–O_1$ 1·645	$O–S–O_1$ 104 $S–O_1–S$ 124

By reacting urea with fluorosulphonic acid a derivative of imidodisulphuric acid can be obtained.

$$OC(NH_2)_2+3FSO_3H \longrightarrow CO_2+FSO_2–NH–SO_2F+NH_4HSO_4+HF$$

The corresponding chloro compound may be similarly prepared; however, a better route is provided by the following method:

$$2PCl_5+H_2NSO_3H \longrightarrow Cl_3P{=}N–SO_2Cl+OPCl_3+HCl$$

$$Cl_3P{=}N–SO_2Cl+ClSO_3H \longrightarrow ClSO_2NH–SO_2Cl+OPCl_3$$ [467]

The latter reaction occurs by protonation of the nitrogen, followed by the cleavage of the PN bond to form the favourable phosphine oxide, $\geqslant P{=}O$. Nucleophilic displacement of the chlorine in $(ClSO_2)_2NH$ by NH_3 results $H_2NSO_2–NH–SO_2NH_2$. Ammonolysis of $S_2O_5Cl_2$ yields the monoamide, which in aqueous media forms an unstable diprotonic acid. Polysulphimido sulphonic acids of the general formula $HO_3S–(HNSO_2)_n–OH$ are obtained from the reaction of NH_3 and SO_3 in a polar solvent such as nitromethane[468]. When sulphamide is heated to 180–200°, a rearrangement occurs:

$$O_2S(NH_2)_2 \longrightarrow [O_2S{=}N][NH_4]$$

This salt can be easily converted into the silver salt, which on treatment with CH_3I yields trimeric $(CH_3NSO_2)_3$. The corresponding parent compound $(HNSO_2)_3$ was found to be stable for a short time in water. It behaves as a diprotonic acid, which readily forms a pyridinium adduct. The latter can also be obtained according to the following scheme:

$$3H_2NSO_2Cl+5C_5H_5N \longrightarrow (O_2SNH)_3.2C_5H_5N+3(C_5H_5NH)Cl$$

[466] G. A. Jeffrey and D. W. Jones, *Acta Cryst.* **9** (1956) 283.
[467] R. Appel, M. Becke-Goehring, M. Eisenhauer and J. Hartenstein, *Chem. Ber.* **95** (1962) 625.
[468] R. Appel and M. Becke-Goehring, *Z. anorg. allg. Chem.* **271** (1953) 171.

$(HNSO_2)_3$ is structurally related to cyanuric acid $(HNCO)_3$. Derivatives of the tetrasulphimide $(HNSO_2)_4$ have also been reported[469]. Hexasulphimid, $(NHSO_2)_6$, is formed by direct interaction $NH(SO_2Cl)_2$ and $SO_2(NH_2)_2$[470].

Nitrides Derived from H_2SO_4

Ammonium nitridotrisulphonate, $N(SO_3NH_4)_3$, is prepared[471] by the reaction of NH_3 vapour and SO_3 in the ratio 1·33 : 1. The compound is also obtained by bubbling SO_2 into an aqueous solution of NH_4NO_2 and NH_4HSO_3. The stability of $[N(SO_3)_3]^{3-}$ in water is not very high since protons formed by hydrolysis autocatalyse the decomposition:

$$[N(SO_3)_3]^{3-} + H_2O \longrightarrow HN(SO_3)_2^{2-} + HSO_4^-$$

Although sulphanuric halides $(XSON)_3$ contain the typical imide and amide but not nitride features they will be discussed within this subsection. Sulphanuric chloride was first prepared by Kirsanov[472]. $ClSO_2{-}N{=}PCl_3$ obtained from the reaction of PCl_5 and H_2NSO_3H (page 913) decomposes thermally to $OPCl_3$ and two products of the composition $(ClSON)_3$. Separation of the α- and β-isomer (m.p. 144–145° and 46–47°) could be achieved by vacuum sublimation. The structure of the α-sulphanuric chloride[473, 474] is known. Since all N–S bonds were found to be of the same distance (1·564 Å), no double bond is localized. This indicates a delocalized π-system. Hydrolysis in acidic media leads to sulphuric acid and imidodisulphamide[475]. The corresponding sulphanuric fluorides are obtained by treating the chloride with KF in carbon tetrachloride. The two isomers differ in the position of their fluorine and oxygen with respect to the non-planar ring, which has been established for α-sulphanuric chloride[473].

TABLE 40. PHYSICAL PROPERTIES OF THE TWO ISOMERS OF TRIMERIC SULPHANURIC FLUORIDES, $(FSNO)_3$

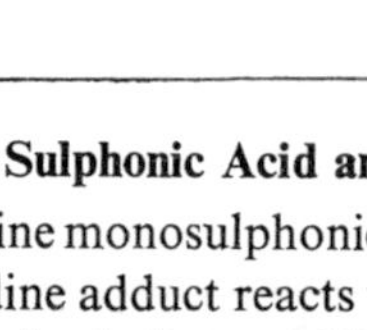

		cis-$(FSNO)_3$	trans-$(FSNO)_3$
	M.p. °C	17.4	−12.5
	B.p. °C	138.4	130.3
	Vap.press. (mm Hg, 25°)	9	10
	H.vapor. (kcal/mol)	9.6	9.8
	Density (g/cm³)	1.92	1.92
cis-isomer	Index of refraction n_D^{25}	1.4166	1.4169
	^{19}F n.m.r.	A-type	AB_2-type

Hydrazine Sulphonic Acid and Its Derivatives

Hydrazine monosulphonic acid is obtained in the form of its hydrazine derivative when SO_3 or its pyridine adduct reacts with anhydrous hydrazine. Potassium fluorosulphonate undergoes hydrazinolysis to yield $KO_3SNH{-}NH_2$. The S–N bond in these compounds is much

469 K. Beucker, G. Leiderer and A. Meuwsen, *Z. anorg. allg. Chem.* **324** (1964) 202.
470 H. A. Lehmann, W. Schneider and R. Hiller, *Z. anorg. allg. Chem.* **365** (1969) 157.
471 N. H. Marsh, U.S. Pat. 2.656.252, Oct. 20, 1953.
472 A. V. Kirsanov, *Zh. Obshch. Khim.* **22** (1952) 93.
473 A. J. Bannister and A. C. Hazell, *Proc. Chem. Soc.* (1962) 282.
474 G. A. Wiegers and A. Vos, *Proc. Chem. Soc.* (1962) 387.
475 M. Goehring, J. Malz and G. Roos, *Z. anorg. allg. Chem.* **273** (1953) 200.

easier to hydrolyse than that of sulphamic acid[476]. In the solid state it exists as a zwitterion, $H_3N\text{-}\overset{+}{N}HSO_3^-$. The infrared spectrum exhibits the characteristic $\overset{+}{N}H_3$-deformation frequences at 1598(w) and 1521(m) cm^{-1}, comparable with those of sulphamic acid, $\overset{+}{H_3}NSO_3^-$, at 1638(w) and 1541(m) cm^{-1}. Hydrazino monosulphonic acid has a pK_a value of 3·85. The reducing properties of hydrazine monosulphonic acid are comparable with hydrazine. Several synthetic approaches have been made for the dihydrazide of sulphuric acid:

$$O_2SCl_2+2N_2H_4 \longrightarrow O_2S(NH\text{-}NH_2)_2+2HCl$$

$$O_2SCl_2+N_2H_3COOH.N_2H_4 \longrightarrow O_2S(NH\text{-}NH_2)_2+2HCl+CO_2$$

Symmetrical hydrazine disulphonic acid can be obtained from chlorosulphonate and hydrazinesulphate[477]. Oxidation of the dipotassium salt with hypochlorite yields $KO_3S\text{—}N\text{=}N\text{—}SO_3K$ (page 912). The asymmetrical hydrazine disulphonic acid has been prepared according to the following scheme:

$$\overset{+}{H_3}N\text{-}OSO_3^- +2HO^- \longrightarrow (NH)+SO_4^{2-}+2H_2O$$

$$HN(SO_3K)_2+(NH) \longrightarrow H_2N\text{-}N(SO_3K)_2$$

In alkaline solution hydroxylamine O-sulphonic acid is a powerful aminating agent, which produces, presumably, the highly reactive NH. Sulphonation of the hydrazine N,N-disulphonic acid with pyridine–sulphur trioxide produces trisulphonate:

$$H_2N\text{-}N(SO_3NH_4)_2+C_5H_5N\text{-}SO_3 \xrightarrow{KOAc} KO_3S\text{-}NH\text{-}N(SO_3K)_2$$

Further sulphonation of the reaction product gives in low yield the tetrasulphonate, $(KO_3S)_2N\text{-}N(SO_3K)_2$. However, a much better yield has been obtained[478] by electrolytic oxidation of $N(SO_3K)_3$.

Hydroxylamides of Sulphuric Acid

Due to three reactive hydrogens in $H_2N\text{-}OH$, four sulphonic acids, all of which have been prepared, may be derived from hydroxylamine.

Hydroxylamine N-Sulphonic Acid

When hydroxylamine N,N-disulphonic acid is hydrolysed, the monosulphonic acid is formed, which is fairly resistant towards further hydrolysis. Alkaline hydrolysis yields sulphite and hyponitrite; the latter is formed via an intermediate NOH[479]. The formation of NO^- in the course of alkaline hydrolysis can be demonstrated by its reaction with $Ni(CN)_4^{2-}$, which produces a red complex.

Hydroxylamine, N,N-Disulphonic Acid

The reaction between SO_2 and nitrite in alkaline solution leads to $HON(SO_3K)_2$[480]:

$$NO_2^- +SO_3H^- +SO_2 \longrightarrow HON(SO_3)^{2-}$$

[476] L. F. Audrieth and R. A. Ogg, *The Chemistry of Hydrazines*, John Wiley and Sons, Inc., New York (1951).
[477] E. Konrad and L. Pellens, *Chem. Ber.* **59** (1926) 135.
[478] R. R. Grinstead, *J. Inorg. Nucl. Chem.* **4** (1957) 287.
[479] R. Nast, K. Nyul and E. Grziwok, *Z. anorg. allg. Chem.* **267** (1952) 304.
[480] F. Seel, E. Degener and H. Knorre, *Z. anorg. allg. Chem.* **299** (1959) 122.

On heating $HON(SO_3H)_2$ splits into $HONH_2$ and H_2SO_4 (Raschig synthesis of hydroxylamine). The reactions between bisulphite and hydroxylamine N-sulphonate and hydroxylamine N-disulphonite yielding imido disulphonate and nitrilo trisulphonate have been studied by Seel[480]:

$$HO\text{–}HN(SO_3)^- + HSO_3^- \longrightarrow HN(SO_3)_2^{2-} + H_2O$$
$$HO\text{–}N(SO_3)_2^{2-} + HSO_3^- \longrightarrow N(SO_3)_3^{3-} + H_2O$$

The free acid $HON(SO_3H)_2$ is not known, but it forms a series of stable salts. Alkaline lead dioxide oxidizes the hydroxylamine N,N-disulphonate to nitrosyl disulphonate (Fremy's salt):

$$HON(SO_3)^{2-} + PbO_2 \longrightarrow ON(SO_3)^{2-}$$

Hydroxylamine O,N-Disulphonic Acid

The salt $K_2(SO_3NHOSO_3)$ was first prepared by Haga[481], and independently by Raschig, by the acid hydrolysis of hydroxylamine trisulphonic acid:

$$O_3SO\text{–}N(SO_3)_2^{3-} + H_2O \xrightarrow{\text{fast}} O_3SO\text{–}NHSO_3^{2-} + HSO_4^-$$

The alkaline hydrolysis of hydroxylamine N,O-disulphonate takes place only slowly in boiling alkali to form sulphate and sulphamate. In contrast, the hydrolysis of its isomer, N,N-disulphonate, proceeds rapidly even in the cold to yield sulphite and nitrite.

Hydroxylamine Trisulphonic Acid

The compound is obtained by the reaction of K_2SO_3 with nitrosodisulphonate. In acidic media the hydrolysis of $KO_3SO\text{–}N(SO_3K)_2$ proceeds fast to give the O,N-disulphonate exclusively. On further hydrolysis the hydroxylamine O-sulphonate is slowly formed.

Nitrosodisulphonic Acid

As described above, the oxidation of hydroxylamine N,N-disulphonate yields nitrosodisulphonate. Fremy first isolated the potassium salt as a dimer (yellow). The monomer exhibits in solution a violet-blue colour. The radical nature of the compound has been studied[482]. Seel[483] has confirmed Raschig's assumption that nitrososulphonic acid played an important role in the lead-chamber process.

2.8. COMPOUNDS COMPOSED OF SULPHUR, NITROGEN AND CARBON

The simplest compound consisting only of S, N and C is dirhodan, N≡C—S—S—C≡N. It is obtained when Ag(SCN) is treated with bromine:

$$2Ag(SCN) + Br_2 \longrightarrow 2AgBr + (SCN)_2$$

$(SCN)_2$ forms light yellow crystals (m.p. −3 to −2°), which decompose on melting under evolution of a yellow smoke and formation of a red amorphous solid. Solutions of $(SCN)_2$ are more stable: however, decomposition to a yellow amorphous product also takes place[484].

[481] T. Haga, *J. Chem. Soc.* **89** (1906) 240; F. Raschig, *Chem. Ber.* **39** (1906) 245.
[482] R. W. Asmussen, *Z. anorg. allg. Chem.* **212** (1933) 317.
[483] F. Seel and H. Meier, *Z. anorg. allg. Chem.* **274** (1953) 197.
[484] H. Remy, *Lehrbuch der Anorganischen Chemie*, p. 571, Akademische Verlagsgesellschaft, Leipzig 1965).

The chemical reactivity of $(SCN)_2$ resembles that of iodine. Depending on the concentrations, an equilibrium is established:

$$I_2+2SCN^- \rightleftharpoons 2I^-+(SCN)_2$$

Hydrogen sulphide reacts according to the following:

$$H_2S+2(SCN)_2 \longrightarrow S(SCN)_2+2HSCN$$

Dirhodano sulphide, $S(SCN)_2$, may also be regarded as dicyano trisulphane, $S_3(CN)_2$. Cyano sulphanes have been prepared from halogen sulphane and mercuric thiocyanate:

$$S_nX_2+Hg(SCN)_2 \longrightarrow S_{n+2}(CN)_2+HgX_2 \ (X = Cl, Br)$$

The fact that a similar synthesis starting with $Hg(CN)_2$ works only with SCl_2, but not with higher chlorosulphanes, is a strong hint that the formation of S–S linkages provides the driving force for the reaction between S_nX_2 and $Hg(SCN)_2$.

$$SCl_2+Hg(CN)_2 \longrightarrow S(CN)_2+HgCl_2$$

The following higher dicyano sulphanes have been prepared[485]: $S_3(CN)_2$, $S_4(CN)_2$, $S_5(CN)_2$, $S_6(CN)_2$, $S_7(CN)_2$ and $S_8(CN)_2$. Rhodanic acid, HSCN, is obtained from $Pb(CSN)_2$ and H_2S:

$$Pb(SCN)_2+H_2S \longrightarrow PbS+2HSCN$$

HSCN is a strong acid, which is completely dissociated in water. Thiocyanates are easily obtained when cyanides are reacted with sulphur (page 819). Ammonium thiocyanate is a technical product, which is prepared from carbon disulphide and ammonia:

$$CS_2+2NH_3 \xrightarrow{100^\circ} NH_4SCN+H_2S$$

The thiocyanate ion forms coloured complexes with various transition metals. This has been studied extensively, because SCN^- may coordinate to the metal atoms in three distinct ways. Formation of a metal nitrogen bond yields the isothiocyanate complexes (preferably with the first row of the transition series). The transition elements of the second and third rows usually form metal sulphur bonds. A third type is found when both nitrogen and sulphur coordinate, thereby yielding bridging complexes. Infrared spectra have been used to distinguish between the various bonding possibilities. In a number of isothiocyanates the carbon nitrogen stretching frequency is higher than it is for the sulphur-bonded forms[486]. Bridging thiocyanate groups also have higher C–N stretches than the corresponding non-bridged forms[487].

The formation of N- or S-bonded thiocyanate complexes depends not only on the metal, but also on the other ligands of the metal. Thus $[Co(-NCS)_4]^{2-}$ is N-bonded, while $[Co(-SCN)_2(Ph_3P)_2]$ is S-bonded[488]. The reverse situation is found in $[Pd(-SCN)_4]^{2-}$ being S-bonded and in $[Pd(NCS)_2(Ph_3P)_2]$ being N-bonded[489].

485 F. Fehér and H. Weber, *Z. Electrochem.* **61** (1957) 285.
486 P. C. H. Mitcheli and R. J. P. Williams, *J. Chem. Soc.* (1960) 1912.
487 D. Forster and D. M. L. Goodgame, *J. Chem. Soc.* (1965) 268.
488 R. Holm and F. A. Cotton, *J. Chem. Phys.* **32** (1960) 1168.
489 A. Turco and C. Pecile, *Nature*, **191** (1961) 66.

2.9. SULPHUR COMPOUNDS OF Vb ELEMENTS

Nitrogen and sulphur form several molecular compounds: S_4N_4, S_2N_2, S_4N_2, $S_{15}N_2$ and $S_{16}N_2$. They are actually nitrides, which have been already discussed in section 2.5. Since P, As, Sb and Bi possess more electropositive character than sulphur, they are able to form sulphides.

Phosphorus Sulphides

Among the sulphides of phosphorus definitely known are the following: P_4S_3, P_4S_5, P_4S_7 and P_4S_{10}. These compounds are prepared by direct combination of the elements. Polymeric phosphorus sulphur compounds of the general formula $(PS_x)_n$ have been synthesized by the reaction of tetrathio phosphoric acid and chlorosulphanes[490]:

$$2xH_3PS_4+3xSCl_2 \longrightarrow (PS_{5.5})_{2x}+6xHCl$$

Disulphur dichloride yields $(PS_7)_x$. The plastic products are light yellow in colour and are stable at room temperature Aqueous alkali decomposes the polymers. A much lower phosphorus content in sulphides is obtained, when phosphorus is incorporated in plastic sulphur. Figure 7 shows the structures[491, 492] of the phosphorus sulphides. Other phosphorus sulphides reported in the literature have been found to be mixtures of the abovementioned four well-characterized compounds. The decomposition of the phosphorus

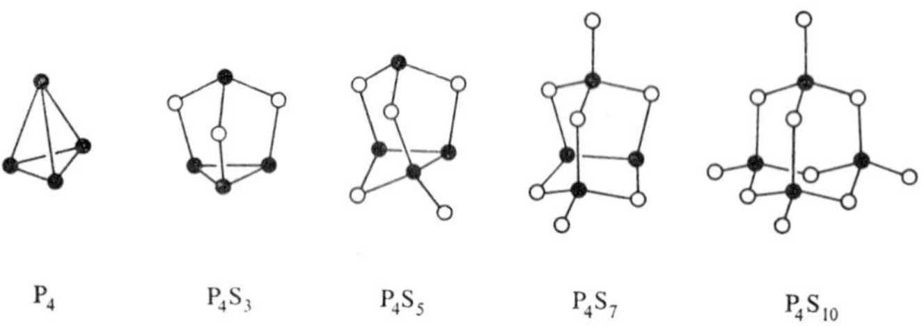

FIG. 7. Phosphorus sulphides.

sulphides by water in acidic and alkaline[493] media results in the formation of H_2S, PH_3, H_3PO_2, H_3PO_3, H_3PO_4 and thiosphoric acid or its salts. Tetraphosphorus trisulphide, P_4S_3, is quite stable towards hydrolysing agents. The reaction of methylamine with phosphorus sulphides has been studied[494].

Tetraphosphorus pentasulphide decomposes to P_4S_3 and P_4S_7 when heated below its melting point. Physical data of the sulphides are noted in Table 41. Because of their inflammability, the phosphorus sulphides (e.g. P_4S_3 starts to burn at $\sim 100°$ in air) have to be prepared in an inert atmosphere (CO_2 or N_2). Technically, these compounds are used for the production of matches, insecticides, lubricants and vulcanized rubber.

Cyclic organyl phosphorus sulphides $(RPS_2)_2$ have been obtained[495] by the reaction of

[490] M. Schmidt, in F. G. A. Stone and W. A. G. Graham (Eds.), *Inorganic Polymers*, p. 143, Academic Press, New York (1962).
[491] Y. C. Leung, J. Waser, S. van Houten, A. Vos and G. A. Wiegers, *Acta Cryst.* **10** (1957) 156.
[492] A. Vos, R. Olthof, F. van Bolhuis and R. Botterweg, *Acta Cryst.* **19** (1965) 864.
[493] W. G. Palmer, *J. Inorg. Nucl. Chem.* **30** (1968) 2367.
[494] E. Fluck and H. Binder, *Z. anorg. allg. Chem.* **359** (1968) 102.
[495] M. Baudler and H. W. Valperts, *Z. Naturforsch.* **22b** (1967) 222.

TABLE 41. PHYSICAL DATA OF PHOSPHORUS SULPHIDES

Compound	P_4S_3	P_4S_5	P_4S_7	P_4S_{10}
M.p. (°C)	171–172·5	170–220	305–310	286–290
B.p. (°C)	407–408	decomp.	523	513–515
Density (g/cm³)	2·03	2·15	2·20	2·09
Colour	yellow	bright yellow	yellowish	yellow
Solubility (g/100 g CS_2)	100	10	0·03	0·2

$RPCl_2$ and H_2S_2 in ether. Surprisingly this reaction occurs already at room temperature, whereas H_2S requires temperatures around 200° to react with $RPCl_2$.

$$2RPCl_2 + H_2S_2 \longrightarrow (R\text{–}\underset{\underset{S}{\|}}{P}\text{–}S)_2 + 2HCl$$

Mixed phosphorus oxysulphides have been prepared[496], e.g. $P_4O_6S_4$, in which the oxygen atoms occupy the bridging positions. Many phosphorus sulphide halides have been reported, of which the tetrahedral[497] thiophosphoryl halides SPX_3 exhibit exceptional hydrolytical stability. The sulphur atom in these compounds possesses a strong nucleophilic character, which is demonstrated by the formation of donor–acceptor compounds with AlX_3, BX_3, $SbCl_5$ and SO_3. The phosphorus sulphides may also react with certain transition metal carbonyls by replacing π-bonded groups. Thus P_4S_3 coordinates with the phosphorus bonded to the three sulphur[498], which has been established by NMR technique for several compounds, e.g. $Ni(P_4S_3)_4$, *cis*-$(P_4S_3)_2M(CO_4)$ (M = Cr, Mo, W).

Structural data are available for the phosphorus halosulphides $(BrSP)_2S_4$[499] (which consists of a six-membered ring with two disulphide bridges) and $P_4S_3I_3$[500] (with S in the bridging position and I in the terminal).

Arsenic Sulphides

Arsenic also forms several sulphides: As_4S_3, As_4S_4, As_2S_3 and possibly As_2S_5. The tetrasulphide is structurally[501] related to S_4N_4. It occurs as mineral realgar. In contrast to the molecular realgar, the crystalline orpiment (As_2S_3) possesses a polymeric layer structure, in which every arsenic atom has three sulphur neighbours in trigonal pyramidal arrangement[502]. In the vapour phase, however, As_4S_6 is found with a structure like that of P_4O_6. Hydrogen sulphide forms with acidic solutions of AsO_4^{3-} a precipitate of the composition As_2S_5, but it is not certain if the compound really exists as As_2S_5 or as a mixture of sulphur in As_2S_3. Acidified arsenites treated with H_2S are readily converted to As_2S_3, which is dissolved in the presence of alkali sulphide or preferably polysulphide ions. The latter acts as an oxidizing agent in the conversion to the oxidation state As(V).

Antimony Sulphides

Antimony forms only one stable sulphide, Sb_2S_3. Orange precipitates of the composition Sb_2S_5 are obtained by acidification of thioantimonate solutions. Similar to As_2S_5, it is not

496 A. J. Stosick, *J. Am. Chem. Soc.* **61** (1939) 1130.
497 M. Baudler, G. Fricke and K. Fichtner, *Z. anorg. allg. Chem.* **327** (1964) 124.
498 R. Jefferson, H. F. Klein and J. F. Nixon, *Chem. Commun.* (1969) 536.
499 F. W. B. Einstein, B. R. Penfold and C. T. Tapsell, *Inorg. Chem.* **4** (1965) 186.
500 D. A. Wright and B. R. Penfold, *Acta Cryst.* **12** (1959) 455.
501 T. Ito, N. Morimoto and R. Sadanaga, *Acta Cryst.* **5** (1952) 775.
502 N. Morimoto, *Mineralog. J. Japan*, **1** (1957) 160.

known whether the product represents Sb_2S_3 and sulphur or Sb_2S_5. Because of Sb_2S_3 dielectrical properties[503], it is used as a semiconductor.

Bismuth Sulphides

Bismuthinite, Bi_2S_3, is the only stable sulphide under ordinary conditions. It exhibits the same structural characteristics as Sb_2S_3[504] and shows semiconducting properties. The bismuth sulphide halides BiSX (X = Cl, Br, I) have orthorhombic structures, in which chains can be distinguished. These chains are linked to each other by additional Bi–S bonds, so that bismuth is surrounded by four sulphur and two chlorine atoms. SbSBr and SbSI have the same structure. The compounds of the general formula MSX have interesting photoelectric, semiconductive and dielectrical properties. Besides the binary sulphides, ternary compounds are known in a large variety, containing the thiophosphate, thioarsenate and thioantimonate moieties. Numerous minerals consisting of sulphur, group V and other elements have been studied. However, only one will be discussed. Surveys of the sulphide minerals of As, Sb and Bi and their structures have been given by Hellner[505], Bokii[506], Nowacki[507] and Berry[508].

The mineral livingstonite of the composition $HgSb_4S_8$ contains two polysulphide ions: S_2^{2-} and S_6^{2-}. The S–S distance in the disulphide amounts to 2·07 Å. The Sb atoms are surrounded by three sulphur in a trigonal pyramidal arrangement. Mercury has two neighbours in linear coordination and four more at larger distances[509].

2.10. SULPHUR COMPOUNDS OF IVb ELEMENTS

The elements of group IV form mono- as well as disulphides. However, they differ greatly in their stabilities. Within the group an increasing stability of the monosulphides is noticed with increasing atomic number of the elements, whereas the disulphides of the lower homologs are favoured.

Carbon Sulphides

CS is only stable when coordinated to metals[510], e.g. in the complex $[SCRhCl(PPh_3)_2]$. The disulphide, CS_2, represents a typical molecular compound, which plays an important role as solvent in many industrial processes. The compound also has been used as reactant for various purposes. The commercial synthesis of CS_2 is carried by reacting sulphur with carbon in a heated reactor. Carbon disulphide and H_2S form in alkaline solution trithiocarbonate, CS_3^{2-}, which decomposes on acidification. In contrast to H_2CO_3, H_2CS_3 has been isolated and studied[511]. Carbon subsulphide, C_3S_2, can be obtained by light or electrical

503 A. P. Grigas and A. S. Karpus, *Kristallografiya*, **12** (1967) 719.
504 W. Hofmann, *Z. Krist.* **86** (1933) 225.
505 E. Hellner, *J. Geol.* **66** (1958) 503.
506 G. B. Bokii and E. M. Romanova, *Kristallografiya*, **6** (1961) 869.
507 W. Nowacki, *Schweiz. Mineral. Petrogr. Mitt.* **44** (1964) 459.
508 L. G. Berry, *Am. Mineralogist*, **50** (1965) 301.
509 N. Niizeki and M. J. Buerger, *Z. Krist.* **109** (1957) 129.
510 J. L. DeBoer, D. Rogers, A. C. Skapski and P. G. H. Troughton, *Chem. Commun.* (1966) 756.

arc induced decomposition of CS_2. Similar to CS, it forms complexes with transition metals. Polymeric carbon sulphur compounds exhibit interesting properties. Thus CS_2 polymerizes under pressure to a black solid[512]. $(CS)_n$ and $(CS_{3+x})_n$ have been prepared[513], the latter according to

$$nS_xCl_2 + nHS\text{–}CS\text{–}SH \longrightarrow (CS_{3+x})_n + 2nHCl$$

The following compounds have been synthesized in excellent yields and in satisfactory purity from SCl_2, S_2Cl_2, S_3Cl_2, S_4Cl_2, S_5Cl_2, S_6Cl_2 and thiocarbonic acid, respectively: $(CS_4)_n$, $(CS_5)_n$, $(CS_6)_n$, $(CS_7)_n$, $(CS_8)_n$ and $(CS_9)_n$. They all form orange-coloured dry powders at room temperature; the colour deepens with increasing chain length. Replacing sulphur in CS_2 by O, Cl or F results in carbon oxisulphide, COS, thiophosgene, $SCCl_2$, and carbon sulphide difluoride, SCF_2, which has been synthesized by chlorine exchange in the $SCCl_2$-dimer:

$$3(SCCl_2)_2 + 4SbF_3 \longrightarrow 4SbCl_3 + 3(F_2CS)_2$$

SCF_2 can be polymerized to $(F_2CS)_n$.

TABLE 42. PHYSICAL DATA OF CARBON SULPHUR COMPOUNDS

Compound	C_3S_2	CS_2	H_2CS_3	COS
M.p. (°C)	0·5	−111·6	−26·9	−138·2
B.p. (°C)	90 dec.	46·3	—	− 50·2
Density (g/cm³)	1·274	1·262 (20°)	1·476 (25°)	1·24 (−87°)

Silicon Sulphides

When mixtures of silicon and sulphur are heated, SiS as a red to black product is formed, which on cooling decomposes to Si and SiS_2. Silicon disulphide has a fibrous structure, consisting of infinite chains, in which SiS_4 tetrahedra share edges[514]. At high pressure, however, the chain structure of SiS_2 is converted into a three-dimensional network[515]. Calcium disilicide has been reacted with chlorosulphanes to form silicon layers containing S_2 bridges[516]. A large number of organosilyl sulphides of the general formula $(R_3Si)_2S_n$ (n = 3 to 9)[517] and $(R_2SiS)_n$ (n = 2, 3) have been described[518]. The structure of disilyl sulphide, $H_3Si\text{–}S\text{–}SiH_3$, has been studied by electron diffraction[519].

Germanium Sulphides

In addition to GeS and GeS_2, a brown compound of the composition Ge_2S_3 is known. It has been obtained by thermal decomposition of GeS_2. Ge_2S_3 might contain Ge^{2+} and Ge^{4+} ions. In GeS_2, a three-dimensional network is built by tetrahedra of GeS_4. The structure of GeS can be explained as a distorted rock salt structure.

511 M. Dräger and G. Gattow, *Angew. Chem.* **80** (1968); G. Gattow and B. Krebs, *Z. anorg. allg. Chem.* **321** (1963) 143.
512 E. Whalley, *Canad. J. Chem.* **38** (1960) 2105.
513 M. Schmidt, in F. G. A. Stone and W. A. G. Graham (Eds.), *Inorganic Polymers*, p. 141.
514 E. Zintl and K. Loosen, *Z. Phys. Chem.* **174** (1935) 301.
515 C. T. Prewitt and H. S. Young, *Science*, **149** (1965) 535.
516 E. Hengge and G. Olbrich, *Z. anorg. allg. Chem.* **365** (1969) 321.
517 F. Fehér and G. Goller, *Z. Naturforsch.* **22b** (1967) 1224.
518 E. W. Abel and D. A. Armitage, in F. G. A. Stone and R. West (Eds.), *Advances in Organometallic Chemistry*, Vol. 5, Academic Press, New York (1967).
519 A. Almenningen, K. Hedberg and R. Seip, *Acta Chem. Scand.* **17** (1963) 2264.

Tin Sulphides

In many respects, tin sulphides are similar to the germanium sulphides. SnS, Sn_2S_3 and SnS_2 have been prepared. Thermal decomposition of SnS_2 yields Sn_2S_3 and presumably Sn_3S_4. In the structure of Sn_2S_3, divalent and tetravalent metal ions can be clearly distinguished. The Sn^{2+} forms with three sulphur neighbours a trigonal pyramid, while Sn^{4+} is octahedrally surrounded[520]. SnS_2 has a layer structure with octahedral coordination of the metal. In the thiostannates Mg_2SnS_4 and $CaSnS_4$ a tetrahedral arrangement is verified.

Lead Sulphides

Lead disulphide has not yet been prepared. All attempts to synthesize it yielded PbS. A large group of sulphide minerals of the composition xPbS.$y$$M_2S_3$ (M = As, Sb, Bi) is known.

2.11. SULPHUR COMPOUNDS OF GROUP IIIb

Boron Sulphur Compounds

Almost all knowledge about boron sulphur chemistry has been obtained within the past two decades. Besides high molecular boron sulphides, molecular compounds attracted the interest of many investigators.

High Molecular Boron Compounds

Three boron sulphides have been studied: $B_{12}S$, B_2S_3 and B_2S_5. Boron sulphide[521], B_2S_3, is formed by passing H_2S, diluted with H_2, over boron at temperatures above 600° and allowing the resulting compound to decompose at 100°. A higher sulphide, B_2S_5, was prepared in a sealed tube at 590°. Passing H_2S over a mixture of boron, sulphur and sodium at 750° yields $NaBS_2$, $NaBS_3$, $Na_4B_2S_5$ and $Na_8B_2S_7$ [522]. When PbS, α-rhombohedral boron and sulphur are heated at 700–900°, $Pb_2B_2S_5$ is formed. CuBS and AgBS were obtained from Cu(Ag), B and S at 600–1150°. A mass spectrometric investigation[523] of compounds obtained by vaporization of a slightly sulphur-rich boron sulphide at 700° showed the existence of two new high molecular boron classes. The ion species of the first class (of which thirty are given) range from $B_2S_3^+$ to $B_{10}H_{17}^+$ and the second series (of which eighteen are given) from $B_3H_3OH^+$ to $B_{11}S_{18}OH^+$. The vaporization and decomposition of metathioboric acid, $(HSBS)_3$, has also been studied by mass spectrometric technique[524]. Besides the parent ions, small amounts of H_2BS_2 and H_2BS_5 have been observed.

Molecular Compounds

Although Stock prepared boron sulphur compounds as early as 1901, the area was not explored until Wiberg[525] published his results concerning the preparation and properties of borthiin. When BBr_3 is reacted with H_2S in CS_2, several products can be obtained, depending on the reaction conditions: $(BrBS)_3$, $(HSBS)_2$ and $(HSBS)_3$.

520 D. Mootz and H. Puhl, *Acta Cryst.* **23** (1967) 471.
521 P. Hagenmuller and F. Chopin, *Compt. Rend.* **255** (1962) 2259.
522 P. Hagenmuller and F. Chopin, *Compt. Rend.* **256** (1963) 5578.
523 F. T. Green and J. L. Margrave, *J. Am. Chem. Soc.* **84** (1962) 3598.
524 J. G. Edwards, H. Wiedemeier and P. W. Gilles, *J. Am. Chem. Soc.* **88** (1966) 2935.
525 E. Wiberg and H. Sturm, *Z. Naturforsch.* **8b** (1953) 529.

$$3BBr + 3H_2S \xrightarrow[-6HBr]{10h} \text{(BrBS)}_3 \xrightarrow[H_2S]{140h} \text{(HSBS)}_3$$

The structure[526] of $(BrBS)_3$ has been elucidated.

The Cl and I derivatives of borthiin have also been prepared. However, all attempts to synthesize $(FBS)_3$ failed. Substitution of the halogens by alkyl, aryl, mercaptanes and amines resulted in the formation of the corresponding compounds. The stability of the borthiin system depends very much on the nature of the substituent. Thus $(ClBS)_3$ decomposes easily at room temperature, whereas $(BrBS)_3$ requires heating up to 80° to yield

TABLE 43. STRUCTURAL DATA OF $(BrBS)_3$

	Bond distance (Å)	Bond angle (°)
$(BrBS)_3$ (six-membered B–S ring, Br on each B)	B–S 1·85 B–Br 1·93	B–S–B 102 S–B–S 138

B_2S_3 and BBr_3. Surprisingly $(IBS)_3$, which contains weak B–I bonds (62 kcal/mole), did not decompose[527] when heated for several hours at 160°. This is in contrast to the properties reported in the literature. The steric effect of the large iodine cannot be very important for the stability, since the corresponding methyl compound (methyl and iodine have similar van der Waals' radii) rearranges to higher molecular species. Hence we have to assume an increase of the B–S bond order by electron back donation from sulphur to boron. By reacting CuBS with iodine, a presumably polymeric $(IBS)_n$ has been obtained[528], which shows a definite melting point (171°). The assumption regarding the polymeric character was based on the observation that the product did not dissolve in organic solvents (e.g. CS_2).

When BBr_3 is allowed to react with H_2S_2, a five-membered rather than a six-membered ring of the composition $(BrB)_2S_3$ is formed in almost quantitative yield[529].

$$2BBr_3 + 2H_2S_2 \longrightarrow \text{Br–B(S–S)(S)B–Br} + S + 4HBr$$

Similarly $(ClB)_2S_3$ may be obtained. However, yields are very low. Boron trichloride is

[526] Z. V. Zvonkava, *Kristallografiya*, **3** (1958) 564.
[527] W. Siebert, Unpublished results.
[528] J. K. Kom, *Ann. Chim.* **9** (1964) 181.
[529] M. Schmidt and W. Siebert, *Chem. Ber.* **102** (1969) 2752.

not electrophilic enough to undergo fast substitution, but it catalyses decomposition of the sulphanes into H_2S and sulphur. This proceeds (schematically) as follows:

$$8H_2S_2 \xrightarrow{[BCl_3]} 4H_2S+4H_2S_3 \longrightarrow 2H_2S+2H_2S_5 \longrightarrow 2H_2S+S_8$$

The catalytic decomposition is also caused by $(BrB)_2S_3$. During the course of this irreversible rearrangement, the formation of higher sulphanes can be easily detected by NMR technique[530]. The instability of $(FB)_2S_3$ is demonstrated by the formation of BF_3, when the bromine exchange in $(BrB)_2S_3$ is attempted even under mild conditions. $(IB)_2S_3$ cannot be prepared from BI_3 and sulphanes, since the resulting HI cleaves S–S bonds. The synthesis of $(IB)_2S_3$ from elemental sulphur and BI_3 represents a new reaction type in inorganic chemistry[529]:

```
                               S—S
                              /   \
2BI3+3/xSx ———————→ I–B         B–I+2I2
                              \   /
                                S
```

This redox reaction under iodine liberation has also been successfully applied to the formation of B–Se, B–O and B–C bonds. Additional BI_3 enlarges the five-membered trithia diborolane system to the borthiin:

$$(IB)_2S_3+BI_3 \longrightarrow (IBS)_3+I_2$$

Heating of the borthiin $(BrBS)_3$ with an excess of sulphur results in ring contraction to $(BrB)_2S_3$, which again indicates that the formation of the B_2S_3 ring is highly favoured.

$$2(BrBS)_3+3S \longrightarrow 3(BrB)_2S_3$$

The rather rare possibility of ring enlargement and ring contraction of inorganic compounds has also been found in the system S_4N_4/S_4N_3Cl (page 899).

A further synthesis for $(BrB)_2S_3$ was established, when tert.-butyl disulphide was treated with BBr_3. Besides the halogen derivatives, alkyl, aryl, mercaptane and amino trithia diborolanes have been prepared. The exceptional thermal stability of all derivatives may be caused by $(p\text{–}p)_\pi$ interaction of the $3p$-electrons of the sulphur atom with the empty p_z-orbital of the boron atom. This can be realized best in a planar system. Since the corresponding oxygen compound, $(HB)_2O_3$, possesses C_{2v} symmetry[531], it is very likely that $(BrB)_2S_3$ is also planar. Additional support for $(B\text{–}S)_\pi$-bonds stems from the observation that dialkyl iodoborane and sulphur form $R_2B\text{–}SS\text{–}BR_2$.

$$2R_2BI+2S \longrightarrow R_2B\text{–}S\text{–}S\text{–}BR_2+I_2$$

When R represents an electron donating group, such as C_6H_5, the redox reaction yields bisboryl sulphide[98]:

$$2(C_6H_5)_2BI+2S \longrightarrow (C_6H_5)_2B\text{–}S\text{–}B(C_6H_5)_2+S+I_2$$

Furthermore, the instability of $R_2B\text{–}S\text{–}BR_2$ (R = alkyl), obtained in solution from the

[530] H. Schmidbaur, M. Schmidt and W. Siebert, *Chem. Ber.* **97** (1964) 3379.
[531] F. A. Grimm and R. F. Porter, *Inorg. Chem.* **8** (1969) 731.

reaction of R_2BX with $(R_3Si)_2S$, indicates that one sulphur is not likely to stabilize two B–S bonds.

Polyhedral thioboranes were first prepared by Muetterties[532]:

$$B_{10}H_{14}+S^{2-}+4H_2O \longrightarrow B_9H_{12}S^- + B(OH)_4^- + 3H_2$$

The reaction of transition metal halides with $B_{10}H_{10}S^{2-}$ ion resulted in the formation of sandwich compounds.

Aluminium Sulphides

Al_2S_3 is the only stable sulphide at room temperature. Similar to B_2S_3, it hydrolyses easily to form H_2S and $Al(OH)_3$. There is some doubt about the composition of a lower sulphide of aluminium, stable above 1000°. No molecular Al–S compounds similar to the B–S compounds have been described. Several ternary sulphides are known, e.g. $ZnAl_2S_4$, $CdAl_2S_4$ and $HgAl_2S_4$.

Gallium Sulphides

The existence of the solid phases Ga_2S, GaS, Ga_4S_5 and Ga_2S_3 has been established. GaS and Ga_2S_3 are semiconductors. GaS has a hexagonal layer structure, which is closely related to that of MoS_2. Ga_2S_3 occurs in three forms and in all the metal is tetrahedrally coordinated. The following sulphide halides, GaSX (X = F, Cl, Br, I), have been prepared.

Indium Sulphides

The indium(III) sulphide, In_2S_3, exists in two forms, of which the α-form is metastable. In_2S has been reported to occur in the gaseous phase. It easily disproportionates on cooling into indium metal and indium trisulphide. An In_6S_7 phase has been detected by X-ray technique, which had been previously called In_4S_5 or In_5S_6. Indium sulphide halides InSX have been prepared[533].

Thallium Sulphides

Several thallium sulphides have been reported. Tl_2S has a layer structure. TlS is tetrahedral, containing Tl^{3+} and Tl^+ ions. In addition to Tl_2S and TlS, a sulphide Tl_4S_3 has been described. A polysulphide of the composition Tl_2S_5 or TlS_2[534] as well as thallium sulphide halogenides has been prepared.

2.12. SULPHUR COMPOUNDS OF GROUP Ia AND IIa ELEMENTS

Monosulphides and Hydrogen Sulphides

The sulphides of the alkali metals and barium easily dissolve in water and are almost completely hydrolysed. In contrast, the higher alkaline earth sulphides are less soluble. Some of the sulphides form hydrates, e.g. $Na_2S.5H_2O$ and $Na_2S.9H_2O$. The latter sulphide is a convenient starting material for the formation of sodium polysulphides, since the containing crystal water on treatment with elemental sulphur is released and then plays

532 E. L. Muetterties, in E. L. Muetterties (Ed.), *The Chemistry of Boron and Its Compounds*, John Wiley & Sons, Inc., New York, London, Sydney (1967); W. R. Herter, F. Klanberg and E. L. Muetterties, *Inorg. Chem.* **6** (1967) 1696.

533 H. Hahn and W. Nickels, *Z. anorg. allg. Chem.* **314** (1962) 303.

534 H. Hahn and W. Klingler, *Z. anorg. allg. Chem.* **260** (1949) 110.

the role of a reaction medium. Several anhydrous sulphides are known: Li_2S, Na_2S, K_2S[535] and Rb_2S, which all have antifluorite structure. The cations are tetrahedrally coordinated; the anions have eight cation neighbours. BeS exists in the zincblende type structure, while MgS, CaS–, SrS and BaS have rocksalt type structure.

Hydrogen sulphides of the type MSH of the alkali metals have been studied, e.g. LiSH possesses tetrahedral arrangement of the ions[536]. NaSH, RbSH and KSH have cubic rocksalt type structures at temperatures above 100–200°.

Polysulphides

When M_2S or MSH is treated with elemental sulphur, various polysulphides are formed depending on the molar ratio of the starting material. It is very convenient to use liquid ammonia as solvent for these reactions. The metal is first dissolved in the ammonia and then sulphur is added in small portions. Of the potassium system the complete row of sulphides is known: K_2S_2, K_2S_3, K_2S_4, K_2S_5 and K_2S_6. Other well-established sulphides are the following: Li_2S_2, Na_2S_2, Na_2S_4, Na_2S_5, Rb_2S_2, Rb_2S_3, Rb_2S_5, Cs_2S_2, Cs_2S_3, Cs_2S_5 and Cs_2S_6. The crystal structure of Cs_2S_6 has been elucidated. Polysulphides of substituted

TABLE 44. STRUCTURAL DATA OF THE S_6^{2-} ION IN Cs_2S_6 AND OF S_4^{2-} IN $BaS_4.H_2O$

		CS_2S_6	$BaS_4 \cdot H_2O$ Ion I	$BaS_4 \cdot H_2O$ Ion II
S_1	*Bond distances* (Å)			
	S(1)–S(2)	1·99	2·02	2·03
S_2	S(2)–S(3)	2·10	2·07	2·07
	S(3)–S(4)	2·03	2·02	2·03
S_3	S(4)–S(5)	2·12		
	S(5)–S(6)	2·03		
S_4	*Bond angles* (°)			
	(1)–(2)–(3)	110·0	104·9	104·1
S_5	(2)–(3)–(4)	106·4	104·9	104·1
	(3)–(4)–(5)	109·7		
S_6	(4)–(5)–(6)	109·2		

ammonium ions have also been prepared up to S_9^{2-}. The S–S distance[537] in Na_2S_2 and K_2S_2 amounts to about 2·15 Å; in MS_2 of the transition metals with pyrite structure the S–S length ranges between 2·07 and 2·21 Å (elemental sulphur: 2·04 Å). VS_2[538] and $Nb_2S_2Cl_2$ possess very short S–S distances in S_2^{2-} of 2·03 to 2·04 Å. The S–S distances[539] in Cs_2S_6 are given in Table 44.

Crystalline SrS_2 and SrS_3 have been prepared by heating the amorphous products. Also known are the following sulphides: BaS_2, BaS_3, SrS_4 and BaS_4. The crystalline lattice of $BaS_4.H_2O$ contains two crystallographically independent S_4^{2-} ions. They possess similar shapes and angles. The S_6^{2-} ions in Cs_2S_6 are non-planary; their structure resembles that of a helix.

[535] W. Klemm, H. Sodomann and P. Langmesser, *Z. anorg. allg. Chem.* **219** (1934) 45.
[536] R. Juza and P. Laurer, *Z. anorg. allg. Chem.* **275** (1954) 79.
[537] H. Föpple, E. Busmann and F. K. Frovath, *Z. anorg. allg. Chem.* **314** (1962) 12.
[538] R. Allmann, J. Baumann, A. Kutoglu, H. Rösch and E. Hellner, *Naturwiss.* **51** (1964) 2264.
[539] S. C. Abrahams and E. Grison, *Acta Cryst.* **6** (1953) 206.

2.13. SULPHUR COMPOUNDS OF TRANSITION ELEMENTS

Since sulphur reacts with almost all transition elements in various proportions, a variety of compounds are found in nature and have been prepared in the laboratories. Within the scope of this section, only the general features of these compounds can be discussed. One of the differences between main group and transition element sulphides is the instability of the former towards hydrolysis. It has been pointed out[540] that with increasing positive charge of the s^2p^6 cation in the main group element sulphide the solubility decreases, which may be credited to the polarizability of the sulphide ion. The sulphides of the cations with d^{10} configurations have smaller solubility products than those of s^2p^6. This is explained in assuming electron back-donation from the *d*-orbitals of the cation to the empty *d*-orbitals of sulphur. From geometrical points, large cations with small electric charges favour these $(d \rightarrow d)_\pi$-interactions; indeed, Cu^+, Ag^+ and Hg^{2+} exhibit a very strong affinity to sulphide. In general, the solubility of the sulphides, in which the cations possess n–d^x configurations, decreases with increasing number of *d*-electrons and with increasing main quantum number *n* of the *d*-shell. These sulphides are obtained from aqueous solutions in amorphous forms, which on standing or heating convert into the more stable crystalline modifications. In these sulphides usually one finds an octahedral coordination of the d^x-cation. The preferred coordination of the cation depends on the size and charge, since geometrical factors play an important role in the spherical symmetry of ions with rare gas configurations. There are exceptions to this: the spherically symmetrical ion Fe^{3+} with the high spin d^5 configuration exhibits usually tetrahedral configuration. A square planar arrangement is observed in Pt^{2+} and Pd^{2+} sulphides and a NiS is known to have a square pyramidal arrangement of the sulphur atoms around Ni^{2+}. The cations of 4*d* and 5*d* elements with one or two electrons often exhibit trigonal prismatic rather than octahedral coordination (e.g. MoS_2, WS_2, NbS_2).

In many disulphides, MS_2, layer structures occur, which may account for the large polarizability of the sulphide ion. The disulphides of group VIII have structures in which at least part of the sulphur is linked to S_2 groups. The pyrite structure (metal octahedrally coordinated) is realized in MnS_2, FeS_2, RuS_2, OsS_2, CoS_2, NiS_2 and in IrS_2, CuS_2, ZnS_2, CdS_2 prepared at high temperature. The sulphur atoms have an approximately tetrahedral arrangement with one sulphur and three metals as neighbours. A large variety of structures is found in compounds ranging in their composition between MS and MS_2. The magnetic and electrical properties of these compounds have received interest[541]. The transition metal sulphides may be divided according to their properties into several classes[542].

1. Semiconductors with ionic magnetism or diamagnetism.
2. Metallic conductors with ionic magnetism (*n* and *p* types).
3. Metallic conductors with temperature-independent paramagnetism or diamagnetism.

If the metal has the d°, d^6 (in octahedral) or d^8 (in square planar) configuration, the compound will exhibit diamagnetism or weak temperature-independent paramagnetism. The platinum metals prefer the diamagnetic configuration.

[540] F. Jellinek, in G. Nickless (Ed.), *Inorganic Sulphur Chemistry*, p. 671, Elsevier, Amsterdam, New York, London (1968).

[541] F. Halliger, *Structure and Bonding*, **4** (1968) 83.

[542] J. B. Goodenough, *Magnetism and the Chemical Bond*, Interscience Publ., New York (1963).

Sulphides of Titanium, Zirconium and Hafnium

The titanium sulphur system is quite complex, since many phases range in the composition TiS to TiS_2. There are also some metal-rich sulphides known. TiS_3, TiS_2 and Ti_2S_3 are semiconductors, while TiS exhibits metallic properties[543]. TiS_3 is diamagnetic, whereas the TiS–TiS_2 sulphides show temperature-independent paramagnetism[544]. Recently an interesting molecular titanium sulphide[545], Cp_2TiS_5, has been prepared from dicyclopentadienyl titanium dichloride, Cp_2TiCl_2, and ammonium polysulphide, $(NH_4)_2S_n$ (n = 2–7), as well as from $Cp_2Ti(SH)_2$ and chlorosulphanes:

$$Cp_2TiCl_2+(NH_4)_2S_5 \longrightarrow Cp_2TiS_5+2[NH_4]Cl$$

The compound has chair conformation with one axial and one equatorial Cp-substituent, which therefore gives rise to two ^1H NMR signals. The sulphides of zirconium and hafnium show structural relation to titanium sulphides. ZrS_3 and HfS_3 are semiconductors, and might be regarded as partial polysulphides $[M^{4+}]S^{2-}(S_2)^{2-}$. Besides these, the following sulphides have been studied: ZrS_2, HfS_2, $ZrS_{1\cdot6}$–$ZrS_{0\cdot9}Zr_2S$, $ZrS_{0\cdot7}$ and Zr_4S_3.

Sulphides of Vanadium, Niobium and Tantalum

Vanadin forms many sulphides within the range VS_4 to V_3S. The most interesting structural features have been found in VS_4, which consists of a fibrous structure[546] with disulphide anions, S_2^{2-}. VS and V_3S_4 are metallic[547]. Of the numerous niobium sulphides, the lower ($NbS_{n\leq 2}$) exhibit metallic character. Tantalum forms a subsulphide Ta_2S. Tantalum disulphide is polymorphic, all forms having a hexagonal or rhombohedral layer lattice. Pentacovalent vanadium, niobium and tantalum form thioanions.

Sulphides of Chromium, Molybdenum and Tungsten

Chromium sulphides occur in the range CrS to Cr_2S_3. The chemical properties and the structure of these compounds are closely related to each other. The preparation of CrBrS and $CrSI_{0\cdot83}$ has been reported[548]. Besides binary chromium sulphides, numerous ternary sulphides, e.g. $NaCrS_2$, $KCrS_2$, $FeCr_2S_4$, have been studied. The following molybdenum sulphides are known: Mo_2S_3, MoS_2, MoS_3, Mo_2S_5 and MoS_4. Molybdenum trisulphide cannot be prepared from the elements, however, by decomposition of $(NH_4)_2MoS_4$. When MoS_3 is heated, it loses sulphur until the composition $Mo_{0\cdot83}S_2$ is reached. MoS_2 may be obtained by heating the elements or by thermal decomposition of MoS_3. It is widely used as a dry lubricant; the lubricating properties in air are comparable to graphite. Another important technical appliance of molybdenum sulphides is their use as catalysts in hydrogenation, hydrocracking and hydrosulphuration reactions.

When Cp_2MoCl_2 is reacted with $(NH_4)_2S_5$, a five-membered rather than a six-membered ring is obtained. Cp_2MoS_4 exhibits only one H–NMR signal, even at −95°. This suggests that the five-membered MoS_4-ring exists at least at low temperature in a conformation of equivalent Cp ligands[549].

543 H. G. Grimmeis, A. Rabenau, H. Hahn and P. Hess, *Z. Electrochem.* **65** (1961) 776.
544 P. Ehrlich, *Z. anorg. allg. Chem.* **260** (1949) 1.
545 H. Köpf, *Chem. Ber.* **102** (1969) 1509.
546 R. Allmann, J. Baumann, A. Kutoglu, H. Rösch and E. Hellner, *Naturwiss.* **51** (1964) 263.
547 S. L. Holt, R. J. Bouchard and A. Wold, *J. Phys. Chem. Solids*, **27** (1966) 755.
548 H. Katscher and H. Hahn, *Naturwiss.* **53** (1966) 361.
549 H. Köpf, *Angew. Chem.* **81** (1969) 332.

Sulphides of Manganese, Technetium and Rhenium

The sulphides of manganese are closely related to Zn and Cd sulphides due to the spherical symmetry of the free-spin d^5 configuration of Mn^{2+}. MnS and MnS_2 are semiconductors and at low temperatures antiferromagnetic. The sulphides of technetium and rhenium differ from those of manganese. They are similar to those of molybdenum and tungsten. Re_2S_7 is obtained from perrhenates and H_2S. Thermal decomposition of Re_2S_7 resembles that of MoS_3. Technetium sulphide, Te_2S_7, is also prepared from pertechnetates and H_2S in acidic solution[550].

Sulphides of Iron, Cobalt and Nickel

The iron sulphide system has been studied extensively. Various sulphides are obtained when the elements are heated together in the proper ratios. FeS_2 occurs in two forms, pyrite and marcarite. The cubic pyrite of the rock salt type exhibits an arrangement of Fe^{2+} and S_2^{2-} ions. Cobalt and nickel also form a pyrite-type disulphide, MS_2. Many polynuclear sulphide complexes of iron, cobalt and nickel have been prepared, in which the sulphide ions are coordinated to two, three and four metal atoms. Usually the metals carry π-accepting terminal ligands (CO, NO, $C_5H_5^-$). Tri-coordinate sulphide ligands are found[551] in $[S_2Fe_3(CO)_9]$, $[S_4Fe_4(C_5H_5)_4]$, and $[SCo_3(CO)_9]$. The complex $[S_2Fe_2(CO)_6]$ contains a disulphide group as bridging ligand (page 930).

Sulphides of the Platinum Elements

The platinum metal sulphides have been studied only sporadically. Rh_2S_3 and Ir_2S_3 have orthorhombic structures with octahedral coordination of the metals[552]. IrS_2 has been prepared at high pressure, but RhS_2 is not known. Several platinum sulphides have been reported, only PtS_2 and PtS are thermally stable compounds. An interesting platinum polysulphide ion, PtS_{15}^{2-}, exists, which consists of three six-membered rings with the joint Pt atom[553]. Palladium also forms sulphides PdS and PdS_2.

Sulphides of Copper, Silver and Gold

Copper disulphide, CuS_2, has a pyrite-type structure, which can be prepared at high temperature. It shows metallic conductivity. Copper prefers to be monovalent in its sulphides, even in copper monosulphide, CuS. Only one binary silver sulphide, Ag_2S, is known, but it exists in several modifications[554]. When $AuCl_3$ is reacted with H_2S, a black product of the composition Au_2S_3 is obtained. Au_2S_3 is decomposed by water and by heat. Au^+ and H_2S yield Au_2S, which is oxidized by polysulphide to form $(AuS_2)^-$.

Sulphides of Zinc, Cadmium and Mercury

These elements are always divalent in their sulphides, even in ZnS_2 and CdS_2. Since ZnS and CdS exhibit sensitivity towards electromagnetic and corpuscular radiations, they are widely used in electronics. By doping them with other metals the compounds become luminescent.

Zinc sulphide exists in several forms, in which both the cation and the anion are tetra-

[550] C. L. Rulfsand, W. W. Meinke, *J. Am. Chem. Soc.* **74** (1952) 235.
[551] C. H. Wei and L. F. Dahl, *Inorg. Chem.* **6** (1967) 1229.
[552] E. Parthe, D. Hohnke and F. Hulliger, *Acta Cryst.* **7** (1967) 832.
[553] P. E. Jones and L. Katz, *Chem. Commun.* (1967) 842.
[554] F. C. Kracek, *Trans. Am. Geophys. Union*, **27** (1946) 364.

hedrally surrounded. The stable CdS crystallizes hexagonal in the wurtzite type. Mercury sulphide is obtained from Hg^{2+} and H_2S usually in the black form, which has a zincblende type of structure (metacinnabar). Upon heating, it is converted to the stable red trigonal form (cinnabar).

2.14. MOLECULAR TRANSITION METAL SULPHIDES

The foregoing section dealt with the transition metal sulphides, which generally exist as polymeric species. Some molecular polysulphides, e.g. Cp_2TiS_5 and Cp_2MoS_4, have been discussed already (page 928). By reacting metal carbonyls with sulphur or sulphur-containing compounds, a large variety of interesting metal carbonyl sulphides have been obtained. Structural investigations using X-ray technique have been mainly carried out by Dahl and his school. Two comprehensive reviews of this area of great research activity have been published[555, 556].

Though the reactions[557, 558] of organic disulphides and mercaptans with metal carbonyls and cyclopentadienyl metal carbonyls have been widely explored, this section will only include the metal sulphides, which have received until recently little attention. Cyclopentadienyl iron dicarbonyl dimer forms[559] with sulphur $(CpFeS)_4$. This compound has also been obtained from $(CpFe(CO)_2)_2$ and cyclohexene sulphide[560]. Its structural characterization as a tetrameric complex was ascertained from a three-dimensional X-ray examination, which showed $(CpFeS)_4$ to form monoclinic crystals. The tetrameric structure can be viewed as formed from two interpenetrating distorted tetrahedra of iron and sulphur atoms. $[(C_5H_5)_2V_2S_5]_n$ has been obtained[559] from $C_5H_5V(CO)_4$ and sulphur. Two ternary iron carbonyl sulphides, $[Fe(CO)_3]_3S_2$ and $[Fe(CO)_3]_2S_2$, have been prepared in a variety of ways[561], e.g. from iron powder, H_2S and CO. These compounds are well characterized and their structures have been elucidated by X-ray studies[562], which showed the presence of a S–S bond between the two bridging atoms, causing an exceptionally low S–Fe–S angle.

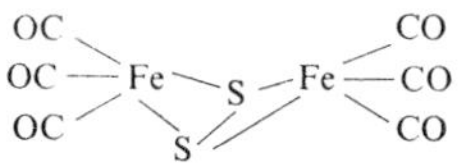

OC, OC, OC — Fe — S — Fe — CO, CO, CO; S (bridging)

The reaction of iron dodecacarbonyl with parathioformaldehyde gives in low yield[563] $[Fe(CO)_3]_3S_2$. A tetra-iron carbonyl sulphide, $[Fe(CO)_3]_4S_2$, obtained from the reaction

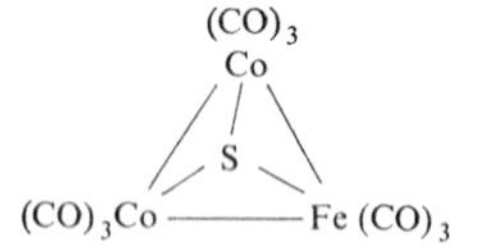

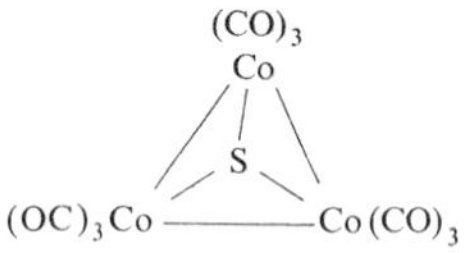

$(CO)_3Co$ / $(CO)_3Co$ — S — $Fe(CO)_3$ $(CO)_3Co$ / $(OC)_3Co$ — S — $Co(CO)_3$

of $Fe_3(CO)_{12}$ with benzyl thiocyanate has been claimed. Replacement of CO in $[Fe(CO)_3]_2$

[555] E. W. Abel and B. C. Crosse, *Organometal. Chem. Rev.* **2** (1967) 443.
[556] L. F. Lindoy, *Coord. Chem. Rev.* **4** (1969) 41.
[557] P. M. Treichel, J. H. Morris and F. G. A. Stone, *J. Chem. Soc.* **2** (1963) 219.
[558] R. B. King, *J. Am. Chem. Soc.* **84** (1962) 2460.
[559] R. A. Schunn, C. J. Fritchie and C. T. Prewitt, *Inorg. Chem.* **5** (1966) 892.
[560] C. H. Wei, G. R. Wilkes, P. M. Treichel and L. F. Dahl, *Inorg. Chem.* **5** (1966) 900.
[561] W. Hieber and J. Gruber, *Z. anorg. allg. Chem.* **296** (1958) 91.
[562] C. H. Wei and L. F. Dahl, *Inorg. Chem.* **4** (1965) 1.
[563] R. Havlin and G. R. Knox, *J. Organometal. Chem.* **4** (1965) 247.

by triphenyl phosphine yields a symmetrically disubstituted product[564] $[Ph_3PFe(CO)_2]_2S_2$. With cobalt carbonyl, $[Fe(CO)_3]_2S_2$ forms a mixed metal carbonyl sulphide, $Co_2Fe(CO)_9S$. This air-stable complex is also formed when sulphur is reacted with a mixture of iron and cobalt carbonyls[565]. Its structure is probably similar to that of $[Co(CO)_3]_3S$.

When cobalt carbonyl and sulphur are reacted in the prescence of CO, several compounds may be obtained[566]: $Co_3(CO)_9S$, $Co_3(CO)_7S_2$ and $[Co_2(CO)_5S]_n$. Treatment of cobalt carbonyl solutions with H_2S gives cobalt sulphide as well as cobalt carbonyl sulphides[567].

$$Co_2(CO)_8 + H_2S \longrightarrow CoS + [Co_2(CO)_5S]_n + Co_3(CO)_9S$$

A three-dimensional X-ray characterization of tricobalt enneacarbonyl sulphide $Co_3(CO)_9S$, obtained by the reaction of dicobalt octacarbonyl with phenylmercaptan, confirms[568] the proposed[566] molecular structure. The geometry of $Co_3(CO)_9S$ with the sulphur symmetrically linked to the $Co_3(CO)_9$ moiety is analogous to those of $CH_3–CCo_3(CO)_9$ and $Ir_4(CO)_{12}$.

$(C_5H_5MoO)_2S_2$ represents the first-known example of an organometallic compound, which contains doubly bridging sulphur atoms linking two transition metals[569]. A substantial π-bonding in the $(MoS)_2$ bridging system of this compound is evidenced by the Mo–S bond being 0·17 Å shorter than the presumed Mo–S distance of 2·049 in $[(C_2H_5OCS_2)_2MoO_2]O$. The $(MoS_2)_2$ in $(C_5H_5MoO)_2S_2$ is rigorously planar. In $[SCo_3(CO)_7]_2S_2$ the disulphide bridging group coordinates symmetrically to four transition metal atoms[570]. When the synthesis of $Ni_3(C_5H_5)_3S$, an electronic equivalent of $Co_3(CO)_9S$, was attempted, the paramagnetic compound tris(cyclopentadienylnickel) disulphide, $Ni_3(C_5H_5)_3S_2$, was obtained[571]. The nickel and sulphur atoms form a regular trigonal bipyramid.

2.15. SULPHUR COMPOUNDS OF LANTHANIDES AND ACTINIDES

Lanthanide Sulphides

An extensive review on the sulphides of the rare-earth metals has been published[572].

In most of the compounds the lanthanides (Ln) are trivalent, forming M^{3+} ions with noble gas configuration. The sesquisulphides M_2S_3 are semiconductors[573], which is due to electrons in the valence band. In monosulphides the electron is retained in the *d*-orbital of the metal, which causes metallic conductivity by overlapping of the *d*-orbitals of the cations. Scandium sulphide, Sc_2S_3, and yttrium sulphide, Y_2S_3, are semiconducting; ScS, Sc_5S_7, YS and Y_5S_7 are metallic. Similarly, Ln_2S and Ln_3S_4 and Ln_5S_7 exhibit metallic properties, but not Ln_2S_3. The divalent ions of Sm^{2+}, Eu^{2+} and Yb^{2+} have configurations $(Xe)4f^6$, $(Xe)4f^7$ and $(Xe)4f^{14}$, respectively. Since these ions lack 5*d*-electrons, their sulphides SmS, EuS and YbS resemble BaS rather than LaS in having no metallic character. In

564 W. Hieber and A. Zeidler, *Z. anorg. allg. Chem.* **329** (1964) 92.
565 S. A. Khattab, L. Marko, G. Bor and B. Marko, *J. Organometal. Chem.* **1** (1964) 373.
566 L. Marko, G. Bor, E. Klumpp, B. Marko and G. Almasy, *Chem. Ber.* **96** (1963) 955.
567 E. Klumpp, L. Marko and G. Bor, *Chem. Ber.* **97** (1964) 926.
568 C. H. Wei and L. F. Dahl, *Inorg. Chem.* **6** (1966) 1229.
569 D. L. Stevenson and L. F. Dahl, *J. Am. Chem. Soc.* **89** (1967) 3721.
570 D. L. Stevenson, V. R. Magnuson and L. F. Dahl, *J. Am. Chem. Soc.* **89** (1967) 3727.
571 H. Vahrenkamp, V. A. Uchtman, L. F. Dahl, *J. Am. Chem. Soc.* **90** (1968) 3272.
572 J. Flahant and P. Laruelle, in L. Eyring (Ed.), *Progress in the Science and Technology of the Rare Earths*, Vol. 3, p. 149, Pergamon Press (1968).
573 F. Jellinek, in G. Nickless (Ed.), *Inorganic Sulphur Chemistry*, p. 729, Elsevier, Amsterdam, New York, London (1968).

Yb_3S_4, Sm_3S_4 and Eu_3S_4 the ions M^{2+}, M_2^{3+} and S_4^{2-} exist according to the magnetic studies. The Y^{3+} ion is octahedrally surrounded, while Yb^2 has a sevenfold coordination. La_3S_4, Ce_3S_4, Pr_3S_4 and Nd_3S_4 resemble in their magnetic moments the trivalent ions. The preferred formation of the d^7 ions Eu^{2+} and Gd^{3+} is demonstrated by the instability of Eu_2S_3 and Gd_3S_4. The lanthanide metals Tb to Tm form monoclinic sulphides Ln_5S_7, which are metallic conductors. Dy_2S_3, Ho_2S_3, Er_2S_3 and Tm_2S_3 are also monoclinic[574] and isotopic with Y_2S_3. The above-mentioned sulphides of the rare-earth metals do not decompose below their melting points. Therefore, some of the sulphides are used as high-temperature semiconductors. In contrast, the disulphides of the rare-earth metals lose sulphur already at about 600°, thereby yielding sesquisulphides. The lanthanides tend to form ternary compounds of the formula Ln_2S_3–MS and Ln_2S_3–M_2S. A rocksalt-type structure is found in the compounds $MLnS_2$ (M = Li, Na, K and Ag). Many mixed sulphides of copper and the lanthanides metals have been prepared. Oxide sulphides of the composition Ln_2O_2S are known of yttrium, lanthanum and all lanthanide metals, which are closely related in the structure to La_2O_3. Sulphide halides of yttrium of the lanthanide elements have been prepared.

Actinium Sulphides

The heavier actinide sulphides have not yet been studied. However, sulphides of thorium[575], uranium, neptunium and plutonium have been reported. The sulphides of plutonium are similar to those of the higher lanthanide elements. Thorium sulphides are diamagnetic; uranium sulphides exhibit paramagnetic properties. The monosulphides ThS, US, NpS and PuS have been studied. Besides U_3S_5, a disulphide US_2 is known which exists in three modifications[576]. ThS_2 has been prepared and it was found to have semi-conducting properties. NpS_2 is unstable. However, the higher sulphides Np_2S_5 and NpS_3 are known. Several oxide sulphides of the actinide elements have been studied: ThOS, PaOS, UOS, Pu_2O_2S, $Pu_4O_4S_3$, NpOS and Np_2O_2S.

2.16. ANALYTICAL USE OF HYDROGEN SULPHIDE[577, 578, 579]

As outlined above, the transition metals readily form sulphides either by heating the elements with sulphur or treating the cations with hydrogen sulphide. Some metals of the Main Groups also exhibit an excellent affinity towards sulphur. However, most of the Main Group element sulphides are hydrolytically unstable and are easily decomposed.

The high tendency for the formation of metal sulphides and their insolubility in aqueous media has been used widely in qualitative and quantitative chemistry. Precipitations by the sulphide ion are of the utmost importance in quantitative analysis, not only for the isolation of single elements, but also for the separation of groups of elements from one another. Hydrogen ion concentration and the character of the solution are very important

574 A. W. Sleight and C. T. Prewit, *Inorg. Chem.* **7** (1968) 2282.

575 J. Graham and F. K. McTaggart, *Austr. J. Chem.* **13** (1960) 67.

576 M. Picon and J. Flahant, *Bull. Soc. Chem. France* (1958) 772.

577 F. Seel, *Grundlagen der Analytischen Chemie und der Chemie in wässrigen Systemen*, Verlag Chemie, Weinheim (1955).

578 W. F. Hillebrand, G. E. F. Lundell, H. A. Bright and J. Hoffman, *Applied Inorganic Analysis*, Wiley, New York (1959).

579 L. G. Sillen and A. E. Martell, *Stability Constants of Metal-ion Complexes*, The Chemical Society, London (1964).

to the precipitations. Thus by varying the acidity, arsenic may be separated from lead, lead from zinc, zinc from nickel and manganese from magnesium. Separation within a group is also possible by conversion of one or more elements into complex anions, e.g. cadmium from copper in cyanide solution. Precipitations with H_2S may be divided into five groups according to the kind of solutions in which the precipitation is made: I, strong acid (pH$<$1); II, dilute acid (pH 2–3); III, very weak acid (pH 5–6); IV, alkaline (pH$>$7); and V, complex anions. The precipitation in strong acid media ranges within a proton concentration of a 0·25 to 13 N hydrochloric acid. Under these conditions the elements of the copper and arsenic group precipitate. The copper group includes copper, silver, mercury, lead, bismuth, cadmium, rhutenium, rhodium, palladium and osmium. Gold, arsenic, platinum, tin, antimony, iridium, germanium, selenium, tellurium and molybdenum belong to the arsenic group. Separation within the group by careful adjustment of the pH does not always give satisfactory results, since the solubility products in some cases are similar. Some of the solubility products are listed below.

TABLE 45. LOGARITHMS OF THE SOLUBILITY PRODUCTS OF SOME SULPHIDES IN WATER AT 25°C

Me_2S		MeS	
Tl_2S	−20	ZnS	−24
Cu_2S	−48	CdS	−26
Ag_2S	−49	HgS	−52
		SnS	−25
Me_2S_3		PbS	−27
$\frac{1}{3}In_2S_3$	−24	PoS	−28
$\frac{1}{3}Sb_2S_3$	−31	MnS	−10
$\frac{1}{3}Bi_2S_3$	−32	FeS	−17
$\frac{1}{3}La_2S_3$	−4	CoS(α)	−20
$\frac{1}{3}Ce_2S_3$	−3	NiS(α)	−19
		CuS	−35

In dilute acid solution (pH 2–3), several sulphides are formed (after having separated the members of the preceding groups). However, the analytical use is not important except for the separation of zinc in this group. Practically the only separations made by the use of hydrogen sulphide in the nearly neutral solution are those of cobalt and nickel from manganese in the absence of iron. Under these conditions, thallium and indium are also completely precipitated. Alkaline sulphide will precipitate a large number of sulphides after prior separation of the sulphides obtained in acid solution. When the solution is treated with tartrate, elements such as aluminium, titanium, chromium and uranium cannot interfere with the precipitation of manganese, iron, zinc, etc.

Certain members of the foregoing groups react with ligands, thereby forming more or less stable complex anions, which are no longer precipitable with the sulphide ion. The elements of the arsenic group are soluble in alkaline sulphide or polysulphide solutions. Usually the sulphides of the copper group are not dissolved by these reagents. Several ions, such as fluoride, oxalate, tetrate and cyanide, allow various elements to separate from each other by precipitation with sulphide ions. In practice, the analytical use of H_2S or sulphide ions in combination with complexing agents depends on the problem to be solved. After removal of the hydrogen sulphide group (from strong acidic media), usually $(NH_4)_2S$ is used for the precipitation of Fe, Cr, Al, Zn, Mn, Ni and Co.

24. SELENIUM, TELLURIUM AND POLONIUM

K. W. Bagnall
The University of Manchester

1. THE ELEMENTS

1.1. INTRODUCTION AND HISTORICAL

The three heaviest elements of the sulphur sub-group—selenium, tellurium and polonium, collectively referred to as the *chalcogens*, a term which includes oxygen and sulphur—have been less intensively studied than either oxygen or sulphur, but even so the main features of their chemistry are well established. The elements have the $ns^2\ np^4$ outer electronic configurations approaching those of the next noble gas (krypton, xenon and radon respectively), so that it is not surprising that their chemistry is predominantly non-metallic, with a gradually increasing metallic character appearing on progressing down the group. Thus selenium, like sulphur, exists in a number of allotropic modifications, only one form being metallic, while with tellurium the metallic form is the more common and polonium is a typical metal. The trend is most clearly shown in the decreasing electronegativity, specific resistivity and first ionization potential (Table 4, p. 944) of the elements as the atomic number increases down the group.

Tellurium was the first of the three to be recognized as an element; Müller von Reichenstein isolated it (and appropriately named it *metallum problematicum*) in 1782 from a Transylvanian ore which had previously been thought to be some kind of alloy of antimony and bismuth, showing that neither was actually present. The name tellurium, derived from the Latin *tellus*, earth, is due to Klaproth.

Although the appearance of deposits of a *red* sulphur, produced when native sulphur was vaporized, was reported in the fourteenth century, this does not appear to have been followed up, but the colour suggests that selenium may have been present. Selenium was finally discovered in 1817 by Berzelius and Gahn when investigating, as a possible source of tellurium, the constitution of the lead chamber mud which was formed when Fahlun copper pyrites was used as the source of sulphur dioxide in this sulphuric acid process. The new element resembled tellurium chemically and was named from the Greek σεληνη, the moon.

In contrast to selenium and tellurium, stable isotopes of polonium are unknown. Its discovery in 1898 by Madame Curie followed from her observation that the radioactivity of certain minerals exceeded that to be expected from their uranium or thorium content. Rightly suspecting that this was due to the presence of one or more unknown, highly radioactive elements, she and her husband processed large quantities of uranium ore residues by conventional chemical group separation methods, ultimately demonstrating that the new element was a higher homologue of tellurium and naming it after her native Poland. This discovery was a most remarkable feat, for this was the first time that invisible quantities of a new element had been identified, separated and investigated solely by making use of its radioactivity, which was, at that time, a newly-discovered phenomenon.

1.2. OCCURRENCE AND DISTRIBUTION

Selenium, tellurium and polonium are comparatively rare, their abundances[1] in the lithosphere being respectively $9 \times 10^{-6}\%$, $2 \times 10^{-7}\%$ and $3 \times 10^{-14}\%$. Native selenium and tellurium are uncommon, usually occurring in conjunction with native sulphur. All of their minerals are rare, the most abundant ones being selenides of lead, copper, silver, mercury and nickel and tellurides of lead, copper, silver, gold and antimony. Some examples are given in Table 1.

The metal selenides and tellurides are nearly all isomorphous with the corresponding sulphides, so explaining the frequent occurrence of sulphides in these minerals. Selenium

TABLE 1. SOME TYPICAL SELENIUM AND TELLURIUM MINERALS

Se:	Clausthalite	$PbSe$
	Crookesite	$(Cu, Tl, Ag)_2Se$
	Eucairite	$(Cu, Ag)_2Se$
	Naumannite	$(Ag, Pb)Se$
Te:	Calaverite	$AuTe_2$
	Nagyagite	$Au_2Sb_2Pb_{10}Te_6S_{15}$
	Petzite	Ag_3AuTe_2
	Sylvanite	$(Au, Ag)Te_2$

is also present in the soil in certain areas of the U.S.A. (the dry plains of Dakota, Wyoming and Kansas) and is taken up by vegetation which then becomes poisonous to animals; their meat is then rendered unfit for human consumption. Selenium is, however, an essential trace element in some animal diets.

Polonium (^{210}Po) is the penultimate number of the radium decay series:

$$\underset{\text{RaD}}{{}^{210}_{82}\text{Pb}} \xrightarrow[22\ \text{yr}]{\beta} \underset{\text{RaE}}{{}^{210}_{83}\text{Bi}} \xrightarrow[5\text{d}]{\beta} \underset{\text{RaF}}{{}^{210}_{84}\text{Po}} \xrightarrow[138{\cdot}4\text{d}]{\alpha} \underset{\text{RaG}}{{}^{206}_{82}\text{Pb}}$$

Uranium ores contain only about 100 μg of polonium per ton.

1.3. PRODUCTION

The methods used for the production of selenium, tellurium[2, 3] and polonium[3, 4] have been described very fully in earlier reviews and only a brief account is given here.

Although selenium was originally obtained mainly from lead chamber mud or from the flue dusts arising from the roasting of seleniferous sulphide ores, and tellurium from lead chamber mud or from silver, gold or lead telluride ores, the main source of both

[1] L. H. Ahrens and S. R. Taylor, *Spectrochemical Analysis*, 2nd ed., p. 93, Pergamon Press, Oxford (1961).
[2] J. W. Mellor, *A Comprehensive Treatise on Inorganic and Theoretical Chemistry*, Vol. X, p. 696, Vol. XI, p. 4, Longmans Green & Co., London (1931).
[3] K. W. Bagnall, *The Chemistry of Selenium, Tellurium and Polonium*, p. 14, Elsevier, Amsterdam (1966).
[4] M. Haissinsky, Polonium, in *Nouveau Traité de Chimie Minérale*, Vol. 13, p. 2041, Masson, Paris (1961).

elements is now the anode slime deposited during the electrolytic refining of copper; this material also contains gold, silver and the platinum metals.

Selenium and tellurium are recovered from the slime by pyrometallurgical methods, and are converted to alkali selenite and tellurite in the process. Selenium is almost completely separated from tellurium at this stage by neutralizing the alkaline selenite and tellurite solution with sulphuric acid. The bulk of the tellurium precipitates as the hydrated dioxide, while the more acidic selenous acid remains in solution, from which 99·5% pure selenium is precipitated by sulphur dioxide[5]:

$$H_2SeO_3+2SO_2+H_2O \longrightarrow Se+2H_2SO_4$$

Until recently, polonium-210 was only available in trace amounts, obtained from uranium ores and radium or its decay products. This nuclide can now be prepared in larger (milligram) amounts by the neutron irradiation of bismuth:

$$^{209}_{83}Bi(n, \gamma)^{210}_{83}Bi \xrightarrow[5d]{\beta} {}^{210}_{84}Po$$

Processes which involve organic materials, such as solvent extraction[6, 7] or ion exchange[8], can be used for the separation of trace, but not for weighable, amounts of polonium from irradiated bismuth. This is because the intense α-radiation from polonium-210 leads to rapid radiation decomposition of all organic materials. Consequently, a direct chemical separation is preferred, involving reduction to elementary polonium by stannous chloride, with tellurium as a carrier if necessary, or electrochemical replacement of the polonium in solution by a less noble metal, such as silver, adjusting the conditions so that bismuth and impurity elements do not plate out with polonium[9, 10]. It should also be possible to separate polonium from metallic bismuth by fractional vacuum sublimation since metallic polonium has a much higher vapour pressure than bismuth at any given temperature, but solutions of polonium in molten bismuth do not behave ideally[11] and the method does not seem to have been used. The sublimation of polonium from bismuth oxide has also been studied[12].

1.4. INDUSTRIAL USES

Selenium has been extensively used in photocells[13], devices in which variations in the intensity of the incident light cause a corresponding variation of an electric current. Grey, crystalline selenium is the only allotrope suitable for this purpose, its conductivity increasing approximately thousandfold when illuminated. The property is also used in xerography, a dry copying process in which a positively charged insulating material is exposed to light, so causing discharge in the exposed areas; a negatively charged powder, which adheres to the

[5] E. E. Elkin and P. R. Tremblay, U.S. Patent 3,127,244 (1964).
[6] F. L. Moore, *Analyt. Chem.* **29** (1957) 1660; *ibid.* **32** (1960) 1048.
[7] J. C. Sheppard and R. Warnock, U.S.A.E.C. unclassified document HW-SA-2632 (1962).
[8] H. Koch and H. Schmidt, *Kernenergie*, **6** (1963) 39.
[9] K. W. Bagnall, J. H. Freeman, D. S. Robertson, P. S. Robinson and M. A. A. Stewart, U.K.A.E.A. unclassified report C/R 2566 (1958).
[10] J. Pauly, *Bull. Soc. chim. France* (1960) 2022.
[11] R. W. Endebrock and P. M. Engle, U.S.A.E.C. unclassified document AECD-4146 (1953).
[12] R. A. Hasty, *J. Inorg. Nuclear Chem.* **29** (1967) 2679.
[13] B. Lange, *Photoelements and their Applications*, Reinhold Publishing Co. Inc., New York, N.Y. (1938).

unexposed, and therefore positively charged, areas is used to "develop" the image, which is transferred to paper and fixed by heat treatment[14].

Grey selenium is used to a greater extent for a.c. rectification, utilizing the property of asymmetric conduction exhibited by thin ("barrier") layers of this allotrope[15]. Less important uses of selenium are in the manufacture of coloured (red or reddish-yellow) glasses or ceramic and enamel pigments, and as bases for phosphors. Tellurium has also been used for these purposes.

Both selenium and tellurium have been used as secondary vulcanizing agents for natural rubber and, in the form of organo-compounds, as oxidation inhibitors in lubricating oils, uses which have been described in more detail elsewhere[3].

Polonium-210 is commonly used for the production of neutron sources[16], in which the polonium is mixed or alloyed with elements for which (α, n) reactions are highly probable, such as beryllium. Another use of polonium is in space satellites, where the radioisotope is used as a heat source (140 W/g) for the production of electrical power by a system which is similar in principle to that of the conventional thermocouple. The attraction of this type of electrical generator is that there are no moving parts to go wrong, but this is offset by the rather short (138·4 day) half-life of polonium.

1.5. PURIFICATION

The purification of selenium or tellurium consists essentially of separating the one element completely from the other, and from traces of sulphur and a few other elements. Consequently a method used to purify selenium can often be used to purify tellurium.

Very pure selenium is obtained by heating the crude material in hydrogen at 650° to form hydrogen selenide which is then passed through a silica tube at 1000° to decompose it[17]. Hydrogen sulphide is more stable to heat than the selenide, and so passes out of the system unchanged; the hydrides of elements which are less stable to heat than hydrogen selenide (tellurium, phosphorus, arsenic and antimony) are not formed at 650°. Tellurium can be purified in a similar manner, by the decomposition at 300° of hydrogen telluride, prepared electrolytically[18], but the overall yield is rather poor ($\sim 24\%$).

The simplest chemical purification method consists in precipitating elementary selenium from a solution of a selenium compound in moderately concentrated (> 8·8 N) hydrochloric acid by reduction with sulphur dioxide. Tellurium(IV) is not reduced at this high acidity, owing to the formation of stable anionic chloro-complex species, but it is precipitated as the element when the acidity is decreased below about 2 N. Sulphur dioxide also reduces selenium(IV), but not tellurium(IV), to the element in concentrated sulphuric acid; hydroxylamine (in citric or tartaric acid solution) and hydriodic acid have also been used for this purpose. In the last case, tellurium(IV) forms the iodo-complex anion TeI_6^{2-}, which is much more stable towards reduction than its selenium analogue. Other chemical purification methods include the dissolution of crude selenium in hot, aqueous alkali sulphite to form the selenosulphate ($SeSO_3^{2-}$), from which selenium is precipitated on acidification, and fusion of the crude material with potassium cyanide, which yields the

[14] C. F. Carlson, U.S. Patents 2,221,776; 2,277,013; 2,297,691 and 2,357,809 (1940–4).
[15] H. K. Henisch, *Metal Rectifiers*, Clarendon Press, Oxford (1945).
[16] H. V. Moyer (Ed.), U.S.A.E.C. unclassified document TID-5221 (1956).
[17] S. Nielsen and R. J. Heritage, *J. Electrochem. Soc.* **106** (1959) 39.
[18] J. Weidel, *Z. Naturforsch.* **9A** (1954) 697.

selenocyanate, KCNSe, and telluride, K_2Te. Tellurium is precipitated from the aqueous solution of the melt on aeration, and selenium is recovered from the filtrate on acidification, although this last step is rather slow.

Sublimation techniques are also used, an effective separation of selenium from tellurium being achieved by repeated sublimation of the more volatile selenium dioxide[19] at 320–350° in air. The purified dioxide is commonly reduced to the element by heating it in ammonia gas:

$$3SeO_2+4NH_3 \longrightarrow 3Se+2N_2+6H_2O$$

Sublimation of elementary selenium does not give as good a separation as is found with the dioxide, but zone refining of selenium appears to be effective[20].

Distillation of selenium tetrabromide from concentrated (> 9 M) hydrobromic acid affords a good separation from tellurium, but not from arsenic or germanium, the bromides of which, if present, distil quantitatively, together with small amounts of tin and antimony[21]. The distillate is usually collected in bromine water at 0° and, after elimination of free bromine by addition of sulphite, selenium is precipitated by boiling the solution with hydroxylamine.

Ion exchange does not appear to be used commonly for the purification of either selenium or tellurium, but sorption of selenous and selenic acids on a strongly basic anion resin (in the acetate form) at pH 2·7–2·8 is a useful way of freeing tellurium from traces of selenium; the separation depends on the fact that orthotelluric acid (Te(VI)) is a much weaker acid than either selenous or selenic acids[22]; selenite and tellurite can also be separated by anion exchange[23]. Although selenium is said[24] to be purified by anion exchange of the hexabromoselenate(IV), $SeBr_6^{2-}$, it is not clear whether tellurium is effectively separated in this procedure.

Solvent extraction is a useful procedure for the purification of tellurium; extraction of tellurium from hydrochloric acid (4–6 N) into a chloroform solution of tribenzylamine leaves any selenium in the aqueous phase, and the tellurium is easily recovered from the chloroform layer by washing it with water[25]. Tellurium is also extracted from its solution in a mixture of iodide, stannous chloride and hydrochloric acid by ethyl acetate[26], from solution in 1 N hydrochloric acid and 0·6 M sodium iodide by a mixture of n-amyl alcohol and diethyl ether[27] and into a solution of dithizone[28] in carbon tetrachloride at pH 1. Tellurium dithizonate is volatile, like its polonium analogue[29], which extracts[30] at pH 0·2 to 5. A simpler method for separating selenium from tellurium, used mainly for the determination of small amounts of selenium, involves the extraction into toluene of diphenylpiazselenol, formed by reaction of selenite with 3,3′-diaminobenzidine[31]. Tellurium

19 D. M. Yukhtanov and N. B. Pleteneva, *Zhur. priklad. Khim.* **33** (1960) 1951.
20 H. Okada, Japan Patent 4306 (1957).
21 W. O. Robinson, H. C. Dudley, K. T. Williams and H. G. Byers, *Ind. and Eng. Chem., Anal. Edn.* **64** (1934) 274.
22 C. R. Veale, *J. Inorg. Nuclear Chem.* **10** (1959) 333.
23 A. Iguchi, *Bull. Chem. Soc. Japan*, **31** (1958) 748.
24 H. Schumann, German (East) Patent (to V.E.B. Chemiewerk Coswig), 32,681 (1964).
25 G. Nakagawa, *Nippon Kagaku Zasshi*, **81** (1960) 1258.
26 R. Vanossi, *Anales Asoc. quim. Argentina*, **38** (1950) 117.
27 C. K. Hanson, *Analyt. Chem.* **29** (1959) 1204.
28 H. Mabuchi, *Bull. Chem. Soc. Japan*, **29** (1956) 842.
29 K. Kimura and H. Mabuchi, *Bull. Chem. Soc. Japan*, **28** (1955) 535.
30 G. Bouissières and C. Ferradini, *Analyt. Chim. Acta*, **4** (1950) 610.
31 K. L. Cheng, *Analyt. Chem.* **28** (1956) 1738.

can also be freed from selenium by electrolytic methods[32], but these have not been extensively investigated.

The purification of weighable amounts of polonium is best achieved by precipitating selenium and tellurium, if present, as the elements from dilute hydrochloric acid with sulphur dioxide, which only reduces polonium to the bipositive state. The sulphide, PoS, is then precipitated and decomposed to the elements at 275° in a vacuum, the final stage being vacuum sublimation of polonium metal[33].

Solvent extraction techniques have been applied to the separation of trace level polonium(IV) from bismuth, thorium, selenium and tellurium; long-chain tetra-alkyl ammonium salts separate the halo-complex anions, probably PoX_6^{2-}, from hydrofluoric or hydrochloric acid solution[34]. Di*iso*propyl and di*iso*butyl carbinol have also been used to extract anionic polonium complexes[35]. Extraction of the polonium complexes with 8-hydroxyquinoline[36], cupferron[37], thenoyl-trifluoroacetone[38], di*iso*propyl ketone[39] and mesityl oxide[40] can be used to purify trace amounts of polonium, while extraction from 6 N hydrochloric acid into tributylphosphate (TBP) in dibutyl ether[41] or dekalin[42] is also reasonably effective for the purification of weighable amounts of polonium, in spite of the radiation damage to the solvent. The extracted complex is probably $PoCl_4.2TBP$, for which the equilibrium constant for the reaction

$$PoCl_4 + 2TBP \rightleftharpoons PoCl_4.2TBP$$

is 0·04 at room temperature.

Ion exchange methods have been used to purify trace level polonium, but are of no practical value for weighable amounts of the element; such methods have been reviewed in detail[43]. Electrodeposition of polonium on to gold[44] or platinum[45], followed by vacuum sublimation of the polonium, is also used widely as a final purification step; the deposition of polonium has been the subject of much study in this connexion and this work has been comprehensively reviewed[10, 43].

1.6. ALLOTROPY

Although the tendency towards ring and chain formation exhibited by sulphur is much less marked with selenium and tellurium, selenium is known to form Se_8 rings, and both selenium and tellurium exist as long-chain species. Polonium does not form either rings or chains.

The allotropy of selenium has not been studied as extensively as that of sulphur, with which there are some analogies, but five modifications have been identified, in contrast to

[32] W. G. Woll and R. T. Gore, U.S. Patent 2,258,963 (1941).
[33] K. W. Bagnall and D. S. Robertson, *J. Chem. Soc.* (1957) 1044.
[34] W. J. Maeck, G. L. Booman, M. E. Kusey and J. E. Rein, *Analyt. Chem.* **33** (1961) 1775.
[35] F. L. Moore, *Analyt. Chem.* **32** (1960) 1048.
[36] K. Kimura and T. Ishimori, *Proc. 2nd Internat. Conf. Peaceful Uses At. Energy, Geneva*, **28** (1958) 151.
[37] A. G. Maddock and G. L. Miles, *J. Chem. Soc.* (1949) S252.
[38] F. Hagemann, *J. Am. Chem. Soc.* **72** (1950) 768.
[39] A. E. Cairo, *Proc. 1st Internat. Conf. Peaceful Uses At. Energy*, **7** (1955) 331.
[40] J. Maréchal-Cornil and E. Picciotto, *Bull. Soc. chim. Belges*, **62** (1953) 372.
[41] D. G. Karracker and D. H. Templeton, *Phys. Rev.* **81** (1951) 510.
[42] K. W. Bagnall and D. S. Robertson, *J. Chem. Soc.* (1957) 509.
[43] K. W. Bagnall, *Chemistry of the Rare Radioelements*, Butterworths, London (1957).
[44] K. W. Bagnall, R. W. M. D'Eye and J. H. Freeman, *J. Chem. Soc.* (1955) 2320.
[45] K. W. Bagnall and R. W. M. D'Eye, *J. Chem. Soc.* (1954) 4295.

tellurium, for which only one form is known for certain, and polonium, where both modifications are simple metal structures. The decreasing complexity of the solid state of the three elements, as compared to sulphur, is reflected in the vapour state. It is apparent from vapour density determinations that Se_8 molecules are present below 550°; the vapour is yellow at the boiling point (685°) and dissociation to Se_6, Se_2 (above 900°) molecules and to atomic selenium occurs with increasing temperature. A mass spectrometric study[45a] of selenium vapour has provided evidence for the existence of Se_4 and Se_7 molecules in the vapour, in addition to those mentioned above, and the enthalpies of vaporization have been measured for each of these species. Tellurium and polonium are less complex in the vapour state, in which only Te_2 and Po_2 molecules are present; although tellurium vapour is known to be golden yellow, there is no report of the colour of polonium vapour.

In the molten state selenium is less complex than sulphur; the viscosity of molten selenium decreases rapidly and uniformly with increasing temperature, in direct contrast to

TABLE 2. THE CRYSTAL STRUCTURES OF THE ALLOTROPES OF SELENIUM, TELLURIUM AND POLONIUM

	Symmetry	Space group	Lattice parameters Å			β
			a_0	b_0	c_0	
Selenium, α	Monoclinic[a]	$P2_1/n$	9·05	9·07	11·61	90°46′
Selenium, β	Monoclinic[b]	$P2_1/a$	12·85	8·07	9·31	93°8′
Selenium, grey	Hexagonal[c] }	$P3_121$ or $P3_221$	4·3662	—	4·9536	—
Tellurium	Hexagonal[d] }		4·4572	—	5·9290	—
Polonium, α	Simple cubic[e]	O_h^1	3·359	—	—	—
Polonium, β	Simple rhombohedral[e]	D_{3d}^5	3·366	(α = 98°5′)		—

[a] R. D. Burbank, *Acta Cryst.* **4** (1951) 140; *ibid.* **5** (1952) 236.
[b] R. E. Marsh, L. Pauling and J. D. McCullough, *Acta Cryst.* **6** (1953) 71.
[c] H. E. Swanson, N. T. Gilfrich and G. M. Ugrinic, *Nat. Bur. Standards, Circular* 539, **5** (1955) 54.
[d] P. Cherin and P. Unger, *Acta Cryst.* **23** (1967) 670.
[e] J. M. Goode, U.S.A.E.C. unclassified document MLM–808, 1953.

the behaviour of sulphur, indicating that the chain length of the selenium polymers in the melt is progressively reduced with increasing temperature; from studies of the decrease in electrical resistivity of molten selenium with increasing temperature, it appears that selenium changes from a semi-conductor to a metallic conductor as the chains in the liquid are thermally destroyed[45b]. Tellurium and polonium are probably even less complex than selenium in the melt, but experimental data are lacking.

Three of the five modifications of selenium are crystalline, two being of monoclinic and the third of hexagonal (or trigonal) symmetry; the last is commonly known as grey metallic selenium. The other modifications are red, amorphous and black, vitreous selenium.

α- and β-monoclinic selenium, originally known collectively as β- or red crystalline selenium, are obtained respectively by the slow and the rapid evaporation of carbon disulphide or benzene solutions of black, vitreous selenium at room temperature. These are unusual polymorphs in that X-ray studies (Table 2) have shown that the unit cell of both modifications contains four Se_8 molecules at very similar intermolecular distances. In both cases each of the Se_8 molecules is non-planar and puckered into the shape of a crown in

[45a] R. Yamdagni and R. F. Porter, *J. Electrochem. Soc.* **115** (1968) 601.
[45b] E. H. Baker, *J. Chem. Soc.* (*A*), (1968) 1089.

which the Se–Se bond distance is 2·34 Å, which is slightly longer than in the sulphur ring structures, as one would expect from the larger selenium atoms. This distance corresponds to a single bond and there is no evidence of double bond character, in contrast to orthorhombic sulphur in which there may be some double bond character involving 3*d* orbitals. The Se–Se–Se angle is about 105·7° as against 107·8° in the S_8 molecule, which is the most stable modification of sulphur. Both α- and β-monoclinic selenium are appreciably soluble in carbon disulphide, giving dark red solutions, and both transform readily to the grey metallic form, but there is no clearly defined transition temperature, nor is there any evidence for an $\alpha \rightarrow \beta$ transformation. The energy of the transformation to grey selenium is very small (~ 180 cal/g atom at 130°).

Grey, or metallic, selenium is thermodynamically the most stable form of the element. It is formed by heating any other modification, by slowly cooling molten selenium or a saturated solution of amorphous selenium in hot aniline, and by condensing selenium vapour at a temperature close to the melting point of the element (~ 220°). It is the only modification which conducts electricity; its conductivity increases at high pressures and on exposure to light (p. 937). The structure consists of infinite helical chains in which branching does not occur, the Se–Se bond distance[46] being 2·373 Å and the bond angle 103·1°. Grey selenium is insoluble in carbon disulphide and its density, 4·82 g/cc, is the highest of any form of the element; the densities of the monoclinic varieties are both 4·46 g/cc and of the amorphous and vitreous varieties, only 4·26 and 4·28 g/cc respectively.

Red, amorphous selenium, originally known as α-selenium, is precipitated when aqueous selenous acid is treated with sulphur dioxide or other reducing agents and when aqueous selenocyanates are acidified; it is also formed by the condensation of selenium vapour on cold surfaces. This amorphous modification has a chain structure rather like that of hexagonal selenium, although it is somewhat deformed[47], an observation confirmed by heat capacity measurements[47a]. In spite of the relationship to hexagonal selenium, this modification is a non-conductor of electricity and is very slightly soluble in carbon disulphide.

Vitreous, or black, selenium, which is somewhat soluble in carbon disulphide, is the ordinary commercial form of the element; it is a somewhat brittle, opaque, dark red-brown to bluish-black lustrous solid and is obtained when molten selenium is cooled suddenly. It does not melt sharply, but softens at about 50° and rapidly transforms to hexagonal selenium at 180–190°; the heat of transformation is about 1·05 kcal/g atom at 125°, which is about 300 cal/g atom less than the heat of transformation of red amorphous to hexagonal selenium. Vitreous selenium is much more complex than any other modification and the polymer rings may contain up to 1000 atoms[48]. Although the transformation to hexagonal selenium occurs to some extent at room temperature, it is catalysed by substances such as halogens and amines, which probably break up the large rings.

Although a so-called amorphous form of tellurium is precipitated when aqueous tellurous acid is treated with reducing agents, such as sulphur dioxide, it is probable that this is merely finely divided crystalline tellurium. Greyish-white, crystalline tellurium has a metallic lustre, but it is a poor conductor of heat and has the electrical properties of a semi-conductor rather than those of a metal. It is quite brittle and can be powdered readily,

46 P. Cherin and P. Unger, *Inorg. Chem.* **6** (1967) 1589.

47 H. Richter, W. Kulcke and H. Specht, *Z. Naturforsch.* 7a (1952) 511.

47a K. K. Manredov, I. G. Kerimov, M. I. Mekhtiev and M. I. Veliev, *Russ. J. Phys. Chem.* **40** (1967) 1655.

48 H. Krebs, *Z. anorg. Chem.* **265** (1951) 156.

and it is insoluble in any liquid with which it does not react. The structure is the same as that of hexagonal selenium, the bond distance being 2·835 Å and the Te–Te–Te angle 103·2°, but the chains are appreciably closer together than are the selenium chains. Hexagonal selenium and tellurium appear to form a continuous range of solid solutions, in which there is fairly random alternation of selenium and tellurium atoms in the infinite chains[49].

Both forms of polonium (Table 2) are metallic and the $\alpha \rightarrow \beta$ phase change appears to occur at about 36°. However, owing to the internal heating arising from the radioactivity, the high temperature α-form is normal at room temperature and the two phases co-exist between 18° and 54°. Both forms have a silvery metallic appearance and are metallic conductors, the temperature coefficient of resistivity being positive[50], in contrast to sulphur, selenium and tellurium for which this coefficient is negative, a characteristic of non-metals.

1.7. NUCLEAR PROPERTIES

Seventeen isotopes of selenium, twenty-one of tellurium and twenty-seven of polonium (Table 3) have been recorded; of these, six selenium and eight tellurium isotopes are stable.

TABLE 3. ISOTOPES OF SELENIUM, TELLURIUM AND POLONIUM[a]

Mass No.[b]	Selenium Mode of decay;[c] half-life; energy of the emitted particle (MeV)	Mass No.[b]	Tellurium Mode of decay;[c] half-life; energy of the emitted particle (MeV)	Mass No.[b]	Polonium Mode of decay;[c] half-life; energy of the emitted particle (MeV)
70	β+ ;~44 min; –	114	β+ or K; 16 min; –	192	α; 0·5 sec; 6·58
71	β+ ; 4·5 min; 3·4	115	β+ or K; 6 min; –	193	α; 4 sec; 6·47
72	K; 9·7d	116	β+ or K; 2·5 hr; –	194	α; 13 sec; 6·38
73	β+ ; 44 min; 1·72	117	β+ 1·7 hr; 2·7	195	α; 30 sec; 6·26
	β+ ; 7·1 hr; 1·32	118	K, β+ ; 6·0d; 2·70	196	α; 1·9 min; 6·14
74	stable (0·87%)	119	K; 4·7d	197	α; 4·0 min; 6·03
75	K; 120d		K, β+ ; 16 hr; 0·63	198	α; 7·0 min; 5·93
76	stable (9·02%)	120	stable (0·089%)	199	α; 12 min; 5·86
77	stable (7·58%)	121	K; 17d	200	α; 10 min; 5·86, 5·77
78	stable (23·52%)	122	stable (2·46%)	201	α, K; 18 min; 5·68
79[d]	β– ; $6{\cdot}5\times10^4$ yr; 0·16	123	stable (0·87%)	202	K, α; 43 min; 5·58
80	stable (49·82%)	124	stable (4·61%)	203	K, α; 42 min; 5·48
81[d]	β– ; 18·6 min; 1·60	125	stable (6·99%)	204	K, α; 3·5 hr; 5·37
82	stable (9·19%)	126	stable (18·71%)	205	K, α; 1·8 hr; 5·22
83[d]	β– ; 69 sec; 3, 4, 1·5	127[d]	β– ; 9·4 hr; 0·70	206	K, α; 8·8d; 5·22
	β– ; 25 min; 0·45, 1·0	128	stable (31·79%)	207	K, β+,α; 5·7 hr; –, 5·10
84[d]	β– ; 3·3 min; –	129[d]	β– ; 74 min; 1·45, 0·99	208	α, K; 2·9 yr; 5·11
85[d]	β– ; 39 sec; –	130	stable (34.48%)	209	α, K; 103 yr; 4·88
87[d]	β– ; 17 sec; –	131[d]	β– ; 25 min; 2·14, 1·69, 1·35	210	α; 138·4d; 5·30
		132[d]	β– ; 78 hr; 0·22	211	α; 25 sec; 7·14
		133[d]	β– ; 2 min; ~2·4		α; 52 sec; 7·44
		134[d]	β– ; 42 min; –	212	α; 0·3 μsec; 8·785
				213	α; 4 μsec; 8·34
				214	α; 164 μsec; 7·69
				215	α, β; 1·8 msec; 7·38, –
				216	α; 0·16 sec; 6·78
				217	α; <10 sec; 6·54
				218	α, β; 3·05 min; 6·00, –

[a] Data from the *Chart of the Nuclides*, 2nd edition, 1961; Kernforschungszentrum Karlsruhe.
[b] Abundance in parentheses.
[c] β+ indicates positron emission, K electron capture and β– negatron emission
[d] Product of the fission of ^{235}U.

Owing to the preponderance of stable heavy isotopes of tellurium, its atomic weight is higher than that of iodine. All isotopes of polonium are radioactive and stable isotopes

[49] E. Grison, *J. Chem. Phys.* **19** (1951) 1109.
[50] C. R. Maxwell, *J. Chem. Phys.* **17** (1949) 1288.

have not been detected in nature; only three isotopes are sufficiently long-lived for macro-chemical studies, ^{208}Po, ^{209}Po and ^{210}Po, the first two being obtained by cyclotron bombardment of bismuth with protons or deuterons [$^{209}Bi(p,2n)^{208}Po$; $^{209}Bi(d,3n)^{208}Po$ using 20–24 MeV deuterons or $^{209}Bi(d,2n)^{209}Po$ with 10–15 MeV deuterons], or of lead with α-particles [$^{207}Pb(\alpha,3n)^{208}Po$]. Polonium-210 occurs naturally, but is usually obtained by neutron bombardment of bismuth (p. 937).

1.8. PHYSICAL PROPERTIES OF THE ELEMENTS

Some physical properties of the elements are given in Table 4, from which it can be seen that the octahedral covalent radii are appreciably greater than the tetrahedral (sp^3) radii,

TABLE 4. SOME PROPERTIES OF SELENIUM, TELLURIUM AND POLONIUM

	Se	Te	Po
Atomic number	34	52	84
Electronic structure	$[Ar]3d^{10}4s^24p^4$	$[Kr]4d^{10}5s^25p^4$	$[Xe]4f^{14}5d^{10}6s^26p^4$
Atomic weight	78·96	127·60	(210)
Density, g/cc	4·82[a]	6·25	9·196(α), 9·398(β)
Atomic radius, Å	1·17	1·37	1·64
Ionic radius, M^{2-}, Å	1·98	2·21	(2·30?)
Ionic radius, M^{4+}, Å		0·89	1·02
Octahedral covalent radius, Å	1·40	1·52	1·52
Tetrahedral radius, Å	1·14	1·32	1·46
Electronegativity[b]	2·48	2·01	1·76
First ionization potential, eV	9·75	9·01	8·43
Melting point, °C	220·5[a]	449·8[c]	254
ΔH fusion, kcal/g atom	1·25	4·27	—
Boiling point, °C	685	1390	962
ΔH vaporization, kcal/mole	22·82	27·26	24·60
Vapour pressure[d]	$A = 8·089$	7·600	7·2345
$\log_{10}p_{mm} = A - B/T$	$B = 4989·5$	5960·2	5377·8
Entropy[e], 298°K, e.u.	10·144	11·88	13
Heat of atomization[f], kcal/g atom (298°K)	49·4	46	—
Specific heat[g], c_p, cal/g	0·0767	0·0481	0·030
Dissociation energies[h] of			
(i) the molecules M_2, kcal/mole	72·94	54·0	—
(ii) the molecules SM, kcal/mole	(90)	(80)	—
(iii) the molecule SeTe	63	—	—
Specific electrical resistivity[i], μohm cm	2×10^{11}	2×10^5	42(α), 44(β)

[a] Grey selenium.
[b] A. L. Allred and E. Rochow, *J. Inorg. and Nuclear Chem.* **5** (1958) 264.
[c] Triple point (A. D. Tevebaugh and E. J. Cairns, *J. Chem. Engng. Data*, **9** (1964) 172).
[d] Data of L. S. Brooks, *J. Am. Chem. Soc.* **74** (1952) 227; *ibid.* **77** (1955) 3211. These equations are reasonably accurate up to 750°C; for a summary of other work see H. Lumbroso in *Nouveau Traité de Chimie Minérale*, Vol. 13, Masson et Cie., Paris, 1961.
[e] U.S.A.E.C. unclassified document ANL–5750.
[f] H. A. Skinner and G. Pilcher, *Quart. Revs.* (*London*), **17** (1963) 264.
[g] Kelly, *Bureau of Mines Bulletin* 592 (1961).
[h] J. Drowart and P. Goldfinger, *Quart. Revs.* (*London*), **20** (1966) 545.
[i] C. R. Maxwell, *J. Chem. Phys.* **17** (1949) 1288.

with the difference becoming less marked with increasing atomic number. The larger octahedral radii, observed in the MBr_6^{2-} ions, for example, are the result of $np^3\,nd^2\,(n+1)s$

bonding, the *ns* orbital being already occupied by an unshared pair of electrons which is truly inert[51].

The known physical properties of polonium, and particularly the low melting and boiling points, are much more in line with those of the elements in the same Period, thallium, lead and bismuth, than with those of tellurium, behaviour which is not uncommon with the heaviest members of the main groups; the horizontal similarity does not, of course, extend to the chemistry of the element.

1.9. CHEMISTRY OF THE ELEMENTS

The preparation of elementary selenium and tellurium by precipitation from aqueous acid solutions with reducing agents such as sulphur dioxide, or by the thermal decomposition of the hydrides, has been mentioned earlier. The elements can also be prepared by reducing the dioxides, for example by heating the selenium compound in gaseous ammonia and the tellurium one with carbon at 800–900°.

Elementary polonium is often obtained by thermally decomposing the sulphide or dioxide in a vacuum, and by reduction of the dioxide in hydrogen at 200°, conditions under which selenium or tellurium dioxide does not react. Earlier methods of producing elementary polonium involved electrodeposition on to gold or platinum from acid solution, or spontaneous deposition on to silver, followed by vacuum sublimation of the polonium from the substrate metal; these procedures, most of which are applicable to weighable amounts of the element, have been reviewed in detail[43]. Polonium is also precipitated from acid solution as the element by reducing agents such as sodium dithionite, tin(II) chloride or titanium(III) chloride; alkaline suspensions of polonium(IV) hydroxide are reduced to the metal by hydrazine, hydroxylamine, sodium dithionite and even concentrated aqueous ammonia, but the reduction by the last, and by anhydrous liquid ammonia (or by primary and secondary amines), is probably due to atomic hydrogen formed as a consequence of the intense α-bombardment[52].

All three elements are very reactive chemically, forming the dioxides slowly in air or oxygen at room temperature and reacting rapidly at higher temperatures, selenium and tellurium burning with a blue flame. Grey selenium is oxidized to some extent by water at 160°, but tellurium is not appreciably affected. Fuming nitric acid or *aqua regia* oxidize selenium and tellurium to the oxyacids of the quadripositive elements, but the more basic character of tellurium is shown by its formation of a basic nitrate. Similarly the even more basic polonium dissolves readily in nitric acid, a variety of nitrates being known (p. 986).

The elements react much less readily with sulphur than they do with oxygen; although selenium reacts with sulphur[53] at 450° to form the red, crystalline sulphide Se_4S_4, and all the species Se_nS_{8-n} have been identified by mass spectrometry[54], the much larger tellurium is less easily incorporated in the S_8 ring and only TeS_7 has been identified[53, 55]. However, high-pressure cocrystallization of sulphur and tellurium appears to yield a phase of composition Te_7S_{10} in which sulphur is presumably incorporated in the tellurium chain[55a].

51 L. Pauling, *The Nature of the Chemical Bond*, p. 251, Cornell University Press, Ithaca, N.Y., 3rd ed. (1960).
52 K. W. Bagnall, P. S. Robinson and M. A. A. Stewart, *J. Chem. Soc.* (1958) 3426.
53 L. L. Hawes, *Nature*, **198** (1963) 1267.
54 R. Cooper and J. V. Culka, *J. Inorg. Nuclear Chem.* **27** (1965) 755; *ibid.* **29** (1967) 1217.
55 R. Cooper and J. V. Culka, *J. Inorg. Nuclear Chem.* **29** (1967) 1877.
55a S. Geller, *Science*, **161** (1968) 290.

Polonium does not react with sulphur but the monosulphide (p. 982) can be obtained by precipitation from acid solution.

The three elements react with halogens in the cold or at moderate temperatures, forming halides (p. 955), the only exception being the failure of selenium to react with iodine; although selenium absorbs iodine, simple iodides have not been recorded.

A notable characteristic of the sulphur sub-group is the reaction of the elements with sulphur or selenium trioxide, or with the corresponding concentrated acids, with the formation of brightly coloured compounds of apparent composition MXO_3 (M = S, Se, Te, Po; X = S, Se) which are described later with the sulphates and selenates (p. 988). Another characteristic is the ease with which the chalcogenides are formed, although the stability of the hydrogen compounds decreases markedly with increasing atomic number. Thus selenium reacts readily with hydrogen above 350° to form hydrogen selenide, H_2Se, but reaction scarcely occurs with tellurium and not at all with polonium. However, all three elements react with most metals and non-metals at moderate temperatures to form selenides, tellurides and polonides (p. 951). Selenium reacts with aqueous solutions of silver salts, silver selenide being precipitated, and tellurium behaves in a similar manner, but polonium seems to be inert. Similarly, both selenium and tellurium dissolve in concentrated aqueous alkali with the formation of a mixture of selenide, or telluride, and selenite, or tellurite, from which the element is reprecipitated when the solution is acidified. In addition, selenium displaces sulphur from a number of sulphides and in anhydrous liquid ammonia displaces tellurium from the potassium tellurides.

Although selenium is not appreciably attacked by dilute hydrochloric acid, tellurium dissolves to some extent in the presence of air and polonium dissolves readily, initially yielding a solution of bipositive polonium which is rapidly oxidized to polonium(IV) by the products of the radiolytic decomposition of the solvent. Selenium also dissolves in antimony pentafluoride, forming the yellow complex $Se(SbF_5)_2$ which is stable to 200°. Whereas selenium is soluble in aqueous or in fused potassium cyanide, forming the selenocyanate KCNSe, tellurium only forms the telluride K_2Te on fusion with the cyanide.

1.10. HEALTH HAZARDS

Polonium is the most dangerous of the three elements because of its intense radioactivity; all three elements appear to be taken up by the kidneys, spleen and liver, the tissues of which undergo irreparable radiation damage in the case of polonium because of the complete absorption of the α-particle energy. Consequently these organs can only tolerate very small amounts of radioactive materials. For the most commonly used isotope of polonium, ^{210}Po, the maximum permissible body burden is only 0·03 μc, equivalent to 2×10^{10} atoms or $6{\cdot}8 \times 10^{-12}$ g of the element. Because of this, the concentration of airborne polonium compounds must be kept at very low levels indeed, the recommended maximum permissible concentration in air being only 4×10^{-11} mg/m³, as compared with concentrations of 0·1 mg/m³ for either selenium or tellurium[56] and 10 mg/m³ for hydrogen cyanide. The ways in which polonium can be safely handled have been fully described in other reviews[3, 43] of the chemistry of the element in which many of the effects of radiation are dealt with in some detail.

[56] I. Sax, *Dangerous Properties of Industrial Materials*, Reinhold Publishing Corporation, New York, N.Y. (1957).

On the basis of their maximum permissible concentrations in air, quoted above, selenium and tellurium are more toxic than hydrogen cyanide by two orders of magnitude. Hydrogen selenide is particularly unpleasant, and when inhaled at very great dilutions in air it causes headache and nausea, while at high concentrations it causes an acute irritation of the mucous membrane. Other selenium compounds, and the organo-compounds in particular, appear to cause eczema and inflammation of the skin, but the most objectionable result of ingesting selenium compounds is the slow release of the element from the body as organo-selenium compounds in the breath and perspiration, both of which develop a very offensive smell. Although hydrogen telluride is as toxic as the selenide, tellurium is rather less unpleasant in this respect, and this may be due to conversion of the ingested tellurium compound to the element in the body; some of it is, however, excreted as foul smelling organotellurium compounds.

It is clear from the above that all three elements should be handled with care and it is certainly better for the well-being of those working with selenium and tellurium to confine as much as possible of the chemistry to well-ventilated glove-boxes, as is the normal practice with radioactive elements such as polonium.

1.11. ANALYTICAL CHEMISTRY

Useful reviews of the analytical determination of selenium[57] and tellurium[58] are available; selenium and tellurium are usually determined gravimetrically as the elements, precipitated from aqueous solution by suitable reducing agents, or volumetrically by oxidation to the sexipositive state. In either case the elements should be in solution in the quadripositive state, although tellurium(VI) can be reduced to the element without prior treatment.

Selenium(VI) is reduced to selenium(IV) by heating with dilute (< 6 N) hydrochloric acid below 100°, the restrictions on acid concentration and temperature being necessary to avoid any loss of selenium as volatile chlorides. Some reduction to the element appears to occur in the dilute acid when the concentration of selenium is low and it is advisable to use solutions more concentrated than 0·2 mg/ml. Tellurium(VI) can be reduced in a similar manner. Elementary selenium and most selenides are oxidized to selenium(IV) by treatment with hydrogen peroxide or a mixture of nitric and hydrochloric acids, excess nitric acid being destroyed by repeated boiling with hydrochloric acid under reflux. Tellurium and the tellurides are best oxidized to tellurium(IV) by the latter procedure. Organoselenium compounds are decomposed, with the formation of selenium(IV), by heating with concentrated nitric acid or fusion with sodium peroxide, in both cases in a sealed tube. Organotellurium compounds are usually decomposed with a mixture of nitric and perchloric acids.

The best reducing agent for the gravimetric estimation of selenium is aqueous sulphur dioxide, preferably with the selenium(IV) in $> 3{\cdot}4$ N hydrochloric acid ($> 8{\cdot}8$ N HCl if tellurium is present as well); reduction with gaseous sulphur dioxide usually results in the loss of some of the selenium as volatile chlorides. Tellurium(IV) and tellurium(VI) are also reduced to the element by sulphur dioxide in dilute (~ 2 N) acid, but a mixture of sulphur dioxide and hydrazine gives a more rapid and complete precipitation. Selenium(IV) and

[57] W. T. Elwell and H. C. J. Saint, in *Comprehensive Analytical Chemistry*, Vol. 1c, p. 296, Classical Analysis, edited by C. L. Wilson and D. W. Wilson, Elsevier, Amsterdam (1962).

[58] S. J. Lyle, *ibid.*, p. 305.

selenium(VI) are both reduced to the element by hydrazine in neutral or acid solution and the nitrogen evolved in the reaction

$$H_2SeO_3 + N_2H_4 \longrightarrow Se + 3H_2O + N_2$$

can be measured instead of weighing the elementary selenium. Other reducing agents (e.g. tin(II) chloride, hypophosphorous acid) tend to give high results because of incorporation of the reducing agent or its oxidation product in the precipitate.

Selenium and tellurium precipitated from cold solution by these procedures are difficult to filter, but, on warming, a more easily filterable precipitate is obtained; it is also advisable to use only a slight excess of reducing agent in the precipitation of tellurium[59]. In both cases the precipitated elements are easily oxidized while they are being dried, but this can be avoided to a considerable extent by eliminating water from the precipitates by washing them with alcohol. Oxidation is also decreased by drying the precipitates in an inert atmosphere or in a vacuum.

Other compounds useful for gravimetry include mercury(II) selenite, precipitated when an excess of aqueous mercury(II) nitrate is added to solutions of selenium(IV) at pH 4–10, and tellurium dioxide, precipitated when bases such as ammonia or pyridine are added to aqueous solutions of tellurium(IV); elements which form insoluble hydroxides must be absent in the latter case. The solubility of barium selenate(VI) in water is about a hundred times that of the sulphate, and this compound is useless for the determination of selenium, but lead selenate and tellurate might be of some use for the determination of these elements, although neither appear to have been used[58].

The best oxidative method for the determination of selenium(IV) is by titration against permanganate in a mixture of dilute sulphuric acid and phosphoric acid, the latter preventing formation of manganese dioxide; the normal procedure consists of adding a known quantity of permanganate followed, after 30 min, by a known excess of standard iron(II) sulphate, the unused portion of which is back-titrated against standard permanganate. Dichromate [or cerium(IV) in the presence of chromium(III), which is equivalent to dichromate] is commonly used for the determination of tellurium(IV), again by a back-titration procedure analogous to that used for selenium(IV). This is a convenient reagent when a mixture of selenium and tellurium is to be analysed, for dichromate does not oxidize selenium(IV), so that the selenium(IV) present can be determined subsequently by a permanganate titration. Tellurium(IV) can, of course, be determined by a permanganate titration in the same way as selenium(IV), or by direct titration in the presence of ruthenium(III) catalyst, in the absence of chloride.

The reduction of selenite to the element by iodide

$$H_2SeO_3 + 4HI \longrightarrow Se + 3H_2O + 2I_2$$

can be used for the iodimetric determination of selenium; the method is precise if the titration is carried out in the presence of carbon tetrachloride or disulphide, or with sufficient starch present to ensure that the elementary selenium remains as a colloid, otherwise rather low results are obtained owing to adsorption of a part of the iodine on the precipitated selenium. Selenites react with thiosulphate in acid solution to form selenopentathionate, $SeS_4O_6^{2-}$,

$$H_2SeO_3 + 4S_2O_3^{2-} + 4H^+ \longrightarrow SeS_4O_6^{2-} + S_4O_6^{2-} + 3H_2O$$

---

[59] O. E. Clauder, *Z. analyt. Chem.* **89** (1932) 270.

so that selenium(IV) can be estimated by adding a known excess of standard thiosulphate and back-titrating against standard iodine. A similar procedure is effective for tellurium(IV). Since thiosulphate reacts more rapidly with selenite than with iodine, a combination of the two methods affords a direct titration procedure in which small amounts of potassium iodide and starch solution are added to the selenite, the blue starch iodide colour remaining visible to the endpoint[60].

Polarographic reduction of selenium and tellurium in ammoniacal solution is also of use for the determination of the two elements, either alone or in mixtures[61], and a polarimetric method for orthotelluric acid, H_6TeO_6, which involves the 1:1 complex with *d*-tartaric acid, has been described[62].

A variety of methods are available for the microscale determination of selenium and tellurium; spectrophotometric methods based on the reaction of selenite with ortho-diamines[63], the selenium(IV) complex with 2,2′-dianthrimide[64] and the selenium(IV) and tellurium(IV) thioglycollates[65] appear to be quite satisfactory, as is the determination of selenium[66] and tellurium[67] by atomic absorption spectrophotometry, while for ultramicro-determinations of selenium[68] and tellurium[69], neutron activation analysis is useful.

The analytical chemistry of polonium seems to be relatively unexplored; the element is usually determined by α-counting or calorimetry, where the heat evolution is due to the stoppage of the disintegration α-particles within the sample or its container. Either procedure requires a knowledge of the half-life of the isotope concerned. No other analytical methods have been reported, but this is hardly surprising in view of the very small (usually submilligram) amounts of polonium available for experiment and the highly radioactive nature of the usable isotopes.

2. COMPOUNDS OF THE ELEMENTS

2.1. INTRODUCTION

In each of the main groups 4–7 of the Periodic Table there is a gradual transition from non-metallic to metallic character with increasing atomic weight; thus, although there is a general similarity between the chemistry of selenium and tellurium and that of the non-metals oxygen and sulphur, there is a little evidence of metallic character appearing for selenium in the formation of one or two basic salts in which the element might be regarded as having some cationic properties. This trend becomes rather more obvious with tellurium, although all its simple salts are still basic, and with polonium the metallic character becomes really pronounced, both normal and basic salts being formed; the latter are, however, formally analogous to those formed by tellurium.

There is a similar trend in the thermal stabilities of the hydrides, H_2X, which decrease

[60] J. S. McNulty, *Analyt. Chem.* **19** (1947) 809.
[61] I. M. Kolthoff and J. J. Lingane, *Polarography*, p. 564, Interscience Publications, New York (1952).
[62] J. G. Lanese and B. Jaselskis, *Analyt. Chem.* **35** (1963) 1878.
[63] E.g. C. A. Parker and L. G. Harvey, *Analyst*, **87** (1962) 558.
[64] I. Dahl and F. J. Langmyhr, *Analyt. Chim. Acta*, **29** (1963) 377; *ibid.* **36** (1966) 24.
[65] G. F. Kirkbright and W. K. Ng, *Analyt. Chim. Acta*, **35** (1966) 116.
[66] C. S. Rann and A. N. Hambly, *Analyt. Chim. Acta*, **32** (1965) 346.
[67] C. L. Chakrabarti, *Analyt. Chim. Acta*, **39** (1967) 293.
[68] H. J. M. Bowen and P. A. Cawse, *Analyst*, **88** (1963) 721; F. J. Conrad and B. T. Kenna, *Analyt. Chem.* **39** (1967) 1001.
[69] H. P. Yule, *Analyt. Chem.* **37** (1965) 129.

with the increasing atomic number of X, the formation of these compounds becoming progressively more endothermic from selenium onwards (Table 6, p. 954). These changes in stability are due to a combination of the increasing radii of X, leading to an increased H–X bond length, and the decreasing electronegativity of X down the group, resulting in the H–Te bond being almost completely non-polar, while the H–Po bond becomes polar but with the polonium atom having some positive charge. These hydrides become progressively more acidic in the same direction, as shown by the increase in their acid dissociation constants in aqueous solution, aqueous hydrogen telluride being almost as acidic as phosphoric acid and more acidic than hydrofluoric acid. The increase in acidity of the Group VI hydrides with atomic number is similar to the trend observed for the halogen acids, HA. As in the halogen case, the trend can be ascribed to the decreasing bond energy H–X, presumably combined with lower hydration enthalpies and entropies of the HX^- ions. The decreasing enthalpy of hydration is, in turn, dependent on the increasing ionic radii of the elements (z^2/r dependence). Derivatives of the hydrides are well known and the numerous compounds formed with metals by selenium, tellurium and polonium are almost invariably isomorphous with their sulphide analogues, so that it is not surprising that selenium and tellurium are frequently found to be constituents of ores containing metallic sulphides.

As well as the formal oxidation state of -2 exhibited in the hydrides and their derivatives, formal oxidation states of $+2$, $+4$ and $+6$ are known for all three elements and $+1$ (X_2A_2) for selenium and tellurium, although the only tellurium halide of this kind is Te_2I_2. Nearly all of these compounds are covalent in character, as in the other groups of non-metallic elements. Thus the halides are generally low-melting, volatile and easily hydrolysed solids, readily forming anionic complex species of the type XA_6^{2-} in aqueous halogen acids. The order of stability of the complexes formed is $I^- > Br^- > Cl^- > F^-$, so demonstrating the Chatt–Ahrland B class character of selenium, tellurium and polonium. However, the simple selenium iodides do not exist, which is explicable on the basis of the appreciably different radii of selenium and iodine, and, more important, the somewhat similar electronegativities[70] of these two elements (Se, 2·48; I, 2·21) which implies almost non-polar bonds, but with some positive charge on the iodine. In view of this, it is somewhat surprising that anionic iodoselenates(IV) have been reported to exist. The atomic radii of tellurium and polonium are, however, similar to that of iodine, and their decreased electronegativities, although implying almost non-polar bonds, do lead to some positive charge on the tellurium or polonium atoms.

Like sulphur, selenium forms dimeric compounds of the type Se_2A_2 (A = Cl, Br, CN, SCN, SeCN), whereas the analogous compounds are not formed by tellurium, apart from the iodide, or polonium; this is partly because the Te–Te and Po–Po bonds will be weaker than the Se–Se bond because of the larger atomic radii of the former, but a more important factor is the decreased electronegativities of the two elements, which will result in a much greater weakening of the Te–Te or Po–Po bond in comparison with the Se–Se bond owing to electron withdrawal by the halide or pseudo-halide ion. Thus it is significant that the only tellurium compound of this type to be reported is Te_2I_2, where the atomic radii are comparable and electron withdrawal by iodide ion is far less effective than it is with the other halide ions.

Selenium, tellurium and polonium all form compounds in which the element is bivalent, with two lone pairs of electrons, the best known examples of which are the organic com-

[70] A. L. Allred and E. Rochow, *J. Inorg. Nuclear Chem.* 5 (1958) 264.

pounds. The halides of selenium and tellurium in this oxidation state are unstable with respect to disproportionation to a compound of the quadrivalent element and the element itself, except when the ion of the bivalent element is stabilized by sulphur donor ligands, but the pseudo-halides (cyanide, etc.) are stable. Polonium appears to form stable dihalides, although oxidation to the quadrivalent state is very easy. Since the outer electronic configuration of the elements is $ns^2\ np^4$, bivalency would only involve pairing of the p electrons, so that one could expect the $+2$ oxidation state to become more stable towards the end of the group where the promotion energies of the np electrons to the $(n+1)d$ shell would be expected to be higher than for earlier members of the group. Quadrivalency would then involve promotion of one p electron to the $(n+1)d$ shell, leaving the ns^2 electrons, and sexivalency the promotion of one s electron to the $(n+1)d$ shell, which would require even more energy. The reason for this increase lies in the poorer shielding of the s electrons from the nucleus by the underlying d shell in selenium, and the additional underlying f shell in polonium, the f shell being an even poorer screener than the d shell, which itself is less effective than the p shell as far as screening is concerned. Unfortunately, thermodynamic and ionization potential data, except for the first ionization, are not available for polonium, so that these arguments cannot be substantiated, but there is at least a qualitative correlation between the Group V and VI elements in the difficulty in oxidizing arsenic and bismuth to the highest, $+5$, state and the corresponding oxidation of selenium and polonium to the $+6$ state, the only reasonably certain example of which for polonium is the hexafluoride.

There is consequently a tendency towards a most stable oxidation state of $+4$, particularly in the oxides and oxo-acids, which is in marked contrast to sulphur, where sexivalency is preferred; in the halides, selenium tetrachloride is much more stable than its sulphur analogue, possibly because of the larger atomic radius of selenium as compared with sulphur. However, selenium tetrabromide is rather unstable with respect to decomposition to bromides of lower oxidation states and the tellurium compound is the most stable to heat of the tetrabromides formed by the three elements.

The bonding in the compounds of the chalcogens is quite straightforward; for compounds of the types XY_2, such as $(CH_3)_2Se$, the configuration is angular, with two σ bonds and two lone pairs of electrons. In the case of trigonal pyramidal species, such as $SeOCl_2$, there is one π and three σ bonds, the bond angles corresponding to those of the tetrahedron (sp^3) formed from the trigonal pyramid and the lone pair of electrons, while with the tetrahedral (sp^3) ions, such as SeO_4^{--}, four σ and two π bonds are involved. The tetrahalides (p. 957) utilize four σ bonds and a lone pair of electrons (sp^3d) and the square pyramidal anions of the type TeY_5^- derived from them involve five σ bonds and one lone pair (sp^3d^2). Octahedral species have been mentioned earlier and in these there are six σ bonds with or without a lone pair, depending on the oxidation state ($+4$ or $+6$).

2.2. ALLOYS

Selenium, tellurium and polonium combine with most metals, and many non-metals, at moderate temperatures (400–1000°C) in the absence of air, forming chalcogenides which are formally derived from the hydrides H_2X. The alkali metal salts (Se, Te) are more conveniently prepared by reaction of the elements in anhydrous liquid ammonia, while some selenides and tellurides can be prepared by reduction of the appropriate oxo-acid salts (e.g. selenites) by ignition with carbon. A number of metal selenides have been prepared by reaction of hydrogen selenide with the anhydrous metal chloride at red heat and a few tellurides have been made by reaction of hydrogen telluride with anhydrous metal chlorides in non-aqueous solvents.

Although aluminium, gold, iron, molybdenum, tantalum and tungsten form both selenides and tellurides, they do not appear to react with polonium. Similarly, polonium and bismuth are said to be miscible in all proportions[71], like selenium and tellurium, but the latter form well-defined compounds Bi_2X_3 (and BiSe) whereas there is no certain evidence for a definite bismuth polonide. Lanthanide compounds of the type M_2X_3 are also known, and a few mono- and ditellurides have been recorded. The lanthanide polonides are particularly stable to heat[72] and are of some use as heat sources for thermo-electric conversion (p. 938). Mercuric polonide is unusual in that it is very volatile[73]; since it is formed from the elements at 200°, and may even be formed when polonium deposits spontaneously

TABLE 5. THE CRYSTAL STRUCTURES OF SOME METAL CHALCOGENIDES

Compound	Symmetry and structure type	Selenides		Tellurides (Lattice parameters, Å)		Polonides	
		a_0	c_0	a_0	c_0	a_0	c_0
CdX	Hexagonal, ZnO	4·30	7·02	—	—	—	—
	Cubic, ZnS	—	—	6·464	—	6·665	—
HgX	Cubic, ZnS	6·07	—	6·434	—	—	—
	Cubic, NaCl	—	—	—	—	6·250	—
MgX	Cubic, NaCl	5·451	—	—	—	—	—
	Hexagonal, ZnO	—	—	4·53	7·38	—	—
	Hexagonal, NiAs	—	—	—	—	4·345	7·077

on to mercury from acid or ketone solutions, care has to be exercised in order to avoid losing polonium by compound formation and subsequent volatilization, particularly in vacuum systems.

Most selenides and tellurides are decomposed by water or dilute acid, with the formation of the hydrogen chalcogenide, but the yields are generally rather poor. There is little information on the behaviour of the analogous polonides, some of which may behave in the same way (p. 953), but aqueous solutions of tellurides are certainly unstable with respect to atmospheric oxidation, which results in the deposition of elementary tellurium, and the polonides should be even less stable in this respect.

Nearly all the known polonides[71, 73] are isomorphous with the analogous selenides and tellurides[74], but in some instances there are structural differences between them. The three commonest structures found for the chalcogenides are the NaCl, ZnO and ZnS types With cations of increasing polarizing power and increasingly easily polarized chalcogenide anions, the ionic NaCl structure gives way to the last two structure types, in both of which the ions are in tetrahedral coordination, the two structures differing in their packing arrangement. In addition to these structure types, the NiAs structure is also observed; here the bonding has become more metallic in character and the cation is now 8-coordinate. Some examples of the four structure types are given in Table 5, including the rather

[71] J. M. Goode, U.S.A.E.C. unclassified document MLM-677 (1952).

[72] G. R. Grove, L. V. Jones and J. F. Eichelberger, U.S.A.E.C. unclassified documents MLM-1139, p. 10 and MLM-1140, p. 16 (1962).

[73] W. G. Wittemann, A. L. Giorgi and D. T. Vier, *J. Phys. Chem.* **64** (1960) 434.

[74] R. W. G. Wyckoff, *Crystal Structures*, Interscience Publishers, New York, N.Y. (1960).

unexpected appearance of the NaCl structure for HgPo, an irregularity which requires further investigation. Many other chalcogenides are known, the lead and alkaline earth compounds, MX (X = Se, Te, Po), having the NaCl structure, whereas the beryllium compounds, BeX, have the ZnS structure and the sodium compounds, Na_2X, the CaF_2 structure. Others are of hexagonal symmetry, such as the platinum compounds, PtX_2 ($Cd(OH)_2$ structure), and the nickel compounds, where there is a range of composition varying continuously between NiX (NiAs structure) and NiX_2 ($Cd(OH)_2$ structure) in the cases of the tellurides and polonides. Nickel selenide, NiSe, is dimorphic, one form having the NiAs structure.

Alkali metal polyselenides, M_2Se_x ($x = 2$ to 4), have been made by dissolving selenium in aqueous alkali metal selenide or, better, by reacting the metal with an excess of selenium in anhydrous liquid ammonia. Analogous polytellurides have been recorded, and alkali metal hydrogen selenides, MHSe (Na, K, Rb, Cs), are also known. All of these compounds are, however, rather unstable and are readily oxidized in air.

2.3. HYDRIDES

Hydrogen selenide, H_2Se, and the telluride, H_2Te, are unpleasant smelling, colourless gases which condense to colourless liquids; as in liquid hydrogen sulphide, there is little association in these. The gases are fairly soluble in water, dissociating to the HX^- and X^{2-} ions, and yield acidic solutions which precipitate selenides or tellurides of many metals from solutions of their salts, but since both these hydrides are easily oxidized, particularly in aqueous solution by air, elementary selenium or tellurium is often precipitated as well. A hexahydrate of hydrogen selenide is known, analogous to the compound formed by hydrogen sulphide. Some physical properties of the hydrides are given in Table 6, the values for H_2Po being extrapolated and not experimental. The decreasing bond angle, M–X–H, in the hydrides is noteworthy and is probably due to the decreasing electronegativity of X, which would lead to polarization of the bonding orbitals, allowing them to come closer together.

Halogens and other oxidizing agents react rapidly with the hydrides, both in the gas phase and in aqueous solution. As in the case of hydrogen sulphide, the reaction of hydrogen selenide with sulphur dioxide in aqueous solution is not straightforward, a 2:1 mixture of sulphur and selenium being precipitated if the selenide is added to aqueous sulphur dioxide:

$$H_2Se + 6SO_2 + 2H_2O \longrightarrow 2S + Se + H_2S_2O_6 + 2H_2SO_4$$
$$H_2Se + 5SO_2 + 2H_2O \longrightarrow 2S + Se + 3H_2SO_4$$

the precipitate is mainly selenium if the sulphur dioxide is added to the hydrogen selenide, a result which might possibly be due to the formation of tetrathionic acid:

$$H_2Se + 6SO_2 + 2H_2O \longrightarrow Se + H_2S_4O_6 + 2H_2SO_4$$

Hydrogen telluride is also oxidized by sulphur dioxide and there is no evidence for the formation of selenium or tellurium analogues of the thionic acids in these reactions.

Hydrogen selenide and telluride can be prepared from the elements at 350° (H_2Se) or 650° (H_2Te) or by the reaction of dilute mineral acid on a suitable selenide or telluride, usually the aluminium compound. The selenide has also been made by heating elementary selenium with lithium or magnesium hydroxides[75] and the telluride by electrolysis of sulphuric acid (15–50%) with a tellurium cathode[18]. Hydrogen diselenide is thought to be formed when selenous acid is reduced with aluminium in hydrochloric acid[76], but little is known about this compound.

Evidence for the existence of hydrogen polonide is somewhat scanty; it seems only to have been prepared on the trace ($\sim 10^{-10}$ g) scale by reduction of polonium in acid solution

TABLE 6. SOME PHYSICAL PROPERTIES[a] OF H_2Se, H_2Te AND H_2Po

	H_2Se	H_2Te	H_2Po
Melting point, °C	−65·73°	−51°	−36°(?)
ΔH fusion, kcal/mole	0·6011	0·970	—
Boiling point/760 mm, °C	−41·3°	−4°	37°(?)
ΔH vap., kcal/mole	4·62	5·6	6·19(?)
Vapour pressure, $\log_{10} p_{mm} =$			
$A - B/T$ (liquid)	A 7·27	A 6·53	—
	B 1030	B 1005	—
$A - B/T$ (solid)	A 8·96	A 7·39	—
	B 1380	B 1220	—
Critical temperature, °C	137°	200°	—
Heat of formation, $\Delta H_{f,\,298}$, kcal/mole[b]	+8(?)	+23·8	—
Bond energy, H—X, kcal/mole[b]	73	64	—
Bond length, H—X, Å	1·46	1·69	—
H—X—H angle[c]	91°0′	89°30′	
Dissociation constant, HX^-, K_1	$1·30\times10^{-4}$	$2·27\times10^{-3}$	
Dissociation constant, X^{2-}, K_2	10^{-11}	$1·59\times10^{-11}$	—

[a] Data from ref. 3, p. 43.
[b] S. R. Gunn, *J. Phys. Chem.* **68** (1964) 949.
[c] HSH = 92·1°.

by magnesium, the volatile product being absorbed in aqueous alkali or silver nitrate solution and followed by its radioactivity. Attempts to prepare it with weighable amounts of the element have been unsuccessful[77].

2.4. CARBONYL COMPOUNDS

Carbonyl selenide and telluride, COSe and COTe, are formed in poor yield by passing carbon monoxide over the elements at 400° (Se[78]) or higher temperatures (Te[79]) and there is some evidence that polonium reacts under these conditions[9]. The selenide is more easily prepared[80] by reaction of aluminium selenide with carbonyl chloride at 219°, and is the best known and most stable of the three compounds; it is a colourless, foul-smelling liquid which boils at −22·9°/725 mm and freezes[78] to a white solid at −122·1°. The vapour pressure is given by

$$\log_{10} p_{mm} = 1149{\cdot}8/T + 7{\cdot}4527$$

and the latent heat of vaporization is 5·26 kcal/mole[80].

[75] J. Datta, *J. Indian Chem. Soc.* **29** (1952) 101, 965.
[76] J. P. Nielsen, S. Maeser and D. S. Jennings, *J. Am. Chem. Soc.* **61** (1939) 440.
[77] Reference 3, p. 46.
[78] T. G. Pearson and P. L. Robinson, *J. Chem. Soc.* (1932) 652.
[79] P. L. Robinson and K. R. Stainthorpe, *Nature*, **153** (1944) 24.
[80] O. Glemser and T. Risler, *Z. Naturforsch.* **3b** (1948) 1.

2.5. HALIDES

Introduction

The only known halides analogous to the sulphur compounds S_2A_2 are the selenium chloride and bromide and the tellurium iodide. The simple selenium dihalides have not been isolated in the solid state, but may be present in the vapour of the tetrahalides, whereas tellurium dichloride and dibromide can be isolated as solids, although they disproportionate rather readily to the metal and the tetrahalide; the polonium dihalides are more stable in this respect. The thermodynamic stability of the tetrahalides increases in passing from selenium to tellurium and in the highest oxidation state the only hexahalides known are the fluorides. These are, as one would expect, the most volatile of all the halides and are predominantly covalent in character, since the central atom could not possibly have as high a charge as +6.

Monohalides, X_2A_2

Fluorides of this composition are unknown, but selenium monochloride and monobromide are well established and are very similar in chemical properties to their sulphur analogues. Both compounds are oily, pungent smelling liquids which decompose at their boiling points (Table 7) to selenium and the tetrahalide

$$2Se_2A_2 = SeA_4+3Se$$

but some selenium dihalide may be present in the vapour of the decomposing monohalides.

The two monohalides are made by reaction of the stoichiometric quantities of the elements or, better, by adding the halogen to a suspension of selenium in carbon disulphide

TABLE 7. PROPERTIES OF THE MONOHALIDES, Se_2A_2

	Se_2Cl_2	Se_2Br_2
Colour	Brownish-red	Blood red
Melting point, °C	−85	—
Boiling point, °C	127/733 mm (d)	225–230 (d)
ΔH vap., kcal/mole	18[a]	—
Conductivity, mhos/cm	2×10^{-7}(18°)	—
Density, g/cc	2·7741 (25°)	3·604 (15°)
Heat of formation $\Delta H_{f,298}$, kcal/mole	−22·2[a]	—
Free energy of formation $\Delta F_{f,298}$, kcal/mole	−18·2[a]	—

[a] A. Glassner, U.S.A.E.C. Report ANL-5750 (1958). d = decomposition.

until dissolution of the element is complete and precipitation of the tetrahalide begins[81]. They can also be made by reducing the tetrahalide with selenium at 120° in a sealed tube and by reaction of selenium, some non-metal selenides or selenium dioxide with a variety of non-metal chlorides or bromides. The compounds are purified either by distillation under reduced pressure in the presence of elementary selenium or by precipitating them

[81] H. Stammreich and R. Forneris, *Spectrochim. Acta*, **8** (1956) 46.

from their solutions with concentrated sulphuric acid, in which the two monohalides are insoluble[82]. Tellurium monoiodide, Te_2I_2, has been identified by X-ray crystallography as a phase formed in the reaction of tellurium with iodine[83], and when a mixture of tellurium and iodine is heated with concentrated hydriodic acid[83a], but little is known about the compound.

The Raman and infrared spectra[84] of the two selenium monohalides indicate that both are of C_2 symmetry; the structure is A–Se–Se–A, with the two Se–A groups at right angles, and there is no evidence for rotation about the Se–Se bond, owing to repulsion between the lone pairs of electrons on the selenium atoms. The structure is therefore similar to that of the hydrogen peroxide molecule. Isomeric sulphur–selenium chlorides, $SSeCl_2$ and $SeSCl_2$, have been reported as formed respectively by reaction of selenium monochloride with sulphur, and sulphur monochloride with selenium[85], but the existence of such isomers is unlikely. However, selenium monochloride, like the sulphur analogue, is a good solvent for sulphur, selenium and iodine[86].

The two monohalides are slowly decomposed by water:

$$2Se_2A_2+2H_2O \longrightarrow SeO_2+3Se+4HA$$

and by ethanol, in which selenium and the tetrahalide are formed. The oxidation of the monochloride by ozone follows a similar pattern to the hydrolysis:

$$2Se_2Cl_2+6O_3 \longrightarrow SeCl_4+3SeO_2+6O_2$$

Both compounds are fairly good halogenating agents, reacting vigorously with quite a wide variety of metals and non-metals. Elementary selenium is formed with anhydrous liquid ammonia[87], but in ammoniacal carbon tetrachloride solution the nitride, Se_4N_4, is formed (p. 982) and in ethereal ammonia at $-80°$ compounds such as Se_2NCl are obtained[87] (p. 982).

Dihalides, XA_2

Selenium and tellurium difluorides are unknown, but the latter exists as the hydrofluoride in the form of stable complexes with sulphur donors such as thiourea (p. 1008). There is, however, some evidence for a polonium(II) fluoride, formed as a bluish-grey precipitate by reduction of polonium(IV) in aqueous fluoride with sulphur dioxide[9]. The compound has not been characterized.

Selenium dichloride is unknown in the solid state, but it is thought to be present in the vapour of the tetrachloride[88]. Solid tellurium dichloride, a black compound which melts to a dark liquid at 208° and boils at 328°, giving a bright red vapour, is obtained only by reaction of dichlorodifluoromethane with fused tellurium[89]. It disproportionates to tellurium(IV) and the element when dissolved in water or organic solvents and it is oxidized to the tetrachloride by chlorine or selenium monochloride and to the dioxide by liquid

82 V. Lenher and C. H. Kao, *J. Am. Chem. Soc.* **47** (1925) 772.
83 W. R. Blackmore, S. C. Abrahams and J. Kalnajs, *Acta Cryst.* **9** (1956) 295.
83a A. Rabenau, H. Rau and P. Eckerlin, *Angew. Chem. Int. Ed. Eng.* **6** (1967) 706.
84 P. J. Hendra and P. J. D. Park, *J. Chem. Soc.* (*A*) (1968) 908.
85 A. Baroni, *Atti Accad. naz. Lincei, Rend. Classe Sci. fis. mat. nat.* **16** (1932) 514; *ibid.* **25** (1937) 719.
86 V. Lenher and C. H. Kao, *J. Am. Chem. Soc.* **48** (1926) 1550.
87 W. Strecker and L. Claus, *Ber.* **56** (1923) 362.
88 D. M. Yost and C. E. Kircher, *J. Am. Chem. Soc.* **52** (1930) 4680.
89 E. E. Aynsley, *J. Chem. Soc.* (1953) 3016.

dinirtogen tetroxide, there being no reaction with the gas, or with gaseous or liquid sulphur dioxide. Tellurium dichloride absorbs dry ammonia gas, possibly forming an amine, but anhydrous liquid ammonia reacts to yield elementary tellurium. A number of complexes are known (p. 1008).

Polonium dichloride[44], a dark ruby-red, hygroscopic solid which sublimes, with decomposition, at 190° in nitrogen, is obtained by reducing the solid tetrachloride with sulphur dioxide (25°), hydrogen sulphide or carbon monoxide (150°) or with hydrogen (200°), but prolonged heating in the last two cases leads to reduction to the element. The compound is also obtained by continued heating of the tetrachloride in a vacuum. Unlike the tellurium compound, it does not disproportionate in water or other solvents, but it is readily oxidized to polonium(IV). Polonium(III) may have a transient existence as an intermediate in this oxidation in hydrochloric acid[90].

Selenium dibromide, like the dichloride, has never been isolated, but it appears to be present in the vapour of the tetrabromide and monobromide, and may be present in equilibrium with the monobromide and tetrabromide in solutions of the latter in carbon tetrachloride[91] or nitrobenzene[92]. Solid tellurium dibromide is known, but it disproportionates readily on heating, so that the melting and boiling point cannot be determined. It is a chocolate-brown solid, prepared by vacuum sublimation on to a cold finger of the solid solution of tellurium in tellurium tetrabromide obtained by reaction of the latter with finely divided tellurium in dry ether in the dark or by reaction of bromotrifluoromethane with fused tellurium[93]. Tellurium dibromide disproportionates rapidly in water, liquid sulphur dioxide or hydrogen cyanide, and more slowly in solution in dry ether or chloroform at room temperature, although disproportionation in the last two solvents is rapid on heating. The compound reacts with gaseous or liquid ammonia in the same way as the dichloride; the known complexes are discussed later (p. 1008).

Polonium dibromide[94] is prepared in a similar way to the dichloride, by reduction of the tetrabromide either with hydrogen sulphide at 25° or by heating in a vacuum at 200°, but reduction with sulphur dioxide is incomplete. It is a purple-brown solid which sublimes, with some decomposition, at 110°/30 μ and disproportionates at the melting point, 270–280° in nitrogen; it is reduced to the metal when heated in dry ammonia and is soluble in water and in a number of ketones, the solutions being purple in colour.

The solid diiodides have never been isolated, but selenium diiodide may be present in solutions of selenium and iodine in carbon disulphide[95]; a few complexes of tellurium diiodide are also known (p. 1008).

Tetrafluorides, XF_4

Selenium tetrafluoride is a white, hygroscopic solid, melting at −9·5° to a colourless liquid which fumes in air; some physical properties are given in Table 8. It is conveniently made by heating selenium tetrachloride, or, better, selenium with silver fluoride[96], but it is more commonly prepared from the elements[97] at 0°, by the action of fluorine on selenium

[90] K. W. Bagnall and J. H. Freeman, *J. Chem. Soc.* (1956) 2770.
[91] N. W. Tideswell and J. D. McCullough, *J. Am. Chem. Soc.* **78** (1956) 3026.
[92] N. Katsaros and J. W. George, *Chem. Comm.* (1968) 662.
[93] E. E. Aynsley and R. H. Watson, *J. Chem. Soc.* (1955) 2603.
[94] K. W. Bagnall, R. W. M. D'Eye and J. H. Freeman, *J. Chem. Soc.* (1955) 3959.
[95] A. F. Kapustinskii and Yu. M. Golutvin, *J. Gen. Chem. (U.S.S.R.)*, **17** (1947) 2010.
[96] O. Glemser, F. Meyer and A. Haas, *Naturwiss.* **52** (1965) 130.
[97] E. E. Aynsley, R. D. Peacock and P. L. Robinson, *J. Chem. Soc.* (1952) 1231.

monochloride[98] or by reaction of sulphur tetrafluoride with selenium dioxide[99] at 100–240°. The compound is then purified by fractional distillation.

The structure[100, 101] of the molecule, obtained by electron diffraction studies is a distorted tetrahedron which results from the replacement of one equatorial bond in a trigonal bipyramid by the lone pair of electrons on the selenium atom, the resulting symmetry[101] being C_{2v}, consistent with the Raman spectrum[102]. Conductivity measurements indicate that there is slight dissociation in the liquid, with the cation being SeF_3^+ and the anion either F^- or SeF_5^-. The known adducts of selenium tetrafluoride (p. 1005) probably involve these ionic species.

Selenium tetrafluoride can be handled in Pyrex when perfectly dry, since it attacks the glass only very slowly. It is a useful fluorinating agent and is completely miscible with diethyl ether, ethanol, iodine pentafluoride and sulphuric acid, and is appreciably soluble in carbon tetrachloride and chloroform. Water hydrolyses it violently and the compound dissolves bromine, iodine, selenium, sulphur and, on heating to 80°, tellurium dioxide but not the trioxide. The tetrafluoride is reduced to selenium by arsine, hydrogen selenide and sulphide, and by potassium iodide, but it reacts with potassium chloride and bromide to form the appropriate selenium tetrahalide.

Tellurium tetrafluoride is also a white, hygroscopic solid, the vapour of which decomposes above 193·8° with the formation of the hexafluoride. It is most simply prepared by reaction of selenium tetrafluoride with tellurium dioxide at 80°, the yield being quantitative[103], and is also formed by the action of nitryl fluoride on tellurium, from the elements at 0° and by the reduction of the hexafluoride with tellurium[104] at 180°.

The crystal symmetry is orthorhombic and the structure in the crystal consists of distorted octahedra linked by *cis* fluorine bridges into endless chains, one apex of each octahedron being occupied by the lone pair of electrons of the selenium atom[105].

Tellurium tetrafluoride is readily hydrolysed and it reacts with glass or silica at 200°. Its chemical behaviour is similar to that of the selenium compound, but it is less useful than the latter as a fluorinating agent.

The involatile product resulting from the radiation decomposition of polonium hexafluoride (p. 964) is probably the tetrafluoride, but the existence of this compound has not been definitely established. A white hydrate (or basic salt) seems to be formed when polonium(IV) hydroxide or the tetrachloride is treated with aqueous hydrofluoric acid[9], but these products have not been investigated.

Tetrachlorides, XCl_4

Selenium, tellurium and polonium tetrachlorides are respectively pale yellow, white and bright yellow, volatile, hygroscopic, readily hydrolysed solids; some physical data are given in Table 8. The selenium compound is almost completely dissociated to lower chlorides and chlorine in the vapour, but the vapour density of the tellurium compound is

98 P. L. Goggin, *J. Inorg. Nuclear Chem.* **28** (1966) 661.
99 A. L. Oppegard, W. C. Smith, E. L. Muetterties and V. A. Engelhardt, *J. Am. Chem. Soc.* **82** (1960) 3835.
100 H. J. M. Bowen, *Nature,* **172** (1953) 171.
101 V. G. Ewing and L. E. Sutton, *Trans. Faraday Soc.* **59** (1963) 1241.
102 J. A. Rolfe, L. A. Woodward and D. A. Long, *Trans. Faraday Soc.* **49** (1953) 1388.
103 R. Campbell and P. L. Robinson, *J. Chem. Soc.* (1956) 785.
104 J. H. Junkins, H. A. Bernhardt and E. J. Barber, *J. Am. Chem. Soc.* **74** (1952) 5749.
105 A. J. Edwards and F. I. Hewaidi, *J. Chem. Soc.* (*A*) (1968) 2977.

normal below 500°, indicating that there is little dissociation in the maroon coloured vapour below this temperature. In contrast, the molten polonium compound, which is straw coloured, becomes scarlet at 350°, probably because of decomposition to the dichloride, and the purple-brown vapour becomes blue-green above 500°, possibly as a result of a progressive decomposition.

The three compounds are commonly prepared from the elements at moderate temperatures or by reaction of the element or its dioxide with chlorinating agents such as carbon tetrachloride, phosphorus pentachloride or thionyl chloride; they are purified by sublimation, usually in an atmosphere of chlorine in order to suppress decomposition. A useful way of preparing the selenium compound consists in reacting chlorine with a solution of the monochloride in ethyl bromide or carbon disulphide, the insoluble tetrachloride precipitating as the reaction proceeds. Residual selenium monochloride is easily removed from the product by washing it with carbon disulphide[81].

TABLE 8. PHYSICAL PROPERTIES OF THE TETRAFLUORIDES AND TETRACHLORIDES[a]

	SeF_4	TeF_4	$SeCl_4$	$TeCl_4$	$PoCl_4$
Melting point, °C	−9·5	129·6	305	224	300
ΔH fusion, kcal/mole[b]	—	6·351	21 (?)[b]	4·51	5·2 (?)
Boiling point, °C	106	193·8d	196 sub.	390	390
ΔH vap., kcal/mole[b]	11·24	8·174	21 (?)*	16·83	19 (?)
Vapour pressure (liquid) $\log_{10} p_{mm} = A - B/T$	A = 9·44 B = 2457	A = 5·6397 B = 1786·4			
Vapour pressure (solid) $\log_{10} p_{mm} = A - B/T$	—	A = 9·0934 B = 3174·3	A = 11·2040 B = 3864		
Heat of formation, $-\Delta H_{f,298}$, kcal/mole[b]	—	205	46·1	77·4	80 (?)
Free energy of formation, $-\Delta F_{f,298}$, kcal/mole[b]	—	185	25·9 (?)	57·6	60 (?)
Density, g/cc	2·75 (18°)				

* ΔH sub.

[a] Data from K. W. Bagnall, *The Chemistry of Selenium, Tellurium and Polonium*, Elsevier, Amsterdam, 1966, unless otherwise stated.

[b] A. Glassner, U.S.A.E.C. Report ANL–5750 (1958).

d = decomp.; sub. = sublimes.

Electron diffraction of the vapour of selenium[106] and tellurium[107] tetrachloride indicates that both have a distorted trigonal bipyramidal structure with one of the equatorial positions occupied by the unshared pair of electrons, a structure consistent with the high dipole moment (2·54 Debye) of the tellurium compound[108]. However, in the solid[109–211] and liquid[109] states the infrared and Raman spectra of tellurium tetrachloride indicate that the ions $TeCl_3^+$ and Cl^- are present, in agreement with the observed electrical conductivity of the molten tetrachloride[113]. Infrared and Raman data for solid selenium tetrachloride[109, 110, 112] are interpreted on the same basis, but no information is available for the polonium compound. X-ray crystallographic data are given in Table 9.

106 R. E. Dodd, L. A. Woodward and H. C. Roberts, *Trans. Faraday Soc.* **52** (1956) 1052.
107 D. P. Stevenson and V. Schomaker, *J. Am. Chem. Soc.* **62** (1940) 1267.
108 C. P. Smyth, A. J. Grossman and S. R. Ginsburg, *J. Am. Chem. Soc.* **62** (1940) 192.
109 H. Gerding and H. Houtgraaf, *Rec. Trav. chim.* **73** (1954) 737.
110 N. N. Greenwood, B. P. Straughan and A. E. Wilson, *J. Chem. Soc.* (*A*) (1966) 1479.
111 G C. Hayward and P. J. Hendra, *J. Chem. Soc.* (*A*) (1967) 643.
112 J. W. George, N. Katsaros and K. J. Wynne, *Inorg. Chem.* **6** (1967) 903.
113 A. Voigt and W. Biltz, *Z. anorg. Chem.* **133** (1924) 298.

Selenium and tellurium tetrachlorides are fairly soluble in non-polar organic solvents and the latter, as well as tellurium tetrabromide and tetraiodide, appear to be trimeric in benzene or toluene; in solvents such as acetone and methyl cyanide these compounds behave as 1:1 electrolytes, possibly in the form $(L_2TeX_3)^+X^-$ where L is the solvent[113a]. The selenium compound is also soluble in hot phosphorus(V) oxochloride; there are few data for the polonium compound, which is apparently soluble in ethanol and in thionyl chloride, and slightly soluble in liquid sulphur dioxide. Selenium tetrachloride is a powerful chlorinating agent, converting, for example, tellurium dioxide to the tetrachloride and selenium dioxide to the oxochloride; it is reduced to the monochloride by sulphur or selenium.

All three tetrachlorides react with anhydrous liquid ammonia, the selenium and tellurium compounds forming the nitride (p. 982) and the polonium compound being

TABLE 9. X-RAY CRYSTALLOGRAPHIC DATA FOR THE TETRAHALIDES

Compound	Symmetry and space group	Lattice parameters, Å			Calc. density g/cc	Ref.
		a_0	b_0	c_0		
$SeCl_4$	Monoclinic, $C2/c$ or Cc	16·46	9·73 $\beta = 117°$	14·93	2·63*	a
$TeCl_4$	Monoclinic, $C2/c$ or Cc	16·91	10·36 $\beta = 117°$	15·25	2·95*	a
$TeBr_4$	Monoclinic, $C2/c$ or Cc	17·75	10·89 $\beta = 116°34'$	15·88	4·33*	a
TeI_4	Orthorhombic, $Pnma\text{-}D_{2h}^{16}$ or $Pn2_1a\text{-}C_{2v}^{9}$	15·54	16·73	14·48	5·145*	b
	or Tetragonal, $I4_1/amd\text{-}D_{4h}^{19}$	16·12	—	11·20	5·7*	b

* 16 molecules/unit cell.

[a] C. B. Shoemaker and S. C. Abrahams, *Acta Cryst.* **18** (1965) 296.

[b] W. R. Blackmore, S. C. Abrahams and J. Kalnajs, *Acta Cryst.* **9** (1956) 295.

reduced to the element (p. 946). The tetrachlorides are also reduced to the element by gaseous ammonia, the tellurium compound at 200–250°, the selenium and polonium compounds at room temperature, the last only slowly. Hydrogen sulphide reduces solid selenium, tellurium and polonium tetrachlorides, the first two to the element at room temperature, but tellurium tetrachloride may form the dichloride at low temperature, and the polonium compound forms the dichloride at 150°; the tetrachlorides react with ketones and β-diketones to form organo-compounds (p. 1002). The complexes of the tetrachlorides are discussed later (p. 1007).

Tetrabromides, XBr_4

The orange-red selenium, yellow tellurium and bright red polonium compounds are hygroscopic, readily hydrolysed solids which are all rather unstable with respect to decomposition with loss of bromine. Thus selenium tetrabromide decomposes appreciably at room temperature and completely at 70°, giving a mixture of the elements and lower selenium bromides. The tellurium compound decomposes above 280° and it melts, in bromine vapour, at 363°, the boiling point lying between 414° and 427°. However, the vapour density indicates almost complete dissociation to bromine and the dibromide.

[113a] N. N. Greenwood, B. P. Straughan and A. E. Wilson, *J. Chem. Soc.* (*A*) (1968) 2209.

The polonium compound seems to be somewhat more stable in this respect; it melts in bromine vapour at 330° and boils at 360°/200 mm.

The three tetrabromides are usually prepared from the elements, using an excess of the halogen, the reaction with polonium requiring heating to 200–250° since metallic polonium is inert to gaseous or liquid bromine at room temperature. The selenium compound is also prepared by the reaction of bromine with the monobromide, either alone or in solution in carbon disulphide, chloroform or ethyl bromide and the tellurium compound can be made by reaction of tellurium with iodine monobromide or by dissolving the dioxide in hydrobromic acid and precipitating the tetrabromide from the solution with moderately concentrated (77%) sulphuric acid. The polonium compound is most easily prepared by heating the dioxide in hydrogen bromide or by evaporating to dryness a solution of the dioxide in hydrobromic acid. Tellurium tetrabromide can be purified by sublimation in bromine vapour or by recrystallization from glacial acetic acid.

The infrared and Raman spectra of the solid selenium[112] and tellurium[110–112] tetrabromides indicate that they are ionic like the tetrachlorides; some crystallographic data are given in Table 9 and the complexes are discussed later (p. 1007).

Selenium tetrabromide is quite soluble in carbon disulphide, carbon tetrachloride, chloroform and ethyl bromide, whereas the tellurium compound is soluble in chloroform and ether, but not in carbon tetrachloride, and polonium tetrabromide is only known to be soluble in ethanol and slightly soluble in liquid bromine. Both the selenium and tellurium compounds react with ammonia to form the nitrides (p. 982), the former in solution in carbon disulphide with the gas and the latter with the anhydrous liquid. Hydrogen sulphide reduces tellurium tetrabromide to the element in solution in chloroform whereas the solid polonium compound is reduced to the dibromide at room temperature.

Tetraiodides, XI_4

Selenium tetraiodide is unknown but anionic iodo-complexes have been isolated; these are discussed later (p. 1006) with the other halo-complexes. The grey-black tellurium and black polonium tetraiodides are appreciably volatile, but decompose on heating, the tellurium compound above 100° and the polonium one above 200°, at which temperature it sublimes in nitrogen, leaving a residue of the metal, possibly because of decomposition to an unidentified lower iodide which subsequently disproportionates. The tellurium compound melts at 280° in a sealed tube, presumably with a considerable degree of decomposition, and two crystal forms are known (Table 9), the orthorhombic one being the more usual. The tetraiodides of the two elements are only slowly hydrolysed by cold water or aqueous alkali, in contrast to the behaviour of the other tetrahalides.

Both tetraiodides are commonly prepared from the elements or by precipitating the compounds from aqueous acid solutions of the quadripositive element, avoiding an excess of hydriodic acid since the compounds redissolve in it to form anionic iodo-complex ions. The tellurium compound is also obtained by heating the element with ethyl iodide or cyanogen iodide and the polonium compound by heating the dioxide in hydrogen iodide at 200°; a black adduct, $PoO_2.xHI$, is formed in the cold[114]. The far infrared spectrum[110] of tellurium tetraiodide shows that it is ionic, like the tetrachloride.

The two tetraiodides are slightly soluble in acetone and ethanol, but insoluble in dilute mineral acids, aliphatic acids and a variety of non-polar solvents. Tellurium tetraiodide

[114] K. W. Bagnall, R. W. M. D'Eye and J. H. Freeman, *J. Chem. Soc.* (1956) 3385.

is also slightly soluble in amyl acetate; although the tellurium compound is unaffected by hydrogen at 100° or hydrogen sulphide at 200°, the latter reduces the polonium one to the element at moderate temperature. However, whereas tellurium tetraiodide reacts with dry ammonia at −80°, apparently forming the nitride, the polonium compound does not react with this reagent nor is it reduced in suspension in dilute (0·1 N) hydriodic acid by hydrazine or sulphur dioxide, even at 100°.

Mixed Halides, $XA_nA'_{4-n}$

Dichlorodibromides, XCl_2Br_2, are known for all three elements; the brown-yellow selenium compound appears to be formed by the action of chlorine on an equimolar mixture of selenium monobromide and tetrabromide[115], whereas the yellow tellurium[89] and salmon-pink polonium[94] compounds are obtained by reaction of the dichloride with bromine. The tellurium compound melts to a ruby-red liquid at 292° and boils at 415°. The dichlorodiiodides, XCl_2I_2, are less well established; tellurium dichloride does not react with iodine, but polonium dichloride, when shaken with a solution of iodine in carbon tetrachloride, appears to form a black, unstable chloroiodide[114]. However, the dibromodiiodides, XBr_2I_2, are more stable with respect to disproportionation; the garnet-red tellurium compound is obtained by evaporating to dryness an ethereal solution of tellurium dibromide and iodine. It melts to a dark red liquid at 323–325° and boils, with decomposition, at 420°, the vapour being purple. The black polonium compound[114] appears to be formed when the dibromide is shaken with iodine in carbon tetrachloride, but the dibromide does not react with iodine vapour.

Two other mixed halides are known, yellow-brown selenium trichlorobromide, $SeCl_3Br$, and the yellow-orange chlorotribromide, $SeClBr_3$. The first is precipitated when chlorine is passed into a solution of selenium monobromide in carbon disulphide and the second when the bromine is added to the monochloride in carbon disulphide[115]. The trichlorobromide decomposes appreciably above 208°, but the vapour density is normal, and the chlorotribromide decomposes at about 200°. Raman and infrared spectra[111] of these two compounds indicate that both have an ionic structure as in tellurium tetrachloride. All of the mixed halides are hygroscopic and are readily hydrolysed.

Hexafluorides, XF_6

The hexafluorides are the only hexahalides known; they are gases at room temperature, the selenium and tellurium compounds being colourless in the vapour state, with a rather unpleasant smell reminiscent of the hydrides. Both of the latter condense to white, volatile solids and some physical data are given in Table 10; neither compound attacks glass. The low boiling points of these compounds are noteworthy and are a reflection of the reduced intermolecular attraction resulting from the sheath of non-polarizable fluorine atoms surrounding each chalcogen atom.

Selenium[116], tellurium[117] and polonium[118] hexafluorides are most conveniently prepared from the elements, but other fluorinating agents, such as bromine trifluoride, can be used to fluorinate the dioxide, for example the selenium compound[119]. The compounds are

[115] F. P. Evans and W. Ramsay, *J. Chem. Soc.* **45** (1884) 62.
[116] D. M. Yost and W. H. Claussen, *J. Am. Chem. Soc.* **55** (1933) 885.
[117] R. Campbell and P. L. Robinson, *J. Chem. Soc.* (1956) 3454.
[118] B. Weinstock and C. L. Chernick, *J. Am. Chem. Soc.* **82** (1960) 4116.
[119] H. J. Emeléus and A. A. Woolf, *J. Chem. Soc.* (1950) 164.

usually purified by repeated sublimation at low temperature. The infrared spectra of selenium[120] and tellurium[121] hexafluorides indicate that both are regular octahedra.

Selenium hexafluoride, like the sulphur compound, is inert to water in which it has a slight but measurable solubility. The tellurium compound, however, is completely hydrolysed in 24 hr at room temperature. This is probably because sulphur and selenium are unable to accept a pair of electrons from water to provide a route for the first stage of the hydrolysis, but tellurium may be able to do so by using either the unoccupied 4*f* orbitals[122] or the unoccupied *d* orbitals and it is significant that tellurium hexafluoride forms fluoro-complexes and adducts with tertiary amines (p. 1007), whereas the selenium and sulphur

TABLE 10. PHYSICAL PROPERTIES[a] OF THE HEXAFLUORIDES AND Te_2F_{10}

	SeF_6	TeF_6	Te_2F_{10}
Melting point, °C	−34·6 (1500 mm)	−37·8	−33·7
Boiling point, °C	−34·8 (945 mm)*	−38·9 (sub.)	59
ΔH vap., kcal/mole	4·38 (at m.p.)	4·5	9·44
Vapour pressure (solid)	A = 9·242	A = 9·161	†A = 9·20
$\log_{10} p_{mm} = A - B/T$	B = 1440·8	B = 1471·4	†B = 2063
Critical temperature, °C	72·4	83·25	170
Heat of formation, $\Delta H_{f,\ 298}$, kcal/mole[b]	−246·0	−315	—
Free energy of formation, $\Delta F_{f,\ 298}$, kcal/mole[b]	−221·8	−292	—
Density, g/cc	3·27 (solid at m.p.)	3·76 (solid at m.p.)	2·9372 (liquid at 0°)
M–F bond length, Å	1·67–1·70	1·84	—

* Sublimes, without fusion, at −46·6°/760 mm.
† V.p. equation for the liquid.
[a] Data from ref. 3, page 91, unless otherwise stated.
[b] A. Glassner, U.S.A.E.C. Report ANL–5750 (1958).

compounds do not. In addition the Te–F bond in the hexafluoride is appreciably longer than the Se–F bond in the selenium analogue (Table 10), so that the bond is more polar in the tellurium compound than in the selenium one, as is also shown in the increased dielectric constant of the former, a factor which also contributes to the increased ease of hydrolysis of the tellurium compound.

Although unaffected by water, aqueous alkali chloride or bromide, selenium hexafluoride is slowly decomposed by aqueous potassium iodide or sodium thiosulphate, a mixture of which can be used to decompose the compound for analysis[123].

The selenium compound reacts slowly with solid potassium iodide at room temperature, with reduction to selenium, and with alkali chloride or bromide forms the tetrachloride or monobromide, respectively at 500° and 300°. It is unaffected by air or sulphur dioxide, but reacts with sodium sulphite or sulphide in the presence of water, forming the selenite; it is also readily decomposed by potassium at 60° and by sodium or lithium above 500°, and reacts with antimony, arsenic and silicon at moderate temperatures. It is reduced to the element by arsine and, more slowly, by ammonia at 200°. Although analogous

120 J. Gaunt, *Trans. Faraday Soc.* **49** (1953) 1122; *ibid.* **50** (1954) 546.
121 J. Gaunt, *Trans. Faraday Soc.* **51** (1955) 893.
122 G. E. Kimball, *J. Chem. Phys.* **8** (1940) 188.
123 C. Dagron, *Compt. rend.* **241** (1955) 418; *ibid.* **242** (1956) 1027.

information does not seem to be available for the tellurium compound, it should be more reactive than the selenium one under similar conditions, and tellurium hexafluoride reacts readily with trimethylsilyl dimethylamide, $(CH_3)_2\,NSi(CH_3)_3$, at −78°, yielding a mixture of pentafluorotellurium dimethylamide, $F_5TeN(CH_3)_2$, and tetrafluorotellurium *bis*dimethylamide, $F_4Te[N(CH_3)_2]_2$; the n.m.r. spectrum of the latter indicates that it is the *cis* isomer[123a]. Both selenium and tellurium hexafluorides are reduced to the tetrafluoride on heating with the element.

Polonium hexafluoride[118] has not been identified with absolute certainty, but the volatile product obtained by fluorination of polonium metal, using the longer-lived isotope ^{208}Po, is probably the hexafluoride. It seems to be stable in the vapour phase, but decomposes to an involatile compound, probably the tetrafluoride, in the solid state, apparently because of α-radiation decomposition.

The only mixed halides derived from the hexafluorides are pentafluorotellurium chloride and bromide, F_5TeY, prepared by reaction of fluorine, diluted with nitrogen, on tellurium tetrachloride or tetrabromide at 25°; both are low boiling liquids[123b].

Ditellurium Decafluoride, Te_2F_{10}

This is a stable, volatile, colourless liquid, prepared[117] by reaction of a mixture of fluorine and oxygen with a mixture of tellurium and its dioxide at 50°–60° or, in poorer yield, from reaction of a mixture of fluorine and nitrogen with tellurium. Some physical properties are given in Table 10.

Its infrared[124] spectrum resembles that of the hexafluoride, and the compound is of the form $F_5Te{-}TeF_5$, with the two octahedral TeF_5 groups staggered (D_{4d} symmetry). It slowly attacks glass, but is scarcely affected by water, acids or dilute alkali and does not react with sulphur or selenium at its boiling point. It is decomposed violently by potassium, less so by sodium and reacts with acetone and organic materials such as tapgrease. The selenium analogue has not been recorded, but the corresponding sulphur compound is known.

Oxohalides

Selenium(IV) oxodifluoride, $SeOF_2$, and oxodichloride, $SeOCl_2$, are colourless, fuming, volatile liquids, and the oxodibromide, $SeOBr_2$, which is appreciably less stable than the sulphur analogue, is an orange solid which decomposes in air at 50°; all three are readily hydrolysed. Some physical properties are given in Table 11.

The three compounds are most conveniently prepared by reaction of selenium dioxide with the appropriate tetrahalide. The oxofluoride is also made by reaction of a mixture of fluorine and nitrogen on selenium dioxide or of fluorine and oxygen on selenium[97] at 200°; the oxodibromide is easily obtained by distilling the oxodichloride from sodium bromide. Other methods of preparing these compounds have been reviewed[3].

The molecular structure of the selenium(IV) oxodihalides is probably pyramidal, like the sulphur analogues, with the selenium atom at the apex, a resemblance shown, for example, by the similarity of the Raman spectra of the oxodifluorides[125].

Selenium(IV) oxodifluoride attacks glass and dissolves selenium and sulphur, the latter

[123a] G. W. Fraser, R. D. Peacock and P. M. Watkins, *Chem. Comm.* (1967) 1248.
[123b] G. W. Fraser, R. D. Peacock and P. M. Watkins, *Chem. Comm.* (1968) 1257.
[124] R. E. Dodd, L. A. Woodward and H. L. Roberts, *Trans. Faraday Soc.* **53** (1957) 1545.
[125] J. A. Rolfe and L. A. Woodward, *Trans. Faraday Soc.* **51** (1955) 778.

reducing the compound to selenium on warming, and it reacts violently with red phosphorus and with powdered silica, but only slowly with silicon. The oxodichloride reacts explosively with potassium and more slowly with a wide variety of metals and non-metals, but it does not react with boron, carbon or silicon and, surprisingly, it does not react with sodium[126] even at 176°. It also dissolves bromine and iodine, and is decomposed by gaseous ammonia on heating, yielding an equimolar mixture of selenium and selenium dioxide. Sulphur reacts with it to form a mixture of sulphur and selenium monochlorides and sulphur dioxide.

Selenium(IV) oxodichloride is a useful solvent for the preparation of metal chloro-complexes, for it has a high dielectric constant, 46·2 at 20°, compared with 9·1 for thionyl

TABLE 11. SOME PHYSICAL PROPERTIES OF THE OXOHALIDES[a]

	$SeOF_2$	$SeOCl_2$	$SeOBr_2$	SeO_2F_2
Melting point, °C	15	10·9	41·6	−99·5
Boiling point, °C	125–6	177·2	217/740 mm (decomp.)	−8·4
ΔH vap., kcal/mole				6·772
Vapour pressure (liquid), $\log_{10} p_{mm} = A - B/T$	$A = 8{\cdot}70$ $B = 2316$	$A = 5{\cdot}8503$*		$A = 8{\cdot}474$ $B = 1480{\cdot}6$
Density, g/cc	2·80 (21·5°)	2·445 (16°)	3·38 (50°)	
Specific conductance, ω^{-1} cm^{-1} at 25°		2×10^{-5}		

* $\log_{10} p_{mm} = 5{\cdot}8503 - 0{\cdot}000219T - 830{\cdot}9/(T-178)$.

[a] Data from ref. 3, pp. 91, 104 and 114.

chloride, and it also has a high dipole moment, 2·62 Debye in benzene. The liquid is appreciably ionized (Table 11):

$$2SeOCl_2 \rightleftharpoons SeOCl^+ + SeOCl_3^-$$

Solvates of a number of metal chlorides have been reported.

Selenium(IV) oxodibromide is less well known, but seems to have rather similar solvent properties and, like the oxodichloride, it is a useful halogenating agent. Unlike the chloride, it reacts explosively with sodium and even more violently with potassium. It is soluble in benzene, xylene and carbon disulphide and dissolves iodine.

Selenium(IV) oxochlorobromide[127], SeOClBr, a dense dark red liquid which boils at 115°/38 mm, is obtained by chlorination of a mixture of selenium monobromide and dioxide at −60°; from n.m.r. studies it has been shown that SeOClBr and SeOClF are present in mixtures of the appropriate oxodihalides[128].

The only tellurium(IV) oxohalides known are the oxodibromide and a rather curious chloride, $Te_6O_{11}Cl_2$, which is said[129] to be formed by reaction of tellurium dioxide with the tetrachloride. This oxochloride apparently melts at 580° and sublimes at 400° in a vacuum. The oxodibromide is apparently obtained by the action of hot, concentrated sulphuric acid on tellurium tetrabromide or by treating a chloroform solution of the

126 V. Lenher, *J. Am. Chem. Soc.* **43** (1921) 29.
127 N. N. Yarovenko, M. A. Raksha and G. B. Gazieva, *Zhur. Obshchei Khim.* **31** (1961) 4006.
128 T. Birchall, R. J. Gillespie and S. L. Vekris, *Canad. J. Chem.* **43** (1965) 1672.
129 P. Khodadad, *Bull. Soc. chim. France*, (1965) 468.

latter with nitric oxide or silver oxide[130], and by thermal decomposition[131] of the compound $TeO_2.2HBr$ at 60–70°. The analogous thiodibromide, $TeSBr_2$, is said[132] to be formed by the action of hydrogen sulphide on solid tellurium tetrabromide, but supporting evidence for these tellurium compounds is lacking. No polonium(IV) oxohalides have been identified with certainty.

A number of oxofluorides of sexavalent selenium and tellurium are known. Selenium(VI) dioxodifluoride, SeO_2F_2, is readily hydrolysed and is a colourless gas at room temperature; some physical properties are given in Table 11. The compound is prepared by reaction of barium selenate(VI) with fluorosulphonic acid under reflux[133] (50°C) or by heating selenium trioxide with potassium tetrafluoroborate[134] at 65–70°, and by reaction of selenium trioxide with selenium tetrafluoride; the product is purified by fractional distillation. Small amounts of the compound are formed in the reaction of selenium dioxide with fluorine[135] and by reaction of selenium trioxide with an excess of arsenic trifluoride[136]. From the Raman and infrared spectra of the dioxodifluoride, the molecule appears to be tetrahedral and of C_{2v} symmetry[137].

Selenium(VI) dioxodifluoride does not attack glass, but reacts violently with ammonia, with reduction to selenium. It is also decomposed by organic matter, such as tap grease[136], and it is slowly reduced to selenium(IV) oxodifluoride by mercury at room temperature; the products of the more rapid reaction with mercury at high temperature have not been identified. The tellurium and polonium analogues are unknown.

Two other selenium(VI) oxofluorides are well established[138], pentafluoroselenium hypofluorite, F_5SeOF, and *bis*pentafluoroselenium peroxide, $F_5SeOOSeF_5$. Both of these are volatile, white solids; the former melts at −54° and boils at −29°, and the latter melts at −62·8° and boils at 76·3°. They are obtained by reaction of selenium dioxide with a mixture of fluorine and nitrogen, the hypofluorite at 80° and the peroxide at 120°, and are purified from each other and from selenium hexafluoride by fractional sublimation or distillation at low pressure. The n.m.r. spectrum of the peroxide confirms the presence of SeF_5 groups. A third oxofluoride, which may have the composition $Se_3O_2F_{14}$ (m.p. −20°) is formed as a by-product in these reactions, all of which also yield much selenium hexafluoride; this third oxofluoride decomposes to selenium hexafluoride and other, unidentified, products at 250°.

The hypofluorite is completely hydrolysed by water or aqueous alkali with the liberation of oxygen:

$$2F_5SeOF + 16KOH \longrightarrow 2K_2SeO_4 + 12KF + 8H_2O + O_2$$

hydrolysis of the sulphur analogue stops at the fluorosulphonate ion. The selenium compound immediately liberates iodine from potassium iodide, whereas the peroxide is much less reactive, being almost inert to water, acid or alkali; it is slowly reduced by potassium iodide and there is only slight attack on glass. The peroxide is decomposed by potassium and reacts rapidly with organic materials; it yields mainly the hexafluoride on heating

130 E. Montignie, *Z. anorg. Chem.* **315** (1962) 102.
131 W. Prandtl and P. Borinski, *Z. anorg. Chem.* **62** (1910) 237.
132 E. Montignie, *Bull. Soc. chim. France*, (1947) 376.
133 A. Engelbrecht and B. Stoll, *Z. anorg. Chem.* **292** (1957) 20.
134 P. Martin, A. Scholer and E. Class, *Chimia* (*Switz.*), **21** (1967) 162.
135 G. Mitra, *J. Indian Chem. Soc.* **37** (1960) 804.
136 H. G. Jerschkewitz, *Angew. Chem.* **69** (1957) 562.
137 T. Birchall and R. J. Gillespie, *Spectrochim. Acta*, **22** (1966) 681.
138 G. Mitra and G. H. Cady, *J. Am. Chem. Soc.* **81** (1959) 2646.

to 200° or with fluorine at 70°, some (18%) hypofluorite also being formed in the latter reaction.

Two tellurium(VI) oxofluorides, of composition $Te_3O_2F_{14}$ and $Te_6O_5F_{26}$, are formed in small yield as by-products in the reaction of tellurium dioxide with fluorine and oxygen[117]; they are probably of the form $F_5Te(OTeF_4)_nOTeF_5$ ($n = 1$ or 4). Very little is known about them. Pentafluorotelluric acid, F_5TeOH, and its derivatives are discussed later (p. 977) with the haloselenic acids and hydrogen halide adducts of selenium dioxide.

2.6. CYANIDES, THIOCYANATES, SELENOCYANATES AND *BIS*TRIFLUOROMETHYL NITROXIDES

These compounds are very similar to the halides but with the compounds of the bi-positive elements much more stable with respect to disproportionation than the dihalides. Although the selenocyanate ion, $SeCN^-$ (p. 939), is well known, the tellurium analogue is unstable and difficult to obtain, only the tetra-ethylammonium salt being known[139].

Selenium monocyanide, $Se_2(CN)_2$, otherwise known as selenocyanogen, is a yellow solid, prepared by oxidation of silver selenocyanate with iodine in carbon tetrachloride, chloroform or ether[140], or of potassium selenocyanate in acetone with iodine pentafluoride[141] or lead tetra-acetate[142]. It is readily hydrolysed and disproportionates when heated in carbon disulphide, yielding the dicyanide, $Se(CN)_2$, and the diselenocyanate, $Se(SeCN)_2$. The corresponding thiocyanate, $Se_2(SCN)_2$, is said[143] to be formed by the action of selenium monochlorlde on mercuric thiocyanate. Very little is known about either of these compounds.

Selenium dicyanide, a colourless solid which melts at 134°, is obtained, mixed with the diselenocyanate, by the oxidation of potassium selenocyanate with dinitrogen tetroxide or iodine pentafluoride, and by the action of selenium monobromide on silver cyanide in ether, probably by way of the disproportionation of selenium monocyanide. The dicyanide sublimes from the mixture at 50° in a vacuum, leaving a residue of the involatile diselenocyanate[141]. The compound is isomorphous with sulphur dicyanide, the crystal symmetry being orthorhombic[144], and both dicyanides are monomeric in benzene. The molecule is V-shaped (C_{2v} symmetry) and its infrared and Raman spectra have been recorded[141]. Tellurium dicyanide is a pale rose solid, the colour possibly being due to traces of elementary tellurium; it is prepared[145] by heating a mixture of tellurium tetrabromide and silver cyanide in dry benzene at 90° for 3 days. It decomposes slowly at 80°, but can be purified by sublimation at 120° in a high vacuum, and is soluble in ether[145]; the infrared spectrum[139, 146] has been recorded. The only known polonium cyanide, a white solid obtained by treating polonium(IV) hydroxide or tetrachloride with aqueous hydrocyanic acid, could be the tetracyanide, but analyses are lacking[147].

Selenium dithiocyanate, $Se(SCN)_2$, a yellow solid which is monomeric in dioxane or acetophenone, crystallizes from aqueous hydrochloric acid solutions of selenium dioxide

139 A. W. Downs, *Chem. Comm.* (1968) 1290.
140 L. Birkenbach and K. Kellermann, *Ber.* **58** (1925) 786, 2377.
141 E. E. Aynsley, N. N. Greenwood and M. J. Sprague, *J. Chem. Soc.* (1964) 704.
142 H. P. Kaufman and F. Kögler, *Ber.* **59** (1926) 178.
143 A. Baroni, *Atti Accad. naz. Lincei, Rend. Classe Sci. fis. mat. nat.* **23** (1936) 139.
144 K. Linke and F. Lemmer, *Z. Naturforsch.* **21*B*** (1966) 192; *Z. anorg. Chem.* **345** (1966) 211.
145 H. E. Cocksedge, *J. Chem. Soc.* **93** (1908) 2175.
146 H. P. Fritz and H. Keller, *Chem. Ber.* **94** (1961) 1524.
147 K. W. Bagnall and J. H. Freeman, *J. Chem. Soc.* (1957) 2161.

and ammonium thiocyanate. It can be recrystallized from anhydrous dioxane, ether being added until a turbidity develops[148]. The crystal symmetry is orthorhombic[149]. The compound is stable below 5° in the absence of water and should be kept in the dark; it is soluble in a variety of organic solvents and, to some extent, in liquid sulphur dioxide. The tellurium compound, about which very little is known, is said[132] to be formed by reaction of tellurium tetrabromide with silver thiocyanate in boiling benzene.

Selenium diselenocyanate is isomorphous[150] with the dithiocyanate; both molecules have C_s symmetry. It is yellow when powdered, red-orange in large crystals, and melts, with decomposition, at 133–134°. The compound is usually prepared by oxidation of an alkali metal or ammonium selenocyanate with chlorine or bromine. Oxidation of potassium selenocyanate with dinitrogen tetroxide is also effective if fuming nitric acid is present to destroy the dicyanide. The product is purified by recrystallization from benzene or chloroform. Selenium diselenocyanate is monomeric in benzene or chloroform and is stable in the dark, being decomposed by visible light or X-radiation.

The selenium and tellurium *bis*trifluoromethyl nitroxides, $[(CF_3)_2NO]_4X$, are both white solids which melt at 25°, obtained by reaction of selenium or tellurium with the *bis*trifluoromethyl nitroxide radical at room temperature[150a]. Very little is known about them as yet, but they are probably best regarded as pseudo-halides.

2.7. OXIDES AND PEROXIDES

Selenium and tellurium monoxides do not exist in the solid state and previous claims for the preparation of the tellurium compound, for example by decomposing the so-called "sulphoxide", $TeSO_3$ (p. 988), at 180° in a vacuum, have been shown to yield a mixture of tellurium and the dioxide[151]. The black polonium compound, an easily oxidized solid, is formed by the spontaneous decomposition of the "sulphoxide", $PoSO_3$ (p. 988), a decomposition which is probably due to the α-radiation of the polonium[44, 152]. There is, however, spectroscopic evidence for the existence of gaseous selenium[153, 154] and tellurium[155] monoxide species in flames at high temperatures; the latter have been reviewed in detail[156], together with the other oxides and oxyacids of tellurium. Physical data for the oxides are given in Table 12.

Selenium, tellurium and polonium dioxides all have some basic character, although this is only very slight for the selenium compound. This last is thermodynamically less stable than either the sulphur or the tellurium compound (Table 12), following the trend in the stability of the halogen oxides of the corresponding periods, where the bromine compounds are very much less stable than the chlorine or iodine compounds.

Selenium dioxide is a white solid which is said to smell like decaying radishes; the melt is pale yellow and the vapour green, but the colour is not due to free selenium. It is usually

[148] S. M. Ohlberg and P. A. van der Meulen, *J. Am. Chem. Soc.* **75** (1953) 997.
[149] S. M. Ohlberg and P. A. Vaughan, *J. Am. Chem. Soc.* **76** (1954) 2649.
[150] O. Aksnes and O. Foss, *Acta Chem. Scand.* **8** (1954) 702, 1787.
[150a] H. G. Ang, J. S. Coombes and V. Sukhoverkhov, *J. Inorg. Nuclear Chem.* **31** (1969) 877.
[151] O. Glemser and W. Poscher, *Z. anorg. Chem.* **256** (1948) 103.
[152] K. W. Bagnall and J. H. Freeman, *J. Chem. Soc.* (1956) 4579.
[153] R. K. Asundi, M. Jan-Khan and R. Samuel, *Proc. Roy. Soc. A* **157** (1936) 28.
[154] L. Bloch, E. Bloch and Shin-Piaw Choong, *Compt. rend.* **201** (1935) 824.
[155] R. F. Barrow and H. J. Hurst, *Nature*, **201** (1964) 699.
[156] W. A. Dutton and W. C. Cooper, *Chem. Rev.* **66** (1966) 657.

prepared from the elements or by dehydration of selenous acid, and is purified by sublimation. The crystal symmetry is tetragonal and the structure[157] consists of infinite chains of selenium and oxygen atoms, whereas discrete molecules are present in the structure of solid sulphur dioxide. The selenium compound is very soluble in water and behaves as a weak base in 100% sulphuric acid, its solution in which is bright green[158]. It is also soluble in selenium(IV) oxodichloride, in which it appears to be a trimer, and, surprisingly, in benzene, but it is only slightly soluble in other organic solvents. The dioxide is reduced

TABLE 12. SOME PHYSICAL PROPERTIES OF THE OXIDES[a]

	SeO_2	TeO_2	SeO_3	TeO_3
Melting point, °C	340 (sealed tube)	732·6[c]	118	—
Heat of fusion, kcal/mole	—	7·0[c]	—	—
Boiling point, °C	315°/760 mm (sublimes)	—	100°/40 mm (sublimes)	—
Heat of sublimation, kcal/mole	24·5[b]	59[c]	9·0*	—
Heat of vaporization, kcal/mole	—	51·7[c]	—	—
Vapour pressure (solid)	A = 12·0267	A = 12·3284[c]	A = 18·159[d]	—
$\log_{10} p_{mm} = A-B/T$	B = 5542·5	B = 13,222[c]	B = 6968[d]	—
Vapour pressure (liquid)	—	A = 10·248	—	—
$\log_{10} p_{mm} = A-B/T$	—	B = 11,300	—	—
Heat of formation, $\Delta H_{f,\,298}$, kcal/mole†	−55·0[b]	−77·75[c]	−44	−83·2 (18°)
Free energy of formation, $\Delta F_{f,\,298}$, kcal/mole	−41·6[b]	−64·5[c]	—	—
Density, g/cc	—	5·75[c]	2·75 (118°)	α, 5·07; β, 6·21
Triple point, °C	—	—	120·9°/5·5 mm	—

* Tetramer.
† The values for SO_2 at 25° and SO_3 at 18° are −71 and −103·2 kcal/mole respectively.
[a] Data from ref. 3, pp. 54–59 unless otherwise stated.
[b] A. Glassner, U.S.A.E.C. Report ANL–5750 (1958).
[c] W. A. Dutton and W. C. Cooper, *Chem. Rev.* **66** (1966) 657.
[d] F. C. Mijlhoff and R. Block, *Rec. Trav. chim.* **83** (1964) 799. The vapour pressure refers to the *partial* pressure of the monomer.

to the element by ammonia, hydrazine and aqueous sulphur dioxide (but not by the anhydrous gas) and is useful as an oxidizing agent in organic chemistry[159].

Tellurium dioxide, a white solid which is yellow when hot, melts to a dark red liquid and is much less volatile than the selenium compound. It is prepared from the elements and by dehydrating tellurous acid or by thermal decomposition of the basic nitrate (p. 986) at 400–430°. The synthetic dioxide is of tetragonal symmetry[160, 161] undergoing a phase change at high (30 kbar) pressure[162], whereas the naturally occurring dioxide (tellurite) is of orthorhombic symmetry[163]. The tetragonal lattice appears to resemble that of rutile, the structure being a distorted square pyramid with the tellurium atom at the apex and

157 J. D. McCullough, *J. Am. Chem. Soc.* **59** (1937) 789.
158 R. H. Flowers, R. J. Gillespie and E. A. Robinson, *J. Inorg. Nuclear Chem.* **9** (1959) 155.
159 G. R. Waitkins and C. W. Clark, *Chem. Rev.* **36** (1945) 235.
160 J. Leciejewicz, *Z. Krist.* **116** (1961) 345.
161 O. Lindqvist, *Acta Chem. Scand.* **22** (1968) 977.
162 S. S. Kabalkina, L. F. Vereschlagin and A. A. Kotilevets, *Zhur. Eksperim. Teor. Fiz.* **51** (1966) 377.
163 H. Beyer, K. Sahl and J. Zemann, *Naturwiss.* **52** (1965) 155.

the pyramids linked by common oxygen atoms[160], an alternative description being a distorted trigonal bipyramid with one equatorial oxygen atom missing[161]. Infrared and Raman data have been recorded[164]. Tellurium dioxide is very soluble in selenium(IV) oxodichloride and in water; it is amphoteric, the solubility being at a minimum at pH 3·8–4·2, the isoelectric point. The compound is reduced to the element by carbon at red heat and by many metals, but hydrogen does not reduce it completely, even at high temperatures.

The Po^{4+} radius calculated from the X-ray data for the dioxide is 1·02–1·04 Å, making the radius ratio Po^{4+}/O^{2-} approximately 0·73, the lower limit of stability for cubic co-ordination[165]. It is therefore not surprising to find that polonium dioxide exists in two crystal modifications, a yellow, low temperature form of face-centred cubic symmetry and a red, high temperature, tetragonal form. The cubic modification has a variable oxygen content[45]; the tetragonal modification darkens at high temperatures and sublimes above 885° in oxygen. Polonium dioxide is prepared from the elements or by heating the sulphate or selenate (p. 987) above 550°; it decomposes to the elements at 500° in a vacuum and is slowly reduced by hydrogen at 200° and by ammonia or hydrogen sulphide at 250°. It does not react with liquid sulphur dioxide and is almost insoluble in water, aqueous ammonia or alkali carbonate.

Pentoxides

Selenium and tellurium pentoxides are probably best regarded as a basic selenate and tellurate of the quadripositive elements and are discussed as such (p. 987). Similarly, the sesquioxide, Se_2O_3, is probably a selenate (p. 988).

Trioxides

Selenium trioxide is a white, hygroscopic solid which decomposes above 165° to the "pentoxide"[166]. It is surprising that the compound is so stable to heat, for the trioxide is thermodynamically unstable with respect to the dioxide, the enthalpy[167] of the reaction

$$SeO_2(s)+\tfrac{1}{2}O_2 \longrightarrow SeO_3(s)$$

being +13 kcal. The trioxide is prepared by reaction of anhydrous potassium selenate with sulphur trioxide under reflux[168] or by dehydrating selenic acid with phosphorus pentoxide[169], and the product is best purified[170] by vacuum sublimation at 120°; it reacts explosively with organic compounds unless purified in this way.

The crystal symmetry of the trioxide is tetragonal[171] and the compound is a tetramer of S_4 symmetry in the solid state[172], but electron diffraction[173] of the vapour indicates that there is appreciable dissociation (30%) to the monomer at 120°, in conformity with the marked curvature of the log (vapour pressure)–reciprocal temperature curve. In the

164 V. P. Cheremisinov and V. P. Zlomanov, *Opt. Spectry.* (*U.S.S.R.*), **12** (1962) 110.
165 A. W. Martin, *J. Phys. Chem.* **58** (1954) 911.
166 H.-G. Jerschkewitz and K. Menning, *Z. anorg. Chem.* **319** (1962) 82.
167 F. C. Mijlhoff, *Rec. Trav. Chim.* **82** (1963) 822.
168 H. A. Lehmann and G. Krüger, *Z. anorg. Chem.* **267** (1952) 315, 324.
169 F. Toul and K. Dostal, *Coll. Czech. Chem. Comm.* **16** (1951) 531.
170 M. Schmidt and I. Wilhelm, *Chem. Ber.* **97** (1964) 872.
171 F. C. Mijlhoff and C. H. MacGillavry, *Acta Cryst.* **15** (1962) 620.
172 F. C. Mijlhoff and R. Block, *Rec. Trav. Chim.* **83** (1964) 799.
173 F. C. Mijlhoff, *Rec. Trav. Chim.* **84** (1965) 74.

molten state, selenium trioxide is probably polymeric, as are the isoelectronic polymetaphosphate ions in the molten salt[172]. The Raman and infrared spectra of the solid and liquid trioxide have been recorded[174].

Selenium trioxide is soluble in ether, dioxane, acetic anhydride and liquid sulphur dioxide; its solution in the last is a useful selenonating agent in organic chemistry[170]. The compound is reduced to the element by sulphur, or hydrogen sulphide in ether, and it dissolves selenium and tellurium, forming the so-called "selenoxides", $XSeO_3$ (p. 988). Reactions leading to the formation of species derived from selenic acid are described later (p. 977).

Tellurium trioxide exists in two modifications, the yellow-orange α-form and the grey β-form. The first decomposes at 406°, forming the "pentoxide", whereas the second is not reduced by hydrogen, even at 400°, but begins to decompose in nitrogen at 430°. The α-trioxide is prepared by dehydrating orthotelluric acid at 300–360°, any tellurium dioxide being removed by washing with hydrochloric acid, to which the trioxide is inert. The β-trioxide is obtained by heating the α-trioxide or orthotelluric acid, usually in the presence of sulphuric acid, at 300–350° for 10–12 hr, preferably in a sealed tube filled with oxygen[174a]; any residual α-trioxide is removed by washing the product with concentrated aqueous potassium hydroxide, in which the β-trioxide is insoluble.

The α-trioxide is a powerful oxidizing agent, reacting violently when heated with a variety of metals and non-metals, but the β-form is much less reactive, and may be a high polymer; no structural data have been recorded for the α-form, but β-TeO_3 is of rhombohedral symmetry[174a]. The α-form, unlike the β-trioxide, is soluble in hot, concentrated alkali, forming tellurates, and is reduced to tellurium(IV) by boiling concentrated hydrochloric acid. The β-form reacts, however, with fused potassium hydroxide and is soluble in concentrated aqueous sodium sulphide. Neither form is soluble in water, dilute acid or dilute alkali[175, 176].

The only evidence for the formation of polonium trioxide is in the anodic deposition of tracer amounts of polonium[177].

2.8. HYDROXIDES AND OXO-ACIDS

The only known compound of the dipositive metals is the dark brown, readily oxidized polonium(II) hydroxide or hydrated oxide, precipitated from solutions of polonium(II) by alkali[44].

Oxo-acids of the Quadrivalent Elements

Selenous and tellurous acids, H_2SeO_3 and H_2TeO_3, are white solids which readily dehydrate to the dioxide, the tellurium compound being the less stable, with a heat of formation, from the solid dioxide and liquid water, of only 1060 cal[178]. The selenium compound is usually prepared by allowing a solution of the dioxide in water to crystallize slowly and the tellurium compound is obtained by hydrolysis of a tetrahalide or by acidifying an aqueous solution of a tellurite, avoiding any heating.

[174] R. Paetzold and H. Amoulong, *Z. anorg. Chem.* **337** (1965) 225.
[174a] D. Dumora and P. Hagenmuller, *Compt. rend.* C**266** (1968) 276.
[175] M. Patry, *Compt. rend.* (1936) 845; *Bull. Soc. chim. France*, 3 (1936) 845.
[176] E. Montignie, *Bull. Soc. chim. France*, (1947) 564.
[177] M. Haissinsky and M. Cottin, *Compt. rend.* **224** (1947) 467.
[178] E. Sh. Ganelina, B. P. Troitskii and V. S. Yakovleva, *Zhur. Priklad. Khim.* **37** (1964) 677.

The crystal symmetry of the selenium compound is orthorhombic and the SeO_3 group is pyramidal[179], but the structure of the tellurium compound is not known; it is, however, a much weaker acid than selenous acid. Both compounds form acid and neutral salts, M^IHXO_3 and $M^I_2XO_3$, by reaction of the appropriate alkali with either selenous acid or tellurium dioxide, and by heating the metal oxide with selenium or tellurium dioxide. The alkali metal salts of both acids are soluble in water; other salts are less soluble.

Selenous acid is very soluble in water and there is some dimerization in concentrated solutions, but there is no dissociation to the dioxide, in contrast to sulphurous acid[180]. It is readily oxidized to selenic acid, for example by ozone in strongly acid solution, a reagent which does not oxidize tellurous acid. However, tellurites are oxidized to tellurates on heating in air or, in solution, by treatment with oxidizing agents such as hydrogen peroxide, and tellurous acid is oxidized to orthotelluric acid by 30% hydrogen peroxide in strong acid under reflux.

The selenium compound is reduced to the element by hydrogen sulphide, sulphur dioxide or iodide ion, and the kinetics of the reaction with the last have been studied[181].

Alkali diselenites, $M^I_2Se_2O_5$, are known; the salts are hydrolysed in aqueous solution. Raman and infrared spectra indicate that the ion contains a non-linear Se–O–Se bridge, the symmetry being C_{2v}, whereas the metabisulphite ion has a sulphur–sulphur bond[182]. The acid, $H_2Se_2O_5$, is present in molten selenous acid[183]. Superacid salts of the type $KH_3(XO_3)_2$ and polyacid salts such as $K_2Te_4O_9$ and $K_2Te_6O_{13}$ have been recorded.

Polonium(IV) hydroxide, a pale yellow solid precipitated by alkali from solutions of polonium(IV), appears to have appreciable acidic character; the colourless solids obtained by heating the dioxide with potassium hydroxide or nitrate are quite soluble in water and are presumably polonites[147].

Oxo-acids of the Sexavalent Elements

The simplest selenium(VI) compound, H_2SeO_4, strongly resembles sulphuric acid in its properties; the ionization constants of the two acids are almost identical, both form alums and the acids form a very similar series of hydrates. In contrast, the two tellurium(VI) compounds, orthotelluric acid, $Te(OH)_6$, and polymetatelluric acid, $(H_2TeO_4)_n$, do not resemble sulphuric or selenic acid in any way and their salts are not isomorphous with the sulphates or selenates. No polonium(VI) oxoacid has been recorded, but fusion of polonium dioxide with potassium chlorate and hydroxide seems to yield[9] a water-soluble polonate(VI).

Anhydrous selenic acid is a white, deliquescent solid (m.p. 58°) which decomposes to selenium dioxide above 260°; it can be distilled at 202°/0·001 mm. Because of supercooling, usually due to impurities in the acid, the compound often remains liquid at room temperature. It is prepared by oxidation of selenium, the dioxide or selenous acid with hydrogen peroxide (> 30%) under reflux, and by oxidation of a selenite with hypochlorous acid or bromine water[184]. The aqueous acid loses water on heating like sulphuric acid.

179 A. F. Wells and M. Bailey, *J. Chem. Soc.* (1949) 1282.
180 M. Falk and P. A. Giguére, *Canad. J. Chem.* **36** (1958) 1689.
181 J. A. Neptune and E. L. King, *J. Am. Chem. Soc.* **75** (1953) 3069.
182 A. Simon and R. Paetzold, *Z. anorg. Chem.* **303** (1960) 39, 46.
183 G. E. Walrafen, *J. Chem. Phys.* **37** (1962) 1468.
184 L. I. Gilbertson and G. B. King, *Inorganic Syntheses*, Vol. 3, p. 137, Ed. L. F. Audrieth, McGraw-Hill Book Co. Inc., New York, N.Y. (1950).

The crystal symmetry of solid selenic acid is orthorhombic and the selenate group is tetrahedral[185]; the infrared and Raman spectra have been published[186, 187].

Selenic acid differs from sulphuric acid in that it is a strong oxidizing agent, although kinetically slow; it is reduced by a number of non-metals, by hydrohalic acids other than hydrogen fluoride, and by hydrogen sulphide at $-10°$, but reduction with sulphurous acid is very slow. It dissolves gold, palladium and silver, but only attacks platinum in the presence of chloride ion. Solutions of sulphur, selenium, tellurium and polonium in the acid are brightly coloured, presumably owing to formation of the "selenoxides" (p. 988), and complex acids with molybdenum(VI), chromium(VI) and chromium(III) are known.

The nitrosyl and nitryl selenates are analogous to the sulphates; the hydrogen selenate, $NOHSeO_4$, a white solid, is prepared by the action of dinitrogen trioxide[188] or tetroxide[189] on selenic acid; mercury reduces it to the blue nitrosiselenic acid, $NO(OH)SeO_3H$, in selenic acid[190]. The analogous nitryl salt, NO_2HSeO_4, is a colourless solid prepared by reaction of nitryl fluoride with selenic acid[191]. The blue dinitrosonium salt, $(NO)_2SeO_4$, is only obtained by the action of dinitrogen tetroxide on selenic acid at low temperatures[192] and is stable below $-13°$.

Like sulphuric acid, selenic acid combines with sulphur trioxide forming $H_2SeO_4.nSO_3$ ($n = 1$ or 2), compounds which decompose on heating and are readily hydrolysed[193]. The potassium salts of these species are formed in the preparation of selenium trioxide from potassium selenate and sulphur trioxide. The selenium trioxide analogues, $H_2Se_2O_7$ (pyroselenic acid), $H_2Se_3O_{10}$ and $H_4Se_3O_{11}$, resemble the compounds formed in the sulphur trioxide–water system. Pyroselenates are prepared by dehydration of the hydrogen selenate and the potassium salts are also obtained by reaction of potassium chloride with selenium trioxide or by pyrolysis of potassium fluoroselenate(VI); the $Se_2O_7^{--}$ ion contains a bent Se–O–Se group[194]. Nitrosyl and nitryl compounds, $NOHSe_2O_7$, $(NO)_2Se_3O_{10}$, $(NO)_2Se_2O_7$ and $(NO_2)_2Se_3O_{10}$ are prepared by reaction of dinitrogen tetroxide with selenium trioxide in nitromethane[195]; with pyroselenic acid or with $H_2Se_3O_{10}$ the product[189] is $(NO)_2Se_2O_7$.

Orthotelluric acid, $Te(OH)_6$, is a white solid which melts in a sealed tube at 136° and decomposes to the trioxide above 300°. It is conveniently prepared by oxidizing tellurium with chloric acid[196], but oxidation of tellurium or the dioxide with either 30% hydrogen peroxide in concentrated sulphuric acid under reflux[197] or potassium permanganate[198] are almost equally effective; the compound is also obtained by the action of a mineral acid on a solution or suspension of a tellurate, an example being sulphuric acid and barium

185 M. Bailey and A. F. Wells, *J. Chem. Soc.* (1951) 968.
186 R. Paetzold and H. Amoulong, *Z. anorg. Chem.* **317** (1962) 288.
187 G. E. Walrafen, *J. Chem. Phys.* **39** (1963) 1479.
188 J. Meyer and W. Wagner, *J. Am. Chem. Soc.* **44** (1922) 1032.
189 K. Dostal, A. Ruzicka and P. Rumisek, *Z. anorg. Chem.* **336** (1965) 219.
190 J. Meyer and W. Gulbins, *Ber.* **59** (1926) 456.
191 G. Hetherington, D. R. Hub and P. L. Robinson, *J. Chem. Soc.* (1955) 4041.
192 V. Lenher and J. H. Mathews, *J. Am. Chem. Soc.* **28** (1906) 516.
193 J. Meyer and V. Stateczny, *Z. anorg. Chem.* **122** (1922)1.
194 R. Paetzold, H. Amoulong and A. Ruzicka, *Z. anorg. Chem.* **336** (1965) 278.
195 G. Kempe, M. Lorenz and H. Thieme-Schneider, *Z. Chem.* **4** (1964) 184; G. Kempe, *Z. Chem.* **8** (1968) 152.
196 J. Meyer and W. Franke, *Z. anorg. Chem.* **193** (1930) 191.
197 L. I. Gilbertson, *J. Am. Chem. Soc.* **55** (1933) 1460.
198 F. C. Mathers, C. M. Rice, H. Broderick and R. Forney, *Inorganic Syntheses*, Vol. 3, p. 145, Ed. L. F. Audrieth, McGraw-Hill Book Co., Inc., New York, N.Y. (1950).

tellurate. The product is purified by recrystallization from water or nitric acid; a tetrahydrate is formed when the acid is crystallized below 10°. The acid oxidizes hydrochloric and hydrobromic acids on heating and slowly liberates iodine from potassium iodide; it is also reduced by hydrazine, hydrogen sulphide and sulphur dioxide.

Although orthotelluric acid was first thought to be the dihydrate of H_2TeO_4, it is correctly $Te(OH)_6$, as was first demonstrated by the preparation of the methyl ester, $Te(OCH_3)_6$. The structure is a slightly distorted octahedron according to its Raman spectrum[199].

The formation of an acid of this form, in contrast to the simple forms of sulphuric and selenic acids, reflects the increase in metallic character on going down the group and is, in part, the result of the increased radius of tellurium(VI), whereby the polarizing power of the central ion decreases so that there is less tendency for a proton to be lost. It is, therefore, not surprising to find that the neighbouring elements in the same period form similar acid species; the isoelectronic ions $Sn(OH)_6^{2-}$ and $Sb(OH)_6^-$ and periodic acid $IO(OH)_5$ are examples, all being close to octahedral arrays of oxygen or hydroxyl groups around the central atom. A few salts of the germanate ion, $Ge(OH)_6^{2-}$, seem to be the only examples of this type of structure in the preceding period, in which the arsenic and germanium oxo-acid ions are usually simple structures, as with selenium, whereas the plumbate(IV) anion is commonly $Pb(OH)_6^{2-}$.

Two crystal modifications[200] of orthotelluric acid are known; the α-form, of cubic crystal symmetry, is obtained by crystallizing the acid from concentrated nitric acid and this is said to transform to the monoclinic β-form when heated. The cubic form appears to be the more stable form at room temperature[156].

Orthotelluric acid is soluble in mineral acids, except nitric acid, and is more soluble in hot water than in cold; it is virtually insoluble in organic solvents. It is a weak dibasic acid, the salts commonly being of the form KH_5TeO_6 or $Li_2H_4TeO_6$ (*not* $Li_2TeO_4.2H_2O$), but salts such as KH_3TeO_5[201] or those containing the TeO_6^{6-} ion are quite well known[202]. The alkali and alkaline earth orthotellurates are soluble in water; they are usually prepared from the acid and the appropriate base. Other salts are at best sparingly soluble and are precipitated from solutions of soluble tellurates on the addition of a soluble metal salt. Complex acid species containing tellurium(VI) are known for arsenic(V), chromium(VI), molybdenum (VI) and tungsten(VI), and a soluble polonium(VI)–chromium(VI) complex acid has been recorded[9].

Polymetatelluric acid, $(H_2TeO_4)_n$, where n is approximately 10, is formed by loss of water when the ortho acid is heated in air at 100–200°; the structure is unknown but the infrared spectrum of the acid suggests that the tellurium atom is 6-coordinate[199]. The compound cannot be obtained pure and always contains some orthotelluric acid, to which the polyacid reverts in solution. Allotelluric acid, obtained by heating the ortho acid in a sealed tube at 305°, is merely a mixture of ortho- and polymetatelluric acids.

Esters of the polyacid are amorphous solids which decompose at about 95° and are soluble in organic solvents; they are usually prepared by evaporating an alcoholic solution of the acid, and the methyl ester can be prepared by heating the orthotellurate above its melting point or by reaction of the polyacid with diazomethane.

[199] H. Siebert, *Z. anorg. Chem.* **301** (1959) 161.
[200] C. Avinens and M. Petit, *Compt. rend.* C**266** (1968) 981.
[201] P. Lammers and J. Zemann, *Z. anorg. Chem.* **334** (1965) 225.
[202] B. N. Kulikovskii, Yu. N. Mikhailov and V. G. Tronev, *Zhur. neorg. Khim.* **10** (1965) 2814; *Russ. J. Inorg. Chem.* **10** (1965) 1526.

Salts of the form $MH_3(TeO_4)_2.1{\cdot}5\ H_2O$, which may be hydrated acid tellurates, are obtained by the action of alkali on ethanolic solutions of the acid. Polytellurates, analogous to the polychromates, such as $K_2Te_4O_{13}$ and $NaHTe_2O_7$, are also known.

Amido derivatives of the Oxo-acids

The only selenium(IV) compounds known are the *bis*dialkylamido derivatives of selenous acid, $SeO(NR_2)_2$; the methyl compound is a colourless solid which melts at 42–43° and decomposes explosively in a vacuum at 50–60°, but can be vacuum distilled at lower temperatures. It is prepared by reaction of selenium(IV) oxodichloride with dimethylamine at low temperature and decomposes slowly at room temperature, losing oxygen to form *bis*dimethylamidoselenide, $[(CH_3)_2N]_2Se$. It also reacts with an excess of the oxodichloride to form the amidochloride, $Cl–SeO–N(CH_3)_2$. The dialkylamido compounds are very soluble in water but are hydrolysed to the alkylamine and selenous acid, and with alcohols the appropriate selenite is formed[203]. The dimethyl compound is more stable than the diethyl one, which decomposes at −20° to yield the amidoselenide, $[(C_2H_5)_2N]_2Se$, and a metabiselenite, $[(C_2H_5)_2NH_2]_2Se_2O_5$.

The simple selenium(VI) analogue, $SeO_2(NH_2)_2$, is formed to some extent in the reaction of dimethyl selenate(VI) with anhydrous liquid ammonia[204]; the methylamido derivatives, $SeO_2(NHCH_3)_2$ and $SeO_2[N(CH_3)_2]_2$, obtained by reaction of dimethyl selenate(VI) with methylamine and dimethylamine respectively, are more readily prepared than the simple

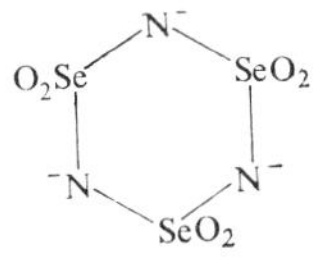

N^-
O_2Se SeO_2
^-N N^-
SeO_2

(I)

amide. In the preparation of the last the major products are the white, water-soluble ammonium salt of the cyclic triselenimide ion (I), $(NH_4NSeO_2)_3$ and a mixture of chain polymers of the type $H_2N[SeO_2N(NH_4)]_nSeO_2NH_2$. These products were first[205] obtained by the reaction of selenium(VI) dioxodifluoride with ammonia gas or, better, with liquid ammonia or with ammonia in strongly cooled carbon tetrachloride solution. The chain polymers are hydrolysed in aqueous solution to the monoamides and are completely decomposed by refluxing with 3–8 N alkali. Water-insoluble triselenimides, such as the potassium, silver and thallium salts, are precipitated from aqueous solutions of the ammonium salt. All of these compounds are explosive, being sensitive to both heat and shock[205].

The white ammonium salt of the selenium analogue of sulphamic acid, $NH_4SeO_3NH_2$, commonly called ammonium amidoselenate, is made by reaction of the selenium trioxide–pyridine complex with anhydrous liquid ammonia at −70°. The compound is very soluble in liquid ammonia and remains in solution, whereas the other product of the reaction, ammonium selenate, is precipitated. The amidoselenate is not formed when gaseous ammonia is used.

203 R. Paetzold and E. Rönsch, *Z. anorg. Chem.* **338** (1965) 22.
204 K. Dostal and L. Zborilova, *Z. Chem.* **4** (1964) 352; *Coll. Czech. Chem. Comm.* **32** (1967) 2809.
205 A. Engelbrecht and F. Clementi, *Monatsh.* **92** (1962) 555, 570.

The ammonium salt is stable in dry air, but explodes if heated rapidly to 120°; decomposition at 100° yields a mixture of tetraselenium tetranitride (p. 982), selenate and selenite. It is very soluble in water but only slightly soluble in methanol, which precipitates the salt from concentrated aqueous solution. The salt reacts with nitrite in dilute acid, evolving nitrogen like the sulphamates[206]. The yellow silver salt, $AgSeO_3NH_2$, is precipitated from a methanol solution of the ammonium salt by silver perchlorate, but in dilute aqueous ammoniacal solution the red-brown, dimorphic $AgSeO_3NAg_2$ is precipitated and in water silver selenate is formed[207]. The infrared spectra of these silver salts have been recorded[208].

Silver amidoselenate, $Ag_3SeO_3NH_2$, reacts exothermically with liquid ammonia, yielding a mixture of the diamine, $Ag(NH_3)_2SeO_3NH_2$, ammonium amidoselenate and the hexamine of silver imidodiselenate, $Ag_3(SeO_3NSeO_3).6NH_3$. The unsolvated ammonium imidodiselenate is prepared from aqueous solutions of the last on addition of ammonium iodide, and the unsolvated silver salt is precipitated from the resulting solution on the addition of silver ion. The infrared spectra of the imidodiselenates resemble those of their sulphur analogues[209].

Ammonium imidodiselenamate, $NH(SeO_3NH_4)_2$, is thought to be formed in the oxidation of selenium in liquid ammonia by oxygen under pressure, but it appears to be difficult to purify this compound[210].

Oxohalo-acids

Potassium fluoroselenite, $KSeO_2F$, a white, hygroscopic, readily hydrolysed solid which melts at 250°, is obtained by melting together equimolar amounts of selenium dioxide and potassium fluoride or by reaction of potassium selenite with selenium(IV) oxodifluoride. The infrared and Raman spectra suggest that there are bridges between neighbouring SeO_2F ions[211]. Potassium trifluoro- and trichloroselenites, $KSeOA_3$, are made by reaction of the selenium(IV) oxodihalide with the appropriate potassium halide; the infrared and Raman spectra of the salts show that the $SeOA_3^-$ anion has a distorted trigonal bipyramidal configuration with two of the halogen atoms in axial positions[212], in reasonable agreement with the crystal structure of the 8-hydroxyquinolinium trichloroselenite in which the environment of the selenium atom is described as a distorted square pyramid[212a]. The parent acids are unknown, but selenium dioxide forms addition compounds with hydrogen halides of the type $SeO_2.2HA$ (A = F, Cl, Br) which appear to behave as the acids $H_2SeO_2A_2$; salts such as $ZnSeO_2F_2.6H_2O$ [213] and $(NH_4)_2SeO_2A_2$ (A = Cl, Br)[214] have been reported, the latter formed from the acid and ammonia in non-aqueous solvents.

The chloro- and bromo-acids, formed from the dioxide and the dry hydrogen halide, are respectively pale yellow and red-brown liquids. The former freezes at −100° and boils, with decomposition, at 174°, and the latter, formed above 30°, is stable to 115°. The chloro-acid can be purified by fractional distillation[215] at 70–75°/15mm; its chemical and

206 K. Dostal and J. Krejci, *Z. anorg. Chem.* **296** (1958) 29.
207 K. Dostal and A. Ruzicka, *Z. anorg. Chem.* **337** (1965) 325.
208 R. Paetzold, K. Dostal and A. Ruzicka, *Z. anorg. Chem.* **348** (1966) 1.
209 K. Dostal and A. Ruzicka, *Z. Chem.* **8** (1968) 153.
210 V. G. Tronev and I. M. Belyakov, *Zhur. neorg. Khim.* **4** (1959) 1932; *Russ. J. Inorg. Chem.* **4** (1959) 875.
211 R. Paetzold and K. Aurich, *Z. anorg. Chem.* **335** (1965) 281.
212 R. Paetzold and K. Aurich, *Z. anorg. Chem.* **348** (1966) 94.
212a A. W. Cordes, *Inorg. Chem.* **6** (1967) 1204.
213 G. Mitra and P. C. Kundu, *Sci. Culture (Calcutta)*, **22** (1956) 119.
214 J. Jackson and G. B. L. Smith, *J. Am. Chem. Soc.* **62** (1940) 543.

solvent properties are very similar to those of selenium(IV) oxodichloride but, unlike the latter, it reacts violently with sodium. Selenium dioxide–hydrogen halide adducts of the type $SeO_2.4HA$ are also known; the chloride, a yellow solid, is stable below 0°, whereas the bromide, a steel grey solid which is formed at or below 30°, is stable to 55°. Both of these are probably present in solutions of selenium dioxide in the appropriate halogen acid[216] and they presumably have some acid character.

Fluoroselenic acid, $HSeO_3F$, is a colourless, viscous, readily hydrolysed liquid which fumes in air; it is a strong oxidizing agent and reacts violently with organic matter, such as filter paper. The acid is prepared from selenium trioxide and anhydrous hydrofluoric acid[217] but much selenium(VI) dioxodifluoride is also formed; the yield of the acid is improved when the reaction is allowed to take place in liquid sulphur dioxide[128].

The potassium salt was first made by distilling selenium(VI) dioxodifluoride on to potassium selenate at −60° and allowing the mixture to warm to room temperature[134]; the alkali metal salts are more easily prepared by reaction of selenium trioxide with the alkali fluoride, either on warming or in liquid sulphur dioxide, and their infrared and Raman spectra have been recorded. The salts are not isostructural[218] with the fluorosulphates, and are decomposed by water with the formation of some selenium(VI) dioxodifluoride. The potassium salt decomposes to the pyroselenate, $K_2Se_2O_7$, at 240° in a vacuum. Esters of fluoroselenic acid, unlike those of fluorosulphuric acid, are associated in the liquid state; they are prepared by reaction of selenium dioxodifluoride with alkyl or aryl selenates. Raman and infrared spectra have been recorded[219].

Chloroselenic acid, $HSeO_3Cl$, is a colourless, readily hydrolysed solid which decomposes, with the evolution of chlorine, above −10°. It is prepared from selenium trioxide and hydrogen chloride at −185° or, in solution in liquid sulphur dioxide, at −40°. The trimethylsilyl ester, made from the chloride and chloroselenic acid or selenium trioxide, appears to be fairly stable[220], but alkali metal salts are not known. An isomer of chloroselenic acid, chloroperoxoselenous acid, ClSeO.OOH, is formed by solvolysis of selenium(IV) oxodichloride with hydrogen peroxide at 0°, but it decomposes readily with the evolution of oxygen and leaves selenium dioxide; the compound cannot be obtained pure[221].

Tellurium(IV) oxohalo-acids are unknown, but a variety of tellurium(VI) oxofluoro-acids and their salts have been isolated. The acids are of the general form $TeF_n(OH)_{6-n}$ ($n = 1$–4[222]; $n = 5$[223]); in addition to their salts, compounds of the type $Te(OH)_6.nM^IF$ ($n = 1$, NaF; $n = 2$, KF; $n = 1{\cdot}5$, NH_4F) are known[224, 225].

Pentafluoro-orthotelluric acid, TeF_5OH, a white, volatile solid which melts at 40° and boils at 60°, is the main product of the reaction of an excess of fluorosulphuric acid with barium tellurate at 160°, the other products being a fluorosulphate, F_5TeOSO_2F (m.p. −89°, b.p. 56°), the sulphuric acid derivative $F_5TeO.SO_3H$ (m.p. −21°, b.p. 133°) and *bis*-

215 T. W. Parker and P. L. Robinson, *J. Chem. Soc.* (1928) 2853.
216 S. Witekowa, *Z. Chem.* **2** (1962) 315.
217 H. Bartels and E. Class, *Helv. Chim. Acta*, **45** (1962) 179.
218 A. J. Edwards, M. A. Mouty and R. D. Peacock, *J. Chem. Soc.* (*A*) (1967) 557.
219 R. Paetzold, R. Kurze and G. Engelhardt, *Z. anorg. Chem.* **353** (1967) 62.
220 M. Schmidt and I. Wilhelm, *Chem. Ber.* **97** (1964) 876.
221 M. Schmidt and P. Bornmann, *Z. anorg. Chem.* **331** (1964) 92.
222 L. Kolditz and I. Fitz, *Z. anorg. Chem.* **349** (1967) 175.
223 A. Engelbrecht and F. Sladky, *Monatsh.* **96** (1965) 159; A. Engelbrecht, W. Loreck and W. Nehoda, *Z. anorg. Chem.* **360** (1968) 88.
224 L. Kolditz and I. Fitz, *Z. anorg. Chem.* **349** (1967) 184.
225 R. F. Weinland and J. Alfa, *Z. anorg. Chem.* **21** (1899) 43.

pentafluorotellurosulphate, $(F_5TeO)_2SO_2$ (m.p. −50°, b.p. 118°). The fluorosulphate and the sulphate are insoluble in water and are relatively stable to hydrolysis compared with the two acids, which are rapidly hydrolysed[223]. All four compounds smell like garlic and they cause transitory asthma when inhaled. The reaction by which these compounds are formed is in marked contrast to the analogous reaction with barium selenate (p. 966), the difference in behaviour presumably resulting from the tendency of tellurium to achieve a higher coordination number than selenium, which forms only the dioxodifluoride.

A number of pentafluoro-orthotellurates, M^IOTeF_5, have been prepared, either from the acid and the base in carbon tetrachloride or by condensing the acid on an alkali chloride, when hydrogen chloride is evolved. The ammonium and pyridinium salts can be sublimed at 140° and 170° respectively, and the alkali metal salts are stable to about 350°. All are very soluble in water, but are hydrolysed to tellurium(VI) hydroxofluoro-acids; some of the salts can be recrystallized from ethanol[226].

The acids $TeF_n(OH)_{6-n}$ ($n = 1$–4) are obtained by reaction of orthotelluric acid with hydrofluoric acid, the four products being separated by paper chromatography; the pyridinium and quinolinium salts are also known. The other alkali metal salts, $Te(OH)_6.nMF$, are prepared from aqueous solutions of the component compounds[224].

Peroxo-acids

Peroxoselenous acid, HOSeO(OOH), an isomer of selenic acid, is made by reaction of selenium dioxide or selenous acid with 100% hydrogen peroxide at 0°. The compound is unstable with respect to loss of oxygen and cannot be obtained pure; it appears to be a stronger oxidizing agent than hydrogen peroxide[221] and is probably an intermediate in the oxidation of selenous to selenic acid by hydrogen peroxide. Permonoselenic acid, H_2SeO_5, is best prepared[227] by hydrolysis of chloroselenic acid with 100% hydrogen peroxide at −40°; it is also obtained in 35% yield by the action of hydrogen peroxide on the selenium trioxide–pyridine complex, SeO_3.py, at −78°. The acid decomposes at room temperature with the loss of oxygen, but is stable at −10°. Perdiselenic acid, $H_2Se_2O_8$, is unknown and cannot be made by anodic oxidation of selenates in selenic acid solution[228].

Potassium peroxo-orthotellurate, $K_2H_4TeO_7$, made[229] by reaction of potassium orthotellurate with hydrogen peroxide at 10°, loses oxygen rapidly in aqueous solution at room temperature. The free acid is known only in aqueous solution and decomposes with loss of oxygen when heated or upon concentration; its aqueous solution is obtained by passing a solution of the potassium salt through a cation exchange column with the resin in the hydrogen form[230]. Peroxoselenic and telluric acids are both stronger oxidizing agents than either hydrogen peroxide or the parent acid.

Oxothio-acids

The configurations of the anions in the isomeric selenosulphates, $M^I_2SO_3Se$, and thioselenates, $M^I_2SeO_3S$, are probably as shown in II and III. The salts are made by boiling an aqueous solution of, respectively, a sulphite with selenium or a selenite with sulphur, in both cases under reflux. Thioselenates isomerize to selenosulphates in aqueous

226 A. Engelbrecht and F. Sladky, *Inorg. Nuclear Chem. Letters*, **1** (1965) 15.
227 M. Schmidt and P. Bornmann, *Z. anorg. Chem.* **330** (1964) 329.
228 L. M. Dennis and J. P. Koller, *J. Am. Chem. Soc.* **41** (1919) 949.
229 E. Montignie, *Z. anorg. Chem.* **253** (1945) 90.
230 M. Schmidt and P. Bornmann, *Naturforsch.* **19***B* (1964) 73.

solution on prolonged boiling, traces of the latter appearing after about 30 min, so that the preparative reaction must be restricted to 20–30 min. Some selenotrithionate (p. 980) is formed in the preparation of selenosulphates, but this is less soluble than the latter and crystallizes out when the solution is cooled. The infrared spectrum of potassium thioselenate has been recorded[231].

Selenosulphates and thioselenates decompose in acid solution with precipitation of selenium and sulphur respectively; thioselenic acid, formed by the action of hydrogen sulphide on chloroselenic acid, and its esters, made by reaction of chloroselenic acid or

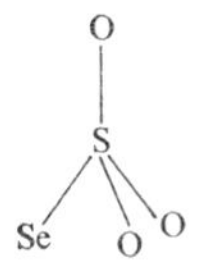

(II) 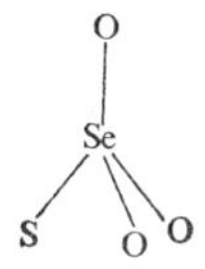 (III)

selenium trioxide with mercaptans and thiophenols, are equally unstable[232]. Aqueous barium hydroxide or chloride decompose selenosulphates in the same way as acids. Solid selenosulphates decompose on heating, yielding a mixture of selenides, polysulphides and sulphate. Silver selenide or sulphide are precipitated respectively when silver salts are added to aqueous selenosulphates or thioselenates; mercury salts react in the same way as silver. Both isomers are oxidized by iodine, presumably with the formation of isomeric diselenotetrathionates[233] (p. 980).

Selenopolythionates of the type $M_2Se_xS_yO_6$ ($x = 1$ or 2; $y = 2$ or 4) are known, but the only tellurium compounds recorded are the telluropentathionates, $M^I_2TeS_4O_6$, and no polonium analogues have been prepared. The selenotrithionates, $M^I_2SeS_2O_6$, are not isomorphous with the corresponding sulphur compounds, $M^I_2S_3O_6$, but the seleno- and telluropentathionates are isomorphous with their sulphur analogues, $M^I_2S_5O_6$. There is, however, a marked difference between the reactions of the selenium and sulphur compounds with sulphite and thiosulphate ions. In the presence of these ions equilibrium is established between selenotrithionate and selenopentathionate in solution:

$$[Se(SO_3)_2]^{2-} + 2S_2O_3^{2-} \rightleftharpoons [Se(S_2O_3)_2]^{2-} + 2SO_3^{2-}$$

Addition of a sulphite ion acceptor, such as formaldehyde, displaces the equilibrium to the right, the sulphite group in the selenotrithionate being replaced by thiosulphate; thus the selenotri- and pentathionates can be regarded as sulphito- and thiosulphato-complexes respectively, and are written as such. The sulphur analogues are also of this type, but addition of sulphite ion to a pentathionate leads to quantitative replacement of thiosulphate ion and no equilibrium is established[234]. Thiosulphate ion is also displaced from selenopentathionates by diethyldithiocarbamate, the selenium complexes of the latter being formed (p. 1008). The anions in the two possible diselenium tetrathionates, $M^I_2Se_2S_2O_6$ and $M^I_2S_2Se_2O_6$, are presumably of the form $[O_3S–Se–Se–SO_3]^{2-}$ and $[O_3Se–S–S–SeO_3]^{2-}$,

231 A. Kozhakova, E. A. Buketov, M. I. Bakeev and A. K. Shokanov, *Zhur. neorg. Khim.* **11** (1966) 1782; *Russ. J. Inorg. Chem.* **11** (1966) 953.
232 M. Schmidt and I. Wilhelm, *Z. anorg. Chem.* **330** (1964) 324.
233 O. M. Baram and M. P. Soldatov, *Zhur. neorg. Khim.* **2** (1957) 1289.
234 O. Foss, *Acta Chem. Scand.* **3** (1949) 435.

analogues to the tetrathionate ion, $[O_3S\text{–}S\text{–}S\text{–}SO_3]^{2-}$, but no structural data are available.

The selenotrithionates, all colourless, crystalline solids, are made by reaction of a sulphite with dimeric selenium acetylacetonate[235] (p. 1002) or of a solution of a selenosulphate with a bisulphite and selenous acid.

$$(C_5H_6O_2{:}Se)_2+4HSO_3^- \longrightarrow 2C_5H_8O_2+2Se(SO_3)_2^{2-}$$

They are also obtained by oxidizing a mixture of selenosulphate and sulphite with iodine or hydrogen peroxide, avoiding an excess in the case of iodine. This is because diselenotetrathionate, $Se_2S_2O_6^{2-}$, is formed in the oxidative procedure and this reacts with sulphite:

$$2SeSO_3^{2-}+I_2 \longrightarrow Se_2S_2O_6^{2-}+2I^-$$
$$Se_2S_2O_6^{2-}+SO_3^{2-} \longrightarrow Se(SO_3)_2^{2-}+SeSO_3^{2-}$$

The overall reaction is:

$$SeSO_3^{2-}+SO_3^{2-}+I_2 \longrightarrow Se(SO_3)_2^{2-}+2I^-$$

The penultimate reaction is reversed on heating owing to the evolution of sulphur dioxide[236].

Although many of the salts decompose slowly on exposure to light or X-rays, the crystal structures of a number of them have been determined[237]. All of the known salts are stable in cold, dilute acid but are decomposed by hot hydrochloric acid. The free acid is best prepared from selenium acetylacetonate in the same way as the salts, but it decomposes when concentrated by evaporation under reduced pressure, with the separation of selenium and evolution of sulphur dioxide[235]. Selenotrithionates decompose slowly in aqueous solution in a similar manner, but some sulphate and selenosulphate are formed as well. The decomposition is catalysed by iodide ion, the first step involving the formation of diselenotetrathionate which subsequently reacts with the sulphurous acid liberated in the first step to regenerate trithionate and, in part, with selenosulphate to form triselenopentathionate[238], $Se_3S_2O_6^{2-}$. A mixture of selenopolythionic acids is formed in the reaction of aqueous sulphurous and selenous acids; the acids $H_2SeS_2O_6$, $H_2Se_2S_2O_6$, $H_2Se_3S_2O_6$ and $H_2Se_4S_2O_6$ have been detected in the mixture[238a], but the reaction appears to be complex and is not well understood. Reaction of selenotrithionates with selenous acid in mineral acid solution also results in the formation of a complex mixture of polyselenothionates, $(Se_nS_2O_6)$, in which $n = 2$ to 6. Nitron salts of these are known, but they decompose in solution, very rapidly at high pH and slowly in neutral solution[239].

Potassium diselenotetrathionate monohydrate, a yellow solid which can be dehydrated at 50–60° and recrystallized from dilute (0·5 N) hydrochloric acid, is sensitive to light and reacts with sulphurous acid as described above. It is obtained by fractional crystallization of the mother liquor from the preparation of the selenotrithionate or by reaction of selenosulphate and sulphite with selenous acid, a reaction in which some triselenopentathionate is formed; this reacts further with sulphite to form selenotrithionate[240]. The acid is apparently

[235] G. T. Morgan and J. D. Main Smith, *J. Chem. Soc.* **119** (1921) 1066.

[236] I. V. Yanitskii and V. I. Zelenkaite, *J. Gen. Chem. U.S.S.R.* **25** (1955) 805.

[237] O. Foss and O. H. Mörch, *Acta Chem. Scand.* **8** (1954) 1169.

[238] V. Zelionkaite, J. Janickis and J. Suliakene, *Lietuvos TSR Aukstuju Mokylu Mokslo Darbai, Chem. ir Chem. Technol.* **4** (1964) 75, 85. (*C.A.* **61** (1964) 9167f; *ibid.* **62** (1965) 62e.)

[238a] E. Blasius and B. Schoenhard, *J. Chromatog.* **28** (1967) 385.

[239] V. Zelionkaite, J. Janickis and D. Kudarauskiene, *Lietuvos TSR Mokslu Akad. Darbai*, Ser. B (1964) 103, 107. (*C.A.* **61** (1964) 5171, 5186.)

[240] I. V. Yanitskii and V. I. Zelenkaite, *Zhur. neorg. Khim.* **2** (1957) 1349.

formed in the hydrolysis of selenium monochloride in acid or neutral sulphite solution by way of the unstable hydroxide, $Se_2(OH)_2$:

$$Se_2Cl_2+2H_2O \longrightarrow Se_2(OH)_2+2HCl$$
$$2Se_2(OH)_2 \longrightarrow 3Se+H_2SeO_3+H_2O$$
$$Se_2(OH)_2+2SO_3^{2-}+2H_2O \longrightarrow H_2Se_2S_2O_6+4OH^-$$

Yellow-green sodium selenopentathionate, $Na_2Se(S_2O_3)_2$, is best prepared by treating an excess of selenous acid or selenium dioxide with sodium thiosulphate in concentrated acetic acid at 0°, the excess being necessary because thiosulphate ion catalyses the decomposition of selenopentathionate. The product is less soluble than sodium tetrathionate, which is

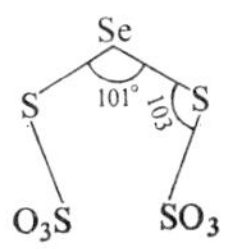

(IV)

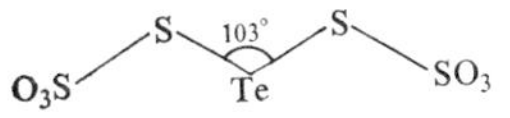

(V)

formed at the same time, and is separated from the latter by recrystallization from 0·2 N hydrochloric acid. The barium, and other alkali metal, salts are prepared from the sodium one by metathesis with the acetates in 0·2 N hydrochloric acid.

All the known selenopentathionates, most of which are hydrated, are of orthorhombic symmetry[242, 243] and the selenopentathionate ion is an unbranched and non-planar chain, with the SO_3 groups *cis* (IV), as in the sulphur analogue, whereas the telluropentathionate ion has the *trans* configuration[244] (V).

Like the sulphur and tellurium compounds, selenopentathionates are decomposed by alkali:

$$2[Se(S_2O_3)_2]^{2-}+6OH^- \longrightarrow 4S_2O_3^{2-}+Se+SeO_3^{2-}+3H_2O$$

The reaction probably involves the unknown selenium(II) hydroxide, $Se(OH)_2$, as an intermediate. Both the sulphur and selenium pentathionates are decomposed by aqueous cyanide, but telluropentathionates are not[245].

The telluropentathionates, $M_2Te(S_2O_3)_2$, are yellow to orange-red solids, prepared in the same way as the selenium compounds[242]. The anion is oxidized by iodine with the formation of tellurium(IV) and tetrathionate and, as with the selenium compounds, thiosulphate groups are displaced by diethyldithiocarbamate or ethyl xanthate[246]. Crystallographic data are available for the alkali metal salts[242, 247] and the anion is known to be an unbranched and non-planar chain (V).

2.9. SULPHIDES

Most of the reported selenium and tellurium sulphides are not true compounds but are merely mixtures of the elements; there is, however, spectroscopic evidence for the existence

[241] J. Janickis and D. Reingardas, *Lietuvos TSR Aukstuju Mokylu Mokslo Darbai, Chem. ir Chem. Technol.* **2** (1962) 47 (*C.A.* **60** (1964) 1324b).
[242] O. Foss and J. Jahr, *Acta Chem. Scand.* **4** (1950) 1560.
[243] O. Foss and O. Tjomsland, *Acta Chem. Scand.* **8** (1954) 1701.
[244] O. Foss, *Acta Chem. Scand.* **7** (1953) 1221.
[245] O. Foss, *Acta Chem. Scand.* **4** (1950) 1241.
[246] O. Foss, *Acta Chem. Scand.* **3** (1949) 708.
[247] O. Foss and P. A. Larssen, *Acta Chem. Scand.* **8** (1954) 1042.

of TeS molecules[248] and examples of replacement of sulphur in S_8 rings by selenium or tellurium are known. The red species of composition Se_4S_4, obtained by fusing equiatomic amounts of the elements, followed by fractional crystallization of the benzene extract, is well defined and other phases[53, 249] identified in similar experiments include Se_2S_6, SeS_7 and TeS_7; evidence for the last has also been obtained by mass spectrometry[250]. A sodium thiotellurate(IV), Na_2TeS_3, is reported[251] to be formed by dissolution of tellurium in solutions of sodium sulphides (Na_2S_2 and Na_2S_3 but not Na_2S) or of sulphur in solutions of sodium telluride, Na_2Te_2, but little is known about the compound.

Polonium monosulphide, PoS, a black solid precipitated by hydrogen sulphide from aqueous solutions of any polonium compound in dilute acid, is also obtained by treating polonium(IV) hydroxide with yellow ammonium sulphide, in which polonium monosulphide is insoluble. The compound decomposes to the elements at 275° in a vacuum and is readily oxidized by aqueous bromine, hypochlorite or *aqua regia*. The solubility product[33] is about $5{\cdot}5 \times 10^{-29}$.

2.10. NITRIDES AND AZIDES

Tetraselenium tetranitride, Se_4N_4, is an orange, hygroscopic solid which is explosive and very sensitive to shock when dry, although it can be stored under chloroform. Like the sulphur analogue, the colour varies with temperature, becoming red at 100° and yellow-orange at liquid air temperature; it decomposes violently at 160°. The compound is prepared by reaction of anhydrous ammonia with selenium halides[252] or dialkyl selenites[253] in solvents such as benzene, carbon disulphide or chloroform, selenium tetrabromide giving

(VI)

the best yield, and by reaction of selenium dioxide with anhydrous liquid ammonia at 70–80° under pressure[254]. Reaction of ammonia with selenium monohalides in ether yields $(Se_2NCl)_n$ or $(Se_2N_2Br)_n$, both of which are probably cyclic species derived from Se_4N_4 [252]. There is also spectroscopic evidence for the formation of the NSe molecule in the reaction of active nitrogen with selenium chlorides[255].

The crystal symmetry of tetraselenium tetranitride is monoclinic and the structure of the molecule is the boat form of a puckered 8-membered ring in which the nitrogen atoms almost form a square and the selenium atoms are at the corners of a slightly deformed tetrahedron (VI), so that the molecule must be subject to considerable strain[256]; the infrared spectrum[254]

[248] H. Mohan and K. Majumdar, *Proc. Phys. Soc.* **77** (1961) 147.
[249] R. Cooper and J. V. Culka, *J. Inorg. Nuclear Chem.* **27** (1965) 755.
[250] R. Cooper and J. V. Culka, *J. Inorg. Nuclear Chem.* **29** (1967) 1217, 1877.
[251] T. N. Greiver and I. G. Zaitseva, *Zhur. priklad. Khim.* **40** (1967) 1920.
[252] W. Strecker and L. Claus, *Ber.* **56** (1923) 362.
[253] W. Strecker and H. E. Schwarzkopf, *Z. anorg. Chem.* **221** (1934) 193.
[254] J. Jander and V. Doetsch. *Angew. Chem.* **70** (1958) 704; *Chem. Ber.* **93** (1960) 561.
[255] P. Goudmand and O. Dessaux, *J. Chim. phys.* **64** (1967) 135.
[256] M. Baernighausen, T. von Volkmann and J. Jander, *Acta Cryst.* **15** (1962) 615; *Angew. Chem.* **77** (1965) 96, *Internat. Edn.* **4** (1965) 72.

is consistent with this structure The enthalpy of formation is +163 kcal/mole and the average N–Se single bond energy is 40 kcal [257].

The nitride is insoluble in most solvents, but is slightly soluble in carbon disulphide and in glacial acetic acid[254]; it is slowly hydrolysed by water or aqueous alkali, with liberation of ammonia, and hydrogen chloride decomposes it, yielding a mixture of ammonium chloride and selenite, hydrogen selenide and selenium tetrachloride. The compound explodes with halogens, hypochlorite or concentrated hydrochloric acid, but it reacts less violently with chlorine diluted with carbon dioxide, the main product being selenium tetrachloride. With dry bromine in carbon disulphide the product is a greenish-brown hygroscopic solid of composition SeN_2Br_4, which is probably a cyclic polymer, but with moist bromine or hydrobromic acid[252] ammonium hexabromoselenate(IV) is formed (p. 1006). Reaction of tetrasulphur tetranitride with selenium monochloride in thionyl chloride yields a mixture of S_4N_3Cl and $(SeS_3N_3)_2SeCl_6$, the average composition being $(SeS_2N_2)Cl_5$; the $(SeS_3N_3)^+$ cation is cyclic[258]. Thionyl chloride reacts with tetraselenium tetranitride to form a pale yellow solid of composition $(NH_4)_2Se(SO)_2Cl_4$, a curious species which is said to exist in two crystal modifications[259].

Tellurium nitride is a citron-yellow solid which is as explosive as the selenium compound; its composition[260] appears to be closer to Te_3N_4 than to Te_4N_4 but its behaviour with halogens, halogen acids and thionyl chloride is very similar to that of the selenium compound. The tellurium compound is best prepared by reaction of tellurium tetrabromide with anhydrous liquid ammonia at −15° to −70°, a reaction in which such species as $Te_3N_2Br_6$ are reported to be formed[261], and by reaction of potassium triimidotellurate(IV), $K_2Te(NH)_3$, with ammonium nitrate in anhydrous liquid ammonia. The imido compound is prepared by the reaction of diphenyl telluride with potassamide in liquid ammonia[262].

The only azide recorded is tellurium azide trichloride, TeN_3Cl_3, prepared by reaction of trimethylsilylazide with tellurium tetrachloride[262a].

2.11. CARBON SELENIDES AND TELLURIDES

Carbon monoselenide, CSe, is obtained by dissociation of the diselenide in a high frequency discharge[263]; it reacts with solid sulphur or tellurium, forming carbon thioselenide, CSSe, or telluroselenide, CSeTe, the products being identified by their infrared spectra at −190°. Similar experiments on the dissociation of carbon disulphide in the presence of solid selenium or tellurium yield the thioselenide, CSSe, and thiotelluride, CSTe, respectively. The thiotelluride is also formed by striking an arc under carbon disulphide between graphite electrodes, one of which is impregnated with tellurium.

Carbon thioselenide, a white solid at −190° which becomes bright yellow at −80°, is a dark yellow liquid at room temperature and boils at 83·9°/749·2 mm. The vapour is lachrymatory and smells of onions at low concentrations in air. It is much less inflammable than the disulphide.

257 C. K. Barker, A. W. Cordes and J. L. Margrave, *J. Phys. Chem.* **69** (1965) 334.
258 A. J. Bannister and J. S. Padley, *J. Chem. Soc.* (*A*) (1967) 1437.
259 H. Garcia-Fernandez, *Compt. rend.* **258** (1964) 2579.
260 S. Strecker and C. Mahr, *Z. anorg. Chem.* **221** (1934) 199.
261 W. Strecker and W. Ebert, *Ber.* **58** (1925) 2527.
262 O. Schmitz-DuMont and B. Ross, *Angew. Chem. Int. Ed.* **6** (1967) 1071.
262a N. Wiberg and K. H. Schmid, *Angew. Chem. Int. Ed.* **6** (1967) 953.
263 R. Stüdel, *Angew. Chem. Int. Ed.* **6** (1967) 635.

The thioselenide is best made[264] by heating ferrous selenide in carbon disulphide vapour at 650°. It is insoluble in water, but soluble in most organic solvents, and it dissolves iodine and, to a small extent, sulphur; it does not dissolve selenium nor does it react with hydrogen sulphide or selenide, but it is decomposed on exposure to light or on heating and by ammonia, bromine or chlorine. A xanthate, $SeCSNaOC_2H_5$, is formed with ethanolic sodium hydroxide.

The analogous thiotelluride, CSTe, a yellowish-red solid, melts at −54° to a bright red liquid which decomposes at room temperature. It is decomposed by light above −50° and apparently smells of garlic[265].

Carbon diselenide, CSe_2, a bright yellow, unpleasant smelling, lachrymatory liquid, boils at 125–126° and freezes at −40° to −45°. It is prepared[266] by heating selenium at 500° in a mixture of nitrogen and dichloromethane vapour, and by heating hydrogen selenide with carbon tetrachloride vapour at 500°.

The diselenide is miscible with carbon disulphide and with most organic solvents; it decomposes slowly at room temperature to form polymers and is rapidly decomposed by chlorine or ammonia. A xanthate, $SeCSeNaOC_2H_5$, is formed with ethanolic potassium hydroxide. The thermochemistry of the diselenide is known in some detail[267].

The violet triselenocarbonates, such as $BaCSe_3$, are precipitated from aqueous or ethanolic solutions of the metal hydrogen selenide (e.g. $Ba(SeH)_2$) by carbon diselenide; the orange dithioselenocarbonates, such as K_2CS_2Se, are prepared in the same way, but with carbon disulphide[268, 269]. All of these salts are very sensitive to oxidation when moist. The reaction of carbon thioselenide with a metal hydrogen selenide or sulphide would presumably yield a thiodiseleno- and a dithioselenocarbonate.

The free acid, H_2CSe_3, is dark red in aqueous solution and is obtained by the action of hydrochloric acid on the barium salt. It is very unstable and its decomposition in aqueous solution is first order[270].

2.12. CARBONATES

The only carbonate recorded for the three elements is the unstable polonium(IV) compound, a white solid obtained by allowing polonium(IV) hydroxide to stand under saturated aqueous carbon dioxide for a long time[9].

2.13. SALTS OF ORGANIC ACIDS

Selenium(II) and tellurium(II) alkyl and aryl thiosulphonates, $(RSO_2S)_2X$, are structural analogues of the seleno- and telluropentathionates (p. 981), all three containing an unbranched S–S–X–S–S chain; they react with thiosulphate to form the pentathionates and the thiosulphate groups are also displaced by diethyldithiocarbamate[271].

264 H. V. A. Briscoe, J. P. Peel and P. L. Robinson, *J. Chem. Soc.* (1929) 59, 1049.
265 A. Stock and P. Praetorius, *Ber.* **47** (1914) 131.
266 D. J. G. Ives, R. W. Pittman and W. Wardlaw, *J. Chem. Soc.* (1947) 1080.
267 G. Gattow and M. Dräger, *Z. anorg. Chem.* **343** (1966) 232.
268 H. Seidel, *Naturwiss.* **52** (1965) 539.
269 H. Hofman-Bang and B. V. Rasmussen, *Naturwiss.* **52** (1965) 660.
270 G. Gattow and M. Dräger, *Z. anorg. Chem.* **348** (1966) 229; *ibid.* **349** (1967) 202.
271 O. Foss, *Acta Chem. Scand.* **5** (1951) 115; *ibid.* **6** (1952) 521.

Selenium(II) methanethiosulphonate is a pale green solid which decomposes above 75°; it is prepared by reaction of selenium monochloride with sodium methanethiosulphonate in ether:

$$Se_2Cl_2+2CH_3SO_2SNa \longrightarrow (CH_3SO_2S)_2Se+Se+2NaCl$$

All three products precipitate and, after removal of the sodium chloride by a water wash, the selenium compound is obtained pure by recrystallization from ethyl acetate or by dissolution in chloroform, from which it is precipitated by carbon disulphide. Tellurium methanethiosulphonate, a yellow solid which decomposes above 120°, is precipitated when the aqueous sodium salt is added to a solution of tellurium dioxide in hydrochloric acid:

$$TeO(OH)^+ +4CH_3SO_2S^- +3H^+ \longrightarrow (CH_3SO_2S)_2Te+(CH_3SO_2S)_2+2H_2O$$

It is recrystallized from chloroform. The analogous benzene- and toluenethiosulphonates have also been made[271]. These compounds are decomposed by aqueous alkali with precipitation of elementary selenium or tellurium.

A basic selenium(IV) acetate, $SeO(CH_3COO)_2$, is obtained by reaction of selenium(IV) oxodichloride with sodium acetate in ethanol; the basic oxalate, $SeO(C_2O_4)$, is prepared by the analogous reaction with silver oxalate in tetrahydrofuran. The structure of the oxalate consists of a pyramidal SeO_3 group with the oxalate group planar[272]. These salts are readily hydrolysed to selenous acid or, with alcohols, to the ester.

Tellurium(IV) acid citrate and tartrate are said to be formed by prolonged (2 months) digestion of the dioxide with the appropriate acid in water at 70°, but the oxalate and succinate cannot be made in this way[273]. The best preparative method for the tellurium(IV) carboxylates is by heating tellurium tetrachloride with the acid in benzene at 60–70°, followed by a final heating of the product above the melting point of the acid until hydrogen chloride is no longer evolved. The salts are slightly hygroscopic, soluble in acetone, ethanol or ethyl acetate, but insoluble in carbon tetrachloride, chloroform or ether[274]. A few polonium(IV) carboxylates have been made by treating polonium tetrachloride or hydroxide with the aqueous acid, but the compositions of these products is unknown[147].

2.14. ALKOXIDES

Selenium and polonium alkoxides have not been reported; the tellurium compounds, $Te(OR)_4$, are all liquids, except for the methoxide, which is a solid. They are prepared by reaction of tellurium tetrachloride with a sodium alkoxide and are all readily hydrolysed[275, 276]. Selenium(IV) dichloride dialkoxides, $Cl_2Se(OR)_2$, all of which are yellow, moderately volatile liquids, are obtained by reaction of alkyl selenites with chlorine in carbon tetrachloride[277]; the analogous tellurium(IV) compounds, $TeCl_x(OR)_{4-x}$, have been made by the action of acetyl chloride on the tetra-alkoxide in benzene and are white solids or colourless liquids; reaction of the tetra-alkoxides with hydrogen chloride yields only

[272] R. Paetzold, *Z. Chem.* **6** (1966) 72.
[273] A. A. Hageman, *J. Am. Chem. Soc.* **41** (1919) 342.
[274] S. Prasad and B. L. Khandelwal, *J. Indian Chem. Soc.* **39** (1962) 67.
[275] P. Dupuy, *Compt. rend.* **240** (1955) 2238.
[276] R. C. Mehrotra and S. N. Mathur, *J. Indian Chem. Soc.* **42** (1965) 1.
[277] R. C. Mehrotra and S. N. Mathur, *Indian J. Chem.* **5** (1967) 375.

$Te(OR)_3Cl$ [278]. Salts of trimethoxyselenous acid, such as $Na[SeO(OCH_3)_3]$, are also known[212].

2.15. NITRATES

Although a tellurium(II) nitrate, $Te(NO_3)_2$, is said[132] to result from the reaction of tellurium tetrabromide with silver nitrate in boiling benzene, its existence is highly unlikely. The only other tellurium nitrate known is the white basic compound, $2TeO_2.HNO_3$, prepared by dissolving tellurium in 8 N nitric acid at 60–70° and evaporating the resulting solution. It decomposes to the dioxide above 190° and is hygroscopic; it is hydrolysed

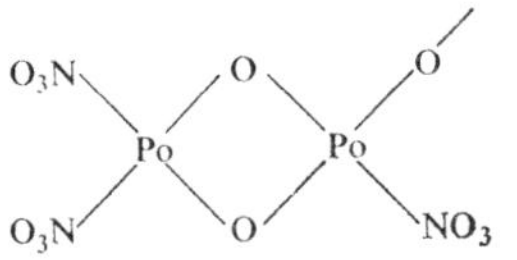

(VII) 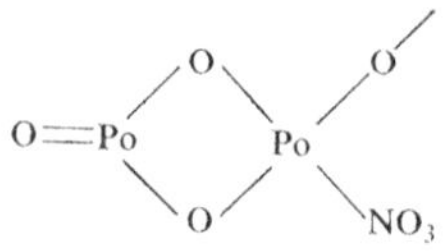 (VIII)

slowly by cold water but rapidly on boiling[279, 280]. The crystal symmetry of the compound is orthorhombic and the structure consists of puckered layers of tellurium and oxygen atoms, with each tellurium atom linked to one other by two oxygen bridges and to two other tellurium atoms by single oxygen bridges; these four Te–O bonds are directed approximately to the two axial and to two of the three equatorial positions of a trigonal bipyramid, as in the tetrachloride structure[281].

Polonium tetranitrate, $Po(NO_3)_4$, a white solid formed by the decomposition of its dinitrogen tetroxide adduct, $Po(NO_3)_4.N_2O_4$, decomposes to the basic salt(VII) in a vacuum. The adduct is the product of the reaction of liquid dinitrogen tetroxide with polonium(IV) hydroxide or chloride; the metal does not react with liquid dinitrogen tetroxide[282].

The white basic nitrate(VII) is prepared by vacuum drying the solid product obtained when polonium(IV) hydroxide or chloride is treated with 0·5 N nitric acid for 12 hr or by exposing polonium metal to a mixture of oxygen and gaseous dinitrogen tetroxide; because oxides of nitrogen are formed by the fixation of nitrogen in air by the α-bombardment, the same end-product is obtained by sealing up polonium metal or its halides in an air-filled container. This basic nitrate decomposes at about 100° to a second basic nitrate(VIII), a yellowish-white solid which is also obtained by evaporating nitric acid solutions of polonium(IV) to a small volume and vacuum drying the residue; it decomposes at 130° to the dioxide[282]. Both of these compounds are probably derived from the unknown $PoO(NO_3)_2$.

2.16. PHOSPHATES

Selenium phosphates are unknown and although a tellurium(IV) phosphate may exist[279], there is no information on its composition or method of preparation. A white, basic

278 R. C. Mehrotra and S. N. Mathur, *Indian J. Chem.* **5** (1967) 206.
279 D. Klein and J. Morel, *Bull. Soc. chim. France*, **43** (1885) 198; *Ann. Chim. Phys.* **5** (1885) 59.
280 J. F. Norris, H. Fay and D. W. Edgerly, *Am. Chem. J.* **23** (1900) 105.
281 L. N. Swink and G. B. Carpenter, *Acta Cryst.* **21** (1966) 578.
282 K. W. Bagnall, D. S. Robertson and M. A. A. Stewart, *J. Chem. Soc.* (1958) 3633.

polonium(IV) phosphate, $2PoO_2.H_3PO_4$, the virtually insoluble gelatinous product of the reaction of 2 M phosphoric acid with polonium(IV) hydroxide or of a soluble phosphate with polonium tetrachloride, appears to be stable to hydrolysis by water or dilute aqueous ammonia, but is decomposed by dilute alkali or mineral acids[9].

2.17. SULPHATES, SELENATES AND TELLURATES

The simple selenium(II) and tellurium(II) sulphates are unknown, although complexes of the latter have been recorded (p. 1008), and polonium(II) sulphate has not been isolated from its aqueous solution, obtained by reducing polonium(IV) disulphate with hydroxylamine in 1–2 N sulphuric acid[152] at 100°.

Selenium(IV) basic sulphate, $SeO_2.SO_3$ or $SeO(SO_4)$, is a deliquescent, colourless solid prepared by dissolving selenous acid in sulphuric acid saturated with sulphur trioxide[283] or by heating selenium dioxide with sulphur trioxide in a sealed tube; the basic selenate, Se_2O_5 or $SeO_2.SeO_3$, is prepared from selenium dioxide in the same way but with selenium trioxide[166, 284]; both compounds are readily hydrolysed. When heated, the sulphate loses sulphur trioxide whereas the selenate melts at 224° and sublimes at $145°/10^{-5}$ mm. The selenate is also formed at 240° in the thermal decomposition of selenium trioxide[285].

The tellurium(IV) and polonium(IV) basic sulphate and selenate are of the form $2XO_2.YO_3$, probably an oxygen bridged structure (IX), so differing markedly from the

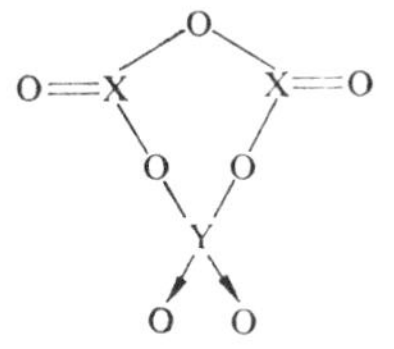

(IX)

selenium compounds, but the tellurium(IV) tellurate has the composition $TeO_2.TeO_3(Te_2O_5)$ and appears to be analogous to the selenium sulphate and selenate. Basic tellurium sulphate, $2TeO_2.SO_3$, and selenate, $2TeO_2.SeO_3$, are white solids, prepared by evaporating solutions of tellurium(IV) with sulphuric or selenic acids[279, 283]. The crystal symmetry of the sulphate is orthorhombic[286], and the compound decomposes above 440° in a vacuum or at 500° in air. It is slowly hydrolysed by cold water and more rapidly on heating. The pale yellow basic tellurate[287] is formed in the decomposition of α-TeO_3 or of orthotelluric acid in air at 406°. It is stable to 430° and is insoluble in water, but soluble, with decomposition, in 30% potassium hydroxide solution. A basic tellurate is also said to be formed by evaporating a solution of tellurium dioxide in orthotelluric acid[283], but details of its composition are lacking.

Basic polonium(IV) sulphate, $2PoO_2.SO_3$, and selenate, $2PoO_2.SeO_3$, are white solids obtained by treating polonium tetrachloride or hydroxide with sulphuric (0·02–0·25 N) or

283 R. Metzner, *Ann. Chim. Phys.* **15** (1898) 203.
284 H. - G. Jerschkewitz, *Angew. Chem.* **69** (1957) 562.
285 E. E. Sidorova, G. I. Blagoveshchenskaya, S. N. Kondrat'ev, K. N. Mochalov and A. G. Salimov, *Zhur. neorg. Khim.* **12** (1967) 1113; *Russ. J. Inorg. Chem.* **12** (1967) 589.
286 J. Loub and H. Hubkova, *Z. Chem.* **5** (1965) 341.
287 J. Rosicky, J. Loub and J. Pavel, *Z. anorg. Chem.* **334** (1965) 312.

selenic (0·15–5·0 N) acids. Both compounds become yellow above 250°, the colour reverting to white on cooling, and both decompose to the dioxide at higher temperatures, the sulphate at 550° and the selenate at 400°. Solubility studies indicate that the basic sulphate is metastable with respect to the disulphate, $Po(SO_4)_2$, in contact with 0·1–0·5 N sulphuric acid.

White, hydrated polonium(IV) disulphate is prepared in the same way as the basic compound, but with more concentrated (> 0·5 N) acid. It is less soluble in dilute sulphuric acid than the basic salt, and its solubility increases very markedly with increasing acid concentration, presumably owing to the formation of sulphato-complexes. The purple anhydrous disulphate is obtained either by heating the hydrate or by washing it with dry ether; water is also lost on long standing. The anhydrous salt decomposes to the dioxide above 550° and is insoluble in organic solvents; it appears to be hydrolysed to some extent by ethanol[152].

2.18. "SULPHOXIDES" AND "SELENOXIDES"

The formation of brightly coloured compounds of selenium, tellurium and polonium with the trioxides of sulphur or selenium, or with the corresponding concentrated acids, has been mentioned earlier (p. 973). These were originally reported to have the composition XSO_3 and $XSeO_3$, but the selenium and tellurium "sulphoxides" are now known to be polysulphates or acid sulphates of tetrameric selenium[288] or tellurium[289] cations. The green solution obtained by dissolving selenium in sulphuric acid contains the Se_8^{2+} ion[288]:

$$8Se + 5H_2SO_4 \longrightarrow Se_8^{2+} + 2H_3O^+ + 4HSO_4^- + SO_2$$

This ion is oxidized by selenium dioxide to the yellow Se_4^{2+} ion and crystalline compounds of the latter, such as yellow $Se_4(S_4O_{13})$ and orange $Se_4(HS_2O_7)_2$, have been isolated from oleum solution, in both of which the Se_4^{2+} cation is square planar[289a,b]. Analogous fluorosulphates and fluoroantimonates can be obtained, and it is probable that the compound $Se(SbF_5)_2$ mentioned earlier (p. 946) also contains a polymeric selenium cation.

The red Te_4^{2+} sulphate is analogous to the Se_4^{2+} compound and a yellow tellurium compound, which may contain the Te_4^{4+} ion, has also been isolated from solutions of tellurium in 100% sulphuric acid[289]. The red polonium compounds, obtained by reaction of the element with concentrated sulphuric or selenic acids[152], may also involve polymeric polonium cations but, unlike the selenium and tellurium compounds, which are decomposed by water with the precipitation of elementary selenium or tellurium, the polonium compounds dissolve completely in water yielding a solution containing polonium(II). The solid polonium compounds decompose spontaneously to polonium monoxide, whereas the tellurium compounds yield a mixture of tellurium, its dioxide and oxides of sulphur or selenium on thermal decomposition.

Solutions of sulphur in selenic acid are indigo, presumably because of the formation of a sulphur selenate, and the latter appears to be formed to some extent in the reaction of sulphur with selenium trioxide in liquid sulphur dioxide at −10°, selenium(IV) selenate

288 J. Barr, R. J. Gillespie, R. Kapoor and K. C. Malhotra, *Canad. J. Chem.* **46** (1968) 149.
289 R. J. Gillespie, R. Kapoor and G. P. Pez, *J. Am. Chem. Soc.* **90** (1968) 6855.
289a J. Barr, D. B. Crump, R. J. Gillespie, R. Kapoor and P. K. Ummet, *Canad. J. Chem.* **46** (1968) 3607.
289b I. D. Brown, D. B. Crump, R. J. Gillespie and D. P. Santry, *Chem. Comm.* (1968) 853.
289c A. Ruzicka and K. Dostal, *Z. Chem.* **7** (1967) 394.
289d N. J. Bjerrum and G. P. Smith, *Progress in Co-ordination Chemistry*, p. 771 Ed. M. Cais (1968).

(p. 987) also being formed[289c]. Similarly, the green selenium selenate is obtained by reaction of selenium with its trioxide[169], although a compound of composition $(Se_2O_3)_2.SO_2$ is formed when this reaction is carried out in liquid sulphur dioxide[289c].

There is also evidence for the formation of the purple Te_4^{2+} ion in the reduction of tellurium tetrachloride with tellurium in a $NaCl/AlCl_3$ eutectic and Te_2AlCl_4 (probably $Te_4(AlCl_4)_2$) is obtained as a dark purple solid by fusing together tellurium, tellurium tetrachloride and aluminium chloride[289d].

2.19. CHROMATES

An orange-yellow polonium(IV) chromate, probably $Po(CrO_4)_2$, is obtained by treating polonium(IV) hydroxide or chloride with aqueous chromium trioxide[9]; it is readily hydrolysed to the dark brown basic salt, $2PoO_2.CrO_3$.

2.20. HALATES AND PERHALATES

The iodates are the only halates known for the three elements; of these tellurium(II) iodate, said[132] to be formed by reaction of tellurium tetrabromide with silver iodate in benzene, is as doubtful as the dinitrate (p. 986), and the selenium(VI) iodates, $I_2O_5.SeO_3$ and $I_2O_5.2SeO_3.H_2O$, colourless solids prepared by the action of selenium trioxide with, respectively, iodine pentoxide or iodic acid, are both polymeric iodyl selenates[290], $[(IO_2)_2SeO_4]_n$ and $[(IO_2)(HSeO_4)]_n$. The white polonium(IV) iodate, probably $Po(IO_3)_4$, is precipitated from nitric acid solutions of polonium(IV) by iodic acid[9]. The compound decomposes slowly at room temperature because of the α-radiation from the polonium and its decomposition is rapid at 350°.

Trihydroxoselenium(IV) perchlorate, $Se(OH)_3ClO_4$, a white, very hygroscopic solid which melts at 33°, is prepared from 70% perchloric acid and selenous acid. Unlike the latter, or selenium dioxide, the compound is soluble in nitromethane in which its conductivity indicates that it is a 1:1 electrolyte[291]; its Raman spectrum indicates that the $Se(OH)_3$ group is pyramidal (C_{3v}), with the Se–O bond somewhat stronger than in the selenite ion, perhaps because of a strong σ bond and some $p\pi d\pi$ bonding[292].

Tellurium(IV) perchlorate[293], $2TeO_2 \cdot HClO_4$ and periodate[229], $2TeO_2 \cdot HIO_4$, are prepared from the dioxide and the perhalic acid; the perchlorate is stable to 300°, so that excess perchloric acid can be removed by heating, and the salt is somewhat more readily hydrolysed than the analogous nitrate. The periodate is also said to be stable to heat; the structures of the two compounds are unknown.

2.21. ORGANO-COMPOUNDS

The only comprehensive reviews of organoselenium and organotellurium chemistry[294,295] are now rather out of date, but a few more recent shorter reviews have appeared in the

[290] G. Kempe and D. Robus, *Z. Chem.* **5** (1965) 394.

[291] E. J. Arlman, *Rec. Trav. Chim.* **58** (1938) 871.

[292] R. Paetzold and M. Garsoffke, *Z. anorg. Chem.* **336** (1965) 52.

[293] F. Fichter and M. Schmid, *Z. anorg. Chem.* **98** (1916) 141.

[294] A. E. Goddard, *Textbook of Inorganic Chemistry*, Ed. J. Newton Friend, Vol. 11, part 4, Chas. Griffin, London, 1937.

[295] E. Krause and A. von Grosse, *Die Chemie der Metallorganischen Verbindungen*, Bornträger, Berlin, 1937.

journals (selenium[296]) or in books[3, 297]. Information on organoselenium chemistry is also often to be found in books on organosulphur chemistry; the volume of literature in these two fields is becoming comparable, but less is known in the organotellurium field and organopolonium chemistry is almost entirely restricted to trace level experiments owing to the charring of the compounds by the α-radiation of the polonium. Identification of the compounds of polonium usually rests on paper chromatographic separations from the tellurium analogues.

Most of the known organo-compounds of the three elements are similar in character to their sulphur analogues; apart from the organometalloid halides, oxides and acids of the type $RSeO_nOH$ ($n = 0$–2), all have unpleasant odours, that of the selenols, RSeH, being particularly repulsive.

The principal classes of organo-compound are given in Table 13, and the relationship between them is expressed schematically for the selenium compounds in Fig. 1. It is clear

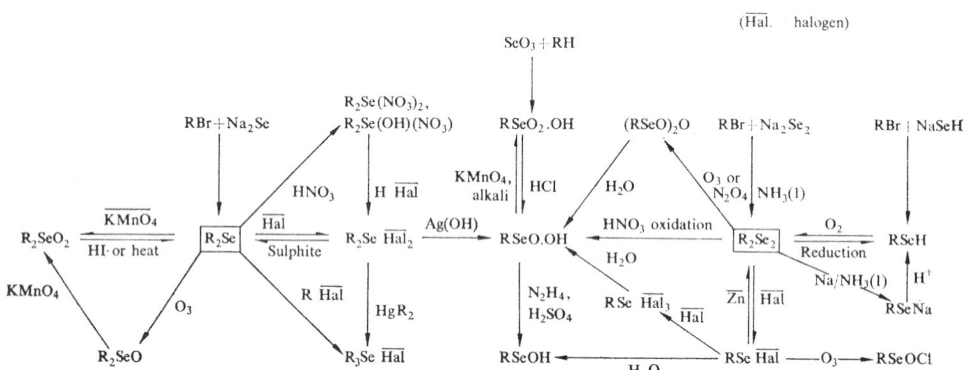

FIG. 1. Reaction scheme for the formation of organo-selenium compounds.

from this scheme that the selenides and diselenides are the two most important types of compound, for the majority of organoselenium compounds can be prepared from them.

By analogy with the other known compounds of the chalcogens, the hydrogen chalcogenides, H_2X, are clearly the parents of the selenols, tellurols, selenides, tellurides and polonides, including the cyclic compounds in which a chalcogen atom is bonded to two carbon atoms. Similarly, the diselenides and ditellurides are formally analogous to the chalcogen monohalides (e.g. to Se_2Cl_2) and the selenoxides and telluroxides are obviously derived from the oxodihalides. Although the selenones and tellurones are evidently derived from the dioxodihalides of the sexavalent elements, it is surprising to find that tellurones exist when the oxo-acids of selenium(VI) and tellurium(VI) are so different in character, a difference which further manifests itself in the non-existence of the telluronic acids. However, the seleninic and tellurinic acids, RXO.OH, are clearly related to the chalcogen(IV) oxo-acids, $XO(OH)_2$. The organochalcogen halides, $RX\overline{hal}_3$, $R_2X\overline{hal}_2$ and $R_3X^+Y^-$, are related to the chalcogen tetrahalides, while the organoselenium monohalides and the selenenic acids, RSeOH, must stem from the unknown selenium dihalides; it is therefore strange to find that the thermodynamically more stable tellurium and polonium dihalides do not appear to give rise to compounds of this type.

296 T. W. Campbell, H. G. Walker and G. M. Coppinger, *Chem. Rev.* **50** (1952) 279.
297 E.g. H. Lumbroso, in *Nouveau Traité de Chimie Minérale*, Vol. 13, Masson, Paris, **1961.**

The preparative and other chemistry of the organo-compounds is described very briefly in the following pages, together with a short account of the known cyclic compounds.

Selenols and Tellurols, RXH

Selenols are colourless when pure, the alkane compounds being volatile liquids and the arene ones solids which have an appreciable vapour pressure. They oxidize readily in air to the corresponding diselenide and are soluble in organic solvents and in aqueous alkali, forming selenolates, but are insoluble in water. The tellurols are scarcely known at all, probably because they are even more readily oxidized than the selenols.

TABLE 13. PRINCIPAL CLASSES OF ORGANO-COMPOUND

General formula	Known for (X)	Nomenclature
RXH	Se, Te (?)	-selenol, -tellurol.
RXR′	Se, Te, Po	-selenide, -telluride, -polonide.
RXXR′	Se, Te	-diselenide, -ditelluride.
RR′XO	Se, Te	-selenoxide, -telluroxide.
$RR'XO_2$	Se, Te	-selenone, -tellurone.
RX Halide	Se	Alkyl/aryl selenium halide.
RX $(Halide)_3$	Se, Te	Alkyl/aryl metalloid trihalide.
R_2X $(Halide)_2$	Se, Te, Po	Dialkyl/aryl metalloid dihalide.
R_3X Halide	Se, Te, Po	Trialkyl/aryl selenonium/telluronium/polononium halide.
RXO_2OH	Se	Alkane/arene selenonic acid.
RXO.OH	Se, Te	Alkane/arene seleninic/tellurinic acid.
RXOH	Se	Alkane/arene selenenic acid.
RR′CX	Se, Te	Seleno-, telluro-carbonyl compounds.
RXCN	Se	Selenocyanates.
$X(CH_2)_nCH_2$ (cyclic)	Se, Te	Selena-, tellura-cycloalkanes.
$X(CH)_nCH$ (cyclic)	Se, Te	Selenophen, tellurophen ($n = 3$).
β-diketone derivatives	Se, Te, Po	Cyclic- 2,4 dioxocyclopentylene chalcogens.
$X(CH_2)_2X'CH_2CH_2$ (cyclic)	Se, Te; X′ = O, S, Se, Te.	X = X′ = Se 1,4 diselenane X = Se, X′ = S, 1,4 selenathiane etc.
Se–N heterocycles	Se	See p. 1004.

Alkane selenols are prepared by reduction of diselenides by dithiothreitol in aqueous solution[298] or with sodium in anhydrous liquid ammonia[299], the alkali selenolate so formed being treated with acid to give the free selenol; they are also obtained by reaction of an alkyl halide or alkali metal alkyl sulphate with a hydroselenide or selenosulphate in a hydrogen atmosphere. The arene compounds are made by reduction of the diselenide or selenocyanate with zinc in acetic acid[300] or by reaction of the appropriate Grignard reagent with selenium in the absence of air[301].

Alkane tellurols are said to be formed in the reaction of alcohols with aluminium telluride at high temperatures in hydrogen[302], a procedure sometimes applied to the preparation of selenols; benzene tellurol may be formed in the reaction of tellurium with phenylmagnesium bromide[303], but even now the existence of the tellurols is still rather uncertain.

298 W. H. H. Guenther, *J. Org. Chem.* **32** (1967) 3931.
299 L. Brandsma and H. Wijers, *Rec. Trav. Chim.* **82** (1963) 68.
300 M. T. Bogert and C. N. Andersen, *Proc. Nat. Acad. Sci. U.S.A.* **11** (1925) 217.
301 D. G. Foster and S. F. Brown, *J. Am. Chem. Soc.* **50** (1928) 1182.
302 A. Baroni, *Atti Accad. naz. Lincei, Rend. Classe Sci. fis. mat. nat.* **27** (1938) 238.
303 M. Guia and F. Cherchi, *Gazzetta*, **50** (1920) 362.

Selenides, Tellurides and Polonides, RXR′

Dialkyl selenides and tellurides are colourless, volatile liquids, any yellow coloration being due to traces of the diselenide or ditelluride; the diaryl compounds, which are almost odourless, are low melting solids, some of which are liquid at room temperature. All are soluble in organic solvents, but insoluble in water; they are decomposed by boiling water, with precipitation of selenium, and form diselenides in the presence of moisture, whereas the tellurides slowly dissolve in cold water owing to oxidation, yielding an alkaline solution containing species of the type $R_2Te(OH)_2$. Diaryl tellurides have some basic character, forming adducts with hydrogen halides such as $(C_6H_5)_2Te.HCl$.

Symmetrical dialkyl selenides or tellurides are prepared by reaction of alkyl halides or dialkyl sulphates with alkali selenosulphate, aluminium telluride or, better, alkali selenide or telluride in anhydrous liquid ammonia[299]. Compounds in which one or both alkyl groups include a carboxylic acid group are prepared in a similar manner, but with the α-halogeno-acid in place of the alkyl halide. Dialkyl and alkaryl selenides are also commonly made by reaction of alkyl halides with alkali selenolates:

$$RSeNa + R'Br \longrightarrow RSeR' + NaBr$$

A wide variety of other methods have been used to prepare dialkyl selenides, but the yields are not usually as good as in those described above. Silyl selenides and tellurides, $(R_3Si)_2X$ and R_3SiXR', germyl tellurides, $(R_3Ge)_2Te$ and $(R_3Ge)TeR$, and the analogous tin and lead derivatives are also known; their infrared spectra have been recorded[304] and a complete assignment of the infrared spectrum of $(CD_3)_2Se$ is available[305].

Trace amounts of dialkyl and diaryl polonides are formed in the decay of radioactive lead(^{210}Pb) tetramethyl[306] and bismuth (^{210}Bi) triaryls[307, 308]. Dimethyl polonide is also formed in the reaction of dimethyl sulphate with sodium telluride/polonide mixtures[309] and, probably, by reaction of methyl radicals formed from acetic acid by the α-bombardment from the polonium[147]; other polonides appear to be formed in the reaction of Grignard reagents with tellurium tetrachloride containing tracer amounts of polonium[310].

Diaryl selenides and tellurides are commonly prepared by heating selenium (or selenium tetrabromide[311]) or tellurium with a mercury diaryl; the selenides are also easily made from a diazonium salt and potassium selenide or selenolate, the latter being used when an unsymmetrical selenide is required[312]:

$$2RN_2^+ + Se^{2-} \longrightarrow (RN_2)_2Se \longrightarrow R_2Se + 2N_2$$

$$RN_2^+ + R'Se^- \longrightarrow RN_2SeR' \xrightarrow{\text{heat}} RSeR' + N_2$$

---

304 H. Schumann and M. Schmidt, *J. Organometallic Chem.* **3** (1965) 485.
305 J. R. Allkins and P. J. Hendra, *Spectrochim. Acta*, **23***A* (1967) 1671.
306 R. A. Mortensen and P. A. Leighton, *J. Am. Chem. Soc.* **56** (1934) 2397.
307 A. N. Murin, V. D. Nefedov, V. M. Zaitsev and S. A. Grachev, *Doklady Akad. Nauk. S.S.S.R.* **133** (1960) 123.
308 V. D. Nefedov, M. Vobetsky and I. Borak, *Radiokhimiya*, **7** (1965) 627, 628; *Soviet Radiochem.* **7** (1965) 623, 625.
309 A. G. Samartzewa, *Trav. Inst. État Radium U.R.S.S.* **4** (1938) 253 (*C.A.* **33** (1939) 4512).
310 V. E. Zhuravlev and N. F. Antipina, *Radiokhimiya*, **9** (1967) 726.
311 H. M. Leicester, *J. Am. Chem. Soc.* **60** (1938) 619.
312 H. M. Leicester and F. W. Bergstrom, *J. Am. Chem. Soc.* **51** (1929) 3587.

Some diaryl selenides can be made from the aryl halide and a selenolate[313], or by heating a diaryl sulphone with selenium, and also by a Friedel–Crafts reaction:

$$(C_6H_5)_2SeCl_2+C_6H_6 \xrightarrow{AlCl_3} (C_6H_5)_2Se+C_6H_5Cl+HCl$$

However, reaction of selenium tetrachloride with benzene under similar conditions is unsatisfactory[312]. Other methods for the preparation of diaryl tellurides include reaction of tellurium with an aryl lithium[314], the method used for the dipentafluorophenyl compound[314a], and, rather less satisfactory, reaction of tellurium or its halides with the appropriate Grignard reagent.

Tellurium tetra-alkyls and aryls are also known; tetraphenyl tellurium, $(C_6H_5)_4Te$, is made by reaction of diphenyl tellurium dichloride with phenyl lithium in ether[315] and the analogous *bis* (2,2′-biphenylene) tellurium[316], a compound in which there is a rapid exchange between the trigonal bipyramidal and tetragonal pyramidal structures, is obtained in a similar manner with dilithium 2,2′-biphenyl or by reaction of the latter with tellurium tetrachloride, tetramethoxide or hexamethylorthotellurate; *bis* (2,2′-biphenylene) tellurium reacts with butyl lithium, as do tellurium tetrachloride and tributyl telluronium iodide, to yield tetrabutyl tellurium[317]. All of these compounds react with electrophiles to yield the telluronium salts (p. 997).

Diselenides and Ditellurides, RXXR′

The yellow diselenides are appreciably less volatile than the corresponding selenols or selenides; they are made by oxidation of selenols in air or with 3% hydrogen peroxide[318], and by reaction of *bis* (methoxymagnesium) diselenide, a compound prepared by reaction of magnesium with selenium in methanol, with halides and γ-lactones[319]. Other methods of preparation include the reduction of aryl seleninic acids (p. 998) with sodium bisulphite or iodide[320], alkaline hydrolysis of selenocyanates, reaction of alkyl halides with alkali diselenide in anhydrous liquid ammonia[299] and, in the case of the dipentafluorophenyl diselenide, reaction of diselenium dichloride, Se_2Cl_2, with the aryl lithium[314a]. Very few ditellurides are known, all of which are red solids; the only aliphatic compounds recorded are derived from acetic and propionic acids, $HOOC(CH_2)_nTeTe(CH_2)_nCOOH$ ($n = 1$ or 2), with the tellurium atoms bonded to the α-carbon atoms. These compounds are prepared by potassium metabisulphite reduction of the *bis* telluritrichlorides, $Cl_3Te(CH_2)_nCO.O.CO(CH_2)_nTeCl_3$, which are made from the acid anhydrides and tellurium tetrachloride under reflux[321, 322]. The only known diarylditelluride is the diphenyl compound, a by-product in the preparation of diphenyl telluride by the Grignard route[323].

313 N. M. Cullinane, A. G. Rees and C. A. J. Plummer, *J. Chem. Soc.* (1939) 151.
314 N. Petragagni and M. de M. Campos, *Chem. Ber.* **96** (1963) 249.
314a S. C. Cohen, M. L. N. Reddy and A. G. Massey, *J. Organometallic Chem.* **11** (1968) 563.
315 G. Wittig and H. Fritz, *Annalen*, **577** (1952) 39.
316 D. Hellwinkel and G. Fahrbach, *Tetrahedron Letters* (1965) 1823; *Annalen*, 712 (1968) 1.
317 D. Hellwinkel and G. Fahrbach, *Chem. Ber.* **101** (1968) 574.
318 H. P. Ward and I. L. O'Donnell, *J. Am. Chem. Soc.* **67** (1945) 883.
319 W. H. H. Guenther, *J. Org. Chem.* **32** (1967) 3929.
320 J. D. McCullough, T. W. Campbell and N. J. Krilanovitch, *Ind. and Eng. Chem., Analyt. Ed.* **18** (1946) 638.
321 G. T. Morgan and H. D. K. Drew, *J. Chem. Soc.* **127** (1925) 531.
322 G. T. Morgan and R. E. Kellett, *J. Chem. Soc.* (1926) 1080.
323 K. Lederer, *Ber.* **48** (1915) 1345.

Selenoxides and Telluroxides, RR′XO

Dialkyl selenoxides are best prepared by the action of ozone on the dialkyl selenide in carbon tetrachloride, chloroform or nitromethane[324] at $-10°$ to $-50°$, or by reaction of the dialkyl selenium dibromide (p. 996) with silver oxide in methanol[324a] at $-15°$; oxidation of the selenide with t-butyl hydroperoxide under anhydrous conditions is also effective, whereas oxidation of the selenide with nitrogen dioxide[324a] yields the adduct $R_2SeO.N_2O_4$. The dialkyl selenoxides are unstable and disproportionate to the selenol and an aldehyde on heating:

$$R(R'CH_2)SeO \longrightarrow RSeH + R'CHO$$

the dinitrogen tetroxide adducts behave in the same way, whereas the diaryl selenoxides lose oxygen, leaving the selenide. The halogenated alkyl derivatives, obtained by reaction of selenium(IV) oxodichloride with unsaturated halocarbons[325], appear to be somewhat more stable to heat:

$$SeOCl_2 + 2CH_2{:}CHCl \longrightarrow (CH_2CHCl_2)_2SeO$$

Hydrolysis of dialkyl selenium dihalides with aqueous alkali is incomplete, leading to basic compounds of the type $R_2SeX_2.R_2SeO$ and oxidation of dialkyl tellurides by air in neutral solution leads to the formation of complexes of the dialkyl telluroxide and the corresponding tellurinic acid, $R_2TeO.RTeO(OH)$. Oxidation of the dialkyl tellurides with nitric acid or alkaline hydrogen peroxide yields the basic nitrate, $R_2Te(OH)NO_3$, or the dihydroxide, $R_2Te(OH)_2$, respectively[326].

The diaryl selenoxides and telluroxides are much more stable entities than their dialkyl analogues; they are colourless solids, soluble in alcohols and slightly soluble in benzene, toluene and water. Their aqueous solutions are alkaline, the selenoxides forming the dihydroxides, $R_2Se(OH)_2$, in moist air and the dihalides, R_2SeY_2, with halogen acids, but the dialkyl compounds form hydrates, $R_2SeO.H_2O$, in which the water is held by hydrogen bonding and with acids form dialkylselenoxonium salts, $R_2Se(OH)X$ ($X = ClO_4^-$, Cl^-, NO_3^-). The infrared spectra of these compounds have been recorded[324a]. The structure of the oxides is probably pyramidal, as in the sulphoxides, but optical isomerism has not been observed[327] with the unsymmetrical selenoxides, RR′SeO. The diaryl telluroxides have high dipole moments, indicating the appreciable polarity of the Te–O bond, as expected from the basicity of the compounds[328].

The diaryl compounds are made by hydrolysis of the selenium[329] or tellurium[330] dihalides or basic nitrates[301], or by oxidation of the selenides with hydrogen peroxide, peracetic acid[331], ozone or t-butyl hydroperoxide[324]. Nitric acid oxidation of the diaryl tellurides yields the dinitrates, $R_2Te(NO_3)_2$, which are hydrolysed to basic salts by water; the product loses water at 150° to form compounds of the type $R_2Te(NO_3)$–O–$Te(NO_3)R_2$. These last yield the telluroxide with alkali[332].

324 G. Ayrey, D. Barnard and D. T. Woodbridge, *J. Chem. Soc.* (1962) 2089.
324a R. Paetzold, V. Lindner, G. Bochmann and P. Reich, *Z. anorg. Chem.* **352** (1967) 295.
325 H. Brintzinger, K. Pfannstiel and H. Vogel, *Z. anorg. Chem.* **256** (1948) 75.
326 M. P. Balfe, C. A. Chaplin and H. P. Phillips, *J. Chem. Soc.* (1938) 341.
327 S. C. Abrahams, *Quart. Rev.* **10** (1956) 431.
328 K. A. Jensen, *Z. anorg. Chem.* **250** (1943) 245.
329 C. K. Banks and C. S. Hamilton, *J. Am. Chem. Soc.* **61** (1939) 2306.
330 G. T. Morgan and F. H. Burstall, *J. Chem. Soc.* (1931) 180.
331 J. D. McCullough, T. W. Campbell and E. S. Gould, *J. Am. Chem. Soc.* **72** (1950) 5753.
332 K. Lederer, *Ber.* **49** (1916) 1082.

Selenones and Tellurones, R_2XO_2

The best known of these compounds are the diaryl selenones, which are white, high melting solids prepared by oxidation of the selenide or selenoxide with potassium permanganate[333]. They are only slightly soluble in organic solvents and are quite powerful oxidizing agents, being reduced to the corresponding selenide by hydriodic acid or when heated with sulphur; they decompose to the selenide and oxygen at 300–400°. The existence of any of the dialkyl compounds seems to be rather doubtful, and the rather unstable cyclic selenones, obtained by direct addition of selenium dioxide to alkyl butadienes in chloroform[334], may be cyclic seleninic acid (p. 998) esters. The analogous tellurones, made by oxidizing a cyclic telluride with hydrogen peroxide, explode when heated[335].

Organoselenium Monohalides, $RSe\overline{hal}$

Only chlorides and bromides are known for this group; these are prepared by oxidation of a selenol or diselenide with the calculated quantity of halogen or by addition of a selenocyanate to the halogen in an inert solvent; addition of the halogen to a selenocyanate yields the selenide[336]. Both types of halide are solid at room temperature, the chlorides being colourless to pale yellow and the bromides orange to deep red. They are soluble in most organic solvents and are reduced to the diselenides by zinc. These halides are appreciably dissociated at room temperature, smelling strongly of the free halogen, and are hydrolysed by water to a mixture of a seleninic acid and a diselenide by way of the selenenic acid[337]:

$$3RSeBr + 2H_2O \longrightarrow RSeO.OH + R_2Se_2 + 3HBr$$

Organometalloid Trihalides, $RX\overline{hal}_3$

These trihalides can be regarded[338] as derivatives of the ortho-form of seleninic or tellurinic acid, $RX(OH)_3$, from which they are obtained by reaction with concentrated halogen acids, except for the

$$RXO.OH + 3H\overline{hal} \longrightarrow RX\overline{hal}_3 + 2H_2O$$

fluorides, of which only one example is known. This is methyltellurium trifluoride, an unstable compound which disproportionates to a mixture of tellurium tetrafluoride and dimethyltellurium difluoride. It is prepared by reaction of the triiodide with silver fluoride in boiling acetone[339]. The selenium trihalides are also obtained by reaction of the monohalide with an excess of halogen and some of the tellurium trichlorides can be made by reaction of cresols[340], alkaryl or diaryl ethers[322] with tellurium tetrachloride, an excess of which is used to avoid formation of the dichloride, R_2TeCl_2. All of these compounds are similar in properties to the monohalides.

333 H. Rheinboldt and E. Geisbrecht, *J. Am. Chem. Soc.* **68** (1946) 2671.
334 H. J. Backer and J. Strating, *Rec. Trav. Chim.* **53** (1934) 1113.
334a W. L. Mock and J. H. McCausland, *Tetrahedron Letters*, (1968) 391.
335 G. T. Morgan and H. Burgess, *J. Chem. Soc.* (1928) 321.
336 O. Behagel and H. Seibert, *Ber.* **66** (1933) 708.
337 O. Behagel and W. Müller, *Ber.* **68** (1935) 1540.
338 D. G. Foster, *J. Am. Chem. Soc.* **55** (1933) 822.
339 H. J. Emeléus and H. G. Heal, *J. Chem. Soc.* (1946) 1126.
340 G. T. Morgan and H. Burgess, *J. Chem. Soc.* (1929) 2214.

Organometalloid Dihalides, $R_2\overline{Xhal_2}$

The tellurium dihalides exist in two forms, the α-dihalides having a simple structure like that of the selenium compounds. This consists of a trigonal bipyramid with the halogen atoms at the apices, and the two organic groups and an unshared pair of electrons in the three equatorial positions, just as in the structures of the tetrachlorides[341] (p. 959). The β-tellurium dihalides are complexes of the type $R_3Te^+RTeI_4^-$; in the methyl compound, the structure consists of trigonal pyramidal Me_3Te^+ ions and square pyramidal anions which are bridged by weak Te–I interactions[342]. This accords with the preparation of the β-dihalides from the telluronium halide and the organotellurium trihalide in acetone. The β-tellurium dichlorides and dibromides can also be prepared from the α-compounds; the latter are converted to the hydroxide by treatment with moist silver oxide, and the product is heated in a vacuum to yield the anhydride, $R_3Te.O.TeR.O$, which forms the β-dihalide with the halogen acid[343]. With hydriodic acid, however, the anhydride yields a mixture of the tellurinic acid and the telluronium iodide.

The selenium and α-tellurium dichlorides and dibromides are obtained either by the reaction of the selenide or telluride with an excess of halogen or by the action of a halogen acid on the nitrate, which is itself formed by the action of nitric acid on the selenide, telluride or the dihydroxide. The rather unstable selenium diiodides are usually made from the dinitrates and aqueous potassium iodide, whereas the α-dialkyltellurium compounds can be obtained by heating tellurium with an alkyl iodide in a sealed tube[344]; the diaryl compounds are obtained from the telluride and iodine in ether. The only known fluorides are dimethylselenium difluoride, a liquid which distils at 25°/0·01 mm, made by reaction of dimethylselenide with silver(II) fluoride[344a] and the white dimethyl and diphenyl tellurium compounds, made from the α-diiodides and silver fluoride in boiling acetone; both tellurium compounds are rather unstable[339] and are presumably α-compounds. Selenium dihalides in which there is a methylene group adjacent to the selenium atom decompose readily to the monohalides:

$$(RCH_2)(R)SeBr_2 \longrightarrow RSeBr + RCH_2Br$$

Telluride dichlorides derived from aliphatic acids are made from tellurium tetrachloride and an excess of the acid anyhdride[321] and a few selenium dichlorides, all derivatives of ketones, can be obtained by reaction of the ketone with selenium tetrachloride in ether[345] or from selenium(IV) oxodichloride and the ketone[346]. Other selenium dichlorides, in which the organic group is halogenated, can be prepared by the action of olefins on selenium monochloride, the primary product being a chloroalkyl selenide; ethylene, for example, yields β,β′-dichlorodiethyl selenide dichloride[347]:

$$2Se_2Cl_2 + 2CH_2 = CH_2 \longrightarrow (ClC_2H_4)_2SeCl_2 + 3Se$$

---

[341] J. D. McCullough, *Acta Cryst.* **6** (1953) 746.
[342] F. Einstein, J. Trotter and C. Williston, *J. Chem. Soc.* (*A*) (1967) 2018.
[343] R. H. Vernon, *J. Chem. Soc.* **119** (1921) 105, 687.
[344] A. Scott, *Proc. Chem. Soc.* **20** (1904) 156.
[344a] K. J. Wynne and J. Puckett, *Chem. Comm.* (1968) 1532.
[345] A. Michaelis and F. Kunckell, *Ber.* **30** (1897) 2823.
[346] R. E. Nelson and R. N. Jones, *J. Am. Chem. Soc.* **52** (1930) 1588.
[347] C. E. Boord and F. F. Cope, *J. Am. Chem. Soc.* **44** (1922) 395.

The β-tellurium diiodides are dark purple, in contrast to the α-compounds, which are brownish-red, and the selenium diiodides, which range from dark red to black; however, the α- and β-dichlorides are both white, and the dibromides are yellow, whereas the selenium dichlorides range from colourless to lemon-yellow and the dibromides are ruby-red.

Dinitrates and basic nitrates can be obtained, and alkoxides, such as $Ph_2Se(OR)_2$, which also have a trigonal bipyramidal structure, with the alkoxide groups at the apices, are easily prepared by reaction of the dibromide or dichloride with the sodium alkoxide. These alkoxides yield diphenyl selenoxide on hydrolysis[348]; this, and other diaryl selenoxides, are converted back to the dihalide with halogen acid.

Although the polonium analogues have never been prepared in weighable amounts, there is evidence for the formation of diaryl polonium dichlorides from trace-scale experiments with the β-decay products of triaryl bismuth-210 dichlorides, and even from the decay of the corresponding triphenyl[307, 308], as well as by halogenation of the polonide, also obtained by decay of the bismuth-210 analogue[349, 350].

Chalcogen-onium Salts, $R_3X^+Y^-$

These are colourless ionic solids which are soluble in polar solvents; the structure of trimethylselenonium iodide is typical for these compounds, consisting of a linear arrangement of Me_3Se^+ and I^- ions, the cation being a slightly distorted pyramid in which a C–Se bond is directed towards the I^- ion[351]. The hydroxides, obtained by treating a halide with moist silver oxide, are basic, the telluronium, compounds being more basic than the selenonium ones.

The trialkyl and alkaryl halides are prepared by reaction of an alkyl iodide[326], or other halide, with the selenide or telluride:

$$RR'X + R''I \longrightarrow RR'R''XI$$

the compounds in which three different groups or atoms are bonded to the selenium[352] or tellurium[353] atom in the cation are optically active and have been resolved. The alkyl sulphates, of the type $RR'R''Se.SO_4CH_3$, are prepared in the same way as the halides, but with a dialkyl sulphate in place of the alkyl halide[354]. Halo-aliphatic acids can be used in the place of alkyl halides; the products are of the type $RR'(CH_2COOH)X^+Y^-$ and are generally known as selenetines or telluretines.

Trialkyl telluronium chlorides have also been made by reaction of zinc alkyls with an excess of tellurium tetrachloride, the excess being necessary in order to avoid formation of the telluride[355]. Other preparative methods include reaction of the dialkyl tellurium diiodide with an alkyl iodide and aqueous sodium sulphite[344] and treatment of the base derived from a β-dialkyl tellurium diiodide with aqueous hydriodic acid[356].

The triaryl selenonium halides are best prepared by reaction of the diaryl selenium dihalide with a mercury diaryl[312] and a few of them have been made by methods resembling

348 R. Paetzold and V. Lindner, *Z. anorg. Chem.* **350** (1967) 295.
349 V. D. Nefedov, M. A. Toropova, V. E. Shuravlev and A. V. Levchenko, *Radiokhimiya*, **7** (1965) 203.
350 V. D. Nefedov, V. E. Zhuravlev, M. A. Toropova, L. N. Grachev and A. V. Levchenko, *Radiokhimiya*, **7** (1965) 245.
351 H. Hope, *Acta Cryst.* **20** (1966) 610.
352 W. J. Pope and A. Neville, *J. Chem. Soc.* **81** (1902) 1552.
353 T. M. Lowry and F. L. Gilbert, *J. Chem. Soc.* (1929) 2867.
354 J. W. Baker and W. G. Moffitt, *J. Chem. Soc.* (1930) 1722.
355 A. Marquardt and A. Michaelis, *Ber.* **21** (1888) 2042.
356 H. D. K. Drew, *J. Chem. Soc.* (1929) 560.

the Friedel–Crafts reaction; for example, triphenyl selenonium chloride is obtained from diphenyl selenium dichloride and benzene in the presence of aluminium chloride[312], and the trianisyl or triphenetyl chlorides result when the ether is heated with selenium dioxide and aluminium chloride[357]. The selenium atom is bonded to a ring carbon atom in these last. The triaryl telluronium halides are obtained by reaction of an aryl Grignard reagent with tellurium tetrachloride or a diaryl tellurium dihalide, but with an excess of the Grignard further reaction occurs, yielding the diaryl telluride. The telluronium ion can be isolated by precipitation of the iodide with hot, aqueous potassium iodide[358]. Triaryl polonium halides have also been prepared by the Grignard route, but only in tracer quantities[349, 350], and there is some evidence for their formation in the β-decay of bismuth-210 triaryls[308].

The preparative methods described above cannot be used to prepare selenonium or telluronium fluorides, the former being obtained when silver fluoride is added to an aqueous solution of a selenonium chloride or bromide and the latter by neutralizing a telluronium hydroxide with hydrofluoric acid[339].

Organochalcogen Acids, RXO_nOH ($n = 0-2$)

These acids can be formally regarded as derived from the 2-, 4- and 6-valent chalcogen. The selenenic acids, RSeOH, orange to violet-red solids, are amphoteric in character and are obtained either by reduction of the aryl seleninic acid with hydrazine sulphate[359] or by hydrolysis of aryl selenium halides in which the electronegativity of the aryl group is low; thus *o*-nitrophenyl selenium bromide yields the selenenic acid, but the *m*-nitrophenyl compound yields the diselenide[337]. Tellurium and polonium analogues of this class of acid are unknown.

The seleninic and tellurinic acids, RXO.OH, are the most stable class of organochalcogen acid. Like the selenenic acids, they are amphoteric and are weaker acids than the analogous carboxylic acids[360]. As mentioned earlier (p. 995), it is probable that the hydrates are really the ortho-acids[338], $RX(OH)_3$, and there is some confirmation of this from the Raman spectra of the aliphatic seleninic acid hydrochlorides, which indicate that they are of the form $RSe(OH)_2Cl$[361]. The structure of the free acid is pyramidal[362].

Alkane[360] and arene[363] seleninic acids, including those in which the organic group is a carboxylate[364], are usually made by nitric acid oxidation of the diselenides; the analogous reaction with diphenyl telluride yields the basic nitrate which is hydrolysed to the tellurinic acid by alkali[323]. Oxidation of selenides with aqueous permanganate or hydrolysis of dialkyl selenium dihalides with moist silver oxide is also satisfactory[365]. Other methods of preparation include hydrolysis of the selenium trihalide (p. 995) with dilute aqueous sodium carbonate[365], nitric acid oxidation of selenocyanates[366] and reduction of aryl selenonic acids with hydrochloric acid[367]. A few alkane tellurinic acids are known, prepared by air or hyperol oxidation of esters of alkyl telluracetic acids ($RTeCH_2COOH$)[326].

[357] T. P. Hilditch and S. Smiles, *J. Chem. Soc.* **93** (1908) 1384.
[358] K. Lederer, *Ber.* **44** (1911) 2287.
[359] M. Rheinboldt and E. Geisbrecht, *Chem. Ber.* **88** (1955) 666.
[360] H. J. Backer and W. van Dam, *Rec. Trav. Chim.* **54** (1935) 531.
[361] R. Paetzold, H.-D. Schumann and A. Simon, *Z. anorg. Chem.* **305** (1960) 98.
[362] J. H. Bryden and J. D. McCullough, *Acta Cryst.* **7** (1954) 833.
[363] M. Stoecker and F. Krafft, *Ber.* **39** (1906) 2197.
[364] H. J. Backer and W. van Dam, *Rec. Trav. Chim.* **48** (1929) 1287; *ibid.* **49** (1930) 482.
[365] M. L. Bird and F. Challenger, *J. Chem. Soc.* (1942) 570.
[366] J. Loevenich, H. Fremdling and M. Fohr, *Ber.* **62** (1929) 2856.
[367] H. W. Doughty, *Am. Chem. J.* **41** (1909) 326.

Seleninic acids are reduced to selenols by zinc and hydrochloric acid, and decompose to selenols and an aldehyde when heated[365], but benzene seleninic acid decomposes to the anhydride when heated in a vacuum[324]. Other alkane and arene seleninic acid anhydrides, $(RSeO)_2O$, in which there is a bent Se–O–Se group in contrast to the S–S bonded sulphur analogues, are obtained by the action of dry nitrogen dioxide[368] or dry ozone[324, 368] on the diselenide in an inert organic solvent. These anhydrides absorb water to form the seleninic acid. 1,2-ethane diseleninic acid anhydride (X) is also known, prepared[369] by

$$\begin{array}{l} CH_2Se(=O) \\ | \qquad\quad \rangle O \\ CH_2Se(=O) \end{array}$$

(X)

nitric acid oxidation of the selenocyanate, $(CH_2SeCN)_2$. The acid chlorides, ReSeOCl, are made by the action of ozone[324, 370] on the selenium monochloride, RSeCl, at $-10°$ to $-50°$. Hydrolysis of the acid chlorides yields the hydrochloride, $RSe(OH)_2Cl$, except in the case of the trifluoromethyl compound, which yields the seleninic acid[370].

The selenonic acids are powerful oxidizing agents, being reduced to the seleninic acid by hydrochloric acid as mentioned above, behaviour which would be expected for a selenium(VI) compound. They are also the strongest acids in this group and are not amphoteric. No tellurium(VI) analogues are known, in conformity with the very different type of structure adopted by telluric acid. Selenonic acids are very similar in properties to the sulphonic acids and are prepared directly from aromatic hydrocarbons and selenic acid[367, 371] or selenium trioxide, preferably in liquid sulphur dioxide[170]. They are also obtained by the oxidation of diaryl selenides with moist chlorine[363] and by oxidation of alkane[365] or arene[372] seleninic acids with alkaline permanganate.

Seleno- and Tellurocarbonyl Derivatives

Four types of compound are known for this group, the ketones, R_2CX, and related selenoketals or mercaptoles, $R_2C(SeR')_2$, selenoaldehydes, RCHSe, selenocarboxylic acids, RCOSeH, and selenoureas, $(NR_2)_2CSe$.

The red selenoketones are dimers, whereas the dark yellow or brown telluroketones are monomers and are the more volatile of the two species; both are oils which are insoluble in water, but soluble in organic solvents, and they are prepared by the action of hydrogen selenide[373] or telluride[374] on a mixture of a ketone and concentrated hydrochloric acid. Selenoketals, which are also red oils, are prepared in a similar manner by condensing a ketone with a selenol in the presence of hydrochloric acid; they are oxidized to seleninic acids by dilute hydrogen peroxide[375].

The white to yellow-orange selenoaldehydes are solids which behave chemically like

368 R. Paetzold, S. Borek and E. Wolfram, *Z. anorg. Chem.* **353** (1967) 53.
369 R. Paetzold and D. Lienig, *Z. anorg. Chem.* **335** (1965) 289.
370 R. Paetzold and E. Wolfram, *Z. anorg. Chem.* **353** (1967) 167.
371 R. Anschutz and F. Teutenberg, *Ber.* **57** (1924) 1018.
372 F. L. Pyman, *J. Chem. Soc.* **115** (1919) 166.
373 W. E. Bradt, *J. Chem. Educ.* **12** (1935) 363.
374 R. E. Lyons and E. D. Scudder, *Ber.* **64** (1931) 530.
375 E. H. Shaw, Jr., and E. E. Reid, *J. Am. Chem. Soc.* **48** (1926) 520.

the thioaldehydes; they are prepared in the same way as the selenoketones[376] or by reaction of the aldehyde with magnesium bromide hydroselenide, BrMgSeH[377]. The selenium analogue of formaldehyde is obtained as a mixture of the trimer, $(CH_2Se)_3$, which is soluble in ethanol, and higher polymers which are insoluble, in the same way as the selenoketones[377a]; the polymer is also obtained by reaction of sodium selenide with formaldehyde in ethyl acetate[377b], the polymer chain being helical in the hexagonal crystalline form[377c]. Trimeric seleno- and telluroformaldehyde have both been obtained by reaction of methylene radicals with thin films of selenium or tellurium[377d]. Selenocarboxylic acids, made from magnesium bromide hydroselenide and the acid chloride[377], are rather unstable and little is known about them.

The selenoureas are white or faintly coloured solids which are obtained by reaction of hydrogen selenide with dicyanamide, or substituted cyanamides, in the presence of concentrated hydrochloric acid[378] and by reaction of aryl isoselenocyanates with primary aromatic amines[379, 380]:

$$RNCSe + NH_2R' \longrightarrow (RNH)(R'NH)CSe$$

Another useful preparative method is the reaction of sodium hydroselenide with an isothiouronium iodide, $(R_2N)_2CSMeI$, in aqueous ethanol[380a]. Aroyl alkyl and aryl selenoureas can be prepared by reaction of the aroyl chloride with potassium selenocyanate and the aliphatic or aromatic primary amine[381], the first step in the reaction being formation of the aroyl isoselenocyanate, identified by infrared spectroscopy[382].

Selenocyanates, RCNSe, and Isoselenocyanates, RNCSe

The alkyl selenocyanates are low melting solids or oils, and the aryl compounds are solids; both are colourless when pure, any colour being due to the presence of diselenides, and are insoluble in water, but soluble in organic solvents. The selenocyanates decompose when heated, the aryl compounds yielding the diaryl selenide and cyanogen.

The selenocyanates are commonly prepared by reaction of potassium selenocyanate with an alkyl[383] or aryl[384] halide, polymethylene dihalide[385], halo-acid, haloketone[386] or diazonium salt[387]; the yield in the last reaction is of the same order as that for the Gattermann–Sandmeyer preparation of aryl halides. Other preparative methods include reaction of selenium selenocyanate with a mercury dialkyl in boiling benzene[388]:

$$R_2Hg + Se(SeCN)_2 \longrightarrow RHgSeCN + RSeCN + Se$$

[376] L. Vanino and A. Schinner, *J. prakt. Chem.* **91** (1915) 116.
[377] Q. Mingoia, *Gazzetta*, **58** (1928) 667.
[377a] L. Mortillaro, L. Credali, M. Russo and C. De Checchi, *J. Polymer Science*, *B*3 (1965) 581.
[377b] M. Prince and B. Bremer, *J. Polymer Science*, *B*5 (1967) 847.
[377c] G. Carazzolo and G. Valle, *J. Polymer Science*, *A*3 (1965) 4013.
[377d] F. D. Williams and F. X. Dunbar, *Chem. Comm.* (1968) 459.
[378] H. J. Backer and H. Bos, *Rec. Trav. Chim.* **62** (1943) 580.
[379] T. Tarantelli and D. Leonesi, *Ann. Chim.* (*Italy*), **53** (1963) 1113.
[380] E. Bulka, K. D. Ahlers and E. Tucek, *Chem. Ber.* **100** (1967) 1459.
[380a] D. L. Klayman and R. J. Shine, *Chem. Comm.* (1968) 372.
[381] I. B. Douglass, *J. Am. Chem. Soc.* **59** (1937) 740.
[382] E. Bulka, K. D. Ahlers and E. Tucek, *Chem. Ber.* **100** (1967) 1367.
[383] W. E. Weaver and W. M. Whaley, *J. Am. Chem. Soc.* **68** (1946) 2115.
[384] E. Fromm and K. Martin, *Annalen*, **401** (1913) 177.
[385] G. T. Morgan and F. H. Burstall, *J. Chem. Soc.* (1931) 173.
[386] G. Hofmann, *Annalen*, **250** (1889) 294.
[387] T. W. Campbell and M. T. Rogers, *J. Am. Chem. Soc.* **70** (1948) 1029.
[388] E. E. Aynsley, N. N. Greenwood and M. J. Sprague, *J. Chem. Soc.* (1965) 2395.

selenium dicyanide reacts in similar manner with mercury diphenyl[388] or triphenyl bismuth[389]. Selenium selenocyanate also reacts directly with aromatic amines, the selenocyanate group entering the ring in the *para* position[389].

The isoselenocyanates are less well known than the selenocyanates; they are low melting, white solids when pure, but decompose on heating, and their vapour is lachrymatory. A characteristic band appears at about 2157 cm^{-1} in their infrared spectra, the analogous selenocyanates absorbing[382] at 2048 and 2127 cm^{-1}. They are prepared by reaction of selenium with an isonitrile in chloroform[382] or petroleum ether[390] under reflux.

Cyclic Selenides and Tellurides, $\overline{CH_2(CH_2)_nX}$

These saturated compounds behave in much the same way as the dialkyl chalcogenides, forming dihalides and selenonium or telluronium salts. The compounds are prepared by reaction of the α,ω-dihaloalkane, preferably the bromide, with aluminium telluride[335], or an alkali metal selenide[391] or telluride[392] in ethanol. Spirocompounds, such as 2,6-diselenaspiro[4]heptane (XI) are obtained in a similar manner, XI being made from potassium selenide and tetrabromoerythritol[393], $C(CH_2Br)_4$. Cyclic selenides are also obtained when the α,ω-dihaloalkane is refluxed with selenium and sodium formaldehyde sulphoxylate[394].

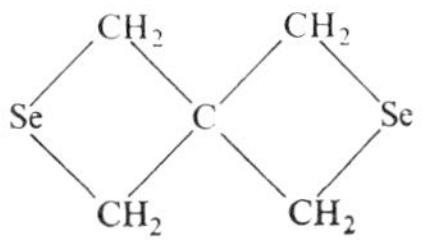

(XI)  (XII) 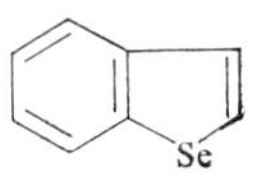 (XIII)

The reaction of 1,5-diiodopentane with tellurium yields cyclotelluropentane diiodide[392], analogous to the formation of the α-dialkyltellurium diiodides (p. 997), and this is reduced to cyclotelluropentane by aqueous sodium sulphite, a reaction common to all dialkyl tellurium dihalides. Cycloselenopropane, the far infrared spectrum of which has been investigated[395], and the tellurium analogue polymerize very readily.

The best known of the unsaturated cyclic selenides is selenophen (XII); the vibrational spectrum of the compound indicates that the molecule is planar, like thiophen, and of C_{2v} symmetry[396]. Selenophen, which melts at $-38°$ and boils at 110°, is a colourless liquid when pure, impurities imparting a pale lemon colour. It smells like a mixture of benzene and carbon disulphide, and is prepared by heating selenium in acetylene at 350–400°, fractionating the product to remove selenonaphthene (XIII) and diselenoheterocycles[397]. Substituted selenophens are obtained in good yield by the reaction of hydrogen selenide with diynes in the presence of Ag^+ or Cu^+ ions as catalyst[397a]; C-substituted selenophens

389 F. Challenger, A. T. Peters and J. Halevy, *J. Chem. Soc.* (1926) 1648.
390 W. J. Franklin and R. L. Werner, *Tetrahedron Letters*, (1965) 3003.
391 G. T. Morgan and F. H. Burstall, *J. Chem. Soc.* (1929) 1096, 1497, 2197; (1930) 1497.
392 W. V. Farrar and J. M. Gulland, *J. Chem. Soc.* (1945) 11.
393 H. J. Backer and H. J. Winter, *Rec. Trav. Chim.* **56** (1937) 492.
394 J. D. McCullough and A. Lefohn, *Inorg. Chem.* **5** (1966) 150.
395 A. B. Harvey, J. R. Durig and A. C. Morrissey, *J. Chem. Phys.* **47** (1967) 4864.
396 A. Trombetti and C. Zauli, *J. Chem. Soc.* (*A*), (1967) 1106.
397 H. V. A. Briscoe, J. B. Peel and P. L. Robinson, *J. Chem. Soc.* (1928) 2628.
397a R. F. Curtis, S. N. Hasnain and J. A. Taylor, *Chem. Comm.* (1968) 365.

are also obtained by heating diketones with phosphorus pentaselenide[398] and 2,5-diacyl-3,4-dihydroxoselenophens (XIV) result from reaction of 1,3-diaroyl or dialkoyl acetones with selenium tetrachloride[399]. More complex species involving two linked selenophen

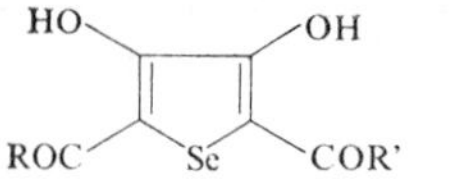

(XIV) 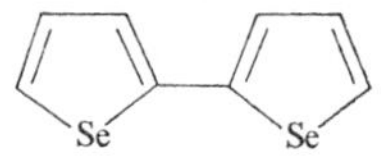 (XV)

groups are also known; 2,2-*bis*(selenophen-yl) or 2,2′-biselenienyl (XV), a solid which melts at 49°, is made by heating 2-iodoselenophen with activated copper[400] in xylene under reflux. Substituted 2,2′-biselenienyls can be made in the same way and a derivative of 3,3′-biselenienyl is also known[401].

Derivatives of β-diketones

These compounds are conveniently discussed with the cyclic chalcogenides since ring compounds are formed in many of the reactions of the chalcogen halides with β-diketones. Acetylacetone has been used as the example throughout, but many other β-diketones react in the same manner; their derivatives have been described in detail elsewhere[294].

Selenium tetrachloride reacts with acetylacetone, or its copper derivative, in chloroform or ether, yielding dimeric selenium acetylacetonate, $Se_2(C_5H_6O_2)_2$, as the main product[402].

$$2SeCl_4+2Cu(C_5H_7O_2)_2 \longrightarrow Se_2(C_5H_6O_2)_2+2C_5H_7O_2Cl+2CuCl_2+2HCl$$

This is a pale yellow solid (m.p. 185°) which, from its n.m.r. and infrared spectra[403], is now known to have the ring structure XVI. This explains the failure of the compound to enolize and indicates that the reaction site in the β-diketone is at the α-carbon atom.

Hydriodic acid reduces XVI to diselenium bisacetylacetone, $Se_2(C_5H_7O_2)_2$ (XVII), which is enolic and can also be prepared by reaction of selenium monochloride with copper

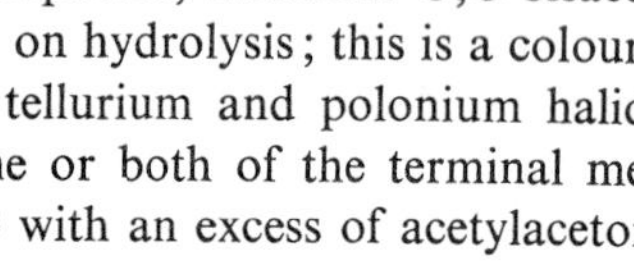

(XVI) 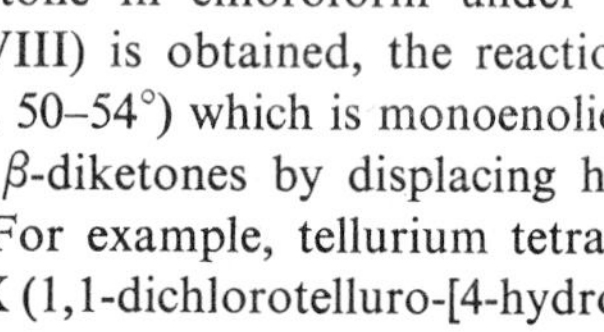 (XVII)

acetylacetonate[404]. By heating XVI with acetylacetone in chloroform under reflux a third compound, selenium O,C-bisacetylacetone (XVIII) is obtained, the reaction being reversed on hydrolysis; this is a colourless solid (m.p. 50–54°) which is monoenolic.

The tellurium and polonium halides react with β-diketones by displacing hydrogen from one or both of the terminal methyl groups. For example, tellurium tetrachloride reacts[402] with an excess of acetylacetone to form XIX (1,1-dichlorotelluro-[4-hydroxypent-

398 M. T. Bogert and C. N. Andersen, *J. Am. Chem. Soc.* **48** (1926) 223.
399 K. Balenovic, A. Deljac, B. Gaspert and Z. Stefanac, *Monatsh.* **98** (1967) 1344.
400 L. Chierici, C. Dell'Erba, A. Guareschi and D. Spinelli, *Ricerca sci.* **8** (1965) 1537.
401 C. Dell'Erba, D. Spinelli, G. Garbarino and G. Leandri, *J. Heterocyclic Chem.* **5** (1968) 45.
402 G. T. Morgan and H. D. K. Drew, *J. Chem. Soc.* **117** (1920) 1456; *ibid.* **121** (1922) 922.
403 D. H. Dewar, J. E. Fergusson, P. R. Hentschel, C. J. Wilkins and P. O. Wilkins, *J. Chem. Soc.* (1964) 688.
404 G. T. Morgan, H. D. K. Drew and T. V. Barker, *J. Chem. Soc.* **121** (1922) 2432.

3en-2one], $(C_5H_7O_2)_2TeCl_2$), but hydrogen chloride is eliminated if the β-diketone is not in excess, the product then being XX (2,4-dioxocyclopentylene-tellurium-1,1-dichloride or 1,1-dichlorotelluracyclohexane-3,5-dione). The colourless solid melts, with decomposition, at 169–173° [402] or 162° [403]. Treatment with alkali or aqueous sulphite eliminates the halogen from XX, yielding the golden yellow 2,4-dioxocyclopentylene-tellurium (m.p. 182°)

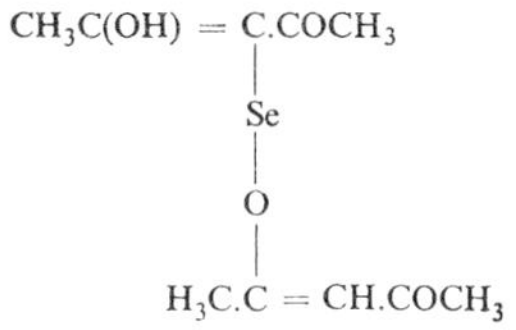

$CH_3C(OH) = C.COCH_3$ | Se | O | $H_3C.C = CH.COCH_3$

(XVIII)

$(CH_3COCH_2COCH_2)_2TeCl_2$

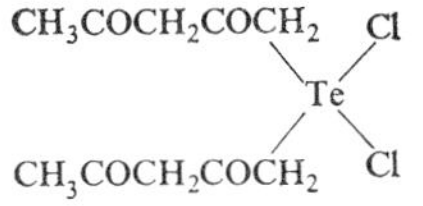

(XIX)

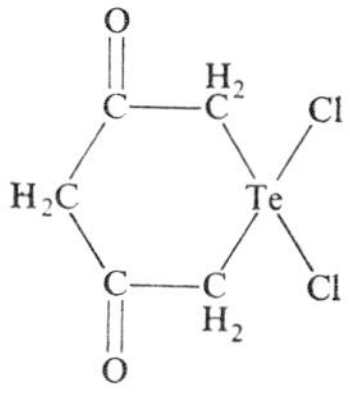

(XX)

which can be sublimed under reduced pressure; this reforms the dihalides when treated with halogen. The polonium tetrahalides react with acetylacetone in the same way as tellurium, the dihalide analogues of XX being yellow-orange solids; 2,4-dioxocyclopentylene polonium is purple-violet, obtained either by reaction of a polonium dihalide with acetylacetone or by removal of halogen from the dihalocompound with aqueous alkali. It reacts with halogen to reform the dihalide and with hydrogen peroxide to form the oxide[9].

Analogues of 1,4-dioxane, $X(CH_2)_2X'CH_2CH_2$

1,4-selenoxane or 1-oxa-4-selenacyclohexane (XXI), a colourless liquid, is prepared by refluxing a β,β′-dihalodiethyl ether with either aqueous sodium selenide under hydrogen[405] (a reaction which, with sodium telluride, yields 1,4-telluroxane[392]) or alkaline sodium formaldehyde sulphoxylate and selenium[394]. The analogous 1,4-selenothiane, or 1-thia-4-selenacyclohexane (m.p. 107°) and 1,4-tellurothiane (m.p. 69·5°) are both volatile

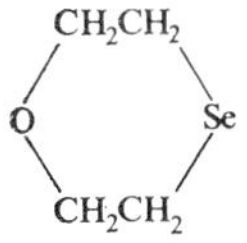

(XXI)

white solids with unpleasant odours, that of the tellurothiane being remarkably repulsive. 1,4-selenothiane is prepared by reaction of β,β′-dichlorodiethyl sulphide under reflux with aqueous[406] or ethanolic[407] sodium selenide, or with selenium and alkaline sodium formaldehyde sulphoxylate[394]; 1,4-tellurothiane is obtained when selenium is replaced by tellurium in the last reaction[408]. 1,4-diselenane, a white solid, is prepared from β,β′-dichlorodiethyl selenide and lithium selenide in acetone under reflux, but the yield is poor; the molecule has the chair form[409].

All of these compounds behave chemically as dialkyl selenides or tellurides, forming

[405] C. S. Gibson and J. D. A. Johnson, *J. Chem. Soc.* (1931) 266.
[406] C. S. Gibson and J. D. A. Johnson, *J. Chem. Soc.* (1933) 1529.
[407] J. D. McCullough and P. Radlick, *Inorg. Chem.* **3** (1964) 924.
[408] J. D. McCullough, *Inorg. Chem.* **4** (1965) 862.
[409] R. E. Marsh and J. D. McCullough, *J. Am. Chem. Soc.* **73** (1951) 1106.

basic nitrates, telluronium salts and Se- or Te-dihalides, the last being reduced to the parent heterocycle by aqueous potassium metabisulphite.

An unsaturated analogue of diselenane, 2,6-(or 2,5-)diphenyl-1,4-diselenacyclohexa-2,5-diene (m.p. 135·5°), which has the boat conformation, is obtained by heating phenylacetylene with red selenium and sodium ethoxide in boiling dixoan[410].

Se–N Organo-compounds

The cyclic 2-aminoselenazoles are obtained by condensing α-haloketones with selenoureas; for example, chloroacetone yields[386] 2-amino-4 methyl-selenazole (XXII). 1-seleno-2,5-diazole (XXIII), which melts at 21° and boils at 138°, is obtained in low yield by reaction of ethylene diamine hydrochloride with selenium monochloride in dimethyl

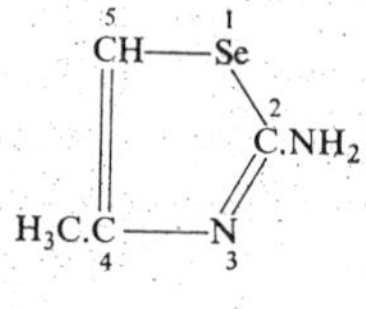

(XXII) 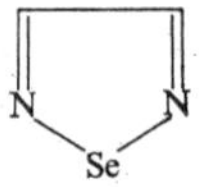 (XXIII)

formamide[411]; the molecule is planar and of C_{2v} symmetry[412]. The 2,3 disubstituted compounds can be made by reaction of α-dioximes with selenium dioxide in dimethyl formamide[413].

Diseleno- and selenothiocarbazic acids are also known; the former, $R_2NNHCSeSe$, are made[414] by reaction of carbon diselenide with a substituted hydrazone at 0°, and the latter by reaction of hydrogen selenide with N-isothiocyanato-dialkylamines. The near infrared spectra of the selenothio-compounds indicate that they have a dipolar structure[415], $R_2NH^+NHCSSe^-$.

Other Selenium Heterocycles

A complete account of the many other types of selenium heterocyclic compound is beyond the scope of this review, and only a few of the representative compounds are

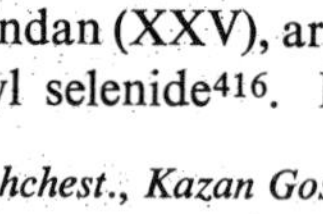

(XXIV)

described below. Selenanthrene (XXIV), which behaves like a diaryl selenide, can be made by treating diphenyl selenide with potassamide in liquid ammonia[262] and indans, such as 2-methyl-2,3-dihydro-1-selenaindan (XXV), are also known; XXV is obtained by the Claisen rearrangement of allyl phenyl selenide[416]. Four-membered ring species such as

[410] Z. S. Titova, *Vestn. Stud. Nauch. Obshchest., Kazan Gos. Univ.* No. 3 (1966) 107 (*C.A.* **68** (1968) 77901f).
[411] L. M. Weinstock, P. David, D. M. Mulvey and J. C. Schaeffer, *Angew. Chem. Int. Edn.* **6** (1967) 364.
[412] E. Benedetti and V. Bertini, *Spectrochim. Acta,* **A24** (1968) 57.
[413] V. Bertini, *Angew. Chem. Int. Edn.* **6** (1967) 563.
[414] U. Anthoni, *Acta Chem. Scand.* **20** (1966) 2742.
[415] U. Anthoni, Ch. Larsen and P. H. Nielsen, *Acta Chem. Scand.* **21** (1967) 2571.
[416] E. G. Kataev, G. A, Chmutova, A. A. Musina and A. P. Anatas'eva, *Zhur. org. Khim.* **3** (1967) 597.

bis-(trifluoromethyl)1,2-diseleneten (XXVI), obtained by reaction of selenium vapour with hexafluorobut-2-yne[417], and 5-membered rings, such as the borolane (XXVII), made

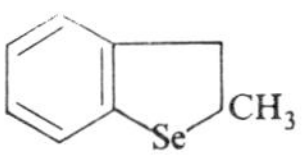

(XXV) 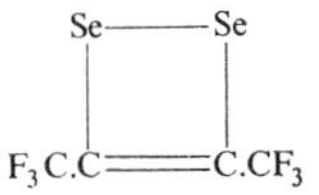 (XXVI) 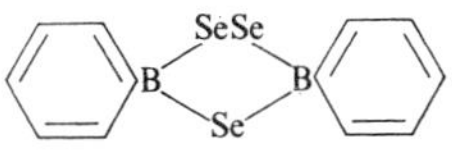 (XXVII) 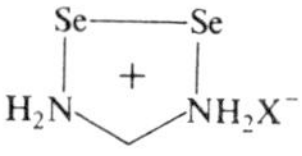 (XXVIII)

by reaction of selenium with phenylboron diiodide[418], or the diselenolylium salt (XXVIII), made by oxidation of ethanolic diselenomalonamide, $CH_2(CSeNH_2)_2$, with ferric chloride or iodine[419], are fairly representative of the wide range of cyclic compounds which can be prepared.

2.22. COMPLEXES

Apart from scattered references to the complexes of the chalcogens in books on their general chemistry[3], there is only one review specifically of the complexes of selenium and tellurium[420]. The complexes formed by the chalcogens are conveniently divided into five groups, comprising the anionic halo-complexes, in which the iodo-complexes are the most stable class for tellurium and polonium, then the anionic complexes formed by oxo-acids, and finally the oxygen, nitrogen and sulphur donor complexes, each type of donor being considered separately. The sulphur donor complexes of tellurium(II) are of particular interest since these are the most stable species known for this oxidation state.

Halo-complexes

The known fluoro-complexes of the quadripositive elements are generally of the form A^IXF_5; the alkali metal and thallous pentafluoroselenates(IV) are white solids obtained by dissolving the fluoride A^IF in selenium tetrafluoride, but the compounds are rather unstable, dissociating appreciably to the component fluorides at room temperature[97]. The analogous pentafluorotellurates(IV) separate as colourless crystals from a solution of the alkali fluoride and tellurium dioxide in hydrofluoric acid[424] or in selenium tetrafluoride[421]. The pyridinium salt is either a pale green[422] or a white[424] solid, made by treating the pyridine complex, TeF_4.py, with aqueous hydrofluoric acid[422] or from a mixture of the dioxide and pyridine in hydrofluoric acid[424]. The nitronium salts, NO_2XF_5, are made by reaction of selenium or tellurium dioxide with nitryl fluoride[423]. The infrared and Raman spectra[424] of the pentafluorotellurates(IV) indicate that the anion is square pyramidal (C_{4v}), confirmed by the X-ray structure determination of the potassium salt[424a]. The solubility of polonium(IV) hydroxide in hydrofluoric acid increases markedly with the concentration of the latter, probably an indication of complex formation, but no salts have been isolated[9].

[417] A. Davison and E. T. Shawl, *Chem. Comm.* (1967) 670.
[418] M. Schmidt, W. Siebert and F. Rittig, *Chem. Ber.* **101** (1968) 281.
[419] K. A. Jensen and U. Kenriksen, *Acta Chem. Scand.* **21** (1967) 1991.
[420] D. I. Ryabchikov and I. I. Nazarenko, *Uspekhi Khim.* **33** (1964) 108. (*Russ, Chem. Rev.* **33** (1964) 55.)
[421] A. J. Edwards, M. A. Mouty, R. D. Peacock and A. J. Suddens, *J. Chem. Soc.* (1964) 4087.
[422] E. E. Aynsley and G. Hetherington, *J. Chem. Soc.* (1953) 2802.
[423] E. E. Aynsley, G. Hetherington and P. L. Robinson, *J. Chem. Soc.* (1954) 1119.
[424] N. N. Greenwood, A. C. Sarma and B. P. Straughan, *J. Chem. Soc.* (1966) 1446.
[424a] A. J. Edwards and M. A. Mouty, *J. Chem. Soc.* (*A*) (1969) 703.

A few rather unstable hexafluoro-complexes are known; the nitrosonium compounds, $(NO)_2XF_6$, are obtained[425] by reaction of nitrososulphuryl fluoride, $NOSO_2F$, with selenium or tellurium tetrachloride:

$$XCl_4+6NOSO_2F \longrightarrow (NO)_2XF_6+4NOCl+6SO_2$$

The selenium compound loses nitrosyl fluoride at room temperature, leaving the pentafluoroselenate(IV), but the tellurium compound is stable to about 100°. The buff pyridinium compound has been reported to be formed from a mixture of pyridine and tellurium dioxide in concentrated hydrofluoric acid[422], but the pentafluoro-complex has been made from the same medium[424], so that the existence of this hexafluoro-complex salt is in doubt.

Tellurium hexafluoride forms complexes with potassium, rubidium and caesium fluorides, the stoichiometry of the compound formed with the last being close to Cs_2TeF_8; the stability of the lattice appears to be inversely related to the polarizing power of the cation and the compounds are only stable in the solid state[426]. Several heptafluorotellurates(VI) have been recorded, made by reaction of tellurium with chlorine trifluoride in anhydrous hydrofluoric acid, followed by addition of the metal fluoride[427].

In contrast to the behaviour of the chalcogen(IV) fluorides, the complexes formed by the other tetrahalides are generally of the type A_2XY_6, in which the anions are octahedral. Crystallographic data are available for many of these salts and are summarized in the literature[3], the general preparative method being precipitation of the complex salt from a solution of the chalcogen dioxide in the concentrated halogen acid, a reaction which is, rather surprisingly, effective for the preparation[428] of hexaiodoselenates(IV); the infrared spectra of the selenium and tellurium hexahaloanions are known[428].

The Raman spectra of solutions of selenium and tellurium dioxides in hydrobromic acid suggest that the pentabromoanions may be present[429] and in non-aqueous solvents it appears that the $TeCl_5^-$ and $TeBr_5^-$ ions predominate[430], although only one salt of this kind, $(C_2H_5)_4NTeCl_5$, has been isolated[431]. A more complex chloro-compound, $(C_2H_5)_4NTe_3Cl_{13}$, isolated from non-aqueous solution, is probably of the type $[(C_2H_5)_4N^+(TeCl_3^+)_3]^{4+}Cl_4$. The infrared and Raman spectra of these species have been recorded[431]. Instability constants for the tellurium halo-complexes[432, 433] and for the polonium iodo-complexes[114], including the PoI_5^- ion, have been published. The vibrational spectra[434] of the adduct $TeCl_4.PCl_5$ are consistent with the formulation $PCl_4^+TeCl_5^-$, whereas those of $TeCl_4.AlCl_3$ suggest that the compound is $TeCl_3^+AlCl_4^-$.

Oxo-acid Complexes

Studies of the solubility of polonium(IV) hydroxide in carbonate[16] and nitrate[282] solution, and the ion exchange behaviour of polonium(IV) at high nitrate ion concentrations[435], indicate that anionic complex species are formed. Tellurium(IV) sulphato-

[425] F. Seel and H. Massat, *Z. anorg. Chem.* **280** (1955) 186.
[426] E. L. Muetterties, *J. Am. Chem. Soc.* **79** (1957) 1004.
[427] A. F. Clifford and A. G. Morris, *J. Inorg. Nuclear Chem.* **5** (1957) 71.
[428] N. N. Greenwood and B. P. Straughan, *J. Chem. Soc.* (*A*) (1966) 962.
[429] P. J. Hendra and Z. Jović, *J. Chem. Soc.* (*A*) (1968) 600.
[430] J. Dobrowolski, *Zesz. Nauk Politech. Gdansk., Chem.* (1966) (11) 3. (*C.A.* **68** (1968) 45817s.)
[431] J. A. Creighton and J. H. S. Green, *J. Chem. Soc.* (*A*) (1968) 808.
[432] R. Korewa and Z. Szponar, *Roczniki Chem.* **39** (1965) 349.
[433] V. I. Murashova, *Zhur. analit. Khim.* **21** (1966) 345. (*J. Analyt. Chem. U.S.S.R.* **21** (1966) 303.)
[434] I. R. Beattie and H. Chudzynska, *J. Chem. Soc.* (*A*) (1967) 984.
[435] E. Orban, U.S.A.E.C. unclassified report MLM-973, 1954.

complexes, of composition $2(2TeO_2.SO_3).MHSO_4.2H_2O$, and the anhydrous compounds obtained by heating the hydrates, have been recorded[283], but the structures are unknown. Carboxylic acids also form anionic complexes with tellurium(IV) and polonium(IV). Examples are the citrato- and tartratotellurates(IV), the silver salts of which are insoluble in water but soluble in nitric acid[436], and evidence of complexing has been obtained from solubility studies with polonium(IV) in formic, acetic, oxalic and tartaric acids, from which it appears that the acetato-complex is more stable than the hexachloroanion[147].

Oxygen Donor Complexes

Very few oxygen donor complexes have been recorded; the infrared spectrum of the 1:1 complex of tellurium tetrafluoride with dioxan indicates that this should be formulated[437] as $[(dioxan)_2TeF_3]^+[TeF_5]^-$ and the 1:1 complex of tellurium tetrachloride with pyridine-N-oxide (PNO) is likewise of the form[438] $[(PNO)_2TeCl_3]^+[TeCl_5]^-$. The nature of the 1:2 complexes of tellurium tetrachloride with acetamide[439], and of polonium tetrachloride with tributyl phosphate[42], is unknown. A polonium(IV) perchlorate complex with tributyl phosphate also seems to be formed in the solvent extraction of polonium from perchloric acid[440]. Selenium trioxide forms 1:1 adducts with ether and dioxan[170], neither of which have been further investigated.

Nitrogen Donor Complexes

The majority of the known nitrogen donor complexes of the chalcogen tetrahalides have the composition $XY_4.2L$, where L is a monodentate ligand, but adducts of selenium or tellurium tetrachlorides with 4 molecules of ammonia or 2 molecules of ethylene diamine are also known[441], some of which may involve 6-coordination [e.g.$(SeCl_2en_2)Cl_2$]. Mono- and bidentate nitrogen donor complexes of composition $TeF_4.L$ and $2TeF_4.L'$ respectively appear[437] to be of the form $[L_2TeF_3]^+[TeF_5]^-$ or $[L'TeF_3]^+[TeF_5]^-$ and the conductivities of the analogous selenium and tellurium tetrachloride complexes appear to indicate that they are 1:1 electrolytes in non-aqueous solvents[438]; similar evidence would indicate that the complexes of composition $XY_4.2L$, where the anion may be a simple halide ion[438], are of the same kind. The 1:1 complexes are probably similar in form to the 1:1 complexes with tellurium tetrafluoride, but it has been suggested[442] that 1:2 complexes, such as $SeCl_4.2py$, contain octahedral $[SeCl_3.2py]^+$ ions, the sixth position being occupied by the lone pair of electrons, as in the structure of the complex $SeOCl_2.2py$, in which the chlorine atoms and pyridine rings occupy *trans* positions in the octahedron and the oxygen atom is *trans* to the lone pair of electrons[443]. However, $SeCl_4.2py$ is isomorphous with $SnCl_4.2py$ which has the *trans* octahedral configuration, and the conductivity data may therefore result from the reaction of the complex with the solvent or impurities in it[443a].

Tellurium hexafluoride is known to form amine complexes of the type $TeF_6.2R_3N$

436 B. Brauner, *J. Chem. Soc.* **55** (1889) 382.
437 N. N. Greenwood, A. C. Sarma and B. P. Straughan, *J. Chem. Soc.* (*A*) (1968) 1561.
438 D. A. Couch, P. S. Elmes, J. E. Fergusson, M. L. Greenfield and C. J. Wilkins, *J. Chem. Soc.* (*A*) (1967) 1813.
439 R. C. Paul and R. Dev, *Indian J. Chem.* **3** (1965) 315.
440 N. I. Ampelogova, *Radiokhimiya*, **5** (1963) 626.
441 V. G. Tronev and A. N. Grigorovich, *Zhur. neorg. Khim.* **2** (1957) 2400.
442 A. W. Cordes and T. V. Hughes, *Inorg. Chem.* **3** (1964) 1640.
443 I. Lindquist and G. Nahringbauer, *Acta Cryst.* **12** (1959) 638.
443a I. R. Beattie, M. Milne, M. Webster, H. E. Blaydon, P. J. Jones, R. C. G. Killean and J. L. Lawrence, *J. Chem. Soc.* (*A*) (1969) 482.

and these appear to 8-coordinate species[444], but no structural data are available for them. Selenium monobromide forms simple nitrogen donor complexes of composition $Se_2Br_2.2L$ with a variety of amines and heterocyclic bases, the compounds precipitating from solutions of the monobromide in carbon disulphide[445], but their structures are unknown.

Sulphur Donor Complexes

Selenium compounds with sulphur donors are scarcely known, but there is evidence[446] for the formation of a selenium(II) species, $[Setu_2]^{2+}$, in the reduction of selenous acid by thiourea (tu) and a selenium tetrachloride complex[441], $SeCl_4.2tu$, is also known. Tellurium tetrachloride and tetrabromide complexes with tetramethyl thiourea (tmu), $Tetmu_2Y_4$, are *trans* octahedral[446a] and crystallize from solutions of the dioxide in the appropriate halogen acid on addition of the ligand and methanol[447], but the primary reaction of tellurium(IV) with thiourea or its N-substituted derivatives is usually reduction to tellurium(II). This oxidation state is stabilized by sulphur donors, so that it has proved possible to isolate a very wide range of complexes of the type $Tetu_2Y_2$ or $Tetu_4Y_2$, where Y is a halide or thiocyanate[448], nitrate[448, 449], sulphate[449], perchlorate[448, 449] or oxalate[449]. All of these complexes are square planar, those of the type $Tetu_2Y_2$ having the *cis* configuration and those of the type $Tetu_4Y_2$ being of the form $[Tetu_4]^{2+}Y_2^-$. However, the thiourea complexes of the methanethiosulphonates (p. 984) have the *trans* configuration. The crystal structures of many of these complexes have been determined by Foss and his co-workers[448]. Most of these tellurium(II) thiourea complexes disproportionate in water but are stable in aqueous thiourea, the fluoride, $Tetu_4(HF_2)_2$, being the least stable of all the thiourea complexes. The infrared and Raman spectra of the tellurium dichloride and dibromide complexes with thiourea have been reported and it appears that the Te–S bond in these compounds is relatively weak[450]. There is trace scale evidence which indicates that polonium is complexed by thiourea, but the composition and nature of the product is not known[451].

Selenium(II) and tellurium(II) alkyl xanthates and dialkyldithiocarbamates are made by reaction of a seleno- or telluropentathionate (p. 981) with an alkali metal alkyl xanthate:

$$XS_4O_6^{2-}+2ROCS_2^- \longrightarrow X(S_2COR)_2+2S_2O_3^{2-}$$

or dialkyldithiocarbamate:

$$XS_4O_6^{2-}+2R_2NCS_2^- \longrightarrow X(S_2CNR_2)_2+2S_2O_3^{2-}$$

The greenish-yellow selenium and red tellurium compounds are not decomposed by water, in which they are insoluble, but the xanthates decompose slowly on keeping[452]. There is also trace scale evidence for the formation of a volatile polonium diethyldithiocarbamate[453].

444 E. L. Muetterties and W. D. Phillips, *J. Am. Chem. Soc.* **79** (1957) 2957.
445 S. Prasad and B. L. Khandelwal, *J. Indian Chem. Soc.* **37** (1960) 645; *ibid.* **38** (1961) 107.
446 E. N. Ovsepyan, G. N. Shaposhnikova and N. G. Galfayan, *Zhur. neorg. Khim.* **12** (1967) 2411, *Russ. J. Inorg. Chem.* **12** (1967) 1271.
446a S. Husebye and J. W. George, *Inorg. Chem.* **8** (1969) 313.
447 O. Foss and W. Johannessen, *Acta Chem. Scand.* **15** (1961) 1939.
448 O. Foss *et al.*, *Acta Chem. Scand.* **13** (1959) 1252, 2155; *ibid.* **15** (1961) 1615, 1616, 1623, 1940, 1947, 1961.
449 J. Vrestal, *Coll. Czech. Chem. Comm.* **25** (1960) 443.
450 P. J. Hendra and Z. Jović, *J. Chem. Soc.* (*A*) (1967) 735.
451 H. Mabuchi, *Bull. Chem. Soc. Japan*, **31** (1958) 245.
452 O. Foss, in *Inorganic Syntheses*, Vol. 4, p. 91, Ed. J. C. Bailar, Jr., McGraw-Hill, New York (1953).
453 K. Kimura and T. Ishimori, *Proc. 2nd Int. Conf. Peaceful Uses At. Energy, U.N.* **28** (1958) 151.

25. FLUORINE

T. A. O'DONNELL
University of Melbourne

For the purposes of organization of data in this treatise as a whole, compounds of fluorine with other elements are regarded formally as fluorides of those other elements. Therefore the detailed physical and chemical properties of fluorine compounds are presented in the appropriate chapters where fluorides of individual or grouped elements are discussed.

Except for section 1, which gives information on the physical and chemical properties of elemental fluorine itself, this chapter is presented as a series of reviews. Hydrogen fluoride and the halogen fluorides are discussed separately as solvent systems, with particular emphasis on acid–base reactions. For main group fluorides those properties are given which result from the peculiar reactivity of fluorine, e.g. its great strength as an oxidant or the ability of fluoro-ligands to stabilize unusual complexes. Any systematic studies of the chemical reactivity of *d*- and *f*-transition metal fluorides, particularly in higher oxidation states, are relatively recent and have not been reviewed comprehensively. This is done in the last section.

Because of its extreme reactivity, elemental fluorine was not isolated until 1886, by which time most of the other naturally occurring elements were known. From that time until about 1940, most study was associated with the ionic fluorides or those either of non-metals or transition elements in lower oxidation states. These were the fluorides which were easy to handle experimentally. The higher fluorides of the elements of Groups VB to VIIB and of the *d*- and *f*-transition elements are, for the most part, extremely reactive and could not be studied adequately until special techniques were developed for their manipulation, although Otto Ruff and his co-workers in Germany did remarkable work over the first third of this century in preparation and initial characterization of many of these fluorides.

Both the inorganic and organic chemistry of fluorine received an enormous impetus during World War II. Separation of the isotopes of natural uranium was carried out by a diffusion process involving uranium hexafluoride vapour and various aspects of fluorine chemistry were basic to the processing of uranium and of plutonium. Fluorinated oils, greases and polymers were needed in the associated technology. While the main effort in this field was concentrated in the war years, there has been continuing activity because of the peaceful application of fission processes. Also, because of the great oxidizing strength of fluorine and of many of its compounds, rocket technologists have been interested in the possibility of using fluorine-containing fuels.

Fluorocarbons have found applications in many industrial processes. Polytetrafluoroethylene and related polymers are widely used for their chemical inertness and because of their very low frictional properties. Low-boiling chlorofluorocarbons are used almost universally in domestic refrigeration and as propellants for pressure-packed commodities.

The large-scale industrial application of inorganic fluorides continues to be in the electrometallurgy of aluminium, where the molten electrolyte is natural or synthetic cryolite Na_3AlF_6; but there has been a growing use of anhydrous hydrogen fluoride in alkylation processes in petroleum refining.

Although the quantity of fluoride involved is relatively small, an aspect of inorganic fluoride chemistry which has had a growing impact on most communities has been the artificial fluoridation of drinking water supplies—a procedure adopted to minimize the incidence of dental caries.

While there are many excellent reviews (which will be referred to in this chapter) on different aspects of fluorine chemistry, the most comprehensive source of information on inorganic and organic fluorine compounds and on biological effects of fluorine and fluorides is the five-volume series "Fluorine Chemistry", edited by J. H. Simons and published by Academic Press between 1950 and 1964.

1. GENERAL PROPERTIES OF FLUORINE AND FLUORIDES

1.1. DISCOVERY OF THE ELEMENT

Although the existence of fluorides had been known for a long time, elemental fluorine was not isolated until 1886 when Moissan electrolysed a dilute solution of potassium fluoride in anhydrous hydrogen fluoride, using platinum electrodes. Because of the high volatility of the electrolyte, he cooled his electrolysis cell to $-24°C$ with boiling methyl chloride.

This first isolation of the element occurred 112 years after Scheele had prepared chlorine and about 70 years after the preparation of elemental bromine and iodine. The long delay in the first preparation of fluorine resulted, of course, from the very great reactivity of the halogen itself. Its compounds are generally thermally stable because of the great strength of the chemical bond between a fluorine and another atom in most situations. Even if fluorine were to be produced relatively easily in a high-temperature reaction, it would immediately react with the materials from which the containing apparatus was made.

Furthermore, fluorine oxidizes water. Therefore no aqueous reaction, or any reaction which produces water, can be used for its preparation. Thus the counterpart of the familiar oxidation in concentrated sulphuric acid of chloride to chlorine by manganese dioxide is not available.

Only non-aqueous reactions at relatively low temperature can be used to prepare fluorine itself. In fact, as will be seen in section 1.3, all of the procedures which have been used to prepare fluorine industrially or in the laboratory are modifications of Moissan's method.

1.2. OCCURRENCE AND INDUSTRIAL APPLICATIONS

A 1962 review paper[1] by G. C. Finger entitled "Fluorine resources and fluorine utilization" gives a detailed account of the minerals in which fluorine is found and the location of the major deposits. An estimate is given there of the reserves of fluorine minerals in the

[1] G. C. Finger, in *Advances in Fluorine Chemistry*, eds. M. Stacey, J. C. Tatlow and A. G. Sharpe, Vol. 2, Butterworths (1962) p. 35.

major countries of the world. Also the way in which these minerals, or chemicals derived from them, are used in industry throughout the world is indicated.

(i) Occurrence

Elemental fluorine is so reactive that it does not occur naturally in the free state. The most reactive form in which fluorine is encountered in nature is as hydrogen fluoride, which can be present in amounts up to 0.03% in gases of volcanic origin. It has been estimated that one such source of volcanic gas in Alaska discharges 200,000 tons of hydrogen fluoride annually into the atmosphere.

In combination, fluorine represents about 0.065% of the earth's crust, being thirteenth of the elements in abundance and occurring there more widely than chlorine, and five to ten times more abundantly than zinc or copper. Finger lists many minor minerals in which fluorine occurs, presumably because of the ease with which the fluoride ion can replace the similarly sized hydroxide ion. For example, in certain deposits, topaz can be represented as having the formula $Al_2SiO_4(F, OH)_2$, where F and OH are interchangeable. The three most important fluorine-containing minerals are fluorspar or fluorite (CaF_2), cryolite (Na_3AlF_6) and fluorapatite which is apatite (calcium hydroxyphosphate) in which some hydroxide has been replaced by fluoride.

The principal producers of fluorspar are the United States, Mexico, USSR and China, with several European countries—Italy, Germany, Spain, France and England—also major producers. Depending on its quality after ore beneficiation, fluorspar is used as a flux in steel-making, as a flux and opacifier in glass and enamel production, or as the raw material from which hydrofluoric acid and anhydrous hydrogen fluoride are produced.

The world's only commercial deposit of cryolite is located in Greenland, and from it 1 to 2 million tons of the mineral in 99% purity have been recovered. Its vast importance is that it is the molten electrolyte from which aluminium metal is produced commercially, although most of the cryolite now used is synthetic rather than natural.

By far, most of the fluorine in the earth's crust is in combined form in fluorapatite, where it occurs to the extent of about 3.5%, whereas it comprises about 50% of fluorspar and of cryolite. The principal producers are the United States, USSR, Morocco and Tunisia, Nauru and other Pacific islands, and Curaçao in the West Indies. However, the mineral is not processed for its fluoride content, but for the phosphate, most of which appears as the fertilizer superphosphate. During treatment of the mineral with sulphuric acid, the fluoride reacts with silica and silicates to form silicon tetrafluoride and fluorosilicates, most of which are lost. Perhaps more effective atmospheric pollution control will lead to the development of recovery processes for the fluorides and the mineral would then become the major source of fluorine.

(ii) Industrial Applications

Fluorspar is separated from silicates and other minerals with which it occurs naturally by density separation processes and by froth flotation. Depending on its quality, it is marketed as one of three grades; when 85% pure it is used as a flux in the steel industry; 85–95% purity is required for use as a flux or an opacifier in the production of glasses and enamels, and a product containing more than 97% of calcium fluoride is needed for hydrogen fluoride production.

Refined fluorspar is reacted in a heated horizontal kiln with sulphuric acid and the resulting vapour, containing about 95% of hydrogen fluoride, is absorbed to give an approximately 70% solution in sulphuric acid. Distillation then gives a product which can be as high as 99.95% in hydrogen fluoride, the most common impurities being water, sulphuric acid, fluorosulphuric acid, sulphur dioxide, silicon tetrafluoride and fluorosilicates. Because the product is stored in steel cylinders, hydrogen can be a major contaminant in the acid as supplied commercially . When aqueous solutions of hydrogen fluoride are required industrially, it is now common practice to add water to the anhydrous product produced as above. A convention adopted by fluorine chemists and engineers is that aqueous solutions of hydrogen fluoride are referred to as hydrofluoric acid whereas the non-aqueous product is called hydrogen fluoride or anhydrous hydrogen fluoride (for which the symbol AHF is widely used).

Most of the hydrogen fluoride produced is used either in the production of synthetic aluminium fluoride or cryolite for the aluminium industry or in the preparation of fluorocarbons, the general name for a wide range of fluoro-organic compounds, polymeric or molecular. The most widely used solid fluorocarbon is polytetrafluoroethylene, for which the best known trade name is Teflon (E. I. du Pont de Nemours & Co.), other trade names being Fluon, etc. Teflon has reasonable elastomeric properties, has excellent resistance to thermal and chemical attack, and exhibits very low values of dielectric constant and coefficient of friction. It is widely used as an insulator, as an inert gasket or gland material, in self-lubricating bearings and as a non-stick surface in food processing. Many liquid fluorocarbons are used as inert dielectrics and lubricants, and the more volatile ones are produced in large quantities to act as the pumping fluid in refrigerators or as propellants in pressure-packed aerosols and similar products. These volatile liquids are the mixed fluoro- and chloro-compounds of methane and ethane, and are best known as the Freons, another du Pont trade name. They are produced by the Swarts reaction, an example of which would be the injection of carbon tetrachloride and hydrogen fluoride into a heated reactor containing antimony trifluoride and a catalyst:

$$3CCl_4+2SbF_3 \longrightarrow 3CCl_2F_2+2SbCl_3$$

Anhydrous hydrogen fluoride acts as a catalyst in alkylation reactions and has application in petroleum refining in the production of high-octane fuels.

The main application of elemental fluorine, obtained by electrolysis of anhydrous hydrogen fluoride as described in the section immediately below, is in the field of nuclear energy. The wanted fissile isotope ^{235}U is normally separated from ^{238}U, which is more abundant in naturally occurring uranium, by a process involving the separation by gaseous diffusion of $^{235}UF_6$ from the slightly denser $^{238}UF_6$. Uranium hexafluoride is volatile, subliming at a pressure of 1 atmosphere at 56°C. While hydrogen fluoride can be used to produce uranium tetrafluoride from the dioxide, the great oxidizing strength of fluorine is needed to convert the tetrafluoride to the hexafluoride. Uranium metal, enriched in ^{235}U, is usually obtained in a two-step reduction of the hexafluoride after recovery from the diffusion process. Hydrogen is used to reduce the hexafluoride to tetrafluoride, which is then further reduced by metals such as calcium or magnesium to give ingots of uranium metal.

1.3. PRODUCTION OF ELEMENTAL FLUORINE

From the turn of this century there was sustained effort on the part of a small number of fluorine chemists, the most notable and productive of whom was Otto Ruff, to find a solid fluoride which could be prepared from hydrogen fluoride or some reagent other than elemental fluorine and which on heating would yield gaseous fluorine. Although there were some reports that some fluoro-complexes, such as of lead(IV), or some binary compounds such as silver(II) fluoride, yielded fluorine on heating, the goal was never reached. The compounds concerned needed elemental fluorine or some other compounds such as the halogen fluorides which were made from fluorine, for their preparation. Furthermore, there is now much evidence that the isolation of a lower fluoride after heating one of these compounds results from complicated hydrolysis reactions involving traces of moisture in the system rather than from thermal decomposition to a lower fluoride and fluorine. Quite recently, Seel and Detmer[2] have reported that fluorine is evolved when potassium fluoride is heated with the adduct $IF_7{\cdot}AsF_5$ to temperatures above 200°C in accordance with the equation

$$IF_7{\cdot}AsF_5+2KF \xrightarrow{>200^\circ C} KIF_6+KAsF_6+F_2$$

However, much fluorine is needed in the formation of the adduct. Therefore this reaction has no value as a preparative method for fluorine.

In fact, there is only one general procedure for the preparation of fluorine—Moissan's method of electrolysis of mixtures of potassium fluoride and hydrogen fluoride. Moissan used a solution in which the mole ratio KF:HF was about 1:13. The vapour pressure of this system was high and, in order to minimize contamination of fluorine with hydrogen fluoride, Moissan had to keep his electrolysis cell at −24°C. Cady[3] measured melting points and vapour pressures over a wide concentration range for the system KF—HF and found that for the mole ratios KF:HF of 1:2 and 1:1 melting points were about 72°C and 240°C and HF vapour pressures were relatively low, being 20–25 mm Hg. In most fluorine generators now in use, the electrolyte is a mixture of potassium fluoride and hydrogen fluoride in an approximate mole ratio of 1:2. The electrolyte is kept molten at 80–90°C, and these are called medium-temperature generators. Some industrial generators, based on the 1:1 electrolyte, have been operated at about 250°C and are referred to as high-temperature generators. In all of them, hydrogen is liberated at the cathode and fluorine at the anode, so that during operation anhydrous hydrogen fluoride must be added to the electrolyte to prevent solidification.

In the development of satisfactory fluorine generators, much of the effort has gone into a study of anode characteristics. Moissan's platinum electrodes were replaced by nickel and subsequently by carbon anodes of varying designs. Corrosion and polarization of the anode have provided the major design problems. Cady has given an excellent account of these problems, of the solutions to them, and of the many resulting designs for fluorine generators[4]. Leech has traced the historical development of generator design[5] and has reviewed the many types of generator in use, particularly in industrial fluorine generation[6]. This latter aspect is covered in great detail in the published account of the American Chemical Society's 1946 Symposium on Fluorine Chemistry[7].

Because hydrogen and fluorine react explosively on mixing, an essential design feature for

[2] F. Seel and O. Detmer, *Z. anorg. u. allgem. Chem.* **301** (1959) 113.
[3] G. H. Cady, *J. Am. Chem. Soc.* **56** (1934) 1431.
[4] G. H. Cady, in *Fluorine Chemistry*, ed. J. H. Simons, Vol. 1, Academic Press (1950) p. 293.
[5] H. R. Leech, *Quart. Revs.* (*London*), **3** (1949) 22.
[6] H. R. Leech, in *Supplement to Mellor's Comprehensive Treatise on Inorganic and Theoretical Chemistry*, Supp. II, Part I, Longman (1956) pp. 15–43.
[7] Symposium Proceedings, *Ind. Eng. Chem.* **39** (1947) 244–288.

a fluorine generator must provide for the separate collection of both products of electrolysis. A simple way of doing this which has proved effective in very small generators in use in some laboratories is to construct a V-shaped cell from copper or nickel[8]. With the electrode in each arm, separate collection is relatively easy.

Most laboratory generators, however, resemble that shown diagrammatically in Fig. 1. The generator is a mild steel pot containing the electrolyte (KF·2HF) with the level of electrolyte shown at *A*. The electrolyte is kept molten at a temperature of 80–90°C by a

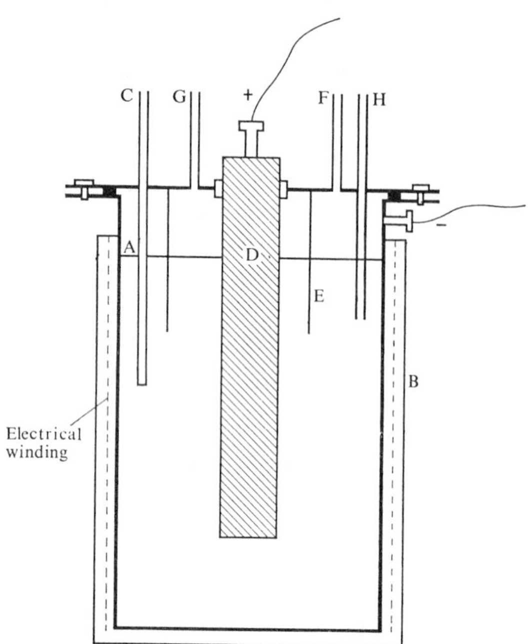

FIG. 1. Medium temperature fluorine generator.

heating jacket *B* controlled by a thermostat in the well at *C*. The pot itself acts as the cathode while the carbon anode *D* is centrally located and insulated from the removable lid of the pot. Hydrogen evolved at the walls of the pot and fluorine at the surface of the anode are kept separate by a metal skirt or diaphragm *E* which dips a few inches into the electrolyte. Hydrogen is piped off to waste at *F*. The fluorine outlet *G* should be provided with positive and negative pressure relief traps so that, if inadvertently a pressure greater or less than atmospheric is applied to the fluorine compartment, fluorine or hydrogen is not forced to pass under the skirt and into the wrong compartment where an explosion would occur. It is also customary to fit to *G* a tower containing solid sodium fluoride which removes much of the hydrogen fluoride contaminant from the fluorine stream by forming sodium fluoride–hydrogen fluoride adducts. A facility is provided at *H* for addition of hydrogen fluoride from a cylinder to enrich the electrolyte after production of hydrogen and fluorine. Laboratory generators are normally operated with d.c. of 10–50 amp. In principle, industrial production involves the use of banks of generators with current capacities of 1000–2000 amp.

[8] L. M. Dennis, J. M. Veeder and E. G. Rochow, *J. Am. Chem. Soc.* **53** (1931) 3263.

The electrolytic generator has limitations in laboratory use in that it cannot be operated to give fluorine pressures greater than one atmosphere and that the fluorine produced is contaminated to a greater or less extent with air and with hydrogen fluoride vapour. Compressed fluorine is now commercially available; but it should be remembered that the hazards of handling the element increase enormously as the working pressure increases. Fortunately several metals, such as mild steel, copper and nickel and its alloys, can be used to handle fluorine because, even though a vigorous exothermic reaction would occur between fluorine and each of the metals if the metal were available for reaction, the surface of the metal is passivated by the formation of a covering layer of the appropriate metal fluoride (see section 1.4). However, if that layer is broken, the metal can burn in fluorine.

A common cause of fluorine fires is reaction at the surface of the metal between the fluorine and adhering grease, even in small amount. The grease burns, the resulting "hot-spot" breaks the protective layer of fluoride, and the hot bulk metal burns. The frightening effect of the deliberate ignition with grease of a fluorine cylinder has been recorded[9] and should be looked at by anyone handling elemental fluorine or very reactive fluorides. *A cardinal rule in using any material of construction with fluorine and oxidizing fluorides is that all exposed surfaces must be rigorously cleansed of grease.* All new components such as valves should be washed with a suitable solvent, and machined metal components should be similarly treated to remove cutting compounds.

However, provided reasonable precautions are taken, fluorine can be handled in safety. Several alloys can be used for fabrication of apparatus (see section 1.4), and suitable valves, pressure regulators and safety barricades are available commercially. Fluorine is available in ½lb and larger cylinders; but it is recommended that the smaller cylinders should normally be used in the laboratory, so that if a fire is started damage is minimized[9a].

Another potential hazard that should be considered by those proposing to use elemental fluorine is that its boiling point is about −188°C, so that if it is at a pressure of 1 atmosphere or greater, it will condense to the liquid in traps cooled by liquid nitrogen. This hazard is probably greatest when opaque traps, made of metal or other materials, are being used.

1.4. MANIPULATION OF FLUORINE AND FLUORIDES IN THE LABORATORY

Unless faced with the necessity of working at high temperatures or pressures or under some other extreme conditions, chemists will generally opt for the convenience of working in glass apparatus. Involatile binary or complex fluorides, such as those of Groups IA and IIA as well as lower fluorides of main group metals and of *d*- and *f*-transition metals, can be handled in conventional glass apparatus providing the temperature is not too high when surface etching will occur.

The relatively inert, and usually involatile, fluorides can be contained in glass or, better, silica for studies of absorption and resonance spectra and for X-ray crystallographic, magnetic susceptibility and similar studies. Also, provided the solvent itself is not a very reactive one, such as hydrogen fluoride or a halogen fluoride, reactions, analyses and physico-

[9] Ref. 7, p. 277.

[9a] Cylinders of fluorine, valves, pressure regulators and safety barricades are available from Matheson, Rutherford, New Jersey, USA. In the interests of safety, these suppliers invite fluorine users to obtain their data sheet on fluorine.

chemical studies in solution can be performed in conventional glass equipment. Aqueous solutions of these fluorides, or of their hydrolysis products, can usually be handled in glass, particularly if the solution is not acidic.

However, glass is etched by fluorine itself, by hydrogen fluoride, by the fluorides of most non-metals—those of carbon and silicon being obvious exceptions—and by the volatile higher fluorides of metals. This etching reaction is a cyclic hydrolysis. Water, adsorbed or loosely chemically bound to the surface of the glass, reacts with a fluoride to form an oxide fluoride and hydrogen fluoride, the latter reacting in turn with glass to give silicon tetrafluoride and water which continues the hydrolysis. Even traces of hydrogen fluoride dissolved in a fluoride will initiate the same hydrolysis cycle. The worst aspect of the etching reaction is the hazard resulting from the build-up of pressure of silicon tetrafluoride. Many serious accidents have occurred with chemical systems that were not regarded as potentially explosive.

For the manipulation of some fluorides, e.g. the hexafluorides of uranium, tungsten and molybdenum, thorough drying of the glass by heating it while the system is evacuated to a pressure of about 10^{-4} to 10^{-5} mm Hg allows use of glass vacuum systems. However, in these situations, grease—even of the halocarbon type—should be avoided. There is usually sufficient moisture in the grease to start hydrolysis and, in addition, outgassing occurs as a result of solubility of the molecular fluoride in the grease.

Usually, however, for handling fluorine, hydrogen fluoride or volatile fluorides, glass vacuum systems are to be avoided, and alternative materials of construction should be selected. Several metals are available. Mild steel and copper are suitable for handling and storing fluorine and most fluorides. For greater inertness nickel and its alloys, e.g. Monel and Inconel, are favoured for apparatus in which careful experimentation is to be carried out. The metals are readily available, easily machined and can be used to relatively high temperatures, because a covering layer of the appropriate fluoride renders them passive to further attack. Steel, copper and nickel can be used safely with fluorine and fluorides to temperatures of approximately 200°, 350° and 650°C respectively. They should be conditioned before use by exposure to an appropriate fluoride to provide the passive layer. The major disadvantage of using metal for vacuum systems is the lack of transparency.

In recent years there has been a marked increase in the fabrication of apparatus for handling fluorides from polymeric materials. Polyethylene has been used for some work with hydrogen fluoride solutions, but its use is not favoured because oxidizing fluorides react readily with it. For chemical inertness, polytetrafluoroethylene (Teflon, Fluon, etc.) is the best material, undergoing virtually no chemical attack to about 250°C, but it is not transparent and, being tough but non-rigid, cannot be machined with precision. Considering the balance of all its properties, the most suitable polymer is polychlorotrifluoroethylene (Kel-F), which is almost as inert as Teflon at room temperature, is transparent and sufficiently rigid that precision equipment such as vacuum valves can be machined from it. Its main limitation is that while, like Teflon, it can be used at liquid nitrogen temperatures, it is more easily pyrolysed than Teflon and cannot be used safely above about 100°C.

In one of the very few reviews dealing with a comparison of several different approaches to the handling of a wide range of reactive fluorides, Canterford and O'Donnell[10] discuss in considerable detail the basis for selection of appropriate materials for construction of apparatus for manipulation of fluorine, hydrogen fluoride and fluorides. They give the

[10] J. H. Canterford and T. A. O'Donnell, in *Technique of Inorganic Chemistry*, eds. H. B. Jonassen and A. Weissberger, Vol. VII, Interscience (1968) p. 273.

circumstances in which glass can be used and those for which metals or polymers or combinations of these materials are most satisfactory. They also describe typical vacuum systems and apparatus for physico-chemical studies of fluorides in the solid, liquid or vapour state, or in solvents such as hydrogen fluoride.

In discussing vacuum systems, Canterford and O'Donnell state that the "backing system", the manifold including valves and protective traps between the vacuum system proper and the pumps, is usually best constructed from glass. They give the relatively unusual conditions under which it is best to construct the vacuum system itself from glass, and they describe typical operations in such a system.

As stated above, certain volatile reactive fluorides can be handled in glass systems provided water that is closely bound to the surface is removed by heating the glass while the system itself is at relatively low pressure; but, even in such vacuum systems, conventional greased glass vacuum valves are unacceptable because volatile fluorides usually react with or dissolve in the grease. Sometimes valves with glass bodies and needles or diaphragms fabricated from Teflon or other inert polymers are used. Alternatively, metal valves can be connected to the glass vacuum systems through compression seals based on Teflon O rings or glands.

Vacuum systems constructed almost entirely from nickel and its alloys have been developed at the Argonne National Laboratory, Illinois, and other USAEC laboratories. Most of the work on characterization of volatile fluorides of transition metals and on the preparation of the binary fluorides of xenon was done in such systems. They have the advantage that all components, valves, traps, all-nickel Bourdon pressure gauges, etc., can withstand very high pressures while being operable under good vacuum conditions; and so they can be used for a wide range of preparative techniques. Another advantage of nickel reaction vessels is that they can be used to temperatures of about 600°C without significant attack by fluorine or fluorides. Canterford and O'Donnell compare these systems with simpler metal ones which are much easier to fabricate and operate, all components being removable from a nickel manifold through compression seals based on Teflon glands[10a].

Provided high pressures are not to be encountered, reaction tubes machined or moulded from Kel-F[10b] can be attached to metal systems so that reactants and products can be observed.

In many situations there are advantages in constructing vacuum systems entirely from Kel-F. It has been observed that, whereas strong Lewis acids, e.g. arsenic and antimony pentafluorides, and anhydrous hydrogen fluoride can each be stored indefinitely in nickel or other metal containers, solutions of such fluoride-acceptors in hydrogen fluoride attack and dissolve metals rapidly. Fortunately such solutions can be handled in Kel-F.

The vacuum system is based on Kel-F tubing and, for normal laboratory-scale operations, $\frac{1}{4}$ in. o.d. tubing is suitable for the manifold and is available commercially from several suppliers. Other components can be fabricated with $\frac{1}{4}$ in. o.d. outlets which can be then connected through unions or T-pieces of the Swagelok type. Alternatively, the Kel-F tubing can be flared with a simple flaring tool and tightened on to the male taper of a component, machined from Kel-F, such as the T-piece shown in Fig. 9 of Canterford and

[10a] The "Swagelok" series of unions, T-pieces, etc., fabricated from brass, Monel, Teflon and other materials, is particularly suitable for demountable vacuum connections. "Whitey" metal needle valves are fitted with "Swagelok" unions. Both series are commercially available from Crawford Fitting Co., Solon, Ohio 44139, USA, or Techmation Ltd., London, NW9, England.

[10b] Moulded Kel-F reaction tubes ($6 \times \frac{3}{4}$ inch) are obtainable from the Argonne National Laboratory, Argonne, Illinois, USA.

O'Donnell's review[10]. Kel-F reaction tubes, machined or moulded, can be connected through suitable heads and valves to the manifold.

The key feature of the Kel-F vacuum system is the vacuum valve, designed in its original form at the Illinois Institute of Technology and modified significantly at the Argonne National Laboratory subsequently. Further modifications have been reported recently[11] and the valve in this form is shown in Fig. 2. The body of the valve is machined from Kel-F

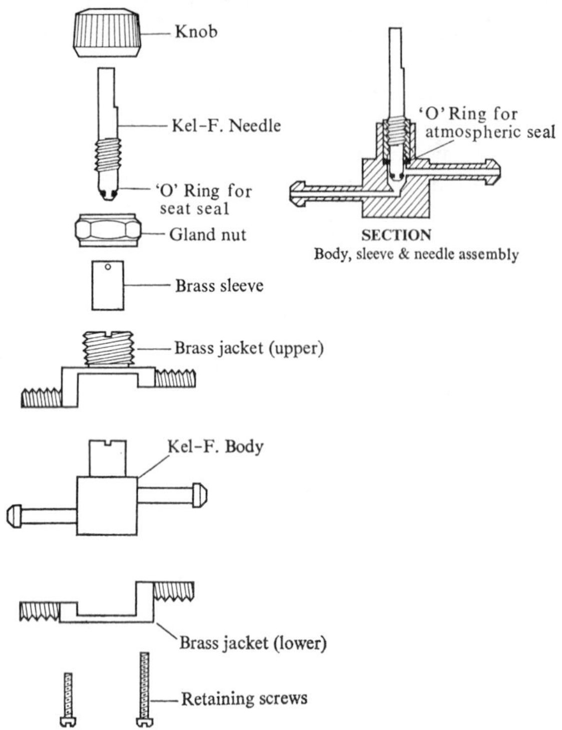

FIG. 2. Kel-F vacuum valve.

but is fitted with metal collars with appropriately located metal threads so that all compression connections to flared or straight Kel-F tubing can be made with mechanically strong threads rather than with Kel-F threads which are easily stripped. The needle of the valve is turned from Kel-F and tipped with Teflon, for good seating. The atmospheric seal between the needle and the valve-body is a Teflon O ring. At room temperature this valve can safely handle fluorine and most fluorides in the liquid or gaseous state because only Kel-F and Teflon are exposed to reactive substances.

While outgassing of Kel-F would prevent attainment of very high vacuum in a Kel-F system, it is easy to achieve a static vacuum sufficiently low to allow distillation and other manipulation of fluorides which have vapour pressures of about 0.1–1 mm Hg and above. One of the working advantages of a Kel-F vacuum system over the other types is that it is physically flexible so that liquids can be poured very easily from one part of the system to another. This is particularly advantageous in the transfer of involatile liquids or in the

[11] T. A. O'Donnell, *Anal. Chem.* **43** (1971) 977.

preparation of a solution of an involatile fluoride in a solvent such as hydrogen fluoride and the subsequent transfer of that solution to a unit such as a conductance or spectral cell.

Canterford and O'Donnell describe typical experimental procedures with fluoride-resistant vacuum systems and also indicate the facilities that now exist for physico-chemical investigation of fluorides. They describe manometers, cells for measurement of conductance and for absorption spectrometry and techniques for X-ray analysis, magnetic measurements and resonance spectrometry. Hyman and Katz, in a review on hydrogen fluoride[12], give considerable detail on techniques for measurement of conductance, and of visible, ultra-violet, infrared and Raman spectra in that solvent.

The safe handling of elemental fluorine has been discussed in section 1.3. The warning given there on the danger of initiating an uncontrollable fire with grease or other combustible material in apparatus that is normally fluorine- or fluoride-resistant applies equally well in this section, particularly if strong oxidants such as the halogen fluorides are being handled. *It is essential to ensure that all grease has been removed from all components with a suitable solvent before fluorides are introduced.* Then the equipment should be "conditioned" by exposure to fluorine, a compound such as chlorine trifluoride or the reactive compound to be used in the ultimate investigation.

Many organic substances, particularly halocarbons, dissolve fairly readily in Kel-F, sometimes causing noticeable swelling and distortion of the polymer on prolonged exposure. For this reason, it is desirable to avoid the use of soluble oils in machining Kel-F. Sufficient oil can be absorbed and retained to initiate chemical reaction with strongly oxidizing fluorides.

Despite all these warnings, the 1947 statement of G. H. Cady[13] is still quite valid: "In my opinion the hazards of working with fluorine and its compounds have been greatly over-rated.... When treated with the respect that is due to it, fluorine is just another substance." In their recent reviews, Hyman and Katz[12] and Canterford and O'Donnell[10] have stressed that suitable techniques are now available for detailed investigation of the physical and chemical properties of most fluorides in the solid, liquid or vapour state or in solution in reactive solvents.

1.5. FLUORINE ISOTOPES

Apart from the stable ^{19}F, the only fluorine isotope with a reasonable existence is ^{18}F with a half-life of 109.5 ± 0.5 min. However, the isotope is not available commercially, and most exchange reactions involving this isotope have been studied at laboratories associated with nuclear energy programmes.

Stranks[14] has described conditions for preparing sodium fluoride containing 0.5 millicurie of ^{18}F at the time of removal from a reactor. He states that chemical tracer work can be carried out with such a sample for a period of 12–24 hr after irradiation. One gram of lithium carbonate is irradiated in a thermal neutron flux of 1×10^{12} neutron cm^{-2} sec^{-1}, leading to $^{16}O(t, n)^{18}F$ reaction with tritons from $^{6}Li(n, \alpha)^{3}H$. Alternatively, fast neutrons can be used in the $^{19}F(n, 2n)^{18}F$ reaction or γ radiation in $^{19}F(\gamma, n)^{18}F$.

12 H. H. Hyman and J. J. Katz, in *Nonaqueous Solvent Systems*, ed. T. C. Waddington, Academic Press (1965) pp. 47–81.
13 Quoted in ref. 5, p. 22.
14 D. R. Stranks, in *Inorganic Synthesis*, ed. J. Kleinberg, Vol. VII, McGraw-Hill (1963) p. 150.

Dove and Sowerby[15] have reviewed the fluorine exchange reactions studied with ^{18}F. They comment on the advantage of *in situ* formation of ^{18}F considering the short half-life of the isotope and report that Sheft and Hyman labelled elemental fluorine directly in the beam of a linear accelerator to study fluorine exchange reactions with the hexafluorides of xenon, sulphur and several transition metals. Sulphur hexafluoride did not exchange up to 350°C, whereas xenon hexafluoride did so readily at 100°C with an estimated energy of activation of 20 kcal mole^{-1} for what appeared to be an associative reaction. Initial studies showed rapid exchange of fluorines below 350°C for the hexafluorides of molybdenum, tungsten, osmium, iridium and uranium.

Elemental fluorine exchanged with halogen fluorides at measurable rates above 190°C with some evidence of surface-catalysed reactions, whereas exchange between hydrogen fluoride and the halogen fluorides was essentially complete at 27°C within the 3 min required for mixing and separation. These latter exchanges were not considered surface reactions because they appeared identical when carried out in nickel or in Kel-F containers. Direct exchange through intermediates was postulated which is not very surprising in light of the ready formation of entities such as ClF_4^- and BrF_6^-, even though these are less easily formed than BrF_4^-, IF_6^- and others.

No exchange was observed to the highest temperatures studied between hydrogen fluoride and either sulphur hexafluoride or fluorocarbons; but silicon tetrafluoride exchanged fluorines with all alkali fluorides before there had been significant formation of hexafluorosilicates, indicating that these complexes are not necessarily intermediates.

A large fraction of the limited number of fluorine exchange reactions has been directed to the technology of handling fluorine and fluorides such as uranium hexafluoride. This last hexafluoride has been studied with fluorine, halogen fluorides and hydrogen fluoride. Also there have been investigations of fluorine-exchange, often with indirect labelling, between fluorine, halogen fluorides and hydrogen fluorides with the fluorides of nickel, copper and zinc, which are metals used in fluoride-resistant alloys.

To date there has been relatively little study of fluorine exchange reactions in aqueous solution. Hydrogen fluoride has been shown to undergo complete exchange with hexafluorosilicate and Anbar and Gutmann investigated the kinetics of exchange of fluorine between alkali metal fluorides and fluoroborates in a study of acid and base hydrolysis of fluoroborates.

1.6. PHYSICAL PROPERTIES OF FLUORINE

Because of the extreme experimental difficulties in handling highly reactive fluorine, quoted physical properties for the element must be regarded as considerably less reliable than for most elements. For example, there have been several vapour pressure equations for the liquid reported, and they give significantly different vapour pressures and derived quantities such as boiling points and enthalpies of vaporization. The values quoted in Table 1 were reported by Johnston and co-workers at Ohio State University and appear to be the most reliable. Most of the thermodynamic data are quoted from the reviews of A. G. Sharpe who wrote an incisive review[16] in 1957 summarizing much of the then known inorganic chemistry of fluorine in terms of energetics of reaction. He included

[15] M. F. A. Dove and B. D. Sowerby, in *Halogen Chemistry*, ed. V. Gutmann, Vol. I, Academic Press (1967) p. 44.

[16] A. G. Sharpe, *Quart. Revs.* (*London*), **11** (1957) 49.

the other halogens, together with later data for fluorine, in a second equally authoritative review[17].

It is now a matter of historical interest only that for a long time the enthalpy of dissociation for molecular fluorine was believed to be in the region 60–70 kcal mole^{-1}. This expected value, "determined" by several different experimentalists between 1920 and

TABLE 1. PHYSICAL PROPERTIES OF FLUORINE

Atomic weight	18.993 (based on ^{12}C = 12.000)
Transition temperature (solid)[a]	45.55°K
Enthalpy of transition[a]	173.90±0.04 cal mole^{-1}
Melting point[a]	53.54°K
Enthalpy of fusion[a]	121.98±0.5 cal mole^{-1}
Boiling point[a]	85.02±0.02°K
Enthalpy of vaporization[a]	1563.98±3 cal mole^{-1} (at 84.71°K and 738 mm)
Vapour pressure (solid)[b]	Equation: $\log_{10}p$(mm) = $(-430.06/T)+8.233$
Vapour pressure (liquid)[a]	Equation (53.54 to 90°K): $\log_{10}p(\text{mm}) = 7.08718 - (357.258)/T - (1.3155\times10^{13})/T^8$
	Values: T (°K) 53.56 60.50 69.57 77.17 81.59 85.05 89.40
	(Exptl.) p(mm) 1.67 12.89 84.30 280.4 504.1 763.1 1219.9
Critical temperature and pressure[b]	144°K and 55 atm
Heat capacity (solid)[a]	T (°K) 13.89 24.16 41.48 52.09
	C_p (cal mole^{-1} deg^{-1}) 1.505 4.406 9.222 17.331
Heat capacity (liquid)[a]	T (°K) 58.14 67.05 76.60 81.32
	C_p (cal mole^{-1} deg^{-1}) 13.685 13.556 13.714 13.797
Density (liquid)[c]	Equation: ρ(g cm^{-3}) = $1.5127+0.00635(T_B-T)$ where T_B = 85.02°K
	Values: T (°K) 66.03 70.00 77.22
	ρ (g cm^{-3}) 1.633 1.608 1.562
Enthalpy of dissociation[d]	37.7±0.1 kcal
Standard entropies[d]	For F_2: 48.4. For F: 37.9 cal mole^{-1} deg^{-1}
Ionization energy[d]	402 kcal
Electronegativity function[e]	4.10 (Allred-Rochow scale) 3.98 (Pauling) 3.91 (Mulliken)
Electron affinity[d]	81.0 kcal
Standard reduction potential[d]	+2.9 volt (calc.)
Enthalpy of hydration (F^-)[d]	121 kcal
Atomic and ionic radii[d]	0.71 and 1.33 Å

[a] J. H. Hu, D. White and H. L. Johnston, *J. Am. Chem. Soc.* **75** (1953) 5642.
[b] H. R. Leech in ref. 6, pp. 46–53.
[c] D. White, J. H. Hu and H. L. Johnston, *J. Am. Chem. Soc.* **76** (1954) 2584.
[d] A. G. Sharpe, ref. 17.
[e] F. A. Cotton and G. Wilkinson, *Advanced Inorganic Chemistry*, 2nd Edn., Interscience (1966) pp. 100–4.

1950, was based on extrapolation of the values of 36.1, 46.1 and 58.2 for iodine, bromine and chlorine. There was considerable surprise, and disbelief, when after 1950 some investigations showed the value to be about 38 kcal.

As will be seen in the section immediately below, this very low value of the fluorine dissociation energy is a vital factor in discussing the chemical reactivity of fluorine. However, there is as yet no generally accepted explanation of the low strength of the F–F bond.

Mulliken[18] noted that single N–N bonds in hydrazines, O–O bonds in peroxides and

[17] A. G. Sharpe, in *Halogen Chemistry*, ed. V. Gutmann, Vol. I, Academic Press (1967) p. 1.
[18] R. S Mulliken, *J. Am. Chem. Soc.* **77** (1955) 884.

the F–F bond were all weaker and longer (relative to the appropriate atomic radii) than the corresponding single P–P, S–S and Cl–Cl linkages. He postulated that partial *pd* hybridization imparts pronounced multiple bond character to the formal P–P, S–S and Cl–Cl single bonds. The low dissociation energy for fluorine relative to that of its neighbouring molecules is then seen in terms of formal multiple bonding in N_2 and O_2 and partial multiple bond character in Cl–Cl. Caldow and Coulson[19], adopting a simple valence bond approach and assuming pure $p\sigma$ bonding, said that the sequence of dissociation energies for the halogens can be explained on the basis of relatively large electron–electron repulsions and relatively less penetration of charge clouds for fluorine than for other halogens. They also commented that coulomb energy terms (electron–nucleus attractions) may be more important than has previously been supposed.

In his direct determination of the enthalpy of dissociation of fluorine by an effusion method, Wise[20] reported values of the equilibrium constant K_p for the dissociation, representative values being 3.88×10^{-5} at 550°K and 2.1×10^{-3} at 790°K.

The standard entropy values quoted in Table 1 are very similar to those of 53.9 and 39.5 for Cl_2 and for Cl and to the corresponding values for bromine and iodine in the gaseous state. The implication of the similarity of entropy values for each of the halogens is that, except at very high temperatures, comparison of reactions involving the halogens themselves can be made quite adequately on the basis of enthalpy values.

For the four halogens fluorine to iodine, the values of the ionization energies 402, 300, 273 and 241 kcal follow the expected order as do the electronegativity functions, 4.10, 2.83, 2.74 and 2.21. Accordingly, there was initial surprise when the electron affinities, calculated from Born–Haber cycles and determined experimentally later, showed a different order with values 81.0, 84.8, 79.0 and 72.1. However, it should be recalled that Pauling defined electronegativity as the power of an atom *in a molecule* to attract electrons to itself, whereas the electron affinity is the energy released when an electron is added to an isolated atom, ion or molecule. An appropriate thermodynamic cycle from which electron affinities for the halogens were calculated is for the halogenation of an alkali metal:

$$M_{(s)} + 1/2X_{2(g)} \longrightarrow MX_{(s)} + \Delta H_f$$

The enthalpy of formation ΔH_f is related to enthalpy of sublimation and the ionization energy for the metal, ΔH_s and $I(M)$, to the dissociation energy and the electron affinity for the halogen, $D(X_2)$ and $A(X)$, and to the lattice energy for the ionic solid $U(MX)$ by the equation

$$\Delta H_f = \Delta H_s + I(M) + 1/2D(X_2) - A(X) - U(MX)$$

The calculated value of the lattice energy could be used with the four previously determined experimental values to give the electron affinity from

$$A(X) = \Delta H_s + I(M) - U(MX) - \Delta H_f + 1/2D(X_2)$$

For the halides of any particular metal, the first two terms are constant. Lattice energies and enthalpies of formation are higher for ionic fluorides than for the corresponding halides, and, in particular, the very low enthalpy of dissociation for fluorine leads to a calculated value for the electron affinity lower than might be expected intuitively or on the basis of electronegativity functions.

Because fluorine reacts with water, no direct measurement of the standard reduction

19 G. L. Caldow and C. A. Coulson, *Trans. Faraday Soc.* **58** (1962) 633.
20 H. Wise, *J. Phys. Chem.* **58** (1954) 389.

potential for the fluorine-hydrated fluoride ion system is possible. However, it can be calculated and compared with that for the other halogen–halide systems, from the cycle

$$1/2X_{2(g)} \longrightarrow X_{(g)} \longrightarrow X^{-}_{(g)} \longrightarrow X^{-}_{aq}$$

The calculated value for E_0 of +2.9 V is very much greater than those of +1.356, +1.065 and +0.535 V determined experimentally for the chlorine, bromine and iodine systems. For the different halogens, two separate effects lead to a greater energy change in the formation of hydrated anions from gaseous molecules in the case of fluorine. The low enthalpy of dissociation for fluorine has already been discussed. In addition the enthalpies of hydration for the four halide ions F^{-}_{aq} to I^{-}_{aq} are 121, 88, 80 and 70 kcal. Comparing the reduction cycle then for fluorine with that for chlorine, much less energy is expended in the dissociation step and, while the electron affinities are very similar for both halogens, an even greater difference in energy occurs for the formation of F^{-}_{aq} than for Cl^{-}_{aq}. The large enthalpy of hydration of fluoride ion is a major factor in considering the relative energetics of any reactions or processes such as dissolution leading to the formation of hydrated halide ions.

The value of 0.71 Å quoted in Table 1 for the atomic radius for fluorine is taken from the fluorine–fluorine distance in molecular fluorine. However, if this value is used with the generally accepted value of the atomic radius of carbon, a carbon–fluorine bond length of 1.48 Å is expected, whereas the measured value is 1.32 Å. It has been suggested that as a working figure 0.64 Å should be taken as the atomic radius for fluorine. The expected order of atomic radii is observed for the halogens, the values for chlorine, bromine and iodine being 0.99, 1.14 and 1.33 Å.

Again, when ionic radii are considered for the halides, an increase is observed with a marked difference between fluoride and chloride. The values determined by X-ray structural studies of lithium halides, in which the halide ions are in contact, are 1.33, 1.81, 1.96 and 2.16 Å for the anions fluoride to iodide. This discontinuity in values can be related also to the marked difference in reactivity of fluorine from other halogens. When ionic solids are formed or react, the size of the fluoride ion leads to high values of lattice energies, and this can materially affect the relative energetics of reaction, as will be shown in the section below.

Another comparison of ionic radii that has interesting chemical implications is between that of the fluoride ion and those for oxide and hydroxide, both about 1.40 Å and both significantly smaller than that of 1.84 Å for sulphide, for example. The easy replacement of hydroxide by fluoride in minerals is then readily explained as is the structural similarity between many fluorides and oxides (e.g. sodium fluoride and magnesium oxide) while different halides of the same metals frequently differ markedly. Thus cadmium and mercuric fluorides have the fluorite structure while cadmium chloride is a layer lattice structure and mercuric chloride is molecular.

There is little to report on spectral properties of fluorine. No absorption bands have been observed, continuous absorption occurring at wavelengths below 4100 Å with a maximum at about 2800 Å.

1.7. CHEMICAL REACTIVITY OF FLUORINE

Fluorine forms compounds with every element except helium, neon and argon, reacting directly with all but nitrogen. Its chemistry is characterized by a unique oxidation state of (−I). With electropositive elements it forms ionic fluorides and, although it never exhibits a formal covalency greater than 1, there is a rapidly growing number of entities which have been shown to contain bridging fluorines, e.g. $Sb_3F_{16}^-$, $As_2F_{11}^-$ [see section 2.4(d)(iii)] and the tetrameric pentafluorides of the transition metals, Mo, W, Nb, Ta, etc. [see section 5.6(b)]. The most spectacular aspect of the chemistry of fluorine is the ability of the strongly oxidizing element to bring out abnormally high oxidation states in elements with which it reacts. Some representative examples of such compounds are BiF_5, IF_7, PtF_6, AgF_2, $KAg^{III}F_4$, TbF_4, $Na_2Pr^{IV}F_6$, PuF_6 and CmF_4.

Bond Energies of Fluorides

The low enthalpy of dissociation of the fluorine molecule relative to the values for the other halogens has been shown in the section above to account in part at least for the greater reactivity of fluorine than of its congeners.

However, equally as important as the weakness of the bond in the reactant fluorine molecule is the strength of the linkage between the fluorine atom and that of the other reactant in the compound formed. Table 2 allows a comparison of bond energies for the four halides of hydrogen, carbon, boron and aluminium. It can be appreciated quite readily that, when halogenation reactions are compared, the enthalpies of fluorination reactions will be much greater than those for other halogenations. Not only is less energy required to form halogen atoms, but so much more energy is evolved in the formation of the halide. Fluorination reactions occur more readily and frequently do occur in situations where other halogenations do not proceed.

TABLE 2. AVERAGE BOND ENERGIES (kcal)

	HX[a]	BX_3[a]	AlX_3[b]	CX_4[a]
F	136	154	139	109
Cl	103	106	102	78
Br	88	88	86	65
I	71	65	68	57

[a] D. A. Johnson, *Some Thermodynamic Aspects of Inorganic Chemistry*, Cambridge U.P. (1968) p. 158.

[b] E. L. Muetterties and C. W. Tullock, in *Preparative Inorganic Reactions* (ed. W. L. Jolly), Vol. 2, Interscience (1965) p. 243.

For these reactions it is valid to make comparisons on the basis of enthalpy changes, or in terms of bond energies, because the entropy values for the different halogens are very similar, as was seen in the section above.

While calculated enthalpy, or free energy changes, usually are reliable in the prediction of reactions involving fluorine and fluorides, the possibility of competing kinetic effects must always be borne in mind. Thus the two compounds carbon tetrafluoride and sulphur hexafluoride are generally regarded as very inert compounds and are not observed to react with water, yet values of −36 and −72 kcal are calculated for the free energy change for the two hydrolyses:

$$CF_4+2H_2O \longrightarrow CO_2+4HF \quad \text{and} \quad SF_6+3H_2O \longrightarrow SO_3+6HF$$

Obviously the activation energies for the two reactions must be very high.

Formation of Covalent Fluorides

Several authors have used a semi-quantitative approach to rationalize the general observation that covalent fluorides are formed more easily than the corresponding other halides, particularly if high oxidation states are involved. Thus while SF_6 is formed readily by direct fluorination of sulphur, the highest chloride is the dichloride, and this difference can be related quantitatively to the energetics of vaporization and atomization of the sulphur in each case, the relative ease of dissociation of each of the halogen molecules to gaseous atoms and the relative strengths of the S–F and S–Cl bonds.

Using a more refined form of this general approach, Bartlett discussed the stability of the compounds of xenon as they were known in 1963 and used his approach predictively.[21] He showed that the xenon fluorides should be expected to be formed from the elements relatively easily and that the chlorides and oxides should not form spontaneously. In the face of uncertain experimental observations, he was able to predict the relative stabilities of the krypton fluorides.

Bartlett considered the following thermodynamic cycle for the formation of a fluoride, chloride or oxide of xenon:

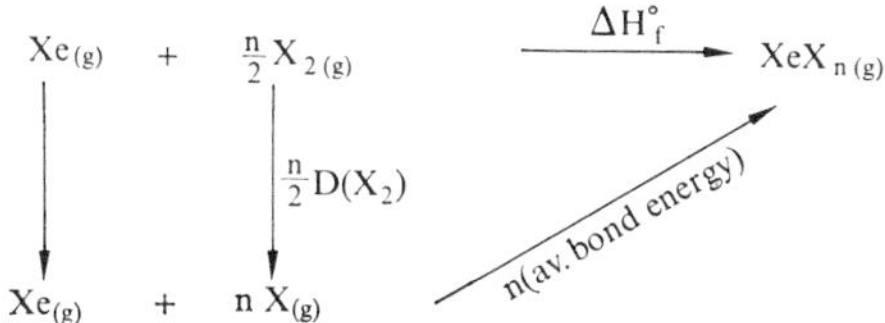

The formation of the compound will be exothermic if the average bond energy of the compound is greater than one-half of the bond dissociation energy for the halogen or oxygen. He quotes experimental bond energy values for the xenon fluorides in the region of 30–35 kcal which are far in excess of 18.9 kcal required to form one fluorine atom

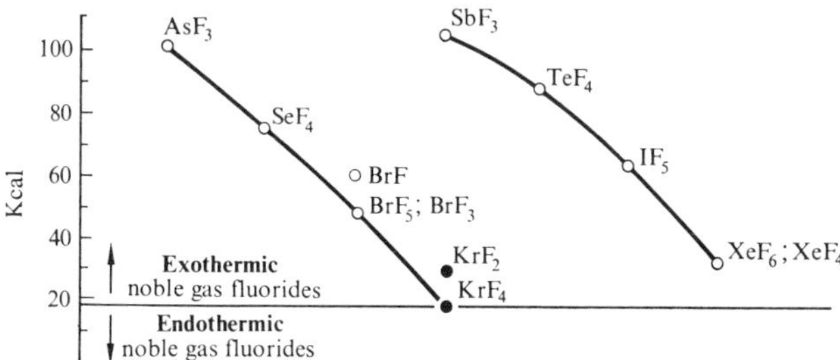

FIG. 3. Average bond energy (kcal) for some fluorides. Open circles, experimental values; closed circles, hypothetical values.

from the molecule, and so accounts for the direct formation of the xenon fluorides from the elements.

In an earlier publication[22] Bartlett had shown that the experimentally determined average bond energies for the xenon fluorides bore the simple relationship to the related fluorides of antimony, tellurium and iodine which is shown in Fig. 3. He used a similar extrapolation to estimate the average bond energies for the krypton fluorides. It is of particular interest to

[21] N. Bartlett, *Endeavour*, **23** (1964) 3.
[22] N. Bartlett, *Chemistry in Canada*, **15** (1963) 33.

note that, at the time Bartlett made these estimates, the preparation of both the difluoride and the tetrafluoride of krypton had been reported. Recognizing that for a noble gas fluoride to have thermodynamic stability its average bond energy must be greater than half the dissociation energy of fluorine (18.9 kcal), Bartlett showed from his extrapolated bond energy values that krypton tetrafluoride should be on the edge of stability (see Fig. 3). However, noting that the bond energy of bromine monofluoride (59.4 kcal) was much greater than those for bromine trifluoride and pentafluoride (48.3 and 47.7 kcal), he suggested that the value for krypton difluoride should be considerably greater than for the tetrafluoride, making the former much more stable thermodynamically. The significant fact is that it is now believed that the early report of the preparation of krypton tetrafluoride was not well-founded and the difluoride is the only stable fluoride.

Referring to his thermodynamic cycle, Bartlett showed for the exothermal formation of both the chlorides and oxides of noble gases, the average single-bond energy must be

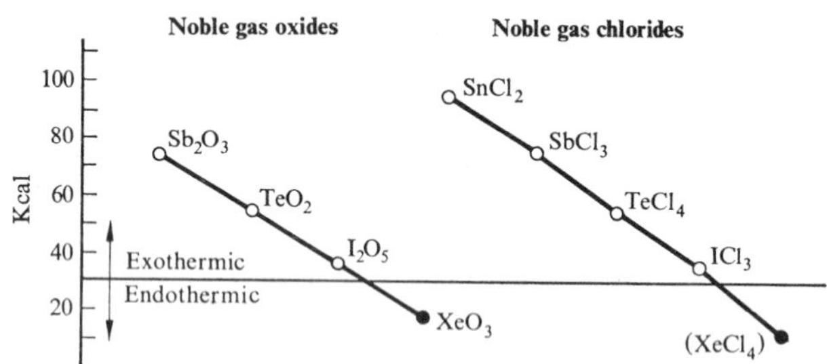

FIG. 4. Average bond energies (kcal) for some oxides and chlorides. Brackets indicate unknown compounds; open circles, experimental values; closed circles, theoretical values.

greater than about 30 kcal, considering a value of 29.1 kcal $mole^{-1}$ for the formation of chlorine atoms from the molecular species and one of 59.2 kcal $mole^{-1}$ for the formation of oxygen atoms from the doubly bonded molecules.

By again extrapolating from values of average bond energies for well-known compounds (as in Fig. 4), he showed that the bond energies for noble gas chlorides and oxides would be expected to be considerably less than the 30 kcal required for thermodynamic stability. The only reported preparation of a chloride, xenon dichloride, involves an indirect preparative route which starts with the difluoride. The well-characterized xenon trioxide must also be prepared indirectly by hydrolysis of higher fluorides of xenon. There is little doubt that the isolation of both compounds depends on kinetic effects rather than on thermodynamic stability of the compounds themselves. In fact, the enthalpy of formation of the trioxide has been shown to be about +95 kcal $mole^{-1}$ and the compound detonates spontaneously.

Formation of Ionic Fluorides

In discussing the relative reactivities of fluorine and the other halogens in the formation of covalent halides, the important factors were the relative bond strengths of the halogens and of the resulting halides in each case. The ease of dissociation of the halogen is, of course, still important in considering the formation of ionic halides; but the overriding factor, particularly for lattices with cations of high positive charge, is the lattice energy which is higher for those solids which contain the small fluoride ion than for the other halides.

The simplest case to consider is the formation of alkali metal halides according to the following cycle:

$$\begin{array}{ccccccc} M_{(s)} & \xrightarrow{\Delta H_1} & M_{(g)} & \xrightarrow{\Delta H_2} & M^+_{(g)} & & \\ & & & & + & \xrightarrow{\Delta H_5} & MX_{(s)} \\ 1/2X_{2(g)} & \xrightarrow{\Delta H_3} & X_{(g)} & \xrightarrow{\Delta H_4} & X^-_{(g)} & & \end{array}$$

For each of the halides of a particular metal the energy required to form gaseous cations ($\Delta H_1+\Delta H_2$), from the solid metal is the same in each case. However, in order to form the gaseous halide ion, energy is absorbed in the dissociation of the halogen (ΔH_3) and evolved in the electron-acceptance step (ΔH_4). For chlorine, bromine and iodine, the difference between the electron affinity and the energy of formation of the atomic species is almost constant at about 55 kcal, while the difference is 62 kcal for fluorine.

However, for this type of reaction, the great difference in driving force comes from the lattice energy term (ΔH_5). Taking the potassium halides as typical examples, the values of lattice energies are 193, 168, 161 and 152 kcal mole^{-1} for the fluoride, chloride, bromide and iodide (see Table 8, p. 1065).

In a much more formal treatment than that given above, Sharpe has considered the energetics of formation of ionic halides containing cations of charge greater than unity and has shown that with increasing charge of the cation, lattice energy considerations become progressively more important[23], and Bartlett has applied similar considerations specifically to the formation of solid fluorides of the platinum metals[24].

An obvious extension from consideration of the formation of ionic halides is to the possibility of a gas–solid reaction between a halogen and a solid ionic halide:

$$MX_{n(s)}+\tfrac{n}{2}X'_{2(g)} \longrightarrow MX'_{n(s)}+\tfrac{n}{2}X_{2(g)}$$

Because of the low dissociation energy for the molecular reactant and the high lattice energy for the ionic product, the reaction as written would proceed most readily for oxidations by fluorine, and iodides would react more readily than chlorides. However, calculation of the relative reactivities would be complicated in the real situation by the formation of interhalogen compounds as products. It will also be seen, from an examination of the appropriate thermodynamic data, why an ionic fluoride does not react with chlorine, bromine or iodine.

The oxidation proposed above is, of course, very different from that in aqueous solution where the relevant equation is

$$X^-_{(aq)}+1/2X_2'_{(g)} \longrightarrow X'^-_{(aq)}+1/2X_{2(g)}$$

Here the standard reduction potential and the hydration energy of the anion are so much greater for the fluorine system than for the other halogens that elemental fluorine would readily liberate chlorine, bromine or iodine from solutions of the other halides. Again the situation would be complicated by a side-reaction, namely the oxidation of water by fluorine.

A reaction situation which is virtually the reverse of those considered up to this point in this section is that in which an ionic fluoride is used as a reagent for introducing fluorine into an organic compound by a halogen–exchange reaction:

$$RCl+MF_{(s)} \longrightarrow RF+MCl_{(s)}$$

---

23 A. G. Sharpe, in ref. 17, pp. 20–23.

24 N. Bartlett, in *Preparative Inorganic Reactions*, ed. W. L. Jolly, Vol. 2, Interscience (1965) pp. 304–306.

The enthalpy for this reaction is obviously related to the difference in lattice energy for the two metal halides, the smaller the difference the more favourable the enthalpy. Since this difference is smaller for caesium fluoride than it is for the other pairs of alkali halides (see Table 8, p. 1065), caesium fluoride is the most powerful fluorinating agent of the alkali metal fluorides. Johnson[25] and Sharpe[17] have extended this treatment to include fluorides of metals other than alkali metals as potential fluorinating agents in halogen-exchange reactions. They make the additional point that if the chloride of the metal is significantly less ionic than the fluoride, the fluorination reaction proceeds even more readily than would be expected from a consideration of lattice energies based on simple electrostatic calculations which assume only ionic interactions in the solid. Thus silver fluoride and mercuric fluoride, for which the corresponding chlorides are markedly covalent, are much more powerful fluorinating agents than the formally analogous sodium and barium fluorides.

1.8. BIOLOGICAL ACTIVITY OF FLUORIDES

Fortunately for those concerned with the chemistry of fluorine, two volumes[26, 27] of the series "Fluorine Chemistry", edited by J. H. Simons, have been devoted to a comprehensive account of biological effects of organic and inorganic fluorides, and no one embarking on a programme of fluorine chemistry should fail to examine these surveys. Although this treatise is devoted to inorganic chemistry, the toxicity of fluoro-organic compounds will be considered briefly.

Toxicity of Fluoro-organic Compounds

Some of the most toxic substances known contain fluorine, many of them developed deliberately in military programmes. On the other hand, some of the least toxic carbon substances contain fluorine, e.g. certain anæsthetics and the fluorocarbons used as refrigerants and pressure-pack propellants.

As stated by Hodge *et al.*[26]: "the toxic effects of the organic fluorine compounds are unlike those of inorganic fluorides; in general, the effects are not consequent on the liberation of fluoride ion. With two or three notable exceptions, the mechanisms of action of these compounds are unknown."

Of the classes of organic compounds likely to be encountered by inorganic chemists, the fluorocarboxylates are the most important and their metabolic activity has been studied extensively. One extraordinary feature of their activity is that those members of the series $F(CH_2)_nCOOH$ for which n is odd are extremely toxic, whereas those for which n is even have little or no toxicity. The first member, monofluoracetic acid, is widely used in the form of its sodium salt to kill rodents and other pests.

The organic fluorophosphates, developed as chemical warfare agents, are less likely to be encountered by inorganic chemists.

Toxicity of Inorganic Fluorides

Here the essential hazard is from the toxic effects on the human or other system of relatively large concentrations of the soluble fluoride ion. Of course, in inorganic chemistry

[25] D. A. Johnson, *Some Thermodynamic Aspects of Inorganic Chemistry*, Cambridge U.P. (1968) p. 47.

[26] H. C. Hodge, F. A. Smith and P. S. Chen, *Fluorine Chemistry*, ed. J H. Simons, Vol. 3, Academic Press (1963).

[27] H. C. Hodge and F. A. Smith, *Fluorine Chemistry*, ed. J. H. Simons, Vol. 4, Academic Press (1965).

a particular fluoride may present an additional or enhanced hazard because of the separate toxicity of the ions or other species associated in the inorganic fluoride.

Hodge and Smith[27] distinguish between acute and chronic effects of inorganic fluorides. They cite 5–10 g of sodium fluoride as a reasonable estimate of a "certainly lethal dose" for a 70 kg man, i.e. 70–140 mg kg^{-1}. They give a table of known fatalities, where sodium or some other soluble fluoride has been taken accidentally or suicidally, with symptoms and the results of autopsies. Of more interest is their table VII in which they give a large number of non-fatal poisonings with symptoms which usually included nausea, vomiting, diarrhoea and acute abdominal pains. They say that ingestion of as little as 150 mg of sodium fluoride can cause serious illness. However, if the poisoning is non-fatal, recovery is very rapid, particularly if the treatment listed on page 34 of their review is followed. Calcium therapy, intravenously and intramuscularly, is the basis of the treatment.

Chronic toxic effects of fluoride (fluorosis) are usually observed as skeletal abnormality or damage, ranging from stiffness and rheumatism to a permanent crippling skeletal rigidity. Men who have ingested large amounts of fluoride (20–80 mg daily for 10–20 years) while working in aluminium or phosphate or other chemical plants in earlier times have shown the extreme symptoms and extraordinarily high fluoride concentrations (up to 13,000 ppm) have been observed in the bones of those affected.

Eye and Skin Burns

The most serious hazard facing anyone using fluorides in the laboratory is the possibility of eye or skin burns through contact with fluorine or fluorides in solid, liquid and gaseous state, as well as in solution.

It is essential to regard skin contact with any water-soluble fluoride—with the possible exception of pure fluorides of the alkali and alkaline earth metals—*as being contact with hydrogen fluoride*. Moisture in the skin will cause hydrolysis to hydrogen fluoride. Even in the case of contact with elemental fluorine, this generalization holds. The immediate effect of a jet of fluorine on the skin is to cause a thermal burn which can be very intense; but hydrogen fluoride is formed in this reaction. In the case of a soluble, and therefore hydrolysable, fluoride, the toxic effects of other entities, e.g. heavy metals, must also be considered.

The reason for instilling the idea that any contact with fluoride must be considered as contact with hydrogen fluoride is that the latter species is highly specific in its deleterious effect on tissue and bone. It is probably the most powerful solvent available for proteins and related compounds. It differs from other mineral acids which cause severe surface burns because it dissolves rapidly in tissue and causes deep-seated burns and damage to underlying bone, in extreme cases. It differs also in that, frequently, pain is not felt immediately and so no damage may be suspected. There is frequently a lag of one or two hours before numbness and then pain are felt. Anyone tempted to underestimate the hazards of fluoride burns to the skin should view pictures that are available[28] showing the destruction of the hand of a woman who suffered hydrofluoric burns and for whom treatment was delayed.

Hodge and Smith give a typical treatment for fluoride burns on page 36 of their review[27]. The affected area should be sluiced with water for several minutes. Then a previously prepared paste of magnesium oxide and glycerol should be applied liberally. This writer disagrees with one feature of the procedure given by Hodge and Smith who recommend

[28] *British Med. Journal* (1951) 728.

injection of a 10% calcium gluconate solution into and around the affected area in the case of severe or extensive burns. It is our practice to have these injections administered by a medical officer whenever skin contact is made with soluble fluorides. After the initial washing and paste treatment, 10% calcium gluconate is injected around and *under* the area of the burn to "freeze" the fluoride as insoluble fluoride. A hard calculus of the insoluble fluoride, which disperses very slowly, has frequently been observed under the surface after such treatment.

Recently, topical application of iced compresses of certain high-molecular-weight quaternary ammonium compounds has been reported[28a] as being superior to calcium gluconate injection, particularly in the case of reasonably severe fluoride burning.

Even greater damage can result from a splash of fluoride in the eye or from inhalation of large concentrations of fluoride vapours. In the former case the eye should be sluiced copiously with water as rapidly as possible, and a prepared eye-wash solution of sodium bicarbonate should be on hand for use. *Specialized* medical attention should be sought immediately in the event of eye contamination or inhalation of large amounts of fluorides.

It should not be necessary to state, of course, that the primary action to be taken when handling fluorides is to wear protective devices. At all times safety spectacles and/or a transparent face visor should be worn. Rubber or plastic gauntlets should be used and, if the amount of fluoride is large, e.g. anhydrous hydrogen fluoride from a cylinder, a plastic apron or other protective clothing which is non-absorbent should be worn.

Beneficial Effects of Fluoride on Bones and Teeth

The deleterious effect on bone of prolonged exposure to high fluoride concentrations has been discussed above. It had been observed that mottling of teeth occurred in communities which had a relatively high concentration of fluoride in their drinking water.

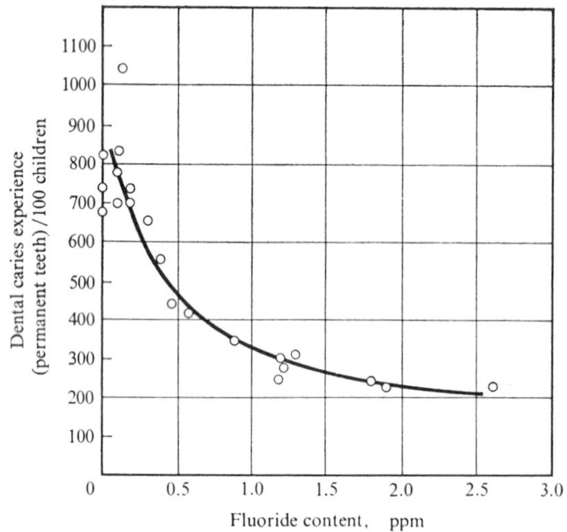

FIG. 5. Relation between dental caries and fluoride content of drinking water. (After Dean, *J. Am. Water Works Assn.*, **35**, 1161, 1943.)

This condition varied from a level which was discernible only to dentists, when the fluoride level was fairly low, to one in which there was unsightly brown staining of the teeth. It

[28a] C. F. Reinhardt, W. G. Hume, A. L. Linch and J. M. Wetherhold, *J. Chem. Educ.* **46** (1969) A171.

was observed further that, in such communities, the incidence of dental caries amongst children was particularly low.

The history of the recognition of the correlation between fluoride level in drinking water and reduction of the incidence of dental decay is given by Hodge and Smith[27]. They comment on the particular importance of the work after 1930 of H. T. Dean and co-workers who put the correlation on a quantitative basis as shown in Fig. 5. Dean's efforts were largely instrumental in the implementation of several programmes in the United States, Canada, the United Kingdom and other countries in which pairs of cities were chosen which were regarded as similar in sociological habits, i.e. in eating, dental care, etc. For each pair, or group, fluoride was added to one community's drinking water supplies at the level of 1 ppm. Dean's work had shown that this was the minimum level to give effective reduction of decay, and that mottling was insignificant at this level. Effectively, mottling occurred only to the extent that it could be detected by clinical inspection. Further, this limited amount of structure change is believed to strengthen the teeth.

A typical control experiment is shown in Fig. 6 for three cities in Ontario, Canada. In Stratford, where the natural level of fluoride in the drinking water is 1.6 ppm, about 50% of children in the age group 9–11 years suffered no dental decay in either 1948 or

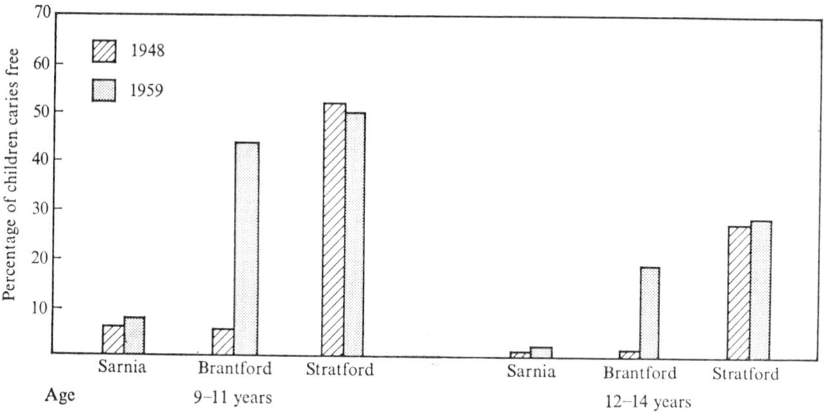

FIG. 6. The percentage of children caries-free in two age groups in a low fluoride area (Sarnia, 0.05 ppm F), in an area with 1.6 ppm in the water (Stratford), and in a third area (Brantford) before and after fluoridation to 1.0 ppm F. (From *Dental Effects of Water Fluoridation*, 7th Report, Dept. of National Health and Welfare, Ottawa, Canada, Sept. 1959.)

1959. Artificial fluoridation (1 ppm) of the Brantford water supply resulted in a six or sevenfold increase in the number of caries-free children over the same time interval. The children of the city of Sarnia, where the fluoride level in the water is very low, showed no significant improvement in dental health over the same period, indicating that changes in social patterns of diet, oral hygiene, etc., did not account for the Brantford results. The same general pattern is retained for older children in all three cities, although, of course, the incidence of dental caries increases with age, with or without fluoridation. It should be appreciated, however, that even in the higher age group, dental health in 1958 was better in Stratford and Brantford than in Sarnia.

Most studies have indicated a reduction of about 60% in the incidence of dental caries amongst children after fluoridation of low-fluoride drinking supplies. Dental clinicians regard a recent Dutch survey[29] as being the most meticulous yet conducted from a clinical point of view. The results suggest that the figure of 60% for reduction of caries is somewhat conservative.

It was stated above that whatever "mottling" occurs when the fluoride level is at 1 ppm is beneficial rather than deleterious. No harmful skeletal or other physical side effect has been established.

Hodge and Smith[27] summarize their section on dental caries by stating: "The total evidence now compiled from areas (a) where fluoride is naturally present in the drinking water and (b) where fluoride has been added establish beyond reasonable doubt the efficiency of fluorides in reducing the incidence of dental caries."

The only valid objection to fluoridation of drinking water supplies appears to be the ethical one based on the undesirability of enforced medication. This and other arguments for and against fluoridation have been examined by several officially constituted commissions set up in Canada, South Africa, Ireland, Tasmania and other countries. Of these reports the most recent, that from Tasmania, Australia[30], deals most exhaustively with chemical implications.

1.9. ANALYTICAL CHEMISTRY OF FLUORINE AND FLUORIDES

A reasonably comprehensive guide to the literature on fluoride analysis is given in this section because most of the volumetric and spectrophotometric methods in use at present are based on non-stoichiometric reactions which are subject to marked and varied chemical interference from other species. In addition, the procedures adopted are frequently tedious and highly empirical and reliable only in the hands of a single practised operator. This comment applies not only to the analytical methods themselves but also to the preliminary separation procedures which frequently must be used to isolate a solution which contains the fluoride for analysis but which has been freed of the many species which can interfere with the analysis. The literature of inorganic fluorine chemistry, as well as that of other fields such as biological and health sciences, is bedevilled by the general unreliability of quoted analyses. In the best of the relatively recent reviews on fluoride analysis, Horton[31] states by way of introduction: "The determination of ionic or elemental fluorine is still very difficult despite the continual appearance of new methods and the variations of older methods described in scientific literature." Wherever possible in this section, Horton's review will be used to provide further detail on methods or iterature references.

(i) Determination of Elemental Fluorine

The two general methods for determination of fluorine are given by Horton on page 287 of his review[31]. The first general procedure is to pass the gas containing fluorine through a sodium chloride solution, the liberated chlorine being collected in sodium hydroxide as hypochlorite and determined volumetrically. In similar fashion, bromine and iodine liberation from bromides or iodides have been reported. The other main approach to the problem is to collect the gas in a metal trap where it is shaken with mercury, forming solid mercuric fluoride. The decrease in volume gives the fluorine content directly and residual gases are determined by conventional gas analytical techniques.

[29] O. B. Dirks, B. Houwink and G. W. Kewant, *Arch. Oral Biol.* **5** (1961) 284.

[30] *Report of the Royal Commission on Fluoridation*, Hobart, Tasmania, Australia (1968).

[31] C. A. Horton in *Treatise on Analytical Chemistry*, eds. I. M. Kolthoff and P. J. Elving, Part II, Vol. 7, Interscience (1961) 207–334.

(ii) Gravimetric Determination of Fluoride[32]

Under ideal conditions, a precipitate of definite stoichiometry is isolated in the gravimetric methods. The methods are reasonably specific for fluoride, although some chemical interferences do occur. However, gravimetric methods are very slow and cannot be instrumented easily.

The gravimetric method widely used in much of the pioneering work on inorganic fluorine chemistry was that in which fluoride was precipitated as calcium fluoride. Unfortunately, the precipitate is frequently quite gelatinous, and phosphate and carbonate tend to precipitate with the fluoride.

A more satisfactory approach involves the precipitation of lead chloride fluoride PbClF from a solution containing excess of both lead and chloride ions. There are some doubts about the true stoichiometry of the precipitate.

(iii) Volumetric Determination of Fluoride[33]

Usually these methods are based on non-stoichiometric reactions and the procedures are highly empirical. The most commonly used procedure is that in which thorium nitrate is used as titrant and an indicator such as Alizarin Red S is added to the fluoride solution. Initially the thorium ions react almost completely to form colourless complexes or precipitates of unknown composition. As the fluoride ion concentration is decreased during the course of the titration, the indicator begins to compete successfully with fluoride for thorium ions to form a pink or red complex. However, there is no easily definable end-point, and the analyst must decide subjectively when the titration is complete. Even the use of a colour match is not easy. Coagulation occurs very slowly during the titration. If one solution is retained after titration in order to attempt to titrate to the same point subsequently, progressive coagulation in the first system makes a real comparison impossible. Some analysts use a homogeneous solution with a suitable pink dye—not the end-point mixture—as a colour match.

Because no fluoro-thorium(IV) compound of definite stoichiometry is formed, the thorium nitrate solution used as titrant must be standardized against a known fluoride solution, i.e. the method is empirical. The titration reaction is extremely pH sensitive, and the solution must be buffered, usually with chloracetate. Several entities interfere markedly, particularly phosphate, arsenate, arsenite and sulphite. Therefore preliminary separation of the fluoride, as discussed in sub-section(vi) below, is usually necessary.

Other titrants, such as zirconium nitrate, and other indicators, such as Chrome Azurol S, have been used but without materially improving the volumetric method. The most encouraging development appears to be the study by Lingane of the use of the lanthanum fluoride membrane electrode (see sub-section(v) below) in determining the end-point in titrations of fluoride against calcium, thorium and lanthanum nitrate solutions[34]. Lingane reports that use of the electrode to define the equivalence point in the lanthanum nitrate titration "makes possible the titration of fluoride with an accuracy hitherto unattainable with visual indicators".

Some reasonably precise indirect volumetric determinations of fluoride have been based on initial precipitation of lead chloride fluoride. If a known excess of chloride is

[32] Ref. 31, pp. 257–259.
[33] Ref. 31, pp. 259–268.
[34] J. J. Lingane, *Anal. Chem.* **39** (1976) 881.

used, residual chloride after precipitation can be determined potentiometrically[35]. Alternatively, as Horton indicates[33], the precipitate can be isolated, dissolved in dilute nitric acid and analysed for lead using EDTA as titrant.

(iv) Spectrophotometric Determination of Fluoride[36]

Early spectrophotometric methods were direct adaptations of the most common of the volumetric methods discussed in the section immediately above. Alizarin Red S was added to a buffered solution of zirconium or thorium nitrate, the former having somewhat wider usage. The extent of bleaching of the resultant coloured solution was related to fluoride concentration by standardizing with known fluoride solutions. The procedures were reasonably rapid and reliable in the hands of a practised analyst; but, as Horton shows for the zirconium–Alizarin Red S reaction in table X[36] of his review, many species caused marked interference. Also, as his table XI shows, a very wide range of experimental conditions have been reported as being optimal for the zirconium–Alizarin Red S and for other reactions, indicating how empirical the whole approach has been. In attempting to make the zirconium–Alizarin Red S method more reliable for determination of fluoride in potable waters, Lim[37] reported what he considered to be optimum acidity of solution, time of development of the colour in the solution for analysis and the zirconium to alizarin ratio. This probably represents a fairly reliable procedure if this reaction is to be used at all.

Many metal cations and complexing chromophores have been investigated and reported as alternatives to the zirconium–alizarin system. Horton has a comprehensive list in his table IX[36]. One chromophore not included in that table but which has been reported as giving good results when used with zirconium solutions is Xylenol Orange[38]. However, Belcher, Leonard and West[39, 40] have introduced a new rigour into the field by designing an alizarin derivative which chelates specifically to cerium(III) to form a red compound, which in turn is complexed by fluoride to give a blue compound containing fluoride and the cerium–alizarin complex in the mole ratio 1 : 1. Details of a spectrophotometric analytical method based on this reaction have been reported[41].

In 1968 a very valuable survey of several spectrophotometric methods for determination of fluoride in potable waters and other aqueous solutions was made by Crosby, Dennis and Stevens[42]. For each method, including the cerium(III)–complexone procedure, they examined the linearity of the calibration graph, stability of the solution of the complex, sensitivity of the method, reproducibility of results in the hands of different analysts and the effect of interfering species. They found the cerium(III)–complexone method to be far superior to the zirconium–alizarin and other spectrophotometric methods studied. However, they also included in their survey analysis using the lanthanum fluoride membrane electrode. This aspect of the survey will be dealt with in the next section.

[35] J. M. Hogan and F. Tortorici, *Anal. Chem.* **39** (1967) 221.
[36] Ref. 31, pp. 268–279.
[37] C. K. Lim, *Analyst*, **87** (1962) 197.
[38] R. Valach, *Talanta*, **8** (1961) 443.
[39] M. A. Leonard and T. S. West, *J. Chem. Soc.* (1960) 4477.
[40] R. Belcher, M. A. Leonard and T. S. West, *J. Chem. Soc.* (1959) 3577.
[41] R. Greenhalgh and J. P. Riley, *Anal. Chim. Acta*, **25** (1961) 179.
[42] N. T. Crosby, A. L. Dennis and J. G. Stevens, *Analyst*, **93** (1968) 643.

(v) Electrometric Determination of Fluoride

Prior to about 1960 there had been many attempts to develop potentiometric, amperometric and polarographic methods of determining fluoride. These were indirect methods based on complexing of various species by fluoride, but none of them proved sufficiently accurate or reliable to be of general use. In 1961 a null-point potentiometric procedure was developed in which, initially, there was zero potential difference between two half-cells, each containing an identical cerium(IV)–cerium(III) solution. The solution containing fluoride for analysis was added to one half-cell, leading to very strong complexing of the cerium(IV) with little if any complexing of cerium(III). As a result, a potential difference was established, and standard sodium fluoride titrant was then added to the other half-cell until the potential difference was again zero. In a general review paper[43], O'Donnell and Stewart have discussed the advantages and disadvantages of their null-point potentiometric method over others then extant.

The simplest, most rapid and probably the most reliable procedure now available, involves use of the fluoride ion activity electrode in conjunction with a standard calomel electrode and a high impedance voltmeter, such as a commercial pH meter. The fluoride electrode is based on a membrane which is a crystal of lanthanum fluoride doped with europium(II). In their original report of the electrode, Frant and Ross[44] said that it gave a reproducible Nernstian response over a wide range of fluoride activities down to about 10^{-5} M fluoride with a non-linear response to about 10^{-6} M. They said further that it was highly specific in its response to fluoride, being unaffected by large concentrations of anions such as chloride, nitrate, sulphate and phosphate or by the actual presence of cations, although, of course, if cations such as aluminium(III) or iron(III) are present they form fluoro complexes and consequently reduce the free fluoride activity.

The one entity which specifically interferes in the analytical use of the fluoride activity electrode is hydroxide ion, which, having the same charge and a size very similar to that of fluoride, probably interacts with the lanthanum fluoride membrane similarly to fluoride. Interference is discernible in solutions in which fluoride and free hydroxide are in comparable concentrations, and a tenfold excess of hydroxide causes the electrode to indicate a fluoride concentration about double the real one. Most analytical work can be carried out free from hydroxide interference as long as the pH is not greater than about 8.5. Although there is no specific interaction with the electrode, moderately large concentrations of hydrogen ion can prevent effective use of the fluoride activity electrode, because the electrode "sees" only the activity of *free* fluoride ions and, since hydrogen fluoride is such a weak acid, protonation of fluoride ions becomes progressively more important with increase of hydrogen ion concentration even at acidities fairly close to neutrality. Effectively, the electrode cannot be used easily for solutions of pH less than 4–5.

In order to overcome some of these problems, especially when using the electrode to determine small concentrations of fluoride, Frant and Ross[45] recommend the use of a total ionic strength adjustment buffer. This is an acetate buffer to give the optimum pH range for the electrode and it has added to it citrate to preferentially complex ions of metals such as aluminium or iron and so leave all fluoride free. In addition, the buffer has a high concentration of sodium chloride to "swamp" the solution with ions. The electrode, which responds to changes in activity of fluoride ion, is then calibrated against known concentrations of sodium fluoride in a solution in which the ionic strength is effectively constant.

43 T. A. O'Donnell and D. F. Stewart in *Electrochemistry*, ed. V. Gutmann, Pergamon Press (1964) p. 332.
44 M. S. Frant and J. W. Ross, Jr., *Science*, **154** (1966) 1553.

Frant and Ross[45], as well as other workers[46], report an accuracy of $\pm 5\%$ in determining fluoride at the level in which it is usually present in potable waters, namely 1 ppm (5×10^{-5} M). The specific fluoride electrode has been used in a wide range of analytical applications and modifications of it have been reported for use in microchemical analysis[46, 47].

As stated above, microdetermination of fluoride in potable waters using the fluoride activity electrode was compared critically with spectrophotometric studies[42]. The conclusion was that "the fluoride electrode surpasses all the colorimetric methods with regard to speed, accuracy and convenience, and is recommended for the routine monitoring of fluoridated water supplies. The indications are that it will be proved satisfactory for the determination of fluoride in a wide range of materials, following a simple pre-treatment to bring the fluoride into solution."

It should be re-emphasized that the fluoride activity electrode also provides the best way of determining the equivalence point in the macrodetermination of fluoride against a titrant such as lanthanum nitrate[34].

It is unfortunate that, at the time of the survey of microanalytical methods for fluoride[42], Bond and O'Donnell's polarographic procedure[48] was not available. Their method applies to the concentration range 2.5×10^{-6} to 10^{-3} M and depends on shift through fluoride complexing of the uranium(V)–uranium(III) half wave potential. The shift is 80 mV at 1 ppm (5×10^{-5} M) and reproducibility at this concentration is $\pm 2\%$. Therefore while this procedure is less simple instrumentally than the fluoride activity electrode, it appears to be somewhat more sensitive and is comparable in specificity. The only entity which interferes at low acidity is phosphate, and this interference can be eliminated, with some loss of sensitivity in fluoride determination, by increasing the acidity.

The polarographic method has a marked advantage over the fluoride activity electrode if it is necessary to work at relatively high acidity where protonation will drastically reduce the activity of free fluoride ions. Such a situation could arise in analysis of solutions containing cations of transition metals and other heavy metals. If, in order to use the fluoride activity electrode easily, the pH of such solutions were maintained at 4 or greater, much fluoride would be lost from solution as fluoro-oxo- or fluoro-hydroxy-species which might well be insoluble. The polarographic method can be used with solutions that are 1 M or higher in acid.

(vi) Elimination of Chemical Interference in Fluoride Analysis

There are many chemical entities which interfere to a greater or less extent with the quantitative determination of fluoride. In those methods which are based on formation of fluoro-complexes in solution, e.g. titrimetric, spectrophotometric or electrometric methods with the exception of the lanthanum fluoride membrane electrode, any species will interfere which forms complexes of stability comparable with the fluoro-complexes. A specific example is phosphate which is present with fluoride in many biological and mineralogical specimens for analysis and which forms soluble complexes or precipitates with zirconium(IV), thorium(IV) and other solutions used in the analysis of fluoride. Even sulphate markedly affects many spectrophotometric analytical procedures. Therefore, especially when these methods are being used, it is customary to carry out a preliminary separation step to remove the fluoride from potentially interfering species. The separation procedure which is the most

45 M. S. Frant and J. W. Ross, Jr., *Anal. Chem.* **40** (1968) 1169.
46 R. A. Durst and J. K. Taylor, *Anal. Chem.* **39** (1967) 1483.
47 R. A. Durst, *Anal. Chem.* **41** (1969) 2089.
48 A. M. Bond and T. A. O'Donnell, *Anal. Chem.* **40** (1968) 1405.

important historically and which is probably the most used is the distillation process proposed by Willard and Winter and described, with its many modifications, by Horton[49]. Fluoride is steam-distilled in glass apparatus from a solution containing sulphuric acid or perchloric acid at about 135°. Powdered glass is frequently added, and the fluoride is recovered in the distillate as hexafluorosilicate which is then usually titrated with thorium nitrate or determined spectrophotometrically. Wade and Yamamura have reported a flask for Willard–Winter distillation in the microdetermination of fluoride[50]. They give a table showing the extent to which several ionic species interfere with their procedure in which microdistillation is followed by spectrophotometric determination with the complexone.

In their survey of microanalytical methods, Crosby, Dennis and Stevens[42] stated that "the distillation procedure is time consuming, potentially hazardous and can only be used by a skilled analyst, as it requires careful control to obtain reliable results".

Preliminary separations based on pyrohydrolysis appear more reliable than the distillation procedures. Two main approaches have been used[49]. Superheated steam has been passed over the fluoride in a platinum or nickel reaction tube at about 1000°. Hydrogen fluoride, dissolved in the condensate, was titrated against standard sodium hydroxide or determined colorimetrically. In subsequent studies a stream of moist oxygen was used in the pyrohydrolysis. Optimum conditions have been reported for pyrohydrolytic separation in the analysis for fluoride in a wide range of ores, slags, rocks and glasses[51].

A separation technique which has proved effective in the analysis of biological samples, including bones and teeth, uses Conway diffusion cells constructed from polypropylene[52] or, better, Teflon[53]. The circular cells have separate concentric compartments. The sample is placed in one of these with perchloric acid. Liberated hydrogen fluoride diffuses into the adjoining compartment and is absorbed in sodium hydroxide after which it is determined spectrophotometrically.

Although in principle cation- or anion-exchange resins should be effective in separating fluoride from possible interfering ions, there has been little satisfactory use of this approach. Sporek has reported a typical separation in the analysis of uranium tetrafluoride[54].

(vii) Comparison of Analytical Procedures

While there is general recognition of the fact that the classical volumetric and spectrophotometric analytical methods, preceded by Willard–Winter distillation or other separation processes, are far from being satisfactory except in the hands of a skilled and practised analyst, there has been surprisingly little exploration of the newer instrumental methods now available. It is relatively rare to find analysts, when faced with the determination of fluoride, who will critically assess the necessity for the preliminary separation, or turn to an analytical method which is not some simple variant of the thorium–alizarin reaction.

There are probably many chemical systems for which preliminary separation of the fluoride is still necessary, particularly if the analytical method is to be volumetric or spectrophotometric. The diffusion method of separation[52] appears to be very effective with samples such as bone, teeth and minerals which are relatively easily dissolved in acid and which contain small concentrations of the heavy metal ions with which fluoride complexes

[49] Ref. 31, pp. 238–244.
[50] M. A. Wade and S. S. Yamamura, *Anal. Chem.* **37** (1965) 1276.
[51] M. J. Nardozzi and L. L. Lewis, *Anal. Chem.* **33** (1961) 1261.
[52] H. W. Wharton, *Anal. Chem.* **34** (1962) 1296.
[53] N. W. Alcock, *Anal. Chem.* **40** (1968) 1397.
[54] K. F. Sporek, *Anal. Chem.* **30** (1958) 1030.

strongly. For more refractory specimens, pyrohydrolysis[51] appears to give reliable recovery.

If macrodetermination based on titration is to be used, Lingane's procedure[34] of titration against lanthanum nitrate using the fluoride activity electrode to indicate the equivalence point appears to be the best method. The cerium(III)–complexone reaction[39–41], or a variation involving lanthanum(III), has been shown[42] to be the most reliable spectrophotometric procedure. However, the same survey indicates that the fluoride activity electrode[44–47] gives the simplest, most rapid and most accurate results in microdeterminations. As shown in sub-section (v), this electrode must be used in a solution for which the pH is between 4 and 8.5, and if it is necessary to analyse for fluoride in a strongly acid solution, the polarographic microdetermination[48] has a definite advantage. These last two electrometric approaches appear to offer the best chance of avoiding preliminary separation of fluoride because they are so specific in their response to fluoride and so little affected by the presence of other soluble species.

2. THE HYDROGEN FLUORIDE SOLVENT SYSTEM

The basis for a physico-chemical approach to study of solution processes in anhydrous hydrogen fluoride appears to have been established in Germany in the decade before World War II. Measurements of conductance and boiling-point elevation were superimposed on the simple determination of solubility of various compounds in hydrogen fluoride. A review[55] in 1939 of the work of K. and H. Fredenhagen, G. Cadenbach and W. Klatt provided a classification of the solution processes as then known; and the general format of this review appears to have been adopted by most authors of textbooks and reviews over the next 25 years. In 1950, Simons[56] reviewed physical evidence for the nature of hydrogen fluoride in the solid, liquid and vapour state, and also described solution processes in the anhydrous solvent. Later[57] he presented the relevant work reported between his first review and 1962.

Apart from giving detailed accounts of the nature of the pure compound hydrogen fluoride, most reviews up to this stage reported qualitatively the way in which the individual members of a large and relatively unrelated range of compounds dissolved in hydrogen fluoride. Solutes, such as water, organic compounds, some fluorides and other halides, cyanides, nitrates, sulphates and other oxy-anions, were described as dissolving with or without ionization or protonation and with or without decomposition. Even in the case of solubilities, there were relatively few quantitative data. The only extensive account of solubility of fluorides of metals was, and still is, the work reported in 1952 by Jache and Cady[58] who measured solubilities of 34 fluorides of metals of the main groups and of the transition series in lower oxidation states over the temperature range from -25 to $+15$°C.

During the last decade or so, very significant advances have been made largely by Kilpatrick, Hyman, Katz and co-workers at the Illinois Institute of Technology and the Argonne National Laboratory, Illinois, in the development of reliable techniques for the observation in hydrogen fluoride solutions of conductances and of visible, ultraviolet, infrared, Raman and resonance spectra.

There are two fairly recent and excellent reviews which give much information on

[55] H. Fredenhagen, *Z. anorg. u. allgem. Chem.* **242** (1939) 23.
[56] J. H. Simons in *Fluorine Chemistry*, ed. J. H. Simons, Vol. 1, Academic Press (1950) pp. 225–289.
[57] J. H. Simons, ref., 56, Vol. 5 (1964) pp. 2–15.
[58] A. W. Jache and G. H. Cady, *J. Phys. Chem.* **56** (1952) 1106.

quantitative physico-chemical investigation of solution processes in hydrogen fluoride. Hyman and Katz[59] give details on determination of conductances and absorption spectra, while Kilpatrick and Jones[60] provide additional information on the use and impact of n.m.r. spectrometry and other recent studies. Most recently Gillespie and co-workers[61, 62] have demonstrated strikingly the value of combining studies of conductance, spectral and colligative properties in determining the nature of dissolved species in hydrogen fluoride. Canterford and O'Donnell[63] have reviewed experimental techniques for handling volatile fluorides, including hydrogen fluoride, and they also report briefly on the application of physico-chemical procedures to hydrogen fluoride solutions.

The collection of quantitative data resulting from the application of improved and modern techniques has led to a rapid growth over the last few years in the amount of reliable information on ionization and other solution processes in hydrogen fluoride. Accordingly, in this review, solution processes will not be described simply in terms of observed solubility of a range of chemical compounds; but, as far as possible, physico-chemical evidence will be used to demonstrate whether particular solutes dissolve with or without ionization, whether solvation or solvolysis occurs and solution processes will be presented in terms of acid–base and oxidation-reduction reactions. Particular emphasis will be placed on Bronsted and Lewis acidity and basicity because of the great acidity of the solvent itself.

2.1. PHYSICAL PROPERTIES OF HYDROGEN FLUORIDE

In their review, Hyman and Katz list an extensive table of physical properties of hydrogen fluoride and they quote the sources of the data. The properties given in Table 3 have been selected on the basis of their relevance to the material in section 2 as a whole and the

TABLE 3. PHYSICAL PROPERTIES OF HYDROGEN FLUORIDE

Freezing point	$-83.55°C$
Cryoscopic constant	$1.55 \pm 0.05°C$ mole^{-1} kg
Boiling point	19.51°C
Ebullioscopic constant	1.9°C mole^{-1} kg
Vapour pressure	Equations: (a) $\log_{10}p(\text{mm}) = 8.38036 - 1952.55/(335.52+t)$ (t = °C)
	(b) $\log_{10}p(\text{mm}) = -1.91173 - 918.24/T + 3.21542 \log_{10}T$ (T = °K)
	Values: t (°C): −50, 0, 25, 100
	p (mm): 34.82, 363.8, 921.4, 7891
Density	Equation: $\rho(\text{g ml}^{-1}) = 1.0020 - (2.2625 \times 10^{-3}t) + (3.125 \times 10^{-6}t^2)$
	Values: t(°C): −50, 0, 25, 100
	ρ(g ml^{-1}): 1.1231, 1.002, 0.9546, 0.796
Dielectric constant	Equation: $\epsilon = 69{,}000/T - 167.5$
	Values: −73°C, 0°C
	175, 83.6
Specific conductance	1×10^{-6} ohm^{-1} cm^{-1} at 0°C
Limiting ionic mobilities	$\lambda_{H^+} = 350$ ohm^{-1} cm^2 $\lambda_{F^-} = 273$ ohm^{-1} cm^2

[59] H. H. Hyman and J. J. Katz in *Nonaqueous Solvent Systems*, ed. T. C. Waddington, Academic Press (1965) pp. 47–81.

[60] M. Kilpatrick and J. G. Jones in *The Chemistry of Nonaqueous Solvents*, ed. J. J. Lagowski, Vol. 2, Academic Press (1967) pp. 43–99.

[61] R. J. Gillespie and K. C. Moss, *J. Chem. Soc. (London)* A (1966) 1170.

[62] R. J. Gillespie and D. A. Humphreys, *J. Chem. Soc. (London)* A (1970) 2311.

[63] J. H. Canterford and T. A. O'Donnell, in *Technique of Inorganic Chemistry*, eds. H. B. Jonassen and A. Weissberger, Vol. VII, Interscience (1968) pp. 273–306.

values quoted are essentially those given by Hyman and Katz. Where they do differ, the sources of the new data are given in the discussion in this section.

It should be noted that there is much confusion in hydrogen fluoride literature because physical measurements, particularly solubilities, have frequently been made at varying temperatures. No fixed point, comparable with that of 25°C for aqueous systems, has been selected. There are great experimental advantages in working at the ice point for conductance, solubility and other studies in hydrogen fluoride; but it should be appreciated that this temperature lies high in the liquid range for the compound. Textbooks frequently use as evidence of the similarity of the two associated liquids, water and hydrogen fluoride, values of the dielectric constant, that for water at 18°C being 81.1 and for hydrogen fluoride at 0°C 83.6. However, at 18° above its freezing point, hydrogen fluoride has a dielectric constant of about 165.

Hyman and Katz refer to a freezing point given by Hu *et al.*, whose value of -83.36°C is based on an experimentally observed point, corrected for a supposed impurity level in their hydrogen fluoride. The value of -83.55°C in Table 3 is based on a triple point temperature of -83.57 determined experimentally by Gillespie and Humphreys[62] and then calculated from the Clausius–Clapeyron equation.

Gillespie and Humphreys demonstrated that hydrogen fluoride is an excellent solvent for cryoscopic studies. They showed freezing points to be reproducible to ± 0.005°C and supercooling to be very small, about 0.06°C. They determined a molar cryoscopic constant of 1.55 ± 0.05°C mole^{-1} kg, using non-ionizing solutes sulphur dioxide, fluorosulphuric acid and xenon difluoride as well as alkali metal fluorides for which a value of 2 was assumed for ν, the number of moles of dissolved particles (molecular or ionic) produced by 1 mole of solute. Their experimental value agrees well with that of 1.52°C mole^{-1} kg calculated from the heat of fusion.

Using water and alkali metal fluorides as solutes, Fredenhagen, Klatt[64] and co-workers measured a molar ebullioscopic constant of 1.9°C mole^{-1} kg. They then observed a value of 2 for ν for ethyl alcohol as a solute. Subsequently, complete protonation of the alcohol was demonstrated by infrared and n.m.r. spectrometry. They also demonstrated the complexity of solution processes for solutes containing oxy-anions, such as sulphate. Protonation of organic solutes and solvolyses of oxy-anions are discussed in greater detail in section 2.4 below.

Several workers have reported equations relating the vapour pressure of hydrogen fluoride to the temperature. In attempting to fit their observed data to an equation, Jarry and Davis[65] were unable to choose between the two equations given in Table 3. While the fit was acceptable in the case of both equations for high values of vapour pressures, there is considerable uncertainty about values for the liquid near the freezing point. Gillespie and Humphreys observed an approximate value of 24 ± 1 mm for the molar depression of the vapour pressure. The value calculated from Raoult's law is 7.13 mm. So their observation suggests an association factor of 3.3 for the solvent at 0°C. This association factor is considerably less than that of 4.7 reported by Jarry and Davis (see p. 1041) for the pure liquid; but the presence of a solute would be expected to modify the degree of association of hydrogen fluoride.

Vapour density measurements in 1924[66] indicated the presence of polymers in the vapour.

[64] W. Klatt, *Z. anorg. u. allgem. Chem.* **232** (1937) 393; **233** (1937) 307; **234** (1937) 189.
[65] R. L. Jarry and W. Davis, *J. Phys. Chem.* **57** (1953) 600.
[66] J. H. Simons and J. H. Hildebrand, *J. Am. Chem. Soc.* **46** (1924) 2183.

It had long been suggested that hexamers were formed in relatively high concentration, and there was considerable controversy over whether these would be zigzag as in the solid (see below) or cyclic. The extent of polymerization of the vapour depends markedly on the temperature and pressure. Jarry and Davis[65] reported values of the average degree of association from 4.72 at 0°C and 3.58 at 25°C to 2.45 at 100°C.

In addition to the vapour density measurements, other physical properties of hydrogen fluoride suggest a high degree of association through hydrogen bonding in the vapour, liquid and solid. The freezing and boiling points are considerably higher than those of the other hydrogen halides, and the liquid range is long. Also the dielectric constant is high as is the heat of fusion, 4.693 kcal mole^{-1}. Recent infrared and n.m.r. studies on the vapour and X-ray structure studies on the solid have confirmed this association through hydrogen bonding.

From an interpretation of infrared spectra, which show bands in the 3 μ region arising from hydrogen stretching fundamentals in hydrogen fluoride polymers, Smith[67] postulated the existence of a predominant equilibrium between hexamers and monomers at temperatures near the boiling point and pressures above 300 mm Hg, although the tetramer was

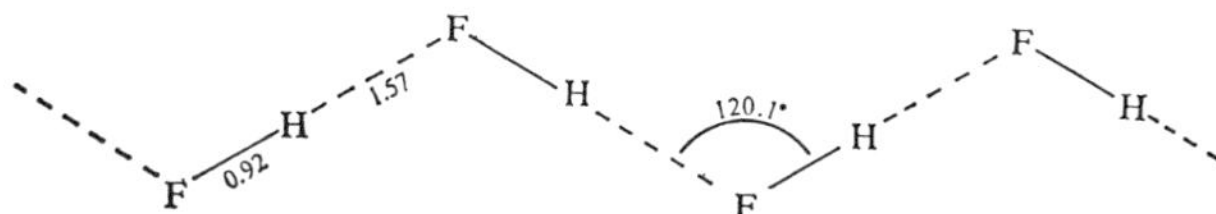

FIG. 7. Structure of solid hydrogen fluoride.

proposed as being present in relatively small amount. At higher temperatures and lower pressures tetramers, dimers and monomers become progressively more important. Low values obtained in dipole moment measurements suggest strongly that the hexamers are cyclic and not linear. The average heat of formation of the hydrogen bond is −6.65 kcal mole^{-1}, about 50% greater than for the hydrogen bond in water.

Intense absorption of infrared radiation makes study of the liquid very difficult. The physical properties of the liquid suggest a high degree of association, but the extent of association is not known. There appears to be a transition from cyclic polymers in the vapour and liquid to the zigzag structure for the solid shown in Fig. 7. The HFH angle has been determined by Atoji and Lipscomb as 120°.1 and the F–F distance as 2.49±0.01 Å. The hydrogen is shown at a distance 0.92 Å from one fluorine because of H–F distances measured in the gas phase. This location is consistent with the experimental observations of Atoji and Lipscomb[68].

Hydrogen fluoride, as produced commercially, contains many impurities which enhance the conductance of the liquid (see section 1.2(ii)). The specific conductance for a reasonable commercial product can be as high as 10^{-3} ohm^{-1} cm^{-1}. Trap to trap distillation in Kel-F traps can reduce this by a factor of 10. Until quite recently, fractional distillation of the liquid rarely gave a product with a specific conductance less than 10^{-5} ohm^{-1} cm^{-1}; but Shamir and Netzer[69] have described a simple modification of the usual distillation procedure which gives hydrogen fluoride with a specific conductance of 1×10^{-6} ohm^{-1} cm^{-1} on a routine basis.

Values of the limiting ionic mobilities for the solvated proton and the solvated fluoride

[67] D. F. Smith, *J. Chem. Phys.* **28** (1958) 1040.
[68] M. Atoji and W. N. Lipscomb, *Acta Cryst.* **7** (1954) 173.
[69] J. Shamir and A. Netzer, *J. Sci. Instr.* Series 2, **1** (1968) 770.

ion are relatively large, 350[61] and 273[70] ohm^{-1} cm^2 respectively. There is good evidence from a comparison of conductances[71] in hydrogen fluoride of solutes such as $HSbF_6$, Et_4NF and Et_4NAsF_6, which contain either one or neither of the ions derived from the solvent, that movement of both the proton and the fluoride ion is by transfer along hydrogen fluoride chains rather than by migration of the solvated ions as separate entities. Transfer of both ions occurs in hydrogen fluoride by a process similar to that in water and in anhydrous sulphuric acid.

2.2 SELF-IONIZATION IN HYDROGEN FLUORIDE

The equilibrium for self-ionization may be written as

$$HF \rightleftharpoons H^+ + F^-$$

However, as in aqueous and other solvent systems, the bare proton has no meaningful separate existence, and both it and the fluoride ion are solvated. Therefore a more acceptable, but still idealized, equation is

$$3HF \rightleftharpoons H_2F^+ + HF_2^-$$

While the hydrogen difluoride ion HF_2^- occurs as the symmetrical anion, with an enthalpy of formation of about -40 kcal $mole^{-1}$, in solids such as KHF_2 the fluoride ion can be considerably more solvated, as indicated by stoichiometric solids such as KH_2F_3, for which crystal structure studies have shown the existence of the separate, non-symmetrical $H_2F_3^-$ anion. In the pure liquid, both anion and cation are solvated to a much greater extent than indicated in the equation above. In fact each ion exists as part of an associated, hydrogen bonded network. If it were not for the convenience of referring to the ions H_2F^+ and HF_2^- in species such as $H_2F^+SbF_6^-$ and $K^+HF_2^-$, it would probably be better to write the autoprotolysis equation as

$$HF \rightleftharpoons H^+_{solv} + F^-_{solv}$$

However, in experimental practice it is not reasonable to assume that the solvated proton and fluoride ion are the only ionic entities even in liquid hydrogen fluoride which has been fractionally distilled. In section 1.2(ii) water, sulphur dioxide, sulphuric and fluorosulphuric acids and silicon tetrafluoride and fluorosilicates were listed as possible impurities in so-called anhydrous hydrogen fluoride. Of these impurities, water makes by far the greatest contribution to the conductance. Hyman and Katz[59] report that a specific conductance of 1.4×10^{-5} ohm^{-1} cm^{-1} corresponds with a water content of 0.0002%. Even dissolution of oxide layers from metal components of apparatus used for handling hydrogen fluoride can introduce protonated water into the liquid during or after distillation.

Assuming that there are no impurities present, that is assuming that for pure anhydrous hydrogen fluoride the specific conductance at 0°C is the lowest observed, 1×10^{-6} ohm^{-1} cm^{-1}, and assuming a value of 623 ohm^{-1} cm^2 for the molar conductance, the concentrations of ionic species in hydrogen fluoride are

$$[H^+] = [F^-] = 1.6 \times 10^{-6} \text{ mole } l^{-1}$$

The ionic product for the liquid is then 2.6×10^{-12} $mole^2$ l^{-2} and the value of pK_a is 13.3. However, some of the conductance for such highly purified hydrogen fluoride must be

[70] M. Kilpatrick and T. J. Lewis, *J. Am. Chem. Soc.* **78** (1956) 5186.
[71] R. J. Gillespie and R. Hulme, unpublished observations.

regarded as being due to the presence of small amounts of water or other impurities, so the real self-dissociation of hydrogen fluoride may be significantly lower than indicated above.

2.3. ACIDITY OF HYDROGEN FLUORIDE

At first sight, considering the relative electronegativities of the halogens themselves, it seems surprising that hydrogen fluoride in aqueous solution is a weak acid, comparable in strength with formic acid, while the other hydrogen halides are very strong. It should be recalled, however, that H_2S, H_2Se and H_2Te, while relatively weak acids in aqueous solution with acidity constants in the range 1.1×10^{-7} to 2.3×10^{-3}, are much stronger than the apparently analogous acid H_2O with an acidity constant of 2×10^{-16}. A brief account of the reasons for the differing relative acidities of the hydrogen halides as solutes in water is given in Chapter 26.

Study of the dissociation of hydrogen fluoride in water is complicated by the formation of the relatively stable aquated species HF_2^-, so that two equilibrium constants K_1 and K_2 must be considered:

$$K_1 = \frac{[H^+][F^-]}{[HF]} \quad \text{and} \quad K_2 = \frac{[HF_2^-]}{[HF][F^-]}$$

A review[72] in 1956 of measurement and values of these constants showed that reported values of K_1 were in the range 2.4 to 7.2×10^{-4} and for K_2 from 5 to 25. Recently, the lanthanum fluoride membrane electrode, which has been described in section 1.9(v) as giving a specific and Nernstian response to fluoride activity, has been used[73] to obtain a value of 6.46×10^{-4} mole l^{-1} for the acidity constant. This is probably the most reliable value reported so far.

Although, as a solute, hydrogen fluoride is a fairly weak acid, it is strongly acidic as a solvent. As the concentration of hydrogen fluoride increases in HF–H_2O mixtures, the system becomes progressively more acidic, with water acting as a very strong base in hydrogen fluoride as solvent. The increase in acidity with increase in hydrogen fluoride concentration is discussed in more detail below. In considering the differing acidity of hydrogen fluoride when it acts as solute or solvent, it should be noted that in dilute aqueous solution, an isolated hydrogen fluoride molecule donates a proton to an aggregate of water molecules and forms an aquated fluoride ion. The situation is very different when only a small amount of water is present in the system. Then a proton is transferred to an isolated water molecule from polymeric hydrogen fluoride and the fluoride ion, now formed, is part of a stable polymeric anionic complex. The differing solvation of the fluoride ion in the two extremes of composition in the H_2O–HF system is probably the major factor in deciding the ease of proton transfer.

Hydrogen fluoride has been studied fairly extensively as a member of a group of extremely acidic solvents, being comparable in acidity with anhydrous sulphuric acid but considerably less acidic than fluorosulphuric acid. For comparison of highly acidic protonic solvents, the Hammett scale has proved useful. Indicators, such as nitroanilines and nitrated hydrocarbons, for which the protonated and non-protonated species show different absorbances,

[72] H. R. Leech, in *Supplement to Mellor's Comprehensive Treatise on Inorganic and Theoretical Chemistry*, Supp. II, Part I, Longman (1956) pp. 111–118.

[73] N. E. Vanderborgh, *Talanta*, **15** (1968) 1009.

are used in spectrophotometric studies on the pure solvents and on solutions. Hammett[74] defined the following relationship for any particular indicator:

$$H_0 = pK_{In} - \frac{\log C_{InH^+}}{\log C_{In}}$$

where K_{In} is the acid dissociation constant for the indicator and C_{InH^+} and C_{In} are the concentrations of the indicator in protonated and non-protonated forms. For solvents of increasing acidity, an overlapping series of indicators of decreasing basicity is studied spectrophotometrically in the visible and ultraviolet region so that values of K_{In} can be evaluated for indicators which are applicable to strongly acidic solvents. In considering a transition to the normal aqueous situation, the Hammett equation would become the conventional pH definition.

Hyman, Kilpatrick and Katz[75] quote a value of -10.2 for the Hammett function for a fractionally distilled sample of hydrogen fluoride having a conductivity of about 10^{-5} $ohm^{-1}\ cm^{-1}$. They have shown that the presence of water markedly reduces values of H_0 for hydrogen fluoride, e.g. a solution 1 M in water has a value of -8.9. Therefore they

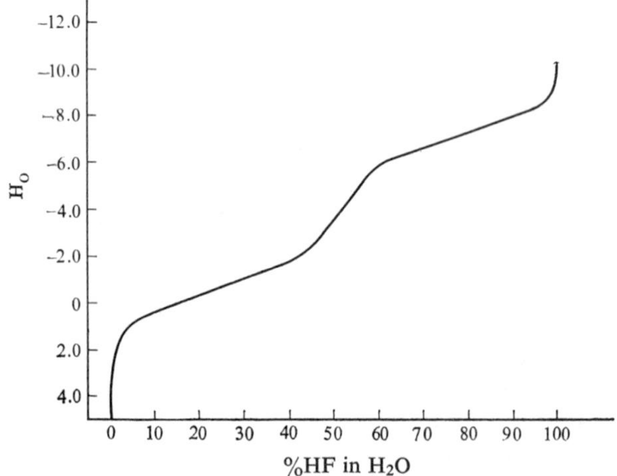

FIG. 8. The H_0 function for the H_2O–HF system

assume that as still drier hydrogen fluoride is studied, values of the Hammett function will increase perhaps to a value of about -11. These workers have studied the increase in acidity of the HF–H_2O system as the concentration of HF increases by measurement of the Hammett function across the whole composition range. Their results are shown in Fig. 8. It is seen that the acidity increases markedly initially, becomes steady with the formation of large concentrations of aquated protons, and then increases rapidly in 1:1 solutions. This has been ascribed to the formation of large concentrations of HF_2^-, $H_2F_3^-$, etc. in aqueous solution. Then there is a steady increase of acidity with H_3O^+ still the predominant cationic species, with another rapid increase near 100% HF, presumably due to the formation of H_2F^+ and polymeric anion chains.

[74] L. P. Hammett and A. J. Deyrup, *J. Am. Chem. Soc.* **54** (1932) 2721.
[75] H. H. Hyman, M. Kilpatrick and J. J. Katz, *J. Am. Chem. Soc.* **79** (1957) 3668.

Bases other than water, e.g. alcohols and fluoride ion, also markedly depress the Hammett function for hydrogen fluoride[75]. A solution 1 M in sodium fluoride has a value for H_0 of -8.4. Conversely acids in hydrogen fluoride markedly increase H_0 values. Kilpatrick and Jones[60] give a value of -15.3 for a 3 M solution of antimony pentafluoride. However, the non-protonic non-electrolyte sulphur dioxide does not significantly affect H_0 values for hydrogen fluoride[75].

2.4. SOLUTION PROCESSES IN HYDROGEN FLUORIDE

As mentioned in the introduction to this section, most textbooks in describing solutions in anhydrous hydrogen fluoride have reported observations, frequently qualitative, on the solubility of a number of unrelated solutes. Frequently, as for sulphates and other oxy-anions, very complex solvolysis reactions occur and attempts have been made to interpret these solvolytic reactions on the basis of very crude physico-chemical measurements. It has not been realized until quite recently that the solubility of some compounds such as silver cyanide can depend markedly on the presence or absence of traces of water in the hydrogen fluoride.

In this review, for simplicity, fluorides will be selected as representative solutes wherever possible. However, protonation of organic and other solutes and solvolysis reactions are so important that they must be included, even though at present there is little reliable information on the course or extent of many solvolyses. As previously indicated, solutes discussed in this section will be classified on the basis that they dissolve with or without ionization or that they undergo solvolytic, acid–base or oxidation-reduction reactions. The term solvolysis will be used here to denote an irreversible chemical interaction of solute and solvent, even though simpler reversible interactions such as solvation and acid–base reactions could be regarded as examples of solvolysis.

(a) Solution Without Ionization

Chemists at the Argonne National Laboratory, Illinois, have measured the solubilities of higher fluorides of transition metals in hydrogen fluoride and have used conductance and Raman spectral studies to investigate possible ionization. Frlec and Hyman[76] report room temperature solubilities in the range 0.5–3 M for the hexafluorides of molybdenum, tungsten, uranium, rhenium and osmium with negligible ionization. They included xenon hexafluoride in this survey and obtained a highly conducting 8.6 M solution, whereas Quarterman and Hyman had shown earlier[77] that xenon difluoride and xenon tetrafluoride did not enhance the conductivity of hydrogen fluoride. The observation by Selig and Frlec[78] that vanadium pentafluoride, although very soluble (3.3 M), gives a non-conducting solution is perhaps a little surprising when comparison is made with the highly conducting solutions obtained from antimony and arsenic pentafluorides (see sub-section [d]). They postulate that vanadium pentafluoride is polymeric in solution and therefore does not accept fluoride ions easily. Selig and Gasner[79] demonstrated that at 25°C rhenium heptafluoride is com-

[76] B. Frlec and H. H. Hyman, *Inorg. Chem.* **6** (1967) 1596.

[77] H. H. Hyman and L. A. Quarterman in *Noble Gas Compounds*, ed. H. H. Hyman, University of Chicago Press (1963) p. 275.

[78] H. Selig and B. Frlec, *J. Inorg. Nucl. Chem.* **29** (1967) 1887.

[79] H. Selig and E. L. Gasner, *J. Inorg. Nucl. Chem* **30** (1968) 658.

parable in solubility with the transition metal hexafluorides, giving a non-conducting 0.62 M solution.

Gillespie and Humphreys[62] have shown from cryoscopic measurements that sulphur hexafluoride is a non-electrolyte and that fluorosulphuric acid and xenon difluoride are not significantly ionized. They have also shown sulphur dioxide to be a non-electrolyte, an observation consistent with the fact that this solute has no effect on the Hammett function for hydrogen fluoride.

Hydrogen chloride has very low solubility in hydrogen fluoride (0.17 M at 0°C), presumably because it is not dissociated, or protonated, to any significant extent. The combination of properties for hydrogen chloride of low solubility in hydrogen fluoride and high volatility provides the basis of an excellent method of preparation of anhydrous fluorides generally. (For example, see preparation of transition metal fluorides, page 1075.) Anhydrous hydrogen fluoride is distilled on to an anhydrous chloride. Hydrogen chloride produced in the metathetical reaction, which is usually energetically favourable, is expelled from solution and the solid fluoride is recovered under anhydrous conditions after allowing excess hydrogen fluoride to evaporate.

(b) Dissociation to Solvated Ions

Primarily because of its very high dielectric constant but also because of its relatively high boiling point, long liquid range and low viscosity anhydrous hydrogen fluoride is an excellent solvent for ionic fluorides. The only reasonably comprehensive report of solubilities

TABLE 4. SOLUBILITY OF IONIC FLUORIDES IN HYDROGEN FLUORIDE

Solute	LiF	NaF	KF	RbF	CsF	NH_4F
Temperature (°C)	12	11	8	20	10	17
Solubility (g per 100 g)	10.3	30.1	36.5	110	199	32.6
Solute	AgF	TlF	MgF_2	CaF_2	SrF_2	BaF_2
Temperature (°C)	12	12	12	12	12	12
Solubility (g per 100 g)	83.2	580	0.025	0.817	14.83	5.60

of inorganic fluorides was given in 1952 by Jache and Cady[58]. Some values from that source are shown in Table 4.

Group IA fluorides show the expected progressive increase in solubility with increase in atomic number of the metal and, as a group, they are much more soluble than the fluorides of Group IIA. Other 1:1 fluorides, of silver and thallium, are much more soluble than those of Group IA. However, whereas Gillespie and Humphreys showed from cryoscopic measurements that Group IA fluorides were effectively completely ionized, silver(I) fluoride gave a value of 1.7 for ν, which corresponds with a dissociation constant at the freezing point of about 0.15. This observation is consistent with Clifford's[80] constant of 0.087, measured potentiometrically at 0°C where the dielectric constant is much less than at the freezing point. The other fluorides studied by Jache and Cady, compounds of transition metals in low oxidation states or of metals of main groups other than IA and IIA, showed very low solubilities by comparison with Group IA and IIA fluorides.

Few, if any, compounds other than fluorides dissolve with simple dissociation in hydrogen fluoride. It has been suggested that potassium perchlorate and periodate do so to some extent, but there is some evidence that the latter, at least, undergoes solvolysis.

[80] A. F. Clifford, W. D. Pardieck and M. W. Wadley, *J. Phys. Chem.* **70** (1966) 3241.

It is customary in reviewing solutions in hydrogen fluoride to list as a separate group those solutes which dissolve by simple dissociation, and the practice has been retained here because the Group IA fluorides in particular represent a very simple class of solute, dissolving according to a well-established pattern and therefore of great value as reference solutes in comparative studies involving measurement of conductance and of colligative properties. However, it should be realized that metal fluorides dissolving in this way might, in a fundamental sense, be classified more effectively as Lewis bases [see sub-section (d)(iv) below].

(c) Solvolysis Reactions

There has been relatively little reliable and comprehensive investigation of solvolysis of solutes in anhydrous hydrogen fluoride. Most of the statements found in books and reviews purporting to describe the behaviour of oxy-anions are based on postulations, such as those in Fredenhagen's review[55], made to explain rudimentary measurements of conductance and colligative properties. While the Fredenhagens and co-workers carried out remarkably good experimental work with quite inadequate materials of construction for their apparatus and showed extraordinary insight in the interpretation of their observations, their work was necessarily exploratory in nature. Careful investigation based on modern techniques is necessary to elucidate the pattern of many solvolytic reactions, which frequently are markedly temperature- and time-dependent.

Carbonates, sulphites and similar compounds obviously undergo solvolysis in the simple way in which they do in acidic aqueous media. Fredenhagen[55] measured the elevation of the boiling point of solutions of potassium sulphate in hydrogen fluoride and, on the basis of an observed value for ν of 6, postulated the following:

$$K_2SO_4+4HF \longrightarrow 2K^++HSO_3F+H_3O^++3F^-$$

This, however, implies that a value of 7 should be obtained for ν. From freezing-point depressions, Gillespie and Humphreys[81] obtained a reliable value of 5 for ν but found that ν increased to about 6 when they observed depressions of vapour pressure of the solution at 0°C. They postulate that, at the freezing point, unionized sulphuric acid is formed:

$$K_2SO_4+4HF \longrightarrow 2K^++2HF_2^-+H_2SO_4$$

but that at higher temperatures solvolysis leads to the formation of unionized fluorosulphuric acid:

$$H_2SO_4+3HF \longrightarrow H_3O^++HSO_3F+HF_2^-$$

They used n.m.r. evidence to demonstrate the presence of fluorosulphuric acid in the warmed solution and showed cryoscopically that, when sulphuric acid was used as the sole solute at the freezing point of hydrogen fluoride, a value of ν close to unity was obtained.

Phosphate undergoes a similar but more extensive solvolysis. ^{19}F and ^{31}P n.m.r. spectrometry[82] in the $HF–P_2O_5–H_2O$ system with measurement of peak areas gave values of equilibrium constants for the successive formation of H_2PO_3F, HPO_2F_2 and $H_3O^+PF_6^-$.

On the basis of some early conductance studies of potassium nitrate dissolved in hydrogen fluoride, it was believed that nitric acid, formed in an initial metathetical reaction, was protonated to $H_2NO_3^+$. However, a more recent infrared and Raman spectral study[83]

[81] R. J. Gillespie and D. A. Humphreys, unpublished observations.

[82] D. P. Ames, S. Ohashi, C. F Callis and J. R. Van Wazer, *J. Am . Chem. Soc.* **81** (1959) 6350.

[83] F. P. Del Greco and J. W. Gryder, *J. Phys. Chem.* **65** (1961) 922.

gives no evidence for the protonated species. Potassium nitrate is now believed to undergo solvolysis according to the equation

$$KNO_3+6HF \longrightarrow K^++NO_2^++H_3O^++3HF_2^-.$$

In relatively dilute solutions, Raman and infrared frequencies characteristic of the nitronium ion were observed. In more concentrated solution a Raman frequency ascribed to molecular HNO_3 was observed. Addition of water to the solution decreased the intensity of the nitronium ion frequencies and increased that for molecular HNO_3. No frequencies which could be assigned to protonated HNO_3 were observed.

Permanganates and chromates are solvolysed to oxide fluorides such as MnO_3F and CrO_2F_2, and there is some evidence that iodates and periodates give a variety of solvolysis products which include oxy-cations or other oxy-anions of iodine. Fredenhagen[55] showed that chlorates and bromates were solvolysed, but the products were not identified unequivocally.

Although there has been little systematic study of solvolysis in hydrogen fluoride, there are many preparative and other reactions that could be classified as solvolytic reactions, e.g. the preparation discussed above of fluorides from anhydrous chlorides. Acetates, and carboxylates generally, dissolve readily because the carboxylic acid formed metathetically is readily protonated [see sub-section (d)(ii) below]. Gillespie, Hulme and Humphreys'[84] cryoscopic investigations show that sodium cyanide reacts with hydrogen fluoride to give non-protonated HCN, Na^+ and HF_2^- and that mercuric cyanide $Hg(CN)_2$ reacts in solution to form the linear cation $Hg_2(CN)_3^+$ as well as HCN and HF_2^-.

(d) Acid–base Reactions

(i) *Bronsted acids.* Perhaps it is not surprising that there has been little study of the possible existence of protonic acids in a solvent as acidic as hydrogen fluoride. Some reviewers have suggested that fluorosulphuric, periodic and perchloric acids can transfer protons to hydrogen fluoride and act as Bronsted acids. Gillespie and Humphreys[62] have demonstrated that fluorosulphuric acid dissolves without appreciably changing the conductance of hydrogen fluoride, and that cryoscopically a value of unity is obtained for ν. They have also indicated that periodates are solvolysed.

The only suggestion that perchloric acid acts as a Bronsted acid comes from Fredenhagen's[55] work on solutions of alkali metal perchlorates in hydrogen fluoride. He showed that potassium perchlorate dissolves to the extent of 0.77 M and he proposed, on the basis of conductance and boiling-point measurements, that it dissociates to simple ions. The separate existence of perchlorate ions would imply that perchloric acid is an acid of significant strength in hydrogen fluoride. However, an alternative explanation of his observations would be that perchlorate ions act as bases, forming an equivalent amount of fluoride ions.

(ii) *Bronsted bases.* Expectedly, many chemical species accept protons from the strongly acidic hydrogen fluoride. The role of water as a base has already been touched upon. Frequently, it is stated or assumed that water is fully protonated in hydrogen fluoride However, boiling point and conductance studies suggest values of about 0.2 for K for the equation

$$2HF+H_2O \rightleftharpoons H_3O^++HF_2^-$$

This may seem surprising when conductance studies on solutes such as alcohols indicate complete protonation. Hyman and Katz suggest[59] that, while for alcohols each molecule

[84] R. J. Gillespie, R. Hulme and D. A. Humphreys, unpublished observations.

may be protonated, hydrogen bonded aggregates may be formed for water as solute, so that the ionization process may be represented as

$$nH_2O+2HF \longrightarrow H(H_2O)_n^+ + HF_2^-$$

More recently, values of ν approaching 2 have been observed cryoscopically for water as a solute[81], indicating that as temperatures approach $-83°C$ protonation of water increases. This would imply that hydrogen fluoride becomes more acidic with decrease in temperature which would be a reflection of its own increasing state of aggregation.

Pioneering conductance and boiling-point investigations by Klatt and Fredenhagen before World War II and more precise conductance measurements later at the Argonne National Laboratory[85] have shown that a wide range of organic acids, alcohols, aldehydes, ketones, ethers and amines act as bases in hydrogen fluoride. These all have one or more electron pairs on an oxygen or a nitrogen atom capable of accepting protons although protonation of hydrocarbons by hydrogen fluoride has also been studied. The protonation of most organic solutes is reversible, the compound being recovered unchanged on removal of the solvent. Hyman and Katz[59] have reviewed conductance studies on solutions in hydrogen fluoride of several different types of organic compound while Kilpatrick and Jones[60] provide considerable detail on the application of several different physical methods to the study of bases in hydrogen fluoride. For organic and other solutes they show how studies of colligative properties, conductances and of absorption and resonance spectra have been used to demonstrate the nature of the species present in solution. For example, early work on alcohols and carboxylic acids had indicated that protonation was essentially complete:

$$R{\cdot}CH_2OH+2HF \longrightarrow R{\cdot}CH_2OH_2^+ + HF_2^-$$

and

$$R{\cdot}COOH+2HF \longrightarrow R{\cdot}C(OH)_2^+ + HF_2^-$$

Subsequently, infrared studies showed that, as hydrogen fluoride is added to alcohols, absorption increases progressively at frequencies assigned to the stretching frequencies for HF_2^-, suggesting protonation of the alcohol. Later again, proton resonance studies showed the presence of the entity $CH_3CH_2OH_2^+$ from the areas of the CH_3,CH_2 and OH_2^+ peaks and from the splitting of the CH_2 resonance.

A dramatic change in basicity of the solute is observed in passing from acetic acid to trifluoroacetic acid, which has frequently been described as weakly basic in hydrogen fluoride. In fact, cryoscopy[62] shows it to be essentially non-protonated. Hyman and Katz[59] warn that it is an over-simplification to try to relate this decrease in basicity too closely to the fact that acetic acid is weakly acidic in water while trifluoroacetic acid is very much stronger. They point out a similar decrease in conductance in hydrogen fluoride for the two solutes CH_3CH_2OH and CF_3CH_2OH, and they say that what should be considered is the molecular environment of the proton-accepting oxygen atom in the base molecule itself. While there are some valid parallels between relative strengths of acids and bases in passing from one solvent to another, it is dangerous to push the concept too far. Thus HCN which is a much weaker acid in water than CH_3COOH is, like HCl, not protonated in hydrogen fluoride.

(iii) *Lewis acids (fluoride acceptors).* Most of the Lewis acids studied in hydrogen fluoride have been pentafluorides which can achieve a stable 6-coordinated structure of

[85] L. A. Quarterman, H. H. Hyman and J. J. Katz, *J. Phys. Chem.* **65** (1961) 90.

fluorines around a central atom by accepting fluoride ions, thereby enhancing the conductance of the solvent by a reaction which, in its simplest form, can be represented as

$$XF_5+2HF \longrightarrow H_2F^+ + XF_6^-.$$

Boron trifluoride and some tetrafluorides have also been investigated to some extent and can be classed as weak Lewis acids. Several metal hexafluorides, as well as rhenium heptafluoride and vanadium pentafluoride, have been shown to have zero strength as acids.

Clifford and co-workers[86] reported a qualitative classification of several pentafluorides, tetrafluorides and trifluorides as Lewis acids in hydrogen fluoride by observing their ability to add fluoride ions and so cause dissolution of sparingly soluble fluorides such as lanthanide trifluorides. Kilpatrick and co-workers at the Illinois Institute of Technology measured conductances of hydrogen fluoride solutions of boron trifluoride and other solutes[87].

The first rigorous and quantitative study was reported in 1961 by Hyman, Quarterman, Kilpatrick and Katz[88] who measured conductances and recorded infrared and Raman

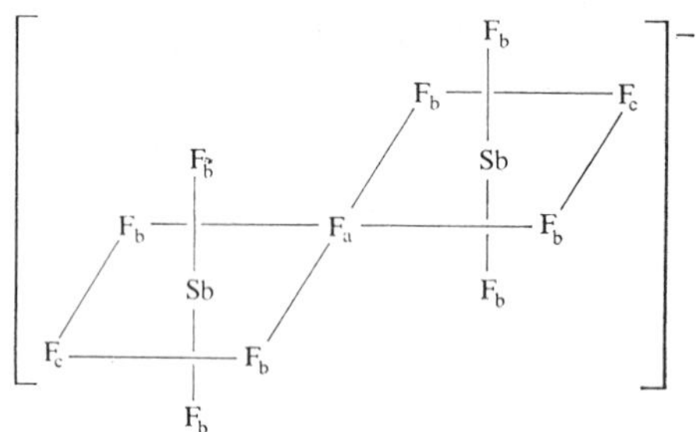

FIG. 9. Structure of the $Sb_2F_{11}^-$ anion.

spectra over the full composition range for the HF–SbF_5 system. They observed that addition of antimony pentafluoride to hydrogen fluoride gave a highly conducting solution, with a maximum conductance at about 10 mole % of the pentafluoride added. In dilute solutions of the pentafluoride, they observed Raman frequencies consistent with the formation of the SbF_6^- anion but the intensity of the frequencies decreased as the pentafluoride concentration increased. They postulated that in relatively dilute solution the ionic species H_2F^+ and SbF_6^- were formed but that ion pair association became increasingly important with increasing pentafluoride concentration. They calculated equilibrium constants for the ionization and for the proposed ion-pair formation.

In 1966 Gillespie and Moss[61] re-investigated the HF–SbF_5 system and obtained values for the conductance over the concentration range which were essentially the same as those of Hyman and co-workers. However, they were able to explain the apparent anomalies in the variation of conductance with increasing concentration of antimony pentafluoride by proposing the step-wise formation of the ions $Sb_2F_{11}^-$, $Sb_3F_{16}^-$, etc., after the initial formation of SbF_6^- at low pentafluoride concentration, and they demonstrated the existence of $Sb_2F_{11}^-$, at least, by ^{19}F n.m.r. spectrometry. The ^{19}F spectrum of a dilute solution (2 mole %) observed at room temperature and at lower temperatures, showed only a broad peak assigned to SbF_6^- in addition to the line for HF. At about 20 mole % the room temperature spectrum showed the broad peak which at lower temperature split into three peaks with

[86] A. F. Clifford, H. C. Beachell and W. M. Jack, *J. Inorg. Nucl. Chem.* **5** (1957) 57.
[87] M. Kilpatrick and F. E. Luborsky, *J. Am. Chem. Soc.* **76** (1954) 5865.
[88] H. H. Hyman, L. A. Quarterman, M. Kilpatrick and J. J. Katz, *J. Phys. Chem.* **65** (1961) 123.

some fine structure. The three new peaks had relative areas in the ratio 1 :8 :2. This spectrum is consistent with that expected for the dimeric anion $Sb_2F_{11}^-$ having the structure shown in Fig. 9 in which there are three different sets of fluorine atoms, a single bridging fluorine (F_a) eight equivalent atoms (F_b) that are *cis* to F_a and two equivalent atoms (F_c) *trans* to F_a.

Gillespie and Moss proposed an ionization pattern which explained the observed conductances and was consistent with their ^{19}F n.m.r. spectra and also with proton resonance spectra which they observed. Monomeric SbF_6^- is formed almost exclusively at low concentration by the strong acid SbF_5. Subsequent cryoscopic investigations have shown ionization to be virtually complete in this system. Over the concentration range 11–22 mole % of SbF_5, monomeric SbF_6^- is gradually replaced by the dimer $Sb_2F_{11}^-$ which in turn is gradually replaced by the trimer $Sb_3F_{16}^-$ over the range 22–33 mole %.

Subsequently Gillespie and co-workers[89] demonstrated by more detailed ^{19}F n.m.r. spectrometric investigation the existence of the anions $Sb_2F_{11}^-$ and $Sb_nF_{5n+1}^-$ in systems such as $CsF–SbF_5$ and $Et_4NSbF_6–SbF_5$. Direct crystallographic evidence is now available for the existence of the entities $Sb_2F_{11}^-$ and $Sb_3F_{16}^-$ in the compounds $XeF_2{\cdot}2SbF_5$ [90] and $Br_2Sb_3F_{16}$ [91].

Of all solutes in hydrogen fluoride, antimony pentafluoride seems to have been the subject of most intensive study. However, there has been some preliminary work on other pentafluorides and related acids or potential acids. Hyman, Lane and O'Donnell[92] used Raman spectrometry to show the presence of fluoroarsenate in solutions of arsenic pentafluoride in hydrogen fluoride and found the solutions to have about half the conductance of equimolar antimony pentafluoride solutions in all but very dilute solutions. Cryoscopic measurements[93] indicate that, even in dilute solution, arsenic pentafluoride forms $As_2F_{11}^-$ rather than AsF_6^-.

Boron trifluoride is considerably less soluble in hydrogen fluoride than arsenic pentafluoride, and phosphorus pentafluoride, also sparingly soluble at 0°C, is weaker again as an acid, causing little enhancement of the conductance of the hydrogen fluoride. Hyman, Lane and O'Donnell[92] showed that molybdenum pentafluoride dissolved in hydrogen fluoride at 25°C only to the extent of 0.0168 M to give a solution for which the molar conductance is only about 3% that of an antimony pentafluoride of corresponding concentration. Considering that the hexafluoromolybdate(V) anion is formed relatively easily in solids such as $KMoF_6$, the reluctance of the pentafluoride when dissolved in hydrogen fluoride to accept fluoride ions probably indicates that it retains in solution the tetrameric structure which X-ray analysis has shown exists in the solid [see section 5.6(b)]. That is, it is already 6-coordinated. Structural factors probably also account for the low solubilities of niobium and tantalum pentafluorides and perhaps also for titanium tetrafluoride. However, the low solubility of silicon tetrafluoride reported by Clifford[81] appears to represent very weak Lewis acidity, when the molecular nature of this tetrafluoride is considered in comparison with the polymeric structure of the transition metal fluorides.

While it is not particularly surprising that the octahedral hexafluorides of transition metals which dissolved readily in hydrogen fluoride [see sub-section (a)] did not enhance the conductance of the solvent[76], the observation by Selig and Frlec[78] that vanadium penta-

[89] J. Bacon, P. A. W. Dean and R. J. Gillespie, *Can. J. Chem.* **47** (1969) 1655.
[90] V. M. McRae, R. D. Peacock and D. R. Russell, *Chem. Communs.* (1969) 62.
[91] A. J. Edwards, G. R. Jones and R. J. C. Sills, *Chem. Communs.* (1968) 1527.
[92] H. H. Hyman, J. Lane and T. A. O'Donnell, 145*th Meeting Amer. Chem. Soc. Abstracts*, p. 63 J.
[93] P. A. W. Dean, R. J. Gillespie, R. Hulme and D. A. Humphreys, *J. Chem. Soc.* (A) (1971) 341.

fluoride dissolved to the extent of 3.3 M without giving a conducting solution or spectral evidence for fluorovanadates was unexpected, and they suggested polymerization of the solute. As stated earlier, rhenium heptafluoride had no acid strength in hydrogen fluoride[79].

It is unfortunate that, at present, there is no comprehensive quantitative survey of Lewis acidity of fluorides in hydrogen fluoride or in any reaction system, although some qualitative observations have been made in solvents such as halogen fluorides, where on the basis of displacement reactions and conductances, an order of acid strengths $SbF_5 > AsF_5 > BF_3 > SnF_4 > TiF_4$ has been suggested[94]. Fox *et al.*[95], in examining the fluoride-donor capabilities of trifluoramine oxide ONF_3, have shown an order of acid strengths $SbF_5 > AsF_5 > BF_3 > PF_5$, because $ONF_2^+SbF_6^-$ is stable at room temperature whereas a vapour pressure of arsenic pentafluoride of 5 mm Hg is in equilibrium with $ONF_2^+AsF_6^-$ at room temperature and $ONF_2^+BF_4^-$ is stable only below −50°C. They were not able to make an adduct with phosphorus pentafluoride. Christie and Pavlath show a similar order in adducts of chlorine trifluoride[96], and Bartlett's[97] observations on the stabilities of compounds containing the XeF^+ and $Xe_2F_3^+$ cations are also consistent with this order. In addition Bartlett's work generally shows iridium and platinum pentafluorides to be very strong Lewis acids. Reviewing the existence and stability of fluoronium cations, Seel and Detmer[98] record that sulphur and selenium tetrafluorides form 1:1 adducts with arsenic, antimony and iridium pentafluorides but not with boron trifluoride, the pentafluorides of phosphorus, niobium and tantalum nor with tin and platinum tetrafluorides.

(*iv*) *Lewis bases* (*fluoride donors*). As stated in sub-section 2.4(b) above, binary fluorides such as the alkali metal fluorides are the simplest source of Lewis base in the hydrogen fluoride system, ionizing to give solvated fluoride ions:

$$MF + HF \longrightarrow M^+ + HF_2^-$$

In general, for essentially ionic fluorides, the solubility decreases as the oxidation state of the metal increases; thus trifluorides are less soluble than difluorides which in turn are less soluble than monofluorides.

Xenon hexafluoride is considerably more soluble in hydrogen fluoride at room temperature (8.6 mole per 1000 g HF) than the reasonably soluble hexafluorides of transition metals, but unlike them, gives a conducting solution. The conductance is about one-third of that of a corresponding solution of sodium fluoride. Considering that the compound $XeF_5^+PtF_6^-$ has been characterized by X-ray structural analysis[99] it seems very likely that xenon hexafluoride ionizes to XeF_5^+ and solvated fluoride ions.

The halogen trifluorides ClF_3 and BrF_3 are reported to dissolve in hydrogen fluoride to give weakly conducting solutions[100] according to the equation

$$XF_3 + HF \longrightarrow XF_2^+ + HF_2^-$$

This dissociation to XF_2^+ is proposed because the resulting solution can be neutralized with a solution of antimony pentafluoride:

$$BrF_2^+HF_2^- + H_2F^+SbF_6^- \longrightarrow BrF_2^+SbF_6^- + 3HF$$

---

[94] A. A. Woolfe, *J. Chem. Soc.* (*London*) (1950) 1053.

[95] W. B. Fox, C. A. Wamser, R. Eibeck, D. K. Huggins, J. S. McKenzie and R. Juurik, *Inorg. Chem.* **8** (1969) 1247.

[96] K. O. Christe and A. E. Pavlath, *Z. anorg. u. allgem. Chem.* **335** (1965) 210.

[97] F. O. Sladky, P. A. Bulliner and N. Bartlett, *J. Chem. Soc.* (London) *A* (1969) 2179

[98] F. Seel and O. Detmer, *Z. anorg. u. allgem. Chem.* **301** (1959) 113.

[99] N. Bartlett, F. Einstein, D. F. Stewart and J. Trotter, *J. Chem. Soc.* (*London*) *A* (1967) 1190.

[100] M. T. Rogers, J. L. Spiers and M. P. Panich, *J. Phys. Chem.* **61** (1957) 366.

Recent conductance measurements[101] show that SF_4 is moderately basic in HF, a value of $4 \pm 2 \times 10^{-2}$ being quoted for the constant at 0°C for the equilibrium

$$SF_4 + HF \rightleftharpoons SF_3^+ + HF_2^-$$

(e) Oxidation–reduction Reactions

There has been little study of oxidation–reduction reactions in hydrogen fluoride. There is little likelihood of chemical oxidation of the solvent to fluorine, but there are some reports of fluorides of elements in very low oxidation states reducing the solvent. Thus chromium difluoride[102] and uranium trifluoride[103] reduce hydrogen fluoride to hydrogen, being oxidized to chromium trifluoride and uranium tetrafluoride. Emeléus observed that hydrogen fluoride reacts with vanadium dichloride to form vanadium trifluoride, hydrogen chloride and hydrogen[104]. This is consistent with the conditions for preparation of vanadium difluoride, namely reduction of vanadium trifluoride with a mixture of hydrogen and hydrogen fluoride at about 1100°C. The fact that attempts to prepare lanthanum difluorides have always given compounds with the lanthanide in an average oxidation state considerably greater than 2 [see section 5.4(a)] probably indicates that these difluorides reduce hydrogen fluoride more readily than the corresponding sulphates reduce water.

As reported in detail in section 5.3(a), O'Donnell and co-workers have studied the ease of reduction of several higher fluorides of transition metals by reductants which are lower

Table 5. Reduction Potentials in Hydrogen Fluoride (at 0°C)

F_2–F^-	+2.71	(+2.87)	volt
Ag(II)–Ag(I)	+2.27	(+1.96)	
Tl(III)–Tl(I)	+1.45	(+1.25)	
Ag^+–Ag	+0.88	(+0.80)	
Hg_2^{2+}–Hg	+0.80	(+0.79)	
Fe^{3+}–Fe^{2+}	+0.58	(+0.77)	
Cu^{2+}–Cu	+0.52	(+0.34)	
H^+–H_2	0.00	(0.00)	
Pb^{2+}–Pb	−0.26	(−0.12)	
Cd^{2+}–Cd	−0.29	(−0.41)	

fluorides of main group elements, usually non-metals. Although direct reaction between the oxidants and reductant was studied in most cases, these reactions could have been carried out in hydrogen fluoride. In fact, the difficult partial reduction of tungsten hexafluoride by phosphorus trifluoride was catalysed in hydrogen fluoride solution. Their work showed that, for the fluorides studied, reductant strength decreased in the order $PF_3 > AsF_3 > SbF_3 \simeq SeF_4 > SF_4$. For the higher transition metal fluorides as oxidants, the order of strengths appears to be $VF_5 \simeq CrF_5 > UF_6 > MoF_6 > ReF_6 > WF_6 > TaF_5 \simeq NbF_5$. On the basis of the small number of reactions available in the chemical literature, it should be possible to include the higher fluorides of the noble metals and of neptunium and plutonium in this list, but there would be considerable uncertainty in some of the placements.

[101] M. Azeem, M. Brownstein and R. J. Gillespie, *Can. J. Chem.* **47** (1969) 4159.
[102] B. J. Sturm, *Inorg. Chem.* **1** (1962) 665.
[103] T. A. O'Donnell and P. W. Wilson, unpublished observations.
[104] H. J. Emeléus and V. Gutmann, *J. Chem. Soc.* (*London*) (1949) 2979.

Although there has been some potentiometry and polarography in aqueous solutions containing high concentrations of hydrogen fluoride, there has been little electrometric work in anhydrous hydrogen fluoride. Krefft is reported to have measured the potential of the cell $H_2(Pt)|KF(HF)|F_2(Pt)$ in Germany in 1939, and in 1954 Koerber and deVries[105] determined the potentials at 0° and 10°C of the cells $M,MF_2(s)|NaF(HF)|Hg_2F_2(s)Hg$ where M = Cd, Cu and Pb. In 1966 Clifford *et al.*[80] carried out e.m.f. measurements on the systems $Cu,CuF_2(s)|TlF(HF)|TlF_3(s)Pt$, $Ag|AgF,TlF(HF)|TlF_3(s)Pt$ and $Ag|AgF(HF)|AgF_2(s)Pt$.

From the results of the three separate projects, Clifford prepared a list of known potentials in hydrogen fluoride. Table 5 lists these data as reduction potentials, the numbers in parentheses being Latimer's values of standard reduction potentials for the corresponding system in water. It is seen that although the two sets of values differ in actual values, the order of strength of oxidants and reductants is the same in each solvent. However, at this stage this must not be accepted as a true generalization because of the limited number and relative simplicity of the systems studied.

3. IONIZATION IN HALOGEN FLUORIDES

Although previously there had been some measurement of the conductances of iodine monochloride and trichloride and of iodine monobromide and some over-simplified speculation about their self-ionization, the first significant work on ionization processes in interhalogen compounds generally was reported in 1949 from the laboratories at Cambridge by Emeléus and his research students. Despite the extreme experimental difficulties involved in preparing, purifying and handling the halogen fluorides, they measured conductances of these compounds and proposed an ionization scheme which has been the basis for all later work and which has been confirmed in all its essential details.

The halogen fluorides generally are powerful, fluorinating oxidants, reacting violently—even explosively—with organic matter and water. They attack many metals, although they can be handled safely in passivated nickel and nickel alloys. Also they etch glass and silica readily unless steps are taken to ensure that these materials are thoroughly dried.

3.1. SELF-IONIZATION AND SOLUTION PROCESSES

The specific conductance was determined by Emeléus and co-workers for chlorine trifluoride, bromine trifluoride and iodine pentafluoride[106]. Values of about 10^{-9}, 8.0×10^{-3} and 5×10^{-6} ohm^{-1} cm^{-1} respectively were obtained for the three compounds at 25°C. Of these, bromine trifluoride with the highest conductivity was studied most intensively. A self-ionization was postulated which in its simplest form can be written as

$$BrF_3 \rightleftharpoons BrF_2^+ + F^-$$

where BrF_2^+ is the Lewis acid of the system and F^- the Lewis base. However, while the degree of solvation of the cation may be uncertain, the fluoride ion is certainly solvated

[105] G. G. Koerber and T. De Vries, *J. Am. Chem. Soc.* **74** (1952) 5008.
[106] A. A. Banks, H. J. Emeléus and A. A. Woolfe, *J. Chem. Soc.* (*London*) (1949) 2861.

at least to the entity BrF_4^-, the structure of which is described below. Hence, the self-ionization is usually given as

$$2BrF_3 \rightleftharpoons BrF_2^+ + BrF_4^-$$

The Cambridge group adduced a considerable amount of evidence to support this proposed self-ionization[107]. The addition of potassium, silver or barium fluorides to bromine trifluoride markedly enhances the conductance. Further, if the solvent is evaporated, solids of stoichiometry $KBrF_4$, $AgBrF_4$ and $Ba(BrF_4)_2$ are recovered. X-ray powder photographs of the potassium compound showed no lines characteristic of potassium bromide or potassium fluoride. It was proposed that enhancement of conductance was caused by formation in solution of the tetrafluorobromate(III) anion and solvated potassium ions. Because bromine trifluoride is such an effective fluorinating agent, the tetrafluorobromates were isolated when compounds other than fluorides, e.g. the appropriate chlorides, were treated with excess bromine trifluoride.

Increased conductance was observed also when antimony(III) oxide, fluoride or oxide chloride was added to bromine trifluoride. On removal of excess solvent, a compound was obtained which was shown by chemical analysis to have the empirical formula $SbBrF_8$. The hexafluoroantimonate(V) anion was well known in alkali metal compounds, and it was suggested that the formulation $BrF_2^+SbF_6^-$ accounted for both the stoichiometry and conductance increase. It will be appreciated that a cleaner method of preparation of the compound would have been the addition of antimony pentafluoride to bromine trifluoride, but, of course, the solvent oxidized and fluorinated each of the antimony(III) compounds to the pentafluoride.

Emeléus and his colleagues did not unambiguously identify the compound $BrF_2^+SbF_6^-$, but they provided additional evidence for their formulation by carrying out a conductometric titration in bromine trifluoride of their compounds $AgBrF_4$ and $SbBrF_8$. They observed a conductivity minimum for a 1 : 1 mole ratio of the two compounds and, on removal of the solvent from the 1 : 1 mixture, the residue was shown by chemical analysis to have the formula $AgSbF_6$. While the fundamental neutralization reaction is

$$BrF_2^+ + BrF_4^- \rightleftharpoons 2BrF_3$$

an overall equation can be written for the reaction as

$$BrF_2^+SbF_6^- + Ag^+BrF_4^- \rightleftharpoons Ag^+SbF_6^- + 2BrF_3$$

When tin, stannous chloride or stannic chloride was treated with excess bromine trifluoride a solution was obtained which on conductometric titration with $KBrF_4$ gave a conductance minimum when the potassium and tin compounds were in the mole ratio 2 : 1. The residue from the solution at the neutralization point was the well-characterized potassium hexafluorostannate(IV). Emeléus proposed that the acidic species $(BrF_2^+)_2SnF_6^{2-}$ reacted as

$$(BrF_2^+)_2SnF_6^{2-} + 2K^+BrF_4^- \rightleftharpoons K_2SnF_6 + 4BrF_3$$

However, the solid compound $(BrF_2^+)_2SnF_6^{2-}$ was not isolated. When bromine trifluoride was pumped at room temperature from a solution to which a tin compound had been added, a product of the composition $SnF_4 1.76BrF_3$ was obtained and at 190°C all of the bromine trifluoride was removed.

There is a high degree of ionization in the pure liquid bromine trifluoride. Woolfe[108]

107 A. A. Woolfe and H. J. Emeléus, *J. Chem. Soc.* (*London*) (1949) 2865.

108 A. A. Woolfe, in *Advances in Inorganic Chemistry and Radiochemistry*, eds. H. J. Emeléus and A. G. Sharpe, Vol. 9, Academic Press (1966) pp. 267–275.

estimates an ionic product $[BrF_2^+][BrF_4^-]$ of about 4×10^{-4}. For solutions 0.05 M in acidic or basic species, the conductance of the solution is only two to three times that of the pure solvent. As a result it is not possible to calculate quantities such as molar conductances and to use the conductance figures rigorously. For this and other reasons, the self-ionization scheme was sometimes considered to be somewhat speculative with regard to both the nature of the ionic species in solution and, in particular, the ionic character of the residue obtained after evaporation of the solvent.

Some doubt was thrown on the existence of ions in the adduct $GeF_4.2BrF_3$, which is stable only below room temperature. When the cold solid adduct was sublimed on to a silver chloride plate and its infrared spectrum observed, there was no strong band at 600 cm^{-1} which has been shown to be characteristic of the hexafluorogermanate anion,

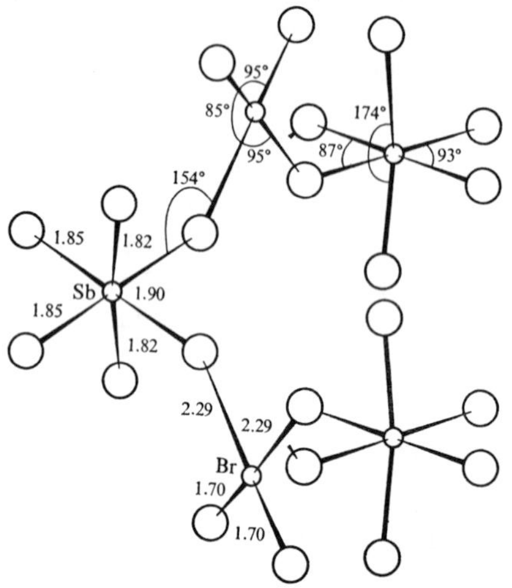

FIG. 10. Structure of the adduct $BrF_3.SbF_5$.

but the spectrum was different from those of either component of the adduct[109]. It was concluded that the compound was not ionic in the solid, but possibly a fluorine-bridged structure. Nevertheless, the possibility of its being ionized on dissolution in bromine trifluoride was not ruled out.

However, fairly recent structural investigations provide convincing support for Emeléus's postulations. There had been some uncertainty from reported X-ray structure analyses as to whether the tetrafluorobromate(III) anion was tetrahedral or square planar. Edwards and Jones have established the latter structure in a neutron diffraction study[110]. The same workers have shown the structure of the BrF_3SbF_5 adduct to be that in Fig. 10.[111] There is slight distortion of the octahedral arrangement of six fluorines around each antimony atom. In the distorted square planar arrangement of four fluorines around each bromine the bromine to fluorine bridge bond (2.29 Å) is considerably longer than the terminal bonds (1.70 Å).

[109] D. H. Brown, K. R. Dixon and D. W. A. Sharp, *Chem. Communs.* (1966) 654.
[110] A. J. Edwards and G. R. Jones, *J. Chem. Soc.* (*London*) *A* (1969) 1936.
[111] A. J. Edwards and G. R. Jones, (a) *Chem. Communs.* (1967) 1304; (b) *J. Chem. Soc.* (*London*) *A* (1969) 1467.

Edwards and Jones say that the ionic formulation $BrF_2^+SbF_6^-$ is a reasonable approximation to the structure, but that there is some contribution from a covalent structure with square-planar BrF_4 units connected by bridging fluorine atoms to SbF_6 octahedra to form chains. They say that, on dissolution in bromine trifluoride, the long bridging bonds would be broken to give solvated BrF_2^+ and SbF_6^- ions.

As stated in the introduction to this section, the work at Cambridge in 1949 was done in the face of enormous experimental difficulties. Polymers such as Kel-F, from which apparatus can be constructed to make easier the handling of volatile reactive fluorides (see section 1.4), were not available. The conductance cells were constructed from silica and, as stated above, compounds from which bromine trifluoride solutions were prepared

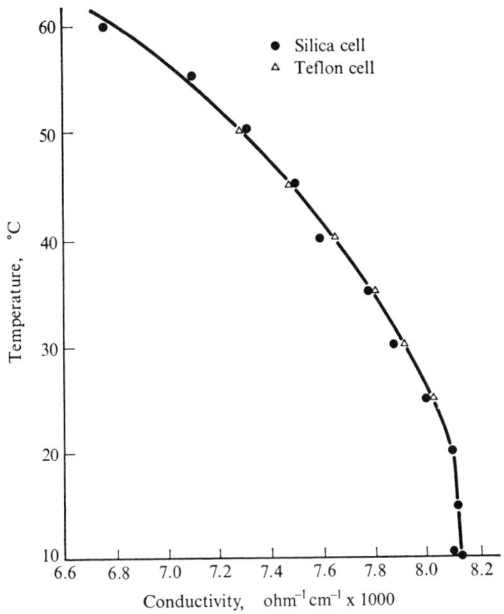

FIG. 11. Variation of conductivity of bromine trifluoride with temperature.

were frequently oxides, chlorides and other relatively unsuitable compounds. It seems likely that water and other impurities which might have been expected to give spurious conductance values were, in fact, eliminated completely from solution because of the extreme reactivity of the solvent. It appears also that relatively simple procedures were used in attempts to purify the solvent by distillation. In a detailed manometric study, Stein[112] has shown that fluorination of bromine does not give pure bromine trifluoride. The product will contain bromine, bromine monofluoride or bromine pentafluoride depending on the relative amounts of the two halogens taken. However Quarterman, Hyman and Katz[113] have demonstrated that the presence of relatively small amounts of bromine or of bromine pentafluoride does not markedly affect the conductance of the trifluoride.

It is significant that, using bromine trifluoride which had been very carefully fractionally

[112] L. Stein, *J. Am. Chem. Soc.* **81** (1959) 1269.
[113] L. A. Quarterman, H. H. Hyman and J. J. Katz, *J. Phys. Chem.* **61** (1957) 912.

distilled and making measurements in a Kel-F conductivity cell, Quarterman, Hyman and Katz obtained conductance data for the trifluoride which did not differ to any appreciable extent from those of Emeléus. They also observed a negative temperature coefficient for the conductance, as reported by Emeléus[106] in 1949. Woolfe[108] showed the agreement between the two sets of conductance data and also the decrease in conductance with increase in temperature in the diagram shown as Fig. 11.

The inability to isolate potassium tetrafluorochlorate(III) $KClF_4$ when potassium fluoride was treated with chlorine trifluoride was related by Emeléus to the very low conductivity of that trifluoride although he did not rule out the possibility of acidic or basic species of that system analogous to those of the bromine trifluoride system.

Although it received much less study than the bromine trifluoride system, iodine pentafluoride yielded evidence for self-ionization along the lines already set out. Woolfe[108] believes that the quoted specific conductance of 5×10^{-6} ohm^{-1} cm^{-1} is probably higher than it should be because of traces of moisture. However, the value supports a postulated equilibrium:

$$2IF_5 \rightleftharpoons IF_4^+ + IF_6^-$$

The addition of potassium fluoride increases the conductance, and Emeléus and Sharpe[114] identified KIF_6 on removal of the solvent. Woolfe[115] found that reaction of iodine pentafluoride with antimony compounds gave a residue of empirical formula $SbIF_{10}$ and that a solution of this compound when mixed with an equivalent amount of KIF_6 in iodine pentafluoride solution gave a residue of solvated potassium hexafluoroantimonate(V) on removal of the solvent. On this evidence, the antimony containing compound was formulated as $IF_4^+SbF_6^-$.

Physical and chemical properties of the halogen fluorides, with emphasis on the ionization equilibria and chemical reactions, have been reviewed by Emeléus[116] and by Sharpe[111]. Stein[118] has produced a very extensive review which is particularly valuable for the detailed information given on vibrational spectra and thermodynamic properties of the halogen fluorides. In a review on heterocations, Woolfe[108] gives a brief but authoritative account of ionization in bromine trifluoride solutions. He is more concerned with the solvent system itself than with preparative and other reactions involving bromine trifluoride.

3.2. PREPARATIVE REACTIONS IN HALOGEN FLUORIDES

Halogen fluorides, particularly chlorine and bromine trifluorides, have been used instead of elemental fluorine for the preparation of binary fluorides such as pentafluorides and hexafluorides of transition metals. However, there is an inherent difficulty in use of these reagents because they form relatively stable adducts with many fluorides. Such adducts have already been described in section 3.1 for antimony and tin compounds reacting with bromine trifluoride. However, in table IX of his review[118], Stein gives an extended list. Chlorine trifluoride forms less stable adducts than the bromine counterpart.

[114] H. J. Emeléus and A. G. Sharpe, *J. Chem. Soc.* (*London*) (1949) 2206.
[115] A. A. Woolfe, *J. Chem. Soc.* (*London*) (1950) 3678.
[116] H. J. Emeléus, in *Fluorine Chemistry*, ed. J. H. Simons, Vol. 2, Academic Press (1959) pp. 39.
[117] A. G. Sharpe, in *Nonaqueous Solvent Systems*, ed. T. C. Waddington, Academic Press (1965) p. 285.
[118] L. Stein, in *Halogen Chemistry*, ed. V. Gutmann, Vol. 1, Academic Press (1967) p. 133.

The real value of the halogen fluorides in preparative reactions is in the formation of ternary compounds through reactions of the type given in section 3.1. Muetterties and Tullock[119] describe a simple procedure for preparation of $AgSbF_6$ and $KTaF_6$ by slow addition of the stoichiometric amounts of silver and antimony or of potassium chloride and metallic tantalum to bromine trifluoride in a Teflon beaker, followed by removal of the liquid by evaporation by warming or by passage of a stream of nitrogen. Preparation of these two compounds and many others had been reported previously by Emeléus and his colleagues. Stein gives a list of about 100 complex fluorides prepared from bromine trifluoride in table X of his review[118], the great majority of these compounds being first isolated at Cambridge.

In some instances, use of bromine trifluoride appears to be the only route to preparation of some compounds. For example, if excess trifluoride is added to metallic gold, the solvent removed and the resulting adduct heated, a pure sample of gold trifluoride is obtained. It has been suggested that the adduct is $BrF_2^+AuF_4^-$ because when a mixture of gold and silver in equivalent amounts is similarly treated, the residue is $AgAuF_4$ which is envisaged as resulting from a neutralization reaction involving the entities $AgBrF_4$ and $BrF_2^+AuF_4^-$.

Emeléus suggests the formation of unstable intermediate acids and bases in reactions leading to the preparation of nitrosonium and nitronium compounds[116]. When equivalent amounts of an antimony compound and either nitrosyl chloride or nitrogen dioxide are treated with bromine trifluoride and excess solvent is removed, the residues have the stoichiometries $NO^+SbF_6^-$ and $NO_2^+SbF_6^-$. It is proposed that the well-characterized compound $BrF_2^+SbF_6^-$ reacts in solution with the unstable bases $NO^+BrF_4^-$ and $NO_2^+BrF_4^-$ which, however, cannot be isolated. Similarly, the isolation of $KAsF_6$, after addition of potassium fluoride and arsenic(III) oxide to bromide trifluoride, is said to result from the intermediate formation of the unstable acid $BrF_2^+AsF_6^-$.

A possible complication in the attempted preparation of a particular binary or complex fluoride using bromide trifluoride is that solvolysis reactions can lead to contamination of the product. Emeléus[116] reports that when a 2 : 1 mixture of potassium bromide and titanium dioxide was treated with bromine trifluoride in a study of the formal neutralization reaction

$$2KBrF_4 + (BrF_2)_2TiF_6 \rightleftharpoons K_2TiF_6 + 4BrF_3$$

the solid product isolated after removal of the solvent was not the pure hexafluorotitanate but a product having the composition $K_2TiF_6 \cdot 0.95BrF_3$ and X-ray analysis showed the presence of $KBrF_4$. Further, when pure K_2TiF_6 was prepared by a reaction using hydrogen fluoride and was then treated with bromine trifluoride, the residue was heavily contaminated with $KBrF_4$ indicating the reversibility of the above reaction. Emeléus cites other examples of solvolysis reactions.

Iodine pentafluoride is the most thermodynamically stable of the halogen fluorides and so is the weakest oxidant. It has been used frequently as a solvent for conductance studies and for fluoride-transfer reactions in studies of Lewis acidity and basicity. Muetterties and Tullock[119] report the specific use of this mild oxidant to react at 200°C with sulphur to give a nearly quantitative yield of very pure sulphur tetrafluoride, which in turn is an extremely important synthetic reagent because it replaces oxygen or sulphur by fluorine in inorganic and organic compounds, as the work of Muetterties's own group has shown.

[119] E. L. Muetterties and C. W. Tullock, in *Preparative Inorganic Reactions*, ed. W. L. Jolly, Vol. 2, Interscience (1965) p. 237.

3.3 FLUOROANIONS OF HALOGENS

The tetrafluorobromate(III) anion BrF_4^- and hexafluoroiodate(V) IF_6^- can be obtained directly from ionization reactions in the solvents BrF_3 and IF_5 and will not be discussed in this section except to point out that there has been considerable study of infrared and Raman spectra of the IF_6^- anion, which is isoelectronic with xenon hexafluoride, to decide the extent to which the structure is different from a regular octahedron.

No compound containing the tetrafluorochlorate(III) anion ClF_4^- has been isolated from chlorine trifluoride solutions. Potassium fluoride enhances the conductance but is recovered unchanged on evaporation of the solvent. Alkali metal compounds $MClF_4$ were obtained for M = K, Rb and Cs but not for M = Li and Na by fluorination of the appropriate metal chlorides[120, 121]; but, even though the reactions were not carried out in the liquid phase, the first preparations which could be regarded as solvolysis reactions were those in which the alkali metal fluorides were reacted with chlorine trifluoride vapour and with bromine pentafluoride at 100°C[112]. Neither lithium nor sodium fluorides formed products, but $MClF_4$ and $MBrF_6$ were formed for M = K, Rb and Cs in about 50%, 90% and 100% yield respectively. This order of reactivity is in accordance with the general rules given by Wiebenga, Havinga and Boswijk that the thermal stability of polyhalide anions increases as the size of the associated cation increases[123]. Subsequently, Christe and Guertin prepared $NO^+ClF_4^-$ by direct combination of nitrosyl fluoride and chlorine trifluoride and also prepared the potassium, rubidium and caesium tetrafluorochlorates. They produced infrared evidence for the ionic nature of the compounds and for the square-planar structure of the anion in the rubidium and caesium compounds[124].

Christe and Guertin also prepared the ClF_2^- anion[125]. They found that nitrosyl fluoride combined directly with chlorine monofluoride at −78°C to give a 1 : 1 compound which was unstable at 25°C. Both the conductance in nitrosyl fluoride and the infrared spectrum of the solid when sublimed on to a silver chloride plate demonstrated the ionic formulation $NO^+ClF_2^-$. Later they combined chlorine monofluoride directly with potassium, rubidium and caesium fluorides to give solids $M^+ClF_2^-$, stable at 25°C, in which the anion was shown by X-ray diffraction techniques to be linear.

Although successful in making and characterizing compounds containing the anions ClF_4^- and ClF_2^-, Christe was unable to make the hexafluorochlorate(V) anion ClF_6^-. He showed that ClF_5 did not form adducts at −78°C with the strong Lewis bases CsF and NOF[126]. When mixed, ClF_5 and NOF exhibited almost ideal vapour pressures at −78°C. Christe comments that the fact that chlorine pentafluoride, with one free localized electron pair, forms adducts with Lewis acids (see section 3.4) but not with Lewis bases suggests that the highest possible coordination number, either for fluorine atoms or for electron pairs, is 6. He says this contention is supported by the fact that the spectra and physical properties indicate little association in the liquid pentafluoride, and he suggests that it seems unlikely that chlorine heptafluoride will be stable.

120 L. B. Asprey, J. L. Margrave and M. E. Silverthorn, *J. Am. Chem. Soc.* **83** (1961) 2955.
121 D. H. Kelly, B. Post and R. W. Mason, *J. Am. Chem. Soc.* **85** (1963) 307.
122 E. D. Whitney, R. O. MacLaren, C. E. Fogle and T. J. Hurley, *J. Am. Chem. Soc.* **86** (1964) 2583.
123 E. H. Wiebanga, E. E. Havinga and K. H. Boswijk, in *Advances in Inorganic Chemistry and Radiochemistry*, eds. H. J. Emeléus and A. G. Sharpe, Vol. 3, Academic Press (1961) pp. 146–148.
124 K. O. Christe and J. P. Guertin, *Inorg. Chem.* **5** (1966) 473.
125 K. O. Christe and J. P. Guertin, *Inorg. Chem.* **4** (1965) 1785.
126 K. O. Christe and D. Pilipovich, *Inorg. Chem.* **8** (1969) 391.

3.4. FLUOROCATIONS OF HALOGENS

Evidence has already been given in section 3.1 for the existence of the cations BrF_2^+ and IF_4^+ derived directly from the appropriate uncharged halogen fluorides; but there are reports of other halogen fluorocations obtained somewhat less directly from solvolysis reactions.

Of the other fluorocations, the difluorochloronium ion ClF_2^+ has received most study. It was one of the cations discussed by Seel and Detmer in a comprehensive review article entitled "On fluoronium compounds"[127]. They reported that after removal of excess chlorine trifluoride from antimony pentafluoride a 1 : 1 adduct remained, which in 0.564 M solution in chlorine trifluoride had a specific conductance of 6.7×10^{-3} ohm^{-1} cm^{-1}, compared with 3×10^{-9} ohm^{-1} cm^{-1} for the solvent. There was a similar 1 : 1 adduct with arsenic pentafluoride formulated as $ClF_2^+AsF_6^-$. Selig and Shamir[128] prepared a 1 : 1 adduct of chlorine trifluoride and boron trifluoride. The molten adduct had a specific conductance of 1.5×10^{-2} ohm^{-1} cm^{-1}, and the conductivity of a chlorine trifluoride solution 1 M in the compound was 2.7×10^{-4} ohm^{-1} cm^{-1}. Further evidence for the ionic formulation $Cl_2F^+BF_4^-$ was an infrared band characteristic of tetrafluoroborate and not present in pure chlorine trifluoride or boron trifluoride.

Christe and Pavlath[129] combined directly equimolar amounts of chlorine trifluoride with boron trifluoride and with phosphorus, arsenic and antimony pentafluorides. They used vapour-pressure measurements to show that the thermal stability of the compounds was in the order $ClF_2^+PF_6^- < ClF_2^+BF_4^- < ClF_2^+AsF_6^- < ClF_2^+SbF_6^-$, indicating an order of strength for the Lewis acids concerned. To show the ionic nature of the compounds, they added to conductance measurements in chlorine trifluorine a conductometric titration of $ClF_2^+AsF_6^-$ against $K^+IF_6^-$ in IF_5. Also the magnitude of the depression of the freezing point of solutions in IF_5 was that for a 1 : 1 electrolyte in solution. Infrared spectral studies showed the ClF_2^+ cation to be bent.

Chlorine pentafluoride was found to form compounds containing the ClF_4^+ cation with the strong Lewis acids SbF_5 and AsF_5 but not with weaker acids such as BF_3. Thus, in the ClF_5–BF_3 system, the total vapour pressure of 570 mm at -95.1°C was only slightly lower than that expected for the two pure compounds in ideal solution, indicating no interaction.

Seel and Detmer[127] reported isolation of the adduct $IF_7.3SbF_5$ which in the melt at 100°C had a specific conductance of 6.65×10^{-3} ohm^{-1} cm^{-1} which they compared with that of 2.8×10^{-3} for molten zinc chloride at 340°C. They tentatively formulated the adduct as $IF_4^{3+}(SbF_6^-)_3$; but, with the existence of the $Sb_3F_{16}^-$ anion well established [see section 2.4(d)(iii)], there seems little doubt that the ionic formulation should be $IF_6^+Sb_3F_{16}^-$. Subsequently, Christe and Sawodny[130] showed that both ions were octahedral in the compound $IF_6^+AsF_6^-$.

Perhaps the least expected fluorocation reported to date is Cl_2F^+ which was formed[131] when 2 : 1 mixtures of chlorine monofluoride and the Lewis acids arsenic pentafluoride or boron trifluoride were allowed to react at -78°C. These compounds are unstable at 25°C. The ionic formulation $Cl_2F^+AsF_6^-$ is supported by infrared spectral evidence;

127 F. Seel and O. Detmer, *Z. anorg u. allgem. Chem.* **301** (1959) 113.
128 H. Selig and J. Shamir, *Inorg. Chem.* **3** (1964) 294.
129 K. O. Christe and A. E. Pavlath, *Z. anorg. u allgem. Chem.* **335** (1965) 210.
130 K. O. Christe and W. Sawodny, *Inorg. Chem.* **6** (1967) 1783.
131 K. O. Christe and W. Sawodny, *Inorg. Chem.* **8** (1969) 212.

but it has been shown recently[132] from a Raman spectral investigation that the bent cation is not the symmetrical Cl–F–Cl$^+$ structure as at first proposed, but the asymmetric Cl–Cl–F$^+$.

4. FLUORIDES OF MAIN GROUP ELEMENTS

As stated earlier, the detailed physical and chemical properties of an individual fluoride are to be found in this treatise in the section devoted to the atom or atoms associated with fluorine in the particular compound.

Furthermore, as recently as 1965, Kemmitt and Sharp[133] have produced a most extensive review giving, for each fluoride, an exhaustive list of the physical properties, the method of preparation, and the stability of the compound and an outline of its chemistry which, while it is very brief, is very useful because it is extensively referenced. Where the informaton is available, their review includes physical properties such as melting, boiling and transition points, critical temperatures and pressures, densities, vapour pressure equations, heats of transition, standard heats and free energies of formation, entropies, structural parameters, infrared, Raman and n.m.r. spectroscopic data and mass spectral observations. Although their emphasis is on binary compounds, Kemmitt and Sharp include in their review mixed halides, oxide fluorides, phosphonitrilic fluorides and sulphur–fluorine compounds which also contain nitrogen or oxygen, as well as hydrides of carbon, nitrogen, phosphorus and sulphur which contain fluorine.

There would be little point in attempting to reproduce a review like that of Kemmitt and Sharp. In this section a general survey will be given of the way in which fluorine brings out special properties or trends in main group fluorides. There will be no particular emphasis on the properties of individual fluorides or even of those of a particular group. Although they will be presented differently, most of the data used are available from Kemmit and Sharp's review, but reference will be made to other papers and specialized reviews. It should be recalled that Simon's "Fluorine chemistry"[134] is a comprehensive source of physical and chemical properties of main group and other fluorides.

4.1. PHYSICAL PROPERTIES AND STRUCTURES OF FLUORIDES

Frequently over-simplified generalizations are made about physical properties and bond types in halides. For example, the boiling points of 704°, 114°, 203° and 346°C for

TABLE 6. MELTING AND BOILING POINTS OF SECOND-ROW CHLORIDES (°C)

	NaCl	$MgCl_2$	$AlCl_3$	$SiCl_4$	PCl_3	SCl_2	Cl_2
M.P.	801	708	190 (2 atm)	−70	−112	−78	−102
B.P.	1413	1412	178 (subl)	57.6	75.5	59 (decomp)	−34

the four tetrahalides of tin from the fluoride to the iodide are presented to show a transition from marked ionic bonding in the tetrafluoride to essentially covalent bonding in the three

132 R. J. Gillespie and M. Morton, *Inorg. Chem.* **9** (1970) 811.

133 R. D. W. Kemmitt and D. W. A. Sharp, in *Advances in Fluorine Chemistry*, eds. M. Stacey, J. C. Tatlow and A. G. Sharpe, Vol. 4, Butterworths (1965), pp. 142–252.

134 *Fluorine Chemistry*, ed. J. H. Simons, Academic Press. For main group fluorides, *see* Vols. I and V.

TABLE 7. MELTING AND BOILING POINTS OF MAIN GROUP FLUORIDES (°C)[a]

LiF	848 1681	BeF_2	542 1096	BF_3	−127·1 −101·0	CF_4	−183·5 −128	NF_3	−206·8 −129·1	OF_2	−223·8 −145·3	F_2	−219·6 −187·99		—
NaF	992 1704	MgF_2	1255 2227	AlF_3	1257 (subl)	SiF_4	−90 −95·1 (subl)	PF_3	−151·5 −101·2	SF_4	−121·0 −40·4	ClF_3	−76·34 11·75		—
								PF_5	−93·8 −84·6	SF_6	−50·8 −63·7 (subl)	ClF_5	< 100 (m.p.)		
KF	858 1502	CaF_2	1418 2513	GaF_3	–	GeF_2	110 160 (decomp)	AsF_3	−5·97 57·13 (742·5 mm)	SeF_4	−9·5 106	BrF_3	8·77 125·75	KrF_2	–
						GeF_4	−15 (3032 mm) −35	AsF_5	−79·8 −53·2	SeF_6	−34·6 −46·6	BrF_5	−60·5 40·76		
RbF	775 1408	SrF_2	1400 2477	InF_3	1170 (m.p.)	SnF_2	213 850	SbF_3	292 319	TeF_4	129·6 (m.p.)	IF_5	9·45 100·5	XeF_2	$\sim$140 (m.p.)
														XeF_4	$\sim$114 (m.p.)
						SnF_4	705 (b.p.)	SbF_5	8·3 141	TeF_6	−37·8 −38·9 (subl)	IF_7	5–6 4·5 (subl)	XeF_6	46 (m.p.)
CsF	703 1253	BaF_2	1320 2277	TlF	327 655	PbF_2	822 1290	BiF_3	727 (m.p.)		—		—		—
				TlF_3	550 (m.p.)	PbF_4	$\sim$600 (m.p.)	BiF_5	151.4 230						

[a] Values from review by Kemmitt and Sharp[133]. Where two values are listed without qualification for a particular fluoride, they are the melting point and boiling point as normally defined. For a single value, its nature is stated. If the value is not a normal melting point or boiling point, the conditions are given, e.g. the pressure at which equilibrium was maintained, whether sublimation (subl) or decomposition (decomp) occurred.

molecular halides which then show the expected increase in boiling point with increase in atomic number of the halogen involved. Similarly, if melting and boiling points for the chlorides of the elements of the second row of the Periodic Classification (Table 6) are considered, those for the ionic compounds sodium and magnesium chloride are relatively high, although, if these are compared with the corresponding fluorides, it is seen that the expected increase in passing from a lattice with a unipositive cation to one containing a dispositive ion is offset by the decrease in the ionic character of the metal–chlorine interaction. Then there is a reasonably smooth transition through the essentially covalent layer lattice structure of aluminium chloride to the volatile molecular chlorides of Group IV and beyond.

However, if the physical properties of the fluorides from sodium to chlorine are examined (Table 7), a marked discontinuity is observed. The fluorides of Groups IV–VII are obviously molecular. However, it is an over-simplification to regard the discontinuity between aluminium and silicon as resulting only from a change in bond type—the difference in electro-negativities is not great enough to account for this. Certainly bonding can be regarded as essentially ionic in aluminium fluoride and covalent in silicon tetrafluoride; but the difference between the halides of aluminium and silicon is greater for the fluorides than for the chlorides because aluminium trichloride is a layer lattice structure which is easily disrupted thermally to yield Al_2Cl_6 dimers, whereas aluminium trifluoride is a three-dimensional arrangement in which the fluorine atoms are approximately close-packed with aluminium atoms in certain of the octahedral holes and much more thermal energy is required to disrupt this structure.

Similar discontinuities, resulting from structural effects superimposed on changes in bond type, are observed in the fluorides within the individual Groups III–V, while not occurring to the same extent for the chlorides. Thus solid boron trifluoride, composed of 3-coordinate molecular units, is very much more volatile than the 6-coordinated solid, aluminium trifluoride. In Group IV, the three tetrafluorides of carbon, silicon and germanium are molecular, low melting and boiling, with the expected decrease in volatility with increase in atomic number of the central element. However, the change from 4-coordination in the germanium fluoride to 6-coordination in stannic fluoride provides another drastic change in melting and boiling points. By comparison the chlorides, bromides and iodides of all the elements of Group IV show a much more gradual increase in physical properties. The trichlorides of nitrogen, phosphorus, arsenic, and antimony boil at +71°, 76°, 130° and 221°C, whereas the decrease in volatility at arsenic trifluoride, and particularly at antimony trifluoride, is very marked. For Groups I and II, where structures of halides are similar, high melting and boiling points do reflect a high degree of ionic binding in the solids. However, for the fluorides of Groups III–V, higher melting and boiling points can result from increase in coordination number as well as increase in ionic bonding.

Conversely, the strong covalent bonds that are formed between fluorine and the less electropositive elements, i.e. those elements which normally form molecular compounds, lead to the more ready formation of molecular entities for the fluorides than for the other halides. For example, phosphorus pentafluoride is molecular under all conditions, whereas in the solid the other two phosphorus pentahalides have the structures $PCl_4^+PCl_6^-$ and $PBr_4^+Br^-$. The halides and mixed halides of Group V provide several such examples. In considering the molecularity of non-metal fluorides, it is of particular interest to note the similarity of the melting points and boiling points for sulphur, selenium and tellurium hexafluorides and the shortness of the liquid range in each case. Here the central atom obviously has little effect on the Van der Waals' interactions between molecules. That this a is real and general effect is shown more clearly by the remarkable similarity in physical properties of the many hexafluorides of a widely differing range of transition metals (see p. 1073).

4.2. IONIC FLUORIDES OF GROUPS I AND II

Effectively the chemistry of the halides of Groups I and II is simply the formation and the disruption through reaction or dissolution of ionic solids. The high melting and boiling points of the fluorides of these two groups, as listed in Table 7, are indicative of the ionic nature of these compounds. However, it is more interesting to make comparisons between the fluorides and other halides by considering lattice energies of the solid compounds. In an extensive review[135], Waddington discusses the methods of calculation of lattice energies and the reliability of the available values for a wide range of ionic

TABLE 8. LATTICE ENERGIES FOR ALKALI METAL HALIDES (KCAL MOLE^{-1})

	F	Cl	Br	I
Li	244	200	190	176
Na	215	184	176	164
K	193	168	161	152
Rb	183	162	156	148
Cs	176	153	150	143

compounds. His recommended values for the halides of Group I are listed in Table 8. A direct comparison of the values for nearly all of the Group I halides is valid because, with the exception of the chloride, bromide and iodide of caesium, they all have the sodium chloride structure. The values in the first column show the decrease which is expected on simple electrostatic grounds in moving from the small cation in lithium fluoride to the large caesium ion. Similarly, the expected decrease occurs with increasing halide ion size, e.g. from the fluoride to the iodide for potassium.

The data avilable for Group II halides are far less complete than for the Group I compounds. However, as expected, the values are much greater for the solids with doubly charged cations. The lattice energies for the four fluorides of magnesium, calcium, strontium and barium are 684, 622, 592, and 561 kcal mole^{-1}. While there may be some reservations about including magnesium fluoride because it has the rutile structure, the other three, with the fluorite lattice, are directly comparable and show the anticipated trend. Similarly strontium chloride, a fluorite structure, has a lattice energy of 496 kcal mole^{-1}, about 100 kcal less than its fluoride counterpart.

4.3. THE "INERT PAIR EFFECT" AND THE FLUORIDES OF GROUPS III–V

The concept of the "inert pair effect" was originally put forward to account for the fact that for some of the main groups, the heavier elements usually exhibited in their compounds oxidation states less by two than that of the formal oxidation state for the group. Thus

[135] T. C. Waddington, in *Advances in Inorganic Chemistry and Radiochemistry*, eds. H. J. Emeléus and A. G. Sharpe, Vol. 1, Academic Press (1959), pp. 157–221.

the stable halides, other than the fluorides, of thallium, lead and bismuth have the general formulae TlX, PbX_2 and BiX_3. It was stated that, with increase in atomic number for the elements of a group, there was progressively less likelihood of involvement in the formation of bonds of the two *s* members of the bonding electrons.

This concept is obviously an over-simplification and an attempt to refine it has been given by Dasent[136]. It can be shown that for the halogens of a particular group, e.g. Group III, there is a monotonic decrease in the strength of the M–Cl bond and, on this basis alone, it can be suggested that the heavier elements do not form sufficiently strong covalent bonds to supply the energy to promote the *s* electrons into bonding orbitals. However, in the earlier and simpler treatments, one fact usually ignored is that with increase in atomic number, the elements of any group become progressively more electropositive and the extent of ionic bonding in the heavier compounds was overlooked. Obviously there is little likelihood of forming a range of halides containing Bi^{5+} or Pb^{4+}. It appears significant that in Groups VI and VII, where the high oxidation state compounds are definitely covalent, the reverse of the "inert pair effect" occurs (see section 4.5), and it may well be that the overriding factor in the whole effect is the ease or difficulty of forming compounds which contain highly charged cations. Unfortunately, sufficient thermodynamic data do not exist to allow a comparison of the possibility of forming for a particular heavy element two ionic halides in each of which the cationic charge differs by 2. Lattice energies would be very difficult to calculate because of the unknown extent of the covalent nature in the bonds of the compounds concerned. Another factor rarely taken into account formally in discussing this effect is the ease of oxidation of the associated halogens. The following progression is particularly significant. There has been no preparation of TlI_3, of $PbBr_4$ or PbI_4 or of any of the halides of Bi(V) other than the fluoride.

The halides of Group III provide a straightforward example of the "inert pair effect". BF_3 is the only stable simple fluoride, B_2F_4 being the only other stable binary compound of boron and fluorine. Extreme conditions are required to prepare aluminium monofluoride, the metal and the trifluoride being heated in vacuum above 700°C, and the monofluoride disproportionates on cooling. It becomes progressively easier to form the monofluoride in passing down the group, with thallous fluoride prepared simply by treating thallous carbonate with hydrofluoric acid. However, thallium trifluoride is also prepared relatively easily by fluorination of the monofluoride, of the metal or any thallium compound. Thallium trichloride is very unstable, beginning to decompose to chlorine and the monochloride at about 40°C and the tribromide is even less stable.

For the halides of carbon, silicon and germanium, excepting the fluorides, there is a unique oxidation state of four. For tin, all four dihalides and all four tetrahalides are stable compounds, easily prepared. As stated previously, no tetrabromide or tetraiodide of lead has been prepared and even the tetrachloride decomposes to chlorine and the dichloride at relatively low temperatures, explosively above 100°C. However, the tetrafluoride is easily prepared by fluorination of the difluoride at 300°C and is thermally stable, its melting point being given as about 600°C.

The difluorides of lead and tin are easily prepared, from aqueous solution if desired. By comparison, the experimental conditions under which the difluorides of germanium, silicon and carbon are formed provide an interesting example of the "inert pair effect" in operation. GeF_2 can be produced easily in gram quantities by the reduction of GeF_4 with the element itself in the range 150–300°C. It is a stable compound, although a

[136] W. E. Dasent, *Nonexistent Compounds*, Edward Arnold, London (1965), pp. 85–96.

vigorous reducing agent, with a measured melting point of 110°C. When SiF_4 is passed over silicon at about 1150°C, the gaseous product approximates to a 1 : 1 mixture of SiF_4 and SiF_2. At 1400° up to 90% of SiF_2 is formed. This difluoride has a half-life at room temperature of about 150 sec. The carbene CF_2 is very unstable and has been prepared much less directly than GeF_2 and SiF_2 by pyrolysis of $(CF_3)_3PF_2$ and of $(CH_3)_3SnCF_3$.

The Group V halides provide a varied and interesting chemistry and have been the subject of many reviews, such as those by Payne[137] and Kolditz[138]. Although there are no halides that are binary and uncharged which contain nitrogen in an oxidation state above III, trifluoramine oxide[139] F_3NO and a compound $NF_4^+AsF_6^-$ containing the tetrafluoronitrogen(V) cation[140] have been reported. Phosphorus and antimony show a wide range of binary halides and mixed halides, both molecular and ionic, in the two oxidation states III and V[133, 137–8] the pentachlorides being formed easily for both elements.

However, arsenic—between phosphorus and antimony—provides an interesting discontinuity. The only simple pentahalide is the fluoride. All the evidence, including a detailed phase study, suggests that there is no stable pentachloride. The cation $AsCl_4^+$ occurs in compounds with the anions PCl_6^- and $SbCl_6^-$. When AsF_3 is chlorinated in the presence of a small amount of water as catalyst, a compound is formed with the empirical formula $AsCl_2F_3$. On the basis of its conductance in liquid AsF_3 and its metathetical reactions[136], the compound is formulated as $AsCl_4^+AsF_6^-$.

The "inert pair effect" situtation is observed with the bismuth halides. Although all four trihalides are formed, the only pentahalide is the fluoride which is obtained fairly easily. Fluorination of metallic bismuth at 2 atm pressure of fluorine and about 550°C yields an essentially ionic pentafluoride which has the α-UF_5 structure and has been observed to melt at 151.4°C.

It should be emphasized that, in this section, stabilities of oxidation states for the elements of Groups III–V have been discussed for halides only. It should be recalled that nitrogen(V), arsenic(V), bismuth(V) and lead(IV), for example, all exist in stable, easily formed oxyanions.

4.4 CATENATION IN GROUP VI FLUORIDES

Catenation in chemical compounds, i.e. extensive bonding between identical or very similar atoms, is found most widely, of course, in Group IV, occurring to an indefinite extent in carbon compounds and becoming progressively less in those of silicon and germanium. Alkylsilanes are catenated compounds with chains of chemically similar atoms.

On a much smaller scale, catenation occurs in Group VI, although it is much less significant in oxygen compounds, e.g. ozone, than with sulphur, where the element itself occurs as S_8 rings; and cations derived from this and related structures have been reported. In addition, there are many compounds with chains of oxygen and sulphur atoms, e.g.

137 D. S. Payne, *Quart. Revs.* (*London*) **15** (1961) 173–189.

138 L. Kolditz, (a) in *Advances in Inorganic Chemistry and Radiochemistry*, eds. H. J. Emeléus and A. G. Sharpe, Vol. 7, Academic Press (1965), pp. 1–26; (b) in *Halogen Chemistry*, ed. V. Gutmann, Vol. 2, Academic Press (1967), pp. 115–168.

139 W. B. Fox, J. S. MacKenzie, E. R. McCarthy, J. R. Holmes, R. F. Stahl and R. Juurik, *Inorg. Chem.* **7** (1968) 2064.

140 J. P. Guertin, K. O. Christe and A. E. Pavlath, *Inorg. Chem.* **5** (1966) 1921.

thiosulphate $[S–S(O)_2–O]^{2-}$, pyrosulphuric acid $HO–S(O)_2–O–S(O)_2–OH$, peroxodisulphuric acid $HO–S(O)_2–O–O–S(O)_2–OH$, dithionic and polythionic acids $HO–S(O)_2–S(O)_2–OH$ and $HO–S(O)_2–S_n–S(O)_2–OH$.

The two oxygen fluorides OF_2 and O_2F_2 have been known for a considerable time. Catenation in oxygen–fluorine compounds appeared to be fairly significant when the compounds O_3F_2 and O_4F_2 were reported over the last few years, and the possible existence of both O_5F_2 and O_6F_2 was proposed. These compounds were all quoted as being formed when a low-pressure mixture of oxygen and fluorine was subjected to an electrical discharge. Turner discusses the evidence for the existence of the compounds[141] and considers that there is considerable doubt that individual compounds more complex than O_2F_2 have separate existence under normal conditions. He proposes that the reported higher compounds may be, in fact, a mixture of O_2F_2 and radicals such as $\cdot O_2F$, the existence of which has been demonstrated using the matrix isolation technique. Further, the radical could be dimerized in the materials isolated. Very recent work[142] indicates that radiolysis of liquid oxygen and fluorine at 77°K yields O_2F_2 and $(O_2F)_n$ with the possibility of very small amounts of higher oxides at this low temperature. However, the reaction product was converted at 195°K to pure O_2F_2, shown to be a yellow diamagnetic solid melting sharply at 119°K. In this paper, it was stated that O_3F_2 was believed to contain the equilibrium mixture

$$n(\cdot O_2F) \rightleftharpoons (O_2F)n$$

in the established compound O_2F_2. Turner[141] surveys the physical evidence for structure and bonding in oxygen fluorides and concludes, on the available evidence, that the O–O bonds in O_2, O_2F and O_2F_2 are very similar and much stronger than in H_2O_2. He also compares the oxygen–hydrogen and oxygen–fluorine radicals.

Sulphur–sulphur links occur in the sulphur difluorides F–S–S–F and $S–SF_2$ and in disulphur decafluoride $F_5S–SF_5$. However, catenation is much more extensive in the many sulphur–oxygen–fluorine compounds listed by Kemmitt and Sharp[133] and reviewed by Cady[143] and by Williamson[144].

The polysulphuryl fluorides from disulphuryl fluoride $S_2O_5F_2$ through $S_3O_8F_2$ and $S_4O_{11}F_2$ to $S_7O_{20}F_2$[133, 144] are all believed to have structures based on that of disulphuryl fluoride $F(O)_2S–O–S(O)_2F$. The introduction of peroxo units provides still further possibilities for catenation in compounds [133, 143–4] such as $F_5S–O–O–SF_5$, $FS(O)_2O–OS(O)_2F$, $F_5S–O–O–S(O)_2F$, $F_5SOSF_4OSF_5$ and $F_5SOSF_4OOSF_4OSF_5$. Fluorine appears to be the only halogen involved in extensively catenated sulphur–oxygen compounds.

4.5. HIGHER OXIDATION STATES IN GROUP VI AND VII FLUORIDES

In section 4.3 it was seen that the more stable halides of the heavier elements of Groups III–V were those in which the oxidation state of the metal concerned was less than the formal one for the appropriate group by two—the so-called "inert pair effect".

[141] J. J. Turner, *Endeavour*, **27** (1968) 42.

[142] C. T. Goetschel, V. A. Campanile, C. D. Wagner and J. N. Wilson, *J. Am. Chem. Soc.* **91** (1969) 4702.

[143] C. H. Cady, in *Advances in Inorganic Chemistry and Radiochemistry*, eds. H. J. Emeléus and A. G. Sharpe, Vol. 2, Academic Press (1960), pp. 105–157.

[144] S. M. Williamson, in *Progress in Inorganic Chemistry*, ed. F. A. Cotton, Vol. 7, Interscience (1966), pp. 39–81.

Proceeding across the Periodic Classification, a transition to the reverse situation is occurring in Group VI fluorides and is complete for those of Group VII. The hexafluorides of sulphur, selenium and tellurium are all prepared easily by direct fluorination of the element concerned. There does not appear to be a marked decrease in thermal stability of the hexafluorides with increasing molecular weight, although it is probably significant that tellurium hexafluoride can be reduced to the tetrafluoride with elemental tellurium. On the other hand, fluorides occur in lower oxidation states for sulphur than for selenium and tellurium. Also, whereas the highest stable chloride of sulphur is the dichloride, stable tetrachlorides can be prepared for both selenium and tellurium, but only tellurium has a stable tetrabromide. For Group VI halides, then, there is a *slight decrease* in stability of the hexafluoride with increase in molecular weight; but there is a *marked increase* in stability of higher chlorides in passing from sulphur to tellurium.

In the Group VII fluorides, the trend to increasing stability of higher oxidation state with increasing atomic number—the reverse of the "inert pair effect"—is well established. The highest fluorides of chlorine and bromine are the pentafluorides, whereas simple fluorination of iodine pentafluoride yields the heptafluoride. Also stability of the monofluorides decreases from the stable chlorine monofluoride through bromine monofluoride which disproportionates to an equilibrium mixture of bromine and bromine trifluoride, to iodine monofluoride which is observed spectroscopically in excited gas mixtures, although the preparation of a very unstable solid which disproportionates to iodine and iodine pentafluoride has been reported. Not even the trifluoride is stable for iodine.

Fluorine, of course, stabilizes higher oxidation states than the other halogens. The highest chlorides of bromine and iodine are the monochloride and the trichloride respectively. It should be noted that in the perhalate anions oxidation state VII occurs for chlorine, iodine and even bromine.

It is interesting, if perhaps incidental, that one of the two recent methods of preparation of the long-elusive perbromate ion involved oxidation of bromate in alkaline solution with elemental fluorine.

4.6. SPECIFICITY OF FLUORINATION IN RARE GAS REACTIVITY

The individual compounds of the rare gases, their methods of preparation and reactions and their structures with proposed bonding schemes have been reviewed several times. Bartlett, who first observed that xenon would enter into chemical reaction with platinum hexafluoride discusses them in Chapter 5 of this treatise. The proceedings of the first Conference on Noble Gas Compounds[145] is a valuable source of information, and reviews and books have been written by Holloway[146], Claassen[147] and Selig[148]. However, in this present review of the particular role of fluorine in forming compounds with main group elements, the key fact is that to date only fluorine has been shown to react *directly* with rare gases.

Photolysis of an equimolar mixture of xenon and fluorine gives pure xenon difluoride. The thermal reaction between the two gaseous elements gives the difluoride, the tetrafluoride or the hexafluoride, depending on the conditions. Higher temperatures and higher

[145] *Noble Gas Compounds*, ed. H. H. Hyman, University of Chicago Press (1963).

[146] J. Holloway, (a) in *Progress in Inorganic Chemistry*, ed. F. A. Cotton, Vol. 7, Interscience (1964), pp. 241–269; (b) in *Noble Gas Chemistry*, Methuen (1968).

[147] H. H. Claassen, *The Noble Gases*, D. C. Heath & Co. (1966).

[148] H. Selig, in *Halogen Chemistry*, ed. V. Gutmann, Vol. 1, Academic Press (1967), pp. 403–429.

fluorine pressures favour the formation of the hexafluoride. An octafluoride of xenon has been reported, but its existence is very doubtful. Electrical discharge in a krypton–fluorine mixture was reported to yield both a difluoride and a tetrafluoride; but it is now believed that the difluoride is the only stable compound. Thermal reaction of radon and fluorine yielded an involatile compound, but its stoichiometry, or whether it was a single compound, was not established.

Many xenon compounds other than binary fluorides and their adducts have been reported; but none of these has been produced directly. All result from the initial production of a fluoride, followed by another reaction. Xenon tetrafluoride disproportionates in aqueous solution to gaseous xenon and an oxy-anion of xenon(VI). Dehydration of the latter species leads to the dangerously explosive xenon trioxide. If an alkaline xenon(VI) solution is oxidized by ozone and then dehydrated, xenon tetroxide can be isolated. Reaction of xenon hexafluoride with excess water gives the same reaction series, while controlled reaction of the hexafluoride with the stoichiometric amount of water vapour yields xenon oxide tetrafluoride $XeOF_4$.

Xenon difluoride has been found to enter into several metathetical reactions. Passage of a discharge through a mixture with silicon tetrachloride is reported to give xenon dichloride and simpler metathetical reactions yield $FXeSO_3F$, $Xe(SO_3F)_2$ and $FXeOTeF_5$.

The details of these and other compounds can be found elsewhere. The important fact here is that no element other than fluorine has been found to react *directly* with the rare gases. The historically important initial reaction of xenon with platinum hexafluoride was, in fact, reaction with fluorine which was formed from the thermal decomposition of the unstable hexafluoride.

4.7. FLUORO-COMPLEXES AND MIXED HALIDES

For complexes of most elements fluoroanions are much more stable than the anions with chlorine ligands which in turn are more stable than the bromo- and iodoanions. This order is reversed, however, for the platinum metals, copper(I), silver, gold, cadmium, mercury, thallium and lead.

Most main group elements, particularly in higher oxidation states, form complex anions containing six fluorine atoms, e.g. AlF_6^{3-}, SiF_6^{2-}, SbF_6^- and IF_6^-. Apart from an obvious case such as BF_4^-, where the tetrahedral arrangement is to be expected in the anion, some interesting exceptions are the 5-coordinate SiF_5^- [149], XeF_7^- [148], XeF_8^{2-} [148] and TeF_8^{2-} and the polymeric anions AlF_5^{2-} and AlF_4^-, which contain chains or layers of AlF_6 octahedra each sharing two or four corners, and $Sb_4F_{16}^{4-}$ [135], the structure of which is shown in Fig. 12 (p. 1072).

As shown in section 2, p. 1052, the pentafluorides of antimony, arsenic and phosphorus are Lewis acids decreasing in strength in that order. Strong Lewis acids form fluoroanions that are so stable that they can stabilize very unusual cations. They usually do this by accepting a fluoride ion and so form a fluorocation. Reactions of trifluoramine oxide, F_3NO, tetrafluorohydrazine, $F_2N \cdot NF_2$, and *cis*-difluorodiazine, $FN \cdot NF$, with arsenic pentafluoride yield the compounds $ONF_2^+ \cdot AsF_6^-$ [150], $N_2F_3^+AsF_6^-$ [151] and $N_2F^+AsF_6^-$ [152],

[149] H. C. Clark, K. R. Dixon and J. G. Nicholson, *Inorg. Chem.* **8** (1969) 450.

[150] W. B. Fox, C. A. Wamser, R. Eibeck, D. K. Huggins, J. S. McKenzie and R. Juurik, *Inorg. Chem.* **8** (1969) 1247.

[151] A. R. Young and D. Moy, *Inorg. Chem.* **6** (1967) 178.

[152] D. Moy and A. R. Young, *J. Am. Chem. Soc.* **87** (1965) 1889.

while the passage of a discharge through a 1 : 1 : 1 mixture of nitrogen trifluoride, fluorine and arsenic pentafluoride gives a white stable crystalline product $NF_4^+AsF_6^-$ [140], containing a cation of nitrogen(V). Earlier, Seel and Detmer[153] had shown that BF_3 and several pentafluorides can accept fluoride to form compounds containing unusual cations, e.g. $SF_3^+BF_4^-$ and $SeF_3^+SbF_6^-$. Bartlett and co-workers have reported fluorocations of xenon in several compounds such as $XeF^+AsF_6^-$ [154], $Xe_2F_3^+IrF_6^-$ [154] and $XeF_5^+PtF_6^-$ [155]. Bartlett was led to react xenon with platinum hexafluoride after estabishing that the hexafluoride could oxidize oxygen and then stabilize the dioxygenyl cation in the compound $O_2^+PtF_6^-$[156]. Similarly, sufficient stability is given by the trimeric fluoroanion of antimony(V) to have allowed a single crystal X-ray structural analysis of $Br_2^+Sb_3F_{16}^-$ (see Fig.12). Many fluorocations and fluoroanions of the halogens have been described in section 3.

Stabilization of the tetrachloroarsenic(V) cation by hexafluoroarsenate(V) occurs in the compound $AsCl_4^+AsF_6^-$. Kolditz[135] cites many similar entities in reviewing the halogen compounds of phosphorus, arsenic and antimony in oxidation state V. An interesting case is that of the compound of empirical formula PCl_2F_3 which is ionic $PCl_4^+PF_6^-$ if formed by the action of arsenic trifluoride on phosphorus pentachloride dissolved in arsenic trichloride. However, if formed by the chlorination of phosphorus trifluoride it is the molecular mixed halide PCl_2F_3 with a trigonal bipyramidal structure with fluorines in equatorial positions and the chlorines at the apices.

There are many molecular mixed halides containing fluorine listed in Kemmitt and Sharp's review. It seems to be a valid generalization that their formation is generally favoured in lower rather than higher oxidations states of the non-halogen element. In a mass spectrometric study of mixed halides of boron[158], all twenty possible ions BX_3^+, BX_2Y^+ and $BXYZ^+$, where X, Y and Z represent F, Cl, Br or I, were identified, although relatively few of them can be isolated as single pure substances. Similarly, there is a wide range of mixed halides of carbon and silicon, as well as the compounds NF_2Cl and $NFCl_2$ and the mixed halides of phosphorus(III). Lead chloride fluoride, PbClF, is of some importance in analytical chemistry (section 1. p. 1033), and there is a corresponding SnClF. Also the similarly formulated compounds SF_5Cl, SF_5Br, TeF_5Cl and TeF_5Br are known.

4.8. FLUORINE BRIDGING

At the present time it seems that fluorine bridging in solids and in species in solution occurs much less frequently than chlorine bridging. Furthermore, it appears to be much more widespread in transition metal fluorides than in main group compounds. Thus the structure of the fluorine-bridged cluster cation $Nb_6F_{12}^{3+}$ has been known for a long time, and more recent work has established that the clusters themselves are bridged through fluorines to form the three-dimensional solid. All of the transition metal pentafluorides are tetramers or chain structures based on an approximately octahedral arrangement of

[153] F. Seel and O. Detmer, *Z. anorg. u. allgem. Chem.* **301** (1959) 113.

[154] F. O. Sladky, P. A. Bulliner and N. Bartlett, *J. Chem. Soc.* (*London*) *A* (1969) 2179.

[155] N. Bartlett, F. Einstein, D. F. Stewart and J. Trotter, *J. Chem. Soc.* (*London*) *A* (1967) 1190.

[156] N. Bartlett and D. H. Lohmann, *J. Chem. Soc.* (*London*), (1962) 5253.

[157] A. J. Edwards, G. R. Jones and R. J. C. Sills, *Chem. Communs.* (1968) 1527.

[158] M. F. Lappert, M. R. Litzow, J. B. Pedley, P. N. K. Riley and A. Tweedale, *J. Chem. Soc.* (*London*) *A* (1968) 3105.

fluorines around each metal atom, with sharing of fluorines by adjacent octahedra. These structures are discussed in detail in section 5. p. 1103. There is also good evidence that polymerization through fluorine bridges occurs in fluoro-anions of transition metals in the formation of entities such as $Nb_2F_{11}^-$, $Ir_2F_{11}^-$, etc.

While there is little evidence at present of extended fluorine bridging in non-ionic solid fluorides of main group elements, complex fluoro-aluminates Tl_2AlF_5 and $TlAlF_4$

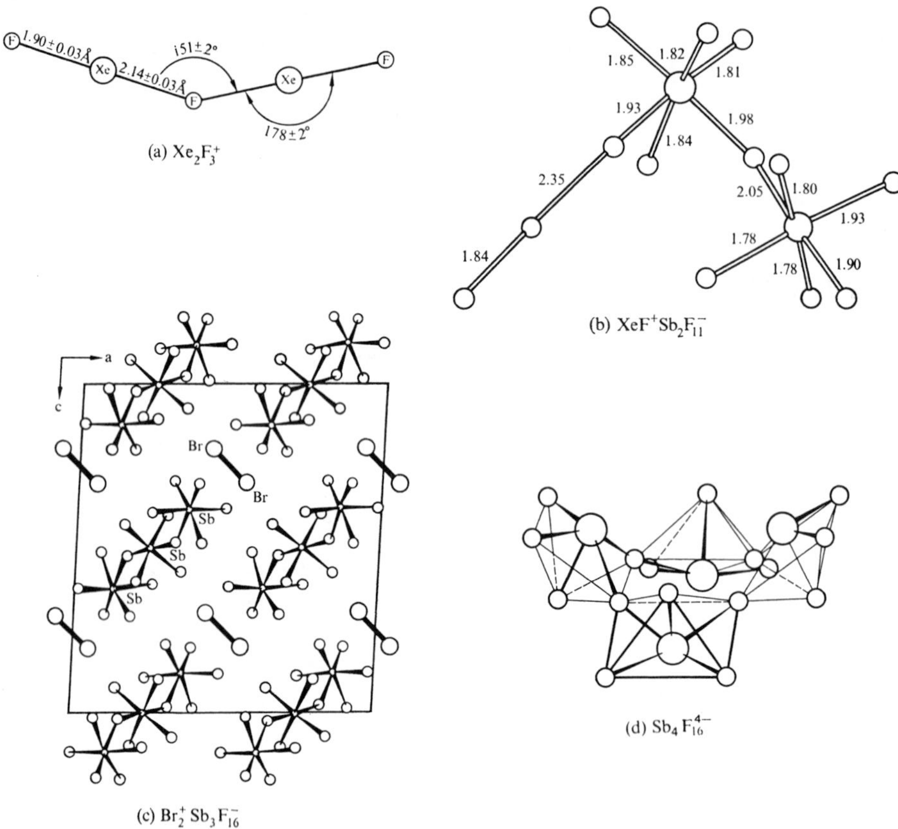

FIG. 12. Fluorine bridging in cations and anions. (a) $Xe_2F_3^+$. N. Bartlett *et al., Chem. Commun.* (1968) 1048. (b) $XeF^+Sb_2F_{11}^-$. R. D. Peacock *et al., Chem. Commun.* (1969) 62. (c) $Br_2^+Sb_3F_{16}^-$. From Ref. 157. (d) $Sb_4F_{16}^{4-}$. From Ref. 137.

are structures in which the anions are chains or layers, each unit of which is an octahedron of six fluorines around an aluminium atom, these units sharing two or four corners[133]. The high viscosity of liquid antimony pentafluoride has long been recognized as a consequence of polymerization. There is a distorted octahedral arrangement of six fluorines about each antimony, the units connected through bridging fluorines.

The cation $Xe_2F_3^+$ is, of course, fluorine bridged and its structure is shown in Fig. 12a. While those who determined the structure of the adduct $XeF_2.SbF_5$ (Fig. 12b) are not prepared to say that it conforms precisely to the ionic formulation $XeF^+Sb_2F_{11}^-$, there is no doubt that fluorine bridging occurs between the two antimony atoms; and the compound

$Br_2^+Sb_3F_{16}^-$ (Fig. 12c) shows more extensive fluorine linkage in the $Sb_3F_{16}^-$ units. The dibromine cations are shown located between these units. One of the fluoro-anions of antimony(III), $Sb_4F_{16}^{4-}$, is also shown (Fig. 12d) as an example of a bridged anion. There is a growing amount of evidence for the existence of other bridged anions such as $As_2F_{11}^-$ and $B_2F_7^-$.

5. CHEMICAL REACTIVITY OF HIGHER FLUORIDES OF *d*- and *f*-TRANSITION ELEMENTS

It appears strange at first sight that the transition metal fluorides on which most investigation of physical and chemical properties has been carried out are those which are the most difficult to handle—the hexafluorides. All the known hexafluorides are very volatile at room temperature. Their boiling points all lie within the range 17.1°C for tungsten hexafluoride to 69.1°C for the uranium compound, with melting points lying between 1.9° and 64.1°C. These properties require effectively that the compounds should be handled in vacuum systems, and yet most of them react readily with glass, the material from which virtually all laboratory vacuum systems were constructed one to two decades ago. This reaction is a cyclic hydrolysis, minute traces of water giving an oxide fluoride of the metal and hydrogen fluoride, which in turn reacts with glass to produce silicon tetrafluoride and to regenerate water to provide further hydrolysis. Obviously, a trace of hydrogen fluoride, a likely contaminant in any volatile fluoride, would initiate the same cycle.

Despite the formidable obstacles to chemical investigation, much work was done with uranium and plutonium hexafluorides under the stimulation of the nuclear energy programmes during and after World War II. Truly heroic work was done on these and other hexafluorides in glass systems, rigorously dried to prevent loss of samples; but the real advances came with the construction of vacuum systems from fluoride-resistant metals and polymers, such as Teflon and Kel-F. Even so, as will be discussed later, most of the emphasis was on preparation, purification and physical characterization of the transition metal hexafluorides through measurement of melting and boiling points and of vapour pressures and through investigation of infrared and Raman spectra. Spectral and diffraction studies had shown the hexafluorides to be octahedral molecular compounds. There was surprisingly little investigation of the chemical properties of hexafluorides before the mid-sixties.

Although the fluorides become easier to handle as the oxidation state of the transition metal is reduced, there has been even less study of fluorides other than hexafluorides. Structures of many pentafluorides have been determined, and they fall into three related polymeric classes. There has been even less chemical study of pentafluorides than of hexafluorides. The only tri- and tetrafluorides of transition metals to receive any detailed study are those of Groups IIIA and IVA and of the lanthanides and actinides. There have been no structural investigations of any consequence of other tri- and tetrafluorides, and chemical studies are fragmentary.

However, despite the enormous gaps in the knowledge of them, the fluorides of *d*- and *f*-transition elements provide an excellent basis for discussion of thermodynamic and chemical stability of oxidation states, particularly the higher ones. Lower oxidation states are not stabilized particularly well by fluorine, other halogens being more effective. Thus, the lowest fluoride reported for titanium is the trifluoride, whereas dihalides occur for the other halogens. Hafnium forms all the trihalides, except the trifluoride, while, for thorium, a formal oxidation state lower than IV has been reported only for the iodide. For the halides generally, fluorine is most effective in bringing out highest oxidation states while iodine stabilizes lowest oxidation states best.

In general, oxygen and fluorine tend to stabilize the highest oxidation states of transition elements. Of the two, oxygen is frequently the more effective, particularly in anionic complexes. It is likely that simpler coordination and multiple bonding in oxo-anions stabilizes higher oxidation states more readily than in the corresponding fluoro-anions. Thus iron(VI) occurs reasonably easily in ferrates whereas no binary or complex fluoride has been prepared with iron in a higher oxidation state than III. Oxidation state VII is

TABLE 9. OXIDATION STATES FOR TRANSITION METAL FLUORIDES[a]

ScF_3	TiF_4	VF_5	CrF_6[b]	MnO_3F[b]	FeF_3	Cs_2CoF_6	K_2NiF_6	Cs_2KCuF_6	ZnF_2
	TiF_3	VF_4	CrF_5	MnF_4[b]	FeF_2	CoF_3	K_3NiF_6	CuF_2	
		VF_3	CrF_4	MnF_3		CoF_2	NiF_2		
		VF_2	CrF_3	MnF_2					
			Cr_2F_5						
			CrF_2						
YF_3	ZrF_4	NbF_5	MoF_6	TcO_3F	RuF_6[b]	RhF_6[b]	PdF_4	Cs_2KAgF_6	CdF_2
	ZrF_3	NbF_4	MoF_5	TcF_6	RuF_5	RhF_5	$Pd^{2+}[PdF_6]^{-2}$	AgF_2	
	ZrF_2	$NbF_{2.5}$	(Mo_2F_9)	TcF_5	RuF_4	RhF_4	PdF_2	AgF	
			MoF_4	K_2TcF_6	RuF_3	RhF_3		Ag_2F	
			MoF_3						
LaF_3	HfF_4	TaF_5	WF_6	ReF_7	OsO_3F_2	IrF_6	PtF_6[b]	AuF_3	HgF_2
			WF_5	ReF_6	OsF_7[b]	IrF_5	PtF_5		Hg_2F_2
			WF_4	ReF_5	OsF_6	K_2IrF_6	PtF_4		
				ReF_4	OsF_5	IrF_3			
					OsF_4				
AcF_3	ThF_4	PaF_5	UF_6	NpF_6	PuF_6[b]	AmO_2F_2	CmF_4	BkF_4[c]	
		PaF_4	UF_5	Rb_2NpF_7	Rb_2PuF_7	AmF_4	CmF_3	BkF_3	
			U_2F_9	NpF_4	Pu_4F_{17}	AmF_3			
			U_4F_{17}	NpF_3	PuF_4				
			UF_4		PuF_3				
			UF_3						
	CeF_4	PrF_4	(Cs_3NdF_7)	(PmF_3)	SmF_3	EuF_3	GdF_3	TbF_4	(Cs_3DyF_7)
	CeF_3	PrF_3	NdF_3		(SmF_2)	(EuF_2)		TbF_3	DyF_3[d]

[a] Where possible, binary fluorides are used to illustrate particular oxidation states. Complex fluorides and oxide fluorides are used only to illustrate oxidation states not exhibited by binary fluorides. Compounds shown in parentheses have not been well characterized.

[b] Compound thermally unstable at or below 20°C.

[c] No fluorides have been reported for actinides beyond berkelium.

[d] Except for the compound $YbF_{2.24}$, only trifluorides have been reported for lanthanides beyond dysprosium.

stable in the permanganate anion and less so in manganese trioxide fluoride MnO_3F; but the highest binary fluoride of manganese is the tetrafluoride. Similarly, chromate and chromyl fluoride are easily formed and stable, while chromium hexafluoride is very difficult to prepare and is reported to be thermally unstable above −100°C. On the other hand, hexafluorides of platinum, iridium and rhodium can all be prepared, although there are no corresponding oxygen compounds. The anions CoF_6^{2-}, NiF_6^{2-}, CuF_6^{3-} and AgF_6^{3-} can all be prepared easily, whereas oxo-anions with the metals in the same oxidation states are ill-defined and it is unlikely that they can be prepared in reasonable purity.

It is worthy of note that a particularly favourable environment for high oxidation states, in addition to that in complex anions, appears to be in oxide fluorides. Thus the compounds MnO_3F, TcO_3F, OsO_3F_2 and AmO_2F_2 have no counterparts in simple halides.

There have been several reviews of the preparation and properties of individual transition metal fluorides. Emeléus, Burg, and later, Simons have described involatile and volatile fluorides in the treatise[159] that still has the most comprehensive coverage of fluorine chemistry. Subsequently Peacock[160] reviewed *d*-transition metal fluorides generally and Bartlett[161] described those of the platinum group metals. More recently, Canterford and Colton[162] have included fluorides in an exhaustive account of halides of *d*-transition metals and, in a book in the same series, Brown[163] has described preparation and properties of halides of lanthanides and actinides. Bagnall[164] has also reviewed actinide fluorides authoritatively quite recently.

The aim of this review is not to describe in detail methods of preparation and properties of individual transition metal fluorides but to demonstrate that there are systematic differences in stability and reactivities of transition metal fluorides, and that these differences can be related to the position of the particular transition metals in the Periodic Classification and to differences in bonding in the fluorides.

Table 9 illustrates the range of oxidation states that occur for transition metals in their fluorine compounds. Most of the discussion in this review will be referred back to this table. Wherever possible, binary fluorides have been selected to demonstrate different oxidation states for each metal. However, where a particular oxidation state is exhibited by a fluoro-complex or by an oxide fluoride and not by a known binary fluoride, such a compound is included in the table. Those fluorides whose existence is not well defined or which have been shown to be thermally unstable at room temperature are indicated appropriately. Actinide fluorides have been placed before those of the lanthanides because the former provide more interesting comparisons with *d*-transition metal fluorides than do the lanthanide fluorides. Because of lack of reliable information, no fluorides are shown for actinide elements beyond berkelium. Simply for convenience of presentation of the table itself, no fluorides of lanthanides beyond dysprosium are included. With the exception of the ill-defined lower fluoride of ytterbium, only trifluorides have been reported for the later lanthanides.

5.1. GENERAL METHODS OF PREPARATION OF TRANSITION METAL FLUORIDES

Detailed references to the preparation of individual fluorides, complex fluorides and oxide fluorides are given in Simons's books[159], by Canterford and Colton[162] and by Brown[163]. Muetterties and Tullock[165] have reviewed types of fluorination procedures rather than the

159 *Fluorine Chemistry*, ed. J. H. Simons, Academic Press: *see* (a) H. J. Emeléus, Vol. 1 (1950), pp. 1–76; (b) A. B. Burg, Vol. 1 (1950), pp. 77–123; (c) J. H. Simons, Vol. 5 (1964), pp. 1–131.

160 R. D. Peacock, in *Progress in Inorganic Chemistry*, ed. F. A. Cotton, Vol. 2, Interscience (1960), pp. 193–249.

161 N. Bartlett, in *Preparative Inorganic Reactions*, ed. W. L. Jolly, Vol. 2, Interscience (1965), pp. 301–339.

162 J. H. Canterford and R. Colton, (a) *Halides of the First Row Transition Metals*, Wiley (1969); (b) *Halides of the Second and Third Row Transition Metals*, Wiley (1968).

163 D. Brown, *Halides of the Lanthanides and Actinides*, Wiley (1968).

164 K. W. Bagnall, in *Halogen Chemistry*, ed. V. Gutmann, Vol. 3, Academic Press (1967), pp. 303–382.

165 E. Muetterties and C. W. Tullock, in *Preparative Inorganic Reactions*, ed. W. L. Jolly, Vol. 2, Interscience (1965), pp. 237–299.

methods of synthesis of individual fluorides, so that the general principles of fluorination might be extended to the preparation of some complex species and of novel structures. They dealt in particular with unusual and potential fluorinating reagants, such as the fluorides of oxygen, sulphur and nitrogen, and they attempted to relate fluorinating strength of reagents to their structure, e.g. in considering their fluoride acceptor strengths. A very generalized account will be given here, designed primarily to indicate stability of transition metals in different oxidation states.

(a) "Stable" or "Normal" Fluorides

For each element it is possible to consider one or more oxidation states which may or may not correspond with the formal oxidation state for the group to which the element belongs, which may be attained without the use of strongly oxidizing or reducing conditions. Thus hydrogen fluoride can be used to prepare fluorides, binary or complex, of elements such as scandium and its congeners including the lanthanides in oxidation state III, of Group IVA elements and thorium in oxidation state IV, of niobium(V) and tantalum(V), of chromium(III) and iron(III), of manganese, nickel, copper, zinc and cadmium in oxidation state II and of silver(I).

In some instances, aqueous hydrofluoric acid or a soluble fluoride can be used as the preparative reagent, as in the case of the lanthanide trifluorides. Usually, however, this procedure introduces the possibility of forming hydrates or oxide fluorides and the use of anhydrous fluoride is to be preferred. At temperatures which are generally in the range 200–300°C, hydrogen fluoride vapour may be passed over a suitable solid, preferably another halide, in a flow system. Alternatively, liquid anhydrous hydrogen fluoride may be condensed on to an anhydrous chloride. For reasons given in section 2.4(a), gaseous hydrogen chloride is evolved and a fluoride in a high state of purity may be recovered by crystallization or after removal of the volatile hydrogen fluoride.

(b) Higher Fluorides

Normally, use is made of the great oxidizing strength of elemental fluorine to produce higher fluorides. Sometimes, trifluorides of bromine and chlorine have been used instead of fluorine because being liquids near room temperature they are somewhat easier to store and to handle than fluorine. However, there is a growing awareness that this advantage can be offset by the difficulty of decomposing many relatively stable adducts formed between the halogen fluorides and the fluorides sought in the preparative reactions.

Generally, oxides and halides may be fluorinated by passing gaseous fluorine over a solid maintained in the temperature range 200–400°C. In order to reduce the risks of contamination of the product, it is often preferable to pass fluorine in a flow system over the appropriate metal in the same temperature range. Sometimes it is necessary to remove surface oxide from a finely divided metallic powder. In this way volatile fluorides such as vanadium and chromium pentafluorides and the hexafluorides of molybdenum, tungsten, osmium, iridium and uranium can be prepared. They are swept from the reaction zone in a stream of fluorine or nitrogen, condensed in traps and purified by distillation or sublimation.

This method also yields involatile solids such as silver difluoride, cobalt trifluoride and cerium and terbium tetrafluorides—all binary fluorides exhibiting high oxidation states for the metal—as well as complexes such as Cs_2KAgF_6, K_3CoF_6 and Cs_2CoF_6,

Cs_2KCuF_6 and the two nickel compounds K_3NiF_6 and K_2NiF_6. It is interesting to note that the only significant difference in experimental conditions for the preparation of the two nickel complexes is that, in the first case, fluorine is passed over a heated 3 : 1 mixture of potassium chloride and anhydrous nickel chloride, whereas, for the second, a 2 : 1 mixture is fluorinated. Stoichiometry of reactants, rather than strength of the oxidant, seems to determine the oxidation state of the metal in the hexafluoronickelate.

Two general approaches have been used for preparation of volatile fluorides which are thermally unstable at room temperature. The first of these appears to have been developed independently in the atomic energy laboratories of the appropriate United States and United Kingdom authorities. It involves heating a metal, either as a filament or by high-frequency induction techniques, in an atmosphere of fluorine with a "cold finger" filled with liquid nitrogen near the heated metal. The fluoride, formed at high temperature, distils from the hot zone and is recovered condensed without significant decomposition. This method was developed for the preparation of plutonium and neptunium hexafluorides and has been applied to the preparation of hexafluorides such as those of platinum and ruthenium.

The alternative approach to the preparation of unstable fluorides involves use of high-pressure fluorine at elevated temperature in a nickel autoclave or "bomb". It has been exploited very successfully by Glemser and co-workers in the preparation of chromium hexafluoride and osmium heptafluoride.

(c) Lower Fluorides

Reduction processes for preparation of lower fluorides include heating a stable transition metal fluoride with hydrogen or with the metal itself or passing a mixture of hydrogen and hydrogen fluoride over a higher fluoride or over a metal. Thus titanium trifluoride may be prepared by heating a mixture of the tetrafluoride and the metal or by passing a hydrogen–hydrogen fluoride mixture over titanium metal or its hydride. Niobium pentafluoride heated with metallic niobium to 400°C yields the tetrafluoride, but on heating to 700°C gives the compound of stoichiometry $NbF_{2\cdot5}$. Vanadium difluoride is prepared by passing hydrogen and hydrogen fluoride over vanadium trifluoride.

Iodine dissolved in iodine pentafluoride has been used as a mild reductant, converting osmium hexafluoride to the pentafluoride and ruthenium pentafluoride to the tetrafluoride. Iodine alone at 150°C reduces ruthenium pentafluoride to the trifluoride. Selenium tetrafluoride reduces the hexafluorides of the platinum group metals. Peacock has used tungsten hexacarbonyl to prepare penta- and tetrafluorides of osmium and rhenium from the appropriate hexafluorides.

5.2. THERMODYNAMIC STABILITY OF *d*-TRANSITION METAL FLUORIDES

(a) Stability by Rows

Reference to Table 9 shows certain general trends in stabilities of transition metal fluorides, the first of these being that for the metals in any row of the Periodic Classification there is a decrease in stability of highest oxidation states with increase in atomic number. The highest possible formal oxidation state is easily attainable for binary fluorides of elements of Groups IIIA, IVA and VA, although there are some experimental difficulties associated with the preparation of the highly reactive vanadium pentafluoride.

However, considering fluorine compounds of later first row metals, manganese shows oxidation state VII only in a thermally unstable oxide fluoride, the highest binary fluoride, also unstable, being the tetrafluoride. Thereafter the only first-row binary fluorides are trifluorides and difluorides, although somewhat higher oxidation states can be reached in fluoro-complexes.

For second-row metals, a similar generalization is apparent. The highest formal oxidation is attained quite easily in binary fluorides through to molybdenum in Group VIA. Despite elegant experimental attempts, no one has yet prepared technetium heptafluoride, although it seems possible that high-pressure high-temperature fluorination could yield a thermally unstable product. In this case, unlike that of manganese, the trioxide fluoride is thermally stable. If special preparative techniques are used, the thermally unstable hexafluorides of ruthenium and rhodium can be isolated, but the highest fluoride of palladium is the tetrafluoride.

The key binary fluoride in the third row is the thermally stable rhenium heptafluoride. By comparison, osmium heptafluoride, prepared at very high fluorine pressures, is thermally unstable above −100°C, although oxidation state VIII occurs in osmium trioxide difluoride. Iridium, unlike its second-row counterpart rhodium, forms a stable hexafluoride, while that of platinum can be prepared by special techniques but is not particularly stable at room temperature. The highest reported fluoride of gold, produced by the action of bromine trifluoride on the metal or on the chloride, is the trifluoride, which, however, decomposes at 500°C to gold and fluorine.

(b) Stability by Groups

Critical assessment of the material presented in the preceding section would bring out one further well-established generalization, namely that within any one group of transition metals, stability of highest oxidation states increases with increasing atomic number.

This is first demonstrable in Group IVA where lower fluorides of titanium and zirconium can be prepared reasonably easily, but only the tetrafluoride of hafnium has been reported. One suspects that titanium difluoride could be made fairly easily. It possibly disproportionates under conditions which might have been tried to date.

Group VA demonstrates the trend even more spectacularly. Vanadium pentafluoride is very easily reduced chemically to the tetrafluoride which disproportionates at 125°C to the trifluoride which in turn can be reduced to the difluoride by a mixture of hydrogen and hydrogen fluoride at 1150°C. Bomb reductions of niobium pentafluoride with metallic niobium are necessary to produce lower niobium fluorides, while there is no satisfactory report of the preparation of a fluoride of tantalum lower than the pentafluoride. The reported trifluoride has been shown to be an oxide fluoride of tantalum(V).

The relative instability of chromium hexafluoride is consistent with the generalization discussed in the preceding sub-section, and all lower chromium fluorides down to the difluoride can be prepared easily. Tungsten forms a very stable hexafluoride, but its pentafluoride and tetrafluoride are difficult to isolate, the former being prepared by igniting a tungsten filament in a limited amount of fluorine while the latter results in small yield from the action of strong reductants on the hexafluoride. The molybdenum compounds, from the hexafluoride to the trifluoride, can be prepared relatively easily.

The binary fluorides of every member of each group preceding Group VIIA can be prepared in the highest formal oxidation state for that group even though one compound, chromium hexafluoride, is thermally unstable. Therefore Group VIIA demonstrates the most clear-cut case of increasing stability of higher fluorides when the thermally unstable manganese tetrafluoride is considered along with technetium hexafluoride and rhenium heptafluoride.

It is convenient to consider fluorides of iron, cobalt and nickel and of the corresponding six noble metals together. Decreasing stability of higher fluorides along each row for these nine elements has already been discussed in the preceding sub-section. Reference to Table 9 shows that in each group the range of existence of higher oxidation states increases with increase in atomic number. The formal oxidation state of VIII is achieved only in osmium trioxide difluoride, the highest binary fluoride being osmium heptafluoride which can be compared with the unstable ruthenium hexafluoride and with iron trifluoride. A similar pattern emerges for the groups headed by cobalt and nickel.

The trends observed in earlier groups become less well defined in Group IB. Oxidation state III occurs for each element, but only in complexes for copper and silver. Comparing it with copper difluoride, there is nothing particularly unusual about silver difluoride which is prepared very easily by fluorination of a compound such as silver chloride. What is perhaps more surprising is that, although the other halides of copper(I) are easily prepared, copper(I) fluoride has never been isolated as a solid, having been detected only in the gas phase. The unusual fluoride in the group is the sub-fluoride Ag_2F produced as lustrous crystals when a saturated solution of silver(I) fluoride is reduced at a silver electrode. At 200°C it disproportionates to silver(I) fluoride and the metal.

Considering the previously reported chemistry and the satisfactory descriptions of bonding, the fluorides of the elements of Group IIB call for no special comment.

(c) Thermal Stability of Intermediate Fluorides

Thermal stability of highest fluorides for each of the transition metals has been indicated in Table 9 and discussed in the preceding sub-sections (a) and (b) above. There is fragmentary evidence in the chemical literature on thermal stabilities of a few intermediate fluorides. Thus at 125°C vanadium tetrafluoride disproportionates to pentafluoride and trifluoride, while niobium tetrafluoride gives the pentafluoride and the compound of stoichiometry $NbF_{2.5}$ at 400°C. These two observations are at least consistent with expected trends in stabilities of higher oxidation states. Molybdenum and platinum pentafluorides both disproportionate to hexafluoride and tetrafluoride, the former at 165°C and rhenium pentafluoride is reported to decompose to the tetrafluoride at 140°.

(d) Unusual Oxidation States

In fluorides listed in Table 9, some *d*-transition elements appear to exhibit unusual oxidation states. The structure of the compound with the formula $NbF_{2.5}$ has been shown by Schaeffer to be based on Nb_6F_{12} cluster units with an additional inter-cluster bridging fluorine atom bonded to each niobium atom to give an infinite three-dimensional array[166].

Chromium trifluoride can be reduced at 1000°C by metallic chromium to the difluoride. Reaction between equimolar amounts of the trifluoride and the difluoride yields the compound $CrF_{2.5}$. A single crystal X-ray study showed this solid to be based on an octahedral arrangement of fluorine atoms around chromium atoms in two different oxidation states.

[166] Ref. 162(b), pp. 94–5.

Peacock has reported isolation of a compound believed to be Mo_2F_9 as a product of fluorination of molybdenum hexacarbonyl, and he has suggested that it might be formulated as $MoF_3^+MoF_6^-$. If it does exist as an ionic solid, it is probably more likely to be similar to the compound $Mo_3Cl_9(MoF_6)_3$ characterized by Stewart and O'Donnell.

Structural analysis of the compound initially reported as palladium trifluoride has shown it to be $Pd^{II}\ Pd^{IV}F_6$.

(e) Summary

Groups VA, VIA and VIIA illustrate adequately the two major trends set out in the previous sub-sections (a) and (b), namely that the highest formal oxidation state decreases in stability in passing across any row of the Periodic Classification. However, stable highest oxidation states are retained further in the third row than in the second row and, in turn, further in the second row than in the first. This last generalization is simply another statement of the one that stability of higher oxidation states increases down any group of transition metals. Thus for binary fluorides of Group VIIA the formal group oxidation state is attained only for the third-row element rhenium. However, proceeding further along the third row, osmium is unable to provide an octafluoride.

The hexafluorides illustrate the overall trends better than any other single series of binary fluorides. To date, chromium is the only first-row metal to have yielded a hexafluoride, and that is unstable. In the second row, the hexafluorides from molybdenum to rhodium show decreasing thermal stability and no hexafluoride of palladium has been prepared. However, in the third row, hexafluorides are stable from tungsten to iridium. That of platinum can be formed but is thermally unstable and highly reactive, being the first compound to be reacted with a rare gas, while gold forms no fluoride higher than the trifluoride.

It appears as a corollary of the two generalizations concerning stabilities of *d*-transition metal fluorides considered by rows and groups that, as stability of higher fluorides decreases, lower fluorides become more significant. This becomes apparent from Table 9 with the fluorides of vanadium and chromium when considering groups, and those of osmium and iridium when considering rows.

Thermodynamic stability of higher fluorides illustrates very well the long-established difference of first row transition metals from those of the second and third rows, especially when binary fluorides are considered, the hexafluorides being particularly good examples. However, there are marked differences between pairs of elements in the second and third rows; but these differences can be illustrated more effectively by a consideration of chemical reactivities than from thermodynamic stabilities. This will be done in section 5.3(e) below.

5.3. CHEMICAL REACTIVITY OF HIGHER FLUORIDES OF *d*-TRANSITION METALS

When the literature describing higher fluorides of transition elements is surveyed, it is seen that much difficult but elegant work was done in the first half of this century by Ruff and other pioneers on preparation, purification and characterization of these compounds. Melting and boiling points and vapour pressures were measured with surprising accuracy. Initially, efforts to characterize fluorides relied heavily on chemical analyses. More recently investigators have used many techniques including infrared, Raman and n.m.r. spectrometry, magnetic susceptibility measurements and X-ray and electron diffraction methods.

Because of the great reactivity of the higher fluorides, particularly the volatile ones, the studies have always been hampered by experimental difficulties. An illustration of the difficulties encountered is that the yellow volatile fluoride described in 1913 as osmium octafluoride was accepted as the only binary compound with eight atoms bonded to a central atom until Weinstock and Malm showed unequivocally in 1957 that it was the hexafluoride.

There is no doubt that the hexafluoride most studied during and soon after World War II was that of the *f*-transition element, uranium. Yet, as Katz and Seaborg[167] stated in 1957, very few chemical reactions had been reported for the compound at that stage. For other higher fluorides of *d*- or *f*-transition metals even less was known about chemical reactivity. The reactions that had been reported were fragmentary. Usually potential reactants appear to have been selected quite unsystematically. Thus there was no basis for a comparison of the reactivities of higher fluorides. It was believed, to quote a specific example, that the hexafluorides of molybdenum and tungsten were very strong oxidants and almost identical in their chemistry[168]. This generalization was probably based on the violence of the hydrolysis reaction for both hexafluorides. However, more recent work has shown that they are quite dissimilar chemically and not particularly reactive—in fact tungsten hexafluoride could be classed as rather inert in all reactions except hydrolysis.

The main reason for the slow development of the chemistry of the higher fluorides is that most of them react with glass, unless it is dried using special procedures, and they react also with most of the components of a conventional vacuum system in which these volatile compounds might be expected to be handled. In the last 10–15 years, construction of vacuum systems from metals such as copper and nickel and their alloys and from polymers such as polytetrafluoroethylene (Teflon, Fluon, etc.) and polychlorotrifluoroethylene (Kel-F) has allowed more reliable study of physical properties of higher fluorides as well as investigation of their chemical reactivities. Typical experimental procedures, based on fluoride-resistant vacuum systems and specially constructed apparatus for physico-chemical measurements, have been described by Canterford and O'Donnell[169].

Some broad generalizations of chemical reactivities of higher metal fluorides have been recognized for some time, while others are quite new. The fairly obvious correlation between thermal instability and high chemical reactivity has been demonstrated in several ways, perhaps the simplest being the series of experiments in which possible reaction of xenon with several hexafluorides was investigated. Xenon was found to react with those hexafluorides which were thermally unstable at room temperature, namely those of platinum, ruthenium, rhodium and plutonium but not with the thermally stable hexafluorides of iridium, molybdenum, osmium and uranium[170].

No direct comparisons of reactivities of related transition metal fluorides appear to have been made before 1960. In 1962 it was reported that nitric oxide reduced molybdenum and uranium hexafluorides but did not react with tungsten hexafluoride. However, no conclusion about the chemical nature of the three fluorides were drawn from these observations. In 1966 O'Donnell and co-workers reported reactions of higher fluorides of uranium and of transition metals of Groups VA and VIA, and they related reactivities of the fluorides studied

[167] J. J. Katz and G. T. Seaborg, *The Chemistry of the Actinide Elements*, Methuen (1957), p. 161.

[168] N. V. Sidgwick, *The Chemical Elements and their Compounds*, Vol. 2, Oxford University Press (1950), p. 1034.

[169] J. H. Canterford and T. A. O'Donnell, in *Technique of Inorganic Chemistry*, eds. H. B. Jonassen and A. Weissberger, Vol. 7, Interscience (1968), pp. 273–306.

[170] C. L. Chernick *et al.*, *Science*, **138** (1962) 136.

to the position of the appropriate metals in the Periodic Classification. This work is discussed below in sections (a) and (d). In the same year, Bartlett discussed reactions of nitric oxide with hexafluorides of third row transition metals and his observations and conclusions are given in sub-sections (b) and (e) below.

Although there are still many and very large gaps in the experimental information available, examination of the order of reactivities of higher fluorides of *d*-transition metals with the wide range of reactants listed in sub-sections (a)–(d) below supports the generalizations put forward in section 5.2(e). Certainly no observation is inconsistent with it. That is, on the basis of reactivities, it is found that stability of higher oxidation states increases with increasing atomic number in any group, but decreases with increasing atomic number in any row.

Study of these reactions supports the long-established generalization that, chemically, the first-row metals differ markedly from those of the second and third rows. However, it has often been suggested, if only implicitly, that in any one subgroup the second- and third-row elements resemble each other closely. A new feature has emerged from the study of chemical reactivities of higher fluorides of second- and third-row elements, namely that in progressing from Group IIIA and IVA through the *d*-transition metals, the second- and third-row elements in each successive group become progressively more dissimilar. This will be illustrated in the sub-sections immediately below, particularly in (d), and will be commented on formally in the summary, i.e. (e).

(a) Oxidation of Lower Fluorides of Non-metals

Until quite recently, molybdenum hexafluoride had been regarded as a strong oxidant[168]. O'Donnell and Stewart[171] used sulphur and selenium tetrafluorides and the trifluorides of

TABLE 10. OXIDATION-REDUCTION REACTIONS OF HIGHER FLUORIDES OF GROUPS VA AND VIA[a]

Reactants[b]	PF_3	AsF_3	SbF_3	CS_2
VF_5	PF_5, VF_4, VF_3[c]	AsF_5, VF_4[c]	SbF_5, VF_4[d]	$(CF_3)_2S_2$, $(CF_3)_2S_3$ SF_4, VF_3
NbF_5	—	—	—	—
TaF_5	—	—	—	—
CrF_5	PF_5, CrF_3	AsF_5, CrF_3	SbF_5, CrF_3[d]	CF_4, SF_6, CrF_3
MoF_6	PF_5, MoF_5[e]	—	—	$(CF_3)_2S_2$, S, MoF_5
WF_6	PF_5, WF_4[d, e]	—	—	—

[a] For purposes of tabulation, the products for some reactions have been simplified. Detailed reaction schemes are available in the original papers.

[b] In each case, products are quoted for excess of the reagent listed horizontally. For many reactions, different products are obtained if the other reagent is in excess.

[c] To greater or less extent, depending on the strength of the product pentafluoride (XF_5) as a fluoride acceptor, complexes of the type $VF_3^+ \cdot XF_6^-$ were obtained.

[d] Reaction very slow unless catalysed by anhydrous HF.

[e] System reaches equilibrium.

phosphorus, arsenic, antimony and bismuth to investigate the ease of reduction of molybdenum hexafluoride and found that only phosphorus trifluoride reacted and even this

[171] T. A. O'Donnell and D. F. Stewart, *Inorg. Chem.* **5** (1966) 1434.

reaction involved an equilibrium because phosphorus pentafluoride partially oxidized molybdenum pentafluoride. They found that phosphorus trifluoride reduced tungsten hexafluoride very slowly and to a very limited extent. Some reduction occurred after several days of shaking the reactants in a Kel-F tube with liquid anhydrous hydrogen fluoride present as catalyst.

Ideally, study of possible reaction of the lower fluorides with chromium hexafluoride should have followed, but this compound is thermally unstable at room temperature and must be presumed to be more chemically reactive than the thermally stable chromium pentafluoride. It was decided to study the pentafluoride which was found to oxidize phosphorus and arsenic trifluorides very readily and to oxidize antimony trifluoride to the pentafluoride if anhydrous hydrogen fluoride was present.

The results of these comparative studies are shown as part of Table 10. The experimentally determined order of reactivities is then $CrF_5 \gg MoF_6 \gg WF_6$. Studies, discussed in sub-section (d), of tungsten hexafluoride reactions with chlorides, established the order $CrF_5 \gg MoF_6 > WF_6$. On the basis of the relative thermal stabilities of the two highest fluorides of chromium, the following order emerges: $CrF_6 \gg MoF_6 > WF_6$. This was consistent with the generalization then established that first-row elements differ markedly from those of the second and third rows; but it indicated a significant difference, brought out much more fully in the sections below, between the reactivities of compounds of second- and third-row elements.

As part of a study of Group VA pentafluorides, Canterford and O'Donnell[172] investigated the possible oxidation of the series of lower fluorides of non-metals by the pentafluorides of vanadium, niobium and tantalum. As Table 10 shows, vanadium pentafluoride proved to be a vigorous oxidant, being reduced in part to vanadium trifluoride by the strongest reductant of the series, phosphorus trifluoride. This work did not discriminate in any way between the pentafluorides of niobium and tantalum as oxidants, since neither pentafluoride reacted with any of the potential reductants.

It has been shown that rhenium hexafluoride is readily reduced to the pentafluoride by phosphorus trifluoride but not by arsenic trifluoride[173]. Thus rhenium hexafluoride is more readily reduced than the preceding third-row compound, tungsten hexafluoride, which is reduced incompletely and only very slowly by phosphorus trifluoride even with the catalyst hydrogen fluoride.

Although sulphur and selenium tetrafluorides do not reduce Group VIA hexafluorides, they do react with some hexafluorides of third-row elements beyond tungsten. Selenium tetrafluoride reduces osmium and iridium hexafluorides to pentafluorides while platinum hexafluoride is reduced to the tetrafluoride. Sulphur tetrafluoride reduces platinum hexafluoride, but not those of osmium and iridium.

Platinum hexafluoride oxidizes chlorine trifluoride to the pentafluoride at temperatures as low as −78°C and also oxidizes bromine trifluoride. Reaction of bromine trifluoride with pentafluorides of osmium, iridium and platinum produces 1 : 1 adducts without reduction in the case of the osmium and iridium compounds but, with platinum pentafluoride, yields platinum tetrafluoride and bromine pentafluoride.

All of these reactions of higher fluorides of osmium, iridium and platinum, when considered with the inertness of tungsten hexafluoride, indicate the trend of decrease in stability of higher fluorides along any transition metal row.

172 J. H. Canterford and T. A. O'Donnell, *Inorg. Chem.* **5** (1966) 1442.
173 J. H. Canterford, T. A. O'Donnell and A. B. Waugh, *Australian J. Chem.* **24** (1971) 243.

(b) Oxidation of Nitric Oxide and of Nitrosyl Fluoride

Several workers, notably Bartlett, have studied reactions of transition metal hexafluorides with nitric oxide and nitrosyl fluoride and these are listed in Table 11. The only significant intra-group relationship is associated with the observation that, while nitric oxide reduces molybdenum hexafluoride, it does not react with the tungsten compound.

For each of the three second-row hexafluorides reacted with nitric oxide, reduction

TABLE 11. REACTIONS OF HEXAFLUORIDES WITH NITRIC OXIDE AND NITROSYL FLUORIDE[a]

	MoF_6	TcF_6	RuF_6		
NO	$NOMoF_6$	$NOTcF_6$	$NORuF_6$		
ONF	$NOMoF_7$	$(NO)_2TcF_8$			
	WF_6	ReF_6	OsF_6	IrF_6	PtF_6
NO	No reaction	$NOReF_6$	$NOOsF_6$	$NOIrF_6(-20°)$ $(NO)_2IrF_6(25°)$	$NOPtF_6(-20°)$ $(NO)_2PtF_6(25°)$
ONF	$(NO)_2WF_8$	$(NO)_2ReF_8$	$NOOsF_7$ $NOOsF_6$	$NOIrF_6$	$NOPtF_6$

[a] Reactions listed are at 25°C unless otherwise stated.

occurs to a hexafluoro-anion with the metal in oxidation state V. For the two second-row cases studied, nitrosyl fluoride forms a 1 : 1 adduct without change in oxidation state.

As Bartlett has stated[174], the reactions of nitric oxide and nitrosyl fluoride with third-row hexafluorides provide the best basis available for comparison of the reactivities of this series of compounds. Decrease in stability of oxidation state VI, i.e. increase in oxidative strength of the hexafluorides, with increase in atomic number is shown for both types of reaction. Nitric oxide does not react with tungsten hexafluoride but reduces rhenium and osmium hexafluorides to oxidation state V. Depending on the temperature of reaction, reduction to oxidation state IV or V may occur for iridium and platinum.

Nitrosyl fluoride does not reduce either tungsten or rhenium hexafluorides, forming adducts, but partially reduces osmium hexafluoride. Reduction of the hexafluoride to the hexafluorometallate(V), with liberation of fluorine, occurs for both strong oxidants iridium and platinum hexafluoride.

(c) Oxidation of Xenon and Carbon Disulphide

Following immediately on Bartlett's report of the reaction of xenon with platinum hexafluoride, Malm and co-workers at the Argonne National Laboratory investigated the possibility of reaction of the rare gas with a series of hexafluorides, including actinide compounds. They found that the *d*-transition metal hexafluorides which reacted were those which were thermally unstable at room temperature, namely those of platinum, ruthenium and rhodium, while molybdenum, osmium and iridium hexafluorides did not react[170].

Although it has not been used widely in such studies, carbon disulphide is particularly effective in indicating differences between fluorides as oxidants, because characteristically different products of reaction are given depending on the oxidizing strength of the fluoride

[174] (a) N. Bartlett, S. P. Beaton and N. K. Jha, *Chem. Communs.* (1966) 168; (b) N. Bartlett, *Angew. Chem. Intern. Ed.* 7 (1968) 433.

concerned. Thus strongly oxidizing fluorides cause rupture of both sulphur–carbon bonds and form carbon tetrafluoride and either sulphur tetrafluoride or sulphur hexafluoride. Milder oxidants, such as iodine pentafluoride, do not rupture both sulphur–carbon linkages and, in addition to sulphur tetrafluoride, form compounds such as bistrifluoromethyldisulphide, $(CF_3)_2S_2$, probably through radicals such as $CF_3S\cdot$. Under very mild fluorinating conditions, e.g. in the electro-chemical oxidation of carbon disulphide in anhydrous hydrogen fluoride, both carbon–sulphur bonds are retained in products such as $CF_2(SF_3)_2$ and $CF_2(SF_5)_2$.

Using phosphorus and arsenic trifluorides as potential reductants, O'Donnell and Stewart[171] observed some differences between the second- and third-row hexafluorides of molybdenum and tungsten, because for the latter it was necessary to carry out reduction in anhydrous hydrogen fluoride to obtain even slight reaction. However, as Table 10 shows, they observed more definite differences in carbon disulphide reactions. Tungsten hexafluoride and carbon disulphide remained without reaction as two miscible liquids, whereas molybdenum hexafluoride was reduced vigorously at room temperature to the pentafluoride and bistrifluoromethyl disulphide was formed. For the first-row element chromium, even the pentafluoride reacted to form carbon tetrafluoride and sulphur hexafluoride. It can be assumed that the thermally unstable chromium hexafluoride would give the same products even more readily. Canterford and O'Donnell[172] observed, as shown also in Table 10, that carbon disulphide is vigorously oxidized by vanadium pentafluoride even at temperatures approaching that of liquid nitrogen, but that this reagent does not differentiate between the pentafluorides of niobium and tantalum since there is no reaction at temperatures up to 20°C. Vanadium pentafluoride is such a strong oxidant that it oxidizes elemental sulphur to the tetrafluoride below room temperature.

(d) Halogen-exchange Reactions with Chlorides

Burg[159] has recorded the observations, usually qualitative, made by Ruff and other pioneers of inorganic fluorine chemistry on reactions of higher transition metal fluorides with chlorides and related compounds. Frequently these observations amounted to little more than the colour which resulted when the fluoride and chloride reacted. Rarely were all of the products of reaction adequately characterized. There does not appear to have been any systematic selection of the chlorides studied and, because of contamination of reactants and products through hydrolysis and other adventitious reactions, the early observations have little significance. Occasionally later workers used reaction with carbon tetrachloride as a measure of reactivity of fluorides.

O'Donnell and co-workers at Melbourne have compared the reactivities of higher fluorides of transition elements of Groups IVA–VIA by studying halogen–exchange reactions of those fluorides with the trichlorides of main Group V, with the tetrachlorides of carbon, silicon and titanium, with boron trichloride and, in a limited study, with the anhydrous chlorides of the alkali and alkaline-earth elements. It was established that phosphorus trichloride entered into halogen-exchange reactions more readily than arsenic trichloride, that titanium tetrachloride was more reactive than carbon and silicon tetrachlorides and that, of the molecular chlorides studied, boron trichloride was the one most likely to enter into this type of reaction—a reflection of the fact that the enthalpy of formation of boron trifluoride is more negative than that for boron trichloride by 172.5 kcal $mole^{-1}$. In many cases, in addition to the halogen-exchange reactions, oxidation-reduction reactions also occurred, either between reactants and products or between the products themselves.

The course and extent of these oxidation-reduction reactions depended on which reactant was present in excess in any particular reaction; but, in all cases, the oxidation-reduction reactions occurring concomitantly with the halogen-exchange reactions were entirely consistent with the oxidation-reduction pattern reported in sub-section (a).

For simplicity of presentation, a selection of the many halogen-exchange reactions has been made for Table 12. The two first-row compounds, vanadium pentafluoride and chromium pentafluoride, are extremely reactive undergoing halogen-exchange and oxidation-reduction. By comparison with them, the pentafluorides of niobium and tantalum are much less reactive, entering into halogen-exchange with some chlorides but no oxidation-reduction occurring. Table 12 could be misleading in suggesting that niobium and tantalum

TABLE 12. HALOGEN- EXCHANGE REACTIONS OF HIGHER FLUORIDES OF GROUPS VA AND VIA[a]

Reactants[b]	PCl_3	$AsCl_3$	BCl_3	$TiCl_4$
VF_5	PF_3, PF_5, $VClF_3$, Cl_2	$VClF_3$, Cl_2	BF_3, VCl_4, Cl_2	TiF_4, VCl_3F, Cl_2
NbF_5	PF_3, $NbCl_5$	No reaction	BF_3, $NbCl_5$	TiF_4, $NbCl_5$
TaF_5	PF_3, $TaCl_5$	No reaction	BF_3, $TaCl_5$[c]	TiF_4, $TaCl_5$
CrF_5	PF_3, PCl_5, $CrCl_3$	AsF_3, $CrCl_3$, Cl_2	BF_3, $CrCl_3$, Cl_2	TiF_4, $CrCl_3$, Cl_2
MoF_6	PF_3, PCl_5, $MoCl_5$	AsF_3, $MoCl_5$, Cl_2	BF_3, $MoCl_5$, Cl_2	TiF_4, $MoCl_5$, Cl_2
WF_6	No reaction (to 150°C)	No reaction (to 150°C)	BF_3, WCl_6	$TiClF_3$, WCl_6

[a] For purposes of tabulation, the products of some reactions have been simplified. Detailed reaction schemes are available in the original papers.

[b] In each case, products are quoted for excess of the reactant listed horizontally. For many reactions, different products are obtained if the other reactant is in excess. Unless otherwise stated, the more volatile reactant was condensed on to the other and the mixture was allowed to equilibrate at room temperature.

pentafluorides react with most chlorides studied. In fact, of the eight chlorides added to these two pentafluorides, the only ones to react were the three for which products are listed in the table. No reaction was observed with antimony and bismuth trichlorides or with the tetrachlorides of carbon, silicon and tin, whereas vanadium pentafluoride reacts vigorously with all eight.

One significant observation was that, where niobium and tantalum pentafluorides entered into halogen-exchange reactions, there were marked kinetic differences in reactivity. Rate of increase of pressure of phosphorus trifluoride or of boron trifluoride was observed to be much greater for the niobium pentafluoride reaction than for the tantalum case. As a result of this observation, the corresponding reactions were studied for zirconium and hafnium tetrafluorides. Kinetic differences did occur but were less marked than for the pair of Group VA pentafluorides.

In Group VIA the first-row compound, chromium pentafluoride, enters into halogen-exchange and oxidation-reduction reactions violently. Molybdenum hexafluoride reacts readily with all chlorides studied. Tungsten hexafluoride differs markedly from its second-row counterpart in reacting only with boron trichloride and titanium tetrachloride. Even at elevated temperatures, it did not react with the other chlorides studied. Phosphorus trichloride, which reacts with all the other higher fluorides involved in this overall study, remains as a liquid immiscible with tungsten hexafluoride at temperatures up to 150°C in a Carius tube. Other chlorides, such as carbon and silicon tetrachlorides, are miscible but do not react on heating.

This series of reactions provided the first systematic evidence that molybdenum and tungsten hexafluorides, previously believed to be very similar chemically as well as physically[168], were quite dissimilar, with tungsten hexafluoride a relatively inert compound. In fact, recently it has been used frequently as an inert solvent for several reactions involving other fluorides.

The implications of the observations that the two pairs of higher fluorides, zirconium and hafnium tetrafluorides and niobium and tantalum pentafluorides, undergo the same reactions but show kinetic differences and that the pair of compounds, molybdenum and tungsten hexafluoride, undergo entirely different reactions will be discussed in sub-section (e) below.

The reaction products quoted in Table 12 are for the cases in which the reacting chloride is in excess. In some of these reactions, particularly when vanadium pentafluoride and molybdenum hexafluoride are involved, quite different products are obtained if the fluoride is in excess. This has been reported and discussed in detail in the relevant papers.

The halogen-exchange reactions of vanadium pentafluoride are extraordinarily complex; but one particular feature of the molybdenum hexafluoride reactions warrants comment. When excess molybdenum hexafluoride reacted with chlorides, the molybdenum-containing product was always analysed as $Mo_2Cl_3F_6$. By showing that the two pure compounds, molybdenum hexafluoride and molybdenum pentachloride, reacted to give the mixed halide and chlorine, Stewart and O'Donnell established that in these halogen-exchange reactions the pentachloride is formed initially, but it then reacts with excess of the hexafluoride. The infrared spectrum of the mixed halide gave evidence for the hexafluoromolybdate(V) anion, and the spectrum in the visible region was remarkably similar to that of the trimeric chloro-complexes of rhenium(III) reported by Cotton and co-workers[162b]. Magnetic susceptibility measurements on the compound itself were consistent with the formulation $Mo_3Cl_9(MoF_6)_3$. A solution of the compound in acetonitrile was electrolysed, and analytical, spectral and magnetic investigations of the separate solutions from the cathode and anode compartments supported the proposed formula.

In most studies of halogen-exchange reactions of higher fluorides, covalent chlorides have been used. O'Donnell and Wilson[175] investigated the reactions of the hexafluorides of molybdenum, tungsten and uranium with anhydrous chlorides of Groups IA and IIA. For the present discussion, this work provided excellent evidence of the marked difference in reactivity between molybdenum and tungsten hexafluorides. The former reacted relatively slowly with all ten chlorides to form the appropriate alkaline or alkaline-earth fluoride and the compound $Mo_3Cl_9(MoF_6)_3$, even when the chloride was in stoichiometric excess. The interpretation of this was that because reaction between the liquid fluoride and ionic chlorides is much slower than the corresponding reactions with molecular chlorides, molybdenum hexafluoride is always *effectively* in excess, so that molybdenum pentachloride reacts as it forms with the hexafluoride.

However, the real value of this work in the present context is that, by comparison with the molybdenum compound, tungsten hexafluoride is very inert, reacting with beryllium chloride—an observation that is consistent with available thermodynamic data. Free energy values are not available; but if enthalpy changes are calculated for the reactions to give the expected products tungsten hexachloride and the appropriate Group IA or IIA fluoride, the only exothermic reaction is that with beryllium chloride. Unfortunately, no correlation with thermodynamic data can be attempted for halogen-exchange reactions

[175] T. A. O'Donnell and P. W. Wilson, *Australian J. Chem.* **21** (1968) 1415.

of molybdenum hexafluoride because one of the products in every case is the compound $Mo_3Cl_9(MoF_6)_3$ on which no thermodynamic studies have been made.

Apart from the work of the Melbourne group, relatively few reactions between higher transition metal fluorides and chlorides have been studied in any detail. Holloway, Rao and Bartlett[176] have described rhodium pentafluoride as an extremely powerful oxidant because it reacts with carbon tetrachloride to form chlorine monofluoride and other products. When viewed against the fact that iridium pentafluoride does not react with carbon tetrachloride, their observation is consistent with the generalization proposed here that reactivity of higher fluorides within any one group of transition elements decreases with increasing atomic number.

In section 5.2(a), the fact that the trifluoride is the highest fluoride of gold was used as evidence that, in any row of transition elements, thermodynamic stability of higher fluorides decreases with increasing atomic number. However, even this fluoride in oxidation state III is considerably more reactive than many higher fluorides, such as tungsten hexafluoride earlier in the same row. Gold trifluoride has been observed by Sharpe to react readily with carbon tetrachloride at 40°C and has been described as a powerful fluorinating agent[177].

By comparison with study of chlorides, there has been very little study of reactions of bromides and iodides with higher transition metal fluorides. O'Donnell and co-workers have included phosphorus tribromide and pentabromide, boron tribromide, the tetrabromide and tetraiodide of carbon and tin tetrabromide in a limited number of their investigations[172, 173]. The general pattern of reactivities is similar to that for chlorides although bromides appear to react more readily than chlorides. For example, tungsten hexafluoride which does not react with phosphorus trichloride undergoes halogen-exchange with the tribromide, although slowly at room temperature. This observation is compatible with the different values of energies of the phosphorus–halogen bond in phosphorus trichloride and tribromide, 79 and 64 kcal $mole^{-1}$ respectively.

(e) Summary: Differences in Reactivity between Second and Third Row Fluorides

The ability, or lack of it, of individual higher transition metal fluorides to oxidize the lower non-metal fluorides, phosphorus and arsenic trifluoride, selenium and sulphur tetrafluoride and bromine and chlorine trifluoride, demonstrates clearly the decrease in stability of higher oxidation states within any one row, e.g. ease of oxidation of selected lower fluorides by hexflauorides increases markedly in the order $WF_6 < ReF_6 < OsF_6 < IrF_6 < PtF_6$.

Bartlett and co-workers have used the ability of third-row hexafluorides from tungsten to platinum to oxidize oxygen, nitric oxide and nitrosyl fluoride to provide a semi-quantitative basis for the increase in oxidizing strength of a hexafluoride with increase in atomic number in any row[174].

Using a calculated value of about -125 kcal $mole^{-1}$ for the lattice energy for $O_2^+PtF_6^-$ and the ionization potential of 281 kcal $mole^{-1}$ for molecular oxygen, they have shown that for the reaction

$$O_{2(g)}+PtF_{6(g)} \longrightarrow O_2^+PtF_6^-{}_{(s)}$$

to occur, the electron affinity for the platinum hexafluoride molecule $A_{(PtF_6)}$ must be greater than -156 kcal $mole^{-1}$.

[176] J. H. Holloway, P. R. Rao and N. Bartlett, *Chem. Communs.* (1965) 306.
[177] A. G. Sharpe, *J. Chem. Soc.* (*London*) (1949) 2901.

Tungsten hexafluoride does not oxidize nitric oxide to the nitrosonium ion whereas rhenium hexafluoride does. Considering a similar lattice energy of about -125 kcal mole^{-1} for NO^+MF^- and the ionization potential of 68 kcal mole^{-1} for nitric oxide, they showed that $A_{(WF_6)}$ must be less negative than -90 kcal mole^{-1}, while that for rhenium hexafluoride must be more so. Using a modified Born–Haber cycle which included the heat of dissociation of nitrosyl fluoride to nitric oxide and fluorine, they calculated that for the observed reaction

$$IrF_6 + ONF \longrightarrow NO^+IrF_6^- + \tfrac{1}{2}F_2$$

to occur, $A_{(IrF_6)}$ must be more negative than -126 kcal mole^{-1}.

They demonstrated that $A_{(MF_6)}$ increases (exothermally) in the series $WF_6 < ReF_6 < OsF_6 < IrF_6 < PtF_6$ with $A_{(ReF_6)} > -90$, $A_{(IrF_6)} > -126$ and $A_{(PtF_6)} > -156$ kcal mole^{-1}. They commented that $A_{(MF_6)}$ appears to increase by about -20 kcal mole^{-1} for unit increase in atomic number of the third row metal M.

Consideration of the chemical reactivities of the *d*-transition element fluorides reinforces the simple and long-established generalization discussed in section 5.2 that the higher oxidation states for fluorides of first-row transition metals are considerably less stable than the corresponding states for the fluorides of the second and third row, i.e. that chemical reactivity of higher first-row fluorides is much greater than for the heavier elements.

However, a survey of more recently studied reactions of higher fluorides brings out the relatively new generalization that, whereas the fluorides of zirconium and hafnium and even those of niobium and tantalum show very similar chemistry, there is a growing dissimilarity in chemical reactivity of fluorides for the pairs of second- and third-row elements which follow.

While oxidation of lower non-metal fluorides showed first-row metal higher fluorides to be much more reactive than those of the second and third rows, they allowed no discrimination in reactivity between pairs of second- and third-row fluorides.

Carbon disulphide does not react with either niobium or tantalum pentafluoride; but it shows clearly that tungsten hexafluoride is much more inert than its formal counterpart, molybdenum hexafluoride, the latter reacting readily to form bistrifluoromethyl-disulphide and other products whereas the former does not react. It should be stressed that the enormous difference in reactivity between higher fluorides of chromium and molybdenum is demonstrated in the carbon disulphide reaction. For chromium, even the pentafluoride oxidizes carbon disulphide to carbon tetrafluoride and sulphur hexafluoride. Chromium hexafluoride would react even more readily.

The covalent chlorides, such as the trichlorides of boron and of Main Group V and the tetrachlorides of titanium and of Main Group IV, demonstrate very specifically the growing chemical differences between pairs of second- and third-row elements in proceeding across the transition metal series. Thus there are discernible kinetic differences in the similar overall reactions of the tetrafluorides of zirconium and hafnium with boron trichloride, with greater kinetic differences when the corresponding reactions of niobium and tantalum are examined. While molybdenum and tungsten hexafluoride each react with both boron trichloride and titanium tetrachloride, molybdenum hexafluoride alone reacts with all of the other covalent chlorides listed in the opening of this paragraph. Also, whereas molybdenum hexafluoride reacts with all chlorides of Groups IA and IIA, tungsten hexafluoride reacts only with beryllium chloride.

Taken together with the highest fluorides of Groups IVA and VA, the transition metal hexafluorides illustrate best the increase in reactivity with increase in atomic number in any row and, more particularly, the increasing differences between second- and third-row elements with increasing atomic number. However, it should be stressed that, while information on pentafluorides and lower fluorides is sketchy and fragmentary, all of the reported observations are consistent with the generalizations set out here.

In summary, *the relatively new generalization that emerges is that, considering reactions of highest fluorides of pairs of second- and third-row transition elements, the tetrafluorides of zirconium and hafnium and the pentafluorides of niobium and tantalum undergo similar reactions but show increasing kinetic differences in reactivity. Molybdenum and tungsten both form highest fluorides of identical stoichiometry, but these undergo substantially different reactions. Technetium and rhenium do not even form the same highest fluorides. In so far as the evidence is available, this widening in the difference between second- and third-row elements persists into the noble metal compounds.*

5.4. THERMODYNAMIC STABILITY OF *f*-TRANSITION METAL FLUORIDES

(a) Lanthanide Fluorides

The chemistry of the lanthanide elements is dominated by the stability of oxidation state III, particularly in aqueous reactions. The extent to which other oxidation states occur for a range of compounds, including oxides, halides and sulphates, has been discussed in a review by Asprey and Cunningham[178].

Considering the fluorides specifically, lanthanide trifluorides are extraordinarily stable and can be prepared simply by precipitation from aqueous solution, by reaction with anhydrous hydrogen fluoride or, for most members of the series, by direct fluorination. However, tetrafluorides are formed easily for cerium and terbium. Fluorine reacts at atmospheric pressure and at relatively low temperatures, e.g. in the range 200–400°C, with the metals, with trifluorides or with the anhydrous chlorides in each case to yield the tetrafluorides. Reaction with the metals themselves can be dangerously vigorous if fluorine is not diluted.

The two dioxides CeO_2 and TbO_2 can be formed by heating lower oxides in oxygen without recourse to high pressures of oxygen, whereas praseodymium dioxide is formed from lower oxides only on heating with high pressure oxygen. Because calculations indicated that it should be thermodynamically stable, the possibility of preparation of praseodymium tetrafluoride by fluorination of the trifluoride was investigated very carefully but without success. Subsequently, fluoro-complexes of praseodymium(IV), such as $NaPrF_5$ and Na_2PrF_6, were prepared relatively easily by fluorination of mixtures of alkali chlorides and praseodymium trichloride. A very elegant preparative approach, in which liquid anhydrous hydrogen fluoride was streamed through the complex suspended on an inert sintered material, yielded the solid tetrafluoride which has been shown by X-ray powder diffraction measurements to be isostructural with other lanthanide and actinide tetrafluorides.

Fluorination at pressures of 1–3 atm and at 400°C of a mixture of caesium chloride and neodymium trichloride or dysprosium trichloride gives evidence for the partial formation of compounds such as Cs_3NdF_7 and Cs_3DyF_7. The evidence for the existence of these complexes is discussed by Asprey and Cunningham[178].

[178] L. B. Asprey and B. B. Cunningham, in *Progress in Inorganic Chemistry*, ed. F. A. Cotton, Vol. 2, Interscience (1960), pp. 267–302.

Whereas tetrafluorides of some lanthanides are formed easily, fluorides of lanthanides in oxidation state II have proved much more difficult to prepare than corresponding sulphates, compounds such as europium(II) sulphate being relatively easily obtained. In all cases reported, the product of a reaction designed to produce the difluoride appears to be a solid solution of the appropriate trifluoride and difluoride. Thus reduction by hydrogen of molten europium trifluoride produces a mixture of the two fluorides. The closest approach to pure ytterbium difluoride has been the formation of a product of stoichiometry $YbF_{2.24}$ and metathetical reactions between samarium(II) sulphate and calcium fluoride have resulted in single phase solid solutions in the range $SmF_{2.2}$ to $SmF_{2.5}$.

(b) Actinide Fluorides

Actinium, although not an *f*-transition element, provides a convenient reference point for considering oxidation states of the fluorides of the elements of the last row of the Periodic Classification. A unique oxidation state of III characterizes all actinium chemistry and the trifluoride can be prepared from aqueous solution, or with hydrogen fluoride, fluorine or halogen fluorides.

It is interesting to compare the stabilities of oxidation states for the metals titanium, zirconium and hafnium in their halogen compounds with those of thorium, for which the ground state electronic configuration of $6d^2 7s^2$ is generally accepted. Fluorides lower than the tetrafluoride occur for titanium and zirconium but not for hafnium and thorium. The comparison can be extended profitably to include halides other than fluorides. With the surprising exception of hafnium di-iodide, which should be formed more readily than the other hafnium dihalides, all of the halides TiX_3, TiX_2, HfX_3 and HfX_2 have been reported for X = Cl, Br and I. For thorium only two lower halides have been reported, the di-iodide and the less satisfactorily characterized tri-iodide. However, the compound having the stoichiometry ThI_2 has been described as $Th^{4+}(e^-)_2(I^-)_2$, i.e. a thorium(IV) compound with delocalized electrons in the solid[179].

Using the facts from the previous paragraph, one could include thorium with titanium, zirconium and hafnium and extend the generalization of section 5.2(b) that for a group of typical *d*-transition elements the stability of higher oxidation states increases regularly with increasing atomic number.

For protactinium, both a penta- and a tetrafluoride can be prepared relatively easily, although oxidation state IV seems somewhat stabler than V in binary fluorides. The pentafluoride is easily reduced to the tetrafluoride by heating with hydrogen but fluorination at about 700°C is required for conversion back. There is no evidence for a trifluoride.

If protactinium were a typical *d*-transition element, it should yield a lower fluoride even less readily than does tantalum, for which no fluoride lower than the pentafluoride has been prepared. The existence and stability of protactinium tetrafluoride is the first evidence in this discussion of the cross-over from *d*- to *f*-transition element character which has been described by Katz and Seaborg[180].

They discuss the relative ease of involvement of *f*-electrons in covalent bonding in compounds of lanthanides and actinides. They state that in the lanthanides the 4*f* electrons do not extend beyond the 5*d* and 6*s* orbitals and, being effectively screened by the 5*d* and 6*s* electrons, are normally not available for bonding. In the actinides, particularly those

[179] L. J. Guggenberger and R. A. Jacobson, *Inorg. Chem.* 7 (1968) 2257.
[180] Ref. 167, pp. 463–470.

immediately following protactinium, the $5f$ electrons are not shielded because they have a greater spatial extension relative to the $6d$ and $7s$ orbitals than in the corresponding lanthanide case. Also, relative to the $6d$ electrons, they have lower binding energies and therefore can be involved relatively easily in bonding. Katz and Seaborg note, however, that for actinium and thorium the binding energies of $6d$ electrons are lower than of $5f$. They postulate, from a consideration of spectral and chemical data, that a cross-over in magnitudes of binding energies for $5f$ and $6d$ electrons occurs at about protactinium. However, they point out that the electronic configuration may differ in a compound from that in a gaseous atom, being dependent on the oxidation state of the element, the type of compound or even the physical state of a particular compound.

Their general hypothesis is quite consistent with the observed thermodynamic stabilities of the fluorides of actinium, thorium and protactinium compared with those of lanthanum, hafnium and tantalum. However, it should be noted that, while protactinium with a stable tetrafluoride and pentafluoride does not resemble tantalum and therefore does not fit into the pattern of a developing series of *d*-transition elements, its chemistry is just as surely not that of a typical lanthanide because it has no trifluoride.

Uranium exhibits a formal oxidation state of VI in the easily prepared hexafluoride; but, again, the cross-over from *d*- to *f*-transition element behaviour can be demonstrated by comparing the fluorides of chromium, molybdenum, tungsten and uranium. As shown in section 5.2(b), there is a marked increase in stability of higher fluorides in passing from chromium through molybdenum to tungsten. If uranium were a typical *d*-transition metal it should form fluorides below the hexafluoride with great difficulty. In fact many fluorides can be formed, the tetrafluoride being particularly stable. The intermediate fluorides UF_5, U_2F_9 and U_4F_{17} are all formed by equilibration of uranium hexafluoride and tetrafluoride, and each disproportionates at elevated temperature to these two fluorides. Katz and Seaborg[181] list the conditions for formation and thermal disproportionation of the intermediate fluorides.

Unlike tungsten, uranium yields a trifluoride by high-temperature reduction of the tetrafluoride with metallic uranium or some other very strong reductant. Even though the trifluoride is easily oxidized, its existence indicates a great difference between tungsten and uranium, a difference which again can be discussed in terms of *f*-element character in uranium. However, this point is demonstrated more effectively in considering the relative chemical reactivities of the fluorides of chromium, molybdenum, tungsten and uranium in section 5.5(d) below.

Proceeding to neptunium, there occurs the most immediately obvious difference between an *f*-element and those which might be considered as the formally analogous Group VII A *d*-elements. For the latter, increasing stability of oxidation state VII with increasing atomic number culminated in the existence of a stable binary heptafluoride ReF_7. However, for neptunium, which if it were a *d*-element should yield the heptafluoride even more readily, the highest binary compound is the hexafluoride and the trifluoride can be prepared easily by reduction of the dioxide with hydrogen and hydrogen fluoride at 500°C. Fluorination gives the tetrafluoride easily; but preparation of the hexafluoride requires special techniques [see section 5.1(b)]. Surprisingly, the pentafluoride has not been prepared, although calculations indicate a large negative value for its enthalpy of formation. However, compounds containing neptunium(V) such as NpF_6^-, NpF_7^{2-} and NpF_8^{3-} are described[163–4].

[181] Ref. 167, p. 157.

In summary, neptunium, with no heptafluoride and easily attained lower fluorides, differs markedly from rhenium, the *d*-element which might be regarded as its formal analogue.

For plutonium there are no fluoro-compounds of oxidation states VIII and VII corresponding to OsO_3F_2 and OsF_7. The hexafluoride can be prepared, but it is thermally unstable and is best prepared at high fluorine pressure or using the "cold finger" technique described in section 5.1(b). The lower fluorides are easily prepared, and the compound Pu_4F_{17}, analogous to the intermediate fluoride of uranium, has been reported. Again, no pentafluoride has been prepared although solids with anions PuF_6^- and PuF_7^{2-} are known[163-4].

Decrease in stability of oxidation state VI in passing from uranium hexafluoride to the thermally unstable plutonium hexafluoride, which is converted completely to the tetrafluoride in 1 hr at 280°C, can be demonstrated by the experimental temperatures required for direct fluorination of lower fluorides to the hexafluorides in each case. Temperatures of 220°, 500° and 750°C are required for smooth conversion to the hexafluoride in the case of the uranium, neptunium and plutonium compounds respectively. Weinstock[182] considers the equilibrium

$$MF_4 + F_2 \rightleftharpoons MF_6$$

at room temperature, and for the equilibrium constant

$$K_p = [MF_6]/[F_2]$$

gives values of 10^{44} for uranium hexafluoride and 4×10^{-5} for the plutonium compound.

Further evidence for decrease in stability of oxidation states VI along the actinide series is that to date none of the many efforts to prepare americium hexafluoride has been successful. Only in the recently prepared and characterized americium dioxide difluoride AmO_2F_2 [183], which is isostructural with the corresponding uranium, neptunium and plutonium dioxide difluorides, is this oxidation state shown for isolated fluoro-compounds of americium. The trifluoride is the most easily prepared fluoride. It is significant that it results from reaction of americium dioxide with a mixture of hydrogen fluoride and oxygen whereas, under the same conditions, plutonium dioxide yields the tetrafluoride. Americium trifluoride can be prepared in the anhydrous state by precipitation from solution, as can its lanthanide counterpart; but americium differs from europium in that the trifluoride can be fluorinated at 500°C to the tetrafluoride. It is noteworthy that for americium no fluoride lower than the trifluoride has been prepared. By comparison with europium, a reasonably stable oxidation state of II might be expected. However, evidence for existence of this oxidation state has been reported only under such drastic conditions as high-temperature dissociation of americium trichloride in liquid plutonium metal.

Another marked difference between actinide and lanthanide character lies in the easy fluorination at 400°C of curium trifluoride to the tetrafluoride, which is isostructural with the tetrafluorides of uranium, neptunium, plutonium and americium. By analogy with

[182] B. Weinstock, *Chem. Eng. News*, **42** (Sept. 21, 1964) 86.
[183] T. K. Keenan, *Inorg. Nucl. Chem. Letters*, **4** (1968) 381.

gadolinium, to which the $4f^75d^16s^2$ ground-state configuration is assigned, a unique oxidation state of III might be expected.

The reported preparation of berkelium tetrafluoride[184] provides a direct parallel between berkelium and terbium. To date there is no evidence for fluorides in oxidation states other than III for elements beyond berkelium; but, of course, the quantities available for study and experimental difficulties in handling the elements make any generalizations tentative in the extreme.

(c) Summary

In very broad outline, the oxidation states other than III exhibited by fluorides of the lanthanides are consistent with the frequently quoted empirical observation that the most stable electronic configuration for these elements in their compounds appears to be one in which the $4f$ orbital is either empty, half-filled or completely filled. That is oxidation state IV occurs easily in fluorides of cerium and terbium and oxidation state II is indicated for europium and ytterbium. However, the rationalization based on extent of filling of f-orbitals is obviously far from perfect when it is recalled that oxidation state IV can be attained for praseodymium and, less easily, for neodymium and dysprosium and that samarium gives evidence of oxidation state II.

The elements following actinium in the last row of the Periodic Classification, the actinides, provide a fascinating series because it is impossible to classify them as precisely as in the case of other series of elements, e.g. the lanthanides or the first, second or third row of d-transition elements.

In representations of the Periodic Classification based on ground-state electronic configurations they are shown almost invariably as though they were a series similar to the lanthanides. If thermodynamic stability of their compounds, particularly the fluorides, is considered, a reasonably strong case for this classification can be made only for the very heavy actinides from about curium onwards.

The halogen chemistry, as well as the other chemistry of actinium and thorium, follows precisely that of the d-elements lanthanum and hafnium. While the chemistry of protactinium has obvious similarities to that of niobium and tantalum, the differences, e.g. the evidence of a tetrafluoride, are sufficiently significant to indicate that there has been an interruption in the development of a simple series of d-elements. As Katz and Seaborg have proposed, a cross-over from d- to f-element character occurs at about this point[180].

From uranium through to americium the chemistry of each element, as characterized by stability of higher and of lower fluorides, becomes progressively less like the d-elements tungsten to iridium. Hexafluorides become less stable thermally from uranium to plutonium while, for americium, oxidation state VI can be achieved in an isolated compound only as the dioxide difluoride. At the same time, oxidation state III, obtained with great experimental difficulty in uranium trifluoride, dominates in the stable americium trifluoride. However, these elements cannot be said to resemble to any degree the lanthanides. As has been suggested above and will be discussed in detail later in section 5.6(c), some rationalization of the chemistry of these elements can be made by postulating the involvement of f-electrons in covalent bonding in their compounds.

Beyond americium, the actinides become progressively more like the corresponding lanthanides with increase in atomic number, although the easily prepared curium tetra-

[184] L. B. Asprey and T. K. Keenan, *Inorg. Nucl. Chem. Letters*, **4** (1968) 537.

fluoride has no gadolinium counterpart. Also there is no americium difluoride to correspond with the europium compound. Overall, the occurrence of a single or dominant oxidation state III is not characteristic of the actinides after actinium. There is growing evidence for a stable oxidation state of II in the very heavy actinides, which at present are available only in micro-amounts.

5.5. CHEMICAL REACTIVITY OF HIGHER FLUORIDES OF *f*-TRANSITION ELEMENTS

There has been virtually no study of chemical reactions into which lanthanide fluorides might enter apart from their ability to form fluoro-complexes and their behaviour in aqueous medium. It is known that the trifluorides are sparingly soluble in water and can be precipitated as relatively pure, stoichiometric compounds. The tetrafluorides, excepting that of cerium, would all oxidize water readily. On thermodynamic grounds, cerium(IV) should also oxidize water, but the kinetics of reaction are so slow that reaction appears negligible. The ill-defined lanthanide difluorides would all reduce water.

As previously stated, Katz and Seaborg commented in 1957[167] that it was strange that despite the technological importance of the uranium fluorides and the great amount of work done on determining their physical properties, so little of their chemistry was known. Their statement applies with even greater validity to chemical study of fluorides of other actinides, even those of elements such as protactinium, neptunium and plutonium which have been readily available for many years. As recently as 1968, Brown was able to comment[185] that the little chemical work which had been done on fluorides applied mostly to uranium hexafluoride.

A small number of reactions was reported in the chemical literature resulting from investigations during World War II, and there have been a small number of other studies, particularly for uranium fluorides, reported since 1960. As far as possible, the reactions of the fluorides of the *f*-transition metals will be presented here on a similar basis to those of the *d*-element fluorides to allow comparisons to be made between the classes of elements. However, for each type of reaction, much less information is available for higher fluorides of actinides than for the *d*-transition metal compounds.

(a) Oxidation of Lower Fluorides of Non-metals

Following a somewhat earlier study[186] in which it was found that phosphorus and arsenic trifluorides reduce uranium hexafluoride to the tetrafluoride at room temperature, O'Donnell and Wilson have shown recently[187] that the fluorides intermediate between uranium hexafluoride and uranium tetrafluoride, namely the pentafluoride and those with formulae U_2F_9 and U_4F_{17}, undergo the same chemical reactions as the hexafluoride. The intermediate fluorides, although they are single compounds, react as though they were stoichiometric mixtures of the hexafluoride and the tetrafluoride.

Uranium hexafluoride has been shown to be stable in the presence of sulphur tetrafluoride at 300°C. In fact it is produced by the action of sulphur tetrafluoride on uranium oxides or on uranyl fluoride at temperatures between 300° and 400°C[188]. It is reduced by

[185] Ref. 163, p. 27.
[186] T. A. O'Donnell, D. F. Stewart and P. W. Wilson, *Inorg. Chem.* **5** (1966) 1438.
[187] T. A. O'Donnell and P. W. Wilson, *Australian J. Chem.* **22** (1969) 1877.
[188] C. E. Johnson, J. Fischer and M. J. Steindler, *J. Am. Chem. Soc.* **83** (1961) 1620.

the tetrafluoride only at temperatures above 500°C, the reduction probably resulting from the disproportionation of sulphur tetrafluoride in that temperature range. In contrast, however, plutonium hexafluoride is reduced readily by sulphur tetrafluoride at temperatures as low as 30°C.

Similarly, bromine trifluoride can be used to prepare uranium hexafluoride from a wide range of uranium-containing compounds; but it is readily oxidized at room temperature to bromine pentafluoride by plutonium hexafluoride which is itself reduced to the tetrafluoride. Neptunium hexafluoride oxidizes bromine trifluoride very slowly at room temperature, but more readily with increase in temperature.

(b) Oxidation of Nitric Oxide, Xenon, Halogens and Carbon Disulphide

Nitric oxide reduces uranium hexafluoride to produce $NO^+UF_6^-$ while nitrosyl fluoride forms a 1 : 1 adduct. By comparison, nitric oxide does not react with tungsten hexafluoride but is oxidized by molybdenum hexafluoride to the nitrosonium ion in $NO^+MoF_6^-$. $NO^+UF_6^-$ is also produced by reaction of uranium hexafluoride with nitrogen dioxide, whereas the dioxide does not react with either tungsten or molybdenum hexafluoride.

In an experimental survey following immediately after Bartlett's announcement of the reaction between platinum hexafluoride and xenon, Malm and co-workers investigated the possibility of reaction between the rare gas and a series of metal hexafluorides. Of the actinide compounds, plutonium hexafluoride reacted but uranium hexafluoride did not[170].

The difference in reactivity of the hexafluorides of uranium and plutonium is demonstrated still further by the fact that the latter oxidizes elemental bromine and iodine but the former does not, and in their reactions with carbon disulphide. At 25°C, uranium hexafluoride reacts with carbon disulphide vapour to form uranium tetrafluoride, sulphur tetrafluoride and the bistrifluoromethylpolysulphides, $(CF_3)_2S_2$ and $(CF_3)_2S_3$, whereas at higher temperatures some carbon tetrafluoride and sulphur hexafluoride are formed. However, the main product of the gas-phase reaction with plutonium hexafluoride is carbon tetrafluoride. As discussed in section 5.3(c), this indicates that plutonium hexafluoride is a stronger oxidant than the corresponding uranium compound because under mild conditions both carbon–sulphur linkages are broken in the plutonium reaction.

The oxidation reactions discussed in the previous section and in this section show that, as an oxidant, uranium hexafluoride is much more reactive than tungsten hexafluoride and somewhat stronger than molybdenum hexafluoride. They also show that, along the actinide series, plutonium hexafluoride is a stronger oxidant than uranium hexafluoride, i.e. stability of oxidation state VI decreases.

(c) Halogen-exchange Reactions with Chlorides

Apart from a report that, in reacting with carbon tetrachloride, plutonium hexafluoride is reduced to the tetrafluoride, the only reported reactions of actinide fluorides with chlorides are those of uranium fluorides. Even so, the literature, drawn in large part from war-time reports, is misleading in that it is usually implied or stated that when uranium hexafluoride reacts with chlorides it is always reduced to the tetrafluoride.

In work published since 1966, O'Donnell and co-workers have made a detailed study of uranium fluorides with a wide range of chlorides, covalent and ionic[186–7, 189]. They have

[189] T. A. O'Donnell and P. W. Wilson, *Australian J. Chem.* **21** (1968) 1415.

shown that uranium hexafluoride reacts very rapidly at room temperature with a stoichiometric excess of a covalent chloride such as boron or aluminium trichloride to form uranium hexachloride and the appropriate trifluoride from the other reactant. This is the thermodynamically favoured reaction if ΔG per mole of uranium hexafluoride is calculated for each of the two possible alternative reactions:

$$UF_6+2XCl_3 \longrightarrow UCl_6+2XF_3$$

or

$$3UF_6+2XCl_3 \longrightarrow 3UF_4+2XF_3+3Cl_2$$

When, however, the hexafluoride was in excess in these two reactions, the products were uranium tetrafluoride, chlorine and boron or aluminium trifluoride. They explained this by showing that the hexafluoride and hexachloride of uranium react spontaneously at room temperature to form uranium tetrafluoride and chlorine, for which reaction ΔG is -29 kcal per mole of uranium hexafluoride. This spontaneous reaction follows the initial formation of uranium hexachloride when the hexafluoride is in excess in the overall reaction.

However, when uranium hexafluoride reacts with arsenic and antimony trichlorides and with silicon tetrachloride, the uranium-containing product is the tetrafluoride, even though

TABLE 13. PRODUCTS OF REACTION BETWEEN IONIC CHLORIDES AND HEXAFLUORIDES

Reactants	MoF_6	WF_6	UF_6
LiCl, NaCl (MCl)	$Mo_3Cl_9(MoF_6)_3$, MF	No reaction	MUF_5, MF
KCl, RbCl, CsCl	$Mo_3Cl_9(MoF_6)_3$, MF	No reaction	No reaction
$BeCl_2$	$Mo_3Cl_9(MoF_6)_3$, BeF_2	WCl_6, BeF_2	UF_4, BeF_2
$MgCl_2$	$Mo_3Cl_9(MoF_6)_3$, MgF_2	No reaction	UF_4, MgF_2
$CaCl_2$, $SrCl_2$, $BaCl_2$	$Mo_3Cl_9(MoF_6)_3$, $M'F_2$	No reaction	UF_4, $M'F_2$

TABLE 14. VALUES OF ΔG (kcal per mole UF_6) FOR POSSIBLE URANIUM HEXAFLUORIDE REACTIONS

Reactants	$UF_6+6MCl^a \rightarrow UCl_6+6MF$	$UF_6+2MCl \rightarrow UF_4+Cl_2+2MF$
LiCl	−38·7	−30·5
NaCl	+20·7	−10·7
KCl	+65·1	+4·1
RbCl	+81·9	+9·7
CsCl	+93·9	+13·7
$BeCl_2$	−138·0	−63·6
$MgCl_2$	−84·9	−45·9
$CaCl_2$	−51·3	−34·7
$SrCl_2$	−24·3	−25·7
$BaCl_2$	+2·4	−16·8

[a] Or $3M'Cl_2$, etc.

in the case of the silicon tetrachloride reaction the formation of uranium hexachloride is favoured thermodynamically.

O'Donnell, Stewart and Wilson postulated[186] that in this instance kinetic factors override thermodynamic considerations. The reaction is relatively slow by comparison with the boron and aluminium trichloride reactions, and *effectively*, uranium hexafluoride is in excess throughout the reaction and reacts with the hexachloride as the latter is formed.

Reactions of uranium hexafluoride with alkali and alkaline-earth chlorides were next studied[189]. Lithium and sodium chlorides reacted to form the alkali fluoride, chlorine and the complexes $LiUF_5$ or $NaUF_5$. The other three alkali chlorides did not react with uranium hexafluoride at room temperature. All of the alkaline-earth chlorides reacted with uranium hexafluoride to give uranium tetrafluoride, chlorine and the appropriate alkaline-earth fluoride. Tables 13 and 14 list these reaction products and values of ΔG for possible reactions which might occur. There is excellent correlation between observed reactivities and ΔG values. Reactions of uranium hexafluoride with the ionic chlorides were found to be much slower than with the covalent chlorides. Therefore, in terms of the postulates given above, formation of uranium hexachloride would not be expected in reactions with alkali or alkaline-earth chlorides. Again, uranium hexafluoride is always in effective excess and reacts with any uranium hexachloride as it forms to give uranium tetrafluoride and chlorine.

As in their oxidizing reactions with lower fluorides of non-metals, the intermediate fluorides UF_5, U_2F_9 and U_4F_{17} were shown by O'Donnell and Wilson[187] to react with chlorides as though they were stoichiometric mixtures of uranium hexafluoride and tetrafluoride, even though structurally they are known to be single compounds. They enter into the same reactions as uranium hexafluoride but they react more slowly than the hexafluoride.

In a brief war-time report, Calkins[190] stated that uranium tetrafluoride reacted at elevated temperature with the trichlorides of boron, aluminium and phosphorus to form uranium tetrachloride, but that it did not react with silicon tetrachloride or phosphorus pentachloride. O'Donnell and Wilson[187] found that reaction between uranium tetrafluoride and beryllium chloride commenced at 440°C, the melting point of beryllium chloride.

(d) Summary

The available data on chemical reactions of higher fluorides of actinide elements support the general proposition put forward in section 5.4(c) that in this series of elements stability of the highest formal oxidation state decreases progressively with increasing atomic number. However, for fluorides of elements beyond uranium studies of chemical reactivities are rare and, where they do exist, fragmentary. For uranium, however, study of the reactivity of its fluorides has allowed a meaningful comparison with the formally corresponding *d*-elements, chromium, molybdenum and tungsten, as will be seen later in this section.

No reactions of actinium trifluoride or thorium tetrafluoride provide any evidence of an oxidation state other than the highest possible formal state. As discussed in section 5.4(c), the ease of reduction of protactinium pentafluoride by hydrogen shows that protactinium is not acting like its formal counterpart tantalum. However, very little other chemistry of protactinium pentafluoride has been reported, apart from the ease of formation of fluoro-complexes or oxide fluorides of protactinium(V).

As with the *d*-transition elements, the hexafluorides of the actinides provide the most useful comparisons of chemical reactivity. As stated in section 5.4(b), increasingly high temperatures are required to fluorinate the appropriate lower fluorides to the hexafluorides of uranium, neptunium and plutonium and, as shown in section 5.5, these three hexafluorides constitute a series of increasing oxidative strength in reactions with compounds such as the halogen fluorides or carbon disulphide. Direct evidence of this trend is given by the fact that plutonium hexafluoride readily oxidizes uranium tetrafluoride to the hexafluoride above 200°C.

The actinide hexafluoride most studied is that of uranium and from its chemical reactivity

190 V. P. Calkins, USAEC Report CD–O. 350. 4 (1945).

emerges evidence, additional to the physical evidence based on spectral and magnetic properties, for *f*-electron involvement in bonding in uranium(VI) compounds. This bonding will be discussed in section 5.6(c); but at this point the appropriate chemical observations will be summarized.

It has been established in section 5.3 that, within any one group of *d*-transition metals, stability of highest oxidation states increases with increasing atomic number, i.e. reactivity of higher fluorides decreases. The systematic studies on the higher fluorides of chromium, molybdenum and tungsten show this conclusively. Further, it is generally accepted that bonding in thorium(IV) compounds involves *s*- and *d*-electrons and the same trend in stabilities is exemplified by the fluorides and other halides of titanium, zirconium, hafnium and thorium. If, then, bonding in uranium hexafluoride were similar to that in the three corresponding fluorides of chromium, molybdenum and tungsten, uranium hexafluoride would be expected to be very inert chemically, less reactive in fact than tungsten hexafluoride.

It should be recalled that tungsten hexafluoride has been shown not to oxidize nitric oxide, carbon disulphide or lower fluorides, excepting phosphorus trifluoride, and to undergo halogen-exchange reactions only with compounds like boron trichloride and titanium tetrachloride which enter these reactions most readily.

However, uranium hexafluoride enters halogen-exchange reactions readily and oxidizes nitric oxide, carbon disulphide and both phosphorus and arsenic trifluorides. By comparison molybdenum hexafluoride oxidizes nitric oxide, carbon disulphide and phosphorus trifluoride, but not arsenic trifluoride. Therefore, while a much weaker oxidant than chromium hexafluoride, uranium hexafluoride, is observably stronger than molybdenum hexafluoride—the most direct evidence of this being that uranium hexafluoride oxidizes molybdenum pentafluoride to hexafluoride.

These observations have been put forward[186] as chemical evidence that the bonding in uranium hexafluoride differs from that in hexafluorides of *d*-elements because *f*-electrons are involved in bonding in the compound. This point will be elaborated on in section 5.6(c).

5.6. BONDING AND STRUCTURE IN HIGHER TRANSITION METAL FLUORIDES

While detailed discussions of bonding in *d*- and *f*-transition metal compounds generally are given in the appropriate chapters of this treatise, the discussion in this section is pertinent to the observed stabilities and reactivities given in this particular review of fluorides of transition metals. Also, detailed structures of individual *d*- and *f*-element fluorides will not be given. General structural trends will be outlined briefly and references given to more detailed sources.

(a) Correlation of Reactivities with Force Constants and Bond Energies for Higher Fluorides

It is generally conceded as unwise to attempt a rigorous correlation of bond strengths and reactivities with force constants calculated from spectral measurements. However, in a chemically related series of compounds of similar structure and bonding, there appears to be good correlation between strength and reactivity of a bond and its stretching frequency ν_1. It is to be appreciated that this frequency is related to vibrational displacement of a terminal atom about an equilibrium position and not to its complete removal, as in consideration of bond energy. However, since in this case neither the symmetry of the molecule as a whole nor the position of the central atom is changed, it appears to be the most suitable frequency to consider.

Values of ν_1 for the hexafluorides all lie in the range 628–771 cm^{-1} and Fig. 13 shows that, for hexafluorides in any row of transition metals, frequencies and hence bond strengths fall with increasing atomic number. However, there is a marked increase in passing from second- to third-row hexafluorides. It is of interest that these values of ν_1 are in the same range as those for sulphur, selenium and tellurium hexafluorides which lie at 773, 708 and 701 cm^{-1} respectively. This suggests that bonding is similar in hexafluorides of *d*-transition metals

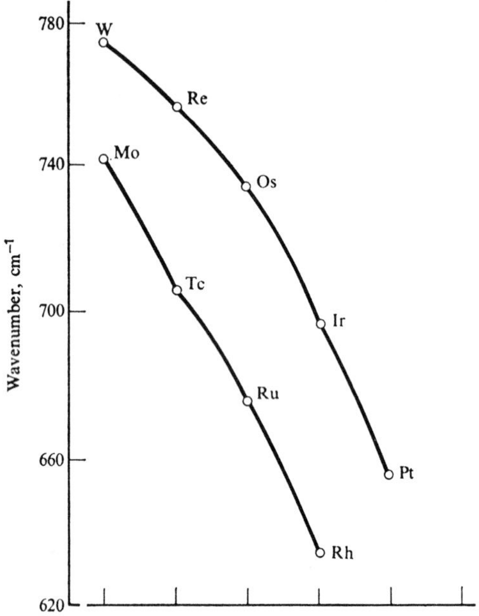

FIG. 13. Metal-fluorine stretching frequencies for hexafluorides.

and of Group VIB elements, although it should be noted that for the main group hexafluorides, frequencies decrease with increasing atomic number.

Table 15 lists bond stretching force constants for second- and third-row transition metal hexafluorides as well as those for actinides. Decrease in bond strength, i.e. increase in reactivity, with increase in atomic number is indicated for hexafluorides in each row. For

TABLE 15. BOND STRETCHING FORCE CONSTANTS (md/Å)

MoF_6 4·84	TcF_6 4·79	RuF_6 4·61	RhF_6 4·27	
WF_6 5·20	ReF_6 5·17	OsF_6 5·14	IrF_6 4·94	PtF_6 4·52
UF_6 3·78	NpF_6 3·71	PuF_6 3·59		

any group, the force constant for a third-row hexafluoride is greater than that for the second row. Also the order $UF_6 < MoF_6 < WF_6$ should be noted. However, although the order of magnitude of force constants is the same as that of reactivities in such a situation, care must be used in applying the arguments too rigorously. The values quoted would suggest that uranium hexafluoride is a very much stronger oxidant than molybdenum hexafluoride, whereas, chemically, it is only marginally so.

If values of force constants were taken as a direct reflection of reactivity of the appropriate hexafluorides, iridium hexafluoride would appear to be less reactive than molybdenum hexafluoride, whereas the former oxidizes nitrosyl fluoride to fluorine and forms the complex $NOIrF_6$ while the latter hexafluoride forms a 1:1 adduct with nitrosyl fluoride without change in oxidation state. On the basis of values of force constants alone, platinum hexafluoride which oxidizes xenon and halogen fluorides would appear less reactive than uranium hexafluoride which does not. The difference in reactivity of platinum and uranium hexafluorides is further illustrated by the fact that platinum hexafluoride can oxidize plutonium tetrafluoride to plutonium hexafluoride, which is itself much more reactive than uranium hexafluoride.

The correlation between force constants and chemical reactivity appears to be valid in a series in which bonding in each member compound is similar. The difference in bonding in the hexafluorides of molybdenum and uranium, which are rather similar chemically, is reflected in the values of bond force constants for each compound. A remarkably similar difference in bond force constants occurs for the two hexafluorides—ruthenium and plutonium. Each has the same number of non-bonding electrons and they are very similar chemically, e.g. each oxidizes xenon. Again, the difference in force constants could well reflect the difference in bonding in the two hexafluorides as a result of involvement of *f*-electrons in the case of the plutonium compound.

TABLE 16. AVERAGE THERMOCHEMICAL BOND ENERGIES (kcal $mole^{-1}$)[a]

TiF_4 140	VF_5[b] 114	CrF_6[c] (78)
ZrF_4 155	NbF_5 136	MoF_6 107
HfF_4 155	TaF_5 144	WF_6 121

[a] All values, except for VF_5 and CrF_6, from W. N. Hubbard, Abs. of Papers, *24th Annual Calorimetry* (*Gordon*) *Conference, Portsmouth, NH, USA, Oct.* 1969, pp. 12-15.

[b] R. G. Cavell and H. C. Clark, *Trans. Faraday Soc.* **59** (1963) 2706.

[c] Estimated value, W. E. Dasent, *Non-existent Compounds*, Edward Arnold, London (1965) p. 138.

A limited number of values of average thermochemical bond energies is available for transition metal fluorides. Some of those listed in Table 16 are described as preliminary data in a programme of fluorine bomb calorimetry initiated by W. N. Hubbard. However, they illustrate the great difference in reactivity between the metal of a first row and its second- and third-row counterparts. More particularly, they bear out the divergence in chemical reactivity between successive pairs of second- and third-row elements.

(b) Bonding and Structures in Higher Fluorides of *d*-Transition Metals

The hexafluorides of the *d*-transition metals are all octahedral arrangements of six fluorines around the central metal atom. Except for four of them which are reported to exhibit some Jahn–Teller distortion, they are quite regular in the vapour and liquid with some tetragonal distortion in the solid state. Weinstock[182] has reviewed the evidence for their structures, including that for the distortion, based on Raman and infrared spectral investigations and on electron diffraction studies. He also discusses the effect of their structures on their volatilities, heat capacities and magnetic properties as well as the apparently anomalous bond lengths in hexafluorides of *d*- and *f*-transition metals.

For the octahedral hexafluoride molecules *as formed*, conventional bonding models are quite adequate, although it is obvious that the availability of *d*-orbitals for octahedral bonding varies by rows, becoming progressively more available further across a row as we

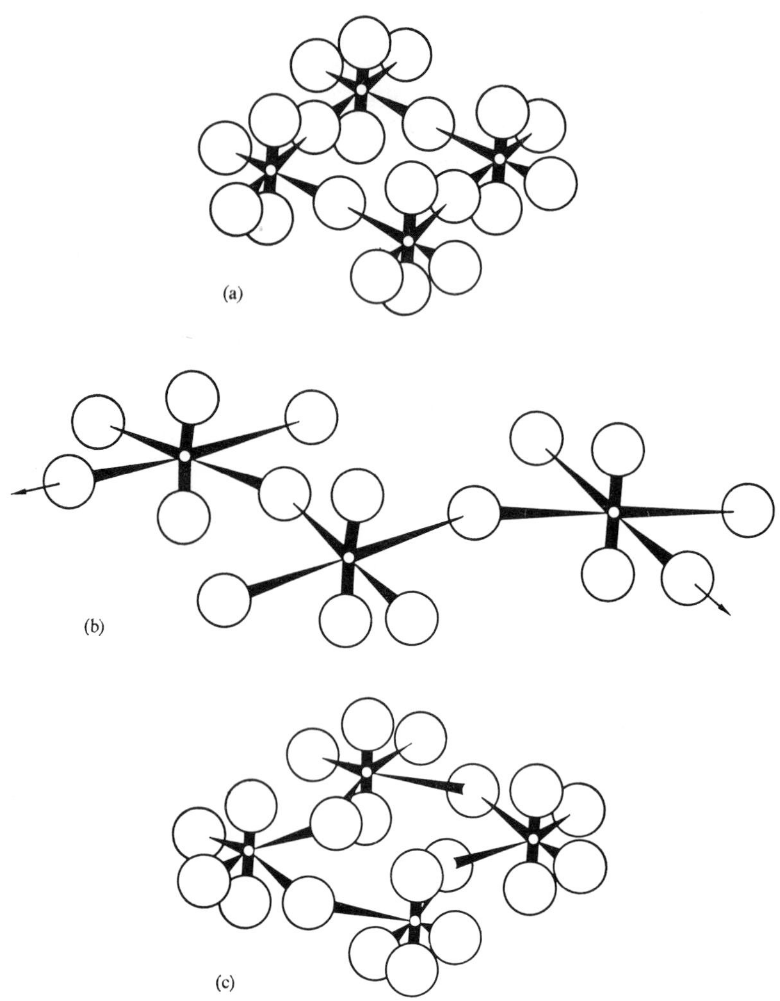

FIG. 14. The structures of transition metal pentafluorides. From J. H. Canterford, R. Colton and T. A. O'Donnell, *Rev. Pure and Appl. Chem.* **17** (1967) 123. (Small circles represent metal, large circles fluorine.)

move in turn through the first, second and third rows. Thus, towards the end of the third row, iridium hexafluoride is quite stable, and platinum hexafluoride is much more stable than chromium hexafluoride early in the first row. The real interest in descriptions of bonding is not in accounting for the existence of the stable hexafluorides but in examining the limits of stability and the chemical reactivity of the higher fluorides. To date any attempts have been quite qualitative and have centred on the number of non-bonding electrons in the members of a series of hexafluorides.

In the papers[174] in which he made semi-quantitative estimates of the values of electron affinities for third row hexafluorides [see section 5.3(e)], Bartlett said that the greater value of the electron affinity for platinum hexafluoride compared with that for tungsten hexafluoride correlates with the greater charge on platinum and the poorer shielding of this charge from the ligands by the non-bonding t_{2g} electrons, the number of which increases from zero in tungsten hexafluoride to four in platinum hexafluoride.

In order to explain both the regular decrease in infrared stretching frequency in the hexafluorides of any one row of transition metals and the fact that stretching frequencies for second-row hexafluorides are lower than those for their third-row counterparts, Canterford and Colton[191] postulated π-bonding in addition to σ-bonding in the hexafluorides by donation of *p*-electrons from fluorine atoms to the metal orbitals. As the number of electrons in the t_{2g} orbitals of the metals in any one row increases, π-bonding will decrease with resulting decrease in stability of the hexafluoride. They state that π-bond effects would be expected to be greater for third-row than for second-row elements. This description would apply also to chemical reactivities which increase along any row and decrease from second to third row.

Apart from the hexafluorides, the only class of *d*-element fluorides for which reliable structures are available are the pentafluorides which are polymeric in the solid state and melt to give liquids which are viscous, as a result presumably of association. With the obvious exception of the niobium and tantalum compounds, the *d*-element pentafluorides are thermally unstable, disproportionating to hexafluoride and tetrafluoride. Edwards, Peacock and co-workers have shown by single crystal X-ray diffraction studies that the pentafluorides of Groups VA, VIA and VIIA fall into two classes of polymers illustrated in Fig. 14a, b. Those of niobium, tantalum, molybdenum and tungsten have four metal atoms at the corners of a square, with linear fluorine bridges. For the four terminal fluorines arranged octahedrally around each metal atom there is some distortion of the out-of-plane fluorines. The remaining pentafluorides of Groups VA–VIIA are infinite chains in which fluorine atoms bridge metal atoms, about each of which four fluorines are arranged octahedrally. In this structure bridging fluorines are in the *cis*-position in each octahedron. The structures of some noble metal pentafluorides have been determined by Edwards, Peacock and colleagues and the remainder by Bartlett and his group. They are also tetrameric in the solid, as in Fig. 14c; but these tetramers are less regular than those in Fig. 14a. The fluorine bridges are non-linear, with the arrangement of the four metal atoms distorted from a square to a rhombus.

Canterford and Colton have reviewed[192] these pentafluoride structures in detail and have suggested that the structures can be rationalized in terms of π-bonding.

Linearity of fluorine bridges is postulated as resulting from significant π-bonding, the extent of such bonding being less, in the chain structures than in the regular square tetramers.

[191] Ref. 162(b), p. 13.
[192] Ref. 162(b), pp. 16–20.

When π-bonding becomes insignificantly small, as in the noble metal pentafluorides, non-linear fluorine bridges occur and these bridging fluorines adopt a tetrahedral structure with two lone pairs and no π-bonding.

Except for the fluorides of metals for which the highest possible oxidation state is III or IV, i.e. those of Groups IIIA and IVA, practically no work has been done on structures—or chemistry—of tetrafluorides or trifluorides of *d*-transition elements. In fact it is now apparent that several compounds previously thought to be lower fluorides are oxide fluorides.

(c) Bonding and Structure in Higher Fluorides of *f*-transition Metals

Regardless of the cited electronic configurations for gaseous lanthanide atoms in their ground states, e.g. $4f^66s^2$ for samarium, it seems apparent that for the lanthanide compounds in the dominant oxidation state, e.g. for the trifluorides, one *d*- and two *s*-electrons are involved in bonding. It seems that the key feature here is that *f*-electrons are not available for bonding.

Katz and Seaborg[180] have discussed the fact that *f*-orbitals generally are energetically favourable for covalent bonding and may well be involved in compounds, such as tellurium hexafluoride, formed by some elements of atomic number less than the lanthanides. However, at atomic numbers about that of cerium, the spatial distribution of *f*-orbitals "shrinks" to the point where they lie well below the 5*s* and 5*p* orbitals. They do not extend sufficiently far to have sufficient overlap with the bonding orbitals of other atoms, although cerium itself with ground-state configuration $4f5d6s^2$ must be a marginal case to account for the comparable stability of compounds of cerium(III) and cerium(IV).

The descriptive chemistry of lanthanide fluorides given in this review supports the recent statement by Brown[193] that the empirical rationalization frequently used to account for oxidation states of IV and II in lanthanide compounds, namely that the preferred states of *f*-orbitals are empty, half-filled or fully filled, is an over-simplification. On this simple basis, only cerium and terbium should form tetrafluorides and only europium and ytterbium should exhibit oxidation state II. However, as shown earlier, praseodymium tetrafluoride has been prepared, fluoro-complexes of neodymium(IV) and dysprosium(IV) are strongly suggested and the evidence for a difluoride of samarium seems no worse than that for ytterbium.

The existence of the exclusive oxidation states of III and IV in fluorides of actinium and thorium, taken with the trends in stabilities of typical *d*-transition metal halides, implies *d*-electron involvement in bonding in these compounds. However, as discussed in detail in the summaries in sections 5.4(c) and 5.5(d), the existence of protactinium tetrafluoride, the thermodynamic stability of lower fluorides of uranium and the significantly greater reactivity of uranium hexafluoride than of tungsten hexafluoride indicate that protactinium and uranium are not typical *d*-elements. Further, if neptunium were a typical *d*-element, a stable heptafluoride would be expected. The decreasing stability of higher fluorides of actinides from uranium onwards is consistent with the qualitative picture put forward by Katz and Seaborg[180] that for elements 89 and 90 binding energies are lower for 6*d* than 5*f* electrons but that for elements beyond protactinium a cross-over occurs and 5*f* electrons have the lower binding energies. Also the 5*f* orbitals are more favourably extended spatially relative to the 7*s* and 7*p* orbitals than in the relative case of 4*f*, 6*s* and 6*p*.

[193] Ref. 163, pp. 1–2.

It would appear that the $5f$ orbitals are available and do in fact participate in covalent bonding in the actinide hexafluorides, in which d^2sf^3 hybrid orbitals could be used. Favourable overlap in hybrids involving f-orbitals has been demonstrated as a general principle[194] and Coulson and Lester have shown that f-orbital involvement in bonding in uranium(VI) compounds would be energetically favourable[195]. However, while the d^2sf^3 hybrid appears to contribute to bonding in actinide hexafluorides, the situation is very complex and it may well be, as Coulson and Lester state, that "in heavy atoms where there are many electrons in the same valency shell, or at comparable distances from the nucleus, the simple language which has been devised to account for bonding between light atoms is no longer applicable. There is—so it would appear—no uniquely compelling description, but rather there are several alternative descriptions which could be employed."

It should be emphasized here that the trends in thermodynamic stabilities and chemical reactivities of higher actinide fluoride do not of themselves indicate f-orbital involvement in bonding in these compounds. They simply show that the bonding in actinide hexafluorides is very different from that in d-transition element hexafluorides where the model based on d^2sp^3 hybridization is used to account for bonding.

There is, of course, a considerable body of physical evidence, based on comparison of spectral and magnetic observations on actinide compounds with similar studies on lanthanide compounds, to indicate population of f-orbitals in the actinide atoms. This evidence has been reviewed by Katz and Seaborg[180], Cunningham[196], Brown[197] and others.

A link between chemical and physical evidence for f-bonding in actinide hexafluorides has already been shown in section 5.6(a). The bond stretching force constants differ markedly and by a constant amount for the two pairs of chemically similar compounds—molybdenum hexafluoride–uranium hexafluoride and ruthenium hexafluoride–plutonium hexafluoride. This constant difference appears to be related directly to bonding differences in each pair—a difference between bonding based on d-electrons and one which will have some f-involvement at least.

It is profitable to discuss in general terms structures of lanthanide and actinide trifluorides together. All these trifluorides show hexagonal symmetry with a 9-coordinate arrangement of fluorine atoms around each metal atom. The trifluorides of some of the heavier lanthanides are dimorphic and show orthorhombic in addition to hexagonal symmetry. Brown[198] illustrates this and gives lattice parameters where they are available.

Similarly, the three lanthanide tetrafluorides of cerium, praseodymium and terbium are isostructural with all of the known tetrafluorides of the actinides, having the structures of zirconium and hafnium tetrafluorides. Again, Brown[199] gives a diagram and lattice dimensions. Penneman originally showed regularity in the change in lattice dimensions with change in atomic number for actinide tetrafluorides and his approach has been refined as a result of more accurate crystallographic measurements by Keenan and Asprey[200].

Of the f-transition metals, only protactinium and uranium have yielded binary pentafluorides to date, the uranium compound being dimorphic. Single crystal X-ray data are not available. Powder patterns indicate that the high temperature α-UF_5 is tetragonal with

[194] J. C. Eisenstein, *J. Chem. Phys.* **25** (1956) 142.
[195] C. A. Coulson and G. R. Lester, *J. Chem. Soc.* (*London*) (1950) 3650.
[196] B. B. Cunningham in *Rare Earth Research*, ed. E. V. Kleber, Macmillan, New York (1961) p. 127.
[197] Ref. 163, pp. 1–9.
[198] Ref. 163, pp. 81–84.
[199] Ref. 163, pp. 54–55.
[200] T. K. Keenan and L. B. Asprey, *Inorg. Chem.* **8** (1969) 235.

an octahedral arrangement of fluorines around each uranium. The low temperature form, β-UF_5, which is isostructural with protactinium pentafluoride, is described as tetragonal with a seven-coordinate arrangement of fluorines around each uranium. The pentafluorides of the *f*-transition metals differ markedly from those of the *d*-elements which have been shown to be regular square tetramers, distorted tetramers or fluorine-bridged chains.

Structures of actinide hexafluorides do not differ in any significant way from those of the *d*-elements and, as stated in the section above, Weinstock[182] has reported the details of the investigation of the structure of these compounds in the solid, liquid and vapour state.

26. CHLORINE, BROMINE, IODINE AND ASTATINE

A. J. Downs and C. J. Adams
University of Oxford

1. INTRODUCTION

1.1. GENERAL ATOMIC PROPERTIES[1–3]

The properties peculiar to non-metals can be attributed more or less directly to the relatively large effective nuclear charge experienced by the valence electrons of the atom, a characteristic reflected in the high ionization potential and electron affinity and the relatively small size of such an atom. Because electrons in the same shell shield one another relatively inefficiently from the nucleus, the incidence of non-metallic properties is a periodic function of atomic number, increase of which in a given period is accompanied by the transition from metallic to non-metallic behaviour. The noble gases, with their unique qualities of electron localization, are the culmination of this development. However, it is the preceding group of typical elements—the halogens—which provides, in physical and chemical terms, the best defined and most homogeneous family of non-metals, described elsewhere[1] as "the most perfect series we have".

The magnitudes of the effective nuclear charges associated with the halogen atoms are indicated by the ionization potentials, which are only 1–4 eV short of those of the corresponding noble gas atoms, and by the relatively small sizes of the atoms. Further, the electron configuration of the halogen atoms in their ground states, ns^2np^5, just one electron short of the corresponding noble gas configuration, causes the atoms to be unusually powerful electron-acceptors, with electron affinities higher than those known for any other atomic species. Inasmuch as that rather elusive quantity electronegativity can be treated as an atomic property, it may be defined in numerous ways, but two useful criteria are (i) the mean of the ionization potential and the electron affinity (Mulliken's scale)[4] and (ii) the effective nuclear charge experienced by an electron at a distance equal to the covalent radius from the nucleus (the Allred–Rochow formulation)[5]. In either case, the combination of the atomic properties outlined clearly implies exceptionally high electronegativity values for the halogen atoms.

[1] N. V. Sidgwick, *The Chemical Elements and their Compounds*, Vol. II, p. 1097. Clarendon Press, Oxford (1950).

[2] A. G. Sharpe, *Halogen Chemistry* (ed. V. Gutmann), Vol. 1, p. 1. Academic Press (1967); see also *Codata Bulletin*, International Council of Scientific Unions, Committee on Data for Science and Technology (1970).

[3] W. Finkelnburg and F. Stern, *Phys. Rev.* **77** (1950) 303; R. W. Kiser, *J. Chem. Phys.* **33** (1960) 1265; M. F. C. Ladd and W. H. Lee, *J. Inorg. Nuclear Chem.* **20** (1961) 163; E. H. Appelman, *J. Amer. Chem. Soc.* **83** (1961) 805.

[4] R. S. Mulliken, *J. Chem. Phys.* **2** (1934) 782.

[5] A. L. Allred and E. G. Rochow, *J. Inorg. Nuclear Chem.* **5** (1958) 264.

These and other numerical properties are summarized in Table 1, while Fig. 1 depicts variations of ionization potential, electron affinity and electronegativity for the halogens and neighbouring atoms in the Periodic Table.

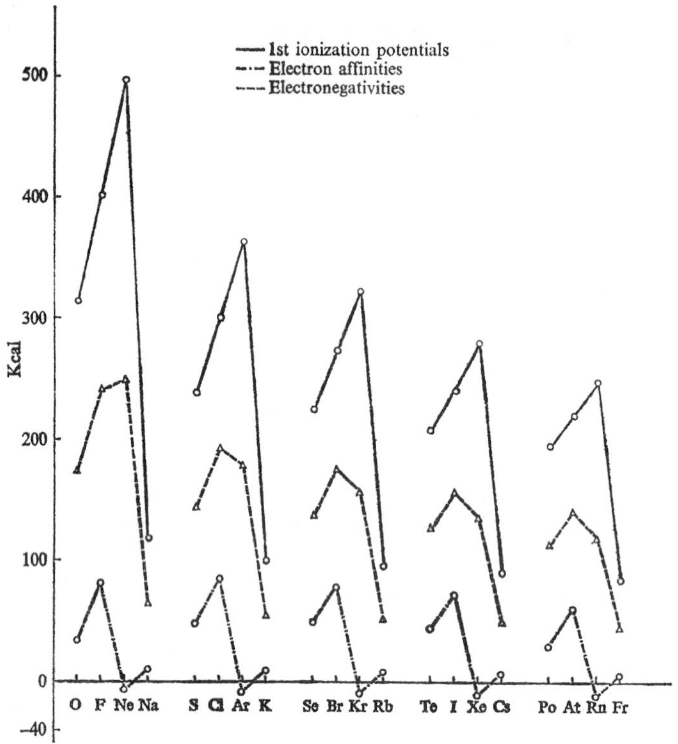

FIG. 1. Ionization potentials, electron affinities and electronegativities of halogen and neighbouring atoms in the Periodic Table.

Formation of Halides

Perhaps the dominant feature of the chemistry of the halogens is the facility with which their atoms acquire an electron to form either the uninegative ion X^- or a single covalent bond –X. As the oxidation potentials of Table 1 and other data suggest, however, this uninegative oxidation state becomes progressively less stable with respect to the free element with increase of atomic number. With the exception of helium, neon and argon, all the elements in the Periodic Table form halides which range in type from ionic aggregates at one extreme to simple molecules at the other. Halides generally are among the most important and common compounds, having played a central role in the historical development of synthetic, structural and interpretative aspects of chemistry. In their capacity as donors, the halogen atoms can be treated formally, like hydrogen and alkyl groups, as one-electron ligands; in common with hydrogen and simple organic groups, e.g. $-CH_3$, the halogens can also function as bridges between two other atoms, as in $(SbF_5)_n$ and Al_2Br_6. Some properties relevant to the formation of halides, including the covalent and ionic radii of the halogens, are shown in Table 1.

TABLE 1. SOME ATOMIC PROPERTIES OF THE HALOGENS[2]

Property	F	Cl	Br	I	At[a]
Atomic number	9	17	35	53	85
Electronic configuration	[He]$2s^2 2p^5$	[Ne]$3s^2 3p^5$	[Ar]$3d^{10} 4s^2 4p^5$	[Kr]$4d^{10} 5s^2 5p^5$	[Xe]$4f^{14} 5d^{10} 6s^2 6p^5$
First ionization potential (kcal)	402	299	273	241	(~220)
Electron affinity at 298°K (kcal)	81·0	84·8	79·0	72·1	(~71)
Electronegativity	4·0	3·0	2·8	2·5	(~2·4)
Dissociation energy of X_2 molecule at 298°K, $D(X_2)$(kcal)	37·7	58·0	46·1	36·1	(~27·7)
$\Delta H_f°[X(g)]$ at 298°K (kcal)	18·88	28·989	26·73	25·517	(~24)
Single-bond covalent radius (Å)[b]	0·71	0·99	1·14	1·33	(~1·4)
Ionic radius of X^- ion (NaCl structure) (Å)	1·33	1·82	1·98	2·20	(~2·3)
$\Delta H_f°[X^-(g)]$ at 298°K (kcal)	−62·1	−55·7	−52·3	−46·6	(~ −47)
$\Delta H^{\circ}_{\text{hydration}}[X^-(g)]$[c] at 298°K (kcal g ion^{-1})	−121	−88	−80	−70	(~ −66)
$E°(\frac{1}{2}X_2/X^-)$, aqueous solution (volts)	+2·87	+1·356	+1·065	+0·535	~ +0·3

[a] Values for astatine given in parentheses are mostly estimated by extrapolation or related calculation (see, for example, ref. 3).

[b] Given by one-half of the interatomic distance in the gaseous X_2 molecules.

[c] Absolute values relative to a hydration enthalpy for the proton of −260·7±2·5 kcal g ion^{-1}.

Elementary Halogens

The diatomic nature of the elementary halogens affords clear evidence of their almost unequivocally non-metallic behaviour: of the other elements, only hydrogen, nitrogen and oxygen share the property of being diatomic under normal conditions. Variations of such properties of the elementary halogens as their melting and boiling points and volatility are compatible with increases in magnitude of the van der Waals' forces with advancing size and polarizability of the atoms or molecules. Nevertheless, that intermolecular forces are significant in the condensed phases is shown (i) by the decrease in band gap for the solid elements with increasing atomic number (thus, with a band gap of 1·3 eV[6], iodine is comparable with various forms of phosphorus, arsenic and selenium), (ii) by the corresponding decrease in the ratio of the distance between next-nearest neighbours to that between nearest neighbours in the elementary solids (1·68 for Cl_2, 1·46 for Br_2 and 1·29 for I_2, see Table 11), (iii) by the degree of change of the vibrational spectrum of the diatomic molecule with the transition from the gas to the solid phase[7], and (iv) by nuclear quadrupole resonance studies of the solid halogens[8]. The general trend of this evidence, together with the

[6] N. N. Kuzin, A. A. Semerchan, L. F. Vereshchagin and L. N. Drozdova, *Doklad. Akad. Nauk S.S.S.R.* **147** (1962) 78; L. F. Vereshchagin and E. V. Zubova, *Fiz. Tverd. Tela*, **2** (1960) 2776; A. S. Balchan and H. G. Drickamer, *J. Chem. Phys.* **34** (1961) 1948.

[7] M. Suzuki, T. Yokoyama and M. Ito, *J. Chem. Phys.* **50** (1969) 3392; **51** (1969) 1929; J. E. Cahill and G. E. Leroi, *ibid.*, p. 4514.

[8] C. H. Townes and B. P. Dailey, *J. Chem. Phys.* **20** (1952) 35; R. Bersohn, *J. Chem. Phys.* **36** (1962) 3445; N. Nakamura and H. Chihara, *J. Phys. Soc. Japan*, **22** (1967) 201; E. A. C. Lucken, *Nuclear Quadrupole Coupling Constants*, Academic Press, London (1969).

increasing stability of positive oxidation states in the series F, Cl, Br, I, makes it all the more intriguing to determine the nature of elementary astatine, about which very little is yet known.

The intermolecular interactions also imply an ambivalence on the part of the halogen molecules which thus have some capacity to act either as donor or acceptor species. The acceptor nature of the heavier halogen molecules of the type XY (where X and Y are the same or different halogens) is amply supported by the very extensive class of complexes formed between XY and a wide range of electron-donors; such complexes range from polyhalide ions like I_3^- to such labile charge-transfer systems as Br_2,C_6H_6. Formally, the halogen molecule is normally regarded as the acceptor partner, but the self-association of these molecules, both in the solid state and in the discrete dimers Br_4 and I_4 recently identified[9] in the gas phase and in solution, suggests that the donor function is not negligible. Intermolecular interactions in these systems extend from those weak enough to be classified as van der Waals' forces, e.g. Br_4, to those recognisable as unambiguous covalent bonds, e.g. I_3^-, a transition that is unusually illuminating in its relevance to some general problems of chemical bonding.

Halogens in Positive Oxidation States

The overall enthalpy changes (in kcal) accompanying the conversion

$$\tfrac{1}{2}X_2(\text{standard state}) \rightarrow X^+(g)$$

are: F, +421; Cl, +328; Br, +300; I, +266·5; while for the process

$$X_2(aq) \rightleftharpoons X^+(aq)+X^-(aq)$$

equilibrium constants of 10^{-40}, 10^{-30} and 10^{-21} have been calculated when X = Cl, Br and I, respectively[10]. Together with other evidence, these data provide some guide to the stability of halogen cations X^+. The removal of one electron from the valence shell of a halogen atom is unlikely to be accompanied by a substantial decrease in size, and it is this, with its consequent effects on lattice and solvation energies, rather than the magnitude of the first ionization potentials, that probably determines the infrequency with which free halogen cations occur under normal chemical conditions[2]. Nevertheless, cationic species such as H_2OX^+ may well participate as intermediates in chemical reactions, whereas, for the heavier halogens, species containing coordinated X^+ units, e.g. BrF_2^+, ICl_2^+ and $[I(NC_5H_5)_2]^+$, have now been characterized (see Section 4). In non-aqueous media such as sulphuric or fluorosulphonic acids, bromine and iodine give coloured solutions whose contents probably include the cations X_2^+ and X_3^+; recent studies of these solutions find, contrary to previous suggestions, no evidence of the species X^+.

In addition to the limited range of cations, there are known, for the heavier halogens but not for fluorine, many derivatives in which a formal positive oxidation state of +1, +3, +5 or +7 is associated with the halogen atom. Such systems invariably include ligands with electronegativities equal to or greater than that of the central halogen atom, the range thereby being restricted to oxygen compounds (e.g. ClO_4^-), and interhalogen systems

[9] L. J. Andrews and R. M. Keefer, *Adv. Inorg. Chem. Radiochem.* **3** (1961) 91; R. S. Mulliken and W. B. Person, *Molecular Complexes*, p. 141, Wiley, New York (1969); A. A. Passchier, J. D. Christian and N. W. Gregory, *J. Phys. Chem.* **71** (1967) 937; A. A. Passchier and N. W. Gregory, *J. Phys. Chem.* **72** (1968) 2697; M. Tamres, W. K. Duerksen and J. M. Goodenow, *ibid.*, p. 966; D. D. Eley, F. L. Isack and C. H. Rochester, *J. Chem. Soc.* (*A*) (1968) 1651.

[10] J. Arotsky and M. C. R. Symons, *Quart. Rev. Chem. Soc.* **16** (1962) 282.

(e.g. BrF_5 and ICl_4^-). The oxidizing nature of the various oxyanions of chlorine, bromine and iodine is evident from the oxidation state diagrams of Fig. 2 which show that under acidic conditions most of the oxidation potentials are close to that of the oxygen couple O_2/H_2O. Other noteworthy features of this figure are the instability of the elementary

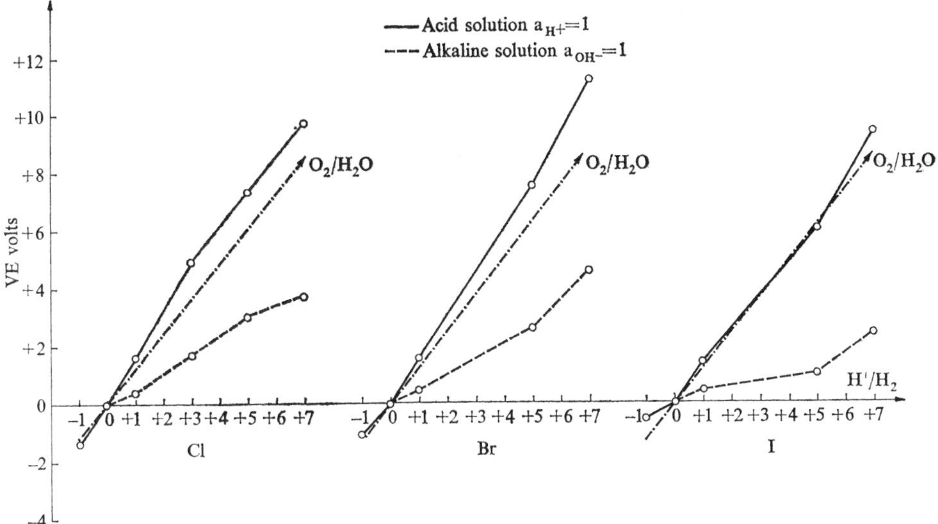

FIG. 2. Oxidation state diagrams for the halogens in aqueous solution.

halogens with respect to disproportionation in alkaline solution, the comparative stability of the IO_3^- ion, the strongly oxidizing character of the recently discovered perbromate ion[11], and the relative instability of the +3 oxidation state for all three halogens.

Stereochemistry and Bonding

The stereochemistries of some typical compounds of chlorine, bromine and iodine, summarized in Table 2, manifest a diversity of structural chemistry which is in striking contrast to that of fluorine compounds. The difference between fluorine and the heavier halogens in this respect arises from the practical inaccessibility to fluorine of oxidation states greater than 0 and of coordination numbers greater than 2 in covalently bonded systems. The formal expansion of the valence shells possible in chlorine, bromine and iodine is not easily rationalized because of the ephemeral nature of those atomic states which the chemist commonly refers to as "valence" states. Nevertheless, atomic properties such as (i) increased size and polarizability, (ii) decreased electronegativity, (iii) smaller energy separations between atomic *ns*, *np*, $(n+1)s$ and $(n+1)p$ states, and (iv) the availability of vacant *nd* orbitals are all likely to subscribe to the increased range of oxidation states and chemical environments open to the heavier halogens.

[11] E. H. Appelman, *J. Amer. Chem. Soc.* **90** (1968) 1900; M. H. Studier, *ibid.*, p. 1901; G. K. Johnson, P. N. Smith, E. H. Appelman and W. N. Hubbard, *Inorg. Chem.* **9** (1970) 119; J. R. Brand and S. A. Bunck, *J. Amer. Chem. Soc.* **91** (1969) 6500.

TABLE 2. STEREOCHEMISTRIES OF COMPOUNDS OF CHLORINE, BROMINE AND IODINE

Coordination number	Geometry	Examples
1	Diatomic unit	Cl_2, Br_2, I_2, BrCl, HCl
2	Linear Angular	I_3^-, $ClICl^-$, $BrICl^-$ ClO_2, ClO_2^-, BrF_2^+
3	Trigonal pyramid T-shaped unit	ClO_3^-, BrO_3^-, IO_3^- ClF_3, BrF_3, $RICl_2$ (R = organic group)
4	Tetrahedral unit Square planar unit Trigonal bipyramid with vacant equatorial site	ClO_4^-, BrO_4^-, IO_4^-, Cl_2O_7, $FClO_3$ ICl_4^-, I_2Cl_6 $IO_2F_2^-$, IF_4^+
5	Square pyramid	ClF_5, BrF_5, IF_5
6	Octahedral unit Distorted octahedron	H_5IO_6, OIF_5 IF_6^-(?), I^vO_6 units, e.g. in NH_4IO_3
7	Pentagonal bipyramid	IF_7

Most of the stereochemistries represented in Table 2 are consistent with the formal disposition around the central halogen atom of 4, 5 or 6 valence electron-pairs at the vertices, respectively, of a tetrahedron, trigonal bipyramid or octahedron. IF_7 apart, clearcut structural information about systems with seven such pairs is generally lacking. Stereochemical details are then explicable on the basis of such factors as (i) the number of σ-bonding and lone-pair electrons on the central atom, (ii) the presence of more than one kind of ligand and hence more than one kind of σ-bond pair, and (iii) the minimization of the repulsive energy of interaction between the different electron pairs[12]. Hence, the structures of species like ClF_3, IF_4^+ and ICl_4^- can be rationalized. Although π-bonding is likely also to be important in derivatives like ClO_4^-, there is no evidence to suggest that this has a major influence on the essential geometry of the unit.

The conventional explanation of the expansion of the valence shells of chlorine, bromine and iodine depends on the *nd* orbitals, which are capable, in principle, of functioning as acceptors of σ- or π-electron density. Historically, this argument derives from the classical concept of the two-centre, two-electron bond, which leads to bonding descriptions of systems such as ClF_3 or IF_5 based on the use of dsp^3 or d^2sp^3 hybrid orbitals by the central halogen atom. Despite the lack of meaningful, quantitative information about the valence states of atoms, the one-electron orbital ionization energies of the free halogen atoms depicted in Fig. 3 furnish a useful comparison. According to such data, the energy separations between the valence *np* states and the excited states *nd*, $(n+1)s$ and $(n+1)p$ range from a maximum of ~370 kcal for fluorine to a minimum of ~170 kcal for iodine. Significantly, at the atomic level the $(n+1)s$ orbitals are appreciably more stable than the *nd* orbitals. Since for effective participation in any bonding scheme it is a fundamental requirement of atomic orbitals that they should be compatible not only in symmetry but also in energy, these large energy separations must cast doubt on the validity of a concept involving the

[12] R. J. Gillespie and R. S. Nyholm, *Quart. Rev. Chem. Soc.* **11** (1957) 339; R. J. Gillespie, *Angew. Chem. Internat. Edn.* **6** (1967) 819.

promotion of electrons to *nd*, $(n+1)s$ or $(n+1)p$ states and the subsequent formation of hybrid orbitals. It is generally acknowledged, however, that the radial functions and energies of *nd* electrons are unusually sensitive both to the formal charge on the atom and to the number of such electrons[13], the result being that the *nd* orbitals do participate in bonding to some,

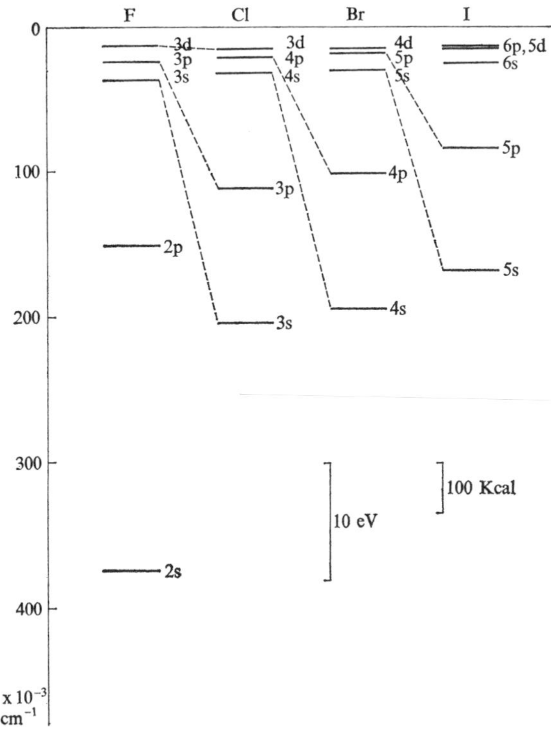

FIG. 3. One-electron orbital ionization energies for the halogen atoms.

as yet undetermined, extent. Equally, it is recognized that the relative involvement of orbitals cannot be precisely prescribed as dsp^3 or d^2sp^3, for bonding may result solely from *s*-orbital or *p*-orbital or *d*-orbital interactions, or from any combination of these.

In general terms, the halogens are electron-rich systems deficient in potentially bonding orbitals. In this respect, there is an obvious analogy between the heavier halogens and noble gases, as illustrated by the resemblance of iodine and xenon in the isoelectronic pairs XeF_6, IF_6^-; XeO_3, IO_3^-; XeO_6^{4-}, IO_6^{5-}. There is a less obvious affinity to the elements like gold or mercury with full or nearly full complements of *d* electrons: as with the halogens, the chemistry of these elements is strongly coloured by the dearth of bonding orbitals. The structural similarity of derivatives such as ICl_2^- and $HgCl_2$ or ICl_4^- and $AuCl_4^-$ suggests an underlying conformity of bonding type. In these, as in noble gas compounds and in species

[13] D. P. Craig and E. A. Magnusson, *J. Chem. Soc.* (1956) 4895; K. A. R. Mitchell, *Chem. Rev.* **69** (1969) 157; C. J. Adams, A. J. Downs and S. Cradock, *Ann. Rep. Chem. Soc.* **65A** (1968) 216, and references cited therein.

such as HF_2^-, the bonding can be described, to a first approximation, in terms of multi-centre molecular orbitals formed exclusively from those *s*- or *p*-orbitals of the central atom that are energetically eligible, the influence of *d*-orbitals being neglected in the first place. For example, according to the simple LCAO-MO treatment developed by Rundle[14] and others[15], the molecular orbitals of the I_3^- ion are represented by a linear combination of the 5*p*-orbitals of the iodine atoms considered to overlap in the direction of the I–I–I axis. The formation of these three-centre molecular orbitals, with the associated energy-level diagram, is illustrated in Fig. 4, which provides a schematic description of so-called three-centre, four-electron bonding[14]. Although the contributions of *s*- and *d*-orbitals and of π-interactions should be incorporated in a more sophisticated account of the bonding, the simplified scheme of Fig. 4 appears more realistic than the "full hybridization" theory which gives

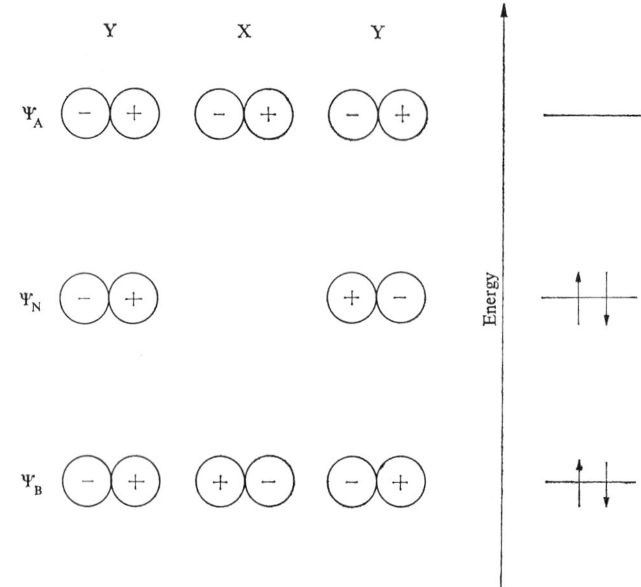

FIG. 4. Formation and energies of three-centre, four-electron molecular orbitals in the I_3^- ion and related systems.

excessive weight to energetically inferior *d*, *s* or *p* functions of the atom. By way of contrast, Fig. 5 illustrates in a purely general way the formation of a single σ-bond between a halogen atom and a second atom having just one available orbital; it is evident that the *ns*-orbital of the halogen makes a relatively greater contribution to the bonding when the second atom is the more electronegative partner. That the extent of *s*-hybridization is usually small ($<10\%$) is suggested by various measurements, notably of the hyperfine coupling tensors for the radical anions XY^- (X, Y = the same or different halogen atoms)[8]. The problems

[14] R. E. Rundle, *Survey of Progress in Chemistry*, **1** (1963) 81; *J. Amer. Chem. Soc.* **85** (1963) 112.

[15] E. E. Havinga and E. H. Wiebenga, *Rec. Trav. Chim.* **78** (1959) 724; E. H. Wiebenga, E. E. Havinga and K. H. Boswijk, *Adv. Inorg. Chem. Radiochem.* **3** (1961) 133; E. H. Wiebenga and D. Kracht, *Inorg. Chem.* **8** (1969) 738.

posed by the bonding in I_3^- illustrate well the sort of impasse that has been reached at the theoretical level, arising partly from limitations of calculated wave-functions and partly from the difficulty of correlating molecular with atomic properties.

The subtlety of problems of molecular structure and stereochemistry is further emphasized by a quite different approach which alludes to the properties not only of the ground but also of excited electronic states. The dependence of the energy of an aggregate upon its geometry

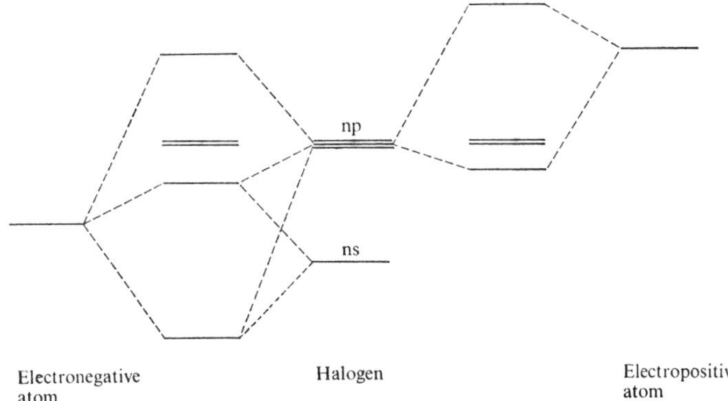

FIG. 5. Correlation diagram for bonds formed by halogens with more electropositive or more electronegative elements.

may be expressed in terms of perturbation theory. If the Hamiltonian operator is expanded[16,17] as a Taylor series in S_i, a symmetry coordinate for molecular deformation:

$$H = H^\circ + H_i' S_i + \tfrac{1}{2} H_{ii}'' S_i^2 + \ldots$$

application of perturbation theory yields for the energy of the system,

$$E = E^\circ + \langle\psi_o \mid H_i' \mid \psi_o\rangle S_i + \left[\tfrac{1}{2}\langle\psi_o \mid H_{ii}'' \mid \psi_o\rangle - \sum_m \frac{\langle\psi_o \mid H_i' \mid \psi_m\rangle^2}{E_m - E_o}\right] S_i^2 + \ldots$$

where the subscripts o and m refer to the ground and to an excited electronic state, respectively. The term in S_i, which describes Jahn–Teller distortions, vanishes for a non-degenerate ground state. However, for a degenerate ground state, such as that of BrF_5 in the form of a trigonal bipyramid[18], the term is finite: this implies instability with respect to a deformation which lifts the degeneracy of the system (achieved for BrF_5, for example, by the transformation trigonal bipyramid → square pyramid). The term in S_i^2 describes the force constant appropriate to S_i; it is evident that a small value of $(E_m - E_o)$, coupled with a non-vanishing matrix element $\langle\psi_o|H_i'|\psi_m\rangle$, can lead to a low or even negative value of the force constant. In the case of a totally symmetric ground state (which is the general rule), $\langle\psi_o|H_i'|\psi_m\rangle$ is non-zero only if H_i' contains the same symmetry species as ψ_m; the characteristics of the force field are then uniquely determined by the lowest-lying excited states[16]. In general, if $(E_m - E_o) \lesssim 4$ eV and a deformation mode contains the same irreducible representation as the product $\psi_o\psi_m$, the force constant appropriate to the appointed deformation is negative,

[16] R. F. W. Bader, *Mol. Phys.* **3** (1960) 137.

[17] D. H. W. DenBoer, P. C. DenBoer and H. C. Longuet-Higgins, *Mol. Phys.* **5** (1962) 387; B. J. Nicholson and H. C. Longuet-Higgins, *ibid.* **9** (1965) 461.

[18] R. S. Berry, M. Tamres, C. J. Ballhausen and H. Johansen, *Acta Chem. Scand.* **22** (1968) 231.

and distortion of the aggregate becomes energetically favourable. The system is then said to suffer a "pseudo-Jahn–Teller effect"[19], for which general symmetry rules have been evolved to predict molecular structure. According to this treatment, molecular structure is not prescribed exclusively by the electronic properties of the individual atoms; rather does it depend also on the symmetry properties and relative energies of the ground and neighbouring electronic states of the aggregate and on the force field operative over the system. These features cannot justifiably be excluded from any meaningful interpretation of the structural chemistry of species like the polyhalide ions and charge-transfer complexes, for which neither the bonding energy nor $(E_m - E_o)$ terms are large, and the frameworks of which are unusually pliable with respect to deformation.

It is also a general precept that the exact structural characteristics of a polyatomic aggregate depend on its environment. That is to say that structure is a function also of the phase in which the unit occurs, or, in the event of solution, of the nature of the solvent, or, in crystalline solids, of features such as the packing of the structural units or the properties of counter-ions. Whenever there are relatively small energy differences between various structural alternatives open to a given system, such constraints, peculiar to the condensed phases, appear to be capable of impressing upon a polyatomic system a structure different from that of the isolated unit. The structure-determining influence of environment is exemplified by numerous structural features of the interhalogens and related species.

1.2. IRREGULARITIES OF BEHAVIOUR OF THE HALOGENS

Although the halogens form a remarkably homologous series, certain discontinuities are evident. Principal among these is the exceptional nature of fluorine, as revealed by the small size of the atom in combination and of the F^- ion, by its high first ionization potential (and electronegativity), and by the relatively low dissociation energy of the F_2 molecule. As a result, there is a much greater divergence of properties between fluorine and chlorine than between the other pairs of neighbouring elements. As in other series of elements, less striking but significant differences are found between the third and fourth members of the Group. These are reflected, for example, in the relatively irregular increases in size of the atoms and ions, such that chlorine and bromine are relatively similar to each other but differ from fluorine on the one side and iodine on the other. Variations in properties like electron affinity, polarizability and standard potential of the couple $\frac{1}{2}X_2/X^-$ (aq) display analogous irregularities. Estimates of the properties of astatine imply[3] that it closely resembles iodine. This kind of variation in properties arises from the irregular increments of nuclear charge of 8, 18, 18 and 32 in the series F, Cl, Br, I and At, attended by the subtle variations in the shielding from the nucleus of the valence p by the core electrons.

At a chemical level we find that iodine differs from chlorine and bromine (i) in its significantly greater capacity to engage in chemical environments with relatively high coordination numbers (6 or even 7), (ii) in the susceptibility of iodides to oxidation, implied by the standard oxidation potentials of Table 1 and by the standard free energies of formation of pure iodides, e.g. HI, and (iii) in the relatively high stability of derivatives of the +5 oxidation state, e.g. IO_3^- and IF_5, with respect both to oxidation and to reduction. The growing importance of the +5 oxidation state with increase of atomic number of the halogens corresponds to a general feature of the B-group elements, that is, the emergence of the

[19] L. S. Bartell, *J. Chem. Educ.* **45** (1968) 754; R. G. Pearson, *J. Amer. Chem. Soc.* **91** (1969) 4947.

oxidation state $N-2$, where N is the group valency. The phenomenon is commonly attributed to the so-called "inert-pair" effect, a term which refers to the formal resistance of the ns^2 electrons to ionization or to participation in covalent bond formation; it is most pronounced for the heavier elements of Groups II–V which are distinguished both by relatively large separations of the *ns* and *np* energy levels of the valence shell and by relatively low energies of chemical combination. For the halogens the "inert-pair" effect plays a less prominent role. This is evident from the fact that the ratio I_s/I_p (where I_s and I_p are the ionization potentials for the removal, respectively, of an *s* and *p* electron) varies neither widely nor systematically; further, the average energies of bonds between a particular halogen and a more electronegative element (e.g. another halogen or oxygen) do not show the steady decrease with atomic number characteristic of the more metallic elements of Groups II–V.

It has been suggested by more than one authority[20] that with respect to combination with a more electronegative element there is an *alternation* of properties in the halogen series such that chlorine resembles iodine whereas bromine shares some of the anomalies exhibited by arsenic and selenium, its horizontal neighbours in the Periodic Table. Arsenic and selenium are noteworthy for the abnormal oxidizing properties associated with their group valence states of 5 and 6, respectively; features such as the failure to isolate arsenic pentachloride and the high oxidation potential of the couple $SeO_4^{2-}(aq)/SeO_3^{2-}(aq)$ give notice of this property. In the past, abortive attempts to prepare either oxides of bromine or perbromates were widely taken to signify an inherent instability of these compounds. With the successful characterization first of bromine oxides[21] and, in 1968, of perbromates[11], reports of the non-existence of these species are seen to be an exaggeration. Much of the mystique of the "alternation" effect, at least in relation to the halogens, has thus evaporated. Accordingly, rationalization of the supposed non-existence of perbromates based, for example, on the spatial properties and energies of the 4*d* orbitals of bromine[22], must now be seen for vain attempts to explain negative experimental evidence. Nothing so devalues the currency of a chemical theory as the discovery that the fact it purports to explain is itself incorrect, and the history of perbromates is a serious indictment of the chemical habit of theorizing about the "non-existence" of compounds involving significant degrees of electron-delocalization[23]. This is not to gainsay the individual behaviour of bromine in many respects; thus, in practical terms, the oxides of bromine are of very low chemical stability[21], while the thermodynamic barrier to the formation of perbromates[11] demands unusually energetic methods of oxidation. However, the exact nature of much of this individuality and the extent to which thermodynamic or kinetic factors are implicated have yet to be properly defined.

1.3. THERMODYNAMIC ASPECTS OF THE CHEMISTRY OF THE HALOGENS

Despite the shortcomings of theoretical accounts, many aspects of halogen chemistry lend themselves unusually well to thermodynamic treatment both at an interpretative and at a predictive level. Two recent accounts[2,24] illustrate clearly the way in which a chemical problem can be treated quantitatively by thermodynamic methods, there being ideally three

[20] See for example C. S. G. Phillips and R. J. P. Williams, *Inorganic Chemistry*, Vol. 1, pp. 431, 663. Clarendon Press, Oxford (1965).

[21] M. Schmeisser and K. Brändle, *Adv. Inorg. Chem. Radiochem.* **5** (1963) 41.

[22] D. S. Urch, *J. Inorg. Nuclear Chem.* **25** (1963) 771.

[23] W. E. Dasent, *Non-existent Compounds*, Edward Arnold and Marcel Dekker (1965).

[24] D. A. Johnson, *Some Thermodynamic Aspects of Inorganic Chemistry*, Cambridge (1968).

stages in the procedure[24]: (i) the statement of the problem in thermodynamic terms; (ii) the restatement of the problem thermodynamically in the one of many possible ways that is the most susceptible to theoretical study; (iii) the theoretical interpretation of the restated problem. Thermodynamic methods cannot of themselves supply explanations for chemical phenomena; rather is their role to provide a precise and self-consistent rephrasing of these phenomena in terms of certain atomic or molecular energy terms. Since most reactions of the halogens and their inorganic compounds are fast (the hydrolysis of some non-metal halides and the reduction of perchlorate ion being among the few noteworthy exceptions), kinetic considerations are seldom of major importance[2]. Further, to the extent that the ionic model provides a useful theoretical basis for numerous, simple derivatives of the halogens, e.g. the crystalline alkali-metal halides and the solvated halide ions, the theoretical interpretation of the thermodynamic data is direct, convincing and often at least semi-quantitative. It is for covalently bonded systems that the theoretical response is less satisfactory.

Schematically the reduction of a free halogen, X_2, in its standard state either (i) to an aquated halide ion X^-, or (ii) to a crystalline ionic solid in conjunction with, say, some univalent cation M^+ is represented in Fig. 6. The free energy change attending the con-

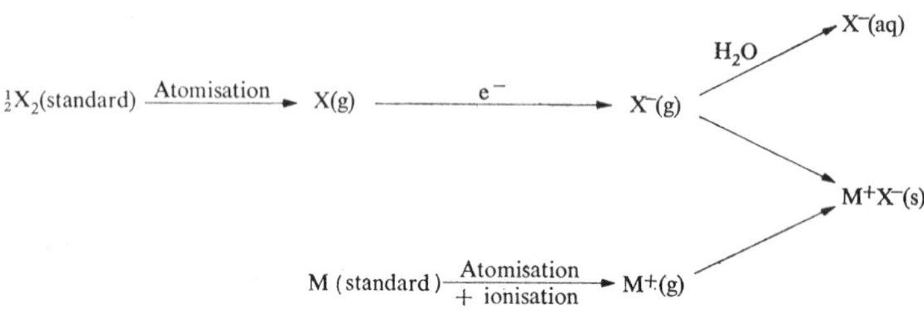

FIG. 6. Schematic representation of the formation (i) of an aquated halide ion and (ii) of a crystalline ionic halide M^+X^-.

version of one atom of the halogen in its standard state to the aquated halide ion under standard conditions of unit activity is seen to be equal to the balance of the free energies (a) of atomization, (b) of electron addition to the gaseous atom, and (c) of solvation of the gaseous anion. Since the entropy contributions are in all cases small compared with the enthalpy terms, and since moreover the entropy changes do not vary greatly from halogen to halogen in a comparative treatment of the problem, the free energy changes can be meaningfully approximated by the corresponding enthalpy terms. On this basis, for example, the standard redox potential $E°$ for the couple $\frac{1}{2}X_2/X^-(aq)$ is given approximately by

$$E° = -\frac{1}{F}[\Delta H(\tfrac{1}{2}X_2(\text{standard}) \rightarrow X^-(\text{aq})) - \Delta H(H^+(\text{aq}) \rightarrow \tfrac{1}{2}H_2(\text{standard}))]$$

$$= -\frac{1}{F}[\Delta H_f°[X(g)] - E_A - \frac{5}{2}RT + \Delta H_{aq}° + 105] \quad (1)$$

where $\Delta H_f°[X(g)]$ is the atomization energy of the halogen, E_A the electron affinity, and $\Delta H_{aq}°$ the enthalpy of hydration of the X^- ion (all in kcal), F is the Faraday and R the gas constant.

Analogous arguments lead to the following expression for the standard enthalpy of formation of the ionic solid M^+X^-, $\Delta H_f°$, variations of which determine, to a good approximation, variations of $\Delta G_f°$, that is, of the thermodynamic stability of the solid with respect to its elements:

$$\Delta H_f^\circ = \Delta H_f^\circ[X(g)] - E_A + I_M - U_{298} - 2RT \quad (2)$$

Here I_M is the first ionization potential of M and U_{298} the lattice energy of M^+X^- at 298°K.

Common to both processes is the conversion

$$\tfrac{1}{2}X_2(\text{standard}) \rightarrow X^-(g)$$

for which the standard enthalpies are given in Table 1; the total range of values in the series F, Cl, Br, I is less than 16 kcal, sufficiently narrow to suggest that variations in this quantity play a relatively minor part in differentiating the chemical properties of the halogens. The other two variables in equations (1) and (2) determined by the nature of the halogen are ΔH_{aq}° and U_{298}, both of which are inverse functions of the size of the ion. Thus, for a singly charged ion, ΔH_{aq}° is given approximately by the basically empirical relationship[25]

$$\Delta H_{aq}^\circ = -\frac{167}{r_{eff}} \text{ kcal g ion}^{-1} \quad (3)$$

wherein r_{eff}, the "effective" radius of the ion, can be equated with the crystal radius of the halide ion. The lattice energy of the ionic crystal M^+X^- at 0° is related very nearly by the expression[2]

$$U_o = \frac{NAe^2}{r_{M^+} + r_{X^-}}\left[1 - \frac{\rho}{r_0}\right] \quad (4)$$

to the electronic charge e, the Avogadro number N, the Madelung constant appropriate to the lattice-type A, the equilibrium interionic separation given, according to the simple model, by $r_{M^+} + r_{X^-} = r_0$, and the compressibility of the ions as determined by the (nearly constant) parameter ρ. It is the sensitivity of both ΔH_{aq}° and U_0 (and the corresponding free energy terms) to variations in the size of the halide ion that accounts for the greatest part of the thermodynamic differences between the halogens with respect (i) to their oxidizing properties under aqueous conditions, and (ii) to the formation of crystalline, ionic halides. Thus, the data of Table 1 reveal variations in the values of ΔH_{aq}° of about 50 kcal g ion^{-1}, while for the sodium halides the values of U_0 (in kcal mol^{-1}) are:[24] NaF, 216; NaCl, 185; NaBr, 176; NaI, 168. In the formation of a purely ionic halide of a metal in any given oxidation state, therefore, if we discount the small effect of possible change of structural type on lattice energy, the fluoride will always have the most negative ΔH_f°.

It can also be shown[2,24] that for the reaction

$$M^{n+}X^-_n(s) + \tfrac{1}{2}X_2(\text{standard}) \rightarrow M^{(n+1)+}X^-_{n+1}(s)$$

the enthalpy term, which is determined primarily by the difference in lattice energies, ΔU_0, of the two ionic solids $M^{n+}X^-_n$ and $M^{(n+1)+}X^-_{n+1}$, becomes increasingly endothermic in the sequence X = F, Cl, Br, I. This is most clearly apparent on the basis of Kapustinskii's approximate expression[26] for the lattice energy of an ionic crystal containing ν ions per chemical formula with charges z_+ and z_- and radii r_+ and r_-:

$$U_0 = \frac{256\nu z_+ z_-}{r_+ + r_-} \text{ kcal mol}^{-1} \quad (5)$$

Hence, if the radii of the M^{n+} and $M^{(n+1)+}$ ions are taken to be approximately equal, being

[25] Ref. 20, p. 163.

[26] A. F. Kapustinskii, *Quart. Rev. Chem. Soc.* **10** (1956) 283.

denoted by r_+, the difference in the lattice energies of MX_{n+1} and MX_n is given roughly by

$$\Delta U_0 = \frac{256(n+2)(n+1)-256(n+1)n}{r_+ + r_{X^-}}$$

$$= \frac{512(n+1)}{r_+ + r_{X^-}} \text{ kcal mol}^{-1}$$

It follows that the lattice energy of MX_{n+1} is invariably greater than that of MX_n, the disparity decreasing as the size of the anion is augmented. Thus the variations in both $-\Delta H_f[X^-(g)]$ and ΔU_0 place the capacities to stabilize the oxidation state $n+1$ in the order F > Cl > Br > I, manifest in the fact that many metals exhibit higher ionic oxidation states in fluorides than in other halides. Conversely, stabilization of the lower halide MX_n with respect to oxidation to MX_{n+1} is best achieved when X = I, consistent with the fact that low oxidation states in iodides are well-known features of the chemistry of many metals, e.g. iron(II), copper(I) and dipositive lanthanides.

Arguments analogous to these serve also to account[2,24] (i) for variations in the efficiency of metal halides as halogen-exchange reagents in reactions such as

$$R\text{–}Cl + M^+F^-(s) \rightarrow R\text{–}F + M^+Cl^-(s)$$

and (ii) for the influence of cation size on the stability of crystalline derivatives of complex halide ions like HCl_2^- and ICl_2^-, and for the mode of decomposition of such ions.

The chemistry of molecular halogen compounds cannot be treated in this way because of the inaccessibility of quantitative information about changes such as

(i) $M^{n+}(g) + nX^-(g) \longrightarrow MX_n(g)$

or (ii) M (standard) ⟶ M(g, ground state) ⟶ M(g,valence state) ⟶ $MX_n(g)$

$n/2\ X_2$(standard) ⟶ nX (g) ⟶ $MX_n(g)$

Nevertheless, certain helpful correlations can be made concerning the standard enthalpy of formation $\Delta H_f°$ of a gaseous molecular halide MX_n in terms of the heats of atomization of M and X_2 and the average bond energy of MX_n, B(M–X) (Fig. 7).

$$\Delta H_f°[MX_n(g)] = \Delta H_f°[M(g)] + n\Delta H_f°[X(g)] - nB(M\text{–}X) \qquad (6)$$

If it is assumed that the enthalpy change associated with any condensation of gaseous MX_n is small and that variations in $T\Delta S°$ terms from halogen to halogen are small compared with

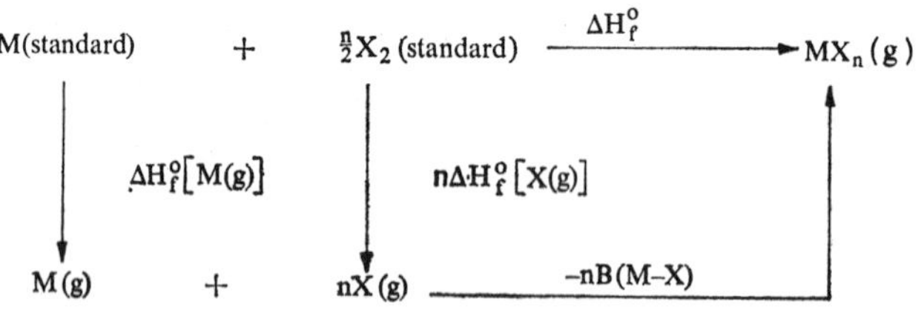

FIG. 7. Thermodynamic cycle for the formation of a gaseous halide MX_n from its elements.

those in $\Delta H°$, variations in $\Delta H_f°[MX_n(g)]$ provide a safe guide to those in $\Delta G_f°[MX_n]$. Bond energy data (see Section 3) show that in all but a very few cases, the order of bond energies in any halide MX_n is F > Cl > Br > I and that the variation is considerable. It follows from this and from the variations in $\Delta H_f°[X(g)]$ displayed in Table 1 that the stability of the halide MX_n, with respect to its constituent elements, decreases in the order $MF_n > MCl_n > MBr_n > MI_n$. The pre-eminent thermodynamic stability of molecular fluorides is a function of the strong bonds that fluorine forms with other elements and the weak bond that it forms with itself; for the other halogens $\Delta H_f°[X(g)]$ varies but little, and the thermodynamic stability of MX_n is primarily a function of B(M–X). The relative stabilities of the halides NX_3, OX_2 and HX and the displacement of a heavier by a lighter halogen from a compound MX_n illustrate this general thermodynamic condition. However, it is difficult to employ the same approach to interpret the relative capacities of the halogens to bring out high oxidation states because of the lack of definite information about the variation of the bond energy of MX_n as n is varied. At best, the interpretative aspect of the bond energy treatment compares unfavourably with that involving the ionic model because bond energies are not theoretically comprehensible in the way that lattice energies are. But such theoretical shortcomings do not impair the value of the thermodynamic correlations, for the light that they may shed on the interdependence of molecular and atomic energy terms.

1.4. SUBSEQUENT TREATMENT OF CHLORINE, BROMINE, IODINE AND ASTATINE

In Sections 2–4 chlorine, bromine and iodine are to be treated side by side in order to underline the homogeneity of much of their chemistry, and to delineate in more detail the resemblance and diversity of these three elements. The arrangement of the sections is illustrated schematically in Fig. 8. In the interests of uniformity within the framework of Comprehensive Inorganic Chemistry, the division of the sections is based on the formal oxidation state of the halogen, the elements (oxidation state 0) being treated first, followed

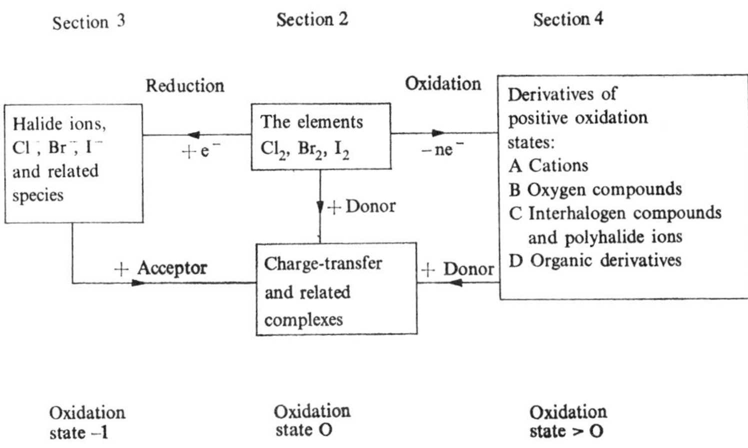

FIG. 8. Organization of the material in Sections 2–4.

initially by the halide ions and derivatives of the halogens with more electropositive elements (oxidation state −1), and then by derivatives with more electronegative elements (oxidation state > 0).

Regrettably, the scheme of things produces, of necessity, some artificial and arbitrary distinctions. For example, the classification according to oxidation state separates the diatomic halogens X_2 (Section 2) from the interhalogens such as ClF and ICl (Section 4), whereas in reality the two classes are closely affiliated; the complication is particularly unfortunate for the apparent heterogeneity it brings to the treatment of donor-acceptor complexes of the two classes, whether these are molecular like Br_2,C_6H_6 or Me_3N,ICl, described mainly in Section 2, or ionic as in the polyhalide species I_3^- and IBr_2^-, presented with the interhalogens in Section 4.

On the other hand, it is inappropriate to consider the heaviest known halogen, astatine, within the context of the scheme of Sections 2–4. Because of the short half-lives of even the most stable isotopes, macroscopic quantities of astatine cannot be accumulated. Accordingly, our knowledge of its chemistry depends entirely on tracer studies. The very limited state of this knowledge, combined with the practical aspects peculiar to its chemistry, requires that a separate section be devoted to astatine.

1.5. PSEUDOHALOGENS[27,28]

Reference must also be made here to the class of pseudohalogens, first defined by Birckenbach and Kellermann[27], comprising such molecules as cyanogen, $(CN)_2$, oxocyanogen, $(OCN)_2$, thiocyanogen, $(SCN)_2$, selenocyanogen, $(SeCN)_2$, and azidocarbondisulphide, $(SCSN_3)_2$. Like the diatomic halogen molecules, these contain two relatively electronegative units (X) directly linked, as in NC–CN; in common with the halogens, they are reduced characteristically to anions X^- or to molecular pseudohalides, e.g. CH_3–X. Examples of pseudohalide anions are CN^-, OCN^-, SCN^-, $SeCN^-$ and $SCSN_3^-$, as well as $TeCN^-$ [29] and N_3^-, for which the pseudohalogen parents are not known. The physical and chemical properties diagnostic of pseudohalogens or pseudohalide ions are, in outline:

1. The parent pseudohalogen is a molecular compound involving a symmetrical combination of two radicals, X–X. With alkalis the reaction is often analogous to that of the halogens, e.g.

$$(CN)_2 + 2OH^- \rightleftharpoons CN^- + OCN^- + H_2O$$

The pseudohalogens also undergo some of the addition reactions typical of halogens, as in

$$CH_2{=}CH_2 + (SCN)_2 \rightarrow NCS\cdot C_2H_4\cdot SCN$$

2. The pseudohalogens react with various metals to give salts containing X^- anions; the salts of silver, mercury(I) and lead(II) are typically sparingly soluble in water.

3. The hydrides HX are acids which are, however, very weak compared with the halogen acids, as illustrated by the following pK_a values: HCl, −7·4; HN_3, 4·4; HCN, 8·9. The difference in acidities is presumably a function primarily of the differences (i) in H–X bond energy and (ii) in hydration energy of the X^- ion.

[27] L. Birckenbach and K. Kellermann, *Chem. Ber.* **58** (1925) 786, 2377.

[28] T. Moeller, *Inorganic Chemistry*, p. 463. Wiley (1952); R. C. Brasted, *Comprehensive Inorganic Chemistry* (ed. M. C. Sneed, J. L. Maynard and R. C. Brasted), Vol. 3, p. 223. Van Nostrand (1954); ref. 20, p. 467; F. A. Cotton and G. Wilkinson, *Advanced Inorganic Chemistry*, 2nd edn., p. 560. Interscience (1966); R. B. Heslop and P. L. Robinson, *Inorganic Chemistry*, 3rd edn., p. 551. Elsevier (1967).

[29] A. W. Downs, *Chem. Comm.* (1968) 1290.

4. The pseudohalogen radicals form compounds among themselves, e.g. $NC \cdot SCN$ and with the halogens, e.g. ClN_3; species analogous to the polyhalide ions are also known, e.g. $(SeCN)_3^-$ and $[I(SCN)_2]^-$ [30].

5. The pseudohalide ions form complexes with metals as do the halide ions, e.g. $[Co(NCS)_4]^{2-}$, though the stabilities of analogous halide and pseudohalide complexes differ widely in many cases.

6. The pseudohalogen forms molecular compounds analogous to molecular halides, e.g. CH_3NCO and $Si(NCO)_4$, though, in contrast with the halogens, the asymmetry of the pseudohalogen radical may lead to isomeric forms depending on the mode of coordination of the radical, as in CH_3–SCN and CH_3–NCS.

7. In common with halide ions, a pseudohalide ion is oxidized to the parent pseudohalogen by suitable oxidizing agents. A typical reaction is

$$2Fe^{3+} + 2SCN^- \rightarrow 2Fe^{2+} + (SCN)_2$$

More detailed aspects of pseudohalogen chemistry are referred to in the context of carbon, nitrogen and appropriate Group VI elements and of transition metal complexes. Except in relation to derivatives of the halogens in positive oxidation states (e.g. interhalogens and polyhalide ions), pseudohalogens are not otherwise treated in this section. Although the classification of materials as pseudohalogens has some practical value, albeit of a limited and largely empirical nature, the affinity of such materials to the halogens themselves must not be exaggerated; even the cursory outline given here suggests a number of significantly divergent properties. Further, it should be appreciated (i) that few of the classical pseudohalogens measure up completely to the idealized behaviour which has been described, and (ii) that, if such deviations are tolerable, then the class is capable of very considerable enlargement[31]; for, in terms of at least some of their properties, species such as $(NO_2)_2$, $(RS)_2$ (where R is an organic group), $(OH)_2$, $(OS(F))O_{22}$ and $(ClO_3)_2$ deserve also to be treated as pseudohalogens.

GENERAL REFERENCES

Brasted, R. C., *Comprehensive Inorganic Chemistry* (ed. M. C. Sneed, J. L. Maynard and R. C. Brasted), Vol. 3, van Nostrand (1954).

Cotton, F. A. and Wilkinson, G., *Advanced Inorganic Chemistry*, 2nd edn., Interscience (1966).

Gmelins Handbuch der Anorganischen Chemie, 8 Auflage: "Chlor", System-nummer 6, Teil A and B, Verlag Chemie, Weinheim/Bergstr. (1968–9); "Brom", System-nummer 7, Verlag Chemie, Berlin (1931); "Iod", System-nummer 8, Verlag Chemie, Berlin (1933).

Gutmann, V. (ed.), *Halogen Chemistry*, Vols. 1–3, Academic Press (1967).

Heslop, R. B. and Robinson, P. L., *Inorganic Chemistry*, 3rd edn., Elsevier (1967).

Jolles, Z. E. (ed.), *Bromine and Its Compounds*, Benn, London (1966).

Mellor, J. W., *A Comprehensive Treatise on Inorganic and Theoretical Chemistry*, Vol. II, Longmans, Green and Co., London (1922); Supplement to *Mellor's Comprehensive Treatise on Inorganic and Theoretical Chemistry*, Supplement II, Part I, Longmans, Green and Co., London (1956).

Pascal, P., *Nouveau Traité de Chimie Minérale*, Vol. XVI, Masson et Cie, Paris (1960).

Phillips, C. S. G. and Williams, R. J. P., *Inorganic Chemistry*, Vol. 1, Clarendon Press, Oxford (1965).

Sidgwick, N. V., *The Chemical Elements and Their Compounds*, Vol. II, Clarendon Press, Oxford (1950).

Wells, A. F., *Structural Inorganic Chemistry*, 3rd edn., Clarendon Press, Oxford (1962).

30 C. Long and D. A. Skoog, *Inorg. Chem.* **5** (1966) 206.

31 See for example L. Birckenbach and K. Huttner, *Chem. Ber.* **62** (1929) 153; *Z. anorg. Chem.* **190** (1930) 1.

2. THE ELEMENTS CHLORINE, BROMINE AND IODINE

2.1. DISCOVERY AND HISTORY[32,33]

Of the halogens, chlorine was the first to be discovered, being prepared by Scheele in 1774 by heating hydrochloric (muriatic) acid with manganese dioxide[34], though the fumes of the gas must have been known from the time of the thirteenth century by all those who made and used aqua regia. The preparation of hydrochloric acid itself (then called *spiritus salis*) was first reported in a fifteenth century Italian manuscript[35], while sodium chloride, known as "salt" from the earliest times, is referred to by Pliny in his *Naturalis Historiae*[36]. In accordance with the views of Lavoisier that all acids contained oxygen, chlorine was first named oxymuriatic acid, a view apparently supported by observations of Berthollet (i) that, if the manganese dioxide is first deprived of some of its oxygen by calcination, it furnishes a smaller quantity of Scheele's gas, and (ii) that oxygen is evolved when the gas reacts with water. After attempts by Gay-Lussac, Thénard and Davy to decompose the so-called oxymuriatic acid had failed, it was Davy who in 1810 first proposed the elementary nature, the name *chlorine* (from the Greek χλωρός, green) and the symbol Cl for the gas[37]. At about the same time, the group name *halogen* (from the Greek ἅλς, sea-salt, and the root γεν –, produce) was first coined to describe the salt-forming tendencies of the individual members of the group.

Next in order of discovery was iodine, first prepared in 1812 by the French chemist Courtois. When an aqueous extract of the calcined ashes of seaweed, that is, kelp, was treated with sulphuric acid, it was noted by Courtois that the black precipitate first produced was converted on heating to liberate the new element in the form of its violet vapour[38]. The name *iode*, the French equivalent of its present name (Greek ἰώδης, violet), was designated in 1813 by Gay-Lussac, who also demonstrated some striking analogies between the new substance and chlorine, and in a famous communication "Mémoire sur l'iode" (1814)[39] described large tracts of iodine's chemical behaviour.

Of the three halogens, bromine had the least eventful history, its elemental nature and its relation to chlorine and iodine being recognized from the very first. The discovery of the element is credited to Balard in 1826 in the course of studying the mother liquor remaining after the crystallization of salt from the water of the Montpellier salt-marshes, which is rich in magnesium bromide[40]. Balard was attracted by the intense yellow coloration which developed when chlorine water was added to the liquor; ether-extraction followed by treatment with potassium hydroxide destroyed the colour, while the residue was shown,

[32] J. W. Mellor, *A Comprehensive Treatise on Inorganic and Theoretical Chemistry*, Vol. II, Longmans, Green and Co., London (1922); Supplement to *Mellor's Comprehensive Treatise on Inorganic and Theoretical Chemistry*, Supplement II, Part I, Longmans, Green and Co., London (1956).

[33] *Gmelins Handbuch der Anorganischen Chemie*, 8 Auflage: "Chlor", System-nummer 6, Teil A, Verlag Chemie, Weinheim/Bergstr. (1968); "Brom", System-nummer 7, Verlag Chemie, Berlin (1931); "Iod", System-nummer 8, Verlag Chemie, Berlin (1933). R. C. Brasted, *Comprehensive Inorganic Chemistry* (ed. M. C. Sneed, J. L. Maynard and R. C. Brasted), Vol. 3, van Nostrand (1954).

[34] C. W. Scheele, *König. Vetens. Akad. Stockholm*, **25** (1774) 89; *Opuscula chimica et physica, Leipzig*, **1** (1788) 232; *The Chemical Essays of C. W. Scheele*, p. 52. London (1901); *Alembic Club Reprints* (1897) 13.

[35] L. Reti, *Chymia*, **10** (1965) 11.

[36] Pliny, *Naturalis Historiae*, Book 33, chapter 25 (first century A.D.).

[37] H. Davy, *Phil. Trans.* **100** (1810) 231; *Alembic Club Reprints* (1894) 9.

[38] B. Courtois, *Ann. Chim.* **88** (1) (1813) 304.

[39] J. L. Gay Lussac, *Ann. Chim.* **91** (1) (1814) 5.

[40] A. J. Balard, *Ann. Chim. Phys.* **32** (2) (1826) 337.

when heated with manganese dioxide and sulphuric acid, to produce red fumes which condensed to a dark brown liquid with an unpleasant smell. There is no question but that the element had been isolated by Joss and by Liebig[41] prior to Balard's discovery; however, neither of these investigators recognized the true nature of their product, Joss mistaking it for selenium and Liebig for iodine chloride. On the other hand, Balard was unquestionably the first to appreciate the elemental nature of the material and its relation to chlorine and iodine. The substance was first called *muride*, but the name *bromine*—from the Greek βρῶμος, a stench—was later preferred.

2.2. NATURAL OCCURRENCE[32,33,42–48]

The halogens are found in nature, not in their highly reactive elemental states, but most commonly in the form of the corresponding halide anions; in the rather exceptional case of iodine, iodate deposits are an important natural source of the element. It has been estimated that igneous rocks, which constitute about 95% of the earth's crust, contain on average the following approximate proportions of the halogens in the combined state: Cl, 0·031%; Br, $1{\cdot}6\times10^{-4}$%; I, 3×10^{-5}%. In accordance with the similarity of radius of the chloride and bromide ions, the mineral chemistry of the two elements is quite closely related, and bromine is known to replace chlorine in numerous minerals. The bulk of the chlorine and bromine present in sedimentary and volcanic rocks takes the place of OH groups in hydroxide-bearing minerals such as hornblendes, micas, clay materials and aluminium hydroxide. There are otherwise known relatively few chlorides or bromides which may be considered as minerals in the strict meaning of the word; the relatively insoluble ores of certain heavy metals (e.g. AgCl, horn silver; AgBr, bromargyrite; Ag(Br,Cl), embolite; Hg_2Cl_2, calomel; CuCl, nantokite) are of negligible commercial importance.

However, the largest natural source of chlorine and bromine is the sea; out of a total average salinity of about 3·4%, sea water contains approximately 1·9% chlorine, mainly as sodium chloride though with smaller amounts of other chlorides, 0.0065% bromine (representing a chlorine:bromine mass ratio of nearly 300:1) and 5×10^{-8}% iodine. Isolated bodies of water in arid regions are frequently found to have a high chloride content; the Great Salt Lake of Utah, for example, contains no less than 23% sodium chloride, while the Dead Sea, with a total salinity of more than 30%, contains near the surface about 3·5% calcium chloride, 8·0% sodium chloride and 13·0% magnesium chloride. Whereas the chloride ion makes up 90% of the total anion content of sea water, the main anion of river water (average salinity 0·01–0·02%) is the bicarbonate ion, the chloride ion amounting to only 2–5%. Both the chloride and bromide ions from weathered rocks are thus dissolved

[41] J. R. Joss, *J. prakt. Chem.* **1** (1834) 129; J. von Liebig, *Liebig's Ann.* **25** (1838) 29.

[42] F. C. Phillips, *Quart. Rev. Chem. Soc.* **1** (1947) 91; D. T. Gibson, *Quart. Rev. Chem. Soc.* **3** (1949) 263; V. M. Goldschmidt, *Geochemistry*, Oxford (1954); V. V. Cherdyntsev, *Abundance of Chemical Elements*, Chicago University Press (1961); G. Gamow, *Biography of the Earth*, Macmillan, London (1962); L. H. Ahrens, *Distribution of the Elements in our Planet*, McGraw-Hill, New York (1965).

[43] C. S. G. Phillips and R. J. P. Williams, *Inorganic Chemistry*, Clarendon Press, Oxford (1965–6).

[44] N. V. Sidgwick, *The Chemical Elements and their Compounds*, Vol. II, p. 1139. Clarendon Press, Oxford (1950).

[45] Z. E. Jolles (ed.), *Bromine and its Compounds*, Benn, London (1966).

[46] *Kirk-Othmer Encyclopedia of Chemical Technology*, 2nd edn., Vols. 1, 3 and 11. Interscience (1963–6).

[47] C. A. Hampel (ed.), *The Encyclopedia of the Chemical Elements*, Reinhold, New York (1968).

[48] J. S. Sconce (ed.), *Chlorine: its Manufacture, Properties and Uses* (A. C. S. Monograph No. 154), Reinhold, New York (1962).

and transported into the hydrosphere. They concentrate in the seas whence they may return to the lithosphere through deposition from parts of the sea separated from the main body of the ocean. The question of whether the primordial ocean contained the same salts in proportions that are the same as, or similar to, those of today's oceans is still a moot point[45], though the discrepancy between the sodium:chlorine ratio in the ocean and that in rocks makes it difficult to subscribe to the view that the composition of the ocean has remained constant. To account, at least in part, for the balance in nature of chlorides and bromides, it has been suggested that carbonates have reacted, and may still be reacting, with hydrogen chloride and bromide present in volcanic gases.

The principal sources of chlorine and bromine of commercial significance are as follows:

Ocean-derived Deposits of Such Minerals as Rock Salt, NaCl; Sylvine, KCl; Carnallite, $MgCl_2$, KCl, $6H_2O$; Kainite, $MgSO_4$, KCl, $3H_2O$

These deposits are widely distributed in the world, many of them being of immense proportions; countries favoured by such natural resources include, for example, the United States, Russia, Germany, Austria, Italy, the United Kingdom and Brazil. For the production of chlorine, sodium chloride is the principal raw material, and where this is obtained from natural rock salt deposits, it may either be mined or pumped to the surface as a saturated aqueous solution. The bromine content of carnallite deposits is no more than 0·1–0·35%, while rock salt contains typically 0·005–0·04% bromine. Minerals richer in bromine, such as bischoffite and tackhydrite ($MgCl_2,6H_2O$ and $CaCl_2,2MgCl_2,12H_2O$, respectively) are not abundant and are less important as sources of bromine than is carnallite. Bromine is not extracted directly from carnallite but from the mother liquor remaining after the extraction of potassium chloride from the mineral.

Natural Brines Derived from Seas and Lakes

These are now a significant source of chlorine and the major source of bromine. Solar evaporation is commonly used to concentrate the brines and to isolate the principal dissolved ingredients. In certain areas, special topographical and climatical conditions greatly facilitate this process; such is the case in the Rann of Cutch in India, in the saline lagoons of the Black Sea and Caspian Sea, in the Sebkha-el-Melah in Tunisia, and on the shores of inland seas and lakes such as the Dead Sea and Great Salt Lake. Although the average bromine content of sea water is only 0·0065%, some isolated bodies of water are much richer in the element: e.g. the Dead Sea, 0·4–0·6%; Sebkha-el-Melah, ~ 0·25%; Sakskoe Ozero (Crimea), 0·28%; Searles Lake (California), 0·085%. At the present time, bromine is extracted commercially from normal ocean water containing no more than 0·0065% bromide ion.

Wells and Springs

Apart from the brine produced artificially by pumping water into underground salt deposits, natural brines from wells and springs are also very widespread; thus, natural waters associated with oil-fields are often relatively rich in halide ions. Although these are most commonly the result of leaching of salt layers, it is possible that some originate from underground pockets left by ocean concentrates. Many such wells and springs are found, for example, in the United States, mainly in Michigan, Ohio and West Virginia, in parts of Russia, Israel and Italy, and many provide important natural sources of common salt, calcium chloride, magnesium chloride and bromine. Whereas chloride is the principal anion of most of the waters, the bromide content varies somewhat unpredictably from zero to proportions (typically 0·1–0·4%) compatible with the commercial production of the element.

Iodine differs in a number of respects from chlorine and bromine. In the first place, it is, by a considerable margin, the rarest of the three halogens. Like chlorine and bromine, it is widely distributed in nature in rocks, soils, underground brines and sea water, but always in low concentration (see Table 36). Unlike chlorine and bromine, iodine occurs naturally not only as iodide but also as iodate, and it is in this form (in the minerals lautarite, $Ca(IO_3)_2$, and dietzeite, $7Ca(IO_3)_2,8CaCrO_4$) that iodine is found in the Caliche beds in Chile, which, with an iodine content of 0·02–1%, remain the major commercial source of iodine. The concentration of iodine in other secondary deposits is insignificant, while the iodide content of sea water is so low as to preclude serious commercial exploitation for the production of iodine. Apart from the Chilean nitrate deposits, some natural brines, commonly derived from oil-well drillings, also contain commercially useful concentrations of iodine (typically 30–40 ppm); such are the natural brines of Michigan and, now on a diminished scale, the oil-well brines of California.

Despite the low concentration of iodine in sea water, some seaweeds, notably the deep-sea varieties, can extract and accumulate the element. Those of the *Laminaria* family are the richest, containing up to 0·45% on a dry basis. The ashes of seaweed provided the first commercial source of iodine. Although largely superseded as the chief raw material for the production of iodine by the Chilean deposits and by natural brines, seaweed continues to be used locally to produce iodine, principally in Japan. A number of other types of marine life, such as oysters, sponges and certain fishes, also concentrate iodine in their systems, and iodides and iodates in sea water enter into the metabolic cycle of most marine flora and fauna.

In the human body, the greatest concentration of iodine is found in the thyroid gland. Iodine appears to be a trace element essential to animal and vegetable life. Bromide ions are found in living organisms, but their biological role is unknown and no precise physiological significance has so far been established. In plants and animals chlorine exists primarily as the chloride ion dissolved in cell liquids, but organo-chlorine compounds, mostly of fungal origin, are also found in nature (see pp. 1336–40).

2.3. FORMATION OF THE ELEMENTARY HALOGENS[32,33,45–48]

General Methods of Production

The formation of free chlorine, bromine or iodine may be effected by one of two chemical methods: (i) oxidation of halide derivatives, and (ii) reduction of compounds of the halogens in positive oxidation states, e.g. ClO^- or IO_3^-. Of these two general methods, the first is very much more important than the second for such practical reasons as the ready availability and cheapness of the starting materials, the comparative facility of selective oxidation rather than selective reduction, and the kinetic inertness exhibited by some of the commoner "positive" halogen derivatives, e.g. ClO_3^- and ClO_4^-. The thermodynamic balance between the two processes is well illustrated by reference to the oxidation state diagrams of Fig. 2 and to the differences between the standard redox potentials of the couples

$XO_3^-/\frac{1}{2}X_2$ and $\frac{1}{2}X_2/X^-$ on the one hand and of the couples $XO_3^-/\frac{1}{2}X_2$ and XO_3^-/X^- on the other. It is thus evident that the majority of reagents capable of reducing the XO_3^- ion to the free halogen in acid solution are also capable of reducing it to the X^- ion.

Oxidation of Halides

Since the free energy of formation of halides, whether pure or in solution, decreases in the series chloride > bromide > iodide, the thermodynamic barrier to the production of free halogen from the corresponding halide decreases in the same order. For chlorine, therefore, the range of chemical agents that will accomplish this oxidation is limited to the more energetic oxidizing agents, whereas for iodine oxidation is easily accomplished and strongly oxidizing conditions are neither necessary nor yet desirable in that they may lead to "positive" iodine compounds.

An example of chemical oxidation to produce chlorine is provided by the reaction of hydrogen chloride with manganese dioxide

$$MnO_2+4HCl \rightarrow MnCl_2+Cl_2+2H_2O$$

Apart from its historical significance[34], this has for long been a method of producing chlorine on a small scale in the laboratory, and was also the basis of the Weldon process formerly used for the manufacture of chlorine. Other effective oxidizing agents of hydrogen chloride or metal chlorides include dichromates, permanganates, lead dioxide, sodium bismuthate, peroxydisulphates and chlorates. Air oxidizes hydrogen chloride at elevated temperatures and in the presence of certain catalysts, e.g. $CuCl_2$, as in the Deacon Process (q.v.). Processes depending on the oxidation of chlorides either by nitric acid or by sulphur trioxide have also been developed on, or up to, commercial proportions. However, electrolytic oxidation of chloride solutions accounts for more than 99% of the chlorine of commerce[32,33,46–48].

In an analogous manner, bromine is liberated from hydrogen bromide, metal bromides or solutions of these by oxidation with reagents such as manganese dioxide, nitric acid or bromates. The Deacon Process of air-oxidation is applicable to the conversion of hydrogen bromide to bromine, as is electrochemical oxidation of bromide ions. But the only methods of importance for the manufacture of bromine are based on the oxidation of bromide-containing solutions by chlorine[32,33,45–47].

Formation of iodine from iodides is accomplished by agents such as chlorine, bromine, manganese dioxide, concentrated sulphuric acid, nitric acid and iodates; even such mild oxidizing agents as Fe^{3+}, $[Fe(CN)_6]^{3-}$, Cu^{2+} and antimonate(V), under appropriate conditions of pH in aqueous solution, oxidize iodide ions quantitatively to free iodine. Because of the ease with which iodine can be estimated volumetrically, a number of these reactions, e.g. that with iodate and copper(II), afford noteworthy methods of quantitative analysis. For the production of iodine on the large scale, oxidation of iodides by manganese dioxide, chlorine, nitrite or dichromate has been exploited. Electrolytic oxidation of iodide solutions provides another potential route to iodine, though it is commercially unrealistic. Conversion of iodides to iodine in commercial quantities is now mostly accomplished by chlorine oxidation in processes that are similar in principle to those used for the extraction of bromine[32,33,46,47].

A modification of detail, though not of principle, is provided by the thermal decomposition of certain metal halides, whereby the elementary halogen is released, though the method is of little practical consequence as a preparative procedure. In thermal decompositions such as

$$PtCl_4 \rightarrow PtCl_2 + Cl_2$$
$$PtCl_2 \rightarrow Pt + Cl_2$$
$$PbCl_4 \rightarrow PbCl_2 + Cl_2$$
$$2AuBr_3 \rightarrow 2Au + 3Br_2$$
$$UO_2Br_2 \rightarrow UO_2 + Br_2$$
$$2EuI_3 \rightarrow 2EuI_2 + I_2$$

the halides range from molecular to predominantly ionic species; the molecular compounds are generally characterized by relatively weak metal–halogen bonds, whereas for the ionic systems decomposition is favoured (i) by the presence of small, highly charged cations, and (ii) by unusually large ionization potentials governing the conversion of the less to the more highly charged form of the metal.

Reduction of Derivatives of the Halogens in Positive Oxidation States

Perhaps the most important reaction of this class is that between halate and the corresponding halide ions in acid solution:

$$5X^- + XO_3^- + 6H^+ \rightleftharpoons 3X_2 + 3H_2O$$

In all cases the balance of this reaction is highly sensitive to pH, and under alkaline conditions the halogens are themselves unstable with respect to disproportionation into XO_3^- and X^- ions. Thus, in one process for the manufacture of bromine from sea water, bromine vapour is absorbed in sodium carbonate solution when it forms bromide and bromate[45,46]; subsequent acidification regenerates free bromine in more concentrated form. Similarly, the naturally occurring iodates found in the Chilean nitrate deposits are converted to iodine by the following processes[46]:

$$IO_3^- + 3HSO_3^- \rightarrow I^- + 3HSO_4^-$$
$$5I^- + IO_3^- + 6H^+ \rightarrow 3I_2 + 3H_2O$$

The reaction between chlorate and hydrochloric acid has also been used for the laboratory preparation of chlorine[32,49].

The oxidation state diagrams of Fig. 2 make it clear that, despite the ready disproportionation of the free halogens under alkaline conditions, the interaction not only of a halate but also of a hypohalite, a chlorite or perhalate with the corresponding halide under acidic conditions is thermodynamically capable of producing the free halogen. The reactions

$$HOCl + Cl^- + H^+ \rightarrow Cl_2 + H_2O$$

and

$$IO_4^- + 7I^- + 8H^+ \rightarrow 4I_2 + 4H_2O$$

afford examples of the practical working of this principle, though the thermodynamic potential of some systems is not realized because of kinetic factors (associated, for example, with the perchlorate ion), or because of side-reactions (e.g. disproportionation). Reaction of both oxyhalogen and interhalogen derivatives with other relatively mild reducing agents

49 E. Dufilho, *Bull. Soc. Pharm. Bordeaux*, **63** (1925) 41.

may also lead to the elementary halogen, as in the following examples:

$$I_2O_5+5CO \longrightarrow I_2+5CO_2$$
$$2ClO_3^-+I_2 \xrightarrow{\text{aq soln}} 2IO_3^-+Cl_2$$
$$5(COOH)_2+2IO_3^-+2H^+ \xrightarrow{\text{aq soln}} 10CO_2+I_2+6H_2O$$
$$6H_5IO_6+15H_2S \xrightarrow{\text{aq soln}} 3I_2+2H_2SO_4+13S+28H_2O$$
$$2NH_3+2ClF_3 \longrightarrow N_2+6HF+Cl_2$$
$$6Nb+10BrF_3 \longrightarrow 6NbF_5+5Br_2$$
$$6ClO_2+2BrF_3 \longrightarrow 6ClO_2F+Br_2$$

Again, thermal decomposition of many oxyhalogen and polyhalide derivatives produces the free halogen. Thus, one mode of decomposition of aqueous hypobromous acid proceeds slowly at room temperature according to the equation:

$$4HOBr \longrightarrow 2Br_2+2H_2O+O_2$$

Other reactions illustrative of this general principle are:

$$2Cl_2O \xrightarrow{>60\,^\circ C} 2Cl_2+O_2$$
$$2ClO_2 \xrightarrow{h\nu} Cl_2+2O_2$$
$$2Cl_2O_7 \xrightarrow{\text{room temp.}} 2Cl_2+7O_2$$
$$2Br_2O \xrightarrow{>-40\,^\circ C} 2Br_2+O_2$$
$$2Mg(XO_3)_2 \longrightarrow 2MgO+5O_2+2X_2 \quad (X = Cl, Br \text{ or } I)$$
$$CsX_3 \longrightarrow CsX+X_2 \quad (X = Br \text{ or } I)$$
$$I_2Cl_6 \longrightarrow 2ICl+2Cl_2$$

Laboratory Preparation and Purification

Because of the ready availability of commercial chlorine, bromine and iodine, it is rarely necessary now to prepare these elements in the laboratory except for didactic reasons, though as recently as 1949 a method was described for the laboratory preparation of chlorine by electrolysis of chloride solutions[50]. Individual halogens can be satisfactorily separated from mixtures of the elements (and other volatile materials) by distillation or by gas-chromatographic[51] methods.

Chlorine, available in conveniently small cylinders, may be of high purity (99·9% or better), but certain commercial samples are liable to contain oxygen, nitrogen, carbon dioxide and monoxide, hydrogen chloride and moisture. A suggested[52] method of purification involves passing the chlorine first through potassium permanganate solution (to remove hydrogen chloride), then through sulphuric acid and over phosphorus pentoxide (to remove moisture), and finally fractionating the material in a vacuum-system, using either low-temperature baths [e.g. acetone–solid CO_2 (−80°C), ethyl bromide (−119°C) and liquid nitrogen (−196°C)] or a low-temperature still. The reaction of chlorine both with mercury and with many kinds of tap-grease must be recognized in all manipulation of the purified material.

Bromine of high quality is readily obtainable. The likely impurities, with the maximum limits imposed for AnalaR grade[53], are as follows: Cl (0·025%), I (0·0001%), sulphate

[50] W. J. Kramers and L. A. Moignard, *Trans. Faraday Soc.* **45** (1949) 903.

[51] J. Janák, M. Nedorost and V. Bubeníková, *Chem. Listy*, **51** (1957) 890; J. F. Ellis and G. Iveson, *Gas Chromatography* (ed. D. H. Desty), p. 300. Butterworths, London (1958); J. F. Ellis, C. W. Forrest and P. L. Allen, *Anal. Chim. Acta*, **22** (1960) 27.

[52] R. E. Dodd and P. L. Robinson, *Experimental Inorganic Chemistry*, Elsevier (1954).

[53] *AnalaR Standards for Laboratory Chemicals*, 6th edn., AnalaR Standards Ltd., London (1967).

(0·005%), arsenic (0·0001%) and non-volatile matter (0·005%). The liquid (b.p. 58·8°C) can be distilled in a glass fractionating apparatus at atmospheric pressure or *in vacuo*[32,52]; like chlorine, bromine attacks mercury, rubber and most tap-greases, which should therefore be excluded from the purification stages. Individual impurities can be removed by the following methods[54]:

Impurity	Treatment
Chlorine	Distillation from, or reaction with, a metal bromide; fractional condensation, low temperature-low pressure sublimation, or chromatographic methods.
Iodine	Extraction with water or conversion to halide followed by treatment with permanganate or bromine.
Water	Concentrated sulphuric acid or phosphorus pentoxide.
Organic matter	(i) Saturation of a 5–30% aqueous bromide solution with crude bromine, followed by distillation, condensation and drying of the bromine.
	(ii) Passage of bromine with oxygen at 1000°C through a furnace packed with fragmented quartz or high-silica (Vycor) glass, followed by removal of any chlorine and drying.

Very pure bromine for atomic weight determinations has been prepared[55] by a lengthy process involving, *inter alia*, reduction of the bromine to bromide; oxidation of the iodide by permanganate; oxidation of the bromide to bromine with halogen-free potassium dichromate and sulphuric acid; distillation, drying over fused calcium oxide and calcium bromide, and finally vacuum-distillation of the bromine.

Iodine of high purity can also be bought. The AnalaR grade has the following maximum limits of impurity[53]: Cl^- and Br^- (0·005%), CN^- (0·005%), sulphate (0·01%) and non-volatile matter (0·01%). Further purification is effected by sublimation[52], after first grinding up the iodine with a small amount of potassium iodide to reduce any free chlorine and bromine and impurities such as ICl, IBr and ICN. Purification has also been accomplished[32,33] by forming an insoluble heavy-metal iodide (AgI or CuI), and either (i) reducing the iodide with hydrogen to the metal and hydrogen iodide, and then oxidizing the hydrogen iodide with nitrite, or (ii) directly oxidizing the metal iodide to the oxide and iodine. Grease and mercury should, ideally, be excluded from the sublimation and other apparatus employed for handling purified iodine.

Commercial Production of Chlorine, Bromine and Iodine

Chlorine[32,33,46–48]

The first patent connected with any industrial use of chlorine, dated 1799, was for its use in bleaching. During the last half of the nineteenth century the demand for chlorine for bleaching grew at a rate that brought about the invention and development of the Weldon and Deacon processes for chlorine production, in both of which hydrogen chloride (mostly from the sulphuric acid treatment of common salt) was the raw material. Towards the end of the nineteenth century, development of electrical generating equipment converted the electrolytic method of chlorine production from a laboratory to a commercial method. At the present time, in practically all countries, more than 99% of the chlorine of commerce is

[54] V. A. Stenger, *Angew. Chem., Internat. Edn.* **5** (1966) 280.
[55] O. Hönigschmid and E. Zintl, *Annalen*, **433** (1923) 201.

produced electrolytically, and chlorine ranks with sulphuric acid, sodium hydroxide, sodium carbonate, ammonia, phosphoric acid and nitric acid as one of the most important tonnage chemicals now made[46]. Chlorine, soda ash (Na_2CO_3) and caustic soda (NaOH) represent the primary products of chemical industries whose basic raw material is common salt[46–48,56].

In the electrolysis of sodium chloride brines, chlorine is produced at the anode, while hydrogen is normally produced at the cathode, together with a simultaneous increase in the local hydroxide ion concentration. The overall reaction may be represented as

$$Cl^- + H_2O \rightarrow OH^- + \tfrac{1}{2}Cl_2 + \tfrac{1}{2}H_2$$

In order to keep the anodic and cathodic products entirely separate, there are two distinct methods of operation: (i) the use of a diaphragm between the two compartments, and (ii) the use of a mercury cathode, whereby sodium, rather than hydrogen, is initially discharged.

In the "diaphragm" process[32,33,46–48], brine is fed continuously to the electrolytic cell, flowing from the anode compartment through an asbestos diaphragm backed by an iron cathode. To minimize back-diffusion and migration, the flow rate is always such that only part of the sodium chloride is converted to the hydroxide, hydrogen and chlorine. The catholyte solution of alkaline brine is evaporated to obtain the sodium hydroxide, in the course of which sodium chloride precipitates, is separated, redissolved and returned to the cell. The assembly consists of an outer steel shell, either cylindrical or rectangular, supporting the cathode, which may take the form of a perforated plate or woven screen inside the shell. The actual cathode surfaces are generally lined with a layer of asbestos, either in the form of paper or of vacuum-deposited fibres. At the minimum practical distance, the anodes are supported with their faces parallel to the diaphragm. An inert, insulating cover, carrying a brine inlet and a chlorine outlet, closes the cell. The choice of materials for the electrodes is important. For economic reasons, not only have corrosion and the consequent need for replacement to be reduced to a minimum, but also the hydrogen and chlorine overvoltages must be kept as low as possible to conserve energy. As cathode material iron or mild steel appears to be in general use, while graphite, with its virtues of moderate price, low apparent density and high overvoltage with respect to oxygen, is usually favoured for the anodes. However, the relatively high overvoltage of graphite with respect to chlorine and its limited mechanical strength and resistance to corrosion, which cause the anodes to disintegrate at a significant rate, have encouraged active interest in alternative anode materials, such as platinized titanium.

In the mercury–cathode type of alkali–chlorine cell[32,33,46–48], there are two main parts, the "electrolyser" or brine cell and the decomposer or "denuder". The "electrolyser" is essentially a closed rectangular trough, long in comparison with its height and breadth. Continuously fed brine is electrolysed between anodes, most commonly of graphite (though, again, platinized titanium offers notable advantages), and a cathode consisting of a sheet of mercury which flows uniformly over the flat bottom of the cell; chlorine gas is formed at the anodes and sodium amalgam at the cathode. Parts of the trough not in contact with mercury are provided with a corrosion-resistant, protective coating such as hard rubber; the bottom of the cell is in some cases bare steel and in others hard rubber, granite or concrete surfaced with a synthetic coating. The anodes are usually horizontal graphite plates suspended on

[56] E. L. Gramse and L. H. Diamond, *Advances in Chemistry Series*, **78** (1968) 1.

rods which extend through the cover of the cell; the anodes, parallel to, and close to, the mercury–brine interface, are perforated or grooved to facilitate the release of chlorine, which is removed from the cell either *via* an outlet in the cover or *via* an enlarged anolyte overflow connection. The sodium amalgam flows continuously to the decomposer, which may be regarded as a secondary cell where the amalgam becomes the anode to a short-circuited graphite (or iron) cathode in an electrolyte of sodium hydroxide solution. There are two general types of amalgam decomposer: (i) a horizontal steel trough mounted parallel to the electrolyser containing graphite grids, and (ii) a vertical steel cylinder packed with pieces of graphite. Purified water is fed to the decomposer, generally in counter-current to the sodium amalgam; hydrogen gas is formed and the electrolyte increases in sodium hydroxide content. A solution containing from 30 to 70% sodium hydroxide at high purity overflows from the decomposer (most cells are operated to give a 50% solution of sodium hydroxide). The denuded mercury is recycled to the electrolyser; typically the amalgam reaching the decomposer contains 0·2% sodium and is returned with less than 0·02% sodium.

For the overall aqueous reactions:

(i) *Diaphragm cell*: $Cl^- + H_2O \longrightarrow OH^- + \frac{1}{2}Cl_2 + \frac{1}{2}H_2$

(ii) *Mercury cathode cell*: $Na^+ + Cl^- \xrightarrow{H_2O/Hg} Na/Hg + \frac{1}{2}Cl_2$

the standard free energies, ΔG°_{298}, are (i) +50·445 kcal mol^{-1} and (ii) approximately +76 kcal mol^{-1} corresponding to standard, reversible cell-voltages at 25°C of (i) 2·19 V and (ii) 3·3 V; the cell-voltages appropriate to the cells under typical operating conditions (given in Table 3) are necessarily somewhat different from these values. The reasons for the much higher voltages in the working cells are as follows: (i) there are high overvoltages of chlorine at the anodes of both cells and of hydrogen at the cathode of the diaphragm

TABLE 3. TYPICAL OPERATING DATA FOR MODERN DIAPHRAGM AND MERCURY CELLS[a–c]

	Unit	Diaphragm cell	Mercury cell
Cell capacity	amp	10,000–60,000	25,000–300,000
Cell capacity	kg Cl_2/day	300–2000	1000–10,000
Anode current density	amp cm^{-2}	~0·1	~0·5
Cathode current density	amp cm^{-2}	~0·1	0·35–0·5
Current efficiency	%	96·5	94–97
Cell voltage	volt	3·75–3·85	4·25–4·50
Energy consumption*	kWh/kg Cl_2	~3·0	~3·5
Energy efficiency	%	~58	~47
Cell temperature	°C	95	65–70
Graphite consumption	kg/1000 kg Cl_2	2·7–3·6	2·0–2·8
Average anode life	days	220–310	150–450
Diaphragm[a] (or mercury[b]) consumption	kg/1000 kg Cl_2	0·6[a]	0·2[b]
Salt consumption	kg NaCl/kg Cl_2	1·8	1·7
NaCl concentration of inlet electrolyte	% by weight	27	26
NaCl concentration of exit electrolyte	% by weight	14–15	23–24

* Electrolysis only

[a] Z. G. Deutsch, C. C. Brumbaugh and F. H. Rockwell, *Kirk-Othmer's Encyclopedia of Chemical Technology*, 2nd edn., Vol. 1, p. 668, Interscience (1963).

[b] C. A. Hampel (ed.), *The Encyclopedia of the Chemical Elements*, Reinhold, New York (1968).

[c] J. S. Sconce (ed.), *Chlorine: its Manufacture, Properties and Uses*, American Chemical Society Monograph No. 154, Reinhold, New York (1962).

cell (sodium has but a small overvoltage at the cathode of the mercury cell); (ii) there are potential drops through the electrolyte gap, through the diaphragm of the diaphragm cell, through the anode and cathode leads, and through the various contacts. The high overvoltage of hydrogen, relative to sodium, at a mercury cathode is one of the major principles embodied in the mercury cell; some 99% of the current brings about discharge of sodium ions and only about 1% liberates hydrogen. In practice, 100% current efficiency is not realized in either type of cell for the following reasons:

(i) There may be current leakages through insulators.

(ii) Secondary reactions occur at the anodes:

(a) $4OH^- \rightarrow 2H_2O+O_2+4e$

(b) $4OH^-+C \rightarrow 2H_2O+CO_2+4e$

(c) $6ClO^-+3H_2O \rightarrow 2ClO_3^-+4Cl^-+6H^++\frac{3}{2}O_2+6e$

at the cathode:

(d) $ClO^-+2H^++2e \rightarrow Cl^-+H_2O$

(e) $ClO_3^-+6H^++6e \rightarrow Cl^-+3H_2O$

(f) $H^++e \rightarrow \frac{1}{2}H_2$ (in the mercury cell)

and in the anolyte or catholyte compartments

(g) $Cl_2+2OH^- \rightleftharpoons Cl^-+ClO^-+H_2O$

(h) $3Cl_2+6OH^- \rightleftharpoons ClO_3^-+5Cl^-+3H_2O$

(i) $2Na(Hg)+Cl_2 \rightarrow 2NaCl+Hg$

The presence of certain impurities is likely also to affect the current efficiency.

(iii) There are always leakages and losses of product.

The relative merits of the two types of cells are nicely balanced, as may be judged in part from the data of Table 3. In practice, the diaphragm cell accounts for more than 70% of chlorine production in the United States, whereas in Canada and Western Europe the mercury cell is favoured. The two most important factors governing the choice between the cells are probably the physical form in which the salt is available and the quality of the sodium hydroxide required. Thus, the mercury cell requires dry salt whereas diaphragm cells are normally operated on saturated brine. In general, therefore, diaphragm cells are more economical where the operation is sited near a natural source of deep-well, saturated brine. On the other hand, the mercury cell produces sodium hydroxide of superior quality, though the factor of mercury-pollution *via* the effluent streams, the dangers of which have only recently, and rather sensationally, come to be appreciated[57], may ultimately weigh against this type of cell.

The auxiliary facilities for plants incorporating diaphragm or mercury cells, indicated in the flow diagrams of Fig. 9, include the following stages: brine purification; direct-current electric power supply; cooling, drying and liquefaction of chlorine; salt recovery (diaphragm cell only); cooling, filtering and concentration of the alkaline (caustic) solution; and cooling and compression of hydrogen, when the latter is utilized. By far the most important co-product is sodium hydroxide, whereas the hydrogen is a relatively minor factor, and much of it is simply burnt for fuel. The chlorine gas emerging from the cells is cooled, either by scrubbing with water or by contact with cooled stoneware, titanium or glass pipes, to remove water vapour and brine spray. It is then dried by a counter-current flow of sulphuric acid.

[57] L. J. Goldwater, *Scientific American*, **224** (5) (1971) 15.

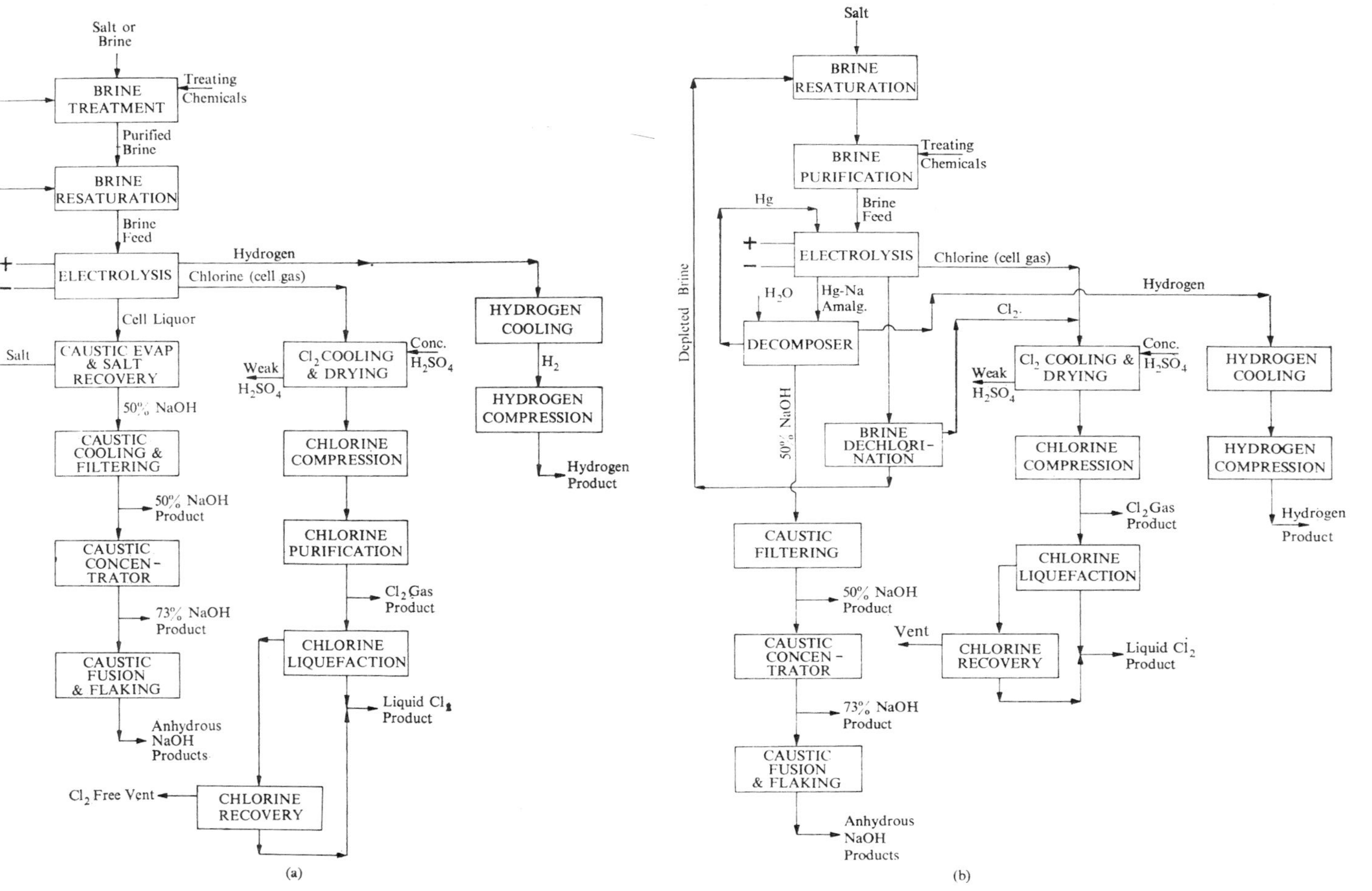

FIG. 9. Flow diagrams (a) of diaphragm cell operation in a chlorine-caustic soda plant and (b) of mercury cathode cell operation in a chlorine-caustic soda plant. [Reproduced with permission from *Kirk-Othmer's Encyclopedia of Chemical Technology*, 2nd edn., Vol. 1, pp. 699, 700, Interscience, John Wiley and Sons, Inc. (1963).]

The concentration of the remaining impurities is further reduced by compression and, ultimately, liquefaction of the chlorine. Very corrosive to all of the common metals, wet chlorine is frequently handled in piping systems constructed of chemical stoneware, glass or some kinds of halogenated plastics. Dry gaseous or liquid chlorine can be handled satisfactorily in steel apparatus at temperatures up to about 120°C.

TABLE 4. MISCELLANEOUS PROCESSES FOR THE MANUFACTURE OF CHLORINE[a–c]

Principle	Details	Comments
Electrolytic oxidation of Cl^-	(i) Aqueous potassium chloride	Operations similar to those involving sodium chloride brines
	(ii) Fused magnesium chloride	Method of producing magnesium; now puts little if any chlorine on the market
	(iii) Fused sodium chloride	Method of producing sodium
	(iv) Hydrochloric acid	Depends economically on the availability of cheap by-product HCl
Chemical methods of oxidizing Cl^-	(i) Reaction of salt and nitric acid	Still practised on a small scale
	(ii) Catalytic air oxidation of HCl	Variations on the Deacon Process; depends on the ready availability of cheap HCl

[a] Supplement to *Mellor's Comprehensive Treatise on Inorganic and Theoretical Chemistry*, Supplement II, Part I, p. 272, Longmans, Green and Co., London (1956).
[b] *Gmelins Handbuch der Anorganischen Chemie*, 8 Auflage, "Chlor", Teil A, p. 2 (1968).
[c] Z. G. Deutsch, C. C. Brumbaugh and F. H. Rockwell, *Kirk-Othmer's Encyclopedia of Chemical Technology*, 2nd edn., Vol. 1, p. 668, Interscience (1963).

Other processes that have been exploited in recent years for the technical production of chlorine[32,33,46] are set out in Table 4. Perhaps the most significant change has been the increasing production of hydrogen chloride from chlorination or dehydrochlorination processes (as in the manufacture of solvents and synthetic resins), which has revived the problem of regenerating chlorine from hydrogen chloride: electrolysis of hydrochloric acid, catalytic oxidation of hydrogen chloride or metallic chlorides using air or oxygen, and the formation and catalytic decomposition of chlorosulphonic acid, $ClSO_3H$, represent, up to the present time, the principal lines of technical development.

Bromine[32,33,45–47]

All methods of bromine production depend on the oxidation of the bromide ion which is found naturally only in relatively low concentrations. In the commercial isolation of bromine from natural brines, bitterns or ocean water, four essential steps may be recognised: (a) oxidation of bromide to bromine; (b) removal of bromine vapour from the solution; (c) condensation of the vapour or fixation in some chemical form; and (d) purification of the product. Chlorine is the only oxidizing agent employed commercially in step (a), though some use has been made of electrolytic methods in the past. Step (b) involves driving out the bromine vapour with a current of air or steam. Steam is suitable when the raw brine is relatively rich in bromine (0·1% or more), but air is more economical when the bromine is extracted from very dilute solutions, such as ocean water[58]. When steam is used, the vapour may be condensed directly; otherwise the bromine must be trapped in an alkaline (sodium carbonate) or reducing (sulphur dioxide) solution. In either case, purification is necessary to remove chlorine.

The so-called "steaming-out" method has been widely employed since the early German development of a continuous process and its further improvement in 1906 by Kubierschky. The Kubierschky process, in one form or another, is operated in Germany, Israel and the United States using brines that are relatively rich in bromide (e.g. mother liquors obtained during the processing of salts from the Stassfurt deposits, or liquors produced by solar evaporation of Dead Sea water). The so-called "blowing-out" process, in which air instead of steam is used to drive out the bromine, was developed in the United States in 1889 by Dow. This represents by far the most important process for obtaining bromine from sea water. It is operated on a large scale by the Dow Chemical Company at Freeport, Texas; other ocean-water plants are found at Hayle (Cornwall), Amlwch (Wales) and Marseilles (France). More detailed accounts of these processes are given elsewhere[45].

In a typical "steaming-out" plant raw brine is preheated to about 90°C and then passes into the chlorinator tower, a tank lined with resistant material and packed with rings. Only a portion of the total chlorine that is to be used is introduced into the bottom of the chlorinator. The brine then passes into the steaming-out tower where it is caused to flow uniformly over a packing made usually of chemical stoneware or porcelain. Steam is injected at the bottom of the tower and the remainder of the chlorine is introduced *via* a separate inlet higher up the tower. The weak acidity of the brine at this stage minimizes the hydrolysis of bromine and increases the efficiency of the steaming-out process; the brine leaving the steaming-out tower is neutralized with lime and finally pumped out through a heat-exchanger. The halogen-laden vapour passes into a condenser and then into a gravity separator. Non-condensable gases bearing some chlorine and bromine are returned to the bottom of the chlorinator tower, while the water layer from the separator, saturated with chlorine and bromine, is returned to the steaming-out tower. Crude bromine collects at the bottom of the separator, whence it passes through a trap to the first of two distillation columns, in which the free halogens are separated from higher-boiling halogenated hydrocarbons. The chlorine from this stage is recycled to the steaming-out tower, while the bromine undergoes a second fractionation for final purification. More than 95% of the bromine is thus recovered from the brine.

By contrast, the following operations are typical of the "blowing-out" process. Sea water is pumped to the top of a brick and concrete blowing-out tower; on the ascent, it is treated with sufficient sulphuric acid to bring the pH down to 3·5 and with a quantity of chlorine 15% in excess of that theoretically required. The pipe through which the water ascends is lined with rubber, as are also the injection pipes for acid and chlorine. At the top of the tower the liquid is distributed so that it descends through parallel chambers filled with wood packing. Here the free bromine is given up to a counter-current stream of air drawn in at the base of the tower. The mixture of air and halogen vapour is then caused to

[58] W. F. McIlhenny, *Mater. Technol.:—Interamer. Approach, Interamer. Conference* (1968) 119; *Chem. Abs.* **70** (1969) 40537g.

react with sulphur dioxide and water vapour:

$$X_2+SO_2+2H_2O \rightarrow 2HX+H_2SO_4 \ (X = Cl \text{ or } Br)$$

The spray of mixed acids is collected and transferred to a steaming-out tower in which the bromine is liberated by the action of chlorine, and collected as in the steaming-out process. The sulphuric and hydrochloric acids formed are used for acidifying the sea water in the first stage of the process. The whole process is automatically controlled, the rate of addition of chlorine being governed by measurement of the oxidation potential of the liquid with a platinum electrode. A high initial temperature of the sea water clearly favours extraction of the bromine by diminishing its solubility in water, and this consideration accounts for the location of some of the plants.

Before the general introduction of sulphur dioxide for the removal of bromine from the air stream, absorption of the halogen was usually carried out using sodium carbonate solution:

$$3CO_3^{2-}+3Br_2 \rightarrow 5Br^-+BrO_3^-+3CO_2$$
$$3CO_3^{2-}+\tfrac{5}{2}Cl_2+\tfrac{1}{2}Br_2 \rightarrow 5Cl^-+BrO_3^-+3CO_2$$

Acidification of the resulting solution of sodium bromide and bromate with sulphuric acid regenerates free bromine which can be steamed out and condensed in the manner already described. This alkaline absorption procedure, in one form or another, is still used in some countries. Some of the other processes which have been used in the extraction of elementary

TABLE 5. MISCELLANEOUS PROCESSES FOR THE EXTRACTION OF BROMINE[a-c]

Principle	Method	Comments
Electrochemical oxidation of aqueous bromide	In general terms as for chloride with either a diaphragm (as in the Wünsche process) or special bipolar carbon electrodes (the Kossuth process)	Now generally uneconomic mainly because of the huge volumes of brine which have to be handled
Chemical oxidation of bromide	Use of manganese dioxide or chlorate under acidic conditions	Early processes for the production of bromine, now obsolete
Alternative methods for the fixation of bromine liberated by blowing-out methods	(i) Reaction with moist scrap iron, ammonia or ferrous bromide solutions	No longer used to produce elementary bromine; may be used to produce bromides
	(ii) Interaction of bromine and phenol to give tribromophenol, followed by fusion of the latter with alkali	A process developed in Russia
	(iii) Interaction of bromine and aniline to give tribromoaniline	Yields relatively poor; tried as a venture by the Ethyl Corporation
	(iv) Solvent extraction; absorption by silica gel or activated charcoal; ion exchange	Methods suggested, e.g. in the patent literature

[a] Z. E. Jolles (ed.), *Bromine and its Compounds*, Benn, London (1966).
[b] V. A. Stenger and G. J. Atchison, *Kirk-Othmer's Encyclopedia of Chemical Technology*, 2nd edn., Vol. 3, p. 750, Interscience (1964).
[c] C. A. Hampel (ed.), *The Encyclopedia of the Chemical Elements*, Reinhold, New York (1968).

bromine are listed in Table 5. Some attention has been given to new ways of isolating bromine, including, for example, the use of ion exchange and solvent extraction[59], but these methods have not come into commercial use. It may be possible to avoid the need for chlorine in the regeneration of bromine from the aqueous solution that results from the interaction of the halogen vapours with sulphur dioxide; a new process entails the functioning of a trace of butyl nitrite or an alkali-metal nitrite in an acidic medium as a catalyst for oxidation by oxygen or air at 0° to 200°C and pressures of 1–4 atm[60].

When the blowing-out process is applied to brine with the object of manufacturing not elementary bromine, but alkali or alkaline earth bromides and bromates, several variations of procedure are advantageous[45–47]. In particular, the air–halogen mixture is usually caused to react with the appropriate alkali carbonate or alkaline–earth hydroxide to give a mixture of bromide and bromate:

$$3Br_2+3H_2O \rightarrow 5Br^-+BrO_3^-+6H^+$$

Subsequent treatment depends upon the compound desired. Thus, bromates of sodium and potassium are less soluble than the bromides and may be crystallized from the mixtures after partial evaporation and cooling. The mother liquors rich in bromide may be heated with scrap iron to reduce the bromate and the bromides crystallized, after removal of the iron oxide precipitate. Other methods of removing bromate include (a) heat treatment, preferably in the presence of charcoal, (b) precipitation of sparingly soluble barium bromate, (c) reduction with hydrogen (or barium) sulphide, and (d) reaction of the bromine with ammonia, whereby bromate formation is suppressed:

$$8NH_3+3Br_2 \rightarrow 6NH_4Br+N_2$$

Equipment coming into contact with wet bromine is generally made of ceramic materials[45–47]. Glass and tile piping are employed most extensively, though some use has been made of tantalum metal, particularly for condensers. The dry halogen may be handled in lead (or lead-lined) or Monel equipment. Brick, granite or concrete construction is favoured for large tanks or other vessels.

Iodine[32,33,46,47]

For the production of iodine from natural brines, one of several possible processes may be used. The first step in any such process is the clarification of the brine to remove oil and other suspended material. In one process, a silver nitrate solution is added in sufficient quantity to precipitate only silver iodide, which is filtered off and treated with clean steel scrap to form metallic silver and a solution of ferrous iodide. The silver is redissolved in nitric acid for use in another cycle and the solution is treated with chlorine or some other oxidizing agent to liberate the iodine.

In another process, the step after clarification is treatment of the brine with chlorine gas to liberate free iodine. The solution is then passed over bales of copper wire with which the iodine reacts to form insoluble cuprous iodide. At intervals the bales are agitated with water to separate the adhering iodide and returned to the precipitators for further use. The cuprous iodide suspended in the water is filtered, dried and stored or transported as such.

[59] R. F. Hein, U.S. Pat. 3,037,845 (1962); 3,174,828 (1965); F. J. Gradishar and R. F. Hein, U.S. Pat. 3,098,716 (1963); L. C. Schoenbeck, U.S. Pat. 3,075,830 (1963); 3,101,250 (1963); E. I. du Pont de Nemours; M. Israel, A. Baniel and O. Schachter, Israel Pat. 15,408 (1962).

[60] W. A. Harding and S. G. Hindin, U.S. Pat. 3,179,498 (1965), Air Products and Chemicals, Inc., and Northern Natural Gas Co.

The largest part of the iodine produced in the United States is derived as a by-product of the Michigan natural brines, from which the iodine is recovered by a process resembling that for the recovery of bromine from sea water. The brine, containing typically 30–40 ppm iodine, is acidified with sulphuric acid and treated with a slight excess of chlorine to liberate the iodine. The liquor is then pumped to a denuding tower in which it is stripped of its iodine by a counter-current of air. The air passes to a second tower where the iodine is absorbed by a solution of hydriodic and sulphuric acids into which sulphur dioxide is passed continuously to reduce the iodine to iodide. This solution, when sufficiently concentrated, is again chlorinated; the iodine precipitated is filtered off, melted under concentrated sulphuric acid and cast into pigs.

Brines may also be treated with a mixture of sulphuric acid and sodium nitrite:

$$2I^- + 2NO_2^- + 4H^+ \rightarrow I_2 + 2NO + 2H_2O$$

The iodine liberated is conveniently extracted, for example, with petroleum or activated charcoal, and converted back to iodide by treatment either with sulphite or with alkali. Oxidation of the iodide concentrate to elementary iodine may be effected with sodium dichromate and sulphuric acid; the use of other oxidizing agents, e.g. bleaching powder, has also been investigated. Electrolytic methods for the extraction of iodine have generally proved to be commercially impractical because of the very low concentrations of iodide present in natural brines. Recently developed ion-exchange techniques have been reported to give a new, more economical extraction process for producing iodine of exceptionally high purity[47].

The process used by the Chilean nitrate industry differs from all the others since the natural source of iodine is here iodate. Sodium iodate is extracted from the caliche, being allowed to accumulate in the mother liquors from crystallization of sodium nitrate until a suitable concentration (about 6 g l^{-1}) has been attained. Part is then drawn off and treated with the exact quantity of sodium bisulphite required to reduce all of the iodate to iodide:

$$2IO_3^- + 6HSO_3^- \rightarrow 2I^- + 6SO_4^{2-} + 6H^+$$

The resulting acidic mixture is then caused to react with just sufficient fresh mother liquor to liberate all the iodine in accordance with the reaction,

$$5I^- + IO_3^- + 6H^+ \rightarrow 3I_2 + 3H_2O$$

The precipitated iodine is filtered off in bag filters and the iodine-free solution returned to the nitrate-leaching cycle after neutralization of any excess acid with sodium carbonate. The iodine cake is washed with water, pressed to reduce the moisture content, broken up and sublimed in concrete-lined iron retorts connected to condensers made of glazed tile. The product is crushed and packed in polythene-lined fibre drums.

In France, Russia and Japan where alginic products are prepared from seaweed, small quantities of iodine can be obtained at low cost as a by-product. The stages in the preparation are: drying and burning the seaweed, leaching of the ash, release of the iodine by chemical reaction (e.g. with MnO_2), and purification.

2.4. USES OF CHLORINE, BROMINE AND IODINE[32,45–48]

Approximate data on the annual production of chlorine, bromine and iodine, together with the principal commercial uses of the elements, are presented in Table 6. Originally the primary commercial outlet for chlorine, the bleaching of textiles now accounts for less than 1% of the total output of chlorine. In fact, chlorine is too destructive for bleaching wool, silk and other products of animal origin, and in recent years hydrogen peroxide has become economically available for these and other purposes. The rapid growth in chlorine usage in recent years (implied by the thirtyfold increase in production in the period 1930–50 and the further tenfold increase in the period 1950–70[32,46,47]) is to be correlated with the growth of the organic chemical industry, and especially with the production of compounds derived from petroleum. Thus, about 65% of the chlorine produced in the United States in 1963 was used for the manufacture of organic chemicals important, for example, as intermediates, solvents, refrigerants, plastics and plasticizers, insecticides and dyes[46]. To complete the pattern of chlorine usage for the United States in 1963, 9% went into the production of inorganic chemicals, e.g. hydrochloric acid, bromine, sodium hypochlorite and various metal chlorides, 17% into the treatment of wood pulp in the production of paper and rayon, 4% into the disinfection of water and the treatment of sewage, and 5% into a variety of minor applications.

Only about 3% of commercial bromine now finds its way into inorganic bromides and bromates, in complete contrast to the early economic history of bromine, which was closely linked to these derivatives[45,46]. By far the most important use of bromine (on a tonnage basis) is in the manufacture of ethylene dibromide for motor fuels; the dibromide serves to convert the lead from the tetraalkyllead antiknock agent to lead(II) bromide, which is volatile at cylinder combustion temperatures and is swept out with the exhaust. The great bulk of commercial bromine is otherwise converted into bromine compounds, as there are relatively few direct applications of liquid bromine outside the laboratory. Organic bromine derivatives find uses, *inter alia*, as industrial organic intermediates, dyestuffs, pharmaceutical chemicals, for fumigation and pest control, as high-density liquids in processes deploying gravity separation, and as fire-extinguishing and fire-proofing agents. Of inorganic bromine compounds, alkali-metal bromides have applications in medicine and (as sources of hypobromite) in bleaching and the desizing of cotton, while silver bromide is a primary agent in photography, and bromates are employed as oxidants for specific purposes.

Unlike chlorine and bromine, iodine has no major commercial outlet, though over half of the iodine produced is consumed industrially, and important quantities also find uses in nutrition, sanitation and medicine. Some of the more significant uses of iodine and its compounds depend upon (a) their function as catalysts, (b) the disinfectant properties of the element and the nutritional properties of iodides, (c) the photosensitive behaviour of silver iodide, a constituent of rapid negative emulsions in photography, and (d) the colour and other properties of iodine-containing dyestuffs.

2.5. ATOMIC CHLORINE, BROMINE AND IODINE[32,44,61]

Formation

In conventional circumstances, halogen atoms are thermodynamically unstable and short-lived with respect to the corresponding diatomic molecules. Nevertheless, there has been a long and sometimes chequered history concerning the formation and characterization of chlorine, bromine and iodine atoms. By making vapour-density measurements at high

61 W. A. Waters, *The Chemistry of Free Radicals*, 2nd edn., Clarendon Press, Oxford (1948).

Table 6. Production and Uses of Chlorine, Bromine and Iodine[a–e]

	Chlorine	Bromine	Iodine
Estimated world production	$1 \cdot 5–2 \cdot 0 \times 10^7$ metric tons per annum	$1 \cdot 5–2 \cdot 0 \times 10^5$ metric tons per annum	$5–9 \times 10^3$ metric tons per annum
Uses	(i) *The production of organic chemicals*, the main uses of which are as (a) solvents, (b) refrigerants and dispersion propellants, (c) plastics, resins, elastomers and plasticizers, (d) intermediates in the production of detergents and other materials, (e) insecticides, (f) dyes, (g) automobile antifreeze and antiknock compounds. Some important organic compounds so produced are: Aliphatic. C_1: CH_3Cl, CH_2Cl_2, $CHCl_3$, CCl_4; C_2: ethylene oxide, ethylene dichloride; di-, tri- and tetra-chloroethylene, vinyl chloride; ethyl chloride, tetraethyllead; monochloroacetic acid, chloral; di(chloroethyl) ether; C_3: glycerol, acrylonitrile; propylene glycol: *C_4 and larger fragments*: butadiene, chloroprene; chloro-derivatives of C_5 and larger fragments. Aromatic: Chlorobenzene, phenol, aniline, dodecylbenzenes, diphenyl, diphenyl ether; chlorotoluenes, *o*- and *p*-dichlorobenzene, DDT; 2,4-dichlorophenoxyacetic acid, hexachlorobenzene, pentachlorophenol. (ii) *The production of inorganic materials*: HCl, metal chlorides, e.g. aluminium trichloride and titanium tetrachloride; sodium hypochlorite and bleaching powder; Br_2 and I_2. (iii) *Treatment of wood pulp* (to remove lignin and other undesirable constituents); *bleaching of pulp and paper*; textile bleaching.	(i) *The manufacture of ethylene dibromide* for antiknock motor fuels. (ii) *The production of organic derivatives* such as methyl bromide, ethylene dibromide and bromochloropropanes for insect control, as space and soil fumigants and fungicides. (iii) *Formation of industrial organic intermediates, dyestuffs, medicinal chemicals and solvents*, e.g. bromoanthraquinones, bromophthaleins, $CF_3CHBrCl$ ("halothane") and mercurochrome. (iv) *As a disinfectant and bleaching agent.* (v) *Production of organic liquids of high density* for gauge fluids and gravity separation processes, e.g. CBr_4, $CHBr_3$, CH_2Br_2, $Br_2CH \cdot CHBr_2$, $BrCH_2 \cdot CH_2Br$ and C_6H_5Br. (vi) *Manufacture of fire-extinguishing agents*, e.g. CH_2ClBr and CF_2Br_2, *and of flame retardants* for incorporation in polymers such as polystyrene, e.g. $CH_2Br[CHBr]_2CH_2Br$, $BrCl_2C \cdot CCl_2Br$, pentabromochlorocyclohexane and tetrabromoisopropylidenebisphenol. (vii) *Production of inorganic bromides and bromates*: AgBr, central to photography; alkali bromides used as mild sedatives; with aqueous hypochlorite bromide gives a solution containing hypobromite useful for bleaching and for desizing cotton; hydrogen bromide, used as a catalyst, e.g. in alkylation reactions; bromates, used as oxidants for	(i) *The element*. Used (a) as a catalyst, e.g. in the conversion of resins, tall oil and other wood products to more stable forms, and of amorphous to black selenium; (b) as a selective absorption agent, e.g. of olefin–paraffin hydrocarbon mixtures; (c) to give charge-transfer complexes with aromatic hydrocarbons which are effective as lubricants; (d) for the control of bacterial growth in cutting-oil emulsions and for disinfection of water and food; (e) "tincture of iodine" and "iodophors"; (f) in quartz-iodine lamps. (ii) *Inorganic iodides*: (a) AgI, used in photography and for the seeding of clouds to induce rainfall; (b) certain iodides used as catalysts, e.g. NiI_2 in the addition of CO to organic compounds and TiI_4 in the production of stereospecific polymers; (c) in the form of iodized salt and protein, which are essential to the health and well-being of man and the higher animals; organic and inorganic derivatives of radioactive ^{131}I have been used in the treatment of certain thyroid and heart conditions; (d) addition of CdI_2 or PbI_2 to electric generator and motor brushes to improve their efficiency and life; (e) the production of high-purity metals or metalloids, e.g. Ti, Zr, Hf and Si, by the van Arkel method. (iii) *Inorganic iodates*: $NaIO_3$, improves the bread-making qualities of certain flours.

(iv) *Sanitation and disinfection*: Treatment of water supplies and sewage; sterilization of process liquors in sugar manufacture and of sea-water cooling circuits. (v) *Extraction and refining of metals*: Ore-beneficiation, fluxing; involved in methods for the extraction of such metals as Ti, Zr, Cu, Pb, Zn, Ni, Au, W and V, and for the refining of metals, e.g. Al, Mg and Pb.	improving the baking quality of wheat flour, in certain hair-wave preparations, and during the malting process in the brewing industry. (All of these probably relate to the oxidation of -SH to -S-S- groups.)	(iv) *Organic iodides*: (a) Dyes, e.g. diiodo-fluorescein, rose bengal and erythrosin, used in photographic processes and in food colouring agents; (b) high-density liquids for gravity separation (mostly in the laboratory), e.g. CH_2I_2; (c) X-ray contrast media, e.g. sodium 3-acetylamine-2,4,6-triiodo-benzoate; (d) germicidal agents, e.g. CHI_3; (e) intermediates, e.g. CH_3I and C_2H_5I.

[a] Supplement to *Mellor's Comprehensive Treatise on Inorganic and Theoretical Chemistry*, Supplement II, Part I, Longmans, Green and Co., London (1956).

[b] Z. E. Jolles (ed.), *Bromine and its Compounds*, Benn, London (1966).

[c] Z. G. Deutsch, C. C. Brumbaugh and F. H. Rockwell, *Kirk-Othmer's Encyclopedia of Chemical Technology*, 2nd edn., Vol. 1, p. 668. Interscience (1963); V. A. Stenger and G. J. Atchison, *ibid.*, Vol. 3, p. 750 (1964); A. W. Hart, M. G. Gergel and J. Clarke, *ibid.*, Vol. 11, p. 847 (1966).

[d] C. A. Hampel (ed.), *The Encyclopedia of the Chemical Elements*, Reinhold, New York (1968).

[e] E. L. Gramse and L. H. Diamond, *Literature of Chemical Technology*, p. 1 (Advances in Chemistry Series Vol. 78), American Chemical Society, Washing. ton D.C. (1968).

temperatures, Victor Meyer[62] showed that the I_2 molecule dissociates reversibly thus: $I_2 \rightleftharpoons 2I$. In 1887 J. J. Thomson reported[63] that bromine and iodine vapours could be dissociated by electrical sparks, though subsequent experiments[64] failed to reproduce the changes of vapour-density which he described. Since 1916 chemists have correctly inferred that some atomic chlorine is produced when chlorine gas is exposed to ultraviolet light, a hypothesis which has subsequently proved a consistent prerequisite to the mechanistic interpretation of chlorination reactions in the gas phase. It was shown in 1928 that traces of atomic chlorine are produced by passing sodium vapour into chlorine gas[61] and that the combination of hydrogen and chlorine gases can thereby be initiated in the absence of light. Appreciable quantities of atomic chlorine and bromine were first obtained in 1933 by the action of an electrical discharge on the appropriate molecular gas[65,66].

The following methods are now generally acknowledged to be effective in the production of halogen atoms.

Thermal Dissociation

The latest thermodynamic data[67] indicate (see, for example, Table 7) that at temperatures

TABLE 7. THERMAL DISSOCIATION OF HALOGEN MOLECULES[a]

Temperature (°K)	Chlorine		Bromine		Iodine	
	log K_p	Degree of dissoc. (%)*	log K_p	Degree of dissoc. (%)*	log K_p	Degree of dissoc. (%)*
298	−36·800	$4·0 \times 10^{-17}$	−28·880	$3·6 \times 10^{-13}$	−24·628	$4·9 \times 10^{-11}$
500	−19·626	$1·5 \times 10^{-8}$	−14·660	$4·7 \times 10^{-6}$	−10·510	$5·6 \times 10^{-4}$
1000	−6·796	$4·0 \times 10^{-2}$	−4·490	0·57	−2·524	4·0
1500	−2·452	5·8	−1·054	25·6	0·172	68·5
2000	−0·256	51·8	0·692	85	1·536	98
3000	1·964	99	2·480	100	2·928	100

* At a pressure of 1 atm.

[a] *JANAF Thermochemical Tables*, The Dow Chemical Company, Midland, Michigan (1960–8).

in the range 600–900°K and pressures near 1 atm the gas-phase dissociation

$$X_2 \rightleftharpoons 2X$$

begins to become appreciable (degree of dissociation > 0·01%): at 1 atm 1% dissociation is achieved at about 1250°, 1050° and 850°K for Cl_2, Br_2 and I_2, respectively.

Rapid, homogeneous heating can be achieved by the use of shock waves[68], following which the rates of dissociation and recombination of the halogens have been investigated by

[62] V. Meyer, *Ber.* **13** (1880) 394.

[63] J. J. Thomson, *Proc. Roy. Soc.* **42** (1887) 343.

[64] E. P. Perman, *Proc. Roy. Soc.* **48** (1890) 45; W. Kropp, *Z. Elektrochem.* **21** (1915) 356.

[65] G.-M. Schwab and H. Friess, *Naturwiss.* **21** (1933) 222; *Z. Elektrochem* **39** (1933) 586.

[66] W. H. Rodebush and W. C. Klingelhoefer, jun., *J. Amer. Chem. Soc.* **55** (1933) 130; G.-M. Schwab *Z. physik. Chem.* **B27** (1934) 452.

[67] *JANAF Thermochemical Tables*, The Dow Chemical Company, Midland, Michigan (1960–1968).

[68] E. F. Greene, *J. Amer. Chem. Soc.* **76** (1954) 2127; E. F. Greene and J. P. Toennies, *Chemical Reactions in Shock Waves*, Edward Arnold, London (1964).

measuring the optical absorption as a function of time. Thus, for mixtures of 5% Cl_2 and 95% Ar at temperatures between 1600° and 2600°K, the rate of dissociation is given[69] by $\log k_D$ ($mol^{-1}\ l\ sec^{-1}$) $= 10{\cdot}66 - 9930/T$, corresponding to an apparent activation energy, E_A, of 45 kcal mol^{-1}; for comparable Br_2/Ar mixtures[70] at temperatures between 1500° and 1900°K, $\log k_D = 10{\cdot}709 - 8240/T$, whence $E_A = 38$ kcal mol^{-1}.

Dissociation by Discharge

The dissociation of halogen molecules into atoms can also be effected in the gas phase by the action of a discharge. Arrangements have been described for the production of atomic chlorine or bromine in which an electrical discharge is maintained either externally[65] or between water-cooled iron electrodes situated inside the reaction vessel[66]. The conditions normally favoured include ambient temperatures and gas pressures < 1 mm Hg, leading to reported degrees of atomization of 10–40%. Relatively high concentrations of chlorine or bromine atoms have also been generated in a flow system by means of microwave[71] or radiofrequency[72] discharges. Curiously, attempts to obtain iodine atoms in this way have been unsuccessful[71]; this difficulty is presumably a function of the hot region which develops immediately after the discharge, because it has been found that iodine atoms can be maintained if they are created outside the discharge.

At $\sim 0{\cdot}1$ mm pressure, the mean lives of chlorine and bromine atoms in a glass tube have been estimated[65,66] to be about 3×10^{-3} and $1{\cdot}8 \times 10^{-3}$ sec, respectively. Reversion of the atoms to the molecular state is a relatively slow and inefficient process because a termolecular collision is necessary to dissipate the energy of combination:

$$X\cdot + X\cdot + M \rightarrow X_2 + M^*$$

and the rate of recombination is largely determined by the effectiveness of the "third body" M as an energy sink, a role which has been investigated for many materials. In effect, recombination normally takes place on the surfaces of the reaction vessel. On a dry quartz surface it has been estimated[73] that the activation energy of recombination of chlorine atoms is ~ 1 kcal mol^{-1}, and that, on average, about 1 in 12 collisions with the surface leads to union of the atoms (cf. the report that only 1 in 10^5–10^9 collisions between pairs of atoms in the gas phase leads to combination). Most metal surfaces are efficient "third bodies" for the recombination of halogen atoms, and a thin silver mirror has been used as a sensitive test for halogen atoms, which, in contrast with the corresponding molecules, cause immediate discoloration[66,71]. However, glass surfaces can be poisoned by certain materials, e.g. H_2SO_4, H_3PO_4, H_3AsO_4, H_3BO_3 and $HClO_4$, which inhibit the rapid recombination of halogen atoms and so permit the achievement of relatively high concentrations of these atoms in a suitable flow system[71].

Optical Dissociation

The absorption spectra of gaseous chlorine, bromine and iodine exhibit in the visible region a series of bands sharply degraded to longer wavelengths with convergence limits at 4795, 5108 and 4991 Å, respectively[74], beyond which the absorption becomes continuous.

69 M. van Thiel, D. J. Seery and D. Britton, *J. Phys. Chem.* **69** (1965) 834; see also R. A. Carabetta and H. B. Palmer, *J. Chem. Phys.* **46** (1967) 1333.
70 C. D. Johnson and D. Britton, *J. Chem. Phys.* **38** (1963) 1455.
71 E. A. Ogryzlo, *Canad. J. Chem.* **39** (1961) 2556.
72 V. Beltran-Lopez and H. G. Robinson, *Phys. Rev.* **123** (1961) 161.
73 G.-M. Schwab, *Z. physik. Chem.* **A178** (1936) 123.
74 A. G. Gaydon, *Dissociation Energies*, 3rd edn., Chapman and Hall, London (1968).

The convergence limit corresponds in each case to the threshold energy of the dissociation

$$X_2(^1\Sigma_g^+) \rightarrow X(^2P_{3/2}) + X(^2P_{1/2})$$

With both bromine and iodine it has been possible to calculate the chemical composition of the photo-stationary state of equilibrium which is reached, under constant illumination, between the dissociation and recombination processes[75]:

$$\text{(a) } X_2 + h\nu \rightarrow X\cdot + X\cdot$$
$$\text{(b) } X\cdot + X\cdot + M \rightarrow X_2 + M^*$$

The energy transmitted to the "third body" M in the recombination reaction is in part converted into heat, with the result that when a photochemically dissociable gas, such as chlorine, bromine or iodine, is first illuminated, there is an increase of pressure, a phenomenon known as the *Budde effect*, arising from the gain in the total kinetic energy of the gas particles[32].

Decisive chemical evidence of the formation of free atoms upon irradiation of the halogens is afforded by a wealth of facts. Thus, the photochemical reactions of chlorine or bromine with hydrogen and of iodine with olefins[76] are satisfactorily explicable only in terms of photodissociation of the halogen molecules. By the methods of flash photolysis[77], halogen atoms at pressures up to several cm Hg have been produced and detected by their absorption spectra; in these circumstances the half-life of chlorine atoms has been estimated to be $\sim 3 \times 10^{-2}$ sec. Flash photolysis has also provided a convenient and elegant method of studying the recombination reactions, notably of iodine atoms[77].

As a rule, the weaker heteronuclear bonds in which a halogen is engaged are susceptible to photolytic dissociation, leading commonly to the formation of free halogen atoms, e.g.

$$R\cdot I \xrightarrow{h\nu} R\cdot + I\cdot \quad ^{61}$$
$$R\cdot O\cdot Cl \xrightarrow{h\nu} RO\cdot + Cl\cdot \qquad (R = \text{organic group})^{78}$$

Neutron Irradiation

Neutron irradiation of halogen-containing species has been shown commonly to proceed with homolytic cleavage of chemical bonds to produce radioactive halogen atoms, which normally react with their chemical environment (see pp. 1169–72)[32,45,79–81].

Formation as Intermediates in Chemical Reactions

The formation of free halogen atoms is believed to be an essential stage in numerous halogenation, dissociation and other reactions, on the evidence usually of detailed studies of the kinetics and conditions of the reactions. Examples are:

$$H\cdot + X_2 \rightarrow HX + X\cdot$$
$$R\cdot + Cl_2 \rightarrow R\text{–}Cl + Cl\cdot \qquad (R = \text{organic group})$$
$$Cl\cdot + XY \rightarrow XCl + Y\cdot \qquad (XY = Br_2, \text{ClBr or ClI})^{82}$$

---

75 E. Rabinowitch and H. L. Lehmann, *Trans. Faraday Soc.* **31** (1935) 689; E. Rabinowitch and W. C. Wood, *ibid.* **32** (1936) 547.

76 G. S. Forbes and A. F. Nelson, *J. Amer. Chem. Soc.* **59** (1937) 693.

77 R. G. W. Norrish, *Angew. Chem.* **64** (1952) 421; G. Porter, *Proc. Roy. Soc.* **200A** (1950) 284; R. G. W. Norrish and B. A. Thrush, *Quart. Rev. Chem. Soc.* **10** (1956) 149.

78 W. A. Pryor, *Free Radicals*, McGraw-Hill, New York (1966).

79 G. Harbottle and N. Sutin, *Adv. Inorg. Chem. Radiochem.* **1** (1959) 267.

80 R. Wolfgang, *Ann. Rev. Phys. Chem.* **16** (1965) 15.

81 R. Wolfgang, *Progress in Reaction Kinetics*, Vol. 3 (ed. G. Porter), p. 97, Pergamon (1965).

82 M. I. Christie, R. S. Roy and B. A. Thrush, *Trans. Faraday Soc.* **55** (1959) 1139, 1149.

$$Br\cdot + XY \rightarrow XBr + Y\cdot \qquad (XY = Cl_2 \text{ or } BrCl)^{82}$$

$$ArN_2Cl \rightarrow Ar\cdot + N_2 + Cl\cdot \qquad (Ar = \text{aryl group})$$

$$CO + Cl\cdot \xleftarrow{\Delta} \cdot COCl \xrightarrow{Cl_2} COCl_2 + Cl\cdot$$

$$NO + Cl_2 \rightarrow NOCl + Cl\cdot \qquad (\text{ref. } 83)$$

$$NOCl \xrightarrow{\Delta} NO + Cl\cdot$$

$$Cl\cdot + ClO + O_2 \xleftarrow{O_3} Cl_2 \xrightarrow{O} ClO + Cl\cdot \qquad (\text{ref. } 84)$$

$$ClO \xrightarrow{O_3} Cl\cdot + 2O_2; \qquad ClO \xrightarrow{O} Cl\cdot + O_2$$

$$H\cdot + HI \rightarrow H_2 + I^*(^2P_{1/2}) \qquad (\text{ref. } 85)$$

$$CuCl_n^{-(n-2)} \rightleftharpoons CuCl_{n-1}^{-(n-2)} + Cl\cdot \qquad (\text{ref. } 86)$$

$$HO\cdot + H_3O^+ + Cl^- \xrightarrow{\text{aq soln}} 2H_2O + Cl\cdot \qquad (\text{ref. } 87)$$

Solvated halogen atoms are also generated, it is believed, by the action of ionizing radiation (neutrons, α-particles, electrons, X- or γ-rays)[88] or of ultraviolet light[89] on aqueous solutions of halide ions, a scheme such as

$$X^-_{aq} \overset{h\nu}{\rightleftharpoons} X^-_{aq}{}^* \rightarrow [X\cdot_{aq} + e^-_{aq}]$$

being proposed to account for the photochemistry of chloride, bromide and iodide ions in aqueous solution[90].

Detection and Characterization

Apart from the qualitative detection of halogen atoms based for example on the discoloration of a metal mirror[71] or on the observed characteristics of a particular reaction, the concentration of the atomic species, and hence the rate of their recombination and other reactions, may be measured by various methods. Notable among these are spectrophotometry, most commonly in the visible region, mass spectroscopy, thermal conductivity measurements, the use of a thermocouple detector (which depends on the heating effect of the recombination of the atoms on a metal wire)[91] and chemical titration procedures involving some rapid reaction such as

$$Cl\cdot + NOCl \rightarrow NO + Cl_2{}^{71}$$

The characterization of atoms of the individual halogens turns for the most part on their spectroscopic properties. Thus, the esr spectra of atoms produced in the gas phase have been observed for chlorine, bromine and iodine in their ground ($^2P_{3/2}$) states[92], as well as for

[83] P. G. Ashmore, *Trans. Faraday Soc.* **49** (1953) 251; P. G. Ashmore and J. Chanmugam, *ibid.* 254.

[84] H. Niki and B. Weinstock, *J. Chem. Phys.* **47** (1967) 3249; P. Huhn, F. Tudos and Z. G. Szabó, *M.T.A. Kém. Oszt. Közl.* **5** (1954) 409.

[85] P. Cadman and J. C. Polanyi, *J. Phys. Chem.* **72** (1968) 3715.

[86] J. K. Kochi, *J. Amer. Chem. Soc.* **84** (1962) 2121.

[87] H. Taube and W. C. Bray, *J. Amer. Chem. Soc.* **62** (1940) 3357.

[88] A. O. Allen, C. J. Hochanadel, J. A. Ghormley and T. W. Davis, *J. Phys. Chem.* **56** (1952) 575; M. Cottin, *J. chim. Phys.* **53** (1956) 903.

[89] J. Jortner, M. Ottolenghi and G. Stein, *J. Phys. Chem.* **68** (1964) 247.

[90] M. F. Fox, *Quart. Rev. Chem. Soc.* **24** (1970) 565.

[91] J. W. Linnett and M. H. Booth, *Nature*, **199** (1963) 1181.

[92] N. Vanderkooi, jun., and J. S. MacKenzie, *Advances in Chemistry Series*, **36** (1962) 98; S. Aditya and J. E. Willard, *J. Chem. Phys.* **44** (1966) 833; E. Wassermann, W. E. Falconer and W. A. Yager, *Ber. Bunsenges. Phys. Chem.* **72** (1968) 248.

chlorine in its excited ($^2P_{1/2}$) state[93], though attempts to detect the atoms trapped in solid, non-polar matrices have failed[94]. Irradiation of crystals of ionic halides does lead formally to the production of trapped halogen atoms, but the properties of the active sites, the so-called "V-centres", leave little doubt that the trapped atoms interact strongly with a neighbouring halide ion to give a unit more aptly described as a $[\text{Hal–Hal}']^-$ molecular ion[94].

Physical Properties of Halogen Atoms

The physical properties of chlorine, bromine and iodine atoms are of two kinds: (i) those which can be precisely identified with the isolated atom, and (ii) those which can be defined or determined for the atom only when it is in the combined state. In practice, category (i) includes the nuclear properties of the different isotopes (Table 8), which are almost immune to variations of chemical environment, and those properties, determined mostly by spectroscopic methods, associated with the different energy states of the atom and derived ions (Table 10(a)); category (ii) includes properties such as atomic radius and electronegativity, whose significance demands that the atom be involved in some form of chemical bonding (Table 10(b)).

Nuclear Properties: Isotopes[32,45,95–97]

Nine well, or reasonably well, authenticated isotopes of chlorine are known, ranging in mass number from 32 to 40. Of these only ^{35}Cl and ^{37}Cl are stable, with natural abundances of 75·77 and 24·23%, respectively; the remainder are radioactive, decaying (i) by positron emission, accompanied by electron capture in one case (^{32}Cl, ^{33}Cl, ^{34}Cl, ^{36}Cl), or (ii) by β-emission (^{36}Cl, ^{38}Cl, ^{39}Cl, ^{40}Cl). Isomers of ^{34}Cl and ^{38}Cl have been described. Isotopes of bromine more-or-less well-defined range in mass number from 74 to 90, but, apart from the very small contribution from unstable isotopes produced in nature by spontaneous fission and nuclear reactions induced by cosmic radiation, naturally occurring bromine consists of a mixture of the two stable isotopes ^{79}Br (50·54%) and ^{81}Br (49·46%). On current evidence, no less than four isobaric pairs of nuclei (mass numbers 77, 79, 80 and 82) have been characterized. Isotopes of bromine with mass numbers 74–78 and 80 decay by simultaneous positron-emission and electron capture, those with mass numbers 80 and 82–90 by β-emission. Naturally occurring iodine is effectively mononuclidic containing the sole, stable isotope ^{127}I. In all, 23 isotopes have been recorded with mass numbers between 117 and 139. In accordance with the usual pattern of behaviour, the lighter (neutron-deficient) radioactive isotopes with mass numbers 117–126 and 128, having a nuclear charge:mass ratio larger than is compatible with stability, undergo positron-emission or electron capture, both processes commonly occurring competitively and each resulting in the reduction of the nuclear charge by one unit. By contrast, the heavier (neutron-rich) isotopes with mass numbers 128 and 130–139, characterized by a nuclear charge:mass ratio which is too small for stability, attempt to redress the balance by β-emission; the three heaviest, and

[93] A. Carrington and D. H. Levy, *J. Phys. Chem.* **71** (1967) 2; A. Carrington, D. H. Levy and T. A. Miller, *J. Chem. Phys.* **47** (1967) 3801.

[94] P. W. Atkins and M. C. R. Symons, *The Structure of Inorganic Radicals*, Elsevier (1967); K. D. J. Root and M. T. Rogers, *Spectroscopy in Inorganic Chemistry* (ed. C. N. R. Rao and J. R. Ferraro), Vol. II, p. 115. Academic Press, New York and London (1971).

[95] C. M. Lederer, J. M. Hollander and I. Perlman, *Table of Isotopes*, 6th edn., Wiley, New York (1968).

[96] G. H. Fuller and V. W. Cohen, *Nuclear Data Tables*, **5A** (1969) 433.

[97] *Handbook of Chemistry and Physics*, The Chemical Rubber Co. (1971–2).

shortest-lived, nuclei, ^{137}I, ^{138}I and ^{139}I, also suffer neutron-emission. The atomic weights recommended by the IUPAC Atomic Weights Commission (1969)[98] are set out in Table 10(a).

Radioactive isotopes of the halogens are produced[32,45,95] by the irradiation of stable isotopes of various elements with neutrons, with charged particles (protons, deuterons or α-particles) or with high-energy photons; bombardment by heavier projectiles, like ^{12}C or ^{14}N, has yielded certain isotopes, access to which is otherwise difficult. The heavier isotopes of bromine and iodine are formed in the fission and spallation–fission of heavier elements. Indeed, isotopes of iodine and, to a lesser extent, of bromine played a prominent part in the early investigations of the fission of uranium and thorium, largely because these two elements were obtained in comparatively high yields and were relatively easy to isolate chemically and to purify; a number of radioactive isotopes of bromine and iodine were thus soon recognized as members of the decay chains which result from the fission process[32].

Neutron-irradiation now affords the commonest route to those isotopes of chlorine and bromine most widely used in radiochemical studies, viz. ^{36}Cl, ^{38}Cl, ^{80}Br, ^{80m}Br and ^{82}Br. In each case a stable nucleus absorbs a neutron and de-excitation of the highly excited composite nucleus thereby formed occurs by γ-ray emission. Of the iodine isotopes of radiochemical significance, ^{128}I is again the outcome of an n,γ reaction (involving naturally occurring ^{127}I), while ^{131}I is produced by the process

$$^{130}Te \xrightarrow{n,\gamma} {}^{131m}Te \xrightarrow[t_{\frac{1}{2}}=30\ hr]{} {}^{131}Te \xrightarrow[t_{\frac{1}{2}}=24\cdot 8\ min]{\beta-} {}^{131}I$$

or by fission of uranium; more recently favoured in radiochemical studies, ^{125}I and ^{132}I are produced thus:

$$^{124}Xe \xrightarrow{n,\gamma} {}^{125}Xe \xrightarrow[t_{\frac{1}{2}}=17\ hr]{\text{Electron-capture}} {}^{125}I \qquad \text{(ref. 99)}$$

$$^{132}Te\ \text{(uranium fission product)} \xrightarrow[t_{\frac{1}{2}}=78\ hr]{\beta-} {}^{132}I \qquad \text{(ref. 100)}$$

It is a characteristic of nuclear reactions of the n,γ-type that the target and product nuclei are chemically identical, a consideration which must limit the specific activity of the chemical product[32]. In other types of nuclear reaction, e.g. n,p, n,α and fission, the nuclear reactant and product are chemically different, making possible, in principle, chemical separation to give so-called "carrier-free" active material. If the product of an n,γ-reaction is itself short-lived with respect to a decay process in which there is a change of atomic number (e.g. β^-, β^+ or electron capture) giving a radioactive halogen nuclide, chemical separation of the product nuclide from the target is again feasible: such a process is exemplified by the production of ^{125}I from ^{124}Xe and of ^{131}I from ^{130}Te. Although some nuclear reactions are thus better suited than others to the production of material of high specific activity, for a variety of reasons the material is seldom truly carrier-free.

If a higher specific activity is required in the product of a simple n,γ-reaction than that permitted by the available neutron flux, recourse may be made to the Szilard–Chalmers process. This depends on the fact that the recoil energy of the product nucleus of an n,γ-reaction (typically $>$ 100 eV) is usually far in excess of ordinary chemical bond energies (1–5 eV), with the result that the nucleus recoils from its immediate chemical environment and may appear in quite a different chemical form. The phenomenon was first observed in 1934 by Szilard and Chalmers, who noted that, when ethyl iodide is irradiated with neutrons,

[98] *Table of Atomic Weights*, 1969. IUPAC Commission on Atomic Weights: *Pure Appl. Chem.* **21** (1970) 95.

[99] P. V. Harper, W. D. Siemens, K. A. Lathrop and H. Endlich, *J. Nucl. Med.* **4** (1963) 277.

[100] Bro. C. Cummiskey, S.M., W. H. Hamill and R. R. Williams, jun., *J. Inorg. Nuclear Chem.* **21** (1961) 205.

most of the resulting radioactive iodine (^{128}I) could be extracted from the ethyl iodide by water[101]. Three conditions must be fulfilled if a useful degree of enrichment of radioactive material is to be realized: formation of the radioactive atom must be attended by its disengagement from the chemical environment (usually a molecule or molecular ion) characteristic of the target; the atom must neither recombine with the fragment from which it has separated nor rapidly interchange with inactive atoms in the target material; and a chemical method must be available for the separation of the target compound from the radioactive material in its new chemical form. Enrichment of radioactive halogen isotopes produced by neutron-capture reactions (e.g. ^{38}Cl, ^{80}Br, ^{82}Br and ^{128}I) has been successfully achieved by the neutron-irradiation of an organic halogen compound such as CCl_4, CH_2X_2, CH_3X, C_2H_5X, C_4H_9X, $C_2H_4Cl_2$ or C_6H_5X (X = Cl, Br or I) and aqueous extraction of the corresponding halide ion. The reactions following rupture of the halide molecule, which are essentially those of the organic radicals and the highly energetic halogen atom formed by recoil, determine to a large extent the efficiency of separation; by contrast, the eventual fate of the active atom is apparently little influenced by the initial recoil energy. Szilard–Chalmers' separations of halogens have also been carried out by neutron-irradiation of solid or dissolved chlorates, perchlorates, bromates, iodates and periodates, from which the active halogen can be removed as silver halide after addition of halide ion carrier.

Quite apart from its importance as a method of isotopic enrichment, the Szilard–Chalmers' process provides a much exploited opportunity to study the chemical behaviour of the highly energetic atoms produced by nuclear recoil. Such atoms are known colloquially as "hot" atoms, and their chemical reactions make up the realm of "hot atom chemistry" (see p. 1169).

For fuller details pertinent to specific radioactive isotopes of the halogens, in respect of the principles and practice of formation, separation, enrichment, handling, detection and estimation, the reader is referred to more comprehensive or specialised texts[32,33,45,102]. Concerning the preparation of labelled chlorine and iodine compounds details have been compiled[102]; synthetic methods appropriate to such bromine compounds have also been outlined[45]. An elegant and effective method of labelling compounds is to inject them on to a gas-chromatography column charged with a radioactive halogen compound. By this method virtually carrier-free propyl bromide[103] and arsenic and germanium chlorides[104] have been obtained. Halogen-labelled molecular chlorides, bromides and iodides can be prepared by heating *in vacuo* an element such as silicon, boron or aluminium with labelled silver or copper(I) halides[105,106].

Of the several methods which have been described for the separation or enrichment of stable isotopes[107], the following have proved most effective for the chlorine isotopes ^{35}Cl and ^{37}Cl [33]: thermal diffusion of gaseous HCl; ultracentrifuge treatment of molecular chlorides; electromagnetic separation of $CuCl^+$ ions in a calutron (a type of cyclotron); migration of chloride ions in solution under the influence of an applied electric field; fractionation of chloride ions by ion-exchange; fractional distillation, e.g. of Cl_2 or HCl; isotopic exchange, e.g. between gaseous Cl_2 or HCl and Cl^- in aqueous solution. Virtually

[101] L. Szilard and T. A. Chalmers, *Nature*, **134** (1934) 462, 494.
[102] R. H. Herber (ed.), *Inorganic Isotopic Syntheses*, Benjamin, New York (1962).
[103] F. Schmidt-Bleek, G. Stöcklin and W. Herr, *Angew. Chem.* **72** (1960) 778.
[104] J. Tadmor, *J. Inorg. Nuclear Chem.* **23** (1961) 158.
[105] K. H. Lieser and H. Elias, *J. Inorg. Nuclear Chem.* **23** (1961) 139.
[106] K. H. Lieser, H. W. Kohlschütter, D. Maulbecker and H. Elias, *Z. anorg. Chem.* **313** (1961) 193.
[107] P. S. Baker, *Survey of Progress in Chemistry*, **4** (1968) 69.

complete separation of $H^{35}Cl$ and $H^{37}Cl$ has been achieved using the thermal gas-diffusion principle. Similar methods have been applied to the naturally occurring bromine isotopes ^{79}Br and ^{81}Br, which are probably best separated by the thermal diffusion of HBr[108], though varying degrees of enrichment have also been brought about in the gas centrifuge[109], by electromagnetic methods[110], and by electrolytic transport in fused zinc or lead bromides[111].

In addition to data about the decay and neutron-capture characteristics of different isotopes, Table 8 also alludes to the magnetic and electrical properties, where these are known. It is evident that, unlike the ^{19}F nucleus, all of the chemically important nuclei of chlorine, bromine and iodine are quadrupolar with nuclear spins $> \frac{1}{2}$. The magnitudes of the nuclear spin, the magnetic moment and the electric quadrupole moment determine, most significantly, the principal characteristics of each nucleus with respect to nmr, esr and nqr experiments, although one or more of these parameters may modulate, *via* second-order interactions, other types of spectroscopic transition, as in the fine structure of microwave spectra or in the hyperfine structure of the atomic spectra. All five of the naturally occurring isotopes of chlorine, bromine and iodine, that is, ^{35}Cl, ^{37}Cl, ^{79}Br, ^{81}Br and ^{127}I, have featured in nmr, esr and nqr measurements. The effect of the quadrupole moment is to provide an efficient mechanism of magnetic relaxation with the result that, in normal chemical environments, the nuclei are invariably characterized by relatively diffuse magnetic resonances, which compare unfavourably with the narrow lines exhibited by ^{19}F ($I = \frac{1}{2}$). Nevertheless, some nmr measurements have been made to explore variations of chemical shift and linewidth, notably in relation to the possible effects of solvation and association of ions in solution (see p. 1239). On the other hand, the very properties which make the nuclei relatively unfavourable for conventional nmr studies make them highly eligible for nqr experiments, which afford a means of investigating the interaction of such quadrupolar nuclei with intramolecular electric fields; nqr measurements have been made for a large number of chlorine, bromine and iodine compounds[112]. With respect to the esr spectra of paramagnetic halogen-containing systems, the simplest examples of which are the halogen atoms (see Table 10), the nuclear properties of the naturally occurring isotopes determine the hyperfine structure of the spectra, there being $2I+1$ equally spaced lines for an individual nucleus with nuclear spin I. Hence, information has been derived, not only about the properties of the halogen nuclei, but about the number of such nuclei in a chemical aggregate and about the density and orbital occupation of the unpaired electrons. The significance of some of the nmr, esr and nqr data is considered in the contexts of the general characteristics of halide species (Section 3) and of individual halogen compounds to which they are relevant.

A feature not evident from the nuclear properties of Table 8 concerns the potential of the different isotopes with respect to the Mössbauer effect, for which the source is a radioactive isotope of reasonable half-life. By radioactive disintegration the isotope populates an excited level which decays to the ground state by emitting low-energy γ-radiation. The only halogen isotopes which fulfil the necessary conditions are ^{127}I and ^{129}I:

[108] H.-U. Hostettler and Kl. Clusius, *Proc. Intern. Symposium on Isotope Separation*, p. 419. Amsterdam, North-Holland Publishing Co. (1957); *Z. Naturforsch.* **12a** (1957) 974.

[109] R. F. Humphreys, *Phys. Rev.* **56** (1939) 684.

[110] C. W. Sheridan, H. R. Gwinn and L. O. Love, U.S. At. Energy Comm. ORNL–3301 (1962); W. Żuk, *Ann. Univ. Mariae Curie-Skłodowska, Lubin-Polonia, Sect. AA*, **12** (1960) 1.

[111] A. Lundén and A. Lodding, *Z. Naturforsch.* **15a** (1960) 320; A. E. Cameron, W. Herr, W. Herzog and A. Lundén, *ibid.* **11a** (1956) 203.

[112] See for example E. A. C. Lucken, *Nuclear Quadrupole Coupling Constants*, Academic Press, London and New York (1969); M. Kubo and D. Nakamura, *Adv. Inorg. Chem. Radiochem.* **8** (1966) 257; H. Sillescu, *Physical Methods in Advanced Inorganic Chemistry* (ed. H. A. O. Hill and P. Day), p. 434. Interscience (1968).

Table 8. Nuclear Properties of the Better Defined Isotopes of Chlorine, Bromine and Iodine[a]

Cl, atomic number 17

Mass number	% natural abundance use as radio-tracer	Principal source	Mode of decay	Half-life	Decay energy (MeV)	Particle energies (MeV)	γ-radiation energies (keV)	Thermal neutron-capture cross-section (b)	Nuclear spin, I ($h/2\pi$)	Magnetic moment, μ (n.m.)	Electric quadrupole moment, Q ($e \times 10^{-24}$)
32		^{32}S–p–n	$\beta^+(\alpha)$	0·31 s	13·2	9·5	Ann. rad. 2210–4770				
33		^{32}S–d–n ^{33}S–p–n	β^+	2·5 s	5·57	4·51	2900				
34m		^{31}P–α–n	β^+, IT	32·0 m		2·5–4·5	Ann. rad. 640–4100				
34		^{34m}Cl decay	β^+	1·56 s	5·482	4·50	Ann. rad.				
35	75·77		STABLE ISOTOPE					44±2	3/2	+0·82183	−0·079
36	Tracer element	^{35}Cl–n–γ	β^- β^+, EC	$3 \cdot 1 \times 10^5$y	0·712 1·14	0·714 (β^-)		100±30	2	+1·285	−0·017
37	24·23		STABLE ISOTOPE					0·430± 0·100 (^{38}Cl) 0·005(^{38m}Cl)	3/2	+0·68411	−0·062
38m		^{37}Cl–n–γ	IT	0·74 s			660				
38	Tracer element		β^-	37·3 m	4·91	1·11–4·81	1600–3760				
39		^{40}Ar–α–α, p ^{40}Ar–γ–p	β^-	55·5 m	3·44	1·91–3·45	246–1520				
40		^{40}Ar–n–p	β^-	1·4 m	7·5	~7·5, ~3·2	1460–5800				

74		^{65}Cu–^{12}C–3*n*	β^+, EC	36 m	~6·8	4·7	Ann. rad. 640				
75		^{65}Cu–^{12}C–2*n* ^{74}Se–*d*–*n* ^{74}Se–*p*–γ	β^+, EC	1·7 h	2·72	0·3–1·70	Ann. rad. 112·5–962·1*				
76		^{75}As–α–3*n*	β^+, EC	16·1 h	4·6	1·2–3·6	Ann. rad. 358·4–4436·7*		1	±0·548	±0·25
77	Tracer element	^{75}As–α–2*n*	β^+, EC	57 h	1·365	0·361	87·86–1005·2*		3/2		
77m		^{76}Se–*p*–γ	IT	4·2 m							
78		^{75}As–α–*n* ^{77}Se–*d*–*n* ^{78}Se–*p*–*n* ^{77}Se–*p*–γ	β^+, EC	6·4 m	3·573	1·937, 2·52	Ann. rad. 614·1				
79	50·54		STABLE ISOTOPE					2·6±0·2 (^{80m}Br) 8·5 (^{80}Br)	3/2	+2·106	+0·31
79m		^{78}Se–*p*–γ ^{79}Br–*n*–*n*′	IT	~5 s							
80m	Tracer element	^{79}Br–*n*–γ	IT	4·4 h	0·085		37, 85		5	+1·317	+0·71
80	Tracer element	^{79}Br–*n*–γ ^{80m}Br decay	β^+, EC β^-	17·6 m	1·871 2·01	0·866 0·70–2·05	616–1257		1	±0·514	±0·18

Table 8 (*cont.*)

Br, atomic number 35 (*cont.*)

Mass number	% natural abundance use as radio-tracer	Principal source	Mode of decay	Half-life	Decay energy (MeV)	Particle energies (MeV)	γ-radiation energies (keV)	Thermal neutron-capture cross-section (b)	Nuclear spin, I ($h/2\pi$)	Magnetic moment, μ (n.m.)	Electric quadrupole moment, Q ($e\times10^{-24}$)
81	49·46		STABLE ISOTOPE					3·0±0·2 (^{82m}Br) 0·26 (^{82}Br)	3/2	+2·270	+0·26
82m		^{81}Br–n–γ	IT β^-		0·046	1·659 2·357	46 (IT), 698·4–1474·8				
82	Tracer element		β^-	35·5 h	3·092	0·257, 0·440	92·3–2056*		5	±1·626	±0·70
83		^{82}Se–n–γ ^{83}Se–β^-	β^-	2·41 h	0·97	0·395, 0·925	32–521				
84		^{87}Rb–n–α fission Th, U, Pu	β^-	31·8 m	4·8	0·89–4·8	270–3930*				
85		Fission U	β^-	3·0 m	2·8	2·8	305·0 (^{85m}Kr)				
86		^{86}Kr–n–p	β^-	54 s	7·1						
87		Fission Th, U, Pu	β^-, n	55 s	6·1	2·6, 8·0	1440–5200*				
88		Fission Th, U, Pu	β^-, n	16 s			760				
89		Fission Th, U, Pu	β^-, n	45 s							
90		Fission Th, U, Pu	β^-, n	1·6 s							

117		La+protons	β^+, EC	7 m			Ann. rad. 160, 340				
118		I+protons	β^+, EC	14 m	7	5·5	Ann. rad. 550–1150				
119		Pd+^{14}N I+protons	β^+, EC	19 m			Ann. rad. 260, 780				
120		I+protons ^{121}Sb–α–5n ^{120}Xe decay	β^+, EC	1·3 h	5·6	2·1, 4·0	Ann. rad. 560–1520				
121		^{121}Sb–α–4n	β^+, EC	2·1 h	2·36	1·2	Ann. rad. 213–740*				
122		^{121}Sb–α–3n ^{122}Te–p–n	β^+, EC	3·5 m	4·14	1·8–3·1	Ann. rad. 560–3450*				
123		^{121}Sb–α–2n	EC	13·3 h	~1·4		159–781·4*		5/2		
124	Tracer element	^{121}Sb–α–n ^{123}Sb–α–3n	β^+, EC	4·2 d	3·17	0·79–2·13	Ann. rad. 602·7–2740*		2		
125	Tracer element	^{123}Sb–α–2n ^{124}Te–d–n ^{125}Xe decay	EC	60 d	0·149		35·48	900±90	5/2	+3·0	−0·89
126		^{123}Sb–α–n ^{125}Te–d–n ^{126}Te–p–n	β^+, EC β^-	13 d	2·150 1·251	β^+ 1·129 β^- 0·385–1·25	Ann. rad. 388·7–1420*		2		
127	100		STABLE ISOTOPE					6·2±0·2	5/2	+2·808	−0·79
128	Tracer element	^{127}I–n–γ	β^+, EC β^-	25·08 m	1·27 2·14	1·13–2·12	442·9–969·5		1		

Table 8 (*cont.*)

I, atomic number 53 (*cont.*)

Mass number	% natural abundance use as radio-tracer	Principal source	Mode of decay	Half-life	Decay energy (MeV)	Particle energies (MeV)	γ-radiation energies (keV)	Thermal neutron-capture cross-section (b)	Nuclear spin, I ($h/2\pi$)	Magnetic moment, μ (n.m.)	Electric quadrupole moment, Q ($e \times 10^{-24}$)
129		Fission U	β^-	$1{\cdot}7 \times 10^7$ y	0·189	0·189	39·58	19±2 (^{130m}I) 9±1 (^{130}I)	7/2	+2·617	−0·55
130m		^{130}Te–d–$2n$ ^{130}Te–p–n ^{129}I–n–γ ^{133}Cs–n–α	IT	8·82 m							
130			β^-	12·3 h	2·99	0·62–1·7	419–1150	18±3	5		
131	Tracer element	^{130}Te–n–γ; spall.–fission Th, U; fission Th, U, Pu	β^-	8·070 d	0·970	0·257–0·806	80·164–722·92*	~0·7	7/2	+2·74	−0·40
132	Tracer element	^{132}Te decay following fission	β^-	2·3 h	3·56	0·72–2·12*	147·10–2395·0*		4	±3·08	±0·08
133		Spall.–fission Pb, U; fission U, Pu; ^{133}Te decay	β^-	20·9 h	1·80	0·7–1·27	151·1–1592·5*		7/2	+2·84	−0·26
134		Spall.–fission U; fission Th, U, Pu	β^-	52 m	4·2	1·10–2·46*	135·4–2467·1*				

135		Spall.–fission U; fission Th, U, Pu	β^-	6·7 h	~2·8	0·5–1·4	135·4–2465·9*		7/2		
136		Fission U, Pu	β^-	83 s	7·0	2·7–7·0	200–3200*				
137		Fission U, Pu	β^-, n	23 s							
138		Fission U	β^-, n	5·9 s							
139		Fission U	β^-, n	2 s							

[a] The major sources of data for this table are:

1. *Handbook of Chemistry and Physics*, 52nd edn., B–253, The Chemical Rubber Co. (1971–72).
2. C. M. Lederer, J. M. Hollander and I. Perlman, *Table of Isotopes*, 6th edn., Wiley, New York (1968).
3. G. H. Fuller and V. W. Cohen, *Nuclear Data Tables*, **5A** (1969) 433.

Column

3 *Source*: Refers to the nuclear features (target element, projectile and outgoing particle, in order) whereby the radioactive isotopes are formed. p = proton; n = neutron; α = α-particle; d = deuteron; γ = γ- or X-rays; spall.–fission = high-energy fission (followed by symbol of target element).

4 *Mode of decay*: IT = isomeric transition; EC = orbital electron capture.

5 *Half-life*: s = seconds; m = minutes; h = hours; d = days; y = years.

7, 8 *Particle and γ-energies*: Ann. rad. refers to the 511·006 keV photon associated with the annihilation of positrons in matter. * = numerous well-defined energies (*Rubber Handbook*) within the limits specified.

9 *Thermal neutron-capture cross-section*: b = barns (10^{-24} cm^2).

10 *Nuclear spin*: units of $h/2\pi$.

11 *Magnetic moment*: units of the nuclear magneton (n.m.), with diamagnetic correction.

12 *Electric quadrupole moment*: units are barns (10^{-24} cm^2).

Mössbauer isotope	γ-ray energy keV	Half-life of Mössbauer transition 10^{-9} s	Parent isotope	Half-life of parent isotope days
^{127}I	57·60	2·68	^{127}Te	105
			^{127}Xe	36
^{129}I	27·72 27·78	16·8	^{129}Te	33

The form of the Mössbauer spectrum is determined by the characteristics of the nuclear transition and of the γ-emission produced thereby, the lifetime of the excited state, the nuclear spins of the excited and ground state, and the field gradient at the nucleus. Nuclear isomer shifts and quadrupole splittings due to both the Mössbauer isotopes of iodine have been measured for a number of solid iodine compounds, e.g. I_2, ICl, IBr, I_2Cl_6, alkali-metal iodides, iodates and periodates[113], and these parameters have been correlated with details of the electron density, bonding and local symmetries of the iodine atoms.

Radioactive halogens were much used in chemical studies even before pile-produced nuclides became generally available; accordingly there is a relatively long history of isotopic exchange and other tracer studies of the courses taken by chemical reactions and of the exploitation of radioactive halogen isotopes in chemical analysis and in various biological studies[114]. The increased availability of radioactive isotopes in more recent years has made possible a notable extension and development of these activities. Of the various uses of radioactive isotopes of chlorine, bromine and iodine, those listed in Table 9 are probably the most significant.

Electronic and Thermodynamic Properties of the Isolated Atoms

Table 10(a) presents for the isolated chlorine, bromine and iodine atoms details relevant to the atomic weight, spectroscopic properties, wave functions, promotion energies, ionization potentials, electron affinity, electronic *g*-factor and thermodynamic properties.

In addition to the references given in the table, there exist substantial reviews of the optical and X-ray spectra of the atoms as reported up to 1956[32]. The separations of the components of the inverted doublet which forms the ground state of each atom imply the following values (in cm^{-1}) for the one-electron spin–own-orbit coupling constants (ζ): F, 269; Cl, 587; Br, 2456; I, 5069. Corresponding with this sequence, whereas the spin–orbit coupling in fluorine is adequately described by the Russell–Saunders laws, the large magnetic interaction in iodine is compatible, not with simple L,S- but with *jj*-coupling. The absence of low-lying excited states is evinced by the relatively high one-electron promotion energies calculated from the mean energies of the appropriate electronic states; these values are sufficiently high to call in question the authenticity of the concept of valence states involving the promotion of one or two *p*-electrons and the subsequent formation of hybrid orbitals.

For a given stage of ionization, the ionization potentials decrease relatively smoothly in the sequence Cl > Br > I after the relatively dramatic decrease in passing from fluorine

113 N. N. Greenwood, *Chem. in Britain*, **3** (1967) 56; M. Pasternak, *Symposia of the Faraday Soc.* **1** (1967) 119; D. W. Hafemeister, *Advances in Chemistry Series*, **68** (1967) 126; R. H. Herber, *Progress in Inorganic Chemistry*, **8** (1967) 1; J. Danon, *Physical Methods in Advanced Inorganic Chemistry* (ed. H. A. O. Hill and P. Day), p. 380, Interscience (1968).

114 J. Kleinberg and G. A. Cowan, U.S. At. Energy Comm. NAS–NS 3005 (1960).

TABLE 9. APPLICATIONS OF RADIOACTIVE HALOGEN ISOTOPES

Application	Examples	Comments
1. (a) *Fundamental studies of halogen-exchange reactions* for information (i) about the incidence of exchange, (ii) about the kinetics of the process, (iii) about its mechanism and (iv) about the equivalence of atoms in a molecule, e.g. PCl_5 or Ph_2I_2.	Studies of exchange[a,b,c] in the following systems: X_2/X^- in aqueous solution; RX/X^- in various solvents; RX/X_2; HX/X_2; XO_3^-/X_2; RX/AlX_3; $POCl_3/Et_4NCl$ in MeCN; PCl_5/Cl_2 in CCl_4; Ph_2I_2/I^- in 50% alcohol; PtX_4^{2-}/X^- and MX_6^{2-}/X^- (M = Re, Os, Pt or Ir) in solution. Possible auto-ionization in molecular halides like $POCl_3$ and $AsCl_3$.	The progress of exchange between two atoms of the same halogen in different chemical environments is followed by "labelling" one of the species with a radioactive halogen isotope, and subsequently separating the two compounds and measuring the variations of activity.
(b) *Investigations of reactions other than those involving isotopic exchange.*	Tracer studies of reactions such as that of Br atoms with aromatic compounds,[d] the iodination of metals,[e] Friedel–Crafts and related reactions,[f] chlorination of hydrocarbons by Bu^tOCl,[g] the I atom-catalysed isomerization of di-iodoethylene,[h] the polymerization of vinyl compounds by iodine-substituted free radicals,[i] the action of bromine-containing inhibitors on the emulsion polymerization of styrene;[j] studies of the stereospecificity of a bromination–debromination sequence starting from 1-bromocyclohexene,[k] and of the formation and reactions of organic free radicals (by trapping with labelled iodine).[l]	Labelled halogen atoms may serve as indicators with respect to the progress, mechanism or the nature of intermediates or products of a chemical reaction.
2. *Investigations of diffusion phenomena and of the distribution of components between two phases.*	Measurements of the self-diffusion of X^-, BrO_3^- or I_3^-[m] in solution, of X^- in crystalline AgX (X = Cl or Br)[n] and of X^- or BrO_3^- in ion-exchange resins;[a,b] determination of residence times in liquid extraction columns;[o] solubility measurements, e.g. of *p*-chloroiodobenzene in ethylene gas,[p] and investigations of precipitation processes, e.g. of AgI[q] and AgBr[b].	Such measurements are significant in the interpretation of solution theory and of the mode of action of ion-exchange resins.
3. *Radiochemical methods of analysis*	Detection and estimation of halogens can be achieved (i) by isotope dilution or related methods: e.g. I^- estimated *via* ^{131}I in conjunction with solvent-extraction[r] or ion-exchange;[s] such methods have also been used in the analysis of fission products for bromine and iodine;[a] (ii) by neutron-activation analysis, used, for example, to estimate the halide content of aqueous media,[b,t] diverse organic systems,[b,u] zinc sulphide phosphors,[b] SiO_2–Al_2O_3 catalysts,[v] and biological material,[b,w] and to evaluate the content and isotopic abundance of halogens in materials such as meteorites.[x]	Such methods are appropriate to the determination of traces (in the order 1 ppm) of the halogens in various systems; neutron-activation analysis is facilitated for chlorine, bromine and iodine (but not fluorine) by the relatively large neutron-capture cross-sections of the naturally occurring isotopes.

Table 9 (*cont.*)

Application	Examples	Comments
4. *Physiological and biochemical applications.*	Radiohalogens have been used to study the transport and distribution of halide in mammalian tissues, e.g. the thyroid gland, central nervous system and bladder, and, to a limited extent, also to trace halogen species in plant physiology.[a,b] In this context numerous compounds of biological interest have been labelled with radiohalogens so that their metabolic fate may be explored. Examples include the use as indicators of an ^{82}Br-labelled analogue of DDT,[a,b] ^{131}I-labelled insulin and antisera,[a] ^{82}Br-labelled proteins and steroids[y] and ^{82}Br- or ^{131}I-labelled growth regulators such as 5-bromouracil and 2-iodo-3-nitrobenzoic acid.[a,b] *Clinical uses*—^{131}I is used for the diagnosis and therapy of thyroid disorders. A method has been developed for the direct irradiation, employing ^{82}Br, of malignant tissue in the bladder.[a,b]	Apart from their practical clinical importance, radioisotopes of the heavier halogens have afforded, *via* tracer experiments, intriguing results concerning the transport of ions in the tissues of living mammals. It has also been shown that iodine entering the thyroid gland as iodide ion is oxidized forming first monoiodo- and then di-iodotyrosine, which suffers oxidative coupling to form thyroxine. Of the major organic constituents of thyroid tissue, di-iodotyrosine and thyroxine, the latter is believed to be the circulating thyroid hormone which governs the metabolic rate of the whole body.[a]
5. *Technological and industrial uses.*	Radioisotopes of chlorine, bromine and iodine, in suitable chemical combination, have found uses in the detection of leaks in process streams,[z] the location of liquid junctions in oil pipelines,[a,b] in studies of the flow of liquids and gases, e.g. atmospheric motions,[aa] in the detection of flaws in the sheathing of telephone cables,[a,b] and in various hydrological investigations, e.g. tracing water movement in soils,[bb] evaluating the recharge-loss balance of ground water,[cc] and studies of sewage dispersion and water pollution.[dd]	These applications depend, for the most part, not on the chemical properties of the radioactive material but on physical properties such as solubility or adsorption.

[a] Supplement to *Mellor's Comprehensive Treatise on Inorganic and Theoretical Chemistry*, Supplement II, Part I, pp. 1013–1063, 1080–1091, Longmans, London (1956).

[b] Z. E. Jolles (ed.), *Bromine and its Compounds*, pp. 425–462, 786–798, Benn, London (1966).

[c] M. F. A. Dove and D. B. Sowerby, *Halogen Chemistry* (ed. V. Gutmann), Vol. 1, p. 41, Academic Press, London and New York (1967).

[d] S. May, M. Roux, Buu-Hoi and R. Daudel, *Compt. rend.* **228** (1949) 1865; G. Gavoret, *J. chim. Phys.* **50** (1953) 183; P. B. D. De la Mare and J. T. Harvey, *J. Chem. Soc.* (1956) 36; (1957) 131.

[e] H. Sugier, *Nukleonika*, **12** (1967) 723 (*Chem. Abs.* **68** (1968) 108367t).

[f] R. M. Roberts and G. J. Fonken, *Friedel-Crafts and Related Reactions* (ed. G. A. Olah), Vol. 1, p. 821, Interscience, New York (1963).

[g] A. A. Zavitsas, *J. Org. Chem.* **29** (1964) 3086.

[h] R. M. Noyes, R. G. Dickinson and V. Schomaker, *J. Amer. Chem. Soc.* **67** (1945) 1319.

[i] K. Ziegler, W. Deparade and H. Kühlhorn, *Annalen*, **567** (1950) 151.

[j] E. J. Meehan, I. M. Kolthoff, N. Tamberg and C. L. Segal, *J. Polymer Sci.* **24** (1957) 215.

[k] C. L. Stevens and J. A. Valicenti, *J. Amer. Chem. Soc.* **87** (1965) 838.

[l] See, for example, G. R. Martin and H. C. Sutton, *Trans. Faraday Soc.* **48** (1952) 812.

[m] K. G. Darrall and G. Oldham, *J. Chem. Soc.* (*A*) (1968) 2584.

[n] J. Nölting, *Z. Physik. Chem.* (*Frankfurt*), **38** (1963) 154; M. Haïssinsky, *Nuclear Chemistry and its Applications* (trans. D. G. Tuck), p. 553, Addison-Wesley (1964).

[o] See, for example, M. Kubin, *Proc. Symp. Radioisotope Tracers Ind. Geophys.*, p. 529, Prague (1966) (*Chem. Abs.* **68** (1968) 60954c).

[p] A. H. Ewald, *Trans. Faraday Soc.* **49** (1953) 1401.

[q] K. Müller and S. Karajannis, *J. Radioanal. Chem.* **2** (1969) 359.

[r] H. G. Richter, *Analyt. Chem.* **38** (1966) 772.

[s] M. Lesigang and F. Hecht, *Mikrochim. Acta*, (1962) 327.

[t] I. F. Yazikov, N. N. Rodin, M. A. Dembrovsky and V. G. Lambrev, *J. Radioanal. Chem.* **3** (1969) 11.

[u] R. Malvano and S. Kwieciński, *J. Radioanal. Chem.* **3** (1969) 257.

[v] P. Bussiere, A. Laurent and E. Junod, *J. Radioanal. Chem.* **2** (1969) 211.

[w] P. Schramel, *J. Radioanal. Chem.* **3** (1969) 29; R. A. Nadkarni and W. D. Ehmann, *ibid.* p. 175.

[x] A. Wyttenbach, H. R. von Gunten and W. Scherle, *Geochimica et Cosmochimica Acta*, **29** (1965) 467, 475.

[y] J. Saroff, R. E. Keenan, A. A. Sandberg and W. R. Slaunwhite, jun., *Steroids*, **10** (1967) 15.

[z] U.S. Pat. 3,370,173 (*Chem. Abs.* **68** (1968) 92298h).

[aa] B. Keisch, R. C. Koch, A. S. Levine, J. Roesmer and W. S. Winnowski, U.S. At. Energy Comm. NSEC–120 (*Nucl. Sci. Abstr.* **21** (1967) 14328).

[bb] A. Hamid and B. P. Warkentin, *Soil Sci.* **104** (1967) 279; E. Wagiel and J. Szymanski, *Pr. Inst. Naft.* (1968) 13 (*Chem. Abs.* **70** (1969) 30558a).

[cc] K. Ubell, *Proc. Symp. Isotop. Hydrol.*, p. 521, Vienna (1966).

[dd] G. E. Eden and R. Briggs, *Proc. Symp. Isotop. Hydrol.*, p. 191, Vienna (1966).

TABLE 10. PROPERTIES OF CHLORINE, BROMINE AND IODINE ATOMS

(a) Isolated atoms

Property	Chlorine	Bromine	Iodine
Atomic number	17	35	53
Mass number of naturally occurring isotopes	35 (75·77%) 37 (24·23%)	79 (50·54%) 81 (49·46%)	127 (100%)
Atomic weight [a]	35·453	79·904	126·9045
Optical spectra	Refs. b, c	Refs. b, d	Refs. b, e
X-ray spectra	Refs. f, g, h	Refs. f, g	Refs. f, g
Wavefunctions for electronic ground state	Refs. i, j	Refs. i, j	Refs. i, j
Electronic configuration and term of ground state	$[Ne]3s^23p^5$ $^2P_{3/2}$	$[Ar]3d^{10}4s^24p^5$ $^2P_{3/2}$	$[Kr]4d^{10}5s^25p^5$ $^2P_{3/2}$
$^2P_{3/2} \rightarrow {}^2P_{1/2}$ promotion energy, cm^{-1} (kcal)	882·36 (2·523)[b]	3,685 (10·54)[b]	7,602·7 (21·76)[b]
One-electron promotion energies, cm^{-1} (kcal)*			
$ns^2np^5 \rightarrow ns^2np^4(n+1)s^1$	72,500 (207)[b]	64,000 (183)[b]	55,400 (159)[b]
$ns^2np^5 \rightarrow ns^2np^4(n+1)p^1$	83,900 (240)[b]	75,100 (215)[b]	65,100 (186)[b]
$ns^2np^5 \rightarrow ns^2np^4nd^1$	~91,000 (258)[c]	~79,000 (225)[d]	~70,000 (200)[e]
Ionization potentials, eV (kcal)[b]			
I_1	12·967 (299·0)	11·84 (273)	10·451 (241·0)
I_2	23·80 (549)	21·6 (498)	19·09 (440)
I_3	39·90 (920)	35·9 (828)	
I_4	53·5 (1234)	47·3 (1091)	
I_5	67·80 (1564)	59·7 (1377)	
Electron affinity at 298°K, eV (kcal)[k]	3·68 (84·8)	3·43 (79·0)	3·13 (72·1)
Esr properties: g-factor (theoretical value assuming Russell–Saunders coupling = $1·3341064$)	1·333923[l]	1·333921[m]	1·333995[n]
$\Delta H°$ for $\frac{1}{2}X_2(g) \rightarrow X(g)$ at 298°K (kcal)[o]	28·989	23·036	18·058
Thermodynamic properties of atoms at 298°K (ref. o)			
$\Delta H_f°$ (kcal)	28·989	26·730	25·517
$\Delta G_f°$ (kcal)	25·170	19·690	16·780
$S°$ (cal deg^{-1})	39·454	41·803	43·182

(b) Bound atoms

Property	Chlorine	Bromine	Iodine
Single-bond covalent radius (Å)[k]	0·99	1·14	1·33
Van der Waals' radius (Å)[p]	1·80	1·95	2·15
Electronegativity[q]	3·0	2·8	2·5
Polarizability, $Å^3$ (ref. r)	2·3	3·2	5·6
Diamagnetic susceptibility, $\times 10^6$ cgs units per g atom[s]	−20·1	−30·6	−44·6

* Calculated from the appropriately weighted mean energies of all the terms of both the ground and excited states.

[a] Based on *Table of Atomic Weights*, 1969 IUPAC Commission on Atomic Weights; *Pure Appl. Chem.* **21** (1970) 95.

[b] C. E. Moore, *Atomic Energy Levels*, Vols. I-III, National Bureau of Standards Circular 467, Washington (1949–58); L. J. Radziemski, jun., and V. Kaufman, *J. Opt. Soc. Amer.* **59** (1969) 424; R. E. Huffman, J. C. Larrabee and Y. Tanaka, *J. Chem. Phys.* **47** (1967) 856.

[c] *Gmelins Handbuch der Anorganischen Chemie*, 8 Auflage, "Chlor", Teil A, p. 123 (1968).

[d] J. L. Tech, *J. Res. Nat. Bur. Stand.* **67A** (1963) 505.

[e] C. C. Kiess and C. H. Corliss, *ibid.* **63A** (1959) 1.

[f] J. A. Bearden, *X-ray Wavelengths*, U.S. Atomic Energy Commission, NYO-10586, Oak Ridge, Tennessee (1964); J. A. Bearden, *Rev. Mod. Phys.* **39** (1967) 78; J. A. Bearden and A. F. Burr, *ibid.* p. 125.

[g] A. E. Sandström, *Experimental Methods of X-ray Spectroscopy: Ordinary Wavelengths, Handbuch der Physik*, **30** (1957) 78.

[h] Ref. c, p. 133.

[i] Functions based on Hartree-Fock-Slater approximation, F. Herman and S. Skillman, *Atomic Structure Calculations*, Prentice-Hall, Englewood Cliffs (1963).

[j] Hartree-Fock wavefunctions, C. Froese, *J. Chem. Phys.* **45** (1966) 1417; J. B. Mann, Contract W-7405-eng-36, Dep. CFSTI; *Nucl. Sci. Abstr.* **22** (1968) 17345.

[k] A. G. Sharpe, *Halogen Chemistry* (ed. V. Gutmann), Vol. 1, p. 1, Academic Press, London and New York (1967); R. S. Berry, *Chem. Rev.* **69** (1969) 533.

[l] V. Beltran-Lopez and H. G. Robinson, *Phys. Rev.* **123** (1961) 161.

[m] J. S. M. Harvey, R. A. Kamper and K. R. Lea, *Proc. Phys. Soc. (London)*, **76** (1960) 979.

[n] K. D. Bowers, R. A. Kamper and C. D. Lustig, *ibid*, **B70** (1957) 1176.

[o] *National Bureau of Standards Technical Note* 270–3, January 1968; *Codata Bulletin*, International Council of Scientific Unions, Committee on Data for Science and Technology (1970).

[p] L. Pauling, *The Nature of the Chemical Bond*, 3rd edn., p. 260, Cornell University Press, Ithaca (1960).

[q] F. A. Cotton and G. Wilkinson, *Advanced Inorganic Chemistry*, 2nd edn., p. 103, Interscience (1966).

[r] R. T. Sanderson, *Inorganic Chemistry*, p. 54, Reinhold (1967).

[s] A. Earnshaw, *Introduction to Magnetochemistry*, p. 6, Academic Press, London (1968).

to chlorine; in this and other respects the variations in ionization potential are normal. Although unquestionably high, the first ionization potentials (I_1) of chlorine, bromine and iodine are nevertheless lower than that of hydrogen (13.6 eV), and I_1 for iodine is not much greater than for some metals—in particular zinc (9·39 eV) and mercury (10·43 eV). Even the sum of the first two potentials for iodine (29·54 eV) is only slightly greater than that for mercury (29·18 eV). Since the removal of one electron from the valence shell of a halogen atom leaves four electrons in the *p*-orbitals, the X^+ cation is unlikely to be substantially smaller than the corresponding atom[115]. It is presumably this size factor, rather than the magnitude of the first ionization potentials, that limits the occurrence of mononuclear halogen cations under normal chemical conditions.

The electron affinities of chlorine, bromine and iodine represent values recently recommended[115]; they are based on an analysis of the results of three distinct methods, viz. (i) determination from computed lattice energies and other quantities in the Born–Haber cycle, (ii) the direct study of the equilibrium $X(g)+e \rightleftharpoons X^-(g)$ at a hot tungsten filament, and (iii) measurements of photochemical electron-detachment from halide ions in shock-heated vapours of alkali-metal halides. The data of Table 10 confirm that the sequence of electron affinities is F < Cl > Br > I. Recently attributed[116] to a destabilization energy amounting to *ca.* 26 kcal g atom^{-1}, which accompanies the interaction of a fluorine atom with an external electron, the anomalous position of fluorine is not really very significant in the comparative chemistry of the halogens. In chemical situations, factors such as lattice energies and solvation energies, which are sensitive functions of the size of the ions produced, invariably outweigh the small differences in electron affinity, and tend to dictate the differences in chemical properties among the halogens.

Table 10(a) also includes values at 298°K for the standard heats of formation, free energies and entropies of the atomic halogens, together with the enthalpies of the reaction

[115] A. G. Sharpe, *Halogen Chemistry* (ed. V. Gutmann), Vol. 1, p. 1, Academic Press (1967).

[116] P. Politzer, *J. Amer. Chem. Soc.* **91** (1969) 6235.

$\frac{1}{2}X_2(g) \rightarrow X(g)$; tables of the usual thermodynamic functions are to be found elsewhere[67]. Variations in the entropies of the gaseous atoms are almost entirely determined by the term $3/2R \ln M + R \ln Q$ in the Sackur–Tetrode equation, M being the atomic weight and Q the electronic multiplicity, i.e. $(2J+1)$, of the ground state. As with the molecular halogens and analogous molecular halides in the vapour phase, variations in the entropy terms are comparatively slight. The principal thermodynamic differences between the halogens arise from enthalpy effects which reflect the different strengths of the X–X bonds, viz. F–F < Cl–Cl > Br–Br > I–I. Several quite different interpretations of this sequence have been advanced, based, for example, on electronic repulsion, which is larger in F_2 than in the other diatomic molecules (in keeping with the destabilization effect[116] already referred to), or on partial multiple bonding in the heavier halogen molecules depending on the use, denied to fluorine, of valence-shell d-orbitals. One analysis[117] concludes that electron repulsion is certainly a very important factor, but that the magnitudes of electron–nucleus attraction and nucleus–nucleus repulsion must also be taken into account; since all of these are a function of the internuclear distance, this quantity is possibly at the root of the somewhat anomalous sequence of bond energies.

Properties of the Bound Atoms

When a halogen atom engages in chemical bonding, its valence electrons are no longer identifiable with localized atomic orbitals. The gross perturbation of the valence electrons by ligand-fields means that most properties of the bound atom are, to some degree, conditioned by its environment. However, the core electrons, being relatively much less responsive to such effects, may be justifiably regarded as occupying individual atomic orbitals. The photoelectron spectrum observed when a halogen atom in a compound is irradiated with X-rays furnishes energies of these core electrons, and it has thus been found, for example, that the energies of the K-electron level and L_1 sub-level vary almost linearly with the oxidation number of chlorine, the total shift being about 9·6 eV as the oxidation number varies from -1 to $+7$[118].

The interatomic distances in the diatomic molecules of the gaseous halogens are conventionally taken to be twice the single-bond covalent radii of the elements, which are set out in Table 10(b). With the uncertainty about the relative contributions of features such as electronic repulsion and d-orbital involvement in the bonding of these molecules, it is by no means clear that this step is sound, but no alternative is available at the present time. From the distances of closest approach of non-bonded halogen atoms, van der Waals' radii have been assigned by Pauling[119]; understandably the van der Waals' radius is much larger than the covalent radius of the halogen atom, being comparable with the ionic radius of the corresponding halide ion. However, van der Waals' radii depend not only on the strength of the attractive forces holding the molecular aggregates together in the crystal, but also on the orientation relative to the covalent bond or bonds formed by the atom. Accordingly undue weight must not be given to the absolute magnitudes of the van der Waals' radii, which, even more than covalent and ionic radii, are the outcome of a highly simplified idea.

The assessment of electronegativity or "the power of an atom in a molecule to attract electrons to itself"[119] remains controversial principally because (i) no element has a unique electronegativity which remains constant throughout the whole range of its compounds,

117 G. L. Caldow and C. A. Coulson, *Trans. Faraday Soc.* **58** (1962) 633.
118 A. Fahlman, R. Carlsson and K. Siegbahn, *Arkiv. Kemi*, **25** (1966) 301.
119 L. Pauling, *The Nature of the Chemical Bond*, 3rd edn., Cornell University Press, Ithaca (1960).

and (ii) the effects of electronegativity cannot be completely extricated from those due to other bonding features. In view of these severe limitations, the electronegativity values given in Table 10(b) are based only on thermochemical data (the Mulliken and Pauling scales) or on estimates of effective nuclear charge and covalent radius (the Allred–Rochow formulation)[120]; no reference has been made to the results of empirical or semi-empirical correlations of electronegativity with dipole moment, chemical shift, nuclear quadrupole coupling constant or vibrational properties of molecules. The electronegativity values are probably meaningful to no better than $\pm 0{\cdot}1$ unit, serving only as rough guides, perhaps as the median numbers in a range for each element. By means of Pauling's expression

$$B(\text{A–X}) = \sqrt{[B(\text{A–A})\cdot B(\text{X–X})]} + 23(\chi_A - \chi_X)^2$$

the energy of a bond between an element A and halogen X, B(A–X), may be estimated roughly in relation to the electronegativities χ_A and χ_X and to the energies of the homonuclear units A–A and X–X. Such estimates, the reliability of which has been tested for some forty halogen-containing bonds[121], are useful as a guide, particularly when experimental data are lacking.

The dipole polarizability, α, gives a measure of the susceptibility to deformation of the electronic charge cloud of an atom under the influence of an externally applied electric field. For a free halogen atom the quantity is amenable neither to experimental determination nor, with the present quality of knowledge about wave functions, to precise calculation[122]. For atoms in chemical combination, however, the molar refractivity of the system provides a means of estimating atomic polarizabilities; since halogen atoms are not commonly found in isotropic environments, values such as those given in Table 10(b) must be regarded as averages of the various components of the atomic polarizability tensor. The polarizability increases in unison with the total number of electrons in the series F < Cl < Br < I; it also increases with the single-bond covalent radius r in accordance with the general empirical rule[123] $\alpha \approx 2{\cdot}3r^3$. As might therefore be expected, the polarizability is smaller for the atom than for the corresponding anion[122] and smaller for a halogen than for preceding atoms in a given row of the Periodic Table[124]. The principal importance of polarizability as an atomic property is in relation to intermolecular binding, notably through the medium of dispersion and Debye interactions, conspicuous, for example, in their influence on the melting and boiling points of molecular halogen compounds, which normally vary thus: $AF_n < ACl_n < ABr_n < AI_n$[43].

Chemical Properties of Halogen Atoms[125]

Although, directly or indirectly, a considerable body of information has been accumulated, no systematic account of the chemical behaviour of the halogen atoms, as distinct from the molecules, has so far been given. In part this reflects the rather heterogeneous nature of experimental enquiries into reactions of the atomic halogens, which have been explored mainly with an eye to their mechanistic interest. Being produced under much less

[120] F. A. Cotton and G. Wilkinson, *Advanced Inorganic Chemistry*, 2nd edn., Interscience (1966).

[121] D. A. Johnson, *Some Thermodynamic Aspects of Inorganic Chemistry*, pp. 158–160 and 167–169, Cambridge (1968).

[122] A. Dalgarno, *Adv. Phys.* **11** (1962) 310.

[123] M. Atoji, *J. Chem. Phys.* **25** (1956) 174.

[124] R. T. Sanderson, *Inorganic Chemistry*, p. 54. Reinhold, New York (1967).

[125] G. C. Fettis and J. H. Knox, *Progress in Reaction Kinetics*, Vol. 2 (ed. G. Porter), p. 1, Pergamon, Oxford (1964); J. G. Calvert and J. N. Pitts, *Photochemistry*, p. 184, Wiley, New York (1966); D. M. Golden and S. W. Benson, *Chem. Rev.* **69** (1969) 125.

forcing conditions than are demanded by atomic hydrogen, nitrogen or oxygen, halogen atoms present fewer technical problems to the study of their reactions; on the other hand, the enhancement in reactivity with respect to the corresponding molecules is less pronounced, and the range of reactions open to the atom but not to the molecule correspondingly more restricted than in the case of hydrogen, nitrogen or oxygen. As already noted, the free atoms are believed to be essential intermediates in many reactions of the halogens and their compounds. Examples of such reactions are given on pp. 1146–7. Inasmuch as the fundamental dissociation reaction $X_2 \rightleftharpoons 2X$ appears to be the precursor of many reactions of the halogen molecules, the reactivity of the atoms is closely associated with the apparent reactivity of the molecules. Thus the reactions characteristic of the molecules—that is, oxidation-reduction, addition and substitution—all have their counterparts in the reactions of the atoms. The thermodynamic properties of the free halogen atoms inevitably make for more favourable free energy and enthalpy balances than in reactions involving the corresponding molecules, while the near-zero activation energies characteristic of many reactions of the atoms contrast strikingly with the substantial activation energies which are the general rule for reactions of the molecules (dissociation of which is commonly implicated). Thus, numerous reactions thermodynamically unattractive to the molecular halogens are feasible for the free atoms, though the products may be short-lived under normal conditions; many other reactions, thermodynamically feasible but kinetically slow for the molecules, proceed very rapidly with the free atoms. Early reports of so-called "active chlorine", produced by the action of an electrical discharge or by irradiation with ultraviolet light, refer to its abnormal chemical reactivity in its direct combination with ozone to produce chlorine monoxide and with sulphur and tellurium to form chlorides[32]. Abnormal reactivity is also illustrated by the report[76] that iodine atoms, produced photochemically, will unite in chloroform solution with olefins at temperatures as low as $-55°C$.

Addition

A primary reaction of the atomic halogens is the recombination $2X \rightarrow X_2$, as previously noted (see p. 1145), a relatively slow process in the gas phase, though highly susceptible to surface-catalysis[126]. Other examples of addition reactions of halogen atoms are:

$$X + \text{>C=C<} \xrightarrow[\text{solution}]{\text{Gas phase or}} \text{>C(X)–C<}\cdot \qquad (1)^{127}$$

$$X\cdot + Y^- \xrightarrow[\text{solution}]{\text{Crystal or}} \cdot XY^- \quad (X,Y = \text{same or different halogen}) \qquad (2)^{94,128}$$

$$X\cdot + YZ \xrightarrow{\text{Gas phase}} \cdot XYZ \quad (X,Y,Z = \text{same or different halogen}) \qquad (3)^{32,129}$$

$$I\cdot + D \xrightarrow{\text{Solution}} \cdot I,D \quad (D = \text{aromatic hydrocarbon, amine, alcohol or alkyl halide}) \qquad (4)^{130,131}$$

$$Cl\cdot + O_2 \xrightarrow{\text{Gas phase}} \cdot ClOO \qquad (5)^{132}$$

$$Cl\cdot + NO \xrightarrow{\text{Gas phase}} NOCl \qquad (6)^{133}$$

$$Cl\cdot + CO \xrightarrow{\text{Gas phase}} \cdot ClCO \qquad (7)^{134}$$

$$Cl\cdot + Hg \xrightarrow{\text{Gas phase}} \cdot HgCl \qquad (8)^{135}$$

Halogenation of many unsaturated organic compounds commonly proceeds by a free radical mechanism[61]. Photochemical chlorination and bromination involve, for example, an atomic chain process such as

$$Cl\cdot + C_2H_4 \rightarrow C_2H_4Cl\cdot \qquad C_2H_4Cl\cdot + Cl_2 \rightarrow C_2H_4Cl_2 + Cl\cdot$$

Similar chains are set up, in the presence of peroxide catalysts, in the reaction of sulphuryl chloride or of hydrogen bromide with unsaturated compounds, e.g.

$$\text{(a) } R\text{–}CO\text{–}O\cdot + H\text{–}Br \rightarrow R\text{–}CO\text{–}OH + Br\cdot$$
$$\text{(b) } Br\cdot + Me\text{–}CH{=}CH_2 \rightarrow Me\text{–}\dot{C}H\text{–}CH_2Br$$
$$\text{(c) } Me\text{–}\dot{C}H\text{–}CH_2Br + H\text{–}Br \rightarrow Me\text{–}CH_2\text{–}CH_2Br + Br\cdot$$

As a result of the reversible addition of halogen atoms, isomerization of olefinic compounds may also be induced; thus, dimethyl maleate is transformed to dimethyl fumarate in the presence of bromine atoms[136]. The action of molecular oxygen as a powerful inhibitor of many reactions of atomic chlorine—photochemical addition no less than chain reactions with hydrogen, hydrocarbons or carbon monoxide—depends on the scavenging action of reactions such as (5).

Addition of a halogen atom X to a halide ion Y^- affords the paramagnetic molecular anion XY^-, which can be identified by its esr, optical or Raman spectrum[94,128]. In this way, the formation of the XY^- anion has been established as a result of γ-irradiation, pulse radiolysis or flash photolysis of aqueous solutions or of low-temperature glasses containing halide ions; the so-called "V-centres" produced by the irradiation of crystalline ionic halides are similarly attributed to species of the type XY^-. Under the appropriate conditions, γ-irradiation of aqueous halide systems has also been found to yield anions identified by their esr spectra as XOH^- (X = Cl, Br or I)[137].

The corresponding addition to a diatomic halogen or interhalogen molecule gives a triatomic radical of the type XYZ (X, Y, Z = the same or different halogen atoms). The formation of such species is supported indirectly by quantum-mechanical calculations, and

126 I. M. Campbell and B. A. Thrush, *Ann. Rep. Chem. Soc.* **62** (1965) 37; S. W. Benson and W. B. DeMore, *Ann. Rev. Phys. Chem.* **16** (1965) 399; J. A. Kerr, *Ann. Rep. Chem. Soc.* **64A** (1967) 75, 121; M. A. A. Clyne, *ibid.* **65A** (1968) 168; J. K. K. Ip and G. Burns, *J. Chem. Phys.* **51** (1969) 3414.

127 J. I. G. Cadogan, *Royal Inst. of Chem. Lecture Series*, No. 6 (1961); J. A. Franklin, G. Huybrechts and C. Cillien, *Trans. Faraday Soc.* **65** (1969) 2094.

128 M. C. R. Symons and W. T. Doyle, *Quart. Rev. Chem. Soc.* **14** (1960) 62; M. Anbar and J. K. Thomas, *J. Phys. Chem.* **68** (1964) 3829; H. C. Sutton, G. E. Adams, J. W. Boag and B. D. Michael, *Pulse Radiolysis* (ed. M. Ebert, J. P. Keene, A. J. Swallow and J. H. Baxendale), p. 61, Academic Press, London (1965); B. Čerček, M. Ebert, C. W. Gilbert and A. J. Swallow, *ibid.* p. 83; J. K. Thomas, *Trans. Faraday Soc.* **61** (1965) 702; R. C. Catton and M. C. R. Symons, *J. Chem. Soc.* (*A*) (1969) 446 and references cited therein; M. Hass and D. L. Griscom, *J. Chem. Phys.* **51** (1969) 5185; J. H. Baxendale and P. L. T. Bevan, *J. Chem. Soc.* (*A*) (1969) 2240.

129 L. Y. Nelson and G. C. Pimentel, *J. Chem. Phys.* **47** (1967) 3671; Y. T. Lee, P. R. LeBreton, J. D. McDonald and D. R. Herschbach, *J. Chem. Phys.* **51** (1969) 455.

130 T. A. Gover and G. Porter, *Proc. Roy. Soc.* **A262** (1961) 476; R. L. Strong, *J. Phys. Chem.* **66** (1962) 2423.

131 R. L. Strong and J. Perano, *J. Amer. Chem. Soc.* **89** (1967) 2535; A. M. Halpern and K. Weiss, *J. Phys. Chem.* **72** (1968) 3863.

132 E. D. Morris, jun., and H. S. Johnston, *J. Amer. Chem. Soc.* **90** (1968) 1918.

133 M. A. A. Clyne and D. H. Stedman, *Trans. Faraday Soc.* **64** (1968) 2698.

134 T. C. Clark, M. A. A. Clyne and D. H. Stedman, *Trans. Faraday Soc.* **62** (1966) 3354.

135 D. G. Horne, R. Gosavi and O. P. Strausz, *J. Chem. Phys.* **48** (1968) 4758.

136 C. Walling, *Free Radicals in Solution*, p. 302. Wiley (1957).

137 R. C. Catton and M. C. R. Symons, *J. Chem. Soc.* (*A*) (1969) 446; I. Marov and M. C. R. Symons, *ibid.* (1971) 201.

some authors have concluded on the basis of kinetic studies that Cl_3 plays a significant part in the recombination of chlorine atoms[32,33,138], though the most recent investigations do not favour this mechanism[133]. More direct evidence of the triatomic radicals comes[129] from molecular-beam studies of atom-recombination reactions and from matrix-isolation, whereby the radical Cl_3, identified by its infrared spectrum, has been trapped in the condensate formed at 20°K following the action of a microwave discharge on a gaseous mixture of chlorine and krypton. $\Delta H_{298}°$ for the reaction $Cl\cdot + Cl_2 \rightarrow \cdot Cl_3$ has been estimated to be *ca.* 4 kcal[139].

Flash photolysis has been used[130,131] to characterize charge-transfer complexes of the halogen atoms and to show, *inter alia*, that the stability constant of the I_{atom}–*o*-xylene complex in solution is larger than that of the I_2–*o*-xylene complex.

Atom Transfer

Atom-transfer reactions involving halogen atoms include

$$X\cdot + H_2 \rightarrow HX + H\cdot \qquad (1)^{32,140}$$

and

$$X\cdot + RH \rightarrow HX + R\cdot \qquad (R = \text{organic group}) \quad (2)^{45,141}$$

which represent the propagation stages of the homolytic reactions between the elementary halogens and either hydrogen or organic compounds RH. The energetics of such reactions are compared in Fig. 10, which illustrates the increasingly endothermic character of

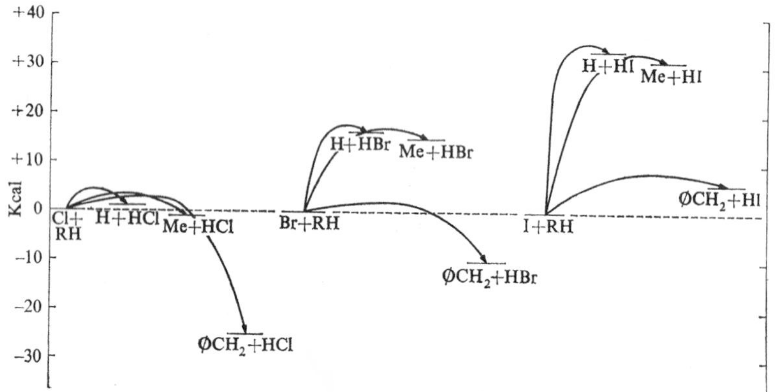

FIG. 10. Energetics of hydrogen atom-transfer reactions of the halogen atoms.

atom-transfer in the series Cl, Br, I, a variation which reflects the relative bond strengths of the hydrides HCl, HBr, HI. Thus, in contrast with the reaction chains with 10^3–10^6 steps exhibited by vapour-phase chlorination reactions, bromination and iodination are characterized by relatively short kinetic chains or by simple radical reactions at all but the highest

[138] E. Hutton and M. Wright, *Trans. Faraday Soc.* **61** (1965) 78.

[139] V. I. Vedeneyev, L. V. Gurvich, V. N. Kondrat'yev, V. A. Medvedev and Ye. L. Frankevich, *Bond Energies, Ionization Potentials and Electron Affinities*, pp. 77, 129. Edward Arnold, London (1966).

[140] M. A. A. Clyne and D. H. Stedman, *Trans. Faraday Soc.* **62** (1966) 2164; A. A. Westenberg and N. de Haas, *J. Chem. Phys.* **48** (1968) 4405; S. W. Benson, F. R. Cruickshank and R. Shaw, *Internat. J. Chem. Kinetics*, **1** (1969) 29.

[141] S. W. Benson and W. B. DeMore, *Ann. Rev. Phys. Chem.* **16** (1965) 405; J. A. Kerr, *Ann. Rep. Chem. Soc.* **64A** (1967) 88, 90; P. Goldfinger, *Acta Chim. Acad. Sci. Hung.* **18** (1959) 17; P. Goldfinger, M. Jeunehomme and G. Martens, *J. Chem. Phys.* **29** (1958) 456.

temperatures. Moreover, compared with chlorine, bromine and iodine atoms are much more selective in their action, and in most cases the reverse action, viz.

$$R\cdot + HX \rightarrow RH + X\cdot \quad (R = H \text{ or organic group})$$

plays a significant role. Intensive studies suggest that atomic chlorine combines with carbon monoxide according to the following chain mechanism[32,61]:

$$\text{(i) } Cl\cdot + CO + M \rightarrow \cdot ClCO + M^* \quad (M \text{ is probably } Cl_2)$$

$$\text{(ii) } \cdot ClCO + Cl_2 \rightarrow OCCl_2 + Cl\cdot$$

The reaction is slow since (a) the initial step (i) requires a three-body collision, and (b) the intermediate radical tends to dissociate:

$$\text{(iii) } \cdot ClCO \rightarrow CO + Cl\cdot$$

on the one hand and to be decomposed by the atom-transfer reaction

$$\text{(iv) } Cl\cdot + \cdot ClCO \rightarrow Cl_2 + CO$$

on the other. Other examples of atom-transfer reactions are:

$\cdot I^* + RI \rightarrow I\cdot + RI^*$ (R = organic group)[142]

$X\cdot + YZ \rightarrow XY + Z\cdot$ (X = Cl, YZ = Br_2, ClBr, ClI; X = Br, YZ = Cl_2, BrCl;[82] X = $I(^2P_{1/2})$, YZ = Cl_2, Br_2, ICl, IBr[143])

$Cl\cdot + NOCl \rightarrow \cdot NO + Cl_2$[71]

$Br\cdot + \cdot ClO_2 \rightarrow \cdot BrO + \cdot ClO$[32]

$Cl\cdot + ClOO\cdot \rightarrow Cl_2 + O_2$[144]

$Cl\cdot + NCl_3 \rightarrow \cdot NCl_2 + Cl_2$[145]

$Cl\cdot + ClN_3 \rightarrow \cdot N_3 + Cl_2$[145]

$Cl\cdot + N_2H_4 \rightarrow \cdot N_2H_3 + HCl$[146]

Apart from the more conventional studies of atoms in their electronic ground states, the behaviour of electronically excited halogen ($^2P_{1/2}$) atoms has also been examined *via* kinetic flash spectroscopy[143]. It has thus been found that, in the presence of Cl_2, Br_2, ICl and IBr, the decay of excited iodine atoms is dominated by reactions with these molecules, whereas, in the presence of CH_3I, quenching of the $I(^2P_{1/2})$ is favoured. It has also been found that hydrogen atom abstraction from paraffins is much more rapid for $I(^2P_{1/2})$ than for $I(^2P_{3/2})$ atoms. The feasibility of such reactions, which are endothermic for the $^2P_{3/2}$ state of iodine, reflects the fact that the $^2P_{1/2} \rightarrow {}^2P_{3/2}$ transition is forbidden, so that the $^2P_{1/2}$ halogen atoms enjoy relatively long radiative lifetimes: F 830 sec, Cl 83 sec, Br 1·1 sec, I 0·13 sec. Provided that a population inversion of the $^2P_{1/2}$ and $^2P_{3/2}$ states is achieved in some primary photolytic step, the lifetimes of the $^2P_{1/2}$ halogen atoms are such as to sustain laser action[143]. Thus, stimulated emission has been observed during flash photolysis of certain alkyl or perfluoroalkyl iodides, though not of molecular iodine:

$$RI \xrightarrow[\text{u.v.}]{h\nu} R + I(^2P_{1/2})$$

$$I(^2P_{1/2}) + h\nu(7603 \text{ cm}^{-1}) \longrightarrow I(^2P_{3/2}) + 2h\nu(7603 \text{ cm}^{-1})$$

Stimulated emission implicating $Br(^2P_{1/2})$ has also been observed following the flash photolysis of IBr. Investigations of such laser systems are potentially important *inter alia* for the light they may shed on primary photochemical processes and on energy-transfer mechanisms.

142 A. F. Trotman-Dickenson, *Free Radicals*, p. 91. Methuen, London (1959).

143 R. J. Donovan, F. G. M. Hathorn and D. Husain, *Trans. Faraday Soc.* **64** (1968) 1228; *J. Chem. Phys.* **49** (1968) 953; D. Husain and R. J. Donovan, *Advances in Photochemistry*, Vol. 8 (ed. J. N. Pitts, jun., G. S. Hammond and W. A. Noyes, jun.), p. 1, Wiley-Interscience (1971).

144 M. A. A. Clyne, *Ann. Rep. Chem. Soc.* **65A** (1968) 183.

145 T. C. Clark and M. A. A. Clyne, *Trans. Faraday Soc.* **65** (1969) 2994.

146 S. N. Foner and R. L. Hudson, *J. Chem. Phys.* **49** (1968) 3724.

Hot-Atom Reactions[32,45,79–81,147–150]

Accelerated or electronically excited halogen atoms or ions, produced for example by an (n,γ) process, differ notably in their reactions from atoms with normal thermal energies, produced typically by photolysis. As distinct from thermal processes, so-called "hot" reactions are characterized by the following general features: (i) products are thereby realized which are not formed by thermal reactions; (ii) the reactions are temperature-independent; (iii) inhibition may be brought about by chemically inert agents which dispel the kinetic, vibrational or electronic energy of the excited species before it undergoes reactive collisions; (iv) the reactions are not affected by low concentrations of reactive species which serve as scavengers for conventional free radicals. The first of these features is prominently displayed in the reactions of halogen atoms with hydrocarbons or organic halides: whereas halogen atoms with normal energies react primarily to abstract hydrogen atoms thereby forming hydrogen halides, "hot" halogen atoms, by virtue of their relatively high kinetic energy, electronic excitation or charge, are able to displace hydrogen atoms, organic radicals or other halogens: e.g.

$$RCH_2Br + Br^* \longrightarrow \begin{cases} RCHBr\cdot + HBr^* \text{ or } RCH_2\cdot + BrBr^* \\ RCH_2Br^* + Br\cdot \\ RCHBrBr^* + H\cdot \\ RBr^* + BrCH_2\cdot \text{ or } CH_2BrBr^* + R\cdot \end{cases}$$

In such a system, radicals may also result from the displacement of two atoms from a molecule: overall it is thus possible to account, at least qualitatively, for the complicated pattern of products formed in the reactions of "hot" bromine atoms with organic bromides[151].

The principal technique for generating and studying "hot" atoms is nuclear recoil, certain transformations, e.g. $^{35}Cl \xrightarrow{n,\gamma} {}^{36}Cl$, $^{80m}Br \longrightarrow {}^{80}Br$ or $^{127}I \xrightarrow{n,\gamma} {}^{128}I$, being capable of producing species of very high energy, which, under properly controlled conditions, become available as "hot" atoms. Although nuclear reactions are normally rare phenomena and nuclear recoil accordingly yields a very small number of "hot" atoms, reactions resulting in the chemical combination of these atoms can readily be traced by means of their radioactivity. A substance, which on nuclear transformation yields the desired "hot" atom, is mixed with the compound with which the "hot" atom is to react; such a mixture must be homogeneous on the scale of the recoil range of the "hot" atom. The sample is exposed to the appropriate nuclear radiation for a period at once long enough to produce the required number of activated atoms, but sufficiently short to avoid appreciable macroscopic decomposition. The products incorporating the "hot" atoms are separated from one another and from the reactant, e.g. by chromatographic methods.

The species initially produced by nuclear transformations usually take the form of very high energy ions. These lose energy in successive interactions with the medium and should become neutralized in the process. Eventually the atoms drop into the range of chemical energies (< 100 eV) and in further collisions may react and combine by a "hot" reaction. However, it is not always clear whether the ion first formed becomes neutralized by charge-exchange processes before it loses enough energy to combine chemically. Whereas there is no evidence to implicate ions in the interaction with methane of ^{80}Br produced by the (n,γ)

147 J. E. Willard, *Ann. Rev. Phys. Chem.* **6** (1955) 141.
148 J. E. Willard, *Nucleonics*, **19**, No. 10 (1961) 61.
149 *Chemical Effects of Nuclear Transformations*, Proceedings of a Symposium, Prague, I.A.E.A. (1961).
150 I. G. Campbell, *Adv. Inorg. Chem. Radiochem.* **5** (1963) 135.
151 M. Milman, *Radiochim. Acta*, **1** (1962) 15.

process from ^{79}Br, ^{128}I from the neutron-irradiation of ^{127}I combines with methane partly as a "hot" atom and partly as a ground-state or excited ion[152].

Many of the current ideas on the mechanism of "hot" atom reactions in liquids and solids derive from the discovery of the Szilard–Chalmers effect in liquid ethyl iodide[101] and from subsequent investigations of halogen interactions in the condensed phase. The very extensive literature relating to such interactions is mainly concerned with attempts to sort out the various types of process that affect the fate of the recoil species. Since it has usually not been possible to define clearly and to separate experimentally these events, comparatively little definite information has so far emerged concerning the detailed mechanism of actual "hot" reactions in the condensed phase. Thus, although considerable effort has been expended on the study of hot-atom reactions in solid halogen-containing compounds, e.g. salts of oxy-anions of the halogens[79,153], most of the interest has focused on the fate of the halogen following activation by radiative neutron-capture. Subsequent annealing, whether induced by thermal or by radiative means, is an important factor in the outcome of such experiments, the effect being in general to restore a part of the initially separable recoil atoms to the form of the parent compound. As with the annealing of radiation damage in solids, this process works upon the defects produced in a matrix of otherwise normal crystal as a result of a sudden nuclear impulse. An informative way of examining the defects consists of following the change in the chemical state of the halogen atom responsible for the defects as a function of the time and temperature of annealing; such experiments supplement studies of radiation damage, which have usually depended on changes in the physical properties of the irradiated solids.

Many features of the reactions of "hot" halogens are quite similar to those of "hot" hydrogen, which have been interpreted by the so-called "impact model"[80,81]. According to this model, the principal mechanism whereby a "hot" atom combines in the gas phase embodies a single-step abstraction reaction. The recoil atom is assumed to enter into stable combination only over a certain energy range; above the upper limit, collisions are too energetic to permit formation of stable products, while the lower limit is determined by the minimum energy required to bring about reaction. By contrast, for all but grazing collisions, energy considerations are not of prime importance in determining the choice of reaction path from the several that are energetically accessible. Instead, the actual course of the reaction is determined (i) by the impact and steric parameters, (ii) by inertial restrictions in relaxation processes required for certain reactions, and (iii) by inductive effects, as yet poorly understood. Apart from differences in steric properties, the reactions of "hot" halogens differ from those of "hot" hydrogen in that, for a given energy, the larger size of the halogen atom makes not only for a lower velocity but also for longer-lived and less localized collisions. There will therefore be more opportunity for the kinetic energy of the incident halogen to spread through the molecule with which it collides, and because the available energy is thus diffused over many bonds, it is less likely that any given bond in the impact area will be broken. Correspondingly there is a greater probability of rupturing two bonds to form a radical incorporating the "hot" atom[151].

The radioisotope produced by thermal neutron irradiation of a compound in dilute solution or in the gas phase is usually obtained in a chemical form other than that of the capturing molecule. However, when pure liquids or solids are irradiated, an appreciable

152 E. P. Rack and A. A. Gordus, *J. Phys. Chem.* **65** (1961) 944; *J. Chem. Phys.* **34** (1961) 1855.

153 See, for example, N. Saito, F. Ambe and H. Sano, *Radiochim. Acta*, **7** (1967) 131; G. E. Boyd and Q. V. Larson, *J. Amer. Chem. Soc.* **90** (1968) 5092.

fraction of the total activity is found in the parent compound, presumably as the result of a secondary re-entry process. To explain this phenomenon, Libby made the general postulate that interactions of "hot" atoms at high energies with individual atoms in a molecule (billiard-ball mechanism), or at lower energies with the whole molecule (epithermal mechanism), could create a radical and leave the "hot" atom with insufficient energy to escape from the cage formed by its chemical environment[154,155]. After further collisional deactivation within the cage, the "hot" atom then combines with the radical previously formed.

The following systems are representative of the reactions of "hot" halogen atoms (X^*) which have been investigated in some detail: hydrocarbons $+X^*$[45,149–151]; $RX+X^*$ (R = organic group)[45,149–151]; olefins $+X^*$[156].

2.6. PHYSICAL PROPERTIES OF THE MOLECULAR HALOGENS

General Characteristics: Thermodynamic Properties

The principal physical properties of the molecular halogens are summarized in Table 11. At ordinary temperatures the elements range from a greenish-yellow gas in chlorine, through a dense mobile liquid with a dark red colour in bromine, to a black, crystalline solid with a slight metallic lustre in iodine. With increasing atomic number, and hence polarizability, there is a strengthening of the van der Waals' forces between the diatomic molecules. The molecules persist throughout the solid, liquid and gaseous phases, though, as discussed below, intermolecular binding in the solid state becomes increasingly important in the series $Cl_2 < Br_2 < I_2$; perhaps the most dramatic evidence of this trend is supplied by the observation that, at high pressures, iodine crystals assume the electrical characteristics of a metal[157]. In their diatomic nature the heavier halogens differ notably from the elements of Groups V and VI, only the lightest members of which are diatomic under conventional conditions. A reflection of this difference is found in the melting and boiling points of chlorine, bromine and iodine, which are much lower than those of the previous members of their respective periods, that is, sulphur, selenium and tellurium.

Table 11 summarizes, *inter alia*, recommended values of the boiling points, melting points, vapour pressures and heat capacities of the elementary halogens. Selected thermodynamic functions, based on the most recent compilations[67,158–159], are also presented for each element in a given phase and for the transitions from one phase to another. Thermodynamic functions have been evaluated for the individual halogens over a wide range of temperatures by applying statistical mechanics to careful analyses of the rotational and vibrational terms of the molecular spectra, with corrections for rotational stretching, vibrational anharmonicity and rotational–vibrational interaction[160]. Where comparisons have been made, the values calculated in this way are in very good agreement with those derived *via* the third-law method from measurements of heat capacity; there is thus no evidence that the solids retain residual entropy at limiting low temperatures. The results of the table confirm the similarity of the standard entropies of the halogens in the vapour phase.

[154] W. F. Libby, *J. Amer. Chem. Soc.* **62** (1940) 1930; *ibid.* **69** (1947) 2523.

[155] M. S. Fox and W. F. Libby, *J. Chem. Phys.* **20** (1952) 487.

[156] See for example C. M. Wai and F. S. Rowland, *J. Amer. Chem. Soc.* **91** (1969) 1053.

[157] A. S. Balchan and H. G. Drickamer, *J. Chem. Phys.* **34** (1961) 1948.

[158] *National Bureau of Standards Circular* 500, U.S. Government Printing Office, Washington (1952).

[159] *National Bureau of Standards Technical Note* 270–3, U.S. Government Printing Office, Washington (1968); *Codata Bulletin*, International Council of Scientific Unions, Committee on Data for Science and Technology (1970).

[160] W. H. Evans, T. R. Munson and D. D. Wagman, *J. Res. Nat. Bur. Stand.* **55** (1955) 147.

TABLE 11. PHYSICAL PROPERTIES OF MOLECULAR CHLORINE, BROMINE AND IODINE

Property	Cl_2	Br_2	I_2
Molecular weight[a]	70·906	159·808	253·8090
Gaseous molecules			
Electronic ground state, configuration $\sigma_g^2\pi_u^4\pi_g^4$	$^1\Sigma_g^+$	$^1\Sigma_g^+$	$^1\Sigma_g^+$
Excitation energies to lowest-lying excited states, T_e(cm^{-1})	$\sigma_g^2\pi_u^4\pi_g^3\sigma_u^1 \leftarrow \sigma_g^2\pi_u^4\pi_g^4$	$\sigma_g^2\pi_u^4\pi_g^3\sigma_u^1 \leftarrow \sigma_g^2\pi_u^4\pi_g^4$	$\sigma_g^2\pi_u^4\pi_g^3\sigma_u^1 \leftarrow \sigma_g^2\pi_u^4\pi_g^4$
$^3\Pi_{1u} \leftarrow {}^1\Sigma_g^+$	~23,500 (continuum)[b]	13,814[c]	11,888[c]
$^3\Pi_{0^+u} \leftarrow {}^1\Sigma_g^+$	17,672·6[c]	15,902·51[d]	15,769·48[d]
Ground state properties			
Internuclear distance, r_e(Å)	1·9881[e]	2·2809[f,g]	2·666[h]
Vibrational frequency, ω_e(cm^{-1})	$^{35}Cl_2$ 559·72[e]	$^{79}Br^{81}Br$ 324·24[f,g]	$^{127}I_2$ 214·52[h]
Anharmonic vibrational constant, $\omega_e x_e$(cm^{-1})	$^{35}Cl_2$ 2·67$_5$[e]	$^{79}Br^{81}Br$ 1·172[f,g]	$^{127}I_2$ 0·60738[h]
Force constant, k_e(mdyne/Å)	3·225[e]	2·475[f,g]	1·720[h]
Dissociation energy:	*cm^{-1}* · *kcal* · *eV*	*cm^{-1}* · *kcal* · *eV*	*cm^{-1}* · *kcal* · *eV*
(i) Spectroscopically determined, D_0°[i]	19,997·2$_5$ ±0·3 · 57·175 · 2·4793 $^{35}Cl_2$	15,894·5 ±0·4 · 45·444 · 1·9706 $^{79}Br_2$ 15,896·6 ±0·5 · 45·451 · 1·9709 $^{81}Br_2$	12,440·9 ±1·2 · 35·570 · 1·5424 $^{127}I_2$
(ii) Recommended thermodynamic value, D°_{298}[j]	— · 57·978 · 2·5141	— · 46·072 · 1·9978	— · 36·115 · 1·5661
Thermodynamic properties			
ΔH_f° at 298°K (kcal)[j]	0	7·388	14·919
ΔG_f° at 298°K (kcal)[j]	0	0·751	4·627
S° at 298°K (cal/deg mol)[j]	53·290	58·640	62·277

Table 11 (*cont.*)

Property	Cl_2	Br_2	I_2
Adiabatic ionization potential, $X_2^+(^2\Pi_{3/2,g}) \leftarrow X_2(^1\Sigma_g^+)$[k]	*kcal* 265·4; *eV* 11·51	*kcal* 242·4; *eV* 10·51	*kcal* 212·6; *eV* 9·22
Properties of the condensed phases	°*K* / °*C*	°*K* / °*C*	°*K* / °*C*
Boiling point	239·10[l] / −34·05[l]	332·62[m] / +59·47[m] 331·93[b] / +58·78[b]	458·39[m] / +185·24[m]
Melting point	172·16[l] / −101·00[l]	265·90[m] / −7·25[m]	386·75[m] / +113·60[m]
Heat of vaporization at boiling point, ΔH_{vap}(kcal/mol)	4·878[n]	7·064[m]	$10{\cdot}025_5$[m]
Entropy of vaporization at boiling point (Trouton constant) (cal/deg mol)	20·40[n]	21·24[m]	21·87[m]
Heat of fusion (kcal/mol)	1·531[n]	2·527[m]	3·709[m]
Heat of sublimation at melting point (kcal/mol)	6·965[n]	10·220[m]	14·477[m]
Entropy of reference state at 298°K, $S°$(cal/deg mol)	53·290(g)[j]	36·379(l)[j]	27·758(c)[j]
Critical temperature	°*K* 417·16° / °*C* 144·0°	°*K* 588[p] / °*C* 315[p]	°*K* 819[p] / °*C* 546[p]
Critical pressure (atm)	76·1°	102[n]	~116[q]
Critical density (g/cm³)	0·573°	1·26[p]	1·64[p]
Vapour pressure, *p* (mm Hg)			
Solid	$\log p = -(2890{\cdot}8/T)+0{\cdot}09914T - 58{\cdot}836 \log T + 132{\cdot}26593$[r]	$\log p = -(11310{\cdot}00/T)+0{\cdot}17483T - 184{\cdot}175 \log T + 444{\cdot}26938$[r]	$\log p = -(3594{\cdot}03/T)+0{\cdot}0004434T - 2{\cdot}9759 \log T + 18{\cdot}80572$[r]
Liquid	$\log p = -(1414{\cdot}8/T)-0{\cdot}01206T + 1{\cdot}34\times10^{-5}T^2 + 10{\cdot}91635$[b]	$\log p = -(2047{\cdot}75/T)-0{\cdot}006100T + 0{\cdot}9589 \log T + 8{\cdot}65047$[r]	$\log p = -(2970/T)-3{\cdot}52 \log T + 18{\cdot}751$[s]
Crystal Structure	Orthorhombic, space group *Cmca*[t]	Orthorhombic, space group *Cmca*[t]	Orthorhombic, space group *Cmca*[u]
Molecules per unit cell	4	4	4
Unit cell dimensions (Å)	*a* 6·24; *b* 4·48; *c* 8·26 (113°K)	*a* 6·67; *b* 4·48; *c* 8·72 (123°K)	*a* 7·136; *b* 4·686; *c* 9·784 (110°K)
Intramolecular distance, *d*(X−X)(Å)	1·980±0·014	2·27±0·10	2·715±0·006
Intermolecular distances, *d*(X · · · X)(Å)	3·32–3·97	3·31–4·14	3·496–4·412

Twice van der Waals' radius of X(Å)	3·60	3·90	4·30
Ratio $\frac{d(X\cdots X)}{d(X-X)}$ next nearest	1·68	1·46	1·29
Nqr data			
Coupling constant, e^2Qq(MHz)	108·95 (20°K)[v]	765·86 (4°K)[v]	2156 (253°K)[v]
Asymmetry parameter, η	≥0·08[v]	0·20[v]	0·175[v]
Thermal properties			
Heat capacity, C_p(cal/mol deg)			
Solid	0·810–13·27 (14·05–172°K)[b,l]	1·725–14·732 (15–265·9°K)[w]	$C_p = -12{\cdot}1098+0{\cdot}059012T + 6{\cdot}686\times10^5T^{-2}$ (10–330°K)[x]
Liquid	16·03–15·70 (172–239°K)[b,l]	18·579–18·077 (265·9–300°K)[w]	19·281 (387–458°K)[m]
Gas	7·576–9·710 (200–6000°K)[m]	8·616–9·562 (298–6000°K)[m]	8·814–9·746 (298–6000°K)[m]
Thermal conductivity (cal/sec/cm²/°C/cm)			
Solid	—	—	$1{\cdot}065\times10^{-3}$ (297–316°K)[b]
Liquid	$\sim4{\cdot}95\times10^{-4}$ (298°K)[y]	$3{\cdot}04–2{\cdot}52\times10^{-4}$ (283–323°K)[z]	—
Gas	$1{\cdot}32–4{\cdot}59\times10^{-5}$ (200–600°K)[z]	$1{\cdot}03–2{\cdot}72\times10^{-5}$ (273–700°K)[z]	$1{\cdot}20–1{\cdot}52\times10^{-5}$ (500–600°K)[z]
Coefficient of cubical expansion (×10⁴ per °C)			
Solid	1·68 (77·4–158·2°K)[aa]	2·47 (167–250°K)[bb]	2·81 (273–387°K)[cc]
Liquid	13·07–54·48 (182–383°K)[l]	11·0 (273–323°K)[b,cc]	8·54[b]
Gas	36·61–38·33 (273–373°K, p = 0–1 atm)[l]	—	—
Mechanical properties			
Density, d(g/cm³)			
Solid	$d = 2{\cdot}098-3{\cdot}5\times10^{-4}T$ (77·4–158·2°K)[l]	4·17–4·05 (0–123°K)[b] $d = 3{\cdot}924-1{\cdot}062\times10^{-3}t$ (t = −23·5 to −106°C)[bb]	5·15–4·886 (78–333°K)[b,cc]

Table 11 (*cont.*)

Property	Cl_2	Br_2	I_2
Density, d (g/cm³) (*cont.*)			
Liquid	1·6552–0·5672 (203–417°K)[b]	3·187–3·009 (273–325°K)[b,cc,dd]	3·960–3·736 (393–453°K)[cc]
Gas	$d = \frac{0{\cdot}86427p}{T}\left[1+\left(\frac{2515-2{\cdot}991(T+199)}{T^2}\right)p+2\left(\frac{2515-2{\cdot}991(T+199)}{T^2}\right)^2p^2+\ldots\right]$ (288–348°K, p up to 2 atm)[b]	0·005480–0·005038 (361–386°K, p = 1 atm)[b]	0·011320–0·007434 at N.T.P. (723–1274°K, p = 0·04–1·05 atm)[q]
Compressibility ($\times 10^6$ per bar)			
Solid	—	—	10–1·25 (0–200 kbar)[ee]
Liquid	116–83 (0–500 bar, 293°K)[ff] 88·9–47·4 (180–240°K)[o]	62·5–49·0 (0–500 bar, 293°K)[ff]	—
Viscosity, η(centipoise)			
Liquid	$\eta = 0{\cdot}385/(1+0{\cdot}005878t - 0{\cdot}00000392t^2)$ (t = −76·5 to −33·8°C)[b]	$\eta = 1{\cdot}241/(1+0{\cdot}012257t + 2{\cdot}721\times10^{-6}t^2)$ (t = −4·3 to +32°C)[b]	2·268–1·414 (389–458°K)[cc]
Gas	0·01294–0·03209 (289–772°K)[l]	0·01526–0·04292 (293–873°K)[b]	0·01785–0·03604 (379–796°K)[b]
Surface tension of liquid, γ(dyne/cm)	31·2–18·4 (213–293°K)[l]	$\gamma = 45{\cdot}5 - 0{\cdot}182t$ (over liquid range, t in °C)[gg]	36·88–34·04 (398–428°K)[hh]
Electrical, optical and magnetic properties			
Dielectric constant, ϵ			
Solid	—	—	10·3 (296°K)[cc] Single crystal at ~300°K: $\epsilon_a = 6$; $\epsilon_b = 3$; $\epsilon_c = 40$[ii]
Liquid	2·142–1·947 (203–283°K)[l]	3·255–3·080 (273–303°K)[dd]	11·08–12·98 (391–441°K)[b]
Gas at N.T.P.	1·00152[b]	1·0047[b]	<1[b]
Specific conductivity, k (ohm^{-1} cm^{-1})			
Solid	—	2·8–8·1 × 10^{-13} (253–266°K)[gg]	1·71–12·0 × 10^{-7} (298–383°K)[cc] Single crystal at room temperature: 5 × 10^{-12} normal, 1·7 × 10^{-8} parallel } to *bc* plane[v] 0·8–1·7 × 10^{4} (350 kbar)[jj]

Liquid	$<1\times10^{-16}$ [l]	$\log k = -11{\cdot}372+0{\cdot}0128t$ (over liquid range, t in °C)[gg]	50–9×10^{-6} (394–413°K)[b,cc]
Refractive index, n			
Solid	—	—	$n_D = 3{\cdot}34$ (293°K)[kk]
Liquid	$n_D = 1{\cdot}3788$ (298°K)[ll]	$n_D = 1{\cdot}6083$ (293°K)[mm]	$n_D = 1{\cdot}98$[kk]
Gas at N.T.P.	$\lambda = 6708$–4800 Å $n = 1{\cdot}00077563$–$1{\cdot}00079166$[nn]	$\lambda = 6708$–5461 Å $n = 1{\cdot}0011525$–$1{\cdot}0011849$[nn]	$\lambda = 6708$–5000 Å $n = 1{\cdot}00197$–$1{\cdot}00212$, exhibits anomalous dispersion[nn]
Molar refraction (cm³/mol) D-line, various temperatures	11·74[mm]	17·61[mm]	28–32[mm]
Diamagnetic susceptibility, $\chi(10^6$ cgs units/mol)			
Solid	—	~ -60[b]	Single crystal: $\chi_a = -84{\cdot}4$; $\chi_b = -89{\cdot}3$; $\chi_c = -75{\cdot}9$[oo]
Liquid	$-40{\cdot}2$[l]	$-53{\cdot}6$[pp]	-99 to -84 (391–432°K)[kk]
Gas	~ -35[qq]	~ -70[b]	—
Standard potential for the system $\frac{1}{2}X_2+e \rightleftharpoons X^-(aq)$, $E°$(V)	$+1{\cdot}356$[rr]	$+1{\cdot}065$[rr]	$+0{\cdot}535$[rr]

[a] *Table of Atomic Weights*, 1969, IUPAC Commission on Atomic Weights; *Pure Appl. Chem.* **21** (1970) 95.

[b] Supplement to *Mellor's Comprehensive Treatise on Inorganic and Theoretical Chemistry*, Supplement II, Part I, Longmans, London (1956).

[c] G. Herzberg, *Molecular Spectra and Molecular Structure. I. Spectra of Diatomic Molecules*, 2nd edn., pp. 512–541. Van Nostrand (1950); W. G. Richards and R. F. Barrow, *Proc. Chem. Soc.* (1962) 297.

[d] J. I. Steinfeld, J. D. Campbell and N. A. Weiss, *J. Mol. Spectroscopy*, **29** (1969) 204; J. A. Coxon, *ibid.* **37** (1971) 39.

[e] A. E. Douglas, Chr. Kn. Møller and B. P. Stoicheff, *Canad. J. Phys.* **41** (1963) 1174; M. A. A. Clyne and J. A. Coxon, *J. Mol. Spectroscopy*, **33** (1970) 381.

[f] J. A. Horsley and R. F. Barrow, *Trans. Faraday Soc.* **63** (1967) 32.

[g] R. J. LeRoy and G. Burns, *J. Mol. Spectroscopy*, **25** (1968) 77.

[h] D. H. Rank and B. S. Rao, *J. Mol. Spectroscopy*, **13** (1964) 34.

[i] R. J. LeRoy and R. B. Bernstein, *J. Mol. Spectroscopy*, **37** (1971) 109.

[j] *National Bureau of Standards Technical Note* 270–3, January 1968; *Codata Bulletin*, International Council of Scientific Unions, Committee on Data for Science and Technology (1970).

[k] D. C. Frost, C. A. McDowell and D. A. Vroom, *J. Chem. Phys.* **46** (1967) 4255; A. W. Potts and W. C. Price, *Trans. Faraday Soc.* **67** (1971) 1242.

[l] *Gmelins Handbuch der Anorganischen Chemie*, 8 Auflage, "Chlor", Teil A (1968).

[m] *JANAF Thermochemical Tables*, The Dow Chemical Company, Midland, Michigan (1960–8).
[n] D. R. Stull and G. C. Sinke, *Advances in Chemistry Series*, **18** (1956).
[o] T. R. Thomson, H. Eyring and T. Ree, *Proc. Nat. Acad. Sci.* **46** (1960) 336.
[p] J. A. Burriel Lluna, C. B. Cragg and J. S. Rowlinson, *An. Quim.* **64** (1968) 1.
[q] M. L. Perlman and G. K. Rollefson, *J. Chem. Phys.* **9** (1941) 362.
[r] A. N. Nesmeyanov, *Vapor Pressure of the Chemical Elements* (ed. R. Gary), Elsevier (1963).
[s] *Landolt-Börnstein Tables*, II Band, 2 Teil, Bandteil a, Springer-Verlag (1960).
[t] J. Donohue and S. H. Goodman, *Acta Cryst.* **18** (1965) 568.
[u] F. van Bolhuis, P. B. Koster and T. Migchelsen, *Acta Cryst.* **23** (1967) 90.
[v] R. Bersohn, *J. Chem. Phys.* **36** (1962) 3445; E. A. C. Lucken, *Nuclear Quadrupole Coupling Constants*, Academic Press, London and New York (1969); N. Nakamura and H. Chihara, *J. Phys. Soc. Japan*, **22** (1967) 201.
[w] D. L. Hildenbrand, W. R. Kramer, R. A. McDonald and D. R. Stull, *J. Amer. Chem. Soc.* **80** (1958) 4129.
[x] D. A. Shirley and W. F. Giauque, *J. Amer. Chem. Soc.* **81** (1959) 4778.
[y] *The Encyclopedia of the Chemical Elements* (ed. C. A. Hampel), Reinhold, New York (1968).
[z] *Landolt-Börnstein Tables*, II Band, 5 Teil, Bandteil b, Springer-Verlag (1968).
[aa] L. L. Hawes and G. H. Cheesman, *Acta Cryst.* **12** (1959) 477.
[bb] L. L. Hawes, *Acta Cryst.* **12** (1959) 34.
[cc] *Kirk-Othmer's Encyclopedia of Chemical Technology*, 2nd edn., Interscience (1963–9).
[dd] G. Fröhlich and W. Jost, *Chem. Ber.* **86** (1953) 1184.
[ee] R. Grover, A. S. Kusubov and H. D. Stromberg, *J. Chem. Phys.* **47** (1967) 4398.
[ff] T. W. Richards and W. N. Stull, *J. Amer. Chem. Soc.* **26** (1904) 408.
[gg] M. S. Chao and V. A. Stenger, *Talanta*, **11** (1964) 271.
[hh] S. Kim and S. Chang, *Daehan Hwahak Hwoejee*, **9** (1965) 110 (*Chem. Abs.* **64** (1966) 5785c).
[ii] M. Simhony, *J. Phys. Chem. Solids*, **24** (1963) 1297.
[jj] A. S. Balchan and H. G. Drickamer, *J. Chem. Phys.* **34** (1961) 1948.
[kk] *Gmelins Handbuch der Anorganischen Chemie*, 8 Auflage, "Iod" (1933).
[ll] A. W. Francis, *J. Chem. Engineering Data*, **5** (1960) 534.
[mm] P. M. Christopher, *Austral. J. Chem.* **20** (1967) 2299.
[nn] C. Cuthbertson and M. Cuthbertson, *Phil. Trans.* **213** (1913) 1.
[oo] J. De Villepin, *J. chim. Phys.* **59** (1962) 901.
[pp] S. Broersma, *J. Chem. Phys.* **17** (1949) 873.
[qq] P. Pascal, *Nouveau Traité de Chimie Minérale*, Vol. XVI, Masson et Cie, Paris (1960).
[rr] A. G. Sharpe, *Halogen Chemistry* (ed. V. Gutmann), Vol. 1, p. 12, Academic Press, London and New York (1967).

The Gaseous Molecules

The bonding in the diatomic halogen molecules can be described approximately in terms of simple molecular-orbital theory (see Fig. 11) as depending on the balance between six electrons in bonding orbitals ($\sigma_g^2\pi_u^4$) and four in anti-bonding orbitals (π_g^4). This represents a

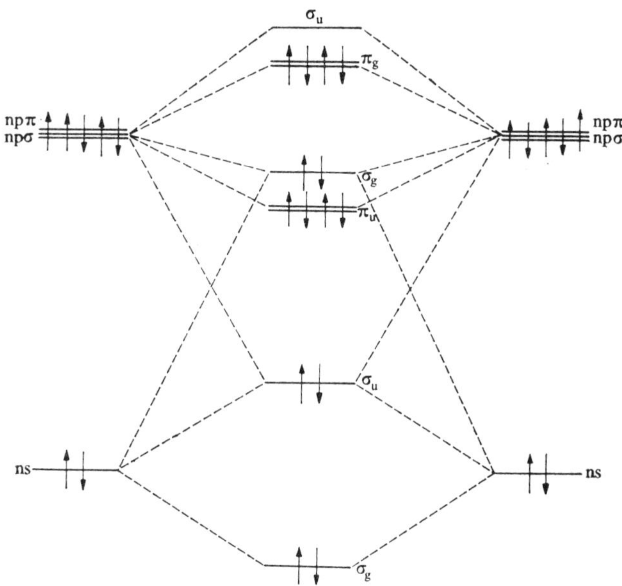

FIG. 11. Schematic energy level diagram showing molecular orbitals for the diatomic halogen molecules.

diamagnetic ground state with a bond order of unity, though a more sophisticated description which takes account of interactions of the *nd*-orbitals would impute at least some $(p\text{-}d)\pi$-character to the X–X bond. The outer π_g and π_u orbitals are mainly atomic, being localized on the halogen atoms.

The electronic spectra of the halogens have attracted much study. The colours of the gaseous systems arise from absorption bands corresponding to the electronic transitions

$$\sigma_g^2\pi_u^4\pi_g^3\sigma_u^1\ (^3\Pi_u \text{ and } ^1\Pi_u) \leftarrow \sigma_g^2\pi_u^4\pi_g^4\ (^1\Sigma_g^+)$$

in which an electron is excited from the anti-bonding π_g to the anti-bonding σ_u orbital. With increase of atomic number there is a decrease in the energy separation of these two orbitals, together with an enhanced probability for the singlet-triplet transition to the lower energy $^3\Pi_u$ state. These features together account for the variations of frequency and intensity exhibited by the visible absorption bands in the series of gaseous molecules Cl_2, Br_2, I_2[32,74,161,162]. However, there is a complication in that the electronic states of Cl_2, Br_2 and

[161] G. Herzberg, *Molecular Spectra and Molecular Structure. I. Spectra of Diatomic Molecules*, 2nd edn., van Nostrand (1950).

[162] R. W. B. Pearse and A. G. Gaydon, *The Identification of Molecular Spectra*, 3rd edn., Chapman and Hall, London (1963).

I_2, unlike those of F_2, approach Hund's coupling case c, wherein the spins and angular momenta of the individual electrons first combine to give separate resultants j_1, j_2, etc., which then yield a final resultant denoted by the quantum number Ω. This situation, corresponding to (j,j) coupling in atoms, renders it impossible to assign a value of the conventional quantum number Λ or of the multiplicity; instead, detailed correlations of Ω with atomic J are necessary.

Chlorine gas shows a strong continuous absorption extending from the blue to about 2500 Å and having a maximum near 3300 Å. With greater path-lengths, there is also seen a weak banded absorption spectrum sharply degraded to the red extending from about 5800 Å to a convergence limit near 4785 Å. Though modified in details of position, intensity and complexity, these characteristics of a continuum in the blue–ultraviolet region and a banded system close to 5000 Å are also found in the absorption spectra of gaseous bromine and iodine. The transition $^3\Pi_{0^+u}(0^+u) \leftarrow {}^1\Sigma_g^+$ is responsible for the system of bands in the green, which shows a well-marked convergence to a continuum at *ca.* 20,880, 19,580 and 20,040 cm^{-1} for Cl_2, Br_2 and I_2 respectively, thresholds corresponding to the energy of the dissociation process

$$X_2 \rightarrow X(^2P_{3/2})+X(^2P_{1/2})$$

Subtraction of the atomic energy of excitation of $X(^2P_{1/2})$ from the energy corresponding to the limit gives the most reliable values at present available for the normal dissociation energy of each of the X_2 molecules; the most recent extrapolation of the convergence limits for Cl_2, Br_2 and I_2 forms the basis of the results in Table 11[163]. The circumstance of absorption in which the observable transitions lead, as the frequency increases, first to bands of higher vibrational quantum numbers and finally to continuous absorption clearly implies the disposition of potential energy curves illustrated in Fig. 12, the shallow minimum

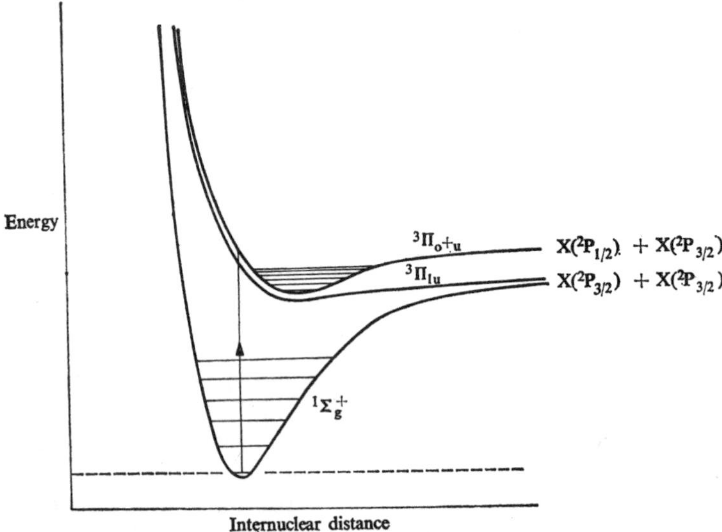

FIG. 12. Schematic form of the potential energy curves of the lowest observed electronic states of the molecules Cl_2, Br_2 and I_2.

163 R. J. LeRoy and R. B. Bernstein, *J. Mol. Spectroscopy*, **37** (1971) 109.

for the upper $^3\Pi_{0^+u}$ state being at an internuclear distance about 0·4 Å greater than that of the ground state.

Detailed vibrational analyses of the banded spectra of the gaseous halogens yield the values given in Tables 11 and 12 for the vibrational frequencies ω_e, the anharmonicity constants $\omega_e x_e$ and $\omega_e y_e$, and the stretching force constants k_e. Where comparisons can be made, good agreement is found between these values and the results derived, for example, from the Raman spectra, as in recent studies of the resonance Raman effect and resonance fluorescence of the gaseous halogens[164]. The vibrational frequencies of liquid chlorine and bromine are not significantly different from those of the gaseous molecules, but the Raman spectra of crystalline chlorine and bromine at low temperatures[165] disclose an appreciable reduction of the vibrational frequencies corresponding to a reduction in k_e of 9% for Cl_2 and 19% for Br_2. A decrease in the X–X stretching frequency is also found commonly to attend the transition from the vapour to the solution phase. The significance of this and related spectroscopic observations is indicated subsequently in the general context of the behaviour of the halogens in solution (pp. 1198–1200).

Detailed rotational analysis of individual vibrational bands of the electronic transition $^3\Pi_{0^+u} \leftarrow {}^1\Sigma_g{}^+$ leads to the most reliable estimates of the rotational constant B_e and hence of the internuclear distance r_e, associated not only with the ground state ($^1\Sigma_g{}^+$) but also with the $^3\Pi_{0^+u}$ excited state of each molecule. Values of these constants and of the rotational constants α_e and D_e, derived from the most recent of such analyses, are collected in Tables 11 and 12.

Various attempts have been made to resolve the continua in the electronic spectra of the halogens into components attributable to distinct transitions. Some of the excited states implicated in these transitions, seldom well-defined and commonly unstable with respect to dissociation, are listed in Table 12; in most cases positive experimental information is lacking and some of the details must therefore be regarded as no more than tentative. Through the action of heat or of an electric discharge on the gaseous halogen or through certain reactions which furnish excited atoms or molecules (e.g. that between hydrogen and the elementary halogen), emission spectra have also been observed; again, despite numerous studies, many features of these spectra have yet to be interpreted satisfactorily.

All three halogens exhibit fluorescence to some degree[32,161]. In this respect iodine has been one of the most widely studied systems, fluorescence occurring in the green as well as in the ultraviolet between *ca.* 1800 and 2600 Å; bromine also exhibits a green fluorescence with an intensity about 1/300th that of the iodine fluorescence; in accordance with this trend, fluorescence in chlorine, only recently reported[164] near 5000 Å, is weaker still. When iodine vapour is illuminated with appropriate monochromatic light, e.g. Hg 5461 Å, it emits a fluorescence spectrum which consists, according to the conditions of excitation, of a series of lines (Wood's resonance series) or of bands. A progression of lines results from the overlapping of a single rotational level of a given electronic transition with the exciting line; further series are developed if the irradiating line is broad enough to cover more than one absorption line. The occurrence of fluorescence implies that the molecule remains in the excited state for periods between 10^{-9} and 10^{-6} sec; measurements with a Kerr cell give a lifetime of $(1{\cdot}0 \pm 0{\cdot}1) \times 10^{-8}$ sec for the excited state leading to the visible fluorescence of iodine. To maintain an excited state which is sufficiently long-lived to promote fluorescence,

164 W. Holzer, W. F. Murphy and H. J. Bernstein, *J. Chem. Phys.* **52** (1970) 399, 469.

165 M. Suzuki, T. Yokoyama and M. Ito, *J. Chem. Phys.* **50** (1969) 3392; **51** (1969) 1929; J. E. Cahill and G. E. Leroi, *ibid.* p. 4514.

Table 12. Spectroscopic Properties of Different States of the Diatomic Halogen Molecules[a]

Molecule	State	T_e (cm^{-1})	ω_e (cm^{-1})	$\omega_e x_e$ (cm^{-1})	$\omega_e y_e$ (cm^{-1})	B_e (cm^{-1})	α_e (cm^{-1})	D_e (cm^{-1})	r_e (Å)	Comments
$^{35}Cl_2$	$E(^1\Pi_g)$	($\sim$75,000)				STABLE				
	$D(^3\Pi_{1g})$	($\sim$67,700)				STABLE				
		$\nu_{max} > 64,100$				UNSTABLE				Numerous continua in the visible and ultraviolet regions
	$C(^3\Sigma_{1u}^+)$	($\sim$58,000)				STABLE				
	$^1\Pi_u$	$\nu_{max} \sim 30,300$				UNSTABLE				
	$^3\Pi_{1u}$, $A^3\Pi_{0^+u}$	$\nu_{max} \sim 23,525$				UNSTABLE				
	$A^3\Pi_{0^+u}$[b]	17,672·6	259·5	5·3		0·1626	$0·0021_2$	$2·36_5 \times 10^{-7}$	$2·43_5$	Convergence limit at 20,880 cm^{-1}
	$X^1\Sigma_g^+$[b]	0	559·72	$2·67_5$	$-0·006_7$	$0·2439_9$	$0·0014_9$	$\sim 1·8_5 \times 10^{-7}$	1·9881	
$^{79}Br^{81}Br$	$F(\Pi_{2,1g})$	(66,500)	(480)							Numerous diffuse emission bands and continua in the visible and near-ultraviolet
	$E(^1\Pi_g)$	(61,444)	(220)							
	$D(^3\Pi_{1g})$	(55,535)	(330)							
	$C(^3\Sigma_{1u}^+)$	(47,000)								
	C?	$\nu_{max} \sim 44,500$								Continua
	$A^1\Pi_u$	$\nu_{max} \sim 24,300$								
	$B^3\Pi_0^+$	$\nu_{max} \sim 20,740$								
	$A^3\Pi_1$	($\nu_{max} \sim 18,630$)								
$^{79}Br^{79}Br$	$B^3\Pi_{0^+u}$[c]	15,902·51	167·55	1·625	$-0·010$	0·059579	0·0004868	$3·09 \times 10^{-8}$	2·6777	Convergence limit at 19,578 cm^{-1}
$^{79}Br^{81}Br$	$A^3\Pi_{1u}$[d]	13,814	153	2·7	$-0·015$	0·0657	0·00155		2·55	
$^{79}Br^{81}Br$	$X^1\Sigma_g^+$[c,e]	0	324·24	1·172	$0·342 \times 10^{-2}$	0·081079	0·00030405	$2·05 \times 10^{-8}$	2·2809	

Table 12 (*cont.*)

Molecule	State	T_e (cm^{-1})	ω_e (cm^{-1})	$\omega_e x_e$ (cm^{-1})	$\omega_e y_e$ (cm^{-1})	B_e (cm^{-1})	α_e (cm^{-1})	D_e (cm^{-1})	r_e (Å)	Comments
$^{127}I_2$	$N(^1\Pi_g)$	(58,620)	(120)							Numerous diffuse emission bands and continua
	$M(\Pi_{2,1g})$	(55,930)	(360)							
	$L(^3\Pi_{1g})$	(51,528)	(215)							
	$K(\Pi_{1,0^+u})$	(62,780)	(260)							
	$J(\Pi_{1,0u})$	(59,218)	(212)							
	$I(\Pi_{1,2u})$	(57,770)	(260)							
	$H(\Pi_{1,2u})$	(56,937)	(206)							
	$G(^1\Sigma_u^+)$	51,708	165·1	0·595	−0·0035					Cordes bands
	$F(^3\Pi_u)$	46,488·7	96·2	0·49						
	$C(^3\Sigma_{1u}^+)$	(44,960)	(90)	(0·17)						
	$E(^1\Sigma_g^+)$	41,407	102·2	0·34						
	$D(^1\Sigma_u^+)$	33,744	104	0·2						
	$B^3\Pi_{0^+u}$ [f]	15,769·48	125·53	0·7339	0·00413	0·028873	0·0001345	$\sim 3{\cdot}5 \times 10^{-9}$	3·033	Convergence limit at 20,037 cm^{-1}
	$A^3\Pi_{1u}$	11,888	44·0	$1{\cdot}0_1$	+0·008					
	$X^1\Sigma_g^+$ [g]	0	214·52	0·60738	$-1{\cdot}307 \times 10^{-3}$	0·037389	0·0001210	$4{\cdot}54 \times 10^{-9}$	2·666	

[a] G. Herzberg, *Molecular Spectra and Molecular Structure. I. Spectra of Diatomic Molecules*, pp. 512–541, van Nostrand (1950); Supplement to *Mellor's Comprehensive Treatise on Inorganic and Theoretical Chemistry*, Supplement II, Part I, Longmans, London (1956).

[b] A. E. Douglas, Chr. Kn. Møller and B. P. Stoicheff, *Canad. J. Phys.* **41** (1963) 1174; M. A. A. Clyne and J. A. Coxon, *J. Mol. Spectroscopy*, **33** (1970) 381.

[c] J. A. Horsley and R. F. Barrow, *Trans. Faraday Soc.* **63** (1967) 32; J. A. Coxon, *J. Mol. Spectroscopy*, **37** (1971) 39.

[d] M. A. A. Clyne and J. A. Coxon, *J. Mol. Spectroscopy*, **23** (1967) 258; J. A. Horsley, *ibid.* **22** (1967) 469.

[e] R. J. LeRoy and G. Burns, *J. Mol. Spectroscopy*, **25** (1968) 77.

[f] J. I. Steinfeld, J. D. Campbell and N. A. Weiss, *J. Mol. Spectroscopy*, **29** (1969) 204.

[g] D. H. Rank and B. S. Rao, *J. Mol. Spectroscopy*, **13** (1964) 34.

Values in parentheses are uncertain.

it is a necessary condition that other modes of deactivation of the excited state do not take place in times shorter than 10^{-9} to 10^{-6} sec. Relaxation of this condition, as by the addition of foreign gas or by the application of a magnetic field, leads ultimately to quenching of the fluorescence.

The absorption spectra of bromine and iodine vapours in the region 2000–3000 Å have recently been found to exhibit significant variations with temperature and concentration[166,167], findings which have been interpreted in terms of the equilibrium

$$2X_2 \rightleftharpoons X_4$$

The temperature-dependence of the equilibrium constant over the range 150–420°C implies that for iodine $\Delta H^\circ_{605^\circ K}$ is $-2{\cdot}9 \pm 0{\cdot}4$ kcal mol^{-1} and $\Delta S^\circ_{605^\circ K}$ is $-14{\cdot}4 \pm 2$ eu, while at 240°C and 2·5 atm pressure 1·4 mol% I_4 is present in the vapour; PVT data for bromine lead to the values $\Delta H^\circ_{400^\circ K} = -2{\cdot}6$ kcal mol^{-1} and $\Delta S^\circ_{400^\circ K} = -15$ eu. The thermodynamic results are consistent with the formation of a complex involving van der Waals' interaction between the X_2 partners which are thus able to move relatively freely with respect to each other. That certain features in the ultraviolet-visible spectra of solutions of iodine in carbon tetrachloride, diethyl ether or 1,2-dichloroethane do not obey Beer's Law has likewise been attributed to the production of the I_4 aggregate with a formation constant ranging from 2·5 (CCl_4) to 0·5 l mol^{-1} (Et_2O)[168–170].

The molecular ions X_2^+ have been variously identified, as by their emission spectra when the gaseous halogen is excited in a discharge tube; by mass-spectrometric analysis of the behaviour of the parent molecule under the action of electron-impact; and by measurements of the photoelectron spectra. This last method provides what are probably the most reliable and precise estimates of the ionization potentials of the diatomic molecules[171]; these values are contained in Tables 11 and 39. The dissociation energy of the molecular ion $D_0(X_2^+)$ is related to $D_0(X_2)$ and the ionization potentials $I(X)$ and $I(X_2)$ by

$$D_0(X_2^+) = D_0(X_2) + I(X) - I(X_2)$$

The fact that $D_0(X_2^+)$ so derived is consistently higher than $D_0(X_2)$ (see Table 39) confirms the anti-bonding character of the π_g orbital from which ionization takes place. Another notable feature is that the dissociation energy of F_2^+ (3·42 eV) is lower than that of Cl_2^+ (3·99 eV), just as the dissociation energy of F_2 is lower than that of Cl_2. Detailed rotational and vibrational analysis[172] of the banded spectrum exhibited in emission by the Cl_2^+ ion has afforded the parameters given in Table 39. Though spectroscopic and magnetic properties of condensed phases containing the Br_2^+ and I_2^+ ions have recently been described, significant perturbation of the ground states of the molecular ions are to be expected as a result of ion–ion or ion–solvent interactions. The influence of chemical environment on the formation and behaviour of halogen cations is more appropriately treated elsewhere (Section 4, p. 1340).

[166] A. A. Passchier, J. D. Christian and N. W. Gregory, *J. Phys. Chem.* **71** (1967) 937.
[167] A. A. Passchier and N. W. Gregory, *J. Phys. Chem.* **72** (1968) 2697; M. Tamres, W. K. Duerksen and J. M. Goodenow, *ibid.* p. 966.
[168] L. J. Andrews and R. M. Keefer, *Adv. Inorg. Chem. Radiochem.* **3** (1961) 91.
[169] R. S. Mulliken and W. B. Person, *Molecular Complexes*, p. 141. Wiley, New York (1969).
[170] D. D. Eley, F. L. Isack and C. H. Rochester, *J. Chem. Soc.* (*A*) (1968) 1651.
[171] A. W. Potts and W. C. Price, *Trans. Faraday Soc.* **67** (1971) 1242; S. Evans and A. F. Orchard, *Inorg. Chim. Acta*, **5** (1971) 81.
[172] F. P. Huberman, *J. Mol. Spectroscopy*, **20** (1966) 29.

The Condensed Phases

The molecular lattices adopted by solid chlorine, bromine and iodine are isostructural. In contrast with the allotropy of neighbouring elements in Group VI, there is no good evidence to suggest the existence of allotropes of solid chlorine, bromine or iodine under normal conditions. The unit cell of the crystals is invariably orthorhombic with the space group *Cmca* containing four molecules and having the dimensions listed in Table 11[173]. The intramolecular X–X distance in solid chlorine or bromine is not significantly different from that of the gaseous molecule, but refinement of the crystal structure of iodine at 110°K shows that at 2·715 Å the I–I bond is appreciably longer than that of the gaseous molecule (2·666 Å). Figure 13 depicts the two-dimensional network formed by nearly linear chains

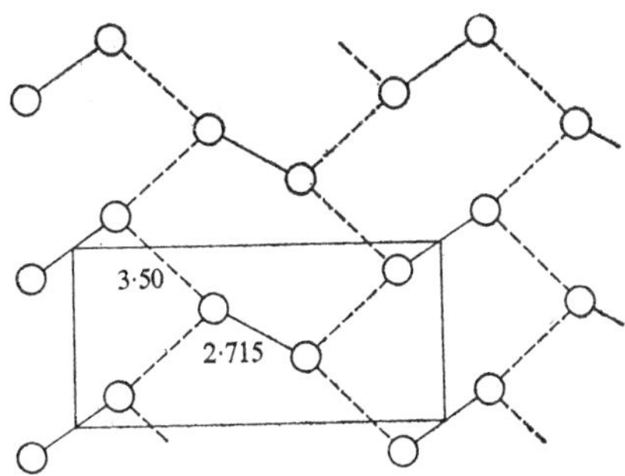

FIG. 13. The two-dimensional network of iodine molecules in the (100) plane of solid iodine. [Reproduced with permission from F. van Bolhuis, P. B. Koster and T. Migchelsen, *Acta Cryst.* **23** (1967) 90.]

I···I–I···I, each iodine atom being involved in two nearly perpendicular chains. The angles of approximately 90° suggest that, to a first approximation, the bonds in intersecting chains are formed by orthogonal 5*p* orbitals of the iodine atom. Each chain can then be described in terms of a model involving four-centre six-electron bonds[174]; alternatively the space group of the unit cell can be exploited to define the matrix integrals for a model in which there is two-dimensional delocalization of *p*-electrons, and which accounts semi-quantitatively for the nqr properties, crystal energy and conductivity of solid iodine[175]. In certain respects, however, an even more apposite description is to be found in the band model, which has been successfully applied to solid iodine on the assumption that electron-exchange between layers is negligible and that only *p* functions are used for binding by the iodine atoms[176].

The relatively short intermolecular X···X distances in solid chlorine and bromine bespeak situations similar in type, if not in degree. Calculations suggest that the orthorhombic structure adopted by the halogens itself testifies to relatively strong intermolecular forces since cubic or hexagonal structures would otherwise seem to be energetically more favourable[177]. That the intermolecular forces assume increasing importance in the series

[173] J. Donohue and S. H. Goodman, *Acta Cryst.* **18** (1965) 568; F. van Bolhuis, P. B. Koster and T. Migchelsen, *ibid.* **23** (1967) 90.

[174] R. J. Hach and R. E. Rundle, *J. Amer. Chem. Soc.* **73** (1951) 4321; G. C. Pimentel, *J. Chem. Phys.* **19** (1951) 446.

[175] J. L. Rosenberg, *J. Chem. Phys.* **40** (1964) 1707.

[176] R. Bersohn, *J. Chem. Phys.* **36** (1962) 3445.

[177] S. C. Nyburg, *J. Chem. Phys.* **40** (1964) 2493; K. Yamasaki, *J. Phys. Soc. Japan*, **17** (1962) 1262.

Cl_2, Br_2, I_2 is well illustrated by the variation in the ratio of the shortest intermolecular X···X to the intramolecular X–X distance (Table 11). Evidence of this same general trend is also to be found in the following features.

(i) As the atomic number advances, so the volatility of the solid halogen decreases, while there is an increase in the melting point and enthalpies of fusion and sublimation; likewise the volatility of the liquid is depressed and the boiling point and heat of volatilization are elevated. These variations follow the general pattern defined by van der Waals' interactions, which are functions primarily of polarizability.

(ii) The vibrational frequency and force constant of the X–X molecule (X = Cl or Br) are significantly reduced in the transition from the gas to the solid phase (see p. 1181). Combined with the observed vibrational frequencies of the lattice and the crystal-field splitting of the intramolecular stretching vibration, the findings indicate stronger intermolecular interaction in solid bromine than in solid chlorine. A similar pattern has also been deduced from studies of the far-infrared spectra of crystalline chlorine, bromine and iodine, whence intermolecular force constants have been calculated[178].

(iii) Measurements of the quadrupole resonance spectra of the ^{35}Cl, ^{37}Cl, ^{79}Br and ^{127}I nuclei in the respective solid halogens afford the results given in Table 11. For solid bromine and iodine, the asymmetry parameter η for nuclear-quadrupole coupling is found to depart substantially from the zero value characteristic of a molecule in a Σ state. This signifies that the electric field at the halogen nucleus is not symmetric about the bound axis, a circumstance presumably dictated by intermolecular binding. The fact that the quadrupole coupling constants found for the crystalline halogens are only slightly less than those of the free atoms gives good grounds for the belief that the bond in the X_2 molecule derives from the overlap of *p*-orbitals with little, if any, admixture of *s*-orbitals. More detailed studies give notice that significant orbital interaction contributes to the intermolecular forces[176,179].

(iv) The layer structure of solid iodine is well characterized by the marked anisotropy of the dielectric constant, electrical conductivity and diamagnetic susceptibility (see Table 11), which have been measured using single crystals of iodine. Monocrystalline iodine thus emerges as a two-dimensional semiconductor having a band-gap estimated to be *ca.* 1·3 eV[180]. It also possesses unique photoconducting properties shown to be the result of hole- rather than electron-conduction, the dominant hole traps being shallow and lying 0·4–0·5 eV above the valence band edge[181].

Under high pressures solid iodine suffers a remarkable increase in electrical conductivity[157,182]; measurements carried out on single crystals indicate a changeover to metallic behaviour in a direction perpendicular to the plane of the molecules at pressures in excess of 160 kbar and a corresponding changeover in directions parallel to this plane at pressures in excess of 220 kbar. At pressures between 160 and 220 kbar the iodine crystal is a semiconductor in one direction and a metal in the other, after the fashion of graphite at atmospheric pressure. These pressures are to be compared with the thresholds for metallic

[178] S. H. Walmsley and A. Anderson, *Mol. Phys.* **7** (1963–4) 411.

[179] C. H. Townes and B. P. Dailey, *J. Chem. Phys.* **20** (1952) 35; N. Nakamura and H. Chihara, *J. Phys. Soc. Japan*, **22** (1967) 201; E. A. C. Lucken, *Nuclear Quadrupole Coupling Constants*, p. 288. Academic Press, London and New York (1969).

[180] N. N. Kuzin, A. A. Semerchan, L. F. Vereshchagin and L. N. Drozdova, *Doklad. Akad. Nauk S.S.S.R.* **147** (1962) 78; L. F. Vereshchagin and E. V. Zubova, *Fiz. Tverd. Tela*, **2** (1960) 2776.

[181] A. Many, M. Simhony, S. Z. Weisz and J. Levinson, *Phys. Chem. Solids*, **22** (1961) 285.

[182] L. F. Vereshchagin, A. A. Semerchan, S. V. Popova and N. N. Kuzin, *Doklad. Akad. Nauk S.S.S.R.* **145** (1962) 757; B. M. Riggleman and H. G. Drickamer, *J. Chem. Phys.* **38** (1963) 2721.

behaviour of 130 and 185 kbar in selenium and stannic iodide, respectively. The development of metallic properties is confirmed by the positive temperature coefficient of the electrical resistance of iodine at high pressures[183]. The effect of such high pressures on the lattice parameters of solid bromine and iodine[184] has also been investigated. Whereas bromine crystals remain orthorhombic, the dimensions of the iodine lattice imply a collapse of the normal structure to a more close-packed one with an increased number of near neighbours surrounding a given iodine atom. Mechanisms for the approach to this condition in and perpendicular to the ac plane have been outlined[184]; whether the I_2 molecules retain their identity and conduction is a result of band overlap or whether the molecules dissociate to give an aggregate of atoms possessing an unfilled conduction band has yet to be resolved. With iodine, therefore, the increased number of relatively low-lying atomic orbitals and the reduced effective nuclear charge experienced by the valence electrons lead to an obscuring of the normal distinction between a molecular and a continuous solid structure. In this respect iodine shows a distinct affinity to elements like selenium and phosphorus, an affinity which is furthered by the relatively high electrical conductivities or semiconductor properties which may be induced in iodine by certain donor moieties, e.g. I_3^- [185], N,N′-diphenyl-*p*-phenylenediamine[186], perylene and pyrene[43], though the precise mechanism of conduction may be open to question in at least some of these systems. Such donor–acceptor interaction encourages the formation of cation radicals representing a source of "holes" and of anion radicals representing a source of electrons. On this basis the electron conductivity of an aggregate is enhanced by a high concentration of the acceptor, viz. iodine, or by incorporation in a polyacceptor network of the type encountered in the black starch–iodine complex[186]. The band gap appears to depend on the separation between the excited charge-transfer and ground states of the donor–acceptor system; in common with the activation energy for conduction, this separation is less than 0·5 eV for some solid complexes of bromine and iodine.

Under normal pressures the liquid phases of chlorine, bromine and iodine each span a temperature range of about 70°C. According to the data of Table 11, liquid chlorine and bromine are both characterized by meagre specific conductivities and low dielectric constants; by contrast, liquid iodine is a much better conductor with a significantly larger dielectric constant. The properties of the liquids suggest that liquid chlorine and bromine are relatively poor solvents, whereas iodine may be expected to function more efficiently in this respect. Extensive studies[32,54,187] have confirmed the validity of this generalization. Thus, whereas alkali-metal halides are, for the most part, sparingly soluble in liquid chlorine or bromine, the iodides of the alkali metals, but not those of the alkaline-earth metals, are readily soluble in liquid iodine. Bromides of large cations, e.g. caesium and tetrabutylammonium, dissolve in bromine to give highly viscous solutions, the equivalent conductance of which increases with increasing concentration; the solubility of caesium bromide provides an efficient process for its separation from the other alkali-metal bromides[54]. On the basis of conductivity measurements on iodides in liquid iodine, the self-ionization

$$nI_2 \rightleftharpoons I^+ + I_{2n-1}^-$$

---

[183] B. M. Riggleman and H. G. Drickamer, *J. Chem. Phys.* **37** (1962) 446.

[184] C. E. Weir, G. J. Piermarini and S. Block, *J. Chem. Phys.* **50** (1969) 2089; R. W. Lynch and H. G. Drickamer, *J. Chem. Phys.* **45** (1966) 1020.

[185] T. Sano, K. Okamoto, S. Kusabayashi and H. Mikawa, *Bull. Chem. Soc. Japan*, **42** (1969) 2505.

[186] V. Hádek, *J. Chem. Phys.* **49** (1968) 5202.

[187] A. G. Sharpe, *Non-aqueous Solvent Systems* (ed. T. C. Waddington), p. 285, Academic Press (1965).

has been postulated[188], the conduction mechanism probably depending primarily on the facile transfer of iodide ions between iodine molecules. According to the proposed scheme for self-ionization, alkali-metal iodides, which readily form polyiodide anions, can be regarded as bases and iodine monohalides as acids. Conductimetric titration of, for example, KI with IBr in molten iodine shows a sharp break at a 1:1 molar ratio, which would correspond to the process

$$I_3^- + IBr \rightarrow Br^- + 2I_2$$

Such a reaction can also be followed potentiometrically. Outside the vapour phase, however, the simple I^+ species has yet to be characterized; on present evidence all that can be said is that the simplest, plausible basis for the self-ionization of iodine appears to involve the formation of I_3^+ and I_3^- ions, but that more complex species may well participate. Solvolytic reactions in liquid iodine, e.g.

$$KCN + I_2 \rightleftharpoons KI + ICN$$

and amphoteric behaviour, e.g.

$$HgI_2 + 2KI \rightleftharpoons K_2HgI_4$$

have also been described. Many molecular halides which do not otherwise react with chlorine, bromine or iodine are soluble in the liquids presumably as the result of relatively strong interactions between the solute, functioning mainly as the donor partner, and the solvent molecules primarily acting in the role of acceptor. Since most studies of such donor-acceptor interactions relate to the solid phase or to conditions in which the donor molecule is present in large excess, detailed discussion of this subject is more aptly taken up in the next subsection. The addition compounds formed by bromine with organic bases such as pyridine, quinoline, acetamide and benzamide form conducting solutions in bromine, probably giving $(\text{base})_n Br^+$ and Br_3^- ions, but the conductivity varies with time, suggesting that bromination of the organic molecule is taking place and at the same time confusing the interpretation of the data[187]. Sulphur, selenium and tellurium are said[187] to dissolve in liquid iodine without chemical change.

2.7. CHEMICAL PROPERTIES OF THE HALOGENS

Redox Properties: Aqueous Chemistry

A considerable degree of order can be found in the reactions of chlorine, bromine and iodine in aqueous solution if full and proper use is made of the standard oxidation potentials appropriate to the elements and their ions. Direct measurements of standard potentials have been made for the couples $\frac{1}{2}Cl_2/Cl^-$, $\frac{1}{2}Br_2/Br^-$ and $\frac{1}{2}I_2/I^-$ with the results given in Table 11, which signify a marked decrease in the oxidizing power of the free halogen in the series $F_2 > Cl_2 > Br_2 > I_2$. There are very few measured potential data for systems involving halogens in other oxidation states, and nearly all the information available[159,189] is derived from thermochemical measurements and estimates or from purely chemical evidence. The doubtful reliability of some of the thermal data is discussed in a paper on the heat of formation and the entropy of the bromate ion in solution[190].

[188] D. J. Bearcroft and N. H. Nachtrieb, *J. Phys. Chem.* **71** (1967) 316.
[189] W. M. Latimer, *The Oxidation States of the Elements and their Potentials in Aqueous Solutions*, 2nd edn., Prentice-Hall, New York (1952).
[190] H. C. Mel, W. L. Jolly and W. M. Latimer, *J. Amer. Chem. Soc.* **75** (1953) 3827.

Most reactions involving oxyhalogen anions also involve hydrogen or hydroxide ions and the position of an equilibrium such as the disproportionation of bromine

$$3Br_2+3H_2O \rightleftharpoons 5Br^-+BrO_3^-+6H^+$$

is very dependent upon the pH of the medium. The potential diagrams originally given by Latimer[189] for the halogens at unit hydrogen ion or at unit hydroxide ion activities are reproduced in a modified form in Fig. 14 (see also Fig. 2), giving what are believed to be the most reliable guides to the aqueous chemistry of the elements at present available.

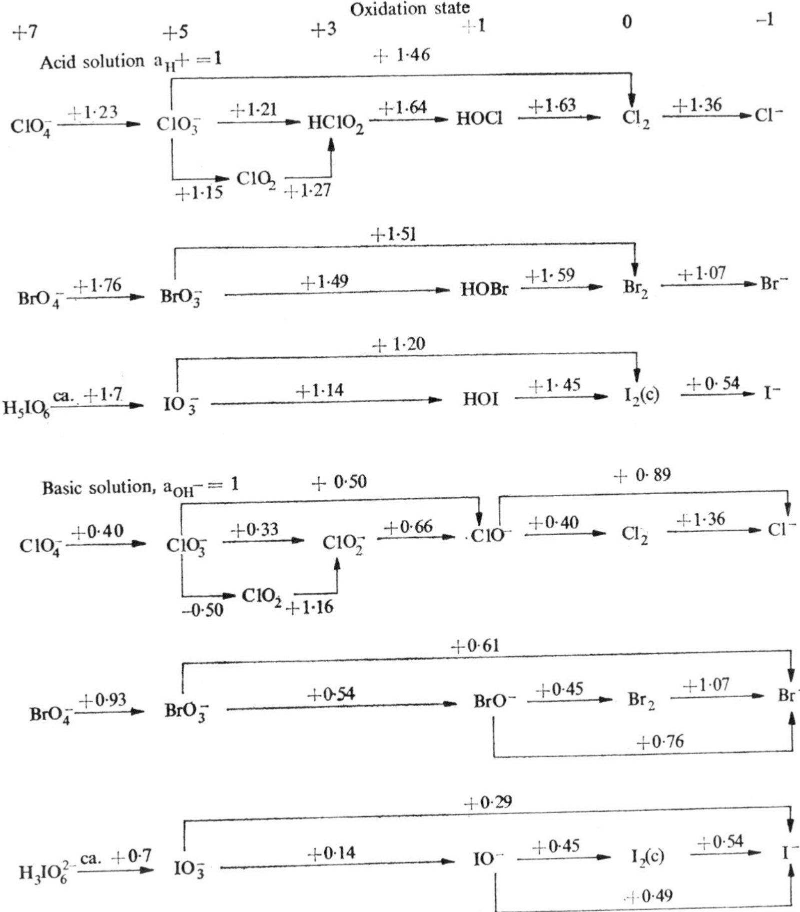

FIG. 14. The standard potentials for systems involving chlorine, bromine and iodine and their ions in aqueous solution, $E°$ in volts.

In the sequence of changes

$$\tfrac{1}{2}X_2(\text{standard}) \rightarrow X(g) \rightarrow X^-(g) \rightarrow X^-(aq)$$

the values of $\Delta H_f°[X^-(g)]$, and consequently of $\Delta G_f°[X^-(g)]$, do not vary widely from halogen to halogen. Accordingly the greatest part of the differences between the redox properties of the couples $\frac{1}{2}X_2/X^-$ arises from the substantial variations in the hydration energies of the anions which follow the sequence $Cl^- > Br^- > I^-$, diminishing as the size of the ion increases. In liquid ammonia the following standard potentials (at 298°K) have been recorded[191]: $\frac{1}{2}Cl_2/Cl^-$, +1·91 V; $\frac{1}{2}Br_2/Br^-$, +1·73 V; $\frac{1}{2}I_2/I^-$, +1·26 V. It is of interest that the increase in the potential in moving from water to liquid ammonia is 0·1 V larger for iodine than for the other halogens. This suggests that, relative to water, the iodide ion coordinates more strongly with liquid ammonia than do the other anions, a conclusion that is in accord with the noticeable solubility of iodides in the latter solvent.

The couple $\frac{1}{2}X_2/X^-$ is reversible for all three halogens, which are characteristically fairly rapid in their oxidizing action, very much faster than molecular oxygen for example. Because of the mechanism of formation of the X_2 molecule by oxidation of the X^- anion, significant activation energies and hence overvoltages must exist; measurements are said to indicate the following order of overvoltages at a platinum electrode: $I_2 > Cl_2 > Br_2$. Were it not for the appreciably higher oxygen overvoltage, chlorine would not be evolved in the electrolysis of aqueous chloride solutions at low current densities; accordingly chlorine overvoltages must necessarily be measured at an electrode which is polarized with respect to oxygen. This phenomenon is of great technical importance in relation to the electrolytic production of chlorine. The mechanism of oxidation is doubtless complex, possibly involving the formation of hypohalite surface compounds which react with halide ions to give the molecular halogen. The relative proportions of oxygen and chlorine evolved in the electrolysis of chloride solutions are known to be influenced by the presence of metal ions, e.g. Mn^{2+}, which preferentially catalyse one or other of the discharge reactions.

The halogens are all to some extent soluble in water. In acid, neutral or alkaline solution the standard potential of the couple $\frac{1}{2}O_2, 2H^+/H_2O$ is +1·23, 0·81 or 0·40 V, respectively. Depending on the precise conditions of pH, all three halogens should therefore be capable of oxidizing water. In contrast with the behaviour of fluorine, the reaction tends to be slow, however, and disproportionation is the initial result:

$$X_2+H_2O \rightleftharpoons HOX+H^++X^-$$

The weak hypohalous acid so produced then undergoes slow decomposition: e.g.

$$HOCl \rightarrow HCl+\tfrac{1}{2}O_2$$

or

$$2HOBr \rightarrow Br_2+H_2O+\tfrac{1}{2}O_2$$

so that oxidation of water is the overall result. However, in the case of bromine and iodine, creation of HOX and HX by disproportionation raises the acidity and eventually stops the reaction for, as the potentials show, the oxidation of water by bromine or iodine does not proceed at unit activity of hydrogen ions.

For saturated solutions of the halogens in water at 298°K the compositions are as shown in Table 13. There is an appreciable concentration of hypochlorous acid in a saturated aqueous solution of chlorine, a smaller concentration of hypobromous acid in a saturated solution of bromine, but only a very meagre concentration of hypoiodous acid in a saturated solution of iodine. It is evident that, because of the unfavourable equilibria, the reaction of the halogen with water does not constitute a suitable method for preparing aqueous solutions of the hypohalous acid. Nevertheless, the yield of hypohalous acid may be improved by judicious choice of conditions whereby the pH is increased, for example by

191 W. L. Jolly, *J. Chem. Educ.* **33** (1956) 512.

TABLE 13. EQUILIBRIUM CONCENTRATIONS IN AQUEOUS SOLUTIONS OF THE HALOGENS AT 298°K[a]

Property	Cl_2	Br_2	I_2
Total solubility (mol l^{-1})	0·0921[b]	0·2141	0·0013
$K_1 = [X_2(aq)]/[X_2(\text{standard})]$	0·062	0·21	0·0013
$K_2 = [H^+][X^-][HOX]/[X_2(aq)]$ (mol^2 l^{-2})	$4·2\times10^{-4}$	$7·2\times10^{-9}$	$2·0\times10^{-13}$
$[X_2(aq)]$(mol l^{-1})	0·062	0·21	0·0013
$[H^+] = [X^-] = [HOX]$(mol l^{-1})	0·030	$1·15\times10^{-3}$	$6·4\times10^{-6}$
Thermodynamic properties of $X_2(aq)$:			
$\Delta G_f°$(kcal mol^{-1})	+1·65	+0·94	+3·92
$\Delta H_f°$(kcal mol^{-1})	−5·6	−0·62	+5·4
$S°$(cal deg^{-1} mol^{-1})	29	31·2	32·8

[a] Ref. 32; F. A. Cotton and G. Wilkinson, *Advanced Inorganic Chemistry*, 2nd edn., p. 569, Interscience (1966); *National Bureau of Standards Technical Note* 270–3, January 1968.
[b] Cl_2 gas at 1 atm pressure.

the use of a suspension of mercuric oxide:

$$2X_2+2HgO+H_2O \rightarrow HgO,HgX_2+2HOX$$

All of the hypohalous acids are rather unstable, being also relatively strong oxidizing agents' especially in acid solution. The analogous sulphur compounds HSX appear to be even less stable on the very limited evidence that is available. Nevertheless, bromine is said[192] to react under rigidly controlled conditions with hydrogen sulphide in chloroform or dichloromethane solution according to the scheme:

$$Br_2+H_2S \rightarrow HBr+HSBr$$

and the isolation of the salt NH_4SBr at low temperature has been reported.

In alkaline solution the halogens are converted to the corresponding hypohalite ions in accordance with the general reaction

$$X_2+2OH^- \rightleftharpoons X^-+XO^-+H_2O$$

The equilibrium constant for this reaction is invariably favourable—$7·5\times10^{15}$ for chlorine, 2×10^8 for bromine and 30 for iodine—and the reaction is rapid. Thus the hydrolysis of chlorine is practically complete in alkaline solution, but the balance is reversed in acid solution. The preparation of bleaching powder and sodium hypochlorite and their action as bleaching agents provide an important illustration of the operation of this equilibrium. However, the situation is complicated by the tendency of the hypohalite ions to disproportionate further in basic solution to produce the corresponding halate ions:

$$3XO^- \rightleftharpoons 2X^-+XO_3^-$$

For this reaction the equilibrium constant is in each case very favourable, viz. 10^{27} for ClO^-, 10^{15} for BrO^- and 10^{20} for IO^-. Thus the actual products obtained when a halogen is dissolved in alkaline solution depend on the rates at which the hypohalite ions initially produced undergo disproportionation. The disproportionation of ClO^- is slow at and below room temperature, but becomes fairly rapid at *ca.* 75°C. By the choice of appropriate conditions good yields of the chlorate ion can be secured, as in the commercial

[192] M. Schmidt and I. Löwe, *Angew. Chem.* **72** (1960) 79.

production of chlorates. The disproportionation of BrO^- is moderately fast even at room temperature, and at temperatures of 50–80°C the BrO_3^- ion is formed quantitatively according to the equation

$$3Br_2+6OH^- \rightarrow 5Br^-+BrO_3^-+3H_2O$$

Thus bromine is thermodynamically stable in acid solution with respect to disproportionation into bromate and bromide, but unstable at a pH of 10; advantage has been taken of this behaviour in an important stage of one process for the extraction of bromine from sea water (see p. 1138). The rate of disproportionation of IO^- is so fast at normal temperatures that the ion has so far eluded proper characterization. With strong base, therefore, iodine reacts quantitatively to give iodate and iodide ions.

Halite ions and halous acids do not feature in the hydrolysis of the halogens. The disproportionation

$$4ClO_3^- \rightleftharpoons Cl^-+3ClO_4^-$$

has an equilibrium constant of 10^{20}, but takes place only very slowly in solution even near 100°C. Such kinetic barriers are an outstanding characteristic of the formation and chemical properties of the perchlorate ion. Likewise, although perbromate emerges as a more powerful oxidant than either perchlorate or periodate, it appears to be sluggish in formation and reaction[193]. Unlike chlorate, iodate is stable with respect to reactions such as

$$7IO_3^-+9H_2O+7H^+ \rightleftharpoons I_2+5H_5IO_6, \quad K = 10^{-85}$$
$$4IO_3^-+3OH^-+3H_2O \rightleftharpoons I^-+3H_3IO_6^{2-}, \quad K = 10^{-44}$$

so that, irrespective of rate, disproportionation with the formation of periodate is not favoured.

Many other redox reactions of the molecular halogens are readily comprehensible on the basis of the thermodynamic data of Fig. 14; some typical reactions are listed in Table 14. Thus the decrease in oxidizing power with increasing atomic number leads to replacement reactions such as

$$Cl_2+2Br^- \rightarrow Br_2+2Cl^-$$

in which a halogen of lower atomic number displaces one of higher atomic number. Likewise redox couples with values of $E°$ less than that of the couple $\frac{1}{2}X_2/X^-$ are usually subject to oxidation by the halogen X_2, though kinetic factors may supervene in some cases. For example, in aqueous solution chlorine, bromine or iodine oxidizes, *inter alia*, nitrite to nitrate, arsenite to arsenate, sulphite to sulphate, hydrogen sulphide to sulphur, Sn^{2+}(aq) to Sn^{4+}(aq) and $[Fe(CN)_6]^{4-}$ to $[Fe(CN)_6]^{3-}$. However, although the balance of the equilibrium

$$2Fe^{2+}(aq)+X_2 \rightleftharpoons 2Fe^{3+}(aq)+2X^-$$

favours the ferric state when X = Cl or Br, in neutral or acid solution ferric ions oxidize iodide to iodine. Again, whereas thiosulphate ions are oxidized to sulphate by chlorine or bromine, the familiar and analytically important reaction

$$2S_2O_3^{2-}+I_2 \rightarrow S_4O_6^{2-}+2I^-$$

takes place with iodine. In fact the standard redox potential of the couple $SO_4^{2-}/\frac{1}{2}S_2O_3^{2-}$ (*ca.* +0·3 V) is such that iodine, in common with the other halogens, should be able to

193 E. H. Appelman, *Inorg. Chem.* **8** (1969) 223.

TABLE 14. SOME REDOX REACTIONS OF THE MOLECULAR HALOGENS IN AQUEOUS SOLUTION[a]

Reagent	Reaction	Comments
Halide ions	$X_2+2Y^- \rightarrow 2X^-+Y_2$ (X = Cl, Y = Br or I; X = Br, Y = I)	Such reactions are important in the extraction of bromine and iodine and in the iodometric determination of free chlorine and bromine.
	The reaction may proceed further, particularly under alkaline conditions: e.g. $I_2+5Cl_2+6H_2O \rightarrow 2IO_3^-+10Cl^-+12H^+$	Bromate and iodate ions are formed.
Nitrite ions	$NO_2^-+\frac{1}{2}X_2+H_2O \rightarrow NO_3^-+X^-+2H^+$ (X = Cl, Br or I)	
Ammonia	$2NH_3+3X_2 \rightarrow N_2+6H^++6X^-$ (X = Cl or Br)	Iodine undergoes a complicated reaction, the course of which depends upon the conditions; products include NH_4I and NI_3.
Hydrazine	$N_2H_4+2X_2 \rightarrow N_2+4H^++4X^-$ (X = Cl, Br or I)	
Azide ions	$2N_3^-+X_2 \rightarrow 3N_2+2X^-$ (X = Br or I)	The reaction is catalysed by traces of SCN^-, $S_2O_3^{2-}$, S^{2-}, thioketones or mercaptans.
Hypophosphite, phosphite and hypophosphate ions	Typically: $X_2+H_2PO_3^-+H_2O \rightarrow H_2PO_4^-+2H^++2X^-$ (X = Cl, Br or I)	The reaction with iodine has been used to estimate phosphites and hypophosphites.
Arsenite	$H_3AsO_3+X_2+H_2O \rightarrow H_3AsO_4+2H^++2X^-$ (X = Cl, Br or I)	The reaction between I_2 and arsenite, which is quantitative at pH 4–9, is an important analytical reaction.
Hydrogen sulphide	$H_2S+X_2 \rightarrow S+2H^++2X^-$ (X = Cl, Br or I)	The sulphur may be further oxidized to SO_4^{2-} by chlorine or bromine.
Sulphite ions	$SO_3^{2-}+X_2+H_2O \rightarrow SO_4^{2-}+2H^++2X^-$ (X = Cl, Br or I)	However, concentrated H_2SO_4 oxidizes Br^- or I^- to the free halogen.
Thiosulphate ions	$S_2O_3^{2-}+4X_2+5H_2O \rightarrow 2HSO_4^-+8H^++8X^-$ (X = Cl or Br)	
	$2S_2O_3^{2-}+I_2 \rightarrow S_4O_6^{2-}+2I^-$	This reaction, normally carried out in acid solution in the presence of iodide ions, is of particular importance in the volumetric estimation of iodine.
Selenite ions	$SeO_3^{2-}+Cl_2+H_2O \rightarrow SeO_4^{2-}+2H^++2Cl^-$	Neither bromine nor iodine is a sufficiently strong oxidizing agent under normal conditions to effect this change.

Table 14 (*cont.*)

Reagent	Reaction	Comments
Carboxylate ions with reducing properties	Examples include: $HCO_2^- + X_2 \rightarrow CO_2 + H^+ + 2X^-$ $C_2O_4^{2-} + X_2 \rightarrow 2CO_2 + 2X^-$ (X = Br or I)	
Tin(II)	$Sn^{2+} + X_2 \rightarrow Sn^{4+} + 2X^-$ (X = Cl, Br or I)	
Iron(II)	$2Fe^{2+}(aq) + X_2 \rightleftharpoons 2Fe^{3+}(aq) + 2X^-$ $2[Fe(CN)_6]^{4-} + X_2 \rightleftharpoons 2[Fe(CN)_6]^{3-} + 2X^-$ (X = Cl, Br or I)	Fe(III) favoured by X = Cl or Br; Fe(II) favoured by X = I. In neutral solution I_2 oxidizes $[Fe(CN)_6]^{4-}$ to $[Fe(CN)_6]^{3-}$, while in strongly acid solution the reverse reaction occurs; this behaviour can be exploited for the estimation both of ferrocyanides and of ferricyanides.
Concentrated nitric acid, persulphate or permanganate	Typically $I_2 + 8H^+ + 10NO_3^- \rightarrow 2IO_3^- + 10NO_2 + 4H_2O$	Preparative reaction for iodic acid.
Oxyhalogen species:		
Hypohalite	A representative reaction is: $X_2 + 5ClO^- + H_2O \rightarrow 2XO_3^- + 5Cl^- + 2H^+$ (X = Br or I)	
Halite	$Cl_2 + 2ClO_2^- \rightarrow 2Cl^- + 2ClO_2$	This represents an important commercial method of producing ClO_2. A tracer study shows that most of the Cl atoms in the ClO_2 are derived from the ClO_2^-; an unsymmetrical intermediate Cl–ClO_2 or Cl–O–Cl–O may be formed.[b]
Halate	$3I_2 + 5ClO_3^- + 3H_2O \rightarrow 6IO_3^- + 6H^+ + 5Cl^-$ $2I_2 + IO_3^- + 10Cl^- + 6H^+ \rightarrow 5ICl_2^- + 3H_2O$	Occurring in strongly acid media, this forms the basis of the Andrews procedure in which potassium iodate is used as an analytical oxidizing agent.
	$Cl_2 + IO_3^- + 2Na^+ + 3OH^- \rightarrow Na_2H_3IO_6 \downarrow + 2Cl^-$	A convenient synthesis of sparingly soluble periodate derivatives; $K_4I_2O_9$ and $Ag_4I_2O_9$ can be prepared in a similar manner.
Perhalate	$I_2 + 5IO_4^- + H_2O \rightarrow 7IO_3^- + 2H^+$	

[a] Refs. 32 and 33.
[b] H. Taube and H. Dodgen, *J. Amer. Chem. Soc.* **71** (1949) 3330.

effect oxidation to the sulphate ion; the fact that it oxidizes thiosulphate quantitatively to tetrathionate therefore implies that this reaction is very much faster than the oxidation to sulphate. It is also to be noted that some redox reactions which take place in aqueous solution do not occur in non-aqueous media. For example, iodine and sulphur dioxide do not react when dissolved together in a mixture of anhydrous methanol and pyridine; only when water is added does oxidation of the sulphur dioxide by the iodine take place (see Table 14). This is the basis of the Karl–Fischer reagent used for the determination of small amounts of water[194].

The oxidation potentials characteristic of oxyhalogen systems show them without exception to be strong oxidizing agents, and the oxidation of the molecular halogens is correspondingly difficult. Nevertheless, although the direct conversion of molecular chlorine to individual oxychlorine species can be achieved with but few reagents, the relative stability of the iodate ion is underlined by the facility with which it is formed from iodine by the action of such oxidizing agents as concentrated nitric acid, chlorate, bromate, chlorine, persulphate or permanganate. It seems probable that the diversity of behaviour exhibited by the halogens with respect to such oxidation depends, at least in part, on kinetic rather than thermodynamic barriers.

Photolysis of aqueous solutions of the halogens induces reactions that are otherwise slow under normal conditions. As early as 1785 Berthollet recorded the action of sunlight on chlorine water[195], while Balard's reports of the properties of bromine water published in 1826[40] likewise alluded to the action of sunlight. The consensus of these and subsequent studies is that photolysis favours oxidation of the water by the halogen to produce oxygen, X^- and XO_3^- ions: e.g.

$$2Cl_2 + 2H_2O \rightarrow 4H^+ + 4Cl^- + O_2$$

$$5Cl_2 + 5H_2O \rightarrow 10H^+ + ClO_3^- + 9Cl^- + O_2$$

Possible mechanisms proposed[32] for these reactions hinge on initiation through the step $X_2 + h\nu \rightarrow 2X\cdot$, and on propagation *via* free radicals such as $\cdot OH$, $\cdot O_2H$ and $ClO\cdot$. The decomposition of hypohalous acids and of hypohalites is also greatly accelerated by photolysis, the relative contributions of the two modes of decomposition,

$$2XO^- \rightarrow 2X^- + O_2$$

$$3XO^- \rightarrow 2X^- + XO_3^-$$

being influenced considerably by temperature, concentration, pH, added salts and exposure to the atmosphere. Such catalytic effects stress once again the importance of kinetic and mechanistic factors in the aqueous chemistry of the halogens. In fact, nearly all the positive oxidation states would be denied existence in aqueous solution but for the slowness of decomposition into the molecular halogen (or halide ion) and oxygen.

We are still far from having a clear picture of the mechanisms of many reactions of the halogens in aqueous media. However, as a general rule, the acceptor character of the halogen molecule with respect to nucleophilic reagents like H_2O, OH^- or X^- leads to labile addition compounds, which may well function as essential reaction intermediates. The formation and characteristics of such compounds provide the theme of the next subsection.

[194] A. I. Vogel, *A Text-book of Quantitative Inorganic Analysis*, 3rd edn., p. 944. Longmans, London (1961).
[195] C. L. Berthollet, *Mém. Acad.* (1785) 276.

Acceptor Functions: Charge–Transfer Interactions[32,33,168,196–204]

In common with molecules like sulphur dioxide, oxygen, the hydrogen halides, quinones and polynitroaromatic systems, halogen molecules of the types X_2 and XY exhibit a primary function as electron-acceptors forming complexes with a wide range of donor species. Not only is this capacity of the halogens at the root of their behaviour as solutes and, in many circumstances, as chemical reagents, its most direct expression, that is in complex-formation, has influenced profoundly the development of our understanding, in particular, of so-called charge-transfer interactions and, in general, of donor-acceptor functions. It is the vacant anti-bonding σ_u orbital of the halogen molecule (see Fig. 11), inevitably more diffuse than the bonding σ_g orbital, which furnishes the acceptor capacity. The halogen molecules are accordingly to be classified as σ-acceptors; further, because the acceptor orbital is anti-bonding in type, with the result that its occupancy must cause a weakening of the X–X or X–Y bond, the molecules are, in Mulliken's language[204], "sacrificial" in action, prone to form relatively weakly bonded complexes. On the basis of semi-empirical considerations, the following vertical electron affinities (in eV) have been deduced for the halogen molecules[205]: Cl_2, $1{\cdot}3 \pm 0{\cdot}4$; Br_2, $1{\cdot}2 \pm 0{\cdot}5$; I_2, $1{\cdot}7 \pm 0{\cdot}5$. Although the acceptor power of iodine thus appears to be superior to those of bromine and chlorine, variations of electron affinity are almost certainly outweighed in normal chemical situations by other factors arising, for example, from charge-transfer (see below), dispersion and other environmental interactions. In addition to the primary σ-acceptor function of a halogen molecule, the occupied π_g orbitals, which afford a π-donor system, exercise a significant secondary influence. This amphoteric behaviour is consistent with the "soft acid" or "class b" character of the halogens in their interactions with species having primarily a donor function; it is also manifest in the relatively strong intermolecular forces in solid iodine and in the aggregates X_4 (X = Br or I) reported to exist in the gas phase and in solution[166–170].

The list of compounds which form recognizable complexes with the halogens encompasses two types of donor:

(i) *σ-donors*, which possess formally non-bonding electrons. These include a wide variety of nitrogen bases (e.g. aliphatic amines, pyridine and its derivatives, nitriles), oxygen bases (e.g. alcohols, ethers, carbonyl compounds), organic sulphides and selenides (e.g. 1,4-dithian, 1,4-diselenacyclohexane), and certain halide derivatives (e.g. alkyl halides and halide ions).

(ii) *π-donors*, wherein the donor function is performed by bonding π-orbitals. These include aromatic systems ranging from benzene to polycyclic hydrocarbons like perylene, as well as molecules containing more localized π-charge clouds, e.g. alkenes.

In Mulliken's terminology the σ-donors are "increvalent" by virtue of their capacity to donate lone-pair electrons, whereas the π-donors are "sacrificial", donation being from a

[196] R. S. Mulliken and W. B. Person, *Ann. Rev. Phys. Chem.* **13** (1962) 107.
[197] G. Briegleb, *Elektronen-Donator-Acceptor-Komplexe*, Springer-Verlag (1961).
[198] O. Hassel and Chr. Rømming, *Quart. Rev. Chem. Soc.* **16** (1962) 1.
[199] J. N. Murrell, *Quart. Rev. Chem. Soc.* **15** (1961) 191.
[200] S. F. Mason, *Quart. Rev. Chem. Soc.* **15** (1961) 353.
[201] J. Rose, *Molecular Complexes*, Pergamon, Oxford (1967).
[202] H. A. Bent, *Chem. Rev.* **68** (1968) 587.
[203] C. K. Prout and J. D. Wright, *Angew. Chem., Internat. Edn.* **7** (1968) 659; C. K. Prout and B. Kamenar, *Molecular Complexes* (ed. R. Foster), Paul Elek (Scientific Books) (1973).
[204] R. S. Mulliken and W. B. Person, *Molecular Complexes*, Wiley, New York (1969).
[205] W. B. Person, *J. Chem. Phys.* **38** (1963) 109.

bonding orbital, with the result that bonding within the donor is weakened by complex-formation. Some systems, e.g. pyridine, are functionally capable of acting both as σ- and π-donors[206], while the presence of energetically accessible vacant orbitals, e.g. *d*- or anti-bonding π-orbitals, may impart amphoteric properties, the donor function being supplemented by a secondary acceptor role. Outside the limits defined by these categories, molecules like cyclohexane and n-heptane possessing only low-energy occupied σ-orbitals may yet experience with the halogen molecules short-lived interactions so weak as to preclude the identification of a distinct complex but sufficiently strong to produce marked perturbation of the electronic spectra of the components[204].

The equilibrium

$$X_2 + D \rightleftharpoons D,X_2$$

involving the interaction of a halogen and donor species D is rapidly established. Evidence concerning the existence, stability and structure of complex species such as D,X_2 has been derived by a variety of physicochemical methods, to be outlined in the following pages.

1. Historical considerations[168]

It has been known for many years that the colour of iodine solutions varies with the nature of the solvent. Thus, the halogen is violet in media such as the aliphatic hydrocarbons and carbon tetrachloride; in other solvents, including the alcohols, ethers and benzene, it is brown or reddish-brown[207]. The visible absorption maximum of the violet solutions is located in the 520–540 mμ region, the overall spectrum being similar to that of iodine in the vapour state. The maximum for brown solutions occurs at shorter wavelengths (460–480 mμ). The solubility of iodine, its heat of solution and the chemical reactivity are generally greater for solvents that give brown solutions than for solvents that give violet solutions. Further, violet solutions often turn brown on the addition of a solvent such as alcohol, whereas brown solutions often turn violet on heating (and brown again on cooling). Although the molecular weight of iodine in both types of solvent corresponds to the diatomic unit I_2, an abnormally small freezing-point depression is observed when a small amount of a solvent forming a brown solution is added to a violet solution.

Several ideas have been invoked to explain the varying colours of iodine solutions: it has been suggested that the brown solutions contain colloidal particles, solvated molecules or associated molecules. A theory of solvent–solute interaction in terms of the cage theory of solutions has also been put forward[208]. However, the formation of molecular complexes, suggested as early as 1930[209], was first demonstrated rather conclusively by the notable studies of Benesi and Hildebrand[210] on the visible and ultraviolet spectra of solutions of iodine in benzene and other aromatic solvents. Most significant was the discovery of an intense ultraviolet absorption band near 300 mμ attributable neither to the solvent nor to iodine. By following the changes in the intensity of absorption with concentration, using an inert solvent, it was shown that the new band arises from a 1:1 complex. Subsequent

[206] R. S. Mulliken, *J. Amer. Chem. Soc.* **91** (1969) 1237; I. D. Eubanks and J. J. Lagowski, *ibid.* **88** (1966) 2425.

[207] A. Lachman, *J. Amer. Chem. Soc.* **25** (1903) 50; J. H. Hildebrand and R. L. Scott, *The Solubility of Non-electrolytes*, 3rd edn., Reinhold, New York (1950).

[208] N. S. Bayliss and A. L. G. Rees, *J. Chem. Phys.* **8** (1940) 377; A. L. G. Rees, *ibid.* p. 429; N. S. Bayliss, *Nature*, **163** (1949) 764; *J. Chem. Phys.* **18** (1950) 292.

[209] M. Chatelet, *Compt. rend.* **190** (1930) 927.

[210] H. A. Benesi and J. H. Hildebrand, *J. Amer. Chem. Soc.* **70** (1948) 2832; **71** (1949) 2703.

studies[168] have disclosed similar behaviour in aromatic solvents on the part of chlorine and bromine as well as interhalogens such as iodine monochloride.

Because of the relatively feeble interaction between the components, complex-formation can often be detected only by studying the physical properties of solutions in which the complexes are in equilibrium with their components. Nevertheless, the persistence in the vapour phase of certain complexes, e.g. I_2,OEt_2, I_2,SEt_2, I_2,C_6H_6 and $C_5H_5N,2I_2$, is suggested by the ultraviolet[211] and mass[212] spectra. Further, numerous crystalline adducts of definite stoichiometry have been isolated; examples include Me_3N,I_2, 4-picoline,I_2, hexamethylenetetramine,$2Br_2$, 1,4-dioxan,X_2 (X = Cl, Br or I), acetone,Br_2, 1,4-dithian,$2I_2$, 1,4-diselenacyclohexane,$2I_2$, C_6H_6,X_2 (X = Cl or Br) and $[p\text{-}Me_2N{\cdot}C_6H_4]_2C{=}CH_2,2X_2$ (X = Br or I). The structures of some of these solid complexes have been determined by X-ray crystallographic analysis carried out by Hassel and others[198], beginning in 1954 with the adduct 1,4-dioxan,Br_2. The results of such analysis are summarized in a subsequent section (see pp. 1201–6 and Table 16).

2. Physicochemical methods of investigation[168,197,201,204]

The most distinctive characteristic associated with the formation of a halogen complex is the appearance of a new intense and broad absorption band in the visible or ultraviolet spectrum, generally accompanied by perturbations of the spectral bands arising from electronic transitions of the component molecules. Several methods have been devised for evaluating the formation constant of such a complex from spectrophotometric measurements of optical density as a function of concentration. Determination of the formation constant at more than one temperature affords values for the enthalpy and entropy changes accompanying complex-formation. Listed in Table 15 are formation constants and enthalpies for some representative complexes of iodine. Formation constants are thus seen to span the range 10^{-1} to 10^4 l mol^{-1}, while enthalpies lie in the range 1–13 kcal mol^{-1}. The new absorption band characteristic of each complex, the so-called "charge-transfer" band, has a molar extinction coefficient of the order 10^4 and a half-width typically of 6000 cm^{-1}.

The interaction of a donor D with a halogen or interhalogen XY implies a certain amount of charge-transfer in the ground state of the complex. Accordingly the vibrational properties of the complex should reflect not only new motions in which D vibrates against XY, but also changes in the strengths of bonds within the D and XY units. Complex-formation is thus attended by the following changes in the infrared and Raman spectra of the component species[213].

(a) New spectral features attributable to vibrations of the D-XY unit may be observed. Thus for Me_3N,I_2 such a band, representing in large part the N–I stretching vibration, has been located at 145 cm^{-1} in infrared absorption.

(b) The frequencies and intensities of vibrational bands associated with internal motions of the D and XY molecules may suffer significant changes. For example, with respect to the

[211] F. T. Lang and R. L. Strong, *J. Amer. Chem. Soc.* **87** (1965) 2345; M. Tamres and J. M. Goodenow, *J. Phys. Chem.* **71** (1967) 1982.

[212] R. Cahay and J. E. Collin, *Nature*, **211** (1966) 1175.

[213] R. F. Lake and H. W. Thompson, *Proc. Roy. Soc.* **A297** (1967) 440; P. Klaboe, *J. Amer. Chem. Soc.* **89** (1967) 3667; J. Gerbier and V. Lorenzelli, *Spectrochim. Acta*, **23A** (1967) 1469; F. Watari, *ibid.* p. 1917; Y. Yagi, A. I. Popov and W. B. Person, *J. Phys. Chem.* **71** (1967) 2439; J. Yarwood and W. B. Person, *J. Amer. Chem. Soc.* **90** (1968) 594, 3930; R. F. Lake and H. W. Thompson, *Spectrochim. Acta*, **24A** (1968) 1321; S. G. W. Ginn, I. Haque and J. L. Wood, *ibid.* p. 1531; K. Yokobayashi, F. Watari and K. Aida, *ibid.* p. 1651; J. P. Kettle and A. H. Price, *J. Chem. Soc., Faraday Trans. II* (1972) 1306.

TABLE 15. PROPERTIES OF SOME HALOGEN COMPLEXES IN SOLUTION[204]

(a) Specific iodine complexes

Donor	Formation constant, K (l mol^{-1})(20°C)	Heat of formation, $-\Delta H$(kcal mol^{-1})	Charge-transfer band		
			λ_{max} (mμ)	ϵ_{max}	$\Delta\nu_{1/2}$(cm^{-1})
Benzene	0·15	1·4	292	16,000	5100
Naphthalene	0·26	1·8	360	7250	4700
Methanol	0·23	3·5	232	13,700	5700
Ethanol	0·26	4·5	230	12,700	6800
Diethyl ether	0·97	4·3	249	5700	6900
Diethyl sulphide	210	7·82	302	29,800	5400
Diethyl disulphide	5·62	4·62	304	15,000	7200
Ammonia	67	4·8	229	23,400	4100
Methylamine	530	7·1	245	21,200	6400
Dimethylamine	6800	9·8	256	26,800	6450
Diethylamine	6320	12·0	278	25,600	8100
Trimethylamine	12,100	12·1	266	31,300	8100
Pyridine	269	7·8	235	50,000	5200
(b) Halogen complexes generally					
π–Donor, e.g. C_6H_6	0·1–2·0	1–4	275–415	5000–15,000	~5000
σ–Donor, e.g. amine	0·5–*ca.*10^3	4–13	225–280	3000–30,000	5000–8000

gas-phase molecules, the following decreases in frequency ($\Delta\nu$ in cm^{-1}) are found for the stretching vibration of XY dissolved in benzene: Cl_2, 31; Br_2, 20; I_2, 10; ICl, 28. These results support the view that the XY molecule is acting as a "sacrificial" σ-acceptor. For weak complexes, the proportional change in stretching force constant for the X–Y bond, $\Delta k/k_0$, is given approximately by $2\Delta\nu/\nu_0$ (where k_0 and ν_0 refer to the unperturbed molecule), and on this basis the donor or acceptor capacities of different molecules have been compared[168,197,201,204]. However, for a relatively strongly bound complex, such as Me_3N,XY, vibrational coupling has been shown to invalidate this simplified treatment of the force field[214]. The methods of vibrational spectroscopy have also been used to follow the course of ionization reactions such as

$$2\text{py,IX} \rightleftharpoons [\text{py}_2\text{I}]^+ + \text{IX}_2^- \qquad (\text{X} = \text{Cl, Br or I; py} = \text{pyridine or } \gamma\text{-picoline})$$

which occur in polar solvents[215].

The enhanced intensity of certain modes with respect to infrared absorption has been the subject of numerous experimental and theoretical enquiries. For the halogen complexes it has been concluded[216] that this effect, manifest in the halogen–halogen stretching vibration, is the outcome of electronic reorientation during the vibration, that is, of vibronic coupling between different electronic states; a similar effect has been described for hydrogen-bonded complexes.

(c) Since the total symmetry is likely to change on complex-formation, vibrational transitions which are forbidden in infrared absorption or Raman scattering in the free molecules may appear in the spectrum of the complex. Thus, the X–X stretching mode of

214 J. N. Gayles, *J. Chem. Phys.* **49** (1968) 1840.

215 I. Haque and J. L. Wood, *Spectrochim. Acta*, **23A** (1967) 959, 2523; S. G. W. Ginn and J. L. Wood, *Trans. Faraday Soc.* **62** (1966) 777.

216 H. B. Friedrich and W. B. Person, *J. Chem. Phys.* **44** (1966) 2161.

chlorine, bromine or iodine, normally inactive in infrared absorption, appears weakly in the infrared spectrum of the halogen in benzene solution. Although, in principle, the operation of the vibrational selection rules should afford a criterion of the geometry of a halogen complex, a simple interpretation is seldom free from ambiguity through the influence of vibronic coupling on band intensities and through other complications.

Other physical properties used for studying complex-formation are as follows.

Conductance[168].

When the product of interaction of a halogen with a donor is appreciably ionic in character, complex-formation may be detected by conductance measurements. Thus, reactions such as that of IX and other halogens with pyridine (see above) give rise to solutions of appreciable conductivity, presumably containing the ions $[py_2X]^+$ and either halide or polyhalide anions.

Solubility Studies[168]

A study of the solubility of iodine in normal solvents (giving a violet colour) was an important part of the work which led Hildebrand to his concept of "regular solutions"[207]. All of these solutions conform reasonably well to the solubility equation for regular solutions

$$RT \ln(a_2/N_2) = V_2\phi_1^2(\delta_2-\delta_1)^2$$

where a_2 denotes the activity of the solid iodine referred to pure liquid iodine, V_2 its liquid molal volume (extrapolated), ϕ_1 the volume fraction of the solvent and δ_2 and δ_1 the "solubility parameters" of iodine and the solvent. The experimental solubility of iodine in a complexing solvent is greater than the value predicted by this equation. Complexing solvents may also be distinguished by plots of log N_2 against $1/T$ which give a family of curves for regular solutions but curves with markedly different slopes for those solutions in which specific interaction occurs. Measurements of the solubility of iodine have also been used to determine formation constants for a number of complexes.

For details of the solubilities of the halogens in aqueous and non-aqueous media, the reader is referred to Mellor's Comprehensive Treatise and Supplement[32], Gmelins Handbuch[33] or Linke's compilation of solubility data[217]. There have also been extensive studies of the adsorption of the halogens on materials such as charcoal, silica gel and magnesium oxide[32]; isotherms have been reported for several systems, and the influence of factors such as the preparation of the solid has also been investigated. Commonly there is but a fine distinction between purely physical processes of solution or adsorption on the one hand and overall chemical reaction on the other. Thus, under appropriate conditions iodine is reversibly adsorbed on certain metals or crystals of metal halides. Likewise, at low concentrations of iodine in aqueous solution the interaction with starch appears to be one of reversible adsorption; only at higher concentrations of iodine does the interaction give rise to a recognizable complex. The adsorption of bromine by graphite affords lamellar compounds by penetration between the carbon layers of the graphite; compounds with bromine contents up to that implied by the compositions C_8Br or $C_{16}Br$ are thus obtained[32,54].

Dielectric Polarization Studies

The product of interaction of a halogen and a donor may be more polar than either

[217] W. F. Linke, *Solubilities: Inorganic and Metal-organic Compounds*, 4th edn., Vol. 1, van Nostrand, Princeton (1958).

reactant. An estimate of the degree of polarization of the complex may then be obtained by measuring the dielectric constant of a solution of the complex in a non-polar medium. In general, the molar polarization of the halogen increases with the strength of the donor. Apparent dipole moments have also been calculated, notably for a number of iodine complexes[168,218]; values of 4–12 *D* have thus been reported for 1:1 complexes of iodine with various amines[219]. However, despite numerous attempts at correlation, no simple relationship appears to exist between the scalar excess moments, i.e. μ(complex)–$\Sigma\mu$(components), and measures of acidic or basic character, e.g. formation constants[219].

Other Methods[168]

Colorimetric measurements have confirmed that the heats of solution of iodine are generally greater in donor than in non-complexing solvents (e.g. cyclohexane < benzene < ethyl acetate < ethyl alcohol < pyridine). Investigations of the colligative properties and of the apparent molar volume of dissolved iodine have also been described. The magnetic susceptibility of benzene solutions of iodine is greater than expected from the normal additivity law, while very pronounced changes in magnetic susceptibility have been reported to accompany the formation of complexes of iodine or bromine with polycyclic aromatic bases such as perylene and pyrene, both in solution and in the solid state. Such adducts probably owe their unusual magnetic properties to the presence of significant concentrations of radical-ions, which would also account for their surprisingly high electrical conductivities.

More recently the ^{1}H nmr spectra of organic sulphide molecules or of methylpyridines have been used as indices to complex-formation with iodine or iodine monochloride[220]. Although the lifetimes of the complexes in solution are too short to permit the observation of more than averaged signals for a given proton, such measurements have been successfully applied to the determination of formation constants. The interaction of olefins and related hydrocarbons with molecular iodine supported on a celite or firebrick column has also been monitored by a gas-solid chromatographic technique; the retention times of the hydrocarbons afford measures of the relative formation constants for the olefin–iodine complexes, which follow a pattern similar to those for olefin–$AgNO_3$ complexes[221].

3. Properties of solid complexes[168,198,201–204]

The theoretical interpretation of donor–acceptor interactions has stimulated numerous structural analyses based on X-ray diffraction studies of crystals of the halogen complexes. It has to be recognized, however, that the configurations found in the crystalline state are not necessarily the same as for individual complexes in solution or in the gas phase; only for especially stable complexes are discrete donor–acceptor units likely to be preserved in the different phases. In weak complexes the crystals typically consist of chains or sheets in which donor and acceptor units alternate in a regular way; a particular orientation of donor and acceptor may thus be favoured over others because it facilitates the attainment of a chain or layer structure, whereas the most stable orientation for a discrete complex in the gas phase may be quite different. Nevertheless, the following details of the crystalline structure are of especial interest:

218 S. N. Bhat and C. N. R. Rao, *J. Amer. Chem. Soc.* **90** (1968) 6008.
219 A. J. Hamilton and L. E. Sutton, *Chem. Comm.* (1968) 460.
220 J. Yarwood, *Chem. Comm.* (1967) 809; E. T. Strom, W. L. Orr, B. S. Snowden, jun., and D. E. Woessner, *J. Phys. Chem.* **71** (1967) 4017.
221 R. J. Cvetanović, F. J. Duncan, W. E. Falconer and W. A. Sunder, *J. Amer. Chem. Soc.* **88** (1966) 1602.

TABLE 16. STRUCTURAL DATA FOR CRYSTALLINE HALOGEN COMPLEXES

Donor atom	Complex	Intermolecular contact (Å)	Van der Waals' radius sum (Å)	Covalent radius sum (Å)	XY intramolecular bond length (Å)	Intramolecular bond length in free XY (Å)	Bond angle D··X–Y (°)	Structural type	Configuration
	(a) *σ–σ complexes*								
N	Me_3N, I_2	N–I, 2·27	3·65	2·03	I–I, 2·83	2·67	179	Isolated mol.	N··I–I linear[a]
	Me_3N, ICl	N–I, 2·30	3·65	2·03	I–Cl, 2·52	2·32	180	Isolated mol.	N··I–Cl ≈ linear (within 3°)[a]
	Pyridine, ICl	N–I, 2·26	3·65	2·03	I–Cl, 2·51	2·32		Isolated mol.	N··I–Cl linear[a]
	Pyridine, IBr	N–I, 2·26	3·65	2·03	I–Br, 2·66	2·47		Isolated mol.	N··I–Br linear[a]
	Pyridine, ICN	N–I, 2·57	3·65	2·03	I–C —	1·99		Isolated mol.	γC··N··I–CN linear[a]
	4-Picoline, I_2	N–I, 2·31	3·65	2·03	I–I, 2·83	2·67	180	Isolated mol.	γC··N··I–I linear[a]
	Phenazine, I_2	N–I, 2·92	3·65	2·03	I–I, 2·75	2·67		Infinite chains	[b]
	Pyridine, $2I_2$	N–I, 2·16 cation	3·65	2·03					Salt $[py_2I]^+I_3^-$, $2I_2$[a]
	Hexamethylene-tetramine, $2Br_2$	N–Br, 2·16	3·45	1·84	Br–Br, 2·43	2·28	180	Isolated mol.	N tetrahedral; N··Br–Br linear[a]
	$2CH_3CN, Br_2$	N–Br, 2·84	3·45	1·84	Br–Br, 2·328	2·28	179·4	Isolated mol.	N··Br–Br··N linear[c]
O	1,4-Dioxan, 2ICl	O–I, 2·57	3·55	1·99	I–Cl, 2·33	2·32	180	Isolated mol.	Cl–I··dioxan··I–Cl units; O··I–Cl linear[a]
	1,4-Dioxan, I_2	O–I, 2·81	3·55	1·99					[d]
	1,4-Dioxan, Br_2	O–Br, 2·71	3·35	1·80	Br–Br, 2·31	2·28		Infinite chains	Chains with equatorial dioxan–Br bonding[a]
	Acetone, Br_2	O–Br, 2·82	3·35	1·80	Br–Br, 2·28	2·28	180	Infinite chains	Chains with O··Br–Br··O linear (see Fig. 15)[a]
	$2CH_3OH, Br_2$	O–Br, 2·80	3·35	1·80	Br–Br, 2·28	2·28	177	Infinite sheets	MeOH··Br_2··MeOH units; linked into H-bonded structure[a]
	1,4-Dioxan, Cl_2	O–Cl, 2·67	3·20	1·65	Cl–Cl, 2·02	1·99		Infinite chains	As in 1,4-dioxan,Br_2[a]
S	Dibenzyl sulphide, I_2	S–I, 2·78	4·00	2·37	I–I, 2·82	2·67	180	Isolated mol.	S··I–I linear[a]
	1,4-Dithian, $2I_2$	S–I, 2·87	4·00	2·37	I–I, 2·79	2·67	177·9	Isolated mol.	I_2 molecule equatorial[a]
	1,4-Dithian, 2IBr	S–I, 2·687	4·00	2·37	I–Br, 2.646	2·47	178·2	Isolated mol.	As in 1,4-dioxan,2ICl[e]
	$2Ph_3PS, 3I_2$	S–I, 2·69	4·00	2·37	I–I, 2·86	2·67	175	Isolated mol.	Crystal contains 2 Ph_3PS··I_2 units linked by a "normal" I_2 molecule (I–I = 2·73 Å)[f]

Se	1,4-Diselenacyclohexane, $2I_2$	Se–I, 2·83	4·15	2·50	I–I, 2·87	2·67	180	Isolated mol.	Chair form with axial I_2[a]
	Selenacyclopentane, I_2	Se–I, 2·76 3·64	4·15	2·50	I–I, 2·91	2·67	179·4	Isolated mol.	Axial I_2 molecule interacting more weakly with a second Se atom also axial[a]
	1-Oxa-4-selenacyclohexane, I_2	Se–I, 2·755 3·708	4·15	2·50	I–I, 2·956	2·67	174·8	Isolated mol.	Structure resembles that of selenacyclopentane, I_2[g]
	1-Oxa-4-selenacyclohexane, ICl	Se–I, 2·630	4·15	2·50	I–Cl, 2·73	2·32	175·8	Isolated mol.	Chair form with ICl bonded to Se in the axial position[h]
Te	Di(*p*-chlorophenyl) telluride, I_2	Te–I, 2·947 2·922	4·35	2·70	I··I, 3·85	2·67		Isolated mol.	Consists of $(p\text{-ClC}_6H_4)_2TeI_2$ molecules with axial I–Te–I bonds but with relatively short I··I intermolecular contacts[i]
	(b) *σ–π complexes*								
	C_6H_6, Br_2	C_6H_6–Br, 3·36	3·65		Br–Br, 2·28	2·28		Infinite chains	Chains of alternate C_6H_6 and X_2 molecules with X_2 perpendicular to the plane of the C_6H_6 ring[a]
	C_6H_6, Cl_2	C_6H_6–Cl, 3·28	3·50		Cl–Cl, 1·99	1·99			

[a] C. K. Prout and J. D. Wright, *Angew. Chem., Internat. Edn.* **7** (1968) 659; C. K. Prout and B. Kamenar, *Molecular Complexes* (ed. R. Foster), Paul Elek (Scientific Books) (1973).

[b] T. Uchida, *Bull. Chem. Soc. Japan*, **40** (1967) 2244.

[c] K.-M. Marstokk and K. O. Strømme, *Acta Cryst.* **B24** (1968) 713.

[d] O. Hassel, *Acta Chem. Scand.* **19** (1965) 2259.

[e] C. Knobler, C. Baker, H. Hope and J. D. McCullough, *Inorg. Chem.* **10** (1971) 697.

[f] W. W. Schweikert and E. A. Meyers, *J. Phys. Chem.* **72** (1968) 1561.

[g] H. Maddox and J. D. McCullough, *Inorg. Chem.* **5** (1966) 522.

[h] C. Knobler and J. D. McCullough, *Inorg. Chem.* **7** (1968) 365.

[i] G. Y. Chao and J. D. McCullough, *Acta Cryst.* **15** (1962) 887; H. A. Bent, *Chem. Rev.* **68** (1968) 587.

(a) The general configuration of the donor–acceptor unit and the factors affecting the relative orientations of the donor and acceptor partners.

(b) Interatomic distances relevant to the donor–acceptor contact and to internal bonds of the donor and acceptor molecules. Hence it may be possible to correlate the magnitude and importance of donor–acceptor forces in the crystal with the donor and acceptor strengths of the component molecules.

Structural data so determined for σ–σ and σ–π complexes of the halogens are summarized in Table 16; the structures of some of the complexes are depicted in Fig. 15. From these results certain general features emerge. Thus, the crystalline σ–σ complexes are characterized by essentially linear D···X–Y units, where D is the donor and X–Y the halogen acceptor; if X and Y are different halogen atoms then D is linked to the heavier atom. The D···X contact is markedly shorter than the sum of the van der Waals' radii for the donor and acceptor atoms; it is invariably greater than the sum of the corresponding covalent radii, though for the strongest donors the margin becomes relatively narrow (as in 1-oxa-4-selenacyclohexane,ICl). With increasing donor strength the D···X distance contracts in relation to the sum of the van der Waals' radii, while, in keeping with the anti-bonding character of the acceptor orbital, the intramolecular bond of the halogen molecule XY is attenuated. The order of donor strength Se > S > O and the order of acceptor strength $I_2 > Br_2 > Cl_2$ may thus be deduced from the variations in interatomic distances given in Table 16. In the extreme case of pyridine with iodine the weakening of the I–I bond by charge transfer is so great that the bond is broken and the final product $C_5H_5N,2I_2$ is best formulated as $[(C_5H_5N)_2I]^+I_3^-,2I_2$; here the cation is centrosymmetric and planar with two equal N–I distances of 2·16 Å[222].

In the blue starch–iodine complex, which has been known for well over a century, some unusual structural details have come to light. The pertinent facts about this bluish-black complex are as follows: (i) it exhibits strong absorption in the 6250 Å region; (ii) only amylose, the linear polymer fraction of the starch, forms the complex; (iii) the complexing unit and chromophore is not normally I_2 but $I_2\cdots I^-$; (iv) the iodine atoms are arranged in linear chains each of which occupies the central channel of a helix formed by the amylose unit; (v) there are about 3·9 glucose units per iodine atom[223,224]. The most recent measurements of adsorption isotherms of the $I_2\cdots I^-$ system in the helical cavity of amylose have been interpreted[225] on the assumption that the bound species can be expressed approximately as I_2,I^-_b, where b varies between 0 and 1. A model for the structure is believed to have been found in the complex $HI_3,2C_6H_5CONH_2$, wherein the benzamide molecules are associated *via* hydrogen-bonding and stacked in such a way as to leave long channels in which nearly linear I_3^- ions are aligned[226]. There is evidence of strong attraction between successive I_3^- ions with $d(I\cdots I) = 3{\cdot}80$ Å as against intramolecular distances of 2·90 and 2.96 Å. There are two such polyiodide chains in each channel probably cross-linked by hydrogen bonds formed by the HI_3 protons; the internal diameter of the amylose helix does not appear to be large enough to accommodate more than one chain. Interaction between the I_3^- ions is believed to be responsible, not only for the blue colour, but also for the esr

222 O. Hassel and H. Hope, *Acta Chem. Scand.* **15** (1961) 407.
223 R. Bersohn and I. Isenberg, *J. Chem. Phys.* **35** (1961) 1640.
224 B. S. Ehrlich and M. Kaplan, *J. Chem. Phys.* **51** (1969) 603.
225 C. L. Cronan and F. W. Schneider, *J. Phys. Chem.* **73** (1969) 3990.
226 J. M. Reddy, K. Knox and M. B. Robin, *J. Chem. Phys.* **40** (1964) 1082.

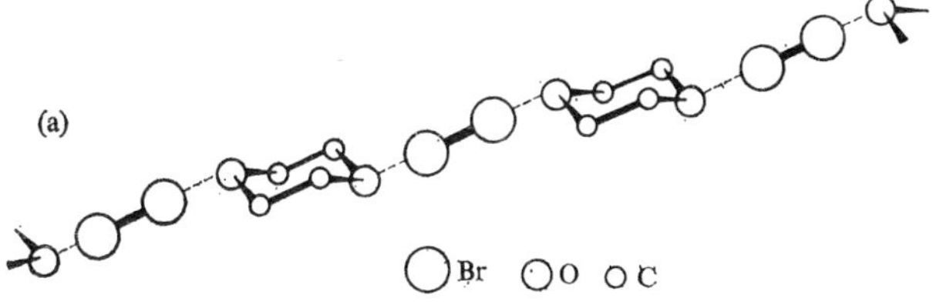

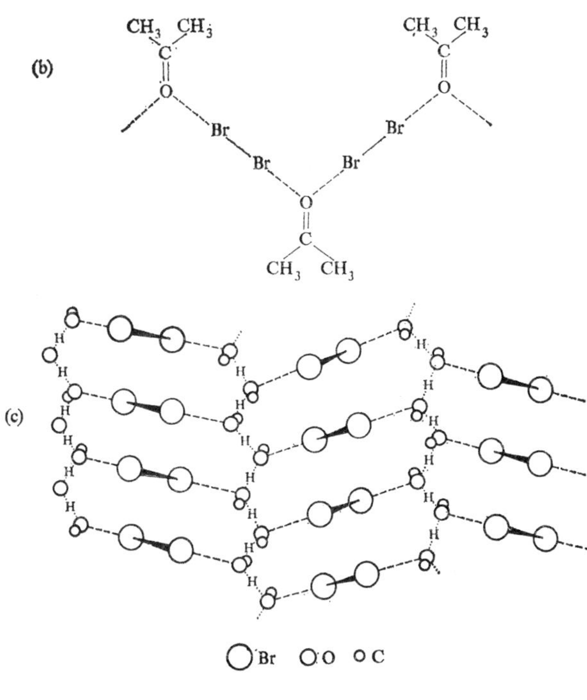

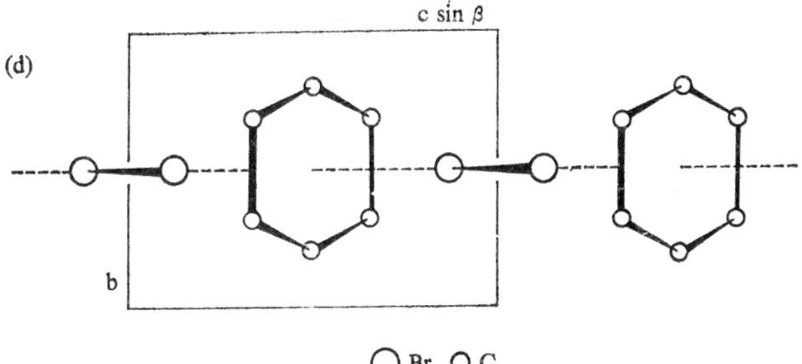

FIG. 15. Schematic diagrams of the structures of crystalline halogen complexes: (a) 1,4-dioxan, Br_2; (b) acetone, Br_2; (c) $2CH_3OH$, Br_2; (d) C_6H_6, Br_2 projected along *a* axis. [Reproduced from *Molecular Complexes* (ed. R. Foster) by kind permission of Paul Elek (Scientific Books) Ltd. and C. K. Prout and B. Kamenar.]

spectrum given by the starch–iodine complex, and on the basis of such evidence the intriguing proposal has been made[223] that the rows of iodine atoms function as one-dimensional metals. However, Mössbauer resonance experiments focused on the ^{129}I nucleus give unambiguous notice that the I_3^- ions, whether in solid CsI_3, $HI_3,2C_6H_5CONH_2$ or amylose-I_3^-, retain non-equivalent iodine atoms[224]. Accordingly the notion of a one-dimensional metal must be treated with some reserve. Other polyhydric compounds such as amylopectin, formed by the hydrolysis of starch (e.g. by enzyme action), also give highly coloured complexes with iodine. Differences in the spectroscopic properties of such complexes have been exploited for analytical measurements.

Charge transfer may ultimately proceed to the reduction of the halogen and simultaneous oxidation of the donor. Thus the interaction of a compound of the type R_2Se or R_2Te (R = organic group) with a halogen may produce, not an adduct, but the corresponding dihalide, a molecular compound with a nearly linear X–Se–X or X–Te–X skeleton. However, the crystal structure of one such compound $(p\text{-}ClC_6H_4)_2TeI_2$ still exhibits certain idiosyncracies suggestive of donor–acceptor interaction: one Te–I bond is significantly longer than the other, while the intermolecular I···I packing distance in the line of the Te–I bonds is 3·85 Å, as against 4·30 Å for the sum of the van der Waals' radii[202]. It appears, then, that by variation of the donor it is possible to move in gradual steps from a well-defined molecular adduct such as dibenzyl sulphide,I_2, where the I–I interaction is relatively strong, through an intermediate stage, represented for example by selenacyclopentane,I_2, where the I–I interaction is relatively weak and some I–Se···I bonding exists, to a compound such as $(p\text{-}ClC_6H_4)_2TeI_2$, where the I···I interaction is very weak and I–Te–I bonding is relatively strong.

In those complexes where the donor–halogen interaction is relatively strong, discrete molecular units are usually identifiable. By contrast, crystals of the weaker complexes frequently contain chains in which donor sites are linked by linear D···X–X···D bridges. In many respects these bridges are comparable with hydrogen bonds, an analogy well displayed in the crystalline complex $Br_2,2CH_3OH$ (Fig. 15): each oxygen atom is here coordinated to three neighbouring oxygens by two hydrogen bonds and one O···Br–Br···O bridge.

Bridging halogen molecules are likewise prominent in the crystalline σ–π complexes X_2,C_6H_6 (X = Cl or Br) on the premises of two-dimensional crystallographic analysis assuming the centrosymmetric space group $C_{2h}{}^3$. The structure consists of infinite chains of alternate benzene and halogen molecules with the latter placed perpendicular to the planes of the benzene rings along a common sixfold axis and equidistant from successive pairs of benzene molecules. The benzene–halogen distances are slightly shorter than the sum of the corresponding van der Waals' radii, but the X–X distances are not measurably different from those in the free halogen molecules. In the face of these conclusions, the infrared spectra of single crystals of Br_2,C_6H_6 are consistent, not with the centrosymmetric disposition of benzene and bromine molecules required by the space group $C_{2h}{}^3$, but rather with a structure having alternate long and short benzene–bromine distances[227]. This apparent contradiction is probably a result of disorder in the crystals. In all probability there are but marginal differences of energy between the different configurations of such a weakly bonded complex; in solution the complex may therefore assume one or more highly labile configurations quite unrelated to the $C_6H_6\cdots X_2$ axial unit characteristic of the crystal.

[227] W. B. Person, C. F. Cook and H. B. Friedrich, *J. Chem. Phys.* **46** (1967) 2521.

Measurements of halogen nqr frequencies[112,228] and of the Mössbauer effect in the ^{129}I nucleus[229] have also been applied to solid complexes of the halogens. Approximate charge distributions calculated from the nqr frequencies of complexes of amines with Br_2, I_2, IBr or ICl clearly signify the transfer of charge from the amine to the halogen molecule in the electronic ground state of the complex. The transferred charge appears to go mainly to the halogen atom Y in the unit D···X–Y, with the result that in complexes of the homonuclear halogens the atom X acquires a partial positive charge. Such a charge distribution correlates well with a scheme of delocalized σ-bonding encompassing the three centres D···X–Y (compare Fig. 4 and p. 1560), and according to which the trihalide ions, amine–halogen complexes and $[(amine)_2halogen]^+$ cations constitute an isoelectronic series. On the other hand, the nqr properties of the solid C_6H_6,Br_2 complex vouch for the implication of the X-ray data that the degree of charge-transfer is here very small[230].

4. Factors influencing the stabilities of halogen complexes[168,196,197,201,204]

Formation constants and related thermodynamic parameters have been measured for a large number of halogen complexes in solution. Combined with the interatomic dimensions of those solid complexes which have been subjected to structural analysis, such results serve as a basis for evaluating donor–acceptor interactions as influenced, for example, by changes in the acceptor, the donor or the environment. The consensus of the evidence is that the relative acceptor strengths of the halogens with respect to a given donor follow the sequence ICl > BrCl > IBr > I_2 > Br_2 > Cl_2. This order has been found to apply to interactions with donors as diverse as aromatic hydrocarbons and halide ions, though it varies somewhat with the choice of donor; certainly the response of the formation constant to changes in the donor varies in magnitude from acceptor to acceptor, being greatest for the strongest acceptor. It is difficult to compare the acceptor strengths of the halogens with those of other Lewis acids because of the wide variation in conditions which have been used in studying different kinds of complex. However, aromatic hydrocarbons form adducts with sulphur dioxide intermediate in stability between the complexes formed with bromine and with chlorine, while iodine and 1,3,5-trinitrobenzene appear to be comparable in strength, if not in electronic action, as acceptors.

Analogous considerations suggest that, with respect to a given halogen, donor capacity increases in the sequence benzene < alkenes < polyalkylbenzenes ≈ alkyl iodides ≈ alcohols ≈ ethers ≈ ketones < organic sulphides < organic selenides < amines. Again some latitude must be allowed for such influences as the acceptor strength of the halogen or steric factors, which may assume importance in many situations. For example, complexes formed by 1,3,5-tri-*tert*-butylbenzene are much less stable than those formed by mesitylene presumably because of the bulk of the alkyl substituents, which inhibits the close approach of the donor and acceptor moieties. The halogens thus emerge as "soft acids". Indeed, iodine is favoured as a reference soft acid for evaluating the relative coordinating power of different donor species, including a wide range of non-aqueous solvents[231]. In this connection

[228] G. A. Bowmaker and S. Hacobian, *Austral. J. Chem.* **22** (1969) 2047.

[229] C. I. Wynter, J. Hill, W. Bledsoe, G. K. Shenoy and S. L. Ruby, *J. Chem. Phys.* **50** (1969) 3872.

[230] H. O. Hooper, *J. Chem. Phys.* **41** (1964) 599; D. F. R. Gilson and C. T. O'Konski, *ibid.* **48** (1968) 2767.

[231] R. S. Drago and K. F. Purcell, *Non-aqueous Solvent Systems* (ed. T. C. Waddington), p. 225, Academic Press (1965).

use has been made of the Drago–Wayland relationship[232],

$$-\Delta H = E_A E_B + C_A C_B$$

which resolves the enthalpy of adduct-formation, ΔH, into parameters E and C representing the susceptibilities of the acid (A) and base (B) to engage in electrostatic interaction (E) and covalent bonding (C). The relationship expresses the generality that all donor–acceptor interactions are composed of some electrostatic and some covalent properties; to this extent so-called "hardness" and "softness" are not mutually exclusive characteristics. The selection of iodine as a reference acid entails the assignment of equal values (unity) of E_A and C_A[233].

The thermodynamic properties designating the formation of halogen complexes show significant variations as the solvent is changed. The enthalpy of complex-formation measured in solution $\Delta H°$(soln) is related to that measured in the gas phase $\Delta H°$(g) by

$$\Delta H°(\text{soln}) = \Delta H°(\text{g}) - \Delta H_{\text{solv}}(\text{X}_2) - \Delta H_{\text{solv}}(\text{D}) + \Delta H_{\text{solv}}(\text{D,X}_2)$$

where ΔH_{solv} denotes the enthalpy of solvation; analogous relationships exist for the free energy and entropy changes attending complex-formation. For a system such as $MeCONMe_2 + I_2 \rightleftharpoons MeCONMe_2,I_2$, only in dilute solution in an inert solvent like carbon tetrachloride or a saturated hydrocarbon is the net solvation energy small compared with the enthalpy of adduct-formation, and only for such solutions is it safe to assume that the enthalpy measured closely corresponds to $\Delta H°$(g). Even with inert solvents the thermodynamic properties of complex-formation vary somewhat from solvent to solvent, a circumstance which can commonly be attributed to minor differences in activity coefficients of the reactants and complex in the different media; variations in the degree of aggregation of one or more components represent another potential source of solvent-dependent thermodynamic properties[234]. A much larger variation results, however, with polar solvents which themselves have some capacity to act either as donors or acceptors. Under these conditions the solvation energies and the balance of such energies may approach, or become relatively large compared with, the measured enthalpies of adduct-formation; this is found to be the case even with a medium as weakly solvating as dichloromethane[232]. Solutions in inert solvents are amenable to "regular" solution theory: for example, one such treatment relates the formation constant K of a complex to the solubility parameter δ_s of the solvent by the equation

$$\log K = a + b\delta_s$$

where a and b depend only on the properties of the donor and acceptor[235]. More generally, however, the effect of solvent perturbation is incompletely understood; it is not even convincingly established that the interactions stabilizing the complex itself are the same in the gas phase as in solution[204].

It is a general rule that the intensity of the charge-transfer band of a complex in the vapour phase is considerably lower than that of the same complex in solution. The reason for this phenomenon is again not clear. One suggestion is that the solvent cage around the complex in solution confines it to the extent that it is under some pressure, resulting in enhanced

232 R. S. Drago, *Chem. in Britain*, **3** (1967) 516.

233 R. S. Drago and B. B. Wayland, *J. Amer. Chem. Soc.* **87** (1965) 3571; O. W. Kolling, *Inorg. Chem.* **8** (1969) 1537.

234 W. Partenheimer, T. D. Epley and R. S. Drago, *J. Amer. Chem. Soc.* **90** (1968) 3886.

235 P. V. Huong, N. Platzer and M. L. Josien, *J. Amer. Chem. Soc.* **91** (1969) 3669.

overlap of donor and acceptor orbitals and hence in an increased transition moment for the charge-transfer absorption. This is consistent with the observed effects of externally applied pressure on the spectra of π–π complexes in solution[204].

5. Nature of the interactions in halogen complexes[168,196–204]

As a rule donor–acceptor interactions are intermediate in character between ordinary van der Waals' contacts and normal covalent bonds. Such interactions, which include hydrogen bonds, may be viewed as the first steps of bimolecular nucleophilic displacement reactions. To this extent there is a close relation between the formation of molecular adducts and certain types of reaction of the elementary halogens. Numerous theories have been developed concerning the nature of donor–acceptor interactions; the views of the early theorists have been surveyed elsewhere in some detail[236,237]. According to the various approaches, the interactions have been described chemically by such phrases as "saturation of residual affinities", "exaltation of valency", "secondary acid–base interactions" or "face-centred bonding"; physically in terms of "adhesion by stray feeler lines of force", "adhesion by attraction of positive and negative patches in molecules", "charge-sharing" or "charge-transfer"; and quantum mechanically by reference to "complex resonance", "no-bond–dative-bond resonance", "filling of anti-bonding orbitals" or "interaction of the highest occupied orbitals (of one component) with the lowest vacant orbitals (of the other)"[202]. These phrases are all useful and complementary, emphasizing different aspects of intermolecular interactions. Some phrases emphasize the directional character of strong intermolecular interactions; or the intermediate length and strength of intermolecular as compared with intramolecular bonds; or the similarity of intramolecular and intermolecular bonds; or the creation of formal charges and "expanded octets"; or the fact that, like intramolecular bonding, intermolecular bonding is essentially an electrostatic phenomenon.

An adequate interpretation of donor–acceptor interactions, applicable to the halogen complexes, must account satisfactorily for the appearance of the intense absorption bands which are characteristic of these adducts, and for the variations in the frequencies and intensities of the bands with changes in the nature of the components. In addition it should be consistent, insofar as it is applicable to the prediction of the orientation of the components with respect to each other, with the thermodynamic properties, vibrational spectra and dipole moments of complexes in solution and with the crystal structures of solid adducts. In this sense the theoretical basis of the interaction of donor and acceptor molecules first given in 1952 by Mulliken, the so-called charge-transfer theory of complex-formation[238], has gained the widest acceptance.

According to Mulliken's model the donor–acceptor complex D,A in its ground state is best represented as a resonance hybrid or combination of a "no-bond" wavefunction $\psi_0(D,A)$ and one or more "dative-bond" functions such as $\psi_1(D^+–A^-)$. The no-bond function includes the electronic energy of the component molecules plus terms representing the effect of dipole interactions, disperson forces, hydrogen-bonding and other intermolecular forces. The dative-bond functions represent states where an electron has been transferred from the donor molecule to the acceptor, introducing electrostatic interactions and forming a weak covalent link between the resulting radical ions. In quantum-mechanical

[236] L. J. Andrews, *Chem. Rev.* **54** (1954) 713.
[237] L. E. Orgel, *Quart. Rev. Chem. Soc.* **8** (1954) 422.
[238] R. S. Mulliken, *J. Amer. Chem. Soc.* **72** (1950) 600; *ibid.* **74** (1952) 811; *J. Phys. Chem.* **56** (1952) 801; *J. Chem. Phys.* **23** (1955) 397; *Rec. trav. chim.* **75** (1956) 845.

terms the wavefunction for the ground state is approximated by

$$\psi_N = a\psi_0(\text{D,A}) + b\psi_1(\text{D}^+ - \text{A}^-) \tag{1}$$

For a weakly bonded complex $a \gg b$, a and b the mixing coefficients being related by the normalization condition $a^2 + b^2 = 1$. Excited states, with a dative structure as the main contributor, have the same form with the coefficients varied to give predominance to the dative-bond contribution. Thus the wavefunction of such a charge-transfer state is given by

$$\psi_E = a^*\psi_1(\text{D}^+ - \text{A}^-) + b^*\psi_0(\text{D,A}) \tag{2}$$

wherein $a^* \gg b^*$. The coefficients a^* and b^* are determined by the quantum-theory requirement that the excited-state wavefunction be orthogonal to the ground-state function: $\int\psi_N\psi_E d\tau = 0$. It follows that $a^* \approx a$ and $b^* \approx b$. The results of the resonance interaction are then:

(i) Admixture of the charge-transfer state gives a ground state for the complex which is more stable than that represented by any of the component wavefunctions.

(ii) Absorption of light induces a transition from the ground to the excited state with transfer of an electron from an orbital largely associated with the donor to an orbital largely associated with the acceptor. This accounts for the charge-transfer band characteristic of the complex.

Intensity of the Charge-Transfer Band

The transition moment for the absorption μ_{EN} is given by

$$\mu_{EN} = -e\int\psi_E \Sigma \mathbf{r}_i \psi_N d\tau \tag{3}$$

$\mathbf{r}_i$ being the position vector of the ith electron. Hence it may be deduced that

$$\mu_{EN} = a^*b(\mu_1 - \mu_0) + (aa^* - bb^*)(\mu_{01} - S\mu_0) \tag{4}$$

where $\mu_1 = -e\int\psi_1\Sigma\mathbf{r}_i\psi_1 d\tau$, $\mu_0 = -e\int\psi_0\Sigma\mathbf{r}_i\psi_0 d\tau$, $\mu_{01} = -e\int\psi_1\Sigma\mathbf{r}_i\psi_0 d\tau$ and $S = \int\psi_0\psi_1 d\tau$. There are therefore two contributions to the transition moment. The first term in $\mu_1 - \mu_0$ is proportional to the dipole moment of the transferred electron and the hole it leaves behind and is related to the stabilization of the ground state through the coefficient a^*b. If this term is to contribute to the intensity of the charge-transfer band, a^*b must be greater than zero, implying some region of overlap between the orbitals of D and those of A; the greater the overlap the more intense the band. However, Mulliken has pointed out that, even if a^*b is very small or zero, so that there is negligible stabilization of the ground state, a charge-transfer band may still be observed through the influence of the second term in equation (4). This provides a plausible explanation of the charge-transfer absorption observed in systems such as iodine in n-heptane or cyclohexane, wherein complex-formation in any sense other than statistical collision-pairing is improbable. In a chance collision encounter between the hydrocarbon and halogen molecules the ground-state potential energy curve is unlikely to exhibit a minimum—that is, no recognizable complex is formed—but the orbitals of the components may overlap adequately to give a mixing of the non-bonded and the charge-transfer states, with a substantial transition moment. The expression "contact charge-transfer" has been coined to describe the corresponding absorption which appears, for example, in iodine–heptane mixtures near 260 mμ. Even in the cases where a definite complex can be identified, collision charge-transfer, as well as complex charge-transfer, contributes to the total intensity of the observed absorption. In a series of related complexes, the portion of the intensity arising from the complexed donor–acceptor pairs progressively increases and that contributed by the collision-pairs decreases as the complexes become more stable. Calculations suggest that as much as three-quarters of the intensity of the observed charge-transfer band of the iodine–benzene system may be derived from collision-pairs[200]. In all probability there is a continuous gradation between the two extremes represented by a true complex and a contact-pair, corresponding to a wide variation in the "stickiness" of collisions responsible for charge-transfer absorption. There may also be some flexibility in respect of those orientations of the components which are required in order that absorption might occur. A mixing of the charge-transfer transition with internal transitions of the component molecules, particularly those of the donor, can also contribute additional intensity to the charge-transfer absorption. According to Murrell[199], such mixing even accounts for the so-called "contact charge-transfer" phenomena.

Energy Terms

Application of the variation principle gives for the ground-state energy E associated with the total wavefunction of the complex

$$(E_0-E)(E_1-E) = (H_{01}-ES)^2 \tag{5}$$

Here $E_0 = \int\psi_0 H\psi_0 d\tau$ is the energy associated with the structure D,A, $E_1 = \int\psi_1 H\psi_1 d\tau$ is the energy associated with the charge-transfer state D^+-A^-, $H_{01} = \int\psi_0 H\psi_1 d\tau$ is the interaction energy due to the mixing of ψ_0 and ψ_1, H is the total exact Hamiltonian for the entire system, while, as before, $S = \int\psi_0\psi_1 d\tau$ denotes the overlap of the functions ψ_0 and ψ_1. From equation (5) it follows that

$$E = E_0-(H_{01}-ES)^2/(E_1-E)$$

There are two solutions of this equation for E. One corresponds to the final ground state (E_N) and the other to the final excited state (E_E). Because the energy of interaction between a halogen and donor molecule is typically small and E_1-E is relatively large, E_N approximates to E_0. Then

$$E_N = E_0-(H_{01}-E_0S)^2/(E_1-E_0) \tag{6a}$$

while the corresponding energy of the excited state is

$$E_E = E_1+(H_{01}-E_1S)^2/(E_1-E_0) \tag{6b}$$

Further, the mixing coefficients a and b of the ground state and a^* and b^* of the excited state are given by

$$\frac{b}{a} = -(H_{01}-E_0S)/(E_1-E_0) \qquad \frac{b^*}{a^*} = -(H_{01}-E_1S)/(E_1-E_0) \tag{7}$$

Equation (6a) emphasizes that the energy of the ground state has contributions both from classical intermolecular forces, electrostatic in origin, through E_0 and from covalent interactions principally embodied in the second term. The partitioning of the energy into electrostatic and covalent parts provides a theoretical justification for the Drago–Wayland relationship (see p. 1208).

The energetics of complex-formation are illustrated schematically in Fig. 16. The stabilization energy of the ground state arising from the mixing of ψ_0 and ψ_1, E_0-E_N, is comparable with, but generally larger than, the heat of formation of the complex ΔH; for the benzene–iodine complex, for example, $E_0-E_N \approx 1{\cdot}3$ kcal mol^{-1}. As a rule S and H_{01} are also relatively small; a and a^* are close to unity and thus large compared with b and b^*,

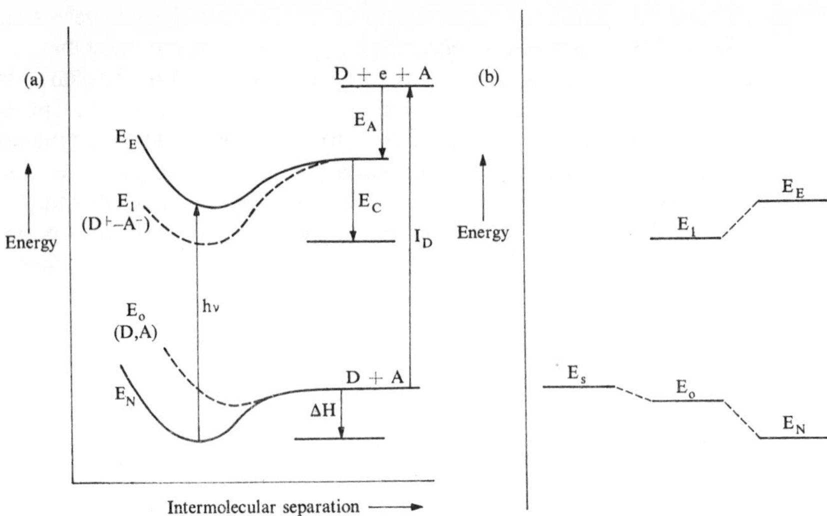

FIG. 16. The relations between the energies of the ground (N), non-bonded (D,A), charge-transfer (D^+-A^-), and excited (E) states and the distance between the donor (D) and acceptor (A) molecules in a charge-transfer complex. E_s is the energy of the infinitely separated donor and acceptor; $E_s - E_o$ = stabilization energy due to classical intermolecular interactions; and $E_o - E_N$ = stabilization energy due to electron delocalization.

while $E_1 - E_0$ is a relatively major energy term, being about 180 kcal mol^{-1} for the benzene–iodine complex, for example.

If I_D is the ionization potential of the donor and E_A the electron affinity of the acceptor, Fig. 16 indicates that these quantities are related to the frequency ν of the charge-transfer transition by

$$h\nu = I_D - E_A - \Delta \tag{8}$$

Δ being the difference between the binding energies of the components in the ground and the excited states. The most important components of Δ are the coulombic energy E_C of the charge-transfer state D^+-A^- and the resonance energies $E_0 - E_N$ and $E_E - E_1$ arising from the mixing of the charge-transfer and non-bonded states. Good linear correlations are found between the charge-transfer frequencies of different iodine complexes and the corresponding ionization potentials of the donor molecules. Similarly, for a particular donor the frequency of the charge-transfer absorption has been shown to be proportional to the electron affinity of the acceptor. However, the slopes of the straight lines relating $h\nu$ to I_D for the iodine complexes are invariably less than unity, implying that Δ varies with the nature of the donor. Account of this is taken in the following approximate relation which can be justified theoretically for a set of closely related weak complexes of a single acceptor:

$$h\nu \approx I_D - C_1 + \frac{C_2}{I_D - C_1} \tag{9}$$

Here $C_1 \approx E_A - E_C + (E_s - E_0)$, the term $(E_s - E_0)$ representing the interaction energy of the donor and acceptor in the formation of the no-bond structure defined by ψ_0, while $C_2 = (H_{01} - E_0 S)^2 + (H_{01} - E_1 S)^2$. For iodine complexes with π-donors, equation (9) implies that $C_1 = 5{\cdot}2$ eV and $C_2 = 1{\cdot}5$ (eV)^2. More generally, however, there is no reason to expect either C_1, with its dependence on the coulombic energy E_C, or the resonance integrals which determine C_2 to remain constant for all donors, even with the same acceptor.

The shift to higher energies in the visible absorption band of iodine as a result of complex-formation corresponds to about $1{\cdot}5\Delta H$. When the iodine molecule, paired off with a donor partner, is excited by the absorption of visible light ($\sigma_u \leftarrow \pi_g$), its suddenly swollen size corresponding to the occupancy of the strongly anti-bonding σ_u orbital introduces an exchange repulsion between it and the donor. It is suggested that the repulsion energy, which should be greater the more intimate the donor–acceptor contact, is added to the usual energy of the excited iodine molecule, thereby increasing the frequency of absorption[204]. Correlation of the visible absorption spectrum of iodine in a wide range of solvents with the ionization potentials of the solvents has recently provided a distinctive classification of the solvents, the variations being attributed to the influences of charge-transfer and contact–charge-transfer interactions[239]. Curiously, however, a recent examination of the spectrum of the strong complex Et_2S,I_2 in the vapour state shows no blue shift in the visible band due to iodine[240]. No good explanation of this behaviour has yet been given.

The energies of the two states represented by ψ_0 and ψ_1 depend upon the relative orientations of the donor and acceptor; only if ψ_1 is much more sensitive to such effects than ψ_0 will the charge-transfer interaction tend to dictate the structure adopted by the complex. Because the energy of the charge-transfer state D^+-A^- may in certain circumstances be a sensitive function of the nature of the solvent, there is not necessarily any relation between the structure of the complex in the solid and solution phases. According to Mulliken's "overlap and orientation" principle, maximum charge-transfer interaction is achieved when the relative orientations of the donor and acceptor molecules provide maximum overlap of the filled donor and vacant acceptor orbitals. This principle is neither proved nor disproved by determinations of individual crystal structures: it is a theoretical principle applying to idealized systems involving no intermolecular forces other than those due to charge-transfer. In reality the configuration of a complex is also affected, and in some cases largely controlled, by electrostatic attraction and exchange-repulsion terms associated with the no-bond structure. Attempts by Mulliken to predict the structures of certain halogen complexes, e.g. C_5H_5N,I_2 and C_6H_6,I_2, on the strength of the overlap and orientation principle, led to conclusions at variance with the structures subsequently identified in the solids; presumably this discord simply reflects the conflicting and subtle requirements of the factors which minimize the total energy of such a system.

More Generalized Treatments

Although the principles of Mulliken's model of charge-transfer interactions have come to be generally accepted, in its simplest form the model is often found to be imperfect. A more general form for the wavefunction of the ground state of a donor-acceptor complex is

$$\psi_N = a\psi_0(D,A) + \Sigma b_i\psi_{1i}(D^+-A^-) + \Sigma c_j\psi_{2j}(D^--A^+) + \cdots \tag{10}$$

wherein more than one charge-transfer state is incorporated. Subject to the overall geometry of the complex, the coefficients b_i and c_j differ from zero only when the corresponding wavefunctions ψ_{1i} and ψ_{2j} belong to the same group-theoretical species as ψ_0. A full treatment should also take account of excited states of the donor and acceptor molecules by inclusion of no-bond functions such as $\psi_0(D^*,A)$ and of charge-transfer functions such as

[239] E. M. Voigt, *J. Phys. Chem.* **72** (1968) 3300.
[240] See for example M. Kroll, *J. Amer. Chem. Soc.* **90** (1968) 1097.

$\psi_1(D^+ - A^{-*})$. In many cases only the first term and one or two terms of the first summation are important, but the possibility of a significant contribution from a charge-transfer state $D^- - A^+$ in which the halogen assumes the donor function cannot be excluded.

There has been considerable debate about the precise nature and relative importance of classical electrostatic forces (e.g. dipole–dipole and dipole–induced-dipole) as against charge-transfer interactions. It has recently been shown, for example, that quadrupole–induced-dipole forces may be comparable in magnitude with charge-transfer interactions in benzene–iodine and related complexes[241]. While some authors have argued the importance, if not preponderance, of coulomb and polarization forces in the formation of weak complexes, others have sought to invoke London dispersion forces as making major contributions to the energies of formation of such complexes. However, although dispersion forces necessarily contribute to the stability of complexes in the vapour phase, their effects are approximately cancelled out in the formation of complexes in solution. A recent analysis concludes that electrostatic, charge-transfer and exchange–repulsion interactions are all important in describing molecular complexes[241]; in the weakest complexes the extent of charge-transfer action may have been overestimated by Mulliken's conventional description, electrostatic forces being pre-eminent. Nevertheless, such considerations do not conflict with the principles of the Mulliken treatment, implying rather the inadequacy of its quantitative use in a highly simplified form.

Alternative descriptions have been based on delocalized molecular orbitals extending over the complex as a whole. A complete account would require all-electron, self-consistent-field MO calculations which have not yet proved tractable for the halogen complexes. However, simplified LCAO-MO approximations have been shown to be compatible with the resonance-structure approach developed by Mulliken, and to underline the similarity between a polyhalide ion like I_3^- and a strong molecular complex like R_3N,I_2 (see p. 1560). Thus, one such approach assigns analogous wavefunctions to the ground states of R_3N,I_2 and of I_3^-, e.g.

$$\psi_N(R_3N,I_2) \approx a\psi_0(R_3N,I_a - I_b) + c\psi(R_3N^+ - I_a,I_b^-) + d\psi(R_3N^+I_a^-I_b) \quad (11)$$

Here the coefficients c and d are unequal; according to nqr measurements[112,228] $c > d$. The distinction between I_3^- and R_3N,I_2 hinges on the relative magnitudes of a, c and d: if d is assumed small, $a = c$ for the symmetric complex I_3^-. A similar situation in R_3N,I_2 would correspond roughly to a structure $R_3N^{+\frac{1}{2}} \cdots I_a \cdots I_b^{-\frac{1}{2}}$ with "half-bonds" between N and I_a and between I_a and I_b. However, the N–I and I–I distances in the solid complex Me_3N,I_2 suggest that c is in fact significantly smaller than a.

6. Intermediacy of halogen complexes in chemical reactions[202,204]

In some complexes two isomeric forms are possible differing strongly in the extent of charge-transfer and often, at the same time, in molecular configuration. Usually only one of the two forms is stable in any one environment, but a change of environment (e.g. from a non-polar to a polar solvent) may favour a change from one to the other form. Such isomeric forms have been designated "outer complexes" (usually loosely bound and with little charge-transfer) and "inner complexes" (usually with more or less complete charge-transfer). The transformation of an outer halogen complex to an inner complex involves dissociation of the halogen molecule, as illustrated in Fig. 17 and by the following examples:

241 M. W. Hanna, *J. Amer. Chem. Soc.* **90** (1968) 285; M. W. Hanna and D. E. Williams, *ibid.* p. 5358; J. L. Lippert, M. W. Hanna and P. J. Trotter, *ibid.* **91** (1969) 4035.

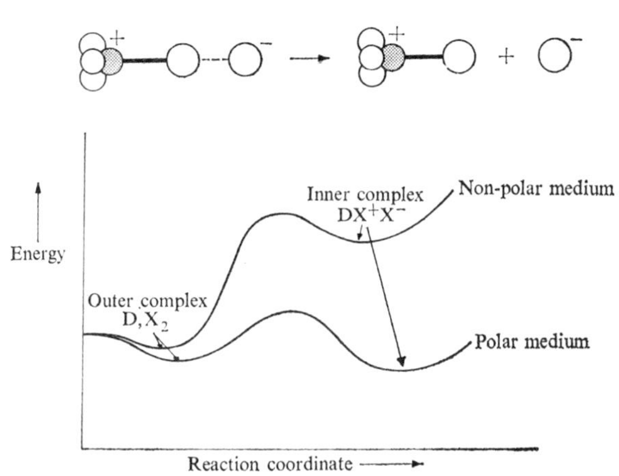

FIG. 17. The formation of "inner" and "outer" complexes of a halogen with a trialkylamine molecule: bimolecular nucleophilic substitution at one of the halogen atoms.

Outer complex — Inner complex

(i) $C_5H_5N + I_2 \rightleftharpoons C_5H_5N, I_2 \underset{\text{Slow}}{\overset{C_5H_5N}{\rightleftharpoons}} [(C_5H_5N)_2I]^+I^-$

(ii) $NH_3 + I_2 \rightleftharpoons H_3N,I_2 \overset{\text{aq. } NH_3}{\rightleftharpoons} [H_3NI]^+I^-$

$[H_3NI]^+I^- \overset{NH_3}{\rightleftharpoons} NH_2I + NH_4^+ + I^-$

(iii) $TMH + I_2 \rightleftharpoons TMH,I_2 \overset{\text{Polar solvent}}{\rightleftharpoons} [\dot{T}MH^+\dot{I}_2^-]_{solv}$

$TMH,I_2 \overset{TMH}{\rightleftharpoons} 2TMH,I_2$

$[\dot{T}MH^+\dot{I}_2^-]_{solv} \rightleftharpoons [T\dot{M}H]^+_{solv} + \dot{I}^-_{2\ solv}$

$(TMH = Me_2N{\cdot}NMe_2)$[242]

Similar reactions with the formation of radical-ions probably occur in the interaction of iodine with donors like perylene or carotene.

(iv) $MR_3 + X_2 \rightleftharpoons R_3M,X_2 \rightleftharpoons [R_3MX]^+X^- \rightleftharpoons R_3MX_2$

$[R_3MX]^+X^- \overset{X_2}{\rightleftharpoons} [R_3MX]^+X_3^-$

$(M = P, \text{As or Sb}; R = CH_3 \text{ or } C_6H_5)$[243]

---

[242] D. Romans, W. H. Bruning and C. J. Michejda, *J. Amer. Chem. Soc.* **91** (1969) 3859.
[243] See for example S. N. Bhat and C. N. R. Rao, *J. Amer. Chem. Soc.* **88** (1966) 3216.

	Outer complex		Inner complex
(v)	$H_2O + X_2 \rightleftharpoons H_2O,X_2$	$\overset{aq.}{\rightleftharpoons}$	$[H_2OX]^+X^-$ $\updownarrow$ aq. $HOX + H_3O^+ + X^-$
(vi)	$H_2S + I_2 \rightleftharpoons H_2S,I_2$	$\rightleftharpoons$	$[H_2SI]^+I^-$ (ref. 244)
(vii)	$R_2Y + X_2 \rightleftharpoons R_2Y,X_2$	$\rightleftharpoons$	$[R_2YX]^+X^- \rightleftharpoons R_2YX_2$

(Y = Se or Te; R = organic group)[202,245]

(viii) $C_6H_6 + X_2 \rightleftharpoons C_6H_6,X_2 \underset{\text{Slow}}{\overset{\text{Polar solvent}}{\rightleftharpoons}} [C_6H_6X]^+ X^-$ (cyclohexadienyl cation bearing H and X)

(ix) $>C{=}C< + Br_2 \rightleftharpoons >C{=}C< \rightarrow Br_2 \overset{Br_2}{\rightleftharpoons} [>C{-}C< \text{ bridged by } Br]^+ Br^-$

The stages involve a bimolecular nucleophilic displacement reaction, comparable with a Walden inversion, at one of the halogen atoms; the formation of a symmetrical complex like I_3^- corresponds to a transition state, sometimes referred to as a "middle complex"[204], intermediate between the outer and inner complexes.

The facility with which such dissociative reactions occur depends upon the natures of the halogen molecule, the donor atom and the substituents borne by the donor atom; the formation of an inner complex is generally a mark of strong donor–acceptor interaction. In contrast with acceptors like the hydrogen halides, which are highly prone to undergo dissociation, the homonuclear halogen molecules are relatively much more reluctant to function in this way. In other words the halogen molecules commonly function simply as sacrificial acceptors, whereas with the hydrogen halide molecules the sacrifice goes virtually to completion; alternatively we may say that the halogen atom X in the molecule X_2 is a relatively poor "leaving group" with respect to nucleophilic displacement at the second atom. However, the relative stabilities of inner and outer complexes are highly susceptible to the influence of environment. In particular, the ionic inner complex is stabilized relative to the outer complex (i) by a polar solvent both through the action of classical electrostatic forces (dependent on properties such as the dielectric constant of the medium) and through the action of the solvent as an auxiliary donor or acceptor, and (ii) by ionic crystal-formation resulting in an increase in coulomb energy per ion-pair as compared with the isolated inner complex. The influence of the solvent is shown qualitatively in Fig. 17: in a non-polar medium the inner complex, while it may have a potential minimum, is typically too high in energy to be formed, while the outer complex is more or less stable as a loose complex; on the other hand, in a sufficiently polar medium the relative stabilities of the two states may well be reversed. Such environmental assistance of the heterolytic dissociation of halogen molecules probably plays a central role in the reactions of the halogens in the condensed phases, and particularly in polar solvents such as water. Some of the more important examples of such reactions are considered briefly in the following sections.

244 J. Jander and G. Türk, *Ber*. **95** (1962) 881.
245 G. C. Hayward and P. J. Hendra, *J. Chem. Soc*. (*A*) (1969) 1760.

Reaction of the Halogens with Water and Hydroxide Ions (see pp. 1188–95 and Table 13)

The rate of hydrolysis of chlorine, bromine and iodine follows the order $Cl_2 < I_2 < Br_2$. In all three cases the intermediate is thought to be the outer complex $[X_2,OH]^-$ or X_2,OH_2 (depending on the conditions), so that the overall reaction rate depends upon the formation of this species and its subsequent transformation to an inner complex:

$$X_2 + OH^- \overset{k_1}{\rightleftharpoons} [X_2,OH]^- \overset{k_2}{\rightarrow} X^- + XOH$$

The stability of the transition state presumably decreases in the order $I_2 > Br_2 > Cl_2$, but the subsequent decomposition depends upon the X–X bond strength, which increases in the order $I_2 < Br_2 < Cl_2$. Understandably this renders unpredictable the overall kinetic pattern. The intermediacy of the species $[I_2,OH]^-$ has been corroborated by studies of alkaline solutions of iodine, which suggest that k_1 is about 100 times greater than k_2[246].

Chlorine hydrate is of historical interest as the solid phase originally thought to be solid chlorine but shown (in 1811) by Davy to contain water[247]; it was later assigned the formula $Cl_2,10H_2O$ by Faraday[248]. Subsequent reports of the crystalline hydrates formed at temperatures close to 0°C gave compositions ranging from $Cl_2,6H_2O$ to $Cl_2,8H_2O$, but, according to a more recent investigation[249], passage of chlorine into dilute solutions of calcium chloride at 0°C gives feathery crystals with the composition $Cl_2,7{\cdot}3H_2O$ and with a melting point above 0°C. The crystal is an example of an ice-like hydrate or clathrate containing polyhedra of hydrogen-bonded water molecules linked in the form characteristic of the ice-I structure. There are 46 water molecules in the unit cell which contains 2 small and 6 rather larger holes. If all the voids are filled, as is probably the case with argon and small molecules like H_2S, this corresponds to the formula $M,5{\cdot}75H_2O$, whereas in chlorine hydrate all the larger holes but only about 20% of the smaller holes are occupied[250]. Bromine forms a similar hydrate with a stoichiometry which is rather ill-defined but which approximates to $Br_2,8H_2O$, melting at about 6°C. Again X-ray diffraction suggests a relatively open ice-like structure with 172 water molecules in the unit cell of the host lattice and a maximum of 20 polyhedral cavities large enough to accommodate Br_2 molecules; occupancy of all these cavities corresponds to a theoretical composition $Br_2,8{\cdot}6H_2O$[251]. In some cases of dramatic proportions, the changes in the solubility of the molecular halogens with the addition of halide ions to an aqueous medium arise from the formation of polyhalide ions such as X_3^- (X = Cl, Br or I), I_2X^- (X = Cl or Br) or Br_2Cl^-. The characteristics of these species are treated more amply in the context of Section 4 (see p. 1534).

Electrophilic Attack of Organic π-Electron Systems[204,252]

Though as yet unproved, the intermediacy of outer and inner complexes is probably essential to the electrophilic action of the molecular halogens (i) on aromatic compounds

[246] J. Sigalla, *J. chim. Phys.* **58** (1961) 602.
[247] H. Davy, *Phil. Trans.* **101** (1811) 155.
[248] M. Faraday, *Quart. J. Science*, **15** (1823) 71.
[249] K. W. Allen, *J. Chem. Soc.* (1959) 4131.
[250] A. F. Wells, *Structural Inorganic Chemistry*, 3rd edn., p. 578. Clarendon Press, Oxford (1962); H. M. Powell, C. K. Prout and S. C. Wallwork, *Ann. Rep. Chem. Soc.* **60** (1963) 576; G. A. Jeffrey, *Dechema Monograph*, **47** (1962) 849; S. Sh. Byk and V. I. Fomina, *Russ. Chem. Rev.* **37** (1968) 469.
[251] K. W. Allen and G. A. Jeffrey, *J. Chem. Phys.* **38** (1963) 2304.
[252] P. B. D. De la Mare and J. H. Ridd, *Aromatic Substitution: Nitration and Halogenation*, Butterworths, London (1959); R. O. C. Norman and R. Taylor, *Electrophilic Substitution in Benzenoid Compounds*, Elsevier (1965); C. K. Ingold, *Structure and Mechanism in Organic Chemistry*, 2nd edn., Bell, London (1969); P. Sykes, *A Guidebook to Mechanism in Organic Chemistry*, 3rd edn., Longmans, London (1970).

which thus undergo substitution reactions (see example (viii) on p. 1216) or (ii) on olefinic or acetylenic compounds which undergo addition (example (ix)). The formation of the ion-pair which constitutes the inner complex is assisted by a polar medium, e.g. water; furthermore, a second acceptor species A may function as a catalyst or so-called "halogen carrier", presumably through the formation of a transition state such as

$$\left[C_6H_6(H)(X)\right]^+ AX^- \quad A = AlX_3, ZnX_2, IX \text{ or } HX \quad (X = Cl, Br \text{ or } I)$$

The catalyst thus assists the action as a "leaving group" of the halogen atom remote from the aromatic nucleus. Under different conditions supplied, for example, by a non-polar medium, relatively high temperatures, the presence of a radical-initiator instead of a halogen-carrier, or irradiation, the interaction may assume a totally different mechanism dominated by homolytic rather than heterolytic fission of the halogen–halogen bond. Thus conditions favouring electrophilic attack lead to substitution at the benzene nucleus rather than addition, which is induced by conditions favouring the formation of halogen atoms.

Detailed studies of the electrophilic halogenation of aromatic molecules by molecular chlorine or bromine are consistent with the reaction sequence

$$C_6H_6 + X_2 \rightleftharpoons C_6H_6 \cdot X^{\delta+}\!-\!X^{\delta-} \underset{\text{slow step}}{\xrightarrow{AlX_3}} C_6H_6 \cdot {}^{+}X \; {}^{-}XAlX_3 \longrightarrow$$

$$\left[C_6H_6(H)(X)\right]^+ + AlX_4^- \longrightarrow C_6H_5X + HX\uparrow + AlX_3$$

$$X = Cl \text{ or } Br$$

the slow step being the formation of a σ-bond between the halogen and the aromatic ring. Molecular iodine is usually unreactive in these circumstances. However, halogenation, including iodination, of aromatic systems can also be achieved by the action of hypohalous acid or a related species (e.g. hypohalite ion or t-butyl hypochlorite) and of interhalogen compounds. Mechanistically the reactions are believed to resemble those involving the molecular halogens. Despite much debate, it seems unlikely that the electrophilic halogen centre is a cation X^+ pre-formed by the heterolytic fission of a halogen molecule; a more plausible view is that the incipient cation is conveyed to the aromatic nucleus in the form of a unit such as $[XOH_2]^+$, X_2, XY, HOX or even perhaps X_3^- (Y = halogen, acetate or alkoxy group).

Olefinic or acetylenic units are readily halogenated by addition of chlorine, bromine or interhalogen compounds. Such reactions are of importance in synthesis, while addition of bromine affords a test of unsaturation, in many cases sufficiently quantitative to be useful for analytical purposes. Addition of iodine is less easily accomplished because the thermodynamic balance is much less favourable, the net difference of bond energies providing little advantage to the forward reaction; in general the reaction also proceeds more slowly. Provided the conditions do not favour the formation of halogen atoms, the addition appears to involve a mechanism analogous to that of electrophilic substitution of aromatic nuclei. The following evidence bears witness to such a mechanism: (a) the addition to olefinic systems is normally found to be stereochemically *trans* and not *cis*; (b) carrying out the addition in the presence of a nucleophile, e.g. halide, NO_3^- or H_2O, results in the formation, in addition to the expected dihalide, of products in which the added nucleophile has become bonded to one of the formerly unsaturated carbon atoms; (c) the addition of a "halogen carrier" accelerates the reaction. Consistent with these observations is the mechanism

$$\mathrm{>C{=}C<} + \mathrm{Br{-}Br} \rightleftharpoons \mathrm{>C{=}C<}\cdots\mathrm{Br}^{\delta+}{-}\mathrm{Br}^{\delta-} \xrightarrow{\text{Slow step}} \mathrm{>\overset{\oplus}{C}{-}C(:Br)<} + \mathrm{Br}^{\ominus} \leftrightarrow \mathrm{>C{-}C<\ (bridged\ Br^{\oplus})} \xrightarrow{\mathrm{Br}^{\ominus}} \mathrm{>CBr{-}CBr<}$$

The halogenation of the α-carbon atom of aliphatic aldehydes, ketones, carboxylic acids and related compounds having some capacity to undergo enolisation probably involves a similar sequence. Addition is facilitated by electron-donating substituents attached to the unsaturated carbon atoms but retarded by electron-withdrawing groups. In extreme cases substituents such as p-$Me_2NC_6H_4$– or p-$MeOC_6H_4$–, which have a powerful fortifying action on the donor properties of ethylene, are reported[253] to favour the formation of inner complexes incorporating relatively stable organic cations: e.g.

Outer complex		Inner complex	
$R_2C{=}CH_2, Br_2 + Br_2$	⟶	$[R_2C\text{-}CH_2Br]^+ Br_3^-$	R = p-$Me_2NC_6H_4$-
$R_2C{=}CR_2, Br_2 + 2Br_2$	⟶	$[R_2C\text{-}CR_2]^{2+} 2Br_3^-$	

Formation of Halogen Cations

First discussed by Noyes and Stieglitz[254] and implicit in many accounts of electrophilic halogenation reactions, the formation of halogen cations of the type X^+ has been reviewed more critically in recent years[255]. However, despite the controversy which still surrounds the existence and significance of species such as H_2OX^+, there is now good evidence of the stable existence of cations such as $[py_2I]^+$ and $[(\gamma\text{-pic})_2I]^+$ (py = pyridine; γ-pic = γ-picoline); once again these may be viewed as the direct outcome of the transformation of an outer into an inner complex. With a polarizable donor partner, e.g. arsenic or tellurium, such a transformation leads characteristically to oxidation of the donor atom (see reactions (iv) and (vii), pp. 1215–6), but with a strong donor resistant to oxidation, the transformation, though formally similar, corresponds to disproportionation of the halogen (reaction (i)). In either case the change may be assisted by the auxiliary action of a second halogen molecule in forming a polyhalide species with the halide ion displaced from the outer complex. The cations $[py_2X]^+$, $[(\gamma\text{-pic})_2X]^+$ and $[(\text{thiourea})_2I]^+$ (X = Br or I) produced in this way are characterized by linear coordination about the halogen atom, a feature substantiated by their vibrational spectra[215,256] and by crystallographic analysis of the complexes $py,2I_2$[222] and $[(\text{thiourea})_2I]^+I^-$[257]. Such cations have also been isolated as salts of the nitrate, perchlorate, tetrafluoroborate or hexafluorophosphate anions.

253 R. Wizinger, *Chimia*, **7** (1953) 273.
254 W. A. Noyes and A. C. Lyon, *J. Amer. Chem. Soc.* **23** (1901) 460; J. Stieglitz, *ibid.* p. 797.
255 R. P. Bell and E. Gelles, *J. Chem. Soc.* (1951) 2734; J. Arotsky and M. C. R. Symons, *Quart. Rev. Chem. Soc.* **16** (1962) 282.
256 I. Haque and J. L. Wood, *J. Mol. Structure*, **2** (1968) 217.
257 H. Hope and G. H.-Y. Lin, *Chem. Comm.* (1970) 169.

Organic donor molecules which are susceptible to oxidation may give rise to radical-ions as the result of the charge-transfer reaction. Such donors include the polyaromatic compounds perylene, pyrene and carotene and the nitrogen donors *p*-phenylenediamine[258], N,N′-diphenyl-*p*-phenylenediamine[186], pyridazine[259] and tetramethylhydrazine (example (iii), p. 1215)[242]. Likewise triphenylamine in propylene carbonate solution is oxidized by iodine to the $[p\text{-}Ph_2N{\cdot}C_6H_4{\cdot}C_6H_4{\cdot}NPh_2\text{-}p]^+$ radical-cation[260]. In cases such as these the distinction between a redox reaction and formation of a charge-transfer complex largely disappears.

Strongly acidic media like sulphuric acid, oleum, fluorosulphonic acid and antimony pentafluoride, individually or in admixture, may favour the formation of cationic species such as X_2^+ and X_3^+ (X = Br or I). For example, when iodine is dissolved in oleum, it gives a deep blue solution containing as a primary ingredient the paramagnetic ion I_2^+[261]. For a review of these and other halogen-containing cations, however, the reader is directed to Section 4 (p. 1340).

Summary of the Reactions of the Halogens

Chemically chlorine, bromine and iodine are well known to be extremely reactive, though the general reactivity is inferior to that of fluorine and decreases in the sequence $Cl_2 > Br_2 > I_2$. With a given reagent the halogens commonly give analogous reaction products. That numerous differences do arise, however, is illustrated by the observations (i) that the ultimate products of halogenation of iron are $FeCl_3$, $FeBr_3$ and FeI_2, of copper are $CuCl_2$, $CuBr_2$ and CuI, and of rhenium are $ReCl_6$, $ReBr_5$ and ReI_4, and (ii) that although chlorine combines with sulphur dioxide, carbon monoxide and nitric oxide to give sulphuryl, carbonyl and nitrosyl chloride respectively, iodine fails to react in this way. The variations in reactivity and in the products of reactions are determined in part by thermodynamic properties, in part by less clearly defined kinetic factors.

Since the enthalpies, and hence free energies, of formation of gaseous halogen atoms or anions do not differ widely for chlorine, bromine and iodine (Tables 1 and 10), the crucial properties which vary considerably and so determine the thermodynamic characteristics of halogenation reactions are (i) the bond energies of molecular halogen compounds or (ii) the sizes of the halide ions through their influence on lattice and solvation energies (see Section 1, pp. 1117–20). With respect to a less electronegative element the halogens appear invariably to follow the bond-energy sequence Cl > Br > I; combination of the halogen with a more electronegative element (oxygen or fluorine) probably leads to a reversal of this sequence. In the reaction with a less electronegative unit M to form a molecular compound MX_n, it follows that iodine is at an energetic disadvantage compared with chlorine because of the weakness of the M–I as compared with the M–Cl bond. The disadvantage may be such that MI_n is unstable with respect to dissociation into molecular iodine and either M or a lower-valent iodide of M. Conversely iodine forms relatively strong bonds with oxygen or fluorine and, compared with chlorine or bromine, it is thus better disposed to give a compound such as IF_m. In these terms it is possible to rationalize, if not to explain, the apparent failure of chlorine or bromine to form a fluoride analogous to IF_7. On the other hand, if consecutive products of halogenation MX_n and MX_{n+1} can be meaningfully described by the ionic model, variations of stability are primarily a function of the lattice energies of the

[258] D. Bargeman and J. Kommandeur, *J. Chem. Phys.* **49** (1968) 4069.
[259] R. J. Hoare and J. M. Pratt, *Chem. Comm.* (1969) 1320.
[260] W. H. Bruning, R. F. Nelson, L. S. Marcoux and R. N. Adams, *J. Phys. Chem.* **71** (1967) 3055.
[261] R. J. Gillespie and J. B. Milne, *Chem. Comm.* (1966) 158; *Inorg. Chem.* **5** (1966) 1236, 1577.

aggregates. For a given oxidation state of M, the stability of the compound with respect to the elements then decreases in the order Cl > Br > I, and the energetic advantage of the oxidation $MX_n \rightarrow MX_{n+1}$ diminishes in the same order (see Section 1, p. 1119). For a given metal of the first transition or lanthanide series, the susceptibilities of the trihalides to the decomposition $MX_3 \rightarrow MX_2 + \frac{1}{2}X_2$ comply with this pattern: accordingly, for M = Eu, Sm or Yb, the decomposition occurs most readily when X = I, while for M = Fe no tri-iodide is known. Similarly the increase in lattice energy of CuX_2 relative to CuX is sufficient to compensate for the (large) second ionization potential of copper in the cases where X = F, Cl or Br but not where X = I.

Kinetically two mechanisms are open to such reactions; these may be formally represented as:

(i) Homolytic fission $$X_2 \xrightarrow[\text{or } \Delta]{h\nu,\text{ radical initiator}} X\cdot + X\cdot$$

(ii) Heterolytic fission $$X_2 \xrightarrow[\text{Halogen carrier}]{\text{Polar solvent}} X^+ + X^-$$

Homolytic fission corresponds to the production of halogen atoms, and the characteristics of reactions in which this mechanism is important have already been discussed (see p. 1165). Reactions in the gas phase or in solution in non-polar solvents generally proceed in this way; the reactions of halogen molecules with hydrogen or hydrocarbons or among themselves come within this category. The foregoing section on complexes of the halogen molecules has provided ample illustration of heterolytic fission, which tends to be the mode of action effective in polar solvents or in the presence of halogen carriers.

Some important general reactions of the halogens are summarized in Table 17. The noble gases and nitrogen apart, the halogens react more or less readily with virtually all the other elements. There are exceptions to this generalization: (i) under the action of a microwave discharge chlorine and xenon react to produce xenon dichloride[262]; (ii) reactions of the halogens with oxygen normally demand forcing conditions in which free atoms and radicals are implicated, though iodine and oxygen are reported to combine directly at high pressure and moderate temperatures[32]; (iii) although interaction of iodine with sulphur or selenium undoubtedly occurs, definite iodides have not as yet been described. Otherwise the readiness with which reaction occurs depends upon the temperature and phase, upon the state of subdivision, density, rigidity, volatility and solubility of any solid materials participating in the reaction, and upon the presence of impurities. Reactions of the halogens with solid elements tend to be slow if the products form coherent surface films which are involatile (in a gas–solid reaction) or insoluble (in a solid–liquid reaction). Depending on whether the solid reactant is in a massive or finely divided form, and on whether the surface area is large or small, the rate of reaction may vary dramatically. For example, whereas finely divided aluminium, antimony or copper burns spontaneously in an atmosphere of bromine, such active metals as sodium and magnesium in the massive state do not react with dry bromine even at temperatures up to 300°C; indeed a magnesium vessel has been used for precise measurements of the heat capacity and vapour pressure of bromine. On the other hand, all the halogens interact readily with metal vapours: with alkali-metal vapours, for example, a luminous flame results and the mechanisms of the accompanying reactions have been the subject of numerous studies[32,61]. Combustion of certain metals (and compounds) in an atmosphere of chlorine lends itself to calorimetric measurements; hence enthalpies of

[262] L. Y. Nelson and G. C. Pimentel, *Inorg. Chem.* **6** (1967) 1758.

TABLE 17. SOME GENERAL REACTIONS OF THE HALOGENS[32,33]

Reagent	Reaction	Comments
1. Donor molecules	Complex-formation with nitrogen-, oxygen-, sulphur-, selenium- or halide-donors, or with π-donors	See pp. 1196–1220.
2. Metals	$2M+nX_2 \rightarrow 2MX_n$	With most metals
3. Hydrogen 4. Molecules containing C–H bonds	$H_2+X_2 \rightarrow 2HX$ $R\text{–}H+X_2 \rightarrow R\text{–}X+HX$ (R = organic group)	Free energy of reaction decreases in the series Cl > Br > I in parallel with the decreasing bond energies of HX or RX (see Fig. 10)
5. Hydrides of B, Si or Ge	Substitution, e.g. $SiH_4+X_2 \rightarrow SiH_3X+HX$ $B_2H_6+6X_2 \rightarrow 2BX_3+6HX$	Products and rates depend markedly upon proportions of reactants, phase, temperature, presence of catalyst
6. Compounds containing $\rangle$N–H bonds	(i) Substitution, e.g. $(CH_2CO)_2NH+Br_2 \rightarrow (CH_2CO)_2NBr+HBr$ $R\cdot CONH_2+Br_2 \rightarrow R\cdot CONHBr+HBr$ Rearrangement and hydrolysis $\downarrow$ RNH_2+CO_2+HBr (ii) Oxidation, e.g. $N_2H_4+2X_2 \rightarrow N_2+4HX$	The relatively low N–X bond energy causes $\rangle$N–X compounds to be strong oxidizing or halogenating agents Hofmann reaction for the synthesis of amines $\rangle$N–X compounds are presumably intermediates
7. Unsaturated organic molecules	Addition, e.g. $\rangle C{=}C\langle + X_2 \rightarrow -\overset{\mid}{\underset{X}{C}}-\overset{X}{\underset{\mid}{C}}-$	Free energy of reaction decreases in the series Cl > Br > I in parallel with the decreasing C–X bond energy
8. Group V element	(i) $2M+3X_2 \rightarrow 2MX_3$ (ii) $MX_3+X_2 \rightarrow MX_5$	M = P, As, Sb or Bi; X = Cl, Br or I M = P, X = Cl or Br; M = Sb, X = Cl
9. Group VI element	(i) $2E+X_2 \rightarrow E_2X_2$ (ii) $E+X_2 \rightarrow EX_2$ (iii) $E+2X_2 \rightarrow EX_4$	E = S or Se, X = Cl or Br; E = Te, X = I E = S, X = Cl; E = Se, Te or Po, X = Cl or Br E = S, X = Cl; E = Se, X = Cl or Br; E = Te or Po, X = Cl, Br or I
10. Other halogens	Formation of interhalogens: $X_2+Y_2 \rightarrow 2XY$ $X_2+3Y_2 \rightarrow 2XY_3$ $X_2+5F_2 \rightarrow 2XF_5$ $I_2+7F_2 \rightarrow 2IF_7$	X, Y = different halogens Y = F, X = Cl, Br or I; Y = Cl, X = Br or I X = Cl, Br or I

Table 17 (*cont.*)

Reagent	Reaction	Comments
11. Halide ions	(i) Formation of polyhalide ions, e.g. $X_2+Y^- \rightarrow X_2Y^-$ (ii) Oxidation: $X_2+2Y^- \rightarrow 2X^- + Y_2$	X, Y = same or different halogen; X = Cl, Y = Br or I; X = Br, Y = I
12. Water	(i) Hydrolysis: $X_2+H_2O \rightarrow H^+ + X^- + HOX$ (ii) Oxidation: $X_2+H_2O \rightarrow 2H^+ + 2X^- + \frac{1}{2}O_2$	Reaction depends on pH. HOX subject to disproportionation. Reaction encouraged by photolysis
13. Metal oxides	Typically, $yX_2+2MO_z+2zC \rightarrow 2MX_y+2zCO$	With many metal oxides at elevated temperatures; carbon not always necessary
14. Metal carbonyls and related compounds	Substitution, e.g. $Mn_2(CO)_{10}+X_2 \rightarrow 2Mn(CO)_5X$ $Ni(CO)_4+X_2 \rightarrow NiX_2+4CO$	X = Cl, Br or I

formation have been directly derived, for example, for VCl_4 [263], $NbCl_5$ [264], $TaCl_5$ [264], $ZrCl_4$ [264] and $HfCl_4$ [263,264].

In their normal (massive) states the following metals are notable for their resistance to attack by the dry halogens at temperatures up to at least 100°C: with respect to gaseous chlorine: nickel, Inconel, Hastelloy, magnesium, Monel, stainless steel, copper, mild steel, cast iron and tantalum; with respect to liquid or gaseous bromine: nickel, magnesium, platinum, lead, tantalum, mild steel and cast iron; with respect to gaseous iodine: magnesium, lead, platinum, gold and bismuth. However, few of these metals offer the same resistance in the presence of moisture, which exacerbates the corrosive properties of the halogens. The pronounced catalytic effect of moisture is probably due in part to hydrolysis of the halogen with the formation of the hydrogen halide and hypohalous acid and in part to the provision of a medium for localized electrochemical action on the surface of the metal.

The reactions with hydrogen range in their thermodynamic and kinetic properties from the explosive, decidedly exothermic reaction of chlorine to the slow, reversible and barely exothermic reaction of iodine (see Table 29). In that all these reactions are now believed to involve the intermediate agency of atoms, certain features have already been discussed in connection with the formation and properties of the halogen atoms. The kinetics of the reaction between gaseous iodine and hydrogen have been studied in great detail, notably in the classical investigations initiated by Bodenstein[32,43]: the reaction is thus found to be bimolecular with an activation energy of *ca.* +39 kcal mol^{-1}. For many years the reaction was taken to be a model of a bimolecular reaction with no more than a single elementary step, namely the formation of the transition state H_2I_2, wherein the bonds of the molecules are weakened while incipient bonds corresponding to the products are beginning to form. Such a mechanism has been widely discussed, elaborated and justified by theoretical analysis[43,265a]. More recently, however, evidence of an atomic mechanism has been found

[263] P. Gross and C. Hayman, *Trans. Faraday Soc.* **60** (1964) 45.

[264] L. A. Reznitskii, *Zhur. fiz. Khim.* **41** (1967) 1482; G. L. Gal'chenko, D. A. Gedakyan and B. I. Timofeev, *Russ. J. Inorg. Chem.* **13** (1968) 159.

[265] (a) K. J. Laidler, *Chemical Kinetics*, 2nd edn., McGraw-Hill, New York (1965); S. W. Benson, *The Foundations of Chemical Kinetics*, McGraw-Hill, New York (1960); G. C. Fettis and J. H. Knox, *Progress in Reaction Kinetics*, Vol. 2 (ed. G. Porter), p. 1, Pergamon (1964); B. A. Thrush, *Progress in Reaction Kinetics*, Vol. 3 (ed. G. Porter), p. 63, Pergamon (1965); (b) J. H. Sullivan, *J. Chem. Phys.* **46** (1967) 73.

on the basis of studies of the photochemical reaction between hydrogen and iodine[265b]. Formally similar mechanisms are thus implied for the interaction of hydrogen with chlorine, bromine and iodine[32,43,265]:

$$\begin{array}{llll} \textit{Initiation}: & X_2 & \xrightarrow[\text{or wall of vessel}]{h\nu,\ \Delta} X+X & (1) \\ \textit{Propagation}: & X+H_2 & \longrightarrow HX+H & (2) \\ & H+X_2 & \longrightarrow HX+X & (3) \\ & H+HX & \longrightarrow H_2+X & (4) \\ \textit{Termination}: & X+X & \longrightarrow X_2 & (5) \\ & H+X & \longrightarrow HX & (6) \\ & H+H & \longrightarrow H_2 & (7) \end{array}$$

Differences arise because of the increasingly endothermic character of reaction (2) in the series Cl, Br, I (Fig. 10). Whereas reaction chains are set up in the hydrogen/chlorine reaction by repetition of processes (2) and (3), such chains are markedly abbreviated in the hydrogen/bromine and hydrogen/iodine reactions at all but the highest temperatures. First proposed in effect by Nernst[266], to account for the photochemical reaction of hydrogen with chlorine, the mechanism implicit in reactions (1)–(7) receives support from numerous findings: for example, (i) the quantum efficiency of the photochemically induced reaction of hydrogen with chlorine is typically 10^5–10^6 (cf. 0·01 for the hydrogen/bromine reaction at room temperature); (ii) assumption of a steady-state condition leads to the rate equations

$$\frac{d}{dt}[\text{HBr}]_{\text{therm}} = 2k_2\left(\frac{k_1}{k_5}\right)^{\frac{1}{2}} \frac{[\text{H}_2][\text{Br}_2]^{\frac{1}{2}}}{1+k_4[\text{HBr}]/k_3[\text{Br}_2]}$$

$$\frac{d}{dt}[\text{HBr}]_{\text{phot}} = 2\frac{k_2}{(k_5)^{\frac{1}{2}}} \frac{[\text{H}_2]I_{\text{abs}}^{\frac{1}{2}}}{1+k_4[\text{HBr}]/k_3[\text{Br}_2]} \qquad (I_{\text{abs}} = \text{number of quanta absorbed})$$

for the thermal and photochemical reactions between hydrogen and bromine; such expressions are identical in their concentration-dependence with those of experimental measurements[32], while the variation with temperature of the term $k_2(k_1/k_5)^{\frac{1}{2}}$ gives an activation energy in satisfactory agreement with the algebraic sum $+\Delta H_2^{\ddagger}+\frac{1}{2}\Delta H_1^{\ddagger}-\frac{1}{2}\Delta H_5^{\ddagger}$; (iii) the presence of free atoms has been detected experimentally. The initiation of the reaction by dissociation of the halogen molecule can be brought about by the action of heat, light or other agents. In the thermal combination of hydrogen with chlorine, dissociation of the chlorine molecules occurs on the walls of the containing vessel, whereas photochemical combination is initiated in the gas phase. In the investigation of these reactions photochemical action has proved important in that it isolates a particular elementary reaction, here the dissociation of the halogen molecule, the rate of which is controlled by the light intensity. Hence the importance of that reaction may be judged, and its elimination allows a closer inspection of the other steps. For this and other reasons, the influence of light on the combination of hydrogen with the halogens has been extensively studied; it is probable that the photochemical reaction of hydrogen with chlorine has been more thoroughly examined than any other, ever since the days of Bunsen. For each system the effects of pressure, temperature and frequency of the irradiating light have been explored in detail. Investigations have also been concerned with the influences of the walls of the reaction vessel, the presence of "third bodies" in the gas phase (e.g. inert gases), and the inhibiting effect of agents such as oxygen or nitric oxide which have the capacity to terminate the reaction chains (see p. 1166). In practical terms, the formation reactions of the hydrogen halides have been suggested as the basis of a fuel cell[267].

[266] W. Nernst, *Z. Elektrochem.* **24** (1918) 335.
[267] W. V. Childs, U.S. Pat. 3,445,292 (1968).

An atomic mechanism is almost certainly responsible for the reactions between the halogens themselves. According to the conditions employed, fluorine reacts (a) with chlorine to form ClF, ClF_3 or ClF_5, (b) with bromine to form BrF, BrF_3 or BrF_5, and (c) with iodine to form IF, IF_3, IF_5 or IF_7. The products of the interaction of chlorine with bromine are $BrCl$ or $BrCl_3$ and related compounds[268]; iodine and chlorine afford ICl or I_2Cl_6; IBr is the only product as yet identified in the reaction between iodine and bromine. Certain reactions require forcing conditions, e.g.

$$Cl_2+5F_2 \xrightarrow[\text{Total pressure 250 atm, 350°C}]{\text{Excess fluorine}} 2ClF_5$$

$$Br_2+3Cl_2 \xrightarrow[\text{Microwave discharge}]{\text{Gas phase}} 2BrCl_3 \qquad \text{(ref. 268)}$$

but others take place with less inducement, e.g.

$$Br_2+Cl_2 \underset{\text{Gas phase or solution}}{\overset{\text{Room temperature}}{\rightleftharpoons}} 2BrCl$$

$$I_2+5F_2 \xrightarrow{\text{Room temperature}} 2IF_5$$

$$I_2+3F_2 \xrightarrow{CCl_3F \text{ suspension, } -45°C} 2IF_3 \qquad \text{(ref. 269)}$$

These reactions and the properties of the interhalogen compounds thereby produced are discussed in Section 4 (p. 1478).

Neither chlorine nor bromine reacts with molecular oxygen under normal conditions, but at 325°C and with a partial pressure of oxygen of *ca.* 1200 atm iodine is reported to give a yield of 2·3% of iodine pentoxide[32]. The action of oxygen atoms on chlorine gives rise to the molecules ClO and ClO_2; the kinetics of the primary reaction

$$O+Cl_2 \rightarrow ClO+Cl$$

have been studied at 300°K by discharge-flow coupled with mass-spectrometric methods of detection[84]. Likewise ozone and chlorine yield the relatively short-lived ClO_3 in a reaction which is susceptible to photolytic action; irradiation by blue light gives detectable amounts of Cl_2O_7. By the action of atomic oxygen or ozone on bromine under suitable conditions bromine oxides are obtained. Chlorine atoms produced, for example, by photolysis combine with oxygen to give, in the first place, the peroxy free-radical ClOO, which then plays a central role in the formation and subsequent decay of the ClO radical according to the mechanism[132,270]:

$$ClO+ClO \rightleftharpoons Cl+ClOO$$
$$ClOO+Cl \rightarrow Cl_2+O_2$$
$$ClOO+M \rightleftharpoons Cl+O_2+M \qquad (M = \text{third body})$$

As produced by flash photolysis of chlorine–oxygen mixtures, ClO has a half-life of a few milliseconds. Nevertheless, it has been possible to explore the formation and fate of species such as ClO and ClOO by measurements of their electronic absorption spectra; ClO has also been characterized by its esr[93] and microwave[271] spectra. Formed by the action on bromine vapour of oxygen atoms produced in a microwave discharge, the gaseous BrO molecule has also been characterized by its esr[93] and microwave[272] spectra, while BrO_2 has been isolated

268 L. Y. Nelson and G. C. Pimentel, *Inorg. Chem.* **7** (1968) 1695.
269 M. Schmeisser, W. Ludovici, D. Naumann, P. Sartori and E. Scharf, *Ber.* **101** (1968) 4214.
270 M. A. A. Clyne, *Ann. Rep. Chem. Soc.* **65A** (1968) 173–174, 183–185.
271 T. Amano, S. Saito, E. Hirota, Y. Morino, D. R. Johnson and F. X. Powell, *J. Mol. Spectroscopy*, **30** (1969) 275.

at low temperatures following the action of an electric discharge on a mixture of bromine vapour and oxygen[32]. The intermediate formation of BrO or other bromine oxides probably accounts for the action of bromine as a "sensitizer" for the photochemical decomposition of ozone, ClO or ClO_2.

Chlorine reacts with many metal oxides to yield anhydrous chlorides or oxychlorides, the presence of carbon commonly being necessary to expedite the reaction. Bromine and iodine are less prone to react with metal oxides, and, where attack does occur, oxyhalogen species such as bromate or iodate tend to be more prominent as products. Thus, when iodine vapour mixed with oxygen is passed over heated alkaline-earth oxides or carbonates, solid periodates are formed: e.g.

$$10CaO+2I_2+7O_2 \rightarrow 2Ca_5(IO_6)_2$$

With mercuric oxide, however, chlorine gives the oxide Cl_2O in a reaction which affords a standard method of preparation; stable only at temperatures below −50°C, Br_2O is likewise formed, but I_2O is not known. Alkali or alkaline-earth metal chlorites are oxidized by chlorine to chlorine dioxide (Table 14). This process, which is important for the industrial production of the dioxide, can be adapted for either continuous or batch operation, and can be carried out under dry or aqueous conditions. Hydrogen peroxide is initially oxidized by chlorine, bromine or iodine according to the equation

$$H_2O_2+X_2 \rightarrow O_2+2HX$$

Hydrogen chloride, once formed, is stable, but oxidation by hydrogen peroxide causes the bromide or iodide to revert to the free halogen. The net result of the interaction with bromine or iodine is therefore the catalyzed decomposition of hydrogen peroxide. The kinetics of these reactions in aqueous solution have been the subject of considerable investigation, on the basis of which possible mechanisms have been discussed[32]. By the formation of halogen atoms, photolysis of the reaction mixture initiates chain reactions in which propagation steps such as

$$X+H_2O_2 \rightarrow HO_2+HX$$

probably participate. By contrast with the behaviour of hydrogen peroxide, peroxydisulphuryl difluoride $FSO_2O{\cdot}OSO_2F$ serves as a source of $FSO_2O{\cdot}$ radicals which oxidize the halogens to produce halogen fluorosulphates. Chlorine is thereby converted to $ClOSO_2F$, bromine to $BrOSO_2F$ or $Br(OSO_2F)_3$ (depending upon the conditions), and iodine to I_3OSO_2F, $IOSO_2F$ or $I(OSO_2F)_3$ (again depending upon the conditions)[273]. These products are to be regarded as derivatives of the halogens in positive oxidation states (see Section 4). Compounds of the type $I(OCOR)_3$ containing formally electropositive iodine atoms are also formed by the action of fuming nitric acid on iodine in the presence of the appropriate acid anhydride $(RCO)_2O$ [R = CH_3, CH_2Cl, $CHCl_2$, CCl_3, CF_3, C_3F_7 or C_6F_5][120,274]. A similar reaction affords the compound IPO_4. Interaction of a halogen with a silver(I) or mercury(II) oxysalt in a non-aqueous medium also gives rise to halogen derivatives of the oxyanion: e.g. $IClO_4$ [120,275], $I(ClO_4)_3$ [120,275], $pyINO_3$ [120] and pyIOCOR[120,274] (py = pyridine; R = organic group). A compound described as "chlorine acetate" and formulated as $ClOCOCH_3$ is likewise the outcome of such a reaction, but the structures of this and related compounds have yet to be elucidated.

[272] F. X. Powell and D. R. Johnson, *J. Chem. Phys.* **50** (1969) 4596.

[273] A. A. Woolf, *New Pathways in Inorganic Chemistry* (ed. E. A. V. Ebsworth, A. G. Maddock and A. G. Sharpe), p. 351, Cambridge (1968).

[274] M. Schmeisser, K. Dahmen and P. Sartori, *Ber.* **100** (1967) 1633.

[275] N. W. Alcock and T. C. Waddington, *J. Chem. Soc.* (1962) 2510.

Compounds containing $\rangle$N–H bonds are susceptible to oxidation by the halogens. For example, under aqueous conditions ammonia and hydrazine are oxidized, at least in part, to nitrogen (see Table 14). However, substitution reactions are also possible, whereby the hydrogen is replaced by a halogen atom. Obtained in this way are compounds such as NH_2Cl, NCl_3, NI_3,NH_3, and N-bromosuccinimide. The $\rangle$N–X bond is characteristically weak with the result that the compounds are powerful oxidizing or halogenating agents; several of them undergo spontaneous decomposition at the slightest provocation, but others, e.g. N-bromosuccinimide, serve as useful halogenating agents for organic systems. The instability of the $\rangle$N–X unit is well illustrated by the Hofmann reaction whereby organic amides are converted to amines:

$$R{\cdot}C(=O){-}NH_2 \xrightarrow{Br_2.OH^-} R{\cdot}C(=O){-}NHBr \xrightarrow{-HBr} [R{\cdot}C(=O){-}N] \xrightarrow{\text{Rearrangement}} R{\cdot}N{=}C{=}O \xrightarrow{\text{Hydrolysis}} R{\cdot}NH_2$$

By the action of an elemental halogen X_2, M–H, M–M and M–C bonds formed by a metal or metalloid M are commonly supplanted by M–X bonds in the formation of binary or mixed halide derivatives of M. A selection of reactions belonging to this category is depicted in Fig. 18[276–278]. The interaction is occasionally violent and difficult to control, as

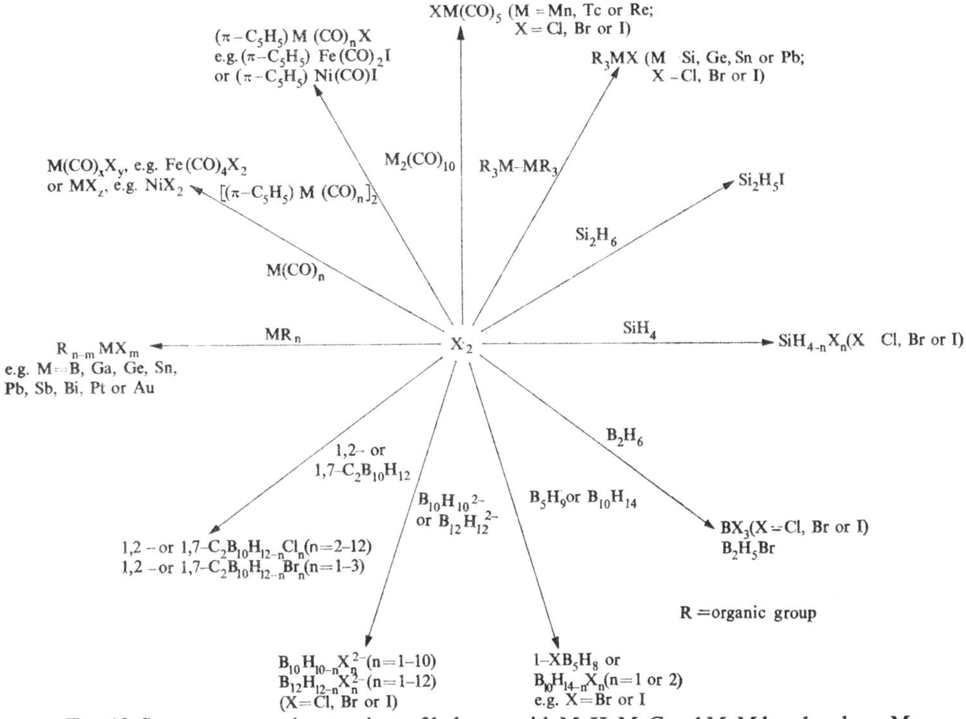

FIG. 18. Some representative reactions of halogens with M–H, M–C and M–M bonds, where M is a metal or metalloid.

[276] E. A. V. Ebsworth, *Volatile Silicon Compounds*, Pergamon, Oxford (1963).

[277] R. M. Adams (ed.), *Boron, Metallo-boron Compounds and Boranes*, Interscience, New York (1964); M. F. Hawthorne, *The Chemistry of Boron and its Compounds* (ed. E. L. Muetterties), p. 223. Wiley, New York (1967).

[278] G. E. Coates, M. L. H. Green and K. Wade, *Organometallic Compounds*, 3rd edn., Methuen, London (1967–8).

between silane and chlorine at room temperature; conversely certain reactions are relatively sluggish at normal temperatures. However, with judiciously chosen conditions of temperature, reaction medium and, where necessary, catalysis, many of the reactions proceed smoothly, affording useful preparative routes, for example, to organometallic halides such as $Fe(CO)_4I_2$ or $(\pi\text{-}C_5H_5)Ni(CO)I$.

In aqueous solution the azide ion is ultimately oxidized to nitrogen, notably by bromine or iodine, in a reaction which is catalyzed by a number of sulphur-containing species, e.g. H_2S, CS_2, SCN^-, $S_nO_6^{2-}$ (n = 4, 5 or 6), N_3SCS^- and cysteine. Advantage is taken of this catalytic action on the iodine–azide reaction to provide a sensitive test for sulphides and sulphur compounds[32,33]. The kinetics of the reactions and their likely mechanisms have been discussed[33,279,280]. Under different conditions the interaction of azides with the molecular halogens has been shown to give the halogen azides XN_3 [281]. The action of the halogens on other metal pseudohalides may likewise lead to compounds such as XCN, though the course of the reaction is profoundly influenced by the choice of conditions: e.g.

(i) $CN^- + X_2 \xrightarrow[\text{dry state}]{\text{Aqueous solution or}} XCN + X^-$ $\quad$ (X = Cl, Br or I)

(ii) $2MSCN + Br_2 \xrightarrow[\text{solvent, e.g. } CHCl_3]{0^\circ C\text{, non-aqueous}} 2MBr + (SCN)_2$
$\qquad\qquad\downarrow$ Dry Cl_2
$\qquad\qquad 2ClSCN$

(M = NH_4, Na, K or $\frac{1}{2}$Pb)[282]

(iii) $SCN^- + 4I_2 + 4H_2O \xrightarrow[\text{pH 7}]{\text{Aqueous soln}} SO_4^{2-} + ICN + 7I^- + 8H^+$

(iv) $SCN^- + 3I_2 + 4H_2O \xrightarrow[\text{pH} < 7]{\text{Aqueous soln}} SO_4^{2-} + 6I^- + HCN + 7H^+$

Apart from its importance as a brominating agent in organic chemistry[45], bromine is a useful oxidizing agent, though under the aqueous conditions commonly employed it is not always certain whether the active agent is Br_2 or a hypobromite species. One facet of these oxidizing properties is presented by the degradation of organic amides (see above), while another consists of the oxidation of aldose sugars to δ-lactones, which are subsequently hydrolysed to aldonic acids. For example, D-glucose can be converted to D-gluconic acid by oxidation with bromine water; the mechanism, which appears to be highly selective, is believed to involve electrophilic attack by bromine on the glycosidic oxygen atom[283]. Extensive use of bromine oxidation was made in the classical studies of Fischer and others on the structure of sugars.

The dissociation of halogen molecules into atoms by the absorption of light is sometimes able to initiate chain reactions between two other substances, the consumption of halogen in the reactions being frequently small or negligible. This property forms the basis of the "sensitizing" action of the halogens on certain reactions. As examples may be cited the chlorine-sensitized combination of carbon monoxide and oxygen, explosion of hydrogen with oxygen, and the oxidation of various organic compounds, the bromine-sensitized decomposition of ozone, and the bromine- or iodine-sensitized decomposition of hydrogen peroxide. It is possible that a similar mechanism operates, at least in

279 J. Weiss, *Trans. Faraday Soc.* **43** (1947) 119.

280 G. Dodd and R. O. Griffith, *Trans. Faraday Soc.* **45** (1949) 546.

281 K. Dehnicke, *Angew. Chem., Internat. Edn.* **6** (1967) 240.

282 A. B. Angus and R. G. R. Bacon, *J. Chem. Soc.* (1958) 774; R. G. R. Bacon and R. S. Irwin, *ibid.* p. 778.

283 D. H. Hutson and D. J. Manners, *Ann. Rep. Chem. Soc.* **61** (1964) 431.

some cases, in the catalytic action of small quantities of iodine on many organic reactions, including halogenation, metalation, dehydration, isomerization and pyrolytic decomposition. Because of this action, iodine is widely employed to accelerate such reactions or to shorten otherwise excessive induction periods. The role of iodine as a catalyst in the hydrogen–chlorine reaction has been discussed, as has its effect on the ignition temperatures or explosion limits of combustion processes involving hydrogen– or hydrocarbon–oxygen mixtures[32].

2.8. ANALYTICAL DETERMINATION OF THE ELEMENTARY HALOGENS[32,284,285]

Chlorine

The detection and determination of free chlorine depend primarily on its oxidizing action. Thus the release of iodine from iodide has been extensively used both as a qualitative test and as a means of quantitative analysis. The familiar starch–iodide test for chlorine may be used to detect as little as 0·1 ppm under aqueous conditions. Chlorine in the atmosphere has been detected by its action on potassium iodide deposited on silver leaf[32]; more commonly the iodine liberated in an aqueous medium is estimated by titration with sodium thiosulphate or arsenite. Another test widely favoured for the detection and estimation of free chlorine involves the use of *o*-tolidine (*o,o'*-dimethylbenzidine); in dilute solution in fairly strong hydrochloric acid this gives a yellow coloration, the intensity of which provides a measure of the concentration of chlorine present. The starch–iodide and *o*-tolidine tests are thought to be about equally sensitive, but the latter is more specific and so less vulnerable to interference from other substances. Even the *o*-tolidine test has considerable shortcomings; not only do other free halogens give a positive reaction, but ions such as Mn^{3+}, NO_2^- and Fe^{3+} also cause severe complications. The search for alternatives to this popular test has yielded other colorimetric procedures involving reagents such as 3,3'-dimethylnaphthidine, which yields a purple-red semiquinoid product and is reported to be superior to *o*-tolidine, *p*-aminodimethylaniline, which gives a red coloration, and methyl orange, which is bleached by chlorine. A useful and rather specific test for chlorine (or bromine) depends upon the König reaction with aqueous cyanide ions to produce ClCN (or BrCN), which gives an intensely coloured di-anil derivative with pyridine and an aromatic amine, e.g. benzidine; the method is applicable to chlorine concentrations up to 2 ppm.

Other techniques for estimating gaseous chlorine involve absorption in alkali in the presence of a reducing agent, the chloride ions thus formed being determined either volumetrically with silver or mercuric nitrate or gravimetrically with silver nitrate. Chlorine gas is assayed commercially by absorption in mercury, with which it reacts quantitatively, followed by volumetric measurement of the residual gases. Alternatively the chlorine may be absorbed in an acetate-buffered aqueous solution of arsenite. In recent years gas-chromatographic methods have been used increasingly for the assay of chlorine gas and the determination of impurities, while infrared spectroscopy has been applied to the detection of organic impurities and moisture.

Bromine[45]

In common with chlorine, free bromine is conveniently estimated by iodometric titration;

[284] G. W. Armstrong, H. H. Gill and R. F. Rolf, *Treatise on Analytical Chemistry* (ed. I. M. Kolthoff, P. J. Elving and E. B. Sandell), Part II, Vol. 7, p. 335. Interscience (1961).

[285] F. D. Snell and L. S. Ettre (eds.), *Encyclopedia of Industrial Chemical Analysis*, Vols. 8, 9 and 14. Interscience (1969–71).

bromine vapour can, for example, be quantitatively absorbed in potassium iodide solution and the liberated iodine determined by titration with thiosulphate. Bromine in the vapour phase can also be absorbed in alkali and either (a) reduced, e.g. with the aid of hydrogen peroxide, to the bromide ion, which as AgBr can be estimated gravimetrically or volumetrically, or (b) oxidized, e.g. by hypochlorite, to the bromate ion, which lends itself to iodometric titration. However, the detection of bromine is often accomplished most easily by one of several colour tests. The colour of free bromine, particularly in carbon tetrachloride or carbon disulphide solution, itself serves as a method of detecting and estimating the element. Various organic reagents which form coloured compounds with bromine are also available, though few are specific; thus most of the reagents which give colorations with chlorine respond similarly to bromine. Reagents recommended for the detection and estimation of free bromine include (a) fluorescein or, better, ethylfluorescein, which affords the red dye eosin; (b) dithizone, which gives an intense red colour in carbon tetrachloride solution; (c) rosaniline (fuchsin), which undergoes bromination to give violet bromo-derivatives; (d) phenol red, which is converted to an indicator of the bromophenol-blue type. Despite problems of interference by other oxidizing agents, several of these colour tests have been successfully applied to the determination of trace quantities of elementary bromine. However, one of the most sensitive procedures for estimating bromine results from its catalytic action on the oxidation of iodine to iodate by acid permanganate; in one such procedure the change in iodine concentration is determined spectrophotometrically using a carbon tetrachloride extract of the aqueous solution.

Various methods have been devised for the evaluation of impurities in liquid bromine. The chlorine content can be determined by measurements of the density of the liquid or of the depression of freezing point, while in amounts of the order 5–10% the halogens can be reduced with sulphur dioxide to the corresponding halides, which are then conveniently estimated by potentiometric titration with silver nitrate. Pure bromine being transparent throughout the normal infrared region, moisture and most organic impurities have been detected with considerable sensitivity by measuring the infrared spectrum of the liquid[54].

Iodine

Free iodine in dilute aqueous solution is readily determined by direct titration with a standard solution of sodium thiosulphate or sodium arsenite, typically using starch as the indicator; the equivalence point may also be determined electrometrically, for example by the so-called "dead-stop" technique. Other methods of volumetric analysis include oxidation by potassium iodate in the presence either of concentrated hydrochloric acid or of cyanide ions, whereby iodine (or iodide) is converted to ICl_2^- or ICN; iodine may also be oxidized to iodate with bromine or permanganate, and the iodate estimated by iodometric titration, six moles of I_2 being thereby released for each mole originally present.

Iodine can be detected as the free element by the violet colours of its vapour and of its solutions in solvents such as carbon tetrachloride, chloroform or carbon disulphide. Perhaps the most widely used test is afforded by the blue-black coloration due to the starch–iodine reaction, the sensitivity of which has been studied as a function of numerous factors. α-Naphthoflavone has been recommended as a test for free iodine more sensitive than starch, 0·1 ppm being readily detected; malachite green has also been shown to be superior to starch in very dilute solutions or in the presence of large amounts of electrolyte or ethanol[286].

[286] J. O. Meditsch, *Analyt. Chim. Acta*, **31** (1964) 286.

Colorimetric methods favoured for the determination of small quantities of free iodine involve the starch–iodine complex, the α-naphthoflavone reagent, the I_3^- ion formed in aqueous solution, or solutions of iodine in organic solvents.

2.9. BIOLOGICAL ACTION OF THE ELEMENTARY HALOGENS[32,33,45,284,285,287]

As vapours, all of the halogens cause similar physiological reactions. Through their oxidizing action, comparatively low concentrations of the vapours are highly irritating to the entire respiratory tract, the mucous membranes and the eyes, producing responses such as coughing and smarting of the eyes; approximate exposure limits are given in Table 18.

TABLE 18. EXPOSURE LIMITS FOR THE HALOGEN VAPOURS[46,284,287]

	Volume concentrations in ppm		
	Chlorine	Bromine	Iodine
Maximum allowable concentration that can be tolerated without disturbance:			
Prolonged exposure (several hours)	0·35–1·0	0·1–1	0·1
Short exposure (½–1 hr)	4	4	0·3
Least detectable odour	3·5	~3·5	
Least amount causing immediate irritation to the throat	10–15	15	
Dangerous to life even on short exposure (½–1 hr)	40–60	40–60	
Rapidly fatal on very short exposure (< ½hr)	1000	1000	

Exposure to high concentrations of the vapours leads to inflammation and congestion of the respiratory system and oedema of the lungs, which in severe cases can be fatal. Because of their greater volatility, chlorine and bromine are much more likely to be encountered in dangerous concentrations than is iodine. Accordingly, few cases of iodine poisoning have been recorded, though this apparently innocuous record should not disguise the lachrymatory action of the vapour or the fact that exposure can evoke pulmonary oedema, eye irritation, rhinitis, chronic pharyngitis and catarrhal inflammation of the mouth. By contrast, the toxic nature of gaseous chlorine is well recognized, as testified by its history as one of the first gases to be used in chemical warfare in World War I. The odours of the halogens give adequate warning against acutely dangerous exposure but not against concentrations which may be harmful as a result of prolonged exposure. However, there have as yet been no reports of cumulative intoxication following several exposures which do not individually exceed the normal safety limits.

Liquid chlorine is a skin irritant causing burns on prolonged contact; it can also cause severe eye damage. Likewise liquid bromine presents a serious hazard in handling. It rapidly attacks the skin and other tissues to produce irritation and necrosis; the burns are not only painful but slow to heal. In short-term contact with the skin, solid iodine causes a discoloration similar to that produced by "tincture of iodine", but prolonged contact can cause burns or allergic dermatitis.

The manipulation of elementary chlorine, bromine and iodine invariably demands the following safety measures: (i) rigorous avoidance of all contact of the halogen with skin,

[287] M. B. Jacobs, *The Analytical Toxicology of Industrial Inorganic Poisons*, p. 635. Interscience (1967).

eyes and clothing, and (ii) adequate ventilation with all possible precautions against inhaling the vapours. Where significant quantities of chlorine or bromine are to be handled, good practice requires that goggles and gloves are worn and that suitable gas masks are available. Preventive health measures and first-aid procedures are given elsewhere[288].

The poisonous action of the free halogens on micro-organisms affords an important means of disinfection. Thus in the treatment of drinking water, the use of small quantities of chlorine has become standard practice, and by using higher halogen concentrations the bacterial count of effluents from industrial plants can be controlled. The activity of chlorine in the destruction of organisms more complex than bacteria, e.g. non-pathogens such as *Crenothrix* formed in pipe-lines, has also been investigated. Likewise dilute aqueous solutions of bromine exert considerable germicidal action; comparative studies suggest that it is commonly intermediate between chlorine and iodine in its activity on bacterial spores, though with respect to some species, e.g. yeast and certain algae, it is said to be more toxic than chlorine. For many years iodine has been used in various forms as a first-aid antiseptic for the treatment of cuts and abrasions, and as routine surgical procedure in the sterilization of operation sites; despite the advent of many antiseptics of higher activity, the use of iodine for these purposes has persisted. The specific action of iodine solutions on a variety of organisms has been widely investigated; the solutions have thus been shown to exhibit both bacteriostatic and bactericidal activity. The halogens are known to react with many materials found in living matter, e.g. unsaturated aliphatic acids (by addition), aromatic amino-acids (by ring-substitution), amino-derivatives (with oxidation or formation of N-halogen compounds), organic donor species (to form molecular complexes which may, for example, be relatively insoluble in water), and thiols and disulphides (with oxidation). While it is not certain that all of these reactions take place *in vivo*, it is conceivable that any one of them could contribute to the biological effects peculiar to the elementary halogens. In specific cases the toxic action of the halogens may well depend upon oxidation of enzyme functional groups such as –SH and –SS–; it is now fairly conclusively proved, for example, that the activity of chlorine towards various micro-organisms is by complete inhibition of the sulphydryl enzymes of bacteria, leading to their death.

3. HALIDE IONS AND RELATED SPECIES: OXIDATION STATE −1

3.1. PROPERTIES OF THE HALIDE IONS

Thermodynamic Properties, Ionic Radii and Polarizabilities

Reduction of a free halogen atom leads to the formation of a halide ion with the closed-shell configuration ns^2np^6 and electronic ground state ^1S; for all the halogens this is a thermodynamically spontaneous process. To the extent that the simple ionic model is applicable to real chemical systems, the halide ions are recognizable entities, to which can be attributed sizes as well as spectroscopic, thermodynamic and magnetic properties. However, even in the most favourable circumstances the simple ionic model, with its implications of a hard sphere bearing unit charge, is unrealistic; in all environments of chemical significance some transfer of charge between the ion and its neighbours is inevitable. Thus, although Table 19 lists some of the better defined physical characteristics of the chloride,

[288] See for example *Chlorine Manual*, 3rd edn., The Chlorine Institute, New York (1959).

TABLE 19. PHYSICAL PROPERTIES OF THE HALIDE IONS

Property	Cl^-		Br^-		I^-		Ref.
Electron configuration and ground state	[Ne]$3s^23p^6$	^1S	[Ar]$3d^{10}4s^24p^6$	^1S	[Kr]$4d^{10}5s^25p^6$	^1S	
Ionic radius (6 : 6-coordination) (Å)	1·81		1·96		2·19		a
Ionization potential (electron affinity of halogen atom) at 298°K	*eV* 3·68	*kcal* 84·8	*eV* 3·43	*kcal* 79·0	*eV* 3·13	*kcal* 72·1	a
Thermodynamic properties at 298°K							a, b
(i) of gaseous anion:							
$-\Delta H_f°$(kcal)	55·7		52·3		46·6		
$S°$(cal deg^{-1} mol^{-1})	36·7		39·1		40·4		
(ii) of the hydrated anion	*rel.* $a_{H^+} = 1$	*rel. gaseous ion*	*rel.* $a_{H^+} = 1$	*rel. gaseous ion*	*rel.* $a_{H^+} = 1$	*rel. gaseous ion*	
$-\Delta H_f°$ (kcal)	39·933	88	29·039	80	13·60	70	
$-\Delta G_f°$ (kcal)	31·383	82	24·900	75	12·44	67	
$S°$ (cal deg^{-1} mol^{-1})	13·57	20	19·91	16	25·60	11	
Spectroscopic properties							
X-ray spectra	Refs. c–e		Refs. d, e		Refs. d, e		
X-ray scattering factors	Ref. g		Ref. g		Ref. g		
Ultraviolet spectra in aqueous solution: wavelength (Å) [frequency (cm^{-1})]; oscillator strength (f)	1820 [55,000]; 0·09		1905 [52,500], 2020 [49,500]; 0·28		1940 [51,500], 2260 [44,300]; 0·47		f
Electrical and magnetic properties							
Polarizability (Å^3)	3·475		4·821		7·216		h
Diamagnetic susceptibility ($\times 10^6$ c.g.s. units per g ion)	−23·4		−34·6		−50·6		i
Standard potential for the aqueous system $\frac{1}{2}X_2/X^-$, $E°$ (V)	+ 1·356		+ 1·065		+ 0·535		a
Ionic mobility (limiting equivalent conductance) in water at 298°K (Λ cm^2 ohm^{-1} g ion^{-1})	76·35		78·14		76·84		j

[a] A. G. Sharpe, *Halogen Chemistry* (ed. V. Gutmann), Vol. 1, p. 1, Academic Press (1967).

[b] *National Bureau of Standards Technical Note* 270–3, U.S. Govt. Printing Office, Washington (1968); *Codata Bulletin*, International Council of Scientific Unions, Committee on Data for Science and Technology (1970).

[c] *Gmelins Handbuch der Anorganischen Chemie*, 8 Auflage, "Chlor", Teil A, p. 133 (1968).

[d] *Mellor's Comprehensive Treatise on Inorganic and Theoretical Chemistry*, Supplement II, Part I, pp. 624–626, 776–778, 910–912, Longmans, Green and Co., London (1956).

[e] J. A. Bearden, *X-ray Wavelengths*, U.S. Atomic Energy Commission, **NYO–10586**, Oak Ridge, Tennessee (1964); J. A. Bearden, *Rev. Mod. Phys.* **39** (1967) 78; J. A. Bearden and A. F. Burr, *ibid.* p. 125; A. E. Sandström, *Experimental Methods of X-ray Spectroscopy: Ordinary Wavelengths, Handbuch der Physik*, **30** (1957) 78.

[f] J. Jortner and A. Treinin, *Trans. Faraday Soc.* **58** (1962) 1503.

[g] *International Tables for X-ray Crystallography*, Vol. III (general ed. K. Lonsdale), p. 201, International Union of Crystallography (1962); H. P. Hanson, F. Herman, J. D. Lea and S. Skillman, *Acta Cryst.* **17** (1964) 1040; D. T. Cromer and J. T. Waber, *Acta Cryst.* **18** (1965) 104.

[h] K. Fajans, *J. Phys. Chem.* **74** (1970) 3407.

[i] A. Earnshaw, *Introduction to Magnetochemistry*, p. 5, Academic Press, London (1968).

[j] R. A. Robinson and R. H. Stokes, *Electrolyte Solutions*, 2nd edn., p. 463, Butterworths, London (1959).

bromide and iodide ions, little absolute significance can be attached to those values, e.g. of ionic radii, which are derived from measurements focused on the halide ion in crystalline solids or in solution. The importance of such results resides not in their accuracy of description but in the success with which they can be applied, through the medium of the ionic model, to the interpretation of the thermodynamic and crystallographic properties of halide systems.

The enthalpy of formation of a gaseous halide anion, $-\Delta H_f[X^-(g)]$, decreases in the sequence of $Cl^- > Br^- > I^-$. The standard entropy of the gaseous anion is lower than that of the corresponding atom because of the difference in electronic multiplicity; according to the Sackur–Tetrode equation, the contribution to the standard entropy of a monatomic gas of electronic multiplicity Q_e is $R \ln Q_e = 0$ or 2·75 cal deg^{-1} for a gaseous halide ion or halogen atom, respectively[289]. Since the standard entropies and entropy changes vary but little from halogen to halogen, variations in $-\Delta G_f[X^-(g)]$ follow the sequence dictated by $-\Delta H_f[X^-(g)]$.

The fact that halides of pre-transition metals have internuclear distances which are approximately additive suggests the feasibility of assigning radii to the individual ions[290]. The radii adopted in Table 19, which refer to a sodium chloride structure, are taken from Sharpe's review of the physical inorganic chemistry of the halogens[289]. In this context, however, two complications are noteworthy. (i) For the very few crystals in which the variation in electron density along a line joining the nuclei has been determined, the minimum in electron density does not occur at the point indicated by the generally accepted ionic radii. In NaCl, for example[291], the minimum occurs at 1·64 Å from the chlorine nucleus. (ii) The ionic radius varies somewhat with the coordination number of the halide ion. This is a consequence of the dependence of the lattice energy of an ionic halide on the Madelung constant A of the lattice-type, viz.

$$U_L = \frac{z_+ e^2 NA}{r}\left[1-\frac{\rho}{r}\right] \tag{1}$$

where z_+ is the charge on the cation, e the electronic charge, N the Avogadro number, r the equilibrium interionic separation and ρ a function of the compressibility (almost constant), which takes into account the interionic repulsion arising from the finite size of the ions. As the number of cation–anion contacts increases, not only A but also r becomes larger. As a result of these mutually compensating changes, there appears to be little difference in energy between different structures, e.g. CsCl and NaCl types, in justification of the empirical formulae

$$U_L = \frac{256\nu z_+}{r_+ + r_-} \quad \text{or} \quad \frac{287\nu z_+}{r_+ + r_-}\left[1-\frac{0{\cdot}345}{r_+ + r_-}\right] \text{kcal mol}^{-1} \tag{2}$$

(ν = number of ions per mole of halide; r_+ and r_- = radii of cation and halide anion respectively) developed by Kapustinskii and Yatsimirskii[292] for the treatment of ionic radii and lattice energies in the absence of structural details. Approximate though the method is, its simplicity makes it highly suitable for comparative purposes, and the conclusions obtained by its use often hold even when the compounds involved are far from ionic[289,293].

289 A. G. Sharpe, *Halogen Chemistry* (ed. V. Gutmann), Vol. 1, p. 1. Academic Press (1967).
290 L. Pauling, *The Nature of the Chemical Bond*, 3rd edn., Cornell University Press, Ithaca (1960).
291 H. Witte and E. Wölfel, *Z. phys. Chem. (Frankfurt)*, **3** (1955) 296.
292 A. F. Kapustinskii, *Quart. Rev. Chem. Soc.* **10** (1956) 283; T. C. Waddington, *Adv. Inorg. Chem. Radiochem.* **1** (1959) 157.
293 D. A. Johnson, *Some Thermodynamic Aspects of Inorganic Chemistry*, Cambridge (1968).

Hence the radius of a single ion has no precise physical basis, the chief justification for values such as those given in the table lying in their practical usefulness in comparative interpretation and in their capacity roughly to reproduce internuclear distances and account for those properties to which such distances are related.

The enhanced susceptibility to charge-delocalization which accompanies the increasing size and nuclear charge of the halide ions is indicated (i) by the polarizabilities and (ii) by the ionization potentials of the ions. Correspondingly the donor capacity of the ions decreases in the order $I^- > Br^- > Cl^-$. Primary interactions of the type $M^{n+} \cdots X^-$ therefore involve a proportion of electron-delocalization (that is, charge-transfer or ultimately covalent bonding), which is greater for X = I than for X = Cl; secondary interactions, e.g. dispersion forces, are also relatively stronger in systems containing the larger halide ions. For this reason, structures which give a good approximation to the ionic model are more commonly formed by fluoride than by iodide ions, the larger size and polarizability of which tend to favour lower coordination numbers, often in conjunction with layer or chain structures. With cations of high electron affinity, formed by non-metals or by metals in high oxidation states, the halide ion cannot survive as a recognizable entity, charge being transferred roughly in accordance with Pauling's Electroneutrality Principle, which requires that, for stability, the charge on any atom of an aggregate shall not greatly exceed $\pm\frac{1}{2}e$ [294]. Such systems may take the form of neutral molecules, e.g. SiX_4 or WCl_6, or of complex ions, e.g. HgX_4^{2-} or PdX_6^{2-}.

Donor Properties of the Halide Ions: Solvation and Charge–Transfer Interactions

Thermodynamic Characteristics

The donor capacity of the halide ions is manifest with respect, not only to cations, but also to appropriate neutral molecules. With halogen or interhalogen molecules the products of this interaction are the chemically recognizable polyhalide ions (see Section 4), but with other, polar molecules, e.g. water, alcohols, amines or acetonitrile, the interaction typically involves a combination of ion–dipole, charge-transfer and dispersion forces, the balance of which governs the power of the liquid to solvate the halide ion.

Most of our knowledge of the quantitative aspects of ion-solvation is restricted to water as the solvent. For the alkali-metal halides, for example, the measured free energy, enthalpy and entropy of the change $M^+(g)+X^-(g) \rightarrow M^+(aq)+X^-(aq)$ indicate that in dilute solutions the thermodynamic properties of one ion are independent of those of another[293,295–297]. This implies that the thermodynamic properties defining hydration can be split up into contributions from the individual ions. All of the numerous attempts which have been made to achieve this end involve some assumption that cannot be thoroughly substantiated, though most yield results which are in fairly close agreement. The absolute values given in Table 19 are based on estimates due to Halliwell and Nyburg[295] and to Harvey and Porter[296]; all of the values are relative to $\Delta G^\circ_{hyd} = -252{\cdot}0$ kcal g-ion^{-1}, $\Delta H^\circ_{hyd} = -260{\cdot}7$ kcal g-ion^{-1} and $S^\circ = -29$ cal g-ion^{-1} deg^{-1} for the proton at 298°K.

294 L. Pauling, *J. Chem. Soc.* (1948) 1461.

295 H. F. Halliwell and S. C. Nyburg, *Trans. Faraday Soc.* **59** (1963) 1126.

296 K. B. Harvey and G. B. Porter, *Introduction to Physical Inorganic Chemistry*, Addison-Wesley, Reading, Mass. (1963).

297 F. D. Rossini, D. D. Wagman, W. H. Evans, S. Levine and I. Jaffe, *Selected Values of Chemical Thermodynamic Properties*, Circular 500, National Bureau of Standards, Washington (1952); National Bureau of Standards Technical Notes 270–1, 270–2, 270–3 and 270–4, U.S. Government Printing Office, Washington (1965–1969); *Codata Bulletin*, International Council of Scientific Unions, Committee on Data for Science and Technology (1970).

The absolute values of the thermodynamic changes accompanying the hydration of the Cl^-, Br^- and I^- ions appear to be primarily a function of ionic size. Thus ΔG°_{hyd}, ΔH°_{hyd} and S° all decrease numerically as the ionic radius increases, corresponding to the attenuation of ion-dipolar forces and the diminishing restriction of the motion of water molecules round the ion. Following a theoretical interpretation first suggested by Born[293], the free energy of electrostatic solvation of an ion of radius r' and charge z in a solvent of dielectric constant ϵ is given by

$$\Delta G^\circ_{solv} = \frac{-Nz^2e^2}{2r'}\left[1-\frac{1}{\epsilon}\right]+1{\cdot}32 \text{ kcal mol}^{-1} \text{ at } 298°\text{K} \quad (3)$$

Of the several shortcomings of this treatment, it may be noted that the radius of an isolated ion in the gas phase or in solution is unlikely to be the same as that (r) in a particular crystal lattice, while the macroscopic value of ϵ is almost certainly inappropriate in the high electric fields which operate close to the ions. If ϵ is equated with the bulk dielectric constant of water, the results of Table 19 imply that for the halide ions r' exceeds r by about 0·2 Å. On the basis of equations (2) and (3) Johnson[293] has shown that the free energy of aqueous dissolution of an ionic halide MX_n may be represented approximately by the relationship

$$\Delta G_s^\circ(\text{kcal}) = \frac{256(n+1)n}{r_+ + r_-} - \frac{164n^2}{r_+ + 0{\cdot}72} + n\Delta G^\circ_{hyd}[X^-] - 7{\cdot}4n - 6{\cdot}1 \quad (4)$$

Expressed as a function of r_+, ΔG_s° passes through a maximum when

$$r_- = r_+[1{\cdot}25\sqrt{(1+1/n)}-1]+0{\cdot}90\sqrt{(1+1/n)} \quad (5)$$

Being very approximate, equations (4) and (5) do not predict accurately the value of the cation radius at which the maximum in ΔG_s° is achieved in a series of salts of given formula

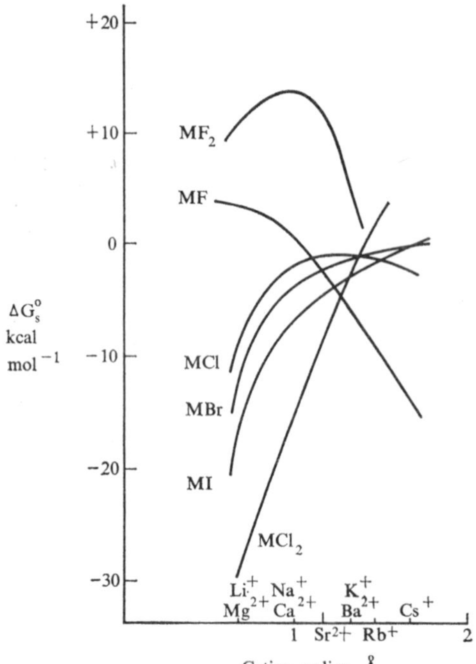

FIG. 19. Variations in the standard free energies of solution, ΔG°_s, of some alkali and alkaline-earth metal halides with the crystal radius of the cation.

type containing the same halide anion. Nevertheless, the following implications with regard to qualitative variations in ΔG_s° are vindicated by experimental data (see Fig. 19): (i) For salts of the same anion, as r_+ is increased, ΔG_s° rises, reaches a maximum and then falls steadily towards a limiting value of $(n\Delta G^\circ_{hyd}[X^-]-7{\cdot}4n-6{\cdot}1)$. (ii) For different series each containing a fixed anion, ΔG_s° reaches a maximum at a larger cation radius as the size of the anion advances. The maximum should occur when the sizes of the cation and anion are suitably matched. (iii) When two series of compounds have different formula types MX_n but contain the same anion, the free energy maximum should occur at a larger cation radius in the series with the higher value of n. Provided that the phase deposited from the saturated solution is the anhydrous compound rather than a hydrate, these generalizations provide a useful and instructive guide to the solubilities of ionic halides, but do not apply well to the halides of metals like silver and mercury whose properties deviate markedly from those expected of ionic compounds.

Optical Spectra

There exist close analogies between the absorption spectra of ionic halides and those of the isoelectronic noble gases[298]. Thus, the first absorption band is attributable to transitions of the type $np^6 \rightarrow np^5(n+1)s^1$; there is some discussion about whether a weak absorption at slightly higher energy is due to a direct ionization of an np-electron to the conduction continuum, or whether it corresponds to the Laporte-forbidden transition $np^6 \rightarrow np^5(n+1)p^1$; further, strong absorption bands at somewhat higher wavenumbers can be ascribed to the transition $np^6 \rightarrow np^5nd^1$. There is little doubt that the description in terms of transitions between localized atomic states is far more appropriate for the alkali-metal halide crystals than is an energy-band description. Because of the relatively localized electronic structure of many simple and complex metal halides it is possible to identify electronic transitions involving charge-transfer from the halide to the metal ion; recent years have witnessed extensive investigations of such transitions[298].

Although ion-dipole interactions undoubtedly play a major role in the solvation of halide ions, some orbital overlap and charge-delocalization are clearly implied by the optical spectra of the ions in solution. The ultraviolet absorption spectra of such solutions show intense ($\epsilon \sim 10^4$) bands, broad and structureless, which are considered to arise from a charge-transfer process:

$$(X^-)\text{solv} \rightarrow (X\cdot + e)\text{solv}$$

and which are thought to represent charge-transfer-to-solvent (c.t.t.s.) transitions[299,300]. Details of the ultraviolet spectra of aqueous solutions of the halide ions are presented in Table 19. Applying Mulliken's theory for donor-acceptor complexes (see pp. 1209–14) leads to the transition energy,

$$E_{max} = I_{X^-} - E_{solv} + \frac{\sigma^2}{I_{X^-} - E_{solv}}(1/2r_i) \tag{6}$$

depending on I_{X^-}, the ionization potential of the anion, E_{solv}, the electron affinity of a group of solvent molecules, σ, an overlap term determined by the overlap integral and polarization terms, and r_i, the crystallographic radius of the ion. Good agreement is observed

298 Chr. K. Jørgensen, *Halogen Chemistry* (ed. V. Gutmann), Vol. 1, p. 265. Academic Press (1967); *Progress in Inorganic Chemistry*, **12** (1970) 101.

299 M. J. Blandamer and M. F. Fox, *Chem. Rev.* **70** (1970) 59.

300 M. F. Fox, *Quart. Rev. Chem. Soc.* **24** (1970) 565.

between predicted and measured transition energy values for halide ions both in acetonitrile and in aqueous solutions.

An alternative approach relates E_{max} to the heats of solvation $\Delta H_{solv}[X^-]$ and $\Delta H_{solv}[X]$ by

$$E_{max} = I_{X^-} + \Delta H_{solv}[X^-] - \Delta H_{solv}[X] + \chi - B \quad (7)$$

χ representing the Franck–Condon strain arising from the immobilization of the solvent dipoles during the electronic transition and B the excited-state electron binding energy. The difficulty of calculating B has been overcome by Franck and Platzman, who assumed a spherically symmetrical excited state centred on the parent anion site: B is then the sum of two contributions, one arising from the polarization of the oriented solvent molecules around the ion, and the other from the electronic polarization of the solvent by the excited-state electron. With these assumptions, equation (7) gives a good account of the observed E_{max} values for the halide ions in aqueous solution but does not reproduce the effect on E_{max} of such variables as change of solvent, added electrolyte, pressure and temperature. Other approaches are[299,300]: (i) a so-called "confined" model of an electron in a potential well, variations in the radius of which determine the changes in E_{max}, and (ii) a "diffuse" model in which the variable radius parameter r_d is combined with the Franck–Platzman treatment to predict both absolute values of E_{max} and environmental effects.

The consensus of the various theories is that the excited state characterized by the ultraviolet absorption bands is either completely or predominantly defined by the solvent and centred on the parent anion site; it typically lies at an energy *ca.* 50 kcal or 18,000 cm^{-1} above that required for ionization of X^-. Charge-transfer is reported to occur even with a supposedly "inert" solvent like carbon tetrachloride[301]. Measurements of optical spectra imply that the halide ions form weak complexes with molecular oxygen[302] but relatively stronger complexes with sulphur dioxide as the acceptor[303]. The action of sulphur dioxide is presumably responsible for the moderate solubility of ionic halides in liquid sulphur dioxide. The excited state of the optical transitions also controls the photolysis of solutions of halide ions, which leads to transient radical-ions such as X_2^- and ultimately to products such as hydrogen and the free halogen[300]. The radicals formed in the primary photolytic process either recombine, dimerise or are scavenged by other solutes, e.g. N_2O, acetone or Brönsted acids.

Charge-transfer spectra are also observed for crystalline and gaseous ionic halides[304]. The ultraviolet absorption spectra of crystalline or gaseous alkali-metal halides consist, for example, of one or more bands attributable to a transition which may be represented roughly as

$$M^+X^- \rightarrow MX$$

---

[301] M. J. Blandamer, T. E. Gough and M. C. R. Symons, *Trans. Faraday Soc.* **62** (1966) 301.

[302] H. Levanon and G. Navon, *J. Phys. Chem.* **73** (1969) 1861.

[303] E. J. Woodhouse and T. H. Norris, *Inorg. Chem.* **10** (1971) 614.

[304] L. E. Orgel, *Quart. Rev. Chem. Soc.* **8** (1954) 422.

[305] Chr. K. Jørgensen, *Orbitals in Atoms and Molecules*, Academic Press, London (1962); *Absorption Spectra and Chemical Bonding in Complexes*, Pergamon, Oxford (1962).

[306] C. Deverell, *Progress in NMR Spectroscopy* (ed. J. W. Emsley, J. Feeney and L. H. Sutcliffe), Vol. 4, p. 235. Pergamon (1969).

[307] C. Hall, *Quart. Rev. Chem. Soc.* **25** (1971) 87.

[308] R. W. Gurney, *Ionic Processes in Solution*, McGraw-Hill, New York (1953); M. Kaminsky, *Discuss. Faraday Soc.* **24** (1957) 171.

[309] O. Ya. Samoilov, *Discuss. Faraday Soc.* **24** (1957) 141; *Structure of Aqueous Electrolyte Solutions and the Hydration of Ions* (transl. by D. J. G. Ives), Consultants Bureau Enterprises Inc., N.Y. (1965).

In terms of the simplest form of the Mulliken theory, the transition energy is given by the equation

$$\left.\begin{aligned} E_{max} &= I_{X^-} - E_{M^+} + \Delta E \\ &= E_X - I_M + \Delta E \end{aligned}\right\} \quad (8)$$

where I signifies the ionization potential and E the electron affinity of the species and ΔE is the change of interaction energy due to the charge-transfer. In keeping with this, the frequency of maximum absorption of the halides of a given metal follows the order of increasing electron affinity, viz. iodine < bromine < chlorine. The general features are reproduced by many crystalline halides, the charge-transfer mechanism providing an effective interpretation of the colours of those halides formed by metals like copper, silver, lead and mercury, which are characterized by relatively high ionization potentials. Crystalline alkali-metal chlorides generally exhibit a single absorption maximum in the ultraviolet, the bromides a pair of bands separated by 3900–4800 cm^{-1} and the iodides a pair of bands separated by 7700–9600 cm^{-1}. The close correspondence between these doublet separations and the energy differences between the two lowest states $^2P_{3/2}$ and $^2P_{1/2}$ of the free halogen atoms (Cl, 882; Br, 3685; I, 7603 cm^{-1}) provides convincing support for the proposed mechanism, as does the fluorescence emission due to excited states of metal atoms, which has been observed for some gaseous halides[305].

As with the photolysis of halide ions in solution, chemical effects may also arise from charge-transfer absorption in a crystalline halide through secondary transformations which may compete with physical processes of deactivation of the excited state. Thus, the formation of metal atoms is responsible for the colour-centres produced by ultraviolet irradiation of alkali halide crystals and for the photosensitivity of the silver halides. Variations in the quantum efficiency of the production of colour-centres in alkali halide crystals indicate that the primary absorption process is virtually independent of temperature, but that the formation of metal atoms depends on the vibrations of the crystal lattice. Whereas the primary process of charge-transfer is completely reversible, only if some specific interaction takes place within the short lifetime of the excited state, is a metal atom produced.

As the ground-state of a system incorporating a halide ion and acceptor species acquires an increasing degree of charge-transfer, the properties of the system become more and more specific, and any purely ionic model becomes progressively less realistic.

Behaviour of Halide Ions in Solution: Nmr and Other Studies [306,307]

The introduction of ions into the lattice characteristic of liquid water gives rise to changes which may be defined by structural and kinetic criteria. In the former case a coordination number may be derived to describe the most probable number of water molecules in the immediate environment of the ion. Although this is a convenient parameter for many transition-metal and other multi-charged ions, for the halide ions, as for the alkali-metal cations, the exchange of water molecules between the first coordination shell and bulk water is so rapid that it is neither meaningful nor expedient to define a separate solvated species in which the central ion has a specific hydration number. In these circumstances it is more appropriate to employ the second criterion and define the ion–solvent interaction in terms of the lifetime of a water molecule in the coordination shell.

In contrast with the F^- ion, Cl^-, Br^- and I^- in aqueous solution all belong to the "structure-breaking" category, tending to break down the water structure in their immediate neighbourhood and increase the mobility of adjacent solvent molecules. Such behaviour is implied by considerations of the entropy of solution, viscosity and temperature-dependence of ionic mobilities in alkali halide solutions. Thus, viscosities of dilute aqueous electrolyte solutions, η, can be represented by the equation[308]

$$\eta/\eta_0 = 1 + Ac^{\frac{1}{2}} + Bc \tag{9}$$

where η_0 is the viscosity of the pure solvent, c is the concentration and A and B are constants. As a function of the interionic forces, A can be calculated theoretically, but the constant B, which takes into account ion–solvent interactions, is purely empirical and may have either a positive or negative sign. It being possible to separate B into individual ion

TABLE 20. PROPERTIES OF THE HALIDE IONS IN AQUEOUS SOLUTION[a,b]

Property	F^-	Cl^-		Br^-		I^-
B coefficient in the Jones–Dole equation $\eta/\eta_0 = 1 + Ac^{\frac{1}{2}} + Bc$ (η = viscosity; c = salt concentration) $(mol/l)^{-1}$ at 298°K		−0·0070		−0·042		−0·0685
Ratio of residence times of water molecules τ_h/τ (τ_h = residence time adjacent to X^- ion; τ = residence time in a site of the water lattice) at 294·7°K		0·63		0·61		0·58
Parameter E in the Samoilov expression $\tau_h/\tau = \exp(E/RT)$ (kcal/mol) at 294·7°K		−0·27		−0·29		−0·32
Ionic molal shift for H_2O molecules, δ (ppm/mol) at 298°K						
^1H resonance	−0·075	+0·026		+0·053		+0·069
^{17}O resonance	−0·92	−1·35		−1·72		−2·35
Ratio of the reorientational correlation times of H_2O molecules τ_d/τ_d° (τ_d = correlation time for molecules close to X^-; τ_d° = correlation time in pure water) at 298°K	2·3	0·9		0·6		0·3
Activation energies for the reorientation of H_2O molecules adjacent to halide ions, E_r (kcal/mol)	4·3	2·6–3·3		2·1–2·9		2·3–2·7
Nmr properties of halide nuclei:	^{19}F	^{35}Cl	^{37}Cl	^{79}Br	^{81}Br	^{127}I
% Natural abundance	100	75·53	24·47	50·54	49·46	100
Sensitivity at constant field relative to ^1H = 1·000	0·833	4·70 $\times 10^{-3}$	2·71 $\times 10^{-3}$	7·86 $\times 10^{-2}$	9·85 $\times 10^{-2}$	9·34 $\times 10^{-2}$
Ground state nuclear spin, $\hbar$	1/2	3/2	3/2	3/2	3/2	5/2
Expectation value $\langle r_i^{-3}\rangle_p$ for the valence-shell p-electrons (a.u.)	6·40	5·74		10·24		12·76
Average excitation energy, Δ (Rydbergs)	0·63	0·56		0·48		0·40
$\frac{1}{\Delta}\langle r_i^{-3}\rangle_p$	10·2	10·2		21·3		31·9
Paramagnetic shift of X^-(aq) relative to X^-(g), σ_{aq}(ppm) (calculated)		−225		−430		−600

[a] C. Deverell, *Progress in NMR Spectroscopy* (ed. J. W. Emsley, J. Feeney and L. H. Sutcliffe), Vol. 4, p. 235, Pergamon (1969).

[b] C. Hall, *Quart. Rev. Chem. Soc.* **25** (1971) 87.

contributions, structure-breaking ions like Cl^-, Br^- and I^- are identified by negative B coefficients, whereas structure-forming ions like F^- are identified by positive coefficients (Table 20). According to Samoilov[309], ion-hydration can be considered in terms of the effect of ions on the translational motion of adjacent water molecules: the mean lifetime (τ_h) of a water molecule in close proximity to an ion is related to that (τ) of a water molecule in an equilibrium position in the water lattice by the expression

$$\tau_h/\tau = \exp(E/RT) \tag{10}$$

Since τ is about $1{\cdot}7 \times 10^{-9}$ sec (at 21·5°C), the values of 0·58–0·63 calculated for τ_h/τ for Cl^-, Br^- and I^- indicate an extremely high frequency of exchange of water molecules from sites adjacent to an ion; further, the parameter E is negative, indicating an increase in the mobility of water molecules in the neighbourhood of the ions (see Table 20). This phenomenon has been termed "negative hydration".

Solutions of halide ions are particularly suited to investigation by nuclear magnetic resonance techniques[306,307]. The resonating nucleus, whether of the halide ion or of an atom in the solvent molecule, undergoes rapid exchange between all possible environments in the solution; accordingly, only one signal is observed at an average frequency determined by the magnetic shielding and lifetime of the nucleus in each of the many allowed sites. The rapidity and randomness of ion–ion and ion–solvent encounters average local magnetic and electric fields to very small values to give relatively narrow resonance lines even for the quadrupolar nuclei of the heavier halogen atoms. Hence it is possible to detect comparatively small differences in magnetic shielding as well as broadening of the resonance line by chemical exchange, hyperfine interaction or quadrupolar effects.

^{1}H and ^{17}O Resonances of Solvent Molecules[306]

Halide ions dissolved in water produce quite small but well-established changes in the ^{1}H resonance and relatively larger changes in the ^{17}O resonance. A linear or near-linear variation of chemical shift with salt concentration is generally observed at concentrations typically up to 2 or 3 molal. The results are normally expressed in molal shifts, that is, the initial slopes of plots of ^{1}H or ^{17}O chemical shift against molality. The ionic contributions appear to be additive at lower concentrations, and accordingly the shifts have been divided into individual ionic contributions. Proton chemical shifts produced by the halide ions other than F^- arise primarily from the rupture of hydrogen bonds, in keeping with their structure-breaking propensities which follow the sequence $Cl^- < Br^- < I^-$[310]. By contrast, direct ion–solvent interactions, rather than disruption of the solvent structure, appear to play the dominant role in determining ^{17}O molal shifts. Whereas all the alkali cations produce similar effects on the ^{17}O resonance of water, the halide ions show significant differences amongst themselves in the effects they have on the ^{17}O chemical shift. A very similar distinction between the behaviour of cations and anions in alkali halide solutions has been found in studies of their Raman spectra, which show the vibrational properties of the solvent molecules to be more strongly perturbed by the halide ions[311]. This result is somewhat surprising, especially since X-ray diffraction data indicate that, for the alkali-metal ions, the ion–dipole interactions with solvent water molecules occur *via* the oxygen atom, whereas it is commonly assumed (from energetic considerations) that the

[310] J. C. Hindman, *J. Chem. Phys.* **36** (1962) 1000.
[311] G. E. Walrafen, *J. Chem. Phys.* **36** (1962) 1035.

halide ions interact *via* the centres of positive charge at the hydrogen atoms. Some Raman studies[312] have been published in support of such a model for halide ion–water interactions. Possible explanations of the unusual response of the solvent molecules to halide ions invoke either a redistribution of charge in the H–O bonds induced by the polarizable halide ions or, alternatively, short-range repulsive forces which operate in collisions between ions and solvent molecules.

Perhaps the most direct evidence of the "negative hydration" of the halide ions in aqueous solution has come from measurements of proton spin-lattice relaxation times[306]. In keeping with the enhanced mobility of water molecules close to Cl^-, Br^- or I^- ions, the proton relaxation rate is reduced as compared with that found in pure water. The variations follow closely the order expected from other estimates of the structure-breaking abilities of the ions; thus, the initial slopes of plots of relaxation rate against molality show an extremely close correlation with the entropy of solution for alkali halides.

Halide Ion Resonances[306,307]

The nuclei ^{35}Cl, ^{79}Br, ^{81}Br and ^{127}I give weaker resonance signals than protons with the result that most investigations are presently confined to solutions of concentration greater than 0·05 molal. On the other hand, a wide range of chemical shifts, spanning as much as 100 ppm, is observed as a result primarily of changes in the paramagnetic contribution to the magnetic shielding. Relaxation times and hence resonance linewidths of the heavier halogen nuclei are determined by processes originating mainly from the quadrupole moments of the nuclei. In consequence, investigations of linewidths can illuminate both the magnitude and time-dependence of electric-field gradients at the nucleus; being determined by the ion–solvent and ion–ion interactions in solution, such gradients give access to information about the times of rotational and translational motions of ions and solvent molecules.

In aqueous solutions of alkali-metal chloride, bromide or iodide, the chemical shift (to low field) of the halogen nucleus varies with the concentration and with the identity of the counter-ion. For a given halide ion, the efficacy of the alkali-metal cations in producing low-field shifts may be ordered $Na^+ < K^+ < Li^+ < Rb^+ < Cs^+$. It has been suggested that the ion shifts so induced originate in the direct binary interactions of the anion with the cation, the contribution from halide–solvent and halide–halide interactions being assumed approximately constant at all but the highest concentrations. Since the larger ions cause greater shifts than the smaller ions, the effect is consistent, not with a direct electrostatic (polarization) interaction, but with the repulsive overlap of the closed-shell orbitals of the ions, a mechanism first suggested by Kondo and Yamashita to account for the shifts of the ions in alkali halide crystals[313]. The experimental shift may be expressed in terms of the fine-structure constant α, the expectation value $\langle r_i^{-3} \rangle_p$ of r_i^{-3} for an outer p-electron, and an average excitation energy Δ as

$$\delta = \frac{-16\alpha^2}{\Delta} \langle r_i^{-3} \rangle_p \left[\sum_j \Lambda_{i-j}{}^{(c)} + (\Lambda_{i-H_2O}{}^{(c)} - \Lambda_{i-H_2O}{}^{\circ}) \right] \qquad (11)$$

where Λ_{i-j} and Λ_{i-H_2O} denote appropriate sums of the squares of overlap integrals for ion–ion and ion–solvent encounters. Table 20 lists for the halide ions values of $\langle r_i^{-3} \rangle_p$ computed from Hartree–Fock wavefunctions and estimates of Δ derived from the ultraviolet absorption spectra of solid alkali halides. Variations in the quotient $\langle r_i^{-3} \rangle_p / \Delta$

[312] G. E. Walrafen, *J. Chem. Phys.* **44** (1966) 1546.
[313] J. Kondo and J. Yamashita, *Phys. Chem. Solids*, **10** (1959) 245.

account for a considerable part of the observed increase in the magnitude of the chemical shifts found with increasing atomic number of the resonating halide ion.

Numerous measurements of the nuclear magnetic relaxation times of the halide ions in solution have been carried out. Though the theory of such relaxation[314] allows us to relate the observed relaxation time to a function of a correlation time (a period of time giving a measure of the timescale of molecular motion), it is a major problem to determine the motion in the liquid to which the correlation time refers. The situation is further complicated by the absence of any knowledge of the quadrupole coupling constant; more or less irrespective of the mechanism, this must itself be an average over many configurations, since the ion experiences many field gradients from various neighbours, all of them random functions of time, notwithstanding at least some degree of correlation. It has been suggested that the solvent molecules contributing to relaxation are mainly those in the rather disorganized "outer-solvation" regions, characterized by structural mismatching between the bulk solvent and the solvent molecules oriented about the ions and by relatively free rotation or reorientation of solvent molecules.

In certain cases the relaxation time of a quadrupolar halogen nucleus can be related to the rates of relatively fast reactions occurring in electrolyte solutions. ^{127}I, which typically exhibits the shortest relaxation time, can thus furnish kinetic information about processes with lifetimes of the order of nanoseconds. Thus, for the aqueous system

$$I^- + I_2 \rightleftharpoons I_3^-$$

the broadening of the ^{127}I resonance gives rate constants which support the view that the equilibrium is established at a diffusionally limited rate[306,307]; that the analogous reaction between Cl^- and Cl_2 exhibits a rate constant at 25°C of $8 \pm 4 \times 10^6$ sec^{-1} is the inference drawn from a more comprehensive evaluation of the ^{35}Cl and ^{37}Cl resonance linewidths[315]. Likewise, analyses of the broadening of the $^{127}I^-$ and $^{81}Br^-$ resonances in the presence of small concentrations of Hg^{2+}, Zn^{2+} or Cd^{2+} ions have led to rate constants for processes such as

$$HgI_4^{2-} + I^{*-} \rightleftharpoons I^- + I^*HgI_3^{2-}$$

which also appear to be diffusion-limited. Having shown that Hg^{2+} ions bind to reactive or exposed sulphydryl groups of proteins, Stengle and Baldeschwieler[316] have developed the so-called "halide ion probe" technique, wherein the mercury tag provides a possible coordination site for halide ions, with the result, in the event of rapid exchange, that the halide ion resonance is broadened. In principle this provides a means of determining the relaxation time of the halide ion nucleus at the binding site, and variation of the conditions, such as pH and temperature or the addition of other substances to the system, reveals changes in the accessibility of the binding site to the halide ions. Furthermore, the broadening of the nmr signal may be used simply to monitor a titration experiment: for example, as mercuric chloride is added to a chloride solution containing small amounts of haemoglobin, the variation of the ^{35}Cl linewidth indicates two reactive –SH groups per haemoglobin to which the mercury binds. Several other studies of this sort have been performed, and the technique has recently been extended to systems other than mercury-labelled molecules, as in the investigation of the binding under various conditions of the Cl^- ion to a zinc metallo-enzyme carbonic anhydrase[317]. A survey of such studies of biomolecules has been given by Hall[307].

[314] H. G. Hertz, *Z. Elektrochem.* **65** (1961) 20; K. A. Valiev, *J. Exp. Theor. Phys.* **10** (1960) 77.
[315] C. Hall, D. W. Kydon, R. E. Richards and R. R. Sharp, *Proc. Roy. Soc.* **A318** (1970) 119.
[316] T. R. Stengle and J. D. Baldeschwieler, *Proc. Nat. Acad. Sci. U.S.A.* **55** (1966) 1020.
[317] R. L. Ward, *Biochemistry*, **8** (1969) 1879.

3.2. GENERAL PROPERTIES OF HALIDES

Attention is now directed to the following general aspects of compounds containing chlorine, bromine or iodine in combination with more electropositive partners: (i) classification; (ii) methods of formation; (iii) physical characteristics, with particular reference to thermodynamic and spectroscopic properties; (iv) nature of bonding. Apart from the hydrogen halides, which form the subject of subsection 3.3, this is not the place to enter into detail about individual halides, the physical and chemical properties of which are presented under the heading of the appropriate electropositive component. It is with the halides as a particularly populous and important class of inorganic system that the present survey is concerned. The approach is dominated by considerations of the ligand function, real or formal, of the halide ions, and of variations in the nature of this function.

Classification[289,293,318–321]

Concerning halides as a class certain generalizations may be remarked at the outset. First, primarily as a result of the small size of the F^- ion and the strong bonds formed by fluorine with other elements, fluorides often differ in stoichiometry or in structure from the other halides. Thus, we note that many metal fluorides have three-dimensional lattices, whereas the corresponding chlorides, bromides and iodides form crystals having layer or

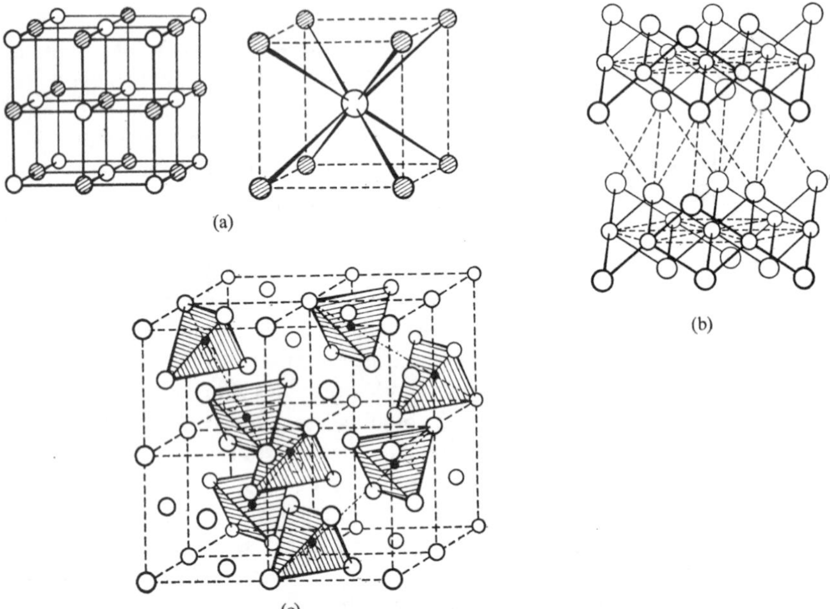

FIG. 20. Representative structures of crystalline halides: (a) the NaCl and CsCl structures; (b) the CdI_2 layer structure [reproduced with permission from C. S. G. Phillips and R. J. P. Williams, *Inorganic Chemistry*, Vol. 1, p. 183, Clarendon Press, Oxford (1965)]; (c) unit cell of the structure of SnI_4 [reproduced with permission from A. F. Wells, *Structural Inorganic Chemistry*, 3rd edn., p. 350, Clarendon Press, Oxford (1962)].

---

318 N. V. Sidgwick, *The Chemical Elements and their Compounds*, Vol. II, Clarendon Press, Oxford (1950).
319 A. F. Wells, *Structural Inorganic Chemistry*, 3rd edn., Clarendon Press, Oxford (1962).
320 C. S. G. Phillips and R. J. P. Williams, *Inorganic Chemistry*, Vol. 1, Clarendon Press, Oxford (1965)
321 F. A. Cotton and G. Wilkinson, *Advanced Inorganic Chemistry*, 2nd edn., Interscience (1966).

chain structures. Secondly, whereas many crystalline fluorides and oxides are isostructural, chlorides, bromides and iodides are often structurally analogous to sulphides, selenides and tellurides of similar formula type, e.g. PbI_2 and SnS_2; CdI_2 and VTe_2. Notwithstanding the substantial differences between the individual halogens, chlorides, bromides and iodides are sufficiently similar to permit a collective classification (see Table 21), though it must be appreciated that there are no clear lines of demarcation between the different classes. Rather is there a uniform gradation from halides which are for all practical purposes ionic, through those of intermediate character, to those which are essentially molecular. Structures representative of the different types of crystalline halide are illustrated in Fig. 20.

Ionic Halides: Three-dimensional Lattices

Ideally an ionic halide exists at normal temperatures as an involatile solid possessing a three-dimensional lattice in which the atoms are regularly disposed. Moreover, although the electrical conductivity of the solid is very low, it is markedly enhanced by fusion. Very careful examination by X-ray techniques has shown for several compounds of this type that the electron density distribution falls to a very low value between adjacent unlike nuclei, and that, if the distribution is divided at this point, each atom possesses an approximately integral net charge[293]. Such measurements disclose further that the internuclear distances are approximately additive. An ionic halide may also be defined as one whose lattice energy can be reproduced, within the limits of experimental error, by the Born–Mayer relation (equation (1)) or more elaborate versions of this expression. According to some or all of these criteria, the class of "ionic" chlorides, bromides and iodides may reasonably be taken to comprise halides of the type MX formed by alkali-metal and other large univalent cations, e.g. Cu^+, Ag^+, NH_4^+ and NR_4^+, and of the types MX_2 and MX_3 formed by metals with large cations—alkaline-earth halides, $PbCl_2$ and UCl_3.

As with hydrides, the closest approach to ionic behaviour is found in the halides formed by elements at the extreme left of the Periodic Table. However, as compared with the hydrides, the pattern is altered because of the more favourable enthalpies of formation of the gaseous anions and because of the greater sizes and polarizabilities of the halide anions. Accordingly "ionic" halides have larger heats of formation and are thermodynamically more stable than the corresponding hydrides, while the elements which form them occupy a wider region of the Periodic Table.

Semi-ionic Halides: Layer or Chain Structures and Systems Involving Metal–Metal Interaction

The capacity of a cation M^{z+} to distort an ionic charge distribution depends on its polarizing power, which can be related to Z^*/r^2 (Z^* = effective nuclear charge; r = ionic radius)[322] or alternatively to the electron affinity of the ion[293]; these properties reflect variations, not only in the ratio z/r, but also in the shielding abilities of the valence and core electrons [cf. the following values of Z^*/r^2 in a.u. (electron affinity in eV): K^+, 0·26 (4·34); Cs^+, 0·20 (3·89); Cu^+, 0·84 (7·72); Au^+, 0·86 (9·22)]. As the polarizing power of the cation increases, non-coulombic interactions make progressively larger contributions to the bonding of metal halides. In the sequence KCl, $CaCl_2$, $ScCl_3$, $TiCl_4$, for example, we pass from a compound well represented by the ionic model to a molecular species most aptly described by some form of bond model. The size and polarizability of the halide ion are also important in determining the character of the halide. Thus, whereas the majority of metal difluorides, which adopt the rutile or the fluorite structure, give a good approximation to the ionic

[322] R. B. Heslop and P. L. Robinson, *Inorganic Chemistry*, 3rd edn., Elsevier (1967).

TABLE 21. PHYSICAL CHARACTERISTICS OF HALIDES MX_n[318–322]

Property	Ionic halides	Macromolecular or 'semi-ionic' halides	Molecular halides
Formed by	Majority of pre-transition metals, lanthanides and actinides in lower charge states (+2, +3)	*d*-Block transition metals in low charge states (+1 to +3) and B-metals	Non-metals and metals in high oxidation states ($\geqslant +3$)
Electronegativity difference between M and X	←	Increasing	
Description	Well represented by the ionic model: regular, three-dimensional lattices involving high coordination numbers; internuclear distances additive; lattice energies well reproduced by the Born–Landé or Born–Mayer equations.	Most conveniently described in terms of departures from the idealized ionic model; some systems amenable to treatment by the band model. Halogen commonly found in relatively unsymmetrical environment and lattices lack the regularity of the idealized systems, e.g. layer or chain structures. Internuclear distances less nearly additive. Non-coulombic interactions contribute significantly to the lattice energy.	Best represented by the bond model involving, for example, valence-bond or molecular-orbital accounts. Stereochemistry of molecules can be treated in terms of valence-shell electron-pair repulsions. Weak intermolecular interactions due to London, dipole–dipole, dipole–induced dipole and quadrupolar forces.
Sublimation energy	Comparable with binding energy of atomic units. Commonly vaporize to give polymer units.	← Increasing May vaporize to give polymer units.	Much smaller than binding energy of atomic units. Molecular units common to the solid and vapour phases.
Boiling point	High ← Generally $MF_n > MCl_n > MBr_n > MI_n$, order dictated by coulombic interactions.	Intermediate	Low Typically $MI_n > MBr_n > MCl_n > MF_n$, order dictated by polarizability.
Melting point	High ← Generally $MF_n > MCl_n > MBr_n > MI_n$, order dictated by lattice energy.	Intermediate	Low Typically $MI_n > MBr_n > MCl_n > MF_n$, order dictated by polarizability.

Colour	Colour, if any, characteristic of internal transitions of cation.	Colour, if any, due to charge–transfer processes of the type $M^+X^- \xrightarrow{h\nu} M\cdot X\cdot$ and/or to internal transitions of cation.	Colour, if any, usually due to charge–transfer transitions.
Charge–transfer properties	Transitions observed as discrete, rather narrow absorption bands of high energy and probably highly localized.	Transitions observed as broad bands of lower energy; in some cases there may be an absorption edge beyond which absorption is continuous. Transitions probably delocalized over the whole crystal. Photoconduction may occur, as in HgI_2.	Transitions observed as discrete, rather narrow absorption bands of relatively low energy.
Electrical conductivity Melt	High conductivity.	Relatively low conductivity.	Very low conductivity presumably due to auto-ionization.
Solid	Low conductivity because of high activation energy for ion-diffusion.	Conduction may be facilitated (i) by ion-diffusion through relatively disordered lattice, e.g. α-AgI, or (ii) by electronic transport, e.g. HgI_2.	Very low conductivity.
Heat of formation, $-\Delta H_{f,298°K}$ per g atom of halogen	← General trend towards higher values		
	ΔH_f well reproduced by calculations based on the simple ionic model. Variations of ΔH_f determined principally by variations of lattice energy.	ΔH_f poorly reproduced by calculations based on the simple ionic model. Deviations increase in the order $MF_n < MCl_n < MBr_n < MI_n$. Variations of ΔH_f reduced, *but not reversed*, by such deviations.	Ionic model qualitatively and quantitatively inappropriate. Variations of ΔH_f determined principally by variations of bond energy which follow the sequence M–F > M–Cl > M–Br > M–I.
Solubility	Favoured by polar, coordinating solvents of relatively high dielectric constant. Dictated by balance between lattice energy and solvation energies of the ions (see p. 1236).		Solubility determined by weak van der Waals' and dipolar forces between molecules. Dissolution commonly favoured by less polar media, e.g. benzene or CCl_4.
	For a given cation, e.g. K^+, solubility in water typically increases in the order $MF_n < MCl_n < MBr_n < MI_n$.	Non-coulombic forces tend to stabilize the solid lattice with respect to dissolution in conventional solvents. For a cation like Ag^+ solubility in water increases in the order $MI_n < MBr_n < MCl_n < MF_n$.	
Hydrolysis	Increasing tendency →		

model, other metal dihalides are mostly characterized by layer structures of the $CdCl_2$ or CdI_2 type, and the discrepancies, approaching 20% in some instances, found between the lattice energies calculated from equations such as (1) and those derived from a Born–Haber cycle allude to considerable non-coulombic interactions[323]. Simultaneously it is found that internuclear distances depart significantly from the additivity rule. These findings are attributable to the peculiarities of the layer structures in which the anion has only three nearest cation neighbours, all of them on the same side of it, an unsymmetrical arrangement clearly incompatible with the requirements of the simple ionic model. The layers are held together by van der Waals' forces between adjacent halide ions. Reliable data concerning the lattice energies of trihalides are sparse, but it is again noteworthy that, in contrast with trifluorides, which favour relatively regular structures incorporating octahedral MF_6 groups, other metal trihalides typically involve the layer structures characteristic of $CrCl_3$ or BiI_3 [319]. Irregular environments are also encountered for metal ions like Cu^{2+} which lack cubic symmetry in their electronic ground states, and for B-metal ions like Tl^+ which possess low-lying excited states of less than cubic symmetry[324].

Whereas hydrogen gives rise to an extensive intermediate class of metal-like or interstitial hydrides, the size and orbital energies of the halogen atoms largely preclude their entry into alloy-like structures. The nearest approach to such structures is found in the iodides of low-valent metals, such as TiI_2 and NbI_4, which, with gross defect structures, behave as semiconductors or metals[320]. Semiconductor behaviour is also found for heavy-metal halides with more regular structures, the conductivity depending either on ion-migration (e.g. AgI) or on electronic transport (e.g. HgI_2). Such halides are amenable to descriptions based, at least qualitatively, on the band model; typical band gaps (in eV) are: CuBr, 2·9 and AgI, 2·8. Further, there are now known many halide species which incorporate clusters of metal atoms held together by strong metal–metal bonding, e.g. $[Re_2Cl_8]^{2-}$, $[Re_3Cl_{12}]^{3-}$, $ReCl_3$, $ReCl_4$, $[M_6X_{12}]^{2+}$ (M = Nb or Ta; X = Cl, Br or I) and $[M'_6X_8]^{4+}$ (M' = Mo or W; X = Cl, Br or I)[325–327]. The formation of these systems, characterized by interatomic distances close to those found in the parent metal, appears to be favoured by metal atoms having relatively unfavourable ionization potentials for the removal of 1–3 electrons, usually coupled with a large heat of atomization.

In fact, most binary metal halides other than those of transition metals in oxidation states $> +3$ are most profitably discussed in terms of the simple ionic model and of deviations from this model. Features of the halides which can thus be rationalized include (a) the thermodynamic functions defining the formation of the compound[289,293], (b) the susceptibility to redox reactions of the type $MX_n + \frac{1}{2}X_2 \rightleftharpoons MX_{n+1}$ (see pp. 1119–20)[289,293], (c) the capacity to undergo halogen-exchange reactions of the type $R-Y+MX(s) \rightarrow R-X+MY(s)$[289,293], (d) the stability of halides containing complex cations susceptible to thermal decomposition, e.g. $PH_4^+X^-$, which commonly decreases in the order I > Br > Cl > F[293], and (e) the solubility in various media (see pp. 1236–7)[293]. In each case the behaviour of the halide is principally a function of the lattice energy, major variations of which reflect variations of ionic charge and size. There is ample evidence that the simple ionic model is

323 D. F. C. Morris, *J. Inorg. Nuclear Chem.* **4** (1957) 8.
324 J. D. Dunitz and L. E. Orgel, *Adv. Inorg. Chem. Radiochem.* **2** (1960) 1.
325 F. A. Cotton, *Quart. Rev. Chem. Soc.* **20** (1966) 389; *Rev. Pure Appl. Chem.* **17** (1967) 25.
326 D. L. Kepert and K. Vrieze, *Halogen Chemistry* (ed. V. Gutmann), Vol. 3, p. 1. Academic Press (1967); M. C. Baird, *Progress in Inorganic Chemistry*, **9** (1968) 1.
327 J. H. Canterford and R. Colton, *Halides of the Second and Third Row Transition Metals*, Wiley-Interscience (1968).

in at least qualitative accord with the energy trends of numerous chemical processes even when there are marked deviations from the properties normally associated with "ionic" behaviour. As pointed out elsewhere[293,320], this is a tribute, not to the descriptive accuracy of the model, but to the self-compensating features embodied in setting up an ionic charge distribution.

As the size of the halide anion changes, the response of the lattice energy varies markedly from one series of halides to another. Thus, with increasing anion size, the lattice energies of the monohalides of copper, silver and gold diminish much less rapidly than do those of the alkali metals, while those of the dihalides of cadmium, mercury and lead fall more slowly than do those of alkaline-earth metals of comparable size[293]. This implies that, *relative to* either the alkali-metal or alkaline-earth series, there is an increase in the strength of the cation–anion binding as the atomic number of the halogen increases. The effect is clearly

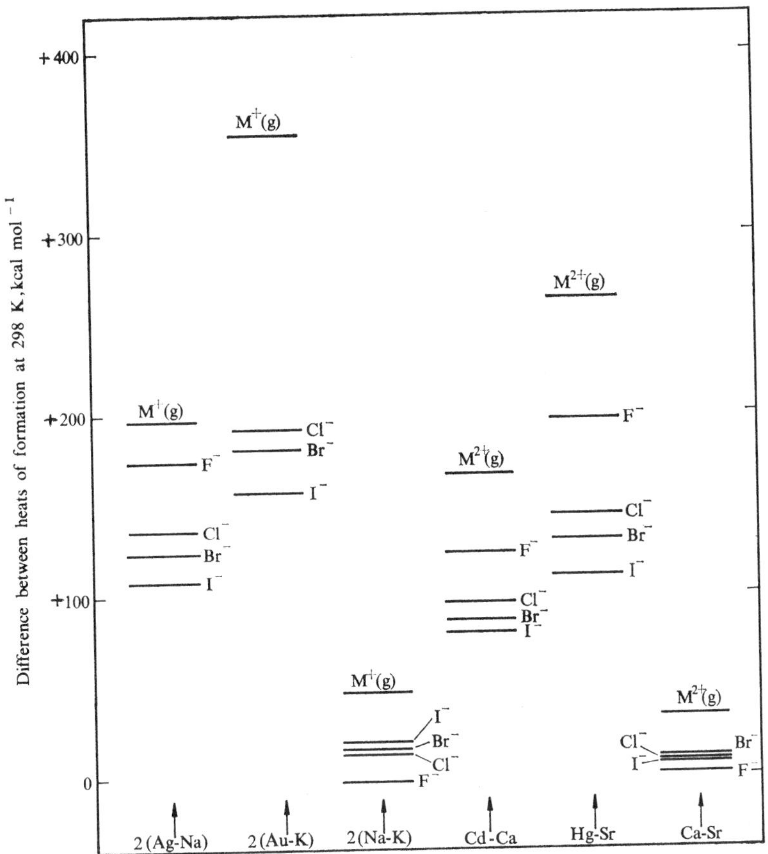

FIG. 21. Differences between the heats of formation of A-metal and B-metal halides for cations of similar size and the same formal charge [after C. S. G. Phillips and R. J. P. Williams, *Inorganic Chemistry*, Vol. II, p. 503, Clarendon Press, Oxford (1966)].

illustrated in Fig. 21 by comparing the heats of formation of A-metal and B-metal halides, M_AX_n and M_BX_n respectively, for cations of the same formal charge and of similar size[320]. According to the simple ionic model, the differences between the heats of formation of two such gaseous cations $[\Delta H_f(M_B^{n+}) - \Delta H_f(M_A^{n+})]$ should be reproduced in the differences between the heats of formation of analogous halides $[\Delta H_f(M_BX_n) - \Delta H_f(M_AX_n)]$. That the latter quantity is invariably more exothermic than the former, the divergence increasing in the sequence $F^- < Cl^- < Br^- < I^-$, betokens the effect of polarization or partial covalent bonding between the B-metal and halide ion. As a result, B-metal ions tend to be class b acceptors[328], forming complexes with stability constants in the order $I^- > Br^- > Cl^- > F^-$ rather than in the reverse sequence characteristic of the metals whose halides approximate better to the ionic model.

Molecular Halides

The halides of non-metals and of metals in high oxidation states usually take the form of discrete molecules which persist throughout the solid, liquid and gaseous phases, e.g. SnI_4, $SbCl_5$, WCl_6 and Al_2Br_6. The molecules are held together by van der Waals' attractions, primarily between the halogen atoms of different molecules, which may be augmented by dipole–dipole and dipole–induced-dipole forces. Accordingly, such halides are characterized by relatively low lattice energies, by low melting and boiling points, by solubility in non-polar solvents and by very meagre electrical conductivities in both the solid and liquid states. For ionic halides, the boiling and melting points tend to fall as the atomic number of the halogen increases, the work required to separate the ions diminishing as they become larger: for molecular halides, the boiling and melting points tend to rise with the atomic number of the halogen (see, for example, the hydrogen halides, subsection 3.3), the polarizability of the halogen here exercising the dominant influence. It must be emphasized, however, that these trends represent the limiting behaviour of each class. Again, whereas most ionic halides dissolve in water to give hydrated metal and halide ions, molecular halides generally suffer ready and irreversible hydrolysis.

Within a halide molecule of the type MX_n, at no point between the M and X nuclei does the electron density approach zero. To account for the electron distribution, we must consider as a primary influence σ-bonding between M and X, and, as a secondary influence of somewhat uncertain importance π-bonding, viz. $M \leftleftarrows X$ (e.g. $p_\pi \rightarrow d_\pi$ interaction, as in $SiCl_4$) or $M \rightrightarrows X$ (e.g. $d_\pi \rightarrow d_\pi$ interaction, as in $PtCl_6^{2-}$). The stereochemistry of the aggregate is best interpreted on the basis of the bond model, whether embodied in localized molecular-orbital methods[329] or in the principle of repulsion between the valence-shell electron pairs of the central atom[330].

For many practical purposes the average bond energy $B(M-X) = 1/n\Delta H$, where ΔH is the enthalpy change for the process $MX_n(g) \rightarrow M(g) + nX(g)$, is most easily related to the thermodynamic properties of MX_n. If the assumption of additivity is a reasonable approximation, it is thus possible qualitatively to account for variations of thermal stability, for example, with respect to a reaction such as

$$MX_n(g) \rightarrow MX_{n-1}(g) + \tfrac{1}{2}X_2(g)$$

or for the occurrence of substitution reactions, e.g.

$$MX_n(g) + \tfrac{1}{2}H_2(g) \rightarrow MX_{n-1}(g) + HX(g)$$

For a given value of n, $B(M-X)$ invariably decreases in the order $X = F > Cl > Br > I$, which is also the sequence of stability of MX_n with respect to the elements. In the

[328] S. Ahrland, J. Chatt and N. R. Davies, *Quart. Rev. Chem. Soc.* **12** (1958) 265.

[329] R. G. Pearson and R. J. Mawby, *Halogen Chemistry* (ed. V. Gutmann), Vol. 3, p. 55. Academic Press (1967).

[330] R. J. Gillespie, *Angew. Chem., Internat. Edn.* **6** (1967) 819.

interpretation of chemical changes, the bond energies of molecular halides fulfil a function analogous to that of the lattice energies of ionic halides. However, the bond-energy approach is marred by the fallibility of the additivity principle and by the fact that bond energies, unlike lattice energies, are not easily accounted for by the theoretical models presently available.

Halide Complexes

In common with fluoride, chloride, bromide and iodide ions function as ligands to form, with the majority of metal ions or molecular halides, complexes such as $AgCl_2^-$, $FeCl_4^-$, $SbBr_4^-$, HgI_4^{2-}, WCl_6^{2-}, etc., as well as a wealth of mixed complexes in company with other ligands; this second class includes neutral or charged species, e.g. $[Co(NH_3)_4Cl_2]^+$, $NbOCl_4^-$, $[Cl_5RuORuCl_5]^{4-}$, $CoBr_2(PHPh_2)_3$, $[CuI(bipyridyl)_2]^+$, $Mn(CO)_5I$ and $[(\pi\text{-}C_2H_4)PtCl_3]^-$.

In the solid state there exist the following possibilities for a compound of the formula A_mBX_n [319]: (a) A, B and X form an infinite three-dimensional array of ions in which no finite or infinite one- or two-dimensional complex can be distinguished, e.g. complex fluorides of the type ABF_3; (b) B and X form infinite two-dimensional or one-dimensional complexes, as in NH_4CdCl_3; (c) B and X form finite complexes, as in K_2SeBr_6, $(NH_4)_2PdCl_4$ or R_4NHX_2 (R = Me, Et or Bu); in some cases finite polynuclear complex ions are encountered, e.g. $Tl_2Cl_9^{3-}$, $[Mo_6X_8]^{4+}$ and $[Nb_6X_{12}]^{2+}$. Sometimes, too, a compound A_mBX_n may contain, not ions BX_n, but ions BX_{n-p} and p separate X^- ions, as in the case of Cs_3CoCl_5, which should accordingly be written $Cs^+_3[CoCl_4]^{2-}Cl^-$. Again, there may be complex ions of two types, as in PCl_5 containing the ions PCl_4^+ and PCl_6^- and $(NH_4)_2SbBr_6$ containing $SbBr_6^{3-}$ and $SbBr_6^-$ ions.

The combination of a univalent metal halide AX and a molecular halide BX_{n-m} to give a finite complex BX_n^{m-} can be represented in the form of a thermodynamic cycle:

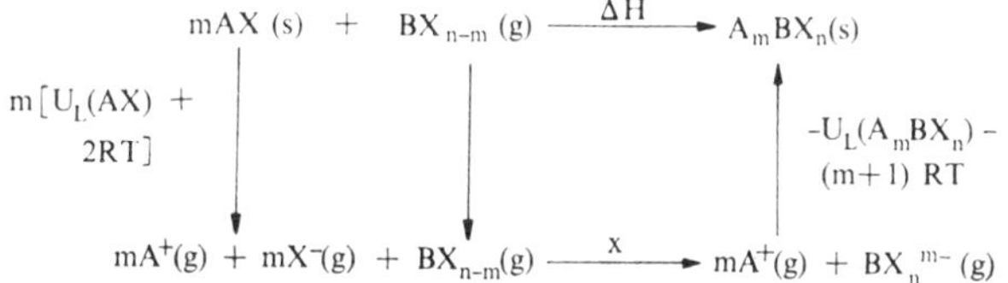

The stability of A_mBX_n then hinges on whether the enthalpy of interaction (x) of gaseous X^- with gaseous BX_{n-m} to give gaseous BX_n^{m-} compensates for the decrease of lattice energy which attends the change from the simple to the complex halide. Since the difference in lattice energy decreases with advancing size of the A^+ cation, it follows that the condition of stability is most likely to be fulfilled when the cation is large, a conclusion in agreement with the observation that caesium salts are the most stable of the alkali-metal salts with respect to thermal decomposition of anions such as HCl_2^-, BCl_4^-, $TiCl_6^{2-}$ and I_3^- [289,331]. The same principle is evident in the extensive use made of quaternary ammonium, phosphonium or arsonium species or of large organic cations derived from amines to stabilize polyhalide, hydrogen dihalide and MX_4^{2-} anions (M = first-row transition metal). Conversely the energy of formation and large size of the complex halogeno-anions can be turned to advantage in the preparation and stabilization of less common cations; examples include Cd_2^{2+} as $[Cd_2][AlCl_4]_2$ [332], NO^+ and NO_2^+ as

[331] F. Basolo, *Coord. Chem. Rev.* **3** (1968) 213.
[332] J. D. Corbett, *Inorg. Chem.* **1** (1962) 700.

$[NO]_2[SnCl_6]$ and $[NO_2][SbCl_6]$[319], ICl_2^+ as $[ICl_2][AlCl_4]$ and $[ICl_2][SbCl_6]$[319], and $AsCl_4^+$ as $[AsCl_4][SbCl_6]$[333]. It is noteworthy that none of these entities forms a salt with simple halide ions, implying that the part played by the energy liberated in forming the complex anion must be a decisive one. In this respect, chloro-complexes appear to be decidedly inferior to fluoro-complexes such as BF_4^-, SbF_6^- and PtF_6^-, which have proved particularly effective in the stabilization of non-metallic cations[289]. However, that the formulation of the solid complex in terms of discrete ions may often be a severe over-simplification is all too clearly demonstrated by the crystal structure of ICl_2SbCl_6 (see p. 1351)[319].

A survey of the relative stabilities of complex halide ions with respect to dissociation into their constituent ions in aqueous solution distinguishes entities such as H^+, Be^{2+}, Ce^{3+}, Fe^{3+} and Ti^{4+} (class a), which form their most stable complexes with F^-, from entities such as Pt^{2+}, Pt^{4+}, Ag^+, Tl^{3+} and Hg^{2+} (class b), which form their least stable complexes with F^- and their most stable complexes with I^- [289,328,334,335]. No rigorous theoretical explanation for either sequence or for the existence of the two classes of acceptor relative to the halide ions has been given. It is likely that the polarizing power Z^*/r^2 (see p. 1245), polarizability and the ability of the metal and ligand to engage in $M \rightarrow X(d_\pi - d_\pi)$ bonding (an opportunity denied to fluorine) are all significant factors, but their relative importance remains unfathomed. However, the stability of a complex ion in solution depends not only on the absolute strength of the M–X bond but also on the relative solvation energies of all the species involved. Hence, the balance of an equilibrium such as

$$MF_6^{2-} + 6Cl^- \rightleftharpoons MCl_6^{2-} + 6F^-$$

turns not only on the balance of M–F and M–Cl bond energies but also on the difference between the hydration energies of MF_6^{2-} and MCl_6^{2-} and six times the difference in hydration energies of F^- and Cl^-. Since the hydration energies of F^- and Cl^- differ by no less than 33 kcal, the order of stability constants is not necessarily that of M–X bond energies in the complex ions. Indeed, the few available results suggest that even for complexes having stability constants which decrease in the sequence I > Br > Cl > F, the actual order of bond energies is F > Cl > Br > I. What seems to be the critical factor about the class b acceptors is that the $M \cdots X$ interaction energies, in common with the lattice energies and heats of formation of related binary halides, decrease less markedly as a function of the atomic number of X (see p. 1249). To this extent, the distinction between the two classes of acceptor is somewhat less fundamental than has been represented elsewhere[289,320,328,336].

When it comes to isolating a complex from solution, solubility considerations play a major, sometimes a dominant, part; there are known many instances in which the species preponderant in solution is not the one to be precipitated from that solution. The presence of complex ions which are very much larger than the counter-ions means that lattice energies are relatively low and vary but little with the size of the counter-ion; by contrast, the major contribution to the solvation energy is made by the counter-ion, being roughly inversely proportional to the radius of that ion. Hence, with increasing size of the counter-ion, the

[333] V. Gutmann, *Z. anorg. Chem.* **266** (1951) 331; F. J. Brinkmann, H. Gerding and K. Olie, *Rec. Trav. Chim.* **88** (1969) 1358.

[334] L. G. Sillén and A. E. Martell (eds.), *Stability Constants of Metal-Ion Complexes*, Chemical Society Special Publication No. 17 (1964); Supplement No. 1, Chemical Society Special Publication No. 25 (1971).

[335] G. P. Haight, jun., *Halogen Chemistry* (ed. V. Gutmann), Vol. 2, p. 351. Academic Press (1967).

[336] A. J. Poë and M. S. Vaidya, *J. Chem. Soc.* (1961) 1023.

situation arises that the sum of the solvation energies of the ions taken separately decreases more rapidly than the lattice energy of the solid complex. Awareness of this generalization is often useful in seeking to isolate a particular halide complex. Thus, of the alkali metals, caesium tends to form the least soluble salts of complex halide anions, a feature which has been exploited in the characterization of the $RhCl_6^{2-}$ ion[293]. However, for complexes of very large cations, e.g. Et_4N^+ and Bu_4N^+, a much larger proportion of the solvation energy is contributed by the anion, with the result that the solubility increases again (cf. Fig. 19). Accordingly, whereas $M_2[RhCl_6]$ (M = Cs or NMe_4) has been isolated from aqueous media, corresponding derivatives of the cations K^+, Rb^+, NEt_4^+ and NBu_4^+ have eluded attempts to synthesize them[293].

Finally it may be mentioned that in effecting the separation of metal ions for analytical or radiochemical purposes, advantage has frequently been taken, in conjunction with ion-exchange or solvent-extraction procedures, of equilibria involving the formation of halide complexes[337–341]. For example, the marked difference in stabilities of the chloride complexes formed by Co^{2+} and Ni^{2+} in aqueous solution makes possible the efficient separation of these two ions by elution on an anion-exchange column with concentrated hydrochloric acid[321]. Likewise, the separation of actinides as a group from the lanthanide ions, together with partial fractionation of the actinides, can be effected by the use of 10 M lithium chloride solution to elute the ions from a suitable anion-exchange column operating at elevated temperatures (up to *ca.* 90°C)[340,342]. Again, the stable halide complexes characteristic of the B-metals are a party to the selective extraction of these metals from strongly acidic aqueous solutions of the halogen acids into organic solvents such as ethers, higher alcohols, ketones or t-butyl phosphate[337,338,341].

General Preparative Methods[320–322,327,343–345]

The general methods of preparing binary chlorides, bromides and iodides, summarized in Table 22, may be divided into two classes, namely wet methods, which are feasible when the halide is not hydrolysed, and dry methods, which are obligatory when such hydrolysis occurs. A further distinction may be made between those methods where oxidizing conditions prevail and those involving reduction of a higher halide, whether by external means or by spontaneous thermal decomposition or disproportionation. Of the oxidizing methods, those involving direct interaction of the elements and halogenation of an oxide are the most widely applicable. In the direct halogenation of an element, it is notable that the chloride produced may contain the element in a higher oxidation state than does the bromide or iodide. However, it is commonly possible, within certain limits, to influence the nature of the product by choice of such reaction conditions as temperature or reactant

337 H. M. N. H. Irving, *Quart. Rev. Chem. Soc.* **5** (1951) 200.

338 D. F. Peppard, *Adv. Inorg. Chem. Radiochem.* **9** (1966) 1.

339 F. Helfferich, *Ion Exchange*, McGraw-Hill (1962).

340 J. Korkisch, *Modern Methods for the Separation of Rarer Metal Ions*, Pergamon (1969).

341 Y. Marcus and A. S. Kertes, *Ion Exchange and Solvent Extraction of Metal Complexes*, Wiley-Interscience (1969).

342 J. J. Katz and G. T. Seaborg, *The Chemistry of the Actinide Elements*, Methuen, London (1957).

343 R. Colton and J. H. Canterford, *Halides of the First Row Transition Metals*, Wiley-Interscience (1969); D. Brown, *Halides of the Lanthanides and Actinides*, Wiley-Interscience (1968); K. W. Bagnall, *Halogen Chemistry* (ed. V. Gutmann), Vol. 3, p. 303. Academic Press (1967).

344 J. D. Corbett, *Preparative Inorganic Reactions* (ed. W. L. Jolly), Vol. 3, p. 1. Interscience, New York (1966).

345 Z. E. Jolles (ed.), *Bromine and its Compounds*, Benn, London (1966).

Table 22. More Important General Methods of Preparing Binary Chlorides, Bromides and Iodides[a-f]

	Method	Examples	Comments
A	DRY METHODS		
1	Direct interaction of the elements	$2Al+3Br_2 \rightarrow Al_2Br_6$ $Sn+2Cl_2 \rightarrow SnCl_4$ $2Ta+5I_2 \rightarrow 2TaI_5$ $Th+2Cl_2 \rightarrow ThCl_4$ $Mo \xrightarrow{Cl_2} MoCl_5$ $Mo \xrightarrow[450°C]{Br_2} MoBr_3 \xrightarrow{650-700°C} MoBr_2$ $Mo \xrightarrow{I_2, 300°C} MoI_3$	Perhaps the most important general method. Elevated temperatures usually required, although with transition metals rapid reaction can often occur with Cl_2 or Br_2 when THF or other ethers are used as the reaction medium, the halide being obtained as a solvate. Where different oxidation states are possible, chlorine, at elevated temperatures, tends to give a higher state than bromine or iodine. By choice of reaction conditions (e.g. temperature, reactant proportions, etc.) it may be possible to preselect the halide produced, e.g. PCl_3 or PCl_5.
2	Metal + hydrogen halide	$Cr+2HCl \xrightarrow{900°C} CrCl_2+H_2$ $Co+2HBr \xrightarrow{\text{Red heat}} CoBr_2+H_2$ $Pu+3HI \xrightarrow{250-300°C} PuI_3+3/2H_2$	Useful method of preparing conventional lower-valent chlorides or bromides (cf. $MoCl_2$ or Ta_6Cl_{14}).
3	Halogenation of metal oxides (i) with molecular halogen (ii) with molecular halogen in the presence of carbon (iii) with carbon tetrahalide or other organic halide	$ZrO_2+2Cl_2 \longrightarrow ZrCl_4$ $WO_2+Br_2 \xrightarrow{460°C} WO_2Br_2$ $TiO_2+C+2Cl_2 \xrightarrow{400-600°C} TiCl_4$ $Ta_2O_5+C+5Br_2 \xrightarrow{>460°C} 2TaBr_5$ $Sc_2O_3+CCl_4 \xrightarrow{600°C} ScCl_3$ $Nb_2O_5+CBr_4 \xrightarrow{370°C} NbBr_5$ $UO_3+Cl_2C{=}CClCCl_3 \xrightarrow{\text{reflux}} UCl_4$	Elevated temperatures invariably required. Oxyhalides may be formed. Useful for chlorides and bromides. Useful for chlorides and bromides. Particularly useful and effective method of making metal chlorides.

	(iv) with aluminium trihalide	$MoO_2 + AlI_3 \xrightarrow{230°C} MoI_2$	AlI_3 has been used extensively for the preparation of binary transition-metal iodides (the metal sulphide has also been used in place of the oxide).[d,e]
	(v) with thionyl chloride Other halogenating agents include the hydrogen halide, NH_4X, $Cl_2 + S_2Cl_2$, $COCl_2$ and PCl_5	$WO_3 + SOCl_2 \xrightarrow{\text{reflux}} WOCl_4$ $4Lu_2O_3 + 9Cl_2 + 3S_2Cl_2 \longrightarrow 8LuCl_3 + 6SO_2$	Final product commonly an oxychloride. Has been used to prepare trichlorides of the lanthanides.
4	Dehydration of hydrated halides		Risk of oxyhalide formation.
	(i) *in vacuo*	$[M(H_2O)_6]Br_3 \xrightarrow{70\text{–}170°C} MBr_3 + 6H_2O$ Controlled vacuum thermal decomposition, M = lanthanide or actinide element.	This method works well for lanthanide and actinide tribromides.[f]
	(ii) with thionyl chloride	$[Cr(H_2O)_6]Cl_3 + 6SOCl_2 \xrightarrow{\text{reflux}} CrCl_3 + 12HCl + 6SO_2$	Efficient method of preparing certain anhydrous chlorides.
	(iii) in presence of anhydrous hydrogen halide or mixed with ammonium halide	$CuCl_2,2H_2O \xrightarrow{HCl, 150°C} CuCl_2 + 2H_2O$	A method which has been applied relatively widely.
	(iv) with 2,2-dimethoxypropane	$MX_n,mH_2O + mMeC(OMe)_2Me \rightarrow MX_n + mMe_2CO + 2mMeOH$	The acetone and/or methanol often becomes coordinated to the halide, but can usually be removed by gentle heating or pumping.[b]
5	Halogen exchange	$FeCl_3 + BBr_3 \longrightarrow FeBr_3 + BCl_3$[g] $MCl_3 + 3HBr \xrightarrow{400\text{–}600°C} MBr_3 + 3HCl$ (M = lanthanide or Pu) $3TaCl_5 + 5AlI_3 \xrightarrow{410°C} 3TaI_5 + 5AlCl_3$	An excess of one reagent is usually required, equilibria normally being established.
6	Reduction of higher halide (i) with the parent element	$AlCl_3 + 2Al \longrightarrow 3AlCl$ $TaI_5 + Ta \xrightarrow{630\text{–}575°C} Ta_6I_{14}$ Thermal-gradient technique particularly effective.[d,e]	

Table 22 (*cont.*)

	Method	Examples	Comments
6	Reduction of higher halide (*cont.*) (ii) with hydrogen	$MX_3+\frac{1}{2}H_2 \longrightarrow MX_2+HX$ (M = Sm, Eu or Yb; X = Cl, Br or I)	These are the principal methods of preparing binary halides of elements in lower valence states which are avoided under oxidizing conditions.
	(iii) with aluminium	$3WBr_5+Al \xrightarrow{475-240°C} 3WBr_4+AlBr_3$ thermal gradient[e]	
7	Thermal decomposition or disproportionation of a halide	$ReCl_5 \xrightarrow{\text{at b.p.}} ReCl_3+Cl_2$ $MoI_3 \xrightarrow{100°C} MoI_2+\frac{1}{2}I_2$ $AuCl_3 \xrightarrow{160°C} AuCl+Cl_2$ $2TaBr_4 \xrightarrow{500°C} TaBr_3+TaBr_5$	
B	WET METHODS		Feasible when the halide is not hydrolysed.
1	Dissolution in aqueous halogen acid (i) of the metal	$Fe+2HCl \rightarrow [Fe(H_2O)_6]Cl_2$	Evaporation of the solution affords the halide, commonly in the form of a hydrated solid. Partial hydrolysis may give oxyhalides, e.g. BiOCl.
	(ii) of the oxide, hydroxide or carbonate	$CoCO_3+2HI \rightarrow [Co(H_2O)_6]I_2$	
2	Precipitation reactions	$Ag^++Cl^- \rightarrow AgCl$ $Cu^{2+}+2I^- \rightarrow CuI+\frac{1}{2}I_2$	
3	Hydrolysis of the molecular halogen	$3X_2+6OH^- \rightarrow 5X^-+XO_3^-+3H_2O$	Used commercially for the production of alkali–metal halides.

[a] *Gmelins Handbuch der Anorganischen Chemie;* P. Pascal, *Nouveau Traité de Chimie Minérale*, Masson et Cie, Paris.
[b] F. A. Cotton and G. Wilkinson, *Advanced Inorganic Chemistry*, 2nd edn., Interscience (1966).
[c] Z. E. Jolles (ed.), *Bromine and its Compounds*, Benn, London (1966).
[d] R. Colton and J. H. Canterford, *Halides of the First Row Transition Metals*, Wiley-Interscience (1969).
[e] J. H. Canterford and R. Colton, *Halides of the Second and Third Row Transition Metals*, Wiley-Interscience (1968).
[f] D. Brown, *Halides of the Lanthanides and Actinides*, Wiley-Interscience (1968).
[g] P. M. Druce, M. F. Lappert and P. N. K. Riley, *Chem. Comm.* (1967) 486.

proportions. Another useful route to certain anhydrous halides of metals in lower oxidation states involves the dehydration of the hydrates readily isolated from aqueous solution. Reduction or decomposition of higher halides provides a route to halides like InCl or Ta_6I_{14} which are avoided by reactions characterized by oxidizing conditions[344]; in practical terms the reaction is often conveniently effected by means of the thermal-gradient technique[327], which is particularly suited to the preparation of halides having a narrow stability range.

For mixed halide species, e.g. oxyhalides like WO_2Br_2 [346] or carbonyl halides like $Mn(CO)_5X$[347], the most widely applicable methods of preparation involve (a) substitution of a binary halide, e.g. with CO, $C_5H_5^-$, CN^-, OH^- or O_2, or (b) halogenation, e.g. of an oxide, sulphide or carbonyl. Halide-exchange, so prominent as a method of producing fluorides, is less important to the preparation of the heavier halides, except where specific properties, e.g. volatility or solubility, facilitate exchange: sodium iodide in acetone, for example, is a good reagent for the replacement of chlorine by iodine mainly because sodium iodide is soluble and sodium chloride insoluble in this medium[289].

For the isolation of solids containing complex halide ions, the most common method involves the treatment of the parent halide MX_n with the halide of a suitable univalent cation, e.g. an alkali-metal or substituted ammonium, phosphonium or arsonium species. Interaction may be conveniently brought about in the melt, in aqueous hydrohalic acid or in a non-aqueous solvent like acetonitrile, ethanol or chloroform[327,343]; in certain cases a non-aqueous solvent which is itself a halogenating agent, e.g. thionyl chloride[348], has been favoured. The nature of the product is often strongly influenced by the relative proportions of the two halides, by the shape as well as the size of the cation, and by the reaction medium employed. In these procedures no deliberate attempt is made to vary the oxidation state of M. Complexes have also been made, however, *via* the reduction or oxidation, by chemical or electrolytic means, of solutions containing halide species, e.g.

$$Ga \xrightarrow[\text{HX, isolated as } Me_4N^+ \text{ salts}]{\text{anodic dissolution in 6 M}} Ga_2X_6^{2-} \quad 349$$

$$TcCl_6^{2-} \xrightarrow{\text{Zn in HCl solution}} Tc_2Cl_8^{3-} \quad 350$$

$$ReOCl_5^{2-} \xrightarrow[\text{solution}]{\text{oxidation by air, HCl}} ReOCl_6^{2-} \quad 351$$

Properties of Halides

Table 21 provides a qualitative summary of some of the physical characteristics of halide systems MX_n, emphasizing how these characteristics vary with the nature of the halide. In keeping with such variations, chemical behaviour also ranges from that attributable to the halide ion X^-, in various conditions of coordination, to the more specific properties diagnostic of the M–X bonds of molecular halides. Accordingly, at one extreme, the chemical properties of ionic halides are determined largely by changes of coulombic energy, through lattice or solvation energy terms, while, at the other, many of the reactions

346 K. Dehnicke, *Angew. Chem., Internat. Edn.* **4** (1965) 22.
347 F. Calderazzo, *Halogen Chemistry* (ed. V. Gutmann), Vol. 3, p. 383. Academic Press (1967).
348 D. M. Adams, J. Chatt, J. M. Davidson and J. Gerratt, *J. Chem. Soc.* (1963) 2189.
349 C. A. Evans and M. J. Taylor, *Chem. Comm.* (1969) 1201.
350 J. D. Eakins, D. G. Humphreys and C. E. Mellish, *J. Chem. Soc.* (1963) 6012.
351 R. Colton, *Austral. J. Chem.* **18** (1965) 435.

of molecular halides can be referred to the strength of the M–X bond, as expressed, for example, by its mean bond energy. To explore more closely the nature of the interaction between M and X and its dependence on the identities of M and X, reference is most commonly made to one or more of the following physical properties, which quantify various aspects of the M · · · X interaction[329]: the mean bond energy; the internuclear distance; the stretching force constant and related vibrational properties; esr, nmr, nqr and Mössbauer parameters; the ligand-field strength, interelectronic repulsion integrals and other terms derived from electronic spectra and magnetic properties; and dipole moments. Brief consideration is now given to these properties.

1. Mean bond energy, *B*(M–X). Defined by ΔH_{at}, the enthalpy of the process $MX_n(g) \rightarrow M(g)+nX(g)$, such that $\Delta H_{at} = nB(M\text{–}X)$, the mean bond energy can be evaluated more or less reliably for many halide molecules[293,297]. An alternative approach, appropriate to metal halides and to complex ions, involves the so-called "coordinate bond energy", which refers to the energy per bond for the heterolytic process $MX_n(g) \rightarrow M^{n+}(g)+nX^-(g)$; compilations of this energy term are to be found elsewhere[329]. In either case, it should be noted, the energy required to break an individual bond may be quite different from the average value.

Table 23 lists bond energy terms for most neutral halide molecules for which experimental data are available. Spectroscopic methods have been used directly to determine the dissociation energy of some of the diatomic halide molecules[352]. More often, however, thermochemical methods furnish the heat of formation of gaseous MX_n; with a knowledge of the heats of atomization of M and X, *B*(M–X) is then easily gained[297,329,353]. Despite the interpretative shortcomings of bond energy terms, some interesting trends may nevertheless be discerned in the results of Table 23. In the first place, for a given element M ($M \neq F$), *B*(M–X) diminishes in the order F > Cl > Br > I. Further, in relation to the typical non-metals M_1, M_2 . . . of one of the vertical Groups IV–VII, the bond energy order is usually $B(M_1\text{–}X) < B(M_2\text{–}X) > B(M_3\text{–}X) > B(M_4\text{–}X)$ for a given halogen; by contrast, for the typical elements of Groups I–III, *B*(M–X) appears to diminish regularly as the atomic number of M increases. Moving from left to right across a given horizontal Period sees *B*(M–X) first rise, reach a maximum, and then fall to relatively low values for the interhalogens and noble-gas fluorides; this trend is modulated by a subsidiary minimum, which is reached with the completion of the *d*-shell at the end of each transition series. Unfortunately, the sparseness of accurate data precludes any meaningful generalization about the bond energies of analogous halides formed by metals of the different transition series, though there is some indication that, where the highest oxidation state is realized, the heaviest metals enjoy the largest bond energies. Where an element forms more than one halide MX_n, it is usually the rule that the bond energy decreases as *n* increases. Compared with the halides formed by typical class a acceptors, e.g. Rb^I or Sr^{II}, halides of class b acceptors like Au^I or Hg^{II} are notable, not only for their comparatively low bond energies, but also, significantly, for the comparatively small changes in bond energy that accompany variations of X, e.g. $B(M\text{–}F) - B(M\text{–}I) = 28$, 29, 39 and 52 kcal for M = Au^I, Hg^{II}, Rb^I and Sr^{II}, respectively.

352 T. L. Cottrell, *The Strengths of Chemical Bonds*, 2nd edn., Butterworths, London (1958); A. G. Gaydon, *Dissociation Energies*, 3rd edn., Chapman and Hall, London (1968).

353 R. C. Feber, *Los Alamos Report, U.S. At. Energy Comm.* **LA-3164** (1965).

TABLE 23. MEAN BOND ENERGIES OF HALIDE MOLECULES (kcal mol^{-1}, 298°K)[a–g]

	H							He
F	135·8^{I}							
Cl	103·3^{I}							
Br	87·5^{I}							
I	71·3^{I}							
	Li	Be	B	C	N	O	F	Ne
F	136^{I}	136^{I} 152^{II}	131^{I} 154·3^{III}	117·0^{IV}	66·5^{III}	51·3^{II}	37·8^{I}	
Cl	113^{I}	92^{I} 110$^{\mathrm{II}h}$	118^{I} 106·1^{III}	78·1^{IV}	45^{III}	49·3^{II}	61·4^{I}	
Br	101^{I}	— 96^{II}	— 88·0^{III}	64·8^{IV}	—	—	59·6^{I}	
I	87^{I}	— 76^{II}	— 64·7^{III}	51^{IV}	—	—	67·0^{I}	
	Na	Mg	Al	Si	P	S	Cl	Ar
F	115^{I}	110^{I} 124^{II}	159^{I} 141^{III}	141$^{\mathrm{III}i}$ 142·6^{IV}	118·7^{III} 111·1^{V}	81·8^{IV} 78·6^{VI}	61·4^{I} 41·7^{III} 36·9$^{\mathrm{V}j}$	
Cl	99^{I}	75^{I} 96^{II}	118^{I} 101·5^{III}	101$^{\mathrm{III}i}$ 95·6^{IV}	78·5^{III} 63·0^{V}	64·7^{II}	58·0^{I}	
Br	88^{I}	— 83^{II}	103^{I} 87^{III}	86$^{\mathrm{III}i}$ 78·8^{IV}	64·4^{III} —	—	52·3^{I}	
I	70^{I}	— 64^{II}	88^{I} 68^{III}	70$^{\mathrm{III}i}$ 59·5^{IV}	44^{III} —	—	50·4^{I}	
	K	Ca	Ga	Ge	As	Se	Br	Kr
F	119^{I}	126^{I} 136^{II}	139^{I} 114^{III}	117^{II} 112·5^{IV}	116·3^{III} 92·5^{VI}	71·6^{VI}	59·6^{I} 48·2^{III} 44·7^{V}	11·7$^{\mathrm{II}m}$
Cl	103^{I}	96^{I} 109^{II}	114^{I} 86·8^{III}	94$^{\mathrm{II}k}$ 81·2^{IV}	73·8^{III} —	57·5^{II}	52·3^{I}	—
Br	92^{I}	— 97^{II}	100^{I} 72·1^{III}	82$^{\mathrm{II}k}$ 67·2^{IV}	61·2^{III} —	53·8^{II}	46·1^{I}	—
I	78^{I}	— 79^{II}	80^{I} 58·9^{III}	65^{II} 51·4^{IV}	49^{III} —	—	42·5^{I}	—
	Rb	Sr	In	Sn	Sb	Te	I	Xe
F	120^{I}	129^{I} 134^{II}	126^{I} 107^{III}	116^{II} 101^{IV}	106^{III} —	79·2^{VI}	67·0^{I} 64·2^{V} 55·4^{VII}	31·8$^{\mathrm{II}n}$ 31·8$^{\mathrm{IV}n}$ 30·6$^{\mathrm{VI}n}$
Cl	103^{I}	96^{I} 111^{II}	103^{I} 78·3^{III}	93^{II} 75·3^{IV}	75·0^{III} 60·4^{V}	—	50·4^{I}	—
Br	92^{I}	— 98^{II}	98·5^{I} 68·6^{III}	80·5^{II} 63·6^{IV}	63·1^{III} —	—	42·5^{I}	—
I	78^{I}	— 82^{II}	80^{I} 54·5^{III}	64^{II} 50^{IV}	48^{III} —	—	36·1^{I}	—

Table 23 (*cont.*)

	Cs	Ba	Tl	Pb	Bi	Po	At	Rn
F	122^{I}	140^{I} 139^{II}	106^{I}	96^{II} 81^{IV}	96^{III}			
Cl	102^{I}	106^{I} 116^{II}	88^{I}	74^{II} 60^{IV}	$66{\cdot}7^{III}$			
Br	92^{I}	— 105^{II}	79^{I}	64^{II} 49^{IV}	56^{IIIo}			
I	76^{I}	— 87^{II}	64^{I}	51^{II} 37^{IV}	43^{IIIo}			

	Sc	Ti	V	Cr	Mn	Fe	Co	Ni	Cu	Zn
F	144^{II} 143^{III}	144^{III} 142^{IV}	141^{II} 134^{III} 109^{V}	114^{II} 111^{III}	111^{II} 102^{III}	117^{II} 110^{III}	113^{II} 102^{III}	112^{II} 90^{III}	102^{I} 91^{II}	99^{II}
Cl	118^{II} 112^{III}	111^{III} 104^{IV}	118^{II} 101^{III} 91^{IVp}	97^{II} 86^{III}	96^{II} 77^{III}	98^{II} 80^{III}	92^{II} 77^{III}	90^{II} 65^{III}	88^{I} 72^{II}	78^{II}
Br	104^{II} 95^{III}	95^{III} 89^{IV}	106^{II} 87^{III}	80^{II} 72^{III}	81^{II} 62^{III}	83^{II} 71^{III}	79^{II} 63^{III}	76^{II} 51^{III}	80^{I} 63^{II}	66^{II}
I	87^{II} 78^{III}	80^{III} 73^{IV}	90^{II} 72^{III}	62^{II} 54^{III}	66^{II} 47^{III}	68^{II} 57^{III}	63^{II} 48^{III}	62^{II} 37^{III}	71^{I} 48^{II}	51^{II}
	Y	**Zr**	**Nb**	**Mo**	**Tc**	**Ru**	**Rh**	**Pd**	**Ag**	**Cd**
F	152^{III}	156^{IV}	136^{V}	$106{\cdot}9^{VI}$		—	—		85^{I}	80^{II}
Cl	120^{III}	117^{IVp}	98^{Vq}	$90{\cdot}9^{IV}$ $81{\cdot}1^{V}$		$80{\cdot}4^{III}$ $70{\cdot}6^{IV}$	$80{\cdot}5^{II}$ $68{\cdot}1^{III}$		78^{I}	69^{II}
Br	105^{III}	103^{IV}	82^{Vq}	—		—	—		73^{I}	59^{II}
I	88^{III}	84^{IV}	68^{V}	88^{II}		—	—		61^{I}	47^{II}

	La		Hf	Ta	W	Re	Os	Ir	Pt	Au	Hg
F	158^{III}		—	—	$121{\cdot}3^{VI}$	—	—			74^{I}	66^{II}
Cl	124^{III}	Rare	124^{IVp}	$102{\cdot}4^{Vq}$	$98{\cdot}1^{IV}$ $89{\cdot}8^{V}$ $83{\cdot}4^{VI}$	$81{\cdot}1^{V}$	$81{\cdot}1^{IV}$			66^{I}.	55^{II}
Br	107^{III}	Earths	—	$86{\cdot}5^{Vq}$	$78{\cdot}4^{V}$	—	—			61^{I}	46^{II}
I	88^{III}		—	—	—	—	—			56^{I}	37^{II}

Superscript Roman numerals refer to the valence state of the element forming the halide. Data derived from:

[a] F. D. Rossini, D. D. Wagman, W. H. Evans, S. Levine and I. Jaffe, *Selected Values of Chemical Thermodynamic Properties*, Circular 500, National Bureau of Standards, Washington (1952); *National Bureau of Standards Technical Notes* 270–1, 270–2, 270–3 and 270–4, U.S. Govt. Printing Office, Washington (1965–9).

[b] T. L. Cottrell, *The Strengths of Chemical Bonds*, 2nd edn., pp. 270–289. Butterworths, London (1958).

[c] R. C. Feber, Los Alamos Report LA–3164 (1965).

[d] C. J. Cheetham and R. F. Barrow, *Adv. High Temperature Chemistry*, **1** (1967) 7; D. L. Hildenbrand, *ibid.* p. 193.

[e] R. G. Pearson and R. J. Mawby, *Halogen Chemistry* (ed. V. Gutmann), Vol. 3, p. 55, Academic Press (1967).

[f] D. A. Johnson, *Some Thermodynamic Aspects of Inorganic Chemistry*, Cambridge (1968).

[g] *Tables of Constants and Numerical Data*, No. 17, *Spectroscopic Data relative to Diatomic Molecules* (general ed. B. Rosen), Pergamon (1970).

[h] H. C. Ko, M. A. Greenbaum, M. Farber and C. C. Selph, *J. Phys. Chem.* **71** (1967) 254.

[i] H. Schäfer, H. Bruderreck and B. Morcher, *Z. anorg. Chem.* **352** (1967) 122.

[j] H. H. Rogers, M. T. Constantine, J. Quaglino, jun., H. E. Dubb and N. N. Ogimachi, *J. Chem. and Eng. Data*, **13** (1968) 307; W. R. Bisbee, J. V. Hamilton, J. M. Gerhauser and R. Rushworth, *ibid*. p. 382.

[k] O. M. Uy, D. W. Muenow and J. L. Margrave, *Trans. Faraday Soc.* **65** (1969) 1296.

[l] P. A. G. O'Hare and W. N. Hubbard, *J. Phys. Chem.* **69** (1965) 4358.

[m] S. R. Gunn, *J. Amer. Chem. Soc.* **88** (1966) 5924.

[n] B. Weinstock, E. E. Weaver and C. P. Knop, *Inorg. Chem.* **5** (1966) 2189.

[o] D. Cubicciotti, *Inorg. Chem.* **7** (1968) 208, 211.

[p] P. Gross and C. Hayman, *Trans. Faraday Soc.* **60** (1964) 45.

[q] P. Gross, C. Hayman, D. L. Levi and G. L. Wilson, *Trans. Faraday Soc.* **58** (1962) 890.

Though it is tempting to relate such fluctuations of bond energy to changes either in the efficiency of orbital-overlap or in the availability of potential donor or acceptor orbitals, it must be appreciated that the *intrinsic* energies of the bonds are inevitably concealed by the general inaccessibility of information about electronic promotion energies separating the ground and valence states. In more practical terms, the bond energy sequences reflect the relative stabilities of halides with respect to thermal decomposition and to substitution reactions. Thus, whereas OF_2 dissociates into its elements only at temperatures above 250°C, OCl_2 is thermodynamically unstable at 25°C, exploding on heating or sparking, OBr_2 begins to decompose above −50°C, and OI_2 is as yet unknown. Again, the dissociation of HI on mild heating contrasts with the explosive formation of HF and HCl from their elements at room temperature, and with the combination of H_2 and Br_2 to give HBr in the presence of a catalyst at 200°C. The lower thermal stability of iodides has been exploited in the preparation of very pure elements, e.g. silicon, boron, titanium and thorium, by pyrolysis of the iodide on a hot wire[293].

2. Bond length, r(M–X)[352,354,355]. Studies of the microwave spectra, of rotational detail superimposed on electronic or vibrational spectra, or of the electron diffraction patterns due to gaseous halide molecules MX_n lead, in principle, to information about molecular dimensions. Determined for the most part by one or more of these methods, the M–X bond lengths of Table 24 refer exclusively to gaseous molecules; in the condensed phases intermolecular interactions may lead to significant changes in these lengths or even to an aggregate in which the molecule ceases to be a recognizable unit. Where the M–X distance has been measured in a number of related molecules, e.g. $MX_{n-m}Y_m$, a mean value is given, though, in common with bond energies, such distances are not strictly independent of the nature and number of the other substituents Y.

For a given element M, r(M–X) increases in the order F < Cl < Br < I in a series of halides of the type MX_n. Likewise, for a given halogen and a particular Group of typical elements, there is an attenuation of r(M–X) as the atomic number of M increases, though, in any horizontal Period, the normal trend with respect to atomic number is in the opposite sense. Evidently there exists no simple correlation between bond length and bond energy. Although short M–X bonds are usually characterized by high bond energies, with an inverse correlation between the length and energy, there are numerous series in which the trend appears to be reversed. Thus, of the diatomic halogen fluorides F_2 exhibits at once the shortest internuclear distance and the lowest bond energy. Deviations of bond distance from the additivity principle (implicit in the designation of covalent radii for atoms) are presumed to reflect the influence of such variables as polarity, π-bonding or intermolecular interactions. Despite the appeals commonly made to bond-contraction as a sign of partial multiple bonding, as in $SiCl_4$, it is difficult to unravel the separate contributions made by such bonding and by the polarity of the bond. The difficulty is exacerbated, moreover, by the fact, manifest in the data of Table 24, that distance is not a very sensitive function of bond character.

[354] L. E. Sutton (ed.), *Tables of Interatomic Distances and Configuration in Molecules and Ions*, The Chemical Society, London (1958); Supplement (1965).

[355] *Tables of Constants and Numerical Data*, No. 17, *Spectroscopic Data relative to Diatomic Molecules* (general ed. B. Rosen), Pergamon (1970).

Table 24. M–X Bond Lengths in Molecular Halides in the Vapour Phase (Å)[a–e]

	H							He
F	0·917^{I}							
Cl	1·275^{I}							
Br	1·415^{I}							
I	1·609^{I}							
	Li	**Be**	**B**	**C**	**N**	**O**	**F**	**Ne**
F	1·564^{I}	1·361^{I} 1·40^{II}	1·262^{I} 1·311^{IIIg}	1·30^{IIi} 1·333^{IV}	1·36^{III}	1·42^{II}	1·418^{I}	
Cl	2·018^{I}	1·77^{If} 1·75^{II}	1·716^{I} 1·74^{III}	1·767^{IV}	1·75^{III}	1·693^{IIj}	1·628^{I}	
Br	2·170^{I}	1·89^{If} 1·91^{II}	1·89^{I} 1·87^{III}	1·938^{IV}	—	—	1·756^{I}	
I	2·392^{I}	2·12^{If} 2·10^{II}	— 2·10^{IIIh}	2·14^{IV}	—	—	1·909^{I}	
	Na	**Mg**	**Al**	**Si**	**P**	**S**	**Cl**	**Ar**
F	1·926^{I}	1·750^{I} 1·77^{II}	1·654^{I} 1·63^{IIIk}	1·591$^{II\,l}$ 1·561^{IV}	1·563^{IIIm} 1·58^{Vn}	1·635^{IIp} 1·60^{IVp} 1·56^{VIp}	1·628^{I} {1·698^{III}, 1·598}	
Cl	2·361^{I}	2·198^{I} 2·18^{II}	2·130^{I} 2·06^{IIIk}	2·019^{IV}	2·03^{III} {2·020^{Vo}, 2·124}	1·99^{II} 2·07^{IV} 1·99^{VI}	1·988^{I}	
Br	2·502^{I}	2·36^{I} 2·34^{II}	2·296^{I} —	2·16^{IV}	2·20^{III} 2·10^{V}	2·24^{II} 2·27^{IV} 2·190^{VI}	2·138^{I}	
I	2·712^{I}	— 2·52^{II}	— 2·44^{IIIk}	2·435^{IV}	2·47^{III} —	—	2·321^{I}	
	K	**Ca**	**Ga**	**Ge**	**As**	**Se**	**Br**	**Kr**
F	2·172^{I}	1·93^{I} 2·10^{II}	1·775^{I} 1·88^{IIIk}	1·68^{IV}	1·712^{III}	{1·638^{IVq}, 1·771} 1·67^{VI}	1·756^{I} {1·721^{III}, 1·810}	1·875^{IIr}
Cl	2·667^{I}	2·439^{I} 2·51^{II}	2·202^{I} —	2·08^{IV}	2·161^{III}	—	2·138^{I}	—
Br	2·821^{I}	— 2·67^{II}	2·353^{I} —	2·32^{IV}	2·33^{III}	—	2·281^{I}	—
I	3·048^{I}	— 2·88^{II}	2·575^{I} 2·44^{IIIk}	2·48^{IV}	2·54^{III}	—	2·485^{I}	—
	Rb	**Sr**	**In**	**Sn**	**Sb**	**Te**	**I**	**Xe**
F	2·270^{I}	2·076^{I} 2·20^{II}	1·985^{I}	2·06^{II} —	—	1·84^{VI}	1·909^{I}	1·977^{IIs} 1·94^{IVs} 1·890^{VIs}
Cl	2·787^{I}	— 2·67^{II}	2·401^{I}	2·43^{II} 2·33^{IV}	2·325^{III}	2·36^{II} 2·33^{IV}	2·321^{I}	—
Br	2·945^{I}	— 2·82^{II}	2·543^{I}	2·55^{II} 2·46^{IV}	2·51^{III}	2·51^{II}	2·485^{I}	—
I	3·177^{I}	— 3·03^{II}	2·754^{I}	2·78^{II} 2·69^{IV}	2·67^{III}	—	2·666^{I}	—

TABLE 24 (*cont.*)

	Cs	Ba	Tl	Pb	Bi	Po	At	Rn
F	2·345^{I}	2·170^{I} 2·32^{II}	2·084^{I}	2·13^{II} —	—			
Cl	2·906^{I}	— 2·82^{II}	2·485^{I}	2·46^{II} 2·43^{IV}	2·48^{III}			
Br	3·072^{I}	— 2·99^{II}	2·618^{I}	2·60^{II} —	2·63^{III}			
I	3·315^{I}	— 3·20^{II}	2·814^{I}	2·81^{II} —	—			

	Sc	Ti	V	Cr	Mn	Fe	Co	Ni	Cu	Zn
F	1·788^{I} 1·91^{III}	—	1·71^{V}	—	1·70^{I} 1·724^{VII}				1·749^{I}	1·81^{II}
Cl	2·23^{I} —	2·185^{IVt}	2·14^{IV} 2·12^{Vt}	2·12^{VIt}	—				2·050^{I}	2·05^{II}
Br	—	2·31^{IVt}	2·30^{IV}	—	—				2·173^{I}	2·21^{II}
I	—	—	—	—	—				2·335^{I}	2·38^{II}
	Y	**Zr**	**Nb**	**Mo**	**Tc**	**Ru**	**Rh**	**Pd**	**Ag**	**Cd**
F	1·925^{I} 2·04^{III}	—	1·88^{V}	1·826^{VIu}			1·83^{VIv}		1·99^{I}	1·97^{II}
Cl	— 2·47^{III}	2·32^{IVu}	2·28^{Vu}	2·23^{IV} 2·27^{Vu}			—		2·281^{I}	2·235^{II}
Br	— 2·63^{III}	2·44^{IVu}	2·45^{Vu}	2·39^{IV}			—		2·392^{I}	2·37^{II}
I	— 2·80^{III}	—	—	—			—		2·544^{I}	2·55^{II}
	La	**Hf**	**Ta**	**W**	**Re**	**Os**	**Ir**	**Pt**	**Au**	**Hg**
F	2·026^{I} 2·22^{III}	—	1·86^{V}	1·833^{VIv}	1·859^{VIIu}	1·831^{VIv}	1·830^{VIv}	1·83^{VIv}		2·05^{II}
Cl	— 2·60^{III} (Rare Earths)	2·33^{IVu}	2·27^{V}	2·26^{VIu}	2·230^{VIIu}	—	—	—		2·29^{II}
Br	— 2·75^{III}	2·43^{IVu}	2·44^{V}	—	—	—	—	—		2·41^{II}
I	— 2·98^{III}	—	—	—	—	—	—	—		2·59^{II}

Superscript Roman numerals refer to valence state of the element forming the halide. Equilibrium bond distances r_e are given wherever the necessary information is available. Distances are derived from the following sources:

[a] T. L. Cottrell, *The Strengths of Chemical Bonds*, 2nd edn., pp. 270–289. Butterworths, London (1958).

[b] *Tables of Interatomic Distances and Configuration in Molecules and Ions* (ed. L. E. Sutton), The Chemical Society, London (1958); Supplement (1965).

[c] C. J. Cheetham and R. F. Barrow, *Adv. High Temperature Chemistry*, **1** (1967) 7.

[d] L. V. Vilkov, N. G. Rambidi and V. P. Spiridonov, *J. Struct. Chem.* **8** (1967) 715.

[e] *Tables of Constants and Numerical Data*, No. 17, *Spectroscopic Data relative to Diatomic Molecules* (general ed. B. Rosen), Pergamon (1970).

[f] A. Snelson, *J. Phys. Chem.* **72** (1968) 250.

[g] K. Kuchitsu and S. Konaka, *J. Chem. Phys.* **45** (1966) 4342.

[h] A. G. Massey, *Adv. Inorg. Chem. Radiochem.* **10** (1967) 1.

[i] F. X. Powell and D. R. Lide, jun., *J. Chem. Phys.* **45** (1966) 1067.

[j] B. Beagley, A. H. Clark and T. G. Hewitt, *J. Chem. Soc. (A)* (1968) 658.

[k] Refers to monomer unit, see ref. d.

[l] V. M. Rao, R. F. Curl, jun., P. L. Timms and J. L. Margrave, *J. Chem. Phys.* **43** (1965) 2557.

[m] E. Hirota and Y. Morino, *J. Mol. Spectroscopy*, **33** (1970) 460.

[n] K. W. Hansen and L. S. Bartell, *Inorg. Chem.* **4** (1965) 1775, 1777; S. B. Pierce and C. D. Cornwell, *J. Chem. Phys.* **48** (1968) 2118.

[o] W. J. Adams and L. S. Bartell, *J. Mol. Structure*, **8** (1971) 23.

[p] H. L. Roberts, *Inorganic Sulphur Chemistry* (ed. G. Nickless), p. 419, Elsevier (1968); *Essays in Structural Chemistry* (ed. A. J. Downs, D. A. Long and L. A. K. Staveley), p. 457, Macmillan (1971).

[q] I. C. Bowater, R. D. Brown and F. R. Burden, *J. Mol. Spectroscopy*, **23** (1967) 272; *ibid.* **28** (1968) 454, 461.

[r] C. Murchison, S. Reichman, D. Anderson, J. Overend and F. Schreiner, *J. Amer. Chem. Soc.* **90** (1968) 5690.

[s] J. H. Holloway, *Noble Gas Chemistry*, Methuen, London (1968); R. M. Gavin, jun., and L. S. Bartell, *J. Chem. Phys.* **48** (1968) 2460, 2466; S. Reichman and F. Schreiner, *J. Chem. Phys.* **51** (1969) 2355.

[t] R. Colton and J. H. Canterford, *Halides of the First Row Transition Metals*, Wiley-Interscience (1969).

[u] J. H. Canterford and R. Colton, *Halides of the Second and Third Row Transition Metals*, Wiley-Interscience (1968).

[v] M. Kimura, V. Schomaker, D. W. Smith and B. Weinstock, *J. Chem. Phys.* **48** (1968) 4001; H. Kim, P. A. Souder and H. H. Claassen, *J. Mol. Spectroscopy*, **26** (1968) 46.

3. Vibrational properties: force constants. Various compilations[345,352,355–359] bear witness to the research effort expended on investigations of the vibrational properties of simple and complex halides. Vibrational frequencies have been derived from the vibrational progressions observed in the electronic spectra of diatomic halides[352,355] or from the infrared and Raman spectra of more complicated species[345,356–359]. Such studies have communicated much valuable information concerning the structure and bonding properties of halide systems.

In a diatomic molecule MX, the frequency of the M–X stretching vibration is determined by two factors, the bond-stretching force constant and the masses of the two atoms. Hence, from the observed frequencies one may readily determine bond-stretching force constants, which are measures of the resistance of the M–X bond to deformation. In contrast with the bond energy, which refers to the process of separation of the M and X atoms to large distances, the force constant gives a measure of bond strength at internuclear distances close to the equilibrium value in terms of the curvature of the potential energy surface. Representative force constants for some diatomic halides are contained in Table 25. In polyatomic systems, the situation is complicated by interactions between formally non-bonded atoms and by the fact that the normal modes involve, not only bond-stretching, but also angular and torsional deformations. Detailed normal coordinate analyses leading to realistic force constants have generally been feasible only for halide molecules or complex ions of relatively high symmetry; even then, there is a need for additional information concerning, typically, isotopically substituted species or Coriolis coupling constants, if the secular equation is to be solved without undue simplification of the molecular force field. If most of the energy of a normal mode is localized in the deformation of a given bond or molecular unit, the frequency of the mode is then characteristic of that bond or unit. The condition whereby a normal mode may be considered as a "group vibration" implies that the frequency of the motion is well separated from others of the same symmetry class. Essentially localized vibrations are assumed for the identification of M–X "stretching frequencies". A popular basis for qualitative analysis and, less reliably, for inferences about chemical bonding, the range and magnitude of such frequencies have been discussed in some detail[345,358,359]; the results have been tabulated or used to construct correlation charts[345,356–359]. However, "group frequency" arguments affecting bonds in which the heavier halogen atoms are engaged are at best approximate. For species of high symmetry, e.g. tetrahedral MX_4 or octahedral MX_6, there exist unique totally symmetric modes localized in the M–X bond coordinates, the frequencies of which can be meaningfully correlated with the M–X stretching force constant[360]. In more complex or less symmetrical units, e.g. $[Mo_6Cl_8Cl_6]^{2-}$ [361], it is more often the rule that the normal modes involve extensive mixing of the various bond vibrations.

Consideration of the mass of data available discloses that the stretching frequencies

[356] K. Nakamoto, *Infrared Spectra of Inorganic and Coordination Compounds*, 2nd edn., Wiley-Interscience, New York (1970).

[357] H. Siebert, *Anwendungen der Schwingungsspektroskopie in der Anorganischen Chemie*, Springer-Verlag (1966).

[358] D. M. Adams, *Metal-Ligand and Related Vibrations*, Edward Arnold, London (1967).

[359] R. J. H. Clark, *Halogen Chemistry* (ed. V. Gutmann), Vol. 3, p. 85. Academic Press (1967).

[360] L. A. Woodward, *Trans. Faraday Soc.* **54** (1958) 1271.

[361] M. J. Ware, *Physical Methods in Advanced Inorganic Chemistry* (ed. H. A. O. Hill and P. Day), p. 241. Interscience (1968).

TABLE 25. STRETCHING FORCE CONSTANTS, k_e (mdyne Å^{-1}), FOR SOME DIATOMIC HALIDE MOLECULES[a,b]

HF 9·66								
HCl 5·16								
HBr 4·11								
HI 3·14								
NaF 1·76	MgF 3·25			AlF 4·22	SiF 4·91	PF 4·98	—	ClF 4·48
NaCl 1·10	MgCl 1·82			AlCl 2·09	SiCl 2·64	PCl 3·24	—	Cl_2 3·23
NaBr 0·96	MgBr 1·51			AlBr 1·70	SiBr 2·22	—	—	ClBr 2·80
NaI 0·76	MgI ~1·16			AlI 1·31	SiI 1·76	—	—	ClI 2·39
KF 1·37	CaF 2·62	MnF 3·18	CuF 3·34	GaF 3·40	GeF 3·92	—	—	BrF 4·09
KCl 0·86	CaCl 1·51	MnCl 1·82	CuCl 2·31	GaCl 1·83	GeCl 2·32	AsCl 2·78	—	BrCl 2·80
KBr 0·70	CaBr 1·27	MnBr 1·57	CuBr 2·05	GaBr 1·52	GeBr 1·97	—	—	Br_2 2·48
KI 0·53	CaI 1·02	MnI ~1·30	CuI 1·74	GaI 1·24	GeI 1·65	—	—	BrI 2·07

[a] T. L. Cottrell, *The Strengths of Chemical Bonds*, 2nd edn., pp. 270–289. Butterworths, London (1958).
[b] *Tables of Constants and Numerical Data*, No. 17, *Spectroscopic Data relative to Diatomic Molecules* (general ed. B. Rosen), Pergamon (1970).

$\nu(MX)$ or, more precisely, the force constants $k(MX)$ of bonds to halogen atoms are functions of the following variables[358,359]:

The ionic character of the bond

The general pattern (see Table 25) is that $\nu(MX)$ and $k(MX)$ diminish as the polarity of the M–X bond increases, reflecting the variation in form of the potential energy curve as the nature of the bond changes.

The nature of M and X

It appears to be a general rule that, for a given element M, $\nu(MX)$ and $k(MX)$ decrease in the order F > Cl > Br > I; in general, too, an inverse correlation with the atomic number of M is found in vertical Groups of the typical elements, but for halides formed by analogous elements of the 4*d* and 5*d* transition series, e.g. MoF_6 and WF_6, $PdCl_6^{2-}$ and $PtCl_6^{2-}$, $CdCl_2$ and $HgCl_2$, there is a significant reversal of this trend. The vibrational properties of transition-metal–halogen bonds also reflect variations of ligand-field stabilization energy and of the bonding or anti-bonding character of the *d*-electron shell[358,359].

Stereochemistry and coordination number

The geometry of a halide aggregate has a profound influence on its vibrational properties in terms, not only of selection rules, but also of the frequencies of the vibrational modes. The general rule has been propounded[358,359] that $\nu(MX)$ decreases with increasing coordination number of M or X: hence, bridging halogen units, as in Al_2Cl_6, are normally characterized by lower $\nu(MX)$ and $k(MX)$ than are terminal M–X bonds.

Oxidation state and charge

An increase in the oxidation state of M is normally accompanied by an increase in $\nu(MX)$ and $k(MX)$. If M is varied in an isoelectronic series, e.g. $AsCl_4^+$, $GeCl_4$, $GaCl_4^-$, $ZnCl_4^{2-}$, the same trend is observed, namely, diminution in $\nu(MX)$ and $k(MX)$ as the negative charge borne by the aggregate increases, an effect which has been related to the radial electronic distribution of M[360].

The presence of other ligands

There is ample evidence that the vibrational properties of M–X bonds are sensitive to the nature and, in some cases, the location of other ligands L in a mixed aggregate such as SnX_nL_{4-n} or PtX_2L_2. Influences such as the electronegativity of L or potential π-bonding between M and L have been weighed in these circumstances[358,359].

Environmental effects

The aggregation of molecules or ions in a crystal is always liable to alter the vibrational properties of these units. Contributory to such changes are the effective symmetry of the molecule or ion, the site of which usually belongs to a symmetry group different from that of the isolated unit, and, in the case of complex ions, the nature of the counter-ion, which may have a marked frequency- or even structure-determining influence. The behaviour of halide species in solution is inevitably dependent on the nature of the solvent, though there have been few systematic studies of the effects of solvent on the vibrations of simple and complex halides[358,359].

4. Esr studies[329,362–368]. Measurements of the electron spin resonance (esr) spectra of paramagnetic transition-metal halides afford information about three parameters, the g-factor, spin-orbit coupling constant λ, and nuclear hyperfine splitting constant A, all of which can be related to the degree of electron-delocalization.

The principles are well exemplified by the classical studies of the $IrCl_6^{2-}$ and $IrBr_6^{2-}$ ions in magnetically dilute mixed crystals[362]. The esr spectra show hyperfine structure lines arising from interaction of the magnetic electrons both with the Ir nucleus ($I = 3/2$) and with the Cl or Br nucleus ($I = 3/2$). From the A- and g-values it has been deduced that the "hole" in the formally non-bonding t_{2g} metal orbitals spends 3–5% of its time on each of the six halogen atoms, and that the hole and five t_{2g} electrons actually occupy molecular orbitals derived from combinations of the metal $5d_{xy}$-, d_{xz}- and d_{yz}-orbitals and suitable p_π-type orbitals of the halogen ligands. Other examples of paramagnetic halides studied in this way include the complex ions CoX_6^{4-}, for which the extent of delocalization increases in the sequence F < Cl < Br < I, and fluorides and chlorides of the lanthanides, in which the delocalization of the 4f-electrons appears to be strictly limited[364,366,367]. Tabulations of esr data relating to bonding in halides are to be found elsewhere[329,363,364,367].

5. Nmr studies. The interactions between ligand nuclei and the magnetic electrons of transition-metal ions, which lead to the hyperfine structure observed in the esr spectra, also result in large shifts in the magnetic resonance of the ligand nucleus. Studied in a number of transition-metal fluorides[329,364,369], these shifts have led to estimates of the fraction of unpaired electrons in fluorine 2s-, $2p_\sigma$- and $2p_\pi$-orbitals, the results being in good agreement with those deduced from esr measurements. Apart from studies of crystalline $FeCl_2$ [369] and of some solutions containing chloride in the presence of a transition-metal ion, however, the influence of electron paramagnetism on the resonances due to the heavier halogen nuclei has so far received little attention.

Measurements of the magnetic resonance of the heavier halogen nuclei in molecular halides have been virtually restricted to ^{35}Cl, though a few relaxation times for other nuclei have been determined indirectly by chemical exchange experiments[307]. Table 26 lists the ^{35}Cl chemical shifts and relaxation times (T_2) of some representative chloride species. Interpretation of the chemical shifts is hampered by the difficulty of rigorous calculation, which requires knowledge of the ground and excited-state wavefunctions. A relatively simple practical equation given by Saika and Slichter for the shift of the ^{19}F resonance in the F_2 molecule relative to the free ion[370] has provided a basis for the discussion of the ^{35}Cl shifts in molecular chlorides, as for example in the case of the Cl_2 molecule[307,371].

362 J. Owen and K. W. H. Stevens, *Nature*, **171** (1953) 836; J. H. E. Griffiths, J. Owen and I. M. Ward, *Proc. Roy. Soc.* **A219** (1953) 526; J. H. E. Griffiths and J. Owen, *Proc. Roy. Soc.* **A226** (1954) 96.

363 B. R. McGarvey, *Transition Metal Chemistry* (ed. R. L. Carlin), Vol. 3, p. 89. Edward Arnold and Marcel Dekker (1966).

364 J. Owen and J. H. M. Thornley, *Rep. Progr. Phys., London*, **29** (1966) 675.

365 A. Carrington and A. D. McLachlan, *Introduction to Magnetic Resonance*, Harper and Row (1967); A. Carrington and H. C. Longuet-Higgins, *Quart. Rev. Chem. Soc.* **14** (1960) 427.

366 A. Abragam and B. Bleaney, *Electron Paramagnetic Resonance of Transition Ions*, Clarendon Press, Oxford (1970).

367 B. A. Goodman and J. B. Raynor, *Adv. Inorg. Chem. Radiochem.* **13** (1970) 135.

368 L. Shields, *J. Chem. Soc.* (*A*) (1971) 1048 and references cited therein.

369 D. R. Eaton and W. D. Phillips, *Adv. Magnetic Resonance*, **1** (1965) 103.

370 A. Saika and C. P. Slichter, *J. Chem. Phys.* **22** (1954) 26.

371 C. Hall, D. W. Kydon, R. E. Richards and R. R. Sharp, *Mol. Phys.* **18** (1970) 711.

TABLE 26. ^{35}Cl CHEMICAL SHIFTS, RELAXATION TIMES AND QUADRUPOLE COUPLING CONSTANTS FOR SOME CHLORINE-CONTAINING SPECIES[a]

Species	Temperature (°K)	Chemical shift* (ppm)	Relaxation time, T_2(msec)	Quadrupole coupling constant, e^2Qq/h(MHz)[b]
ClO_3^-(aq)	Room	−1050	0·024	—
ClO_4^-(aq)	299	−946	5·1	—
$TiCl_4$	299	−840	0·40	12·2
$VOCl_3$	299	−791	0·161	23·1
SO_2Cl_2	299	−760	0·030	75·4
CrO_2Cl_2	299	−603	0·107	31·4
CCl_4	299	−500	0·025	81·2
S_2Cl_2	299	−480	0·022	71·6
$POCl_3$	299	−430	0·027	57·9
$CHCl_3$	Room	−410	0·025	76·7
Cl_2	298	−370	0·031	109
PCl_3	299	−370	0·060	52·2
CH_2Cl_2	Room	−220	0·042	72·0
$SiCl_4$	299	−174	0·079	40·8
$GeCl_4$	299	−170	0·041	51·4
Cl_3^-(aq)	298	~ −150	0·003†	116†
			0·015‡	51‡
$AsCl_3$	299	−150	0·023	50·3
$SnCl_4$	299	−120	0·022	48·2
CH_3Cl	Room	−40	0·099	68·1
Cl^-(aq), dil soln*	Room	0	—	—

* External reference. † Central atom. ‡ Terminal atom.

[a] C. Hall, *Quart. Rev. Chem. Soc.* **25** (1971) 87.

[b] E. A. C. Lucken, *Nuclear Quadrupole Coupling Constants*, Academic Press, London and New York (1969).

With ^{35}Cl, as with ^{19}F, the nuclear magnetic shielding appears to be dominated by the paramagnetic contribution[370–372], which has been related by the expression

$$\sigma_{par} = \frac{-2}{3}\left(\frac{e^2\hbar}{m^2c^2}\right)\Delta E^{-1}\left\langle \frac{1}{r^3} \right\rangle_{np} \tag{12}$$

to the mean excitation energy ΔE, effectively expressing the accessibility of electronic states above the ground state, and to the expectation value of r^{-3} for the valence *p*-electron. The periodic dependence of the *range* of chemical shifts on the atomic number of the nucleus has been shown[372] to follow the pattern dictated by the term $\langle 1/r^3 \rangle_{np}$; for the halogens, these ranges are (in ppm): F, 625; Cl, 1000; Br, ~ 1650. More recently, Johnson, Hunt and Dodgen[373] have sought to express the ^{35}Cl chemical shifts in terms of the quadrupole coupling constant e^2Qq, an accurately known, independent experimental parameter, obtaining

$$\sigma_{par} \propto (e^2Qq/h)\Delta E^{-1}\langle 1/r^3 \rangle_{np}$$

Although the mean excitation energy is not clearly defined, transition-metal halides which

[372] C. J. Jameson and H. S. Gutowsky, *J. Chem. Phys.* **40** (1964) 1714.

[373] K. J. Johnson, J. P. Hunt and H. W. Dodgen, *J. Chem. Phys.* **51** (1969) 4493.

have relatively low-lying excited states, but rather small quadrupole coupling constants, are mostly characterized by large chemical shifts. Furthermore, there is a general trend in the direction of larger shifts in compounds with large quadrupole coupling constants. However, there are also distinct anomalies, e.g. the wide variation of chemical shift for the series CH_3Cl, CH_2Cl_2, $CHCl_3$, CCl_4.

There exist many data relating to the chemical shifts[374–379] and coupling constants[374,379,380] of nuclei such as ^{1}H[374], ^{11}B[374,378], ^{13}C[374,377], ^{14}N[374,376], ^{19}F[374], ^{31}P[374,375], or ^{195}Pt[379] in halogen-containing molecules. In the interpretation of chemical shifts, the shielding of a nucleus is conveniently expressed as the algebraic sum of the local diamagnetic (σ_d) and paramagnetic (σ_p) contributions. For nuclei other than hydrogen, variations in shielding are largely dictated by variations in σ_p, which, being sensitive to changes in the wavefunction of the valence electrons, reflects the bonding properties of the resonant atom. Hence the ^{195}Pt chemical shifts of the platinum hydrides *trans*-$[Pt(PEt_3)_2HX]$ have been correlated with changes in the σ- and π-character of the Pt–X bond[379]. However, recent analyses[376,377] of carbon and nitrogen compounds have revealed that changes in σ_d can be quite large and that errors of interpretation can follow the neglect of these terms. In general, the complexity of factors which influence the magnetic shielding and interaction of nuclei tends to frustrate attempts to reach unambiguous conclusions about chemical bonding, though there are to be found numerous correlations, commonly of an empirical nature, with properties such as bond length, bond angle, hybridization and electronegativity[374].

6. Nqr studies[329,365,381–386]. The energy of a non-spherical atomic nucleus in an inhomogeneous electric field varies according to the orientation of the nucleus about some fixed axis. The interaction between the nucleus and field gives rise to so-called nuclear quadrupole coupling; an inhomogeneous electric field is provided, for example, by the incompletely filled *p*-shell found in halogen atoms. In a nuclear quadrupole resonance (nqr) experiment, radiation in the radiofrequency region is employed to effect transitions among the various orientations of a quadrupolar nucleus in an asymmetric intramolecular field; the frequencies of the transitions depend upon both the field gradient q produced by the valence electrons and

[374] J. A. Pople, W. G. Schneider and H. J. Bernstein, *High-resolution Nuclear Magnetic Resonance*, McGraw-Hill (1959); J. W. Emsley, J. Feeney and L. H. Sutcliffe, *High Resolution Nuclear Magnetic Resonance Spectroscopy*, Pergamon (1965–6).

[375] M. M. Crutchfield, C. H. Dungan, J. H. Letcher, V. Mark and J. R. Van Wazer, *P^{31} Nuclear Magnetic Resonance*, Interscience (1967).

[376] R. Grinter and J. Mason, *J. Chem. Soc.* (*A*) (1970) 2196.

[377] J. Mason, *J. Chem. Soc.* (*A*) (1971) 1038.

[378] See for example H. Nöth and H. Vahrenkamp, *Ber.* **99** (1966) 1049; M. F. Lappert, M. R. Litzow, J. B. Pedley and A. Tweedale, *J. Chem. Soc.* (*A*) (1971) 2426.

[379] R. R. Dean and J. C. Green, *J. Chem. Soc.* (*A*) (1968) 3047.

[380] W. McFarlane, *Quart. Rev. Chem. Soc.* **23** (1969) 187.

[381] W. J. Orville-Thomas, *Quart. Rev. Chem. Soc.* **11** (1957) 162.

[382] T. P. Das and E. L. Hahn, *Nuclear Quadrupole Resonance Spectroscopy*, Academic Press, New York (1958).

[383] M. Kubo and D. Nakamura, *Adv. Inorg. Chem. Radiochem.* **8** (1966) 257.

[384] H. Sillescu, *Physical Methods in Advanced Inorganic Chemistry* (ed. H. A. O. Hill and P. Day), p. 434. Interscience (1968).

[385] E. A. C. Lucken, *Nuclear Quadrupole Coupling Constants*, Academic Press (1969).

[386] E. A. C. Lucken, *Structure and Bonding*, **6** (1969) 1.

the quadrupole moment eQ of the nucleus. Structural information about a compound can be obtained by considering how different structural and electronic effects influence q, which is an index to the asymmetry of the electron environment. The direct measurement of quadrupole coupling by absorption of radiofrequency radiation is restricted to the crystalline phase, where the axes of q are fixed in space; quadrupole coupling constants for species in the gas phase are most easily derived from the fine structure in the pure-rotational (microwave) spectrum. Nuclear quadrupole coupling in solids has also been extensively investigated by the following techniques: nuclear magnetic resonance, electron spin resonance, electron-nuclear double resonance (ENDOR) and Mössbauer spectroscopy.

The naturally occurring quadrupolar nuclei ^{35}Cl, ^{37}Cl, ^{79}Br, ^{81}Br and ^{127}I have been a mainstay of nqr experiments, many of which have been directed to exploring the properties of halogen-bearing bonds. An analysis of the factors affecting the field gradient experienced by a halogen atom in a molecule indicates that the major contribution arises from the way in which the p-orbitals are occupied by the valence electrons; by contrast, electron density in the spherically symmetrical s-orbital does not give rise to a field gradient. At one extreme, represented by the free halogen atom, the p-electron "hole" produces a large field gradient, e.g. $e^2Qq = -109{\cdot}746$ MHz for ^{35}Cl: at the other, represented by the free halide ion, the spherical symmetry of the closed electron shell produces a vanishing field gradient at the nucleus and $e^2Qq = 0$. The measured nqr frequencies therefore give a relatively direct indication of the ionic character of a halide species (see Table 27), though considerable difficulties arise when a more quantitative treatment is attempted.

Most interpretations have been based on the approximate valence-bond approach pioneered by Townes and Dailey[387], which leads to the following general equation relating e^2Qq_{exp}, the measured quadrupole coupling constant for the halogen atoms in a halide species MX_n, to e^2Qq_{at}, the corresponding constant for the isolated atom, by[388]

$$e^2Qq_{\text{exp}} = [1-s-d-i(1-s-d)]e^2Qq_{\text{at}} \tag{13}$$

Here s and d represent the fractions of s- and d-character in the σ-bonding orbitals of the halogen atom, and i denotes the degree of ionic character in the M–X bond. A modified form of the equation, namely,

$$e^2Qq_{\text{exp}} = [1-s+d-i-\pi]e^2Qq_{\text{at}} \tag{14}$$

has also been employed[389] to give an account of the extent of π-interaction in M–X bonds. In view of the discrepancy between the number of measurable and unknown parameters, a rigorous solution is not possible without additional information. Whereas a σ-bond to a halogen should produce an axially symmetric field gradient, π-bonding leads to asymmetry, the extent of such bonding being related to the asymmetry parameter η. For example, this has formed the basis of a detailed study of π-bonding in halides of the Group IV elements[390], and also affords cogent evidence of extensive π-bonding in the boron trihalide molecules[385].

[387] C. H. Townes and B. P. Dailey, *J. Chem. Phys.* **17** (1949) 782; *ibid.* **23** (1955) 118; B. P. Dailey, *Discuss. Faraday Soc.* **19** (1955) 255.

[388] R. S. Drago, *Physical Methods in Inorganic Chemistry*, Reinhold, New York (1965).

[389] B. P. Dailey, *J. Chem. Phys.* **33** (1960) 1641.

[390] R. Bersohn, *J. Chem. Phys.* **22** (1954) 2078; J. D. Graybeal and P. J. Green, *J. Phys. Chem.* **73** (1969) 2948.

Table 27. Nuclear Quadrupole Coupling Constants and Ionic Character in Some Halide Species

(a) *Binary halides*[a]

Halide	Quadrupole coupling constant, e^2Qq/h (MHz)	Ionic character	Halide	Quadrupole coupling constant, e^2Qq/h (MHz)	Ionic character
$K^{35}Cl$	0·04	1·000	$F^{35}Cl$	146·0	0·259
$Cs^{35}Cl$	3	0·968	$Cl^{127}I$	2944	0·229
$Na^{79}Br$	58	0·911	$D^{79}Br$	533	0·186
$Na^{127}I$	259·87	0·867	$Cl^{79}Br$	876·8	0·110
$Tl^{35}Cl$	15·8	0·831	$D^{127}I$	1823·38	0·065
$F^{79}Br$	1089·0	0·329	$^{35}Cl_2$	108·95	0

e^2Qq/h atom (MHz): ^{35}Cl, −109·746; ^{37}Cl, −86·510; ^{79}Br, 769·756; ^{81}Br, 643·032; ^{127}I, −2292·712.

(b) *Complex halides*[b]

Hexahalide	X = ^{35}Cl		X = ^{79}Br		X = ^{127}I	
	Quadrupole coupling constant, e^2Qq/h (MHz)	Ionic character*	Quadrupole coupling constant, e^2Qq/h (MHz)	Ionic character*	Quadrupole coupling constant, e^2Qq/h (MHz)	Ionic character*
K_2WX_6	20·4	0·78 (0·43)	—	—	—	—
K_2ReX_6	27·9	0·72 (0·45)	231	0·66 (0·39)	822	0·57 (0·32)
K_2OsX_6	33·8	0·62 (0·47)	—	—	—	—
K_2IrX_6	41·7	0·55 (0·47)	—	—	—	—
K_2PtX_6	52·0	0·44	406	0·38	1360	0·30
K_2PdX_6	53·5	0·43	411	0·37	—	—
K_2SnX_6	31·5	0·66	261	0·60	884	0·55
$(NH_4)_2PbX_6$	34·5	0·63	—	—	—	—
K_2SeX_6	41·2	0·56	346	0·47	—	—
K_2TeX_6	30·3	0·68	271	0·58	1020	0·48

* The numbers in brackets stand for the total ionic character where π-bonding has been considered.

[a] C. H. Townes and B. P. Dailey, *J. Chem. Phys.* **17** (1949) 782; *ibid.* **23** (1955) 118.

[b] H. Sillescu, *Physical Methods in Advanced Inorganic Chemistry* (ed. H. A. O. Hill and P. Day), p. 456. Interscience (1968).

By assuming the d- and π-contributions to be unimportant and the extent of s-hybridization to be 15%, Townes and Dailey[387] have calculated the ionic character i for a large number of diatomic halides (see, for example, Table 27). Alternatively, the experimental quadrupole coupling constants have been used to calculate the s-character of halogen bonds, the ionic character being defined through its relation to valence-state energies[391]. That the interpretation of the results for polyatomic halides[385] has been less successful is indicated by the divergence of opinion expressed by various authors regarding the estimates of the quantities i, π, d and s in equation (14). A recent analysis[392] suggests that there is a distinct change in the s-character of the chlorine σ-orbital when chlorides pass from the solid to the gaseous state, and that the s-character depends upon the atom to which the chlorine is bonded. However, the general conclusion, here and elsewhere[386], that in very few cases do the results provide clear evidence of outer d-orbital participation, e.g. in Si–X, P–X or S–X bonds, must be tempered by the recognition that approximate methods are necessarily involved in the analysis of the experimental data. The problem of ionic character in transition-metal halides and hexahalo-complexes of heavier atoms has also been investigated by direct measurements of the nqr spectra[383–385]; some of the results are reproduced in Table 27. In some of these systems the relative contributions of σ- and π-bonding have been evaluated, but, in view of the burden of assumptions made, the conclusions must be regarded as open to doubt[327,384,388]. A more general treatment of quadrupole coupling constants in terms of a molecular-orbital approach[393] may serve to remove some of the uncertainties, but radical progress must be inhibited by the paucity of observable parameters.

7. Mössbauer spectra[394–397]. An alternative source of information about nuclear quadrupole coupling, Mössbauer spectroscopy is based on the phenomenon of recoilless emission and resonance absorption of low energy γ-radiation[394]. Chemical applications of this technique depend on hyperfine interactions between the nuclear energy levels and the extranuclear electrons, which give rise (i) to nuclear isomer shifts δ, (ii) to quadrupole splittings, and (iii) to magnetic hyperfine Zeeman splittings. To achieve recoilless conditions for the nuclei under examination, studies of Mössbauer, as of nqr, spectra are essentially confined to solids. The quadrupole splitting pattern of a Mössbauer spectrum gives direct access to the quadrupole coupling constant e^2Qq, and if $I > 3/2$ for one of the nuclear states implicated in the γ-transition, the pattern may also disclose the asymmetry parameter η. The interpretation of such results in terms of site symmetries and field gradients within a crystal and of the imbalance of p- and d-electrons then follows the pattern outlined in the previous section; hence, features such as ionic character, sp-hybridization and π-bonding have been analysed.

[391] M. A. Whitehead and H. H. Jaffé, *Theoret. Chim. Acta*, **1** (1963) 209.
[392] M. Kaplansky and M. A. Whitehead, *Mol. Phys.* **16** (1969) 481.
[393] F. A. Cotton and C. B. Harris, *Proc. Nat. Acad. Sci. U.S.A.* **56** (1966) 12.
[394] E. Fluck, *Adv. Inorg. Chem. Radiochem.* **6** (1964) 433; N. N. Greenwood, *Chem. in Britain*, **3** (1967) 56; V. I. Goldanskii, *Angew. Chem., Internat. Edn.* **6** (1967) 830; R. H. Herber, *Progress in Inorganic Chemistry*, **8** (1967) 1; J. Danon, *Physical Methods in Advanced Inorganic Chemistry* (ed. H. A. O. Hill and P. Day), p. 380, Interscience (1968); V. I. Goldanskii and R. H. Herber (eds), *Chemical Applications of Mössbauer Spectroscopy*, Academic Press (1968).
[395] M. Pasternak, *Symposia of the Faraday Soc.* **1** (1967) 119.
[396] D. W. Hafemeister, *Advances in Chemistry Series*, **68** (1967) 126.
[397] G. J. Perlow, *Chemical Applications of Mössbauer Spectroscopy* (ed. V. I. Goldanskii and R. H. Herber), p. 377. Academic Press (1968).

Of particular interest is the isomer shift parameter δ, which is unique to the Mössbauer experiment. This derives from the fact that an s-electron has a finite probability of being found within the nuclear volume. Since the nucleus alters its radius r by a small amount δr during a γ-decay, there is a change in the fraction of the electron within the nucleus, with a concomitant change in electrostatic energy. If the source and absorber in a Mössbauer experiment are chemically distinct, first-order perturbation theory gives the shift in the resonance line as

$$\delta = \frac{4\pi}{5} Ze^2 r^2 \frac{\delta r}{r} \left[|\psi(0)|^2_{\text{absorber}} - |\psi(0)|^2_{\text{source}} \right] \quad (15)$$

where Z is the nuclear charge, e the electronic charge and $|\psi(0)|^2$ the total s-electron density at the nucleus. The isomer shift is therefore the product of a nuclear term, which is a physical constant for a given isotope and can be determined, and a chemical term relating to the change in s-electron density. By setting up suitable calibration scales, it is possible to monitor the s-electron density at a particular nucleus in a series of compounds, e.g. halide derivatives of ^{57}Fe, ^{119}Sn and ^{197}Au. As a measure of s-orbital occupancy, the isomer shift depends upon the valence state and bonding orbitals of the atom, upon the shielding effects of p-, d- or f-electrons, upon the ionic-covalent balance of interactions involving the atom, and upon the donor or acceptor capacity of neighbouring atoms through σ- or π-bonding.

Of the halogen nuclei, only the two isotopes of iodine ^{127}I and ^{129}I are suitable for Mössbauer experiments (see p. 1158)[395–397]. Experimental measurements affecting these

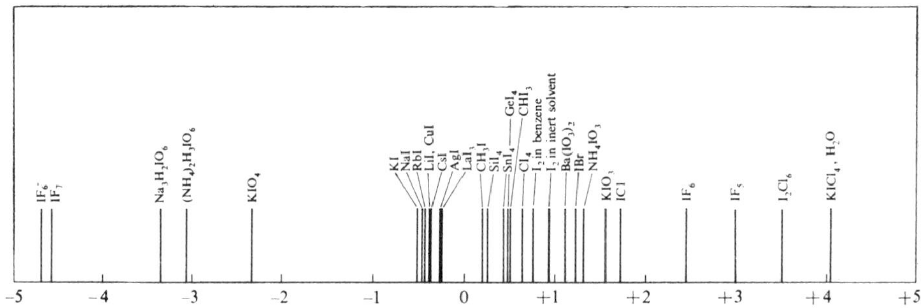

FIG. 22. ^{129}I isomer shifts (measured with respect to a ZnTe source) of iodine compounds.

isotopes have been focused, *inter alia*, on the following compounds: alkali-metal iodides[396], CH_nI_{4-n} (n = 0–3)[398], MI_4 (M = C, Si, Ge or Sn)[395], IX (X = Cl, Br, I or CN)[395,399], IF_5, IF_7 and related ions[400], I_2Cl_6[399], and derivatives of polyhalide, iodate and periodate ions[395–397]. Figure 22 illustrates the range of isomer shifts for ^{129}I; the spectra of many compounds also exhibit quadrupole splitting, whence the parameters e^2Qq and η have been derived. Through their dependence on the 5s-electron density at the nucleus, the isomer

398 B. S. Ehrlich and M. Kaplan, *J. Chem. Phys.* **50** (1969) 2041.

399 M. Pasternak and T. Sonnino, *J. Chem. Phys.* **48** (1968) 1997, 2009.

400 S. Bukshpan, C. Goldstein, J. Soriano and J. Shamir, *J. Chem. Phys.* **51** (1969) 3976; S. Bukshpan, J. Soriano and J. Shamir, *Chem. Phys. Letters*, **4** (1969) 241.

shifts in different iodine compounds provide an index to the relative participation of the 5*s*- and 5*p*-electrons in the bonds formed by the iodine atom. In alkali-metal iodides, the isomer shifts are linearly dependent on the number of *p* "holes" in the closed-shell configuration $5s^25p^6$, calculated independently from nmr and nqr data[396]; this result has been used to calibrate the isomer shift in terms of the relative change in 5*s* density. Of the factors influencing the 5*s* density—electrostatic interactions, covalency, electronic shielding and overlap deformation of the free-ion orbitals—it appears that shielding and overlap make the major contributions. Where only *p*-electrons are involved in the bonding, the removal of *p*-electron density for bonding decreases the shielding of the *s*-electrons and therefore increases the *s*-electron density at the nucleus, with the result that the isomer shift relative to I^- is positive (see Fig. 22). To a first approximation, the isomer shift is a linear function of the number of 5*p* "holes", h_p[395,399]. On this basis, it has been concluded[395–397,399] that the bonds in I_2, ICl, IBr, ICN, I_2Cl_6, ICl_2^- and ICl_4^- are derived almost exclusively from 5*p*-electrons of the iodine with negligible *sp*-hybridization. By contrast, where *sp*-hybridization is involved, e.g. in IO_4^- and CH_nI_{4-n} ($n = 0$–3), *s*-electron density is removed from the neighbourhood of the iodine nucleus into the bonding orbitals, and, as indicated in Fig. 22, the isomer shift relative to I^- is negative. In this case the isomer shift is empirically expressed by an equation of the type

$$\delta = -ah_s + bh_p + c$$

where h_s is the number of 5*s* "holes" and *a*, *b* and *c* are constants, which must be determined experimentally[395,399,400].

8. Electronic and magnetic properties[298,305,329,401–405]. Knowledge about the differences between the ground and excited electronic states can be gained from the electronic spectra of halide species; most widely and profitably studied have been spectroscopic transitions involving internal rearrangements of the d^n electrons of transition-metal derivatives. For details of the ground and very low-lying excited states of transition-metal halides, such studies have been augmented by measurements of the magnetic properties. The energies of the *d*-states depend upon the balance of the following interdependent factors: (i) the interaction of the *d*-electrons with the nucleus from which they are incompletely screened; (ii) the electron–electron repulsion, variations of which for different states of a given d^n configuration may be expressed in terms of the two Racah parameters *B* and *C*; (iii) the magnetic coupling of the spin and orbital momenta of the electrons, which may be defined in terms of the one-electron parameter ζ; and (iv) the field due to the ligands, which is usually referred to the ligand- or crystal-field parameter *Dq*. In the rearrangement of the d^n electrons it is commonly assumed, as a first approximation, that the energy due

401 T. M. Dunn, *Modern Coordination Chemistry* (ed. J. Lewis and R. G. Wilkins), p. 229. Interscience, New York (1960).

402 H.-H. Schmidtke, *Physical Methods in Advanced Inorganic Chemistry* (ed. H. A. O. Hill and P. Day), p. 107. Interscience (1968).

403 A. B. P. Lever, *Inorganic Electronic Spectroscopy*, Elsevier (1968); J. Ferguson, *Progress in Inorganic Chemistry*, **12** (1970) 159.

404 B. N. Figgis and J. Lewis, *Progress in Inorganic Chemistry*, **6** (1964) 37; B. N. Figgis, *Introduction to Ligand Fields*, Interscience, New York (1966).

405 M. Gerloch and J. R. Miller, *Progress in Inorganic Chemistry*, **10** (1968) 1.

to the first term is unchanged, so that the study of *d–d* spectroscopic transitions bears witness to the effects of electron–electron repulsion, spin-orbit coupling[406–408] and the ligand field.

Interelectron Repulsion: the Nephelauxetic Effect

Comparison of the spectra of compounds formed by a given transition-metal ion with that of the gaseous metal ion shows that the Racah parameters are smaller for the compounds than for the free ion, implying that coordination decreases the repulsions between *d*-electrons on the transition-metal ion. Decrease of the effective nuclear charge on the metal *d*-electrons through covalent σ-bonding with the ligands causes an expansion of the *d*-orbitals, an effect which may be enhanced by metal-to-ligand π-bonding. Accordingly, the reduction in the Racah parameters, represented, for example, by the ratio $B_{\text{compound}}/B_{\text{free ion}} = \beta$, can be used as an index to covalency in the metal–ligand bond. Ligands have been ordered on the basis of their increasing "nephelauxetic" effect (decreasing β) with respect to a given metal ion, e.g.

$$F^- < H_2O < NH_3 < H_2NCH_2CH_2NH_2 < SCN^- < Cl^- \sim CN^- < Br^- < S^{2-} \sim I^-$$

which is also the order of increasing polarizability of the ligand atom.

Ligand-field Effects

The ligand-field parameter *Dq* is well known to be a function of the metal ion, the ligand and the stereochemistry of the system. For a given metal and stereochemistry, the nature of the ligand causes *Dq* to increase in accordance with the familiar spectrochemical series, an abbreviated version of which follows:

$$I^- < Br^- < Cl^- < S^{2-} < F^- < H_2O < SCN^- < NH_3 < NO_2^- < CH_3^- < CN^-$$

Such an order cannot be rationalized on the basis of the simple point-charge model of crystal-field theory. In fact, the following factors all subscribe, in varying degrees, to the magnitude of *Dq*: interactions due to electrostatic perturbation, the metal–ligand σ-bond, metal-to-ligand π-bonding, ligand-to-metal π-bonding, and metal electron–ligand electron repulsions. For certain systems a scheme has been devised to resolve these various contributions[409]: the following order of σ-bonding interaction with a given metal ion is thus found:

$$NH_3 > H_2O > F^- > Cl^- > Br^- > I^-$$

while the π-donor capacity follows the sequence

$$I^- > Br^- > Cl^- > F^- > NH_3$$

---

406 B. N. Figgis, J. Lewis, F. E. Mabbs and G. A. Webb, *J. Chem. Soc.* (*A*) (1966) 1411.
407 J. Owen, *Proc. Roy. Soc.* **A227** (1955) 183; *Discuss. Faraday Soc.* **19** (1955) 127.
408 T. M. Dunn, *J. Chem. Soc.* (1959) 623.
409 D. S. McClure, *Advances in the Chemistry of the Coordination Compounds* (ed. S. Kirschner), p. 498. Macmillan (1961).

9. Dipole Moments[329,410,411]. A neutral halide molecule in which the bonds to the halogen atoms are not symmetrically disposed inevitably possesses a dipole moment. For polyatomic halides, a bond moment can be defined for each bond, provided that the geometry of the molecule is known, though, as vector quantities, such moments are seldom strictly additive. In fact, the dipole moment of a unit M–X depends, not only upon the ionic character, but also upon additional factors: namely, (i) the overlap moment arising from the fact that the orbitals of two atoms of dissimilar size tend to overlap closer to the smaller atom; (ii) the hybridization moment due to the possible asymmetry of the atomic orbitals involved in the bond, as in PX_3; and (iii) the induced moment arising from the polarization of valence electrons. Only when allowance has been made for these other effects does the dipole moment provide a measure of the ionic character of diatomic halides which is substantially in accord with estimates based on nqr or electronegativity parameters[387]. More generally, however, the properties of M–X bonds in polyatomic halide species are not easily assessed in terms of dipole moments, though evidence of π-bonding has more than once been adduced in this way[412]. Dipole moments of halide molecules have been compiled elsewhere[410,411].

Nature of Bonding: a Summary

The electronegativity of an atom being a measure of the energy of the valence electrons in relation to that of other atoms, the high electronegativity commonly associated with the halogens signifies the relatively low energy (in absolute terms) of the valence orbitals of these, compared with other, atoms. The binding of a diatomic halide is then determined by the relative energies of the valence orbitals of M and X, by the number and type of such orbitals, and by the number of valence electrons carried by M. In most circumstances only the outer *ns*- and *np*-orbitals of the halogen appear to make significant bonding contributions, and, with respect to more electropositive elements, even the *ns*-orbital is predominantly non-bonding in character; the participation of the vacant *nd*-orbitals of chlorine, bromine or iodine cannot be excluded, but the extent is now conceded by most theoretical studies to be small[413]. Relatively direct evidence that the "lone-pair" electrons of the halogen atom commonly occupy orbitals which are almost atomic in character comes from the recently described photoelectron spectra of halide molecules[414]; in simple monohalide derivatives of saturated hydrocarbons such orbitals have thus been shown to be virtually non-bonding, but in certain organic polyhalides and in numerous inorganic halides there is good reason to believe that these orbitals assume varying degrees of π-bonding character.

For a typical element M in the same horizontal Period as X, the energy level diagram of MX, illustrated schematically in Fig. 23, varies from that characteristic of a highly polar alkali-metal halide molecule, with minimal perturbation of the appropriate atomic orbitals, to that characteristic of the exclusively covalent halogen molecule, the formation of which implies marked perturbation of the two separate sets of atomic orbitals. The bonding

[410] B. Lakatos and J. Bohus, *Acta Chim. Acad. Sci. Hung.* **20** (1959) 115; B. Lakatos, J. Bohus, and G. Medgyesi, *ibid.* p. 1.

[411] A. L. McClellan, *Tables of Experimental Dipole Moments*, Freeman, San Francisco (1963).

[412] See for example E. A. V. Ebsworth, *Volatile Silicon Compounds*, pp. 57–58, 162. Pergamon, Oxford (1963); J. Lorberth and H. Nöth, *Ber.* **98** (1965) 969.

[413] See for example B. M. Deb and C. A. Coulson, *J. Chem. Soc.* (*A*) (1971) 958.

[414] D. W. Turner, C. Baker, A. D. Baker and C. R. Brundle, *Molecular Photoelectron Spectroscopy*, p. 214. Wiley-Interscience (1970).

orbitals of the MX molecule having a maximum capacity of 10 electrons, the highest dissociation energies are exhibited by the Group III monohalides BF, AlCl, etc.

The acceptor function of the halogen orbitals in the process of compound-formation is a characteristic of all halide species. Consequences of this electron-transfer include the

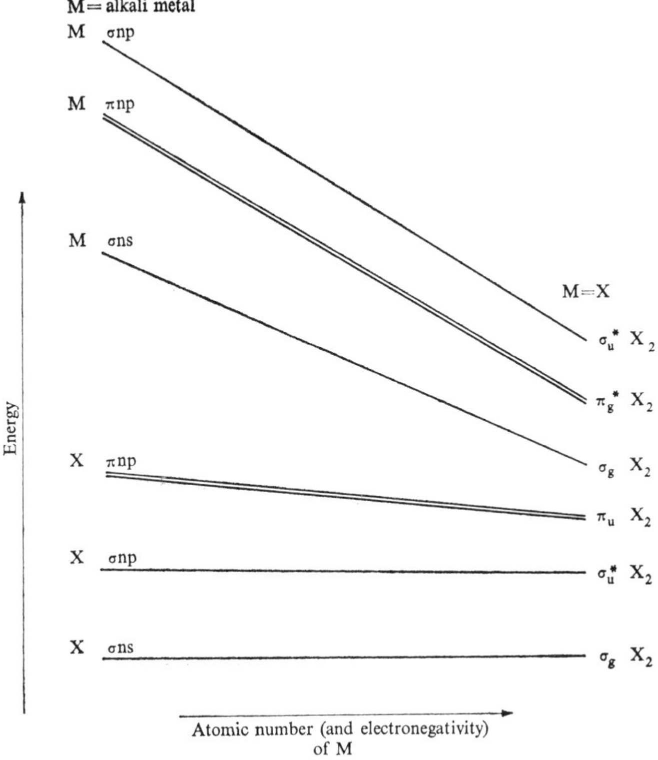

FIG. 23. Schematic energy level diagram for the diatomic halide MX of a typical element M belonging to the same period as X.

contraction and stabilization of outer orbitals on the atom partnered by the halogen; hence, vacant *d*-orbitals on atoms such as silicon or phosphorus may be induced to play a significant part in σ- or π-interactions[415]. Even molecular-orbital accounts of electron-rich species like $XeCl_2$ or ICl_4^-, which deny *d*-orbital involvement in the first instance, accentuate the importance of charge-transfer to the halogen ligand, since it is here that the non-bonding molecular orbital of the three-centre–four-electron model (see Fig. 4) is largely localized.

Theoretical accounts of the bonding in metal halides have been reviewed by Pearson and Mawby[329]. Total coordinate bond energies corresponding to the process $MX_n(g) \rightarrow M^{n+}(g) + nX^-(g)$ have been calculated on the basis of the following models:

[415] D. P. Craig and E. A. Magnusson, *J. Chem. Soc.* (1956) 4895; D. P. Craig, *Chemical Society Special Publication* No. 12, p. 343. The Chemical Society, London (1958); K. A. R. Mitchell, *Chem. Rev.* **69** (1969) 157.

(i) an ionic model assuming spherical, non-polarizable ions; (ii) an ionic model incorporating polarizable ions; (iii) a localized molecular-orbital model in which a two-centre molecular orbital is constructed for each M–X bond; and (iv) a modified version of the semi-empirical Wolfsberg–Helmholz method, an LCAO-MO approach using the one-electron approximation. The hard-sphere model gives a good account of the alkali-metal halides, somewhat less satisfactory results for the alkaline-earth halides, and understandably poor results for compounds like $AlCl_3$ and $TiCl_4$. The polarizable ion model works comparatively well for a wide range of metal halides, but gives poor results for some, e.g. $BeCl_2$ and $TiCl_4$. The energies so calculated are generally in fairly close correspondence with those derived from the localized molecular-orbital model, showing that both are approximate ways of evaluating the same effect, namely the distortion of the electron cloud of the anion in the field of the cation. The modified Wolfsberg–Helmholz method gives coordinate bond energies in good agreement with experiment for halides as widely different as NaCl and CCl_4, but some of the assumptions of this treatment appear rather arbitrary and lacking in physical significance. Though the complexity of the Hartree–Fock method at present restricts its application to simple molecules containing atoms of relatively low atomic number, halides of Groups I–IV have been successfully treated by the extended Hückel method utilizing single-exponent, one-electron Slater orbitals[416]. Properties deduced in this way include the electronic structure, charge distribution, ionization potential and bond angle. It has even been suggested that the ionic nature of the halides favours such calculations by permitting only a minimum of orbital overlap; relatively crude atomic orbitals are not, therefore, incompatible with a reasonable molecular-orbital scheme.

3.3. THE HYDROGEN HALIDES

Historical Background

About AD 77, Pliny, in his *Naturalis Historiae*[417], described the purification of gold by heating it with salt, "misy" (iron or copper sulphate) and "schistos" (clay). Although this mixture would give off fumes of hydrogen chloride, attention was focused on the effect of the treatment on the metal and not on the nature of the effluvia. There is reference in the "Alchimia Geberi"—possibly written as early as the thirteenth century—to the preparation of aqua regia[418], while the preparation of hydrochloric acid itself was first reported in a fifteenth-century Italian manuscript[419]. Although there is no clear record, it is almost inconceivable that aqua regia or hydrochloric acid was not made earlier than this, since all the necessary ingredients were in the hands of early chemists. In Lavoisier's nomenclature, hydrochloric acid was designated "muriatic acid", a name still used in American industry[420], but, following Davy's investigations on chlorine[421], the name "hydrochloric acid" came into general usage. Reports of the preparation of hydrogen bromide and hydrogen iodide followed hard upon the identification of the corresponding elements. Thus, Balard showed

[416] J. W. Hastie and J. L. Margrave, *J. Phys. Chem.* **73** (1969) 1105.

[417] Pliny, *Naturalis Historiae*, Book 33, chapter 25 (first century A.D.).

[418] J. W. Mellor, *A Comprehensive Treatise on Inorganic and Theoretical Chemistry*, Vol. II, Longmans, Green and Co., London (1922).

[419] L. Reti, *Chymia*, **10** (1965) 11.

[420] W. R. Kleckner and R. C. Sutter, *Kirk-Othmer Encyclopedia of Chemical Technology*, 2nd edn., Vol. 11, p. 307. Interscience (1966).

[421] H. Davy, *Phil. Trans.* **100** (1810) 231; *Alembic Club Reprints* (1894) 9.

that the passage of a mixture of hydrogen and bromine vapour through a red-hot tube containing iron turnings gives rise to hydrogen bromide[422]. Courtois first prepared hydrogen iodide without recognizing its nature, but, in his historic memoir on iodine, Gay-Lussac showed that, although hydrogen and iodine do not react at ordinary temperatures, combination does occur in the vapour phase at elevated temperatures[423].

Preparation

1. Outline of methods[345,418,424–428]. The methods available for the preparation of the hydrogen halides, whether as gases or as aqueous solutions, are compared in Table 28. As indicated in the table, the methods fall into three general categories.

The Direct Combination of the Elements

The characteristics of this important and much studied reaction have been outlined in Section 2 (pp. 1223–4); more detailed accounts are to be found in references 345, 418 and 424–426. It has also been pointed out in preceding discussions that the relative ease of direct synthesis of the hydrogen halides varies with the bond energy of the HX molecule. Thus, the burning of hydrogen in chlorine, which takes place without catalytic agency, is a commercially important process for the production of hydrogen chloride. The less facile combination of hydrogen and bromine is most conveniently brought about in the presence of a catalyst (e.g. platinized asbestos, platinized silica gel or activated charcoal) at temperatures between 200° and 400°C. Energetically least favoured is the formation from its elements of hydrogen iodide, though, with the aid of a platinum catalyst typically heated to at least 300°C, this probably affords the best method of preparing hydrogen iodide in other than small quantities.

Reduction of the Parent Halogen by Agents other than Hydrogen

Several of these reactions have been referred to incidentally in Section 2. From the preparative standpoint the most important reducing agents are as follows. *Red phosphorus and water*, which furnishes a useful method of producing HBr or HI on the laboratory scale. *Various hydrocarbons*, the chlorination of which now provides much of the hydrogen chloride of commerce; the reaction of bromine or iodine with suitable unsaturated hydrocarbons—notably tetrahydronaphthalene—represents a valuable laboratory route to the corresponding hydrogen halide. *Sulphur dioxide*: this is used to convert aqueous bromine to hydrobromic acid in one of the stages leading to the commercial production of elemental bromine (see pp. 1137–8). *Hydrogen sulphide*: the reaction of this with an aqueous suspension of iodine has been used to prepare aqueous hydriodic acid.

[422] A. J. Balard, *Ann. Chim. Phys.* **32** (2) (1826) 337.

[423] B. Courtois, *Ann. Chim.* **88** (1) (1813) 304, 311; J. L. Gay-Lussac, *ibid.* **91** (1) (1814) 5.

[424] *Gmelins Handbuch der Anorganischen Chemie*, 8 Auflage: "Chlor", System-nummer 6, Verlag Chemie, Berlin (1927); "Brom", System-nummer 7, Verlag Chemie, Berlin (1931); "Iod", System-nummer 8, Verlag Chemie, Berlin (1933).

[425] *Gmelins Handbuch der Anorganischen Chemie*, 8 Auflage, "Chlor", System-nummer 6, Teil B—Lieferung 1, Verlag Chemie, Weinheim/Bergstr. (1968).

[426] Supplement to *Mellor's Comprehensive Treatise on Inorganic and Theoretical Chemistry*, Supplement II, Part I, Longmans, Green and Co., London (1956).

[427] G. Brauer, *Handbook of Preparative Inorganic Chemistry*, 2nd edn., Vol. 1, Academic Press (1963).

[428] *Inorganic Syntheses*, Vols. 1, 3 and 7, McGraw-Hill (1939–63).

TABLE 28. PREPARATION OF THE HYDROGEN HALIDES[a–f]

Method of preparation	HCl	HBr	HI
A. *Direct synthesis from the elements*:			
(a) without catalyst:			
slow combination at	100°C	250°C	500°C*
rapid combination at	200°C	500°C	—
(b) with catalyst:			
slow combination at	—	150°C	300°C*
rapid combination at	20°C	250°C	600°C*
(c) under the influence of visible light:			
rapid combination at	20°C	200°C	In ultraviolet only
Comments	Important method of synthesis on the large scale	Principal method of synthesis on the large scale	Used as a method of synthesis, usually on a relatively small scale
B. *Reduction of the parent halogen by agents other than hydrogen*:			
(a) with water vapour, rapid reaction at	500°C, over charcoal	500°C, over charcoal	700–800°C, over charcoal. HI solutions also produced by heating an aqueous suspension of I_2 with activated charcoal
(b) with sulphur dioxide	15–90°C, in aqueous solution	One of the stages in the commercial production of bromine (in aqueous solution)	Reaction occurs but has been little exploited
(c) with hydrogen sulphide	Occurs in the gas phase or in solution; S_2Cl_2 also tends to form	Reaction tends to give S_2Br_2 as well as HBr	Has been used to prepare hydriodic acid, H_2S being passed into an aqueous suspension of I_2
(d) with red phosphorus and water	Reaction violent and inconvenient	Provides a well-known laboratory method of preparing HBr	Provides a convenient laboratory method of preparation
(e) with hydrocarbons	HCl produced as a by-product of the chlorination of hydrocarbons, e.g. CH_4, C_6H_6. Commercially a major source of HCl	Reaction of Br_2 with benzene, toluene, naphthalene or petroleum (in the presence of a catalyst) gives HBr. Rapid reaction with tetrahydronaphthalene (tetralin) at 20°C is a useful method of preparing HBr in the laboratory	Few hydrocarbons react readily but the reaction of I_2 with *boiling* tetrahydronaphthalene has been used to prepare HI in the laboratory

C. *From inorganic or organic halides*			
(a) action of concentrated H_2SO_4 or H_3PO_4 on metal halide	"Salt-cake" process using H_2SO_4 and NaCl still of considerable industrial importance	H_2SO_4 causes some decomposition through oxidation of the HBr; H_3PO_4 gives satisfactory results	H_2SO_4 causes extensive oxidation of HI; H_3PO_4 can be used satisfactorily
(b) action of concentrated H_2SO_4 or other dehydrating agents on a concentrated aqueous solution of HX	A useful method of preparing gaseous HCl in the laboratory	H_2SO_4, $CaBr_2$ or P_2O_5 may be used as dehydrating agents	P_2O_5 may be used as a dehydrating agent
(c) by hydrolysis	E.g. PCl_3, $TiCl_4$. Hydrolysis of $MgCl_2$, $CaCl_2$ and NaCl has been utilized commercially. Hydrolysis of organic acid chloride useful for the preparation of DCl	E.g. PBr_3 (see above), $AlBr_3$. The hydrolysis of $AlBr_3$ has been used to prepare HBr of high purity	E.g. PI_3, SiI_4
(d) by reaction with H_2, hydrocarbons or H_2S	Under suitable conditions, H_2 or simple hydrocarbons react with many metal chlorides to give HCl; of little practical use as methods of preparation	Heavy-metal bromides reduced to anhydrous HBr by either H_2 or a hydrocarbon at elevated temperatures. H_2S may be used, as in $CdBr_2+H_2S \rightarrow CdS+2HBr$; of negligible practical significance	H_2S may be used, as in $2CuI+H_2S \rightarrow Cu_2S+2HI$; of negligible practical significance

* Extensive thermal decomposition of HI occurs at temperatures above 300°C, so that an equilibrium mixture of HI, H_2 and I_2 is obtained.

[a] *Gmelins Handbuch der Anorganischen Chemie*, 8 Auflage: System-nummer 6, "Chlor" (1927); "Chlor", Teil B, Lieferung 1 (1968); System-nummer 7, "Brom" (1931); System-nummer 8, "Iod" (1933).

[b] J. W. Mellor, *A Comprehensive Treatise on Inorganic and Theoretical Chemistry*, Vol. II, Longmans, Green and Co., London (1922).

[c] Supplement to *Mellor's Comprehensive Treatise on Inorganic and Theoretical Chemistry*, Supplement II, Part I, Longmans, Green and Co., London (1956).

[d] Z. E. Jolles (ed.), *Bromine and its Compounds*, Benn, London (1966).

[e] G. Brauer, *Handbook of Preparative Inorganic Chemistry*, 2nd edn., Vol. 1, Academic Press (1963).

[f] *Inorganic Syntheses*, Vols. 1, 3 and 7, McGraw-Hill (1939–63).

From Inorganic Halides

The displacement of the hydrogen halide from a metal halide by the action of a less volatile protic acid forms the basis of the classical method of producing hydrogen chloride on both the laboratory and the commercial scale. Thus, the reactions

$$NaCl + H_2SO_4 \rightarrow NaHSO_4 + HCl$$

and

$$NaHSO_4 + NaCl \rightarrow Na_2SO_4 + HCl$$

make up the so-called "salt-cake" process, which, from its inception as part of the Leblanc process, has been used for over a century in the manufacture of hydrogen chloride. By contrast, the oxidizing action of hot, concentrated sulphuric acid militates against the preparative success of analogous reactions involving a metal bromide or iodide, since the hydrogen halide tends to be oxidized to the parent halogen. Nevertheless, the use of syrupy phosphoric acid in place of sulphuric acid obviates this difficulty. The dehydration of concentrated aqueous solutions of the hydrohalic acids, either by chemical agents like sulphuric acid or phosphorus pentoxide or by thermal stripping, also serves as a convenient source of the gaseous hydrogen halide. The hydrolysis of certain halides has likewise been turned to advantage, as in the preparation of pure hydrogen bromide by the hydrolysis of aluminium bromide, or in the production of hydrogen chloride as a by-product resulting from the hydrolysis of magnesium chloride (in the Dow process for extracting magnesium from sea water). The hydrolysis of organic acid chlorides, e.g. C_6H_5COCl, with deuterium oxide affords one of the most convenient methods of preparing deuterium chloride[427].

2. Laboratory methods[345,424–430]. Although the aqueous hydrohalic acids are readily available from commercial sources, pure samples of a gaseous hydrogen halide are less easily obtained in this way. Of the various methods which have been suggested for the production of the gaseous compounds, the following are probably the most effective:

Hydrogen chloride: (i) the action of concentrated or pure sulphuric acid on sodium chloride; (ii) the action of concentrated sulphuric acid on concentrated (e.g. constant-boiling) hydrochloric acid.

Hydrogen bromide: (i) direct union of the gaseous elements in the presence of a catalyst; (ii) the action of bromine on a mixure of red phosphorus and water represented mainly by the equation

$$2P + 6H_2O + 3Br_2 \rightarrow 6HBr + 2H_3PO_3$$

(iii) the reaction of bromine with tetrahydronaphthalene at room temperature

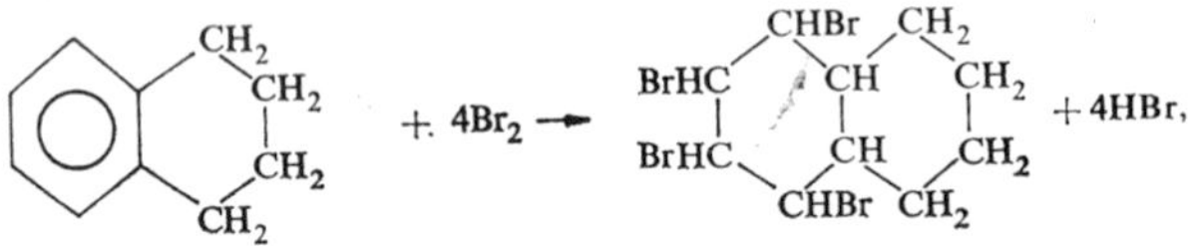

$$C_6H_4(CH_2)_4 + 4Br_2 \rightarrow (BrHC)_2(CH)_2(CHBr)_2(CH_2)_4 + 4HBr,$$

though this involves the loss of at least half of the bromine taken; (iv) dehydration of a concentrated aqueous solution of hydrobromic acid, e.g. with phosphorus pentoxide, concentrated sulphuric acid[431] or calcium bromide.

429 R. E. Dodd and P. L. Robinson, *Experimental Inorganic Chemistry*, pp. 197–200. Elsevier (1954).
430 R. H. Herber (ed.), *Inorganic Isotopic Syntheses*, Benjamin, New York (1962).
431 A. D. B. Sloan, *Chem. and Ind.* (1964) 574.

Hydrogen iodide: (i) direct union of the gaseous elements in the presence of a catalyst; (ii) the action of iodine on a mixture of red phosphorus and water; (iii) the iodination of boiling tetrahydronaphthalene; (iv) dehydration of a concentrated aqueous solution of hydriodic acid.

Practical details are given in references 424 and 426–431. Gaseous hydrogen chloride is freed from sulphur dioxide by scrubbing with sulphuric acid, which also serves as a drying agent. Hydrogen bromide is stripped of elemental bromine by passage over moist red phosphorus or through tetrahydronaphthalene; phosphorus acids and bromides are absorbed in a small quantity of water, while the best drying agents appear to be anhydrous calcium bromide and activated alumina. Iodine is removed from hydrogen iodide by condensation or by reaction with a metal iodide, either as a solid or as a saturated aqueous solution; anhydrous calcium iodide and phosphorus pentoxide are recommended drying agents. Samples of hydrogen halide of high purity, suitable, for example, for liquid-phase studies of their solvent properties[432], are obtained by repeated fractionation *in vacuo*. A guide to the purity of the material is provided by its vapour pressure, though the most sensitive test is said to be afforded by the specific conductivity of the liquid[432]. Of the three compounds, hydrogen iodide presents the most taxing problems of manipulation since it is decomposed by light, by many organic materials, e.g. tap-grease, and also by extensive glass surfaces, particularly in the presence of traces of water.

An aqueous solution of a hydrogen halide free from impurity is readily obtained by absorption of the purified gas in water. For exceptional purposes, very pure hydrochloric acid has been prepared by isothermal distillation of a concentrated solution of the acid in a desiccator containing a sample of scrupulously purified water[425]; an alternative procedure involves boiling the acid with small amounts of potassium permanganate to remove traces of bromine and iodine, followed by distillation *via* a quartz condenser[427]. Constant-boiling hydrobromic acid is conveniently prepared from potassium bromide, water and concentrated sulphuric acid, while constant-boiling hydriodic acid can be made by the reduction of iodine in aqueous suspension with either hydrogen sulphide or hypophosphorous acid[428]. Concentrated solutions of hydrobromic acid are somewhat susceptible to oxidation, and, unless protective measures are taken, the formation of bromine and polybromide ions causes pronounced darkening over a period of time. The tendency to suffer oxidation is even more pronounced in the case of hydriodic acid, the decomposition of which is favoured by the action of light and of impurities; storage in a well-stoppered, dark-glass bottle is recommended to preserve the acid in a satisfactory condition for extended periods. The addition of small amounts of hypophosphorous acid or of red phosphorus inhibits the oxidation of hydriodic acid. Ion-exchange methods have proved effective in the purification of hydrobromic and hydriodic acids[433].

Methods used to prepare isotopic variants of the hydrogen halides include[345,425–430] (i) the direct interaction of deuterium or tritium with the halogen; (ii) hydrolysis with deuterium oxide or tritium-enriched water of a suitable halide, examples being PX_3 (X = Cl, Br or I, either as the pure compound or formed *in situ* by the reaction of the elements), PCl_5, $SiCl_4$, $SOCl_2$, $MgCl_2$, $AlCl_3$ and C_6H_5COCl; (iii) the reaction of dry sodium chloride with deuterosulphuric acid; (iv) the reduction with deuterium or tritium of a metal halide,

432 M. E. Peach and T. C. Waddington, *Non-aqueous Solvent Systems* (ed. T. C. Waddington), p. 83. Academic Press (1965).

433 H. Irving and P. D. Wilson, *Chem. and Ind.* (1964) 653.

e.g. $CoCl_2$ and AgCl; and (v) exchange reactions, as between concentrated aqueous HCl and tritium-enriched water or sulphuric acid, or between tritium and HCl at elevated temperatures. Labelling by introduction of a radioactive halogen isotope has also been accomplished by one or other of these methods, e.g.

$$D_2SO_4+Na^{36}Cl \rightarrow NaDSO_4+D^{36}Cl$$ [428]

while thermal diffusion of the gaseous hydrogen halide affords the most effective means of separating the naturally occurring species $H^{35}Cl$ and $H^{37}Cl$ and likewise $H^{79}Br$ and $H^{81}Br$ (see p. 1150).

Manufacture[420,434,435]

Hydrogen chloride and hydrochloric acid are commodities of much commercial importance, world production being in excess of 3,000,000 metric tons of hydrogen chloride per annum. By contrast, hydrogen bromide and iodide are of small commercial significance, whether in the gaseous or in the aqueous states. The commercial production of hydrogen chloride awaited the Leblanc process for the manufacture of sodium carbonate, in which the gas was a co-product with salt-cake (sodium sulphate) of the first step, the reaction between salt and sulphuric acid. For a time, the gas was merely vented to the atmosphere, but legislation was enacted prohibiting its indiscriminate discharge, and thus necessitating its recovery. Although the Solvay process has supplanted the Leblanc process, the reaction between salt and sulphuric acid is still exploited because of the industrial demand for salt-cake (e.g. by the paper and glass industries) and hydrochloric acid. At the present time, four major processes are used commercially to produce hydrogen chloride and hydrochloric acid: (i) the salt–sulphuric acid process, (ii) the Hargreaves process, (iii) the direct synthetic method, and (iv) organic chlorination processes, which yield hydrogen chloride as a by-product. Usually two or more of these processes are operated in every major industrialized country of the world, depending upon the availability of raw materials and the size and diversity of the chemical industry of that country.

Salt–sulphuric Acid Process

The endothermic reactions

$$NaCl+H_2SO_4 \rightarrow NaHSO_4+HCl$$

and

$$NaCl+NaHSO_4 \rightarrow Na_2SO_4+HCl$$

take place at temperatures in the order of 150 and 540–600°C, respectively. In practice, temperatures are usually kept below 650°C to prevent the salt-cake from fusing. The reactions are carried out in various types of equipment: for example, a cast-iron retort; the so-called Mannheim furnace, a mechanical furnace consisting of a stationary circular muffle in the form of a basal concave pan and a domed cover, separated by a cylindrical mantle; the Laury-type furnace, which has a mobile oil-fired combustion chamber and, for the reaction vessel, a horizontal two-chambered rotating cylinder; or the Cannon fluid-bed

[434] V. A. Stenger and G. J. Atchison, *Kirk-Othmer Encyclopedia of Chemical Technology*, 2nd edn., Vol. 3, p. 767. Interscience (1964).

[435] A. W. Hart, M. G. Gergel and J. Clarke, *Kirk-Othmer Encyclopedia of Chemical Technology*, 2nd edn., Vol. 11, p. 857. Interscience (1966).

reactor, wherein sulphuric acid vapours are injected with the gases from a gas-fired combustion chamber into a fluid bed of salt. Many references in the patent literature relate to improvements in the efficiency of the process, by modifications of temperature, relative quantities of reactants, and design of reaction chambers. In some cases, potassium or calcium chloride has been employed in place of sodium chloride.

Hargreaves Process

Here the reactants are salt, sulphur dioxide, air and water vapour; the products are the same as in the salt–sulphuric acid process. The essential reaction

$$4NaCl+2SO_2+O_2+2H_2O \rightarrow 2Na_2SO_4+4HCl$$

is exothermic, and sufficient heat is evolved to maintain the process once the reactants have been brought to temperatures in the order of 430–540°C. In practical terms, a mixture of sulphur dioxide, steam and air is passed through stacks of salt briquets lying on perforated trays within a vertical chamber. Introduced in England in the last half of the nineteenth century, the Hargreaves process promised for a time to become a major source of salt-cake and hydrochloric acid, but various factors have led to its decline in more recent times.

Direct Synthetic Process

Hydrogen chloride in high concentration and of high purity is synthesized by the combustion of a controlled mixture of hydrogen and chlorine. The chlorine is derived, either wet or dry, from the electrolysis of brine; the hydrogen may come likewise from the electrolysis of brine, or from other sources, e.g. the hydrocarbon–steam reaction. The equipment varies in detail with the qualities of the raw materials and of the product desired, but the essential parts comprise a burner and combustion chamber, necessary control and safety devices, and facilities for processing the hydrogen chloride. The burner consists of a nozzle that injects the reactants into a vertical or horizontal combustion chamber; materials favoured for construction are silica, brick-lined steel, water-jacketed steel or water-cooled graphite. The reaction is initiated by igniting the hydrogen in a stream of air with a retractable torch burning either hydrogen or coal-gas; chlorine is then introduced into the burner to establish a hydrogen–chlorine flame. To ensure the formation of chlorine-free hydrogen chloride, a slight excess (typically 2–5%) of hydrogen is employed.

By-product Hydrogen Chloride

The chlorination of many organic materials produces hydrogen chloride as a by-product. In recent years the scale of chlorination processes of this sort has reached such proportions that in the United States, for example, more than 75% of all the hydrochloric acid of commerce is so derived. The chlorination of methane or benzene is typical of the reactions which generate hydrogen chloride as a by-product; the gas is also produced by pyrolysis of certain chlorinated organic compounds, as in the cracking of 1,2-dichloroethane or 1,1,2,2-tetrachloroethane:

$$ClCH_2{\cdot}CH_2Cl \longrightarrow CH_2{=}CHCl+HCl$$

$$Cl_2CH{\cdot}CHCl_2 \xrightarrow{\text{catalyst}} CCl_2{=}CHCl+HCl$$

The hydrogen chloride evolved from these reactions is liable to be contaminated with chlorine, air, organic chloro-compounds, excess reactants and moisture, depending upon the individual process. Accordingly, extensive treatment is commonly necessary to strip the hydrogen chloride of its impurities. As noted previously, hydrochloric acid is also a by-product of the Dow process for extracting magnesium from sea water.

Gas Treatment

After leaving the generating plant, the hydrogen chloride is treated in several steps. These may include three or more of the following: (i) *Removal of suspended solids* in settling chambers, centrifugal traps or cyclone-type separators. (ii) *Cooling* by heat-exchange methods. (iii) *Absorption* in water, usually achieved by counter-current flow of gas and liquid. (iv) *Desorption* of gaseous hydrogen chloride from the concentrated aqueous acid, usually brought about by thermal stripping at 75–130°C, though chemical agents, such as concentrated sulphuric acid or calcium chloride, have also been employed. (v) *Purification.* Methods used to purify by-product hydrogen chloride typically involve scrubbing of the gas with one or more non-aqueous solvents; alternatively the hydrogen chloride is removed from the gaseous products and purified by an absorption–desorption process using the aqueous acid as solvent. Activated charcoal has also been used to remove organic impurities. (vi) *Liquefaction* by compression and cooling of the anhydrous gas. The exact treatment depends on the composition and temperature of the raw gas and on the composition and nature of the end-product.

Materials of Construction

The choice of materials for the fabrication of equipment allowing the manipulation, storage and transport of hydrochloric acid is conditioned by the highly corrosive nature of the solution. Even with relatively dilute solutions, the corrosion problem is severe, and the situation is further complicated by the presence of contaminating materials, originating in the preparative process, which accelerate corrosion rates and shorten the lifetime of equipment. In the production of reagent-grade acid, the choice of materials is practically limited to tantalum, glass and impervious graphite. Where contamination can be tolerated, an economic balance must be struck between the life of the material and its cost. The following materials have found extensive use: rubber-lined carbon steel, high-silicon iron alloys (e.g. "Durichlor"), nickel alloys (e.g. "Chlorimet" and "Hastelloy"), copper alloys, tantalum, titanium, acid-proof brick, chemical stoneware or porcelain, glass, certain rubbers, plastics (e.g. polyvinyl chloride, polyethylene, polystyrene, polytetrafluoroethylene and glass-reinforced polyesters), baked carbon, graphite or impregnated carbon and graphite. The limitations on the use of metals do not apply to the anhydrous gas; only in the presence of moisture do the corrosion problems become severe. In fact, no ignition point has been found with anhydrous hydrogen chloride in contact with steel even at 760°C, and the gas or liquid is usually stored in steel vessels.

Uses

Hydrochloric acid is widely used industrially in such diverse fields as the pickling of metals for scale removal, the reactivation of bone charcoal and carbon in sugar-refining operations, the production of glucose and corn sugar from starch, the production of chloroprene and vinyl chloride—important intermediates in the manufacture of synthetic rubber—and the activation of petroleum wells. Other applications are found in the production of alumina for the manufacture of aluminium; in the Dow process for isolating magnesium from sea water; in many extractive metallurgical processes for treating high-grade ores, among which are those yielding radium, vanadium, tungsten, tantalum, manganese and germanium; in the production of numerous organic chloro-derivatives and metal chlorides; in the production of phosphoric acid from phosphate rock; as a catalyst in organic syntheses; and as an analytical reagent. In 1963 it has been estimated that, of the hydrochloric acid consumed, nearly 50% went to the synthesis of organic chemicals, 17% to the production of metals, 18% to the activation of petroleum wells, 7% to metal and industrial cleaning operations, 5% to the production of inorganic chemicals, and 4% to food-processing. However, oversupply in recent years, combined with the increasing pressure to reduce pollution, has stimulated the search for new industrial outlets for the acid. Two developments which may assume importance in this respect are the production of chlorine from hydrochloric acid, either by electrolysis or by chemical oxidation (see Section 2), and "oxyhydrochlorination" processes, whereby hydrogen chloride, air or oxygen, and a hydrocarbon give rise to chlorinated organic products and water, e.g.

$$C_6H_6+HCl+\frac{1}{2}O_2 \rightarrow C_6H_5Cl+H_2O$$

Hydrobromic and Hydriodic Acids

Hydrogen bromide gas is prepared commercially by burning a mixture of hydrogen and bromine vapour. The gas is passed through hot activated charcoal to remove free bromine and is either liquefied or absorbed in water. The major use of aqueous hydrobromic acid is in the manufacture of inorganic bromides and of certain organic bromo-derivatives. Neither hydrogen iodide nor hydriodic acid appears to be produced on the large scale or to find significant use outside the laboratory.

Physical Properties[289,345,424–426,432,436]

1. General physical characteristics. Quantitative physical properties of the anhydrous hydrogen halides are summarized in Table 29, which illustrates the relatively uniform pattern of behaviour in the series HCl, HBr, HI; in many respects, however, this uniformity is not shared by hydrogen fluoride (see Chapter 25). Thus, unlike hydrogen fluoride, which boils at 19·5°C, hydrogen chloride, bromide and iodide all boil well below room temperature, the order of the boiling points being HF > HCl < HBr < HI. The strongly hydrogen-bonded fluoride apart, the relative volatility of the compounds reflects the strengthening of the van der Waals' forces which accompanies the increasing number of electrons, and hence the polarizability, in the series Cl, Br, I. Since the dipole moment of the molecules decreases progressively from HF to HI, dipole–dipole interactions presumably play a less important part than do dispersion or dipole–induced-dipole interactions in the intermolecular binding of the heavier hydrogen halides. On the evidence of the boiling points, Trouton constants and spectroscopic and other properties, hydrogen chloride, bromide and iodide differ from the fluoride in showing no signs of appreciable association in the liquid or gaseous states. The HX molecules persist throughout the solid, liquid and gaseous phases, though, as discussed below, hydrogen-bonding probably makes a significant contribution to the intermolecular binding of the solid. The pattern of behaviour thus exhibited by the hydrogen halides finds a close parallel in the properties of the hydrides H_2O, H_2S, H_2Se and H_2Te and, to a lesser degree, in those of the corresponding Group V hydrides, each series being distinguished by the anomalous properties occasioned by the associative tendencies of the first member.

[436] T. C. Waddington, *MTP International Review of Science: Inorganic Chemistry Series One*, Vol. 3 (ed. V. Gutmann), p. 85, Butterworths and University Park Press (1972).

TABLE 29. PHYSICAL PROPERTIES OF THE HYDROGEN HALIDES

Property	HCl	HBr	HI
Molecular weight	36·461[a]	80·912[a]	127·912$_5$ [a]
Gaseous molecules			
Electronic ground state configuration $\sigma^2\pi^4$	$^1\Sigma^+$	$^1\Sigma^+$	$^1\Sigma^+$
Observed transitions in the ultraviolet,[b] $\nu_{0,0}$(cm^{-1})	$\sigma^2\pi^3\sigma^* \leftarrow \sigma^2\pi^4$	$\sigma^2\pi^3\sigma^* \leftarrow \sigma^2\pi^4$	$\sigma^2\pi^3\sigma^* \leftarrow \sigma^2\pi^4$
$^1\Pi \leftarrow {}^1\Sigma^+$(X) C ← X	77,540	70,542	62,320
B ← X	75,134	67,084	56,750
A ← X	Continuous absorption starting at 44,000 cm^{-1} with maximum at ~65,000 cm^{-1} (ϵ_{max} 890)	Continuous absorption starting at ~35,000 cm^{-1} with maximum at ~55,000 cm^{-1} (ϵ_{max} 525)	Continuous absorption starting at 27,500 cm^{-1} with maximum at ~48,000 cm^{-1} (ϵ_{max} 174)
Ground state properties			
Internuclear distance, r_e(Å)	1·2746[c,d]	1·4145[d]	1·6090[d]
Dipole moment, D	1·07[e]	0·828[e]	0·448[e]
Molecular quadrupole moment (× 10^{26} esu cm^2)	5·8[f]	5·5[f]	—
Vibrational frequency, ω_e(cm^{-1})	H^{35}Cl, 2990·7027;[c] H^{37}Cl, 2988·4799[c] D^{35}Cl, 2144·77;[d] D^{37}Cl, 2140·20[c]	H^{79}Br, 2649·3876;[g] H^{81}Br, 2648·9752[g] D^{79}Br, 1885·33;[h] D^{81}Br, 1884·75[h]	H^{127}I, 2308·901[i] D^{127}I, 1639·655[i]
Anharmonic vibrational constant, $\omega_e x_e$(cm^{-1})	H^{35}Cl, 52·5707;[c] H^{37}Cl, 52·5166[c] D^{35}Cl, 26·92;[c,d] D^{37}Cl, 26·99[c,d]	H^{79}Br, 45·2316;[g] H^{81}Br, 45·2175[g] D^{79}Br, 22·73;[h] D^{81}Br, 22·72[h]	H^{127}I, 38·9810[i] D^{127}I, 19·873[i]
Force constant, k_e(mdyne/Å)	5·161[c]	4·114[g]	3·138[i]
Dissociation energy:	*kcal* / *eV*	*kcal* / *eV*	*kcal* / *eV*
D_0°	102·24[j] / 4·433[j]	86·64[j] / 3·757[j]	70·43[j] / 3·054[j]
$D^\circ_{298^\circ K}$	103·16[j] / 4·473[j]	87·54[j] / 3·796[j]	71·32[j] / 3·093[j]
Thermodynamic properties			
ΔH_f° at 298°K (kcal mol^{-1})	−22·063[j]	− 8·70[j]	+6·30[j]
ΔG_f° at 298°K (kcal mol^{-1})	−22·778[j]	−12·77[j]	+0·38[j]
S° at 298°K (cal deg^{-1} mol^{-1})	44·643[j]	47·463[j]	49·350[j]

Ionization potentials:[k]	*kcal*	*eV*	*kcal*	*eV*	*kcal*	*eV*
$HX^+(^2\Pi_{3/2}) \leftarrow HX(^1\Sigma^+)$	293·8	12·74	269·1	11·67	239·4	10·38
$HX^+(^2\Pi_{1/2}) \leftarrow HX(^1\Sigma^+)$	295·7	12·82	276·7	12·00	254·8	11·05
$HX^+(^2\Sigma^+) \leftarrow HX(^1\Sigma^+)$	374·3	16·23	352·6	15·29	319·4	13·85
Properties of the condensed phases	°*K*	°*C*	°*K*	°*C*	°*K*	°*C*
Boiling point[l]	188·11	− 85·05	206·43	−66·73	237·80	−35·36
Melting point[l]	158·94	−114·22	186·28	−86·88	222·36	−50·80
Liquid range at normal pressures (°C)	29·17[l]		20·15[l]		15·44[l]	
Heat of vaporization at boiling point, ΔH_{vap}(kcal mol^{-1})	3·860[l]		4·210[l]		4·724[l]	
Entropy of vaporization at boiling point (Trouton constant) (cal deg^{-1} mol^{-1})	20·5[l]		20·39[l]		19·86[l]	
Heat of fusion (kcal mol^{-1})	0·476[l]		0·575[l]		0·686[l]	
	°*K*	°*C*	°*K*	°*C*	°*K*	°*C*
Critical temperature	324·7[c]	51·54[c]	363·1[m]	89·9[m]	424·1[m]	150·9[m]
Critical pressure (atm)	82·07[c]		84[m]		82[m]	
Critical density (g cm^{-3})	0·424[c]		—		—	
Vapour pressure, p (mm Hg)						
Solid	$\log p = -(1114/T) - 1{\cdot}285 \log T - 0{\cdot}0009467T + 12{\cdot}005$ (130–160°K)[h]		$\log p = -(1250/T) - 0{\cdot}2766 \log T - 0{\cdot}004084T + 10{\cdot}488$[n]		$\log p = -(1406/T) - 0{\cdot}377 \log T - 0{\cdot}003167T + 10{\cdot}493$[h]	
Liquid	$\log p = -(905{\cdot}53/T) + 1{\cdot}75 \log T - 0{\cdot}005077T + 4{\cdot}65739$[h]		$\log p = -(1338/T) - 4{\cdot}672 \log T + 20{\cdot}179$[n]		$\log p = -(1636/T) - 7{\cdot}111 \log T + 0{\cdot}002293T + 26{\cdot}119$[h]	

TABLE 29 (*cont.*)

Property	HCl	HBr	HI
Crystal structure Each hydrogen halide exists in polymorphic crystalline forms, of which that stable at low temperatures has been most fully characterized. The incomplete and not always concordant evidence of various physical measurements on this form points to the presence of zigzag planar or nearly planar chains: X X H H H H X X	*α-phase* (stable below 98·38°K)[o] Face-centred orthorhombic lattice, space group $Bb2_1m$. 4 molecules per unit cell. Unit cell dimensions for DCl at 92·4°K (Å) *a* *b* *c* 5·082 5·410 5·826 d(D–Cl) = 1·25 ± 0·02 Å d(Cl · · · Cl) = 3·688 ± 0·001 Å ∠(Cl · · Cl · · Cl) = 93° 3′ Ferroelectric[p] *β-phase* (stable above 98·38°K)[q] Face-centred cubic lattice, space group $Fm3m$–O_h^5. 4 molecules per unit cell. Lattice parameter a_0 = 5·482 Å for DCl at 118·5°K. d(D–Cl) = 1·17 ± 0·04 Å	*Low-temperature form(s)* (stable below *ca.* 110°K)[r] Face-centred orthorhombic lattice. 4 molecules per unit cell. Unit cell dimensions for DBr (Å) *Temp.* *a* *b* *c* 84°K 5·44 5·614 6·120* 107°K 5·559 5·649 6·120† d(D–Br) ~ 1·38 ± 0·03 Å ∠(Br · · Br · · Br) = 91° 41′* or 90° 56′† * Phase isomorphous with α-phase of DCl, space group $Bb2_1m$. Ferroelectric.[p] † Disordered phase stable between 86° and 118°K, space group *Bbcm* *High-temperature form* (stable above *ca.* 110°K)[h] Face-centred cubic lattice 4 molecules per unit cell Lattice parameter a_0 = 5·76–5·78 Å	*Low-temperature form* (stable below *ca.* 120°K)[s] Face-centred tetragonal lattice. 4 molecules per unit cell. Unit cell dimensions at 4·2°K (Å) *a* *c/a* 5·987 1·082 *High-temperature form* (stable above *ca.* 120°K)[s] Face-centred cubic lattice. 4 molecules per unit cell. Lattice parameter at 130°K, a_0 = 6·246 Å
Halogen nqr data[t] Coupling constants, e^2Qq(MHz)	$H^{35}Cl$, 67·51 $D^{35}Cl$, 67·41 $D^{37}Cl$, 53·14 $T^{35}Cl$, 67·21 $T^{37}Cl$, 52·96	$H^{79}Br$, 535·44 $H^{81}Br$, 447·76 $D^{79}Br$, 530·58 $D^{81}Br$, 443·39 $T^{79}Br$, 528·75 $T^{81}Br$, 441·26	$H^{127}I$, 1831·07 $D^{127}I$, 1823·38 $T^{127}I$, 1822·61
Vibrational spectra	Infrared; [u1–u5] Raman[u4,u6]	Infrared;[u1–u4,u7] Raman[u4,u6,u7]	Infrared[u1]
Thermal properties Heat capacity, C_p(cal mol^{-1} deg^{-1}) Solid Liquid Gas	 0·340–11·69 (10·76–147·4°K)[e] 14·41–14·44 (163·0–173·4°K)[e] 6·959–9·388 (100–6000°K)[x]	 1·831–12·32 (15·72–182·09°K)[v] 14·20–14·31 (189·93–205·11°K)[v] 6·959–9·422 (100–6000°K)[x]	 2·745–12·04 (17·11–218·00°K)[w] 14·34–14·15 (227·05–236·15°K)[w] 6·959–9·511 (100–6000°K)[x]

Thermal conductivity (cal sec^{-1} cm^{-2} deg^{-1} cm^{-1})			
Solid	—	0·625–1·321 × 10^{-3} (78·0–142·2°K)[y]	—
Liquid	8·71–1·66 × 10^{-4} (199·2–337·2°K; pressure 30–200 bar)[z]	—	—
Gas	2·18–6·73 × 10^{-5} (200–600°K; p = 210–260 mm Hg)[aa,bb]	1·24–4·29 × 10^{-5} (200–600°K; p = 130–250 mm Hg)[aa]	—
Coefficient of cubical expansion (× 10^4 per °C)			
Solid	6·2[c]	5·4[cc]	5 (4·2–180°K)[s]
Liquid	27·8–24·0 (159–188°K)[c]	21·19 (206·1–227·2°K)[h]	14·4 (237·8°K)[dd]
Gas	37·69–36·61 (273–373°K, p = 0–1 atm)[c]	—	—
Mechanical properties			
Density, d(g cm^{-3})			
Solid	1·54–1·47 (0–107°K)[h]	2·96–2·55 (0–150°K)[h,ee]	3·67–3·05 (4·2–222°K)[s,hh]
Liquid	1·267–0·630 (160·01–319·35°K)[c]	$d = 2{\cdot}757 - 1{\cdot}238 \times 10^{-3}T - 7{\cdot}01 \times 10^{-6}T^2$ (T = 193–333°K)[ff]	2·860–2·7862 (223·3–240·4°K)[hh]
Gas at N.T.P.	1·6392 × 10^{-3} [c]	3·6445 × 10^{-3} [gg]	5·7888 × 10^{-3} [hh]
Compressibility (× 10^6 per bar)			
Solid	36·0–7·0 (0–20 kbar, 130°K)[ee]	36·5–7·4 (0–20 kbar, 150°K)[ee]	—
Liquid	380–53 (100–2000 bar, 293°K)[c]	—	—
Viscosity, η(centipoise)			
Liquid	0·566–0·058 (160·8–318·2°K)[ii]	0·874–0·816 (186·8–199·4°K)[ii]	1·418–1·298 (223·2–236·4°K)[ii]
Gas	0·01313–0·02530 (273·2–523·2°K)[h]	0·01835–0·02365 (291·9–373·4°K)[h]	0·01731–0·03228 (273·2–524·2°K)[ii]
Surface tension of liquid, γ(dyne cm^{-1})	26·912–22·409 (168·5–192·6°K)[jj]	30·191–25·399 (182–204°K)[gg]	29·06–26·96 (225·3–236·5°K)[hh]

TABLE 29 (*cont.*)

Property	HCl	HBr	HI
Electrical, optical and magnetic properties			
Dielectric constant, ϵ			
Solid	*α-phase*: 2·6–4·3.[c] Slight dielectric absorption. *β-phase*: Static dielectric constant ϵ_0 = 24·7–15·2 (98·4–158·9°K)[c] Dielectric absorption gives complex dielectric constant.	Static dielectric constant ϵ_0 = 200–8·23 (70–187°K; 20–1·3 × 10^8 Hz).[kk,ll] Dielectric absorption gives complex dielectric constants; behaviour very sensitive to phase transitions.	Static dielectric constant ϵ_0 = 28–3 (63–222°K; 20–8·6 × 10^7 Hz).[kk,ll] Dielectric absorption gives complex dielectric constants; behaviour very sensitive to phase transitions.
Liquid	14·30–4·6 (158–301°K)[c] Some dielectric absorption	7·33–3·82 (188–298°K)[ll]	3·55–2·90 (222·6–295°K)[ll]
Gas	ϵ = 1·007452–1·001182 (201·4–588·8°K, p = 1 atm)[c,h] ϵ = 1·00526–1·24474 (p = 0·993–39·936 atm, T = 298°K)[c] ϵ = 1·00476–1·18271 (p = 0·995–40·210 atm, T = 323°K)[c]	ϵ = 1·004545–1·000943 (217·5–599·1°K, p = 1 atm)[h]	ϵ = 1·002687–1·0009706 (244·5–612·2°K, p = 1 atm)[hh]
Specific conductivity, k (ohm^{-1} cm^{-1})			*Conductivity* × 10^{10} *Temp. range*, °K | ν = 60 *kc/s* | ν = 300 *c/s*
Solid	0·13–1·3 × 10^{-9} (87·1–99·1°K)[c] <10^{-10} (99·1–159·0°K)[c]	—	85·7–118·5 | 6·8–1·7 | <0·1 126·3–200·2 | <1 | <0·1 209·6–222·4 | 3·4–4·2 | <0·1–0·6
Liquid	3·5 × 10^{-9} (188°K)[c,h] Temp. coefficient = −6·5 × 10^{-11} (ohm^{-1} cm^{-1} deg^{-1})	1·4 × 10^{-9} (189°K)[mm]	222·6–236·3 | 4·2–13 | 1·0–3·4 (ref. hh)
Refractive index, n			
Liquid	n_D = 1·254–1·252 (283–290°K)[jj]	n_D = 1·325 (283°K)[gg]	n_D = 1·466 (289·7°K)[hh]
Gas at N.T.P.	λ = 6562·9–2302·83 Å n = 1·0004434–1·000523525[c,h]	λ = 6707·85–4799·9 Å n = 1·00060752–1·00062160[gg]	λ = 6707·5–4799·9 Å n = 1·00091087–1·00093900[hh]
Molar refraction (cm^3 mol^{-1}), D-line	6·666[h]	9·161[nn]	13·73[nn]

Diamagnetic susceptibility, $\chi(\times 10^6$ c.g.s. units $mol^{-1})$			
Solid	—	—	$-47{\cdot}2$ (195°K)[oo]
Liquid	$-23{\cdot}6$ to $-22{\cdot}6$ (195–273°K)[oo]	$-33{\cdot}9$ to $-32{\cdot}9$ (195–273°K)[oo]	$-48{\cdot}3$ to $-47{\cdot}7$ (233–281°K)[oo]
Gas	-25[c]	—	—
^1H nmr chemical shift of gas (ppm relative to CH_4)	$+0{\cdot}45$[pp]	$+4{\cdot}35$[pp]	$+13{\cdot}25$[pp]

[a] *Table of Atomic Weights*, 1969, IUPAC Commission on Atomic Weights; *Pure Appl. Chem.* **21** (1970) 95.

[b] G. Herzberg, *Molecular Spectra and Molecular Structure*. I. *Spectra of Diatomic Molecules*, 2nd edn., pp. 534–540. Van Nostrand (1950).

[c] *Gmelins Handbuch der Anorganischen Chemie*, 8 Auflage, "Chlor", Teil B (1968).

[d] *Tables of Constants and Numerical Data*, No. 17, *Spectroscopic Data relative to Diatomic Molecules* (general ed. B. Rosen), Pergamon (1970).

[e] A. L. McClellan, *Tables of Experimental Dipole Moments*, Freeman, San Francisco (1963); F. A. Van Dijk and A. Dymanus, *Chem. Phys. Letters*, **5** (1970) 387.

[f] S. Weiss and R. H. Cole, *J. Chem. Phys.* **46** (1967) 644.

[g] D. H. Rank, U. Fink and T. A. Wiggins, *J. Mol. Spectroscopy*, **18** (1965) 170.

[h] *Mellor's Comprehensive Treatise on Inorganic and Theoretical Chemistry*, Supplement II, Part I, Longmans, Green and Co. (1956).

[i] S. C. Hurlock, R. M. Alexander, K. N. Rao, N. Dreska and L. A. Pugh, *J. Mol. Spectroscopy*, **37** (1971) 373.

[j] *National Bureau of Standards Technical Note* 270–3, U.S. Govt. Printing Office, Washington (1968); *Codata Bulletin*, International Council of Scientific Unions, Committee on Data for Science and Technology (1970).

[k] H. J. Lempka, T. R. Passmore and W. C. Price, *Proc. Roy. Soc.* **A304** (1968) 53.

[l] F. D. Rossini, D. D. Wagman, W. H. Evans, S. Levine and I. Jaffe, *Selected Values of Chemical Thermodynamic Properties*, Circular 500, National Bureau of Standards, Washington (1952).

[m] G. Woolsey, *J. Amer. Chem. Soc.* **59** (1937) 1577.

[n] J. R. Bates, J. O. Halford and L. C. Anderson, *J. Chem. Phys.* **3** (1935) 531.

[o] E. Sándor and R. F. C. Farrow, *Nature*, **213** (1967) 171.

[p] S. Shimaoka, N. Niimura and S. Hoshino, *Acta Cryst.* **A25** (1969) S50; S. Hoshino, S. Shimaoka and N. Niimura, *Phys. Rev. Letters*, **19** (1967) 1286.

[q] E. Sándor and R. F. C. Farrow, *Nature*, **215** (1967) 1265.

[r] E. Sándor and M. W. Johnson, *Nature*, **217** (1968) 541.

[s] F. A. Mauer, C. J. Keffer, R. B. Reeves and D. W. Robinson, *J. Chem. Phys.* **42** (1965) 1465.

[t] T. Tokuhiro, *J. Chem. Phys.* **47** (1967) 109.

[u] (1) D. F. Hornig and W. E. Osberg, *J. Chem. Phys.* **23** (1955) 662; G. L. Hiebert and D. F. Hornig, *ibid.* **26** (1957) 1762; **27** (1957) 752, 1216; **28** (1958) 316; (2) A. Anderson, H. A. Gebbie and S. H. Walmsley, *Mol. Phys.* **7** (1964) 401; (3) L.-C. Brunel and M. Peyron, *Compt. rend.* **264C** (1967) 821, 930; (4) R. Savoie and A. Anderson, *J. Chem. Phys.* **44** (1966) 548; (5) J. Blanchard, L.-C. Brunel and M. Peyron, *Compt. rend.* **272B** (1971) 366; (6) M. Ito, M. Suzuki and T. Yokoyama, *J. Chem. Phys.* **50** (1969) 2949; (7) E. L. Pace, *Spectrochim. Acta*, **27A** (1971) 491.

[v] W. F. Giauque and R. Wiebe, *J. Amer. Chem. Soc.* **50** (1928) 2193.
[w] W. F. Giauque and R. Wiebe, *J. Amer. Chem. Soc.* **51** (1929) 1441.
[x] *JANAF Thermochemical Tables*, The Dow Chemical Company, Midland, Michigan (1960–8).
[y] A. Eucken and E. Schröder, *Ann. Phys.* **36**(5) (1939) 609.
[z] H. Ziebland and D. P. Needham, *Proceedings of the 4th Symposium on Thermophysical Properties* (ed. J. R. Moszynski), p. 296, American Society of Mechanical Engineers, New York (1968).
[aa] E. U. Franck, *Z. Elektrochem.* **55** (1951) 636.
[bb] A. K. Barua, A. Manna and P. Mukhopadhyay, *J. Chem. Phys.* **49** (1968) 2422; C. E. Baker and R. S. Brokaw, *ibid.* **43** (1965) 3519.
[cc] W. Biltz and A. Lemke, *Z. anorg. Chem.* **203** (1932) 321.
[dd] S. J. Yosim, *J. Chem. Phys.* **40** (1964) 3069.
[ee] J. W. Stewart, *J. Chem. Phys.* **36** (1962) 400.
[ff] W. G. Strunk and W. H. Wingate, *J. Amer. Chem. Soc.* **76** (1954) 1025.
[gg] *Gmelins Handbuch der Anorganischen Chemie*, 8 Auflage, "Brom" (1931).
[hh] *Gmelins Handbuch der Anorganischen Chemie*, 8 Auflage, "Iod" (1933).
[ii] *Landolt–Börnstein Tables*, II Band, *Eigenschaften der Materie in ihren Aggregatzuständen*, 5 Teil, Bandteil a, "Transportphänomene I" (1969).
[jj] *Gmelins Handbuch der Anorganischen Chemie*, 8 Auflage, "Chlor" (1927).
[kk] P. P. M. Groenewegen and R. H. Cole, *J. Chem. Phys.* **46** (1967) 1069; R. H. Cole and S. Havriliak, jun., *Discuss. Faraday Soc.* **23** (1957) 31.
[ll] *Landolt–Börnstein Tables*, II Band, *Eigenschaften der Materie in ihren Aggregatzuständen*, 6 Teil, "Elektrische Eigenschaften I" (1959).
[mm] M. E. Peach and T. C. Waddington, *Non-aqueous Solvent Systems* (ed. T. C. Waddington), p. 85. Academic Press (1965).
[nn] N. Bauer and K. Fajans, *J. Amer. Chem. Soc.* **64** (1942) 3023.
[oo] *Landolt–Börnstein Tables*, II Band, *Eigenschaften der Materie in ihren Aggregatzuständen*, 10 Teil, "Magnetische Eigenschaften II" (1967).
[pp] J. A. Pople, W. G. Schneider and H. J. Bernstein, *High-resolution Nuclear Magnetic Resonance*, p. 90, McGraw-Hill (1959).

All the hydrogen halides are colourless. The gases fume strongly in moist air and have penetrating, acrid odours. In the absence of a polar solvent, the properties are exclusively those of molecular compounds, but dissolution in a solvent like water brings about dissociation to give the solvated proton and halide ion. The hydrogen halides are thus acceptor molecules with a marked capacity, in Mulliken's terminology[437], to form "inner complexes":

$$\underset{\text{Polar solvent}}{S + HX} \rightleftharpoons \underset{\text{Outer complex}}{S,HX_{solv}} \rightleftharpoons \underset{\text{Inner complex}}{[SH]^{+}X^{-}{}_{solv}}$$

This distinctive property is related to the relative proximity to the electronic ground state of excited states of the type $H^{+}X^{-}$ and to the large solvation energy accruing to the proton. By contrast with hydrogen fluoride, hydrogen chloride, bromide and iodide are all strong acids in aqueous solution.

Table 29 also includes recommended values of spectroscopic properties, vapour pressures, heat capacities and thermodynamic functions for the hydrogen halides. Satisfactory agreement is found between the thermodynamic functions derived *via* the third-law method from measurements of heat capacity and those calculated from spectroscopic data, so that there is no evidence that the solids retain residual entropy at limiting low temperatures. Whereas the standard entropies vary but little from one gaseous molecule to the other, the bond energy sequence HCl > HBr > HI reflects the marked decline in $-\Delta H_f$. It is this enthalpy term which dictates the relative stabilities of the gaseous hydrogen halides with respect to thermal dissociation; the results of Table 30 demonstrate the dramatic

TABLE 30. THERMAL STABILITY OF THE GASEOUS HYDROGEN HALIDES[a]

Temperature (°K)	HCl		HBr		HI	
	% dissoc.*	log K_p	% dissoc.*	log K_p	% dissoc.*	log K_p
100	—	48·709	—	19·520	—	−9·920
200	—	24·624	—	12·011	—	−2·599
300	—	16·596	—	9·290	—	−0·247
500	$8{\cdot}4\times10^{-4}$	10·151	0·11	5·927	22·9	+1·055
1000	0·23	5·265	2·58	3·155	30·1	+0·732
1500	1·53	3·615	7·27	2·212	33·1	+0·613
2000	3·89	2·785	11·9	1·737	34·6	+0·554
3000	9·58	1·950	19·0	1·260	36·1	+0·496
4000	14·7	1·529	23·6	1·020	36·9	+0·466
6000	22·0	1·097	29·1	0·774	37·8	+0·431

* At a pressure of 1 atm.
[a] *JANAF Thermochemical Tables*, The Dow Chemical Company, Midland, Michigan (1960–8).

change of these stabilities, from hydrogen chloride, which is dissociated to the extent of no more than 9·6% even at 3000°K, to hydrogen iodide, 23% of which is dissociated at 500°K. Only the slowness of the decomposition, in the absence of a catalyst, keeps hydrogen iodide from being about 50% dissociated at room temperature.

Details of the physical properties of the deuterium halides have not generally been included in Table 29, but are to be found in references 425 and 426.

[437] R. S. Mulliken and W. B. Person, *Molecular Complexes*, Wiley-Interscience, New York (1969).

2. Physical properties of the gaseous molecules. The $^1\Sigma^+$ electronic ground state of the hydrogen halide molecules involves the configuration $(ns\sigma)^2(np\sigma)^2(np\pi)^4$ for the valence electrons[438]. Molecular-orbital calculations indicate that the $(ns\sigma)^2$ and $(np\pi)^4$ electrons are mainly atomic and that the formation of the H–X bond is principally due to the $(np\sigma)^2$ electrons (see Fig. 24), though a small contribution probably derives from interaction with vacant hydrogen orbitals (e.g. $2p_\pi$). The electronic states of the molecules have been studied theoretically in great detail by Mulliken[438].

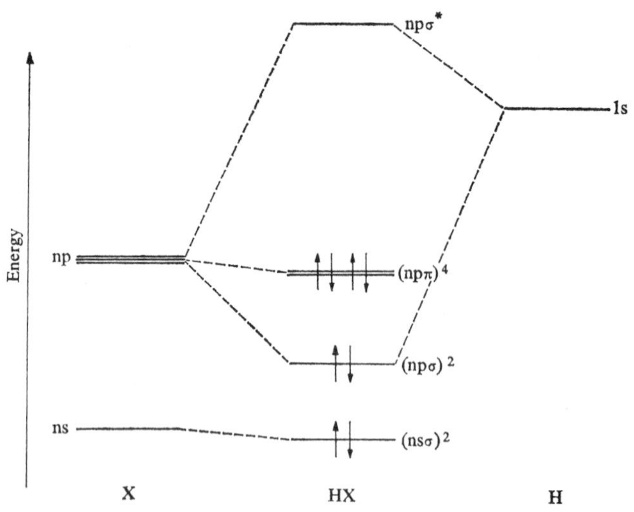

FIG. 24. Energy level diagram showing the molecular orbitals of a hydrogen halide molecule.

The electronic spectra of the molecules, studied in the ultraviolet region both in absorption and emission[425,426,439], exhibit the following features: (i) a continuous absorption band at low frequencies attributed to the transition $^3\Pi,^1\Pi\ (np\sigma)^2(np\pi)^3(np\sigma^*) \leftrightarrow {}^1\Sigma^+\ (np\sigma)^2(np\pi)^4$; (ii) banded systems represented as $^3\Pi,^1\Pi\ (np\sigma)^2(np\pi)^3(n+1)s\sigma \leftrightarrow {}^1\Sigma^+\ (np\sigma)^2(np\pi)^4$, where the excited states involve JJ-like coupling between the outer σ-electron and the HX^+ core; (iii) band systems ascribed to transitions of the type $^1\Sigma^+\ (np\sigma)(np\pi)^4(n+1)s\sigma \leftrightarrow {}^1\Sigma^+\ (np\sigma)^2(np\pi)^4$; and (iv) bands arising from transitions involving low-lying states of the molecular ions HX^+. An analysis of such measurements and of the photoelectron spectra[439] of the molecules has led to the construction of potential energy diagrams such as that given in Fig. 25 for HCl. The disposition and form of the potential energy curves do not favour a banded spectrum at low frequencies analogous to that observed for Cl_2, Br_2 or I_2. Accordingly, no accurate value for the dissociation energy of the HX molecule can be obtained directly from spectroscopic measurements. Instead, the most reliable value of this parameter, quoted in Table 29, is derived from thermochemical data. Reference to the electronic and photoelectron spectra associated with the molecular ions HX^+ has provided the information about these species which is presented in Table 31.

438 R. S. Mulliken, *Rev. Mod. Phys.* **4** (1932) 1; *Phys. Rev.* **50** (1936) 1017; *ibid.* **51** (1937) 310.

439 H. J. Lempka, T. R. Passmore and W. C. Price, *Proc. Roy. Soc.* **A304** (1968) 53 and references cited therein.

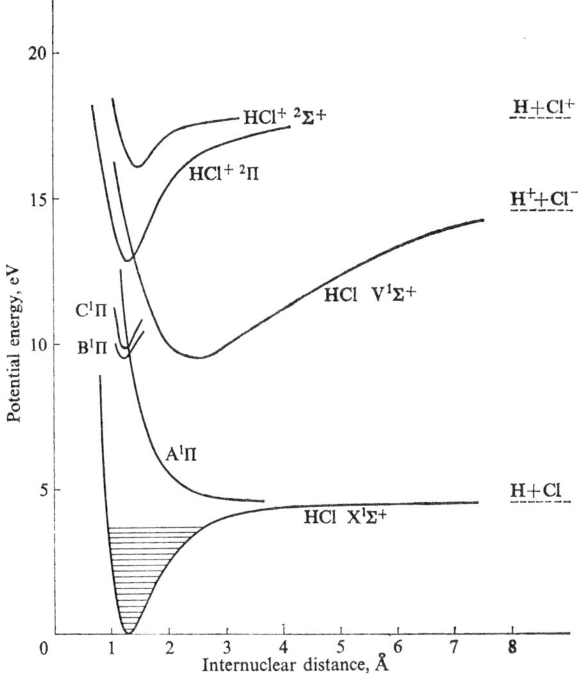

FIG. 25. Potential energy curves for the ground and excited states of HCl and HCl^+.

TABLE 31. PROPERTIES OF THE MOLECULAR IONS HX^+ [a,b]

Molecular ion	Electronic state	Adiabatic ionization potential relative to $^1\Sigma^+$ ground state of HX(eV)	Dissociation energy, $D_0°$		Internuclear distance, r_e(Å)	Vibrational frequency, $\Delta G_{\frac{1}{2}}$ (cm^{-1})
			kcal	*eV*		
HCl^+	$^2\Pi_{3/2}$	12·74	108·3	4·70	1·3153	2568
	$^2\Pi_{1/2}$	12·82	106·5	4·62		
	$^2\Sigma^+$	16·23	41·5	1·80	1·5140	1527
HBr^+	$^2\Pi_{3/2}$	11·67	90·6	3·93	1·448	2348
	$^2\Pi_{1/2}$	12·00	82·9	3·60		
	$^2\Sigma^+$	15·29	47·5	2·06	1·684	1329
HI^+	$^2\Pi_{3/2}$	10·38	72·1	3·13	1·62	2170
	$^2\Pi_{1/2}$	11·05	56·7	2·46		
	$^2\Sigma^+$	13·85	64·8	2·81	1·90	1040

[a] *Tables of Constants and Numerical Data*, No. 17, *Spectroscopic Data relative to Diatomic Molecules* (general ed. B. Rosen), Pergamon (1970).

[b] H. J. Lempka, T. R. Passmore and W. C. Price, *Proc. Roy. Soc.* **A304** (1968) 53.

It is noteworthy that the internuclear distance is somewhat longer, and the vibrational frequency somewhat lower, for the molecular ion in its ground state than for the parent molecule. Despite this evidence of weaker binding in the molecular ion, the energy of dissociation of the ion is slightly greater than that of the molecule. However, the general tenor of the results gives clear notice that the $p\pi$ electrons of the HX molecule, which are the first to ionize, are virtually non-bonding. The ionization energies of the $p\pi$ electrons bear a linear relationship to the ionization energies of the corresponding noble gases, whence the hydrogen halides are formally derived by withdrawing a proton from the nucleus[439].

Detailed analyses of the vibrational–rotational and microwave spectra of the gaseous hydrogen halides yield the most reliable estimates of the internuclear distances r_e (see Table 29). Recent measurements of the Stark effect in the microwave spectra of hydrogen bromide and iodide give precise values for the dipole moments of these molecules[440]; with the corresponding parameters recommended on the basis of earlier measurements for the other hydrogen halides, these confirm the striking decrease in dipole moment along the series HF, HCl, HBr, HI. Quadrupole coupling constants have also been obtained from the microwave spectra of hydrogen, deuterium and tritium halides[441]. The significance of these parameters has been discussed in subsection 3.2 (pp. 1271–4), and estimates of the ionic character of the H–Br and H–I bonds, based on the classical Townes–Dailey approximation[387], are included in Table 27. Nmr measurements on the gaseous hydrogen halides show that the proton resonance moves progressively to higher field in the sequence HCl < HBr < HI[374]; the increased shielding of the proton thus implied is consistent with the variations of electron density to be expected as the electronegativity of the halogen decreases.

The infrared and Raman spectra of the gaseous hydrogen halides give the values listed in Table 29 for the vibrational frequencies ω_e, the anharmonicity constants $\omega_e x_e$, and the stretching force constants k_e. The influence of change of physical state on the infrared and Raman spectra of the hydrogen halides has been the subject of several investigations, though a full interpretation of the results has not always been possible[356,426]. The low-frequency shift of the vibrational fundamental on going from the gaseous to the condensed phases provides clear evidence of intermolecular interaction involving the hydrogen halide molecule, notably in the solid phase (q.v.) or when dissolved in solvents which do not promote dissociation. The infrared spectra of hydrogen halide molecules isolated in a variety of inert matrices at 20°K signify that diffusion occasions the formation of cyclic dimers and trimers and various multimeric species; in the presence of N_2, CO_2, CO or C_2H_4 (D) in argon matrices at 20°K, evidence is found for interactions of the type HX–D, HX–D–HX and $(HX)_2$–D, while the HX–C_2H_4 system is believed to give rise to a specific complex[442,443].

3. The solid phase. The solid hydrogen halides are noteworthy for their polymorphism. Thus, solid hydrogen chloride undergoes a first-order transition at 98·4°K; solid hydrogen bromide undergoes three λ-type transitions at 89·75°, 113·62° and 116·86°K, while for solid hydrogen iodide transition points have been detected at *ca.* 25°, 70° and

440 F. A. Van Dijk and A. Dymanus, *Chem. Phys. Letters*, **5** (1970) 387.
441 T. Tokuhiro, *J. Chem. Phys.* **47** (1967) 109.
442 A. J. Barnes and H. E. Hallam, *Quart. Rev. Chem. Soc.* **23** (1969) 398.
443 A. J. Barnes, H. E. Hallam and G. F. Scrimshaw, *Trans. Faraday Soc.* **65** (1969) 3150, 3159, 3172.

125°K[444]. Of the different forms, those stable at low temperatures have been most fully characterized. On the evidence of neutron diffraction studies of deuterium chloride[445] and bromide[446] at low temperatures, the crystals adopt a face-centred orthorhombic lattice with a unit cell belonging to the space group $Bb2_1m$ and containing four molecules, the dimensions being as listed in Table 29. Within the generous limits of error, the intramolecular H–X distance is not significantly different from that of the gaseous molecule. The structure of linear zigzag chains of HX molecules is confirmed by the vibrational spectra of the solids[447–450], though recent interpretations[449,450] of these spectra favour non-planar rather than the planar chains implicit in the space group of the unit cell deduced by neutron diffraction. Qualitatively, therefore, the structure resembles that found in gaseous and crystalline hydrogen fluoride, and there are several reasons for believing that it is based, not on a simple dipolar array, but on hydrogen-bonding between adjacent HX molecules. Such hydrogen-bonding is, however, relatively weak. At low temperatures hydrogen iodide differs from the chloride and bromide in adopting an ordered face-centred tetragonal lattice[451]. At higher temperatures, all the solids appear to assume disordered face-centred cubic lattices, again with four molecules per unit cell[426,451,452]. A model has been proposed[452] for this phase of deuterium chloride, in which the deuterium atom has twelve equally probable equilibrium positions situated along the lines connecting the chlorine atom of the DCl molecule with its twelve nearest neighbours; the reorientational motions of the hydrogen halide molecules have also been discussed[450] in relation to the observed Raman spectra of the high-temperature phases. The relative mobility of molecules in the solid hydrogen halides is also indicated by the ferroelectric character reported for the orthorhombic phase of hydrogen chloride[453], and by the complicated picture presented by measurements of the dielectric properties of the solids[454].

Properties of the Liquids: Function as Non-aqueous Solvents[432,436,455]

The anhydrous hydrogen halides HCl, HBr and HI have some interest as solvent systems, both in their own right and also because of the comparison thus allowed with liquid hydrogen fluoride (see Chapter 25)[456]. The first experiments using liquid hydrogen chloride as a solvent were reported in 1865 by Gore[457], who, on the basis of qualitative visual observations of solubility, concluded that the liquid "has but a feeble solvent power for solid bodies in

444 G. L. Hiebert and D. F. Hornig, *J. Chem. Phys.* **26** (1957) 1762.

445 E. Sándor and R. F. C. Farrow, *Nature*, **213** (1967) 171.

446 E. Sándor and M. W. Johnson, *Nature*, **217** (1968) 541.

447 D. F. Hornig and W. E. Osberg, *J. Chem. Phys.* **23** (1955) 662; G. L. Hiebert and D. F. Hornig, *ibid.* **27** (1957) 752, 1216; *ibid.* **28** (1958) 316.

448 A. Anderson, H. A. Gebbie and S. H. Walmsley, *Mol. Phys.* **7** (1964) 401.

449 R. Savoie and A. Anderson, *J. Chem. Phys.* **44** (1966) 548; L.-C. Brunel and M. Peyron, *Compt. rend.* **264C** (1967) 821.

450 M. Ito, M. Suzuki and T. Yokoyama, *J. Chem. Phys.* **50** (1969) 2949.

451 F. A. Mauer, C. J. Keffer, R. B. Reeves and D. W. Robinson, *J. Chem. Phys.* **42** (1965) 1465.

452 E. Sándor and R. F. C. Farrow, *Nature*, **215** (1967) 1265.

453 K. Shimaoka, N. Niimura and S. Hoshino, *Acta Cryst.* **A25** (1969) S50.

454 P. P. M. Groenewegen and R. H. Cole, *J. Chem. Phys.* **46** (1967) 1069; R. H. Cole and S. Havriliak, jun., *Discuss. Faraday Soc.* **23** (1957) 31.

455 T. C. Waddington, *Non-aqueous Solvents*, Nelson, London (1969); R. A. Zingaro, *Nonaqueous Solvents*, Raytheon Educ. Co., Boston (1968).

456 H. H. Hyman and J. J. Katz, *Non-aqueous Solvent Systems* (ed. T. C. Waddington), p. 47. Academic Press (1965).

457 G. Gore, *Phil. Mag.* **29** (4) (1865) 541.

general". Nevertheless, the study of hydrogen chloride, bromide and iodide as non-aqueous solvents was resumed just after the turn of the century when Steele, McIntosh and Archibald described physical and thermodynamic constants of the liquids, as well as solubility, conductivity, transport number and ebullioscopic measurements[458]; little correlation could be found between dissolving power, dielectric constant, conductivity and the molecular weights measured for solute species. More recently, there have been extensive studies of the liquids, particularly hydrogen chloride, as reaction media[432].

Of the physical properties of the hydrogen halides recorded in Table 29, the relatively narrow liquid range (15–30°), with its implied experimental difficulties, and low dielectric constant (3–14) are of especial significance in relation to the exploitation of the liquids as solvents. By contrast, hydrogen fluoride has a liquid range of 100° and a dielectric constant of 175 at −73°C. The low dielectric constants of hydrogen chloride, bromide and iodide mean that only salts with low lattice energies, e.g. the tetra-alkylammonium halides, are appreciably soluble, and that extensive ion-pairing occurs in all but the most dilute solutions. The electrical conductivities of the liquids, though lower than those of hydrogen fluoride and water, suggest some form of self-ionization such as

$$3HX \rightleftharpoons H_2X^+ + HX_2^-$$

with solvation of both the proton and halide ion produced by the primary dissociation process. The only reasonably well authenticated example of an H_2X^+ ion in solution is that where X = F, which has been detected in the HF–SbF_5 system. The postulated H_2Cl^+ ion is known to exist in the gas phase[459]; it may also be present in the compounds $HCl,HClO_4$, HCl,H_2SO_4 or HCl,HBr, though there is no convincing evidence to this effect. The ions H_2Br^+ and H_2I^+ have likewise eluded detection in the condensed phases. On the other hand, the existence of the hydrogen dihalide anions HX_2^- is now well established (see pp. 1313–21); the presence of more extensively solvated halide ions, e.g. $Cl(HCl)_n^-$ ($n > 1$), has also been considered[460]. If the self-ionization is correctly represented by the above equation, two definitions of acids and bases are applicable, based upon a difference of emphasis rather than of principle; this arises from the fact that either halide ion- or proton-transfer can be regarded as the primary step in the equilibrium. Accordingly, acids can be defined as either proton-donors or halide ion-acceptors, and bases as either proton-acceptors or halide ion-donors. Thus, hydrogen chloride shows some of the characteristics of a "chloridotropic" solvent like arsenic trichloride and some of an acidic solvent like sulphuric acid. Because of the highly acidic nature of the solvents, it has been impossible to find any reasonably strong solvo-acids, though the following compounds appear to function as weak acids: boron(III) halides, tin(IV) halides, and, in liquid hydrogen chloride, PF_5 (the strongest acid yet found), XCl and HX (X = Br or I). Conversely, there are very many solvo-bases, some of which act as strong electrolytes. Such bases comprise (a) salts which ionize readily to give a free halide ion, e.g. tetra-alkylammonium halides, and (b) compounds containing atoms with a lone pair of electrons, e.g. amines, phosphines, ethers

[458] B. D. Steele, D. McIntosh and E. H. Archibald, *Phil. Trans.* **205** (1905) 99; *Z. physik. Chem.* **55** (1906) 129.

[459] D. O. Schissler and D. P. Stevenson, *J. Chem. Phys.* **24** (1956) 926; F. H. Field and F. W. Lampe, *J. Amer. Chem. Soc.* **80** (1958) 5583.

[460] C. J. Ludman and T. C. Waddington, *2nd International Raman Conference*, Oxford (1970); T. C. Waddington, *6th International Symposium on Fluorine Chemistry*, Durham (1971).

or sulphides, or a π-bonded system which can easily be protonated, e.g. aromatic olefins and organic compounds containing the groups $-C{\equiv}N$, $-N{=}N-$, $\rangle C{=}O$ or $\ni P{=}O$.

Attention has been directed, not only to acid–base behaviour, but also to solvolysis and redox reactions in liquid hydrogen halide solutions[432]. Solvolysis corresponds to the replacement of a ligand Y by a halogen atom X, a type of reaction which has been observed[432,461] in solutions in hydrogen chloride when Y is phenyl, hydroxyl or fluorine, e.g.

$$Ph_3SnCl + HCl \rightarrow Ph_2SnCl_2 + PhH$$
$$Ph_3COH + 3HCl \rightarrow Ph_3C^+HCl_2^- + H_3O^+Cl^-$$
$$SbF_3 + 3HCl \rightarrow SbCl_3 + 3HF$$

By means of conductimetric titrations, it has been possible to establish, *inter alia*, the following oxidation reactions of chlorine, bromine and iodine monochloride in liquid hydrogen chloride[432]:

$$I^- + 2Cl_2 \rightarrow ICl_4^-$$
$$Br^- + Cl_2 \rightarrow BrCl_2^-$$

$$I^- + ICl + HCl \xrightarrow{\text{oxidation}} I_2\downarrow + HCl_2^-$$
$$HCl_2^- \xrightarrow[\text{acid-base}]{ICl} HCl + ICl_2^-$$

$$PCl_3 + X_2 + HCl \longrightarrow PCl_3X^+ + HClX^- \quad (X = Cl \text{ or } Br)$$

Experimental studies of the solvent properties of the hydrogen halides normally demand that the liquids be handled at low temperatures in an enclosed vacuum system. The methods principally exploited to investigate the behaviour of solutes and to monitor reactions in solution are as follows.

Conductimetric Measurements

These have been the mainstay of many recent investigations[432]. The mechanism of the conduction of acids and bases is not known, but probably involves halide ion-transfer, at least in basic solution, e.g.

$$Cl{-}H{-}Cl^- \cdots\cdots H{-}Cl$$
$$\downarrow$$
$$Cl{-}H \cdots\cdots Cl{-}H{-}Cl^-$$

The variation of the molar conductance Λ_m with the concentration c of strongly basic solutions is noteworthy, for plots of log Λ_m against log c or of Λ_m against $\sqrt{c}$ tend to exhibit a minimum, a behaviour characteristic of solvents of low dielectric constant[432]. To account for these results, two modes of ionization have been suggested for an ion-pair A^+B^-:

$$A^+B^- \rightleftharpoons A^+ + B^- \quad \text{very dilute solutions}$$
$$\left.\begin{array}{l} A^+B^- + A^+ \rightleftharpoons A_2B^+ \\ A^+B^- + B^- \rightleftharpoons AB_2^- \end{array}\right\} \quad \text{more concentrated solutions}$$

The variation of the conductance of weak electrolytes is more complicated, and cannot

461 M. E. Peach, *Inorg. Nuclear Chem. Letters*, 7 (1971) 75.

easily be explained. Reactions involving changes in the ionic species present in solution can be monitored conductimetrically. The processes of salt-formation, adduct-formation and ionization, or a combination of these, give characteristic conductimetric curves. Hence, it has been possible to follow the course of acid–base reactions such as

$$Me_4N^+HCl_2^- + BCl_3 \rightarrow Me_4N^+BCl_4^- + HCl$$

or of redox reactions such as

$$PCl_3 + Cl_2 + HCl \rightarrow PCl_4^+HCl_2^-$$

Spectroscopic Measurements

The spectroscopic properties of solutions in hydrogen halides have been very little studied, mainly because of the technical difficulties of maintaining the sample either at low temperatures or at relatively high pressures. However, nmr spectra have confirmed the presence of $Ph_2C{\cdot}CH_3^+$ in solutions of $Ph_2C{=}CH_2$ and of $PhC{\cdot}CH_3^{2+}$ in solutions of $PhC{\equiv}CH$, both in liquid hydrogen chloride at room temperature[432]. The weakness and simplicity of the Raman spectrum of the pure liquid facilitate the application of Raman spectroscopy to solutions in hydrogen chloride; according to preliminary reports[460], such spectra serve to identify certain solute species in solutions at room temperature.

Phase Diagrams

While not revealing much about the nature of the solutions, phase diagrams in which one of the hydrogen halides is a component give indications of some of the compounds that may be formed in solution, and may also yield valuable information about the function of solvo-acids and solvo-bases. Thus, the fact that the system BCl_3–HCl gives no sign of compound-formation suggests that BCl_3 must be a very weak solvo-acid in hydrogen chloride. The diagrams for several compounds that are solvo-bases with hydrogen chloride or bromide have also been studied.

Cryoscopic and Ebullioscopic Measurements

These have been made on some solutions in hydrogen chloride, bromide and iodide, but the results are commonly difficult to interpret because of the complex mode of ionization[432].

Preparative Methods

Because of their low boiling points and consequent easy removal, the liquid hydrogen halides are useful media for certain preparations. Convenient routes to the following species or their derivatives have thus been devised: BX_4^- [432], BF_3Cl^- [432], $B_2Cl_6^{2-}$ [432], NO_2Cl[432], $Al_2Cl_7^-$ [462], R_2SCl^+ and $RSCl_2^+$ [463], PCl_3Br^+ [432], $Ni_2Cl_4(CO)_3$ and $Ni(NO)_2Cl_2$ (products of the reactions of $Ni(CO)_4$ with Cl_2 and NOCl, respectively)[464], $[Fe(CO)_5H]^+$ and $[(\pi{-}C_5H_5)Fe(CO)_2]_2H^+$ (formed by the protonation of $Fe(CO)_5$ and $[(\pi{-}C_5H_5)Fe(CO)_2]_2$, respectively[465]).

Solutions of the Hydrogen Halides

In common with hydrogen fluoride, the heavier hydrogen halides are notable for their profuse solubility in water; in no case do the solutions comply even approximately with Henry's law. Details of the solubility and of the density of the resulting solutions are presented in Table 32. Each of the aqueous systems gives rise to a maximum-boiling azeotrope,

462 M. E. Peach, V. L. Tracy and T. C. Waddington, *J. Chem. Soc. (A)* (1969) 366.
463 M. E. Peach, *Canad. J. Chem.* **47** (1969) 1675.
464 Z. Iqbal and T. C. Waddington, *J. Chem. Soc. (A)* (1968) 2958; *ibid.* (1969) 1092.
465 D. A. Symon and T. C. Waddington, *J. Chem. Soc. (A)* (1971) 953.

Table 32. Behaviour of the Hydrogen Halides in Aqueous Solution[a–f]

Property	HF	HCl	HBr	HI
Solubility in water (g/100 g soln at 1 atm)	Miscible in all proportions	45·15 at 0°C 42·02 at 20°C 37·34 at 50°C	68·85 at 0°C 65·88 at 25°C 63·16 at 50°C	~71 at 0°C — —
Density of aqueous solution (g cm^{-3} at 20°C) concentration:				
10 g/100 g solution	1·035	1·047	1·073	1·072
20 g/100 g solution	1·072	1·091	1·158	1·167
saturated	—	1·205	1·79	1·99
Constant-boiling solution at 1 atm				
boiling point (°C)	112	108·58	124·3	126·7
concentration (g/100 g solution)	38	20·22	47·63	56·7
density (g cm^{-3} at 25°C)	1·138	1·096	1·482	1·708
Thermodynamic functions for the process $HX(g) + \infty H_2O \rightarrow H^+X^-(\infty H_2O)$ at 25°C:	*undissociated* / *ionized*			
$\Delta H°$ (kcal mol^{-1})	−11·70 / −14·70	−17·890	−20·35	−19·52
$\Delta G°$ (kcal mol^{-1})	− 5·65 / − 1·34	− 8·595	−12·08	−12·74
$\Delta S°$ (cal deg^{-1} mol^{-1})	−20·3 / −44·8	$-31{\cdot}1_5$	−27·8	$-22{\cdot}7_5$
Free energy changes for ionization of HX molecules in water at 25°C, $\Delta G°$ (kcal mol^{-1})				
1. $HX(aq) \rightarrow HX(g)$	+ 5·65	− 1·2	− 0·7	− 0·5
2. $HX(g) \rightarrow H(g) + X(g)$	+128·6	+ 96·6	$+ 81{\cdot}0_5$	+ 65·0
3. $H(g) + X(g) \rightarrow H^+(g) + X^-(g)$	+232·9	+229·1	+234·9	+241·8
4. $H^+(g) + X^-(g) \rightarrow H^+(aq) + X^-(aq)$	−362·3	−334·0	−327·1	−318·2
$HX(aq) \rightarrow H^+(aq) + X^-(aq)$, $\Delta G°$ calc.	+ 4·85	− 9·5	− 11·9	− 11·9

TABLE 32 (*cont.*)

Property	HF	HCl	HBr	HI
$HX(aq) \rightarrow H^+(aq) + X^-(aq)$, ΔH° calc. (kcal mol^{-1})	$-0{\cdot}5$	$-13{\cdot}7$	$-15{\cdot}2$	$-14{\cdot}1$
$HX(aq) \rightarrow H^+(aq) + X^-(aq)$, ΔS° calc. (cal deg^{-1} mol^{-1})	-18	-14	-11	$-7{\cdot}4$
Calculated pK_a for acid $= \Delta G^\circ/1{\cdot}36$	3·6	$-7{\cdot}0$	$-8{\cdot}8$	$-8{\cdot}8$
Observed pK_a for acid	3·2	~ -7	< -7	< -7

[a] J. W. Mellor, *A Comprehensive Treatise on Inorganic and Theoretical Chemistry*, Vol. II, Longmans, Green and Co., London (1922); Supplement II, Part I (1956).

[b] *Gmelins Handbuch der Anorganischen Chemie*, 8 Auflage, "Chlor" (1927); "Brom" (1931); "Iod" (1933).

[c] Z. E. Jolles (ed.), *Bromine and its Compounds*, Benn, London (1966).

[d] A. G. Sharpe, *Halogen Chemistry* (ed. V. Gutmann), Vol. 1, p. 1. Academic Press (1967).

[e] J. C. McCoubrey, *Trans. Faraday Soc.* **51** (1955) 743.

[f] *National Bureau of Standards Technical Note* 270–3, U.S. Govt. Printing Office, Washington (1968).

the characteristics of which are listed in the table. The stability of composition provided by constant-boiling hydrochloric acid has led to its widespread use as a standard for analytical purposes; samples of the constant-boiling acid prepared by distillation have been found, after storage for more than three years, to differ in composition by less than 0·1% from freshly prepared samples[426].

Unlike hydrogen fluoride, the other hydrogen halides undergo virtually complete ionization in aqueous solution at all but the highest concentrations. Thermodynamic functions defining the dissolution and dissociation of the hydrogen halides in water are included in Table 32, wherein the thermodynamic parameters for the process

$$HX(aq) \rightleftharpoons H^+(aq) + X^-(aq)$$

are related to the sum of the appropriate parameters for the following stages:

1. $HX(aq) \rightarrow HX(g)$[297,466]
2. $HX(g) \rightarrow H(g) + X(g)$[297]
3. $H(g) + X(g) \rightarrow H^+(g) + X^-(g)$[289,297]
4. $H^+(g) + X^-(g) \rightarrow H^+(aq) + X^-(aq)$[296]

For reactions (1), (2) and (4), free energy and enthalpy changes have been derived from published measurements or estimates; for reaction (3), the free energy has been estimated from the heats of ionization of the hydrogen and halogen atoms[289,297], together with an entropy contribution of $-R \log_e 4 - R \log_e 2 = -4{\cdot}13$ cal deg^{-1} mol^{-1}, due to the change of electronic multiplicity. Such an analysis reproduces, within the limits of error, the observed pattern of ionization, even though a relatively small change of free energy, namely < 10 kcal, is implied by the distinction between a strong acid like HCl and a weak acid like HF. The results demonstrate that, in acid strength, there is little to choose between aqueous hydrochloric, hydrobromic and hydriodic acids. The difference between hydrogen fluoride and the other hydrogen halides resides mainly in the more endothermic enthalpy of ionization; this, in turn, arises primarily from the higher bond energy of the HF molecule, but partly also from the enhanced hydration energy of the undissociated HF molecule (the consequence of hydrogen-bonding to water molecules) and the reduced electron affinity of fluorine. These factors together more than compensate for the high hydration energy of the fluoride ion.

The physical properties of the aqueous hydrohalic acids, and particularly hydrochloric acid, have been the subject of many investigations. Included in more comprehensive accounts of the halogens[418,424–426] are details of the following properties for one or more of the aqueous hydrohalic acids: density, vapour pressure, specific heat, heats of solution and neutralization, activity coefficients, viscosity, surface tension, compressibility, diffusion coefficients, distribution coefficients between water and other liquids, molal volume, electrical conductivity, transport numbers of the ions, dielectric constant, refractive index and magnetic susceptibility.

Cooling aqueous solutions of the hydrogen halides produces a variety of solid hydrates, the properties of which are summarized in Table 33. The vibrational spectra and recent X-ray analyses leave little doubt that these hydrates are to be formulated as $[(H_2O)_nH]^+X^-$. The compounds are of interest in the opportunity they provide for the study of the hydrated

[466] J. C. McCoubrey, *Trans. Faraday Soc.* **51** (1955) 743.

TABLE 33. SOLID HYDRATES OF THE HYDROGEN HALIDES

Formula	X = Cl[a]	X = Br[b,c]	X = I[c]
HX,H_2O	M.p. −15·35°C. Indicated by i.r. spectrum[d] and confirmed by X-ray analysis to be $H_3O^+Cl^-$ with O–H · · Cl = 2·95 Å and O–H = 0·96 ± 0·08 Å (see Fig. 26).[e]	Stable between −3·3° and −15·5°C under pressure. I.r. spectrum consistent with the formulation $H_3O^+Br^-$.[d]	I.r. spectrum consistent with the formulation $H_3O^+I^-$.[d]
$HX,2H_2O$	M.p. −17·7°C. Shown by X-ray analysis to be $(H_2O)_2H^+Cl^-$; the bonding arrangement around one end of the $(H_2O)_2H^+$ ion is almost planar and pyramidal around the other.[f] 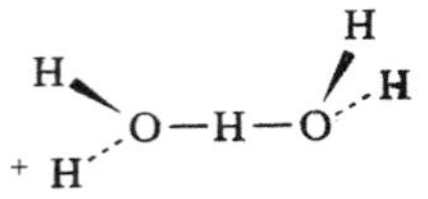 Central O–H distances = 2·41 Å.	M.p. −11·3°C.	M.p. *ca.* −43°C.
$HX,3H_2O$	M.p. −24·9°C. Shown by X-ray analysis to be $H_5O_2{}^+Cl^-,H_2O$ with nearly eclipsed $(H_2O)_2H^+$ units.[g] 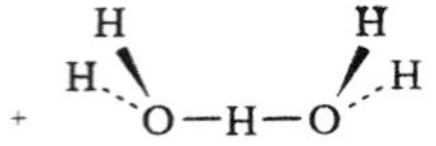 Central O–H distances = 2·43 Å.	Decomposes −47·9°C.	M.p. *ca.* −48°C.
$HX,4H_2O$	Not known.	M.p. −55·8°C. Shown by X-ray analysis to be $[(H_2O)_3H]^+[(H_2O)_4H]^+2Br^-,H_2O$ with H–O bond lengths of 2·465–2·75 Å. The aggregates $[(H_2O)_3H]^+$ and $[(H_2O)_4H]^+$ may be regarded as $H_3O^+,2H_2O$ and $H_3O^+,3H_2O$, respectively.[h]	M.p. −36·5°C.
$HX,6H_2O$	Very unstable, m.p. −70°C.	Decomposes −88·2°C.	

[a] *Gmelins Handbuch der Anorganischen Chemie*, 8 Auflage, System-nummer 6, "Chlor", Teil B, Lieferung 1 (1968).
[b] Z. E. Jolles (ed.), *Bromine and its Compounds*, Benn, London (1966).
[c] J. W. Mellor, *A Comprehensive Treatise on Inorganic and Theoretical Chemistry*, Vol. II, Longmans, Green and Co., London (1922).
[d] C. C. Ferriso and D. F. Hornig, *J. Chem. Phys.* **23** (1955) 1464.
[e] Y. K. Yoon and G. B. Carpenter, *Acta Cryst.* **12** (1959) 17.
[f] J.-O. Lundgren and I. Olovsson, *Acta Cryst.* **23** (1967) 966.
[g] J.-O. Lundgren and I. Olovsson, *Acta Cryst.* **23** (1967) 971.
[h] J.-O. Lundgren and I. Olovsson, *J. Chem. Phys.* **49** (1968) 1068.

proton. Thus, X-ray analysis of single crystals at low temperatures has identified the aggregates H_3O^+, $[(H_2O)_2H]^+$, $[(H_2O)_3H]^+$ and $[(H_2O)_4H]^+$ [467], while some of the earliest direct evidence of the H_3O^+ ion was gained from the infrared spectra of the monohydrates at −195°C[468].

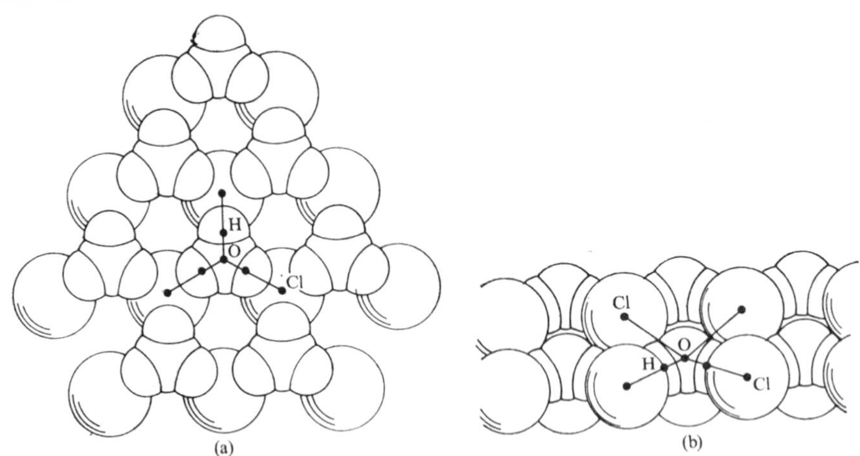

FIG. 26. Crystal structure of HCl,H_2O: (a) [111] plane; (b) [$1\bar{2}1$] plane. [Reproduced with permission from Y. K. Yoon and G. B. Carpenter, *Acta Cryst.* **12** (1959) 20.]

Solutions of the hydrogen halides in non-aqueous solvents have also been the focus of some attention; notwithstanding the dearth of systematic measurements of solubility, a significant weight of information, both qualitative and quantitative, has accumulated[418,424–426,469]. Though limited by the reactivity of the solute, a wide range of solvents has thus been identified, extending from non-polar to highly polar media, e.g. hydrocarbons, carbon tetrachloride, chloroform and other organo-halogen compounds, alcohols, diethyl ether, sulphur dioxide, pyridine, aniline, N-methylacetamide, acetonitrile, nitrobenzene, acetic acid and hydrogen fluoride. The form of the solute varies correspondingly from HX molecules to solvated H^+ and X^- ions, as may be judged by the electrical conductivity. It has been concluded that the tendency of the hydrogen halides to dissociate in solution is largely determined by the capacity of the solvent molecule to act as a proton acceptor; the dielectric constant of the medium is thought to have a much smaller effect. Heats of solution of hydrogen chloride and hydrogen bromide in solvents like carbon tetrachloride, chloroform and various hydrocarbons[297], falling typically in the range −2·6 to −4·2 kcal, are relatively more exothermic than those of the parent halogens. The vibrational spectra of solutions of the hydrogen halides, like those of the parent halogens, in organic solvents show variations attributable to specific interactions between the solute and solvent molecules; the extent of hydrogen-bonding between HX and solvent molecules has thus been gauged in a qualitative manner. The very sparing electrical conductivity of many solutions, e.g. in benzene or sulphur dioxide, is greatly increased by the presence of traces of water,

467 Y. K. Yoon and G. B. Carpenter, *Acta Cryst.* **12** (1959) 17; J.-O. Lundgren and I. Olovsson, *J. Chem. Phys.* **49** (1968) 1068.

468 C. C. Ferriso and D. F. Hornig, *J. Chem. Phys.* **23** (1955) 1464.

469 W. F. Linke, *Solubilities: Inorganic and Metal-organic Compounds*, 4th edn., Vol. 1, van Nostrand, Princeton (1958).

as a result of the reaction

$$HX + H_2O \rightarrow H_3O^+ + X^-$$

In an acidic solvent like anhydrous acetic acid, both hydrogen chloride and hydrogen bromide behave as weak acids, the pK values of 8·85 and 6·40, respectively, giving clear notice that hydrogen bromide is the stronger acid under these conditions[426]. The order of acid strengths HI > HBr > HCl is reported to prevail in acetonitrile, various alcohols, pyridine and aniline.

Chemical Behaviour[424–426,436]

The hydrogen halide molecules share with diatomic halogen or interhalogen molecules the primary function of electron acceptors. However, relative to the ground state, interaction of the HX molecule with a donor species D, like water, ammonia or an organic base, has a markedly greater stabilizing influence on excited states represented approximately by the formulation H^+X^-, as a result of which the "outer complex" initially formed is very much more prone to transformation into an "inner complex":

$$\underset{\text{outer complex}}{HX + D \rightleftharpoons D, HX} \rightleftharpoons \underset{\text{inner complex}}{[DH]^+X^-}$$

the net result being the heterolytic fission of the H–X bond. This mechanism is characteristic of many solution reactions of the hydrogen halides, which, in their action as proton-donors, behave as acids in the more restricted sense defined by Lowry and Brønsted; it is as proton-donors, for example, that the hydrogen halides react with metals and metal oxides, hydroxides, carbonates and sulphides. The halide ions released by heterolytic fission may suffer various possible fates: thus, they may be stabilized as such by solvation or by incorporation in a solid lattice; they may give rise to complex ions, e.g. I_3^-, $GaBr_4^-$ or $PtCl_6^{2-}$; or they may undergo oxidation to the parent halogen or even, in some circumstances, to oxy-halogen species like IO_3^-.

Homolytic fission of the HX molecule

$$HX \rightarrow H + X$$

which occurs with increasing readiness in the series HCl < HBr < HI, represents an alternative mechanism for reaction, being favoured at elevated temperatures in the gas phase or in solution in non-polar solvents. This process is facilitated by photolysis or radiolysis or by the agency of suitable catalysts.

Because of these different mechanisms, with their dependence on the reaction medium, the chemical behaviour of the hydrogen halides is unusually sensitive to the nature of this medium. Thus, the kinetic barrier to both homolytic and heterolytic fission is such for the anhydrous materials that they are relatively inert, e.g. with respect to most metals and their oxides. By contrast, the reactivity of solutions in highly polar media like water has for long been a familiar feature of inorganic chemistry.

The reactions of the hydrogen halides may also be classified, according to their outcome, as either addition or substitution. Addition reactions may be further sub-divided into those wherein the HX bond remains intact, as in the formation of hydrogen dihalide anions, and those wherein this bond is broken, as when a hydrogen halide adds to the $>C=C<$ or

—C≡C— units of organic molecules. Substitution reactions are exemplified by protonation, e.g.

$$M+nH_3O^+ + nX^- \rightarrow M^{n+} + nX^- + nH_2O + n/2H_2$$

and

$$O^{2-} + 2H_3O^+ + 2X^- \rightarrow 2H_2O + 2X^-$$

by oxidation, e.g.

$$XO_3^- + 6H_3O^+ + 5X^- \rightarrow 3X_2 + 9H_2O$$

and by exchange, e.g.

$$HX + D_2 \rightleftharpoons DX + HD$$

Reactions representative of addition and substitution processes are also indicated schematically in the flowchart of Fig. 27.

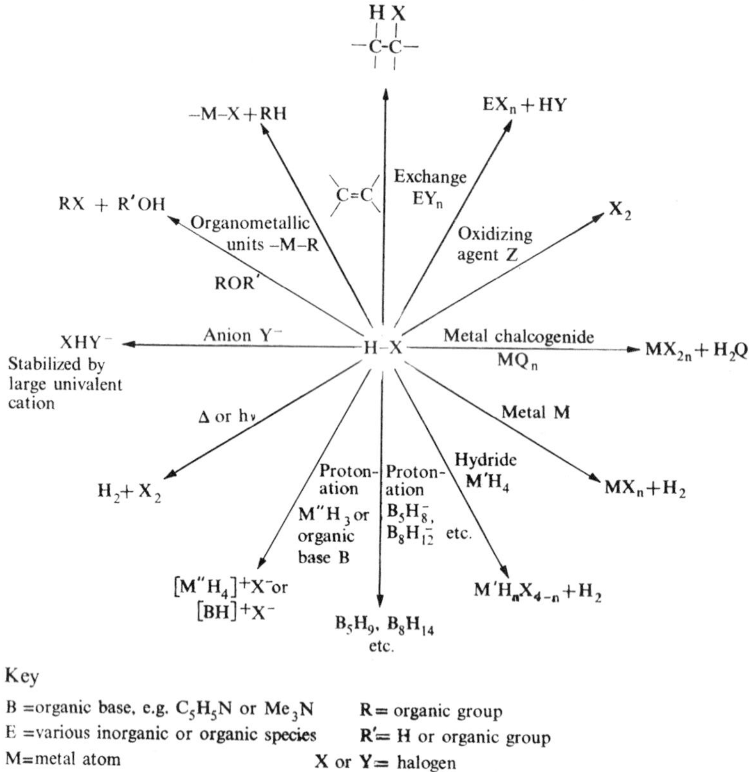

FIG. 27. Some representative reactions of the hydrogen halides.

1. Addition reactions in which the H–X bond remains intact. Numerous derivatives of the hydrogen halides are known in which it is likely, though seldom certain, that the H–X bond survives addition. Such compounds are formally analogous to the charge-transfer complexes of the molecular halogens, but, with the exception of anionic species of the type HXY^- (see below), they remain relatively ill-characterized. Detailed studies of numerous HX–solvent systems reveal the formation of distinct compounds, which are commonly low-melting and stable with respect to dissociation only at low temperatures, though salt-like derivatives containing anions of the type $X(HX)_n{}^-$ (see below) of appreciable stability may result from the interaction of hydrogen halides with certain organic bases. Spectroscopic measurements on the systems Me_2O,HCl[470] and R_3P,HX (R = Me or Ph; X = Cl, Br or I)[471] signify the presence of a hydrogen-bonded molecular species, as distinct from ions. By contrast, materials most realistically represented by an ionic composition, e.g. $[MeCONH_3]^+Cl^-$, $[Et_2OH]^+[Cl(HCl)_n]^-$ or $[MeC{=}NH]^+[Cl(HCl)_n]^-$ [432], are formed by various ethers, nitriles and other organic bases susceptible to protonation; heterolytic cleavage of at least some H–X bonds is clearly implicit in such formulations. Neutral addition compounds which probably belong to the molecular category include HCl,HBr (somewhat implausibly formulated as $[H_2Cl]^+Br^-$)[426], $2H_2S,3HBr$[345], Me_2O,HCl[470], olefin,nHCl (n = 1 or 2)[472], acetylene,nHCl (n = 1, 2 or 4)[472], Ar,HCl (Ar = aromatic hydrocarbon)[472], R_3P,HX[471], and clathrate compounds with phenols[345], though the precise nature of many of these systems remains obscure.

Addition of a halogen atom to a hydrogen halide molecule gives rise to a free radical short-lived under normal conditions. However, evidence has been obtained for the formation of the complex I · · · I–H on photolysis of ethyl iodide trapped in a hydrocarbon matrix at 77°K[473]. By contrast, the action of a discharge on a gaseous mixture of a hydrogen halide HX with the corresponding molecular halogen X_2 is believed to induce the reaction

$$X+H\text{–}X \rightarrow X\text{–}H\text{–}X \quad (X = Cl, Br \text{ or } I)$$

According to the infrared spectrum attributed by Pimentel, Noble and Bondybey to the HX_2 radical in the matrix-isolated condition, the stretching force constant of the H–X bond sees little change with the transition $HX_2^- \rightarrow HX_2+e$. Although this finding gives appealing support to the non-bonding character of the highest occupied molecular orbital of the HX_2^- ion (see below), the conclusions are clouded by circumstantial evidence pointing to the possibility that the trapped species is actually HX_2^- and *not* HX_2[474].

Hydrogen Dihalide Anions and Related Species[436,475]

All the hydrogen halides HX have the capacity to function as acceptors with respect to a univalent anionic donor Y^-, which may be a halide, pseudohalide or oxyanion, and so form relatively well-defined species of the type YHX^-. It is in the affinity of these species to trihalide ions like ICl_2^- that the analogy between the hydrogen halide and diatomic halogen or interhalogen molecules as acceptor species is most clearly manifest. The existence of the hydrogen dihalide anions requires hydrogen-bridging to exercise a primary rather than the secondary bonding function evident, for example, in the stabilization of crystal structures and in the association of liquids like ammonia, water and hydrogen fluoride[476,477]. However, the anions formed by the heavier halogens are notably less stable

470 G. Govil, A. D. H. Clague and H. J. Bernstein, *J. Chem. Phys.* **49** (1968) 2821.
471 M. Van den Akker and F. Jellinek, *Rec. Trav. Chim.* **86** (1967) 275.
472 D. Cook, Y. Lupien and W. G. Schneider, *Canad. J. Chem.* **34** (1956) 957, 964.
473 D. Timm, *Acta Chem. Scand.* **20** (1966) 2219.
474 P. N. Noble and G. C. Pimentel, *J. Chem. Phys.* **49** (1968) 3165; V. Bondybey, G. C. Pimentel and P. N. Noble, *ibid.* **55** (1971) 540; P. N. Noble, *ibid.* **56** (1972) 2088; but see D. E. Milligan and M. E. Jacox, *ibid.* **53** (1970) 2034.
475 D. G. Tuck, *Progress in Inorganic Chemistry*, **9** (1968) 161.
476 G. C. Pimentel and A. L. McClellan, *The Hydrogen Bond*, Freeman, San Francisco (1960).
477 W. C. Hamilton and J. A. Ibers, *Hydrogen Bonding in Solids*, Benjamin, New York (1968).

than HF_2^- as regards thermal decomposition; unlike HF_2^-, they are not easily stabilized in the crystalline phase by simple monatomic cations. For this and other reasons, the anions derived from hydrogen chloride, bromide and iodide have attracted less attention than HF_2^-, though recent research has gone some way towards redressing this balance[475]. In representing the simplest framework within which hydrogen-bonding can be studied in isolation, such anions are of particular significance. By contrast, the difficulty of distinguishing between the effects of hydrogen-bonding and those of other interactions necessitates a relatively pragmatic approach to such bonding in more complicated networks.

Inevitably the earliest correct identification of HXY^- species must be a matter of debate, especially since the authors who first reported the existence of salts containing an extra molecule or more of hydrogen halide did not always formulate the materials as derivatives of the HXY^- anion. Within these limitations, it appears that the earliest report was by Dilthey[478], who described the compounds $[Si(acac)_3]Cl,HCl$ and $[Si(dibenz)_3]Cl,HCl$ (acac = acetylacetonate; dibenz = dibenzoylacetonate). Similar compounds with tetra-alkylammonium, pyridinium and quinolinium cations were reported shortly afterwards by Kaufler and Kunz[479], who prepared and correctly formulated derivatives of the HCl_2^-, HBr_2^- and HI_2^- anions, and also gave notice of chloride and bromide species of the types $H_2X_3^-$ and $H_3X_4^-$. Ephraim established the reversible nature of the formation and dissociation processes[480], but few investigations of these and related compounds were otherwise undertaken until more recent times, which have witnessed considerable research activity in this area.

Preparation[475]

Preparative routes leading to derivatives of the HX_2^- ions are essentially independent of the nature of X, provided that the appropriate cation is selected. The earliest researchers[478,479] prepared these salts by the action of the dry, gaseous hydrogen halide on amines or substituted ammonium halides, a technique which has also been used in more recent studies. A solid tetra-alkylammonium halide generally takes up excess hydrogen halide to form products, which are often liquid at room temperature, of the type $R_4NH_nX_{n+1}$; excess hydrogen halide can be removed by pumping, and, in the absence of moisture, the final product is usually R_4NHX_2. An alternative technique[432,475] is to dissolve the R_4NX salt in the liquid hydrogen halide and then evaporate the solution. The use of an organic solvent as a reaction medium has also been reported. Thus, treatment of tropenyl methyl ether with excess hydrogen halide in ether affords crystalline tropenium$^+$$HX_2^-$ (X = Cl or Br), while the reaction of gaseous hydrogen iodide with tetrabutylammonium iodide in dichloromethane gives Bu_4NHI_2; with chloro-tri-*p*-methoxyphenylmethane or 9-chloro-9-phenylxanthin in benzene solution, hydrogen chloride gives the HCl_2^- salt of the appropriate carbonium cation. As with HF_2^- and related species $H_nF_{n+1}^-$, the composition of the anions in crystalline materials depends markedly on the nature of the cation, and, to some extent, on the temperature of crystallization.

Few purely inorganic salts containing HX_2^- ions have been reported, and even some of these are of questionable authenticity. As long ago as 1881, Berthelot reported[481] that ammonium bromide and hydrogen bromide could "possibly" combine, but there is no

478 W. Dilthey, *Ber.* **36** (1903) 923; *Annalen*, **344** (1906) 300.
479 F. Kaufler and E. Kunz, *Ber.* **42** (1909) 385, 2482.
480 F. Ephraim, *Ber.* **47** (1914) 1828.
481 M. Berthelot, *Ann. Chim. Phys.* **23** (5) (1881) 98.

record that this observation has subsequently been put to the test. On the other hand, there is unambiguous evidence that caesium salts of the anions HCl_2^-, $HClBr^-$ and $HClI^-$ are formed by the direct interaction of hydrogen chloride and the appropriate caesium halide at low temperatures[482,483]; high dissociation pressures are reported for $CsHCl_2$ at room temperature[483]. Deuterium analogues of these and other salts have also been prepared. The precipitate formed when hydrogen chloride is bubbled through a concentrated aqueous solution of caesium chloride has been the subject of some controversy[475,484]. Nevertheless, recent investigations indicate that two distinct crystalline phases are produced, and three-dimensional X-ray studies of single crystals of the hexagonal phase establish it as $CsCl,1/3[H_3O.HCl_2]$, containing the ions Cs^+, H_3O^+, Cl^- and HCl_2^- [484]; the analogous bromide compound $CsBr,1/3[H_3O.HBr_2]$ has likewise been prepared and characterized. No sign of HCl_2^- species could be detected in the Raman spectra of aqueous solutions of hydrochloric acid of various concentrations up to 8 M in lithium chloride[485].

Various mixed anions of the type HXY^- have been obtained as crystalline salts, usually by preparative methods in which a solid is treated with hydrogen halide gas, or in which a salt R_4NY is dissolved in the liquid hydrogen halide HX and crystallized as R_4NHXY. Deuterium analogues have also been obtained in a number of cases. The following anions have been reported to date:[475]

X	Y
F^-	Cl^-, Br^-, I^-
Cl^-	Br^-, I^-, CN^-, NO_3^-, formate, acetate
Br^-	I^-, CN^-, formate, acetate

and there is good reason to believe that this range is capable of considerable expansion.

Among the species analogous to the halogen-containing anions are the following, which have been characterized only in recent years[475]: $H(NCS)_2^-$, $H(CN)_2^-$, $H(NO_2)_2^-$, $H(NO_3)_2^-$, $H(IO_3)_2^-$, $H(carboxylate)_2^-$, $H[M(CO)_5]_2^-$ (M = Cr, Mo or W), $H(OH_2)_2^+$, H(α-picoline oxide)$_2^+$, and $H[(\pi\text{-}C_5H_5)Fe(CO)_2]_2^+$ [465].

Physical Properties[436,475,482–484]

The main interest in the study of salts of HX_2^- and HXY^- anions has centred on their physical rather than chemical properties. Some of these physical properties, determined with somewhat variable degrees of certainty, are presented in Table 34. Although not included in the table, vibrational and nqr properties of some deuterated derivatives have also been examined.

Whereas salts of the HF_2^- and $H_2F_3^-$ anions have been the focus of a number of structural investigations using X-ray or neutron-diffraction methods, very little definitive structural information has so far been accumulated about salts containing other halogen-bearing anions of the type HX_2^- or HXY^-. According to three-dimensional X-ray studies, single crystals of the compounds $CsX,1/3[H_3O^+HX_2^-]$ (X = Cl or Br)[484] contain strings of HX_2^- ions parallel to the *c*-axis. The X · · · X distance is $3{\cdot}14 \pm 0{\cdot}02$ Å in the HCl_2^- ion

482 J. W. Nibler and G. C. Pimentel, *J. Chem. Phys.* **47** (1967) 710.
483 G. C. Stirling, C. J. Ludman and T. C. Waddington, *J. Chem. Phys.* **52** (1970) 2730.
484 L. W. Schroeder and J. A. Ibers, *Inorg. Chem.* **7** (1968) 594.
485 A. G. Maki and R. West, *Inorg. Chem.* **2** (1963) 657.

TABLE 34. PHYSICAL PROPERTIES OF HYDROGEN DIHALIDE ANIONS HX_2^- AND HXY^- [a,b]

Anion	Counter-ion in crystalline salt	Thermodynamics of hydrogen-bond formation			X · · · X distance in HX_2^- anion (Å)	Vibrational frequencies (cm^{-1})*			Reference
		$-\Delta H°$ (kcal mol^{-1})	$-\Delta G°$ (kcal mol^{-1})	$-\Delta S°$ (eu)		ν_1	ν_2	ν_3	
HF_2^-	K^+, Rb^+, Cs^+, NH_4^+, NMe_4^+	37	—	—	2·27 (NH_4HF_2)	600	1233	1473	a,b
HCl_2^- **	Cs^+	10·2	—	—	3·14[d] ($CsCl,1/3[H_3O^+HCl_2^-]$)	199	602,660	1670	a–d
	NMe_4^+	12·0	2·31	29·4	3·22[e] ($NMe_4^+HCl_2^-$)	210	?	1575	a,e,f
	NEt_4^+	13·7	4·28	28·6	—	—	?	730	a,f
	$NBu^n_4{}^+$	12·6–14·2	—	—	—	219	?	~1530	a,f
	$[Si(acac)_3]^+$	14·7	—	—	—	—	—	—	a
HBr_2^-	Cs^+	—	—	—	3·35[d] ($CsBr,1/3[H_3O^+HBr_2^-]$)	—	—	—	d
	NMe_4^+	9·4	1·80	27·8	—	126	?	1420	a,g
	NEt_4^+	11·6	—	—	—	160	?	—	a,g
	$NBu^n_4{}^+$	12·8	—	—	—	—	—	700	a,g
	$NPent^n_4{}^+$	—	—	—	—	150	?	770	g
HI_2^-	NEt_4^+	7·3	0·77	23·9	—	—	—	—	a
	$NBu^n_4{}^+$	12·4	1·60	32·8	—	—	?	1650–1700	a,h
$HFCl^-$	NEt_4^+	—	—	—	—	275	823,863	2710	i
$HFBr^-$	NEt_4^+	—	—	—	—	220	740	~2900	i
HFI^-	$NBu^n_4{}^+$	—	—	—	—	180	635	3145	i
$HClBr^-$	Cs^+	—	—	—	—	—	508	1705	b
	NMe_4^+	—	—	—	—	145	?	1890	j
	NEt_4^+	9·1†	1·54†	22·7†	—	170	?	1570,1650	a,j
	$NBu^n_4{}^+$	9·2†	—	—	—	—	?	1650,1730	a,j
$HClI^-$	Cs^+	—	—	—	—	—	485	2200	b
	$NBu^n_4{}^+$	8·6††	—	—	—	—	?	~2025	a,h

acac = acetylacetonate.

* Assignments are given on the basis of the general conclusions reached in reference b.

** ^{35}Cl nqr frequencies (in MHz) for solid derivatives of the HCl_2^- ion (temperature in °K): $CsHCl_2$, 20·47 (294); $CsCl,1/3[H_3O^+HCl_2^-]$, $11{\cdot}89_5$ (294); NMe_4HCl_2, 19·51 (294); NEt_4HCl_2, 11·89 (294)[k].

† NR_4Br + HCl.

†† NR_4I + HCl.

[a] D. G. Tuck, *Progress in Inorganic Chemistry*, **9** (1968) 161.

[b] J. W. Nibler and G. C. Pimentel, *J. Chem. Phys.* **47** (1967) 710.

[c] G. C. Stirling, C. J. Ludman and T. C. Waddington, *J. Chem. Phys.* **52** (1970) 2730.

[d] L. W. Schroeder and J. A. Ibers, *Inorg. Chem.* **7** (1968) 594.

[e] J. S. Swanson and J. M. Williams, *Inorg. Nuclear Chem. Letters*, **6** (1970) 271.

[f] J. C. Evans and G. Y.-S. Lo, *J. Phys. Chem.* **70** (1966) 11.

[g] J. C. Evans and G. Y.-S. Lo, *J. Phys. Chem.* **71** (1967) 3942.

[h] J. A. Salthouse and T. C. Waddington, *J. Chem. Soc. A* (1966) 28.

[i] J. C. Evans and G. Y.-S. Lo, *J. Phys. Chem.* **70** (1966) 543.

[j] J. C. Evans and G. Y.-S. Lo, *J. Phys. Chem.* **70** (1966) 20.

[k] C. J. Ludman, T. C. Waddington, J. A. Salthouse, R. J. Lynch and J. A. S. Smith, *Chem. Comm.* (1970) 405.

and 3·35±0·02 Å in the HBr_2^- ion; *ca.* 0·47 Å less than the radius sum of the appropriate X^- ions, these values imply the formation of very strong hydrogen bonds. The symmetrical structure of the HX_2^- ion in these salts, indicated by the presence of a mirror plane perpendicular to the X···X axis, is supported by the nqr[486] and vibrational[487] spectra. By contrast, the absence of such a mirror plane implies an unsymmetrical HCl_2^- anion in crystalline Me_4NHCl_2 [488]; in this case the Cl···Cl distance is 3·22±0·02 Å. In general, the results of X-ray[484,488], neutron-scattering[483], ^{35}Cl nqr[385,486], and vibrational spectroscopic[475,482,483,487] measurements give good grounds for believing that HX_2^- anions may exist either in a centrosymmetric or in a non-centrosymmetric form, the environment being the determining factor. The anions in the salts $CsHCl_2$, Me_4NHCl_2 and $Bu^n{}_4NHCl_2$ are thus presumed to be unsymmetrical, those in the salts Et_4NHCl_2, Pr_4NHCl_2 and $Pent_4NHCl_2$ to be symmetrical. Evidently, through its influence on the optimum energy of crystal-packing, the nature of the cation imposes a major constraint on the structure of the anion. Recent investigations[482,483] suggest that the HCl_2^- ion in $CsHCl_2$ is non-linear with the symmetry C_{2v} or C_s, while the conclusion that the salt Me_4NHCl_2, like KHF_2, has virtually no residual entropy at, or approaching, absolute zero[489] implies a potential energy curve for the anion having a single minimum, and so favours C_{2v} symmetry. For salts containing centrosymmetric HCl_2^- ions, the small shift in ^{35}Cl nqr frequency accompanying deuteration points to a flat or nearly flat potential well[486].

The analysis of the vibrational spectra of HX_2^- and HXY^- anions[475,482,483] has proved to be unexpectedly difficult, partly because most of the cations used to stabilize the anions themselves contribute a rich assortment of bands, and partly because the bands attributable to the anions are usually very broad at room temperature. Nevertheless, considerable clarification has been achieved by Nibler and Pimentel[482], who have developed an experimental technique for preparing, and obtaining the infrared spectra of, the caesium salts of HCl_2^-, DCl_2^-, $HClBr^-$, $HClI^-$ and $DClI^-$ at 20°K; under these conditions, the width of the absorption bands is considerably reduced, even to the point where some fine structure becomes apparent. Analysis of the results, leading to assignments incorporated in Table 34, indicates frequencies for the bending mode ν_2 of the anions which are about one-half the previously accepted values. The high intensity in infrared absorption of the overtone $2\nu_2$ (previously assigned as ν_2) is attributed to the abnormally large second derivative of the transition-dipole characteristic of hydrogen bonds, and arising from the asymmetry of the potential function[482]. Vibrational assignments and force constants, reported elsewhere[356,475] for ions of the types HX_2^- (X = Cl, Br or I), HFY^- (Y = Cl, Br or I) and $HClY^-$ (Y = Br, I or NO_3), and for some of their deuterated derivatives, are mostly subject to modification in the light of these findings. In no case do the results lend themselves to unambiguous deductions about the structures of the ions or about the form of the potential well in which the proton moves.

Salts of the HX_2^- and HXY^- anions have mostly been described as white crystalline compounds. Somewhat exceptional, therefore, are the coloured tropenium (Tr) derivatives $TrHX_2$, the absorption spectra of which (in dichloromethane solution) have been related to anion-cation charge-transfer processes (X = Cl or Br) or to internal transitions of the

[486] C. J. Ludman, T. C. Waddington, J. A. Salthouse, R. J. Lynch and J. A. S. Smith, *Chem. Comm.* (1970) 405.

[487] L. W. Schroeder, *J. Chem. Phys.* **52** (1970) 1972, 6453.

[488] J. S. Swanson and J. M. Williams, *Inorg. Nuclear Chem. Letters*, **6** (1970) 271.

[489] S.-S. Chang and E. F. Westrum, jun., *J. Chem. Phys.* **36** (1962) 2571.

anion (X = I). However, such studies are complicated by interference from absorptions originating from the cation, and also by the tendency of the anion to dissociate in solution[475]:

$$HX_2^- + S \rightleftharpoons S, H^+ + 2X^-$$

or

$$HX_2^- + S \rightleftharpoons S, HX + X^- \quad (S = \text{solvent molecule})$$

Thus, in contrast with HF_2^-, other HX_2^- or HXY^- ions appear to dissociate completely in aqueous solution. The ^1H nmr spectrum of a solution of the HCl_2^- or HBr_2^- anion in a basic solvent like dimethyl sulphoxide or acetonitrile exhibits only a single resonance line attributable to the acidic proton, indicating a rapid exchange of hydrogen between HX_2^- and either S,H^+ or S,HX.

In common with other thermally unstable complex halides (e.g. the polyhalides), solid derivatives of the HX_2^- or HXY^- anions gain in stability as the size of the cation M^+ increases; according to arguments of the type invoked elsewhere (see pp. 1251–2)[289,293], this behaviour can be correlated with the difference between the lattice energies (U) of the salts MHX_2 (or MHXY) and MX. Estimates of the free energy and enthalpy (ΔH_1) changes accompanying the reaction

$$MX(s) + HX(g) \rightleftharpoons MHX_2(s)$$

have been derived either from pressure-composition isotherms over a series of temperatures or from direct calorimetric measurements of the heat output. Hence, through the thermodynamic cycle[475]

$$\begin{array}{ccc} MX(s)+HX(g) & \xrightarrow{\Delta H_1} & MHX_2(s) \\ U(MX)+2RT \downarrow & & \uparrow -U(MHX_2)-2RT \\ M^+(g)+X^-(g)+HX(g) & \xrightarrow{\Delta H_2} & M^+(g)+HX_2^-(g) \end{array}$$

access has been gained to the enthalpy change ΔH_2 of the process

$$HX(g) + X^-(g) \rightarrow HX_2^-(g)$$

Used as an index to the strength of the H–X bond in HX_2^-, ΔH_2 is loosely termed the "hydrogen bond energy", though it is really a net enthalpy term since the H–X distance is almost certainly greater in the HX_2^- ion than in the parent HX molecule. A sufficiently large cation sees a convergence of the lattice energies U(MX) and $U(MHX_2)$, and hence of the enthalpy changes ΔH_1 and ΔH_2. In keeping with this, $-\Delta H_1$ for the reaction of hydrogen chloride with a tetra-alkylammonium chloride follows the order $Bu^n{}_4N^+ > Et_4N^+ > Me_4N^+$, and graphical arguments suggest that the values obtained for ΔH_1 are essentially at a maximum for the tetrabutylammonium cation. Minimal values have thus been deduced for the "hydrogen bond energies" in HCl_2^-, HBr_2^- and HI_2^-. The results of such experiments are quoted in Table 34, together with corresponding values of ΔG° and ΔS°, where these have been determined. For the HF_2^- ion, the same approach implies a "hydrogen bond energy" not much in excess of 37 kcal, whereas, it should be noted, estimates of the lattice energies of the salts MHF_2 (M = K, Rb or Cs) give a value of 58 ± 5 kcal. Since no entirely adequate explanation of this discrepancy has yet been found, doubts about the reliability of both calculations must prevail. Reference to salts including cations like tropylium and pyridinium, as well as tetra-alkylammonium, has shown, not unexpectedly, that ΔH_1 is influenced to some extent by factors other than the bulk of the cation.

A further application of calculations of this type is in the prediction or rationalization of decomposition processes affecting salts of HXY^- anions[289,475]. If we consider the alternative decomposition paths and construct the cycle

$$\begin{array}{ccccc}
MX(s) + HY(g) & \longrightarrow & M^+(g) + X^-(g) + HY(g) & \longrightarrow & M^+(g) + X(g) + HY(g) + e \\
\uparrow \Delta H_3 & & & & \downarrow \\
\boxed{MHXY(s)} & & & & M^+(g) + Y(g) + H(g) + X(g) + e \\
\downarrow \Delta H_4 & & & & \uparrow \\
MY(s) + HX(g) & \longrightarrow & M^+(g) + Y^-(g) + HX(g) & \longrightarrow & M^+(g) + Y(g) + HX(g) + e
\end{array}$$

it is evident that

$$\Delta H_3 - \Delta H_4 = \Delta U + \Delta E - \Delta D$$

where ΔU is the difference in lattice energies of MX and MY, ΔE the difference in electron affinities of X and Y, and ΔD the difference in the dissociation energies of HX and HY[289]. For $HBrCl^-$, ΔE is 6 kcal and ΔD 16 kcal. If, therefore, ΔU is less than 10 kcal and we ignore entropy factors, the products should be MBr and HCl. Since, even for CsCl and CsBr, ΔU is only 5 kcal, this condition is certain to be fulfilled in tetra-alkylammonium salts. Likewise, it can be reasoned that the decomposition of the salt $R_4NHClNO_3$ should afford R_4NNO_3 and HCl, as is actually observed. The determining factor in the decomposition appears to be the wide spread of bond energies in the HX molecules. By contrast, it has generally been taken for granted that the thermal decomposition of polyhalides always produces the simple halide having the smallest anion[289].

Chemical Properties[475]

Relatively little attention has been paid to the chemical reactions of salts of HX_2^- and HXY^- anions; information about the reactions is therefore sparse and provides little opportunity for systematic correlation.

The anions are generally unstable with respect to moisture, tending to decompose in moist air with the production of HX gas. This is a special case of the general reaction between the anions and bases B (q.v.):

$$HX_2^- + B \rightleftharpoons BH^+ + 2X^-$$

in which the bases B and X^- are in competition for the proton. The balance of such a process depends on the HX_2^- bond strength, the base strength of B and the states of the various components. Even washing with a base as mild as acetone causes $Bu^n{}_4NHI_2$ to lose HI. There appears to have been no report of the application of HX_2^- and HXY^- salts in synthetic chemistry, though it is possible that they may be useful as anhydrous sources of the appropriate hydrogen halide. It has also been suggested that the anions may be significant as intermediates in certain reactions.

Theoretical Aspects of the Bonding in HX_2^- *and* HXY^- *Anions*[475-477]

Despite the simplicity of the HX_2^- and HXY^- anions as isolated hydrogen-bonded entities, a generalized, accurate treatment of the bonding is still lacking. Although numerous models based on electrostatic or molecular-orbital treatments have been proposed, the qualitative predictive power of these models is strictly limited, and few quantitative predictions are possible.

The first theory proposed for hydrogen-bonding in various complexes was founded on the electrostatic model, which relates the strength of the hydrogen bond to the polarity of the H–X bond. The advantages and disadvantages of this model have been well summarized by Pimentel and McClellan[476]. Published calculations of hydrogen bond strength in HX_2^- anions have referred only to HF_2^-; although appealingly close agreement between calculated and experimental values is found, this is probably somewhat fortuitous, and cannot be taken as proof of the model's accuracy. A simple qualitative approach which assumes that the bonds in HX_2^- can be treated by an LCAO-MO description has been advanced by Pimentel[475,476]. The atomic orbitals involved are the hydrogen $1s$ and the halogen np_σ orbitals, the appropriate combinations of which yield bonding, non-bonding and anti-bonding molecular orbitals in a three-centre scheme formally very similar to that of Fig. 4. In the ground state of the HX_2^- ion the two electron pairs originally accommodated in the X^- orbitals occupy the bonding and the non-bonding molecular orbitals, giving the equivalent of two H–X bonds each of order 0·5. Alternatively, in the formalism suggested by Linnett[490], the structure can be represented in terms of two one-electron bonds

$$\circ\,|\overline{\underline{X}}\times H\circ\overline{\underline{X}}|\times$$

Calculations based on a more sophisticated molecular-orbital description and carried out on the HF_2^- ion suggest, *inter alia*, that the hydrogen $2p_\pi$ orbitals are quite important in the bonding scheme; the potential importance of π-bonding in HX_2^- has likewise been demonstrated by calculations of overlap integrals in HF_2^- and HCl_2^-.

Attempts have also been made to estimate the covalent contribution to hydrogen-bonding by considering a number of idealized formulations and attempting to deduce a wave equation which represents the appropriate mixing of the individual wavefunctions in such a way as to give an accurate description of the bond character. A comprehensive review of these and other calculations on the hydrogen bond is given elsewhere[491]. Of particular significance, however, are the findings of a recent analysis of the contributions to hydrogen-bonding made by coulombic interactions, electron-exchange, polarization of lone-pair electrons, dispersion forces and charge-transfer[492]: in summary, these are (a) that the hydrogen atom is unique because it has no inner shells and therefore the exchange energy in a hydrogen-bond is low; (b) that the coulombic energy is the largest attractive term but is not adequately represented by the dipole–dipole approximation; (c) that delocalization effects, which may be introduced into the model by charge-transfer terms, make an important contribution to the energy only for moderately strong hydrogen bonds, as in HX_2^- or HXY^-; and (d) that the coulombic, exchange and charge-transfer energies are all enhanced by a low s-character for the lone-pair electrons, the cumulative effect of all three being responsible for the general feature that orbitals low in s-character are much better acceptors with respect to hydrogen-bonding than those rich in s-character.

2. Addition reactions in which the H–X bond is broken. Examples of such reactions are the protonation of the Group V hydrides MH_3 (M = N, P or As) and of a wide range of organic bases: e.g.

$$MH_3 + HX \rightarrow [MH_4]^+X^-$$

---

[490] J. W. Linnett, *The Electronic Structure of Molecules: A New Approach*, Methuen, London (1964).
[491] S. Bratož, *Adv. Quant. Chem.* 3 (1967) 209.
[492] J. N. Murrell, *Chem. in Britain*, 5 (1969) 107.

and the addition to olefins, acetylenes, epoxides and related organic compounds: e.g.

$$\gt C{=}C\lt \; + \; HX \rightarrow -\overset{|}{\underset{H}{C}}-\overset{|}{\underset{X}{C}}-$$

$$\gt \underset{C}{\overset{C}{|}}\!>\!O \; + \; HX \rightarrow \; \begin{matrix} \gt C-OH \\ | \\ \gt C-X \end{matrix}$$

as well as to certain multiply bonded or low-valent inorganic compounds: e.g.

$$SO_3 + HCl \rightarrow ClSO_2OH$$

$$(HBNH)_3 + 3HX \rightarrow (HXB\text{–}NH_2)_3$$

(though the exact nature of this product is still open to question)[493]

$$MX_2 + HX \rightarrow HMX_3 \; (M = Si \text{ or } Ge)^{424,494}$$

The reaction of a gaseous base B with the gaseous hydrogen halide to form a solid salt BH^+X^- may be discussed in terms of the following cycle[289,293]:

$$\begin{array}{ccc} B(g) + HX(g) & \xrightarrow[\text{H}^+\text{and X}^-,\ \Delta H^\circ_{diss}]{\text{Dissociation of HX into}} & B(g) + H^+(g) + X^-(g) \\ \downarrow \Delta H^\circ & & \downarrow \text{Combination of B and H}^+ \\ BH^+X^-(s) & \xleftarrow[-U_L[BH^+X^-]-2RT]{} & BH^+(g) + X^-(g) \end{array}$$

The differences in entropy changes for processes involving different hydrogen halides are small, and accordingly the relative stability of the salt BH^+X^- is primarily a function of the standard enthalpy change ΔH°. For halides of the same cation,

$$\Delta H^\circ = -\,U_L[BH^+X^-] + \Delta H^\circ_{diss} + \text{constant}$$

At 298°K ΔH°_{diss} = 369·8, 333·5, 323·6 and 314·3 kcal for X = F, Cl, Br and I, respectively. If, therefore, the solid fluoride BH^+F^- is to be more stable with respect to dissociation than the solid iodide BH^+I^-, the lattice energy of BH^+F^- must exceed that of BH^+I^- by at least 55·5 kcal. The difference in lattice energies depends, however, on the effective radius of the cation BH^+, $r(BH^+)$; according to the Kapustinskii approximation, the preceding condition requires that

$$512\left[\frac{1}{r(BH^+)+1{\cdot}33} - \frac{1}{r(BH^+)+2{\cdot}19}\right] > 55{\cdot}5$$

or that $r(BH^+) <$ *ca.* 1·1 Å. Extension of this reasoning to other pairs of halides shows that the iodide should be the salt most stable with respect to dissociation into gaseous B and HX for any cation BH^+ with a radius greater than *ca.* 1·6 Å. Setting aside complications which may arise from the precise shape of the BH^+ cation, or from hydrogen-bonding in one or more of the halides, we find that the predictions are almost entirely fulfilled in practice. Thus, for the NH_4^+ ion with a thermochemical radius of 1·45 Å, ΔH° = −35·1, −42·1, −45·0 and −43·5 kcal for X = F, Cl, Br and I, respectively, while for the larger

[493] E. K. Mellon, jun., and J. J. Lagowski, *Adv. Inorg. Chem. Radiochem.* **5** (1963) 259; K. Niedenzu and J. W. Dawson, *The Chemistry of Boron and its Compounds* (ed. E. L. Muetterties), p. 377. Wiley (1967).

[494] O. M. Nefedov and M. N. Manakov, *Angew. Chem., Internat. Edn.* **5** (1966) 1021.

PH_4^+ ion, no fluoride of which has been isolated, the values of $\Delta H°$ (in kcal) are: PH_4Cl, $-14{\cdot}0$; PH_4Br, $-23{\cdot}1$; PH_4I, $-24{\cdot}3$. Hence, it is not surprising that experiments involving the co-condensation of arsine and a hydrogen halide at 110°K yielded infrared evidence for the formation of AsH_4^+ in the case of the bromide and iodide only, and suggested that the latter is the more stable system[495]. Here, as in other aspects of the acidity of the hydrogen halides, it is the relative weakness of the H–Br and H–I bonds that is the most important single factor in determining the relative magnitudes of the overall energy changes.

All the hydrogen halides add to $\rangle C{=}C\langle$ and —C≡C— units in a great variety of organic compounds, including conjugated dienes, where both 1,2- and 1,4-additions are possible[496,497]. Under normal conditions, the addition is presumed to take place by an electrophilic mechanism; the rate-determining step is protonation of the multiply bonded system, possibly *via* an initially formed π-complex, the addition being completed by subsequent attack of the nucleophile X^-.

$$\rangle C{=}C\langle \xrightarrow[\text{slow}]{H^+} H{-}\overset{|}{\underset{|}{C}}{-}\overset{|}{\underset{|}{C}}^{\oplus} \xrightarrow{X^-} H{-}\overset{|}{\underset{|}{C}}{-}\overset{|}{\underset{|}{C}}{-}X$$

In support of this, it is found that the ease of addition increases in the series HF < HCl < HBr < HI, which reflects the increasing acid strength of the hydrogen halides. Studies of the stereochemistry of the process reveal no fixed pattern; whereas some reactions are stereospecific, others are not. Foreseeably, since the rate-determining step involves electrophilic attack, the reaction is assisted by electron-repelling substituents and retarded by halogens or other electron-withdrawing groups attached to the π-bonded carbon atoms. The orientation of the hydrogen and halogen atoms in the product is usually defined by the empirical Markownikov rule, the halogen attaching itself to the site of lower electron density; this course is determined by the relative stabilities of the intermediate carbonium ions. In solution in water or hydroxylic solvents, acid-catalysed hydration

$$\rangle C{=}C\langle \longrightarrow H{-}\overset{|}{\underset{|}{C}}{-}\overset{|}{\underset{|}{C}}{-}OH$$

constitutes a competing reaction. Less polar solvents encourage radical-formation, and, in the presence of peroxide catalysts, hydrogen bromide has the capacity to add to unsaturated molecules by a free-radical mechanism, leading to a reversal of the normal

495 A. Heinemann, *Naturwiss.* 48 (1961) 568.

496 See for example J. Hine, *Physical Organic Chemistry*, 2nd edn., McGraw-Hill (1962); R. T. Morrison and R. N. Boyd, *Organic Chemistry*, 2nd edn., Allyn and Bacon, Boston (1966); J. March, *Advanced Organic Chemistry: Reactions, Mechanisms and Structure*, McGraw-Hill (1968); P. Sykes, *A Guidebook to Mechanism in Organic Chemistry*, 3rd edn., Longmans, London (1970); C. K. Ingold, *Structure and Mechanism in Organic Chemistry*, 2nd edn., Bell, London (1969).

497 B. Capon, M. J. Perkins and C. W. Rees (eds.), *Organic Reaction Mechanisms*, Interscience (1965–7); B. Capon and C. W. Rees (eds.), *Organic Reaction Mechanisms*, Interscience (1968–70).

orientation of the added atoms.

$$1 \quad \underset{\substack{\text{Radical} \\ \text{initiator}}}{RO\cdot} + HX \longrightarrow ROH + X\cdot$$

$$2 \quad >C{=}C< + X\cdot \longrightarrow \cdot\overset{|}{\underset{|}{C}}-\overset{|}{\underset{|}{C}}-X$$

$$3 \quad \cdot\overset{|}{\underset{|}{C}}-\overset{|}{\underset{|}{C}}-X + HX \longrightarrow H-\overset{|}{\underset{|}{C}}-\overset{|}{\underset{|}{C}}-X + X\cdot$$

(see Section 2, p. 1167). In this case, attack is initiated by a halogen atom. The virtually complete control of orientation which can be achieved in the addition of hydrogen bromide to unsaturated molecules by introducing radicals or radical-acceptors has been turned to advantage in organic synthesis. Hydrogen bromide is unique among the hydrogen halides in this respect because the steps of the free-radical mechanism are all exothermic. With hydrogen fluoride, stage (1) is strongly endothermic, and, though with hydrogen iodide this stage is energetically favoured, the iodine atoms formed are not sufficiently reactive to promote the later stages. With hydrogen chloride, radical-addition has been observed only in a few cases, but the reaction chains are usually so short at ordinary temperatures as to make this path less attractive than the electrophilic mechanism.

It is likely, but by no means certain, that addition of hydrogen chloride to the multiple bonds of CO or RCN (R = H or an organic group), aided by a chloride ion-acceptor, is an essential prelude to reactions with aromatic compounds (ArH), as in the so-called Gatterman–Koch, Gatterman or Hoesch reactions:

$$CO + HCl \xrightarrow{AlCl_3} [HCO]^+AlCl_4^- \xrightarrow[CuCl]{ArH} ArCHO$$

$$RCN + HCl \xrightarrow{ZnCl_2} [R{\cdot}C{=}NH]^+Cl^- \xrightarrow[\text{hydrolysis}]{ArH} ArCOR$$

and with other compounds, e.g.

$$RCN + R'OH \xrightarrow[\substack{\text{anhydrous} \\ \text{conditions}}]{HCl} \underset{\substack{\text{hydrochloride of} \\ \text{imino ester}}}{[R{\cdot}C(OR'){=}NH_2]^+Cl^-} \xrightarrow{\text{hydrolysis}} R{\cdot}C(OR'){=}O$$

3. Substitution reactions. Such reactions inevitably entail cleavage of the H–X bond; depending on the subsequent fate of the fragments, the hydrogen halide exercises in any given reaction at least two of the following possible functions: oxidation, reduction, protonation or halogenation.

$$HX \longrightarrow \underset{\text{Protonating agent}}{H^+} + \underset{\text{Halogenating agent}}{X^-}$$

Protonating agent H^+ — Reducing agent, e.g. metal → $\frac{1}{2}H_2$

Halogenating agent X^- — Oxidizing agent → $\frac{1}{2}X_2$

Oxidation: Reaction with Metals

For the reaction of a gaseous hydrogen halide with an element M to proceed

$$M + nHX \rightarrow MX_n + n/2H_2$$

it is necessary but not sufficient that $\Delta G_f°[MF_n] < -65{\cdot}3n$, $\Delta G_f°[MCl_n] < -22{\cdot}78n$,

$\Delta G_f°[MBr_n] < -12{\cdot}77n$ or $\Delta G_f°[MI_n] < +0{\cdot}38n$ kcal mol^{-1}, thermodynamic conditions which imply that most metals should react[436]. In practice, the facility of reaction between metals and the hydrogen halides varies markedly with the nature of the metal and with its physical state, though, with most metals in the massive state, reaction is slow at all but elevated temperatures. As noted in Table 22, reactions such as

$$Co + 2HBr \xrightarrow{\text{Red heat}} CoBr_2 + H_2$$

$$Pu + 3HI \xrightarrow{250\text{–}300°C} PuI_3 + 3/2H_2$$

provide expedient methods of preparing anhydrous metal halides, while the analogous process

$$Si + 2HX \xrightarrow{\sim 350°C} SiX_2 + H_2$$

is probably the first stage of the reaction of a hydrogen halide with silicon to produce (mainly) $HSiX_3$ and SiX_4. The formation of a relatively volatile metal halide, through the action of a gaseous hydrogen halide, has been exploited to effect vapour-phase transport of the metal (e.g. iron or nickel) at temperatures well below those required to cause significant vaporization of the pure metal[425].

Aqueous hydrohalic acids attack most metals in accordance with the equation

$$M + nH_3O^+ \rightarrow M^{n+} + nH_2O + n/2H_2$$

However, the thermodynamic feasibility of the reaction depends not only on the standard electrode potential of M, but also upon the concentration of the acid, the solubility of the halide MX_n, and the stability of potential complex species. The importance of this last feature is illustrated by the observation that, whereas copper is not readily attacked by hydrochloric acid, the presence of thiourea, which complexes with the Cu^+ ion, causes the metal to dissolve in the 1M acid with a brisk evolution of hydrogen. The thermodynamic and kinetic readiness of reaction may also depend upon the presence of an oxidizing agent, e.g. air or an added agent such as nitric acid. Detailed studies of the dissolution of metals in hydrochloric acid have shown that the rate of reaction depends on the following variables: temperature, concentration of acid, the physical form of the metal (including the mechanical or thermal treatment it may have undergone), access of the solution species to the metal surface, the nature, distribution and concentration of impurities in the metal, and the presence of complexing, oxidizing or reducing agents in the liquid phase. Of these factors, the rate at which the reactive species in solution can diffuse to the metal surface is influenced by the viscosity of the acid solution, by the presence of an oxide film on the metal surface, and by the rate at which the acid solution is stirred or the metal sample rotated. The formation of a coherent, insoluble film of halide on the surface of the metal inevitably inhibits continued dissolution. The action of impurities in the metal or aqueous phase probably depends, at least in many cases, on the effect they have on the localized electrochemical cells set up on the surface of the metal. Metals notable for their resistance to attack by hydrochloric acid have been identified in connection with the commercial production and handling of the acid. The degree of corrosion of metals by the acid may be effectively reduced in the presence of dissolved inhibitors, e.g. phenolic compounds or quinoline.

Reducing Action of the Hydrogen Halides

In keeping with the standard potential of the couple $\frac{1}{2}X_2/X^-$, the hydrogen halides become increasingly strong reducing agents in the series HCl < HBr < HI. Illustrative of the redox reactions common to all three compounds are the following[418,424–426,436]:

$$R\cdot + HX \rightarrow R{\cdot}H + X\cdot \quad (R = \text{H or an organic group})$$
$$O_2 + 4HX \rightarrow 2X_2 + 2H_2O$$
$$2P + 8HX \rightarrow 2PH_4X + 3X_2$$
$$MO_2 + 4HX \rightarrow MX_2 + X_2 + 2H_2O \qquad (M = \text{Mn or Pb})$$
$$H_2O_2 + 2H_3O^+ + 2X^- \rightarrow X_2 + 4H_2O$$
$$XO_3^- + 6H_3O^+ + 5X^- \rightarrow 3X_2 + 9H_2O$$
$$UF_6 + 2HX \rightarrow UF_4 + 2HF + X_2$$

Other agents with the capacity to oxidize each hydrogen halide to the corresponding molecular halogen include atomic nitrogen and oxygen, ozone, fluorine, hypochlorite, vanadate, selenate and tellurate, persulphate, permanganate, periodate and ruthenium and osmium tetroxide. On the other hand, only hydrogen bromide and iodide are oxidized by hot, concentrated sulphuric acid or by chromates or chlorine, while the power of hydrogen iodide as a reducing agent is indicated by its reactions with sulphur, sulphur chlorides, interhalogens, bromine, iron(III), copper(II), oxy-nitrogen compounds, phosphorus(V), arsenic(V) and antimony(V) derivatives, tetrasulphur tetranitride and organo-halogen compounds, e.g.[418,424–426,436]

$$2X^- + Cl_2 \xrightarrow{\text{aq soln}} X_2 + 2Cl^-$$

(X = Br or I; primary step in the manufacture of bromine and iodine—see Section 2, pp. 1136–40).

$$2HI + S \underset{\text{aq soln}}{\overset{\text{anhydrous}}{\rightleftharpoons}} I_2 + H_2S$$

$$SOCl_2 + 4H_3O^+ + 6I^- \longrightarrow H_2S + 2Cl^- + 5H_2O + 3I_2$$
$$N_2O + 10H_3O^+ + 8I^- \longrightarrow 2NH_4^+ + 11H_2O + 4I_2$$
$$NO_2^- + 2H_3O^+ + I^- \longrightarrow NO + 3H_2O + \tfrac{1}{2}I_2$$
$$H_3AsO_4 + 2H_3O^+ + 2I^- \underset{\text{alkali}}{\overset{\text{acid}}{\rightleftharpoons}} H_3AsO_3 + 3H_2O + I_2$$
$$S_4N_4 + 24H_3O^+ + 20I^- \longrightarrow 4H_2S + 4NH_4^+ + 24H_2O + 10I_2$$
$$RI + H_3O^+ + I^- \longrightarrow RH + H_2O + I_2$$

Nitric acid is reduced by excess concentrated hydrohalic acid, but the product varies from nitrosyl chloride,

$$NO_3^- + 3Cl^- + 4H_3O^+ \rightarrow NOCl + Cl_2 + 6H_2O$$

believed to be the principal active agent of the mixture well known as aqua regia, to nitric oxide:

$$NO_3^- + 3I^- + 4H_3O^+ \rightarrow NO + 3/2I_2 + 6H_2O$$

The reaction of a hydrogen halide molecule with atomic hydrogen or an organic radical R· represents a relatively well-established propagation stage of the chain reaction between the parent halogen and either H_2 or RH; as such, it has been referred to in Section 2 (pp. 1168–9). The oxidation of the hydrogen halides by molecular oxygen is accelerated by photolytic action or by the agency of various catalysts. Detailed studies suggest that the homogeneous reaction of hydrogen bromide with oxygen proceeds by the following mechanism:

$$HBr + O_2 \rightarrow HOOBr$$
$$HOOBr + HBr \rightarrow 2HOBr$$
$$HOBr + HBr \rightarrow H_2O + Br_2$$

Added inert gases decrease the rate, probably by accelerating the decomposition of the intermediate HOOBr. By contrast, the pathway of the photochemical oxidation of hydrogen iodide is believed to be

Initiation	HI	$\xrightarrow{h\nu} H\cdot + I\cdot$
Propagation	$H\cdot + HI$	$\rightarrow H_2 + I\cdot$
	$H\cdot + O_2$	$\rightarrow HO_2\cdot$
	$HO_2\cdot + HI$	$\rightarrow H_2O_2 + I\cdot$
Termination	$I\cdot + I\cdot$	$\rightarrow I_2$

As in the hydrogen–oxygen reaction, $HO_2\cdot$ plays a central role in the mechanism.

Under forcing conditions, oxidation of the halide ions may give rise to oxyhalogen species: thus, depending on the exact conditions of concentration and current density, anodic oxidation of aqueous hydrochloric acid may yield either chloric or perchloric acid, while, in the presence of potassium persulphate, a mixture of hydriodic acid and silver nitrate is oxidized to the sparingly soluble periodate Ag_3IO_5. Generally, however, oxidation of halide to oxyhalogen ions by chemical means is more easily accomplished in alkaline media (see Fig. 2).

Protonation and Halogenation

The protonating action of aqueous hydrohalic acids is familiar through reactions such as

$$O^{2-} + 2H_3O^+ \rightarrow 3H_2O$$

$$S^{2-} + 2H_3O^+ \rightarrow H_2S + 2H_2O$$

and

$$CO_3^{2-} + 2H_3O^+ \rightarrow CO_2 + 3H_2O$$

which are significant as methods of bringing metal ions into solution (for example, in qualitative analysis), as wet methods of producing metal halides or halide complexes (see Table 22), and as laboratory routes to hydrogen sulphide and carbon dioxide. Anhydrous halides or oxyhalides are formed by reaction of the gaseous hydrogen halide—most commonly the chloride—with metal oxides at elevated temperatures[436]; for example, the reaction of HCl with Sb_2O_3 to form $SbCl_3$ is complete in 45 min at 300°C. Transport reactions have been described[425] whereby a metal initially in the form of an involatile oxide, e.g. BeO, Al_2O_3 or TiO_2, is converted at elevated temperatures into a relatively volatile chloride or oxychloride by a stream of gaseous hydrogen chloride. Crystals of a material like FeOCl have thus been prepared.

Nitrides, borides, silicides, germanides and certain carbides are also susceptible to protonation by the hydrogen halides in solution or in the gaseous phase, usually with the formation of the corresponding hydrides, e.g.

$$Mg_3N_2 + 6HX \rightarrow 3MgX_2 + 2NH_3$$

though agents less volatile and less prone to undergo side-reactions are generally preferred for the preparation of these hydrides. In the chemistry of boranes, however, anhydrous hydrogen chloride is widely favoured as a means of protonating anionic derivatives to produce neutral molecules, e.g.

$$B_8H_{14}{}^{498a} \xleftarrow{B_8H_{12}^-} HCl \xrightarrow{B_5H_8^-} B_5H_9{}^{498d}$$

$$HCl \xrightarrow{B_9H_{14}^-} B_9H_{15}{}^{498b}$$

$$HCl \xrightarrow{B_{10}H_{10}^{2-},\ Et_2S} B_{10}H_{12}(SEt_2)_2{}^{498c}$$

The halogenating action of gaseous hydrogen halides is uppermost in reactions such as

$$MH_4+HX \rightarrow MH_3X+H_2$$

and

$$MH_3X+HX \rightarrow MH_2X_2+H_2 \quad (M = \text{Si or Ge but not C})$$

Catalysed by the appropriate aluminium halide, these provide a useful method of synthesizing halogen-substituted silanes and germanes. Likewise, the gaseous hydrogen halides halogenate diborane and other neutral boranes to give, for example, the terminally substituted derivatives B_2H_5X (X = Br or I)[499]. Probably as a sequel to protonation, metal–carbon bonds in many organometallic compounds, C–O bonds in alcohols and ethers, and C–N bonds in diazoketones and tertiary aromatic amines are also subject to halogenation by the hydrogen halides under various conditions, e.g.

$$SnPh_4+HX \xrightarrow{\text{aq soln}} Ph_3SnX+PhH^{500}$$

$$2(\pi\text{-}CH_2CHCH_2)_2Ni+2HCl \xrightarrow[\text{conditions}]{\text{anhydrous}} [(\pi\text{-}CH_2CHCH_2)NiCl]_2+2CH_2{=}CHCH_3{}^{501}$$

$$(R_3P)_2PtMe_2+HCl \xrightarrow[\text{conditions}]{\text{anhydrous}} (R_3P)_2PtMeCl+MeH^{501}$$

$$ROH+HX \xrightarrow[\text{conditions}]{\text{anhydrous}} RX+H_2O^{496}$$

$$ROR'+HX \xrightarrow[\text{acid}]{\text{conc aqueous}} ROH+R'X \quad (X = \text{Br or I};\ R' = \text{alkyl group})^{496}$$

$$ArNR_2+2HX \xrightarrow[\text{acid}]{\text{conc aqueous}} RX+[ArNH_2R]^+X^- \quad (X = \text{Br or I})^{496}$$

By these means, certain organometallic halides and organo-halogen compounds are expediently prepared, while the cleavage of methoxy groups by constant-boiling hydriodic acid forms the basis of the Zeisel method of estimating such groups in aromatic ethers. Further, reactions of this kind are probably involved in the degradation of naturally occurring organic materials, e.g. cellulose, starch and gelatin, by the hydrogen halides, either in the gaseous or concentrated aqueous phase.

Typical of the hydrogen- or halogen-exchange reactions involving the molecular hydrogen halides are the following:

[498] (a) J. Dobson and R. Schaeffer, *Inorg. Chem.* **7** (1968) 402; (b) J. Dobson, P. C. Keller and R. Schaeffer, *Inorg. Chem.* **7** (1968) 399; (c) M. D. Marshall, R. M. Hunt, G. T. Hefferan, R. M. Adams and J. M. Makhlouf, *J. Amer. Chem. Soc.* **89** (1967) 3361; (d) A. J. Downs, G. M. Sheldrick and J. J. Turner, *Ann. Rep. Chem. Soc.* **64A** (1967) 234.

[499] M. F. Hawthorne, *The Chemistry of Boron and its Compounds* (ed. E. L. Muetterties), p. 223. Wiley (1967).

[500] C. S. G. Phillips and R. J. P. Williams, *Inorganic Chemistry*, Vol. II, p. 563. Clarendon Press, Oxford (1966).

[501] G. E. Coates, M. L. H. Green and K. Wade, *Organometallic Compounds*, 3rd edn., Vol. II, Methuen, London (1968).

$$D_2+HX \xrightleftharpoons{\text{gas phase}} HD+DX^{426}$$

$$B_{10}H_{14}+6DCl \xrightleftharpoons{AlCl_3,CS_2} B_{10}H_8D_6+6HCl^{425}$$

$$RH+DCl \xrightleftharpoons{\text{gas phase}} RD+HCl^{425,426}$$

$$RX+HY \xrightleftharpoons{\text{gas phase}} RY+HX^{502}$$

$$X_2+HY \xrightleftharpoons{\text{gas phase}} XY+HX$$

(used as bases for chemical lasers)[436,502]

$$MX_n+mHY \xrightleftharpoons[\text{phase}]{\text{gas or liquid}} MX_{n-m}Y_m+mHX^{502}$$

[R = organic group; M = B, Al, C, Si, Sn, P or As; X, Y = same or different halogen.]

In addition, there have been numerous qualitative or quantitative accounts of halide-exchange implicating complex halide or organo-halogen species in polar media[502]. The kinetic and thermodynamic properties of some of these reactions, together with the effects of chemical or photochemical catalysis, have been explored, notably with the aid of isotopically labelled species, in attempts to elucidate their mechanisms. The advantage taken of exchange reactions for isotopic substitution is illustrated by the reaction between decaborane(14) and deuterium chloride (whereby deuteration of specific sites of the B_{10} framework is achieved), by the preparation of deuterated benzenes through the action of deuterium chloride on benzene in the presence of aluminium chloride, and by some of the methods which have been employed to produce tritium chloride. Similarly, catalyzed exchange reactions involving hydrogen bromide afford a practical means of converting chloro- to corresponding bromo-alkanes[345], though, in analogous situations including hydrogen iodide, reduction commonly prevails over halogen-exchange.

3.4. DETECTION AND ANALYTICAL DETERMINATION OF THE HYDROGEN HALIDES AND HALIDE IONS[345,426,503,504]

Hydrogen Halides

In commercial practice, the concentration of hydrochloric acid is commonly measured in terms of its specific gravity. In the laboratory, however, it is normal to assay hydrochloric and the other hydrohalic acids either volumetrically, e.g. by titration with standard base, or gravimetrically by precipitation of the silver halide. Hydrochloric acid is in common use as a primary or secondary standard in chemical analysis, a context in which various methods have been described for the preparation of solutions of accurately defined concentration. Impurities in the reagent-grade concentrated acid may include free halogens ($< 10^{-4}$%), sulphate and sulphite (each $< 10^{-4}$%), bromide ($< 5\times10^{-3}$%), ammonium ions ($< 3\times10^{-4}$%), arsenic ($< 10^{-6}$%), iron ($< 2\times10^{-5}$%), heavy metals ($< 10^{-4}$%), extractable organic materials ($< 5\times10^{-4}$%) and involatile matter ($< 5\times10^{-4}$%). Tests of these specifications and of the optical transparency of the acid are described elsewhere[420,504]; the nature of the manufacturing process determines which tests are most relevant. Hydrogen halides in the vapour phase, irrespective of concentration, are first trapped by absorption in a suitable medium, e.g. water, standard sodium hydroxide or

[502] M. F. A. Dove and D. B. Sowerby, *Halogen Chemistry* (ed. V. Gutmann), Vol. 1, p. 41. Academic Press (1967).

[503] G. W. Armstrong, H. H. Gill and R. F. Rolf, *Treatise on Analytical Chemistry* (ed. I. M. Kolthoff, P. J. Elving and E. B. Sandell), Part II, Vol. 7, p. 335. Interscience (1961).

[504] V. A. Stenger, *Encyclopedia of Industrial Chemical Analysis* (ed. F. D. Snell and L. S. Ettre), Vol. 8, p. 1. Interscience (1969); G. Oplinger, *ibid.* Vol. 9, p. 333. Interscience (1970).

sodium carbonate solution, which is subsequently analysed by acid-base titration or by treatment with silver nitrate solution.

Detection and Separation of Halide Ions

Classically, detection of the halide ions in aqueous solution is normally accomplished by precipitation of the sparingly soluble silver halide in acidic solution. The individual ions may be distinguished by the colour of the silver halide and by its solubility in ammoniacal solution, which decreases in the order AgCl > AgBr > AgI. In admixture with the other silver halides, silver chloride can be identified by its solubility in sodium arsenite solution; subsequent acidification of the solution again yields a positive test for chloride when silver nitrate is added. However, the most distinctive tests for individual halide ions are based on their redox properties. Thus, in dilute aqueous solution chloride ions are oxidized to chlorine only by the strongest oxidizing agents. Even with a dichromate and concentated sulphuric acid, a solid chloride yields, not the elemental halogen (as does a bromide or iodide), but chromyl chloride; this affords one of the more distinctive positive tests for chloride ions. By contrast, the presence of bromide, even in a solution rich in chloride ions, can be established by treatment of the acidified solution with hypochlorite, permanganate or hydrogen peroxide; the bromine liberated is then characterized by its yellow-brown colour in carbon tetrachloride solution or by the coloration it produces with a dyestuff like fluorescein, fuchsin or *o*-naphthoflavone (see Section 2, p. 1230); alternatively bromide is detected by oxidation to bromate by hypochlorite in hot alkaline solution. Iodides are oxidized to iodine by relatively mild agents, e.g. iron(III) or nitrite; the iodine is typically detected by its action on starch solution or by the colour of its solution in an organic solvent (see Section 2, p. 1230). Iodides, if present, must be removed before the test for bromine is carried out. This may be accomplished by oxidizing the iodide with iron(III) or nitrite and boiling the solution until the free iodine has been expelled. Mixtures of all three halide ions have been identified by selective oxidation, for example, with either persulphate or permanganate. Thus, in acetic acid solution, only iodide is oxidized by persulphate; acidification with sulphuric acid and the addition of some persulphate then oxidizes bromide but not chloride.

Efficient separation of the halide ions in solution is achieved by the use of a strong-base anion-exchange resin, a solution of sodium nitrate typically being used as the eluent. After separation, the eluted halide ions can be estimated by normal methods (see below). The anions have likewise been separated by the techniques of paper and thin-layer chromatography. In either case, an individual halide can be identified by its relative retention time (R_f value), while a semi-quantitative measure of its concentration is gained from the intensity of the developed spot. Mixtures of volatile halides are usually fractionated by distillation or by the techniques of gas-liquid or gas-solid chromatography. Thus, one of the most rigorous procedures devised for the separation of the halide ions[505] exploits the selective oxidation principle in conjunction with distillation of the parent halogen. First, iodide is oxidized with hydrogen peroxide, and then bromide with 50% nitric acid; at each stage, the free halogen is distilled from the mixture, condensed and trapped in scrubbing bottles containing hydrazine sulphate solution, and then estimated as the halide either by potentiometric or by turbidimetric measurements.

Estimation of the Halide Ions[345,426,503,504]

The principal methods applicable to the quantitative analysis of chloride, bromide and iodide ions are summarized in Table 35. In practice, the anions are usually determined

[505] T. J. Murphy, W. S. Clabaugh and R. Gilchrist, *J. Res. Nat. Bur. Stand.* **53** (1954) 13.

Table 35. Methods Available for the Quantitative Estimation of Chloride, Bromide and Iodide Ions[a–d]

Halide ions	Method	Reaction type	Reagent	Comments
1. Cl^-, Br^-, I^-	Gravimetric	Precipitation	Silver nitrate	Still probably the most accurate method of determining halide ions.
2. Cl^-, Br^-, I^-	Volumetric	Precipitation	Silver nitrate	(a) End-point detected using K_2CrO_4 as indicator, pH 5–7 (Mohr titration); unsatisfactory for iodides. (b) End-point detected using an adsorption indicator, e.g. fluorescein, dichlorofluorescein, eosin, diphenylcarbazone or *p*-ethoxychrysoidine[e] (Fajans method). (c) End-point found by adding excess $AgNO_3$ and then back-titrating with a standard thiocyanate solution using iron(III) as the indicator (Volhard method); widely used to estimate total halide concentration. (d) End-point determined potentiometrically (may be used to estimate mixtures of halide ions). Differential electrolytic potentiometry[f] enables the end-point to be located with enhanced precision; this method has been used to determine nanogram amounts of halide at extreme dilution. (e) End-point determined by amperometric methods, e.g. the dead-stop technique.
3. Cl^-, Br^-, I^-	Volumetric	Complex-formation	Mercury(II) nitrate or perchlorate	First excess of Hg^{2+} ions detected typically using diphenylcarbazide or diphenylcarbazone as indicator; conductimetric methods have also been used.[g]
4. Br^-	Volumetric	Oxidation-reduction	Hypochlorite	Basis of van der Meulen method and numerous variations of this method. Br^- oxidized to BrO_3^-, which is then estimated by iodimetric titration. Widely used for the determination of bromide in the presence of chloride.
			Potassium permanganate, chromic acid and other agents in presence of CN^-	Br^- oxidized to BrCN. The reaction can be monitored potentiometrically or the BrCN estimated by iodimetric titration; to estimate traces of bromide in the presence of moderate amounts of chloride, the BrCN is distilled and determined potentiometrically by a sensitive null-point method.[d]

TABLE 35 (*cont.*)

Halide ions	Method	Reaction type	Reagent	Comments
5. I^-	Volumetric	Oxidation-reduction	Potassium iodate, nitrite and other agents in acid solution	I^- oxidized to I_2, which is estimated with standard thiosulphate or arsenite solution.
			Potassium iodate in a strongly acid medium	I^- oxidized to species such as ICl_2^-. End-point determined typically by the disappearance of free iodine from an organic solvent in contact with the reaction mixture.
			Chlorine or bromine water, hypochlorite or potassium permanganate	I^- oxidized to IO_3^-, which is then estimated by iodimetric titration.
6. Cl^-, Br^-, I^-	Spectrophotometric	Oxidation-reduction	Oxidation, e.g. with hypochlorite or nitrite	Free halogen (usually Br_2 or I_2) estimated either dissolved in an organic solvent or following reaction with a suitable reagent, e.g. rosaniline, phenol red, starch or palladium iodide (see p. 1230).
		Complex-formation	Miscellaneous	Optical density due to halide complex determined at a suitable frequency. A variation of this method is involved in the estimation of Cl^- with $Hg(SCN)_2$, whereby the SCN^- liberated is determined colorimetrically with Fe^{3+}.
7. Cl^-, Br^-, I^-	Nephelometric or turbidimetric	Precipitation	Silver nitrate	Favoured for the determination of chloride in trace amounts.
8. Cl^-, Br^-, I^-	Potentiometric	Oxidation-reduction	—	Potential difference measured between reference electrode and sensing electrode which is reversible with respect to halide ions, e.g. Ag/AgX. Method easily automated and can be used to monitor halide ion concentrations, for example, in industrial wastes.
9. Cl^-, Br^-, I^-	Coulometric	Electrolysis	—	Constant-current and constant-potential versions of the method have been used. Electrolytically generated reagents, e.g. Ag^+ or Hg_2^{2+}, have been employed as the titrant to determine halide ions in this way. Such methods have the advantage of precision, convenience and rapidity of determination, and ease of automation; applicable to low concentrations and to halide ions in admixture.

10. Cl^-, Br^-, I^-	Polarographic	Electrolytic oxidation or reduction	—	Most extensively used following quantitative oxidation of X^- to XO_3^- ($X = Br$ or I), which is then reduced irreversibly back to X^- at a dropping mercury electrode; microgram quantities of halide can thus be estimated. Anodic waves have also been used to estimate halide ions, e.g. Br^- in blood.[b]
11. Cl^-, Br^-, I^-	X-ray fluorescence or absorption	—	—	Applied principally to the determination of halogens in organic materials. Suitability of the method depends on the properties of the matrix provided by the sample. Used to determine bromine in petrol and other hydrocarbons, in blood serum, urine and tissue. A recent innovation involves analysis of chlorine and bromine in aromatic hydrocarbons by absorption of monochromatic X-rays (K-capture spectroscopy).[h]
12. Cl^-, Br^-, I^-	Spectrographic	—	—	Spectroscopic lines suitable for analytical work occur in the ultraviolet; these are produced in emission, for example, by the "copper spark" technique. The method is simple, rapid and applicable to mixtures of halides, but is of limited precision.[a,b]
13. Cl^-, Br^-, I^-	Radiochemical: (a) isotope dilution	Miscellaneous	—	Radioactive halogen isotopes, e.g. ^{131}I, have been used for the low-level detection and estimation of halogens, in conjunction with precipitation, solvent-extraction or ion-exchange procedures. Hence the iodine content of the thyroid gland has been measured.
	(b) neutron activation analysis	Miscellaneous or no change	—	Used to estimate microgram or sub-microgram quantities of halogens, e.g. bromide in water or biological material. Chemical separation is necessary if mixtures of halides are to be analysed in this way.

TABLE 35 (*cont.*)

Halide ions	Method	Reaction type	Reagent	Comments
14. Br^-, I^-	Chronometric	Catalytic action on redox reaction	Miscellaneous	The catalytic actions of iodide on the rate of reaction between Ce(IV) and As(III) and of bromide on that between MnO_4^- and I_2 in acid solution[d] provide very sensitive procedures for the estimation of the halide ion, e.g. 0·005–0·45 μg of I^- in natural waters or common salt can thus be analysed.[e,i] The progress of the reaction can be monitored spectrophotometrically. The catalytic action on thiocyanate oxidations[j] and on the polarographic reduction of In(III)[k] has also been used to determine traces of iodide.
15. Cl^-, Br^-, I^-	Chromatographic	Distribution between two phases: (a) liquid–solid	Stationary phase may be: Amberlite anion-exchange resin, paper or silica gel	The halide ions can be separated by ion-exchange, thin-layer or paper chromatography. Separation may be followed by quantitative analysis using $AgNO_3$ or by semi-quantitative evaluation of the intensities of coloured spots.
		(b) gas–liquid or gas–solid	silica gel or celite	Used to separate, identify and estimate volatile halides, notably organo–halogen derivatives.

[a] *Mellor's Comprehensive Treatise on Inorganic and Theoretical Chemistry*, Supplement II, Part I, Longmans, Green and Co. (1956).

[b] Z. E. Jolles (ed.), *Bromine and its Compounds*, Benn, London (1966).

[c] G. W. Armstrong, H. H. Gill and R. F. Rolf, *Treatise on Analytical Chemistry* (ed. I. M. Kolthoff, P. J. Elving and E. B. Sandell), Part II, Vol. 7, p. 335. Interscience (1961).

[d] V. A. Stenger, *Encyclopedia of Industrial Chemical Analysis* (ed. F. D. Snell and L. S. Ettre), Vol. 8, p. 1. Interscience (1969); G. Oplinger, *ibid.* Vol. 9, p. 333 (1970).

[e] K. N. Tandon and R. C. Mehrotra, *Analyt. Chim. Acta*, **27** (1962) 15.

[f] E. Bishop and R. G. Dhaneshwar, *Analyt. Chem.* **36** (1964) 726.

[g] C. L. Wilson, D. W. Wilson and C. R. N. Strouts, *Comprehensive Analytical Chemistry*, Vol. IIA, p. 206. Elsevier (1964).

[h] W. Scaman, H. C. Lawrence and H. C. Craig, *Analyt. Chem.* **29** (1957) 1631.

[i] H. V. Malmstadt and T. P. Hadjiioannou, *Analyt. Chem.* **35** (1963) 2157; T. P. Hadjiioannou, *Analyt. Chim. Acta*, **30** (1964) 488, 537.

[j] K. B. Yatsimirsky, L. I. Budarin, N. A. Blagoveshchenskaya, R. V. Smirnova, A. P. Fedorova and V. K. Yatsimirsky, *Zhur. analit. Khim.* **18** (1963) 103.

[k] A. J. Engel, J. Lawson and D. A. Aikens, *Analyt. Chem.* **37** (1965) 203.

volumetrically, though measurements of the highest precision still rely on the classical gravimetric procedures. The equivalence point of the titration of a halide with standard silver nitrate solution may be detected by the use of an indicator which forms a sparingly soluble silver salt, e.g. silver chromate, whose solubility product nevertheless exceeds those of the silver halides (the Mohr method), or by the use of a dyestuff like dichlorofluorescein, which is adsorbed by the silver halide and whose colour responds to the presence of excess silver ions in the reaction mixture. Various indirect methods of titration also exist, of which the best known (the Volhard method) involves the precipitation of the halide with a measured excess of silver ions, followed by back-titration of the excess with a standard thiocyanate solution. The very limited ionization of mercury(II) halides makes possible the titration of halide ions with a solution of an ionized mercury(II) salt, e.g. the nitrate or perchlorate; diphenylcarbazide or diphenylcarbazone is typically used as an indicator to detect the first excess of mercury(II) ions.

Numerous volumetric methods have been described in which bromide is subjected to oxidation to bromine, hypobromite, bromine cyanide or bromate. Of these, one of the most widely employed is based on the so-called van der Meulen method, whereby bromide is oxidized to bromate with alkaline hypochlorite; excess hypochlorite is removed by the action of formate, and the bromate is then determined iodimetrically. Such a procedure is particularly advantageous for the analysis of bromide as a minor constituent in the presence of chloride, e.g. in natural brines and similar salt solutions, though a more rapid procedure involves oxidation by hypochlorite or other agents in acidic solution, together with spectrophotometric analysis of the bromine thereby liberated[504]. A method has also been devised whereby traces of bromide can be separated from moderate amounts of chloride by oxidation with chromic acid in the presence of cyanide ions; following distillation from the reaction mixture, the bromine cyanide thus formed is determined potentiometrically by a sensitive null-point method[504].

Iodide ion is oxidized comparatively easily to iodate by chlorine, bromine, hypochlorite or bleaching powder; after removal of the excess of oxidizing agent, the iodate is determined by conventional iodimetric titration. Hence, iodide can be determined in low concentrations, e.g. in natural brines, even in the presence of chloride and bromide. Appropriate also to samples containing small amounts of iodide is the liberation, by the action, for example, of iron(III), nitrite or hydrogen peroxide, of free iodine, which then lends itself to spectrophotometric analysis.

The titration of halide ions with silver nitrate can be accurately monitored by potentiometric measurements; in this way it is also possible to analyse mixtures of the halide ions in a single operation. In a development of the potentiometric principle, a very small current —below the polarographic diffusion limit—is passed through a pair of indicator electrodes immersed in the solution; silver ions are thus generated electrolytically, and the concentration of these ions determines the potential assumed by the electrodes. This method of so-called "differential electrolytic potentiometry", which has the virtues of both sensitivity and precision, has been applied to the determination of nanogram amounts of halide at extreme dilution[345,506]. Direct potentiometric measurements using a sensing electrode which is reversible with respect to halide ions—usually silver–silver halide—have been extensively employed in the analysis of halide ions, forming a convenient basis for the continuous and automatic monitoring of the ions in solution. One of the most sensitive

[506] E. Bishop and R. G. Dhaneshwar, *Analyt. Chem.* **36** (1964) 726.

procedures for evaluating bromide or iodide depends upon the catalytic action of the ions on certain redox reactions, e.g. between iodine and permanganate or between cerium(IV) and arsenic(III); in a typical experiment the progress of the redox reaction is followed spectrophotometrically. Hence, minute amounts of bromide or iodide have been analysed, for example, in blood serum, tissues or natural waters. Other physical methods which have been applied to the estimation of halide ions (see Table 35) include polarography, X-ray fluorescence and absorption, and radiochemical procedures.

Prior to analysis, organically bound halogen atoms are converted to the corresponding anions by hydrolysis, oxidation or reduction; the halide ions are then determined by one of the methods of Table 35. Relating to the decomposition stage, many methods have been described, the choice often depending on the nature of the compound to be analysed. The most commonly used methods are based on: (i) hydrolysis of labile C–X bonds; (ii) reduction, e.g. with sodium metal or sodium biphenyl in a suitable solvent; (iii) wet oxidation, e.g. with fuming nitric acid (the Carius method) or chromic acid; (iv) oxidation with sodium peroxide in a sealed bomb; (v) combustion in a stream of oxygen either with (the Pregl method) or without a catalyst; (vi) combustion in an enclosed atmosphere of oxygen (the Schöniger oxygen-flask method). Either wet oxidation or alkaline incineration is favoured for the destruction of organic matter when the halogen content of biological material is to be determined; this treatment follows the precipitation or extraction of proteins in the isolation of protein-bound halogen. Many of the recent determinations of iodine in biological material have been based on the use of radioactive isotopes of the element, combined with radio-counting techniques employing either sodium iodide scintillation or Geiger-Müller counters; with the ^{131}I isotope, for example, protein-bound iodine has thus been analysed[503].

3.5. BIOLOGICAL ACTION OF THE HYDROGEN HALIDES AND HALIDE IONS[345,426,507–509]

The principal hazards of concentrated solutions of the hydrohalic acids are associated with the chemical burns or dermatitis arising from contact with the skin, although the action of the fumes as a respiratory irritant must not be overlooked. Gaseous hydrogen chloride is a corrosive poison, for the fumes it produces in air are actually composed of minute droplets of concentrated hydrochloric acid. It exerts a destructive action on the mucous membrane and skin. Inhalation of excessive concentrations of the gas produces a severe irritation of the upper respiratory tract. The oedema or spasm of the larynx and inflammation of the respiratory system which follow the destructive action of the acid may ultimately prove fatal. No evidence of chronic systemic effects has been found. Concentrations of 0·13–0·2% are lethal to human beings in exposures lasting only a few minutes, while the maximum concentration which can be tolerated for exposures of 60 min is in the range 0·005–0·01%. The maximum concentration in the atmosphere permissible for normal working conditions has been set at 5 ppm. Although there are few details concerning the toxicities of hydrogen bromide and hydrogen iodide, as gases or as aqueous acids, practical experience suggests that the physiological effects resemble those due to hydrogen chloride.

The principal features of the biochemistry and geochemistry of the halogens concern the halide anions and organo-halogen compounds; some of these features are summarized in Table 36[508]. Both chlorine, as one of the more abundant elements, and iodine, as a trace

[507] M. B. Jacobs, *The Analytical Toxicology of Industrial Inorganic Poisons*, p. 640. Interscience (1967).
[508] H. J. M. Bowen, *Trace Elements in Biochemistry*, Academic Press (1966).
[509] E. J. Underwood, *Trace Elements in Human and Animal Nutrition*, 3rd edn., Academic Press (1971).

element, are now known to be essential ingredients of biochemical processes. On the other hand, there is as yet no unequivocal evidence that bromine performs any such vital function, despite the fact that all animal tissues other than the thyroid, where the position is reversed, contain 50–100 times more bromine than iodine. The halogens are taken up by animals mainly as halide anions present, in varying proportions, in foodstuffs, table salt and water. In common with other anions, the halide ions suffer diffusion in soft tissues and enter the bloodstream, whence they are removed and actively pumped into the urine by the action of the kidney. Next to urea, chlorides are in fact the most abundant constituents of the urine. The anions are also excreted by mammals *via* the hair and nails, which are fed by the bloodstream, and in which relatively high concentrations of the anions are found.

Chloride ions exhibit certain unique functions, their concentration being regulated by active transport. They are found mainly, but not exclusively, in the intracellular space, and make up the bulk of the anions in blood plasma. The best known active secretion of chloride in vertebrates takes place in the lining of the stomach where specialized oxyntic cells are able to secrete approximately 0·17 M hydrochloric acid, which aids the digestion of food. The oxyntic cells are able to pump bicarbonate ions into the bloodstream and hydrogen ions into the stomach; chloride ions pass from the blood to the stomach contents to maintain electrical neutrality. It is also found that the distribution of Cl^- ions between plasma and the red cells of blood is similar and intimately related to that of the HCO_3^- ions: as the HCO_3^- concentration varies causing the ions to diffuse into or out of the red cells, electrolytic equilibrium is maintained by a counterbalancing migration of chloride ions. Such shifts of Cl^- and HCO_3^- ions between the plasma and the red cells play an important role in stabilizing the pH of the blood. Likewise, changes in the concentrations of Na^+ and K^+ ions are often accompanied by alterations in the concentrations of Cl^- and HCO_3^- ions in the extracellular fluid. These compensating movements of the diffusable Cl^- ion through cell membranes appear to be a major feature of the chloride metabolism, which is therefore closely related to the metabolism of ions such as HCO_3^- and Na^+. In common with other ions whose concentration within a cell differs from that in the medium in which the cell lives, Cl^- ions exhibit an essential electrochemical function. The most important aspect of this function appears to be the availability of the ions as a source of free energy during cell stimulation, but the stabilization of emulsions formed by highly charged colloidal particles also contained within the cell may represent a significant secondary influence of the ions. Moreover, nearly all the essential elements, both major and minor, are believed to have one or more catalytic functions in the cell. In this context, it is noteworthy that chloride ions are reported to activate the enzyme α-amylase, but, in general, little is yet known about the chloride ion as a biochemical catalyst.

Organo-chlorine compounds are also found in nature. Most of them are of fungal origin, e.g. griseofulvin ($C_{17}H_{17}O_6Cl$) derived from *Penicillium griseofulvum*, and some are powerful antibiotics. Their function is presumably that of restricting bacterial growth in the neighbourhood of fungal hyphae. Organo-chlorine compounds are also present in some red algae, and their participation in photosynthesis has been mooted.

Studies with ^{82}Br indicate that bromine is retained for only short periods in the body tissues of mammals, and that it is excreted mostly in the urine; claims that it is concentrated in the thyroid and pituitary glands have not been substantiated[508]. Bromide and chloride ions readily interchange to some degree in the body tissues, so that administration of bromide results in some displacement of body chloride and *vice versa*. Bromides exert a prolonged depressant action upon all cerebrospinal centres, with the exception of those in the medulla,

TABLE 36. BIOGEOCHEMISTRY OF CHLORINE, BROMINE AND IODINE[a]

Location	Cl	Br	I
Igneous rocks (ppm)	130	2·5	0·5
Shales (ppm)	180	4	2·2
Sandstones (ppm)	10	1	1·7
Limestones (ppm)	150	6·2	1·2
Fresh water (ppm)	7·8	0·2	0·002
Sea water (ppm)	19,000	65	0·06
Air (μg m^{-3})	1·2	—	—
Soils (ppm)	100	5	5
	Much higher in alkaline soils near the sea and in salt deserts. A major exchangeable anion in many soils.	Said to be enriched in soil organic matter.	Said to be strongly absorbed by humus. Extensive areas of soil deficiency, resulting in mammalian goitre, exist.
Marine plants (ppm)	4700	740	30–1500
		Highest in brown algae.	Accumulated by brown algae and some diatoms.
Land plants (ppm)	2000	15	0·42
	Much higher in maritime and salt desert plants.	Higher in some species of Cucurbitaceae and Chenopodiaceae.	Accumulated by Feijoa sellowiana.
Marine animals (ppm)	5000–90,000	60–1000	1–150
	Highest in soft coelenterates, lowest in calcareous tissues.	Accumulated in scleroproteins of many sponges and by several corals and molluscs.	Accumulated by some sponges and corals and in the tunics of ascidians.
Land animals (ppm)	2800	6	0·43
	Highest in mammalian hair and skin.		Accumulated by mammalian thyroid, also by hair.
Mammalian blood (ppm)	2900	4·6	0·063
plasma (ppm)	3950	3·9	0·077
red cells (ppm)	1890	(5·6)	(0·044)
atoms/red cell	$2·5 \times 10^9$	$3·3 \times 10^6$	16,000

Functions	Essential for angiosperms and mammals; has electrochemical and probably also catalytic functions. Present in some antibiotics and pigments from fungi. Also present in compounds in red algae.	Several brominated amino acids have been isolated from marine organisms. Unconfirmed evidence for essentiality to mammals.	Essential to red algae, brown algae and mammals. Thyroxine and other iodinated amino acids are found in sponges, corals, ascidians and vertebrate thyroids.
Reviews	Plants,[b] animals[c]	Terrestrial organisms,[d] marine organisms[e]	Algae,[d,f] plants,[b,d] animals[d,g]
Toxicity	Relatively harmless to organisms.	Relatively harmless to organisms, but prolonged administration may cause "bromism".	Scarcely toxic, but excessive dosages lead to the condition of "iodism".

Calculated values in brackets.

[a] H. J. M. Bowen, *Trace Elements in Biochemistry*, Academic Press (1966).

[b] W. Stiles, *Encyclopedia of Plant Physiology* (ed. W. Ruhland), Vol. 4, p. 558. Springer, Berlin (1958).

[c] E. Cotlove and C. A. M. Hogben, *Mineral Metabolism* (ed. C. L. Comar and F. Bronner), Vol. 2B, p. 109. Academic Press (1962).

[d] E. J. Underwood, *Trace Elements in Human and Animal Nutrition*, 3rd edn., Academic Press (1971).

[e] J. Roche, M. Fontaine and J. Leloup, *Comparative Biochemistry* (ed. M. Flovkin and H. S. Mason), Vol. 5, p. 493. Academic Press (1963).

[f] T. I. Shaw, *Proc. Roy. Soc.* **B150** (1959) 356.

[g] J. Gross, *Mineral Metabolism* (ed. C. L. Comar and F. Bronner), Vol. 2B, p. 221. Academic Press (1962).

and have been used in medicine since 1835 as mild sedatives, particularly in the treatment of nervous disorders[345]. However, the mechanism whereby bromide depresses nervous activity has still to be elucidated. Excessive doses of bromide can produce toxic effects, manifest in the condition known as "bromism"; symptoms commonly include dermatitis, but mental and emotional disturbances are the most prominent and serious features. Notable amongst the few naturally occurring organo-bromine compounds are the pigment Tyrian purple (6,6′-dibromo-indigo) produced by some molluscs, a brominated phenol extracted from certain species of red algae, and the scleroproteins of certain corals and horny sponges which incorporate the molecules 3-bromotyrosine and 3,5-dibromotyrosine. However, the place of bromine in the economy of such invertebrates is still uncertain.

The ancient Greeks are reputed to have used burnt sponges successfully but quite empirically in the treatment of human goitre[426,509]. Knowledge of this fact and the discovery of iodine in abundance in sponges as early as 1819 led the French physician Coindet to use salts of iodine therapeutically in the treatment of goitre. The presence of iodine in the alimentary intake of mammals is now known to be essential; a deficiency causes the thyroid to become enlarged—that is, the condition characteristic of goitre—in an attempt to secure more iodine by increasing the volume of blood traversing the gland. To counteract the iodine deficiency, prevalent in the natural foodstuffs and water supplies of many areas, it has become common practice deliberately to add iodide *via* the water supplies and *via* "iodized" table salt. However, excessive dosage of iodide may induce the condition of "iodism", typically marked by the symptoms of headache, catarrh and dermatitis ("iodine rash"). Iodine occurs in body tissues as iodide ions and as organo-iodine compounds. In the thyroid it exists as iodide ions, mono- and di-iodotyrosine, thyroxine, tri-iodothyroxine, polypeptides containing thyroxine, thyroglobulin, and probably other iodinated compounds. The iodinated amino-acids are bound with other amino-acids in peptide linkage to form thyroglobulin, the unique iodinated protein of the thyroid gland. The chief constituent of the colloid filling the follicular lumen, thyroglobulin is a glycoprotein with a molecular weight of 650,000; it constitutes the storage form of the thyroid hormone, believed to be thyroxine, and normally accounts for some 90% of the total iodine of the gland. Details of the metabolism of the thyroid function are treated elsewhere[509]. Iodinated amino-acids, e.g. 3-iodotyrosine and 2-iodohistidine, are also incorporated in the scleroproteins of certain corals and horny sponges, but the functions of these compounds and of the iodine in brown algae remain obscure.

4. DERIVATIVES OF CHLORINE, BROMINE AND IODINE IN POSITIVE OXIDATION STATES

A. HALOGEN CATIONS[510]

1. INTRODUCTION

Although in some ways the choice is arbitrary, it is nevertheless difficult to decide exactly which species and which phenomena should be discussed under the heading of "halogen cations". In a review bearing this title published in 1962[510], the authors restricted

[510] J. Arotsky and M. C. R. Symons, *Quart. Rev. Chem. Soc.* **16** (1962) 282; R. J. Gillespie and M. J. Morton, *ibid.* **25** (1971) 553.

themselves to the species Cl^+, Br^+ and I^+; however, most of the evidence then construed in terms of X^+ has since been reinterpreted in favour of X_2^+, so that the circumstantial basis for such an exclusive account no longer exists. On the other hand, it would not be feasible to cite all the instances when positively-charged halogen-containing molecules have been invoked to rationalize experimental observations. It is therefore intended that this part of Section 4 shall cover the chemistry of cationic derivatives of the positive halogens generally, with particular reference to the cations X^+ and X_2^+ (X = Cl, Br, I) and their complexes with neutral donors (e.g. water, pyridine); the species X_3^+, being but special examples of the triatomic polyhalogen cations, are here compared with the ions XY_2^+ (X = F, Y = Cl; X = Cl, Y = F; X = Br, Y = F, I; X = I, Y = F, Cl, Br), and accounts are also given of the pentatomic cations I_5^+ and XF_4^+ (X = Cl, Br, I), of IF_6^+, and of the oxohalogen cations. Organohalogen cations are discussed in detail in Section 4D.

While it is only recently that many of the cations described below have been properly characterized, cationic halogen species have featured more or less plausibly in the literature over many years. Earlier reviewers[510] documented (but perhaps did not evade) the "obsessive desire" to crown the increasingly electropositive character of the heavier halogens with the simple iodine cation I^+; for example, the discovery that iodine and iodine monochloride formed conducting solutions in ethanol[511] engendered the belief that I^+ was present in these solutions. Since many deductions were made from purely chemical evidence, positive halogen compounds were often and understandably confused with cationic species: the true nature of any cations actually discovered was even less well understood. In the nomenclature of this section, cations or positively-*charged* species are referred to as such, to distinguish them from other "positive" halogen compounds. The criteria of cation-formation are taken largely from structural and physicochemical measurements rather than from more qualitative chemical details.

Before proceeding to discuss the various cations, it will be useful to outline the areas of halogen chemistry wherein evidence for cationic behaviour has been found. Since the ions are powerful electrophiles, their existence as distinct species is to be expected only in media of low nucleophilicity, i.e. in strong acids. Here they may be formed either by oxidation of a halogen or by abstraction of halide ion from an interhalogen. The ionization of halogens, interhalogens, and positive-halogen derivatives in common non-aqueous solvents of higher nucleophilicity (e.g. EtOH, MeCN) has been intimated, while kinetic measurements also imply that the species X^+(aq) and H_2OX^+ may be intermediates in the halogenation of aromatic compounds. In keeping with the increasingly electropositive character of the elements, the incidence of cation-formation increases from fluorine, for which FCl^+ is the only likely candidate, to iodine, which has many cationic derivatives.

2. THE HALOGENS IN OXIDIZING ACIDIC MEDIA

Like the neighbouring elements antimony and tellurium, iodine dissolves in oxidizing acidic media to form highly coloured polyatomic cations: the identity of the species produced depends on the oxidizing and acidic strengths of the medium, as is nicely exemplified in the iodine–sulphuric acid system.

Iodine is but slightly soluble ($< 10^{-4}$ M) in concentrated sulphuric acid, yielding a pink solution whose absorption spectrum is akin to that of iodine vapour. Oleums are both

[511] P. Walden, *Z. phys. Chem.* **43** (1903) 385.

stronger acids and stronger oxidizing agents than the parent sulphuric acid. The solubility limit of iodine in 30% oleum is *ca.* 0·5 M and the red-brown solutions contain I_3^+ and possibly I_5^+: concentrations of iodine in excess of 10 M may be obtained in 65% oleum, when the deep blue paramagnetic solutions contain I_2^+ and SO_2, but in disulphuric acid (45% oleum) oxidation of iodine to I_2^+ is incomplete. In 30% oleum I_3^+ is readily converted to I_2^+ by iodate, persulphate or peroxide; further oxidation of I_2^+ by these reagents produces iodine(III), usually in the form of solvated IO^+. Concentrated sulphuric acid itself is too weakly acidic to sustain I_2^+, so that oxidation of iodine with, for example, iodate produces only I_3^+ and IO^+.

That strongly electrophilic iodine species are present in the brown and blue solutions in oleum was early recognized by the respective iodination of such inert aromatic compounds as nitrobenzene and pyridine. The identity of the cations present and the equilibria relating them in oleums and other solvents, including the "magic acid" systems and aprotic media such as IF_5 and SbF_5, have been determined by investigation of the magnetic, colligative, conductimetric and spectroscopic properties of the solutions, as listed in Table 37: solid compounds isolated from solution also reveal the character of the halogen cations. An elegant profile of the sulphuric acid systems has been presented by Gillespie and his co-workers[512], who controlled the formal oxidation state of the iodine in solution by using iodine–iodic acid mixtures; in fluorosulphuric acid and related solvents $S_2O_6F_2$ is an expedient oxidant[513].

Cationic derivatives of chlorine and bromine are stable only in media considerably more acidic than those which support the corresponding iodine cations. Both chlorine and bromine dissolve unchanged in 65% oleum, and on further oxidation give tervalent compounds, in the case of bromine *via* Br_3^+. Br_2^+ and Br_3^+ have been characterized spectroscopically in the solvent $HSO_3F–SbF_5–3SO_3$ following the dissolution of bromine fluorosulphates (or $Br_2–S_2O_6F_2$ mixtures)[514], and are also known in the red solids $[Br_2]^+[Sb_3F_{16}]^-$ [515] and $[Br_3]^+[AsF_6]^-$ [516] made from bromine, bromine pentafluoride and, respectively, antimony or arsenic pentafluoride. The order of electrophilicity $Br_2^+ > Br_3^+ \sim I_2^+ > I_3^+$ is indicated by the acidic strengths of the media which stabilize these species.

Solutions of chlorine or chlorine fluorides in SbF_5 and related solvents apparently contain two paramagnetic species in temperature-dependent equilibrium with one another; by one account[517] the species predominant at −80°C is Cl_2^+ and that at +60°C ClF^+, although the respective identities Cl_2O^+ and $ClOF^+$ have also been proposed[518]. Since the only solids to have been isolated from such solutions contain diamagnetic cations like Cl_2F^+ and Cl_3^+, it seems probable that the paramagnetic moieties are indeed Cl_2O^+ and $ClOF^+$, produced by reaction with the cell walls or with impurities.

[512] R. A. Garrett, R. J. Gillespie and J. B. Senior, *Inorg. Chem.* **4** (1965) 563; see also references in Table 37.

[513] F. Aubke and G. H. Cady, *Inorg. Chem.* **4** (1965) 269; R. J. Gillespie and J. B. Milne, *ibid.* **5** (1966) 1236, 1577.

[514] R. J. Gillespie and M. J. Morton, *Chem. Comm.* (1968) 1565.

[515] A. J. Edwards, G. R. Jones and R. J. C. Sills, *Chem. Comm.* (1968) 1527; A. J. Edwards and G. R. Jones, *J. Chem. Soc.* (*A*) (1971) 2318.

[516] O. Glemser and A. Šmalc, *Angew. Chem., Internat. Edn.* **8** (1969) 517.

[517] G. A. Olah and M. B. Comisarow, *J. Amer. Chem. Soc.* **91** (1969) 2172.

[518] R. S. Eachus, T. P. Sleight and M. C. R. Symons, *Nature*, **222** (1969) 769.

TABLE 37. INVESTIGATIONS OF IODINE-CONTAINING CATIONS IN STRONG ACIDS

Solvent	Substrate	Oxidant	Species found in solution	Solids prepared	Techniques[a]	Reference
H_2SO_4	I_2	HIO_3	I_5^+, I_3^+, IO^+		cryoscopy, cond.	R. A. Garrett, R. J. Gillespie and J. B. Senior, *Inorg. Chem.* **4** (1965) 563.
	ICl	HIO_3	I_2Cl^+, ICl_2^+			
	IBr	HIO_3	I_2Br^+			
30% oleum	I_2	solvent	I_5^+, I_3^+		u.v., cond.	J. Arotsky, H. C. Mishra and M. C. R. Symons, *J. Chem. Soc.* (1961) 12.
		HIO_3	I_2^+, IO^+			
$H_2S_2O_7$ (45% oleum)	I_2	solvent	I_3^+, I_2^+		cryoscopy, cond., u.v.	R. J. Gillespie and K. C. Malhotra, *Inorg. Chem.* **8** (1969) 1751.
65% oleum	I_2	solvent	I_2^+, IO^+	$[IO]^+[HS_2O_7]^-$	cryoscopy, cond., u.v., Raman.	R. J. Gillespie and K. C. Malhotra, *Inorg. Chem.* **8** (1969) 1751; R. J. Gillespie, J. B. Milne and M. J. Morton, *ibid.* **7** (1968) 2221.
	KI	solvent	I_2^+			
	ICl	solvent	I_2^+			
HSO_3F	I_2	$S_2O_6F_2$	I_3^+, I_2^+, I_4^{2+}, $I(SO_3F)_3$		cond., m.w., mag., u.v., Raman.	F. Aubke and G. H. Cady, *Inorg. Chem.* **4** (1965) 269; R. J. Gillespie and J. B. Milne, *ibid.* **5** (1966) 1236, 1577; R. J. Gillespie, J. B. Milne and M. J. Morton, *ibid.* **7** (1968) 2221.
IF_5	I_2	solvent	I_3^+		mag., u.v., cond., i.r. (solids)	R. D. W. Kemmitt, M. Murray, V. M. McRae, R. D. Peacock, M. C. R. Symons and T. A. O'Donnell, *J. Chem. Soc.* (*A*) (1968) 862.
	I_2	MF_5 (M = P, As, Sb, Nb, Ta)	I_2^+	$[I_2]^+[M_2F_{11}]^-$ (M = Sb, Ta)		
	I_2	air	I_2^+			E. E. Aynsley, N. N. Greenwood and D. H. W. Wharmby, *J. Chem. Soc.* (1963) 5369.
SbF_5	I_2	solvent	I_2^+ (with excess SbF_5)	$I(SbF_5)_2$[b]; $[I_2]^+[Sb_2F_{11}]^-$; $ISbF_5$[c]	mag., u.v., cond., i.r. (solids), Raman.	R. D. W. Kemmitt, M. Murray, V. M. McRae, R. D. Peacock, M. C. R. Symons and T. A. O'Donnell, *J. Chem. Soc.* (*A*) (1968) 862; R. J. Gillespie and M. J. Morton, *J. Mol. Spectroscopy*, **30** (1969) 178.

[a] cond., conductivity measurements; mag., magnetic susceptibility measurements; m.w., molecular weight determinations; i.r., infrared spectroscopy; u.v., ultraviolet-visible absorption spectroscopy.

[b] Solid contains I_2^+, Sb(III) and Sb(V).

[c] Solid contains I_3^+, Sb(III) and Sb(V).

3. THE INTERHALOGENS WITH HALIDE ION-ACCEPTORS

The interhalogens interact with Lewis acids such as BF_3, SbF_5 or $SbCl_5$, to give crystalline ionic complexes: typical reactions include

$$BrF_5+2SbF_5 \rightarrow [BrF_4]^+[Sb_2F_{11}]^-$$
$$ClF_3+BF_3 \rightarrow [ClF_2]^+[BF_4]^-$$
$$ICl_3+SbCl_5 \rightarrow [ICl_2]^+[SbCl_6]^-$$
$$IF_7+AsF_5 \rightarrow [IF_6]^+[AsF_6]^-$$

There is insufficient thermodynamic data about derivatives of a given Lewis acid with different interhalogens usefully to allow comparison of the relative stabilities of the different cations; for one particular cation the thermal stabilities of the complexes vary according to the acceptor strength of the Lewis acid.

Some compounds have been prepared which are formally derived from unknown interhalogens such as Cl_2F_2, Cl_3F and Br_3F. Thus ClF and SbF_5 afford, not the 1:1 complex ClF,SbF_5 (or $Cl^+ SbF_6^-$), but the 2:1 adduct formulated as $[Cl_2F]^+[SbF_6]^-$ [519]: AsF_5 combines with equimolar mixtures of Cl_2 and ClF to give $[Cl_3]^+[AsF_6]^-$ [519], and with Br_2/BrF_3 or Br_2/BrF_5 mixtures forming $[Br_3]^+[AsF_6]^-$ [516].

The adducts function as electrolytes in solvents such as AsF_3 or anhydrous HF. X-ray and spectroscopic studies of the solids suggest that, while the best model for these compounds is an ionic lattice, there are relatively strong interactions involving halogen-bridging between the cationic and anionic units.

The significant electrical conductance displayed by some liquid interhalogens (Section 4C) is consistent with self-ionization involving interhalogen cations, e.g.[520]

$$2BrF_3 \rightleftharpoons BrF_2^+ + BrF_4^-$$

4. HALOGEN CATIONS IN AROMATIC HALOGENATION

It has already been remarked that the cations I_2^+, I_3^+ and I_5^+, found in solutions of iodine in oleum, are exceedingly potent agents for iodinating even unreactive aromatic compounds. There is also evidence from kinetic studies that cationic halogen species may be involved in halogenation under milder aqueous conditions: the reader is referred to a recent review for a detailed and critical survey of this work[521]. It appears that the kinetic terms $[HOCl][H^+]$ and $[HOBr][H^+]$ which are observed in halogenations effected by acidified hypochlorous and hypobromous acids represent the distinct species H_2OCl^+ and H_2OBr^+. With the free halogen as reagent in uncatalyzed reactions the active agents are the undissociated molecules Cl_2, Br_2 and I_2. The intervention of the aquated Cl^+ ion in zeroth-order chlorinations by acidified hypochlorous acid, though eloquently supported by the kinetic data, is unlikely in view of the calculated equilibrium concentration of this species (p. 1345). In many experiments silver salts have been added to limit the concentration of X^-, and in these cases the possibility that the complex AgX_2^+ is the halogenating agent cannot be discounted[510].

519 R. J. Gillespie and M. J. Morton, *Inorg. Chem.* **9** (1970) 811.

520 See for example A. G. Sharpe, *Non-aqueous Solvent Systems* (ed. T. C. Waddington), p. 285, Academic Press (1965).

521 E. Berliner, *J. Chem. Educ.* **43** (1966) 124.

5. POSITIVE HALOGEN DERIVATIVES IN NON-AQUEOUS SOLVENTS

The chemical or electrochemical behaviour of many positive halogen derivatives dissolved in the common non-aqueous solvents implies ionization involving halogen-containing cations, although there is little justification for the belief, formerly widespread, that simple species such as I^+ occur in these solutions. Ethanolic solutions of INO_3 may exchange up to 70% of their iodine for protons on contact with a protonated cation-exchange resin, but the active agent is probably $EtOHI^+$, the conjugate acid of ethyl hypoiodite[522]. When saturated solutions of $I(OCOCH_3)_3$ in acetic anhydride are electrolyzed, the amount of silver iodide formed at a silvered platinum gauze cathode is consistent with Faraday's law calculations based on I^{3+},

$$I^{3+}+Ag+3e^- \rightarrow AgI$$

The nature of the cation is still obscure.

The important role of nucleophilic solvents in stabilizing halogen cations is demonstrated by pyridine, wherein I_2, ICl and IBr dissolve with the formation of $[py_2I]^+$ cations[523]: numerous derivatives of this and similar cations have been isolated and characterized (see below). The moderate electrical conductivity of solutions of ICl in acetonitrile may similarly be explained by a partial ionization such as

$$ICl+2MeCN \rightleftharpoons [(MeCN)_2I]^+ + Cl^-$$

6. MONATOMIC CATIONS X+

The cations F^+, Cl^+, Br^+ and I^+ have the ground state ns^2np^4 and are highly endothermic and electrophilic species (Table 38), so that their existence as distinct chemical species would depend critically on the provision of a suitably polar, acidic environment. It was formerly supposed[510] that the paramagnetic entity present in the blue solutions of iodine in strong acids was I^+, but later work (Table 37) has shown that I_2^+ is in fact responsible for the characteristic magnetic and spectroscopic properties of these solutions.

TABLE 38. THERMODYNAMIC DATA FOR THE MONATOMIC HALOGEN CATIONS

Property	F	Cl	Br	I
$\Delta H_{f,0}°[X^+(g)]$(kcal mol^{-1})[a]	422·14	330·74	301·41	268·16
Electron affinity of X^+(kcal mol^{-1})[a]	402	299	273	241
$X_2(aq) \rightleftharpoons X^+(aq)+X^-(aq)$[b]				
$\Delta G°$(kcal mol^{-1})		+56	+39	+27
K		10^{-40}	10^{-30}	10^{-21}

[a] *Selected Values of Chemical Thermodynamic Properties*, N.B.S. Technical Note 270–3 (1968).

[b] J. Arotsky and M. C. R. Symons, *Quart. Rev. Chem. Soc.* **16** (1962) 282.

---

[522] H. Brusset and T. Kikindai, *Compt. rend.* **232** (1951) 1840.

[523] S. G. W. Ginn and J. L. Wood, *Trans. Faraday Soc.* **62** (1966) 777.

In media such as fluorosulphuric acid and the various oleums, iodine(I) compounds undergo, not simple ionization, e.g.

$$IOSO_2F \rightleftharpoons I^+ + SO_3F^-,$$

but complete disproportionation to I_2^+ and an iodine(III) derivative.

$$10IOSO_2F \rightleftharpoons 4I_2^+ + 2I(OSO_2F)_3 + 4SO_3F^-$$

Similarly equimolar mixtures of bromine and $S_2O_6F_2$ dissolved in HSO_3F–SbF_5–$3SO_3$ produce Br_2^+ and $Br(OSO_2F)_3$ rather than Br^+ [519].

Many kinetic studies have suggested that the aquated X^+ cations (X = Cl, Br, I) are intermediates in aromatic halogenations conducted in aqueous solution using the free halogen as reagent. However, the spontaneous ionic dissociation

$$X_2(aq) \rightleftharpoons X^+(aq) + X^-(aq)$$

is not favoured thermodynamically (p. 1345), even allowing that the aquated X^+ cations, having the electronic configuration ns^2np^4, enjoy a ligand field stabilization of <28 kcal mol^{-1} [510]: despite the immense uncertainty in the solvation energies assigned to the cations, the calculated equilibrium constants (Table 38) show that only $I^+(aq)$ may reasonably be considered as an intermediate in aqueous reactions[521].

The suggestion that protonated hypohalous acids H_2OX^+ (X = Cl, Br, I) intervene in aromatic halogenations is approved on both kinetic and thermodynamic grounds[510,521]. Structurally there is a nice distinction between $X^+(aq)$ and H_2OX^+: in the former, ion–dipole interactions with the surrounding water molecules should leave the halogen atom with its unpaired electrons; in the latter, the localized covalent bond between H_2O and X^+ produces a diamagnetic, spin-paired species.

Despite the lack of evidence for the discrete ions Cl^+, Br^+ and I^+, stable complexes of these cations with donors such as aromatic amines have been known for a long time. As was mentioned above, I_2, IBr and ICl dissolve in pyridine with the formation of the $[py_2I]^+$ cation[523]. Solid derivatives are generally prepared by the action of a silver salt and the amine on the halogen in an inert solvent[524].

$$AgNO_3 + I_2 + 2py \xrightarrow{CHCl_3} AgI + [py_2I]^+NO_3^-$$

$$[Ag(py)_2]SbF_6 + Br_2 \xrightarrow{MeCN} AgBr + [py_2Br]^+SbF_6^-$$

$$AgOCOCH_3 + I_2 + py \longrightarrow AgI + pyIOCOCH_3$$

Addition of base to solutions of a halogen salt is also effective[525].

$$ClNO_3 + 2py \xrightarrow{EtOH} [py_2Cl]^+NO_3^-$$

In proof of the presence of monovalent halogen, the iodine compounds release iodine quantitatively on reaction with aqueous iodide.

$$L_nIA + I^- \rightarrow I_2 + A^- + nL$$

The di-ligand complexes behave as 1:1 electrolytes in solution and are best formulated as e.g. $[py_2I]^+NO_3^-$, while compounds like $pyIOCOCH_3$ are essentially covalent. To some extent the stoichiometry of the complex depends on the anion and it appears that the halogen cations X^+ have a strong tendency towards two-coordination: thus the weakly

524 H. Schmidt and H. Meinert, *Angew. Chem.* **71** (1959) 126.
525 M. Schmeisser and K. Brändle, *Angew. Chem.* **73** (1961) 388.

coordinating anions ClO_4^- and SbF_6^- give only $[py_2I]^+ClO_4^-$ and $[py_2I]^+SbF_6^-$, nitrate affords both $pyIONO_2$ and $[py_2I]^+NO_3^-$, while with carboxylate anions the compounds are of the type [pyIOCOR].

The solid "py_2I_2" prepared directly from pyridine and iodine has been shown by X-ray studies to contain, in addition to I_3^- and I_2 molecules, centrosymmetric planar $[py_2I]^+$ cations with linear N—I—N units [r(N—I) = 2·16 Å][526]. Infrared and Raman spectroscopic measurements have found essentially similar $[py_2Br]^+$ and $[py_2I]^+$ ions both in the crystal and in solution[527]. The $\{[(H_2N)_2CS]_2I\}^+$ cation detected crystallographically in $[(H_2N)_2CS]_2I_2$ has a linear symmetrical S—I—S unit [r(I—S) = 2·629 Å], but the ion as a whole is not planar[528].

7. THE DIATOMIC CATIONS X_2^+

Of the three cations X_2^+ (X = Cl, Br, I), I_2^+ is familiar as the origin of the paramagnetism and deep blue colour of solutions of iodine in oleum, SbF_5 and similar media, where it has been defined by physicochemical studies (Table 37), including a striking resonance Raman experiment; structural details are not yet available for the solids $[I_2]^+[M_2F_{11}]^-$ (M = Sb, Ta)[529]. The red cation Br_2^+ has been characterized spectroscopically in solution in the "super-acid" SbF_5—HSO_3F—$3SO_3$[514] and crystallographically in $[Br_2]^+[Sb_3F_{16}]^-$[515], which is prepared from bromine, bromine pentafluoride and antimony pentafluoride. The claim that Cl_2^+ is found following the dissolution of chlorine or chlorine fluorides in SbF_5 and is in temperature-dependent equilibrium with ClF^+ is based mainly on esr measurements[517], and has been seriously disputed[518]: the solid complex Cl_2IrF_6 has also been reported[530]. In the vapour phase, however, Cl_2^+ has been thoroughly characterized by analysis of its electronic band spectrum.

Because the ionization of the molecular halogens involves removal of an electron from an antibonding orbital, the bonds in the X_2^+ cations are stronger than those in the neutral molecules: this is reflected in the higher bond dissociation energies, shorter interatomic distances and increased vibrational frequencies found in the cations (Table 39). In each case the unpaired electron produces a paramagnetic moment, measured for Br_2^+ and I_2^+, which is in accord with the expected $^2\Pi_{3/2}$ ground state: while Cl_2^+ is reputedly detectable by its esr absorption[517], the failure to record esr spectra for Br_2^+ and I_2^+, even at 4°K, is attributed to a large spread in the g- and hyperfine tensors, coupled with efficient spin-relaxation[530]. A fully satisfactory explanation of the ultraviolet-visible spectra of Br_2^+ and I_2^+ is still being sought.

In concentrated solution I_2^+ disproportionates to I_3^+ and an iodine(III) species; e.g. in HSO_3F the equilibrium

$$8I_2^+ + 3SO_3F^- \rightleftharpoons I(SO_3F)_3 + 5I_3^+$$

is satisfied by the equilibrium constant[531]

$$K = \frac{[I(SO_3F)_3]^{\frac{1}{8}}[I_3^+]^{\frac{5}{8}}}{[I_2^+][SO_3F^-]^{\frac{3}{8}}} = 2·0$$

---

[526] O. Hassel and H. Hope, *Acta Chem. Scand.* **15** (1961) 407.
[527] I. Haque and J. L. Wood, *J. Mol. Structure*, **2** (1968) 217.
[528] H. Hope and G. H.-Y. Lin, *Chem. Comm.* (1970) 169.
[529] R. D. W. Kemmitt, M. Murray, V. M. McRae, R. D. Peacock, M. C. R. Symons and T. A. O'Donnell, *J. Chem. Soc.* (*A*) (1968) 862.
[530] N. Bartlett, cited in ref. 529.
[531] R. J. Gillespie and J. B. Milne, *Inorg. Chem.* **5** (1966) 1577.

TABLE 39. SOME PROPERTIES OF THE DIATOMIC CATIONS Cl_2^+, Br_2^+ AND I_2^+

Property	$^{35}Cl_2^+$		Br_2^+		I_2^+		Reference
Ionization potentials:	*eV*	*kcal*	*eV*	*kcal*	*eV*	*kcal*	a
$X_2^+(^2\Pi_{3/2,g}) \leftarrow X_2(^1\Sigma_g^+)$	11·51	265·4	10·51	242·4	9·22	212·6	
$X_2^+(^2\Pi_{1/2,g}) \leftarrow X_2(^1\Sigma_g^+)$							
$X_2^+(^2\Pi_{3/2,u}) \leftarrow X_2(^1\Sigma_g^+)$	13·96	321·9	12·41	286·2	10·74	247·7	
$X_2^+(^2\Pi_{1/2,u}) \leftarrow X_2(^1\Sigma_g^+)$							
$X_2^+(^2\Sigma_g^+) \leftarrow X_2(^1\Sigma_g^+)$	15·72	362·5	14·28	329·3	12·66	292·0	
$\Delta H_{f,0}°[X_2^+(g)]$ (kcal mol^{-1})	265·4		253·3		228·3		a,b
Dissociation energy, $D_0°(X_2^+)$	*eV*	*kcal*	*eV*	*kcal*	*eV*	*kcal*	a–c
$^2\Pi_{3/2,g}$	3·98	91·8	3·30	76·1	2·77	63·9	
$^2\Pi_{1/2,g}$							
$^2\Pi_{3/2,u}$	1·53	35·3	1·40	32·3	1·25	28·8	
$^2\Pi_{1/2,u}$							
Dissociation energy, $D_0°(X_2)$	2·479	57·18	1·971	45·45	1·542	35·57	
Vibrational frequency (cm^{-1})	$^{35}Cl_2^+(g)$[d] $^2\Pi_{3/2,g}$ $\omega_e = 645·61$ $^2\Pi_{1/2,g}$ $\omega_e = 644·77$ $^2\Pi_u$ $\omega_e = \sim 310$		$Br_2^+(g)$[e] ω_e $^2\Pi_g$ 376·0 $^2\Pi_{3/2,u}$ 190·0 $^2\Pi_{1/2,u}$ 152·0	*Br_2^+ in solid $[Br_2][Sb_3F_{16}]$*[f] 368 *Br_2^+ in solutions in SbF_5–HSO_3F–$3SO_3$*[h] 360	*I_2^+ in solutions in oleum, SbF_5 or HSO_3F*[g] $^2\Pi_g$ 238		d–h
Anharmonic vibrational constants (cm^{-1})	$^{35}Cl_2^+(g)$		$Br_2^+(g)$		$I_2^+(g)$		d,e
$^2\Pi_{3/2,g}$	$\omega_e x_e = 3·015$, $\omega_e y_e = 0·007$		$\omega_e x_e = 1·25$		—		
$^2\Pi_{1/2,g}$	$\omega_e x_e = 2·988$						
$^2\Pi_{3/2,u}$	—		$\omega_e x_e = 1·0$		—		
$^2\Pi_{1/2,u}$	—		$\omega_e x_e = 0·35$		—		

TABLE 39 (*cont.*)

Property	$^{35}Cl_2^+$	Br_2^+	I_2^+	Reference
Rotational constant, B_e (cm^{-1})				
$^2\Pi_{3/2,g}$	0·26950	—	—	d
$^2\Pi_{1/2,g}$	0·2697	—	—	
$^2\Pi_u$	*ca.* 0·19	—	—	
	α_e D_e			
$^2\Pi_{3/2,g}$	0·00164 $0{\cdot}18\times10^{-6}$	—	—	
$^2\Pi_{1/2,g}$	0·00167 —	—	—	
Internuclear distance (Å)	$Cl_2^+(g)$[d] $^2\Pi_{3/2,g}$ $r_e = 1{\cdot}892$ $^2\Pi_{1/2,g}$ $r_e = 1{\cdot}891$ $^2\Pi_u$ $r_e = ca.\ 2{\cdot}25$	Br_2^+ *in* $[Br_2]^+[Sb_3F_{16}]^-$ [f] 2·15±0·01	—	d,f
Magnetic moment (B.M.)	—	Br_2^+ *in* $[Br_2]^+[Sb_3F_{16}]^-$ [f] 1·6	I_2^+ *in* $[I_2]^+[Sb_2F_{11}]^-$ [i] 2·25 I_2^+ *in solution*[j] 2·00±0·1	f,i,j
Esr spectrum	Ref. k			k
Ultraviolet-visible spectrum	Ref. d,e	Ref. e,f,h	Ref. i,j	d–f, h–j

[a] A. W. Potts and W. C. Price, *Trans. Faraday Soc.* **67** (1971) 1242.

[b] *Selected Values of Chemical Thermodynamic Properties*, N.B.S. Technical Note 270–3, U.S. Government Printing Office, Washington (1968).

[c] *Tables of Constants and Numerical Data*, No. 17, *Spectroscopic Data relative to Diatomic Molecules* (general ed. B. Rosen), Pergamon (1970).

[d] F. P. Huberman, *J. Mol. Spectroscopy*, **20** (1966) 29.

[e] G. Herzberg, *Molecular Spectra and Molecular Structure*. I. *Spectra of Diatomic Molecules*, p. 520, van Nostrand (1950); *Supplement to Mellor's Comprehensive Treatise on Inorganic and Theoretical Chemistry*, Supplement II, Part I, Longmans, London (1956).

[f] A. J. Edwards, G. R. Jones and R. J. C. Sills, *Chem. Comm.* (1968) 1527; A. J. Edwards and G. R. Jones, *J. Chem. Soc.* (*A*) (1971) 2318.

[g] R. J. Gillespie and M. J. Morton, *J. Mol. Spectroscopy*, **30** (1969) 178.

[h] R. J. Gillespie and M. J. Morton, *Chem. Comm.* (1968) 1565.

[i] R. D. W. Kemmitt, M. Murray, V. M. McRae, R. D. Peacock, M. C. R. Symons and T. A. O'Donnell, *J. Chem. Soc.* (*A*) (1968) 862.

[j] See references in Table 37.

[k] G. A. Olah and M. B. Comisarow, *J. Amer. Chem. Soc.* **91** (1969) 2172, but see R. S. Eachus, T. P. Sleight and M. C. R. Symons, *Nature*, **222** (1969) 769.

At low temperatures the dimerization $2I_2^+ \rightleftharpoons I_4^{2+}$ has been detected in HSO_3F solution[532]; the heat of association is -10 kcal mol^{-1}.

8. TRIATOMIC CATIONS X_3^+ AND XY_2^+

The known triatomic halogen and interhalogen cations are listed below: the central atom is placed first.

ClF_2^+	BrF_2^+	IF_2^+
Cl_2F^+	Br_3^+	ICl_2^+
Cl_3^+		$[IBr_2^+]$
		$[I_2Br^+]$
		I_3^+

The bracketed ions have so far been found only in solution, while the others are also known in more-or-less well characterized solid derivatives, usually combined with complex halogeno-anions such as BF_4^-, AsF_6^- or $SbCl_6^-$. Cl_3^+ is known in the adduct $[Cl_3]^+[AsF_6]^-$ formed from Cl_2, ClF and AsF_5 [519], while $[Br_3]^+[AsF_6]^-$ is prepared either from Br_2, BrF_5 and AsF_5 or from $[O_2]^+[AsF_6]^-$ and bromine[516]. Br_3^+ is also known in solution in oleum[510] and in fluorosulphuric acid. First detected in oleum by its iodinating properties, I_3^+ is contained in the solid $I(SbF_5)_2$, which liberates the cation on dissolution in AsF_3 [529]. The black fluorosulphate I_3OSO_2F is prepared from stoichiometric amounts of I_2 and $S_2O_6F_2$ [513].

In these cations the formally tripositive central halogen is bonded to two formally uninegative halides, giving an average oxidation state of $+\frac{1}{3}$. Structural data are available for ClF_2^+, Cl_2F^+, Cl_3^+, BrF_2^+ and ICl_2^+, which are all found to be bent, with the heaviest atom at the apex and with angles of 90–100°; the atoms of the other triatomic cations are presumably similarly disposed. X-ray investigations show a close affinity between the structures of the solids $[ClF_2]^+[SbF_6]^-$, $[BrF_2]^+[SbF_6]^-$ and $[ICl_2]^+[SbCl_6]^-$. The octahedral anions and angular cations are linked by bridging-halogen bonds, giving much distorted square-planar arrangements about the heavy atom (Table 40) in forming infinite helical chains (Fig. 28): in both $[BrF_2]^+[SbF_6]^-$ and $[ICl_2]^+[SbCl_6]^-$ the bridging utilizes *cis*-halogen atoms on the anion, although in $[ClF_2]^+[SbF_6]^-$ the anion exercises the rarer *trans*-bridging option, possibly because of a smaller covalent contribution to the structure. Interionic chlorine bridges are also found in $[ICl_2]^+[AlCl_4]^-$ [533]. While the ionic model best represents the structures of these salts, the importance of interionic interactions in fixing structural details is amply demonstrated by the variations in the fundamental frequencies recorded for BrF_2^+ in a series of solids[534] and in HF–BrF_3 solution[535]. As indexed by changes in the vibrational frequencies of AsF_6^-, the extent of fluorine-bridging in the hexafluoroarsenates decreases in the order $BrF_2^+ > ClF_2^+ > Cl_2F^+ > Cl_3^+$ [519,535], consonant with the decreasing polarizing power of the cations.

9. PENTATOMIC CATIONS

No structural data are available for I_5^+, the cation formed in sulphuric acid from HIO_3 and an excess of iodine[512]. There has been speculation about possible shapes and electronic structures.

[532] R. J. Gillespie, J. B. Milne and M. J. Morton, *Inorg. Chem.* **7** (1968) 2221.

[533] C. G. Vonk and E. H. Wiebenga, *Rec. Trav. Chim.* **78** (1959) 913; *Acta Cryst.* **12** (1959) 859.

[534] K. O. Christe and C. J. Schack, *Inorg. Chem.* **9** (1970) 2296.

[535] T. Surles, H. H. Hyman, L. A. Quarterman and A. I. Popov, *Inorg. Chem.* **10** (1971) 611.

[536] K. O. Christe and W. Sawodny, *Inorg. Chem.* **8** (1969) 212; H. Meinert, U. Gross and A.-R. Grimmer, *Z. Chem.* **10** (1970) 226.

TABLE 40. COORDINATION ABOUT THE TRIVALENT HALOGEN IN $[ClF_2]^+[SbF_6]^-$, $[BrF_2]^+[SbF_6]^-$ and $[ICl_2]^+[SbCl_6]^-$

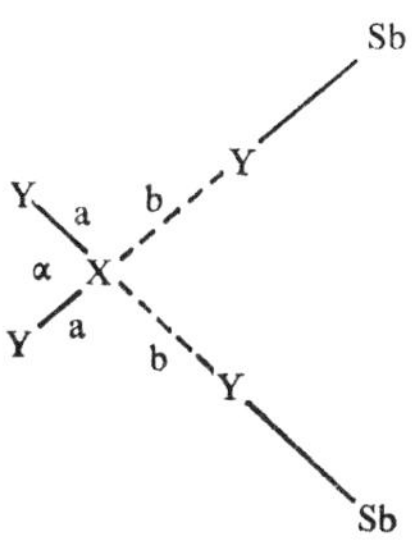

X	Y	*a* (Å)	*b* (Å)	α (°)	Reference
Cl	F	1·58	2·43 2·33	95·9	A. J. Edwards and R. J. C. Sills, *J. Chem. Soc.* (*A*) (1970) 2697.
Br	F	1·69	2·29	93·5	A. J. Edwards and G. R. Jones, *J. Chem. Soc.* (*A*) (1969) 1467.
I	Cl	2·33 2·29	3·00 2·85	92·5	C. G. Vonk and E. H. Wiebenga, *Acta Cryst.* **12** (1959) 859.

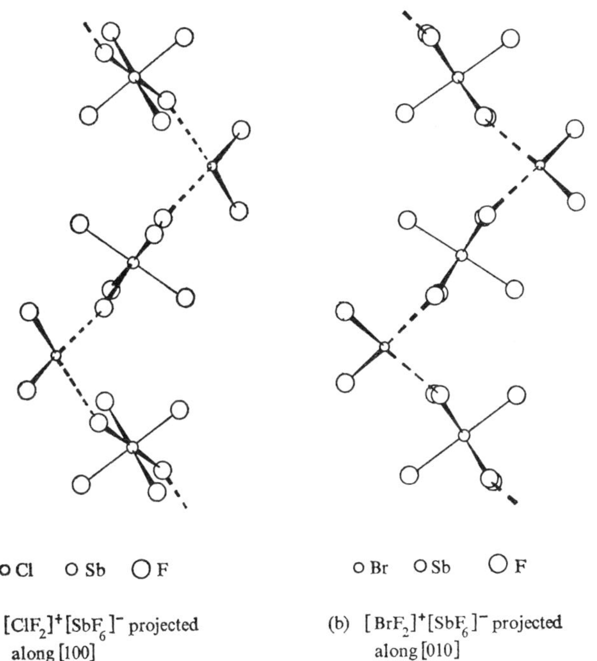

(a) $[ClF_2]^+[SbF_6]^-$ projected along [100]

(b) $[BrF_2]^+[SbF_6]^-$ projected along [010]

FIG. 28. Crystal structures of $[ClF_2]^+[SbF_6]^-$ and $[BrF_2]^+[SbF_6]^-$.

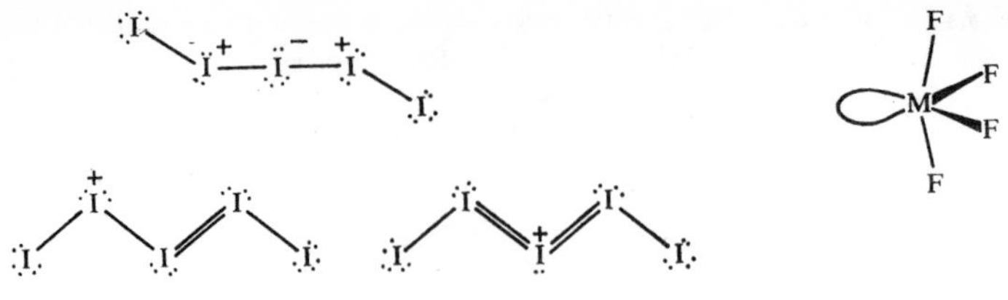

(a) (b)

FIG. 29. Structures of pentatomic halogen cations: (a) possible structures of I_5^+; (b) C_{2v} unit of MF_4^+ (M = Cl, Br, I).

In so far as they can be recognized in the adducts of the halogen pentafluorides with SbF_5, the ions ClF_4^+, BrF_4^+ and IF_4^+ exhibit the C_{2v} structure expected of systems in which the valence-shell of the central atom consists of four bonding electron pairs and one lone pair, and established for the isoelectronic chalcogen tetrafluorides[536]. Extensive fluorine-bridging in the solids increases the coordination number of the halogen (Cl, 6; Br, 6; I, 8), but "axial" and "equatorial" halogen–fluorine bonds may be distinguished from the longer interionic linkages. In $[IF_4]^+[SbF_6]^-$ the "axial" and "equatorial" I–F bond lengths are 1·79 Å and 1·83 Å, respectively, with "axial" and "equatorial" F–I–F angles of 148° and 107° [537]; the crystal structure of the complex $[BrF_4]^+[Sb_2F_{11}]^-$ has likewise been determined[538].

10. IF_6^+

The discrete octahedral IF_6^+ cation is found in complexes of IF_7 with acceptors such as AsF_5 and SbF_5; ^{19}F nmr[539], vibrational[540] and ^{129}I Mössbauer[541] spectroscopic studies have been reported. The molecular force field is unusual in that ν_2 (e_g) has a higher frequency than ν_1 (a_{1g}). The cation is also formed by the oxidation of IF_5 with KrF^+ salts.

$$[KrF]^+[Sb_2F_{11}]^- + IF_5 \rightarrow Kr + [IF_6]^+[Sb_2F_{11}]^-$$

Analogous attempts to synthesize BrF_6^+ derivatives by oxidation of BrF_5 have been unsuccessful.

11. HALOGEN OXOCATIONS

Four halogen oxocations have been reasonably well established[542], viz. ClO^+, ClO_2^+, IO^+ and IO_2^+; like other halogen cations, they are stable only in highly electrophilic media. They are formed (i) in the reaction of a halogen oxide with an acid or an acid anhydride, e.g.

$$2ClO_2 + 3SO_3 \rightarrow [ClO][ClO_2][S_3O_{10}]$$
$$I_2O_3 + H_2SeO_4 \rightarrow [IO]_2SeO_4 + H_2O$$

or (ii) by the interaction of an oxyfluoride and a Lewis acid, e.g.

$$[ClO_2][SO_3F] \xleftarrow{SO_3} ClO_2F \xrightarrow{AsF_5} [ClO_2][AsF_6]$$

537 H. W. Baird and H. F. Giles, *Acta Cryst.* **A25** (1969) S115.
538 M. D. Lind and K. O. Christe, *Inorg. Chem.* **11** (1972) 608.
539 J. F. Hon and K. O. Christe, *J. Chem. Phys.* **52** (1970) 1960.
540 K. O. Christe, *Inorg. Chem.* **9** (1970) 2801.
541 S. Bukshpan, J. Soriano and J. Shamir, *Chem. Phys. Letters*, **4** (1969) 241.
542 M. Schmeisser and K. Brändle, *Adv. Inorg. Chem. Radiochem.* **5** (1963) 41.

There is some evidence that ClO_3^+ may be an intermediate (i) in the preparation of ClO_3F from ClO_4^- and acids such as HF, SbF_5 or HSO_3F, or (ii) in the formation of aromatic perchloryl derivatives (p. 1573) from arenes, ClO_3F and Friedel–Crafts catalysts: claims that IO_3F forms unstable and presumably ionic adducts with BF_3 and AsF_5 are not unambiguously proven. No oxocations of bromine have been observed.

In solution in HSO_3F or $H_2S_2O_7$ chloryl salts function as electrolytes of strength comparable with analogous nitrosyl compounds. The vibrational spectrum of the bent discrete ClO_2^+ cation in the white solids $[ClO_2]^+[SbF_6]^-$ and $[ClO_2]^+[BF_4]^-$ has been reported[543], although in other solids extensive cation–anion interactions occur, as witnessed by the orange colour of $[ClO_2][SO_3F]$ and the blood-red hue of $[ClO_2]_2[S_3O_{10}]$. The volatility and thermal stability of the complexes $[ClO_2]^+[MF_n]^-$ are definite functions of the acceptor strength of the parent Lewis acid: ClO_2F,VF_5 decomposes below $-78°C$, while ClO_2F,SbF_5 is stable up to $+230°C$. The chloryl component is readily displaced by nitrogen oxides,

$$[NO]^+[AsF_6]^- + ClO_2 \xleftarrow{NO} [ClO_2]^+[AsF_6]^- \xrightarrow{NO_2} [NO_2]^+[AsF_6]^- + ClO_2$$

although the simple substitution is frequently complicated by other redox reactions.

While the chlorine oxocations are relatively discrete species, IO^+ and IO_2^+ are found in solution only in solvated and associated forms: in dilute solution in oleum I_2O_5 is present as undissociated IO_2HSO_4, but at high concentrations polymerization occurs and a white solid I_2O_5,SO_3 separates[544]. Similarly, conductimetric studies of IO_2SO_3F in HSO_3F show that the solute is not completely dissociated, even at high dilution[545]. Spectroscopic measurements suggest that solids which are formally derivatives of IO^+ and IO_2^+ [e.g. $(IO)_2SO_4$, $(IO)IO_3$, $(IO_2)SbF_6$, $(IO_2)AsF_6$] in fact contain polymeric cationic networks with I–O–I bridges, cross-linked by the anions; in solid $(IO_2)SO_3F$, however, isolated IO_2^+ cations are joined by bridging fluorosulphate groups.

B. THE OXYGEN COMPOUNDS OF THE HALOGENS

1. THE HALOGEN–OXYGEN BOND

Introduction

Concern for the preservation of the octet of electrons about the halogen atom restricted early accounts of the bonding in species such as ClO_4^- to "dative-coordinate" (1) or "semi-ionic" (2) electron-pair bonds. At a later stage the idea of double-bonding (3) between the halogen and oxygen atoms was mooted, but at the expense of the halogen octet. Canonical forms devised for halogen–oxygen molecules utilized all three types of bond. The application of more sophisticated valence theory, however, prompts two important and not unconnected questions about the bonding in halogen–oxygen molecules:

(1) Which orbitals on the halogen atom are involved in the bonding; specifically, is it necessary to invoke participation by halogen *nd* orbitals (or for iodine even 4*f* orbitals) in order adequately to account for observed molecular properties?

(2) To what extent is π-bonding encountered in these compounds; is it p_π–p_π or d_π–p_π in character?

543 K. O. Christe, C. J. Schack, D. Pilipovich and W. Sawodny, *Inorg. Chem.* **8** (1969) 2489.
544 R. J. Gillespie and J. B. Senior, *Inorg. Chem.* **3** (1964) 440, 972.
545 H. A. Carter and F. Aubke, *Inorg. Chem.* **10** (1971) 2296.

Since the problem is crucial to the subsequent discussion, consideration is given at the outset to the ways in which *nd* orbitals may be involved in bonding; the results of experiment and calculation are reviewed later.

$$\underset{(1)}{X \rightarrow O} \qquad \underset{(2)}{X^+ - O^-} \qquad \underset{(3)}{X = O}$$

It should be appreciated, however, that any discussion of this sort—and, indeed, the distinction between *s*, *p* and *d* electrons in many-electron systems—is an implicit acknowledgement of the one-electron approach to atomic and molecular orbitals; in this sense, many of the problems of bonding are themselves imposed by the limitations of the theoretical models currently at our disposal.

nd Orbitals

The role of valence-shell *d* orbitals in the ground states of molecules formed by the typical elements has attracted considerable comment and attention during the past decade[546–548]. Although the *nd* functions undoubtedly have symmetry properties appropriate for bonding, there has been much debate about their potency in overlapping with ligand orbitals. The *nd* orbitals of the free atom are distinctly outer functions, both in size and energy. Calculations for 2nd row elements (Si, P, S, Cl) have suggested, however, that for an atom in a high valence state, or combined with highly electronegative ligands, the *nd* orbitals may be so stabilized and contracted that their effective participation in bonding is possible: such an effect would, in principle, lead to the involvement of *nd* orbitals in both σ- and π-interactions. The particular subjects of this section are oxyhalogen molecules, but many remarks made below are of a general nature and apply to interhalogen species and to related derivatives of the heavier elements of other Groups.

The contribution of *nd* orbitals to σ-bonding is probably unimportant. The valence-bond notion of octet expansion by means of the full hybridization of one or more *nd* orbitals in σ-bonding frameworks (suggested to account for coordination numbers >4, and allowing sp^3d schemes for pentacoordinate, sp^3d^2 schemes for hexacoordinate systems) is largely defunct, retreating before the realization that there is no causal relationship between the number of ligands and the number of bonding orbitals[549]. The molecular orbital approach to this problem of hypervalency[550], exemplified by Rundle's three-centre–four-electron bond[551], initially ignores *nd* orbitals as energetically inferior to the valence-shell *s* and *p* functions; the contributions of *nd* orbitals and of π-interactions (see below) are incorporated in a more sophisticated account, but may have only slight significance[552–554].

There is, however, a good *a priori* case for d_π–p_π bonding to oxygen superimposed on the molecular framework of σ-bonds[546,547,555]. A σ-bond between oxygen and a less electronegative halogen atom (Cl, Br, I) invariably embodies some charge separation

546 K. A. R. Mitchell, *Chem. Rev.* **69** (1969) 157.
547 D. W. J. Cruickshank, *J. Chem. Soc.* (1961) 5486.
548 L. Pauling, *The Nature of the Chemical Bond*, 3rd. edn., Cornell University Press, Ithaca (1960).
549 A. J. Downs, *Unusual Coordination Numbers*, in *New Pathways in Inorganic Chemistry* (ed. E. A. V. Ebsworth, A. G. Maddock and A. G. Sharpe), p. 15, Cambridge (1968).
550 J. I. Musher, *Angew. Chem., Internat. Edn.* **8** (1969) 54.
551 R. E. Rundle, *Survey Progr. Chem.* **1** (1963) 81.
552 L. S. Bartell, *Inorg. Chem.* **5** (1966) 1635.
553 B. M. Deb and C. A. Coulson, *J. Chem. Soc.* (*A*) (1971) 958.
554 R. S. Berry, M. Tamres, C. J. Ballhausen and H. Johansen, *Acta Chem. Scand.* **22** (1968) 231.
555 H. H. Jaffé, *J. Phys. Chem.* **58** (1954) 185.

$X^{\delta+}O^{\delta-}$, which is formally complete in the case of a terminal halogen–oxygen bond (2). Electron donation from a filled oxygen $2p_\pi$ orbital to a vacant halogen *nd* orbital of suitable symmetry (Fig. 30) presents an attractive means of neutralizing the charge separation and presumably strengthening the bond. The extent of such interactions is expected to depend on the other groups attached to the halogen and oxygen atoms, as well as on the oxidation state and coordination number of the halogen. The number of *nd* orbitals able effectively to participate in d_π–p_π interactions may be restricted by the geometry of the molecule[555]. In the tetrahedral XO_4^- anions (X = Cl, Br, I) all the *nd* functions possess symmetry characteristics which would, in principle, allow bonding with the two filled $2p_\pi$ orbitals of each

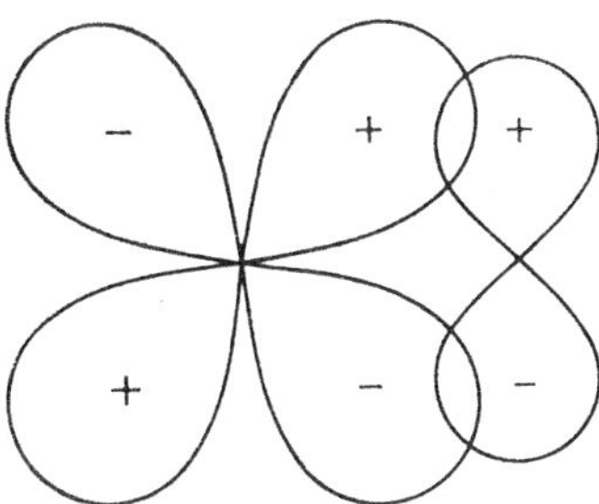

FIG. 30. Overlap for a d_π–p_π bond.

oxygen atom oriented perpendicular to the X–O bond axes. As was persuasively argued by Cruickshank[547], however, just the two *d* orbitals of *E* symmetry (d_{z^2} and $d_{x^2-y^2}$) have good overlap with the oxygen orbitals, forming two strongly bonding molecular orbitals having local π symmetry with respect to the X–O bonds; the T_2 MOs are regarded as being only weakly π-bonding. A closely related idea, expressed in valence-bond terminology, is the duodecet rule: that the valency shell of a second row element tends to be occupied by twelve electrons[556].

Experimental Data

The experimentally determinable parameters which bear most directly on the strength of bonding in molecules are bond lengths, valence force constants and bond energies (both dissociation energies and bond energy terms)[557]. Interatomic distances and valence force constants have been measured for many halogen–oxygen molecules; in Table 41 we have collated these two parameters for species containing only halogen atoms and oxygen atoms, and also for some other molecules. Bond dissociation energies, which have been less widely investigated, are listed in Table 42; bond energy terms for the chlorine oxides will be found in Table 45. The bond properties are markedly responsive to the oxidation state and coordination number of the halogen.

Bonding in halogen(I) oxycompounds is effectively described by conventional electron-pair bonds between the halogen and oxygen atoms. Unlike the oxygen fluorides, wherein the oxygen–fluorine bond is polarized $O^{\delta+}$–$F^{\delta-}$, single bonds between oxygen and the heavier halogens are polarized $X^{\delta+}$–$O^{\delta-}$, a difference typified in the distinctive chemistries of

[556] R. J. Gillespie and E. A. Robinson, *Canad. J. Chem.* **42** (1964) 2496.
[557] T. L. Cottrell, *The Strengths of Chemical Bonds*, 2nd edn., Butterworths, London (1958).

TABLE 41. BOND LENGTHS AND VALENCE FORCE CONSTANTS FOR HALOGEN–OXYGEN MOLECULES

		Cl		Br		I	
		r(Cl–O)[a] (Å)	f_r[b] (mdyne Å^{-1})	r(Br–O)[a] (Å)	f_r[b] (mdyne Å^{-1})	r(I–O)[a] (Å)	f_r[b] (mdyne Å^{-1})
Calculated single bond	"normal"[c]	1·69		1·82		1·99	
X_2O		1·70[h]	2·9[h]		2·4[s]		
XO^-		3·3[i]		3·0[i]			
XO		1·57[j]	6·4[j]	1·72[j]	4·0[j]	1·87[j]	3·9[j]
XO_2^-		1·57[w]	4·3[k]		4·2[k]		
XO_2		1·47[l]	7·2[l]				
O_2XOXO_2	terminal					1·79[m]	
	bridging					1·93[m]	
XO_2^+			9·0[d]				
XO_3^-		1·46–1·48[x]	5·9[n]	1·64–1·68[x]	5·3[n]	1·82[x]	5·5[n]
O_3XOXO_3	terminal	1·41[o]	9·3[o]				
	bridging	1·71[o]	3·2[o]				
XO_4^-		1·42–1·47[p]	8·2[q]	1·61[p]	6·0$_5$[q]	1·78[p]	5·9[q]
$O_3XO_3XO_3^{4-}$	terminal					1·77[r]	5·3[g]
	bridging					2·01[r]	3·0[g]
HOX		1·70[v]	3·9[v]		3·6[v]		
$HOXO_2$	terminal					1·80[e]	
	bridging					1·90[e]	
$HOXO_3$	terminal	1·41[t]	9·20[t]				
	bridging	1·64[t]	3·85[t]				
$(HO)_5XO$	terminal					1·78[r]	5·4[g]
	bridging					1·89[r]	3·0[g]
FXO_3		1·40[u]	9·4[u]				
NXO_3^{2-}		1·50[f]					

[a] Rounded to nearest 0·01 Å.
[b] Rounded to nearest 0·1 mdyne Å^{-1}.
[c] L. Pauling, *The Nature of the Chemical Bond*, 3rd edn., pp. 221–230, Cornell University Press, Ithaca (1960).
[d] K. O. Christe, C. J. Schack, D. Pilipovich and W. Sawodny, *Inorg. Chem.* **8** (1969) 2489.
[e] For molecule in anhydroiodic acid HIO_3,I_2O_5; Y. D. Feikema and A. Vos, *Acta Cryst.* **20** (1966) 769.
[f] N. I. Golovina, G. A. Klitskaya and L. O. Atovmyan, *J. Struct. Chem.* **9** (1968) 817.
[g] H. Siebert, *Fortschr. Chem. Forsch.* **8** (1967) 470.
[h–x] For references see: (h) Table 46; (i) Table 62; (j) Table 55; (k) Table 66; (l) Table 47; (m) Table 52; (n) Table 69; (o) Table 49; (p) Table 76; (q) Table 75; (r) Table 83; (s) Table 50; (t) Table 77; (u) Table 59; (v) Table 61; (w) Table 67; (x) Table 70.

TABLE 42. BOND DISSOCIATION ENTHALPIES OF OXYHALOGEN COMPOUNDS

Molecule	Process	$\Delta H^\circ_{298°K}$ (kcal mol^{-1})	Method
Cl_2O	$Cl_2O \rightarrow ClO + Cl$	35·0	Calc.[a]
HOCl	$HOCl \rightarrow HO + Cl$	32·3±2	Mass spec.[b]
OCl^-	$OCl^- \rightarrow Cl^- + O$	60±10	Calc.[c]
ClO	$ClO \rightarrow Cl + O$	64·29	Calc.[d]
		63·31±0·03(D_0°)	Spectroscopic[e]
ClO^+	$ClO^+ \rightarrow Cl^+ + O$	111±5	Calc.[c]
ClOO	$ClOO \rightarrow Cl + O_2$	8±2	Estimated[f]
OClO	$OClO \rightarrow ClO + O$	58·7	Calc.[a]
		66·5	Spectroscopic[a]
ClO_3	$ClO_3 \rightarrow OClO + O$	47·6	Calc.[a]
Cl_2O_7	$O_3ClOClO_3 \rightarrow ClO_3 + ClO_4$	30±4	Mass spec.[b]
$HOClO_3$	$HOClO_3 \rightarrow HO + ClO_3$	47·6	Calc.[g]
		46	Mass spec.[h]
$FClO_3$	$FClO_3 \rightarrow FClO_2 + O$	57	Mass spec.[i]
HOBr	$HOBr \rightarrow HO + Br$	56±10	Calc.[a]
BrO	$BrO \rightarrow Br + O$	56·23	Calc.[d]
		55·2±0·6 (D_0°)	Spectroscopic[e]
OBrO	$OBrO \rightarrow BrO + O$	≥70	Estimated[a]
HOI	$HOI \rightarrow HO + I$	56±10	Calc.[a]
IO	$IO \rightarrow I + O$	43·25	Calc.[d]
		42±5 (D_0°)	Spectroscopic[e]

[a] V. I. Vedeneyev, L. V. Gurvich, V. N. Kondrat'yev, V. A. Medvedev and Ye. L. Frankevich, *Bond Energies, Ionization Potentials and Electron Affinities*, Edward Arnold, London (1966).
[b] I. P. Fisher, *Trans. Faraday Soc.* **64** (1968) 1852.
[c] P. A. G. O'Hare and A. C. Wahl, *J. Chem. Phys.* **54** (1971) 3770.
[d] *Selected Values of Chemical Thermodynamic Properties*, N.B.S. Technical Note 270–3 (1968).
[e] R. A. Durie and D. A. Ramsey, *Canad. J. Phys.* **36** (1958) 45; R. A. Durie, F. Legay and D. A. Ramsey, *ibid.* **38** (1960) 444.
[f] S. W. Benson and J. H. Buss, *J. Chem. Phys.* **27** (1957) 1382.
[g] J. B. Levy, *J. Phys. Chem.* **66** (1962) 1092.
[h] G. A. Heath and J. R. Majer, *Trans. Faraday Soc.* **60** (1964) 1783.
[i] V. H. Dibeler, R. M. Reese and D. E. Mann, *J. Chem. Phys.* **27** (1957) 176.

hypofluorites and hypochlorites. The three measured O–Cl(I) bond lengths (Cl_2O, 1·70 Å; HOCl, 1·69 Å; CH_3OCl, 1·67 Å) are all close to the "normal" single bond length (1·69 Å) calculated from the covalent radii of the elements with due allowance for the electronegativity correction proposed by Schomaker and Stevenson[548,558]. In terms of valence force constant, however, Cl–O single bonds are less homogeneous, the values extending from 3·9 mdyne Å^{-1} (HOCl) to 2·6 mdyne Å^{-1} ($ClOSO_2F$). Investigations of Cl_2O using photoelectron spectroscopy and *ab initio* LCAO–SCF calculations imply that there is minimal (≤ 1%) *d*-orbital participation in the Cl_2O molecular orbitals[559].

The contraction of many halogen–oxygen bonds relative to the "normal" single-bond lengths and the concomitant increase in valence force constant are commonly attributed to multiple bonding[546,547]. The occurrence of shortening is independent of the coordination

[558] V. Schomaker and D. P. Stevenson, *J. Amer. Chem. Soc.* **63** (1941) 37.
[559] A. B. Cornford, D. C. Frost, F. G. Herring and C. A. McDowell, *J. Chem. Phys.* **55** (1971) 2820.

number of the halogen, but is restricted almost exclusively to bonds with terminal oxygen atoms; thus the bridging Cl–O distances in Cl_2O_7 are *longer* than the bonds in Cl_2O, although the terminal Cl–O distances in Cl_2O_7 are very short. [An exception to this rule is the Cl–OH distance in anhydrous perchloric acid (1·64 Å).] It is arguable to what extent the strengthening of the Cl–O bond implied by the changes in bond length and force constant, for example in going from ClO^- to ClO_4^-, is reflected (i) in related bond dissociation enthalpies (Table 42) or (ii) in the bond energy terms of the chlorine oxides (Table 45). Though most evidence is available for oxychlorine molecules, what data there are suggest similar diminutions in bromine–oxygen and iodine–oxygen distances as the oxidation number of the halogen rises, although the accompanying increases in stretching force constant are less than those for chlorine–oxygen bonds.

The issues of interatomic distances and stretching force constants apart, evidence for *nd* orbital participation and multiple bonding is spectroscopic in origin, and confusing in interpretation.

(i) It is claimed that the X-ray fluorescence spectra of oxyanions of second row elements (P, S and Cl) exhibit features directly testifying to $3d_\pi$–$2p_\pi$ bonding[560]: for oxychlorine anions the occupancy of 3*d* orbitals increases in the expected sequence $ClO^- < ClO_2^- < ClO_3^- < ClO_4^-$ [561]. However, X-ray spectroscopic evidence also places high positive charges on the Cl atoms of ClO_3^- and ClO_4^- (Table 44)[562], implying little, if any, charge neutralization, either by π-donation or σ-polarization.

(ii) Bond polarizabilities derived from the Raman intensities of the fundamentals in totally symmetric vibrations of the halogen oxyanions imply nd_π–$2p_\pi$ bonding in ClO_3^- and ClO_4^-, but not in bromine and iodine systems[563].

(iii) The esr parameters of oxyhalogen radicals are usually interpreted with the assumption of negligible spin-density in the halogen *nd* orbitals, although this does not necessarily exclude the participation of these functions in bonding molecular orbitals.

(iv) Halogen nuclear quadrupole spectroscopy yields results which are equivocal in respect of *nd* orbital-participation because of the severe approximations necessary in their analysis (see pp. 1271–4). Similar reservations prevail about the interpretation of ^{129}I Mössbauer spectra[564].

Calculations

That π-bonding occurs in oxygen derivatives of intermediate and higher oxidation states of the halogens is supported, at least for oxychlorine compounds, by molecular orbital calculations of varying degrees of complexity. Wagner[565] used an internally consistent LCAO–MO method to obtain π-bond orders and atomic charges for a number of species; the results, summarized in Table 43, show an increase in π-bonding in the order $ClO^- < Cl_2O < ClO_2^- < ClO < ClO_3^- < ClO_2 < ClO_4^-$. No p_π character is predicted for the anions ClO_2^-, ClO_3^- and ClO_4^-, the chlorine 3*p* functions being pre-empted in σ-bonding orbitals and in inert pair orbitals tetrahedrally disposed about the central atom. ClO_2 has strong d_π–p_π and p_π–p_π interactions, in keeping with the short interatomic distances and

560 D. S. Urch, *J. Chem. Soc.* (*A*) (1969) 3026.
561 V. I. Nefedov, *J. Struct. Chem.* **8** (1967) 919.
562 W. Nefedow, *Phys. Status Solidi*, **2** (1962) 904.
563 G. W. Chantry and R. A. Plane, *J. Chem. Phys.* **32** (1960) 319; *ibid.* **34** (1961) 1268.
564 E. A. C. Lucken, *Structure and Bonding*, **6** (1969) 1.
565 E. L. Wagner, *J. Chem. Phys.* **37** (1962) 751.

TABLE 43. LCAO-MO CALCULATIONS FOR OXYCHLORINE MOLECULES[a]

Molecule	Atomic charges		π-bond order	%$3d_\pi$ character
	Q_{Cl}	Q_O		
ClO^-	0	−1·00	0	—
Cl_2O	0	0	0	—
ClO_2^-	+0·429	−0·715	0·132	100
ClO[b]	+0·509	−0·509	0·500	0
ClO_3^-	+0·293	−0·431	0·608	100
ClO_2	+0·769	−0·385	0·737	52
ClO_4^-	+0·437	−0·359	0·908	100

[a] E. L. Wagner, *J. Chem. Phys.* **37** (1962) 751.
[b] *Ab initio* Hartree–Fock–Roothaan SCF-MO calculations indicate "not insignificant" *d*-orbital participation; P. A. G. O'Hare and A. C. Wahl, *J. Chem. Phys.* **54** (1971) 3770.

high force constant; allowing only p_π–p_π interactions would give a balance of one π-electron between two bonds and much weaker bonding. Detailed *ab initio* calculations for ClO[566] also suggest that d_π–p_π bonding in this molecule is "not insignificant", though Wagner predicted only p_π–p_π interactions.

It has been shown that the inclusion of chlorine's $3d$ orbitals in MO calculations for the oxyanions results in lower positive charges on the Cl atoms than are assigned without the $3d$ functions (Table 44). While chemical intuition suggests that the halogen atom in these assemblies should not bear a charge $> +1$, X-ray spectra of the oxychlorine anions[561,562] are interpreted in terms of much higher charges; infrared[567] and Raman[563] intensity measurements are consistent with lower charges on the Cl atoms. Comparative studies of ClO_4^-, BrO_4^- and IO_4^- [using Hartree–Fock–Slater atomic wave functions and self-consistent charge and configuration (extended Hückel) MO calculations] conclude[568] that the radial nodes on the bromine $4d$ and iodine $5d$ functions do not have an adverse effect on d_π–$2p_\pi$ overlap relative to the nodeless chlorine $3d$ orbitals. No attempt was made to quantify the extent of *d*-orbital participation in the bonding in perhalate anions.

Bond Orders

Several authors have tried to correlate the lengths r and force constants f_r of halogen–oxygen bonds with their bond orders n. Pauling[548] gives the following bond orders: ClO_2^-, 1·37; ClO_3^-, 1·64; ClO_4^-, 2·10.

Robinson[569] discovered an empirical relation of the form

$$\log_{10} f_r(\text{Cl–O}) = -a \log_{10} r(\text{Cl–O}) + b \quad (1)$$

whence he derived an equation

$$n(\text{Cl–O}) = c f_r(\text{Cl–O}) + d \quad (2)$$

566 P. A. G. O'Hare and A. C. Wahl, *J. Chem. Phys.* **54** (1971) 3770.
567 G. N. Krynauw and C. J. H. Schutte, *Z. physik. Chem., N.F.* **55** (1967) 121.
568 M. M. Cox and J. W. Moore, *J. Phys. Chem.* **74** (1970) 627.
569 E. A. Robinson, *Canad. J. Chem.* **41** (1963) 3021.

TABLE 44. CALCULATED AND EXPERIMENTAL ATOMIC CHARGES IN OXYCHLORINE ANIONS

ClO_2^-		ClO_3^-		ClO_4^-		Method	Reference
Q_{Cl}	Q_O	Q_{Cl}	Q_O	Q_{Cl}	Q_O		
+0·35	a	+0·34	a	+0·34	−0·33	Estimated	L. Pauling, *The Nature of the Chemical Bond*, 3rd edn., pp. 320–324, Cornell (1960).
+0·43	−0·72	+0·29	−0·43	+0·44	−0·36	LCAO-MO calculations[b]	E. L. Wagner, *J. Chem. Phys.* **37** (1962) 751.
				+0·64[b]	−0·41[b]	Extended Hückel	M. M. Cox and J. W. Moore, *J. Phys. Chem.* **74** (1970) 627.
				+0·57[c]	−0·39[c]	MO calculations	
				+2·02[d]	−0·76[d]		
+0·36	a	+1·02	a	+2·10	−0·78	LCAO-MO calculations[d,e]	R. Manne, *J. Chem. Phys.* **46** (1967) 4645.
		a	−0·26	+0·65	−0·41	I.r. intensities	G. N. Krynauw and C. J. H. Schutte, *Z. phys. Chem. N.F.* **55** (1967) 121.
		+1·00	a	+1·00	−0·50	Raman intensities	G. W. Chantry and R. A. Plane, *J. Chem. Phys.* **34** (1961) 1268.
		+1·72	a	+2·27	−0·82	X-ray spectroscopy	W. Nefedow, *Phys. Status Solidi*, **2** (1962) 904.

[a] Not estimable because of uncertainty in "lone-pair charges".
[b] Includes chlorine 3*d* orbitals.
[c] Includes chlorine 4*s* orbitals.
[d] Neglects chlorine 3*d* orbitals.
[e] Consistent with E.S.C.A. data.

based on assumed (valence-bond) bond orders for ClO_4^- (1·5) and ClO_3F (1·67) and a calculated force constant for the "normal" single bond (f_r = 3·3 mdyne Å^{-1}). Robinson evaluated the constants: *a*, 6·45; *b*, 2·0; *c*, 0·102; and *d*, 0·66; using up-to-date data, the constants are: *a*, 6·3$_0$; *b*, 1·94; *c*, 0·107; and *d*, 0·63. Wagner[565] also arrived at an expression of the type (2), having calculated bond orders by the LCAO–MO method; unhappily he used a set of force constants different from those of Robinson. Siebert, too, has correlated I–O bond lengths and force constants[570].

Conclusions

The deductions made about π-bonding solely on the basis of the bond lengths and force constants in halogen–oxygen compounds are inconclusive, being based on a concept of doubtful validity, viz. the "normal" single bond; while spectroscopic (X-ray, Raman etc.) measurements endorse d_π–p_π bonding in Cl–O systems, there is no direct physical support for π-interactions in oxybromine or oxyiodine molecules. The variations observed in bond lengths and force constants may be attributable for I–O, for Br–O, and (at least in part) for Cl–O bonds to changes in the oxidation state and coordination number of the halogen, in the ionic charge on the assembly, and in ligation of the oxygen. In particular, the radius of the halogen atom may be very sensitive to the charge which the atom bears and to its coordination number.

Scepticism may also be necessary in considering quantitative estimates of multiple-bond character. Empirically derived bond orders may be useful for comparing different halogen–oxygen bonds, although this purpose might better be served by a plain statement of the physical parameters which define the bond order. Bond orders obtained in MO calculations may be more reliable, although the same calculations often produce charge distributions in the molecule very different from those determined experimentally (Table 44).

In discussing oxyhalogen molecules in this section use will be made of the formulations (1), (2) and (3), since they facilitate the depiction of redox and other reactions; to do so does not imply the correctness of each or any of the formulations.

2. THE OXIDES OF THE HALOGENS

INTRODUCTION

Fourteen halogen oxides have been isolated and reasonably well characterized: they are

Cl_2O	Br_2O	
Cl_2O_3		
ClO_2	BrO_2	
$ClOClO_3$		"I_2O_4"
		I_4O_9
	Br_2O_5	I_2O_5
	Br_3O_8	
Cl_2O_6	BrO_3	
Cl_2O_7		

570 H. Siebert, *Anwendungen der Schwingungsspektroskopie in der Anorganischen Chemie*, p. 106, Springer-Verlag (1966).

Claims have also been made for the existence of I_2O_7 and ClO_4, but the evidence is not compelling. It is noteworthy that the halogen perchlorates $BrOClO_3$ and $I(ClO_4)_3$ are formally mixed oxides.

At room temperature the chlorine oxides are gases or fairly volatile liquids; despite their explosive nature, they have been rather fully investigated, and chlorine dioxide is used commercially as an oxidant. By contrast, the known bromine oxides are all very unstable above −40°C, and the iodine oxides are solids which decompose into iodine and oxygen on heating. Such are the disparities (both in chemical character and in the extent of our available knowledge) between formally analogous derivatives of chlorine, bromine and iodine that the ensuing discussion is best organized in the sequence: chlorine oxides, bromine oxides, iodine oxides, rather than by comparison of compounds of the same empirical formula.

CHLORINE OXIDES

Introduction

In keeping with the endothermic nature of the chlorine oxides, chlorine and oxygen do not combine under ordinary conditions: chlorine reacts with monatomic oxygen to form Cl_2O_6, but the oxides are generally prepared by less direct chemical methods. Chlorine dioxide was discovered early in the nineteenth century during several independent investigations of the action of sulphuric acid on potassium chlorate. In 1834 the oxidation of chlorine with mercuric oxide (yielding Cl_2O) was reported, while in 1843 Millon produced by the photolysis of ClO_2 a red oil—presumably chlorine trioxide but analyzed then as Cl_6O_{17}. The isolation of Cl_2O_7 awaited the dehydration of perchloric acid in 1900. Two other oxides have only recently been characterized—Cl_2O_3 in 1967 and $ClOClO_3$ in 1970; the unknown compound chlorine tetroxide has a colourful record of alleged preparations.

The dangers inherent in the manipulation of chlorine oxides have been recognized since the earliest days of their chemistry: in 1815 Davy cautioned that the reaction of $KClO_3$ with sulphuric acid be performed with only very small quantities of chemicals. The need to employ adequate safety procedures when working with these unpredictable endothermic materials cannot too often be stressed. All the compounds, but especially ClO_2, are liable to detonate in the event of thermal or physical shock, or even change of phase. They are strong oxidizing agents, causing organic matter to ignite spontaneously and sometimes explosively. The reader is referred to general texts on laboratory safety[571], and should be fully conversant with the literature before attempting to prepare or use any of these materials.

Physical and thermochemical data for the oxides of chlorine are itemized in Table 45. The account of their chemistry presented here summarizes and supplements the most recent reviews[572–575], which together provide an exhaustive coverage of the preparations,

[571] *Handbook of Laboratory Safety*, 2nd edn. (ed. N. V. Steere), Chemical Rubber Co., Cleveland (1971).

[572] *Gmelins Handbuch der Anorganischen Chemie*, 8 Auflage, "Chlor", System-nummer 6, Teil B, Lieferung 2, Verlag Chemie, Weinheim/Bergstr. (1969).

[573] C. C. Addison, *Mellor's Comprehensive Treatise on Inorganic and Theoretical Chemistry*, Supplement II, Part I, pp. 514–544, Longmans, London (1956).

[574] H. L. Robson, *Kirk-Othmer Encyclopedia of Chemical Technology*, 2nd edn., Vol. 5, pp. 7–50, Interscience, New York (1964).

[575] (a) M. Schmeisser and K. Brändle, *Adv. Inorg. Chem. Radiochem.* **5** (1963) 41; (b) R. J. Brisdon, *MTP International Review of Science: Inorganic Chemistry Series One*, Vol. 3 (ed. V. Gutmann), p. 215, Butterworths and University Park Press (1972).

TABLE 45. PHYSICAL PROPERTIES OF THE OXIDES OF CHLORINE

	Cl_2O	ClO_2	$ClOClO_3$	Cl_2O_6	Cl_2O_7
Physical properties					
Colour: solid	brown	red	pale yellow	yellow (−180°C)	colourless
liquid	red-brown	red	pale yellow	[dark red]	colourless
vapour	yellow-green	yellow-green	pale yellow	[red-brown]	colourless
Melting point (°C)	−120·6[a]	−59[e]	−117±2[i]	3·5[k]	−91·5[n]
Boiling point (°C)	2·0[b]	9·7[f]	44·5[i]	203[k]	81[n]
Vapour pressure, $\log p$(mm) =	$7{\cdot}87-1373/T$[b] (173°K $<T<$ 288°K)	$7{\cdot}7427-1275{\cdot}1/T$[f] (227°K $<T<$ 312°K)	$7{\cdot}82-1586/T$[i] (226°K $<T<$ 296°K)	liq: $7{\cdot}1-2070/T$[k] solid: $9{\cdot}3-2690/T$ (233°K $<T<$ 293°K)	$8{\cdot}03-1818/T$[n] (270°K $<T<$ 303°K)
ΔH_{vap} at b.p. (kcal mol^{-1})	6·20[b]	6·29[f]	7·17[i]	9·5[k]	8·29[n]
Trouton's constant (cal deg^{-1} mol^{-1})	22·5[b]	22·23[f]	22·6[i]	21[k]	23·4[n]
Thermodynamic properties of the gaseous molecule					
$\Delta H_f°$ at 0°K (kcal mol^{-1})	19·71[c]	25·09[c]			
$\Delta H_f°$ at 298°K (kcal mol^{-1})	19·2[c]	24·5[c]	*ca.* 43[i]	[37][c,l]	65·0[c]
$\Delta G_f°$ at 298°K (kcal mol^{-1})	23·4[c]	28·8[c]			
$S°$ at 298°K (cal deg^{-1} mol^{-1})	63·60[c]	61·36[c]	78·21[j]		
$C_p°$ at 298°K (cal deg^{-1} mol^{-1})	10·85[c]	10·03[c]	20·56[j]		
Thermochemical properties					
Bond energy term (kcal mol^{-1})	49[c]	62[c]		[57·2][c,l]	
Ionization potential (kcal mol^{-1})	254[d]	256[g]		[270][l,m]	> 300[g]
Electron affinity (kcal mol^{-1})		79·1[h]		[91·4][l,m]	

[a] C. H. Secoy and G. H. Cady, *J. Amer. Chem. Soc.* **62** (1940) 1036.
[b] C. F. Goodeve, *J. Chem. Soc.* (1930) 2733.
[c] *Selected Values of Chemical Thermodynamic Properties*, N.B.S. Technical Note 270–3 (1968).
[d] Measured by photoelectron spectroscopy: A. B. Cornford, D. C. Frost, F. G. Herring and C. A. McDowell, *J. Chem. Phys.* **55** (1971) 2820.
[e] F. E. King and J. R. Partington, *J. Chem. Soc.* (1926) 925.
[f] H. Grubitsch and E. Suppan, *Monatsh. Chem.* **93** (1962) 246.
[g] Measured by mass spectroscopy: V. H. Dibeler, R. M. Reese and D. E. Mann, *J. Chem. Phys.* **27** (1957) 176.
[h] Estimated from the heat of hydration of ClO_2^-: V. I. Vedeneyev, L. V. Gurvich, V. N. Kondrat'yev, V. A. Medvedev and Ye. L. Frankevich, *Bond Energies, Ionization Potentials and Electron Affinities*, p. 196, Edward Arnold, London (1966).
[i] C. J. Schack and D. Pilipovich, *Inorg. Chem.* **9** (1970) 1387.
[j] K. O. Christe, C. J. Schack and E. C. Curtis, *Inorg. Chem.* **10** (1971) 1589.
[k] C. F. Goodeve and F. D. Richardson, *J. Chem. Soc.* (1937) 294.
[l] Bracketed values refer to ClO_3 molecules.
[m] Estimated from the lattice energies of chlorates: ref. h.
[n] C. F. Goodeve and J. Powney, *J. Chem. Soc.* (1932) 2078.

properties and uses of these compounds. Further historical information is best sought in the original edition of Mellor[576].

Dichlorine Monoxide

Preparation

The name "dichlorine monoxide" is adopted for the compound Cl_2O; the use of "chlorine monoxide", common in the older literature, invites confusion with the diatomic radical ClO.

In the laboratory Cl_2O is best obtained by treating freshly prepared yellow mercuric oxide either with chlorine gas (diluted with dry air) or with a solution of chlorine in carbon tetrachloride[577].

$$2Cl_2+2HgO \rightarrow HgCl_2.HgO+Cl_2O$$

Large-scale preparations may also follow this method. Dichlorine monoxide is very soluble in water and is partially converted to hypochlorous acid, of which it is the anhydride.

$$Cl_2O+H_2O \rightleftharpoons 2HOCl$$

Concentrated solutions of hypochlorous acid contain significant amounts of dichlorine monoxide, and the vapour over such solutions also contains Cl_2O: the oxide may be extracted either with CCl_4 or by passing air through the solution. On dissolution in alkali, Cl_2O generates hypochlorite. Much of the dichlorine monoxide produced industrially is used directly in the manufacture of hypochlorous acid and hypochlorites.

Structure and reactions

The compound is a yellowish-green gas at room temperature. The angular Cl–O–Cl molecule is well defined in the vapour phase by spectroscopic and diffractometric techniques (Table 46); the Cl–O distance is consistent with a normal single bond. Comparison of the infrared spectra of the vapour and solid suggests that a similar C_{2v} unit prevails in the condensed phase.

Cl_2O is liable to explode on heating or on impact, although very pure samples are allegedly quite stable in this respect. The photolytic decomposition of the gas into Cl_2 and O_2 proceeds with a quantum yield of 2 by the mechanism

$$Cl_2O \xrightarrow{h\nu} ClO+Cl$$

$$Cl+Cl_2O \rightarrow ClO+Cl_2$$

$$ClO+ClO \rightarrow Cl_2+O_2$$

Photolysis of carbon tetrachloride solutions or controlled thermal degradation (60°C < T < 140°C) produces significant amounts of ClO_2. Photolysis of Cl_2O isolated at 14°K in an inert gas matrix[578] affords, in addition to $(ClO)_2$, the radical ClClO, characterized by its infrared spectrum as containing a very weak Cl–Cl bond and a Cl–O linkage not much different from that in the diatomic ClO molecule (see Table 55).

576 J. W. Mellor, *A Comprehensive Treatise on Inorganic and Theoretical Chemistry*, Vol. 2, pp. 240–418, Longmans, Green and Co., London (1922).

577 G. H. Cady, *Inorganic Syntheses*, Vol. 5 (ed. T. Moeller), p. 156, McGraw-Hill (1957).

578 M. M. Rochkind and G. C. Pimentel, *J. Chem. Phys.* **46** (1967) 4481; W. G. Alcock and G. C. Pimentel, *ibid.* **48** (1968) 2373.

TABLE 46. SPECTROSCOPIC AND RELATED INVESTIGATIONS OF DICHLORINE MONOXIDE

Microwave spectrum[a,b] Measured for $^{35}Cl^{16}O^{35}Cl$, $^{35}Cl^{16}O^{37}Cl$, $^{37}Cl^{16}O^{37}Cl$ C_{2v} symmetry
r_0(Cl–O) = 1·70038 ± 0·00069 Å; $\theta_{Cl-O-Cl}$ = 110·96 ± 0·08°[a]
r_s(Cl–O) = 1·70038 ± 0·00043 Å; $\theta_{Cl-O-Cl}$ = 110·86 ± 0·04°[a]
Force constants (mdyne Å^{-1}):[a] f_r = 2·88 ± 0·08; f_{rr} = 0·31 ± 0·05
f_θ/r^2 = 0·432 ± 0·002; $f_{r\theta}/r$ = 0·17 ± 0·05
Quadrupole coupling data:[b] e^2Qq = −140 MHz; η = 0·06
Electron diffraction[c] C_{2v} symmetry
$r_g(1)$(Cl–O) = 1·693 ± 0·003 Å; $\theta_{Cl-O-Cl}$ = 111·2 ± 0·3°
Infrared spectrum[d] Measured for vapour, solid and matrix-isolated molecule; includes isotopic data (^{16}O, ^{18}O, ^{35}Cl, ^{37}Cl).
Fundamental frequencies (cm^{-1}), solid $^{35}Cl_2{}^{16}O$ at 77°K:
$\nu_1(a_1)$ 630·7, $\nu_2(a_1)$ 296·4, $\nu_3(b_1)$ 670·8
Force constants (mdyne Å^{-1}):[d,e] f_r = 2·75 ± 0·02; f_{rr} = 0·40 ± 0·01
f_θ/r^2 = 0·46 ± 0·01; $f_{r\theta}/r$ = 0·15 ± 0·01
Ultraviolet-visible spectrum[f] Measured for vapour.
Continuous absorption with three maxima
Mass spectrum[g] Measured at −78°C

Positive ion	O_2	^{35}Cl	^{37}Cl	^{35}ClO	^{37}ClO	$^{35}Cl^{35}Cl$
Relative intensity	0·07	0·7	0·7	100·0	33·4	5·4

Positive ion	$^{35}Cl^{37}Cl$	$^{37}Cl^{37}Cl$	$^{35}ClO^{35}Cl$	$^{35}ClO^{37}Cl$	$^{37}ClO^{37}Cl$
Relative intensity	3·6	0·5	43·5	28·3	4·7

Photoelectron spectrum[h]
Vertical i.p. (eV): 11·02, 12·37, 12·65, 12·79, 15·90, 16·65, 17·68, 20·64
Dissociation enthalpies (kcal mol^{-1})
ClOCl → ClO + Cl, $\Delta H°_{298}$ = 34·2 (th/d); 32·3 ± 2 (mass spec.)[g]
ClOCl → Cl_2 + O, $\Delta H°_{298}$ = 40·4 (th/d)
Dipole moment[i] Measured for CCl_4 solution: 0·78 ± 0·08*D*

[a] G. E. Herberich, R. H. Jackson and D. J. Millen, *J. Chem. Soc.* (*A*) (1966) 336.
[b] R. H. Jackson and D. J. Millen, *Proc. Chem. Soc.* (1959) 10.
[c] B. Beagley, A. H. Clark and T. G. Hewitt, *J. Chem. Soc.* (*A*) (1968) 658.
[d] M. M. Rochkind and G. C. Pimentel, *J. Chem. Phys.* **42** (1965) 1361.
[e] B. Beagley, A. H. Clark and D. W. J. Cruickshank, *Chem. Comm.* (1966) 458.
[f] C. F. Goodeve and J. I. Wallace, *Trans. Faraday Soc.* **26** (1930) 254; W. Finkelnburg, H. J. Schumacher and G. Steiger, *Z. phys. Chem.* **B15** (1931) 127.
[g] I. P. Fisher, *Trans. Faraday Soc.* **64** (1968) 1852.
[h] A. B. Cornford, D. C. Frost, F. G. Herring and C. A. McDowell, *J. Chem. Phys.* **55** (1971) 2820.
[i] D. Sundhoff and H. J. Schumacher, *Z. phys. Chem.* **B28** (1935) 17.

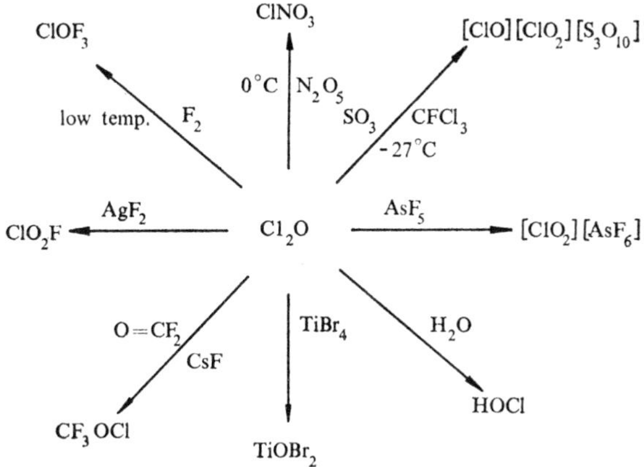

SCHEME 1. Reactions of dichlorine monoxide.

Chlorination of organic compounds by Cl_2O,

$$Cl_2O + 2RH \rightarrow 2RCl + H_2O$$

follows a mixed chain reaction with OCl and Cl as chain-propagating species; its selectivity is not as great as that of other chlorinating agents[579].

Dichlorine monoxide has found several minor synthetic applications, particularly where it can act as a substitute for the more dangerous ClO_2. Its reaction with N_2O_5 is a very convenient route to $ClNO_3$ [580], which is also formed from Cl_2O and other nitrogen oxides: in each case the exact stoichiometry of the reaction depends on the phase in which it is conducted[575], but the step leading to chlorine nitrate is

$$ClO + NO_2 \rightarrow ClNO_3$$

The low-temperature fluorination of Cl_2O gives $ClOF_3$, while AgF_2 at 70°C readily converts it to ClO_2F, probably *via* ClO_2 [581]. The reactions whereby metal halides are transformed into oxyhalides probably involve metal hypochlorite intermediates; boron halides similarly afford anhydrous boric oxide[575]. Both SO_3 and AsF_5 oxidize Cl_2O in the formation of ionic complexes[582].

The contrasting polarizations of the halogen–oxygen bonds in Cl_2O and F_2O are marked in their reactions with $O{=}CF_2$, which are catalysed by CsF: CF_3O^- attacks the positively charged chlorine atom forming CF_3OCl, whereas with F_2O, CF_3OOF is the initial product[583].

Dichlorine Trioxide

Early reports of the preparation of Cl_2O_3 by the partial reduction of chloric acid with reagents such as As_2O_3 or undecylenic acid are unfounded; the products were mixtures of chlorine and chlorine dioxide[576]. However, the concept of Cl_2O_3 as a transient intermediate accounts satisfactorily (i) for the induction time found in the photolytic decomposition of ClO_2 in a static system, and (ii) for some aspects of the bleaching action of chlorous acid.

Careful photolysis of chlorine dioxide at −78°C affords, in addition to chlorine and oxygen, two products[584]: orange chlorine trioxide, and a dark brown solid identified as Cl_2O_3. The latter compound, though stable at −78°C, explodes on vaporizing. It is assigned the structure O–Cl–Cl(=O)(=O) (O–Cl–Cl with two O atoms on the second Cl), containing a very weak chlorine–chlorine bond, and its heat of formation is estimated to be +45 kcal mol^{-1}.

Chlorine Dioxide

Precautions

Discussion of the chemistry of chlorine dioxide (which was once called "chlorine peroxide") is prefaced by a warning of the risks involved in manipulating this capricious compound. The deep red liquid is explosive above −40°C, and detonations can occur at temperatures as low as −100°C; concentrations of the yellow-green vapour of > 10% in air

579 D. D. Tanner and N. Nychka, *J. Amer. Chem. Soc.* **89** (1967) 121.
580 M. Schmeisser, *Inorganic Syntheses*, Vol. 9 (ed. S. Y. Tyree, jun.), p. 127, McGraw-Hill (1967).
581 Y. Macheteau and J. Gillardeau, *Bull. Soc. chim. France* (1967) 4075.
582 C. J. Schack and D. Pilipovich, *Inorg. Chem.* **9** (1970) 387.
583 D. E. Gould, L. R. Anderson, D. E. Young and W. B. Fox, *J. Amer. Chem. Soc.* **91** (1969) 1310.
584 E. T. McHale and G. von Elbe, *J. Amer. Chem. Soc.* **89** (1967) 2795; *J. Phys. Chem.* **72** (1968) 1849.

may explode under the action of heat, light or shock; spontaneous combustion can occur with organic or other readily oxidizable materials. Dilution with polyatomic gases stabilizes ClO_2 to some extent. The compound is normally generated ready for immediate use, although small amounts may be stored in the liquid phase at temperatures below $-50°C$ or in CCl_4 solution in the dark: the storage of large quantities is strongly discouraged.

Preparation

Despite subsequent improvements in experimental conditions, the reaction between concentrated sulphuric acid and a solid chlorate, whereby chlorine dioxide was first discovered, holds great risks of explosion; the essential process is the disproportionation of chloric acid.

$$3HClO_3 \rightarrow 2ClO_2 + HClO_4 + H_2O$$

Most of the many alternative syntheses since devised involve either reduction of chlorates or oxidation of chlorites; only those methods affording convenient access to the compound are mentioned here.

The oxidation of sodium chlorite with chlorine

$$2NaClO_2 + Cl_2 \rightarrow 2ClO_2 + 2NaCl$$

is effected industrially by mixing concentrated aqueous solutions of the reagents[574] while on a small scale chlorine gas, diluted with dry air, is passed through columns packed with the solid chlorite[585].

Reduction of $KClO_3$ in the laboratory is conveniently accomplished with moist oxalic acid, since the simultaneously evolved carbon oxides act as diluents for ClO_2 [575].

$$2ClO_3^- + C_2O_4^{2-} + 4H^+ \rightarrow 2ClO_2 + 2CO_2 + 2H_2O$$

Industrial processes for the reduction of chlorate depend on two major reagents: chloride ion and sulphur dioxide[574]. Using Cl^-, the ultimate reaction

$$Cl^- + 5ClO_3^- + 6H^+ \rightarrow 6ClO_2 + 3H_2O$$

is not feasible thermodynamically; under optimum conditions the reaction approximates to

$$2Cl^- + 2ClO_3^- + 4H^+ \rightarrow Cl_2 + 2ClO_2 + 2H_2O$$

and the product is contaminated with chlorine. 90% yields of ClO_2 may be obtained, however, if sulphur dioxide is used to reduce solutions of sodium chlorate in 6–9 M sulphuric acid.

The reaction of silver chlorate with chlorine at 90°C provided samples of pure ClO_2 for the measurement of physical properties[586].

$$2AgClO_3 + Cl_2 \rightarrow 2ClO_2 + O_2 + 2AgCl$$

Structural chemistry

The ClO_2 molecule, with 19 electrons in its valence shell, has C_{2v} symmetry in its $^2B_1$ ground state. The bond length of 1·471 Å is some 0·22 Å shorter than the calculated single-bond length, while the O–Cl–O angle is only slightly smaller (*ca.* 2°) than the bond angle of sulphur dioxide (119·3°)[587]. That the chlorine 3*d* orbitals are greatly involved in

585 R. I. Derby and W. S. Hutchinson, *Inorganic Syntheses*, Vol. 4 (ed. J. C. Bailar, jun.), p. 152, McGraw-Hill (1953).
586 F. E. King and J. R. Partington, *J. Chem. Soc.* (1926) 925.
587 Y. Morino, Y. Kikuchi, S. Saito and E. Hirota, *J. Mol. Spectroscopy*, **13** (1964) 95.

TABLE 47. SPECTROSCOPIC AND RELATED INVESTIGATIONS OF CHLORINE DIOXIDE

Microwave spectrum[a,b] Measured for $^{35}Cl^{16}O_2$, $^{37}Cl^{16}O_2$, $^{35}Cl^{16}O^{18}O$ C_{2v} symmetry
r_s(Cl–O) = 1·473±0·01 Å; θ_{O-Cl-O} = 117·6±1°[a]
Quadrupole coupling constants:[b] $e^2Qq_{aa} = -51·90$ MHz
$e^2Qq_{bb} = +2·28$ MHz
$e^2Qq_{cc} = +49·62$ MHz

Electron diffraction[c] C_{2v} symmetry
r_g(1)(Cl–O) = 1·475±0·003 Å; θ_{O-Cl-O} = 117·7±1·7°
Infrared spectrum Measured for vapour[d] and matrix-isolated molecule[h]
Fundamental frequencies (cm^{-1}), vapour:

	$\nu_1(a_1)$	$\nu_2(a_1)$	$\nu_3(b_1)$
$^{35}ClO_2$	945·2	447·3	1110·8
$^{37}ClO_2$	940·4	444·6	1098·1

Force constants (mdyne $Å^{-1}$):[e] $f_r = 7·23$; $f_{rr} = -0·02$
$f_\theta/r^2 = 0·63$; $f_{r\theta}/r = 0·25$
Rotational analysis of $\nu_3(b_1)$[f]
Raman spectrum Measured for solution in water[g]
Ultraviolet-visible spectrum (absorption)
Solution in H_2O:[i] λ_{max} = 345 mμ, $\epsilon \sim 10^3$
Vapour: band system 2500–5000 Å, $^2A_2 \leftarrow {}^2B_1$; vibration-rotation analysis yields excited state parameters[d,e] r(Cl–O) = 1·619±0·016 Å, θ_{O-Cl-O} = 107·0±0·28°, $f_r = 3·69$, $f_{rr} = 0·08$, $f_\theta/r^2 = 0·29$, $f_{r\theta}/r = 0·10$ mdyne $Å^{-1}$.
In the vacuum-ultraviolet region three band systems are observed: at 1829 Å, 1628 Å and 1568 Å.[j]
Mass spectrum[k] Measured at −78°C

Positive ion	O_2	^{35}Cl	^{37}Cl	^{35}ClO	^{37}ClO	$^{35}ClO_2$	$^{37}ClO_2$
Relative intensity	2·1	1·9	0·6	31·3	10·1	100·0	32·0

Ionization potential, by electron impact 10·7±0·1 eV
Bond dissociation enthalpy (kcal mol^{-1})[l]
OClO → OCl+O, $\Delta H°_{298}$ 58·7 (th/d); 66·5 (spectroscopic)
OClO → Cl+O_2, $\Delta H°_{298}$ 4·0 (th/d); 4·6 (spectroscopic)
Esr spectrum Measured for solutions and for the matrix-isolated molecule.
In H_2SO_4 at 300°K[m] $g_{av} = 2·0093$, A_{iso}[u] = +16·5 G
In H_2SO_4 at 77°K[m] $g_{av} = 2·0102$, A_{iso}[u] = +18·0 G
In $KClO_3$ at 295°K[m] $g_{av} = 2·0101$, A_{iso}[u] = +18·0 G
Other studies: n–r.
Paramagnetic susceptibility[s] 1310×10^{-6} cgs units
Dipole moment[t] Measured for CCl_4 solutions; 1·69±0·09*D*

[a] R. F. Curl, jun., J. L. Kinsey, J. G. Baker, J. C. Baird, G. R. Bird, R. F. Heidelberg, T. M. Sugden, D. R. Jenkins and C. N. Kenney, *Phys. Rev.* **121** (1961) 1119; R. F. Curl, jun., R. F. Heidelberg and J. L. Kinsey, *ibid.* **125** (1962) 1993.
[b] R. F. Curl, jun., *J. Chem. Phys.* **37** (1962) 1779.
[c] A. H. Clark and B. Beagley, *J. Chem. Soc.* (*A*) (1970) 46.
[d] A. W. Richardson, R. W. Redding and J. C. D. Brand, *J. Mol. Spectroscopy*, **29** (1969) 93.
[e] J. C. D. Brand, R. W. Redding and A. W. Richardson, *J. Mol. Spectroscopy*, **34** (1970) 399.
[f] A. W. Richardson, *J. Mol. Spectroscopy*, **35** (1970) 43.
[g] T. G. Kujumzelis, *Physik. Z.* **39** (1938) 665.
[h] A. Arkell and I. Schwager, *J. Amer. Chem. Soc.* **89** (1967) 5999.
[i] N. Konopik, J. Derkosch and E. Berger, *Monatsh. Chem.* **84** (1953) 214.
[j] C. M. Humphries, A. D. Walsh and P. A. Warsop, *Discuss. Faraday Soc.* **35** (1963) 137, 230.
[k] I. P. Fisher, *Trans. Faraday Soc.* **63** (1967) 684.
[l] V. I. Vedeneyev, R. V. Gurvich, V. N. Kondrat'yev, V. A. Medvedev and Ye. L. Frankevich, *Bond Energies, Ionization Potentials and Electron Affinities*, p. 78, Edward Arnold, London (1966).
[m] R. S. Eachus, P. R. Edwards, S. Subramanian and M. C. R. Symons, *J. Chem. Soc.* (*A*) (1968) 1704.
[n] P. W. Atkins, J. A. Brivati, N. Keen, M. C. R. Symons and P. A. Trevalion, *J. Chem. Soc.* (1962) 4785.
[o] J. R. Byberg, S. J. K. Jensen and L. T. Muus, *J. Chem. Phys.* **46** (1967) 131.
[p] J. C. Fayet, C. Pariset and B. Thiéblemont, *Compt. rend.* **268B** (1969) 1317.
[q] J. E. Bennett and D. J. E. Ingram, *Phil. Mag.* **1** (1956) 109; J. E. Bennett, D. J. E. Ingram and D. Schonland, *Proc. Phys. Soc.* **69A** (1956) 556.
[r] T. Cole, *Proc. Nat. Acad. Sci.* **46** (1960) 506.
[s] N. W. Taylor, *J. Amer. Chem. Soc.* **48** (1926) 854.
[t] D. Sundhoff and H. J. Schumacher, *Z. phys. Chem.* **B28** (1935) 17.
[u] Hyperfine data for ^{35}Cl.

the bonding is supported by molecular orbital calculations[565,588]; the short bond length and high valence force constant similarly testify to the strength of the chlorine–oxygen bond. The spectroscopic properties of the molecule (Table 47) are consistent with the delocalization of the unpaired electron over all three atoms in a $4b_1$ antibonding orbital comprising the chlorine $3p$ and oxygen $2p$ orbitals perpendicular to the plane of the molecule; esr measurements indicate a spin density of 64% on chlorine[589].

Chlorine dioxide presents an interesting example of an odd-electron molecule which is stable towards dimerization; the physical and spectroscopic properties of the compound show no evidence of aggregation in the vapour, liquid or solid phases or in solution. This reluctance to dimerize probably stems from the delocalization of the unpaired electron. The energy expended in reorganizing the ClO_2 unit to permit localization of the electron would be insufficiently offset by chlorine–chlorine overlap, which is notably weak in ClClO and Cl_2O_3. Curiously, though, the isoelectronic thionite ion (SO_2^-) exists as $S_2O_4^{2-}$ in crystalline $Na_2S_2O_4$ (but with a long S–S bond, 2·389 Å)[590], and chlorine trioxide is associated in the liquid state despite the fact that the unpaired electron is here more delocalized than in ClO_2. There is also some evidence that the BrO_2 radical dimerizes in aqueous solution (p. 1377).

Reactions

Photochemical or thermal decomposition of ClO_2 commences with the reaction

$$ClO_2 \rightarrow ClO + O \qquad \Delta H^\circ_{298} = 66{\cdot}5 \text{ kcal mol}^{-1}$$

If the molecule is photolysed with uv light when isolated at low temperature in a rigid inert matrix[589,591] (whether a solidified noble gas, a solution frozen at 77°K, or a host perchlorate lattice), the major product is the radical ClOO formed in a cage back-reaction; some ClO may also be able to flee the cage.

After a short induction period, photolysis of the dry gas at room temperature proceeds with the formation of chlorine, oxygen and some chlorine trioxide which adheres to the walls of the vessel and is broken down on prolonged illumination[573].

$$ClO_2 + O \rightarrow ClO_3$$
$$ClO + ClO \rightarrow Cl_2 + O_2$$
$$2ClO_3 \rightarrow Cl_2O_6 \qquad \text{(a)}$$
$$2ClO_3 \rightarrow Cl_2 + 3O_2 \qquad \text{(b)}$$

Since step (b) has the higher heat of activation, more chlorine trioxide is formed at lower temperatures.

Photolysis of solid ClO_2 at −78°C produces some Cl_2O_3 as well as Cl_2O_6 (p. 1366)[584]; the formation and decomposition of the sesquioxide are held to be responsible for the initial period of induction in the decomposition of ClO_2.

$$ClO_2 + ClO \rightleftharpoons Cl_2O_3$$

Dark green aqueous solutions containing up to 8 g ClO_2 per litre (heat of solution = 6·6 kcal mol^{-1}) decompose only very slowly in the dark; the crystalline hydrate of approximate

588 J. L. Gole and E. F. Hayes, *Internat. J. Quantum Chem.*, Symposium No. 3, p. 519 (1969–70).
589 P. W. Atkins, J. A. Brivati, N. Keen, M. C. R. Symons and P. A. Trevalion, *J. Chem. Soc.* (1962) 4785.
590 J. D. Dunitz, *Acta Cryst.* **9** (1956) 579.
591 A. Arkell and I. Schwager, *J. Amer. Chem. Soc.* **89** (1967) 5999.

composition $ClO_2,8H_2O$, formed when the gas is passed into cold water, is a clathrate of the 8-gas, 46-water moles type with incomplete occupation of the cells by chlorine dioxide[592]. Illumination of neutral aqueous solutions inaugurates rapid decomposition to a mixture of chloric and hydrochloric acids,

$$\begin{aligned} ClO_2 &\xrightarrow{h\nu} ClO+O \\ ClO+H_2O &\rightarrow H_2ClO_2 \\ H_2ClO_2+ClO &\rightarrow HCl+HClO_3 \end{aligned}$$

while photodecomposition of moist ClO_2 gas produces a mist of droplets containing HOCl, $HClO_2$, $HClO_3$ and $HClO_4$ *via* hydrolysis of the intermediates Cl_2O_3 and Cl_2O_6. In alkaline solution hydrolysis rapidly (and possibly explosively) generates a mixture of chlorite and chlorate anions, and for this reason ClO_2 is regarded as the mixed anhydride of chlorous and chloric acids.

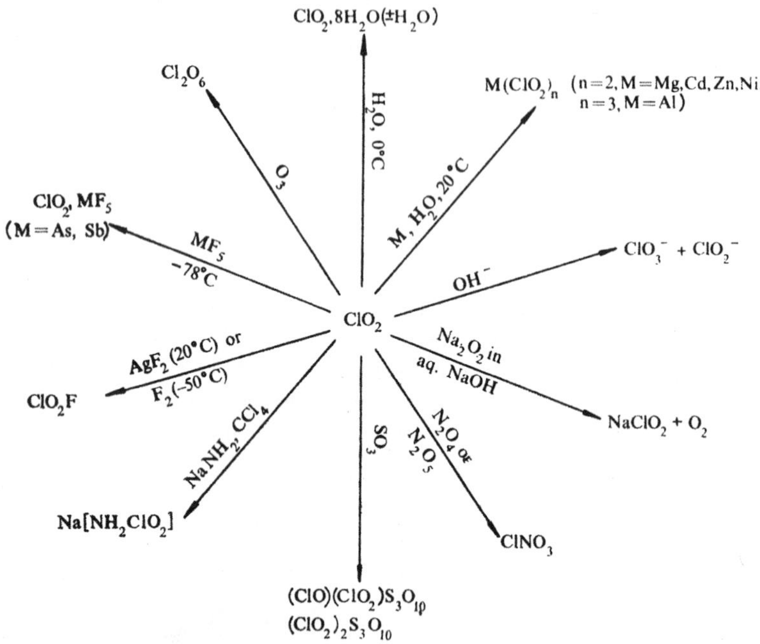

SCHEME 2. Reactions of chlorine dioxide.

Some reactions of chlorine dioxide are summarized in Scheme 2. The compound is a strong oxidizing agent, combining readily with organic matter and numerous inorganic materials, including phosphorus, sulphur, phosphorus halides and potassium borohydride. The reaction of ClO_2 with iodide ion involves ClO_2I^- as intermediate[593]; the stoichiometry of the reaction depends on the pH of the solution.

Acid solution: $2ClO_2+10I^-+8H^+ \rightarrow 2Cl^-+4H_2O+5I_2$

Neutral solution: $2ClO_2+2I^- \rightarrow 2ClO_2^-+I_2$

[592] M. Bigorgne, *Compt. rend.* **236** (1953) 1966.
[593] H. Fukutomi and G. Gordon, *J. Amer. Chem. Soc.* **89** (1967) 1362.

The latter reaction suggests that, in the presence of a suitable reductant, the disproportionative hydrolysis of ClO_2 in alkaline solution may be diverted to the production of chlorites; the most convenient reagents for this purpose are alkaline peroxides.

$$2ClO_2+O_2^{2-} \rightarrow 2ClO_2^-+O_2$$

The oxidation of powdered metals to chlorites by cold aqueous solutions of ClO_2 probably involves direct electron transfer; for the redox potential of the process

$$ClO_2(aq)+e^-(aq) \rightleftharpoons ClO_2^-(aq)$$

electrochemical measurements[572] give a value of +0·936 V at 25°C (cf. Fig. 14 and ref. 289).

Fluorination of ClO_2 with fluorine at −50°C or AgF_2 at 20°C readily affords chloryl fluoride[575]. The reaction of gaseous ClO_2 with fluorine is homogeneous and bimolecular at low partial pressures and low temperatures[594], the rate-determining step being

$$ClO_2+F_2 \rightarrow ClO_2F+F$$

At room temperature only spontaneous decomposition of ClO_2 is observed when fluorination is attempted. Little is known about the 1:1 adducts with arsenic or antimony pentafluoride, which are stable at temperatures up to 80°C[595]; the nature of sodium amidochlorate $Na[NH_2ClO_2]$ is equally obscure[596].

Analysis[574]

The most convenient method of determining chlorine dioxide is by liberating iodine from neutral or acidic iodide solution, followed by thiosulphate titration. The presence of halogen oxyanions causes complications, since these species also oxidize I^- to I_2; chlorine dioxide may be blown out of solution in a current of nitrogen, and absorbed in neutral KI solution. ClO_2 can be estimated in the presence of chlorine by treatment with acid KI and back-titration to determine the acid consumed.

Uses[574,597]

Chlorine dioxide, manufactured by the reduction of chlorates, is used in making sodium chlorite (subsequently employed for bleaching), and is itself used for the bleaching of paper pulp on the scale of 10^5 tons per annum; the bleaching and oxidizing powers of the compound have also been applied to oils, fats, waxes, flour and textiles, to treating leather prior to tanning (by decomposing the keratin), to improving the viscosity of rubber-based varnishes and glues, and to sterilizing foodstuffs such as cottage cheese.

When used in the purification of water, chlorine dioxide is normally generated by the action of chlorine on sodium chlorite. ClO_2 enjoys several advantages over chlorine in its ability to destroy ill-tasting phenols and to convert iron and manganese rapidly to insoluble forms.

Chlorine Perchlorate (see also p. 1473)

This unusual chlorine oxide[598] is prepared by the reaction

$$MClO_4+ClOSO_2F \xrightarrow{-45°C} MSO_3F+ClOClO_3 \qquad (M = Cs \text{ or } NO_2)$$

594 P. J. Aymonino, J. E. Sicre and H. J. Schumacher, *J. Chem. Phys.* **22** (1954) 756.
595 M. Schmeisser and W. Fink, *Angew. Chem.* **69** (1957) 780.
596 G. Beck, *Z. anorg. Chem.* **233** (1937) 155.
597 W. Masschelein, *Chim. Ind., Genie Chim.* **97** (1967) 49, 346.
598 C. J. Schack and D. Pilipovich, *Inorg. Chem.* **9** (1970) 1387.

It is thermodynamically and kinetically less stable than chlorine dioxide, decomposing at ambient temperatures to Cl_2, O_2 and Cl_2O_6. The formally positive chlorine atom is quantitatively replaced by bromine in forming the mixed oxide $BrOClO_3$ [599]:

$$Br_2+2ClOClO_3 \xrightarrow{-45°C} Cl_2+2BrOClO_3$$

Dichlorine Hexoxide

Preparation

The best method of making Cl_2O_6 is probably the ozonolysis of chlorine dioxide[600].

$$2ClO_2+2O_3 \rightarrow Cl_2O_6+2O_2$$

The compound is also produced in the photochemical combination of chlorine and ozone, in the photolysis of chlorine dioxide, and (with 80% yield) in the decomposition of chlorine perchlorate[598]. A neat and original synthesis involves the thermal degradation of the diperchlorate of xenon[601].

$$Xe(ClO_4)_2 \xrightarrow{20°} Xe+O_2+Cl_2O_6$$

Structure

In the vapour phase the compound probably exists as the pyramidal ClO_3 molecule[602]; this paramagnetic species has also been studied by esr and ultraviolet-visible absorption spectroscopy, both in frozen solutions and trapped in crystalline lattices (Table 48). Along the isoelectronic series PO_3^{2-}, SO_3^-, ClO_3, the esr parameters indicate increasing delocalization of the odd electron onto the oxygen atoms; there is an accompanying increase in bond angle: PO_3^{2-}, 110°; SO_3^-, 111°; ClO_3, 112° [603]. Photolysis of the isolated ClO_3 molecule readily produces ClO and oxygen[604].

$$ClO_3 \xrightarrow{h\nu} ClO+O_2$$

It should be noted that the ultraviolet-visible spectrum attributed to gaseous ClO_3 [605a] (Table 48) bears little resemblance to that for ClO_3 produced by the γ-radiolysis of 12·5 M $HClO_4$ frozen at -196°C[604]; the vapour-phase spectrum, however, is distinctly like that of ClO (p. 1381).

The molecular weight of the compound, measured cryoscopically in carbon tetrachloride, is consistent with a dimeric formulation Cl_2O_6, as are the properties of solutions in oleum and H_3PO_4. γ-Radiolysis of solutions of Cl_2O_6 frozen at -196°C produces ClO_3; photolysis of such solutions gives ClO as the principal decomposition fragment[604].

In the liquid and solid phases the oxide is said to exist *exclusively* as the dimer Cl_2O_6, although the high melting point and boiling point could indicate a polymeric formulation. Early measurements of the paramagnetic susceptibility of the liquid[605a] were interpreted in terms of the equilibrium

$$Cl_2O_6 \rightleftharpoons 2ClO_3$$

---

599 C. J. Schack, K. O. Christe, D. Pilipovich and R. D. Wilson, *Inorg. Chem.* **10** (1971) 1078.
600 H. J. Schumacher and G. Stieger, *Z. anorg. Chem.* **184** (1929) 272.
601 N. Bartlett, M. Wechsberg, F. O. Sladky, P. A. Bulliner, G. R. Jones and R. D. Burbank, *Chem. Comm.* (1969) 703.
602 C. F. Goodeve and F. A. Todd, *Nature*, **132** (1933) 514.
603 A. Begum, S. Subramanian and M. C. R. Symons, *J. Chem. Soc.* (*A*) (1970) 918.
604 V. N. Belevskii and L. T. Bugaenko, *Russ. J. Inorg. Chem.* **12** (1967) 1203.
605 (a) J. Farquharson, C. F. Goodeve and F. D. Richardson, *Trans. Faraday Soc.* **32** (1936) 790; (b) A. Pavia, J. L. Pascal and A. Potier, *Compt. rend.* **272C** (1971) 1495.

TABLE 48. SPECTROSCOPIC AND RELATED INVESTIGATIONS OF CHLORINE TRIOXIDE/DICHLORINE HEXOXIDE

Ultraviolet-visible spectrum
ClO_3 Vapour:[a] $\lambda_{max} \sim 2780$ Å, $\epsilon \sim 1{\cdot}2\times10^3$ (see text)
In 12·5 M $HClO_4$ at −196°C:[b] $\lambda_{max} \sim 4300$ Å, $\epsilon \sim 5\times10^3$
Cl_2O_6 Liquid;[a] CCl_4 solution;[c] oleum solution.[b]
Magnetic susceptibility[a] Measured using Gouy method as a function of temperature between −40°C and +10°C. For the dissociation $Cl_2O_6 \rightleftharpoons 2ClO_3$, $\log K = -0{\cdot}974 - 1730/2{\cdot}3RT$ (see text)
Esr spectrum Measured for ClO_3 in frozen solutions or isolated in crystalline lattices; the radical is produced by γ-radiolysis of an oxychlorine substrate.
In 12·5 M $HClO_4$ at 77°K:[d] $g_{av} = 2{\cdot}009$, A_{iso}[e] = 128·6 G
In $KClO_4$ at 300°K:[f] $g_{av} = 2{\cdot}011$, A_{iso}[e] = 122 G
In NH_4ClO_4 at 300°K:[g] $g_{av} = 2{\cdot}008$, A_{iso}[e] = 128 G
Also for ClO_3 in $LiClO_4$ (195°K),[f] $NaClO_4$ (195°K),[f] $Mg(ClO_4)_2$ (195°K),[f] aq. $HClO_3$ (77°K),[d] Cl_2O_6 (77°K),[b] oleum (77°K),[b] $FClO_4$ (77°K),[b] $NOClO_4$ (77°K),[b] $NaClO_3$ (298°K).[h,i]

[a] C. F. Goodeve and F. D. Richardson, *Trans. Faraday Soc.* **32** (1936) 790.
[b] V. N. Belevskii and L. T. Bugaenko, *Russ. J. Inorg. Chem.* **12** (1967) 1203.
[c] M. H. Kalina and J. W. T. Spinks, *Canad. J. Res.* **B16** (1938) 381.
[d] V. N. Belevskii and L. T. Bugaenko, *Zhur. Fiz. Khim.* **41** (1967) 144.
[e] Hyperfine data for ^{35}Cl.
[f] P. W. Atkins, J. A. Brivati, N. Keen, M. C. R. Symons and P. A. Trevalion, *J. Chem. Soc.* (1962) 4785.
[g] T. Cole, *J. Chem. Phys.* **35** (1961) 1169.
[h] F. T. Gamble, *J. Chem. Phys.* **42** (1965) 3542.
[i] J. C. Fayet and B. Thiéblemont, *Compt. rend.* **261** (1965) 1501.

but recent esr investigations of the condensed phases could not detect the presence of ClO_3[604]; the sole paramagnetic species was the impurity ClO_2 in 0·01 M concentration. Little credence can therefore be given to the much-quoted heat of dissociation, viz. 1·73 kcal mol^{-1}.

No structural data are available for the dimer, but there are two outstanding possibilities, viz.

$O_3Cl{-}ClO_3$ (4) $\quad$ $O_2Cl(\mu\text{-}O)_2ClO_2$ (5)

Structure (4) is attractive in that little rearrangement of the ClO_3 pyramids is needed to pair the odd electrons, although the bridged dimer (5) comes closer to the ionic formulation $ClO_2^+ClO_4^-$, which is judged to be compatible with the vibrational spectrum attributed to Cl_2O_6 at 87°K[605b].

Reactions

The dark red liquid is allegedly the least explosive of the chlorine oxides. In many of its reactions it appears to behave as chloryl perchlorate $[ClO_2]^+[ClO_4]^-$ [575]; similar behaviour is exhibited, for example, by diborane in some of its reactions. Nitrogen oxides and their derivatives displace ClO_2 with formation of nitrosyl or nitryl perchlorates though the reactions are not reversible. With anhydrous HF, an equilibrium is set up and ClO_2F and $HClO_4$ are formed; Cl_2O_6 may also be used to synthesize metal perchlorates. With AsF_5 an unusual ClO_3 adduct of unknown structure is formed. Hydrolysis gives a mixture of chloric and perchloric acids.

$$Cl_2O_6 + H_2O \rightarrow HClO_3 + HClO_4$$

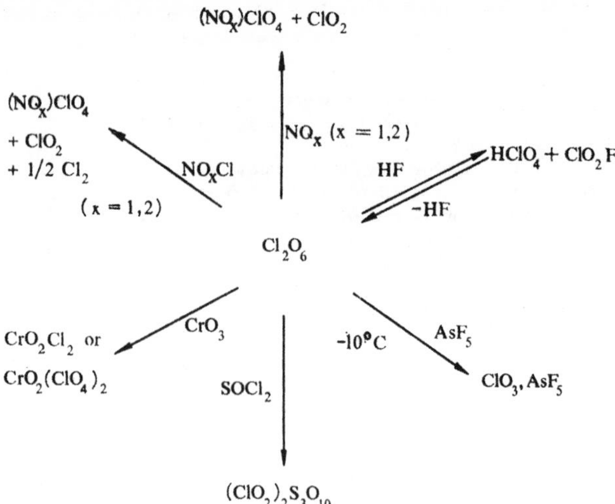

SCHEME 3. Reactions of dichlorine hexoxide.

Dichlorine Heptoxide

Preparation

The oily liquid is best obtained by the careful dehydration of perchloric acid with phosphoric acid at −10°C, followed by equally careful distillation at −35°C and 1 mm pressure[606]. It is also formed when chlorine and ozone react in blue light.

Structure

The gaseous molecule has a $O_3Cl–O–ClO_3$ structure of C_2 symmetry, the ClO_3 groups being 15° from the symmetrically staggered C_{2v} configuration. The dimensions of the bridging system (Table 49) are consistent with "normal" single bonds; repulsion between the ClO_3 groups accounts for the opening of the Cl–O–Cl angle from the expected tetrahedral value. Whether the shortness of the terminal Cl–O bonds evinces d_π–p_π bonding was discussed in Section 4B1.

Reactions

Cl_2O_7 is relatively stable for a chlorine oxide; though exploding when heated or subjected to shock, it does not ignite organic material at room temperature. Its synthesis shows it to be the true anhydride of perchloric acid, and it regenerates ClO_4^- on dissolution in water or alkali.

Thermal decomposition of dichlorine heptoxide[607] to chlorine and oxygen in both the liquid and vapour phase is initiated by rupture of a bridging Cl–O bond:

$$Cl_2O_7 \rightarrow ClO_3 + ClO_4$$

The activation energy for this process is 32·9 kcal mol^{-1} for the vapour and 32·1 kcal mol^{-1} for the liquid.

606 C. F. Goodeve and J. Powney, *J. Chem. Soc.* (1932) 2078.
607 I. P. Fisher, *Trans. Faraday Soc.* **64** (1968) 1852.

TABLE 49. SPECTROSCOPIC AND RELATED INVESTIGATIONS OF DICHLORINE HEPTOXIDE

Electron diffraction[a]
Molecule has $O_3ClOClO_3$ structure of C_2 symmetry
$r_g(1)[Cl–O_{term}] = 1{\cdot}405 \pm 0{\cdot}002$ Å; $r_g(1)[Cl–O_{br}] = 1{\cdot}709 \pm 0{\cdot}004$ Å
$\theta_{Cl–O_{br}–Cl} = 118{\cdot}6 \pm 0{\cdot}7°$; $\theta_{O_t–Cl–O_t} = 115{\cdot}2 \pm 0{\cdot}2°$
ClO_3 groups are oriented 15° from C_{2v} (staggered) configuration.
$Cl–O_{br}$ bonds are inclined at 4·7° to threefold axis of ClO_3 groups.
Infrared spectrum[b] Measured for vapour and solid.
Raman spectrum[c] Measured for liquid.
Force constants[d] $f_r(Cl–O_{term}) = 9{\cdot}32$ mdyne Å^{-1}
$f_r(Cl–O_{br}) = 3{\cdot}2$ mdyne Å^{-1}
Ultraviolet absorption spectrum[e] Measured for vapour.
Dipole moment[f] Measured for CCl_4 solution: $0{\cdot}72 \pm 0{\cdot}02D$.
Mass spectrum[g]

Positive ion	O_2	^{35}Cl	^{37}Cl	^{35}ClO	^{37}ClO	$^{35}ClO_2$	$^{37}ClO_2$
Relative intensity	0·27	1·08	0·34	9·01	2·88	31·2	10·4
Positive ion	$^{35}Cl_2$	$^{35}Cl^{37}Cl$	$^{37}Cl_2$	$^{35}ClO_3$	$^{37}ClO_3$	$^{35}ClO_4$	$^{37}ClO_4$
Relative intensity	0·99	0·54	0·11	100·0	33·4	0·67	0·23
Positive ion	$^{35}Cl_2O_7$	$^{35}Cl^{37}ClO_7$		$^{37}Cl_2O_7$			
Relative intensity	0·67	0·45		0·09			

Dissociation enthalpy
$O_3Cl–OClO_3 \rightarrow O_3Cl + ClO_4$; $\Delta H = 30 \pm 4$ kcal mol^{-1}

[a] B. Beagley, *Trans. Faraday Soc.* **61** (1965) 1821.
[b] R. Savoie and P. A. Giguère, *Canad. J. Chem.* **40** (1962) 991.
[c] J. D. Witt and R. M. Hammaker, *Chem. Comm.* (1970) 667.
[d] Based on incorrect assignment: E. A. Robinson, *Canad. J. Chem.* **41** (1963) 3021; R. J. Gillespie and E. A. Robinson, *ibid.* **42** (1964) 2496.
[e] C. F. Goodeve and B. A. M. Windsor, *Trans. Faraday Soc.* **32** (1936) 1518.
[f] R. Fonteyne, *Natuurw. Tijdschr.* **20** (1938) 112, 275.
[g] I. P. Fisher, *Trans. Faraday Soc.* **64** (1968) 1852.

Chlorine Tetroxide

Gomberg claimed that the reaction of iodine with silver perchlorate in anhydrous ether produced chlorine tetroxide[608].

$$I_2 + 2AgClO_4 \rightarrow 2AgI + (ClO_4)_2$$

Subsequent investigations of this reaction have shown the products to be iodine perchlorates (p. 1473)[609].

The ClO_4 molecule is a probable intermediate in the thermal decomposition of Cl_2O_7 [607]. A species of this stoichiometry is also formed in the γ-radiolysis of potassium chlorate at 77°K, but probably has the structure

$$\begin{matrix} O \\ & \diagdown \\ & & Cl–O–O \\ & \diagup \\ O \end{matrix}$$

[610].

The electron affinity of the hypothetical molecule is calculated to be 134 kcal mol^{-1} (from the lattice energies of perchlorates)[611].

608 M. Gomberg, *J. Amer. Chem. Soc.* **45** (1923) 398.
609 N. W. Alcock and T. C. Waddington, *J. Chem. Soc.* (1962) 2510.
610 R. S. Eachus, P. R. Edwards, S. Subramanian and M. C. R. Symons, *J. Chem. Soc.* (*A*) (1968) 1704.
611 V. I. Vedeneyev, L. V. Gurvich, V. N. Kondrat'yev, V. A. Medvedev and Ye. L. Frankevich, *Bond Energies, Ionization Potentials and Electron Affinities*, Edward Arnold, London (1966).

BROMINE OXIDES[575,612,613]

Dibromine Monoxide

In accord with the synthesis of Cl_2O, Br_2O is formed in the reaction of mercuric oxide with bromine vapour or with cold solutions of bromine in carbon tetrachloride.

$$2Br_2+2HgO \rightarrow HgBr_2.HgO+Br_2O$$

Yields are low, particularly for the vapour-phase reaction, and pure dibromine monoxide is best prepared by the decomposition of BrO_2 *in vacuo*[614]; pumping off other products at −60°C leaves Br_2O.

TABLE 50. PROPERTIES OF DIBROMINE MONOXIDE

Colour of solid:[a] brown-black
Melting point:[a] −17·5°C (with decomposition)
Infrared spectrum:[b] Measured for solid at −196°C
Fundamental frequencies (cm^{-1}) and assignments in C_{2v} symmetry.
$\nu_1(a_1)$ 504; $\nu_2(a_1)$ 197; $\nu_3(b_1)$ 587.
Force constants (mdyne $Å^{-1}$) $f_r = 2{\cdot}4 \pm 0{\cdot}2$; $f_{rr} = 0{\cdot}4 \pm 0{\cdot}2$; $f_\theta/r^2 = 0{\cdot}4 \pm 0{\cdot}1$; $f_{r\theta}/r = 0{\cdot}2 \pm 0{\cdot}1$.
Ultraviolet-visible spectrum[c] Measured for solutions in CCl_4.

[a] M. Schmeisser and K. Brändle, *Adv. Inorg. Chem. Radiochem.* **5** (1963) 41.
[b] C. Campbell, J. P. M. Jones and J. J. Turner, *Chem. Comm.* (1968) 888.
[c] W. Benschede and H. J. Schumacher, *Z. anorg. Chem.* **226** (1936) 370.

The compound is unstable above −40°C with respect to decomposition into bromine and oxygen. The infrared spectrum of the brown–black solid at −196°C is consistent with a bent C_{2v} structure for the BrOBr molecule; the estimated bond angle is 111°[614]. Few reactions of the compound have been studied.

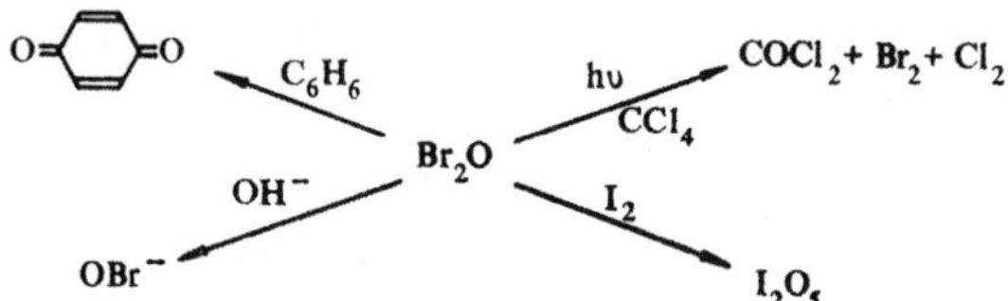

SCHEME 4. Reactions of dibromine monoxide.

Bromine Dioxide[575,613]

The quantitative ozonolysis of bromine in a fluorocarbon solvent at low temperatures produces bromine dioxide as a light yellow crystalline solid.

$$Br_2+4O_3 \xrightarrow[-78°C]{CF_3Cl} 2BrO_2+4O_2$$

[612] A. G. Sharpe, *Mellor's Comprehensive Treatise on Inorganic and Theoretical Chemistry*, Supplement II, Part I, pp. 747–749, Longmans, London (1956).
[613] P. J. M. Radford and M. Schmeisser, *Bromine and its Compounds* (ed. Z. E. Jolles), p. 147, Benn, London (1966).
[614] C. Campbell, J. P. M. Jones and J. J. Turner, *Chem. Comm.* (1968) 888.

The compound also results from the action of oxygen atoms on bromine, e.g. during glow-discharge of bromine–oxygen mixtures.

The solid is believed to have a polymeric structure, and is unstable much above −40°C. Calorimetric study of the violent decomposition which occurs on rapid warming to 0°C,

$$2BrO_2 \rightarrow Br_2 + 2O_2$$

implies that $\Delta H_f°[BrO_2(s)]$ is $+12{\cdot}5 \pm 0{\cdot}7$ kcal mol^{-1}; the compound is *less* endothermic than ClO_2 ($\Delta H_f° = 25{\cdot}1$ kcal mol^{-1}). Slow warming of BrO_2 under vacuum evolves Br_2O and a white solid—presumably a higher oxide.

Hydrolysis of bromine dioxide, which proceeds in 5 M alkali, reflects the instability of Br(III) in aqueous solution.

$$6BrO_2 + 6OH^- \rightarrow 5BrO_3^- + Br^- + 3H_2O$$

Reactions with N_2O_4 and fluorine have also been investigated.

$$[NO_2]^+[Br(NO_3)_2]^- \xleftarrow{N_2O_4} BrO_2 \xrightarrow{F_2} BrO_2F$$

About the discrete BrO_2 radical little is known. It may be generated in aqueous solution by the flash photolysis or pulse radiolysis of BrO_3^- (pp. 1380–5); the kinetics of its hydrolysis by OH^- suggest that it may exist in equilibrium with the dimer Br_2O_4 [615].

$$2BrO_2 \rightleftharpoons Br_2O_4 \qquad K = 19\ \text{M}^{-1}$$

Higher Oxides

The higher oxides of bromine have not been well characterized. While the ozonolysis of bromine to BrO_2 proceeds smoothly at −78°C in an inert solvent, execution of the reaction in the vapour phase at higher temperatures has yielded at least three products, depending on the material of the reaction vessel and the pressure of the reactants: only approximate elemental analyses have commonly been given. Any (or all) of these white solids may also be formed in the controlled decomposition of BrO_2. No structural studies of these compounds have been reported, and attempts to dehydrate perbromic acid to Br_2O_7 have been unsuccessful.

TABLE 51. THE HIGHER OXIDES OF BROMINE[a,b]

Composition	Br_2O_5	Br_3O_8	BrO_3
Colour	white	white	white
Preparation	Ozonolysis of Br_2 at 0°C in pyrex: $5 < \frac{[O_3]}{[Br_2]} < 15$	Ozonolysis of Br_2 at 0°C in quartz: $5 < \frac{[O_3]}{[Br_2]} < 15$	[a] Ozonolysis of Br_2 at 0°C: $\frac{[O_3]}{[Br_2]} > 25$ [b] Glow discharge on 2% mixtures of Br_2 in O_2
Hydrolysis product	$HBrO_3$	"$H_4Br_3O_{10}$"	$HBrO_3 + O_2$

[a] M. Schmeisser and K. Brändle, *Adv. Inorg. Chem. Radiochem.* **5** (1963) 72.
[b] P. J. M. Radford and M. Schmeisser, *Bromine and its Compounds* (ed. Z. E. Jolles), p. 147, Benn, London (1966).

[615] G. V. Buxton and F. S. Dainton, *Proc. Roy. Soc.* **A304** (1968) 427.

The BrO_3 radical, produced by the γ-radiolysis of $KBrO_3$ at 77°K, has been studied by esr spectroscopy (Table 54); the O–Br–O angle is estimated to be 114° [603].

IODINE OXIDES[575,576,616]

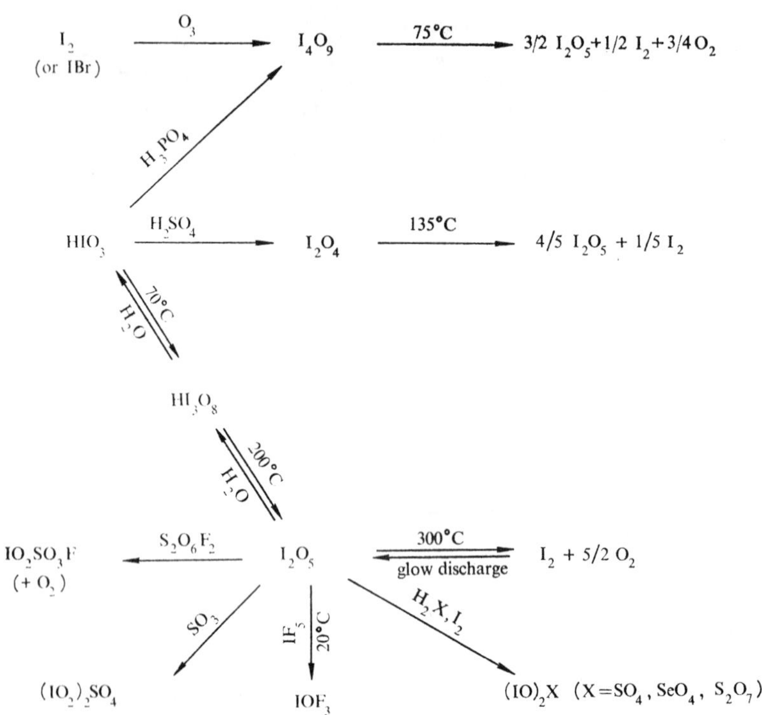

SCHEME 5. The oxides of iodine.

Iodine Pentoxide[617]

I_2O_5 is the most important and thermally the most stable of the iodine oxides. In the most convenient preparation, iodic acid is dehydrated at 200°C in a stream of dry air. The hygroscopic white solid absorbs water readily from the atmosphere, giving HI_3O_8 (formulated I_2O_5,HIO_3); commercial samples of "iodine pentoxide" may contain copious amounts of this hydrate[618]. On dissolution in water iodine pentoxide regenerates HIO_3.

The crystal structure of the pentoxide[619] contains I_2O_5 molecules (Fig. 31), consisting of two IO_3 pyramids joined through one corner: the bridging I–O distances correspond to single bonds, whereas the terminal bonds are considerably shorter. That intermolecular iodine–oxygen contacts ($\geqslant$2·23 Å; represented by broken lines in Fig. 31) contribute significantly to the bonding is also evident from the vibrational spectra of the solid. HI_3O_8

[616] G. J. Hills, *Mellor's Comprehensive Treatise on Inorganic and Theoretical Chemistry*, Supplement II, Part I, pp. 870–873, Longmans, London (1956).
[617] A. W. Hart, M. G. Gergel and J. Clarke, *Kirk-Othmer Encyclopedia of Chemical Technology*, 2nd edn., Vol. 11, p. 859, Interscience, New York (1966).
[618] K. Selte and A. Kjekshus, *Acta Chem. Scand.* **22** (1968) 3309.
[619] K. Selte and A. Kjekshus, *Acta Chem. Scand.* **24** (1970) 1912.

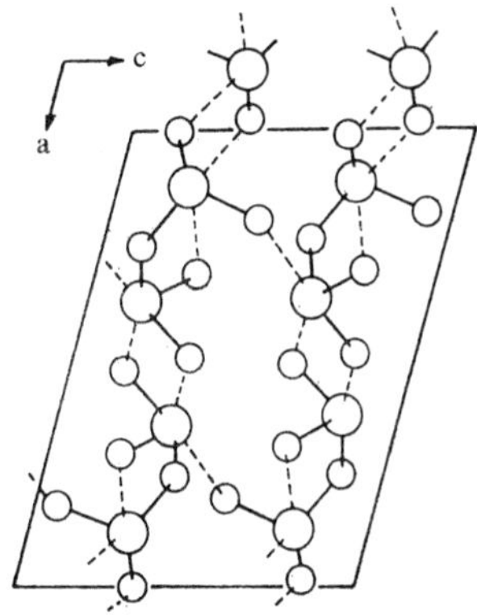

FIG. 31. Crystal structure of I_2O_5.

TABLE 52. SOME PROPERTIES OF IODINE PENTOXIDE

$\Delta H_f°$(s) at 298°K	−37·78 kcal mol^{-1} [a]			
Crystal structure[b]	*Intramolecular distances* (Å)			
	I_1–O_1 1·78(3)		I_2–O_3	1·83(3)
	I_1–O_2 1·77(3)		I_2–O_4	1·79(3)
	I_1–O_5 1·92(2)		I_2–O_5	1·95(3)
	Intramolecular angles (°)			
	O_1–I_1–O_2	99·5(1·3)		
	O_1–I_1–O_5	96·5(1·2)		
Intermolecular interactions are depicted in Fig. 31.	O_2–I_1–O_5	101·9(1·0)		
	O_3–I_2–O_4	94·8(1·1)		
	O_3–I_2–O_5	93·1(1·1)		
	O_4–I_2–O_5	97·5(1·0)		
	I_1–O_5–I_2	139·2(1·4)		
Infrared and Raman spectra[c]	Measured for solid I_2O_5			
Mass spectrum[d]				
^{127}*I Quadrupole resonance spectrum*[e]	Measured at 77°K			
Quadrupole coupling constant (77°K)	e^2Qq = 1068 MHz			
	η = 0·33			

[a] *Selected Values of Chemical Thermodynamic Properties*, p. 37, N.B.S. Technical Note 270–3 (1968).

[b] K. Selte and A. Kjekshus, *Acta Chem. Scand.* **24** (1970) 1912.

[c] P. M. A. Sherwood and J. J. Turner, *Spectrochim. Acta*, **26A** (1970) 1975.

[d] M. H. Studier and J. L. Huston, *J. Phys. Chem.* **71** (1967) 457.

[e] S. Kojima, K. Tsukada, S. Ogawa and A. Shimauchi, *J. Chem. Phys.* **23** (1955) 1963.

contains a network of interacting I_2O_5 and HIO_3 molecules[620]; the I_2O_5 bond lengths match those found in the pure pentoxide, but the I–O–I bridging angle is much larger in I_2O_5 (139·2°) than in anhydriodic acid, and the relative orientations of the IO_2 groups are also different.

I_2O_5 oxidizes many common materials such as NO, C_2H_4, H_2S and CO: the last of these reactions proceeds quantitatively at ambient temperatures when CO is passed over the powdered oxide[617].

$$I_2O_5 + 5CO \rightarrow I_2 + 5CO_2$$

The ease with which iodine may be determined makes this a very useful method for the estimation of CO in the atmosphere or in other gaseous mixtures. Fluorination of I_2O_5 (with F_2, BrF_3, ClF_3, SF_4 or ClO_2F) affords IF_5, which itself reacts with I_2O_5 to give IOF_3. With SO_3 or $S_2O_6F_2$, salts of the iodyl cation IO_2^+ are formed[621], while in concentrated acids (H_2SO_4, $H_2S_2O_7$, H_2SeO_4) iodine pentoxide is reduced by iodine to iodosyl derivatives $(IO)_2X$ (X = SO_4, S_2O_7, SeO_4)[622].

Other Oxides

The IO_2 radical (half-life 50 μsec) has been detected in flash photolysis of aqueous iodate solutions (p. 1382), but on the testimony of its infrared and Mössbauer spectra the yellow diamagnetic crystalline solid "I_2O_4" is considered to be iodosyl iodate $[IO]^+[IO_3]^-$ [623]; the iodosyl cation presumably forms a polymeric chain $[(I{-}O{-})^+]$, as in other iodosyl derivatives, and is cross-linked by iodate anions. Alkaline hydrolysis of I_2O_4 produces iodate and iodide.

$$3I_2O_4 + 6OH^- \rightarrow 5IO_3^- + I^- + 3H_2O$$

Hydrochloric acid degrades it to iodine monochloride.

$$I_2O_4 + 8HCl \rightarrow 2ICl + 3Cl_2 + 4H_2O$$

The commonly held view that the yellow hygroscopic material I_4O_9 is in fact iodine triiodate $I(IO_3)_3$ is based solely on the apparent stoichiometry of its reactions with water and with hydrogen chloride[575].

The existence of other iodine oxides is uncertain. By analogy with the synthesis of Cl_2O_7, attempts to prepare I_2O_7 have centred on the dehydration of periodic acid[624]: the most convincing claim is that 65% oleum produces a highly reactive orange solid with approximately the correct composition[625]. Thermal dehydration of paraperiodic acid at *ca.* 130°C produces a solid I_2O_5,I_2O_7. Diiodine trioxide is unknown, but the iodosyl cation (p. 1352) and various iodosyl salts are derivatives of this hypothetical oxide.

Oxyhalogen Radicals

As well as the oxides which may be obtained as pure materials, a number of molecular species have been identified, either as transients in the vapour phase or in solution, or trapped in solid matrices.

620 Y. D. Feikema and A. Vos, *Acta Cryst.* **20** (1966) 769.
621 F. Aubke, G. H. Cady and C. H. L. Kennard, *Inorg. Chem.* **3** (1964) 1799.
622 G. Daehlie and A. Kjekshus, *Acta Chem. Scand.* **18** (1964) 144.
623 J. H. Wise and H. H. Hannan, *J. Inorg. Nuclear Chem.* **23** (1961) 31.
624 M. Drátovský and L. Pačesová, *Russ. Chem. Rev.* **37** (1968) 243.
625 H. C. Mishra and M. C. R. Symons, *J. Chem. Soc.* (1962) 1194.

TABLE 53. OXYHALOGEN RADICALS

ClClO[a]		
ClO[a,b,c]	BrO[b,c]	IO[b,c]
[OClO][d]	OBrO[c]	OIO[c]
ClOO[a,b]	BrOO[b]	IOO[b]
[ClO_3][d]	BrO_3[a,c]	IO_3[c]
ClO_4[a]		

[a] Characterized by matrix-isolation.
[b] Characterized in the vapour phase.
[c] Characterized in solution.
[d] Stable species under normal conditions.

ClO, BrO and IO

The radicals ClO, BrO and IO are produced in the vapour phase

(i) by the introduction of the halogen or its compounds into an oxyhydrogen flame,
(ii) by ozonolysis of the halogen, or
(iii) by the action of flash photolysis or microwave discharge on halogen–oxygen mixtures.

The most important reactions are:

$$O+X_2 \rightarrow XO+X$$

and

$$X+O_2 \rightarrow XO+O$$

ClO is an efficient carrier in chain reactions, and its role as an intermediate in the synthesis and decomposition of other oxychlorine molecules is well exemplified in preceding sections.

Detailed spectroscopic studies of the vapour-phase radicals have yielded the ground state parameters listed in Table 55. The high dissociation energy and short bond lengths of ClO are evidence of a strong bond, as is the fundamental frequency of 995 cm^{-1} attributed to ^{35}ClO isolated in an argon matrix, with the derived force constant of 6·4 mdyne $Å^{-1}$ [626]; Hartree–Fock SCF–MO calculations predict a stretching frequency of 975 cm^{-1} [566]. (However, analyses of the electronic and microwave spectra of the gaseous molecule suggest lower stretching frequencies—868 and 610 ± 60 cm^{-1} respectively.)

The reactions of ClO, BrO and IO in the vapour phase have been studied using flow systems[627]. In the absence of competing species, the gaseous XO radicals decompose according to the scheme(s):

$$XO+XO \rightarrow X+XOO$$
$$XOO \rightarrow X+O_2 \quad (X = Cl, Br \text{ or } I)$$
$$XOO+X \rightarrow X_2+O_2 \quad (X = Cl)$$

Dissociation of the unstable peroxybromine and peroxyiodine molecules is immediate; the more stable ClOO may also degrade, however, *via* reaction with chlorine atoms. At high pressures decomposition of ClO may proceed through the dimer $(ClO)_2$. This dimer is also found in inert gas matrices following the photolysis of Cl_2O isolated at 14°K; in

[626] L. Andrews and J. I. Raymond, *J. Chem. Phys.* **55** (1971) 3087.
[627] M. A. A. Clyne and J. A. Coxon, *Trans. Faraday Soc.* **62** (1966) 1175; M. A. A. Clyne and H. W. Cruse, *ibid.* **66** (1970) 2214, 2227.

TABLE 54. SPECTROSCOPIC STUDIES OF SOME OXYHALOGEN RADICALS

Molecule	Medium	Temp. (°K)	Method of production	Technique	Ref.	Comments
ClClO	Solid nitrogen	14	photolysis of dichlorine monoxide	ir	a	Weak $(p-\pi^*)\sigma$ Cl–Cl bond
ClO[h]	Solid argon	4	$Cl_2O + Li \rightarrow ClO + LiCl$	ir	b	Fundamental frequency: ^{35}ClO 994·8 cm^{-1}; ^{37}ClO 986·0 cm^{-1}. Force constant = 6·41 mdyne $Å^{-1}$
	Solid H_3PO_4 and solid H_2SO_4	77	u.v. photolysis of chlorine dioxide	esr	c	$g_{xx} = 1·9909$, $g_{yy} = 2·0098$, $g_{zz} = 1·9909$, $g_{av} = 1·9972$; $A_{xx} = 5·7$, $A_{yy} = -11·4$, $A_{zz} = 5·7$, $A_{iso} = -5·7$ G°
	Solid H_2SO_4	77	u.v. photolysis of chlorine dioxide	uv	d	λ_{max} *ca.* 2650 Å
ClOO	Vapour	*ca.* 300	photolysis of Cl_2 in presence of O_2	ir, uv, kinetics	e	Broad absorption at 2400 Å
	Solid argon	4	photolysis of chlorine dioxide	ir	f	Fundamental frequencies (cm^{-1}): 1441, 407, 373. Force constants: f_r(O–O), 9·65 mdyne $Å^{-1}$; f_r(O–Cl), 1·29 mdyne $Å^{-1}$.
	$KClO_4$	77	γ-radiolysis of ClO_4^-	esr	g	
ClO_4	$KClO_4$	77	γ-radiolysis of ClO_4^-	esr	h	Probable structure $(O)_2$Cl—O—O
BrO[i]	aq. NaOH	*ca.* 295	pulse γ-radiolysis of OBr^-, BrO_2^-	uv, kinetics	j	$\lambda_{max} = 3500$ Å; half-life *ca.* 2μsec
	aq. NaOH	*ca.* 295	flash photolysis of BrO_2^-	uv, kinetics	k	
OBrO	aq. NaOH	*ca.* 295	pulse radiolysis of BrO_2^-, BrO_3^-	uv, kinetics	j	$\lambda_{max} = 4750$ Å
	aq. NaOH	*ca.* 295	flash photolysis of BrO_2^-, BrO_3^-	uv, kinetics	k,l	
BrO_3	$KBrO_3$	*ca.* 130	γ-radiolysis of BrO_3^- (at 77°K)	esr	m	$g_{av} = 2·070$; $A_{iso} = 614$ G
IO[i]	aq. NaOH	*ca.* 295	formed in secondary reactions after pulse γ-radiolysis and flash photolysis of iodine oxyanions.	uv, kinetics	n	$\lambda_{max} = 4900$ Å
IO_2	aq. NaOH	*ca.* 295	pulse γ-radiolysis and flash photolysis of IO_3^-	uv, kinetics	n	$\lambda_{max} = 7150$ Å
IO_3	aq. NaOH	*ca.* 295	pulse γ-radiolysis and flash photolysis of IO_3^-	uv, kinetics	n	$\lambda_{max} = 3800$ Å

[a] M. M. Rochkind and G. C. Pimentel, *J. Chem. Phys.* **46** (1967) 4481.
[b] L. Andrews and J. I. Raymond, *J. Chem. Phys.* **55** (1971) 3087.
[c] P. W. Atkins, J. A. Brivati, N. Keen, M. C. R. Symons and P. A. Trevalion, *J. Chem. Soc.* (1962) 4785.
[d] I. Norman and G. Porter, *Proc. Roy. Soc.* **A230** (1955) 399.

[e] H. S. Johnston, E. D. Morris, jun. and J. Van den Bogaerde, *J. Amer. Chem. Soc.* **91** (1969) 7712.
[f] A. Arkell and I. Schwager, *J. Amer. Chem. Soc.* **89** (1967) 5999.
[g] R. S. Eachus, P. R. Edwards, S. Subramanian and M. C. R. Symons, *J. Chem. Soc.* (*A*) (1968) 1704.
[h] J. R. Morton, *J. Chem. Phys.* **45** (1966) 1800.
[i] For information about vapour-phase molecule see Table 55.
[j] G. V. Buxton, F. S. Dainton and F. Wilkinson, *Chem. Comm.* (1966) 320.
[k] O. Amichai, G. Czapski and A. Treinin, *Israel J. Chem.* **7** (1969) 351.
[l] F. Barat, L. Gilles, B. Hickel and J. Sutton, *Chem. Comm.* (1969) 1485.
[m] A. Begum, S. Subramanian and M. C. R. Symons, *J. Chem. Soc.* (*A*) (1970) 918.
[n] O. Amichai and A. Treinin, *J. Phys. Chem.* **74** (1970) 830.
[o] ^{35}Cl hyperfine interactions.

TABLE 55. PROPERTIES OF THE VAPOUR-PHASE SPECIES ClO, BrO and IO

	ClO	BrO	IO
Thermodynamic properties at 298°K			
ΔH_f° (kcal mol^{-1})[a]	24·34	30·06	41·84
ΔG_f° (kcal mol^{-1})[a]	23·45	25·87	35·80
S° (cal deg^{-1} mol^{-1})[a]	54·14	56·75	58·65
C_p° (cal deg^{-1} mol^{-1})[a]	7·52	7·67	7·86
Electronic spectrum	b	b	b
Ground state $^2\Pi_{3/2}$			
Microwave spectrum	c	d	
Esr spectrum	e	f	f
r_e(Å)	1·569±0·001[c]	1·721±0·001[d]	[r_0 = 1·868±0·001[f]]
ω_e(cm^{-1})	[see text][i]	713[b]	681·4[b]
Dissociation energy, D_0° (kcal mol^{-1}) determined spectroscopically[b]	63·31±0·03	55·2±0·6	42±5
Ionization energy (kcal mol^{-1})	239·8[g]		
Electron affinity (kcal mol^{-1})	67·09[h]	61[j]	
Dipole moment, *D*	1·239[c]	1·55[d]	
Force constant (mdyne Å^{-1})	6·41[i]	3·99[b]	3·87[b]

[a] *Selected Values of Chemical Thermodynamic Properties*, N.B.S. Technical Note 270–3 (1968).
[b] R. A. Durie and D. A. Ramsey, *Canad. J. Phys.* **36** (1958) 45; R. A. Durie, F. Legay and D. A. Ramsey, *ibid.* **38** (1960) 444.
[c] T. Amano, S. Saito, E. Hirota, Y. Morino, D. R. Johnson and F. X. Powell, *J. Mol. Spectroscopy*, **30** (1969) 275.
[d] F. X. Powell and D. R. Johnson, *J. Chem. Phys.* **50** (1969) 4596.
[e] A. Carrington and D. H. Levy, *J. Chem. Phys.* **44** (1966) 1298.
[f] A. Carrington, P. N. Dyer and D. H. Levy, *J. Chem. Phys.* **52** (1970) 309.
[g] Measured by mass spectroscopy: V. H. Dibeler, R. M. Reese and D. E. Mann, *J. Chem. Phys.* **27** (1957) 176.
[h] Calculated from the heat of hydration of OCl^-: V. I. Vedeneyev, L. V. Gurvich, V. N. Kondrat'yev, V. A. Medvedev and Ye. L. Frankevich, *Bond Energies, Ionization Potentials and Electron Affinities*, Edward Arnold, London (1966).
[i] L. Andrews and J. I. Raymond, *J. Chem. Phys.* **55** (1971) 3087.
[j] G. V. Buxton and F. S. Dainton, *Proc. Roy. Soc.* **A304** (1968) 427.

contrast with O_2F_2, which is noted for its strong O–O bond, $(ClO)_2$ is but a weakly associated species, resembling $(NO)_2$ [578]. A molecule of composition Cl_2O_2 (probably Cl–Cl(O)(O)) is postulated as an intermediate in the reaction of ClO_2^- with chlorine or hypochlorites.

BrO has been detected by its ultraviolet-visible absorption spectrum following the flash photolysis[628] or pulse radiolysis[615] of aqueous bromite or hypobromite solutions: similar treatment of IO_2^- is impossible, but IO is formed in the reactions

$$IO_2 + IO_3^- \rightarrow IO + IO_4^-$$

and

$$I_2^- + IO^- \rightarrow IO + 2I^-$$

which occur after the flash photolysis or pulse radiolysis of, respectively, aqueous iodate and aqueous hypoiodite solutions[629]. The spectra and reactions of both radicals have been

[628] O. Amichai, G. Czapski and A. Treinin, *Israel J. Chem.* **7** (1969) 351.
[629] O. Amichai and A. Treinin, *J. Phys. Chem.* **74** (1970) 830.

studied in solution. BrO oxidizes bromite

$$BrO + BrO_2^- \rightarrow BrO^- + BrO_2$$

and disproportionates during hydrolysis[615]:

$$2BrO + H_2O \rightarrow BrO^- + BrO_2^- + 2H^+$$

ClOO

The peroxychlorine radical is an important species in the chain reactions of oxychlorine compounds, e.g.

$$ClO + ClO \rightarrow ClOO + Cl$$

and as such has been identified by its uv spectrum, using the technique of "modulation kinetic spectroscopy"[630]. Since its dissociation into chlorine atoms and oxygen molecules is slightly endothermic ($\Delta H = 8 \pm 2$ kcal mol^{-1})[631], ClOO is sufficiently long-lived to be a very useful intermediate; for example, it is responsible for the high efficiency shown by oxygen in promoting the recombination of chlorine atoms.

$$Cl + O_2 + M \rightarrow ClOO + M$$

$$ClOO + Cl \rightarrow Cl_2 + O_2$$

The heat of formation of the gaseous molecule is estimated to be 21 ± 2 kcal mol^{-1}; ClOO is therefore somewhat *less* endothermic than the more familiar isomer chlorine dioxide.

ClOO is also formed when chlorine dioxide is photolysed at low temperature in rigid inert matrices[591].

$$ClO_2 \xrightarrow{h\nu} \underset{\text{cage}}{[ClO + O]} \rightarrow ClOO$$

Esr and infrared studies show the isolated molecule (Table 54) closely to resemble OOF: the odd electron occupies a π^* orbital; the O–O bond is strong, and the O–Cl bond weak. The formation of chlorine dioxide in the γ-radiolysis of $KClO_4$ involves ClOO as an intermediate[610].

BrO_2 and IO_2

The radicals XO_2 are produced in aqueous solution (i) by flash photolysis of XO_3^- (X = Cl, Br or I)[632]:

$$XO_3^-,H_2O \xrightarrow{h\nu} [XO_3^-,H_2O]^* \rightarrow XO_2 + OH + OH^-$$

(ii) by γ-radiolysis of XO_3^- (X = Br or I)[615,629]

$$BrO_3^- + e^-(aq) \rightarrow [BrO_3^{2-}] \rightarrow BrO_2 + O^{2-}$$

BrO_2 and IO_2 may also be formed in secondary processes following photolysis or radiolysis of other anions. The radicals have been characterized by their ultraviolet-visible absorption spectra (Table 54).

Whereas ClO_2 may be recovered unchanged from aqueous solution, BrO_2 and IO_2 have very short lifetimes. The rate of hydrolysis of BrO_2 has been explored (see p. 1377), while it has been reported that IO_2 oxidizes iodate to periodate, itself being reduced to IO (see above)[629].

630 H. S. Johnston, E. D. Morris, jun. and J. Van den Bogaerde, *J. Amer. Chem. Soc.* **91** (1969) 7712.
631 S. W. Benson and J. H. Buss, *J. Chem. Phys.* **27** (1957) 1382.
632 F. Barat, L. Gilles, B. Hickel and J. Sutton, *Chem. Comm.* (1969) 1485.

BrO_3 [603]

γ-Radiolysis of $KBrO_3$ at 77°K or of frozen alkaline bromate solutions produces a paramagnetic centre identified by its esr spectrum as BrO_3. The derived O–Br–O bond angle of 114° indicates considerable flattening of the pyramid relative to the parent anion, and is consistent with trends along the series PO_3^{2-} (110°), SO_3^- (111°), ClO_3 (112°) and AsO_3^{2-} (110°), SeO_3^- (112°), BrO_3 (114°).

3. THE OXYFLUORIDES OF THE HALOGENS

The twelve halogen oxyfluorides which have so far been identified contain pentavalent or heptavalent chlorine, bromine or iodine; derivatives of the trivalent halogens have not yet been synthesized.

TABLE 56. OXYFLUORIDES OF THE HALOGENS

$ClO_2F^{a,b}$	BrO_2F	$IO_2F^{a,b}$
$ClOF_3^{a,b}$		IOF_3
$ClO_3F^{a?}$	$BrO_3F^{a?}$	$IO_3F^{a?}$
$ClO_2F_3^{a,c}$		$IO_2F_3^{b}$
		IOF_5
ClO_3OF		

[a] Compound functions as donor of F^-.
[b] Compound functions as acceptor of F^-.
[c] Two isomers known.

The compounds are generally obtained by fluorination of an appropriate halogen oxide, oxyacid or oxysalt. As might have been expected by interpolation from the corresponding oxides and fluorides, the compounds are oxidizing and fluorinating agents; the chlorine oxyfluorides in particular have been extensively investigated as potential oxidants for rocket fuels. They also tend to function as donors or acceptors of F^- in forming complexes with, respectively, Lewis acids and fluoride ion-donors, and hydrolyse to an oxyacid maintaining the oxidation state of the halogen. The literature up to 1962 has been reviewed[575a], while the oxyfluorides of chlorine have been the subject of a more recent, comprehensive survey[572].

Definitive structural characterization is restricted to gaseous ClO_3F and the solids IOF_3 and KIO_2F_2. However, vibrational spectra and valence force constants have been determined for many chlorine oxyfluorides, for the most part by Christe and his coworkers; chlorine–oxygen and chlorine–fluorine stretching force constants are listed in Table 57. Both neutral and charged species contain strong Cl–O bonds (force constants $\geqslant$ that of ClO_4^-), but considerable weakening of the Cl–F bonds occurs with coordination by F^-.

Halogenyl Fluorides

Preparation

The three compounds ClO_2F, BrO_2F and IO_2F are obtained by fluorination of an oxide

TABLE 57. VALENCE FORCE CONSTANTS OF SOME CHLORINE OXYFLUORIDES AND RELATED SPECIES

Molecule	f_r(Cl−O) (mdyne Å^{-1})	f_r(Cl−F) (mdyne Å^{-1})	Reference
$ClOF_2^+$	11·21	3·44	K. O. Christe, E. C. Curtis and C. J. Schack, *Inorg.Chem.* **11** (1972) 2212.
$ClOF_3$	9·37	3·16(eq) 2·34(ax)	K. O. Christe and E. C. Curtis, *Inorg. Chem.* **11** (1972) 2196.
$ClOF_4^-$	9·13	1·79	K. O. Christe and E. C. Curtis, *Inorg. Chem.* **11** (1972) 2209.
ClO_2^+	8·96		K. O. Christe, C. J. Schack, D. Pilipovich and W. Sawodny, *Inorg. Chem.* **8** (1969) 2489.
ClO_2F	9·07	2·53	D. F. Smith, G. M. Begun and W. H. Fletcher, *Spectrochim. Acta*, **20** (1964) 1763.
$ClO_2F_2^-$	8·3	1·6	K. O. Christe and E. C. Curtis, *Inorg. Chem.* **11** (1972) 35.
ClO_3F	9·41	3·91	W. Sawodny, A. Fadini and K. Ballein, *Spectrochim. Acta*, **21** (1965) 995.
ClF_2^+		4·74	K. O. Christe and C. J. Schack, *Inorg. Chem.* **9** (1970) 2296.
ClF_3		4·2(eq) 2·7(ax)	R. A. Frey, R. L. Redington and A. L. K. Aljibury, *J. Chem. Phys.* **54** (1971) 344.
ClF_4^-		2·11	K. O. Christe and W. Sawodny, *Z. anorg. Chem.* **374** (1970) 306.
ClF_5		3·47(ap) 2·67(bas)	K. O. Christe, E. C. Curtis, C. J. Schack and D. Pilipovich, *Inorg. Chem.* **11** (1972) 1679.

or oxysalt of the tetravalent or pentavalent halogen. The most convenient syntheses are

$$ClO_2 + AgF_2 \xrightarrow{20°C} ClO_2F + AgF$$

$$KBrO_3 + BrF_5 \xrightarrow{-50°C} BrO_2F + KBrF_4 + \tfrac{1}{2}O_2$$

$$I_2O_5 + F_2 \xrightarrow[\text{liq HF}]{20°C} 2IO_2F + \tfrac{1}{2}O_2$$

but in each case other oxyhalogen substrates and/or fluorinating agents may be used[575,581]. Synthesis of ClO_2F by fluorination of Cl_2O or Cl_2O_6 probably involves initial conversion to ClO_2. IO_2F is formed by dismutation of IOF_3.

$$2IOF_3 \xrightarrow{110°C} IO_2F + IF_5$$

Properties

Table 58 lists some properties of the halogenyl fluorides. The vibrational spectra of chloryl fluoride confirm the presence of discrete $FCl\langle^{O}_{O}$ molecules of C_s symmetry in the vapour and liquid; IO_2F is likewise revealed as a polymeric solid.

Reactions

Thermolysis of ClO_2F in a Monel system is appreciable only above 250°C, producing ClF and oxygen[633]. Bromyl fluoride, however, is unstable above its melting point, and decomposes vigorously at 56°C, probably according to the equation

$$3BrO_2F \rightarrow BrF_3 + Br_2 + 3O_2$$

TABLE 58. PROPERTIES OF THE HALOGENYL FLUORIDES

Property	ClO_2F	BrO_2F	IO_2F
Colour	colourless	yellow	colourless
Melting point (°C)	−115[a]	−9[b]	>200[b] (decomposition)
Boiling point (°C)	−6[a]		
Decomposition temp. (°C)	>250[c]	56[b]	*ca.* 200[b]
Vapour pressure, log p(mm) =	$8{\cdot}23-1412/T$[a]		
Trouton's constant (cal deg^{-1} mol^{-1})	23·2[a]		
ΔH_{vap} (kcal mol^{-1})	6·2[a]		
Infrared spectrum	gas[f]		solid[d,e]
Raman spectrum	liq.[f]		solid[e]
Force constants (mdyne Å^{-1})			
f_r(Cl−O)	9·07[f]		
f_r(Cl−F)	2·53[f]		

[a] H. Schmitz and H. J. Schumacher, *Z. anorg. chem.* **249** (1942) 238.
[b] M. Schmeisser and K. Brändle, *Adv. Inorg. Chem. Radiochem.* **5** (1963) 41.
[c] Y. Macheteau and J. Gillardeau, *Bull. Soc. chim. France* (1969) 1819.
[d] P. W. Schenk and D. Gerlatzek, *Z. Chem.* **10** (1970) 153.
[e] H. A. Carter and F. Aubke, *Inorg. Chem.* **10** (1971) 2296.
[f] D. F. Smith, G. M. Begun and W. H. Fletcher, *Spectrochim. Acta*, **20** (1964) 1763.

Both compounds attack glass at room temperature, the former slowly, the latter rapidly.

Hydrolysis of ClO_2F, BrO_2F or IO_2F produces the appropriate halate ion:

$$XO_2F+2OH^- \rightarrow XO_3^-+F^-+H_2O$$

Hydrolysis of BrO_2F can proceed with explosive violence, while in the hydrolysis of chloryl fluoride ClO_2 has been detected as an intermediate.

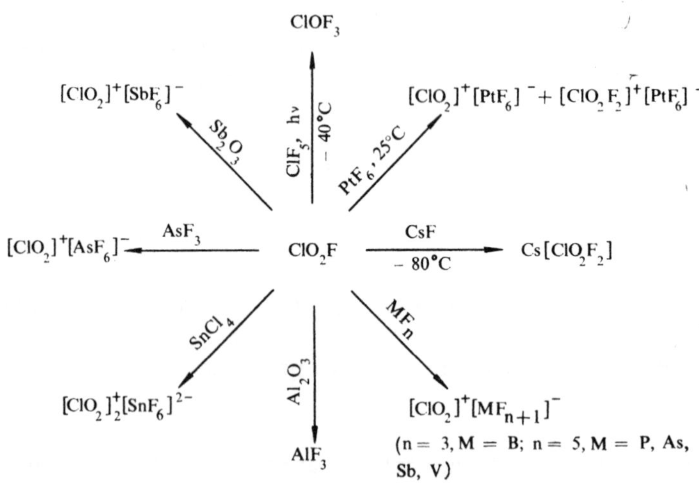

SCHEME 6. Reactions of chloryl fluoride.

633 Y. Macheteau and J. Gillardeau, *Bull. Soc. chim. France* (1969) 1819.

The stabilities of the complexes formed by ClO_2F with Lewis acids parallel the strengths of the acceptors, viz. $SbF_5 > AsF_5 > PF_5 > BF_3 > VF_5$; on the basis of their vibrational spectra, the adducts are formulated as salts containing discrete $[ClO_2]^+$ cations. The heat of dissociation of the BF_3 complex, that is, for the reaction

$$[ClO_2]^+[BF_4]^-(s) \rightarrow ClO_2F(g)+BF_3(g)$$

is 24 kcal mol^{-1} [634]. Chloryl fluoro-complexes are commonly products of oxidation or fluorination reactions effected by ClO_2F (Scheme 6). While bromyl fluoride apparently does not react with BF_3, AsF_5 or SbF_5, iodyl fluoride combines with acceptor molecules in the presence of a suitable solvent.

$$[IO_2]^+[SO_3F]^- \xleftarrow[\text{reflux}]{SO_3} IO_2F \xrightarrow[\text{liq HF}]{AsF_5} [IO_2]^+[AsF_6]^-$$

The structural chemistry of chloryl and iodyl salts was reviewed in Section 4A (p. 1352).

ClO_2F and PtF_6 combine together at 25° in two competing reactions[635]:

$$2ClO_2F+2PtF_6 \rightarrow [ClO_2]^+[PtF_6]^- + [ClO_2F_2]^+[PtF_6]^-$$

and

$$2ClO_2F+2PtF_6 \rightarrow 2[ClO_2]^+[PtF_6]^- + F_2$$

The chlorine(VII) salt comprises about 10% of the solid product.

Acceptor properties are evident in the formation of $Cs[ClO_2F_2]$[636] (from CsF and ClO_2F) and $K[IO_2F_2]$[637] (from KF and IO_2F in anhydrous HF). The spectroscopically characterized $ClO_2F_2^-$ anion[638] has C_{2v} symmetry, the framework being derived from a trigonal–bipyramid with two weak axial Cl–F bonds (force constant 1·6 mdyne Å^{-1}) and two strong equatorial Cl–O bonds (force constant 8·3 mdyne Å^{-1}); the remaining equatorial position is occupied by the lone pair of electrons. A similar $IO_2F_2^-$ anionic unit has been defined by X-ray crystallography[639] and infrared and Raman spectroscopy[545] in KIO_2F_2 [Fig. 32(a)].

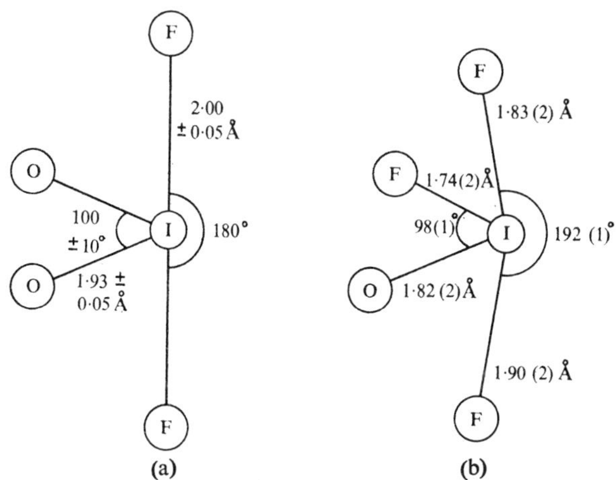

FIG. 32. Molecular structures of (a) $IO_2F_2^-$ (in KIO_2F_2) and (b) IOF_3.

634 K. O. Christe, C. J. Schack, D. Pilipovich and W. Sawodny, *Inorg. Chem.* **8** (1969) 2489.
635 K. O. Christe, *Inorg. Nuclear Chem. Letters*, **8** (1972) 453.
636 D. K. Huggins and W. B. Fox, *Inorg. Nuclear Chem. Letters*, **6** (1970) 337.
637 J. J. Pitts, S. Kongpricha and A. W. Jache, *Inorg. Chem.* **4** (1965) 257.
638 K. O. Christe and E. C. Curtis, *Inorg. Chem.* **11** (1972) 35.
639 L. Helmholz and M. T. Rogers, *J. Amer. Chem. Soc.* **62** (1940) 1537.

Halogen Oxide Trifluorides

$ClOF_3$[640,641] is prepared (i) by direct low-temperature fluorination of covalent inorganic hypochlorites such as Cl_2O or $ClONO_2$, or (ii) photochemically from gaseous mixtures, e.g. ClO_3F/F_2, $Cl_2/O_2/F_2$, ClF/IOF_5 or ClO_2F/ClF_5.

The pure compound is a colourless gas at room temperature, is an excellent oxidizing fluorinating agent, and is allegedly more corrosive than ClF_3. Measured physical properties are: m.p., -42°C; b.p., 29°C; vapour pressure, $\log p(\text{mm}) = 8{\cdot}433 - 1680/T$; $\Delta H_{vap} = 7{\cdot}7$ kcal mol^{-1}; Trouton's constant $= 25{\cdot}4$ cal deg^{-1} mol^{-1}. Defined spectroscopically in the vapour phase[642], the $ClOF_3$ molecule has a C_s structure analogous to that of IOF_3; the Cl–O bond is equatorial and fairly strong (force constant 9·37 mdyne Å^{-1}). The ^{19}F nmr spectrum of the vapour features a single line at -327 ppm relative to CCl_3F (external), which shifts to -272 ppm and broadens for the neat liquid; no splitting is observed on cooling the sample. Molecular association *via* the axial fluorine atoms is indicated for the condensed phases.

The oxyfluoride reacts readily with the Lewis acids BF_3, AsF_5 and SbF_5 to give complexes containing the $ClOF_2^+$ cation, structurally akin to SOF_2[643], and with alkali-metal fluorides forms salts $M^+[ClOF_4]^-$ (M = K or Cs) in which the anion has C_{4v} symmetry[644].

Iodine pentoxide dissolves in boiling IF_5; white hygroscopic needles of IOF_3 separate on cooling the solution.

$$I_2O_5 + 3IF_5 \rightarrow 5IOF_3$$

The crystal structure of the solid contains molecular species (Fig. 32) linked by weak I–F–I bridges[645]. Vibrational spectra have also been recorded[646]. On warming to 110°C, IOF_3 dismutates reversibly into IF_5 and iodyl fluoride.

$$2IOF_3 \rightarrow IF_5 + IO_2F$$

Perhalogenyl Fluorides

The compounds ClO_3F and IO_3F were first recognized in the early 1950s; not synthesized until after the discovery of perbromates (p. 1451), BrO_3F still awaits detailed investigation.

Synthesis

The most convenient syntheses involve fluorination of the perhalic acid or its salts.

$$KClO_4 \xrightarrow{SbF_5 \text{ or } HSO_3F} ClO_3F$$

$$KBrO_4 \xrightarrow{SbF_5} BrO_3F$$

$$HIO_4 \xrightarrow[\text{liquid HF}]{F_2} IO_3F$$

The reactions possibly proceed *via* the intermediate formation of the XO_3^+ cation (X = Cl, Br or I).

640 R. Bougon, J. Isabey and P. Plurien, *Compt. rend.* **271C** (1970) 1366.
641 D. Pilipovich, C. B. Lindahl, C. J. Schack, R. D. Wilson and K. O. Christe, *Inorg. Chem.* **11** (1972) 2189; D. Pilipovich, H. H. Rogers and R. D. Wilson, *ibid.* **11** (1972) 2192.
642 K. O. Christe and E. C. Curtis, *Inorg. Chem.* **11** (1972) 2196.
643 K. O. Christe, E. C. Curtis and C. J. Schack, *Inorg. Chem.* **11** (1972) 2212.
644 K. O. Christe and E. C. Curtis, *Inorg. Chem.* **11** (1972) 2209.
645 J. W. Viers and H. W. Baird, *Chem. Comm.* (1967) 1093.
646 H. A. Carter and F. Aubke, *Inorg. Chem.* **10** (1971) 2296.

Perchloryl Fluoride

Properties

The interesting and useful compound ClO_3F has been the subject of three comprehensive reviews[572,647,648], and its physical properties have been measured in some detail (Table 59). Electron diffraction has defined the C_{3v} molecule in the gas phase (Fig. 33), while microwave measurements imply a dipole moment of no more than 0·023 *D*[649]. The symmetry of the electric field experienced by the central chlorine atom is also demonstrated by the measured ^{35}Cl and ^{37}Cl nuclear quadrupole coupling constants and by the ^{35}Cl–F and ^{37}Cl–F spin–spin coupling observed in the ^{19}F nmr spectrum. The ^{19}F chemical shift of ClO_3F (−287 ppm relative to CCl_3F) compares with that of molecular fluorine (−428·7 ppm relative to CCl_3F). ClO_3F offers the highest resistance to electrical breakdown known for any gas, and has been used as an insulator in high-voltage systems.

TABLE 59. PROPERTIES OF PERCHLORYL FLUORIDE, ClO_3F

Physical properties and thermodynamic parameters					
Melting point (°C)	−147·74[a]		$\Delta H_f°$ at 25°C (kcal mol^{-1})		−5·7[c]
Boiling point (°C)	−46·67[a]		$\Delta G_f°$ at 25°C (kcal mol^{-1})		11·5[c]
T_{crit} (°C)	95·17[b]		$S°$ at 25°C (cal deg^{-1} mol^{-1})		66·65[c]
P_{crit} (atm)	53·0[b]		$C_p°$ at 25°C (cal deg^{-1} mol^{-1})		15·52[c]
ΔH_{fusion} (kcal mol^{-1})	0·9163[a]				
ΔH_{vap} at b.pt. (kcal mol^{-1})	4·619[a]				
Trouton's constant (cal deg^{-1} mol^{-1})	20·4[a]				
Vapour pressure equation:[a] $\log p(\text{mm}) = -1652·37/T - 8·62625 \log T + 0·0046098T + 29·44780$ for T = 164–229°K.					
Liquid density, ρ (g cm^{-3}):[d] $\rho = 2·266 - 1·603\times10^{-3}T - 4·080\times10^{-6}T^2$ for T = 131–234°K.					
Viscosity, η (centipoise):[e] $\log \eta = 299/T - 1·755$ for T = 196–327°K.					
Surface tension (dyne cm^{-1}):[e] 24·1–21·3 for T = 198–218°K.					
Molecular spectra, etc.					
Infrared:	gas[f]				
Raman:	gas[g]	liquid[f]			
^{19}F nmr:		liquid[h]			
Mass spectrum:	gas[i]				
Microwave spectrum:	gas[j]				
Electron diffraction:	gas[k]				
Molecular parameters					
Bond dissociation energies (kcal mol^{-1}):[i]	*D*(Cl–O), 57;	*D*(Cl–F), 60			
Vibrational frequencies (cm^{-1}):[g]	ν_1 (a_1)	1062	ν_4 (e)	1314	
	ν_2 (a_1)	716	ν_5 (e)	573	
	ν_3 (a_1)	549	ν_6 (e)	414	
Valence force constants (mdyne Å^{-1}):[l]	f_r(Cl–O)	9·41;	f_r(Cl–F)	3·91	
Bond lengths (Å):[k,m]	r_g(1) (Cl–O)	1·404±0·002			
	r_g(1) (Cl–F)	1·619±0·004			

647 J. F. Gall, *Kirk-Othmer Encyclopedia of Chemical Technology*, 2nd edn., Vol. 9, p. 598, Interscience, New York (1966).

648 V. M. Khutoretskii, L. V. Okhlobystina and A. A. Fainzil'berg, *Russ. Chem. Rev.* **36** (1967) 145.

649 A. A. Maryott and S. J. Kryder, *J. Chem. Phys.* **27** (1957) 1221; D. R. Lide, jun. *ibid.* **43** (1965) 3767.

Table 59 (*cont.*)

^{19}F chemical shift (ppm relative to CCl_3F as internal standard):[h]			
	−287		
J(Cl–F) (Hz):[h]	278±5		
Nuclear quadrupole coupling constants, e^2Qq (MHz):[j]			
	^{35}Cl, −19·2±0·5;	^{37}Cl,	−15·4±1·5
Ionization potential (kcal mol^{-1}):[i]	314±4		
Dipole moment, D:[n]	0·023±0·003		

[a] J. K. Koehler and W. F. Giauque, *J. Amer. Chem. Soc.* **80** (1958) 2659.
[b] A. Engelbrecht and H. Atzwanger, *J. Inorg. Nuclear Chem.* **2** (1956) 348.
[c] *Selected Values of Chemical Thermodynamic Properties*, N.B.S. Technical Note 270–3 (1968).
[d] R. L. Jarry, *J. Phys. Chem.* **61** (1957) 498.
[e] J. Simkin and R. L. Jarry, *J. Phys. Chem.* **61** (1957) 503.
[f] D. R. Lide, jun., and D. E. Mann, *J. Chem. Phys.* **25** (1956) 1128; F. X. Powell and E. R. Lippincott, *ibid.* **32** (1960) 1883.
[g] H. H. Claassen and E. H. Appelman, *Inorg. Chem.* **9** (1970) 622.
[h] S. Brownstein, *Canad. J. Chem.* **38** (1960) 1597; H. Agahigian, A. P. Gray and G. D. Vickers, *ibid.* **40** (1962) 157; J. Bacon, R. J. Gillespie and J. W. Quail, *ibid.* **41** (1963) 3063; A. A. Maryott, T. C. Farrar and M. S. Malmberg, *J. Chem. Phys.* **54** (1971) 64.
[i] V. H. Dibeler, R. M. Reese and D. E. Mann, *J. Chem. Phys.* **27** (1957) 176.
[j] D. R. Lide, jun., *J. Chem. Phys.* **43** (1965) 3767.
[k] A. H. Clark, B. Beagley, D. W. J. Cruickshank and T. G. Hewitt, *J. Chem. Soc.* (*A*) (1970) 872.
[l] W. Sawodny, A. Fadini and K. Ballein, *Spectrochim. Acta*, **21** (1965) 995.
[m] For bond angles see Fig. 33.
[n] A. A. Maryott and S. J. Kryder, *J. Chem. Phys.* **27** (1957) 1221.

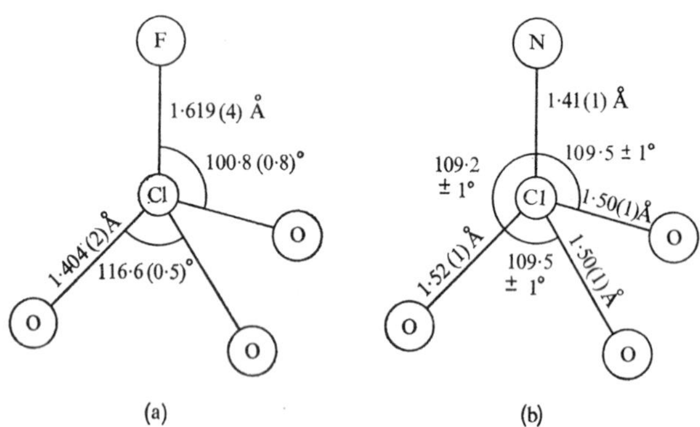

FIG. 33. Molecular structures of (a) ClO_3F and (b) $[NClO_3]^{2-}$ (in K_2NClO_3).

Reactions

Perchloryl fluoride is often regarded as an inert substance, although in fact many of its reactions occur under mild conditions. While $\Delta G_f°$ is slightly positive, $\Delta H_f°$ is negative, so that spontaneous decomposition at room temperature is virtually impossible, and the compound is stable towards explosion; thermal decomposition is noticeable only above 400°C. The oxidizing powers of ClO_3F depend strongly on the nature of the reaction medium and on the temperature. In 0·1 M acid, iodide ion is quantitatively oxidized to iodine.

$$ClO_3F+8I^-+6H^+ \rightarrow Cl^-+F^-+4I_2+3H_2O$$

Many materials towards which it is inert at 20°C are rapidly attacked at 150–200°C. Perchloryl fluoride is used in large quantities, either alone or admixed with halogen fluorides, as an oxidant for rocket fuels; mixtures of ClO_3F with ClF_5 yield $ClOF_3$ on ultraviolet photolysis[641].

ClO_3F is highly susceptible to nucleophilic attack at the chlorine atom. Under aqueous conditions hydrolysis occurs only on heating the compound to 200°C in a sealed tube with concentrated alkali, but in alcoholic potash hydrolysis is rapid and quantitative at room temperature. ClO_3F reacts smoothly with liquid or aqueous ammonia:

$$ClO_3F+3NH_3 \rightarrow NH_4[NHClO_3]+NH_4F$$

From aqueous solutions of this ammonium salt other amidoperchlorates may be prepared, e.g. M^INHClO_3, $M^I_2NClO_3$ and $M^{II}NClO_3$ (M^I = Na, K, Cs or Ag; M^{II} = Sr, Ba or Pb); many of the solids are explosive. Structural investigations[650] of these amidoperchlorates include an X-ray study of crystalline K_2NClO_3; the salt is isomorphous with K_2SO_4, and the anion approximates to C_{3v} symmetry (Fig. 33)[651]. Although the unstable explosive liquid N-perchloropiperidine has been prepared, other organic amines undergo oxidative decomposition with ClO_3F.

In organic chemistry ClO_3F has been widely used as a mild fluorinating agent. While phenyllithium displaces F^- from ClO_3F in a straightforward manner (Scheme 7), reaction with carbanions involves C–F bond-formation in many cases where the anion is stabilized by resonance with one or more neighbouring electronegative centres, as in diesters and polynitro compounds.

$$R_3C^-+ClO_3F \rightarrow R_3CF+ClO_3^-$$

The probable mechanism involves attack on ClO_3F by the more nucleophilic of the anionic sites, followed by rapid formation of the strong C–F bond[652].

$$EtO_2C\text{-}C(R)\text{=}C(\text{-}O^-)\text{-}OEt + F\text{-}ClO_3 \longrightarrow EtO_2C\text{-}C(R)(F)\text{-}C(=O)\text{-}OEt \cdots ClO_3^- \longrightarrow EtO_2C\text{-}C(R)(F)\text{-}C(=O)\text{-}OEt + ClO_3^-$$

A related mechanism is proposed for the oxofluorination of carbon–carbon bonds.

Unlike ClO_2F, perchloryl fluoride does not form complexes with Lewis acids such as SbF_5 or SO_3; neither is it soluble in liquid HF. In the presence of Friedel–Crafts catalysts, however, ClO_3F inserts the ClO_3 group into aromatic nuclei (p. 1573), a reaction which may involve the intermediate agency of ClO_3^+.

Perbromyl Fluoride[653]

BrO_3F is a colourless reactive gas, solidifying at about −110°C to a white solid. At

[650] J. Goubeau, E. Kilcioglu and E. Jacob, *Z. anorg. Chem.* **357** (1968) 190; A. I. Karelin, Yu. Ya. Kharitonov and V. Ya. Rosolovskii, *Zhur. priklad. Spektroskopii*, **8** (1968) 256, 458; *Russ. J. Inorg. Chem.* **13** (1968) 1234.
[651] N. I. Golovina, G. A. Klitskaya and L. O. Atovmyan, *J. Struct. Chem.* **9** (1968) 817.
[652] W. A. Sheppard, *Tetrahedron Letters* (1969) 83.

room temperature it slowly decomposes in metal or Kel-F containers. As with ClO_3F, the infrared and Raman spectra of the gaseous compound imply a molecule of C_{3v} symmetry[654]. It is hydrolysed by aqueous alkali.

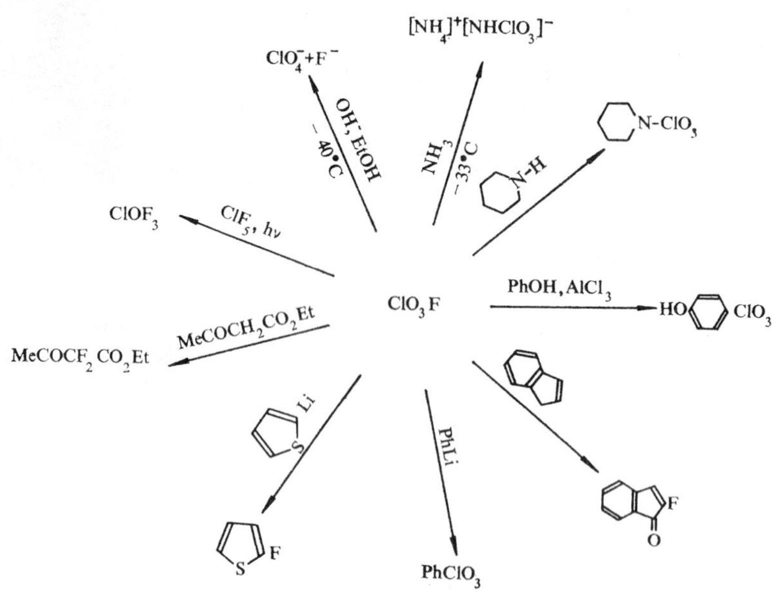

SCHEME 7. Reactions of perchloryl fluoride.

Periodyl Fluoride[575a]

IO_3F is a white crystalline solid, stable in glass, and decomposing at 100°C to IO_2F and oxygen. It is soluble in liquid HF, reacting therein with BF_3 and AsF_5 to give materials which may be formulated as $[IO_3][BF_4]$ and $[IO_3][AsF_6],10HAsF_6$, respectively. SO_3 reduces IO_3F to iodyl fluorosulphate:

$$IO_3F+SO_3 \rightarrow IO_2SO_2F+O_2$$

Like ClO_3F, periodyl fluoride reacts with ammonia, but the product, NH_3,IO_3NH_2, has not been characterized.

Iodine Oxypentafluoride

Iodine oxypentafluoride, IOF_5, is obtained by the reaction of IF_7 with water, silica, glass or I_2O_5[655]; though not susceptible to hydrolysis, the colourless compound, m.p. 4·5°C, is difficult to purify. Spectroscopic investigations are consistent with a structure of C_{4v} symmetry (6)[656].

653 E. H. Appelman and M. H. Studier, *J. Amer. Chem. Soc.* **91** (1969) 4561.
654 H. H. Claassen and E. H. Appelman, *Inorg. Chem.* **9** (1970) 622.
655 R. J. Gillespie and J. W. Quail, *Proc. Chem. Soc.* (1963) 278; N. Bartlett and L. E. Levchuk, *ibid.* p. 342; J. H. Holloway, H. Selig and H. H. Claassen, *J. Chem. Phys.* **54** (1971) 4305.
656 D. F. Smith and G. M. Begun, *J. Chem. Phys.* **43** (1965) 2001.

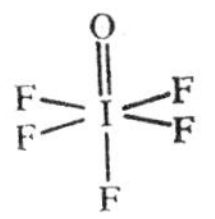

(6)

Normal coordinate analysis of the vibrational frequencies implies little difference between the axial and equatorial I–F bonds (force constants 4·60 and 4·42 mdyne Å^{-1}, respectively), though the ^{19}F chemical shifts clearly discriminate between axial and equatorial fluorine atoms (relative to SiF_4, the values are -272 and -236 ppm, respectively)[657]. The microwave spectrum has not been analysed to yield interatomic distances, but the dipole moment is given as $1{\cdot}08 \pm 0{\cdot}1\,D$[658].

Halogen Dioxide Trifluorides

ClO_2F_3

Two materials of composition ClO_2F_3 have been described in the literature.

(1) Treatment of $[ClO_2]^+[PtF_6]^-$ containing about 10% $[ClO_2F_2]^+[PtF_6]^-$ (p. 1389) with NOF in a sapphire reactor at -78°C displaces, among other volatile products, a material of composition ClO_2F_3, colourless in the solid, liquid and vapour phases, and apparently stable at 25°C[659].

$$[ClO_2F_2]^+[PtF_6]^- + NOF \rightarrow ClO_2F_3 + [NO]^+[PtF_6]^-$$

Its infrared spectrum is consistent with a molecule of C_{2v} symmetry (cf. IO_2F_3 below) having equatorial oxygen atoms. The existence of $ClO_2F_2^+$ indicates fluoride-donor abilities which have yet to be directly tested by experiment.

(2) A violet solid O_2ClF_3 [660], a vigorous oxidizing agent even at low temperature, has been prepared by two routes:

$$ClF_3 + O_2 \xrightarrow[195°K]{h\nu} O_2ClF_3 \xleftarrow[120°K]{} ClF + O_2F_2$$

The compound is formulated on spectroscopic grounds as $FClOOClF_2$; with excess ClF, O_2F_2 gives a blue compound believed to be $F_2ClOOClF_2$.

IO_2F_3

Iodine dioxide trifluoride has been prepared by the following route[661].

$$Ba_3H_4(IO_6)_2 \xrightarrow{HSO_3F} [HIO_2F_4] \xrightarrow{SO_3} IO_2F_3$$

The yellow sublimable solid melts at 41°C; ^{19}F nmr studies of the melt are consistent with the presence of both C_s (7) and C_{2v} (8) isomers.

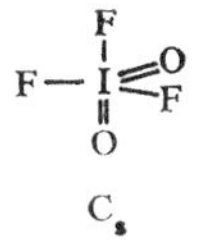

C_s (7) C_{2v} (8)

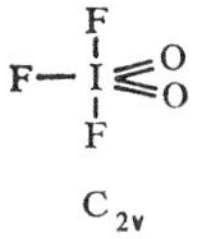

[657] N. Bartlett, S. Beaton, L. W. Reeves and E. J. Wells, *Canad. J. Chem.* **42** (1964) 2531.
[658] S. B. Pierce and C. D. Cornwell, *J. Chem. Phys.* **47** (1967) 1731.
[659] K. O. Christe, *Inorg. Nuclear Chem. Letters*, **8** (1972) 457.
[660] A. V. Grosse and A. G. Streng, *Chem. Abs.* **66** (1967) 39492z; *ibid.* **67** (1967) 75007z.
[661] A. Engelbrecht and P. Peterfy, *Angew. Chem., Internat. Edn.* **8** (1969) 768.

Like IOF_5, the compound is quite resistant to hydrolysis. On exposure to sunlight it liberates oxygen (and some ozone).

$$IO_2F_3 \xrightarrow{h\nu} IOF_3 + \tfrac{1}{2}O_2$$

Fluorine Perchlorate

While fluorine perchlorate has the empirical composition of a halogen oxyfluoride, it has no halogen–fluorine bond, and is considered below (p. 1473) alongside the other halogen perchlorates.

4. OXYACIDS AND OXYSALTS OF THE HALOGENS

(A) INTRODUCTION

Listed in Table 60 are the known oxyacids of the halogens. Many of these acids and their salts were first discovered, or at least recognized for themselves, during the great advance in synthetic and descriptive inorganic chemistry which marked the period 1770–1830. Only recently has the list been completed, however, when the successful syntheses of hypofluorous acid[662], perbromic acid[663] and perbromate salts[663] exorcised a considerable accumulation of conjecture about their "non-existence". Many of the acids are known only in solution (Table 60), but, except for iodites and hypoiodites, reasonably stable crystalline salts have been isolated; the existence of HIO_2 and IO_2^- as more than transient species in solution is doubtful. The molecules HOX (X = F, Cl or Br), HIO_3 and HXO_4 (X = Cl, Br or I) are sufficiently stable in the vapour phase to permit their investigation by mass spectroscopy and other techniques. A series of hydrates of perchloric acid has been characterized, while iodic acid forms a 1:1 compound with iodine pentoxide, HIO_3,I_2O_5.

TABLE 60. THE OXYACIDS OF THE HALOGENS

HOF[a]	HOCl[b,e]	HOBr[b,e]	HOI[e]
	$HClO_2$ [e]	$HBrO_2$(?)[e]	HIO_2(??)[e]
	$HClO_3$ [e]	$HBrO_3$ [e]	HIO_3 [b,d,e]
	$HClO_4$ [b,c,d,e]	$HBrO_4$ [b,e]	HIO_4 [b,d]
			$H_7I_3O_{14}$ [d]
			H_5IO_6 [d,e]

[a] Stable in solid and vapour phases at low temperature; rapidly oxidizes water.
[b] Known in vapour phase.
[c] Known as pure liquid.
[d] Known in solid phase.
[e] Known in aqueous solution.

The anhydrous molecular acids H_mXO_n (X = F, Cl, Br or I) all contain O–H rather than X–H bonds, and similar structures are assumed for the undissociated acids in solution. Periodic acid H_5IO_6 (formally $HIO_4,2H_2O$) exists as the ortho-acid $(HO)_5IO$ in the solid state and solution, although the solid hydrates of perchloric acid consist of ClO_4^- anions

[662] M. H. Studier and E. H. Appelman, *J. Amer. Chem. Soc.* **93** (1971) 2349.
[663] E. H. Appelman, *J. Amer. Chem. Soc.* **90** (1968) 1900.

balanced by aquated protons. While H_5IO_6 behaves as a weak tribasic acid (pK_1, 3·29; pK_2, 8·3; pK_3, 11·6), the strengths of the monobasic acids rise as the oxidation state of the halogen rises; the chlorine acids in particular closely follow Pauling's rules[664], as shown by the approximate pK_as: HOCl, 7·52; $HClO_2$, 1·94; $HClO_3$, −3; $HClO_4$, −10. The pK_as of related acids increase in the sequence: Cl < Br < I. Iodic acid and (especially) periodic acid and their derivatives undergo polymerization in solution, and salts containing oligomeric anions have been prepared.

The halogen–oxygen bond lengths in the acid H_mXO_n or anion XO_n^- contract as the oxidation state of the halogen rises, and the valence force constant undergoes a simultaneous increase. These effects, discussed in Section 4B1, may be explained either in terms of d_π–p_π bonding in the X–O bond, or by an increasing charge separation $X^{\delta+}$–$O^{\delta-}$ in the bond. The charges on the atoms in the anions have been investigated by calculation and by spectroscopic methods (Table 44), but without producing consistent results.

Another noteworthy aspect of the structural chemistry of the oxyacids and their derivatives is that higher coordination numbers are found for iodine than for chlorine or bromine; this must be due, at least in part, to the greater size of the iodine atom. While chlorine and bromine are restricted to coordination numbers $\leqslant 4$ in the oxyacids and their salts, many periodates contain six iodine–oxygen bonds, and the crystal structures ofiodates (including HIO_3) display significant iodine–oxygen contacts at less than the sum of the van der Waals' radii (3·55 Å), which increase the number of neighbours to 6, 7 or 8. The ease with which iodine expands its coordination shell has been used to rationalize the observation that redox and exchange reactions involving iodates and periodates occur much more quickly than those of their chlorine and bromine relatives. The readiness with which iodates and periodates ligate metal ions (forming iodato- and periodato-complexes) probably reflects a fairly high charge separation in the I–O bond, producing polarizable oxygen atoms. While chlorates, bromates, (presumably) perbromates and (especially) perchlorates are weaker complexing agents than iodine anions (ClO_4^- being so weak that perchlorates are commonly used to establish media of constant ionic strength in studies of the complexation equilibria of other species), it is clear from spectroscopic and crystallographic data that these anions (even ClO_4^-) will attach themselves to cations if stronger donors are absent.

Both thermodynamic and kinetic factors are important in the chemistry of the halogen oxyacids, and particularly in the interrelationships of the various oxidation states of the halogens in the condensed phases; moreover, both the thermodynamic and kinetic parameters which control the reactions of the acids and anions in solution are critically sensitive to pH. The examples in the subsequent paragraphs illustrate the importance of thermodynamic, kinetic and environmental features in discussing the chemistry of these systems; many additional examples will be found in the remainder of the section.

The thermodynamic properties of oxyacids and oxysalts have already been summarized in connection with the oxidation state diagram and redox potentials for halogen derivatives in aqueous solution (Figs. 2 and 14) and similar patterns are encountered for the solid derivatives; some relevant redox potentials are also tabulated below amongst the properties of the various species. In general, the ease of oxidation of the halogens increases in the order F $\ll$ Br < Cl < I; for the aqueous halate ions $\Delta G_f^\circ = +4{\cdot}55$ (BrO_3^-), $-1{\cdot}90$ (ClO_3^-) and $-30{\cdot}2$ kcal g-ion^{-1} (IO_3^-). The ready oxidation of iodine is expected on the basis of

[664] L. Pauling, *The Nature of the Chemical Bond*, 3rd edn., pp. 324–328, Cornell University Press, Ithaca (1960).

established trends in the Periodic Table. The observation that bromine compounds are thermodynamically less stable than analogous chlorine and iodine compounds is an example of the "alternation effect"[665], which rationalizes the "anomalous" behaviour of the elements of the third row in terms of the irregular increase in nuclear charge on descending a Group, with particular emphasis on the shielding characteristics of different orbitals and on the resultant variation in promotion energies. Similar effects are seen in the instability of bromine oxides, fluorides and oxyfluorides.

Though the oxyacids and their salts are generally stable thermodynamically with respect to formation from the elements (the notable exceptions with $\Delta G_f^\circ > 0$ being the aqueous BrO_3^- and BrO_4^- ions), other decomposition processes are possible. The reaction

$$XO_n^- \rightarrow X^- + n/2O_2$$

is more or less exothermic for chlorine and bromine anions, though for kinetic reasons (rupture of the X–O bond is necessary) it is observed at room temperature only for the hypohalites. The oxidizing powers of the compounds are also noted in many reactions in solution, the mechanisms usually involving oxygen-atom transfer. In acid solution only HIO_3 is stable with respect to liberation of oxygen from water, although the reactions of the potentially unstable species are very slow unless suitably catalysed. The oxidizing powers of these compounds are exploited in the laboratory in numerous preparative and analytical processes, on a large scale in bleaching and sterilizing operations, and in the manufacture of rocket fuels and explosives. The same oxidizing properties also bring risks of fire and explosion in manipulation, especially in the presence of inflammable organic matter; perchlorate explosions have caused injury or death to many workers.

The closeness of successive redox potentials accords with the experimental observation that disproportionation reactions are common. For chlorine, bromine and iodine the $+1$ and $+3$ oxidation states are unstable with respect to formation of -1 and $+5$ (or $+7$) states, while in the chlorine systems evolution of ClO_2 is a possibility, especially from acid solution. In general, several different disproportionation processes may be open to each species; the choice of reaction then turns on kinetic rather than thermodynamic factors.

The evident effect of pH on the redox potentials (Fig. 14) is best translated into chemical terms by an example. The disproportionation of the halogens in hot alkaline solution,

$$3X_2 + 6OH^- \rightarrow 5X^- + XO_3^- + 3H_2O$$

is used to prepare halates in high yield; the equilibrium constants are very favourable (Cl, 4×10^{74}; Br, 2×10^{37}; I, 1×10^{23}), and the reaction proceeds (*via* XO^- and XO_2^-) to completion in a matter of hours for chlorate, of minutes for iodate. However, in acid solution the reverse reaction,

$$XO_3^- + 5X^- + 6H^+ \rightarrow 3X_2 + 3H_2O$$

has favourable equilibrium constants (Cl, 2×10^9; Br, 1×10^{38}; I, 6×10^{44}). In the case of bromate and iodate, the reaction is instantaneous and is used in the analysis of these anions; chlorate is reduced to ClO_2 (in addition to chlorine) by Cl^- in acid solution.

For chlorine and bromine the rates of decomposition, exchange and redox reactions[666] generally decrease as the oxidation state of the halogen rises, e.g. $ClO^- > ClO_2^- > ClO_3^- > ClO_4^-$. Perchlorates and perbromates are noted for being sluggish oxidizing agents at

[665] C. S. G. Phillips and R. J. P. Williams, *Inorganic Chemistry*, Vol. 1, p. 663, Clarendon Press, Oxford (1965); R. S. Nyholm, *Proc. Chem. Soc.* (1961) 273.

[666] A. G. Sykes, *Kinetics of Inorganic Reactions*, pp. 199, 269, Pergamon (1966).

room temperature, and exchange oxygen only very slowly with water: the half-life of ClO_4^- in 6 M $HClO_4$ with respect to oxygen exchange is more than 100 years at 25°C. However, periodates are vigorous oxidizing agents and trade oxygen rapidly with water.

The rates of reactions are usually faster in acid solution than in neutral or alkaline solution, although it may be hard to distinguish acid catalysis from other changes brought about by decreasing the pH—especially changes in the thermodynamic parameters; it has also been shown that in solutions containing both an acid and its anion, the acid is usually more reactive. Transition metal cations catalyse many decomposition reactions of the oxyacids and oxyanions in solution, probably *via* intermediate complexes which labilize a halogen–oxygen bond.

Finally, the rates of reaction increase as the size of the halogen increases. The equilibrium constants for the disproportionation of the hypohalites in alkaline solution

$$3OX^- \rightleftharpoons XO_3^- + 2X^-$$

are all sufficiently large (Cl, 3×10^{26}; Br, 8×10^{14}; I, 5×10^{23}) to discount thermodynamic influences; while OCl^- persists for days in solution, OBr^- decomposes in hours, and OI^- disappears almost instantaneously. Similar observations have been made for the exchange of oxygen between water and XO_3^-, which proceeds at a measurable rate for ClO_3^- at 100°C in acid solution, for BrO_3^- at 30°C in acid solution, and for IO_3^- at 20°C in neutral solution.

(B) HYPOHALOUS ACIDS AND HYPOHALITES

Introduction

The story of halogen(I) oxyacids[576] begins in 1774, when Scheele remarked that chlorine water was able to bleach vegetable colours. That this property might be useful commercially was suggested in 1785 by Berthollet, who also noted that solutions of chlorine in potash lye were more concentrated and more powerful bleaches than aqueous solutions and did not have the deleterious effects on workers and materials caused by excess chlorine; patents for this bleaching process were taken out in 1789. Influenced by the high cost of alkalis, Tennant in 1789 prepared bleaching solutions by dissolving chlorine in aqueous suspensions of lime, strontia or baryta. He subsequently (1798) patented a process for the manufacture of "bleaching powder" by saturating dry calcium hydroxide with chlorine gas.

The chemical constitution of these bleaching solutions was first truly recognized in 1834 by A. J. Balard, who prepared an aqueous solution of hypochlorous acid and isolated the anhydride Cl_2O; he had earlier (1821) obtained solutions of hypobromous acid by the gradual addition of yellow mercuric oxide to bromine water. Solutions of hypoiodous acid were similarly generated, but not until 1897. HOI is seen in the mass spectrum of H_5IO_6.

Of the acids, only hypofluorous acid has been obtained in the pure form, milligram amounts being produced in the fluorination of ice[662]; it is unstable at ambient temperatures. Solutions of HOCl, HOBr, HOI or their salts are unstable, as are the solid salts themselves. Nevertheless their oxidizing powers have been widely used in commercial and household bleaching and disinfecting operations, as well as in analytical and preparative procedures in the laboratory. The acids and their derivatives (especially esters) are important halogenating agents in organic chemistry.

The following reviews of the chemistry of the halogen(I) oxyacids and their derivatives have recently appeared: hypochlorous acid and hypochlorites[572,574,575b,667]; hypobromous acid and hypobromites[575b,668,669]; hypoiodous acid and hypoiodites[575b,616].

Preparation

Hypohalous Acids

The most convenient methods for obtaining aqueous solutions of the acids HOCl, HOBr and HOI involve perturbing the equilibria in the disproportionative hydrolysis of the halogens (p. 1188)

$$X_2+H_2O \rightleftharpoons H^+ + X^- + HOX$$

(which normally lies well to the left), by removal of halide ion in the form of an insoluble or sparingly dissociated salt. Although many silver(I) and mercury(II) salts have been used for this purpose, Ag_2O and HgO are probably the best reagents, affording the metal halide without introducing extraneous anions into solution. The stability of acid solutions depends on the nature of other ions present; for HOCl and HOBr, some purification may be effected by distillation of the solution under reduced pressure.

Concentrated (>5 M) solutions of hypochlorous acid may be prepared by treating dichlorine monoxide (either the pure liquid or solutions in CCl_4) with water at 0°C[577]; on a large scale the acid is produced by passing Cl_2O gas into water[574].

Hypohalites

Disproportionation of the halogens in alkaline solution is thermodynamically more favourable than in neutral media (p. 1191), and is rapid at room temperature; aqueous solutions containing ClO^-, BrO^- or IO^- result from dissolving the halogen in a *cold* solution or suspension of the appropriate base.

$$X_2+2OH^- \rightleftharpoons X^- + OX^- + H_2O$$

Made by this method, the solutions are contaminated by halide ion, and further by the subsequent disproportionation of the hypohalites themselves (see below). The presence of excess base stabilizes the solutions to some extent.

Methods have been reported for the electrochemical oxidation of halides to hypohalites in cold dilute solution[576]. Chemical oxidation of halide ions is also possible: in alkaline solution hypochlorites oxidize bromides to hypobromites, and hypoiodites can be generated from iodide with either hypochlorite or hypobromite.

Hypochlorite solutions have been prepared by the neutralization of hypochlorous acid or dichlorine monoxide[577]. Separation of hypochlorites from chlorides may be effected by treatment with an amine base (or an alcohol), which with OCl^- generates the chloramine (or hypochlorite ester) (p. 1410); the organic derivative is readily separated, and the hypochlorite released with alkali. The hydrolysis of N-chloro or N-bromo compounds, e.g. "chloramine-T" ($Na^+[CH_3C_6H_4SO_2NCl]^-$), is an expedient method for the *in situ* generation of hypohalite ions.

The preparation of pure solid hypochlorites is difficult, if not impossible; procedures have been described for obtaining materials with purity <90%. Hypochlorite solutions are

[667] C. C. Addison, *Mellor's Comprehensive Treatise on Inorganic and Theoretical Chemistry*, Supplement II, Part I, pp. 544–569, Longmans, London (1956).

[668] B. Cox, *Mellor's Comprehensive Treatise on Inorganic and Theoretical Chemistry*, Supplement II, Part I, pp. 750–752, Longmans, London (1956).

[669] P. J. M. Radford, *Bromine and its Compounds* (ed. Z. E. Jolles), pp. 154–158, Benn, London (1966).

sufficiently stable in the absence of light to allow evaporation or crystallization at room temperature[577] (at higher temperatures disproportionation dominates), but the solids so derived are more or less contaminated with chloride, chlorite and chlorate; hypochlorites of Li, Na, K, Mg, Ca, Sr and Ba have been reported in both hydrated and anhydrous forms[667]. Basic hypochlorites of the alkaline earths have been synthesized (and manufactured) by the chlorination of metal hydroxides (either the solid or an aqueous suspension); such salts frequently contain large amounts of chloride.

Solid yellow sodium and potassium hypobromites [$NaOBr,xH_2O$, where $x = 5$ or 7, and $KOBr,3H_2O$] can be crystallized from the supersaturated solutions which result from the addition of bromine to cold ($<0°C$) concentrated ($\geqslant 40\%$) aqueous solutions of NaOH or KOH; the compounds are slightly impure, and are unstable above 0°C[669].

There are no substantiated accounts of the isolation of solid metal hypoiodites.

Structure and Properties

Microwave studies (Table 61) have confirmed the HOX structures (as distinct from HXO) of the gaseous hypofluorous acid and hypochlorous acid molecules. The bond angle in HOF is smaller than that in either H_2O (104·7°) or F_2O (103·2°); it is equally consistent with the essential neutrality of the fluorine atom in HOF that the O–F bond is longer and weaker than in F_2O (1·412 Å). By contrast, the angle in HOCl is only slightly less than that of the water molecule; within the experimental errors, the O–H and Cl–O bond lengths are identical with those of the parent oxides H_2O (0·96 Å) and Cl_2O (1·70 Å). The O–Cl bond in CH_3OCl is slightly shorter (1·67 Å; Table 64).

The stretching force constants of the O–X bonds in HOX (X = Cl or Br) are much greater than those of the oxides X_2O (Table 41); the values for OCl^- and OBr^- (Table 62) fall between the two extremes. The hypohalite ions are isoelectronic with the corresponding diatomic halogen fluorides, but have smaller stretching force constants. The ultraviolet-visible spectra of the aqueous ions display weak maxima (listed in Table 62) associated with the transitions $^3\Pi_0{}^+ \leftarrow {}^1\Sigma^+$, and stronger absorptions at shorter wavelengths identified with charge-transfer-to-solvent transitions. Dissociation in the $^3\Pi_0{}^+$ excited state [giving $X^-(^1S)$ and $O(^3P)$] probably accounts for the high photochemical sensitivity of the OX^- anions.

The hypohalous acids are weak acids; this factor, combined with their extreme instability, makes the accurate determination of dissociation constants a difficult undertaking. The values for HOCl and HOBr given in Table 61 summarize many individual studies listed elsewhere (HOCl[667,670] and HOBr[669]). Because the acids are so weak, solutions of the salts display an alkaline reaction, the equilibrium

$$OX^- + H_2O \rightleftharpoons HOX + OH^-$$

lying well to the right. Except at high pH, hypohalite solutions contain significant quantities of the free acid.

Concentrated solutions of hypochlorous acid contain large amounts of the anhydride, dichlorine monoxide, which can form a separate liquid layer; the solution at the eutectic

670 J. C. Morris, *J. Phys. Chem.* **70** (1966) 3798.
671 C. H. Secoy and G. H. Cady, *J. Amer. Chem. Soc.* **62** (1940) 1036.

TABLE 61. PROPERTIES OF THE HYPOHALOUS ACIDS

	HOF	HOCl	HOBr	HOI
Molecular HOX				
$\Delta H_f°$(g) (kcal mol^{-1})[a]		−21·5	−20 ± 10	−21 ± 10
Microwave spectrum	b	c		
r_e(O–X) (Å)	1·442 ± 0·001	1·693 ± 0·005		
r_e(O–H) (Å)	0·964 ± 0·01	0·97 ± 0·01		
∠ H–O–X	97·2 ± 0·6°	103 ± 3°		
Infrared spectrum		vapour[e,f]		
	matrix-isolated[d,h]	matrix-isolated[g]	matrix-isolated[g]	
Fundamental frequencies (cm^{-1})[i]				
ν_1 O–H stretch	3537·1	3581	3590	
ν_2 Bend	1359·0	1239	1164	
ν_3 O–X stretch	886·0	729	626	
Force constants				
f_r(O–H) (mdyne Å^{-1})		7·35[g]	7·14[g]	
f_R(O–X) (mdyne Å^{-1})		3·86	3·59	
f_θ (mdyne Å rad^{-1})		0·77	0·70	
$f_{r\theta}$ (mdyne rad^{-1})		0·45	0·64	
Ultraviolet-visible spectrum		vapour[j]		
HOX in aqueous solution				
Colour		colourless	pale yellow	
$\Delta H_f°$ (undiss.) (kcal mol^{-1})[k]		−28·9	−27·0	−33·0
$\Delta G_f°$ (undiss.) (kcal mol^{-1})[k]		−19·1	−19·7	−23·7
$S°$ (undiss.) (cal deg^{-1} mol^{-1})[k]		34	34	22·8
Dissociation constant, $K_{a,298}$		*ca.* 3×10^{-8} [l]	*ca.* 5×10^{-9} [l]	*ca.* 1×10^{-11} [m] $4·5 \times 10^{-13}$ [n]
Reduction potentials, in acid solution[o]				
$E°(XO_3^-/HOX)$ (volts)		+1·43	+1·49	+1·14
$E°(HOX/\frac{1}{2}X_2)$ (volts)		+1·63	+1·59	+1·45
$E°(HOX/X^-)$ (volts)		+1·50	+1·33	+1·00
Ultraviolet-visible spectrum		p	q	
Kinetic studies of decomposition		r	s	t

[a] V. I. Vedeneyev, L. V. Gurvich, V. N. Kondrat'yev, V. A. Medvedev and Ye. L. Frankevich, *Bond Energies, Ionization Potentials and Electron Affinities*, pp. 78, 130, Edward Arnold, London (1966).

[b] Measured for HOF and DOF; H. Kim, E. F. Pearson and E. H. Appelman, *J. Chem. Phys.* **56** (1972) 1.

[c] Measured for $HO^{35}Cl$, $HO^{37}Cl$, $DO^{35}Cl$, $DO^{37}Cl$; D. C. Lindsay, D. G. Lister and D. J. Millen, *Chem. Comm.* (1969) 950.

[d] Measured for HOF; J. A. Goleb, H. H. Claassen, M. H. Studier and E. H. Appelman, *Spectrochim. Acta*, **28A** (1972) 65.

[e] K. Hedberg and R. M. Badger, *J. Chem. Phys.* **19** (1951) 508.

[f] High resolution study of ν_1 for $HO^{35}Cl$, $HO^{37}Cl$, $DO^{35}Cl$, $DO^{37}Cl$ gives r(O–Cl) = $1{\cdot}689 \pm 0{\cdot}006$ Å, r(O–H) = $0{\cdot}97_4 \pm 0{\cdot}02$ Å, $\angle$ H–O–Cl = $104{\cdot}8 \pm 5°$; R.A. Ashby, *J. Mol. Spectroscopy*, **23** (1967) 439.

[g] Includes isotopic data (H, D; ^{16}O, ^{18}O); HOX produced by photolysis of Ar–HX–O_3 mixtures; I. Schwager and A. Arkell, *J. Amer. Chem. Soc.* **89** (1967) 6006.

[h] Measured for HF · · · HOF; P. N. Noble and G. C. Pimentel, *Spectrochim. Acta*, **24A** (1968) 797.

[i] Data for HOF isolated in N_2; HOCl, HOBr isolated in Ar.

[j] W. C. Fergusson, L. Slotin and D. W. G. Style, *Trans. Faraday Soc.* **32** (1936) 956.

[k] *Selected Values of Chemical Thermodynamic Properties*, N.B.S. Technical Note 270–3 (1968).

[l] See text.

[m] Determined from kinetic studies; Y. T. Chia, U.S. Atomic Energy Commission Rept. UCRL-8311 (1958); *Chem. Abs.* **53** (1958) 2914e.

[n] Measured analytically; A. Skrabal, *Z. Elektrochem.* **28** (1922) 57; *Ber.* **75B** (1942) 1570.

[o] A. G. Sharpe, *Halogen Chemistry* (ed. V. Gutmann), Vol. 1, p. 13, Academic Press (1967).

[p] Maxima at 235 nm ($\epsilon \sim 100\ M^{-1}$) and 290 nm ($\epsilon \sim 27\ M^{-1}$); J. C. Morris, *J. Phys. Chem.* **70** (1966) 3798.

[q] L. Farkas and F. S. Klein, *J. Chem. Phys.* **16** (1948) 886.

[r] M. W. Lister, *Canad. J. Chem.* **30** (1953) 879.

[s] P. J. M. Radford, *Bromine and its Compounds* (ed. Z. E. Jolles), p. 154, Benn, London (1966).

[t] M. L. Josien and C. Courtial, *Bull. Soc. chim. France* (1949) 374.

TABLE 62. PROPERTIES OF HYPOHALITE IONS IN AQUEOUS SOLUTION

	OCl^-	OBr^-	OI^-
Thermodynamic functions			
$\Delta H_f°$ (kcal mol^{-1})[a]	−25·6	−22·5	−25·7
$\Delta G_f°$ (kcal mol^{-1})[a]	−8·8	−8·0	−9·2
$S°$(cal deg^{-1} mol^{-1})[a]	10	10	−1·3
Reduction potentials, in basic solution[b]			
$E°(XO_3^-/OX^-)$(volts)	+0·50	+0·54	+0·14
$E°(OX^-/\frac{1}{2}X_2)$(volts)	+0·40	+0·45	+0·45
$E°(OX^-/X^-)$(volts)	+0·89	+0·76	+0·49
Raman frequency (cm^{-1})	713[c]	620[d]	
Force constant (mdyne Å^{-1})	3·29[c]	3·0[d]	
Ultraviolet-visible spectrum			
λ_{max}(nm)	292[e]	331[f]	365[g]
ϵ_{max}(M^{-1})	350	326	31
Disproportionation, $3OX^- \rightleftharpoons 2X^- + XO_3^-$, **in basic solution**			
K[h]	3×10^{26}	8×10^{14}	5×10^{23}
Rate	Slow at 20°C; fairly rapid at 75°C	Moderately fast at 20°C	Very fast at all temperatures
Kinetic studies	i	j	k

[a] *Selected Values of Chemical Thermodynamic Properties*, N.B.S. Technical Note 270–3 (1968).

[b] A. G. Sharpe, *Halogen Chemistry* (ed. V. Gutmann), Vol. 1, p. 13, Academic Press (1967).

[c] T. G. Kujumzelis, *Physik. Z.* **39** (1938) 665.

[d] J. C. Evans and G. Y.-S. Lo, *Inorg. Chem.* **6** (1967) 1483.

[e] J. C. Morris, *J. Phys. Chem.* **70** (1966) 3798.

[f] C. H. Cheek and V. J. Linnenbom, *J. Phys. Chem.* **67** (1963) 1856.

[g] O. Haimovich and A. Treinin, *Nature*, **207** (1965) 185.

[h] Calculated from electrode potentials (ref. b).

[i] M. W. Lister, *Canad. J. Chem.* **34** (1956) 465, 479.

[j] P. Engel, A. Oplatka and B. Perlmutter-Hayman, *J. Amer. Chem. Soc.* **76** (1954) 2010.

[k] C. H. Li and C. F. White, *J. Amer. Chem. Soc.* **65** (1943) 335; M. L. Josien and C. Courtial, *Bull. Soc. chim. France* (1949) 374; M. W. Lister and P. Rosenblum, *Canad. J. Chem.* **41** (1963) 3013; O. Haimovich and A. Treinin, *J. Phys. Chem.* **71** (1967) 1941.

point (−39·6°C) contains 47% HOCl and the solid phase is then $HOCl,2H_2O$[671]. Hypobromous acid, however, is stable only in dilute solution (<7% HOBr); hypoiodous acid is unstable at all concentrations.

Decomposition

The manner and rate of decomposition of hypohalous acids or hypohalite ions in solution are much influenced by the concentration, pH and temperature of the solution, by added salts, and by photolysis. The two fundamental, competing styles of decomposition are

$$2HOX \rightarrow 2H^+ + 2X^- + O_2 \qquad (\text{or } 2OX^- \rightarrow 2X^- + O_2)$$

and

$$3HOX \rightarrow 3H^+ + 2X^- + XO_3^- \qquad (\text{or } 3OX^- \rightarrow 2X^- + XO_3^-)$$

The free acids react more readily than the anions, so that hypohalites are most stable in basic solution. Because of the commercial importance (in bleaching and sterilizing processes) of the decomposition reactions of hypochlorites, there have been extensive studies of the effect of various additives as catalysts, promoters and activators; catalysis increases the rate of oxygen-evolution but not of disproportionation.

For the hypohalite ions in basic solution disproportionation is the prominent method of decomposition. All the equilibrium constants are very favourable (Table 62), and the rate of the reaction increases in the sequence $ClO^- < BrO^- < IO^-$. The disproportionation of hypochlorite ion is bimolecular in OCl^-, and involves the intermediate agency of the ClO_2^- ion; the mechanism is

$$OCl^- + OCl^- \xrightarrow{\text{slow}} ClO_2^- + Cl^- \quad (1)$$

$$OCl^- + ClO_2^- \xrightarrow{\text{fast}} ClO_3^- + Cl^- \quad (2)$$

Disproportionation of OBr^- and OI^- is autocatalysed by the halide ion formed; halite ion is again an intermediate, and solid bromites can be prepared by the controlled decomposition of hypobromite solutions (p. 1413). Halite ion arises directly from hypohalite [as in eqn. (1) above], and also in a halide-catalysed reaction

$$XO^- + X^- + H^+ \rightarrow [X_2OH]^- \xrightarrow[\text{slow}]{XO^-} 2X^- + HXO_2$$

The bromite and iodite ions probably degrade *via* reaction with hypohalite [cf. eqn. (2)], though there is also some evidence for the alternative fast reaction

$$XO_2^- + XO_2^- \rightarrow XO_3^- + XO^-$$

Photolysis of alkaline hypochlorite solutions produces chloride, chlorate and oxygen, together with traces of chlorine dioxide (from photolysis of the chlorite intermediate)[672]. Evolution of oxygen from hypohalite solutions is also catalysed by cobalt, nickel and copper ions[673].

The hypohalous acids decompose more rapidly than the anions, and are more prone to yield oxygen, especially in the presence of light or catalytic amounts of metal ions; oxygen-evolution represents the exclusive mode of decomposition of hypoiodous acid. Solutions of hypobromous and hypoiodous acids rapidly discolour through formation of the free halogens and trihalide ions in subsequent reactions of the disproportionation or decomposition products.

$$BrO_3^- + 5Br^- + 6H^+ \rightarrow 3Br_2 + 3H_2O$$

$$HOI + I^- + H^+ \rightarrow I_2 + H_2O$$

$$X_2 + X^- \rightarrow X_3^- \qquad (X = Br \text{ or } I)$$

The rate of disproportionation of hypochlorous acid is proportional to the square of the HOCl concentration; the proposed mechanism has chlorous acid as an intermediate,

$$2HOCl \xrightarrow{\text{slow}} 2H^+ + Cl^- + ClO_2^-$$

$$HOCl + ClO_2^- \xrightarrow{\text{fast}} H^+ + Cl^- + ClO_3^-$$

Oxygen-evolution from hypochlorous acid is unimolecular in HOCl.

Like their solutions, solid hypohalites decompose more or less readily at room temperature either by disproportionation or by oxygen-evolution; the reactions are encouraged by

672 G. V. Buxton and R. J. Williams, *Proc. Chem. Soc.* (1962) 141.
673 M. W. Lister, *Canad. J. Chem.* **34** (1956) 479.

heating and by irradiation. The stability of solid hypochlorites is an important factor in determining their usefulness in the storage and transport of "active chlorine" for bleaching processes (see below).

Reactions

The hypohalites exchange oxygen rapidly with labelled water[674]. Hypochlorites oxidize many inorganic substrates (Scheme 8); kinetic studies reveal that either OCl^- or HOCl is the oxidant, both species rarely being active in the same reaction. Most of the reactions involve oxygen atom-transfer to the substrate; isotopic studies have shown that hypochlorite oxidation of NO_2^- involves quantitative transfer of oxygen from chlorine to nitrogen[674], but with SO_3^{2-} some sulphate is formed *via* $ClSO_3^-$.

$$SO_3^{2-} + HOCl \rightarrow ClSO_3^- + OH^-$$
$$ClSO_3^- + 2OH^- \rightarrow SO_4^{2-} + Cl^- + H_2O$$

While hypobromites and hypoiodites are thermodynamically weaker oxidizing agents, their reactions, which follow pathways similar to those of OCl^-, are more rapid. Hypochlorites convert iodates to periodates, but are reluctant to oxidize chlorates[675]; hypoiodite rapidly brings about both reactions, but none of the hypohalites oxidizes BrO_3^- [676].

While the halogens disproportionate in alkaline solution to halide and hypohalite (at least initially), in acid solution hypohalites combine with halides to regenerate the halogen:

$$HOX + X^- + H^+ \rightarrow X_2 + H_2O$$

The equilibria are favourably disposed (Cl, $2{\cdot}4 \times 10^3$; Br, $1{\cdot}4 \times 10^8$; I, 5×10^{12}), and rapidly established in the case of bromine and iodine, the reactions occurring during the decomposition of HOBr and HOI. The reaction of HOCl with Cl^- is slow, but the acid oxidizes Br^- to Br_2 and I^- to I_2.

Hypohalite anions react with basic nitrogen by forming N–X bonds. Thus, hypochlorites convert ammonia and related compounds to chloramines; in dilute equimolar solutions NH_3 and hypochlorite generate the unstable chloramine NH_2Cl,

$$NH_3 + OCl^- \rightarrow NH_2Cl + OH^-$$

while with excess hypochlorous acid nitrogen trichloride is the product[677]. The familiar "chlorine" odour of water which has been sterilized with hypochlorite is due to chloramines produced from bacteria. However, hypobromites oxidize amines quantitatively to nitrogen, a facility exploited in the analysis of urea:

$$3OBr^- + 2OH^- + CO(NH_2)_2 \rightarrow 3Br^- + CO_3^{2-} + N_2 + 3H_2O$$

With primary amides the intermediate bromoamide loses CO_2 in rearranging to a primary amine. The reaction of hypohalites with cyanide involves the halogen cyanide as an intermediate:

$$CN^- + OCl^- + H_2O \rightarrow ClCN + 2OH^-$$
$$ClCN + 2OH^- \rightarrow NCO^- + Cl^- + H_2O$$

---

674 M. Anbar and H. Taube, *J. Amer. Chem. Soc.* **80** (1958) 1073.

675 M. W. Lister and P. Rosenblum, *Canad. J. Chem.* **39** (1961) 1645.

676 O. Haimovich and A. Treinin, *J. Phys. Chem.* **71** (1967) 1941.

677 D. G. Nicholson, *Kirk-Othmer Encyclopedia of Chemical Technology*, 2nd edn., Vol. 5, p. 2, Interscience, New York (1964).

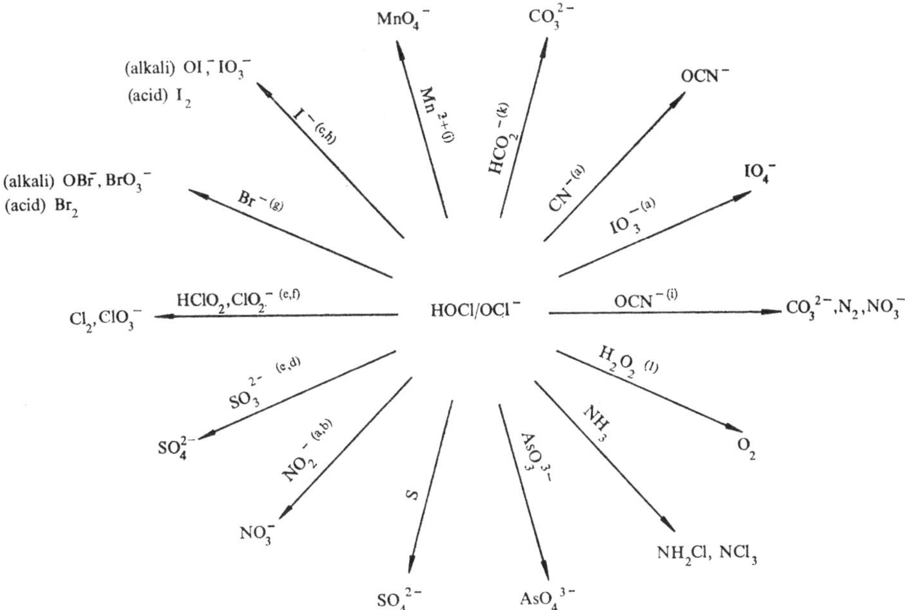

SCHEME 8. Reactions of hypochlorous acid and hypochlorites with inorganic substrates.

Notes for Scheme 8.

Kinetic studies of hypochlorite reactions:

[a] M. W. Lister and P. Rosenblum, *Canad. J. Chem.* **39** (1961) 1645.
[b] M. Anbar and H. Taube, *J. Amer. Chem. Soc.* **80** (1958) 1073.
[c] M. W. Lister and P. Rosenblum, *Canad. J. Chem.* **41** (1963) 3013.
[d] J. Halperin and H. Taube, *J. Amer. Chem. Soc.* **74** (1952) 380.
H. Taube and H. Dodgen, *J. Amer. Chem. Soc.* **71** (1949) 3330.
[f] F. Emmenegger and G. Gordon, *Inorg. Chem.* **6** (1967) 633.
[g] L. Farkas, M. Lewin and R. Bloch, *J. Amer. Chem. Soc.* **71** (1949) 1988.
[h] Y. T. Chia and R. E. Connick, *J. Phys. Chem.* **63** (1959) 1518.
[i] M. W. Lister, *Canad. J. Chem.* **34** (1956) 489.
[j] M. W. Lister, *Canad. J. Chem.* **34** (1956) 465, 479.
[k] M. W. Lister and P. Rosenblum, *Canad. J. Chem.* **41** (1963) 2727.
[l] R. E. Connick, *J. Amer. Chem. Soc.* **69** (1947) 1509.

The hypohalous acids are used in organic chemistry as aromatic and aliphatic halogenating agents; HOBr and HOI are normally generated *in situ*. The ease of aromatic halogenation increases in the order hypochlorite < hypobromite < hypoiodite, and is encouraged by lead or silver salts. In acid solution the active species may be the protonated hypohalous acids H_2OX^+ (p. 1344)[678]; phenol is converted rapidly to 2,4,6-tribromo- or 2,4,6-tri-iodophenol in alkaline solution. Chlorination of aliphatic compounds with HOCl proceeds by a free radical mechanism[579].

$$\begin{aligned} &HOCl && \rightarrow HO\cdot + Cl\cdot \\ &RH + Cl\cdot && \rightarrow R\cdot + HCl \\ &HCl + HOCl && \rightleftharpoons H_2O + Cl_2 \\ &R\cdot + Cl_2 && \rightarrow RCl + Cl\cdot \end{aligned}$$

[678] E. Berliner, *J. Chem. Educ.* **43** (1966) 124.

Hypohalites cleave methyl ketones, forming carboxylate anions and a haloform.

$$RCOCH_3+3OX^- \rightarrow RCO_2^- +CHX_3+2OH^-$$

The iodoform test has been used for many years to identify and enumerate CH_3CO-groups in organic compounds, but direct titration against hypobromites using Bordeaux indicator has recently been used to determine methyl ketones[679].

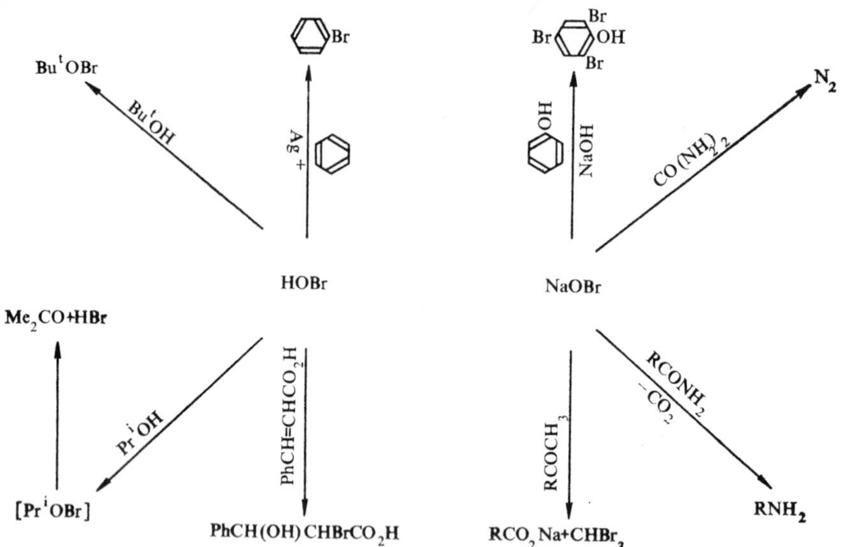

SCHEME 9. Reactions of hypobromous acid and hypobromites with organic substrates.

The acids generally add across ethylenic double bonds as though they were the ion-pair X^+OH^-, though the reaction of HOCl with hindered olefins may follow a radical pathway. Hydrolysis of chlorohydrins (formed from olefins and HOCl) is a useful route to α-glycols. HOBr and especially HOI (i.e. Br_2 or I_2 with HgO) have been used to oxidize alcohols *via* decomposition of the hypohalite ester.

Bleaching and Other Uses

Hypochlorites are widely used as relatively cheap agents for bleaching fabrics and wood pulp, although they are more likely to damage fabrics than more expensive bleaches (e.g. acid chlorite). Hypobromites are faster and stronger bleaching agents than hypochlorites[680], and are conveniently generated *in situ* by adding small amounts of alkali bromide to hypochlorite solutions. The bleaching action of both systems depends on three main reactions:

(1) Disruptive oxidation of coloured molecules.
(2) Addition of HOX across olefinic functions.
(3) Halogenation of saturated compounds.

The effects of various activators, promoters and catalysts on hypochlorite bleaching processes have been investigated.

[679] M. H. Hashmi and A. A. Ayaz, *Anal. Chem.* **36** (1964) 384.
[680] M. Lewin, *Bromine and its Compounds* (ed. Z. E. Jolles), pp. 704–713, Benn, London (1966).

TABLE 63. SOME HYPOCHLORITE PREPARATIONS USED IN BLEACHING, ETC.

Name and composition	Approximate "available chlorine" content	Special uses
"Liquid bleach"; sodium hypochlorite solution, pH $\geqslant$ 11	5%	Domestic bleaching and sanitation.
	10%	Small-scale commercial bleaching (e.g. in laundries).
"Bleach liquor"; solution of $Ca(OCl)_2$ and $CaCl_2$	85 g l^{-1}	Pulp and paper bleaching.
Calcium hypochlorite; normally dried $Ca(OCl)_2,2H_2O$; water $<1\%$	70%	Swimming pool sanitation.
"Bleaching powder"; $Ca(OCl)_2,CaCl_2,Ca(OH)_2,2H_2O$	35%	General bleaching and sanitation.
"Tropical bleaching powder" } $Ca(OCl)_2,CaCl_2,Ca(OH)_2,2H_2O + CaO$	34%	} More stable than ordinary "bleaching powder", especially in hot climates.
"Supertropical bleaching powder" } $Ca(OCl)_2,CaCl_2,Ca(OH)_2,2H_2O + CaO$	30%	}
Water sterilizing powder } $Ca(OCl)_2,CaCl_2,Ca(OH)_2,2H_2O + CaO$	25%	}
Chlorinated trisodium phosphate dodecahydrate	3·5%	Detergent manufacture.
Lithium hypochlorite; normally diluted with sulphates	40%	Processes where calcium is undesirable; sanitation of hard water; some dairy applications.

The oxidizing powers of oxychlorine bleaches are indexed by the "available chlorine" content. This concept may be illustrated by considering the (possibly hypothetical) liberation of iodine from hydriodic acid. Comparison of the reactions

$$Cl_2+2HI \rightarrow 2HCl+I_2$$

and

$$LiOCl+2HI \rightarrow LiCl+I_2+H_2O$$

shows that 1 mol of I_2 is liberated by 70·92 g of chlorine and also by 58·4 g of anhydrous lithium hypochlorite. The "available chlorine" content of LiOCl is defined as the weight of chlorine which liberates the same amount of I_2 as a given weight of the compound; expressed as a percentage, the "available chlorine" content of lithium hypochlorite is (70·92/58·4) × 100, or 121%.

Some of the hypochlorite preparations used for domestic and industrial bleaching and sterilization operations are listed in Table 63. Basic sodium hypochlorite solutions (pH *ca.* 10·5) are sold for domestic use, but solid hypochlorites, being more stable and more concentrated in "available chlorine", are more convenient in terms of storage and transport. The most widely used materials are the hydrated and basic forms of calcium hypochlorite (including the traditional "bleaching powder") made by chlorination of slaked lime; "bleaching powder" has now been largely superseded by products which have either enhanced stability ("tropical" or "supertropical bleaching powder") or a higher "available chlorine" content. The stability of bleaching powders is adversely affected by carbonates, transition metal ions and organic impurities; calcium oxide increases the stability by taking up excess water.

Esters of Hypohalous Acids

Primary, secondary and tertiary alkyl hypohalites are produced by the reaction of an alcohol with a hypohalous acid (which may be generated *in situ* by hydrolysis of a halogen):

$$ROH+HOX \rightarrow ROX+H_2O$$

Primary and secondary derivatives are unstable, readily expelling hydrogen halide while forming respectively an aldehyde or a ketone,

$$RR'CHOX \rightarrow RR'C{=}O+HX$$

TABLE 64. SPECTROSCOPIC INVESTIGATIONS OF METHYL HYPOCHLORITE

Microwave spectrum[a] Measured for $^{12}CH_3O^{35}Cl$, $^{12}CH_3O^{37}Cl$, $^{13}CH_3O^{35}Cl$, $^{12}CH_2DO^{35}Cl$, $^{12}CD_3O^{35}Cl$, $^{12}CD_3O^{37}Cl$

r_{av}(C–H) = 1·099 ± 0·018 Å; θ_{av}(H–C–H) = 109·4 ± 1·8°
r(O–Cl) = 1·674 ± 0·019 Å; θ(C–O–Cl) = 112·8 ± 2·1°
r(O–C) = 1·389 ± 0·028 Å

Quadrupole coupling constants ($CH_3O^{35}Cl$):

$e^2Qq = -84{\cdot}34$ MHz
$\eta = 0{\cdot}408$

Barrier to internal rotation = 3060 ± 150 cal mol^{-1}

Infrared spectrum[b] Recorded for vapour

[a] J. S. Rigden and S. S. Butcher, *J. Chem. Phys.* **40** (1964) 2109.
[b] R. Fort, J. Favre and L. Denivelle, *Bull. Soc. chim. France*, (1955) 534.

and are intermediates in the oxidation of alcohols by hypohalous acids (or by the halogen in the presence of Ag_2O or HgO). Methyl hypochlorite has been characterized spectroscopically in the vapour phase (Table 64).

Tertiary-alkyl hypochlorites[667] (and hypobromites[681]) are yellow liquids (and orange solids) at ambient temperatures, at least in the absence of light; illumination induces decomposition to a ketone and an alkyl halide:

$$RR'R''COX \xrightarrow{h\nu} RR'C{=}O + R''X$$

The ease of fragmentation has been studied as a function of the alkyl substituents[682]. Tertiary-butyl hypochlorite and hypobromite are efficient halogenating agents for aliphatic and allylic C–H bonds; the radical mechanism involves the t-butoxy radical as a chain-carrier[683]:

$$Me_3COX \xrightarrow{h\nu} Me_3CO\cdot + X\cdot$$
$$Me_3CO\cdot + RH \rightarrow Me_3COH + R\cdot$$
$$R\cdot + Me_3COX \rightarrow RX + Me_3CO\cdot$$

Inorganic chlorination by Me_3COCl has also been reported[684]:

$$HNF_2 + Bu^tOCl \rightarrow Bu^tOH + ClNF_2$$

Heterolysis of the O–X bond is a possible alternative in some reactions. Thus, tertiary-alkyl hypohalites add across olefinic double bonds in a polar fashion, e.g.

$$\text{cyclohexene} + Bu^tOBr \xrightarrow{MeOH} \text{2-bromo-1-methoxycyclohexane (OMe, Br)} + Bu^tOH$$

and can act as sources of positive halogen for substitution into aromatic compounds: Bu^tOCl is an excellent reagent for the oxidation of sulphides to sulphoxides without concomitant formation of sulphones[685].

$$R_2S \xrightarrow{Bu^tOCl} [R_2SCl]^+[Bu^tO]^- \xrightarrow[-Bu^tOH]{R'OH} R_2S(OR')Cl \xrightarrow[-R'Cl]{\Delta} R_2S{=}O$$

Fluoroalkyl Hypochlorites

These compounds are considerably more stable than the parent aliphatic esters; they are prepared either (1) by the elimination of HF between an alcohol and ClF[686],

$$ROH + ClF \xrightarrow{25°C} ROCl + HF$$

$$[R = (CF_3)_3C,\ CH_3(CF_3)_2C,\ (CF_3)_2CH \text{ or } CF_3CH_2]$$

or (2) by the addition at −20°C of ClF to a carbonyl compound in a reaction catalysed by CsF[583,687], BF_3 or AsF_5 [688]:

$$RR'C{=}O + ClF \rightarrow RR'FCOCl$$

$$[R = F,\ R' = F \text{ or } CF_3;\ R = CF_3,\ R' = CF_3 \text{ or } CF_2Cl]$$

---

681 C. Walling and A. Padwa, *J. Org. Chem.* **27** (1962) 2976.
682 C. Walling and A. Padwa, *J. Amer. Chem. Soc.* **85** (1963) 1593.
683 C. Walling and J. A. McGuinness, *J. Amer. Chem. Soc.* **91** (1969) 2053.
684 K. O. Christe, *Inorg. Chem.* **8** (1969) 1539.
685 C. R. Johnson and J. J. Rigau, *J. Amer. Chem. Soc.* **91** (1969) 5398.
686 D. E. Young, L. R. Anderson, D. E. Gould and W. B. Fox, *J. Amer. Chem. Soc.* **92** (1970) 2313.
687 C. J. Schack and W. Maya, *J. Amer. Chem. Soc.* **91** (1969) 2902.
688 D. E. Young, L. R. Anderson and W. B. Fox, *Inorg. Chem.* **9** (1970) 2602.

The action of $F_4S{=}O$ on ClF in the presence of CsF generates the related compound F_5SOCl[583,689]. CF_3OCl is a by-product in the fluorination of OCF_2 with ClF_3 adsorbed on γ-alumina[690], and may also be obtained from OCF_2 and Cl_2O[583].

Various spectroscopic techniques have been applied **to** characterize the compounds. Their thermal stabilities decrease in the order $CF_3OCl > C_2F_5OCl > (CF_3)_2CFOCl > SF_5OCl$; CF_3OCl decomposes only slowly at 150°C following a bipartite reaction path which involves the CF_3O radical:

$$CF_3OCl \rightarrow CF_3O\cdot + Cl\cdot$$
$$2CF_3O\cdot \rightarrow CF_3OOCF_3 \qquad (91\%)$$
$$CF_3O\cdot \rightarrow OCF_2 + F\cdot \qquad (9\%)$$

By contrast, photolysis of CF_3OCl affords OCF_2 as the only detectable fragment[691]. The compounds combine readily with free-radical sources (e.g. N_2F_4), and add to carbon monoxide and sulphur dioxide to form respectively chloroformates and chlorosulphates.

$$CF_3OCl + \tfrac{1}{2}N_2F_4 \xrightarrow{h\nu} CF_3ONF_2 + \tfrac{1}{2}Cl_2$$
$$CF_3OCl + CO \rightarrow CF_3OC(=O)Cl$$
$$CF_3OCl + SO_2 \rightarrow CF_3OSO_2Cl$$

Thiohypohalous Acids

Under carefully controlled conditions bromine reacts with H_2S in chloroform or dichloromethane[692]:

$$Br_2 + H_2S \rightarrow HBr + HSBr$$

HSBr can be "fixed" with NH_3 in the form of an ammonium salt, which is stable at low temperature but decomposes rapidly in aqueous solution,

$$NH_4SBr \rightarrow S + NH_4^+ + Br^-$$

or on warming to room temperature. HSI has been similarly obtained, and SBr^- is believed to be an intermediate in the oxidation by bromine of NaHS in ethereal solution.

(C) HALOUS ACIDS AND HALITES

Introduction

The acids HXO_2 and the anions XO_2^- (X = Cl, Br or I) are intermediates in the disproportionation of HOX and OX^- respectively, as well as in the oxidation of halides by halates. Chlorous acid, itself the least stable of the oxyacids of chlorine and known only in aqueous solution, is the best characterized of the halous acids; the existence of $HBrO_2$ and HIO_2 as other than transient species is still questionable. ClO_2^- and BrO_2^- persist in neutral or alkaline solution if impurities are absent, and stable salts containing these ions may be isolated; IO_2^- decomposes rapidly under all conditions.

689 C. J. Schack, R. D. Wilson, J. S. Muirhead and S. N. Cohz, *J. Amer. Chem. Soc.* **91** (1969) 2907.
690 R. Veyre, M. Quenault and C. Eyraud, *Compt. rend.* **268C** (1969) 1480.
691 K. O. Christe and D. Pilipovich, *J. Amer. Chem. Soc.* **93** (1971) 51.
692 M. Schmidt and I. Löwe, *Angew. Chem.* **72** (1960) 79.

Sodium chlorite is manufactured on a large scale, being extensively used as a bleaching agent (in acid solution) and as a source of chlorine dioxide for water purification and tallow bleaching.

Both in the solid state and in solution the decomposition of chlorites produces chlorine dioxide. Under normal conditions chlorite solutions do not evolve ClO_2 in dangerous amounts; however, explosive concentrations of ClO_2 may result if acid is dropped or spilled on to solid chlorites, or if the salts are heated. Appropriate precautions should be taken.

For reviews of the chemistry of the halogen(III) oxyacids and their salts the reader is referred to the following: chlorous acid and the chlorites[572,574,575b,576,693,694]; bromous acid and the bromites[575b,576,668,695]; iodous acid and the iodites[575b,576,616].

Preparation

Halous Acids

Chlorous acid is formed (together with $HClO_3$) in the decomposition of aqueous solutions of chlorine dioxide. Solutions of the acid were first obtained by reducing chloric acid with tartaric acid[576], but the best method of preparation is to treat suspensions of barium chlorite with sulphuric acid:

$$Ba(ClO_2)_2+H_2SO_4 \rightarrow BaSO_4+2HClO_2$$

Unstable solutions containing appreciable amounts of $HBrO_2$ [695] or HIO_2 [616] allegedly result from the reaction of aqueous silver salts with an excess of halogen; other acids (HX, HOX and HXO_3) are also present in these solutions, among which the halous acids are apparently recognized by their characteristic oxidizing properties.

Chlorites

Solutions from which alkali or alkaline earth chlorites may be crystallized are best formed by reducing ClO_2: both industrially[574] and in the laboratory, peroxides are convenient reducing agents (usually H_2O_2 in conjunction with a solution or suspension of the metal hydroxide or carbonate).

$$2ClO_2+O_2^{2-} \rightarrow 2ClO_2^-+O_2$$

Other reductants which have been used or suggested include organic matter, amines, nitrites, sulphur compounds, iodides and sodium amalgams.

Some powdered metals of intermediate electropositive character (e.g. Al, Cd or Zn) unite directly with chlorine dioxide in aqueous solution (p. 1370); many more chlorites may be made metathetically from $Ba(ClO_2)_2$ and the appropriate metal sulphate.

Bromites

Aqueous solutions containing BrO_2^- can be obtained by the controlled disproportionation of cold concentrated alkaline hypobromite solutions[695-697]; when the optimum bromite concentration is achieved, residual OBr^- is destroyed by the addition of ammonia

693 C. C. Addison, *Mellor's Comprehensive Treatise on Inorganic and Theoretical Chemistry*, Supplement II, Part I, pp. 569–576, Longmans, London (1956).

694 A. S. Chernyshev, V. V. Shtutser and N. G. Semenova, *Uspekhi Khim.* **25** (1956) 91.

695 P. J. M. Radford, *Bromine and its Compounds* (ed. Z. E. Jolles), pp. 159–162, Benn, London (1966).

696 H. Fuchs and R. Landsberg, *Z. anorg. Chem.* **372** (1970) 127.

697 M. Sédiey, Fr. Addn. 92,474 and 92,491 (1968); *Chem. Abs.* **71** (1969) 126609q, 126610h.

or acetone. From such solutions bromites of lithium, sodium, potassium and barium may be crystallized. Anhydrous $LiBrO_2$ [698] and $Ba(BrO_2)_2$ [699] have been prepared by dry reactions:

$$2LiBrO_3 + LiBr \xrightarrow{190^\circ} 3LiBrO_2$$

$$Ba(BrO_3)_2 \xrightarrow{250^\circ} Ba(BrO_2)_2 + O_2$$

Properties

Chlorous Acid

Chlorous acid is a moderately strong acid (Table 65). In concentrations of only a few grams per litre it soon decomposes at room temperature, discolouring as chlorine dioxide is formed (see below). Acid solutions of chlorite are potentially stronger oxidizing agents than are alkaline solutions.

TABLE 65. PROPERTIES OF AQUEOUS CHLOROUS ACID

ΔH_f° for $HClO_2$ (undissoc.)	$-12{\cdot}4$ kcal mol^{-1} [a]
ΔG_f° for $HClO_2$ (undissoc.)	$+1{\cdot}4$ kcal mol^{-1} [a]
S° for $HClO_2$ (undissoc.)	45·0 cal deg^{-1} mol^{-1} [a]
Dissociation constant, K_a	$1{\cdot}1 \times 10^{-2}$ [b]
Reduction potentials in acid solution[c]	
$E^\circ(ClO_3^-/HClO_2)$	$+1{\cdot}21$ volts
$E^\circ(ClO_2/HClO_2)$	$+1{\cdot}27$ volts
$E^\circ(HClO_2/HOCl)$	$+1{\cdot}64$ volts
$E^\circ(HClO_2/Cl^-)$	$+1{\cdot}57$ volts
Ultraviolet-visible absorption spectrum	d

[a] *Selected Values of Chemical Thermodynamic Properties*, N.B.S. Technical Note 270–3 (1968).
[b] Consensus of the following individual studies: (i) Potentiometric titration of ClO_2^- with HCl; $K_a = 1{\cdot}07 \times 10^{-2}$; G. F. Davidson, *J. Chem. Soc.* (1954) 1649. (ii) Spectrophotometric; $K_a = 1{\cdot}15 \times 10^{-2}$; D. Leonesi and G. Piantoni, *Ann. Chim. (Rome)*, **55** (1965) 668. (iii) pH titration of $Ba(ClO_2)_2$ with H_2SO_4; $K_a = 1{\cdot}10 \times 10^{-2}$; B. Barnett, Ph.D. Thesis, University of California (1935). (iv) Unspecified method; $K_a = 1{\cdot}01 \times 10^{-2}$; M. W. Lister, *Canad. J. Chem.* **30** (1952) 879.
[c] A. G. Sharpe, *Halogen Chemistry* (ed. V. Gutmann), Vol. 1, p. 13, Academic Press (1967).
[d] D. Leonesi and G. Piantoni, *Ann. Chim. (Rome)*, **55** (1965) 668.

Halites

The chlorite and bromite ions have been studied spectroscopically in solution (Table 66) and in crystalline salts (Table 67); $AgClO_2$ and NH_4ClO_2 have been investigated crystallographically. The O–Cl–O angle (111°) is close to the expected tetrahedral value (cf. ClO_3^-, 110° and ClO_2, 118°).

There are large and unexplained discrepancies between the vibrational frequencies attributed to BrO_2^- in the infrared spectra of the solids $LiBrO_2$ and $Ba(BrO_2)_2$ and those observed in the Raman and infrared spectra of BrO_2^- in aqueous solution and in $NaBrO_2,3H_2O$; the latter study is more plausible, since the frequencies observed for the lithium and barium salts are high in comparison with those of ClO_2^-, and yield on analysis an unusually large valence force constant for Br–O stretching (4·75 mdyne Å^{-1})[699].

Decomposition of Chlorites

Alkaline solutions of sodium chlorite remain unchanged for up to one year if light is

[698] P. Hagenmuller and B. Tanguy, *Compt. rend.* **260** (1965) 3974.
[699] B. Tanguy, B. Frit, G. Turrell and P. Hagenmuller, *Compt. rend.* **264C** (1967) 301.

TABLE 66. PROPERTIES OF AQUEOUS HALITE IONS

	ClO_2^-	BrO_2^-
Colour	Colourless	Yellow
Thermodynamic properties at 298°K		
$\Delta H_f°$ (kcal mol^{-1})	$-15{\cdot}9$[a]	
$\Delta G_f°$ (kcal mol^{-1})	$+4{\cdot}1$[a]	
$S°$ (cal deg^{-1} mol^{-1})	24·2[a]	
Reduction potentials in alkaline solution[b]		
$E°(XO_3^-/XO_2^-)$ (volts)	+0·33	
$E°(XO_2/XO_2^-)$ (volts)	+1·16	
$E°(XO_2^-/XO^-)$ (volts)	+0·66	
$E°(XO_2^-/X^-)$ (volts)	+0·77	
Raman spectrum	c	d
Fundamental frequencies (cm^{-1})		
$\nu_1(a_1)$	790	709
$\nu_2(a_1)$	400	324
$\nu_3(b_1)$	[840][e]	680
Force constants (mdyne $Å^{-1}$)	f	d
f_r	4·26	4·2
f_{rr}	0·11	
f_θ/r^2	0·52	
$f_{r\theta}/r$	0·02	
Infrared spectrum		d
Ultraviolet-visible absorption spectrum	g	h
λ_{max}(nm)	260	290

[a] *Selected Values of Chemical Thermodynamic Properties*, N.B.S. Technical Note 270–3 (1968).
[b] A. G. Sharpe, *Halogen Chemistry* (ed. V. Gutmann), Vol. 1, p. 13, Academic Press (1967).
[c] J. P. Mathieu, *Compt. rend.* **234** (1952) 2272.
[d] J. C. Evans and G. Y.-S. Lo, *Inorg. Chem.* **6** (1967) 1483.
[e] Estimated from spectra of solid chlorites.
[f] H. Siebert, *Anwendungen der Schwingungsspektroskopie in der Anorganischen Chemie*, p. 50, Springer-Verlag (1966).
[g] W. Buser and H. Hänisch, *Helv. Chim. Acta*, **35** (1952) 2547; H. L. Friedman, *J. Chem. Phys.* **21** (1953) 319; D. Leonesi and G. Piantoni, *Ann. Chim. (Rome)*, **55** (1965) 668.
[h] O. Amichai and A. Treinin, *J. Phys. Chem.* **74** (1970) 830.

excluded; even with boiling no decomposition occurs. The stability decreases as the pH falls, cold neutral solutions being stable in the dark, but degrading slowly if heated. Acid solutions decompose at measurable rates, especially if the pH is less than 4.

The decomposition of chlorous acid may be expressed by three equations, none of which on its own accurately describes the reaction:

(i) $$3HClO_2 \rightarrow 3H^+ + 2ClO_3^- + Cl^- \qquad \Delta G° = -33{\cdot}2 \text{ kcal mol}^{-1}$$

(ii) $$5HClO_2 \rightarrow 4ClO_2 + Cl^- + H^+ + 2H_2O \qquad \Delta G° = -34{\cdot}5 \text{ kcal mol}^{-1}$$

(iii) $$HClO_2 \rightarrow Cl^- + H^+ + O_2 \qquad \Delta G° = -29{\cdot}5 \text{ kcal mol}^{-1}$$

Oxygen-evolution (iii) is a minor process, accounting for less than 3% of the chlorite consumed. The simple disproportionation to chlorate and chloride (i) is not observed, and in 2 M perchloric acid the stoichiometry of the disproportionation of $HClO_2$ initially

TABLE 67. PROPERTIES OF SOLID HALITES

X-ray diffraction

NH_4ClO_2[a] (at −35°C) r(Cl–O) = 1·57±0·03 Å; θ_{O-Cl-O} = 110·5±1·4°

$AgClO_2$[b] r(Cl–O) = 1·55±0·05 Å; θ_{O-Cl-O} = 111±3°

Vibrational spectra

		$\nu_1(a_1)$ (cm^{-1})	$\nu_2(a_1)$ (cm^{-1})	$\nu_3(b_1)$ (cm^{-1})
$NaClO_2,3H_2O$	Raman[c]	786	402	844
	Infrared[f]			860
$NaBrO_2,3H_2O$	Raman[d]	720	~330	680
$LiBrO_2$	Infrared[e]	775	400	800
$Ba(BrO_2)_2$	Infrared[e]	775	400	800

^{35}Cl quadrupole resonance spectra[g]

Resonance frequencies: $AgClO_2$, 54·08 MHz; $NaClO_2$, 51·82 MHz.

X-ray spectra: $NaClO_2$ Emission[h] Fluorescence[j]
E.S.C.A.[i]

[a] R. B. Gillespie, R. A. Sparks and K. N. Trueblood, *Acta Cryst.* **12** (1959) 867.
[b] J. Cooper and R. E. Marsh, *Acta Cryst.* **14** (1961) 202.
[c] J. P. Mathieu, *Compt. rend.* **234** (1952) 2272.
[d] J. C. Evans and G. Y.-S. Lo, *Inorg. Chem.* **6** (1967) 1483.
[e] B. Tanguy, B. Frit, G. Turrell and P. Hagenmuller, *Compt. rend.* **264C** (1967) 301.
[f] Includes data for $AgClO_2$ and $Pb(ClO_2)_2$; C. Duval, J. Lecomte and J. Morandat, *Bull. Soc. chim. France* (1951) 745.
[g] J. L. Ragle, *J. Chem. Phys.* **32** (1960) 403.
[h] V. I. Nefedov, *J. Struct. Chem.* **8** (1967) 919.
[i] A. Fahlman, R. Carlsson and K. Siegbahn, *Arkiv. Kemi*, **25** (1966) 301.
[j] J. A. Bearden, *Rev. Mod. Phys.* **39** (1967) 86; D. S. Urch, *J. Chem. Soc.* (*A*) (1969) 3026.

approximates to equation (iv)[700]:

(iv) $$4HClO_2 \rightarrow 2ClO_2+ClO_3^-+Cl^-+2H^++H_2O$$

However, reaction (ii) is catalysed by Cl^-, so that the rate of formation of ClO_3^- relative to ClO_2 decreases as the reaction proceeds; if appreciable amounts of chloride are present, reaction (ii) predominates. A very small amount of ClO_2 also results from the interaction of chlorous acid with chlorate[701], viz.

$$HClO_2+H^++ClO_3^- \rightarrow 2ClO_2+H_2O$$

The overall rate law has the form

$$\frac{-d[HClO_2]}{dt} = k_1[HClO_2]^2+\frac{k_2[HClO_2][Cl^-]^2}{K+[Cl^-]}$$

and the following mechanisms have been proposed.

(a) *Uncatalysed reaction*

$$HClO_2+HClO_2 \rightarrow HOCl+H^++ClO_3^- \text{ (rate-determining)}$$

$$HOCl+HClO_2 \rightarrow \left[Cl{-}Cl\genfrac{}{}{0pt}{}{\diagup O}{\diagdown O}\right]+H_2O$$

$$HOCl+H^++Cl^- \rightleftharpoons Cl_2+H_2O$$

$$Cl_2+HClO_2 \rightarrow \left[Cl{-}Cl\genfrac{}{}{0pt}{}{\diagup O}{\diagdown O}\right]+H^++Cl^-$$

$$\left[Cl{-}Cl\genfrac{}{}{0pt}{}{\diagup O}{\diagdown O}\right]+H_2O \rightarrow Cl^-+ClO_3^-+2H^+$$

$$2\left[Cl{-}Cl\genfrac{}{}{0pt}{}{\diagup O}{\diagdown O}\right] \rightarrow Cl_2+2ClO_2$$

[700] R. G. Kieffer and G. Gordon, *Inorg. Chem.* **7** (1968) 235, 239.
[701] H. Taube and H. Dodgen, *J. Amer. Chem. Soc.* **71** (1949) 3330.

(b) *Chloride ion catalysed reaction*

$$HClO_2 + Cl^- \rightleftharpoons [HCl_2O_2^-]$$
$$[HCl_2O_2^-] + Cl^- \rightarrow \text{products (rate-determining)}$$

At low hydrogen ion concentrations the following path may also assume importance[701]:

$$HClO_2 + ClO_2^- \rightarrow [HClO_2\text{–}ClO_2^-] \rightarrow \text{products}$$

Chlorous acid solutions can be stabilized by the addition of H_2O_2, which regenerates ClO_2^- from chlorine dioxide.

Pyrolysis of sodium chlorite proceeds with disproportionation to chlorate and chloride[702]:

$$3NaClO_2 \xrightarrow{260°C} 2NaClO_3 + NaCl$$

While disproportionation is also the process initially observed on heating chlorites of heavier metals, many of these salts decompose explosively at higher temperatures or on rapid heating, probably *via* the evolution of ClO_2 [703].

Decomposition of Bromites and Iodites

The mechanisms by which aqueous BrO_2^- and IO_2^- ions decompose have been much less fully studied than those of the chlorite reactions; though the primary processes are presumably analogous to those in the disproportionation of chlorite, there is some evidence for the occurrence in alkaline media of the simple reaction[676,704]

$$XO_2^- + XO_2^- \rightarrow XO_3^- + XO^- \qquad (X = Br \text{ or } I)$$

There are no reports that heating solid bromites occasions disproportionation. $LiBrO_2$ (at 225°C)[705] and $Ba(BrO_2)_2$ (at 220°C)[699] lose oxygen in forming the metal bromide, whereas after dehydration $Mg(BrO_2)_2,6H_2O$ decomposes to magnesium oxide[706]. Crystalline bromites are stable at room temperature, even if illuminated.

Reactions

Fluorination of sodium chlorite produces low yields of $ClOF_3$ [641]. Chlorine and hypochlorous acid oxidize chlorites to ClO_2 or ClO_3^- or to a mixture of both. The reaction of chlorine with solid sodium chlorite comprises an expedient synthesis of ClO_2.

$$2NaClO_2 + Cl_2 \rightarrow 2NaCl + 2ClO_2$$

In acidic solution chlorine reacts more rapidly and produces less chlorate than does hypochlorous acid, although the product distribution also depends on the ratio of the reactants; excess chlorite encourages the formation of ClO_2 [707]. The reaction

$$HOCl + HClO_2 \rightarrow ClO_3^- + Cl^- + 2H^+$$

is the second stage in the disproportionation of hypochlorites (p. 1405), and similar reactions have been postulated between the other halites and hypohalous acids. Chlorous acid is oxidized by permanganate to chloric acid

702 F. Solymosi, T. Bansagi and E. Berenyi, *Magy. Kem. Foly.* **74** (1968) 23; *Chem. Abs.* **68** (1968) 51465s.
703 F. Solymosi, *Z. physik. Chem., N.F.* **57** (1968) 1.
704 C. H. Cheek and V. J. Linnenbom, *J. Phys. Chem.* **67** (1963) 1856.
705 R. Diament, *Compt. rend.* **264C** (1967) 589.
706 R. Diament and M. Sédiey, *Compt. rend.* **268C** (1969) 1243.
707 F. Emmenegger and G. Gordon, *Inorg. Chem.* **6** (1967) 633.

In acid solution chlorites and bromites oxidize iodide ion to iodine, which is further oxidized to IO_3^-. Bromites rapidly oxidize cold alkaline arsenite solutions, with which chlorites react only slowly. The oxidation of transition metal ions by chlorite has also been explored[708].

Uses

Sodium chlorite is widely used in acid solution as a bleaching agent for textiles[574]; typically pH values of 3–5 have been used. The nature of the bleaching reactions is uncertain; Cl_2O_3 is a possible active intermediate, although oxidation of coloured impurities by oxygen and chlorine dioxide (decomposition products of $HClO_2$) is also important. Recent improvements in the efficiency of chlorite bleaching have involved the use of solutions of relatively high pH (to hinder the escape of ClO_2) and of various activators.

Alkaline bromite solutions are used in desizing cellulose fabrics; this treatment removes starches which are added to fabrics to facilitate weaving but which interfere with the penetration of reagents used in subsequent bleaching and dyeing processes.

(D) HALIC ACIDS AND HALATES

Introduction

In 1658 J. R. Glauber claimed that he could convert hydrochloric acid into "nitric acid", and common salt into "saltpetre"; the substances he produced were almost certainly chloric acid and sodium chlorate[576]. Towards the end of the eighteenth century several workers investigated crystalline solids generated by the interaction of the recently discovered chlorine with alkaline materials; the chemical constitution of these salts (chlorates) was recognized by Berthollet. In 1814 Gay Lussac obtained solutions of chloric acid by treating barium chlorate with sulphuric acid.

The oxidation of iodine, suspended in water, to iodic acid was reported in 1813 by Davy (who used ClO_2) and by Gay Lussac (who used chlorine); the latter author also investigated the disproportionation of iodine (forming iodide and iodate) in alkaline media. Barium bromate and bromic acid were described by Balard in 1826. Calcium iodate occurs naturally in the form of the minerals lautarite and dietzeite, found, for example, in the Caliche deposits of Chile (see p. 1127).

The halates are strong oxidizing agents in acid solution, and in so acting are usually reduced to the corresponding halide ions. As judged by the redox potentials (Table 69), their oxidizing powers decrease in the order bromate $\gtrsim$ chlorate $>$ iodate, but the rates of reaction follow the sequence iodate $>$ bromate $>$ chlorate. The couples (BrO_3^-/Br^-) and ($BrO_3^-/\frac{1}{2}Br_2$) have approximately the same potentials in acid solution as the (MnO_4^-/Mn^{2+}) couple (+1·51 volts). The solid compounds form spontaneously inflammable or explosive mixtures with oxidizable material, and acidified chlorates are liable to release ClO_2 in dangerously high local concentrations. Chlorates are important precursors in the manufacture of chlorine dioxide and perchlorates, and are used in making matches and fireworks; bromates and iodates have only minor commercial applications, but are useful analytical reagents.

[708] G. Gordon and F. Feldman, *Inorg. Chem.* 3 (1964) 1728; B. Z. Shakhashiri and G. Gordon, *J. Amer. Chem. Soc.* **91** (1969) 1103.

There have been several reviews of chloric acid and the chlorates[572,575b,576,709,710], bromic acid and the bromates[575b,576,711,712] and iodic acid and the iodates[575b,576,624,713].

Formation and Preparation

The best routes to many chlorates, bromates and iodates (not excluding the acids) involve metathetical or neutralization reactions in aqueous solution. Lists of individual reactions will be found elsewhere[572,576,709,711,713], but a general method involves the action of a sulphate on the appropriate barium halate:

$$Ba(XO_3)_2+M_2SO_4 \rightarrow 2MXO_3+BaSO_4$$

The anions themselves are usually produced either (i) by disproportionation of the parent halogen in alkaline solution or (ii) by chemical or electrolytic oxidation of the parent halogen or of the corresponding halide ion. Only the most important processes are described below.

Halic Acids

Treating $Ba(ClO_3)_2$ or $Ba(BrO_3)_2$ with the stoichiometric amount of sulphuric acid remains the best means for obtaining solutions of $HClO_3$ or $HBrO_3$.

$$Ba(XO_3)_2+H_2SO_4 \rightarrow BaSO_4+2HXO_3$$

Although amenable to the same mode of preparation, iodic acid is more readily synthesized by the oxidation of an aqueous suspension of iodine, either electrolytically or chemically (with fuming nitric acid); crystallization of moderately acid solutions deposits α-HIO_3, but HI_3O_8 emerges from solution in concentrated (> 5 M) acid.

Laboratory Synthesis of Halates

The disproportionation of the halogens in basic solution offers a facile route to halate ions:

$$3X_2+6OH^- \rightarrow XO_3^- +5X^- +3H_2O$$

Disproportionation is speeded at higher temperatures, so that the free halogen is added to a warm (50–80°C) solution of a hydroxide or a carbonate; in the case of bromates, boiling of the solution is recommended to hasten the conversion of intermediate hypobromites and bromites to bromate. Fractional crystallization readily separates the less soluble halate from the accompanying halide. It is also possible to chlorinate suspensions of metal oxides or hydroxides.

To synthesize bromates and iodates, oxidation of the halogen or halide is often more economical than disproportionation of the halogen. Thus bromides are transformed into bromates by hypochlorites in aqueous solution (conveniently executed by passing chlorine into an alkaline bromide solution), or by molten chlorates. Aqueous chlorate oxidizes bromine only slowly.

[709] C. C. Addison, *Mellor's Comprehensive Treatise on Inorganic and Theoretical Chemistry*, Supplement II, Part I, pp. 576–596, Longmans, London (1956).
[710] T. W. Clapper and W. A. Gale, *Kirk-Othmer Encyclopedia of Chemical Technology*, 2nd edn., Vol. 5, pp. 50–61, Interscience, New York (1964).
[711] B. Cox, *Mellor's Comprehensive Treatise on Inorganic and Theoretical Chemistry*, Supplement II, Part I, pp. 753–773, Longmans, London (1956).
[712] P. J. M. Radford, *Bromine and its Compounds* (ed. Z. E. Jolles), pp. 162–178, Benn, London (1966).
[713] G. J. Hills, *Mellor's Comprehensive Treatise on Inorganic and Theoretical Chemistry*, Supplement II, Part I, pp. 874–895, Longmans, London (1956).

Iodates may be prepared directly by heating alkali-metal iodides to *ca.* 600°C under high pressures of oxygen. Chlorates (or bromates) oxidize iodine to the corresponding iodates, e.g.

$$I_2+2NaClO_3 \rightarrow 2NaIO_3+Cl_2$$

Hypochlorite may not be used since this causes further oxidation of iodine(V) to iodine-(VII)[714]. Iodates also result from the thermal decomposition of periodates (p. 1457).

Manufacture of Chlorates and Bromates

Chemical methods for manufacturing chlorates and bromates involve the halogenation of an alkali (as described above). For chlorate this process has largely been superseded by electrolytic oxidation of chloride, but some bromate is still produced by dissolving bromine in a warm alkaline solution or suspension, since the bromide simultaneously formed is a desirable by-product.

Brine is the common electrolyte in cells for chlorate production[710,715]. No diaphragm separates the electrodes (a steel cathode and, usually, a graphite anode) which are situated close together. Under optimum conditions present cells have a current efficiency of 80–90% for the overall cell process

$$Cl^- + 3H_2O \rightarrow ClO_3^- + 3H_2$$

The primary electrode reactions are oxidation of Cl^- to Cl_2 at the anode:

$$Cl^- \rightarrow \tfrac{1}{2}Cl_2 + e$$

and production of hydroxide ions at the cathode:

$$H_2O + e \rightarrow \tfrac{1}{2}H_2 + OH^-$$

Diffusion in the cell allows the chlorine to disproportionate:

$$Cl_2 + 2OH^- \rightarrow Cl^- + OCl^- + H_2O$$

Chlorate then arises either by disproportionation of hypochlorite (p. 1405) or by direct oxidation of OCl^- at the anode. Losses occur through extraneous oxidation and reduction processes at the electrodes, by decomposition of hypochlorite (to chloride and oxygen), and by consumption of the anode.

The electrochemical production of bromates follows similar lines[712]. However, the electrolyte is now a solution of bromine in alkali, so that some OBr^- is initially present, and anodes coated with PbO_2 are employed; reduction of OBr^- at the cathode is prevented by the addition of small amounts of dichromate.

Properties and Structural Chemistry

Halic Acids

$HClO_3$ and $HBrO_3$ can be obtained only in aqueous solution. Dilute solutions decompose on boiling, but they may be concentrated to syrupy liquids under vacuum at room temperature. The limiting concentrations for stability are: $HClO_3$ *ca.* 40%, $HBrO_3$ *ca.* 50%; decomposition paths are discussed below. Iodic acid solutions crystallize on evaporation, and heating solid HIO_3 occasions only dehydration, producing ultimately its anhydride, I_2O_5.

[714] H. H. Willard, *Inorganic Syntheses*, Vol. I (ed. H. S. Booth), p. 168, McGraw-Hill (1939).
[715] J. C. Schumacher, *J. Electrochem. Soc.* **116** (1969) 68C.

Both chloric and bromic acids are strong acids ($pK_a \lesssim 0$); by contrast, iodic acid is only moderately strong (pK_a, 0·8), and the presence of the undissociated molecule may be detected in solution by spectroscopic methods (Table 68).

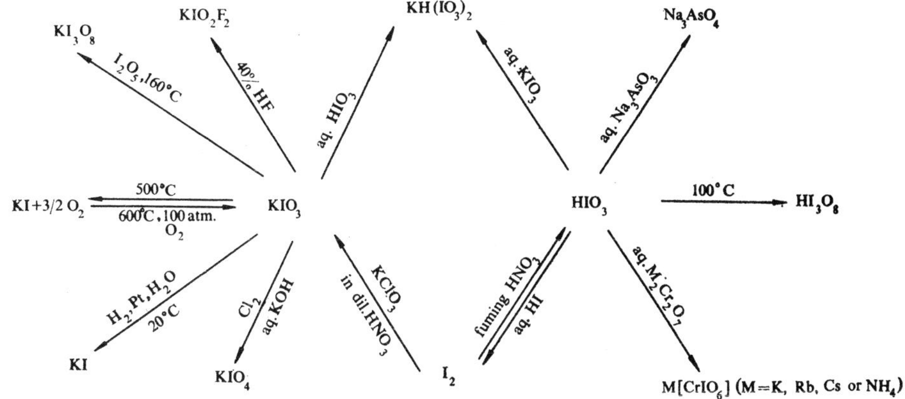

SCHEME 10. Reactions of some iodates.

TABLE 68. PROPERTIES OF IODIC ACID AND ANHYDRIODIC ACID HI_3O_8

	HIO_3	HI_3O_8
Melting point (°C)[a]	*ca.* 110	
Dehydration temperature (°C)[a]	*ca.* 100 (→ HI_3O_8)	*ca.* 200 (→ I_2O_5)
Thermodynamic functions at 298°K[b]		
$\Delta H_f°$(c) (kcal mol^{-1})	−55·4[c]	−92·2[d]
$\Delta H_f°$(undissoc,aq) (kcal mol^{-1})	−50·5[d]	
$\Delta G_f°$(undissoc,aq) (kcal mol^{-1})	−31·7[d]	
$S°$(undissoc,aq) (cal deg^{-1} mol^{-1})	39·9[d]	
Acidity		
Dissociation constant, K_a	0·157[e]	
Hammett acidity function, H_0	expt[f]	
	calc[g]	
Crystal structure		
X-ray diffraction	α-HIO_3[h]	[i]
Neutron diffraction	α-HIO_3[j]	
	Contains $HOIO_2$ molecules	Contains $HOIO_2$ and I_2O_5 molecules
Dimensions of $HOIO_2$ molecule	[h]	[i]
r(I=O) (Å)	1·80, 1·82	1·79, 1·81
r(I−OH) (Å)	1·89	1·90
θ_{O-I-OH}	95·7°, 98·2°	94·2°, 95·4°
$\theta_{O=I=O}$	101·4°	99°
Spectroscopic investigations		
Raman spectrum	solid[k,q]	solid[k]
	aq soln[l]	
Infrared spectrum	solid[a,k,m,n]	solid[a,k]
^{127}I quadrupole resonance spectrum	solid[o]	
e^2Qq(MHz)	1126·9	
η	0·4505	
Mass spectrum	vapour[p]	

Table 68 (*cont.*)

[a] K. Selte and A. Kjekshus, *Acta Chem. Scand.* **22** (1968) 3309.

[b] For redox potentials see Table 69.

[c] Estimated from the heat of reaction of HIO_3 with aqueous N_2H_4; P. B. Howard and H. A. Skinner, *J. Chem. Soc.* (*A*) (1967) 269.

[d] *Selected Values of Chemical Thermodynamic Properties*, p. 40, N.B.S. Technical Note 270–3 (1968).

[e] Conductimetric, potentiometric and kinetic-salt effect investigations of solutions $<0{\cdot}1$M; A. D. Pethybridge and J. E. Prue, *Trans. Faraday Soc.* **63** (1967) 2019; includes critical survey of earlier measurements of K_a for HIO_3.

[f] Measured for solutions <6 M using *o*- and *p*-nitroaniline as indicators; J. G. Dawber, *J. Chem. Soc.* (1965) 4111.

[g] J. G. Dawber, *J. Chem. Soc.* (*A*) (1968) 1532.

[h] Error limits for distances $\pm 0{\cdot}04$ Å, for angles $\pm 5°$; M. T. Rogers and L. Helmholz, *J. Amer. Chem. Soc.* **63** (1941) 278.

[i] Error limits for distances $\pm 0{\cdot}01$ Å, for angles $\pm 0{\cdot}6°$; Y. D. Feikema and A. Vos, *Acta Cryst.* **20** (1966) 769.

[j] r(I—OH) = $1{\cdot}899 \pm 0{\cdot}011$ Å, r(I=O) = $1{\cdot}780 \pm 0{\cdot}010$ Å, $1{\cdot}816 \pm 0{\cdot}010$ Å; B. S. Garrett, U.S. Atomic Energy Comm. ORNL-1745 (1954); *Chem. Abs.* **49** (1955) 5064b.

[k] P. M. A. Sherwood and J. J. Turner, *Spectrochim. Acta*, **26A** (1970) 1975.

[l] G. C. Hood, A. C. Jones and C. A. Reilly, *J. Phys. Chem.* **63** (1959) 101.

[m] Includes data for HIO_3 and DIO_3; W. E. Dasent and T. C. Waddington, *J. Chem. Soc.* (1960) 2429.

[n] L. Couture, M. Krauzman and J. P. Mathieu, *Compt. rend.* **269B** (1969) 1278.

[o] Single crystal study, including Zeeman effect; R. Livingston and H. Zeldes, *J. Chem. Phys.* **26** (1957) 351.

[p] M. H. Studier and J. L. Huston, *J. Phys. Chem.* **71** (1967) 457.

[q] Includes data for HIO_3 and DIO_3; J. R. Durig, O. D. Bonner and W. H. Breazeale, *J. Phys. Chem.* **69** (1965) 3886.

Investigations of the system I_2O_5–H_2O have shown two modifications of iodic acid and the presence of an intermediate phase, anhydriodic acid HI_3O_8 [618]. Both α-HIO_3 and β-HIO_3 are orthorhombic, differing only in the ratio of the crystallographic axes. Single-crystal diffractometric studies (summarized in Table 68) reveal $HOIO_2$ molecules in α-HIO_3, and both $HOIO_2$ and I_2O_5 molecules in HI_3O_8 (which is therefore to be formulated as HIO_3,I_2O_5); in both compounds there is evidence of hydrogen-bonding and of significant I···O intermolecular contacts (at 2·5–3·0 Å) which result in distorted octahedral coordination about the unique iodine atom in α-HIO_3 and about two of the iodine atoms in HI_3O_8. The third iodine atom in HI_3O_8 has 4 intermolecular I···O contacts, giving irregular hepta-coordination.

Polymerization of Iodic Acid

The physical (conductimetric and potentiometric) properties of aqueous iodic acid solutions are consistent with the formation in moderately concentrated solution of the anion $[H(IO_3)_2]^-$ [716], having a stability constant of 4 l mol^{-1} as defined by the equation

$$HIO_3 + IO_3^- \rightleftharpoons [H(IO_3)_2]^-$$

The behaviour of the acidity function of iodic acid in concentrated solution indicates further association to furnish $[IO_3(HIO_3)_n]^-$ ($n > 1$)[717].

Hydrogen biiodate and dihydrogen triiodate salts, $M^IH(IO_3)_2$ and $M^IH_2(IO_3)_3$, may be crystallized from solutions containing M^IIO_3 and excess HIO_3 [718]. $KH(IO_3)_2$, which is used as an alkalimetric standard, suffers dehydration at 105°C; infrared studies of its

716 A. D. Pethybridge and J. E. Prue, *Trans. Faraday Soc.* **63** (1967) 2019.

717 J. G. Dawber, *J. Chem. Soc.* (1965) 4111; *ibid.* (*A*) (1968) 1532.

718 S. B. Smith, *J. Amer. Chem. Soc.* **69** (1947) 2285.

structure have proved inconclusive[719], while preliminary crystallographic data have been reported for the α- and γ-modifications[720]. Heating these hydrogen iodates produces salts such as $K_2H_2I_6O_{17}$, which contains I–O–I bonds[721].

Chlorates, Bromates and Iodates

Alkali-metal chlorates and bromates crystallize with distorted NaCl-type lattices; however, the corresponding iodates adopt perovskite structures and display piezoelectric properties. The solubility of these compounds in water decreases in the order chlorate > bromate > iodate. The iodates of some tetravalent metal ions (e.g. Ce, Zr, Hf and Th) are sufficiently insoluble to afford a useful means of separation. Double salts between alkali-metal iodates and halides have also been investigated [e.g. $K_3Cl(IO_3)_2$]. Iodates form complex polyanions with the oxyanions of other metallic and non-metallic elements, e.g. sulphates, selenates, tellurates, molybdates, tungstates, chromates and vanadates, and also iodato-complexes with some metal ions, e.g. $[M(IO_3)_6]^{2-}$ (M = Mn, Ti and Pb).

As expected with the presence of a non-bonding electron pair on the halogen atom, the XO_3^- anions defined by diffractometric techniques in the salts listed in Table 70 are pyramidal, and usually have C_{3v} symmetry within the experimental error. Owing to the difficulty encountered in fixing oxygen atom positions, many early studies of bromates and iodates yielded only imprecise parameters which have not been included in Table 70. The ClO_3^- and BrO_3^- ions are seen to be rather flat pyramids with O–X–O angles of *ca.* 106°; however, the iodate ion has smaller bond angles (*ca.* 97°), possibly indicating that there is less *s*-character in the I–O bonds. As might be anticipated, the X–O bond lengths increase from chlorate to bromate to iodate. The $HOIO_2$ molecules in α-iodic acid and HI_3O_8 have two I–O bonds close in length to those of the iodate ion (*ca.* 1·80 Å), but the I–OH bonds are longer (1·90 Å); the bond angles are all close to 100° (Table 68).

Significant intermolecular I···O interactions are found in crystalline iodates and iodic acid, the distances (2·5–3·3 Å) being appreciably less than the sum of the van der Waals' radii of iodine and oxygen (3·55 Å); such interactions give rise to a trigonally distorted octahedral environment about the iodine in α-HIO_3, $LiIO_3$, NH_4IO_3 and $Ce(IO_3)_4,H_2O$ and about two of the iodine atoms in HI_3O_8, to hepta-coordination in $Sr(IO_3)_2,H_2O$ and (about one of the iodine atoms) in HI_3O_8, and a square-antiprism of eight oxygens about the iodine in $NaIO_3$ and $Ce(IO_3)_4$. The high viscosity of aqueous iodate solutions and the low mobility found for the IO_3^- ion imply that the ion is strongly solvated in solution[722].

Infrared and Raman spectroscopic results confirm that the XO_3^- ions have in solution essentially the same pyramidal configuration as in solid derivatives; the spectroscopic features of the XO_3^- anion in simple halates are somewhat modified by the crystal environment of the anion. However, the spectra of some bromates and iodates [including $M(IO_3)_2$ (M = Pb or Hg), $K_2M(IO_3)_6$ (M = Pb, Ti or Mn), $Fe(IO_3)_3$, $[Co(NH_3)_3(H_2O)(IO_3)_2]IO_3$ and $M(BrO_3)_3,4H_2O$ (M = La, Pr or Nd)] display bands attributable to stretching motions of $M–O–XO_2$ units rather than to the four fundamentals characteristic of the isolated XO_3^- anion. The salt NH_4CrIO_6 contains the anion $[O_3CrOIO_2]^-$, composed of a CrO_4 tetrahedron and an IO_3 pyramid sharing one corner; its dimensions are: $r(Cr–O_{term})$ = 1·64 Å; $r(Cr–O_{bridge})$ = 1·90 Å; $r(I–O_{term})$ = 1·83 Å; $r(I–O_{bridge})$ = 2·06 Å; ∠I–O–Cr = 118° (see Table 70, ref. m).

719 W. E. Dasent and T. C. Waddington, *J. Chem. Soc.* (1960) 2429.
720 G. Argay, I. Naray-Szabo and E. Peter, *J. Therm. Anal.* **1** (1969) 413.
721 T. Dupuis, *Mikrochim. Acta,* (1962) 289.
722 H. A. Bent, *Chem. Rev.* **68** (1968) 615.

TABLE 69. PROPERTIES OF HALATE IONS IN AQUEOUS SOLUTION

	ClO_3^-	BrO_3^-	IO_3^-
Thermodynamic functions at 298°K			
$\Delta H_f°$(kcal mol^{-1})[a]	−24·87	−15·95	−52·51
$\Delta G_f°$(kcal mol^{-1})[a]	−1·90	+4·55	−30·20
$S°$(cal deg^{-1} mol^{-1})[a]	38·8	38·6	28·3
Reduction potentials			
Acid solution			
$E°(XO_4^-/XO_3^-)$(volts)[a]	+1·23	+1·76	
$E°(H_5XO_6/XO_3^-)$(volts)[b]			ca. +1·7
$E°(XO_3^-/XO_2)$(volts)[b]	+1·15		
$E°(XO_3^-/\frac{1}{2}X_2)$(volts)[a,b]	+1·46	+1·51	+1·20
$E°(XO_3^-/X^-)$(volts)[a,b]	+1·44	+1·44	+1·09
Alkaline solution			
$E°(XO_4^-/XO_3^-)$(volts)[a,b]	+0·40	+0·93	+0·81
$E°(XO_3^-/\frac{1}{2}X_2)$(volts)[a,b]	+0·47	+0·52	+0·20
$E°(XO_3^-/X^-)$(volts)[a,b]	+0·62	+0·61	+0·29
Spectra, etc.			
Raman spectrum	c,d,e	d,f,g	d,e,h
Infrared spectrum	f		
Fundamental frequencies (cm^{-1})	c,d	d,g	e
$\nu_1(a_1)$	935	806	779
$\nu_2(a_1)$	610	421	390
$\nu_3(e)$	982	836	826
$\nu_4(e)$	479	356	330
Force constants[i]			
f_r(mdyne Å^{-1})	5·87	5·28	5·48
f_θ/r^2(mdyne Å^{-1})	1·01	0·64	0·55
Ultraviolet-visible spectrum[j]			
λ_{max}(nm)	<200	195	195
Nmr spectrum			
^{17}O chemical shift (ppm with respect to H_2O)	−287[k]	−297[k]	
^{35}Cl chemical shift (ppm with respect to Cl^-)	−1050[l]		
Ionic mobility at 25°C, Λ_0 (cm^2 ohm^{-1} g-ion^{-1})	64·58[m]	55·8[m]	40·54[n]

[a] G. K. Johnson, P. N. Smith, E. H. Appelman and W. N. Hubbard, *Inorg. Chem.* **9** (1970) 119.

[b] A. G. Sharpe, *Halogen Chemistry* (ed. V. Gutmann), Vol. 1, p. 13, Academic Press (1967).

[c] C. Rocchiccioli, *Ann. Chim.* Series 13, **5** (1960) 999.

[d] G. W. Chantry and R. A. Plane, *J. Chem. Phys.* **34** (1961) 1268.

[e] S. T. Shen, Y. T. Yao and T. Wu, *Phys. Rev.* **51** (1937) 235.

[f] M. Rolla, *Gazz. Chim. Ital.* **69** (1939) 779.

[g] C. Rocchiccioli, *Compt. rend.* **249** (1959) 236.

[h] G. C. Hood, A. C. Jones and C. A. Reilly, *J. Phys. Chem.* **63** (1959) 101.

[i] H. Siebert, *Anwendungen der Schwingungsspektroskopie in der Anorganischen Chemie*, p. 58, Springer-Verlag (1966).

[j] A. Treinin and M. Yaacobi, *J. Phys. Chem.* **68** (1964) 2487.

[k] B. N. Figgis, R. G. Kidd and R. S. Nyholm, *Proc. Roy. Soc.* **269A** (1962) 469.

[l] Y. Saito, *Canad. J. Chem.* **43** (1965) 2530.

[m] C. B. Monk, *J. Amer. Chem. Soc.* **70** (1948) 3281.

[n] M. Spiro, *J. Phys. Chem.* **60** (1956) 976.

Table 70. Average Interatomic Distances and Interbond Angles in Crystalline Chlorates, Bromates and Iodates

Compound	X-ray or neutron diffraction	Coordination number of halogen	Intraionic dimensions		Intermolecular $r(I \cdots OIO_2)$ (Å)	I ··· O distances $r(I \cdots OH_2)$ (Å)	Reference
			r(X–O) (Å)	∠ O–X–O (°)			
NH_4ClO_3	X-ray	3	1·45 ±0·03	106			R. B. Gillespie, P. K. Gantzel and K. N. Trueblood, *Acta Cryst.* **15** (1962) 1271.
$Ba(ClO_3)_2,H_2O$	Neutron	3	1·485±0·004	106·3±0·2			S. K. Sikka, S. N. Momin, H. Rajagopal and R. Chidambaram, *J. Chem. Phys.* **48** (1968) 1883.
NH_4BrO_3	X-ray	3	1·64 ±0·02	Not quoted			A. Santoro and S. Siegel, *Acta Cryst.* **13** (1960) 1017.
$Sm(BrO_3)_3,9H_2O$	Neutron	3	1·67 ±0·02 (twice) 1·54 ±0·02	105·7±0·7 107·3±1·3 (twice)			S. K. Sikka, *Acta Cryst.* **A25** (1969) 621.
$LiIO_3$	X-ray	6	1·817±0·017	98·65±0·6	2·873±0·017		A. Rosenzweig and B. Morosin, *Acta Cryst.* **20** (1966) 758.
NH_4IO_3	X-ray	6	1·765±0·008 1·806±0·008 1·836±0·012	102·3 – 105·6	2·778, 2·819, 2·830		E. T. Keve, S. C. Abrahams and J. L. Bernstein, *J. Chem. Phys.* **54** (1971) 2556
$Ce(IO_3)_4,H_2O$[a]	X-ray	6 6 6 6	1·83 1·82 1·84 1·80	97·1±0·3	2·93, 2·99, 3·00 2·56, 2·78, 2·99 2·51, 2·73 2·55, 2·66	 3·10 3·00	J. A. Ibers, *Acta Cryst.* **9** (1956) 225.
$Ca(IO_3)_2,6H_2O$	X-ray	6	1·78 ±0·03 1·85 ±0·03 1·90 ±0·03	97·8±1·5	2·85±0·02	2·86±0·07 2·89±0·03	A. Braibanti, A. M. M. Lanfredi, M. A. Pellinghelli and A. Tiripicchio, *Inorg. Chim. Acta*, **5** (1971) 590.
$Sr(IO_3)_2,H_2O$	X-ray	7	1·79 ±0·01	99·2±1	3·17±0·01 2·85±0·01 (twice) 3·22±0·01		A. M. M. Lanfredi, M. A. Pellinghelli, A. Tiripicchio and M. T. Camellini, *Acta Cryst.* **B28** (1972) 679.
$Zr(IO_3)_4$	X-ray	8	1·83 ±0·02	99±1 93±1 (twice)	2·55±0·02 2·83±0·02 2·94±0·02 (twice) 3·11±0·02		A. C. Larson and D. T. Cromer, *Acta Cryst.* **14** (1961) 128.

TABLE 70 (*cont.*)

Other studies: $NaClO_3$[b,c]; $KClO_3$[c]; $NaBrO_3$[d]; $KBrO_3$[e]; $Zn(BrO_3)_2,6H_2O$[f]; $Hg(OH)BrO_3$[g]; $Nd(BrO_3)_3,9H_2O$[h]; $NaIO_3$[i]; $Cu(OH)IO_3$[j]; $Ce(IO_3)_4$[k]; $K_2H(IO_3)_2Cl$[l]; NH_4CrIO_6[m].

[a] Contains four unique iodine atoms.
[b] W. H. Zachariasen, *Z. Krist.* **71** (1929) 517.
[c] J. G. Bower, R. A. Sparks and K. N. Trueblood, *U.S. Govt. Research Rept.* **32** (1959) 119.
[d] W. H. Zachariasen, *Norske Videnskapsselsk Skr.* **4** (1928) 90.
[e] J. E. Hamilton, *Z. Krist.* **100** (1938) 104.
[f] S. H. Yü and C. A. Beevers, *Z. Krist.* **95** (1936) 426.
[g] G. Bjornlund, *Acta Chem. Scand.* **25** (1971) 1645.
[h] L. Helmholz, *J. Amer. Chem. Soc.* **61** (1939) 1544.
[i] C. H. MacGillavry and C. L. P. van Eck, *Rec. Trav. Chim.* **62** (1943) 729.
[j] S. Ghose, *Acta Cryst.* **15** (1962) 1105.
[k] D. T. Cromer and A. C. Larson, *Acta Cryst.* **9** (1956) 1015.
[l] A. Braibanti, A. Tiripicchio and A. M. M. Lanfredi, *Chem. Comm.* (1967) 1128.
[m] K.-A. Wilhelmi and P. Löfgren, *Acta Chem. Scand.* **15** (1961) 1413.

TABLE 71. SPECTROSCOPIC AND OTHER INVESTIGATIONS OF CRYSTALLINE CHLORATES, BROMATES AND IODATES

Measurements	Chlorates	Bromates	Iodates
Spectroscopic measurements			
Infrared spectra	1–6	2, 5, 7–10	2, 11–15
Raman scattering	1, 4, 16	16	11, 17
Nuclear quadrupole resonance	18–21 (^{35}Cl)	22–24 (^{79}Br, ^{81}Br)	25 (^{127}I)
Mössbauer spectrum		26	27, 28
X-ray spectra			
emission	29		
E.S.C.A.	30		31
fluorescence	32		
Thermal decomposition			
Calorimetry	33–35	35	
Thermal gravimetric analysis; differential thermal analysis	36–39	39–42	39
Other thermochemical studies	43	43	43–45

1 Infrared (absorption and reflection) and Raman study of $NaClO_3$ single crystal; J. L. Hollenberg and D. A. Dows, *Spectrochim. Acta*, **16** (1960) 1155.

2 Infrared spectra of $NaClO_3$, $KClO_3$, $Ba(ClO_3)_2,H_2O$, $NaBrO_3$, $KBrO_3$, $AgBrO_3$, $NaIO_3$, KIO_3, $Ca(IO_3)_2,6H_2O$; F. A. Miller and C. H. Wilkins, *Anal. Chem.* **24** (1952) 1253; F. A. Miller, G. L. Carlson, F. F. Bentley and W. H. Jones, *Spectrochim. Acta*, **16** (1960) 135.

3 Infrared spectra of alkali halides doped with ClO_3^-; W. E. Klee, *Z. anorg. Chem.* **370** (1969) 1; G. N. Krynauw and C. J. H. Schutte, *Z. phys. Chem., N.F.* **55** (1967) 8, 121.

4 Infrared and Raman spectra of crystalline $NaClO_3$; Raman spectra of molten $LiClO_3$, $NaClO_3$, $KClO_3$ and $AgClO_3$; D. W. James and W. H. Leong, *Austral. J. Chem.* **23** (1970) 1087.

5 Infrared reflection spectra of $NaClO_3$ and $NaBrO_3$ single crystals; R. Duverney, *Compt. rend.* **254** (1962) 1954; M. Debeau, *ibid.* **256** (1963) 109.

6 Infrared spectra of molten alkali-metal chlorates; J. K. Wilmshurst, *J. Chem. Phys.* **36** (1962) 2415.

7 Infrared spectra of bromates of Na, K, Ag, Ba, Sr, Pb, Co, Ni, Mg, Zn and Cu; C. Rocchiccioli, *Compt. rend.* **249** (1959) 236.

8 Infrared spectra of isotopically labelled (^{16}O, ^{18}O) $CsBrO_3$; C. Campbell and J. J. Turner, *J. Chem. Soc.* (*A*) (1967) 1241.

9 Infrared spectra of rare-earth bromates; G. M. Yakunina, S. E. Kharzeeva and V. V. Serebrennikov, *Russ. J. Inorg. Chem.* **14** (1969) 1541.

10 Infrared reflection spectrum of $NaBrO_3$; R. Duverney, C. Deloupy and R. Lalauze, *Compt. rend.* **260** (1965) 5749; M. Galtier, J. Barcelo and C. Deloupy, *ibid.* **265B** (1967) 1322.

11 Infrared and Raman spectra of $NaIO_3$; P. M. A. Sherwood and J. J. Turner, *Spectrochim. Acta*, **26A** (1970) 1975.

12 Infrared spectrum of NH_4IO_3; P. W. Schenk and D. Gerlatzek, *Z. Chem.* **10** (1970) 153.

13 Infrared spectra of 20 iodates and iodato-complexes, including $KH(IO_3)_2$; W. E. Dasent and T. C. Waddington, *J. Chem. Soc.* (1960) 2429.

14 Infrared spectra of cobalt iodates and iodato-complexes; H. E. LeMay, jun., and J. C. Bailar, jun., *J. Amer. Chem. Soc.* **89** (1967) 5577.

15 Infrared spectra of polyiodates; T. Dupuis, *Mikrochim. Acta* (1962) 289.

16 Raman spectra of $NaClO_3$ and $NaBrO_3$; J. P. Mathieu, *J. Chim. Phys.* **46** (1949) 58.

17 Raman spectrum of $NaIO_3$; J. R. Durig, O. D. Bonner and W. H. Breazeale, *J. Phys. Chem.* **69** (1965) 3886.

18 ^{35}Cl nqr spectra of 16 chlorates; R. W. Moshier, *Inorg. Chem.* **3** (1964) 199.

Table 71 (*cont.*)

[19] ^{35}Cl nqr spectrum of $NaClO_3$; T. Wang, *Phys. Rev.* **99** (1955) 566.

[20] ^{35}Cl nqr spectrum of $KClO_3$; T. Kushida, G. B. Benedek and N. Bloembergen, *Phys. Rev.* **104** (1956) 1364; D. B. Utton, *J. Chem. Phys.* **47** (1967) 371.

[21] ^{35}Cl nqr spectrum of $AgClO_3$; L. Anderson, *Bull. Amer. Phys. Soc.* **5** (1960) 412.

[22] ^{79}Br nqr spectra of $NaBrO_3$ and $KBrO_3$; R. F. Tipsword, J. T. Allender, E. A. Stahl, jun. and C. D. Williams, *J. Chem. Phys.* **49** (1968) 2464.

[23] ^{79}Br nqr spectra of bromates of Li, Na, K, Ag, Ca, Sr and Ba; K. Shimomura, T. Kushida and N. Inoue, *J. Chem. Phys.* **22** (1954) 350.

[24] ^{81}Br nqr spectra of $NaBrO_3$ single crystal; R. Fusaro and J. W. Doane, *J. Chem. Phys.* **47** (1967) 5446.

[25] ^{127}I nqr spectra of KIO_3 and $Ca(IO_3)_2,6H_2O$; G. W. Ludwig, *J. Chem. Phys.* **25** (1956) 159.

[26] ^{83}Br Mössbauer spectrum of $KBrO_3$; M. Pasternak and T. Sonnino, *Phys. Rev.* **164** (1967) 384.

[27] ^{127}I Mössbauer spectra (ZnTe source) of $NaIO_3$ and $Na_3Mn(IO_3)_6$; P. Jung and W. Triftshäuser, *Phys. Rev.* **175** (1968) 512.

[28] ^{129}I Mössbauer spectra of KIO_3, NH_4IO_3 and $Ba(IO_3)_2$; D. W. Hafemeister, G. de Pasquali and H. de Waard, *Phys. Rev.* **135B** (1964) 1089.

[29] X-ray emission spectrum of $NaClO_3$; V. I. Nefedov, *J. Struct. Chem.* **8** (1967) 919; W. Nefedow, *Phys. Stat. Solidi*, **2** (1962) 904.

[30] $NaClO_3$; A. Fahlman, R. Carlsson and K. Siegbahn, *Arkiv. Kemi*, **25** (1966) 301.

[31] KIO_3; C. S. Fadley, S. B. M. Hagstrom, M. P. Klein and D. A. Shirley, *J. Chem. Phys.* **48** (1968) 3779.

[32] Spectrum of $NaClO_3$; J. A. Bearden, *Rev. Mod. Phys.* **39** (1967) 86; D. S. Urch, *J. Chem. Soc.* (*A*) (1969) 3026.

[33] $KClO_3$ decomposition; A. A. Gilliland and D. D. Wagman, *J. Res. Nat. Bur. Stand.* **69A** (1965) 1.

[34] $NaClO_3$ and $KClO_3$ decomposition; A. F. Vorob'ev, N. M. Privalova, A. S. Monaenkova and S. M. Skuratov, *Doklad. Akad. Nauk S.S.S.R.* **135** (1960) 1388; A. F. Vorob'ev, N. M. Privalova and L. T. Huang, *Vestn. Mosk. Univ. Ser. II, Khim.* **18** (1963) 27.

[35] $KClO_3$ and $KBrO_3$ decomposition; G. K. Johnson, P. N. Smith, E. H. Appelman and W. N. Hubbard, *Inorg. Chem.* **9** (1970) 119.

[36] Decomposition of alkali-metal chlorates; M. M. Markowitz, D. A. Boryta and H. Stewart, jun., *J. Phys. Chem.* **68** (1964) 2282.

[37] $AgClO_3$ decomposition; F. Solymosi, *Z. phys. Chem., N.F.* **57** (1968) 1.

[38] NH_4ClO_3 decomposition; F. Solymosi and T. Bansagi, *Combust. Flame*, **13** (1969) 262.

[39] Decomposition of chlorates, bromates and iodates of K, Rb and Cs; O. N. Breusov, N. I. Kashina and T. V. Revzina, *Russ. J. Inorg. Chem.* **15** (1970) 316.

[40] Decomposition of 17 alkali-metal, alkaline earth, transition metal and rare earth bromates; G. M. Bancroft and H. D. Gesser, *J. Inorg. Nuclear Chem.* **27** (1965) 1545.

[41] Decomposition of hydrated bromates of divalent metals; B. Dušek and F. Petru, *Russ. J. Inorg. Chem.* **13** (1968) 307.

[42] Decomposition of hydrated bromates of rare-earth elements; I. Mayer and Y. Glasner, *J. Inorg. Nuclear Chem.* **29** (1967) 1605.

[43] Calculation of lattice energies; A. Finch and P. J. Gardner, *J. Phys. Chem.* **69** (1965) 384.

[44] Reaction of $KIO_3(c)$ with aqueous hydrazine; P. B. Howard and H. A. Skinner, *J. Chem. Soc.* (*A*) (1967) 269.

[45] Reaction of $KIO_3(c)$ with aqueous HI; C. Wu, M. M. Birky and L. G. Hepler, *J. Phys. Chem.* **67** (1963) 1202.

TABLE 72. HALOGEN ATOM CHARGES IN CHLORATE, BROMATE AND IODATE IONS

Technique	Charge on halogen atom in		
	ClO_3^-	BrO_3^-	IO_3^-
Infrared intensity measurements	a	$+1{\cdot}72$[b]	
Raman intensity measurements	*ca.* $+1$[j]	*ca.* $+1$[j]	*ca.* $+1$[j]
Mössbauer spectroscopy			$+0{\cdot}83$[c]
X-ray photoelectron spectroscopy	$+1{\cdot}02$[h]		$+1{\cdot}10$[d]
Nuclear quadrupole resonance spectroscopy[e]	$+1{\cdot}21$[f]	$+0{\cdot}87$[f]	$+0{\cdot}69$[f]
	$+1{\cdot}49$[g]	$+0{\cdot}72$[g]	$+2{\cdot}73$[g]
X-ray emission spectroscopy	$+1{\cdot}72$[i]		

[a] Charge on oxygen estimated to be $-0{\cdot}25$; chlorine charge not estimated; G. N. Krynauw and C. J. H. Schutte, *Z. phys. Chem., N.F.* **55** (1967) 121.
[b] R. Duverney, C. Deloupy and R. Lalauze, *Compt. rend.* **260** (1965) 5749.
[c] D. W. Hafemeister, G. de Pasquali and H. de Waard, *Phys. Rev.* **135B** (1964) 1089.
[d] C. S. Fadley, S. B. M. Hagstrom, M. P. Klein and D. A. Shirley, *J. Chem. Phys.* **48** (1968) 3779.
[e] E. A. C. Lucken, *Nuclear Quadrupole Coupling Constants*, p. 285, Academic Press (1969).
[f] Calculated assuming tetrahedral bond angles.
[g] Calculated using experimental bond angles.
[h] R. Manne, *J. Chem. Phys.* **46** (1967) 4645; A. Fahlman, R. Carlsson and K. Siegbahn, *Arkiv. Kemi*, **25** (1966) 301.
[i] W. Nefedow, *Phys. Stat. Solidi*, **2** (1962) 904.
[j] G. W. Chantry and R. A. Plane, *J. Chem. Phys.* **32** (1960) 319.

The results of various spectroscopic investigations have been used to calculate atomic charges for the halate ions; some of these values are reproduced in Table 72. In view of the assumptions necessary in analysing the data, the discrepancies between the different values are hardly surprising; although all the calculations imply considerable polarity in the X–O bond, there is no hint of any trend in the derived charges.

Reactions

Decomposition Reactions

Dilute solutions of $HClO_3$ and $HBrO_3$ decompose if heated, but the cold solutions may be concentrated *in vacuo*. While the decomposition of bromic acid is described by the equation

$$4HBrO_3 \rightarrow 2Br_2 + 5O_2 + 2H_2O$$

chloric acid disproportionates, possibly with explosive violence. Dilute solutions evolve chlorine and oxygen in a reaction approximating to

$$8HClO_3 \rightarrow 4HClO_4 + 2H_2O + 3O_2 + 2Cl_2$$

but in more concentrated solution chlorine dioxide is one of the products. Large amounts of the dioxide result from the action of concentrated sulphuric acid on solid chlorates (p. 1367).

$$3HClO_3 \rightarrow HClO_4 + H_2O + 2ClO_2$$

Three decomposition processes are possible for the halate ions:

(1) $$4XO_3^- \rightarrow X^- + 3XO_4^-$$

(2) $$2XO_3^- \rightarrow 2X^- + 3O_2$$

(3) $$4XO_3^- \rightarrow 2X_2 + 5O_2 + 2O^{2-}$$

At temperatures near to 25°C reaction (1) is thermodynamically feasible only for chlorates, as witnessed by the thermodynamic data for the potassium salts (Table 73) and by the equilibrium constants in aqueous solution (ClO_3^-, 10^{22}; BrO_3^-, 10^{-30}; IO_3^-, 10^{-53}). Disproportionation proceeds very slowly even in boiling chlorate solutions; it is observed, however, as an exothermic process in the pyrolysis of metal chlorates. For the molten alkali-metal salts reaction (1) comprises the major decomposition route at temperatures below 500°C, the melt becoming viscid as perchlorate is formed; at higher temperatures the perchlorates are themselves unstable. For other chlorates, though observed initially, disproportionation is superseded by other decomposition processes. While there is no evidence for formation of BrO_4^- or IO_4^- during the pyrolysis of bromates or iodates, the anhydrous alkaline earth iodates decompose at high temperatures according to the reaction

$$5M(IO_3)_2 \rightarrow M_5(IO_6)_2 + 4I_2 + 9O_2$$

Perhalate formation is supposedly an internal oxygenation process.

Disproportionation apart, whether decomposition of metal halates occurs by reaction (2) (with formation of the metal halide) or by reaction (3) (with oxide-formation) is determined mainly on thermodynamic grounds, by the relative stabilities of the metal halide and oxide, but also kinetically, by the relative activation energies of the two processes[723]; on both counts oxide-formation is favoured by the presence of a strongly polarizing cation. The effects of particle size and rate of heating are also important kinetically. Halide-formation is observed for alkali-metal, alkaline earth and silver salts, while magnesium, transition metal and rare earth salts favour oxide-formation, occasionally with explosive evolution of chlorine dioxide from chlorates, e.g. from $Cu(ClO_3)_2$. Barium bromite may be prepared by the controlled pyrolysis of barium bromate[699].

The thermal stability of the salts falls in the sequence iodate > chlorate > bromate, and also decreases as the polarizing power of the cation rises. Oxygen-evolution from alkali chlorates is catalysed by transition metal salts, which depress the temperature of decomposition and preclude disproportionation. Heating mixtures of potassium chlorate and manganese dioxide has long been exploited, particularly for didactic purposes, as a method of producing oxygen on the small scale.

While the initial step in the thermal decomposition of alkali-metal halates is rupture of the X–O bond, which occurs only at high temperature (> 300°C), decomposition of the ammonium halates begins at much lower temperatures (NH_4ClO_3, 50°C; NH_4BrO_3, −5°C; NH_4IO_3, *ca.* 100°C), and the compounds explode on further heating. It is believed that, as in the case of NH_4ClO_4, the reaction commences with proton transfer,

$$\begin{array}{ccccc} NH_4^+XO_3^-(s) & \longrightarrow & NH_3(s) & + & HXO_3(s) \\ & & \updownarrow & & \updownarrow \\ & & NH_3(g) & & HXO_3(g) \end{array}$$

723 G. M. Bancroft and H. D. Gesser, *J. Inorg. Nuclear Chem.* **27** (1965) 1545.

TABLE 73. THERMODYNAMIC PROPERTIES OF POTASSIUM HALATES AND PERHALATES AT 25°C

X =	Chlorine	Bromine	Iodine
$\Delta H_f°$(c), KXO_3(kcal mol^{-1})[a]	−95·1	−86·0	−119·5
$\Delta G_f°$(c), KXO_3(kcal mol^{-1})[a]	−70·8	−64·7	−99·6
$\Delta H_f°$(c), KXO_4(kcal mol^{-1})[a]	−103·2	−68·7	−110·1
$\Delta G_f°$(c), KXO_4(kcal mol^{-1})[a]	−72·2	−40·7	−83·5
$KXO_3(c) \rightarrow KX(c)+3/2\ O_2(g)$[a,b]			
$\Delta H°$(kcal mol^{-1})	−9·1	−7·8	+41·2
$\Delta G°$(kcal mol^{-1})	−26·8	−25·9	+22·6
$KXO_4(c) \rightarrow KX(c)+2\ O_2(g)$[a,b]			
$\Delta H°$(kcal mol^{-1})	−1·0	−25·1	+31·8
$\Delta G°$(kcal mol^{-1})	−25·4	−49·9	+6·5
$KXO_3(c) \rightarrow 3/4\ KXO_4(c)+1/4\ KX(c)$[a,b]			
$\Delta H°$(kcal mol^{-1})	−8·4	+10·0	+17·3
$\Delta G°$(kcal mol^{-1})	−7·7	+11·5	+17·7
$KXO_n(s) \rightarrow 1/2\ K_2O(s)+1/2\ X_2(g)+(2n-1)/4\ O_2(g)$[a,b]			
$n = 3$; $\Delta H°$(kcal mol^{-1})	+51·9	+46·5	+83·8
$n = 4$; $\Delta H°$(kcal mol^{-1})	+60·0	+25·6	+69·2
Decomposition temperature (°C)	c	a	d
KXO_3	472 ($\rightarrow 3/4\ KClO_4+1/4\ KCl$)	390 ($\rightarrow KBr+3/2\ O_2$)	525 ($\rightarrow KI+3/2\ O_2$)
KXO_4	534 ($\rightarrow KCl+2O_2$)	275 ($\rightarrow KBrO_3+1/2\ O_2$)	303 ($\rightarrow KIO_3+1/2\ O_2$)

[a] G. K. Johnson, P. N. Smith, E. H. Appelman and W. N. Hubbard, *Inorg. Chem.* **9** (1970) 119.
[b] *Selected Values of Chemical Thermodynamic Properties*, N.B.S. Technical Note 500, Washington (1952).
[c] M. M. Markowitz, D. A. Boryta and H. Stewart, jun., *J. Phys. Chem.* **68** (1964) 2282; M. M. Markowitz, D. A. Boryta and R. F. Harris, *J. Phys. Chem.* **65** (1961) 261.
[d] O. N. Breusov, N. I. Kashina and T. V. Revzina, *Russ. J. Inorg. Chem.* **15** (1970) 316.

followed by rapid decomposition of the acid and oxidation of ammonia to give the products, which include a residue of NH_4NO_3.

Flash photolysis of the XO_3^- anions[632] in aqueous solution (Table 54) produces XO_2 radicals by the reaction

$$XO_3^-,H_2O \rightarrow (XO_3^-,H_2O)^* \rightarrow XO_2+OH+OH^-$$

Depending on the temperature of irradiation and subsequent annealing processes, the action on crystalline chlorates[610,724] and bromates[603,725] of X-rays or γ-rays can produce many species, including perhalate ions, oxyhalogen radicals (Table 54) and radical anions (p. 1460).

$$XO_3^- \xrightarrow{\text{X or } \gamma} XO_4^-, XO_2^-, XO^-, X^-, XO_3, XO_2, X\text{–}XO_3^-, XO_3^{2-}, O_2 \text{ and } O_3 \ (X = \text{Cl or Br})$$

Similar decay follows the capture of thermal neutrons by the halates[726].

Oxidation by Chlorate, Bromate and Iodate

Both thermodynamically and kinetically the oxidizing powers of the halates in aqueous solution are marked functions of the hydrogen ion concentration. The potentials for the conversion of halate to halide (the usual reduction process) are more favourable in acid solution (1·1–1·5 volts as against 0·3–0·7 volt in alkaline solution), and the reactions are more rapid. In alkaline media chlorate exhibits no oxidizing capabilities, and its reduction in weakly acidic solution may be very slow, requiring catalysis by transition metal derivatives, e.g. of Os(VIII), V(V) or Mn(VII). In concentrated solution chlorates are often reduced to chlorine dioxide rather than to chloride.

Kinetic differences apart, the three halate anions are quite similar in their oxidizing properties; some reactions of bromates are summarized in Scheme 11. Bromates and iodates undergo some redox reactions in alkaline solution; thus, reaction with hydroxylamine or hydrazine (as their sulphates) produces nitrogen (alkaline chlorates not reacting), and iodate oxidizes vanadyl salts to the pentavalent state:

$$6VO^{2+}+IO_3^-+18OH^- \rightarrow 6VO_3^-+I^-+9H_2O$$

However, many transformations which are rapidly wrought by the alkaline halites or hypohalites at room temperature, e.g. the oxidation of As(III) to As(V), take place only slowly with the halates, even in acid solution.

Halates and halides react in acid solution in all nine possible combinations: iodides are oxidized quantitatively to iodine, and bromides to bromine (setting aside subsequent polyhalide-formation): the liberation of iodine is a common procedure in analysing halates.

$$XO_3^-+5X^-+6H^+ \rightarrow 3X_2+3H_2O \quad (X = \text{Cl, Br or I})$$

$$XO_3^-+6Y^-+6H^+ \rightarrow X^-+3Y_2+3H_2O \quad (X = \text{Cl};\ Y = \text{I or Br};\ X = \text{Br};\ Y = \text{I})$$

$$XO_3^-+5Y^-+6H^+ \rightarrow 2Y_2+XY+3H_2O \quad (X = \text{I};\ Y = \text{Br})$$

$$2ClO_3^-+2Cl^-+4H^+ \rightarrow Cl_2+2ClO_2+2H_2O$$

Chlorides produce both Cl_2 and ClO_2 with chlorate, bromine and chlorine with bromate, and chlorine and iodine chlorides with iodate. The reactions share common kinetic features with the (halide-catalysed) exchange of oxygen between the halate ions and water, which is

724 H. G. Heal, *Canad. J. Chem.* **37** (1959) 979.

725 L. C. Brown, G. M. Begun and G. E. Boyd, *J. Amer. Chem. Soc.* **91** (1969) 2250.

726 T. Andersen, H. E. L. Madsen and K. Olesen, *Trans. Faraday Soc.* **62** (1966) 2409.

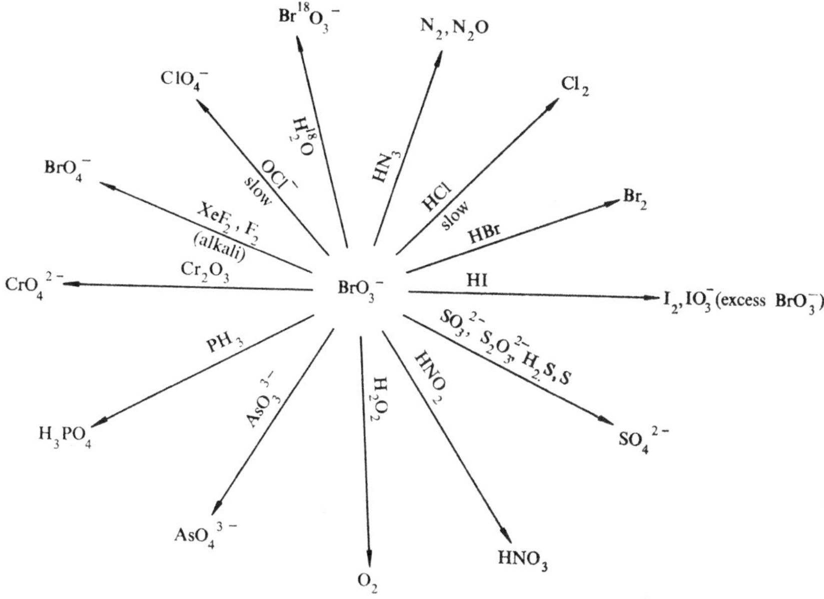

SCHEME 11. Reactions of aqueous bromates.

also acid-catalysed. The general rate equations have the form[701,727,728]

$$\text{rate} = k[XO_3^-][H^+]^2[Y^-]$$

and the proposed mechanism involves an intermediate of the type YXO_2:

$$XO_3^- + 2H^+ \underset{}{\overset{\text{fast}}{\rightleftharpoons}} [H_2XO_3^+]$$

$$[H_2XO_3^+] + Y^- \longrightarrow H_2O + [YXO_2] \xrightarrow{Y^-} Y_2 + XO_2^-$$

$$[YXO_2] \xrightarrow{H_2O} [H_2XO_3^+] + Y^- \qquad [YXO_2] \longrightarrow Y_2 + 2XO_2 \quad (\text{with } YXO_2;\ X,Y = Cl)$$

Subsequent reaction of XO_2^- with Y^- is assumed to be rapid. It is uncertain whether the YXO_2 intermediate has the structure Y–X(=O)(=O) or O–X–O–Y. The former is favoured by comparison with known compounds (the halogenyl fluorides), but the latter structure could also account for the kinetically related oxidation of NO_2^- and SO_3^{2-}, which involves quantitative transfer of oxygen from the halate ion to the substrate. Halide ion catalysis has been found for other oxidation reactions of the halate ions.

Iodine replaces chlorine and bromine in ClO_3^- and BrO_3^-:

$$I_2 + 2XO_3^- \rightarrow X_2 + 2IO_3^-$$

---

727 A. F. M. Barton and G. A. Wright, *J. Chem. Soc.* (*A*) (1968) 1747.
728 M. Anbar and S. Guttmann, *J. Amer. Chem. Soc.* **83** (1961) 4741.

The bromate–iodine reaction has the following stages at pH 1·5–2·5 [729]:

(1) An induction period in which a catalyst is produced (probably HOBr)
(2) $I_2 + BrO_3^- \rightarrow IBr + IO_3^-$
(3) $3IBr + 2BrO_3^- + 3H_2O \rightarrow 5Br^- + 3IO_3^- + 6H^+$
(4) $5Br^- + BrO_3^- + 6H^+ \rightarrow 3Br_2 + 3H_2O$

The reaction between iodate and sulphite in acid solution (Landolt's reaction) is interesting: three different steps are involved which give rise to a periodic appearance and disappearance of iodine in solution.

(1) $IO_3^- + 3SO_3^{2-} \rightarrow I^- + 3SO_4^{2-}$
(2) $5I^- + IO_3^- + 6H^+ \rightarrow 3I_2 + 3H_2O$
(3) $3I_2 + 3SO_3^{2-} + 3H_2O \rightarrow 6I^- + 6H^+ + 3SO_4^{2-}$

The oxidation of HN_3 by BrO_3^- in perchloric acid solution affords N_2, N_2O, OBr^- and Br_2; the oxygen in N_2O is derived mostly from the solvent[730].

Solid chlorates readily oxidize organic material and non-metals such as sulphur, selenium, tellurium, phosphorus and arsenic. Dry mixtures are generally stable, but traces of moisture may induce ignition or explosive decomposition (*via* the formation of $HClO_3$). Bromates and iodates also form unstable mixtures with combustible or oxidizable substances. Molten chlorates have been used to oxidize iodide to iodate.

Oxidation of Halates

Oxidation of halogen(V) to halogen(VII) is thermodynamically more facile in alkaline solution than in acid solution. Perchlorates arise in the disproportionation of chlorates, from which they may also be obtained by electrolytic or chemical oxidation, e.g. with persulphate ($E^\circ = +2{\cdot}01$ volts). Iodates are more readily oxidized, since chlorine ($E^\circ = +1{\cdot}36$ volts), bromine (+1·07 volts), hypochlorite (+0·89 volt), permanganate (+1·23 volts) or persulphate effect the change; anodic oxidation is hindered by reduction of iodate at the cathode.

The problem of oxidizing bromates has perplexed chemists for many years. Though the synthesis of perbromates has now been accomplished[663,731] using F_2, XeF_2 or electrical means to oxidize aqueous BrO_3^-, the reduction potential of the BrO_4^-/BrO_3^- couple (+1·76 volts in acid solution), though greater than those for the other halogen(VII)/halogen(V) couples, is not sufficiently great to explain why ozone or persulphate does not effect the oxidation.

A detailed kinetic study has been made of the oxidation of halates by XeF_2 in aqueous solution[732]; the optimum yields are ClO_4^-, 93%; BrO_4^-, 12%; IO_4^-, 93%. The oxidation of ClO_3^- and BrO_3^- is accomplished by an intermediate in the reaction of XeF_2 with water, producing a second intermediate which either goes on to give perhalate or goes back to halate; the latter reaction predominates for bromate. In dilute solution iodate-oxidation entails a similar mechanism, but in concentrated iodate solutions there is a direct reaction *via* a xenon iodate:

$$XeF_2 + IO_3^- \longrightarrow [FXeOIO_2] \xrightarrow{H_2O} Xe + HF + H^+ + IO_4^-$$

Fluorination of solid potassium chlorate affords ClO_3F (p. 1391).

729 D. E. C. King and M. W. Lister, *Canad. J. Chem.* **46** (1968) 279.
730 R. C. Thompson, *Inorg. Chem.* **8** (1969) 1891.
731 E. H. Appelman, *Inorg. Chem.* **8** (1969) 223.
732 E. H. Appelman, *Inorg. Chem.* **10** (1971) 1881.

Uses

Sodium chlorate is manufactured in very large amounts for conversion into (i) chlorine dioxide (by reduction with chloride or sulphur dioxide), and (ii) perchloric acid and perchlorates (by electrolysis in acid solution). Quantities are also used as oxidizers in the making of matches and in fireworks, and form a common ingredient of weed-killing preparations.

The commercial applications of bromates and iodates depend on their ability to oxidize thiolic residues in proteins to S–S linkages, a characteristic which is utilized in improving the baking properties of flours, and in preparations for waving hair. Potassium bromate and iodate are important primary analytical standards[733]; bromates are used especially in the determination of trivalent arsenic and antimony, by oxidation from the trivalent to the pentavalent state, while iodate is used in numerous procedures for the estimation of a variety of metals.

(E) PERHALIC ACIDS AND PERHALATES

Introduction

Discovery

Potassium perchlorate was discovered in 1816 by von Stadion, who treated fused potassium chlorate with sulphuric acid; chlorine dioxide was evolved and potassium perchlorate was crystallized from the mixture. Subsequent investigations by the same worker led to the isolation of perchloric acid as a result of distilling $KClO_3$–H_2SO_4 mixtures, and to the electrolytic oxidation of $KClO_3$ in aqueous solution. The nature of the disproportionation of chlorates was first truly recognized by Serullas (in 1830) when he isolated $KClO_4$ from pyrolysed $KClO_3$ without recourse to treatment with sulphuric acid; he also prepared perchloric acid solutions by the reaction

$$2KClO_4 + H_2SiF_6 \rightarrow K_2SiF_6 \downarrow + 2HClO_4$$

Perchlorates have since been found in natural nitrate and saline deposits, and in sea water.

Interest in perchlorates waned until about 1894, when manufacture by electrolysis of chlorate solutions was undertaken in Sweden for use in explosives. Since that date, the level of commercial production of perchlorate has fluctuated in inverse proportion to the extent of world peace; current world production is in excess of 50,000 tons per annum, much of which is put to use as oxidizers in rocket fuels. The natural interest of national governments in these applications has meant that the physicochemical properties of perchloric acid and its derivatives have been investigated in great detail. There have been several comprehensive reviews of the chemistry of perchloric acid and the perchlorates[572,575b,576,734–738]; two of these articles give special emphasis to the physical chemistry of perchloric acid[736,737].

[733] J. H. Thompson, *Comprehensive Analytical Chemistry* (ed. C. L. Wilson and D. W. Wilson), Vol. IB, p. 259, Elsevier (1960).

[734] J. C. Schumacher (ed.), *Perchlorates*, A.C.S. Monograph No. 146, Reinhold, New York (1960).

[735] J. C. Schumacher and R. D. Stewart, *Kirk-Othmer Encyclopedia of Chemical Technology*, 2nd edn., Vol. 5, pp. 61–84, Interscience, New York (1964).

[736] G. S. Pearson, *Adv. Inorg. Chem. Radiochem.* **8** (1966) 177.

[737] A. A. Zinov'ev, *Russ. Chem. Rev.* **32** (1963) 268.

[738] C. C. Addison, *Mellor's Comprehensive Treatise on Inorganic and Theoretical Chemistry*, Supplement II, Part I, pp. 596–620, Longmans, London (1956).

Periodic acid and its salts[575b,576,624,739,740] have been known for many years; thus, the oxidation of iodates by chlorine was reported as early as 1833. By contrast, perbromates were not discovered until 1968 during studies of the β-decay of $^{82}SeO_4^{2-}$, an event rapidly followed by the disclosure of chemical syntheses of perbromates. The difficulties encountered in oxidizing bromates were described in the preceding section, although the inaccessibility of perbromates is due not so much to their intrinsic properties as to the high-energy kinetic barrier which opposes the conversion of Br(V) to Br(VII).

A Note of Caution

It must be emphasized that, despite their kinetic inertness at ambient temperatures, perchloric acid and the perchlorates are potentially very strong oxidizing agents. Anhydrous perchloric acid is subject to spontaneous explosive decomposition on storage or in the presence of contaminants, and many of the perchlorates have highly exothermic decomposition processes open to them, e.g.

$$AgClO_4 \rightarrow AgCl + 2O_2 \qquad \Delta H \sim -25 \text{ kcal mol}^{-1}$$

which bring considerable risks of explosion to operations such as drying and grinding, especially if the cation is readily oxidizable. Though 72% perchloric acid is an extremely useful and safe analytical reagent, the dangers involved in mixing 100% perchloric acid with organic material would be hard to exaggerate: in 1947 an explosion caused by the introduction of some plastic into a tank containing 220 U.S. gallons of a mixture of perchloric acid and acetic acid resulted in 17 deaths and the destruction of 116 buildings. Organic salts of perchloric acid have a bad record of laboratory explosions; arene diazonium perchlorates were once thought to be the most explosive compounds known. The preparation of anhydrous perchloric acid may be prohibited by local regulations, and should in any case be undertaken only in small amounts; the literature contains detailed accounts of appropriate safety precautions[571,734].

Comparison of Perchlorates, Perbromates and Periodates

Some aspects of the chemistry of the perhalic acids and their derivatives are summarized in Table 74. Though data for the perbromates are still sparse, recent results show that they have much in common with the perchlorates, whereas the chemistry of the periodates is rather different and resembles that of the tellurates(VI). Many of the differences between periodates and the corresponding chlorine and bromine derivatives are attributable, at least in part, to the greater size of the iodine atom.

While the perchlorates and perbromates are found only with tetra-coordinated halogen atoms, hexa-coordination about iodine is the rule rather than the exception in periodates. A much-quoted example concerns the phases having the composition H_5XO_6, where X = Cl or I; these are crystalline at room temperature with the structure $[H_5O_2]^+ClO_4^-$ in the one case or $(HO)_5IO$ in the other, whereas anhydrous $HClO_4$ is liquid under similar conditions. HIO_4 may be made by controlled dehydration of $(HO)_5IO$, and IO_4^- is the principal constituent of alkaline periodate solutions; the anion enters into a rapid hydration–dehydration equilibrium in solution:

$$IO_4^- + 2H_2O \rightleftharpoons (HO)_4IO_2^-$$

---

[739] G. J. Hills, *Mellor's Comprehensive Treatise on Inorganic and Theoretical Chemistry*, Supplement II, Part I, pp. 896–907, Longmans, London (1956).

[740] H. Siebert, *Fortschritte Chem. Forsch.* **8** (1967) 470.

TABLE 74. GENERAL COMPARISON OF THE HALOGEN(VII) OXYACIDS

Property	Halogen		
	Chlorine	Bromine	Iodine
Structure of the acid	$HOClO_3$	$HOBrO_3$ probably	$(HO)_5IO$
Physical state at 20°C	liquid	—	solid
Acid character	very strong, monobasic	very strong, monobasic	weak, polybasic
Anions found in solution	ClO_4^-	BrO_4^-	$H_4IO_6^-$ $H_3IO_6^{2-}$ $H_2IO_6^{3-}$ $H_2I_2O_{10}^{4-}$ IO_4^-
Oxidizing character of aqueous solution	absent at 20°C strong at 100°C	sluggish at 20°C strong at 100°C	potent and rapid at 20°C
Oxygen-exchange between H_2O and XO_4^-	not observed	not observed	extremely rapid

which is not observed for the other perhalates. The hydrated periodates persist in acid solution and are isolable as salts. Periodates also form polymeric anions with I–O–I bridges, as well as very stable complexes and heteropolyanions with I–O–M bridges; no analogous polymerization of perchlorates (except to give the anhydride Cl_2O_7) or perbromates has been reported, and perchlorato-complexes contain only weakly bound ClO_4^- ions.

The lability of the groups attached to iodine is also responsible for the rapidity with which periodates oxidize many materials. While perchlorates and especially perbromates are also strong oxidizing agents, their reactions are hampered at room temperature by the absence of suitable intermediates of low energy; nucleophilic attack on the small halogen atom to give a five-coordinate intermediate is difficult because of the relatively congested tetrahedron of negatively charged oxygen atoms surrounding it, while the alternative pathway—rupture of a Cl–O or Br–O bond—is also expensive of energy.

Simple perchlorates and perbromates M^IXO_4, like many other salts containing tetrahedral anions, crystallize at room temperature with the orthorhombic Barite ($BaSO_4$) structure; at higher temperatures (>200°C) perchlorates transform to a cubic lattice, a change associated with the onset of free rotation of the ClO_4^- ions. Metaperiodates M^IIO_4 adopt the tetragonal Scheelite structure. The anions XO_4^- have been characterized by spectroscopic and single-crystal diffractometric studies (Tables 75, 76, 79 and 83); in simple salts the anions are tetrahedral within the limits of accuracy of X-ray diffraction experiments, although the vibrational spectra of the solids show effects, such as the appearance of a weak infrared absorption due to the formally forbidden symmetric stretching mode $\nu_1(a_1)$, arising out of the occupation by the anion of a site of symmetry lower than T_d. The bond lengths increase from *ca.* 1·45 Å in perchlorates to 1·61 Å in $KBrO_4$ and 1·78 Å in $NaIO_4$. While the valence force constants decrease from 8·2 mdyne Å^{-1} for ClO_4^- to *ca.* 6·0 mdyne Å^{-1} for BrO_4^- and IO_4^-, the decreases in the deformation and interaction force

TABLE 75. PROPERTIES OF THE AQUEOUS PERHALATE IONS XO_4^-

	ClO_4^-	BrO_4^-	IO_4^-
Thermodynamic functions at 298°K			
$\Delta H_f°$(kcal mol^{-1})[a]	−30·70	3·19	−34·56
$\Delta G_f°$(kcal mol^{-1})[a]	−1·84	29·18	−11·09
$S°$(cal deg^{-1} mol^{-1})[a]	43·6	44·7	48·8
Reduction potentials			
Acid solution			
$E°(XO_4^-/XO_3^-)$(volts)[a]	+1·23	+1·76	b
$E°(XO_4^-/\frac{1}{2}X_2)$(volts)[a,c]	+1·39	+1·58	
$E°(XO_4^-/X^-)$(volts)[a,c]	+1·39	+1·52	
Alkaline solution			
$E°(XO_4^-/XO_3^-)$(volts)[a,c]	+0·40	+0·93	+0·81
$E°(XO_4^-/\frac{1}{2}X_2)$(volts)[a,c]	+0·45	+0·64	+0·37
$E°(XO_4^-/X^-)$(volts)[a,c]	+0·56	+0·69	+0·40
Spectroscopic properties			
Raman spectrum	d,e,f	g,h	d,i
Infrared spectrum	j		
Fundamental frequencies (cm^{-1})			
$\nu_1(a_1)$ (R)	931	801	791
$\nu_2(e)$ (R)	462	331	256
$\nu_3(f_2)$ (R, i.r.)	1107	878	853
$\nu_4(f_2)$ (R, i.r.)	631	410	325
Force constants (mdyne Å^{-1})[g]			
f_r	8·24	6·05	5·90
f_θ/r^2	0·87	0·48	0·30
$f_{\theta\theta}/r^2$	−0·21	−0·12	−0·09
$(f_{r\theta}-f_{r\theta}')/r$	0·78	0·38	0·07
Ultraviolet-visible absorption spectrum	q	h	k
λ_{max}(nm)	<185	187	222·5
ϵ_{max}(M^{-1})		$3·9\times10^3$	$10·7\times10^3$
Nmr spectrum			
^{17}O chemical shift (ppm with respect to H_2O)	−288±5[l]		
^{35}Cl chemical shift (ppm with respect to Cl^-)	−1000[m]		
^{127}I spectrum			n
Ionic mobility (25°C) (cm^2 ohm^{-1} g-ion^{-1})	67·4[o]		54·6[p]

R, Raman-active; i.r., infrared-active.

[a] G. K. Johnson, P. N. Smith, E. H. Appelman and W. N. Hubbard, *Inorg. Chem.* **9** (1970) 119.

[b] See Table 80 for periodate potentials in acid solution.

[c] A. G. Sharpe, *Halogen Chemistry* (ed. V. Gutmann), Vol. 1, p. 13, Academic Press (1967).

[d] Includes absolute intensity measurements; G. W. Chantry and R. A. Plane, *J. Chem. Phys.* **32** (1960) 319; *ibid.* **34** (1961) 1268.

[e] Includes effect of metal ions on the spectrum of ClO_4^-; M. M. Jones, E. A. Jones, D. F. Harmon and R. T. Semmes, *J. Amer. Chem. Soc.* **83** (1961) 2038; R. E. Hester and R. A. Plane, *Inorg. Chem.* **3** (1964) 769.

[f] J. W. Akitt, A. K. Covington, J. G. Freeman and T. H. Lilley, *Trans. Faraday Soc.* **65** (1969) 2701.

[g] L. C. Brown, G. M. Begun and G. E. Boyd, *J. Amer. Chem. Soc.* **91** (1969) 2250.

[h] E. H. Appelman, *Inorg. Chem.* **8** (1969) 223.

[i] H. Siebert and G. Willfahrt, cited in H. Siebert, *Fortschritte Chem. Forsch.* **8** (1967) 470.

[j] J. D. S. Goulden and D. J. Manning, *Spectrochim. Acta*, **23A** (1967) 2249.

[k] C. E. Crouthamel, H. V. Meek, D. S. Martin and C. V. Banks, *J. Amer. Chem. Soc.* **71** (1949) 3031.

[l] B. N. Figgis, R. G. Kidd and R. S. Nyholm, *Proc. Roy. Soc.* **269A** (1962) 469.

Table 75 (*cont.*)

[m] Y. Saito, *Canad. J. Chem.* **43** (1965) 2530.
[n] R. M. Kren, H. W. Dodgen and C. J. Nyman, *Inorg. Chem.* **7** (1968) 446.
[o] J. H. Jones, *J. Amer. Chem. Soc.* **67** (1945) 855.
[p] C. B. Monk, *J. Amer. Chem. Soc.* **70** (1948) 3281.
[q] L. J. Heidt, G. F. Koster and A. M. Johnson, *J. Amer. Chem. Soc.* **80** (1959) 6471.

TABLE 76. INVESTIGATIONS OF SOLID PERHALATES

Measurements	Perchlorates	Perbromates	Periodates
X-ray crystallography	1–7	8	9
Spectroscopic investigations			
Infrared and Raman spectra	1, 10–14	15, 16	17–21
Mössbauer spectra			22, 23
X-ray spectra:			
E.S.C.A.	24		25
emission	26		
fluorescence	27		
Thermal decomposition			
Calorimetry	28	29	
Thermal gravimetric analysis; differential thermal analysis	30–32		32, 33

[1] See also Table 79.

[2] $LiClO_4$, $r_{av}(Cl-O) = 1{\cdot}44 \pm 0{\cdot}01$ Å; $KClO_4$, $r_{av}(Cl-O) = 1{\cdot}43 \pm 0{\cdot}02$ Å; R. Prosen, Ph.D. Thesis, University of California (1955), cited in ref. 4.

[3] NH_4ClO_4, $r_{av}(Cl-O) = 1{\cdot}46 \pm 0{\cdot}03$ Å; K. Venkatesan, *Proc. Indian Acad. Sci.* **A56** (1957) 134.

[4] $[NO_2]ClO_4$, $r_{av}(Cl-O) = 1{\cdot}464 \pm 0{\cdot}007$ Å; M. R. Truter, D. W. J. Cruickshank and G. A. Jeffrey, *Acta Cryst.* **13** (1960) 855.

[5] $[N_2H_5]ClO_4,\frac{1}{2}H_2O$, $r_{av}(Cl-O) = 1{\cdot}447 \pm 0{\cdot}004$ Å; R. Liminga, *Acta Chem. Scand.* **21** (1967) 1217.

[6] $HClO_4$(mesitaldehyde)$_2$, $r_{av}(Cl-O) = 1{\cdot}458 \pm 0{\cdot}007$ Å; C. D. Fisher, L. H. Jensen and W. M. Schubert, *J. Amer. Chem. Soc.* **87** (1965) 33.

[7] $Cu(C_6H_5N_3)_2(ClO_4)_2$, $r_{av}(Cl-O) = 1{\cdot}422 \pm 0{\cdot}01$ Å; J. J. Bonnet and Y. Jeannin, *Acta Cryst.* **26B** (1970) 318.

[8] $KBrO_4$, $r_{av}(Br-O) = 1{\cdot}610 \pm 0{\cdot}016$ Å; S. Siegel, B. Tani and E. H. Appelman, *Inorg. Chem.* **8** (1969) 1190.

[9] See also Table 83.

[10] Infrared spectra of alkali halides doped with ClO_4^-; W. E. Klee, *Z. anorg. Chem.* **370** (1969) 1; G. N. Krynauw and C. J. H. Schutte, *Z. phys. Chem., N.F.* **55** (1967) 8, 121.

[11] Infrared spectra of $MClO_4$ (M = Li, Na, K, Rb, Cs, Tl and Ag); A. Hezel and S. D. Ross, *Spectrochim. Acta*, **22** (1966) 1949.

[12] Infrared spectra of $MClO_4$ (M = Na, K and NH_4) and $M(ClO_4)_2,xH_2O$ (M = Mg, Ca, Ba, Fe, Co, Ni, Cu and Zn); S. D. Ross, *Spectrochim. Acta*, **18** (1962) 225.

[13] Infrared spectrum of $KClO_4$ single crystal at low temperature; R. A. Schroeder, E. R. Lippincott and C. E. Weir, *J. Inorg. Nuclear Chem.* **28** (1966) 1397.

[14] Infrared spectra of $M(ClO_4)_3,n$DMSO ($n = 7$, M = Sm, Gd and Y; $n = 8$, M = La, Ce, Pr and Nd); Y. N. Krishnamurthy and S. Soundararajan, *J. Inorg. Nuclear Chem.* **29** (1967) 517; *Z. anorg. Chem.* **349** (1967) 220.

[15] Infrared and Raman spectra of $KBrO_4$; E. H. Appelman, *Inorg. Chem.* **8** (1969) 223.

[16] Infrared and Raman spectra of $CsBrO_4$ and Ph_4AsBrO_4; L. C. Brown, G. M. Begun and G. E. Boyd, *J. Amer. Chem. Soc.* **91** (1969) 2250.

Table 76 (*cont.*)

[17] See also Table 84.

[18] Infrared and Raman spectra of MIO_4 (M = NH_4, Na, K, Rb, Cs and Ag); R. G. Brown, J. Denning, A. Hallett and S. D. Ross, *Spectrochim. Acta*, **26A** (1970) 963.

[19] Infrared spectra of M_5IO_6 (M = Li and Na) and $M_5(IO_6)_2$ (M = Ca, Sr and Ba); J. Hauck, *Z. Naturforsch.* **25b** (1970) 647.

[20] Infrared spectra of $Na_3H_2IO_6$, $Na_2H_3IO_6$, $NaIO_4,3H_2O$, $K_4I_2O_9$, $K_4I_2O_9,9H_2O$, $AgIO_4$, Ag_3IO_5, Ag_5IO_6, Ag_2HIO_5 and $Ag_2H_3IO_6$; J. R. Kyrki, *Suomen Kemistilehti*, **B38** (1965) 192.

[21] Infrared spectra of many alkali-metal periodates; H. Siebert, *Z. anorg. Chem.* **272** (1953) 1; *ibid.* **303** (1960) 162; *ibid.* **304** (1960) 266; *Fortschritte Chem. Forsch.* **8** (1967) 470.

[22] ^{127}I Mössbauer spectra of $NaIO_4$ and $Na_3H_2IO_6$; P. Jung and W. Triftshäuser, *Phys. Rev.* **175** (1968) 512.

[23] ^{129}I Mössbauer spectrum of KIO_4; D. W. Hafemeister, G. de Pasquali and H. de Waard, *Phys. Rev.* **135B** (1964) 1089.

[24] Spectrum of $NaClO_4$; A. Fahlman, R. Carlsson and K. Siegbahn, *Arkiv. Kemi*, **25** (1966) 301.

[25] Spectrum of KIO_4; C. S. Fadley, S. B. M. Hagstrom, M. P. Klein and D. A. Shirley, *J. Chem. Phys.* **48** (1968) 3779.

[26] Spectrum of $NaClO_4$; V. I. Nefedov, *J. Struct. Chem.* **8** (1967) 919; W. Nefedow, *Phys. Stat. Solidi*, **2** (1962) 904.

[27] Spectrum of $NaClO_4$; J. A. Bearden, *Rev. Mod. Phys.* **39** (1967) 86; D. S. Urch, *J. Chem. Soc.* (*A*) (1969) 3026.

[28] Decomposition of $KClO_4$; W. H. Johnson and A. A. Gilliland, *J. Res. Nat. Bur. Stand.* **65A** (1961) 63.

[29] Decomposition of $KBrO_4$; G. K. Johnson, P. N. Smith, E. H. Appelman and W. N. Hubbard, *Inorg. Chem.* **9** (1970) 119.

[30] Decomposition of alkali-metal perchlorates; M. M. Markowitz and D. A. Boryta, *J. Phys. Chem.* **69** (1965) 1114.

[31] Decomposition of perchlorates of Mg, Ba, Cd, Zn and Pb; F. Solymosi, *Magy. Kem. Foly.* **74** (1968) 155.

[32] Decomposition of perchlorates and metaperiodates of K, Rb and Cs; O. N. Breusov, N. I. Kashina and T. V. Revzina, *Russ. J. Inorg. Chem.* **15** (1970) 316.

[33] For a review of this subject see M. Drátovský and L. Pačesová, *Russ. Chem. Rev.* **38** (1968) 243.

constants, reflecting the decrease in crowding around the central atom, are much more marked (Table 75). The frequencies of the first fully allowed electronic transitions $^1T_2(t_1{}^5t_2{}^1) \leftarrow {}^1A_1(t_1{}^6)$ increase in the order $IO_4^- < BrO_4^- < ClO_4^-$.

Perbromates are thermodynamically less stable than comparable perchlorates and periodates, and perbromic acid is potentially as strong an oxidizing agent as periodic acid (Tables 75 and 80). Comparison of the electronic structures and stabilities of ClO_4^-, BrO_4^- and IO_4^- using Hartree–Fock–Slater atomic wave functions and self-consistent charge and configuration (extended Hückel) MO calculations[568] shows that, although the total overlap population in BrO_4^- is slightly less than that in ClO_4^- or IO_4^-, none of the earlier ideas about the "non-existence" of perbromates (high *s*–*p* promotion energy, poor *d*–*p* overlap or high inner-shell repulsions) was well founded; neither is extra stability conferred on the IO_4^- ion by coulombic interactions or 4*f*-orbital participation. Inclusion of $(n+1)s$ and *nd* orbitals gives charges for the halogen atoms of +0·57 for Cl, +0·65 for Br and +0·75 for I; for ClO_4^- this is somewhat lower than the value obtained from spectroscopic techniques (Table 44), while investigations of IO_4^- (Table 76) have yielded charges for the iodine atom of +1·24 (E.S.C.A.) and +1·44 (Mössbauer effect).

Perchloric Acid and the Perchlorates

1. Preparation. The production of perchlorates as primary products may be accomplished in a number of ways[734]:

(1) Electrolytic oxidation of chloride or chlorate to perchlorate in aqueous solution.
(2) Action of a strong mineral acid (e.g. H_2SO_4) on a chlorate, to give perchlorate, chloride and chlorine dioxide.
(3) Thermal disproportionation of an alkali-metal chlorate, e.g. $KClO_3$ at *ca.* 500°C.
(4) Chemical oxidation of chlorates to perchlorates with, for example, ozone, peroxydisulphates or lead dioxide.

While the chemical methods are useful in the laboratory, only the electrolytic oxidation of chlorate is important commercially. The cells have a steel cathode and platinum anode, and the pH of the electrolyte is maintained at *ca.* 6·7; losses by reduction are minimized by the addition of chromate, which supposedly forms an insoluble film on the cathode. The predominant practice is to make sodium perchlorate first and thence other perchlorates by metathesis (Scheme 12).

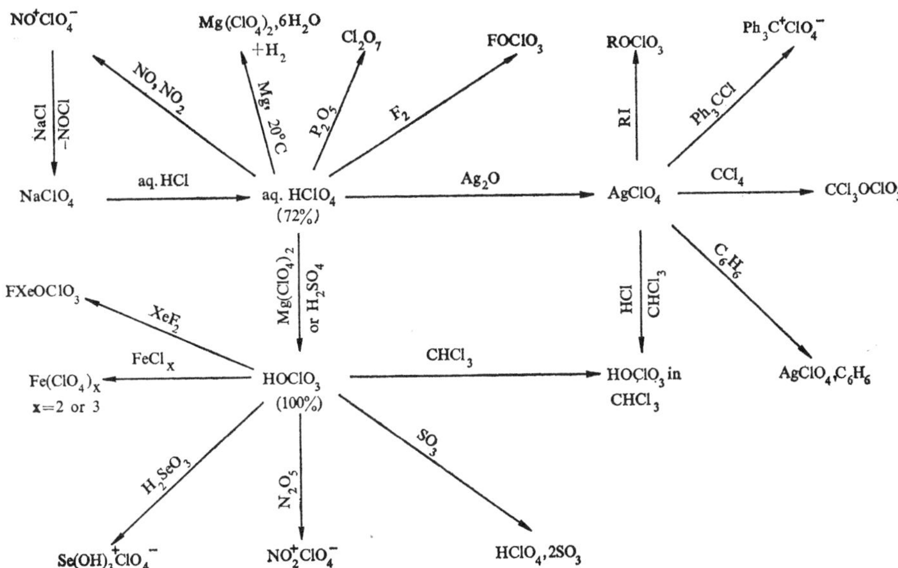

SCHEME 12. Interconversions of perchlorates.

Because of the size and limited coordinating power of the anion, many metal perchlorates crystallize from solution with hydrated cations; the synthesis of anhydrous perchlorates has been attempted by several routes. The low solubility of metal salts in anhydrous $HClO_4$ restricts one possible method. Some success has been secured in the reaction of $NO^+ClO_4^-$ with metal salts:

$$MX_n + nNO^+ClO_4^- \rightarrow M(ClO_4)_n + nNOX \qquad (X = F, Cl, Br \text{ or } NO_3)$$

The high solubility of $AgClO_4$ in organic solvents has been utilized in synthesizing perchlorates of non-metallic cations, and also in preparing esters of perchloric acid.

2. Perchloric Acid

Preparation[736]

Aqueous solutions of perchloric acid are obtained by treating anhydrous $NaClO_4$ or $Ba(ClO_4)_2$ with concentrated hydrochloric acid, filtering off the precipitated metal chloride, and concentrating the filtrate by distillation. Depending on the reaction conditions, dehydration of the constant boiling acid (72·4% $HClO_4$) affords either Cl_2O_7 (p. 1374) or anhydrous $HClO_4$ as the major product; contamination of $HClO_4$ with Cl_2O_7 may be minimized, but not eliminated, by using a modest ratio (*ca.* 3:1) of dehydrating agent (oleum or $Mg(ClO_4)_2$) to acid, and by immediately distilling the mixture at room temperature and very low pressure. Pure $HClO_4$ is best prepared by distilling the monohydrate $H_3O^+ClO_4^-$, obtained by treating crude anhydrous $HClO_4$ with the stoichiometric amount of the 72% acid (*not* with water). Deuteroperchloric acid has been obtained by distilling mixtures of D_2SO_4 with $NaClO_4$ or $Mg(ClO_4)_2$.

Properties

Some properties of anhydrous perchloric acid are listed in Table 77; more extensive tabulations will be found elsewhere[734,736,737]. The stability of the acid is adversely affected by impurities, especially Cl_2O_7, and many early studies of its chemistry are unreliable because the presence of Cl_2O_7 went unrecognized.

Molecular $HOClO_3$ has been defined in the vapour phase by electron diffraction (Table 77); infrared and Raman spectroscopic results affirm that the anhydrous acid retains this molecular form in the solid and liquid phases, although there is evidence of association *via* O–H···O hydrogen bonds. Recent investigations of some physical properties of liquid $HClO_4$ (vapour pressure, viscosity, electrical conductivity and electrolysis)[741,742] suggest that the dissociation

$$3HClO_4 \rightleftharpoons Cl_2O_7 + H_3O^+ + ClO_4^-$$

occurs with an equilibrium constant at 25°C of $0{\cdot}68\times10^{-6}$.

Anhydrous perchloric acid is an extremely powerful oxidizing agent; it reacts explosively with most organic substances, ignites hydrogen iodide and thionyl chloride, and rapidly oxidizes gold and silver. Some metal salts dissolve with formation of the anhydrous metal perchlorate, e.g.

$$Al_2Cl_6 + 6HClO_4 \rightarrow 2Al(ClO_4)_3 + 6HCl$$

$$Mn(NO_3)_2 + 6HClO_4 \rightarrow Mn(ClO_4)_2 + 2NO_2^+ClO_4^- + 2H_3O^+ClO_4^-$$

Solutions of $HClO_4$ in non-oxidizable organic liquids (e.g. $CHCl_3$, MeCN or $MeNO_2$) are best obtained by passing dry HCl gas into the appropriate solution of silver perchlorate. In contrast with the hypohalous acids, $HOClO_3$ does not add across olefinic double bonds.

Aqueous Solutions

Perchloric acid forms a constant boiling mixture with water which contains 72·4% by weight of the acid and boils at 203°C with slight decomposition. At room temperature aqueous solutions have the properties only of a very strong acid, dissolving active metals with liberation of hydrogen and forming substituted ammonium salts with amines. 79% perchloric acid has an indicator activity equivalent to that of 98% sulphuric acid. Spectroscopic studies (Table 78) show dissociation of the acid to be complete at concentrations < 6M,

[741] V. Ya. Rosolovskii, *Russ. J. Inorg. Chem.* **11** (1966) 1158.
[742] N. Bout and J. Potier, *Rev. Chim. Minérale*, **4** (1967) 621.

TABLE 77. PROPERTIES OF ANHYDROUS PERCHLORIC ACID

Thermodynamic and other physical properties	
$\Delta H_f°$ for $HClO_4$(l) at 25°C	−9·70 kcal mol^{-1} [a]
$S°$ for $HClO_4$(g) at 25°C	68·2 cal deg^{-1} mol^{-1} [b]
$C_p°$ for $HClO_4$(l) at 25°C	28·8 cal deg^{-1} mol^{-1} [b]
Colour	colourless when freshly distilled
Melting point	−101°C[b]
ΔH_{fusion}	1·657 kcal mol^{-1} [b]
Boiling point	120·5°C[c]
ΔH_{vap} at the boiling point	8·13 kcal mol^{-1} [c]
Trouton's constant	20·6 cal deg^{-1} mol^{-1} [c]
Vapour pressure	Ref. c
Density of the liquid at 25°C	1·761 g cm^{-3} [d]
Electrical conductivity	Ref. e
Molecular structure	
Electron diffraction:	Measured at 35°C[f]
r_g(1)[Cl=O]	1·408(2) Å
r_g(1)[Cl—OH]	1·635(7) Å
∠ O=Cl=O	112·8(5)°
∠ O=Cl—OH	105·8(7)°
Spectroscopic properties	
Infrared spectrum	Measured for $HClO_4$ and $DClO_4$ in the solid, liquid and vapour phases[g]
Raman spectrum	Measured for the liquid[h] and solid[i]
Vibrational assignment of skeletal fundamentals	$\nu_{symm}(ClO_3)$, 1036; ν(Cl—OH), 742; $\delta_{symm}(ClO_3)$, 577; $\nu_{asymm}(ClO_3)$, 1230; $\delta_{asymm}(ClO_3)$, 577; $\rho(ClO_3)$, 430 cm^{-1}
Force constants: f_r(Cl=O)	9·20 mdyne $Å^{-1}$
f_r(Cl—OH)	3·85 mdyne $Å^{-1}$
Mass spectrum	Refs. j, k, l
Ionization potential	13·05±0·1 eV, measured by electron impact[k]
Bond dissociation energy: $D(HO—ClO_3)$	46 kcal mol^{-1} [l]

[a] *Selected Values of Chemical Thermodynamic Properties*, N.B.S. Technical Note 270–3, Washington (1968).
[b] J. C. Trowbridge and E. F. Westrum, jun., *J. Phys. Chem.* **68** (1964) 42.
[c] D. J. Sibbett and I. Geller, cited in G. S. Pearson, *Adv. Inorg. Chem. Radiochem.* **8** (1966) 177.
[d] G. Mascherpa, *Bull. Soc. chim. France* (1961) 1259.
[e] N. Bout and J. Potier, *Rev. Chim. Minérale*, **4** (1967) 621.
[f] A. H. Clark, B. Beagley, D. W. J. Cruickshank and T. G. Hewitt, *J. Chem. Soc.* (*A*) (1970) 1613.
[g] P. A. Giguère and R. Savoie, *Canad. J. Chem.* **40** (1962) 495.
[h] A. Simon and M. Weist, *Z. anorg. Chem.* **263** (1952) 301.
[i] A. J. Dahl, J. C. Trowbridge and R. C. Taylor, *Inorg. Chem.* **2** (1963) 654.
[j] H. F. Cordes and S. R. Smith, *J. Chem. Eng. Data*, **15** (1970) 158.
[k] I. P. Fisher, *Trans. Faraday Soc.* **63** (1967) 684.
[l] G. A. Heath and J. R. Majer, *Trans. Faraday Soc.* **60** (1964) 1783.

and not less than 90% at 12 M. In 6 M solution there are four water molecules for every ionic species in solution, and incomplete ionization at higher concentration is a result of the limited number of water molecules available to solvate the proton. In accord with these observations, perchloric acid is normally extracted from water into ether as $[(H_2O)_4H]^+[ClO_4]^-$.

Perchloric acid solutions exhibit very little oxidizing power in the cold; they are slowly reduced by $SnCl_2$, V_2O_3 and V(II), V(III) and Ti(III) ions. Rates of reaction with V_2O_3 fall in the sequence $IO_4^- > BrO_3^- > IO_3^- > ClO_4^-$. The hot concentrated acid is a vigorous oxidizing agent, however, evolving hydrogen chloride with metals and decomposing organic materials; the "wet fire" method (used for ashing organic material prior to the determination of inorganic residues) utilizes perchloric acid alone or in mixtures with nitric or periodic acid.

TABLE 78. PROPERTIES OF AQUEOUS PERCHLORIC ACID

Thermodynamic properties at 298°K	
$\Delta H_f°$	−30·91 kcal mol^{-1} [a]
$\Delta G_f°$	−2·06 kcal mol^{-1} [a]
$S°$	43·5 cal deg^{-1} mol^{-1} [a]
$\Delta H_{dilution}$ ($HClO_4 \rightarrow HClO_4,\infty H_2O$)	−21·21 kcal mol^{-1} [a]
Electrical conductivity	Ref. b
Acidity	
Spectroscopic studies of dissociation for $[HClO_4] < 12$ M	Raman studies [c,d]
	^1H nmr studies [c,e,f]
	^{35}Cl nmr studies [c]
Hammett acidity function, H_0	Experiment [g]
	Calculated [h]

[a] *Selected Values of Chemical Thermodynamic Properties*, N.B.S. Technical Note 270–3, Washington (1968).
[b] L. H. Brickwedde, *J. Res. Nat. Bur. Stand.* **42A** (1949) 309.
[c] J. W. Akitt, A. K. Covington, J. G. Freeman and T. H. Lilley, *Trans. Faraday Soc.* **65** (1969) 2701.
[d] K. Heinzinger and R. E. Weston, jun., *J. Chem. Phys.* **42** (1965) 272.
[e] R. W. Duerst, *J. Chem. Phys.* **48** (1968) 2275.
[f] G. C. Hood and C. A. Reilly, *J. Chem. Phys.* **32** (1960) 127.
[g] Measured using substituted aniline indicators; K. Yates and H. Wai, *J. Amer. Chem. Soc.* **86** (1964) 5408.
[h] J. G. Dawber, *Chem. Comm.* (1966) 3.

Hydrates of Perchloric Acid

The melting point diagram of the system Cl_2O_7–H_2O indicates the existence of at least nine intermediate phases[743], several of which have been characterized by diffractometric and spectroscopic techniques (Table 79) as containing ClO_4^- ions hydrogen-bonded to $[H(OH_2)_x]^+$ cations. The monohydrate $H_3O^+ClO_4^-$ has been particularly well studied. The phase transition at 243°K is associated with a change in the rotation of the H_3O^+ ion in the lattice; proton spin-relaxation times for the polycrystalline solid demonstrate that below 243°K the H_3O^+ ion is rotating about a threefold axis, but that above 243°K the reorientation is essentially isotropic. The ClO_4^- anion in the orthorhombic high-temperature form is essentially tetrahedral, but the monoclinic low-temperature modification contains slightly distorted ClO_4^- anions such as have been found in the other hydrates; Cl–O bonds longer than the mean are involved in hydrogen-bonding. The $H_5O_2^+$ cation in $HClO_4,2H_2O$ is centrosymmetric.

Decomposition of Anhydrous Perchloric Acid

In contrast with its hydrates and aqueous solutions—72% $HClO_4$ boils at 203°C with

[743] G. Mascherpa, *Rev. Chim. Minérale*, **2** (1965) 379.

Table 79. Properties of the Hydrates of Perchloric Acid

Property	Composition					
	$(HClO_4)_4,H_2O$	$HClO_4,H_2O$	$HClO_4,2H_2O$	$HClO_4,5/2H_2O$	$HClO_4,3H_2O$	$HClO_4,7/2H_2O$
Structure	—	$[H_3O]^+[ClO_4]^-$	$[H_5O_2]^+[ClO_4]^-$	$[H_3O]^+{}_2[ClO_4]^-{}_2,3H_2O$	$[H_7O_3]^+[ClO_4]^-$	—
Melting point (°C)	Decomposes at −73·1[a]	+49·905[b]	−20·65±0·1[c]	−33·1±0·1[c]	−40·2±0·1[c]	−45·9±0·1[c]
Boiling point (°C)	—	—	203	—	—	—
Transition temperature (if any) (°C)	−100[a]	*ca.* −30[d]	—	—	—	—
$\Delta H_f°$ at 25°C (kcal mol^{-1})	—	−91·35[e]	−162·04[e]	—	—	—
X-ray diffraction	—	at 25°C[f,g] at −80°C[h]	at −196°C[i]	at −188°C[j]	at −188°C[k]	—
Raman spectrum	—	at temperatures in the range −50° to 25°C[l]	at −180°C[m]	—	—	—
Infrared spectrum	—	at −180°C[n] and 25°C	at −180°C[m]	—	—	—
^1H nmr spectrum	—	at temperatures in the range −180° to 25°C[d]	—	—	—	—
^{35}Cl nqr spectrum	—	at temperatures in the range −180° to 25°C[d]	—	—	—	—

[a] G. Mascherpa, *Compt. rend.* **252** (1961) 1800.
[b] G. F. Smith and O. E. Goehler, *Ind. Eng. Chem. Anal. Edn.* **3** (1931) 61.
[c] G. Mascherpa, A. C. Pavia and A. Potier, *Compt. rend.* **254** (1962) 3215.
[d] D. E. O'Reilly, E. M. Peterson and J. M. Williams, *J. Chem. Phys.* **54** (1971) 96.
[e] *Selected Values of Chemical Thermodynamic Properties*, N.B.S. Technical Note 270–3, Washington (1968).
[f] Average r(Cl–O) = 1·452(5) Å; M. R. Truter, *Acta Cryst.* **14** (1961) 318.
[g] Average r(Cl–O) = 1·42(1) Å; F. S. Lee and G. B. Carpenter, *J. Phys. Chem.* **63** (1959) 279.
[h] Average r(Cl–O) = 1·464(6) Å; C. E. Nordman, *Acta Cryst.* **15** (1962) 18.
[i] Average r(Cl–O) = 1·438(3) Å; I. Olovsson, *J. Chem. Phys.* **49** (1968) 1063.
[j] Average r(Cl–O) = 1·434(2) Å; 1·439(2) Å; J. Ahmlöf, J.-O. Lundgren and I. Olovsson, *Acta Cryst.* **B27** (1971) 898.
[k] Average r(Cl–O) = 1·437(2) Å; J. Ahmlöf, *Acta Cryst.* **B28** (1972) 481.
[l] R. C. Taylor and G. L. Vidale, *J. Amer. Chem. Soc.* **78** (1956) 5999.
[m] A. C. Pavia and P. A. Giguère, *J. Chem. Phys.* **52** (1970) 3551.
[n] R. Savoie and P. A. Giguère, *J. Chem. Phys.* **41** (1964) 2698.

only slight decomposition—anhydrous $HClO_4$ is apparently unstable even at $-78°C$[744]. The explosive reputation of $HClO_4$ itself is probably exaggerated, at least for very pure samples; however, the acid ages badly, and the presence of impurities, especially Cl_2O_7, greatly facilitates decomposition.

Pyrolysis of perchloric acid vapour

$$4HClO_4 \rightarrow 2H_2O + 2Cl_2 + 7O_2$$

has been studied in the temperature range 150–439°C[745,746]. Above 310°C the reaction is homogeneous and first-order in $HClO_4$; the suggested rate-determining step is rupture of the Cl–OH bond,

$$HOClO_3 \rightarrow HO + ClO_3$$

followed by a fast reaction between a hydroxyl radical and another perchloric acid molecule

$$HO + HOClO_3 \rightarrow H_2O + ClO_4$$

and the decomposition of ClO_3 and ClO_4 to chlorine and oxygen *via* ClO_2 and ClO. The experimental activation energy (45·1 kcal mol^{-1}) accords well with the dissociation energy of the Cl–OH bond measured by electron impact (46 kcal mol^{-1})[747] or calculated from thermodynamic data (47·6 kcal mol^{-1}). Below 310°C the decomposition is heterogeneous and second-order in $HClO_4$, while at temperatures well above 450°C hydrogen chloride is produced by reaction between chlorine and water:

$$Cl_2 + H_2O \rightarrow 2HCl + \tfrac{1}{2}O_2$$

On standing at room temperature, the initially colourless, mobile, hygroscopic liquid discolours, passing through yellow and red to become dark brown in a matter of 2 days; at this point evolution of oxygen and chlorine oxides sets in, and after 4 days the liquid, now colourless again, deposits white crystals of $H_3O^+ClO_4^-$. Again the initial step is cleavage of the Cl–OH bond. The instability of anhydrous $HClO_4$ compared with its solutions is caused by the relative weakness of the Cl–OH bond, which is of course absent in the dissociated acid. The presence of only small amounts of dichlorine heptoxide significantly alters the character of the decomposition.

Other Systems Involving Perchloric Acid

Perchloric acid is sufficiently strong an acid to protonate many molecules which are themselves acidic in aqueous solution: selenous acid and perchloric acid form the compound $[Se(OH)_3]^+ClO_4^-$, while mixtures of anhydrous nitric and perchloric acids at $-40°C$ give evidence for the equilibrium

$$HNO_3 + 2HClO_4 \rightleftharpoons NO_2^+ClO_4^- + H_3O^+ClO_4^-$$

Physical properties have been reported for systems involving perchloric acid and the following acids[734,735,736]: $CH_nCl_{3-n}CO_2H$ (n = 0, 1, 2 or 3), H_2SO_4, HNO_2, H_3PO_4 and CF_3CO_2H[748].

744 A. A. Zinov'ev, *Zhur. Neorg. Khim.* **3** (1958) 1205.
745 J. B. Levy, *J. Phys. Chem.* **66** (1962) 1092.
746 D. J. Sibbett and I. Geller, cited in G. S. Pearson, *Adv. Inorg. Chem. Radiochem.* **8** (1966) 177.
747 G. A. Heath and J. R. Majer, *Trans. Faraday Soc.* **60** (1964) 1783.

Perchloric acid is apparently completely dissociated in solvents such as acetonitrile. In organic acids, however, it is only moderately strong; for example, the pK_a value is *ca.* 0·5 in formic acid, 2·7 in acetic acid and *ca.* 1·3 in trifluoroacetic acid. Perchloric acid has been used in acetic acid (or acetic acid–acetic anhydride mixtures) for the potentiometric or conductimetric titration of organic bases[749,750], such as amines or phosphine oxides, which give only diffuse end-points in aqueous solution. The method has been extended to the analysis of ketones by titration as semicarbazone or hydrazone derivatives[751], and other solvents have been used, including chlorobenzene and acetonitrile.

Infrared and Raman investigations of solutions of $HClO_4$ in acetic anhydride have shown[752,753] that the equilibrium

$$(CH_3CO)_2O + HOClO_3 \rightleftharpoons CH_3C(O)OClO_3 + CH_3CO_2H$$

is established. Acetyl perchlorate has also been prepared from silver perchlorate and acetyl chloride; it is a vigorous acetylating agent. The benzoyl analogue is also known.

3. Ionic Perchlorates. The many perchlorate salts which have been isolated contain both inorganic and organic cations. The latter compounds, although unstable thermodynamically, are isolable at room temperature because of the kinetic inertness of the perchlorate ion; they may be dangerously explosive if subjected to shock or heating. Perchloric acid forms salts $BH^+ClO_4^-$ by protonating weak organic bases B (amines, amine oxides or ketones); carbonium and diazonium perchlorates are also known. Non-metallic inorganic cations include NO^+, NO_2^+ and $N_2H_5^+$.

The large size of the perchlorate ion regulates many of the properties and uses of the compounds. Thus, many metal perchlorates crystallize with hydrated cations, and ammines of perchlorates have been reported. Anhydrous perchlorates, e.g. $Mg(ClO_4)_2$, are frequently deliquescent, and absorb not only water but many other polar molecules from the vapour phase. Perchlorates of large unipositive cations tend to be insoluble in water; that $KClO_4$ is far less soluble than $NaClO_4$ has been utilized in the separation of these alkali metals, while complex cations are often conveniently crystallized as perchlorates for subsequent investigation by X-ray diffraction. By contrast, the perchlorates of lithium, sodium, the alkaline earths and especially silver are extremely soluble in organic media. Silver perchlorate is used in many metathetical preparations of other perchlorates (Scheme 12), and forms crystalline complexes $AgClO_4$,L, $AgClO_4$,2L and $4AgClO_4$,L (L = aromatic hydrocarbon) in which the organic molecule is bonded to the cation; some of these compounds have been studied crystallographically, e.g. $AgClO_4,C_6H_6$ [754], $AgClO_4$,2(*m*-xylene)[755] and $4AgClO_4$,L (L = naphthalene or anthracene)[756]. The relation between the relative sizes of counter-ions, stoichiometry and the solubility of salts has been discussed in terms of the variation in lattice and solvation energies[757].

748 J. Bessière, *Bull. Soc. chim. France*, (1969) 3356.

749 A. H. Beckett and E. H. Tinley, *Titration in Non-aqueous Solvents*, 3rd edn., B.D.H., Poole (1960); I. M. Kolthoff and S. Bruckenstein, *Treatise on Analytical Chemistry* (ed. I. M. Kolthoff, P. J. Elving and E. B. Sandell), Part I, Vol. I, p. 475, Interscience, New York (1959).

750 V. Vajgand and T. Pastor, *J. Electroanal. Chem.* **8** (1964) 40.

751 D. B. Cowell and B. D. Selby, *Analyst*, **88** (1963) 974.

752 A.-M. Avedikian and A. Commeyras, *Bull. Soc. chim. France*, (1970) 1258.

753 E. S. Sorokin, N. I. Geidel'man and V. Ya. Bytenskii, *Zhur. priklad. Khim.* **43** (1970) 1595.

754 H. G. Smith and R. E. Rundle, *J. Amer. Chem. Soc.* **80** (1958) 5075.

755 I. F. Taylor, jun., E. A. Hall and E. L. Amma, *J. Amer. Chem. Soc.* **91** (1969) 5745.

756 E. A. Hall and E. L. Amma, *J. Amer. Chem. Soc.* **91** (1969) 6538.

757 D. A. Johnson, *Some Thermodynamic Aspects of Inorganic Chemistry*, pp. 97–115, Cambridge (1968).

Structural Aspects

Essentially tetrahedral ClO_4^- anions have been defined crystallographically in the salts listed in Table 76, though rather more precise dimensions have been obtained for the crystalline hydrates of perchloric acid (Table 79). The bond lengths, uncorrected for thermal motion, are *ca.* 1·44 Å; including the correction increases the bond lengths to *ca.* 1·46 Å. Many early studies of the perchlorates were inaccurate, and these have not been listed; likewise excluded from the list are some more recent crystallographic investigations of perchlorates containing complex cations, which achieved their main aim of characterizing the cation at the expense of reporting chemically unrealistic parameters for the ClO_4^- unit.

Coordination of perchlorate groups to metal ions has been detected by vibrational spectroscopy: while the Raman spectra of concentrated aqueous solutions show no effects not attributable to ion-pairing[758], the infrared spectra of some complex transition metal perchlorates evince coordination of the ClO_4^- unit to the cation. The discrete tetrahedral ClO_4^- ion has four characteristic Raman lines (Table 75), two of which coincide with distinctive infrared absorptions at *ca.* 1050–1150 cm^{-1} (ν_3) and 630 cm^{-1} (ν_4); the totally symmetric stretching fundamental, although formally forbidden, is often observed as a weak band at *ca.* 930 cm^{-1} in the infrared spectra of the solids. If the perchlorate ion becomes coordinated to a metal ion by one oxygen atom, the local symmetry of the $-OClO_3$ group is reduced to C_{3v} (with the assumption of a linear M–O–Cl unit). As a result, the broad degenerate ν_3 band splits into two well-defined features between 1000 and 1200 cm^{-1}; stretching of the coordinated Cl–O bond (formally related to ν_1 of ClO_4^-) is seen as an intense infrared absorption at 925–950 cm^{-1}, while the degenerate bending mode ν_4 also splits into two bands. Such coordination has been found for perchlorates of some divalent metals, including *trans*-Co[*o*-phenylenebisdimethylarsine]$_2(ClO_4)_2$ [759], $Ni(MeCN)_x(ClO_4)_2$ ($x = 2$ or 4)[760] and $NiL_2(ClO_4)_2$ (where L = N-methylethylenediamine or N,N′-dimethylethylenediamine)[761]. The magnitude of the band splittings observed is apparently consistent with a force constant of *ca.* 6 mdyne $Å^{-1}$ for the coordinated Cl–O bond, as opposed to *ca.* 8 mdyne $Å^{-1}$ for the terminal Cl–O bonds[762].

Bidentate perchlorate groups, characterized by further changes in the infrared spectrum, have been suggested for the compounds $NiL_2(ClO_4)_2$ (L = N,N,N′-trimethylethylenediamine)[761], Ni(ethylenediamine)$_2(ClO_4)_2$ [763] and $UO_2(ClO_4)_2$ [764].

Thermal Decomposition

Depending on the cation, metal perchlorates decompose to yield the corresponding oxide or chloride; in most cases the reaction achieves a measurable rate below 600°C.

$$M(ClO_4)_n \rightarrow MCl_n + 2nO_2$$
$$M(ClO_4)_n \rightarrow MO_{n/2} + n/2Cl_2 + 7n/4O_2$$

The manner of decomposition may be rationalized by considering the relative equivalent free energies of the possible products[765]. The perchlorates of the alkali metals, silver, calcium, barium, cadmium and lead decompose to afford the corresponding chlorides; for

758 R. E. Hester and R. A. Plane, *Inorg. Chem.* **3** (1964) 769.
759 G. A. Rodley and P. W. Smith, *J. Chem. Soc.* (*A*) (1967) 1580.
760 A. E. Wickenden and R. A. Krause, *Inorg. Chem.* **4** (1965) 404.
761 S. F. Pavkovic and D. W. Meek, *Inorg. Chem.* **4** (1965) 1091.
762 H. Brintzinger and R. E. Hester, *Inorg. Chem.* **5** (1966) 980.
763 M. E. Farago, J. M. James and V. C. G. Trew, *J. Chem. Soc.* (*A*) (1967) 820.
764 V. M. Vdovenko, L. G. Mashirov and D. N. Suglobov, *Soviet Radiochemistry*, **9** (1967) 37.
765 M. M. Markowitz, *J. Inorg. Nuclear Chem.* **25** (1963) 407.

these elements, $\Delta G_f°$(chloride) $- \frac{1}{2}\Delta G_f°$(oxide) < -20 kcal mol^{-1}. Aluminium and ferric perchlorates, on the other hand, produce the corresponding oxides (here $\Delta G_f°$(chloride) $- \frac{1}{2}\Delta G_f°$(oxide) > 0 kcal mol^{-1}), while for magnesium and zinc both reactions are observed. Dehydration of hydrated perchlorates may be difficult, commonly being accompanied by decomposition, in the event of which hydrolysis reactions may also occur.

Decomposition of the alkali-metal perchlorates involves little heat change:

$$MClO_4 \rightarrow MCl + 2O_2 \qquad \Delta H° \sim 0$$

While the corresponding chlorates have long liquid ranges, rapid decomposition of the heavier alkali-metal perchlorates begins at *ca.* 580°C and apparently causes melting; the exact temperature depends on particle size and on the previous history of the sample. Only $LiClO_4$ has a congruent melting point (at 247°C). There is no appreciable accumulation of intermediate chlorates, which are, of course, highly unstable at the temperatures involved; by contrast, formation of the thermally more stable halate is a well-defined step in the pyrolysis of alkali-metal perbromates and metaperiodates (Table 73). The activation energy for perchlorate decomposition (corresponding to rupture of a Cl–O bond) decreases from 60–70 kcal mol^{-1} for the alkali-metal perchlorates to 40 kcal mol^{-1} for $Zn(ClO_4)_2$.

The thermal decomposition of ammonium perchlorate has been extensively studied, not only for its chemical interest, but also because of its application in the oxidation of rocket fuels; there have been three recent reviews[766,767,768]. The decomposition was first observed in 1869 when the simple reaction $NH_4ClO_4 \rightarrow NH_4Cl + 2O_2$ was proposed; it has since been realized, however, that three fairly distinct stages are involved:

(1) "Low-temperature" decomposition in the range *ca.* 200–300°C, which is characterized by an induction period, an acceleratory region, a rate maximum and a deceleratory region, the reaction stopping before all the material is consumed: the approximate stoichiometry is

$$4NH_4ClO_4 \rightarrow 2Cl_2 + 8H_2O + 2N_2O + 3O_2$$

(2) "High-temperature" decomposition, between 350°C and 400°C, with immeasurably fast initiation, the stoichiometry being

$$2NH_4ClO_4 \rightarrow Cl_2 + 4H_2O + 2NO + O_2$$

(3) Deflagration at *ca.* 450°C, for which two limiting equations have been found:

(a) at low pressure:

$$2NH_4ClO_4 \rightarrow Cl_2 + 4H_2O + 2NO + O_2$$

(b) at high pressure:

$$4NH_4ClO_4 \rightarrow 4HCl + 6H_2O + 2N_2 + 5O_2$$

Other minor products reported in individual studies include N_2O, N_2O_3, NO_2, N_2O_4, N_2, HCl, ClO_2, NOCl, NO_2Cl, HNO_3 and $HClO_4$.

Careful investigation has shown that the reaction is first-order and that the activation energy (*ca.* 30 kcal mol^{-1}) is the same in all three regions; the rate-determining step is believed to be proton-transfer from NH_4^+ to ClO_4^- in the crystal, followed by evaporation into the gas phase. Decomposition of $HClO_4$ and subsequent oxidation of NH_3 by the fragmentation products in vapour-phase reactions account for the majority of the observed

[766] A. G. Keenan and R. F. Siegmund, *Quart. Rev. Chem. Soc.* **23** (1969) 430.
[767] V. V. Boldyrev, *Dokl. Akad. Nauk S.S.S.R.* **181** (1968) 1406.
[768] P. W. M. Jacobs and H. M. Whitehead, *Chem. Rev.* **69** (1969) 551.

products, whose distribution depends on the reaction conditions.

$$\begin{array}{ccccccc} NH_4^+ClO_4^-(s) & \rightleftharpoons & NH_3(ads.) & + & HClO_4(ads.) & \rightsquigarrow & \text{products} \\ & & \updownarrow & & \updownarrow & & \\ \text{sublimate} & \longleftarrow & NH_3(g) & + & HClO_4(g) & \longrightarrow & \text{products} \end{array}$$

Radiolytic Decomposition

The action of X-rays and γ-rays on the alkali-metal and alkaline earth perchlorates has been comprehensively studied[769]. Chemically identifiable fragments include the metal oxide (or superoxide), ClO_3^-, ClO_2, ClO_2^-, ClO^-, Cl^- and O_2; O_3^-, ClOO, ClO_3 and ClO_4 have been detected by esr spectroscopy as paramagnetic centres stable at low temperature[610]. Similar results have been obtained in the radiolysis of frozen or liquid aqueous solutions of perchloric acid, although these studies are complicated by the reaction of radical intermediates with water[736]. The mechanism is complex, but probably involves several of the possible primary decomposition processes as well as excitation of ClO_4^-.

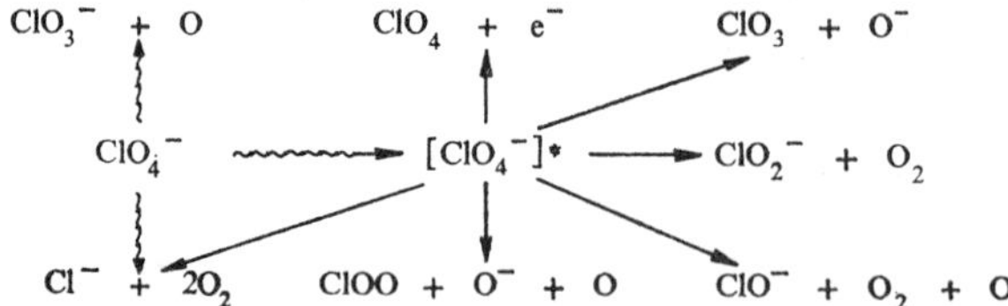

It is interesting to note that ClOO is a *precursor* of chlorine dioxide, and that ClO_4 probably has the structure O_2ClOO[610]; Cl^- is a primary product in radiolysis of the solid, whereas in solution it arises from the attack on ClO_3^- by various free radicals.

4. Esters of Perchloric Acid

Alkyl Perchlorates

Ever since their discovery in 1841 the alkyl perchlorates have been recognized as treacherously explosive compounds; their sensitivity to shock decreases as the molecular weight rises, although the higher temperatures needed to distil the compounds of higher molecular weight introduce a compensatory hazard. The perchlorate esters have been stabilized in the form of urea inclusion compounds[770].

Although the oily liquids may be obtained from the anhydrous acid and the alcohol at low temperature, the best preparative procedure involves metathesis between silver perchlorate and an alkyl halide in an organic solvent. The infrared spectra of the liquids are consistent with the structure $ROClO_3$ [770]; features at *ca.* 705, 1035, 1230 and 1260 cm^{-1} are attributed to vibrations of the $-OClO_3$ group.

Studies have been made of the substitution reactions of methyl perchlorate. Hydrolysis involves attack at both the chlorine and carbon atoms, while methanolysis is catalysed by ClO_4^- [771]:

$$MeOClO_3 + MeOH \rightarrow Me_2O + HOClO_3$$

---

[769] L. A. Prince and E. R. Johnson, *J. Phys. Chem.* **69** (1965) 359, 377.

[770] J. Radell, J. W. Connolly and A. J. Raymond, *J. Amer. Chem. Soc.* **83** (1961) 3958.

[771] D. N. Kevill and H. S. Posselt, *Chem. Comm.* (1967) 438; J. Koskikallio, *Suomen Kemistilehti*, **40B** (1967) 199.

Extremely explosive materials have been made by partial or complete reaction of anhydrous perchloric acid with polyfunctional alcohols including glycol, glycerol and pentaerythritol; explosions have occurred merely on pouring the liquids from one container to another (cf. nitroglycerine). The compound CCl_3OClO_3 explodes on contact with organic material.

Inorganic Molecular Perchlorates

Inorganic analogues of the esters of perchloric acid include the halogen perchlorates $FOClO_3$, $ClOClO_3$, $BrOClO_3$ and $I(OClO_3)_3$, which are discussed fully in Section 4B5, and the xenon compounds $FXeOClO_3$ and $Xe(OClO_3)_2$, prepared from xenon difluoride and anhydrous perchloric acid[601], and involving linear coordination about the xenon atom.

5. Uses

In Explosives and Propellants[734,768]

$KClO_4$ and NH_4ClO_4 are important oxygen-carriers for solid explosives and propellants; more powerful explosives contain ammonium perchlorate, since with suitably combustible material this gives entirely gaseous products:

$$NH_4ClO_4 + 2C \xrightarrow{\text{induced explosion}} NH_3 + HCl + 2CO_2$$

Perchlorate explosives are not sensitive to shock and are capable of wide modification to suit the particular purpose in hand. Perchlorates have also been included in slower-burning mixtures used for fireworks or signal flares; heavy metal salts are used to introduce colours.

Typical rocket propellants, whose combustion might be described as a slow explosion, consist of 75% NH_4ClO_4, 20% fuel (which also serves to bind the material into pellets) and 5% additives to provide desired effects on physical properties, storage or burning capabilities. Lithium perchlorate has recently been used as an oxidizer since it has twice as much oxygen per unit volume as NH_4ClO_4 and gives a higher combustion temperature.

Other Uses

Several analytical uses of perchloric acid—as an acid in non-aqueous titrimetry and as an oxidant in the wet ashing of organic material—have been mentioned in the preceding text. Perchloric acid has also been used for the digestion of chromium, including its quantitative separation from other metals, as an electrolyte (together with acetic acid) in the electro-polishing of aluminium, and as a catalyst for esterification reactions. Miscellaneous uses are listed elsewhere[736].

Perchlorates have been used as drying agents: anhydrous $Mg(ClO_4)_2$ can absorb up to 60% of its own weight of water, and does not become sticky upon handling; regeneration can be achieved *in vacuo* at 200°C. The minimal complexing ability of the ClO_4^- anion in aqueous solution has led to the widespread adoption of perchlorates to provide media of constant ionic strength for the study of the kinetics and equilibrium constants of many aqueous processes.

Perbromic Acid and the Perbromates[575b]

First obtained in studies of the β-decay of $^{83}SeO_4^{2-}$ [663], perbromates are also formed in the γ-radiolysis of crystalline bromates[725]. Oxidation of bromate ions in aqueous solution is effected electrolytically[663], with XeF_2 [663], or, most expeditiously, by passing fluorine into strongly basic solutions containing bromate[731].

Perbromic acid is a strong acid bearing one proton per heptavalent bromine atom; in aqueous solution it is completely dissociated into aquated protons and tetrahedral BrO_4^- anions, which do not exchange oxygen with $H_2{}^{18}O$. Solutions more than 6 M (55%) in $HBrO_4$ are unstable in air, undergoing an apparently autocatalytic decomposition which is complete for 80% concentrations: decomposition is also catalysed by metal ions, e.g. Ag^+ and Ce^{4+}. Despite this decomposition, some perbromic acid may be distilled, and the mass spectrum of $HBrO_4$ has been reported[772]. On very rapid evaporation, crystallization of perbromic acid solutions occurs (possibly to give $HBrO_4,2H_2O$) just before decomposition sets in. Attempts to dehydrate the acid have met only with decomposition.

The vibrational spectra reported for the tetrahedral BrO_4^- anion vary but little, irrespective of whether the ion is in solution or in various crystalline salts. $KBrO_4$ crystallizes with the orthorhombic Barite structure[773], being isomorphous with the room-temperature modifications of the alkali-metal perchlorates; the BrO_4^- unit is tetrahedral within the limits of experimental error, having a Br–O bond length of 1·61 Å. $KBrO_4$ is converted by SbF_5 into BrO_3F (p. 1393)[653].

The exothermic reaction

$$KBrO_4(c) \rightarrow KBr(c) + 2O_2(g) \qquad \Delta H^\circ_{298} = -25{\cdot}38 \pm 0{\cdot}10 \text{ kcal mol}^{-1}$$

proceeds in two stages: (i) production of $KBrO_3$ at *ca.* 275°C, and (ii) its subsequent decomposition at *ca.* 390°C[774]. From the calorimetric study of this decomposition, and of the dissolution of $KBrO_4$ in water, were derived the thermodynamic properties of $KBrO_4$ and BrO_4^-(aq) listed in Tables 73 and 75[774]. The order of thermal stability $KBrO_3 > KBrO_4$ resembles that for the iodine systems rather than that for the analogous chlorine compounds ($KClO_4 > KClO_3$). Unlike NH_4ClO_4, NH_4BrO_4 [775] is neither shock- nor friction-sensitive; in its thermal decomposition (producing N_2, Br_2, O_2 and H_2O), it resembles NH_4BrO_3 rather than NH_4ClO_4 (p. 1449).

Despite the large potential (+1·76 volts) of the BrO_4^-/BrO_3^- couple in acidic solution, perbromic acid is described as a sluggish oxidizing agent at room temperature, though somewhat more powerful than perchloric acid: dilute solutions oxidize Br^- and I^- only slowly, but the 12 M acid rapidly oxidizes chloride. At 100°C 6 M perbromic acid fairly rapidly converts Mn^{2+} to MnO_4^-. Reduction to bromide may be accomplished with $SnCl_2$.

Periodic Acid and the Periodates

Nomenclature

Despite efforts[740] to establish a logical system for naming derivatives of H_5IO_6 in various stages of deprotonation, dehydration and aggregation, the older nomenclature is still most commonly used, based on four acids: H_5IO_6, *para-* or *ortho*periodic acid; HIO_4, *meta*periodic acid; (hypothetical) H_2IO_5, *meso*periodic acid; and $H_7I_3O_{14}$, *tri*periodic acid. Protonated anions are described thus: $H_3IO_6^{2-}$, trihydrogen paraperiodate; but degrees of aggregation are difficult to render unambiguously using this system. In discussing periodates it is least confusing and cumbersome to use formulae whenever possible.

[772] M. H. Studier, *J. Amer. Chem. Soc.* **90** (1968) 1901.
[773] S. Siegel, B. Tani and E. H. Appelman, *Inorg. Chem.* **8** (1969) 1190.
[774] G. K. Johnson, P. N. Smith, E. H. Appelman and W. N. Hubbard, *Inorg. Chem.* **9** (1970) 119.
[775] J. N. Keith and I. J. Solomon, *Inorg. Chem.* **9** (1970) 1560.

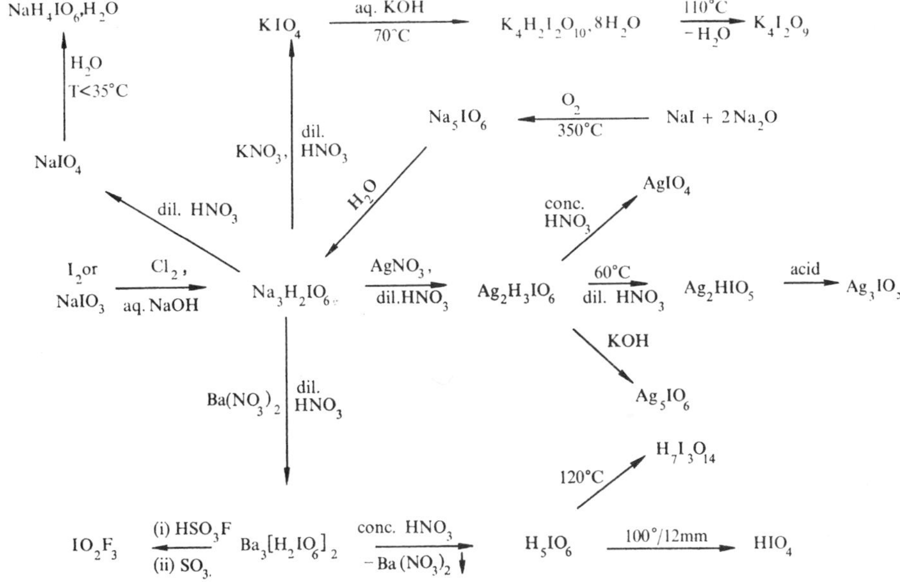

SCHEME 13. Interconversion of periodates.

Synthesis

The synthesis of periodates rests mainly on the oxidation of iodide, iodine or iodate in aqueous solution. Industrial processes employ electrochemical oxidation with a PbO_2 anode. A starting point for the laboratory preparation of many periodates[714] is the sodium salt $Na_3H_2IO_6$ (Scheme 13), conveniently produced by passing chlorine into aqueous alkaline solutions of iodine or sodium iodate[714]. The paraperiodates of the alkaline earth metals can be obtained by disproportionation of the iodates, e.g.

$$5Ba(IO_3)_2 \rightarrow Ba_5(IO_6)_2 + 4I_2 + 9O_2$$

The thermal stability of paraperiodates also allows their synthesis by the oxidation of solid mixtures of iodides and oxides with oxygen.

Periodic Acid

Aqueous solutions of periodic acid are best obtained by treating barium paraperiodate with concentrated nitric acid: the only solid phase stable in contact with such a solution is paraperiodic acid, H_5IO_6 (Table 80). Structural investigations of the white crystalline solid confirm that it is a genuine *ortho* acid $(HO)_5IO$ rather than a hydrate $HIO_4,2H_2O$ or an oxonium salt $[H_5O_2]^+IO_4^-$ analogous to the dihydrate of perchloric acid (Table 79). Dehydration of H_5IO_6 on heating to *ca.* 120°C produces $H_7I_3O_{14}$, while heating at 100°C *in vacuo* gives HIO_4, which has also been characterized in the mass spectrum of periodic acid; at higher temperatures evolution of oxygen sets in, producing initially a solid I_2O_5,I_2O_7 and, at 150°C, just I_2O_5. The reaction of H_5IO_6 with 60% oleum is said to produce an orange solid of approximate composition I_2O_7 [625].

Periodic acid is a fairly weak acid, and in acid solution exists in the undissociated form $(HO)_5IO$. The difference between the apparent (pK 1·6) and true (pK 3·29) first dissociation constants of H_5IO_6 is caused by hydration equilibria between CO_2 and the undissociated

TABLE 80. PROPERTIES OF PERIODIC ACID, H_5IO_6

Melting point	128·5°C (with decomposition)[a]
$\Delta H_f°$	−199·3 kcal mol^{-1} [b,c]
$\Delta H_f°$(aq)	−188·9 kcal mol^{-1} [b]
Dissociation constants	pK_1 3·29, pK_2 8·3, pK_3 11·6[c]
Reduction potential, in acid solution	
$E°(H_5IO_6/HIO_3)$	*ca.* +1·7 volts[d]
Crystal structure determination	$(HO)_5IO$ molecules
X-ray diffraction	[e]
Neutron diffraction	[f]
Bond lengths	[f]
r(I=O)	1·78(2) Å
r(I—OH)	1·89(2) Å
r_{av}(O—H)	0·96 Å
Infrared spectrum	solid[g,h]
Raman spectrum	solution[h,i]
Force constants[j]	
f_r(I=O)	5·4 mdyne $Å^{-1}$
f_r(I−OH)	3·0 mdyne $Å^{-1}$
Mass spectrum[k]	
Ions: HIO_4^+, HIO_3^+, HIO_2^+, IO_2^+, HIO^+, IO^+, HI^+, I^+	

[a] L. Pačesova and Z. Hauptman, *Z. anorg. Chem.* **325** (1963) 325.
[b] E. E. Mercer and D. T. Farrar, *Canad. J. Chem.* **46** (1968) 2679; J. H. Stern and J. J. Jasnosz, *J. Chem. Eng. Data*, **9** (1964) 534.
[c] See Tables 81 and 82.
[d] A. G. Sharpe, *Halogen Chemistry* (ed. V. Gutmann), Vol. 1, p. 13, Academic Press (1967).
[e] Y. D. Feikema, *Acta Cryst.* **14** (1961) 315.
[f] Y. D. Feikema, *Acta Cryst.* **20** (1966) 765.
[g] H. Siebert, *Z. anorg. Chem.* **303** (1960) 162.
[h] See also Table 84.
[i] H. Siebert, *Z. anorg. Chem.* **273** (1953) 21.
[j] H. Siebert, *Fortschritte Chem. Forsch.* **8** (1967) 470.
[k] M. H. Studier and J. L. Huston, *J. Phys. Chem.* **71** (1967) 457.

acid. In very strongly acidic media (e.g. 10 M $HClO_4$) there is evidence for the protonated cation $[I(OH)_6]^+$ [625]. The principal anions present in aqueous solutions of higher pH are tetrahedral IO_4^-, the hydrated species $[(HO)_4IO_2]^-$, $[(HO)_3IO_3]^{2-}$ and $[(HO)_2IO_4]^{3-}$, and the dimer $[(HO)_2I_2O_8]^{4-}$. The equilibria which link the various species (Tables 81 and 82) are apparently established rapidly and are pH-dependent. Hydration of the metaperiodate ion

$$IO_4^- + 2H_2O \rightleftharpoons H_4IO_6^-$$

is not catalysed by hydrogen ions, and is among the most rapid reactions of its kind; the specific (pseudo-first-order) rate constant is $(1·9 \pm 0·2) \times 10^2$ sec^{-1} at 20°C (ionic strength 0·1 M, pH range 3·4–5·0). The mechanism is still uncertain: both one-step and two-step paths (Scheme 14) are consistent with the kinetic data, though for various reasons the two-step mechanism (involving a five-coordinate intermediate) is considered more likely[776].

Solutions of periodic acid decompose slowly into iodic acid and ozonized oxygen, a reaction which is catalysed by the presence of colloidal platinum.

[776] K. Kustin and E. C. Lieberman, *J. Phys. Chem.* **68** (1964) 3869.

TABLE 81. INVESTIGATIONS OF PERIODATE EQUILIBRIA IN AQUEOUS SOLUTION

Medium	Temperature (°C)	Technique(s)[a]	Reference
$0 \lesssim pH \lesssim 12$	0–70	u.v.; pot. tit. of periodate with OH^-	C. E. Crouthamel, A. M. Hayes and D. S. Martin, *J. Amer. Chem. Soc.* **73** (1951) 82.
aq $HClO_4$ ($\leq$ 10 M)	25	u.v.	H. C. Mishra and M. C. R. Symons, *J. Chem. Soc.* (1962) 1194.
$0 \lesssim pH \lesssim 11$	5–75	solubility of Ph_4AsIO_4	S. H. Laurie, J. M. Williams and C. J. Nyman, *J. Phys. Chem.* **68** (1964) 1311.
$3{\cdot}4 < pH < 5{\cdot}0$	20	kinetics	K. Kustin and E. C. Lieberman, *J. Phys. Chem.* **68** (1964) 3869.
		dil. tit. of H_5IO_6 with OH^-	K. F. Jahr and E. Gegner, *Angew. Chem., Internat. Edn.* **6** (1967) 707.
	25	calorimetry	E. E. Mercer and D. T. Farrar, *Canad. J. Chem.* **46** (1968) 2679.
$1{\cdot}3 < pH < 5{\cdot}8$	0–40	solubility of $CsIO_4$; ^{127}I nmr	R. M. Kren, H. W. Dodgen and C. J. Nyman, *Inorg. Chem.* **7** (1968) 446.
$pH > 7$	25	pot. tit.; ultracent.; Raman	J. Aveston, *J. Chem. Soc.* (*A*) (1969) 273.
$6 < pH < 12$	1–70	u.v.; Raman	G. J. Buist, W. C. P. Hipperson and J. D. Lewis, *J. Chem. Soc.* (*A*) (1969) 307.

[a] u.v., ultraviolet absorption spectroscopy; pot. tit., potentiometric titration; dil. tit., dilatometric titration; ultracent., ultracentrifuge studies.

TABLE 82. PERIODATE EQUILIBRIA IN AQUEOUS SOLUTION

Equilibrium	K_{298}	ΔH° (kcal mol^{-1})
$H_6IO_6^+ \rightleftharpoons H_5IO_6 + H^+$	6·3[a]	
$H_5IO_6 \rightleftharpoons H_4IO_6^- + H^+$	$5{\cdot}1 \times 10^{-4}$ [b]	$+2{\cdot}55 \pm 0{\cdot}7$[e]
$H_4IO_6^- \rightleftharpoons H_3IO_6^{2-} + H^+$	$4{\cdot}9 \times 10^{-9}$ [c]	~0[b]
$H_3IO_6^{2-} \rightleftharpoons H_2IO_6^{3-} + H^+$	$2{\cdot}5 \times 10^{-12}$ [c]	
$H_4IO_6^- \rightleftharpoons IO_4^- + 2H_2O$	40;[b] 29[d]	$+11{\cdot}9$[e]
$2H_3IO_6^{2-} \rightleftharpoons H_2I_2O_{10}^{4-} + 2H_2O$	*ca.* 820[c]	

[a] H. C. Mishra and M. C. R. Symons, *J. Chem. Soc.* (1962) 1194.
[b] C. E. Crouthamel, A. M. Hayes and D. S. Martin, *J. Amer. Chem. Soc.* **73** (1951) 82.
[c] G. J. Buist, W. C. P. Hipperson and J. D. Lewis, *J. Chem. Soc.* (*A*) (1969) 307.
[d] R. M. Kren, H. W. Dodgen and C. J. Nyman, *Inorg. Chem.* **7** (1968) 446.
[e] E. E. Mercer and D. T. Farrar, *Canad. J. Chem.* **46** (1968) 2679.

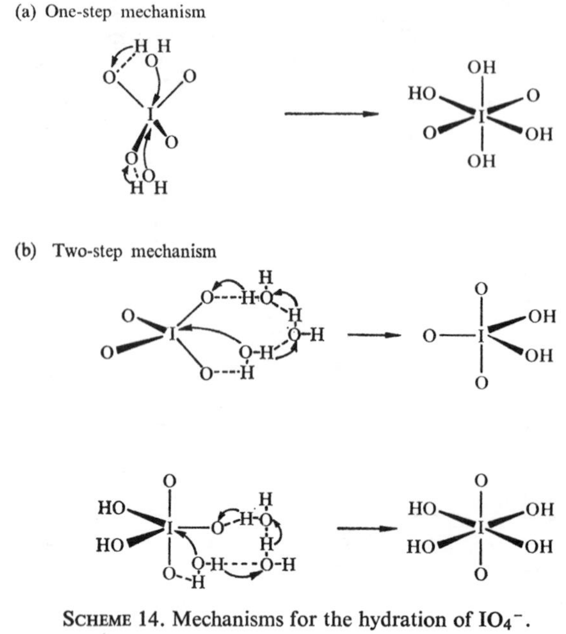

(a) One-step mechanism

(b) Two-step mechanism

SCHEME 14. Mechanisms for the hydration of IO_4^-.

Periodates[624,740]

Salts of periodic acid have been isolated which contain, or purport to contain, the following anions:

metaperiodates:	IO_4^-
mesoperiodates:	$I_2O_9^{4-}$, $I_2O_{10}^{6-}$, $HI_2O_{10}^{5-}$, $H_2I_2O_{10}^{4-}$, $H_3I_2O_{10}^{3-}$
orthoperiodates:	IO_6^{5-}, $H_2IO_6^{3-}$, $H_3IO_6^{2-}$, $H_4IO_6^-$, $I_2O_{11}^{8-}$
triperiodates:	$H_2I_3O_{14}^{5-}$, $H_4I_3O_{14}^{3-}$

together with various amounts of water of crystallization. Most cations generate several compounds which are often interconvertible *via* aqueous solution (Scheme 13) by changing the concentration of the solution, its pH or the temperature of crystallization; some conversions are effected by controlled thermal dehydration, e.g.

$$3LiIO_4,3H_2O \rightarrow Li_3H_4I_3O_{14}+7H_2O$$

Definitive X-ray investigations have been few (Table 83), but extensive use has been made of infrared and Raman spectroscopy in characterizing the salts (Table 84), all of which, except the metaperiodates, involve octahedral coordination about iodine. Derivatives of mesoperiodic acid ("H_3IO_5") invariably possess dimeric anions, as found for $I_2O_9^{4-}$ (9) and $H_2I_2O_{10}^{4-}$ (10), which are formed from two IO_6 octahedra sharing respectively a common face and a common edge.

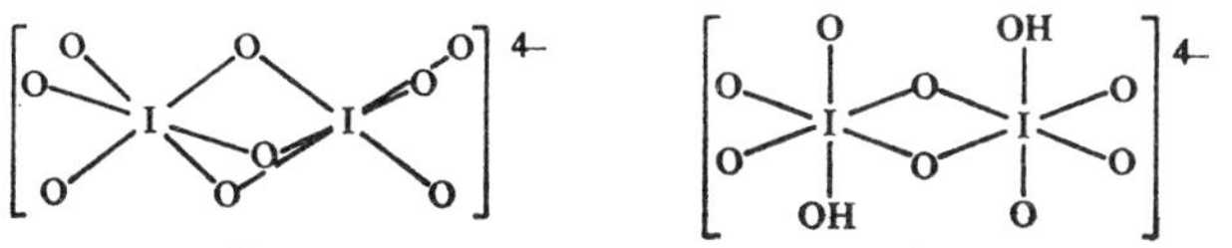

(9) (10)

The thermal decomposition of periodates[624] of various stoichiometries is accompanied by loss of water, oxygen and iodine (in some cases only one or two of these components). Reactions include partial or complete dehydration, conversion to a more stable periodate of a different composition,

$$5Ba(IO_4)_2 \rightarrow Ba_5(IO_6)_2+4I_2+14O_2$$

reduction to a salt of the anions IO_3^- or IO_4^{2-} (see below),

$$Na_2H_3IO_6 \xrightarrow{180°C} Na_2IO_4+3/2H_2O+1/4O_2$$

$$NaIO_4 \xrightarrow{300°C} NaIO_3+\tfrac{1}{2}O_2$$

or formation of the metal iodide (for electropositive metals), the oxide or the metal itself.

Reactions

The activity of periodic acid as an oxidizing agent varies greatly as a function of pH and is capable of subtle control. In acid solution it is one of the most powerful oxidizing agents known, quantitatively and rapidly converting manganese(II) salts to permanganate, while in alkaline solution it is slightly less oxidizing than hypochlorite. Despite the ease with which they perform all the oxidations effected by iodate, periodates have not been much used in inorganic preparative and analytical operations; titration with $NaIO_4$ has been suggested as a method of analysing mixtures of S^{2-}, SO_3^{2-}, HSO_3^-, $S_2O_4^{2-}$ and $S_2O_3^{2-}$ [777].

However, there are numerous reports of the oxidation of organic compounds by periodic acid or periodates[778], not merely in conjunction with perchloric acid in the "wet fire" destructive oxidation of organic compounds[779], but as reagents able to execute a number of well-defined reactions. The specific cleavage of 1,2-diols (and the related compounds α-diketones, α-ketols, α-aminoalcohols and α-diamines) has been widely exploited in the realm of carbohydrates and nucleic acids, and the mechanism has been well established

777 R. L. Kaushik and R. Prosad, *J. Indian Chem. Soc.* **46** (1969) 405.
778 B. Sklarz, *Quart. Rev. Chem. Soc.* **21** (1967) 3.
779 G. F. Smith and H. Diehl, *Talanta*, **4** (1960) 185.

TABLE 83. AVERAGE INTERNUCLEAR DISTANCES IN PERIODATES

Unit	Compound	Average internuclear distances in Å				References
		Terminal I–O units	I–OH units	Bridging I–OI units	O–H···O units	
IO_4^-	$NaIO_4$	1·775±0·007				A. Kálmán and D. W. J. Cruickshank, *Acta Cryst.* **B26** (1970) 1782.
$I_2O_9^{4-}$	$K_4I_2O_9$	1·77		2·01		B. Brehler, H. Jacobi and H. Siebert, *Z. anorg. Chem.* **362** (1968) 301.
$(HO)_5IO$	H_5IO_6	1·78	1·89		2·60–2·78	Y. D. Feikema, *Acta Cryst.* **20** (1966) 765.
$(HO)_3IO_3^{2-}$	$CdH_3IO_6,3H_2O$	1·86	1·95		2·62–3·00	A. Braibanti, A. Tiripicchio, F. Bigoli and M. A. Pellinghelli, *Acta Cryst.* **B26** (1970) 1069.
	$[Mg(OH_2)_6]H_3IO_6$	1·86	1·98		2·56–2·86	F. Bigoli, A. M. M. Lanfredi, A. Tiripicchio and M. T. Camellini, *Acta Cryst.* **B26** (1970) 1075.
$(HO)_2I_2O_8^{4-}$	$K_4H_2I_2O_{10},8H_2O$	1·807±0·011	1·980±0·016	1·992±0·012 2·017±0·007	2·79–2·97	H. Siebert and H. Wedemeyer, *Angew. Chem., Internat. Edn.* **4** (1965) 523; A. Ferrari, A. Braibanti and A. Tiripicchio, *Acta Cryst.* **19** (1965) 629.

TABLE 84. CHARACTERISTIC VIBRATIONAL FREQUENCIES OF PERIODATES[a]

Frequency (cm^{-1})	Character of vibration
3100–3400	O–H stretching: weakly hydrogen-bonded
2200–2900	O–H stretching: strongly hydrogen-bonded
1600, 3400	Water of crystallization
1070–1310	I–OH deformation
840–950	I–OH torsion (found only with strong hydrogen bonds)
600–850	I–O stretching
500–600	I–O–I stretching
250–500	OIO deformation

[a] H. Siebert, *Fortschritte Chem. Forsch.* **8** (1967) 470; H. Siebert and G. Wieghardt, *Spectrochim. Acta*, **27A** (1971) 1677; see also Table 76.

(Scheme 15). In rigid systems only *cis*-functional groups are oxidized, the specificity being due to the cyclic intermediate.

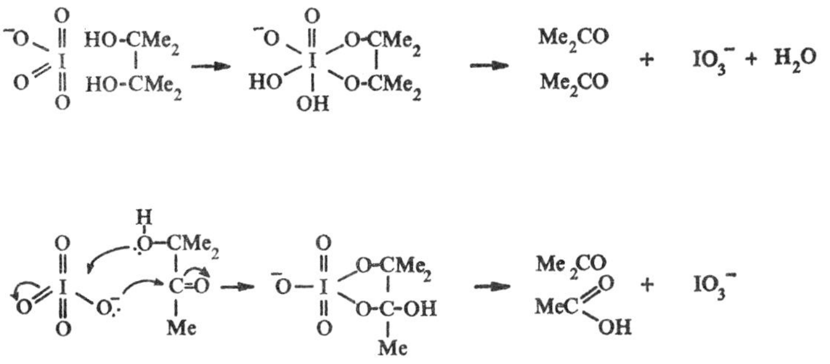

$$[IO_4]^- + HO\text{-}CMe_2\text{-}CMe_2\text{-}OH \longrightarrow [O(HO)_2I(O\text{-}CMe_2)_2]^- \longrightarrow 2\,Me_2CO + IO_3^- + H_2O$$

$$[IO_4]^- + HO\text{-}CMe_2\text{-}C(=O)Me \longrightarrow [O_2I(O\text{-}CMe_2)(O\text{-}C(OH)Me)]^- \longrightarrow Me_2CO + MeC(=O)OH + IO_3^-$$

SCHEME 15. Periodate fission of glycols and α-ketols.

Periodate has also been applied to the oxidation of phenols (to quinones), sulphides (to sulphoxides) and hydrazine derivatives:

$$R_3CNH{\cdot}NH_2 \xrightarrow{IO_4^-} R_3CH + N_2$$

Periodate Complexes and Heteropolyanions

By mixing solutions of alkali-metal periodates with solutions of appropriate transition metal salts (accompanied, if necessary, by chemical or electrochemical oxidation), it has been possible to isolate salts containing the periodate complexes and heteropolyanions listed in Table 85; contact with a protonic cation-exchange resin affords, in some cases, the free acid in solution, e.g. $H_2[FeIO_6]$ and $H_5[I(MoO_4)_6]$. The iodine and metal atoms in these compounds are linked *via* I–O–M bridges; the iodine atom is invariably surrounded by an octahedron of oxygens, but the coordination of the metal atom is determined by the crystal-field and other demands peculiar to its oxidation state.

The periodate complexes exhibit high formation constants, e.g. $[Cu(IO_6)_2]^{7-}$ *ca.* 10^{10} and $[Co(IO_6)_2]^{7-}$ *ca.* 10^{18}, and may contain metal atoms in unusually high formal oxidation states, e.g. Cu(III) and Ag(III); they may be broken down in strongly acid solution. Like the

TABLE 85. CONSTITUTION OF PERIODATE COMPLEXES AND HETEROPOLYANIONS[a]

Anion	n	Metal
Periodate complexes		
$[M_4(IO_6)_3]^{n-}$	3	Fe(III), Co(III)
$[MIO_6]^{n-}$	1	Ni(IV), Mn(IV)
	2	Fe(III), Co(III)
$[M(IO_6)_2]^{n-}$	6	Pd(IV), Pt(IV), Ce(IV)
	7	Co(III), Fe(III), Cu(III), Ag(III), Au(III)
$[M(IO_6)_3]^{n-}$	11	Mn(IV)
Heteropolyanions		
$[I(MO_4)_6]^{n-}$	5	Mo(VI), W(VI)
$[IO_2(MO_4)_4]^{n-}$	5	Mo(VI)
$[IO_5(MO_4)]^{n-}$	5	Mo(VI), W(VI)

[a] For references see: H. Siebert, *Fortschritte Chem. Forsch.* **8** (1967) 470; M. Drátovský and L. Pačesová, *Russ. Chem. Rev.* **37** (1968) 243.

simple periodates, the complex anions often contain protons (attached to I–O bonds) which ionize only with great difficulty if at all: the green acid $H_7[Co(IO_6)_2]$ shows only five protons which can be titrated (pK_1, small; $pK_2 = 1{\cdot}95$; $pK_3 = 7{\cdot}1$; $pK_4 = 8{\cdot}0$; $pK_5 = 12{\cdot}1$)[780], while $H_{11}[Mn(IO_6)_3]$ is apparently heptabasic (pK_1 and $pK_2 \lesssim 0$; $pK_3 = 2{\cdot}75$; $pK_4 = 4{\cdot}35$; $pK_5 = 5{\cdot}45$; $pK_6 = 9{\cdot}55$; $pK_7 = 10{\cdot}45$)[780]. Crystal-structure determinations show the salts $Na_3KH_3[Cu(IO_6)_2],14H_2O$[781] and $Na_7H_4[Mn(IO_6)_3],17H_2O$[782] to contain the anions (11) and (12), in both of which the periodate functions as a bidentate ligand. The metal enjoys square-planar coordination in the diamagnetic copper salt and octahedral geometry in the paramagnetic manganese(IV) compound.

The heteropolyanions (Table 85) have received comparatively little attention.

(11)

(12)

(F) PARAMAGNETIC OXYANIONS

The action of ionizing radiation on chlorates and bromates produces defect centres, some of which have been identified spectroscopically at low temperatures as radical anions; these include $[Cl{-}ClO_2]^-$ and ClO_3^{2-} in γ-irradiated $KClO_3$[610], $[Br{-}BrO_3]^-$, $[BrO{-}BrO_3]^-$

[780] M. W. Lister, *Canad. J. Chem.* **39** (1961) 2330.
[781] I. Hadinec, L. Jenšovský, A. Línek and V. Syneček, *Naturwiss.* **47** (1960) 377.
[782] A. Línek, *Czech. J. Phys.* **13** (1963) 398.
[783] R. C. Catton and M. C. R. Symons, *Chem. Comm.* (1968) 1472.

and BrO_3^{2-} in $KBrO_3$ treated with thermal neutrons[726], and $ClOH^-$ in γ-irradiated $BaCl_2,2H_2O$[783].

While these transient species trapped in crystalline lattices are interesting intermediates in the radiolytic decomposition of the halogen oxyanions, macroscopic quantities of several paramagnetic materials have been isolated which are apparently derivatives of iodine(VI)[624]. White or yellow solids with the stoichiometries $M^I_2IO_4$ (M^I = Li or Na), $Be_3I_2O_9,4H_2O$, $M^{II}IO_4,nH_2O$ (n = 1, M = Ca or Ba; n = 0·5, M = Mg; n = 0, M = Pb) and $Al_2(I_2O_7)_3,2H_2O$ are formed on heating certain hydrated periodates to *ca.* 200°C, and themselves decompose at *ca.* 400°C.

$$Ba_2I_2O_9,3H_2O \xrightarrow{220^\circ C} 2BaIO_4,H_2O+H_2O+\tfrac{1}{2}O_2$$

The amorphous solids are stable in air at room temperature, but hydrolyse in water to give iodine(V) and iodine(VII) anions. Their mass magnetic susceptibilities lie in the range $0{\cdot}2$–$5{\cdot}2\times10^{-6}$ cgs units, giving moments in the range 0·5–2·1 B.M. There is no chemical evidence for the presence of peroxides, and the possibility that impurities are responsible for the paramagnetism has apparently been excluded[624]. It is claimed that crystalline barium iodate(VI) can be obtained by heating the amorphous $BaIO_4,H_2O$ in a eutectic mixture of sodium and potassium nitrates at 250°C; the solid gives rise not only to a sharp X-ray diffraction pattern quite different from that of barium iodate(V), but also to a measurable esr signal[784]. Tentatively identified by their ultraviolet-visible absorption spectra, the ion IO_4^{2-} and the radical $\cdot IO_4$ are believed to be formed as transients during the pulse radiolysis of aqueous periodate[785].

Up to the present time there has been no report of long-lived oxy-chlorine or oxy-bromine salts analogous to the iodate(VI) derivatives, though the parent oxide ClO_3 can be prepared and handled on the macroscopic scale (see p. 1372). However, there is an ostensible likeness between iodate(VI) and the oxy-selenium(V) and oxy-tellurium(V) compounds which appear to be formed in the thermal decomposition of SeO_3, TeO_3, $Te(OH)_6$ and certain tellurates[624].

(G) ANALYTICAL CHEMISTRY OF THE OXYANIONS[572,786–789]

The limitations of space preclude any attempt to itemize the many methods which have been used or suggested to detect, separate and estimate the oxyanions of the halogens. The intention is rather to describe the way in which the chemical properties of these species have been turned to advantage in their quantitative analysis. No references will be given for procedures which are already well-documented in standard texts on analytical chemistry[786,787], or in previous reviews of the analysis of oxyanions formed by chlorine[788], bromine[788,789] and iodine[788]; the reader is also referred to these sources for details of empirical, qualitative tests for the presence of the anions. Discussion is restricted to methods in current use.

[784] M. Drátovský, J. Julák and V. Trkal, *Coll. Czech. Chem. Comm.* **32** (1967) 3977.

[785] F. Barat, L. Gilles, B. Hickel and B. Lesigne, *Chem. Comm.* (1971) 847.

[786] A. I. Vogel, *A Text-book of Quantitative Inorganic Analysis*, 3rd edn., Longmans, London (1961).

[787] L. A. Haddock, *Comprehensive Analytical Chemistry* (ed. C. L. Wilson and D. W. Wilson), Vol. IC, pp. 341–366, Elsevier (1962).

[788] L. A. Haddock, *Mellor's Comprehensive Treatise on Inorganic and Theoretical Chemistry*, Supplement II, Part I, pp. 660–685, 800–812, 937–956, Longmans, London (1956).

[789] P. G. W. Scott and M. L. Parker, *Bromine and its Compounds* (ed. Z. E. Jolles), pp. 768–772, Benn, London (1966).

The methods outlined here may also be applied to other "positive halogen" derivatives (e.g. interhalogens, halogen esters of oxyacids, etc.), since on hydrolysis in aqueous solution these compounds commonly yield oxyanions.

Apart from gravimetric and spectrophotometric techniques, the methods used in determining the oxyhalogen anions involve their reduction. Polarographic analysis has been achieved for all the anions except ClO_4^- and BrO_4^- (Table 86), and, in principle, a wide variety of chemical reducing agents could also be used; the literature abounds with suggestions. In practice, however, only a small number of reductants is routinely employed, and only three methods are of major significance:

(1) Direct titration with a reducing agent, the end-point being located with indicators, or by potentiometry, amperometry, coulometry or conductimetric measurements. It is plainly important for titrimetry that the reaction be rapid (although back-titration of excess reductant is always possible), so that, for less reactive species like ClO_3^-, acid solution and a catalyst may be necessary. Sodium arsenite reduces all the anions except ClO_4^- and BrO_4^-, and is by far the best titrant (see below); starch–potassium iodide is a common chemical indicator, and back-titration, when necessary, is conveniently carried out with standard bromate solution.

(2) Reduction to halide, which is subsequently determined by a standard procedure (p. 1331), is a method open to all the oxyanions, but is used especially for the chlorine species. The strength of the reducing agent and the reaction conditions needed depend on the anion being reduced. Perchlorate requires vigorous methods such as fusion or reduction with Ti^{3+}; other reagents which have been used include stannous chloride, ferrous sulphate and sulphurous acid.

(3) With the exception of perchlorate, all the anions (and ClO_2) have been estimated by analysis of the iodine which they liberate (directly or indirectly) from acidified potassium iodide solution. Reaction of hypochlorite with iodide is rapid at pH 8, but chlorates and bromates in strongly acidic solution require catalysis (by ferrous salts and molybdates respectively) for a speedy reaction. In some cases the Dietz method may be used, whereby reaction of the oxyanion with bromide in strongly acid solution liberates bromine, which is subsequently allowed to react with iodide ions. The iodine has been determined spectrophotometrically as I_3^- or by titration, for example, with thiosulphate.

Other reduction procedures are discussed below in connection with the analysis of mixtures of the oxyanions; gravimetric and physical methods of analysis are listed in Table 86.

Because ClO_4^- is so hard to reduce at room temperature, the analysis of perchlorates has presented some difficulties. Many workers have reduced perchlorate to chloride under forcing conditions (e.g. by fusion with sodium carbonate in a platinum crucible or by combustion with an ammonium salt) and subsequently estimated the chloride formed. Reducwith an iron(III) solution), but the method requires an inert atmosphere; reaction with vanadium(II) sulphate produces vanadium(IV), which may be determined spectrophotometrically. Recent developments have involved gravimetric analysis by precipitation of insoluble perchlorates using large organic cations, e.g. Ph_4M^+ (M = P, As or Sb), though other large anions are liable to interfere; potentiometric, conductimetric and amperometric titration of perchlorate solutions using these reagents has also been proposed. Colorimetric reagents (mainly complicated quaternary ammonium salts such as methylene blue)

TABLE 86. SPECTROPHOTOMETRIC, GRAVIMETRIC AND POLAROGRAPHIC METHODS OF ANALYSING OXYHALOGEN SPECIES

Oxyhalogen species	Spectrophotometry		Gravimetric analysis	Polarography
	Direct	With a chromophoric agent		
ClO^-	At 290 mμ[1]			23
ClO_2^-	At 261 mμ[1]			24, 25
	At 250 mμ[2]			
ClO_2	At 360 mμ[2]	4		24, 25
ClO_3^-		5		26
ClO_4^-		6–10	19–21	
BrO^-				27, 28
BrO_2^-				27–30
BrO_3^-		11–13		27, 31
BrO_4^-		14		
IO^-				32
IO_3^-		15		31
IO_4^-	At 222·5 mμ[3]	16–18	22	33

[1] T. Chen, *Anal. Chem.* **39** (1967) 804.

[2] $HClO_2$ and ClO_2^- have the same extinction coefficient at 250 mμ; C. C. Hong and W. H. Rapson, *Canad. J. Chem.* **46** (1968) 2061.

[3] G. O. Aspinall and R. J. Ferrier, *Chem. and Ind.* (1957) 1216.

[4] With tyrosine; P. Kerenyi and P. Kuba, *Chem. Zveste*, **17** (1963) 146.

[5] With *o*-tolidine; P. Urone and E. Bonde, *Anal. Chem.* **32** (1960) 1666.

[6] Methylene blue-perchlorate complex extracted into $ClCH_2CH_2Cl$; I. Swasaki, S. Utsumi and C. Kang, *Bull. Chem. Soc. Japan*, **36** (1963) 325; complex precipitated; G. M. Nobar and C. R. Ramachandran, *Anal. Chem.* **31** (1959) 263.

[7] Crystal violet-perchlorate complex extracted into C_6H_5Cl; S. Uchikawa, *Bull. Chem. Soc. Japan*, **40** (1966) 798.

[8] Ferroin perchlorate extracted into C_3H_7CN; J. Fritz, J. E. Abbink and P. A. Campbell, *Anal. Chem.* **36** (1964) 2123; ferroin perchlorate precipitated; Z. Gregorowicz, F. Bull and Z. Klima, *Mikrochim. Ichnoanal. Acta* (1963) 116.

[9] With α-furildioxime; N. L. Trautwein and J. C. Guyon, *Anal. Chem.* **40** (1968) 639.

[10] Neutral red-perchlorate complex extracted into $C_6H_5NO_2$; M. Tsubouchi and Y. Yuroko, *Bunseki Kagaku*, **19** (1970) 966.

[11] With *o*-arsanilic acid; J. C. MacDonald and J. H. Yoe, *Anal. Chim. Acta*, **28** (1963) 383.

[12] With *o*-aminobenzoic acid; M. H. Hashmi, H. Ahmad, A. Rashid and A. A. Ayaz, *Anal. Chem.* **36** (1964) 2028.

[13] With 1,2,3-tris-(2-diethylaminoethoxy)benzene; I. Odler, *Anal. Chem.* **41** (1969) 116.

[14] Crystal violet-perbromate complex extracted into C_6H_5Cl; L. C. Brown and G. E. Boyd, *Anal. Chem.* **42** (1970) 291.

[15] With 1-naphthylamine; A. S. Vorob'ev, *Uch. Zap. Udmurtsk. Gos. Ped. Inst.* **11** (1957) 150; *Chem. Abs.* **54** (1960) 6386d.

[16] Crystal violet-periodate complex extracted into benzene; C. E. Hedrick and D. A. Berger, *Anal. Chem.* **38** (1966) 791.

[17] With benzhydrazide; A. M. Escarilla, P. F. Maloney and P. M. Maloney, *Anal. Chim. Acta*, **45** (1969) 199.

[18] With *o*-dianisidine; M. Guernet, *Compt. rend.* **254** (1962) 3688.

[19] As $[(n\text{-}C_5H_{11})_4N]ClO_4$; R. G. Dosch, *Anal. Chem.* **40** (1968) 829; $[(n\text{-}C_5H_{11})_4N]Br$ used in the potentiometric titration of ClO_4^-.

[20] As $[Ph_4P]ClO_4$ or $[Ph_4Sb]ClO_4$; H. H. Willard and L. R. Perkins, *Anal. Chem.* **25** (1953) 1634; $[Ph_4Sb]_2SO_4$ used in the amperometric titration of ClO_4^-; M. D. Morris, *Anal. Chem.* **37** (1965) 977.

Table 86 (*cont.*)

[21] As $[Ph_4As]ClO_4$; T. Okuba, A. Fumio and T. Teraoka, *Nippon Kagaku Zasshi*, **89** (1968) 432; $[Ph_4As]Cl$ used in amperometric and conductimetric titrations of ClO_4^-; G. M. Smith, *Ind. Eng. Chem., Anal. Edn.* **11** (1939) 186; R. J. Baczuk and W. T. Bolleter, *Anal. Chem.* **39** (1967) 93.

[22] As $[Ph_4As]IO_4$; S. H. Laurie, J. M. Williams and C. J. Nyman, *J. Phys. Chem.* **68** (1964) 1311.

[23] E. N. Jenkins, *J. Chem. Soc.* (1951) 2627.

[24] O. Schwarzer and R. Landsberg, *J. Electroanal. Chem.* **14** (1967) 339.

[25] G. Raspi and F. Pergola, *J. Electroanal. Chem.* **20** (1969) 419.

[26] Requires OsO_4 as a catalyst; L. Meites and H. Fofsass, *Anal. Chem.* **31** (1959) 119.

[27] H. Fuchs and R. Landsberg, *Anal. Chim. Acta*, **45** (1969) 505.

[28] F. Pergola, G. Raspi and A. Massagei, *J. Electroanal. Chem.* **22** (1969) 175.

[29] A. F. Krivis and G. R. Supp, *Anal. Chem.* **40** (1968) 2063.

[30] N. Velghe and A. Claeys, *Bull. Soc. Chim. Belges*, **75** (1966) 498.

[31] J. Proszt, V. Cieleszby and K. Gyorbiro, *Polarographie*, p. 332, Akademie Kiado, Budapest (1967).

[32] O. Souchay, *Anal. Chim. Acta*, **2** (1948) 17.

[33] R. D. Corlett, W. C. Breck and G. W. Hay, *Canad. J. Chem.* **48** (1970) 2474.

have also been used, but most require a precipitation or extraction step before measurement, and the method is subject to interference from other ions.

In the absence of interfering species, the determination of individual anions is straightforward using any of the methods described in the foregoing paragraphs or listed in Table 86. However, the successful analysis of mixtures of the anions may require considerable ingenuity. In some cases, differences in reaction rates may be exploited, and several species may be successively titrated by careful pH control and the use of catalysts. It may be possible selectively to reduce or otherwise destroy some components of the mixture, so that a series of determinations of the total oxidizing power may be made with different components rendered inactive. Again, a combination of several techniques may be needed. The remainder of this section is devoted to some specific problems of this sort and to some of the solutions which have been proposed.

ClO^-, ClO_2^-, ClO_2, ClO_3^- *and* ClO_4^-

Hypochlorite, chlorite and chlorate may be determined by potentiometric titration with arsenite in, respectively, alkaline solution, alkaline solution with OsO_4 as a catalyst, and acid solution with OsO_4 as a catalyst; a similar procedure but using amperometric titrations has been described for mixtures containing Cl_2, chloramine, ClO_2 and ClO_2^-.

Recent investigations of these systems have utilized spectroscopic procedures. Thus, ClO_2, having been determined spectrophotometrically at 360 mμ, may be removed by passing an inert gas through the solution (simultaneously sweeping out some HOCl), whereupon chlorite and chlorous acid can be estimated together at 250 mμ, where they have the same absorbance; the total oxyanion concentration is determined iodometrically[790]. Alternative procedures involve the spectrophotometric measurement of I_3^- for ClO^- and ($ClO_2^-+ClO^-$) (following oxidation of KI at pH 8·3 and pH 2·3 respectively) and of Fe^{3+} for chlorate (following oxidation of ferrous sulphate)[791,792]. In either case, perchlorate may be determined as the difference between the chloride formed on total reduction of the mixture, e.g. with Ti^{3+}, and that formed on partial reduction of the mixture, e.g. with hot

[790] C. C. Hong and W. H. Rapson, *Canad. J. Chem.* **46** (1968) 2061.

[791] T. Chen, *Anal. Chem.* **39** (1967) 804.

[792] L. A. Prince, *Anal. Chem.* **36** (1964) 613.

H_2SO_3, which reduces only chlorates, chlorites and hypochlorites. The presence of ClO_2 interferes with iodometric measurements since it too oxidizes I^- to I_2; in analysing such solutions, the iodine liberated at pH 8·3 and pH 2·3 corresponds to $(ClO^- + 1/2ClO_2)$ and $(ClO^- + 2ClO_2^- + 5/2ClO_2)$ respectively.

Mixtures Containing BrO^-, BrO_2^-, BrO_3^- and BrO_4^-

The +1, +3 and +5 oxidation states of bromine are readily determined by potentiometric titration with sodium arsenite: hypobromite is so determined in cold alkaline solution, bromite in alkaline solution with OsO_4 as a catalyst, and bromate in acid solution[793,794]. Hypobromite may be determined in the presence of bromite by potentiometric titration with ferrocyanide[794]. The colorimetric estimation of BrO_4^- with crystal violet is apparently insensitive to the presence of other oxy-bromine anions.

Mixtures Containing IO^-, IO_3^- and Periodates

Hypoiodites may be measured in the presence of other oxy-iodine species by their reaction with phenol which produces a precipitate of 2,4,6-triiodophenol. Two portions of the solution are acidified and one is treated with phenol; both are then treated with potassium iodide, and the difference in the amounts of iodine liberated corresponds to the hypoiodite content of the solution. Selective iodometry allows iodates and periodates to be estimated: the iodine liberated from KI at pH 4·4–7 corresponds to iodine(VII)'s being reduced to iodine(V), but at lower pH iodate is itself reduced with the liberation of further iodine. Similarly, in a solution buffered with sodium carbonate, periodate is reduced only to iodate. Periodate is stabilized against reaction with I^- in the presence of molybdate, when a heteropolyanion is formed; under these conditions iodate liberates iodine, which may be estimated with thiosulphate, and periodate is then regenerated by treatment with oxalic acid[795].

Chlorates, Bromates and Iodates

These species can be separated by ion-exchange chromatography on a strongly basic resin[796], by paper electrophoresis, and by paper chromatography; their characteristic infrared spectra have been used as a basis for quantitative determinations[797]. Chemical methods for their estimation in the presence of one another depend on the increase in reactivity as the atomic number of the halogen increases. Thus, in dilute acid in the presence of chlorate, bromate is reduced to Br^- by oxalic acid, and oxidizes I^- to I_2; chlorate undergoes neither reaction. In a very weakly acid solution containing BrO_3^-, IO_3^- and F^-, only IO_3^- reacts with I^-. Alternatively, a simple colorimetric procedure[798] for analysing BrO_3^- and IO_3^- individually or in admixture uses a 1:1 mixture of isonicotinic acid hydrazide and 2,3,5-triphenyltetrazolium chloride, which in dilute hydrochloric acid gives a pink colour with IO_3^- at 20°C, whereas the bromate colour develops only on heating; chlorate apparently does not interfere. Polarography can also distinguish BrO_3^- or IO_3^- in the presence of ClO_3^-, since in the absence of a catalyst the oxy-chlorine anion is only slowly reduced.

[793] P. Norkus and S. Stulgiene, *Zhur. Anal. Khim.* **24** (1969) 1565.
[794] T. Andersen and H. E. L. Madsen, *Anal. Chem.* **37** (1965) 49.
[795] R. Belcher and A. Townshend, *Anal. Chim. Acta*, **41** (1968) 395.
[796] M. Kikindai-Cassel, *Ann. Chim.* **3** (13) (1958) 5.
[797] M. W. Miller, R. H. Philp, jun. and A. L. Underwood, *Talanta*, **10** (1963) 763.
[798] M. H. Hashmi, H. Ahmad, A. Rashid and F. Azam, *Anal. Chem.* **36** (1964) 2471.

5. HALOGEN DERIVATIVES OF OXYACIDS

Introduction

It has been known for some forty years that in polar non-aqueous solvents chlorine, bromine and iodine undergo reactions with silver salts[799], e.g.

$$Cl_2+AgNO_3 \xrightarrow{EtOH} AgCl+ClNO_3$$

That the resulting solutions conducted electricity and liberated the elemental halogen at the cathode during electrolysis convinced earlier researchers that the reactions could be described as metatheses, and the products as salts of halogen cations, e.g.

$$Cl^+Cl^- + Ag^+NO_3^- \rightarrow Ag^+Cl^- + Cl^+NO_3^-$$

It proved impossible to isolate the pure halogen derivatives from these solutions, although adducts with organic bases, e.g. $C_5H_5N,IOCOCH_3$ and $[Cl(C_8H_{11}N)_2]NO_3$, were readily obtained; the chemistry of these adducts was discussed in Section 4A (p. 1345). More recent synthetic work has shown the pure halogen "salts" to be covalent compounds with halogen–oxygen bonds, and it is the chemistry of these materials—formally esters—with which the present section is concerned.

Halogen(I) nitrates, fluorosulphates, perchlorates and carboxylates have structures related to that of the parent acid, the hydroxylic hydrogen being replaced by fluorine, chlorine, bromine or iodine. The thermal stability of these compounds decreases as the atomic number of the halogen increases, although the estimation of stabilities is complicated by the susceptibility of the materials to decomposition in the presence of trace impurities. In common with esters of the hypohalous acids (which have similar R–O–Hal skeletons), they add across olefinic double bonds with rupture of the oxygen–halogen linkage:

$$Bu^tOCH_2{\cdot}CH_2Cl \xleftarrow{Bu^tOCl} CH_2{=}CH_2 \xrightarrow{ClONO_2} ClCH_2{\cdot}CH_2ONO_2$$

They are distinguished from covalent hypochlorites by their reactions with metal halides:

$$\xleftarrow[\not\;]{F_5SOCl} CsF \xrightarrow{ClOSO_2F} CsSO_3F+ClF$$

That comparable hypochlorites, like F_5SOCl and CF_3OCl, fail to react with CsF is attributed to the instability of the anion formed on heterolysis of the Cl–O bond. Further indication of the "positive" character of the halogen in oxyacid derivatives is the *replacement* of chlorine in reactions with bromine or iodine, e.g.

$$2ClOClO_3+Br_2 \rightarrow 2BrOClO_3+Cl_2$$

No chlorine(III) oxyacid derivatives are known, but bromine and iodine are readily converted to the tripositive state, even by silver salts in non-aqueous solution:

$$2Br_2+3AgNO_3 \rightarrow Br(NO_3)_3+3AgBr$$

The trivalent bromine and iodine compounds are usually more stable than the corresponding monovalent species (whose instability is comparable with the ready disproportionation of BrF and IF). Few derivatives of this kind containing pentavalent halogen atoms have been reported, and none is known for the heptavalent elements.

[799] M. I. Ushakov, *J. Gen. Chem.* (*U.S.S.R.*) **1** (1931) 1258.

Halogen Fluorosulphates

Fluorosulphate derivatives of the halogens have featured in a recent general review of fluorosulphates[800]. The most important compounds are:

$ClOSO_2F$	$BrOSO_2F$	$IOSO_2F$
	$Br(OSO_2F)_3$	$I(OSO_2F)_3$
	$M^I[Br(OSO_2F)_4]$	$M^I[I(OSO_2F)_4]$
		ICl_2OSO_2F
	Br_3OSO_2F (?)	I_3OSO_2F
ClO_2OSO_2F		IO_2OSO_2F
		$IF_3(OSO_2F)_2$

That much of the chemistry of interhalogen compounds—exchange, addition, displacement and complexation reactions—can be simulated by these materials is evident from the reactions of the iodine fluorosulphates. Thus, in many respects, the fluorosulphate group is behaving as a pseudohalogen (see p. 1122).

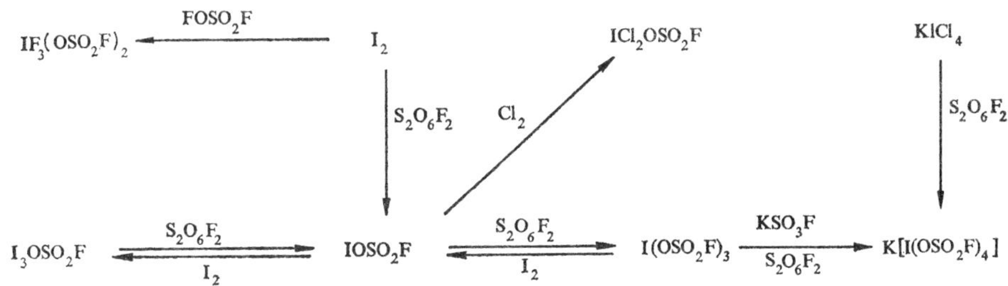

SCHEME 16. Iodine fluorosulphates.

Most of the halogen fluorosulphates are highly sensitive to moisture, and decompose slowly on standing at room temperature. The simpler compounds are formed by mixing the halogen and peroxydisulphuryl difluoride, $FSO_2O{\cdot}OSO_2F$, in requisite proportions under appropriate conditions. The equilibria established when various proportions of halogen and peroxydisulphuryl difluoride are dissolved in fluorosulphuric acid (or related solvents) have been studied spectroscopically, conductimetrically and by measurements of colligative properties; this topic has been alluded to in Section 4A (p. 1341) in connection with the formation and characterization of halogen cations.

Halogen(I) Fluorosulphates

Direct combination of the halogen with an equimolar quantity of peroxydisulphuryl difluoride affords $ClOSO_2F$, $BrOSO_2F$ or $IOSO_2F$ (Table 87). $ClOSO_2F$ may also be prepared in high yield from sulphur trioxide and chlorine monofluoride[801], while $BrOSO_2F$ and

[800] A. A. Woolf, *New Pathways in Inorganic Chemistry* (ed. E. A. V. Ebsworth, A. G. Maddock and A. G. Sharpe), p. 327, Cambridge (1968).

[801] W. P. Gilbreath and G. H. Cady, *Inorg. Chem.* **2** (1963) 496.

TABLE 87. PROPERTIES OF HALOGEN(I) FLUOROSULPHATES

	$FOSO_2F$	$ClOSO_2F$	$BrOSO_2F$	$IOSO_2F$
Reaction conditions, $X_2+S_2O_6F_2 \rightarrow 2XOSO_2F$	a	125°C, pressure[b]	0–50°C[c]	20–60°C[d]
Physical properties				
Colour	colourless[e]	yellow[b]	red-brown[c]	black[d]
Melting point (°C)	−158·5[e]	−84·3[b]	−31·5[c]	51·5[d]
Boiling point (°C)	−31·3[e]	45·1[b]	117·3[c]	
ΔH_{vap} (kcal mol^{-1})	5·35[e]	7·66[b]	9·94[c]	
Trouton's constant (cal deg^{-1} mol^{-1})	22·1[e]	24·0[b]	25·8[c]	
Molecular spectra and parameters				
Infrared spectrum	gas[e]	gas[f] solid[f]		
Raman spectrum	liquid[g]	liquid[g]	liquid[g]	
^{19}F nmr spectrum	liquid[h]	liquid[h]		
Force constant, f_r(O–X) (mdyne Å^{-1})		2·6[f]		
^{19}F chemical shift, δ(S–F) (ppm relative to CCl_3F)[h]	−36·3	−33·9	−41·3	

[a] Prepared from SO_3 and excess F_2.
[b] W. P. Gilbreath and G. H. Cady, *Inorg. Chem.* **2** (1963) 496.
[c] F. Aubke and R. J. Gillespie, *Inorg. Chem.* **7** (1968) 599.
[d] F. Aubke and G. H. Cady, *Inorg. Chem.* **4** (1965) 269.
[e] F. B. Dudley, G. H. Cady and D. F. Eggers, jun., *J. Amer. Chem. Soc.* **78** (1956) 290.
[f] K. O. Christe, C. J. Schack and E. C. Curtis, *Spectrochim. Acta*, **26A** (1970) 2367.
[g] A. M. Qureshi, L. E. Levchuk and F. Aubke, *Canad. J. Chem.* **49** (1971) 2544; F. Aubke and A. M. Qureshi, *Inorg. Chem.* **10** (1971) 1116.
[h] F. A. Hohorst and J. M. Shreeve, *Inorg. Chem.* **5** (1966) 2069.

$IOSO_2F$ are formed in the thermal decomposition of the respective trisfluorosulphates. The vibrational spectrum of $ClOSO_2F$ is consistent with a molecular structure of C_s symmetry (analogous to those of $HOSO_2F$ and $FOSO_2F$) in the solid, liquid and vapour phases. $BrOSO_2F$ is a viscous liquid which somewhat resembles BrF_3; the specific conductivity ($7{\cdot}21 \times 10^{-4}$ ohm^{-1} cm^{-1} at 25°C) signifies some self-ionization, possibly according to the equation[802]

$$3BrOSO_2F \rightleftharpoons [Br_2OSO_2F]^+ + [Br(OSO_2F)_2]^-$$

In fluorosulphuric acid solution $IOSO_2F$ disproportionates[803]:

$$5IOSO_2F \rightleftharpoons 2I_2^+ + I(OSO_2F)_3 + 2SO_3F^-$$

Disproportionation of $BrOSO_2F$ occurs only in more acidic media, e.g. the "super-acid" HSO_3F–SbF_5–SO_3 [804].

Halogen(I) fluorosulphates are extremely reactive compounds. Organic materials (including Kel-F) are attacked at ambient temperatures, and hydrolysis of $ClOSO_2F$ proceeds violently with evolution of some oxygen. Photolysis of $ClOSO_2F$ at low temperature produces Cl_2 and $FSO_2O{\cdot}OSO_2F$, while the ready addition of $XOSO_2F$ (X = Cl, Br or I) across olefinic double bonds also involves rupture of the halogen–oxygen bond.

$$CF_2{=}CF_2 + BrOSO_2F \rightarrow CF_2Br{\cdot}CF_2OSO_2F$$

$BrOSO_2F$ reacts with molecular chlorides according to the general equation

$$MCl_x + yBrOSO_2F \rightarrow MCl_{x-y}(OSO_2F)_y + yBrCl$$

Compounds thus derived include $[C(O)OSO_2F]_2$, $C(OSO_2F)_4$ and $Cl_2P(O)OSO_2F$[805]. $ClOSO_2F$ similarly attacks alkali-metal salts to produce chlorine(I) derivatives of the anion together with the alkali-metal fluorosulphate:

$$ClOClO_3 + M^ISO_3F \xleftarrow{M^IClO_4} ClOSO_2F \xrightarrow{M^IF} ClF + M^ISO_3F$$

Halogen(III) Fluorosulphates

Whereas IF_3 readily disproportionates to iodine and IF_5, both the trisfluorosulphates $Br(OSO_2F)_3$ and $I(OSO_2F)_3$ (prepared from the halogen and an excess of $FSO_2O{\cdot}OSO_2F$) are in this respect quite stable at room temperature. At 80°C $I(OSO_2F)_3$ decomposes slowly *in vacuo* to give $IOSO_2F$, SO_3, $IF(OSO_2F)_2$ and possibly $I(OSO_2F)_5$ [806]. The Raman spectra of the pale yellow solids $Br(OSO_2F)_3$ (m.p. 59°C) and $I(OSO_2F)_3$ (m.p. 32°C) contain lines attributable to both terminal and bridging fluorosulphate groups[807]. Ampholytic behaviour is noted in fluorosulphuric acid solution, wherein cryoscopic and conductimetric measurements have identified the equilibria[808]

$$I(OSO_2F)_3 \rightleftharpoons [I(OSO_2F)_2]^+ + SO_3F^-; \qquad K_b \sim 10^{-5}\ \text{mol kg}^{-1}$$

$$I(OSO_2F)_3 + SO_3F^- \rightleftharpoons [I(OSO_2F)_4]^-; \qquad K_a \sim 10\ \text{mol}^{-1}\ \text{kg}$$

---

802 F. Aubke and R. J. Gillespie, *Inorg. Chem.* **7** (1968) 599.
803 R. J. Gillespie and J. B. Milne, *Inorg. Chem.* **5** (1966) 1577.
804 R. J. Gillespie and M. J. Morton, *Chem. Comm.* (1968) 1565.
805 D. D. Des Marteau, *Inorg. Chem.* **7** (1968) 434.
806 F. Aubke and G. H. Cady, *Inorg. Chem.* **4** (1965) 269.
807 H. A. Carter, S. P. L. Jones and F. Aubke, *Inorg. Chem.* **9** (1970) 2485.
808 R. J. Gillespie and J. B. Milne, *Inorg. Chem.* **5** (1966) 1236.

The Raman spectra of the ions $[Br(OSO_2F)_4]^-$ and $[I(OSO_2F)_4]^-$ in alkali-metal salts imply square-planar MO_4 skeletons (compare BrF_4^- and ICl_4^-), as well as suggesting that the Br–O bonds involve a greater degree of covalency than the I–O bonds[807].

Oxidation of $IOSO_2F$ with chlorine gives a product of approximate composition $ICl_2(OSO_2F)$, while reaction between stoichiometric amounts of I_2 and $IOSO_2F$ affords the dark brown, hygroscopic solid I_3OSO_2F, which melts at 92°C with the liberation of iodine and forms the cation I_3^+ on dissolution in fluorosulphuric acid[806]. Br_3OSO_2F has not been isolated as such, but its probable presence as an impurity in "$BrOSO_2F$" prepared from $FSO_2O{\cdot}OSO_2F$ and a slight excess of bromine is indicated[802,809] by discrepancies between the appearance, physical properties and reactivity of such samples and those reported more recently for pure specimens. In the "super-acid" solvent $HSO_3F–SbF_5–3SO_3$, 3:1 mixtures of Br_2 and $FSO_2O{\cdot}OSO_2F$ produce Br_3^+ and SO_3F^-[810].

Halogen(V) Fluorosulphates

While there is no evidence for the direct formation of $I(OSO_2F)_5$ from iodine and a large excess of $FSO_2O{\cdot}OSO_2F$, the compound is possibly one of the products of the thermal decomposition of $I(OSO_2F)_3$.

Two oxyhalogen(V) fluorosulphates have been reported, viz. ClO_2SO_3F[810] and IO_2SO_3F[646,805]; these have already been cited in the general context of cationic oxyhalogen species (p. 1352).

Halogen Nitrates

The established nitrate derivatives of chlorine, bromine and iodine are

$ClONO_2$	$BrONO_2$	$IONO_2$
	$M^I[Br(ONO_2)_2]$	$M^I[I(ONO_2)_2]$
	$Br(ONO_2)_3$	$I(ONO_2)_3$
		$M^I[I(ONO_2)_4]$
	BrO_2NO_3	

The chemistry of these compounds as known up to 1961 has been reviewed by Schmeisser and Brändle[811], while the status of bromine nitrates (up to 1966)[812] and of chlorine nitrate (up to 1968)[572] has been the burden of more recent reviews.

The action of silver nitrate on an alcoholic solution of chlorine, bromine or iodine gives rise to the halogen(I) nitrate; with excess silver nitrate, bromine and iodine give $Br(NO_3)_3$ and $I(NO_3)_3$ respectively. While the pure materials cannot be isolated from these solutions, adducts with organic nitrogen bases, e.g. $[(C_5H_5N)_2Br]NO_3$, are readily formed. Electrolysis of alcoholic solutions of INO_3 releases iodine at the cathode.

Chlorine(I) nitrate is a product of almost any reaction between a chlorine oxide and an oxide of nitrogen, being formed by the combination of ClO and NO_2 radicals. Preparatively, the best synthesis is depicted by the equation[580]

$$Cl_2O + N_2O_5 \xrightarrow{0°C} 2ClONO_2$$

---

[809] J. E. Roberts and G. H. Cady, *J. Amer. Chem. Soc.* **82** (1960) 352.

[810] R. J. Gillespie and M. J. Morton, *Chem. Comm.* (1968) 1565.

[811] M. Schmeisser and K. Brändle, *Angew. Chem.* **73** (1961) 388.

[812] M. Schmeisser and E. Schuster, *Bromine and its Compounds* (ed. Z. E. Jolles), pp. 209–213, Benn, London (1966).

TABLE 88. HALOGEN(I) NITRATES

	$FONO_2$	$ClONO_2$	$BrONO_2$	$IONO_2$
Colour	colourless[a]	colourless[c]	yellow[d]	yellow[d]
Melting point (°C)	−175[a]	−107[c]	−42[d]	
Boiling point (°C)	−45·9[a]	18[c]		
Decomposition temperature (°C)	ambient[b]	ambient[b]	< 0[d]	< 0[d]
ΔH_{vap} (kcal mol^{-1})	4·726[a]	7·3[c]		
Trouton's constant (cal deg^{-1} mol^{-1})	20·8[a]	25·1[c]		
$\Delta H^\circ_{f,\,298}[XONO_2(g)]$ (kcal mol^{-1})[b]	+2·5	+6·97		
$\Delta G^\circ_{f,\,298}[XONO_2(g)]$ (kcal mol^{-1})[b]	+17·56	+22·08		
Infrared spectrum	gas[b]	gas[b]	gas[e]	
	solid[b]	solid[b]		
^{14}N nmr spectrum		liquid[f]		
Bond dissociation energies (kcal mol^{-1})[b]				
D(O–X)	32·95	38·55		
$D(O_2N–OX)$	46·59	25·31		
Force constant, f_r(O–X) (mdyne Å^{-1})[b]	3·25	2·83		

[a] G. H. Cady, *J. Amer. Chem. Soc.* **56** (1934) 2635; O. Ruff and W. Kwasnik, *Angew. Chem.* **48** (1935) 238.
[b] R. H. Miller, D. L. Bernitt and I. C. Hisatsune, *Spectrochim. Acta*, **23A** (1967) 223.
[c] H. Martin, *Angew. Chem.* **70** (1958) 97.
[d] M. Schmeisser and K. Brändle, *Angew. Chem.* **73** (1961) 388.
[e] C. J. Schack, cited in C. J. Schack, K. O. Christe, D. Pilipovich and R. D. Wilson, *Inorg. Chem.* **10** (1971) 1078.
[f] A. M. Qureshi, J. A. Ripmeester and F. Aubke, *Canad. J. Chem.* **47** (1969) 4247.

although the action of ClF on nitric acid produces $ClONO_2$ in 90% yield[813]. The nitrate has been used to make other halogen nitrates (Scheme 17). Furthermore, its reaction with metal chlorides (conveniently executed at −78°C in liquid $ClONO_2$) provides a route to anhydrous metal nitrates[814]. Other reactions are summarized in Scheme 17. Like $ClONO_2$, $IONO_2$ adds across olefinic linkages, and has also been used to oxidize alcohols[815].

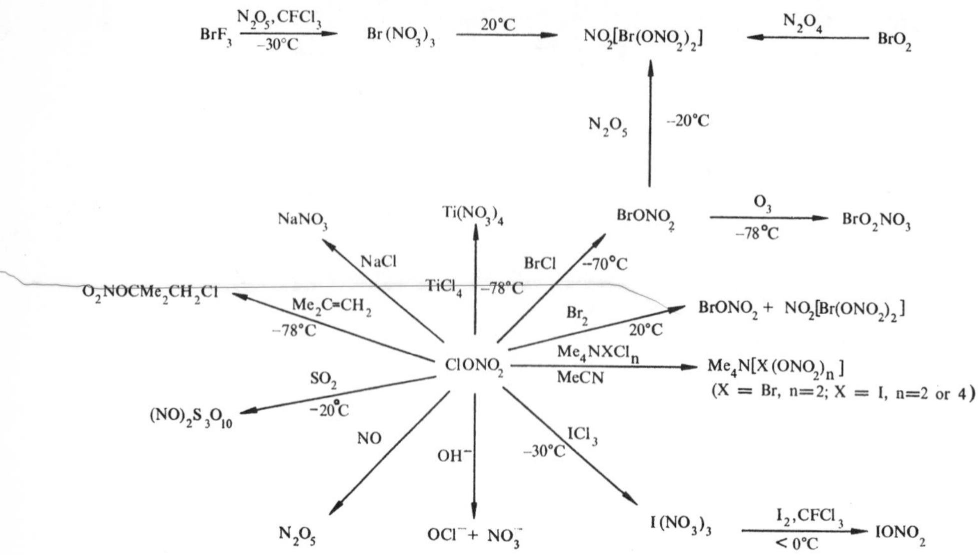

SCHEME 17. Halogen nitrates.

Detailed investigations of its infrared spectrum establish that $ClONO_2$ is structurally akin to $FONO_2$. In keeping with the differences in halogen–oxygen and nitrogen–oxygen bond energies, however, the first step in the decomposition of fluorine nitrate is

$$FONO_2 \rightarrow F + ONO_2$$

but chlorine nitrate disintegrates *via* the reaction

$$ClONO_2 \rightarrow ClO + NO_2$$

for which an activation energy of 30 kcal mol^{-1} has been derived experimentally[816].

While $I(NO_3)_3$ loses nitrogen dioxide at 0°C, the yellow compound $Br(NO_3)_3$ is more stable, melting with decomposition at 48°C; orange bromyl nitrate, BrO_2NO_3, does not persist much above −78°C. Infrared spectroscopic investigations of the complex nitratohalogen anions $[X(NO_3)_n]^-$ ($X = Br$, $n = 2$; $X = I$, $n = 2$ or 4) could not determine whether the nitrate groups function as mono- or bidentate ligands[817].

[813] C. J. Schack, *Inorg. Chem.* **6** (1967) 1938.

[814] C. C. Addison and N. Logan, *Preparative Inorganic Reactions* (ed. W. L. Jolly), Vol. 1, p. 141, Interscience (1964).

[815] U. E. Diner and J. W. Lown, *Chem. Comm.* (1970) 333.

[816] L. F. R. Cafferata, J. E. Sicre and H. J. Schumacher, *Z. physik. Chem., N.F.* **29** (1961) 188.

[817] M. Lustig and J. K. Ruff, *Inorg. Chem.* **5** (1966) 2124.

Halogen Perchlorates

The interaction of iodine and silver perchlorate in organic solvents (which Gomberg believed to produce ClO_4 [608]) is in fact rather complicated. The various studies of this reaction have been reviewed by Alcock and Waddington[609], whose own careful scrutiny led to the findings summarized in Table 89. The course of the reaction depends on the solvent and on the reaction conditions, and in no case can a pure product be isolated.

TABLE 89. REACTIONS OF IODINE WITH SILVER PERCHLORATE IN NON-AQUEOUS SOLVENTS

Solvent	Conditions	Reaction
EtOH	−85°C	$I_2 + AgClO_4 \rightarrow AgI + IClO_4$
Et_2O	−85°C; I_2 added to $AgClO_4$	$I_2 + 2AgClO_4 \rightarrow AgI + Ag[I(ClO_4)_2]$
	−85°C; $AgClO_4$ added to I_2	$I_2 + AgClO_4 \rightarrow AgI + IClO_4$ $I_2 + 2AgClO_4 \rightarrow AgI + Ag[I(ClO_4)_2]$ $Ag[I(ClO_4)_2] + IClO_4 \rightarrow AgI + I(ClO_4)_3$
EtOH or Et_2O	room temperature	rapid reaction with solvent

TABLE 90. PROPERTIES OF HALOGEN(I) PERCHLORATES

	$FOClO_3$	$ClOClO_3$ [a]	$BrOClO_3$
Colour	colourless[b]	pale yellow[e]	red[g]
Melting point (°C)	−167·3[b]	−117±2[e]	< −78[g]
Boiling point (°C)	−15·9/755 mm[b]	44·5[e]	
Decomposition temperature (°C)	~100[c]	ambient[e]	−20[g]
Infrared spectrum	gas[c]	gas[f] matrix-isolated[f] solid[f]	gas[f] matrix-isolated[f]
Raman spectrum		liquid[f]	
Force constants (mdyne Å^{-1})			
f_r(Cl=O)	9·58[h]	8·8[f]	8·8[f]
f_r(Cl—O)		2·65[f]	2·65[f]
f_r(O—X)		2·65[f]	2·65[f]
^{19}F nmr spectrum	d		

[a] See also Table 45.
[b] G. H. Rohrback and G. H. Cady, *J. Amer. Chem. Soc.* **69** (1947) 677.
[c] Y. Macheteau and J. Gillardeau, *Bull. Soc. chim. France*, (1969) 1819.
[d] H. Agahigian, A. P. Gray and G. D. Vickers, *Canad. J. Chem.* **40** (1962) 157.
[e] C. J. Schack and D. Pilipovich, *Inorg. Chem.* **9** (1970) 1387.
[f] K. O. Christe, C. J. Schack and E. C. Curtis, *Inorg. Chem.* **10** (1971) 1589.
[g] C. J. Schack, K. O. Christe, D. Pilipovich and R. D. Wilson, *Inorg. Chem.* **10** (1971) 1078.
[h] E. A. Robinson, *Canad. J. Chem.* **41** (1963) 3021.

The action of fluorine on concentrated perchloric acid generates the unstable compound fluorine perchlorate[818], which has a pronounced tendency to explode, e.g. on freezing. It

[818] G. H. Rohrback and G. H. Cady, *J. Amer. Chem. Soc.* **69** (1947) 677; *ibid.* **70** (1948) 2603.

decomposes thermally by two routes:

$$ClF+2O_2 \leftarrow FOClO_3 \rightarrow ClO_2F+O_2$$

and readily oxidizes iodide ions:

$$FOClO_3+2I^- \rightarrow I_2+ClO_4^-+F^-$$

Chlorine(I) perchlorate, which has been briefly treated in connection with the oxides of chlorine (p. 1371), has been used to synthesize bromine[599] and iodine perchlorates (Scheme 18). All the halogen perchlorates are hazardous compounds, liable to explode if subjected to thermal or physical shock.

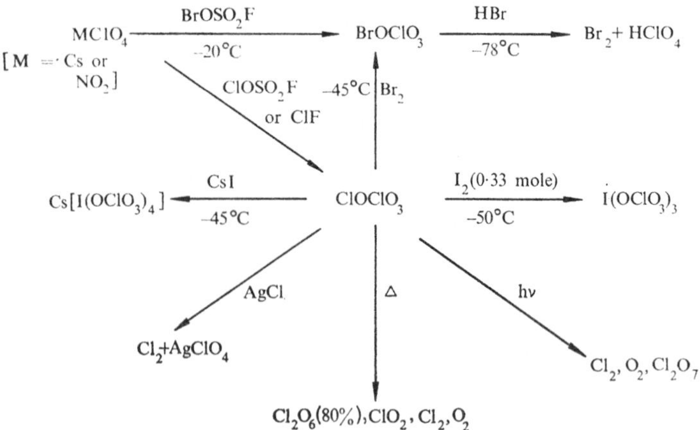

SCHEME 18. Halogen perchlorates.

Vibrational spectroscopic studies confirm that the halogen(I) perchlorates have structures directly related to that of perchloric acid, the hydrogen being replaced by fluorine, chlorine or bromine. While $FOClO_3$ is apparently the most explosive member of the set, *thermal* stability decreases in the sequence $FOClO_3 > ClOClO_3 > BrOClO_3$ as the polarizability of the terminal halogen increases. Attempts to make $IOClO_3$ from equimolar amounts of I_2 and $ClOClO_3$ have been unsuccessful, the iodine being oxidized to the trivalent state[819].

The reaction of iodine, ozone and anhydrous perchloric acid allegedly affords a material of empirical composition $I(ClO_4)_3,2H_2O$[820]. Pure $I(ClO_4)_3$[819] is a white solid, which is polymeric (according to Raman data) and unstable at temperatures much above -45°C; the mode of decomposition depends on the rate of warming, but the products include chlorine oxides (predominantly Cl_2O_7), an iodine(V) perchlorate (probably IO_2ClO_4) and I_2O_5. The pale yellow salt $Cs[I(OClO_3)_4]$ is stable at ambient temperatures: the anion apparently contains a square-planar IO_4 unit.

Halogen Carboxylates

With the exception of perfluoroacyl hypofluorites, halogen(I) carboxylates are unstable and cannot be isolated in the pure state. They are intermediates in the Hunsdiecker reaction,

[819] K. O. Christe and C. J. Schack, *Inorg. Chem.* **11** (1972) 1682.
[820] F. Fichter and H. Kappeler, *Z. anorg. Chem.* **91** (1915) 134.

whereby equimolar amounts of the halogen and a silver carboxylate, mixed in an inert solvent, produce an alkyl halide *via* decarboxylation of the initially formed halogen carboxylate[821]:

$$RCO_2Ag + X_2 \longrightarrow AgX + [RCO_2X] \xrightarrow{-CO_2} RX \text{ (X = Cl, Br or I)}$$

This intermediacy has been verified by the spectroscopic detection of O–Cl bonds in carbon tetrachloride solutions containing CH_3CO_2Ag and chlorine[822]. Aliphatic halogen(I) carboxylates lose carbon dioxide at room temperature, but perfluoro- derivatives undergo decarboxylation only at 100°C. Iodine trifluoroacetate is stabilized as its pyridine complex $C_5H_5N,IOCOCF_3$, and analogous derivatives of other iodine(I) carboxylates have also been reported; the compounds give only very weakly conducting solutions in acetone. The halogen acetates $ClOCOCH_3$ and $BrOCOCH_3$ are believed to be the effective halogenating agents in mixtures of acetic anhydride with chlorine and bromine respectively[823].

Oxidation of iodine with fuming nitric acid in the presence of a carboxylic acid or anhydride affords the iodine(III) compounds $I(OCOR)_3$ [R = CH_nCl_{3-n}, where $n = 0, 1, 2$ or 3; CF_3, C_3F_7 and C_6F_5][824]. The infrared spectra of the solid perfluoro- compounds are consistent with the presence of covalent I–O–C(=O)R units, although in electrolysis of iodine triacetate the quantity of silver iodide formed at a silvered platinum gauze cathode is in good agreement with Faraday's Law calculations based on the equation

$$I^{3+} + Ag + 3e \rightarrow AgI$$

which assumes that I^{3+} is present in solution. The iodine(III) carboxylates are quite sensitive to moisture and possess limited thermal stability; thus, the colourless crystals of iodine(III) acetate are fairly stable in the cold, but begin to decompose at 100°C and explode at 140°C[825]. The compound reacts with methanesulphonic acid to give $(CH_3SO_3)_3I$.

Other Halogen Oxysalts

On oxidation of iodine with concentrated nitric acid in the presence of acetic anhydride and phosphoric acid, the compound IPO_4 is obtained. Its nature remains a matter for conjecture, but hydrolysis causes the iodine(III) to disproportionate:

$$5IPO_4 + 9H_2O \rightarrow I_2 + 3HIO_3 + 5H_3PO_4$$

Moreover, oxyhalogen derivatives of the acids H_2SO_4, $H_2S_2O_7$, $H_2S_3O_{10}$ and H_2SeO_4 have been prepared by reactions such as

$$2ClO_2 + 3SO_3 \rightarrow [ClO][ClO_2][S_3O_{10}]$$

and

$$HIO_3 + H_2SO_4 \rightarrow IO_2HSO_4 + H_2O$$

The nature of the oxyhalogen unit in these and related compounds has been considered in Section 4A (p. 1352).

821 R. G. Johnson and R. K. Ingham, *Chem. Rev.* **56** (1956) 219.
822 M. Anbar and I. Dostrovsky, *J. Chem. Soc.* (1954) 1105.
823 P. B. D. de la Mare, I. C. Hilton and C. A. Vernon, *J. Chem. Soc.* (1960) 4039; P. B. D. de la Mare and J. L. Maxwell, *Chem. and Ind.* (1961) 553.
824 M. Schmeisser, K. Dahmen and P. Sartori, *Ber.* **100** (1967) 1633.
825 N. V. Sidgwick, *The Chemical Elements and their Compounds*, Vol. II, p. 1244, Clarendon Press, Oxford (1950).

C. INTERHALOGENS AND POLYHALIDE ANIONS[826-841]

1. INTRODUCTION

The existence of compounds formed by union of the halogens with one another came to be recognized soon after the identification of the halogens themselves. Thus, the mono- and trichloride of iodine were discovered by both Gay Lussac[842] and Davy[843] in 1813–14. Balard[844] first made iodine monobromide in 1826; he also noticed that when bromine is mixed with chlorine the intensity of the colour is much diminished, though a century elapsed before spectroscopic evidence established beyond doubt the existence of bromine monochloride[845]. First of the halogen fluorides to be identified was iodine pentafluoride, which may have been prepared by Kämmerer in 1862 when he heated dry silver fluoride with iodine[846]; however, the true nature of the reaction was not established until 1870:

$$5AgF + 3I_2 \rightarrow 5AgI + IF_5$$

and it fell to Moissan first to describe the preparation from its elements and characterization of the pentafluoride[846]. Working independently, Lebeau and Prideaux discovered bromine trifluoride in 1905 and remarked upon its extreme reactivity[847]. With the exception of iodine mono- and trifluoride and chlorine pentafluoride, the remaining halogen fluorides were first prepared by Ruff and his collaborators at Breslau between 1925 and 1934: chlorine monofluoride[848] and trifluoride[849] were made from chlorine and fluorine at 250°C; the

826 J. W. Mellor, *A Comprehensive Treatise on Inorganic and Theoretical Chemistry*, Vol. II, Longmans, Green and Co., London (1922).

827 *Supplement to Mellor's Comprehensive Treatise on Inorganic and Theoretical Chemistry*, Supplement II, Part I, Longmans, Green and Co., London (1956).

828 *Gmelins Handbuch der Anorganischen Chemie*, 8 Auflage: "Chlor", System-nummer 6, Verlag Chemie, Berlin (1927); "Brom", System-nummer 7, Verlag Chemie, Berlin (1931); "Iod", System-nummer 8, Verlag Chemie, Berlin (1933).

829 *Gmelins Handbuch der Anorganischen Chemie*, 8 Auflage, "Chlor", System-nummer 6, Teil **B**, Lieferung 2, p. 543. Verlag Chemie, Weinheim/Bergstr. (1969).

830 A. G. Sharpe, *Quart. Rev. Chem. Soc.* **4** (1950) 115.

831 N. N. Greenwood, *Rev. Pure Appl. Chem.* **1** (1951) 84.

832 R. C. Brasted, *Comprehensive Inorganic Chemistry* (ed. M. C. Sneed, J. L. Maynard and R. C. Brasted), Vol. 3, p. 179, van Nostrand (1954).

833 H. C. Clark, *Chem. Rev.* **58** (1958) 869.

834 E. E. Havinga and E. H. Wiebenga, *Rec. Trav. Chim.* **78** (1959) 724.

835 E. H. Wiebenga, E. E. Havinga and K. H. Boswijk, *Adv. Inorg. Chem. Radiochem.* **3** (1961) 133.

836 A. G. Sharpe, *Non-aqueous Solvent Systems* (ed. T. C. Waddington), p. 285, Academic Press (1965).

837 Z. E. Jolles (ed.), *Bromine and its Compounds*, Benn, London (1966).

838 L. Stein, *Halogen Chemistry* (ed. V. Gutmann), Vol. 1, p. 133. Academic Press (1967).

839 H. Meinert, *Z. Chem.* **7** (1967) 41.

840 A. I. Popov, *Halogen Chemistry* (ed. V. Gutmann), Vol. 1, p. 225, Academic Press (1967); *MTP International Review of Science: Inorganic Chemistry Series One*, Vol. 3 (ed. V. Gutmann), p. 53, Butterworths and University Park Press (1972).

841 A. A. Opalovskii, *Russ. Chem. Rev.* **36** (1967) 711.

842 J. L. Gay Lussac, *Ann. Chim.* [1] **91** (1814) 5.

843 H. Davy, *Phil. Trans.* **104** (1814) 487.

844 A. J. Balard, *Ann. Chim. Phys.* [2] **32** (1826) 337.

845 A. J. Balard, *Ann. Chim. Phys.* [2] **32** (1826) 371; N. V. Sidgwick, *The Chemical Elements and their Compounds*, p. 1149. Clarendon Press, Oxford (1950).

846 H. Kämmerer, *J. prakt. Chem.* **85** (1862) 452; G. Gore, *Proc. Roy. Soc.* **19** (1870) 235; H. Moissan, *Compt. rend.* **135** (1902) 563.

847 P. Lebeau, *Compt. rend.* **141** (1905) 1018; E. B. R. Prideaux, *J. Chem. Soc.* **89** (1906) 316.

848 O. Ruff, E. Ascher, J. Fischer and F. Laass, *Z. anorg. Chem.* **176** (1928) 258.

849 O. Ruff and H. Krug, *Z. anorg. Chem.* **190** (1930) 270.

unstable bromine monofluoride[850] was prepared from the elements at 10°C; and bromine pentafluoride[851] and iodine heptafluoride[852] were identified as the products of the fluorination of bromine trifluoride (at 200°C) and iodine pentafluoride (at 280°C), respectively. In common with bromine monofluoride, iodine monofluoride cannot be isolated as a pure substance at room temperature since it disproportionates rapidly to iodine and iodine pentafluoride; nevertheless, in 1951 Durie[853] identified the IF molecule by its emission spectrum (4350–6900 Å) in the bright greenish-yellow flame which accompanies the reaction of fluorine with iodine crystals. As formed by the action of fluorine on iodine suspended in trichlorofluoromethane, iodine trifluoride has been represented by Schmeisser and his co-workers[854] as a yellow material stable up to −28°C, while most recently characterized of the interhalogen compounds is chlorine pentafluoride, first reported by Smith[855] in 1963 as the issue of heating to 350°C a 14:1 mixture of fluorine and chlorine at a total pressure of 250 atm.

That the donor action of a halide ion on either a halogen or interhalogen molecule gives rise to a polyhalide anion, e.g. I_3^-, ICl_4^- or BrF_4^-, has likewise been established through a long and sometimes contentious history. Shortly after the discovery of iodine, Pelletier and Caventou[856] reported the preparation of a crystalline addition compound of strychnine and iodine, which they called "hydroiodure iodurê", and which was undoubtedly strychninium tri-iodide, $C_{21}H_{22}O_2N_2,HI_3$. The solubility of iodine in potassium iodide solutions and the formation of compounds between metal halides and iodine or an iodine halide attracted early attention to the polyhalides, to which many publications in the nineteenth and early twentieth century bear witness. Cremer and Duncan[857] made a careful and thorough re-examination of the methods of preparation, physical properties and reactions in solution and in the solid state of the polyhalides; their systematic survey summarizes most of the early work and adds a considerable weight of new material. In recent years, the range of polyhalide complexes has been substantially enlarged, while a wealth of detail concerning stability and structure has also been brought to light; the outstanding experimental advances have come with the crystallographic analysis of numerous solid complexes, with the development of physicochemical techniques for monitoring reactions in solution, and with the increasing use of non-aqueous solvents as reaction media. It has also become evident that certain halogen and interhalogen molecules are susceptible either to oxidation or to halide ion-abstraction, whereby polyhalogen cations like Cl_2F^+, BrF_2^+ or I_3^+ are produced. Such species are the burden of Section 4A and will receive only incidental mention here.

Because of their unusual compositions, polyhalogen compounds are not easily described in terms of classical valence rules, and they offer to this day an intriguing test of modern theories of structure and bonding; more than once they have been the subject of theoretical polemics. In this context, there exist numerous analogies between polyhalogen species and, on the one hand, noble-gas halides or, on the other, halide derivatives of the elements

[850] O. Ruff and A. Braida, *Z. anorg. Chem.* **214** (1933) 81.
[851] O. Ruff and W. Menzel, *Z. anorg. Chem.* **202** (1931) 49.
[852] O. Ruff and R. Keim, *Z. anorg. Chem.* **193** (1930) 176.
[853] R. A. Durie, *Proc. Roy. Soc.* **A207** (1951) 388; *Canad. J. Phys.* **44** (1966) 337.
[854] M. Schmeisser and E. Scharf, *Angew. Chem.* **72** (1960) 324; M. Schmeisser, W. Ludovici, D. Naumann, P. Sartori and E. Scharf, *Ber.* **101** (1968) 4214.
[855] D. F. Smith, *Science*, **141** (1963) 1039.
[856] P. Pelletier and J. B. Caventou, *Ann. Chim. Phys.* [2] **10** (1819) 164.
[857] H. W. Cremer and D. R. Duncan, *J. Chem. Soc.* (1931) 1857, 2243; *ibid.* (1932) 2031; *ibid.* (1933) 181.

of Groups V and VI. The striking structural and chemical affinities revealed by the members of formally isoelectronic sequences such as SbX_6^{3-}, TeX_6^{2-}, IF_6^-, XeF_6; SbX_5^{2-}, TeX_5^-, IF_5, XeF_5^+; and SbX_4^-, TeX_4 and IF_4^+ clearly imply that the polyhalogens are representative of a more general class of molecular aggregate. Typically this class is distinguished (i) by the presence of a central atom coordinated by more ligands than are formally compatible with the achievement of an 8-electron valence-shell, and (ii) by a surplus of valence *p*-electrons relative to the number of bonding molecular orbitals which are available through the interaction of valence *s*- or *p*-functions of the central atom. The various theoretical treatments which have been applied to the interpretation of the physical properties—and notably of the stereochemistries—of polyhalogen species will be dealt with by way of conclusion to the present section. Apart from their intrinsic interest, however, interhalogen compounds are also noteworthy for their action as halogenating agents, a characteristic which has been turned to considerable synthetic advantage, and for the solvent function exhibited by several members, e.g. ICl, BrF_3 and IF_5 [836].

2. INTERHALOGEN COMPOUNDS

Introduction

In composition, an interhalogen compound stable under conventional conditions belongs to one of the classes XY, XY_3, XY_5 or XY_7; the compounds which have so far been characterized with some degree of assurance are listed in Table 91. To these may

TABLE 91. KNOWN INTERHALOGEN COMPOUNDS

Type XY	Type XY_3	Type XY_5	Type XY_7
ClF[a]	ClF_3 [a]	ClF_5 [a]	IF_7 [a]
BrF[b]	BrF_3 [a]	BrF_5 [a]	
IF[b]	IF_3 [b]	IF_5 [a]	
BrCl[c]	I_2Cl_6 [a]		
ICl[a]			
IBr[a]			

[a] Pure material long-lived at room temperature; physical properties at least reasonably well defined.
[b] Rapidly disproportionates at room temperature into a mixture of the element and higher fluorides; few reliable physical properties known.
[c] Rapidly dissociates at room temperature into a mixture of the elements; few reliable physical properties known.

be added the recently identified bromine chlorides produced by the action of a microwave discharge on mixtures of chlorine and bromine and trapped in an inert matrix at 20°K[858]; analysis of the infrared spectra suggests that the trapped molecules include several with T-shaped frameworks possessing either Cl–Br–Cl or Cl–Br–Br "linear" units, e.g. Br_2Cl_2, $BrCl_3$ and Br_3Cl. The general stoichiometry of a stable interhalogen compound, that is XY_n, is such that n is an odd number, whence it follows that the molecule contains an even number of valence electrons, and the observed diamagnetism is to be anticipated. With the notable exception of the newly discovered Br_2Cl_2 and Br_3Cl[858], molecules XY_n with n greater than 1 are characterized by a central atom X the atomic

[858] L. Y. Nelson and G. C. Pimentel, *Inorg. Chem.* 7 (1968) 1695.

number of which exceeds that of Y. The greater the atomic number, and hence the size, of X, the more substituents Y it can carry, while the smaller the substituents Y, the more of these can be accommodated around a given X atom: for n to exceed 1, stability appears to demand that Y be either fluorine or chlorine, and for n to exceed 3 that Y be limited to fluorine. Where compounds with higher values of n are known, it is generally the case that the stability is enhanced as the atomic number of X increases. Whether or not a particular interhalogen XY_n can be prepared, however, depends on its stability with respect, not only to dissociation into the parent elements X_2 and Y_2, but also to disproportionation into X_2 and XY_m, where $m > n$.

The molecules Br_4 and I_4 identified by spectroscopic measurements (see Section 2, p. 1184) formally belong to the interhalogen class XY_3; however, in that there are as yet no grounds for believing that the X_4 units embody more than loose association between the X_2 molecules, these appear to have little in common with species like ClF_3 or I_2Cl_6. More closely related to the interhalogens are derivatives, mostly of the type XY' and XY'_3, formed by pseudohalogens Y′: examples include XCN, XSCN, XN_3 and $I(NCS)_3$ (X = Cl, Br or I). In many of their properties, e.g. their action as Lewis acids, these pseudohalide compounds resemble conventional interhalogens. Reliable physical properties have been described for but few of the pseudohalides, and, in the absence of sufficient information for a systematic account, such compounds will be represented here, not in detail, but mainly for the comparison they afford with the better defined interhalogen compounds.

Radicals of the type XYZ (where X, Y and Z are the same or different halogen atoms), which are short-lived under normal conditions, have been postulated as intermediates in certain gas-phase reactions of halogen and interhalogen molecules. Representative of this class is the radical Cl_3, which has also been identified in the matrix-isolated condition by its infrared spectrum (see Section 2, p. 1168)[859]. Molecular-beam studies of atom-recombination reactions also imply the formation of triatomic species of this kind[860]; according to estimates of dissociation energy and mean lifetime, the most stable configuration of radicals containing different atoms is that with the least electronegative atom in the central position. The thermal or photolytic dissociation of ClF_3 is thought to proceed *via* the formation of ClF_2, for which ΔH_f° is estimated to be *ca.* -19 kcal mol^{-1} [861]. Substantiation of this view is provided by the infrared spectrum reported for ClF_2 as produced by the photolysis of chlorine trifluoride or mixtures of chlorine monofluoride and fluorine isolated in an inert matrix at 16°K[861]; the vibrational characteristics indicate that, whereas Cl_3 is linear, ClF_2 is bent with an apex angle of $136 \pm 15°$. Such experiments apart, however, interhalogen radicals are the subjects more of speculation than of first-hand evidence.

Preparation, Purification and Manipulation[826–829,838–840,862]

All the known interhalogen compounds may be (and most commonly are) obtained by direct combination of the elements. A survey of more specific details for individual compounds, which appears in Table 92, emphasizes that the product formed by the direct

[859] L. Y. Nelson and G. C. Pimentel, *J. Chem. Phys.* **47** (1967) 3671.

[860] Y. T. Lee, P. R. LeBreton, J. D. McDonald and D. R. Herschbach, *J. Chem. Phys.* **51** (1969) 455.

[861] J. A. Blauer, H. G. McMath and F. C. Jaye, *J. Phys. Chem.* **73** (1969) 2683; G. Mamantov, E. J. Vasini, M. C. Moulton, D. G. Vickroy and T. Maekawa, *J. Chem. Phys.* **54** (1971) 3419.

[862] G. Brauer, *Handbook of Preparative Inorganic Chemistry*, 2nd edn., Vol. 1, Academic Press (1963); *Inorganic Syntheses*, Vols. 1, 3 and 9, McGraw-Hill (1939–67); R. E. Dodd and P. L. Robinson, *Experimental Inorganic Chemistry*, Elsevier (1954).

TABLE 92. THE PREPARATION OF INDIVIDUAL INTERHALOGEN COMPOUNDS

Compound	Method	Conditions	Comments	References
ClF	1. $Cl_2+F_2 \rightarrow 2ClF$ 2. $ClF_3+Cl_2 \rightarrow 3ClF$	Gas phase, 220–250°C. Gas phase, 250–350°C.	Copper reaction vessel typically employed. Both methods are effective and useful. ClF is purified by distillation, micro-sublimation or gas chromatography.	a–e
BrF	1. $Br_2+F_2 \rightarrow 2BrF$ 2. $BrF_3+Br_2 \rightarrow 3BrF$ or $BrF_5+2Br_2 \rightarrow 5BrF$	Gas phase. In suitable circumstances the reaction gives rise to a flame. BrF favoured at elevated temperatures.	BrF detected by its electronic spectrum; further fluorination or disproportionation of BrF leads to BrF_3 or BrF_5. BrF detected spectroscopically but cannot be isolated as a pure compound because of disproportionation at normal temperatures.	a, b, f
IF	1. $I_2+F_2 \rightarrow 2IF$ or $RI+F_2 \rightarrow IF+RF$ (R = Me or Et) 2. $IF_3+I_2 \rightarrow 3IF$ 3. $I_2+AgF \rightarrow AgI+IF$	Combustion of I_2 or RI with F_2 produces a flame; also reaction of I_2 with F_2 in CCl_3F at −45°C[g]. In CCl_3F at −78°C[g,h]. −10° to +30°C[i].	IF detected by its electronic spectrum. It cannot be isolated as a pure compound at room temperature because it disproportionates rapidly: $5IF \rightarrow 2I_2+IF_5$ However, adducts IF,L (L = 2,2′-bipyridyl, quinoline or pyridine) have been reported[g].	b, g–i
BrCl	$Br_2+Cl_2 \rightleftharpoons 2BrCl$	In the gas phase[a,j] or in solution, e.g. in CCl_4 [a,j] or water[k], at normal temperatures.	BrCl characterized by spectrophotometric or pressure measurements. The pure compound cannot be isolated because of the facility of its dissociation into Br_2 and Cl_2 under normal conditions.	a, e, j–l
ICl	1. $I_2+Cl_2 \rightarrow 2ICl$	The reaction may be carried out with or without a solvent and at or below room temperature; methods using either liquid or gaseous Cl_2 have been described.	ICl is normally prepared by direct combination of the elements and best purified by fractional crystallization of the melt or by low-temperature distillation.	a, c, d, f, l

	2.	$I^- + Cl^- \rightarrow ICl + 2e$	Aqueous acidic solution at room temperature; oxidizing agents which have been used include $KMnO_4$, chlorine water and KIO_3.	A procedure of analytical rather than preparative importance.	
IBr		$I_2 + Br_2 \rightarrow 2IBr$	Direct interaction of the pure elements at or near room temperature.	Material purified by fractional crystallization of the melt. Dissociation into I_2 and Br_2 in the vapour, even at room temperature, makes it difficult to obtain the compound completely free from I_2 and Br_2.	a, c, l
ClF_3	or	$Cl_2 + 3F_2 \rightarrow 2ClF_3$ $ClF + F_2 \rightarrow ClF_3$	Gas phase, 200–300°C.	Copper, nickel, Monel, Inconel or Kel-F apparatus typically employed. Manufactured commercially as a technical-grade chemical. Purification commonly involves treatment with NaF (to remove HF) followed by fractional distillation.	a–e
BrF_3	1. 2.	$Br_2 + 3F_2 \rightarrow 2BrF_3$ $Br_2 + 2ClF_3 \rightarrow 2BrF_3 + Cl_2$ [m]	Reaction between gaseous F_2 and Br_2 vapour, with or without a diluent, at or near room temperature; alternatively, the F_2 is bubbled into or over liquid Br_2. ClF_3 is passed into liquid Br_2 at 0°C.	Copper, nickel, Monel or Kel-F apparatus employed. BrF_3 is manufactured commercially as a technical-grade chemical. Purification normally effected by fractional distillation.	a–d, f, m
IF_3	1. 2.	$I_2 + 3F_2 \rightarrow 2IF_3$ $I_2 + 3XeF_2 \rightarrow 2IF_3 + 3Xe$ [n]	Action of F_2 on I_2 suspended in CCl_3F at −45°C.	IF_3 is described as a yellow solid stable up to −28°C; above this temperature it decomposes rapidly to I_2, IF and IF_5. Derivatives of IF_3 are known, e.g. IF_3,MF_5 (M = As or Sb), MIF_4 (M = alkali metal or NO) and IF_3,L (L = aromatic nitrogen-base).	b, h, n, o

TABLE 92 (*cont.*)

Compound	Method	Conditions	Comments	References
$BrCl_3$, Br_2Cl_2 and Br_3Cl	$Br_2+3Cl_2 \rightarrow 2BrCl_3$ etc.	Action of microwave discharge on gaseous mixtures of Cl_2 and Br_2.	Species trapped in solid inert matrices at 20°K.	p
I_2Cl_6	1. $I_2+3Cl_2 \rightarrow I_2Cl_6$	Reaction best carried out between I_2 and an excess of liquid Cl_2 at −80°C.	Solid I_2Cl_6 obtained by evaporation of the excess chlorine; cannot be purified by crystallization or vaporization because of its ready dissociation into ICl and Cl_2.	a, c, f, l
	2. $I_2+5Cl^-+ClO_3^-+6H^+ \rightarrow I_2Cl_6+3H_2O$	Reaction carried out in strongly acidic, aqueous solution at <40°C [c].	Reported as a convenient method of preparing I_2Cl_6.	
ClF_5	1. $Cl_2+5F_2 \xrightarrow{\Delta} 2ClF_5$	Gas phase with excess fluorine, total pressure 250 atm, temperature ~ 350°C.	Compound first prepared by this method.	b, e, m, q–t
	2. $ClF_3+F_2 \xrightarrow{h\nu} ClF_5$	Photochemical reaction carried out at room temperature and 1 atm pressure [q].	A convenient preparative route to ClF_5.	
	3. $MClF_4(s)+F_2 \rightarrow MF(s)+ClF_5$	80–150°C, conversion up to 90% [r].	Probably the most expedient routes yet devised to ClF_5.	
	or $MCl(s)+3F_2 \rightarrow MF(s)+ClF_5$ (M = alkali metal)	100–300°C, conversion 50–70% [s].		
	4. Electrolytic oxidation of Cl_2 or ClF_3 in an HF medium [t].			
BrF_5	1. $Br_2+5F_2 \rightarrow 2BrF_5$	Gas phase with excess F_2, temperature > 150°C.	Copper, nickel, Monel, Inconel or Kel-F apparatus typically employed. BrF_5 is manufactured commercially as a technical-grade chemical. Purification usually involves fractional distillation.	a–d, f, m, s
	2. $BrF_3+F_2 \rightarrow BrF_5$	Gas phase, temperature 200°C.		
	3. $KBr(s)+3F_2 \rightarrow KF(s)+BrF_5$	25°C, yield ~ 50% [s].		
IF_5	1. $I_2+5F_2 \rightarrow 2IF_5$	Reaction of gaseous F_2 with solid I_2 at room temperature.	Method (1) is most generally used for the preparation of IF_5, though fluorination of I_2O_5 with ClF_3, BrF_3 or SF_4 provides an expedient method of preparation on the small scale. The reaction vessel is typically	a–d, m, u
	2. Fluorination of I_2 with AgF, ClF_3, BrF_3 or RuF_5.	Various conditions [b].		
	3. HI or metal iodide + F_2.	Various conditions [b].		

	4. I_2O_5 or metal iodate + F_2, ClF_3, BrF_3 or SF_4.	Various conditions[b,m].	of copper or nickel; purified IF_5 may subsequently be manipulated in Kel-F, silica or (at room temperature) glass apparatus. IF_5 is manufactured commercially as a technical-grade chemical. Rigorous purification involves treatment (i) with F_2 to oxidize I_2 and (ii) with NaF to remove HF, followed by fractionation[u].	
IF_7	1. $I_2 + 7F_2 \rightarrow 2IF_7$ or $IF_5 + F_2 \rightarrow IF_7$ 2. $KI(s) + 4F_2 \rightarrow KF(s) + IF_7$ or $MI_2(s) + 8F_2 \rightarrow MF_2(s) + 2IF_7$	Gas phase, 250–300°C[b]. Gas phase, 150°C[v]. 250°C[w]. e.g. M = Pd[x]. Method (2) recommended for the preparation of pure IF_7 because of the difficulty of drying I_2.	IF_7 reacts with silica, glass, I_2O_5 or traces of water to form OIF_5, from which it can be separated only with difficulty. It is normally handled in Monel or nickel apparatus and purified by fractional sublimation.	a–c, m, v–x

[a] *Mellor's Comprehensive Treatise on Inorganic and Theoretical Chemistry*, Supplement II, Part I, Longmans, Green and Co., London (1956).
[b] L. Stein, *Halogen Chemistry* (ed. V. Gutmann), Vol. 1, p. 133, Academic Press (1967).
[c] G. Brauer, *Handbook of Preparative Inorganic Chemistry*, 2nd edn., Vol. 1, Academic Press (1963).
[d] R. E. Dodd and P. L. Robinson, *Experimental Inorganic Chemistry*, pp. 223–226, Elsevier (1954).
[e] *Gmelins Handbuch der Anorganischen Chemie*, 8 Auflage, "Chlor", System-nummer 6, Teil B, Lieferung 2, Verlag Chemie (1969).
[f] *Inorganic Syntheses*, Vols. 1, 3 and 9, McGraw-Hill (1939–67).
[g] M. Schmeisser, P. Sartori and D. Naumann, *Chem. Ber.* **103** (1970) 590, 880.
[h] M. Schmeisser and E. Scharf, *Angew. Chem.* **72** (1960) 324.
[i] H. Schmidt and H. Meinert, *Angew. Chem.* **72** (1960) 109.
[j] N. N. Greenwood, *Rev. Pure Appl. Chem.* **1** (1951) 84.
[k] H. Gutmann, M. Lewin and B. Perlmutter-Hayman, *J. Phys. Chem.* **72** (1968) 3671.
[l] *Gmelins Handbuch der Anorganischen Chemie*, 8 Auflage: "Brom", System-nummer 7, Verlag Chemie (1931); "Iod", System-nummer 8, Verlag Chemie (1933).
[m] H. Meinert, *Z. Chem.* **7** (1967) 41.
[n] N. Bartlett, personal communication.
[o] M. Schmeisser, W. Ludovici, D. Naumann, P. Sartori and E. Scharf, *Chem. Ber.* **101** (1968) 4214.
[p] L. Y. Nelson and G. C. Pimentel, *Inorg. Chem.* **7** (1968) 1695.
[q] R. Gatti, R. L. Krieger, J. E. Sicre and H. J. Schumacher, *J. Inorg. Nuclear Chem.* **28** (1966) 655; R. L. Krieger, R. Gatti and H. J. Schumacher, *Z. phys. Chem.* **51** (1966) 240.
[r] D. Pilipovich, W. Maya, E. A. Lawton, H. F. Bauer, D. F. Sheehan, N. N. Ogimachi, R. D. Wilson, F. C. Gunderloy, jun., and V. E. Bedwell, *Inorg. Chem.* **6** (1967) 1918.
[s] G. A. Hyde and M. M. Boudakian, *Inorg. Chem.* **7** (1968) 2648.
[t] E. A. Lawton and H. H. Rogers, U.S. Pat. 3,373,096 (1968).
[u] D. W. Osborne, F. Schreiner and H. Selig, *J. Chem. Phys.* **54** (1971) 3790.
[v] C. J. Schack, D. Pilipovich, S. N. Cohz and D. F. Sheehan, *J. Phys. Chem.* **72** (1968) 4697.
[w] H. Selig, C. W. Williams and G. J. Moody, *J. Phys. Chem.* **71** (1967) 2739; H. H. Claassen, E. L. Gasner and H. Selig, *J. Chem. Phys.* **49** (1968) 1803.
[x] N. Bartlett and L. E. Levchuk, *Proc. Chem. Soc.* (1963) 342.

interaction of elemental halogens is acutely dependent on the conditions under which the reaction is carried out. For the formation of certain halogen fluorides, advantage has sometimes been taken of fluorinating agents other than elemental fluorine; possible agents include other halogen fluorides, e.g. ClF_3 or BrF_3, as well as the noble-gas fluorides KrF_2 and XeF_2[838], platinum and plutonium hexafluorides[838,863], and dioxygen difluoride[838]. Another variant of the normal convention involves fluorination of a suitable metal halide, usually by elemental fluorine; chlorine and bromine pentafluorides and iodine heptafluoride are expediently produced by methods of this sort. Thus, for the preparation of pure iodine heptafluoride, fluorination of potassium iodide has been recommended[864] in preference to the reaction between iodine and excess fluorine, primarily because iodine is difficult to dry, with the result that the product of the latter reaction tends to be contaminated with iodine oxypentafluoride, OIF_5. Reactions of the type

$$X_2+XF_3 \rightarrow 3XF$$

have also been exploited in the formation of the halogen monofluorides. Chlorine monofluoride, free from the trifluoride, is thereby efficiently produced at 250°C, and there are good reasons for believing that the monofluorides of bromine and iodine are formed in a similar fashion, albeit fleetingly under normal conditions.

Without exception, the interhalogen compounds are highly reactive materials, seldom easy to purify or to manipulate in the pure condition. The destructive action on organic materials also makes the compounds highly injurious to living matter. The toxicity of the vapours depends upon their virulent action on the eyes, nose and throat, the bronchi and mucous membrane being especially vulnerable to attack; the liquids likewise have a drastic physiological effect on the skin, giving rise to painful burns and necrosis. Precise details of the biological effects are sparse, though qualitative reports of the action of certain halogen fluorides have been published[838,862,865], and procedures have been described for the safe handling and monitoring of toxic levels of chlorine trifluoride[838,866].

With respect to dissociation, the compounds BrCl, ICl, IBr and I_2Cl_6 are of relatively low stability at normal temperatures. Such dissociation renders impossible the production of bromine monochloride free from elemental chlorine and bromine, and inhibits the use of distillation for the purification of the other compounds. Following their preparation, iodine monochloride and monobromide are commonly refined by fractional crystallization of the melt[827,862], but the dissociation pressure of iodine trichloride is such as to frustrate even crystallization as a method of purification. Accordingly, iodine trichloride is normally produced under conditions which ensure that contamination is restricted to materials like chlorine, which can be removed simply by vaporization[827,862]. All of these interhalogen compounds attack not only metals, but also cork, rubber and many other organic materials. They are best handled *in vacuo* or in an inert atmosphere using glass apparatus which excludes grease or moisture, storage being accomplished preferably in sealed glass ampoules.

By contrast, the fluorides BrF, IF and IF_3 are susceptible to disproportionation which is rapid at room temperature; on the evidence presently available, it is doubtful whether

[863] F. P. Gortsema and R. H. Toeniskoetter, *Inorg. Chem.* **5** (1966) 1925.

[864] H. Selig, C. W. Williams and G. J. Moody, *J. Phys. Chem.* **71** (1967) 2739.

[865] H. C. Hodge and F. A. Smith, *Fluorine Chemistry* (ed. J. H. Simons), Vol. IV. Academic Press, New York and London (1965).

[866] L. M. Vincent and J. Gillardeau, Comm. Energ. At. (France), Rapt. CEA No. 2360 (1963).

pure samples of any of these compounds have yet been prepared. Although the well-defined fluorides ClF, ClF_3, ClF_5, BrF_3, BrF_5, IF_5 and IF_7 are prone neither to dissociate nor to disproportionate at normal temperatures, their susceptibility to reaction with moisture, glass, organic materials and most metals is such that their preparation, manipulation and storage calls for much the same stringency of operation that is demanded by fluorides like XeF_2, ReF_7 and PtF_6 [867,868]. According to numerous reports, several of the halogen fluorides, e.g. ClF_3, BrF_3 and IF_7, attack Pyrex glass or quartz even at room temperature, while this becomes the general pattern of behaviour at elevated temperatures. It is probable, however, that impurities like hydrogen fluoride initiate or accelerate this reaction, as it has also been reported, for example, that very pure chlorine trifluoride has no effect on Pyrex or quartz at normal temperatures[838]. Materials found to offer the best resistance to attack include Monel, Inconel, nickel, copper, stainless steel and sapphire, together with Kel-F and Teflon plastics, which, despite their instability at elevated temperatures, are frequently used for the fabrication of gaskets, valve-seatings and both reaction and storage vessels. The handling of materials like BrF_3 and IF_7 requires, ideally, a vacuum-system having a working manifold, valves and a Bourdon or bellows manometer constructed in Monel, nickel, copper or stainless steel; access is thence gained to traps or reaction vessels made either in metal or Kel-F, depending on the severity of treatment to be applied to the fluoride. For investigations of the infrared absorption of a halogen fluoride, a cell having a nickel body and bearing, typically, AgCl, CaF_2 or polythene windows has been effectively employed[868], while, to examine the Raman spectrum and certain other properties of the vapour or liquid, silica vessels have been extensively used, though baking *in vacuo* and preliminary exposure to the fluoride are recommended for the pre-seasoning of the equipment. The electrical conductivity of the pure liquid or of solutions of other materials in the liquid is probably best measured with the aid of a conductivity cell made in Kel-F[869], though earlier investigations relied, with varying degrees of success, on glass or silica cells.

Impurities likely to be contained in freshly prepared or commercial samples of the halogen fluorides include the elemental halogens, hydrogen fluoride, other halogen fluorides and oxyhalogen derivatives, e.g. OIF_5. To purify a particular halogen fluoride, the two measures most commonly taken are (i) treatment with an anhydrous alkali-metal fluoride to remove hydrogen fluoride and certain of the halogen fluoride impurities in the form of involatile salts, e.g. $NaHF_2$, and (ii) fractional distillation or sublimation, either from trap to trap or *via* a suitable low-temperature column. For example, the samples of iodine pentafluoride used for recent calorimetric and vapour-pressure measurements were derived from the commercial product in the following stages: first, treatment at room temperature, in a Kel-F container, with a small amount of fluorine to oxidize elemental iodine to the pentafluoride; second, removal of traces of hydrogen fluoride by heating to 150°C in the presence of finely divided and thoroughly dried sodium fluoride; and finally, fractionation of the volatile material to separate the pentafluoride from the more volatile heptafluoride[870]. Similarly, for the most recent vapour-pressure measurements on iodine heptafluoride[871],

[867] D. F. Shriver, *The Manipulation of Air-sensitive Compounds*, p. 105, McGraw-Hill (1969); B. Weinstock, *Record of Chemical Progress*, **23** (1962) 23; H. H. Hyman (ed.), *Noble-gas Compounds*, The University of Chicago Press, Chicago (1963).

[868] J. H. Canterford and T. A. O'Donnell, *Technique of Inorganic Chemistry* (ed. H. B. Jonassen and A. Weissberger), Vol. 7, p. 273. Interscience, New York (1968).

[869] L. A. Quarterman, H. H. Hyman and J. J. Katz, *J. Phys. Chem.* **61** (1957) 912.

[870] D. W. Osborne, F. Schreiner and H. Selig, *J. Chem. Phys.* **54** (1971) 3790.

[871] C. J. Schack, D. Pilipovich, S. N. Cohz and D. F. Sheehan, *J. Phys. Chem.* **72** (1968) 4697.

the sample was prepared by the fluorination of iodine pentafluoride and purified (i) by treatment with anhydrous potassium fluoride to abstract both hydrogen fluoride and iodine pentafluoride (as KHF_2 and KIF_6, respectively), and (ii) by fractional sublimation *in vacuo*. The use of gas chromatography has also been reported[872] for the separation of mixtures of halogen fluorides, fluorine and other halogens, hydrogen fluoride and other volatile fluorides (e.g. UF_6); a typical assembly for this purpose incorporates a nickel column which is packed with Kel-F powder coated with Kel-F oils.

Pseudohalide analogues of the interhalogen compounds are commonly prepared by the action of an elemental halogen or an interhalogen compound on a derivative of the pseudohalide anion[827,832,862,873,874], e.g.

$$MCN + X_2 \rightarrow XCN + MX$$

(M = alkali metal or $\frac{1}{2}$Hg; X = Cl, Br or I; aqueous or non-aqueous media)

$$MN_3 + X_2 \rightarrow XN_3 + MX$$

(M = alkali metal, Ag or $\frac{1}{2}Hg_2$; X = Cl, Br or I; various conditions)

$$3ICl + 3MECN \rightarrow 3MCl + I(NCE)_3 + I_2$$

(M = K, NH_4, Ag or $\frac{1}{2}$Pb; E = O or S)

The products range from the toxic but comparatively stable cyanides XCN to the highly explosive azides XN_3. A tendency to suffer polymerization appears to be a common characteristic, e.g.

$$\underset{\text{Halogen cyanide}}{3XCN} \rightarrow \underset{\text{Cyanuric trihalide}}{[XCN]_3}$$

$$2XNCO \rightarrow X_2N{\cdot}CO{\cdot}NCO \qquad (X = Cl \text{ or } Br)$$

Physical Properties[827,829,831,835,838–840]

1. **General characteristics.** In listing the principal physical properties of the interhalogen compounds, Table 93 alludes to structural, thermodynamic and spectroscopic parameters of the molecules, as well as to characteristics of the condensed phases, notably melting and boiling points, vapour pressure, density, viscosity and electrical conductivity. However, it should be emphasized that the data are not all equally reliable. In some instances, e.g. that of bromine monochloride, thermal dissociation precludes the possibility of accurate measurements of properties like the melting and boiling points, and the figures quoted merely represent a very approximate estimate of the range of existence of the liquid phase. The facility of disproportionation of the monofluorides BrF and IF likewise frustrates measurements of meaningful physical properties for the condensed phases, though the molecules have been characterized spectroscopically in the vapour phase. In the light of this, the significance of the physical properties originally reported for bromine monofluoride by Ruff and Braida[850] must be regarded as dubious. For iodine trifluoride, numerical physical properties have yet to be described.

[872] J. F. Ellis, C. W. Forrest and P. L. Allen, *Anal. Chim. Acta*, **22** (1960) 27; J. G. Million, C. W. Webber and P. R. Kuehn, U.S. Atomic Energy Commission Rept. K-1639 (1966).

[873] K. Dehnicke, *Angew. Chem., Internat. Edn.* **6** (1967) 240.

[874] A. Hassner, M. E. Lorber and C. Heathcock, *J. Org. Chem.* **32** (1967) 540; F. W. Fowler, A. Hassner and L. A. Levy, *J. Amer. Chem. Soc.* **89** (1967) 2077; A. Hassner and F. Boerwinkle, *ibid.* **90** (1968) 216.

Table 93. Physical Properties of the Interhalogens[1–9]

(a) Type XY

Property	ClF	BrF	IF	BrCl	ICl	IBr
Molecular weight[10]	54·451	98·902	145·9029	115·357	162·357$_5$	206·808$_5$
Gaseous molecules						
Electronic ground state configuration $\sigma^2\pi^4\pi^{*4}$, $X^1\Sigma^+$						
Electronic transitions, T_e(cm^{-1}), relative to the $X^1\Sigma^+$ state (excited state in parentheses)[11]		61,615 (II) 57,900 (I)		61,563 (D) 59,318 (C)	58,168* (D) 53,457* (C) 37,741 (E)	56,369* (G) 51,701* (F) 39,126 (E) 38,713 (D) 35,427 (C)
	18,956 ($^3\Pi_{0^+}$)	18,281·2 ($^3\Pi_{0^+}$) 17,385 ($^3\Pi_1$)	19,053·7$_5$ ($^3\Pi_{0^+}$)	16,795 ($^3\Pi_{0^+}$)	~18,000* [B′(0$^+$)] ~17,344* ($^3\Pi_{0^+}$) 13,556·21 ($^3\Pi_1$)	16,814 [B′(0$^+$)] 16,165 ($^3\Pi_{0^+}$) 12,213 ($^3\Pi_1$)
Ground state properties						
Internuclear distance, r_e(Å)[11]	1·6281	1·756	1·908$_9$	2·138	2·3207	2·485
Dipole moment, D	0·881[11]	1·29[11]	—	0·57[11]	0·65[12]	1·21[12]
Vibrational frequency, ω_e(cm^{-1})	786·34 (^{35}ClF)[11]	672·6[11]	608·1$_9$[11]	440[11]	384·293 (I^{35}Cl)[11]	268·71 (I^{79}Br)[11]
Anharmonic vibrational constant, $\omega_e x_e$(cm^{-1})	6·23 (^{35}ClF)[11]	4·5[11]	2·4$_8$[11]	1·6[11]	1·501 (I^{35}Cl)[11]	0·83 (I^{79}Br)[11]
Force constant, k_e (mdyne/Å)	4·484	4·089	3·600	2·80	2·386	2·071
Dissociation energy: $D_0°$	*kcal* 60·35[8] *eV* 2·617[8]	*kcal* 59·42[8] *eV* 2·577[8]	*kcal* ?66·2[8] *eV* ?2·87[8]	*kcal* 51·4[13] *eV* 2·23[13]	*kcal* 49·63[13] *eV* 2·152[13]	*kcal* 41·91[13] *eV* 1·817[13]
Thermodynamic properties						
$\Delta H_f°$[XY(g)] at 298°K (kcal mol^{-1})	−13·5[8]	−14·0[8]	−22·6[8]	+3·50[14]	+4·18[15,16]	+9·76[14]
$\Delta G_f°$[XY(g)] at 298°K (kcal mol^{-1})	−13·8[8]	−17·6[8]	−28·1[8]	−0·23[14]	−1·37[15,16]	+0·89[14]
$S°$[XY(g)] at 298°K (cal deg^{-1} mol^{-1})	52·06[8]	54·70[8]	56·45[8]	57·36[14]	59·140[14]	61·822[14]
$\Delta H_f°$[XY(s)] at 298°K (kcal mol^{-1})	—	—	—	—	−8·44(α)[15,16]	−2·5[14]

TABLE 93 (*cont.*)

(a) Type XY

Property	ClF	BrF	IF	BrCl	ICl	IBr
Thermodynamic properties (cont.)						
$\Delta G_f°$[XY(s)] at 298°K (kcal mol^{-1})	—	—	—	—	−3·334(α)[15,16]	—
$S°$[XY(s)] at 298°K (cal deg^{-1} mol^{-1})	—	—	—	—	23·405(α)[15,16]	—
$\Delta H_f°$[XY(l)] at 298°K (kcal mol^{-1})	—	—	—	—	−5·69[15,16]	—
$\Delta G_f°$[XY(l)] at 298°K (kcal mol^{-1})	—	—	—	—	−3·314[15,16]	—
$S°$[XY(l)] at 298°K (cal deg^{-1} mol^{-1})	—	—	—	—	32·3[14]	—
Heat capacity, $C_p°$ (cal deg^{-1} mol^{-1})						
Gas (100–6000°K)[17]	6·646–9·722	6·970–9·704	6·983–10·146	7·101–9·540	7·213–9·580	7·623–9·620
Liquid (303·68–317·76°K)[15]	—	—	—	—	24·63–24·61	—
Solid (17·71–295·04°K)[15]	—	—	—	—	1·91–13·74(α)	—
(263–273°K)[5]	—	—	—	—	*ca.* 13·76(β)	—
Ionization potentials:	*kcal* *eV*	*kcal* *eV*	*kcal* *eV*	*kcal* *eV*	*kcal* *eV*	*kcal* *eV*
$XY^+(^2\Pi_{3/2}) \leftarrow XY(^1\Sigma^+)$	} 293[19] 12·7[19]	272[19] 11·8[19]	242[19] 10·5[19]	256[19] 11·1[19]	226·0[18] 9·80[18]	220·9[18] 9·58[18]
$XY^+(^2\Pi_{1/2}) \leftarrow XY(^1\Sigma^+)$	}				241·0[18] 10·45[18]	235·7[18] 10·22[18]
$XY^+(^2\Pi) \leftarrow XY(^1\Sigma^+)$	—	—	—	—	284·8[18] 12·35[18]	265·2[18] 11·50 $(^2\Pi_{3/2})$[18]
$XY^+(^2\Sigma^+) \leftarrow XY(^1\Sigma^+)$	—	—	—	—	320·5[18] 13·90[18]	309·0[18] 13·40[18]
Properties of the condensed phases	°*K* °*C*	°*K* °*C*	°*K* °*C*	°*K* °*C*	°*K* °*C*	°*K* °*C*
Boiling point	173·0[8] −100·1[8]	~ 293[2] ~ 20[2]	Disproportionates	~ 278[2] ~ 5[2]	370–373[7] 97–100[7]	~ 389[9] ~ 116[9]
		Disproportionates		Dissociates	Some dissociation	Dissociates
Melting point	117·5[8] −155·6[8]	~ 240[2] ~ −33[2]	— —	~ 207[2] ~ −66[2]	300·53[15] 27·38[15](α)	314[9] 41[9]
					287·0[9] 13·9[9](β)	
Heat of vaporization (kcal mol^{-1})	4·80 (173·0°K)[8]	—	—	—	9·876 (liquid at 298·15°K)[15,16] 12·622 (solid, α at 298·15°K)[15,16]	—

Entropy of vaporization at boiling point (Trouton constant) (cal deg^{-1} mol^{-1})	28·0	—	—	—	~ 26·5	—
Heat of fusion (kcal mol^{-1})	—	—	—	—	2·773(α)[15] 2·27(β)[9]	—
Vapour pressure, p(mm Hg)	$\log p = 15{\cdot}738 - 3109/T + 1{\cdot}538 \times 10^5/T^2$ (liquid, 123–168°K)[8]	—	—	—	Partial pressure ICl(l): 38·52–135·5 (303·60–330°K)[16] ICl(s): 0·44–32·62 (250–300·53°K)[16]	—
Crystal structure					Monoclinic, space group $P2_1/c$ (α[20] and β[21])	Orthorhombic, space group $Ccm2_1$(no. 36)[22]
Molecules per unit cell					8 (α and β)	4
Unit cell dimensions (Å)					α β	
a					12·60 8·883	4·903
b					4·38 8·400	6·993
c					11·90 7·568	8·931
β					119·5° 91·35°	
Intramolecular distance (Å)					2·37 2·35 2·44 2·44	2·52
Intermolecular distance (Å)					3·00 2·94 3·08 3·06 †	3·18 3·76 ††

† Structures consist of zigzag chains composed of two types of ICl molecule: (i) with both I and Cl atoms engaged in short-range intermolecular interactions along the chain, and (ii) with the Cl atom branching off the chain and not involved in such interactions (Fig. 36). The closest approach between atoms in different molecular chains implies normal van der Waals' interaction. *α-form*: chains puckered and branching Cl atoms *cis* to each other[20]. *β-form*: chains essentially planar and branching Cl atoms *trans* to each other[21].

†† Molecules form a herring-bone pattern similar to that in crystalline I_2[22].

TABLE 93 (*cont.*)

(a) Type XY

Property	ClF	BrF	IF	BrCl	ICl	IBr
Spectroscopic properties						
Vibrational spectra of condensed phases	Refs. 8, 23	—	—	Ref. 23	Ref. 23	Ref. 23
Quadrupole coupling constants, e^2Qq(MHz):						
Gaseous molecules[24]	^{35}Cl, −146·0	^{79}Br, 1089·0	—	^{35}Cl, −103·6 ^{79}Br, 876·8	^{35}Cl, −82·5 ^{127}I, −2944	^{79}Br, 722 ^{127}I, −2731
Solid	—	—	—	—	^{127}I, −3046 (77°K)[25]	^{127}I, −2892[26]
Asymmetry parameter for the solid, η	—	—	—	—	0·0298[25]	~0·06[26]
^{129}I Mössbauer spectrum: isomer shift relative to standard ZnTe source (cm sec^{-1})	—	—	—	—	0·173[26]	0·123[26]
Nmr spectra	^{19}F resonance studied[8]	—	—	—	—	—
Mechanical properties						
Density (g cm^{-3}):						
Solid	—	—	—	—	3·85 (α, 0°C)[9] 3·66 (β, 0°C)[9]	4·4157–4·4135 (273–283°K)[5]
Liquid	1·62 (173°K)[2]	—	—	—	$3{\cdot}18223/[1+9{\cdot}1590\times10^{-4}t+8{\cdot}3296\times10^{-7}t^2+2{\cdot}7501\times10^{-9}t^3]$, t in °C[5]	3·7616–3·7343 (315–323°K)[5]
Viscosity of liquid (centipoise)	—	—	—	—	4·19 (28°C)[7]	—
Electrical and magnetic properties						
Specific conductivity of liquid (ohm^{-1} cm^{-1})	$1{\cdot}9\times10^{-7}$ (145°K)[8]	—	—	—	$4{\cdot}403\text{–}1{\cdot}817\times10^{-3}$ (26·75–122·0°C)[2]	$3{\cdot}02\text{–}4{\cdot}0\times10^{-4}$ (40–55°C)[7,9]
Magnetic susceptibility, $\chi(\times10^6$ cgs units)	—	—	—	—	−54·5 (s, 17°C)[27]	—

TABLE 93 (*cont.*)

(b) Type XY_3

Property	ClF_3	BrF_3	IF_3	I_2Cl_6
Molecular weight[10]	92·448	136·899	183·899$_5$	466·527
Ground state properties of gaseous XY_3 molecule				
Structure: T-shaped molecule, C_{2v} symmetry				
Dimensions: r(Å)	1·698[8]	1·810[8]	—	—
R(Å)	1·598[8]	1·721[8]	—	—
θ	87°29′[8]	86°12·6′[8]	—	—
Dipole moment, D	0·557[8]	1·19[8]	—	—
Vibrational frequencies (cm^{-1}):	Refs. 28, 29	Refs. 28, 29		
ν_1 (a_1)	752·1	675	—	—
ν_2 (a_1)	529·3	552	—	—
ν_3 (a_1)	328	242	—	—
ν_4 (b_1)	702	612	—	—
ν_5 (b_1)	442	350	—	—
ν_6 (b_2)	328	242	—	—
Valence stretching force constants (mdyne/Å)[28,29]:				
f_r	2·704	3·009	—	—
f_R	4·193	4·084	—	—
Thermodynamic properties:	*ClF_3(g)* *$[ClF_3]_2$(g)*	*BrF_3(g)*	*IF_3(g)*	
$\Delta H_f°$ at 298°K (kcal mol^{-1})	−39·35[30] −81·1[14]	−61·09[14]	~ −116*	—
$\Delta G_f°$ at 298°K (kcal mol^{-1})	−29·75[30] −56·7[14]	−54·84[14]	~ −110*	—
$S°$ at 298°K (cal deg^{-1} mol^{-1})	67·28[14] 117[14]	69·89[14]	~72*	—
Mean X–Y bond energy of XY_3 molecule at 298°K (kcal)	41·7	48·2	~66*	—
Heat capacity of gas, $C_p°$ (cal deg^{-1} mol^{-1}) (100–6000°K)[17]	9·394–19·857	9·347–19·859	—	—
Ionization potential[19]	*kcal* 300 *eV* 13·0	*kcal* 297 *eV* 12·9	*kcal* 265 *eV* 11·5	—

Table 93 (*cont.*)

(b) Type XY_3

Property	ClF_3	BrF_3	IF_3	I_2Cl_6
Properties of the condensed phases	°K / °C	°K / °C		°K / °C
Boiling point	284·90[8] / 11·75[8]	398·90[8] / 125·75[8]	Described as a yellow solid which decomposes at −28°C[31]	Dissociates
Melting point	196·83[8] / −76·32[8]	281·92[8] / 8·77[8]		384[9] / 101 (16 atm)[9]
Transition temperature (solid I to II)	190·49[8] / −82·66[8]	— / —		— / —
Heat of vaporization at boiling point, ΔH_{vap} (kcal mol^{-1})	6·580[8]	10·235[8]	—	—
Entropy of vaporization at boiling point (Trouton constant) (cal deg^{-1} mol^{-1})	23·10[8]	25·7[8]	—	—
Heat of fusion (kcal mol^{-1})	1·8193[8]	2·875[8]	—	—
Vapour pressure of liquid, p(mm Hg)	$\log p = 7{\cdot}36711 - 1096{\cdot}917/(t+232{\cdot}75)$; $t = -46{\cdot}97$ to $+29{\cdot}55$°C[8]	$\log p = 7{\cdot}74853 - 1685{\cdot}8/(t+220{\cdot}57)$; $t = 38{\cdot}72$ to $154{\cdot}82$°C[8]	—	—
Thermodynamic properties:	ClF_3(l)	BrF_3(l)		ICl_3(s)
$\Delta H_f°$ at 298°K (kcal mol^{-1})	−45·65[30]	−71·9[14]	—	−21·340[16]
$\Delta G_f°$ at 298°K (kcal mol^{-1})	—	−57·5[14]	—	−5·413[16]
$S°$ at 298°K (cal deg^{-1} mol^{-1})	—	42·6[14]	—	40·395[16]
Heat capacity of solid, $C_p°$ (cal deg^{-1} mol^{-1})	—	—	—	1·90–24·50 (15–330°K)[16]
Crystal structure	Orthorhombic, space group *Pnma* (153°K)[32]	Orthorhombic, space group $Cmc2_1$ (148°K)[33]		Triclinic, space group $P\bar{1}$[34]
Molecules per unit cell	4	4		1 I_2Cl_6 molecule
Unit cell dimensions	$a = 8{\cdot}82_5$ Å; $b = 6{\cdot}09$ Å; $c = 4{\cdot}52$ Å	$a = 5{\cdot}34$ Å; $b = 7{\cdot}35$ Å; $c = 6{\cdot}61$ Å		$a = 5{\cdot}71$ Å, $\alpha = 130°50'$; $b = 10{\cdot}88$ Å, $\beta = 80°51'$; $c = 5{\cdot}48$ Å, $\gamma = 108°30'$
Molecular geometry and dimensions	T-shaped ClF_3 molecules, point group C_{2v}; $d(Cl\text{–}F_{ax}) = 1{\cdot}716$ Å; $d(Cl\text{–}F_{eq}) = 1{\cdot}621$ Å; $\angle F_{ax}\text{–}Cl\text{–}F_{eq} = 86°59'$	T-shaped BrF_3 molecules, point group C_{2v}; $d(Br\text{–}F_{ax}) = 1{\cdot}85$ Å; $d(Br\text{–}F_{eq}) = 1{\cdot}72$ Å; $\angle F_{ax}\text{–}Br\text{–}F_{eq} = 82{\cdot}0, 88{\cdot}4°$		Planar I_2Cl_6 molecules, point group D_{2h}; $d(I\text{–}Cl_{terminal}) = 2{\cdot}38, 2{\cdot}39$ Å; $d(I\text{–}Cl_{bridge}) = 2{\cdot}68, 2{\cdot}72$ Å; $\angle Cl_{terminal}\text{–}I\text{–}Cl_{terminal} = 94°$; $\angle Cl_{bridge}\text{–}I\text{–}Cl_{bridge} = 84°$

Spectroscopic properties				
Vibrational spectra of the condensed phases	Refs. 8, 28 and 29	Refs. 8, 28 and 29	—	Ref. 35
Halogen nqr spectra of the solid:				
e^2Qq(MHz)	^{35}Cl, 150·26[24]	—	—	$^{35}Cl_{terminal}$, −68·34[36] $^{35}Cl_{bridge}$, +27·48[36] ^{127}I, 3035 (77°K)[25]
Asymmetry parameter, η	—	—	—	^{127}I, 0·077[25]
^{129}I Mössbauer spectrum of the solid: isomer shift relative to standard ZnTe source (cm sec^{-1})	—	—	—	0·350[26]
^{19}F nmr spectrum:				
Chemical shift of liquid (ppm rel. CF_3COOH as an external standard)	−192·2, −81·0 (210°K)[37]	−54·3 (~300°K)[38]	—	—
J(F–F) (Hz)	422·2 (210°K)[37]	—	—	—
	At higher temperatures the multiplets coalesce ultimately giving a single broad resonance	Single resonance at normal temperatures		
^{19}F chemical shift of gas (ppm rel. SiF_4)	−313·24, −187·32 (300°K)[37]	—	—	—
J(F–F) (Hz)	441 (300°K)[37]	—	—	—
Mechanical properties				
Density, d(g cm^{-3}):				
Solid	2·53 (153°K)[32]	3·23 (292°K)[8]	—	$d_4^{-40} = 3·203$; $d_4^{15} = 3·1107$[5]
Liquid	$d_4^t = 1·8853 - 2·942 \times 10^{-3}t - 3·79 \times 10^{-6}t^2$ ($t = -5°$ to 46°C)[8]	2·8030 (298°K)[38] 2·7351 (323°K)[38]	—	—
Surface tension of liquid (dyne cm^{-1})	26·6–18·7 (273·2–323·1°K)[39]	37·1–33·8 (285·2–318·2°K)[40]	—	—

TABLE 93 (*cont.*)

(b) Type XY_3

Property	ClF_3	BrF_3	IF_3	I_2Cl_6
Mechanical properties (cont.)				
Viscosity of liquid (centipoise)	0·448–0·316 (290·5–323·2°K)[39]	3·036–1·775 (286·4–312·8°K)[40]	—	—
Electrical, magnetic and optical properties				
Dielectric constant, ϵ:				
Liquid	$\epsilon_t = 4{\cdot}754 - 0{\cdot}018t$ (t = 0–42°C)[41]	—	—	—
Gas	1·002825–1·001929 (319·3–413·3°K)[42]	1·003748–1·002964 (415·5–448·2°K)[42]	—	—
Specific conductivity (ohm^{-1} cm^{-1}):				
Solid	—	—	—	$0{\cdot}019\times10^{-3} - 8{\cdot}35\times10^{-3}$ (343–373°K)[2]
Liquid	$4{\cdot}9\times10^{-9}$ (298°K)[8]	$8{\cdot}34\times10^{-3} - 7{\cdot}31\times10^{-3}$ (283·7–323·2°K)[2,43]	—	$8{\cdot}60\times10^{-3} - 0{\cdot}36\times10^{-3}$ (375–420°K)[2]
Magnetic susceptibility, χ ($\times10^6$ cgs units)	−26·5 (l, 300°K)[44]	−33·9 (l, 300°K)[44]	—	−90·2 (ICl_3 solid, 290°K)[27]
Molar refraction (cm^3 mol^{-1})	10·34 (g, 299°K)[8]	12·92 (g, 326°K)[8] 13·22 (l, 298°K)[8]	—	—

TABLE 93 (*cont.*)

(c) Types XY_5 and XY_7

Property	ClF_5	BrF_5	IF_5	IF_7
Molecular weight[10]	130·445	174·896	221·8964	259·8933
Ground state properties of gaseous XY_n molecule				
Structure	Square pyramid, C_{4v} symmetry[8] $d(Cl–F_{apical}) = \sim 1·62$ Å $d(Cl–F_{basal}) = \sim 1·72$ Å	Square pyramid, C_{4v} symmetry[45] $d(Br–F_{apical}) = 1·689$ Å $d(Br–F_{basal}) = 1·774$ Å $\angle F_{apical}–Br–F_{basal} = 84·8°$	Square pyramid, C_{4v} symmetry[45] $d(I–F_{apical}) = 1·844$ Å $d(I–F_{basal}) = 1·869$ Å $\angle F_{apical}–I–F_{basal} = 81·9°$	Pentagonal bipyramid approximating to D_{5h} symmetry[46] $d(I–F_{ax}) = 1·786$ Å $d(I–F_{eq}) = 1·858$ Å Deformation from D_{5h} symmetry on the average by 7·5° ring-puckering displacements (e_2'' symmetry) and 4·5° axial bending displacements (e_1' symmetry)
Dipole moment, *D*	—	1·51[8]	2·18[8]	0[47] Some evidence of polar conformations at low temperatures
Vibrational frequencies (cm^{-1})	Refs. 48, 49 $\nu_1 (a_1)$ = 709 (R, i.r.) $\nu_2 (a_1)$ = 539 (R, i.r.) $\nu_3 (a_1)$ = 495 (R, i.r.) $\nu_4 (b_1)$ = 480 (R)[1] $\nu_5 (b_1)$ = 346* (R) $\nu_6 (b_2)$ = 375 (R)[1] $\nu_7 (e)$ = 725 (R, i.r.) $\nu_8 (e)$ = 484 (R, i.r.) $\nu_9 (e)$ = 299 (R, i.r.) [1] Refers to liquid-phase. Other frequencies refer to gas-phase	Ref. 48 $\nu_1 (a_1)$ = 683 (R, i.r.) $\nu_2 (a_1)$ = 587 (R, i.r.) $\nu_3 (a_1)$ = 369 (R, i.r.) $\nu_4 (b_1)$ = 535 (R)[1] $\nu_5 (b_1)$ = 281* (R) $\nu_6 (b_2)$ = 312 (R)[1] $\nu_7 (e)$ = 644 (R, i.r.) $\nu_8 (e)$ = 415 (R, i.r.) $\nu_9 (e)$ = 237 (R, i.r.)[1] [1] Refers to liquid-phase. Other frequencies refer to gas-phase	Refs. 50, 51 $\nu_1 (a_1)$ = 710 (R, i.r.) $\nu_2 (a_1)$ = 614 (R, i.r.) $\nu_3 (a_1)$ = 318 (R, i.r.) $\nu_4 (b_1)$ = 602 (R) $\nu_5 (b_1)$ = 181* (R) $\nu_6 (b_2)$ = 274 (R) $\nu_7 (e)$ = 631, 635 (R, i.r.) $\nu_8 (e)$ = 372 (R, i.r.) $\nu_9 (e)$ = 189 (R, i.r.) All refer to gas-phase	Ref. 52 $\nu_1 (a_1')$ = 676 (R) $\nu_2 (a_1')$ = 635 (R) $\nu_3 (a_2'')$ = 670 (i.r.) $\nu_4 (a_2'')$ = 365 (i.r.) $\nu_5 (e_1')$ = 746 (i.r.) $\nu_6 (e_1')$ = 425 (i.r.) $\nu_7 (e_1')$ = 257 (i.r.) $\nu_8 (e_1'')$ = 510 (R) $\nu_9 (e_1'')$ = 352 (R) $\nu_{10} (e_1'')$ = 310 (R) $\nu_{11} (e_2'')$ inactive All refer to gas-phase

TABLE 93 (*cont.*)

(c) Types XY_5 and XY_7

Property	ClF_5	BrF_5	IF_5	IF_7
Valence stretching force constants (mdyne/Å):				
Apical X–F bonds	3·47[49]	4·03[49]	4·82[49]	$4{\cdot}09_5$[52]
Basal or equatorial X–F bonds	2·67[49]	3·24[49]	3·82[49]	3·01[52]
Thermodynamic properties of gaseous molecules:				
$\Delta H_f°$ at 298°K (kcal mol^{-1})	−60·9±4·5[53]	−102·5[14]	−200·8[51]	−229·9[8,51]
$\Delta G_f°$ at 298°K (kcal mol^{-1})	−39·0±4·5[53,54]	−83·8[14]	−184·6[51]	−201·3[8,51]
$S°$ at 298°K (cal deg^{-1} mol^{-1})	74·3[54]	76·50[14]	80·45[51]	87·40[8]
Mean X–F bond energy of XF_n molecule at 298°K (kcal)	36·9	44·7	64·2	55·4
Heat capacity of gas, $C_p°$ (cal deg^{-1} mol^{-1})	20·92–30·85 (250–1000°K)[54]	10·839–31·770 (100–6000°K)[17]	11·830–31·767 (100–6000°K)[17]	13·213–43·682 (100–6000°K)[17]
Ionization potential[19]	—	—	311 kcal (13·5 eV)	—
Properties of the condensed phases	*°K* / *°C*	*°K* / *°C*	*°K* / *°C*	*°K* / *°C*
Boiling point	260·05[53] / −13·1[53]	314·45[8] / 41·30[8]	377·63[51] / 104·48[51]	277·92[8] / 4·77[8] (sublimation temperature)
Melting point	170±4[55] / −103±4[55]	212·65[8] / −60·5[8]	282·571[51] / 9·421[51]	— / —
Triple point	— / —	— / —	282·553[51] / 9·403[51]	279·60[8] / 6·45[8]
Transition temperature (solid I to II)	— / —	— / —	— / —	~153[8] / ~−120[8]
Heat of vaporization at boiling point, ΔH_{vap} (kcal mol^{-1})	5·313[53]	7·31[8]	8·595[51]	5·91[56] ΔH_{subl} = 6·0[57]
Entropy of vaporization at boiling point (Trouton constant) (cal deg^{-1} mol^{-1})	20·4[53]	23·3[8]	22·76[51]	21·1[56]
Heat of fusion (kcal mol^{-1})	—	—	2·682[51]	—
Vapour pressure, p(mm Hg):				
Liquid	$\log p = 7{\cdot}4837-1197/T$ (192·95–391·05°K)[53]	$\log p = 6{\cdot}4545+0{\cdot}001101t-895/(t+206)$[8] ($t$ in °C)	$\log p = 29{\cdot}02167-3090{\cdot}14/T-6{\cdot}96834 \log T$[51]	$\log p = 7{\cdot}4967-1291{\cdot}58/T$[56]
Solid	—	0·1–1·7 (185–208°K)[2]	$\log p = 11{\cdot}764-3035/T$[8]	$\log p = 11{\cdot}2319-3046{\cdot}93/T+197{,}769/T^2$ (195–273°K)[57]

Property				
Critical properties	Ref. 53	—	—	—
Thermodynamic properties:		*BrF_5*(l)	*IF_5*(l)	
$\Delta H_f°$ at 298°K (kcal mol^{-1})	—	−109·6[14]	−210·8[51]	—
$\Delta G_f°$ at 298°K (kcal mol^{-1})	—	−84·1[14]	−186·6[51]	—
$S°$ at 298°K (cal deg^{-1} mol^{-1})	—	53·8[14]	53·74[51]	—
Heat capacity, $C_p°$ (cal deg^{-1} mol^{-1}):				
Liquid	—	—	41·92–41·55 (282·571–350°K)[51]	—
Solid	—	—	0·243–32·75 (6–282·571°K)[51]	—
Crystal structure		Orthorhombic, space group $Cmc2_1$[33]	Monoclinic, space group *Cc* or *C*2/*c*[33]	*Solid I* (<153°K): orthorhombic, space group *Aba*2[58] *Solid II* (153–273°K): cubic, space group $I\bar{4}3m$, *I*432 or *Im*3*m*[58]
				Solid I / *Solid II*
Molecules per unit cell		4	20	4 / 2
Unit cell dimensions		$a = 6·422$ Å $b = 7·245$ Å $c = 7·846$ Å	$a = 15·16$ Å $b = 6·86$ Å $c = 18·20$ Å $\beta = 93°14'$	*Solid I*: $a = 8·74$ Å, $b = 8·87$ Å, $c = 6·14$ Å at 128°K *Solid II*: $a = 6·28$ Å at 165°K
Molecular geometry and dimensions		Square-pyramidal BrF_5 molecules: $d(Br–F_{apical}) = 1·68$ Å $d(Br–F_{basal}) = 1·75–1·82$ Å $\angle F_{apical}–Br–F_{basal} = 80·5–86·5°$	More detailed structural analysis has not been attempted. Structure of IF_5 molecule in crystalline XeF_2,IF_5: square-pyramid with C_{4v} symmetry[59] $d(I–F_{apical}) = 1·862$ Å $d(I–F_{basal}) = 1·892$ Å $\angle F_{apical}–I–F_{basal} = 80·9°$	The data are compatible with the presence of IF_7 pentagonal bipyramids with $d(I–F) = 1·80$ Å

Table 93 (*cont.*)

(c) Types XY_5 and XY_7

Property	ClF_5	BrF_5	IF_5	IF_7
Spectroscopic properties				
Nuclear quadrupole coupling constants, e^2Qq(MHz):				
Gas (microwave spectrum)[60]	—	^{79}Br, $-280{\cdot}9$ ^{81}Br, $-233{\cdot}3$	^{127}I, $+1056{\cdot}6$	—
Solid (Mössbauer spectrum)[61]	—	—	^{127}I, $+1073$	^{127}I, -148
^{129}I Mössbauer spectrum of the solid: isomer shift relative to standard ZnTe source (cm sec^{-1})	—	—	$+3{\cdot}00$[61]	$-4{\cdot}56$[61]
^{19}F nmr spectrum:				
Chemical shift of liquid (ppm)				
Basal or equatorial F atoms	-247 rel $CFCl_3$[55]	-219 rel CF_3CO_2H (ext)[62]	$-173{\cdot}6$ rel SiF_4[63]	-334 rel SiF_4[63]
Apical F atoms	-412 rel $CFCl_3$[55]	-349 rel CF_3CO_2H (ext)[62]	$-222{\cdot}2$ rel SiF_4[63]	-334 rel SiF_4[63]
J(F–F) (Hz)	130[55]	110[62]	$85{\cdot}0$[63]	—
Mechanical properties				
Density, d(g cm^{-3}):				
Solid	—	$3{\cdot}09$ (212°K)[8]	$3{\cdot}961$–$3{\cdot}678$ ($77{\cdot}15$–$273{\cdot}15$°K)[51]	$3{\cdot}62$ (128°K)[58]
Liquid	$d = 3{\cdot}553 - 1{\cdot}396\times10^{-2}T + 4{\cdot}565\times10^{-5}T^2 - 6{\cdot}311\times10^{-8}T^3$ (193–372°K)[53]	$d = 2{\cdot}5509 - 3{\cdot}484\times10^{-3}t - 3{\cdot}45\times10^{-6}t^2$ ($t = -15°$ to 76°C)[8]	$3{\cdot}263$–$3{\cdot}031$ ($283{\cdot}40$–$343{\cdot}96$°K)[51]	$d = 2{\cdot}7918 - 0{\cdot}0049t$ ($t = 7{\cdot}13$–$24{\cdot}32$°C)[56]
Surface tension of liquid (dyne cm^{-1})	—	$24{\cdot}3$–$21{\cdot}6$ ($282{\cdot}4$–$305{\cdot}8$°K)[40]	$30{\cdot}8$–$28{\cdot}2$ ($291{\cdot}6$–$311{\cdot}4$°K)[40]	—
Viscosity of liquid (centipoise)	—	$0{\cdot}824$–$0{\cdot}590$ ($275{\cdot}5$–$302{\cdot}1$°K)[40]	$2{\cdot}490$–$1{\cdot}686$ ($292{\cdot}1$–$313{\cdot}2$°K)[40]	—
Electrical, magnetic and optical properties				
Dielectric constant, ϵ:				
Liquid	$\epsilon = 3{\cdot}08 - 0{\cdot}015t$ ($t = -80°$ to -17°C)[53]	$\epsilon = 8{\cdot}20 - 0{\cdot}0117t$ ($t = -11{\cdot}7°$ to $24{\cdot}5$°C)[64]	$\epsilon = 41{\cdot}09 - 0{\cdot}198t$ ($t = 0$–42°C)[41]	$1{\cdot}75$ (298°K)[56]
Gas	$1{\cdot}00279$ ($297{\cdot}05$°K)[53]	$1{\cdot}006320$–$1{\cdot}004378$ ($345{\cdot}6$–$430{\cdot}8$°K)[64]	$1{\cdot}009108$–$1{\cdot}007135$ ($392{\cdot}7$–$446{\cdot}0$°K)[41]	—

Specific conductivity of the liquid (ohm^{-1} cm^{-1})	$0{\cdot}37-1{\cdot}25\times10^{-9}$ ($193-256$°K)[53]	$7{\cdot}8-9{\cdot}91\times10^{-8}$ ($213-298$°K)[65]	$5{\cdot}4\times10^{-6}$ (298°K)[8]	$<10^{-9}$ (~ 298°K)[56]
Magnetic susceptibility of the liquid, χ ($\times10^6$ cgs units)	—	$-45{\cdot}1$ (298°K)[44]	$-58{\cdot}1$ (~ 298°K)[44]	—
Molar refraction (cm^3 mol^{-1})	—	15·41 (l, 298°K)[8] 15·48 (g, 298°K)[8]	19·17 (g, 298°K)[8]	17·46 (l, 298°K)[56]

*Calculated value. R, Raman-active. i.r., infrared-active.

1 J. W. Mellor, *A Comprehensive Treatise on Inorganic and Theoretical Chemistry*, Vol. II, Longmans, Green and Co., London (1922).
2 *Supplement to Mellor's Comprehensive Treatise on Inorganic and Theoretical Chemistry*, Supplement II, Part I, Longmans, Green and Co., London (1956).
3 *Gmelins Handbuch der Anorganischen Chemie*, 8 Auflage: "Brom", System-nummer 7, Verlag Chemie (1931); "Iod", System-nummer 8, Verlag Chemie (1933).
4 *Gmelins Handbuch der Anorganischen Chemie*, 8 Auflage: "Chlor", System-nummer 6, Teil B, Lieferung 2, p. 543. Verlag Chemie (1969).
5 N. N. Greenwood, *Rev. Pure Appl. Chem.* **1** (1951) 84.
6 E. H. Wiebenga, E. E. Havinga and K. H. Boswijk, *Adv. Inorg. Chem. Radiochem.* **3** (1961) 148.
7 A. G. Sharpe, *Non-aqueous Solvent Systems* (ed. T. C. Waddington), p. 285. Academic Press (1965).
8 L. Stein, *Halogen Chemistry* (ed. V. Gutmann), Vol. 1, p. 133. Academic Press (1967).
9 H. Meinert, *Z. Chem.* **7** (1967) 41.
10 *Table of Atomic Weights*, 1969, IUPAC Commission on Atomic Weights: *Pure Appl. Chem.* **21** (1970) 95.
11 *Tables of Constants and Numerical Data*, No. 17, *Spectroscopic Data relative to Diatomic Molecules* (general ed. B. Rosen), Pergamon (1970).
12 A. L. McClellan, *Tables of Experimental Dipole Moments*, Freeman, San Francisco (1963).
13 A. G. Gaydon, *Dissociation Energies*, 3rd edn., Chapman and Hall, London (1968).
14 *National Bureau of Standards Technical Note* 270–3, U.S. Government Printing Office, Washington (1968).
15 G. F. Calder and W. F. Giauque, *J. Phys. Chem.* **69** (1965) 2443.
16 R. H. Lamoreaux and W. F. Giauque, *J. Phys. Chem.* **73** (1969) 755.
17 *JANAF Thermochemical Tables*, The Dow Chemical Company, Midland, Michigan (1960–8).
18 Adiabatic ionization potentials: S. Evans and A. F. Orchard, *Inorg. Chim. Acta*, **5** (1971) 81; A. W. Potts and W. C. Price, *Trans. Faraday Soc.* **67** (1971) 1242.
19 A. P. Irsa and L. Friedman, *J. Inorg. Nuclear Chem.* **6** (1958) 77.
20 K. H. Boswijk, J. van der Heide, A. Vos and E. H. Wiebenga, *Acta Cryst.* **9** (1956) 274.
21 G. B. Carpenter and S. M. Richards, *Acta Cryst.* **15** (1962) 360.
22 L. N. Swink and G. B. Carpenter, *Acta Cryst.* **B24** (1968) 429.
23 H. Stammreich, R. Forneris and Y. Tavares, *Spectrochim. Acta*, **17** (1961) 1173.
24 E. A. C. Lucken, *Nuclear Quadrupole Coupling Constants*, p. 288. Academic Press (1969).
25 R. S. Yamasaki and C. D. Cornwell, *J. Chem. Phys.* **30** (1959) 1265.

TABLE 93 (*Footnotes cont.*)

[26] M. Pasternak and T. Sonnino, *J. Chem. Phys.* **48** (1968) 1997.
[27] F. W. Gray and J. Dakers, *Phil. Mag.* [7] **11** (1931) 81.
[28] H. Selig, H. H. Claassen and J. H. Holloway, *J. Chem. Phys.* **52** (1970) 3517.
[29] R. A. Frey, R. L. Redington and A. L. K. Aljibury, *J. Chem. Phys.* **54** (1971) 344.
[30] R. C. King and G. T. Armstrong, *J. Res. Nat. Bur. Stand.* **74A** (1970) 769.
[31] M. Schmeisser, W. Ludovici, D. Naumann, P. Sartori and E. Scharf, *Chem. Ber.* **101** (1968) 4214.
[32] R. D. Burbank and F. N. Bensey, *J. Chem. Phys.* **21** (1953) 602.
[33] R. D. Burbank and F. N. Bensey, jun., *J. Chem. Phys.* **27** (1957) 982.
[34] K. H. Boswijk and E. H. Wiebenga, *Acta Cryst.* **7** (1954) 417.
[35] R. Forneris, J. Hiraishi, F. A. Miller and M. Uehara, *Spectrochim. Acta*, **26A** (1970) 581.
[36] J. C. Evans and G. Y.-S. Lo, *Inorg. Chem.* **6** (1967) 836.
[37] L. G. Alexakos and C. D. Cornwell, *J. Chem. Phys.* **41** (1964) 2098.
[38] J. W. Emsley, J. Feeney and L. H. Sutcliffe, *High Resolution Nuclear Magnetic Resonance Spectroscopy*, Vol. 2, p. 871. Pergamon (1966).
[39] A. A. Banks, A. Davies and A. J. Rudge, *J. Chem. Soc.* (1953) 732.
[40] M. T. Rogers and E. E. Garver, *J. Phys. Chem.* **62** (1958) 952.
[41] M. T. Rogers, H. B. Thompson and J. L. Speirs, *J. Amer. Chem. Soc.* **76** (1954) 4841.
[42] M. T. Rogers, R. D. Pruett and J. L. Speirs, *J. Amer. Chem. Soc.* **77** (1955) 5280.
[43] H. H. Hyman, T. Surles, L. A. Quarterman and A. I. Popov, *J. Phys. Chem.* **74** (1970) 2038.
[44] M. T. Rogers, M. B. Panish and J. L. Speirs, *J. Amer. Chem. Soc.* **77** (1955) 5292.
[45] A. G. Robiette, R. H. Bradley and P. N. Brier, *Chem. Comm.* (1971) 1567.
[46] W. J. Adams, H. B. Thompson and L. S. Bartell, *J. Chem. Phys.* **53** (1970) 4040.
[47] E. W. Kaiser, J. S. Muenter, W. Klemperer and W. E. Falconer, *J. Chem. Phys.* **53** (1970) 53.
[48] G. M. Begun, W. H. Fletcher and D. F. Smith, *J. Chem. Phys.* **42** (1965) 2236.
[49] K. O. Christe, E. C. Curtis, C. J. Schack and D. Pilipovich, *Inorg. Chem.* **11** (1972) 1679.
[50] H. Selig and H. Holzman, *Israel J. Chem.* **7** (1969) 417; L. E. Alexander and I. R. Beattie, *J. Chem. Soc.* (*A*) (1971) 3091.
[51] D. W. Osborne, F. Schreiner and H. Selig, *J. Chem. Phys.* **54** (1971) 3790.
[52] H. H. Claassen, E. L. Gasner and H. Selig, *J. Chem. Phys.* **49** (1968) 1803; R. K. Khanna, *J. Mol. Spectroscopy*, **8** (1962) 134.
[53] H. H. Rogers, M. T. Constantine, J. Quaglino, jun., H. E. Dubb and N. N. Ogimachi, *J. Chem. Eng. Data*, **13** (1968) 307; W. R. Bisbee, J. V. Hamilton, J. M. Gerhauser and R. Rushworth, *ibid.* p. 382.
[54] R. Bougon, J. Chatelet and P. Plurien, *Compt. rend.* **264C** (1967) 1747.
[55] D. Pilipovich ,W. Maya, E. A. Lawton, H. F. Bauer, D. F. Sheehan, N. N. Ogimachi, R. D. Wilson, F. C. Gunderloy, jun., and V. E. Bedwell, *Inorg Chem.* **6** (1967) 1918.
[56] H. Selig, C. W. Williams and G. J. Moody, *J. Phys. Chem.* **71** (1967) 2739.
[57] C. J. Schack, D. Pilipovich, S. N. Cohz and D. F. Sheehan, *J. Phys. Chem.* **72** (1968) 4697.
[58] R. D. Burbank and F. N. Bensey, jun., *J. Chem. Phys.* **27** (1957) 981; R. D. Burbank, *Acta Cryst.* **15** (1962) 1207; J. Donohue, *Acta Cryst.* **18** (1965) 1018.
[59] G. R. Jones, R. D. Burbank and N. Bartlett, *Inorg. Chem.* **9** (1970) 2264.
[60] M. J. Whittle, R. H. Bradley and P. N. Brier, *Trans. Faraday Soc.* **67** (1971) 2505; R. H. Bradley, P. N. Brier and M. J. Whittle, *Chem. Phys. Letters*, **11** (1971) 192.
[61] S. Bukshpan, C. Goldstein, J. Soriano and J. Shamir, *J. Chem. Phys.* **51** (1969) 3976.
[62] E. L. Muetterties and W. D. Phillips, *J. Amer. Chem. Soc.* **81** (1959) 1084; *Adv. Inorg. Chem. Radiochem.* **4** (1962) 245.
[63] N. Bartlett, S. Beaton, L. W. Reeves and E. J. Wells, *Canad. J. Chem.* **42** (1964) 2531.
[64] M. T. Rogers, R. D. Pruett, H. B. Thompson and J. L. Speirs, *J. Amer. Chem. Soc.* **78** (1956) 44.
[65] M. T. Rogers, J. L. Speirs and M. B. Panish, *J. Amer. Chem. Soc.* **78** (1956) 3288.

The details of Table 93 disclose that iodine monochloride, monobromide and trichloride are solids at room temperature, whereas bromine trifluoride and pentafluoride and iodine pentafluoride are liquids, and the fluorides of chlorine and iodine heptafluoride are gases. The following sequences of volatility are apparent:

$$ClF > ClF_3 < ClF_5$$
$$BrF_3 < BrF_5$$
$$IF_5 < IF_7$$
$$ClF_5 > BrF_5 > IF_5$$

It is noteworthy that the trifluorides ClF_3 and BrF_3 are significantly less volatile than the corresponding pentafluorides, while IF_7 represents the most volatile interhalogen derivative of iodine. In the solid or liquid states, the materials exhibit electrical conductivities in the range 10^{-2} to 10^{-9} ohm^{-1} cm^{-1}; the significance of the comparatively high conductivities of some of the liquids is discussed subsequently in connection with the function of these interhalogens as solvent systems. In all cases, however, the electrical conductivity falls below the range characteristic of fused salts. Thus, all of the physical properties of the interhalogens are compatible with their formulation as molecular compounds. Accordingly, variations in the boiling and melting points and in the volatility of the compounds follow the pattern set by molecular polarizability (which is also a function of molecular geometry) and dipole moment (which spans the range from virtually zero for IF_7 to 2·18 D for IF_5).

The properties of the interhalogens signify that the mononuclear unit XY_n is commonly the only recognizable molecular aggregate in the solid, liquid or gaseous phases. A notable exception is provided, however, by iodine trichloride, crystals of which are composed of planar I_2Cl_6 molecules (see below); the mononuclear unit ICl_3 has not been characterized. Again, polymeric networks, which denote appreciable interaction between the IX units, have been established for crystalline iodine(I) chloride and bromide. According to the Trouton constants given in the table, chlorine(I) fluoride and iodine(I) chloride would appear also to be associated in the liquid state, but the experimental results may well be substantially in error. The non-ideality of the vapours of chlorine and bromine trifluorides have been attributed to the equilibrium

$$2XF_3 \rightleftharpoons X_2F_6$$

for which equilibrium constants have been reported[838]; thus, for the chlorine compound, $K_p(= p^2_{ClF_3}/p_{Cl_2F_6})$ is 35·4 atm at 24·2°C. When chlorine or bromine trifluoride is isolated at comparatively high concentrations in solid inert matrices maintained at 5–25°K, the infrared spectrum bears direct witness that dimers and other polymeric species exist as distinct, though weakly associated, aggregates[875]. Some degree of aggregation is also indicated by the comparatively large Trouton constants reported for the trifluorides. Likewise, a comparison of the Raman spectra of gaseous and liquid iodine pentafluoride points to a measure of association in the liquid[876]. Contrary to earlier reports, however, the Trouton constant of iodine pentafluoride has now been established as well within the range observed for normal liquids[870]; evidently, therefore, the degree of association at the boiling point is slight, though this does not invalidate the evidence for association at lower temperatures.

[875] R. A. Frey, R. L. Redington and A. L. K. Aljibury, *J. Chem. Phys.* **54** (1971) 344.
[876] H. Selig and H. Holzman, *Israel J. Chem.* **7** (1969) 417.

The only pseudohalide derivatives for which detailed physical constants, thermodynamic properties and structural data have been described are the cyanides ClCN, BrCN and ICN[832,837,862,877–879]. According to the vibrational-rotational and microwave spectra of the gaseous materials, the molecules are invariably linear with the following dimensions: ClCN, $r(\text{C–Cl}) = 1{\cdot}629$ Å, $r(\text{C}{\equiv}\text{N}) = 1{\cdot}163$ Å; BrCN, $r(\text{C–Br}) = 1{\cdot}790$ Å, $r(\text{C}{\equiv}\text{N}) = 1{\cdot}158$ Å; ICN, $r(\text{C–I}) = 1{\cdot}995$ Å, $r(\text{C}{\equiv}\text{N}) = 1{\cdot}158$ Å. The molecular crystals are composed of linear chains $\cdots \text{X–C}{\equiv}\text{N} \cdots \text{X–C}{\equiv}\text{N} \cdots$, in which the halogen and nitrogen atoms of each XCN molecule engage in relatively short-range intermolecular interactions[879]. The vibrational spectra of the species XCN[880], XSCN[880] and XN_3[881] have also been reported, and those of the cyanides have been exploited in investigations of the coordinating properties of the molecules, e.g. with respect to various inorganic halides[880,882].

2. **Thermodynamic properties.** Selected thermodynamic functions, based on the most recent estimates[838,878,883], appear for the majority of the interhalogen compounds listed in Table 93; calorimetric data are available only for ICl[884,885], I_2Cl_6[885], ClF_3[827], BrF_3[827] and IF_5[870]. The results of the table show that the standard entropies of all the gaseous diatomic molecules form a well-graded series, with the values for the interhalogen compounds falling between those of the parent halogens. The entropy of the unsymmetrical molecule XY exceeds the mean of those for the corresponding symmetrical molecules by a small margin represented, in the ideal case, by $R \log_e 2 = 1{\cdot}4$ eu. It follows that the entropy change accompanying the gaseous reaction

$$X_2 + Y_2 \rightarrow 2XY$$

is small and that the stability of XY with respect to its elements is effectively determined by the appropriate enthalpy change. However, a reaction of the type

$$X_2 + nY_2 \rightarrow 2XY_n \quad (n = 3, 5 \text{ or } 7)$$

is possible only in the event of a significant negative enthalpy change, since the decrease in the number of molecules occasions a marked decrease in entropy.

The most precise estimates of the heats of formation of the diatomic molecules ClF, BrF, IF, ICl and IBr derive from the convergence limits of the band systems observed in the visible region of the spectrum and thought to arise from the transition $A\,^3\Pi_{0+} \leftarrow X\,^1\Sigma^+$ (ClF, BrF or IF) or $A\,^3\Pi_1 \leftarrow X\,^1\Sigma^+$ (ICl or IBr), X referring to the ground state of the molecule[827,831,838]. The convergence limits thus found are (in cm^{-1}): ClF, 21,512;

[877] *Tables of Interatomic Distances and Configuration in Molecules and Ions*, Chemical Society Special Publication No. 11, London (1958); Supplement, Chemical Society Special Publication No. 18, London (1965).

[878] F. D. Rossini, D. D. Wagman, W. H. Evans, S. Levine and I. Jaffe, *Selected Values of Chemical Thermodynamic Properties*, Circular 500, National Bureau of Standards, Washington (1952); National Bureau of Standards Technical Note 270–3, U.S. Government Printing Office, Washington (1968).

[879] A. F. Wells, *Structural Inorganic Chemistry*, 3rd edn., Clarendon Press, Oxford (1962).

[880] H. Siebert, *Anwendungen der Schwingungsspektroskopie in der Anorganischen Chemie*, Springer-Verlag (1966); K. Nakamoto, *Infrared Spectra of Inorganic and Coordination Compounds*, 2nd edn., Wiley-Interscience, New York (1970).

[881] D. E. Milligan and M. E. Jacox, *J. Chem. Phys.* **40** (1964) 2461.

[882] K. Kawai and I. Kanesaka, *Spectrochim. Acta*, **25A** (1969) 263.

[883] W. H. Evans, T. R. Munson and D. D. Wagman, *J. Res. Nat. Bur. Stand.* **55** (1955) 147; *JANAF Thermochemical Tables*, The Dow Chemical Company, Midland, Michigan (1960–8).

[884] G. V. Calder and W. F. Giauque, *J. Phys. Chem.* **69** (1965) 2443.

[885] R. H. Lamoreaux and W. F. Giauque, *J. Phys. Chem.* **73** (1969) 755.

BrF, 21,190; IF, 23,570; ICl, 17,357; IBr, 14,660. The behaviour resembles closely that of the parent halogen molecules, except that the correlations are rather different as the *g* or *u* symmetries of the states do not have to be taken into account. The assumption that the products of dissociation of the fluoride molecules are $X(^2P_{3/2})+F(^2P_{1/2})$ (X = Cl, Br or I), whereas those of ICl and IBr are normal halogen atoms in their electronic ground states, then leads to the dissociation energies given in Table 93. Some ambiguity still clothes the products of dissociation of the halogen fluorides, for the alternative combination $X(^2P_{1/2})$ $+F(^2P_{3/2})$ cannot be excluded; the corresponding dissociation energies are then consistently lower than those given in the table, viz. ClF, 59·0; BrF, 50·05; IF, 45·6 kcal mol^{-1}. However, from a comparison of bond energies, Slutsky and Bauer[886] conclude that the higher dissociation energy is the logical choice for each of the diatomic halogen fluorides. Where thermochemical data are available, as for ClF, BrF, ICl and IBr, the results provide satisfactory support for the suggested modes of dissociation and for the spectroscopic values of the dissociation energy. Combination of the heat of formation of ClF, determined by direct calorimetric measurements, with the spectroscopic value for $D_0°(ClF)$ implies that $D_0°(F_2)$ = *ca.* 36·5 kcal mol^{-1}, a conclusion of some importance in that it provided the first cogent evidence against the values of 60–70 kcal at one time favoured for this quantity[830,831]. By contrast, no corresponding band system has been observed for BrCl; in this case, the thermodynamic parameters recommended by the National Bureau of Standards[878] are founded on the experimentally measured equilibrium constants and enthalpy change characterizing the process

$$Br_2(g)+Cl_2(g) \rightleftharpoons 2BrCl(g)$$

these data being supplemented by statistically calculated entropy values.

The enthalpies of the reactions

$$ClF_3(g)+2H_2(g)+100H_2O(l) \rightarrow [HCl,3HF,100H_2O](l)$$

and

$$\tfrac{1}{2}Cl_2(g)+\tfrac{1}{2}H_2(g)+[3HF,100H_2O](l) \rightarrow [HCl,3HF,100H_2O](l)$$

have recently been measured directly in a flame calorimeter at 1 atm and 303·5°K[887]; together with data from previous investigations, the values yield $\Delta H°_{f,298}[ClF_3(g)]$ $= -39·35 \pm 1·23$ kcal mol^{-1}. As applied to the reactions of gaseous ClF_5 with hydrogen and ammonia[888], the methods of reaction calorimetry furnish $\Delta H°_{f,298}[ClF_5(g)] = -60·9$ $\pm 4·5$ kcal mol^{-1}, while measurements of the equilibrium constant for the reaction

$$ClF_3(g)+F_2(g) \rightleftharpoons ClF_5(g)$$

at elevated temperatures suggest a value of $-57·7 \pm 1$ kcal mol^{-1} [889]. For BrF_3 and BrF_5, the enthalpies of formation have been determined by direct combination of the elements in an adiabatic calorimeter, with the results given in Table 93. The corresponding term favoured for IF_5 is that measured by Settle, O'Hare, Hubbard and Jeffes and referred to in the context of recent calorimetric studies of the pentafluoride[870]. From equilibrium-pressure measurements of the dissociation

$$IF_7(g) \rightleftharpoons IF_5(g)+F_2(g)$$

---

886 L. Slutsky and S. H. Bauer, *J. Amer. Chem. Soc.* **76** (1954) 270.

887 R. C. King and G. T. Armstrong, *J. Res. Nat. Bur. Stand.* **74A** (1970) 769.

888 W. R. Bisbee, J. V. Hamilton, J. M. Gerhauser and R. Rushworth, *J. Chem. and Eng. Data*, **13** (1968) 382.

889 R. Bougon, J. Chatelet and P. Plurien, *Compt. rend.* **264C** (1967) 1747.

Bernstein and Katz calculate that $\Delta H° = 28{\cdot}5 \pm 2{\cdot}0$ kcal mol^{-1} and $\Delta S° = 43{\cdot}5 \pm 3{\cdot}0$ eu in the temperature range 450–550°C[838]. On the basis of their measurements and of the enthalpy of formation of IF_5, $\Delta H°_{f,298}[IF_7(g)]$ is $-229{\cdot}9$ kcal mol^{-1}.

Mean bond energies for the interhalogen molecules appear in Table 93, while, for the halogen fluorides XF_n, the variation of bond energy as a function of X and n is depicted in Fig. 34. For a given molecular type XF_n, the general sequence of bond energies is

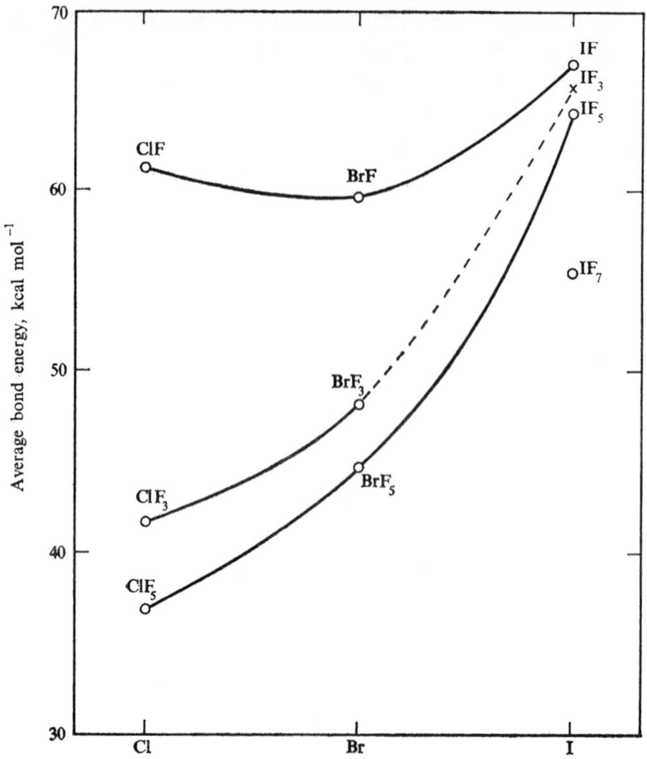

FIG. 34. Average bond energies of halogen fluorides.

$ClF_n < BrF_n < IF_n$, the energy increasing as the X–F bond becomes more polar; ClF and BrF form the only exceptions to this rule. The bond energies of the diatomic interhalogen molecules are reproduced satisfactorily by the Pauling relationship between electronegativity X and bond energy B[890]:

$$(\chi_X - \chi_Y)^2 = \frac{1}{23}[B(X\text{–}Y) - \sqrt{B(X\text{–}X)B(Y\text{–}Y)}]$$

For a given halogen X, the X–F bond energy of XF_n diminishes as n increases, as in the series ClF, ClF_3, ClF_5, but the rate of decrease of bond energy with respect to n is notably less when X = I than when X = Cl. To reproduce the bond energies of polyatomic interhalogen molecules XF_n by means of the Pauling relationship, it is necessary to assign

[890] D. A. Johnson, *Some Thermodynamic Aspects of Inorganic Chemistry*, p. 153, Cambridge (1968).

to X an effective electronegativity which increases in step with n. From the bond energies of the other halogen fluorides (Fig. 34), the average bond energy of IF_3 is estimated to be *ca.* 66 kcal; the enthalpy of formation of the gaseous compound from the elements in their standard states at 25°C is then *ca.* -116 kcal mol^{-1}.

The stability of the gaseous diatomic molecules XY with respect to the gaseous elements is determined almost exclusively by the difference between the bond energy of the XY molecule and the mean of those of the X_2 and Y_2 molecules; in unison with this energy term, which reflects the difference in electronegativity between X and Y, the stability decreases in the order IF > BrF > ClF > ICl > IBr > BrCl. The gain in energy with the formation of an interhalogen XY may thus be ascribed to the polar character of the X–Y bond. With respect to dissociation into the gaseous elements, all of the interhalogen molecules are thermodynamically stable at 25°C, though only marginally so in the case of BrCl.

However, the existence of a particular interhalogen compound XY_n, even when it contains a strong X–Y bond, depends also on its thermodynamic stability with respect to other compounds containing the atoms X and Y, and on the speed of whatever disproportionation processes may be open to it. A plot of $\Delta G^\circ_{f,298}$ for gaseous halogen fluoride molecules XF_n as a function of n (Fig. 35) gives clear notice, through the relative slopes $\Delta G^\circ/\Delta n$, that, whereas ClF is stable, BrF and IF are thermodynamically vulnerable to disproportionation

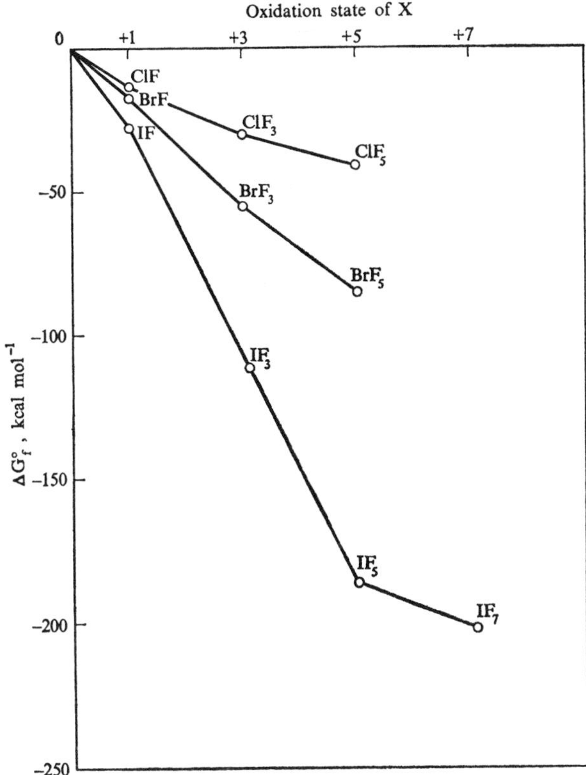

FIG. 35. Free energies of formation (at 298°K) of gaseous halogen fluorides XF_n.

processes of the type

$$3XF \rightarrow X_2+XF_3$$

and

$$5XF \rightarrow 2X_2+XF_5$$

None of the other diatomic interhalogen molecules, BrCl, ICl or IBr, shows any tendency to undergo reactions analogous to these. The qualitative evidence presently available[854] suggests that, in the condensed phase, IF_3 disproportionates at low temperatures in accordance with the reaction

$$2IF_3 \rightarrow IF+IF_5$$

or, at room temperature, in accordance with the reaction

$$5IF_3 \rightarrow I_2+3IF_5$$

In general, disproportionation of intermediate interhalogen compounds XY_n is likely to occur in those situations where the bond energy decreases but little with increase of n (cf. IF, IF_3 and IF_5 in Fig. 34).

A notable feature of the thermodynamic properties of the interhalogen compounds is the unique stability of iodine pentafluoride (see Table 93 and Fig. 35), which is not disposed to disproportionate and may be expected to dissociate significantly only at temperatures in excess of 1000°C, e.g.

$$IF_5(g) \xrightarrow{\sim 1400°C} IF_3(g)+2F(g)$$ [838]

In this respect there is evidently a close parallel in thermodynamic status between IF_5 and the oxy-species I_2O_5 and IO_3^-. At elevated temperatures, the marked increase of entropy accompanying reactions such as

$$ClF_5(g) \xrightleftharpoons{350°C} ClF_3(g)+F_2(g)$$ [889]

$$BrF_5(g) \xrightarrow{>550°C} BrF_3(g)+F_2(g)$$ [838]

and

$$IF_7(g) \xrightarrow{>450°C} IF_5(g)+F_2(g)$$ [838]

promotes the dissociation of certain polyatomic halogen fluorides, while even at relatively low temperatures the reaction

$$ICl_3(g) \rightarrow ICl(g)+Cl_2(g)$$

appears to proceed effectively to completion.

3. **Structural properties**[829,835,838–840,877,879,891,892]. As is evident from the details of Table 93 and Fig. 36, the structures of almost all the interhalogen compounds are now known, the only exceptions being ClF_5, for which no definitive structural parameters are yet available, and the ill-defined IF_3. The most precise estimates of the interatomic distances in the diatomic molecules come from the microwave spectra and from the rotational detail observed at high resolution in the electronic spectra of the gaseous materials[891]. For

[891] *Tables of Constants and Numerical Data*, No. 17, *Spectroscopic Data relative to Diatomic Molecules* (general ed. B. Rosen), Pergamon (1970).

[892] H. A. Bent, *Chem. Rev.* **68** (1968) 599.

ClF_3[893] and BrF_3[894] microwave studies of the vapours again provide definitive structural details, while the combination of microwave and electron-diffraction measurements has recently enabled the structures of BrF_5 and IF_5 to be determined with some confidence[895]. Electron diffraction also forms the basis of the most detailed structural analysis so far carried out on the IF_7 molecule in the vapour phase[896]. The geometry of ClF_5 has been deduced from its vibrational and ^{19}F nmr spectra; though consistent with the vibrational properties, the approximate dimensions given in the table are prompted largely by analogy with related molecules.

X-ray diffraction measurements have been reported for numerous interhalogen compounds in the crystalline condition, but only in the cases of ICl[897], IBr[898] and I_2Cl_6[899] have the results been refined to obtain closely defined structural parameters; less well characterized, with inferior structure factors, are crystalline ClF_3[900], BrF_3[901], BrF_5[901] and IF_7[901,902]. Crystalline IF_5 is monoclinic with a structure so complicated as to militate against significant refinement of the raw crystallographic data; however, the recently determined crystal structure of the molecular addition compound XeF_2,IF_5 discloses the presence of more-or-less discrete IF_5 molecules, whose structure has thus been established with some precision[903].

The measured interatomic distance in a diatomic interhalogen molecule XY is generally somewhat shorter than the sum of the covalent radii of X and Y (Fig. 36), the margin increasing as the electronegativities of X and Y diverge. Such a pattern complies at once with the variation in bond energies for this group of molecules, and with the general implications of the empirical Schomaker–Stevenson equation relating bond length to electronegativity difference[879]. It is now established beyond dispute that the ClF_3 and BrF_3 molecules have planar skeletons of C_{2v} symmetry with an unusual T-shaped configuration. Each molecule contains two nearly collinear X–F bonds, which are, to a first approximation, perpendicular to the third, though the idealized T-form is distorted in such a way that the angle between the bonds is slightly, but significantly, less than 90°. The unique X–F bond is somewhat shorter, whereas the other two bonds are somewhat longer, than that in the corresponding diatomic molecule. The fluorine atoms of the BrF_5 and IF_5 molecules assume the form of a square-based pyramid belonging to the symmetry group C_{4v}, the bromine or iodine atom being situated somewhat below the basal plane of the pyramid, so that the angle between the apical and basal X–F bonds is rather less than 90°. In each of these molecules, the unique, apical bond is shorter than those of the basal XF_4 unit, but, whereas the limits defined by the two Br–F distances in BrF_5 span that in diatomic BrF, all the bonds in IF_5 are appreciably shorter than that in IF. Curiously, too, with respect to idealized octahedral coordination with angles of 90° at the central atom, IF_5 experiences the greater angular distortion ($\angle F_{apical}$–I–F_{basal} = 81·9°; $\angle F_{apical}$–Br–F_{basal} = 84·8°), but with a

893 D. F. Smith, *J. Chem. Phys.* **21** (1953) 609.
894 D. W. Magnuson, *J. Chem. Phys.* **27** (1957) 223.
895 A. G. Robiette, R. H. Bradley and P. N. Brier, *Chem. Comm.* (1971) 1567.
896 W. J. Adams, H. B. Thompson and L. S. Bartell, *J. Chem. Phys.* **53** (1970) 4040.
897 K. H. Boswijk, J. van der Heide, A. Vos and E. H. Wiebenga, *Acta Cryst.* **9** (1956) 274; G. B. Carpenter and S. M. Richards, *ibid.* **15** (1962) 360.
898 L. N. Swink and G. B. Carpenter, *Acta Cryst.* **B24** (1968) 429.
899 K. H. Boswijk and E. H. Wiebenga, *Acta Cryst.* **7** (1954) 417.
900 R. D. Burbank and F. N. Bensey, *J. Chem. Phys.* **21** (1953) 602.
901 R. D. Burbank and F. N. Bensey, jun., *J. Chem. Phys.* **27** (1957) 982.
902 J. Donohue, *J. Chem. Phys.* **30** (1959) 1618; *Acta Cryst.* **18** (1965) 1018.
903 G. R. Jones, R. D. Burbank and N. Bartlett, *Inorg. Chem.* **9** (1970) 2264.

(a) Type XY

Molecule	r_e, Å	$r_{calc.}$, Å	$r_e - r_{calc.}$, Å
ClF	1.62811	1.703	−0.075
BrF	1.756	1.851	−0.095
IF	1.908_9	2.042	− 0.133
BrCl	2.138	2.136	+ 0.002
ICl	2.3207	2.327	− 0.006
IBr	2.485	2.475	+ 0.010

(b) Type XY_3

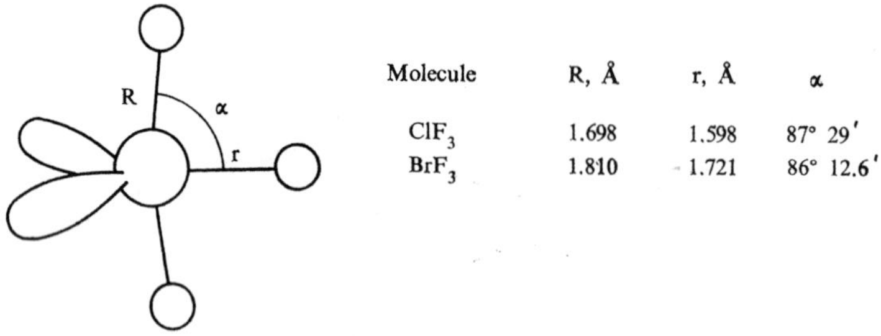

Molecule	R, Å	r, Å	α
ClF_3	1.698	1.598	87° 29′
BrF_3	1.810	1.721	86° 12.6′

(c) Type X_2Y_6

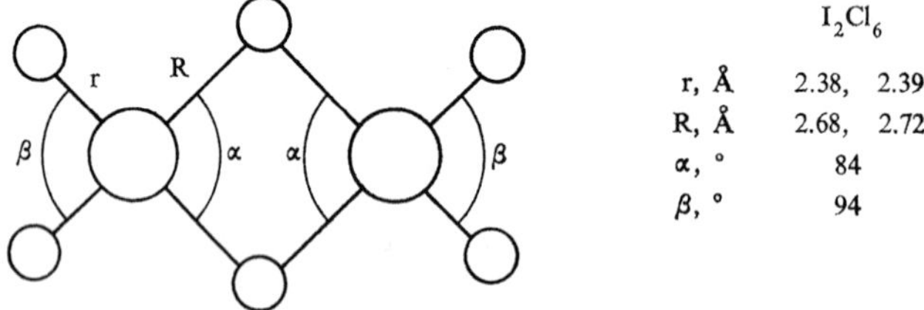

	I_2Cl_6
r, Å	2.38, 2.39
R, Å	2.68, 2.72
α, °	84
β, °	94

FIG. 36. Structures of interhalogen molecules of the types XY, XY_3 and X_2Y_6.

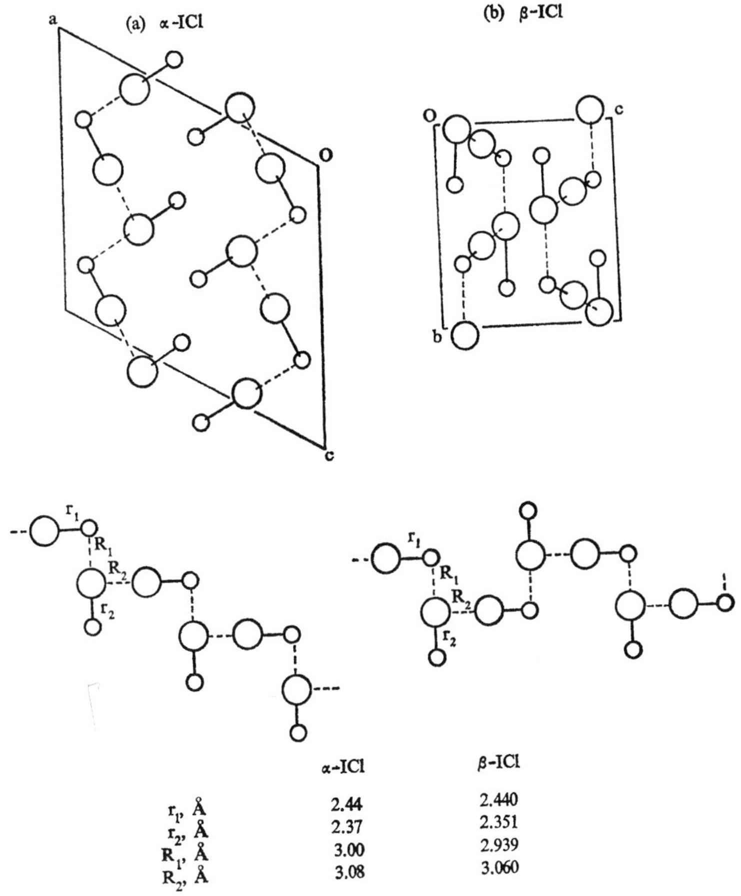

	α-ICl	β-ICl
r_1, Å	2.44	2.440
r_2, Å	2.37	2.351
R_1, Å	3.00	2.939
R_2, Å	3.08	3.060

cf. r_e(I-Cl) = 2·321 Å; sum of van der Waals radii: r(I) + r(Cl) = 3·95 Å; 2 x r(I) = 4·30 Å.

FIG. 36. Structures of α- and β-forms of crystalline ICl.

(a) IBr

Arrangement within each mirror plane of the crystal

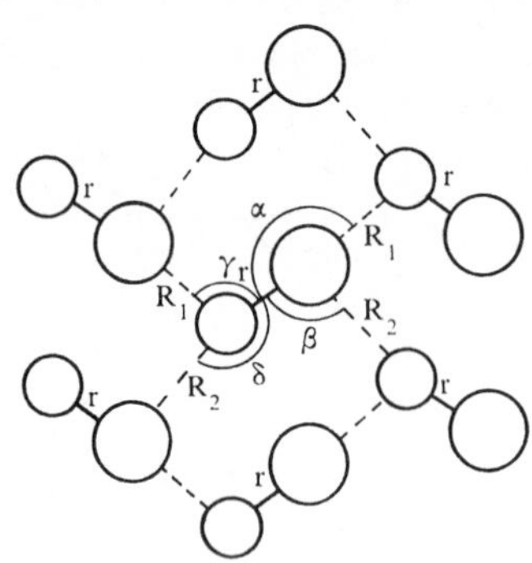

r, Å	2.524	α, °	175.2
R_1, Å	3.181	β, °	94.6
R_2, Å	3.764	γ, °	104.4
		δ, °	165.5

cf. r_e(I-Br) = 2·485 Å; sum of van der Waals radii: r(I) + r(Br) = 4·10 Å; 2 x r(I) = 4·30 Å.

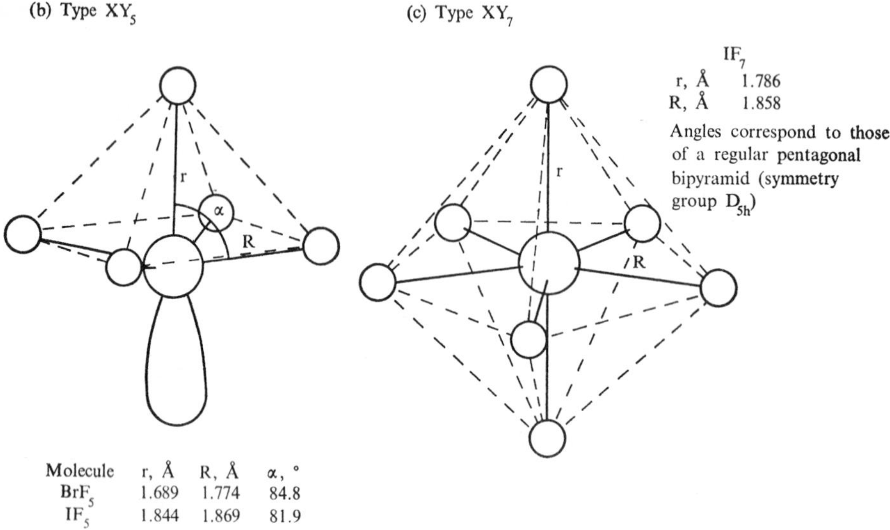

Molecule	r, Å	R, Å	α, °
BrF_5	1.689	1.774	84.8
IF_5	1.844	1.869	81.9

FIG. 36. Structures (a) of solid IBr, (b) of molecular interhalogens of the type XY_5, and (c) of molecular IF_7.

smaller disparity between apical and basal bond lengths[895]. The isolated IF_7 molecule has the form of a pentagonal bipyramid belonging to the symmetry group D_{5h}, in which the bonds of the apical F–I–F unit are shorter than those of the more congested equatorial IF_5 unit; the mean I–F distance is shorter than that observed in IF_5, despite the increased crowding. However, the details of the electron-diffraction pattern[896] imply that the molecule is deformed from D_{5h} symmetry on the average by puckering of the equatorial ring of fluorine atoms and by bending of the axial F–I–F unit, the angular displacements about the central atom being 7·5° and 4·5°, respectively. The simplest interpretation of the diffraction intensities is that the molecule undergoes essentially free pseudorotation, involving facile intramolecular rearrangement, along a pathway (predominantly described by e_2'' displacement coordinates) connecting 10 equivalent structures of C_2 symmetry *via* intermediates of C_s symmetry. The implication of deformed, polar configurations in this mechanism accounts for the fact that at low temperatures, but not at room temperature, a molecular beam of IF_7 is deflected by an inhomogeneous electric field, the molecules behaving as if they possess electric dipole moments[904]. The extraordinary flexibility characteristic of the IF_7 molecule appears also to be reproduced, with heightened amplitudes of distortion, in ReF_7, the only other molecule of the type XY_7 which has been analysed in this way[905].

On the basis of available crystallographic data, the molecules ClF_3, BrF_3 and BrF_5 suffer comparatively minor changes in the transition from the vapour to the solid phase; likewise, the dimensions reported for the IF_5 molecule in crystalline XeF_2,IF_5 are in close agreement with those of the gaseous molecule. There has been considerable controversy over the interpretation of X-ray diffraction measurements for IF_7, the issue being complicated by the apparent disorder in the crystal. Inevitably, it can now be appreciated, the ease with which the free molecule can be deformed from one configuration to another causes complications when the molecules pack together in the crystal. The most recent analyses conclude that the X-ray diffraction data are consistent with, but do not prove, the presence of IF_7 molecules with D_{5h} symmetry; a distance of 1·80 Å has thus been deduced for each of the seven I–F bonds[902]. In both ICl[897] and IBr[898] in the crystalline state, the IX molecules are incorporated in polymeric networks (Fig. 36); that there is relatively strong interaction between the molecules is evident from the observations (i) that the intermolecular distances are significantly shorter than the sum of the van der Waals' radii of the atoms concerned, and (ii) that the intramolecular I–X distances are somewhat longer than those in the gaseous IX molecules. ICl crystallizes in two forms, both of which are characterized by zigzag chains containing two types of ICl molecule: in molecules of one type, both atoms participate in short-range intermolecular interactions along the chain; in molecules of the other type, the chlorine atoms branch off the chain and are not involved in such interactions. With respect to one another, the branching chlorine atoms are disposed *trans* in β-ICl, but *cis* in α-ICl. Further, the chains in β-ICl are essentially planar, while in α-ICl they are puckered. The structures suggest that, in the intermolecular interactions along the chains, the chlorine atoms act chiefly as electron donors, while the iodine atoms act either as electron-acceptors or simultaneously as donors and acceptors, after the fashion of the iodine atoms in solid iodine[892]. Whereas there are thus only half as many strong intermolecular interactions per molecule in crystalline ICl as in iodine, the molecules of IBr crystals are linked to form planar sheets in a herring-bone pattern analogous to that in iodine. Of the

904 E. W. Kaiser, J. S. Muenter, W. Klemperer and W. E. Falconer, *J. Chem. Phys.* **53** (1970) 53.
905 E. J. Jacob and L. S. Bartell, *J. Chem. Phys.* **53** (1970) 2235.

interhalogen compounds iodine trichloride appears to be unique in that the crystals are composed of planar, symmetrical I_2Cl_6 molecules separated from one another by normal van der Waals' distances. Within each I_2Cl_6 molecule, the terminal I–Cl bonds are of about the same length as in ICl, but the bridging I–Cl bonds are considerably longer.

4. **Spectroscopic properties.** In contrast with a homonuclear halogen molecule, a diatomic interhalogen molecule XY lacks a centre of symmetry. Thus, although the valence-shell structure is similar to that of a symmetrical halogen molecule, orbitals may no longer be labelled with a parity suffix. Corresponding atomic orbitals still mix with one another to produce σ- or π-molecular orbitals, though the coefficients designating the contributions made by the individual atoms, viz. a and b in $a\psi_X \pm b\psi_Y$, are no longer equal. Hence, while the valence shell of an XY molecule may be written as

$$(1\sigma Y)^2(2\sigma X)^2(3\sigma Y)^2(1\pi Y)^4(2\pi X)^4 \qquad {}^1\Sigma^+$$

where X is the halogen of higher atomic number, mixing interaction increases the extent to which the 2π-orbital should be associated with the Y atom. The 1σ and 2σ molecular orbitals derive mainly from the valence s-orbitals of the X and Y atoms, while the remaining high-energy molecular orbitals are functions primarily of the corresponding p-orbitals. Promotion of a $(1\pi Y)$- or $(2\pi X)$-electron to the anti-bonding $(4\sigma X)$-orbital gives rise to the excited Π-states

$$B(1\sigma Y)^2(2\sigma X)^2(3\sigma Y)^2(1\pi Y)^3(2\pi X)^4(4\sigma X)^1 \qquad {}^{1,3}\Pi$$

$$A(1\sigma Y)^2(2\sigma X)^2(3\sigma Y)^2(1\pi Y)^4(2\pi X)^3(4\sigma X)^1 \qquad {}^{1,3}\Pi$$

Transitions implicating these and other excited states have been observed, either in absorption or in emission, for all of the diatomic interhalogen molecules, and physical constants, such as transition energies, vibrational frequencies and internuclear distances, have been deduced for some of the states thus identified. For fuller details the reader is referred elsewhere[827,891,906].

Although absorption spectra in the visible and ultraviolet regions have also been described for some of the polyatomic interhalogen compounds[838], no proper interpretation of the results has yet been feasible. In common with the parent halogens, gaseous BrCl, ICl and IBr have been shown to exhibit fluorescence, and for all three molecules resonance Raman and resonance fluorescence spectra have lately been reported[907]. The electronic absorption spectra of the species BrCl, ICl and IBr dissolved in a variety of solvents have also been investigated, and, as with the parent halogen molecules, evidence has thus been adduced for charge–transfer interactions between the interhalogen and donor molecules. In the acceptor action thus manifest, the uniformity of behaviour of diatomic halogen molecules is such that it would be unduly artificial to separate homonuclear X_2 from heteronuclear XY species; accordingly, this particular aspect of the behaviour of interhalogen compounds has already been treated in Section 2 (see p. 1196).

Recent measurements of the photoelectron spectra of the gaseous molecules ICl and IBr[908] have led to the ionization potentials recorded in Table 93. The spectra show obvious analogies to those of the parent halogen molecules, the observed bands being consistent with the anticipated energy sequence $(3\sigma Y)^2 < (1\pi Y)^4 < (2\pi I)^4$ (Y = Cl or Br). Upper

906 G. Herzberg, *Molecular Spectra and Molecular Structure. I. Spectra of Diatomic Molecules*, 2nd edn., p. 501, van Nostrand (1950).

907 W. Holzer, W. F. Murphy and H. J. Bernstein, *J. Chem. Phys.* **52** (1970) 399.

limits for the ionization potential have also been based on the appearance potentials of the appropriate molecular ions observed in mass-spectrometric studies of these and other interhalogen compounds[909].

Numerous studies testify to the interest occasioned by the vibrational properties of the interhalogen molecules[838,910,911]. The earlier measurements of the vibrational spectra of polyatomic systems had as a primary objective the elucidation of the molecular geometries. Thus, one of the first positive indications of the C_{4v} symmetry of the ClF_5, BrF_5 and IF_5 molecules was afforded by their infrared and Raman spectra[911]. However, with the more definitive conclusions of microwave, electron-diffraction and X-ray investigations, the main ambitions of the most recent experiments have been (i) to determine for an individual molecule reliable assignments of the vibrational frequencies, (ii) to explore the nature of the force field encompassing the atoms of the molecule, and (iii) to monitor chemical changes affecting the compound. Evaluated, for the most part, by reference to vibrational progressions in the electronic spectra, the vibrational frequencies of the isolated diatomic molecules signify that the stretching force constant decreases uniformly in the sequence ClF > BrF > IF > BrCl > ICl > IBr, which complies with the pattern of increasing internuclear distance, but not with the somewhat irregular pattern of bond energies (see Table 93). The infrared and Raman spectra observed for several of these compounds confirm the findings based on the electronic spectra[907,910]; further, variations of the frequencies, or of other spectroscopic properties, which accompany changes of phase or solvent or the formation of distinct molecular adducts, serve as an index to the intermolecular interactions experienced by the molecules in the condensed phases. The reduction both of the vibrational frequency and of the stretching force constant defining the X–Y unit gives clear notice of the anti-bonding character of the acceptor orbitals in molecules like ICl. The recently reported Raman spectra of gaseous ClF_3 [912], BrF_3 [912], IF_5 [876] and IF_7 [913], together with the infrared spectra of ClF_3, BrF_3 and BrF_5 in the matrix-isolated condition[875], have brought new facts to light about the vibrational spectra of these molecules, as well as eliminating some of the uncertainties and errors in earlier accounts[838]. The vibrational assignments thus deduced are presented in Table 93. The force fields for the polyatomic halogen fluorides are as yet poorly determined, though normal coordinate analyses of the molecules XF_3 (X = Cl or Br)[875], XF_5 (X = Cl, Br or I)[911] and IF_7 [913] yield valence force constants, which clearly discriminate between the two types of X–F bond found in each of these molecules; in every case, the more attenuated X–F bonds are characterized by smaller force constants. It is noteworthy that the longer bonds in BrF_3 have higher force constants than the corresponding bonds in ClF_3, whereas corresponding parameters for molecules of the type XF_5 increase in the order $ClF_5 < BrF_5 < IF_5$. As previously noted, the vibrational spectra of the molecules ClF_3 [875], BrF_3 [875] and IF_5 [876] under various conditions allude to significant degrees of polymerization in the condensed phases at low temperatures. However, whereas the dimers X_2F_6 isolated in solid inert matrices at 5–25°K appear to be of C_{2h} symmetry and to involve unsymmetrical X · · · F–X bridges, the infrared and Raman

908 A. W. Potts and W. C. Price, *Trans. Faraday Soc.* **67** (1971) 1242; S. Evans and A. F. Orchard, *Inorg. Chim. Acta*, **5** (1971) 81.

909 A. P. Irsa and L. Friedman, *J. Inorg. Nuclear Chem.* **6** (1958) 77.

910 H. Stammreich, R. Forneris and Y. Tavares, *Spectrochim. Acta*, **17** (1961) 1173.

911 G. M. Begun, W. H. Fletcher and D. F. Smith, *J. Chem. Phys.* **42** (1965) 2236.

912 H. Selig, H. H. Claassen and J. H. Holloway, *J. Chem. Phys.* **52** (1970) 3517.

913 H. H. Claassen, E. L. Gasner and H. Selig, *J. Chem. Phys.* **49** (1968) 1803; R. K. Khanna, *J. Mol. Spectroscopy*, **8** (1962) 134.

spectra of solid iodine(III) chloride substantiate the presence of symmetrical I–Cl–I bridges in the planar I_2Cl_6 molecules[914]. Although geometrically analogous to Au_2Cl_6, I_2Cl_6 differs from this molecule in its abnormally long I–Cl bridge bonds, which are also uncommonly weak, as judged by their stretching force constants.

The ^{19}F nmr spectra of all the halogen fluorides except BrF, IF and IF_3 have been recorded, in some instances in both the liquid and vapour states[838]; the relevant chemical shifts and coupling constants are collected in Table 93. Although the chemical shifts are relatively widely scattered, the resonances invariably occur at low fields (on a scale which has the F_2 and HF molecules as, respectively, the low- and high-field limits), implying, as with the noble-gas fluorides[915], that the fluorine atoms are relatively poorly shielded. The general trend in a series of fluorides XF_n is for the shielding of the fluorine to decline as n increases or as the atomic number of X decreases. Because of the inherent difficulties in direct application of Ramsey's theory of the chemical shift, a full interpretation of the factors contributing to the shielding is not yet accessible, though the results can be rationalized in terms of a substantial degree of ionic character in the X–F bonds[915]; numerical estimates of ionic character and of the charge densities at the fluorine atoms have thus been pronounced. It is noteworthy that the apical fluorine atoms of ClF_5, BrF_5 and IF_5 are less well shielded than the basal atoms; similarly, the two fluorine atoms of the nearly linear F–Cl–F unit of ClF_3 are less well shielded than the third fluorine. This reduction in shielding of certain fluorine atoms is attributable to the virtually exclusive use of atomic *p*-functions in the bonding of such atoms. Under suitable conditions, the ^{19}F nmr spectra of the molecules ClF_3, ClF_5, BrF_5 and IF_5 give unequivocal notice of the geometries of the molecules, distinct multiplet resonances due to magnetically non-equivalent fluorines being observed. The temperature-dependence of the liquid-phase spectra indicates that intermolecular fluorine-exchange becomes rapid at *ca.* 60°C for ClF_3 and at *ca.* 200°C for IF_5, the estimated activation energies being 4·8 and 13 kcal mol^{-1}, respectively. Fluorine-exchange in ClF_3 is much accelerated by the presence of hydrogen fluoride, even at very low concentrations. Thus, although the exchange may well proceed, as first suggested by Muetterties and Phillips[916], *via* transient fluorine-bridged dimers of the type identified by matrix-isolation, ionization as well as association may also provide a pathway for exchange. An analogous mechanism can be postulated for the fluorine-exchange in IF_5. No sign of rapid fluorine-exchange has been found with BrF_5 at temperatures up to 180°C, implying that the activation barrier to such exchange is significantly greater in this case, but, to judge by the single ^{19}F resonance observed for the liquid, fluorine-exchange between BrF_3 molecules is rapid even at room temperature. On the nmr timescale, equivalence is preserved for each of the seven fluorine atoms of IF_7, irrespective of the phase or temperature, but in this instance the rapid *intra*molecular rearrangement inferred from other measurements (q.v.) is presumed to be responsible for the averaging of the magnetic environments experienced by the fluorine atoms. The observed temperature-dependence of the spectrum of liquid IF_7 is then attributed, not to intermolecular exchange, but to the effects of coupling between the ^{19}F and quadrupolar ^{127}I nuclei[838].

[914] H. Stammreich and Y. Kawano, *Spectrochim. Acta*, **24A** (1968) 899; R. Forneris, J. Hiraishi, F. A. Miller and M. Uehara, *ibid.* **26A** (1970) 581.

[915] H. H. Hyman, *Science*, **145** (1964) 773; J. C. Hindman and A. Svirmickas, *Noble-gas Compounds* (ed. H. H. Hyman), p. 251, University of Chicago Press, Chicago (1963); C. J. Jameson and H. S. Gutowsky, *J. Chem. Phys.* **40** (1964) 2285.

[916] E. L. Muetterties and W. D. Phillips, *J. Amer. Chem. Soc.* **79** (1957) 322.

Quadrupole coupling of the heavier halogen nuclei in interhalogen compounds has also been investigated through the agency of the microwave spectra of the vapours or through measurements of the nqr or ^{129}I Mössbauer spectra of the solids[917,918]. The quadrupole coupling constants and, where known, the asymmetry parameters are included in Table 93, as are the ^{129}I isomer shifts characterizing the Mössbauer spectra of ICl, IBr, I_2Cl_6, IF_5 and IF_7[918]. The significance of these parameters has been discussed in Section 3.2 (pp. 1271–6 and Fig. 22), and estimates of the ionic character of certain interhalogen bonds, based on the classical Townes–Dailey approximation, appear in Table 27. The general tenor of the data supports the view that bonding in interhalogen molecules arises primarily through the interactions of valence *p*-electrons of the halogen atoms, with but a meagre contribution from the atomic *d*-functions; to judge, for example, by the ^{129}I isomer shifts, *s*-orbital participation appears also to be minimal in all but the IF_7 molecule[918].

Chemical Properties[826–840]

1. **General characteristics.** In general, the chemical properties of the interhalogen compounds XY_n resemble those of the parent halogens. The two principal categories of their reactions are:

Donor–acceptor (or Acid–base) Interactions

These range from the relatively weak intermolecular interactions involving XY_n molecules to interactions which may be strong enough, in effect, to bring about the heterolytic dissociation of an X–Y bond.

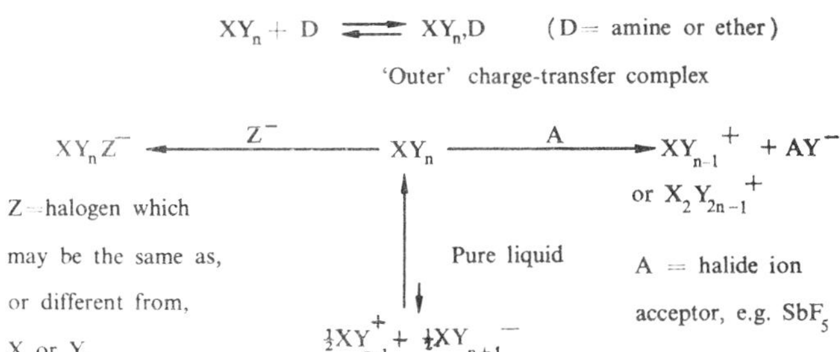

Thus, with donor molecules such as those of amines or ethers, which resist oxidation, an interhalogen molecule like BrCl, ICl or IBr experiences weak intermolecular binding. The properties of many of the solutions formed by the interhalogen compound in different solvents and of the low-melting addition compounds afforded by some of these systems are explicable in terms of charge–transfer interactions between the molecular components (see Section 2, p. 1196)[826–829,919]. Presumably to be formulated thus are the crystalline complexes formed by neutral oxygen- or nitrogen-bases, not only with a diatomic interhalogen molecule, e.g. BrCl,py (py = pyridine or substituted pyridine)[920] and

917 E. A. C. Lucken, *Nuclear Quadrupole Coupling Constants*, p. 288, Academic Press (1969).

918 M. Pasternak and T. Sonnino, *J. Chem. Phys.* **48** (1968) 1997; S. Bukshpan, C. Goldstein, J. Soriano and J. Shamir, *ibid.* **51** (1969) 3976.

919 L. J. Andrews and R. M. Keefer, *Adv. Inorg. Chem. Radiochem.* **3** (1961) 91; R. S. Mulliken and W. B. Person, *Molecular Complexes*, Wiley, New York (1969).

920 T. Surles and A. I. Popov, *Inorg. Chem.* **8** (1969) 2049.

IF,quinoline[921], but also with bromine pentafluoride[922], iodine tri-[839,923] and pentafluoride[838], and iodine trichloride[827,839], e.g. BrF_5,pyridine, IF_3,pyrazine, IF_5,dioxan, IF_5,$HCONMe_2$ and ICl_3,pyridine, though it is possible that at least some of these materials are better represented by an ionic constitution, e.g. $[(\text{quinoline})_2I]^+IF_2^-$ or $[(\text{pyridine})_2I]^+$ IF_6^-[839]. Charge-transfer spectra have recently been reported even for iodine heptafluoride in n-hexane or cyclohexane solutions[924]. According to the crystallographic evidence concerning certain of the crystalline adducts (Table 16), the X–Y bond of the interhalogen molecule is commonly attenuated by the interaction, but the integrity of the molecule is preserved. Compared with a parent halogen molecule, an interhalogen is, by virtue of its polarity, a more potent acceptor; it is also more susceptible to heterolytic cleavage. Potential cationic products of such a process which have lately been identified with some degree of assurance include XF_2^+ and XF_4^+ (X = Cl, Br or I), Cl_2F^+, ICl_2^+ and IF_6^+ (see Section 4A)[925], while numerous polyhalide anions, for example of the type XY_nZ^- (Z = a halogen which may be the same as, or different from, X or Y), are now well authenticated (see p. 1534)[835,840,841]. In accordance with the behaviour outlined in the scheme above, most interhalogen compounds are ambivalent with respect to halide ions, which they have the capacity to accept in forming a polyhalide anion or to relinquish in forming a cationic derivative. Heterolytic cleavage of an interhalogen bond is encouraged by a powerful halide ion-acceptor, e.g. BF_3, AsF_5 or SbX_5 (X = F or Cl), and by conditions tending to stabilize the highly electrophilic interhalogen cation thus formed, which can exist only in the presence of molecules and anions of very low basicity. Conversely, polyhalide anions are stabilized by environments low in acidity.

Halogenation

Here heterolytic or homolytic cleavage of an interhalogen bond effects halogenation of a reagent. Thus, diatomic interhalogen molecules add to unsaturated units, e.g.

$$XY + \;>C=C<\; \longrightarrow -\underset{X}{\overset{|}{C}}-\underset{Y}{\overset{|}{C}}- \quad {}^{827,926,927}$$

$$ClF + SF_4 \longrightarrow SF_5Cl \quad {}^{928}$$

Substitution reactions, often with simultaneous oxidation of the reagent, form a chemical common denominator for interhalogen compounds. Typical of such behaviour is the fluorinating action of halogen fluorides on a wide range of elements and their binary derivatives (see p. 1525)[838].

2. **Acid–base properties: solvent functions**[836,838]. Because of their great reactivity, the interhalogens dissolve only a limited number of substances without causing them to

[921] M. Schmeisser, P. Sartori and D. Naumann, *Ber.* **103** (1970) 880.
[922] H. Meinert and U. Gross, *Z. Chem.* **9** (1969) 190.
[923] M. Schmeisser, K. Dahmen and P. Sartori, *Ber.* **103** (1970) 307; M. Schmeisser, P. Sartori and D. Naumann, *ibid.* pp. 312, 590.
[924] P. R. Hammond, *J. Phys. Chem.* **74** (1970) 647.
[925] R. J. Gillespie and M. J. Morton, *Quart. Rev. Chem. Soc.* **25** (1971) 553.
[926] W. K. R. Musgrave, *Adv. Fluorine Chemistry*, **1** (1960) 1.
[927] R. D. Chambers, W. K. R. Musgrave and J. Savory, *Proc. Chem. Soc.* (1961) 113; *J. Chem. Soc.* (1961) 3779; P. Sartori and A. J. Lehnen, *Ber.* **104** (1971) 2813.
[928] F. Nyman and H. L. Roberts, *J. Chem. Soc.* (1962) 3180

undergo reactions more drastic than ion-formation or solvation. Most investigations of these substances as solvents have been purely qualitative in nature, and, even where quantitative data are available, their interpretation is often obscure. There are many practical obstacles to such investigations: BrF_3, the only interhalogen solvent to have the status of a reagent of some importance in preparative inorganic chemistry, explodes with water and most organic matter, reacts with asbestos with incandescence, and can be manipulated in quartz apparatus only because of the sluggishness of the reaction with silica in this form at normal temperatures; the iodine chlorides, though they are less violent in their reactions, nevertheless possess the property, inconvenient in electrochemical studies, of readily dissolving gold and platinum. Measurements of electrical conductivity provide some indication of the number and mobility of the ions present, but give little information as to their nature. Identification of products liberated at electrodes may throw some light on this problem, but too often the relationship between what carries the current and what is discharged at the electrodes is open to question. Even if, as is sometimes the case, the structures of solid adducts derived from the solvent system are known, far-reaching changes may occur on dissolution, and halogen-exchange processes are often very rapid. Transport and emf measurements, which have contributed so much to our knowledge of the nature of aqueous solutions, have not been performed to any profitable extent in interhalogen solvents. The picture presented by experimental findings is therefore not a very clear one, and much further work will be needed before the nature of the solutions is well understood.

Intimately related to the solvent properties of an interhalogen compound are the reactions whereby a halide ion is formally gained or lost (q.v.), since these represent acid–base functions characteristic of the compound. Of particular relevance therefore is the behaviour (i) with respect to a Lewis acid like BF_3 or $SbCl_5$ endowed with the power to abstract a halide ion, and (ii) with respect to a base which can serve, directly or indirectly, as a source of halide ions. Where solid complexes arise from such interactions (Table 94), characterization has been achieved, sometimes definitively by X-ray crystallographic investigations, more often in terms of the vibrational or nmr spectra of the materials or of their electrical conductivities in solution. However, the acid or base strength of an interhalogen molecule is only one of several factors that contribute to the effectiveness of the liquid as a solvent, and clear signs of halide-transfer have been found for some compounds with but slight or poorly authenticated solvent properties. It is appropriate, nevertheless, that the aspects of acid–base and solvent properties should be considered in the same context. In the following survey, the interhalogen compounds are treated in turn according to formula type, viz. XY, XY_3, XY_5 and XY_7.

Compounds of Formula XY[827,830,831,836,839]

Of the compounds of this general formula only ICl and IBr have been described as solvents. One special difficulty attends the interpretation of data on the fused compounds: that is their dissociation into free halogens. For ICl the degree of dissociation is 0·4% at 25°C and 1·1% at 100°C; for IBr the corresponding figures are 8·8 and 13·4%. Meaningful physical properties of the liquids are thus sparsely documented (see Table 93).

Irrespective of the fact that the potential of chlorine monofluoride as a solvent appears not to have been explored seriously, its ampholytic behaviour is clearly implied by the solid complexes formed through the reactions

$$Cl_2F^+AF^- \xleftarrow{\text{A}} 2ClF \xrightarrow{\text{2MF}} 2M^+ClF_2^-$$

($A = BF_3$ or AsF_5)[925] (M = alkali metal or NO)[838,841]

Table 94. Solid Complexes Formed by Interhalogen Compounds: (a) with Halide Ion-Acceptors and (b) with Ionic Halides[a–c]

Interhalogen	Complexes formed with halide ion-acceptors		Complexes formed with ionic halides	
	Composition	Formulation	Composition	Formulation
ClF	$BF_3,2ClF$; $AsF_5,2ClF$	$[Cl_2F]^+[BF_4]^-$; $[Cl_2F]^+[AsF_6]^-$	MF,ClF; NOF,ClF	$M^+[ClF_2]^-$; $NO^+[ClF_2]^-$
BrF	—	—	—	—
IF	—	—	$[NEt_4]F,IF$	$[NEt_4]^+[IF_2]^-$
$BrCl$	—	—	$MCl,BrCl$; $[NR_4]Cl,BrCl$ (R = Me or Et)	$M^+[BrCl_2]^-$; $[NR_4]^+[BrCl_2]^-$
ICl	$AlCl_3,2ICl$; $SbCl_5,2ICl$; $SbCl_5,3ICl$	$[I_2Cl]^+[AlCl_4]^-$; $[I_2Cl]^+[SbCl_6]^-$	MCl,ICl	$M^+[ICl_2]^-$*
IBr	—	—	MBr,IBr	$M^+[IBr_2]^-$*
ClF_3	ClF_3,BF_3; ClF_3,MF_5 (M = P, As, Sb or Pt)	$[ClF_2]^+[BF_4]^-$; $[ClF_2]^+[MF_6]^-$*	MF,ClF_3; NOF,ClF_3	$M^+[ClF_4]^-$; $NO^+[ClF_4]^-$
BrF_3	BrF_3,AuF_3; BrF_3,MF_5 (M = Sb, Bi, Nb, Ta or Ru); $MF_4,2BrF_3$ (M = Ge, Sn, Pd or Pt)	$[BrF_2]^+[AuF_4]^-$; $[BrF_2]^+[MF_6]^-$*; $[BrF_2{}^+]_2[MF_6]^{2-}$	MF,BrF_3; AgF,BrF_3; NOF,BrF_3; $BaF_2,2BrF_3$	$M^+[BrF_4]^-$*; $Ag^+[BrF_4]^-$; $NO^+[BrF_4]^-$; $Ba^{2+}[BrF_4{}^-]_2$
IF_3	IF_3,MF_5 (M = As or Sb)	$[IF_2]^+[MF_6]^-$	MF,IF_3; NOF,IF_3	$M^+[IF_4]^-$; $NO^+[IF_4]^-$
I_2Cl_6	$ICl_3,AlCl_3$; $ICl_3,SbCl_5$	$[ICl_2]^+[AlCl_4]^-$*; $[ICl_2]^+[SbCl_6]^-$*	MCl,ICl_3; $[NR_4]Cl,ICl_3$	$M^+[ICl_4]^-$*; $[NR_4]^+[ICl_4]^-$
ClF_5	ClF_5,MF_5 (M = As, Sb or Pt)	$[ClF_4]^+[MF_6]^-$?	—	—
BrF_5	$BrF_5,2SbF_5$; BrF_5,SO_3	$[BrF_4]^+[Sb_2F_{11}]^-$*; $[BrF_4]^+[SO_3F]^-$?	MF,BrF_5	$M^+[BrF_6]^-$
IF_5	IF_5,MF_5 (M = Sb or Pt); $IF_5,1{\cdot}17SO_3$	$[IF_4]^+[MF_6]^-$*; $[IF_4]^+[SO_3F]^-$?	MF,IF_5; NOF,IF_5; $[NR_4]F,IF_5$	$M^+[IF_6]^-$; $NO^+[IF_6]^-$; $[NR_4]^+[IF_6]^-$
IF_7	IF_7,BF_3; IF_7,AsF_5; $IF_7,3SbF_5$	$[IF_6]^+[BF_4]^-$; $[IF_6]^+[AsF_6]^-$*; $[IF_6]^+[Sb_3F_{16}]^-$	—	—

M = alkali metal or NH_4; R = organic group. * Crystallographic evidence available.

[a] L. Stein, *Halogen Chemistry* (ed. V. Gutmann), Vol. 1, p. 133, Academic Press (1967).

[b] H. Meinert, *Z. Chem.* **7** (1967) 41.

[c] R. J. Gillespie and M. J. Morton, *Quart. Rev. Chem. Soc.* **25** (1971) 553.

the identity of the ions Cl_2F^+ and ClF_2^- having been established primarily by their vibrational spectra. Such behaviour suggests that the meagre electrical conductivity of liquid chlorine monofluoride may arise from the auto-ionization

$$3ClF \rightleftharpoons Cl_2F^+ + ClF_2^-$$

though the Cl_2F^+ ion appears to be unstable in solution, disproportionating completely, for example, in an antimony pentafluoride–hydrogen fluoride medium even at −76°C[925].

$$2Cl_2F^+ \rightarrow ClF_2^+ + Cl_3^+$$

The slight but significant conductivities of the liquid iodine monohalides have commonly been attributed to auto-ionization schemes of the type

$$2IX \rightleftharpoons I^+ + IX_2^-$$

resembling that suggested for liquid iodine. Altogether more plausible, however, are schemes such as

$$3IX \rightleftharpoons I_2X^+ + IX_2^-$$

or

$$4IX \rightleftharpoons I_2X^+ + I_2X_3^-$$ [840]

whereby an interhalogen cation, rather than an isolated monatomic cation, is produced, though it is possible that cations of the type I_2X^+ are extensively disproportionated to give the species I_3^+ and IX_2^+ [925]. If the auto-ionization is correctly represented in this way, a halide ion-acceptor, which promotes the formation of the interhalogen cation, can be regarded as an acid, whereas a source of halide ions behaves as a base in the liquid iodine monohalide. In that potential acids include the appropriate aluminium(III), tin(IV) or antimony(V) halides, it is noteworthy that phase studies of the systems $AlCl_3$–ICl and $SbCl_5$–ICl reveal the formation of the adducts $AlCl_3$,2ICl, $SbCl_5$,2ICl and $SbCl_5$,3ICl. A reasonable formulation of these involves the I_2Cl^+ ion, in keeping with the structures of the materials $AlCl_3$,ICl_3 and $SbCl_5$,ICl_3, which, it is known, are better represented as $ICl_2^+AlCl_4^-$ and $ICl_2^+SbCl_6^-$ (see below). Stable acids belonging to the IBr solvent system have not yet been isolated. On the other hand, the existence of the anions ICl_2^- and IBr_2^-, for example in crystalline alkali-metal salts and in $PCl_4^+ICl_2^-$, is well established by X-ray diffraction and other measurements; complexes of the type MX,2IX may well contain pentahalide anions $I_2X_3^-$, as in the case where M = K and X = Cl[840].

The electrochemical and cryoscopic properties of ICl, IBr and their binary mixtures with organic and inorganic liquids, investigated in detail by Fialkov and his collaborators[929,930], are also consistent with the postulated auto-ionization of ICl and IBr. Thus, the monohalides give conducting solutions when dissolved in arsenic trichloride, sulphur dioxide, sulphuryl chloride, nitrobenzene, glacial acetic acid or diethyl ether[827,830,831], and it has been shown that, on electrolysis of ICl, for example, in nitrobenzene or acetic acid, both iodine and chlorine appear at the anode in amounts commensurate with the discharge of the ICl_2^- anion.

Iodine(I) chloride was first shown to be an ionizing solvent by Cornog and Karges[931], who found that the electrical conductivity of the pure liquid is increased considerably when

[929] Ya. A. Fialkov and K. Ya. Kaganskaya, *J. Gen. Chem.* (*U.S.S.R.*) **18** (1948) 289.

[930] Ya. A. Fialkov and I. D. Muzyka, *J. Gen. Chem.* (*U.S.S.R.*) **18** (1948) 802, 1205; *ibid.* **19** (1949) 1416; *ibid.* **20** (1950) 385; Ya. A. Fialkov and O. I. Shor, *ibid.* **19** (1949) 1787.

[931] J. Cornog and R. A. Karges, *J. Amer. Chem. Soc.* **54** (1932) 1882; J. Cornog, R. A. Karges and H. W. Horrabin, *Proc. Iowa Acad. Sci.* **39** (1932) 159.

potassium or ammonium chloride is dissolved in it; the equivalent conductance of the solute increases with progressive dilution, the values at infinite dilution at 35°C being about 32 and 26 ohm^{-1} cm^2, respectively. Other compounds which are appreciably soluble to yield conducting solutions include RbCl, CsCl, KBr, KI, $AlCl_3$, $AlBr_3$, PCl_5, $SbCl_5$, pyridine, acetamide and benzamide. Compounds that are but slightly soluble include LiCl, NaCl, AgCl and $BaCl_2$, while the molecular halides $SiCl_4$, $TiCl_4$, $NbCl_5$ and $SOCl_2$ dissolve but have little or no effect on the conductivity. The interpretation of the conductivity data is complicated by changes in density, viscosity and degree of dissociation which occur on dilution. Acid–base reactions have been monitored by conductimetric titration. Thus, the titration of RbCl with $SbCl_5$ or of KCl with $NbCl_5$ shows a break corresponding to 1:1 molar proportions, whilst that of NH_4Cl with $SnCl_4$ shows a break corresponding to a molar ratio of 2:1, which implies the formation of the ion $SnCl_6^{2-}$. The compound PCl_5,ICl, which in the solid state is $PCl_4^+ICl_2^-$, behaves as an acid towards $KICl_2$ and as a base towards $SbCl_5$. From the preparative viewpoint, reactions in fused iodine(I) chloride are limited by the prevalence of solvolysis and by the frequent necessity to isolate reaction products by extraction techniques rather than by precipitation; accordingly, pure products are seldom isolable. Nevertheless, it is possible that the molten reagent may prove to be useful in stabilizing chloride complexes of elements in high oxidation states.

Fused iodine(I) bromide as a solvent closely resembles the chloride. Alkali-metal bromides furnish polyhalides of formula M^IBr_2, and phosphorus pentabromide gives the compound PBr_5,IBr, which, by analogy with the chloro- compound, is probably $PBr_4^+IBr_2^-$; in IBr this behaves exclusively as a base. Tin(IV) bromide acts as an acid, which may be shown by conductimetric titration to undergo neutralization reactions such as

$$2IBr_2^- + SnBr_4 \rightarrow SnBr_6^{2-} + 2IBr$$

Compounds of Formula XY_3[827,830,831,836,838,839]

In this category, studies of solvent behaviour have been restricted to the compounds chlorine trifluoride, bromine trifluoride and iodine trichloride. Physical properties relevant to the solvent functions of the liquids, e.g. dielectric constant, specific conductivity, Trouton constant and viscosity, are contained in Table 93.

Chlorine trifluoride is characterized by a very low electrical conductivity and dielectric constant; in general, the signs do not augur well for its capacity as an ionizing solvent. Although no studies of ionic reactions in the liquid have been reported, it is noteworthy that derivatives recently prepared and characterized contain severally the ions ClF_2^+ [838,925] and ClF_4^- [838,841], which, together, are the most likely outcome of the auto-ionization reaction

$$2ClF_3 \rightleftharpoons ClF_2^+ + ClF_4^-$$

Thus, compounds of the type $MClF_4$ (M = alkali metal) are produced, not only by the action of fluorine on alkali-metal chlorides at elevated temperatures, but also by the direct interaction of chlorine trifluoride with alkali-metal fluorides. That these solid complexes contain more-or-less discrete ClF_4^- anions is implied by their vibrational spectra[932]. Chlorine trifluoride also forms 1 : 1 complexes with strong Lewis acids like BF_3, AsF_5 and SbF_5, which, on the premises of their vibrational spectra, are best formulated as salts of the ClF_2^+ cation, e.g. $ClF_2^+AsF_6^-$, a conclusion recently confirmed by the experimentally

932 K. O. Christe and W. Sawodny, *Z. anorg. Chem.* **374** (1970) 306.

determined crystal structure of ClF_3,SbF_5 (Fig. 28)[933]. Thermodynamic properties have been estimated for some of these complexes on the basis of their dissociation pressures[838]; for $ClF_2^+BF_4^-$ and $ClF_2^+PF_6^-$, for example, $\Delta H°_{dissoc}$ = 23·6 and 16·4 kcal mol^{-1}, respectively. The existence of the ion ClF_2^+ in the liquid phase is also denoted by measurements of the Raman and infrared spectra and electrical conductivities of liquid mixtures of the trifluorides ClF_3 and BrF_3; the fluoride ion-transfer process

$$ClF_3 + BrF_3 \rightleftharpoons ClF_2^+ + BrF_4^-$$

with an equilibrium constant of *ca.* 10^{-4} at 25°C is thereby implied[934]. Evidently, therefore, chlorine trifluoride is a significantly weaker fluoride ion-acceptor (i.e. acid) than bromine trifluoride.

Of the interhalogens, bromine trifluoride is probably the most widely used as a laboratory reagent, chiefly for the preparation of complex fluorides. The compound fluorinates everything which dissolves in it, and a discussion of solubilities is therefore restricted to inorganic fluorides. These fall into two groups: I. Alkali-metal fluorides, silver(I) fluoride and barium fluoride; II. Fluorides of gold(III), boron, titanium(IV), silicon, germanium(IV), vanadium(V), niobium(V), tantalum(V), phosphorus(V), arsenic(V), antimony(V), platinum(IV), ruthenium(V) and a few other metals[836,838]. Members of both groups have been shown to enhance the electrical conductivity of bromine trifluoride, and many form thermally stable adducts with the solvent. To judge by their vibrational spectra[934–936] and conductivities in liquid bromine trifluoride, solid complexes of the type MF,BrF_3 (M = alkali metal, Ag, ½Ba or NO) derived from fluorides of category I are most aptly formulated as $M^+BrF_4^-$; the presence of a square-planar BrF_4^- anion in the tetragonal crystals formed by $KBrF_4$ has ultimately been confirmed by neutron-diffraction studies[937]. By contrast, solid complexes derived from the interaction of bromine trifluoride with the Lewis acids forming category II are best represented as $BrF_2^+AF^-$ (e.g. A = AuF_3, AsF_5, SbF_5, NbF_5 or RuF_5) or $[BrF_2^+]_2AF_2^{2-}$ (e.g. A = GeF_4, SnF_4 or PtF_4). Again, these formulations find support in the infrared and Raman spectra of the solids[934,936] and in the crystal structure established for the adduct BrF_3,SbF_5 (Fig. 28)[938]. According to the measured dissociation pressures, the enthalpies of the reactions

$$KBrF_4(s) \rightarrow KF(s) + BrF_3(g)$$

and

$$BrF_2SbF_6(l) \rightarrow SbF_5(g) + BrF_3(g)$$

are +4·1 and +27·8 kcal mol^{-1}, respectively[838]. That the ions BrF_2^+ and BrF_4^- also exist in the liquid mixtures bromine trifluoride–hydrogen fluoride and bromine trifluoride–chlorine trifluoride is intimated, moreover, by the vibrational spectra and electrical conductivities of the liquids[934]. Accordingly, the consensus of the experimental evidence is that the relatively high specific conductivity of pure liquid bromine trifluoride (> 8·01 × 10^{-3} ohm^{-1} cm^{-1} at 25°C)[939] is attributable to the self-ionization

$$2BrF_3 \rightleftharpoons BrF_2^+ + BrF_4^-$$

---

933 A. J. Edwards and R. J. C. Sills, *J. Chem. Soc.* (*A*) (1970) 2697.

934 T. Surles, H. H. Hyman, L. A. Quarterman and A. I. Popov, *Inorg. Chem.* **9** (1970) 2726; *ibid.* **10** (1971) 611, 913.

935 J. Shamir and I. Yaroslavsky, *Israel J. Chem.* **7** (1969) 495.

936 K. O. Christe and C. J. Schack, *Inorg. Chem.* **9** (1970) 1852, 2296.

937 A. J. Edwards and G. R. Jones, *J. Chem. Soc.* (*A*) (1969) 1936.

938 A. J. Edwards and G. R. Jones, *J. Chem. Soc.* (*A*) (1969) 1467.

939 H. H. Hyman, T. Surles, L. A. Quarterman and A. Popov, *J. Phys. Chem.* **74** (1970) 2038.

It was first pointed out by Woolf and Emeléus[940] that dissociation according to this equation implies the possibility of acid–base neutralization reactions in liquid bromine trifluoride. The fact that any fluoride in class I reacts with a member of class II to yield a complex halide is then easily understood in terms, first, of the formation of derivatives of the BrF_2^+ and BrF_4^- ions, and, subsequently, of the interaction of these according to an equation such as

$$BrF_2^+SbF_6^- + Ag^+BrF_4^- \rightarrow Ag^+SbF_6^- + 2BrF_3$$

In a few instances, such as this one, it has been shown by measuring the conductivities of solutions containing different molar proportions of the acid and base that the conductivity exhibits a well-defined minimum at the equivalence point, which corresponds to a molar ratio of 1 : 1. Similar measurements signify that the complex $SnF_4,2BrF_3$ reacts with $KBrF_4$ in accordance with the equation

$$[BrF_2^+]_2SnF_6^{2-} + 2K^+BrF_4^- \rightarrow K^+{}_2SnF_6^{2-} + 4BrF_3$$

Such reactions form the basis of an extremely useful method for the preparation of complex fluorides, which may be isolated, following neutralization, by evaporation of the excess solvent. Among a very large number of preparations accomplished by this method, the following suffice to illustrate its scope[836,838]:

$$Ag + Au \xrightarrow{BrF_3} AgBrF_4 + BrF_2AuF_4 \rightarrow AgAuF_4$$

$$NOCl + SnF_4 \xrightarrow{BrF_3} [NO]_2SnF_6$$

$$N_2O_4 + Sb_2O_3 \xrightarrow{BrF_3} [NO_2]SbF_6$$

$$VF_5 + LiF \xrightarrow{BrF_3} LiVF_6$$

$$Ru + KCl \xrightarrow{BrF_3} KRuF_6$$

Often it is necessary only to treat equivalent amounts of the acid- and base-forming elements (or compounds of the elements) with bromine trifluoride, which thus functions both as a fluorinating agent and as a reaction medium; the excess of the interhalogen is subsequently removed *in vacuo* in order to obtain the complex. However, the evaporation stage is not always easily completed: thus, complications may arise through solvolysis or through incomplete interaction of the acid and base according to the neutralization equation

$$BrF_2^+ + BrF_4^- \rightleftharpoons 2BrF_3$$

High valence states of elements are commonly stabilized in bromine trifluoride. Thus, although palladium(IV) fluoride has not yet been isolated, salts such as K_2PdF_6 are readily obtained by treatment of alkali-metal chloropalladates(II) with bromine trifluoride; complexes of chromium(V), e.g. $KCrOF_4$, have also been made from the corresponding dichromates under analogous conditions. In a similar vein, bromine trifluoride has recently emerged as a solvent for xenon tetrafluoride[941], and the formation of complexes of the latter with the acids PF_5, AsF_5 and SbF_5 has thus been monitored by conductimetric

[940] A. A. Woolf and H. J. Emeléus, *J. Chem. Soc.* (1949) 2865.
[941] D. Martin, *Compt. rend.* **268C** (1969) 1145.

titration; the solid $XeF_4,4SbF_5$ is said to precipitate from solutions containing XeF_4 and SbF_5. Bromine trifluoride has lately found a further use as a medium for reaction calorimetry[942].

Despite the fact that iodine trifluoride has virtually eluded physical characterization, it is the parent of certain derivatives which are more amenable to investigation, and which testify to fluoride-transfer processes analogous to those described for bromine and chlorine trifluorides. Thus, complexes of the type MF,IF_3 (M = alkali metal or NO) result from the reaction of MF with the trifluoride at low temperatures, or from the action of iodine pentafluoride or fluorine on MI[838,854]. Presumably to be formulated as tetrafluoroiodates(III), i.e. $M^+IF_4^-$, these compounds are typically white solids which are thermally stable but decompose readily in the presence of moisture. The abstraction of a fluoride ion from iodine trifluoride also appears to be realized by the action of AsF_5 or SbF_5 at $-70°C$. That the 1 : 1 adduct thus formed from SbF_5 is plausibly represented as a derivative of the IF_2^+ ion is substantiated by the ^{19}F nmr spectrum of the solid[854].

Iodine trichloride in the molten state is appreciably dissociated into the monochloride and chlorine, a circumstance which, with other factors, has militated against investigations of its use as a solvent. The measured electrical conductivity of the molten material may well be due in part to the reaction

$$I_2Cl_6 \rightleftharpoons ICl_2^+ + ICl_4^-$$

though there is as yet no consistent evidence concerning the nature of the ions produced under these conditions. Approximately square-planar in form, the ICl_4^- ion has been identified by crystallographic studies of the complex $KICl_4,H_2O$[943], while the crystal structures of the complexes $ICl_3,AlCl_3$ and $ICl_3,SbCl_5$ (Table 40)[944] indicate that these are derivatives of the ICl_2^+ ion. Hence, in its acid–base characteristics, if not in its thermal stability, iodine trichloride evidently has much in common with the halogen trifluorides.

Compounds of Formula XY_5 [827,830,831,836,838,839]

The dielectric constants, degrees of association and electrical conductivities of the three pure liquids (see, for example, Table 93) clearly mark iodine pentafluoride as having the greatest potential as a solvent; the self-ionization implied by the conductivities, which increase in the sequence $ClF_5 < BrF_5 < IF_5$, is probably best represented by the equation

$$2XF_5 \rightleftharpoons XF_4^+ + XF_6^-$$

though the species ClF_6^- has yet to be characterized. With the fluoride ion-acceptors AsF_5 and SbF_5, though not with BF_3, chlorine pentafluoride certainly forms 1 : 1 adducts, the vibrational spectra of which, it has been argued, support the formulation $[ClF_4]^+[MF_6]^-$[945]. On the other hand, no complex attributable to the formation of the ClF_6^- ion has yet come to light *via* the interaction of the pentafluoride with either alkali-metal or nitrosyl fluorides. Of the complexes formed by bromine pentafluoride with the Lewis acids SbF_5 and SO_3, $BrF_5,2SbF_5$ is plausibly assigned the constitution $[BrF_4]^+[Sb_2F_{11}]^-$ on the basis of its crystal structure as well as its infrared and ^{19}F nmr

942 G. W. Richards and A. A. Woolf, *J. Chem. Soc.* (*A*) (1969) 1072.
943 R. J. Elema, J. L. de Boer and A. Vos, *Acta Cryst.* **16** (1963) 243.
944 C. G. Vonk and E. H. Wiebenga, *Acta Cryst.* **12** (1959) 859.
945 K. O. Christe and D. Pilipovich, *Inorg. Chem.* **8** (1969) 391; D. V. Bantov, B. É. Dzevitskii, Yu. S. Konstantinov, V. F. Sukhoverkhov and Yu. A. Ustynyuk, *Doklady Chem.* **180** (1968) 491.

spectra[946a], while BrF_5,SO_3 is presumably either $[BrF_4]^+[SO_3F]^-$ or molecular FSO_2OBrF_4. There is also evidence that the 1 : 1 complexes formed by bromine pentafluoride with MF (M = K, Rb or Cs) contain the BrF_6^- anion: thus, the vibrational spectra of the rhombohedral crystals testify to the presence of such a unit, the bromine atom apparently occupying a site with D_{3d} symmetry[946b]. The neutralization reaction

$$[BrF_4]^+[Sb_2F_{11}]^- + 2Cs^+BrF_6^- \rightarrow 3BrF_5 + 2Cs^+SbF_6^-$$

is reported to occur in the molten complex $BrF_5,2SbF_5$ [946a]. Iodine pentafluoride forms 1 : 1 adducts with SbF_5 and PtF_5, while giving with SO_3 a constant-boiling mixture of composition $IF_5,1{\cdot}17SO_3$; several oxides and salts of oxy-acids, e.g. NO_2, MoO_3 and KIO_4, also yield definite adducts, the nature of which is as yet obscure. Preliminary reports of the crystal structure[947] and vibrational spectrum[935] of IF_5,SbF_5 sustain the representation $[IF_4]^+[SbF_6]^-$, the cation being structurally analogous to the SF_4 molecule; a similar constitution is to be expected for IF_5,PtF_5, while $[IF_4]^+[SO_3F]^-$ is possibly an ingredient of the constant-boiling mixture derived from SO_3. Solid derivatives of the type MF,IF_5 have been described for a variety of univalent cations (M = alkali metal[838,839], Ag[838,839], NO[838,839] or tetra-alkylammonium[948]). Although no definitive structural analysis has yet been reported, the vibrational[949] and ^{129}I Mössbauer[950] spectra of the solids support the formulation $M^+IF_6^-$; the presence of the IF_6^- ion is also implied by the electrical conductivities and vibrational and ^{19}F nmr spectra of solutions of the salts. In conjunction with a bulky cation such as R_4N^+ (where R = Me or Et) or $[(pyridine)_2X]^+$ (where X = I, ICl_2, Br or Cl), the related anion $[IF_5Cl]^-$ may well exist in the newly disclosed complexes R_4NCl,IF_5 and $[(pyridine)_2X]Cl,IF_5$ [951].

A method reported for the preparation of the complex K_2PtF_6 has exploited bromine pentafluoride as a medium for the reaction of $[BrF_2^+]_2[PtF_6]^{2-}$ with potassium fluoride, but the solvent properties of chlorine and bromine pentafluoride appear otherwise to have received little attention. On the other hand, the conductivity of liquid iodine pentafluoride is reported to be increased by the dissolution either of a base such as KIF_6 or of one of the Lewis acids SbF_5, SO_3, BF_3, HF and KIO_3. Neutralization reactions between an acid and base, analogous to those which take place in bromine trifluoride solution, have been established in a few cases, e.g.

$$[IF_4]^+SbF_6^- + K^+IF_6^- \rightarrow K^+SbF_6^- + 2IF_5$$
$$[IF_4]^+BF_4^- + K^+IF_6^- \rightarrow K^+BF_4^- + 2IF_5$$

Hence, the complex salts $KSbF_6$, KBF_4, $KPtF_6$ and $CsRhF_6$ have been obtained, but various factors, e.g. incomplete interaction and solvation of the product, render iodine pentafluoride inferior to bromine trifluoride as a reaction medium for the preparation of such complexes. Iodine pentafluoride–hydrogen fluoride mixtures appear to offer significantly greater solvent power than pure iodine pentafluoride, and the formation of various complex salts in such mixtures has been described[838]. According to recent reports,

[946] (a) M. D. Lind and K. O. Christe, *Inorg. Chem.* **11** (1972) 608; H. Meinert, U. Gross and A.-R. Grimmer, *Z. Chem.* **10** (1970) 226; (b) R. Bougon, P. Charpin and J. Soriano, *Compt. rend.* **272C** (1971) 565.

[947] H. W. Baird and H. F. Giles, *Acta Cryst.* **A25** (1969) S115; N. Bartlett, private communication.

[948] H. Meinert and H. Klamm, *Z. Chem.* **8** (1968) 195.

[949] K. O. Christe, J. P. Guertin and W. Sawodny, *Inorg. Chem.* **7** (1968) 626; S. P. Beaton, D. W. A. Sharp, A. J. Perkins, I. Sheft, H. H. Hyman and K. Christe, *ibid.* p. 2174; H. Klamm, H. Meinert, P. Reich and K. Witke, *Z. Chem.* **8** (1968) 393, 469.

[950] S. Bukshpan, J. Soriano and J. Shamir, *Chem. Phys. Letters*, **4** (1969) 241.

[951] H. Klamm and H. Meinert, *Z. Chem.* **10** (1970) 270.

iodine pentafluoride dissolves xenon difluoride and xenon tetrafluoride[952], and claims have been laid to a number of distinct adducts derived from these systems[953]. Comparatively weak intermolecular association is clearly implied by the finding that the crystalline compound XeF_2,IF_5 contains discrete XeF_2 and IF_5 molecules not perceptibly perturbed by the interaction, which is presumed to be essentially coulombic in origin[903]. Certain redox reactions have also been reported to occur in liquid iodine pentafluoride. For example, alkali-metal iodides dissolve in the pentafluoride at room temperature with the liberation of elemental iodine and the formation of the tetrafluoroiodate(III) anion, derivatives of which may be isolated from the resulting solution[836,838]. Iodine dissolves in iodine pentafluoride in the presence of a Lewis acid like SbF_5 to form blue solutions until recently supposed to contain the I^+ cation[836,838,839]. However, with the characterization of the I_2^+ ion[925], the absorption spectra and other properties of the blue solutions clearly proclaim[954] this to be the predominant coloured constituent.

Compounds of Formula XY_7 [836,838,839]

Since the triple point is at 6·45°C and the vapour pressure of the solid reaches 760 mm at 4·77°C, iodine heptafluoride is manifestly characterized by a very restricted liquid range. In contrast with previously quoted values, the Trouton constant of the liquid is normal; in view of this and of the meagre values of the dielectric constant and specific conductivity, the liquid offers little promise as a solvent. Of the products of the hypothetical fluoride ion-transfer reaction

$$2IF_7 \rightleftharpoons IF_6^+ + IF_8^-$$

the IF_6^+ cation has been characterized with some assurance. Thus, the heptafluoride has been shown to form the complexes IF_7,AsF_5, IF_7,BF_3 and $IF_7,3SbF_5$, and for the first of these the formulation $[IF_6]^+[AsF_6]^-$ has been established by spectroscopic[950,955,956] and X-ray crystallographic[957] measurements. By contrast, there is no sign of interaction of iodine heptafluoride with alkali-metal fluorides, which do not dissolve in the liquid. Although no complexes of the type MF,IF_7 are known, it is noteworthy that IF_7, IF_7,AsF_5 and $IF_7,3SbF_5$ give off fluorine when heated with potassium fluoride to 200–250°C: in these circumstances, neutralization reactions such as

$$[IF_6]^+[AsF_6]^- + K^+F^- \rightarrow K^+[AsF_6]^- + IF_7$$

are believed to precede the decomposition.

3. **Halogenating action of interhalogen compounds**[827,829–832,838,839]. Except in situations where electrophilic attack is encouraged, the halogenating action of an interhalogen compound XY_n typically involves the addition or substitution of the more electronegative atom Y with simultaneous reduction of the less electronegative atom X, for example to the parent halogen X_2. The general pattern of behaviour is illustrated by the reactions

952 F. O. Sladky and N. Bartlett, *J. Chem. Soc.* (*A*) (1969) 2188.

953 H. Meinert and G. Kauschka, *Z. Chem.* **9** (1969) 35; V. A. Legasov, V. B. Sokolov and B. B. Chaivanov, *Zhur. fiz. Khim.* **43** (1969) 2935; A. V. Nikolaev, A. A. Opalovskii, A. S. Nazarov and G. V. Tret'yakov, *Doklady Chem.* **189** (1969) 982; *Doklady Phys. Chem.* **191** (1970) 262.

954 R. D. W. Kemmitt, M. Murray, V. M. McRae, R. D. Peacock and M. C. R. Symons, *J. Chem. Soc.* (*A*) (1968) 862.

955 K. O. Christe and W. Sawodny, *Inorg. Chem.* **6** (1967) 1783; K. O. Christe, *ibid.* **9** (1970) 2801.

956 J. F. Hon and K. O. Christe, *J. Chem. Phys.* **52** (1970) 1960; M. R. Barr and B. A. Dunell, *Canad. J. Chem.* **48** (1970) 895.

957 S. P. Beaton, Ph.D. Thesis, University of British Columbia, Vancouver, British Columbia (1966).

of iodine monochloride or monobromide with a wide range of elements E:

$$E + nIX \rightarrow EX_n + n/2 I_2 \quad (X = Cl \text{ or } Br)$$

wherein chlorination or bromination, but never iodination, of E is the rule. Likewise the halogen fluorides are noted for their activity as fluorinating agents. Although, in most cases, only qualitative data on the rates and products of the reactions are available, all the halogen fluorides are moderate-to-vigorous fluorinating agents, the approximate order of reactivity being $ClF_3 > BrF_5 > IF_7 > ClF > BrF_3 > IF_5 > BrF > IF_3 > IF$[839]: our rather ill-furnished knowledge of the reactions of ClF_5[958] suggests that it is somewhat less reactive than ClF_3. That such a sequence reflects only in part the thermodynamic potential of the individual fluorides is shown by the molar free energies of reduction listed in Table

TABLE 95. MOLAR FREE ENERGIES OF REDUCTION OF THE HALOGEN AND CERTAIN OTHER FLUORIDES

Reaction	ΔG°_{298} (kcal mol^{-1})
$ClF_5(g) \rightarrow ClF_3(g) + F_2(g)$	+9·25[a]
$IF_7(g) \rightarrow IF_5(l) + F_2(g)$	+14·7[a]
$ClF_3(g) \rightarrow ClF(g) + F_2(g)$	+15·95[a]
$2ClF(g) \rightarrow Cl_2(g) + F_2(g)$	+27·6[a]
$BrF_5(l) \rightarrow BrF_3(l) + F_2(g)$	+27·6[a]
$2BrF(g) \rightarrow Br_2(l) + F_2(g)$	+35·2[a]
$BrF_3(l) \rightarrow BrF(g) + F_2(g)$	+39·9[a]
$2IF(g) \rightarrow I_2(s) + F_2(g)$	+56·2[a]
$IF_5(l) \rightarrow IF_3(g) + F_2(g)$	~ +79*[b]
$IF_3(g) \rightarrow IF(g) + F_2(g)$	~ +80*[b]
$XeF_6(g) \rightarrow XeF_4(g) + F_2(g)$	+8·1[c]
$XeF_4(g) \rightarrow XeF_2(g) + F_2(g)$	+15·2[c]
$XeF_2(g) \rightarrow Xe(g) + F_2(g)$	+17·9[c]
$2CoF_3(s) \rightarrow 2CoF_2(s) + F_2(g)$	~ +44*[d]
$AsF_5(g) \rightarrow AsF_3(l) + F_2(g)$	+62·9[d,e]
$2AgF_2(s) \rightarrow 2AgF(s) + F_2(g)$	~ +69*[d]
$SbF_5(l) \rightarrow SbF_3(s) + F_2(g)$	~ +90*[b]
$PF_5(g) \rightarrow PF_3(g) + F_2(g)$	+161·8[d]

* Estimated values.
[a] See Table 93.
[b] L. Stein, *Science*, **168** (1970) 362.
[c] H. Selig, *Halogen Chemistry* (ed. V. Gutmann), Vol. 1, p. 403, Academic Press (1967).
[d] *National Bureau of Standards Technical Notes* 270–3 and 270–4, U.S. Government Printing Office, Washington (1968–9).
[e] P. A. G. O'Hare and W. N. Hubbard, *J. Phys. Chem.* **69** (1965) 4358.

95, which signify that at 25°C the fluorinating capacity should decrease in the sequence $ClF_5(g) > IF_7(g) > ClF_3(g) > ClF(g) \sim BrF_5(l) > BrF(g) > BrF_3(l) > IF(g) > IF_5(l) \sim IF_3(g)$. With the changes of phase suffered by the oxidized or reduced forms at different temperatures, some modification of this order is to be expected, but the following general conclusions seem to emerge: (a) in common with the noble-gas fluorides, halogen fluorides

[958] D. Pilipovich, W. Maya, E. A. Lawton, H. F. Bauer, D. F. Sheehan, N. N. Ogimachi, R. D. Wilson, F. C. Gunderloy, jun., and V. E. Bedwell, *Inorg. Chem.* **6** (1967) 1918.

like ClF_3 and IF_7 are energetically not far removed as fluorinating agents from elemental fluorine; and (b) kinetic factors, including the catalytic action of surfaces and of auxiliary Lewis acids like HF, modify the characteristics of the fluorination reactions to a significant, though as yet largely unspecified, extent.

In the gas phase or in solution in a non-polar solvent, homolytic fission of an interhalogen bond is presumably the primary step of halogenation: in a polar solvent, electrophilic attack, either by the electron-deficient site of the undissociated interhalogen molecule or by a cation like ClF_2^+, is favoured, particularly with the catalytic assistance of a halide ion-acceptor like HF or CoF_3. The intermediate agency of radicals has been substantiated in the decomposition of gaseous chlorine trifluoride[861], while several reports testify to the intervention of surface-reactions in processes that purport to occur in the gas phase, but, for the most part, the mechanistic details of such reactions are obscure. No less speculative are the mechanisms of halogenation reactions belonging to the second category, which are accompanied by heterolytic dissociation of an interhalogen bond. In this case, it is the capacity of the interhalogen XY_n to act as a donor or acceptor of Y^- ions, rather than the strength of the X–Y bond, which determines the reactivity of the compound. It is likely, though by no means certain, that the fluorination of elements, oxides and salts of oxy-acids by liquid bromine trifluoride proceeds *via* fluoride ion-transfer processes of this sort, a view supported by the fluorinating action of derivatives like $[BrF_2]^+[SbF_6]^-$ and $KBrF_4$ originating from the interaction of suitable acids or bases with bromine trifluoride[839].

Cogent evidence of the influence of the environment on the course of halogenation stems from a systematic study of the reaction between iodine monochloride and salicylic acid or phenol[959]. Thus, the main reaction between the organic compound and the interhalogen vapour involves chlorination since homolytic fission of the I–Cl molecule leads to chlorination by Cl_2 rather than iodination by the less reactive I_2; in carbon tetrachloride solution (low dielectric constant) iodination predominates, accompanied to a small extent by chlorination, heterolytic fission of the I–Cl bond now providing the more facile means of attack; in a solvent of high dielectric constant, such as nitrobenzene, heterolytic fission eclipses the homolytic process, and iodination occurs exclusively. Likewise, with respect to aromatic compounds, iodine monobromide normally acts as a brominating agent as a result of the high degrèe of thermal dissociation which it experiences and of the relative rates of halogenation of the free bromine and iodine thus generated. Only in a solvent of high dielectric constant, like water, does electrophilic iodination occur with this reagent, homolytic dissociation of which can, nevertheless, be suppressed by the addition of elemental iodine[839]. In general, the interhalogen compounds are less reactive than the parent halogen molecules with respect to homolytic dissociation, but more reactive with respect to heterolytic dissociation. Correspondingly, fluorination reactions which would be difficult to control with elemental fluorine sometimes proceed relatively smoothly when a reagent like chlorine trifluoride, bromine trifluoride or iodine pentafluoride is used as the fluorinating agent. Conversely, the utilization of an interhalogen compound is sometimes advantageous in situations where electrophilic attack by the parent halogen is unduly sluggish. For instance, with respect to electrophilic iodination of organic materials, iodine monochloride is significantly more reactive than iodine itself.

The most notable applications of the halogen fluorides stem from their action as fluorinating agents. Thus, chlorine trifluoride and bromine triflouride have been exploited for the

[959] F. W. Bennett and A. G. Sharpe, *J. Chem. Soc.* (1950) 1383.

preparation of binary and complex fluorides—notably of transition metals—from the corresponding elements or their oxides, oxy-salts or halides; reference to some representative reactions has already been made in the context of the solvent action of bromine trifluoride (see p. 1521). Further, as judged by its reactions with sulphur, selenium, molybdenum and tungsten and several of their derivatives[928,960–962], chlorine monofluoride has lately emerged as a useful, more moderate fluorinating agent. On the large scale, the halogen fluorides have been widely used in atomic energy installations for the preparation of uranium hexafluoride, and processes have thus been developed for separating uranium from plutonium and fission products in spent nuclear fuels. As attested by a very extensive project literature, chlorine trifluoride and bromine trifluoride appear to be the most serviceable of the fluorides for such recovery processes. Uranium metal and oxides are converted by these agents to volatile uranium hexafluoride, anhydrous hydrogen fluoride being used to accelerate the reaction with chlorine trifluoride. Hence, plutonium is converted to the involatile tetrafluoride, and, tellurium, iodine and molybdenum apart, most fission products also form involatile fluorides, which remain in the reaction vessel when the uranium hexafluoride is volatilized. The uranium hexafluoride is then separated from other volatile components by distillation. The parent halogens and lower-valent halogen fluorides formed by reduction of the trifluoride are treated, in a typical procedure, with fluorine to regenerate the trifluoride. An experimental reactor has even been operated with uranium hexafluoride fuel stabilized with respect to reduction to lower fluorides by the presence of chlorine trifluoride[838].

Several halogen fluorides act upon a wide range of oxygen-bearing compounds to liberate oxygen quantitatively, a reaction which has been turned to account for the determination of oxygen in metal oxides and derivatives of oxy-anions, e.g. carbonates, silicates and phosphates. Individual reagents used for this purpose include bromine trifluoride, bromine pentafluoride, $KBrF_4$, $[BrF_2]^+[SbF_6]^-$ and a gaseous mixture of chlorine trifluoride with hydrogen fluoride[838].

Extensive use has been made of chlorine trifluoride, bromine trifluoride and iodine pentafluoride for the fluorination of organic compounds, though the reactions are commonly difficult to control. Moreover, the exceedingly energetic character of many of the reactions of halogen fluorides with hydrogen-containing compounds has prompted investigations of the feasibility of chlorine trifluoride, bromine trifluoride and bromine pentafluoride as oxidizers for such rocket fuels as hydrazine and its derivatives. The system hydrazine–chlorine trifluoride has been studied most extensively, since it combines high performance with the advantage over cryogenic systems that both the fuel and oxidizer may be stored as liquids at ambient temperature. The fluorides have also been employed as hypergollic fluids for the ignition of hydrogen–oxygen or solid propellants. Of the three materials so tested, chlorine trifluoride clearly emerges as the most effective, but no development beyond the experimental stage has yet been reported[838].

Reactions with Elements

The reactivity of the interhalogen compounds is indicated by the facts that few elements

[960] J. J. Pitts and A. W. Jache, *Inorg. Chem.* **7** (1968) 1661.
[961] C. J. Schack and R. D. Wilson, *Inorg. Chem.* **9** (1970) 311.
[962] R. Veyre, M. Quenault and C. Eyraud, *Compt. rend.* **268C** (1969) 1480; C. J. Schack and W. Maya, *J. Amer. Chem. Soc.* **91** (1969) 2902; D. E. Gould, L. R. Anderson, D. E. Young and W. B. Fox, *ibid.* p. 1310; C. J. Schack, R. D. Wilson, J. S. Muirhead and S. N. Cohz, *ibid.* p. 2907; D. E. Young, L. R. Anderson and W. B. Fox, *Inorg. Chem.* **9** (1970) 2602.

withstand their action at elevated temperatures, and that many elements react violently, sometimes with spontaneous ignition or explosion, even at or below ambient temperatures. Details of such reactions are to be found in more comprehensive accounts of the compounds[826–829,832,838,839]. The rate of reaction is a function, not only of the nature of the reagents and of the temperature, but also of the physical states of the reagents and products, of various catalytic agencies (e.g. solid surfaces, foreign materials or, possibly, one or more of the products of reaction), and, where a solvent is employed, of the nature of that solvent and of the solubilities therein of all the species participating in the reaction. Reaction is likely to be inhibited by the formation of a coherent coating of halide on the surface of a solid element which is either involatile or insoluble in whatever medium surrounds it. Thus, the resistance of nickel and copper and their alloys to attack by the halogen fluorides depends upon the protection afforded by such a superficial coating of the corresponding fluoride. In the interaction of an interhalogen compound with an element E exhibiting more than one valence state, oxidation of E commonly proceeds, in conjunction with the halogen of lower atomic number, to the highest valence state, whereas the halogen of higher atomic number is reduced either to the element or to a lower-valent interhalogen derivative. The oxidizing capacity of the interhalogens varies substantially from compound to compound. At one extreme, we find in chlorine trifluoride, for example, an oxidizing agent of unusual power: with but few exceptions, it reacts vigorously with virtually every element at room or elevated temperatures; as in the corresponding reaction with elemental fluorine, the highest known valence state of the element is usually realized in the fluoride thereby produced. With hydrogen, potassium, phosphorus, arsenic, antimony, sulphur, selenium, tellurium, powdered molybdenum, tungsten, rhodium, iridium and iron, spontaneous ignition occurs; bromine and iodine also inflame in the trifluoride, forming mixtures of bromine and iodine fluorides (cf. Table 95), while even the noble metals platinum, palladium and gold are attacked at elevated temperatures. Much milder, by contrast, is the action of iodine pentafluoride. Although alkali metals, molybdenum, tungsten, phosphorus, arsenic, antimony and boron react energetically at room temperature or when heated, hydrogen, silver, magnesium, copper, iron and chromium are not disposed to react at normal temperatures. The variation in oxidizing power of the halogen fluorides is clearly delineated by the observation that only certain of them react with the heaviest noble gases xenon and radon. Thus, chlorine trifluoride and iodine heptafluoride, but not iodine pentafluoride, oxidize xenon to xenon fluorides[838,963], while in hydrogen fluoride solutions microgram quantities of radon are oxidized at temperatures between $-195°$ and 25°C by all the stable halogen fluorides other than iodine pentafluoride[964].

While remaining highly reactive materials, iodine monochloride and monobromide lie, nevertheless, at the lower end of the scale of oxidizing capacity. Investigations of their behaviour with respect to a large number of elements[965] reveal that moderate-to-vigorous reactions are common, each affording the corresponding chloride or bromide of the element in question, but that numerous elements, including boron, carbon, cadmium, lead, zirconium, niobium, molybdenum and tungsten, fail to react even with the molten interhalogen compound. Where reaction occurs, the highest valence state of the element is less frequently realized with these reagents: for example, although phosphorus gives the

963 N. Bartlett, private communication.

964 L. Stein, *J. Amer. Chem. Soc.* **91** (1969) 5396; *Yale Scientific Magazine*, **44** (1970) 2; *Science*, **168** (1970) 362.

965 V. Gutmann, *Z. anorg. Chem.* **264** (1951) 169; *Monatsh.* **82** (1951) 280.

corresponding pentahalide, vanadium is converted by iodine monochloride, not to the tetrachloride, but to the trichloride, which is conveniently prepared in this way[827].

Reactions with Oxides and Salts of Oxy-acids

The range of halogenating power of the interhalogen compounds is also prominent in their reactions with oxides and salts of oxy-acids. Powerfully oxidizing halogen fluorides, such as chlorine trifluoride or bromine trifluoride, bromine pentafluoride and iodine heptafluoride, typically react with these species to form fluoride or oxyfluoride derivatives, together with oxygen or oxyhalogen species such as ClO_2F or OIF_5. With chlorine trifluoride the reaction of many metal oxides is sufficiently vigorous to promote ignition, while bromine trifluoride, bromine pentafluoride or the complexes $[BrF_2]^+[SbF_6]^-$ and $KBrF_4$ liberate oxygen quantitatively from numerous oxides and oxysalts, e.g.[838]

$$6Sb_2O_3+32BrF_3 \xrightarrow{\text{Liquid}} 12[BrF_2][SbF_6]+10Br_2+9O_2$$

$$3MReO_4+4BrF_3 \xrightarrow{\text{Liquid}} 3MReO_2F_4+2Br_2+3O_2$$

$$(M = K, Rb, Cs, Ag, \tfrac{1}{2}Ca, \tfrac{1}{2}Sr \text{ or } \tfrac{1}{2}Ba)$$

$$KAlSi_3O_8+8BrF_5 \xrightarrow{450^\circ C} KF+AlF_3+3SiF_4+4O_2+8BrF_3$$

On the other hand, milder reactions are the rule with iodine pentafluoride: it has been found, for example, that V_2O_5, Sb_2O_5, MoO_3, WO_3 and CrO_3 form the following compounds in reactions with the hot or boiling pentafluoride: $2VOF_3,3IOF_3$, $SbF_5,3IO_2F$, $2MoO_3,3IF_5$, $WO_3,2IF_5$ and CrO_2F_2; while potassium permanganate and perrhenate react thus:

$$KMO_4+IF_5 \rightarrow MO_3F+IOF_3+KF \quad (M = Mn \text{ or } Re)$$

Included within the same general category are the reactions of glass and quartz, which occur more or less readily with all the halogen fluorides, e.g.

$$2B_2O_3+4BrF_3 \rightarrow 4BF_3+2Br_2+3O_2$$

$$3SiO_2+4BrF_3 \rightarrow 3SiF_4+2Br_2+3O_2$$

$$SiO_2+2IF_7 \rightarrow SiF_4+2OIF_5$$

It is possible that attack is usually initiated by traces of hydrogen fluoride and sustained by a continuous cycle of reactions such as

$$4HF+SiO_2 \rightarrow SiF_4+2H_2O$$

$$IF_7+H_2O \rightarrow OIF_5+2HF$$

This mechanism is consistent with the observations that very pure samples of chlorine trifluoride or iodine pentafluoride are slow to etch glass or quartz apparatus.

Again, the action of water illustrates well the diversity of oxidizing properties of the interhalogen compounds. Strongly oxidizing halogen fluorides like chlorine or bromine trifluoride interact violently with water unless precautions are taken to moderate the process, for example by the use of low temperatures and by dilution; the products formed depend partly upon the proportions of the reactants, but may include HF, HX, HOX, HXO_3, O_2 and X_2, as well as halogen oxides (e.g. ClO_2 and OF_2) and oxyhalogen fluorides (e.g. ClO_2F) in certain cases. Whereas iodine pentafluoride also suffers vigorous hydrolysis, the iodine does not undergo reduction:

$$3H_2O+IF_5 \rightarrow HIO_3+5HF \qquad \Delta H_{291} = -22{\cdot}05 \text{ kcal mol}^{-1}$$

Altogether more moderate are the hydrolytic reactions of the diatomic interhalogens BrCl, ICl and IBr and of iodine trichloride. The primary steps in the hydrolysis of the diatomic molecules in the absence of added halide ions are:

$$IX+H_2O \rightarrow H^++X^-+HOI \ (X = Cl \text{ or } Br)$$

and

$$BrCl+H_2O \rightarrow H^++Cl^-+HOBr$$

The reactions occurring during the hydrolysis of iodine trichloride are complicated, but have been formally represented by the equation

$$2ICl_3+3H_2O \rightarrow IO_3^-+ICl_2^-+4Cl^-+6H^+$$

However, in acidic solutions containing excess of halide ions, hydrolysis of the interhalogen yields place to the formation of relatively stable polyhalide ions such as ICl_2^- and IBr_2^-. In the form of the ICl_2^- ion, present in strongly acid solution, iodine monochloride features in volumetric analysis, being for example the reduction product of iodate in the well-known Andrews titration. Iodine trichloride gives as possible products Cl^-, ICl_4^-, ICl_2^- and IO_3^- in proportions depending on the precise conditions of reaction[839]. Although Br–Cl, I–Cl and I–Br bonds can thus survive hydrolysis, halogen–fluorine bonds appear invariably to suffer hydrolysis, this despite the enhanced energy of such bonds. Under aqueous conditions, therefore, an electropositive halogen atom has the characteristics of a class "b" acceptor[966], though, as noted elsewhere (Section 3, p. 1252), this property hinges upon *differences* of bond energy and of hydration energy, being related only indirectly to intrinsic properties of the interhalogen bonds.

With certain compounds containing multiple bonds between carbon and oxygen or between sulphur and oxygen, the diatomic molecules ClF and BrF undergo addition reactions. In some instances, the addition is localized on the carbon or sulphur atom, e.g.

$$CO + ClF \longrightarrow FClC{=}O \quad [838]$$

$$SO_2 + XF \longrightarrow XS(=O)_2F \qquad (X = Cl^{961} \text{ or } Br^{839}).$$

but in others addition occurs across one of the multiple bonds, e.g.

$$R_1R_2C{=}O+ClF \rightarrow R_1R_2CFOCl$$

(R_1, R_2 = F, CF_3 or other perhalogenoalkyl groups)[962]

$$F_4S{=}O+ClF \rightarrow SF_5OCl \quad [962]$$

$$SO_3+ClF \rightarrow FSO_2OCl \quad [961]$$

---

966 S. Ahrland, J. Chatt and N. R. Davies, *Quart. Rev. Chem. Soc.* **12** (1958) 265.

Reactions of the latter class provide an expeditious route to molecular hypochlorites, being catalysed either by a base like caesium fluoride or by an acid like hydrogen fluoride[962].

Reactions with Halides

Fluorination, frequently allied to oxidation, is innate to the reactions of the halogen fluorides with many other halides derived from metals or non-metals. For example, the chlorides $NiCl_2$, AgCl and $CoCl_2$ are converted to NiF_2, AgF_2 and CoF_3 by chlorine trifluoride at 250°C[838]. Arsenic trichloride is oxidized by chlorine mono- or trifluoride to the compound formulated as $[AsCl_4]^+[AsF_6]^-$, while with antimony pentachloride, chlorine monofluoride gives $SbCl_4F$. The addition of chlorine monofluoride to sulphur tetrafluoride at 350°C affords one of the most efficient methods yet devised for the preparation of pentafluorosulphur chloride[928]; at 180°C chlorine trifluoride converts sulphur tetrafluoride to a mixture of pentafluorosulphur chloride and sulphur hexafluoride[838]. Liquid bromine trifluoride acts upon many metallic chlorides, bromides and iodides to produce binary or complex fluorides in which the highest accessible valence state of the metal is often attained, though the formation of insoluble surface films of fluoride may prevent the reaction from going to completion, as with lead(II), thallium(I) and cobalt(II) halides, which yield mixtures of lower- and higher-valent fluorides. However, when the product is volatile (e.g. UF_6) or soluble in the reagent (e.g. the fluorides of the alkali metals), conversion into the fluoride is quantitative.

Reactions with Hydrogen-containing and Organic Compounds

Many hydrogen-containing compounds, both organic and inorganic, inflame or explode when mixed with a halogen fluoride as a result of highly energetic reactions such as[838]

$$2NH_3+2ClF_3 \rightarrow N_2+6HF+Cl_2$$

and

$$3N_2H_4+4ClF_3 \rightarrow 3N_2+12HF+2Cl_2$$

On the other hand, some of the reactions may be moderated by diluting the interhalogen with an inert gas, by dissolving the reagent in a relatively inert solvent such as carbon tetrachloride or a fluorocarbon, or by lowering the temperature. For example, all the chlorine fluorides convert difluoroamine to chlorodifluoroamine, e.g.

$$ClF_3+3HNF_2 \rightarrow ClNF_2+N_2F_4+3HF$$

in reactions which can be moderated by the use of complexes of either the chlorine fluoride or of the difluoroamine[967]. Chlorodifluoroamine is also the product of hazardous reactions involving chlorine trifluoride and ammonium fluoride or bifluoride[838].

The reactions of the halogen fluorides with organic compounds have been reviewed by Musgrave[926], and will be described here only in outline. Even where moderating conditions are used, substitution reactions in which the chlorine, bromine or iodine of organo-halogen compounds is replaced by fluorine are generally less exothermic and easier to control than reactions involving C–H bonds. However, cleavage of C–C bonds can also occur, particularly with the more energetic reagents, and violent explosions have been reported as a result of contact between chlorine trifluoride and relatively inert hydrocarbons, e.g. Kel-F oil[838]. Even with iodine pentafluoride, the mildest of the halogen fluorides, compounds rich in hydrogen inflame at room temperature. Nevertheless, numerous organo-halogen

[967] D. Pilipovich and C. J. Schack, *Inorg. Chem.* 7 (1968) 386.

compounds undergo smooth substitution reactions with iodine pentafluoride: for example, CCl_4 is slowly converted to CCl_3F together with some CCl_2F_2, CHI_3 gives mainly CHF_3, while CF_3I is the principal product of the reaction with CI_4[830]. Other organic compounds whose reactions with one or more of the halogen fluorides have been investigated include carbon tetrabromide, carbon disulphide, trichloroacetic acid, polychlorobutadienes, acetonitrile and related compounds, aliphatic amines, various ketones, as well as benzene, hexachlorobenzene and other aromatic compounds[827,830,838,926]. Substitution reactions are the general rule, but in some instances, as in the vapour-phase fluorination of benzene and toluene by chlorine trifluoride, both addition- and substitution-products are formed. Many of the reactions have been the subjects of patents; certain of them, e.g. between iodine pentafluoride and carbon tetraiodide or carbon disulphide, played an important part in providing, for the first time, routes to perfluoroalkyl derivatives[926], but, in the main, these have now been supplanted by procedures more amenable to control and less severe in their technical demands.

As noted previously, the course of substitution reactions involving the compounds BrCl, ICl and IBr varies with the nature of the medium employed. Most extensively studied has been the electrophilic halogenation of C–H bonds in aromatic systems. In particular, the reactions of iodine monochloride with, *inter alia*, the following compounds have been studied in some detail: phenols, aromatic carboxylic acids, aromatic ethers, acetanilide and other N- or ring-substituted aniline derivatives, aromatic amino-acids and amides, and various thiazole derivatives. Iodine trichloride also reacts under appropriate conditions with various aromatic compounds, including thiophen, to give chloro-substituted products, usually together with no more than small amounts of the corresponding iodo-substituted derivatives. With certain organometallic derivatives, substitution at the metal–carbon bond may take a somewhat different course. Thus, iodine monochloride is reported to act upon organo-mercury compounds in various solvents to give organo-iodine, rather than organo-chlorine, compounds, unless secondary reactions intervene, whereas iodine trichloride converts aryl–tin or aryl–mercury compounds into the corresponding diaryliodonium derivatives (q.v.), e.g.

$$ICl_3 + 2PhSnCl_3 \rightarrow Ph_2ICl + 2SnCl_4$$

A characteristic reaction of the diatomic interhalogen molecules is addition to the units $\gt C{=}C \lt$ or $-C{\equiv}C-$ of organic compounds. Interaction of this kind has been reported, not only for the compounds ClF, BrCl, ICl and IBr, but also for the less well characterized fluorides BrF and IF. As produced *in situ* by the reaction of the parent halogen either with silver(I) fluoride[968] or with a higher-valent halogen fluoride, e.g. BrF_3 or IF_5[926,927], the monofluorides add to the double bonds of various fluoro-olefins[926,927] or of unsaturated carbohydrate derivatives like D-glucal triacetate[968]. Numerous addition reactions with olefinic and acetylenic compounds have also been reported for bromine and iodine monochloride; indeed the products of such reactions afforded some of the earliest positive evidence for the existence of bromine monochloride as a distinct compound. The reactions with iodine monochloride have been successfully exploited to determine the degree of unsaturation (the "iodine number") in natural and synthetic rubbers. The orientation of addition and the kinetic characteristics of such reactions are consistent with a mechanism

968 L. D. Hall and J. F. Manville, *Canad. J. Chem.* 47 (1969) 361, 379.

analogous to that proposed for the addition of a homonuclear halogen molecule (see p. 1218), electrophilic attack being initiated by the more electropositive halogen atom which forms a cationic organo-halogen intermediate. Despite the instability of the pure compounds, the propensity of BrN_3, IN_3 and INCO to add to olefinic species has also been established, the pseudohalide molecule typically being generated *in situ* by the reaction of the parent halogen with a salt of the pseudohalide anion[874]; it is reported that, with appropriately chosen conditions, BrN_3 can undergo either electrophilic or free-radical addition, with corresponding variations in the orientation of the bromine and azide substituents, but that IN_3 acts exclusively as an electrophile. Iodine trichloride adds to acetylene apparently to give the compound $ClCH{=}CHICl_2$, and also to other hydrocarbons containing one or two isolated double bonds or two conjugated double bonds, though the nature of the products is not well defined[827].

3. POLYHALIDE ANIONS[826–829,832,834,835,840,841,857]

Introduction and Classification

A polyhalide anion may be defined as an addition product of a halide ion acting as a Lewis base with one or more halogen or interhalogen molecules acting as Lewis acids. Such an ion can be assigned the generalized formula $X_mY_nZ_p^-$, where X, Y and Z represent either identical or different halogen atoms. For all the well-defined species of this type, $m+n+p$ is an odd number that can be 3, 5, 7 or 9. Hence, it is clear that the anion contains an even number of electrons. The apparent anomaly posed by the complex "CsI_4" is resolved by the discovery that the material is diamagnetic and contains the I_8^{2-} ion. Other complex species have been alluded to in which the sum $m+n+p$ appears either to exceed 9 or to be an even number, but, with very few exceptions, convincing evidence of the formulation or structure is lacking.

TABLE 96. POLYHALIDE ANIONS $X_mY_nZ_p^-$ FOUND IN CRYSTALLINE SALTS[a]

$m+n+p=$				Other species
3	5	7	9	
I_3^-	I_5^-	I_7^-	I_9^-	I_8^{2-}
I_2Br^-	I_4Cl^-	I_6Br^-		
I_2Cl^-	I_4Br^-	IF_6^-		
IBr_2^-	$I_2Br_3^-$	Br_6Cl^-		
ICl_2^-	$I_2Br_2Cl^-$	BrF_6^-		
$IBrCl^-$	$I_2BrCl_2^-$			
$IBrF^-$	$I_2Cl_3^-$			
IF_2^- [b]	$IBrCl_3^-$			
Br_3^-	ICl_4^-			
Br_2Cl^-	ICl_3F^-			
$BrCl_2^-$	IF_4^-			
Cl_3^-	BrF_4^-			
ClF_2^-	ClF_4^-			

[a] A. I. Popov, *Halogen Chemistry* (ed. V. Gutmann), Vol. 1, p. 225, Academic Press (1967).

[b] H. Meinert, *Z. Chem.* 7 (1967) 41.

Table 96 lists the polyhalide anions identified more or less convincingly in the form of crystalline complexes; the catalogue includes neither the anions of compounds whose identification remains dubious nor those solution species whose identification rests solely on physicochemical measurements of stability constants. Crystalline derivatives of the anions, as of the hydrogen dihalide anions (Section 3, p. 1313), are formed most easily with large univalent cations, e.g. Cs^+, NR_4^+ (R = alkyl group) or $AsPh_4^+$. Many crystalline polyhalides also contain solvent or other foreign donor molecules, which are apparently essential to the stability of the compound, since the material suffers irreversible decomposition when attempts are made to remove the molecules. Examples of these solvated polyhalides include KI_3,H_2O, KI_7,H_2O, $KICl_4,H_2O$, $MI_3,2PhCN$ (M = Na or K) and $KI_3,2MeHNCHO$. Such complexes should not be confused with the polyhalide salts formed by many organic bases which suffer either protonation or coordination to a cationic halogen-centre to accommodate the formation and stabilization of the polyhalide anion, e.g.

$$\text{2,2'-bipyridyl,2ICl} \xrightarrow{\text{protonation}} [\text{2,2'-bipyridylH}]^+ICl_2^-$$

$$2C_5H_5N,ICl \xrightarrow{\text{polar solvent}} [(C_5H_5N)_2I]^+[ICl_2]^-$$

Preparation

The general method for the preparation of polyhalides simply involves direct combination between an ionic halide and the appropriate halogen or interhalogen, the experimental conditions being dictated by the reactivity of the reactants and products. This may be brought about (i) by exposing the solid halide to the gaseous or liquid halogen or interhalogen; (ii) by mixing the parent halogen or interhalogen compound with a halide or polyhalide in a suitable solvent, e.g. water, an alcohol or acetonitrile; (iii) by producing the interhalogen compound and effecting combination *in situ*; (iv) by displacement of a halogen of higher atomic number in an existing polyhalide by one of lower atomic number. These general methods have been comprehensively described elsewhere[840,841,857,969]. Similar methods have been used to prepare pseudohalide analogues of the polyhalide ions: for example, oxidation of solutions containing iodine and thiocyanate ions gives rise to the anion $[I(SCN)_2]^-$[970]. In various circumstances, polyhalide ions may well have an intermediate agency; thus, there are spectroscopic reasons for believing that, when silver iodide crystals are subjected to high pressures, I_3^- ions are formed, presumably as a stage in the evolution of metallic properties[971].

The gas-solid reactions are, for the most part, comparatively slow and inefficient, often failing to produce pure products. On the other hand, many polyhalide complexes (excluding those containing fluorine) can be successfully prepared in methanol or ethanol solutions. With the more reactive polyhalides, a solvent less susceptible to halogenation, e.g. 1,2-dichloroethane, has to be selected. As previously noted (p. 1515), derivatives of fluorine-containing polyhalide anions can commonly be gained by the direct action of the liquid halogen fluoride on a suitable ionic fluoride; alternatively, the fluorinating action of the halogen fluoride or of elemental fluorine on chloride, bromide or iodide derivatives is

969 A. I. Popov and R. E. Buckles, *Inorganic Syntheses*, Vol. 5 (ed. T. Moeller), p. 167, McGraw-Hill, New York (1957).
970 C. Long and D. A. Skoog, *Inorg. Chem.* **5** (1966) 206.
971 M. J. Moore and D. W. Skelly, *J. Chem. Phys.* **46** (1967) 3676.

turned to account in reactions such as[838,841]

$$MCl + 2F_2 \xrightarrow{90\text{–}250^\circ C} MClF_4 \text{ (M = K, Rb or Cs)}$$

and

$$6MCl + 8BrF_3 \xrightarrow{\text{liquid } BrF_3} 6MBrF_4 + 3Cl_2 + Br_2 \text{ (M = K, Ag or } \tfrac{1}{2}\text{Ba)}$$

Contrary to earlier reports[857], it now appears that not only halogen, but also interhalogen, molecules add to appropriate polyhalide ions, e.g.

$$ICl_2^- + XCl \rightarrow IXCl_3^- \text{ (X = Cl or I)}$$

It has thus been possible to prepare a variety of mixed pentahalide salts, e.g. $M[I_2X_3]$ and $M[I_2X_2Y]$, where $M^+ = NR_4^+$, $[C_5H_5NH]^+$, [2,2′-bipyridylH]$^+$ or [1,10-phenanthrolineH]$^+$ and X, Y = Cl or Br[840].

It is relatively simple to prepare polyhalide salts of anions like I_3^-, ICl_2^- and IBr_2^-, which are comparatively stable with respect both to dissociation and to halogenation reactions, but the preparative difficulties increase rapidly as the chemical stability of the anion deteriorates. For example, the Cl_3^- ion is comparatively unstable with respect to the dissociation $Cl_3^- \rightleftharpoons Cl_2 + Cl^-$, whether in the solid state or in solution; as a result, salts of this anion are scarce and rather poorly characterized. Some of the salts of fluorine-containing anions can be formed only at low temperature, and the difficulty of preparation is further aggravated by their vigorous activity as fluorinating agents and by their susceptibility to hydrolysis.

In the formation of polyhalide complexes, the final product is usually determined more by the composition of the reaction mixture than by the precise nature of the reagents. Thus, potassium dichloroiodate(I) can be obtained by the reaction (i) of potassium iodide with chlorine, (ii) of potassium chloride with iodine monochloride, or (iii) of potassium tetrachloroiodate(III) with potassium iodide. When a polyhalide ion engages in a replacement reaction, the terminal atoms are typically displaced by more electronegative substituents. For example, the reaction of bromine with the I_2Br^- or I_3^- ions invariably leads to the formation of IBr_2^-, but reaction does not proceed beyond this stage to give Br_3^-. Likewise the reaction

$$Et_4NICl_2 + 2AgF \xrightarrow{\text{acetonitrile}} Et_4NIF_2 + 2AgCl$$

has been reported[839].

Phase studies of binary systems, such as halogen– or interhalogen–alkali halide, and of ternary systems, such as halogen– or interhalogen–alkali halide–water, have been used by numerous investigators for the identification and characterization of polyhalide salts. In a number of cases the experimental results have clearly established the existence of individual polyhalides, but, on occasion, claims have been made for compounds which have not been substantiated and whose existence or nature still remains open to question[840]. The tendency of polyhalides to form solvated solids by interaction with water and other donor molecules evidently becomes more pronounced as the bulk of the cation decreases. Thus, whereas rubidium and caesium readily form solvent-free tri-iodides, the corresponding derivatives of lithium, sodium and potassium are known only as solvated materials, e.g. LiI_3,4PhCN and KI_3,H_2O; similarly, solvent-free salts of the ICl_4^- anion are formed by all the alkali-metal cations other than lithium. In the formation of solvent-free polyhalides, there is ample experimental evidence that the stability of the solid complex is governed by the nature of the cation. Thus, to judge by the efforts made to prepare such complexes, small or multiply charged cations are generally ill-adapted, while bulky univalent organic cations such as NR_4^+, $[C_5H_5NH]^+$, SR_3^+ and $AsPh_4^+$ are most conducive, to the stabilization of polyhalide anions. This generalization appears to be valid in spite of certain irregularities imposed by the nature of the polyhalide anion: for example, tetrachloroiodate(III) derivatives of the alkaline-earth and divalent transition metals have been prepared, as have $Ni(ClF_2)_2$[972], $Ni(ClF_4)_2$[972], $Ba(BrF_4)_2$, $[Ni(NH_3)_6][I_7]_2$ and $[Co(NH_3)_6][I_3]_3$. With a suitable cation, crystalline polyhalides containing up to 9 halogen atoms, as in NMe_4I_9, have been characterized. Curiously, only the I_3^-, I_5^- and I_9^- salts can be obtained with the cation NMe_4^+ and the I_3^-, I_5^- and I_7^- salts with NEt_4^+. However, the complex derived by precipitation or crystallization is determined not only by the formation constant of the anion, but also by the relative solubilities of the different components, which bear a subtle and largely obscure relation to the properties of the ions. Hence, preparative success is not necessarily a guide to the thermodynamic stabilities of complex polyhalides in the solvent-free condition. Moreover, there are good reasons for believing that bulk and charge are not the only properties of the cation which affect the stability of solid polyhalides; due weight must also be given to the precise shape, polarizability and charge-distribution of the cation, though this has seldom been feasible at more than a qualitative, empirical level in studies up to the present time.

The reaction of a halogen or interhalogen compound with a molecular halide sometimes entails the formation of a crystalline addition compound. Depending on the relative power of the reagents to accept halide ions, there are two ways of formulating the product, viz.

$$MX_n + 2XY \rightarrow [XY_2]^+[MX_{n+1}]^-$$

or

$$MX_n + XY \rightarrow [MX_{n-1}]^+[X_2Y]^-$$

which may thus contain either a polyhalogen cation or a polyhalide anion; in practice, the difference between these two conditions may well be clouded by the formation of a polymeric network of the type found in ClF_3,SbF_5 (see Fig. 28). The situation is further complicated in certain cases where amphoteric behaviour appears to be possible, as with the complex ICl,PCl_5, and, depending on experimental conditions, halide ion-transfer can take place in either direction. Formed from phosphorus tri- and pentahalides are complexes such as IX,PCl_5, IX,PBr_5 and Br_2,PBr_5. That the solids contain trihalide anions has been verified in the cases of ICl,PCl_5 [836,840] and Br_2,PBr_5 [973], which are to be formulated, according to crystallographic analysis, as $[PCl_4]^+[ICl_2]^-$ and $[PBr_4]^+[Br_3]^-$, respectively. By contrast, the power of antimony(V) halides as halide ion-acceptors appears to favour the formation of cationic polyhalogen derivatives.

The protonic acids corresponding to the polyhalide ions have not been prepared in the pure state, and it is doubtful whether the stabilities of the materials are compatible with isolation under conventional conditions. The acids are presumably formed in solutions whenever a halogen or interhalogen is added to a hydrohalic acid; the limited evidence available suggests that their strength as proton-donors is superior to that of the hydrohalic acids. Aqueous solutions containing up to 94% of $HIBr_2$ and 80% of HIBrCl or $HICl_2$ have been prepared, but attempts to isolate crystalline phases containing the acid have been unsuccessful. Orange-yellow crystals of the hydrate $HICl_4,4H_2O$ result on cooling the

[972] L. Stein, J. M. Neil and G. R. Alms, *Inorg. Chem.* **8** (1969) 1779.
[973] G. L. Breneman and R. D. Willett, *Acta Cryst.* **23** (1967) 467.

solution formed by passing chlorine through a suspension of iodine in concentrated hydrochloric acid; the material decomposes when attempts are made to dehydrate it. Representative of the solids which have been identified as complexes of a protonic acid with an appropriate nitrogen-base are HI_2Cl_3,C_5H_5N, $HI_3,2PhCONH_2$ and $HI_3,4PhCN$; on the evidence available the proton is accommodated either as a discrete cation, e.g. $[C_5H_5NH]^+$, or within a hydrogen-bonded network of some kind. Spectrophotometric investigation of the hydrogen iodide–iodine system in carbon tetrachloride suggests an equilibrium constant for the reaction $HI+I_2 \rightleftharpoons HI_3$ in the range 25–400[840].

Physical Properties

The physical characterization of polyhalide species has assumed three principal aims: (i) qualitatively or quantitatively the thermodynamic stability with respect to dissociation has been assessed for many polyhalides either as solids or in solution; (ii) the structures of numerous solid complexes have been determined; and (iii) spectroscopic properties of the complexes in the solid or solution phases have been evaluated.

1. Stability of the Polyhalides

Solids

A solid polyhalide derivative of the type $M^+[X_mY_nZ_p]^-$ tends to dissociate spontaneously into a halogen or interhalogen compound and a simple monohalide derivative of the cation M^+. In many cases, the dissociation pressure is already appreciable at room temperature. In accordance with this behaviour, the chemical reactions of the polyhalides are mostly those of the dissociation products. The relative thermal stabilities of crystalline trihalides have been assessed by comparing the relative magnitudes of the dissociation pressure at a given temperature or the temperatures at which the dissociation pressure reaches a particular value (e.g. 1 atm). An alternative approach has consisted of suspending the polyhalide salt in an organic solvent like carbon tetrachloride and determining the equilibrium concentration of free halogen or interhalogen in the liquid phase; since both the poly- and monohalide salts are virtually insoluble in non-polar organic liquids, this concentration is a direct index to the degree of dissociation of the polyhalide. More recently, the free energy change accompanying a reaction of the type

$$MI_{x+2y}(s) \rightarrow MI_x(s)+yI_2$$

$$(M = Rb, Cs, NH_4, NMe_4 \text{ or } NEt_4;\ x = 1, 3 \text{ or } 5 \text{ and } y = 1 \text{ or } 2)$$

has been determined[974] from emf measurements of the solid-state galvanic cell

$$Ag \mid AgI \mid C, MI_x+MI_{x+2y}$$

$$Ag \mid AgI \mid C, I_2(s)$$

over the temperature range 25–113·5°C; hence, the results presented in Table 97 have been derived.

On the basis of the various measurements, the following generalizations may be made:

(i) For several series of polyhalides containing different cations and the same anion, thermal stability is promoted by increasing the size of the cation.

[974] L. E. Topol, *Inorg. Chem.* **7** (1968) 451; *ibid.* **10** (1971) 736.

TABLE 97. VALUES OF ΔG°, ΔH° AND ΔS° AT 25°C FOR THE REACTION $MI_x(s)+yI_2(s) \rightarrow MI_{x+2y}(s)$[a]

Reaction	Cation	$-\Delta G^\circ$ (kcal mol^{-1})	$-\Delta H^\circ$ (kcal mol^{-1})	ΔS° (cal deg^{-1} mol^{-1})
$MI(s)+I_2(s) \rightarrow MI_3(s)$	NH_4	1·8	2·1	−1·1
	Rb	2·4	3·1	−2·3
	Cs	3·5	3·7	−0·9
	NMe_4	3·5	1·7	+6·7
	NEt_4	5·5	—	—
$MI_3(s)+\frac{1}{2}I_2(s) \rightarrow MI_4(s)$	Cs	0·63	0·80	−0·6
$MI_3(s)+I_2(s) \rightarrow MI_5(s)$	NMe_4	3·21	2·2	+3·2
$MI_3(s)+2I_2(s) \rightarrow MI_7(s)$	NEt_4	5·27	5·5	−0·9
$MI_5(s)+2I_2(s) \rightarrow MI_9(s)$	NMe_4	2·14	0·3	+6·2

[a] L. E. Topol, *Inorg. Chem.* **7** (1968) 451; *ibid.* **10** (1971) 736.

(ii) On dissociation, the solid monohalide afforded by the smallest halide ion results. Thus, $CsICl_2$ gives $CsCl+ICl$ rather than $CsI+Cl_2$.

(iii) For polyhalides containing the same cation but different trihalide ions, the precise stability sequence varies according to the method used to assess the extent of dissociation, but measurements of dissociation pressure imply the following order of decreasing stability: $I_3^- > IBr_2^- > ICl_2^- > I_2Br^- > Br_3^- > BrCl_2^- > Br_2Cl^-$. Evidently the most stable anions are those containing iodine as the central atom, while the stability of symmetrical anions, e.g. IBr_2^-, is superior to that of related but unsymmetrical anions, e.g. I_2Br^-.

Solutions

The behaviour of the polyhalide ions in solution has been the subject of many investigations, though in comparatively few cases have reliable thermodynamic properties been established for the solution species. The solvents employed include, in addition to water, chloroalkanes, e.g. $ClCH_2{\cdot}CH_2Cl$ or CCl_3Br, alcohols, alcohol–water mixtures, anhydrous or aqueous acetic acid, trichloroacetic acid, acetonitrile, acetone and nitromethane. Studies of the solutions are complicated by the fact that the properties of polyhalide ions are drastically influenced by the nature of the solvent; in particular, the possibility of solvolysis or of halogenation of the solvent must be taken into account. Water is a far from perfect medium because of the susceptibility to hydrolysis of the polyhalide anions and of the halogens or interhalogens formed by the dissociation reaction. While hydrolysis can be more-or-less suppressed by the addition of an acid in fairly high concentration, this not only increases the ionic strength, but also introduces another reactive species into the system.

The following physicochemical methods have provided the experimental mainstay for the characterization of the polyhalide ions in solution: spectrophotometry (mostly in the ultraviolet and visible regions), electrical conductance, potentiometry, polarographic and voltammetric techniques, liquid–liquid distribution experiments, and measurements of solubilities or colligative properties. Hence, it has been possible to establish the existence in aqueous solutions of trihalide ions corresponding to all the possible combinations of chlorine, bromine and iodine atoms. Even the highly unstable Cl_3^- ion exists in concentrated chloride solutions saturated with chlorine. By contrast, if fluorine-containing polyhalide ions are capable of existing in aqueous media, convincing evidence of the fact has still to be found. In media where the opportunity of solvolysis is denied, e.g. bromine trifluoride[934] or acetonitrile[949], the presence of fluoro-anions like BrF_4^- and IF_6^- is clearly intimated by the vibrational spectra or electrical conductivities of the solutions; of the stability of such species with respect to dissociation, however, little is known.

TABLE 98. THERMODYNAMIC CONSTANTS FOR THE FORMATION OF POLYHALIDE ANIONS IN AQUEOUS SOLUTION[a–e]

Ion	Formed from	Formation constant at 298°K (K_f)	$-\Delta G°_{298}$ (kcal mol^{-1})	$-\Delta H°_{298}$ (kcal mol^{-1})	$\Delta S°_{298}$ (cal deg^{-1} mol^{-1})
I_3^-(aq)	I_2(aq)+I^-(aq)	710[e]	3·9	4·5	−2·2
IBr_2^-(aq)	IBr(aq)+Br^-(aq)	480[d]	3·7	—	—
ICl_2^-(aq)	ICl(aq)+Cl^-(aq)	166[a]	3·0	—	—
$IBrCl^-$(aq)	IBr(aq)+Cl^-(aq)	34[d]	2·1	—	—
Br_3^-(aq)	Br_2(aq)+Br^-(aq)	16·3[a]	1·68	1·50	+0·6
I_2Br^-(aq)	I_2(aq)+Br^-(aq)	10·5[d]	1·4	1·4	−0·3
I_2Cl^-(aq)	I_2(aq)+Cl^-(aq)	1·66[a]	0·3	—	—
Br_2Cl^-(aq)	Br_2(aq)+Cl^-(aq)	1·14[c]	0·08	0	+0·4
Cl_3^-(aq)	Cl_2(aq)+Cl^-(aq)	0·19[a]	−1·0	—	—
$BrCl_2^-$(aq)	Br_2(aq)+Cl^-(aq)	$7·2\times10^{-3}$*[c]	—	—	—
I_2Cl^-(aq)	ICl(aq)+I^-(aq)	3×10^{8} [e]	11·6	—	—
I_2Br^-(aq)	IBr(aq)+I^-(aq)	$2·6\times10^{6}$ [d]	8·8	—	—
I_5^-(aq)	I_2(aq)+I_3^-(aq)	~ 9[a]	1·3	—	—
Br_5^-(aq)	Br_2(aq)+Br_3^-(aq)	1·45[a]	0·2	2·2	−7·0

* $K_f = [BrCl_2^-][Br^-]/[Br_2][Cl^-]^2$.

[a] A. I. Popov, *Halogen Chemistry* (ed. V. Gutmann), Vol. 1, p. 225, Academic Press (1967).

[b] *National Bureau of Standards Technical Note* 270–3, U.S. Government Printing Office, Washington (1968).

[c] R. P. Bell and M. Pring, *J. Chem. Soc.* (*A*) (1966) 1607.

[d] R. Guidelli and F. Pergola, *J. Inorg. Nuclear Chem.* **31** (1969) 1373.

[e] G. Piccardi and R. Guidelli, *J. Phys. Chem.* **72** (1968) 2782.

In Table 98 are listed some thermodynamic constants for the formation of polyhalide ions in aqueous solution; with but few exceptions, however, the limitations of the physico-chemical methods of analysis tend to impair the reliability of these data. For example, in calculations of the formation constant K_f of a trihalide ion, it has commonly been assumed that the activity coefficients of the mono- and trihalide ions are equal and therefore cancel out in the expression for K_f; in fact, this simplifying assumption is not valid. Still more fallible are the thermodynamic properties deduced for polyhalide ions containing more than three atoms, since the existence of successive equilibria is typically characterized by comparatively small departures from the pattern expected of the initial process $XY+Z^- \rightleftharpoons XYZ^-$. Hence, the stabilities of most pentahalide species are still poorly defined, while the existence of other species remains problematic; falling within the second category, for example, are the ions I_6^{2-} and the somewhat implausible $BrCl_6^{5-}$, ICl_6^{5-} and IBr_4^{3-}, which have been variously postulated as ingredients of aqueous equilibria[840]. Studies of some of the mixed polyhalide ions are further complicated by their tendency to disproportionate in favour of symmetrical species, e.g.

$$2BrICl^- \rightarrow IBr_2^- + ICl_2^-$$ [840]

$$2Br_2Cl^- \rightarrow Br_3^- + BrCl_2^-$$ [975]

As with solid polyhalide salts, dissociation in solution invariably takes the course which yields the smallest monohalide ion. Moreover, the stabilities of the polyhalide ions also follow a trend which is similar in outline to that established for the solid derivatives, though, not surprisingly, the intervention of a solvent appears to produce some differences of detail between the two stability sequences. The transition from water to a non-aqueous solvent is commonly attended by a dramatic increase in the stability constant of the polyhalide ion. Set against a value of 710 in aqueous solution, for example, the formation constant of the I_3^- ion at 25°C increases by several orders of magnitude when water is replaced as the solvent by 1,2-dichloroethane ($K_f = 10^7$), nitromethane ($K_f = 10^{6\cdot 7}$), acetone ($K_f = 10^{8\cdot 3}$) or acetonitrile ($K_f = 10^{6\cdot 6}$); likewise K_f for the Br_3^- ion is reported to take the following values in different solvents at 25°C: 16 (water), 177 (methanol) and $1\cdot 66 \times 10^4$ (bromotrichloromethane). The enhanced stability of polyhalide ions in non-aqueous media presumably accounts also for the claim laid to the existence of the Cl_5^- anion in acetonitrile solutions for which the ratio $[Cl_2]/[Cl^-]$ exceeds unity[976]; as yet, this ion is unknown in aqueous solution. There is evidence, too, that the *relative* stabilities of different polyhalide ions varies from solvent to solvent: for example, when nitromethane, acetone or acetonitrile is the solvent, the trihalide ions appear to *gain* in stability in the sequence $I_3^- < Br_3^- < Cl_3^-$, which is the reverse of that found in aqueous solution. Similarly, the stability of the ICl_4^- ion in solution depends on the nature of the solvent: in acetonitrile or 1,2-dichloroethane it dissociates to ICl_2^- and Cl_2 ($K_f = ca.$ 7000 or 5400, respectively), whereas in trichloroacetic acid it dissociates completely to ICl, Cl_2 and Cl^-[840]. Variations of stability of a particular anion have been correlated with the dielectric constant of the medium, although, since the formation of a polyhalide ion is not accompanied by the separation of charges, there is no *prima facie* reason for such a correlation. It seems more likely that the variations are attributable, at least in the first instance, to the protean effects of solvation and to the competition introduced by the acidic or basic functions of the solvent molecules.

Discussion[977]

The mode of decomposition of solid polyhalide salts and the dependence of the thermal stability on the size of the cation can be interpreted by thermodynamic reasoning analogous to that presented, in general, for other thermally unstable complex halides (Section 3, p. 1251) and, in particular, for solid derivatives of the hydrogen dihalide anions (Section 3, p. 1319). Thus, the factors which influence the mode of decomposition of a mixed polyhalide M^+XYZ^- may be established by considering the alternative decomposition paths and constructing the cycle depicted in Fig. 37.

975 R. P. Bell and M. Pring, *J. Chem. Soc.* (*A*) (1966) 1607.

976 J. C. Evans and G. Y.-S. Lo, *J. Chem. Phys.* **44** (1966) 3638.

977 A. G. Sharpe, *Halogen Chemistry* (ed. V. Gutmann), Vol. 1, p. 29. Academic Press (1967); D. A. Johnson, *Some Thermodynamic Aspects of Inorganic Chemistry*, p. 59, Cambridge (1968).

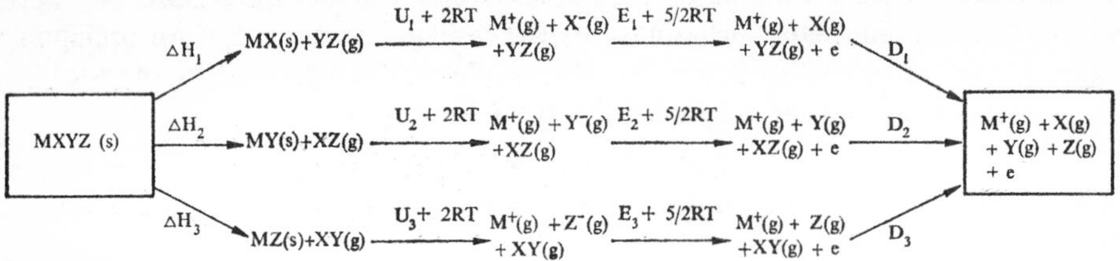

FIG. 37. Modes of decomposition of a mixed polyhalide M^+XYZ^-.

That the overall change in entropy varies but little for the different paths is evident from the fact that a reaction such as $KI(s)+\frac{1}{2}Cl_2(g) \rightarrow KCl(s)+\frac{1}{2}I_2(g)$ experiences an entropy change of less than 1 cal deg^{-1}. Variations of free energy are therefore determined almost exclusively by variations in the enthalpy terms ΔH_1, ΔH_2 and ΔH_3. Reference to the cycle then shows that it is the interplay of the differences (a) between the lattice energies of the monohalides (U_1, U_2 or U_3), (b) between the electron affinities of the halogen atoms (E_1, E_2 or E_3), and (c) between the dissociation energies of the diatomic halogen molecules (D_1, D_2 or D_3) that determines the course of the reaction. In fact, the contributions of (b) and (c) are mutually compensating, so that, overall, the determining factor in the decomposition is the lattice energy of the solid monohalide, which increases as the anion becomes smaller. Thus, for the complex CsIBrCl, the situation is as follows: (i) CsI(s)+BrCl(g): $D_1+E_1=$ 124 kcal and $U_1 = 144$ kcal; (ii) CsBr(s)+ICl(g): $D_2+E_2 = 129$ kcal and $U_2 = 151$ kcal; and (iii) CsCl(s)+IBr(g): $D_3+E_3 = 127$ kcal and $U_3 = 156$ kcal. Hence, the sequence $\Delta H_1 > \Delta H_2 > \Delta H_3$ is implied, the favoured products being CsCl(s)+IBr(g). Even if the decomposition products include the diatomic halogen in the condensed phase, the additional enthalpy terms are still transcended by the differences in the lattice energies of the monohalides; moreover, the effects of condensation are further mitigated by an unfavourable entropy change. Thus, in contrast with solid derivatives of the hydrogen dihalide anions, which decompose to give the hydrogen halide with the maximum bond energy, solid polyhalides decompose to give the monohalide having the smallest anion. The difference in behaviour between complexes of the two classes reflects the circumstance that changes in the nature of the halogen evoke substantial variations in the bond energies of the hydrogen halides but only modest variations in those of diatomic halogen or interhalogen molecules. An interpretation of the mode of dissociation of polyhalides in solution follows similar lines, except that the production of the smallest anion is then dictated by the solvation energies rather than the lattice energies of the different species.

The effect of cation size on the stability of solid derivatives of a particular polyhalide anion, e.g. ICl_2^-, may be discussed in terms of the following thermodynamic cycle:

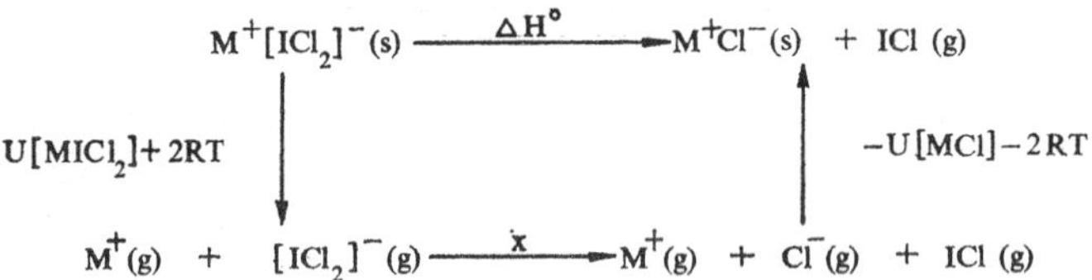

If variations in the nature of the cation M^+ are assumed, to a first approximation, not to affect the entropy term associated with the decomposition reaction, the pattern of thermodynamic stability is inevitably determined by the enthalpy change $\Delta H°$. According to the cycle,

$$\Delta H° = U[MICl_2] - U[MCl] + x$$

where x is a constant for a given polyhalide anion. Using Kapustinskii's approximation to express the lattice energies of the simple and complex halides in terms of the appropriate ionic radii r, we have (in kcal)

$$\Delta H° = 512\left[\frac{1}{r(M^+)+r(ICl_2^-)} - \frac{1}{r(M^+)+r(Cl^-)}\right]+x$$

Since the effective thermochemical radius of the $[ICl_2]^-$ anion must exceed that of the Cl^- anion, it follows that the contribution to $\Delta H°$ made by the lattice energies is intrinsically exothermic. However, the incentive to decomposition represented by this term clearly decreases as the radius of the cation expands. The ionic model therefore accounts successfully for the observation that solid derivatives of a particular polyhalide anion become progressively more stable as the size of the cation increases. The limitations of multivalent cations M^{n+} in stabilizing polyhalide anions depend, not only on the smaller size of such cations, but also on the heightened differences between the lattice energies of the simple and complex halides, since for the decomposition of 1 g-ion of ICl_2^- in the salt $M(ICl_2)_n$

$$\Delta H° = 256(n+1)\left[\frac{1}{r(M^{n+})+r(ICl_2^-)} - \frac{1}{r(M^{n+})+r(Cl^-)}\right]+x$$

Another contributory factor is probably the destabilizing effect of anion–anion contact, the impact of which is most severe on derivatives of multivalent cations; in fact, it is likely that the anions in complexes like $Ba(BrF_4)_2$ take the form of extended polymeric networks.

Although the effectiveness of large tetra-alkylammonium ions like NMe_4^+ in forming stable polyhalides appears to comply with these principles, the recently measured free energies of decomposition of some polyhalide salts $[NR_4]I_n$ show that the corresponding entropy changes are both opposite in sign and larger in magnitude than those observed for simpler cations like Rb^+, Cs^+ and NH_4^+ [974]. According to the results of Table 97, the disordering which accompanies the formation of a solid tetra-alkylammonium polyiodide makes a major contribution to the stability of the complex. On the basis of these and other thermodynamic parameters, the following enthalpy terms (at 25°C) have been deduced: $\Delta H° = -24{\cdot}_0$ kcal for the reaction $I_2(g)+I^-(g) \rightarrow I_3^-(g)$; $\Delta H_f°[I_3^-(g)] = -56{\cdot}_4$ kcal; and $\Delta H° = -44{\cdot}_0$ kcal for the hydration of the gaseous I_3^- ion[974]. Although strictly a net enthalpy change, the first of these terms is analogous to the so-called "hydrogen bond energy" of an HX_2^- anion[977], testifying indirectly to the strength of the bonds in the isolated tri-iodide anion.

2. **Structures of solid polyhalides**[834,835,840,978]. Our present knowledge of the structures of polyhalide anions has been gained pre-eminently from X-ray diffraction studies of crystalline salts, though these have lately been augmented by recourse to neutron diffraction and vibrational, nqr and Mössbauer spectroscopy. Summarized in Table 99 are the structural details presently available for anionic polyhalide derivatives; the geometries of the anions are also depicted in Fig. 38.

[978] R. E. Rundle, *Acta Cryst.* **14** (1961) 585.

1. Trihalide species

(a) 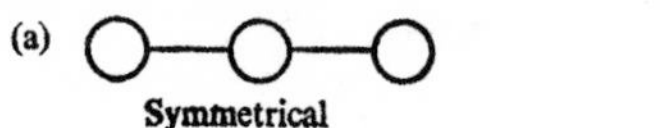

Symmetrical

(b) 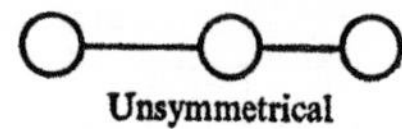

Unsymmetrical

2. Pentahalide species

(a)

2·82 Å
3·17 Å
175°
95°
3·17 Å
175°
2·82 Å

I_5^- in Me_4NI_5

(b) 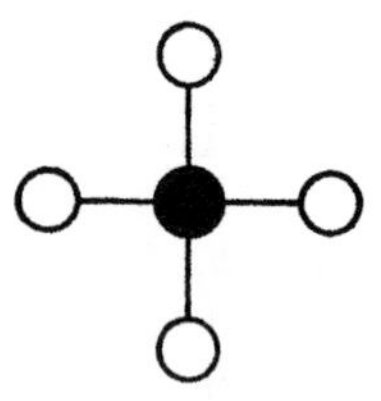

BrF_4^- in $KBrF_4$ and
ICl_4^- in $KICl_4,H_2O$

3. Heptahalide species

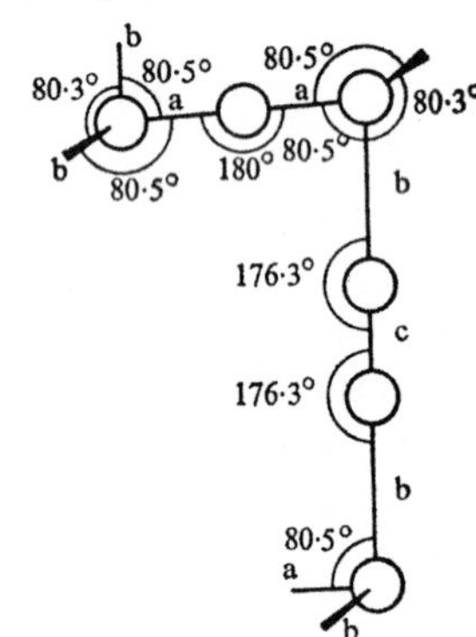

a=2.904 Å
b=3.435 Å
c=2.735 Å

Structure of anionic network in Et_4NI_7: I_3^- ions and I_2 molecules are linked to form a three-dimensional array

4. Enneahalide species

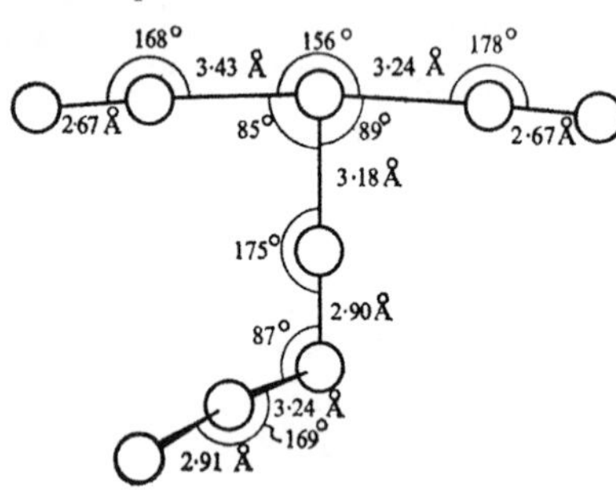

Structure of anionic network in Me_4NI_9

5. Other species

2·85 Å
177°
3·00 Å
175°
2·80 Å
3·42 Å
3·42 Å
81°
175°
3·00 Å
177°
2·85 Å

Structure of the I_8^{2-} anion in Cs_2I_8

FIG. 38. Structures of anionic polyhalide aggregates.

Table 99. Structural Details for Anionic Polyhalide Complexes

Polyhalide anion	Solid phase: Cation	Shape of anion	Bond lengths (Å)	Bond length in corresponding diatomic molecule (Å)	Bond angles	Spectroscopic studies
Triatomic Anions						
Cl_3^-	NEt_4^+, $NPr^n_4^+$ or $NBu^n_4^+$	Linear, symmetrical (V)	—	—	—	V[1]
Br_3^-	Cs^+	Nearly linear, unsymmetrical (X)[2]	Br–Br = 2·440 Br–Br = 2·698	2·281 2·281	∠Br–Br–Br = 177·5°	V[3,4] NQR[5]
	PBr_4^+	Nearly linear, unsymmetrical (X)[6]	Br–Br = 2·39 Br–Br = 2·91	2·281 2·281	∠Br–Br–Br = 177·3°	
	Me_3NH^+ in $[Me_3NH^+]_2Br^-Br_3^-$	Nearly linear, symmetrical (X)[7]	Br–Br = 2·53, 2·54	2·281	∠Br–Br–Br = 171°	
I_3^-	Cs^+	Nearly linear, unsymmetrical (X)[7]	I–I = 2·83 I–I = 3·03	2·666 2·666	∠I–I–I = 176°	V[8–11] NQR[5,12] M[13]
	NH_4^+	Nearly linear, unsymmetrical (X)[7]	I–I = 2·82 I–I = 3·10	2·666 2·666	∠I–I–I = 177°	
	$AsPh_4^+$	Nearly linear, symmetrical (X)[7]	I–I = 2·90	2·666	∠I–I–I = 176°	
	$[PhCONH_2]_2H^+$ in $2PhCONH_2,HI_3$	Nearly linear, slightly unsymmetrical I_3^- ions linked in chains (X)[14]	I–I = 2·910 I–I = 2·951	2·666 2·666	∠I–I–I = 176·8°	
	NEt_4^+	*Modification I*: contains two independent centrosymmetric I_3^- ions (X)[15]	(a) I–I = 2·928 (b) I–I = 2·943	2·666 2·666	∠I–I–I = 180°	
		Modification II: contains two independent nearly linear but unsymmetrical I_3^- ions (X)[15]	(a) I–I = 2·912, 2·961 (b) I–I = 2·892, 2·981	2·666 2·666	∠I–I–I = 179·5° ∠I–I–I = 177·7°	
	K^+ in $KI,KI_3,$-$6MeCONHMe$	The structure contains I_3^- ions which appear to be linear and symmetrical, alternating with I^- ions along the trigonal axis of the unit cell (X)[16]	I–I = 2·945	2·666	∠I–I–I = 180°	

Table 99 (*cont.*)

Polyhalide anion	SOLID PHASE					Spectroscopic studies
	Cation	Shape of anion	Bond lengths (Å)	Bond length in corresponding diatomic molecule (Å)	Bond angles	
ClF_2^-	NO^+	Linear, symmetrical $[F–Cl–F]^-$ ion (V)[17]	—	—	—	V[17]
	Rb^+ and Cs^+	Unsymmetrical $[F–Cl–F]^-$ ion (V)[17]	—	—	—	
$BrCl_2^-$	Cs^+, NMe_4^+, NEt_4^+, $NPr^n_4{}^+$ and $NBu^n_4{}^+$	Linear, symmetrical $[Cl–Br–Cl]^-$ ion (V)[3,18]	—	—	—	V[3,18]
ICl_2^-	NMe_4^+	Linear, symmetrical $[Cl–I–Cl]^-$ ion (X)[19]	I–Cl = 2·55	2·321	∠Cl–I–Cl = 180°	V[3,8] NQR[5]
	PCl_4^+	Linear, symmetrical $[Cl–I–Cl]^-$ ion (X)[7]	I–Cl = 2·36 though probably in error	2·321	∠Cl–I–Cl = 180°	
	piperazinium in $[C_4H_{12}N_2][ICl_2]_2$	Linear, unsymmetrical $[Cl–I–Cl]^-$ ion (X)[20]	I–Cl = 2·47 I–Cl = 2·69	2·321 2·321	∠Cl–I–Cl = 180°	
	triethylenediammonium in $[HN(C_2H_4)_3NH][ICl_2]_2$	Structure contains two non-equivalent $[Cl–I–Cl]^-$ ions, both linear and unsymmetrical (X)[21]	(a) I–Cl = 2·54, 2·67 (b) I–Cl = 2·53, 2·63	2·321 2·321	∠Cl–I–Cl = 180° ∠Cl–I–Cl = 180°	
IBr_2^-	Cs^+	Nearly linear, unsymmetrical $[Br–I–Br]^-$ ion (X)[22]	I–Br = 2·62 I–Br = 2·78	2·485 2·485	∠Br–I–Br = 178·0°	V[8,9] NQR[12]
I_2Br^-	Cs^+	Nearly linear $[I–I–Br]^-$ ion (X)[23]	I–I = 2·777 I–Br = 2·906	2·666 2·485	∠I–I–Br = 178·0°	V[8,10]
$IBrCl^-$	NH_4^+	Nearly linear $[Cl–I–Br]^-$ ions statistically distributed in the crystal (studied at 140°K, X)[24]	I–Cl = ~2·91 I–Br = ~2·51	2·321 2·485	∠Br–I–Cl = 179°	V[8]
Penta-atomic Anions						
Cl_5^-	—	—	—	—	—	V[1]
Br_5^-	—	—	—	—	—	V[4]

I_5^-	NMe_4^+	Nearly planar V-shaped I_5^- anions (X)[7] (see Fig. 38)	I_{apex}–I = 3·17 $I_{terminal}$–I = 2·82	2·666 2·666	∠I–I_{apex}–I = 95° ∠I_{ap}–I–I_{term} = 175°	V[10]
ClF_4^-	Rb^+ and Cs^+	Square-planar ClF_4^- anion with D_{4h} symmetry (V)[25]	—	—	—	V[25]
BrF_4^-	K^+	Square-planar BrF_4^- anion with D_{4h} symmetry (N)[26]	Br–F = 1·89	1·756	∠F–Br–F(adj) = 90·9, 89·1 ± 2·0°	V[27–29]
IF_4^-	Cs^+	C_{2v} symmetry implied for the anion (V)[29]	—	—	—	V[29]
ICl_4^-	K^+ in $KICl_4,H_2O$	Approximately square-planar ICl_4^- anion (X)[30]	I–Cl = 2·42, 2·47, 2·53 and 2·60	2·321	∠Cl–I–Cl(adj) = 90·6, 90·7, 89·2 and 89·5°	V[3] NQR[5] M[31]
HEPTA-ATOMIC ANIONS						
I_7^-	NEt_4^+	Structure contains centrosymmetric I_3^- anions linked to I_2 molecules in a 3-dimensional network (X)[7] (see Fig. 38)	I–I = 2·904 (I_3^-), 2·735 (I_2) and 3·435 for shortest $I_2 \cdots I_3^-$ distance	2·666	∠I–I–I = 180° (I_3^-), 80·3, 80·5 and 176·3°	V[10]
BrF_6^-	K^+, Rb^+ or Cs^+	D_{3d} symmetry implied for the BrF_6^- anion (V)[32]; rhombohedral crystal system (X)	—	—	—	V[32]
IF_6^-	K^+, Cs^+, NMe_4^+ and NEt_4^+	Non-octahedral symmetry implied for the IF_6^- anion (V and M)[33,34]	—	—	—	V[33] M[34]
ENNEA-ATOMIC ANIONS						
I_9^-	NMe_4^+	The structure can be considered to arise from the association of I_7^- units with additional I_2 molecules (X)[7] (see Fig. 38)	I–I = 2·67, 2·90, 3·18, 3·24 and 3·43	2·666	∠I–I–I = 85, 87 and 89°; 156, 168, 169, 175 and 178°	V[10]
OTHER SPECIES						
I_8^{2-}	Cs^+	Planar Z-shaped I_8^{2-} anions present, having two arms consisting of asymmetric I_3^- anions linked by an I_2 molecule (X)[7]	I–I = 2·85 and 3·00 (I_3^-) I–I = 2·80 (I_2) and 3·42 for shortest $I_2 \cdot\cdot I_3^-$ distance	2·666 2·666	∠I–I–I = 177° (I_3^-) ∠I–I–$I^- \cdot\cdot$I = 81° ∠I··I–I = 175°	NQR[12]

TABLE 99 (*cont.*)

M = Mössbauer spectra; N = neutron diffraction; NQR = nqr spectra; V = vibrational spectra; X = X-ray diffraction.

[1] J. C. Evans and G. Y.-S. Lo, *J. Chem. Phys.* **44** (1966) 3638.
[2] G. L. Breneman and R. D. Willett, *Acta Cryst.* **B25** (1969) 1073.
[3] W. B. Person, G. R. Anderson, J. N. Fordemwalt, H. Stammreich and R. Forneris, *J. Chem. Phys.* **35** (1961) 908.
[4] J. C. Evans and G. Y.-S. Lo, *Inorg. Chem.* **6** (1967) 1483.
[5] E. A. C. Lucken, *Nuclear Quadrupole Coupling Constants*, p. 289. Academic Press (1969).
[6] G. L. Breneman and R. D. Willett, *Acta Cryst.* **23** (1967) 467.
[7] E. H. Wiebenga, E. E. Havinga and K. H. Boswijk, *Adv. Inorg. Chem. Radiochem.* **3** (1961) 148.
[8] A. G. Maki and R. Forneris, *Spectrochim. Acta*, **23A** (1967) 867.
[9] G. C. Hayward and P. J. Hendra, *Spectrochim. Acta*, **23A** (1967) 2309.
[10] F. W. Parrett and N. J. Taylor, *J. Inorg. Nuclear Chem.* **32** (1970) 2458.
[11] S. G. W. Ginn and J. L. Wood, *Chem. Comm.* (1965) 262.
[12] G. A. Bowmaker and S. Hacobian, *Austral. J. Chem.* **21** (1968) 551.
[13] B. S. Ehrlich and M. Kaplan, *J. Chem. Phys.* **51** (1969) 603.
[14] J. M. Reddy, K. Knox and M. B. Robin, *J. Chem. Phys.* **40** (1964) 1082.
[15] T. Migchelsen and A. Vos, *Acta Cryst.* **23** (1967) 796.
[16] K. Toman, J. Honzl and J. Ječný, *Acta Cryst.* **18** (1965) 673.
[17] K. O. Christe, W. Sawodny and J. P. Guertin, *Inorg. Chem.* **6** (1967) 1159.
[18] J. C. Evans and G. Y.-S. Lo, *J. Chem. Phys.* **44** (1966) 4356.
[19] G. J. Visser and A. Vos, *Acta Cryst.* **17** (1964) 1336.
[20] Chr. Rømming, *Acta Chem. Scand.* **12** (1958) 668.
[21] S. Gran and Chr. Rømming, *Acta Chem. Scand.* **22** (1968) 1686.
[22] J. E. Davies and E. K. Nunn, *Chem. Comm.* (1969) 1374.
[23] G. B. Carpenter, *Acta Cryst.* **20** (1966) 330.
[24] T. Migchelsen and A. Vos, *Acta Cryst.* **22** (1967) 812.
[25] K. O. Christe and W. Sawodny, *Z. anorg. Chem.* **374** (1970) 306 and references cited therein.
[26] A. J. Edwards and G. R. Jones, *J. Chem. Soc.* (*A*) (1969) 1936.
[27] K. O. Christe and C. J. Schack, *Inorg. Chem.* **9** (1970) 1852 and references cited therein.
[28] T. Surles, H. H. Hyman, L. A. Quarterman and A. I. Popov, *Inorg. Chem.* **9** (1970) 2726; *ibid.* **10** (1971) 913.
[29] J. Shamir and I. Yaroslavsky, *Israel J. Chem.* **7** (1969) 495.
[30] R. J. Elema, J. L. de Boer and A. Vos, *Acta Cryst.* **16** (1963) 243.
[31] D. W. Hafemeister, *The Mössbauer Effect and its Application in Chemistry*, p. 126. Advances in Chemistry Series No. 68, American Chemical Society (1967).
[32] R. Bougon, P. Charpin and J. Soriano, *Compt. rend.* **272C** (1971) 565.
[33] K. O. Christe, J. P. Guertin and W. Sawodny, *Inorg. Chem.* **7** (1968) 626; S. P. Beaton, D. W. A. Sharp, A. J. Perkins, I. Sheft, H. H. Hyman and K. Christe, *ibid.* p. 2174; H. Klamm, H. Meinert, P. Reich and K. Witke, *Z. Chem.* **8** (1968) 393, 469.
[34] S. Bukshpan, J. Soriano and J. Shamir, *Chem. Phys. Letters*, **4** (1969) 241.

On the basis of this information, polyhalide anions are evidently characterized by the following general features:

(a) Heteronuclear anions, e.g. ICl_2^-, $IBrCl^-$ or ICl_4^-, invariably adopt the form in which the halogen of highest atomic number is the central "polyvalent" atom.

(b) Interatomic distances are always greater, by some tenths of an ångström, than those in the corresponding diatomic halogen molecules. Apparent exceptions to this rule have now been diagnosed as the outcome of inaccuracies in earlier crystallographic studies of polyhalide salts containing the anions ICl_2^- [979], ICl_4^- [943] and $BrICl^-$ [980]. Though still substantially less than twice the van der Waals' radius of iodine, some of the $I \cdots I$ distances in the polyiodide salts NEt_4I_7, NMe_4I_9 and Cs_2I_8 are not much shorter than those *between* iodine molecules in crystalline iodine.

(c) Bond angles close to 90° or 180° predominate in the anionic aggregates, though values as low as 80° or 156° have been encountered.

(d) A trihalide ion varies its symmetry and dimensions according to the demands of its environment, the individual bond distances changing in a systematic way with the overall length of the ion[978,981]. This stereochemical flexibility is probably inherent, to some degree, in all polyhalide anions.

Trihalides

All of the trihalide anions which have been structurally characterized in the crystalline phase comply with the general rules (a)–(c): the anions are invariably linear, or nearly linear, with bond angles at the central atom in the range 171–180°. The characteristic (d) has been shown to apply to the anions Br_3^-, I_3^- and ICl_2^-, each of which has been studied in three or more salts. Trihalide ions may be either symmetrical or unsymmetrical in form. With the I_2Br^- and $BrICl^-$ anions, found, respectively, in the salts CsI_2Br[982] and NH_4BrICl[980], the dissymmetry is a necessary consequence of the composition of the ion or of the operation of generalization (a). However, other anions, e.g. Br_3^-, I_3^- and ICl_2^-, have the capacity to assume either a symmetrical or an unsymmetrical form, depending on the precise nature of the salt in which they occur. Thus, the Br_3^- anion appears to be symmetrical in the mixed salt $[Me_3NH^+]_2Br^-[Br_3]^-$ [835] but unsymmetrical in $CsBr_3$[981] and $[PBr_4]^+[Br_3]^-$ [973]; the I_3^- anion appears to be symmetrical in crystalline $[AsPh_4]^+[I_3]^-$ [835], $[NEt_4]I_7$[835], the complex $KI,KI_3,6MeCONHMe$[983] and modification I of $[NEt_4]^+[I_3]^-$ [984], but unsymmetrical in CsI_3[835], NH_4I_3[835], Cs_2I_8[835], the complex HI_3,2benzamide[985] and modification II of $[NEt_4]^+[I_3]^-$ [984]; whereas the ICl_2^- anion is reported to be symmetrical in $[NMe_4]^+[ICl_2]^-$ [979] but unsymmetrical in the piperazinium[986] and triethylenediammonium[987] salts. The IBr_2^- anion is unsymmetrical in the one salt whose structure has been determined, viz. $CsIBr_2$[988]. In some cases, e.g. that of the complex $KI,KI_3,6MeCONHMe$, it is impossible to rule out the circumstance that the appearance of centrosymmetry is determined by statistical disordering of the trihalide anions which are

979 G. J. Visser and A. Vos, *Acta Cryst.* **17** (1964) 1336.
980 T. Migchelsen and A. Vos, *Acta Cryst.* **22** (1967) 812.
981 G. L. Breneman and R. D. Willett, *Acta Cryst.* **B25** (1969) 1073.
982 G. B. Carpenter, *Acta Cryst.* **20** (1966) 330.
983 K. Toman, J. Honzl and J. Ječný, *Acta Cryst.* **18** (1965) 673.
984 T. Migchelsen and A. Vos, *Acta Cryst.* **23** (1967) 796
985 J. M. Reddy, K. Knox and M. B. Robin, *J. Chem. Phys.* **40** (1964) 1082.
986 Chr. Rømming, *Acta Chem. Scand.* **12** (1958) 668.
987 S. Gran and Chr. Rømming, *Acta Chem. Scand.* **22** (1968) 1686.
988 J. E. Davies and E. K. Nunn, *Chem. Comm.* (1969) 1374.

individually unsymmetrical. The general pattern of behaviour is that centrosymmetric trihalide anions are favoured by the presence of bulky cations such as NR_4^+ or $AsPh_4^+$, though the recognition of two structural modifications of the salt $[NEt_4][I_3]$[984], one containing symmetrical and the other unsymmetrical I_3^- anions, implies that the size of the cation is not the only factor critical to the shape of the anion. In general, the variations of structure are attributable to the electrostatic field developed by ions in the neighbourhood of a given trihalide anion[978,984,989]. Molecular-orbital calculations which take account of this crystal field for the I_3^- ion[989] imply that one equilibrium position is available to the central iodine atom when the distance between the two terminal iodine atoms D becomes equal to or less than a certain critical value D_c, whereas there are two such positions for this atom when D exceeds D_c (see Fig. 39). When D is large, the $I_2 \cdots I^-$ interaction is essentially that of an

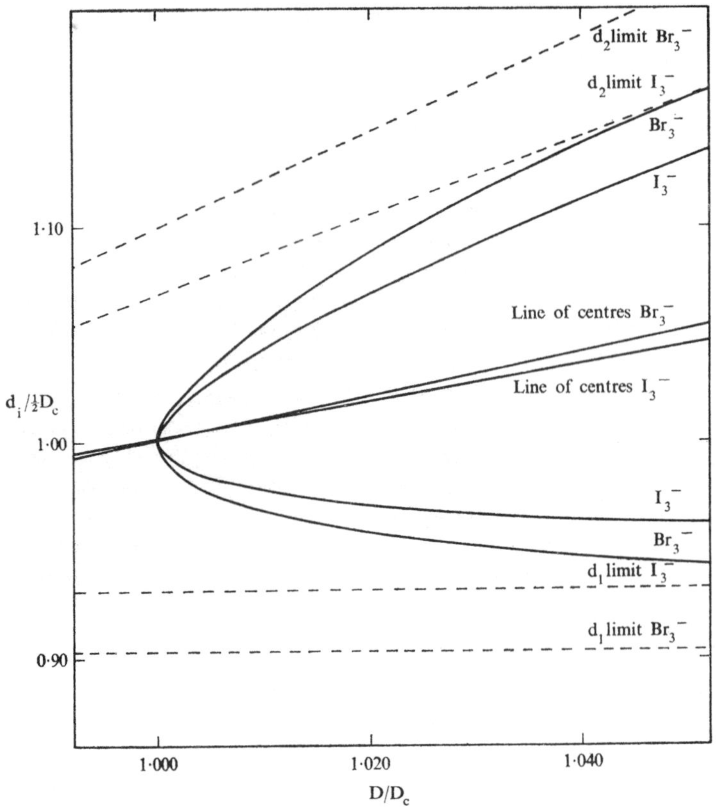

FIG. 39. Comparison of the configurations of the Br_3^- and I_3^- anions (after ref. 981). D = total length of the X_3^- ion; D_c = critical value of D where the ion becomes symmetrical; $d_i = X_{central}-X_{terminal}$ distance with i = 1 for the short bond and i = 2 for the long bond.

ion with a polarizable molecule, but as D approaches D_c, orbital-overlap and covalent bonding become increasingly important. These general features disclose a striking resemblance between the triatomic systems X_2–X^- and H_2–H[978,981], though with the major difference

989 R. D. Brown and E. K. Nunn, *Austral. J. Chem.* **19** (1966) 1567.

that the magnitude of *D* for the former is acutely responsive to the influence of neighbouring ions. In a salt containing relatively small counter-ions, e.g. Cs^+, the crystal field experienced by the I_3^- ions is unsymmetrical, with the result that one of the two equilibrium positions open to the central iodine atom is preferentially stabilized; consequently the I_3^- assumes an unsymmetrical form. By contrast, the relatively weak, symmetrical crystal field created by an environment of large cations, e.g. $AsPh_4^+$, produces minimal perturbation of the short, centrosymmetric unit which characterizes the isolated I_3^- ion. Somewhat less plausible is the suggestion of Slater[990] that *D* for the isolated I_3^- ion is comparable with the largest values observed in crystals, but that the "pressure" of certain environments forces the ion to become shorter and, ultimately, centrosymmetric.

Unfortunately, no clearcut decision is possible concerning the shape of trihalide ions in solution. Whereas the infrared spectrum of the salt $[NBu_4][I_3]$ in benzene, nitrobenzene and pyridine is consistent with the presence of centrosymmetric I_3^- ions[991], the Raman spectra of solutions containing Br_3^-, I_3^- or IBr_2^- anions do not comply with such an assumption[992]. In view of these ambiguities, it is impossible, as yet, to assess the effect of the anisotropic environment which the ions are likely to experience in solution.

The kinship of the trihalide ions with charge-transfer complexes formed between diatomic halogen or interhalogen molecules and neutral donor species is underlined by reference to the "effective radii" R_1 and R_2 of the central iodine atom in units of the type X–I–Y (X = halogen; Y = halide ion or donor atom of neutral molecule), these being evaluated by subtracting the accepted covalent radii of the ligand atoms X and Y from the measured interatomic distances[993]. To a good approximation, a linear relationship is found to connect R_1 and R_2, viz. $R_2 = -1{\cdot}97R_1 + 4{\cdot}59$ Å, irrespective of whether X–I–Y corresponds to a trihalide anion or to a neutral complex. Quantitative expression is thus given to the generalization that the strengths of the bonds formed by the central atom are complementary, the intermolecular I–Y bond gaining in strength at the expense of the intramolecular I–X bond until, as with the centrosymmetric I_3^- anion, the distinction between the two bonds is lost.

Higher Polyhalides

The structures of pentahalide ions and of more complicated aggregates do not show the same regularity as those of the trihalide anions. The three pentahalide anions for which definitive results are available adopt one of two structures: (i) a square-planar array, or (ii) a nearly planar V-shaped unit (see Fig. 38). Following earlier vicissitudes, the square-planar structure of the BrF_4^- anion in the salt $KBrF_4$ has now been established by neutron diffraction[937], while refinement of the crystal structure of the salt $KICl_4,H_2O$[943] reveals a similar, though somewhat distorted, structure for the ICl_4^- ion. By contrast, the I_5^- ion in the complex $[NMe_4][I_5]$ belongs to the second category: each arm of the nearly planar V-shaped unit corresponds to an unsymmetrical I_3^- anion, the apex angle being 95°; the ion may thus be formulated as $[I(I_2)_2]^-$ with two iodine molecules coordinated to a single iodide ion. On the premises of the vibrational spectra, the BrF_4^-[934–936] and ICl_4^-[994] ions retain their square-planar geometry in solution, and are isostructural with the

990 J. C. Slater, *Acta Cryst.* **12** (1959) 197.

991 S. G. W. Ginn and J. L. Wood, *Chem. Comm.* (1965) 262.

992 A. G. Maki and R. Forneris, *Spectrochim. Acta*, **23A** (1967) 867.

993 O. Hassel and Chr. Rømming, *Acta Chem. Scand.* **21** (1967) 2659.

994 W. B. Person, G. R. Anderson, J. N. Fordemwalt, H. Stammreich and R. Forneris, *J. Chem. Phys.* **35** (1961) 908.

ClF_4^- ion in its rubidium or caesium salt[932]. On the other hand, a V-shaped skeleton analogous to that of I_5^- may well be assumed by the species Cl_5^- [976], Br_5^- [995], $I_2Cl_3^-$ [996] and $I_2Cl_2Br^-$[996], a view consistent with the incompletely defined vibrational spectra attributed to the anions, but as yet unsubstantiated by more compelling evidence.

The complex $[NEt_4]I_7$ is the only representative of the group of heptahalides which has been the subject of detailed structural analysis[835,840]. The crystal of this compound contains, not discrete I_7^- ions, but a three-dimensional network composed of centrosymmetric I_3^- ions and I_2 molecules; adjacent to each I_3^- ion are four I_2 molecules, and to each I_2 molecule are two I_3^- ions (Fig. 38). Hence, the complex is more realistically formulated as $[NEt_4]^+[I_3(I_2)_2]^-$. At 3·435 Å the closest approach of the I_3^- and I_2 units differs but little from the shortest intermolecular distance (3·50 Å) in crystalline iodine; that it is substantially less than twice the van der Waals' radius of iodine (4·30 Å) indicates, nonetheless, comparatively strong interaction between the I_3^- and I_2 units. The structures adopted by the heptahalide anions BrF_6^- and IF_6^- are of especial interest in view of their relationship to xenon hexafluoride[997], with the implication of a valence shell for the central atom having a complement of 14 electrons. Vibrational[946b,949] and ^{129}I Mössbauer[950] spectra signify non-octahedral structures for both anions whether in solution or in the crystalline phase; it has not been possible to specify the precise geometry, though the vibrational spectra of the crystalline complexes $MBrF_6$ (M = K, Rb or Cs) imply that the octahedron of fluorine atoms is distorted about a threefold axis to impart D_{3d} symmetry to the environment of the bromine atom[946b].

The crystal structure of the salt $[NMe_4]I_9$ reveals aggregates of iodine atoms which may be considered as I_9^- anions, though one of the interatomic distances (3·43 Å) is scarcely shorter than the closest approach between iodine atoms of different I_9^- units (3·49 Å)[835,840]. It is perhaps more realistic to consider the structure in terms of non-planar Z-shaped I_7^- ions each linked by relatively strong interaction with an additional I_2 molecule and with other I_7^- entities. A more distinct anionic unit is observed in the crystal structure of the salt formulated on the strength of its observed diamagnetism as Cs_2I_8[835,840]. Here it is possible to distinguish planar Z-shaped I_8^{2-} aggregates composed of two unsymmetrical I_3^- ions linked by a common I_2 molecule. To be represented therefore as $[I_2(I_3)_2]^{2-}$, each aggregate involves interaction between the I_3^- ions and I_2 molecule (I · · · I distance 3·42 Å) which is substantially stronger than that between adjacent I_8^{2-} aggregates (shortest I · · · I approach 3·88 Å).

3. **Spectroscopic properties.** The thermodynamic and structural investigations outlined in the preceding sections have been augmented by measurements relating to the vibrational[992,994], electronic[840,998–1000], nqr[917,1001] and ^{127}I or ^{129}I Mössbauer[950,1002] spectra of polyhalide species.

[995] J. C. Evans and G. Y.-S. Lo, *Inorg. Chem.* **6** (1967) 1483.
[996] Y. Yagi and A. I. Popov, *J. Inorg. Nuclear Chem.* **29** (1967) 2223.
[997] R. M. Gavin, jun. and L. S. Bartell, *J. Chem. Phys.* **48** (1968) 2460, 2466.
[998] M. B. Robin, *J. Chem. Phys.* **40** (1964) 3369.
[999] Chr. K. Jørgensen, *Halogen Chemistry* (ed. V. Gutmann), Vol. 1, p. 360. Academic Press (1967).
[1000] D. Meyerstein and A. Treinin, *Trans. Faraday Soc.* **59** (1963) 1114.
[1001] G. A. Bowmaker and S. Hacobian, *Austral. J. Chem.* **21** (1968) 551.
[1002] D. W. Hafemeister, *Advances in Chemistry Series*, **68** (1967) 126; G. J. Perlow and M. R. Perlow, *J. Chem. Phys.* **45** (1966) 2193.

Thus, the infrared and Raman spectra of polyhalides have been employed, not only to investigate the structure of the anions in different environments, but also to determine the properties of the potential field experienced by the atoms within the anion. Valence force constants calculated from the vibrational frequencies of polyhalide anions known or presumed to be symmetrical are listed in Table 100; particular interest attaches to the

TABLE 100. FORCE CONSTANTS FOR SYMMETRICAL POLYHALIDE ANIONS

Anion	Principal stretching force constant, f_r(YXY) (mdyne/Å)	Interaction constant between collinear bonds, f_{rr}(YXY) (mdyne/Å)	Interaction constant between perpendicular bonds, f_{rr}'(YXY) (mdyne/Å)	Bending force constant for linear YXY unit, f_θ/r^2 (mdyne/Å)	$\frac{f_{rr}(\text{YXY})}{f_r(\text{YXY})}$	$\frac{f_r(\text{YXY})}{f_r(\text{XY})}$ [f_r(XY) = stretching force constant of XY molecule]
Cl_3^- [a]	0·96	0·55	—	—	0·57	0·30
Br_3^- [b]	0·91	0·32	—	—	0·36	0·38
I_3^- [c]	0·71	0·23	—	—	0·32	0·42
ClF_2^- [d]	2·35	0·17	—	—	0·07	0·55
$BrCl_2^-$ [e]	1·02	0·47	—	—	0·46	0·36
ICl_2^- [b,f]	1·00	0·36	—	0·13	0·36	0·46
IBr_2^- [f,g]	0·91	0·30	—	0·10	0·33	0·42
ClF_4^- [h]	2·11	0·23	0·29	—	0·11	0·48
BrF_4^- [h]	2·38	0·20	0·28	—	0·08	0·58
ICl_4^- [b]	1·25	0·33	0·08	—	0·26	0·53

[a] J. C. Evans and G. Y.-S. Lo, *J. Chem. Phys.* **44** (1966) 3638.
[b] W. B. Person, G. R. Anderson, J. N. Fordemwalt, H. Stammreich and R. Forneris, *J. Chem. Phys.* **35** (1961) 908.
[c] S. G. W. Ginn and J. L. Wood, *Chem. Comm.* (1965) 262.
[d] K. O. Christe, W. Sawodny and J. P. Guertin, *Inorg. Chem.* **6** (1967) 1159.
[e] J. C. Evans and G. Y.-S. Lo, *J. Chem. Phys.* **44** (1966) 4356.
[f] A. G. Maki and R. Forneris, *Spectrochim. Acta*, **23A** (1967) 867.
[g] G. C. Hayward and P. J. Hendra, *Spectrochim. Acta*, **23A** (1967) 2309.
[h] K. O. Christe and C. J. Schack, *Inorg. Chem.* **9** (1970) 1852.

comparison of the principal stretching force constant f_r(YXY) of each anion with the corresponding interaction constant between collinear bonds f_{rr}(YXY) and with the stretching force constant of the corresponding diatomic halogen molecule f_r(XY). Analysis reveals, firstly, that the force constant f_r(YXY) is only 30–58% of the magnitude of f_r(XY), and, secondly, that the interaction constant f_{rr}(YXY) commonly takes unusually large values in proportion to f_r(YXY); as the ratio $f_{rr}(\text{YXY})/f_r(\text{YXY})$ increases over the range 0·07–0·57, so $f_r(\text{YXY})/f_r(\text{XY})$ diminishes. These findings confirm the analogy between the polyhalide and hydrogen dihalide anions, both being characterized by unusually weak bonds; further, in that the term $[f_r(\text{YXY})-f_{rr}(\text{YXY})]$ defines the force constant for vibration of the central atom within the one-dimensional potential well bounded by the ligands, the curvature of the well is singularly flat at the equilibrium position of the central atom. That the central atom of such an anion is thus relatively loosely confined to its equilibrium position accounts for the susceptibility to structural changes of the anion under the influence of different crystal or solvation forces. The marked inferiority of the valence force constants of the polyhalide anions, compared with corresponding diatomic halogen molecules or with polyhalogen cations such as ClF_2^+, supports the molecular-orbital scheme (see Fig. 4) first developed by Pimentel and Hach and Rundle[1003], in which only the valence *p*-orbitals of the halogen atoms are presumed to subscribe to the bonding. The results may then be reconciled with the three-centre, four-electron character of the bonds in polyhalide anions; approximate bond orders of 0·5 or less may be divined *via* the ratio $f_r(YXY)/f_r(XY)$.

Variously exploited for the measurement of stability constants, the optical spectra of polyhalide anions in aqueous or non-aqueous media have been described by several investigators[840,998–1000], but with the interpretation of the details little headway has been made. With the assumption of valence-shell molecular orbitals composed exclusively of atomic *p*-functions, the molecular orbital scheme for a centrosymmetric trihalide anion takes the form depicted in Fig. 40, whence the two transitions of lowest energy are seen to be

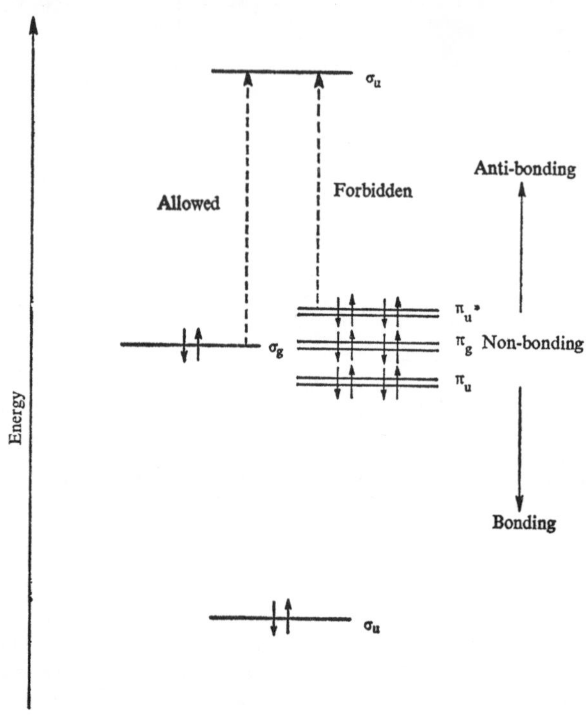

FIG. 40. Molecular-orbital scheme and electronic transitions for a linear trihalide anion.

$\sigma_u^* \leftarrow \pi_u^*$ and $\sigma_u^* \leftarrow \sigma_g$. As with the parent halogen molecules, the first of these is formally forbidden in a centrosymmetric unit; that it appears relatively weakly in absorption at 23,000–30,000 cm^{-1} [1000] may be due partly to spin–orbit coupling phenomena and partly to the distortion impressed upon the ion by crystal or solvent interactions. The transition occurs, in effect, within the *p*-shell of the central atom, and the associated absorption has some of the characteristics of a ligand-field band; the following frequencies (in cm^{-1}) show signs of the evolution of a spectrochemical series: I_3^-, 23,000; IBr_2^-, 27,000; $IBrCl^-$,

[1003] G. C. Pimentel, *J. Chem. Phys.* **19** (1951) 446; R. J. Hach and R. E. Rundle, *J. Amer. Chem. Soc.* **73** (1951) 4321.

28,100; ICl_2^-, 29,800. On the other hand, the transition $\sigma_u^* \leftarrow \sigma_g$, which is allowed irrespective of the geometry of the trihalide anion, entails electron-transfer from one of the terminal atoms to the central halogen atom:

$$[Y—X—Y]^- \xrightarrow{h\nu} [Y—X^- \quad Y]$$

a process plainly akin to the charge-transfer which characterizes neutral complexes of the diatomic halogen molecules. Analogies with the spectra of such complexes, combined with thermodynamic arguments[1000], favour the assignment to this transition of the intense absorption bands observed at 28,000–44,000 cm^{-1}. Spin–orbit splitting is then presumed to account for the intense doublet feature (at 28,300 and 34,800 cm^{-1}), to which the aqueous I_3^- anion owes its red colour. Alternatively the bands may have their origin in the transitions $\sigma_u^* \leftarrow \pi_u^*$ and $\sigma_u^* \leftarrow \pi_g$, having probabilities much enhanced by the mixing of σ- and π-orbitals which would accompany the departure from linearity of the I_3^- skeleton[998]. If the latter interpretation is extended to a one-dimensional chain of I_3^- ions, subject to exciton-like coupling, a semi-quantitative explanation is found for the blue colour of the starch–iodine and related complexes[998].

The quadrupole coupling constants derived from the nqr[917,1001] or iodine Mössbauer[950,1002] spectra have been used to explore the electron distribution in the valence orbitals of the atoms forming polyhalide species. That the total p-σ electron population of a trihalide anion is invariably close to 4, within the limitations of the Townes–Dailey approximation, vindicates the four-electron, three-centre scheme of bonding; on similar grounds, a bonding scheme of this sort is also favoured by the quadrupole coupling constants for the pentahalide anion ICl_4^-. The essential p-character of the bonding is endorsed, moreover, by the measured isomer shifts of the few Mössbauer spectra[950,1002] which have been reported for iodine-containing polyhalide salts.

β-Decay of the ^{129}I nucleus in the salts $KICl_4,H_2O$ and $KICl_2,H_2O$ has provided an ingenious method of synthesizing *in situ* the previously unknown compounds $XeCl_4$ and $XeCl_2$, which are short-lived under conventional conditions; since the decay populates the excited nuclear level of ^{129}Xe, the products are conveniently detected and characterized by their Mössbauer spectra[1004]. The properties deduced for the molecules $XeCl_4$ and $XeCl_2$ stress their close resemblance, in bonding and structure, to the respective isoelectronic species ICl_4^- and ICl_2^-, from which they issue.

Chemical Properties[840]

The chemical reactions of a polyhalide anion are largely those of the dissociation products, and therefore reflect the stability of the complex with respect to dissociation. The function of polyhalide derivatives as sources of the corresponding halogen or interhalogen compounds is illustrated by the fluorinating action of solid tetrafluorochlorate(III) and tetrafluorobromate(III) salts[838,841]; thus, although bromine trifluoride is commonly presumed to be the active principle, it is notable that the fluorination of a number of metal oxides proceeds to completion more quickly with potassium tetrafluorobromate(III) than with the parent trifluoride. Polyhalide salts dissolved in non-aqueous media are also known to act as halogenating agents. For example, polybromides have been used in the bromination of β-naphthol, styrene, aniline and phenol and in the preparation of 2,2-dibromopropanol[840].

[1004] G. J. Perlow and M. R. Perlow, *Rev. Mod. Phys.* **36** (1964) 353; *Chemical Effects of Nuclear Transformations*, Vol. II, p. 443, Int. Atomic Energy Agency, Vienna (1965); *J. Chem. Phys.* **41** (1964) 1157.

Likewise tetrabutylammonium tetrachloroiodate(III) has been shown to be a chlorinating agent in 1,2-dichloroethane, since, upon illumination, the solution yields the (1-chlorobutyl)tributylammonium ion[840].

Apart from dissociation, the polyhalide anions are also susceptible to solvolysis, disproportionation, oxidation and exchange. Thus, in aqueous solution the anions all suffer some degree of hydrolysis, the tri-iodide being most resistant but such resistance diminishing as iodine is replaced by the more electronegative halogens. Further, on the evidence furnished, for example, by the voltammetric behaviour of halide–halogen or halide–interhalogen systems, unsymmetrical trihalide anions are prone to disproportionate in favour of symmetrical species (see p. 1541). The oxidation of a polyhalide anion, notably by a halogen or interhalogen molecule, leads either to a new polyhalide anion richer in the number of halogen atoms it contains or, where such a species is unstable, to the corresponding interhalogen compound: e.g.

$$ICl_2^- + ICl \rightarrow I_2Cl_3^- \quad {}^{840}$$

$$MClF_4 + F_2 \rightarrow MF + ClF_5 \quad {}^{958}$$

Polarographic and chronopotentiometric studies show that in solvents which stabilize the I_3^- ion, e.g. acetonitrile, nitromethane, acetone or 1,2-dimethoxyethane, anodic oxidation of the iodide ion occurs in two steps, viz.

$$3I^- \xrightarrow{-2e} I_3^- \xrightarrow{-e} 3/2 I_2 \quad {}^{840}$$

However, the intermediate formation of the I_3^- ion is not apparent in the one-step electro-oxidation of iodide which is observed with solutions in water, alcohols or acetic acid. For certain redox couples implicating stable polyhalide species, e.g. I_2/I_3^- and I_3^-/I^-, standard potentials have been determined. That exchange processes such as

$$*I^- + I_3^- \overset{k_2}{\rightleftharpoons} *I_3^- + I^-$$

occur very rapidly in solution has been established by experiments using radioactive halogen isotopes[1005]; for the bimolecular reaction between $*I^-$ and I_3^-, ^{127}I nmr measurements signify that k_2 (at 27°C), $\Delta H_2\ddagger$ and $\Delta S_2\ddagger$ are $2{\cdot}2 \times 10^9$ sec^{-1} mol^{-1} l, 4·5 kcal mol^{-1} and 0·7 eu respectively (see p. 1243).

Despite the fact that polyhalide anions enter into many of the reactions of halogen and interhalogen molecules and are potentially significant as intermediates, for example, in reactions involving electrophilic attack by the neutral molecules, details about specific reactions of the polyhalides, other than dissociation, are still comparatively sparse. Accordingly, a clearer definition of the chemical functions and reactivity of the anions must await the findings of more comprehensive studies.

4. BONDING IN NEUTRAL AND ANIONIC POLYHALOGEN SPECIES[834,835,978,1003,1006,1007]

An adequate interpretation of the electronic structure and bonding within an interhalogen molecule or polyhalide anion must take account of the geometry and dimensions, as well as the thermodynamic and spectroscopic properties, which have been described in

[1005] M. F. A. Dove and D. B. Sowerby, *Halogen Chemistry* (ed. V. Gutmann), Vol. 1, p. 41, Academic Press (1967).
[1006] E. H. Wiebenga and D. Kracht, *Inorg. Chem.* **8** (1969) 738.
[1007] B. M. Deb and C. A. Coulson, *J. Chem. Soc.* (*A*) (1971) 958.

the preceding sections. Several distinct approaches to the problem have been essayed. The first model to be proposed invoked essentially electrostatic interactions between ions and polarized molecules to account for the formation of polyhalide anions[834,835]. To describe the valency state of the central halogen atom in species such as ClF_3 or ICl_4^-, a second approach assumed that electrons are promoted from the *s*- or *p*-orbitals to the vacant *d*-orbitals of the central atom; hybridization of the valence orbitals, represented, for example, as sp^3d or sp^3d^2, then gives rise, *via* the principle of maximum overlap, to localized electron-pair bonds[834,835]. An alternative view does not specify the involvement of *d*-orbitals, but takes the localized electron-pairs to be a consequence of the repulsions between electrons of the same spin implicit in the Pauli exclusion principle[1008]. In yet another approach, delocalization of the *p*-electrons of the halogen atoms *via* multicentre molecular orbitals is considered to be the primary stabilizing influence in the formation of a neutral or anionic polyhalogen aggregate[834,835,978,1003,1006]. With this premise, simple and modified Hückel procedures have been used in an attempt to explain the structures, stabilities and spectroscopic properties of polyhalogen systems[1006]. The most sophisticated calculations so far undertaken, those of Deb and Coulson[1007], employ all the valence-shell atomic *s*- and *p*-orbitals as basis functions, in conjunction with CNDO/2 and INDO methods, to deduce ground-state properties of interhalogen molecules.

The Electrostatic Model

According to the electrostatic model, the formation of a trihalide anion is the outcome of ion–dipole interaction between the halide ion and a point dipole induced by it in the associated halogen molecule. Hence, the stability of the trihalide complex is evidently a function of the polarizability and radius of the constituent halogen atoms. It has been shown that the combination of ion–dipole attraction, Born repulsion and van der Waals' attraction does lead to the right order of magnitude for the energy of formation of the I_3^- ion[835]. However, the significance of the calculations is questionable in view of several empirical factors which they include and of the very crude approximation of an induced *point* dipole, which is certainly far from valid. Electrostatic interaction between an I^- ion and an I_2 molecule inevitably results in an unsymmetrical I_3^- ion because the two bonds are, by their nature, non-equivalent. It follows that the centrosymmetric I_3^- ion found, for example, in $[AsPh_4][I_3]$ and $[NEt_4]I_7$ is not compatible with the electrostatic model, unless it is assumed that the interaction involves an I^+ ion and two I^- ions or that the pressure of the environment induces a centrosymmetric ion. Although the model gives a satisfactory qualitative account of the structures adopted by the I_5^- and I_8^{2-} ions and by the anionic aggregate of $[NMe_4]I_9$, it is not directly amenable to other polyhalide anions or to neutral interhalogen molecules. It must be concluded, therefore, that a satisfactory explanation of the bonding in polyhalogen systems demands at least some degree of electron-delocalization.

Localized Covalent Bonds

Covalent bonding in a diatomic interhalogen molecule XY presents no problems, being analogous to that in the homonuclear molecules X_2 and Y_2. Similar molecular-orbital schemes apply to both types of molecule, which owe their stability to the bonding capacity of the doubly occupied σ-orbital derived almost exclusively from the *p*-orbitals

[1008] R. J. Gillespie and R. S. Nyholm, *Quart. Rev. Chem. Soc.* **11** (1957) 339; R. J. Gillespie, *Angew. Chem., Internat. Edn.* **6** (1967) 819.

of the halogen atoms (see Fig. 11). In order to accommodate the additional bonds formed by one or more halogen atoms in all the other polyhalogen complexes, some form of hybridization involving *d*-orbitals has commonly been presumed. As the promotion of a valence electron to a *d*-orbital in a halogen atom requires a very large amount of energy (see Table 10), it has to be assumed that this energy can be recouped by the formation of strongly directed hybridized bonds and by the improved separation of the charge clouds of the lone-pair electrons when they go into the hybrid orbitals. Polarization of the *d*-orbitals of the central halogen atom by the surrounding electronegative ligands is likely to be an important factor in facilitating the participation of these orbitals in bond formation[1009]. The promotion of one electron in the ground-state configuration ns^2np^5 of the halogen atom to an *nd*-orbital furnishes three unpaired electrons in the valence state of the atom, and thus provides for the formation of up to three bonds with other halogen atoms. In order to account for the observed shapes of the molecules or ions which result, the central atom is assumed to employ five sp^3d hybrid orbitals; according to experimental experience, the d_{z^2} orbital is thus implicated in hybridization and the orbitals are directed to the corners of a trigonal bipyramid. In a molecule such as ClF_3 or BrF_3, two of the hybrid orbitals are used to accommodate the lone-pair electrons on the central atom, while in a trihalide ion such as ICl_2^-, there are three lone-pairs to be accommodated in this way. Promotion of two electrons to *nd*-orbitals and the assumption of octahedrally disposed sp^3d^2 hybrid orbitals by the central atom are the corresponding prerequisites to the formation of interhalogen molecules of the type XY_5 or of pentahalide anions such as BrF_4^- or ICl_4^-. The five X–Y bonds of the XY_5 molecule account for five bonding-pairs of electrons, which, with a lone-pair, complete the valence shell of X, whereas in an anion of the type XY_4^- the six hybrid orbitals centred on X are occupied by four bonding- and two lone-pairs of electrons. Finally, the structure of IF_7 can be rationalized in terms of sp^3d^3 hybridization, requiring the promotion of three electrons to *d*-orbitals. Hence, all the valence electrons of the iodine atom are used in bond-formation.

Depending on the precise proportions of *s*-, *p*- and *d*-functions which make up the hybrid orbitals, the bonds in individual polyhalogen aggregates may be equivalent or non-equivalent. Hence, it is possible qualitatively to understand the variations of interatomic distance within a particular aggregate. For instance, the disparity in bond lengths in the ClF_3 or BrF_3 molecule (Table 93) can be rationalized by the assumption that the heavy atom draws on some form of *pd*-hybrid orbital for the two nearly collinear bonds and on *sp*-mixing for the orbitals which accommodate the remaining bonding and non-bonding valence electrons. On this basis, the nearly collinear bonds coincide approximately with the principal axis, while the third bond and the two lone-pairs of electrons occupy the equatorial belt of the trigonal bipyramid defined by the 10-electron valence shell (Fig. 36). More generally, however, the concept of hybridization does not form a useful basis for interpreting the finer stereochemical points exhibited by polyhalogen systems.

A model which has been successfully applied to stereochemical arguments is that in which each electron pair in the valence shell of the central atom is regarded as a point on the surface of a sphere surrounding that atom[1008]. The distribution of the electron pairs is then conditioned by the Pauli exclusion principle so as to keep them as far apart as possible. Alternatively, the problem may be considered in terms of the packing of spheres

[1009] D. P. Craig, *Chemical Society Special Publication* No. 12, p. 343. The Chemical Society, London (1958).

around a central nucleus, each sphere representing the orbital of a pair of electrons[1010]. In reality, electron clouds separated by a mean distance r do not suffer purely coulombic interactions with an associated repulsive force $F = e^2/r^2$, nor are they strictly to be treated as hard objects such that $F = e^2/r^\infty$; instead a force law with $F = e^2/r^n$, where n commonly takes values between 6 and 12, is likely to give a more realistic account of the interactions. Hence, the most stable arrangement of 5 electron-pairs is at the vertices of a trigonal bipyramid for $2 \leqslant n < \infty$, though the square pyramid is inferior by a comparatively small margin. By the same token, 6 electron-pairs should be octahedrally disposed for $2 \leqslant n \leqslant \infty$. For 7 electron-pairs, however, it has been found that, as n decreases, the most stable arrangement changes from 1:3:3 to 1:4:2 and finally to 1:5:1 corresponding to the pentagonal bipyramid (see Fig. 41); calculation shows that the energies of all three arrangements are

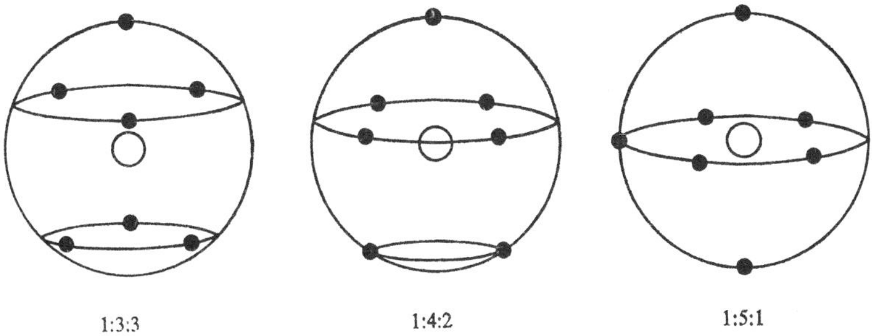

FIG. 41. Possible stereochemistries of IF_7 and related species in which the valence shell of the central atom contains seven electron pairs[1008].

very similar. The susceptibility of a molecule like IF_7 to facile intramolecular exchange ("pseudorotation") thus finds a ready explanation. It is also evident that the trigonal–bipyramid and pentagonal–bipyramid arrangements of electron pairs present non-equivalent apical and equatorial sites. For example, each apical electron-pair in the former arrangement has three nearest neighbours at 90°, while each equatorial pair has only two such nearest neighbours at 90° and two at 120°. For $n \geqslant 4$, the total force experienced by the apical pairs exceeds that experienced by the equatorial pairs, and equilibrium is reached only if the apical pairs are at a greater distance from the nucleus of the central atom than are the equatorial pairs. If it is also accepted that the charge clouds of lone-pair (LP) electrons are larger and occupy more of the "surface" of the central atom than the bonding electrons (BP), then the repulsive forces must decrease in the sequence LP–LP > LP–BP > BP–BP, and it follows that the replacement of bonding-pairs by lone-pairs is attended by a decrease in the angles between the bonding-pairs. With these assumptions, it is possible to explain the geometries of ClF_3 and BrF_3 and of symmetrical trihalide anions. Here the lone-pairs of electrons occupy the more spacious equatorial sites in the trigonal bipyramid so as to minimize the interactions between electron-pairs mutually disposed at 90°. As with other molecules in which the central atom has 5 electron-pairs in its valence-shell, the apical bonds of ClF_3 and BrF_3 are longer than the equatorial bond, the presence of the lone-pairs in the equatorial belt having the effect of amplifying the disparity. Likewise, a simple explanation is found for the geometries of molecules of the type XY_5 and of

[1010] H. A. Bent, *J. Chem. Educ.* **40** (1963) 446.

anions such as BrF_4^-. The enhanced repulsion between a lone-pair and adjacent electron-pairs then accounts for the elongation of the basal relative to the apical bonds of XF_5 molecules and for the *trans*-arrangement of the lone-pairs in the pentahalide anions. Moreover, the different spatial demands of bonding and non-bonding electron-pairs in molecules of the types XF_3 and XF_5 lead to interbond angles significantly less than the ideal values of 90° and 180°. Of the 7 electron-pairs which formally make up the valence-shell of the central atom in the anions BrF_6^- and IF_6^-, one is a lone-pair, and it has been argued that the most stable arrangement is the 1:3:3 structure (Fig. 41), in which the lone-pair occupies the unique axial position and so has the minimum number of nearest neighbours[1008]. Such a structure probably represents the most stable configuration of the molecule XeF_6 [997], but the facility with which both XeF_6 and IF_7 [896] undergo intramolecular rearrangement may mean that crystal or solvation forces ultimately determine the structures adopted by BrF_6^- and IF_6^-. Analogous to these species are the ions SeX_6^{2-} and TeX_6^{2-} (X = Cl, Br or I), which are characterized by regular octahedral structures, although the isoelectronic complexes SbX_6^{3-} appear to vary their stereochemistry to suit the environment[1011].

Against the successes of those models which invoke localized covalent bonding must be set a number of weaknesses or failures. In general, the burden of assumptions denies any quantitative treatment, and the models leave unexplained some of the finer structural and energetic points. The consensus of recent opinion is that the contribution of *d*-orbitals is greatly exaggerated in the simple hybridization model; although the principle of electron-pair repulsion does not specify whether the *d*-orbitals of the central atom are involved, its basis becomes obscure unless such participation is assumed[1006]. Systems poorly served by the models of localized covalent bonding include solid ICl, IBr and I_2Cl_6, the trihalide anions, and the salts $[NMe_4][I_5]$, $[NEt_4]I_7$, $[NMe_4]I_9$ and Cs_2I_8. Thus, localized electron-pairs are not easily reconciled with the shapes of the I_5^- and I_8^{2-} anions or with the structural mutability of the I_3^- ion; more generally, there is no satisfactory basis for explaining the wide range of halogen–halogen distances within these aggregates.

Delocalized Bonding

To counter the shortcomings of earlier models, molecular-orbital schemes have been devised for polyhalogen systems to provide, in effect, for delocalization of the valence *p*-electrons of the halogen atoms engaged in bonding[834,835,978,1003]. The treatment typically entails, in its simplest form, the following assumptions[834,835]:

(i) The molecular orbitals ψ^k ($k = 1 \ldots n$) are represented by linear combinations of the valence *np*-orbitals ϕ_i ($i = 1 \ldots n$) of the halogen atoms; completely neglected as bonding agents are the low-energy *ns*- and high-energy *nd*-functions of each atom.

(ii) Overlap integrals S_{ij} for neighbouring and more remote pairs of atoms are neglected.

(iii) Bond or interaction integrals $H_{ij} = \int \phi_i^* H \phi_j d\tau$ (where i and j refer to atomic orbitals of different atoms) are taken to be zero, except for those orbitals of adjacent atoms which interact in the direction defined by the bond-axis, and with which a common value of $H_{ij} = \beta$ is associated.

(iv) To evaluate the coulomb integrals, different approximations have been made. As a zero approximation, all the integrals α are supposed to have the same value regardless of the nature of the halogen atoms and their formal charge. In a first approximation, the nature and formal charge of the halogen atoms are acknowledged by the following rules:

[1011] C. J. Adams and A. J. Downs, *Chem. Comm.* (1970) 1699.

(a) For I, Br, Cl and F the values are taken to be α, $\alpha+0{\cdot}2\beta$, $\alpha+0{\cdot}4\beta$ and $\alpha+1{\cdot}0\beta$, respectively, the increments being approximately proportional to the corresponding differences in the electronegativities of the atoms.

(b) A formal charge Q, calculated for a given atom in the zero approximation, is supposed to increase the absolute value of its coulomb integral by $0{\cdot}2Q\beta$. Because these numerical factors are relatively arbitrary, the first approximation cannot be expected to give truly quantitative results, though it should indicate the sense in which the results obtained in the zero approximation must be modified in a more realistic account.

The application of this relatively naïve Hückel theory leads therefore to the molecular orbitals

$$\psi^k = \sum_{i=1}^{n} C_i^k \phi_i$$

where C_i^k is the coefficient of the ith atomic orbital in the kth molecular orbital. According to the variation principle,

$$\sum C_i^k (H_{ij} - E_k S_{ij}) = 0$$

so that the electronic energy E_k corresponds to the eigenvalues determined by the secular equation

$$|H_{ij} - E_k S_{ij}| = 0$$

Of the different possible configurations of a polyhalogen aggregate, that corresponding to the lowest calculated E_k should be the most stable. Thus, for a symmetrical trihalide anion, the idealized energy-level diagram takes the form illustrated in Fig. 4; if account is taken of lateral overlap of p-orbitals to implement π-bonding, however, a more realistic diagram is that of Fig. 40.

A fundamental prerequisite to such an approach is that the atoms must be sufficiently close together to allow the non-zero bond integrals H_{ij} to assume appreciable values. For example, to form I_3^-, the I^- ion must first be brought close to the I_2 molecule before an appreciable delocalization energy can be realized. In this respect, there is a nice distinction between the formally similar triatomic systems $H+H_2$ and I^-+I_2 (see p. 1550). With $H+H_2$ the Born repulsion tends to keep the components well separated because the van der Waals' attraction between them is very weak; accordingly the combined action of the various forces leads to a very loosely bound complex, having a trifling delocalization energy. However, the approach of I_2 and I^- is attended by relatively strong van der Waals' and electrostatic attractions, which tend to bring the species quite close together. Furthermore, the overlap of $5p$-functions starts to become effective at interatomic distances relatively large compared with those appropriate to the $1s$-functions which provide for delocalization in the system $H+H_2$. In general, it appears that appreciable stabilization by delocalization of p-electrons, whether in interhalogen molecules, polyhalide anions or noble-gas compounds, is promoted by the longer-range forces of van der Waals' and electrostatic interactions. Because of the complexity of a system such as I_3^-, however, there is no predicting *a priori* whether the anion will be symmetrical or unsymmetrical.

The simple LCAO-MO method outlined here leads to the following conclusions[834,835]:

(a) The scheme accounts in a reasonable way for the predominance of bond angles close to 90° or 180° in polyhalogen systems.

(b) Where different arrangements of the atoms are possible, the configuration with the lowest calculated energy corresponds with that observed for the polyhalogen molecule or ion. For instance, in the zero approximation the delocalization energy of a trihalide anion is zero when the bond angle at the central atom is 90° but 0·83β when the angle is 180°; the reverse order of stability is found for a trihalogen cation like ICl_2^+. The energy of the linear $IBrCl^-$ ion has also been shown to be at a minimum when the iodine atom occupies the central position. Again, the V-shaped configuration emerges with the lowest energy for a homonuclear pentahalide anion like I_5^-, but gives place to the square-planar configuration for the ICl_4^- anion. However, while these calculations serve as guides to the configuration favoured by a given number of atoms, they give little hint of how the stabilities of complexes vary with the nature or number of halogen atoms: for example, the calculated delocalization energies of the molecules ClF_3 and I_4 make little distinction of stability.

(c) For most of the known polyhalogen systems the calculations have given estimates of bond orders and of the net charges borne by the atoms. It appears that the bond orders, which fall in the range 0·6–1·0, invariably comply with the patterns established by measured interatomic distances and stretching force constants. If the asymmetry commonly associated with trihalide anions like I_3^- is correctly assumed to reflect the influence of crystal forces, then adjustment of the coulomb integrals again provides qualitative agreement between the calculated bond orders and observed bond lengths. The net charges on the halogen atoms are found to vary widely, even in homonuclear aggregates. In the symmetrical I_3^- ion, for instance, the net charge on the terminal atoms is approximately −0·5e, whereas that on the central atom is close to zero. Analogous calculations for ICl_4^- imply that the net charges are −0·52e for each of the chlorine atoms and +1·07e for the central iodine atom. Although undoubtedly exaggerated by the simple model, these are in tune with the measured nqr parameters of polyhalogen compounds[917,1001].

(d) The calculated energies for the three possible modes of dissociation of the $BrICl^-$ anion show that the process $BrICl^-(g) \rightarrow Cl^-(g) + IBr(g)$ offers the minimum thermodynamic barrier. This factor may well augment the decrease in lattice energy in accounting for the observation that crystalline salts of the $BrICl^-$ anion dissociate on heating to give IBr and the corresponding monochloride (see p. 1541). Qualitative agreement between theory and experiment has also been obtained for the energies of formation of diatomic interhalogen molecules in the gas phase[835].

Various refinements and extensions of the simple molecular-orbital approach have been essayed. Thus, a more sophisticated view of the electronic structure of polyhalogen systems must take account of $(p\text{–}p)\pi$-orbitals associated with individual halogen–halogen bonds and of their capacity to mix with the $(p\text{–}p)\sigma$-orbitals when the interbond angles deviate from the limiting values of 90° or 180° [998]. Some weight must also be given to the bonding contributions of *ns*- and *nd*-orbitals. Hence, the σ-bonding orbitals of the ICl_4^- anion are undoubtedly augmented through interactions involving the $5d_{x^2-y^2}$ orbital of the central iodine atom, while $(p \rightarrow d)\pi$-bonding may well provide an important mechanism for the transfer of charge between halogen atoms[978]. The most recent refinement of the Hückel approach[1006] retains the exclusive use of valence *p*-orbitals as basis functions, but takes into account (a) the electrostatic interaction of the net charges on the atoms, and (b) the change of energy of each atom, including its core electrons, as a function of its charge. The iterative procedures of this modified Hückel theory have been used to recalculate the net charges, bond orders and energy terms of polyhalogen compounds. In their general tenor, the results resemble those derived from the simple model, but the implied reduction in the net charges borne by the atoms makes for greater realism and for improved quantitative agreement with structural, thermodynamic and spectroscopic findings.

In a more elaborate description of the electronic properties of interhalogen molecules, Deb and Coulson[1007] have considered as basis functions all the valence-shell atomic orbitals other than *d*. They have applied separately the approximate CNDO/2 and INDO methods in a consistent and non-empirical fashion to determine the energies of the different configurations open to individual molecules, and then used the theoretical equilibrium geometries to calculate dipole moments, orbital energies and ionization potentials, electronic transition energies, harmonic force constants and other molecular parameters. If the calculations have failed sometimes correctly to assess the relative stabilities of alternative configurations, they have enjoyed more success in explaining the equilibrium shape of a given configuration and in reproducing experimental trends in certain properties. Accordingly, the neglect of *d*-orbitals would appear to be vindicated.

Jahn–Teller and "Pseudo-Jahn–Teller" Effects[1012,1013]

A quite different approach, outlined in Section 1 (p. 1115), employs perturbation theory to relate the energy of a polyhalogen aggregate to the deformation of its ground-state geometry. On this basis, the structural characteristics are a function of the symmetry properties and relative energies of the ground and excited electronic states, which in turn bear upon the force field operative over the system. The possibility then exists that the structure of the aggregate may be susceptible to spontaneous distortion (a) if an orbitally degenerate ground state is implicated (the Jahn–Teller effect), or (b) if there exists an excited state ψ_m separated from the ground state ψ_0 by a comparatively small energy gap $E_m - E_0$ (the "pseudo-Jahn–Teller" effect)[1012]. Hence, calculations relating to the electronic and geometric structures of the molecules PF_5, AsF_5 and BrF_5 not only support the assumption that *d*-orbitals of the central atom play only a small part in the bonding, but rationalize the structural differences in terms of Jahn–Teller distortion of the trigonal–bipyramid model of BrF_5 [1013]. Repulsions between the lone-pair and bonding-pairs of electrons in this molecule then manifest the nodal properties of the highest occupied molecular orbital. The ease with which certain polyhalogen units are deformed, whether reversibly *via* rapid intramolecular exchange, as with IF_7, or irreversibly under the influence of their environment, as with I_3^-, may be associated with the proximity of ground and excited electronic states (e.g. < 4 eV for I_3^-). In general, the mode of deformation is then appointed by the symmetry properties of the excited state. Thus, the excited electronic state Σ_u^+ of the centrosymmetric I_3^- ion dictates the off-centre displacement of the central atom, which is accommodated by the σ_u^+ vibration ν_3. The anomalously small force constant associated with the deformation thus finds an explanation (see p. 1553), while the additional stabilization liable to accrue through the mixing-in of the excited state allowed by the deformation ultimately favours a permanently unsymmetrical I_3^- unit. Similar arguments apply to the anions BrF_6^- and IF_6^-, which may be expected to share the stereochemical flexibility of the isoelectronic species XeF_6[997], and consequently to be unusually responsive to the influence of their surroundings.

[1012] L. S. Bartell, *J. Chem. Educ.* **45** (1968) 754; R. G. Pearson, *J. Amer. Chem. Soc.* **91** (1969) 4947.
[1013] R. S. Berry, M. Tamres, C. J. Ballhausen and H. Johansen, *Acta Chem. Scand.* **22** (1968) 231.

D. ORGANIC POLYVALENT HALOGEN DERIVATIVES

1. INTRODUCTION

Despite its title, the present section is restricted mainly to the chemistry of organic compounds containing iodine(III) and iodine(V). The obvious limitations of space and context preclude any treatment *per se* of conventional univalent organo-halogen compounds, e.g. the alkyl, aryl and acyl halides. Moreover, by comparison with the iodine systems, the array of known organic polyvalent chlorine and bromine compounds is very small, consisting almost entirely of some chloronium and bromonium derivatives which are discussed below alongside their iodine(III) analogues. No iodine(VII) compounds containing carbon–iodine bonds have as yet been prepared, but some aromatic perchloryl derivatives are known.

Organic polyvalent iodine compounds were first described by Willgerodt[1014] and have been the subject of two detailed reviews[1015,1016]. They may be classified according to the oxidation state of the iodine atom and the number of carbon–iodine bonds on the same iodine atom.

Iodine(III) Compounds

(a) with one carbon–iodine bond, viz. iodo disalts RIX_2 and iodoso compounds RIO;
(b) with two carbon–iodine bonds on the same iodine atom, viz. iodonium compounds RR′IX, including iodolium compounds (1);
(c) with three carbon–iodine bonds on the same iodine atom, e.g. Ph_3I.

(1)

Iodine(V) Compounds

(a) with one carbon–iodine bond, e.g. iodoxy compounds RIO_2 and "iodoso disalts" $RIOX_2$;
(b) with two carbon–iodine bonds on the same iodine atom, e.g. iodosyl salts R_2IOX.

In each class the majority of stable compounds contain aromatic organic residues; Scheme 19 illustrates the major reactions by means of which the various aryl derivatives may be interconverted. Kinetically stable non-aromatic derivatives are known in which the organic groups are resistant to nucleophilic attack at the 1-carbon atom, and the current development of imaginative synthetic procedures should increase the number of these compounds in the near future.

1014 C. Willgerodt, *J. prakt. Chem.* **33** (1886) 154.
1015 F. M. Beringer and E. M. Gindler, *Iodine Abstr. and Revs.* **3** (1956).
1016 D. F. Banks, *Chem. Rev.* **66** (1966) 243.

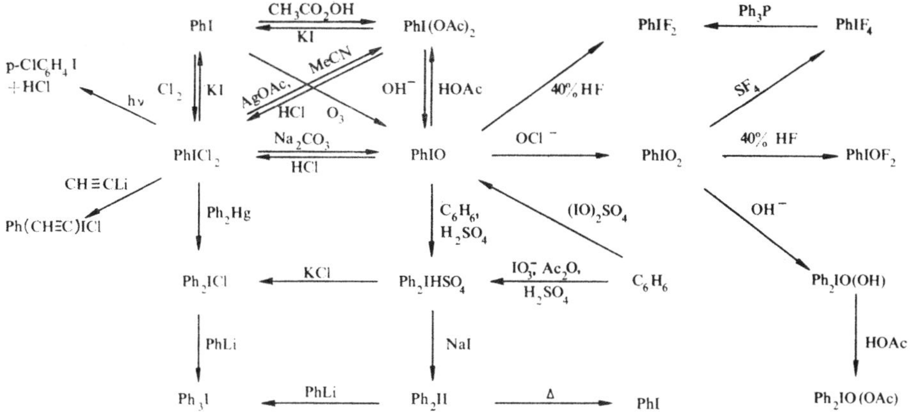

SCHEME 19. Interconversions of aromatic polyvalent iodine compounds.

Although cyclic chloronium, bromonium and iodonium ions (2) have long been postulated as intermediates in organic substitution and addition reactions[1017], stable organochloronium and organobromonium compounds were first isolated only in 1952[1018]. The chemistry of these derivatives has been developed largely by the Russian school of Nesmeyanov.

X^+ bridging C----C

(X=Cl,Br,I)

(2)

The problems of structure and bonding presented by organic derivatives of polyvalent halogens are closely akin to those of the interhalogens and polyhalide ions discussed above in Section 4C. Chemically the compounds are noticeable for the ease with which one or more bonds to the positive halogen may be fractured; whether this cleavage is heterolytic or homolytic in nature depends on the system in question. There is insufficient evidence

TABLE 101. CARBON–IODINE BOND LENGTHS

Compound	C–I bond length (Å)	Reference
p-IC_6H_4NO	2·00[a]	M. J. Webster, *J. Chem. Soc.* (1956) 2841
$C_6H_5ICl_2$	2·00±0·05	E. M. Archer and T. G. D. van Schalkwyk, *Acta Cryst.* **6** (1953) 88
$(C_6H_5)_2ICl$	2·08[a]	T. L. Khotsyanova, *Doklad. Akad. Nauk, S.S.S.R.*, **110** (1956) 71
$(C_6H_5)_2IBF_4$	2·02±0·03	Yu. T. Struchkov and T. L. Khotsyanova, *Bull. Acad. Sci., U.S.S.R.* (1960) 771
p-$ClC_6H_4IO_2$	1·93[a]	E. M. Archer, *Acta Cryst.* **1** (1948) 64

[a] No limits of error quoted.

1017 For a review of this subject see: J. G. Traynham, *J. Chem. Educ.* **40** (1963) 392; B. Capon, *Quart. Rev. Chem. Soc.* **18** (1964) 45.

1018 R. B. Sandin and A. S. Hay, *J. Amer. Chem. Soc.* **74** (1952) 274.

accurately to chart the variation in the properties of the carbon–halogen bond throughout this series of compounds; such carbon–iodine bond lengths as have been determined crystallographically are listed in Table 101.

2. IODO DISALTS $ArIX_2$

Aromatic iodo disalts $ArIX_2$ are derivatives of the hypothetical acids $ArI(OH)_2$; in solution they are readily and reversibly converted into the iodoso compound.

$$ArIX_2 \underset{HX}{\overset{OH^-}{\rightleftarrows}} ArIO$$

They may be prepared by two general routes, involving either oxidation of an aryl iodide or substitution of X^- into some other organo-iodine(III) derivative. Combination of ICl_3 with an organometallic compound is not a good method for the synthesis of $ArICl_2$, since only in special cases does the reaction stop at this stage; iodonium compounds typically constitute the major product.

Iodo dichlorides, probably the most important members of this class, are best obtained by passing dry chlorine into a chloroform solution of the parent iodo compound[1019].

$$ArI + Cl_2 \rightleftharpoons ArICl_2$$

The reaction is reversible and the dichlorides dissociate more or less readily on heating or on dissolution in polar media; electrophilic substituents in the aromatic group facilitate this decomposition, as indexed by the variation both of the dissociation constant in solution[1020] and of the thermal stability of the solids[1021]. On standing, the compounds evolve hydrogen chloride in a photo-catalysed reaction, e.g.

$$C_6H_5ICl_2 \xrightarrow{h\nu} p\text{-}ClC_6H_4I + HCl$$

Iodobenzene dichloride is used as a mild chlorinating agent in organic synthesis. Chlorination of olefinic functions[1022] proceeds preferentially by a free radical mechanism [Scheme 20(a)] giving predominantly the *trans*-product; in the presence of water or of radical inhibitors, reaction proceeds more slowly by an ionic mechanism [20(b)]. By contrast, iodobenzene dichloride acts in aromatic chlorination reactions subject to metal halide catalysis merely as a source of molecular chlorine[1023], the rate-determining step being dissociation of the dichloride.

Other disalts crystallize (i) following dissolution of the iodoso compound in an acid [e.g. $ArIF_2$, $ArI(OAc)_2$], (ii) after addition of acid to a solution of $ArIO$ or $ArICl_2$ in acetic acid [e.g. $ArI(ONO_2)_2$][1024], or (iii) after treatment of $ArICl_2$ with a silver salt in acetonitrile [e.g. $ArI(OCOCF_3)_2$][1025]. About the white unstable solids $ArIF_2$ little is known; iodo dibromides, diiodides and dipseudohalides are apparently unstable at

[1019] H. J. Lucas and E. R. Kennedy, *Organic Syntheses*, Collective Vol. III (editor-in-chief E. C. Horning), p. 482, Wiley, New York (1955).
[1020] R. M. Keefer and L. J. Andrews, *J. Amer. Chem. Soc.* **82** (1960) 4547.
[1021] D. A. Bekoe and R. Hulme, *Nature*, **177** (1956) 1230.
[1022] D. D. Tanner and G. C. Gidley, *J. Org. Chem.* **33** (1968) 38.
[1023] R. M. Keefer and L. J. Andrews, *J. Amer. Chem. Soc.* **79** (1957) 4348.
[1024] R. B. Sandin, *Chem. Rev.* **32** (1943) 249.

ambient temperatures[1025].

$$PhICl_2 + 2AgX \xrightarrow[-40°C]{MeCN} PhI + 2AgCl + X_2 \quad (X = Br, I, CN, NCO)$$

Derivatives of weak oxyacids are unstable with respect to formation of the iodoso compound[1025].

$$PhICl_2 + 2AgNO_2 \xrightarrow[-40°C]{MeCN} PhIO + 2AgCl + N_2O_3$$

Iodo dicarboxylates, which may also be produced directly from ArI and percarboxylic acids, are used as oxidizing agents in organic synthesis. Like lead tetraacetate, $ArI(OAc)_2$ acts to cleave 1,2-glycols or to substitute methyl groups into aromatic molecules.

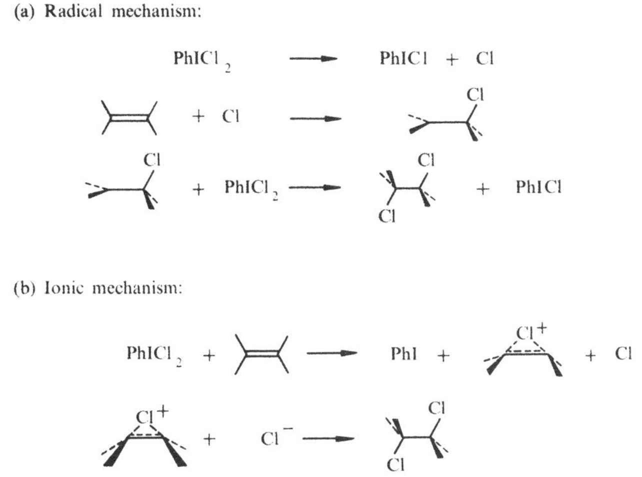

(a) Radical mechanism:

$PhCl_2 \longrightarrow PhICl + Cl$

alkene + Cl ⟶ chloroalkyl radical

chloroalkyl radical + $PhICl_2$ ⟶ 1,2-dichloroalkane + PhICl

(b) Ionic mechanism:

$PhICl_2$ + alkene ⟶ PhI + chloronium ion (Cl^+) + Cl

chloronium ion (Cl^+) + Cl^- ⟶ 1,2-dichloroalkane

SCHEME 20. Mechanisms of olefinic chlorination using iodobenzene dichloride.

There is no evidence that iodo disalts undergo ionic dissociation in solution. The $PhICl_2$ molecule revealed in the crystal structure of this compound[1026] is T-shaped, containing a linear symmetrical Cl–I–Cl group arranged almost perpendicular to the plane of the ring; the structure may be described as a trigonal bipyramid with the phenyl group and two lone pairs in the equatorial positions and chlorine atoms at the apices (Fig. 42). There are, however, significant intermolecular I–Cl contacts (3·40 Å), close to those found in $ICl_3,SbCl_5$ (p. 1351). In p-$ClC_6H_4ICl_2$ the iodine atom enjoys a similar environment, although the molecule as a whole is planar[1021]. The Mössbauer[1027] and ^{35}Cl nqr[1028] spectra of $PhICl_2$ are consistent with a bonding scheme involving only p-functions of the iodine atom. Similarly the influence of substituents at the aromatic ring on the stability and spectra of iodo disalts is explicable solely in terms of inductive effects, and there is no cause to invoke resonance interactions involving the iodine $5d$-orbitals and the aromatic π-cloud.

[1025] N. W. Alcock and T. C. Waddington, *J. Chem. Soc.* (1963) 4103.
[1026] E. M. Archer and T. G. D. van Schalkwyk, *Acta Cryst.* **6** (1953) 88.
[1027] B. S. Ehrlich and M. Kaplan, *J. Chem. Phys.* **54** (1971) 612.
[1028] J. C. Evans and G. Y.-S. Lo, *J. Phys. Chem.* **71** (1967) 2730.

Regarding non-aromatic derivatives, simple alkyl iodo dichlorides are formed by reaction between iodoalkanes and chlorine, but are unstable even at low temperature[1029]; iodomethane dichloride, the most stable member of the series, decomposes at $-30°C$ to

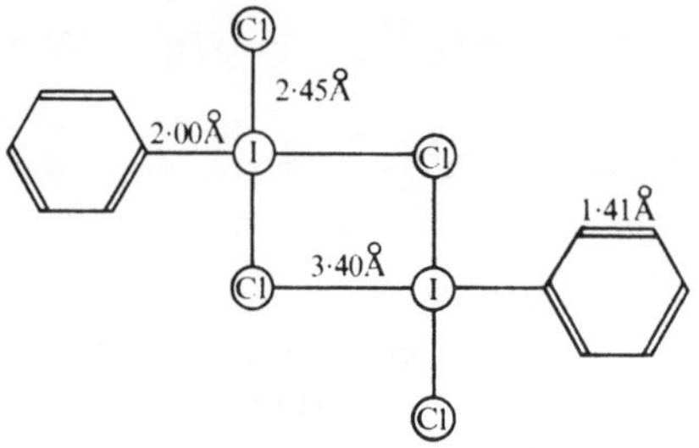

2·45Å
2·00Å
1·41Å
3·40Å

FIG. 42. Structure of iodobenzene dichloride.

methyl chloride and iodine monochloride. On the other hand, bridgehead iodides (e.g. 1-norbornyl)[1030], α-iodomethylsulphones RSO_2CH_2I[1031], and simple substituted iodoalkenes (e.g. ClCH=CHI)[1029] afford on chlorination reasonably stable dichlorides which resemble their aryl counterparts. Perfluoroalkyl iodo difluorides R_fIF_2 have also been described[1032] as low-melting solids obtained by fluorinating R_fI ($R_f = C_2F_5$, n-C_3F_7, n-C_4F_9) with fluorine or halogen fluorides; the compounds are monomeric in the liquid phase.

3. IODOSO COMPOUNDS

Aryl iodoso compounds ArIO are pale yellow solids which disproportionate slowly at room temperature, rapidly at 100°C.

$$2ArIO \rightarrow ArI + ArIO_2$$

They may decompose explosively on melting. Methods for their preparation include:

(i) the basic hydrolysis of iodo disalts in aqueous solution;

(ii) direct oxidation of ArI, e.g. with ozone, fuming nitric acid or potassium permanganate;

(iii) the action of iodosyl sulphate, $(IO)_2SO_4$ (p. 1352), on ArH in the absence of excess sulphuric acid.

Chemically, iodoso compounds behave as anhydrides of the hypothetical dibasic acids $ArI(OH)_2$. Exchange studies using H_2O^{18} disclose that iodosobenzene undergoes 30% exchange in neutral solution and 60% in basic solution[1033].

$$C_6H_5IO + H_2O^{18} \rightleftharpoons \left[C_6H_5I \begin{matrix} \diagup OH \\ \diagdown O^{18}H \end{matrix} \right] \rightleftharpoons C_6H_5IO^{18} + H_2O$$

The protonated species $C_6H_5IOH^+$ is believed to be the active intermediate in the oxidation of 1,2-glycols to 1,2-diketones, which is readily effected by iodosobenzene.

1029 J. Thiele and W. Peter, *Annalen*, **369** (1909) 119; J. Thiele and H. Haakh, *ibid.* p. 131.
1030 J. B. Dence and J. D. Roberts, *J. Org. Chem.* **33** (1968) 1251.
1031 O. Exner, *Coll. Czech. Chem. Comm.* **24** (1959) 3567.
1032 C. S. Rondestvedt, jun., *J. Amer. Chem. Soc.* **91** (1969) 3054.
1033 I. P. Gragerov and A. F. Levit, *J. Gen. Chem. (U.S.S.R.)* **33** (1963) 536.

The iodoso function is highly susceptible to nucleophilic attack, which may be intramolecular if the aromatic ring contains a nucleophilic group in the *ortho*-position. Thus the oxidation of 2-iodoisophthalic acid produces the lactone (3) rather than 2-iodosoisophthalic acid[1034]. The reaction of equivalent amounts of $ArIO_2$ and ArIO in the presence of silver oxide,

$$ArIO_2 + ArIO + AgOH \rightarrow Ar_2I^+OH^- + AgIO_3$$

probably proceeds *via* attack of $ArIO_3H^-$ on the iodoso derivative.

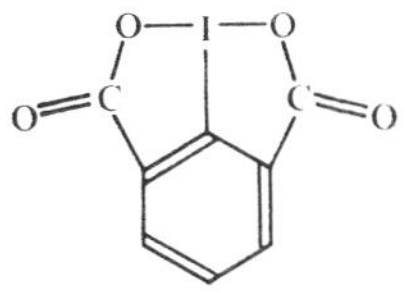

(3)

Structurally little is known about iodoso compounds, since no definitive studies have been made. The Mössbauer spectrum of solid iodosobenzene[1027] is consistent with a bent C–I–O unit ($\angle$ C–I–O *ca.* 90°), the iodine atom drawing almost exclusively on its *p*-orbitals for the purposes of bonding. Intense absorptions at 600–700 cm^{-1} in the infrared spectra of iodoso compounds are attributed to fundamentals in which I–O stretching plays the major part[1035].

4. DIORGANOHALONIUM COMPOUNDS

Diaryliodonium compounds ArAr′IX are known containing a wide variety of anionic groups (X) and both carbocyclic and heterocyclic aromatic residues. The anions may be readily interchanged by standard methods, and two general routes are available for synthesis of the cations.

(i) The acid-catalysed condensation of iodoso compounds with aromatic hydrocarbons gives good yields of iodonium compounds.

$$Ar'IO + ArH \xrightarrow{HX} ArAr'IX + H_2O$$

Intramolecular coupling processes are particularly fruitful, e.g.[1036]

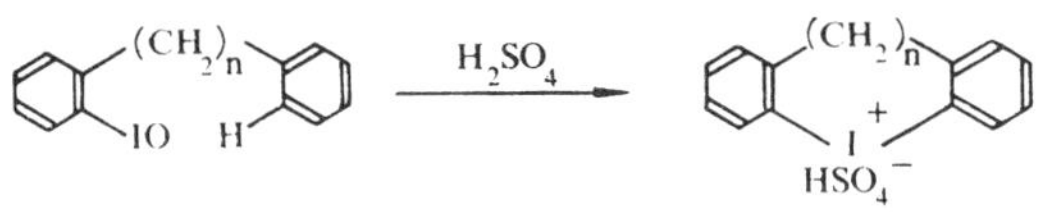

[n 0, 1, 2, 3]

Oxidation of aryl iodides to iodoso compounds in the presence of hydrocarbons affords an opportunity for this reaction, while symmetric iodonium bisulphates $Ar_2I^+HSO_4^-$ result from the treatment of ArH with iodosyl sulphate in the presence of *excess* sulphuric acid.

1034 W. C. Agosta, *Tetrahedron Letters*, (1965) 2681.

1035 C. Furlani and G. Sartori, *Ann. Chim.* (*Rome*), **47** (1957) 124; M. V. King, *J. Org. Chem.* **26** (1961) 3323.

1036 J. Collette, D. McGreer, R. Crawford, F. Chubb and R. B. Sandin, *J. Amer. Chem. Soc.* **78** (1956) 3819.

(ii) Iodine trichloride combines directly with many organometallic reagents (e.g. Ar_2Hg, $ArSnCl_3$) to give Ar_2ICl. A neat variation of this method exploits the instability of tri-organoiodine(III) derivatives[1037],

$$ClCH{=}CHICl_2 \xrightarrow[\text{toluene}]{\text{ArLi}} [ClCH{=}CHIAr_2] \rightarrow Ar_2ICl + CH{\equiv}CH$$

Diarylchloronium, diarylbromonium and diaryliodonium salts have been isolated following the thermolysis of diazonium compounds in the presence of the appropriate aryl halide[1038].

$$PhN_2^+BF_4^- + PhX \rightarrow Ph_2X^+BF_4^- + N_2 \text{ (X = Cl, Br or I)}$$

Only exceptionally do yields exceed 10% of the diarylhalonium salt, although intramolecular coupling is more successful[1018].

$$\text{(2-}N_2^+\text{-C}_6\text{H}_4\text{)–Y–(2-X-C}_6\text{H}_4\text{)}\ HSO_4^- \longrightarrow \text{cyclic } X^+ \text{ (Y-bridged) } HSO_4^- + N_2$$

(X = Cl, Br, I; Y = S, SO, SO_2)

The bromine(III) ylide (4) has also been reported[1039].

$$O{=}C_6H_2(Bu^t)_2{=}\!^{-}Br^+\text{–}C_6H_2(Pr^i)_2\text{–}OH$$

(4)

Diarylhalonium compounds are potent arylating agents, undergoing thermal or photo-chemical cleavage under mild conditions[1038].

$$Ph_2O + PhI + HCl \xleftarrow{\text{PhOH}} Ph_2ICl \xrightarrow{C_5H_5N} \text{(N-phenylpyridinium)} + PhI$$

$$Ph_3Se^+BF_4^- \xleftarrow{Ph_2Se} Ph_2IBF_4 \xrightarrow{C_5H_5N} C_5H_5N^+\text{—Ph } BF_4^-$$

$$Ph_4Pb \xleftarrow{\text{Pb}} Ph_2BrBF_4 \xrightarrow{\text{Tl}} Ph_2Tl^+BF_4^-$$

$$Ph_2ISbCl_4 \xrightarrow{\text{Sb}} Ph_2SbCl + Ph_2SbCl_3 + Ph_3SbCl_2$$

Reactions of diarylhalonium halides proceed largely *via* homolysis of the halogen–carbon bond and follow radical pathways. By contrast, the intervention of Ar^+ cations is implicated in the arylation effected by the tetrafluoroboronates on the double salts with metal halides. Chlorine and bromine derivatives react more readily than their iodine analogues, although their relative inaccessibility precludes their extensive use in this respect.

[1037] F. M. Beringer and R. A. Nathan, *J. Org. Chem.* **34** (1969) 685.
[1038] A. N. Nesmeyanov, L. G. Makarova and T. P. Tolstaya, *Tetrahedron*, **1** (1957) 145.
[1039] W. H. Pirkle and G. F. Koser, *J. Amer. Chem. Soc.* **90** (1968) 3598.

In the structural chemistry of halonium compounds Ar_2XY, attention has centred on the nature of the X–Y bond. In polar solvents ionic dissociation is favoured, but the influence of other solvents has not been adequately studied. Evidence for the solid state rests on two-dimensional crystal-structure determinations of three diphenyliodonium salts. Discrete angular Ph_2I^+ cations are recognizable in Ph_2IBF_4[1040]. Ph_2I units of very similar geometry feature in diphenyliodonium chloride and iodide[1041], although in these halides dimeric molecules are found (Fig. 43), which are obviously closely related to the I_2Cl_6 dimer (Section 4C). Ph_2ClX and Ph_2BrX (X = Cl, I, BF_4) are isomorphous with their iodonium analogues[1042]. In these systems spectroscopic techniques may not be a sensitive index to structure[1043].

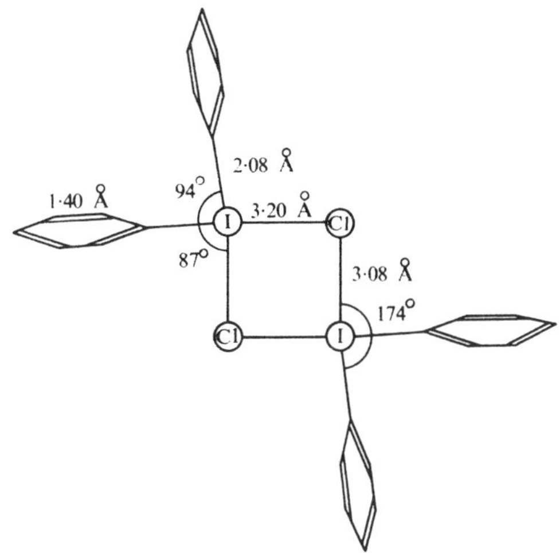

FIG. 43. Structure of diphenyliodonium chloride.

Although some iodonium compounds have been prepared containing one non-aromatic group, e.g. Ph(CH≡C)ICl, true alkylhalonium derivatives are not well known. On the basis of kinetic and stereochemical evidence, cyclic halonium ions (2) have long been indicted as intermediates in organic substitution and addition reactions. More recently, nmr

Br^+ Br_3^-

(5)

studies of solutions of aliphatic halides in "super-acids", such as SbF_5–SO_2 mixtures, have revealed the formation of cyclic ions $[(CH_2)_nX]^+$ (n = 2, 4; X = Cl, Br, I) by abstraction of X^- from $X(CH_2)_nX$, and of open-chain dialkylhalonium ions $RR'X^+$ (R, R' = Me, Et,

1040 Yu. T. Struchkov and T. L. Khotsyanova, *Bull. Acad. Sci., U.S.S.R.* (1960) 771.
1041 T. L. Khotsyanova, *Doklad. Akad. Nauk, S.S.S.R.* **110** (1956) 71.
1042 T. L. Khotsyanova and Yu. T. Struchkov, *Soviet Phys. Crystallog.* **2** (1957) 378.
1043 A. N. Nesmeyanov, L. M. Épshtein, L. S. Isaeva, T. P. Tolstaya and L. A. Kazitsyna, *Bull. Acad. Sci., U.S.S.R.* (1964) 574.

Pr; X = Cl, Br, I) from the corresponding alkyl halide(s)[1044]; the open-chain ions possibly play a part in Friedel–Crafts alkylation and isomerization reactions[1045]. Curiously, the only stable alkyl derivative so far isolated is (5), obtained as a crystalline solid during the bromination of adamantylideneadamantane in CCl_4 solution [1046]. Attempts to synthesize dialkyliodonium salts R_2IX by the direct methods of iodoso coupling or organometallic exchange have not been successful, even for cyclopropyl, 1-norbornyl or 9-triptycyl derivatives[1030].

5. TRIARYLIODINE COMPOUNDS

These compounds, which result from combination in an inert solvent of an organolithium compound with an iodonium salt, are nearly all unstable towards spontaneous decomposition into free radicals. Two relatively stable derivatives are known, namely triphenyliodine (from PhLi and $Ph_2I^+I^-$) and (6) (from PhLi and dibenziodolium iodide)[1016,1047].

(6)

6. IODINE(V) COMPOUNDS

Iodoxy compounds are formed in the slow disproportionation of iodoso derivatives:

$$2ArIO \rightarrow ArI + ArIO_2$$

the simultaneously produced iodoarene is readily separated by steam distillation. Oxidation of iodoso to iodoxy compounds is also effected by aqueous hypochlorite solution or by Caro's acid.

The white solids $ArIO_2$ are subject to detonation on impact or on heating. They are slightly soluble in water, where they behave as weak monobasic acids.

$$ArIO_2 + 2H_2O \rightleftharpoons H_3O^+ + ArIO_2(OH)^-$$

The pK_a of iodoxybenzene is 10·4[1048]. On the other hand, amphoteric character is indicated by protonation in acidic media.

In cold alkaline solution the conjugate base $PhIO_2(OH)^-$ is quantitatively and irreversibly converted to diphenyliodyl hydroxide, a strong oxidizing agent.

$$2PhIO_2(OH)^- \rightarrow Ph_2IO^+OH^- + IO_3^- + OH^-$$

Addition of acetic acid or trifluoroacetic acid produces acetate or trifluoroacetate salts from which other diphenyliodyl compounds, $Ph_2IO^+X^-$ (X = F, Cl, Br, IO_3), may be

[1044] G. A. Olah and J. M. Bollinger, *J. Amer. Chem. Soc.* **89** (1967) 4744; *ibid.* **90** (1968) 947; G. A. Olah, J. M. Bollinger and J. Brinich, *ibid.* **90** (1968) 2587, 6988.

[1045] G. A. Olah and J. R. DeMember, *J. Amer. Chem. Soc.* **91** (1969) 2113.

[1046] J. Strating, J. H. Wieringa and H. Wynberg, *Chem. Comm.* (1969) 907.

[1047] K. Clauss, *Ber.* **88** (1955) 268.

[1048] I. Masson, E. Race and F. E. Pounder, *J. Chem. Soc.* (1935) 1669

obtained metathetically[1049]. The infrared spectra of the carboxylates indicate an ionic formulation, viz. $[Ph_2IO]^+[RCO_2]^-$.

Reaction of hot 40% hydrofluoric acid with an iodoxy compound gives an aryliodoso difluoride[1050].

$$ArIO_2 + 2HF \rightarrow ArIOF_2 + H_2O$$

With sulphur tetrafluoride, iodoxybenzene is converted to $PhIF_4$[1051], while the related perfluoroalkyl derivatives R_fIF_4 ($R_f = C_2F_5$, C_3F_7 or n-C_4F_9) result from fluorination of R_fI with excess of ClF_3, BrF_3 or BrF_5[1032]. To judge by their ^{19}F nmr spectra, the tetrafluoroiodo- molecules have square-pyramidal structures, related to that of IF_5, with apical organic groups.

In keeping with the behaviour of inorganic iodates and iodine pentoxide, the infrared spectra of iodoxy, iodoso and iodyl derivatives are marked by strong absorptions in the region 650–800 cm^{-1}, presumably arising from I–O bond-stretching vibrations. The two I–O linkages in *p*-$ClC_6H_4IO_2$ are unequal in length (1·60 and 1·65 Å; $\angle$ O–I–O = 103°; $\angle$ O–I–C = 94·5°), and probably entail some sort of multiple bonding; short intermolecular I · · · O interactions (2·87 Å) must also represent a major binding force within the crystal[1052]. The Mössbauer spectrum of iodoxybenzene itself indicates some involvement of the iodine 5*s*-orbital in the bonding[1027].

7. PERCHLORYL COMPOUNDS

Perchloryl aromatic compounds[1053] are prepared from perchloryl fluoride by two routes:

$$C_6H_5Li \xrightarrow{ClO_3F} C_6H_5ClO_3 + LiF$$

$$C_6H_5F \xrightarrow[AlCl_3]{ClO_3F} FC_6H_4ClO_3 + FC_6H_4ClO_3 \text{ (}o\text{-}O_3Cl\text{)}$$

They are stable compounds which can be distilled in steam; the $-ClO_3$ group resembles $-NO_2$ in the way that it deactivates the aromatic nucleus. Hydrolysis in concentrated alkali replaces $-ClO_3$ by $-OH$. Hydrogenation on a platinum catalyst proceeds with rupture of the C–Cl bond, but the ClO_3 group is unaffected by reductants such as $LiAlH_4$, $SnCl_2$ or zinc and hydrochloric acid.

5. ASTATINE[1054–1061]

5.1. INTRODUCTION

The element of atomic number 85, the ultimate member of the Halogen Group, was

1049 F. M. Beringer and P. Bodlaender, *J. Org. Chem.* **33** (1968) 2981.
1050 R. F. Weinland and W. Stille, *Ber.* **34** (1901) 2631.
1051 L. M. Yagupol'skii, V. V. Lyalin, V. V. Orda and L. A. Alekseeva, *J. Gen. Chem.* (*U.S.S.R.*) **38** (1968) 2714.
1052 E. M. Archer, *Acta Cryst.* **1** (1948) 64
1053 V. M. Khutoretskii, L. V. Okhlobystina and A. A. Fainzil'berg, *Russ. Chem. Rev.* **36** (1967) 145; J. F. Gall, *Kirk-Othmer Encyclopedia of Chemical Technology*, 2nd edn., Vol. 9, p. 598. Interscience, New York (1966).
1054 A. G, Maddock, *Mellor's Comprehensive Treatise on Inorganic and Theoretical Chemistry*, Supplement II, Part I, p. 1064. Longmans, Green and Co., London (1956).
1055 K. W. Bagnall, *Chemistry of the Rare Radioelements*, p. 97. Butterworths, London (1957).
1056 E. K. Hyde, *J. Chem. Educ.* **36** (1959) 15.
1057 E. Anders, *Ann. Rev. Nucl. Sci.* **9** (1959) 203.
1058 A. H. W. Aten, jun., *Adv. Inorg. Chem. Radiochem.* **6** (1964) 207.
1059 V. D. Nefedov, Yu. V. Norseev, M. A. Toropova and V. A. Khalkin, *Russ. Chem. Rev.* **37** (1968) 87.
1060 D. R. Corson, K. R. MacKenzie and E. Segrè, *Phys. Rev.* **57** (1940) 459, 1087; *Nature*, **159** (1947) 24.
1061 (a) E. H. Appelman, U.S. Atomic Energy Commission Reports UCRL-9025 and NAS-NS 3012 (1960); (b) E. H. Appelman, *MTP International Review of Science*: *Inorganic Chemistry Series One*, Vol. 3 (ed. V. Gutmann), p. 181, Butterworths and University Park Press (1972).

named astatine by Corson, MacKenzie and Segrè, who in some experiments reported in 1940 were the first to prepare and identify unambiguously an isotope of this element and to explore some of its chemical properties[1060]. The name, which was derived from the Greek word αστατοσ meaning "unstable", emphasizes the instability of all the known isotopes of the element with respect to radioactive decay. For a number of reasons astatine is one of the most difficult elements to investigate from a chemical point of view.

(i) The first isotopes to be discovered, ^{211}At and ^{210}At, with half-lives of 7·2 and 8·3 hr, respectively, are still, and will probably remain, the longest-lived isotopes. The short half-life of even the most suitable isotopes makes it virtually impossible to work with the element in a laboratory which does not itself have the facilities to produce the isotopes. Accordingly the larger part of our present knowledge concerning astatine depends on the activities of only a small number of investigators; following the pioneering work of Corson, MacKenzie and Segrè, the elegant studies of Appelman[1061], also at the Berkeley Radiation Laboratory, have been pre-eminent in bringing some qualitative and semi-quantitative sense to the seemingly erratic behaviour of the element.

(ii) Astatine being chemically available only as a result of nuclear synthesis, its preparation is made more difficult because none of its isotopes is accessible by neutron irradiation. Instead of the more conventional and convenient nuclear reactor, an accelerator is required to produce the high-energy α, ^{12}C or ^{14}N particles that are the necessary reagents for the synthesis of the longest-lived astatine isotopes.

(iii) Chemical operations with milligram amounts of these isotopes would be quite impossible, even if such quantities were available, the specific activity of ^{210}At being, for example, about 2000 C/mg, a level which would elicit enormous heating and radiation effects. For this reason, all information on the chemistry of astatine has necessarily been derived from radiochemical studies of trace concentrations, typically 10^{-11} to 10^{-15} M; the most concentrated aqueous solution of astatine recorded by Appelman[1061] was no more than 10^{-8} M. The specific activity of more concentrated solutions is such as to induce radiolysis of the water, leading to the formation of hydrogen peroxide and oxygen and either masking or influencing the behaviour of the astatine.

(iv) At the dilutions which must be employed to explore the chemistry of astatine, concentration makes a major contribution to the thermodynamic behaviour, and the extrapolation of results to deduce macroscopic behaviour is correspondingly uncertain. These uncertainties will be most prominent in the event of dissociation or ionization or if the reaction involves a change in the number of astatine-containing species. The need for caution is underlined by observations implying that the chemistry of iodine on a tracer scale occasionally differs notably from its macrochemistry[1062]: not only do new ionic species appear at low concentrations, but the situation is further complicated by reactions with ever-present impurities, by adsorption effects, or by radiocolloid formation.

(v) As with other tracers, the chemistry of astatine has been studied mainly by extraction and by coprecipitation, neither of which provides results easy to interpret. Although astatine should follow iodine in many of its chemical reactions, so that iodine is the obvious carrier element (cf. the use of caesium as a carrier for francium), the analogy between the two elements often leaves much to be desired, and in many situations the astatine tracer

[1062] M. Kahn and A. C. Wahl, *J. Chem. Phys.* **21** (1953) 1185; M. L. Good and R. R. Edwards, *J. Inorg. Nuclear Chem.* **2** (1956) 196.

does not follow the iodine carrier in the way one would expect. For example, ions or molecules of astatine compounds often do not fit well into the host lattice of the corresponding iodine compound; astatine compounds are often more strongly adsorbed than the corresponding iodine compounds; again, following redox reactions with certain reagents, it is possible that iodine and astatine appear in different oxidation states. Any one of these complications may lead to apparent inconsistencies in the behaviours of the astatine tracer and the iodine carrier.

Although investigations of the chemistry of astatine were initiated more than 30 years ago, much of the behaviour of the element is still in doubt, and even some of the conclusions drawn from the known experimental results may well turn out to be erroneous on closer acquaintance. In particular, very little is yet known about the physical chemistry of astatine, though interesting estimates have been made of the lattice energies of alkali-metal astatides, the dissociation energy of the At_2 molecule and the electron affinity of the At atom[1063], and, on the basis of chemical studies, an approximate potential diagram has been deduced for the chemistry of the element in aqueous solution[1064]. From its position in the Periodic Table, one may expect the element to resemble iodine fairly closely, though some significant differences should also exist. For example, the +7 and −1 oxidation states of astatine should be somewhat less stable, and the electropositive character of the lower oxidation states more pronounced, than in the case of iodine. In evidence of these differences, perastatates emerge, reputedly, only under the most energetic conditions of oxidation, the astatide ion appears to be a stronger aqueous reducing agent than the iodide ion, and polyhalide complexes such as $AtCl_2^-$ have significantly greater formation constants than their iodine counterparts. The incipient metallic character of solid iodine should be more highly developed in solid astatine, in keeping with the transition from semiconductor to metal in the neighbouring elements tellurium and polonium. As yet the At_2 molecule has not been characterized, and it is perhaps suggestive that, whereas the molecules HAt, MeAt, AtI, AtBr and AtCl have been identified by mass spectroscopy, the At_2 molecule has eluded detection[1065]. Inevitably more polarizable and probably having lower ionization potentials, astatine atoms and molecules are likely to experience relatively stronger intermolecular or charge–transfer interactions than do the atoms and molecules of the lighter halogens. To this extent the At_2 molecule is likely to be at once a better donor and a better acceptor than the other halogen molecules; it would therefore be interesting to know whether the molecule retains its identity in the solid element or whether, in the interests of long-range stability, astatine, like polonium, adopts a three-dimensional lattice, wherein atoms rather than molecules are the only recognizable subunits. It has even been suggested, albeit on the basis of very limited evidence, that in its properties astatine shows a closer resemblance to polonium and bismuth, its horizontal neighbours in the Periodic Table, than to iodine.

5.2. HISTORY, DISCOVERY AND NATURAL OCCURRENCE[1054–1059]

The discovery and interpretation of atomic numbers by Moseley and the establishment of the physical basis of the Periodic Classification of the elements by Bohr in the early 1920's indicated that a fifth halogen might exist[1066]. Consideration of the properties of the neighbouring elements suggested that it would probably prove to be radioactive, though no

[1063] M. F. C. Ladd and W. H. Lee, *J. Inorg. Nuclear Chem.* **20** (1961) 163.
[1064] E. H. Appelman, *J. Amer. Chem. Soc.* **83** (1961) 805.
[1065] E. H. Appelman, E. N. Sloth and M. H. Studier, *Inorg. Chem.* **5** (1966) 766.
[1066] E. Wagner, *Z. Elektrochem.* **26** (1920) 260.

basis for the estimation of its radioactive properties was known at that time. The subsequent search for the element followed two principal lines: one involved chemical studies, usually combined with a sensitive physicochemical method of detection, especially X-ray emission spectroscopy; the other depended on radioactive studies. It was realized that, unless the element possessed a stable or very long-lived isotope, it must occur in association with one of the three natural decay series, so that attention was focused on uranium and thorium minerals, though some investigators, heedless of the likely radioactive character of the element, explored the possibility of its natural occurrence in macroscopic quantities.

Several reports of the discovery of element 85 appeared as a result of these investigations. Names such as Alabamine[1067], Helvetium[1068], Anglo-Helvetium[1069], Leptine[1070] and Daccine[1071] were variously assigned to the element, as were definite chemical properties. However, most of these early claims were rejected by a critical reinvestigation of the methods of separation and identification[1072]. Hevesy[1073] and Andersen[1074] were unable to detect astatine in nature, though the methods used by the latter have been shown to be inapplicable to the separation of astatine, and the results are therefore inconclusive[1060]. On investigating the possible radiochemical derivation of astatine, Hahn[1075] observed that the currently recognized radioactive decay processes could produce isotopes of element 85 only by α-decay of an isotope of the unknown element 87 or by β-decay of an isotope of polonium. The possible genesis of element 85 by α-decay of element 87, francium, linked the searches for these two elements, whose histories have several features in common.

From the vantage point of our present, more sophisticated knowledge of nuclear stabilities, we know that the search for stable or moderately long-lived isotopes of astatine was foredoomed to failure. However, with the development of a sufficiently large cyclotron, Corson, MacKenzie and Segrè had access to beams of α-particles of sufficiently high energy to surmount the coulomb barrier of the heavy elements and so reach an energy range where the cross-section for α-induced nuclear reaction becomes appreciable. By bombardment of a bismuth target with high-energy α-particles they succeeded in producing astatine according to the nuclear scheme given below[1060]:

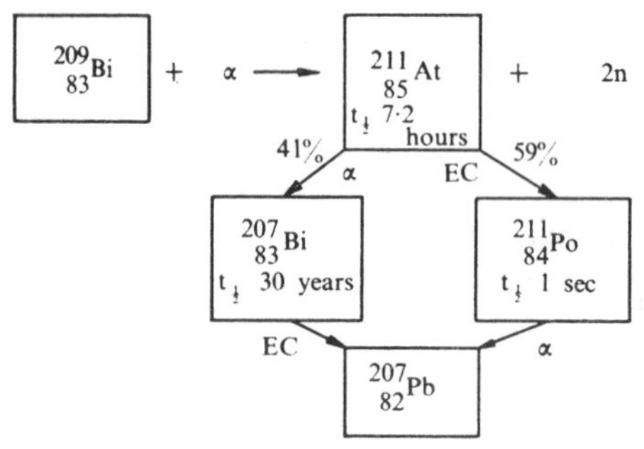

1067 F. Allison, E. R. Bishop and A. L. Sommer, *J. Amer. Chem. Soc.* **54** (1932) 616.
1068 W. Minder, *Helv. Phys. Acta*, **13** (1940) 144.
1069 A. Leigh-Smith and W. Minder, *Nature*, **150** (1942) 767.
1070 C. W. Martin, *Nature*, **151** (1943) 309.
1071 R. De, *Separate*, Bani Press, Dacca (1937).
1072 B. Karlik and T. Bernert, *Sitzungsber. Akad. Wiss. Wien, Abt. IIa*, **151** (1942) 255; *Naturwiss.* **30** (1942) 685; B. Karlik, *Monatsh.* **77** (1947) 348.
1073 G. v. Hevesy and R. Hobbie, *Z. anorg. Chem.* **208** (1932) 107.
1074 E. B. Andersen, *K. Danske Vidensk. Selsk.* **16** (1938) No. 5, 1.
1075 O. Hahn, *Naturwiss.* **14** (1926) 158.

The ^{211}At isotope thereby produced exhibited α-activity of two kinds decaying with a half-life of about 7 hr; the α-particles of lower energy (*ca.* 40%) correspond to direct α-decay of the isotope to ^{207}Bi, while those of higher energy (*ca.* 60%) correspond to orbital electron capture in the first place to produce a polonium isotope ^{211}Po, whose α-decay to ^{207}Pb has a very short half-life. The α-activity characteristic of ^{211}At could be separated from the target by distillation and was also shown to be separable from mercury, lead, thallium, bismuth and polonium. The possibility of fission was excluded by both chemical and physical considerations. Although the chemical properties described in the first publication were subsequently proved to be rather inaccurate, that the α-activity was correctly attributed to the isotope ^{211}At is now beyond dispute.

Astatine can be said to occur in nature only in the widest sense of the term, being a short-lived, rare branch product in the decay of the uranium and thorium radioactive series[1056,1076]. In accordance with the genetic relationship between polonium and astatine, Karlik and Bernert[1077] and later Walen[1078] observed, in studying the radioactivity of the daughter products of radon, that about 0·02% of Ra-A (^{210}Po) disintegrates by β-decay to ^{218}At with a half-life of about 2 sec. Following this discovery of β-branching in decay of Ra-A, which served as an impetus to the study of other natural radioactive series, Karlik and Bernert[1079] found that Ac-A (^{215}Po) also undergoes β-decay to the extent of 5×10^{-4}% to form ^{215}At, which undergoes α-decay with a half-life of about 10^{-4} sec, and which is

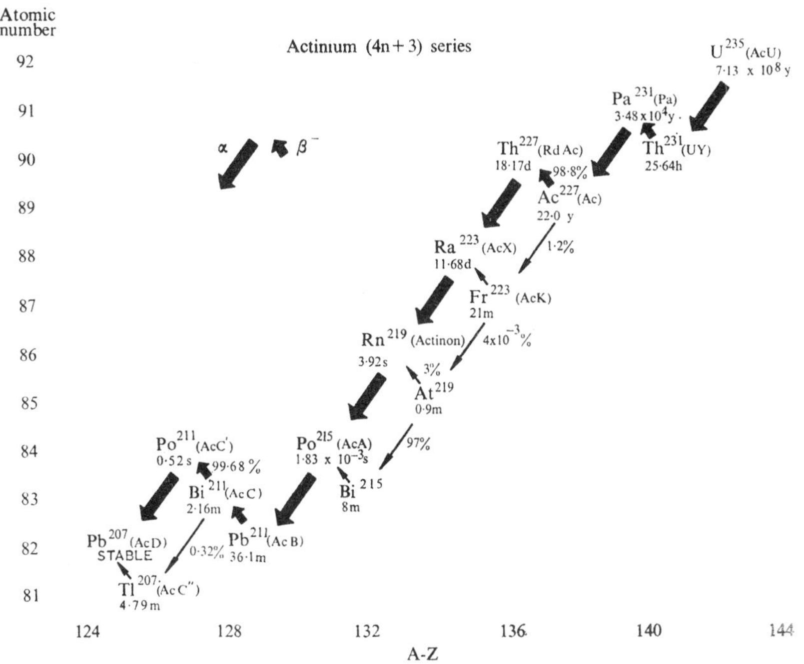

FIG. 44. Actinium (4n+3) series showing the relationships of the rare branch products ^{223}Fr, ^{219}At and ^{215}Bi. [Reproduced with permission from E. K. Hyde, *J. Chem. Educ.* **36** (1959) 16.]

1076 E. K. Hyde and A. Ghiorso, *Phys. Rev.* **90** (1953) 267.
1077 B. Karlik and T. Bernert, *Naturwiss.* **32** (1944) 44.
1078 R. J. Walen, *J. Phys. Radium*, **10** (1949) 95.
1079 B. Karlik and T. Bernert, *Z. Physik*, **123** (1944) 51.

also produced in the successive α-disintegration of ^{227}Pa[1080]. On the basis of the regularities in the α- and β-decays in the natural radioactive series, it was suggested that, as a result of rare instances of α-decay, ^{227}Ac forms a minor decay chain parallel to the main series, which avoids element 85 (see Fig. 44)[1070]. In verification of this suggestion, a careful scrutiny of the actinium ($4n+3$) series led to the isolation from the ^{227}Ac source of ^{223}Fr (half-life 21 min) and hence of an α-emitting ^{219}At isotope (half-life 0·9 min) formed by α-branching to the extent of $4 \times 10^{-3}\%$. ^{219}At was identified by chemical means, this being the only case of *chemical* isolation of astatine from a natural source[1076].

The situation with respect to the natural occurrence of astatine may be summarized as follows: (i) minute amounts of ^{218}At ($t_{\frac{1}{2}} = 2$ sec) and possibly of ^{215}At ($t_{\frac{1}{2}} = 10^{-4}$ sec) formed in uranium ores as a result of rare β-branching of the A-products; (ii) minute amounts of ^{217}At ($t_{\frac{1}{2}} = 0{\cdot}03$ sec) from the steady production of very small amounts of ^{237}Np and ^{233}U in uranium ores through natural neutron processes; (iii) minute amounts of ^{219}At ($t_{\frac{1}{2}} = 0{\cdot}9$ min) present in ^{235}U ore samples because of a very minor α-branching of ^{223}Fr, which itself is a rare product. ^{219}At is the longest-lived isotope of astatine likely to be discovered in natural sources. Estimates suggest that the outermost mile of the earth's crust contains no more than about 70 mg of astatine, as compared with 24·5 g of francium or the relatively abundant polonium (4000 tons) and actinium (11,300 tons)[1081]. Astatine thus appears to be the rarest element in nature.

5.3. ISOTOPES OF ASTATINE[1056–1059]

Some 21 isotopes of astatine have now been more or less well defined. The half-lives, modes of decay and decay energies are listed in Table 102, together with details of the nuclear reactions whereby the isotopes have been obtained and identified. Isotopes possessing 126 neutrons or less (e.g. ^{210}At and ^{211}At) are produced by bombarding bismuth with α-particles having energies in excess of 21 MeV. More recently some of the lighter isotopes of astatine have been identified in nuclear reactions involving the bombardment of gold targets with accelerated ^{12}C, ^{14}N or ^{16}O nuclei. Most of the isotopes having more than 126 neutrons are descendants of the neutron-deficient protactinium isotopes prepared in the high-energy bombardment of thorium[1082]. The heaviest isotope for which data are available is ^{219}At, the daughter of ^{223}Fr; heavier isotopes are certainly β^- unstable with half-lives of less than a minute. Almost all the known isotopes of astatine are also formed by fission of thorium and uranium by high-energy protons[1083,1084].

The half-lives and decay energies of the isotopes both for α-emission and for electron capture are of great interest in relation to the closed configurations of 82 protons and 126 neutrons, which exercise a major influence on the stability of the heaviest nuclei[1056]. Figure 45 indicates the general trends in nuclear stability in the region of the elements following lead in the form of an idealized sketch of the mass-energy surface. In general, the surface takes the form of a trough sloping upwards in the direction of higher masses, and the bottom of the trough defines the region of β-stability. The steepness of the slope determines the

1080 W. W. Meinke, A. Ghiorso and G. T. Seaborg, *Phys. Rev.* **75** (1949) 314.
1081 I. Asimov, *J. Chem. Educ.* **30** (1953) 616.
1082 W. W. Meinke, A. Ghiorso and G. T. Seaborg, *Phys. Rev.* **81** (1951) 782.
1083 F. F. Momyer, jun. and E. K. Hyde, *J. Inorg. Nuclear Chem.* **1** (1955) 274.
1084 M. Lefort, G. Simonoff and X. Tarrago, *Bull. Soc. chim. France*, (1960) 1726; B. N. Belyaev, Yung-Yu Wang, E. N. Sinotova, L. Nemet and V. A. Khalkin, *Radiokhimiya*, **2** (1960) 603.

TABLE 102. ISOTOPES OF ASTATINE[a]

Isotope	Observed mode of decay Mass excess (Δ MeV)	Energy of α-radiations (MeV)	Half-life	Genetic relationships		Principal means of production
				Daughter	Parent	
^{200}At	α	6·42 (~60%) 6·47 (~40%)	0·9 min		^{200}Po	^{197}Au–^{12}C–9*n*
^{201}At	α, EC	6·35	1·5 min	^{205}Fr	^{201}Po	^{197}Au–^{12}C–8*n*
^{202}At	α(12%), EC(88%) Δ −10	6·12 (64%) 6·23 (36%)	3·0 min		^{202}Po	^{197}Au–^{12}C–7*n*
^{203}At	α(14%), EC(86%) Δ −11	6·09	7·4 min		^{203}Po	^{197}Au–^{12}C–6*n*
^{204}At	α(4·52%), EC(95·5%) Δ −11	5·95	8·9–9·3 min		^{204}Po	^{197}Au–^{12}C–5*n* ^{209}Bi–α–9*n*
^{205}At	α(18%), EC(82%) Δ −13	5·90	26·2 min	^{209}Fr	^{205}Po	^{197}Au–^{12}C–4*n* ^{209}Bi–α–8*n* ^{197}Au–^{14}N–6*n*
^{206}At	α(~88%), EC(~12%) Δ −12	5·70	29·5–32·8 min		^{206}Po	^{197}Au–^{12}C–3*n* ^{197}Au–^{14}N–5*n* ^{209}Bi–α–7*n*
^{207}At	α(~10%), EC(~90%) Δ −13·41	5·76	1·8 hr	^{207}Rn ^{211}Fr	^{207}Po ^{203}Bi	^{209}Bi–α–6*n* ^{197}Au–^{14}N–4*n*
^{208}At	α(0·5%), EC(>99%) Δ −12	5·65	1·6 hr	^{212}Fr	^{208}Po	^{209}Bi–α–5*n*
^{209}At	α(~5%), EC(~95%) Δ −12·89	5·65	5·5 hr	^{209}Rn ^{213}Fr	^{209}Po ^{205}Bi	^{209}Bi–α–4*n*
^{210}At	α(0·17%), EC(>99%) Δ −12·12	5·519 (32%) 5·437 (31%) 5·355 (37%)	8·3 hr	^{210}Rn	^{210}Po ^{206}Bi	^{209}Bi–α–3*n*

TABLE 102 (*cont.*)

Isotope	Observed mode of decay Mass excess (Δ MeV)	Energy of α-radiations (MeV)	Half-life	Genetic relationships		Principal means of production
				Daughter	Parent	
^{211}At $I = 9/2$[b]	α(40·9%), EC(59·1%) Δ −11·64	5·868	7·21 hr	^{211}Rn	^{207}Bi ^{211}Po	^{209}Bi–α–2*n*
^{212}At	α Δ −8·64	7·60 (20%) 7·66 (80%)	0·30 sec			^{209}Bi–α–*n*
^{212m}At	α Δ −8·42	7·82 (80%) 7·88 (20%)	0·12 sec			^{209}Bi–α–*n*
^{213}At	α Δ −6·5	9·2	⩽1 sec			Descendant ^{225}Pa
^{214}At	α Δ −3·42	8·78	~2×10^{-6} sec	^{218}Fr		Descendant ^{226}Pa
^{215}At	α Δ −1·25	8·01	~10^{-4} sec	^{219}Fr ^{215}Po	^{211}Bi	Descendant ^{227}Pa
^{216}At	α Δ 2·25	7·80	~3×10^{-4} sec	^{220}Fr	^{212}Bi	Descendant ^{228}Pa
^{217}At	α Δ 4·38	7·07	0·0323 sec	^{221}Fr	^{213}Bi	Descendant ^{225}Ac
^{218}At	α(>99%), β$^-$(0·1%) Δ 8·11	6·70 (94%) 6·65 (6%)	1·5–2·0 sec	^{218}Po	^{214}Bi	Daughter ^{218}Po
^{219}At	α(~97%), β$^-$(~3%) Δ 10·5	6·28	0·9 min	^{223}Fr	^{219}Rn ^{215}Bi	Descendant ^{227}Ac natural source
>^{219}At	Unknown β$^-$ emitters with high decay energies and very short half-lives					

$\alpha = {}^4_2He$; *n* = neutron; EC = orbital electron capture.
The means of production by nuclear reactions are represented as: target element–projectile–outgoing particle.
[a] C. M. Lederer, J. M. Hollander and I. Perlman, *Table of Isotopes*, 6th edn. (1968); E. K. Hyde, *J. Chem. Educ.* **36** (1959) 15; V. D. Nefedov, Yu. V. Norseev, M. A. Toropova and V. A. Khalkin, *Russ. Chem. Rev.* **37** (1968) 87.
[b] G. H. Fuller and V. W. Cohen, *Nuclear Data Tables*, **5A** (1969) 459.

amount of energy available for α-decay. In the region of lead and bismuth the surface is relatively flat so that the decay energy for α-emission is very low. However, immediately after lead, in the region appropriate to isotopes of francium and astatine, the surface rises steeply, with the result that all the isotopes have very short lives. With the move to still

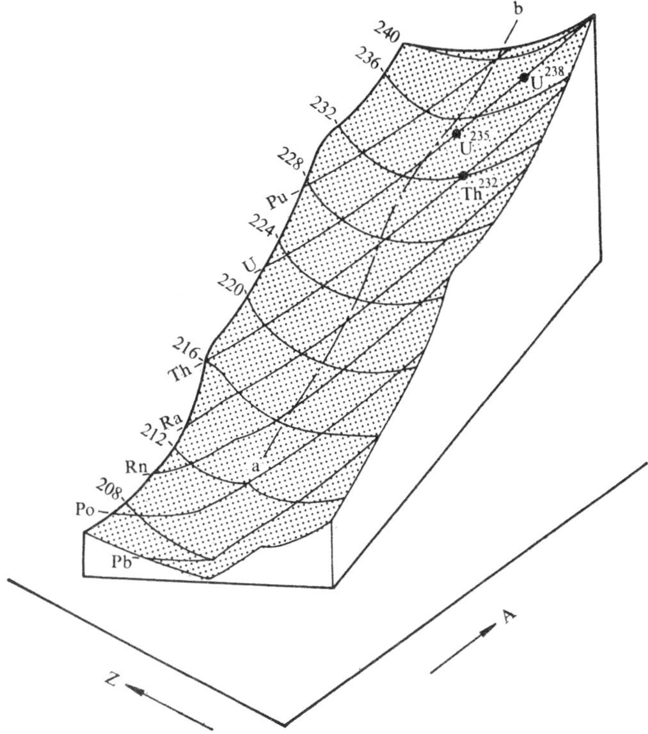

FIG. 45. Mass-energy surface for the heaviest elements. Atomic masses displayed on an idealized mass-energy surface to show general trends in nuclear instability. Vertical height is proportional to mass. Coordinates in the base plane are the atomic number Z and the mass number A. [Reproduced with permission from E. K. Hyde, *J. Chem. Educ.* **36** (1959) 15.]

heavier elements, the surface levels off somewhat, and it is in this region that the long-lived isotopes ^{232}Th, ^{235}U and ^{238}U are found. The general features of nuclear binding which underlie the gross changes on this energy surface are rather well correlated by the independent particle or shell model of the nucleus: as nuclei are built up by the addition of neutrons and protons, there occur favoured closed shells of neutrons and protons in a manner reminiscent of the occurrence of especially stable electron configurations in the building up of the chemical elements. ^{208}Pb is particularly tightly bound because it contains such closed configurations or "magic numbers" of 82 protons and 126 neutrons. The additional nucleons which must be added to make the nuclides immediately succeeding lead are not as firmly bound, a feature reflected in the steepness of the mass-energy surface in the region of element 85.

Though all of the isotopes of astatine with less than 126 neutrons (mass number < 211) undergo α-decay, the half-life of this process is sometimes quite long: e.g. 5·5 years for ^{210}At.

However, all such isotopes lie well to the neutron-deficient side of β-stability and the decay energies for electron capture are several MeV. By contrast, isotopes of astatine having more than 126 neutrons are characterized by marked instability with respect to α-decay; as with francium, the discontinuity in atomic masses which occurs at 126 neutrons causes a related discontinuity in the α-decay energies. Instead of decreasing as the mass number increases in accordance with the usual pattern, these energies first decrease and then rise abruptly to a maximum once the 126-neutron shell has been completed, corresponding to the sharp decrease in binding energy of the nucleons. The heaviest astatine isotopes show a modest increase in stability with respect to α-decay, but susceptibility to β-decay finally denies the existence of long-lived species with mass numbers in excess of 211. The β-stability of astatine has been explored by the method of closed decay-energy cycles, which suggest that all isotopes of astatine are unstable with respect to β-decay processes[1085].

The ^{210}At isotope ($t_{\frac{1}{2}} = 8{\cdot}3$ hr) is the longest-lived form of astatine at present known, ^{209}At and ^{211}At having comparable lifetimes ($t_{\frac{1}{2}} = 5{\cdot}5$ and $7{\cdot}2$ hr, respectively). There remains the possibility that an isomer of one of the known isotopes will ultimately be discovered with an appreciably longer half-life. One isomer ^{212m}At has been identified with some assurance; others, less convincing, have been reported. The consensus of opinion is that ^{210}At is the longest-lived form of astatine likely ever to be encountered and that chemical studies of astatine are likely always to be circumscribed by the constraints and ambiguities of tracer methods.

5.4. PREPARATION, SEPARATION AND ESTIMATION OF ASTATINE[1054–1059,1061]

The chemical properties of astatine are most conveniently explored using the relatively long-lived isotopes with mass numbers 209, 210 and 211; although not quite the longest-lived, ^{211}At is, from the point of view of the chemist, the most widely favoured. All three isotopes are prepared by the α-bombardment of bismuth. The threshold α-energy for astatine production is about 21 MeV, and the cross-section for the formation of ^{211}At rises to about 0·9 barn at 26 MeV; if the α-particles are accelerated to more than 29 MeV, the threshold of the (α, 3n) reaction is passed and ^{210}At is produced, while ^{209}At is formed with a fairly large cross-section by α-particles of 60 MeV. To produce ^{211}At of the highest radiochemical purity, therefore, the particle energy should not exceed 29 MeV. Astatine is the main product of the α-bombardment of bismuth; the primary reaction does not give significant yields of other elements.

The bismuth is irradiated either as the metal or as the oxide. The metal is normally applied to a support of gold, silver or aluminium by fusion or by sputtering; the oxide is most readily introduced into holes drilled in a metal slab. To avoid loss of astatine by volatilization during the irradiation, the target must be cooled efficiently.

Separation of the astatine from the target is normally effected either (i) by distillation or (ii) by "wet" extraction methods.

(i) To separate astatine from metallic bismuth, advantage is taken of the relatively high volatility of the astatine. Distillation is typically performed in a current of nitrogen (at a pressure of a few mm Hg) or *in vacuo*, the target being heated to 300–600°C. The astatine is condensed on a suitable cold surface, offered either by a glass cold finger or by a cooled platinum disc. Further purification is achieved by redistillation. The astatine is

[1085] F. Everling, L. A. König, J. H. E. Mattauch and A. H. Wapstra, *Nucl. Phys.* **18** (1960) 529; A. H. Wapstra, *ibid.* **28** (1961) 29.

finally washed from the glass or metal surface on which it has been collected with a suitable aqueous solution, e.g. dilute nitric acid. If the distillation is carried out at too high a temperature, the astatine is contaminated with polonium and possibly even with weighable quantities of bismuth, and tends to form radiocolloids which may seriously interfere with many experiments. The proportion of astatine effectively transferred in this way is generally low (5–30%) and irreproducible, probably because much of it is lost by adsorption or by chemical reactions on the walls of the vessels used or on other materials inevitably present.

(ii) Wet methods for the isolation of astatine can be employed either for the preparation of carrier-free solutions or of solutions in an excess of iodine. In a typical procedure the irradiated bismuth or bismuth oxide is dissolved in perchloric acid containing a little iodine as a carrier for the astatine. The bismuth and some of the polonium are then precipitated as phosphates. For many purposes this liquid can be used as an aqueous solution of AtI, though it still contains some polonium, which may give rise to difficulties in certain situations. Purification can be achieved by extracting the I_2 containing AtI into carbon tetrachloride or chloroform, whence it may be re-extracted into a reducing aqueous phase. Alternatively, the astatine, initially in a hydrochloric acid solution in the presence of iron(II) salts, is extracted into di-isopropyl ether. This method has the advantage that, after the addition of a suitable quantity of tributyl phosphate, polonium and bismuth can be quantitatively returned to an aqueous phase containing nitric and hydrochloric acids. By the use of such wet methods losses of astatine have been restricted so as not to exceed 5%.

Methods have also been devised for separating astatine from lead, bismuth, thorium and polonium following the fission of thorium under the action of high-energy protons[1059]. These depend upon the dissolution of the target in acid, followed by selective redox and precipitation reactions, together with solvent-extraction or chromatographic techniques.

The preparation of astatine for estimation is largely dependent on the composition of the solution and the method of analysis. If only volatile materials are present in appreciable quantities, and if one is dealing with aqueous solutions, the liquid may be evaporated to dryness on silver or platinum. However, quantitative deposition of astatine on evaporation is difficult to achieve from organic systems. Alternatively astatine may be precipitated from aqueous solutions on a silver foil after the fashion of the technique used to study polonium. Another approach is to coprecipitate the astatine with other materials: astatine(0) is coprecipitated quantitatively with the sulphides of bismuth, mercury, silver and antimony, rather less completely with HgO, $Fe(OH)_3$, $Al(OH)_3$ and $La(OH)_3$. Tellurium-precipitation carries astatine satisfactorily when tellurous acid is reduced with sulphur dioxide in a hydrochloric acid medium. In the presence of iodine, astatine is best isolated by the precipitation of an iodide from a reducing solution. For this purpose the precipitation of palladium iodide is reported to be more reliable than that of silver iodide.

The characteristic α-spectrum of a ^{211}At sample serves as a distinctive autograph of the isotope; the 40 : 60 ratio of the 5·87 MeV α-particles of ^{211}At and the 7·43 MeV α-particles of ^{211}Po, and the controlling half-life of 7·2 hr for both, furnish conclusive proof of the identity and purity of a stock solution. This spectrum can be obtained with a gridded ion chamber whose output is subjected to linear amplification and pulse height analysis. If the radiochemical purity of the sample is assured, it is possible to follow the fate of astatine through tracer experiments simply by employing, for example, a liquid scintillation counter. If the purity is in question, there is the danger that ^{210}Po may interfere with such measurements. The α-activity may be measured either on a very thin or on a very thick sample; the former arrangement is more accurate for counting but the preparation is more adventitious, whereas the latter arrangement is generally quite easy to prepare but the assay may produce considerable errors. However, numerous investigators prefer to determine ^{211}At by counting the 80 keV K–X-rays originating in the electron-capture branch of the astatine decay. For example, Appelman has analysed astatine solutions by dilution to a standard volume and by counting the 80 keV X-rays with a counter containing a well-type crystal of sodium iodide. The α-emission of ^{210}At being insignificant, this nuclide is estimated by the X-rays or γ-rays produced by decay; likewise ^{209}At decays to the extent of only *ca.* 5% by α-emission and is most conveniently determined by X-ray or γ-ray counting. For chemical purposes it does not, as a rule, matter very much if the astatine activity observed is due partly, say, to ^{210}At and partly to ^{211}At, though the X- and γ-radiations emitted by ^{209}At, ^{210}At and ^{211}At differ widely in energy, and discrimination can be effected by a crystal with good resolution. Various methods have been described whereby absolute determinations of the activity of astatine can be carried out, though α-standardization of ^{211}At by liquid scintillation counting appears to be most reliable. Autoradiography of astatine has also been used successfully in biological work, the activity measured being the α-emission of ^{211}At.

5.5. PHYSICAL PROPERTIES OF ASTATINE AND ITS COMPOUNDS

Since the half-lives of known astatine isotopes are so short as to preclude the isolation of the element or its compounds in macroscopic quantities, few extranuclear physical properties of astatine-containing systems can be measured. Most of the properties described have been interpolated or extrapolated by various theoretical or empirical treatments from the data available for neighbouring or homologous elements; some of these properties are listed in Table 103. For example, a first ionization potential of *ca.* 9·5 eV has been estimated[1086], while extrapolation of the almost linear relationship between I_1 for O, S, Se and Te and I_2 for F, Cl, Br and I implies that I_2 = 17·0 eV (rather than 18·2 eV as suggested elsewhere[1086]). Extrapolation of the vibrational frequencies ω_e of the familiar diatomic halogen molecules points to a value of ω_e of *ca.* 160 cm^{-1} for the At_2 molecule. Use of the empirical relationships

$$\log \omega_e = a - b \log n^2 I$$
$$\log D = c + d \log I$$

(n = principal quantum number of the valence shell; a, b, c and d = constants or near-constants)

between ω_e and I and D, the ionization potential and dissociation energy, respectively, of the At_2 molecule then leads to the estimates of I and D given in the table[1087]. These results, combined with a probable value of *ca.* 20 kcal mol^{-1} for $\Delta H_f[At_2(g)]$, imply that astatine should be better disposed than iodine to form cations of the type At^+ or At_2^+ as chemically recognizable entities.

As evaluated by several investigators[1088], the conventional ionic radius of the At^- ion is close to 2·3 Å. Radius-ratio considerations suggest that CsAt is therefore likely to have a caesium chloride-type structure, LiAt a zinc blende lattice and the remaining alkali-metal astatides to be isostructural with sodium chloride; together with compressibilities and other factors estimated by reference to the known alkali halides, these assumptions have been

1086 W. Finkelnburg and F. Stern, *Phys. Rev.* **77** (1950) 303.

1087 R. W. Kiser, *J. Chem. Phys.* **33** (1960) 1265.

1088 W. H. Zachariasen, *The Actinide Elements* (ed. G. T. Seaborg and J. J. Katz), p. 775, McGraw-Hill, New York (1954); L. Genov, *J. Gen. Chem.* (*U.S.S.R.*) **29** (1959) 683; G. A. Krestov, *Radiokhimiya*, **4** (1962) 690.

TABLE 103. ESTIMATED PHYSICAL PROPERTIES OF ASTATINE AND ITS DERIVATIVES

Property	Value	Reference
Atomic properties		
Atomic number	85	
Electronic configuration of neutral At atom in its ground state	$[Xe]4f^{14}5d^{10}6s^26p^5$ ($^2P_{3/2}$)	
	eV *kcal*	
Ionization potentials, I_1	9·5 220	a
I_2	17·0 392	Estimated from I_1(Po)
Electron affinity at 298°K	3·1 71	b*
Electronegativity	2·4	Based on Mulliken scale of electronegativity
Absorption spectrum of atomic astatine	2 lines observed	c
Properties of elementary astatine		
Dissociation energy of At_2 molecule, $D(At_2)_{298°K}$	*eV* *kcal* 1·2 27·7	d
$\Delta H^\circ_{f,298°K}$[At(g)]	1·0 24	d, e
Internuclear distance, r_e(At–At)	2·9 Å	Estimated by extrapolation
	eV *kcal*	
Ionization potential, $I(At_2)$	8·3 191	d
Vibrational frequency, ω_e	160 cm^{-1}	d
Melting point	~300°C	f
Boiling point	~335°C	f
Properties of the astatide ion		
Ionic radius of At^- ion	2·3 Å	g
	eV *kcal*	
$-\Delta H^\circ_{f,298°K}[At^-(g)]$	2·0 47	b
$-\Delta G^\circ_{hyd}[At^-]$ at 298°K	64 kcal	e, estimated from ionic radius
$-\Delta H^\circ_{hyd}[At^-]$ at 298°K	66 kcal	e, estimated from ionic radius
Reduction potential, $E^\circ(\frac{1}{2}At_2/At^-)$	+0·3 V	h

[a] W. Finkelnburg and F. Stern, *Phys. Rev.* **77** (1950) 303.

[b] M. F. C. Ladd and W. H. Lee, *J. Inorg. Nuclear Chem.* **20** (1961) 163; * also involves estimate of $\Delta H_f[At_2(g)]$.

[c] R. McLaughlin, *J. Opt. Soc. Amer.* **54** (1964) 965.

[d] R. W. Kiser, *J. Chem. Phys.* **33** (1960) 1265.

[e] See for example A. G. Sharpe, *Halogen Chemistry* (ed. V. Gutmann), Vol. 1, pp. 10–12, Academic Press (1967); D. A. Johnson, *Some Thermodynamic Aspects of Inorganic Chemistry*, pp. 70–71, 101–105, Cambridge (1968).

[f] R. E. Honig and D. A. Kramer, *Vapour Pressure Curves of the Elements*, Sheet C, RCA Laboratories (1969).

[g] W. H. Zachariasen, *The Actinide Elements* (ed. G. T. Seaborg and J. J. Katz), p. 775, McGraw-Hill, New York (1954); L. Genov, *J. Gen. Chem.* (*U.S.S.R.*) **29** (1959) 683; G. A. Krestov, *Radiokhimiya*, **4** (1962) 690.

[h] E. H. Appelman, *J. Amer. Chem. Soc.* **83** (1961) 805.

shown[1063] to furnish the following lattice energies (U_0, kcal mol^{-1}): LiAt, 172; NaAt, 157; KAt, 147; RbAt, 142; CsAt, 140. Heats of formation of these astatides have also been computed by an extrapolation technique, whence $\Delta H_f[At^-(g)]$ has been evaluated. On the basis of this and the estimated values of $D(At_2)$ and $\Delta H_f[At_2(g)]$, the electron affinity of astatine emerges as *ca.* 71 kcal mol^{-1}. Other derivations of thermodynamic parameters for astatine and some of its compounds have been given[1059], while the ionic radius of At^- has been applied here to the estimation of absolute values of $\Delta G°$ and $\Delta H°$ for the hydration of the ion (see Table 103).

The bulk of the At^- ion clearly implies that the lattice and hydration energies of ionic astatides are likely to be significantly less than those of corresponding chlorides, bromides or iodides. Accordingly the At^- ion, even more than the I^- ion, is liable to be unstable (with respect to redox processes) in solids containing metal ions in relatively high charge states, e.g. Fe^{3+} or Cu^{2+}, but stable in the presence of metal ions in relatively low charge states, e.g. Cu^+ or Eu^{2+} (see p. 1119). Likewise the solubilities of ionic astatides should follow a pattern similar to that of the corresponding iodides, though for cations of a given charge the free energy of solution should reach a maximum (corresponding to low solubility) at a larger cation radius. With B-metal ions like Ag^+ or Hg^{2+}, however, charge-transfer and electron delocalization are likely to exercise an even greater influence in solid astatide derivatives than in the corresponding iodides. In all probability the energies of covalent bonds engaging a halogen to a more electropositive atom follow the sequence F > Cl > Br > I > At, though with respect to a more electronegative partner astatine may well form the strongest bonds.

5.6. CHEMICAL PROPERTIES OF ASTATINE[1056–1059,1061]

As already indicated, information about the chemical properties of astatine is available almost exclusively from tracer experiments, which are notoriously liable to misinterpretation. The following factors are typical of those which must be weighed in any attempt to interpret the behaviour of astatine under these conditions: (i) the very low concentration of the species under investigation, (ii) adsorption phenomena, (iii) interference from impurities, (iv) formation of radiocolloids, (v) photosensitivity of reactions, and (vi) differences in behaviour between astatine and the carrier element iodine.

Notwithstanding these difficulties, at least four oxidation states of astatine are known at present: a -1 state that coprecipitates with AgI or PdI_2; a zero state which can be extracted into organic solvents; a $+5$ state that is carried by insoluble iodates; and an intermediate state ($+1$ or $+3$?) that is neither extracted into organic solvents nor carried from solution by insoluble iodates. The redox equilibria among these oxidation states have been studied qualitatively in acidic solution using reversible redox couples and approaching the equilibria from both sides[1064]. On the basis of these chemical tests, the reduction potential scheme shown in Fig. 46 has been tentatively proposed; the behaviour of iodine and astatine is compared in the accompanying oxidation state diagram. Figure 47 summarizes the principal features of the aqueous chemistry of astatine that appear to have been established beyond the realm of dispute.

Oxidation State 0

When astatine is isolated from a bismuth target by volatilization, it shows many properties expected of the zero oxidation state. It is volatile from glass surfaces at room temperature, but is adsorbed strongly by metals such as silver, gold and platinum and weakly by others, e.g. nickel and copper. Adsorption by silver remains high at 325°C indicating the formation of a stable compound. Zerovalent astatine is soluble in nitric acid and is commonly used in this form for tracer studies. Evaporation of the solution causes some volatilization of the astatine. The retention of the halogen upon various surfaces during this

(a) *Standard potentials E° in volts*

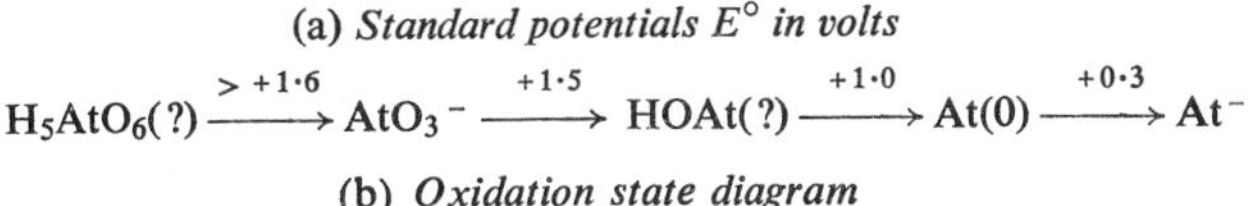

(b) *Oxidation state diagram*

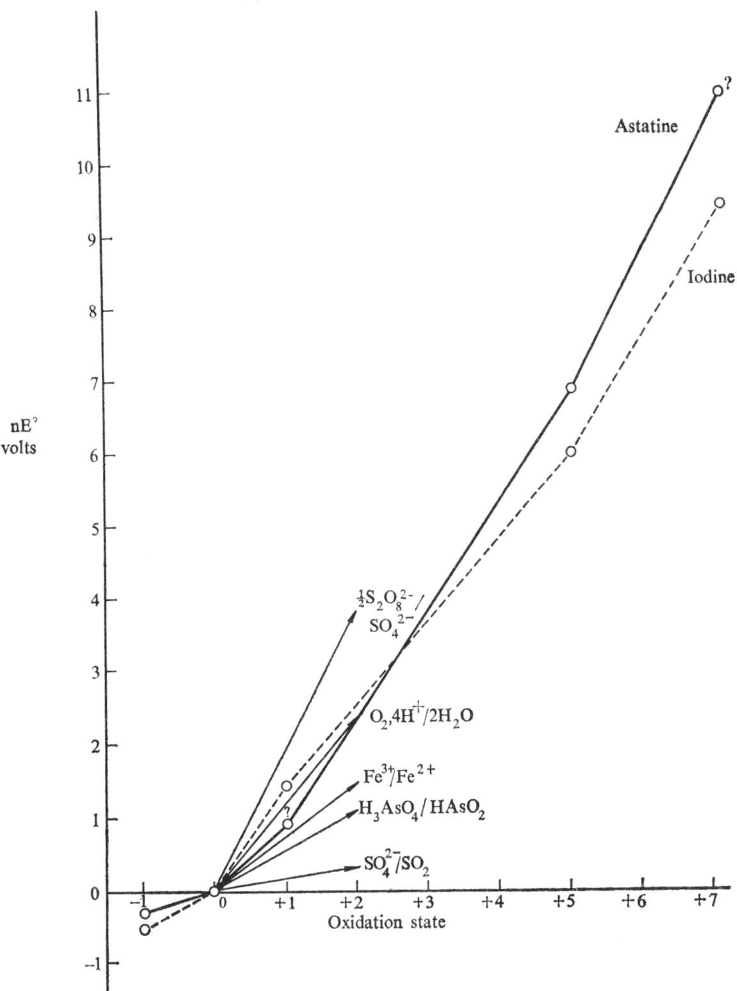

FIG. 46. Suggested oxidation states and approximate reduction potentials of astatine in aqueous solution, $a_{H^+} = 1{\cdot}0$.

evaporation depends upon the nature of the surface, the composition of the aqueous solution and the conditions of evaporation, but the volatility losses are not as high as those for similar solutions containing iodine. There exists a considerable fund of coprecipitation data relevant to the aqueous chemistry of the lowest oxidation states of astatine. Thus, astatine is carried quantitatively by insoluble sulphides precipitated from strong hydrochloric acid and by elemental tellurium deposited by the action of sulphur dioxide, zinc or

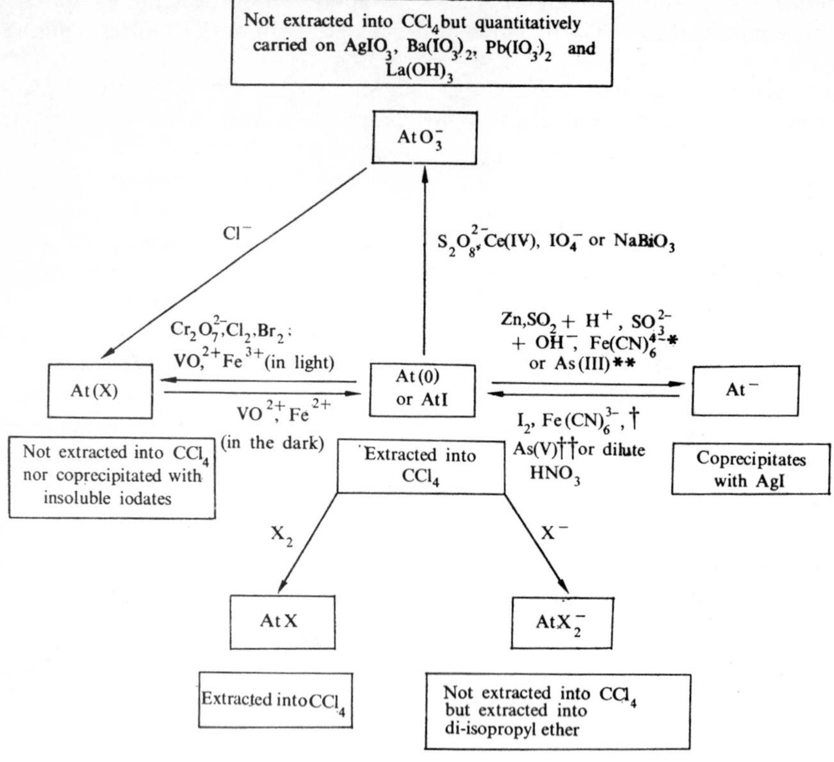

Conditions

* Low acidity and ionic strength, $[Fe(CN)_6^{4-}] \gg [Fe(CN)_6^{3-}]$ (E < +0·4 V)

** Low acidity and ionic strength, $[As(III)] \gg [As(V)]$ (E < +0·3 V)

† 1:1 $Fe(CN)_6^{3-}$/ $Fe(CN)_6^{4-}$ mixture at pH< 3 (E > +0·4 V)

†† pH< 4 (E > +0·3 V)

FIG. 47. Aqueous reactions of astatine.

other strong reducing agents, but only partially by insoluble hydroxides. It is also deposited from dilute nitric acid solutions on such surfaces as copper, bismuth and silver. In this respect it closely resembles the neighbouring element polonium.

The zero oxidation state of astatine resembles that of iodine in the freedom with which it is extracted from aqueous media into organic solvents, which include carbon tetrachloride, benzene and other hydrocarbons, as well as donor solvents such as ethers and tributyl phosphate. With the exception of astatine(0) and interhalogen compounds such as AtI, none of the other oxidation states is extracted into a non-polar organic solvent like carbon tetrachloride. It follows that, in principle, such solvent-extraction procedures afford an excellent means of deriving quantitative information about astatine chemistry. By observing changes in extraction coefficients with astatine concentration, with acidity, with the concentration of potential complexing agents or with oxidation-reduction conditions, it should be possible to specify the formula of the extracted species, the extent of complexing and the oxidation potentials. In practice, only crude estimates of these quantities have generally been forthcoming because of the difficulty of realizing reproducible, non-drifting results. The general characteristic of solvent-extraction studies is that the nominal distribution coefficients achieved when a dilute acid solution is repeatedly put in contact with an organic solvent decrease quickly to a very low value, while the distribution coefficient for the repeated "back-washing" of the first solvent fraction rises steadily towards some limiting value of perhaps 50–100, which is not well reproduced with a different stock solution of astatine.

Presumably the solvent-extraction behaviour of an element as reactive as iodine or astatine at extremely low concentrations is conditioned by adsorption on, or reaction with, impurities, colloidal particles or the surfaces in contact with the solutions. Under these handicaps only large changes in distribution coefficients are meaningful, and it has not been possible to decide on the basis of present evidence whether the species extracted by the organic phase is At_2 or At. Nevertheless, our knowledge of astatine chemistry has been substantially enlarged by qualitative studies of this sort. Thus, a *decrease* of the partition coefficient of astatine between the organic and aqueous layers has been observed with the following reagents[1056–1059,1064]:

Alkali. This suggests that in aqueous alkaline solution astatine, like iodine, disproportionates to the -1 and a higher oxidation state. Increasing the acidity of the solution restores the astatine to its initial state and it is again extracted by the organic phase.

Reducing Agents such as Sulphur Dioxide, Arsenic(III) or Metallic Zinc. Evidently astatine(0) is reduced to the astatide ion.

Halide Ions, e.g. Cl^-, Br^- or I^-. Presumably polyhalide ions incorporating astatine are formed. The distribution coefficient diminishes markedly in the presence of a non-polar solvent like carbon tetrachloride but remains high in the presence of a donor solvent like di-isopropyl ether.

Oxidizing Agents of Intermediate Power, e.g. Cl_2, Br_2, Vanadate, Concentrated Nitric Acid and Fe^{3+}. These oxidize the astatine to some oxidation state X intermediate between 0 and +5. Very little is known about this state. Several of the reactions are further complicated, moreover, by their photosensitivity. The reaction with chlorine, bromine or iodine gives the interhalogen compounds AtX (X = Cl, Br or I)[1065].

Powerful Oxidizing Agents, e.g. $S_2O_8^{2-}$, Periodate, HOCl, Ce(IV) or Sodium Bismuthate. These convert astatine to a form which coprecipitates with insoluble iodates, presumably AtO_3^-.

It is also significant that Fe^{2+} reduces higher oxidation states of astatine to astatine(0) but not to the At^- ion[1064]. Distillation experiments from a variety of media (e.g. nitric, hydrochloric, perchloric and sulphuric acids) indicate that the involatile states of astatine present in most of these solutions can be reduced to volatile astatine(0) by Fe^{2+}. Solvent-extraction and coprecipitation studies show that astatine in the +X oxidation state is reduced to At(0) in the dark by the VO_2^+/VO^{2+} couple ($E^\circ = +0{\cdot}95$ V) and by the Fe^{3+}/Fe^{2+} couple ($E^\circ = +0{\cdot}76$ V) even when the ratio $[Fe^{3+}]/[Fe^{2+}]$ is large ($E = 0{\cdot}89$ V). Under the influence of light, however, the vanadium couple oxidizes At(0) quantitatively to At(+X) provided some iodine is present. The reaction of astatine with the ferric–ferrous couple is not sensitive to light unless $[Fe^{3+}]/[Fe^{2+}] > 100$ ($E > +0{\cdot}87$ V). When the ratio is greater than this, the astatine is nearly completely photo-oxidized to At(+X) whether or not iodine is present. The photochemical reactions thus appear to involve a gross shift of a chemical equilibrium under the influence of light. Such an effect is likely only if the reaction under investigation involves a species present at trace concentrations, but the primary photochemical process involves only species present at macro concentrations. The information about these reactions is insufficient to justify speculation as to their mechanisms, but the results make clear the hazards of interpretation that beset tracer studies of astatine.

Astatide is oxidized to At(0) by the $\frac{1}{2}I_2/I^-$ couple ($E^\circ = +0{\cdot}54$ V), by a 1:1 ferrocyanide–ferricyanide mixture at pH < 3 ($E > +0{\cdot}4$ V) and by the As(V)/As(III) couple at pH < 4 ($E > +0{\cdot}3$ V). On the other hand, the ferrocyanide–ferricyanide couple reduces At(0) to At^- at low acidity and ionic strength if the concentration of ferrocyanide ions is several times that of ferricyanide ($E < +0{\cdot}4$ V), and the As(V)/As(III) couple behaves similarly at pH > 4 ($E < +0{\cdot}3$ V). These results serve to define a value of *ca.* +0·3 V for $E^\circ[\mathrm{At(0)/At^-}]$.

In summary, astatine(0) has been characterized as volatile, soluble in organic solvents and susceptible to adsorption on various surfaces; unfortunately it is also noted for its severe irreproducibility of behaviour.

Oxidation State −1

The formation of the astatide ion by reduction of astatine(0) with a sufficiently powerful reducing agent is indicated (i) by electromigration experiments and (ii) by coprecipitation with insoluble iodides such as AgI, TlI or PdI_2, irrespective of pH and of other anions which may be present. Unlike astatine(0), which may also be carried from solution, the astatide ion is not removed from the insoluble iodides by washing with acetone, nor is it extracted from aqueous solutions by carbon tetrachloride. Although the astatide ion probably represents the best defined chemical state of astatine, having a striking resemblance to the iodide ion, it is not retained very well by some iodide precipitates like AgI, though initially it coprecipitates quite efficiently. The ratio of the diffusion coefficients $D_{I^-}/D_{At^-} = 1{\cdot}41$ in 1% sodium chloride solution containing $1{\cdot}2\times10^{-3}$ M KI and 4×10^{-3} M Na_2SO_3 has been determined[1058], a value implying that the diffusion coefficients of the halide ions pass through a maximum and then decrease as a function of atomic number. The result can also be used to evaluate the mobility of the At^- ion in aqueous solution, which likewise conforms to the pattern $F^- < Cl^- \lesssim Br^- \gtrsim I^- > At^-$.

The earliest tracer studies showed signs of the formation of a volatile hydride in acid solutions containing the astatide ion. More recently the hydride has been identified by its mass spectrum[1065], though whether it is produced by the hydrolysis of astatine by traces of water or by substitution reactions with organic impurities is not clear.

Interhalogens and Polyhalide Ions

Whereas the behaviour of astatine(0) is irreproducible, by mixing astatine(0) with other halogens or halide ions it is often possible to obtain reproducible behaviour characteristic of interhalogen species. In the presence of a large excess of elemental iodine, for example, the so-called "zero" oxidation state is represented exclusively by AtI molecules. Under these circumstances the astatine is protected against adsorption and a fairly accurate and reliable value of *ca.* 5·5 for the distribution ratio between carbon tetrachloride and water has been obtained; for the more polar AtBr the corresponding value is 0·04. The coprecipitation of AtI with I_2 from chloroform solutions has also been studied. Mass spectrometric studies confirm that the direct addition of excess iodine, bromine or chlorine to a freshly sublimed sample of astatine converts a large part of the astatine into the corresponding monohalide[1065]. Although higher astatine chlorides and bromides might be expected to exist, they would probably not be formed under the rather mild halogenating conditions employed in this experiment. Curiously the action of ClF_3 on astatine fails to give a volatile product.

The formation of polyhalide species has been studied by means of solvent-extraction experiments. One such study showed that astatine is extracted by di-isopropyl ether from chlorine-treated hydrochloric acid solutions, affording strong evidence of the presence of a polyhalide ion, though whether this should be formulated as $AtCl_2^-$ or $AtCl_4^-$ could not be determined. Appelman[1089] has exploited the distribution of astatine between water and carbon tetrachloride to investigate the reactions of the element with I_2, I^-, IBr, Br^- and Cl^-. Hence the species AtI, AtBr, AtI_2^-, $AtIBr^-$, $AtICl^-$, $AtBr_2^-$ and $AtCl_2^-$ have been characterized and the equilibrium constants interrelating them have been evaluated. These constants correlate well with analogous data relating to the lighter halogens, and it is interesting to note that astatine is always complexed a little more strongly than iodine in the corresponding circumstances, the ratios of the formation constants amounting to a factor of 2·5–8. The formation of AtCl and $AtCl_2^-$ has likewise been confirmed more recently by measurements of the distribution of astatine between an aqueous oxidizing solution and metallic platinum coated with an oxide film, and stability constants for these species have thus been derived[1090]. The $AtCl_2^-$ ion has also been characterized by paper electromigration experiments and is adsorbed on sulphonic-cation-exchange resins, e.g. Dowex 50.

An alternative approach to the investigation of astatine-containing polyhalide species is to produce solid derivatives. For example, CsI_3 containing $CsAtI_2$ is easily prepared by adding an excess of solid iodine to an aqueous solution of CsI containing some AtI. The thermal decomposition of $CsAtI_2$ obeys the general rule for the decomposition of polyhalides, namely that the halogen of lowest atomic number is retained to form the involatile monohalide; hence, during the decomposition of $CsAtI_2$ in the CsI_3 host lattice, astatine in the form of AtI is volatilized and involatile CsI remains[1091].

Oxidation State +X

In aqueous solution there appears to exist an oxidation state intermediate between 0 and +5. This may represent either AtO^- or AtO_2^-, but no convincing evidence of its chemical nature is yet available, and it must be borne in mind that so-called "At(+X)" need not represent a single species or even a single oxidation state. Indeed the nature of the state may vary with the mode of preparation. Formed by the oxidation of aqueous astatine(0) with oxidants such as bromine or dichromate or, in the presence of light, vanadate or iron(III), astatine(X) is characterized by not being extracted into carbon tetrachloride and by its failure to coprecipitate with either insoluble iodates or iodides. The oxidation state also results when astatate is reduced with chloride ions [$E°(\frac{1}{2}Cl_2/Cl^-) = +1·36$ V]; the periodate–iodate couple [$E° = ca. +1·6$ V] effects the reverse change, although the reaction is sometimes slow and incomplete at room temperature. A potential of +1·5 V is thus indicated[1064] for the couple AtO_3^-/At(X). On the evidence of such redox behaviour,

1089 E. H. Appelman, *J. Phys. Chem.* **65** (1961) 325.
1090 Yu. V. Norseev and V. A. Khalkin, *J. Inorg. Nuclear Chem.* **30** (1968) 3239.
1091 G. A. Brinkman, J. Th. Veenboer and A. H. W. Aten, jun., *Radiochim. Acta*, **2** (1963) 48.

the +X state appears to be significantly more stable than either IO^-, which is known, or IO_2^-, which is not.

Recent experimental results[1059,1061b] on the electrodeposition of astatine from oxidizing solutions imply the presence of an astatine cation having a charge of +1. This species—possibly At^+ or AtO^+—is adsorbed on monofunctional sulphonic-cation-exchang eresins, on certain precipitates containing large univalent cations, e.g. caesium tungstophosphate and silver iodate, and on metallic platinum coated with an oxide film (see above). The astatine deposited on platinum is held strongly on the metal surface; adsorption is increased by raising the temperature and decreased by reducing the pH; desorption is easily accomplished by anodic polarization of the platinum. Evidence of a cationic species analogous to $[(C_5H_5N)_2I]^+$ is also forthcoming from tracer studies of the formation of $[(C_5H_5N)_2I]^+X^-$ ($X = NO_3$ or ClO_4) in the presence of astatine[1092]. The ratio of astatine to iodine is higher in these derivatives than in the iodine used for the synthesis, a finding concordant with the enhanced ability of astatine to form cationic species, e.g. according to the reaction

$$X_2 \xrightleftharpoons{\text{Donor species}} X^+_{\text{coord}} + X^-_{\text{coord}}$$

In all probability the free energy of this process in strongly coordinating solvents like water is most favourably disposed towards ionization in the case where X = At.

Oxidation States +5 and +7

Treatment of astatine(0) with strong oxidizing agents converts it to a form which is not extracted into carbon tetrachloride but is quantitatively coprecipitated with insoluble iodates, e.g. $AgIO_3$, $Ba(IO_3)_2$ and $Pb(IO_3)_2$, as well as with lanthanum hydroxide. On this basis, the existence of an astatate ion, presumably formulated as AtO_3^-, has been inferred, the negative charge on the ion being demonstrated by means of electromigration experiments. Curiously it has been reported that although, in the presence of strong oxidizing agents, astatine coprecipitates well with $Pb(IO_3)_2$, a large part of the activity due to the astatine may be eluted from the precipitate by acetone.

After the most vigorous oxidation with agents such as periodate or cerium(IV) in 6 M perchloric acid, the astatine appears to be entirely in the form of AtO_3^-; precipitation of potassium periodate fails to carry any of the activity with it. Lately, however, there have been reports[1093] testifying that aqueous perastatate is generated either through the agency of such potent oxidants as XeF_2, $K_2S_2O_8$ or KOCl or by anodic oxidation; the identity of the AtO_4^- ion is sustained both by electrophoretic analysis and by coprecipitation with MIO_4 salts (M = K or Cs) from solutions at pH 6.

Organic Derivatives of Astatine[1058,1059,1061b]

The method of producing and studying organic derivatives of astatine which has so far proved most practicable involves the use of iodine as a carrier and investigation of the subsequent fate of the activity due to the traces of astatine. Identification of the astatine compounds is achieved most conveniently with the aid of chromatographic techniques. However, the difficulties encountered in working with the inorganic forms of astatine are exacerbated in the study of organic derivatives of the element. Irregular extraction and

[1092] J. J. C. Schats and A. H. W. Aten, jun., *J. Inorg. Nuclear Chem.* **15** (1960) 197.

[1093] V. A. Khalkin, Yu. V. Norseev, V. D. Nefedov, M. A. Toropova and V. I. Kuzin, *Doklady Chem.* **195** (1970) 855; G. A. Nagy, P. Groz, V. A. Hakin, Do Kim Tuong and Yu. V. Norseev, *Kozp. Fiz. Kut. Intez. Kozlem.* **18** (1970) 173.

coprecipitation, combined with the relative instability of many astatine compounds, often make the interpretation of observations quite uncertain. If, for instance, iodoform is synthesized from iodine containing AtI, the first precipitate of CHI_3 shows a very appreciable activity, which disappears rapidly in the course of a series of recrystallizations. It is not clear whether this signifies that the lattice did not contain $CHAtI_2$ at all or whether this compound, once formed, is very easily decomposed, though the first explanation seems more plausible. This situation probably explains the early rather conflicting reports made about the synthesis of organic compounds of astatine[1058].

A general consideration of the chemical properties of astatine suggests that the best chance for incorporation of astatine into organic molecules may exist in those cases where the astatine atom carries a partial positive charge, both in the final product and in any intermediate stage through which it passes during the reaction. In keeping with this general observation, methods have been described[1059,1094] for the synthesis of astatine-labelled compounds, not only of the type RAt, but also of the types $RAtCl_2$, R_2AtCl and $RAtO_2$ (where R = phenyl or *p*-tolyl); the details are outlined in the following scheme:

$$R_2ICl + KI(At) \longrightarrow R_2I{\cdot}I(At) \xrightarrow{170\text{-}190^\circ C} RI(At) \xrightarrow{Cl_2} RI(At)Cl_2 \begin{cases} \xrightarrow{R_2Hg} R_2I(At)Cl \\ \xrightarrow[70\text{-}100^\circ C]{NaOCl} RI(At)O_2 \end{cases}$$

A variety of methods has been exploited lately to synthesize astatobenzene[1095], e.g.

$$Ph_2II \xrightarrow{At^-} Ph_2IAt \xrightarrow{175^\circ C} PhI + PhAt$$

$$PhI \xrightarrow{AtI} PhAt \xleftarrow{AtI_2^-,\, I_3^-} PhNHNH_2$$

$$PhN_2Cl \xrightarrow{At^-} PhAt \xleftarrow[130\text{–}200^\circ C]{At^-} PhI$$

Reference has also been made to hot-atom reactions which apparently afford the opportunity of generating carrier-free astatobenzene: in one such procedure triphenylbismuth is irradiated with α-particles, and in another the electron-capture decay of ^{211}Rn to ^{211}At is engineered in the presence of benzene. Successful syntheses have been claimed, moreover, for *p*-astatobenzoic and *p*-astatobenzenesulphonic acid[1058]; the latter is of interest as providing a means of incorporating astatine into different proteins, a process which can also be accomplished by means of diazo-coupling of benzidine to protein. Elsewhere there are good grounds for believing that the interaction of the At^- ion, under appropriate conditions, with an alkyl iodide[1096] or with ICH_2COOH[1097] furnishes the corresponding organo-astatine compound.

Separation and characterization of organic derivatives have commonly entailed the use of thin-layer, paper or gas chromatography, the last of these techniques inviting

1094 V. D. Nefedov, Yu. V. Norseev, Kh. Savlevich, E. N. Sinotova, M. A. Toropova and V. A. Khalkin, *Doklady Chem.* **144** (1962) 507.

1095 G. Samson, *Chem. Weekbl.* **65** (1969) 27; G. Samson and A. H. W. Aten, jun., *Radiochim. Acta*, **13** (1970) 220; V. D. Nefedov, M. A. Toropova, V. A. Khalkin, Yu. V. Norseev and V. I. Kuzin, *Radiokhimiya*, **12** (1970) 194; V. I. Kuzin, V. D. Nefedov, Yu. V. Norseev, M. A. Toropova and V. A. Khalkin, *Radiokhimiya*, **12** (1970) 414.

1096 G. Samson and A. H. W. Aten, jun., *Radiochim. Acta*, **12** (1969) 55; M. Gesheva, A. Kolachkovsky and Yu. V. Norseev, Preprint, Joint Institute for Nuclear Research, Dubna, P6-5683 (1971).

1097 G. Samson and A. H. W. Aten, jun., *Radiochim. Acta*, **9** (1968) 53.

estimates of the boiling points of several volatile astatine compounds[1095,1096]. In aqueous solution the acid $AtCH_2COOH$ appears, on the evidence of solvent-extraction experiments, to be marginally weaker than ICH_2COOH, the pK_a values at 22°C being 3·70 and 3·12, respectively[1097].

5.7. BIOLOGICAL BEHAVIOUR OF ASTATINE[1058,1059]

Preparations of astatine for biological purposes typically involve distillation *in vacuo* from a bismuth target and condensation in a trap on a thin ice surface; after melting the ice, an aqueous solution of astatine in minimal volume is obtained. Solutions used for physiological studies are prepared by the addition of the appropriate amount of sodium chloride to the aqueous solution of astatine. For such investigations very small quantities of astatine ($< 10\ \mu C$) are usually employed, and to obtain precise information about the distribution of astatine, a precise analytical method of separating it from biological specimens has been developed by Hamilton and his coworkers[1098]. The organic substance is destroyed by wet combustion with a mixture of perchloric and nitric acids. The solution is subsequently evaporated or diluted with distilled water to bring the concentration of perchloric acid to 3 M; the astatine is then separated from this solution for counting by coprecipitation with metallic thallium or by adsorption on a silver disc, the latter being more suitable for the serial analysis of biological materials. The distribution of astatine can also be observed in autoradiograms, in which the ^{211}At is recognized by its α-tracks.

It has been observed that, if astatine is administered as a radiocolloid, it tends to accumulate in the liver, a tendency common to most radiocolloids. However, if administered as a true solution, astatine, like iodine, accumulates in the thyroid gland, as was first demonstrated in 1940 by Hamilton and Soley[1098] when comparing the metabolism of radioiodine and astatine. It is not clear whether the astatine is carried by the blood as At^- or as At(0). For some time its chemical condition in the thyroid gland was also in doubt, but it has now been shown[1098] that a large fraction of the astatine in the thyroid follows the protein fraction as it is precipitated by trichloroacetic acid or by concentrated ammonium sulphate. This suggests that astatine can take up positions in the thyroid protein similar to those normally occupied by iodine. The uptake of astatine by the thyroid gland is reduced by the administration of thiocyanate, a behaviour quite analogous to that of iodine. On the other hand, the influence of propylthiouracil enhances the uptake of astatine but reduces that of iodine. Relatively small amounts of astatine are capable of provoking profound changes in thyroid tissue without inducing any noticeable alterations of structure in the parathyroid gland or other peritrachial tissues.

Astatine is superior to radioiodine for destruction of abnormal thyroid tissue because of the localized action of the α-particles, which dissipate 5·9 MeV within a range of 70 microns of tissue, whereas the much weaker β-rays of radioiodine have a maximum range of approximately 2000 microns in tissue. Detailed investigations have been made in guinea pigs, rats and monkeys[1098]. On this basis, the use of astatine to treat hyperthyroidism in humans has been proposed, though the hazards from deleterious side-effects have not been fully explored; reports that mammary and pituitary tumours can be induced in rats with a single injection of astatine do not breed confidence in the future of such treatment.

[1098] J. G. Hamilton and M. H. Soley, *Proc. Nat. Acad. Sci. U.S.A.* **26** (1940) 483; J. G. Hamilton, *Radiology*, **39** (1942) 541; J. G. Hamilton, P. W. Durbin, C. W. Asling and M. E. Johnston, *Proc. Intern. Conference on Peaceful Uses of Atomic Energy, Geneva*, **10** (1956) 175.

INDEX